W9-AXW-420

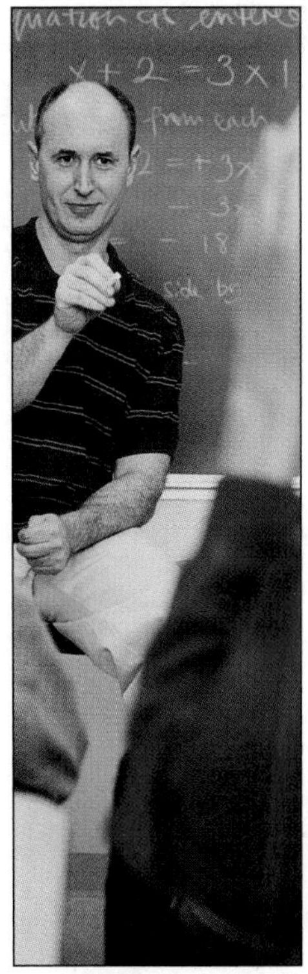

WileyPLUS is built around the activities you perform

Prepare & Present

Create outstanding class presentations using a wealth of resources, such as PowerPoint™ slides, image galleries, interactive learningware, and more. Plus you can easily upload any materials you have created into your course, and combine them with the resources *WileyPLUS* provides.

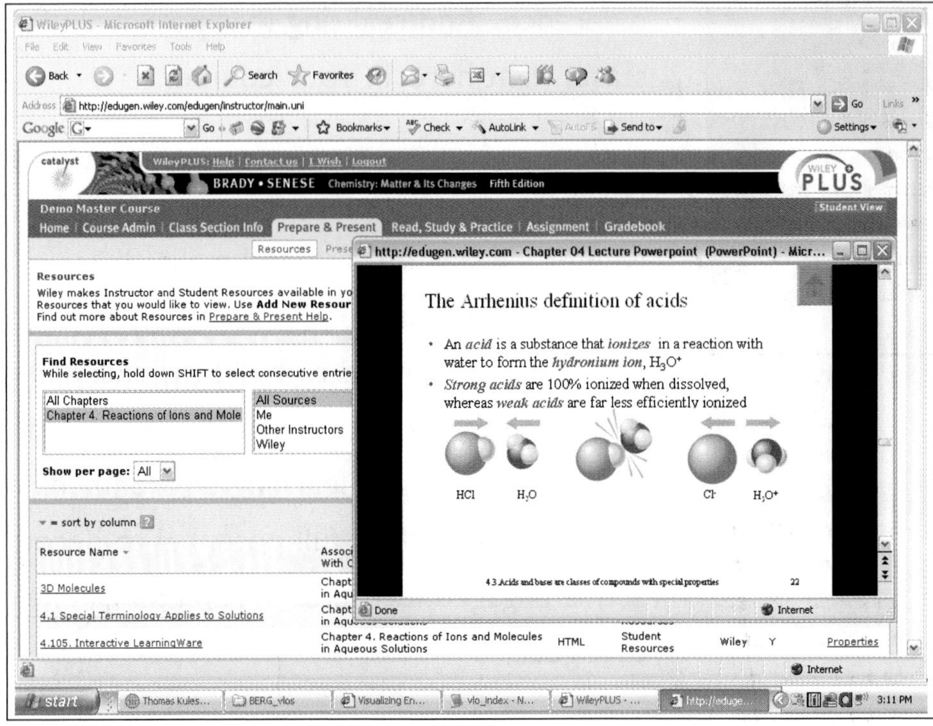

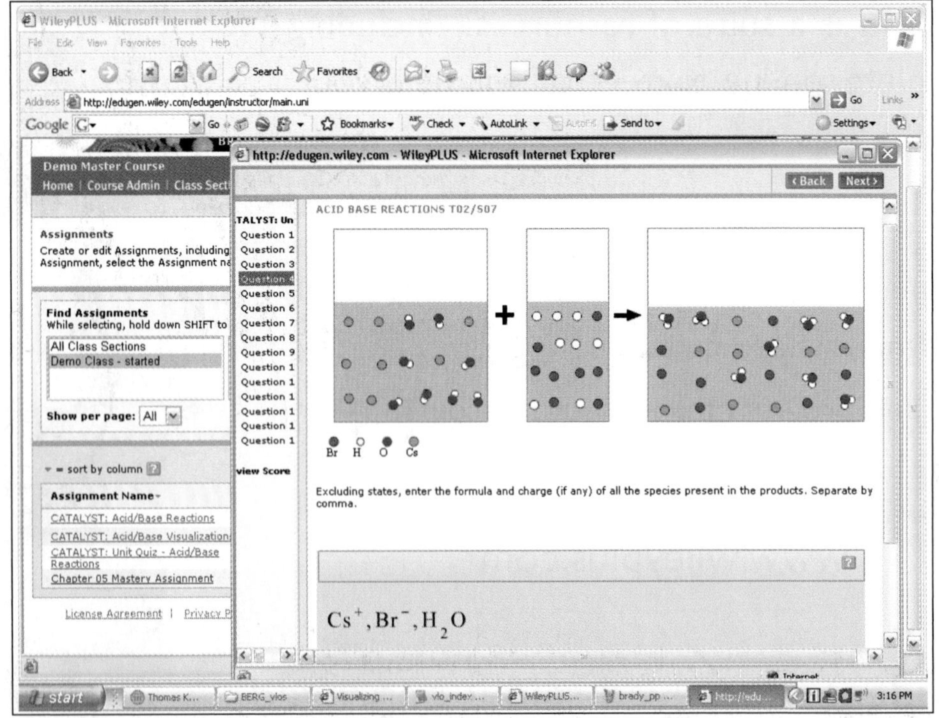

catalyst

CATALYST

With the 5th edition of Brady/Senese, we are introducing an innovative on-line learning program called CATALYST. The CATALYST assignments ask students to consider the key concepts and topics at hand from different perspectives; with different givens and desired responses required each time a new question is presented.

Create Assignments

Automate the assigning and grading of homework or quizzes by using the provided question banks. Student results will be automatically graded and recorded in your gradebook. *WileyPLUS* also links homework problems to relevant sections of the online text, hints, or solutions—context-sensitive help where students need it most!

*Based upon 7,000 survey responses from student users of *WileyPLUS* in academic year 2006-2007.

in your class each day. With WileyPLUS you can:

Track Student Progress

Keep track of your students' progress via an instructor's gradebook, which allows you to analyze individual and overall class results. This gives you an accurate and realistic assessment of your students' progress and level of understanding.

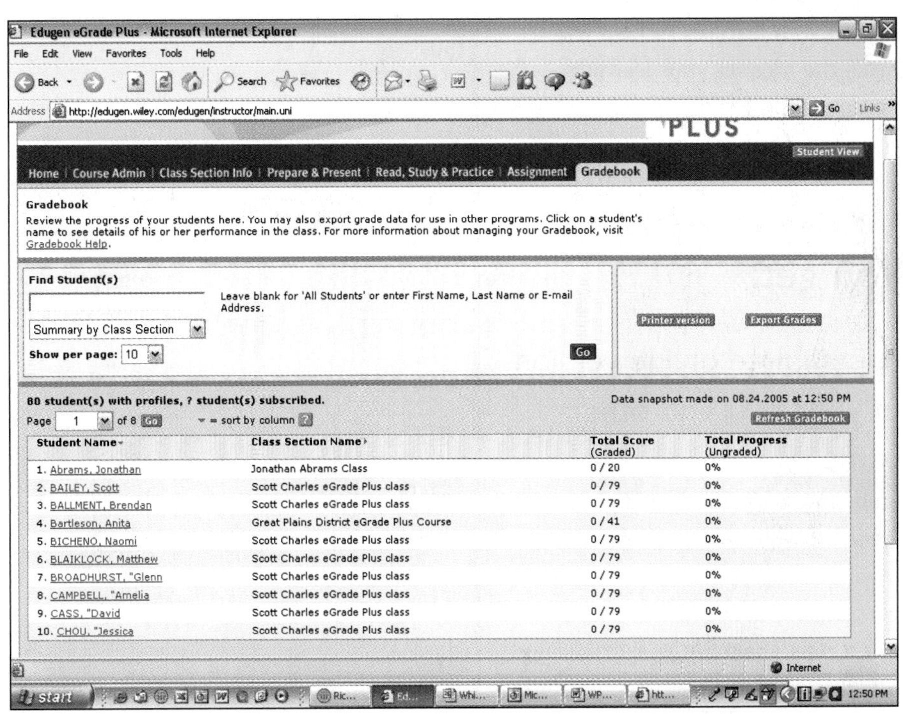

Now Available with WebCT and eCollege!

Now you can seamlessly integrate all of the rich content and resources available with *WileyPLUS* with the power and convenience of your WebCT or eCollege course. You and your students get the best of both worlds with single sign-on, an integrated gradebook, list of assignments and roster, and more. If your campus is using another course management system, contact your local Wiley Representative.

"I studied more for this class than I would have without *WileyPLUS*."

Melissa Lawler, *Western Washington Univ.*

For more information on what *WileyPLUS* can do to help your students reach their potential, please visit

www.wileyplus.com/experience

84% of students would recommend *WileyPLUS* to their next instructors *

TO THE STUDENT

You have the potential to make a difference!

WileyPLUS is a powerful online system packed with features to help you make the most of your learning potential, and get the best grade you can!

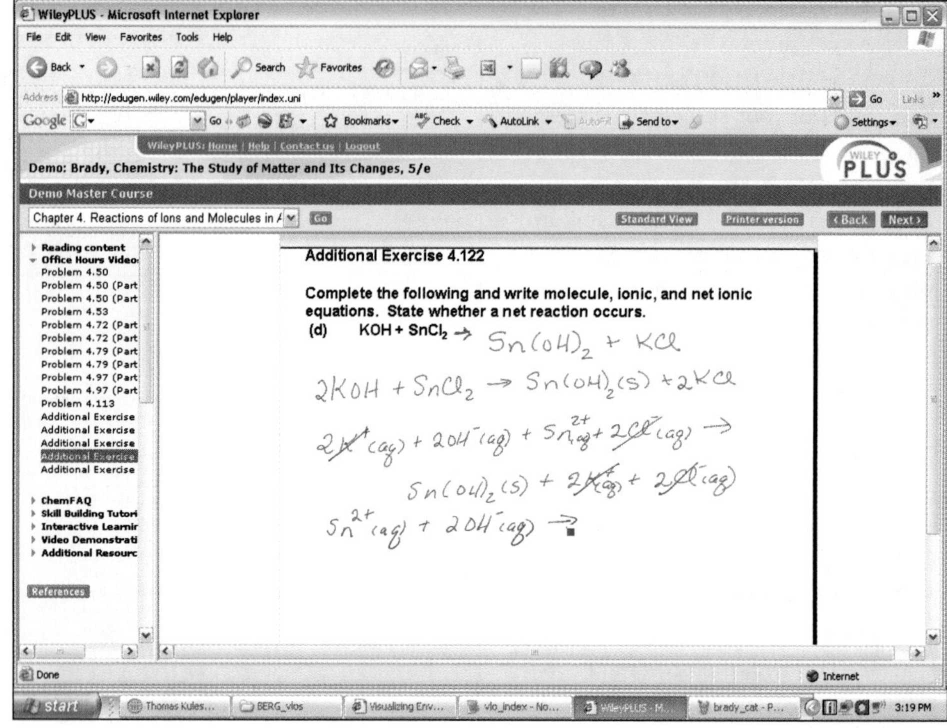

With Wiley**PLUS** you get:

A complete online version of your text and other study resources

Study more effectively and get instant feedback when you practice on your own. Resources like self-assessment quizzes, interactive learningware, video clips, chem FAQs, skill building tutorials, and office hours videos bring the subject to life, and help you master the material.

Problem-solving help, instant grading, and feedback on your homework and quizzes

You can keep all of your assigned work in one location, making it easy for you to stay on task. Plus, many homework problems contain direct links to the relevant portion of your text to help you deal with problem-solving obstacles at the moment they come up.

The ability to track your progress and grades throughout the term.

A personal gradebook allows you to monitor your results from past assignments at any time. You'll always know exactly where you stand.

If your instructor uses *WileyPLUS*, you will receive a URL for your class. If not, your instructor can get more information about *WileyPLUS* by visiting www.wileyplus.com

"It has been a great help, and I believe it has helped me to achieve a better grade."
Michael Morris, *Columbia Basin College*

74% of students surveyed said it helped them get a better grade.*

th EDITION

CHEMISTRY
MATTER AND ITS CHANGES

AMES E. BRADY

St. John's University, New York

RED SENESE

Frostburg State University, Maryland

In collaboration with

NEIL D. JESPERSEN

St. John's University, New York

JOHN WILEY & SONS, INC.

EXECUTIVE EDITOR Stuart Johnson
PROJECT EDITOR Jennifer Yee
EXECUTIVE MARKETING MANAGER Amanda Wainer
SENIOR PRODUCTION EDITOR Elizabeth Swain
SENIOR MEDIA EDITOR Thomas Kulesa
SENIOR DESIGNER Madelyn Lesure
SENIOR ILLUSTRATION EDITOR Anna Melhorn
SENIOR PHOTO EDITOR Jennifer MacMillan

FRONT COVER PHOTO: © Chris Ewels
BACK COVER PHOTOS: (from top to bottom): Laguna Design/Photo Researchers, Inc.; Courtesy Dr. Ernst Richter; Courtesy Dr. Mark McClure, University of North Carolina at Pembroke; Courtesy Wikimedia Commons.

This book was set in 10.5/12 Adobe Garamond by Prepare and printed and bound by Courier Kendallville. The cover was printed by Courier Kendallville.

This book is printed on acid free paper. ∞

To order books or for customer service please, call 1-800-CALL WILEY (225-5945).

Library of Congress Cataloging-in-Publication Data:
Brady, James E.
Chemistry: matter and its changes.—5th ed./James E. Brady, Fred Senese.
p. cm.
Includes index.
ISBN 978-0-470-12094-1 (cloth)
Instructor's edition: ISBN 978-0-470-28644-9
1. Chemistry. I. Senese, Frederick. II. Title.
QD33.2.B73 2009
540—dc22

2007033355

Printed in the United States of America
10 9 8 7 6 5 4 3 2 1

PREFACE

The goal of this textbook and its supporting material is to address the needs within the entire range of student ability in the general chemistry course. Our approach provides tutorial help and instruction precisely when students need it - without burdening them when they do not. This format has reached its present form through a process of evolution. Over the course of numerous editions, we have responded to suggestions by instructors and students who have used the text. Their responses to the innovations that we have introduced over time have allowed us to polish our approach, each time making this text a more effective teaching tool.

Refining the Problem-Solving Approach

We are among those who firmly believe that problem solving reinforces concepts, and that learning to become a good problem solver is essential for anyone studying the sciences. This belief has served as a guiding principle through previous editions and for this edition as well.

One of the strengths of the 4th edition teaching package was its integrated system of support, and our intent in this edition was to refine our approach to make it even more effective. Toward this end, we've made the following changes, refinements, and additions.

- We have thoroughly reviewed and edited the worked examples to be sure they follow our three-step approach to solving problems: *Analysis*, in which the nature of the problem is analyzed and the method of solution is developed; *Solution*, in which the actual solution to the problem is reached; and *Is the Answer Reasonable?*, in which we describe how we check to be sure the answer makes sense.

- We've added more examples in the *Is the Answer Reasonable?* discussions that illustrate how simple chemical logic and/or approximate arithmetic can be used to judge whether an answer is "in the right ballpark."

- We've increased the number of Practice Exercises following the worked examples. The first Practice Exercise in a set includes a hint designed to get the student's thinking started in the right direction.

- We've retained the Chemical Tools concept and reorganized the summary of the Tools at the end of each chapter to make it easier for students to see how they are applied to problem solving. Where appropriate, we include equations or figures.

- In the worked examples, we've made an effort to improve the connection between the problem solving "Tools" described in the body of the text and their application to solving problems.

- In addition to the end-of-chapter Review Questions, Review Problems, and Additional Exercises, we've added a new set of exercises called Exercises in Critical Thinking, which offer more open-ended problems for students with advanced problem-solving skills. These have been thoroughly reviewed.

- As part of our Web-based support, we offer a new feature called Office Hours, in which an experienced teacher provides detailed explanations for the solution of problems chosen from among the Review Problems and Exercises in each chapter. Using screen capture technology with accompanying voiceover instruction, our Office Hours videos offer students a virtual tutorial session with a chemistry instructor.

- We've expanded the number of questions and problems in the periodic review segments previously called *Test of Facts and Concepts*, now more aptly named *Bringing it Together.*

Significant Changes in This Edition

In addition to fine tuning our already robust approach to teaching problem solving, a principal goal in this revision was to produce a shorter textbook that focused on topics most often taught by chemistry teachers.

Chemistry textbooks, including ours, have grown by a process of accretion to the point where it has become very difficult to cover their contents within a two-semester course. To address this issue, we carefully reviewed the topics included in the last edition and selected for removal content that we - and our reviewers - felt was not essential. We also rewrote some discussions to both reduce their length and to improve their clarity. For example,

- As part of a reorganization and streamlining of the beginning chapters in the text, we moved the topics of measurement, units, and the factor label method to Chapter 1. They are no longer in a separate chapter. We also eliminated the discussion of specific gravity.
- We rewrote, shortened, and clarified topics in the chapter dealing with stoichiometry.
- We rewrote and shortened somewhat the chapters dealing with reactions in aqueous solutions and oxidation-reduction.
- Topics in the thermochemistry chapter have been rewritten to improve clarity and decrease length.
- We returned the discussion of crystal structures to the chapter dealing with the properties of solids, liquids, and changes of state.
- We removed discussions of liquid crystals and high-tech ceramics. Although these topics are interesting, few instructors felt they had enough time to teach or assign them.
- The discussion of polymers has been moved to the chapter on organic chemistry.
- Carbon-14 dating is now included as a topic in the chapter on kinetics.
- The nuclear chemistry chapter has been shortened somewhat. Nuclear fusion has been added to the chapter.
- Descriptive chemistry chapters have been condensed and some topics have been removed.

Organization and Content Development

For those looking at our text for the first time, we provide here a more detailed overview of the book. As noted earlier, it is designed for a mainstream university-level general chemistry course for science majors (e.g., chemistry, biology, pre-med). As in previous editions, we employ a relaxed writing style and student-friendly attitude while providing clear and thorough explanations of difficult concepts. So as not to lose the less-prepared student, we do not assume students have had a previous course in chemistry, and mastery of only basic algebra is expected.

In structuring the text we have sought to provide a logical progression of topics arranged to yield the maximum flexibility for the teacher in organizing his or her course. As much as possible, chapters have been written so they can easily be presented out of order if the instructor wishes to alter the topic sequence to suit his or her course. For example, the chapter dealing with the properties of gases could easily be moved to an earlier point if so desired.

Chapter content is based on our conviction that a general chemistry course serves a variety of goals in the education of a student. First, of course, it must provide a sound foundation in the basic facts and concepts of chemistry upon which theoretical models can be constructed. The general chemistry course should also give the student an appreciation of the central role that chemistry plays among the sciences as well as the importance of chemistry in society and day-to-day living. In addition, it should enable the student to develop skills in analytical thinking and problem solving.

To assist students in previewing chapter contents and in reviewing key concepts, we use descriptive phrases both for section headings as well as subheadings. This enables us to use the **Chapter Outline** at the start of each chapter to provide a meaningful overview of the chapter contents. This is followed by a section titled **This Chapter in Context** where we describe the nature of the chapter contents and where they fit within the broad scope of the course. The sequence of chapters described below gives an overview of the development of concepts in the text.

Foundations in reaction chemistry and stoichiometry

Chapters 1 through 6 develop a foundation in reaction chemistry, the importance of measurement and units, stoichiometry, and thermochemistry, along with a basic introduction to the structure of matter and the periodic table.

To enable students from the outset to obtain a feel for the nature of chemistry, Chapters 1 and 2 cover the basic concepts of atoms, molecules, elements, and compounds, with a brief treatment of the nature of chemical reactions. These chapters include discussions of measurement, units, significant figures, and unit conversions and introduce the periodic table and the nature of ionic and molecular compounds. The naming of chemical compounds is presented on an "as needed" basis. In Chapter 2, methods of naming ionic and molecular compounds are discussed, while naming acids and bases is postponed until Chapter 4 when these compounds are introduced.

Chapter 3 provides a careful and thorough discussion of the mole concept and chemical stoichiometry. Chapter 4 deals with simple acid-base chemistry, metathesis reactions, and solution stoichiometry. Chapter 5 focuses on redox reactions and includes discussions of the activity series of metals and oxidation reactions involving oxygen. Chapters 4 and 5 provide students with a foundation in the basic descriptive chemistry of solution reactions and gives them a knowledge base that serves as a foundation for theoretical concepts developed in Chapters 7–9 (which deal with atomic structure and bonding).

Chapter 6 has been partially rewritten to give students a better understanding of the nature of heat and how it is measured. The kinetic molecular theory is presented here as well as the first law of thermodynamics.

Electronic structure and bonding

Chapters 7 through 9 cover electronic structures of atoms and bonding in compounds. In Chapter 7 (The Quantum Mechanical Atom), the introduction to quantum theory has been improved. We show with minimal mathematics how the concept of standing waves and the de Broglie hypothesis can be combined to yield the energy levels for a confined electron. Discussions of irregularities in periodic trends in ionization energy and electron affinity are covered as a special topic ("Facets of Chemistry"). This enables teachers who do not wish to dwell on these finer details to easily omit them. Yet they are available for teachers who wish to discuss them.

The discussion of bonding is divided between two chapters. The first treats the topic at a relatively elementary level, describing the principal features of ionic and covalent bonds using Lewis structures. The second bonding chapter deals with molecular structure (VSEPR theory) and the valence bond and molecular orbital theories (including some simple heteronuclear diatomic molecules).

Physical properties and the states of matter

Chapters 10 through 12 focus on the properties of the states of matter and solutions. In Chapter 10, dealing with gases, we have expanded the discussion of real gases to include an explanation of the origin of the correction terms in the van der Waals equation.

Chapter 12 examines the physical properties of solutions and presents students with a preview of the concept of entropy in the discussion of the factors that influence the solubilities of substances in various solvents.

Kinetics and equilibrium

Chapters 13 and 14 examine rates of chemical reactions and chemical equilibrium. In the kinetics chapter we include calculations involving carbon-14 dating so that teachers who find it difficult to find time to discuss nuclear reactions can include some treatment of this important subject in their course. Chapter 15 (Acids and Bases: A Second Look) brings together the various views of acids and bases, including the Brønsted-Lowry and Lewis acid-base concepts. The pH concept is introduced in Chapter 15 and applied to solutions of strong acids and bases.

Chapters 16 and 17 cover equilibria in aqueous solutions. Equilibria involving weak acids and bases, including polyprotic acids and their salts, are discussed in Chapter 16. As in the last edition, treatment of problems requiring the quadratic equation or the method of successive approximations are placed in a separate section. Chapter 17 deals with solubility and complex ion equilibria.

Thermodynamics and electrochemistry

The discussion of thermodynamics in Chapter 18 places this topic after the chapters on equilibria, so we are able to incorporate the treatment of thermodynamic equilibrium constants. We also include a discussion of the calculation of bond energies from thermodynamic data.

Chapter 19 (Electrochemistry) ties together concepts of thermodynamics and equilibrium as well as practical applications of electrolysis and galvanic cells. We begin with the discussion of galvanic cells, including standard reduction potentials and the Nernst equation. We have an up-to-date treatment of practical galvanic cells that includes nickel-metal hydride and lithium ion batteries, which are used extensively in modern electronics, and we have updated our treatment of fuel cells. By placing galvanic cells at the beginning of the chapter, discussion of electrolysis reactions proceeds with less mystery.

Nuclear, inorganic, and organic chemistry

Chapter 20 presents an overview of nuclear reactions and the role they play in chemistry and society. This chapter, which could actually be presented earlier in the course should the teacher elect to do so, has been revised to keep it up to date. Nuclear fission and fusion reactions are included in our discussions.

Recognizing the fact that most teachers do not have time to teach a great deal of descriptive chemistry, we provide one chapter that deals with highlights or this material. Here we take a unique approach that looks at trends in properties that can be explained using the principles of bonding, thermodynamics, and kinetics developed in earlier chapters. We also examine trends in properties and structure that extend across periods and down groups in the periodic table. Included in Chapter 21 are discussions of the structures, nomenclature, and bonding involving complex ions, particularly those of the transition metals.

Chapter 22, the final chapter of the text, serves as an introduction to organic and biochemistry. In this chapter we include a discussion of some important organic polymers.

LEARNING FEATURES THAT ENHANCE PROBLEM-SOLVING AND CRITICAL THINKING SKILLS

Aware of possible student difficulties with problem solving and analytical thinking, we have adopted a unique approach to developing thinking skills. We distinguish three types of learning aids: those that enhance problem-solving skills; those that further comprehension and learning; and those that extend the breadth and knowledge of the student.

Many students entering college today lack experience in analytical thinking. A course in chemistry should provide an ideal opportunity to help students sharpen their reasoning skills because problem solving in chemistry operates on two levels. Because of the nature of the subject, in addition to mathematics, many problems also involve the application of theoretical concepts. Students often have difficulty at both levels, and one of the goals of this text has been to develop a unified approach that addresses each level.

Chemical tools approach to problem analysis

Students are taught a variety of basic skills, such as finding the number of grams in a mole of a substance or writing the Lewis structure of a molecule. Problem solving often involves bringing together a sequence of such simple tasks. Therefore, if we are to teach problem solving, we must teach students how to seek out the necessary relationships required to obtain solutions to problems.

We use an innovative approach to problem solving that makes an analogy between the abstract tools of chemistry and the concrete tools of a mechanic. Students are encouraged to think of simple skills as tools that can be used to solve more complex problems. When faced with a new problem, the student is urged to examine the tools that have been taught and to select those that bear on the problem at hand.

To foster this approach to thinking through problems, we present a comprehensive program of reinforcement and review:

TOOLS FOR PROBLEM SOLVING

In this chapter you learned to apply the following concepts as tools in solving problems dealing with reactions in aqueous solutions. Study each one carefully so that you know what each is used for. When faced with solving a problem, recall what each tool does and consider whether it will be helpful in finding a solution. This will aid you in selecting the tools you need.

Criteria for a balanced ionic or net ionic equation *(page 135)* For an equation that includes the formulas of ions to be balanced, it must satisfy two criteria. The number of atoms of each kind must be the same on both sides, and the total net electrical charge shown on both sides must be the same.

Equation for the ionization of an acid in water *(page 137)* Equation 4.2 describes how acids react with water to form hydronium ion plus an anion.

$$HA + H_2O \longrightarrow H_3O^+ + A^-$$

Use this tool to write equations for ionizations of acids and to determine the formula of the anion formed when the acid molecule loses an H^+. The equation also applies to acid anions such as HSO_4^- which gives SO_4^{2-} when it loses an H^+. Often H_2O is omitted from the equation and the hydronium ion is abbreviated as H^+.

Equation for the ionization of a molecular base in water *(page 139)* Equation 4.3 describes how molecules of molecular bases acquire H^+ from H_2O to form a cation plus a hydroxide ion.

$$B + H_2O \longrightarrow BH^+ + OH^-$$

Use this tool to write equations for ionizations of bases and to determine the formula of the cation formed when the base molecule gains an H^+. *Molecular bases are weak and are not completely ionized.*

Table of strong acids *(page 140)* Formulas of the most common strong acids are given here. If you learn this list and encounter an acid that's *not* on the list, you can assume it to be a weak acid. The most common strong acids are HCl, HNO_3, and H_2SO_4. *Remember that strong acids are completely ionized in water.*

Predicting the existence of a net ionic equation *(page 146)* A net ionic equation will exist and a reaction will occur when:

- A precipitate is formed from a mixture of soluble reactants.
- An acid reacts with a base. *This includes strong or weak acids reacting with strong or weak bases or insoluble metal hydroxides or oxides.*
- A weak electrolyte is formed from a mixture of strong electrolytes.
- A gas is formed from a mixture of reactants.

These criteria are tools to determine whether or not a net reaction will occur in a solution.

Solubility rules *(page 147)* The rules in Table 4.1 are the tool we use to determine whether a particular salt is soluble in water. (If a salt is soluble, it's completely dissociated into ions.) They also serve as a tool to help predict the course of metathesis reactions.

Substances that form gases in metathesis reactions *(page 152)* Use Table 4.2 as a tool to help predict the outcome of metathesis reactions. The most common gas formed in such reactions is CO_2, which comes from the reaction of an acid with a carbonate or bicarbonate.

Molarity *(page 154)* Molarity provides the connection between moles of a solute and the volume of its solution. The definition

TOOLS ICON. The Tools icon in the margin calls attention to each chemical tool when it is first introduced and is accompanied by a brief statement that identifies the tool. Following the Summary at the end of the chapter, the tools are reviewed under the heading **Tools for Problem Solving,** preparing students for the exercises that follow.

WORKED EXAMPLES, as noted earlier, follow a three-step process. Each begins with an Analysis that describes the thought processes and critical links involved in selecting the tools needed to solve the problem, as well as how information will be assembled to achieve the solution. In many of the example problems in this edition, the analysis step has been expanded to include greater detail. Next comes the *Solution* step in which the problem is solved according to the plan developed in the *Analysis*. Examples conclude with a section titled *Is the Answer Reasonable?* in which the answer is studied to see whether it "makes sense." Here, students are taught to perform approximate arithmetic to get ballpark estimates of the answers to numerical problems. They are also warned of common errors and other ways to check their work.

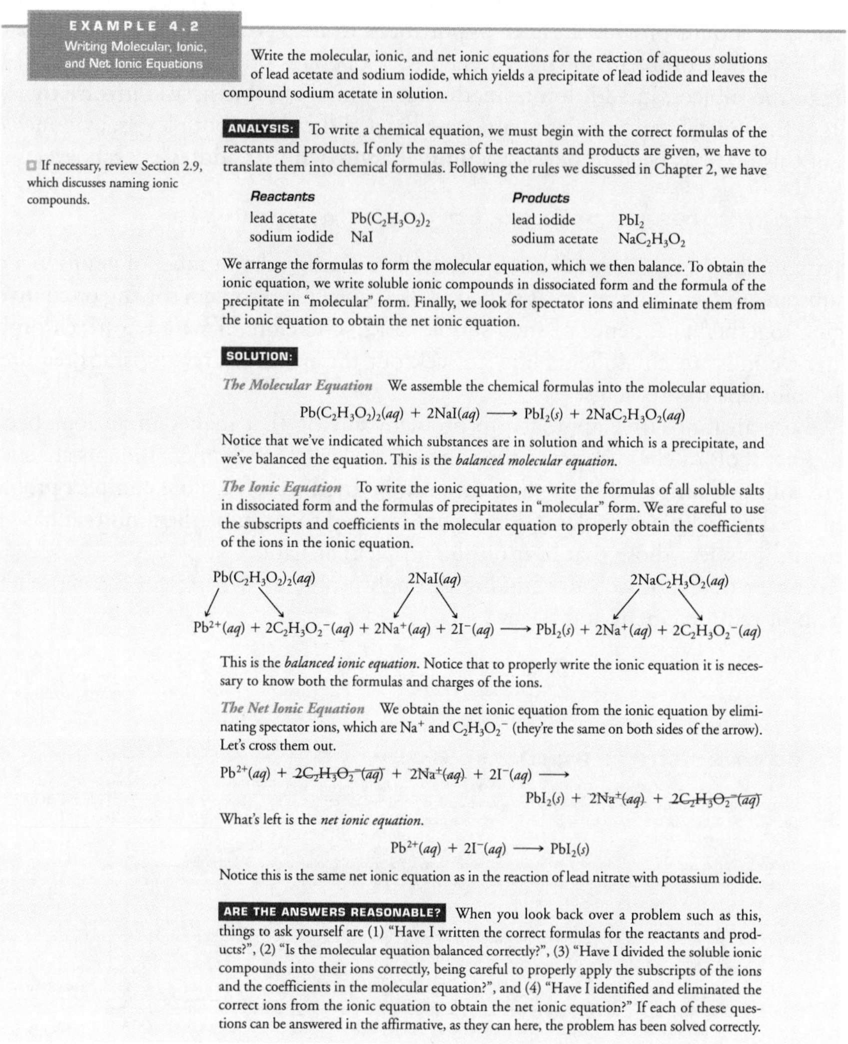

EXAMPLE 4.2
Writing Molecular, Ionic, and Net Ionic Equations

Write the molecular, ionic, and net ionic equations for the reaction of aqueous solutions of lead acetate and sodium iodide, which yields a precipitate of lead iodide and leaves the compound sodium acetate in solution.

ANALYSIS: To write a chemical equation, we must begin with the correct formulas of the reactants and products. If only the names of the reactants and products are given, we have to translate them into chemical formulas. Following the rules we discussed in Chapter 2, we have

If necessary, review Section 2.9, which discusses naming ionic compounds.

Reactants		*Products*	
lead acetate	$Pb(C_2H_3O_2)_2$	lead iodide	PbI_2
sodium iodide	NaI	sodium acetate	$NaC_2H_3O_2$

We arrange the formulas to form the molecular equation, which we then balance. To obtain the ionic equation, we write soluble ionic compounds in dissociated form and the formula of the precipitate in "molecular" form. Finally, we look for spectator ions and eliminate them from the ionic equation to obtain the net ionic equation.

SOLUTION:

The Molecular Equation We assemble the chemical formulas into the molecular equation.

$$Pb(C_2H_3O_2)_2(aq) + 2NaI(aq) \longrightarrow PbI_2(s) + 2NaC_2H_3O_2(aq)$$

Notice that we've indicated which substances are in solution and which is a precipitate, and we've balanced the equation. This is the *balanced molecular equation.*

The Ionic Equation To write the ionic equation, we write the formulas of all soluble salts in dissociated form and the formulas of precipitates in "molecular" form. We are careful to use the subscripts and coefficients in the molecular equation to properly obtain the coefficients of the ions in the ionic equation.

$Pb(C_2H_3O_2)_2(aq)$ $2NaI(aq)$ $2NaC_2H_3O_2(aq)$

$$Pb^{2+}(aq) + 2C_2H_3O_2^-(aq) + 2Na^+(aq) + 2I^-(aq) \longrightarrow PbI_2(s) + 2Na^+(aq) + 2C_2H_3O_2^-(aq)$$

This is the *balanced ionic equation.* Notice that to properly write the ionic equation it is necessary to know both the formulas and charges of the ions.

The Net Ionic Equation We obtain the net ionic equation from the ionic equation by eliminating spectator ions, which are Na^+ and $C_2H_3O_2^-$ (they're the same on both sides of the arrow). Let's cross them out.

$$Pb^{2+}(aq) + \cancel{2C_2H_3O_2^-(aq)} + \cancel{2Na^+(aq)} + 2I^-(aq) \longrightarrow PbI_2(s) + \cancel{2Na^+(aq)} + \cancel{2C_2H_3O_2^-(aq)}$$

What's left is the *net ionic equation.*

$$Pb^{2+}(aq) + 2I^-(aq) \longrightarrow PbI_2(s)$$

Notice this is the same net ionic equation as in the reaction of lead nitrate with potassium iodide.

ARE THE ANSWERS REASONABLE? When you look back over a problem such as this, things to ask yourself are (1) "Have I written the correct formulas for the reactants and products?", (2) "Is the molecular equation balanced correctly?", (3) "Have I divided the soluble ionic compounds into their ions correctly, being careful to properly apply the subscripts of the ions and the coefficients in the molecular equation?", and (4) "Have I identified and eliminated the correct ions from the ionic equation to obtain the net ionic equation?" If each of these questions can be answered in the affirmative, as they can here, the problem has been solved correctly.

Practice Exercise 3: When solutions of $(NH_4)_2SO_4$ and $Ba(NO_3)_2$ are mixed, a precipitate of $BaSO_4$ forms, leaving soluble NH_4NO_3 in the solution. Write the molecular, ionic, and net ionic equations for the reaction. (Hint: Remember that polyatomic ions do not break apart when ionic compounds dissolve in water.)

Practice Exercise 4: Write molecular, ionic, and net ionic equations for the reaction of aqueous solutions of cadmium chloride and sodium sulfide to give a precipitate of cadmium sulfide and a solution of sodium chloride.

PRACTICE EXERCISES follow most worked examples to enable the student to apply what has just been studied to a similar problem. The first member of each Practice Exercise set includes a hint designed to focus the student's thinking in the right direction. All of the Practice Exercises have answers in Appendix B at the back of the book.

END-OF-CHAPTER EXERCISES are divided into four sections. **Review Questions,** classified according to topic, enable students to gauge their progress in learning concepts presented in a chapter. **Review Problems,** also classified according to topic, are presented in pairs of similar problems, with the first member of each pair having its answer in Appendix B. These problems provide routine practice in the use of basic tools as well as opportunities to incorporate the tools in the solution of more complex problems. **Additional Exercises** at the end of the problem sets are unclassified. Many problems are cumulative, requiring two or more concepts, and several in later chapters require skills learned in earlier chapters. The range of problem difficulty provides the instructor flexibility is assigning homework. **Exercises in Critical Thinking** are sets of questions that require students to "think outside the box" and often go beyond explanations provided in the text, frequently requiring exploration on the internet. Many are quite open ended.

CHAPTERS 4-6

BRINGING IT TOGETHER

We pause again to allow you to test your understanding of concepts, your knowledge of scientific terms, and your skills at solving chemistry problems. Read through the following questions carefully, and answer each as fully as possible. When necessary, review topics you are uncertain of. If you can answer these questions correctly, you are ready to go on to the next group of chapters.

1. What is the difference between a strong electrolyte and a weak electrolyte? Formic acid, $HCHO_2$, is a weak acid. Write a chemical equation showing its reaction with water.
2. Write an equation showing the reaction of water with itself to form ions.
3. Methylamine, CH_3NH_2, is a weak base. Write a chemical equation showing its reaction with water.
4. Write molecular, ionic, and net ionic equations for the reaction that occurs when a solution containing hydrochloric acid is added to a solution of the weak base methylamine (CH_3NH_2).
5. According to the solubility rules, which of the following salts would be classified as soluble?

(a) $Ca_3(PO_4)_2$	(f) $Au(ClO_4)_3$	(k) $ZnSO_4$
(b) $Ni(OH)_2$	(g) $Cu(C_2H_3O_2)_2$	(l) Na_2S
(c) $(NH_4)_2HPO_4$	(h) $AgBr$	(m) $CoCO_3$
(d) $SnCl_2$	(i) KOH	(n) $BaSO_3$
(e) $Sr(NO_3)_2$	(j) Hg_2Cl_2	(o) MnS

6. What are the two criteria that must be met for an ionic equation to be balanced correctly?
7. Write molecular, ionic, and net ionic equations for any reactions that would occur between the following pairs of compounds. If no reaction occurs, write "N.R."
   (a) $CuCl_2(aq)$ and $(NH_4)_2CO_3(aq)$
   (b) $HCl(aq)$ and $MgCO_3(s)$
   (c) $ZnCl_2(aq)$ and $AgC_2H_3O_2(aq)$
   (d) $HClO_4(aq)$ and $NaCHO_2(aq)$
   (e) $MnO(s)$ and $H_2SO_4(aq)$
   (f) $FeS(s)$ and $HCl(aq)$
8. Write a chemical equation for the complete neutralization of H_3PO_4 by NaOH.
9. Which ion exists in abundance in all solutions of strong acids?
10. Which ion makes a solution basic?
11. Define *monoprotic acid, diprotic acid,* and *polyprotic acid.* What is the general definition of a *salt*?
12. Which of the following oxides are acidic and which are basic: P_4O_6, Na_2O, SeO_3, CaO, PbO, and SO_2?
13. Write the formulas of any acid salts that could be formed by the reaction of the following acids with potassium hydroxide.
   (a) sulfurous acid
   (d) lithium hydrogen sulfate
   (e) bromic acid
16. How many milliliters of 0.200 M $BaCl_2$ must be added to 27.0 mL of 0.600 M Na_2SO_4 to give a complete reaction between their solutes?
17. What mass of $Mg(OH)_2$ will be formed when 30.0 mL of 0.200 M $MgCl_2$ solution is mixed with 25.0 mL of 0.420 M NaOH solution? What will be the molar concentrations of the ions remaining in solution?
18. How many milliliters of 6.00 M HNO_3 must be added to 200 mL of water to give 0.150 M HNO_3?
19. How many grams of CO_2 must be dissolved in 300 mL of 0.100 M Na_2CO_3 solution to change the solute entirely into $NaHCO_3$?
20. A certain toilet cleaner uses $NaHSO_4$ as its active ingredient. In an analysis, 0.500 g of the cleaner was dissolved in 30.0 mL of distilled water and required 24.60 mL of 0.105 M NaOH for complete neutralization in a titration. What was the percentage by weight of $NaHSO_4$ in the cleaner?
21. A volume of 28.50 mL of a freshly prepared solution of KOH was required to titrate 50.00 mL of 0.0922 M HCl solution. What was the molarity of the KOH solution?
22. To neutralize the acid in 10.0 mL of 18.0 M H_2SO_4 that was accidentally spilled on a laboratory bench top, solid sodium bicarbonate was used. The container of sodium bicarbonate was known to weigh 155.0 g before this use and out of curiosity its mass was measured as 144.5 g afterward. The reaction forms sodium sulfate. Was sufficient sodium bicarbonate used? Determine the limiting reactant and calculate the maximum yield in grams of sodium sulfate.
23. How many milliliters of concentrated sulfuric acid (18.0 M) are needed to prepare 125 mL of 0.144 M H_2SO_4?
24. The density of concentrated phosphoric acid solution is 1.689 g solution/mL solution at 20 °C. It contains 144 g H_3PO_4 per 1.00×10^2 mL of solution.
   (a) Calculate the molar concentration of H_3PO_4 in this solution.
   (b) Calculate the number of grams of this solution required to hold 50.0 g H_3PO_4.
25. A mixture consists of lithium carbonate (Li_2CO_3) and potassium carbonate (K_2CO_3). These react with hydrochloric acid

BRINGING IT TOGETHER are problem sets placed between chapters at strategic intervals. These enable the student to review and gauge their progress. Many of the problems in these sets incorporate concepts developed over two or more chapters.

FEATURES THAT FURTHER COMPREHENSION AND LEARNING

MACRO-TO-MICRO ILLUSTRATIONS. To help students make the connection between the macroscopic world we see and events that take place at the molecular level, we have a substantial number of illustrations that combine both views. A photograph, for example, will show a chemical reaction as well as an artist's rendition of the chemical interpretation of what is taking place between the atoms, molecules, or ions involved. We include a variety of illustrations that visualize reactions at the molecular level. The goal is to show how models of nature enable chemists to better understand their observations and to get students to visualize events at the molecular level.

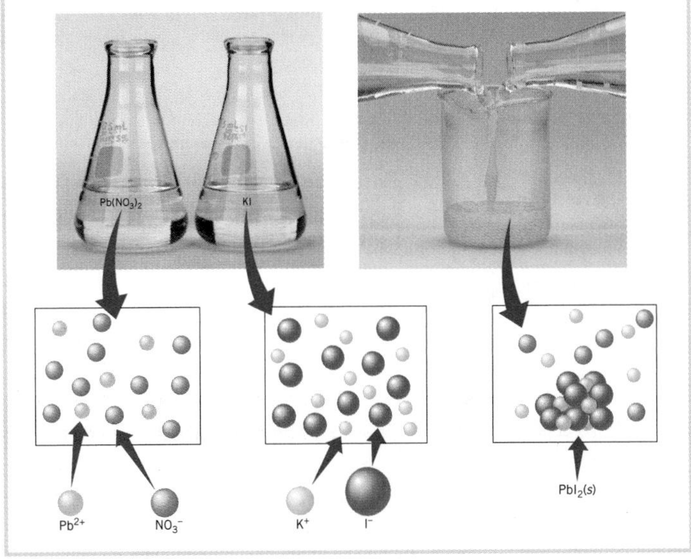

MARGIN COMMENTS make it easy to enrich a discussion, without carrying the aura of being essential. Some margin comments jog the student's memory concerning a definition of a term.

PERIODIC TABLE CORRELATIONS One of our goals was to call particular attention to the usefulness of the periodic table in correlating chemical and physical properties of the elements.

BOLDFACE TERMS alert the student to "must-learn" items. Especially important equations are highlighted with a beige background. Definitions of boldfaced terms are included in the Glossary.

PROBLEM ANALYSIS AT A GLANCE Where appropriate, figures contain flow-charts that summarize the relationships involved in solving problems, the approach used to analyze the method of attack on problems, or the approach to applying rules of chemical nomenclature.

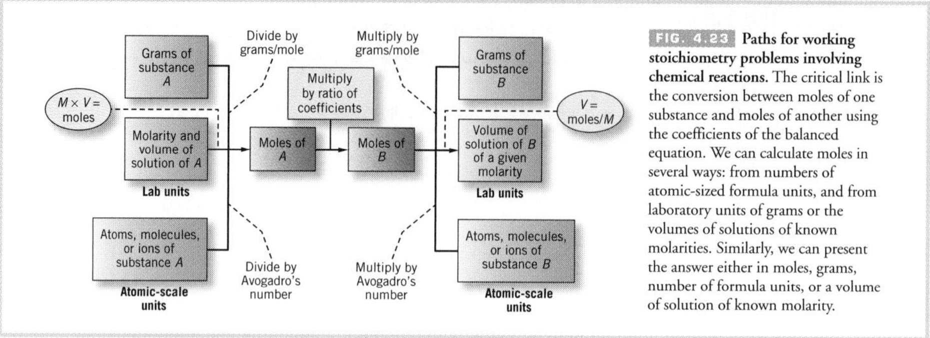

FIG. 4.23 **Paths for working stoichiometry problems involving chemical reactions.** The critical link is the conversion between moles of one substance and moles of another using the coefficients of the balanced equation. We can calculate moles in several ways: from numbers of atomic-sized formula units, and from laboratory units of grams or the volumes of solutions of known molarities. Similarly, we can present the answer either in moles, grams, number of formula units, or a volume of solution of known molarity.

CHAPTER SUMMARIES use the boldface terms to show how the terms fit into statements that summarize concepts.

FEATURES THAT EXTEND THE BREADTH OF KNOWLEDGE OF THE STUDENT

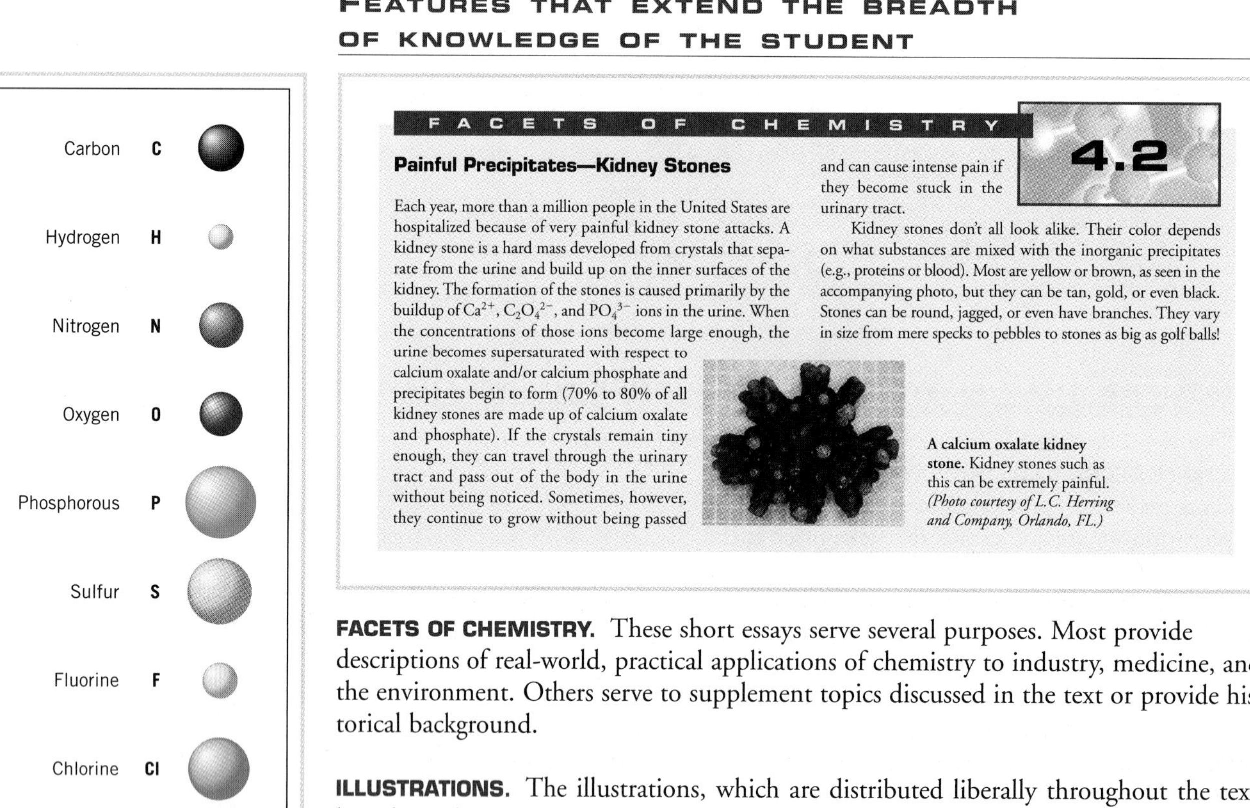

FACETS OF CHEMISTRY 4.2

Painful Precipitates—Kidney Stones

Each year, more than a million people in the United States are hospitalized because of very painful kidney stone attacks. A kidney stone is a hard mass developed from crystals that separate from the urine and build up on the inner surfaces of the kidney. The formation of the stones is caused primarily by the buildup of Ca^{2+}, $C_2O_4^{2-}$, and PO_4^{3-} ions in the urine. When the concentrations of those ions become large enough, the urine becomes supersaturated with respect to calcium oxalate and/or calcium phosphate and precipitates begin to form (70% to 80% of all kidney stones are made up of calcium oxalate and phosphate). If the crystals remain tiny enough, they can travel through the urinary tract and pass out of the body in the urine without being noticed. Sometimes, however, they continue to grow without being passed and can cause intense pain if they become stuck in the urinary tract.

Kidney stones don't all look alike. Their color depends on what substances are mixed with the inorganic precipitates (e.g., proteins or blood). Most are yellow or brown, as seen in the accompanying photo, but they can be tan, gold, or even black. Stones can be round, jagged, or even have branches. They vary in size from mere specks to pebbles to stones as big as golf balls!

A calcium oxalate kidney stone. Kidney stones such as this can be extremely painful. *(Photo courtesy of L.C. Herring and Company, Orlando, FL.)*

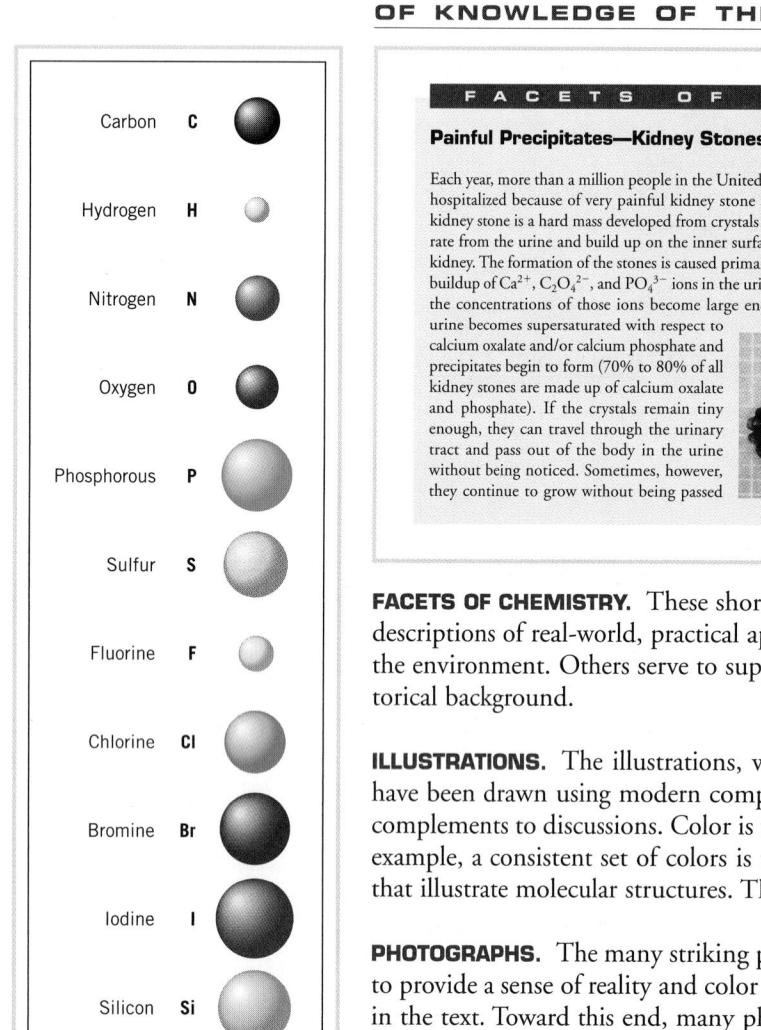

Carbon C
Hydrogen H
Nitrogen N
Oxygen O
Phosphorous P
Sulfur S
Fluorine F
Chlorine Cl
Bromine Br
Iodine I
Silicon Si
Boron B

FACETS OF CHEMISTRY. These short essays serve several purposes. Most provide descriptions of real-world, practical applications of chemistry to industry, medicine, and the environment. Others serve to supplement topics discussed in the text or provide historical background.

ILLUSTRATIONS. The illustrations, which are distributed liberally throughout the text, have been drawn using modern computer techniques to provide accurate, eye-appealing complements to discussions. Color is used constructively rather than for its own sake. For example, a consistent set of colors is used to identify atoms of the elements in drawings that illustrate molecular structures. These are shown in the margin.

PHOTOGRAPHS. The many striking photographs in the book serve two purposes. One is to provide a sense of reality and color to the chemical and physical phenomena described in the text. Toward this end, many photographs of chemicals and chemical reactions are included. The photographs also serve to illustrate how chemistry relates to the world outside of the laboratory. The chapter-opening photos, for example, call the students' attention to the relationship between the chapter's content and common (and often not-so-common) things. Similar photos within the chapters illustrate practical examples and applications of chemical reactions and physical phenomena.

SUPPLEMENTS

A comprehensive package of supplements has been created to assist both the teacher and the student and includes the following:

Study Guide by Neil Jespersen of St. John's University. This guide has been written to further enhance understanding of concepts. It is an invaluable tool for students and contains chapter overviews, additional worked-out problems giving detailed steps involved in solving them, alternate problem-solving approaches, as well as extensive review exercises.

Solutions Manual by Alison Hyslop of St. John's University. The manual contains worked-out solutions for text problems whose answers appear in Appendix B.

Laboratory Manual for Principles of General Chemistry, Eighth Edition by Jo Beran of Texas A&M University, Kingsville. This comprehensive laboratory manual is for use in the general chemistry course. This manual is known for its broad selection of topics and experiments, and for its clear layout and design. Contai-ning enough material for two or three terms, this lab manual emphasizes techniques, helping students learn the time and situation for their correct use. The accompanying Instructor's Manual (IM) presents the details of each experiment, including overviews, an instructor's lecture outline, and teaching hints. The IM also contains answers to the pre-lab assignment and laboratory questions.

Office Hours Videos by Dixie Goss, of Hunter College. These video clips with accompanying audio voiceover walk students through the problem solving process and provide students with a virtual office hour experience with a chemistry instructor. The videos focus on selected end of chapter problems from the text.

Instructor's Manual by Mark A. Benvenuto of University of Detroit–Mercy. In addition to lecture outlines, alternate syllabi, and chapter overviews, this manual contains suggestions for small group active-learning projects, class discussions, and short writing projects, and contains relevant web links for each chapter.

Test Bank by Jason D'Acchioli of the University of Wisconsin – Stevens Point. The Test Bank contains over 1,800 questions including; multiple choice, true-false, short answer questions, and critical thinking problems.

Computerized Test Bank IBM and Macintosh versions of the entire Test Bank are available with full editing features to help the instructor customize tests.

Instructor's Solutions Manual by Alison Hyslop of St. John's University. Contains worked-out solutions to all end of chapter problems.

Digital Image Archive Text web site includes downloadable files of text images in JPG format. Instructors may use these images to customize their presentations and to provide additional visual support for quizzes and exams.

Power Point Lecture Slides by Nancy Mullins of Jacksonville Community College. Featuring images from the text, the slides are customizable to fit your course.

Personal Response Systems/"Clicker" Questions. A bank of questions is available for anyone using personal response systems technology in their classroom.

WileyPLUS with CATALYST

WileyPlus with CATALYST partners with the instructor to teach students how to think their way through a problem, rather than rely on a list of memorized equations, by placing a strong emphasis on developing problem-solving skills and conceptual understanding. *WileyPlus* with CATALYST incorporates an online learning system designed to facilitate dynamic learning and retention of learned concepts. CATALYST was developed by Dr. Patrick Wegner (California State University, Fullerton) to promote conceptual understanding and visualization of chemical phenomena in undergraduate chemistry courses.

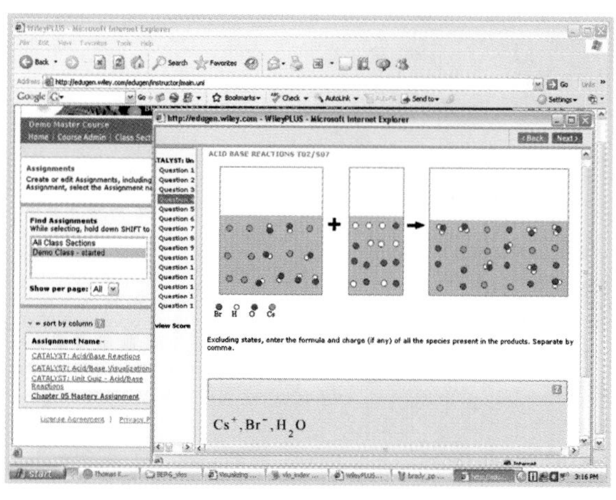

CATALYST assignments have multiple levels of parameterization and test on key concepts from multiple points of view (visual, symbolic, graphical, quantitative). Hundreds of end-of-chapter problems are available for assignment, and all are available with multiple forms of problem-solving support.

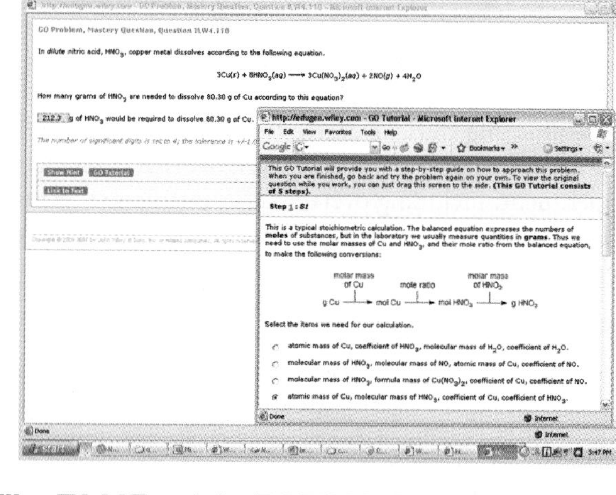

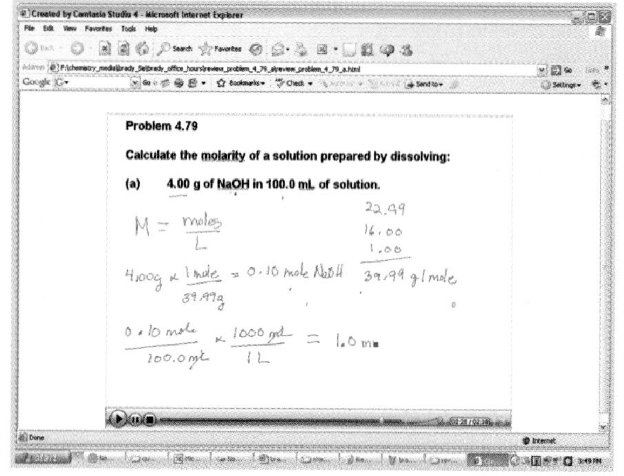

WileyPLUS with CATALYST resources

For Students

Office Hours Videos are video worked examples based on end-of-chapter problems.

Chem FAQs give alternate presentations for topics in the book using interactive examples, virtual experiments, animations, and learning games. They guide students to construct knowledge themselves through a series of questions.

Skill Building Tutorials are animated presentations of key concepts, problem-solving strategies, and fundamental tools for approaching general chemistry

Interactive LearningWare are interactive exercises that allow students to work through the solutions to key end of chapter problems

Study Guide. The Study Guide further enhances understanding of text concepts and contains additional worked examples giving detailed steps involved in solving them, additional problems and questions, chapter overviews, alternate problem-solving approaches, and extensive review exercises.

Video Demonstrations are short video clips illustrating chemical principles and laboratory procedures.

Conceptual Self Quizzes feature detailed feedback based on common errors students make.

Additional Resources • 3D Molecules • Audio Pronunciation Guide • Key Tables

For Instructors

A description of each of the following resources can be found on page xv.

- Instructor's Solutions Manual
- Test Bank
- Digital Image Archive
- PowerPoint Lecture Slides

ACKNOWLEDGMENTS

We begin with words of welcome to Neil Jespersen, a respected analytical chemist, colleague at St. John's University, and close friend for many years. Neil's collaboration on this edition has added a fresh perspective on teaching general chemistry. His contributions have enhanced the quality of presentations in many places throughout the book.

We express our fond thanks to our wives, June Brady, Marilyn Jespersen, and Lori Senese, and our children, Mark and Karen Brady, Lisa Fico and Kristen Pierce, and Kai and Keiran Senese, for their constant support, understanding, and patience. They have been, and continue to be a constant source of inspiration for us all.

We deeply appreciate the contributions of others who have helped in preparing materials for this edition, in particular Alison Hyslop of St. John's University for her diligence in preparing the Answer Appendix, and Conrad Bergo of East Stroudsburg University for checking the answers for accuracy. We thank Mike Borut for his thoughtful suggestions regarding various aspects of the text. We thank Dixie Goss, of Hunter College, for her skillful creation of the Office Hours videos.

It is with particular pleasure that we thank the staff at Wiley for their careful work, encouragement, and sense of humor, particularly our editors, Jennifer Yee and Stuart Johnson. We are also grateful for the efforts of Marketing Manager Amanda Wainer, our Editorial Program Assistant, Catherine Donovan, Senior Media Editor, Thomas Kulesa, our Photo Editor, Jennifer MacMillan, our Designer, Madelyn Lesure, our Illustration Editor, Anna Melhorn, the entire production team, and especially Elizabeth Swain for her tireless attention to getting things right. Our thanks also go to Pietro Paolo Adinolfi and others at Preparé (the Compositor) for their unflagging efforts toward changing a manuscript into a book.

We express gratitude to the colleagues whose careful reviews, helpful suggestions, and thoughtful criticism of previous editions as well as the current edition manuscript have been so important in the development of this book. Additional thanks go to those who participated in the media development by creating content and reviewing extensively. Our thanks go out to the reviewers of previous editions and of the current edition, and to authors and reviewers of the supporting media package:

Hugh Akers
Lamar University

Robert D. Allendoerfer
State University of New York, Buffalo

Patricia Amateis
Virginia Polytechnic Institute

Mark Amman
Alfred State College

David Anderson
University of Colorado

Dale Arlington
South Dakota School of Mines and Technology

George C. Bandik
University of Pittsburgh

Wesley Bentz
Alfred University

Conrad Bergo
East Stroudsburg University

Mark A. Benvenuto
University of Detroit–Mercy

Keith O. Berry
Oklahoma State University

William Bitner
Alvin Community College

Neal Boehnke
Jacksonville University

Simon Bott
University of North Texas

Donald Brandvold
New Mexico Institute of Mining and Technology

Timothy Brewer
Eastern Michigan University

Brian P. Buffin
University of Michigan, Flint

Robert F. Bryan
University of Virginia

Steven W. Buckner
Columbus State University

C. Eugene Burchill
University of Manitoba

Barbara A. Burke
California State Polytechnic University-Pomona

Jerry Burns
Pellissippi State Technical College

Jidhyt Burstyn
University of Wisconsin

Sheila Cancella
Raritan Valley Community College

Deborah Carey
Stark Learning Center

Tara S. Carpenter
University of Maryland, Baltimore County

Charles Carraher
Florida Atlantic University

Jefferson D. Cavalieri
Dutchess Community College

Laura Chaudhury
Broward Community College

Michael Chetcuti
University of Notre Dame

Ronald J. Clark
Florida State University

Wendy Clevenger
University of Tennessee at Chattanooga

Paul S. Cohen
The College of New Jersey

Kathleen Crago
Loyola University

Jason D'Acchioli
University of Wisconsin, Stevens Point

Henry Daley
Bridgewater State College

Diana Daniel
General Motors Institute

John E. Davidson
Eastern Kentucky University

William Davies
Emporia State University

William Deese
Louisiana Tech University

David F. Dever
Macon College

Gregg R. Diekmann
University of Texas at Dallas

Bonnie Dixon
University of Maryland

David Dobberpuhl
Creighton University

Joseph Dreisbach
University of Scranton

Barbara Drescher
Middlesex County College

Wendy Elcesser
Indiana University of Pennsylvania

William B. Euler
University of Rhode Island

James Farrar
University of Rochester

Dongling Fei
Manatee Community College

John H. Forsberg
St. Louis University

David Frank
Ferris State University

Donna Jean A. Fredeen
Southern Connecticut State University

Donna Friedman
Florrisant Valley Community College

Linda Galang
Madison Area Technical College

Ronald A. Garber
California State University, Long Beach

Paul Gaus
College of Wooster

John I. Gelder
Oklahoma State University

Jim Giles
Mesa Community College

David B. Green
Pepperdine University

Thomas Greenbowe
Iowa State University

Tammy S. Gummersheimer
Schenectady County Community College

Michael Guttman
Miami-Dade Community College

Peter Hambright
Howard University

Paul Hanson
University of New Orleans

Henry Harris
Armstrong State College

Mark Harris
Lebanon Valley College

Daniel T. Haworth
Marquette University

Harlon J. Hawthorne
Broward Community College

Sherell Hickman
Brevard Community College

Craig Hoag
SUNY Plattsburgh

Carl A. Hoeger
University of California, San Diego

Jason Hofstein
Sienna College

Paul A. Horton
Indian River Community College

Thomas Huang
East Tennessee State University

Dan Huchital
Florida Atlantic University

Alison Hyslop
St. John's University

Peter Iyere
Tennessee State University

Denley Jacobson
Purdue University

Andrew Jorgensen
University of Toledo

George Kaminski
Central Michigan University

Wendy L. Keeney-Kennicutt
Texas A & M University

Janice Kelland
Memorial University of Newfoundland

Henry C. Kelly
Texas Christian University

Reynold Kero
Saddleback College

Ernest Kho
University of Hawaii-Hilo

Laura Kibler-Herzog
Georgia State University

Louis Kirschenbaum
University of Rhode Island

Nina Klein
Montana Tech

Larry Krannich
University of Alabama-Birmingham

Robert M. Kren
University of Michigan-Flint

Chandrika Kulatilleke
Baruch College

Russell D. Larsen
Texas Technical University

Gerald Lesley
Southern Connecticut State University

Melvin Lesley
Southern Connecticut University

Shari J. Lillard
California State University, Northridge

Patrick Lloyd
Kingsborough Community College

Ken Loach
SUNY-Plattsburgh

Glen Loppnow
University of Alberta

Steve Lower-Simon
Fraser University

Sunil Malapati
Ferris State University

David Marten
Westmont College

Barbara McGoldrick
Union County College

Cathy MacGowan
Armstrong Atlanta State University

Garrett J. McGowan
Alfred University

Michael McIntire
University of Wisconsin, Green Bay

Sara Leslie McIntosh
Rensselaer Polytechnic Institute

Jeanette Medina
SUNY-Geneseo

William A. Meena
Rock Valley College

Patrick Meyer
Grand Valley State University

Stephen Mezyk
California State, Long Beach

Jalal U. Mondal
University of Texas, Pan American

Chad Morris
University of Vermont

Barbara Mowery
Thomas Nelson Community College

Patricia Moyer
Phoenix College

Nancy Mullins
Florida Community College, Jacksonville

Robert Nakon
West Virginia University

Alex Nazarenko
Buffalo State University

Edward Neth
University of Connecticut

Anne-Marie Nickel
Milwaukee School of Engineering

James Niewahner
Northern Kentucky University

Brian Nordstrom
Embry-Riddle Aeronautical University

Sabrina Godfrey Novick
Hofstra University

Robert H. Paine
Rochester Institute of Technology

Naresh Pandya
Kapiolani Community College

Cynthia Peck
Delta College

Lee Pedersen
University of North Carolina, Chapel Hill

James Penner Hahn
University of Michigan

Les Pesterfield
Western Kentucky University

Giuseppe Petrucci
University of Vermont

Casey Raymond
SUNY Oswego

Jason Ribbett
Ball State University

Michelle Richards-Babb
West Virginia University

Nina Rokainen
Benedictine University

Richard J. Rosso
St. John's University

Alan Sadurski
Ohio Northern University

Jerry L. Sarquis
Miami University

Lisa Seagraves
Broward Community College

Paula Secondo
Western Connecticut State University

Ronald See
St. Louis University

Karl Seff
University of Hawaii

Edward Senkbeil
Salisbury State University

Venkatesh Shanbhag
Mississippi State University

Ralph W. Sheets
Southwest Missouri State University

John Sheriden
Rutgers University – Newark

David Shinn
University of Hawaii

Anton Shurpik
U.S. Merchant Marine Academy

John W. Sibert
University of Texas at Dallas

Reuben Simoyi
West Virginia University

Mary Sohn
Florida Institute of Technology

Thomas E. Sorensen
University of Wisconsin, Milwaukee

S. Paul Steed
Sacramento City College

Darel Straub
University of Pittsburgh

Agnes Tenney
University of Portland

Wayne Tikkanen
California State, Los Angeles

Roselin Wagner
Hofstra University

David White
State University of New York, New Paltz

S.D.Worley
Auburn University

Warren Yeakel
Henry Ford Community College

David Young
Ohio University

Jose Zambrana
Queens College

E. Peter Zurbach
St. Joseph's University

The college level course in chemistry that you are about to begin will be one of the most challenging in your academic career. Successfully meeting this challenge will have its own rewards. One of these rewards is that you will have earned an essential foundation for any one of a large selection of technical career paths including art conservation, basic chemical research, environmental science, medicine, and veterinary science. Another will be the confidence you gain as you understand the chemical world around you and learn how you can make informed decisions. Finally, the discipline you gain in the study of chemistry, regular reading before class, working problems in a logical manner and thinking through the larger concepts, will carry over and have a positive effect on all of your studies and beyond. Let's take a look at some of the specific attributes of this book so that you can utilize them fully.

The language of chemistry

Every job description, from plumber to pharmacist, comes with its own vocabulary that helps the practitioner communicate with colleagues quickly and efficiently. Chemistry is no different. There is a large body of new words and phrases that need to be remembered. This will make it easier to learn the concepts of chemistry. To help all users of the text, even those who have never had a previous chemistry course, we highlight in bold text the first appearance of all new terms. These terms are defined in the glossary and can be accessed on-line by clicking on the word or phrase. Looking up a term in the glossary will lead you to the section(s) that use the term in a significant way. Learning the vocabulary of chemistry and using it appropriately is an important step in understanding and applying the ideas and concepts of this fundamental science.

Problem solving reinforces concepts

You may hear that chemistry involves solving many problems, and that is true. The reason for all of the problem solving is that chemistry is built on certain fundamental concepts describing how the physical world works. Problems are designed to reinforce these concepts and to illustrate how, in many instances, several concepts can be combined to obtain useful results. As a modern student you have a calculator and access to a computer to do mathematical calculations. Some of these take mere seconds to perform compared to hours that students previously spent on the same calculations. This just emphasizes the fact that problem solving in chemistry is NOT just mathematical calculation. In fact, we present the calculation part of a problem as a small segment of an overall thought process.

The thought process involves three steps, ***analysis, solution,*** and deciding ***is the answer reasonable?***. The first step is an ***analysis*** of the problem. What specifically does the problem ask us to do? What concepts and laws are involved in the question? If a calculation is involved, what data is given, and where can additional facts be found? What tools for problem solving have we learned that can help us? What sequence of logical steps are needed for our calculation? Then there is the ***solution*** step, where the information in the analysis step is put together in a logical fashion to develop an answer. A reasoned solution includes understanding how you arrived at the answer. For math problems it includes the correct use of significant figures. Finally, it is important to ask ***is the answer reasonable?*** By the time you finish with the solution, it is a natural tendency to feel that the problem is finished. However, if you start with one drop of reactant and end up with a swimming pool of product, something is obviously wrong. That is a much different error from one where you calculate 2.65 g of product and the actual answer is 2.68 g. Each problem in this book illustrates how you can check your answers to avoid large errors that will take you off course. We provide a variety of ways to check your answers as illustrations throughout the book so that you can see which method might be best for you. The three-step method of ***analysis, solution,*** and ***is the answer reasonable?*** works well in chemistry and also in many other subjects.

TOOLS FOR PROBLEM SOLVING

Earlier we mentioned that within the plumbing profession there is a specialized vocabulary. Plumbers also have specialized tools used to perform their jobs. In developing proficiency in their craft, the professional plumber must learn to use the proper tool for the job at hand. We can visualize the concepts, laws, and equivalencies of chemistry as tools to do our job of solving problems.

Throughout this book we point out (with a tool symbol as shown in the margin) the tools that should be thoroughly understood in order to solve problems. These tools will be especially helpful for problems within the chapter and will often pop up again in additional problems in the rest of the book. To help with the tools, we collect them at the end of each chapter so you can review them quickly to determine which tools will apply to your problem. If you don't find a tool in a chapter, look in previous chapters. We have made a special effort to never ask questions before you have all of the needed information to solve the problem.

We place special emphasis on problem solving in this text. Along with laboratory work, it is the preferred method for learning and applying the concepts of chemistry. By applying concepts to a variety of situations and often combining two or more concepts in a single problem we see the true richness of the science of chemistry.

Problem solving is a skill that needs to be developed in a logical progression. Each chapter has many ***Examples*** that give detailed solutions using our three-step process. Often, the step-by-step algebraic process is given and the cancellation of units is clearly shown in color. Immediately after the examples are the ***Practice Exercises.*** These reinforce the just-solved example and the first practice exercise usually includes a hint to help you get started. You will find answers to all the Practice Exercises in Appendix B at the back of the book. Finally, we have the end-of-chapter ***Questions, Problems, and Exercises.*** The questions often ask non-numerical questions about the material in the chapter. The problems tend to be numerical and they are presented in pairs of similar problems, one of which has the answer in Appendix B. Finally, the exercises present more difficult problems and open-ended ***Critical Thinking*** questions that often require use of reference material outside this book or on the internet.

Certain end-of-chapter problems are designated with an asterisk (*) to indicate that they are more difficult. Other questions have the symbol **ILW**, indicating that they are available as an Interactive Learningware problem, which can be analyzed and solved online within a stepped-out tutorial. Other questions are marked with the symbol **OH**, to indicate that they are available with an Office Hours video to help guide you through the problem. Within Office Hours, problems are solved by an instructor giving details and insights that are not possible in the printed text. Finally, we hope that your instructor will set up on-line problem sets using the WileyPlus with CATALYST system. Here you will be able to solve end-of-chapter problems and get instant feedback on whether or not your answers are correct. Problem-solving assistance and support is available with each question. With WileyPlus with CATALYST, you will be able to test your problem-solving skills by tackling a variety of unique questions each time you access the system.

FACETS OF CHEMISTRY

BRIEF CONTENTS

CONTENTS

1 FUNDAMENTAL CONCEPTS AND UNITS OF MEASUREMENT

Mike Peterson #54 of the Jacksonville Jaguars sacks Houston quarterback David Carr #8. Athletes such as these find their sports safer than ever before thanks to high tech materials made possible through chemical research. Most of the materials used in their uniforms, helmets, and protective pads do not occur naturally in our world and would not exist without discoveries made by observant chemists. The fruits of chemical science touch all our lives every day in ways most of us rarely think of. *(Lisa Blumenfeld/Getty Images/NewsCom.)*

CHAPTER OUTLINE

THIS CHAPTER IN CONTEXT Chemistry is a science that has impacted every aspect of our lives. We have come to take for granted so many of the materials, discovered by chemists, that make us comfortable, provide for our entertainment, and ensure that the foods we place on our tables are fresh and wholesome. Most of the medicines to cure disease and relieve pain, and nearly all of the objects used by doctors in hospitals, would not exist if chemists had not synthesized the materials from which they are made. As we guide you through the study of chemistry, we will provide numerous examples of how this subject relates to the world in which we live. Our aim is to give you an appreciation of the significant role that chemistry plays in modern society.

This chapter has three principal goals. The first is to provide you with an appreciation of the central role that chemistry plays among the sciences. The second is to have you understand the way scientists approach the study of nature and how they construct mental pictures of the microscopic world to explain the results of experimental observations. And third, we will begin to discuss the principal substances that serve as building blocks for all the materials we encounter in our daily lives.

If you've had a prior course in chemistry, perhaps in high school, you're likely to be familiar with many of the topics that we cover in this chapter. Nevertheless, it is important to be sure you have a mastery of these subjects, because if you don't start this course with a firm understanding of the basics, you may find yourself in trouble later on.

1.1 CHEMISTRY IS IMPORTANT FOR ANYONE STUDYING THE SCIENCES

In our discussions, we do not assume that you have had a prior course in chemistry. However, we do urge you to study this chapter thoroughly, because the concepts developed here will be used in later chapters.

Chemistry[1] is the study of the composition and properties of matter, which includes all of the chemicals that make up tangible things, from rocks to people to pizza. Chemists search for answers to fundamental questions about the effect of a substance's composition on its properties. They also seek to learn the way substances change, often dramatically, when they interact with each other in *chemical reactions*. And permeating all of this is a search for knowledge about the basic underlying structure of matter and the forces that determine the properties we observe through our senses. From these studies has come the ability to create materials never before found on earth, materials with especially desirable properties that fulfill specific needs of society. This knowledge has also enabled biologists to develop a fundamental understanding of many of the processes taking place in living organisms.

Although you may not plan to be a chemist, some knowledge of chemistry will surely be valuable to you. In fact, the involvement of chemistry among the various branches of science is evidenced by the names of some of the divisions of the American Chemical Society, the largest scientific organization in the world (see Table 1.1).

TABLE 1.1 Names of Some of the Divisions of the American Chemical Society

Agricultural & Food Chemistry	Computers in Chemistry
Agrochemicals	Environmental Chemistry
Biochemical Technology	Fuel Chemistry
Biological Chemistry	Geochemistry
Business Development & Management	Industrial & Engineering Chemistry
Carbohydrate Chemistry	Medicinal Chemistry
Cellulose and Renewable Materials	Nuclear Chemistry & Technology
Chemical Health & Safety	Petroleum Chemistry
Chemical Toxicology	Polymer Chemistry
Chemistry & the Law	Polymeric Materials: Science & Engineering
Colloid & Surface Chemistry	Rubber

[1] Important terms will be set in bold type to call them to your attention. Be sure you learn their meanings.

1.2 THE SCIENTIFIC METHOD HELPS US BUILD MODELS OF NATURE

Scientists who work in university, industrial, and government laboratories follow a general approach to their work called the **scientific method.** In very simple terms, it is a cyclical process in which we gather and assemble information about nature, formulate explanations for what we've observed, and then test the explanations with new experiments.

In the sciences, we usually gather information by performing experiments in laboratories under controlled conditions so observations we make are reproducible (Figure 1.1). An **observation** *is a statement that accurately describes something we see, hear, taste, feel, or smell.*

Observations gathered during an experiment often lead us to make conclusions. A **conclusion** *is a statement that's based on what we think about a series of observations.* For example, consider the following statements about the fermentation of grape juice to make wine:

1. Before fermentation, grape juice is very sweet and contains no alcohol.
2. After fermentation, the grape juice is no longer as sweet and it contains a great deal of alcohol.
3. In fermentation, sugar is converted into alcohol.

Statements 1 and 2 are observations because they describe properties of the grape juice that can be tasted and smelled. Statement 3 is a conclusion because it *interprets* the observations that are available.

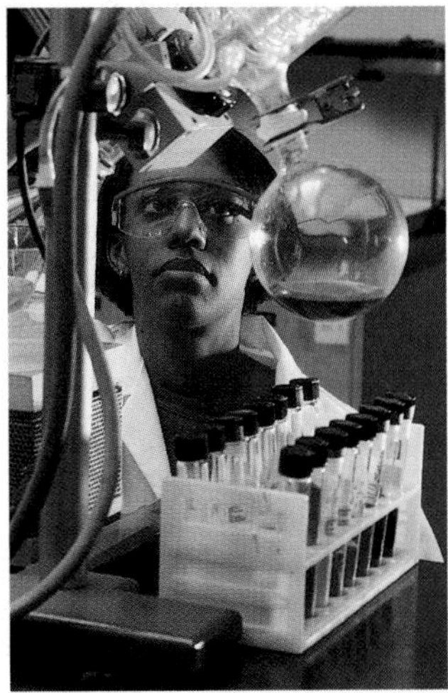

FIG. 1.1 A scientist working in a chemical research laboratory. Reproducible conditions in a laboratory permit experiments to yield reliable results. *(Index Stock.)*

Experimental observations lead to scientific laws

Observations we make while performing experiments are referred to as **data.** For example, if we study the behavior of gases, such as the air we breathe, we soon discover that the volume of a gas depends on a number of factors, including the mass of the gas, its temperature, and its pressure. The observations we record relating these factors are our data.

One of the goals of science is to organize facts so that relationships or generalizations among the data can be established. For instance, one generalization we would make from our observations is that when the temperature of a gas is held constant, squeezing the gas into half its original volume causes the pressure of the gas to double. If we were to repeat our experiments many times with numerous different gases, we would find that this generalization is uniformly applicable to all of them. Such a broad generalization, based on the results of many experiments, is called a **law** or **scientific law.**

We would say that the pressure of the gas is inversely proportional to its volume; the smaller the volume, the larger the pressure.

We often express laws in the form of mathematical equations. For example, if we represent the pressure of a gas by the symbol P and its volume by V, the inverse relationship between pressure and volume can be written as

$$P = \frac{C}{V}$$

where C is a proportionality constant. (We will discuss gases and the laws relating to them in greater detail in Chapter 10.)

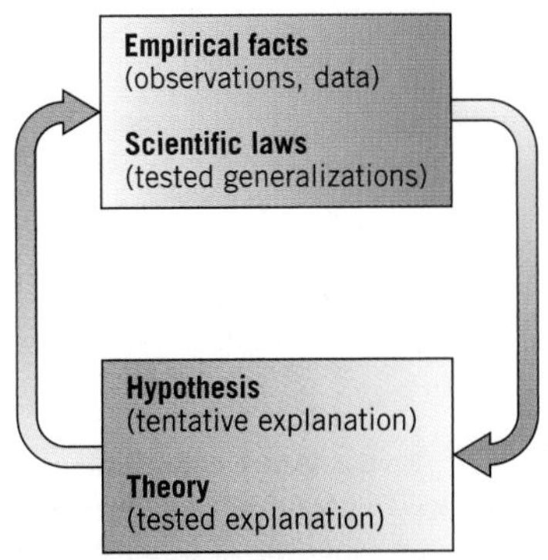

FIG. 1.2 The scientific method is cyclical. Observations suggest explanations, which suggest new experiments, which suggest new explanations, and so on.

Hypotheses and theories are models of nature

As useful as they may be, laws only state what happens; they do not provide explanations. *Why*, for example, are gases so easily compressed to a smaller volume? More specifically, *what must gases be like at the most basic, elementary level for them to behave as they do?* Answering such questions when they first arise is no simple task and requires much speculation. But over time scientists build mental pictures, called **theoretical models,** that enable them to explain observed laws.

In the development of a theoretical model, researchers form tentative explanations called **hypotheses** (Figure 1.2). They then perform experiments that test predictions derived from the model. Sometimes the results show the model is wrong. When this happens, the model must be abandoned or modified to account for the new data. Eventually, if the model survives repeated testing, it achieves the status of a theory. *A* **theory** *is a tested explanation of the behavior of nature*. You should keep in mind, however, that it is impossible to perform every test that might show a theory to be wrong, so we can never prove *absolutely* that a theory is correct.

Science doesn't always proceed in the orderly stepwise fashion described above. Luck sometimes plays an important role. For example, in 1828 Frederick Wöhler, a German chemist, was testing one of his theories and obtained an unexpected material when he heated a substance called ammonium cyanate. Out of curiosity he analyzed it and found it to be urea (a component of urine). This was exciting because it was the first time anyone had knowingly made a substance produced only by living creatures from a chemical not having a life origin. The fact that this could be done led to the beginning of a whole branch of chemistry called *organic chemistry*. Yet, had it not been for Wöhler's curiosity and his application of the scientific method to his unexpected results, the importance of his experiment might have gone unnoticed.

Many breakthrough discoveries in science have come about by accident.

As a final note, it is significant that the most spectacular and dramatic changes in science occur when major theories are proved to be wrong. Although this happens only rarely, when it occurs, scientists are sent scrambling to develop new theories, and exciting new frontiers are opened.

The atomic theory is a model of nature

Virtually every scientist would agree that the most significant theoretical model of nature ever formulated is the atomic theory. According to this theory, which we will discuss further in Chapter 2, all chemical substances are composed of tiny particles that we call **atoms.** Individual atoms combine in diverse ways to form more complex particles called **molecules.** Consider, for example, the substance water. Experimental evidence suggests that water molecules are each composed of two atoms of hydrogen and one of oxygen. To aid in our understanding and to help visualize how atoms combine, we often use drawings such as Figure 1.3. According to what we wish to emphasize, a variety of ways are used to describe the structures of molecules, as illustrated in Figure 1.4 for molecules of methane (the combustible fuel in natural gas).

Today we know a great deal about atoms and how they combine to form more complex materials. In coming chapters you will learn how we've come to apply this knowledge to making connections between what we physically observe in our large, *macroscopic* world and what we believe takes place in the tiny, submicroscopic world of atoms and molecules.

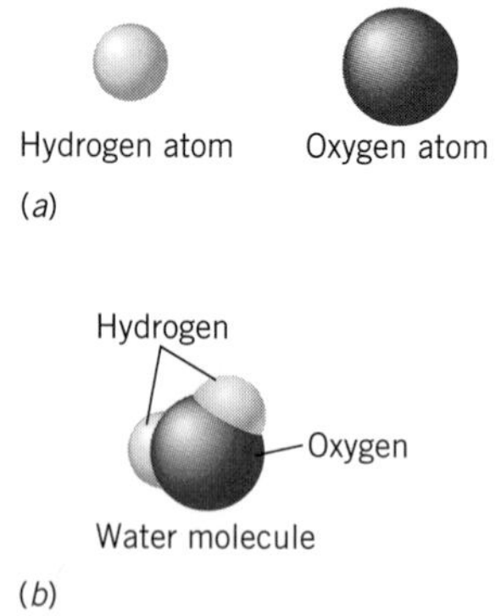

Hydrogen atom Oxygen atom
(*a*)

Hydrogen
Oxygen
Water molecule
(*b*)

FIG. 1.3 Atoms combine to form molecules. Illustrated here is a molecule of water, which consists of one atom of oxygen and two atoms of hydrogen. (*a*) Colored spheres are used to represent individual atoms, white for hydrogen and red for oxygen. (*b*) A drawing that illustrates the shape of a water molecule.

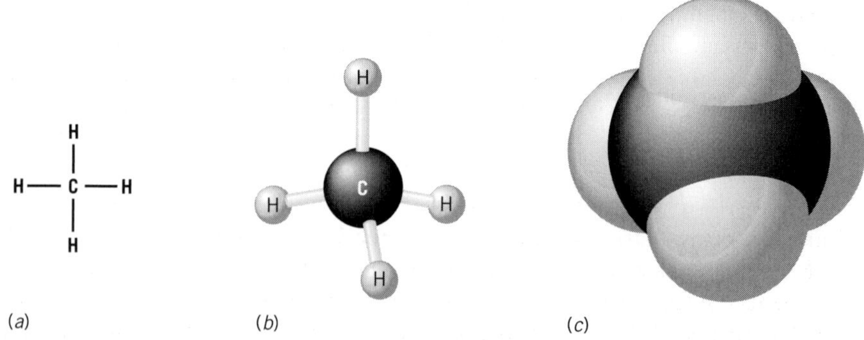

FIG. 1.4 Some of the different ways that the structures of molecules are represented. (*a*) A structure using chemical symbols to stand for atoms and dashes to indicate how the atoms are connected to each other. The molecule is methane, the substance present in natural gas that fuels stoves and Bunsen burners. A methane molecule is composed of one atom of carbon (C) and four atoms of hydrogen (H). (*b*) A *ball-and-stick model* of methane. The black ball is the carbon atom and the white balls are hydrogen atoms. (*c*) A *space-filling model* of methane that shows the relative sizes of the C and H atoms. Ball-and-stick and space-filling models are used to illustrate the three-dimensional shapes of molecules.

Learning to appreciate how chemists interpret behavior on a macroscopic level in terms of the composition of substances on an atomic scale should be one of your major goals in studying this course.

Macroscopic refers to objects large enough to be observed with the naked eye. Here we use the term to mean the things we observe with our senses, whether it be in the laboratory or in the world we encounter in our day-to-day living.

1.3 MATTER IS COMPOSED OF ELEMENTS, COMPOUNDS, AND MIXTURES

Earlier we described chemistry as being concerned with the properties and transformations of matter. **Matter** is *anything that occupies space and has mass.* It is the stuff our universe is made of, and all of the chemicals that make up tangible things, from rocks to pizza to people, are examples of matter.

In this definition, we've used the term *mass* rather than *weight.* The words mass and weight are often used interchangeably even though they refer to different things. **Mass** *refers to how much matter there is in a given object,*[2] *whereas* **weight** *refers to the force with which the object is attracted by gravity.* For example, a golf ball contains a certain amount of matter and has a certain mass, which is the same regardless of the golf ball's location. However, a golf ball on earth weighs about six times more than on the moon because the gravitational attraction of the earth is six times that of the moon. Because mass does not vary from place to place, we use mass rather than weight when we specify the amount of matter in an object. Mass is measured with an instrument called a balance, which we will discuss in Section 1.5.

Elements cannot be decomposed into simpler substances by chemical reactions

Chemistry is especially concerned with **chemical reactions,** which are *transformations that alter the chemical compositions of substances.* An important type of chemical reaction is **decomposition** in which one substance is changed into two or more others. For example, if we pass electricity through molten (melted) sodium chloride (salt), the silvery metal sodium and the pale green gas, chlorine, are formed. This change has decomposed sodium chloride into two simpler substances. No matter how we try, however, sodium and chlorine cannot be decomposed further by chemical reactions into still simpler substances that can be stored and studied.

In chemistry, *substances that cannot be decomposed into simpler materials by chemical reactions are called* **elements.** Sodium and chlorine are two examples. Others you may be familiar with include iron, aluminum, sulfur, and carbon (as in charcoal). Some elements are gases at room temperature. Examples include chlorine, oxygen, hydrogen, nitrogen, and helium. Elements are the simplest forms of matter that chemists work with directly. All more complex substances are composed of elements in various combinations.

Chemical symbols are used to identify elements

So far, scientists have discovered 90 existing elements in nature and have made 27 more, for a total of 117. Each element is assigned a unique **chemical symbol,** which can be used as an abbreviation for the name of the element. Chemical symbols are also used to stand for atoms of elements when we write *chemical formulas* such as H_2O (water) and CO_2 (carbon dioxide). We will have a lot more to say about formulas later.

In most cases, an element's chemical symbol is formed from one or two letters of its English name. For instance, the symbol for carbon is C, for bromine it is Br, and for silicon it is Si. For some elements, the symbols are derived from the non-English names given to those elements long ago. Table 1.2 contains a list of elements whose symbols come to us in that way.[3] Regardless of the origin of the symbol, the first letter is always capitalized and the second letter, if there is one, is always written lowercase. The names and chemical symbols of the elements are given on the inside front cover of the book.

[2] Mass is a measure of an object's momentum, or resistance to a change in motion. Something with a large mass, such as a truck, contains a lot of matter and is difficult to stop once it's moving. An object with less mass, such as a baseball, is much easier to stop.

[3] The symbol for tungsten is W, from the German name *wolfram.* This is the only element whose symbol is neither related to its English name nor derived from its Latin name.

TABLE 1.2 Elements That Have Symbols Derived from Their Latin Names

Element	Symbol	Latin Name	Element	Symbol	Latin Name
Sodium	Na	Natrium	Gold	Au	Aurum
Potassium	K	Kalium	Mercury	Hg	Hydrargyrum
Iron	Fe	Ferrum	Antimony	Sb	Stibium
Copper	Cu	Cuprum	Tin	Sn	Stannum
Silver	Ag	Argentum	Lead	Pb	Plumbum

Compounds are composed of two or more elements in fixed proportions

By means of chemical reactions, elements combine in various *specific proportions* to give all of the more complex substances in nature. Thus, hydrogen and oxygen combine to form water (H_2O), and sodium and chlorine combine to form sodium chloride (NaCl, common table salt). Water and sodium chloride are examples of compounds. A **compound** *is a substance formed from two or more* ***different elements*** *in which the elements are always combined in the same fixed (i.e., constant) proportions by mass.* For example, if any sample of pure water is decomposed, the mass of oxygen obtained is *always* eight times the mass of hydrogen. Similarly, when hydrogen and oxygen react to form water, the mass of oxygen consumed is always eight times the mass of hydrogen, never more and never less.

Mixtures can have variable compositions

Elements and compounds are examples of **pure substances.**[4] The composition of a pure substance is always the same, regardless of its source. Pure substances are rare, however. Usually, we encounter mixtures of compounds or elements. Unlike elements and compounds, **mixtures** *can have variable compositions.* For example, Figure 1.5 shows three mixtures that contain sugar. They have different degrees of sweetness because the amount of sugar in a given size sample varies from one to the other.

Mixtures can be either homogeneous or heterogeneous. A **homogeneous mixture** *has the same properties throughout the sample.* An example is a thoroughly stirred mixture of sugar in water. We call such a homogeneous mixture a **solution.** Solutions need not be liquids, just homogeneous. For example, the alloy used in the U.S. 5 cent coin is a solid solution of copper and nickel, and clean air is a gaseous solution of oxygen, nitrogen, and a number of other gases.

FIG. 1.5 Orange juice, Coca-Cola, and pancake syrup are mixtures that contain sugar. The amount of sugar varies from one to another because mixtures can have variable compositions. *(Thomas Brase/Stone/Getty Images; Andy Washnik; Andy Washnik.)*

[4] We have used the term *substance* rather loosely until now. Strictly speaking, **substance** really means *pure substance.* Each unique chemical element and compound is a *substance;* a mixture consists of two or more substances.

A **heterogeneous mixture** *consists of two or more regions, called* **phases,** *that differ in properties.* A mixture of olive oil and vinegar in a salad dressing, for example, is a two-phase mixture in which the oil floats on the vinegar as a separate layer (Figure 1.6). The phases in a mixture don't have to be chemically different substances like oil and vinegar, however. A mixture of ice and liquid water is a two-phase heterogeneous mixture in which the phases have the same chemical composition but occur in different *physical states* (a term we will discuss further in the next section).

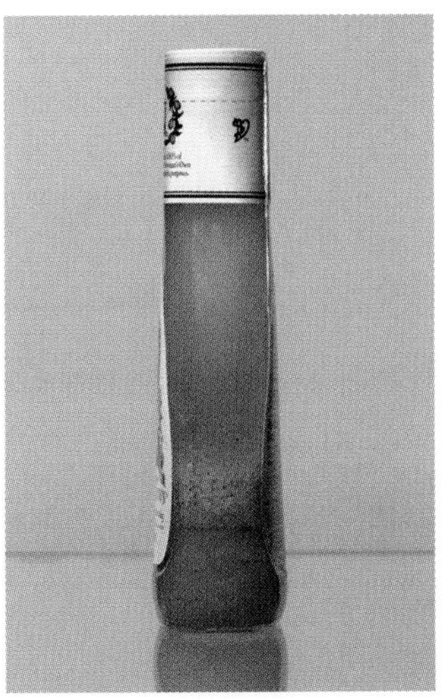

FIG. 1.6 A heterogeneous mixture. The salad dressing shown here contains vinegar and vegetable oil (plus assorted other flavorings). Vinegar and oil do not dissolve in each other; instead, they form two layers. The mixture is heterogeneous because each of the separate phases (oil, vinegar, and other solids) has its own set of properties that differ from the properties of the other phases. *(Andy Washnik.)*

(a)

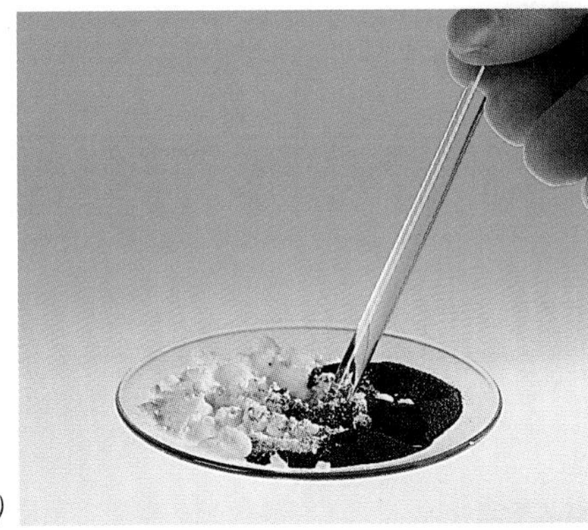

(b)

FIG. 1.7 Formation of a mixture of iron and sulfur. (*a*) Samples of powdered sulfur and powdered iron. (*b*) A mixture of sulfur and iron is made by stirring the two powders together. *(Michael Watson.)*

The process we use to create a mixture is said to involve a **physical change,** because *no new chemical substances form.* This is illustrated in Figure 1.7 for powdered samples of the elements iron and sulfur. By simply dumping them together and stirring, the mixture forms, but both elements retain their original properties. To separate the mixture, we could similarly use just physical changes. For example, we could remove the iron by stirring the mixture with a magnet—a physical operation. The iron powder sticks to the magnet as we pull it out, leaving the sulfur behind (Figure 1.8). The mixture also could be separated by treating it with a liquid called carbon disulfide, which is able to dissolve the sulfur but not the iron. Filtering the sulfur solution from the solid iron, followed by evaporation of the liquid carbon disulfide from the sulfur solution, gives the original components, iron and sulfur, separated from each other.

The formation of a compound involves a **chemical change** (chemical reaction) because *the chemical makeup of the substances involved are changed.* Iron and sulfur, for example, combine to form a compound often called "fool's gold" because of its appearance (Figure 1.9). In this compound the elements no longer have the same properties they had before they combined, and they cannot be separated by physical means. The decomposition of fool's gold into iron and sulfur is also a chemical reaction.

The relationships among elements, compounds, and mixtures are shown in Figure 1.10.

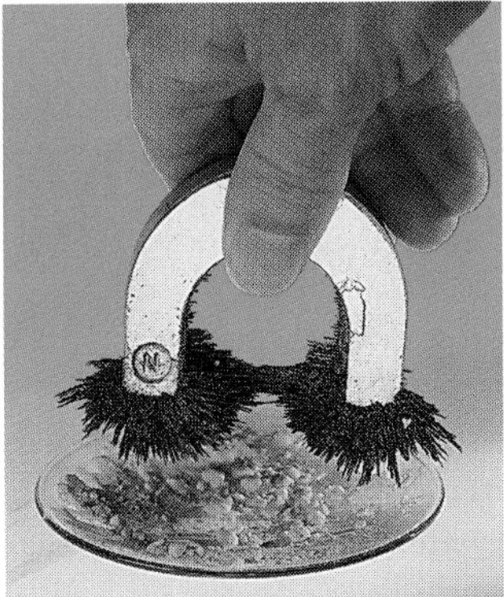

FIG. 1.8 Formation of a mixture is a physical change. Here we see that forming the mixture has not changed the iron and sulfur into a compound of the two elements. The mixture can be separated by pulling the iron out with a magnet. *(Michael Watson.)*

FIG. 1.9 "Fool's gold." The mineral pyrite (also called iron pyrite) has an appearance that caused some miners to mistake it for real gold. *(D. Harms/ Peter Arnold, Inc.)*

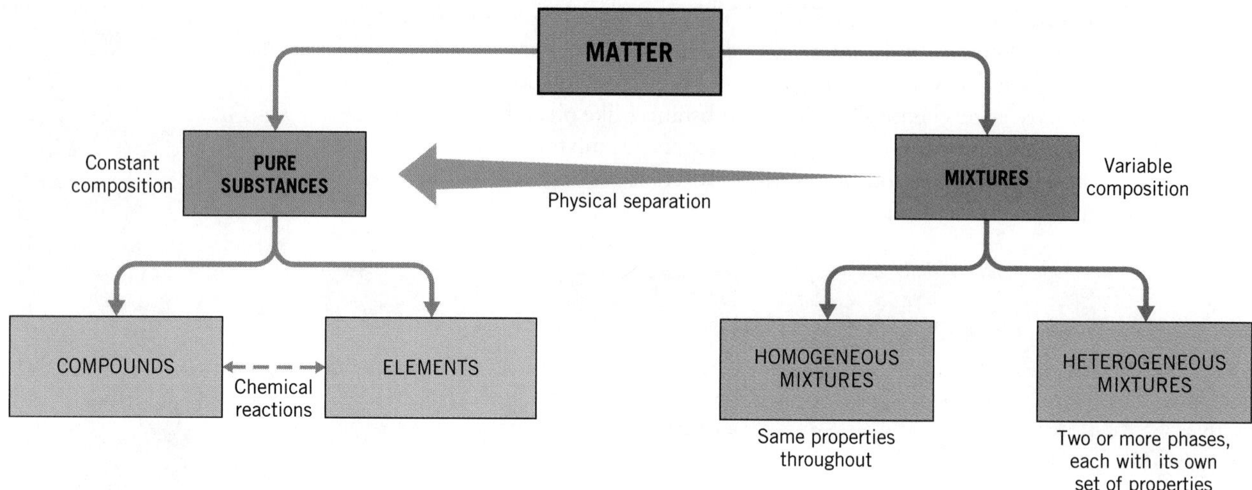

FIG. 1.10 Classification of matter.

1.4 Properties of Matter Can Be Classified in Different Ways

In chemistry we use **properties** (characteristics) of materials to identify them and to distinguish one kind from another. To help organize our thinking, we classify properties into different types.

Properties can be classified as physical or chemical

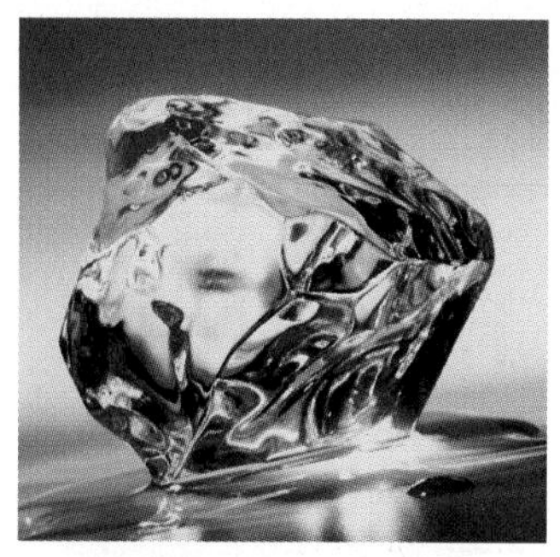

FIG. 1.11 Liquid water and ice are both composed of water molecules. Melting the ice cube doesn't change the chemical composition of the molecules. *(Susumu Sato/Corbis Images.)*

One way to classify properties is based on whether or not the chemical composition of an object is changed by the act of observing the property. *A* **physical property** *is one that can be observed without changing the chemical makeup of a substance.* For example, a physical property of gold is that it is yellow. The act of observing this property (color) doesn't change the chemical makeup of the gold. Neither does observing that gold conducts electricity, so color and electrical conductivity are physical properties.

Sometimes, observing a physical property does lead to a physical change. To measure the melting point of ice, for example, we observe the temperature at which the solid begins to melt (Figure 1.11). This is a physical change because it does not lead to a change in chemical composition; both ice and liquid water are composed of water molecules.

Solids, liquids, and gases are physical states of matter

Although ice, liquid water, and steam have quite different appearances and physical properties, they are just different forms of the same substance, water. **Solid, liquid,** and **gas** are the most common **states of matter.** As with water, most substances are able to exist in all three of these states, and the state we observe generally depends on the temperature. The obvious properties of solids, liquids, and gases can be interpreted at a submicroscopic level according to the different ways the individual atomic-size particles are organized (Figure 1.12). For a given substance, a change from one state to another is a physical change.

A chemical property describes a chemical change

A **chemical property** *describes a chemical change (chemical reaction) that a substance undergoes.* When a chemical reaction takes place, chemicals interact to form entirely *different* substances with different properties. An example is the rusting of iron, which involves a chemical reaction between iron, oxygen, and water. When the substances react, the product, rust, no longer looks like iron, oxygen, or water. It's a brown solid that isn't at all like a metal and it is not attracted by a magnet (Figure 1.13).

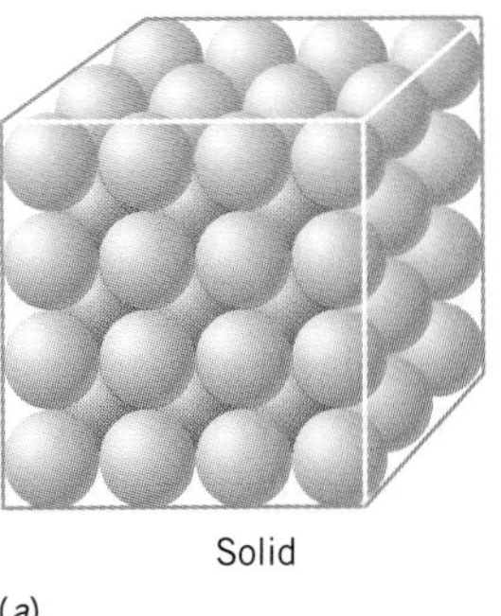

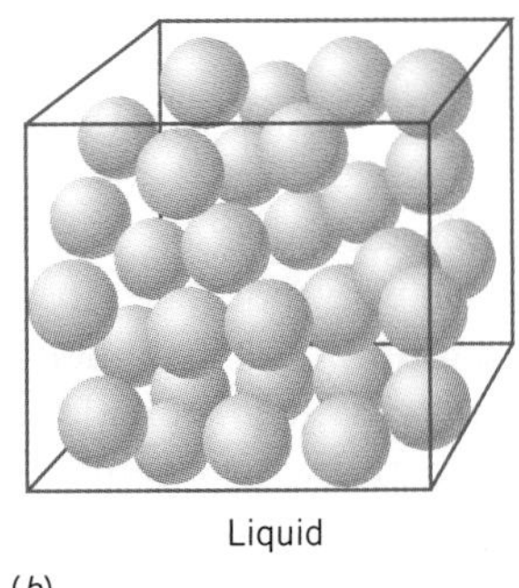

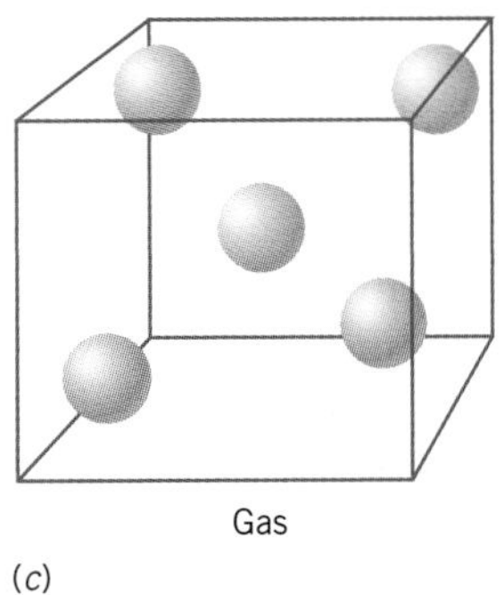

(*a*) (*b*) (*c*)

FIG. 1.12 **Solid, liquid and gaseous states of matter as viewed by the atomic model of matter.** (*a*) In a solid, the particles are tightly packed and cannot move easily. (*b*) In a liquid, the particles are still close together but can move past one another. (*c*) In a gas, the particles are far apart with much empty space between them.

The ability of iron to form rust in the presence of oxygen and moisture is a chemical property of iron. When we observe this property, the reaction changes the iron, oxygen, and water into rust, so after we've made the observation we no longer have the same substances as before.

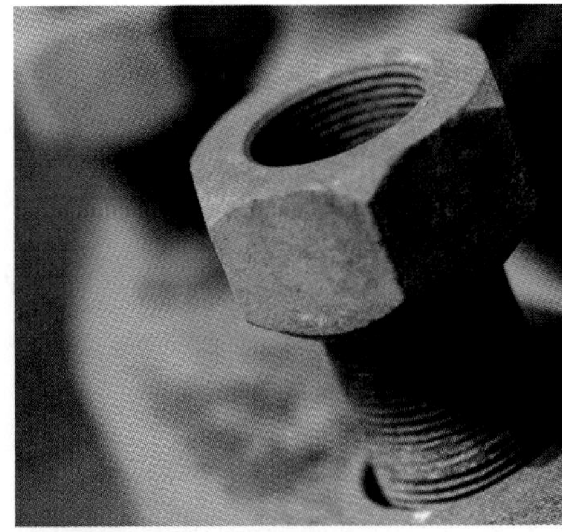

FIG. 1.13 **Chemical reactions cause changes in composition.** Here we see a coating of rust that has formed on an iron object. The properties and chemical composition of the rust are entirely different from those of the iron. *(George B. Diebold/ Corbis Images.)*

Properties can also be classified as intensive or extensive

Another way of classifying a property is according to whether or not it depends on the size of the sample under study. For example, two different pieces of gold can have different volumes, but both have the same characteristic shiny yellow color and both will begin to melt if heated to the same temperature. Volume is said to be an **extensive property**—*a property that depends on sample size.* Color and melting point (and boiling point, too) are examples of **intensive properties**—*properties that are independent of sample size.*

Some kinds of properties are better than others for identifying substances

A job chemists often perform is *chemical analysis.* They're asked, "What is a particular sample composed of?" To answer such a question, the chemist relies on the properties of the chemicals that make up the sample. For identification purposes, intensive properties are more useful than extensive ones because every sample of a given substance exhibits the same set of intensive properties.

Color, freezing point, and boiling point are examples of intensive physical properties that can help us identify substances. Chemical properties are also intensive properties and also can be used for identification. For example, gold miners were able to distinguish between real gold and fool's gold, a mineral also called pyrite (Figure 1.9, page 7), by heating the material in a flame. Nothing happens to the gold, but the pyrite sputters, smokes, and releases bad-smelling fumes because of its ability, when heated, to react chemically with oxygen in the air.

1.5 MEASUREMENTS ARE ESSENTIAL TO DESCRIBE PROPERTIES

Observations can be qualitative or quantitative

Earlier you learned that an important step in the scientific method is observation. In general, observations fall into two categories, qualitative and quantitative. **Qualitative observations,** such as the color of a chemical or that a mixture becomes hot when a reaction occurs, do not involve numerical information and are usually of limited value. More important are **quantitative observations,** or **measurements,** which do yield numerical data. You make such observations in everyday life, for example, when you glance at your watch, or step onto a bathroom scale. In chemistry, we make various measurements that aid us in describing both chemical and physical properties.

Measurements always include units

Measurements involve numbers, but they differ from the numbers used in mathematics in two crucial ways.

First, measurements always involve a comparison. When you say that a person is six feet tall, you're really saying that the person is six times taller than a reference object that is 1 foot high, where *foot* is an example of a **unit.** Both the number and the unit are essential parts of the measurement, because the unit gives the reported value a sense of size. For example, if you were told that the distance between two points is 25, you would naturally ask "25 what?" The distance could be 25 inches, 25 feet, 25 miles, or 25 of any other unit that's used to express distance. A number without a unit is really meaningless. *Writing down a measurement without a unit is a common and serious mistake, and one you should avoid.*

The second important difference is that measurements always involve uncertainty; they are *inexact.* The act of measurement involves an estimation of one sort or another, and both the observer and the instruments used to make the measurement have inherent physical limitations. As a result, measurements always include some uncertainty, which can be minimized but never entirely eliminated. We will say more about this topic in Section 1.6.

SI units are standard in science

A standard system of units is essential if measurements are to be made consistently. In the sciences, and in every industrialized nation on earth, metric-based units are used. The advantage of working with metric units is that converting to larger or smaller units can be done simply by moving a decimal point, because metric units are related to each other by simple multiples of ten.

In 1960, a simplification of the original metric system was adopted by the General Conference on Weights and Measures (an international body). It is called the **International System of Units,** abbreviated **SI** from the French name, *Le Système International d'Unités.* The SI is now the dominant system of units in science and engineering, although there is still some usage of older metric units.

The SI has as its foundation a set of **base units** (Table 1.3) for seven measured quantities. For now, we will focus on the base units for length, mass, time, and temperature. We will discuss the unit for amount of substance, the mole, at length in Chapter 3. The unit for electrical current, the ampere, will be discussed briefly when we study electrochemistry in Chapter 19. The unit for luminous intensity, the candela, will not be important to us in this book.

Most of the base units are defined in terms of reproducible physical phenomena. For instance, the meter is defined as exactly the distance light travels in a vacuum in 1/299,792,458 of a second. Everyone has access to this standard because light and a vacuum are available to all. Only the base unit for mass is defined by an object made by human hands—a carefully preserved platinum–iridium alloy block stored at the International Bureau of Weights and Measures in France (Figure 1.14). This block serves indirectly as the calibrating standard for all "weights" used for scales and balances in the world.[5]

FIG. 1.14 The international standard kilogram. This standard for mass in the SI is made of a platinum–iridium alloy and is kept at the International Bureau of Weights and Measures in France. Other nations such as the United States maintain their own standard masses that have been carefully calibrated against this international standard. *(Courtesy Bureau International des Poids et Mesures.)*

TABLE 1.3 The SI Base Units

Measurement	Unit	Symbol
Length	meter	m
Mass	kilogram	kg
Time	second	s
Electric current	ampere	A
Temperature	kelvin	K
Amount of substance	mole	mol
Luminous intensity	candela	cd

[5] Scientists are working on a method of accurately counting atoms whose masses are accurately known. Their goal is to develop a new definition of the kilogram that doesn't depend on an object that can be stolen, lost, or destroyed.

All SI units are built from the base units

The SI units for *any* physical quantity can be built from these seven base units. For example, there is no SI base unit for area, but we know that to calculate the area of a rectangular room we multiply its length by its width. Therefore, the *unit* for area is derived by multiplying the *unit* for length by the *unit* for width. Length (or width) is a base measured quantity in the SI and has the *meter* (m) as its base unit.

$$\text{length} \times \text{width} = \text{area}$$
$$(\text{meter}) \times (\text{meter}) = (\text{meter})^2$$
$$\text{m} \times \text{m} = \text{m}^2$$

The SI **derived unit** for area is therefore m^2 (read as *meters squared*, or *square meter*).

In deriving SI units, we employ a very important concept that we will use repeatedly throughout this book when we perform calculations: *Units undergo the same kinds of mathematical operations that numbers do.* We will see how this fact can be used to convert from one unit to another in Section 1.7.

EXAMPLE 1.1
Deriving SI Units

Linear momentum is a measure of the "push" a moving object has, equal to the object's mass times its velocity. What is the SI derived unit for linear momentum?

A Word about Problem Solving This is the first of many encounters you will have with solving problems in chemistry. Helping you learn how to approach and solve problems is one of the major goals of this textbook. We view problem solving as a three-step process. The first step is figuring out what has to be done to solve the problem, which is the function of the *Analysis* step described below. The second is actually performing whatever is required to obtain the answer (the *Solution* step). And finally, we examine the answer to determine whether it seems to be *reasonable.* For more information on the aids that are available to assist you in problem solving, we recommend that you read the "To the Student" section at the beginning of the book.

ANALYSIS: To derive a unit for a quantity we must first express it in terms of simpler quantities. We're told that linear momentum is mass times velocity. Therefore, the SI unit for linear momentum will be the SI unit for mass times the SI unit for velocity. The SI unit for mass is the kilogram (kg). Velocity is distance traveled (length) per unit time, so it has derived SI units of meters per second, m/s. Multiplying these units should give the derived unit for linear momentum.

SOLUTION:

$$\text{mass} \times \text{velocity} = \text{linear momentum}$$
$$\text{mass} \times \text{length/time} = \text{linear momentum}$$
$$\text{kilogram} \times \text{meter/second} = \text{kilogram meter/second}$$
$$\text{kg} \times \text{m/s} = \text{kg m/s}$$

IS THE ANSWER REASONABLE? *Before leaving a problem, it is always wise to examine the answer to see whether it makes sense.* For numerical calculations, ask yourself, "Is the answer too large, or too small?" Judging the answers to such questions serves as a check on the arithmetic as well as on the method of obtaining the answer and can help you find obvious errors. In this problem, the check is simple. The derived unit for linear momentum should be the product of units for mass and velocity, and this is obviously true. Therefore, our answer is correct.

Practice Exercise 1:[6] The volume of a sphere is given by the formula $V = \frac{4}{3}\pi r^3$, where r is the radius of the sphere. From this equation, determine the SI unit for volume. (Hint: r is a distance, so it must have a distance unit.)

[6] Answers to the Practice Exercises are found at the back of the book.

Practice Exercise 2: When you "step hard on the gas" in a car you feel an invisible force pushing you back in your seat. This force equals the product of your mass, m, times the acceleration, a, of the car. In equation form, this is $F = ma$. Acceleration is the change in velocity, v, with time, t:

$$a = \frac{\text{change in } v}{\text{change in } t}$$

Therefore, the units of acceleration are those of velocity divided by time. Velocity is a ratio of distance divided by time, $v = d/t$, so the units of velocity are those of distance divided by time. What is the SI derived unit for force expressed in SI base units?

We can construct SI units of any convenient size using decimal multipliers

Sometimes the basic units are either too large or too small to be used conveniently. For example, the meter is inconvenient for expressing the size of very small things such as bacteria. The SI solves this problem by forming larger or smaller units by applying **decimal multipliers** to the base units. Table 1.4 lists the most commonly used decimal multipliers and the prefixes used to identify them.

When the name of a unit is preceded by one of these prefixes, the size of the unit is modified by the corresponding decimal multiplier. For instance, the prefix *kilo-* indicates a multiplying factor of 10^3, or 1000. Therefore, a *kilo*meter is a unit of length equal to 1000 meters.[7] The symbol for kilometer (km) is formed by applying the symbol meaning kilo (k) as a prefix to the symbol for meter (m). Thus 1 km = 1000 m (or alternatively, 1 km = 10^3 m). Similarly a decimeter (dm) is 1/10 of a meter, so 1 dm = 0.1 m (1 dm = 10^{-1} m).

The symbols and multipliers listed in colored, boldface type in Table 1.4 are the ones most commonly encountered in chemistry.

SI prefixes

TABLE 1.4 Decimal Multipliers That Serve as SI Prefixes[a]

Prefix	Meaning	Symbol	Multiplication factor (fraction)	Multiplication factor (power of ten)
exa		E		10^{18}
peta		P		10^{15}
tera		T		10^{12}
giga		G		10^{9}
mega	**millions of**	**M**	**1,000,000**	$\mathbf{10^{6}}$
kilo	**thousands of**	**k**	**1000**	$\mathbf{10^{3}}$
hecto		h		10^{2}
deka		da		10^{1}
deci	**tenths of**	**d**	**0.1**	$\mathbf{10^{-1}}$
centi	**hundredths of**	**c**	**0.01**	$\mathbf{10^{-2}}$
milli	**thousandths of**	**m**	**0.001**	$\mathbf{10^{-3}}$
micro	**millionths of**	**μ**	**0.000001**	$\mathbf{10^{-6}}$
nano	**billionths of**	**n**	**0.000000001**	$\mathbf{10^{-9}}$
pico	**trillionths of**	**p**	**0.000000000001**	$\mathbf{10^{-12}}$
femto		f		10^{-15}
atto		a		10^{-18}

[a]Be sure you learn the prefixes shown in bold colored type.

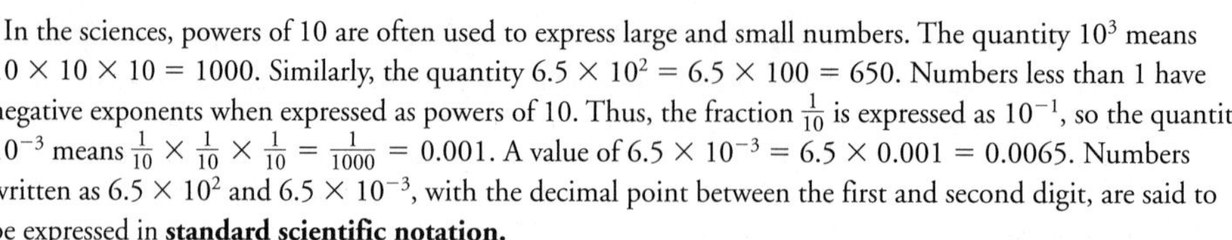
[7] In the sciences, powers of 10 are often used to express large and small numbers. The quantity 10^3 means $10 \times 10 \times 10 = 1000$. Similarly, the quantity $6.5 \times 10^2 = 6.5 \times 100 = 650$. Numbers less than 1 have negative exponents when expressed as powers of 10. Thus, the fraction $\frac{1}{10}$ is expressed as 10^{-1}, so the quantity 10^{-3} means $\frac{1}{10} \times \frac{1}{10} \times \frac{1}{10} = \frac{1}{1000} = 0.001$. A value of $6.5 \times 10^{-3} = 6.5 \times 0.001 = 0.0065$. Numbers written as 6.5×10^2 and 6.5×10^{-3}, with the decimal point between the first and second digit, are said to be expressed in **standard scientific notation.**

Non-SI units are still in common use

Some older metric units that are not part of the SI system are still used in the laboratory and in the scientific literature. Some of these units are listed in Table 1.5; others will be introduced as needed in upcoming chapters.

The United States is the only large nation still using the *English system* of units, which measures distance in inches, feet, and miles; volume in ounces, quarts, and gallons; and mass in ounces and pounds. However, a gradual transition to metric units is occurring. Beverages, food packages, tools, and machine parts are often labeled in both English and metric units (Figure 1.15). Common conversions between the English system and the SI are given in Table 1.6 and inside the rear cover of the book.[8]

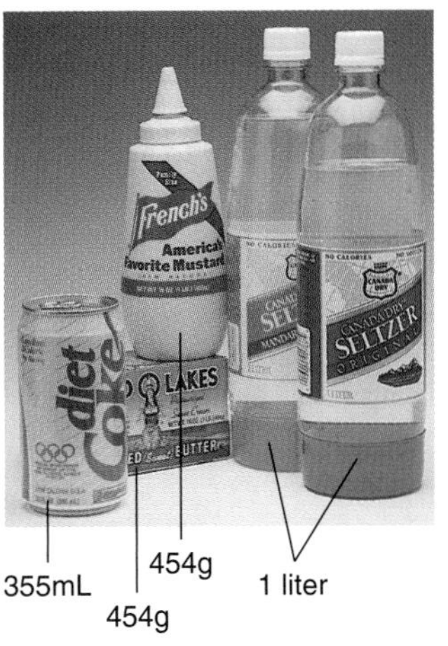

FIG. 1.15 Metric units are becoming commonplace on many consumer products. *(Michael Watson.)*

TABLE 1.5 Some Non-SI Metric Units Commonly Used in Chemistry

Measurement	Name	Symbol	Value in SI units
Length	angstrom	Å	1 Å = 0.1 nm = 10^{-10} m
Mass	atomic mass unit	u (amu)	1 u = 1.66054×10^{-27} kg, approximately
	metric ton	t	1 t = 10^3 kg
Time	minute	min	1 min = 60 s
	hour	h (hr)	1 h = 60 min = 3600 s
Temperature	degree Celsius	°C	Add 273.15 to obtain the Kelvin temperature
Volume	liter	L	1 L = 1000 cm^3

TABLE 1.6 Some Useful Conversions

Measurement	English to Metric	Metric to English
Length	1 in. = 2.54 cm	1 m = 39.37 in.
	1 yd = 0.9144 m	1 km = 0.6215 mi
	1 mi = 1.609 km	
Mass	1 lb = 453.6 g	1 kg = 2.205 lb
	1 oz = 28.35 g	
Volume	1 gal = 3.785 L	1 L = 1.057 qt
	1 qt = 946.4 mL	
	1 oz (fluid) = 29.6 mL	

We use several common units in laboratory measurements

Units for laboratory measurements

The most common measurements you will make in the laboratory will be those of length, volume, mass, and temperature.

Length

The SI base unit for length, the **meter (m),** is too large for most laboratory purposes. More convenient units are the **centimeter (cm)** and the **millimeter (mm).** They are related to the meter as follows.

$$1 \text{ cm} = 10^{-2} \text{ m} = 0.01 \text{ m}$$
$$1 \text{ mm} = 10^{-3} \text{ m} = 0.001 \text{ m}$$

It is also useful to know the relationships

$$1 \text{ m} = 100 \text{ cm} = 1000 \text{ mm}$$
$$1 \text{ cm} = 10 \text{ mm}$$

An older non-SI unit called the angstrom (Å) is often used to describe the dimensions of atomic and molecular sized particles: 1 Å = 0.1 nm = 10^{-10} m

[8] Originally, these conversions were established by measurement. For example, if a metric ruler is used to measure the length of an inch, it is found that 1 in. equals 2.54 cm. Later, to avoid confusion about the accuracy of such measurements, it was agreed that these relationships would be taken to be exact. For instance, 1 in. is now defined as *exactly* 2.54 cm. Exact relationships also exist for the other quantities, but for simplicity many have been rounded off. For example, 1 lb = 453.59237 g, *exactly*.

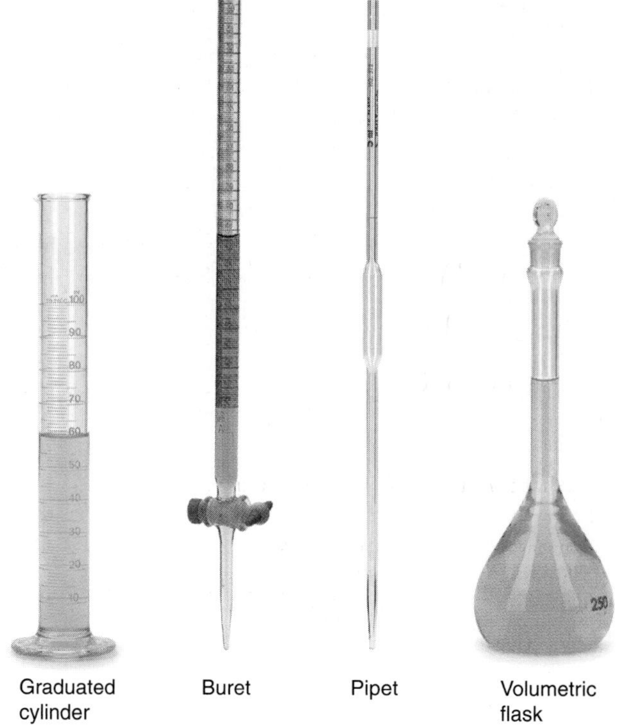

FIG. 1.16 Common laboratory glassware used for measuring volumes. Graduated cylinders are used to measure volumes to the nearest milliliter. Precise measurements of volume are made using burets, pipets, and volumetric flasks. *(Andy Washnik.)*

Volume

Volume is a derived unit with dimensions of (length)^3. With these dimensions expressed in meters, the derived SI unit for volume is the **cubic meter, m^3.**

In chemistry, measurements of volume usually arise when we measure amounts of liquids. The traditional metric unit of volume used for this is the **liter (L).** In SI terms, a liter is defined as exactly 1 cubic decimeter.

$$1 \text{ L} = 1 \text{ dm}^3$$

However, even the liter is too large to conveniently express most volumes measured in the lab. The glassware we normally use, such as that illustrated in Figure 1.16, is marked in **milliliters (mL).**[9]

$$1 \text{ L} = 1000 \text{ mL}$$

Because 1 dm = 10 cm, then 1 dm^3 = 1000 cm^3. Therefore, 1 mL is exactly the same as 1 cm^3.

$$1 \text{ cm}^3 = 1 \text{ mL}$$

$$1 \text{ L} = 1000 \text{ cm}^3 = 1000 \text{ mL}$$

Sometimes you may see cm^3 abbreviated cc (especially in medical applications), although the SI frowns on this symbol. Figure 1.17 compares the cubic meter, liter, and milliliter.

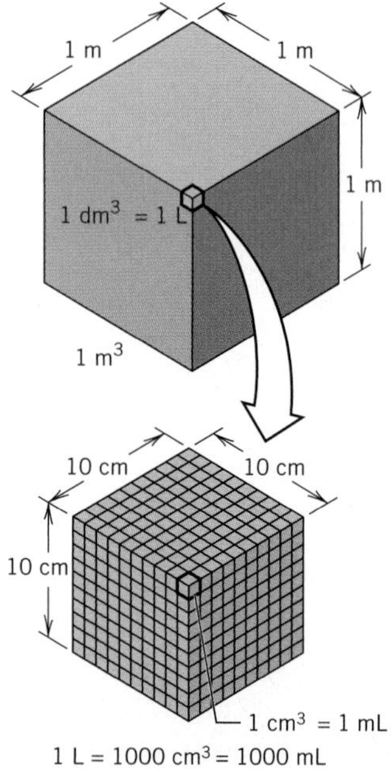

FIG. 1.17 Comparing volume units. A cubic meter (m^3) is approximately equal to a cubic yard, 1000 cm^3 is approximately a quart, and 1 cm^3 is approximately $\frac{1}{30}$ of a fluid ounce.

Mass

In the SI, the base unit for mass is the **kilogram (kg),** although the **gram (g)** is a more conveniently sized unit for most laboratory measurements. One gram, of course, is $\frac{1}{1000}$ of a kilogram (1 kilogram = 1000 g, so 1 g must equal 0.001 kg).

Mass is measured by comparing the weight of a sample with the weights of known standard masses. The operation is called **weighing,** and the apparatus used is called a **balance** (Figure 1.18). For the balance in Figure 1.18*a*, we would place our sample on the left pan and then add standard masses to the other. When the weight of the sample and the total weight of the standards are in balance (when they match), their masses are then equal. Figure 1.19 gives the masses of some common objects in SI units.

[9] Use of the abbreviations L for liter and mL for milliliter is rather recent. Confusion between the printed letter 1 and the number 1 prompted the change from l for liter and ml for milliliter. You may encounter the abbreviation ml in other books or on older laboratory glassware.

(a)

(b)

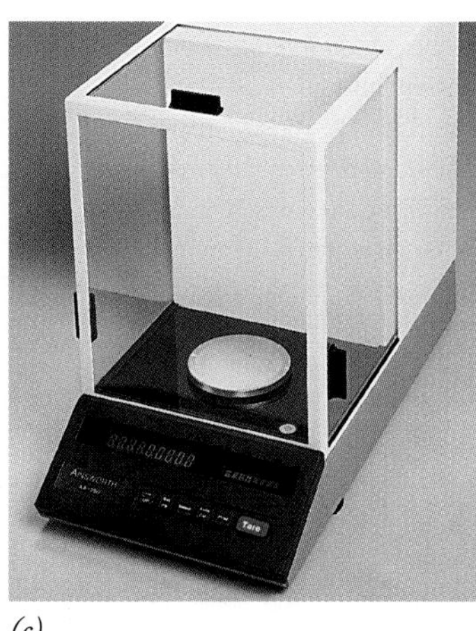

(c)

FIG. 1.18 Typical laboratory balances. (*a*) A traditional two-pan analytical balance capable of measurements to the nearest 0.0001 g. (*b*) A modern top-loading balance capable of mass measurements to the nearest 0.001 g (fitted with a cover to reduce the effects of air currents and thereby improve precision). (*c*) A modern analytical balance capable of measurements to the nearest 0.0001 g. *(Michael Watson; Courtesy Central Scientific Co.; Courtesy Cole-Parmer Instrument Co.)*

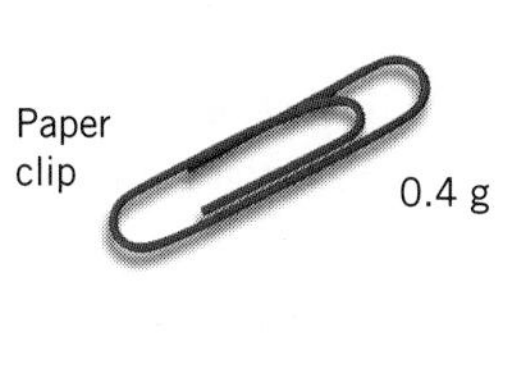

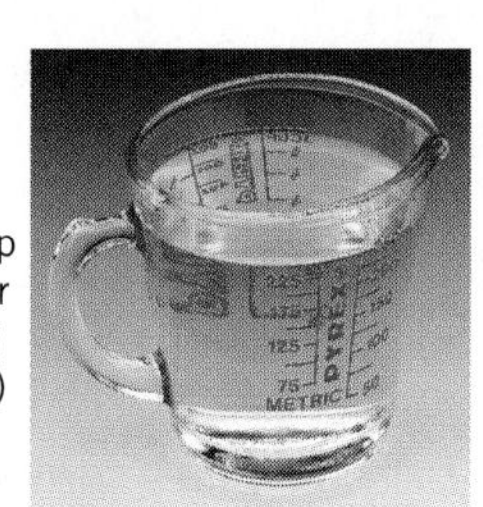

FIG. 1.19 Masses of several common objects in metric and English units. *(Coco McCoy/Rainbow; Coco McCoy/Rainbow; Andy Washnik; Jim Cummins/Taxi/Getty Images.)*

Temperature

Temperature is usually measured with a thermometer (Figure 1.20). Thermometers are graduated in *degrees* according to one of two temperature scales. Both scales use as reference

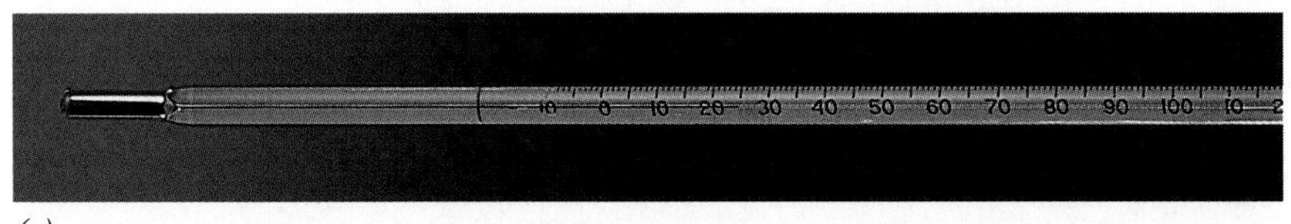

(a)

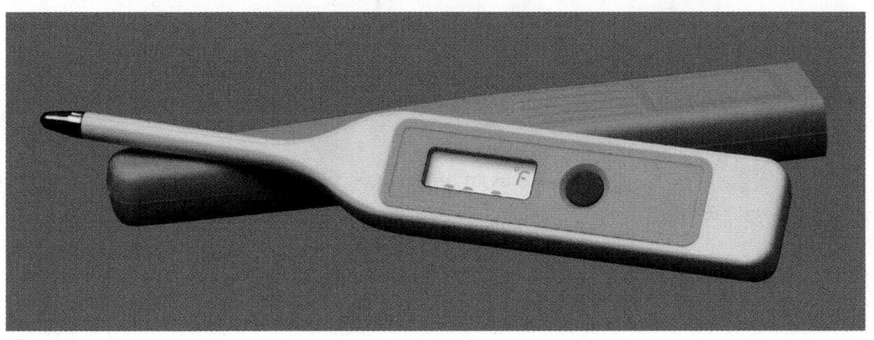
(b)

FIG. 1.20 Typical laboratory thermometers. (*a*) A traditional mercury thermometer. (*b*) An electronic thermometer. *(Michael Watson/Corbis Images.)*

In chemistry, reference data are commonly tabulated at 25 °C, which is close to room temperature. Biologists often carry out their experiments at 37 °C because that is normal human body temperature.

points the temperature at which water freezes[10] and the temperature at which it boils. On the **Fahrenheit scale** water freezes at 32 °F and boils at 212 °F. If you've been raised in the United States, this is probably the scale you're most familiar with. In recent times, however, you have probably noticed an increased use of the Celsius scale, especially in weather broadcasts. This is the scale we use most often in the sciences. On the **Celsius scale** water freezes at 0 °C and boils at 100 °C. (See Figure 1.21.)

As you can see in Figure 1.21, on the Celsius scale there are 100 degree units between the freezing and boiling points of water, while on the Fahrenheit scale this same temperature range is spanned by 180 degree units. Consequently, 5 Celsius degrees are the same as 9 Fahrenheit degrees. We can use the following equation as a tool to convert between these temperature scales.

TOOLS
Celsius to Fahrenheit conversions

$$t_F = \left(\frac{9\ °\text{F}}{5\ °\text{C}}\right)t_C + 32\ °\text{F} \quad (1.1)$$

In this equation, t_F is the Fahrenheit temperature and t_C is the Celsius temperature. As noted earlier, units behave like numbers in calculations, and we see in Equation 1.1 that °C "cancels out" to leave only °F. The 32 °F is added to account for the fact that the freezing point of water (0 °C) occurs at 32 °F on the Fahrenheit scale. Equation 1.1 can easily be rearranged to permit calculating °C from °F.

We will use a capital T to stand for the Kelvin temperature and a lowercase t (as in t_C) to stand for the Celsius temperature. This conforms to usage described by the International Bureau of Weights and Measures in Sèvres, France, and the National Institute of Standards and Technology (NIST) in Gaithersburg, MD.

The SI unit of temperature is the **kelvin (K),** which is the degree unit on the **Kelvin temperature scale.** Notice that the temperature unit is K, not °K (the degree symbol, °, is omitted). Also notice that the name of the unit, kelvin, is not capitalized. Equations that include temperature as a variable sometimes take on a simpler form when Kelvin temperatures are used. We will encounter this situation many times throughout the book.

The name of the temperature scale, the Kelvin scale, is capitalized, but the name of the unit, the kelvin, is not. However, the symbol for the kelvin is the capital letter K.

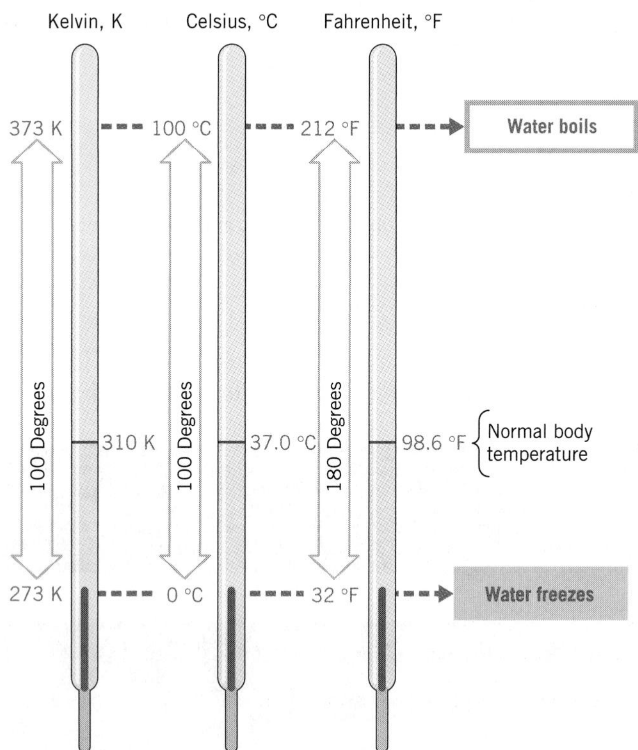

FIG. 1.21 **Comparison among Kelvin, Celsius, and Fahrenheit temperature scales.**

[10] Water freezes and ice melts at the same temperature, and a mixture of ice and liquid water will maintain a constant temperature of 32 °F or 0 °C. If heat is added, some ice melts; if heat is removed, some liquid water freezes, but the temperature doesn't change. This constancy of temperature is what makes the "ice point" convenient for calibrating thermometers.

Figure 1.21 shows how the Kelvin, Celsius, and Fahrenheit temperature scales relate to each other. Notice that the kelvin is *exactly* the same size as the Celsius degree. *The only difference between these two temperature scales is the zero point.* The zero point on the Kelvin scale is called **absolute zero** and corresponds to nature's coldest temperature. It is 273.15 degree units below the zero point on the Celsius scale, which means that 0 °C equals 273.15 K, and 0 K equals −273.15 °C. Thermometers are never marked with the Kelvin scale, so to convert from Celsius to Kelvin temperatures the following equation applies.

$$T_K = (t_C + 273.15\ °\text{C})\left(\frac{1\ \text{K}}{1\ °\text{C}}\right) \quad (1.2)$$

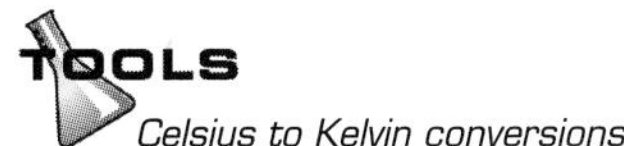

Celsius to Kelvin conversions

This amounts to simply adding 273.15 to the Celsius temperature to obtain the Kelvin temperature. Often we are given Celsius temperatures rounded to the nearest degree, in which case we round 273.15 to 273. Thus, 25 °C equals (25 + 273) K or 298 K.

EXAMPLE 1.2
Converting among Temperature Scales

Thermal pollution, the release of large amounts of heat into rivers and other bodies of water, is a serious problem near power plants and can affect the survival of some species of fish. For example, trout will die if the temperature of the water rises above approximately 25 °C. (a) What is this temperature in °F? (b) Rounded to the nearest whole degree unit, what is this temperature in kelvins?

ANALYSIS: *Usually, the first job in solving a problem is determining which tools are required to do the work.* Both parts of the problem here deal with temperature conversions. Therefore, we ask ourselves, "What tools do we have that relate temperature scales to each other?" Let's write them:

Equation 1.1 $$t_F = \left(\frac{9\ °\text{F}}{5\ °\text{C}}\right) t_C + 32\ °\text{F}$$

Equation 1.2 $$T_K = (t_C + 273.15\ °\text{C})\left(\frac{1\ \text{K}}{1\ °\text{C}}\right)$$

Equation 1.1 relates Fahrenheit temperatures to Celsius temperatures, so this is the tool we need to answer part (a). Equation 1.2 relates Kelvin temperatures to Celsius temperatures, and this is the tool we need for part (b). Now that we have what we need, the rest follows.

SOLUTION:

(a) We substitute the value of the Celsius temperature (25 °C) for t_C.

$$t_F = \left(\frac{9\ °\text{F}}{5\ \cancel{°\text{C}}}\right)(25\ \cancel{°\text{C}}) + 32\ °\text{F}$$

$$= 77\ °\text{F}$$

Therefore, 25 °C = 77 °F. (Notice that we have canceled the unit °C in the equation above. As noted earlier, units behave the same as numbers do in calculations.)

(b) Once again, we have a simple substitution. Since $t_C = 25\ °\text{C}$, the Kelvin temperature (rounded) is

$$T_K = (25\ °\text{C} + 273\ °\text{C})\left(\frac{1\ \text{K}}{1\ °\text{C}}\right)$$

$$= 298\ \cancel{°\text{C}}\left(\frac{1\ \text{K}}{1\ \cancel{°\text{C}}}\right) = 298\ \text{K}$$

Thus, 25 °C = 298 K.

ARE THE ANSWERS REASONABLE? For part (a), we know that a Fahrenheit degree is about half the size of a Celsius degree, so 25 Celsius degrees should be about 50 Fahrenheit degrees. The positive value for the Celsius temperature tells us we have a temperature *above* the freezing point of water. Since water freezes at 32 °F, the Fahrenheit temperature should be approximately 32 °F + 50 °F = 82 °F. The answer of 77 °F is quite close.

For part (b), we recall that 0 °C = 273 K. A temperature above 0 °C must be higher than 273 K. Our calculation, therefore, appears to be correct.

Practice Exercise 3: What Fahrenheit temperature corresponds to a Celsius temperature of 86 °C? (Hint: What tool relates the two temperature scales?)

Practice Exercise 4: What Celsius temperature corresponds to 50 °F? What Kelvin temperature corresponds to 68 °F (expressed to the nearest whole kelvin unit)?

1.6 MEASUREMENTS ALWAYS CONTAIN SOME UNCERTAINTY

We noted in the preceding section that measurements are inexact; they contain **uncertainties** (also called **errors**). One source of uncertainty is associated with limitations in our ability to read the scale of the measuring instrument. Uncontrollably changing conditions at the time of the measurement can also cause errors that are more important than scale reading errors. For example, if you are measuring a length of wire with a ruler, you may not be holding the wire perfectly straight every time.

If we were to take an enormous number of measurements using appropriately adjusted instruments, statistically half of the measurements should be larger and half smaller than the true value of the measured quantity. And, in fact, we do observe that a series of measurements tend to cluster around some central value, which we generally assume is close to the true value. We can estimate the central value quite simply by reporting the **average,** or **mean,** of the series of measurements. This is done by summing the measurements and then dividing by the number of measurements we made. Although making repeated measurements is tedious, the more measurements we make, the more confident we can be that the average is close to the true value that all measurements would be grouped around.

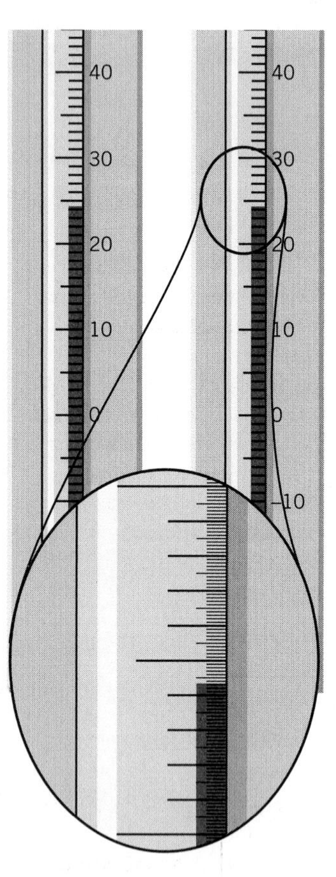

FIG. 1.22 Thermometers with different scales give readings with different precision. The thermometer on the left has marks that are one degree apart, allowing the temperature to be estimated to the nearest tenth of a degree. The thermometer on the right has marks every 0.1 °C. This scale permits estimation of the hundredths place.

Uncertainties in measurements are a natural part of reading a scale

One kind of error that can't be eliminated arises when we attempt to obtain a measurement by reading the scale on an instrument. Consider, for example, reading the same temperature from each of the two thermometers in Figure 1.22.

The marks on the left thermometer are one degree apart, and we can see that the temperature lies between 24 °C and 25 °C. When reading a scale, we always record the last digit to the nearest tenth of the smallest scale division. Looking closely, therefore, we might estimate that the fluid column falls about 3/10 of the way between the marks for 24 and 25 degrees, so we can report the temperature to be 24.3 °C. However, it would be foolish to say that the temperature is *exactly* 24.3 °C. The last digit is only an estimate, and the left thermometer might be read as 24.2 °C by one observer or 24.4 °C by another. Because different observers might obtain values that differ by 0.1 °C, there is an uncertainty of ±0.1 °C in the measured temperature. We can express this, if we wish, by writing the temperature as 24.3 ± 0.1 °C.

The thermometer on the right has marks that are 1/10 of a degree apart, which allows us to estimate the temperature as 24.32 °C. In this case, we are estimating the hundredths place and the uncertainty is ±0.01 °C. We could write the temperature as 24.32 ± 0.01 °C. Notice that because the thermometer on the right is more finely graduated, we are able to obtain measurements with smaller uncertainties. We would have more confidence in temperatures read from the thermometer on the right in Figure 1.22 because it has more digits and a smaller amount of uncertainty. *The reliability of a piece of data is indicated by the number of digits used to represent it.*

By convention in science, *all digits in a measurement up to and including the first estimated digit are recorded.* If a reading measured with the thermometer on the right seemed exactly on the 24 °C mark, we would record the temperature as 24.00 °C, not 24 °C, to show that the thermometer can be read to the nearest 1/100 of a degree.

Measurements are written using the significant figures convention

The concepts discussed above are so important that we have special terminology to describe numbers that come from measurements.

> *Digits that result from measurement such that only the digit farthest to the right is not known with certainty are called* **significant figures** (*or* **significant digits**).

The number of significant figures in a measurement is equal to the number of digits known for sure *plus* one that is estimated. Let's look at our two temperature measurements.

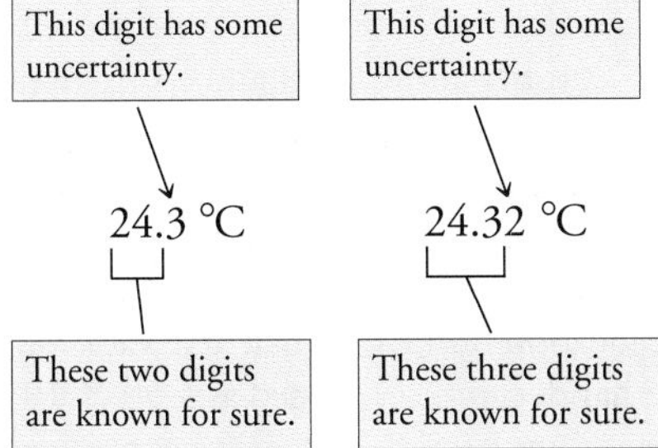

This digit has some uncertainty.	This digit has some uncertainty.
24.3 °C	24.32 °C
These two digits are known for sure.	These three digits are known for sure.

The first measurement, 24.3 °C, has three significant figures; the second, 24.32 °C, has four significant figures.

Accuracy is correctness; precision is reproducibility

Two words often used in reference to measurements are *accuracy* and *precision*. **Accuracy** refers to how close a measurement is to the true or correct value. **Precision** refers to how closely repeated measurements of a quantity come to each other and to the average. Notice that the two terms are not equivalent, because the average doesn't always correspond to the true or correct value. A practical example of how accuracy and precision differ is illustrated in Figure 1.23.

For measurements to be **accurate,** the measuring device must be carefully calibrated (adjusted) so it gives correct values when a standard reference is used with it. For example, to calibrate an electronic balance, a known reference mass is placed on the balance and a calibration routine within the balance is initiated. Once calibrated, the balance will give correct readings, the accuracy of which is determined by the accuracy of the standard mass used. Standard reference masses (also called "weights") can be purchased from scientific supply companies.

Precision refers to how closely repeated measurements of the same quantity come to each other. In general, the smaller the uncertainty (i.e., the "plus or minus" part of the measurement), the more precise the measurement. This translates as: *the more significant figures in a measured quantity, the more precise the measurement.*

We usually assume that a very precise measurement is also of high accuracy. We can be wrong, however, if our instruments are improperly calibrated. For example, the improperly marked ruler in Figure 1.24 might yield measurements which vary by a hundredth of a centimeter (±0.01 cm), but all the measurements would be too large by 1 cm—a case of good precision but poor accuracy.

When are zeros significant digits?

Usually, it is simple to determine the number of significant figures in a measurement; we just count the digits. Thus 3.25 has three significant figures and 56.205 has five of them. When zeros come at the beginning or the end of a number, however, they sometimes cause confusion.

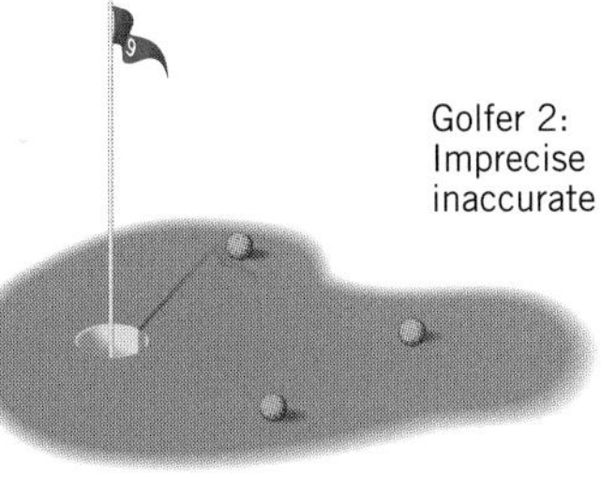

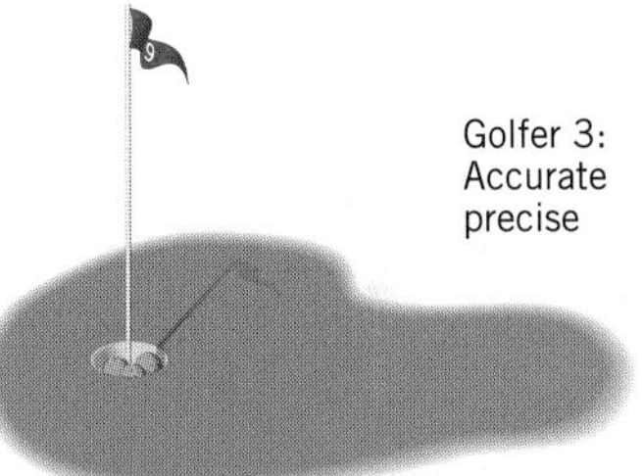

FIG. 1.23 The difference between precision and accuracy in the game of golf. Golfer 1 hits shots that are precise (because they are tightly grouped) but the accuracy is poor because the balls are not near the target (the "true" value). Golfer 2 needs help. His shots are neither precise nor accurate. Golfer 3 wins the prize with shots that are precise (tightly grouped) and accurate (in the hole).

How accurate would measurements be with this ruler?

FIG. 1.24 An improperly marked ruler. This improperly marked ruler will yield measurements that are each wrong by one whole unit. The measurements might be precise, but the accuracy would be very poor.

Counting significant figures

Zeros to the right of a decimal point are always counted as significant. Thus, 4.500 m has four significant figures because the zeros would not be written unless those digits were known to be zeros.

Zeros to the left of the first nonzero digit are never counted as significant. For instance, a length of 2.3 mm is the same as 0.0023 m. Since we are dealing with the same measured value, its number of significant figures cannot change when we change the units. Both quantities have two significant figures.

Zeros on the end of a number without a decimal point are assumed not to be significant. For example, suppose you were told that a protest march was attended by 45,000 people. If this was just a rough estimate, it might be uncertain by as much as several thousand, in which case the value 45,000 represents just two significant figures, since the "5" is the uncertain digit. None of the zeros would then count as significant figures. On the other hand, suppose the protesters were carefully counted using an aerial photograph, so that the count could be reported to be 45,000 give or take about 100 people. In this case, the value represents 45,000 $\pm$ 100 protesters and contains three significant figures, with the uncertain digit being the zero in the hundreds place. So a simple statement such as "there were 45,000 people attending the march" is ambiguous. We can't tell how many significant digits the number has from the number alone. We can be sure that the nonzero digits are significant, though. The best we can do is to say, "45,000 has *at least* two significant figures."

We can avoid confusion by using *scientific notation* when we report a measurement. For example, if we want to report the number of protesters as 45,000 give or take a thousand, we can write the rough estimate as 4.5×10^4. The 4.5 shows the number of significant figures and the 10^4 tells us the location of the decimal. The value obtained from the aerial photograph count, on the other hand, can be expressed as 4.50×10^4. This time the 4.50 shows three significant figures and an uncertainty of $\pm$100 people.

Measurements limit the precision of results calculated from them

When several measurements are obtained in an experiment they are usually combined in some way to calculate a desired quantity. For example, to determine the area of a rectangular carpet we require two measurements, length and width, which are then multiplied to give the answer we want. To get some idea of how precise the area really is, we need a way to take into account the precision of the various values used in the calculation. To make sure this happens, we follow certain rules according to the kinds of arithmetic being performed.

Multiplication and division

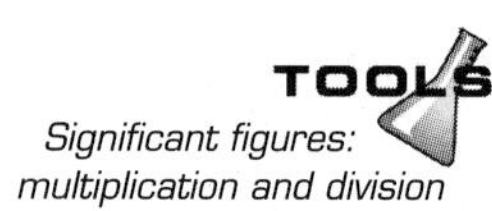

Significant figures: multiplication and division

For multiplication and division, the number of significant figures in the answer should not be greater than the number of significant figures in the least precise measurement. Let's look at a typical problem involving some measured quantities.

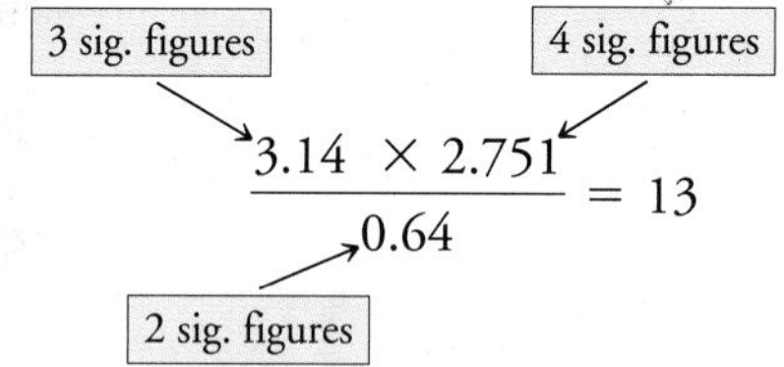

$$\frac{3.14 \times 2.751}{0.64} = 13$$

The result displayed on a calculator[11] is 13.49709375. However, the least precise factor, 0.64, has only two significant figures, so the answer should have only two. The correct answer, 13, is obtained by rounding off the calculator answer.[12]

[11] Calculators usually give too many significant figures. An exception is when the answer has zeros at the right that are significant figures. For example, an answer of 1.200 would be displayed on most calculators as 1.2. If the zeros belong in the answer, be sure to write them down.

[12] When we wish to round off a number at a certain point, we simply drop the digits that follow if the first of them is less than 5. Thus, 8.1634 rounds to 8.16, if we wish to have only two decimal places. If the first digit after the point of round off is larger than 5, or if it is 5 followed by other nonzero digits, then we add 1 to the preceding digit. Thus 8.167 and 8.1653 both round to 8.17. Finally, when the digit after the point of round off is a 5 and no other digits follow the 5, then we drop the 5 if the preceding digit is even and add 1 if it is odd. Thus, 8.165 rounds to 8.16 and 8.175 rounds to 8.18.

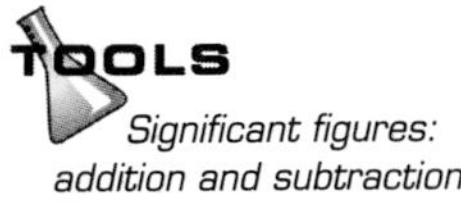

Addition and subtraction

For addition and subtraction, the answer should have the same number of decimal places as the quantity with the fewest number of decimal places. As an example, consider the following addition of measured quantities.

$$\begin{array}{r} 3.247 \\ 41.36 \\ +\ 125.2 \\ \hline 169.8 \end{array}$$

+ 125.2 ← (This number has only 1 decimal place.)

169.8 ← (The answer has been rounded to 1 decimal place.)

In this calculation, the digits beneath the 6 and the 7 are unknown; they could be anything. (They're not necessarily zeros because if we *knew* they were zeros, then zeros would have been written there.) Adding an unknown digit to the 6 or 7 will give an answer that's also unknown, so for this sum we are not justified in writing digits in the second and third places after the decimal point. Therefore, we round the answer to the nearest tenth.

Exact numbers contain no uncertainty

Numbers that come from definitions, such as 12 in. = 1 ft, and those that come from a direct count, such as the number of people in a room, have no uncertainty. We can assume that such **exact numbers** have an infinite number of significant figures. Therefore, we ignore exact numbers when applying the rules described above.

Practice Exercise 5: Perform the following calculations involving measurements and round the results so they have the correct number of significant figures and proper units. (Hint: Apply the rules for significant figures described in this section and keep in mind that units behave like numbers do in calculations.)

(a) 21.0233 g + 21.0 g

(b) 10.0324 g/11.7 mL

(c) $\dfrac{14.25 \text{ cm} \times 12.334 \text{ cm}}{(2.223 \text{ cm} - 1.04 \text{ cm})}$

Practice Exercise 6: Perform the following calculations involving measurements and round the results so that they are written to the correct number of significant figures and have the correct units.

(a) 32.02 mL − 2.0 mL

(b) 54.183 g − 0.0278 g

(c) 10.0 g + 1.03 g + 0.243 g

(d) $43.4 \text{ in.} \times \dfrac{1 \text{ ft}}{12 \text{ in.}}$

(e) $\dfrac{1.03 \text{ m} \times 2.074 \text{ m} \times 3.9 \text{ m}}{12.46 \text{ m} + 4.778 \text{ m}}$

1.7 UNITS CAN BE CONVERTED USING THE FACTOR-LABEL METHOD

After analyzing a problem and assembling the necessary information to solve it, the next step is working the problem to obtain an answer. For numerical problems, scientists usually use a system called the **factor-label method** (also called **dimensional analysis**) to help them perform the correct arithmetic. As you will see, often this method also helps in analyzing the problem and selecting the tools needed to solve it.

In the factor-label method we treat a numerical problem as one involving a conversion of units from one kind to another. To do this we use one or more *conversion factors* to change the units of the given quantity to the units of the answer.

(Given quantity) × (conversion factor) = (desired quantity)

A **conversion factor** *is a fraction formed from a valid relationship or equality between units and is used to switch from one system of measurement and units to another.* To illustrate, suppose we want to express a person's height of 72.0 inches in centimeters. To do this we need the relationship between the inch and the centimeter. We can obtain this from Table 1.6.

$$2.54 \text{ cm} = 1 \text{ in. (exactly)} \tag{1.3}$$

To construct a valid conversion factor, the relationship between the units must be true. For example, the statement

3 ft = 41 in.

is false. Although you might make a conversion factor out of it, any answers you would calculate are sure to be incorrect. *Correct answers require correct relationships between units.*

If we divide both sides of this equation by 1 in., we obtain a conversion factor.

$$\frac{2.54 \text{ cm}}{1 \text{ in.}} = \frac{\cancel{1} \cancel{\text{in.}}}{\cancel{1} \cancel{\text{in.}}} = 1$$

Notice that we have canceled the units from both the numerator and denominator of the center fraction, leaving the first fraction equaling 1. As mentioned earlier, *units behave just as numbers do in mathematical operations;* this is a key part of the factor-label method. Let's see what happens if we multiply 72.0 inches, the height that we mentioned, by this fraction.

The relationship between the inch and the centimeter is exact, so the numbers in 1 in. = 2.54 cm have an infinite number of significant figures.

$$72.0 \cancel{\text{in.}} \times \frac{2.54 \text{ cm}}{1 \cancel{\text{in.}}} = 183 \text{ cm}$$

$$\begin{pmatrix}\text{given}\\ \text{quantity}\end{pmatrix} \times \begin{pmatrix}\text{conversion}\\ \text{factor}\end{pmatrix} = \begin{pmatrix}\text{desired}\\ \text{quantity}\end{pmatrix}$$

Because we have multiplied 72.0 in. by something that is equal to 1, we know we haven't changed the magnitude of the person's height. We have, however, changed the units. Notice that we have canceled the unit inches. The only unit left is centimeters, which is the unit we want for the answer. The result, therefore, is the person's height in centimeters.

One of the benefits of the factor-label method is that it often lets you know when you have done the *wrong* arithmetic. From the relationship in Equation 1.3, we can actually construct two conversion factors:

$$\frac{2.54 \text{ cm}}{1 \text{ in.}} \quad \text{and} \quad \frac{1 \text{ in.}}{2.54 \text{ cm}}$$

We used the first one correctly, but what would have happened if we had used the second by mistake?

$$72.0 \text{ in.} \times \frac{1 \text{ in.}}{2.54 \text{ cm}} = 28.3 \text{ in.}^2/\text{cm}$$

In this case, none of the units cancel. We get units of in.^{2}/cm because inches times inches is inches squared. Even though our calculator may be very good at arithmetic, we've got the wrong answer. *The factor-label method lets us know we have the wrong answer because the units are wrong!*

We will use the factor-label method extensively throughout this book to aid us in setting up the proper arithmetic in problems. In fact, we will see that it also helps us assemble the information we need to solve the problem. The following examples illustrate the method.

EXAMPLE 1.3
Applying the Factor-Label Method

Convert 3.25 m to millimeters (mm).

ANALYSIS: To clearly identify the problem, let's write the given quantity (with its units) on the left and the *units* of the desired answer on the right.

$$3.25 \text{ m} = ? \text{ mm}$$

To solve this problem our tool will be a conversion factor that relates the unit meter to the unit millimeter. From the table of decimal multipliers, the prefix "milli" means "$\times\ 10^{-3}$," so we can write

$$1 \text{ mm} = 10^{-3} \text{ m}$$

Notice that this relationship connects the units given to the units desired.

We now have all the information we need to solve the problem.

SOLUTION: From the relationship above, we can form two conversion factors.

$$\frac{1\text{ mm}}{10^{-3}\text{ m}} \quad \text{and} \quad \frac{10^{-3}\text{ m}}{1\text{ mm}}$$

We know we have to cancel the unit meter, so we need to multiply by a conversion factor with this unit in the denominator. Therefore, we select the one on the left as our tool. This gives

$$3.25\ \cancel{\text{m}} \times \frac{1\text{ mm}}{10^{-3}\ \cancel{\text{m}}} = 3.25 \times 10^{3}\text{ mm}$$

Notice we have expressed the answer to three significant figures because that is how many there are in the given quantity, 3.25 m.

IS THE ANSWER REASONABLE? We know that millimeters are much smaller than meters, so 3.25 m must represent a lot of millimeters. Our answer, therefore, makes sense.

EXAMPLE 1.4
Applying the Factor-Label Method

A liter, which is slightly larger than a quart, is defined as 1 cubic decimeter (1 dm^3). How many liters are there in 1 cubic meter (1 m^3)?

ANALYSIS: Let's begin once again by stating the problem in equation form.

$$1\text{ m}^3 = ?\text{ L}$$

Next, we assemble the tools. What relationships do we know that relate these various units? We are given the relationship between liters and cubic decimeters,

$$1\text{ L} = 1\text{ dm}^3 \qquad (1.4)$$

From the table of decimal multipliers, we also know the relationship between decimeters and meters,

$$1\text{ dm} = 0.1\text{ m}$$

but we need a relationship between cubic units. Since units undergo the same kinds of operations numbers do, we simply cube each side of this equation (being careful to cube *both* the numbers and the units).

$$(1\text{ dm})^3 = (0.1\text{ m})^3$$

$$1\text{ dm}^3 = 0.001\text{ m}^3 \qquad (1.5)$$

Notice how Equations 1.4 and 1.5 provide a path from the given units to those we seek. Such a path is always a necessary condition when we apply the factor-label method.

$$\text{m}^3 \xrightarrow[\text{Equation 1.5}]{} \text{dm}^3 \xrightarrow{\text{Equation 1.4}} \text{L}$$

Now we are ready to solve the problem.

SOLUTION: The first step is to eliminate the units m^3. We use Equation 1.5.

$$1\ \cancel{\text{m}^3} \times \frac{1\text{ dm}^3}{0.001\ \cancel{\text{m}^3}} = 1000\text{ dm}^3$$

Then we use Equation 1.4 to take us from dm^3 to L.

$$1000\ \cancel{\text{dm}^3} \times \frac{1\text{ L}}{1\ \cancel{\text{dm}^3}} = 1000\text{ L}$$

Thus, 1 m^3 = 1000 L.

Usually, when a problem involves the use of two or more conversion factors, they can be "strung together" in a "chain calculation" to avoid having to compute intermediate results. For example, this problem can be set up as follows.

$$1\ \cancel{m^3} \times \frac{1\ \cancel{dm^3}}{0.001\ \cancel{m^3}} \times \frac{1\ L}{1\ \cancel{dm^3}} = 1000\ L$$

IS THE ANSWER REASONABLE? One liter is about a quart. A cubic meter is about a cubic yard. Therefore, we expect a large number of liters in a cubic meter, so our answer seems reasonable. (Notice here that in our analysis we have approximated the quantities in the calculation in units of quarts and cubic yards, which may be more familiar than liters and cubic meters if you've been raised in the United States. We get a feel for the approximate magnitude of the answer using our familiar units and then relate this to the actual units of the problem.)

EXAMPLE 1.5
Applying the Factor-Label Method

Some mountain climbers are susceptible to high altitude pulmonary edema (HAPE), a life-threatening condition that causes fluid retention in the lungs. It can develop when a person climbs rapidly to heights greater than 2,500 meters (2.5×10^3 m). What is this distance expressed in feet?

ANALYSIS: The problem can be stated as

$$2500\ m = ?\ ft$$

We are converting a metric unit of length (the meter) into an English unit of length (the foot). The critical link between the two will be a metric-to-English length conversion. One of several sets of tools we can use is

$$1\ cm = 10^{-2}\ m \qquad \text{(from Table 1.4)}$$
$$1\ in. = 2.54\ cm \qquad \text{(from Table 1.6)}$$
$$1\ ft = 12\ in.$$

Notice how they provide a path from meters to centimeters to inches to feet.

SOLUTION: Now we apply the factor-label method by eliminating unwanted units to bring us to the units of the answer.

$$2.5 \times 10^3\ \cancel{m} \times \frac{1\ \cancel{cm}}{10^{-2}\ \cancel{m}} \times \frac{1\ \cancel{in.}}{2.54\ \cancel{cm}} \times \frac{1\ ft}{12\ \cancel{in.}} = 8.2 \times 10^3\ ft$$

Notice that if we were to stop after the first conversion factor, the units of the answer would be centimeters; if we stop after the second, the units would be inches, and after the third we get feet—the units we want. This time the answer has been rounded to two significant figures because that's how many there were in the measured distance. Notice that the numbers 12 and 2.54 do not affect the number of significant figures in the answer because they are exact numbers derived from definitions.

This is not the only way we could have solved this problem. Other sets of conversion factors could have been chosen. For example, we could have used 1 yd = 0.9144 m and 3 ft = 1 yd. Then the problem would have been set up as

$$2500\ \cancel{m} \times \frac{1\ \cancel{yd}}{0.9144\ \cancel{m}} \times \frac{3\ ft}{1\ \cancel{yd}} = 8200\ ft \qquad \text{(rounded correctly)}$$

Many problems that you meet, just like this one, have more than one path to the answer. There isn't necessarily any *one* correct way to set up the solution. *The important thing is for you to be able to reason your way through a problem and find some set of relationships that can take you from the given information to the answer.* The factor-label method can help you search for these relationships if you keep in mind the units that must be eliminated by cancellation.

IS THE ANSWER REASONABLE? Let's do some approximate arithmetic to get a feel for the size of the answer. A meter is slightly longer than a yard, so let's approximate the given distance, 2500 m, as 2500 yd. In 2500 yd, there are $3 \times 2500 = 7500$ ft. Since the meter is a bit longer than a yard, our answer should be a bit longer than 7500 ft, so the answer of 8200 ft seems to be reasonable.

Practice Exercise 7: Use the factor-label method to convert an area of 124 ft^2 to square meters. (Hint: What relationships would be required to convert feet to meters?)

Practice Exercise 8: Use the factor-label method to perform the following conversions: (a) 3.00 yd to inches, (b) 1.25 km to centimeters, (c) 3.27 mm to feet, and (d) 20.2 miles/gallon to kilometers/liter.

1.8 DENSITY IS A USEFUL INTENSIVE PROPERTY

In our earlier discussion of properties we noted that intensive properties are useful for identifying substances. One of the interesting things about extensive properties is that if you take the ratio of two of them, the resulting quantity is usually independent of sample size. In effect, the sample size cancels out and the calculated quantity becomes an intensive property. A useful property obtained this way is **density,** *which is defined as the ratio of an object's mass to its volume.* Using the symbols d for density, m for mass, and V for volume, we can express this mathematically as

$$d = \frac{m}{V} \tag{1.6}$$

Density

Notice that to determine an object's density we make two measurements, mass and volume.

EXAMPLE 1.6
Calculating Density

A sample of blood completely fills an 8.20 cm^3 vial. The empty vial has a mass of 10.30 g. The vial has a mass of 18.91 g after being filled with blood. What is the density of blood?

ANALYSIS: This problem asks you to connect the mass and volume of blood with its density. The critical link between these quantities is the definition of density, given by Equation 1.6. (Without knowing this definition, you cannot solve the problem.) Equation 1.6 becomes the tool we'll use to obtain the answer.

SOLUTION: The volume of the blood equals the volume of the vial, 8.20 cm^3. The mass of the blood is the difference between the masses of the full and empty vials:

$$\text{Mass of blood} = 18.91\text{ g} - 10.30\text{ g} = 8.61\text{ g}$$

To determine the density we simply take the ratio of mass to volume.

$$\text{Density} = \frac{m}{V} = \frac{8.61\text{ g}}{8.20\text{ cm}^3} = 1.05\text{ g/cm}^3$$

This could also be written as

$$\text{Density} = 1.05\text{ g/mL}$$

because $1\text{ cm}^3 = 1\text{ mL}$.

IS THE ANSWER REASONABLE? First, the answer has the correct units, so that's encouraging. In the calculation we are dividing 8.61 by 8.20, a number that is slightly smaller. The answer should be slightly larger than one, which it is, so a density of 1.05 g/cm^3 seems reasonable.

There is more mass in 1 cm^3 of gold than in 1 cm^3 of iron.

Although the density of water varies slightly with temperature, it is useful to remember the value 1.00 g/cm^3. It can be used if the water is near room temperature and only three (or fewer) significant figures are required.

Each pure substance has its own characteristic density (Table 1.7). Gold, for instance, is much more dense than iron. Each cubic centimeter of gold has a mass of 19.3 g, so its density is 19.3 g/cm^3. By comparison, the density of water is 1.00 g/cm^3 and the density of air at room temperature is about 0.0012 g/cm^3.

Most substances, such as the mercury in the bulb of a thermometer, expand slightly when they are heated, so the amount of matter packed into each cubic centimeter is less. Therefore, density usually decreases slightly with increasing temperature.[13] For solids and liquids the size of this change is small, as you can see from the data for water in Table 1.8. When only two or three significant figures are required, we can often ignore the variation of density with temperature.

TABLE 1.7

Densities of Some Common Substances in g/cm^3 at Room Temperature

Water	1.00
Aluminum	2.70
Iron	7.86
Silver	10.5
Gold	19.3
Glass	2.2
Air	0.0012

TABLE 1.8 Density of Water as a Function of Temperature

Temperature (°C)	Density (g/cm^3)
10	0.999700
15	0.999099
20	0.998203
25	0.997044
30	0.995646

Use density to relate a material's mass to its volume

A useful property of density is that it provides a way to convert between the mass and volume of a substance. It defines a relationship, which we will call an **equivalence**, between the amount of mass and its volume. For instance, the density of gold (19.3 g/cm^3) tells us that 19.3 g of the metal is equivalent to a volume of 1.00 cm^3. We express this relationship symbolically as

$$19.3 \text{ g gold} \Leftrightarrow 1.00 \text{ cm}^3 \text{ gold}$$

where we have used the symbol $\Leftrightarrow$ to mean "is equivalent to." (We can't really use an equals sign in this expression because grams can't *equal* cubic centimeters; one is a unit of mass and the other is a unit of volume.)

In setting up calculations by the factor-label method, an equivalence can be used to construct conversion factors just as equalities can. From the equivalence we have just written we can form two conversion factors:

$$\frac{19.3 \text{ g gold}}{1.00 \text{ cm}^3 \text{ gold}} \quad \text{and} \quad \frac{1.00 \text{ cm}^3 \text{ gold}}{19.3 \text{ g gold}}$$

The following example illustrates how we use density in calculations.

EXAMPLE 1.7
Calculations Using Density

Seawater has a density of about 1.03 g/mL. (a) What mass of seawater would fill a sampling vessel to a volume of 225 mL? (b) What is the volume, in milliliters, of 45.0 g of seawater?

ANALYSIS: For both parts of this problem, we are relating the mass of a material to its volume. Density is the critical link that we need between these two quantities. The given density tells us that *1.03 g of seawater is equivalent to 1.00 mL of seawater*, which we write as

$$1.03 \text{ g seawater} \Leftrightarrow 1.00 \text{ mL seawater}$$

[13] Liquid water behaves oddly. Its maximum density is at 4 °C, so when water at 0 °C is warmed, its density increases until the temperature reaches 4 °C. As the temperature is increased further the density of water gradually decreases.

From this relationship we can construct two conversion factors. These will be the tools we use to obtain the answers.

$$\frac{1.03 \text{ g seawater}}{1.00 \text{ mL seawater}} \quad \text{and} \quad \frac{1.00 \text{ mL seawater}}{1.03 \text{ g seawater}}$$

SOLUTION: (a) The question can be restated as 225 mL seawater ⇔ ? g seawater. We need to eliminate the unit *mL seawater,* so we choose the conversion factor on the left as our tool.

$$225 \cancel{\text{mL seawater}} \times \frac{1.03 \text{ g seawater}}{1.00 \cancel{\text{mL seawater}}} \Leftrightarrow 232 \text{ g seawater}$$

Thus, 225 mL of seawater has a mass of 232 g.

(b) The question is, 45.0 g seawater ⇔ ? mL seawater. This time we need to eliminate the unit *g seawater,* so we use the conversion factor on the right as our tool.

$$45.0 \cancel{\text{g seawater}} \times \frac{1.00 \text{ mL seawater}}{1.03 \cancel{\text{g seawater}}} \Leftrightarrow 43.7 \text{ mL seawater}$$

Thus, 45.0 g of seawater has a volume of 43.7 mL.

ARE THE ANSWERS REASONABLE? Notice that the density tells us that 1 mL of seawater has a mass of slightly more than 1 g. So for part (a), we might expect that 225 mL of seawater should have a mass slightly more than 225 g. Our answer, 232 g, is reasonable. For part (b), 45 g of seawater should have a volume not too far from 45 mL, so our answer of 43.7 mL is the right size.

Practice Exercise 9: A gold-colored metal object has a mass of 365 g and a volume of 22.12 cm^3. Is the object composed of pure gold? (Hint: How does the density of the object compare with that of pure gold?)

Practice Exercise 10: A certain metal alloy has a density of 12.6 g/cm^3. How many pounds would 0.822 ft^3 of this alloy weigh? (Hint: What is the density of the alloy in units of lb/ft^3?)

Practice Exercise 11: An ocean-dwelling dinosaur was estimated to have had a body volume of 1.38×10^6 cm^3. The animal's mass when alive was estimated at 1.24×10^6 g. What is its density?

Practice Exercise 12: The density of diamond is 3.52 g/cm^3. What is the volume in cubic centimeters of a 1 carat diamond, which has a mass of 200 mg? (Assume three significant figures.)

Conclusions must be drawn from reliable measurements

We saw earlier that substances can be identified by their properties. If we are to rely on properties such as density for identification of substances, it is very important that our measurements be reliable. We must have some idea of what the measurement's accuracy and precision are.

The importance of accuracy is obvious. If we have no confidence that our measured values are close to the true values, we certainly cannot trust any conclusions that are based on the data we have collected.

Precision of measurements can be equally important. For example, suppose we had a gold wedding ring and we wanted to determine whether or not the gold was 24 carat. We could determine the mass of the ring, and then its volume, and compute the density of the ring. We could then compare our experimental density with the density of 24 carat gold (which is 19.3 g/mL). Suppose the ring had a volume of 1.0 mL and the ring had a mass of 18 g, as measured using a graduated cup measure and a kitchen scale. The density of the ring would then be 18 g/mL, to the correct number of significant figures. Could we conclude that the ring was made of 24 carat gold? We know the density to only two significant figures. The experimental density could be as low as 17 g/mL or as high as 19 g/mL, which means the ring *could* be 24 carat gold—or it could be 22 carat gold (which has a density of around 17.7 to 17.8 g/mL) or maybe even 18 carat gold (which has a density up to 16.9 g/mL).

Suppose we now measure the mass of the ring with a laboratory balance capable of measurements to the nearest ±0.001 g and obtain a mass of 18.153 g. We measure the volume using volumetric glassware and find a volume of 1.03 mL. The density is 17.6 g/mL to the correct number of significant figures. The difference between this density and the density of 24 carat gold is 19.3 g/mL − 17.6 g/mL = 1.7 g/mL. This is considerably larger than the uncertainty in the experimental density (which is about ±0.1 g/mL). We can be reasonably confident that the ring is not 24 carat gold, and in fact the measurements point toward the ring being composed of 22 carat gold.

To trust conclusions drawn from measurements, we must be sure the measurements are accurate and that they are of sufficient precision to be meaningful. This is a key consideration in designing experiments.

SUMMARY

Chemistry and the Scientific Method. **Chemistry** is a science that studies the properties and composition of **matter,** which is anything that has **mass** and occupies space. It employs the **scientific method** in which **observations** are used to collect **empirical facts,** or **data,** that can be summarized in **scientific laws. Models** of nature begin as **hypotheses** that mature into **theories** when they survive repeated testing. According to the **atomic theory,** matter is composed of **atoms** that combine to form more complex substances, many of which consist of **molecules** composed of two or more atoms.

Elements, Compounds, and Mixtures. An **element,** which is identified by its **chemical symbol,** cannot be decomposed into something simpler by a **chemical reaction.** Elements combine in fixed proportions to form **compounds.** Elements and compounds are **pure substances** that may be combined in *varying* proportions to give **mixtures.** If a mixture has two or more **phases,** it is **heterogeneous.** A one-phase **homogeneous** mixture is called a **solution.** Formation or separation of a mixture into its components can be accomplished by a **physical change,** which doesn't alter the chemical composition of the substances involved. Formation or decomposition of a compound takes place by a **chemical change** that changes the chemical makeup of the substances involved.

Properties of Materials. **Physical properties** are measured without changing the chemical composition of a sample. **Solid, liquid,** and **gas** are the most common **states of matter.** Their properties can be related to the different ways the individual atomic-size particles are organized. A **chemical property** describes a chemical reaction a substance undergoes. **Intensive properties** are independent of sample size; **extensive properties** depend on sample size.

Units of Measurement. **Qualitative observations** lack numerical information, whereas **quantitative observations** require numerical measurements. The units used for scientific measurements are based on the set of seven **SI base units** which can be combined to give various **derived units.** These all can be scaled to larger or smaller sized units by applying **decimal multiplying factors.** In the laboratory we routinely measure length, volume, mass, and temperature. Convenient units for length and volume are, respectively, **centimeters** or **millimeters,** and **liters** or **milliliters. Mass** is a measure of the amount of matter in an object and differs from weight. Mass is measured with a **balance** and is expressed in units of **kilograms** or **grams.** Temperature is measured in units of **degrees Celsius** (or **Fahrenheit**) using a thermometer. For many calculations, temperature must be expressed in **kelvins (K).** The zero point on the **Kelvin temperature scale** is called **absolute zero.**

Significant Figures. The **precision** of a measured quantity is revealed by the number of **significant figures** that it contains, which equals the number of digits known for sure plus the first one that possesses some uncertainty. Measured values are **precise** if they contain many significant figures and therefore differ from each other by small amounts. A measurement is **accurate** if its value lies very close to the true value. When measurements are combined in calculations, rules help us determine the correct number of significant figures in the answer (see below). **Exact numbers** are considered to have an infinite number of significant figures.

Factor-Label Method. The **factor-label method** is based on the ability of units to undergo the same mathematical operations as numbers. **Conversion factors** are constructed from *valid relationships* between units. These relationships can be either equalities or **equivalencies** (indicated by the symbol ⇔) between units. Unit cancellation serves as a guide to the use of conversion factors and aids us in correctly setting up the arithmetic for a problem.

Density. **Density** is an intensive property equal to the ratio of a sample's mass to its volume. Besides serving as a means for identifying substances, density provides a conversion factor that relates mass to volume.

TOOLS FOR PROBLEM SOLVING

In this chapter you learned to apply the following concepts as tools in solving problems. Study each one carefully so that you know what each is used for. When faced with solving a problem, recall what each tool does and consider whether it will be helpful in finding a solution. This will aid you in selecting the tools you need.

SI Prefixes *(Table 1.4, page 12)* We use the prefixes to create larger and smaller units. They are also used to create conversion factors for converting between differently sized units. Be sure you are familiar with the ones in bold colored type in Table 1.4.

Units in laboratory measurements *(pages 13-17)* Often we must convert among units commonly used for laboratory measurements.

Length: 1 m = 100 cm = 1000 mm

Volume: 1 L = 1000 mL = 1000 cm^3

Temperature conversions *(pages 16-17)* Use Equations 1.1 and 1.2 to convert between temperature scales.

$$t_F = \left(\frac{9\ °F}{5\ °C}\right)t_C + 32\ °F \qquad T_K = (t_C + 273.15\ °C)\left(\frac{1\ K}{1\ °C}\right)$$

Add 273.15 to Celsius temperature to obtain the Kelvin temperature.

Rules for counting significant figures in a number *(page 19)* To gauge the quality of a measurement, we must know the number of significant figures it contains:

- All nonzero digits are significant.
- Zeros to the right of the decimal are significant if they follow a nonzero digit.
- Zeros between significant digits are significant.
- Zeros that are to the *left* of the first nonzero digit are not significant.
- Zeros on the end of a number without a decimal point are assumed not to be significant. (To avoid confusion, scientific notation should be used.)

Rules for arithmetic and significant figures *(pages 20-21)* We use these rules in almost every numerical problem to obtain the correct number of significant figures in the answer.

Multiplication and division: Round the answer to the same number of significant figures as the least precise factor.

Addition and subtraction: Round the answer to match the same number of decimal places as the quantity with the fewest number of decimal places.

Exact numbers: Exact numbers, such as those that arise from definitions, do not affect the number of significant figures in the result of a calculation.

Density *(page 25)* The density, *d*, relates mass, *m*, and volume, *V*, for a substance.

$$d = \frac{m}{V}$$

Density provides an equivalence between mass and volume, from which we can construct conversion factors to convert between mass and volume for a substance.

QUESTIONS, PROBLEMS, AND EXERCISES

Answers to problems whose numbers are printed in color are given in Appendix B. More challenging problems are marked with asterisks. ILW = Interactive Learningware solution is available at *www.wiley.com/college/brady*. OH = an Office Hours video is available for this problem.

REVIEW QUESTIONS

Introduction; The Scientific Method

1.1 After some thought, give two reasons why a course in chemistry will benefit *you* in the pursuit of your particular major.

1.2 What steps are involved in the scientific method?

1.3 What is the difference between (a) a law and a theory, (b) an observation and a conclusion, (c) an observation and data?

Properties of Substances

OH **1.4** Define *matter.* Which of the following are examples of matter? (a) air, (b) a pencil, (c) a cheese sandwich, (d) a squirrel, (e) your mother

1.5 What is *a physical property?* What is *a chemical property?* What is the chief distinction between physical and chemical properties? Define the terms *intensive property* and *extensive property.* Give two examples of each.

1.6 "A sample of calcium (an electrically conducting white metal that is shiny, relatively soft, melts at 850 °C, and boils at 1440 °C) was placed into liquid water that was at 25 °C. The calcium reacted slowly with the water to give bubbles of gaseous hydrogen and a solution of the substance calcium hydroxide." In this description, what physical properties and what chemical properties are described?

OH **1.7** In places like Saudi Arabia, freshwater is scarce and is recovered from seawater. When seawater is boiled, the water evaporates and the steam can be condensed to give pure water that people can drink. If all the water is evaporated, solid salt is left behind. Are the changes described here chemical or physical?

1.8 Name the three states of matter.

Elements, Compounds, and Mixtures

1.9 Define (a) element, (b) compound, (c) mixture, (d) homogeneous, (e) heterogeneous, (f) phase, and (g) solution.

1.10 What is the chemical symbol for each of the following elements? (a) chlorine, (b) sulfur, (c) iron, (d) silver, (e) sodium, (f) phosphorus, (g) iodine, (h) copper, (i) mercury, (j) calcium

1.11 What is the name of each of the following elements? (a) K, (b) Zn, (c) Si, (d) Sn, (e) Mn, (f) Mg, (g) Ni, (h) Al, (i) C, (j) N

SI Units

1.12 Why must measurements always be written with units?

1.13 What is the only SI base unit that includes a decimal prefix?

1.14 What is the meaning of each of the following prefixes? (a) centi-, (b) milli-, (c) kilo-, (d) micro-, (e) nano-, (f) pico-, (g) mega-

1.15 What abbreviation is used for each of the prefixes named in Question 1.14?

1.16 What reference points do we use in calibrating the scale of a thermometer? What temperature on the Celsius scale do we assign to each of these reference points?

OH **1.17** In each pair, which is larger: (a) A Fahrenheit degree or a Celsius degree? (b) A Celsius degree or a kelvin? (c) A Fahrenheit degree or a kelvin?

Significant Figures; the Factor-Label Method

1.18 Define the term *significant figures.*

1.19 What is the difference between *accuracy* and *precision?*

1.20 Suppose a length had been reported to be 31.24 cm. What is the minimum uncertainty implied in this measurement?

1.21 Suppose someone suggested using the fraction $\frac{3\text{ yd}}{1\text{ ft}}$ as a conversion factor to change a length expressed in feet to its equivalent in yards. What is wrong with this conversion factor? Can we construct a valid conversion factor relating centimeters to meters from the equation 1 cm = 1000 m? Explain your answer.

1.22 In 1 hour there are 3600 seconds. By what conversion factor would you multiply 250 seconds to convert it to hours? By what conversion factor would you multiply 3.84 hours to convert it to seconds?

1.23 If you were to convert the measured length 4.165 ft to yards by multiplying by the conversion factor (1 yd/3 ft), how many significant figures should the answer contain? Why?

Density

1.24 Write the equation that defines density. Identify the symbols in the equation.

1.25 Silver has a density of 10.5 g cm^{-3}. Express this as an equivalence between mass and volume for silver. Write two conversion factors that can be formed from this equivalence for use in calculations.

REVIEW PROBLEMS

SI Prefixes

1.26 What number should replace the question mark in each of the following?

(a) 1 cm = ? m
(b) 1 km = ? m
(c) 1 m = ? pm
(d) 1 dm = ? m
(e) 1 g = ? kg
(f) 1 cg = ? g

1.27 What numbers should replace the question marks below?

(a) 1 nm = ? m
(b) 1 μg = ? g
(c) 1 kg = ? g
(d) 1 Mg = ? g
(e) 1 mg = ? g
(f) 1 dg = ? g

Temperature Conversions

1.28 Perform the following conversions.

(a) 50 °C to °F
(b) 10 °C to °F
(c) 25.5 °F to °C
(d) 49 °F to °C
(e) 60 °C to K
(f) −30 °C to K

1.29 Perform the following conversions.

(a) 96 °F to °C
(b) −6 °F to °C
(c) −55 °C to °F
(d) 273 K to °C
(e) 299 K to °C
(f) 40 °C to K

1.30 A healthy dog has a temperature ranging from 37.2 to 39.2 °C. Is a dog with a temperature of 103.5 °F within normal range?

1.31 The coldest permanently inhabited place on earth is the Siberian village of Oymyakon in Russia. In 1964 the temperature reached a shivering −96 °F! What is this temperature in °C?

1.32 Estimates of the temperature at the core of the sun range from 10 megakelvins to 25 megakelvins. What is this range in °C and °F?

1.33 Natural gas is mostly methane, a substance that boils at a temperature of 111 K. What is its boiling point in °C and °F?

1.34 Helium has the lowest boiling point of any liquid. It boils at 4 K. What is its boiling point in °C?

1.35 The atomic bomb detonated over Hiroshima, Japan, at the end of World War II raised the temperature on the ground below to about 6000 K. Is this hot enough to melt concrete? (Concrete melts at 2000 °C.)

Significant Figures

1.36 How many significant figures do the following measured quantities have?

(a) 37.53 cm
(b) 37.240 cm
(c) 202.0 g
(d) 0.00024 kg
(e) 0.07080 m
(f) 2400 mL

1.37 How many significant figures do the following measured quantities have?

(a) 0.0230 g
(b) 105.303 m
(c) 0.007 kg
(d) 614.00 mg
(e) 10 L
(f) 3.8105 mm

OH **1.38** Perform the following arithmetic and round off the answers to the correct number of significant figures. Include the correct units with the answers.
(a) 0.0023 m × 315 m
(b) 84.25 kg – 0.01075 kg
(c) (184.45 g − 94.45 g)/(31.4 mL − 9.9 mL)
(d) (23.4 g + 102.4 g + 0.003 g)/(6.478 mL)
(e) (313.44 cm – 209.1 cm) × 8.2234 cm

1.39 Perform the following arithmetic and round off the answers to the correct number of significant figures. Include the correct units with the answers.
(a) 3.58 g/1.739 mL
(b) 4.02 mL + 0.001 mL
(c) (22.4 g − 8.3 g)/(1.142 mL − 0.002 mL)
(d) (1.345 g + 0.022 g)/(13.36 mL – 8.4115 mL)
(e) (74.335 m – 74.332 m)/(4.75 s × 1.114 s)

Unit Conversions by the Factor-Label Method

OH **1.40** Perform the following conversions.
(a) 32.0 dm/s to km/hr
(b) 8.2 mg/mL to μg/L
(c) 75.3 mg to kg
(d) 137.5 mL to L
(e) 0.025 L to mL
(f) 342 pm^2 to dm^2

1.41 Perform the following conversions.
(a) 92 dL to μm^3
(b) 22 ng to μg
(c) 83 pL to nL
(d) 230 km^3 to m^3
(e) 87.3 cm s^{-2} to km hr^{-2}
(f) 238 mm^2 to nm^2

1.42 Perform the following conversions. If necessary, refer to Tables 1.4 and 1.6.
(a) 36 in. to cm
(b) 5.0 lb to kg
(c) 3.0 qt to mL
(d) 1 cup (8 oz) to mL
(e) 55 mi/hr to km/hr
(f) 50.0 mi to km

1.43 Perform the following conversions. If necessary, refer to Tables 1.4 and 1.6.
(a) 250 mL to qt
(b) 3.0 ft to m
(c) 1.62 kg to lb
(d) 1.75 L to fluid oz
(e) 35 km/hr to mi/hr
(f) 80.0 km to mi

1.44 Perform the following conversions.
(a) 8.4 ft^2 to cm^2 (b) 223 mi^2 to km^2 (c) 231 ft^3 to cm^3

1.45 Perform the following conversions.
(a) 2.4 yd^2 to m^2 (b) 8.3 $in.^2$ to mm^2 (c) 9.1 ft^3 to L

1.46 The human stomach can expand to hold up to 4.2 quarts of food. A pistachio nut has a volume of about 0.9 mL. Use this information to estimate the maximum number of pistachios that can be eaten in one sitting.

1.47 In the movie *Cool Hand Luke* (1967), Luke wagers that he can eat 50 eggs in one hour. The prisoners and guards bet against him, saying, "Fifty eggs gotta weigh a good six pounds. A man's gut can't hold that." A chewed, peeled chicken egg has a volume of approximately 53 mL. If Luke's stomach has a volume of 4.2 quarts, does he have any chance of winning the bet?

ILW **1.48** The winds in a hurricane can reach almost 200 miles per hour. What is this speed in meters per second? (Assume three significant figures.)

1.49 A bullet is fired at a speed of 2435 ft/s. What is this speed expressed in kilometers per hour?

1.50 A bullet leaving the muzzle of a pistol was traveling at a speed of 2230 feet per second. What is this speed in miles per hour?

1.51 On average, water flows over Niagara Falls at a rate of 2.05×10^5 cubic feet per second. One cubic foot of water weighs 62.4 lb. Calculate the rate of water flow in tons of water per day. (1 ton = 2000 lb.)

1.52 The brightest star in the night sky in the northern hemisphere is Sirius. Its distance from earth is estimated to be 8.7 light-years. A light-year is the distance light travels in one year. Light travels at a speed of 3.00×10^8 m/s. Calculate the distance from earth to Sirius in miles. (1 mi = 5280 ft.)

1.53 One degree of latitude on the earth's surface equals 60.0 nautical miles. One nautical mile equals 1.151 statute miles. (A *statute mile* is the distance over land that we normally associate with the unit mile.) Calculate the circumference of the earth in statute miles.

1.54 The deepest point in the earth's oceans is found in the Mariana Trench, a deep crevasse located about 1000 miles southeast of Japan beneath the Pacific Ocean. Its maximum depth is 6033.5 fathoms. One fathom is defined as 6 feet. Calculate the depth of the Mariana Trench in meters.

1.55 At sea level, our atmosphere exerts a pressure of about 14.7 lb/in.^{2}, which means that each square inch of your body experiences a force of 14.7 lb from the air that surrounds you. As you descend below the surface of the ocean, the pressure produced by the seawater increases by about 14.7 lb/in.^{2} for every 10 meters of depth. In the preceding problem you calculated the maximum depth of the Mariana Trench, located in the Pacific Ocean. What is the approximate pressure in pounds per square inch and in tons per square inch exerted by the sea at the deepest point of the trench? (1 ton = 2000 lb.)

Density

1.56 A sample of kerosene weighs 36.4 g. Its volume was measured to be 45.6 mL. What is the density of the kerosene?

1.57 A block of magnesium has a mass of 14.3 g and a volume of 8.46 cm^3. What is the density of magnesium in g/cm^3?

OH **1.58** Acetone, the solvent in some nail polish removers, has a density of 0.791 g/mL. What is the volume of 25.0 g of acetone?

1.59 A glass apparatus contains 26.223 g of water when filled at 25 °C. At this temperature, water has a density of 0.99704 g/mL. What is the volume of this apparatus?

1.60 Chloroform, a chemical once used as an anesthetic, has a density of 1.492 g/mL. What is the mass in grams of 185 mL of chloroform?

1.61 Gasoline has a density of about 0.65 g/mL. How much does 34 L (approximately 18 gallons) weigh in kilograms? In pounds?

ILW **1.62** A graduated cylinder was filled with water to the 15.0 mL mark and weighed on a balance. Its mass was 27.35 g. An object made of silver was placed in the cylinder and completely submerged in the water. The water level rose to 18.3 mL. When reweighed, the cylinder, water, and silver object had a total mass of 62.00 g. Calculate the density of silver.

1.63 Titanium is a metal used to make golf clubs. A rectangular bar of this metal measuring 1.84 cm × 2.24 cm × 2.44 cm was found to have a mass of 45.7 g. What is the density of titanium?

1.64 The space shuttle uses liquid hydrogen as its fuel. The external fuel tank used during takeoff carries 227,641 lb of hydrogen with a volume of 385,265 gallons. Calculate the density of liquid hydrogen in units of g/mL. (Express your answer to three significant figures.)

1.65 Some time ago, a U.S. citizen traveling in Canada observed that the price of regular gasoline was 0.959 Canadian dollars per liter. The exchange rate at the time was 1.142 Canadian dollars per one U.S. dollar. Calculate the price of the Canadian gasoline in units of U.S. dollars per gallon. (Just the week before, the traveler had paid $2.249 per gallon in the United States.)

ADDITIONAL EXERCISES

OH **1.66** You are the science reporter for a daily newspaper and your editor asks you to write a story based on a report in the scientific literature. The report states that analysis of the sediments in Hausberg Tarn (elevation 4350 m) on the side of Mount Kenya (elevation 4600–4700 m) shows that the average temperature of the water rose by 4.0 °C between 350 BC and AD 450. Your editor tells you that she wants all the data expressed in the English system of units. Make the appropriate conversions.

1.67 An astronomy website states that neutron stars have a density of 1.00×10^8 tons per cubic centimeter. The site does not specify whether "tons" means metric tons (1 metric ton = 1000 kg) or English tons (1 English ton = 2000 pounds). How many grams would one teaspoon of a neutron star weigh, if the density were in metric tons per cm^3? How many grams would the teaspoon weigh if the density were in English tons per cm^3? (One teaspoon is approximately 4.93 mL.)

1.68 The star Arcturus is 3.50×10^{14} km from the earth. How many days does it take for light to travel from Arcturus to earth? What is the distance to Arcturus in light-years? One light-year is the distance light travels in one year (365 days); light travels at a speed of 3.00×10^8 m/s.

1.69 A pycnometer is a glass apparatus used for accurately determining the density of a liquid. When dry and empty, a certain pycnometer had a mass of 27.314 g. When filled with distilled water at 25.0 °C, it weighed 36.842 g. When filled with chloroform (a liquid once used as an anesthetic before its toxic properties were known), the apparatus weighed 41.428 g. At 25.0 °C, the density of water is 0.99704 g/mL. (a) What is the volume of the pycnometer? (b) What is the density of chloroform?

1.70 Radio waves travel at the speed of light, 3.00×10^8 m/s. If you were to broadcast a question to an astronaut on the moon, which is 239,000 miles from earth, what is the minimum time that you would have to wait to receive a reply?

1.71 Suppose you have a job in which you earn $4.50 for each 30 minutes that you work.

(a) Express this information in the form of an equivalence between dollars earned and minutes worked.

(b) Use the equivalence defined in (a) to calculate the number of dollars earned in 1 hr 45 min.

(c) Use the equivalence defined in (a) to calculate the number of minutes you would have to work to earn $17.35.

1.72 When an object floats in water, it displaces a volume of water that has a weight equal to the weight of the object. If a ship has a weight of 4255 tons, how many cubic feet of seawater will it displace? Seawater has a density of 1.025 g cm^{-3}; 1 ton = 2000 lb.

1.73 Aerogel or "solid smoke" is a novel material that is made of silicon dioxide, like glass, but is a thousand times less dense than glass because it is extremely porous. Material scientists at NASA's Jet Propulsion Laboratory created the lightest aerogel ever in 2002, with a density of 0.00011 pounds per cubic inch. The material was used for thermal insulation in the 2003 Mars Exploration Rover. If the maximum space for insulation in the spacecraft's hull was 2510 cm^3, what mass (in grams) did the aerogel insulation add to the spacecraft?

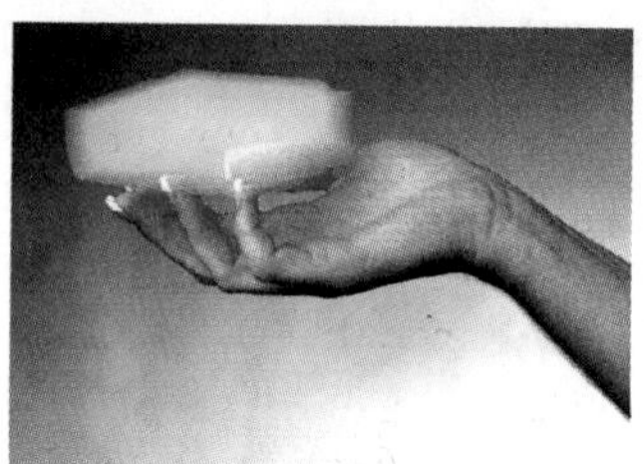

Aerogel. *(NASA/JPL.)*

1.74 A liquid known to be either ethanol (ethyl alcohol) or methanol (methyl alcohol) was found to have a density of 0.798 ± 0.001 g/mL. Consult the *Handbook of Chemistry and Physics* to determine which liquid it is. What other measurements could help to confirm the identity of the liquid?

1.75 An unknown liquid was found to have a density of 69.22 lb/ft^3. The density of ethylene glycol (the liquid used in antifreeze) is 1.1088 g/mL. Could the unknown liquid be ethylene glycol?

1.76 When an object is heated to a high temperature, it glows and gives off light. The color balance of this light depends on the temperature of the glowing object. Photographic lighting is described, in terms of its color balance, as a temperature in kelvins. For example, a certain electronic flash gives a color balance (called color temperature) rated at 5800 K. What is this temperature expressed in °C?

OH ***1.77** There exists a single temperature at which the value reported in °F is numerically the same as the value reported in °C. What is that temperature?

***1.78** In the text, the Kelvin scale of temperature is defined as an absolute scale in which one Kelvin degree unit is the same size as one Celsius degree unit. A second absolute temperature scale exists called the Rankine scale. On this scale, one Rankine degree unit (°R) is the same size as one Fahrenheit degree unit. (a) What is the only temperature at which the Kelvin and Rankine scales possess the same numerical value? Explain your answer. (b) What is the boiling point of water expressed in °R?

***1.79** Density measurements can be used to analyze mixtures. For example, the density of solid sand (without air spaces) is about 2.84 g/mL. The density of gold is 19.3 g/mL. If a 1.00 kg sample of sand containing some gold has a density of 3.10 g/mL (without air spaces), what is the percentage of gold in the sample?

***1.80** An artist's statue has a surface area of 14.6 ft^2. The artist plans to apply gold plate to the statue and wants the coating to be 2.50 μm thick. If the price of gold were $625.10 per troy ounce, how much would it cost to give the statue its gold coating? (1 troy ounce = 31.1035 g; the density of gold is 19.3 g/mL.)

***1.81** A cylindrical metal bar has a diameter of 0.753 cm and a length of 2.33 cm. It has a mass of 8.423 g. Calculate the density of the metal in the units lb/ft^3.

1.82 What is the volume in cubic millimeters of a 3.54 carat diamond, given that the density of the diamond is 3.51 g/mL? (1 carat = 200 mg.)

1.83 Because of the serious consequences of lead poisoning, the Federal Centers for Disease Control in Atlanta has set a threshold of concern for lead levels in children's blood. This threshold was based on a study that suggested that lead levels in blood as low as *10 micrograms of lead per deciliter of blood* can result in subtle effects of lead toxicity. Suppose a child had a lead level in her blood of 2.5×10^{-4} grams of lead per liter of blood. Is this person in danger of exhibiting the effects of lead poisoning?

***1.84** Gold has a density of 19.31 g cm^{-3}. How many grams of gold are required to provide a gold coating 0.500 mm thick on a ball bearing having a diameter of 2.000 mm?

***1.85** A Boeing 747 jet airliner carrying 568 people burns about 5.0 gallons of jet fuel per mile. What is the rate of fuel consumption in units of miles per gallon per person? Is this better or worse than the rate of fuel consumption in an automobile carrying two people that gets 21.5 miles per gallon? If the airliner were making the 3470 mile trip from New York to London, how many pounds of jet fuel would be consumed? (Jet fuel has a density of 0.803 g/mL.)

EXERCISES IN CRITICAL THINKING

1.86 A homogeneous solution is defined as a uniform mixture consisting of a single phase. With our vastly improved abilities to "see" smaller and smaller particles, down to the atomic level, present an argument for the proposition that all mixtures are heterogeneous. Present the argument that the ability to observe objects as small as an atom has no effect on the definitions of heterogeneous and homogeneous.

1.87 Find two or more websites that give the values for each of the seven base SI units. Keeping in mind that not all websites provide reliable information, which website do you believe provides the most reliable values? Justify your answer.

1.88 Reference books such as the *Handbook of Chemistry and Physics* report the specific gravities of substances instead of their densities. Find the definition of specific gravity and discuss the relative merits of specific gravity and density in terms of their usefulness as a physical property.

1.89 A student used a graduated cylinder having volume markings every 2 mL to carefully measure 100 mL of water for an experiment. A fellow student said that by reporting the volume as "100 mL" in her lab notebook, she was only entitled to one significant figure. She disagreed. Why did her fellow student say the reported volume had only one significant figure? Considering the circumstances, how many significant figures are in her measured volume? Justify your answer.

1.90 Download a table of data for the density of water between its freezing and boiling points. Use a spreadsheet program to plot (a) the density of water versus temperature and (b) the volume of a kilogram of water versus temperature. Interpret the significance of these plots.

1.91 List the physical and chemical properties mentioned in this chapter. What additional physical and chemical properties can you think of to extend the list?

ELEMENTS, COMPOUNDS, AND CHEMICAL REACTIONS

2

From a safe distance lightning illuminates the sky and puts on a splendid show in a springtime thunderstorm. In addition to the dazzling show, lightning has the energy to cause chemical reactions to occur between the atmosphere's oxygen and nitrogen molecules. Some of these compounds are fertilizers, essential for life below, and others are compounds that we call pollutants. In this chapter we see how to use chemical reactions to summarize the interactions of elements and compounds. *(Scott Stulberg/Corbis.)*

CHAPTER OUTLIN

THIS CHAPTER IN CONTEXT Students sometimes say that to them "chemistry is a foreign language." The statement is not far from the truth. We can consider the elements in the periodic table to be our new alphabet; the formulas for compounds are the words of chemistry; and balanced equations that show how those compounds react with each other are the sentences of this new language. Learning the language of chemistry will help you succeed because you will be able to concentrate on new concepts that depend on being fluent in our new language.

In Chapter 1 we learned about the broad scope and nature of the subject of chemistry. Importantly we learned that precise and accurate measurements and calculations are central to all sciences, especially chemistry. Now that you've learned these introductory concepts we turn our attention to atoms, elements, chemical compounds, and chemical reactions.

Our study begins with *Dalton's atomic theory,* a theory that has its roots in two basic laws of nature. The law of conservation of mass and the law of definite proportions are the foundation of one of the most important scientific theories.

The atomic theory leads us to the study of the atom's basic parts, *electrons, protons,* and *neutrons,* and how one atom and its isotopes differ from another. *Mendeleev's periodic table* on the other hand is a storehouse of relationships, trends, and similarities between the elements. When elements combine to form a compound, they do so in two broad general ways, either by the sharing of electrons between atoms to make *molecular compounds* or by the transfer of one or more electrons from one atom to another forming *ions* and *ionic compounds.* Compounds are described by *chemical formulas* and their corresponding names. How compounds and elements react with each other is described by a *balanced chemical equation.*

This chapter has three principal goals. The first is to teach you about the atomic theory so you develop an appreciation of scientific theories in general. The second is to help you understand the nature of ionic and molecular substances, how they are formed, and some of their properties. This includes understanding how to interpret chemical formulas and the basics of balancing equations. The third is to introduce you to *chemical nomenclature*—the system used to name chemical compounds. Being able to describe compounds by name is essential for communication among scientists. Therefore, we urge you to make the effort to learn how to name compounds and how to translate chemical names into chemical formulas.

As with the preceding chapter, you may already be familiar with some of the topics we discuss here. Nevertheless, be sure to study them thoroughly. By doing so you begin to build your store of factual knowledge that will enable you to more easily interpret and understand advanced topics as we get to them in later chapters.

2.1 Elements and Atoms Are Described by Dalton's Atomic Theory

In our discussion of elements in the preceding chapter, no reference was made to the atomic nature of matter. In fact, the distinction between elements and compounds had been made even before the atomic theory of matter was formulated. In this section we will examine how the atomic theory began and take a closer look at elements in terms of our modern view of atomic structure.

Dalton's atomic theory explained chemical laws

In modern science, we have come to take for granted the existence of atoms and molecules. In fact, we've already used the atomic theory to explain some of the properties of materials. However, scientific evidence for the existence of atoms is relatively recent, and chemistry did not progress very far until that evidence was found.

The concept of atoms began nearly 2500 years ago when certain Greek philosophers expressed the belief that matter is ultimately composed of tiny indivisible particles, and it is from the Greek word *atomos,* meaning "not cut," that the word *atom* is derived. The philosophers' conclusions, however, were not supported by any evidence; they were derived simply from philosophical reasoning.

Laws of chemical combination evolved from experimental observations

The concept of atoms remained a philosophical belief, having limited scientific usefulness, until the discovery of two quantitative laws of chemical combination: the *law of conservation of mass* and the *law of definite proportions.* The evidence that led to the discovery of these laws came from the experimental observations of many scientists in the eighteenth and early nineteenth centuries.

TOOLS
Law of conservation of mass and law of definite proportions

Law of Conservation of Mass. No detectable gain or loss of mass occurs in chemical reactions. Mass is *conserved.*

Law of Definite Proportions. In a given chemical compound, the elements are always combined in the same proportions by mass.

Notice that both of these laws refer to the masses of substances because the balance was one of the few chemical instruments in those times. Earlier, in our definition of matter, we noted that mass is a measure of the amount of matter in an object. Recall that mass and weight are not the same and you should be careful to use these terms correctly.

The **law of conservation of mass** means that if a chemical reaction takes place in a sealed vessel that permits no matter to enter or escape, the mass of the vessel and its contents after the reaction will be identical to its mass before. Although this may seem quite obvious to us now, it wasn't quite so clear in the early history of modern chemistry when colorless gases could easily be overlooked. When scientists were able to make sure that *all* substances, including any gaseous reactants and/or products, were included when masses were measured, the law of conservation of mass could be truly tested.

We actually used the **law of definite proportions** on page 6 when we defined a compound as a substance in which two or more elements are chemically combined in a *definite fixed proportion by mass.* Thus, if we decompose samples of water (a compound) into the elements oxygen and hydrogen, we always find that the ratio of oxygen to hydrogen, *by mass,* is 8 to 1. In other words, the mass of oxygen obtained is always eight times the mass of hydrogen.

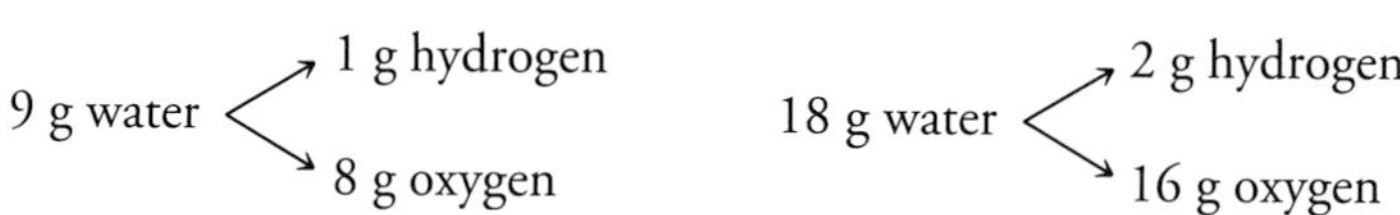

In any sample of water, the mass of oxygen is always eight times the mass of hydrogen.

Similarly, if we form water from oxygen and hydrogen, the mass of oxygen consumed will always be eight times the mass of hydrogen that reacts. This is true even if there's a large excess of one of them. For instance, if 100 g of oxygen is mixed with 1 g of hydrogen and the reaction to form water is initiated, all the hydrogen would react but only 8 g of oxygen would be consumed; there would be 92 g of oxygen left over. No matter how we try, we can't alter the chemical composition of the water formed in the reaction.

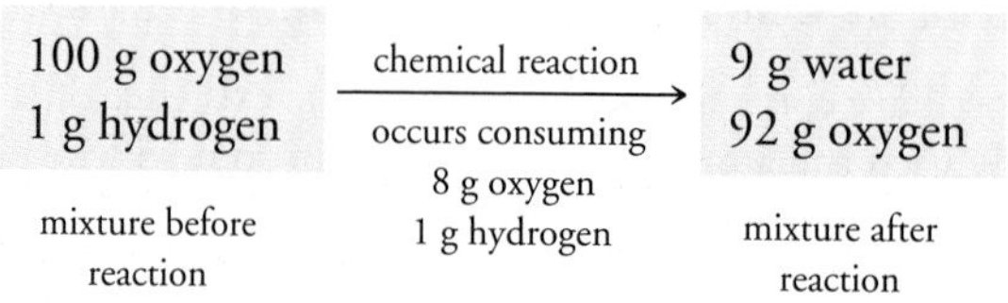

Let's look at a sample calculation that shows how we might use the law of definite proportions.

EXAMPLE 2.1
Applying the Law of Definite Proportions

The element molybdenum (Mo) combines with sulfur (S) to form a compound commonly called molybdenum disulfide that is useful as a dry lubricant, similar to graphite. It is also used in specialized lithium batteries. A sample of this compound contains 1.50 g of Mo for each 1.00 g of S. If a different sample of the compound contains 2.50 g of S, how many grams of Mo does it contain?

ANALYSIS: As you learned in Chapter 1, much of the effort in solving a chemistry problem is devoted to determining which concepts have to be applied. We view these concepts as *tools*, each with its specific uses when applied to problem solving. Our goal in this Analysis step is to identify which tools we need and how we will apply them.

Let's begin by examining the problem and asking a question: What have we learned that relates the masses of elements in two samples of the same compound? We've described two tools relating to masses, the laws of conservation of mass and definite proportions. The law of conservation of mass concerns only the total mass of the chemicals in a reaction, so it doesn't seem to help us here. The law of definite proportions does seem to apply since it concerns the mathematical relationships of elements within a compound no matter where the sample came from. It states that the proportions of the elements by mass must be the same in both samples; so the law of definite proportions is the tool we need to apply. The law tells us that the ratio of grams of Mo to grams of S must be the same in both samples. To solve the problem, then, we will set up the mass ratios for the two samples. In the ratio for the second sample, the mass of molybdenum will be an unknown quantity. We'll equate the two ratios and solve for the unknown quantity.

SOLUTION: Now that we've determined what we need to do to solve the problem, the rest is pretty easy. The first sample has a Mo to S mass ratio of

$$\frac{1.50 \text{ g Mo}}{1.00 \text{ g S}}$$

In the second sample, we know the mass of S (2.50 g) and we want to find the mass of Mo (the unknown is *mass of Mo*). The mass ratio of Mo to S in the second sample is therefore

$$\frac{\textit{mass of Mo}}{2.50 \text{ g S}}$$

Now we equate them, because the two ratios must be equal.

$$\frac{\textit{mass of Mo}}{2.50 \text{ g S}} = \frac{1.50 \text{ g Mo}}{1.00 \text{ g S}}$$

Solving for the *mass of Mo* gives

$$\textit{Mass of Mo} = 2.50 \cancel{\text{g S}} \times \frac{1.50 \text{ g Mo}}{1.00 \cancel{\text{g S}}} = 3.75 \text{ g Mo}$$

IS THE ANSWER REASONABLE? To avoid errors, it's always wise to do a rough check of the answer. Usually, some simple reasoning is all we need to see if the answer is "in the right ball park." This is how we might do such a check here: Notice that the amount of sulfur in the second sample is more than twice the amount in the first sample. Therefore, we should expect the amount of Mo in the second sample to be somewhat more than twice what it is in the first. The answer we obtained, 3.75 g Mo, is more than twice 1.50 g Mo, so our answer seems to be reasonable. In addition we can check that the units "g S" cancel, as shown, to leave the desired units "g Mo."

Practice Exercise 1: Cadmium sulfide is a yellow compound that is used as a pigment in artist's oil colors. A sample of this compound is composed of 1.25 g of cadmium and 0.357 g of sulfur. If a second sample of the same compound contains 3.50 g of sulfur, how many grams of cadmium does it contain? (Hint: Identify the law and write its mathematical form as it applies to this problem.)

Practice Exercise 2: Several samples of compounds containing only iron and sulfur were analyzed by taking the sample and heating it strongly to produce gaseous sulfur oxides and metallic iron. Which of the following compounds are the same and which are different?

Sample	Mass of compound before heating	Mass of iron after heating
A	25.36 g	16.11 g
B	15.42 g	8.28 g
C	7.85 g	4.22 g
D	11.87 g	7.54 g

The atomic theory was proposed by John Dalton

The laws of conservation of mass and definite proportions served as the *experimental foundation* for the atomic theory. They prompted the question: "What must be true about the nature of matter, given the truth of these laws?" In other words, what is matter made of?

At the beginning of the nineteenth century, John Dalton (1766–1844), an English scientist, used the Greek concept of atoms to make sense out of the laws of conservation of mass and definite proportions. Dalton reasoned that if atoms really exist, they must have certain properties to account for these laws. He described such properties, and the list constitutes what we now call **Dalton's atomic theory.**

Dalton's Atomic Theory

1. *Matter consists of tiny particles called atoms.*
2. *Atoms are indestructible. In chemical reactions, the atoms rearrange but they do not themselves break apart.*
3. *In any sample of a pure element, all the atoms are identical in mass and other properties.*
4. *The atoms of different elements differ in mass and other properties.*
5. *When atoms of different elements combine to form compounds, new and more complex particles form. However, in a given compound the constituent atoms are always present in the same fixed* ***numerical*** *ratio.*

Dalton's theory easily explained the law of conservation of mass. According to the theory, a chemical reaction is simply a reordering of atoms from one combination to another. If no atoms are gained or lost and if the masses of the atoms can't change, then the mass after the reaction must be the same as the mass before. This explanation of the law of conservation of mass works so well that it serves as the reason for balancing chemical equations, which we will discuss in the next chapter.

The law of definite proportions is also easy to explain. According to the theory, a given compound always has atoms of the same elements in the same numerical ratio. Suppose, for example, that two elements, A and B, combine to form a compound in which the number of atoms of A equals the number of atoms of B (i.e., the *atom ratio* is 1 to 1). If the mass of a B atom is twice that of an A atom, then every time we encounter a sample of this compound, the mass ratio (A to B) would be 1 to 2. This same mass ratio would exist regardless of the size of the sample, so in samples of this compound the elements A and B are always present in the same proportion by mass.

The atomic theory led to the discovery of the law of multiple proportions

Strong support for Dalton's theory came when Dalton and other scientists studied elements that are able to combine to give two (or more) compounds. For example, sulfur and oxygen form two different compounds, which we call sulfur dioxide and sulfur trioxide. If we decompose a 2.00 g sample of sulfur dioxide, we find it contains 1.00 g of S and 1.00 g of O. If we decompose a 2.50 g sample of sulfur trioxide, we find it also contains 1.00 g of S, but this time the mass of O is 1.50 g. This is summarized in the following table.

Compound	Sample Size	Mass of Sulfur	Mass of Oxygen
Sulfur dioxide	2.00 g	1.00 g	1.00 g
Sulfur trioxide	2.50 g	1.00 g	1.50 g

First, notice that sample sizes aren't the same; they were chosen so that each has the *same mass of sulfur.* Second, the ratio of the masses of oxygen in the two samples is one of small whole numbers.

$$\frac{\text{mass of oxygen in sulfur trioxide}}{\text{mass of oxygen in sulfur dioxide}} = \frac{1.50\text{ g}}{1.00\text{ g}} = \frac{3}{2}$$

Similar observations are made when we study other elements that form more than one compound with each other, and these observations form the basis of the **law of multiple proportions.**

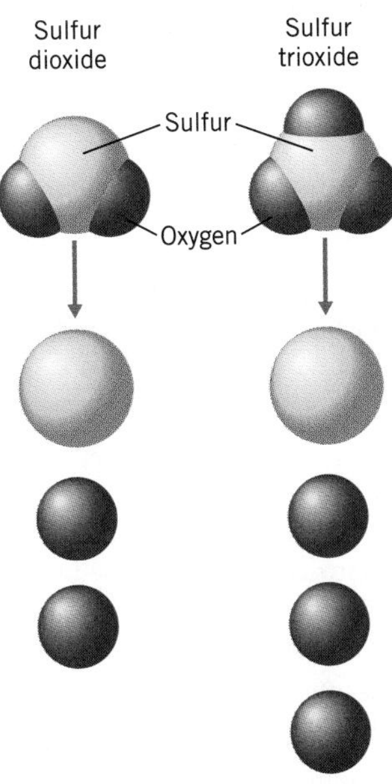

FIG. 2.1 Oxygen compounds of sulfur demonstrate the law of multiple proportions. Illustrated here are molecules of sulfur trioxide and sulfur dioxide. Each has one sulfur atom, and therefore the same mass of sulfur. The oxygen ratio is 3 to 2, both by atoms and by mass.

Law of Multiple Proportions. Whenever two elements form more than one compound, the different masses of one element that combine with the same mass of the other element are in the ratio of small whole numbers.

Dalton's theory explains the law of multiple proportions in a very simple way. Suppose a molecule of sulfur trioxide contains one sulfur and three oxygen atoms, and a molecule of sulfur dioxide contains one sulfur and two oxygen atoms (Figure 2.1). If we had just one molecule of each, then our samples each would have one sulfur atom and therefore the same mass of sulfur. Then, comparing the oxygen atoms, we find they are in a numerical ratio of 3 to 2. But because oxygen atoms all have the same mass, the mass ratio must also be 3 to 2.

The law of multiple proportions was not known before Dalton presented his theory, and its discovery demonstrates the scientific method in action. Experimental data suggested to Dalton the existence of atoms, and the atomic theory suggested the relationships that we now call the law of multiple proportions. Repeated experimental tests have uncovered no instances where the law of multiple proportions fails. These successful tests added great support to the atomic theory. In fact, for many years the law was one of the strongest arguments in favor of the existence of atoms.

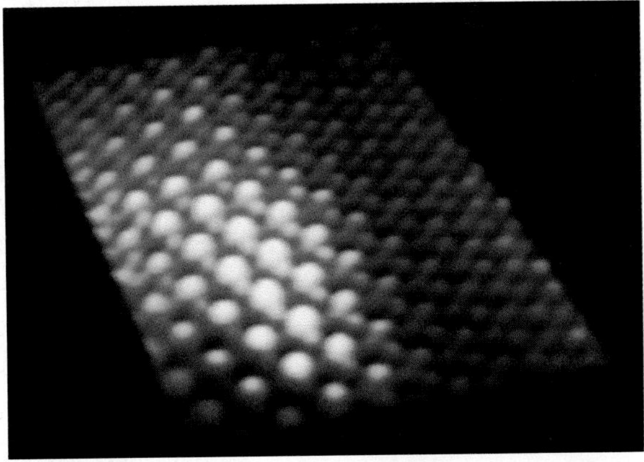

FIG. 2.2 Individual atoms can be imaged using a scanning tunneling microscope. This STM micrograph reveals the pattern of individual atoms of palladium deposited on a graphite surface. Palladium is a silvery white metal used in alloys such as white gold and dental crowns. *(Eurelios/Phototake.)*

Modern experimental evidence exists for atoms

Atoms are so incredibly tiny that even the most powerful optical microscopes are unable to detect them. In recent times, though, scientists have developed very sensitive instruments that are able to map the surfaces of solids with remarkable resolution. One such instrument is called a **scanning tunneling microscope.** It was invented in the early 1980s by Gerd Binnig and Heinrich Rohrer and earned them the 1986 Nobel Prize in Physics. With this instrument, the tip of a sharp metal probe is brought very close to an electrically conducting surface and an electric current bridging the gap is begun. The flow of current is extremely sensitive to the distance between the tip of the probe and the sample. As the tip is moved across the surface, the height of the tip is continually adjusted to keep the current flow constant. By accurately recording the height fluctuations of the tip, a map of the hills and valleys on the surface is obtained. The data are processed using a computer to reveal images such as that shown in Figure 2.2.

2.2 ATOMS ARE COMPOSED OF SUBATOMIC PARTICLES

The earliest theories about atoms imagined them to be indestructible and totally unable to be broken into smaller pieces. However, as you probably know, atoms are not quite as indestructible as Dalton had thought. During the late 1800s and early 1900s, experiments were performed that demonstrated that atoms are composed of **subatomic particles.** (For some of the details about these experiments, see Facets of Chemistry 2.1.) From this work the current theoretical model of atomic structure evolved. We will examine it in general terms in this chapter. A more detailed discussion of atomic structure will follow in Chapter 7.

Protons, neutrons, and electrons are subatomic particles

Physicists have discovered a large number of subatomic particles, but protons, neutrons, and electrons are the only ones that will concern us at this time.

Protons are in all nuclei. Except for ordinary hydrogen, all nuclei also contain neutrons.

Experiments have shown that atoms are composed of three principal kinds of subatomic particles: **protons, neutrons,** and **electrons.** Experiments also revealed that at the center of an atom there exists a very tiny, extremely dense core called the **nucleus,** which is where an atom's protons and neutrons are found. Because they are found in nuclei, protons and neutrons are sometimes called **nucleons.** The electrons in an atom surround the nucleus and fill the remaining volume of the atom. (*How* the electrons are distributed around the nucleus is the subject of Chapter 7.) The properties of the subatomic particles are summarized in Table 2.1, and the general structure of the atom is illustrated in Figure 2.3.

Notice that two of the subatomic particles carry electrical charges. Protons carry a single unit of **positive charge** and electrons carry one unit of the opposite charge, a **negative charge.** Two particles that have the same electrical charge repel each other and two particles that have opposite charges will experience an attractive force. In an atom the negatively charged electrons are attracted to positively charged protons. In fact, it is this attraction that holds the electrons around the nucleus. Neutrons have no charge and are said to be electrically neutral (hence the name *neutron*).

Because of their identical charges, electrons repel each other. The repulsions between the electrons keep them spread out throughout the volume of the atom, and it is the *balance* between the attractions the electrons feel toward the nucleus and the repulsions they feel toward each other that controls the sizes of atoms.

The binding energy of nucleons is discussed in Chapter 20.

Protons also repel each other, but they are able to stay together in the small volume of the nucleus because their repulsions are apparently offset by powerful nuclear forces that involve other subatomic particles we will not study.

Matter as we generally find it in nature appears to be electrically neutral, which means that it contains equal numbers of positive and negative charges. Therefore, *in a neutral atom, the number of electrons must equal the number of protons.*

The proton and neutron are much more massive than the electron, so in any atom almost all of the atomic mass is contributed by the particles that are found in the nucleus. (The mass of an electron is only about 1/1800 of that of a proton or neutron.) It is also interesting to note, however, that the diameter of an atom is approximately 10,000 times the diameter of its nucleus, so almost all the *volume* of an atom is occupied by its electrons, which fill the space around the nucleus. (To place this on a more meaningful scale, if the nucleus were 1 ft in diameter, it would lie at the center of an atom with a diameter of approximately 1.9 miles!)

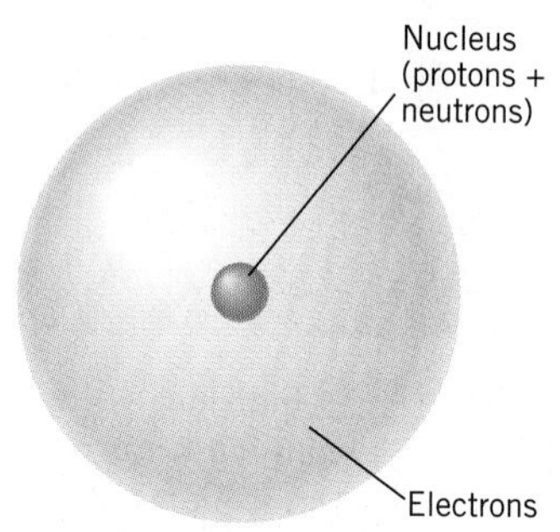

FIG. 2.3 The internal structure of an atom. An atom is composed of a tiny nucleus that holds all the protons and neutrons; the electrons fill the space outside the nucleus.

TABLE 2.1 Properties of Subatomic Particles

Particle	Mass (g)	Electrical Charge	Symbol
Electron	9.109383×10^{-28}	1−	${}_{-1}^{0}e$
Proton	$1.6726217 \times 10^{-24}$	1+	${}_{1}^{1}H^{+}$, ${}_{1}^{1}p$
Neutron	$1.6749273 \times 10^{-24}$	0	${}_{0}^{1}n$

FACETS OF CHEMISTRY

Experiments Leading to the Discovery of Subatomic Particles

Our current knowledge of atomic structure was pieced together from facts obtained from experiments by scientists that began in the nineteenth century. In 1834, Michael Faraday discovered that the passage of electricity through aqueous solutions could cause chemical changes, which was the first hint that matter was electrical in nature. Later in that century, scientists began to experiment with *gas discharge tubes* in which a high-voltage electric current was passed through a gas at low pressure in a glass tube (Figure 1). Such a tube is fitted with a pair of metal *electrodes* and when the electricity begins to flow between them, the gas in the tube glows. This flow of electricity is called an *electric discharge*, which is how the tubes got their names. (Modern neon signs work this way.)

The physicists who first studied this phenomenon did not know what caused the tube to glow, but tests soon revealed that negatively charged particles were moving from the negative electrode (the *cathode*) to the positive electrode (the *anode*). According to legend, it was Benjamin Franklin who decided which electrode was positive and which was negative. The physicists called these emissions *rays*, and because the rays came from the cathode, they were called *cathode rays*.

In 1897, the British physicist J. J. Thomson constructed a special gas discharge tube to make quantitative measurements of the properties of cathode rays. In some ways, the *cathode ray tube* he used was similar to a TV picture tube, as Figure 2 shows. In Thomson's tube, a beam of cathode rays was focused on a glass surface coated with a phosphor that glows when the cathode rays strike it (point 1). The cathode ray beam passed between the poles of a magnet and between a pair of metal electrodes that could be given electrical charges. The magnetic field tends to bend the beam in one direction (to point 2) whereas the charged electrodes bend the beam in the opposite direction (to point 3). By adjusting the charge on the electrodes, the two effects can be made to cancel, and from the amount of charge on the electrodes required to balance the effect of the magnetic field, Thomson was able to calculate the first bit of quantitative information about a cathode ray particle—the ratio of its charge to its mass (often expressed as e/m, where e stands for charge and m stands for mass). The charge-to-mass ratio has a value of -1.76×10^8 coulombs/gram, where the coulomb (C) is a standard unit of electrical charge and the negative sign reflects the negative charge on the particle.

Many experiments were performed using the cathode ray tube and they demonstrated that cathode ray particles are in all matter. They are, in fact, *electrons*.

Measuring the Charge and Mass of the Electron. In 1909, a researcher at the University of Chicago, Robert Millikan, designed a clever experiment that enabled him to measure the electron's charge (Figure 3). During an experiment he would spray a fine mist of oil droplets above a pair of parallel metal plates, the top one of which had a small hole in it. As the oil drops settled, some would pass through this hole into the space between the plates, where he would irradiate them briefly with X rays. The X rays knocked electrons off molecules in the air, and the electrons became attached to the oil drops, which thereby were given an electrical charge. By observing the rate of fall of the charged drops both when the metal plates were electrically charged and when they were not, Millikan was able to calculate the amount of charge carried by each drop. When he examined his results, he found that all the values he obtained were whole-number multiples of -1.60×10^{-19} C. He reasoned that since a drop could only pick up whole numbers of electrons, this value must be the charge carried by each individual electron.

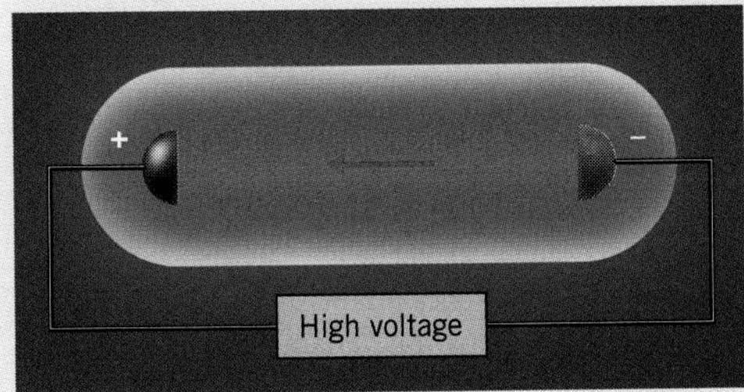

FIG. 1 **A gas discharge tube.** Cathode rays flow from the negatively charged cathode to the positively charged anode.

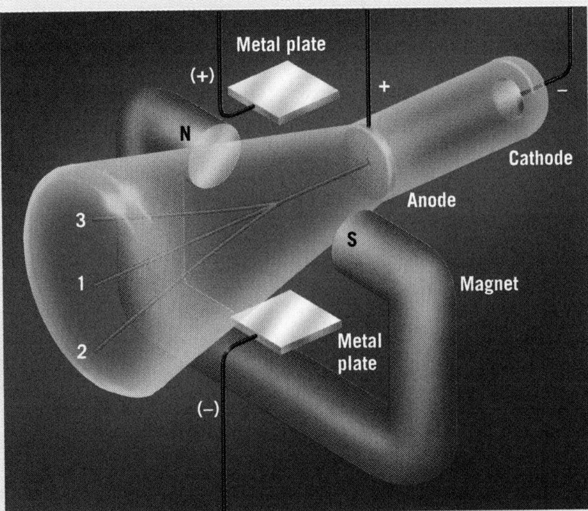

FIG. 2 **Thomson's cathode ray tube.** This device was used to measure the charge-to-mass ratio for the electron.

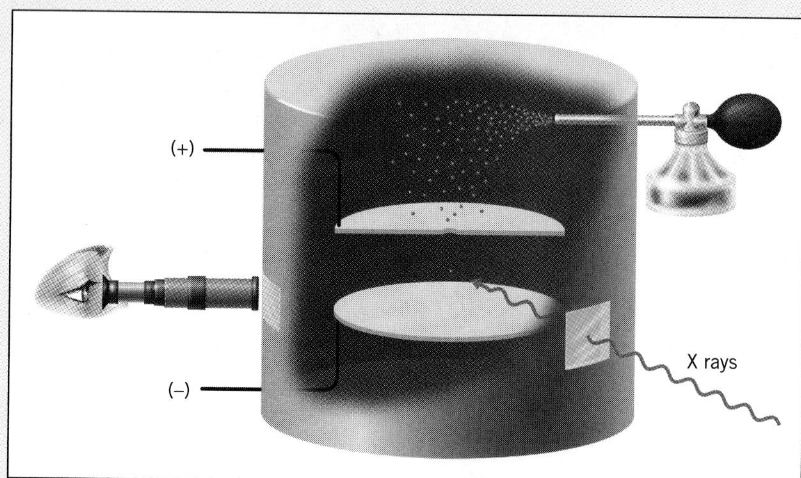

FIG. 3 Millikan's oil drop experiment. Electrons, which are ejected from air molecules by the X rays, are picked up by very small drops of oil falling through the tiny hole in the upper metal plate. By observing the rate of fall of the charged oil drops, with and without electrical charges on the metal plates, Millikan was able to calculate the charge carried by an electron.

Once Millikan had measured the electron's charge, its mass could then be calculated from Thomson's charge-to-mass ratio. This mass was calculated to be 9.09×10^{-28} g. More precise measurements have since been made, and the mass of the electron is currently reported to be 9.109383×10^{-28} g.

Discovery of the Proton. The removal of electrons from an atom gives a positively charged particle (called an *ion*). To study these, a modification was made in the construction of the cathode ray tube to produce a new device called a *mass spectrometer*. This apparatus is described in Facets of Chemistry 2.2 on page 47 and was used to measure the charge-to-mass ratios of positive ions. These ratios were found to vary, depending on the chemical nature of the gas in the discharge tube, showing that their masses also varied. The lightest positive particle observed was produced when hydrogen was in the tube, and its mass was about 1800 times as heavy as an electron. When other gases were used, their masses always seemed to be whole-number multiples of the mass observed for hydrogen atoms. This suggested the possibility that clusters of the positively charged particles made from hydrogen atoms made up the positively charged particles of other gases. The hydrogen atom, minus an electron, thus seemed to be a fundamental particle in all matter and was named the *proton*, after the Greek word *proteios,* meaning "of first importance."

Discovery of the Atomic Nucleus

Early in the twentieth century, Hans Geiger and Ernest Marsden, working under Ernest Rutherford at Great Britain's Manchester University, studied what happened when *alpha rays* hit thin metal foils. Alpha rays are composed of particles having masses four times those of the proton and bearing two positive charges; they are emitted by certain unstable atoms in a phenomenon called *radioactivity*. Most of the alpha particles sailed right on through as if the foils were virtually empty space (Figure 4). A significant number of alpha particles,

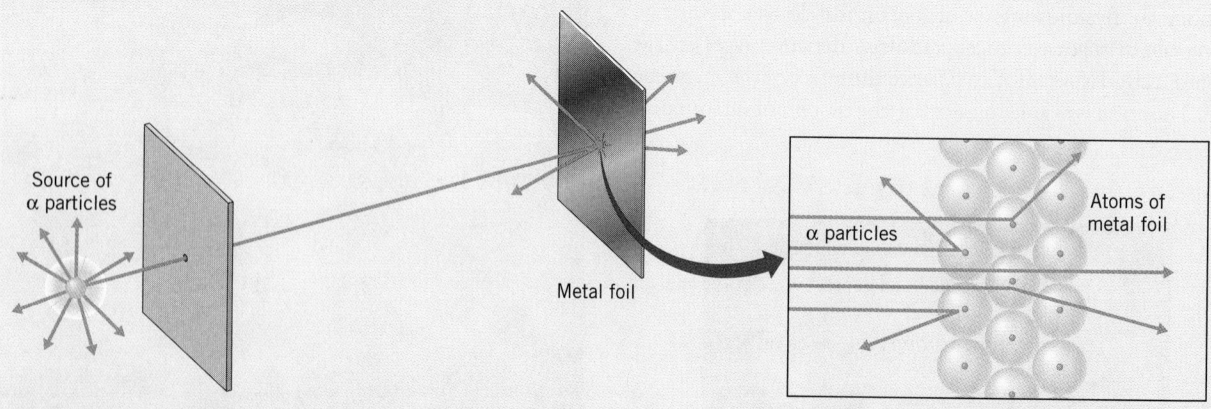

FIG. 4 Rutherford's alpha-particle experiment. Alpha particles are scattered in all directions by a thin metal foil. Some hit something very massive head-on and are deflected backward. Many sail through. Some, making near misses with the massive "cores" (nuclei), are still deflected, because alpha particles have the same kind of charge (+) as these cores.

however, were deflected at very large angles. Some were even deflected backward, as if they had hit stone walls. Rutherford was so astounded that he compared the effect to that of firing a 15 in. artillery shell at a piece of tissue paper and having it come back and hit the gunner! From studying the angles of deflection of the particles, Rutherford reasoned that only something extraordinarily massive and positively charged could cause such an occurrence. Since most of the alpha particles went straight through, he further reasoned that the metal atoms in the foils must be mostly empty space. Rutherford's ultimate conclusion was that virtually all of the mass of an atom must be concentrated in a particle having a very small volume located in the center of the atom. He called this massive particle the atom's *nucleus.*

Discovery of the Neutron

From the way alpha particles were scattered by a metal foil, Rutherford and his students were able to estimate the number of positive charges on the nucleus of an atom of the metal. This had to be equal to the number of protons in the nucleus, of course. But when they computed the nuclear mass based on this number of protons, the value always fell short of the actual mass. In fact, Rutherford found that only about half of the nuclear mass could be accounted for by protons. This led him to suggest that there were other particles in the nucleus that had a mass close to or equal to that of a proton, but with no electrical charge. This suggestion initiated a search that finally ended in 1932 with the discovery of the *neutron* by Sir James Chadwick, a British physicist.

Atomic numbers define elements and mass numbers describe isotopes

TOOLS *The number of electrons, protons, and neutrons in atoms*

What distinguishes one element from another is the number of protons in the nuclei of its atoms, because *all the atoms of a particular element have an identical number of protons.* In fact, this allows us to redefine an **element** as *a substance whose atoms all contain the identical number of protons.* Thus, each element has associated with it a unique number, which we call its **atomic number** (***Z***), that equals the number of protons in the nuclei of any of its atoms.

$$\text{Atomic number } (Z) = \text{number of protons}$$

Most elements exist in nature as mixtures of similar atoms called *isotopes* that differ only in mass. What makes isotopes of the same element different are the numbers of neutrons in their nuclei. *The* **isotopes** *of a given element have atoms with the same number of protons but different numbers of neutrons.* The numerical sum of the protons and neutrons in the atoms of a particular isotope is called the **mass number** (***A***) of the isotope.

$$\text{Isotope mass number } (A) = (\text{number of protons}) + (\text{number of neutrons})$$

Therefore, every isotope is fully defined by two numbers, its atomic number and its mass number. Sometimes these numbers are added to the left of the chemical symbol as a subscript and a superscript, respectively. Thus, if X stands for the chemical symbol for the element, an isotope of X is represented as

$$^{A}_{Z}X$$

The isotope of uranium used in nuclear reactors, for example, can be symbolized as follows:

mass number (protons + neutrons) → $^{235}_{92}\text{U}$

atomic number (number of protons) →

uranium-235

TOOLS *Atomic symbols for isotopes*

As indicated, the name of this isotope is uranium-235 or U-235. Each neutral atom contains 92 protons and $(235 - 92) = 143$ neutrons as well as 92 electrons. In writing the symbol for the isotope, the atomic number is often omitted because it is redundant. Every atom of uranium has 92 protons, and every atom that has 92 protons is an atom of uranium. Therefore, this uranium isotope can be represented simply as ^{235}U.

It is useful to remember that for a neutral atom, the atomic number equals both the number of protons and the number of electrons.

In naturally occurring uranium, a more abundant isotope is ^{238}U. Atoms of this isotope also have 92 protons, but the number of neutrons is 146. Thus, atoms of ^{235}U and ^{238}U have the identical number of protons but differ in the numbers of neutrons.

EXAMPLE 2.2
Counting Protons, Neutrons, and Electrons

How many electrons, protons, and neutrons does the Cr-52 isotope have?

ANALYSIS: This problem asks for all three of the major subatomic particles in the chromium isotope that has a nominal mass of 52. Therefore, all three of the tools for subatomic particles must be used:

$$\text{Protons} = \text{atomic number} = Z$$

$$\text{Electrons} = \text{atomic number} = Z$$

$$\text{Neutrons} = \text{mass number} - \text{atomic number} = A - Z$$

SOLUTION: Applying the tools we find that $Z = 24$ and $A = 52$ and we conclude

$$\text{Protons} = 24 \qquad \text{Electrons} = 24 \qquad \text{Neutrons} = 52 - 24 = 28$$

IS THE ANSWER REASONABLE? One check is to be sure that the sum of the number of protons and neutrons is the mass number. A second check is that the number of any of the particles is not larger than the mass (the largest number given in the problem) and in most cases the number of electrons, protons, or neutrons is usually close to half of the mass. Our answers fulfill these conditions.

Practice Exercise 3: Write the symbol for the isotope of plutonium (Pu) that contains 146 neutrons. How many electrons does it have? (Hint: Review the tools for writing isotope symbols and counting electrons.)

Practice Exercise 4: How many protons, neutrons, and electrons are in each atom of $^{35}_{17}Cl$?

Practice Exercise 5: In the previous exercise, can we discard the 35 or the 17 or both from the symbol without losing the ability to solve the problem? Explain your reasoning.

Relative atomic masses of elements can be found

One of the most useful concepts to come from Dalton's atomic theory is that atoms of an element have a constant, characteristic **atomic mass** (or **atomic weight**). This concept opened the door to the determination of chemical formulas and ultimately to one of the most useful devices chemists have for organizing chemical information, the periodic table of the elements. But how can the masses of atoms be measured?

Individual atoms are much too small to weigh in the traditional manner. However, the *relative masses* of the atoms of elements can be determined *provided we know the ratio in which the atoms occur in a compound.* Let's look at an example to see how this could work.

Hydrogen (H) combines with the element fluorine (F) to form the compound hydrogen fluoride. Each molecule of this compound contains one atom of hydrogen and one atom of fluorine, which means that in *any* sample of this substance the fluorine-to-hydrogen *atom ratio* is always 1 to 1. It is also found that when a sample of hydrogen fluoride is decomposed, the mass of fluorine obtained is always 19.0 times larger than the mass of hydrogen, so the fluorine-to-hydrogen *mass ratio* is always 19.0 to 1.00.

F-to-H atom ratio: 1 to 1

F-to-H mass ratio: 19.0 to 1.00

How could a 1-to-1 atom ratio give a 19.0-to-1.00 mass ratio? *Only if each fluorine atom is 19.0 times heavier than each H atom.*

Notice that even though we haven't found the actual masses of F and H atoms, we do now know how their masses compare (i.e., we know their *relative masses*). Similar procedures with other elements in other compounds are able to establish relative mass relationships among the other elements as well. What we need next is a way to place all these masses on the same mass scale.

TOOLS *Atomic masses are relative*

Carbon-12 is the standard on the atomic mass scale

To establish a uniform mass scale for atoms it is necessary to select a standard against which the relative masses can be compared. Currently, the agreed-upon reference is the most abundant isotope of carbon, called carbon-12 and symbolized ^{12}C. One atom of this isotope is assigned *exactly* 12 units of mass, which are called **atomic mass units.** Some prefer to use the symbol **amu** for the atomic mass unit. The internationally accepted symbol is **u,** which is the symbol we will use throughout the rest of the book. By assigning 12 u to the mass of one atom of ^{12}C, the size of the atomic mass unit is established to be $\frac{1}{12}$ of the mass of a single carbon-12 atom:

The atomic mass unit is sometimes called a dalton, 1 u = 1 dalton.

1 atom of ^{12}C has a mass of 12 u (exactly)
1 u equals $\frac{1}{12}$ the mass of 1 atom of ^{12}C (exactly)

In modern terms, the atomic mass of an element is the average mass of the element's atoms (as they occur in nature) relative to an atom of carbon-12, which is assigned a mass of 12 units. Thus, if an average atom of an element has a mass twice that of a ^{12}C atom, its atomic mass would be 24 u.

The definition of the size of the atomic mass unit is really quite arbitrary. It could just as easily have been selected to be $\frac{1}{24}$ of the mass of a carbon atom, or $\frac{1}{10}$ of the mass of an iron atom, or any other value. Why $\frac{1}{12}$ of the mass of a ^{12}C atom? First, carbon is a very common element, available to any scientist. Second, and most important, by choosing the atomic mass unit of this size, the atomic masses of nearly all the other elements are almost whole numbers, with the lightest atom (hydrogen) having a mass of approximately 1 u.

Chemists generally work with whatever *mixture* of isotopes comes with a given element as it occurs naturally. Because the composition of this isotopic mixture is very nearly constant regardless of the source of the element, we can speak of an *average atom* of the element—average in terms of mass. For example, naturally occurring hydrogen is a mixture of two isotopes in the relative proportions given in the margin. The "average atom" of the element hydrogen, as it occurs in nature, has a mass that is 0.083992 times that of a ^{12}C atom. Since $0.083992 \times 12.000 \text{ u} = 1.0079 \text{ u}$, the average atomic mass of hydrogen is 1.0079 u. Notice that this average value is just a little larger than the atomic mass of ^{1}H because naturally occurring hydrogen also contains a little ^{2}H as shown in Table 2.2.

Even the smallest laboratory sample of an element has so many atoms that the relative proportions of the isotopes is constant.

In general, the mass number of an isotope differs slightly from the atomic mass of the isotope. For instance, the isotope ^{35}Cl has an atomic mass of 34.968852 u. In fact, the *only* isotope that has an atomic mass equal to its mass number is ^{12}C; *by definition* the mass of this atom is exactly 12 u.

TABLE 2.2 Abundance of Hydrogen Isotopes

Hydrogen Isotope	Mass	Percentage Abundance
^{1}H	1.007825 u	99.985
^{2}H	2.0140 u	0.015

Average atomic masses can be calculated from isotopic abundances

Originally, the relative atomic masses of the elements were determined in a way similar to that described for hydrogen and fluorine in our earlier discussion. A sample of a compound was analyzed and from the formula of the substance the relative atomic masses were calculated. These were then adjusted to place them on the unified atomic mass scale. In modern times, methods have been developed to measure very precisely both the relative abundances of the isotopes of the elements and their atomic masses. (See Facets of Chemistry 2.2.) This kind of information has permitted the calculation of more precise values of the average atomic masses, which are found in the table on the inside front cover of the book. Example 2.3 illustrates how this calculation is done.

EXAMPLE 2.3

Calculating Average Atomic Masses from Isotopic Abundances

Naturally occurring chlorine is a mixture of two isotopes. In every sample of this element, 75.77% of the atoms are ^{35}Cl and 24.23% are atoms of ^{37}Cl. The accurately measured atomic mass of ^{35}Cl is 34.9689 u and that of ^{37}Cl is 36.9659 u. From these data, calculate the average atomic mass of chlorine.

ANALYSIS: In any natural sample containing many atoms of chlorine, 75.77% of the mass is contributed by atoms of ^{35}Cl and 24.23% is contributed by atoms of ^{37}Cl. This means that when we calculate the mass of the "average atom," we have to proportion it according to both the masses of the isotopes and their relative abundances. It is convenient to imagine an "average atom" to be composed of 75.77% of ^{35}Cl and 24.23% of ^{37}Cl. (Keep in mind, of course, that such an atom doesn't really exist.) We also recall that when we need to use percentages in calculations, we must divide the percentage by 100 to obtain a decimal number. This decimal number, when multiplied by the isotope mass, will tell us how much of the average mass is contributed by that isotope. All we need to do is add the contributions for all isotopes to obtain the average mass.

SOLUTION: We will calculate 75.77% of the mass of an atom of ^{35}Cl, which is the contribution of this isotope to the "average atom":

$$\text{contribution of } ^{35}\text{Cl} = \frac{75.77\%\ ^{35}\text{Cl} \times 34.9689 \text{ u}}{100\ \%} = 26.496 \text{ u}$$

and for the ^{37}Cl, its contribution is

$$\text{contribution of } ^{37}\text{Cl} = \frac{24.23\%\ ^{37}\text{Cl} \times 36.9659 \text{ u}}{100\%} = 8.9568 \text{ u}$$

Then we add these contributions to give us the total mass of the "average atom."

$$26.496 \text{ u} + 8.957 \text{ u} = 35.453 \text{ u rounded to } 35.45 \text{ u}$$

Notice that in a two-step problem we kept one extra significant figure until the final rounding to four significant figures.

IS THE ANSWER REASONABLE? Once again, the final step is a check to see if the answer makes sense. Here is how we might do such a check: First, from the masses of the isotopes, we know the average atomic mass is somewhere between approximately 35 and 37. If the abundances of the two isotopes were equal, the average would be nearly 36. But there is more ^{35}Cl than ^{37}Cl, so a value closer to 35 than 37 seems reasonable; therefore, we can feel pretty confident our answer is correct.

Practice Exercise 6: Aluminum atoms have a mass that is 2.24845 times that of an atom of ^{12}C. What is the atomic mass of aluminum? (Hint: Recall that we have a tool that gives the relationship between the atomic mass unit and ^{12}C.)

Practice Exercise 7: How much heavier than an atom of ^{12}C is the average atom of naturally occurring copper? Refer to the table inside the front cover of the book for the necessary data.

Practice Exercise 8: Naturally occurring boron is composed of 19.8% of ^{10}B and 80.2% of ^{11}B. Atoms of ^{10}B have a mass of 10.0129 u and those of ^{11}B have a mass of 11.0093 u. Calculate the average atomic mass of boron.

FACETS OF CHEMISTRY 2.2

The Mass Spectrometer and the Experimental Measurement of Atomic Masses

When a spark is passed through a gas, electrons are knocked off the gas molecules. Because electrons are negatively charged, the particles left behind carry positive charges; they are called *positive ions*. These positive ions have different masses, depending on the masses of the molecules from which they are formed. Thus, some molecules have large masses and give heavy ions, while others have small masses and give light ions.

The device that is used to study the positive ions produced from gas molecules is called a *mass spectrometer* (illustrated in the figure at the right). In a mass spectrometer, positive ions are created by passing an electrical spark (called an *electric discharge*) through a sample of the particular gas being studied. As the positive ions are formed, they are attracted to a negatively charged metal plate that has a small hole in its center. Some of the positive ions pass through this hole and travel onward through a tube that passes between the poles of a powerful magnet.

One of the properties of charged particles, both positive and negative, is that their paths become curved as they pass through a magnetic field. This is exactly what happens to the positive ions in the mass spectrometer as they pass between the poles of the magnet. However, the extent to which their paths are bent depends on the masses of the ions. This is because the path of a heavy ion, like that of a speeding cement truck, is difficult to change, but the path of a light ion, like that of a motorcycle, is influenced more easily. As a result, heavy ions emerge from between the magnet's poles along different lines than the lighter ions. In effect, an entering beam containing ions of different masses is sorted by the magnet into a number of beams, each containing ions of the same mass. This spreading out of the ion beam thus produces an array of different beams called a *mass spectrum*.

In practice, the strength of the magnetic field is gradually changed, which sweeps the beams of ions across a detector located at the end of the tube. As a beam of ions strikes the detector, its intensity is measured and the masses of the particles in the beam are computed based on the strength of the magnetic field and the geometry of the apparatus.

Among the benefits derived from measurements using the mass spectrometer are very accurate isotopic masses and relative isotopic abundances. These serve as the basis for the very precise values of the atomic masses that you find in the periodic table.

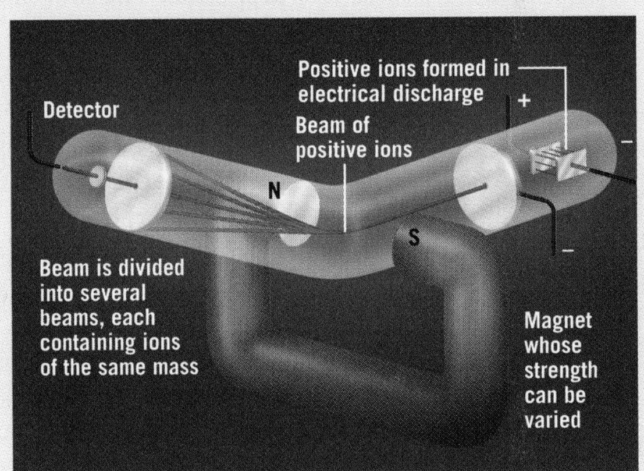

2.3 THE PERIODIC TABLE IS USED TO ORGANIZE AND CORRELATE FACTS

When we study different kinds of substances, we find that some are elements and others are compounds. Among compounds, some are composed of discrete molecules. Others, as you will learn, are made up of atoms that have acquired electrical charges. For elements such as sodium and iron we mentioned their metallic properties. On the other hand, chlorine and oxygen are not classified as metals. If we were to continue on this way, without attempting to build our subject around some central organizing structure, it would not be long before we became buried beneath a mountain of information in the form of seemingly unconnected facts.

The need for organization was recognized by many early chemists, and there were numerous attempts to discover relationships among the chemical and physical properties of the elements. A number of different sequences of elements were tried in the search for some sort of order or pattern. A few of these arrangements came quite close, at least in some respects, to our current periodic table, but either they were flawed in some way or they were presented to the scientific community in a manner that did not lead to their acceptance.

Mendeleev created the first periodic table

The periodic table we use today is based primarily on the efforts of a Russian chemist, Dmitri Ivanovich Mendeleev (1834–1907), and a German physicist, Julius Lothar Meyer (1830–1895). Working independently, these scientists developed similar periodic tables only a few months apart in 1869. Mendeleev is usually given the credit, however, because he had the good fortune to publish first.

Mendeleev was preparing a chemistry textbook for his students at the University of St. Petersburg. Looking for some pattern among the properties of the elements, he found that when he arranged them in order of increasing atomic mass, similar chemical properties were repeated over and over again at regular intervals. For instance, the elements lithium (Li), sodium (Na), potassium (K), rubidium (Rb), and cesium (Cs) are soft metals that are very reactive toward water. They form compounds with chlorine that have a 1-to-1 ratio of metal to chlorine. Similarly, the elements that immediately follow each of these also constitute a set with similar chemical properties. Thus, beryllium (Be) follows lithium, magnesium (Mg) follows sodium, calcium (Ca) follows potassium, strontium (Sr) follows rubidium, and barium (Ba) follows cesium. All of these elements form a water-soluble chlorine compound with a 1-to-2 metal to chlorine atom ratio. Mendeleev used such observations to construct his **periodic table,** which is illustrated in Figure 2.4.

Periodic refers to the recurrence of properties at regular intervals.

The elements in Mendeleev's table are arranged in order of increasing atomic mass. When the sequence is broken at the right places and stacked, the elements fall naturally into columns, called *groups,* in which the elements of a given group have similar chemical properties. The rows themselves are called *periods.*

Mendeleev's genius rested on his placing elements with similar properties in the same group even when this left occasional gaps in the table. For example, he placed arsenic (As) in Group V under phosphorus because its chemical properties were similar to those of phosphorus, even though this left gaps in Groups III and IV. Mendeleev reasoned, correctly, that the elements that belonged in these gaps had simply not yet been discovered. In fact, on the basis of the location of these gaps Mendeleev was able to predict with remarkable accuracy the properties of these yet-to-be-found substances. His predictions helped serve as a guide in the search for the missing elements.

FIG. 2.4 **The first periodic table.** Mendeleev's periodic table roughly as it appeared in 1871. The numbers next to the symbols are atomic masses.

Periods	Group I	Group II	Group III	Group IV	Group V	Group VI	Group VII	Group VIII
1	H 1							
2	Li 7	Be 9.4	B 11	C 12	N 14	O 16	F 19	
3	Na 23	Mg 24	Al 27.3	Si 28	P 31	S 32	Cl 35.5	
4	K 39	Ca 40	— 44	Ti 48	V 51	Cr 52	Mn 55	Fe 56, Co 59 Ni 59, Cu 63
5	(Cu 63)	Zn 65	— 68	— 72	As 75	Se 78	Br 80	
6	Rb 85	Sr 87	?Yt 88	Zr 90	Nb 94	Mo 96	— 100	Ru 104, Rh 104 Pd 105, Ag 108
7	(Ag 108)	Cd 112	In 113	Sn 118	Sb 122	Te 128	I 127	
8	Cs 133	Ba 137	?Di 138	?Ce 140	—	—	—	— —
9	—	—	—	—	—	—	—	
10	—	—	?Er 178	?La 180	Ta 182	W 184	—	Os 195, Ir 197 Pt 198, Au 199
11	(Au 199)	Hg 200	Tl 204	Pb 207	Bi 208	—		
12	—	—	—	Th 231	—	U 240	—	— — — —

The elements tellurium (Te) and iodine (I) caused Mendeleev some problems. According to the best estimates at that time, the atomic mass of tellurium was greater than that of iodine. Yet if these elements were placed in the table according to their atomic masses, they would not fall into the proper groups required by their properties. Therefore, Mendeleev switched their order and in so doing violated his ordering sequence. (Actually, he believed that the atomic mass of tellurium had been incorrectly measured, but this wasn't so.)

The table that Mendeleev developed is in many ways similar to the one we use today. One of the main differences, though, is that Mendeleev's table lacks the column containing the elements helium (He) through radon (Rn). In Mendeleev's time, none of these elements had yet been found because they are relatively rare and because they have virtually no tendency to undergo chemical reactions. When these elements were finally discovered, beginning in 1894, another problem arose. Two more elements, argon (Ar) and potassium (K), did not fall into the groups required by their properties if they were placed in the table in the order required by their atomic masses. Another switch was necessary and another exception had been found. It became apparent that atomic mass was not the true basis for the periodic repetition of the properties of the elements. To determine what the true basis was, however, scientists had to await the discoveries of the atomic nucleus, the proton, and atomic numbers.

The modern periodic table arranges elements by atomic number

When atomic numbers were discovered, it was soon realized that the elements in Mendeleev's table are arranged in precisely the order of increasing atomic number. In other words, if we take atomic numbers as the basis for arranging the elements in sequence, no annoying switches are required and the elements Te and I or Ar and K are no longer a problem. The fact that it is the atomic number—the number of protons in the nucleus of an atom—that determines the order of elements in the table is very significant. We will see later that this has important implications with regard to the relationship between the number of electrons in an atom and the atom's chemical properties.

The modern periodic table is shown in Figure 2.5 and also appears on the inside front cover of the book. We will refer to the table frequently, so it is important for you to become familiar with it and with some of the terminology applied to it.

Special terminology is associated with the periodic table

Rows are periods

As in Mendeleev's table, the elements are arranged in rows that we call **periods**, but here they are arranged in order of increasing atomic number. For identification purposes the periods are numbered. We will find these numbers useful later on. Below the main body of the table are two long rows of 14 elements each. These actually belong in the main body of the table following La ($Z = 57$) and Ac ($Z = 89$), as shown in Figure 2.6. They are almost always placed below the table simply to conserve space. If the fully spread-out table is printed on one page, the type is so small that it's difficult to read. Notice that in the fully extended form of the table, with all the elements arranged in their proper locations, there is a great deal of empty space. An important requirement of a detailed atomic theory, which we will get to in Chapter 7, is that it must explain not only the repetition of properties, but also why there is so much empty space in the table.

Recall that the symbol Z stands for atomic number.

Columns are groups or families

Again, as in Mendeleev's table, the vertical columns are called **groups.** However, there is not uniform agreement among chemists on how they should be numbered. In an attempt to standardize the table, the International Union of Pure and Applied Chemistry (the IUPAC), an international body of scientists responsible for setting standards in chemistry, officially adopted a system in which the groups are simply numbered sequentially, 1 through 18, from left to right using Arabic numerals. Chemists in North America favor the system where the longer groups are labeled IA to VIIIA and the shorter groups are labeled IB to VIIIB in the sequence depicted in Figure 2.5. Note that Group VIIIB encompasses three columns; moreover, the sequence of the B-group elements is unique and will make sense when we learn more about the structure of the atom in Chapter 7. European chemists favor a third numbering system with roman numerals and the designation of A and B groups but with a different sequence from the North American table. In Figure 2.5 and on the

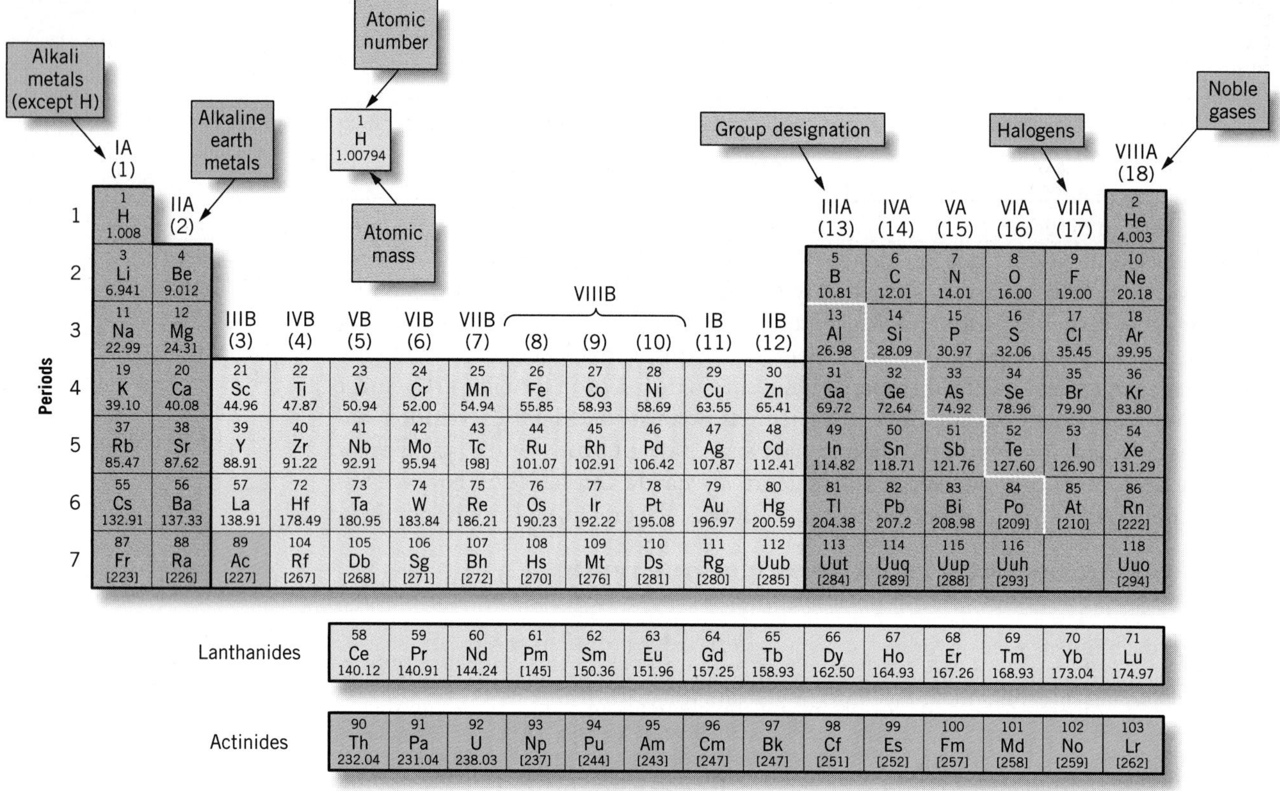

FIG. 2.5 **The modern periodic table.** At room temperature, mercury and bromine are liquids. Eleven elements are gases including the noble gases and the diatomic gases of hydrogen, oxygen, nitrogen, fluorine, and chlorine. The remaining elements are solids.

inside front cover of the book, we have used both the North American labels as well as those preferred by the IUPAC. Because of the lack of uniform agreement among chemists on how the groups should be specified, we will use the North American A-group/B-group designations in Figure 2.5 when we wish to specify a particular group.

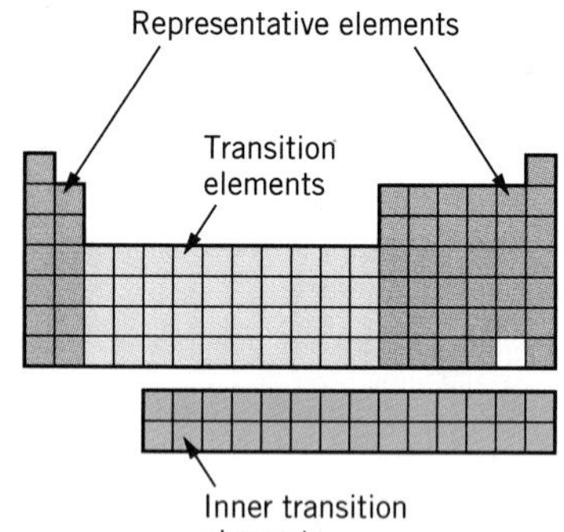

As we have already noted, the elements in a given group bear similarities to each other. Because of such similarities, groups are sometimes referred to as **families of elements.** The elements in the longer columns (the A groups) are known as the **representative elements** or **main group elements.** Those that fall into the B groups in the center of the table are called **transition elements.** The elements in the two long rows below the main body of the table are the **inner transition elements,** and each row is named after the element that it follows in the main body of the table. Thus, elements 58–71 are called the **lanthanide elements** because they follow lanthanum ($Z = 57$), and elements 90–103 are called the **actinide elements** because they follow actinium ($Z = 89$). The lanthanides are also called **rare earth metals.**

Some of the groups have acquired common names. For example, except for hydrogen, the Group IA elements are metals. They form compounds with oxygen that dissolve in water to give solutions that are strongly alkaline, or caustic. As a result, they are called the **alkali metals** or simply the *alkalis.* The Group IIA elements are also metals. Their oxygen compounds are alkaline, too, but many compounds of the Group IIA elements are unable to dissolve in water and are found in deposits in the ground. Because of their properties and where they occur in nature, the Group IIA elements became known as the **alkaline earth metals.**

On the right side of the table, in Group VIIIA, are the **noble gases.** They used to be called the **inert gases** until it was discovered that the heavier members of the group show a small degree of chemical reactivity. The term *noble* is used when we wish to suggest a very limited degree of chemical reactivity. Gold, for instance, is often referred to as a noble metal because so few chemicals are capable of reacting with it.

1 H																															2 He
3 Li	4 Be																									5 B	6 C	7 N	8 O	9 F	10 Ne
11 Na	12 Mg																									13 Al	14 Si	15 P	16 S	17 Cl	18 Ar
19 K	20 Ca	21 Sc															22 Ti	23 V	24 Cr	25 Mn	26 Fe	27 Co	28 Ni	29 Cu	30 Zn	31 Ga	32 Ge	33 As	34 Se	35 Br	36 Kr
37 Rb	38 Sr	39 Y															40 Zr	41 Nb	42 Mo	43 Tc	44 Ru	45 Rh	46 Pd	47 Ag	48 Cd	49 In	50 Sn	51 Sb	52 Te	53 I	54 Xe
55 Cs	56 Ba	57 La	58 Ce	59 Pr	60 Nd	61 Pm	62 Sm	63 Eu	64 Gd	65 Tb	66 Dy	67 Ho	68 Er	69 Tm	70 Yb	71 Lu	72 Hf	73 Ta	74 W	75 Re	76 Os	77 Ir	78 Pt	79 Au	80 Hg	81 Tl	82 Pb	83 Bi	84 Po	85 At	86 Rn
87 Fr	88 Ra	89 Ac	90 Th	91 Pa	92 U	93 Np	94 Pu	95 Am	96 Cm	97 Bk	98 Cf	99 Es	100 Fm	101 Md	102 No	103 Lr	104 Rf	105 Db	106 Sg	107 Bh	108 Hs	109 Mt	110 Ds	111 Rg	112 Uub	113 Uut	114 Uuq	115 Uup	116 Uuh		118 Uuo

FIG. 2.6 Extended form of the periodic table. The two long rows of elements below the main body of the table in Figure 2.5 are placed in their proper places in this table.

Finally, the elements of Group VIIA are called the **halogens,** derived from the Greek word meaning "sea" or "salt." Chlorine (Cl), for example, is found in familiar table salt, a compound that accounts in large measure for the salty taste of seawater. The other groups of the representative elements have less frequently used names and we will name those groups based on the first element in the family; for example, Group VA is the **nitrogen family** and Group VIA is the **oxygen family.**

2.4 Elements can be metals, nonmetals, or metalloids

TOOLS Periodic table metals, nonmetals, and metalloids

The periodic table organizes all sorts of chemical and physical information about the elements and their compounds. It allows us to study systematically the way properties vary with an element's position within the table and, in turn, makes the similarities and differences among the elements easier to understand and remember.

Even a casual inspection of samples of the elements reveals that some are familiar metals and that others, equally well known, are not metals. Most of us recognize metals such as lead, iron, or gold and nonmetals such as oxygen or nitrogen. A closer look at the nonmetallic elements, though, reveals that some of them, silicon and arsenic to name two, have properties that lie between those of true metals and true nonmetals. These elements are called **metalloids.** The elements are not evenly divided into the categories of metals, nonmetals, and metalloids. (See Figure 2.7.) Most elements are metals, slightly over a dozen are nonmetals, and only a handful are metalloids.

Notice that the metalloids are grouped around the bold stair-step line that is drawn diagonally from boron (B) to astatine (At).

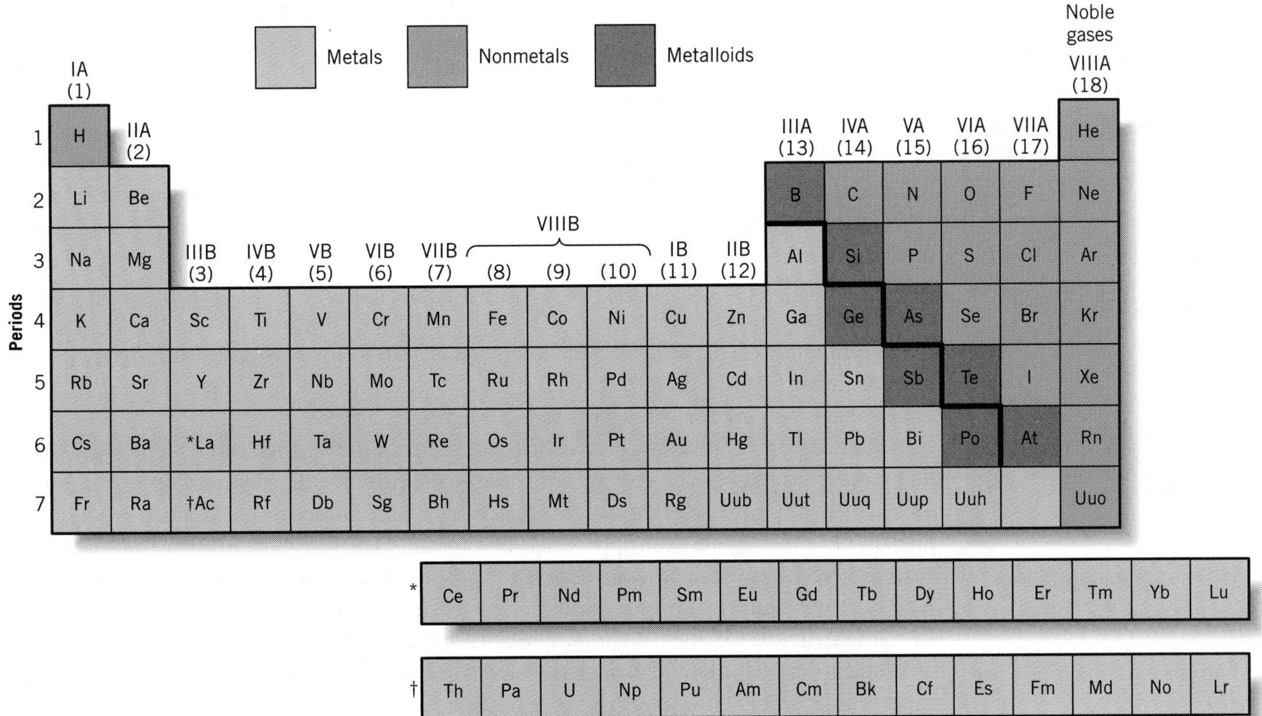

FIG. 2.7 Distribution of metals, nonmetals, and metalloids among the elements in the periodic table.

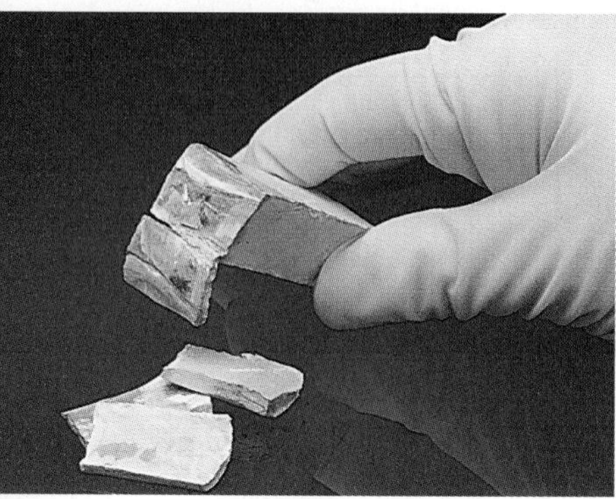

FIG. 2.8 **Sodium is a metal.** The freshly exposed surface of a bar of sodium reveals its shiny metallic luster. The metal reacts quickly with moisture and oxygen to form a white coating. Its high reactivity makes it dangerous to touch with bare skin. *(Michael Watson.)*

Metals have distinctive physical properties

You probably know a **metal** when you see one. Metals tend to have a shine so unique that it's called a *metallic luster*. For example, the silvery sheen of the freshly exposed surface of sodium in Figure 2.8 would most likely lead you to identify sodium as a metal even if you had never seen or heard of it before. We also know that metals conduct electricity. Few of us would hold an iron nail in our hand and poke it into an electrical outlet. In addition, we know that metals conduct heat very well. On a cool day, metals always feel colder to the touch than do neighboring nonmetallic objects because metals conduct heat away from your hand very rapidly. Nonmetals seem less cold because they can't conduct heat away as quickly and therefore their surfaces warm up faster.

Other properties that metals possess, to varying degrees, are **malleability**—the ability to be hammered or rolled into thin sheets—and **ductility**—the ability to be drawn into wire. The ability of a blacksmith to fashion horseshoes from a bar of iron (Figure 2.9) depends on the malleability of iron and steel, and the manufacture of electrical wire is based on the ductility of copper.

Hardness is another physical property that we usually think of for metals. Some, such as chromium or iron, are indeed quite hard; but others, like copper and lead, are rather soft. The alkali metals such as sodium (Figure 2.8) are so soft they can be cut with a knife, but they are also so chemically reactive that we rarely get to see them as free elements.

Thin lead sheets are used for sound-deadening because the easily deformed lead absorbs the sound vibrations.

All the metallic elements, except mercury, are solids at room temperature (Figure 2.10). Mercury's low freezing point (−39 °C) and fairly high boiling point (357 °C) make it useful as a fluid in thermometers. Most of the other metals have much higher melting points, and some are used primarily because of this. Tungsten, for example, has the highest melting point of any metal (3400 °C, or 6150 °F), which explains its use as filaments that glow white-hot in electric lightbulbs.

The chemical properties of metals vary tremendously. Some, such as gold and platinum, are very unreactive toward almost all chemical agents. This property, plus their natural beauty and rarity, makes them highly prized for use in jewelry. Other metals, however, are so reactive that few people except chemists and chemistry students ever get to see them in their "free" states. For instance, the metal sodium reacts very quickly with oxygen or moisture in the air, and its bright metallic surface tarnishes almost immediately. In contrast, compounds of sodium, such as table salt and baking soda, are quite stable and very common.

We use the term *free element* to mean an element that is not chemically combined with any other element.

Nonmetals lack the properties of metals

Substances such as plastics, wood, and glass that lack the properties of metals are said to be *nonmetallic*, and an element that has nonmetallic properties is called a **nonmetal** or **nonmetallic element.** Most often, we encounter the nonmetals in the form of compounds or mixtures of compounds. There are some nonmetals, however, that are very important to us in their elemental forms. The air we breathe, for instance, contains mostly nitrogen and oxygen. Both are gaseous, colorless, and odorless nonmetals. Since we can't see, taste, or smell them, however, it's difficult to experience their existence. (Although if you step into an atmosphere without oxygen, your body will soon tell you that something is missing!) Probably the most commonly *observed* nonmetallic element is carbon. We find it as the graphite in pencils, as coal, and as the charcoal used for barbecues. It also occurs in a more valuable form as diamond (Figure 2.11). Although diamond and graphite differ in appearance, each is a form of elemental carbon.

FIG. 2.9 **Malleability of iron.** A blacksmith uses the malleability of hot iron to fashion horseshoes from an iron bar. *(Stone/Getty Images.)*

Many of the nonmetals are solids at room temperature and atmospheric pressure, while many others are gases. Photographs of some of the nonmetallic elements appear in Figure 2.12. Their properties are almost completely opposite those of metals. Each of these elements lacks the characteristic appearance of a metal. They are poor conductors of heat and, with the exception of the graphite form of carbon, are also poor conductors of electricity. The electrical conductivity of graphite appears to be an accident of molecular structure, since the structures of metals and graphite are completely different.

The nonmetallic elements lack the malleability and ductility of metals. A lump of sulfur crumbles when hammered and breaks apart when pulled on. Diamond cutters rely on

FIG. 2.10 Mercury droplet. The metal mercury (once known as quicksilver) is a liquid at room temperature, unlike other metals, which are solids. *(OPC, Inc.)*

FIG. 2.11 Diamond. This gem is simply another form of the element carbon. *(Charles D. Winters/ Photo Researchers, Inc.)*

FIG. 2.12 Some nonmetallic elements. In the bottle on the left is dark-red liquid bromine, which vaporizes easily to give a deeply colored orange vapor. Pale green chlorine fills the round flask in the center. Solid iodine lines the bottom of the flask on the right and gives off a violet vapor. Powdered red phosphorus occupies the dish in front of the flask of chlorine, and black powdered graphite is in the watch glass. Also shown are lumps of yellow sulfur. *(Michael Watson.)*

the brittle nature of carbon when they split a gem-quality stone by carefully striking a quick blow with a sharp blade.

As with metals, nonmetals exhibit a broad range of chemical reactivity. Fluorine, for instance, is extremely reactive. It reacts readily with almost all the other elements. At the other extreme is helium, the gas used to inflate children's balloons and the blimps seen at major sporting events. This element does not react with anything, a fact that chemists find useful when they want to provide a totally *inert* (unreactive) atmosphere inside some apparatus.

Metalloids have physical properties between metals and nonmetals

The properties of metalloids lie between those of metals and nonmetals. This shouldn't surprise us since the metalloids are located between the metals and the nonmetals in the periodic table. In most respects, metalloids behave as nonmetals, both chemically and physically. However, in their most important physical property, electrical conductivity, they somewhat resemble metals. Metalloids tend to be **semiconductors;** they conduct electricity, but not nearly so well as metals. This property, particularly as found in silicon and germanium, is responsible for the remarkable progress made during the last five decades in the field of solid-state electronics. The operation of every computer, audio system, TV receiver, DVD or CD player, and AM–FM radio relies on transistors made from semiconductors. Perhaps the most amazing advance of all has been the fantastic reduction in the size of electronic components that semiconductors have allowed (Figure 2.13). To it, we owe the development of small and versatile cell phones, cameras, MP3 players, handheld calculators,

FIG. 2.13 Modern electronic circuits rely on the semiconductor properties of silicon. The silicon wafer shown here contains more electronic components (10 billion) than there are people on our entire planet (about 6.5 billion)! *(Courtesy Sematech; Courtesy NASA)*

and microcomputers. The heart of these devices is an integrated circuit that begins as a wafer of extremely pure silicon (or germanium) that is etched and chemically modified into specialized arrays of thousands of transistors.

FIG. 2.14 **Hydrogen reacts with oxygen to form water.** The reaction of hydrogen with oxygen provides the thrust for the main rocket engines of the space shuttle shown here just after liftoff. *(Corbis Images.)*

Metallic and nonmetallic character is related to an element's position in the periodic table

The occurrence of the metalloids between the metals and the nonmetals is our first example of trends in properties within the periodic table. We will frequently see that as we move from position to position across a period or down a group, chemical and physical properties change in a more or less regular way. There are few abrupt changes in the characteristics of the elements as we scan across a period or down a group. The location of the metalloids can be seen, then, as an example of the gradual transition between metallic and nonmetallic properties. From left to right across Period 3, we go from aluminum, an element that has every appearance of a metal; to silicon, a semiconductor; to phosphorus, an element with clearly nonmetallic properties. A similar gradual change is seen going down Group IVA. Carbon is certainly a nonmetal, silicon and germanium are metalloids, and tin and lead are metals. Trends such as these are useful to spot because they help us remember properties.

2.5 Formulas and Equations Describe Substances and Their Reactions

A property possessed by nearly every element is the ability to combine with other elements to form compounds, although not all combinations appear to be possible. For example, iron reacts with oxygen to form a compound that we commonly call rust, but no compound is formed between sodium and iron.

In Chapter 1 we noted that during chemical reactions, the properties of the substances change, often dramatically, when a reaction takes place. This is certainly true in the reactions of elements to form compounds. An example is the reaction between hydrogen and oxygen to form ordinary water.

At room temperature both hydrogen and oxygen are clear colorless gases. When they are mixed and ignited, these elements combine explosively to form the familiar compound water, which of course is a liquid at room temperature. As with nearly all chemical reactions, the properties of the substances present prior to the reaction differ quite a lot from the properties of those present afterwards.

The reaction between hydrogen and oxygen is not one we would expect to encounter in our daily lives, but it does have applications. When hydrogen and oxygen are cooled to sufficiently low temperatures, they condense to form liquids that serve as the fuel for the main rocket engines of the space shuttle (Figure 2.14). Hydrogen has also been used to power nonpolluting vehicles in which its reaction with oxygen (from air) yields an exhaust containing only water vapor.

A chemical formula describes the composition of a substance

To describe chemical substances, both elements and compounds, we commonly use **chemical formulas,** in which chemical symbols are used to represent atoms of the elements that are present. For a **free element** (*one that is not combined with another element in a compound*) we often simply use the chemical symbol. Thus, the element sodium is represented by its symbol, Na, which is interpreted to mean one atom of sodium.

Except for the noble gases, all the free nonmetallic elements exist as molecules that contain two or more atoms. Many of those we encounter frequently occur as **diatomic molecules** (molecules composed of two atoms each). Among them are the gases hydrogen, oxygen, and nitrogen and the halogens (fluorine, chlorine, bromine, and iodine). We represent these elements with chemical formulas in which subscripts indicate the number of atoms in a molecule. Thus, the **formula** for molecular hydrogen is H_2, and those for oxygen, nitrogen, and chlorine are O_2, N_2, and Cl_2, respectively. (See Figure 2.15.) The elements that occur as diatomic molecules are listed in Table 2.3. This would be a good time to learn them because

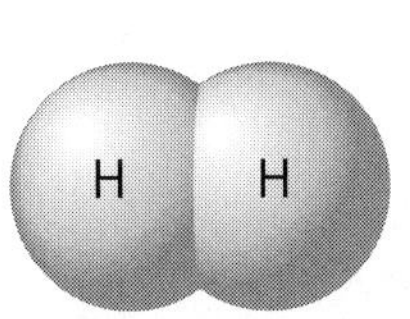

Hydrogen molecule, H_2

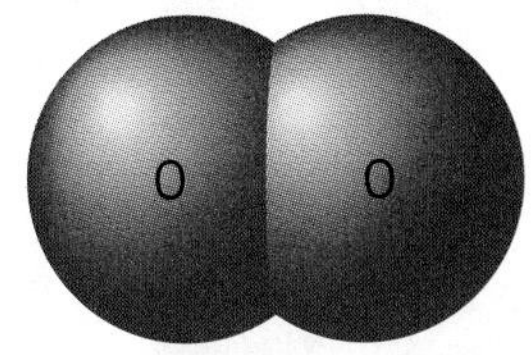

Oxygen molecule, O_2

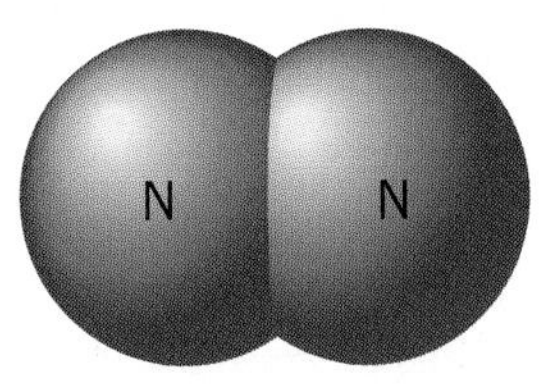

Nitrogen molecule, N_2

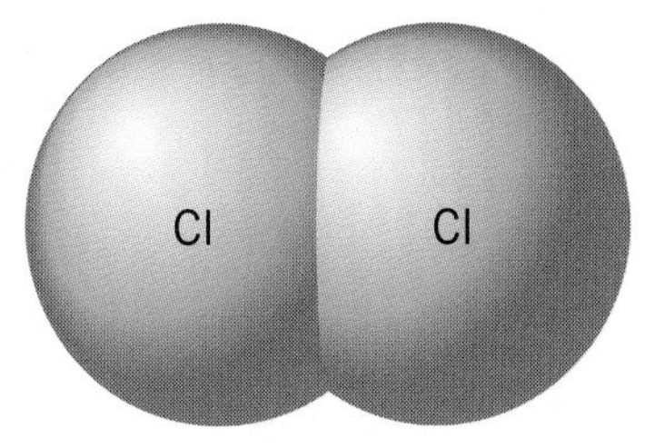

Chlorine molecule, Cl_2

FIG. 2.15 Models that depict the diatomic molecules of hydrogen, oxygen, nitrogen, and chlorine. Each contains two atoms per molecule; their different sizes reflect differences in the sizes of the atoms that make up the molecules. The atoms are shaded by color to indicate the element (hydrogen, white; oxygen, red; nitrogen, blue; and chlorine, green).

TABLE 2.3 Elements That Occur Naturally as Diatomic Molecules

Hydrogen	H_2	Fluorine	F_2
Nitrogen	N_2	Chlorine	Cl_2
Oxygen	O_2	Bromine	Br_2
		Iodine	I_2

you will come upon them often throughout the course. Other nonmetals have their atoms arranged in even more complex combinations. Elemental sulfur, for example, contains molecules of S_8 and one form of phosphorus has molecules of P_4.

Just as chemical symbols can be used as shorthand notations for the names of elements, a chemical formula is a shorthand way of writing the name for a compound. However, *the most important characteristic of a compound's formula is that it specifies the composition of the substance.*

In the formula of a compound, each element present is identified by its chemical symbol. Table salt, for example, has the chemical formula NaCl which indicates it is composed of the elements sodium (Na) and chlorine (Cl). When more than one atom of an element is present, the number of atoms is given by a **subscript** after the symbol. For instance, the iron oxide in rust has the formula Fe_2O_3, which tells us that the compound is composed of iron (Fe) and oxygen (O), and that in this compound there are two atoms of iron for every three atoms of oxygen. When no subscript is written, we assume it to be 1, so in NaCl we find one atom of sodium (Na) for each atom of chlorine (Cl). Similarly, the formula H_2O tells us that in water there are two hydrogen atoms for every one oxygen atom, and the formula for chloroform, $CHCl_3$, indicates that one atom of carbon, one atom of hydrogen, and three atoms of chlorine have combined (Figure 2.16).

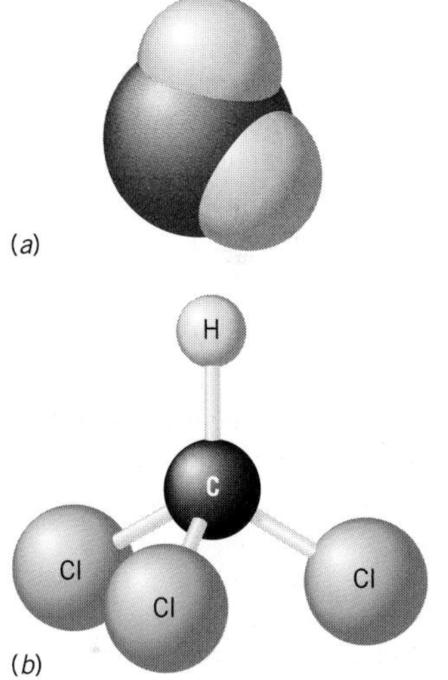

FIG. 2.16 Molecules of water and chloroform. (*a*) A space-filling model of H_2O. (*b*) A ball-and-stick model of $CHCl_3$. (Hydrogen, white; oxygen, red; carbon, black; and chlorine, green.)

For more complicated compounds, we sometimes find formulas that contain parentheses. An example is the formula for urea, $CO(NH_2)_2$, which tells us that the group of atoms within the parentheses, NH_2, occurs twice. (The formula for urea also could be written as CON_2H_4, but there are good reasons for writing certain formulas with parentheses, as you will see later.)

Hydrates are crystals that contain water in fixed proportions

Certain compounds form crystals that contain water molecules. An example is ordinary plaster—the material often used to coat the interior walls of buildings. Plaster consists of crystals of calcium sulfate, $CaSO_4$, that contain two molecules of water for each $CaSO_4$. These water molecules are not held very tightly and can be driven off by heating the crystals. The dried crystals absorb water again if exposed to moisture, and the amount of water absorbed always gives crystals in which the H_2O-to-$CaSO_4$ ratio is 2 to 1. Compounds whose crystals contain water molecules in fixed ratios are quite common and are called **hydrates**. The formula for this hydrate of calcium sulfate is written $CaSO_4 \cdot 2H_2O$ to show that there are two molecules of water per $CaSO_4$. The raised dot is used to indicate that the water molecules are not bound too tightly in the crystal and can be removed.

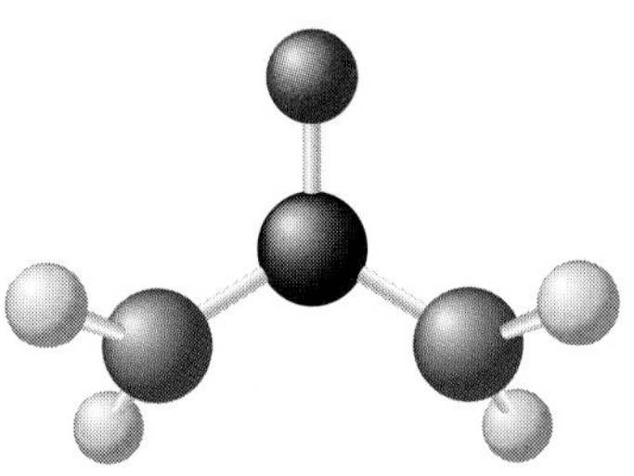
Ball-and-stick model of the urea molecule, $CO(NH_2)_2$.

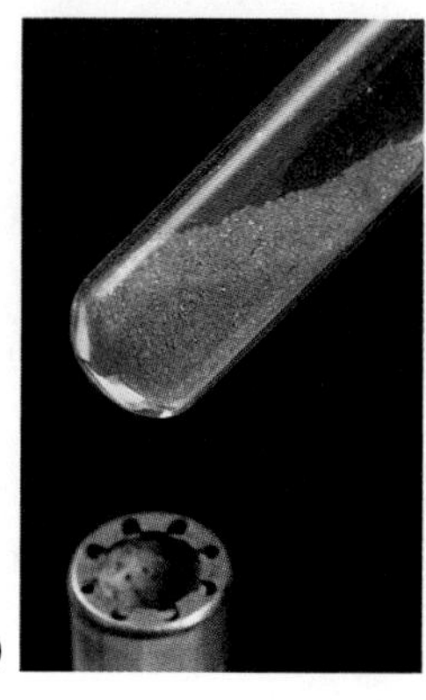
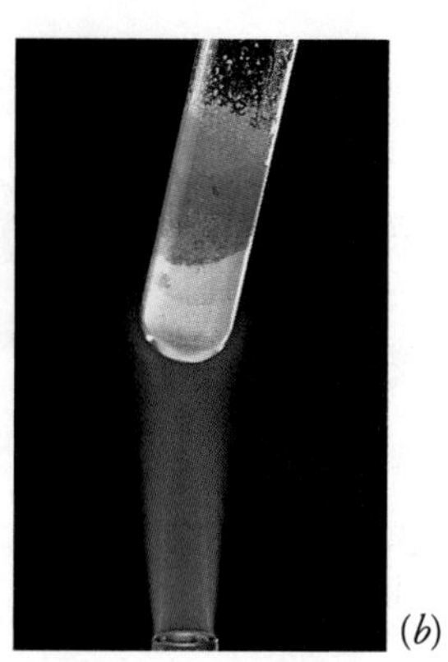

FIG. 2.17 Water can be driven from hydrates by heating. (*a*) Blue crystals of copper sulfate pentahydrate, $CuSO_4{\cdot}5H_2O$, about to be heated. (*b*) The hydrate readily loses water when heated. The light colored solid observed in the lower half of the test tube is pure $CuSO_4$. *(Richard Megna/Fundamental Photographs; Michael Watson)*

Sometimes the *dehydration* (removal of water) of hydrate crystals produces changes in color. An example is copper sulfate, which is sometimes used as an agricultural fungicide. Copper sulfate forms blue crystals with the formula $CuSO_4{\cdot}5H_2O$ in which there are five water molecules for each $CuSO_4$. When these blue crystals are heated, most of the water is driven off and the solid that remains, now nearly pure $CuSO_4$, is almost white (Figure 2.17). If left exposed to the air, the $CuSO_4$ will absorb moisture and form blue $CuSO_4{\cdot}5H_2O$ again.

When all the water is removed, the solid is said to be **anhydrous**, meaning "without water."

Counting atoms in formulas is a necessary skill

Counting the number of atoms of the elements in a chemical formula is an operation you will have to perform many times, so let's look at an example.

EXAMPLE 2.4
Counting Atoms in Formulas

How many atoms of each element are represented by the formulas (a) $Al_2(SO_4)_3$ and (b) $CoCl_2{\cdot}6H_2O$?

ANALYSIS: The essential tool to use here is the set of principles governing how we count atoms in a chemical formula in the preceding section. To review, the subscript following an element indicates how many of that element are part of the formula; a subscript of 1 is implied if there is no subscript. We also must recall that a quantity within parentheses is repeated a number of times equal to the subscript that follows, and a raised dot in a formula indicates the substance is a hydrate in which the number preceding H_2O specifies how many water molecules are present.

SOLUTION: (a) Here we must recognize that all the atoms within the parentheses occur three times.

Subscript 3 indicates three SO_4 units.

$$Al_2(SO_4)_3$$

Each SO_4 contains one S and four O atoms, so three of them contain three S and twelve O atoms. The subscript for Al tells us there are two Al atoms. Therefore, the formula $Al_2(SO_4)_3$ shows

2 Al 3 S 12 O

(b) This is a formula for a hydrate, as indicated by the raised dot. It contains six water molecules, each with two H and one O, for every $CoCl_2$.

The 6 indicates there are six molecules of H_2O.

$$CoCl_2{\cdot}6H_2O$$

Dot indicates the compound is a hydrate.

Therefore, the formula $CoCl_2 \cdot 6H_2O$ represents

1 Co   2 Cl   12 H   6 O

ARE THE ANSWERS REASONABLE? The only way to check the answer here is to perform a recount.

Practice Exercise 9: How many atoms of each element are expressed by the formulas below? (Hint: Pay special attention to counting elements within parentheses.)

(a) $NiCl_2$ (b) $FeSO_4$ (c) $Ca_3(PO_4)_2$ (d) $Co(NO_3)_2 \cdot 6H_2O$

Practice Exercise 10: How many atoms of each element are present in each of the formulas that follow? Consult the table inside the front cover to write the full name of each element as well as its symbol.

(a) NH_4NO_3 (b) $FeNH_4(SO_4)_2$ (c) $Mo(NO_3)_2 \cdot 5H_2O$ (d) $C_6H_4ClNO_2$

Chemical equations describe what happens in chemical reactions

A **chemical equation** *describes what happens when a chemical reaction occurs.* It uses chemical formulas to provide a before-and-after picture of the chemical substances involved. Consider, for example, the reaction between hydrogen and oxygen to give water. The chemical equation that describes this reaction is

$$2H_2 + O_2 \longrightarrow 2H_2O$$

The two substances that appear to the left of the arrow are the **reactants;** they are the substances present before the reaction begins. To the right of the arrow we find the formula for the **product** of the reaction, water. In this example, only one substance is formed in the reaction, so there is only one product. As we will see, however, in most chemical reactions there is more than one product. The products are the substances that are formed and that exist after the reaction is over. The arrow means "reacts to yield." Thus, this equation tells us that *hydrogen and oxygen react to yield water.*

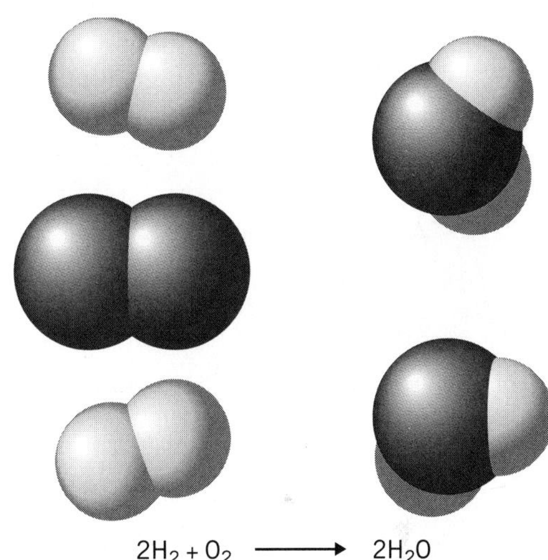

FIG. 2.18 The reaction between molecules of hydrogen and oxygen. The reaction between two molecules of hydrogen and one molecule of oxygen gives two molecules of water.

Coefficients are written in front of formulas to satisfy the law of conservation of mass

In the equation for the reaction of hydrogen and oxygen, you'll notice that the number 2 precedes the formulas of hydrogen and water. Numbers in front of the formulas are called **coefficients,** and they indicate the number of molecules of each kind among the reactants and products. Thus, $2H_2$ means two molecules of H_2, and $2H_2O$ means two molecules of H_2O. When no number is written, the coefficient is assumed to be 1 (so the coefficient of O_2 equals 1).

Coefficients are needed to have the equation conform to the law of conservation of mass, as illustrated in Figure 2.18. Because atoms cannot be created or destroyed in a chemical reaction, we must have the same number of atoms of each kind present before and after the reaction (that is, on both sides of the arrow). When this condition is met, we say the equation is **balanced.** Another example is the equation for the combustion of butane, C_4H_{10}, the fluid in disposable cigarette lighters (Figure 2.19).

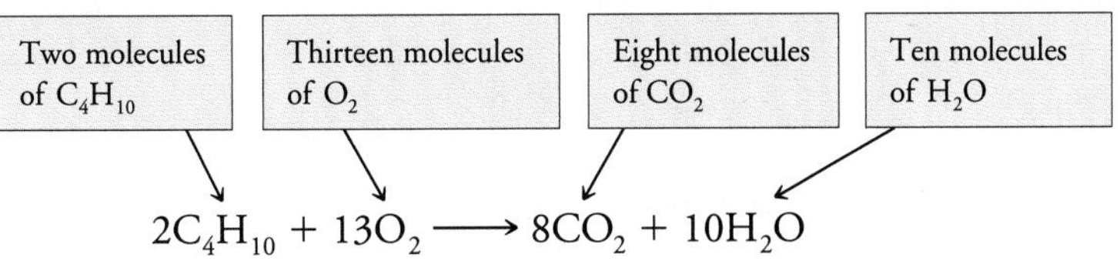

FIG. 2.19 The combustion of butane, C_4H_{10}. The products are carbon dioxide and water vapor. *(Robert Capece.)*

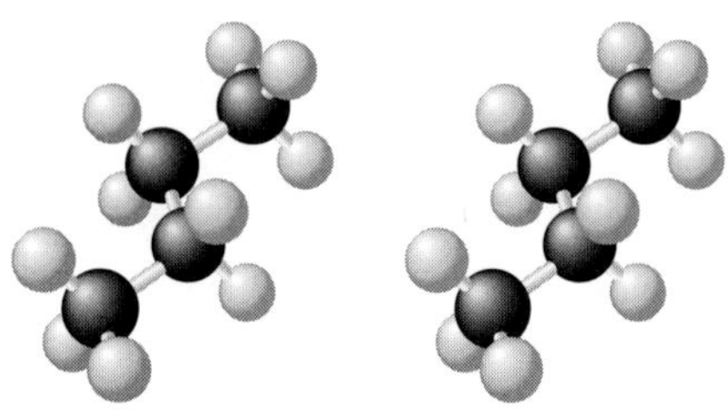

Two molecules of butane contain 8 atoms of C and 20 atoms of H.

FIG. 2.20 Understanding coefficients in an equation. The expression $2C_4H_{10}$ describes two molecules of butane, each of which contains 4 carbon and 10 hydrogen atoms. This gives a total of 8 carbon and 20 hydrogen atoms.

The 2 before the C_4H_{10} tells us that two molecules of butane react. This involves a total of 8 carbon atoms and 20 hydrogen atoms, as we see in Figure 2.20. Notice we have multiplied the numbers of atoms of C and H in one molecule of C_4H_{10} by the coefficient 2. On the right we find 8 molecules of CO_2, which contain a total of 8 carbon atoms. Similarly, 10 water molecules contain 20 hydrogen atoms. Finally, we can count 26 oxygen atoms on both sides of the equation. You will learn to balance equations such as this in Chapter 3.

The states of the reactants and products can be specified in a chemical equation

In a chemical equation we sometimes find it useful to specify the physical states of the reactants and products, that is, whether they are solids, liquids, or gases. This is done by writing *s* for solid, *l* for liquid, or *g* for gas in parentheses after the chemical formulas. For example, the equation for the combustion of the carbon in a charcoal briquette can be written as

$$C(s) + O_2(g) \longrightarrow CO_2(g)$$

At times we will also find it useful to indicate that a particular substance is dissolved in water. We do this by writing *aq*, meaning "*aqueous* solution," in parentheses after the formula. For instance, the reaction between stomach acid (an aqueous solution of HCl) and the active ingredient in Tums, $CaCO_3$, is

$$2HCl(aq) + CaCO_3(s) \longrightarrow CaCl_2(aq) + H_2O(l) + CO_2(g)$$

Remember:

s = solid
l = liquid
g = gas
aq = aqueous

EXAMPLE 2.5
Determining if an Equation Is Balanced

Determine whether or not the following chemical equations are balanced. Support your conclusions by writing how many atoms of each element are on each side of the arrow.

(a) $Fe(OH)_3 + 2HNO_3 \longrightarrow Fe(NO_3)_3 + 2H_2O$

(b) $BaCl_2 + H_2SO_4 \longrightarrow BaSO_4 + 2HCl$

(c) $C_6H_{12}O_6 + 6O_2 \longrightarrow 6CO_2 + 6H_2O$

ANALYSIS: The statement of the problem asks if the equations are balanced. You can prove an equation is balanced if each element has the same number of atoms on each side of the arrow. Again, in this example, we use the tool that tells us how to use subscripts to count atoms in each formula. Also, a given atom, such as oxygen, in these equations may appear in both reactants or both products and we need to be sure to account for all of them.

SOLUTION: (a) Reactants: 1 Fe, 9 O, 5 H, 2 N. Products: 1 Fe, 11 O, 4 H, 3 N. Only Fe has the same number of atoms on each side of the arrow. This is *not* balanced.
(b) Reactants: 1 Ba, 2 Cl, 2 H, 1 S, 4 O. Products: 1 Ba, 2 Cl, 2 H, 1 S, 4 O. This equation *is* balanced.
(c) Reactants: 6 C, 12 H, 18 O. Products: 6 C, 18 O, 12 H. This equation *is* balanced.

ARE THE ANSWERS REASONABLE? The appropriate way to check this is to recount the atoms. Try counting the atoms in the reverse direction this time.

Practice Exercise 11: How many atoms of each element appear on each side of the arrow in the following equation? (Hint: Recall that coefficients multiply the elements in the entire formula.)

$$Mg(OH)_2 + 2HCl \longrightarrow MgCl_2 + 2H_2O$$

Practice Exercise 12: Rewrite the equation in Practice Exercise 11 to show that $Mg(OH)_2$ is a solid, HCl and $MgCl_2$ are dissolved in water, and H_2O is a liquid.

Practice Exercise 13: Count each of the atoms in the following equation to determine if it is balanced.

$$2(NH_4)_3PO_4 + 3Ba(C_2H_3O_2)_2 \longrightarrow Ba_3(PO_4)_2 + 6NH_4C_2H_3O_2$$

2.6 MOLECULAR COMPOUNDS CONTAIN NEUTRAL PARTICLES CALLED MOLECULES

The concept of molecules dates to the time of Dalton's atomic theory, where a part of his theory was that atoms of elements combine in fixed numerical ratios to form "molecules" of a compound. By our modern definition *a* **molecule** *is an electrically neutral particle consisting to two or more atoms.* Accordingly, the term molecule applies to many elements such as H_2 and O_2 as well as to **molecular compounds.**

Experimental evidence exists for molecules

One phenomenon that points to the existence of molecules is called **Brownian motion** [named after Robert Brown (1773–1858), the Scottish botanist who first observed it]. When very small particles such as tiny grains of pollen are suspended in a liquid and observed under a microscope, the tiny particles are seen to be constantly jumping and jiggling about. It appears as though they are continually being knocked back and forth by collisions with something. An explanation is that this "something" is *molecules* of the liquid. The microscopic particles are constantly bombarded by molecules of the liquid, but because the suspended particles are so small, the collisions are not occurring equally on all sides. The unequal numbers of collisions cause the lightweight particles to jerk about.

There is additional evidence for molecules, and today scientists accept the existence of molecules as fact. Looking more closely, within molecules atoms are held to each other by attractions called **chemical bonds,** which are electrical in nature. In molecular compounds chemical bonds arise from the sharing of electrons between one atom and another. We will discuss such bonds at considerable length in Chapters 8 and 9. What is important to know about molecules now is that *the group of atoms that make up a molecule move about together and behave as a single particle,* just as the various parts that make up a car move about as one unit. The chemical formulas that we write to describe the compositions of molecules are called **molecular formulas,** which specify the actual numbers of atoms of each kind that make up a single molecule.

Molecular compounds form when nonmetals combine

As a general rule, *molecular compounds are formed when nonmetallic elements combine.* For example, you learned that H_2 and O_2 combine to form molecules of water. Similarly, carbon and oxygen combine to form either carbon monoxide, CO, or carbon dioxide, CO_2. (Both are gases that are formed in various amounts as products in the combustion of fuels such as gasoline and charcoal.) Although molecular compounds can be formed by the direct combination of elements, often they are the products of reactions between compounds. You will encounter many such reactions in your study of chemistry.

Carbon monoxide is a poisonous gas found in the exhaust of automobiles.

Although there are relatively few nonmetals, the number of molecular substances formed by them is huge. This is because of the variety of ways in which they combine as well as the varying degrees of complexity of their molecules. Variety and complexity reach a maximum with compounds in which carbon is combined with a handful of other elements such as hydrogen, oxygen, and nitrogen. There are so many of these compounds, in fact, that their study encompasses the chemical specialties called organic chemistry and biochemistry.

Molecules vary in size from small to very large. Some contain as few as two atoms (diatomic molecules). Most molecules are more complex, however, and contain more atoms.

Molecules of water (H_2O), for example, have three atoms and those of ordinary table sugar ($C_{12}H_{22}O_{11}$) have 45. There also are molecules that are very large, such as those that occur in plastics and in living organisms, some of which contain millions of atoms.

At this early stage we can only begin to look for signs of order among the vast number of nonmetal–nonmetal compounds. To give you a taste of the subject, we will look briefly at some simple compounds that the nonmetals form with hydrogen, as well as some simple compounds of carbon.

Hydrogen forms compounds with many nonmetals

Compounds that elements form with hydrogen are often called **hydrides,** and the formulas of the simple hydrides of the nonmetals are given in Table 2.4.[1] These compounds provide an opportunity to observe how we can use the periodic table as an aid in remembering factual information, in this case, the formulas of the hydrides. Notice that the number of hydrogen atoms combined with the nonmetal atom equals *the number of spaces to the right that we have to move in the periodic table to get to a noble gas.* (You will learn *why* this is so in Chapter 8, but for now we can just use the periodic table to help us remember the formulas.)

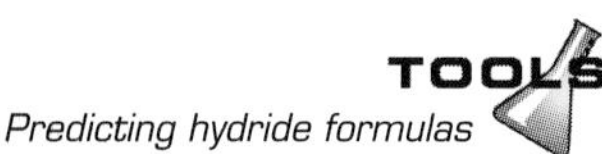

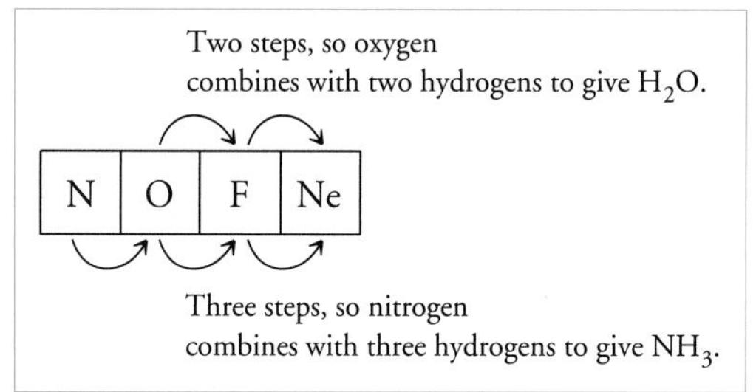

Many of the nonmetals form more complex compounds with hydrogen, but we will not discuss them here.

Also note in Table 2.4 that the formulas of the simple hydrides are similar for nonmetals within a given group of the periodic table. If you know the formula for the hydride of the top member of the group, then you know the formulas of all of them in that group.

We live in a three-dimensional world, and this is reflected in the three-dimensional shapes of molecules. The shapes of the simple nonmetal hydrides of nitrogen, oxygen, and fluorine are illustrated as space-filling models in Figure 2.21. Our understanding of the geometric shapes of molecules is described in Chapter 9.

TABLE 2.4 Simple Hydrogen Compounds of the Nonmetallic Elements

	Group			
Period	IVA	VA	VIA	VIIA
2	CH_4	NH_3	H_2O	HF
3	SiH_4	PH_3	H_2S	HCl
4	GeH_4	AsH_3	H_2Se	HBr
5		SbH_3	H_2Te	HI

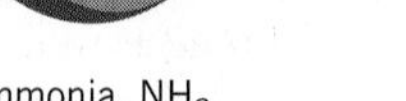

Ammonia, NH_3

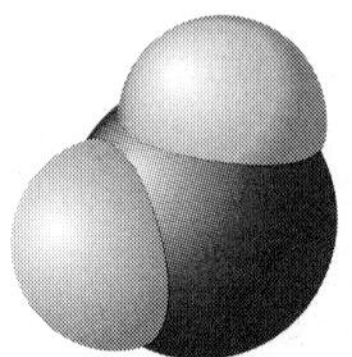

Water, H_2O

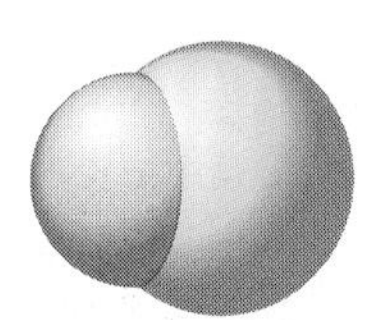

Hydrogen fluoride, HF

FIG. 2.21 Nonmetal hydrides of nitrogen, oxygen, and fluorine.

[1] Table 2.4 shows how the formulas are normally written. The order in which the hydrogens appear in the formula is not of concern to us now. Instead, we are interested in the *number* of hydrogens that combine with a given nonmetal.

Compounds of carbon form the basis for organic chemistry

Among all the elements, carbon is unique in the variety of compounds it forms with elements such as hydrogen, oxygen, and nitrogen. As a consequence, the number and complexity of such **organic compounds** is enormous, and their study constitutes the major specialty called **organic chemistry**. The term *organic* here comes from an early belief that these compounds could only be made by living organisms. We now know this isn't true, but the name organic chemistry persists nonetheless.

Our goal at this time is to acquaint you with some of the important kinds of organic compounds we encounter regularly, so our discussion here is brief.

Organic compounds are around us everywhere and we will frequently use such substances as examples in our discussions. Therefore, it will be helpful if you can begin to learn some of them now.

The study of organic chemistry begins with **hydrocarbons** (compounds of carbon and hydrogen). The simplest hydrocarbon is methane, CH_4, which is a member of a series of hydrocarbons with the general formula C_nH_{2n+2}, where n is an integer (i.e., a whole number). The first six members of this series, called the **alkane** series, are given in Table 2.5 along with their boiling points. Notice that as the molecules become larger, their boiling points increase.

TABLE 2.5 Hydrocarbons Belonging to the Alkane Series

Compound	Name	Boiling Point (°C)
CH_4	Methane[a]	−161.5
C_2H_6	Ethane[a]	−88.6
C_3H_8	Propane[a]	−42.1
C_4H_{10}	Butane[a]	−0.5
C_5H_{12}	Pentane	36.1
C_6H_{14}	Hexane	68.7

[a] Gases at room temperature (25 °C) and atmospheric pressure.

Molecular formulas can be written in different ways depending on what information is needed. A condensed formula such as C_2H_6 for ethane or C_3H_8 for propane simply indicates the number of each type of atom in the molecule. Structural formulas indicate how the carbon atoms are connected. Ethane is written as CH_3CH_3 and propane is $CH_3CH_2CH_3$ in the structural format. Line structures and a ball-and-stick representation are illustrated in Figure 1.3 for methane, while Figure 2.22. illustrates ethane and propane in space-filling models.

The alkanes are common substances. They are the principal constituents of petroleum from which most of our useful fuels are produced. Methane itself is the major component of natural gas that is often used for home heating and cooking. Gas-fired barbecues and some homes use propane as a fuel, and butane is the fuel in inexpensive cigarette lighters.[2] Hydrocarbons with higher boiling points are found in gasoline, kerosene, paint thinners, diesel fuel, and even candle wax.

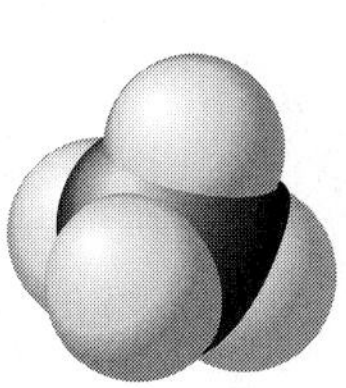

Methane, CH_4

Ethane, C_2H_6

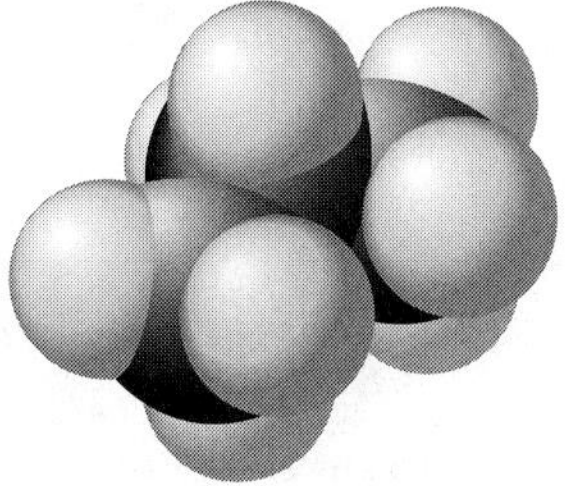

Propane, C_3H_8

FIG. 2.22 The first three members of the alkane series of hydrocarbons. White atoms represent hydrogen and black atoms represent carbon in these space-filling models that demonstrate the shapes of the molecules.

[2] Propane and butane are gases when they're at the pressure of the air around us, but they become liquids when compressed. When you purchase these substances, they are liquids with pressurized gas above them. The gas can be drawn off and used by opening a valve to the container.

The chemically correct names for ethylene and acetylene are ethene and ethyne, respectively.

Alkanes are not the only class of hydrocarbons. For example, there are three two-carbon hydrocarbons. In addition to ethane, C_2H_6, there are ethylene, C_2H_4 (an alkene from which polyethylene is made), and acetylene, C_2H_2 (an alkyne, which is the fuel used in *acetylene* welding torches).

The hydrocarbons serve as the foundation for organic chemistry. Derived from them are various other classes of organic compounds. An example is the class of compounds called **alcohols,** in which the atoms OH replace a hydrogen in the hydrocarbon. Thus, *methanol*, CH_3OH (also called *methyl alcohol*), is related to methane, CH_4, by removing one H and replacing it with OH (Figure 2.23). Methanol is used as a fuel and as a raw material for making other organic chemicals. Another familiar alcohol is *ethanol* (also called *ethyl alcohol*), C_2H_5OH. Ethanol, known as grain alcohol because it is obtained from the fermentation of grains, is in alcoholic beverages. It is also mixed with gasoline to reduce petroleum consumption. A 10% ethanol/90% gasoline mixture is known as gasohol; an 85% mixture of ethanol and gasoline is called E85.

Methanol is also known as wood alcohol because it was originally made by distilling wood. It is quite poisonous. Ethanol in high doses is also a poison.

Alcohols constitute just one class of compounds derived from hydrocarbons. We will discuss some others after you've learned more about how atoms bond to each other and about the structures of molecules.

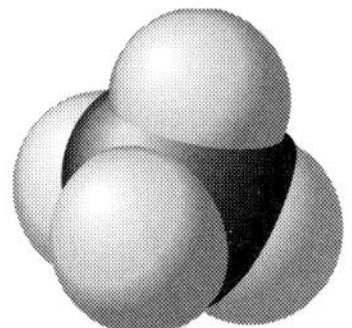

Methane

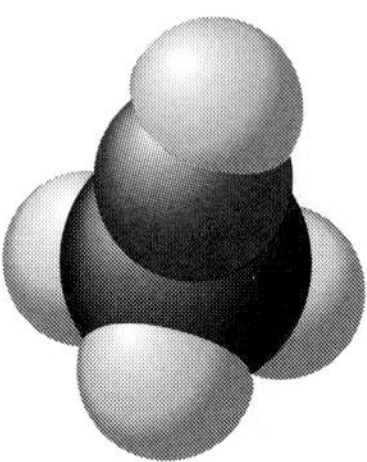

Methanol

FIG. 2.23 Relationship between an alkane and an alcohol. The alcohol methanol is derived from methane by replacing one H by OH. (Color code: carbon is black, hydrogen is white, and oxygen is red.)

Practice Exercise 14: Gasoline used in modern cars is a complex mixture of hundreds of different organic compounds. Less than 1% of gasoline is actually octane. Write the formula for octane using the condensed and structural format. (Hint: The prefix "octa-" stands for the number 8.)

Practice Exercise 15: What is the formula of the alkane hydrocarbon having 10 carbon atoms, decane? Write the formula in the condensed form and in the structural form. Download the space-filling molecule from the Internet.

Practice Exercise 16: On the basis of the discussions in this section, what are the formulas of (a) propanol and (b) butanol? Write the formula in the condensed form and in the structural form. Download the space-filling molecules from the Internet.

2.7 IONIC COMPOUNDS ARE COMPOSED OF CHARGED PARTICLES CALLED IONS

Formation of an ionic compound from its elements involves electron transfer

Under appropriate conditions, atoms are able to transfer electrons between one another when they react. This is what happens, for example, when the metal sodium combines with the nonmetal chlorine. As shown in Figure 2.24, sodium is a typical shiny metal and chlorine is

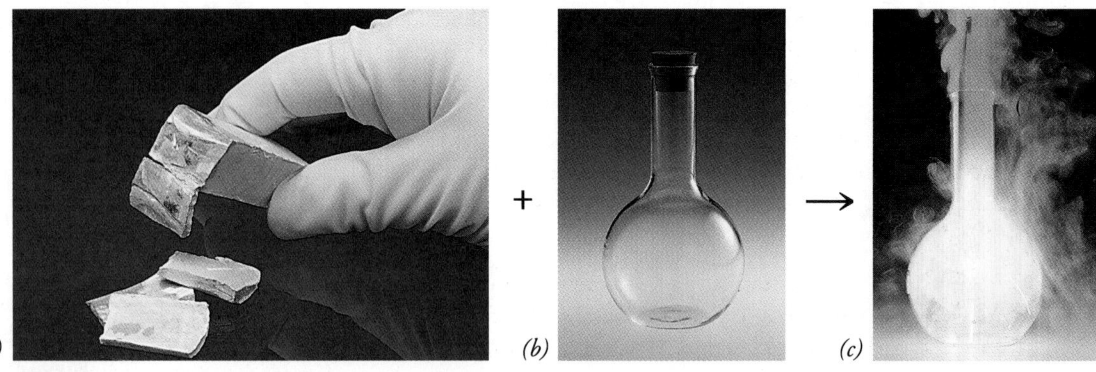

FIG. 2.24 Sodium reacts with chlorine to give the ionic compound sodium chloride. (*a*) Freshly cut sodium has a shiny metallic surface. The metal reacts with oxygen and moisture, so it cannot be touched with bare fingers. (*b*) Chlorine is a pale green gas. (*c*) When a small piece of sodium is melted in a metal spoon and thrust into the flask of chlorine, it burns brightly as the two elements react to form sodium chloride. The smoke coming from the flask is composed of fine crystals of salt. *(Michael Watson; Richard Megna/Fundamental Photographs; Richard Megna/Fundamental Photographs.)*

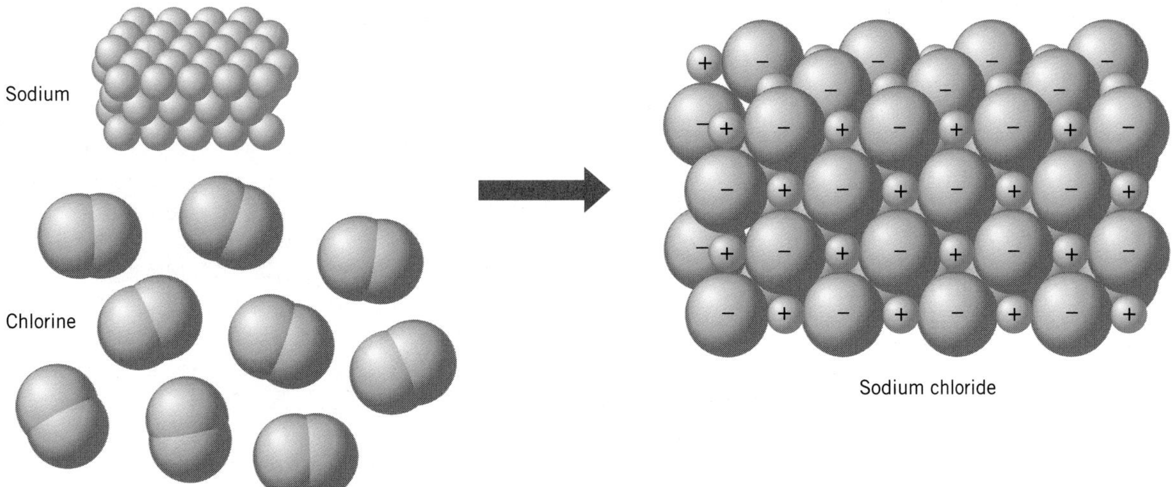

FIG. 2.25 **The reaction of sodium with chlorine viewed at the atomic level.** Electrically neutral atoms and molecules react to yield positive and negative ions which are held to each other by electrostatic attractions (attractions between opposite electrical charges).

a pale green gas. When a piece of heated sodium is thrust into the chlorine, a vigorous reaction takes place yielding a white powder, salt (NaCl). The equation for the reaction is

$$2\text{Na}(s) + \text{Cl}_2(g) \longrightarrow 2\text{NaCl}(s)$$

The changes that take place at the atomic level are illustrated in Figure 2.25.

The formation of the **ions** in salt results from the transfer of electrons between the reacting atoms. Specifically, each sodium atom gives up one electron to a chlorine atom. We can diagram the changes in equation form by using the symbol e^- to stand for an electron.

$$\overset{e^-}{\text{Na} + \text{Cl}} \longrightarrow \text{Na}^+ + \text{Cl}^-$$

Here we are concentrating on what happens to the individual atoms, so we have not shown chlorine as diatomic Cl_2 molecules.

The electrically charged particles formed in this reaction are a sodium ion (Na^+) and a chloride ion (Cl^-). The sodium ion has a positive 1+ charge, indicated by the superscript plus sign, because the loss of an electron leaves it with one more proton in its nucleus than there are electrons outside. Similarly, by gaining one electron the chlorine atom has added one more negative charge, so the chloride ion has a single negative charge indicated by the minus sign. Solid sodium chloride is composed of these charged sodium and chloride ions and is said to be an **ionic compound.**

A neutral sodium atom has 11 protons and 11 electrons; a sodium ion has 11 protons and 10 electrons, so it carries a unit positive charge. A neutral chlorine atom has 17 protons and 17 electrons; a chloride ion has 17 protons and 18 electrons, so it carries a unit negative charge.

As a general rule, *ionic compounds are formed when metals react with nonmetals.* In the electron transfer, however, not all atoms gain or lose just one electron; some gain or lose more. For example, when calcium atoms react, they lose two electrons to form Ca^{2+} ions and when oxygen atoms form ions they each gain two electrons to give O^{2-} ions. (We will have to wait until a later chapter to study the reasons why certain atoms gain or lose one electron each, whereas other atoms gain or lose two or more electrons.)

Practice Exercise 17: For each of the following atoms or ions, give the number of protons and the number of electrons in one particle. (a) an Fe atom, (b) an Fe^{3+} ion, (c) an N^{3-} ion, (d) an N atom. (Hint: Recall that electrons have a negative charge and ions that have a negative charge are atoms must have gained electrons.)

Practice Exercise 18: For each of the following atoms or ions, give the number of protons and the number of electrons in one particle. (a) an O atom, (b) an O^{2-} ion, (c) an Al^{3+} ion, (d) an Al atom.

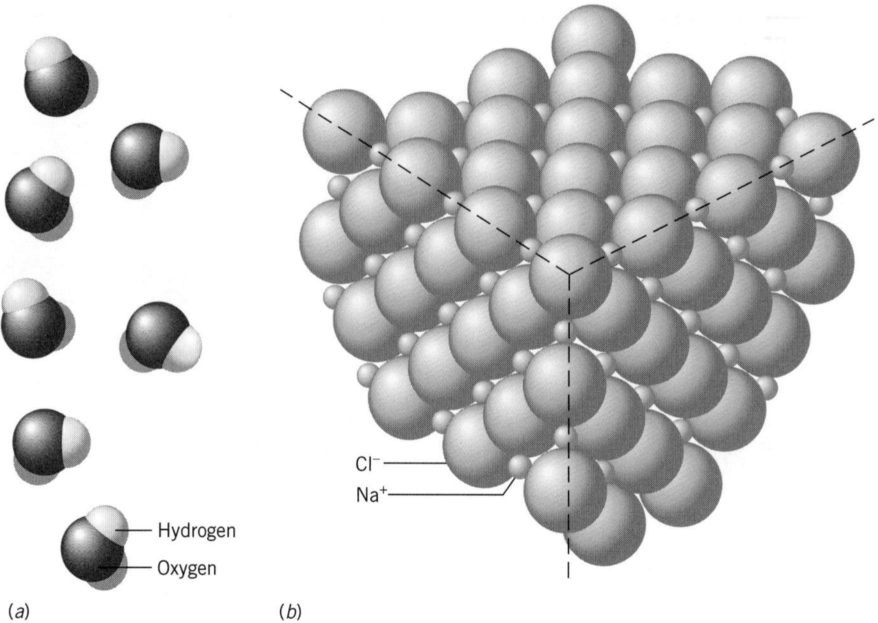

FIG. 2.26 **Molecular and ionic substances.** (*a*) In water there are discrete molecules that each consist of one atom of oxygen and two atoms of hydrogen. Each particle has the formula H_2O. (*b*) In sodium chloride, ions are packed in the most efficient way. Each Na^+ is surrounded by six Cl^-, and each Cl^- is surrounded by six Na^+. Because individual molecules do not exist, we simply specify the ratio of ions as NaCl.

Notice that the charges on the ions are omitted when writing formulas for compounds. This is because compounds are electrically neutral overall.

Figure 2.26 compares the structures of water and sodium chloride and demonstrates an important difference between molecular and ionic compounds. In water it is safe to say that two hydrogen atoms "belong" to each oxygen atom in a particle having the formula H_2O. However, in NaCl it is impossible to say that a particular Na^+ ion belongs to a particular Cl^- ion. The ions in a crystal of NaCl are simply packed in the most efficient way, so that positive ions and negative ions can be as close to each other as possible. In this way, the attractions between oppositely charged ions, which are responsible for holding the compound together, can be as strong as possible.

Because molecules don't exist in ionic compounds, the subscripts in their formulas are always chosen to specify the smallest whole-number ratio of the ions. This is why the formula of sodium chloride is given as NaCl rather than Na_2Cl_2 or Na_3Cl_3. Although the smallest unit of an ionic compound can't be called a molecule, the idea of "smallest unit" is still quite often useful. Therefore, we take the smallest unit of an ionic compound to be whatever is represented in its formula and call this unit a **formula unit.** Thus, one formula unit of NaCl consists of one Na^+ and one Cl^-, whereas one formula unit of the ionic compound $CaCl_2$ consists of one Ca^{2+} and two Cl^- ions. (In a broader sense, we can use the term *formula unit* to refer to whatever is represented by a formula. Sometimes the formula specifies a set of ions, as in NaCl; sometimes it is a molecule, as in O_2 or H_2O; sometimes it can be just an ion, as in Cl^- or Ca^{2+}; and sometimes it might be just an atom, as in Na.)

Experimental evidence exists for ions in compounds

We know that metals conduct electricity because electrons can move from one atom to the next in a wire when connected to a battery. Solid ionic compounds are poor conductors of electricity as are molecular substances such as water. However, if an ionic compound is dissolved in water or is heated to a high temperature, so that it melts, the resulting liquids are able to conduct electricity easily. These observations suggest that ionic compounds are composed of charged ions rather than neutral molecules and these ions when made mobile by dissolving or melting can conduct electricity. Figure 2.27 illustrates how the electrical conductivity can be tested.

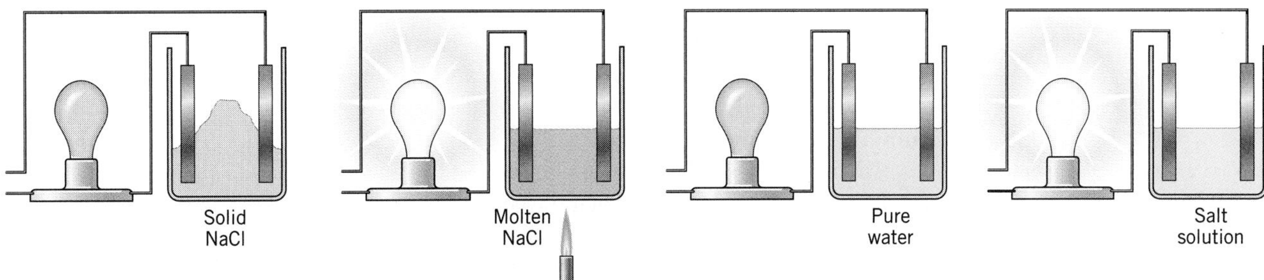

FIG. 2.27 An apparatus to test for electrical conductivity. The electrodes are dipped into the substance to be tested. If the lightbulb glows when electricity is applied, the sample is an electrical conductor. Here we see that solid sodium chloride does not conduct electricity, but when the solid is melted it does conduct. Liquid water, a molecular compound, is not a conductor of electricity because it does not contain electrically charged particles. An aqueous salt solution contains ions of the salt and will conduct electricity.

2.8 THE FORMULAS OF MANY IONIC COMPOUNDS CAN BE PREDICTED

In the preceding section we noted that metals combine with nonmetals to form ionic compounds. In such reactions, metal atoms lose one or more electrons to become positively charged ions and nonmetal atoms gain one or more electrons to become negatively charged ions. In referring to these particles, we will frequently call a positively charged ion a **cation** (pronounced *CAT-i-on*) and a negatively charged ion an **anion** (pronounced *AN-i-on*).[3] Thus, solid NaCl is composed of sodium cations and chloride anions.

Ions formed by representative metals and nonmetals can be remembered using the periodic table

In Section 2.6 you saw that the periodic table can be helpful in remembering the formulas of the nonmetal hydrides. It can also help us remember the kinds of ions formed by many of the representative elements (elements in the A groups of the periodic table). For example, except for hydrogen, the neutral atoms of the Group IA elements always lose one electron each when they react, thereby becoming ions with a charge of 1+. Similarly, atoms of the Group IIA elements always lose two electrons when they react; so these elements always form ions with a charge of 2+. In Group IIIA, the only important positive ion we need consider now is that of aluminum, Al^{3+}; an aluminum atom loses three electrons when it reacts to form the ion.

Positive ions, called **cations**, are usually metals that have lost one or more electrons.

All these ions are listed in Table 2.6. *Notice that the number of positive charges on each of the cations is the same as the group number when we use the North American numbering of groups in the periodic table.* Thus, sodium is in Group IA and forms an ion with a 1+

TABLE 2.6 Some Ions Formed from the Representative Elements

Group Number						
IA	IIA	IIIA	IVA	VA	VIA	VIIA
Li^+	Be^{2+}		C^{4-}	N^{3-}	O^{2-}	F^-
Na^+	Mg^{2+}	Al^{3+}	Si^{4-}	P^{3-}	S^{2-}	Cl^-
K^+	Ca^{2+}				Se^{2-}	Br^-
Rb^+	Sr^{2+}				Te^{2-}	I^-
Cs^+	Ba^{2+}					

[3] The names *cation* and *anion* come from the way the ions behave when electrically charged metal plates called electrodes are dipped into a solution that contains them. We will discuss this in detail in Chapter 19.

charge, barium (Ba) is in Group IIA and forms an ion with a 2+ charge, and aluminum is in Group IIIA and forms an ion with a 3+ charge. Although this generalization doesn't work for all the metallic elements (it doesn't work for the transition elements, for instance), it does help us remember what happens to the metallic elements of Groups IA and IIA and aluminum when they react.

Negative ions, called **anions**, are monatomic nonmetals that have gained one or more electrons.

Among the nonmetals on the right side of the periodic table we also find some useful generalizations. For example, when they combine with metals, the halogens (Group VIIA) form ions with one negative charge (written as 1−) and the nonmetals in Group VIA form ions with two negative charges (written as 2−). Notice that *the number of negative charges on the anion is equal to the number of spaces to the right that we have to move in the periodic table to get to a noble gas.*

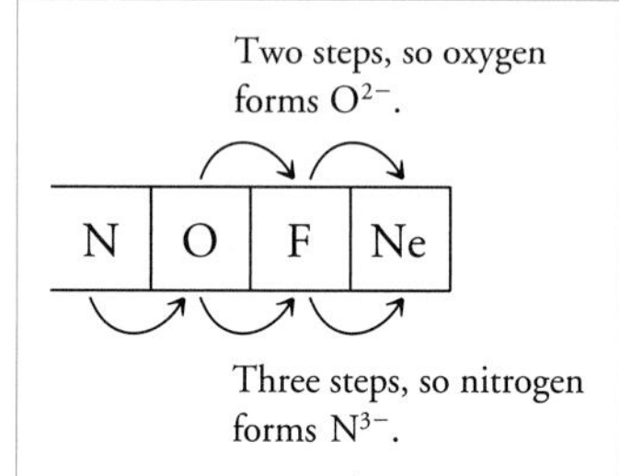

You've probably also noticed that the number of negative charges on an anion is the same as the number of hydrogens in the simple nonmetal hydride of the element, as shown in Table 2.4.

Writing formulas for ionic compounds follows certain rules

All chemical compounds are electrically neutral, so the ions in an ionic compound always occur in a ratio such that the total positive charge is equal to the total negative charge. This is why the formula for sodium chloride is NaCl; the 1-to-1 ratio of Na^+ to Cl^- gives electrical neutrality. In addition, as we've already mentioned, discrete molecules do not exist in ionic compounds, so we always use the smallest set of subscripts that specify the correct ratio of the ions. The following, therefore, are the rules we use in writing the formulas of ionic compounds.

A substance is electrically neutral, with a net charge of zero, if the total positive charge equals the total negative charge.

Rules for Writing Formulas of Ionic Compounds

1. The positive ion is given first in the formula. (This isn't required by nature, but it is a custom we always follow.)
2. The subscripts in the formula must produce an electrically neutral formula unit. (Nature *does* require electrical neutrality.)
3. The subscripts should be the smallest set of whole numbers possible. For instance, if all subscripts are even, divide them by 2. (You may have to repeat this simplification step several times.)
4. The charges on the ions are not included in the finished formula for the substance. When a subscript is 1 it is left off; no subscript implies a subscript of 1.

EXAMPLE 2.6
Writing Formulas for Ionic Compounds

Write the formulas for the ionic compounds formed from (a) Ba and S, (b) Al and Cl, and (c) Al and O.

ANALYSIS: To correctly write the formula, we have to apply our tool that summarizes the rules for ionic compounds listed above. First, we need to figure out the charges of the ions and since we're working with representative elements, we can use the periodic table to do this. Then we need to assemble the ions so that the formula unit is electrically neutral.

SOLUTION: (a) The element Ba is in Group IIA, so the charge on its ion is 2+. Sulfur is in Group VIA, so its ion has a charge of 2−. Therefore, the ions are Ba^{2+} and S^{2-}. Since the charges are equal but opposite, a 1-to-1 ratio will give a neutral formula unit. Therefore, the formula is BaS. Notice that we have not included the charges on the ions in the finished formula.

(b) By using the periodic table, the ions of these elements are Al^{3+} and Cl^-. We can obtain a neutral formula unit by combining one Al^{3+} with three Cl^-. (The charge on Cl^- is 1−; the 1 is understood.)

$$1(3+) + 3(1-) = 0$$

The formula is $AlCl_3$.

(c) For these elements, the ions are Al^{3+} and O^{2-}. In the formula we seek there must be the same number of positive charges as negative charges. This number must be a whole-number multiple of both 3 and 2. The smallest number that satisfies this condition is 6, so there must be two Al^{3+} and three O^{2-} in the formula.

$$\begin{array}{ll} 2Al^{3+} & 2(3+) = 6+ \\ 3O^{2-} & \underline{3(2-) = 6-} \\ & \text{sum} = 0 \end{array}$$

The formula is Al_2O_3.

A "trick" you may have seen before is to use the *number* of positive charges for the subscript of the anion and the *number* of negative charges as the subscript for the cation as shown in the diagram.

$$Al^{③+} \quad O^{②-}$$

When using this method, always be sure to check that the subscripts cannot be reduced to smaller numbers.

ARE THE ANSWERS REASONABLE? In writing a formula, there are two things to check. First, be sure you've correctly written the formulas of the ions. (This is where students make a lot of mistakes.) Then check that you've combined them in a ratio that gives electrical neutrality.

Practice Exercise 19: Write formulas for ionic compounds formed from (a) Na and F, (b) Na and O, (c) Mg and F, and (d) Al and C. (Hint: One element must form a cation and the other will form an anion based on its position in the periodic table.)

Practice Exercise 20: Write the formulas for the compounds made from (a) Ca and N, (b) Al and Br, (c) Na and P, and (d) Cs and Cl.

Many of our most important chemicals are ionic compounds. We have mentioned NaCl, common table salt, and $CaCl_2$, which is a substance often used to melt ice on walkways in the winter. Other examples are sodium fluoride, NaF, used by dentists to give fluoride treatments to teeth, and calcium oxide, CaO, an important ingredient in cement.

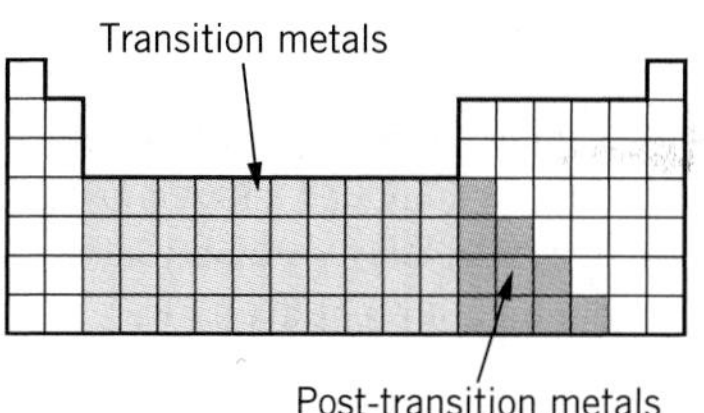

Distribution of transition and post-transition metals in the periodic table.

Transition and post-transition metals form more than one cation

The transition elements are located in the center of the periodic table, from Group IIIB on the left to Group IIB on the right (Groups 3 to 12 using the IUPAC system). All of them lie to the left of the metalloids, and they all are metals. Included here are some of our most familiar metals, including iron, chromium, copper, silver, and gold.

TABLE 2.7 Ions of Some Transition Metals and Post-transition Metals

Transition Metals	
Chromium	Cr^{2+}, Cr^{3+}
Manganese	Mn^{2+}, Mn^{3+}
Iron	Fe^{2+}, Fe^{3+}
Cobalt	Co^{2+}, Co^{3+}
Nickel	Ni^{2+}
Copper	Cu^{+}, Cu^{2+}
Zinc	Zn^{2+}
Silver	Ag^{+}
Cadmium	Cd^{2+}
Gold	Au^{+}, Au^{3+}
Mercury	$Hg_2{}^{2+}$, Hg^{2+}
Post-transition Metals	
Tin	Sn^{2+}, Sn^{4+}
Lead	Pb^{2+}, Pb^{4+}
Bismuth	Bi^{3+}

Most of the transition metals are much less reactive than the metals of Groups IA and IIA, but when they react they also transfer electrons to nonmetal atoms to form ionic compounds. However, the charges on the ions of the transition metals do not follow as straightforward a pattern as do those of the alkali and alkaline earth metals. One of the characteristic features of the transition metals is the ability of many of them to form more than one positive ion. Iron, for example, can form two different ions, Fe^{2+} and Fe^{3+}. This means that iron can form more than one compound with a given nonmetal. For example, with the chloride ion, Cl^-, iron forms two compounds, with the formulas $FeCl_2$ and $FeCl_3$. With oxygen, we find the compounds FeO and Fe_2O_3. As usual, we see that the formulas contain the ions in a ratio that gives electrical neutrality. Some of the most common ions of the transition metals are given in Table 2.7. Notice that one of the ions of mercury is diatomic $Hg_2{}^{2+}$. It consists of two Hg^+ ions joined by the same kind of bond found in molecular substances. The simple Hg^+ ion does not exist.

The prefix *post* means "after."

The **post-transition metals** are those metals that occur in the periodic table immediately following a row of transition metals. The two most common and important ones are tin (Sn) and lead (Pb). Except for bismuth, post-transition metals have the ability to form two different ions, and therefore two different compounds with a given nonmetal. For example, tin forms two oxides, SnO and SnO_2. Lead also forms two oxides that have similar formulas (PbO and PbO_2). The ions that these metals form are also included in Table 2.7.

Practice Exercise 21: Write formulas for the chlorides and oxides formed by (a) chromium and (b) copper. (Hint: There are more than one chloride and oxide for each of these transition metals.)

Practice Exercise 22: Write the formulas for the sulfides and nitrides of (a) gold and (b) tin.

Ions may be composed of more than one element

A substance is **diatomic** if it is composed of molecules that contain only two atoms. It is a **binary compound** if it contains two different elements, regardless of the number of each. Thus, BrCl is a binary compound and is also diatomic; CH_4 is a binary compound but is not diatomic.

The metal compounds that we have discussed so far have been **binary compounds**—compounds formed from *two* different elements. There are many other ionic compounds that contain more than two elements. These substances usually contain **polyatomic ions,** which are ions that are themselves composed of two or more atoms linked by the same kinds of bonds that hold molecules together. Polyatomic ions differ from molecules, however, in that they contain either too many or too few electrons to make them electrically neutral. Table 2.8 lists some important polyatomic ions. It is very important that you learn the formulas, charges, and names of all of these ions.

The formulas of compounds formed from polyatomic ions are determined in the same way as are those of binary ionic compounds; the ratio of the ions must be such that the formula unit is electrically neutral, and the smallest set of whole-number subscripts is used. One difference in writing formulas with polyatomic ions is that parentheses are needed around the polyatomic ion if a subscript is required.

Polyatomic ions

TABLE 2.8 Formulas and Names of Some Polyatomic Ions

Ion	Name (Alternate Name in Parentheses)
NH_4^+	ammonium ion
H_3O^+	hydronium ion[a]
OH^-	hydroxide ion
CN^-	cyanide ion
NO_2^-	nitrite ion
NO_3^-	nitrate ion
ClO^- or OCl^-	hypochlorite ion
ClO_2^-	chlorite ion
ClO_3^-	chlorate ion
ClO_4^-	perchlorate ion
MnO_4^-	permanganate ion
$C_2H_3O_2^-$	acetate ion
$C_2O_4^{2-}$	oxalate ion
CO_3^{2-}	carbonate ion
HCO_3^-	hydrogen carbonate ion (bicarbonate ion)[b]
SO_3^{2-}	sulfite ion
HSO_3^-	hydrogen sulfite ion (bisulfite ion)[b]
SO_4^{2-}	sulfate ion
HSO_4^-	hydrogen sulfate ion (bisulfate ion)[b]
SCN^-	thiocyanate ion
$S_2O_3^{2-}$	thiosulfate ion
CrO_4^{2-}	chromate ion
$Cr_2O_7^{2-}$	dichromate ion
PO_4^{3-}	phosphate ion
HPO_4^{2-}	monohydrogen phosphate ion
$H_2PO_4^-$	dihydrogen phosphate ion

[a]You will encounter this ion only in aqueous solutions.
[b]You will often see and hear the alternate names for these ions.

In general, polyatomic ions are not formed by the direct combination of elements. They are the products of reactions between compounds.

EXAMPLE 2.7
Formulas That Contain Polyatomic Ions

One of the minerals responsible for the strength of bones is the ionic compound calcium phosphate, which is formed from Ca^{2+} and PO_4^{3-}. Write the formula for this compound.

ANALYSIS: The essential tool for solving this problem is the identity of the formula, including the charge, of the polyatomic ion. We have related much information about ions to the periodic table. Unfortunately, the polyatomic ions must be memorized. Knowledge of the polyatomic ions is required for this problem.

SOLUTION: As before, if the number of positive charges on the cation is equal to the number of negative of charges on the anion, the formula unit will contain one of each. If the number of charges are not equal then we use the number of positive charges as the subscript for the anion and the number of negative charges as the subscript for the cation. We will need three calcium ions to give a total charge of 6+ and two phosphate ions to give a charge of 6− so that the total charge is $+6 - 6 = 0$. The formula is written with parentheses to show that the PO_4^{3-} ion occurs two times in the formula unit.

$$Ca_3(PO_4)_2$$

IS THE ANSWER REASONABLE? We double-check to see that electrical neutrality is achieved for the compound. We have six positive charges from the three Ca^{2+} ions and six negative charges from the two PO_4^{3-} ions. The sum is zero and our compound is electrically neutral as required.

Practice Exercise 23: Write the formula for the ionic compound formed from (a) potassium ion and acetate ion, (b) strontium ion and nitrate ion, and (c) Fe^{3+} and acetate ion. (Hint: See if you remember these polyatomic ions before looking at the table.)

Practice Exercise 24: Write the formula for the ionic compound formed from (a) Na^+ and CO_3^{2-}, (b) NH_4^+ and SO_4^{2-}.

Polyatomic ions are found in a large number of very important compounds. Examples include $CaSO_4$ (calcium sulfate, in plaster of Paris), $NaHCO_3$ (sodium bicarbonate, also called baking soda), NaOCl (sodium hypochlorite, in liquid laundry bleach), $NaNO_2$ (sodium nitrite, a meat preservative), $MgSO_4$ (magnesium sulfate, also known as Epsom salts), and $NH_4H_2PO_4$ (ammonium dihydrogen phosphate, a fertilizer).

2.9 Molecular and Ionic Compounds Are Named Following a System

In conversation, chemists rarely use formulas to describe compounds. Instead, names are used. For example, you already know that water is the name for the compound having the formula H_2O and that sodium chloride is the name of NaCl.

At one time there was no uniform procedure for assigning names to compounds, and those who discovered compounds used whatever method they wished. Today, we know of more than 15 million different chemical compounds, and it is necessary to have a logical system for naming them. Chemists around the world now agree on a systematic method for naming substances that is overseen by the International Union of Pure and Applied Chemistry, IUPAC. By using the **IUPAC rules** we are able to write the correct formula given the name for the many compounds we will encounter. Additionally we will be able to take a formula and correctly name it.

In this section we discuss the **nomenclature** (naming) of simple molecular and ionic inorganic compounds. In general, **inorganic compounds** are substances that would *not* be considered to be derived from hydrocarbons such as methane (CH_4), ethane (C_2H_6), and other carbon–hydrogen compounds. As we noted earlier, the hydrocarbons and compounds that can be thought of as coming from them are called organic compounds. We will have more to say about naming organic compounds later.

Even if we exclude organic compounds, the number and variety of molecular substances is quite enormous. To introduce you to the naming of them, we will restrict ourselves to binary compounds.

Binary compounds composed of two nonmetals are named using Greek prefixes

Our goal is to be able to translate a chemical formula into a name that contains information that would enable someone else, just looking at the name, to reconstruct the formula. For a binary molecular compound, therefore, we must indicate which two elements are present and the number of atoms of each in a molecule of the substance.

To identify the first element in a formula, we just specify its English name. Thus, for HCl the first word in the name is "hydrogen" and for PCl_5 the first word is "phosphorus." To identify the second element, we append the suffix *-ide* to the stem of the element's English name. Here are some examples:

Element	*Stem*	*Name as second element*
oxygen	ox-	oxide
sulfur	sulf-	sulfide
nitrogen	nitr-	nitride
phosphorus	phosph-	phosphide
fluorine	fluor-	fluoride
chlorine	chlor-	chloride
bromine	brom-	bromide
iodine	iod-	iodide

To form the name of the compound, we place the two parts of the name one after another. Therefore, the name of HCl is hydrogen chloride. However, to name PCl_5, we need a way to specify the number of Cl atoms bound to the phosphorus in the molecule. This is done using the following Greek prefixes:

mono- = 1 (often omitted)	hexa- = 6
di- = 2	hepta- = 7
tri- = 3	octa- = 8
tetra- = 4	nona- = 9
penta- = 5	deca- = 10

To name PCl_5, therefore, we add the prefix *penta-* to chloride to give the name phosphorus pentachloride. Notice how easily this allows us to translate the name back into the formula.

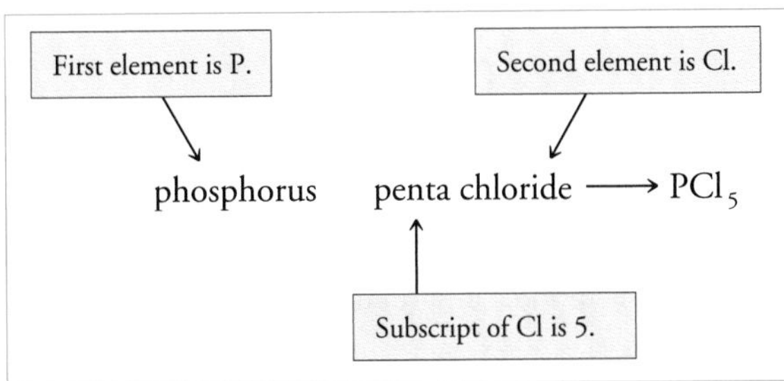

TOOLS *Naming binary molecular compounds*

The prefix *mono-* is used when we want to emphasize that only one atom of a particular element is present. For instance, carbon forms two compounds with oxygen, CO and CO_2. To clearly distinguish between them, the first is called carbon monoxide (one of the o's is omitted to make the name easier to pronounce) and the second is carbon dioxide.

As indicated above, the prefix *mono-* is often omitted from a name. Therefore, in general, if there is no prefix before the name of an element, we take it to mean there is only one atom of that element in the molecule. An exception to this is in the names of binary compounds of nonmetals with hydrogen. An example is hydrogen sulfide. The name tells us the compound contains the two elements hydrogen and sulfur. We don't have to be told how many hydrogens are in the molecule because, as you learned earlier, we can use the periodic table to determine the number of hydrogen atoms in molecules of the simple nonmetal hydrides. Sulfur is in Group VIA, so to get to the noble gas column we have to move two steps to the right; the number of hydrogens combined with the atom of sulfur is two. The formula for hydrogen sulfide is therefore H_2S.

EXAMPLE 2.8 Naming Compounds and Writing Formulas

(1) What is the name of $AsCl_3$? (2) What is the formula for dinitrogen tetraoxide?

ANALYSIS: (1) In naming compounds, the first step is to determine what type of compound is involved. Looking at the periodic table, we see that $AsCl_3$ is made up of two nonmetals, so we conclude that it is a molecular compound. Once we've done this, we apply the tool for naming molecular compounds described above.

(2) To write the formula from the name, we convert the prefixes to numbers and apply them as subscripts to the chemical symbols of the elements.

☐ After we've discussed ionic compounds this first step in the analysis will be particularly important, because different rules apply depending on the type of compound being named.

SOLUTION: (1) In $AsCl_3$, As is the symbol for arsenic and, of course, Cl is the symbol for chlorine. The first word in the name is just arsenic and the second will contain chloride with an appropriate prefix to indicate number. There are three Cl atoms, so the prefix is tri-. Therefore, the name of the compound is arsenic trichloride.

(2) As we did earlier for phosphorus pentachloride, we convert the prefixes to numbers and apply them as subscripts.

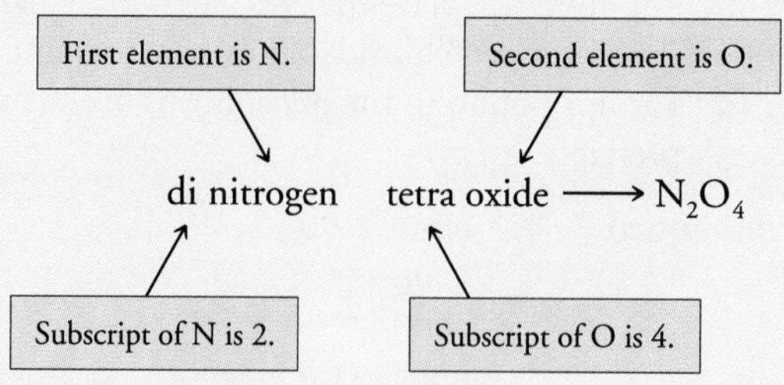

Sometimes we drop the *a* before an *o* for ease of pronunciation. N_2O_4 would then be named dinitrogen tetroxide.

ARE THE ANSWERS REASONABLE? To feel comfortable with the answers, be sure to double-check for careless errors. Next, take your answers and reverse the process. Does arsenic trichloride result in the original formula, $AsCl_3$? Does N_2O_4 have a name of dinitrogen tetraoxide? We can say yes to both and have confidence in our work.

Practice Exercise 25: Name the following compounds using Greek prefixes when needed: (a) PCl_3, (b) SO_2, and (c) Cl_2O_7. (Hint: See the list of prefixes above.)

Practice Exercise 26: Write formulas for the following compounds: (a) arsenic pentachloride, (b) sulfur hexachloride, and (c) disulfur dichloride.

Common names exist for many molecular compounds

Not every compound is named according to the systematic procedure described above. Many familiar substances were discovered long before a systematic method for naming them had been developed, and they acquired common names that are so well known that no attempt has been made to rename them. For example, following the scheme described above we might expect that H_2O would have the name hydrogen oxide (or even dihydrogen monoxide). Although this isn't wrong, the common name water is so well known that it is always used. Another example is ammonia, NH_3, whose odor you have no doubt experienced while using household ammonia solutions or the glass cleaner Windex. Common names are used for the other hydrides of the nonmetals in Group VA as well. The compound PH_3 is called phosphine and AsH_3 is called arsine.

Common names are also used for very complex substances. An example is sucrose, which is the chemical name for table sugar, $C_{12}H_{22}O_{11}$. The structure of this compound is pretty complex, and its name assigned following the systematic method is equally complex. It is much easier to say the simple name sucrose, and be understood, than to struggle with the cumbersome systematic name for this common compound.

For ionic compounds, the name of the cation precedes that of the anion

In naming ionic compounds, our goal is the same as in naming molecular substances—we want a name that someone else could use to reconstruct the formula. The system we use here, however, is somewhat different than for molecular compounds.

For ionic compounds, the name of the cation is given first, followed by the name of the anion. This is the same as the sequence in which the ions appear in the formula. If the metal in the compound forms only one cation, such as Na^+ or Ca^{2+}, the cation is specified by just giving the English name of the metal. The anion in a binary compound is formed from a nonmetal and its name is created by adding the suffix *-ide* to the stem of the name for the nonmetal just as we did for molecular compounds. An example is KBr, potassium bromide. Table 2.9 lists some common **monatomic** (one-atom) negative ions and their names. It is also useful to know that the *-ide* suffix is usually used only for monatomic ions, with just two common exceptions—*hydroxide ion* (OH^-) and *cyanide ion* (CN^-).[4]

[4] If the name of a compound ends in *-ide* and it isn't either a hydroxide or a cyanide, you can feel confident the substance is a binary compound.

TABLE 2.9 Monatomic Negative Ions

H^-	hydride	N^{3-}	nitride	O^{2-}	oxide	F^-	fluoride
C^{4-}	carbide	P^{3-}	phosphide	S^{2-}	sulfide	Cl^-	chloride
Si^{4-}	silicide	As^{3-}	arsenide	Se^{2-}	selenide	Br^-	bromide
				Te^{2-}	telluride	I^-	iodide

To form the name of an ionic compound, we simply specify the names of the cation and anion. *Use of prefixes to identify the number of cations and anions would be redundant and therefore prefixes are never used when naming ionic compounds.* The reason is that once we know what the ions are, we can assemble the formula correctly just by taking them in a ratio that gives electrical neutrality.

To keep the name as simple as possible, we give the minimum amount of information necessary to be able to reconstruct the formula. To write the formula of an ionic compound, we only need the formulas of the ions.

EXAMPLE 2.9
Naming Compounds and Writing Formulas

(a) What is the name of $SrBr_2$? (b) What is the formula for aluminum selenide?

ANALYSIS: The first step in naming a compound is to determine the type of compound it is. As you've already seen, there are slightly different rules for ionic and molecular substances. What clues do we have there that these are ionic?

(a) The element Sr is a metal and Br is a nonmetal. Compounds of a metal and nonmetal are ionic, so we use the rules for naming ionic compounds.

(b) Aluminum is a metal. The only compounds of metals that we've discussed are ionic, so we'll proceed on that assumption. The *-ide* ending of selenide suggests the anion is composed of a single atom of a nonmetal. The only one that begins with the letters "selen-" is selenium, Se. (See the table inside the front cover.)

SOLUTION: (a) The compound is composed of the ions Sr^{2+} (an element from Group IIA) and Br^- (an element from Group VIIA). The cation simply takes the name of the metal, which is strontium. The anion's name is derived from bromine by replacing *-ine* with *-ide*; it is the bromide ion. The name of the compound is strontium bromide.

(b) The name tells us that the cation is the aluminum ion, Al^{3+}. The anion is formed from selenium (Group VIA), and its formula is Se^{2-}. The correct formula must represent an electrically neutral formula unit. Using the number of charges on one ion as the subscript of the other, the formula is Al_2Se_3.

ARE THE ANSWERS REASONABLE? First, we review the analysis and check to be sure we've applied the correct rules, which we have. Next we can reverse the process to be sure our name strontium bromide does mean $SrBr_2$, and that it is reasonable to call Al_2Se_3 aluminum selenide.

Practice Exercise 27: Give the correct formulas for (a) potassium oxide, (b) barium bromide, (c) sodium nitride, and (d) aluminum sulfide. (Hint: Recall what the ending *-ide* means.)

Practice Exercise 28: Give the correct names for (a) $AlCl_3$, (b) BaS, (c) NaBr, and (d) CaF_2.

When a metal can form more than one ion, the charge is indicated with a Roman numeral

Many of the transition metals and post-transition metals are able to form more than one positive ion. Iron, a typical example, forms ions with either a 2+ or a 3+ charge (Fe^{2+} or Fe^{3+}). Compounds that contain these different iron ions have different formulas, so in their names it is necessary to specify which iron ion is present.

FACETS OF CHEMISTRY

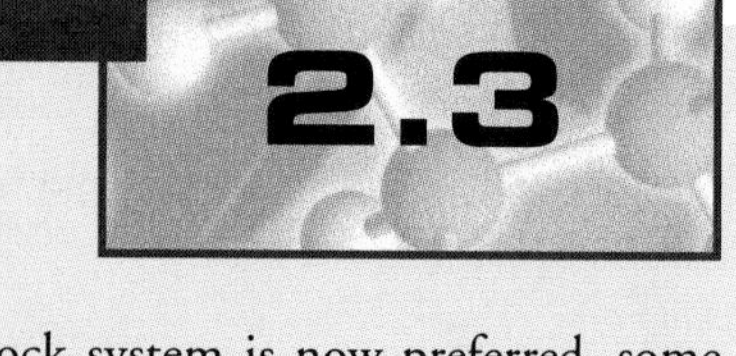

Another System for Naming Ionic Compounds

The Stock system, which we use when naming compounds of metals able to form more than one positive ion, is a relatively recent development in inorganic nomenclature. A slightly different method existed before it.

In the older system, the suffix *-ous* is used to specify the ion with the lower charge and the suffix *-ic* is used to specify the ion with the higher charge. With this method, we also use the Latin stem for elements whose symbols are derived from their Latin names. Some examples are

Fe^{2+}	ferrous ion	$FeCl_2$	ferrous chloride
Fe^{3+}	ferric ion	$FeCl_3$	ferric chloride
Cu^{+}	cuprous ion	$CuCl$	cuprous chloride
Cu^{2+}	cupric ion	$CuCl_2$	cupric chloride

One difficulty with this method is that it does not specify what the charges on the metal ions are, so it becomes necessary to memorize the ions formed by the metals. Additional examples are given in the table below. Notice that mercury is an exception; we use the English stem when naming its ions.

Even though the Stock system is now preferred, some chemical companies still label bottles of chemicals using the old system. These old names also appear in the older scientific literature, which still holds much excellent data. This means that you may need to learn both systems.

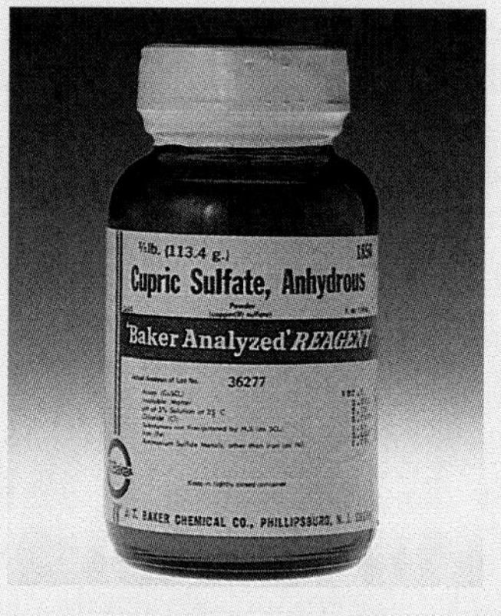

The older system of nomenclature is still found on the labels of many laboratory chemicals. This bottle contains copper(II) sulfate, which according to the older system of nomenclature is called cupric sulfate. *(Michael Watson.)*

Cr^{2+}	chromous	Mn^{2+}	manganous	Fe^{2+}	ferrous	Co^{2+}	cobaltous
Cr^{3+}	chromic	Mn^{3+}	manganic	Fe^{3+}	ferric	Co^{3+}	cobaltic
Au^{+}	aurous	Hg_2^{2+}	mercurous	Sn^{2+}	stannous	Pb^{2+}	plumbous
Au^{3+}	auric	Hg^{2+}	mercuric	Sn^{4+}	stannic	Pb^{4+}	plumbic

Originally the cations with different charges were given names that distinguished the higher charge from the lower one. That system is described in Facets of Chemistry 2.3 for those who are interested.

Alfred Stock (1876–1946), a German inorganic chemist, was one of the first scientists to warn the public of the dangers of mercury poisoning.

The currently preferred method for naming ions of metals that can have more than one charge in compounds is called the **Stock system.** Here we use the English name followed, *without a space*, by the numerical value of the charge written as a Roman numeral in parentheses.[5] Examples of using the Stock system are shown below.

Fe^{2+}	iron(II)	$FeCl_2$	iron(II) chloride
Fe^{3+}	iron(III)	$FeCl_3$	iron(III) chloride
Cr^{2+}	chromium(II)	CrS	chromium(II) sulfide
Cr^{3+}	chromium(III)	Cr_2S_3	chromium(III) sulfide

Using the Stock system

Remember that *the Roman numeral equals the positive charge on the metal ion*; it is not necessarily a subscript in the formula. For example, copper forms two oxides, one containing the Cu^{+} ion and the other containing the Cu^{2+} ion. Their formulas are Cu_2O and CuO and their names are as follows[6]:

[5] Silver and nickel are almost always found in compounds as Ag^{+} and Ni^{2+}, respectively. Therefore, AgCl and $NiCl_2$ are almost always called simply silver chloride and nickel chloride.

[6] For some metals, such as copper and lead, one of their ions is much more commonly found in compounds than any of their others. For example, most common copper compounds contain Cu^{2+} and most common lead compounds contain Pb^{2+}. For compounds of these metals, if the charge is not indicated by a Roman numeral, we assume the ion present has a 2+ charge. Thus, it not unusual to find $PbCl_2$ called lead chloride, or for $CuCl_2$ to be called copper chloride.

Cu^+	copper(I)	Cu_2O	copper(I) oxide
Cu^{2+}	copper(II)	CuO	copper(II) oxide

These copper compounds illustrate that in deriving the formula from the name, you must figure out the formula from the ionic charges, as discussed in Section 2.8 and illustrated in the preceding example.

EXAMPLE 2.10
Naming Compounds and Writing Formulas

The compound $MnCl_2$ has a number of commercial uses, including disinfecting, the manufacture of batteries, and purifying natural gas. What is the name of the compound?

ANALYSIS: The first step is to determine the kind of compound involved so we can apply the appropriate rules. The compound here is made up of a metal and a nonmetal, so we use the rules for naming ionic compounds.

To name the cation, we need to know whether the metal is one that forms more than one positive ion. Manganese (Mn) is a transition element, and transition elements often do form more than one cation, so we should apply the Stock method as our tool. To do this, we need to determine the charge on the manganese cation. We can figure this out because the sum of the charges on the Mn and Cl ions must equal zero, and because the only ion chlorine forms has a single negative charge.

SOLUTION: The anion of chlorine (the chloride ion) is Cl^-, so a total of two negative charges are supplied by the two Cl^- ions. Therefore, for $MnCl_2$ to be electrically neutral, the Mn ion must carry two positive charges, 2+. The cation is named as manganese(II), and the name of the compound is manganese(II) chloride.

IS THE ANSWER REASONABLE? Performing a quick check of the arithmetic assures us we've got the correct charges on the ions. Everything appears to be okay.

EXAMPLE 2.11
Naming Compounds and Writing Formulas

What is the formula for cobalt(III) fluoride?

ANALYSIS: To answer this question, we first need to determine the charges on the two ions using the tool above. Then we assemble them into a chemical formula being sure to achieve an electrically neutral formula unit.

SOLUTION: Cobalt(III) corresponds to Co^{3+}. The fluoride ion is F^-. To obtain an electrically neutral substance, we must have three F^- ions for each Co^{3+} ion, so the formula is CoF_3.

IS THE ANSWER REASONABLE? We can check to see that we have the correct formulas of the ions and that we've combined them to achieve an electrically neutral formula unit. This will tell us we've obtained the correct answer.

Practice Exercise 29: Name the compounds (a) K_2S, (b) Mg_3P_2, (c) $NiCl_2$, and (d) Fe_2O_3. Use the Stock system where appropriate. (Hint: Determine which metals can have more than one charge.)

Practice Exercise 30: Write formulas for (a) aluminum sulfide, (b) strontium fluoride, (c) titanium(IV) oxide, and (d) gold(III) oxide.

Similar rules apply to naming ionic compounds that contain polyatomic ions

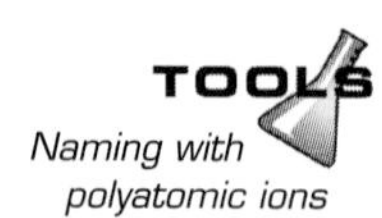

Naming with polyatomic ions

The extension of the nomenclature system to include ionic compounds containing polyatomic ions is straightforward. Most of the polyatomic ions listed in Table 2.8 are anions and their names are used as the second word in the name of the compound. For example, Na_2SO_4 contains the sulfate ion, SO_4^{2-}, and is called sodium sulfate. Similarly, $Cr(NO_3)_3$ contains the nitrate ion, NO_3^-. Chromium is a transition element, and in this compound its charge must be 3+ to balance the negative charges of three NO_3^- ions. Therefore, $Cr(NO_3)_3$ is called chromium(III) nitrate.

Among the ions in Table 2.8, the only cation that forms compounds which can be isolated is ammonium ion, NH_4^+. It forms ionic compounds such as NH_4Cl (ammonium chloride) and $(NH_4)_2SO_4$ (ammonium sulfate), even though NH_4^+ is not a metal cation. Notice that the latter compound is composed of two polyatomic ions.

It is important that you learn the formulas (including charges) and the names of the polyatomic ions in Table 2.8. You will encounter them frequently throughout your chemistry course.

EXAMPLE 2.12
Naming Compounds and Writing Formulas

What is the name of $Mg(ClO_4)_2$, a compound used commercially for removing moisture from gases?

ANALYSIS: To answer this question, it is essential that you recognize that the compound contains a polyatomic ion. If you've learned the contents of Table 2.8, you know that "ClO_4" is the formula (without the charge) of the perchlorate ion. With the charge, the ion's formula is ClO_4^-. We also have to decide whether we need to apply the Stock system in naming the metal Mg. We use the naming tool for polyatomic ion compounds to complete the exercise.

SOLUTION: Magnesium is in Group IIA, and forms only the ion Mg^{2+}. Therefore, we don't need to use the Stock system in naming the cation; it is named simply as "magnesium." The anion is perchlorate, so the name of $Mg(ClO_4)_2$ is magnesium perchlorate.

IS THE ANSWER REASONABLE? We can check to be sure we've named the anion correctly, and we have. The metal is magnesium, which only forms Mg^{2+}. Therefore, the answer seems to be correct.

Practice Exercise 31: What are the names of (a) Li_2CO_3 and (b) $Fe(OH)_3$? (Hint: Recall the names of the polyatomic ions and the positions of Li and Fe in the periodic table.)

Practice Exercise 32: Write the formulas for (a) potassium chlorate and (b) nickel(II) phosphate.

Hydrates are named using Greek prefixes

Earlier we discussed compounds called hydrates, such as $CuSO_4 \cdot 5H_2O$. Usually, hydrates are ionic compounds whose crystals contain water molecules in fixed proportions relative to the ionic substance. To name them, we provide two pieces of information: the name of the ionic compound and the number of water molecules in the formula. The number of water molecules is specified using the Greek prefixes mentioned earlier (mono-, di-, tri-, etc.), which precede the word "hydrate." Thus, $CuSO_4 \cdot 5H_2O$ is named as "copper sulfate *pentahydrate*." Similarly, $CaSO_4 \cdot 2H_2O$ is named calcium sulfate dihydrate, and $FeCl_3 \cdot 6H_2O$ is iron(III) chloride hexahydrate.[7]

[7] Chemical suppliers (who do not always follow current rules of nomenclature) sometimes indicate the number of water molecules using a number and a dash. For example, one supplier lists $Ca(NO_3)_2 \cdot 4H_2O$ as "Calcium nitrate, 4-hydrate."

Naming a compound requires that we select the rules that apply

In this chapter we've discussed how to name two classes of compounds, molecular and ionic, and you saw that slightly different rules apply to each. To name chemical compounds successfully we need to make a series of decisions based on the rules we just covered. At this point we can summarize this decision process in a flowchart such as the one shown in Figure 2.28. The example below illustrates how to use the flowchart, and when working on the Review Problems, you may want to refer to Figure 2.28 until you are able to develop the skills that will enable you to work without it.

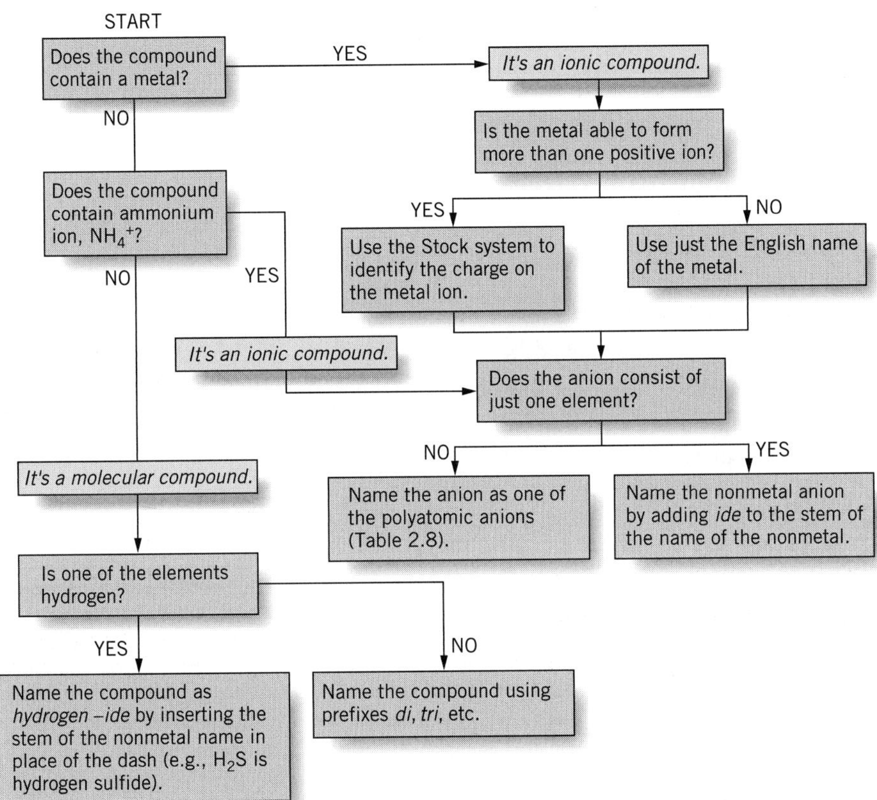

FIG. 2.28 Flowchart for naming molecular and ionic compounds.

EXAMPLE 2.13
Applying the Rules for Naming Compounds

What is the name of (a) $CrCl_3$, (b) P_4S_3, and (c) NH_4NO_3?

ANALYSIS: For each compound, we use the tools summarized in Figure 2.28 and proceed through the decision processes to arrive at the way to name the compound. As you read through the solution below, be sure to refer to Figure 2.28 so you can see how we decide which rules we need to apply.

SOLUTION: (a) Starting at the top of Figure 2.28, we first determine that the compound contains a metal (Cr), so it's an ionic compound. Next, we see that the metal is a transition element, and chromium is one of those that forms more than one cation, so we have to apply the Stock method. To do this, we need to know what the charge is on the metal ion. We can figure this out using the charge on the anion and the fact that the compound must be electrically neutral overall. The anion is formed from chlorine, so its charge must be 1− (the anion is Cl^-). Therefore, the metal ion must be Cr^{3+}; we name the metal as *chromium(III)*. Next, we see that there is only one nonmetallic element in the compound, Cl, so the name of the anion ends in *-ide*; it's the *chloride* ion. The compound $CrCl_3$ is therefore named *chromium(III) chloride*.

(b) Once again, we start at the top. First we determine that the compound doesn't contain a metal. It also doesn't contain NH_4, so the compound is molecular. It doesn't contain hydrogen, so we are led to the decision that we must use Greek prefixes to specify the numbers of atoms of each element. Applying the procedure on page 71, the name of the compound P_4S_3 is *tetraphosphorus trisulfide.*

(c) We begin at the top. Studying the formula, we see that it does not contain the symbol for a metal, so we proceed down the left side of the figure. The formula does contain NH_4 which indicates the compound contains the *ammonium* ion, NH_4^+ (it's an ionic compound). The rest of the formula is NO_3, which consists of more than one atom. This suggests the polyatomic anion, NO_3^- (*nitrate* ion). The name of the compound NH_4NO_3 is *ammonium nitrate.*

ARE THE ANSWERS REASONABLE? To check the answers in a problem of this kind, review the decision processes which led you to the names. In part (a), you can check to be sure you've calculated the charge on the chromium ion correctly. Also, check to be sure you've used the correct names of any polyatomic ions. Doing these things will show we've named the compounds correctly.

Practice Exercise 33: The compound I_2O_5 is used in respirators where it serves to react with highly toxic carbon monoxide to give the much less toxic gas, carbon dioxide. What is the name of I_2O_5? (Hint: Is this a molecular or ionic compound?)

Practice Exercise 34: The compound $Cr(C_2H_3O_2)_3$ is used in the tanning of leather. What is the name for this compound?

SUMMARY

Elements and Atoms. When accurate masses of all the reactants and products in a reaction are measured and compared, no observable changes in mass accompany chemical reactions (the **law of conservation of mass**). The mass ratios of the elements in any compound are constant regardless of the source of the compound or how it is prepared (the **law of definite proportions**). **Dalton's atomic theory** explained the laws of chemical combination by proposing that matter consists of indestructible atoms with masses that do not change during chemical reactions. During a chemical reaction, atoms may change partners, but they are neither created nor destroyed. After Dalton had proposed his theory, it was discovered that whenever two elements form more than one compound, the different masses of one element that combine with a fixed mass of the other are in a ratio of small whole numbers (the **law of multiple proportions**). Using modern instruments such as the **scanning tunneling microscope,** scientists are able to "see" atoms on the surfaces of solids.

Atomic Mass. An element's **atomic mass** (**atomic weight**) is the relative mass of its atoms on a scale in which atoms of carbon-12 have a mass of exactly 12 u (**atomic mass units**). Most elements occur in nature as uniform mixtures of a small number of **isotopes,** whose masses differ slightly. However, all isotopes of an element have very nearly identical chemical properties and the percentages of the isotopes that make up an element are generally so constant throughout the world that we can say that the average mass of their atoms is a constant.

Atomic Structure. Atoms can be split into **subatomic particles,** such as **electrons, protons,** and **neutrons. Nucleons** are particles that make up the atomic **nucleus** and include the protons, each of which carries a single unit of **positive charge** (charge = 1+), and neutrons (no charge). The number of protons is called the **atomic number** (***Z***) of the element. Each element has a different atomic number. The electrons, each with a unit of **negative charge** (charge = 1−) are found outside the nucleus; their number equals the atomic number in a neutral atom. Isotopes of an element have identical atomic numbers but different numbers of neutrons. In more modern terms, an **element** can be defined as a substance whose atoms all have the same number of protons in their nuclei.

The Periodic Table. The search for similarities and differences among the properties of the elements led Mendeleev to discover that when the elements are placed in (approximate) order of increasing atomic mass, similar properties recur at regular, repeating intervals. In the modern **periodic table** the elements are arranged in rows, called **periods,** in order of increasing atomic number. The rows are stacked so that elements in the columns, called **groups** or **families,** have similar chemical and physical properties. The A-group elements (IUPAC Groups 1, 2, and 13–18) are called **representative elements;** the B-group elements (IUPAC Groups 3–12) are called **transition elements.** The two long rows of **inner transition elements** located below the main body of the table consist of the **lanthanides,** which follow La ($Z = 57$), and the **actinides,** which follow Ac ($Z = 89$). Certain groups are given family names: Group IA (Group 1), except for hydrogen, are the **alkali metals** (the alkalis); Group IIA (Group 2), the **alkaline earth metals;** Group VIIA (Group 17), the **halogens;** Group VIIIA (Group 18), the **noble gases.**

Metals, Nonmetals, and Metalloids. Most elements are **metals;** they occupy the lower left-hand region of the periodic table (to the left of a line drawn approximately from boron, B, to astatine, At). **Nonmetals** are found in the upper right-hand region of the table. **Metalloids** occupy a narrow band between the metals and nonmetals.

Metals exhibit a **metallic luster,** tend to be **ductile** and **malleable,** and conduct electricity. Nonmetals tend to be brittle, lack metallic luster, and are nonconductors of electricity. Many nonmetals are gases. Bromine (a nonmetal) and mercury (a metal) are the two elements that are liquids at ordinary room temperature. Metalloids have properties intermediate between those of metals and nonmetals and are **semiconductors** of electricity.

Reactions of Elements to Form Compounds. Nearly every element has the ability to form compounds, although not all combinations are possible. When a chemical reaction takes place, the properties of the substances present at the start disappear and are replaced by the properties of the substances formed in the reaction. Chemical symbols are used to write **chemical formulas,** both for **free elements** that occur as molecules (e.g., **diatomic molecules** of elements such as H_2, O_2, N_2, and Cl_2) and for chemical compounds. **Subscripts** are used to specify how many atoms of each element are present. Some compounds form solids called **hydrates,** which contain water molecules in definite proportions. Heating a hydrate usually drives off the water.

Chemical Equations. A **chemical equation** presents a before-and-after description of a chemical reaction. When **balanced,** an equation contains **coefficients** that make the number of atoms of each kind the same among the **reactants** and the **products.** In this way, the equation conforms to the law of conservation of mass. The physical states of the reactants and products can be specified in an equation by placing the following symbols within parentheses following the chemical formulas: *s*, *l*, *g*, and *aq*, which stand for solid, liquid, gas, and aqueous solution (dissolved in water), respectively.

Molecules and Molecular Compounds. Molecules are electrically neutral particles consisting of two or more atoms. The erratic movements of microscopic particles suspended in a liquid (**Brownian motion**) can be interpreted to be caused by collisions with molecules of the liquid. Molecules are held together by **chemical bonds** that arise from the sharing of electrons between atoms. Formulas we write for molecules are **molecular formulas.** Molecular compounds are formed when nonmetals combine with each other. The simple nonmetal hydrides have formulas that can be remembered by the position of the nonmetal in the periodic table. **Organic compounds** are **hydrocarbons,** or compounds considered to be derived from hydrocarbons by replacing H atoms with other atoms.

Ions and Ionic Compounds. Binary ionic compounds are formed when metals react with nonmetals. In the reaction, electrons are transferred from a metal to a nonmetal. The metal atom becomes a positive ion (a **cation**); the nonmetal atom becomes a negative ion (an **anion**). The formula of an ionic compound specifies the smallest whole-number ratio of the ions. The smallest unit of an ionic compound is called a **formula unit,** which specifies the smallest whole-number ratio of the ions that produces electrical neutrality. Many ionic compounds also contain **polyatomic ions**—ions that are composed of two or more atoms.

Naming Molecular Compounds. The system of **nomenclature** for **binary** molecular **inorganic compounds** uses a set of **Greek prefixes** to specify the numbers of atoms of each kind in the formula of the compound. The first element in the formula is specified by its English name; the second element takes the suffix *-ide,* which is added to the stem of the English name. For simple nonmetal hydrides, it is not necessary to specify the number of hydrogens in the formula. Many familiar substances as well as very complex molecules are usually identified by **common names.**

Naming Ionic Compounds. In naming an ionic compound, the cation is specified first, followed by the anion. Metal cations take the English name of the element, and when more than one positive ion can be formed by the metal, the **Stock system** is used to identify the amount of positive charge on the cation. This is done by placing a Roman numeral equal to the positive charge in parentheses following the name of the metal. Simple **monatomic** anions are formed by nonmetals and their names are formed by adding the suffix *-ide* to the stem of the nonmetal's name. Only two common polyatomic anions (cyanide and hydroxide) end in the suffix *-ide.*

Properties of Molecular and Ionic Compounds. We found that it is often possible to distinguish ionic compounds from molecular compounds by their ability to **conduct electricity.** Molecular compounds are generally poor electrical conductors, whereas ionic compounds when melted into the liquid state or dissolved in water will conduct electricity readily.

TOOLS FOR PROBLEM SOLVING

We have learned the following concepts which can be applied as tools in solving problems. Study each one carefully so that you know what each is used for. When faced with solving a problem, recall what each tool does and consider whether it will be helpful in finding a solution. This will aid you in selecting the tools you need. If necessary, refer to these tools when working on the exercises in the chapter and the review questions and problems that follow. Remember that tools from Chapter 1 may be needed at times to solve problems in this chapter.

Law of definite proportions *(page 36)* If we know the **mass ratio** of the elements in one sample of a compound, we know the ratio will be the same in a different sample of the same compound.

Law of conservation of mass *(page 36)* The total mass of chemicals present before a reaction starts equals the total mass after the reaction is finished. We can use this law to check whether we have accounted for all the substances formed in a reaction.

Subatomic particles *(page 43* There are three important relationships between the numbers of **protons, neutrons (nucleons)**, and **electrons.** These relationships are

For a neutral atom: number of electrons = number of protons
Atomic number (Z) = number of protons
Mass number (A) = number of protons + number of neutrons

Relative atomic masses *(page 45)* Atomic masses are relative to the mass of a ^{12}C atom that has a mass of exactly 12 atomic mass units (u). Therefore the atomic mass of ^{12}C is exactly 12 u.

Chemical formula *(page 54)* **Subscripts** in a formula specify the number of atoms of each element in one formula unit of the substance. This gives us *atom ratios* that we will find useful when we deal with the compositions of compounds in Chapter 3.

Rules for writing formulas of ionic compounds *(page 66)* The rules permit us to write correct chemical formulas for ionic compounds. You will need to learn to use the periodic table (see below) to remember the charges on the cations and anions of the representative metals and nonmetals.

Polyatomic ions *(page 69)* Certain groups of atoms arrange themselves into stable configurations that we call polyatomic ions. It is very important that you commit to memory the names, formulas, and charges of these ions (which are given in Table 2.8).

Monatomic anion names *(page 73)* The list on this page gives the common names of anions that must be remembered.

Greek prefixes *(page 71)* This page has a list of the Greek prefixes from one to ten that you should know for naming molecular compounds and hydrates.

Rules for naming molecular compounds *(page 71)* These rules give us a logical system for naming binary molecular compounds by specifying the number of each type of atom using Greek prefixes.

Rules for naming ionic compounds *(page 72)* These rules give us a systematic method for naming ionic compounds. The name of the cation is combined with the name of a monatomic anion as given on page 73. These rules are used with slight modification for cations that can have more than one possible charge (see using the **Stock system** below) and for situations where a polyatomic ion is involved (see naming with **polyatomic ions** below)

Using the Stock system *(page 74)* The Stock system specifies the charge of a cation by placing a Roman numeral in parentheses just after the name of the cation. The Stock system and its Roman numerals are only used for cations that can have more than one possible charge.

Naming with polyatomic ions *(page 76)* Naming compounds that contain polyatomic anions is done by specifying the cation name, using the Stock system if needed, and then specifying the polyatomic anion name as given in Table 2.8. The one polyatomic cation, the ammonium ion (NH_4^+), uses its name and then the appropriate name of the anion.

Periodic table The periodic table has several tool icons in this chapter illustrating its use in a variety of different ways. The periodic table is used to find group numbers and period numbers of the elements *(page 49)*. The table will help us recall group names and we can obtain atomic numbers and average masses of the elements *(page 50)*. From an element's position in the periodic table, we can tell whether it's a metal, nonmetal, or metalloid *(page 51)*. From a nonmetal's position in the periodic table we can write the formula of its simple hydride *(page 60)* and predict the charge of monatomic anions *(page 65)*. For the metals in Groups IA and IIA, we can use the elements' positions in the periodic table to obtain the charges on their ions *(page 65)*.

QUESTIONS, PROBLEMS, AND EXERCISES

Answers to problems whose numbers are printed in color are given in Appendix B. More challenging problems are marked with asterisks. ILW = Interactive Learningware solution is available at www.wiley.com/college/brady. OH = an Office Hours video is available for this problem.

REVIEW QUESTIONS

Laws of Chemical Combination and Dalton's Theory

2.1 Name and state the two laws of chemical combination discussed in this chapter.

2.2 Why didn't the existence of isotopes affect the apparent validity of the atomic theory?

2.3 In your own words, describe how Dalton's theory explains the law of conservation of mass and the law of definite proportions.

2.4 Which of the laws of chemical combination is used to define the term *compound*?

2.5 Describe what you need to do in the laboratory to test (a) the law of conservation of mass, (b) the law of definite proportions, and (c) the law of multiple proportions.

Atomic Masses and Atomic Structure

2.6 What are the names, symbols, and electrical charges of the three subatomic particles introduced in this chapter?

2.7 Where in an atom is nearly all of its mass concentrated? Explain your answer in terms of the particles that contribute to this mass.

2.8 What is a *nucleon*? Which ones have we studied?

2.9 Define the terms *atomic number* and *mass number*. What symbols are used to designate these terms?

2.10 Consider the symbol ${}_{b}^{a}X$, where X stands for the chemical symbol for an element. What information is given in locations (a) a and (b) b?

2.11 Write the symbols of the isotopes that contain the following. (Use the table of atomic masses and numbers printed inside the front cover for additional information, as needed.)
(a) An isotope of iodine whose atoms have 78 neutrons.
(b) An isotope of strontium whose atoms have 52 neutrons.
(c) An isotope of cesium whose atoms have 82 neutrons.
(d) An isotope of fluorine whose atoms have 9 neutrons.

The Periodic Table

2.12 In the compounds formed by Li, Na, K, Rb, and Cs with chlorine, how many atoms of Cl are there per atom of the metal? In the compounds formed by Be, Mg, Ca, Sr, and Ba with chlorine, how many atoms of Cl are there per atom of metal? How did this kind of information lead Mendeleev to develop his periodic table?

2.13 On what basis did Mendeleev construct his periodic table? On what basis are the elements arranged in the modern periodic table?

2.14 On the basis of their positions in the periodic table, why is it not surprising that strontium-90, a dangerous radioactive isotope of strontium, replaces calcium in newly formed bones?

2.15 In the refining of copper, sizable amounts of silver and gold are recovered. Why is this not surprising?

2.16 Why would you reasonably expect cadmium to be a contaminant in zinc but not in silver?

2.17 Using the symbol for nitrogen, N, indicate what information is conveyed by the superscripts before and after the symbol and by subscripts before and after the symbol.

2.18 Make a rough sketch of the periodic table and mark off those areas where you would find (a) the representative elements, (b) the transition elements, and (c) the inner transition elements.

2.19 Which of the following is
(a) an alkali metal: Ca, Cu, In, Li, S?
(b) a halogen: Ce, Hg, Si, O, I?
(c) a transition element: Pb, W, Ca, Cs, P?
(d) a noble gas: Xe, Se, H, Sr, Zr?
(e) a lanthanide element: Th, Sm, Ba, F, Sb?
(f) an actinide element: Ho, Mn, Pu, At, Na?
(g) an alkaline earth metal: Mg, Fe, K, Cl, Ni?

Physical Properties of Metals, Nonmetals, and Metalloids

2.20 Name five physical properties that we usually observe for metals.

2.21 Why is mercury used in thermometers? Why is tungsten used in lightbulbs?

2.22 Which nonmetals occur as monatomic gases (gases whose particles consist of single atoms)?

2.23 Which two elements exist as liquids at room temperature and pressure?

2.24 Which physical property of metalloids distinguishes them from metals and nonmetals?

2.25 Sketch the shape of the periodic table and mark off those areas where we find (a) metals, (b) nonmetals, and (c) metalloids.

2.26 Most periodic tables have a heavy line that looks like a staircase starting from boron down to polonium. What information does this line convey?

2.27 Which metals can you think of that are commonly used to make jewelry? Why isn't iron used to make jewelry? Why isn't potassium used?

2.28 What trends (regular changes in physical or chemical properties) in the periodic table have been mentioned in this chapter?

2.29 Find a periodic table on the Internet that lists physical properties of the elements. Can you distinguish trends in the periodic table based on (a) melting point, (b) boiling point, or (c) density?

Chemical Formulas

2.30 What are two ways to interpret a chemical symbol?

2.31 What is the difference between an atom and a molecule?

2.32 Write the formulas and names of the nonmetallic elements that exist in nature as diatomic molecules.

Chemical Equations

2.33 What do we mean when we say a chemical equation is *balanced*? Why do we balance chemical equations?

2.34 For a chemical reaction, what do we mean by the term *reactants*? What do we mean by the term *products*?

2.35 The combustion of a thin wire of magnesium metal (Mg) in an atmosphere of pure oxygen produces the brilliant light of a flashbulb, once commonly used in photography. After the reaction, a thin film of magnesium oxide is seen on the inside of the bulb. The equation for the reaction is

$$2Mg + O_2 \longrightarrow 2MgO$$

(a) State in words how this equation is read.
(b) Give the formula(s) of the reactants.
(c) Give the formula(s) of the products.
(d) Rewrite the equation to show that Mg and MgO are solids and O_2 is a gas.

2.36 The chemical equation for the combustion of octane (C_8H_{18}), a component of gasoline, is

$$2C_8H_{18} + 25O_2 \longrightarrow 16CO_2 + 18H_2O$$

Rewrite the equation so that it specifies octane and water as liquids and oxygen and carbon dioxide (CO_2) as gases.

Molecular Compounds of Nonmetals

2.37 Which are the only elements that exist as free, individual atoms when not chemically combined with other elements?

2.38 Write chemical formulas for molecules of elemental sulfur and phosphorus mentioned in this chapter.

2.39 Which kind of elements normally combine to form molecular compounds?

2.40 Without referring to Table 2.4, but using the periodic table, write chemical formulas for the simplest hydrogen compounds of (a) carbon, (b) nitrogen, (c) tellurium, and (d) iodine.

2.41 The simplest hydrogen compound of phosphorus is phosphine, a highly flammable and poisonous compound with an odor of decaying fish. What is the formula for phosphine?

2.42 Astatine, a radioactive member of the halogen family, forms a compound with hydrogen. Predict its chemical formula.

2.43 Under appropriate conditions, tin can be made to form a simple molecular compound with hydrogen. Predict its formula.

2.44 Write the chemical formulas for (a) methane, (b) ethane, (c) propane, and (d) butane. Give one practical use for each of these hydrocarbons.

2.45 What are the formulas of (a) methanol and (b) ethanol?

2.46 What is the formula for the alkane that has 10 carbon atoms?

2.47 Candle wax is a mixture of hydrocarbons, one of which is an alkane with 23 carbon atoms. What is the formula for this hydrocarbon?

2.48 The formula for a compound is correctly given as $C_6H_{12}O_6$. State two reasons why we expect this to be a molecular compound, rather than an ionic compound.

2.49 Explore the Internet and find a reliable source of structures for molecular compounds. For questions 44 to 47 print out the ball-and-stick and space-filling models of the compounds mentioned.

Ionic Compounds

2.50 Describe what kind of event must occur (involving electrons) if the atoms of two different elements are to react to form (a) an ionic compound or (b) a molecular compound.

2.51 With what kind of elements do metals react?

2.52 With what kind of elements do nonmetals react?

2.53 Why are nonmetals found in more compounds than are metals, even though there are fewer nonmetals than metals?

2.54 What is an ion? How does it differ from an atom or a molecule?

2.55 Why do we use the term *formula unit* for ionic compounds instead of the term *molecule*?

2.56 Most compounds of aluminum are ionic, but a few are molecular. How do we know that Al_2Cl_6 is molecular?

2.57 Consider the sodium atom and the sodium ion.
(a) Write the chemical symbol of each.
(b) Do these particles have the same number of nuclei?
(c) Do they have the same number of protons?
(d) Could they have different numbers of neutrons?
(e) Do they have the same number of electrons?

2.58 Define *cation, anion,* and *polyatomic ion.*

2.59 How many electrons has a titanium atom lost if it has formed the ion Ti^{4+}? What are the total numbers of protons and electrons in a Ti^{4+} ion?

2.60 If an atom gains an electron to become an ion, what kind of electrical charge does the ion have?

2.61 How many electrons has a nitrogen atom gained if it has formed the ion N^{3-}? How many protons and electrons are in an N^{3-} ion?

2.62 What is wrong with the formula $RbCl_3$? What is wrong with the formula SNa_2?

2.63 A student wrote the formula for an ionic compound of titanium as Ti_2O_4. What is wrong with this formula? What should the formula be?

2.64 What are the formulas of the ions formed by (a) iron, (b) cobalt, (c) mercury, (d) chromium, (e) tin, and (f) copper?

2.65 Which of the following formulas are incorrect? (a) NaN_2, (b) RbCl, (c) K_2S, (d) Al_2Cl_3, (e) MgO_2

2.66 What are the formulas (including charges) for (a) cyanide ion, (b) ammonium ion, (c) nitrate ion, (d) sulfite ion, (e) chlorate ion, and (f) sulfate ion?

2.67 What are the formulas (including charges) for (a) hypochlorite ion, (b) bisulfate ion, (c) phosphate ion, (d) dihydrogen phosphate ion, (e) permanganate ion, and (f) oxalate ion?

2.68 What are the names of the following ions? (a) $Cr_2O_7^{2-}$, (b) OH^-, (c) $C_2H_3O_2^-$, (d) CO_3^{2-}, (e) CN^-, (f) ClO_4^-

2.69 From what you have learned in Section 2.8, write correct balanced equations for the reactions between (a) calcium and chlorine, (b) magnesium and oxygen, (c) aluminum and oxygen, and (d) sodium and sulfur.

2.70 Write the balanced equations (including phases) for the reactions described below: (a) Solid iron(III) hydroxide reacts with gaseous hydrogen chloride forming water and iron(III) chloride. (b) Aqueous silver nitrate is reacted with aqueous barium chloride to form solid silver chloride and aqueous barium nitrate.

2.71 Write the balanced equations (including phases) for the reactions described below: (a) Gaseous propane reacts with gaseous oxygen to form carbon dioxide and water. (b) Sodium metal is added to water and the products are aqueous sodium hydroxide and hydrogen gas.

Naming Compounds

2.72 What is the difference between a binary compound and one that is diatomic? Give examples that illustrate this difference.

2.73 In naming the compounds discussed in this chapter, why is it important to know whether a compound is molecular or ionic?

2.74 In naming ionic compounds of the transition elements, why is it essential to know the charge on the anion?

2.75 Describe (a) the three situations in which Greek prefixes are used and (b) when Roman numerals are used.

REVIEW PROBLEMS

Laws of Chemical Combination

OH **2.76** Ammonia is composed of hydrogen and nitrogen in a ratio of 9.33 g of nitrogen to 2.00 g of hydrogen. If a sample of ammonia contains 6.28 g of hydrogen, how many grams of nitrogen does it contain?

2.77 A compound of phosphorus and chlorine used in the manufacture of a flame retardant treatment for fabrics contains 1.20 grams of phosphorus for every 4.12 g of chlorine. Suppose a sample of this compound contains 6.22 g of chlorine. How many grams of phosphorus does it contain?

2.78 Refer to the data about ammonia in Problem 2.76. If 4.56 g of nitrogen combined completely with hydrogen to form ammonia, how many grams of ammonia would be formed?

2.79 Refer to the data about the phosphorus–chlorine compound in Problem 2.77. If 12.5 g of phosphorus combined completely with chlorine to form this compound, how many grams of the compound would be formed?

2.80 Molecules of a certain compound of nitrogen and oxygen contain one atom each of N and O. In this compound there are 1.143 g of oxygen for each 1.000 g of nitrogen. Molecules of a different compound of nitrogen and oxygen contain one atom of N and two atoms of O. How many grams of oxygen would be combined with each 1.000 g of nitrogen in the second compound?

2.81 Tin forms two compounds with chlorine. In one of them (compound 1), there are two Cl atoms for each Sn atom; in the other (compound 2), there are four Cl atoms for each Sn atom. When combined with the same mass of tin, what would be the ratio of the masses of chlorine in the two compounds? In compound 1, 0.597 g of chlorine is combined with each 1.000 g of tin. How many grams of chlorine would be combined with 1.000 g of tin in compound 2?

Atomic Masses and Isotopes

2.82 The chemical substance in natural gas is a compound called methane. Its molecules are composed of carbon and hydrogen, and each molecule contains four atoms of hydrogen and one atom of carbon. In this compound, 0.33597 g of hydrogen is combined with 1.0000 g of carbon-12. Use this information to calculate the atomic mass of the element hydrogen.

OH **2.83** A certain element *X* forms a compound with oxygen in which there are two atoms of *X* for every three atoms of O. In this compound, 1.125 g of *X* is combined with 1.000 g of oxygen. Use the average atomic mass of oxygen to calculate the average atomic mass of *X*. Use your calculated atomic mass to identify the element *X*.

2.84 If an atom of carbon-12 had been assigned a relative mass of 24.0000 u, what would be the average atomic mass of hydrogen relative to this mass?

2.85 One atom of ^{109}Ag has a mass that is 9.0754 times that of a ^{12}C atom. What is the atomic mass of this isotope of silver expressed in atomic mass units?

Atomic Structure

ILW **2.86** Naturally occurring copper is composed of 69.17% of ^{63}Cu, with an atomic mass of 62.9396 u, and 30.83% of ^{65}Cu, with an atomic mass of 64.9278 u. Use these data to calculate the average atomic mass of copper.

2.87 Naturally occurring magnesium (one of the elements in milk of magnesia) is composed of 78.99% of ^{24}Mg (atomic mass, 23.9850 u), 10.00% of ^{25}Mg (atomic mass, 24.9858 u), and 11.01% of ^{26}Mg (atomic mass, 25.9826 u). Use these data to calculate the average atomic mass of magnesium.

ILW **2.88** Give the numbers of neutrons, protons, and electrons in the atoms of each of the following isotopes. (Use the table of atomic masses and numbers printed inside the front cover for additional information, as needed.) (a) radium-226, (b) ^{206}Pb, (c) carbon-14, (d) ^{23}Na

2.89 Give the numbers of electrons, protons, and neutrons in the atoms of each of the following isotopes. (As necessary, consult the table of atomic masses and numbers printed inside the front cover.) (a) cesium-137, (b) ^{238}U, (c) iodine-131, (d) ^{197}Au

Chemical Formulas

2.90 The compound $Cr(C_2H_3O_2)_3$ is used in the tanning of leather. How many atoms of each element are given in this formula?

2.91 Asbestos, a known cancer-causing agent, has as a typical formula, $Ca_3Mg_5(Si_4O_{11})_2(OH)_2$. How many atoms of each element are given in this formula?

2.92 Epsom salts is a hydrate of magnesium sulfate, $MgSO_4 \cdot 7H_2O$. What is the formula of the substance that remains when Epsom salts is completely dehydrated?

2.93 Rochelle salt is the tetrahydrate of $KNaC_4H_4O_6$. Write the formula for Rochelle salt.

2.94 How many atoms of each element are represented in each of the following formulas? (a) $K_2C_2O_4$, (b) H_2SO_3, (c) $C_{12}H_{26}$, (d) $HC_2H_3O_2$, (e) $(NH_4)_2HPO_4$

2.95 How many atoms of each kind are represented in the following formulas? (a) Na_3PO_4, (b) $Ca(H_2PO_4)_2$, (c) C_4H_{10}, (d) $Fe_3(AsO_4)_2$, (e) $C_3H_5(OH)_3$

2.96 How many atoms of each kind are represented in the following formulas? (a) $Ni(ClO_4)_2$, (b) $CuCO_3$, (c) $K_2Cr_2O_7$, (d) CH_3CO_2H, (e) $(NH_4)_2HPO_4$

2.97 How many atoms of each kind are represented in the following formulas? (a) $CH_3CH_2CO_2C_3H_7$, (b) $MgSO_4 \cdot 7H_2O$, (c) $KAl(SO_4)_2 \cdot 12H_2O$, (d) $Cu(NO_3)_2$, (e) $(CH_3)_3COH$

2.98 How many atoms of each element are represented in each of the following expressions? (a) $3N_2O$, (b) $4NaHCO_3$, (c) $2CuSO_4 \cdot 5H_2O$

OH **2.99** How many atoms of each element are represented in each of the following expressions? (a) $7CH_3CO_2H$, (b) $2(NH_2)_2CO$, (c) $5K_2Cr_2O_7$

Chemical Equations

2.100 Consider the balanced equation

$$2Fe(NO_3)_3 + 3Na_2CO_3 \longrightarrow Fe_2(CO_3)_3 + 6NaNO_3$$

(a) How many atoms of Na are on each side of the equation?
(b) How many atoms of C are on each side of the equation?
(c) How many atoms of O are on each side of the equation?

2.101 Consider the balanced equation for the combustion of octane, a component of gasoline:

$$2C_8H_{18} + 25O_2 \longrightarrow 16CO_2 + 18H_2O$$

(a) How many atoms of C are on each side of the equation?
(b) How many atoms of H are on each side of the equation?
(c) How many atoms of O are on each side of the equation?

Ionic Compounds

2.102 Use the periodic table, but not Table 2.6, to write the symbols for the ions of (a) K, (b) Br, (c) Mg, (d) S, and (e) Al.

2.103 Use the periodic table, but not Table 2.6, to write the symbols for ions of (a) barium, (b) oxygen, (c) fluorine, (d) strontium, and (e) rubidium.

OH **2.104** Write formulas for ionic compounds formed between (a) Na and Br, (b) K and I, (c) Ba and O, (d) Mg and Br, and (e) Ba and F.

2.105 Write the formulas for the ionic compounds formed by the following transition metals with the chloride ion, Cl^-: (a) chromium, (b) iron, (c) manganese, (d) copper, and (e) zinc.

2.106 Write formulas for the ionic compounds formed from (a) K^+ and nitrate ion, (b) Ca^{2+} and acetate ion, (c) ammonium ion and Cl^-, (d) Fe^{3+} and carbonate ion, and (e) Mg^{2+} and phosphate ion.

2.107 Write formulas for the ionic compounds formed from (a) Zn^{2+} and hydroxide ion, (b) Ag^+ and chromate ion, (c) Ba^{2+} and sulfite ion, (d) Rb^+ and sulfate ion, and (e) Li^+ and bicarbonate ion.

2.108 Write formulas for two compounds formed between O^{2-} and (a) lead, (b) tin, (c) manganese, (d) iron, and (e) copper.

2.109 Write formulas for the ionic compounds formed from Cl^- and (a) cadmium ion, (b) silver ion, (c) zinc ion, and (d) nickel ion.

Naming Compounds

2.110 Name the following molecular compounds.
(a) SiO_2 (b) XeF_4 (c) P_4O_{10} (d) Cl_2O_7

2.111 Name the following molecular compounds.
(a) ClF_3 (b) S_2Cl_2 (c) N_2O_5 (d) $AsCl_5$

2.112 Name the following ionic compounds.
(a) CaS (b) $AlBr_3$ (c) Na_3P (d) Ba_3As_2 (e) Rb_2S

2.113 Name the following ionic compounds.
(a) NaF (b) Mg_2C (c) Li_3N (d) Al_2O_3 (e) K_2Se

2.114 Name the following ionic compounds using the Stock system.
(a) FeS (b) CuO (c) SnO_2 (d) $CoCl_2 \cdot 6H_2O$

2.115 Name the following ionic compounds using the Stock system.
(a) Mn_2O_3 (b) Hg_2Cl_2 (c) PbS (d) $CrCl_3 \cdot 4H_2O$

2.116 Name the following. If necessary, refer to Table 2.8 on page 69.
(a) $NaNO_2$ (b) $KMnO_4$ (c) $MgSO_4 \cdot 7H_2O$ (d) KSCN

2.117 Name the following. If necessary, refer to Table 2.8 on page 69.
(a) K_3PO_4
(b) $NH_4C_2H_3O_2$
(c) $Fe_2(CO_3)_3$
(d) $Na_2S_2O_3 \cdot 5H_2O$

2.118 Identify each of the following as molecular or ionic and give its name:
(a) $CrCl_2$
(b) S_2Cl_2
(c) $NH_4C_2H_3O_2$
(d) SO_3
(e) KIO_3
(f) P_4O_6
(g) $CaSO_3$
(h) AgCN
(i) $ZnBr_2$
(j) H_2Se

2.119 Identify each of the following as molecular or ionic and give its name:
(a) $V(NO_3)_3$
(b) $Co(C_2H_3O_2)_2$
(c) Au_2S_3
(d) Au_2S
(e) $GeBr_4$
(f) K_2CrO_4
(g) $Fe(OH)_2$
(h) I_2O_4
(i) I_4O_9
(j) P_4Se_3

OH **2.120** Write formulas for the following.
(a) sodium monohydrogen phosphate
(b) lithium selenide
(c) chromium(III) acetate
(d) disulfur decafluoride
(e) nickel(II) cyanide
(f) iron(III) oxide
(g) antimony pentafluoride

2.121 Write formulas for the following.
(a) dialuminum hexachloride
(b) tetraarsenic decaoxide
(c) magnesium hydroxide
(d) copper(II) bisulfate
(e) ammonium thiocyanate
(f) potassium thiosulfate
(g) diiodine pentaoxide

2.122 Write formulas for the following.
(a) ammonium sulfide
(b) chromium(III) sulfate hexahydrate
(c) silicon tetrafluoride
(d) molybdenum(IV) sulfide
(e) tin(IV) chloride
(f) hydrogen selenide
(g) tetraphosphorus heptasulfide

2.123 Write formulas for the following.
(a) mercury(II) acetate
(b) barium hydrogen sulfite
(c) boron trichloride
(d) calcium phosphide
(e) magnesium dihydrogen phosphate
(f) calcium oxalate
(g) xenon tetrafluoride

2.124 The compounds Se_2S_6 and Se_2S_4 have been shown to be antidandruff agents. What are their names?

2.125 The compound P_2S_5 is used to manufacture safety matches. What is the name of this compound?

ADDITIONAL EXERCISES

2.126 An element has 25 protons in its nucleus.
(a) Is the element a metal, a nonmetal, or a metalloid?
(b) On the basis of the average atomic mass, write the symbol for the element's most abundant isotope.
(c) How many neutrons are in the isotope you described in part (b)?
(d) How many electrons are in atoms of this element?
(e) How many times heavier than ^{12}C is the average atom of this element?

*2.127 Elements *X* and *Y* form a compound in which there is one atom of *X* for every four atoms of *Y*. When these elements react, it is found that 1.00 g of *X* combines with 5.07 g of *Y*. When 1.00 g of *X* combines with 1.14 g of O, it forms a compound containing two atoms of O for each atom of *X*. Calculate the atomic mass of *Y*.

2.128 An iron nail is composed of four isotopes with the percentage abundances and atomic masses given in the following table. Calculate the average atomic mass of iron.

Isotope	Percentage Abundance	Atomic Mass (u)
^{54}Fe	5.80	53.9396
^{56}Fe	91.72	55.9349
^{57}Fe	2.20	56.9354
^{58}Fe	0.28	57.9333

OH *2.129 Bromine (shown in Figure 2.12, page 53) is a dark red liquid that vaporizes easily and is very corrosive to the skin. It is used commercially as a bleach for fibers and silk. Naturally occurring bromine is composed of two isotopes: ^{79}Br, with a mass of 78.9183 u, and ^{81}Br, with a mass of 80.9163 u. Use this information and the average atomic mass of bromine given in the table on the inside front cover of the book to calculate the percentage abundances of the two isotopes.

OH *2.130 Rust contains an iron–oxygen compound in which there are three oxygen atoms for each two iron atoms. In this compound, the iron to oxygen mass ratio is 2.325 g Fe to 1.000 g O. Another compound of iron and oxygen contains these elements in the ratio of 2.616 g Fe to 1.000 g O. What is the ratio of iron to oxygen atoms in this other iron–oxygen compound?

2.131 One atomic mass unit has a mass of $1.6605389 \times 10^{-24}$ g. Calculate the mass, in grams, of one atom of magnesium. What is the mass of one atom of iron, expressed in grams? Use these two answers to determine how many atoms of Mg are in 24.31 g of magnesium and how many atoms of Fe are in 55.85 g of iron. Compare your answers. What conclusions can you draw from the results of these calculations? Without actually performing any calculations, how many atoms do you think would be in 40.08 g of calcium?

OH 2.132 What are the formulas for mercury(I) nitrate dihydrate and mercury(II) nitrate monohydrate?

2.133 Consider the following substances: Cl_2, CaO, HBr, $CuCl_2$, AsH_3, $NaNO_3$, and NO_2.
(a) Which are binary substances?
(b) Which is a triatomic molecule?
(c) In which do we find only electron sharing?
(d) Which are diatomic?
(e) In which do we find only attractions between ions?
(f) Which are molecular?
(g) Which are ionic?

2.134 Using the old system of nomenclature, write the names for the following compounds. (See Facets of Chemistry 2.3.)
(a) gold(III) sulfate
(b) gold(III) nitrate
(c) lead(IV) oxide
(d) mercury(I) chloride
(e) copper(II) sulfate
(f) mercury(II) chloride
(g) cobalt(II) hydroxide
(h) tin(II) chloride
(i) tin(IV) sulfide

2.135 Write the names of the following compounds using the Stock system of nomenclature. Also write their formulas. (See Facets of Chemistry 2.3.)
(a) cupric bromide
(b) cuprous iodide
(c) ferrous sulfate
(d) chromous chloride
(e) mercurous nitrate
(f) manganous sulfate
(g) plumbous acetate

2.136 A student needed a sample of $Fe(NO_3)_3 \cdot 9H_2O$, but when she went to the latest edition of the catalog of a major chemical supplier, she could not find it listed alphabetically under "iron." Knowing that suppliers of laboratory chemicals still often list chemicals under their older names, what name should she look for in the catalog?

2.137 Write the balanced chemical equation for the reaction between elements with atomic numbers of (a) 20 and 35, (b) 6 and 17, (c) 13 and 16. For each of these determine the ratio of the mass of the heavier element to the lighter element in the compound.

OH 2.138 Write the balanced gas phase chemical equation for the reaction of dinitrogen pentoxide with sulfur dioxide to form sulfur trioxide and nitrogen oxide. What small, whole-number ratios are expected for oxygen in the nitrogen oxides and the sulfur oxides?

EXERCISES IN CRITICAL THINKING

2.139 Imagine a world where, for some reason, hydrogen and helium have not been discovered. Would Mendeleev have had enough information to predict their existence?

2.140 Around 1750 Benjamin Franklin knew of two opposite types of electric charge, produced by rubbing a glass rod or amber rod with fur. He decided that the charge developed on the glass rod should be the "positive" charge and from there on charges were defined. What would have changed if Franklin decided the amber rod acquired the positive charge?

2.141 Explore the Internet and find for yourself a reliable source of physical properties of elements and compounds. Justify how you decided the site was reliable.

2.142 Spreadsheet applications such as Microsoft Excel can display data in a variety of ways; some of these are shown throughout this book. What method of displaying periodic trends (line graphs, tables, bar graphs, 3-D views, etc.) is most effective for your learning style? Explain your answer by stating why your chosen display is better than the others.

2.143 Scientists often validate measurements such as measuring the circumference of the earth by using two independent methods to measure the same value. Describe two independent methods for determining the atomic mass of an element. Explain how these methods are truly independent.

THE MOLE: RELATING THE MICROSCOPIC WORLD OF ATOMS TO LABORATORY MEASUREMENTS

3

The rich and famous often fly to exotic vacation spots using private airplanes such as this. Even now you may be dreaming about your upcoming vacation after the final exam. Here, the plane is prepared for takeoff with the necessary mass (weight) of aviation fuel to reach the planned destination with some excess to spare. Professionals use complex calculations that include the mole concept described in this chapter to determine the mass of fuel needed to transport vacationers safely to their next destination. *(Alice M. Prescott/Unicorn Stock Photos)*

CHAPTER OUTLIN

3.1 The mole conveniently links mass to number of atoms or molecules

3.2 Chemical formulas relate amounts of substances in a compound

3.3 Chemical formulas can be determined from experimental mass measurements

3.4 Chemical equations link amounts of substances in a reaction

3.5 The reactant in shortest supply limits the amount of product that can form

3.6 The predicted amount of product is not always obtained experimentally

THIS CHAPTER IN CONTEXT In Chapters 1 and 2 we reviewed the basics of the mathemat language of chemistry. In this chapter we combine the material from Chapters 1 and 2 to learn the fundame chemical calculations. These calculations are important for success in laboratory work in this course. You find this chapter to be important for future courses in organic chemistry, biochemistry, and almost any other a laboratory course in the sciences.

Our chemical calculations are called **stoichiometry** (stoy-kee-AH-meh-tree), which loosely translates as "th ure of the elements." Stoichiometry involves converting chemical formulas and equations that represent ir atoms, molecules, and formula units to the laboratory scale that uses milligrams, grams, and even kilograms substances. To do this we introduce the *mole concept.* The mole concept allows the chemist to scale up fr atomic and molecular level to the laboratory scale much as the baker in Figure 3.1 scales up the amount of ingr

FIG. 3.1 **Manufacture on small and large scales.** Making a single gingerbread man requires 10 currants for eyes and buttons, 1/10 of a cup of spiced cookie dough, and a teaspoon of glacé icing. To manufacture gingerbread men by the million, you will have to order raw materials by the ton. To scale up the recipe, you will need to know the masses of raw materials required to manufacture some fixed number of gingerbread men.

from a single gingerbread man to a mass-production scale. Chemical conversions are similar in nature to the gingerbread man example but of course they answer different questions. Some of these questions are

How many grams of product can be made if we react *x* grams of *A* and *y* grams of *B*?

How can we be sure to get the most product from an expensive reactant?

If a reaction has a 70% yield, how many grams of reactants are needed to produce the amount of product needed?

How many grams of each reactant are needed so that there will be no reactants left over (i.e., no waste)?

Our stoichiometric calculations are almost always factor-label conversions from one set of units to another as we saw in Chapter 1. To be successful at factor-label calculations we need two things, a knowledge of the equalities that can be made into conversion factors and a logical sequence of steps to apply the conversion factors to our problem. Figure 3.6 at the end of this chapter organizes the equalities and sequence of steps in a simple flowchart. As you read this chapter, you might want to look at Figure 3.6 to see how all the parts fit together neatly.

3.1 THE MOLE CONVENIENTLY LINKS MASS TO NUMBER OF ATOMS OR MOLECULES

We can tell from the fundamental measurements of the mass of the proton, neutron, and electron that even the largest of the atoms must have extremely small masses and correspondingly small sizes. Any sample of matter that is observable by the naked eye must have very large numbers of atoms or molecules. The methods of calculation developed in Chapter 1 and the mole concept allow us to count by weighing and then use that information to solve some very interesting problems.

Counting by weighing is familiar to everyone even if you are not aware of it. A pound of chocolate chips counts out the needed number of chocolate chips for your cookies. A quarter pound of rice counts out the correct number of rice grains to accompany your meal. Weighing a bag of dimes, knowing that each dime weighs 2.27 grams, will allow you to calculate the number of coins. Similarly, the mass of a chemical substance can be used to determine the number of atoms or molecules in the sample. This last conversion is possible because of the mole concept.

The SI unit for the *amount of substance* is the mole

The **mole** (abbreviated as mol) is the SI unit for the amount of substance. The amount of substance does not refer to the mass or volume of your sample but it does refer to the number of atoms, molecules, or formula units, etc., in your sample. **One mole is defined**

Mole is a Latin word with several meanings, including: a shapeless mass; a large number; or trouble or difficulty.

as the number of atoms in exactly 12 grams of ^{12}C atoms. Based on this definition and the fact that the average atomic masses in the periodic table are relative values, we can deduce that we will have a mole of any element if we weigh an amount equal to the atomic mass in gram units (this is often called the **gram atomic mass**). For example, the atomic mass of sodium is 22.99 u, so one mole of sodium has a mass of 22.99 g and contains as many atoms as there are in a 12.00 g sample of carbon-12.

TOOLS *One mole of any element is equal to the atomic mass with gram units*

$$1 \text{ mole of element } X = \text{gram atomic mass of } X$$

Figure 3.2 is a photo showing one mole of some common elements: iron, mercury, copper, and sulfur.

The mole concept also applies to compounds

Many chemists use the terms *molecular weight* and *atomic weight* for molecular mass and atomic mass.

Molecules and ionic compounds discussed in Chapter 2 have definite formulas. For molecular compounds and elements, adding the atomic masses of all atoms in the formula results in the **molecular mass (molecular weight)**. The gram molecular mass of a molecular substance (a mass in grams numerically equal to the molecular mass) is also equal to one mole.

$$1 \text{ mole of molecule } X = \text{gram molecular mass of } X$$

For example, a molecule of H_2O consists of 2 atoms of H (with a total mass of 2×1.00 u $= 2.00$ u) and one atom of O (with a mass of 16.00 u) for a total of 18.00 u. Therefore, the molecular mass of H_2O is 18.00 u. The gram molecular mass of H_2O is 18.00 g, so 1 mol $H_2O = 18.00$ g H_2O.

TOOLS ***Molecular mass** or **formula mass** is the sum of the masses of all atoms in a chemical formula*

Similarly, the **formula mass** of an ionic compound is the sum of the masses of all the atoms in the formula of an ionic compound.

$$1 \text{ mole of ionic compound } X = \text{gram formula mass of } X$$

For example, the ionic compound calcium chloride, $CaCl_2$, has a formula mass that is the sum of the atomic masses of one calcium atom and two chlorine atoms. One calcium atom has an atomic mass of 40.08 u. Two chlorine atoms each with an atomic mass of 35.45 u have a total mass of 70.90 u. Adding these together gives us 110.98 u for the formula mass of $CaCl_2$. Therefore, 1 mol $CaCl_2$ has a mass of 110.98 g. There is a distinct similarity between all three equations above. To simplify discussions we will often use the following relationship between moles and mass unless one of the other, equivalent, definitions provides more clarity.

TOOLS ***Molar mass** is the mass in grams of one mole of any substance*

$$1 \text{ mole of } X = \text{gram molar mass of } X$$

The gram **molar mass** is simply the mass of one mole of the substance under consideration. Figure 3.3 depicts one mole of four different compounds.

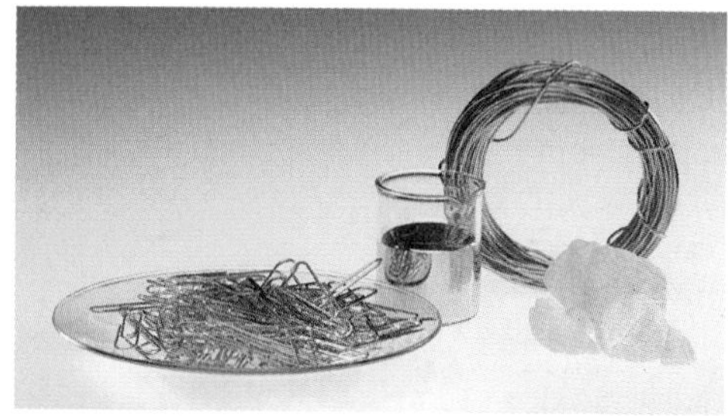

FIG. 3.2 **Moles of elements.** Each sample of these elements contains the same number of atoms. *(Michael Watson.)*

FIG. 3.3 **Moles of compounds.** One mole of four different compounds: water, sodium chloride, copper sulfate pentahydrate, and sodium chromate. Each sample contains the identical number of formula units or molecules. *(Michael Watson.)*

Convert mass to moles and moles to mass using molar masses

At this point we recognize the above relationships or equalities as the necessary information for doing conversion problems similar to those in Chapter 1. Now, however, the problems will be couched in chemical terms. We will also use all of the principles in Section 1.6 to end up with the maximum, and correct, number of significant figures in our answers. To do this we will need to round our atomic or molar masses to at least one more significant figure than the data given in the example.

The term *molar mass* has been coined as a general term to cover all items previously called atomic masses, atomic weights, molecular masses, molecular weights, formula masses, and formula weights.

EXAMPLE 3.1
Converting from Grams to Moles

How many moles of sulfur are there in a 23.5 g sample of sulfur?

ANALYSIS: We see that the problem starts with a certain mass of sulfur and asks us to convert it to moles. This uses the tool we just described that equates moles and grams of an element. The equality shows that there is a one-step conversion possible using 1 mol S = 32.06 g S.

SOLUTION: Let's begin by expressing the question in the form of an equation.

$$\underset{\text{start}}{23.5 \text{ g S}} = \underset{\text{end}}{? \text{ mol S}}$$

Now use the equality to cancel the grams of sulfur as shown below.

$$23.5 \cancel{\text{g S}} \times \left(\frac{1 \text{ mol S}}{32.06 \cancel{\text{g S}}}\right) = 0.733 \text{ mol S}$$

Solving a stoichiometry problem is much like giving directions to get from your house to your college. You need to know both the starting and ending points. Setting up the problem in this way gives you those reference points. The factor-label ratios get us from one to the other.

IS THE ANSWER REASONABLE? First review the math to be sure that the units cancel properly, and they do. Second, round off all numbers to one significant figure and calculate an estimated answer. This gives 20/30 = 2/3, and our answer is not very different (very different is a factor of 10 or more) from 2/3. We are justified in being confident in our answer.

EXAMPLE 3.2
Converting from Moles to Grams

We need 0.254 mol of iron(III) chloride for a certain experiment. How many grams do we need to weigh?

ANALYSIS: The formula for iron(III) chloride is $FeCl_3$. As in the last example we need the tool for the conversion between mass and moles. We write the equality as

$$1 \text{ mol } FeCl_3 = \text{gram molar mass } FeCl_3$$

This tells us we need the gram molar mass of $FeCl_3$ that is the sum of the gram atomic masses of one iron atom and three chlorine atoms.

$$\text{gram molar mass} = 55.85 \text{ g Fe mol}^{-1} + (3 \times 35.45 \text{ g Cl mol}^{-1}) = 162.20 \text{ g } FeCl_3 \text{ mol}^{-1}$$

The equality can be written as 1 mol $FeCl_3$ = 162.20 g $FeCl_3$.

SOLUTION: The problem starts with

$$0.254 \text{ mol } FeCl_3 = ? \text{ g } FeCl_3$$

Use the factor label to perform the conversion:

$$0.254 \cancel{\text{mol } FeCl_3} \left(\frac{162.20 \text{ g } FeCl_3}{1 \cancel{\text{mol } FeCl_3}}\right) = 41.2 \text{ g } FeCl_3$$

IS THE ANSWER REASONABLE? First we verify that the units cancel properly, and they do. Next we do an approximate calculation. If we round 0.254 to 0.25 and 162.20 to 160, the arithmetic becomes 0.25 × 160 = 40, which gives a result that is very close to the calculated value. We also could have estimated the answer by rounding to one significant figure, which gives 0.3 × 200 = 60, and still have concluded that the math was correct. Remember, this is just an estimate, but it tells us our more precise answer is correct.

Practice Exercise 1: How many moles of aluminum are there in a 3.47 gram sheet of aluminum foil used to wrap your sandwich for lunch today? (Hint: Recall the tool that relates the mass of an element to moles of that element.)

Practice Exercise 2: Your laboratory balance can weigh samples to three decimal places. If the uncertainty in your weighing is ±0.002 g, what is the uncertainty in moles if the sample being weighed is pure silicon?

The number of particles in a mole is called Avogadro's number

Avogadro's number was named for Amedeo Avogadro (1776–1856), an Italian scientist who was one of the pioneers of stoichiometry.

Avogadro's number

The definition of the mole refers to a number equal to the number of atoms in exactly 12 g of ^{12}C. Just what is that number? After much experimentation the scientific community agrees that the value, to four significant figures, is 6.022×10^{23}. In honor of Amedeo Avogadro, this value has been named **Avogadro's number.** Now we can write a very important relationship between the atomic scale and the laboratory scale as

$$1 \text{ mole of } X = 6.022 \times 10^{23} \text{ units of } X$$

Avogadro's number is a link between moles of substance and elementary units of substance in a stoichiometry problem. If a problem does not mention atoms or molecules at all, you don't need to use Avogadro's number in the calculation!

The units of our chemicals can be atoms, molecules, formula units, etc.

We use Avogadro's number to relate the macroscopic and microscopic worlds

The relationships developed above allow us to connect the laboratory scale with the atomic scale using our standard factor-label calculations as shown in the next two examples.

EXAMPLE 3.3
Converting from the Laboratory Scale to the Atomic Scale

How many atoms of copper are there in a piece of pure copper wire that weighs 14.3 grams?

ANALYSIS: Here we do not have any tool that directly converts grams of copper to atoms of copper. However, we do have one tool that relates mass to moles (the atomic mass of copper) and another tool that relates moles to atoms (Avogadro's number). We need to start with the grams of copper and use the appropriate conversion factors derived from these tools in sequence.

$$\text{grams copper} \longrightarrow \text{moles copper} \longrightarrow \text{atoms copper}$$

SOLUTION: We start by mathematically stating the problem:

$$14.3 \text{ g Cu} = ? \text{ atoms Cu}$$

The first step will be to convert grams of copper to moles of copper. The atomic mass gives us

$$1 \text{ mol Cu} = 63.546 \text{ g Cu}$$

Estimates are done without calculators. For exponential numbers the numerical part of the calculation is separated from the exponents. The numerical part, to one significant figure, is usually easy to evaluate. The exponent is easier since it involves simple addition and subtraction.

This will allow us to construct a conversion factor to take us from grams to moles. To go from moles of copper to the number of copper atoms, we need Avogadro's number, which gives the relationship

$$1 \text{ mol Cu} = 6.02 \times 10^{23} \text{ atom Cu}$$

This will provide a conversion factor to take us from moles to atoms.

To assemble the solution, we begin with the given amount of copper (14.3 g) and apply conversion factors that eliminate units we don't want and take us to the units of the answer. This could be done in two steps or in one complete step as shown here. Note how the units cancel.

$$14.3 \cancel{\text{g Cu}} \times \left(\frac{1 \cancel{\text{mol Cu}}}{63.546 \cancel{\text{g Cu}}}\right) \times \left(\frac{6.022 \times 10^{23} \text{ atoms Cu}}{1 \cancel{\text{mol Cu}}}\right) = 1.35 \times 10^{23} \text{ atoms Cu}$$

Also notice that the first factor converts grams of copper to moles of copper and the second takes us from moles of copper to atoms of copper.

IS THE ANSWER REASONABLE? The most common mistake students make in this kind of calculation is using Avogadro's number incorrectly, or not using it at all. Think about the answer for a moment. We know copper atoms are very small, so in 14.3 g Cu there will be an enormous number of atoms. Our answer is a very large number, so it appears to be reasonable.

EXAMPLE 3.4
Calculating the Mass of a Molecule

What is the mass in grams of one molecule of carbon tetrachloride?

ANALYSIS: Using the nomenclature tools of Chapter 2, the formula of the molecule in question is CCl_4. In this problem we're asked to express the mass of a single CCl_4 molecule (composed of just 5 atoms) in a laboratory sized unit, grams. This tells us we need to use Avogadro's number, which is a tool that relates molecules to moles. Then we can use the molecular mass as a tool to relate moles to grams.

SOLUTION: Let's begin by expressing the question in the form of an equation.

$$1 \text{ molecule } CCl_4 = ?\text{ g } CCl_4$$

Avogadro's number lets us write

$$1 \text{ mol } CCl_4 = 6.02 \times 10^{23} \text{ molecules } CCl_4$$

This allows us to calculate the number of moles of CCl_4

$$1 \cancel{\text{molecule } CCl_4} \times \left(\frac{1 \text{ mol } CCl_4}{6.02 \times 10^{23} \cancel{\text{molecules } CCl_4}}\right) = 1.661 \times 10^{-24} \text{ mol } CCl_4$$

To find grams, we need the molecular mass of CCl_4, which is 153.823. Therefore, 1 mol CCl_4 = 153.823 g CCl_4. Now we can calculate grams of CCl_4.

$$1.661 \times 10^{-24} \cancel{\text{mol } CCl_4} \times \left(\frac{153.823 \text{ g } CCl_4}{1 \cancel{\text{mol } CCl_4}}\right) = 2.56 \times 10^{-22} \text{ g } CCl_4$$

In stepwise calculations such as this we normally keep at least one extra significant figure until the final result is determined. The calculated mass of one molecule of CCl_4 is 2.56×10^{-22} g.

IS THE ANSWER REASONABLE? We expect a single molecule to have a very small mass. Our answer is a very small number, so it appears to be reasonable.

Practice Exercise 3: Would you be able to use a balance capable of weighing to the nearest 0.001 g to weigh 5.64×10^{18} formula units of calcium nitrate? [Hint: What is the mass of this amount of $Ca(NO_3)_2$?]

Practice Exercise 4: If the uncertainty in weighing a sample in the lab is ±0.002 g, what is this uncertainty in terms of molecules of sucrose, $C_{12}H_{22}O_{11}$?

3.2 CHEMICAL FORMULAS RELATE AMOUNTS OF SUBSTANCES IN A COMPOUND

Consider the chemical formula for water, H_2O:

- One molecule of water contains 2 H atoms and 1 O atom.
- Two molecules of water contain 4 H atoms and 2 O atoms.
- A dozen molecules of water contain 2 dozen H atoms and 1 dozen O atoms.
- A mole of molecules of water contains 2 moles of H atoms and 1 mole of O atoms.

Whether we're dealing with atoms, dozens of atoms, or moles of atoms, the chemical formula tells us that the ratio of H atoms to O atoms is always 2 to 1. In addition we can write the following **Stoichiometric equivalencies** concerning the water molecule:

$$1 \text{ mol } H_2O \Leftrightarrow 2 \text{ mol H} \qquad 1 \text{ mol } H_2O \Leftrightarrow 1 \text{ mol O} \qquad 1 \text{ mol O} \Leftrightarrow 2 \text{ mol H}$$

We recall that the symbol ⇔ means "is chemically equivalent to" and it is treated mathematically as an equal sign (see page 26 in Chapter 1).

TOOLS

Subscripts tell us the number of atoms in a formula

Within chemical compounds, moles of atoms always combine in the same ratio as the individual atoms themselves.

This fact lets us prepare mole-to-mole conversion factors involving elements in compounds as we need them. For example, in the formula P_4O_{10}, the subscripts mean that there are 4 moles of P for every 10 moles of O in this compound. We can relate P and O within the compound using the following conversion factors.

$$4 \text{ mol P} \Leftrightarrow 10 \text{ mol O} \quad \text{from which we write} \quad \frac{4 \text{ mol P}}{10 \text{ mol O}} \quad \text{or} \quad \frac{10 \text{ mol O}}{4 \text{ mol P}}$$

The formula P_4O_{10} also implies other equivalencies, each with its two associated conversion factors.

$$1 \text{ mol } P_4O_{10} \Leftrightarrow 4 \text{ mol P} \quad \text{or} \quad \frac{1 \text{ mol } P_4O_{10}}{4 \text{ mol P}} \quad \text{and} \quad \frac{4 \text{ mol P}}{1 \text{ mol } P_4O_{10}}$$

$$1 \text{ mol } P_4O_{10} \Leftrightarrow 10 \text{ mol O} \quad \text{or} \quad \frac{1 \text{ mol } P_4O_{10}}{10 \text{ mol O}} \quad \text{and} \quad \frac{10 \text{ mol O}}{1 \text{ mol } P_4O_{10}}$$

EXAMPLE 3.5

Calculating Amount of a Compound by Analyzing One Element

Calcium phosphate is widely found in nature in the form of natural minerals. It is also found in bones and some kidney stones. In many instances if we determine one element in a compound we can find out how much of the compound is present. In one case a sample is found to contain 0.864 mole of phosphorus. How many moles of $Ca_3(PO_4)_2$ will that represent?

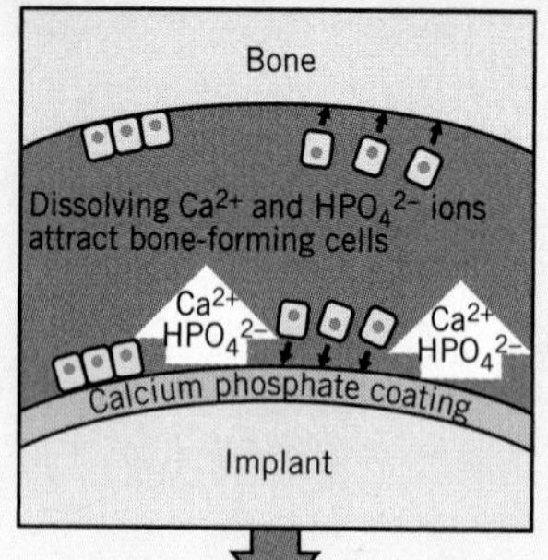

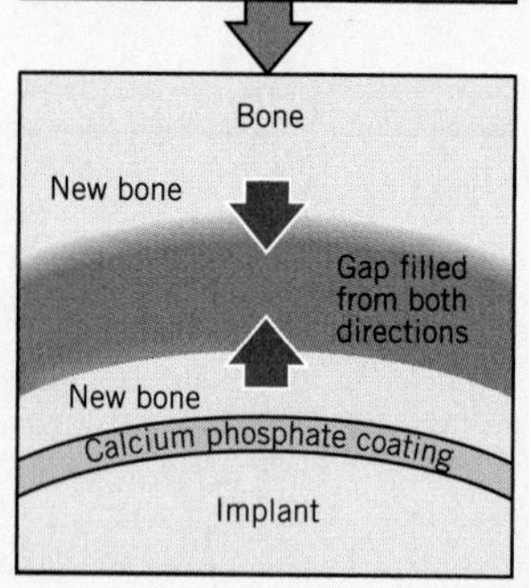

Some surfaces on bone implants are coated with calcium phosphate to permit bone to actually bond to the surface.

ANALYSIS: We need to use the mole ratio tool that gives the relationships between the compound's formula and the individual elements in that formula. Specifically we need to use the equivalence 2 mol P ⇔ 1 mol $Ca_3(PO_4)_2$ to construct the conversion factor needed to convert moles of P into moles of $Ca_3(PO_4)_2$.

SOLUTION: The initial question is written as an equation:

$$0.864 \text{ mol P} = ? \text{ mol } Ca_3(PO_4)_2$$

Seeing that mol P must cancel and mol $Ca_3(PO_4)_2$ must remain, the correct conversion factor must be 1 mol $Ca_3(PO_4)_2$/2 mol P and we solve the equation as

$$0.864 \cancel{\text{mol P}} \left(\frac{1 \text{ mol } Ca_3(PO_4)_2}{2 \cancel{\text{mol P}}} \right) = 0.432 \text{ mol } Ca_3(PO_4)_2$$

IS THE ANSWER REASONABLE? For a quick check, you can round 0.864 to 1 and divide by 2 to get 0.5. There is little difference between our estimate, 0.5, and the calculated answer, 0.432.

Whenever a problem asks you to convert an amount of one substance into an amount of a different substance, the most important conversion factor in the problem is usually a mole-to-mole relationship between the two substances.

Practice Exercise 5: Aluminum sulfate is analyzed and it is determined that the sample contains 0.0774 mole of sulfate ions. How many moles of aluminum does the sample contain? (Hint: Construct the correct formula for aluminum sulfate, then use the tool in this section.)

Practice Exercise 6: How many moles of nitrogen atoms are combined with 8.60 mol of oxygen atoms in dinitrogen pentoxide, N_2O_5?

One common use of stoichiometry in the lab occurs when we must relate the masses of two raw materials that are needed to make a compound. These calculations are summarized by the following sequence of steps to convert the mass of compound *A* to the mass of compound *B*.

$$\text{mass of } A \longrightarrow \text{moles of } A \longrightarrow \text{moles of } B \longrightarrow \text{mass of } B$$

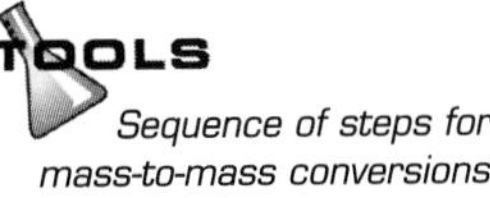

Sequence of steps for mass-to-mass conversions

In the following example we see how this is applied.

EXAMPLE 3.6
Calculating Amounts of One Element from Amounts of Another in a Compound

Chlorophyll a, the green pigment in leaves, has the formula $C_{55}H_{72}MgN_4O_5$. If 0.0011 g of Mg is available to a plant cell for chlorophyll a synthesis, how many grams of carbon will be required to completely use up the magnesium?

ANALYSIS: Let's begin, as usual, by restating the problem as follows.

$$0.0011 \text{ g Mg} \Leftrightarrow ? \text{ g C} \qquad \text{(for chlorophyll a only)}$$

A mole-to-mole ratio is the tool to use in problems that convert the moles of one substance into the moles of another. The formula of chlorophyll a, $C_{55}H_{72}MgN_4O_5$, relates Mg to C within the compound. We use the formula's subscripts as the tool we need to establish the relationship between moles of Mg and moles of C.

$$1 \text{ mol Mg} \Leftrightarrow 55 \text{ mol C}$$

We know we'll need to use this relationship to solve the problem. Let's drop it into the middle of our calculation sequence:

1 mol Mg ⇔ 55 mol C

$$0.0011 \text{ g Mg} \longrightarrow \text{mol Mg} \longrightarrow \text{mol C} \longrightarrow ? \text{ g C}$$

By placing this tool between our given information (0.0011 g Mg) and our desired units (g C) sequence, we've cut a difficult problem into two simpler ones. All we have to do now to complete the conversion of units is relate grams of Mg to moles of Mg, and moles of C to grams of C. The atomic mass of C links moles of C to grams of C, and the atomic mass of Mg links moles of Mg to grams of Mg. Rounding them to three significant figures, we can write

$$1 \text{ mol Mg} = 24.3 \text{ g Mg}$$
$$1 \text{ mol C} = 12.0 \text{ g C}$$

Our complete sequence for the problem is

1 mol Mg = 24.31 g Mg | 1 mol Mg ⇔ 55 mol C | 1 mol C = 12.0 g C

$$0.0011 \text{ g Mg} \longrightarrow \text{mol Mg} \longrightarrow \text{mol C} \longrightarrow \text{g C}$$

SOLUTION: We now set up the solution by forming conversion factors so the units cancel:

$$0.0011 \cancel{\text{g Mg}} \times \left(\frac{1 \cancel{\text{mol Mg}}}{24.3 \cancel{\text{g Mg}}}\right) \times \left(\frac{55 \cancel{\text{mol C}}}{1 \cancel{\text{mol Mg}}}\right) \times \left(\frac{12.0 \text{ g C}}{1 \cancel{\text{mol C}}}\right) = 0.030 \text{ g C}$$

A plant cell must supply 0.030 g C for every 0.0011 g Mg to completely use up the magnesium in the synthesis of chlorophyll a.

IS THE ANSWER REASONABLE? After checking that our units cancel properly, a quick estimate of the answer can be made by rounding all numbers to one significant figure. One way to do this results in the following expression (without units):

$$\frac{0.001 \times 1 \times 50 \times 10}{20 \times 1 \times 1} = \frac{0.5}{20} = \frac{0.05}{2} = 0.025$$

This value is close to the answer we got and gives us confidence that it is correct. (Note that if we rounded the 55 up to 60 our estimate would have been 0.030, which would still confirm our conclusion.)

Our answer was 0.030 g C from 0.0011 g Mg, so the mass of carbon we obtained is about 30 times the mass of the magnesium we started with. This seems reasonable because there are 55 times as many C atoms as Mg but a Mg atom weighs twice as much as a carbon atom.

Practice Exercise 7: How many grams of iron are needed to combine with 25.6 g of O to make Fe_2O_3? (Hint: Determine the mole ratios that the formula provides.)

Practice Exercise 8: An important iron ore called hematite contains iron(III) oxide. How many grams of iron are in a 15.0 g sample of hematite?

Practice Exercise 9: How many grams of iron will combine with 12.0 g of oxygen to form iron(III) oxide? Hematite, mentioned above, is often highly polished and used as a semiprecious gemstone.

3.3 CHEMICAL FORMULAS CAN BE DETERMINED FROM EXPERIMENTAL MASS MEASUREMENTS

In pharmaceutical research, chemists often synthesize entirely new compounds, or isolate new compounds from plant and animal tissues. They must then determine the formula and structure of the new compound. This is usually accomplished using mass spectroscopy, which gives an experimental value for the molecular mass. The compound can also be decomposed chemically to find the masses of elements within a given amount of compound. Let's see how experimental mass measurements can be used to determine a compound's formula.

Percentage composition describes the relative masses of the elements in a compound

The usual form for describing the relative masses of the elements in a compound is a list of *percentages by mass* called the compound's **percentage composition.** The **percentage by mass** of an element is the number of grams of the element present in 100 g of the compound. In general, a percentage by mass is found by using the following equation.

TOOLS *Percentage composition*

$$\% \text{ by mass of element} = \frac{\text{mass of element}}{\text{mass of whole sample}} \times 100\% \tag{3.1}$$

EXAMPLE 3.7
Calculating a Percentage Composition from Chemical Analysis

A sample of a liquid with a mass of 8.657 g was decomposed into its elements and gave 5.217 g of carbon, 0.9620 g of hydrogen, and 2.478 g of oxygen. What is the percentage composition of this compound?

ANALYSIS: We must use the tool expressed in Equation 3.1 and apply it to each element. The "mass of whole sample" here is 8.657 g, so we take each element in turn and perform the calculations.

SOLUTION:

For C: $\dfrac{5.217 \text{ g C}}{8.657 \text{ g sample}} \times 100\% = 60.26\% \text{ C}$

For H: $\dfrac{0.9620 \text{ g H}}{8.657 \text{ g sample}} \times 100\% = 11.11\% \text{ H}$

For O: $\dfrac{2.478 \text{ g O}}{8.657 \text{ g sample}} \times 100\% = 28.62\% \text{ O}$

Sum of percentages: 99.99%

One of the useful things about a percentage composition is that it tells us the mass of each of the elements in 100 g of the substance. For example, the results in this problem tell us that in 100.00 g of the liquid there are 60.26 g of carbon, 11.11 g of hydrogen, and 28.62 g of oxygen.

IS THE ANSWER REASONABLE? The "check" is that the percentages must add up to 100%, allowing for small differences caused by rounding. We can also check the individual results by rounding all the numbers to one significant figure to estimate the results. For example, the percentage C would be estimated as $5/9 \times 100$, which is a little over 50% and agrees with our answer.

Practice Exercise 10: An organic compound weighing 0.6672 g is decomposed giving 0.3481 g carbon and 0.0870 g hydrogen. What are the percentages of hydrogen and carbon in this compound? Is it likely that this compound contains another element? (Hint: Recall the tool concerning the conservation of mass.)

Practice Exercise 11: From 0.5462 g of a compound there was isolated 0.2012 g of nitrogen and 0.3450 g of oxygen. What is the percentage composition of this compound? Are any other elements present?

Experimental percentage compositions can help identify an unknown compound

Elements can combine in many different ways. Nitrogen and oxygen, for example, form all of the following compounds: N_2O, NO, NO_2, N_2O_3, N_2O_4, and N_2O_5. To identify an unknown sample of a compound of nitrogen and oxygen, one might compare the percentage composition found by experiment with the calculated, or theoretical, percentages for each possible formula. Which formula, for example, fits the percentage composition calculated in Practice Exercise 11? A strategy for matching empirical formulas with mass percentages is outlined in the following example.

EXAMPLE 3.8
Calculating a Theoretical Percentage Composition from a Chemical Formula

Do the mass percentages of 25.94% N and 74.06% O match the formula N_2O_5?

ANALYSIS: To calculate the theoretical percentages by mass of N and O in N_2O_5, we need the masses of N and O in a specific sample of N_2O_5. *If we choose 1 mol of the given compound to be this sample, calculating the percentages will be simple using the tool in Equation 3.1.*

SOLUTION: We know that 1 mol of N_2O_5 must contain 2 mol N and 5 mol O. The corresponding number of grams of N and O are found as follows.

2 N: $2 \cancel{\text{mol N}} \times \dfrac{14.01 \text{ g N}}{1 \cancel{\text{mol N}}} = 28.02 \text{ g N}$

5 O: $5 \cancel{\text{mol O}} \times \dfrac{16.00 \text{ g O}}{1 \cancel{\text{mol O}}} = 80.00 \text{ g O}$

$1 \text{ mol } N_2O_5 = 108.02 \text{ g } N_2O_5$

Now we can calculate the percentages.

$$\text{For \% N:} \quad \frac{28.02 \text{ g N}}{108.02 \text{ g sample}} \times 100\% = 25.94\% \text{ N in } N_2O_5$$

$$\text{For \% O:} \quad \frac{80.00 \text{ g O}}{108.02 \text{ g sample}} \times 100\% = 74.06\% \text{ O in } N_2O_5$$

Thus the experimental values do match the theoretical percentages for the formula N_2O_5.

IS THE ANSWER REASONABLE? We can easily check our math. Since the denominator in each fraction is close to 100, the numerator is a simple estimate of the percentage. For nitrogen the numerator of 28 compares well to our 25.94% answer to satisfy us that the calculation was done correctly.

Practice Exercise 12: Calculate the theoretical percentage composition of N_2O_4. (Hint: Recall the definition of percentage composition.)

Practice Exercise 13: Calculate the theoretical percentage compositions for N_2O, NO, NO_2, N_2O_3, N_2O_4, and N_2O_5. Which of these compounds produced the data in Practice Exercise 11?

An empirical formula can be determined from the masses of the different elements in a sample of a compound

The compound that forms when phosphorus burns in oxygen consists of molecules with the formula P_4O_{10}. When a formula gives the composition of one *molecule*, it is called a **molecular formula.** Notice, however, that both the subscripts 4 and 10 are divisible by 2, so the *smallest* numbers that tell us the *ratio* of P to O are 2 and 5. We can write a simpler (but less informative) formula that expresses this ratio, P_2O_5. This is called the **empirical formula** because it can be obtained from an experimental analysis of the compound.

Empirical formula

The empirical formula expresses the simplest whole number ratio of the atoms of each element in a compound.

We already know that the ratio of atoms in a compound is the same as a ratio of the moles of those atoms in the compound. We will determine the simplest ratio of moles from experimental data. The experimental data we need is any information that allows us to determine the moles of each element in the compound. We will investigate three types of data that can be used to determine empirical formulas. They are (a) masses of the elements, (b) percentage composition, and (c) combustion data. In all three, the goal is to obtain the simplest ratio of moles of each element in the formula.

The next four examples illustrate how we can calculate empirical formulas. We will then look at what additional data are required to obtain a compound's molecular formula.

EXAMPLE 3.9
Calculating an Empirical Formula from Mass Data

A 2.57 g sample of a compound composed of only tin and chlorine was found to contain 1.17 g of tin. What is the compound's empirical formula?

ANALYSIS: The subscripts in an empirical formula give the relative number of moles of elements in a compound. If we can find the *mole* ratio of Sn to Cl, we will have the empirical formula. The first step, therefore, is to convert the numbers of grams of Sn and Cl to the numbers of moles of Sn and Cl using the tool that relates mass to moles. Then we convert these numbers into their simplest *whole-number* ratio.

The problem did not give the mass of chlorine in the 2.57 g sample. But there are only two elements present in the compound, tin and chlorine. We know the mass of the tin, and we know the total mass of compound. The law of conservation of mass, one of our tools from the previous chapter, requires that the mass of Cl is the difference between 2.57 g of compound and 1.17 g of Sn.

SOLUTION: First, we find the mass of Cl in 2.57 g of compound:

$$\text{Mass of Cl} = 2.57 \text{ g compound} - 1.17 \text{ g Sn} = 1.40 \text{ g Cl}$$

Now we use the atomic masses to convert the mass data for tin and chlorine into moles.

$$1.17 \cancel{\text{g Sn}} \times \frac{1 \text{ mol Sn}}{118.71 \cancel{\text{g Sn}}} = 0.00986 \text{ mol Sn}$$

$$1.40 \cancel{\text{g Cl}} \times \frac{1 \text{ mol Cl}}{35.45 \cancel{\text{g Cl}}} = 0.0395 \text{ mol Cl}$$

We could now write a formula: $Sn_{0.00986}Cl_{0.0395}$, which does express the mole ratio, but subscripts also represent atom ratios and need to be integers. To convert the decimal subscripts to integers we begin by dividing each by the smallest number in the set. *This is always the way to begin the search for whole-number subscripts; pick the smallest number of the set as the divisor.* It's guaranteed to make at least one subscript a whole number, namely, 1. Here, we divide both numbers by 0.00986.

$$Sn_{\frac{0.00986}{0.00986}}Cl_{\frac{0.0395}{0.00986}} = Sn_{1.00}Cl_{4.01}$$

We may round 4.01 to 4, because even if the third significant digit is uncertain, we do know the second significant digit, the zero, with certainty; so the empirical formula is $SnCl_4$.

IS THE ANSWER REASONABLE? If $SnCl_4$ is the right formula, one mole of the compound contains 118.7 grams of tin and 142 grams of chlorine, or a little more chlorine than tin. The statement of the problem also gives us slightly more chlorine than tin. You should also recall from Chapter 2 that tin forms either the Sn^{2+} or the Sn^{4+} ion and that chlorine forms only the Cl^- ion. Therefore either $SnCl_2$ or $SnCl_4$ are reasonable compounds and one of them was our answer.

□ We cannot forget that we now have a storehouse of reasonable chemical formulas that were developed in Chapter 2.

Practice Exercise 14: A 1.525 g sample of a compound between nitrogen and oxygen contains 0.712 g of nitrogen. Calculate its empirical formula. (Hint: How many grams of oxygen are there?)

Practice Exercise 15: A 1.525 g sample of a compound between sulfur and oxygen was prepared by burning 0.7625 g of sulfur in air and collecting the product. What is the empirical formula for the compound formed?

Sometimes our strategy of using the lowest common divisor does not give whole numbers. Let's see how to handle such a situation.

EXAMPLE 3.10
Calculating an Empirical Formula from Mass Composition

One of the compounds of iron and oxygen, "black iron oxide," occurs naturally in the mineral magnetite. When a 2.448 g sample was analyzed it was found to have 1.771 g of Fe. Calculate the empirical formula of this compound.

ANALYSIS: To calculate the empirical formula, we need to know the masses of both iron and oxygen, but we've only been given the mass of iron. We use the law of conservation of mass as a tool (page 36) to calculate the mass of oxygen. Next we will use our tools for converting masses to moles, recalling that one mole of an element is equal to its atomic mass in gram units.

The mineral magnetite, like any magnet, is able to affect the orientation of a compass needle. *(Paul Silverman/Fundamental Photographs.)*

Then we'll write a trial formula and see whether we can adjust the subscripts to their smallest whole numbers by the strategy learned in Example 3.9.

SOLUTION: The mass of O is, as we said, found by difference.

$$2.448 \text{ g compound} - 1.771 \text{ g Fe} = 0.677 \text{ g O}$$

The moles of Fe and O in the sample can now be calculated.

$$1.771 \text{ g Fe} \times \frac{1 \text{ mol Fe}}{55.845 \text{ g Fe}} = 0.03171 \text{ mol Fe}$$

$$0.677 \text{ g O} \times \frac{1 \text{ mol O}}{16.00 \text{ g O}} = 0.0423 \text{ mol O}$$

These results let us write the formula as $Fe_{0.03171}O_{0.0423}$.

Our first effort to change the ratio of 0.03171 to 0.0423 into whole numbers is to divide both by the smaller, 0.03171.

$$Fe_{\frac{0.03171}{0.03171}}O_{\frac{0.0423}{0.03171}} = Fe_{1.000}O_{1.33}$$

This time we cannot round 1.33 to 1.0. The subscript for oxygen has three significant figures, so we can be sure that the digits 1.3 in 1.33 are known with certainty. Therefore the subscript for O, 1.33, is much too far from a whole number to round off and retain the required precision. In a *mole* sense, the ratio of 1 to 1.33 is correct; we just need a way to restate this ratio in whole numbers. Let's look at a simple strategy that will give us whole-number subscripts.

TOOLS *Techniques for finding whole-number subscripts for empirical formulas*

Trial and error is an easy way to obtain integer subscripts. We try multiplying the subscripts by 2, then 3, 4, 5, 6, and so on. The lowest multiplier that results in integer subscripts is the correct one to use. For example, let's multiply each subscript in $Fe_{1.000}O_{1.33}$ by a whole number, 2. *Since we multiply all subscripts by 2 we do not change the ratio*; it changes only the size of the numbers used to state it.

$$Fe_{(1.000 \times 2)}O_{(1.33 \times 2)} = Fe_{2.000}O_{2.66}$$

This didn't work either; 2.66 is also too far from a whole number (based on the allowed precision) to be rounded off. Let's try using 3 instead of 2 *on the original ratio of 1.000 to 1.33.*

$$Fe_{(1.000 \times 3)}O_{(1.33 \times 3)} = Fe_{3.000}O_{3.99}$$

We are now justified in rounding; 3.99 is acceptably close to 4. The empirical formula of the oxide of iron is Fe_3O_4. A different method is noted in the margin, where Table 3.1 shows us which whole number multiplier should be used to convert a decimal that cannot be rounded to a whole number.

TABLE 3.1

Decimal Numbers and Their Rational Fractions

Decimal	Fraction[a]
0.20	1/5
0.25	1/4
0.33	1/3
0.40	2/5
0.50	1/2
0.60	3/5
0.66	2/3
0.75	3/4
0.80	4/5

[a]Use the denominator of the fraction as a multiplier to create whole-number subscripts in empirical formulas.

IS THE ANSWER REASONABLE? First, the fact that our calculation gives whole number subscripts is a good indicator that we've solved the problem correctly. Second, another way to check our answer is to estimate the percentage of iron from the given data and from our result. The given data are 1.771 g Fe and 2.448 g of sample and the percentage iron is estimated as

$$\frac{1.771 \text{ g}}{2.448 \text{ g}} \times 100 \approx \frac{\mathbf{1.8\ g}}{\mathbf{2.4\ g}} \times \mathbf{100} = \frac{3}{4} \times 100 = \text{approximately } 75\%$$

In one mole of the compound Fe_3O_4, the mass of iron is $3 \times 55.8 = 167.4$ and the molar mass is 231.4. The percentage of iron is estimated as

$$\frac{167.4 \text{ g}}{231.4 \text{ g}} \times 100 \approx \frac{170 \text{ g}}{230 \text{ g}} \times 100 = \frac{\mathbf{1.7\ g}}{\mathbf{2.3\ g}} \times \mathbf{100} = \text{approximately } 75\%$$

We don't have to do any calculations because we can see that mathematical expressions from both calculations are almost the same, and the answers will be very close to each other. Let's compare the bold term from each equation,

$$\frac{1.8 \text{ g}}{2.4 \text{ g}} \times 100 \approx \frac{1.7 \text{ g}}{2.3 \text{ g}} \times 100$$

We are able to conclude that our percentages of iron are the same and our empirical formula is reasonable.

Practice Exercise 16: When aluminum is produced on an industrial scale 5.68 tons of aluminum and 5.04 tons of oxygen are obtained. What is the empirical formula of the compound used to produce aluminum? (Hint: 1 ton = 2000 lb and 1 lb = 454 g.)

Practice Exercise 17: A 2.012 g sample of a compound of nitrogen and oxygen has 0.522 g of nitrogen. Calculate its empirical formula.

Empirical formulas can be determined from mass percentages

Only rarely is it possible to obtain the masses of every element in a compound by the use of just one weighed sample. Two or more analyses carried out on different samples are often needed. For example, suppose an analyst is given a compound known to consist exclusively of calcium, chlorine, and oxygen. The mass of calcium in one weighed sample and the mass of chlorine in another sample would be determined in separate experiments. Then the mass data for calcium and chlorine would be converted to percentages by mass *so that the data from different samples relate to the same sample size, namely, 100 g of the compound.* The percentage of oxygen would be calculated by difference because % Ca + % Cl + % O = 100%. Each mass percentage represents a certain number of grams of the element, which is next converted into the corresponding number of moles of the element. The mole proportions are converted to whole numbers in the way we just studied, giving us the subscripts for the empirical formula. Let's see how this works.

TOOLS

Percentage composition helps to correlate results from different experiments

EXAMPLE 3.11

Calculating an Empirical Formula from Percentage Composition

A white powder used in paints, enamels, and ceramics has the following percentage composition: Ba, 69.6%; C, 6.09%; and O, 24.3%. What is its empirical formula? What is the name of this compound?

ANALYSIS: Consider having 100 grams of this compound. The percentages of the elements given in the problem are numerically the same as the masses of these elements in our 100 g sample. We see a general principle that by assuming a 100 g sample, all percent signs can be changed to gram units. Now that we have the masses of the elements we can use the procedures and tools used in Examples 3.9 and 3.10.

SOLUTION: Assuming a 100 g sample of the compound we quickly convert 69.6% Ba to 69.6 g Ba, 6.09% C to 6.09 g C, and 24.3% O to 24.3 g O. Now we convert these to moles.

$$\text{Ba:} \quad 69.6 \cancel{\text{g Ba}} \times \frac{1 \text{ mol Ba}}{137.33 \cancel{\text{g Ba}}} = 0.507 \text{ mol Ba}$$

$$\text{C:} \quad 6.09 \cancel{\text{g C}} \times \frac{1 \text{ mol C}}{12.01 \cancel{\text{g C}}} = 0.507 \text{ mol C}$$

$$\text{O:} \quad 24.3 \cancel{\text{g O}} \times \frac{1 \text{ mol O}}{16.00 \cancel{\text{g O}}} = 1.52 \text{ mol O}$$

Our preliminary empirical formula is then

$$Ba_{0.507}C_{0.507}O_{1.52}$$

We next divide each subscript by the smallest, 0.507.

$$Ba_{\frac{0.507}{0.507}}C_{\frac{0.507}{0.507}}O_{\frac{1.52}{0.507}} = Ba_{1.00}C_{1.00}O_{3.00}$$

The subscripts are whole numbers, so the empirical formula is $BaCO_3$, representing barium carbonate.

IS THE ANSWER REASONABLE? The fact that our calculations led to whole number subscripts is a strong clue that the answer is correct. In addition, our knowledge of ionic compounds and the polyatomic carbonate ion from Chapter 2 tell us that the barium ion is Ba^{2+} and the carbonate ion is CO_3^{2-}; $BaCO_3$ is a reasonable formula.

Practice Exercise 18: A white solid used to whiten paper has the following percentage composition: Na, 32.4%; S, 22.6%. The unanalyzed element is oxygen. What is the compound's empirical formula? (Hint: What law allows you to calculate the % oxygen?)

Practice Exercise 19: Cinnamon gets some of its flavor from cinnamaldehyde that is 81.79% C, 6.10% H, and the rest oxygen. Determine the empirical formula for this compound.

Percentage composition is important since it allowed us to determine the amount of three substances using only two experiments. This in itself is a considerable saving in time and effort. Additionally, it is often difficult to analyze a sample for certain elements, oxygen for example, and using percentage measurements helps avoid this problem.

Empirical formulas can be determined from indirect analyses

In practice, a compound is seldom broken down completely to its *elements* in a quantitative analysis. Instead, the compound is changed into other *compounds*. The reactions separate the elements by capturing each one entirely (quantitatively) in a *separate* compound *whose formula is known*.

In the following example, we illustrate the indirect analysis of a compound made entirely of carbon, hydrogen, and oxygen. Such compounds burn completely in pure oxygen—it is called a *combustion reaction*—and the sole products are carbon dioxide and water. (This particular kind of indirect analysis is sometimes called a *combustion analysis*.) The complete combustion of methyl alcohol (CH_3OH), for example, occurs according to the following equation.

Organic compounds react with a stream of pure oxygen to give CO_2 and H_2O. The flowing gases pass through a preweighed tube of $CaSO_4$ to absorb the water and then through a tube of NaOH deposited on a binder to absorb the carbon dioxide. The increase in mass of the $CaSO_4$ and NaOH tubes represents the mass of water and CO_2, respectively.

$$2CH_3OH + 3O_2 \longrightarrow 2CO_2 + 4H_2O$$

The carbon dioxide and water can be separated and are individually weighed. Notice that all of the carbon atoms in the original compound end up among the CO_2 molecules, and all of the hydrogen atoms are in H_2O molecules. In this way at least two of the original elements, C and H, are quantitatively measured.

We will calculate the mass of carbon in the CO_2 collected, which equals the mass of carbon in the original sample. Similarly, we will calculate the mass of hydrogen in the H_2O collected, which equals the mass of hydrogen in the original sample. When added together, the mass of C and mass of H are less than the total mass of the sample because part of the sample is composed of oxygen. The law of conservation of mass allows us to subtract the sum of the C and H masses from the original sample mass to obtain the mass of oxygen in the sample of the compound.

EXAMPLE 3.12
Empirical Formula from Indirect Analysis

A 0.5438 g sample of a liquid consisting of only C, H, and O was burned in pure oxygen, and 1.039 g of CO_2 and 0.6369 g of H_2O were obtained. What is the empirical formula of the compound?

ANALYSIS: There are two parts to this problem. First we need to calculate the mass of the elements, C and H, by determining the number of grams of C in the CO_2 and the number of grams of H in the H_2O. (This kind of calculation was illustrated in Example 3.6 and uses the tools that tell us how to create conversion factors from grams to moles and from moles of one substance to moles of another.) These values represent the number of grams of C and H in the original sample. Adding them together and subtracting the sum from the mass of the original sample will give us the mass of oxygen in the sample.

In the second half of the solution, we use the masses of C, H, and O to calculate the empirical formula as in Example 3.9.

SOLUTION: First we find the number of grams of C in the CO_2 and of H in the H_2O. We use the normal conversion sequence from grams of compound, to moles of compound, to moles of element, and then to grams of element as shown in the next two equations.

$$1.039 \, \cancel{\text{g CO}_2} \times \frac{1 \, \cancel{\text{mol CO}_2}}{44.009 \, \cancel{\text{g CO}_2}} \times \frac{1 \, \cancel{\text{mol C}}}{1 \, \cancel{\text{mol CO}_2}} \times \frac{12.011 \text{ g C}}{1 \, \cancel{\text{mol C}}} = 0.2836 \text{ g C}$$

$$0.6369 \, \cancel{\text{g H}_2\text{O}} \times \frac{1 \, \cancel{\text{mol H}_2\text{O}}}{18.015 \, \cancel{\text{g H}_2\text{O}}} \times \frac{2 \, \cancel{\text{mol H}}}{1 \, \cancel{\text{mol H}_2\text{O}}} \times \frac{1.0079 \text{ g H}}{1 \, \cancel{\text{mol H}}} = 0.07125 \text{ g H}$$

The total mass of C and H is therefore the sum of these two quantities.

$$\text{Total mass of C and H} = 0.2836 \text{ g C} + 0.07125 \text{ g H} = 0.3548 \text{ g}$$

The difference between this total and the 0.5438 g in the original sample is the mass of oxygen (the only other element).

$$\text{Mass of O} = 0.5438 \text{ g} - 0.3548 \text{ g} = 0.1890 \text{ g O}$$

Now we can convert the masses of the elements to an empirical formula.

$$\text{For C:} \quad 0.2836 \, \cancel{\text{g C}} \times \frac{1 \text{ mol C}}{12.011 \, \cancel{\text{g C}}} = 0.02361 \text{ mol C}$$

$$\text{For H:} \quad 0.07125 \, \cancel{\text{g H}} \times \frac{1 \text{ mol H}}{1.0079 \, \cancel{\text{g H}}} = 0.07068 \text{ mol H}$$

$$\text{For O:} \quad 0.1890 \, \cancel{\text{g O}} \times \frac{1 \text{ mol O}}{15.999 \, \cancel{\text{g O}}} = 0.01181 \text{ mol O}$$

Our preliminary empirical formula is thus $C_{0.02361}H_{0.07068}O_{0.01181}$. We divide all of these subscripts by the smallest number, 0.01181.

$$C_{\frac{0.02361}{0.01181}}H_{\frac{0.07068}{0.01181}}O_{\frac{0.01181}{0.01181}} = C_{1.998}H_{5.985}O_1$$

The results are acceptably close to integers to say that the empirical formula is C_2H_6O.

IS THE ANSWER REASONABLE? Our checks on problems need to be quick and efficient. In previous examples we have been able to use our knowledge of ionic compounds to see if formulas are reasonable. In the future, with some knowledge of organic chemistry you will know that the formula C_2H_6O is reasonable. Beyond that, the fact that we obtained whole number subscripts after all these calculations is also a good indicator that we've solved the problem correctly.

Practice Exercise 20: A sample containing only sulfur and carbon is completely burned in air. The analysis produced 0.640 g SO_2 and 0.220 g of CO_2. What is the empirical formula? (Hint: Use the tools for relating grams of a compound to grams of an element.)

Practice Exercise 21: The combustion of a 5.048 g sample of a compound of C, H, and O gave 7.406 g CO_2 and 3.027 g H_2O. Calculate the empirical formula of the compound.

As we said earlier, more than one sample of a substance must be analyzed whenever more than one reaction is necessary to separate the elements. This is also true in indirect analyses. For example, if a compound contains C, H, N, and O, combustion converts the C and H to CO_2 and H_2O, which are separated by special techniques and weighed. The mass of C in the CO_2 sample and the mass of H in the H_2O sample are then calculated in the usual way. A second reaction with a different sample of the compound can be used to obtain the nitrogen, either as N_2 or as NH_3. Because different size samples are used, the masses of C and H from one sample and the mass of N from another cannot be used to calculate the empirical formula. So we convert the masses of the elements found in their respective samples into percentages of the element by mass in the compound. We add up the percentages, subtract from 100 to get the percentage of O, and then calculate the empirical formula from the percentage composition as described earlier.

Molecular formulas are determined from empirical formulas and molecular masses

The empirical formula is the accepted formula unit for ionic compounds. For molecular compounds, however, chemists prefer *molecular* formulas because they give the number of atoms of each type in a molecule.

Molecular masses can sometimes be obtained using mass spectroscopy, discussed in Chapter 1. Other methods will be discussed in Chapters 10, 11, and 12.

Sometimes an empirical formula and a molecular formula are the same. Two examples are H_2O and NH_3. Usually, however, the subscripts of a molecular formula are whole-number multiples of those in the empirical formula. The subscripts of the molecular formula P_4O_{10}, for example, are each two times those in the empirical formula, P_2O_5, as you saw earlier. The molecular mass of P_4O_{10} is likewise two times the formula mass of P_2O_5. This observation provides us with a way to find out the molecular formula for a compound provided we have a way of determining experimentally the molecular mass of the compound. If the experimental molecular mass *equals* the calculated empirical formula mass, the empirical formula itself is also a molecular formula. Otherwise, the experimental molecular mass will be some whole-number multiple of the value calculated from the empirical formula. Whatever the whole number is, it's a common multiplier for the subscripts of the empirical formula.

EXAMPLE 3.13
Determining a Molecular Formula from an Empirical Formula and a Molecular Mass

Styrene, the raw material for polystyrene foam plastics, has an empirical formula of CH. Its molecular mass is 104 g mol^{-1}. What is its molecular formula?

ANALYSIS: The molecular mass of styrene, 104 g mol^{-1}, is some simple multiple of the formula mass of the empirical formula, CH. When we compute that multiple it will tell us how many CH units make up the styrene molecule. We can compute this multiple by dividing the molecular mass by the empirical formula mass.

SOLUTION: For the empirical formula, CH, the formula mass is

$$12.01 + 1.008 = 13.02$$

To find how many CH units weighing 13.02 are in a mass of 104, we divide.

$$\frac{104}{13.02} = 7.99$$

Rounding this to 8, we see that 104 is 8 times larger than 13.02, so the correct molecular formula of styrene must have subscripts 8 times those in CH. Styrene, therefore, is C_8H_8.

IS THE ANSWER REASONABLE? The molecular mass of C_8H_8 is approximately $(8 \times 12) + (8 \times 1) = 104$ g mol^{-1}, which is consistent with the molecular mass we started with.

Practice Exercise 22: After determining that the empirical formulas of two different compounds were CH_2Cl and $CHCl$ a student mixed up the data for the molecular masses. However, the student knew that one compound had a molecular mass of 100 and the other had a molecular mass of 289 g mol^{-1}. What are the likely molecular formulas of the two compounds? (Hint: Recall the relationship between the molecular and empirical formula.)

Practice Exercise 23: The empirical formula of hydrazine is NH_2, and its molecular mass is 32.0 g mol^{-1}. What is its molecular formula?

3.4 CHEMICAL EQUATIONS LINK AMOUNTS OF SUBSTANCES IN A REACTION

Balancing an equation involves adjusting coefficients

We learned in Chapter 2 that a *chemical equation* is a shorthand, quantitative description of a chemical reaction. An equation is *balanced* when all atoms present among the reactants are also somewhere among the products. As we learned, coefficients, the numbers in front of formulas, are multiplier numbers for their respective formulas, and the values of the coefficients determine whether an equation is balanced.

Always approach the balancing of an equation as a two-step process.

Step 1. *Write the unbalanced "equation."* Organize the formulas in the pattern of an equation with plus signs and an arrow. Use *correct* formulas. (You learned to write many of them in Chapter 2, but until we have studied more chemistry, you will usually be given formulas.)

Step 2. *Adjust the coefficients to get equal numbers of each kind of atom on both sides of the arrow.* When doing step 2, make no changes in the formulas, either in the atomic symbols or their subscripts. If you do, the equation will involve different substances from those intended. You may still be able to balance it, but the equation will not be for the reaction you want.

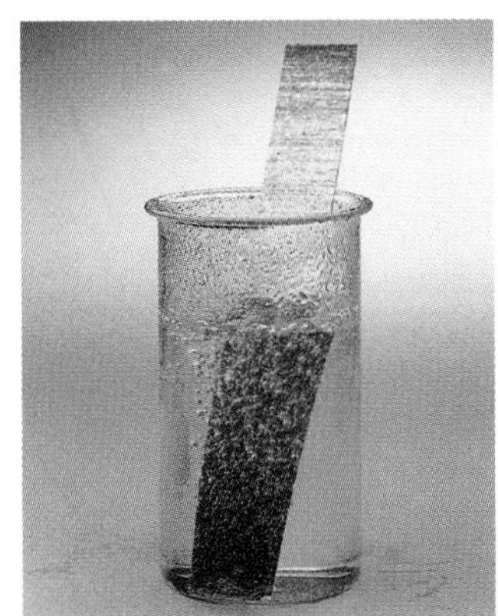

Zinc metal reacts with hydrochloric acid. *(Richard Megna/Fundamental Photographs.)*

We'll begin with simple equations that can be balanced easily by inspection. An example is the reaction of zinc metal with hydrochloric acid (margin photo). First, we need the correct formulas, and this time we'll include the physical states because they are different. The reactants are zinc, $Zn(s)$, and hydrochloric acid, an aqueous solution of the gas hydrogen chloride, HCl, symbolized as $HCl(aq)$. We also need formulas for the products. Zn changes to a water-soluble compound, zinc chloride, $ZnCl_2(aq)$, and hydrogen gas, $H_2(g)$, bubbles out as the other product. (Recall that hydrogen occurs naturally as a *diatomic molecule*, not as an atom.)

Step 1. Write an unbalanced equation.

$$Zn(s) + HCl(aq) \longrightarrow ZnCl_2(aq) + H_2(g) \qquad \text{(unbalanced)}$$

Step 2. Adjust the coefficients to get equal numbers of each kind of atom on both sides of the arrow.

There is no simple set of rules for adjusting coefficients. Experience is the greatest help, and experience has taught chemists that the following guidelines often get to the solution most directly when they are applied in the order given.

Guidelines for balancing chemical equations

Some Guidelines for Balancing Equations

1. Balance elements other than H and O first.
2. Elements should be balanced last (e.g., Zn and H_2 in our example).
3. Balance as a group those polyatomic ions that appear unchanged on both sides of the arrow.

Using the guidelines given here, we'll look at Cl first in our example. Because there are two Cl to the right of the arrow but only one to the left, we put a 2 in front of the HCl on the left side. The result is

$$Zn(s) + 2HCl(aq) \longrightarrow ZnCl_2(aq) + H_2(g)$$

We then balance the hydrogen and zinc and find that no additional coefficient changes are needed. Everything is now balanced. On each side we find 1 Zn, 2 H, and 2 Cl. One complication is that an infinite number of *balanced* equations can be written for any given reaction! We might, for example, have adjusted the coefficients so that our equation came out as follows.

$$2Zn(s) + 4HCl(aq) \longrightarrow 2ZnCl_2(aq) + 2H_2(g)$$

This equation is also correctly balanced. For simplicity, we prefer the *smallest* whole-number coefficients when writing balanced equations.

EXAMPLE 3.14
Writing a Balanced Equation

Sodium hydroxide and phosphoric acid, H_3PO_4, react as aqueous solutions to give sodium phosphate and water. The sodium phosphate remains in solution. Write the balanced equation for this reaction.

ANALYSIS: First, we need to write an unbalanced equation that includes the reactant formulas on the left-hand side and the product formulas on the right. We are given the formula only for phosphoric acid. We need to use the tools in Chapter 2 to determine that sodium hydroxide is NaOH, water is H_2O, and sodium phosphate has a formula of Na_3PO_4. Next we write the unbalanced equation placing all the reactants to the left of the arrow and all products to the right. Finally we use the procedures suggested in the tool for balancing equations to adjust stoichiometric coefficients (*never change subscripts!*) until there are the same number of each type of atom on the left and right sides of the equation.

SOLUTION: We include the designation (*aq*) for all substances dissolved in water (except H_2O itself; we'll usually not give it any designation when it is in its liquid state).

$$NaOH(aq) + H_3PO_4(aq) \longrightarrow Na_3PO_4(aq) + H_2O \qquad \text{(unbalanced)}$$

There are several things not in balance, but our guidelines suggest that we work with Na first rather than with O, H, or PO_4. There are 3 Na on the right side, so we put a 3 in front of NaOH on the left, as a trial.

$$3NaOH(aq) + H_3PO_4(aq) \longrightarrow Na_3PO_4(aq) + H_2O \qquad \text{(unbalanced)}$$

Now the Na are in balance. The unit of PO_4 is balanced also. Not counting the PO_4, we have on the left 3 O and 3 H in 3NaOH plus 3 H in H_3PO_4, for a net of 3 O and 6 H on the left. On the right, in H_2O, we have 1 O and 2 H. The ratio of 3 O to 6 H on the left is equivalent to the ratio of 1 O to 2 H on the right, so we write the multiplier (coefficient) 3 in front of H_2O.

$$3NaOH(aq) + H_3PO_4(aq) \longrightarrow Na_3PO_4(aq) + 3H_2O \qquad \text{(balanced)}$$

We now have a balanced equation.

IS THE ANSWER REASONABLE? On each side we have 3 Na, 1 PO_4, 6 H, and 3 O besides those in PO_4, and since the coefficients for $H_3PO_4(aq)$ and $Na_3PO_4(aq)$ are 1 our coefficients cannot be reduced to smaller whole numbers.

Practice Exercise 24: Write the balanced chemical equation that describes what happens when a solution containing aluminum chloride is mixed with a solution containing sodium phosphate and the product of the reaction is solid aluminum phosphate and a solution of sodium chloride. (Hint: Write the correct formulas based on information in Chapter 2.)

Practice Exercise 25: When aqueous solutions of calcium chloride, $CaCl_2$, and potassium phosphate, K_3PO_4, are mixed, a reaction occurs in which solid calcium phosphate, $Ca_3(PO_4)_2$, separates from the solution. The other product is $KCl(aq)$. Write the balanced equation.

The strategy of balancing whole units of polyatomic ions, like PO_4, as a group is extremely useful. Using this method we have less atom counting to do and balancing equations is often easier.

Coefficients in a balanced equation provide mole-to-mole ratios among reactants and products

So far we have focused on relationships between elements within a single compound. We have seen that the essential conversion factor between substances within a compound is the mole-to-mole ratio obtained from the compound's formula. In this section, we'll see that the same techniques can be used to relate substances involved in a chemical reaction. The critical link between substances involved in a reaction is a mole-to-mole ratio obtained from the coefficients in the chemical equation that describes the reaction.

To see how chemical equations can be used to obtain mole-to-mole relationships, consider the equation that describes the burning of octane (C_8H_{18}) in oxygen (O_2) to give carbon dioxide and steam:

> The chemical equation gives relative amounts of molecules of each type that participate in the reaction. It does *not* mean that 2 octane molecules actually collide with 25 O_2 molecules. The reaction occurs in many steps, which the chemical equation does not show.

$$2C_8H_{18}(l) + 25O_2(g) \longrightarrow 16CO_2(g) + 18H_2O(g)$$

This equation can be interpreted on a *microscopic* (molecular) scale as follows:

For every two molecules of liquid octane that react with twenty-five molecules of oxygen gas, sixteen molecules of carbon dioxide gas and eighteen molecules of steam are produced.

This statement immediately suggests many equivalence relationships that can be used to build conversion factors in stoichiometry problems:

$$2 \text{ molecules } C_8H_{18} \Leftrightarrow 25 \text{ molecules } O_2$$

$$2 \text{ molecules } C_8H_{18} \Leftrightarrow 16 \text{ molecules } CO_2$$

$$2 \text{ molecules } C_8H_{18} \Leftrightarrow 18 \text{ molecules } H_2O$$

$$25 \text{ molecules } O_2 \Leftrightarrow 16 \text{ molecules } CO_2$$

$$25 \text{ molecules } O_2 \Leftrightarrow 18 \text{ molecules } H_2O$$

$$16 \text{ molecules } CO_2 \Leftrightarrow 18 \text{ molecules } H_2O$$

Any of these microscopic relationships can be scaled up to the macroscopic level by multiplying both sides of the equivalency by Avogadro's number, which effectively allows us to replace "molecules" with "moles" or "mol":

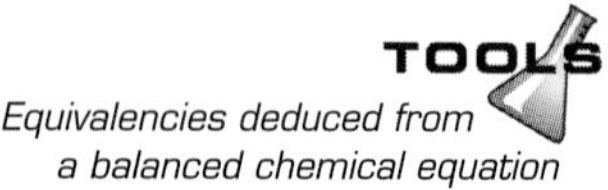

Equivalencies deduced from a balanced chemical equation

$$2 \text{ mol } C_8H_{18} \Leftrightarrow 25 \text{ mol } O_2$$

$$2 \text{ mol } C_8H_{18} \Leftrightarrow 16 \text{ mol } CO_2$$

$$2 \text{ mol } C_8H_{18} \Leftrightarrow 18 \text{ mol } H_2O$$

$$25 \text{ mol } O_2 \Leftrightarrow 16 \text{ mol } CO_2$$

$$25 \text{ mol } O_2 \Leftrightarrow 18 \text{ mol } H_2O$$

$$16 \text{ mol } CO_2 \Leftrightarrow 18 \text{ mol } H_2O$$

We can interpret the equation on a macroscopic (mole) scale as follows:

> *Two moles of liquid octane react with twenty-five moles of oxygen gas to produce sixteen moles of carbon dioxide gas and eighteen moles of steam.*

To use these equivalencies in a stoichiometry problem, the equation must be **balanced.** That means that every atom found in the reactants must also be found somewhere in the products. You must always check to see whether this is so for a given equation before you can use the coefficients in the equation to build equivalencies and conversion factors.

First, let's see how mole-to-mole relationships obtained from a balanced chemical equation can be used to convert moles of one substance to moles of another when both substances are involved in a chemical reaction.

EXAMPLE 3.15
Stoichiometry of Chemical Reactions

How many moles of sodium phosphate can be made from 0.240 mol of sodium hydroxide by the following unbalanced reaction?

$$NaOH(aq) + H_3PO_4(aq) \longrightarrow Na_3PO_4(aq) + H_2O$$

ANALYSIS: The question asks us to relate amounts of two different substances. *A mole-to-mole relationship is the tool that defines the relationship between two different substances in a stoichiometry problem.* The balanced equation is the tool to use because it gives the coefficients we need. We are given an unbalanced equation and need to balance this equation to use this tool.

$$3NaOH(aq) + H_3PO_4(aq) \longrightarrow Na_3PO_4(aq) + 3H_2O$$

From the coefficients, we now know that

$$3 \text{ mol NaOH} \Leftrightarrow 1 \text{ mol } Na_3PO_4$$

This enables us to prepare the conversion factor that we need.

SOLUTION: We write the question in equation form as

$$0.240 \text{ mol NaOH} \Leftrightarrow ? \text{ mol } Na_3PO_4$$

and convert 0.240 mol NaOH to the numbers of moles of Na_3PO_4 equivalent to it in the reaction as follows.

$$0.240 \cancel{\text{mol NaOH}} \times \frac{1 \text{ mol } Na_3PO_4}{3 \cancel{\text{mol NaOH}}} = 0.0800 \text{ mol } Na_3PO_4$$

Thus we can make 0.0800 mol Na_3PO_4 from 0.240 mol NaOH.

IS THE ANSWER REASONABLE? The equation tells us that 3 mol NaOH $\Leftrightarrow$ 1 mol Na_3PO_4, so the actual number of moles of Na_3PO_4 (0.0800 mol) should be one-third the actual number of moles of NaOH (0.240 mol), and it is. We can also check that the units cancel correctly.

Practice Exercise 26: In the reaction $2SO_2(g) + O_2(g) \longrightarrow 2SO_3(g)$ how many moles of O_2 are needed to produce 6.76 moles of SO_3? (Hint: Write the equivalence that relates O_2 to SO_3.)

Practice Exercise 27: How many moles of sulfuric acid, H_2SO_4, are needed to react with 0.366 mol of NaOH by the following reaction?

$$2NaOH(aq) + H_2SO_4(aq) \longrightarrow Na_2SO_4(aq) + 2H_2O$$

Mole-to-mole ratios link masses of different substances in chemical reactions

The most common stoichiometric calculation the chemist does is to relate grams of one substance with grams of another in a chemical reaction. For example, glucose ($C_6H_{12}O_6$) is one of the body's primary energy sources. The body combines glucose and oxygen to give carbon dioxide and water. The balanced equation for the overall reaction is

$$C_6H_{12}O_6(aq) + 6O_2(aq) \longrightarrow 6CO_2(aq) + 6H_2O(l)$$

How many grams of oxygen must the body take in to completely process 1.00 g of glucose? The problem can be expressed as

$$1.00 \text{ g } C_6H_{12}O_6 \Leftrightarrow ? \text{ g } O_2$$

The first thing we should notice about this problem is that we're relating *two different substances* in a reaction. The equivalence that relates the substances is the mole-to-mole relationship between glucose and O_2 given by the chemical equation. In this case, the equation tells us that

$$1 \text{ mol } C_6H_{12}O_6 \Leftrightarrow 6 \text{ mol } O_2$$

If we insert that mole-to-mole conversion between our starting point (1.00 g $C_6H_{12}O_6$) and the desired quantity (g O_2) we have cut the problem into three simple steps:

TOOLS *Sequence of steps for mass-to-mass calculations using balanced chemical equations*

$$\boxed{1 \text{ mol } C_6H_{12}O_6 \Leftrightarrow 6 \text{ mol } O_2}$$

$$1.00 \text{ g } C_6H_{12}O_6 \longrightarrow \text{mol } C_6H_{12}O_6 \longrightarrow \text{mol } O_2 \longrightarrow \text{g } O_2$$

We convert grams of $C_6H_{12}O_6$ to moles of $C_6H_{12}O_6$ (using the molecular mass of $C_6H_{12}O_6$). We then convert mol $C_6H_{12}O_6$ to mol O_2 using the equivalence relationship from the balanced equation. Finally, we convert mol O_2 into g O_2 using the molecular mass of O_2.

Figure 3.4 outlines this flow for *any* stoichiometry problem that relates reactant or product masses. If we know the *balanced equation* for a reaction and the *mass* of any reactant or product, we can calculate the required or expected mass of *any* other substance in the equation. Example 3.16 shows how it works.

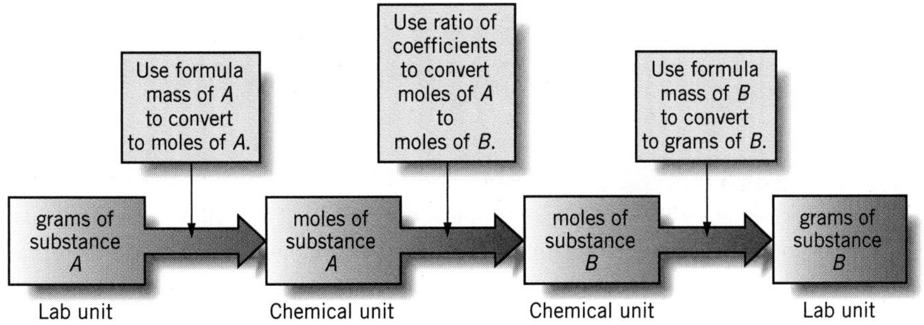

FIG. 3.4 The sequence of calculations for solving stoichiometry problems. This sequence applies to all calculations that start with the mass of one substance (*A*) and require the mass of a second substance (*B*) as the answer.

EXAMPLE 3.16
Stoichiometric Mass Calculations

Portland cement is a mixture of the oxides of calcium, aluminum, and silicon. The raw material for its calcium oxide is calcium carbonate, which occurs as the chief component of a natural rock, limestone. When calcium carbonate is strongly heated it decomposes. One product, carbon dioxide, is driven off to leave the desired calcium oxide as the only other product.

A chemistry student is to prepare 1.50×10^2 g of calcium oxide in order to test a particular "recipe" for Portland cement. How many grams of calcium carbonate should be used, assuming that all will be converted?

ANALYSIS: We begin by obtaining a balanced chemical reaction. From the nomenclature tools we find that calcium carbonate is $CaCO_3$. Using the tools from Chapter 2, the formulas for carbon dioxide and calcium oxide are CO_2 and CaO, respectively. Our balanced equation must be

$$CaCO_3(s) \xrightarrow{\text{heat}} CaO(s) + CO_2(g)$$

Special reaction conditions are often indicated with words or symbols above the arrow. In this reaction temperatures above 2000 °C are needed and this is indicated with the word, *heat*, above the arrow.

Now we can state the problem in mathematical form as

$$1.50 \times 10^2 \text{ g CaO} \Leftrightarrow ? \text{ g } CaCO_3$$

In problems that convert an amount of one substance to amount of a different substance in a chemical reaction, the tool we use is a mole-to-mole conversion factor. From the balanced chemical equation, we have

$$1 \text{ mol CaO} \Leftrightarrow 1 \text{ mol } CaCO_3$$

In our road map, this is the central conversion between CaO and $CaCO_3$ as shown below.

$$\boxed{1 \text{ mol CaO} \Leftrightarrow 1 \text{ mol } CaCO_3}$$

$$1.50 \times 10^2 \text{ g CaO} \longrightarrow \text{mol CaO} \longrightarrow \text{mol } CaCO_3 \longrightarrow \text{g } CaCO_3$$

For the complete calculation we must convert g CaO to mol CaO. *The equality tool relating mass and moles uses the molar mass:*

$$56.08 \text{ g CaO} = 1 \text{ mol CaO}$$

Next, the mole ratio tool converts moles of CaO to moles of $CaCO_3$. Finally, we must also convert mol $CaCO_3$ to g $CaCO_3$. Again, the tool defining the equality between mass and moles uses the molar mass:

$$100.09 \text{ g } CaCO_3 = 1 \text{ mol } CaCO_3$$

Putting this all together, our overall strategy will be as follows.

$$\boxed{1 \text{ mol CaO} \Leftrightarrow 1 \text{ mol } CaCO_3}$$

$$1.50 \times 10^2 \text{ g CaO} \longrightarrow \text{mol CaO} \longrightarrow \text{mol } CaCO_3 \longrightarrow \text{g } CaCO_3$$

$$\boxed{56.08 \text{ g CaO} = 1 \text{ mol CaO}} \qquad \boxed{1 \text{ mol } CaCO_3 = 100.09 \text{ g } CaCO_3}$$

SOLUTION: We assemble conversion factors so the units cancel correctly:

$$1.50 \times 10^2 \cancel{\text{g CaO}} \times \left(\frac{1 \cancel{\text{mol CaO}}}{56.08 \cancel{\text{g CaO}}}\right) \times \left(\frac{1 \cancel{\text{mol } CaCO_3}}{1 \cancel{\text{mol CaO}}}\right) \times \left(\frac{100.09 \text{ g } CaCO_3}{1 \cancel{\text{mol } CaCO_3}}\right)$$

$$= 268 \text{ g } CaCO_3$$

Notice how the calculation flows from grams of CaO to moles of CaO, then to moles of $CaCO_3$ (using the equation), and finally to grams of $CaCO_3$. We cannot emphasize too much that *the key step in all calculations of reaction stoichiometry is the use of the balanced equation.*

IS THE ANSWER REASONABLE? In a mass-to-mass calculation like this the first check is the magnitude of the answer compared to the starting mass. In the majority of reactions the calculated mass is not less than 1/5 of the starting mass nor is it larger than five times the starting mass. Our result is reasonable on this criterion. We can make a more detailed check by first making sure that the units cancel properly. We can also round all numbers to one or two significant figures and estimate the answer. We would estimate $\frac{150 \times 100}{50} = 300$ and this value is close to our answer of 268, giving us confidence in the calculation. Alternately we may round the denominator to 60 instead of 50 to get an equally good estimate of 250.

EXAMPLE 3.17
Stoichiometric Mass Calculations

The thermite reaction is one of the most spectacular reactions of aluminum with iron(III) oxide by which metallic iron and aluminum oxide are made. So much heat is generated that the iron forms in the liquid state (Figure 3.5).

A certain welding operation requires at least 86.0 g of iron each time a weld is made. What is the minimum mass, in grams, of iron(III) oxide that must be used for each weld? Also calculate how many grams of aluminum are needed.

FIG. 3.5 **The thermite reaction.** Pictured here is a device for making white-hot iron by the reaction of aluminum with iron oxide and letting the molten iron run down into a mold between the ends of two steel railroad rails. The rails are thereby welded together. *(Courtesy Orgo-Thermit.)*

ANALYSIS: First we need to write a balanced chemical equation. From the information given and our nomenclature tools (Chapter 2) we determine that Al and Fe_2O_3 are the reactants and that Fe and Al_2O_3 are the products. The equation is written and then balanced to obtain

$$2Al(s) + Fe_2O_3(s) \longrightarrow Al_2O_3(s) + 2Fe(l)$$

Now, we state the problem in mathematical form:

$$86.0 \text{ g Fe} \Leftrightarrow ? \text{ g } Fe_2O_3$$

Remember that all problems in reaction stoichiometry must be solved at the mole level because an equation's coefficients disclose *mole* ratios, not mass ratios. So we use our tool that relates mass to moles to convert the number of grams of Fe to moles. Then we can use the tool that gives us the mole-to-mole relationship indicated by the coefficients in the balanced equation,

$$1 \text{ mol } Fe_2O_3 \Leftrightarrow 2 \text{ mol Fe}$$

to see how many *moles* of Fe_2O_3 are needed. Finally we use our tools to convert this answer into grams of Fe_2O_3. The other calculations follow the same pattern.

SOLUTION: We'll set up the first calculation as a chain; the steps are summarized below the conversion factors.

$$86.0 \cancel{\text{g Fe}} \times \frac{1 \cancel{\text{mol Fe}}}{55.85 \cancel{\text{g Fe}}} \times \frac{1 \cancel{\text{mol } Fe_2O_3}}{2 \cancel{\text{mol Fe}}} \times \frac{159.70 \text{ g } Fe_2O_3}{1 \cancel{\text{mol } Fe_2O_3}} = 123 \text{ g } Fe_2O_3$$

$$\text{grams Fe} \rightarrow \text{moles Fe} \rightarrow \text{moles } Fe_2O_3 \rightarrow \text{grams } Fe_2O_3$$

A minimum of 123 g of Fe_2O_3 is required to make 86.0 g of Fe.

Next, we calculate the number of grams of Al needed, but we know that we must first find the number of *moles* of Al required. Only from this can the grams of Al be calculated. The relevant mole-to-mole relationship, again using the balanced equation, is

$$2 \text{ mol Al} \Leftrightarrow 2 \text{ mol Fe}$$

☐ We could simplify the mole ratio to 1 mol Al ⇔ 1 mol Fe. Leaving the 2-to-2 ratio maintains the relationship with the balanced equation coefficients.

Employing another chain calculation to find the mass of Al needed to make 86.0 g of Fe, we have (using 26.98 as the atomic mass of Al)

$$86.0 \cancel{\text{g Fe}} \times \frac{1 \cancel{\text{mol Fe}}}{55.85 \cancel{\text{g Fe}}} \times \frac{2 \cancel{\text{mol Al}}}{2 \cancel{\text{mol Fe}}} \times \frac{26.98 \text{ g Al}}{1 \cancel{\text{mol Al}}} = 41.5 \text{ g Al}$$

$$\text{grams Fe} \rightarrow \text{moles Fe} \rightarrow \text{moles Al} \rightarrow \text{grams Al}$$

ARE THE ANSWERS REASONABLE? The estimate that our answers in a mass-to-mass calculation should be within 1/5 to 5 times the initial mass is true for both Al and Fe_2O_3. Rounding the numbers to one or two significant figures and estimating the answer (after rechecking that the units cancel properly) results in

$$\frac{90 \times 160}{60 \times 2} = 120 \text{ g } Fe_2O_3 \quad \text{and} \quad \frac{90 \times 30}{60} = 45 \text{ g Al}$$

Both estimates are close to our calculated values and give us confidence that we calculated correctly.

Practice Exercise 28: Using the information in Example 3.17 calculate the mass of Al_2O_3 formed under the conditions specified. (Hint: Recall the law of conservation of mass.)

Practice Exercise 29: How many grams of carbon dioxide are also produced by the reaction described in Example 3.16?

3.5 THE REACTANT IN SHORTEST SUPPLY LIMITS THE AMOUNT OF PRODUCT THAT CAN FORM

We've seen that balanced chemical equations can tell us how to mix reactants together in just the right proportions to get a certain amount of product. For example, ethanol, C_2H_5OH, is prepared industrially as follows:

$$\underset{\text{ethylene}}{C_2H_4} + H_2O \longrightarrow \underset{\text{ethanol}}{C_2H_5OH}$$

The equation tells us that one mole of ethylene will react with one mole of water to give one mole of ethanol. We can also interpret the equation on a molecular level: Every molecule of ethylene that reacts requires one molecule of water to produce one molecule of ethanol:

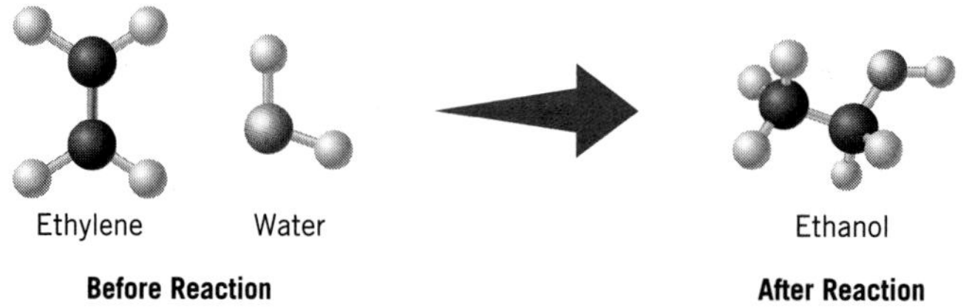

Notice that in both the "before" and "after" views of the reaction, the numbers of carbon, hydrogen, and oxygen atoms are the same.

If we have three molecules of ethylene reacting with three molecules of water, then three ethanol molecules are produced:

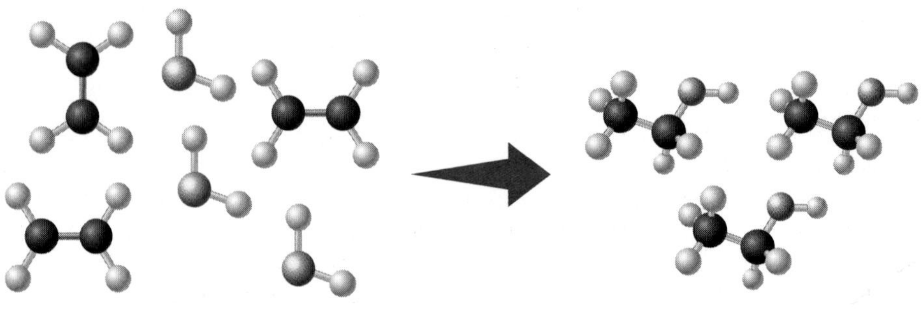

What happens if we mix 3 molecules of ethylene with 5 molecules of water? The ethylene will be completely used up before all the water is, and the product will contain two unreacted water molecules:

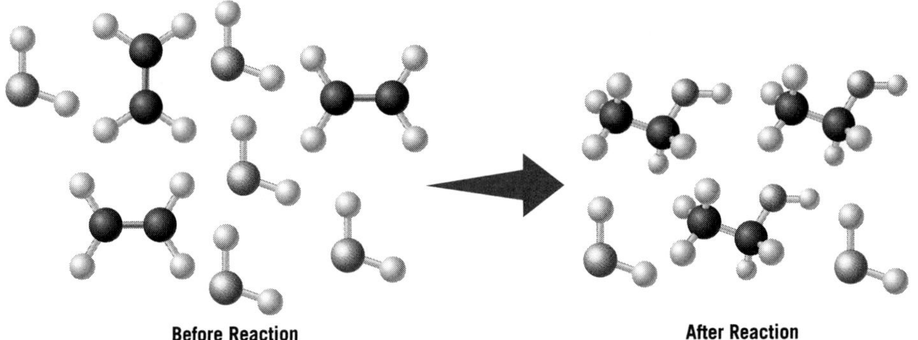

We don't have enough ethylene to use up all the water. The excess water remains after the reaction stops. This situation can be a problem in the manufacture of chemicals because not only do we waste one of our reactants (water, in this case), but we also obtain a product that is contaminated with unused reactant.

In this reaction mixture, ethylene is called the **limiting reactant** because it limits the amount of product (ethanol) that forms. The water is called an **excess reactant,** because we have more of it than is needed to completely consume all the ethylene.

To predict the amount of product we'll actually obtain in a reaction, we need to know which of the reactants is the limiting reactant. In the last example above, we saw that we needed only 3 H_2O molecules to react with 3 C_2H_4 molecules, but we had 5 H_2O molecules, so H_2O is present in excess and C_2H_4 is the limiting reactant. We could also have reasoned that 5 molecules of H_2O would require 5 molecules of C_2H_4, and since we have only 3 molecules of C_2H_4, it must be the limiting reactant.

Once we have identified the limiting reactant, it is possible to compute the amount of product that will actually form, and the amount of excess reactant that will be left over after the reaction stops. We must use the amount of the limiting reactant given in the problem for these calculations.

Example 3.18 shows how to solve a typical limiting reactant problem when the amounts of the reactants are given in mass units.

EXAMPLE 3.18

Limiting Reactant Calculation

Gold(III) hydroxide is used for electroplating gold onto other metals. It can be made by the following reaction.

$$2KAuCl_4(aq) + 3Na_2CO_3(aq) + 3H_2O \longrightarrow 2Au(OH)_3(aq) + 6\,NaCl(aq) + 2KCl(aq) + 3CO_2(g)$$

To prepare a fresh supply of $Au(OH)_3$, a chemist at an electroplating plant has mixed 20.00 g of $KAuCl_4$ with 25.00 g of Na_2CO_3 (both dissolved in a large excess of water). What is the maximum number of grams of $Au(OH)_3$ that can form?

ANALYSIS: The clue that tells us this is a limiting reactant question is that *the quantities of two reactants are given.* Once we determine which reactant limits the product, we use its mass in our calculation. We will need to use a combination of our stoichiometry tools to solve this problem.

To find the limiting reactant, we arbitrarily pick one of the reactants ($KAuCl_4$ or Na_2CO_3) and calculate whether it would all be used up. If so, we've found the limiting reactant. If not, the other reactant limits. (We were told that water is in excess, so we know that it does not limit the reaction.)

SOLUTION: We will show, in the two boxes below, the calculations needed to determine which is the limiting reactant. In solving a limiting reactant problem you will need to do only one of these calculations.

We start with $KAuCl_4$ as the reactant to work with and calculate how many grams of Na_2CO_3 *should* be provided to react with 20.00 g of $KAuCl_4$. The formula masses are 377.88 for $KAuCl_4$ and 105.99 for Na_2CO_3. We'll set up a chain calculation using the following mole- to-mole relationship.

$$2 \text{ mol KAuCl}_4 \Leftrightarrow 3 \text{ mol Na}_2\text{CO}_3$$

$$\text{grams KAuCl}_4 \rightarrow \text{moles KAuCl}_4 \rightarrow \text{moles Na}_2\text{CO}_3 \rightarrow \text{grams Na}_2\text{CO}_3$$

$$20.00 \cancel{\text{ g KAuCl}_4} \times \frac{1 \cancel{\text{ mol KAuCl}_4}}{377.88 \cancel{\text{ g KAuCl}_4}} \times \frac{3 \cancel{\text{ mol Na}_2\text{CO}_3}}{2 \cancel{\text{ mol KAuCl}_4}} \times \frac{105.99 \text{ g Na}_2\text{CO}_3}{1 \cancel{\text{ mol Na}_2\text{CO}_3}}$$
$$= 8.415 \text{ g Na}_2\text{CO}_3$$

We find that 20.00 g of $KAuCl_4$ needs 8.415 g of Na_2CO_3. The 25.00 g of Na_2CO_3 taken is therefore more than enough to let the $KAuCl_4$ react completely. The Na_2CO_3 is the excess reactant, so **$KAuCl_4$ is the limiting reactant.**

We start with Na_2CO_3 as the reactant to work with and calculate how many grams of $KAuCl_4$ *should* be provided to react with 25.00 g of Na_2CO_3. The formula masses are 377.88 for $KAuCl_4$ and 105.99 for Na_2CO_3. We'll set up a chain calculation using the following mole-to-mole relationship.

$$3 \text{ mol Na}_2\text{CO}_3 \Leftrightarrow 2 \text{ mol KAuCl}_4$$

$$\text{grams Na}_2\text{CO}_3 \rightarrow \text{moles Na}_2\text{CO}_3 \rightarrow \text{moles KAuCl}_4 \rightarrow \text{grams KAuCl}_4$$

$$25.00 \cancel{\text{ g Na}_2\text{CO}_3} \times \frac{1 \cancel{\text{ mol Na}_2\text{CO}_3}}{105.99 \cancel{\text{ g Na}_2\text{CO}_3}} \times \frac{2 \cancel{\text{ mol KAuCl}_4}}{3 \cancel{\text{ mol Na}_2\text{CO}_3}} \times \frac{377.88 \text{ g KAuCl}_4}{1 \cancel{\text{ mol KAuCl}_4}}$$
$$= 59.42 \text{ g KAuCl}_4$$

We find that 25.00 g Na_2CO_3 would require much more $KAuCl_4$ than provided, so we conclude that **$KAuCl_4$ is the limiting reactant.**

□ After determining which reactant is the limiting reactant, return to the statement of the problem and use the amount of the limiting reactant stated in the problem to perform further calculations.

The result from either calculation above is sufficient to designate $KAuCl_4$ as the limiting reactant. From here on, we have a routine calculation converting the mass of the limiting reactant, $KAuCl_4$, to mass of product, $Au(OH)_3$. We know from the equation's coefficients that

$$1 \text{ mol KAuCl}_4 \Leftrightarrow 1 \text{ mol Au(OH)}_3$$

Using this, and the conversion factors constructed from the formula masses, we set up the following chain calculation.

$$\text{grams KAuCl}_4 \rightarrow \text{moles KAuCl}_4 \rightarrow \text{moles Au(OH)}_3 \rightarrow \text{grams Au(OH)}_3$$

$$20.00 \cancel{\text{ g KAuCl}_4} \times \left(\frac{1 \cancel{\text{ mol KAuCl}_4}}{377.88 \cancel{\text{ g KAuCl}_4}}\right) \times \left(\frac{1 \cancel{\text{ mol Au(OH)}_3}}{1 \cancel{\text{ mol KAuCl}_4}}\right) \times \left(\frac{247.99 \text{ g Au(OH)}_3}{1 \cancel{\text{ mol Au(OH)}_3}}\right)$$
$$= 13.13 \text{ g Au(OH)}_3$$

Thus from 20.00 g of $KAuCl_4$ we can make a maximum of 13.13 g of $Au(OH)_3$.

In this synthesis, some of the initial 25.00 g of Na_2CO_3 is left over. Since one of our calculations showed that 20.00 g of $KAuCl_4$ requires only 8.415 g of Na_2CO_3 out of 25.00 g Na_2CO_3, the difference, (25.00 g − 8.415 g) = 16.58 g of Na_2CO_3, remains unreacted. It is possible that the chemist used an excess to ensure that every last bit of the very expensive $KAuCl_4$ would be changed to $Au(OH)_3$.

IS THE ANSWER REASONABLE? First, the resulting mass is within the range of 1/5 to 5 times the starting mass and is not unreasonable. Again, we check that our units cancel properly and then we estimate the answer as $\frac{20 \times 250}{400} = 12.5$ g $Au(OH)_3$. That estimate is close to our answer, which assures us the calculation was done correctly.

Practice Exercise 30: A Kipp generator is an old device for making carbon dioxide as needed. It consists of an enclosed flask that contains limestone, $CaCO_3$, and has a valve to add hydrochloric acid, $HCl(aq)$, as needed. The reaction between the limestone and hydrochloric acid produces carbon dioxide as shown in the reaction

$$CaCO_3(s) + 2HCl(aq) \longrightarrow CO_2(g) + CaCl_2(aq) + H_2O$$

How many grams of CO_2 can be made by reacting 125 g of $CaCO_3$ with 125 g of HCl? How many grams of which reactant are left over? (Hint: Find the limiting reactant.)

Practice Exercise 31: In an industrial process for making nitric acid, the first step is the reaction of ammonia with oxygen at high temperature in the presence of a platinum gauze. Nitrogen monoxide forms as follows.

$$4NH_3 + 5O_2 \longrightarrow 4NO + 6H_2O$$

How many grams of nitrogen monoxide can form if a mixture initially contains 30.00 g of NH_3 and 40.00 g of O_2?

3.6 The Predicted Amount of Product Is Not Always Obtained Experimentally

In most experiments designed for chemical synthesis, the amount of a product actually isolated falls short of the calculated maximum amount. Losses occur for several reasons. Some are mechanical, such as materials sticking to glassware. In some reactions, losses occur by the evaporation of a volatile product. In others, a product is a solid that separates from the solution as it forms because it is largely insoluble. The solid is removed by filtration. What stays in solution, although relatively small, contributes to some loss of product.

One of the common causes of obtaining less than the stoichiometric amount of a product is the occurrence of a **competing reaction** (or **side reaction**). It produces a **by-product,** a substance made by a reaction that competes with the **main reaction.** The synthesis of phosphorus trichloride, for example, gives some phosphorus pentachloride as well, because PCl_3 can react further with Cl_2.

$$\text{Main reaction:} \quad 2P(s) + 3Cl_2(g) \longrightarrow 2PCl_3(l)$$

$$\text{Competing reaction:} \quad PCl_3(l) + Cl_2(g) \longrightarrow PCl_5(s)$$

The competition is between newly formed PCl_3 and still unreacted phosphorus for still unchanged chlorine.

The **actual yield** of desired product is simply how much is isolated, stated in mass units or moles. The **theoretical yield** of the product is what must be obtained if no losses occur. When less than the theoretical yield of product is obtained, chemists generally calculate the *percentage yield* of product to describe how well the preparation went. The **percentage yield** is the actual yield calculated as a percentage of the theoretical yield.

TOOLS
Theoretical, actual, and percentage yields

$$\text{Percentage yield} = \frac{\text{actual yield}}{\text{theoretical yield}} \times 100\%$$

Both the actual and theoretical yields must be in the same units, of course.

It is important to realize that the actual yield is an experimentally determined quantity. It cannot be calculated. The theoretical yield is always a calculated quantity based on a chemical equation and the amounts of the reactants available.

Let's now work an example that combines the determination of the limiting reactant with a calculation of percentage yield.

EXAMPLE 3.19
Calculating a Percentage Yield

A chemist set up a synthesis of phosphorus trichloride by mixing 12.0 g of phosphorus with 35.0 g of chlorine gas and obtained 42.4 g of liquid phosphorus trichloride. Calculate the percentage yield of this compound.

ANALYSIS: We start by determining the formulas for the reactants and products and then balancing the chemical equation. Phosphorus is represented as $P(s)$, chlorine gas is $Cl_2(g)$, and the product is $PCl_3(l)$. The balanced equation is

$$2P(s) + 3Cl_2(g) \longrightarrow 2PCl_3(l)$$

Now we notice that the masses of *both* reactants are given, so this must be a limiting reactant problem. The first step is to figure out which reactant, P or Cl_2, is the limiting reactant, because we must base all calculations on the limiting reactant. Our basic tools to use are the mass-to-moles relationship and the mole ratio expressed by the balanced equation.

SOLUTION: In any limiting reactant problem, we can arbitrarily pick one reactant and do a calculation to see whether it can be entirely used up. We'll choose phosphorus and see whether there is enough to react with 35.0 g of chlorine. The following calculation gives us the answer.

$$12.0 \cancel{\text{g P}} \times \frac{1 \cancel{\text{mol P}}}{30.97 \cancel{\text{g P}}} \times \frac{3 \cancel{\text{mol Cl}_2}}{2 \cancel{\text{mol P}}} \times \frac{70.90 \text{ g Cl}_2}{1 \cancel{\text{mol Cl}_2}} = 41.2 \text{ g Cl}_2$$

Thus, with 35.0 g of Cl_2 provided but 41.2 g of Cl_2 needed, there is not enough Cl_2 to react with all 12.0 g of P. The Cl_2 will be all used up before the P is used up, so Cl_2 is the limiting reactant. We therefore base the calculation of the theoretical yield of PCl_3 on Cl_2. (Be careful to use the 35.0 g of Cl_2 given in the problem, *not* the 41.2 g calculated while we determined the limiting reactant.)

To find the *theoretical yield* of PCl_3, we calculate how many grams of PCl_3 could be made from 35.0 g of Cl_2 if everything went perfectly according to the equation given.

$$35.0 \cancel{\text{g Cl}_2} \times \frac{1 \cancel{\text{mol Cl}_2}}{70.90 \cancel{\text{g Cl}_2}} \times \frac{2 \cancel{\text{mol PCl}_3}}{3 \cancel{\text{mol Cl}_2}} \times \frac{137.32 \text{ g PCl}_3}{1 \cancel{\text{mol PCl}_3}} = 45.2 \text{ g PCl}_3$$

$$\text{grams Cl} \rightarrow \text{moles Cl}_2 \rightarrow \text{moles PCl}_3 \rightarrow \text{grams PCl}_3$$

The actual yield was 42.4 g of PCl_3, not 45.2 g, so the percentage yield is calculated as follows.

$$\text{Percentage yield} = \frac{42.4 \text{ g PCl}_3}{45.2 \text{ g PCl}_3} \times 100\% = 93.8\%$$

Thus 93.8% of the theoretical yield of PCl_3 was obtained.

IS THE ANSWER REASONABLE? The obvious check is that the calculated or theoretical yield can never be *less* than the actual yield. Second, our answer is within the range of 1/5 to 5 times the starting amount. Finally an estimated answer, after checking that all units cancel properly, is $\frac{35 \times 2 \times 140}{70 \times 3} \approx 50$ g PCl_3, which is close to the 45.2 g we calculated.

Practice Exercise 32: In the synthesis of aspirin we react salicylic acid with acetic anhydride. The balanced chemical equation is

$$\underset{\text{salicylic acid}}{2HOOCC_6H_4OH} + \underset{\text{acetic anhydride}}{C_4H_6O_3} \longrightarrow \underset{\text{acetyl salicylic acid}}{2HOOCC_6H_4O_2C_2H_3} + \underset{\text{water}}{H_2O}$$

If we mix together 28.2 grams of salicylic acid with 15.6 grams of acetic anhydride in this reaction we obtain 30.7 grams of aspirin. What are the theoretical and percentage yields of our experiment? (Hint: What is the limiting reactant?)

Practice Exercise 33: Ethanol, C_2H_5OH, can be converted to acetic acid (the acid in vinegar), $HC_2H_3O_2$, by the action of sodium dichromate in aqueous sulfuric acid according to the following equation.

$$3C_2H_5OH(aq) + 2Na_2Cr_2O_7(aq) + 8H_2SO_4(aq) \longrightarrow 3HC_2H_3O_2(aq) + 2Cr_2(SO_4)_3(aq) + 2Na_2SO_4(aq) + 11H_2O$$

In one experiment, 24.0 g of C_2H_5OH, 90.0 g of $Na_2Cr_2O_7$, and an excess of sulfuric acid were mixed, and 26.6 g of acetic acid ($HC_2H_3O_2$) was isolated. Calculate the theoretical and percentage yields of $HC_2H_3O_2$.

SUMMARY

Mole Concept and Formula Mass or Molecular Mass. In the SI definition, one mole of any substance is an amount with the same number, **Avogadro's number** (6.022×10^{23}), of atoms, molecules, or formula units as there are atoms in 12 g (exactly) of carbon-12. For monatomic elements, the **atomic mass** in grams is one mole of that element. The sum of the atomic masses of all of the atoms appearing in a chemical formula gives the **formula mass** or the **molecular mass.**

Molar Mass. This is a general term for the mass in grams numerically equal to the atomic mass, molecular mass, or formula mass. Like an atomic mass, a formula mass, or a molecular mass, the **molar mass** is a tool for grams-to-moles or moles-to-grams conversions.

Chemical Formulas. The actual composition of a molecule is given by its **molecular formula.** An **empirical formula** gives the ratio of atoms, but in the smallest whole numbers, and it is generally the *only* formula we write for ionic compounds. In the case of a molecular compound, the molecular mass is a small whole-number multiple of the empirical formula mass.

Empirical Formulas. An empirical formula may be experimentally determined if there is some way to determine the small whole-number ratio of the atoms in the substance. This calculation can be done if we know the mass or percentage of each element in the compound.

Formula Stoichiometry. A chemical formula is a tool for stoichiometric calculations, because its subscripts tell us the mole ratios in which the various elements are combined.

Balanced Equations and Reaction Stoichiometry. A balanced equation is a tool for reaction stoichiometry because its coefficients disclose the stoichiometric equivalencies. When balancing an equation, only the coefficients can be adjusted, never the subscripts. All problems of reaction stoichiometry must be solved by first converting to moles.

Yields of Products. A reactant taken in a quantity less than required by another reactant, as determined by the reaction's stoichiometry, is called the **limiting reactant.** The **theoretical yield** of a product can be no more than permitted by the limiting reactant. Sometimes **competing reactions** (side reactions) producing by-products reduce the **actual yield.** The ratio of the actual to the theoretical yields, expressed as a percentage, is the **percentage yield.**

Stoichiometric Calculations. These are generally problems where units are converted. The sequence of steps typically used in stoichiometric calculations is shown in Figure 3.6. Conversion factors in this sequence are found in the molar mass, Avogadro's number, the chemical formula, or the balanced chemical reaction.

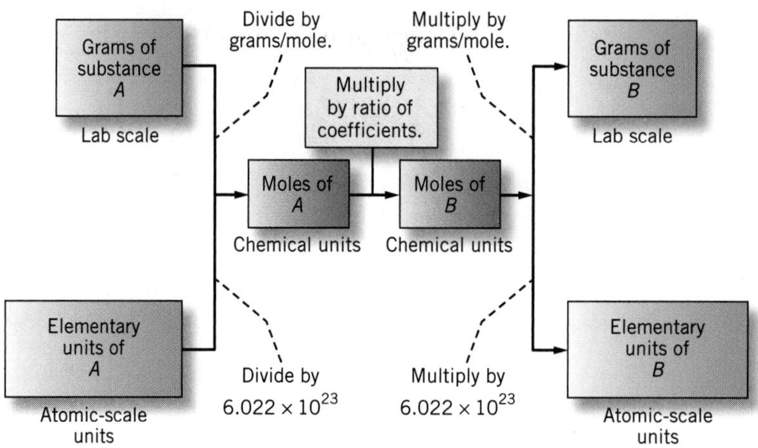

FIG. 3.6 Stoichiometry pathways. This summarizes all of the possible stoichiometric calculations encountered in this chapter. The boxes represent either grams, moles, or elementary units (atoms, molecules, formula units, or ions). Problems will give starting information representing one of the boxes. The question asked will tell you where to end. Perform conversions as noted in the instructions between each box.

TOOLS FOR PROBLEM SOLVING

In this chapter you learned to apply the following concepts as tools in solving problems. Study each one carefully so that you know what each is used for. When faced with solving a problem, recall what each tool does and consider whether it will be helpful in finding a solution. This will aid you in selecting the tools you need. In this chapter we see that many of the tools from Chapters 1 and 2 must be used with the new tools from this chapter.

Atomic mass *(page 88)* Atomic masses are used to form a conversion factor to calculate mass from moles of an element, or moles from the mass of an element.

$$\text{Gram atomic mass of } X = 1 \text{ mole } X$$

Formula mass, molecular mass *(page 88)* The formula mass or molecular mass is used to form a conversion factor to calculate mass from moles of a compound, or moles from the mass of a compound.

$$\text{Gram molecular mass of } X = 1 \text{ mole } X$$

$$\text{Gram formula mass of } X = 1 \text{ mole } X$$

Molar mass *(page 88)* This is a general term encompassing atomic, molecular, and formula masses. All are the sum of the masses of the elements in the chemical formula.

$$\text{Gram molar mass of } X = 1 \text{ mole of } X$$

Avogadro's number *(page 90)* This relates macroscopic lab-sized quantities (e.g., moles) to numbers of individual atomic-sized particles such as atoms, molecules, or ions.

$$1 \text{ mole } X = 6.022 \times 10^{23} \text{ particles of } X$$

Chemical formula, subscripts *(page 92)* Subscripts in a formula establish atom ratios and mole ratios between the elements in the substance.

Conversion sequence *(page 93)* This is the logical sequence of steps required for a mass-to-mass conversion problem using a chemical formula; also see Figure 3.6.

Percentage composition *(page 94 and 99)* The percentage composition is used to represent the composition of a compound and can be the basis for computing the empirical formula. Comparing experimental and theoretical percentage compositions can help establish the identity of a compound. Percentage composition also helps correlate information from different experiments.

$$\text{Percentage of } X = \frac{\text{mass of } X \text{ in the sample}}{\text{mass of the entire sample}} \times 100\%$$

Determination of an empirical formula *(page 96)* The simplest ratio of elements in a molecule or in the formula for an ionic compound can be calculated when the mass of each element of a compound is experimentally determined.

Methods for finding integer subscripts *(page 98)* Dividing all molar amounts by the smallest value often normalizes subscripts to integers. If decimals remain, multiplication by a small whole number can result in integer subscripts.

Guidelines for balancing chemical equations *(page 104)* Balancing equations means setting coefficients so that equal numbers of each atom will be reactants as well as products. A logical sequence for balancing equations by inspection is presented.

Equivalencies obtained from balanced equations *(page 106)* Balanced chemical equations give us relationships between all reactants and products that can be used in factor-label calculations.

Sequence of conversions using balanced equations *(page 107)* As with chemical formulas, a logical sequence of conversions allows calculation of amounts of all components of a chemical reaction. See Figure 3.6.

Limiting reactant calculations *(page 111)* When the amounts of at least two reactants are known, solution of problems requires identifying the limiting reactant. All calculations must be based only on the amount of the limiting reactant available.

Theoretical, actual, and percentage yields *(page 113)* The theoretical yield is calculated from the limiting reactant. The actual yield must be determined by experiment, and the percentage yield relates the magnitude of the actual yield to the percentage yield.

$$\text{Percentage yield} = \frac{\text{actual mass by experiment}}{\text{theoretical mass by calculation}} \times 100\%$$

QUESTIONS, PROBLEMS, AND EXERCISES

Answers to problems whose numbers are printed in color are given in Appendix B. More challenging problems are marked with asterisks. ILW = Interactive Learningware solution is available at www.wiley.com/college/brady. OH = an Office Hours video is available for this problem.

REVIEW QUESTIONS

Mole Concept

3.1 How would you estimate the number of atoms in a gram of iron, using the mass in grams of an atomic mass unit?

3.2 What is the definition of the mole?

3.3 Why are moles used when all stoichiometry problems could be done using only the mass in grams of an atomic mass unit?

3.4 Which contains more molecules: 2.5 mol of H_2O or 2.5 mol of H_2?

Chemical Formulas

3.5 How many moles of iron atoms are in one mole of Fe_2O_3? How many iron atoms are in one mole of Fe_2O_3?

3.6 Write all the mole-to-mole conversion factors that can be written based on the following chemical formulas.
(a) SO_2 (b) As_2O_3 (c) K_2SO_4 (d) Na_2HPO_4

3.7 Write all the mole-to-mole conversion factors that can be written based on the following chemical formulas.
(a) Mn_3O_4 (b) Sb_2S_5 (c) $(NH_4)_2SO_4$ (d) Hg_2Cl_2

3.8 What information is required to convert grams of a substance into moles of that same substance?

3.9 Why is the expression "1.0 mol of oxygen" ambiguous? Why doesn't a similar ambiguity exist in the expression "64 g of oxygen?"

3.10 The atomic mass of aluminum is 26.98. What specific conversion factors does this value make available for relating a mass of aluminum (in grams) and a quantity of aluminum given in moles?

Empirical Formulas

3.11 In general, what fundamental information, obtained from experimental measurements, is required to calculate the empirical formula of a compound?

3.12 Why is the changing of subscripts not allowed when balancing a chemical equation?

3.13 Under what circumstances can we change, or assign, subscripts in a chemical formula?

3.14 How many distinct empirical formulas are shown by the following models for compounds formed between elements *A* and *B*? Explain. (Element *A* is represented by a black sphere and element *B* by a light gray sphere.)

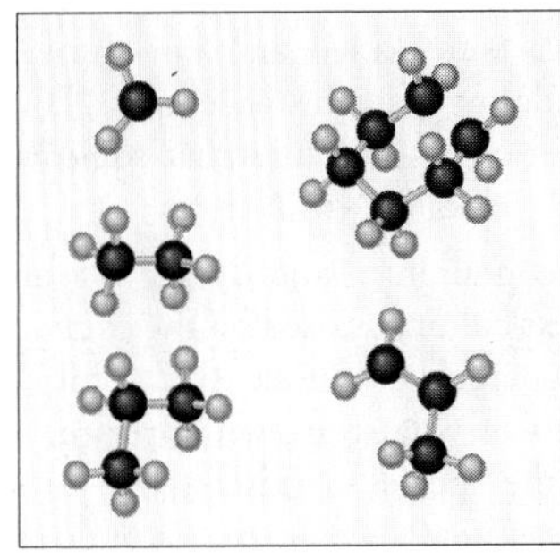

Avogadro's Number

3.15 How would Avogadro's number change if the atomic mass unit were to be redefined as 2×10^{-27} kg, exactly?

3.16 What information is required to convert grams of a substance into molecules of that same substance?

Stoichiometry with Balanced Equations

3.17 When given the *unbalanced* equation

$$Na(s) + Cl_2(g) \longrightarrow NaCl(s)$$

and asked to balance it, student *A* wrote

$$Na(s) + Cl_2(g) \longrightarrow NaCl_2(s)$$

and student *B* wrote

$$2Na(s) + Cl_2(g) \longrightarrow 2NaCl(s)$$

Both equations are balanced, but which student is correct? Explain why the other student's answer is incorrect.

3.18 Give a step-by-step procedure for estimating the number grams of *A* required to completely react with 10 moles of *B*, given the following information:

A and *B* react to form A_5B_2.
A has a molecular mass of 100.0.
B has a molecular mass of 200.0.
There are 6.022×10^{23} molecules of *A* in a mole of *A*.

Which of these pieces of information weren't needed?

3.19 If two substances react completely in a 1-to-1 ratio *both* by mass and by moles, what must be true about these substances?

3.20 What information is required to determine how many grams of sulfur would react with a gram of arsenic?

3.21 A mixture of 0.020 mol of Mg and 0.020 mol of Cl_2 reacted completely to form $MgCl_2$ according to the equation

$$Mg + Cl_2 \longrightarrow MgCl_2$$

What information describes the *stoichiometry* of this reaction? What information gives the *scale* of the reaction?

3.22 In a report to a supervisor, a chemist described an experiment in the following way: "0.0800 mol of H_2O_2 decomposed into 0.0800 mol of H_2O and 0.0400 mol of O_2." Express the chemistry and stoichiometry of this reaction by a conventional chemical equation.

3.23 On April 16, 1947, in Texas City, Texas, two cargo ships, the *Grandcamp* and the *High Flier*, were each loaded with approximately 2000 tons of ammonium nitrate fertilizer. The *Grandcamp* caught fire and exploded, followed by the *High Flier*. Over 600 people were killed and one-third of the city was destroyed. Considering a much smaller mass, how would you calculate the number of N_2 molecules that could be produced after the explosion of 1.00 kg of NH_4NO_3?

3.24 Molecules containing *A* and *B* react to form *AB* as shown at the top of the next column. Based on the equations and the contents of the boxes labeled "Initial," sketch for each reaction the molecular models of what is present after the reaction is over. (In both cases, the species *B* exists as B_2. In reaction 1, *A* is monatomic; in reaction 2, *A* exists as diatomic molecules A_2.)

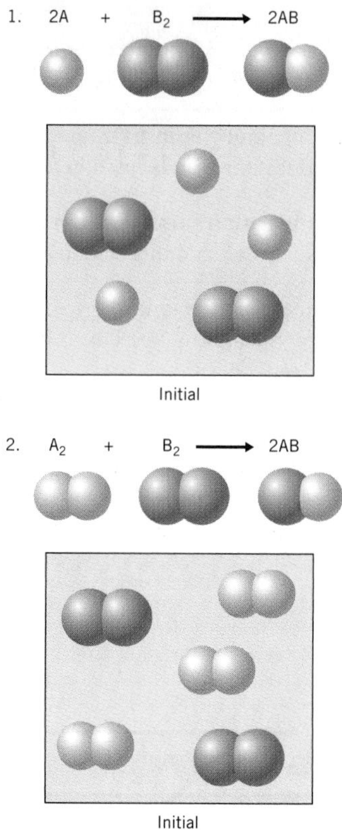

REVIEW PROBLEMS

The Mole Concept and Stoichiometric Equivalencies

3.25 In what smallest whole-number ratio must N and O atoms combine to make dinitrogen tetroxide, N_2O_4? What is the mole ratio of the elements in this compound?

3.26 In what atom ratio are the elements present in methane, CH_4 (the chief component of natural gas)? In what mole ratio are the atoms of the elements present in this compound?

3.27 How many moles of tantalum atoms correspond to 1.56×10^{21} atoms of tantalum?

3.28 How many moles of iodine molecules correspond to 1.80×10^{24} molecules of I_2?

3.29 Sucrose (table sugar) has the formula $C_{12}H_{22}O_{11}$. In this compound, what is the
(a) atom ratio of C to H?
(b) mole ratio of C to O?
(c) atom ratio of H to O?
(d) mole ratio of H to O?

3.30 Nail polish remover is sometimes the volatile liquid ethyl acetate, $CH_3COOC_2H_5$. In this compound, what is the
(a) atom ratio of C to O?
(b) mole ratio of C to O?
(c) atom ratio of C to H?
(d) mole ratio of C to H?

3.31 How many moles of Bi atoms are needed to combine with 1.58 mol of O atoms to make bismuth oxide, Bi_2O_3?

3.32 How many moles of vanadium atoms, V, are needed to combine with 0.565 mol of O atoms to make vanadium pentoxide, V_2O_5?

3.33 How many moles of Cr are in 2.16 mol of Cr_2O_3?

3.34 How many moles of O atoms are in 4.25 mol of calcium carbonate, $CaCO_3$, the chief constituent of seashells?

3.35 Aluminum sulfate, $Al_2(SO_4)_3$, is a compound used in sewage treatment plants.
(a) Construct a pair of conversion factors that relate moles of aluminum to moles of sulfur for this compound.
(b) Construct a pair of conversion factors that relate moles of sulfur to moles of $Al_2(SO_4)_3$.
(c) How many moles of Al are in a sample of this compound if the sample also contains 0.900 mol S?
(d) How many moles of S are in 1.16 mol $Al_2(SO_4)_3$?

3.36 Magnetite is a magnetic iron ore. Its formula is Fe_3O_4.
(a) Construct a pair of conversion factors that relate moles of Fe to moles of Fe_3O_4.
(b) Construct a pair of conversion factors that relate moles of Fe to moles of O in Fe_3O_4.
(c) How many moles of Fe are in 2.75 mol of Fe_3O_4?
(d) If this compound could be prepared from Fe_2O_3 and O_2, how many moles of Fe_2O_3 would be needed to prepare 4.50 mol Fe_3O_4?

3.37 How many moles of H_2 and N_2 can be formed by the decomposition of 0.145 mol of ammonia, NH_3?

3.38 How many moles of S are needed to combine with 0.225 mol Al to give Al_2S_3?

ILW 3.39 How many moles of UF_6 would have to be decomposed to provide enough fluorine to prepare 1.25 mol of CF_4? (Assume sufficient carbon is available.)

3.40 How many moles of Fe_3O_4 are required to supply enough iron to prepare 0.260 mol Fe_2O_3? (Assume sufficient oxygen is available.)

3.41 How many atoms of carbon are combined with 4.13 moles of hydrogen in a sample of the compound propane, C_3H_8? (Propane is used as the fuel in gas barbecues.)

3.42 How many atoms of hydrogen are found in 2.31 mol of propane, C_3H_8?

3.43 What is the total number of C, H, and O atoms in 0.260 moles of glucose, $C_6H_{12}O_6$?

3.44 What is the total number of N, H, and O atoms in 0.356 mol of ammonium nitrate, NH_4NO_3, an important fertilizer?

Measuring Moles of Elements and Compounds

3.45 How many atoms are in 6.00 g of carbon-12?

OH 3.46 How many atoms are in 1.50 mol of carbon-12? How many grams does this much carbon-12 weigh?

3.47 Determine the mass in grams of each of the following:
(a) 1.35 mol Fe (b) 24.5 mol O (c) 0.876 mol Ca

3.48 Determine the mass in grams of each of the following:
(a) 0.546 mol S (b) 3.29 mol N (c) 8.11 mol Al

3.49 What is the mass, in grams, of 2×10^{12} atoms of potassium?

3.50 What is the mass, in grams, of 4×10^{17} atoms of sodium?

3.51 How many moles of nickel are in 17.7 g of Ni?

3.52 How many moles of chromium are in 85.7 g of Cr?

3.53 Calculate the formula mass of each of the following to the maximum number of significant figures possible using the periodic table inside the front cover.
(a) $NaHCO_3$
(b) $(NH_4)_2CO_3$
(c) $CuSO_4 \cdot 5H_2O$
(d) potassium dichromate
(e) aluminum sulfate

3.54 Calculate the formula mass of each of the following to the maximum number of significant figures possible using the periodic table inside the front cover.
(a) calcium nitrate
(b) $Pb(C_2H_5)_4$
(c) $Na_2SO_4 \cdot 10H_2O$
(d) $Fe_4[Fe(CN)_6]_3$
(e) magnesium phosphate

3.55 Calculate the mass in grams of the following.
(a) 1.25 mol $Ca_3(PO_4)_2$
(b) 0.625 mmol iron(III) nitrate
(c) 0.600 μmol C_4H_{10}
(d) 1.45 mol ammonium carbonate

3.56 What is the mass in grams of the following?
(a) 0.754 mol zinc chloride
(b) 0.194 μmol potassium chlorate
(c) 0.322 mmol $POCl_3$
(d) 4.31×10^{-3} mol $(NH_4)_2HPO_4$

3.57 Calculate the number of moles of each compound in the following samples.
(a) 21.5 g calcium carbonate
(b) 1.56 ng NH_3
(c) 16.8 g strontium nitrate
(d) 6.98 μg Na_2CrO_4

3.58 Calculate the number of moles of each compound in the following samples.
(a) 9.36 g calcium hydroxide
(b) 38.2 kg lead(II)sulfate
(c) 4.29 g H_2O_2
(d) 4.65 mg $NaAuCl_4$

ILW 3.59 One sample of CaC_2 contains 0.150 mol of carbon. How many moles and how many grams of calcium are also in the sample? [Calcium carbide, CaC_2, was once used to make signal flares for ships. Water dripped onto CaC_2 reacts to give acetylene (C_2H_2), which burns brightly.]

OH 3.60 How many moles of iodine are in 0.500 mol of $Ca(IO_3)_2$? How many grams of calcium iodate are needed to supply this much iodine? [Iodized salt contains a trace amount of calcium iodate, $Ca(IO_3)_2$, to help prevent a thyroid condition called goiter.]

3.61 How many moles of nitrogen, N, are in 0.650 mol of ammonium carbonate? How many grams of this compound supply this much nitrogen?

3.62 How many moles of nitrogen, N, are in 0.556 mol of ammonium nitrate? How many grams of this compound supply this much nitrogen?

3.63 How many kilograms of a fertilizer made of pure $(NH_4)_2CO_3$ would be required to supply 1.00 kilogram of nitrogen to the soil?

3.64 How many kilograms of a fertilizer made of pure P_2O_5 would be required to supply 1.00 kilogram of phosphorus to the soil?

Percentage Composition

3.65 Calculate the percentage composition by mass for each of the following:
(a) sodium dihydrogenphosphate
(b) $NH_4H_2PO_4$
(c) $(CH_3)_2CO$
(d) calcium sulfate dihydrate
(e) $CaSO_4 \cdot 2H_2O$

3.66 Calculate the percentage composition by mass of each of the following:
(a) $(CH_3)_2N_2H_2$
(b) $CaCO_3$
(c) iron(III) nitrate
(d) C_3H_8
(e) aluminum sulfate

3.67 Which has a higher percentage of oxygen: morphine ($C_{17}H_{19}NO_3$) or heroin ($C_{21}H_{23}NO_5$)?

3.68 Which has a higher percentage of nitrogen: carbamazepine ($C_{15}H_{12}N_2O$) or carbetapentane ($C_{20}H_{31}NO_3$)?

3.69 Freon is a trade name for a group of gaseous compounds once used as propellants in aerosol cans. Which has a higher percentage of chlorine: Freon-12 (CCl_2F_2) or Freon-141b ($C_2H_3Cl_2F$)?

3.70 Which has a higher percentage of fluorine: Freon-12 (CCl_2F_2) or Freon 113 ($C_2Cl_3F_3$)?

OH **3.71** It was found that 2.35 g of a compound of phosphorus and chlorine contained 0.539 g of phosphorus. What are the percentages by mass of phosphorus and chlorine in this compound?

3.72 An analysis revealed that 5.67 g of a compound of nitrogen and oxygen contained 1.47 g of nitrogen. What are the percentages by mass of nitrogen and oxygen in this compound?

3.73 Phencyclidine ("angel dust") is $C_{17}H_{25}N$. A sample suspected of being this illicit drug was found to have a percentage composition of 84.71% C, 10.42% H, and 5.61% N. Do these data acceptably match the theoretical data for phencyclidine?

3.74 The hallucinogenic drug LSD has the molecular formula $C_{20}H_{25}N_3O$. One suspected sample contained 74.07% C, 7.95% H, and 9.99% N.
(a) What is the percentage O in the sample?
(b) Are these data consistent for LSD?

3.75 How many grams of O are combined with 7.14×10^{21} atoms of N in the compound dinitrogen pentoxide?

3.76 How many grams of C are combined with 4.25×10^{23} atoms of H in the compound C_5H_{12}?

Empirical Formulas

3.77 Write empirical formulas for the following compounds.
(a) S_2Cl_2 (b) $C_6H_{12}O_6$ (c) NH_3 (d) As_2O_6 (e) H_2O_2

3.78 What are the empirical formulas of the following compounds?
(a) $C_2H_4(OH)_2$
(b) $H_2S_2O_8$
(c) C_4H_{10}
(d) B_2H_6
(e) C_2H_5OH

3.79 Quantitative analysis of a sample of sodium pertechnetate with a mass of 0.896 g found 0.111 g of sodium and 0.477 g of technetium. The remainder was oxygen. Calculate the empirical formula of sodium pertechnetate. (Radioactive sodium pertechnetate is used as a brain-scanning agent in medicine.)

3.80 A sample of Freon was found to contain 0.423 g C, 2.50 g Cl, and 1.34 g F. What is the empirical formula of this compound?

3.81 A dry-cleaning fluid composed of only carbon and chlorine was found to be composed of 14.5% C and 85.5% Cl (by mass). What is the empirical formula of this compound?

3.82 One compound of mercury with a formula mass of 519 g mol^{-1} contains 77.26% Hg, 9.25% C, and 1.17% H (with the balance being O). Calculate the empirical and molecular formulas.

3.83 Cinnamic acid, a compound related to the flavor component of cinnamon, is 72.96% carbon, 5.40% hydrogen, and the rest is oxygen. What is the empirical formula of this acid?

3.84 Vanillin, a compound used as a flavoring agent in food products, has the following percentage composition: 63.2% C, 5.26% H, and 31.6% O. What is the empirical formula of vanillin?

ILW **3.85** When 0.684 g of an organic compound containing only carbon, hydrogen, and oxygen was burned in oxygen, 1.312 g CO_2 and 0.805 g H_2O were obtained. What is the empirical formula of the compound?

3.86 Methyl ethyl ketone (often abbreviated MEK) is a powerful solvent with many commercial uses. A sample of this compound (which contains only C, H, and O) weighing 0.822 g was burned in oxygen to give 2.01 g CO_2 and 0.827 g H_2O. What is the empirical formula for MEK?

3.87 When 6.853 mg of a sex hormone was burned in a combustion analysis, 19.73 mg of CO_2 and 6.391 mg of H_2O were obtained. What is the empirical formula of the compound?

3.88 When a sample of a compound in the vitamin D family was burned in a combustion analysis, 5.983 mg of the compound gave 18.490 mg of CO_2 and 6.232 mg of H_2O. What is the empirical formula of the compound?

Molecular Formulas

3.89 The following are empirical formulas and the masses per mole for three compounds. What are their molecular formulas?
(a) NaS_2O_3; 270.4 g/mol
(b) C_3H_2Cl; 147.0 g/mol
(c) C_2HCl; 181.4 g/mol

3.90 The following are empirical formulas and the masses per mole for three compounds. What are their molecular formulas?
(a) Na_2SiO_3; 732.6 g/mol
(b) $NaPO_3$; 305.9 g/mol
(c) CH_3O; 62.1 g/mol

3.91 The compound described in Problem 3.87 was found to have a molecular mass of 290. What is its molecular formula?

3.92 The compound described in Problem 3.88 was found to have a molecular mass of 399 g mol^{-1}. What is the molecular formula of this compound?

ILW **3.93** A sample of a compound of mercury and bromine with a mass of 0.389 g was found to contain 0.111 g bromine. Its molecular mass was found to be 561 g mol^{-1}. What are its empirical and molecular formulas?

3.94 A 0.6662 g sample of "antimonal saffron" was found to contain 0.4017 g of antimony. The remainder was sulfur. The formula mass of this compound is 404 g mol^{-1}. What are the empirical and molecular formulas of this pigment? (This compound is a red pigment used in painting.)

3.95 A sample of a compound of C, H, N, and O, with a mass of 0.6216 g was found to contain 0.1735 g C, 0.01455 g H, and 0.2024 g N. Its formula mass is 129 g mol^{-1}. Calculate its empirical and molecular formulas.

3.96 Strychnine, a deadly poison, has a formula mass of 334 g mol^{-1} and a percentage composition of 75.42% C, 6.63% H, 8.38% N, and the balance oxygen. Calculate the empirical and molecular formulas of strychnine.

Balancing Chemical Equations

3.97 How many moles of hydrogen are part of the expression "$2Ba(OH)_2 \cdot 8H_2O$," taken from a balanced equation?

3.98 How many moles of oxygen are part of the expression "$3Ca_3(PO_4)_2$," taken from a balanced equation?

OH **3.99** Write the equation that expresses in acceptable chemical shorthand the following statement: "Iron can be made to react with molecular oxygen to give iron(III) oxide."

3.100 The conversion of one air pollutant, nitrogen monoxide, produced in vehicle engines, into another, nitrogen dioxide, occurs when nitrogen monoxide reacts with molecular oxygen in the air. Write the balanced equation for this reaction.

3.101 Balance the following equations.
(a) Calcium hydroxide reacts with hydrogen chloride to form calcium chloride and water.
(b) Silver nitrate and calcium chloride react to form calcium nitrate and silver chloride.
(c) Lead nitrate reacts with sodium sulfate to form lead sulfate and sodium nitrate.
(d) Iron(III) oxide and carbon react to form iron and carbon dioxide.
(e) Butane reacts with oxygen to form carbon dioxide and water.

3.102 Balance the following equations.
(a) $SO_2 + O_2 \longrightarrow SO_3$
(b) $NaHCO_3 + H_2SO_4 \longrightarrow Na_2SO_4 + H_2O + CO_2$
(c) $P_4O_{10} + H_2O \longrightarrow H_3PO_4$
(d) $Fe_2O_3 + H_2 \longrightarrow Fe + H_2O$
(e) $Al + H_2SO_4 \longrightarrow Al_2(SO_4)_3 + H_2$

3.103 Balance the following equations.
(a) $Mg(OH)_2 + HBr \longrightarrow MgBr_2 + H_2O$
(b) $HCl + Ca(OH)_2 \longrightarrow CaCl_2 + H_2O$
(c) $Al_2O_3 + H_2SO_4 \longrightarrow Al_2(SO_4)_3 + H_2O$
(d) $KHCO_3 + H_3PO_4 \longrightarrow K_2HPO_4 + H_2O + CO_2$
(e) $C_9H_{20} + O_2 \longrightarrow CO_2 + H_2O$

3.104 Balance the following equations.
(a) $CaO + HNO_3 \longrightarrow Ca(NO_3)_2 + H_2O$
(b) $Na_2CO_3 + Mg(NO_3)_2 \longrightarrow MgCO_3 + NaNO_3$
(c) $(NH_4)_3PO_4 + NaOH \longrightarrow Na_3PO_4 + NH_3 + H_2O$
(d) $LiHCO_3 + H_2SO_4 \longrightarrow Li_2SO_4 + H_2O + CO_2$
(e) $C_4H_{10}O + O_2 \longrightarrow CO_2 + H_2O$

3.105 Chemical reactions can be used to change the charge of ions. Fe^{3+} is converted to Fe^{2+} when iron(III) chloride reacts with tin(II) chloride to make iron(II) chloride and tin(IV) chloride. Write and balance the equation that represents this reaction.

3.106 A precipitation reaction is one where soluble reactants form an insoluble product. A common precipitation reaction involves the reaction of the soluble salt aluminum chloride with soluble silver nitrate to form the insoluble silver chloride and soluble aluminum nitrate. Write and balance the equation for this reaction along with the appropriate states indicated in parentheses.

Stoichiometry Based on Chemical Equations

3.107 Chlorine is used by textile manufacturers to bleach cloth. Excess chlorine is destroyed by its reaction with sodium thiosulfate, $Na_2S_2O_3$, as follows.

$$Na_2S_2O_3(aq) + 4Cl_2(g) + 5H_2O \longrightarrow 2NaHSO_4(aq) + 8HCl(aq)$$

(a) How many moles of $Na_2S_2O_3$ are needed to react with 0.12 mol of Cl_2?
(b) How many moles of HCl can form from 0.12 mol of Cl_2?
(c) How many moles of H_2O are required for the reaction of 0.12 mol of Cl_2?
(d) How many moles of H_2O react if 0.24 mol HCl is formed?

3.108 The octane in gasoline burns according to the following equation.

$$2C_8H_{18} + 25O_2 \longrightarrow 16CO_2 + 18H_2O$$

(a) How many moles of O_2 are needed to react fully with 6 mol of octane?
(b) How many moles of CO_2 can form from 0.5 mol of octane?
(c) How many moles of water are produced by the combustion of 8 mol of octane?
(d) If this reaction is used to synthesize 6.00 mol of CO_2, how many moles of oxygen are needed? How many moles of octane?

3.109 The following reaction is used to extract gold from pretreated gold ore:

$$2Au(CN)_2^-(aq) + Zn(s) \longrightarrow 2Au(s) + Zn(CN)_4^{2-}(aq)$$

(a) How many grams of Zn are needed to react with 0.11 mol of $Au(CN)_2^-$?
(b) How many grams of Au can form from 0.11 mol of $Au(CN)_2^-$?
(c) How many grams of $Au(CN)_2^-$ are required for the reaction of 0.11 mol of Zn?

OH **3.110** Propane burns according to the following equation.

$$C_3H_8 + 5O_2 \longrightarrow 3CO_2 + 4H_2O$$

(a) How many grams of O_2 are needed to react fully with 3 mol of propane?
(b) How many grams of CO_2 can form from 0.1 mol of propane?
(c) How many grams of water are produced by the combustion of 4 mol of propane?

3.111 The incandescent white of a fireworks display is caused by the reaction of phosphorus with O_2 to give P_4O_{10}.
(a) Write the balanced chemical equation for the reaction.
(b) How many grams of O_2 are needed to combine with 6.85 g. of P?
(c) How many grams of P_4O_{10} can be made from 8.00 g of O_2?
(d) How many grams of P are needed to make 7.46 g of P_4O_{10}?

3.112 The combustion of butane, C_4H_{10}, produces carbon dioxide and water. When one sample of C_4H_{10} was burned, 4.46 g of water was formed.
(a) Write the balanced chemical equation for the reaction.
(b) How many grams of butane were burned?
(c) How many grams of O_2 were consumed?
(d) How many grams of CO_2 were formed?

ILW **3.113** In *dilute* nitric acid, HNO_3, copper metal dissolves according to the following equation.

$$3Cu(s) + 8HNO_3(aq) \longrightarrow 3Cu(NO_3)_2(aq) + 2NO(g) + 4H_2O$$

How many grams of HNO_3 are needed to dissolve 11.45 g of Cu according to this equation?

3.114 The reaction of hydrazine, N_2H_4, with hydrogen peroxide, H_2O_2, has been used in rocket engines. One way these compounds react is described by the equation

$$N_2H_4 + 7H_2O_2 \longrightarrow 2HNO_3 + 8H_2O$$

According to this equation, how many grams of H_2O_2 are needed to react completely with 852 g of N_2H_4?

3.115 Oxygen gas can be produced in the laboratory by decomposition of hydrogen peroxide (H_2O_2):

$$2H_2O_2(aq) \longrightarrow 2H_2O + O_2(g)$$

How many kilograms of O_2 can be produced from 1.0 kg of H_2O_2?

3.116 Oxygen gas can be produced in the laboratory by decomposition of potassium chlorate ($KClO_3$):

$$2KClO_3(s) \longrightarrow 2KCl(s) + 3O_2(g)$$

How many kilograms of O_2 can be produced from 1.0 kg of $KClO_3$?

Limiting Reactant Calculations

3.117 The reaction of powdered aluminum and iron(III) oxide,

$$2Al + Fe_2O_3 \longrightarrow Al_2O_3 + 2Fe$$

produces so much heat the iron that forms is molten. Because of this, railroads use the reaction to provide molten steel to weld steel rails together when laying track. Suppose that in one batch of reactants 4.20 mol of Al was mixed with 1.75 mol of Fe_2O_3.
(a) Which reactant, if either, was the limiting reactant?
(b) Calculate the number of grams of iron that can be formed from this mixture of reactants.

3.118 Ethanol (C_2H_5OH) is synthesized for industrial use by the following reaction, carried out at very high pressure:

$$C_2H_4(g) + H_2O(g) \longrightarrow C_2H_5OH(l)$$

What is the maximum amount of ethanol that can be produced when 1.0 kg of ethylene (C_2H_4) and 0.010 kg of steam are placed into the reaction vessel?

ILW **3.119** Silver nitrate, $AgNO_3$, reacts with iron(III) chloride, $FeCl_3$, to give silver chloride, AgCl, and iron(III) nitrate, $Fe(NO_3)_3$. A solution containing 18.0 g of $AgNO_3$ was mixed with a solution containing 32.4 g of $FeCl_3$. How many grams of which reactant remains after the reaction is over?

3.120 Chlorine dioxide, ClO_2, has been used as a disinfectant in air-conditioning systems. It reacts with water according to the equation

$$6ClO_2 + 3H_2O \longrightarrow 5HClO_3 + HCl$$

If 142.0 g of ClO_2 is mixed with 38.0 g of H_2O, how many grams of which reactant remain if the reaction is complete?

3.121 Some of the acid in acid rain is produced by the following reaction:

$$3NO_2(g) + H_2O(l) \longrightarrow 2HNO_3(aq) + NO(g)$$

If a falling raindrop weighing 0.050 g comes into contact with 1.0 mg of $NO_2(g)$, how much HNO_3 can be produced?

3.122 Phosphorus pentachloride reacts with water to give phosphoric acid and hydrogen chloride according to the following equation.

$$PCl_5 + 4H_2O \longrightarrow H_3PO_4 + 5HCl$$

In one experiment, 0.360 mol of PCl_5 was slowly added to 2.88 mol of water.
(a) Which reactant, if either, was the limiting reactant?
(b) How many grams of HCl were formed in the reaction?

Theoretical Yield and Percentage Yield

3.123 Barium sulfate, $BaSO_4$, is made by the following reaction.

$$Ba(NO_3)_2(aq) + Na_2SO_4(aq) \longrightarrow BaSO_4(s) + 2NaNO_3(aq)$$

An experiment was begun with 75.00 g of $Ba(NO_3)_2$ and an excess of Na_2SO_4. After collecting and drying the product, 64.45 g of $BaSO_4$ was obtained. Calculate the theoretical yield and percentage yield of $BaSO_4$.

3.124 The Solvay process for the manufacture of sodium carbonate begins by passing ammonia and carbon dioxide through a solution of sodium chloride to make sodium bicarbonate and ammonium chloride. The equation for the overall reaction is

$$H_2O + NaCl + NH_3 + CO_2 \longrightarrow NH_4Cl + NaHCO_3$$

In the next step, sodium bicarbonate is heated to give sodium carbonate and two gases, carbon dioxide and steam.

$$2NaHCO_3 \longrightarrow Na_2CO_3 + CO_2 + H_2O$$

What is the theoretical yield of sodium carbonate, expressed in grams, if 120 g NaCl was used in the first reaction? If 85.4 g of Na_2CO_3 was obtained, what was the percentage yield?

ILW **3.125** Aluminum sulfate can be made by the following reaction.

$$2AlCl_3(aq) + 3H_2SO_4(aq) \longrightarrow Al_2(SO_4)_3(aq) + 6HCl(aq)$$

It is quite soluble in water, so to isolate it the solution has to be evaporated to dryness. This drives off the volatile HCl, but the residual solid has to be heated to a little over 200 °C to drive off all of the water. In one experiment, 25.0 g of $AlCl_3$ was mixed with 30.0 g of H_2SO_4. Eventually, 28.46 g of pure $Al_2(SO_4)_3$ was isolated. Calculate the percentage yield.

3.126 The combustion of methyl alcohol in an abundant excess of oxygen follows the equation

$$2CH_3OH + 3O_2 \longrightarrow 2CO_2 + 4H_2O$$

When 6.40 g of CH_3OH was mixed with 10.2 g of O_2 and ignited, 6.12 g of CO_2 was obtained. What was the percentage yield of CO_2?

*3.127 The potassium salt of benzoic acid, potassium benzoate ($KC_7H_5O_2$), can be made by the action of potassium permanganate on toluene (C_7H_8) as follows.

$$C_7H_8 + 2KMnO_4 \longrightarrow KC_7H_5O_2 + 2MnO_2 + KOH + H_2O$$

If the yield of potassium benzoate cannot realistically be expected to be more than 71%, what is the minimum number of grams of toluene needed to produce 11.5 g of potassium benzoate?

*3.128 Manganese trifluoride, MnF_3, can be prepared by the following reaction.

OH

$$2MnI_2(s) + 13F_2(g) \longrightarrow 2MnF_3(s) + 4IF_5(l)$$

If the percentage yield of MnF_3 is always approximately 56%, how many grams of MnF_3 can be expected if 10.0 grams of each reactant is used in an experiment?

ADDITIONAL EXERCISES

3.129 Mercury is an environmental pollutant because it can be converted by certain bacteria into the very poisonous substance methyl mercury, $(CH_3)_2Hg$. This compound ends up in the food chain and accumulates in the tissues of aquatic organisms, particularly fish, which renders them unsafe to eat. It is estimated that in the United States 263 tons of mercury are released into the atmosphere each year. If only 1.0 percent of this mercury is changed to $(CH_3)_2Hg$, how many pounds of this compound are formed annually?

*3.130 Lead compounds are often highly colored and are toxic to mold, mildew, and bacteria, properties that in the past were useful for paints used before 1960. Today we know lead is very hazardous and it is not used in paint; however, old paint is still a problem. If a certain lead-based paint contains 14.5% $PbCr_2O_7$ and 73% of the paint evaporates as it dries, what mass of lead will be in a paint chip that weighs 0.15 g?

3.131 A superconductor is a substance that is able to conduct electricity without resistance, a property that is very desirable in the construction of large electromagnets. Metals have this property if cooled to temperatures a few degrees above absolute zero, but this requires the use of expensive liquid helium (boiling point 4 K). Scientists have discovered materials that become superconductors at higher temperatures, but they are ceramics. Their brittle nature has so far prevented them from being made into long wires. A recently discovered compound of magnesium and boron, which consists of 52.9% Mg and 47.1% B, shows special promise as a high-temperature superconductor because it is inexpensive to make and can be fabricated into wire relatively easily. What is the formula of the compound?

*3.132 A 0.1246 g sample of a compound of chromium and chlorine was dissolved in water. All of the chloride ion was then captured by silver ion in the form of AgCl. A mass of 0.3383 g of AgCl was obtained. Calculate the empirical formula of the compound of Cr and Cl.

*3.133 A compound of Ca, C, N, and S was subjected to quantitative analysis and formula mass determination, and the following data were obtained. A 0.250 g sample was mixed with Na_2CO_3 to convert all of the Ca to 0.160 g of $CaCO_3$. A 0.115 g sample of the compound was carried through a series of reactions until all of its S was changed to 0.344 g of $BaSO_4$. A 0.712 g sample was processed to liberate all of its N as NH_3, and 0.155 g NH_3 was obtained. The formula mass was found to be 156. Determine the empirical and molecular formulas of the compound.

3.134 Ammonium nitrate will detonate if ignited in the presence of certain impurities. The equation for this reaction at a high temperature is

$$2NH_4NO_3(s) \xrightarrow{>300\ °C} 2N_2(g) + O_2(g) + 4H_2O(g)$$

Notice that all of the products are gases and so must occupy a vastly greater volume than the solid reactant.

(a) How many moles of *all* gases are produced from 1 mol of NH_4NO_3?

(b) If 1.00 ton of NH_4NO_3 exploded according to this equation, how many moles of *all* gases would be produced? (1 ton = 2000 lb.)

3.135 A lawn fertilizer is rated as 6.00% nitrogen, meaning 6.00 g of N in 100 g of fertilizer. The nitrogen is present in the form of urea, $(NH_2)_2CO$. How many grams of urea are present in 100 g of the fertilizer to supply the rated amount of nitrogen?

*3.136 Nitrogen is the "active ingredient" in many quick acting fertilizers. You are operating a farm of 1500 acres to produce soybeans. Which of the following fertilizers will you choose as the most economical for your farm? (a) NH_4NO_3 at \$625 for 25 kg, (b) $(NH_4)_2HPO_4$ at \$55 for 1 kg, (c) urea, CH_4ON_2, at \$60 for 5 kg, (d) ammonia, NH_3, at \$128 for 50 kg

3.137 Based solely on the amount of available carbon, how many grams of sodium oxalate, $Na_2C_2O_4$, could be obtained from 125 g of C_6H_6? (Assume that no loss of carbon occurs in any of the reactions needed to produce the $Na_2C_2O_4$.)

3.138 According to NASA, the space shuttle's external fuel tank for the main propulsion system carries 1,361,936 lb of liquid oxygen and 227,641 lb of liquid hydrogen. During takeoff, these chemicals are consumed as they react to form water. If the reaction is continued until all of one reactant is gone, how many pounds of which reactant are left over?

*3.139 For a research project, a student decided to test the effect of the lead(II) ion (Pb^{2+}) on the ability of salmon eggs to hatch. This ion was obtainable from the water-soluble salt, lead(II) nitrate, $Pb(NO_3)_2$, which the student decided to make by the following reaction. (The desired product was to be isolated by the slow evaporation of the water.)

$$PbO(s) + 2HNO_3(aq) \longrightarrow Pb(NO_3)_2(aq) + H_2O$$

Losses of product for various reasons were anticipated, and a yield of 86.0% was expected. In order to have 5.00 g of product at this yield, how many grams of PbO should be taken? (Assume that sufficient nitric acid, HNO_3, would be used.)

3.140 Chlorine atoms cause chain reactions in the stratosphere that destroy ozone that protects the earth's surface from ultraviolet radiation. The chlorine atoms come from chlorofluorocarbons, compounds that contain carbon, fluorine, and chlorine, which were used for many years as refrigerants. One of these compounds is Freon-12, CF_2Cl_2. If a sample contains 1.0×10^{-9} g of Cl, how many grams of F should be present if all of the F and Cl atoms in the sample came from CF_2Cl_2 molecules?

*3.141 Lime, CaO, can be produced in two steps shown in the equations below. If the percentage yield of the first step is 83.5% and the percentage yield of the second step is 71.4%, what is the expected overall percentage yield for producing CaO from $CaCl_2$?

$$CaCl_2(aq) + CO_2(g) + H_2O \longrightarrow CaCO_3(s) + 2HCl(aq)$$

$$CaCO_3(s) \xrightarrow{\text{heat}} CaO + H_2O$$

EXERCISES IN CRITICAL THINKING

3.142 A newspaper story describing the local celebration of Mole Day on October 23 (selected for Avogadro's number, 6.022×10^{23}) attempted to give the readers a sense of the size of the number by stating that a mole of M&Ms would be equal to 18 tractor trailers full. Assuming that an M&M occupies a volume of about 0.5 cm^3, calculate the dimensions of a cube required to hold one mole of M&Ms. Would 18 tractor trailers be sufficient?

3.143 Suppose you had one mole of pennies and that you were going to spend 500 million dollars each and every second until you spent your entire fortune. How many years would it take you to spend all this cash? (Assume 1 year = 365 days.)

3.144 Using the above two exercises as examples, devise a creative way to demonstrate the size of the mole, or Avogadro's number.

3.145 List the different ways in which a chemist could use the information used to determine empirical formulas.

3.146 Calculate the percentage carbon in $C_{20}H_{42}$ and $C_{21}H_{44}$. If you were to burn one gram of each compound and quantitatively collect the carbon dioxide and water produced, would you be able to discern the difference between these two compounds with the equipment in your laboratory? In your evaluation consider the difference in masses expected from each sample and the number of decimal places your balance must read to. Also, list, in order of importance, all factors that can create error in this experiment.

CHAPTERS 1-3 BRINGING IT TOGETHER

Many of the fundamental concepts and problem-solving skills that were developed in the preceding chapters will carry forward into the rest of this book. Therefore, we recommend that you pause here to see how well you have grasped the concepts, how familiar you are with important terms, and how able you are at working chemical problems. Don't be discouraged if some of the problems seem to be difficult at first. Where necessary, take some time to review the concepts and problem-solving tools required.

Some of the problems here require data or other information found in tables in this book, including those inside the covers. Freely use these tables as needed. For problems that require mathematical solutions, we recommend that you first assemble the necessary information in the form of equivalencies and then use them to set up appropriate conversion factors needed to obtain the answers.

1. A rectangular box was found to be 24.6 cm wide, 0.35140 m high, and 7,424 mm deep.
  (a) How many significant figures are in each measurement?
  (b) Calculate the volume of the box in units of cm^3. Be sure to express your answer to the correct number of significant figures.
  (c) Use the answer in part (b) to calculate the volume of the box in cubic feet.
  (d) Suppose the box was solid and composed entirely of zinc, which has a density of 7.140 g/cm^3. What would be the mass of the box in kilograms?

2. What is the difference between an atom and a molecule? What is the difference between a molecule and a mole?

3. If a 10 g sample of element *X* contains twice as many atoms as a 10 g sample of element *Y*, how does the atomic mass of *X* compare with the atomic mass of *Y*?

4. How did Dalton's atomic theory account for the law of conservation of mass? How did it explain the law of definite proportions?

5. If atom *A* has the same number of neutrons as atom *B*, must *A* and *B* be atoms of the same element? Explain.

6. Construct a conversion factor that would enable you to convert a volume of 3.14 ft^3 into cubic centimeters (cm^3).

7. The atoms of an isotope of plutonium, Pu, each contain 94 protons, 150 neutrons, and 94 electrons. Write a symbol for this element that incorporates its mass number and atomic number. Write the symbol for a different isotope of plutonium.

8. An atom of an isotope of nickel has a mass number of 60. How many protons, neutrons, and electrons are in this atom?

9. A solution was found to contain particles consisting of 12 neutrons, 10 electrons, and 11 protons. Write the chemical symbol for this particle, consulting the periodic table as needed.

10. For each of the following, indicate whether it is possible to see the item specified with the naked eye. If not, explain.
  (a) A molar mass of iron
  (b) An atom of iron
  (c) A molecule of water
  (d) A mole of water
  (e) An ion of sodium
  (f) A formula unit of sodium chloride

11. Make a sketch of the general shape of the modern periodic table and mark off those areas where we find the metals, metalloids, and nonmetals.

12. Which of the following elements would most likely be found together in nature: Ca, Hf, Sn, Cu, Zr?

13. Match an element on the left with a description on the right.

Calcium	Halogen
Iron	Noble gas
Helium	Alkali metal
Gadolinium	Alkaline earth metal
Iodine	Transition metal
Sodium	Inner transition metal

14. Define *ductile* and *malleable.*

15. Which metal is a liquid at room temperature? Which metal has the highest melting point?

16. What is the most important property that distinguishes a metalloid from a metal or a nonmetal?

17. Give the symbols of the post-transition metals.

18. Give chemical formulas for the following.
  (a) potassium nitrate
  (b) calcium carbonate
  (c) cobalt(II) phosphate
  (d) magnesium sulfite
  (e) iron(III) bromide
  (f) magnesium nitride
  (g) aluminum selenide
  (h) copper(II) perchlorate
  (i) bromine pentafluoride
  (j) dinitrogen pentaoxide
  (k) strontium acetate
  (l) ammonium dichromate
  (m) copper(I) sulfide

19. Give chemical names for the following.

(a) $NaClO_3$	(g) K_2CrO_4
(b) $Ca_3(PO_4)_2$	(h) $Ca(CN)_2$
(c) $NaMnO_4$	(i) $MnCl_2$
(d) AlP	(j) $NaNO_2$
(e) ICl_3	(k) $Fe(NO_3)_2$
(f) PCl_3	

20. Why do we always write empirical formulas for ionic compounds?

21. Which of the following are binary substances: Al_2O_3, Cl_2, MgO, NO_2, $NaClO_4$?

22. A sample of a compound with a mass of 204 g consists of 1.00×10^{23} molecules. What is its molar mass?

23. Calculate the mass in grams of one formula unit of $K_4Fe(CN)_6$.

24. How many grams of copper(II) nitrate trihydrate, $Cu(NO_3)_2 \cdot 3H_2O$, are present in 0.118 mol of this compound?

25. A sample of 0.5866 g of nicotine was analyzed and found to consist of 0.4343 g C, 0.05103 g H, and 0.1013 g N. Calculate the percentage composition of nicotine.

26. A compound of potassium had the following percentage composition: K, 37.56%; H, 1.940%; P, 29.79%. The rest was oxygen. Calculate the empirical formula of this compound (arranging the atomic symbols in the order K H P O).

27. How many molecules of ethyl alcohol, C_2H_5OH, are in 1.00 fluid ounce of the liquid? The density of ethyl alcohol is 0.798 g/mL (1 oz = 29.6 mL).

28. What volume in liters is occupied by a sample of ethylene glycol, $C_2H_6O_2$, that consists of 5.00×10^{24} molecules. The density of ethylene glycol is 1.11 g/mL.

29. If 2.56 g of chlorine, Cl_2, will be used to prepare dichlorine heptoxide, how many moles and how many grams of molecular oxygen are needed?

30. Balance the following equations.

(a) $Fe_2O_3 + HNO_3 \longrightarrow Fe(NO_3)_3 + H_2O$

(b) $C_{21}H_{30}O_2 + O_2 \longrightarrow CO_2 + H_2O$

31. How many moles of nitric acid, HNO_3, are needed to react with 2.56 mol of Cu in the following reaction?

$$3Cu + 8HNO_3 \longrightarrow 3Cu(NO_3)_2 + 2NO + 4H_2O$$

32. Under the right conditions, ammonia can be converted to nitrogen monoxide, NO, by the following reaction.

$$4NH_3 + 5O_2 \longrightarrow 4NO + 6H_2O$$

How many moles and how many grams of O_2 are needed to react with 56.8 g of ammonia by this reaction?

33. Dolomite is a mineral consisting of calcium carbonate and magnesium carbonate. When dolomite is strongly heated, its carbonates decompose to their oxides (CaO and MgO) and carbon dioxide is expelled.

(a) Write separate equations for the decompositions of calcium carbonate and magnesium carbonate.

(b) When a dolomite sample with a mass of 5.78 g was heated strongly, the residue had a mass of 3.02 g. Calculate the masses in grams and the percentages of calcium carbonate and magnesium carbonate in this sample of dolomite.

34. Adipic acid, $C_6H_{10}O_4$, is a raw material for making nylon, and it can be prepared in the laboratory by the following reaction between cyclohexene, C_6H_{10}, and sodium dichromate, $Na_2Cr_2O_7$, in sulfuric acid, H_2SO_4.

$$3C_6H_{10}(l) + 4Na_2Cr_2O_7(aq) + 16H_2SO_4\,(aq) \longrightarrow$$
$$3C_6H_{10}O_4(s) + 4Cr_2(SO_4)_3\,(aq) + 4Na_2SO_4(aq) + 16H_2O$$

There are side reactions. These plus losses of product during its purification reduce the overall yield. A typical yield of purified adipic acid is 68.6%.

(a) To prepare 12.5 g of adipic acid in 68.6% yield requires how many grams of cyclohexene?

(b) The only available supply of sodium dichromate is its dihydrate, $Na_2Cr_2O_7 \cdot 2H_2O$. (Since the reaction occurs in an aqueous medium, the water in the dihydrate causes no problems, but it does contribute to the mass of what is taken of this reactant.) How many grams of this dihydrate are also required in the preparation of 12.5 g of adipic acid in a yield of 68.6%?

35. One of the ores of iron is hematite, Fe_2O_3, mixed with other rock. One sample of this ore is 31.4% hematite. How many tons of this ore are needed to make 1.00 ton of iron if the percentage recovery of iron from the ore is 91.5% (1 ton = 2000 lb)?

36. Gold occurs in the ocean in a range of concentration of 0.1 to 2 mg of gold per ton of seawater. Near one coastal city the gold concentration of the ocean is 1.5 mg/ton.

(a) How many tons of seawater have to be processed to obtain 1.0 troy ounce of gold if the recovery is 65% successful? (The troy ounce, 31.1 g, is the standard "ounce" in the gold trade.)

(b) If gold can be sold for $625.10 per troy ounce, what is the breakeven point in the dollar-cost per ton of processed seawater for extracting gold from the ocean at this location?

37. *C.I. Pigment Yellow 45* ("sideran yellow") is a pigment used in ceramics, glass, and enamel. When analyzed, a 2.164 g sample of this substance was found to contain 0.5259 g of Fe and 0.7345 g of Cr. The remainder was oxygen. Calculate the empirical formula of this pigment. What additional data are needed to calculate the molecular mass of this compound?

38. When 6.584 g of one of the hydrates of sodium sulfate was heated so as to drive off all of its water of hydration, the residue of anhydrous sodium sulfate had a mass of 2.889 g. What is the formula of the hydrate?

39. In an earlier problem we described the reaction of ammonia with oxygen to form nitrogen monoxide, NO.

$$4NH_3 + 5O_2 \longrightarrow 4NO + 6H_2O$$

How many moles and how many grams of NO could be formed from a mixture of 45.0 g of NH_3 and 58.0 g of O_2? How many grams of which reactant would remain unreacted?

40. A sample of 14.0 cm^3 of aluminum, in powdered form, was mixed with an excess of iron(III) oxide. A reaction between them was initiated that formed aluminum oxide and metallic iron. How many cubic centimeters of metallic iron were formed?

4 REACTIONS OF IONS AND MOLECULES IN AQUEOUS SOLUTIONS

The ability of molecules and ions to come into intimate contact when a substance is dissolved in a liquid forms the basis for the chemical reaction that takes place when Alka-Seltzer tablets are dropped into a glass of water. In this chapter we explore a variety of types of chemical reactions that occur in aqueous solutions. *(Gusto Productions/ Photo Researchers, Inc.)*

CHAPTER OUTLINE

THIS CHAPTER IN CONTEXT Ionic compounds are common, and many of them are soluble in water where they break apart into individual ions. Examples of such solutions include seawater and the fluids that surround cells in our bodies. In this chapter our goal is to teach you what happens when ionic substances dissolve in water, the nature of the chemical reactions they undergo, and the products that form.

We will also introduce you to another important class of compounds called acids and bases. These are also common substances that include many household products as well as compounds found in all living creatures. In this chapter we will examine the kinds of substances that are acids and bases and their reactions in aqueous solutions.

In the laboratory, liquid solutions in general (and aqueous solutions in particular) serve as a medium for many chemical reactions. This is because, for a reaction to occur, the particles of the reactants must make physical contact. The particles need freedom of motion, which is made possible when all of the reactants are in one fluid phase. When possible, therefore, solutions of reactants are combined to give a fluid reaction mixture in which chemical changes can occur swiftly and smoothly. To deal quantitatively with such reactions, we will extend the principles of stoichiometry you learned in Chapter 3 to deal with chemical reactions in solution.

4.1 SPECIAL TERMINOLOGY APPLIES TO SOLUTIONS

When water is a component of a solution, it is usually considered to be the solvent even when it is present in small amounts.

Before we get to the meat of our subject, we first must define some terms. A **solution** is a homogeneous mixture in which the molecules or ions of the components freely intermingle. When a solution forms, at least two substances are involved. One is the *solvent* and all of the others are *solutes*. The **solvent** is the medium into which the solutes are mixed or dissolved. In this chapter we deal with *aqueous solutions*, so the solvent will be liquid water.[1] A **solute** is any substance dissolved in the solvent. It might be a gas, like the carbon dioxide dissolved in carbonated beverages. Some solutes are liquids, like ethylene glycol dissolved in water to protect a vehicle's radiator against freezing. Solids, of course, can be solutes, like the sugar dissolved in lemonade or the salt dissolved in seawater.

To describe the composition of a solution, we often specify a **concentration,** which is the *ratio* of the amount of solute either to the amount of solvent or to the amount of solution. A **percentage concentration,** for example, is the number of grams of solute per 100 g of *solution*, a "solute-to-solution" ratio. Thus, the concentration of salt in seawater is often given as 3 g salt/100 g seawater.

The *relative* amounts of solute and solvent are often loosely given without specifying actual quantities. In a **dilute solution** the ratio of solute to solvent is small, for example, a few crystals of salt in a glass of water. In a **concentrated solution,** the ratio of solute to solvent is large. Syrup, for example, is a very concentrated solution of sugar in water.

Concentrated and dilute are relative terms. For example, a solution of 100 g of sugar in 100 mL of water is concentrated compared to one with just 10 g of sugar in 100 mL of water, but the latter solution is more concentrated than one that has 1 g of sugar in 100 mL of water.

Usually there is a limit to the amount of a solute that can dissolve in a given amount of solvent at a given temperature. When this limit is reached, we have a **saturated solution** and any excess solute that's added simply sits at the bottom of the solution. The **solubility** of a solute is the amount required to give a saturated solution, usually expressed as grams dissolved in 100 g of *solvent* at a given temperature. The temperature must be specified because solubilities vary with temperature. A solution having less solute than required for saturation is called an **unsaturated solution.** It is able to dissolve more solute.

In most cases, the solubility of a solute increases with temperature, so more solute can be dissolved by heating a saturated solution in the presence of excess solute. If the temperature of such a warm, saturated solution is subsequently lowered, the additional solute should separate from the solution, and indeed, this tends to happen spontaneously.

[1] Liquid water is a typical and very common solvent, but the solvent can actually be in any physical state, solid, liquid, or gas.

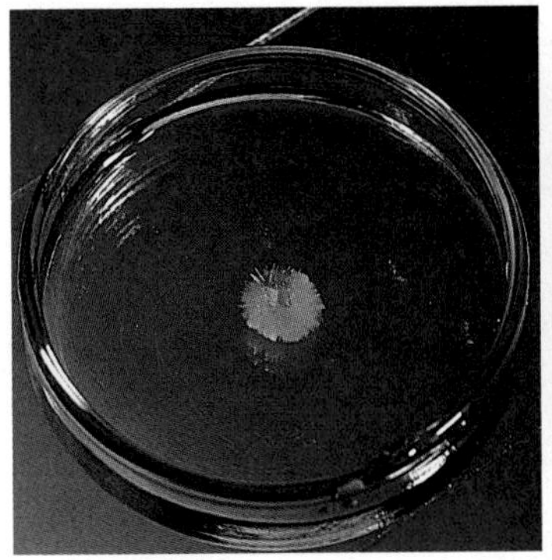

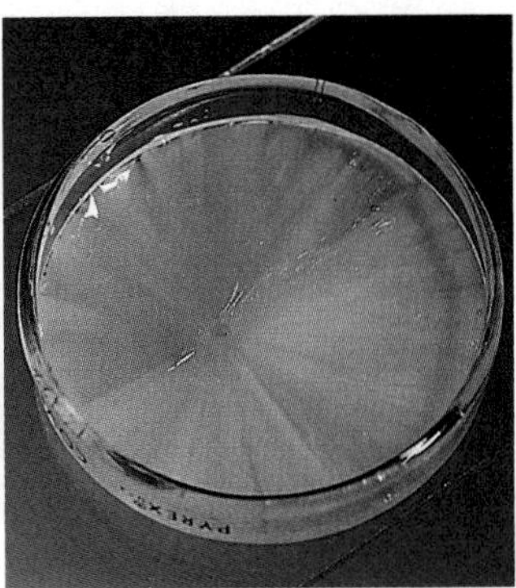

FIG. 4.1 **Crystallization.** When a small seed crystal of sodium acetate is added to a supersaturated solution of the compound, excess solute crystallizes rapidly until the solution is just saturated. The crystallization shown in this sequence took less than 10 seconds! *(Andy Washnik.)*

However, sometimes the solute doesn't separate, leaving us with a **supersaturated solution,** a solution that actually contains more solute than required for saturation. Supersaturated solutions are unstable and can only be prepared if there are no traces of undissolved solute. If even a tiny crystal of the solute is present or is added, the extra solute crystallizes (Figure 4.1). A solid that forms in a solution is called a **precipitate,** and a chemical reaction that produces a precipitate is called a **precipitation reaction.**

4.2 IONIC COMPOUNDS CONDUCT ELECTRICITY WHEN DISSOLVED IN WATER

Water itself is a very poor electrical conductor because it consists of electrically neutral molecules that are unable to transport electrical charges. However, as we noted in Chapter 2, when an ionic compound dissolves in water the resulting solution conducts electricity well. This is illustrated in Figure 4.2*a* for a solution of copper sulfate, $CuSO_4$.

Solutes such as $CuSO_4$, which yield electrically conducting aqueous solutions, are called **electrolytes.** Their ability to conduct electricity suggests the presence of electrically charged particles that are able to move through the solution. The generally accepted reason is that when an ionic compound dissolves in water, the ions separate from each other and enter the solution as more or less independent particles that are surrounded by molecules of the solvent. This change is called the **dissociation** of the ionic compound, and is

Solutions of electrolytes conduct electricity in a way that's different from metals. This is discussed more completely in Chapter 19.

(a)

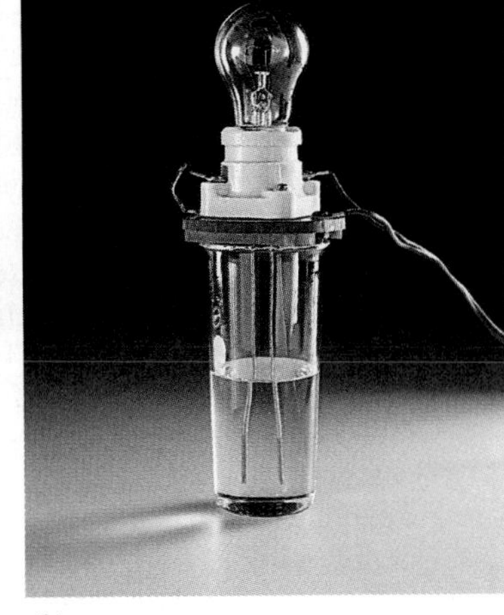

(b)

FIG. 4.2 **Electrical conductivity of solutions of electrolytes versus nonelectrolytes.** *(a)* The copper sulfate solution is a strong conductor of electricity, and $CuSO_4$ is a strong electrolyte. *(b)* Neither sugar nor water is an electrolyte, and this sugar solution is a nonconductor. *(Michael Watson.)*

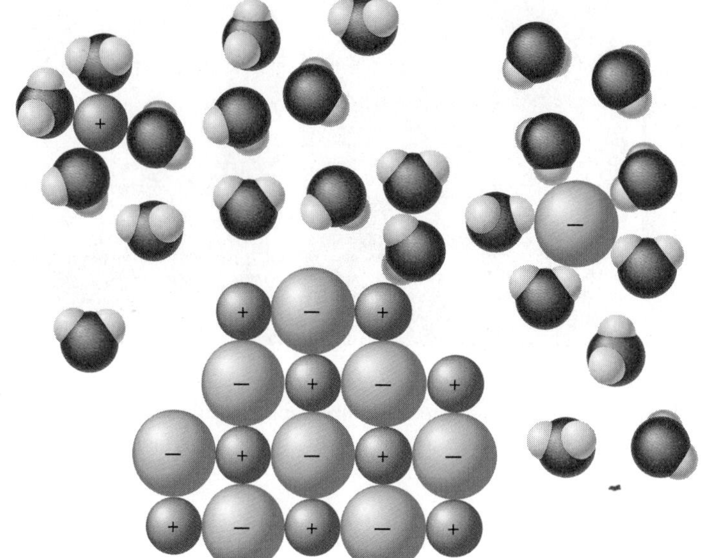

FIG. 4.3 Dissociation of an ionic compound as it dissolves in water. Ions separate from the solid and become surrounded by molecules of water. The ions are said to be hydrated. In the solution, the ions are able to move freely and the solution is able to conduct electricity.

illustrated in Figure 4.3. In general, *we will assume that in water the dissociation of any* ***salt*** *(i.e., ionic compound) is complete* and that the solution contains no undissociated formula units of the salt. Thus, an aqueous solution of $CuSO_4$ is really a solution that contains Cu^{2+} and SO_4^{2-} ions, with virtually no undissociated formula units of $CuSO_4$. Because these solutions contain so many ions, they are strong conductors of electricity, and salts are said to be **strong electrolytes.**

Keep in mind that a strong electrolyte is 100% dissociated in an aqueous solution.

Many ionic compounds have low solubilities in water. An example is AgBr, the light sensitive compound in most photographic film. Although only a tiny amount of this compound dissolves in water, all of it that does dissolve is completely dissociated. However, because of the extremely low solubility, the number of ions in the solution is extremely small and the solution doesn't conduct electricity well. Nevertheless, it is still convenient to think of AgBr as a strong electrolyte because it serves to remind us that salts are completely dissociated in aqueous solution.

Aqueous solutions of most molecular compounds do not conduct electricity, and such solutes are called **nonelectrolytes.** Examples are sugar (Figure 4.2*b*) and ethylene glycol (the solute in antifreeze solutions). Both consist of uncharged molecules that stay intact and simply intermingle with water molecules when they dissolve.

Ethylene glycol, $C_2H_4(OH)_2$, is a type of alcohol. Other alcohols, such as ethanol and methanol, are also nonelectrolytes.

Equations for dissociation reactions show the ions

A convenient way to describe the dissociation of an ionic compound is with a chemical equation. Thus, for the dissociation of calcium chloride in water we write

$$CaCl_2(s) \longrightarrow Ca^{2+}(aq) + 2Cl^-(aq)$$

We use the symbol *aq* after a charged particle to mean that it is surrounded by water molecules in the solution. We say it is **hydrated.** By writing the formulas of the ions separately, we mean that they are essentially independent of each other in the solution. Notice that each formula unit of $CaCl_2(s)$ releases three ions, one $Ca^{2+}(aq)$ and two $Cl^-(aq)$.

Often, when the context is clear that the system is aqueous, the symbols (*s*) and (*aq*) are omitted. They are "understood." You should not be disturbed, therefore, when you see an equation such as

$$CaCl_2 \longrightarrow Ca^{2+} + 2Cl^-$$

Polyatomic ions generally remain intact as dissociation occurs. When copper sulfate dissolves, for example, both Cu^{2+} and SO_4^{2-} ions are released.

$$CuSO_4(s) \longrightarrow Cu^{2+}(aq) + SO_4^{2-}(aq)$$

Be sure you know the formulas and charges on the polyatomic ions listed in Table 2.8 on page 69.

EXAMPLE 4.1
Writing the Equation for the Dissociation of an Ionic Compound

Ammonium sulfate is used as a fertilizer to supply nitrogen to crops. Write the equation for the dissociation of this compound when it dissolves in water.

ANALYSIS: To write the equation correctly, we need to know the formulas of the ions that make up the compound. In this case, the cation is NH_4^+ (ammonium ion) and the anion is SO_4^{2-} (sulfate ion). The correct formula of the compound is therefore $(NH_4)_2SO_4$, which means there are *two* NH_4^+ ions for each SO_4^{2-} ion. We have to be sure to indicate this in the equation.

SOLUTION: We write the formula for the solid on the left of the equation and indicate its state by (*s*). The ions are written on the right side of the equation and are shown to be in aqueous solution by the symbol (*aq*) following their formulas.

$$(NH_4)_2SO_4(s) \longrightarrow 2NH_4^+(aq) + SO_4^{2-}(aq)$$

The subscript 2 becomes the coefficient for NH_4^+.

IS THE ANSWER REASONABLE? There are two things to check when writing equations such as this. First, be sure you have the correct formulas for the ions, including their charges. Second, be sure you've indicated the number of ions of each kind that comes from one formula unit when the compound dissociates. Performing these checks here confirms we've solved the problem correctly.

Practice Exercise 1: Write equations that show the dissociation of the following compounds in water: (a) $FeCl_3$ and (b) potassium phosphate. (Hint: Identify the ions present in each compound.)

Practice Exercise 2: Write equations that show what happens when the following solid ionic compounds dissolve in water: (a) $MgCl_2$, (b) $Al(NO_3)_3$, and (c) sodium carbonate.

Equations for ionic reactions can be written in different ways

Often, ionic compounds react with each other when their aqueous solutions are combined. For example, when solutions of lead nitrate, $Pb(NO_3)_2$, and potassium iodide, KI, are mixed, a bright yellow precipitate of lead iodide, PbI_2, forms (Figure 4.4). The chemical equation for the reaction is

$$Pb(NO_3)_2(aq) + 2KI(aq) \longrightarrow PbI_2(s) + 2KNO_3(aq) \quad (4.1)$$

where we have noted the insolubility of PbI_2 by writing (*s*) following its formula. This is called a **molecular equation** because all the formulas are written with the ions together, as if the substances in solution consist of neutral "molecules." Equation 4.1 is fine for performing stoichiometric calculations, but let's look at other ways that we might write the chemical equation.

Soluble ionic compounds are fully dissociated in solution, so $Pb(NO_3)_2$, KI, and KNO_3 are not present in the solution as intact units or "molecules." To show this, we can write the formulas of all soluble strong electrolytes in "dissociated" form to give the **ionic equation** for the reaction.

$$Pb(NO_3)_2(aq) + 2KI(aq) \longrightarrow PbI_2(s) + 2KNO_3(aq)$$

$$Pb^{2+}(aq) + 2NO_3^-(aq) + 2K^+(aq) + 2I^-(aq) \longrightarrow PbI_2(s) + 2K^+(aq) + 2NO_3^-(aq)$$

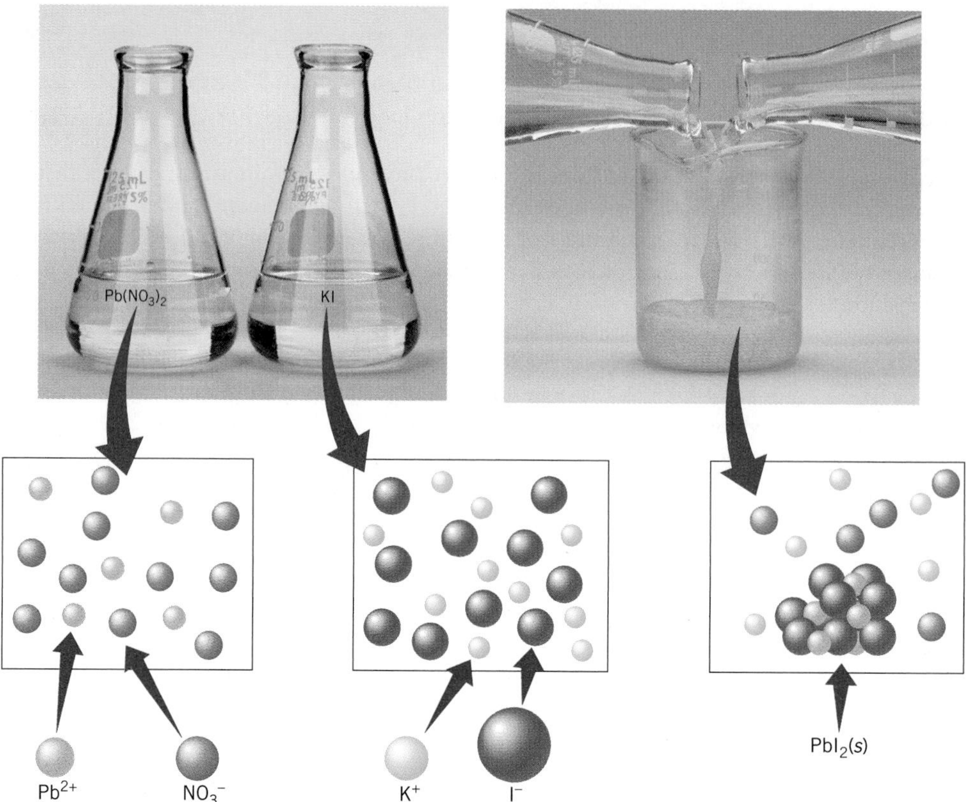

FIG. 4.4 **The reaction of $Pb(NO_3)_2$ with KI.** On the left are flasks containing solutions of lead nitrate and potassium iodide. These solutes exist as separated ions in their respective solutions. On the right, we observe that when the solutions of the ions are combined, there is an immediate reaction as the Pb^{2+} ions join with the I^- ions to give a precipitate of small crystals of solid, yellow PbI_2. The reaction is so rapid that the yellow color develops where the two streams of liquid come together. If the $Pb(NO_3)_2$ and KI are combined in a 1-to-2 mole ratio, the solution would now contain only K^+ and NO_3^- ions (the ions of KNO_3). *(Andy Washnik.)*

Notice that we have *not* separated PbI_2 into its ions in this equation. This is because PbI_2 has an extremely low solubility in water; it is essentially insoluble. When the Pb^{2+} and I^- ions meet in the solution, insoluble PbI_2 forms and separates as a precipitate. Therefore, after the reaction is over, the Pb^{2+} and I^- ions are no longer able to move independently. They are trapped in the insoluble product.

The ionic equation gives a clearer picture of what is actually going on in the solution during the reaction. The Pb^{2+} and I^- ions come together to form the product, while the other ions, K^+ and NO_3^-, are unchanged by the reaction. *Ions that do not actually take part in a reaction are sometimes called* **spectator ions;** in a sense, they just "stand by and watch the action."

To emphasize the actual reaction that occurs, we can write the **net ionic equation,** *which is obtained by eliminating spectator ions from the ionic equation.* Let's cross out the spectator ions, K^+ and NO_3^-.

$$Pb^{2+}(aq) + \cancel{2NO_3^-(aq)} + \cancel{2K^+(aq)} + 2I^-(aq) \longrightarrow PbI_2(s) + \cancel{2K^+(aq)} + \cancel{2NO_3^-(aq)}$$

What remains is the net ionic equation,

$$Pb^{2+}(aq) + 2I^-(aq) \longrightarrow PbI_2(s)$$

Notice how it calls our attention to the ions that are actually participating in the reaction as well as the change that occurs.

FACETS OF CHEMISTRY

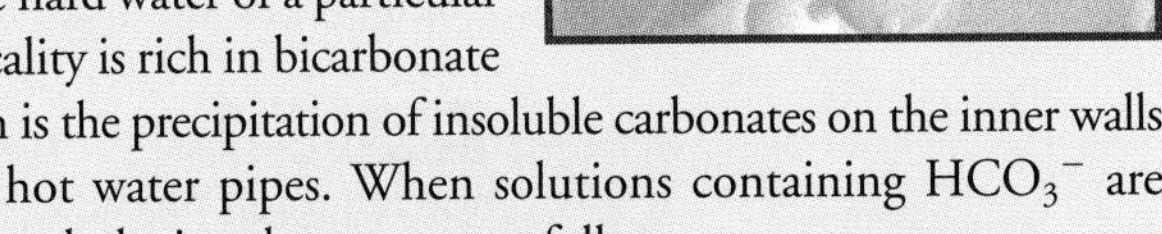

Boiler Scale and Hard Water

Precipitation reactions occur around us all the time and we hardly ever take notice until they cause a problem. One common problem is caused by **hard water**—groundwater that contains the "hardness ions," Ca^{2+}, Mg^{2+}, Fe^{2+}, or Fe^{3+}, in concentrations high enough to form precipitates with ordinary soap. Soap normally consists of the sodium salts of organic acids derived from animal fats or oils (so-called *fatty acids*). An example is sodium stearate, $NaC_{18}H_{35}O_2$. The negative ion of the soap forms an insoluble "scum" with hardness ions, which reduces the effectiveness of the soap for removing dirt and grease.

Hardness ions can be removed from water in a number of ways. One way is to add hydrated sodium carbonate, $Na_2CO_3 \cdot 10H_2O$, often called washing soda, to the water. The carbonate ion forms insoluble precipitates with the hardness ions; an example is $CaCO_3$.

$$Ca^{2+}(aq) + CO_3^{2-}(aq) \longrightarrow CaCO_3(s)$$

Once precipitated, the hardness ions are not available to interfere with the soap.

Another problem when the hard water of a particular locality is rich in bicarbonate ion is the precipitation of insoluble carbonates on the inner walls of hot water pipes. When solutions containing HCO_3^- are heated, the ion decomposes as follows.

$$2HCO_3^-(aq) \longrightarrow H_2O + CO_2(g) + CO_3^{2-}(aq)$$

Like most gases, carbon dioxide becomes less soluble as the temperature is raised, so CO_2 is driven from the hot solution and the HCO_3^- is gradually converted to CO_3^{2-}. As the carbonate ions form, they are able to precipitate the hardness ions. This precipitate, which sticks to the inner walls of pipes and hot water boilers, is called *boiler scale*. In locations that have high concentrations of Ca^{2+} and HCO_3^- in the water supply, boiler scale is a very serious problem, as illustrated in the accompanying photograph.

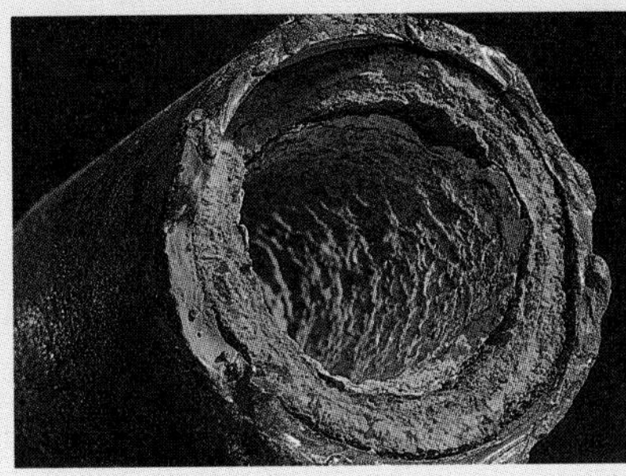

Boiler scale built up on the inside of a water pipe. *(Courtesy of Betz Company.)*

The net ionic equation is especially useful because it permits us to *generalize*. It tells us that if we combine *any* solution that contains Pb^{2+} with *any* other solution that contains I^-, we ought to expect a precipitate of PbI_2. And this is exactly what happens if we mix aqueous solutions of lead acetate, $Pb(C_2H_3O_2)_2$, and sodium iodide, NaI. A yellow precipitate of PbI_2 forms immediately (Figure 4.5). Example 4.2 demonstrates how we construct the molecular, ionic, and net ionic equations for the reaction.

FIG. 4.5 Another reaction that forms lead iodide. The net ionic equation tells us that any soluble lead compound will react with any soluble iodide compound to give lead iodide. This prediction is borne out here as a precipitate of lead iodide is formed when a solution of sodium iodide is added to a solution of lead acetate. *(Andy Washnik.)*

EXAMPLE 4.2
Writing Molecular, Ionic, and Net Ionic Equations

Write the molecular, ionic, and net ionic equations for the reaction of aqueous solutions of lead acetate and sodium iodide, which yields a precipitate of lead iodide and leaves the compound sodium acetate in solution.

ANALYSIS: To write a chemical equation, we must begin with the correct formulas of the reactants and products. If only the names of the reactants and products are given, we have to translate them into chemical formulas. Following the rules we discussed in Chapter 2, we have

If necessary, review Section 2.9, which discusses naming ionic compounds.

Reactants		***Products***	
lead acetate	$Pb(C_2H_3O_2)_2$	lead iodide	PbI_2
sodium iodide	NaI	sodium acetate	$NaC_2H_3O_2$

We arrange the formulas to form the molecular equation, which we then balance. To obtain the ionic equation, we write soluble ionic compounds in dissociated form and the formula of the precipitate in "molecular" form. Finally, we look for spectator ions and eliminate them from the ionic equation to obtain the net ionic equation.

SOLUTION:

The Molecular Equation We assemble the chemical formulas into the molecular equation.

$$Pb(C_2H_3O_2)_2(aq) + 2NaI(aq) \longrightarrow PbI_2(s) + 2NaC_2H_3O_2(aq)$$

Notice that we've indicated which substances are in solution and which is a precipitate, and we've balanced the equation. This is the *balanced molecular equation.*

The Ionic Equation To write the ionic equation, we write the formulas of all soluble salts in dissociated form and the formulas of precipitates in "molecular" form. We are careful to use the subscripts and coefficients in the molecular equation to properly obtain the coefficients of the ions in the ionic equation.

$$Pb(C_2H_3O_2)_2(aq) \qquad 2NaI(aq) \qquad 2NaC_2H_3O_2(aq)$$

$$Pb^{2+}(aq) + 2C_2H_3O_2^-(aq) + 2Na^+(aq) + 2I^-(aq) \longrightarrow PbI_2(s) + 2Na^+(aq) + 2C_2H_3O_2^-(aq)$$

This is the *balanced ionic equation.* Notice that to properly write the ionic equation it is necessary to know both the formulas and charges of the ions.

The Net Ionic Equation We obtain the net ionic equation from the ionic equation by eliminating spectator ions, which are Na^+ and $C_2H_3O_2^-$ (they're the same on both sides of the arrow). Let's cross them out.

$$Pb^{2+}(aq) + \cancel{2C_2H_3O_2^-(aq)} + \cancel{2Na^+(aq)} + 2I^-(aq) \longrightarrow PbI_2(s) + \cancel{2Na^+(aq)} + \cancel{2C_2H_3O_2^-(aq)}$$

What's left is the *net ionic equation.*

$$Pb^{2+}(aq) + 2I^-(aq) \longrightarrow PbI_2(s)$$

Notice this is the same net ionic equation as in the reaction of lead nitrate with potassium iodide.

ARE THE ANSWERS REASONABLE? When you look back over a problem such as this, things to ask yourself are (1) "Have I written the correct formulas for the reactants and products?", (2) "Is the molecular equation balanced correctly?", (3) "Have I divided the soluble ionic compounds into their ions correctly, being careful to properly apply the subscripts of the ions and the coefficients in the molecular equation?", and (4) "Have I identified and eliminated the correct ions from the ionic equation to obtain the net ionic equation?" If each of these questions can be answered in the affirmative, as they can here, the problem has been solved correctly.

Practice Exercise 3: When solutions of $(NH_4)_2SO_4$ and $Ba(NO_3)_2$ are mixed, a precipitate of $BaSO_4$ forms, leaving soluble NH_4NO_3 in the solution. Write the molecular, ionic, and net ionic equations for the reaction. (Hint: Remember that polyatomic ions do not break apart when ionic compounds dissolve in water.)

Practice Exercise 4: Write molecular, ionic, and net ionic equations for the reaction of aqueous solutions of cadmium chloride and sodium sulfide to give a precipitate of cadmium sulfide and a solution of sodium chloride.

FIG. 4.6 An acid–base indicator. Litmus paper, a strip of paper impregnated with the dye litmus, becomes blue in aqueous ammonia (a base) and pink in lemon juice (which contains citric acid). *(Ken Karp.)*

In a balanced ionic or net ionic equation, both atoms and charge must balance

In the ionic and net ionic equations we've written, not only are the atoms in balance, but so is the net electrical charge, which is the same on both sides of the equation. Thus, in the ionic equation for the reaction of lead nitrate with potassium iodide, the sum of the charges of the ions on the left (Pb^{2+}, $2NO_3^-$, $2K^+$, and $2I^-$) is zero, which matches the sum of the charges on all of the formulas of the products (PbI_2, $2K^+$, and $2NO_3^-$).[2] In the net ionic equation the charges on both sides are also the same: on the left we have Pb^{2+} and $2I^-$, with a net charge of zero, and on the right we have PbI_2, also with a charge of zero. We now have an additional requirement for an ionic equation or net ionic equation to be balanced: *the net electrical charge on both sides of the equation must be the same.*

Criteria for a balanced ionic equation

Criteria for Balanced Ionic and Net Ionic Equations

1. **Material balance.** There must be the same number of atoms of each kind on both sides of the arrow.
2. **Electrical balance.** The *net* electrical charge on the left must equal the *net* electrical charge on the right (although this charge does not necessarily have to be zero).

4.3 Acids and Bases Are Classes of Compounds with Special Properties

Acids and bases constitute a class of compounds that include some of our most familiar chemicals and important laboratory reagents. Vinegar, lemon juice, and the liquid in an automobile battery contain acids. The white crystals of lye in some drain cleaners, the white substance that makes milk of magnesia opaque, and household ammonia are all bases.

There are some general properties that are common to aqueous solutions of acids and bases. For example, **acids** generally have a tart (sour) taste, whereas **bases** have a somewhat bitter taste and have a soapy "feel." (However, taste is ***never*** used as a laboratory test for acids or bases; some are extremely corrosive to animal tissue. ***Never taste chemicals in the laboratory!***)

Acids and bases also affect the colors of certain dyes we call **acid–base indicators.** An example is litmus (Figure 4.6), which has a pink or red color in an acidic solution and a blue color in a basic solution.[3]

One of the most important properties of acids and bases is their reaction with each other, a reaction referred to as **neutralization.** For example, when solutions of hydrochloric acid, $HCl(aq)$, and the base sodium hydroxide, $NaOH(aq)$, are mixed the following reaction occurs.

$$HCl(aq) + NaOH(aq) \longrightarrow NaCl(aq) + H_2O$$

Acids and bases should be treated with respect because of their potential for causing bodily injury if spilled on the skin. If you spill an acid or base on yourself in the lab, be sure to wash it off immediately and notify your instructor at once.

[2] There is no charge written for the formula of a compound such as PbI_2, so as we add up charges, we take the charge on PbI_2 to be zero.

[3] Litmus paper, commonly found among the items in a locker in the general chemistry lab, consists of strips of absorbent paper that have been soaked in a solution of litmus and dried. Red litmus paper is used to test if a solution is basic. A basic solution turns red litmus blue. To test if the solution is acidic, blue litmus paper is used. Acidic solutions turn blue litmus red.

When the reactants are combined in a 1-to-1 ratio by moles, the acidic and basic properties of the solutes disappear and the resulting solution is neither acidic nor basic. We say an *acid–base neutralization* has occurred. Svante Arrhenius,[4] a Swedish chemist, was the first to suggest that an acid–base neutralization is simply the combination of a hydrogen ion with a hydroxide ion to produce a water molecule, thus making H^+ ions and OH^- ions disappear.

Today we know that in aqueous solutions hydrogen ions, H^+, attach themselves to water molecules to form **hydronium ions,** H_3O^+. However, for the sake of convenience, we often use the term *hydrogen ion* as a substitute for *hydronium ion*, and in many equations, we use $H^+(aq)$ to stand for $H_3O^+(aq)$. In fact, whenever you see the symbol $H^+(aq)$, you should realize that we are actually referring to $H_3O^+(aq)$.

Even the formula H_3O^+ is something of a simplification. In water the H^+ ion is associated with more than one molecule of water, but we use the formula H_3O^+ as a simple representation.

For most purposes, we find that the following modified versions of Arrhenius' definitions work satisfactorily when we deal with aqueous solutions.

> ***Arrhenius Definition of Acids and Bases***
>
> An **acid** is a substance that reacts with water to produce hydronium ion, H_3O^+.
>
> A **base** is a substance that produces hydroxide ion in water.

In general, the reaction of an acid with a base produces an ionic compound as one of the products. In the reaction of HCl(*aq*) with NaOH(*aq*), the compound is sodium chloride, or salt. This reaction is so general, in fact, that *we use the word* **salt** *to mean any ionic compound that doesn't contain either hydroxide ion, OH^-, or oxide ion, O^{2-}.* (Ionic compounds that contain OH^- or O^{2-} are bases, as described below.)

In aqueous solutions, acids give H_3O^+

In general, **acids** are molecular substances that react with water to produce ions, one of which is the hydronium ion, H_3O^+. Thus, when gaseous molecular HCl dissolves in water, a hydrogen ion (H^+) transfers from the HCl molecule to a water molecule. The reaction is depicted in Figure 4.7 using space-filling models, and is represented by the chemical equation

If gaseous HCl is cooled to about −85 °C, it condenses to a liquid that doesn't conduct electricity. No ions are present in pure liquid HCl.

$$HCl(g) + H_2O \longrightarrow H_3O^+(aq) + Cl^-(aq)$$

This is an **ionization reaction** because ions form where none existed before. Because the solution contains ions, it conducts electricity, so acids are electrolytes.

Sometimes acids also contain hydrogen atoms that are not able to form H_3O^+. An example is acetic acid, $HC_2H_3O_2$, the acid that gives vinegar its sour taste. This acid reacts with water as follows.

$$HC_2H_3O_2(aq) + H_2O \longrightarrow H_3O^+(aq) + C_2H_3O_2^-(aq)$$

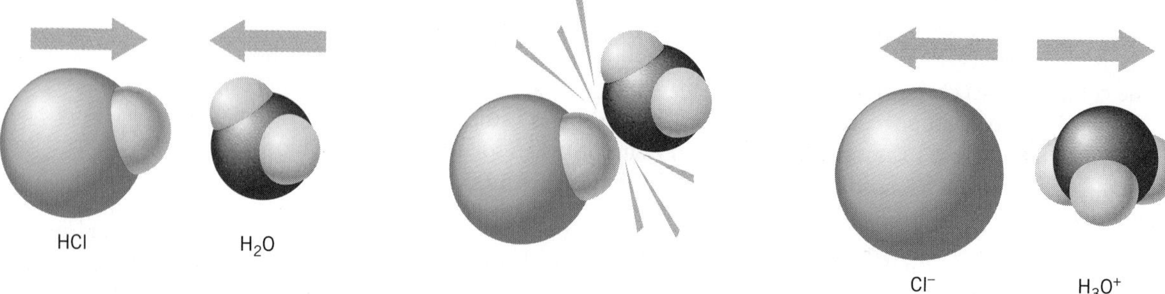

FIG. 4.7 **Ionization of HCl in water.** Collisions between HCl molecules and water molecules lead to a transfer of H^+ from HCl to H_2O, giving Cl^- and H_3O^+ as products.

[4] Arrhenius proposed his theory of acids and bases in 1884 in his Ph.D. thesis. He won the Nobel Prize in Chemistry for his work in 1903.

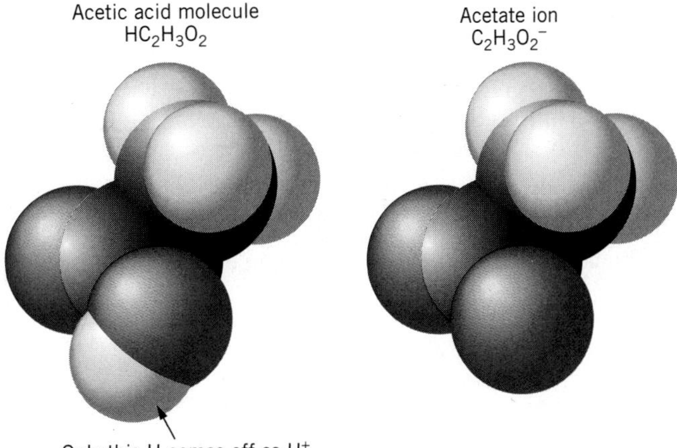

FIG. 4.8 Acetic acid and acetate ion. The structures of acetic acid and acetate ion are illustrated here. In acetic acid, only the hydrogen attached to an oxygen can come off as H^+.

Notice that only the hydrogen written first in the formula is able to transfer to H_2O to give hydronium ions. The structures of the acetic acid molecule and the acetate ion are shown in Figure 4.8, with the hydrogen that can be lost by the acetic acid molecule indicated.

Hydrogens that are able to be transferred to water molecules to form hydronium ions are usually written first in the formula for the acid.

As noted earlier, the "active ingredient" in the hydronium ion is H^+, which is why $H^+(aq)$ is often used in place of $H_3O^+(aq)$ in equations. Using this simplification, the ionization of HCl and $HC_2H_3O_2$ in water can be represented as

$$HCl(g) \xrightarrow{H_2O} H^+(aq) + Cl^-(aq)$$

and

$$HC_2H_3O_2(aq) \xrightarrow{H_2O} H^+(aq) + C_2H_3O_2^-(aq)$$

In the ionization reactions of HCl and $HC_2H_3O_2$, an anion is formed when the acid transfers an H^+ to the water molecule. If we represent the acid molecule by the general formula *HA*, we might represent the ionization of an acid in general terms by the equation

$$HA + H_2O \longrightarrow H_3O^+ + A^- \qquad (4.2)$$

Ionization of an acid in water

The molecules HCl and $HC_2H_3O_2$ are said to be **monoprotic acids** because they are capable of furnishing only *one* H^+ per molecule of acid. **Polyprotic acids** can furnish more than one H^+ per molecule. They undergo reactions similar to those of HCl and $HC_2H_3O_2$, except that the loss of H^+ by the acid occurs in two or more steps. Thus, the ionization of sulfuric acid, a **diprotic acid,** takes place by two successive steps.

$$H_2SO_4(aq) + H_2O \longrightarrow H_3O^+(aq) + HSO_4^-(aq)$$

$$HSO_4^-(aq) + H_2O \longrightarrow H_3O^+(aq) + SO_4^{2-}(aq)$$

Triprotic acids ionize in three steps, as illustrated in Example 4.3.

EXAMPLE 4.3
Writing Equations for Ionization Reactions of Acids

Phosphoric acid, H_3PO_4, is a triprotic acid found in some soft drinks such as Coca-Cola where it adds a touch of tartness to the beverage. Write equations for its stepwise ionization in water.

ANALYSIS: We are told that H_3PO_4 is a triprotic acid, which is also indicated by the three hydrogens at the beginning of the formula. Because there are three hydrogens to come off the molecule, we expect there to be three steps in the ionization. Each step removes one H^+, and we can use that knowledge to deduce the formulas of the products. Let's line them up so we can see the progression.

$$H_3PO_4 \xrightarrow{-H^+} H_2PO_4^- \xrightarrow{-H^+} HPO_4^{2-} \xrightarrow{-H^+} PO_4^{3-}$$

Notice that loss of H^+ decreases the number of hydrogens by one and increases the negative charge by one unit. Also, the product of one step serves as the reactant in the next step. We'll use Equation 4.2 for the ionization of an acid as a tool in writing the chemical equation for each step.

SOLUTION: The first step is the reaction of H_3PO_4 with water to give H_3O^+ and $H_2PO_4^-$.

$$H_3PO_4(aq) + H_2O \longrightarrow H_3O^+(aq) + H_2PO_4^-(aq)$$

The second and third steps are similar to the first.

$$H_2PO_4^-(aq) + H_2O \longrightarrow H_3O^+(aq) + HPO_4^{2-}(aq)$$
$$HPO_4^{2-}(aq) + H_2O \longrightarrow H_3O^+(aq) + PO_4^{3-}(aq)$$

IS THE ANSWER REASONABLE? Check to see whether the equations are balanced in terms of atoms and charge. If any mistakes were made, something would be out of balance and we would discover the error. In this case, all the equations are balanced, so we can feel confident we've written them correctly.

Practice Exercise 5: Write the equation for the ionization of $HCHO_2$ (formic acid) in water. Formic acid is used industrially to remove hair from animal skins prior to tanning. (Hint: Formic acid and acetic acid are both examples of organic acids.)

Practice Exercise 6: Write equations for the stepwise ionization in water of citric acid, $H_3C_6H_5O_7$, the acid in citrus fruits.

Nonmetal oxides can be acids

The acids we've discussed so far have been molecules containing hydrogen atoms that can be transferred to water molecules. Nonmetal oxides form another class of compounds that yield acidic solutions in water. Examples are SO_3, CO_2, and N_2O_5 whose aqueous solutions contain H_3O^+ and turn litmus red. These oxides are called **acidic anhydrides,** where *anhydride* means "without water." They react with water to form molecular acids containing hydrogen, which are then able to undergo reaction with water to yield H_3O^+.

$$SO_3(g) + H_2O \longrightarrow H_2SO_4(aq) \qquad \text{sulfuric acid}$$
$$N_2O_5(g) + H_2O \longrightarrow 2HNO_3(aq) \qquad \text{nitric acid}$$
$$CO_2(g) + H_2O \longrightarrow H_2CO_3(aq) \qquad \text{carbonic acid}$$

Although carbonic acid is too unstable to be isolated as a pure compound, its solutions in water are quite common. Carbon dioxide from the atmosphere dissolves in rainwater and the waters of lakes and streams where it exists partly as carbonic acid and its ions (HCO_3^- and CO_3^{2-}). This makes these waters naturally slightly acidic. Carbonic acid is also present in carbonated beverages.

Not all nonmetal oxides are acidic anhydrides, only those that are able to react with water. For example, carbon monoxide doesn't react with water, so its solutions in water are not acidic; carbon monoxide, therefore, is not classified as an acidic anhydride.

Bases are substances that give OH^- in water

Bases fall into two categories: ionic compounds that contain OH^- or O^{2-}, and molecular compounds that react with water to give hydroxide ions. Because solutions of bases contain ions, they conduct electricity. Therefore, bases are electrolytes.

Ionic bases are metal hydroxides and oxides

Ionic bases include metal hydroxides, such as NaOH and $Ca(OH)_2$. When dissolved in water, they dissociate just like other soluble ionic compounds.

$$NaOH(s) \longrightarrow Na^+(aq) + OH^-(aq)$$
$$Ca(OH)_2(s) \longrightarrow Ca^{2+}(aq) + 2OH^-(aq)$$

Soluble metal oxides are **basic anhydrides** because they react with water to form the hydroxide ion as one of the products. Calcium oxide is typical.

$$CaO(s) + H_2O \longrightarrow Ca(OH)_2(aq)$$

This reaction occurs when water is added to dry cement or concrete because calcium oxide or "quicklime" is an ingredient in those materials. In this case it is the oxide ion, O^{2-}, that actually forms the OH^-.

$$O^{2-} + H_2O \longrightarrow 2OH^-$$

Continual contact of your hands with fresh Portland cement can lead to irritation because the mixture is quite basic.

Even insoluble metal hydroxides and oxides are basic because they are able to neutralize acids. We will study these reactions in Section 4.5.

Many nitrogen compounds are molecular bases

The most common molecular base is the gas ammonia, NH_3, which dissolves in water and reacts to give a basic solution by an ionization reaction.

$$NH_3(aq) + H_2O \longrightarrow NH_4^+(aq) + OH^-(aq)$$

Organic compounds called amines, in which fragments of hydrocarbons are attached to nitrogen in place of hydrogen, are similar to ammonia in their behavior toward water. An example is methylamine, CH_3NH_2, in which a **methyl group,** CH_3, replaces a hydrogen in ammonia.

$$CH_3NH_2(aq) + H_2O \longrightarrow CH_3NH_3^+(aq) + OH^-(aq)$$

The hydrogen taken from the H_2O molecule becomes attached to the nitrogen atom of the amine. This is how nitrogen-containing bases behave, which is why we've included the H^+ with the other two hydrogens on the nitrogen.

Notice that when a molecular base reacts with water, an H^+ is lost by the water molecule and gained by the base. (See Figure 4.9.) One product is a cation that has one more H and one more positive charge than the reactant base. Loss of H^+ by the water gives the other product, the OH^- ion, which is why the solution is basic. We might represent this by the general equation

$$\text{base} + H_2O \longrightarrow \text{baseH}^+ + OH^-$$

If we signify the base by the symbol *B*, this becomes

$$B + H_2O \longrightarrow BH^+ + OH^- \quad (4.3)$$

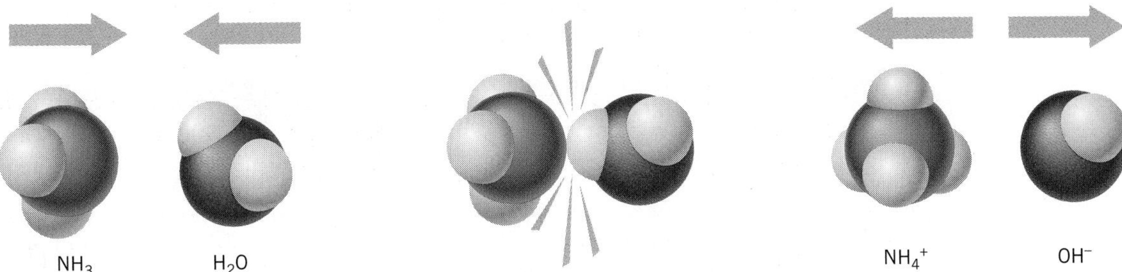

FIG. 4.9 Ionization of ammonia in water. Collisions between NH_3 molecules and water molecules lead to a transfer of H^+ from H_2O to NH_3, giving NH_4^+ and OH^- ions.

EXAMPLE 4.4
Writing the Equation for the Ionization of a Base

Dimethylamine, $(CH_3)_2NH$, is used as an attractant for boll weevils so they can be destroyed. This insect has caused more than a $14 billion loss to the yield of cotton in the United States since it arrived from Mexico in 1892. The compound is a base in water. Write an equation for its ionization.

ANALYSIS: The reactants in the equation are $(CH_3)_2NH$ and H_2O. To write the equation, we need to know the formulas of the products. We've been told that $(CH_3)_2NH$ is a base, so Equation 4.3 is the tool we will use to write the chemical equation.

SOLUTION: When a base reacts with water, it takes an H^+ from H_2O, leaving OH^- behind. Therefore, when an H^+ is picked up by $(CH_3)_2NH$, the product will be $(CH_3)_2NH_2^+$. The equation for the reaction is

$$(CH_3)_2NH(aq) + H_2O \longrightarrow (CH_3)_2NH_2^+(aq) + OH^-(aq)$$

IS THE ANSWER REASONABLE? Compare the equation we've written with the general equation for reaction of a base with water. Notice that the formula for the product has one more H and a positive charge, and that the H^+ has been added to the nitrogen. Also, notice that the water has become OH^- when it loses H^+. The equation is therefore correct.

Practice Exercise 7: Triethylamine, $(C_2H_5)_3N$, is a base in water. Write an equation for its reaction with the solvent. (Hint: How do nitrogen-containing bases react toward water?)

Practice Exercise 8: Hydroxylamine, $HONH_2$, is a base in water. Write an equation for its reaction with the solvent.

Acids and bases are classified as strong or weak

Ionic compounds such as NaCl and $CaCl_2$ break up essentially 100% into ions in water. No "molecules" of either NaCl or $CaCl_2$ are detectable in their aqueous solutions. Because these solutions contain so many ions, they are strong conductors of electricity, so ionic compounds are said to be **strong electrolytes.**

All strong acids are strong electrolytes.

Hydrochloric acid is also a strong electrolyte. Its ionization in water is essentially complete; its solutions are strongly acidic, and it is said to be a *strong acid.* In general, *acids that are strong electrolytes are called* **strong acids.** There are relatively few strong acids; the most common ones are listed below.

List of strong acids

Strong Acids

$HClO_4(aq)$	perchloric acid
$HCl(aq)$	hydrochloric acid
$HBr(aq)$	hydrobromic acid
$HI(aq)$	hydroiodic acid[5]
$HNO_3(aq)$	nitric acid
$H_2SO_4(aq)$	sulfuric acid

Metal hydroxides are ionic compounds, so they are also strong electrolytes. Those that are soluble are the hydroxides of Group IA and the hydroxides of calcium, strontium, and barium of Group IIA. Solutions of these compounds are strongly basic, so these substances are considered to be **strong bases.** The hydroxides of other metals have very low solubilities in water. They are strong electrolytes in the sense that the small amounts of them that dissolve in solution are completely dissociated. However, because of their low solubility in water, their solutions are very weakly basic.

[5] Sometimes the first "o" in the name of HI(*aq*) is dropped for ease of pronunciation to give *hydriodic acid.*

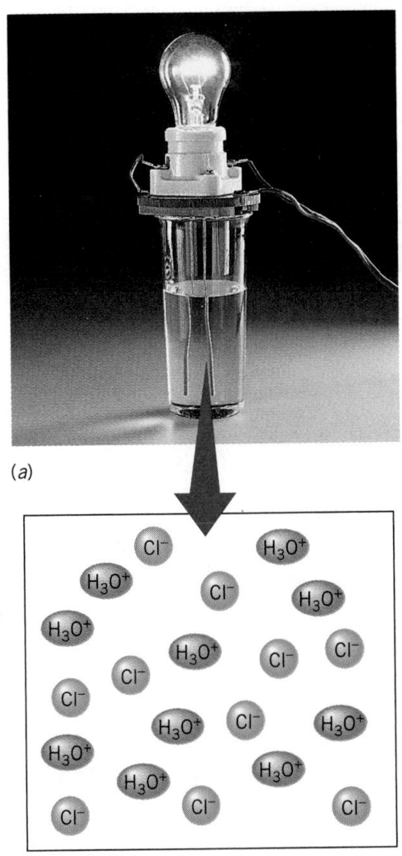

All the HCl is ionized in the solution, so there are many ions present.

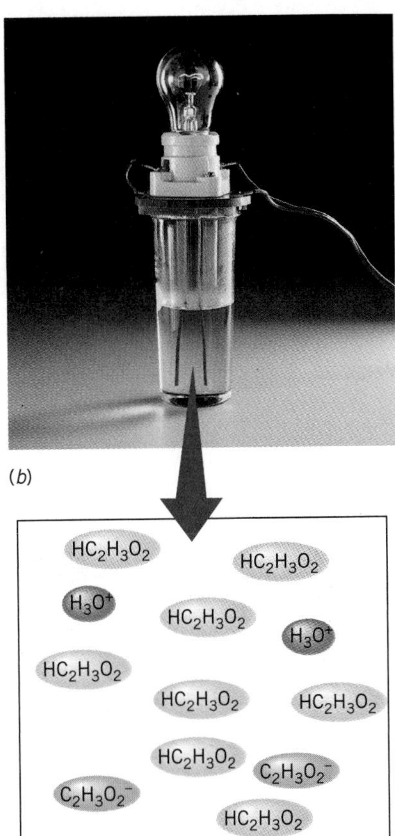

Only a small fraction of the acetic acid is ionized, so there are few ions to conduct electricity. Most of the acetic acid is present as neutral molecules of $HC_2H_3O_2$.

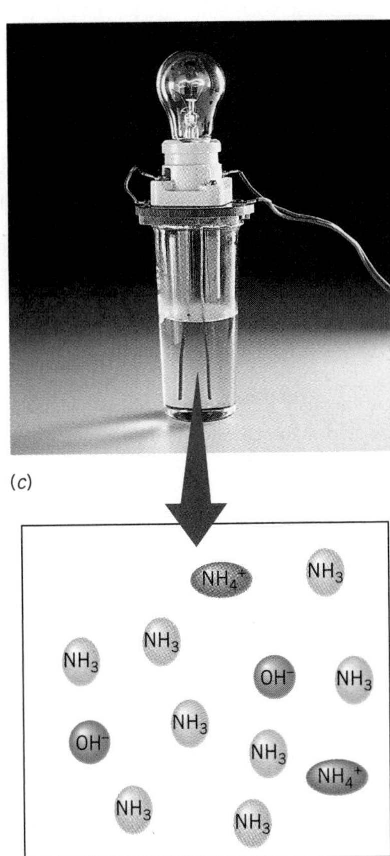

Only a small fraction of the ammonia is ionized, so few ions are present to conduct electricity. Most of the ammonia is present as neutral molecules of NH_3.

FIG. 4.10 Electrical conductivity of solutions of strong and weak acids and bases at equal concentrations. *(a)* HCl is 100% ionized and is a strong conductor, enabling the light to glow brightly. *(b)* $HC_2H_3O_2$ is a weaker conductor than HCl because the extent of its ionization is far less, so the light is dimmer. *(c)* NH_3 also is a weaker conductor than HCl because the extent of its ionization is low, and the light remains dim. *(Michael Watson.)*

Weak acids and bases are weak electrolytes

Most acids are not completely ionized in water. For instance, a solution of acetic acid, $HC_2H_3O_2$, is a relatively poor conductor of electricity compared to a solution of HCl with the same concentration (Figure 4.10), so acetic acid is classified as a **weak electrolyte** and is a **weak acid.**

The reason an acetic acid solution is a poor conductor is because in the solution only a small fraction of the acid exists as H_3O^+ and $C_2H_3O_2^-$ ions. The rest is present as molecules of $HC_2H_3O_2$. This is because $C_2H_3O_2^-$ ions have a strong tendency to react with H_3O^+ when the ions meet in the solution. As a result, there are two opposing reactions occurring simultaneously (Figure 4.11). One involves the formation of the ions,

$$HC_2H_3O_2(aq) + H_2O \longrightarrow H_3O^+(aq) + C_2H_3O_2^-(aq)$$

and the other removes ions

$$H_3O^+(aq) + C_2H_3O_2^-(aq) \longrightarrow HC_2H_3O_2(aq) + H_2O$$

A balance is reached when ions form and disappear at the same rate, and for acetic acid this happens when only a small percentage of the $HC_2H_3O_2$ is ionized.

Acetic acid molecule collides with a water molecule.
Transfer of a proton yields an acetate ion and a hydronium ion.
Water
Acetic acid
Acetate ion
Hydronium ion
Acetate ion collides with a hydronium ion.
Transfer of a proton yields an acetic acid molecule and water.
Hydronium ion
Acetate ion
Acetic acid
Water

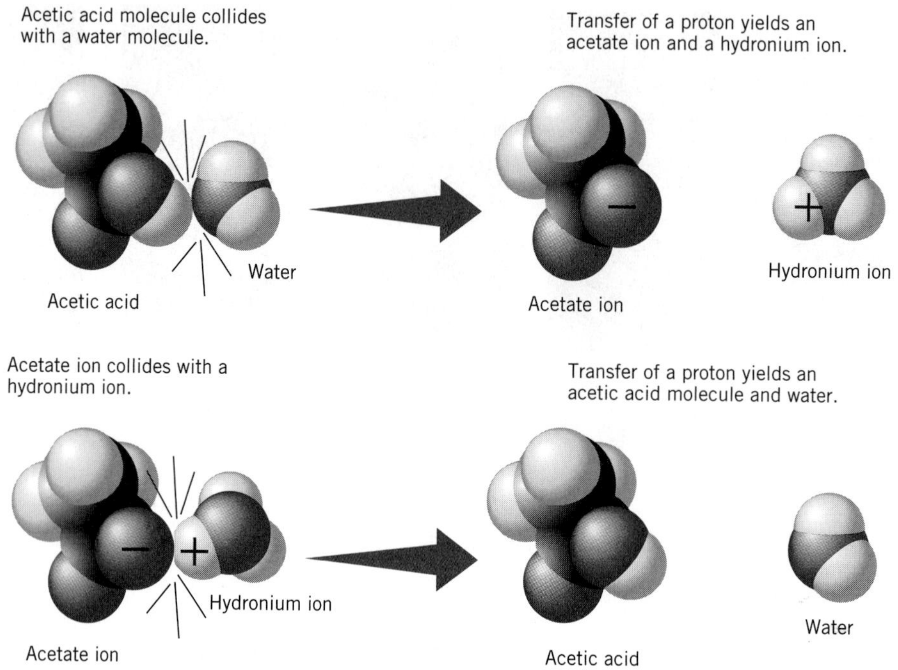

FIG. 4.11 Equilibrium in a solution of acetic acid. Two opposing reactions take place simultaneously in a solution of acetic acid. Molecules of acid collide with molecules of water and form acetate ions and H_3O^+ ions. Meanwhile, acetate ions collide with H_3O^+ ions to give acetic acid molecules and water molecules. (The usual colors are used: white = H, red = O, black = C.)

The condition we've just described, with two opposing reactions occurring at the same rate, is called a **chemical equilibrium** or **dynamic equilibrium.** It is an *equilibrium* because the concentrations of the substances present in the solution do not change with time; it is *dynamic* because the opposing reactions continue endlessly.

The two opposing processes in a dynamic equilibrium are usually represented in a single equation by using double arrows, $\rightleftharpoons$. For acetic acid, we write

$$HC_2H_3O_2(aq) + H_2O \rightleftharpoons H_3O^+(aq) + C_2H_3O_2^-(aq)$$

The **forward reaction** (read from left to right) forms the ions; the **reverse reaction** (from right to left) removes them from the solution.

Molecular bases, such as ammonia and methylamine, are also weak electrolytes and have a low percentage ionization. They are classified as **weak bases.** (See Figure 4.10*c*.) In a solution of ammonia, only a small fraction of the solute is ionized to give NH_4^+ and OH^- because the ions have a strong tendency to react with each other. This leads to the dynamic equilibrium (Figure 4.12)

$$NH_3(aq) + H_2O \rightleftharpoons NH_4^+(aq) + OH^-(aq)$$

in which most of the base is present as NH_3 molecules.

Let's briefly summarize the results of our discussion.

Weak acids and weak bases are weak electrolytes.

Strong acids and strong bases are strong electrolytes.

In describing equilibria such as those above, we will often talk about the **position of equilibrium.** By this we mean the extent to which the forward reaction proceeds toward completion. If very little of the products are present at equilibrium, the forward reaction has not gone far toward completion and we say "the position of equilibrium lies to the left," toward the reactants. On the other hand, if large amounts of the products are present at equilibrium, we say "the position of equilibrium lies to the right."

For any weak electrolyte, only a small percentage of the solute is actually ionized at any instant after equilibrium is reached, so the position of equilibrium lies to the left. To call acetic acid a *weak* acid, for example, is just another way of saying that the forward reaction in this equilibrium is far from completion.

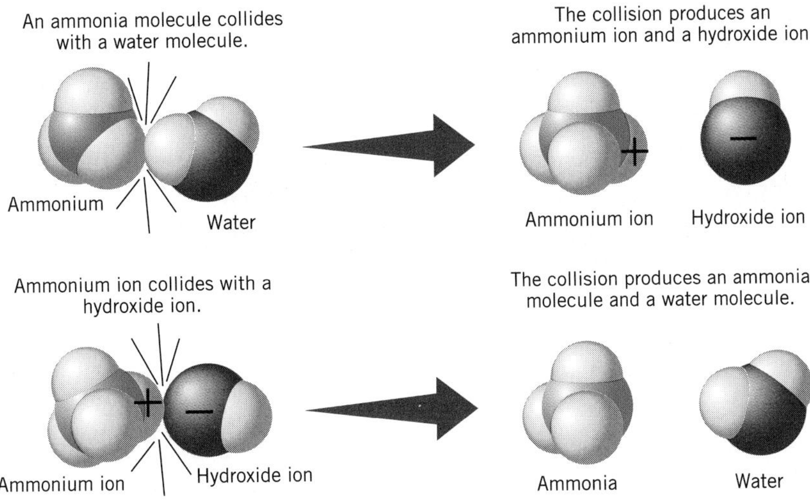
An ammonia molecule collides with a water molecule.
Ammonium
Water
The collision produces an ammonium ion and a hydroxide ion.
Ammonium ion
Hydroxide ion
Ammonium ion collides with a hydroxide ion.
Ammonium ion
Hydroxide ion
The collision produces an ammonia molecule and a water molecule.
Ammonia
Water

FIG. 4.12 Equilibrium in a solution of the weak base ammonia. Collisions between water and ammonia molecules produce ammonium and hydroxide ions. The reverse process, which involves collisions between ammonium ions and hydroxide ions, removes ions from the solution and forms ammonia and water molecules.

Strong acids do not participate in equilibria because they are fully ionized

With molecular compounds that are strong electrolytes, the tendency of the forward ionization reaction to occur is very large, while the tendency of the reverse reaction to occur is extremely small. In aqueous HCl, for example, there is little tendency for Cl^- and H_3O^+ to react to form molecules of HCl and H_2O. As a result, all of the HCl molecules dissolved in water become converted to ions—the acid becomes 100% ionized. For this reason, *we do not use double arrows in describing what happens when HCl(g) or any other strong electrolyte undergoes ionization or dissociation.*

Practice Exercise 9: Earlier you learned that methylamine, CH_3NH_2 (a fishy smelling substance found in herring brine), is a base in water. Write the equation that shows that methylamine is a weak base. (Hint: How do we show an equilibrium exists in the solution?)

Practice Exercise 10: Nitrous acid, HNO_2, is a weak acid thought to be responsible for certain cancers of the intestinal system. Write the chemical equation that shows that HNO_2 is a weak acid in water.

4.4 Naming Acids and Bases Follows a System

Although at first there seems to be little order in the naming of acids, there are patterns that help organize names of acids and the anions that come from them when the acids are neutralized.

Hydrogen compounds of nonmetals can be acids

The binary compounds of hydrogen with many of the nonmetals are acidic, and in their aqueous solutions they are referred to as **binary acids.** Some examples are HCl, HBr, and H_2S. In naming these substances as acids, we add the prefix *hydro-* and the suffix *-ic* to the stem of the nonmetal name, followed by the word *acid.* For example, aqueous solutions of hydrogen chloride and hydrogen sulfide are named as follows:

Name of the molecular compound		*Name of the binary acid in water*	
HCl(*g*)	hydrogen chloride	HCl(*aq*)	*hydro*chlor*ic acid*
$H_2S(g)$	hydrogen sulfide	$H_2S(aq)$	*hydro*sulfur*ic acid*

Notice that the gaseous molecular substances are named in the usual way as binary compounds. *It is their aqueous solutions that are named as acids.*

When an acid is neutralized, the salt that is produced contains the anion formed by removing a hydrogen ion, H^+, from the acid molecule. Thus HCl yields salts containing the chloride ion, Cl^-. Similarly, HBr gives salts containing the bromide ion, Br^-. In general, then, neutralization of a binary acid yields the simple anion of the nonmetal.

In the name of an acid, the prefix *hydro-* tells us it is a binary acid. If the prefix *hydro-* is absent, it tells us the substance is not a binary acid.

Oxoacids contain hydrogen, oxygen, and another element

Acids that contain hydrogen, oxygen, plus another element are called **oxoacids.** Examples are H_2SO_4 and HNO_3. These acids do not take the prefix *hydro-*. Many nonmetals form two or more oxoacids that differ in the number of oxygen atoms in their formulas. When there are two oxoacids, the one with the larger number of oxygens takes the suffix *-ic* and the one with the fewer number of oxygens takes the suffix *-ous*.

H_2SO_4	sulfur*ic acid*	HNO_3	nitr*ic acid*
H_2SO_3	sulfur*ous acid*	HNO_2	nitr*ous acid*

The halogens can occur in as many as four different oxoacids. The oxoacid with the most oxygens has the prefix *per-*, and the one with the least has the prefix *hypo-*.

$HClO_4$	*per*chlor*ic acid*	$HClO_2$	chlor*ous acid*
$HClO_3$	chlor*ic acid*	$HClO$	*hypo*chlor*ous acid* (usually written HOCl)

This relationship between name of the acid and name of the anion carries over to other acids that end in the suffix *-ic*. For example, acetic acid gives the anion acetate, and citric acid gives the anion citrate.

The neutralization of oxoacids produces negative polyatomic ions. The name of the polyatomic ion is related to that of its parent acid.

(1) *-ic* acids give *-ate* anions: HNO_3 (nitr*ic acid*) $\longrightarrow$ NO_3^- (nitr*ate* ion)

(2) *-ous* acids give *-ite* anions: H_2SO_3 (sulfur*ous acid*) $\longrightarrow$ SO_3^{2-} (sulf*ite* ion)

In naming polyatomic anions, the prefixes *per-* and *hypo-* carry over from the name of the parent acid. Thus perchloric acid, $HClO_4$, gives perchlorate ion, ClO_4^-, and hypochlorous acid, HClO, gives hypochlorite ion, ClO^-.

EXAMPLE 4.5
Naming Acids and Their Salts

Bromine forms four oxoacids, similar to those of chlorine. What is the name of the acid $HBrO_2$ and what is the name of the salt $NaBrO_3$?

ANALYSIS AND SOLUTION: Let's review the acids formed by chlorine and then reason by analogy. For chlorine we have

$HClO_4$	perchloric acid	$HClO_2$	chlorous acid
$HClO_3$	chloric acid	$HClO$	hypochlorous acid

The acid $HBrO_2$ is similar to chlorous acid, so to name it we will use the stem of the element name bromine (brom-) in place of chlor-. Therefore, the name of $HBrO_2$ is *bromous acid.*

To find the name of $NaBrO_3$, let's begin by asking "What acid would give this salt by neutralization?" Neutralization involves removing an H^+ from the acid molecule and replacing it with a cation, in this case Na^+. Therefore, the salt $NaBrO_3$ would be obtained by neutralizing the acid $HBrO_3$. This acid has one more oxygen than bromous acid, $HBrO_2$, so it would have the ending *-ic*. Thus, $HBrO_3$ is named bromic acid. Neutralizing an acid that has a name that ends in *-ic* gives an anion with a name that ends in *-ate*, so the anion BrO_3^- is the bromate ion. Therefore, the salt $NaBrO_3$ is *sodium bromate.*

ARE THE ANSWERS REASONABLE? There's really not much we can do to check the answers here. For the salt, if $HClO_3$ is chloric acid, then it seems reasonable that $HBrO_3$ would be bromic acid, which would mean that BrO_3^- is the bromate ion and $NaBrO_3$ is sodium bromate.

Practice Exercise 11: The formula for arsenic acid is H_3AsO_4. What is the name of the salt Na_3AsO_4? (Hint: Recall how the name of the anion is related to the name of the acid.)

Practice Exercise 12: Formic acid is $HCHO_2$. What is the name of the salt $Ca(CHO_2)_2$?

Practice Exercise 13: Name the water solutions of the following acids: HF, HBr. Name the sodium salts formed by neutralizing the acids with NaOH.

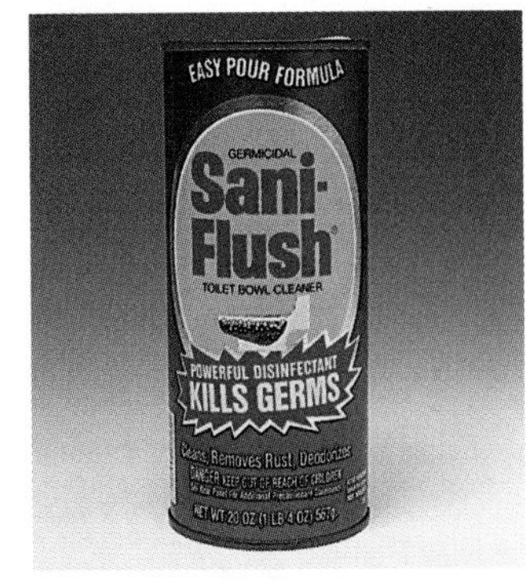

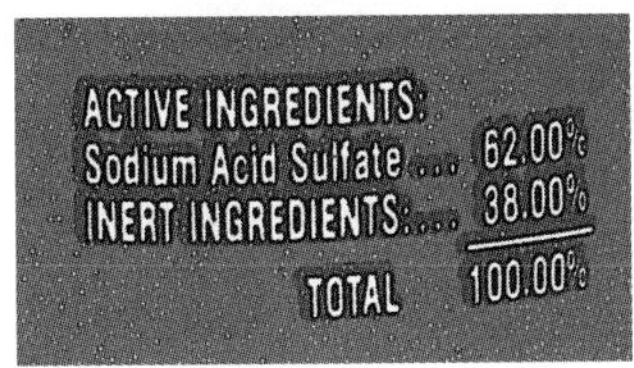

FIG. 4.13 **Many acid salts have useful applications.** As its active ingredient, this familiar product contains sodium hydrogen sulfate (sodium bisulfate), which the manufacturer calls "sodium acid sulfate." *(Robert Capece.)*

Acid salts can be formed by polyprotic acids

Monoprotic acids such as HCl and $HC_2H_3O_2$ have only one hydrogen that can be removed by neutralization and these acids form only one anion. However, polyprotic acids can be neutralized stepwise and the neutralization can be halted before all the hydrogens have been removed. For example, partial neutralization of H_2SO_4 gives the HSO_4^- ion, which forms salts such as $NaHSO_4$. This compound is called an **acid salt** because its anion, HSO_4^-, is capable of furnishing additional H^+.

In naming ions such as HSO_4^-, we specify the number of hydrogens that can still be neutralized if the anion were to be treated with additional base. Thus, HSO_4^- is called the hydrogen sulfate ion; it's the active ingredient in Sani-Flush (Figure 4.13). Similarly, $H_2PO_4^-$ is named as the dihydrogen phosphate ion. These anions give the following salts with Na^+:

$NaHSO_4$ sodium hydrogen sulfate
NaH_2PO_4 sodium dihydrogen phosphate

For acid salts of diprotic acids, the prefix ***bi-*** is still often used.

$NaHCO_3$ sodium bicarbonate
or
sodium hydrogen carbonate

Notice that the prefix bi- does *not* mean "two"; it means that there is an acidic hydrogen in the compound.

Practice Exercise 14: What is the formula for sodium bisulfite? What is the chemically correct name for this compound? (Hint: What information do we get from the prefix bi- and the suffix -ite?)

Practice Exercise 15: Write molecular equations for the stepwise neutralization of phosphoric acid by sodium hydroxide. What are the names of the salts that are formed?

Bases are named as hydroxides or molecules

Metal compounds that contain the ions OH^- or O^{2-}, such as NaOH and Na_2O, are ionic and are named just like any other ionic compound. Thus, NaOH is sodium hydroxide and Na_2O is sodium oxide.

Molecular bases such as NH_3 (ammonia) and CH_3NH_2(methylamine) are specified by just giving the name of the molecule.[6] There is nothing special in their names that tells us they are bases.

[6] Solutions of ammonia are sometimes called *ammonium hydroxide*, although there is no evidence that the species NH_4OH actually exists.

4.5 Ionic Reactions Can Often Be Predicted

In our discussion of the reaction of KI with $Pb(NO_3)_2$ (page 132), you saw that the net ionic equation reveals a change in the number of ions in solution when the reaction takes place. Such changes characterize ionic reactions in general. In this section you will learn how we can use the existence or nonexistence of a net ionic equation as a criterion to determine whether or not an ionic reaction occurs in a solution of mixed solutes.

TOOLS
Predicting net ionic equations

In general, a net ionic equation will exist (and a reaction will occur) under the following conditions:

A precipitate is formed from a mixture of soluble reactants.
An acid reacts with a base.
A weak electrolyte is formed from a mixture of strong electrolytes.
A gas is formed from a mixture of reactants.

It is also important to note that *no net reaction will occur if all the substances in the ionic equation cancel.* There will be no net ionic equation, and therefore no net reaction!

Predicting precipitation reactions

The reaction between $Pb(NO_3)_2$ and KI,

$$Pb(NO_3)_2(aq) + 2KI(aq) \longrightarrow PbI_2(s) + 2KNO_3(aq)$$

is just one example of a large class of ionic reactions in which cations and anions change partners. The technical term we use to describe them is **metathesis,** but they are also sometimes called **double replacement reactions.** (In the formation of the products, PbI_2 and KNO_3, the I^- replaces NO_3^- in the lead compound and NO_3^- replaces I^- in the potassium compound.) Metathesis reactions in which a precipitate forms are sometimes called **precipitation reactions.**

FACETS OF CHEMISTRY 4.2

Painful Precipitates—Kidney Stones

Each year, more than a million people in the United States are hospitalized because of very painful kidney stone attacks. A kidney stone is a hard mass developed from crystals that separate from the urine and build up on the inner surfaces of the kidney. The formation of the stones is caused primarily by the buildup of Ca^{2+}, $C_2O_4^{2-}$, and PO_4^{3-} ions in the urine. When the concentrations of those ions become large enough, the urine becomes supersaturated with respect to calcium oxalate and/or calcium phosphate and precipitates begin to form (70% to 80% of all kidney stones are made up of calcium oxalate and phosphate). If the crystals remain tiny enough, they can travel through the urinary tract and pass out of the body in the urine without being noticed. Sometimes, however, they continue to grow without being passed and can cause intense pain if they become stuck in the urinary tract.

Kidney stones don't all look alike. Their color depends on what substances are mixed with the inorganic precipitates (e.g., proteins or blood). Most are yellow or brown, as seen in the accompanying photo, but they can be tan, gold, or even black. Stones can be round, jagged, or even have branches. They vary in size from mere specks to pebbles to stones as big as golf balls!

A calcium oxalate kidney stone. Kidney stones such as this can be extremely painful. *(Photo courtesy of L.C. Herring and Company, Orlando, FL.)*

TABLE 4.1 Solubility Rules for Ionic Compounds in Water

Soluble Compounds

1. All compounds of the alkali metals (Group IA) are soluble.
2. All salts containing NH_4^+, NO_3^-, ClO_4^-, ClO_3^-, and $C_2H_3O_2^-$ are soluble.
3. All chlorides, bromides, and iodides (salts containing Cl^-, Br^-, or I^-) are soluble *except* when combined with Ag^+, Pb^{2+}, and Hg_2^{2+} (note the subscript "2").
4. All sulfates (salts containing SO_4^{2-}) are soluble *except* those of Pb^{2+}, Ca^{2+}, Sr^{2+}, Hg_2^{2+}, and Ba^{2+}.

Insoluble Compounds

5. All metal hydroxides (ionic compounds containing OH^-) and all metal oxides (ionic compounds containing O^{2-}) are insoluble *except* those of Group IA and those of Ca^{2+}, Sr^{2+}, and Ba^{2+}.

   When metal oxides do dissolve, they react with water to form hydroxides. The oxide ion, O^{2-}, does not exist in water. For example,

   $$Na_2O(s) + H_2O \longrightarrow 2NaOH(aq)$$

6. All salts that contain PO_4^{3-}, CO_3^{2-}, SO_3^{2-}, and S^{2-} are insoluble *except* those of Group IA and NH_4^+.

TOOLS
Solubility rules

Lead nitrate and potassium iodide react because one of the products is insoluble. This is what leads to a net ionic equation. Such reactions can be predicted if we know which substances are soluble and which are insoluble. To help us, we can use a set of **solubility rules** (Table 4.1) to tell us, in many cases, whether an ionic compound is soluble or insoluble. To make the rules easier to remember, they are divided into two categories. The first includes compounds that are soluble, with some exceptions. The second describes compounds that are generally insoluble, with some exceptions. Some examples will help clarify their use.

Rule 1 states that all compounds of the alkali metals are soluble in water. This means that you can expect *any* compound containing Na^+ or K^+, or any of the Group IA metal ions, *regardless of the anion*, to be soluble. If one of the reactants in a metathesis is Na_3PO_4, you now know from Rule 1 that it is soluble. Therefore, you would write it in *dissociated* form in the ionic equation. Similarly, Rule 6 states, in part, that all carbonate compounds are *insoluble* except those of the alkali metals and the ammonium ion. If one of the products in a metathesis reaction is $CaCO_3$, you'd expect it to be insoluble, because the cation is not an alkali metal or NH_4^+. Therefore, you would write its formula in undissociated form as $CaCO_3(s)$ in the ionic equation.

Let's look at an example that illustrates how we can use the rules to predict the outcome of a reaction.

EXAMPLE 4.6
Predicting Reactions and Writing Their Equations

Predict whether a reaction will occur when aqueous solutions of $Fe_2(SO_4)_3$ and $Pb(NO_3)_2$ are mixed. Write molecular, ionic, and net ionic equations for it.

ANALYSIS: We know the molecular equation will take the form

$$Fe_2(SO_4)_3 + Pb(NO_3)_2 \longrightarrow$$

To complete the equation we have to determine the makeup of the products. We begin, therefore, by predicting what a double replacement (metathesis) might produce. Then we proceed to expand the molecular equation into an ionic equation, and finally we drop spectator ions to obtain the net ionic equation. The existence of a net ionic equation tells us that a reaction does indeed take place. To accomplish all of this we need to know solubilities, and here our tool is the solubility rules.

The critical step in determining whether a reaction occurs is obtaining a net ionic equation.

SOLUTION: The reactants, $Pb(NO_3)_2$ and $Fe_2(SO_4)_3$, contain the ions Pb^{2+} and NO_3^-, and Fe^{3+} and SO_4^{2-}, respectively. To write the formulas of the products, we interchange anions. We combine Pb^{2+} with SO_4^{2-}, and for electrical neutrality, we must use one ion of each. Therefore, we write $PbSO_4$ as one possible product. For the other product, we combine Fe^{3+} with NO_3^-. Electrical neutrality now demands that we use *three* NO_3^- to *one* Fe^{3+} to make $Fe(NO_3)_3$. The correct formulas of the products, then, are $PbSO_4$ and $Fe(NO_3)_3$. The unbalanced molecular equation at this point is

$$Fe_2(SO_4)_3 + Pb(NO_3)_2 \longrightarrow Fe(NO_3)_3 + PbSO_4 \qquad \text{(unbalanced)}$$

Always write equations in two steps: First write correct formulas for the reactants and products, then adjust the coefficients to balance the equation.

Next, let's determine solubilities. The reactants are ionic compounds and we are told that they are in solution, so we know they are water soluble. Solubility Rules 2 and 4 tell us this also. For the products, we find that Rule 2 says that all nitrates are soluble, so $Fe(NO_3)_3$ is soluble; Rule 4 tells us that the sulfate of Pb^{2+} is *insoluble*. This means that a precipitate of $PbSO_4$ will form. Writing (*aq*) and (*s*) following appropriate formulas, the unbalanced molecular equation is

$$Fe_2(SO_4)_3(aq) + Pb(NO_3)_2(aq) \longrightarrow Fe(NO_3)_3(aq) + PbSO_4(s) \qquad \text{(unbalanced)}$$

When balanced, we obtain the *molecular equation*.

$$Fe_2(SO_4)_3(aq) + 3Pb(NO_3)_2(aq) \longrightarrow 2Fe(NO_3)_3(aq) + 3PbSO_4(s)$$

Next, we expand this to give the *ionic equation* in which soluble compounds are written in dissociated (separated) form as ions, and insoluble compounds are written in "molecular" form. Once again, we are careful to apply the subscripts of the ions and the coefficients.

$$2Fe^{3+}(aq) + 3SO_4^{2-}(aq) + 3Pb^{2+}(aq) + 6NO_3^-(aq) \longrightarrow 2Fe^{3+}(aq) + 6NO_3^-(aq) + 3PbSO_4(s)$$

By removing spectator ions (Fe^{3+} and NO_3^-), we obtain

$$3Pb^{2+}(aq) + 3SO_4^{2-}(aq) \longrightarrow 3PbSO_4(s)$$

Finally, we reduce the coefficients to give us the correct *net ionic equation.*

$$Pb^{2+}(aq) + SO_4^{2-}(aq) \longrightarrow PbSO_4(s)$$

The existence of the net ionic equation confirms that a reaction does take place between lead nitrate and iron(III) sulfate.

IS THE ANSWER REASONABLE? One of the main things we have to check in solving a problem such as this is that we've written the correct formulas of the products. For example, in this problem some students might be tempted (without thinking) to write $Pb(SO_4)_2$ and $Fe_2(NO_3)_3$, or even $Pb(SO_4)_3$ and $Fe_2(NO_3)_2$. *This is a common error.* Always be careful to figure out the charges on the ions that must be combined in the formula. Then take the ions in a ratio that gives an electrically neutral formula unit.

Once we're sure the formulas of the products are right, we check that we've applied the solubility rules correctly, which we have. Then we check that we've properly balanced the equation (We have.), that we've correctly divided the soluble compounds into their ions (We have.), and that we've eliminated the spectator ions to obtain the net ionic equation (We have.).

Practice Exercise 16: Show that in aqueous solutions there is no net reaction between $Zn(NO_3)_2$ and $Ca(C_2H_3O_2)_2$. (Hint: Write molecular, ionic, and net ionic equations.)

Practice Exercise 17: Predict the reaction that occurs on mixing the following solutions. Write molecular, ionic, and net ionic equations for the reactions that take place. (a) $AgNO_3$ and NH_4Cl, (b) sodium sulfide and lead acetate.

Predicting acid-base reactions

Earlier we discussed neutralization as one of the key properties of acids and bases. Many such reactions can be viewed as metathesis. An example is the reaction between HCl and NaOH.

$$\mathrm{HCl}(aq) + \mathrm{NaOH}(aq) \longrightarrow \mathrm{NaCl}(aq) + \mathrm{H_2O}$$

Writing this as an ionic equation gives

$$\mathrm{H^+}(aq) + \mathrm{Cl^-}(aq) + \mathrm{Na^+}(aq) + \mathrm{OH^-}(aq) \longrightarrow \mathrm{Na^+}(aq) + \mathrm{Cl^-}(aq) + \mathrm{H_2O}$$

where we have used H^+ as shorthand for H_3O^+. The net ionic equation is obtained by removing spectator ions.

$$\mathrm{H^+}(aq) + \mathrm{OH^-}(aq) \longrightarrow \mathrm{H_2O}$$

In this case, a net ionic equation exists because of the formation of a very weak electrolyte, H_2O, instead of a precipitate. In fact, we find this same net ionic equation for any reaction between a strong acid and a soluble strong base.

The formation of water in a neutralization reaction is such a strong driving force for reaction that it will form even if the acid is weak or if the base is insoluble, or both. Here are some examples.

Reaction of a weak acid with a strong base

Molecular equation:

$$\underset{\text{weak acid}}{\mathrm{HC_2H_3O_2}(aq)} + \underset{\text{strong base}}{\mathrm{NaOH}(aq)} \longrightarrow \mathrm{NaC_2H_3O_2}(aq) + \mathrm{H_2O}$$

Net ionic equation:

$$\mathrm{HC_2H_3O_2}(aq) + \mathrm{OH^-}(aq) \longrightarrow \mathrm{C_2H_3O_2^-}(aq) + \mathrm{H_2O}$$

This reaction is illustrated in Figure 4.14.

Reaction of a strong acid with an insoluble base

Figure 4.15 shows the reaction of hydrochloric acid with milk of magnesia, which contains $Mg(OH)_2$.

Molecular equation:

$$\underset{\text{strong acid}}{2\mathrm{HCl}(aq)} + \underset{\text{insoluble base}}{\mathrm{Mg(OH)_2}(s)} \longrightarrow \mathrm{MgCl_2}(aq) + 2\mathrm{H_2O}$$

Net ionic equation:

$$2\mathrm{H^+}(aq) + \mathrm{Mg(OH)_2}(s) \longrightarrow \mathrm{Mg^{2+}}(aq) + 2\mathrm{H_2O}$$

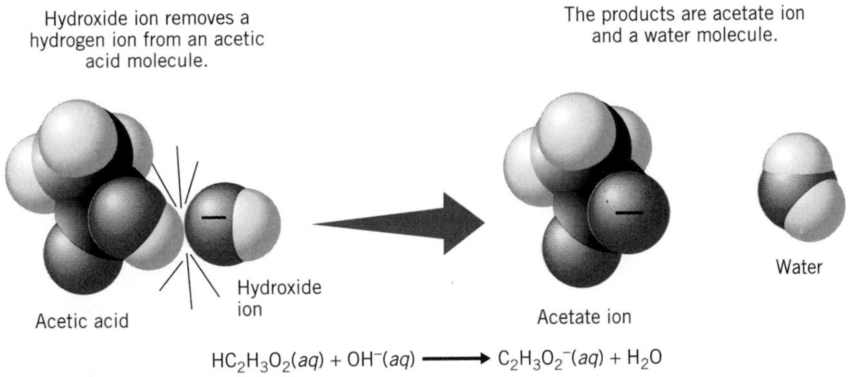

Hydroxide ion removes a hydrogen ion from an acetic acid molecule.

The products are acetate ion and a water molecule.

Acetic acid

Hydroxide ion

Acetate ion

Water

$$\mathrm{HC_2H_3O_2}(aq) + \mathrm{OH^-}(aq) \longrightarrow \mathrm{C_2H_3O_2^-}(aq) + \mathrm{H_2O}$$

FIG. 4.14 Net reaction of acetic acid with a strong base. The neutralization of acetic acid by hydroxide ion occurs primarily by the removal of H^+ from acetic acid molecules by OH^- ions.

FIG. 4.15 Hydrochloric acid is neutralized by milk of magnesia. A solution of hydrochloric acid is added to a beaker containing milk of magnesia. The thick white solid in milk of magnesia is magnesium hydroxide, $Mg(OH)_2$, which is able to neutralize the acid. The mixture is clear where some of the solid $Mg(OH)_2$ has already reacted and dissolved. *(Andy Washnik.)*

Reaction of a weak acid with an insoluble base

Molecular equation:

$$2HC_2H_3O_2(aq) + Mg(OH)_2(s) \longrightarrow Mg(C_2H_3O_2)_2(aq) + 2H_2O$$

Net ionic equation:

$$2HC_2H_3O_2(aq) + Mg(OH)_2(s) \longrightarrow Mg^{2+}(aq) + 2C_2H_3O_2^-(aq) + 2H_2O$$

Notice that in the last two examples, the formation of water drives the reaction, even though one of the reactants is insoluble. To correctly write the ionic and net ionic equations, it is important to know both the solubility rules and which acids are strong and weak. If you've learned the list of strong acids, you can expect that any acid *not* on the list will be a weak acid. (Unless specifically told otherwise, you should assume weak acids to be water soluble.)

Reaction of an acid with a weak base

Acid–base neutralization doesn't always involve the formation of water. We see this in the reaction of an acid with a weak base such as NH_3. For a strong acid such as HCl, we have

Molecular equation:

$$HCl(aq) + NH_3(aq) \longrightarrow NH_4Cl(aq)$$

Net ionic equation (using H^+ as shorthand for H_3O^+):

$$H^+(aq) + NH_3(aq) \longrightarrow NH_4^+(aq)$$

Figure 4.16 depicts the transfer of H^+ from H_3O^+ to NH_3.

With a weak acid such as $HC_2H_3O_2$, we have

Molecular equation:

$$HC_2H_3O_2(aq) + NH_3(aq) \longrightarrow NH_4C_2H_3O_2(aq)$$

Ionic and net ionic equation:

$$HC_2H_3O_2(aq) + NH_3(aq) \longrightarrow NH_4^+(aq) + C_2H_3O_2^-(aq)$$

Even though solutions of $HC_2H_3O_2$ contain some H^+, and solutions of NH_3 contain some OH^-, when these solutions are mixed the predominant reaction is between molecules of acid and base. This is illustrated in Figure 4.17.

Practice Exercise 18: Write the molecular, ionic, and net ionic equations for the neutralization of $HNO_3(aq)$ by $Ca(OH)_2(aq)$. (Hint: First determine whether the acid and base are strong or weak.)

Practice Exercise 19: Write molecular, ionic, and net ionic equations for the reaction of (a) HCl with KOH, (b) $HCHO_2$ with LiOH, and (c) N_2H_4 with HCl.

Practice Exercise 20: Write molecular, ionic, and net ionic equations for the reaction of the weak base methylamine, CH_3NH_2, with formic acid, $HCHO_2$ (a weak acid).

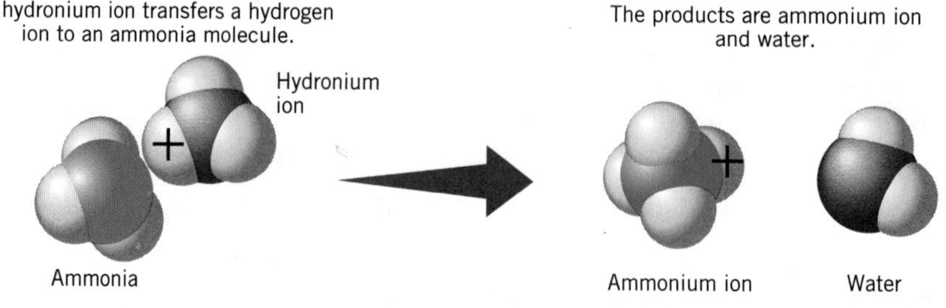
A hydronium ion transfers a hydrogen ion to an ammonia molecule.
The products are ammonium ion and water.
Hydronium ion
Ammonia
Ammonium ion
Water
$NH_3(aq) + H_3O^+(aq) \longrightarrow NH_4^+(aq) + H_2O$

FIG. 4.16 Reaction of ammonia with a strong acid. The reaction occurs primarily by the direct attack of H_3O^+ on NH_3 molecules. Transfer of a proton to the ammonia molecule produces an ammonium ion and a water molecule.

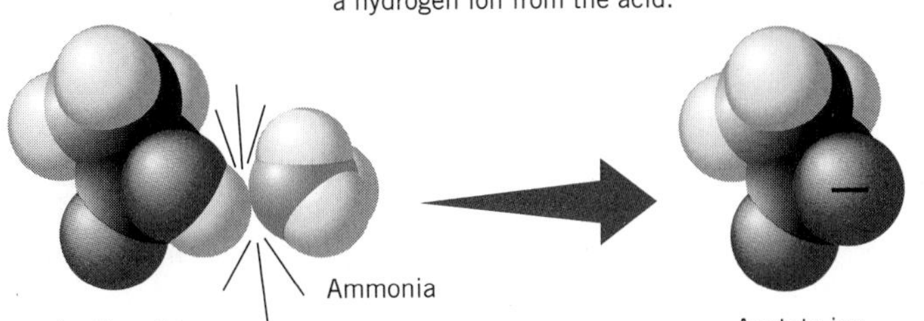

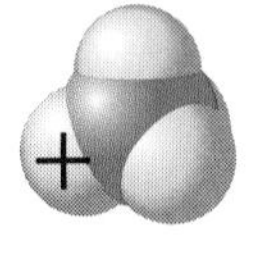

$$HC_2H_3O_2(aq) + NH_3(aq) \longrightarrow C_2H_3O_2^-(aq) + NH_4^+(aq)$$

FIG. 4.17 Reaction of acetic acid with ammonia. The collision of an ammonia molecule with an acetic acid molecule leads to a transfer of H^+ from the acetic acid to ammonia and the formation of ions.

Predicting reactions in which a gas is formed

Sometimes a product of a metathesis reaction is a substance that normally is a gas at room temperature and is not very soluble in water. The most common example is carbon dioxide. This product forms when an acid reacts with either a bicarbonate or carbonate. For example, as a sodium bicarbonate solution is added to hydrochloric acid, bubbles of carbon dioxide are released (Figure 4.18). This is the same reaction that occurs if you take sodium bicarbonate to soothe an upset stomach. Stomach acid is HCl and its reaction with the $NaHCO_3$ both neutralizes the acid and produces CO_2 gas (burp!). The molecular equation for the metathesis reaction is

$$HCl(aq) + NaHCO_3(aq) \longrightarrow NaCl(aq) + H_2CO_3(aq)$$

FIG. 4.18 The reaction of sodium bicarbonate with hydrochloric acid. The bubbles contain the gas carbon dioxide. *(Michael Watson.)*

Carbonic acid, H_2CO_3, is too unstable to be isolated in pure form. When it forms in appreciable amounts as a product in a metathesis reaction, it decomposes into its anhydride (the gas CO_2) and water. Carbon dioxide is only slightly soluble in water, so most of the CO_2 bubbles out of the solution. The decomposition reaction is

$$H_2CO_3(aq) \longrightarrow H_2O + CO_2(g)$$

Therefore, the overall molecular equation for the reaction is

$$HCl(aq) + NaHCO_3(aq) \longrightarrow NaCl(aq) + H_2O + CO_2(g)$$

The ionic equation is

$$H^+(aq) + Cl^-(aq) + Na^+(aq) + HCO_3^-(aq) \longrightarrow Na^+(aq) + Cl^-(aq) + H_2O + CO_2(g)$$

and the net ionic equation is

$$H^+(aq) + HCO_3^-(aq) \longrightarrow H_2O + CO_2(g)$$

Similar results are obtained if we begin with a carbonate instead of a bicarbonate. In this case, hydrogen ions combine with carbonate ions to give H_2CO_3, which subsequently decomposes to water and CO_2.

$$2H^+(aq) + CO_3^{2-}(aq) \longrightarrow H_2CO_3(aq) \longrightarrow H_2O + CO_2(g)$$

The net reaction is

$$2H^+(aq) + CO_3^{2-}(aq) \longrightarrow H_2O + CO_2(g)$$

The release of CO_2 by the reaction of a carbonate with an acid is such a strong driving force for reaction that it enables insoluble carbonates to dissolve in acids (strong and weak). The reaction of limestone, $CaCO_3$, with hydrochloric acid is shown in Figure 4.19. The molecular and net ionic equations for the reaction are as follows:

$$CaCO_3(s) + 2HCl(aq) \longrightarrow CaCl_2(aq) + CO_2(g) + H_2O$$
$$CaCO_3(s) + 2H^+(aq) \longrightarrow Ca^{2+}(aq) + CO_2(g) + H_2O$$

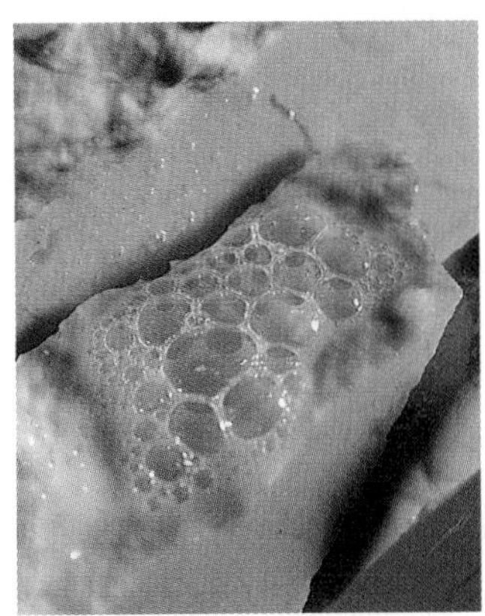

FIG. 4.19 Limestone reacts with acid. Bubbles of CO_2 are formed in the reaction of limestone ($CaCO_3$) with hydrochloric acid. *(Andy Washnik.)*

Carbon dioxide is not the only gas formed in metathesis reactions. Table 4.2 lists others and the reactions that form them.

TOOLS
Gases formed in metathesis reactions

TABLE 4.2 Gases Formed in Metathesis Reactions

Gas	Formed by Reaction of Acids with:	Equation for Formation[a]
H_2S	Sulfides	$2H^+ + S^{2-} \longrightarrow H_2S$
HCN	Cyanides	$H^+ + CN^- \longrightarrow HCN$
CO_2	Carbonates	$2H^+ + CO_3^{2-} \longrightarrow (H_2CO_3) \longrightarrow H_2O + CO_2$
	Bicarbonates (hydrogen carbonates)	$H^+ + HCO_3^- \longrightarrow (H_2CO_3) \longrightarrow H_2O + CO_2$
SO_2	Sulfites	$2H^+ + SO_3^{2-} \longrightarrow (H_2SO_3) \longrightarrow H_2O + SO_2$
	Bisulfites (hydrogen sulfites)	$H^+ + HSO_3^- \longrightarrow (H_2SO_3) \longrightarrow H_2O + SO_2$
Gas	**Formed by Reaction of Bases with:**	**Equation for Formation**
NH_3	Ammonium salts[b]	$NH_4^+ + OH^- \longrightarrow NH_3 + H_2O$

[a]Formulas in parentheses are of unstable compounds that break down according to the continuation of the sequence.
[b]In writing a metathesis reaction, you may be tempted sometimes to write NH_4OH as a formula for "ammonium hydroxide." That compound does not exist. In water, it is nothing more than a solution of NH_3.

EXAMPLE 4.7
Predicting Reactions and Writing Their Equations

What reaction (if any) occurs when solutions of ammonium carbonate, $(NH_4)_2CO_3$, and propionic acid, $HC_3H_5O_2$, are mixed?

ANALYSIS: Our tools for working a problem such as this are the list of strong acids (page 140), the solubility rules (Table 4.1, page 147), and the list of gases formed in metathesis reactions (Table 4.2). We begin by writing a potential metathesis equation in molecular form. Then we examine the reactants and products to see if any are weak electrolytes or substances that give gases. We also look for soluble or insoluble ionic compounds. Then we form the ionic equation and search for spectator ions which we eliminate to obtain the net ionic equation.

SOLUTION: We begin by constructing a molecular equation, treating the reaction as a metathesis. For the acid, we take the cation to be H^+ and the anion to be $C_3H_5O_2^-$. Therefore, after exchanging cations between the two anions we can obtain the following balanced molecular equation.

$$(NH_4)_2CO_3 + 2HC_3H_5O_2 \longrightarrow 2NH_4C_3H_5O_2 + H_2CO_3$$

In the statement of the problem we are told that we are working with a *solution* of $HC_3H_5O_2$, so we know it's soluble. Also, it is not on the list of strong acids, so we expect it to be a weak acid; we will write it in molecular form in the ionic equation.

Next, we recognize that H_2CO_3 decomposes into $CO_2(g)$ and H_2O. (This information is also found in Table 4.2.) Let's rewrite the molecular equation taking this into account.

$$(NH_4)_2CO_3 + 2HC_3H_5O_2 \longrightarrow 2NH_4C_3H_5O_2 + CO_2(g) + H_2O$$

Next, we need to determine which of the ionic substances are soluble. The solubility rules tell us that all ammonium salts are soluble, and we know that all salts are strong electrolytes. Therefore, we will write $(NH_4)_2CO_3$ and $NH_4C_3H_5O_2$ in dissociated form. Now we are ready to expand the molecular equation into the ionic equation.

$$2NH_4^+(aq) + CO_3^{2-}(aq) + 2HC_3H_5O_2(aq) \longrightarrow 2NH_4^+(aq) + 2C_3H_5O_2^-(aq) + CO_2(g) + H_2O$$

The only spectator ion is NH_4^+. Dropping this gives the net ionic equation.

$$CO_3^{2-}(aq) + 2HC_3H_5O_2(aq) \longrightarrow 2C_3H_5O_2^-(aq) + CO_2(g) + H_2O$$

IS THE ANSWER REASONABLE? There are some common errors that people make in working problems of this kind, so it is important to double-check. First, *proceed carefully.* Be sure you've written the formulas of the products correctly. (If you need review, you might look at Example 4.6 on page 147.) Look for weak acids. (You need to know the list of strong ones; if an acid isn't on the list, it's a weak acid.) Look for gases, or substances that decompose into gases. (Be sure you've studied Table 4.2.) Check for insoluble compounds. (You need to know the solubility rules in Table 4.1.) If you've learned what is expected of you, and checked each step, it is likely your result is correct.

EXAMPLE 4.8
Predicting Reactions and Writing Their Equations

What reaction (if any) occurs in water between potassium nitrate and ammonium chloride?

ANALYSIS: First, we have to convert the names of the compounds into chemical formulas. Here we use the principles of nomenclature rules from Chapter 2. The ions in potassium nitrate are K^+ and NO_3^-, so the salt has the formula KNO_3. In ammonium chloride, the ions are NH_4^+ and Cl^-, so the salt is NH_4Cl. Now we can proceed to writing molecular, ionic, and net ionic equations as in the preceding example.

SOLUTION: First we write the molecular equation, being sure to construct correct formulas for the products.

$$KNO_3 + NH_4Cl \longrightarrow KCl + NH_4NO_3$$

Looking over the substances in the equation, we don't find any that are weak acids or that decompose to give gases. Next, we check solubilities.

Solubility Rule 2 tells us that both KNO_3 and NH_4Cl are soluble. By solubility Rules 1 and 2, both products are also soluble in water. The anticipated molecular equation is therefore

$$\underset{\text{soluble}}{KNO_3(aq)} + \underset{\text{soluble}}{NH_4Cl(aq)} \longrightarrow \underset{\text{soluble}}{KCl(aq)} + \underset{\text{soluble}}{NH_4NO_3(aq)}$$

and the ionic equation is

$$K^+(aq) + NO_3^-(aq) + NH_4^+(aq) + Cl^-(aq) \longrightarrow$$
$$K^+(aq) + Cl^-(aq) + NH_4^+(aq) + NO_3^-(aq)$$

Notice that the right side of the equation is the same as the left side except for the order in which the ions are written. When we eliminate spectator ions, everything goes. There is no net ionic equation, which means there is no net reaction.

IS THE ANSWER REASONABLE? Once again, we perform the same checks here as in Example 4.7, and they tell us our answer is right.

Practice Exercise 21: Knowing that salts of the formate ion, CHO_2^-, are water soluble, predict the reaction between $Co(OH)_2$ and formic acid, $HCHO_2$. Write molecular, ionic, and net ionic equations. (Hint: Apply the tools we used in Example 4.7.)

Practice Exercise 22: Predict whether a reaction will occur in aqueous solution between the following pairs of substances. Write molecular, ionic, and net ionic equations. (a) $KCHO_2$ and HCl, (b) $CuCO_3$ and $HC_2H_3O_2$, (c) calcium acetate and silver nitrate, and (d) sodium hydroxide and nickel(II) chloride.

4.6 THE COMPOSITION OF A SOLUTION IS DESCRIBED BY ITS CONCENTRATION

As you learned earlier, the composition of a solution is specified by giving its *concentration*. Percentage concentration (grams of solute per 100 g of solution) was used as an example. To deal with the stoichiometry of reactions in solution, however, percentage concentration is not a convenient way to express concentrations of solutes. Instead, we express the amount of solute in moles and the amount of solution in liters.

The **molar concentration,** or **molarity** (abbreviated *M*), of a solution is defined as *the number of moles of solute per liter of solution*. It is a ratio of moles of solute to the volume of the solution expressed in liters.

$$\text{Molarity } (M) = \frac{\text{moles of solute}}{\text{liters of solution}} \tag{4.4}$$

Thus, a solution that contains 0.100 mol of NaCl in 1.00 L has a molarity of 0.100 *M*, and we would refer to the solution as 0.100 *molar* NaCl or as 0.100 *M* NaCl. The same concentration would result if we dissolved 0.0100 mol of NaCl in 0.100 L (100 mL) of solution, because the *ratio* of moles of solute to volume of solution is the same.

$$\frac{0.100 \text{ mol NaCl}}{1.00 \text{ L NaCl soln}} = \frac{0.0100 \text{ mol NaCl}}{0.100 \text{ L NaCl soln}} = 0.100\ M \text{ NaCl}$$

Molarity is a conversion factor relating moles of solute and volume of a solution

Whenever we have to deal with a problem that involves an amount of a chemical and a volume of a solution of that substance, you can expect that solving the problem will involve molarity.

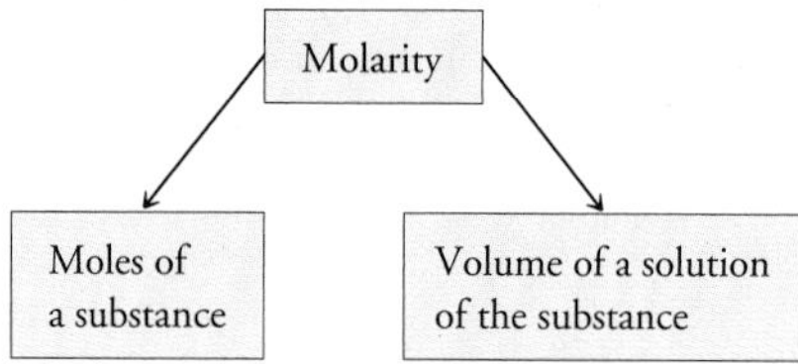

Molarity is a tool that provides the conversion factors we need to convert between moles and volume (either in liters or milliliters). Consider, for example, a solution labeled 0.100 *M* NaCl. The unit *M* always translates to mean "moles per liter," so we can write

$$0.100\ M \text{ NaCl} = \frac{0.100 \text{ mol NaCl}}{1.00 \text{ L soln}}$$

This gives us an equivalence relationship between "mol NaCl" and "L soln" that we can use to form two conversion factors.[7]

$$0.100 \text{ mol NaCl} \Leftrightarrow 1.00 \text{ L soln}$$

$$\frac{0.100 \text{ mol NaCl}}{1.00 \text{ L NaCl soln}} \qquad \frac{1.00 \text{ L NaCl soln}}{0.100 \text{ mol NaCl}}$$

[7] Some students find it easier to translate the "1.00 L" part of these factors into the equivalent 1000 mL here rather than convert between liters and milliliters at some other stage of the calculation. Factors such as the two above, therefore, can be rewritten as follows whenever it is convenient. (Remember that "1000") in the following is regarded as having an infinite number of significant figures because, standing as it does for 1 L, it is part of the definition of molarity and is an exact number.)

$$\frac{0.100 \text{ mol NaCl}}{1000 \text{ mL NaCl soln}} \text{ and } \frac{1000 \text{ mL NaCl soln}}{0.100 \text{ mol NaCl}}$$

EXAMPLE 4.9
Calculating the Molarity of a Solution

To study the effect of dissolved salt on the rusting of an iron sample, a student prepared a solution of NaCl by dissolving 1.461 g of NaCl in a total volume of 250.0 mL. What is the molarity of this solution?

ANALYSIS: The tool we'll use to solve this problem is Equation 4.4, which defines molarity as the ratio of *moles of solute* to *liters of solution*. If we can find these two pieces of information, we can arrange them as a ratio:

$$\text{Molarity} = \frac{?\ \text{mol NaCl}}{?\ \text{L soln}}$$

Therefore, we have to convert 1.461 g of NaCl to moles of NaCl and 250.0 mL to liters. Then we simply divide one by the other to find the molarity.

SOLUTION: The number of moles of NaCl is found using the formula mass of NaCl, 58.443 g mol^{-1}.

$$1.461\ \cancel{\text{g NaCl}} \times \frac{1\ \text{mol NaCl}}{58.443\ \cancel{\text{g NaCl}}} = 0.02500\ \text{mol NaCl}$$

To find the volume of the solution in liters, we move the decimal three places to the left, so 250.0 mL equals 0.2500 L.

The ratio of moles to liters, therefore, is

$$\frac{0.02500\ \text{mol NaCl}}{0.2500\ \text{L}} = 0.1000\ M\ \text{NaCl}$$

If necessary, practice converting between liters and milliliters. It's a task you will have to perform frequently.

IS THE ANSWER REASONABLE? Let's use our answer to do a rough calculation of the amount of NaCl in the solution. If our answer is right, we should find a value not too far from the amount given in the problem (1.461 g). If we round the formula mass of NaCl to 60, and use 0.1 *M* as an approximate concentration, then one liter of the solution contains 0.1 mol of NaCl, or approximately 6 g of NaCl (one-tenth of 60 g). But 250 mL is only 1/4 of a liter, so the mass of NaCl will be approximately 1/4 of 6 g, or about 1.5 g. This is pretty close to the amount that was given in the problem, so our answer is probably correct.

Practice Exercise 23: A certain solution contains 16.9 g of HNO_3 dissolved in 125 mL of solution. Water is added until the volume is 175 mL. What is the molarity of HNO_3 in the final solution? (Hint: Does the amount of HNO_3 change when the water is added?)

Practice Exercise 24: Suppose 1.223 g of NaCl is added to the 250.0 mL of NaCl solution described in Example 4.9. If there is no change in the total volume of the solution, what is the molarity of the new NaCl solution?

EXAMPLE 4.10
Using Molar Concentrations

How many milliliters of 0.250 *M* NaCl solution must be measured to obtain 0.100 mol of NaCl?

ANALYSIS: We can restate the problem as follows:

$$0.100\ \text{mol NaCl} \Leftrightarrow ?\ \text{mL soln}$$

To relate moles and volume, the tool we use is the molarity.

$$0.250\ M\ \text{NaCl} = \frac{0.250\ \text{mol NaCl}}{1\ \text{L NaCl soln}}$$

The fraction on the right relates moles of NaCl to liters of the solution, which we can express as an equivalence.

$$0.250 \text{ mol NaCl} \Leftrightarrow 1 \text{ L NaCl soln}$$

The equivalence allows us to construct two conversion factors.

$$\frac{0.250 \text{ mol NaCl}}{1 \text{ L NaCl soln}} \quad \text{and} \quad \frac{1.00 \text{ L NaCl soln}}{0.250 \text{ mol NaCl}}$$

To obtain the answer, we select the one that will allow us to cancel the unit "mol NaCl."

SOLUTION: We operate with the second factor on 0.100 mol NaCl.

$$0.100 \cancel{\text{mol NaCl}} \times \frac{1.00 \text{ L NaCl soln}}{0.250 \cancel{\text{mol NaCl}}} = 0.400 \text{ L of } 0.250 \, M \text{ NaCl}$$

Because 0.400 L corresponds to 400 mL, 400 mL of 0.250 *M* NaCl provides 0.100 mol of NaCl.

IS THE ANSWER REASONABLE? The molarity tells us one liter contains 0.250 mol NaCl, so we need somewhat less than half of a liter (500 mL) to obtain just 0.100 mol. The answer, 400 mL, is reasonable.

Practice Exercise 25: A student measured 175 mL of 0.250 *M* HCl solution into a beaker. How many moles of HCl were in the beaker? (Hint: Molarity gives the equivalence between moles of solute and volume of solution in liters.)

Practice Exercise 26: How many milliliters of 0.250 *M* HCl solution contain 1.30 g of HCl?

Moles of solute can always be obtained from the volume and molarity of a solution

If you worked Practice Exercise 25 you learned that we can use the volume and molarity of a solution to calculate the number of moles of solute in it. This is such a useful relationship that it warrants special attention. Solving Equation 4.4 for *moles of solute* gives

Molarity times volume gives moles

$$\text{molarity} \times \text{volume (L)} = \text{moles of solute} \qquad (4.5)$$

$$\frac{\text{mol solute}}{\cancel{\text{L soln}}} \times \cancel{\text{L soln}} = \text{mol solute}$$

Thus, *any time you know both the molarity and volume of a solution, you can easily calculate the number of moles of solute in it.* As you will see, this concept will be very useful in solving a variety of problems.

One situation in which Equation 4.5 is useful is when we must prepare some specific volume of a solution having a desired molarity (for example, 250 mL of 0.0800 *M* Na_2CrO_4). To proceed, we have to calculate the amount of solute that will be in the solution after it's made. Thus, in 250 mL of 0.0800 *M* Na_2CrO_4 there are

In the laboratory, 250 mL is easily measured to a precision equal to or greater than ±1 mL, so we will take 250 mL to have three significant figures.

$$\frac{0.0800 \text{ mol } Na_2CrO_4}{1 \cancel{\text{L soln}}} \times 0.250 \cancel{\text{L soln}} = 0.0200 \text{ mol } Na_2CrO_4$$

Figure 4.20 shows how we would use a 250 mL volumetric flask to prepare such a solution. (A *volumetric flask* is a narrow-necked flask having an etched mark high on its neck. When filled to the mark, the flask contains precisely the volume given by the flask's label.)

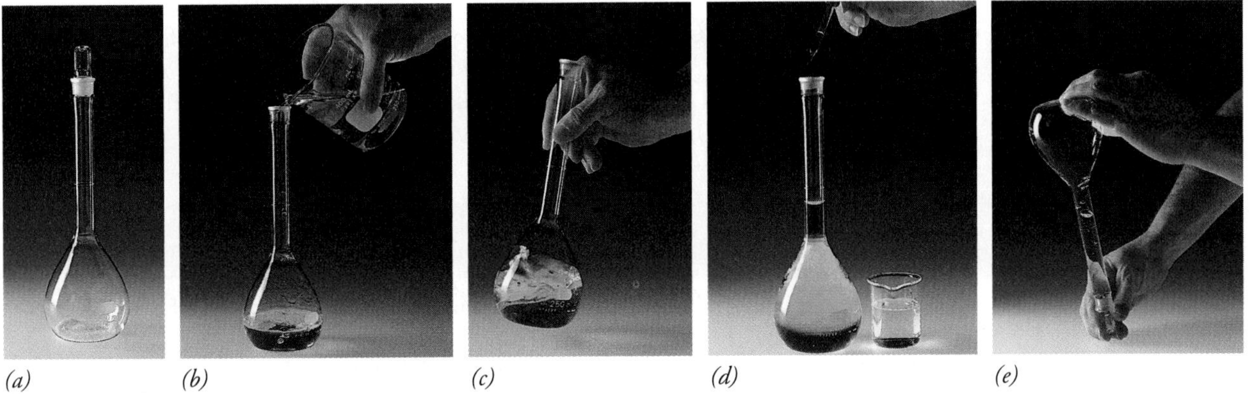

(a) (b) (c) (d) (e)

FIG. 4.20 **The preparation of a solution having a known molarity.** *(a)* A 250 mL volumetric flask, one of a number of sizes available for preparing solutions. When filled to the line etched around its neck, this flask contains exactly 250 mL of solution. The flask here already contains a weighed amount of solute. *(b)* Water is being added. *(c)* The solute is brought completely into solution before the level is brought up to the narrow neck of the flask. *(d)* More water is added to bring the level of the solution to the etched line. *(e)* The flask is stoppered and then inverted several times to mix its contents thoroughly. *(Michael Watson.)*

EXAMPLE 4.11
Preparing a Solution with a Known Molarity

Strontium nitrate, $Sr(NO_3)_2$, is used in fireworks to produce brilliant red colors. Suppose we need to prepare 250.0 mL of 0.100 *M* $Sr(NO_3)_2$ solution. How many grams of strontium nitrate are required?

ANALYSIS: The critical link in solving this problem is realizing that we know both the volume and molarity of the final solution, which permits us to calculate the number of moles of $Sr(NO_3)_2$ that will be in it. Once we know the number of moles of $Sr(NO_3)_2$, we can calculate its mass using the molar mass of the salt.

SOLUTION: You've learned that the product of molarity and volume equals moles of solute, so Equation 4.5 is the tool we need. The volume 250.0 mL converts to 0.2500 L. Therefore, multiplying the molarity by the volume in liters takes the following form.

$$\underbrace{\frac{0.100 \text{ mol Sr(NO}_3)_2}{1.00 \cancel{\text{ L Sr(NO}_3)_2 \text{ soln}}}}_{\text{molarity}} \times \underbrace{0.2500 \cancel{\text{ L Sr(NO}_3)_2 \text{ soln}}}_{\text{volume (L)}} = \underbrace{0.0250 \text{ mol Sr(NO}_3)_2}_{\text{moles of solute}}$$

Finally, we convert from moles to grams using the molar mass of $Sr(NO_3)_2$, which is 211.63 g mol^{-1}.

$$0.0250 \cancel{\text{ mol Sr(NO}_3)_2} \times \frac{211.63 \text{ g Sr(NO}_3)_2}{1 \cancel{\text{ mol Sr(NO}_3)_2}} = 5.29 \text{ g Sr(NO}_3)_2$$

Thus, to prepare the solution we need to dissolve 5.29 g of $Sr(NO_3)_2$ in a total volume of 250.0 mL.

We could also have set this up as a chain calculation as follows, with the conversion factors strung together.

$$0.2500 \cancel{\text{ L Sr(NO}_3)_2 \text{ soln}} \times \frac{0.100 \cancel{\text{ mol Sr(NO}_3)_2}}{1.00 \cancel{\text{ L Sr(NO}_3)_2 \text{ soln}}} \times \frac{211.63 \text{ g Sr(NO}_3)_2}{1 \cancel{\text{ mol Sr(NO}_3)_2}} = 5.29 \text{ g Sr(NO}_3)_2$$

IS THE ANSWER REASONABLE? If we were working with a full liter of this solution, it would contain 0.1 mol of $Sr(NO_3)_2$. The molar mass of the salt is 211.63 g mol^{-1}, so 0.1 mol is slightly more than 20 g. However, we are working with just a quarter of a liter (250 mL), so the amount of $Sr(NO_3)_2$ needed is slightly more than a quarter of 20 g, or 5 g. The answer, 5.29 g, is close to this, so it makes sense.

Practice Exercise 27: Suppose you wished to prepare 50 mL of 0.2 *M* $Sr(NO_3)_2$ solution. Using the kind of approximate arithmetic we employed in the Is the Answer Reasonable step in the preceding example, estimate the number of grams of $Sr(NO_3)_2$ required. [Hint: How many moles of $Sr(NO_3)_2$ would be in one liter of the solution?]

Practice Exercise 28: How many grams of $AgNO_3$ are needed to prepare 250.0 mL of 0.0125 *M* $AgNO_3$ solution?

Diluting a solution reduces the concentration

Laboratory chemicals are usually purchased in concentrated form and must be *diluted* (made less concentrated) before being used. This is accomplished by adding more solvent to the solution, which spreads the solute through a larger volume and causes the concentration (the amount per unit volume) to decrease.

During dilution, the amount of solute remains constant. This means that the product of molarity and volume, which equals the moles of solute, must be the same for both the concentrated and diluted solution.

$$\underbrace{\begin{pmatrix}\text{Volume of}\\ \text{dilute solution}\\ \textit{to be prepared}\end{pmatrix} \times M_{\text{dilute}}}_{\text{moles of solute in the dilute solution}} = \underbrace{\begin{pmatrix}\text{Volume of}\\ \text{concentrated solution}\\ \textit{to be used}\end{pmatrix} \times M_{\text{conc}}}_{\text{moles of solute in the concentrated solution}}$$

Or,

Dilution of solutions

$$V_{\text{dil}} \cdot M_{\text{dil}} = V_{\text{conc}} \cdot M_{\text{conc}} \tag{4.6}$$

Any units can be used for volume in Equation 4.6 provided that the volume units are the same on both sides of the equation. We thus normally solve dilution problems using *milliliters* directly in Equation 4.6.

EXAMPLE 4.12
Preparing a Solution of Known Molarity by Dilution

How can we prepare 100.0 mL of 0.0400 *M* $K_2Cr_2O_7$ from 0.200 *M* $K_2Cr_2O_7$?

ANALYSIS: This is the way such a question comes up in the lab, but what it is really asking is, "How many milliliters of 0.200 *M* $K_2Cr_2O_7$ (the more concentrated solution) must be diluted to give a solution with a final volume of 100.0 mL and a final molarity of 0.0400 *M*?" Once we see the question this way, we realize that Equation 4.6 is the tool we need to solve the problem.

SOLUTION: It's a good idea to assemble the data first, noting what is missing (and therefore what has to be calculated).

$$V_{\text{dil}} = 100.0 \text{ mL} \qquad M_{\text{dil}} = 0.0400\ M$$
$$V_{\text{conc}} = ? \qquad M_{\text{conc}} = 0.200\ M$$

Next, we use Equation 4.6 ($V_{dil} \times M_{dil} = V_{conc} \times M_{conc}$):

$$100.0 \text{ mL} \times 0.0400\ M = V_{conc} \times 0.200\ M$$

Solving for V_{conc} gives

$$V_{conc} = \frac{100.0 \text{ mL} \times 0.0400\ M}{0.200\ M}$$
$$= 20.0 \text{ mL}$$

Therefore, the answer to the question as asked is, We would withdraw 20.0 mL of 0.200 *M* $K_2Cr_2O_7$, place it in a 100 mL volumetric flask, and then add water until the final volume is exactly 100 mL. (See Figure 4.21.)

IS THE ANSWER REASONABLE? Notice that the concentrated solution is 5 times as concentrated as the dilute solution ($5 \times 0.04 = 0.2$). To reduce the concentration by a factor of 5 requires that we increase the volume by a factor of 5, and we see that 100 mL is 5 times 20 mL. The answer appears to be correct.

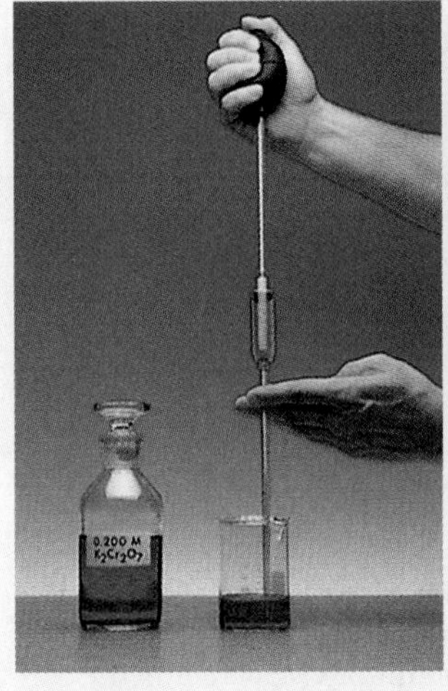

(a)

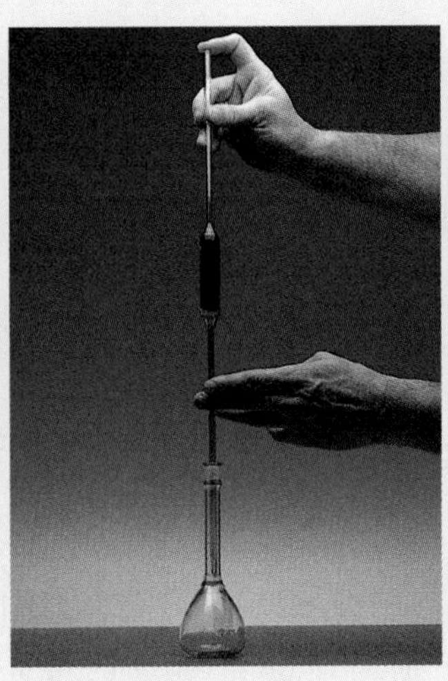
(b)

(c)

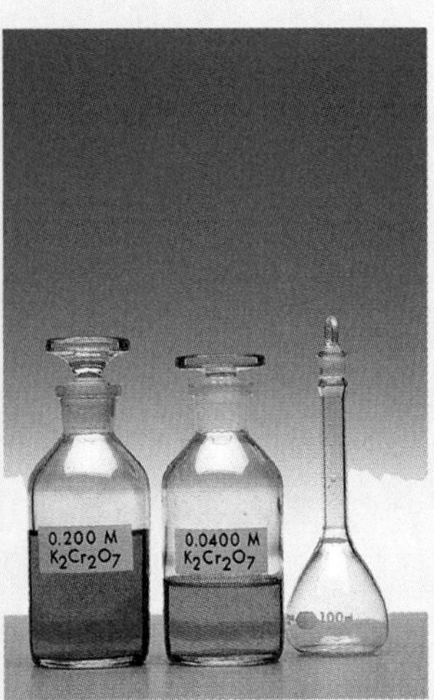

(d)

FIG. 4.21 Preparing a solution by dilution. *(a)* The calculated volume of the more concentrated solution is withdrawn from the stock solution by means of a volumetric pipet. *(b)* The solution is allowed to drain entirely from the pipet into the volumetric flask. *(c)* Water is added to the flask, the contents are mixed, and the final volume is brought up to the etch mark on the narrow neck of the flask. *(d)* The new solution is put into a labeled container. *(OPC, Inc.)*

Practice Exercise 29: To what final volume must 100.0 mL of 0.125 *M* H_2SO_4 solution be diluted to give a 0.0500 *M* H_2SO_4 solution? (Hint: Write the equation we used for dilution problems.)

Practice Exercise 30: How many milliliters of water have to be *added* to 150 mL of 0.50 *M* HCl to reduce the concentration to 0.10 *M* HCl?

4.7 MOLARITY IS USED FOR PROBLEMS IN SOLUTION STOICHIOMETRY

When we deal quantitatively with reactions in solution, we often work with volumes of solutions and molarity.

EXAMPLE 4.13
Stoichiometry Involving Reactions in Solution

One of the solids present in photographic film is silver bromide, AgBr. One way to prepare it is to mix solutions of silver nitrate and calcium bromide. Suppose we wished to prepare AgBr by the following precipitation reaction.

$$2AgNO_3(aq) + CaBr_2(aq) \longrightarrow 2AgBr(s) + Ca(NO_3)_2(aq)$$

How many milliliters of 0.125 *M* $CaBr_2$ solution must be used to react with the solute in 50.0 mL of 0.115 *M* $AgNO_3$?

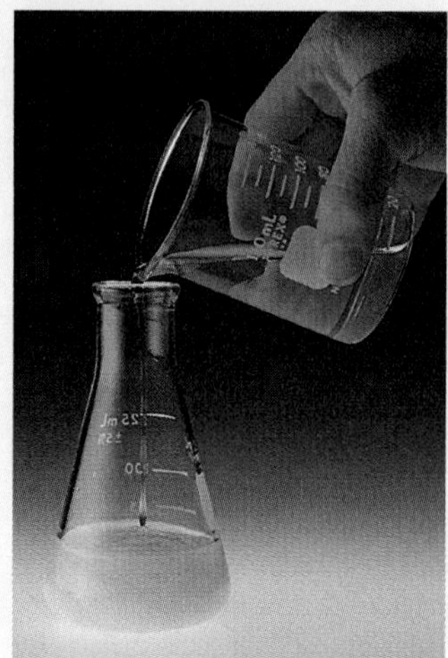

Silver bromide (AgBr) precipitates when solutions of calcium bromide and silver nitrate are mixed. *(Michael Watson.)*

ANALYSIS: As you've learned, when we have a stoichiometry problem dealing with a chemical reaction, the tool that relates the amounts of the substances is their coefficients in the equation. For this problem, therefore, we can write

$$2 \text{ mol } AgNO_3 \Leftrightarrow 1 \text{ mol } CaBr_2$$

However, we're not given moles directly. Instead we have molarities and the volume of the $AgNO_3$ solution. The critical link in solving the problem is recognizing that the molarity and volume of the $AgNO_3$ solution provides a path to finding the number of moles of $AgNO_3$.

Knowing that we will use molarity as a tool in working the problem, let's outline the path to the answer. We can calculate the moles of $AgNO_3$ by multiplying the volume and molarity of the $AgNO_3$ solution. We then use the coefficients in the equation to translate to moles of $CaBr_2$. Finally, we use the molarity of the $CaBr_2$ solution as a conversion factor to find the volume of the solution needed. The calculation flow will look like the following.

mol $AgNO_3$ —(coefficients of balanced equation)→ mol $CaBr_2$

volume and molarity of $AgNO_3$ solution (from **$AgNO_3$ soln** to mol $AgNO_3$)

moles of $CaBr_2$ and molarity of $CaBr_2$ solution (from mol $CaBr_2$ to **$CaBr_2$ soln**)

	$AgNO_3$ soln		**$CaBr_2$ soln**
Given:	50.0 mL of 0.115 M $AgNO_3$	To find:	? mL of 0.125 M $CaBr_2$

SOLUTION: First, we find the moles of $AgNO_3$ taken. Changing 50.0 mL to 0.0500 L,

$$\underbrace{0.0500 \text{ L } \cancel{AgNO_3 \text{ soln}}}_{\text{volume (L)}} \times \underbrace{\frac{0.115 \text{ mol } AgNO_3}{1.00 \text{ L } \cancel{AgNO_3 \text{ soln}}}}_{\text{molarity}} = 5.75 \times 10^{-3} \text{ mol } AgNO_3$$

Next, we use the coefficients of the equation to calculate the amount of $CaBr_2$ required.

$$5.75 \times 10^{-3} \cancel{\text{mol } AgNO_3} \times \frac{1 \text{ mol } CaBr_2}{2 \cancel{\text{mol } AgNO_3}} = 2.88 \times 10^{-3} \text{ mol } CaBr_2$$

Finally we calculate the volume (mL) of 0.125 M $CaBr_2$ that contains this many moles of $CaBr_2$. Here we use the fact that the molarity of the $CaBr_2$ solution, 0.125 M, gives two possible conversion factors:

$$\frac{0.125 \text{ mol } CaBr_2}{1.00 \text{ L } CaBr_2 \text{ soln}} \quad \text{and} \quad \frac{1.00 \text{ L } CaBr_2 \text{ soln}}{0.125 \text{ mol } CaBr_2}$$

We use the one that cancels the unit "mol $CaBr_2$."

$$2.88 \times 10^{-3} \cancel{\text{mol } CaBr_2} \times \frac{1.00 \text{ L } CaBr_2 \text{ soln}}{0.125 \cancel{\text{mol } CaBr_2}} = 0.0230 \text{ L } CaBr_2 \text{ soln}$$

Thus 0.0230 L, or 23.0 mL, of 0.125 M $CaBr_2$ has enough solute to combine with the $AgNO_3$ in 50.0 mL of 0.115 M $AgNO_3$.

IS THE ANSWER REASONABLE? The molarities of the two solutions are about the same, but only 1 mol of $CaBr_2$ is needed for each 2 mol $AgNO_3$. Therefore, the volume of $CaBr_2$ solution needed (23.0 mL) should be about half the volume of $AgNO_3$ solution taken (50.0 mL), which it is.

Practice Exercise 31: How many milliliters of 0.0475 *M* H_3PO_4 could be completely neutralized by 45.0 mL of 0.100 *M* KOH? The balanced equation for the reaction is

$$H_3PO_4(aq) + 3KOH(aq) \longrightarrow K_3PO_4(aq) + 3H_2O$$

(Hint: Outline the path of the calculations as in the preceding example.)

Practice Exercise 32: How many milliliters of 0.124 *M* NaOH contain enough NaOH to react with the H_2SO_4 in 15.4 mL of 0.108 *M* H_2SO_4 according to the following equation?

$$2NaOH(aq) + H_2SO_4(aq) \longrightarrow Na_2SO_4(aq) + 2H_2O$$

Net ionic equations can be used in stoichiometric calculations

In the preceding problem, we worked with a molecular equation in solving a stoichiometry problem. Ionic and net ionic equations can also be used, but this requires that we work with the concentrations of the ions in solution.

Calculating concentrations of ions in solutions of electrolytes

The concentrations of the ions in a solution of an electrolyte are obtained from the formula and molar concentration of the solute. For example, suppose we are working with a solution labeled "0.10 *M* $CaCl_2$." In 1.0 L of this solution there is 0.10 mol of $CaCl_2$, which is fully dissociated into Ca^{2+} and Cl^- ions.

$$CaCl_2 \longrightarrow Ca^{2+} + 2Cl^-$$

From the stoichiometry of the dissociation, we see that 1 mol Ca^{2+} and 2 mol Cl^- are formed from each 1 mol of $CaCl_2$. Therefore, 0.10 mol $CaCl_2$ will yield 0.10 mol Ca^{2+} and 0.20 mol Cl^-. In 0.10 *M* $CaCl_2$, then, the concentration of Ca^{2+} is 0.10 *M* and the concentration of Cl^- is 0.20 *M*. Thus,

The solution doesn't actually contain any $CaCl_2$, even though this is the solute used to prepare the solution. Instead, the solution contains Ca^{2+} and Cl^- ions.

The concentration of a particular ion equals the concentration of the salt multiplied by the number of ions of that kind in one formula unit of the salt.

TOOLS *Molarity of ions in a salt solution*

EXAMPLE 4.14
Calculating the Concentrations of Ions in a Solution

What are the molar concentrations of the ions in 0.20 *M* aluminum sulfate?

ANALYSIS: First, we need the formula for the solute. Aluminum forms the Al^{3+} ion and sulfate ion is SO_4^{2-}. For electrical neutrality, the formula of the salt must be $Al_2(SO_4)_3$. In the solution, the concentrations of the ions are determined by the stoichiometry of the salt. Therefore, we determine the number of ions of each kind formed from one formula unit of $Al_2(SO_4)_3$. These values are then used along with the given concentration of the salt to calculate the ion concentrations.

SOLUTION: When $Al_2(SO_4)_3$ dissolves, it dissociates as follows:

$$Al_2(SO_4)_3(s) \longrightarrow 2Al^{3+}(aq) + 3SO_4^{2-}(aq)$$

Each formula unit of $Al_2(SO_4)_3$ yields two Al^{3+} ions and three SO_4^{2-} ions. Therefore, 0.20 mol $Al_2(SO_4)_3$ yields 0.40 mol Al^{3+} and 0.60 mol SO_4^{2-}, and we conclude that the solution contains 0.40 *M* Al^{3+} and 0.60 *M* SO_4^{2-}.

IS THE ANSWER REASONABLE? The answers here have been obtained by simple mole reasoning. We could have found the answers in a more formal manner using the factor-label method. For example, for Al^{3+}, we have

$$\frac{0.20 \text{ } \cancel{\text{mol } Al_2(SO_4)_3}}{1.0 \text{ L soln}} \times \frac{2 \text{ mol } Al^{3+}}{1 \text{ } \cancel{\text{mol } Al_2(SO_4)_3}} = \frac{0.40 \text{ mol } Al^{3+}}{1.0 \text{ L soln}} = 0.40 \text{ } M \text{ } Al^{3+}$$

A similar calculation would give the concentration of SO_4^{2-} as 0.60 *M*. Study both methods. With just a little practice, you will have little difficulty with the reasoning approach we used first.

EXAMPLE 4.15
Calculating the Concentration of a Salt from the Concentration of One of Its Ions

A student found that the sulfate ion concentration in a solution of $Al_2(SO_4)_3$ was 0.90 *M*. What was the concentration of $Al_2(SO_4)_3$ in the solution?

ANALYSIS: Once again, we use the formula of the salt to determine the number of ions released when it dissociates. This time we use the information to work backward to find the salt concentration.

SOLUTION: Let's set up the problem using the factor label method to be sure of our procedure. We will use the fact that 1 mol $Al_2(SO_4)_3$ yields 3 mol SO_4^{2-} in solution.

$$1 \text{ mol } Al_2(SO_4)_3 \Leftrightarrow 3 \text{ mol } SO_4^{2-}$$

Therefore,

$$\frac{0.90 \text{ } \cancel{\text{mol } SO_4^{2-}}}{1.0 \text{ L soln}} \times \frac{1 \text{ mol } Al_2(SO_4)_3}{3 \text{ } \cancel{\text{mol } SO_4^{2-}}} = \frac{0.30 \text{ mol } Al_2(SO_4)_3}{1.0 \text{ L soln}} = 0.30 \text{ } M \text{ } Al_2(SO_4)_3$$

The concentration of $Al_2(SO_4)_3$ is 0.30 *M*.

IS THE ANSWER REASONABLE? We'll use the reasoning approach to check our answer. We know that 1 mol $Al_2(SO_4)_3$ yields 3 mol SO_4^{2-} in solution. Therefore, the number of moles of $Al_2(SO_4)_3$ is only one-third the number of moles of SO_4^{2-}. So the concentration of $Al_2(SO_4)_3$ must be one-third of 0.90 *M*, or 0.30 *M*.

Practice Exercise 33: What are the molar concentrations of the ions in 0.40 *M* $FeCl_3$? (Hint: How many ions of each kind are formed when $FeCl_3$ dissociates?)

Practice Exercise 34: In a solution of Na_3PO_4, the PO_4^{3-} concentration was determined to be 0.250 *M*. What was the sodium ion concentration in the solution?

Net ionic equations can be used in stoichiometry calculations

You have seen that a net ionic equation is convenient for describing the net chemical change in an ionic reaction. Let's study an example that illustrates how such equations can be used in stoichiometric calculations.

EXAMPLE 4.16
Stoichiometric Calculations Using a Net Ionic Equation

How many milliliters of 0.100 *M* $AgNO_3$ solution are needed to react completely with 25.0 mL of 0.400 *M* $CaCl_2$ solution? The net ionic equation for the reaction is

$$Ag^+(aq) + Cl^-(aq) \longrightarrow AgCl(s)$$

ANALYSIS: In many ways, this problem is similar to Example 4.13. However, to use the net ionic equation, we will need to work with the concentrations of the ions. Therefore, the tools we will use to solve this problem are the formulas of the salts (to find the molar concentrations of the ions) and the coefficients of the equation (to relate moles of Ag^+ and Cl^-).

To solve the problem, the first step will be to calculate the molarities of the ions in the solutions being mixed. Next, using the volume and molarity of the Cl^- solution, we will calculate the moles of Cl^- available. Then we'll use the coefficients of the equation to find the moles of Ag^+ that react. Finally, we'll use the molarity of the Ag^+ solution to determine the volume of the 0.100 M $AgNO_3$ solution needed.

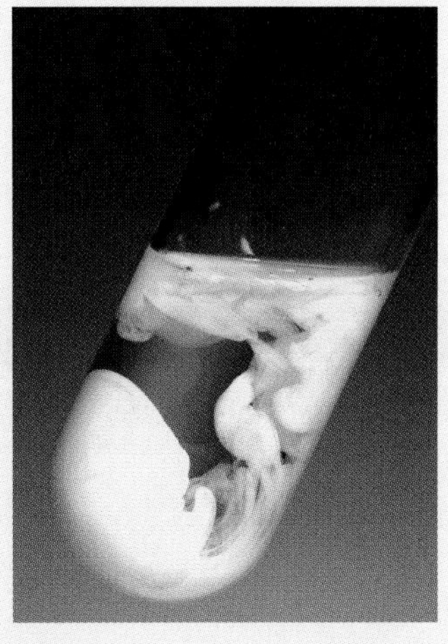

A solution of $AgNO_3$ is added to a solution of $CaCl_2$, producing a precipitate of AgCl. *(Andy Washnik.)*

SOLUTION: We begin by finding the concentrations of the ions in the reacting solutions:

0.100 M $AgNO_3$ contains 0.100 M Ag^+ and 0.100 M NO_3^-
0.400 M $CaCl_2$ contains 0.400 M Ca^{2+} and 0.800 M Cl^-

We're only interested in the Ag^+ and Cl^-; the Ca^{2+} and NO_3^- are spectator ions. For our purposes, then, the solution concentrations are 0.100 M Ag^+ and 0.800 M Cl^-. Having these values, we can now restate the problem: How many milliliters of 0.100 M Ag^+ solution are needed to react completely with 25.0 mL of 0.800 M Cl^- solution?

$$25.0 \text{ mL } Cl^- \text{ soln} \Leftrightarrow ? \text{ mL } Ag^+ \text{ soln}$$

The moles of Cl^- available for reaction are obtained from the molarity and volume (0.0250 L) of the Cl^- solution.

$$0.0250 \cancel{\text{L } Cl^- \text{ soln}} \times \frac{0.800 \text{ mol } Cl^-}{1.00 \cancel{\text{L } Cl^- \text{ soln}}} = 0.0200 \text{ mol } Cl^-$$

The coefficients of the equation tell us the Ag^+ and Cl^- combine in a 1-to-1 mole ratio, so 0.0200 mol $Cl^- \Leftrightarrow$ 0.0200 mol Ag^+. Finally, we calculate the volume of the Ag^+ solution using its molarity as a conversion factor. As we've done earlier, we use the factor that makes the units cancel correctly.

Simple reasoning works well here. Since the coefficients of Ag^+ and Cl^- are the same, the numbers of moles that react must be equal.

$$0.0200 \cancel{\text{mol } Ag^+} \times \frac{1.00 \text{ L } Ag^+ \text{ soln}}{0.100 \cancel{\text{mol } Ag^+}} = 0.200 \text{ L } Ag^+ \text{ soln}$$

Our calculations tell us that we must use 0.200 L or 200 mL of the $AgNO_3$ solution. We could also have used the following chain calculation, of course.

$$0.0250 \cancel{\text{L } Cl^- \text{ soln}} \times \frac{0.800 \cancel{\text{mol } Cl^-}}{1.00 \cancel{\text{L } Cl^- \text{ soln}}} \times \frac{1 \cancel{\text{mol } Ag^+}}{1 \cancel{\text{mol } Cl^-}} \times \frac{1.00 \text{ L } Ag^+ \text{ soln}}{0.100 \cancel{\text{mol } Ag^+}} = 0.200 \text{ L } Ag^+ \text{ soln}$$

IS THE ANSWER REASONABLE? The silver ion concentration is one-eighth as large as the chloride ion concentration. Since the ions react one-for-one, we will need eight times as much silver ion solution as chloride solution. Eight times the amount of chloride solution, 25 mL, is 200 mL, which is the answer we obtained. Therefore, the answer appears to be correct.

Practice Exercise 35: Suppose 18.4 mL of 0.100 M $AgNO_3$ solution was needed to react completely with 20.5 mL of $CaCl_2$ solution. What is the molarity of the $CaCl_2$ solution? Use the net ionic equation in the preceding example to work the problem. (Hint: How can you calculate molarity from moles and volume, and how can you calculate the molarity of the $CaCl_2$ solution from the molarity of Cl^-?)

Practice Exercise 36: How many milliliters of 0.500 M KOH are needed to react completely with 60.0 mL of 0.250 M $FeCl_2$ solution to precipitate $Fe(OH)_2$? The net ionic equation is $Fe^{2+}(aq) + 2OH^-(aq) \longrightarrow Fe(OH)_2(s)$.

4.8 CHEMICAL ANALYSIS AND TITRATION ARE APPLICATIONS OF SOLUTION STOICHIOMETRY

Chemical analyses fall into two categories. In a **qualitative analysis** we simply determine which substances are present in a sample without measuring their amounts. In a **quantitative analysis,** our goal is to measure the amounts of the various substances in a sample.

When chemical reactions are used in a quantitative analysis, a useful strategy is to capture *all* of a desired chemical species in a compound with a known formula. From the amount of this compound obtained, we can determine how much of the desired chemical species was present in the original sample. The calculations required for these kinds of problems are not new; they are simply applications of the stoichiometric calculations you've already learned.

EXAMPLE 4.17
Calculation Involving a Quantitative Analysis

A certain insecticide is a compound known to contain carbon, hydrogen, and chlorine. Reactions were carried out on a 0.134 g sample of the compound that converted all of its chlorine to chloride ion dissolved in water. This aqueous solution required 37.80 mL of 0.0500 *M* $AgNO_3$ to precipitate all the chloride ion as AgCl. What was the percentage by mass of Cl in the original insecticide sample? The precipitation reaction was

$$Ag^+(aq) + Cl^-(aq) \longrightarrow AgCl(s)$$

ANALYSIS: The percentage Cl in the sample will be calculated as follows:

$$\%\text{ Cl by mass} = \frac{\text{mass of Cl in sample}}{\text{mass of sample}} \times 100\%$$

So, we need to determine the mass of Cl. We can use Equation 4.5 (page 156) and the volume and molarity of the $AgNO_3$ solution to find the number of moles of Ag^+ that reacts with the Cl^- in the solution. The ionic equation tells us Ag^+ and Cl^- react in a 1-to-1 mole ratio, so the moles of Ag^+ equals the moles of Cl^- that react. This is the same as the number of moles of Cl in the sample. Then we'll use the molar mass of Cl as a tool to change moles of Cl to grams of Cl.

The molar mass of a chlorine atom and a chloride ion differ by such a small amount that we can use them interchangeably.

SOLUTION: In 0.0500 *M* $AgNO_3$, the molarity of Ag^+ is 0.0500 *M*. Applying Equation 4.5,

$$0.03780 \cancel{\text{L Ag}^+\text{ soln}} \times \frac{0.0500 \text{ mol Ag}^+}{1.00 \cancel{\text{L Ag}^+\text{ soln}}} = 1.89 \times 10^{-3} \text{ mol Ag}^+$$

From the stoichiometry of the equation, when 1.89×10^{-3} mol Ag^+ reacts, 1.89×10^{-3} mol Cl^- reacts. This is the amount of Cl^- that came from the sample, so the sample must have contained 1.89×10^{-3} mol Cl. The atomic mass of Cl is 35.45, so 1 mol Cl = 35.45 g Cl. Therefore, the mass of Cl in the sample was

$$1.89 \times 10^{-3} \cancel{\text{mol Cl}} \times \frac{35.45 \text{ g Cl}}{1 \cancel{\text{mol Cl}}} = 6.70 \times 10^{-2} \text{ g Cl}$$

The percentage by mass of Cl in the sample is

$$\%\text{ Cl} = \frac{6.70 \times 10^{-2} \text{ g Cl}}{0.134 \text{ g sample}} \times 100\%$$

$$= 50.0\%$$

The insecticide was 50.0% Cl by mass.

IS THE ANSWER REASONABLE? Although we could do some approximate arithmetic to check our calculation, that's more easily done with the calculator. Therefore, let's look over the reasoning and calculations to see if they make sense. We've used the molarity and volume of the $AgNO_3$ solution to calculate the number of moles of Ag^+ that reacted (1.89×10^{-3} mol Ag^+). This has to be the same as the moles of Cl^- that reacted, and because the Cl^- came from the sample, the sample must have contained 1.89×10^{-3} mol of Cl. The mass of this Cl was calculated in the usual way using the molar mass of Cl. The value we obtained, 0.067 g, is half the sample weight of 0.134 g, so half (50%) of the sample weight was Cl. The answer, therefore, seems to be correct.

Practice Exercise 37: A solution containing Na_2SO_4 was treated with 0.150 *M* $BaCl_2$ solution until all the sulfate ion had reacted to form $BaSO_4$. The net reaction

$$Ba^{2+}(aq) + SO_4^{2-}(aq) \longrightarrow BaSO_4(s)$$

required 28.40 mL of the $BaCl_2$ solution. How many grams of Na_2SO_4 were in the solution? (Hint: How do moles of SO_4^{2-} relate to moles of Na_2SO_4?)

Practice Exercise 38: A sample of a mixture containing $CaCl_2$ and $MgCl_2$ weighed 2.000 g. The sample was dissolved in water and H_2SO_4 was added until the precipitation of $CaSO_4$ was complete.

$$CaCl_2(aq) + H_2SO_4(aq) \longrightarrow CaSO_4(s) + 2HCl(aq)$$

The $CaSO_4$ was filtered, dried completely, and weighed. A total of 0.736 g of $CaSO_4$ was obtained.

(a) How many moles of Ca^{2+} were in the $CaSO_4$?
(b) How many moles of Ca^{2+} were in the original 2.000 g sample?
(c) How many moles of $CaCl_2$ were in the 2.000 g sample?
(d) How many grams of $CaCl_2$ were in the 2.000 g sample?
(e) What was the percentage by mass of $CaCl_2$ in the original mixture?

Acid–base titrations are useful in chemical analyses

Titration is an important laboratory procedure used in performing chemical analyses. The apparatus is shown in Figure 4.22. The long tube is called a **buret**, which is marked for volumes, usually in increments of 0.10 mL. The valve at the bottom of the buret is called a **stopcock,** and it permits the analyst to control the amount of **titrant** (the solution in the buret) that is delivered to the receiving vessel (the beaker shown in the drawing).

In a typical titration, a solution containing one reactant is placed in the receiving vessel. Carefully measured volumes of a solution of the other reactant are then added from the buret. (One of the two solutions is of a precisely known concentration and is called a **standard solution.**) This addition is continued until something (usually a visual effect, like a color change) signals that the two reactants have been combined in just the right proportions to give a complete reaction.

In acid–base titrations, an **acid–base indicator** is used to detect the completion of the reaction by a change in color. Indicators are dyes that have one color in an acidic solution and a different color in a basic solution. Litmus was mentioned earlier. Phenolphthalein is

The theory of acid–base indicators is discussed in Chapter 16.

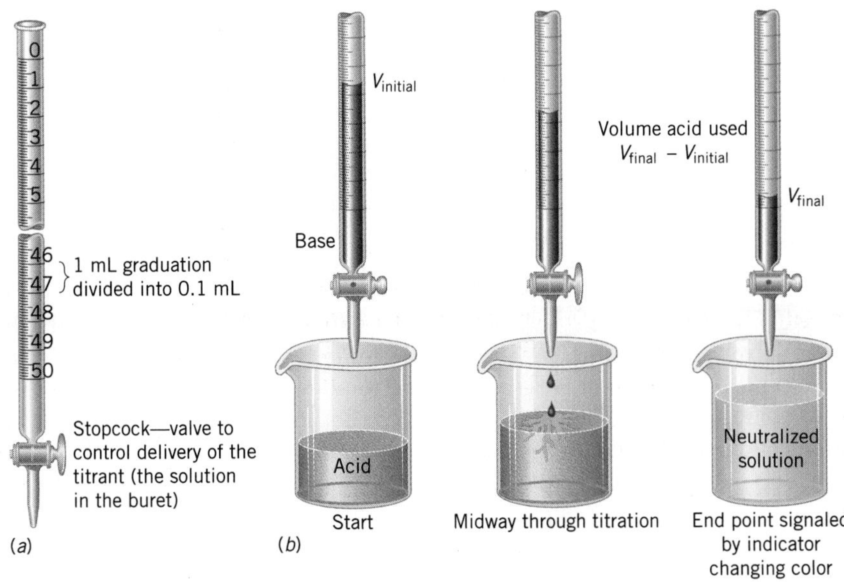

FIG. 4.22 Titration. *(a)* A buret. *(b)* The titration of an acid by a base in which an acid–base indicator is used to signal the end point, which is the point at which all of the acid has been neutralized and addition of the base is halted.

a common indicator for titrations; it changes from colorless to pink when a solution changes from acidic to basic. This color change is very abrupt, and occurs with the addition of only one final drop of the titrant just as the end of the reaction is reached. When we observe the color change, the **end point** has been reached and the addition of titrant is stopped. We then record the total volume of the titrant that's been added to the receiving flask.

EXAMPLE 4.18

Calculation Involving Acid–Base Titration

A student prepares a solution of hydrochloric acid that is approximately 0.1 *M* and wishes to determine its precise concentration. A 25.00 mL portion of the HCl solution is transferred to a flask, and after a few drops of indicator are added, the HCl solution is titrated with 0.0775 *M* NaOH solution. The titration requires exactly 37.46 mL of the standard NaOH solution to reach the end point. What is the molarity of the HCl solution?

A solution of HCl is titrated with a solution of NaOH using phenolphthalein as the acid–base indicator. The pink color that phenolphthalein has in a basic solution can be seen where a drop of the NaOH solution has entered the HCl solution, to which a few drops of the indicator had been added. *(Michael Watson.)*

ANALYSIS: This is really a straightforward stoichiometry calculation involving a chemical reaction. The first step in solving it is to write the balanced equation, which will give us the stoichiometric equivalency between HCl and NaOH. The reaction is an acid–base neutralization, so the product is a salt plus water. Following the procedures developed earlier, the reaction is

$$\text{HCl}(aq) + \text{NaOH}(aq) \longrightarrow \text{NaCl}(aq) + \text{H}_2\text{O}$$

Solving the problem will follow the same route as in Example 4.13. Our tools will be the molarity and volume of the NaOH solution (to calculate moles of NaOH that react), the coefficients of the equation (to obtain moles of HCl), and the definition of molarity (to calculate the molarity of the HCl solution from moles HCl and the volume of the HCl sample taken).

SOLUTION: From the molarity and volume of the NaOH solution, we calculate the number of moles of NaOH consumed in the titration.

$$0.03746 \, \cancel{\text{L NaOH soln}} \times \frac{0.0775 \text{ mol NaOH}}{1.00 \, \cancel{\text{L NaOH soln}}} = 2.90 \times 10^{-3} \text{ mol NaOH}$$

The coefficients in the equation tell us that NaOH and HCl react in a 1-to-1 mole ratio,

$$2.90 \times 10^{-3} \, \cancel{\text{mol NaOH}} \times \frac{1 \text{ mol HCl}}{1 \, \cancel{\text{mol NaOH}}} = 2.90 \times 10^{-3} \text{ mol HCl}$$

so in this titration, 2.90×10^{-3} mol HCl was in the flask. To calculate the molarity of the HCl, we simply apply the definition of molarity. We take the ratio of the number of moles of HCl that reacted (2.90×10^{-3} mol HCl) to the volume (in liters) of the HCl solution used (25.00 mL, or 0.02500 L).

$$\text{Molarity of HCl soln} = \frac{2.90 \times 10^{-3} \text{ mol HCl}}{0.02500 \text{ L HCl soln}}$$

$$= 0.116 \, M \, \text{HCl}$$

The molarity of the hydrochloric acid is 0.116 *M*.

IS THE ANSWER REASONABLE? If the concentrations of the NaOH and HCl were the same, the volumes used would have been equal. However, the volume of the NaOH solution used is larger than the volume of HCl solution. This must mean that the HCl solution is more concentrated than the NaOH solution. The value we obtained, 0.116 *M*, is larger than 0.0775 *M*, so our answer makes sense.

Practice Exercise 39: In a titration, a sample of H_2SO_4 solution having a volume of 15.00 mL required 36.42 mL of 0.147 *M* NaOH solution for *complete* neutralization. What is the molarity of the H_2SO_4 solution? (Hint: Check to be sure your chemical equation is written correctly and balanced.)

Practice Exercise 40: "Stomach acid" is hydrochloric acid. A sample of gastric juice having a volume of 5.00 mL required 11.00 mL of 0.0100 *M* KOH solution for neutralization in a titration. What was the molar concentration of HCl in this fluid? If we assume a density of 1.00 g mL^{-1} for the fluid, what was the percentage by weight of HCl?

SUMMARY

Solution Vocabulary. A **solution** is a homogeneous mixture in which one or more **solutes** are dissolved in a **solvent.** A solution may be **dilute** or **concentrated,** depending on the amount of solute dissolved in a given amount of solvent. **Concentration** (e.g., **percentage concentration**) is a ratio of the amount of solute to either the amount of solvent or the amount of solution. The amount of solute required to give a **saturated** solution at a given temperature is called the solute's **solubility. Unsaturated** solutions will dissolve more solute, but **supersaturated** solutions are unstable and tend to give a **precipitate.**

Electrolytes. Substances that **dissociate** or **ionize** in water to produce cations and anions are **electrolytes;** those that do not are called **nonelectrolytes.** Electrolytes include salts and metal hydroxides as well as molecular acids and bases that ionize by reaction with water. In water, ionic compounds are completely dissociated into ions and are **strong electrolytes.**

Ionic and Net Ionic Equations. Reactions that occur in solution between ions and are called **ionic reactions.** Solutions of soluble strong electrolytes often yield an insoluble product which appears as a **precipitate.** Equations for these reactions can be written in three different ways. In **molecular equations,** complete formulas for all reactants and products are used. In an **ionic equation,** soluble strong electrolytes are written in dissociated (ionized) form; "molecular" formulas are used for solids and weak electrolytes. A **net ionic equation** is obtained by eliminating **spectator ions** from the ionic equation, and such an equation allows us to identify other combinations of reactants that give the same net reaction. An ionic or net ionic equation is balanced only if both atoms *and* net charge are balanced.

Acids and Bases as Electrolytes. An **acid** is a substance that produces hydronium ions, H_3O^+, when dissolved in water, and a **base** produces hydroxide ions, OH^-, when dissolved in water. The oxides of nonmetals are generally **acidic anhydrides** and react with water to give acids. Metal oxides are usually **basic anhydrides** because they tend to react with water to give metal hydroxides or bases.

Strong acids and bases are also strong electrolytes. **Weak acids and bases** are **weak electrolytes,** which are incompletely ionized in water. In a solution of a weak electrolyte there is a **chemical equilibrium (dynamic equilibrium)** between the nonionized molecules of the solute and the ions formed by the reaction of the solute with water.

Predicting Metathesis Reactions. Metathesis or **double replacement** reactions take place when anions and cations of two salts change partners. A metathesis reaction will occur if there is a net ionic equation. This happens if (1) a precipitate forms from soluble reactants, (2) an acid–base neutralization occurs, (3) a gas is formed, or (4) a weak electrolyte forms from soluble strong electrolytes. You should learn the **solubility rules** (Table 4.1), and remember that all salts are strong electrolytes. Remember that all strong acids and bases are strong electrolytes, too. Strong acids react with strong bases in neutralization reactions to produce a salt and water. Acids react with insoluble oxides and hydroxides to form water and the corresponding salt. Many **acid–base neutralization** reactions can be viewed as a type of metathesis reaction in which one product is water. Be sure to learn the reactions that produce gases in metathesis reactions, which are found in Table 4.2.

Molar Concentration, Dilution, and Solution Stoichiometry. Molarity is the ratio of moles of solute to liters of solution. Molarity provides two conversion factors relating moles of solute and the volume of a solution.

$$\frac{\text{mol solute}}{\text{1 L soln}} \quad \text{and} \quad \frac{\text{1 L soln}}{\text{mol solute}}$$

Concentrated solutions of known molarity can be diluted quantitatively using volumetric glassware such as pipets and volumetric flasks. When a solution is diluted by adding solvent, the amount of solute doesn't change but the concentration decreases.

In ionic reactions, the concentrations of the ions in a solution of a salt can be derived from the molar concentration of the salt, taking into account the number of ions formed per formula unit of the salt.

Titration is a technique used to make quantitative measurements of the amounts of solutions needed to obtain a complete reaction. The apparatus is a long tube called a **buret** that has a **stopcock** at one end, which is used to control the flow of **titrant.** In an acid–base titration, the **end point** is normally detected visually using an **acid–base indicator.** A color change indicates complete reaction, at which time addition of titrant is stopped and the volume added is recorded.

TOOLS FOR PROBLEM SOLVING

In this chapter you learned to apply the following concepts as tools in solving problems dealing with reactions in aqueous solutions. Study each one carefully so that you know what each is used for. When faced with solving a problem, recall what each tool does and consider whether it will be helpful in finding a solution. This will aid you in selecting the tools you need.

Criteria for a balanced ionic or net ionic equation *(page 135)* For an equation that includes the formulas of ions to be balanced, it must satisfy two criteria. The number of atoms of each kind must be the same on both sides, and the total net electrical charge shown on both sides must be the same.

Equation for the ionization of an acid in water *(page 137)* Equation 4.2 describes how acids react with water to form hydronium ion plus an anion.

$$HA + H_2O \longrightarrow H_3O^+ + A^-$$

Use this tool to write equations for ionizations of acids and to determine the formula of the anion formed when the acid molecule loses an H^+. The equation also applies to acid anions such as HSO_4^- which gives SO_4^{2-} when it loses an H^+. Often H_2O is omitted from the equation and the hydronium ion is abbreviated as H^+.

Equation for the ionization of a molecular base in water *(page 139)* Equation 4.3 describes how molecules of molecular bases acquire H^+ from H_2O to form a cation plus a hydroxide ion.

$$B + H_2O \longrightarrow BH^+ + OH^-$$

Use this tool to write equations for ionizations of bases and to determine the formula of the cation formed when the base molecule gains an H^+. *Molecular bases are weak and are not completely ionized.*

Table of strong acids *(page 140)* Formulas of the most common strong acids are given here. If you learn this list and encounter an acid that's *not* on the list, you can assume it to be a weak acid. The most common strong acids are HCl, HNO_3, and H_2SO_4. *Remember that strong acids are completely ionized in water.*

Predicting the existence of a net ionic equation *(page 146)* A net ionic equation will exist and a reaction will occur when:

- A precipitate is formed from a mixture of soluble reactants.
- An acid reacts with a base. *This includes strong or weak acids reacting with strong or weak bases or insoluble metal hydroxides or oxides.*
- A weak electrolyte is formed from a mixture of strong electrolytes.
- A gas is formed from a mixture of reactants.

These criteria are tools to determine whether or not a net reaction will occur in a solution.

Solubility rules *(page 147)* The rules in Table 4.1 are the tool we use to determine whether a particular salt is soluble in water. (If a salt is soluble, it's completely dissociated into ions.) They also serve as a tool to help predict the course of metathesis reactions.

Substances that form gases in metathesis reactions *(page 152)* Use Table 4.2 as a tool to help predict the outcome of metathesis reactions. The most common gas formed in such reactions is CO_2, which comes from the reaction of an acid with a carbonate or bicarbonate.

Molarity *(page 154)* Molarity provides the connection between moles of a solute and the volume of its solution. The definition provided by Equation 4.4 serves as a tool for calculating molarity from values of moles of solute and volume of solution (in liters).

$$\text{Molarity } (M) = \frac{\text{moles of solute}}{\text{liters of solution}}$$

Molarity is the tool we use to write an equivalence between moles of solute and volume of solution, from which appropriate conversion factors can be formed.

Product of molarity and volume gives moles *(page 156)* For any problem in which you're given both molarity and volume of a solution of a substance, you can always calculate the number of moles of the substance using Equation 4.5.

$$\text{Molarity} \times \text{volume (L)} = \text{moles of solute}$$

Recognizing this relationship is very important when working stoichiometry problems involving solutions.

Dilution problems *(page 158)* Equation 4.6 is the tool we use for working dilution problems.

$$V_{dil} \cdot M_{dil} = V_{conc} \cdot M_{conc}$$

The volume units must be the same on both sides of the equation.

Concentrations of ions in a solution of a salt *(page 161)* When using a net ionic equation to work stoichiometry problems, we need the concentrations of the ions in the solutions. *The concentration of a particular ion equals the concentration of the salt multiplied by the number of ions of that kind in one formula unit of the salt.*

Overview of stoichiometry problems In this chapter you encountered another way that the data in stoichiometry problems are presented. Figure 4.23 gives an overview of the various paths through problems that involve chemical reactions. All funnel through the coefficients of the equation as the means to convert from moles of one substance to moles of another. The starting and finishing quantities can be moles, grams, or volumes of solutions of known molarity.

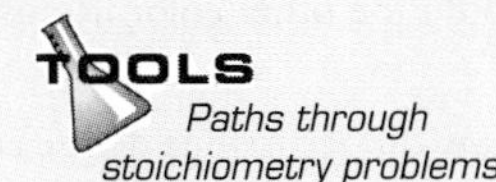

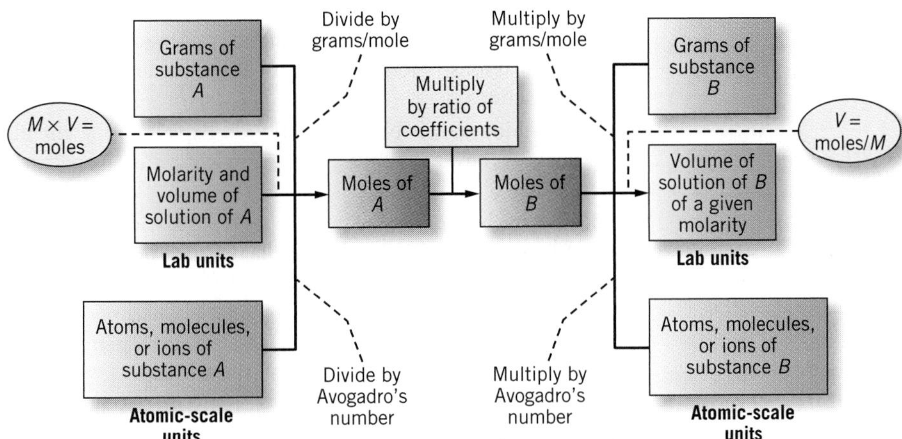

FIG. 4.23 Paths for working stoichiometry problems involving chemical reactions. The critical link is the conversion between moles of one substance and moles of another using the coefficients of the balanced equation. We can calculate moles in several ways: from numbers of atomic-sized formula units, and from laboratory units of grams or the volumes of solutions of known molarities. Similarly, we can present the answer either in moles, grams, number of formula units, or a volume of solution of known molarity.

QUESTIONS, PROBLEMS, AND EXERCISES

Answers to problems whose numbers are printed in color are given in Appendix B. More challenging problems are marked with asterisks. ILW = Interactive Learningware solution is available at *www.wiley.com/college/brady*. OH = an Office Hours video is available for this problem.

REVIEW QUESTIONS

Solution Terminology

4.1 Define the following: (a) solvent, (b) solute, (c) concentration.

4.2 Define the following: (a) concentrated, (b) dilute, (c) saturated, (d) unsaturated, (e) supersaturated, (f) solubility.

4.3 Why are chemical reactions often carried out using solutions?

4.4 Describe what will happen if a crystal of sugar is added to (a) a saturated sugar solution, (b) a supersaturated solution of sugar, and (c) an unsaturated solution of sugar.

4.5 What is the meaning of the term *precipitate*? What condition must exist for a precipitate to form spontaneously in a solution?

Electrolytes

4.6 What is an *electrolyte*? What is a *nonelectrolyte*?

4.7 Why is an electrolyte able to conduct electricity while a nonelectrolyte cannot? What does it mean when we say that an ion is "hydrated?"

4.8 Define *dissociation* as it applies to ionic compounds that dissolve in water.

4.9 Write equations for the dissociation of the following in water: (a) $CaCl_2$, (b) $(NH_4)_2SO_4$, (c) $NaC_2H_3O_2$.

Ionic Reactions

4.10 How do molecular, ionic, and net ionic equations differ? What are spectator ions?

4.11 The following equation shows the formation of cobalt(II) hydroxide, a compound used to improve the drying properties of lithographic inks.

$$Co^{2+}(aq) + 2Cl^-(aq) + 2Na^+(aq) + 2OH^-(aq) \longrightarrow Co(OH)_2(s) + 2Na^+(aq) + 2Cl^-(aq)$$

Which are the spectator ions? Write the net ionic equation.

4.12 What two conditions must be fulfilled by a balanced ionic equation? The following equation is not balanced. How do we know? Find the errors and fix them.

$$3Co^{3+}(aq) + 2HPO_4^{2-}(aq) \longrightarrow Co_3(PO_4)_2(s) + 2H^+(aq)$$

Acids, Bases, and Their Reactions

4.13 Give two general properties of an acid. Give two general properties of a base.

4.14 If you believed a solution was basic, which color litmus paper (blue or pink) would you use to test the solution to see if you were correct? What would you observe if you've selected correctly? Why would the other color litmus paper not lead to a conclusive result?

4.15 How did Arrhenius define an acid and a base?

4.16 Which of the following undergo dissociation in water? Which undergo ionization? (a) NaOH, (b) HNO_3, (c) NH_3, (d) H_2SO_4.

4.17 Which of the following would yield an acidic solution when they react with water? Which would give a basic solution? (a) P_4O_{10}, (b) K_2O, (c) SeO_3, (d) Cl_2O_7.

4.18 What is a *dynamic equilibrium*? Using acetic acid as an example, describe why all the $HC_2H_3O_2$ molecules are not ionized in water.

4.19 Why don't we use double arrows in the equation for the reaction of a strong acid with water?

4.20 Which of the following are strong acids? (a) HCN, (b) HNO_3, (c) H_2SO_3, (d) HCl, (e) $HCHO_2$, (f) HNO_2.

4.21 Which of the following produce a strongly basic solution when dissolved in water? (a) C_5H_5N, (b) $Ba(OH)_2$, (c) KOH, (d) $C_6H_5NH_2$, (e) Cs_2O, (f) N_2O_5.

4.22 Methylamine, CH_3NH_2, reacts with hydronium ion in very much the same manner as ammonia.

$$CH_3NH_2(aq) + H_3O^+(aq) \longrightarrow CH_3NH_3^+(aq) + H_2O$$

On the basis of what you have learned so far in this course, sketch the molecular structures of CH_3NH_2 and $CH_3NH_3^+$ (the methylammonium ion).

Nomenclature of Acids and Bases

4.23 Name the following: (a) $H_2Se(g)$, (b) $H_2Se(aq)$

4.24 Iodine, like chlorine, forms several acids. What are the names of the following? (a) HIO_4, (b) HIO_3, (c) HIO_2, (d) HIO, (e) HI.

4.25 For the acids in the preceding question, (a) write the formulas and (b) name the ions formed by removing a hydrogen ion (H^+) from each acid.

4.26 Write the formula for (a) chromic acid, (b) carbonic acid, and (c) oxalic acid. (Hint: Check the table of polyatomic ions.)

4.27 Name the following acid salts: (a) $NaHCO_3$, (b) KH_2PO_4, (c) $(NH_4)_2HPO_4$.

4.28 Write the formulas for all the acid salts that could be formed from the reaction of NaOH with the acid H_3PO_4.

4.29 Name the following oxoacids and give the names and formulas of the salts formed from them by neutralization with NaOH: (a) HOCl, (b) HIO_2, (c) $HBrO_3$, (d) $HClO_4$.

4.30 The formula for the arsenate ion is AsO_4^{3-}. What is the formula for arsenous acid?

4.31 Butyric acid, $HC_4H_7O_2$, gives rancid butter is bad odor. What is the name of the salt $NaC_4H_7O_2$?

4.32 Calcium propionate, $Ca(C_3H_5O_2)_2$ is used in baked foods as a preservative and to prevent the growth of mold. What is the name of the acid $HC_3H_5O_2$?

Predicting Ionic Reactions

4.33 What factors lead to the existence of a net ionic equation in a reaction between ions?

4.34 What is another name for *metathesis reaction*?

4.35 Silver bromide is "insoluble." What does this mean about the concentrations of Ag^+ and Br^- in a saturated solution of AgBr? Explain why a precipitate of AgBr forms when solutions of the soluble salts $AgNO_3$ and NaBr are mixed.

4.36 If a solution of trisodium phosphate, Na_3PO_4, is poured into seawater, precipitates of calcium phosphate and magnesium phosphate are formed. (Magnesium and calcium ions are among the principal ions found in seawater.) Write net ionic equations for the reactions.

4.37 Washing soda is $Na_2CO_3{\cdot}10H_2O$. Explain, using chemical equations, how this substance is able to remove Ca^{2+} ions from "hard water."

4.38 With which of the following will the weak acid $HCHO_2$ react? Where there is a reaction, write the formulas of the products. (a) KOH, (b) MgO, (c) NH_3

4.39 Suppose you suspected that a certain solution contained ammonium ions. What simple chemical test could you perform that would tell you whether your suspicion was correct?

4.40 What gas is formed if HCl is added to (a) $NaHCO_3$, (b) Na_2S, and (c) K_2SO_3?

Molarity and Dilution

4.41 What is the definition of molarity? Show that the ratio of millimoles (mmol) to milliliters (mL) is equivalent to the ratio of moles to liters.

4.42 A solution is labeled 0.25 *M* HCl. Construct two conversion factors that relate moles of HCl to the volume of solution expressed in liters.

4.43 When the units *molarity* and *liter* are multiplied, what are the resulting units?

4.44 When a solution labeled 0.50 *M* HNO_3 is diluted with water to give 0.25 *M* HNO_3, what happens to the number of moles of HNO_3 in the solution?

4.45 Two solutions, A and B, are labeled "0.10 *M* $CaCl_2$" and "0.20 *M* $CaCl_2$," respectively. Both solutions contain the same number of moles of $CaCl_2$. If solution A has a volume of 50 mL, what is the volume of solution B?

Chemical Analyses and Titrations

4.46 What is the difference between a qualitative analysis and a quantitative analysis?

4.47 Describe each of the following: (a) buret, (b) titration, (c) titrant, and (d) end point.

4.48 What is the function of an indicator in a titration? What color is phenolphthalein in (a) an acidic solution and (b) a basic solution?

REVIEW PROBLEMS

Ionic Reactions

4.49 Write ionic and net ionic equations for these reactions.
(a) $(NH_4)_2CO_3(aq) + MgCl_2(aq) \longrightarrow 2NH_4Cl(aq) + MgCO_3(s)$
(b) $CuCl_2(aq) + 2NaOH(aq) \longrightarrow Cu(OH)_2(s) + 2NaCl(aq)$
(c) $3FeSO_4(aq) + 2Na_3PO_4(aq) \longrightarrow Fe_3(PO_4)_2(s) + 3Na_2SO_4(aq)$
(d) $2AgC_2H_3O_2(aq) + NiCl_2(aq) \longrightarrow 2AgCl(s) + Ni(C_2H_3O_2)_2(aq)$

OH **4.50** Write balanced ionic and net ionic equations for these reactions.
(a) $CuSO_4(aq) + BaCl_2(aq) \longrightarrow CuCl_2(aq) + BaSO_4(s)$
(b) $Fe(NO_3)_3(aq) + LiOH(aq) \longrightarrow LiNO_3(aq) + Fe(OH)_3(s)$
(c) $Na_3PO_4(aq) + CaCl_2(aq) \longrightarrow Ca_3(PO_4)_2(s) + NaCl(aq)$
(d) $Na_2S(aq) + AgC_2H_3O_2(aq) \longrightarrow NaC_2H_3O_2(aq) + Ag_2S(s)$

Acids and Bases as Electrolytes

4.51 Pure $HClO_4$ is molecular. In water it is a strong acid. Write an equation for its ionization in water.

4.52 HBr is a molecular substance that is a strong acid in water. Write an equation for its ionization in water.

OH **4.53** Hydrazine is a toxic substance that can form when household ammonia is mixed with a bleach such as Clorox. Its formula is N_2H_4, and it is a weak base. Write a chemical equation showing the reaction of hydrazine with water.

4.54 Pyridine, C_5H_5N, is a fishy smelling compound used as an intermediate in making insecticides. It is a weak base. Write a chemical equation showing its reaction with water.

4.55 Nitrous acid, HNO_2, is a weak acid that can form when sodium nitrite, a meat preservative, reacts with stomach acid (HCl). Write an equation showing the ionization of HNO_2 in water.

4.56 Pentanoic acid, $HC_5H_9O_2$, is found in a plant called valerian, which cats seem to like almost as much as catnip. Also called valeric acid, it is a weak acid. Write an equation showing its reaction with water.

4.57 Carbonic acid, H_2CO_3, is a weak diprotic acid formed in rainwater as it passes through the atmosphere and dissolves carbon dioxide. Write chemical equations for the equilibria involved in the stepwise ionization of H_2CO_3 in water.

4.58 Phosphoric acid, H_3PO_4, is a weak acid found in some soft drinks. It undergoes ionization in three steps. Write chemical equations for the equilibria involved in each of these reactions.

Metathesis Reactions

4.59 Write *balanced* ionic and net ionic equations for these reactions.
(a) $FeSO_4(aq) + K_3PO_4(aq) \longrightarrow Fe_3(PO_4)_2(s) + K_2SO_4(aq)$
(b) $AgC_2H_3O_2(aq) + AlCl_3(aq) \longrightarrow AgCl(s) + Al(C_2H_3O_2)_3(aq)$

4.60 Write *balanced* ionic and net ionic equations for these reactions.
(a) $Fe(NO_3)_3(aq) + KOH(aq) \longrightarrow KNO_3(aq) + Fe(OH)_3(s)$
(b) $Na_3PO_4(aq) + SrCl_2(aq) \longrightarrow Sr_3(PO_4)_2(s) + NaCl(aq)$

4.61 Aqueous solutions of sodium sulfide and copper(II) nitrate are mixed. A precipitate of copper(II) sulfide forms at once. The solution that remains contains sodium nitrate. Write the molecular, ionic, and net ionic equations for this reaction.

4.62 If an aqueous solution of iron(III) sulfate (a compound used in dyeing textiles and also for etching aluminum) is mixed with a solution of barium chloride, a precipitate of barium sulfate forms and the solution that remains contains iron(III) chloride. Write the molecular, ionic, and net ionic equations for this reaction.

4.63 Use the solubility rules to decide which of the following compounds are *soluble* in water.
(a) $Ca(NO_3)_2$ (d) silver nitrate
(b) $FeCl_2$ (e) barium sulfate
(c) $Ni(OH)_2$ (f) copper(II) carbonate

4.64 Predict which of the following compounds are *soluble* in water.
(a) $HgBr_2$ (d) ammonium phosphate
(b) $Sr(NO_3)_2$ (e) lead(II) iodide
(c) Hg_2Br_2 (f) lead(II) acetate

Acid–Base Neutralization Reactions

4.65 Complete and balance the following equations. For each, write the molecular, ionic, and net ionic equations. (All the products are soluble in water.)
(a) $Ca(OH)_2(aq) + HNO_3(aq) \longrightarrow$
(b) $Al_2O_3(s) + HCl(aq) \longrightarrow$
(c) $Zn(OH)_2(s) + H_2SO_4(aq) \longrightarrow$

4.66 Complete and balance the following equations. For each, write the molecular, ionic, and net ionic equations. (All the products are soluble in water.)
(a) $HC_2H_3O_2(aq) + Mg(OH)_2(s) \longrightarrow$
(b) $HClO_4(aq) + NH_3(aq) \longrightarrow$
(c) $H_2CO_3(aq) + NH_3(aq) \longrightarrow$

4.67 How would the electrical conductivity of a solution of $Ba(OH)_2$ change as a solution of H_2SO_4 is added slowly to it? Use a net ionic equation to justify your answer.

4.68 How would the electrical conductivity of a solution of $HC_2H_3O_2$ change as a solution of NH_3 is added slowly to it? Use a net ionic equation to justify your answer.

Ionic Reactions That Produce Gases

4.69 Write balanced net ionic equations for these reactions:
(a) $HNO_3(aq) + K_2CO_3(aq)$
(b) $Ca(OH)_2(aq) + NH_4NO_3(aq)$

4.70 Write balanced net ionic equations for these reactions:
(a) $H_2SO_4(aq) + NaHSO_3(aq)$
(b) $HNO_3(aq) + (NH_4)_2CO_3(aq)$

Predicting Ionic Reactions

4.71 Explain why the following reactions take place.
(a) $CrCl_3 + 3NaOH \longrightarrow Cr(OH)_3 + 3NaCl$
(b) $ZnO + 2HBr \longrightarrow ZnBr_2 + H_2O$

OH **4.72** Explain why the following reactions take place.
(a) $MnCO_3 + H_2SO_4 \longrightarrow MnSO_4 + H_2O + CO_2$
(b) $Na_2C_2O_4 + 2HNO_3 \longrightarrow 2NaNO_3 + H_2C_2O_4$

4.73 Complete and balance the molecular, ionic, and net ionic equations for the following reactions.
(a) $HNO_3 + Cr(OH)_3 \longrightarrow$
(b) $HClO_4 + NaOH \longrightarrow$
(c) $Cu(OH)_2 + HC_2H_3O_2 \longrightarrow$
(d) $ZnO + H_2SO_4 \longrightarrow$

4.74 Complete and balance molecular, ionic, and net ionic equations for the following reactions.
(a) $NaHSO_3 + HBr \longrightarrow$
(b) $(NH_4)_2CO_3 + NaOH \longrightarrow$
(c) $(NH_4)_2CO_3 + Ba(OH)_2 \longrightarrow$
(d) $FeS + HCl \longrightarrow$

ILW **4.75** Write balanced molecular, ionic, and net ionic equations for the following pairs of reactants. If all ions cancel, indicate that no reaction (N.R.) takes place.
(a) sodium sulfite and barium nitrate
(b) formic acid ($HCHO_2$) and potassium carbonate
(c) ammonium bromide and lead(II) acetate
(d) ammonium perchlorate and copper(II) nitrate

4.76 Write balanced molecular, ionic, and net ionic equations for the following pairs of reactants. If all ions cancel, indicate that no reaction (N.R.) takes place.
(a) ammonium sulfide and sodium hydroxide
(b) chromium(III) sulfate and potassium carbonate
(c) silver nitrate and chromium(III) acetate
(d) strontium hydroxide and magnesium chloride

***4.77** Choose reactants that would yield the following net ionic equations. Write molecular equations for each.
(a) $HCO_3^-(aq) + H^+(aq) \longrightarrow H_2O + CO_2(g)$
(b) $Fe^{2+}(aq) + 2OH^-(aq) \longrightarrow Fe(OH)_2(s)$
(c) $Ba^{2+}(aq) + SO_3^{2-}(aq) \longrightarrow BaSO_3(s)$
(d) $2Ag^+(aq) + S^{2-}(aq) \longrightarrow Ag_2S(s)$
(e) $ZnO(s) + 2H^+(aq) \longrightarrow Zn^{2+}(aq) + H_2O$

***4.78** Suppose that you wished to prepare copper(II) carbonate by a precipitation reaction involving Cu^{2+} and CO_3^{2-}. Which of the following pairs of reactants could you use as solutes?
(a) $Cu(OH)_2 + Na_2CO_3$
(b) $CuSO_4 + (NH_4)_2CO_3$
(c) $Cu(NO_3)_2 + CaCO_3$
(d) $CuCl_2 + K_2CO_3$
(e) $CuS + NiCO_3$

Molar Concentration

OH **4.79** Calculate the molarity of a solution prepared by dissolving
(a) 4.00 g of sodium hydroxide in 100.0 mL of solution
(b) 16.0 g of calcium chloride in 250.0 mL of solution

4.80 Calculate the molarity of a solution that contains
(a) 3.60 g of sulfuric acid in 450.0 mL of solution
(b) 2.0×10^{-3} mol iron(II) nitrate in 12.0 mL of solution

4.81 How many milliliters of 0.265 *M* $NaC_2H_3O_2$ are needed to supply 14.3 g $NaC_2H_3O_2$?

4.82 How many milliliters of 0.615 *M* HNO_3 contain 1.67 g HNO_3?

ILW **4.83** Calculate the number of grams of each solute that has to be taken to make each of the following solutions.
(a) 125 mL of 0.200 *M* NaCl
(b) 250.0 mL of 0.360 *M* $C_6H_{12}O_6$ (glucose)
(c) 250.0 mL of 0.250 *M* H_2SO_4

4.84 How many grams of solute are needed to make each of the following solutions?
(a) 250.0 mL of 0.100 *M* potassium sulfate
(b) 100.0 mL of 0.250 *M* iron(III) chloride
(c) 500.0 mL of 0.400 *M* barium acetate

Dilution of Solutions

4.85 If 25.0 mL of 0.56 *M* H_2SO_4 is diluted to a volume of 125 mL, what is the molarity of the resulting solution?

4.86 A 150 mL sample of 0.45 *M* HNO_3 is diluted to 450 mL. What is the molarity of the resulting solution?

ILW **4.87** To what volume must 25.0 mL of 18.0 *M* H_2SO_4 be diluted to produce 1.50 *M* H_2SO_4?

4.88 To what volume must 50.0 mL of 1.50 *M* HCl be diluted to produce 0.200 *M* HCl?

4.89 How many milliliters of water must be added to 150.0 mL of 2.50 *M* KOH to give a 1.00 *M* solution? (Assume volumes are additive.)

4.90 How many milliliters of water must be added to 120.0 mL of 1.50 *M* HCl to give 1.00 *M* HCl?

Concentrations of Ions in Solutions of Electrolytes

4.91 Calculate the number of moles of each of the ions in the following solutions.
(a) 32.3 mL of 0.45 *M* $CaCl_2$
(b) 50.0 mL of 0.40 *M* $AlCl_3$

4.92 Calculate the number of moles of each of the ions in the following solutions.
(a) 18.5 mL of 0.40 *M* $(NH_4)_2CO_3$
(b) 30.0 mL of 0.35 *M* $Al_2(SO_4)_3$

4.93 Calculate the concentrations of each of the ions in (a) 0.25 *M* $Cr(NO_3)_2$, (b) 0.10 *M* $CuSO_4$, (c) 0.16 *M* Na_3PO_4, and (d) 0.075 *M* $Al_2(SO_4)_3$.

4.94 Calculate the concentrations of each of the ions in (a) 0.060 *M* $Ca(OH)_2$, (b) 0.15 *M* $FeCl_3$, (c) 0.22 *M* $Cr_2(SO_4)_3$, and (d) 0.60 *M* $(NH_4)_2SO_4$.

4.95 In a solution of $Al_2(SO_4)_3$ the Al^{3+} concentration is 0.12 *M*. How many grams of $Al_2(SO_4)_3$ are in 50.0 mL of this solution?

4.96 In a solution of $NiCl_2$, the Cl^- concentration is 0.055 *M*. How many grams of $NiCl_2$ are in 250 mL of this solution?

Solution Stoichiometry

OH **4.97** How many milliliters of 0.25 *M* $NiCl_2$ solution are needed to react completely with 20.0 mL of 0.15 *M* Na_2CO_3 solution? How many grams of $NiCO_3$ will be formed? The reaction is

$$Na_2CO_3(aq) + NiCl_2(aq) \longrightarrow NiCO_3(s) + 2NaCl(aq)$$

4.98 How many milliliters of 0.100 *M* NaOH are needed to completely neutralize 25.0 mL of 0.250 *M* $H_2C_4H_4O_6$? The reaction is

$$2NaOH(aq) + H_2C_4H_4O_6(aq) \longrightarrow Na_2C_4H_4O_6(aq) + 2H_2O$$

4.99 What is the molarity of an aqueous solution of potassium hydroxide if 21.34 mL is exactly neutralized by 20.78 mL of 0.116 *M* HCl? Write and balance the molecular equation for the reaction.

4.100 What is the molarity of an aqueous phosphoric acid solution if 12.88 mL is completely neutralized by 26.04 mL of 0.1024 *M* NaOH? Write and balance the molecular equation for the reaction.

4.101 Aluminum sulfate, $Al_2(SO_4)_3$, is used in water treatment to remove fine particles suspended in the water. When made basic, a gel-like precipitate forms that removes the fine particles as it settles. In an experiment, a student planned to react $Al_2(SO_4)_3$ with $Ba(OH)_2$. How many grams of $Al_2(SO_4)_3$ are needed to react with 85.0 mL of 0.0500 *M* $Ba(OH)_2$?

4.102 How many grams of baking soda, $NaHCO_3$, are needed to react with 162 mL of stomach acid having an HCl concentration of 0.052 *M*?

4.103 How many milliliters of 0.150 *M* $FeCl_3$ solution are needed to react completely with 20.0 mL of 0.0450 *M* $AgNO_3$ solution? How many grams of AgCl will be formed? The net ionic equation for the reaction is

$$Ag^+(aq) + Cl^-(aq) \longrightarrow AgCl(s)$$

4.104 How many grams of cobalt(II) chloride are needed to react completely with 60.0 mL of 0.200 *M* KOH solution? The net ionic equation for the reaction is

$$Co^{2+}(aq) + 2OH^-(aq) \longrightarrow Co(OH)_2(s)$$

ILW 4.105 Consider the reaction of aluminum chloride with silver acetate. How many milliliters of 0.250 *M* aluminum chloride would be needed to react completely with 20.0 mL of 0.500 *M* silver acetate solution? The net ionic equation for the reaction is

$$Ag^+(aq) + Cl^-(aq) \longrightarrow AgCl(s)$$

4.106 How many milliliters of ammonium sulfate solution having a concentration of 0.250 *M* are needed to react completely with 50.0 mL of 1.00 *M* sodium hydroxide solution? The net ionic equation for the reaction is

$$NH_4^+(aq) + OH^-(aq) \longrightarrow NH_3(g) + H_2O$$

*4.107 Suppose that 4.00 g of solid Fe_2O_3 is added to 25.0 mL of 0.500 *M* HCl solution. What will the concentration of the Fe^{3+} be when all the HCl has reacted? What mass of Fe_2O_3 will not have reacted?

*4.108 Suppose 3.50 g of solid $Mg(OH)_2$ is added to 30.0 mL of 0.500 *M* H_2SO_4 solution. What will the concentration of Mg^{2+} be when all of the acid has been neutralized? How many grams of $Mg(OH)_2$ will not have dissolved?

*4.109 Suppose that 25.0 mL of 0.440 *M* NaCl is added to 25.0 mL of 0.320 *M* $AgNO_3$.
(a) How many moles of AgCl would precipitate?
(b) What would be the concentrations of each of the ions in the reaction mixture after the reaction?

*4.110 A mixture is prepared by adding 25.0 mL of 0.185 *M* Na_3PO_4 to 34.0 mL of 0.140 *M* $Ca(NO_3)_2$.
(a) What mass of $Ca_3(PO_4)_2$ will be formed?
(b) What will be the concentrations of each of the ions in the mixture after the reaction?

Titrations and Chemical Analyses

4.111 In a titration, 23.25 mL of 0.105 *M* NaOH was needed to react with 21.45 mL of HCl solution. What is the molarity of the acid?

4.112 A 12.5 mL sample of vinegar, containing acetic acid, was titrated using 0.504 *M* NaOH solution. The titration required 20.65 mL of the base.
(a) What was the molar concentration of acetic acid in the vinegar?
(b) Assuming the density of the vinegar to be 1.0 g mL^{-1}, what was the percentage (by mass) of acetic acid in the vinegar?

ILW OH 4.113 Lactic acid, $HC_3H_5O_3$, is a monoprotic acid that forms when milk sours. An 18.5 mL sample of a solution of lactic acid required 17.25 mL of 0.155 *M* NaOH to reach an end point in a titration. How many moles of lactic acid were in the sample?

4.114 Ascorbic acid (vitamin C) is a diprotic acid having the formula $H_2C_6H_6O_6$. A sample of a vitamin supplement was analyzed by titrating a 0.1000 g sample dissolved in water with 0.0200 *M* NaOH. A volume of 15.20 mL of the base was required to completely neutralize the ascorbic acid. What was the percentage by mass of ascorbic acid in the sample?

4.115 Magnesium sulfate forms a hydrate known as *Epsom salts*. A student dissolved 1.24 g of this hydrate in water and added a barium chloride solution until the precipitation reaction was complete. The precipitate was filtered, dried, and found to weigh 1.174 g. Determine the formula for Epsom salts.

4.116 A sample of iron chloride weighing 0.300 g was dissolved in water and the solution was treated with $AgNO_3$ solution to precipitate the chloride as AgCl. After precipitation was complete, the AgCl was filtered, dried, and found to weigh 0.678 g. Determine the empirical formula of the iron chloride.

4.117 A certain lead ore contains the compound $PbCO_3$. A sample of the ore weighing 1.526 g was treated with nitric acid, which dissolved the $PbCO_3$. The resulting solution was filtered from undissolved rock and required 29.22 mL of 0.122 *M* Na_2SO_4 to completely precipitate all the lead as $PbSO_4$. What is the percentage by mass of lead in the ore?

4.118 An ore of barium contains $BaCO_3$. A 1.542 g sample of the ore was treated with HCl to dissolve the $BaCO_3$. The resulting solution was filtered to remove insoluble material and then treated with H_2SO_4 to precipitate $BaSO_4$. The precipitate was filtered, dried, and found to weigh 1.159 g. What is the percentage by mass of barium in the ore? (Assume all the barium is precipitated as $BaSO_4$.)

4.119 To a mixture of NaCl and Na_2CO_3 with a mass of 1.243 g was added 50.00 mL of 0.240 *M* HCl (an excess of HCl). The mixture was warmed to expel all of the CO_2 and then the unreacted HCl was titrated with 0.100 *M* NaOH. The titration required 22.90 mL of the NaOH solution. What was the percentage by mass of NaCl in the original mixture of NaCl and Na_2CO_3?

4.120 A mixture was known to contain both KNO_3 and K_2SO_3. To 0.486 g of the mixture, dissolved in enough water to give 50.00 mL of solution, was added 50.00 mL of 0.150 *M* HCl (an excess of HCl). The reaction mixture was heated to drive off all of the SO_2, and then 25.00 mL of the reaction mixture was titrated with 0.100 *M* KOH. The titration required 13.11 mL of the KOH solution to reach an end point. What was the percentage by mass of K_2SO_3 in the original mixture of KNO_3 and K_2SO_3?

ADDITIONAL EXERCISES

4.121 Classify each of the following as a strong electrolyte, weak electrolyte, or nonelectrolyte.
(a) KCl
(b) $C_3H_5(OH)_3$ (glycerin)
(c) NaOH
(d) $C_{12}H_{22}O_{11}$ (sucrose, or table sugar)
(e) $HC_2H_3O_2$ (acetic acid)
(f) CH_3OH (methyl alcohol)
(g) H_2SO_4
(h) NH_3

OH **4.122** Complete the following and write molecular, ionic, and net ionic equations. State whether a net reaction occurs in each case.
(a) $CaCO_3 + HNO_3 \longrightarrow$
(b) $CaCO_3 + H_2SO_4 \longrightarrow$
(c) $FeS + HBr \longrightarrow$
(d) $KOH + SnCl_2 \longrightarrow$

OH ***4.123** Aspirin is a monoprotic acid called acetylsalicylic acid. Its formula is $HC_9H_7O_4$. A certain pain reliever was analyzed for aspirin by dissolving 0.250 g of it in water and titrating it with 0.0300 *M* KOH solution. The titration required 29.40 mL of base. What is the percentage by weight of aspirin in the drug?

***4.124** In an experiment, 40.0 mL of 0.270 *M* barium hydroxide was mixed with 25.0 mL of 0.330 *M* aluminum sulfate.
(a) Write the net ionic equation for the reaction that takes place.
(b) What is the total mass of precipitate that forms?
(c) What are the molar concentrations of the ions that remain in the solution after the reaction is complete?

***4.125** Qualitative analysis of an unknown acid found only carbon, hydrogen, and oxygen. In a quantitative analysis, a 10.46 mg sample was burned in oxygen and gave 22.17 mg CO_2 and 3.40 mg H_2O. The molecular mass was determined to be 166 g mol^{-1}. When a 0.1680 g sample of the acid was titrated with 0.1250 *M* NaOH, the end point was reached after 16.18 mL of the base had been added.
(a) Calculate the percentage composition of the acid.
(b) What is its empirical formula?
(c) What is its molecular formula?
(d) Is the acid mono-, di-, or triprotic?

***4.126** How many milliliters of 0.10 *M* HCl must be added to 50.0 mL of 0.40 *M* HCl to give a final solution that has a molarity of 0.25 *M*?

EXERCISES IN CRITICAL THINKING

4.127 Compare the advantages and disadvantages of performing a titration using the mass of the sample and titrant rather than the volume.

4.128 What kinds of experiments could you perform to measure the solubility of a substance in water? Describe the procedure you would use and the measurements you would make. What factors would limit the precision of your measurements?

4.129 Describe experiments, both qualitative and quantitative, that you could perform to show that lead chloride is more soluble in water than lead iodide.

4.130 How could you check the accuracy of a 100 mL volumetric flask?

4.131 Suppose a classmate doubted that an equilibrium really exists between acetic acid and its ions in an aqueous solution. What argument would you use to convince that person that such an equilibrium does exist?

4.132 When Arrhenius originally proposed that ions exist in solution, his idea was not well received. Propose another explanation for the conduction of electricity in molten salts and aqueous salt solutions.

4.133 Carbon dioxide is one obvious contributor to excessive global warming. What is your plan for controlling CO_2 emissions? What are the advantages and disadvantages of your plan?

5 OXIDATION-REDUCTION REACTIONS

This athlete is able to perform because her body derives energy from food she consumed. The metabolism of foods provides the energy we need for all sorts of activities, from simply thinking to participating in sports. The chemical reactions involved can be viewed as occurring by electron transfer. In this chapter we study such reactions, how to balance equations that describe them, and how these reactions are useful for laboratory work. *(Tim Pannell/Corbis)*

CHAPTER OUTLINE

'HIS CHAPTER IN CONTEXT In the preceding chapter you learned about some important reactions at take place in aqueous solutions. This chapter expands on that knowledge with a discussion of a class of reacns that can be viewed as involving the transfer of one or more electrons from one reactant to another. Our goal to teach you how to recognize and analyze the changes that occur in these reactions and how to balance equans for reactions that involve electron transfer. You will also learn how to apply the principles of stoichiometry to ese reactions and how to predict a type of reaction called single replacement. The reactions discussed here contute a very broad class with many practical examples that range from combustion to batteries to the metabolism foods by our bodies. Although we will introduce you to some of those reactions in this chapter, we will have to wait til a later time to discuss some of the others.

5.1 OXIDATION–REDUCTION REACTIONS INVOLVE ELECTRON TRANSFER

Among the first reactions studied by early scientists were those that involved oxygen. The combustion of fuels and the reactions of metals with oxygen to give oxides were described by the word *oxidation*. The removal of oxygen from metal oxides to give the metals in their elemental forms was described as *reduction*.

In 1789, the French chemist Antoine Lavoisier discovered that *combustion* involves the reaction of chemicals in various fuels, like wood and coal, not just with air but specifically with the oxygen in air. Over time, scientists came to realize that such reactions were actually special cases of a much more general phenomenon, one in which electrons are transferred from one substance to another. Collectively, electron transfer reactions came to be called **oxidation–reduction reactions,** or simply **redox reactions.** The term **oxidation** was used to describe the loss of electrons by one reactant, and **reduction** was used to describe the gain of electrons by another. For example, the reaction between sodium and chlorine to yield sodium chloride involves a loss of electrons by sodium (*oxidation* of sodium) and a gain of electrons by chlorine (*reduction* of chlorine). We can write these changes in equation form including the electrons, which we represent by the symbol e^-.

Notice that when we write equations of this type, the electron appears as a "product" if the process is oxidation and as a "reactant" if the process is reduction.

$$Na \longrightarrow Na^+ + e^- \qquad \text{(oxidation)}$$

$$Cl_2 + 2e^- \longrightarrow 2Cl^- \qquad \text{(reduction)}$$

We say that sodium (Na) is oxidized and chlorine (Cl_2) is reduced.

Oxidation and reduction always occur together. No substance is ever oxidized unless something else is reduced, and the total number of electrons lost by one substance is always the same as the total number gained by the other. If this were not true, electrons would appear as a product of the overall reaction, and this is never observed.[1] In the reaction of sodium with chlorine, for example, the overall reaction is

$$2Na + Cl_2 \longrightarrow 2NaCl$$

When two sodium atoms are oxidized, two electrons are lost, which is exactly the number of electrons gained when one Cl_2 molecule is reduced.

For a redox reaction to occur, one substance must accept electrons from the other. The substance that accepts the electrons is called the **oxidizing agent;** it is the agent that allows the other substance to lose electrons and be oxidized. Similarly, the substance that supplies the electrons is called the **reducing agent** because it helps something else to be reduced. In our example above, sodium is serving as a reducing agent when it supplies electrons to chlorine. In the process, sodium is oxidized. Chlorine is an oxidizing agent when it accepts electrons from the sodium, and when that happens, chlorine is reduced to chloride ion. One way to remember this is by the following summary:

The *oxidizing agent* causes oxidation to occur by accepting electrons (it gets reduced). The *reducing agent* causes reduction by supplying electrons (it gets oxidized).

TOOLS
Oxidizing and reducing agents

> The reducing agent is the substance that is oxidized.
> The oxidizing agent is the substance that is reduced.

Redox reactions are very common. They occur in batteries, which are built so that the electrons transferred can pass through some external circuit where they are able to light a flashlight or power an iPod. The metabolism of foods, which supplies our bodies with energy, also occurs by a series of redox reactions. And ordinary household bleach works by oxidizing substances that stain fabrics, making them colorless or easier to remove from the fabric (see Figure 5.1).

[1] If electron loss didn't equal electron gain, it would also violate the law of conservation of mass, and that doesn't happen in chemical reactions.

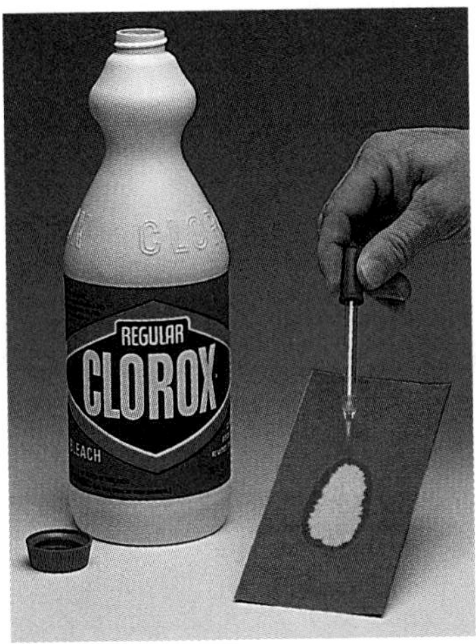

FIG. 5.1 **"Chlorine" bleach.** This common household product is a dilute aqueous solution of sodium hypochlorite, NaOCl, which destroys fabric stains by oxidizing them to colorless products. *(OPC, Inc.)*

EXAMPLE 5.1
Identifying Oxidation–Reduction

The bright light produced by the reaction between magnesium and oxygen often is used in fireworks displays. The product of the reaction is magnesium oxide, an ionic compound. Which element is oxidized and which is reduced? What are the oxidizing and reducing agents?

Fireworks display over Washington, D.C. *(Pete Saloutos/Corbis Images.)*

ANALYSIS: This example asks us to apply the definitions presented above, so we need to know how the electrons are transferred in the reaction. As a first step, let's write the chemical equation for the reaction.

Magnesium oxide is a compound of a metal and a nonmetal, so it's an ionic substance. The locations of the elements in the periodic table tell us that the ions are Mg^{2+} and O^{2-}, so the formula for magnesium oxide is MgO. (You probably already knew the formulas of these ions, but we're showing how you can apply previous knowledge to the problem at hand.) The reactants are magnesium, Mg, and molecular oxygen, O_2. The equation for the reaction is therefore

$$2Mg + O_2 \longrightarrow 2MgO$$

Now that we have the equation and the formulas of the ions we can determine how electrons are exchanged.

SOLUTION: When a magnesium atom becomes a magnesium ion, it must lose two electrons. Because electrons are lost in this process, they appear on the right when we express it as an equation.

$$Mg \longrightarrow Mg^{2+} + 2e^-$$

By losing electrons, *magnesium is oxidized*. This also means that *Mg must be the reducing agent.*

When oxygen reacts to yield O^{2-} ions, each oxygen atom must gain two electrons, so an O_2 molecule must gain four electrons. Because electrons are gained, they appear on the left in the equation.

$$O_2 + 4e^- \longrightarrow 2O^{2-}$$

By gaining electrons, *O_2 is reduced and must be the oxidizing agent.*

ARE THE ANSWERS REASONABLE? There are two things we can do to check our answers. First, we can check to be sure that we've placed electrons on the correct sides of the equations. As with ionic equations, *the number of atoms of each kind and the net charge must be the same on both sides.* We see that this is true for both equations. (If we had placed the electrons on the wrong side, the charges would not balance.) By observing the locations of the electrons in the equations (on the right for oxidation; on the left for reduction), we come to the same conclusions that Mg is oxidized and O_2 is reduced.

Another check is noting that we've identified one substance as being oxidized and the other as being reduced. If we had made a mistake, we might have concluded that both were oxidized, or both were reduced. But that's impossible, because in every reaction in which there is oxidation, there must also be reduction.

Practice Exercise 1: When sodium reacts with molecular oxygen, O_2, the product is sodium peroxide, Na_2O_2, which contains the peroxide ion, O_2^{2-}. In this reaction, is O_2 oxidized or reduced? (Hint: Which reactant gains electrons, and which loses electrons?)

Practice Exercise 2: Identify the substances oxidized and reduced and the oxidizing and reducing agents in the reaction of aluminum and chlorine to form aluminum chloride.

Oxidation numbers are used to follow redox changes

Unlike the reaction of magnesium with oxygen, not all reactions with oxygen produce ionic products. For example, sulfur reacts with oxygen to give sulfur dioxide, SO_2, which is molecular. Nevertheless, it is convenient to also view this as a redox reaction, but this requires that we change the way we define oxidation and reduction. To do this, chemists developed a bookkeeping system called **oxidation numbers,** which provides a way to keep tabs on electron transfers.

The oxidation number of an element in a particular compound is assigned according to a set of rules, which are described below. For simple, monatomic ions in a compound such as NaCl, the oxidation numbers are the same as the charges on the ions, so in NaCl the oxidation number of Na^+ is $+1$ and the oxidation number of Cl^- is -1. The real value of oxidation numbers is that they can also be assigned to atoms in molecular compounds. In such cases, it is important to realize that the oxidation number does not actually equal a charge on an atom. To be sure to differentiate oxidation numbers from actual electrical charges, we will specify the sign *before* the number when writing oxidation numbers, and *after* the number when writing electrical charges. Thus, a sodium ion has a charge of 1+ and an oxidation number of +1.

A term that is frequently used interchangeably with *oxidation number* is **oxidation state**. In NaCl, sodium has an oxidation number of +1 and is said to be "in the +1 oxidation state." Similarly, the chlorine in NaCl is said to be in the −1 oxidation state. There are times when it is convenient to specify the oxidation state of an element when its name is written out. This is done by writing the oxidation number as a Roman numeral in parentheses after the name of the element. For example, "iron(III)" means iron in the +3 oxidation state.

Oxidation numbers are sometimes written as a superscript using Roman numerals. Thus S^{VI} stands for sulfur in the +6 oxidation state.

We can use these new terms now to redefine a redox reaction.

A **redox reaction** is a chemical reaction in which changes in oxidation numbers occur.

We will find it easy to follow redox reactions by taking note of the changes in oxidation numbers. To do this, however, we must be able to assign oxidation numbers to atoms in a quick and simple way.

Rules permit assignment of oxidation numbers to almost any atom

We can use some basic knowledge learned earlier plus the set of rules below to determine the oxidation numbers of the atoms in almost any compound.

Assigning oxidation numbers

Rules for Assigning Oxidation Numbers

1. The oxidation number of any *free element* (an element not combined chemically with a different element) is zero, regardless of how complex its molecules are.
2. The oxidation number for any simple, monatomic ion (e.g., Na^+ or Cl^-) is equal to the charge on the ion. The charge on a polyatomic ion can be viewed as the *net* oxidation number of the ion.
3. The sum of all the oxidation numbers of the atoms in a molecule or polyatomic ion must equal the charge on the particle.
4. In its compounds, fluorine has an oxidation number of −1.
5. In its compounds, hydrogen has an oxidation number of +1.
6. In its compounds, oxygen has an oxidation number of −2.

Rule 1 means that O in O_2, P in P_4, and S in S_8 all have oxidation numbers of zero.

As you will see shortly, occasionally there is a conflict between two of these rules. When this happens, *we apply the rule with the lower number and ignore the conflicting rule.*

In addition to these basic rules, there is some other chemical knowledge you will need to remember. Recall that we can use the periodic table to help us remember the charges on certain ions of the elements. For instance, all the metals in Group IA form ions with a 1+ charge, and all those in Group IIA form ions with a 2+ charge. This means that when we find sodium in a compound, we can assign it an oxidation number of +1 because its simple ion, Na^+, has a charge of 1+. (We have applied Rule 2.) Similarly, calcium in a compound exists as Ca^{2+} and has an oxidation number of +2.

In binary ionic compounds with metals, the nonmetals have oxidation numbers equal to the charges on their anions (Table 2.6 on page 65). For example, the compound Fe_2O_3 contains the oxide ion, O^{2-}, which is assigned an oxidation number of −2. Similarly, Mg_3P_2 contains the phosphide ion, P^{3-}, which has an oxidation number of −3.

The numbered rules given above usually come into play when an element is capable of having more than one oxidation state. For example, you learned that transition metals can form more than one ion. Iron, for example, forms Fe^{2+} and Fe^{3+} ions, so in an iron compound we have to use the rules to figure out which iron ion is present. Similarly, when nonmetals are combined with hydrogen and oxygen in compounds or polyatomic ions, their oxidation numbers can vary and must be calculated using the rules.

With this as background, let's look at some examples that illustrate how we apply the rules, especially when they don't explicitly cover all the atoms in a formula.

EXAMPLE 5.2
Assigning Oxidation Numbers

Titanium dioxide, TiO_2, is a white pigment used in making paint. A now outmoded process of making TiO_2 from its ore involved $Ti(SO_4)_2$ as an intermediate. What is the oxidation number of titanium in $Ti(SO_4)_2$?

ANALYSIS: The key here is recognizing SO_4 as the sulfate ion, SO_4^{2-}. Once we know this, the rest is straightforward. Because we're only interested in the oxidation number of titanium, we use the charge on SO_4^{2-} to equal the net oxidation number of the ion (Rule 2).

SOLUTION: The sulfate ion has a charge of 2−, which we can take to be its net oxidation number. Then we apply the summation rule (Rule 3) to find the oxidation number of titanium, which we will represent by x.

$$\begin{array}{lrl} \text{Ti} & (1\text{ atom}) \times (x) = x & \\ SO_4^{2-} & (2\text{ ions}) \times (-2) = -4 & \\ \hline & \text{Sum} = 0 & \text{(Rule 3)} \end{array}$$

Obviously, the oxidation number of titanium is +4 so that the sum can be zero.

IS THE ANSWER REASONABLE? We have the sum of oxidation numbers adding up to zero, the charge on $Ti(SO_4)_2$, so our answer is okay.

EXAMPLE 5.3
Assigning Oxidation Numbers

Determine the oxidation numbers of the atoms in hydrogen peroxide, H_2O_2, a common antiseptic purchased in pharmacies.

ANALYSIS: The compound is molecular, so there are no ions to work with. Scanning the rules, we see we have one for hydrogen and another for oxygen. However, we begin to see we have a conflict between two of the rules. Rule 5 tells us to assign an oxidation number of +1

Hydrogen peroxide destroys bacteria by oxidizing them. *(Robert Capece.)*

to hydrogen, and Rule 6 tells us to give oxygen an oxidation number of −2. Both of these can't be correct because the sum must be zero (Rule 3).

$$\begin{array}{lrl} \text{H} & (2 \text{ atoms}) \times (+1) = +2 & \text{(Rule 5)} \\ \text{O} & (2 \text{ atoms}) \times (-2) = -4 & \text{(Rule 6)} \\ \hline & \text{Sum} \neq 0 & \text{(violates Rule 3)} \end{array}$$

As mentioned above, when there is a conflict between the rules, we ignore the higher numbered rule that causes the conflict.

SOLUTION: Rule 6 is the higher numbered rule causing the conflict, so we have to ignore it this time and just apply Rules 3 (the sum rule) and 5 (the rule that tells us the oxidation number of hydrogen is +1). Because we don't have a rule that applies to oxygen in this case, we'll represent the oxidation number of O by x.

$$\begin{array}{lrl} \text{H} & (2 \text{ atoms}) \times (+1) = +2 & \text{(Rule 5)} \\ \text{O} & (2 \text{ atoms}) \times (x) = 2x & \\ \hline & \text{Sum} = 0 & \text{(Rule 3)} \end{array}$$

For the sum to be zero, $2x = -2$, so $x = -1$. Therefore, in this compound, oxygen has an oxidation number of −1.

$$\text{H} = +1 \qquad \text{O} = -1$$

IS THE ANSWER REASONABLE? A conflict between the rules occurs only rarely, but when it happens the conflict becomes apparent because it causes a violation of the sum rule. However, *the sum rule always applies,* so the answer is reasonable.

Sometimes, oxidation numbers calculated by the rules have fractional values, as illustrated in the next example.

EXAMPLE 5.4
Assigning Oxidation Numbers

The air bags used as safety devices in modern autos are inflated by the very rapid decomposition of the ionic compound sodium azide, NaN_3. The reaction gives elemental sodium and gaseous nitrogen. What is the average oxidation number of the nitrogen in sodium azide?

The explosive decomposition of sodium azide releases nitrogen gas, which rapidly inflates an air bag during a crash. *(Corbis-Bettmann.)*

ANALYSIS: We're told that NaN_3 is ionic, so we know there is a cation and an anion. The cation is the sodium ion, Na^+, which means that the remainder of the formula unit, "N_3," must carry one unit of negative charge. However, there could not be three nitrogen anions each with a charge of $-\frac{1}{3}$ because it would mean that each nitrogen atom had acquired one-third of an electron. Whole numbers of electrons are always involved in electron transfers. Therefore, the anion must be a single particle with a negative one charge, namely, N_3^-. For this anion, the sum of the oxidation numbers of the three nitrogens must be –1. We can use this information, now, to solve the problem.

SOLUTION: The sum of the oxidation numbers of the nitrogen must be equal to −1. If we let x stand for the oxidation number of just one of the nitrogens, then

$$(3 \text{ atoms}) \times (x) = 3x = -1$$
$$x = -\tfrac{1}{3}$$

We could have also tackled the problem just using the rules for assigning oxidation numbers. We know that sodium exists as the ion Na^+ and that the compound is neutral overall. Therefore,

$$\begin{array}{lrl} \text{Na} & (1 \text{ atom}) \times (+1) = +1 & \text{(Rule 2)} \\ \text{N} & (3 \text{ atoms}) \times (x) = 3x & \\ \hline & \text{Sum} = 0 & \text{(Rule 3)} \end{array}$$

The sum of the oxidation numbers of the three nitrogen atoms in this ion must add up to -1, so each nitrogen must have an oxidation number of $-\frac{1}{3}$.

IS THE ANSWER REASONABLE? We've solved this example two ways, so we can certainly feel confident the answer is correct.

Practice Exercise 3: Chlorite ion, ClO_2^-, has been shown to be a potent disinfectant, and solutions of it are sometimes used to disinfect air-conditioning systems in cars. What is the oxidation number of chlorine in this ion? (Hint: The sum of the oxidation numbers is not zero.)

Practice Exercise 4: Assign oxidation numbers to each atom in (a) $NiCl_2$, (b) Mg_2TiO_4, (c) $K_2Cr_2O_7$, (d) HPO_4^{2-}, and (e) $V(C_2H_3O_2)_3$.

Practice Exercise 5: Iron forms a magnetic oxide with the formula Fe_3O_4 that contains both Fe^{2+} and Fe^{3+} ions. What is the *average* oxidation number of iron in this oxide?

Oxidation numbers are used to identify oxidation and reduction

Oxidation numbers can be used in several ways. One is to define oxidation and reduction in the most comprehensive manner, as follows:

Oxidation is an increase in oxidation number.
Reduction is a decrease in oxidation number.

Let's see how these definitions apply to the reaction of hydrogen with chlorine. To avoid ever confusing oxidation numbers with actual electrical charges, we will write oxidation numbers directly above the chemical symbols of the elements.

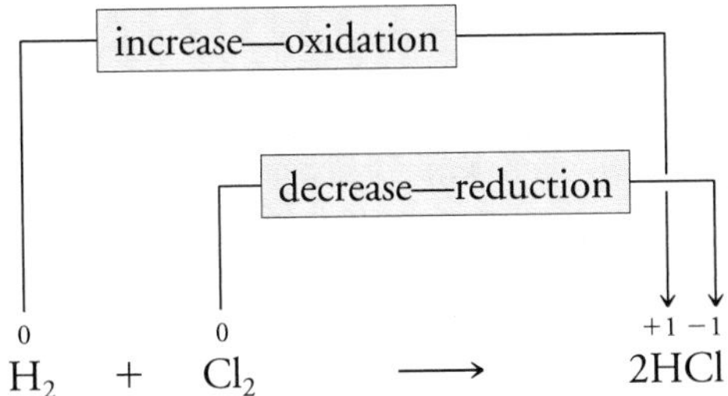

$$\overset{0}{H_2} + \overset{0}{Cl_2} \longrightarrow 2\overset{+1}{H}\overset{-1}{Cl}$$

Notice that we have assigned the atoms in H_2 and Cl_2 oxidation numbers of zero, in accord with Rule 1. The changes in oxidation number tell us that hydrogen is oxidized and chlorine is reduced.

EXAMPLE 5.5
Using Oxidation Numbers to Follow Redox Reactions

Is the following a redox reaction? If so, identify the substance oxidized and the substance reduced as well as the oxidizing and reducing agents.

$$2KCl + MnO_2 + 2H_2SO_4 \longrightarrow K_2SO_4 + MnSO_4 + Cl_2 + 2H_2O$$

ANALYSIS: To determine whether the reaction is redox and to give the specific answers requested, our tools will be the rules for assigning oxidation numbers and the revised definitions of oxidation and reduction. This will tell us whether redox is occurring, and if so, what is oxidized and reduced. Then we recall that the substance oxidized is the reducing agent, and the substance reduced is the oxidizing agent.

SOLUTION: To determine whether redox is occurring here, we first must assign oxidation numbers to each atom on both sides of the equation. Following the rules, we get

$$\overset{+1-1}{2KCl} + \overset{+4\ -2}{MnO_2} + \overset{+1+6-2}{2H_2SO_4} \longrightarrow \overset{+1+6-2}{K_2SO_4} + \overset{+2+6-2}{MnSO_4} + \overset{0}{Cl_2} + \overset{+1\ -2}{2H_2O}$$

Next we look for changes, keeping in mind that an increase in oxidation number is oxidation and a decrease is reduction.

A change from −1 to 0 is going up the number scale, so it's an increase in oxidation number

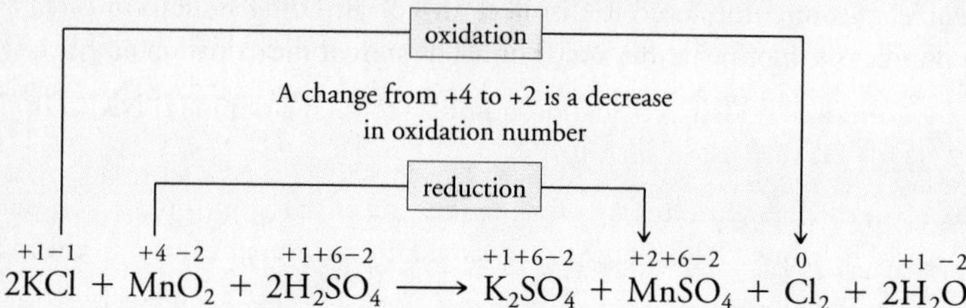

Thus the Cl in KCl is oxidized and the Mn in MnO_2 is reduced. The reducing agent is KCl and the oxidizing agent is MnO_2. (Notice that when we identify the oxidizing and reducing agents, we give the entire formulas for the substances that contain the atoms with oxidation numbers that change.)

IS THE ANSWER REASONABLE? There are lots of things we could check here. We have found changes that lead us to conclude that redox is happening. Assigning oxidation numbers also allows us to identify one change as oxidation and the other as reduction, which gives us confidence that we've done the rest of the work correctly.

Practice Exercise 6: Consider the following reactions:

$$N_2O_5 + 2NaHCO_3 \longrightarrow 2NaNO_3 + 2CO_2 + H_2O$$

$$KClO_3 + 3HNO_2 \longrightarrow KCl + 3HNO_3$$

Which one is a redox reaction? For the redox reaction, which compound is oxidized and which is reduced? (Hint: How is redox defined using oxidation numbers?)

Practice Exercise 7: Chlorine dioxide, ClO_2, is used to kill bacteria in the dairy industry, meat industry, and other food and beverage industry applications. It is unstable, but can be made by the following reaction.

$$Cl_2 + 2NaClO_2 \longrightarrow 2ClO_2 + 2NaCl$$

Identify the substances oxidized and reduced as well as the oxidizing and reducing agents in the reaction.

Practice Exercise 8: When hydrogen peroxide is used as an antiseptic, it kills bacteria by oxidizing them. When the H_2O_2 serves as an oxidizing agent, which product might be formed from it, O_2 or H_2O? Why?

5.2 THE ION-ELECTRON METHOD CREATES BALANCED NET IONIC EQUATIONS FOR REDOX REACTIONS

Many redox reactions take place in aqueous solution and many of these involve ions; they are ionic reactions. An example is the reaction of laundry bleach with substances in the wash water. The active ingredient in the bleach is hypochlorite ion, OCl^-, which is the oxidizing agent in these reactions. To study redox reactions, it is often helpful to write ionic and net ionic equations, just as we did in our analysis of metathesis reactions earlier in Chapter 4. Balancing net ionic equations for redox reactions is especially easy if we follow a procedure called the ion–electron method.

The ion–electron method uses a divide and conquer approach

In the **ion–electron method,** we divide the oxidation and reduction processes into individual equations called **half-reactions** that are balanced separately. Each half-reaction is made to obey *both* criteria for a balanced ionic equation: *both atoms and charge have to balance.* Then we combine the balanced half-reactions to obtain the fully balanced net ionic equation.

Equations for redox reactions are often more complex than those for metathesis reactions and can be difficult to balance by inspection. The ion–electron method is a systematic procedure for balancing redox equations.

In balancing the half-reactions, we must take into account that for many redox reactions in aqueous solutions, H^+ or OH^- ions play an important role, as do water molecules. For example, when solutions of $K_2Cr_2O_7$ and $FeSO_4$ are mixed, the acidity of the mixture decreases as dichromate ion, $Cr_2O_7^{2-}$, oxidizes Fe^{2+} (Figure 5.2). This is because the reaction uses up H^+ as a reactant and produces H_2O as a product. In other reactions, OH^- is consumed, while in still others H_2O is a reactant. Another fact is that in many cases the products (or even the reactants) of a redox reaction will differ depending on the acidity of the solution. For example, in an acidic solution MnO_4^- is reduced to Mn^{2+} ion, but in a neutral or slightly basic solution, the reduction product is insoluble MnO_2.

Because of these factors, redox reactions are generally carried out in solutions containing a substantial excess of either acid or base, so before we can apply the ion–electron method, we have to know whether the reaction occurs in an acidic or a basic solution. (This information will always be given to you in this book.)

FIG. 5.2 A redox reaction taking place in an acidic solution. A solution of $K_2Cr_2O_7$ oxidizes Fe^{2+} to Fe^{3+} in an acidic solution. At the same time, the orange dichromate ion is reduced to Cr^{3+}. *(Peter Lerman.)*

H^+ and H_2O help balance redox equations for acidic solutions

As you just learned, $Cr_2O_7^{2-}$ reacts with Fe^{2+} in an acidic solution to give Cr^{3+} and Fe^{3+} as products. This information permits us to write the **skeleton equation,** which shows only the ions (or sometimes molecules, too) involved in the redox changes.

$$Cr_2O_7^{2-} + Fe^{2+} \longrightarrow Cr^{3+} + Fe^{3+}$$

We then proceed through the steps described below to find the balanced equation. As you will see, the ion–electron method will tell us how H^+ and H_2O are involved in the reaction; we don't need to know this in advance.

Step 1. ***Divide the skeleton equation into half-reactions.*** We choose one of the reactants, let's say $Cr_2O_7^{2-}$, and write it at the left of an arrow. On the right, we write what $Cr_2O_7^{2-}$ changes to, which is Cr^{3+}. This gives us the beginnings of one half-reaction. For the second half-reaction, we write the other reactant, Fe^{2+}, on the left and the other product, Fe^{3+}, on the right. *Notice we are careful to use the complete formulas for the ions that appear in the skeleton equation.* Except for hydrogen and oxygen, the same elements must appear on both sides of a given half-reaction.

$$Cr_2O_7^{2-} \longrightarrow Cr^{3+}$$
$$Fe^{2+} \longrightarrow Fe^{3+}$$

Step 2. ***Balance atoms other than H and O.*** There are two Cr atoms on the left and only one on the right, so we place a coefficient of 2 in front of Cr^{3+}. The second half-reaction is already balanced in terms of atoms, so we leave it "as is."

$$Cr_2O_7^{2-} \longrightarrow 2Cr^{3+}$$
$$Fe^{2+} \longrightarrow Fe^{3+}$$

Many students tend to forget this step. If you do, you may end up in trouble later on.

Step 3. ***Balance oxygen by adding H_2O to the side that needs O.*** There are seven oxygen atoms on the left of the first half-reaction and none on the right. Therefore, we add $7H_2O$ to the right side of the first half-reaction to balance the oxygens. (There is no oxygen imbalance in the second half-reaction, so there's nothing to do there.)

$$Cr_2O_7^{2-} \longrightarrow 2Cr^{3+} + 7H_2O$$
$$Fe^{2+} \longrightarrow Fe^{3+}$$

We use H_2O, not O or O_2, to balance oxygen atoms because H_2O is what is actually present in the solution.

The oxygen atoms now balance, but we've created an imbalance in hydrogen. That issue is addressed next.

Step 4. ***Balance hydrogen by adding H^+ to the side that needs H.*** After adding the water, we see that the first half-reaction has 14 hydrogens on the right and none on the left. To balance hydrogen, we add $14H^+$ to the left side of the half-reaction. When you do this step (or others) *be careful to write the charges on the ions.* If they are omitted, you will not obtain a balanced equation in the end.

$$14H^+ + Cr_2O_7^{2-} \longrightarrow 2Cr^{3+} + 7H_2O$$

$$Fe^{2+} \longrightarrow Fe^{3+}$$

Now each half-reaction is balanced in terms of atoms. Next we will balance the charge.

Step 5. ***Balance the charge by adding electrons.*** First we compute the net electrical charge on each side. For the first half-reaction we have

$$\underbrace{14H^+ + Cr_2O_7^{2-}}_{\text{Net charge} = (14+) + (2-) = 12+} \longrightarrow \underbrace{2Cr^{3+} + 7H_2O}_{\text{Net charge} = 2(3+) + 0 = 6+}$$

The algebraic difference between the net charges on the two sides equals the number of electrons that must be added to the more positive (or less negative) side. In this instance, we must add $6e^-$ to the left side of the half-reaction.

$$6e^- + 14H^+ + Cr_2O_7^{2-} \longrightarrow 2Cr^{3+} + 7H_2O$$

This half-reaction is now complete; it is balanced in terms of both atoms and charge. (We can check this by recalculating the charge on both sides.)

To balance the other half-reaction, we add one electron to the right.

$$Fe^{2+} \longrightarrow Fe^{3+} + e^-$$

Step 6. ***Make the number of electrons gained equal to the number lost and then add the two half-reactions.*** At this point we have the two balanced half-reactions

$$6e^- + 14H^+ + Cr_2O_7^{2-} \longrightarrow 2Cr^{3+} + 7H_2O$$

$$Fe^{2+} \longrightarrow Fe^{3+} + e^-$$

Six electrons are gained in the first, but only one is lost in the second. Therefore, before combining the two equations we multiply all of the coefficients of the second half-reaction by 6.

Because we know the electrons will cancel, we really don't have to carry them down into the combined equation. We've done so here just for emphasis.

$$6e^- + 14H^+ + Cr_2O_7^{2-} \longrightarrow 2Cr^{3+} + 7H_2O$$

$$6(Fe^{2+} \longrightarrow Fe^{3+} + e^-)$$

$$\text{(Sum)} \quad 6e^- + 14H^+ + Cr_2O_7^{2-} + 6Fe^{2+} \longrightarrow 2Cr^{3+} + 7H_2O + 6Fe^{3+} + 6e^-$$

Step 7. ***Cancel anything that is the same on both sides.*** This is the final step. Six electrons cancel from both sides to give the final balanced equation.

$$14H^+ + Cr_2O_7^{2-} + 6Fe^{2+} \longrightarrow 2Cr^{3+} + 7H_2O + 6Fe^{3+}$$

Notice that both the charge and the atoms balance.

In some reactions, after adding the two half-reactions you may have H_2O or H^+ on both sides—for example, $6H_2O$ on the left and $2H_2O$ on the right. Cancel as many as you can. Thus,

$$\ldots + 6H_2O \ldots \longrightarrow \ldots + 2H_2O \ldots$$

reduces to

$$\ldots + 4H_2O \ldots \longrightarrow \ldots$$

The following is a summary of the steps we've followed for balancing an equation for a redox reaction in an acidic solution. If you don't skip any steps and you perform them in the order given, you will always obtain a properly balanced equation.

Ion–Electron Method—Acidic Solution

Step 1. Divide the equation into two half-reactions.
Step 2. Balance atoms other than H and O.
Step 3. Balance O by adding H_2O.
Step 4. Balance H by adding H^+.
Step 5. Balance net charge by adding e^-.
Step 6. Make e^- gain equal e^- loss; then add half-reactions.
Step 7. Cancel anything that's the same on both sides.

Ion–electron method for acidic solutions

EXAMPLE 5.6
Using the Ion–Electron Method

Balance the following equation. The reaction occurs in an acidic solution.

$$MnO_4^- + H_2SO_3 \longrightarrow SO_4^{2-} + Mn^{2+}$$

ANALYSIS: In using the ion–electron method, there's not much to analyze. It's necessary to know the steps and to follow them in sequence.

SOLUTION: We follow the steps given above.

Step 1. Divide the skeleton equation into half-reactions.

$$MnO_4^- \longrightarrow Mn^{2+}$$
$$H_2SO_3 \longrightarrow SO_4^{2-}$$

Step 2. There is nothing to do for this step. All the atoms except H and O are already in balance.

Step 3. Add H_2O to balance oxygens.

$$MnO_4^- \longrightarrow Mn^{2+} + 4H_2O$$
$$H_2O + H_2SO_3 \longrightarrow SO_4^{2-}$$

Step 4. Add H^+ to balance H.

$$8H^+ + MnO_4^- \longrightarrow Mn^{2+} + 4H_2O$$
$$H_2O + H_2SO_3 \longrightarrow SO_4^{2-} + 4H^+$$

Step 5. Balance charge by adding electrons to the more positive side.

$$5e^- + 8H^+ + MnO_4^- \longrightarrow Mn^{2+} + 4H_2O$$
$$H_2O + H_2SO_3 \longrightarrow SO_4^{2-} + 4H^+ + 2e^-$$

Step 6. Make electron loss equal to electron gain, then add half-reactions

$$2(5e^- + 8H^+ + MnO_4^- \longrightarrow Mn^{2+} + 4H_2O)$$
$$5(H_2O + H_2SO_3 \longrightarrow SO_4^{2-} + 4H^+ + 2e^-)$$

$$10e^- + 16H^+ + 2MnO_4^- + 5H_2O + 5H_2SO_3 \longrightarrow$$
$$2Mn^{2+} + 8H_2O + 5SO_4^{2-} + 20H^+ + 10e^-$$

Step 7. Cancel $10e^-$, $16H^+$, and $5H_2O$ from both sides. The final equation is

$$2MnO_4^- + 5H_2SO_3 \longrightarrow 2Mn^{2+} + 3H_2O + 5SO_4^{2-} + 4H^+$$

IS THE ANSWER REASONABLE? The check involves *two* steps. First, we check that each side of the equation has the same number of atoms of each element, which it does. Second, we check to be sure that the net charge is the same on both sides. On the left we have $2MnO_4^-$ with a net charge of 2−. On the right we have $2Mn^{2+}$ and $4H^+$ (total charge = 8+) along with $5SO_4^{2-}$ (total charge = 10−), so the net charge is also 2−. Having both atoms *and* charge in balance makes it a balanced equation and confirms that we've worked the problem correctly.

Practice Exercise 9: Explain why the following equation is not balanced. Balance it using the ion–electron method. (Hint: What are the criteria for a balanced ionic equation?)

$$Al(s) + Cu^{2+}(aq) \longrightarrow Al^{3+}(aq) + Cu(s)$$

Practice Exercise 10: The element technetium (atomic number 43) is radioactive and one of its isotopes, ^{99}Tc, is used in medicine for diagnostic imaging. The isotope is usually obtained in the form of the pertechnetate anion, TcO_4^-, but its use sometimes requires the technetium to be in a lower oxidation state. Reduction can be carried out using Sn^{2+} in an acidic solution. The skeleton equation is

$$TcO_4^- + Sn^{2+} \longrightarrow Tc^{4+} + Sn^{4+} \qquad \text{(acidic solution)}$$

Balance the equation by the ion–electron method.

Practice Exercise 11: What is the balanced net ionic equation for the following reaction in an acidic solution?

$$Cu + NO_3^- \longrightarrow Cu^{2+} + N_2O$$

Additional steps produce balanced equations for basic solutions

In a basic solution, the concentration of H^+ is very small; the dominant species are H_2O and OH^-. Strictly speaking, these should be used to balance the half-reactions. However, the simplest way to obtain a balanced equation for a basic solution is to first *pretend* that the solution is acidic. We balance the equation using the seven steps just described, and then we use a simple three-step procedure described below to convert the equation to the correct form for a basic solution. The conversion uses the fact that H^+ and OH^- react in a 1-to-1 ratio to give H_2O.

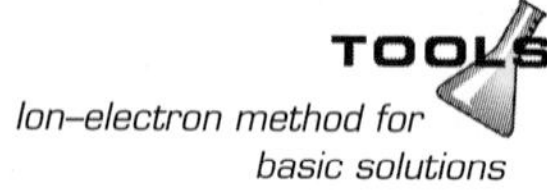

Ion–electron method for basic solutions

Additional Steps in the Ion–Electron Method for Basic Solutions

Step 8. Add to *both* sides of the equation the same number of OH^- as there are H^+.
Step 9. Combine H^+ and OH^- to form H_2O.
Step 10. Cancel any H_2O that you can.

As an example, suppose we wanted to balance the following equation for a basic solution.

$$SO_3^{2-} + MnO_4^- \longrightarrow SO_4^{2-} + MnO_2$$

Following Steps 1 through 7 for acidic solutions gives

$$2H^+ + 3SO_3^{2-} + 2MnO_4^- \longrightarrow 3SO_4^{2-} + 2MnO_2 + H_2O$$

Conversion of this equation to one appropriate for a basic solution proceeds as follows.

Step 8. ***Add to both sides of the equation the same number of OH^- as there are H^+.*** The equation for acidic solution has $2H^+$ on the left, so we add $2OH^-$ to *each* side. This gives

$$2OH^- + 2H^+ + 3SO_3^{2-} + 2MnO_4^- \longrightarrow 3SO_4^{2-} + 2MnO_2 + H_2O + 2OH^-$$

Step 9. ***Combine H^+ and OH^- to form H_2O.*** The left side has $2OH^-$ and $2H^+$, which become $2H_2O$. So in place of $2OH^- + 2H^+$ we write $2H_2O$.

$$\underbrace{2OH^- + 2H^+}_{\downarrow} + 3SO_3^{2-} + 2MnO_4^- \longrightarrow 3SO_4^{2-} + 2MnO_2 + H_2O + 2OH^-$$

$$2H_2O + 3SO_3^{2-} + 2MnO_4^- \longrightarrow 3SO_4^{2-} + 2MnO_2 + H_2O + 2OH^-$$

Step 10. ***Cancel any H_2O that you can.*** In this equation, one H_2O can be eliminated from both sides. The final equation, balanced for basic solution, is

$$H_2O + 3SO_3^{2-} + 2MnO_4^- \longrightarrow 3SO_4^{2-} + 2MnO_2 + 2OH^-$$

Practice Exercise 12: Consider the following half-reaction balanced for an acidic solution:

$$2H_2O + SO_2 \longrightarrow SO_4^{2-} + 4H^+ + 2e^-$$

What is the balanced half-reaction for a basic solution? (Hint: Remember that H^+ and OH^- react to form H_2O.)

Practice Exercise 13: Balance the following equation for a basic solution.

$$MnO_4^- + C_2O_4^{2-} \longrightarrow MnO_2 + CO_3^{2-}$$

5.3 METALS ARE OXIDIZED WHEN THEY REACT WITH ACIDS

Earlier you learned that one of the properties of acids is that they react with bases. Another important property is their ability to react with certain metals. These are redox reactions in which the metal is oxidized and the acid is reduced. But in these reactions, the part of the acid that's reduced depends on the composition of the acid itself as well as on the metal.

When a piece of zinc is placed into a solution of hydrochloric acid, bubbling is observed and the zinc gradually dissolves (Figure 5.3). The chemical reaction is

$$Zn(s) + 2HCl(aq) \longrightarrow ZnCl_2(aq) + H_2(g)$$

for which the net ionic equation is

$$Zn(s) + 2H^+(aq) \longrightarrow Zn^{2+}(aq) + H_2(g)$$

In this reaction, zinc is oxidized and hydrogen ions are reduced. Stated another way, *the* H^+ *of the acid is the oxidizing agent.*

Many metals react with acids just as zinc does—by being oxidized by hydrogen ions. In these reactions a metal salt and gaseous hydrogen are the products.

Metals that are able to react with acids such as HCl and H_2SO_4 to give hydrogen gas are said to be *more active* than hydrogen (H_2). For other metals, however, hydrogen ions are not powerful enough to cause their oxidation. Copper, for example, is significantly less reactive than zinc or iron, and H^+ cannot oxidize it. Copper is an example of a metal that is *less active* than H_2.

FIG. 5.3 Zinc reacts with hydrochloric acid. Bubbles of hydrogen are formed when a solution of hydrochloric acid comes in contact with metallic zinc. The same reaction takes place if hydrochloric acid is spilled on galvanized (zinc coated) steel. *(Andy Washnik.)*

The oxidizing power of an acid depends on the nature of the anion

Hydrochloric acid contains H_3O^+ ions (which we abbreviate as H^+) and Cl^- ions. The hydrogen ion in hydrochloric acid is able to be an oxidizing agent by being reduced to H_2. However, the Cl^- ion in the solution has no tendency at all to be an oxidizing agent, so in a solution of HCl the only oxidizing agent is H^+. The same applies to dilute solutions of sulfuric acid, H_2SO_4.

The hydrogen ion in water is actually a rather poor oxidizing agent, so hydrochloric acid and sulfuric acid have rather poor oxidizing abilities. They are often called **nonoxidizing acids,** even though their hydrogen ions are able to oxidize certain metals. (When we call something a nonoxidizing acid, we mean the *anion* of the acid is a weaker oxidizing agent than H^+ and that the anion of the acid is more difficult to reduce than H^+.) Some acids contain anions that are stronger oxidizing agents than H^+ and are called **oxidizing acids.** (See Table 5.1.)

The strongest oxidizing agent in a solution of a "nonoxidizing" acid is H^+.

Nitric acid is a powerful oxidizing agent

Nitric acid, HNO_3, ionizes in water to give H^+ and NO_3^- ions. In this solution the nitrate ion is a more powerful oxidizing agent than the hydrogen ion. This makes it able to oxidize metals that H^+ cannot, such as copper and silver. For example, the molecular equation for the reaction of concentrated HNO_3 with copper, shown in Figure 5.4, is

$$Cu(s) + 4HNO_3(aq) \longrightarrow Cu(NO_3)_2(aq) + 2NO_2(g) + 2H_2O$$

Oxidizing and nonoxidizing acids

TABLE 5.1 Nonoxidizing and Oxidizing Acids

Nonoxidizing Acids	
$HCl(aq)$	
$H_2SO_4(aq)$[a]	
$H_3PO_4(aq)$	
Most organic acids (e.g., $HC_2H_3O_2$)	
Oxidizing Acids	**Reduction reaction**
HNO_3	(conc.) $NO_3^- + 2H^+ + e^- \longrightarrow NO_2(g) + H_2O$
	(dilute) $NO_3^- + 4H^+ + 3e^- \longrightarrow NO(g) + 2H_2O$
	(very dilute, with strong reducing agent) $NO_3^- + 10H^+ + 8e^- \longrightarrow NH_4^+ + 3H_2O$
H_2SO_4	(hot, conc.) $SO_4^{2-} + 4H^+ + 2e^- \longrightarrow SO_2(g) + 2H_2O$
	(hot, conc., with strong reducing agent) $SO_4^{2-} + 10H^+ + 8e^- \longrightarrow H_2S(g) + 4H_2O$

[a] H_2SO_4 is a nonoxidizing acid when cold and dilute.

$$Cu + 4HNO_3 \longrightarrow Cu(NO_3)_2 + 2NO_2 + 2H_2O$$

FIG. 5.4 The reaction of copper with concentrated nitric acid. A copper penny reacts vigorously with concentrated nitric acid, as this sequence of photographs shows. The dark red brown vapors are nitrogen dioxide, the same gas that gives smog its characteristic color. *(Michael Watson.)*

If we convert this to a net ionic equation and then assign oxidation numbers, we can see that NO_3^- is the oxidizing agent and Cu is the reducing agent.

oxidation (0 → +2); reduction (+5 → +4)

$$\overset{0}{Cu(s)} + 2\overset{+5}{N}O_3^-(aq) + 4H^+(aq) \longrightarrow \overset{+2}{Cu^{2+}}(aq) + 2\overset{+4}{N}O_2(g) + 2H_2O$$

Cu: reducing agent; NO_3^-: oxidizing agent

Notice that in this reaction, *no hydrogen gas is formed*. The H^+ ions of the HNO_3 are an essential part of the reaction, but they just become part of water molecules without a change in oxidation number.

The nitrogen-containing product formed in the reduction of nitric acid depends on the concentration of the acid and the reducing power of the metal. With *concentrated* nitric acid, nitrogen dioxide, NO_2, is often the reduction product. With *dilute* nitric acid the product is often nitrogen monoxide (also called nitric oxide), NO, instead. Copper, for example, reacts as follows:

Concentrated HNO_3

$$Cu(s) + 4H^+(aq) + 2NO_3^-(aq) \longrightarrow Cu^{2+}(aq) + 2NO_2(g) + 2H_2O$$

Dilute HNO_3

$$3Cu(s) + 8H^+(aq) + 2NO_3^-(aq) \longrightarrow 3Cu^{2+}(aq) + 2NO(g) + 4H_2O$$

Nitric acid is a very effective oxidizing acid. All metals except the most unreactive ones, such as platinum and gold, are attacked by it. Nitric acid also does a good job of oxidizing organic compounds, so it is wise to be especially careful when working with this acid in the laboratory. Very serious accidents have occurred when inexperienced people have used concentrated nitric acid around organic substances.

Nitric acid causes severe skin burns, so be careful when you work with it in the laboratory. If you spill any on your skin, wash it off immediately and seek the help of your lab teacher.

Hot concentrated sulfuric acid is an oxidizing acid

In a dilute solution, the sulfate ion of sulfuric acid has little tendency to serve as an oxidizing agent. However, if the sulfuric acid is both concentrated and hot, it becomes a fairly potent oxidizer. For example, copper is not bothered by cool dilute H_2SO_4, but it is attacked by hot concentrated H_2SO_4 according to the following equation:

$$Cu + 2H_2SO_4(\text{hot, conc.}) \longrightarrow CuSO_4 + SO_2 + 2H_2O$$

Because of this oxidizing ability, hot concentrated sulfuric acid can be very dangerous. The liquid is viscous and can stick to the skin, causing severe burns.

Practice Exercise 14: Write the balanced half-reactions for the reaction of zinc with hydrogen ions. (Hint: Be sure to place the electrons on the correct sides of the half-reactions.)

Practice Exercise 15: When very dilute nitric acid reacts with a strong oxidizing agent such as magnesium, the nitrate ion is reduced to ammonium ion. Write a balanced net ionic equation for the reaction.

Practice Exercise 16: Write balanced molecular, ionic, and net ionic equations for the reaction of hydrochloric acid with (a) magnesium and (b) aluminum. (Both metals are oxidized by hydrogen ions.)

5.4 A MORE ACTIVE METAL WILL DISPLACE A LESS ACTIVE ONE FROM ITS COMPOUNDS

The formation of hydrogen gas in the reaction of a metal with an acid is a special case of a more general phenomenon—one element displacing (pushing out) another element from a compound by means of a redox reaction. In the case of a metal–acid reaction, it is the metal that displaces hydrogen from the acid, changing $2H^+$ to H_2.

Another reaction of this same general type occurs when one metal displaces another metal from its compounds, and is illustrated by the experiment shown in Figure 5.5. Here we see a brightly polished strip of metallic zinc that is dipped into a solution of copper sulfate. After the zinc is in the solution for a while, a reddish brown deposit of metallic copper forms on the zinc, and if the solution were analyzed, we would find that it now contains zinc ions, as well as some remaining unreacted copper ions.

The results of this experiment can be summarized by the following net ionic equation.

$$Zn(s) + Cu^{2+}(aq) \longrightarrow Cu(s) + Zn^{2+}(aq)$$

Sulfate ion is a spectator ion in this reaction.

Metallic zinc is oxidized as copper ion is reduced. In the process, Zn^{2+} ions have taken the place of the Cu^{2+} ions, so a solution of copper sulfate is changed to a solution of zinc sulfate. An atomic-level view of what's happening at the surface of the zinc during the reaction is depicted in Figure 5.6. A reaction such as this, in which one element replaces another in a compound, is sometimes called a **single replacement reaction.**

The activity series arranges metals according to their ease of oxidation

In the reaction of zinc with copper ion, the more "active" zinc displaces the less "active" copper in a compound, where we have used the word *active* to mean "easily oxidized." This is actually a general phenomenon: *an element that is more easily oxidized will displace one that is less*

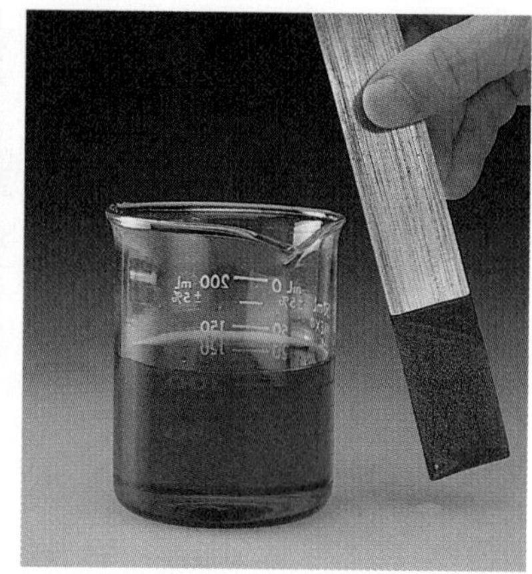

FIG. 5.5 **The reaction of zinc with copper ion.** (*Left*) A piece of shiny zinc next to a beaker containing a copper sulfate solution. (*Center*) When the zinc is placed in the solution, copper ions are reduced to the free metal while the zinc dissolves. (*Right*) After a while the zinc becomes coated with a red-brown layer of copper. Notice that the solution is a lighter blue than before, showing that some of the copper ions have left the solution. *(Michael Watson.)*

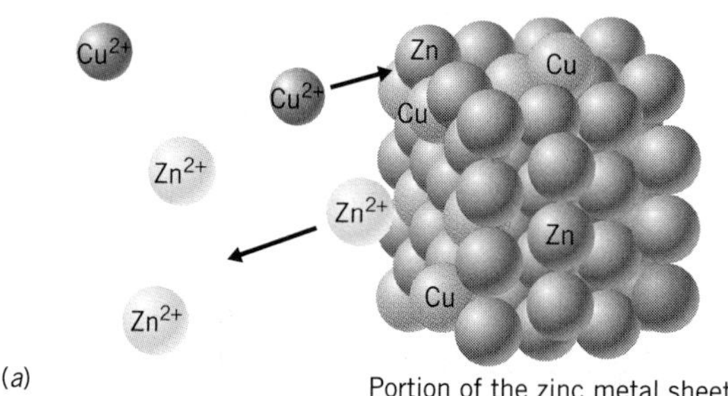

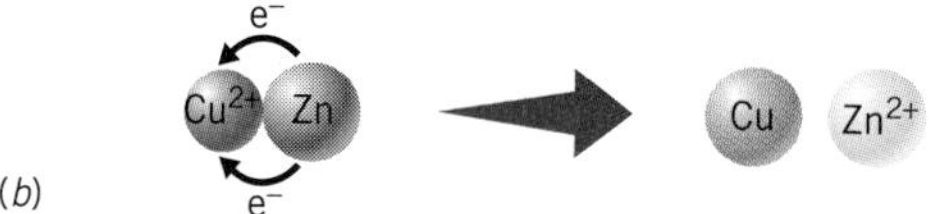

FIG. 5.6 **The reaction of copper ions with zinc, viewed at the atomic level.** (*a*) Copper ions (blue) collide with the zinc surface where they pick up electrons from zinc atoms (gray). The zinc atoms become zinc ions (yellow) and enter the solution. The copper ions become copper atoms (red-brown) and stick to the surface of the zinc. (For clarity, the water molecules of the solution and the sulfate ions are not shown.) (*b*) A close-up view of the exchange of electrons that leads to the reaction.

easily oxidized from its compounds. By comparing the relative ease of oxidation of various metals using experiments like the one pictured in Figure 5.5, we can arrange metals in order of their ease of oxidation. This yields the **activity series** shown in Table 5.2. In this table, metals at the bottom are more easily oxidized (are more active) than those at the top. *This means that a given element will be displaced from its compounds by any metal below it in the table.*

Notice that we have included hydrogen in the activity series. Metals below hydrogen in the series can displace hydrogen from solutions containing H^+. These are the metals that are capable of reacting with nonoxidizing acids. On the other hand, metals above hydrogen in the table do not react with acids having H^+ as the strongest oxidizing agent.

Metals at the very bottom of the table are very easily oxidized and are extremely strong reducing agents. They are so reactive, in fact, that they are able to reduce the hydrogen in water molecules. Sodium, for example, reacts vigorously (see Figure 5.7).

$$2Na(s) + 2H_2O \longrightarrow H_2(g) + 2NaOH(aq)$$

For metals below hydrogen in the activity series, there's a parallel between the ease of oxidation of the metal and the speed with which it reacts with H^+. For example, in

TABLE 5.2 Activity Series for Some Metals (and Hydrogen)

	Element	Oxidation Product
Least Active	Gold	Au^{3+}
	Mercury	Hg^{2+}
	Silver	Ag^{+}
	Copper	Cu^{2+}
	HYDROGEN	H^{+}
	Lead	Pb^{2+}
	Tin	Sn^{2+}
	Cobalt	Co^{2+}
	Cadmium	Cd^{2+}
	Iron	Fe^{2+}
	Chromium	Cr^{3+}
	Zinc	Zn^{2+}
	Manganese	Mn^{2+}
	Aluminum	Al^{3+}
	Magnesium	Mg^{2+}
	Sodium	Na^{+}
	Calcium	Ca^{2+}
	Strontium	Sr^{2+}
	Barium	Ba^{2+}
	Potassium	K^{+}
	Rubidium	Rb^{+}
Most Active	Cesium	Cs^{+}

Increasing ease of oxidation of the metal (↓)

Increasing ease of reduction of the ion (↑)

FIG. 5.7 Metallic sodium reacts violently with water. The heat of the reaction ignites the sodium metal, which can be seen burning and sending sparks from the surface of the water. In the reaction, sodium is oxidized to Na^{+} and water molecules are reduced to give hydrogen gas and hydroxide ions. When the reaction is over, the solution contains sodium hydroxide. *(OPC, Inc.)*

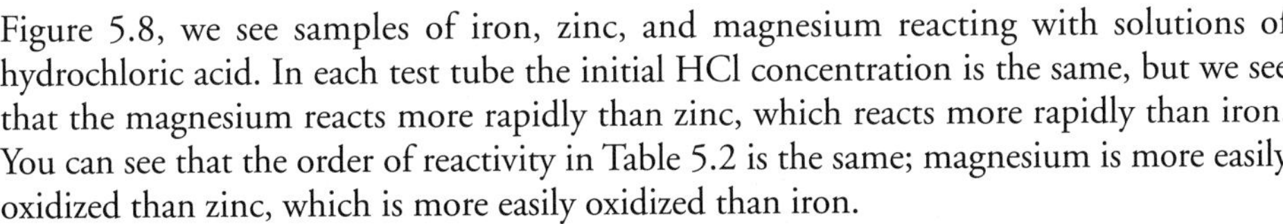

Figure 5.8, we see samples of iron, zinc, and magnesium reacting with solutions of hydrochloric acid. In each test tube the initial HCl concentration is the same, but we see that the magnesium reacts more rapidly than zinc, which reacts more rapidly than iron. You can see that the order of reactivity in Table 5.2 is the same; magnesium is more easily oxidized than zinc, which is more easily oxidized than iron.

The activity series can be used to predict reactions

The activity series in Table 5.2 permits us to make predictions of the outcome of single replacement redox reactions, as illustrated in the following examples.

$2HCl(aq) + Fe(s) \longrightarrow FeCl_2(aq) + H_2(g)$

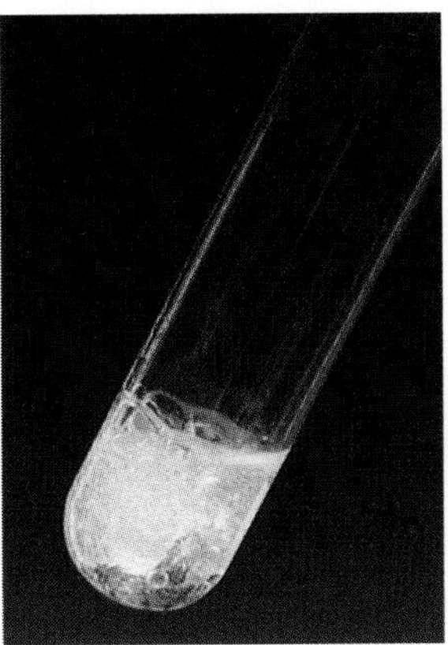

$2HCl(aq) + Zn(s) \longrightarrow ZnCl_2(aq) + H_2(g)$

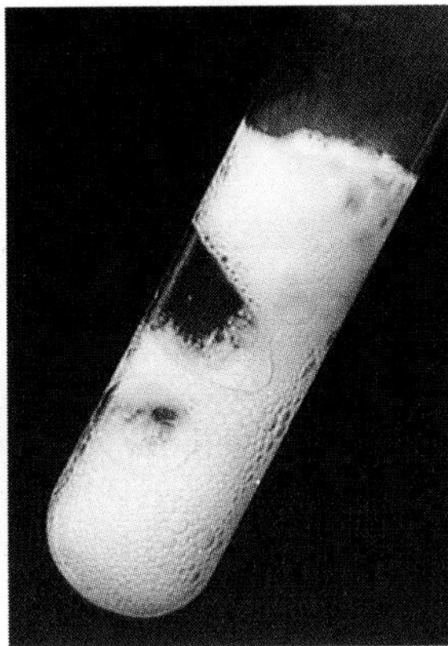

$2HCl(aq) + Mg(s) \longrightarrow MgCl_2(aq) + H_2(g)$

FIG. 5.8 The relative ease of oxidation of metals parallels their rates of reaction with hydrogen ions of an acid. The products are hydrogen gas and the metal ion in solution. All three test tubes contain HCl(*aq*) at the same concentration. The first also contains pieces of iron, the second, pieces of zinc, and the third, pieces of magnesium. Among these three metals, the ease of oxidation increases from iron to zinc to magnesium. *(OPC, Inc.)*

EXAMPLE 5.7
Using the Activity Series

What will happen if an iron nail is dipped into a solution containing copper(II) sulfate? If a reaction occurs, write its molecular equation.

ANALYSIS: Reading the question, we have to ask, what *could* happen? If a chemical reaction were to occur, iron would have to react with the copper sulfate. A *metal* possibly reacting with the *salt of another metal?* This suggests the possibility of a single replacement reaction. The tool we use to predict such reactions is the activity series.

Examining Table 5.2 we see that iron is below copper. This means iron is more easily oxidized than copper, so we expect metallic iron to displace copper ions from the solution. *A reaction will occur.* (We've answered one part of the question.) The formula for copper(II) sulfate is $CuSO_4$. To write an equation for the reaction, we have to know the final oxidation state of the iron. In the table, this is indicated as +2, so the Fe atoms change to Fe^{2+} ions. To write the formula of the salt in the solution we pair Fe^{2+} with SO_4^{2-} to give $FeSO_4$. Copper(II) ions are reduced to copper atoms.

SOLUTION: Our analysis told us that a reaction *will* occur and it also gave us the products, so the equation is

$$Fe(s) + CuSO_4(aq) \longrightarrow Cu(s) + FeSO_4(aq)$$

IS THE ANSWER REASONABLE? We can check the activity series again to be sure we've reached the correct conclusion, and we can check to be sure the equation we've written has the correct formulas and is balanced correctly. Doing this confirms that we've got the right answers.

EXAMPLE 5.8
Using the Activity Series

What happens if an iron nail is dipped into a solution of aluminum sulfate? If a reaction occurs, write the molecular equation.

ANALYSIS: Once again, we have to realize that we're looking for a potential single replacement reaction. Scanning the activity series, we see that aluminum metal is *more* easily oxidized than iron metal. This means that aluminum atoms would be able to displace iron ions from an iron compound. But it also means that iron atoms cannot displace aluminum ions from its compounds, and iron *atoms* plus aluminum *ions* are what we're given.

SOLUTION: Our analysis has told us that iron atoms will not reduce aluminum ions, so we must conclude that no reaction can occur.

$$Fe(s) + Al_2(SO_4)_3(aq) \longrightarrow \text{no reaction}$$

IS THE ANSWER REASONABLE? Checking the activity series again, we are confident we've come to the correct answer. In writing the equation, we've also been careful to correctly write the formula of aluminum sulfate.

Practice Exercise 17: Suppose a mixture is prepared containing magnesium sulfate ($MgSO_4$), copper(II) sulfate ($CuSO_4$), metallic magnesium, and metallic copper. What reaction, if any, will occur? (Hint: What reactions could occur?)

Practice Exercise 18: Write a chemical equation for the reaction that will occur, if any, when (a) aluminum metal is added to a solution of copper(II) chloride and (b) silver metal is added to a solution of magnesium sulfate. If no reaction will occur, write "no reaction" in place of the products.

5.5 Molecular Oxygen Is a Powerful Oxidizing Agent

Oxygen is a plentiful chemical; it's in the air and available to anyone who wants to use it, chemist or not. Furthermore, O_2 is a very reactive oxidizing agent, so its reactions have been well studied. When they are rapid, with the evolution of light and heat, we call them **combustion**. The products of reactions with oxygen are generally oxides, *molecular oxides* when oxygen reacts with nonmetals and *ionic oxides* when oxygen reacts with metals.

Organic compounds burn in oxygen

Experience has taught you that certain materials burn. For example, if you had to build a fire to keep warm, you no doubt would look for combustible materials like twigs, logs, or other pieces of wood to use as fuel. When you drive a car, it is probably powered by the combustion of gasoline. Wood and gasoline are examples of substances or mixtures of substances that chemists call *organic compounds*—compounds whose structures are determined primarily by the linking together of carbon atoms. When organic compounds burn, the products of the reactions are usually easy to predict.

Hydrocarbons are important fuels

Fuels such as natural gas, gasoline, kerosene, heating oil, and diesel fuel are examples of *hydrocarbons*—compounds containing only the elements carbon and hydrogen. Natural gas is composed principally of methane, CH_4. Gasoline is a mixture of hydrocarbons, the most familiar of which is octane, C_8H_{18}. Kerosene, heating oil, and diesel fuel are mixtures of hydrocarbons in which the molecules contain even more atoms of carbon and hydrogen.

TOOLS
Hydrocarbon combustion with plentiful supply of O_2

When hydrocarbons burn in a ***plentiful*** *supply of oxygen, the products of combustion are always carbon dioxide and water.* Thus, methane and octane combine with oxygen according to the equations

$$CH_4 + 2O_2 \longrightarrow CO_2 + 2H_2O$$

$$2C_8H_{18} + 25O_2 \longrightarrow 16CO_2 + 18H_2O$$

Many people don't realize that water is one of the products of the combustion of hydrocarbons, even though they have seen evidence for it. Perhaps you've seen clouds of condensed water vapor coming from the exhaust pipes of automobiles on cold winter days, or you may have noticed that shortly after you first start a car, drops of water fall from the exhaust pipe. This is water that has been formed during the combustion of the gasoline. Similarly, the "smokestacks" of power stations release clouds of condensed water vapor (Figure 5.9), which is often mistaken for smoke from fires used to generate power to make electricity. Actually, many of today's power stations produce very little smoke because they burn clean natural gas instead of coal.

FIG. 5.9 **Water is a product of the combustion of hydrocarbons.** Here we see clouds of condensed water vapor coming from the stacks of an oil-fired electric generating plant during the winter. *(Sandra Baker/Liaison Agency, Inc./Getty Images.)*

TOOLS
Hydrocarbon combustion with limited supply of O_2

When the supply of oxygen is somewhat restricted during the combustion of a hydrocarbon, some of the carbon is converted to carbon monoxide. The formation of CO is a pollution problem associated with the use of gasoline engines, as you may know.

$$2CH_4 + 3O_2 \longrightarrow 2CO + 4H_2O \qquad \text{(in a limited oxygen supply)}$$

TOOLS
Hydrocarbon combustion with extremely limited supply of O_2

When the oxygen supply is extremely limited, only the hydrogen of a hydrocarbon mixture is converted to the oxide (water). The carbon atoms emerge as elemental carbon. For example, when a candle burns, the fuel is a high-molecular-weight hydrocarbon (e.g., $C_{20}H_{42}$) and incomplete combustion forms tiny particles of carbon that glow brightly. If a cold surface is held in the flame, the unburned carbon deposits, as seen in Figure 5.10.

An important commercial reaction is the incomplete combustion of methane in a very limited oxygen supply, which follows the equation

$$CH_4 + O_2 \longrightarrow C + 2H_2O \qquad \text{(in a very limited oxygen supply)}$$

This finely divided form of carbon is also called *carbon black.*

The carbon that forms is very finely divided and would be called *soot* by almost anyone observing the reaction. Nevertheless, such soot has considerable commercial value when collected and marketed under the name *lampblack.* This sooty form of carbon is used to manufacture inks and much of it is used in the production of rubber tires, where it serves as a binder and a filler. When soot from incomplete combustion is released into air, its tiny particles constitute a component of air pollution referred to as *particulates,* which contribute to the haziness of smog.

Combustion of organic compounds that contain oxygen also produces CO_2 and H_2O

The formula for cellulose can be expressed as $(C_6H_{10}O_5)n$, which indicates that the molecule contains the $C_6H_{10}O_5$ unit repeated some large number n times.

Earlier we mentioned that you might choose wood to build a fire. The chief combustible ingredient in wood is cellulose, a fibrous material that gives plants their structural strength. Cellulose is composed of the elements carbon, hydrogen, and oxygen. Each cellulose molecule consists of many small, identical groups of atoms that are linked together to form a very long molecule, although the lengths of the molecules differ. For this reason we cannot specify a molecular formula for cellulose. Instead, we use the empirical formula, $C_6H_{10}O_5$, which represents the small, repeating "building block" units in large cellulose molecules. When cellulose burns, the products are also carbon dioxide and water. The only difference between its reaction and the reaction of a hydrocarbon with oxygen is that some of the oxygen in the products comes from the cellulose.

TOOLS
Combustion of organic compounds containing C, H, and O

$$C_6H_{10}O_5 + 6O_2 \longrightarrow 6CO_2 + 5H_2O$$

The complete combustion of all other organic compounds containing only carbon, hydrogen, and oxygen produces the same products, CO_2 and H_2O, and follows similar equations.

Burning organic compounds that contain sulfur gives SO_2 as one of the products

TOOLS
Combustion of organic compounds containing sulfur

A major pollution problem in industrialized countries is caused by the release into the atmosphere of sulfur dioxide formed by the combustion of fuels that contain sulfur or its compounds. *The products of the combustion of organic compounds of sulfur are carbon dioxide, water, and sulfur dioxide.* A typical reaction is

$$2C_2H_5SH + 9O_2 \longrightarrow 4CO_2 + 6H_2O + 2SO_2$$

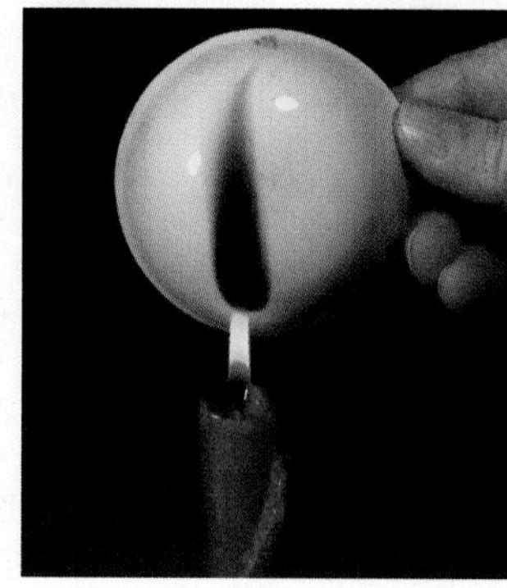

FIG. 5.10 Incomplete combustion of a hydrocarbon. The bright yellow color of a candle flame is caused by glowing particles of elemental carbon. Here we see that a black deposit of carbon is formed when the flame contacts a cold porcelain surface. *(Andy Washnik.)*

A solution of sulfur dioxide in water is acidic, and when rain falls through polluted air it picks up SO_2 and becomes "acid rain." Some SO_2 is also oxidized to SO_3, which reacts with moisture to give H_2SO_4, making the acid rain even more acidic.

Practice Exercise 19: Write a balanced chemical equation for the combustion of candle wax, $C_{20}H_{42}$, in a very limited supply of oxygen. (Hint: What happens to methane under these conditions?)

Practice Exercise 20: Write a balanced equation for the combustion of butane, C_4H_{10}, in an abundant supply of oxygen. Butane is the fuel used in disposable cigarette lighters.

Practice Exercise 21: Ethanol, C_2H_5OH, is now mixed with gasoline, and the mixture is sold under the name *gasohol.* Write a chemical equation for the combustion of ethanol.

Many metals react with oxygen

We don't often think of metals as undergoing combustion, but have you ever seen an old-fashioned flashbulb fired to take a photograph? The source of light is the reaction of the metal magnesium with oxygen (see Figure 5.11). A close look at a fresh flashbulb reveals a fine web of thin magnesium wire within the glass envelope. The wire is surrounded by an atmosphere of oxygen, a colorless gas. When the flashbulb is used, a small electric current surges through the thin wire, causing it to become hot enough to ignite, and it burns rapidly in the oxygen atmosphere. The equation for the reaction is

$$2Mg + O_2 \longrightarrow 2MgO$$

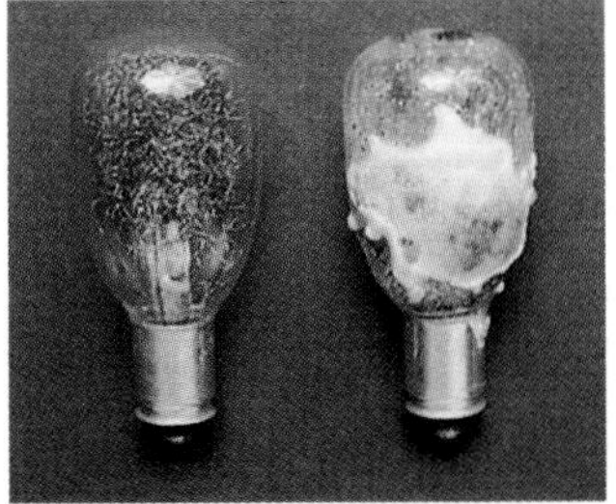

FIG. 5.11 A flashbulb, before and after firing. Fine magnesium wire in an atmosphere of oxygen fills the flashbulb at the left. After being used (*right*), the interior of the bulb is coated with a white film of magnesium oxide. *(Robert Capece.)*

Most metals react directly with oxygen, although not so spectacularly, and usually we refer to the reaction as **corrosion** or **tarnishing** because the oxidation products dull the shiny metal surface. Iron, for example, is oxidized fairly easily, especially in the presence of moisture. As you know, under these conditions the iron corrodes—it rusts. Rust is a form of iron(III) oxide, Fe_2O_3, that also contains an appreciable amount of absorbed water. The formula for rust is therefore normally given as $Fe_2O_3{\cdot}xH_2O$ to indicate its somewhat variable composition. Although the rusting of iron is a slow reaction, the combination of iron with oxygen can be speeded up if the metal is heated to a very high temperature under a stream of pure O_2 (see Figure 5.12).

An aluminum surface, unlike that of iron, is not noticeably dulled by the reaction of aluminum with oxygen. Aluminum is a common metal found around the home in uses ranging from aluminum foil to aluminum window frames, and it surely appears shiny. Yet, aluminum is a rather easily oxidized metal, as can be seen from its position in the activity series (Table 5.2). A *freshly* exposed surface of the metal does react very quickly with oxygen and becomes coated with a very thin film of aluminum oxide, Al_2O_3, so thin that it doesn't obscure the shininess of the metal beneath. Fortunately, the oxide coating adheres very tightly to the surface of the metal and makes it very difficult for additional oxygen to combine with the aluminum. Therefore, further oxidation of aluminum occurs very slowly.

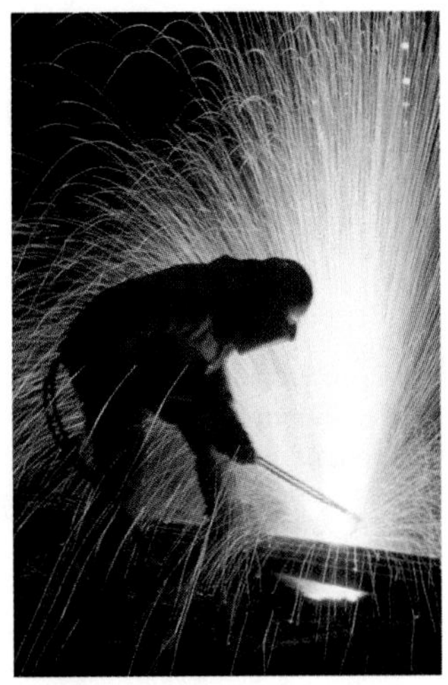

FIG. 5.12 Cutting steel with an oxyacetylene torch. An oxygen–acetylene flame is used to heat steel until its glowing red hot. Then the acetylene is turned off and the steel is cut by a stream of pure oxygen whose reaction with the hot metal produces enough heat to melt the steel and send a shower of burning steel sparks flying. *(Scott T. Smith/Corbis Images.)*

Practice Exercise 22: Write a balanced chemical equation for the reaction of molecular oxygen with strontium metal to form the oxide. (Hint: Strontium, Sr, is in the same group in the periodic table as calcium.)

Practice Exercise 23: The oxide formed in the reaction shown in Figure 5.12 is iron(III) oxide. Write a balanced chemical equation for the reaction.

Most nonmetals react with oxygen directly

Most nonmetals combine as readily with oxygen as do the metals, and their reactions usually occur rapidly enough to be described as combustion. To most people, the most important nonmetal combustion reaction is that of carbon because the reaction is a source of heat. Coal and charcoal, for example, are common carbon fuels. Coal is used worldwide in large amounts to generate electricity, and charcoal is a popular fuel for broiling hamburgers. If plenty of oxygen is available, the combustion of carbon gives CO_2, but when the supply

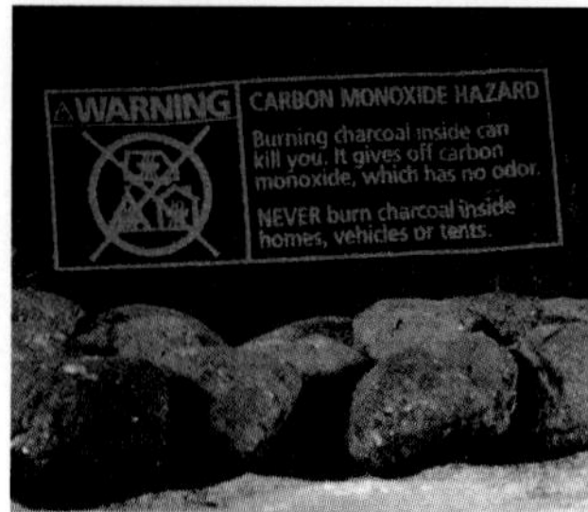

The label on a bag of charcoal displays a warning about carbon monoxide. *(Andy Washnik.)*

of O_2 is limited, some CO forms as well. Manufacturers that package charcoal briquettes, therefore, print a warning on the bag that the charcoal shouldn't be used indoors for cooking or heating.

Sulfur is another nonmetal that burns readily in oxygen. In the manufacture of sulfuric acid, the first step is the combustion of sulfur to produce sulfur dioxide. As mentioned earlier, sulfur dioxide also forms when sulfur compounds burn, and the presence of sulfur and sulfur compounds as impurities in coal and petroleum is a major source of air pollution. Power plants that burn coal are making strides to remove the SO_2 from their exhausts and it is being used to make sulfuric acid. As we noted earlier, when SO_2 does enter the atmosphere, it can dissolve in rainwater and become one of the components of acid rain. Some SO_2 is also oxidized slowly to SO_3, which gives the strong acid H_2SO_4 when it dissolves in rainwater.

5.6 Redox Reactions Follow the Same Stoichiometric Principles as Other Reactions

In general, working stoichiometry problems involving redox reactions follows the same principles we've applied to other reactions. The principal difference is that the chemical equations are more complex. Nevertheless, once we have a balanced equation, moles of substances involved in the reaction are related by the coefficients in the balanced equation.

Because so many reactions involve oxidation and reduction, it should not be surprising that they have useful applications in the lab. Some redox reactions are especially useful in chemical analyses, particularly in titrations. Unlike in acid–base titrations, however, there are no simple indicators that can be used to conveniently detect the end points in redox titrations, so we have to rely on color changes among the reactants themselves.

One of the most useful reactants for redox titrations is potassium permanganate, $KMnO_4$, especially when the reaction can be carried out in an acidic solution. Permanganate ion is a powerful oxidizing agent, so it oxidizes most substances that are capable of being oxidized. That's one reason why it is used. Especially important, though, is the fact that the MnO_4^- ion has an intense purple color and its reduction product in acidic solution is the almost colorless Mn^{2+} ion. Therefore, when a solution of $KMnO_4$ is added from a buret to a solution of a reducing agent, the chemical reaction that occurs forms a nearly colorless product. This is illustrated in Figure 5.13, where we see a solution of $KMnO_4$ being poured into an acidic solution containing Fe^{2+}. As the $KMnO_4$ solution is added, the purple color continues to be destroyed as long as there is any reducing agent left. In a titration, after the last trace of the reducing agent has been consumed, the MnO_4^- ion in the next drop of titrant has nothing to react with, so it colors the solution pink. This signals the end of the titration. In this way, permanganate ion serves as its own indicator in redox titrations. The next example illustrates a typical analysis using $KMnO_4$ in a redox titration.

In concentrated solutions, MnO_4^- is purple, but dilute solutions of the ion appear pink.

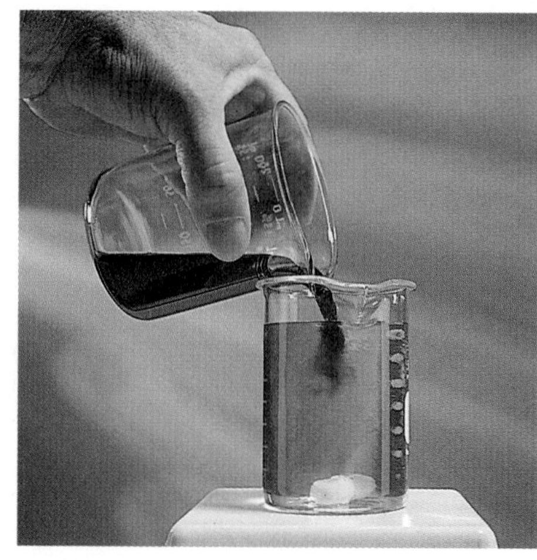

FIG. 5.13 Reduction of MnO_4^- by Fe^{2+}. A solution of $KMnO_4$ is added to a stirred acidic solution containing Fe^{2+}. The reaction oxidizes the pale blue-green Fe^{2+} to Fe^{3+} while the MnO_4^- is reduced to the almost colorless Mn^{2+} ion. The purple color of the permanganate will continue to be destroyed until all of the Fe^{2+} has reacted. Only then will the iron-containing solution take on a pink or purple color. This ability of MnO_4^- to signal the completion of the reaction makes it especially useful in redox titrations, where it serves as its own indicator. *(Andy Washnik.)*

EXAMPLE 5.9
Redox Titrations in Chemical Analysis

All the iron in a 2.000 g sample of an iron ore was dissolved in an acidic solution and converted to Fe^{2+}, which was then titrated with 0.1000 M $KMnO_4$ solution. In the titration the iron was oxidized to Fe^{3+}. The titration required 27.45 mL of the $KMnO_4$ solution to reach the end point.

(a) How many grams of iron were in the sample?

(b) What was the percentage iron in the sample?

(c) If the iron was present in the sample as Fe_2O_3, what was the percentage by mass of Fe_2O_3 in the sample?

ANALYSIS: We're dealing with a chemical reaction, so the first step will be to write the balanced equation using the ion–electron method covered earlier in this chapter. With this as a start, let's look over the tools we'll use to answer the first two questions. We'll tackle the last part afterward.

- Molarity and volume of the $KMnO_4$: These will give us moles of $KMnO_4$ used in the titration (remember: volume (L) × molarity = moles).
- Coefficients of the equation: These permit us to find moles of iron from the moles of $KMnO_4$ used.
- Molar mass of iron: This lets us calculate the mass of iron in the sample from moles of iron.
- Equation to calculate the percentage of iron: $\% \text{ Fe} = \dfrac{\text{g Fe}}{\text{g sample}} \times 100\%$.

SOLUTION: The skeleton equation for the reaction is

$$Fe^{2+} + MnO_4^- \longrightarrow Fe^{3+} + Mn^{2+}$$

Balancing it by the ion–electron method for acidic solutions gives

$$5Fe^{2+} + MnO_4^- + 8H^+ \longrightarrow 5Fe^{3+} + Mn^{2+} + 4H_2O$$

The number of moles of $KMnO_4$ consumed in the reaction is calculated from the volume of the solution used in the titration and its concentration.

$$0.02745 \cancel{\text{L KMnO}_4 \text{ soln}} \times \frac{0.1000 \text{ mol KMnO}_4}{1.000 \cancel{\text{L KMnO}_4 \text{ soln}}} \Leftrightarrow 0.002745 \text{ mol KMnO}_4$$

Next, we use the coefficients of the equation to calculate the number of moles of Fe^{2+} that reacted. The chemical equation tells us five moles of Fe^{2+} react per mole of MnO_4^- consumed.

$$0.002745 \cancel{\text{mol KMnO}_4} \times \frac{5 \text{ mol Fe}^{2+}}{1 \cancel{\text{mol KMnO}_4}} \Leftrightarrow 0.01372 \text{ mol Fe}^{2+}$$

This is the number of moles of iron in the ore sample, so the mass of iron in the sample is

$$0.01372 \cancel{\text{mol Fe}} \times \frac{55.845 \text{ g Fe}}{1 \cancel{\text{mol Fe}}} = 0.7662 \text{ g Fe}$$

This is the answer to part (a) of the problem. Next we calculate the percentage of iron in the sample, which is the mass of iron divided by the mass of the sample, all multiplied by 100%.

$$\% \text{ Fe} = \frac{\text{mass of Fe}}{\text{mass of sample}} \times 100\%$$

Substituting gives

$$\% \text{ Fe} = \frac{0.7662 \text{ g Fe}}{2.000 \text{ g sample}} \times 100\% = 38.31\% \text{ Fe}$$

The answer to part (b) is that the sample is 38.31% iron.

ARE THE ANSWERS REASONABLE? We can use some approximate arithmetic to estimate the answer. In the titration we used approximately 30 mL, or 0.030 L, of the $KMnO_4$ solution, which is 0.10 *M*. Multiplying these tells us we've used approximately 0.003 mol of $KMnO_4$. From the coefficients of the equation, five times as many moles of Fe^{2+} react, so the amount of Fe in the sample is approximately $5 \times 0.003 = 0.015$ mol. The atomic mass of Fe is about 55 g/mol, so the mass of Fe in the sample is approximately $0.015 \times 55 \approx 0.8$ g. Our answer (0.7662 g) is reasonable. Calculating % Fe is then straightforward.

ANALYSIS CONTINUED: Now we can work on the last part of the question. Earlier in the problem we determined the number of moles of iron that reacted, 0.01372 mol Fe. How many moles of Fe_2O_3 would have contained that number of moles of iron? That's the critical question we have to answer. Once we know this, we can calculate the mass of the Fe_2O_3 and the percentage of Fe_2O_3 in the original sample. The tools we'll use in solving this part of the problem are

- The chemical formula, Fe_2O_3: The formula relates moles of iron to moles Fe_2O_3.
- Molar mass of Fe_2O_3: This lets us calculate the mass of Fe_2O_3.
- Formula for calculating percentage of Fe_2O_3: $\% \; Fe_2O_3 = \dfrac{g \; Fe_2O_3}{g \; sample} \times 100\%$.

SOLUTION CONTINUED: The chemical formula for the iron oxide gives us

$$1 \text{ mol } Fe_2O_3 \Leftrightarrow 2 \text{ mol Fe}$$

This provides the conversion factor we need to determine how many moles of Fe_2O_3 were present in the sample. Working with the number of moles of Fe,

$$0.01372 \cancel{\text{mol Fe}} \times \frac{1 \text{ mol } Fe_2O_3}{2 \cancel{\text{mol Fe}}} \Leftrightarrow 0.006860 \text{ mol } Fe_2O_3$$

This is the number of moles of Fe_2O_3 in the sample. The formula mass of Fe_2O_3 is 159.69 g mol^{-1}, so the mass of Fe_2O_3 in the sample was

$$0.006860 \cancel{\text{mol } Fe_2O_3} \times \frac{159.69 \text{ g } Fe_2O_3}{1 \cancel{\text{mol } Fe_2O_3}} = 1.095 \text{ g } Fe_2O_3$$

Finally, the percentage of Fe_2O_3 in the sample was

$$\% \; Fe_2O_3 = \frac{1.095 \text{ g } Fe_2O_3}{2.000 \text{ g sample}} \times 100\% = 54.75\% \; Fe_2O_3$$

The ore sample contained 54.75% Fe_2O_3.

IS THE ANSWER REASONABLE? We've noted that the amount of Fe in the sample is approximately 0.015 mol. The amount of Fe_2O_3 that contains this much Fe is 0.0075 mol. The formula mass of Fe_2O_3 is about 160, so the mass of Fe_2O_3 in the sample was approximately $0.0075 \times 160 = 1.2$ g, which isn't too far from the mass we obtained (1.095 g). Since 1.095 g is about half of the total sample mass of 2.000 g, the sample was approximately 50% Fe_2O_3, in agreement with the answer we obtained.

Practice Exercise 24: A 15.00 mL sample of a solution containing oxalic acid, $H_2C_2O_4$, was titrated with 0.02000 *M* $KMnO_4$. The titration required 18.30 mL of the $KMnO_4$ solution. What was the molarity of the $H_2C_2O_4$ solution? In the reaction, oxalate ion ($C_2O_4^{2-}$) is oxidized to CO_2 and MnO_4^- is reduced to Mn^{2+}. (Hint: You will need to calculate number of moles of $H_2C_2O_4$ in the sample and then apply the definition of molarity.)

Practice Exercise 25: A researcher planned to use chlorine gas in an experiment and wished to trap excess chlorine to prevent it from escaping into the atmosphere. To accomplish this, the reaction of sodium thiosulfate ($Na_2S_2O_3$) with chlorine gas in an acidic aqueous solution to give sulfate ion and chloride ion would be used. How many grams of $Na_2S_2O_3$ are needed to trap 4.25 g of chlorine?

Practice Exercise 26: A sample of a tin ore weighing 0.3000 g was dissolved in an acid solution and all the tin in the sample was changed to tin(II). In a titration, 8.08 mL of 0.0500 *M* $KMnO_4$ solution was required to oxidize the tin(II) to tin(IV).
(a) What is the balanced equation for the reaction in the titration?
(b) How many grams of tin were in the sample?
(c) What was the percentage by mass of tin in the sample?
(d) If the tin in the sample had been present in the compound SnO_2, what would have been the percentage by mass of SnO_2 in the sample?

SUMMARY

Oxidation–Reduction. **Oxidation** is the loss of electrons or an algebraic increase in oxidation number; **reduction** is the gain of electrons or an algebraic decrease in oxidation number. Both always occur together in **redox reactions.** The substance oxidized is the **reducing agent;** the substance reduced is the **oxidizing agent. Oxidation numbers** are a bookkeeping device that we use to follow changes in redox reactions. They are assigned according to the rules on page 178. The term **oxidation state** is equivalent to oxidation number.

Ion–Electron Method. In a balanced redox equation, the number of electrons gained by one substance is always equal to the number lost by another substance. This fact forms the basis for the **ion–electron method** (page **183**), which provides a systematic method for deriving a net ionic equation for a redox reaction in aqueous solution. According to this method the *skeleton* net ionic equation is divided into two **half-reactions,** which are balanced separately before being recombined to give the final balanced net ionic equation. For reactions in basic solution, the equation is balanced as if it occurred in an acidic solution, and then the balanced equation is converted to its proper form for basic solution by adding an appropriate number of OH^-.

Metal–Acid Reactions. In **nonoxidizing acids,** the strongest oxidizing agent is H^+ (Table 5.1). The reaction of a metal with a nonoxidizing acid gives hydrogen gas and a salt of the acid. Only metals more active than hydrogen react this way. These are metals that are located below hydrogen in the **activity series** (Table 5.2). **Oxidizing acids,** like HNO_3, contain an anion that is a stronger oxidizing agent than H^+, and they are able to oxidize many metals that nonoxidizing acids cannot.

Metal-Displacement Reactions. If one metal is more easily oxidized than another, it can displace the other metal from its compounds by a redox reaction. Such reactions are sometimes called **single replacement reactions.** Atoms of the more active metal become ions; ions of the less active metal generally become atoms. In this manner, any metal in the **activity series** can displace any of the others above it in the series from their compounds.

Oxidations by Molecular Oxygen. **Combustion** is the rapid reaction of a substance with oxygen accompanied by the evolution of heat and light. Combustion of a hydrocarbon in the presence of excess oxygen gives CO_2 and H_2O, two molecular oxides. When the supply of oxygen is limited, some CO also forms, and in a very limited supply of oxygen the products are H_2O and very finely divided, elemental carbon (as soot or lampblack). The combustion of organic compounds containing only carbon, hydrogen, and oxygen also gives the same products, CO_2 and H_2O. Sulfur burns to give SO_2, which also forms when sulfur-containing fuels burn. Most nonmetals also burn in oxygen to give molecular oxides.

Many metals combine with oxygen in a process often called **corrosion,** but only sometimes is the reaction rapid enough to be considered combustion. The products are ionic metal oxides.

Redox Titrations. Potassium permanganate is often used in redox titrations because it is a powerful oxidizing agent and serves as its own indicator. In acidic solutions, the purple MnO_4^- ion is reduced to the nearly colorless Mn^{2+} ion.

TOOLS FOR PROBLEM SOLVING

In this chapter you learned to apply the following concepts as tools in solving problems. Study each one carefully so that you know what each is used for. When faced with solving a problem, recall what each tool does and consider whether it will be helpful in finding a solution. This will aid you in selecting the tools you need.

Identifying oxidizing and reducing agents *(page 176)* The substance reduced is the oxidizing agent; the substance oxidized is the reducing agent.

Rules for assigning oxidation numbers *(page 178)* The rules permit us to assign oxidation numbers to elements in compounds and ions. You use changes in oxidation numbers to identify oxidation and reduction. Remember that when there is a conflict between two rules, the rule with the lower number is followed and the rule with the higher number is ignored.

Ion-electron method *(For acidic solutions, page 185; for basic solutions, page 186)* Use this method when you need to obtain a balanced net ionic equation for a redox reaction. Be sure to follow the steps in the order given and do not skip steps. Also, be sure to include charges on all ions.

Table of oxidizing and nonoxidizing acids *(Table 5.1, page 188)* Refer to this table to identify oxidizing and nonoxidizing acids, which enables you to anticipate the products of reactions of metals with acids. Nonoxidizing acids will react with metals below hydrogen in Table 5.2 to give H_2 and the metal ion.

Activity series of metals *(Table 5.2, page 191)* When a question deals with the possible reaction of one metal with the salt of another, refer to the activity series to determine the outcome. A metal in the table will reduce the ion of any metal above it in the table, leading to a single replacement reaction.

Combustion reactions of hydrocarbons with oxygen *(pages 193 and 194)* The products of the reaction do not depend on the identity of the hydrocarbon, but they do depend on the availability of oxygen.

$$\text{hydrocarbon} + O_2 \longrightarrow CO_2 + H_2O \qquad \text{(plentiful supply of } O_2\text{)}$$

$$\text{hydrocarbon} + O_2 \longrightarrow CO + H_2O \qquad \text{(limited supply of } O_2\text{)}$$

$$\text{hydrocarbon} + O_2 \longrightarrow C + H_2O \qquad \text{(very limited supply of } O_2\text{)}$$

Combustion of compounds containing C, H, and O *(page 194)* Complete combustion gives CO_2 and H_2O.

$$\text{(C,H,O compound)} + O_2 \longrightarrow CO_2 + H_2O \qquad \text{(complete combustion)}$$

Organic compounds containing sulfur give SO_2 when burned *(page 194)* If an organic compound contains sulfur, SO_2 is formed in addition to CO_2 and H_2O when the compound is burned.

QUESTIONS, PROBLEMS, AND EXERCISES

Answers to problems whose numbers are printed in color are given in Appendix B. More challenging problems are marked with asterisks. ILW = Interactive Learningware solution is available at www.wiley.com/college/brady. OH = an Office Hours video is available for this problem.

REVIEW QUESTIONS

Oxidation–Reduction

5.1 Define *oxidation* and *reduction* (a) in terms of electron transfer and (b) in terms of oxidation numbers.

5.2 In the reaction $2Mg + O_2 \longrightarrow 2MgO$, which substance is the oxidizing agent and which is the reducing agent? Which substance is oxidized and which is reduced?

5.3 Why must both oxidation and reduction occur simultaneously during a redox reaction? What is an oxidizing agent? What happens to it in a redox reaction? What is a reducing agent? What happens to it in a redox reaction?

5.4 In the compound As_4O_6, arsenic has an *oxidation number* of +3. What is the *oxidation state* of arsenic in this compound?

5.5 Is the following a redox reaction? Explain.

$$2NO_2 \longrightarrow N_2O_4$$

5.6 Is the following a redox reaction? Explain.

$$2CrO_4^{2-} + 2H^+ \longrightarrow Cr_2O_7^{2-} + H_2O$$

5.7 If the oxidation number of nitrogen in a certain molecule changes from +3 to −2 during a reaction, is the nitrogen oxidized or reduced? How many electrons are gained (or lost) by each nitrogen atom?

Ion–Electron Method

5.8 The following equation is not balanced. Why? Use the ion–electron method to balance it.

$$Ag + Fe^{2+} \longrightarrow Ag^+ + Fe$$

5.9 Use the ion–electron method to balance the following equation.

$$Cr^{3+} + Zn \longrightarrow Cr + Zn^{2+}$$

5.10 What are the net charges on the left and right sides of the following equations? Add electrons as necessary to make each of them a balanced half-reaction.
(a) $NO_3^- + 10H^+ \longrightarrow NH_4^+ + 3H_2O$
(b) $Cl_2 + 4H_2O \longrightarrow 2ClO_2^- + 8H^+$

5.11 In the preceding question, which half-reaction represents oxidation? Which represents reduction?

Reactions of Metals with Acids and the Activity Series

5.12 What is a *single replacement reaction*?

5.13 What is a nonoxidizing acid? Give two examples. What is the oxidizing agent in a nonoxidizing acid?

5.14 What is the strongest oxidizing agent in an aqueous solution of nitric acid?

5.15 If a metal is able to react with a solution of HCl, where must the metal stand relative to hydrogen in the activity series?

5.16 Where in the activity series (Table 5.2) do we find the best reducing agents? Where do we find the best oxidizing agents?

5.17 Which metals in Table 5.2 will not react with nonoxidizing acids?

5.18 Which metals in Table 5.2 will react with water? Write chemical equations for each of these reactions.

5.19 When manganese reacts with silver ion, is manganese oxidized or reduced? Is it an oxidizing agent or a reducing agent?

Oxygen as an Oxidizing Agent

5.20 Define *combustion*.

5.21 Why is "loss of electrons" described as oxidation?

5.22 What products are produced in the combustion of $C_{10}H_{22}$ (a) if there is an excess of oxygen available? (b) If there is a slightly limited oxygen supply? (c) If there is a very limited supply of oxygen?

5.23 If one of the impurities in diesel fuel has the formula C_2H_6S, what products will be formed when it burns? Write a balanced chemical equation for the reaction.

5.24 Burning ammonia in an atmosphere of oxygen produces stable N_2 molecules as one of the products. What is the other product? Write the balanced equation for the reaction.

REVIEW PROBLEMS

Oxidation–Reduction; Oxidation Numbers

5.25 Assign oxidation numbers to the atoms in the following: (a) S^{2-}, (b) SO_2, (c) P_4, and (d) PH_3.

OH **5.26** Assign oxidation numbers to the atoms in the following: (a) ClO_4^-, (b) $CrCl_3$, (c) SnS_2, and (d) $Au(NO_3)_3$.

5.27 Assign oxidation numbers to each atom in the following: (a) NaOCl, (b) $NaClO_2$, (c) $NaClO_3$, and (d) $NaClO_4$.

5.28 Assign oxidation numbers to the elements in the following. (a) $Ca(VO_3)_2$, (b) $SnCl_4$, (c) MnO_4^{2-}, (d) MnO_2

5.29 Assign oxidation numbers to the elements in the following. (a) PbS, (b) $TiCl_4$, (c) CsO_2, (d) O_2F_2

5.30 Assign oxidation numbers to the elements in the following. (a) $Sr(IO_3)_2$, (b) Cr_2S_3, (c) OF_2, (d) HOF

5.31 When chlorine is added to drinking water to kill bacteria, some of the chlorine is changed into ions by the following equilibrium:

$$Cl_2(aq) + H_2O \rightleftharpoons H^+(aq) + Cl^-(aq) + HOCl(aq)$$

In the forward reaction (the reaction going from left to right), which substance is oxidized and which is reduced? In the reverse reaction, which is the oxidizing agent and which is the reducing agent?

5.32 A pollutant in smog is nitrogen dioxide, NO_2. The gas has a reddish brown color and is responsible for the red-brown color associated with this type of air pollution. Nitrogen dioxide is also a contributor to acid rain because when rain passes through air contaminated with NO_2, it dissolves and undergoes the following reaction:

$$3NO_2(g) + H_2O \longrightarrow NO(g) + 2H^+(aq) + 2NO_3^-(aq)$$

In this reaction, which element is reduced and which is oxidized?

5.33 For the following reactions, identify the substance oxidized, the substance reduced, the oxidizing agent, and the reducing agent.
(a) $2HNO_3 + 3H_3AsO_3 \longrightarrow 2NO + 3H_3AsO_4 + H_2O$
(b) $NaI + 3HOCl \longrightarrow NaIO_3 + 3HCl$
(c) $2KMnO_4 + 5H_2C_2O_4 + 3H_2SO_4 \longrightarrow 10CO_2 + K_2SO_4 + 2MnSO_4 + 8H_2O$
(d) $6H_2SO_4 + 2Al \longrightarrow Al_2(SO_4)_3 + 3SO_2 + 6H_2O$

5.34 For the following reactions, identify the substance oxidized, the substance reduced, the oxidizing agent, and the reducing agent.
(a) $Cu + 2H_2SO_4 \longrightarrow CuSO_4 + SO_2 + 2H_2O$
(b) $3SO_2 + 2HNO_3 + 2H_2O \longrightarrow 3H_2SO_4 + 2NO$
(c) $5H_2SO_4 + 4Zn \longrightarrow 4ZnSO_4 + H_2S + 4H_2O$
(d) $I_2 + 10HNO_3 \longrightarrow 2HIO_3 + 10NO_2 + 4H_2O$

Ion–Electron Method

ILW **5.35** Balance the following equations for reactions occurring in an acidic solution.
(a) $S_2O_3^{2-} + OCl^- \longrightarrow Cl^- + S_4O_6^{2-}$
(b) $NO_3^- + Cu \longrightarrow NO_2 + Cu^{2+}$
(c) $IO_3^- + AsO_3^{3-} \longrightarrow I^- + AsO_4^{3-}$
(d) $SO_4^{2-} + Zn \longrightarrow Zn^{2+} + SO_2$
(e) $NO_3^- + Zn \longrightarrow NH_4^+ + Zn^{2+}$
(f) $Cr^{3+} + BiO_3^- \longrightarrow Cr_2O_7^{2-} + Bi^{3+}$
(g) $I_2 + OCl^- \longrightarrow IO_3^- + Cl^-$
(h) $Mn^{2+} + BiO_3^- \longrightarrow MnO_4^- + Bi^{3+}$
(i) $H_3AsO_3 + Cr_2O_7^{2-} \longrightarrow H_3AsO_4 + Cr^{3+}$
(j) $I^- + HSO_4^- \longrightarrow I_2 + SO_2$

5.36 Balance these equations for reactions occurring in an acidic solution.
(a) $Sn + NO_3^- \longrightarrow SnO_2 + NO$
(b) $PbO_2 + Cl^- \longrightarrow PbCl_2 + Cl_2$
(c) $Ag + NO_3^- \longrightarrow NO_2 + Ag^+$

(d) $Fe^{3+} + NH_3OH^+ \longrightarrow Fe^{2+} + N_2O$
(e) $HNO_2 + I^- \longrightarrow I_2 + NO$
(f) $C_2O_4^{2-} + HNO_2 \longrightarrow CO_2 + NO$
(g) $HNO_2 + MnO_4^- \longrightarrow Mn^{2+} + NO_3^-$
(h) $H_3PO_2 + Cr_2O_7^{2-} \longrightarrow H_3PO_4 + Cr^{3+}$
(i) $VO_2^+ + Sn^{2+} \longrightarrow VO^{2+} + Sn^{4+}$
(j) $XeF_2 + Cl^- \longrightarrow Xe + F^- + Cl_2$

5.37 Balance equations for these reactions occurring in a basic solution.
(a) $CrO_4^{2-} + S^{2-} \longrightarrow S + CrO_2^-$
(b) $MnO_4^- + C_2O_4^{2-} \longrightarrow CO_2 + MnO_2$
(c) $ClO_3^- + N_2H_4 \longrightarrow NO + Cl^-$
(d) $NiO_2 + Mn(OH)_2 \longrightarrow Mn_2O_3 + Ni(OH)_2$
(e) $SO_3^{2-} + MnO_4^- \longrightarrow SO_4^{2-} + MnO_2$

5.38 Balance equations for these reactions occurring in a basic solution.
(a) $CrO_2^- + S_2O_8^{2-} \longrightarrow CrO_4^{2-} + SO_4^{2-}$
(b) $SO_3^{2-} + CrO_4^{2-} \longrightarrow SO_4^{2-} + CrO_2^-$
(c) $O_2 + N_2H_4 \longrightarrow H_2O_2 + N_2$
(d) $Fe(OH)_2 + O_2 \longrightarrow Fe(OH)_3 + OH^-$
(e) $Au + CN^- + O_2 \longrightarrow Au(CN)_4^- + OH^-$

5.39 Laundry bleach such as Clorox is a dilute solution of sodium hypochlorite, NaOCl. Write a balanced net ionic equation for the reaction of NaOCl with $Na_2S_2O_3$. The OCl^- is reduced to chloride ion and the $S_2O_3^{2-}$ is oxidized to sulfate ion.

OH **5.40** Calcium oxalate is one of the minerals found in kidney stones. If a strong acid is added to calcium oxalate, the compound will dissolve and the oxalate ion will be changed to oxalic acid (a weak acid). Oxalate ion is a moderately strong reducing agent. Write a balanced net ionic equation for the oxidation of $H_2C_2O_4$ by $K_2Cr_2O_7$ in an acidic solution. The reaction yields Cr^{3+} and CO_2 among the products.

5.41 Ozone, O_3, is a very powerful oxidizing agent, and in some places ozone is used to treat water to kill bacteria and make it safe to drink. One of the problems with this method of purifying water is that if there is any bromide ion in the water, it becomes oxidized to bromate ion, which has shown evidence of causing cancer in test animals. Assuming that ozone is reduced to water, write a balanced chemical equation for the reaction. (Assume an acidic solution.)

5.42 Chlorine is a good bleaching agent because it is able to oxidize substances that are colored to give colorless reaction products. It is used in the pulp and paper industry as a bleach, but after it has done its work, residual chlorine must be removed. This is accomplished using sodium thiosulfate, $Na_2S_2O_3$, which reacts with the chlorine, reducing it to chloride ion. The thiosulfate ion is changed to sulfate ion, which is easily removed by washing with water. Write a balanced chemical equation for the reaction of chlorine with thiosulfate ion, assuming an acidic solution.

Reactions of Metals with Acids

5.43 Write balanced molecular, ionic, and net ionic equations for the reactions of the following metals with hydrochloric acid to give hydrogen plus the metal ion in solution.
(a) Manganese (gives Mn^{2+})
(b) Cadmium (gives Cd^{2+})
(c) Tin (gives Sn^{2+})

5.44 Write balanced molecular, ionic, and net ionic equations for the reaction of each of the following metals with dilute sulfuric acid.
(a) Nickel (gives Ni^{2+})
(b) Chromium (gives Cr^{3+})
(c) Aluminum (gives Al^{3+})

5.45 On the basis of the discussions in this chapter, suggest chemical equations for the oxidation of metallic silver to Ag^+ ion with (a) dilute HNO_3 and (b) concentrated HNO_3.

OH **5.46** When hot and concentrated, sulfuric acid is a fairly strong oxidizing agent. Write a balanced net ionic equation for the oxidation of metallic copper to copper(II) ion by hot concentrated H_2SO_4, in which the sulfur is reduced to SO_2. Write a balanced molecular equation for the reaction.

Single Replacement Reactions and the Activity Series

5.47 Use Table 5.2 to predict the outcome of the following reactions. If no reaction occurs, write N.R. If a reaction occurs, write a balanced chemical equation for it.
(a) $Fe + Mg^{2+} \longrightarrow$
(b) $Cr + Pb^{2+} \longrightarrow$
(c) $Ag^+ + Fe \longrightarrow$
(d) $Ag + Au^{3+} \longrightarrow$

OH **5.48** Use Table 5.2 to predict the outcome of the following reactions. If no reaction occurs, write N.R. If a reaction occurs, write a balanced chemical equation for it.
(a) $Mn + Fe^{2+} \longrightarrow$
(b) $Cd + Zn^{2+} \longrightarrow$
(c) $Mg + Co^{2+} \longrightarrow$
(d) $Cr + Sn^{2+} \longrightarrow$

5.49 The following reactions occur spontaneously.

$$Pu + 3Tl^+ \longrightarrow Pu^{3+} + 3Tl$$
$$Ru + Pt^{2+} \longrightarrow Ru^{2+} + Pt$$
$$2Tl + Ru^{2+} \longrightarrow 2Tl^+ + Ru$$

List the metals Pu, Pt, and Tl in order of increasing ease of oxidation.

5.50 The following reactions occur spontaneously.

$$2Y + 3Ni^{2+} \longrightarrow 2Y^{3+} + 3Ni$$
$$2Mo + 3Ni^{2+} \longrightarrow 2Mo^{3+} + 3Ni$$
$$Y^{3+} + Mo \longrightarrow Y + Mo^{3+}$$

List the metals Y, Ni, and Mo in order of increasing ease of oxidation.

5.51 It is found that the following reaction occurs spontaneously.

$$Ru^{2+}(aq) + Cd(s) \longrightarrow Ru(s) + Cd^{2+}(aq)$$

What reaction will occur if a mixture is prepared containing the following: $Cd(s)$, $Cd(NO_3)_2(aq)$, $Tl(s)$, $TlCl(aq)$? (Refer to the information in Problem 5.49 above.)

5.52 It is observed that when magnesium metal is dipped into a solution of nickel(II) chloride, some of the magnesium dissolves and nickel metal is deposited on the surface of the magnesium. Referring to Problem 5.50, can you tell which of the following reactions will occur spontaneously? Explain the reason for your answer.
(a) $2Mo^{3+} + 3Mg \longrightarrow 3Mg^{2+} + 2Mo$
(b) $2Mo + 3Mg^{2+} \longrightarrow 2Mo^{3+} + 3Mg$

Reactions of Oxygen

5.53 Write balanced chemical equations for the complete combustion (in the presence of excess oxygen) of the following:
(a) C_6H_6 (benzene, an important industrial chemical and solvent)
(b) C_3H_8 (propane, a gaseous fuel used in many stoves)
(c) $C_{21}H_{44}$ (a component of paraffin wax)

5.54 Write balanced chemical equations for the complete combustion (in the presence of excess oxygen) of the following:
(a) $C_{12}H_{26}$ (a component of kerosene)
(b) $C_{18}H_{36}$ (a component of diesel fuel)
(c) C_7H_8 (toluene, a raw material in the production of TNT)

5.55 Write balanced equations for the combustion of the hydrocarbons in Problem 5.53 in (a) a slightly limited supply of oxygen and (b) a very limited supply of oxygen.

5.56 Write balanced equations for the combustion of the hydrocarbons in Problem 5.54 in (a) a slightly limited supply of oxygen and (b) a very limited supply of oxygen.

5.57 Methanol, CH_3OH, has been suggested as an alternative to gasoline as an automotive fuel. Write a balanced chemical equation for its complete combustion.

OH **5.58** Metabolism of carbohydrates such as glucose, $C_6H_{12}O_6$, produces the same products as complete combustion. Write a chemical equation representing the metabolism (combustion) of glucose.

5.59 Write the balanced equation for the combustion of dimethylsulfide, $(CH_3)_2S$, in an abundant supply of oxygen.

5.60 Thiophene, C_4H_4S, is an impurity in crude oil and is a source of pollution if not removed. Write an equation for the combustion of thiophene.

5.61 Write chemical equations for the reaction of oxygen with (a) zinc, (b) aluminum, (c) magnesium, and (d) iron to form iron(III) oxide.

5.62 Write chemical equations for the reaction of oxygen with (a) beryllium, (b) lithium, (c) barium, and (d) bismuth to form bismuth(III) oxide.

Redox Reactions and Stoichiometry

5.63 Iodate ion reacts with sulfite ion to give sulfate ion and iodide ion.
(a) Write a balanced net ionic equation for the reaction.
(b) How many grams of sodium sulfite are needed to react with 5.00 g of sodium iodate?

5.64 Potable water (drinking water) should not have manganese concentrations in excess of 0.05 mg/mL. If the manganese concentration is greater than 0.1 mg/mL, it imparts a foul taste to the water and discolors laundry and porcelain surfaces. Manganese(II) ion is oxidized to permanganate ion by bismuthate ion, BiO_3^-, in an acidic solution. In the reaction, BiO_3^- is reduced to Bi^{3+}.
(a) Write a balanced net ionic equation for the reaction.
(b) How many milligrams of $NaBiO_3$ are needed to oxidize the manganese in 18.5 mg of manganese(II) sulfate?

OH **5.65** How many grams of copper must react to displace 12.0 g of silver from a solution of silver nitrate?

5.66 How many grams of aluminum must react to displace all the silver from 25.0 g of silver nitrate? The reaction occurs in aqueous solution.

5.67 In an acidic solution, permanganate ion reacts with tin(II) ion to give manganese(II) ion and tin(IV) ion.
(a) Write a balanced net ionic equation for the reaction.
(b) How many milliliters of 0.230 *M* potassium permanganate solution are needed to react completely with 40.0 mL of 0.250 *M* tin(II) choride solution?

5.68 In an acidic solution, bisulfite ion reacts with chlorate ion to give sulfate ion and chloride ion.
(a) Write a balanced net ionic equation for the reaction.
(b) How many milliliters of 0.150 *M* sodium chlorate solution are needed to react completely with 30.0 mL of 0.450 *M* sodium bisulfite solution?

5.69 Sulfites are used worldwide in the wine industry as antioxidant and antimicrobial agents. However, sulfites have also been identified as causing certain allergic reactions suffered by asthmatics, and the FDA mandates that sulfites be identified on the label if they are present at levels of 10 ppm (parts per million) or higher. The analysis of sulfites in wine uses the "Ripper method" in which a standard iodine solution, prepared by the reaction of iodate and iodide ions, is used to titrate a sample of the wine. The iodine is formed in the reaction

$$IO_3^- + 5I^- + 6H^+ \longrightarrow 3I_2 + 3H_2O$$

The iodine is held in solution by adding an excess of I^- which combines with I_2 to give I_3^-. In the titration, the SO_3^{2-} is converted to SO_2 by acidification and the reaction during the titration is

$$SO_2 + I_3^- + 2H_2O \longrightarrow SO_4^{2-} + 3I^- + 4H^+$$

Starch is added to the wine sample to detect the end point, which is signaled by the formation of a dark blue color when excess iodine binds to the starch molecules. In a certain analysis, 0.0421 g of $NaIO_3$ was dissolved in dilute acid and excess NaI was added to the solution, which was then diluted to a total volume of 100.0 mL. A 50.0 mL sample of wine was then acidified and titrated with the iodine-containing solution. In the titration, 2.47 mL of the iodine solution was required.
(a) What was the molarity of the iodine (actually, I_3^-) in the standard solution?
(b) How many grams of SO_2 were in the wine sample?
(c) If the density of the wine was 0.96 g/mL, what was the percentage of SO_2 in the wine?
(d) Parts per million (ppm) is calculated in a manner similar to percent (which is equivalent to *parts per hundred*).

$$\text{ppm} = \frac{\text{grams of component}}{\text{grams of sample}} \times 10^6 \text{ ppm}$$

What was the concentration of sulfite in the wine, expressed as parts per million SO_2?

5.70 Methylbromide, CH_3Br, is used in agriculture to fumigate soil to rid it of pests such as nematodes. It is injected directly into the soil, but over time it has a tendency to escape before it can undergo natural degradation to innocuous products. Soil chemists have found that ammonium thiosulfate, $(NH_4)_2S_2O_3$, a nitrogen and sulfur fertilizer, drastically reduces methylbromide emissions by causing it to degrade.

In a chemical analysis to determine the purity of a batch of commercial ammonium thiosulfate, a chemist first prepared a standard solution of iodine following the procedure in the preceding problem. First, 0.462 g of KIO_3 was dissolved in 100 mL of water. The solution was made acidic and treated with excess potassium iodide, which caused the following reaction to take place:

$$IO_3^- + 8I^- + 6H^+ \longrightarrow 3I_3^- + 3H_2O$$

The solution containing the I_3^- was then diluted to exactly 250 mL in a volumetric flask. Next, the chemist dissolved 0.218 g of the fertilizer in water, added starch indicator, and titrated it with the standard I_3^- solution. The reaction was

$$2S_2O_3^{2-} + I_3^- \longrightarrow S_4O_6^{2-} + 3I^-$$

The titration required 27.99 mL of the I_3^- solution.
(a) What was the molarity of the I_3^- solution used in the titration?
(b) How many grams of $(NH_4)_2S_2O_3$ were in the fertilizer sample?
(c) What was the percentage by mass of $(NH_4)_2S_2O_3$ in the fertilizer?

ILW **5.71** A sample of a copper ore with a mass of 0.4225 g was dissolved in acid. A solution of potassium iodide was added, which caused the reaction

$$2Cu^{2+}(aq) + 5I^-(aq) \longrightarrow I_3^-(aq) + 2CuI(s)$$

The I_3^- that formed reacted quantitatively with exactly 29.96 mL of 0.02100 M $Na_2S_2O_3$ according to the following equation.

$$I_3^-(aq) + 2S_2O_3^{2-}(aq) \longrightarrow 3I^-(aq) + S_4O_6^{2-}(aq)$$

(a) What was the percentage by mass of copper in the ore?
(b) If the ore contained $CuCO_3$, what was the percentage by mass of $CuCO_3$ in the ore?

5.72 A 1.362 g sample of an iron ore that contained Fe_3O_4 was dissolved in acid and all the iron was reduced to Fe^{2+}. The solution was then acidified with H_2SO_4 and titrated with 39.42 mL of 0.0281 M $KMnO_4$, which oxidized the iron to Fe^{3+}. The net ionic equation for the reaction is

$$5Fe^{2+} + MnO_4^- + 8H^+ \longrightarrow 5Fe^{3+} + Mn^{2+} + 4H_2O$$

(a) What was the percentage by mass of iron in the ore?
(b) What was the percentage by mass of Fe_3O_4 in the ore?

5.73 Hydrogen peroxide (H_2O_2) solution can be purchased in drug stores for use as an antiseptic. A sample of such a solution weighing 1.000 g was acidified with H_2SO_4 and titrated with a 0.02000 M solution of $KMnO_4$. The net ionic equation for the reaction is

$$6H^+ + 5H_2O_2 + 2MnO_4^- \longrightarrow 5O_2 + 2Mn^{2+} + 8H_2O$$

The titration required 17.60 mL of $KMnO_4$ solution.
(a) How many grams of H_2O_2 reacted?
(b) What is the percentage by mass of the H_2O_2 in the original antiseptic solution?

5.74 Sodium nitrite, $NaNO_2$, is used as a preservative in meat products such as frankfurters and bologna. In an acidic solution, nitrite ion is converted to nitrous acid, HNO_2, which reacts with permanganate ion according to the equation

$$H^+ + 5HNO_2 + 2MnO_4^- \longrightarrow 5NO_3^- + 2Mn^{2+} + 3H_2O$$

A 1.000 g sample of a water-soluble solid containing $NaNO_2$ was dissolved in dilute H_2SO_4 and titrated with 0.01000 M $KMnO_4$ solution. The titration required 12.15 mL of the $KMnO_4$ solution. What was the percentage by mass of $NaNO_2$ in the original 1.000 g sample?

5.75 A sample of a chromium-containing alloy weighing 3.450 g was dissolved in acid, and all the chromium in the sample was oxidized to CrO_4^{2-}. It was then found that 3.18 g of Na_2SO_3 was required to reduce the CrO_4^{2-} to CrO_2^- in a basic solution, with the SO_3^{2-} being oxidized to SO_4^{2-}.
(a) Write a balanced equation for the reaction of CrO_4^{2-} with SO_3^{2-} in a basic solution.
(b) How many grams of chromium were in the alloy sample?
(c) What was the percentage by mass of chromium in the alloy?

5.76 Solder is an alloy containing the metals tin and lead. A particular sample of the alloy weighing 1.50 g was dissolved in acid. All the tin was then converted to the +2 oxidation state. Next, it was found that 0.368 g of $Na_2Cr_2O_7$ was required to oxidize the Sn^{2+} to Sn^{4+} in an acidic solution. In the reaction the chromium was reduced to Cr^{3+} ion.
(a) Write a balanced net ionic equation for the reaction between Sn^{2+} and $Cr_2O_7^{2-}$ in an acidic solution.
(b) Calculate the number of grams of tin that were in the sample of solder.
(c) What was the percentage by mass of tin in the solder?

5.77 Both calcium chloride and sodium chloride are used to melt ice and snow on roads in the winter. A certain company was marketing a mixture of these two compounds for this purpose. A chemist, wishing to analyze the mixture, dissolved 2.463 g of it in water and precipitated calcium oxalate by adding sodium oxalate, $Na_2C_2O_4$. The calcium oxalate was carefully filtered from the solution, dissolved in sulfuric acid, and titrated with 0.1000 M $KMnO_4$ solution. The reaction that occurred was

$$6H^+ + 5H_2C_2O_4 + 2MnO_4^- \longrightarrow 10CO_2 + 2Mn^{2+} + 8H_2O$$

The titration required 21.62 mL of the $KMnO_4$ solution.
(a) How many moles of $C_2O_4^{2-}$ were present in the calcium oxalate precipitate?
(b) How many grams of calcium chloride were in the original 2.463 g sample?
(c) What was the percentage by mass of calcium chloride in the sample?

5.78 A way to analyze a sample for nitrite ion is to acidify a solution containing NO_2^- and then allow the HNO_2 that is formed to react with iodide ion in the presence of excess I^-. The reaction is

$$2HNO_2 + 2H^+ + 3I^- \longrightarrow 2NO + 2H_2O + I_3^-$$

Then the I_3^- is titrated with $Na_2S_2O_3$ solution using starch as an indicator.

$$I_3^- + 2S_2O_3^{2-} \longrightarrow 3I^- + S_4O_6^{2-}$$

In a typical analysis, a 1.104 g sample that was known to contain $NaNO_2$ was treated as described above. The titration required 29.25 mL of 0.3000 M $Na_2S_2O_3$ solution to reach the end point.
(a) How many moles of I_3^- had been produced in the first reaction?
(b) How many moles of NO_2^- had been in the original 1.104 g sample?
(c) What was the percentage by mass of $NaNO_2$ in the original sample?

ADDITIONAL EXERCISES

5.79 What is the oxidation number of sulfur in the tetrathionate ion, $S_4O_6^{2-}$?

***5.80** In Practice Exercise 7 (page 182), some of the uses of chlorine dioxide were described along with a reaction that could be used to make ClO_2. Another reaction that is used to make this substance is

$$HCl + NaOCl + 2NaClO_2 \longrightarrow 2ClO_2 + 2NaCl + NaOH$$

Which element is oxidized? Which element is reduced? Which substance is the oxidizing agent and which is the reducing agent?

5.81 What is the average oxidation number of carbon in (a) C_2H_5OH (grain alcohol), (b) $C_{12}H_{22}O_{11}$ (sucrose—table sugar), (c) $CaCO_3$ (limestone), and (d) $NaHCO_3$ (baking soda)?

5.82 The following chemical reactions are *observed to occur* in aqueous solution.

$$2Al + 3Cu^{2+} \longrightarrow 2Al^{3+} + 3Cu$$

$$2Al + 3Fe^{2+} \longrightarrow 3Fe + 2Al^{3+}$$

$$Pb^{2+} + Fe \longrightarrow Pb + Fe^{2+}$$

$$Fe + Cu^{2+} \longrightarrow Fe^{2+} + Cu$$

$$2Al + 3Pb^{2+} \longrightarrow 3Pb + 2Al^{3+}$$

$$Pb + Cu^{2+} \longrightarrow Pb^{2+} + Cu$$

Arrange the metals Al, Pb, Fe, and Cu in order of increasing ease of oxidation.

5.83 In the preceding problem, were all the experiments described actually necessary to establish the order?

5.84 According to the activity series in Table 5.2, which of the following metals react with nonoxidizing acids? (a) silver, (b) gold, (c) zinc, (d) magnesium

5.85 In each pair below, choose the metal that would most likely react more rapidly with a nonoxidizing acid such as HCl. (a) aluminum or iron, (b) zinc or nickel, (c) cadmium or magnesium

5.86 In June 2002, the Department of Health and Children in Ireland began a program to distribute tablets of potassium iodate to households as part of Ireland's National Emergency Plan for Nuclear Accidents. Potassium iodate provides iodine, which when taken during a nuclear emergency, works by "topping off" the thyroid gland to prevent the uptake of radioactive iodine that might be released into the environment by a nuclear accident.

To test the potency of the tablets, a chemist dissolved one in 100 mL of water, made the solution acidic, and then added excess potassium iodide, which caused the following reaction to occur.

$$IO_3^- + 8I^- + 6H^+ \longrightarrow 3I_3^- + 3H_2O$$

The resulting solution containing I_3^- was titrated with 0.0500 *M* $Na_2S_2O_3$ solution, using starch indicator to detect the end point. (In the presence of iodine, starch turns dark blue. When the $S_2O_3^{2-}$ has consumed all the iodine, the solution becomes colorless.) The titration required 22.61 mL of the thiosulfate solution to reach the end point. The reaction during the titration was

$$I_3^- + 2S_2O_3^{2-} \longrightarrow 3I^- + S_4O_6^{2-}$$

How many milligrams of KIO_3 were in the tablet?

5.87 Use Table 5.2 to predict whether the following displacement reactions should occur. If no reaction occurs, write N.R. If a reaction does occur, write a balanced chemical equation for it.

(a) $Zn + Sn^{2+} \longrightarrow$
(b) $Cr + H^+ \longrightarrow$
(c) $Pb + Cd^{2+} \longrightarrow$
(d) $Zn + Co^{2+} \longrightarrow$
(e) $Mn + Pb^{2+} \longrightarrow$

5.88 Sucrose, $C_{12}H_{22}O_{11}$, is ordinary table sugar. Write a balanced chemical equation representing the metabolism of sucrose. (See Review Problem 5.58.)

***5.89** Balance the following equations by the ion–electron method.

(a) $NBr_3 \longrightarrow N_2 + Br^- + HOBr$ (basic solution)
(b) $Cl_2 \longrightarrow Cl^- + ClO_3^-$ (basic solution)
(c) $H_2SeO_3 + H_2S \longrightarrow S + Se$ (acidic solution)
(d) $MnO_2 + SO_3^{2-} \longrightarrow Mn^{2+} + S_2O_6^{2-}$ (acidic solution)
(e) $XeO_3 + I^- \longrightarrow Xe + I_2$ (acidic solution)
(f) $(CN)_2 \longrightarrow CN^- + OCN^-$ (basic solution)

5.90 Lead(IV) oxide reacts with hydrochloric acid to give chlorine. The unbalanced equation for the reaction is

$$PbO_2 + Cl^- \longrightarrow PbCl_2 + Cl_2$$

How many grams of PbO_2 must react to give 15.0 g of Cl_2?

***5.91** A solution contains $Ce(SO_4)_3^{2-}$ at a concentration of 0.0150 *M*. It was found that in a titration, 25.00 mL of this solution reacted completely with 23.44 mL of 0.032 *M* $FeSO_4$ solution. The reaction gave Fe^{3+} as a product in the solution. In this reaction, what is the final oxidation state of the Ce?

***5.92** A copper bar with a mass of 12.340 g is dipped into 255 mL of 0.125 *M* $AgNO_3$ solution. When the reaction that occurs has finally ceased, what will be the mass of unreacted copper in the bar? If all the silver that forms adheres to the copper bar, what will be the total mass of the bar after the reaction?

5.93 A solution containing 0.1244 g of $K_2C_2O_4$ was acidified, changing the $C_2O_4^{2-}$ ions to $H_2C_2O_4$. The solution was then titrated with 13.93 mL of a $KMnO_4$ solution to reach a faint pink end point. In the reaction, $H_2C_2O_4$ was oxidized to CO_2 and MnO_4^- was reduced to Mn^{2+}. What was the molarity of the $KMnO_4$ solution used in the titration?

***5.94** It was found that a 20.0 mL portion of a solution of oxalic acid, $H_2C_2O_4$, requires 6.25 mL of 0.200 *M* $K_2Cr_2O_7$ for complete reaction in an acidic solution. In the reaction, the oxidation product is CO_2 and the reduction product is Cr^{3+}. How many milliliters of 0.450 *M* NaOH are required to completely neutralize the $H_2C_2O_4$ in a separate 20.00 mL sample of the same oxalic acid solution?

***5.95** A mixture is made by combining 300 mL of 0.0200 *M* $Na_2Cr_2O_7$ with 400 mL of 0.060 *M* $Fe(NO_3)_2$. Initially, the H^+ concentration in the mixture is 0.400 *M*. Dichromate ion oxidizes Fe^{2+} to Fe^{3+} and is reduced to Cr^{3+}. After the reaction in the mixture has ceased, how many milliliters of 0.0100 *M* NaOH will be required to neutralize the remaining H^+?

***5.96** A solution with a volume of 500.0 mL contained a mixture of SO_3^{2-} and $S_2O_3^{2-}$. A 100.0 mL portion of the solution was found to react with 80.00 mL of 0.0500 *M* CrO_4^{2-} in a basic solution to give CrO_2^-. The only sulfur-containing product was SO_4^{2-}. After the reaction, the solution was treated with excess 0.200 *M* $BaCl_2$ solution, which precipitated $BaSO_4$. This solid was filtered from the solution, dried, and found to weigh 0.9336 g.

Explain in detail how you can determine the molar concentrations of SO_3^{2-} and $S_2O_3^{2-}$ in the original solution.

*5.97 An organic compound contains carbon, hydrogen, and sulfur. A sample of it with a mass of 1.045 g was burned in oxygen to give gaseous CO_2, H_2O, and SO_2. These gases were passed through 500.0 mL of an acidified 0.0200 *M* $KMnO_4$ solution, which caused the SO_2 to be oxidized to SO_4^{2-}. Only part of the available $KMnO_4$ was reduced to Mn^{2+}. Next, 50.00 mL of 0.0300 *M* $SnCl_2$ was added to a 50.00 mL portion of this solution, which still contained unreduced $KMnO_4$. There was more than enough added $SnCl_2$ to cause all of the remaining MnO_4^- in the 50 mL portion to be reduced to Mn^{2+}. The excess Sn^{2+} that still remained after the reaction was then titrated with 0.0100 M $KMnO_4$, requiring 27.28 mL of the $KMnO_4$ solution to reach the end point. What was the percentage of sulfur in the original sample of the organic compound that had been burned?

*5.98 A bar of copper weighing 32.00 g was dipped into 50.0 mL of 0.250 *M* $AgNO_3$ solution. If all the silver that deposits adheres to the copper bar, how much will the bar weigh after the reaction is complete? Write and balance any necessary chemical equations.

EXERCISES IN CRITICAL THINKING

5.99 The ion $OSCN^-$ is found in human saliva. Discuss the problems in assigning oxidation numbers to the atoms in this ion. Suggest a reasonable set of oxidation numbers for the atoms in $OSCN^-$.

5.100 We described the ion–electron method for balancing redox equations. Can you devise an alternate method using oxidation numbers?

5.101 Assuming that a chemical reaction with DNA could lead to damage causing cancer, would a very strong or a weak oxidizing agent have a better chance of being a carcinogen? Justify your answer.

5.102 Would you expect atomic oxygen and chlorine to be better or worse oxidizing agents than molecular oxygen and molecular chlorine? Justify your answer.

5.103 Do we live in an oxidizing or reducing environment? What effect might our environment have on chemistry we do in the laboratory? What effect might the environment have on the nature of the chemicals (minerals, etc.) we find on earth?

6 ENERGY AND CHEMICAL CHANGE

The power for this "top fuel dragster" comes from the combustion of fuel, but the amount of energy produced depends on the type of fuel and what is used to burn it. In this chapter we study the nature of energy, how it is measured, and how it relates to chemical reactions. *(© 2006 Jason Ellis)*

CHAPTER OUTLINE

THIS CHAPTER IN CONTEXT Energy is a term we often see in the news. We hear reports on the cost of energy and what's happening to the world's energy supplies. One might come away with the notion that energy is something you can hold in your hands and place in a bottle. Energy is not like matter, however. Rather, it is something that matter can possess, something that enables objects to move or cause other objects to move. Understanding what energy *really* is should be one of your goals in studying this chapter.

Nearly every chemical and physical change is accompanied by a *change* in energy. The energy changes associated with the evaporation and condensation of water drive global weather systems. The combustion of fuels produces energy changes we use to power cars and generate electricity, and our bodies use energy released in the metabolism of foods to drive biochemical processes. In this chapter we study **thermochemistry,** the branch of chemistry that deals specifically with the energy absorbed or released by chemical reactions. Thermochemistry has many practical applications, but it is also of great theoretical importance, because it provides an important link between laboratory measurements (such as temperature changes) and events on the molecular level that occur when molecules form or break apart.

Thermochemistry is part of the science of **thermodynamics,** the study of energy transfer and energy transformation. Thermodynamics allows scientists to predict whether a proposed physical change or chemical reaction can occur under a given set of conditions. It is an essential part of chemistry (and all the natural sciences). We'll continue our study of thermodynamics in Chapter 18.

6.1 AN OBJECT HAS ENERGY IF IT IS CAPABLE OF DOING WORK

Work is done by an object when it causes something to move. A moving car has energy because it can move another car in a collision.

As mentioned in the introduction above, energy is intangible; you can't hold it in your hand to study it and you can't put it in a bottle. **Energy** *is something an object has if the object is able to do work.* It can be possessed by the object in two different ways, as kinetic energy and as potential energy.

Kinetic energy (KE) *is the energy an object has when it is moving.* It depends on the object's mass and velocity; the larger its mass and the greater its velocity, the more kinetic energy it has. A simple equation relates kinetic energy (KE) to these quantities:

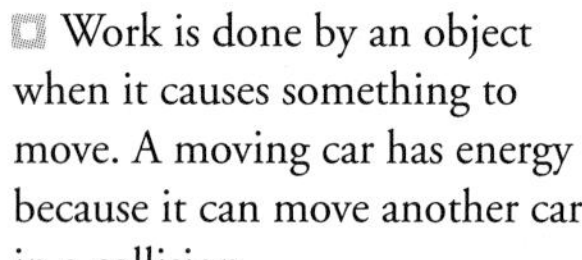

$$KE = \frac{1}{2}mv^2 \tag{6.1}$$

where m is the mass and v is the velocity.

Unlike kinetic energy, there is no single, simple equation that can be used to calculate the amount of potential energy an object has.

Potential energy (PE) *is energy an object has that can be changed to kinetic energy; it can be thought of as* **stored energy.** For example, when you wind an alarm clock, you transfer energy to a spring. The spring holds this stored energy (potential energy) and gradually releases it, in the form of kinetic energy, to make the clock's mechanism work.

Chemicals also possess potential energy, which is sometimes called **chemical energy.** When chemical reactions occur, the chemical energy possessed by the substances involved changes, leading to either an absorption or release of energy (as heat or light, for instance). For example, the explosive reaction between hydrogen and oxygen in the main engines of the space shuttle, shown in Figure 6.1, produces light, heat, and the expanding gases that help lift the space vehicle from its launchpad. One of the practical uses of chemical reactions such as this is to satisfy the energy needs of society.

Potential energy depends on position

An important aspect of potential energy is the way it depends on the positions of objects that experience attractions or repulsions toward other objects. For example, a book has potential energy because it experiences a gravitational attraction toward the earth. Lifting the book, which changes its position, increases the potential energy. This energy is supplied by the person doing the lifting. Letting the book fall allows the potential energy to decrease. The lost potential energy is changed to kinetic energy as the book gains speed during its descent.

How potential energy varies with position for objects that attract or repel is illustrated in Figure 6.2 for two balls connected by a spring. In the center we see the spring in its relaxed state and the two balls have the minimum potential energy. Stretching the spring, illustrated in the top two drawings, opposes a force that tends to pull the balls toward each other. This requires work and the energy supplied is stored as potential energy in the stretched spring. Similarly, compressing the spring as illustrated in the bottom two drawings also requires work, and the energy supplied is stored as potential energy in the compressed spring. These observations lead to two very important relationships we will use often.

Factors that affect potential energy

- Potential energy increases when objects that attract move apart, and decreases when they move toward each other.
- Potential energy increases when objects that repel move toward each other, and decreases when they move apart.

If you ever have trouble remembering this, think back to the balls connected by a spring.

In chemical systems we don't have springs between particles, but we do have attractions and repulsions between electrical charges. Electrons are attracted to protons because of their opposite electrical charges. Electrons repel electrons and nuclei repel other nuclei because they have the same kind of electrical charge. Changes in the relative positions of these particles as atoms join to form molecules or break apart when molecules decompose lead to changes in potential energy. These are the kinds of potential energy changes that lead to the release or absorption of energy by chemical systems during chemical reactions.

FIG. 6.1 Liquid hydrogen and oxygen serve as fuel for the space shuttle. Almost invisible points of blue flame come from the main engines of the space shuttle, which consume hydrogen and oxygen in the formation of water. *(Corbis Images.)*

Energy cannot be created or destroyed

One of the most important facts about energy is that *it cannot be created or destroyed; it can only be changed from one form to another.* This fact was established by many experiments and observations, and is known today as the **law of conservation of energy.** (Recall that when we say in science that something is "conserved," we mean that it is unchanged or remains constant.) You observe this law whenever you toss something—a ball, for instance—into the air. You give the ball some initial amount of kinetic energy when you throw it. As it rises, its potential energy increases. Because energy cannot come from nothing, the rise in potential energy comes at the expense of the ball's kinetic energy. Therefore, the ball's $\frac{1}{2}mv^2$ becomes smaller, and because the mass of the ball cannot change, the

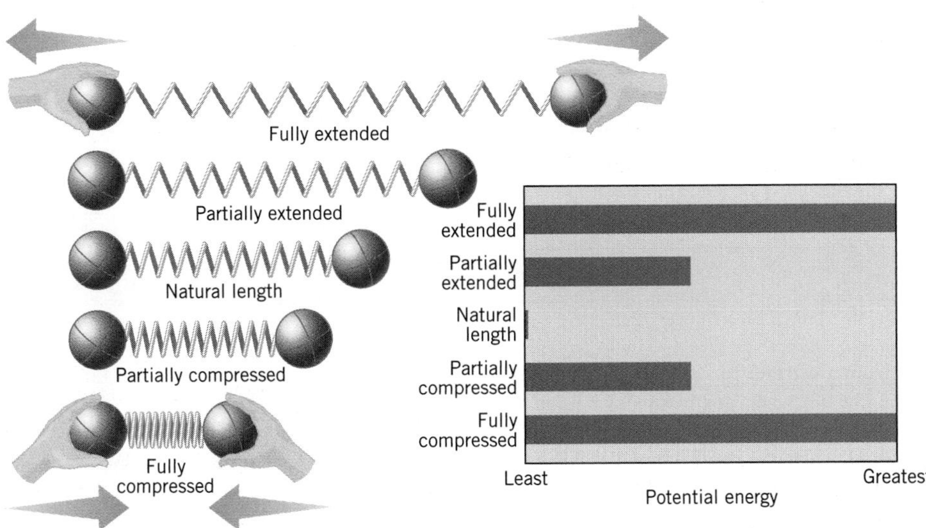

FIG. 6.2 The potential energy of a spring depends on its length. Either stretching or squeezing the spring raises the potential energy. The potential energy is at its lowest when the spring is at its natural length.

velocity (v) becomes less—the ball slows down. When all the kinetic energy has changed to potential energy, the ball can go no higher; it has stopped moving and its potential energy is at a maximum. The ball then begins to fall, and its potential energy is changed back to kinetic energy.

Heat and temperature are not the same

In any object, the atoms, molecules, or ions are constantly moving and colliding with each other, which gives them varying amounts of kinetic energy. *The temperature of an object is proportional to the **average** kinetic energy of its particles—the higher the average kinetic energy, the higher the temperature.* What this means is that when the temperature of an object is raised, the molecules move faster. (Recall that KE $= \frac{1}{2}mv^2$. Increasing the average kinetic energy doesn't increase the masses of the atoms, so it must increase their speeds.)

Heat is energy (also called **thermal energy**) that is transferred between objects caused by differences in their temperatures, and as you know, heat always passes spontaneously from a warmer object to a cooler one. This energy transfer continues until both objects come to the same temperature.

When a hot object is placed in contact with a cold one, the faster atoms of the hot object collide with and lose kinetic energy to the slower atoms of the cold object (Figure 6.3). This decreases the average kinetic energy of the particles of the hot object, causing its temperature to drop. At the same time, the average kinetic energy of the particles in the cold object is raised, causing the temperature of the cold object to rise. Eventually, the average kinetic energies of the atoms in both objects become the same and the objects reach the same temperature. Thus, the transfer of heat is interpreted as a transfer of kinetic energy between two objects.

All forms of energy (potential and kinetic) can be transformed to heat energy. For example, when you "step on the brakes" to stop your car, the kinetic energy of the car is changed to heat energy by friction between the brake shoes and the wheels.

The SI derived unit for energy is the joule

The SI unit of energy is a derived unit called the **joule** (symbol **J**) and corresponds to the amount of kinetic energy possessed by a 2 kilogram object moving at a speed of 1 meter per second. Using the equation for kinetic energy, KE $= \frac{1}{2}mv^2$,

$$1\text{ J} = \frac{1}{2}(2\text{ kg})\left(\frac{1\text{ m}}{1\text{ s}}\right)^2$$

$$1\text{ J} = 1\text{ kg m}^2\text{ s}^{-2}$$

The joule is actually a rather small amount of energy and in most cases we will use the larger unit, the **kilojoule (kJ);** $1\text{ kJ} = 1000\text{ J} = 10^3\text{ J}$.

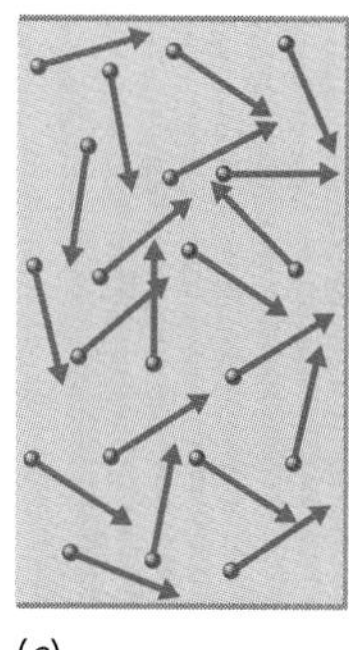
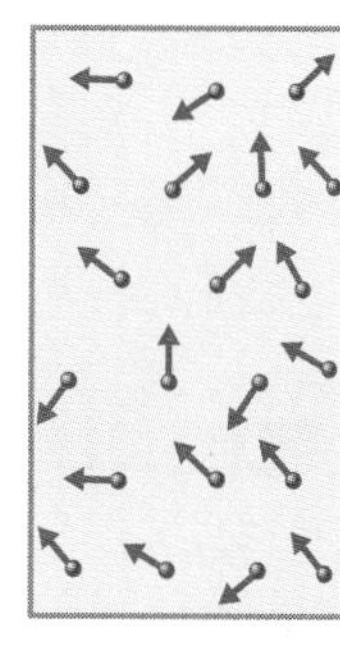
(a)

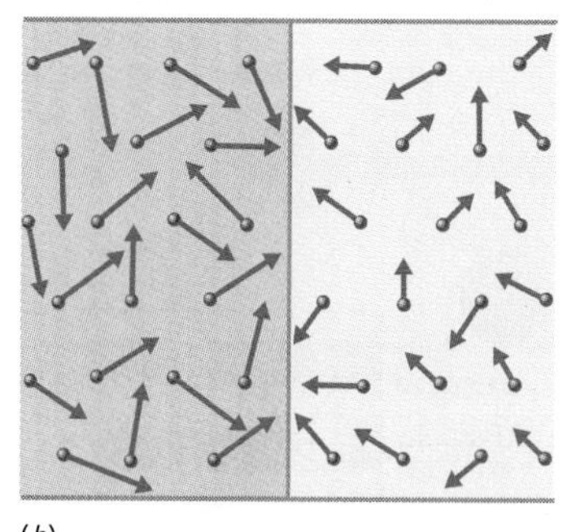
(b)

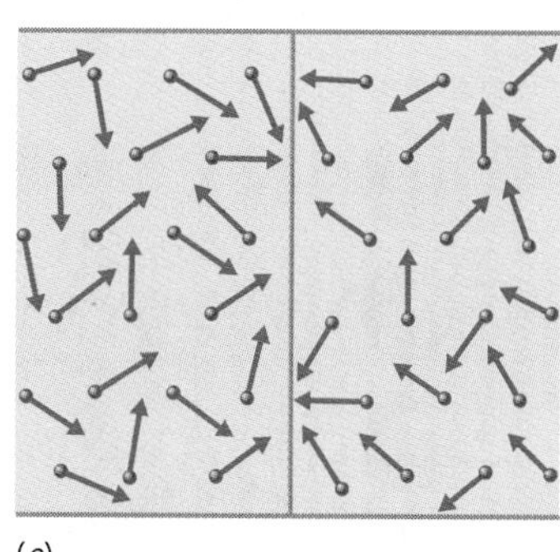
(c)

FIG. 6.3 **Energy transfer from a warmer to a cooler object.** (*a*) The longer arrows on the left denote higher kinetic energies and so something warmer, like hot water just before it's poured into a cooler coffee cup. (*b*) The hot water and the cup's inner surface are now in thermal contact. Collisions between the fast moving water molecules and the slower molecules in the cup's material cause the water molecules to slow down and the cup's molecules to speed up. In this way, kinetic energy is transferred from the water to the cup. (*c*) Thermal equilibrium is established: the temperatures of the water and the cup wall are now equal.

Another energy unit you may be familiar with is called the **calorie (cal).** Originally, it was defined as the energy needed to raise the temperature of 1 gram of water by 1 degree Celsius. (Temperature units are discussed below.) With the introduction of the SI, the calorie has been redefined as follows:

$$1 \text{ cal} = 4.184 \text{ J (exactly)} \tag{6.2}$$

The larger unit **kilocalorie (kcal),** which equals 1000 calories, can also be related to the kilojoule.

$$1 \text{ kcal} = 4.184 \text{ kJ}$$

The nutritional or dietary Calorie (note the capital), Cal, is actually one kilocalorie.

$$1 \text{ Cal} = 1 \text{ kcal} = 4.184 \text{ kJ}$$

While joules and kilojoules have been accepted worldwide as the standard units of energy, calories and kilocalories are still in common use, so you will need to be able to convert joules into calories and vice versa.

6.2 INTERNAL ENERGY IS THE TOTAL ENERGY OF AN OBJECT'S MOLECULES

In the preceding section we introduced *heat* as a transfer of energy that occurs between objects with different temperatures. For example, heat will flow from a hot cup of coffee into the cooler surroundings. Eventually the coffee and surroundings come to the same temperature and we say they are in **thermal equilibrium** with each other. The temperature of the coffee has dropped and the temperature of the surroundings has increased a bit.

Energy that is transferred as heat comes from an object's fund of internal energy. **Internal energy (*E*)** is the sum of energies for all of the individual particles in a sample of matter. All of the particles within any object are in constant motion. For example, in a sample of air at room temperature, oxygen and nitrogen molecules travel faster than rifle bullets, constantly colliding with each other and with the walls of their container. The molecules spin as they move, and the atoms within the molecules jiggle and vibrate; these internal molecular motions also contribute to the kinetic energy of the molecule and so to the internal energy of the sample. We'll use the term **molecular kinetic energy** for the energy associated with such motions. Each particle has a certain value of molecular kinetic energy at any given moment. Molecules are continually exchanging energy with each other during collisions, but as long as the sample is isolated, the total kinetic energy of all the molecules remains constant.

Recall that kinetic energy is energy of motion and is given by KE = $1/2\ mv^2$, where m is the mass of an object and v is its velocity.

Internal energy is often given the symbol E.[1] In studying both chemical and physical changes, we will be interested in the *change* in internal energy that accompanies the process. This is defined as the **internal energy change (ΔE)**, where the symbol Δ (Greek letter delta) signifies a change.

$$\Delta E = E_{\text{final}} - E_{\text{initial}}$$

The symbol Δ denotes a *change between some initial and final state.*

For a chemical reaction, E_{final} corresponds to the internal energy of the products, so we'll write it as E_{products}. Similarly, we'll use the symbol $E_{\text{reactants}}$ for E_{initial}. So for a chemical reaction the change in internal energy is given by

$$\Delta E = E_{\text{products}} - E_{\text{reactants}} \tag{6.3}$$

Notice carefully an important convention illustrated by this equation. Changes in something like temperature (Δt) or in internal energy (ΔE) are always figured by taking "final minus initial" or "products minus reactants." This means that if a system *absorbs* energy from its surroundings during a change, its final energy is greater than its initial energy and ΔE is positive. This is what happens, for example, when photosynthesis occurs or when something in the surroundings supplies energy to charge a battery. As the system (the battery) absorbs the energy, its internal energy increases and is then available for later use elsewhere.

[1] Sometimes the symbol U is used for internal energy.

Temperature is related to average molecular kinetic energy

Temperature is related to average molecular kinetic energy, but it is not equal to it. Temperature is ***not*** an energy.

The notion, introduced in Section 6.1, that atoms and molecules are in constant random motion forms the basis for the **kinetic molecular theory.** It is this theory that tells us in part that *temperature* is related to the *average kinetic energy* of the atoms and molecules of an object. Be sure to keep in mind the distinction between temperature and internal energy; the two are quite different. Temperature is related to the *average* molecular kinetic energy, whereas internal energy is related to the *total* molecular kinetic energy.[2]

The concept of an *average* kinetic energy implies that there is a distribution of kinetic energies among the molecules in an object. Let's examine this further. At any particular moment, the individual particles in an object are moving at different velocities, which gives them a broad range of kinetic energies. An extremely small number of particles will be standing still as a result of balanced collisions, so their kinetic energies will be zero. Another small number will have very large velocities acquired through successive "rear end" collisions; these will have very large kinetic energies. Between these extremes there will be many molecules with intermediate amounts of kinetic energy.

Figure 6.4 shows graphs that describe the kinetic energy distributions among molecules in a sample at two different temperatures. The vertical axis represents the *fraction* of molecules with a given kinetic energy (i.e., the number of molecules with a given KE divided by the total number of molecules in the sample). Each curve in Figure 6.4 describes how the fraction of molecules with a given KE varies with KE.

Each curve starts out with a fraction equal to zero when KE equals zero. This is because the fraction of molecules with zero KE (corresponding to molecules that are motionless) is essentially zero, regardless of the temperature. Moving to higher values of KE, we find greater fractions. For very large values of KE, the fraction drops off again because very few molecules have very large velocities.

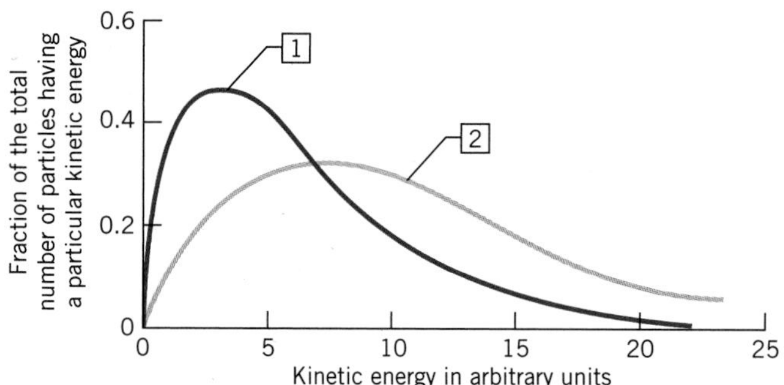

FIG. 6.4 The distribution of kinetic energies among gas particles. The distribution of individual kinetic energies changes in going from a lower temperature, curve 1, to a higher temperature, curve 2. The highest point on each curve is the most probable value of kinetic energy for that temperature. It's the value of the molecular kinetic energy that we would most frequently find, if we could get inside the system and observe and measure the kinetic energy of each molecule. The most probable value of molecular kinetic energy is less at the lower temperature. At the higher temperature, more molecules have high speeds and fewer molecules have low speeds, so the maximum shifts to the right and the curve flattens.

[2] When we're looking at heat flow, the total molecular kinetic energy is the part of the internal energy we're most interested in. However, internal energy is the total energy of all the particles in the object, so molecular potential energies (from forces of attraction and repulsion that operate between and within molecules) can and do make a large contribution to the internal energy. This is especially true in liquids and solids.

Each curve in Figure 6.4 has a characteristic peak or maximum corresponding to the most frequently experienced values for molecular KE. Because the curves are not symmetrical, the *average* values of molecular KE lie slightly to the right of the maxima. Notice that when the temperature increases (going from curve 1 to 2), the curve flattens and the maximum shifts to a higher value, as does the average molecular KE. In fact, *if we double the Kelvin temperature, the average KE also doubles, so the Kelvin temperature is directly proportional to the average KE.* The reason the curve flattens is because the area under each curve corresponds to the *sum* of all the fractions, and when all the fractions are added, they must equal one. (No matter how you cut up a pie, if you add all the fractions, you must end up with *one* pie!)

We will find the graphs illustrated in Figure 6.4 very useful later when we discuss the effect of temperature on such properties as the rates of evaporation of liquids and the rates of chemical reactions.

Internal energy is a state function

Equation 6.3 defines what we mean by a change in the internal energy of a chemical system during a reaction. This energy change can be made to appear entirely as heat, and as you will learn soon, we can measure heat. However, there is no way to actually measure either E_{products} or $E_{\text{reactants}}$. Fortunately, we are more interested in ΔE than we are in the absolute amounts of energy that the reactants or products have.

If we can measure heat, but not internal energy itself, why talk about energy at all? It turns out that *the energy of an object depends only on the object's current condition.* It doesn't matter how the object acquired that energy, or how it will lose it. This simple fact vastly simplifies thermochemical calculations.

The complete list of properties that specify an object's current condition is known as the **state** of the object. In chemistry, it is usually enough to specify the object's pressure, temperature, volume, and chemical composition (numbers of moles of all substances present).

"State," as used in thermochemistry, doesn't have the same meaning as when it's used in terms such as "solid state" or "liquid state."

Any property that, like energy, depends *only* on an object's current state is called a **state function.** Pressure, temperature, and volume are themselves examples of state functions and it's easy to understand why. A system's current temperature, for example, does not depend on what it was yesterday. Nor does it matter *how* the system acquired it, that is, the path to its current value. If it's now 25 °C, we know all we can or need to know about its temperature. We do not have to specify how it got there or where it's going. Also, if the temperature were to increase, say, to 35 °C, the change in temperature, Δt, is simply the difference between the final and the initial temperatures.

$$\Delta t = t_{\text{final}} - t_{\text{initial}} \tag{6.4}$$

To make the calculation of Δt, we do not have to know what caused the temperature change—exposure to sunlight, heating by an open flame, or any other mechanism. All we need are the initial and final values. *This independence from the method or mechanism by which a change occurs is the important feature of all state functions.* As we'll see later, the advantage of recognizing that some property is a state function is that many calculations are then much, much easier.

6.3 HEAT CAN BE DETERMINED BY MEASURING TEMPERATURE CHANGES

By measuring the amount of heat that is absorbed or released by an object, we are able to quantitatively study nearly any type of energy transfer. For example, if we want to measure the energy transferred by an electrical current, we can force the current through something with high electrical resistance, like the heating element in a toaster, and the energy transferred by the current becomes heat.

In our study it is very important to specify the **boundary** across which heat flows. The boundary might be visible (like the walls of a beaker) or invisible (like the boundary that separates warm air from cold air along a weather front). The boundary encloses the **system,** which is the object we are interested in studying. Everything outside the system is called the **surroundings.** The system and the surroundings together are called the **universe.**

Three types of systems are possible, depending on whether matter or energy can cross the boundary.

- **Open systems** can gain or lose mass and energy across their boundaries. The human body is an example of an open system.
- **Closed systems** can absorb or release energy, but not mass, across the boundary. The mass of a closed system is constant, no matter what happens inside. A lightbulb is an example of a closed system.
- **Isolated systems** cannot exchange matter or energy with their surroundings. Because energy cannot be created or destroyed, the energy of an isolated system is constant, no matter what happens inside. Processes that occur within an isolated system are called **adiabatic,** from the Greek *a* + *diabatos*, meaning "not passable." A stoppered Thermos bottle is a good approximation of an isolated system.

The heat an object gains or loses is directly proportional to its temperature change

There is no instrument available that directly measures heat. Instead, we measure temperature changes, and then use them to calculate heat. Common sense and experience tell you that the more heat you add to an object, the more its temperature will rise. In fact, experiments show that the temperature change, Δt, is *directly proportional* to the amount of heat absorbed, which we will identify by the symbol q. This can be expressed in the form of an equation as

Heat capacity

$$q = C\,\Delta t \tag{6.5}$$

where C is a proportionality constant called the **heat capacity.** The units of heat capacity are usually J °C^{-1}, expressing the amount of energy needed to raise the temperature of an object by 1 °C.

Heat capacity depends on two factors. One is the size of the sample; if we double the size of the sample, we need twice as much heat to cause the same temperature increase. Heat capacity also depends on what the sample is made of. For example, it's found that it takes more heat to raise the temperature of one gram of water by 1 °C than to cause the same temperature change in one gram of iron.

Specific heat is an intensive property related to heat capacity

From the preceding discussion, we expect that different samples of a given substance will have different heat capacities, and this is what is observed. For example, if we have 10.0 g of water, it must absorb 41.8 J of heat energy to raise its temperature by 1.00 °C. The heat capacity of the sample is found by solving Equation 6.5 for C.

$$C = \frac{q}{\Delta t} = \frac{41.8\text{ J}}{1.00\text{ °C}} = 41.8\text{ J/°C}$$

On the other hand, a 100 g sample of water must absorb 418 J of heat to have its temperature raised by 1.00 °C. The heat capacity of this sample is

$$C = \frac{q}{\Delta t} = \frac{418\text{ J}}{1.00\text{ °C}} = 418\text{ J/°C}$$

Notice that a 10-fold increase in sample size has produced a 10-fold increase in the heat capacity. This indicates that the heat capacity is directly proportional to the mass of the sample,

$$C = m \times s \tag{6.6}$$

where m is the mass of the sample and s is a constant called the **specific heat capacity** (or simply the **specific heat**). If the heat capacity has units of J/°C, and if the mass is in grams, the specific heat capacity has units of (J/g °C) or J g^{-1} °C^{-1}. The units tell us that the specific heat capacity is the amount of heat required to raise the temperature of 1 gram of a substance by 1 °C.

Because C depends on the size of the sample, *heat capacity is an extensive property.* When we discussed density in Section 1.8 you learned that if we take the ratio of two extensive properties, we can obtain an intensive property—one that is independent of the size of the sample. Because heat capacity and mass both depend on sample size, their ratio obtained by solving Equation 6.6 for s yields a property that is the same for any sample of a substance.

If we substitute Equation 6.6 into Equation 6.5, we can obtain an equation for q in terms of specific heat.

Specific heat

$$q = ms\,\Delta t \tag{6.7}$$

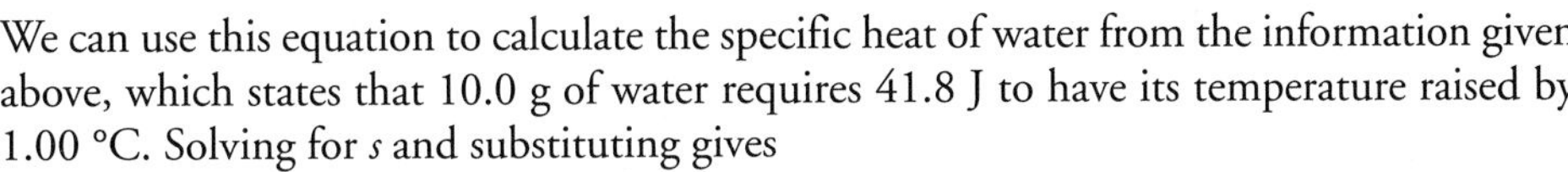

We can use this equation to calculate the specific heat of water from the information given above, which states that 10.0 g of water requires 41.8 J to have its temperature raised by 1.00 °C. Solving for s and substituting gives

$$s = \frac{q}{m\,\Delta t} = \frac{41.8\text{ J}}{10.0\text{ g} \times 1.00\,°\text{C}}$$

$$= 4.18\text{ J g}^{-1}\,°\text{C}^{-1}$$

Recall that the older energy unit calorie (cal) is currently defined as 1 cal = 4.184 J. Therefore, the specific heat of water expressed in calories is

$$s = 1.00\text{ cal g}^{-1}\,°\text{C}^{-1} \qquad (\text{for } H_2O)$$

You should learn these values for water:

$s = 1.00$ cal g^{-1} °C^{-1}
$s = 4.18$ J g^{-1} °C^{-1}

In measuring heat, we often use an apparatus containing water, so the specific heat of water is a quantity we will use in working problems. Every substance has its own characteristic specific heat, and some are given in Table 6.1.

TABLE 6.1 Specific Heats

Substance	Specific Heat, J g^{-1} °C^{-1} (25 °C)
Carbon (graphite)	0.711
Copper	0.387
Ethyl alcohol	2.45
Gold	0.129
Granite	0.803
Iron	0.4498
Lead	0.128
Olive oil	2.0
Silver	0.235
Water (liquid)	4.18

When comparing or working with mole-sized quantities of substances, we can use the **molar heat capacity,** which is the amount of heat needed to raise the temperature of 1 mol of a substance by 1 °C. Molar heat capacity equals the specific heat times the molar mass and has units of J mol^{-1} °C^{-1}.

FACETS OF CHEMISTRY

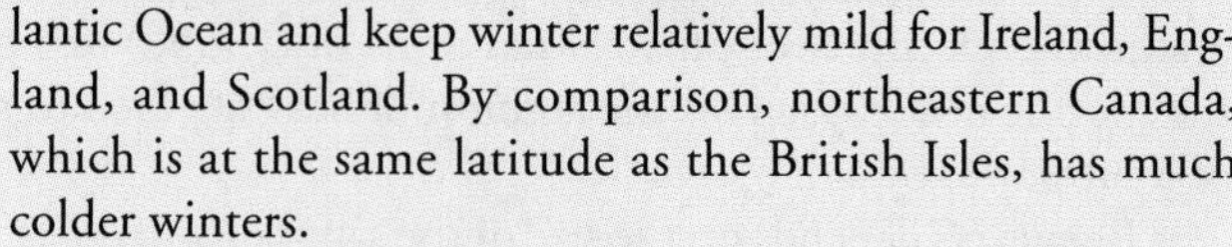

Water, Climate, and the Body's "Thermal Cushion"

Compared with most substances, water has a very high specific heat. A body of water can therefore gain or lose a substantial amount of heat without undergoing a large change in temperature. Because of this, the oceans of the world have a very significant moderating effect on climate. This is particularly apparent when we compare temperature extremes of locales near the oceans with those inland away from the sea and away from large lakes (such as the Great Lakes in the upper United States). Places near the sea tend to have cooler summers and milder winters than places located inland because the ocean serves as a thermal "cushion," absorbing heat in the summer and giving some of it back during the winter.

Warm and cool ocean currents also have global effects on climate. For example, the warm waters of the Gulf of Mexico are carried by the Gulf Stream across the Atlantic Ocean and keep winter relatively mild for Ireland, England, and Scotland. By comparison, northeastern Canada, which is at the same latitude as the British Isles, has much colder winters.

Water also serves as a thermal cushion for the human body. The adult body is about 60% water by mass, so it has a high heat capacity. This makes it relatively easy for the body to maintain a steady temperature of 37 °C, which is vital to survival. In other words, the body can exchange considerable energy with the environment but experience only a small change in temperature. With a substantial thermal cushion, the body adjusts to large and sudden changes in outside temperature while experiencing very small fluctuations of its core temperature.

The algebraic sign of *q* is used to indicate the direction of heat flow

Heat is energy that's transferred from one object to another. This means that heat is lost by one object and the same amount of heat is gained by the other. To indicate the direction of heat flow, we assign a positive sign to q if the heat is gained and a negative sign if the heat is lost. For example, if a piece of warm iron is placed into a beaker of cool water and the iron loses 10.0 J of heat, the water gains 10.0 J of heat. For the iron, $q = -10.0$ J and for the water, $q = +10.0$ J.

The relationship between the algebraic signs of q in a transfer of heat can be stated in a general way by the equation

Heat transfer

$$q_1 = -q_2 \tag{6.8}$$

where 1 and 2 refer to the objects between which the heat is transferred.

EXAMPLE 6.1
Determining the Heat Capacity of an Object

Central processing chips in computers generate a tremendous amount of heat—enough to damage themselves permanently if the chip is not cooled somehow. Aluminum "heat sinks" are often attached to the chips to carry away excess heat. Suppose that a heat sink at 71.3 °C is dropped into a Styrofoam cup containing 100.0 g of water at 25.0 °C. The temperature of the water rises to 27.4 °C. What is the heat capacity of the heat sink, in J/°C?

ANALYSIS: Heat capacity is related to temperature change by Equation 6.5, so that's the tool we need to solve the problem. To find C, we divide both sides by the temperature change, Δt:

$$C = q/\Delta t$$

Let's find q first, *being very careful about algebraic signs.* We can assume that the heat lost by the heat sink, $q_{\text{(heat sink)}}$, will be gained by the much cooler water, q_{H_2O}. Because the water is gaining heat, q_{H_2O} will be a positive quantity, and because the heat sink loses heat, $q_{\text{(heat sink)}}$ will be a negative quantity *equal in size* to q_{H_2O}. This lets us apply Equation 6.8 and write

$$q_{\text{(heat sink)}} = -q_{H_2O}$$

The negative sign on the right assures us that when we substitute the positive value for q_{H_2O} the sign of $q_{\text{(heat sink)}}$ will be negative.

To calculate q_{H_2O}, our tool will be Equation 6.7, which enables us to use the specific heat of water (4.18 J g^{-1} °C^{-1}), the mass of water, and the temperature change for water. Once we've found q_{H_2O}, we change its algebraic sign to obtain $q_{(\text{heat sink})}$. Then we use the temperature change for the heat sink to calculate C for the heat sink.

SOLUTION: The temperature of the water rises from 25.0 °C to 27.4 °C, so for the water,

$$\Delta t_{H_2O} = t_{\text{final}} - t_{\text{initial}} = 27.4\ °\text{C} - 25.0\ °\text{C} = 2.4\ °\text{C}$$

The specific heat of water is 4.18 J g^{-1} °C^{-1} and its mass is 100.0 g. Therefore, for water, the heat absorbed is

$$q_{H_2O} = ms\Delta t = 100.0\ \text{g} \times 4.18\ \text{J g}^{-1}\ °\text{C}^{-1} \times 2.4\ °\text{C}$$

$$= +1.0 \times 10^3\ \text{J}$$

Therefore, $q_{(\text{heat sink})} = -1.0 \times 10^3$ J

The temperature of the heat sink decreases from 71.3 °C to 27.4 °C, so

$$\Delta t_{(\text{heat sink})} = t_{\text{final}} - t_{\text{initial}} = 27.4\ °\text{C} - 71.3\ °\text{C} = -43.9\ °\text{C}$$

The heat capacity of the heat sink is then

$$C = \frac{q_{(\text{heat sink})}}{\Delta t} = \frac{-1.0 \times 10^3\ \text{J}}{-43.9\ °\text{C}} = 23\ \text{J/°C}$$

IS THE ANSWER REASONABLE? In any calculation involving energy transfer, we first check to see that all quantities have the correct signs. Heat capacities are positive for common objects, so the fact that we've obtained a positive value for C tells us we've handled the signs correctly.

For the transfer of a given amount of heat, the larger the heat capacity, the smaller the temperature change. The heat capacity of the water is $C = ms = 100\ \text{g} \times 4.18\ \text{J g}^{-1}\ °\text{C}^{-1} =$ 418 J °C^{-1} and the water changes temperature by 2.4 °C. The size of the temperature change for the heat sink, 43.9 °C, is almost 20 times as large as that for the water, so the heat capacity should be about 1/20 that of the water. Dividing 418 J °C^{-1} by 20 gives 20.9 J °C^{-1}, which is not too far from our answer, so our calculations seem to be reasonable.

EXAMPLE 6.2
Calculating Heat from a Temperature Change, Mass, and Specific Heat

If a gold ring with a mass of 5.50 g changes in temperature from 25.00 to 28.00 °C, how much heat has it absorbed?

ANALYSIS: The question asks us to connect the heat absorbed by the ring with its temperature change, Δt. Equations 6.5 or 6.7 could provide this connection. However, we don't know the heat capacity of the ring. We do know the mass of the ring, and the fact that it is made of gold, so we can look up the specific heat and use Equation 6.7 as our tool to solve the problem. Table 6.1 gives the specific heat of gold as 0.129 J g^{-1} °C^{-1}.

SOLUTION: The mass m of the ring is 5.50 g, the specific heat s is 0.129 J g^{-1} °C^{-1}, and the temperature increases from 25.00 to 28.00 °C, so Δt is 3.00 °C. Using these values in Equation 6.7 gives

$$q = ms\,\Delta t$$

$$= (5.50\ \text{g}) \times (0.129\ \text{J g}^{-1}\,°\text{C}^{-1}) \times (3.00\ °\text{C})$$

$$= 2.13\ \text{J}$$

The sign of Δt will determine the sign of the heat exchanged. Δt is positive because ($t_{final} - t_{initial}$) is positive.

Thus, only 2.13 J raises the temperature of 5.50 g of gold by 3.00 °C. Because Δt is positive, so is q, 2.13 J. Thus the *sign* of the energy change is in agreement with the fact that the ring *absorbs* heat.

IS THE ANSWER REASONABLE? If the ring had a mass of only 1 g and its temperature increased by 1 °C, we'd know from the specific heat of gold (let's round it to 0.13 J g^{-1} $°C^{-1}$) that the ring would absorb 0.13 J. For a 3 °C increase, the answer would be three times as much, or 0.39 J, nearly 0.4 J. For a ring a little heavier than 5 g, the heat absorbed would be five times as much, or about 2.0 J. So our answer (2.13 J) is clearly reasonable.

Practice Exercise 1: A ball bearing at 220 °C is dropped into a cup containing 250 g of water at 20.0 °C. The water and ball bearing come to a temperature of 30.0 °C. What is the heat capacity of the ball bearing, in J/°C? (Hint: How are the algebraic signs of q related for the ball bearing and the water?)

Practice Exercise 2: The temperature of 250 g of water is changed from 25.0 to 30.0 °C. How much energy was transferred into the water? Calculate your answer in joules, kilojoules, calories, and kilocalories.

6.4 ENERGY IS ABSORBED OR RELEASED DURING MOST CHEMICAL REACTIONS

Almost every chemical reaction involves the absorption or release of energy. When this happens, the potential energy (also called chemical energy) of the substances involved in the reaction changes. To understand the origin of this energy change we need to explore the origin of potential energy in chemical systems.

In Chapter 2 we introduced you to the concept of *chemical bonds*, which are the attractive forces that bind atoms to each other in molecules, or ions to each other in ionic compounds. In this chapter you learned that when particles experience attractions or repulsions, potential energy changes occur when the particles come together or move apart. We can now bring these concepts together to understand the origin of energy changes in reactions.

Exothermic reactions release heat; endothermic reactions absorb heat

Chemical reactions generally involve *both* the breaking and making of chemical bonds. In most reactions, when bonds *form*, things that attract each other move closer together, which tends to decrease the potential energy of the reacting system. When bonds *break*, on the other hand, things that normally attract each other are forced apart, which increases the potential energy of the reacting system. Every reaction, therefore, has a certain net overall potential energy change, a net balance between the "costs" of breaking bonds and the "profits" from making them.

In many reactions, the products have *less* potential energy (chemical energy) than the reactants. When the gas methane burns in a Bunsen burner, for example, molecules of CH_4 and O_2 with large amounts of chemical energy but relatively low amounts of molecular kinetic energy change into products (CO_2 and H_2O) that have less chemical energy but much more molecular kinetic energy. Thus some chemical energy changes into molecular *kinetic* energy. The increase in molecular kinetic energy leads to a temperature increase in the reaction mixture, and if the reaction is occurring in an uninsulated system, some of this energy can be transferred to the surroundings as heat. The net result, then, is that the drop in chemical energy appears as heat that's transferred to the surroundings. In the chemical equation, therefore, we can write heat as a product.

$$CH_4(g) + 2O_2(g) \longrightarrow CO_2(g) + 2H_2O(g) + \text{heat}$$

exo means "out" *endo* means "in" *therm* means "heat"

Any reaction in which heat is a product is said to be **exothermic.**

In some reactions the products have *more* chemical energy than the reactants. For example, green plants make energy-rich molecules of glucose ($C_6H_{12}O_6$) and oxygen from carbon dioxide and water by a multistep process called *photosynthesis.* It requires a continuous supply of energy, which is provided by the sun. The plant's green pigment, chlorophyll, is the solar energy absorber for photosynthesis.

$$6CO_2 + 6H_2O + \text{solar energy} \xrightarrow[\text{many steps}]{\text{chlorophyll}} C_6H_{12}O_6 + 6O_2$$

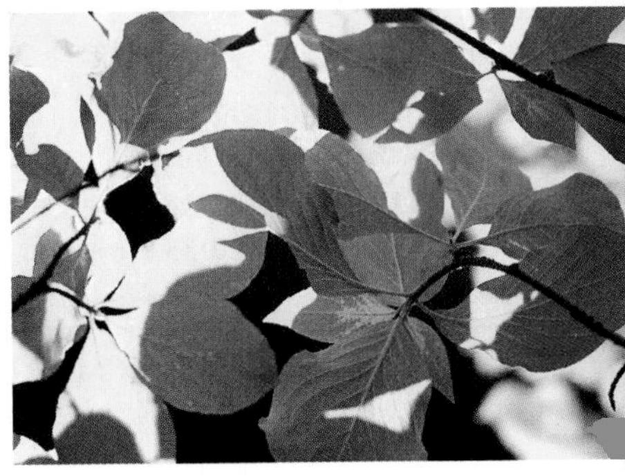

By an endothermic reaction, photosynthesis converts solar energy, trapped by the green pigment chlorophyll, into plant compounds that are rich in chemical energy. (© *Index Stock.*)

Reactions that consume energy are said to be **endothermic.** Usually such reactions change kinetic energy into potential energy (chemical energy), so the temperature of the system tends to drop as the reaction proceeds. If the reaction is occurring in an uninsulated vessel, the temperature of the surroundings will drop as heat flows into the system.

Energy can be released by breaking weak bonds to form stronger ones

The strength of a chemical bond is measured by how much energy is needed to separate the atoms that the bond holds together, or by how much energy is released when the bond forms. The larger this amount of energy, the stronger the bond.

Breaking weak bonds requires relatively little energy compared to the amount of energy released when strong bonds form. This is key to understanding why burning fuels such as CH_4 produce heat. Weaker bonds between carbon and hydrogen within the fuel break so that stronger bonds in the water and carbon dioxide molecules can form. Burning a hydrocarbon fuel produces one molecule of carbon dioxide for each carbon atom and one molecule of water for each pair of hydrogen atoms in the fuel. Generally, the more carbon and hydrogen atoms a molecule of the fuel contains, the more strong bonds can form when it burns—and the more heat it will produce, per molecule.

We will have a lot more to say about chemical bonds and their strengths in later chapters.

6.5 Heats of Reaction Are Measured at Constant Volume or Constant Pressure

The amount of heat absorbed or released in a chemical reaction is called the **heat of reaction.** Heats of reaction are determined by measuring the change in temperature they cause in their surroundings, using an apparatus called a **calorimeter.** The calorimeter is often just a container with a known heat capacity in which the reaction is carried out. We can calculate the heat of reaction by measuring the temperature change the reaction causes in the calorimeter. The science of using a calorimeter for determining heats of reaction is called **calorimetry.**

There is no instrument that *directly* measures energy. A calorimeter does not do this; its thermometer is the only part to provide raw data, the temperature change. We *calculate* the energy change.

Calorimeter design is not standard, varying according to the kind of reaction and the precision desired. Calorimeters are usually designed to measure heats of reaction under conditions of either constant volume or constant pressure. We have constant-volume conditions if we run the reaction in a closed, rigid container. Running the reaction in an open container imposes constant-pressure conditions.

Pressure is an important variable in calorimetry, as we'll see in a moment. If you have ever blown up a balloon or worked with an auto or bicycle tire, you have already learned something about *pressure.* You know that when a tire holds air at a relatively high pressure, it is hard to dent or push in its sides because "something" is pushing back. That something is the air pressure inside the tire.

Pressure is the amount of *force* acting on a unit of area; it's the ratio of force to area.

$$\text{Pressure} = \frac{\text{force}}{\text{area}}$$

You can increase the pressure in a tire by pushing or forcing air into it. Its pressure, when you quit, is whatever pressure you have added *above the initial pressure,* namely, the atmospheric pressure. We describe the air in the tire as being *compressed,* meaning that it's at a higher pressure than the atmospheric pressure.

Atmospheric pressure is the pressure exerted by the mixture of gases in our atmosphere. In English units this ratio of force to area is approximately 14.7 lb in.$^{-2}$ at sea level. The ratio does vary a bit with temperature and weather, and it varies a lot with altitude. The value of 14.7 lb in.$^{-2}$ is very close to two other common pressure units, the **standard atmosphere** (abbreviated **atm**), and the **bar,** which will be discussed in Chapter 10. A container that is open to the atmosphere is under a constant pressure of about 14.7 lb in.$^{-2}$, which is approximately 1 bar or 1 atm.

More precisely, 14.696 lb in.$^{-2}$ = 1.0000 atm = 1.0133 bar.

We've used the symbol q for heat; the symbols q_v and q_p are often used to show heats measured at constant volume or constant pressure, respectively. We must distinguish q_v from q_p. For reactions that involve big changes in volume (such as consumption or production of a gas) the difference between the two can be quite large.

EXAMPLE 6.3
Computing q_v and q_p Using Calorimetry

A gas-phase chemical reaction occurs inside an apparatus similar to that shown in Figure 6.5. The reaction vessel is a cylinder topped by a piston. The piston can be locked in place with a pin. The cylinder is immersed in an insulated bucket containing a precisely weighed amount of water. A separate experiment determined that the calorimeter (which includes the piston, cylinder, bucket, and water) has a heat capacity of 8.101 kJ/°C. The reaction was run twice with identical amounts of reactants each time. The following data were collected.

Run	Pin position	Initial bucket temperature (°C)	Final bucket temperature (°C)
1	*a* (piston locked)	24.00	28.91
2	*b* (piston unlocked)	27.32	31.54

Determine q_v and q_p for this reaction.

ANALYSIS: The key to solving the problem is realizing that from the heat capacity and temperature change, we can calculate the heat absorbed by the calorimeter. The tool is Equation 6.5. Run 1 will give the heat at constant volume, q_v, since the piston cannot move. Run 2 will give the heat at constant pressure, q_p, because with the piston unlocked the entire reaction will run under atmospheric pressure.

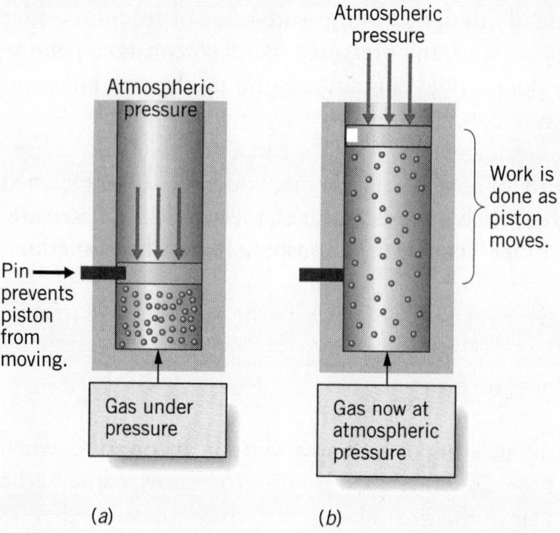

FIG. 6.5 Pressure–volume work. (*a*) A gas is confined under pressure in a cylinder fitted with a piston that is held in place by a sliding pin. (*b*) When the piston is released, the gas inside the cylinder expands and pushes the piston upward against the opposing pressure of the atmosphere. As it does so, the gas does some pressure–volume work on the surroundings.

SOLUTION: For Run 1, Equation 6.5 gives the heat absorbed by the calorimeter as

$$q = C\,\Delta t = (8.101\text{ kJ/°C}) \times (28.91\text{ °C} - 24.00\text{ °C}) = 39.8\text{ kJ}$$

Because the calorimeter gains heat, this amount of heat is released by the reaction, so *q for the reaction* must be negative. Therefore, $q_v = -39.8$ kJ.

For Run 2,

$$q = C\,\Delta t = (8.101\text{ kJ/°C}) \times (31.54\text{ °C} - 27.32\text{ °C}) = 34.2\text{ kJ}$$

so $q_p = -34.2$ kJ.

IS THE ANSWER REASONABLE? The arithmetic is straightforward, but in any calculation involving heat, always check to see that the signs of the heats in the problem make sense. The calorimeter absorbs heat, so its heats are positive. The reaction releases heat, so its heats must be negative.

Why are q_v and q_p different for reactions that involve a significant volume change? The system in this case is the reacting mixture. If the system expands against atmospheric pressure, it is doing work. Some of the energy that would otherwise appear as heat is used up when the system pushes back the atmosphere. In the example, the work done to expand the system against atmospheric pressure is equal to amount of "missing" heat in the constant-pressure case:

$$\text{Work} = (-39.8\text{ kJ}) - (-34.2\text{ kJ}) = -5.6\text{ kJ}.$$

The minus sign indicates that energy is leaving the system. This is called **expansion work** (or, more precisely, **pressure–volume work** or ***P–V* work**). A common example of pressure–volume work is the work done by the expanding gases in a cylinder of a car engine as they move a piston. Another example is the work done by expanding gases to lift a rocket from the ground. The amount of expansion work w done can be computed from atmospheric pressure and the volume change that the system undergoes[3]:

$$w = -P\,\Delta V \tag{6.9}$$

The sign of w confirms that the system loses energy by doing work pushing back the atmosphere.

P is the *opposing pressure* against which the piston pushes, and ΔV is the change in the volume of the system (the gas) during the expansion, i.e., $\Delta V = V_{\text{final}} - V_{\text{initial}}$. Because V_{final} is greater than V_{initial}, ΔV must be positive. This makes the expansion work negative.

Heat and work are both ways to transfer energy

In chemistry, *a minus sign on an energy transfer always means that the system loses energy.* Consider what happens whenever work can be done or heat can flow in the system shown in Figure 6.5*b*. If work is negative (as in an expansion) the system loses energy and the surroundings gain it. We say work is done *by* the system. If heat is negative (as in an

[3] $P\,\Delta V$ must have units of energy if it is referred to as work. Work is accomplished when an opposing force, F, is pushed through some distance or length, L. The amount of work done is equal to the strength of the opposing force multiplied by the distance the force is moved.

$$\text{Work} = F \times L$$

Because pressure is force (F) per unit area, and area is simply length squared, L^2, we can write the following equation for pressure.

$$P = \frac{F}{L^2}$$

Volume (or a volume change) has dimensions of length cubed, L^3, so pressure times volume change is

$$P\,\Delta V = \frac{F}{L^2} \times L^3 = F \times L$$

But, from above, $F \times L$ also equals work.

exothermic reaction) the system loses energy and the surroundings gain it. Either event should cause a drop in internal energy. On the other hand, if work is positive (as in a compression), the system gains energy and the surroundings lose it. We say that work is done *on* the system. If heat is positive (as in an endothermic reaction), again, the system will gain energy and the surroundings will lose it. Either positive work or positive heat should cause a positive change in internal energy.

Work and heat are simply alternative ways to transfer energy. Using this sign convention, we can relate the work w and the heat q that go into the system to the internal energy change (ΔE) the system undergoes:

TOOLS
First law of thermodynamics

$$\Delta E = q + w \tag{6.10}$$

In Section 6.2 we pointed out that the internal energy depends only on the current state of the system; we said that *internal energy is a state function.* This statement (together with Equation 6.10, which is a definition of internal energy change) is a statement of the **first law of thermodynamics.** The first law is one of the most subtle and most powerful principles ever devised by science. It implies that we can move energy around in various ways, but we cannot create energy, or destroy it.

ΔE is independent of how a change takes place; it depends only on the state of the system at the beginning and the end of the change. But the values of q and w depend on what happens *between* the initial and final states. Thus, neither q nor w is a state function. Their values depend on the *path* of the change. For example, consider the discharge of an automobile battery by two different paths (see Figure 6.6). Both paths take us between the same two states, one being the fully charged state and the other the fully discharged state. Because ΔE is a state function and because both paths have the same initial and final states, *ΔE must be the same for both paths.* But how about q and w?

In path 1, we simply short the battery by placing a heavy wrench across the terminals. If you have ever done this, even by accident, you know how violent the result can be. Sparks fly and the wrench becomes very hot as the battery quickly discharges. Lots of heat is given off, but *the system does no work* ($w = 0$). All of ΔE appears as heat.

In path 2, we discharge the battery more slowly by using it to operate a motor. Along this path, much of the energy represented by ΔE appears as work (running the motor) and

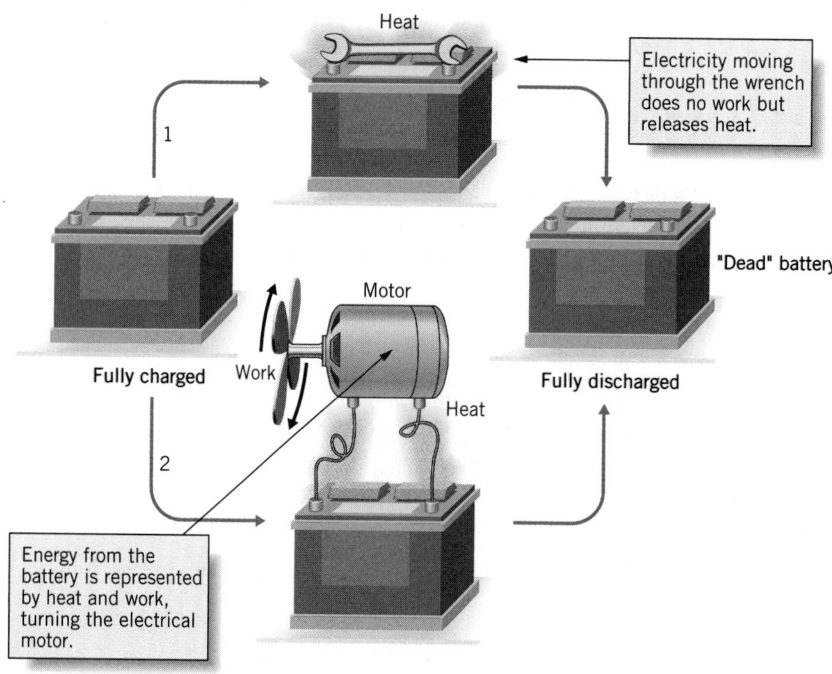

FIG. 6.6 Energy, heat, and work. The complete discharge of a battery along two different paths yields the same total amount of energy, ΔE. However, if the battery is simply shorted with a heavy wrench, as shown in path 1, this energy appears entirely as heat. Path 2 gives part of the total energy as heat, but much of the energy appears as work done by the motor.

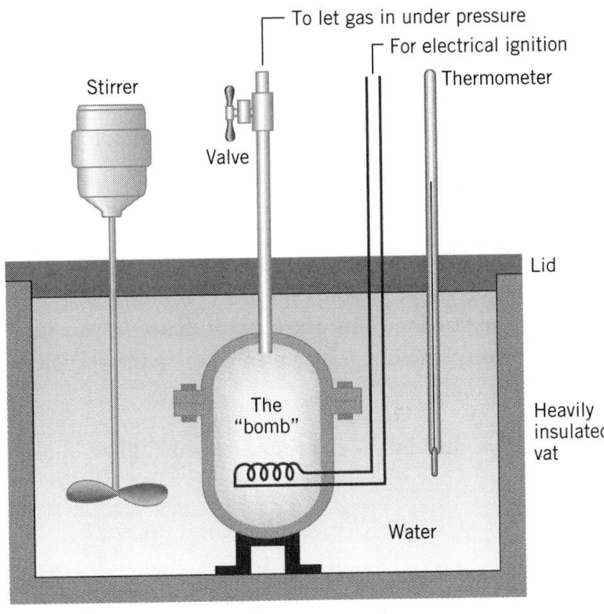

FIG. 6.7 A bomb calorimeter. The water bath is usually equipped with devices for adding or removing heat from the water, thus keeping its temperature constant up to the moment when the reaction occurs in the bomb. The reaction chamber is of fixed volume, so $P\,\Delta V$ must equal zero for reactions in this apparatus.

only a relatively small amount appears as heat (from the friction within the motor and the electrical resistance of the wires).

There are two vital lessons here. The first is that neither q nor w is a state function. Their values depend *entirely* on the path between the initial and final states. Yet, their sum, ΔE, as we said, *is* a state function.

Heats of combustion are determined using constant-volume calorimetry

The heat produced by a combustion reaction is called a **heat of combustion.** Because combustion reactions require oxygen and produce gaseous products, we have to measure heats of combustion in a closed container. Figure 6.7 shows the apparatus that is usually used to determine heats of combustion. The instrument is called a *bomb calorimeter* because the vessel holding the reaction itself resembles a small bomb. The "bomb" has rigid walls, so the change in volume, ΔV, is zero when the reaction occurs. This means, of course, that $P\,\Delta V$ must also be zero, and no expansion work is done, so w in Equation 6.10 is zero. Therefore, the heat of reaction measured in a bomb calorimeter is the **heat of reaction at constant volume, q_v,** and corresponds to ΔE.

$$\Delta E = q_v$$

Food scientists determine dietary calories in foods and food ingredients by burning them in a bomb calorimeter. The reactions that break down foods in the body are complex, but they have the same initial and final states as the combustion reaction for the food.

EXAMPLE 6.4 Bomb Calorimetry

(a) When 1.000 g of olive oil was completely burned in pure oxygen in a bomb calorimeter like the one shown in Figure 6.7, the temperature of the water bath increased from 22.000 °C to 26.049 °C. How many dietary Calories are in olive oil, per gram? The heat capacity of the calorimeter is 9.032 kJ/°C.
(b) Olive oil is almost pure glyceryl trioleate, $C_{57}H_{104}O_6$. The equation for its combustion is

$$C_{57}H_{104}O_6 + 80O_2 \longrightarrow 57CO_2 + 52H_2O$$

What is the change in internal energy, ΔE, for the combustion of one mole of glyceryl trioleate? Assume the olive oil burned in part (a) was pure glyceryl trioleate.

ANALYSIS: In part (a), Equation 6.5 is the tool to compute the heat absorbed by the calorimeter from its temperature change and heat capacity. Then we place a minus sign in front of the result to get the heat released by the combustion reaction.

Since the heat capacity is in kilojoules, the heat released will be in kilojoules, too. We'll need to convert kilojoules to dietary Calories, which are actually kilocalories (kcal). We can use the relationship

$$1 \text{ kcal} = 4.184 \text{ kJ}$$

Part (b) asks for the change in internal energy, ΔE. Bomb calorimetry measures q_v, which is equal to the internal energy change for the reaction. Thus the heat calculated in part (a) is equal to ΔE for combustion of 1.000 g of glyceryl trioleate. The molar mass of glyceryl trioleate is the tool to convert ΔE per gram to ΔE per mole.

SOLUTION: First, let's compute the heat absorbed by the calorimeter when 1.000 g of olive oil is burned, using Equation 6.5:

$$q_{\text{calorimeter}} = C\,\Delta t = (9.032 \text{ kJ/°C}) \times (26.049 \text{ °C} - 22.000 \text{ °C}) = 36.57 \text{ kJ}$$

Changing the algebraic sign gives the heat of combustion of 1.000 g of olive oil, $q_v = -36.57$ kJ. Part (a) asks for the heat in dietary Calories. We convert to kilocalories (kcal), which are equivalent to dietary Calories (Cal):

$$\frac{-36.57 \text{ kJ}}{1.000 \text{ g oil}} \times \frac{1 \text{ kcal}}{4.184 \text{ kJ}} \times \frac{1 \text{ Cal}}{1 \text{ kcal}} = -8.740 \text{ Cal/g oil}$$

We can say "8.740 dietary Calories are released when one gram of olive oil is burned." For part (b), we'll convert the heat produced per gram to the heat produced per mole, using the molar mass of $C_{57}H_{104}O_6$, 885.39 g/mol:

$$\frac{-36.57 \text{ kJ}}{1.000 \text{ g } C_{57}H_{104}O_6} \times \frac{885.39 \text{ g } C_{57}H_{104}O_6}{1 \text{ mol } C_{57}H_{104}O_6} = -3.238 \times 10^4 \text{ kJ/mol}$$

Since this is heat at constant volume, we have $\Delta E = q_v = -3.238 \times 10^4$ kJ for the combustion of 1 mol of $C_{57}H_{104}O_6$.

ARE THE ANSWERS REASONABLE? Let's do some simple approximate arithmetic. In part (a) the temperature change is about 4 °C, so the product $C\,\Delta t$ is about 9 kJ/°C × 4 °C ≈ 36 kJ. We must be careful to give q a negative sign to show that heat is released, so $q_v \approx -36$ kJ. To find q_v in kcal, we have to divide −36 kJ by about 4 (1 kcal ≈ 4 kJ), so q_v is approximately −9 kcal or −9 Cal. The answer obtained above, −8.740 Cal/g, is close to this, so we can be confident it is correct.

For part (b), to calculate q_v in units of kJ/mol we multiply q_v per gram (about −36 kJ) by the molar mass, which is somewhat smaller than 1000 g mol^{-1}. The answer should be somewhat smaller than −36,000 kJ/mol or -3.6×10^4 kJ/mol. Our answer, -3.238×10^4 kJ/mol, agrees with this analysis, so we can be confident it is correct.

Practice Exercise 3: The heat of combustion of methyl alcohol, CH_3OH, is −715 kJ mol^{-1}. When 2.85 g of CH_3OH was burned in a bomb calorimeter, the temperature of the calorimeter changed from 24.05 °C to 29.19 °C. What is the heat capacity of the calorimeter in units of kJ/°C? (Hint: Which equation relates temperature change, energy, and heat capacity?)

Practice Exercise 4: A 1.50 g sample of carbon is burned in a bomb calorimeter which has a heat capacity of 8.930 kJ/°C. The temperature of the water jacket rises from 20.00 °C to 25.51 °C. What is ΔE for the combustion of 1 mole of carbon?

Heats of reactions in solution are determined by constant-pressure calorimetry

Most reactions that are of interest to us do not occur at constant volume. Instead, they run in open containers like test tubes, beakers, and flasks, where they experience the constant pressure of the atmosphere. When reactions run under constant pressure, they may transfer energy as the **heat of reaction at constant pressure, q_p,** *and* as expansion work, w, so to calculate ΔE we need the equation

> Biochemical reactions also occur at constant pressure.

$$\Delta E = q_p + w \tag{6.11}$$

This is inconvenient. If we want to calculate the internal energy change for the reaction, we'll have to measure its volume change and then use Equation 6.9. To avoid this problem, scientists have defined a "corrected" internal energy called **enthalpy,** or ***H*.** Enthalpy is defied by the equation

> From the Greek *en* + *thalpein*, meaning "to heat" or "to warm."

$$H = E + PV$$

At constant pressure,

$$\Delta H = \Delta E + P\,\Delta V$$
$$= (q_p + w) + P\,\Delta V$$

From Equation 6.9, $P\,\Delta V = -w$, so

$$\Delta H = (q_p + w) + (-w)$$
$$\Delta H = q_p \tag{6.12}$$

Like E, H is a state function.

As with internal energy, an **enthalpy change, ΔH,** is defined by the equation

$$\Delta H = H_{\text{final}} - H_{\text{initial}}$$

For a chemical reaction, this can be rewritten as follows.

$$\Delta H = H_{\text{products}} - H_{\text{reactants}} \tag{6.13}$$

Positive and negative values of ΔH have the same interpretation as positive and negative values of ΔE.

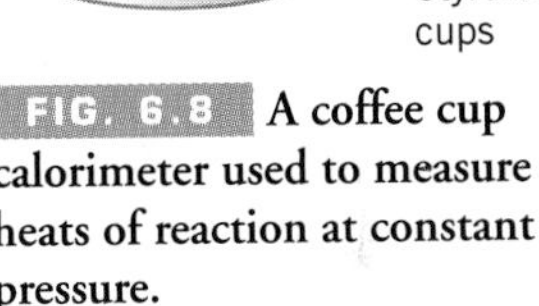

FIG. 6.8 A coffee cup calorimeter used to measure heats of reaction at constant pressure.

Significance of the sign of ΔH:

For an endothermic change, ΔH is positive.

For an exothermic change, ΔH is negative.

TOOLS *Algebraic sign of ΔH*

The difference between ΔH and ΔE for a reaction equals $P\,\Delta V$. This difference can be fairly large for reactions that produce or consume gases, because these reactions can have very large volume changes. For reactions that involve only solids and liquids, though, the values of ΔV are tiny, so ΔE and ΔH for these reactions are nearly identical.

A very simple constant-pressure calorimeter, dubbed the coffee cup calorimeter, is made of two nested and capped cups made of Styrofoam, a very good insulator (Figure 6.8). A reaction occurring in such a calorimeter exchanges very little heat with the surroundings, particularly if the reaction is fast. The temperature change is rapid and easily measured. We can use Equation 6.5 to find the heat of reaction, if we have determined the heat capacity of the calorimeter and its contents before the reaction. The Styrofoam cup and the thermometer absorb only a tiny amount of heat, and we can usually ignore them in our calculations.

> Research-grade calorimeters have greater accuracy and precision than the coffee cup calorimeter.

EXAMPLE 6.5
Constant-Pressure Calorimetry

The reaction of hydrochloric acid and sodium hydroxide is very rapid and exothermic. The equation is

$$HCl(aq) + NaOH(aq) \longrightarrow NaCl(aq) + H_2O$$

In one experiment, a student placed 50.0 mL of 1.00 *M* HCl at 25.5 °C in a coffee cup calorimeter. To this was added 50.0 mL of 1.00 *M* NaOH solution also at 25.5 °C. The mixture was stirred, and the temperature quickly increased to a maximum of 32.2 °C. What is ΔH expressed in kJ per mole of HCl? Because the solutions are relatively dilute, we can assume that their specific heats are close to that of water, 4.18 J g^{-1} °C^{-1}. The density of 1.00 *M* HCl is 1.02 g mL^{-1} and that of 1.00 *M* NaOH is 1.04 g mL^{-1}. (We will neglect the heat lost to the Styrofoam itself, to the thermometer, or to the surrounding air.)

ANALYSIS: The reaction is taking place at constant pressure, so the heat we're calculating is q_p, which equals ΔH. We're given the specific heat of the solutions, so to calculate q_p we also need the system's total mass and the temperature change. (The tool used here will be Equation 6.7, $q = ms\,\Delta t$.) The mass here refers to the *total* grams of the combined solutions, but we've been given volumes. So we have to use their densities as a tool to calculate their masses, which you learned to do in Chapter 1.

SOLUTION: For the HCl solution, the density is 1.02 g mL^{-1} and we have

$$\text{Mass (HCl)} = \frac{1.02 \text{ g}}{1.00 \cancel{\text{mL}}} \times 50.0 \cancel{\text{mL}} = 51.0 \text{ g}$$

Similarly, for the NaOH solution, the density is 1.04 g mL^{-1} and

$$\text{Mass (NaOH)} = \frac{1.04 \text{ g}}{1.00 \cancel{\text{mL}}} \times 50.0 \cancel{\text{mL}} = 52.0 \text{ g}$$

The mass of the final solution is thus the sum, 103.0 g.

The reaction changes the system's temperature by ($t_{\text{final}} - t_{\text{initial}}$), so

$$\Delta t = 32.2 \text{ °C} - 25.5 \text{ °C} = 6.7 \text{ °C}$$

Now we can calculate the heat absorbed by the solution using Equation 6.7.

$$\begin{aligned} \text{Heat absorbed by the solution} &= \text{mass} \times \text{specific heat} \times \Delta t \\ &= 103.0 \text{ g} \times 4.18 \text{ J g}^{-1} \text{ °C}^{-1} \times 6.7 \text{ °C} \\ &= 2.9 \times 10^3 \text{ J} = 2.9 \text{ kJ} \end{aligned}$$

Changing the sign gives the heat evolved by the reaction, $q_p = -2.9$ kJ. But this is q_p specifically for the mixture prepared; the problem calls for kilojoules *per mole* of HCl. The tool we use to calculate the number of moles of HCl is the molarity, which provides a conversion factor connecting volume and moles. In 50.0 mL of HCl solution (0.0500 L) we have

$$0.0500 \cancel{\text{L HCl soln}} \times \frac{1.00 \text{ mol HCl}}{1.00 \cancel{\text{L HCl soln}}} = 0.0500 \text{ mol HCl}$$

The neutralization of 0.0500 mol of acid has $q_p = -2.9$ kJ. To calculate the heat released per mole, ΔH, we simply take the ratio of joules to moles.

$$\Delta H \text{ per mole of HCl} = \frac{-2.9 \text{ kJ}}{0.0500 \text{ mol HCl}} = -58 \text{ kJ mol}^{-1} \text{ HCl}$$

Thus, ΔH for neutralizing HCl by NaOH is -58 kJ mol^{-1} HCl.

IS THE ANSWER REASONABLE? Let's first review the logic of the steps we used. *Notice how the logic is driven by definitions, which carry specific units.* Working backward, knowing that we want units of kilojoules per mole in the answer, we must calculate separately the number of moles of acid neutralized and the number of kilojoules that evolved. The latter will emerge when we multiply the solution's mass (g) and specific heat ($J\ g^{-1}\ °C^{-1}$) by the degrees of temperature increase (°C). A simple change from joules to kilojoules is also required.

We can check the answer with some simplified arithmetic. Let's start with the mass of the solution. Because the densities are close to $1\ g\ mL^{-1}$, each solution has a mass of about 50 g, so the mixture weighs about 100 g.

The heat evolved will equal the specific heat times the mass times the temperature change. The specific heat is around $4\ J\ g^{-1}\ °C^{-1}$, the mass is about 100 g, and the temperature change is about 7 °C. The heat evolved will be approximately 4×700, or 2800 J. That's equal to 2.8×10^3 J or 2.8 kJ so $q_p \approx -2.8$ kJ. Our answer of -2.9 kJ is certainly reasonable. This much heat is associated with neutralizing 0.05 mol HCl, so the heat per mole is -2.9 kJ divided by 0.05 mol, or $-58\ kJ\ mol^{-1}$.

Practice Exercise 5: For the preceding worked example, calculate ΔH per mole of NaOH. (Hint: How many moles of NaOH were neutralized in the reaction?)

Practice Exercise 6: When pure sulfuric acid dissolves in water, much heat is given off. To measure it, 175 g of water was placed in a coffee cup calorimeter and chilled to 10.0 °C. Then 4.90 g of sulfuric acid (H_2SO_4), also at 10.0 °C, was added, and the mixture was quickly stirred with a thermometer. The temperature rose rapidly to 14.9 °C. Assume that the value of the specific heat of the solution is $4.18\ J\ g^{-1}\ °C^{-1}$, and that the solution absorbs all the heat evolved. Calculate the heat evolved in kilojoules by the formation of this solution. (Remember to use the *total* mass of the solution, the water plus the solute.) Calculate also the heat evolved *per mole* of sulfuric acid.

6.6 THERMOCHEMICAL EQUATIONS ARE CHEMICAL EQUATIONS THAT QUANTITATIVELY INCLUDE HEAT

The amount of heat a reaction produces or absorbs depends on the number of moles of reactants we combine. It makes sense that if we burn two moles of carbon, we're going to get twice as much heat as we would if we had burned one mole. For heats of reaction to have meaning, we must describe the **system** completely. Our description must include amounts and concentrations of reactants, amounts and concentrations of products, temperature, and pressure, because all of these things can influence heats of reaction.

Unless we specify otherwise, whenever we write ΔH we mean ΔH for the *system*, not the surroundings.

Chemists have agreed to a set of **standard states** to make it easier to report and compare heats of reaction. Most thermochemical data are reported for a pressure of 1 bar, or (for substances in aqueous solution) a concentration of 1 *M*. A temperature of 25 °C (298 K) is often specified as well, although temperature is not part of the definition of standard states in thermochemistry.

$\Delta H°$ is the enthalpy change for a reaction with all reactants and products at standard state

The **standard heat of reaction ($\Delta H°$)** is the value of ΔH for a reaction occurring under standard conditions and involving the actual numbers of *moles* specified by the coefficients of the equation. To show that ΔH is for *standard* conditions, a degree sign is added to ΔH to make $\Delta H°$ (pronounced "delta H naught" or "delta H zero"). The units of $\Delta H°$ are normally kilojoules.

To illustrate clearly what we mean by $\Delta H°$, let us use the reaction between gaseous nitrogen and hydrogen that produces gaseous ammonia.

$$N_2(g) + 3H_2(g) \longrightarrow 2NH_3(g)$$

When specifically 1.000 mol of N_2 and 3.000 mol of H_2 react to form 2.000 mol of NH_3 at 25 °C and 1 bar, the reaction releases 92.38 kJ. Hence, for the reaction *as given by the above equation*, $\Delta H° = -92.38$ kJ. Often the enthalpy change is given immediately after the equation; for example,

$$N_2(g) + 3H_2(g) \longrightarrow 2NH_3(g) \qquad \Delta H° = -92.38 \text{ kJ}$$

Thermochemical equations

An equation that also shows the value of $\Delta H°$ is called a **thermochemical equation.** It always gives the physical states of the reactants and products, and *its $\Delta H°$ value is true only when the coefficients of the reactants and products are taken to mean moles of the corresponding substances.* The above equation, for example, shows a release of 92.38 kJ if *two* moles of NH_3 form. If we were to make twice as much or 4.000 mol of NH_3 (from 2.000 mol of N_2 and 6.000 mol of H_2), then twice as much heat (184.8 kJ) would be released. On the other hand, if only 0.5000 mol of N_2 and 1.500 mol of H_2 were to react to form 1.000 mole of NH_3, then only half as much heat (46.19 kJ) would be released. To describe the various *sizes* of the reactions just described, we write the following thermochemical equations.

$$N_2(g) + 3H_2(g) \longrightarrow 2NH_3(g) \qquad \Delta H° = -92.38 \text{ kJ}$$
$$2N_2(g) + 6H_2(g) \longrightarrow 4NH_3(g) \qquad \Delta H° = -184.8 \text{ kJ}$$
$$\tfrac{1}{2}N_2(g) + \tfrac{3}{2}H_2(g) \longrightarrow NH_3(g) \qquad \Delta H° = -46.19 \text{ kJ}$$

Because the coefficients of a thermochemical equation always mean *moles*, not molecules, we may use fractional coefficients. (In the kinds of equations you've seen up till now, fractional coefficients were not allowed because we cannot have fractions of *molecules*, but we can have fractions of moles in a thermochemical equation.)

You must write down physical states for all reactants and products in thermochemical equations. The combustion of 1 mol of methane, for example, has different values of $\Delta H°$ if the water produced is in its liquid or its gaseous state.

$$CH_4(g) + 2O_2(g) \longrightarrow CO_2(g) + 2H_2O(l) \qquad \Delta H° = -890.5 \text{ kJ}$$
$$CH_4(g) + 2O_2(g) \longrightarrow CO_2(g) + 2H_2O(g) \qquad \Delta H° = -802.3 \text{ kJ}$$

(The difference in $\Delta H°$ values for these two reactions is the amount of energy that would be released by the physical change of 2 mol of water vapor at 25 °C to 2 mol of liquid water at 25 °C.)

EXAMPLE 6.6
Writing a Thermochemical Equation

The following thermochemical equation is for the exothermic reaction of hydrogen and oxygen that produces water.

$$2H_2(g) + O_2(g) \longrightarrow 2H_2O(l) \qquad \Delta H° = -571.8 \text{ kJ}$$

What is the thermochemical equation for this reaction when it is conducted to produce 1.000 mol H_2O?

ANALYSIS: The given equation is for 2.000 mol of H_2O, and any changes in the coefficient for water must be made identically to all other coefficients, *as well as to the value of* $\Delta H°$.

SOLUTION: We divide everything by 2, to obtain

$$H_2(g) + \tfrac{1}{2}O_2(g) \longrightarrow H_2O(l) \qquad \Delta H° = -285.9 \text{ kJ}$$

IS THE ANSWER REASONABLE? Compare the equation just found with the initial equation to see that the coefficients and the value of $\Delta H°$ are all divided by 2.

Practice Exercise 7: The combustion of methane follows the thermochemical equation

$$CH_4(g) + 2O_2(g) \longrightarrow CO_2(g) + 2H_2O(l) \qquad \Delta H° = -890.5 \text{ kJ}$$

Write the thermochemical equation for the combustion of methane when 1/2 mol of $H_2O(l)$ is formed. [Hint: What do you have to do to the coefficient of $H_2O(l)$ to give $\frac{1}{2}H_2O(l)$?]

Practice Exercise 8: What is the thermochemical equation for the formation of 2.500 mol of $H_2O(l)$ from $H_2(g)$ and $O_2(g)$?

6.7 Thermochemical Equations Can Be Combined because Enthalpy Is a State Function

We mentioned earlier that enthalpy is a state function. This important fact permits us to calculate heats of reaction for reactions we cannot actually carry out in the laboratory. You will see soon that to accomplish this, we will combine known thermochemical equations, which usually requires that we manipulate them in some way.

Both the coefficients and direction of thermochemical equations can be adjusted

In the last section you learned that if we change the size of a reaction by multiplying or dividing the coefficients of a thermochemical equation by some factor, the value of $\Delta H°$ is multiplied by the same factor. Another way to manipulate a thermochemical equation is to change its direction. For example, the thermochemical equation for the combustion of carbon in oxygen to give carbon dioxide is

$$C(s) + O_2(g) \longrightarrow CO_2(g) \qquad \Delta H° = -393.5 \text{ kJ}$$

The reverse reaction, which is extremely difficult to carry out, would be the decomposition of carbon dioxide to carbon and oxygen. The law of conservation of energy requires that its value of $\Delta H°$ equals +393.5 kJ.

$$CO_2(g) \longrightarrow C(s) + O_2(g) \qquad \Delta H° = +393.5 \text{ kJ}$$

In effect, these two thermochemical equations tell us that the combustion of carbon is exothermic (as indicated by the negative sign of $\Delta H°$) and that the reverse reaction is endothermic. The same amount of energy is involved in both reactions; it is just the direction of energy flow that is different. The lesson to learn here is that *we can reverse any thermochemical equation as long as we change the sign of its* $\Delta H°$.

Enthalpy changes depend only on initial and final states, not on the path between them

What we're leading up to is a method for combining known thermochemical equations in a way that will allow us to calculate an unknown $\Delta H°$ for some other reaction. Let's revisit the combustion of carbon to see how this works.

We can imagine two paths leading from 1 mole each of carbon and oxygen to 1 mol of carbon dioxide.

One-Step Path. Let C and O_2 react to give CO_2 directly.

$$C(s) + O_2(g) \longrightarrow CO_2(g) \qquad \Delta H° = -393.5 \text{ kJ}$$

Two-Step Path. Let C and O_2 react to give CO, and then let CO react with more O_2 to give CO_2.

$$\text{Step 1:} \quad C(s) + \tfrac{1}{2}O_2(g) \longrightarrow CO(g) \qquad \Delta H° = -110.5 \text{ kJ}$$

$$\text{Step 2:} \quad CO(g) + \tfrac{1}{2}O_2(g) \longrightarrow CO_2(g) \qquad \Delta H° = -283.0 \text{ kJ}$$

Overall, the two-step path consumes 1 mol each of C and O_2 to make 1 mol of CO_2, just like the one-step path. The initial and final states for the two routes to CO_2, in other words, are identical.

Because $\Delta H°$ is a state function dependent only on the initial and final states and is independent of path, the values of $\Delta H°$ for both routes should be identical. We can see that this is exactly true simply by adding the equations for the two-step path and comparing the result with the equation for the single step.

$$\text{Step 1:} \quad C(s) + \tfrac{1}{2}O_2(g) \longrightarrow CO(g) \qquad \Delta H° = -110.5 \text{ kJ}$$

$$\text{Step 2:} \quad CO(g) + \tfrac{1}{2}O_2(g) \longrightarrow CO_2(g) \qquad \Delta H° = -283.0 \text{ kJ}$$

$$CO(g) + C(s) + O_2(g) \longrightarrow CO_2(g) + CO(g) \qquad \Delta H° = -110.5 \text{ kJ} + (-283.0 \text{ kJ})$$

$$\Delta H° = -393.5 \text{ kJ}$$

The equation resulting from adding Steps 1 and 2 has "CO(g)" appearing *identically* on opposite sides of the arrow. We can cancel them to obtain the net equation. *Such a cancellation is permitted only when both the formula and the physical state of a species are identical on opposite sides of the arrow.* The net thermochemical equation for the two-step process, therefore, is

$$C(s) + O_2(g) \longrightarrow CO_2(g) \qquad \Delta H° = -393.5 \text{ kJ}$$

The results, chemically and thermochemically, are thus identical for both routes to CO_2.

Enthalpy diagrams show alternative pathways between initial and final states

The energy relationships among the alternative pathways for the same overall reaction are clearly seen using a graphical construction called an **enthalpy diagram.** Figure 6.9 is an enthalpy diagram for the formation of CO_2 from C and O_2. Each horizontal line corresponds to a certain total amount of enthalpy, which we can't actually measure. We *can* measure differences in enthalpy, however, and that's what we use the diagram for. Lines higher up the enthalpy scale represent larger amounts of enthalpy, so going from a lower line to a higher line corresponds to an increase in enthalpy and a positive value for $\Delta H°$ (an endothermic change). The size of $\Delta H°$ is represented by the vertical distance between the two lines. Likewise, going from a higher line to a lower one represents a decrease in enthalpy and a negative value for $\Delta H°$ (an exothermic change).

Notice in Figure 6.9 that the line for the absolute enthalpy of $C(s) + O_2(g)$, taken as the sum, is above the line for the final product, CO_2. A horizontal line represents the *sum*

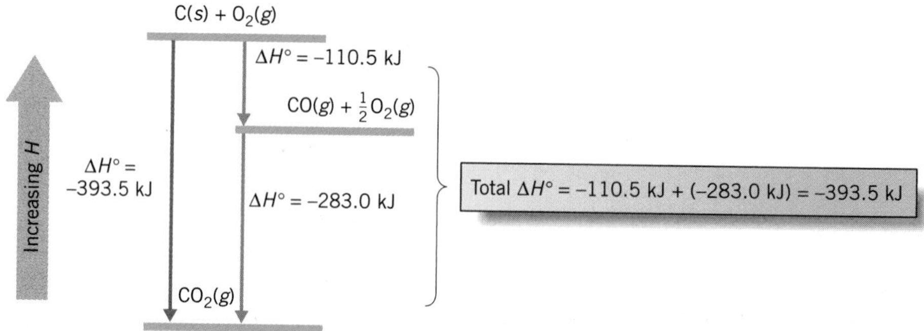

FIG. 6.9 An enthalpy diagram for the formation of $CO_2(g)$ from its elements by two different paths. On the left is path 1, the direct conversion of $C(s)$ and $O_2(g)$ to $CO_2(g)$. On the right, path 2 shows two shorter, downward pointing arrows. The first step of path 2 takes the elements to $CO(g)$, and the second step takes $CO(g)$ to $CO_2(g)$. The overall enthalpy change is identical for both paths, as it must be, because enthalpy is a state function.

of the enthalpies of all of the substances on the line, *in the physical states specified.* Near the left side of Figure 6.9, we see a long downward arrow that connects the enthalpy level for the reactants, $C(s) + O_2(g)$, to that of the final product, $CO_2(g)$. This is for the one-step path, the direct path.

On the right we have the two-step path. Here, Step 1 brings us to an intermediate enthalpy level corresponding to the intermediate products, $CO(g) + \frac{1}{2}O_2(g)$. These include the mole of $CO(g)$ made in the first step plus $\frac{1}{2}$ mol of O_2 which has not reacted. Then Step 2 occurs to give the final product. What the enthalpy diagram shows is that the total decrease in enthalpy is the same regardless of the path.

EXAMPLE 6.7
Preparing an Enthalpy Diagram

Hydrogen peroxide, H_2O_2, decomposes into water and oxygen by the following equation.

$$H_2O_2(l) \longrightarrow H_2O(l) + \tfrac{1}{2}O_2(g)$$

Construct an enthalpy diagram for the following two reactions of hydrogen and oxygen, and use the diagram to determine the value of $\Delta H°$ for the decomposition of hydrogen peroxide.

$$H_2(g) + O_2(g) \longrightarrow H_2O_2(l) \qquad \Delta H° = -188 \text{ kJ}$$

$$H_2(g) + \tfrac{1}{2}O_2(g) \longrightarrow H_2O_2(l) \qquad \Delta H° = -286 \text{ kJ}$$

ANALYSIS: The two given reactions are exothermic, so their values of $\Delta H°$ will be associated with downward pointing arrows. The highest enthalpy level, therefore, must be for the elements themselves. The lowest level must be for the product of the reaction with the largest negative $\Delta H°$, the formation of water by the second equation. The enthalpy level for the H_2O_2, formed by a reaction with the less negative $\Delta H°$, must be in between.

SOLUTION: First, let's diagram the two reactions having known values of $\Delta H°$. Notice that the first reaction requires an entire mole of O_2, whereas the second needs only 1/2 mol of O_2. On the top line we'll include enough O_2 for the first reaction. This leaves $\frac{1}{2}O_2(g)$ unreacted when we form $H_2O(l)$, as indicated on the bottom line.

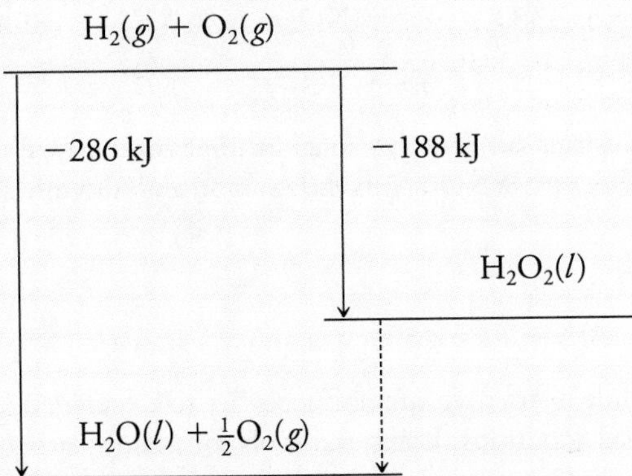

Notice that the gap on the right side (represented by the dotted arrow) corresponds exactly to the change of $H_2O_2(l)$ into $H_2O(l) + \frac{1}{2}O_2(g)$, the reaction for which $\Delta H°$ is sought. This enthalpy separation corresponds to the difference between the enthalpy changes -286 kJ and -188 kJ. Thus for the decomposition of $H_2O_2(l)$ into $H_2O(l)$ and $\frac{1}{2}O_2(g)$,

$$\Delta H° = [-286 \text{ kJ} - (-188 \text{ kJ})] = -98 \text{ kJ}$$

Our completed enthalpy diagram is shown on the next page, indicating that $\Delta H°$ for the decomposition of 1 mol of $H_2O_2(l)$ is -98 kJ.

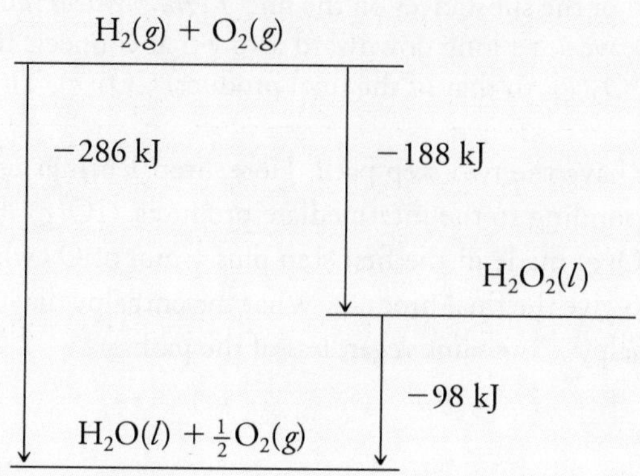

IS THE ANSWER REASONABLE? First be sure that the chemical formulas are arrayed properly on the horizontal lines in the enthalpy diagram. The arrows pointing downward should indicate negative values of $\Delta H°$. These arrows must be consistent *in direction* with the given equations (in other words, they must point from reactants to products). Finally, notice that the total amount of energy associated with the two-step process, namely, (−188 kJ) plus (−98 kJ) equals −286 kJ, the energy for the one-step process.

Practice Exercise 9: Two oxides of copper can be made from copper by the following reactions.

$$2Cu(s) + O_2(g) \longrightarrow 2CuO(s) \qquad \Delta H° = -310 \text{ kJ}$$
$$2Cu(s) + \tfrac{1}{2}O_2(g) \longrightarrow Cu_2O(s) \qquad \Delta H° = -169 \text{ kJ}$$

Using these data, construct an enthalpy diagram that can be used to find $\Delta H°$ for the reaction $Cu_2O(s) + \frac{1}{2}O_2(g) \rightarrow 2CuO(s)$. Is the reaction endothermic or exothermic? (Hint: Remember, an arrow points down for a negative $\Delta H°$ and up for a positive $\Delta H°$.)

Practice Exercise 10: Consider the following thermochemical equations:

$$\tfrac{1}{2}N_2(g) + \tfrac{1}{2}O_2(g) \longrightarrow NO(g) \qquad \Delta H° = +90.4 \text{ kJ}$$
$$NO(g) + \tfrac{1}{2}O_2(g) \longrightarrow NO_2(g) \qquad \Delta H° = -56.6 \text{ kJ}$$

Using these data, construct an enthalpy diagram that can be used to find $\Delta H°$ for the reaction $\frac{1}{2}N_2(g) + O_2(g) \longrightarrow NO_2(g)$. Is the reaction endothermic or exothermic?

Predict any heat of reaction using Hess's law

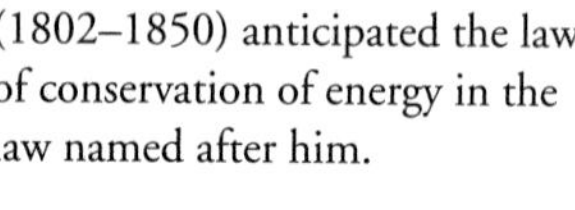
Germain Henri Hess (1802–1850) anticipated the law of conservation of energy in the law named after him.

Enthalpy diagrams, while instructive, are not necessary to calculate $\Delta H°$ for a reaction from known thermochemical equations. Using the tools for manipulating equations, we ought to be able to calculate $\Delta H°$ values simply by algebraic summing. G. H. Hess was the first to realize this, so the associated law is called **Hess's law of heat summation** or, simply, **Hess's law.**

Hess's Law. The value of $\Delta H°$ for any reaction that can be written in steps equals the sum of the values of $\Delta H°$ of each of the individual steps.

For example, suppose we add the two equations in Practice Exercise 10 and then cancel anything that's the same on both sides.

$$\tfrac{1}{2}N_2(g) + \tfrac{1}{2}O_2(g) \longrightarrow NO(g)$$

$$NO(g) + \tfrac{1}{2}O_2(g) \longrightarrow NO_2(g)$$

$$\tfrac{1}{2}N_2(g) + \cancel{NO(g)} + \underbrace{\tfrac{1}{2}O_2(g) + \tfrac{1}{2}O_2(g)}_{O_2(g)} \longrightarrow \cancel{NO(g)} + NO_2(g)$$

Notice that we've combined the two $\frac{1}{2}O_2(g)$ to give $O_2(g)$. Rewriting the equation gives

$$\tfrac{1}{2}N_2(g) + O_2(g) \longrightarrow NO_2(g)$$

This is the equation for which we were asked to determine $\Delta H°$. According to Hess's law, we should be able to obtain $\Delta H°$ for this reaction by simply adding the $\Delta H°$ values for the equations we've added.

$$\Delta H° = (+90.4\text{ kJ}) + (-56.6\text{ kJ}) = +33.8\text{ kJ}$$

This was the answer to Practice Exercise 10. Thus, we've added the two given equations to obtain the desired equation and we've added their $\Delta H°$ values to obtain the desired $\Delta H°$.

The chief use of Hess's law is to calculate the enthalpy change for a reaction for which such data cannot be determined experimentally or are otherwise unavailable. Often this requires that we manipulate equations, so let's recapitulate the few rules that govern these operations.

Manipulating thermochemical equations

Rules for Manipulating Thermochemical Equations

1. When an equation is reversed—written in the opposite direction—the sign of $\Delta H°$ must also be reversed.[4]
2. Formulas canceled from both sides of an equation must be for the substance in identical physical states.
3. If all the coefficients of an equation are multiplied or divided by the same factor, the value of $\Delta H°$ must likewise be multiplied or divided by that factor.

EXAMPLE 6.8
Using Hess's Law

Carbon monoxide is often used in metallurgy to remove oxygen from metal oxides and thereby give the free metal. The thermochemical equation for the reaction of CO with iron(III) oxide, Fe_2O_3, is

$$Fe_2O_3(s) + 3CO(g) \longrightarrow 2Fe(s) + 3CO_2(g) \qquad \Delta H° = -26.7\text{ kJ}$$

Use this equation and the equation for the combustion of CO,

$$CO(g) + \tfrac{1}{2}O_2(g) \longrightarrow CO_2(g) \qquad \Delta H° = -283.0\text{ kJ}$$

to calculate the value of $\Delta H°$ for the following reaction.

$$2Fe(s) + \tfrac{3}{2}O_2(g) \longrightarrow Fe_2O_3(s)$$

ANALYSIS: The tools will be Hess's law and the rules for manipulating thermochemical equations. In this problem, we cannot simply add the two given equations, because this will not produce the equation we want. We first have to manipulate the equations so that when we add them we will get the target equation. In performing these adjustments, we have to keep our eye

[4] To illustrate, the reverse of the equation

$$C(s) + O_2(g) \longrightarrow CO_2(g) \qquad \Delta H° = -394\text{ kJ}$$

is the following equation:

$$CO_2(g) \longrightarrow C(s) + O_2(g) \qquad \Delta H° = +394\text{ kJ}$$

on our target—the final desired equation. When we've done this, we can then add the adjusted $\Delta H°$ values to obtain the desired $\Delta H°$.

SOLUTION: We can manipulate the two given equations as follows.

Step 1. We begin by trying to get the iron atoms to come out right. The target equation must have 2Fe on the *left*, but the first equation above has 2Fe to the *right* of the arrow. To move it to the left, we reverse the *entire* equation, remembering also to reverse the sign of $\Delta H°$. This puts Fe_2O_3 to the right of the arrow, which is where it has to be after we add the adjusted equations. After these manipulations, and reversing the sign of $\Delta H°$, we have

$$2Fe(s) + 3CO_2(g) \longrightarrow Fe_2O_3(s) + 3CO(g) \qquad \Delta H° = +26.7 \text{ kJ}$$

Step 2. There must be $\frac{3}{2}O_2$ on the left, and we must be able to cancel *three* CO and *three* CO_2 when the equations are added. If we multiply the second of the equations given above by 3, we will obtain the necessary coefficients. We must also multiply the value of $\Delta H°$ of this equation by 3, because three times as many moles of substances are now involved in the reaction. When we have done this, we have

$$3CO(g) + \tfrac{3}{2}O_2(g) \longrightarrow 3CO_2(g) \qquad \Delta H° = 3 \times (-283.0 \text{ kJ}) = -849.0 \text{ kJ}$$

Let's now put our two equations together and find the answer.

$$2Fe(s) + \cancel{3CO_2(g)} \longrightarrow Fe_2O_3(s) + \cancel{3CO(g)} \qquad \Delta H° = +26.7 \text{ kJ}$$

$$\cancel{3CO(g)} + \tfrac{3}{2}O_2(g) \longrightarrow \cancel{3CO_2(g)} \qquad \Delta H° = -849.0 \text{ kJ}$$

$$\text{Sum} \qquad 2Fe(s) + \tfrac{3}{2}O_2(g) \longrightarrow Fe_2O_3(s) \qquad \Delta H° = -822.3 \text{ kJ}$$

Thus, the value of $\Delta H°$ for the oxidation of 2 mol Fe(*s*) to 1 mol $Fe_2O_3(s)$ is -822.3 kJ. (The reaction is *very* exothermic.)

IS THE ANSWER REASONABLE? There is no quick "head check." But for each step, double-check that you have heeded the rules for manipulating thermochemical equations. Also, be careful with the algebraic signs when adding the $\Delta H°$ values.

Practice Exercise 11: Consider the thermochemical equation

$$H_2(g) + \tfrac{1}{2}O_2(g) \longrightarrow H_2O(l) \qquad \Delta H° = -285.9 \text{ kJ}$$

What is the value of $\Delta H°$ for the following reaction? (Hint: How does altering the coefficients and changing direction change $\Delta H°$?)

$$3H_2O(l) \longrightarrow 3H_2(g) + \tfrac{3}{2}O_2(g)$$

Practice Exercise 12: Given the following thermochemical equations:

$$2NO(g) + O_2(g) \longrightarrow 2NO_2(g) \qquad \Delta H° = -113.2 \text{ kJ}$$

$$2N_2O(g) + 3O_2(g) \longrightarrow 4NO_2(g) \qquad \Delta H° = -28.0 \text{ kJ}$$

calculate $\Delta H°$ for this reaction: $N_2O(g) + \frac{1}{2}O_2(g) \longrightarrow 2NO(g)$

Practice Exercise 13: Ethanol, C_2H_5OH, is made industrially by the reaction of water with ethylene, C_2H_4. Calculate the value of $\Delta H°$ for the reaction

$$C_2H_4(g) + H_2O(l) \longrightarrow C_2H_5OH(l)$$

given the following thermochemical equations.

$$C_2H_4(g) + 3O_2(g) \longrightarrow 2CO_2(g) + 2H_2O(l) \qquad \Delta H° = -1411.1 \text{ kJ}$$

$$C_2H_5OH(l) + 3O_2(g) \longrightarrow 2CO_2(g) + 3H_2O(l) \qquad \Delta H° = -1367.1 \text{ kJ}$$

6.8 TABULATED STANDARD HEATS OF REACTION CAN BE USED TO PREDICT ANY HEAT OF REACTION USING HESS'S LAW

Enormous databases of thermochemical equations have been compiled to allow the calculation of any heat of reaction, using Hess's law. The most frequently tabulated reactions are combustion reactions, phase changes, and formation reactions. We'll discuss the enthalpy changes that accompany phase changes in Chapter 11.

The **standard heat of combustion, ΔH_c°,** of a substance is the amount of heat released when one mole of a fuel substance is completely burned in pure oxygen gas, with all reactants and products brought to 25 °C and 1 bar of pressure. All carbon in the fuel becomes carbon dioxide gas, and all the fuel's hydrogen becomes liquid water. *Combustion reactions are always exothermic*, so ΔH_c° is always negative.

EXAMPLE 6.9
Writing an Equation for a Standard Heat of Combustion

How many moles of carbon dioxide gas are produced by a gas-fired power plant for every 1.00 MJ (megajoule) of energy it produces? The plant burns methane, $CH_4(g)$, for which ΔH_c° is −890 kJ/mol.

ANALYSIS: We can restate the problem as an equivalency relation:

$$1.00\text{ MJ released} \Leftrightarrow ?\text{ mol } CO_2(g)$$

We need to link moles of carbon dioxide with megajoules of heat produced. The tool we will use is the balanced thermochemical equation. This is a combustion reaction, and you learned in Chapter 5 that when a hydrocarbon burns, the products are CO_2 and H_2O. The first step in the solution will be to write a balanced chemical equation for the reaction of CH_4 with O_2. Because ΔH_c° gives the heat evolved *per mole* of CH_4, we will have to be sure the coefficient of CH_4 is one. In the equation CH_4, O_2, and CO_2 will be gases and H_2O will be a liquid (because that's the standard form of H_2O at 25 °C.)

Once we have the thermochemical equation, the coefficients will let us relate moles of CO_2 to kilojoules of heat released. To relate megajoules with kilojoules, our tools will be the SI prefixes *kilo-* and *mega-*.

$$1\text{ MJ} = 10^6\text{ J}$$
$$1\text{ kJ} = 10^3\text{ J}$$

SOLUTION: First, we write and balance the thermochemical equation, remembering that ΔH° is ΔH_c°.

$$CH_4(g) + 2O_2(g) \longrightarrow CO_2(g) + 2H_2O(l) \qquad \Delta H_c^\circ = -890\text{ kJ}$$

So, 1 mol CO_2 is formed for every 890 kJ of heat that is released. The number of moles of CO_2 released for production of 1.0 MJ of energy is

$$1.00\text{ MJ} \times \frac{10^6\text{ J}}{1\text{ MJ}} \times \frac{1\text{ kJ}}{10^3\text{ J}} \times \frac{1\text{ mol } CO_2}{890\text{ kJ}} = 1.12\text{ mol } CO_2$$

IS THE ANSWER REASONABLE? When 1 mol CH_4 burns, 890 kJ is released. This is just a little less than 1000 kJ (or 1 MJ), so the amount of CH_4 needed to release 1 MJ should be a bit larger than 1 mol. Because 1 mol $CH_4 \Leftrightarrow$ 1 mol CO_2, our answer, 1.12 mol CO_2, is reasonable.

Practice Exercise 14: The heat of combustion, ΔH_c°, of acetone, C_3H_6O, is 1790.4 kJ/mol. How many kilojoules of heat are evolved in the combustion of 12.5 g of acetone? (Hint: Calculate the molar mass of acetone.)

Practice Exercise 15: n-Octane, $C_8H_{18}(l)$, has a standard heat of combustion of 5450.5 kJ/mol. A 15 gallon automobile fuel tank could hold about 480 moles of n-octane. How much heat could be produced by burning a full tank of n-octane?

Forming one mole of a compound from its elements involves the heat of formation

The **standard enthalpy of formation, ΔH_f°,** of a substance, also called its **standard heat of formation,** is the amount of heat absorbed or evolved when specifically *one mole* of the substance is formed at 25 °C and 1 bar from its elements in their *standard states.* An element is in its **standard state** when it is at 25 °C and 1 bar and in its most stable form and physical state (solid, liquid, or gas). Oxygen, for example, is in its standard state only as a gas at 25 °C and 1 bar and only as O_2 molecules, not as O atoms or O_3 (ozone) molecules. Carbon must be in the form of graphite, not diamond, to be in its standard state, because the graphite form of carbon is the more stable form under standard conditions.

Older thermochemical data used a standard pressure of 1 atm, not 1 bar. Because 1 atm = 1.01325 bar, the new definition made little difference in tabulated heats of reaction.

Standard enthalpies of formation for a variety of substances are given in Table 6.2, and a more extensive table of standard heats of formation can be found in Appendix C. Notice in particular that *all values of ΔH_f° for the elements in their standard states are zero.* (Forming

TABLE 6.2 Standard Enthalpies of Formation of Typical Substances

Substance	ΔH_f° (kJ mol^{-1})	Substance	ΔH_f° (kJ mol^{-1})
$Ag(s)$	0	$H_2O_2(l)$	−187.6
$AgBr(s)$	−100.4	$HBr(g)$	−36
$AgCl(s)$	−127.0	$HCl(g)$	−92.30
$Al(s)$	0	$HI(g)$	26.6
$Al_2O_3(s)$	−1669.8	$HNO_3(l)$	−173.2
$C(s, \text{graphite})$	0	$H_2SO_4(l)$	−811.32
$CO(g)$	−110.5	$HC_2H_3O_2(l)$	−487.0
$CO_2(g)$	−393.5	$Hg(l)$	0
$CH_4(g)$	−74.848	$Hg(g)$	60.84
$CH_3Cl(g)$	−82.0	$I_2(s)$	0
$CH_3I(g)$	14.2	$K(s)$	0
$CH_3OH(l)$	−238.6	$KCl(s)$	−435.89
$CO(NH_2)_2(s)$ (urea)	−333.19	$K_2SO_4(s)$	−1433.7
$CO(NH_2)_2(aq)$	−391.2	$N_2(g)$	0
$C_2H_2(g)$	226.75	$NH_3(g)$	−46.19
$C_2H_4(g)$	52.284	$NH_4Cl(s)$	−315.4
$C_2H_6(g)$	−84.667	$NO(g)$	90.37
$C_2H_5OH(l)$	−277.63	$NO_2(g)$	33.8
$Ca(s)$	0	$N_2O(g)$	81.57
$CaBr_2(s)$	−682.8	$N_2O_4(g)$	9.67
$CaCO_3(s)$	−1207	$N_2O_5(g)$	11
$CaCl_2(s)$	−795.0	$Na(s)$	0
$CaO(s)$	−635.5	$NaHCO_3(s)$	−947.7
$Ca(OH)_2(s)$	−986.59	$Na_2CO_3(s)$	−1131
$CaSO_4(s)$	−1432.7	$NaCl(s)$	−411.0
$CaSO_4 \cdot \frac{1}{2}H_2O(s)$	−1575.2	$NaOH(s)$	−426.8
$CaSO_4 \cdot 2H_2O(s)$	−2021.1	$Na_2SO_4(s)$	−1384.5
$Cl_2(g)$	0	$O_2(g)$	0
$Fe(s)$	0	$Pb(s)$	0
$Fe_2O_3(s)$	−822.2	$PbO(s)$	−219.2
$H_2(g)$	0	$S(s)$	0
$H_2O(g)$	−241.8	$SO_2(g)$	−296.9
$H_2O(l)$	−285.9	$SO_3(g)$	−395.2

an element *from itself,* of course, would yield no change in enthalpy.) In most tables, values of ΔH_f° for the elements are not included for this reason.

It is important to remember the meaning of the subscript f in the symbol ΔH_f°. It is applied to a value of ΔH° only when *one mole* of the substance is formed *from its elements in their standard states.* Consider, for example, the following four thermochemical equations and their corresponding values of ΔH°.

$$H_2(g) + \tfrac{1}{2}O_2(g) \longrightarrow H_2O(l) \qquad \Delta H_f^\circ = -285.9 \text{ kJ/mol}$$

$$2H_2(g) + O_2(g) \longrightarrow 2H_2O(l) \qquad \Delta H^\circ = -571.8 \text{ kJ}$$

$$CO(g) + \tfrac{1}{2}O_2(g) \longrightarrow CO_2(g) \qquad \Delta H^\circ = -283.0 \text{ kJ}$$

$$2H(g) + O(g) \longrightarrow H_2O(l) \qquad \Delta H^\circ = -971.1 \text{ kJ}$$

Only in the first equation is ΔH° given the subscript f. It is the only reaction that satisfies both of the conditions specified above for standard enthalpies of formation. The second equation shows the formation of *two* moles of water, not one. The third involves a *compound* as one of the reactants. The fourth involves the elements as *atoms,* which are *not standard states* for these elements. Also notice that the units of ΔH_f° are kilojoules *per mole,* not just kilojoules, because the value is for the formation of 1 mol of the compound (from its elements). We can obtain the enthalpy of formation of two moles of water (ΔH° for the second equation) simply by multiplying the ΔH_f° value for 1 mol of H_2O by the factor, 2 mol $H_2O(l)$.

$$\left(\frac{-285.9 \text{ kJ}}{\cancel{\text{mol } H_2O(l)}}\right) \times 2 \cancel{\text{mol } H_2O(l)} = -571.8 \text{ kJ}$$

EXAMPLE 6.10

Writing an Equation for a Standard Heat of Formation

What equation must be used to represent the formation of nitric acid, $HNO_3(l)$, when we want to include its value of ΔH_f°?

ANALYSIS: Answering the question requires that we know the definition of ΔH_f°, which demands that the equation must show only *one mole* of the product. We begin with its formula and take whatever fractions of moles of the elements are required to make it. We must also be careful to include the physical states. Table 6.2 gives the value of ΔH_f° for $HNO_3(l)$, -173.2 kJ mol^{-1}.

SOLUTION: The three elements, H, N, and O, all occur as diatomic molecules in the gaseous state, so the following fractions of moles supply exactly enough of each to make one mole of HNO_3.

$$\tfrac{1}{2}H_2(g) + \tfrac{1}{2}N_2(g) + \tfrac{3}{2}O_2(g) \longrightarrow HNO_3(l) \qquad \Delta H_f^\circ = -173.2 \text{ kJ mol}^{-1}$$

IS THE ANSWER REASONABLE? The answer correctly shows only 1 mol of HNO_3, and this governs the coefficients for the reactants. So simply check to be sure everything is balanced.

Practice Exercise 16: Write the thermochemical equation that would be used to represent the standard heat of formation of $NH_4Cl(s)$. (Hint: Be sure you show substances in their standard states.)

Practice Exercise 17: Write the thermochemical equation that would be used to represent the standard heat of formation of sodium bicarbonate, $NaHCO_3(s)$.

Standard enthalpies of formation are useful because they provide a convenient method for applying Hess's law without having to manipulate thermochemical equations. To see how this works, let's look again at Example 6.7 where we calculated $\Delta H°$ for the decomposition of H_2O_2.

$$H_2O_2(l) \longrightarrow H_2O(l) + \tfrac{1}{2}O_2(g)$$

The two equations we used were

$$(1) \quad H_2(g) + O_2(g) \longrightarrow H_2O_2(l) \qquad \Delta H_f^\circ(1) = -188 \text{ kJ}$$

$$(2) \quad H_2(g) + \tfrac{1}{2}O_2(g) \longrightarrow H_2O(l) \qquad \Delta H_f^\circ(2) = -286 \text{ kJ}$$

This time we have numbered the equations and identified the enthalpy changes as ΔH_f° because in each case we're forming one mole of the compound from elements in their standard states. To combine these equations to find $\Delta H°$ for the decomposition of H_2O_2, we have to reverse equation (1), which means we change the sign of its ΔH.

$$\text{(1 reversed)} \quad H_2O_2(l) \longrightarrow H_2(g) + O_2(g) \qquad \Delta H° = -\Delta H_f^\circ(1) = +188 \text{ kJ}$$

$$(2) \quad H_2(g) + \tfrac{1}{2}O_2(g) \longrightarrow H_2O(l) \qquad \Delta H_f^\circ(2) = -286 \text{ kJ}$$

$$\text{Sum} \quad H_2O_2(l) \longrightarrow H_2O(l) + \tfrac{1}{2}O_2(g) \qquad \Delta H° = \Delta H_f^\circ(2) + [-\Delta H_f^\circ(1)]$$

$$= -286 \text{ kJ} + (+188 \text{ kJ})$$

$$= -98 \text{ kJ}$$

We can use either heats of combustion or heats of formation for the reactants and products in Hess's law, but don't mix them. Use heats of combustion for all reactants and all products, *or* use heats of formation for all reactants and all products.

Thus, the desired $\Delta H°$ was obtained by subtracting the ΔH_f° for the reactant (H_2O_2) from the ΔH_f° for the product (H_2O). In fact, this works for *any* reaction where we know the heats of formation of all the reactants and products. So if we're working with heats of formation, Hess's law can be restated as follows: the net $\Delta H^\circ_{\text{reaction}}$ equals the sum of the heats of formation of the products minus the sum of the heats of formation of the reactants, each ΔH_f° value multiplied by the appropriate coefficient given by the thermochemical equation. In other words, we can express Hess's law in the form of the **Hess's law equation.**

Hess's law equation

$$\Delta H^\circ_{\text{reaction}} = \begin{pmatrix}\text{sum of } \Delta H_f^\circ \text{ of all} \\ \text{of the products}\end{pmatrix} - \begin{pmatrix}\text{sum of } \Delta H_f^\circ \text{ of all} \\ \text{of the reactants}\end{pmatrix} \qquad (6.14)$$

As long as we have access to a table of standard heats of formation, using Equation 6.14 avoids having to manipulate a series of thermochemical equations. Finding $\Delta H^\circ_{\text{reaction}}$ reduces to performing arithmetic, as we see in the next example.

EXAMPLE 6.11

Using Hess's Law and Standard Enthalpies of Formation

Some chefs keep baking soda, $NaHCO_3$, handy to put out grease fires. When thrown on the fire, baking soda partly smothers the fire and the heat decomposes it to give CO_2, which further smothers the flame. The equation for the decomposition of $NaHCO_3$ is

$$2NaHCO_3(s) \longrightarrow Na_2CO_3(s) + H_2O(l) + CO_2(g)$$

Use the data in Table 6.2 to calculate the $\Delta H°$ for this reaction in kilojoules.

ANALYSIS: When values of ΔH_f° are available, Hess's law equation (Equation 6.14) is our basic tool for computing values of $\Delta H°$. We will calculate the sum of ΔH_f° for the products, and then the sum of ΔH_f° for the reactants. Then we will subtract the total ΔH_f° for the reactants from the total ΔH_f° for the products to calculate $\Delta H^\circ_{\text{reaction}}$. For each of the reactants and products, we compute its $\Delta H°$ by multiplying its ΔH_f° (from Table 6.2) by its coefficient in the balanced chemical equation.

SOLUTION: First, let's set up the calculation using symbols for the quantities we will use.

$$\Delta H^\circ = [1 \text{ mol } Na_2CO_3(s) \times \Delta H^\circ_{f\, Na_2CO_3(s)} + 1 \text{ mol } H_2O(l) \times \Delta H^\circ_{f\, H_2O(l)} + 1 \text{ mol } CO_2(g) \times \Delta H^\circ_{f\, CO_2(g)}] - [2 \text{ mol } NaHCO_3(s) \times \Delta H^\circ_{f\, NaHCO_3(s)}]$$

We now go to Table 6.2 to find the values of ΔH°_f for each substance *in its proper physical state.*

$$\Delta H^\circ = \left[1 \text{ mol } Na_2CO_3 \times \left(\frac{-1131 \text{ kJ}}{\text{mol } Na_2CO_3}\right) + 1 \text{ mol } H_2O \times \left(\frac{-285.9 \text{ kJ}}{\text{mol } H_2O}\right) + 1 \text{ mol } CO_2 \times \left(\frac{-393.5 \text{ kJ}}{\text{mol } CO_2}\right)\right] - \left[2 \text{ mol } NaHCO_3 \times \left(\frac{-947.7 \text{ kJ}}{\text{mol } NaHCO_3}\right)\right]$$

This reduces to

$$\Delta H^\circ = (-1810 \text{ kJ}) - (-1895 \text{ kJ})$$
$$= +85 \text{ kJ}$$

Thus, under standard conditions, the reaction is endothermic by 85 kJ. (Notice that we did not have to manipulate any equations.)

IS THE ANSWER REASONABLE? Double-check that all of the coefficients found in the chemical equation are correctly applied as multipliers on the ΔH°_f quantities. Be sure that the *signs* of the values of ΔH°_f have all been carefully used. Keeping track of the algebraic signs requires particular care and is the greatest source of error in these calculations. Also check to be sure you've followed the correct order of subtraction specified in Equation 6.14. In other words, have you subtracted the ΔH° of the *reactants* from the ΔH° of the *products*?

Practice Exercise 18: Use heats of formation data from Table 6.2 to calculate ΔH° for this reaction: $CaCl_2(s) + H_2SO_4(l) \longrightarrow CaSO_4(s) + 2HCl(g)$. (Hint: Be sure to follow the correct order of subtraction.)

Practice Exercise 19: Write thermochemical equations corresponding to ΔH°_f for $SO_3(g)$ and $SO_2(g)$ and show how they can be manipulated to calculate ΔH° for the following reaction:

$$SO_3(g) \longrightarrow SO_2(g) + \tfrac{1}{2}O_2(g) \qquad \Delta H^\circ = ?$$

Use Equation 6.14 to calculate ΔH° for the reaction using the heats of formation of $SO_3(g)$ and $SO_2(g)$. How do the answers compare?

Practice Exercise 20: Calculate ΔH° for the following reactions.

(a) $2NO(g) + O_2(g) \longrightarrow 2NO_2(g)$
(b) $NaOH(s) + HCl(g) \longrightarrow NaCl(s) + H_2O(l)$

SUMMARY

Introduction. **Thermochemistry,** which deals with energy absorbed or released in chemical reactions, is a part of **thermodynamics,** the study of energy transfer and energy transformation. **Heat** is energy **(thermal energy)** that transfers between objects having different temperatures. Heat continues to flow until the two objects come to the same temperature and **thermal equilibrium** is established.

Energy Is the Ability to Do Work or Supply Heat. An object has **kinetic energy (KE)** if it is moving (KE $= \frac{1}{2}mv^2$) and **potential energy (PE)** when it experiences attractions or repulsions toward other objects. When objects that attract are moved apart, PE increases. Similarly, when objects that repel are pushed together, PE increases. Because electrical attractions and repulsions occur within atoms, molecules, and ions, substances have **chemical energy,** a form of **potential energy.** The particles that make up matter are in constant random motion, so they also possess kinetic energy, specifically, **molecular kinetic energy.** According to the **kinetic molecular theory,** the average molecular kinetic energy is directly proportional to the Kelvin temperature. The **law**

of conservation of energy states that energy can be neither created nor destroyed, but only changed from one form to another. The SI unit of energy is the **joule;** 4.184 J = 1 cal (calorie). Larger units are the **kilojoule (kJ)** and **kilocalorie (kcal,** which is the same as the nutritional Calorie).

Internal Energy and Temperature. The **state** of a system is the list of properties that describe its current condition. The **internal energy** of a system, E, is a **state function,** which is a property that depends only on the current state of the system. E equals the sum of the system's molecular kinetic and potential energies. A change in E is defined as $\Delta E = E_{\text{final}} - E_{\text{initial}}$, although absolute amounts of E cannot be measured or calculated. For a chemical reaction, the definition translates to $\Delta E = E_{\text{products}} - E_{\text{reactants}}$. A positive value for ΔE, which can be measured, means a system absorbs energy from its surroundings during a change.

Measuring Heat. The boundary across which heat is transferred encloses the **system** (the object we're interested in). Everything else in the universe is the system's **surroundings.** The heat flow q is related to the temperature change Δt by $q = C\,\Delta t$, where C is the **heat capacity** of the system (the heat needed to change the temperature of the system by one degree Celsius). The heat capacity for a pure substance can be computed from its mass, m, using the equation $C = ms$, where s is the **specific heat** of the material (the heat needed to change the temperature of 1 g of a substance by 1 °C). Water has an unusually high specific heat. We can compute a heat flow when we know the mass and specific heat of an object using the equation $q = ms\,\Delta t$. The heat, q, is given a positive sign when it flows into a system and a negative sign when it flows out.

Energy Changes when Bonds Are Formed or Broken. Bond breaking increases potential energy (chemical energy); bond formation decreases potential energy (chemical energy). In an **exothermic** reaction, chemical energy is changed to molecular kinetic energy. If the system is **adiabatic** (no heat leaves it), the internal temperature increases. Otherwise, the heat has a tendency to leave the system. In **endothermic** reactions, molecular kinetic energy of the reactants is converted into potential energy of the products. This tends to lower the system's temperature and lead to a flow of heat into the system.

Heats of Reaction. The change in chemical potential energy in a reaction is the **heat of reaction, *q*,** which can be measured at constant volume or constant pressure. Pressure is the ratio of force to the area over which the force is applied. **Atmospheric pressure** is the pressure exerted by the mixture of gases in our atmosphere. When a volume change, ΔV, occurs at constant opposing pressure, P, the associated **pressure–volume work (expansion work)** is given by $w = -P\,\Delta V$. The energy expended in doing this pressure–volume work causes heats of reaction measured at constant volume (q_v) to differ numerically from heats measured at constant pressure (q_p).

The **first law of thermodynamics** says that no matter how the change in energy accompanying a reaction may be allocated between q and w, their sum, ΔE, is the same: $\Delta E = q + w$. The algebraic sign for q and w is negative when the system gives off heat to or does work on the surroundings. The sign is positive when the system absorbs heat or receives work energy done to it. When the volume of a system cannot change, as in a **bomb calorimeter,** w is zero, and q_v is the **heat of reaction at constant volume,** ΔE.

When the system is under conditions of constant pressure, the energy of the system is called its **enthalpy, *H*.** At constant pressure, $\Delta H = \Delta E + P\,\Delta V$. The **heat of reaction at constant pressure,** q_p, is the **enthalpy change, ΔH.** ΔH is a state function. Its value differs from that of ΔE by the work involved in interacting with the atmosphere when the change occurs at constant atmospheric pressure. In general, the difference between ΔE and ΔH is quite small. Exothermic reactions have negative values of ΔH; endothermic changes have positive values.

Thermochemical Equations. A balanced chemical equation that includes both the enthalpy change and the physical states of the substances is called a **thermochemical equation.** Coefficients in a thermochemical equation represent mole quantities of reactants and products. Such equations can be added, reversed (reversing also the sign of ΔH), or multiplied by a constant multiplier (doing the same to ΔH). If formulas are canceled or added, they must be of substances in identical physical states.

The reference conditions for thermochemistry, called **standard conditions,** are 25 °C and 1 bar of pressure. An enthalpy change measured under those conditions is called the **standard enthalpy of reaction** or the **standard heat of reaction,** given the symbol **$\Delta H°$.**

Hess's law of heat summation is possible because enthalpy is a state function. Values of $\Delta H°$ can be determined *by the manipulation of any combination of thermochemical equations that add up to the final net equation. The units for $\Delta H°$ are generally joules or kilojoules.* An **enthalpy diagram** provides a graphical description of the enthalpy changes for alternative paths from reactants to products.

When the enthalpy change is for the complete combustion of one mole of a pure substance under standard conditions in pure oxygen, $\Delta H°$ is called the **standard heat of combustion** of the compound, symbolized as ΔH_c°.

When the enthalpy change is for the formation of *one* mole of a substance under standard conditions from its *elements in their standard states,* $\Delta H°$ is called the **standard heat of formation** of the compound, symbolized as ΔH_f° (usually in units of kilojoules per mole, kJ mol^{-1}). The value of $\Delta H°$ for a reaction can be calculated from tabulated values of ΔH_f° using the **Hess's law equation.**

TOOLS FOR PROBLEM SOLVING

In this chapter you learned to apply the following concepts as tools in solving problems. Study each one carefully so that you know what each is used for. When faced with solving a problem, recall what each tool does and consider whether it will be helpful in finding a solution. This will aid you in selecting the tools you need.

Kinetic energy *(page 208)* Kinetic energy (KE) can be calculated from an object's mass, m, and its speed or velocity, v, using Equation 6.1: $KE = \frac{1}{2}mv^2$.

Factors that affect potential energy *(page 209)* You should know how potential energy varies when the distance changes between objects that attract or repel.

Heat capacity *(page 214)* Heat capacity, C, provides a way of determining heat by measuring a temperature change. Its units are energy divided by temperature (e.g., J °C^{-1}).

$$q = C\,\Delta t$$

The value of C depends on sample size.

Specific heat capacity *(page 215)* Also called specific heat, s, specific heat capacity is an intensive property. When mass, m, and temperature change, Δt, are known, q is calculated by the equation

$$q = ms\,\Delta t$$

Heat transfer *(page 216)* When heat is transferred between two objects, the *size* of q is identical for both objects, but the algebraic signs of q are opposite.

$$q_1 = -q_2$$

First law of thermodynamics *(page 222)* This law relates energy transfer in the forms of heat, q, and work, w, to the internal energy change, ΔE.

$$\Delta E = q + w$$

The algebraic sign of ΔH *(page 225)* The sign of ΔH indicates the direction of energy flow.

For an endothermic change, ΔH is positive.

For an exothermic change, ΔH is negative.

Thermochemical equations *(page 228)* Thermochemical equations show enthalpy changes for reactions where the amounts of reactants and products in moles are represented by the coefficients. They can be manipulated and combined to determine heats of reactions for other reactions.

Enthalpy diagrams *(page 230)* We use enthalpy diagrams to provide a graphical picture of the enthalpy changes associated with a set of thermochemical equations that are combined to give some net reaction.

Hess's law *(page 232)* This law allows us to combine thermochemical equations to give a final desired equation and its associated $\Delta H°$. We adjust the coefficients and directions of the given equations, and make appropriate adjustments to their $\Delta H°$ values, so the equations add to give the desired equation. Adding the adjusted $\Delta H°$ values gives the $\Delta H°$ for the desired equation.

Manipulating thermochemical equations *(page 233)* Changing the direction of a reaction changes the sign of $\Delta H°$. When the coefficients are multiplied by a factor, the value of $\Delta H°$ is multiplied by the same factor. Formulas can be canceled only when the substances are in the same physical state. These rules are used in applying Hess's law.

Hess's law equation *(page 238)* We use this equation with standard heats of formation, $\Delta H_f°$, to calculate $\Delta H°$ for some desired equation.

$$\Delta H°_{\text{reaction}} = \begin{pmatrix}\text{sum of } \Delta H_f° \text{ of all} \\ \text{of the products}\end{pmatrix} - \begin{pmatrix}\text{sum of } \Delta H_f° \text{ of all} \\ \text{of the reactants}\end{pmatrix}$$

QUESTIONS, PROBLEMS, AND EXERCISES

Answers to problems whose numbers are printed in color are given in Appendix B. More challenging problems are marked with asterisks. ILW = Interactive Learningware solution is available at www.wiley.com/college/brady. OH = an Office Hours video is available for this problem.

REVIEW QUESTIONS

Kinetic and Potential Energy

6.1 Give definitions for (a) *energy* and (b) *work.*

6.2 Define (a) *kinetic energy* and (b) *potential energy.*

6.3 State the equation used to calculate an object's kinetic energy. Define the symbols used in the equation.

6.4 If a car increases its speed from 30 mph to 60 mph, by what factor does the kinetic energy of the car increase?

6.5 What is meant by the term *chemical energy*?

6.6 How does the potential energy change (increase, decrease, or no change) for each of the following?
(a) Two electrons come closer together.
(b) An electron and a proton become farther apart.
(c) Two atomic nuclei approach each other.
(d) A ball rolls downhill.

6.7 State the *law of conservation of energy*. Describe how it explains the motion of a child on a swing.

6.8 Define *heat.* How do *heat* and *temperature* differ?

6.9 On a molecular level, how is thermal equilibrium achieved when a hot object is placed in contact with a cold object?

6.10 What is the SI unit of energy? How much energy (in joules and in calories) does a 75 kg object have if it's moving at 45 m s^{-1}?

Internal Energy and the Kinetic Theory of Matter

6.11 How is internal energy related to molecular kinetic and potential energy? How is a *change* in internal energy defined for a chemical reaction?

6.12 Consider the distribution of molecular kinetic energies shown in the diagram below for a gas at 25 °C.

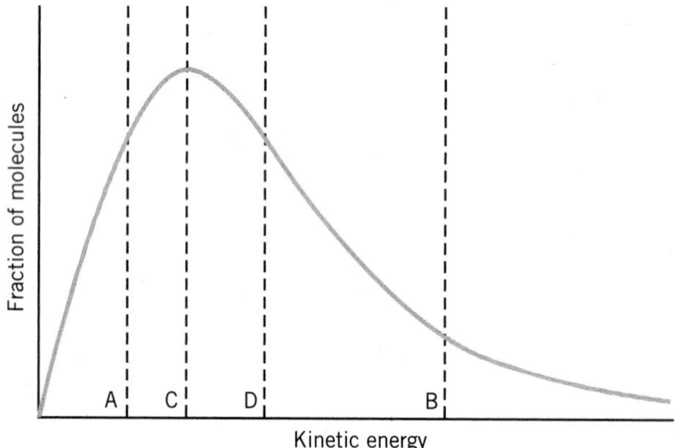

(a) Which point corresponds to the most frequently occurring (also called the most probable) molecular kinetic energy?
(b) Which point corresponds to the average molecular kinetic energy?
(c) If the temperature of the gas is raised to 50 °C, how will the height of the curve at B change?
(d) If the temperature of the gas is raised to 50 °C, how will the height of the curve at A change? How will the maximum height of the curve change?

6.13 Suppose the temperature of an object is raised from 100 °C to 200 °C by heating it with a Bunsen burner. Which of the following will be true?
(a) The average molecular kinetic energy will increase.
(b) The total kinetic energy of all the molecules will increase.
(c) The number of fast-moving molecules will increase.
(d) The number of slow-moving molecules will increase.
(e) The chemical potential energy will decrease.

6.14 A quart of boiling water will cause a more severe burn if it's spilled on you than just a drop of boiling water. Explain this in terms of the amounts of kinetic energy possessed by the water molecules in the two samples.

6.15 How can the *state* of a system be specified? What is a *state function*?

Experimental Measurement of Heat

6.16 What do the terms *system* and *surroundings* mean? What is the difference between an isolated system and a closed system?

6.17 What is the name of the thermal property whose values can have the following units?
(a) J g^{-1} $°C^{-1}$ (b) J mol^{-1} $°C^{-1}$ (c) J $°C^{-1}$

6.18 For samples with the same mass, which kind of substance needs more energy to undergo an increase of 5 °C, something with a *large* or with a *small* specific heat? Explain.

6.19 How do heat capacity and specific heat differ?

6.20 What is the meaning of a negative value for q?

6.21 Suppose object A has twice the specific heat and twice the mass of object B. If the same amount of heat is applied to both objects, how will the temperature change of A be related to the temperature change in B?

Energy Changes in Chemical Reactions

6.22 In a certain chemical reaction, there is a decrease in the potential energy (chemical energy) as the reaction proceeds.
(a) How does the total kinetic energy of the particles change?
(b) How does the temperature of the reaction mixture change?

6.23 What term do we use to describe a reaction that liberates heat to its surroundings? How does the chemical energy change during such a reaction? What is the algebraic sign of q for such a reaction?

6.24 What term is used to describe a reaction that absorbs heat from the surroundings? How does the chemical energy change during such a reaction? What is the algebraic sign of q for such a reaction?

6.25 Some instant cold packs sold in pharmacies contain a bag of water inside a pouch of ammonium nitrate, NH_4NO_3. When the bag of water is broken and the two substances mixed, the ammonium nitrate dissolves and the solution becomes quite cold. What happens to the average kinetic energy of the substances in the cold pack? What happens to the chemical energy of the water and ammonium nitrate?

6.26 When gasoline burns, it reacts with oxygen in the air and forms hot gases consisting of carbon dioxide and water vapor. How does the potential energy of the gasoline and oxygen compare with the potential energy of the carbon dioxide and water vapor?

6.27 Describe how the potential energy of the system of atomic-sized particles changes when each of the following events occurs.
(a) The wax of a candle burns in air, giving a yellow flame. (The system consists of the wax and O_2 in the air.)
(b) Ammonium nitrate dissolves in water to produce a cooling effect.

Internal Energy and Enthalpy

6.28 Write the equation that states the first law of thermodynamics. In your own words, what does this statement mean in terms of energy exchanges between a system and its surroundings?

6.29 How is *enthalpy* defined?

6.30 What is the *sign* of ΔH for an exothermic change?

6.31 If the enthalpy of a system increases by 100 kJ, what must be true about the enthalpy of the surroundings? Why?

6.32 When we measure the heat of combustion of glucose, $C_6H_{12}O_6$, in a bomb calorimeter, what products are formed in the reaction? Is the heat we measure equal to ΔE or ΔH?

6.33 If a system containing gases expands and pushes back a piston against a constant opposing pressure, what equation describes the work done on the system?

6.34 Consider the reaction

$$C_{12}H_{22}O_{11}(s) + 12O_2(g) \longrightarrow 12CO_2(g) + 11H_2O(l)$$

Are the values of ΔE and ΔH expected to be appreciably different?

Enthalpy and Heats of Reaction

6.35 Why do standard reference values for temperature and pressure have to be selected when we consider and compare heats of reaction for various reactions? What are the values for the standard temperature and standard pressure?

6.36 What distinguishes a *thermochemical* equation from an ordinary chemical equation?

6.37 Why are fractional coefficients permitted in a balanced thermochemical equation? If a formula in a thermochemical equation has a coefficient of $\frac{1}{2}$, what does it signify?

Hess's Law

6.38 What fundamental fact about ΔH makes Hess's law possible?

6.39 What *two* conditions must be met by a thermochemical equation so that its standard enthalpy change can be given the symbol ΔH_f°?

6.40 What is Hess's law expressed in terms of standard heats of formation?

REVIEW PROBLEMS

First Law of Thermodynamics

6.41 If a system does 45 J of work and receives 28 J of heat, what is the value of ΔE for this change?

6.42 If a system absorbs 48 J of heat and does 22 J of work, what is the value of ΔE for this change?

6.43 An automobile engine converts heat into work via a cycle. The cycle must finish exactly where it started, so the energy at the start of the cycle must be exactly the same as the energy at the end of the cycle. If the engine is to do 100 J of work per cycle, how much heat must it absorb?

6.44 If the engine in the previous problem absorbs 250 joules of heat per cycle, how much work can it do per cycle?

Thermal Properties, Measuring Energy Changes

6.45 How much heat, in joules and in calories, must be removed from 1.75 mol of water to lower its temperature from 25.0 to 15.0 °C?

OH **6.46** How many grams of water can be heated from 25.0 °C to 35.0 °C by the heat released from 85.0 g of iron that cools from 85.0 °C to 30.0 °C?

6.47 A 5.00 g mass of a metal was heated to 100.0 °C and then plunged into 100.0 g of water at 24.0 °C. The temperature of the resulting mixture became 28.0 °C.
(a) How many joules did the water absorb?
(b) How many joules did the metal lose?
(c) What is the heat capacity of the metal sample?
(d) What is the specific heat of the metal?

6.48 A sample of copper was heated to 120.00 °C and then thrust into 200.0 g of water at 25.00 °C. The temperature of the mixture became 26.50 °C.
(a) How much heat in joules was absorbed by the water?
(b) The copper sample lost how many joules?
(c) What was the mass in grams of the copper sample?

OH **6.49** Calculate the molar heat capacity of iron in J mol^{-1} $°C^{-1}$. Its specific heat is 0.4498 J g^{-1} $°C^{-1}$.

6.50 What is the molar heat capacity of ethyl alcohol, C_2H_5OH, in units of J mol^{-1} $°C^{-1}$, if its specific heat is 0.586 cal g^{-1} $°C^{-1}$?

Calorimetry

6.51 A vat of 4.54 kg of water underwent a decrease in temperature from 60.25 to 58.65 °C. How much energy in kilojoules left the water? (For this range of temperature, use a value of 4.18 J g^{-1} $°C^{-1}$ for the specific heat of water.)

6.52 A container filled with 2.46 kg of water underwent a temperature change from 25.24 °C to 27.31 °C. How much heat, measured in kilojoules, did the water absorb? (The specific heat of water is 4.18 J g^{-1} $°C^{-1}$.)

OH **6.53** Nitric acid neutralizes potassium hydroxide. To determine the heat of reaction, a student placed 55.0 mL of 1.3 *M* HNO_3 in a coffee cup calorimeter, noted that the temperature was 23.5 °C, and added 55.0 mL of 1.3 *M* KOH, also at 23.5 °C. The mixture was stirred quickly with a thermometer, and its temperature rose to 31.8 °C. Write the balanced equation for the reaction. Calculate the heat of reaction in joules. Assume that the specific heats of all solutions are 4.18 J g^{-1} $°C^{-1}$ and that all densities are 1.00 g mL^{-1}. Calculate the heat of reaction per mole of acid (in units of kJ mol^{-1}).

6.54 A dilute solution of hydrochloric acid with a mass of 610.29 g and containing 0.33183 mol of HCl was exactly neutralized in a calorimeter by the sodium hydroxide in 615.31 g of a dilute NaOH solution. The temperature increased from 16.784 to 20.610 °C. The specific heat of the HCl solution was 4.031 J g^{-1} $°C^{-1}$; that of the NaOH solution was 4.046 J g^{-1} $°C^{-1}$. The heat capacity of the calorimeter was 77.99 J $°C^{-1}$. Write the balanced equation for the reaction. Use the data above to calculate the heat evolved. What is the heat of neutralization per mole of HCl? Assume that the original solutions made independent contributions to the total heat capacity of the system following their mixing.

6.55 A 1.00 mol sample of propane, a gas used for cooking in many rural areas, was placed in a bomb calorimeter with excess oxygen and ignited. The initial temperature of the calorimeter was 25.000 °C and its total heat capacity was 97.1 kJ $°C^{-1}$. The reaction raised the temperature of the calorimeter to 27.282 °C.
(a) Write the balanced chemical equation for the reaction in the calorimeter.
(b) How many joules were liberated in this reaction?
(c) What is the heat of reaction of propane with oxygen expressed in kilojoules per mole of C_3H_8 burned?

6.56 Toluene, C_7H_8, is used in the manufacture of explosives such as TNT (trinitrotoluene). A 1.500 g sample of liquid toluene was placed in a bomb calorimeter along with excess oxygen. When the combustion of the toluene was initiated, the temperature of the calorimeter rose from 25.000 °C to 26.413 °C. The products of the combustion are $CO_2(g)$ and $H_2O(l)$, and the heat capacity of the calorimeter was 45.06 kJ $°C^{-1}$.
(a) Write the balanced chemical equation for the reaction in the calorimeter.
(b) How many joules were liberated by the reaction?
(c) How many joules would be liberated under similar conditions if 1.00 mol of toluene were burned?

Enthalpy Changes and Heats of Reaction

6.57 One thermochemical equation for the reaction of carbon monoxide with oxygen is

$$3CO(g) + \tfrac{3}{2}O_2(g) \longrightarrow 3CO_2(g) \qquad \Delta H° = -849 \text{ kJ}$$

(a) Write the thermochemical equation for the reaction using 2 mol of CO.
(b) What is $\Delta H°$ for the formation of 1 mol of CO_2 by this reaction?

OH **6.58** Ammonia reacts with oxygen as follows.

$$4NH_3(g) + 7O_2(g) \longrightarrow 4NO_2(g) + 6H_2O(g) \qquad \Delta H° = -1132 \text{ kJ}$$

(a) Calculate the enthalpy change for the combustion of 1 mol of NH_3.
(b) Write the thermochemical equation for the reaction in which one mole of H_2O is formed.

6.59 Magnesium burns in air to produce a bright light and is often used in fireworks displays. The combustion of magnesium follows the thermochemical equation

$$2Mg(s) + O_2(g) \longrightarrow 2MgO(s) \qquad \Delta H° = -1203 \text{ kJ}$$

How much heat (in kilojoules) is liberated by the combustion of 6.54 g of magnesium?

6.60 Methanol is the fuel in "canned heat" containers (e.g., Sterno) that are used to heat foods at cocktail parties. The combustion of methanol follows the thermochemical equation

$$2CH_3OH(l) + 3O_2(g) \longrightarrow 2CO_2(g) + 4H_2O(g) \qquad \Delta H° = -1199 \text{ kJ}$$

How many kilojoules are liberated by the combustion of 46.0 g of methanol?

Hess's Law

6.61 Construct an enthalpy diagram that shows the enthalpy changes for a one-step conversion of germanium, Ge(*s*), into $GeO_2(s)$, the dioxide. On the same diagram, show the two-step process, first to the monoxide, GeO(*s*), and then its conversion to the dioxide. The relevant thermochemical equations are the following.

$$Ge(s) + \tfrac{1}{2}O_2(g) \longrightarrow GeO(s) \qquad \Delta H° = -255 \text{ kJ}$$

$$Ge(s) + O_2(g) \longrightarrow GeO_2(s) \qquad \Delta H° = -534.7 \text{ kJ}$$

Using this diagram, determine $\Delta H°$ for the following reaction.

$$GeO(s) + \tfrac{1}{2}O_2(g) \longrightarrow GeO(s)$$

6.62 Construct an enthalpy diagram for the formation of $NO_2(g)$ from its elements by two pathways: first, from its elements and, second, by a two-step process, also from the elements. The relevant thermochemical equations are

$$\tfrac{1}{2}N_2(g) + O_2(g) \longrightarrow NO_2(g) \qquad \Delta H° = +33.8 \text{ kJ}$$

$$\tfrac{1}{2}N_2(g) + \tfrac{1}{2}O_2(g) \longrightarrow NO(g) \qquad \Delta H° = +90.37 \text{ kJ}$$

$$NO(g) + \tfrac{1}{2}O_2(g) \longrightarrow NO_2(g) \qquad \Delta H° = ?$$

Be sure to note the signs of the values of $\Delta H°$ associated with arrows pointing up or down. Using the diagram, determine the value of $\Delta H°$ for the third equation.

6.63 Show how the equations

$$N_2O_4(g) \longrightarrow 2NO_2(g) \qquad \Delta H° = +57.93 \text{ kJ}$$

$$2NO(g) + O_2(g) \longrightarrow 2NO_2(g) \qquad \Delta H° = -113.14 \text{ kJ}$$

can be manipulated to give $\Delta H°$ for the following reaction.

$$2NO(g) + O_2(g) \longrightarrow N_2O_4(g)$$

6.64 We can generate hydrogen chloride by heating a mixture of sulfuric acid and potassium chloride according to the equation

$$2KCl(s) + H_2SO_4(l) \longrightarrow 2HCl(g) + K_2SO_4(s)$$

Calculate $\Delta H°$ in kilojoules for this reaction from the following thermochemical equations.

$$HCl(g) + KOH(s) \longrightarrow KCl(s) + H_2O(l) \qquad \Delta H° = -203.6 \text{ kJ}$$

$$H_2SO_4(l) + 2KOH(s) \longrightarrow K_2SO_4(s) + 2H_2O(l) \qquad \Delta H° = -342.4 \text{ kJ}$$

6.65 Calculate $\Delta H°$ in kilojoules for the following reaction, the preparation of the unstable acid nitrous acid, HNO_2.

$$HCl(g) + NaNO_2(s) \longrightarrow HNO_2(l) + NaCl(s)$$

Use the following thermochemical equations.

$$2NaCl(s) + H_2O(l) \longrightarrow 2HCl(g) + Na_2O(s) \qquad \Delta H° = +507.31 \text{ kJ}$$

$$NO(g) + NO_2(g) + Na_2O(s) \longrightarrow 2NaNO_2(s) \quad \Delta H° = -427.14 \text{ kJ}$$

$$NO(g) + NO_2(g) \longrightarrow N_2O(g) + O_2(g) \quad \Delta H° = -42.68 \text{ kJ}$$

$$2HNO_2(l) \longrightarrow N_2O(g) + O_2(g) + H_2O(l) \quad \Delta H° = +34.35 \text{ kJ}$$

6.66 Calcium hydroxide reacts with hydrochloric acid by the following equation.

$$Ca(OH)_2(aq) + 2HCl(aq) \longrightarrow CaCl_2(aq) + 2H_2O(l)$$

Calculate $\Delta H°$ in kilojoules for this reaction, using the following equations as needed.

$$CaO(s) + 2HCl(aq) \longrightarrow CaCl_2(aq) + H_2O(l) \quad \Delta H° = -186 \text{ kJ}$$

$$CaO(s) + H_2O(l) \longrightarrow Ca(OH)_2(s) \quad \Delta H° = -65.1 \text{ kJ}$$

$$Ca(OH)_2(s) \xrightarrow[\text{dissolving in water}]{} Ca(OH)_2(aq) \quad \Delta H° = -12.6 \text{ kJ}$$

6.67 Given the following thermochemical equations:

$$CaO(s) + Cl_2(g) \longrightarrow CaOCl_2(s) \quad \Delta H° = -110.9 \text{ kJ}$$

$$H_2O(l) + CaOCl_2(s) + 2NaBr(s) \longrightarrow 2NaCl(s) + Ca(OH)_2(s) + Br_2(l) \quad \Delta H° = -60.2 \text{ kJ}$$

$$Ca(OH)_2(s) \longrightarrow CaO(s) + H_2O(l) \quad \Delta H° = +65.1 \text{ kJ}$$

calculate the value of $\Delta H°$ (in kilojoules) for the reaction

$$\tfrac{1}{2}Cl_2(g) + NaBr(s) \longrightarrow NaCl(s) + \tfrac{1}{2}Br_2(l)$$

6.68 Given the following thermochemical equations:

$$2Cu(s) + S(s) \longrightarrow Cu_2S(s) \quad \Delta H° = -79.5 \text{ kJ}$$

$$S(s) + O_2(g) \longrightarrow SO_2(g) \quad \Delta H° = -297 \text{ kJ}$$

$$Cu_2S(s) + 2O_2(g) \longrightarrow 2CuO(s) + SO_2(g) \quad \Delta H° = -527.5 \text{ kJ}$$

calculate the standard enthalpy of formation (in kilojoules per mole) of $CuO(s)$.

6.69 Given the following thermochemical equations:

$$4NH_3(g) + 7O_2(g) \longrightarrow 4NO_2(g) + 6H_2O(g) \quad \Delta H° = -1132 \text{ kJ}$$

$$6NO_2(g) + 8NH_3(g) \longrightarrow 7N_2(g) + 12H_2O(g) \quad \Delta H° = -2740 \text{ kJ}$$

calculate the value of $\Delta H°$ (in kilojoules) for the reaction

$$4NH_3(g) + 3O_2(g) \longrightarrow 2N_2(g) + 6H_2O(g)$$

OH **6.70** Given the following thermochemical equations:

$$3Mg(s) + 2NH_3(g) \longrightarrow Mg_3N_2(s) + 3H_2(g) \quad \Delta H° = -371 \text{ kJ}$$

$$\tfrac{1}{2}N_2(g) + \tfrac{3}{2}H_2(g) \longrightarrow NH_3(g) \quad \Delta H° = -46 \text{ kJ}$$

calculate $\Delta H°$ (in kilojoules) for the following reaction.

$$3Mg(s) + N_2(g) \longrightarrow Mg_3N_2(s)$$

Hess's Law and Standard Heats of Formation

6.71 Write the thermochemical equations, including values of $\Delta H_f°$ in kilojoules per mole (from Table 6.2), for the formation of each of the following compounds from their elements, everything in standard states.

(a) $HC_2H_3O_2(l)$, acetic acid
(b) $C_2H_5OH(l)$, ethyl alcohol
(c) $CaSO_4 \cdot 2H_2O(s)$, gypsum

6.72 Write the thermochemical equations, including values of $\Delta H_f°$ in kilojoules per mole (from Appendix C), for the formation of each of the following compounds from their elements, everything in standard states.

(a) $MgCl_2 \cdot 2H_2O(s)$
(b) $(NH_4)_2Cr_2O_7(s)$
(c) $POCl_3(g)$

OH **6.73** Using data in Table 6.2, calculate $\Delta H°$ in kilojoules for the following reactions.

(a) $2H_2O_2(l) \longrightarrow 2H_2O(l) + O_2(g)$
(b) $HCl(g) + NaOH(s) \longrightarrow NaCl(s) + H_2O(l)$

6.74 Using data in Table 6.2, calculate $\Delta H°$ in kilojoules for the following reactions.

(a) $CH_4(g) + Cl_2(g) \longrightarrow CH_3Cl(g) + HCl(g)$
(b) $2NH_3(g) + CO_2(g) \longrightarrow CO(NH_2)_2(s) + H_2O(l)$

6.75 The enthalpy change for the combustion of *one* mole of a compound under standard conditions is called the standard heat of combustion, and its symbol is $\Delta H_c°$. The value for sucrose, $C_{12}H_{22}O_{11}$, is -5.65×10^3 kJ mol^{-1}. Write the thermochemical equation for the combustion of 1 mol of sucrose and calculate the value of $\Delta H_f°$ for this compound. The sole products of combustion are $CO_2(g)$ and $H_2O(l)$. Use data in Table 6.2 as necessary.

6.76 The thermochemical equation for the combustion of acetylene gas, $C_2H_2(g)$, is

$$2C_2H_2(g) + 5O_2(g) \longrightarrow 4CO_2(g) + 2H_2O(l) \quad \Delta H° = -2599.3 \text{ kJ}$$

Using data in Table 6.2, determine the value of $\Delta H_f°$ for acetylene gas.

ADDITIONAL EXERCISES

***6.77** A 2.00 kg piece of granite with a specific heat of 0.803 J g^{-1} °C^{-1} and a temperature of 95.0 °C is placed into 2.00 L of water at 22.0 °C. When the granite and water come to the same temperature, what will that temperature be?

6.78 In the recovery of iron from iron ore, the reduction of the ore is actually accomplished by reactions involving carbon monoxide. Use the following thermochemical equations:

$$Fe_2O_3(s) + 3CO(g) \longrightarrow 2Fe(s) + 3CO_2(g) \quad \Delta H° = -28 \text{ kJ}$$

$$3Fe_2O_3(s) + CO(g) \longrightarrow 2Fe_3O_4(s) + CO_2(g) \quad \Delta H° = -59 \text{ kJ}$$

$$Fe_3O_4(s) + CO(g) \longrightarrow 3FeO(s) + CO_2(g) \quad \Delta H° = +38 \text{ kJ}$$

to calculate $\Delta H°$ for the reaction

$$FeO(s) + CO(g) \longrightarrow Fe(s) + CO_2(g)$$

6.79 Use the results of Problem 6.78 and data in Table 6.2 to calculate the value of $\Delta H_f°$ for FeO. Express the answer in units of kilojoules per mole.

6.80 The amino acid glycine, $C_2H_5NO_2$, is one of the compounds used by the body to make proteins. The equation for its combustion is

$$4C_2H_5NO_2(s) + 9O_2(g) \longrightarrow 8CO_2(g) + 10H_2O(l) + 2N_2(g)$$

For each mole of glycine that burns, 973.49 kJ of heat is liberated. Use this information plus values of ΔH_f° for the products of combustion to calculate ΔH_f° for glycine.

6.81 The value of ΔH_f° for HBr(g) was first evaluated using the following standard enthalpy values obtained experimentally. Use these data to calculate the value of ΔH_f° for HBr(g).

$$Cl_2(g) + 2KBr(aq) \longrightarrow Br_2(aq) + 2KCl(aq) \quad \Delta H^\circ = -96.2 \text{ kJ}$$

$$H_2(g) + Cl_2(g) \longrightarrow 2HCl(g) \quad \Delta H^\circ = -184 \text{ kJ}$$

$$HCl(aq) + KOH(aq) \longrightarrow KCl(aq) + H_2O(l) \quad \Delta H^\circ = -57.3 \text{ kJ}$$

$$HBr(aq) + KOH(aq) \longrightarrow KBr(aq) + H_2O(l) \quad \Delta H^\circ = -57.3 \text{ kJ}$$

$$HCl(g) \xrightarrow[\text{dissolving in water}]{} HCl(aq) \quad \Delta H^\circ = -77.0 \text{ kJ}$$

$$Br_2(l) \xrightarrow[\text{dissolving in water}]{} Br_2(aq) \quad \Delta H^\circ = -4.2 \text{ kJ}$$

$$HBr(g) \xrightarrow[\text{dissolving in water}]{} HBr(aq) \quad \Delta H^\circ = -79.9 \text{ kJ}$$

***6.82** Acetylene, C_2H_2, is a gas commonly used in welding. It is formed in the reaction of calcium carbide, CaC_2, with water. Given the thermochemical equations below, calculate the value of ΔH_f° for acetylene in units of kilojoules per mole.

$$CaO(s) + H_2O(l) \longrightarrow Ca(OH)_2(s) \quad \Delta H^\circ = -65.3 \text{ kJ}$$

$$CaO(s) + 3C(s) \longrightarrow CaC_2(s) + CO(g) \quad \Delta H^\circ = +462.3 \text{ kJ}$$

$$CaCO_3(s) \longrightarrow CaO(s) + CO_2(g) \quad \Delta H^\circ = +178 \text{ kJ}$$

$$CaC_2(s) + 2H_2O(l) \longrightarrow Ca(OH)_2(s) + C_2H_2(g) \quad \Delta H^\circ = -126 \text{ kJ}$$

$$2C(s) + O_2(g) \longrightarrow 2CO(g) \quad \Delta H^\circ = -220 \text{ kJ}$$

$$2H_2O(l) \longrightarrow 2H_2(g) + O_2(g) \quad \Delta H^\circ = +572 \text{ kJ}$$

OH **6.83** The reaction for the metabolism of sucrose, $C_{12}H_{22}O_{11}$, is the same as for its combustion in oxygen to yield $CO_2(g)$ and $H_2O(l)$. The standard heat of formation of sucrose is -2230 kJ mol^{-1}. Use data in Table 6.2 to compute the amount of energy (in kJ) released by metabolizing 1 oz (28.3 g) of sucrose.

***6.84** For ethanol, C_2H_5OH, which is mixed with gasoline to make the fuel gasohol, $\Delta H_f^\circ = -277.63$ kJ/mol. Calculate the number of kilojoules released by burning completely 1 gallon of ethanol. The density of ethanol is 0.787 g cm^{-3}. Use data in Table 6.2 to help in the computation.

6.85 Consider the following thermochemical equations:

$$(1)\ CH_3OH(l) + O_2(g) \longrightarrow HCHO_2(l) + H_2O(l) \quad \Delta H^\circ = -411 \text{ kJ}$$

$$(2)\ CO(g) + 2H_2(g) \longrightarrow CH_3OH(l) \quad \Delta H^\circ = -128 \text{ kJ}$$

$$(3)\ HCHO_2(l) \longrightarrow CO(g) + H_2O(l) \quad \Delta H^\circ = -33 \text{ kJ}$$

Suppose Equation (1) is reversed and divided by 2, Equations (2) and (3) are multiplied by $\frac{1}{2}$, and then the three adjusted equations are added. What is the net reaction, and what is the value of ΔH° for the net reaction?

6.86 Chlorofluoromethanes (CFMs) are carbon compounds of chlorine and fluorine and are also known as Freons. Examples are Freon-11 ($CFCl_3$) and Freon-12 (CF_2Cl_2), which have been used as aerosol propellants. Freons have also been used in refrigeration and air-conditioning systems. It is feared that as these Freons escape into the atmosphere they will lead to a significant depletion of ozone from the upper atmosphere where ozone protects the earth's inhabitants from harmful ultraviolet radiation. In the stratosphere CFMs absorb high-energy radiation from the sun and split off chlorine atoms that hasten the decomposition of ozone, O_3. Possible reactions are

$$(1)\ O_3(g) + Cl(g) \longrightarrow O_2(g) + ClO(g) \quad \Delta H^\circ = -126 \text{ kJ}$$

$$(2)\ ClO(g) + O(g) \longrightarrow Cl(g) + O_2(g) \quad \Delta H^\circ = -268 \text{ kJ}$$

$$(3)\ O_3(g) + O(g) \longrightarrow 2O_2(g)$$

The O atoms in Equation 2 come from the breaking apart of O_2 molecules caused by radiation from the sun. Use Equations 1 and 2 to calculate the value of ΔH° (in kilojoules) for Equation 3, the net reaction for the removal of O_3 from the atmosphere.

6.87 Suppose a truck with a mass of 14.0 tons (1 ton = 2000 lb) is traveling at a speed of 45.0 mi/hr. If the truck driver slams on the brakes, the kinetic energy of the truck is changed to heat as the brakes slow the truck to a stop. How much would the temperature of 5.00 gallons of water increase if all the heat could be absorbed by the water?

EXERCISES IN CRITICAL THINKING

6.88 Suppose we compress a spring, tie it up tightly, and then dissolve the spring in acid. What happens to the potential energy contained in the compressed spring? How could you measure it?

6.89 Carefully and precisely describe the difference between H, ΔH, and ΔH°.

6.90 Why do we usually use ΔH° rather than ΔE° when we discuss energies of reaction? Develop an argument for using ΔE° instead.

6.91 Explain why we can never determine values for E or H. Make a list of the factors we would need to know to determine E or H.

6.92 Find the heats of formation of some compounds that are explosives. What do they have in common? Compare them to the ΔH_f° values of stable compounds such as H_2O, NaCl, and $CaCO_3$. Formulate a generalization to summarize your observations.

CHAPTERS 4-6

BRINGING IT TOGETHER

We pause again to allow you to test your understanding of concepts, your knowledge of scientific terms, and your skills at solving chemistry problems. Read through the following questions carefully, and answer each as fully as possible. When necessary, review topics you are uncertain of. If you can answer these questions correctly, you are ready to go on to the next group of chapters.

1. What is the difference between a strong electrolyte and a weak electrolyte? Formic acid, $HCHO_2$, is a weak acid. Write a chemical equation showing its reaction with water.

2. Write an equation showing the reaction of water with itself to form ions.

3. Methylamine, CH_3NH_2, is a weak base. Write a chemical equation showing its reaction with water.

4. Write molecular, ionic, and net ionic equations for the reaction that occurs when a solution containing hydrochloric acid is added to a solution of the weak base methylamine (CH_3NH_2).

5. According to the solubility rules, which of the following salts would be classified as soluble?

(a) $Ca_3(PO_4)_2$ (f) $Au(ClO_4)_3$ (k) $ZnSO_4$
(b) $Ni(OH)_2$ (g) $Cu(C_2H_3O_2)_2$ (l) Na_2S
(c) $(NH_4)_2HPO_4$ (h) AgBr (m) $CoCO_3$
(d) $SnCl_2$ (i) KOH (n) $BaSO_3$
(e) $Sr(NO_3)_2$ (j) Hg_2Cl_2 (o) MnS

6. What are the two criteria that must be met for an ionic equation to be balanced correctly?

7. Write molecular, ionic, and net ionic equations for any reactions that would occur between the following pairs of compounds. If no reaction occurs, write "N.R."

(a) $CuCl_2(aq)$ and $(NH_4)_2CO_3(aq)$
(b) $HCl(aq)$ and $MgCO_3(s)$
(c) $ZnCl_2(aq)$ and $AgC_2H_3O_2(aq)$
(d) $HClO_4(aq)$ and $NaCHO_2(aq)$
(e) $MnO(s)$ and $H_2SO_4(aq)$
(f) $FeS(s)$ and $HCl(aq)$

8. Write a chemical equation for the complete neutralization of H_3PO_4 by NaOH.

9. Which ion exists in abundance in all solutions of strong acids?

10. Which ion makes a solution basic?

11. Define *monoprotic acid, diprotic acid,* and *polyprotic acid.* What is the general definition of a *salt*?

12. Which of the following oxides are acidic and which are basic: P_4O_6, Na_2O, SeO_3, CaO, PbO, and SO_2?

13. Write the formulas of any acid salts that could be formed by the reaction of the following acids with potassium hydroxide.

(a) sulfurous acid
(b) nitric acid
(c) hypochlorous acid
(d) phosphoric acid
(e) carbonic acid

14. Name the following

(a) HIO_3 (d) $Ca(H_2PO_4)_2$
(b) HOBr (e) $Fe(HSO_4)_3$
(c) HNO_2

15. Write formulas for the following:

(a) bromous acid
(b) hypoiodous acid
(c) sodium dihydrogen phosphate
(d) lithium hydrogen sulfate
(e) bromic acid

16. How many milliliters of 0.200 *M* $BaCl_2$ must be added to 27.0 mL of 0.600 *M* Na_2SO_4 to give a complete reaction between their solutes?

17. What mass of $Mg(OH)_2$ will be formed when 30.0 mL of 0.200 *M* $MgCl_2$ solution is mixed with 25.0 mL of 0.420 *M* NaOH solution? What will be the molar concentrations of the ions remaining in solution?

18. How many milliliters of 6.00 *M* HNO_3 must be added to 200 mL of water to give 0.150 *M* HNO_3?

19. How many grams of CO_2 must be dissolved in 300 mL of 0.100 *M* Na_2CO_3 solution to change the solute entirely into $NaHCO_3$?

20. A certain toilet cleaner uses $NaHSO_4$ as its active ingredient. In an analysis, 0.500 g of the cleaner was dissolved in 30.0 mL of distilled water and required 24.60 mL of 0.105 *M* NaOH for complete neutralization in a titration. What was the percentage by weight of $NaHSO_4$ in the cleaner?

21. A volume of 28.50 mL of a freshly prepared solution of KOH was required to titrate 50.00 mL of 0.0922 *M* HCl solution. What was the molarity of the KOH solution?

22. To neutralize the acid in 10.0 mL of 18.0 *M* H_2SO_4 that was accidentally spilled on a laboratory bench top, solid sodium bicarbonate was used. The container of sodium bicarbonate was known to weigh 155.0 g before this use and out of curiosity its mass was measured as 144.5 g afterward. The reaction forms sodium sulfate. Was sufficient sodium bicarbonate used? Determine the limiting reactant and calculate the maximum yield in grams of sodium sulfate.

23. How many milliliters of concentrated sulfuric acid (18.0 *M*) are needed to prepare 125 mL of 0.144 *M* H_2SO_4?

24. The density of concentrated phosphoric acid solution is 1.689 g solution/mL solution at 20 °C. It contains 144 g H_3PO_4 per 1.00×10^2 mL of solution.

(a) Calculate the molar concentration of H_3PO_4 in this solution.
(b) Calculate the number of grams of this solution required to hold 50.0 g H_3PO_4.

25. A mixture consists of lithium carbonate (Li_2CO_3) and potassium carbonate (K_2CO_3). These react with hydrochloric acid as follows.

$$Li_2CO_3(s) + 2HCl(aq) \longrightarrow 2LiCl(aq) + H_2O + CO_2(g)$$

$$K_2CO_3(s) + 2HCl(aq) \longrightarrow 2KCl(aq) + H_2O + CO_2(g)$$

When 4.43 g of this mixture was analyzed, it consumed 53.2 mL of 1.48 *M* HCl. Calculate the number of grams of each carbonate and their percentages.

26. One way to prepare iodine is to react sodium iodate, $NaIO_3$, with hydroiodic acid, HI. The following reaction occurs.

$$NaIO_3 + 6HI \longrightarrow 3I_2 + NaI + 3H_2O$$

Calculate the number of moles and the number of grams of iodine that can be made this way from 16.4 g of $NaIO_3$.

27. A white solid was known to be the anhydrous form of either sodium carbonate or sodium bicarbonate. Both react with hydrochloric acid to give sodium chloride, water, and carbon dioxide, but the mole proportions are not the same.
(a) Write the balanced equation for each reaction.
(b) It was found that 0.5128 g of the solid reacted with 47.80 mL of 0.2024 *M* HCl, and that the addition of more acid caused the formation of no more carbon dioxide. Perform the calculations that establish which substance the unknown solid was.

28. Assign oxidation numbers to the atoms in the following formulas: (a) As_4, (b) $HClO_2$, (c) $MnCl_2$, and (d) $V_2(SO_3)_3$.

29. For the following unbalanced equations, write the reactants and products in the form they should appear in an ionic equation. Then write balanced net ionic equations by applying the ion–electron method.
(a) $K_2Cr_2O_7 + HCl \longrightarrow KCl + Cl_2 + H_2O + CrCl_3$
(b) $KOH + SO_2(aq) + KMnO_4 \longrightarrow K_2SO_4 + MnO_2 + H_2O$

30. Balance the following equations by the ion–electron method for *acidic solutions.*
(a) $Cr_2O_7^{2-} + Br^- \longrightarrow Br_2 + Cr^{3+}$
(b) $H_3AsO_3 + MnO_4^- \longrightarrow H_2AsO_4^- + Mn^{2+}$

31. Balance the following equations by the ion–electron method for *basic solutions.*
(a) $I^- + CrO_4^{2-} \longrightarrow CrO_2^- + IO_3^-$
(b) $SO_2 + MnO_4^- \longrightarrow MnO_2 + SO_4^{2-}$

32. In the previous two questions, identify the oxidizing agents and reducing agents.

33. Complete and balance the following equations if a reaction occurs.
(a) $Sn(s) + HCl(aq) \longrightarrow$
(b) $Cu(s) + HNO_3$ (conc.) $\longrightarrow$
(c) $Zn(s) + Cu^{2+}(aq) \longrightarrow$
(d) $Ag(s) + Cu^{2+}(aq) \longrightarrow$

34. Write a balanced chemical equation for the combustion of cetane, $C_{16}H_{34}$, a hydrocarbon present in diesel fuel, (a) in the presence of excess oxygen, (b) in a somewhat limited supply of oxygen, and (c) in a severely limited supply of oxygen.

35. Stearic acid, $C_{17}H_{35}CO_2H$, is derived from animal fat. Write a balanced chemical equation for the combustion of stearic acid in an abundant supply of oxygen.

36. Methanethiol, CH_3SH, is a foul-smelling gas produced in the intestinal tract by bacteria acting on albumin in the absence of air. Write a chemical equation for the combustion of CH_3SH in an excess supply of oxygen.

37. Why is $KMnO_4$ such a useful laboratory oxidizing agent? Write half-reactions for the reduction of MnO_4^- in (a) acidic solution and (b) basic solution.

38. Write molecular equations for the reaction of O_2 with (a) magnesium, (b) aluminum, (c) phosphorus, and (d) sulfur.

39. *Bordeaux mixture* is traditionally prepared by mixing copper sulfate and calcium hydroxide in water. The resulting suspension of copper hydroxide is sprayed on trees and shrubs to fight fungus diseases. This fungicide is also available in commercial preparations. In an analysis of one such product, a sample weighing 0.238 g was dissolved in hydrochloric acid. Excess KI solution was then added, forming copper (I) iodide and iodine, and the iodine that was formed was titrated with 0.01669 *M* $Na_2S_2O_3$ solution using starch as an indicator. The titration required 28.62 mL of the thiosulfate solution. What was the percentage by weight of copper in the sample of Bordeaux mixture?

40. An aluminum ball 0.500 cm in diameter is to be dissolved in hydrochloric acid. What is the minimum volume of 0.500 *M* HCl needed to dissolve the aluminum?

41. Suppose we have a system that consists of 225 g of pure water at 25.00 °C and 1 atm of pressure. What value do we use for the heat capacity of this system? What value do we use for the specific heat of the water? (For the energy part of the unit, use the joule.)

42. In what specific ways do a standard enthalpy of reaction and a standard enthalpy of formation differ?

43. Write the chemical equation for which the $\Delta H°$ is equal to the standard heat of formation of nitric acid.

44. Why was the concept of a *standard state* introduced into chemistry?

45. Under what conditions is an element in its standard state? Which form of oxygen can exist in the standard state of oxygen?

46. Why isn't the value of ΔH_f° equal to zero for ozone, O_3?

47. The specific heat of helium is 5.19 J/g °C and that of nitrogen is 1.04 J/g °C. How many joules can one mole of each gas absorb when its temperature increases by 1.00 °C?

48. A calorimeter vat in which a stirring motor, a thermometer, and a "bomb" are immersed absorbed the heat released by the combustion of 0.514 g of benzoic acid. The thermochemical equation for its combustion is as follows.
$$2C_7H_6O_2(s) + 15O_2(g) \longrightarrow 14CO_2(g) + 6H_2O(l) \qquad \Delta H° = -7048 \text{ kJ}$$
The temperature of the calorimeter rose from 24.112 °C to 24.866 °C. What is the heat capacity of this calorimeter in kJ/°C?

49. A sample of 10.1 g of ammonium nitrate, NH_4NO_3, was dissolved in 125 g of water in a coffee cup calorimeter. The temperature changed from 24.5 °C to 18.8 °C. Calculate the *heat of solution* of ammonium nitrate in kJ/mol. Assume that the energy exchange involves only the solution and that the specific heat of the solution is 4.18 J/g °C.

50. When 0.6484 g of cetyl palmitate, $C_{32}H_{64}O_2$ (a fruit wax), was burned in a bomb calorimeter with a heat capacity of 11.99 kJ/°C, the temperature of the calorimeter rose from 24.518 °C to 26.746 °C. Calculate the molar heat of combustion of cetyl palmitate in kJ/mol.

51. When we say that the value of ΔH for a chemical reaction is a *state function*, what do we mean?

52. The combustion of methane (the chief component of natural gas) follows the equation
$$CH_4(g) + 2O_2(g) \longrightarrow CO_2(g) + 2H_2O(g)$$
$\Delta H°$ for this reaction is −802.3 kJ. How many grams of methane must be burned to provide enough heat to raise the temperature of 250 mL of water from 25.0 °C to 50.0 °C?

53. Label the following thermal properties as intensive or extensive.
(a) specific heat
(b) heat capacity
(c) ΔH_f°
(d) $\Delta H°$
(e) molar heat capacity

54. What is the definition of *internal energy*? State the first law of thermodynamics.

55. What is *pressure–volume work*? Give the equation that could be used to calculate it.

56. The change in the internal energy, ΔE, is equal to the "heat of reaction at constant volume." Explain the reason for this.

57. Suppose that a gas is produced in an exothermic chemical reaction between two reactants in an aqueous solution. Which quantity will have the larger magnitude, ΔE or ΔH?

58. The thermochemical equation for the combustion reaction of half a mole of carbon monoxide is as follows.

$$\tfrac{1}{2}CO(g) + \tfrac{1}{4}O_2(g) \longrightarrow \tfrac{1}{2}CO_2(g) \quad \Delta H^\circ = -141.49 \text{ kJ}$$

Write the thermochemical equation for
(a) the combustion of 2 mol of $CO(g)$
(b) the decomposition of 1 mol of $CO_2(g)$ to $O_2(g)$ and $CO(g)$

59. The standard heat of combustion of eicosane, $C_{20}H_{42}(s)$, a typical component of candle wax, is 1.332×10^4 kJ/mol, when it burns in pure oxygen and the products are cooled to 25 °C. The only products are $CO_2(g)$ and $H_2O(l)$. Calculate the value of the standard heat of formation of eicosane (in kJ/mol) and write the corresponding thermochemical equation.

60. Using data in Table 6.2 calculate values for the standard heats of reaction (in kilojoules) for the following reactions.
(a) $H_2SO_4(l) \longrightarrow SO_3(g) + H_2O(l)$
(b) $C_2H_6(g) \longrightarrow C_2H_4(g) + H_2(g)$

61. Calculate the standard heat of formation of calcium carbide, $CaC_2(s)$, in kJ/mol using the following thermochemical equations.

$$Ca(s) + 2H_2O(l) \longrightarrow Ca(OH)_2(s) + H_2(g) \quad \Delta H^\circ = -414.79 \text{ kJ}$$

$$2C(s) + O_2(g) \longrightarrow 2CO(g) \quad \Delta H^\circ = -221.0 \text{ kJ}$$

$$CaO(s) + H_2O(l) \longrightarrow Ca(OH)_2(s) \quad \Delta H^\circ = -65.19 \text{ kJ}$$

$$2H_2(g) + O_2(g) \longrightarrow 2H_2O(l) \quad \Delta H^\circ = -571.8 \text{ kJ}$$

$$CaO(s) + 3C(s) \longrightarrow CaC_2(s) + CO(g) \quad \Delta H^\circ = +462.3 \text{ kJ}$$

THE QUANTUM MECHANICAL ATOM

7

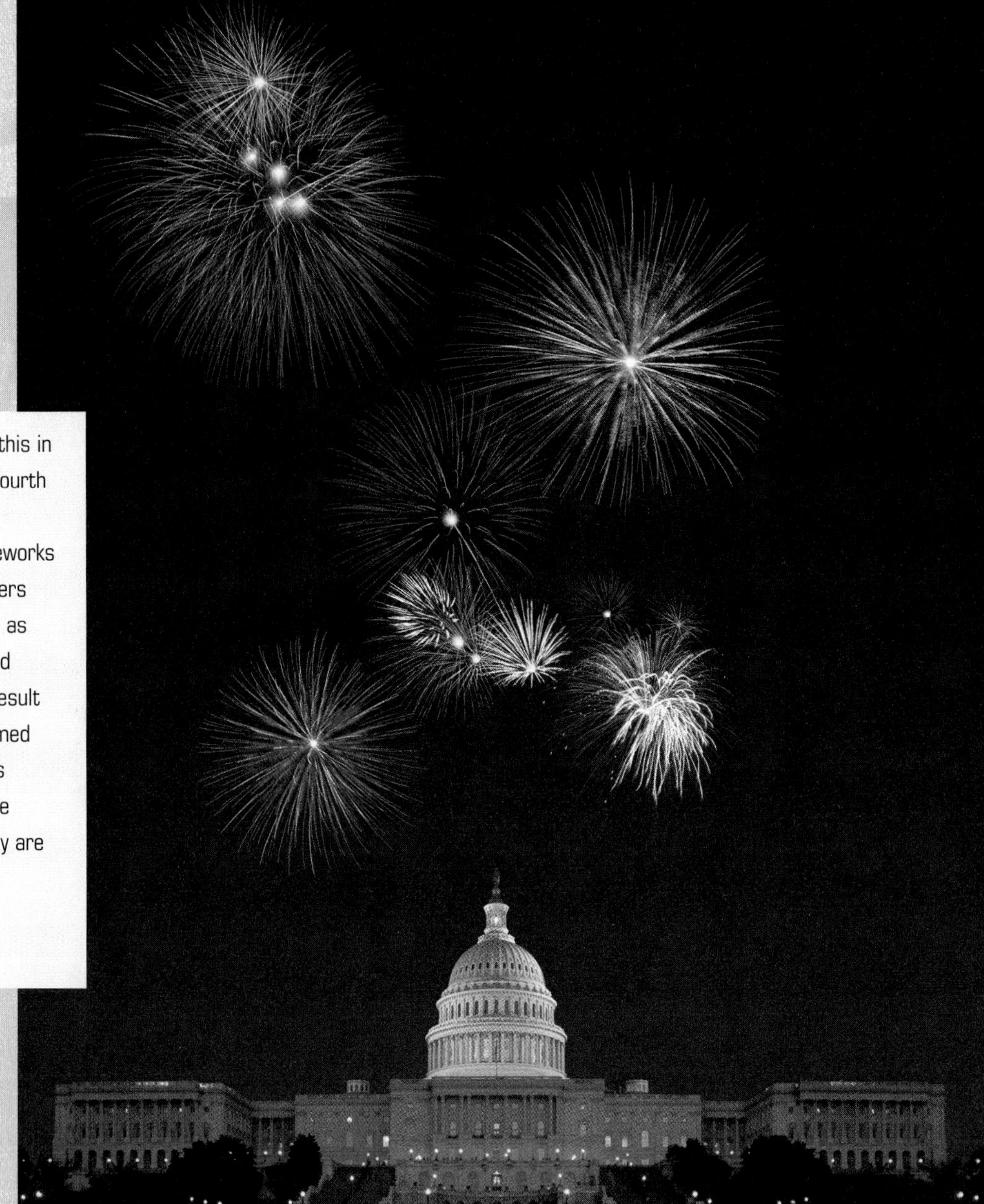

Colorful fireworks such as this in Washington, D.C., on the fourth of July are popular events worldwide. Many of the fireworks displays resemble the flowers they are named after, such as chrysanthemum, peony, and palm. Each display is the result of two or more carefully timed chemical explosions. In this chapter we will see that the colors in a fireworks display are easily explained by modern quantum mechanics. *(Bill Ross/Corbis)*

CHAPTER OUTLINE

THIS CHAPTER IN CONTEXT Nature presents the observer with very complex and often large chemical structures that are essential for life and also make up the nonliving world around us. Understanding these structures starts at the atomic level. If we know how the atoms are put together and their properties we can understand how they bond to each other (Chapter 8) and then we will understand the basic three-dimensional geometries (Chapter 9) that then allow us to understand the larger structures and the interactions of matter. This chapter introduces us to the most modern concepts of the atom itself.

In previous chapters we have described how the structure of the atom was deduced from a variety of experimental evidence. Each experiment added more detail to the make-up of the atom. It started with the indivisible atom and grew to the nuclear model proposed by Rutherford where negatively charged electrons surrounded an extremely dense positively charged nucleus composed of protons and neutrons. This model explained the mass relationships between the elements and the nature of isotopes. However, there were many unanswered questions. Why do metals tend to form cations and why do nonmetals tend to form anions? Why are certain combinations of elements very common while other combinations are never observed? Why do two nonmetals combine to form compounds while no similar process exists for two metals to form a compound? Why are the noble gases virtually inert? Why does the periodic table have the arrangement and shape it has? Many more questions could also be posed, but there was one major question.

This fundamental question concerned the fact that classical physics predicted that atoms simply could not exist. The physics of the late 1800s predicted that electrons surrounding the nucleus must quickly lose energy and crash into the nucleus. We call this the **collapsing atom paradox.** At the same time other unsolvable problems arose for classical physics. Heated objects should have emitted vast quantities of ultraviolet (UV) light. In fact they emit very little UV radiation. This was known as the **ultraviolet catastrophe.** In addition, emerging studies of particles, such as electrons, passing through very small openings gave results (diffraction patterns) that could only be explained if we regarded particles as waves. This led to the concept of the **wave/particle duality** of matter and energy.

The above problems with classical physics made it clear that an entirely new set of concepts would be needed to define modern physics and chemistry. These concepts are commonly called **wave mechanics, quantum mechanics,** or **quantum theory** and they are now a cornerstone of modern chemistry.

Light emitted from atoms gives us clues about how electrons are arranged within an atom. At the same time an understanding of standing waves develops insights to explain why atoms do not collapse. Therefore we must begin our introduction to quantum mechanics by describing the nature of electromagnetic radiation.

7.1 Electromagnetic Radiation Provides the Clue to the Electronic Structures of Atoms

Electromagnetic radiation can be described as a wave or as a stream of photons

You've learned that objects can have energy in only two ways, as kinetic energy and as potential energy. You also learned that energy can be transferred between things, and in Chapter 6 our principal focus was on the transfer of heat. Energy can also be transferred between atoms and molecules in the form of light or electromagnetic energy. This is a very important form of energy in chemistry. For example, many chemical systems emit visible light as they react (see Figure 7.1).

Many experiments show that electromagnetic radiation carries energy through space by means of **waves.** Waves are an oscillation that moves outward from a disturbance (think of ripples moving away from a pebble dropped into a pond). In the case of electromagnetic radiation, the disturbance can be a vibrating electric charge. When the charge oscillates, it produces a pulse in the electric field around it. As the electric field pulses, it creates a pulse in the magnetic field. The magnetic field pulse gives rise to yet another electric field pulse

Electricity and magnetism are closely related to each other. A moving charge creates an electric current, which in turn creates a magnetic field around it. This is the fundamental idea behind electric motors. A moving magnetic field creates an electric field or current. This is the idea behind electrical generators and turbines.

(a)

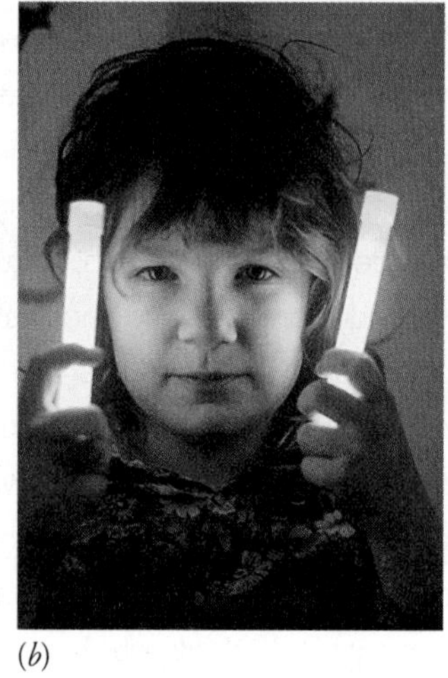
(b)

(c)

FIG. 7.1 Light is given off in a variety of chemical reactions. (*a*) Combustion. (*b*) Cyalume light sticks. (*c*) A lightning bug. *(Peter Arnold, Inc.; IPA/The Image Works; Edward Degginger/Bruce Coleman, Inc.)*

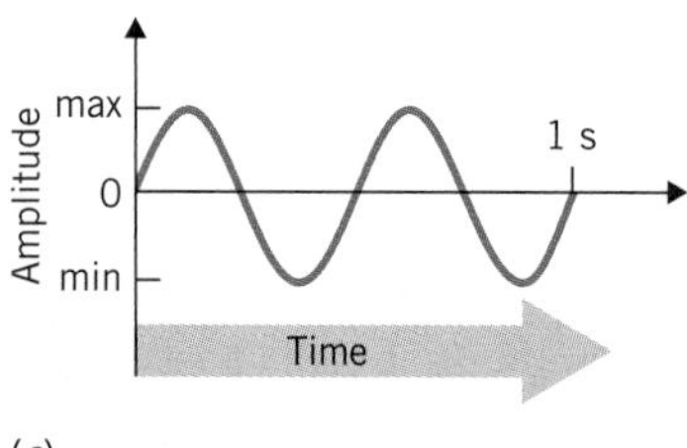

(a)

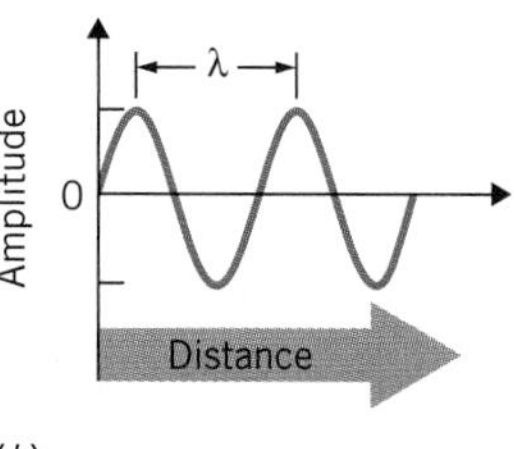

(b)

FIG. 7.2 Two views of electromagnetic radiation. (*a*) The frequency, ν, of a light wave is the number of complete oscillations each second. Here two cycles span a one-second time interval, so the frequency is 2 cycles per second, or 2 Hz. (*b*) Electromagnetic radiation frozen in time. This curve shows how the amplitude varies along the direction of travel. The distance between two peaks is the wavelength, λ, of the electromagnetic radiation.

further away from the disturbance. The process continues, with a pulse in one field giving rise to a pulse in the other, and the resulting train of pulses in the electric and magnetic fields is called an **electromagnetic wave.** This wave ripples away from the source at extremely high speeds.

Electromagnetic waves don't need a medium to travel through, as water and sound waves do. They can cross empty space. The speed of the electromagnetic wave in a vacuum is the same no matter how the radiation is created (about 3.00×10^8 m/s).

An electromagnetic wave is often depicted as a sine wave that has an amplitude, wavelength, and frequency. Figure 7.2 shows how the amplitude or intensity of the wave varies with time and with distance as the wave travels through space. **Amplitude** of the wave is related to the intensity or brightness of the radiation. In Figure 7.2*a*, we see two complete oscillations or *cycles* of the wave during a one-second interval. The number of cycles per second is called the **frequency** of the electromagnetic radiation, and its symbol is $\boldsymbol{\nu}$ (the Greek letter *nu*, pronounced "new"). In the SI, the unit of time is the second (s), so frequency is given the unit "per second," which is $\frac{1}{\text{second}}$, or $(\text{second})^{-1}$. This unit is given the special name **hertz (Hz).**

$$1 \text{ Hz} = 1 \text{ s}^{-1}$$

The SI symbol for the second is s.

$$\text{s}^{-1} = \frac{1}{\text{s}}$$

As electromagnetic radiation moves away from its source, the positions of maximum and minimum amplitude (peaks and troughs) are regularly spaced. The peak-to-peak distance is called the radiation's **wavelength,** symbolized by $\boldsymbol{\lambda}$ (the Greek letter *lambda*). See Figure 7.2*b*. Because wavelength is a distance, it has distance units (for example, meters).

If we multiply the wavelength by frequency, the result is the speed of the wave. We can see this if we analyze the units.

For any wave, the product of its wavelength and its frequency equals the speed of the wave.

(In SI units)

$$\text{meters} \times \frac{1}{\text{second}} = \frac{\text{meters}}{\text{second}} = \text{speed}$$

$$\text{m} \times \frac{1}{\text{s}} = \frac{\text{m}}{\text{s}} = \text{m s}^{-1}$$

The speed of electromagnetic radiation in a vacuum is a constant and is commonly called the **speed of light.** Its value to three significant figures is 3.00×10^8 m/s (or m s^{-1}). This important physical constant is given the symbol $\boldsymbol{c}$.

The speed of light is one of our most carefully measured constants, because the meter is defined in terms of it. The precise value of the speed of light in a vacuum is 2.99792458×10^8 m/s, and a meter is defined as exactly the distance traveled by light in 1/299,792,458 of a second.

$$c = 3.00 \times 10^8 \text{ m s}^{-1}$$

From the preceding discussion we obtain a very important relationship that allows us to convert between wavelength, λ and frequency, ν.

$$\lambda \times \nu = c = 3.00 \times 10^8 \text{ m s}^{-1} \tag{7.1}$$

Wavelength–frequency relationship

EXAMPLE 7.1
Calculating Frequency from Wavelength

Mycobacterium tuberculosis, the organism that causes tuberculosis, can be completely destroyed by irradiation with ultraviolet light with a wavelength of 254 nm. What is the frequency of this radiation?

ANALYSIS: To convert between wavelength and frequency we use the tool expressed by Equation 7.1. However, we must be careful about the units.

SOLUTION: To calculate the frequency, we solve Equation 7.1 for ν.

$$\nu = \frac{c}{\lambda}$$

Next, we substitute for c (3.00×10^8 m s^{-1}) and use 254 nm for the wavelength. However, to cancel units correctly, we must have the wavelength in meters. Recall from Chapter 1 that nm means nanometer and the prefix nano implies the factor "$\times 10^{-9}$."

$$1 \text{ nm} = 1 \times 10^{-9} \text{ m}$$

Therefore, 254 nm equals 254×10^{-9} m. Substituting gives

$$\begin{aligned} \nu &= \frac{3.00 \times 10^8 \cancel{\text{m}}\,\text{s}^{-1}}{254 \times 10^{-9} \cancel{\text{m}}} \\ &= 1.18 \times 10^{15} \text{ s}^{-1} \\ &= 1.18 \times 10^{15} \text{ Hz} \end{aligned}$$

IS THE ANSWER REASONABLE? One way to test if our answer is correct is to see if the given wavelength, multiplied by our calculated frequency, gives us the speed of light. We will round the 254 nm to 250×10^{-9} m and round our answer to 1×10^{15} s^{-1}

$$(250 \times 10^{-9} \text{ m}) \times (1 \times 10^{15} \text{ s}^{-1}) = 250 \times 10^{6} \text{ m s}^{-1} = 2.5 \times 10^8 \text{ m s}^{-1}$$

which is close to the actual speed of light, within one significant figure. Using the actual data and a calculator gives us the exact value for the speed of light.

EXAMPLE 7.2
Calculating Wavelength from Frequency

Radio station WGBB on Long Island, New York, broadcasts its AM signal, a form of electromagnetic radiation, at a frequency of 1240 kHz. What is the wavelength of the radio waves expressed in meters?

ANALYSIS: This question will utilize the same equation as Example 7.1. This time we solve for wavelength in the units asked for in the problem.

SOLUTION: Solving Equation 7.1 for the wavelength gives

$$\lambda = \frac{c}{\nu} = \frac{3.00 \times 10^8 \text{ m s}^{-1}}{1240 \text{ kHz}}$$

In order for the units to work out we recall that the prefix "k" in kHz means kilo and stands for "$\times 10^3$" and Hz means s^{-1}. Replacing k with $\times 10^3$ and Hz with s^{-1} results in the frequency written as $1240 \times 10^3 \text{ s}^{-1}$. Substituting gives

$$\lambda = \frac{3.00 \times 10^8 \text{ m } \cancel{\text{s}^{-1}}}{1240 \times 10^3 \cancel{\text{s}^{-1}}}$$

$$= 242 \text{ m}$$

IS THE ANSWER REASONABLE? If this wavelength is correct, we should be able to multiply it by the original frequency (in Hz) and get the speed of light, as we did in the previous example. Another approach is to test our answer by dividing the speed of light by the wavelength to get back the frequency. Rounding to one significant figure we get

$$\frac{3 \times 10^8 \cancel{\text{m}}/\text{s}}{2 \times 10^2 \cancel{\text{m}}} = 1.5 \times 10^6 \text{ Hz} = 1500 \text{ kHz}$$

which is reasonably close to the original frequency of 1240 kHz.

Practice Exercise 1: Helium derives it name from the Latin name for the sun. Helium was discovered when spectroscopists found that the 588 nm wavelength (among others) was missing from the sun's spectrum. What is the frequency of this radiation? (Hint: Recall the metric prefixes so units will cancel correctly.)

Practice Exercise 2: The most intense radiation emitted by the earth has a wavelength of about 10.0 μm. What is the frequency of this radiation in hertz?

Practice Exercise 3: An FM radio station in West Palm Beach, Florida, broadcasts electromagnetic radiation at a frequency of 104.3 MHz (megahertz). What is the wavelength of the radio waves, expressed in meters?

Electromagnetic waves are categorized by frequency

Remember that there is an inverse relationship between wavelength and frequency. The lower the frequency, the longer the wavelength.

Electromagnetic radiation comes in a broad range of frequencies called the **electromagnetic spectrum,** illustrated in Figure 7.3. Some portions of the spectrum have popular names. For example, radio waves are electromagnetic radiations having very low frequencies (and therefore very long wavelengths). Microwaves, which also have low frequencies, are emitted by radar instruments such as those the police use to monitor the speeds of cars. In microwave ovens, similar radiation is used to heat water in foods, causing the food to cook quickly. Infrared radiation is emitted by hot objects and consists of the range of frequencies that can make molecules of most substances vibrate internally. You can't see infrared radiation, but you can feel how your body absorbs it by holding your hand near a hot radiator; the absorbed radiation makes your hand warm. Gamma rays (γ rays) are at the high-frequency end of the electromagnetic spectrum. They are produced by certain

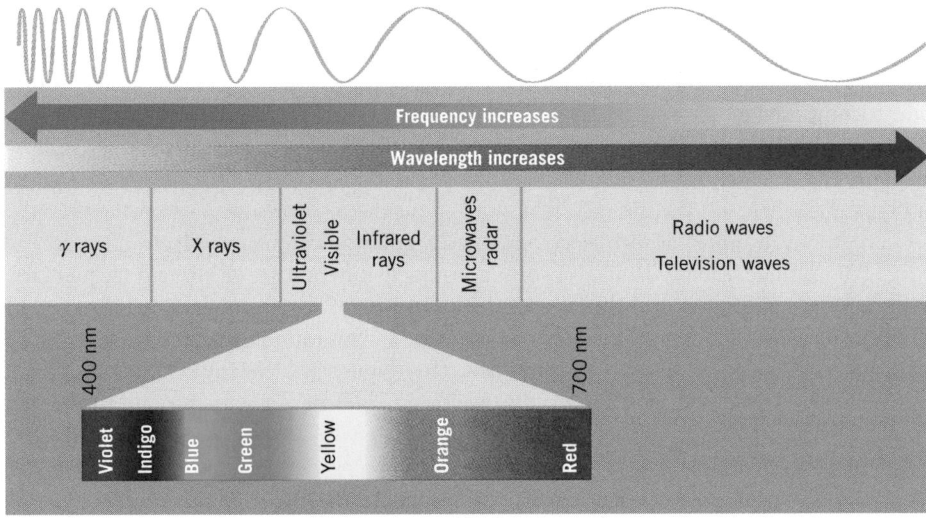

(*a*)

(*b*)

(*c*)

FIG. 7.3 The electromagnetic spectrum. (*a*) The electromagnetic spectrum is divided into regions according to the wavelengths of the radiation. (*b*) The visible spectrum is composed of wavelengths that range from about 400 to 700 nm. (*c*) The production of a visible spectrum by splitting white light into its rainbow of colors. (*From "The Gift of Color," Eastman Kodak Company.*)

elements that are radioactive. X rays are very much like gamma rays, but they are usually made by special equipment. Both X rays and gamma rays penetrate living things easily.

Most of the time, you are bombarded with electromagnetic radiation from all portions of the electromagnetic spectrum. Radio and TV signals pass through you; you feel infrared radiation when you sense the warmth of a radiator; X rays and gamma rays fall on you from space; and light from a lamp reflects into your eyes from the page you're reading. Of all these radiations, your eyes are able to sense only a very narrow band of wavelengths ranging from about 400 to 700 nm. This band is called the **visible spectrum** and consists of all the colors you can see, from red through orange, yellow, green, blue, and violet. White light is composed of all these colors and it can be separated into them by focusing a beam of white light through a prism, which spreads the various wavelengths apart. This is illustrated in Figure 7.3*b*. A photograph showing the production of a visible spectrum is given in Figure 7.3*c*.

The way substances absorb electromagnetic radiation often can help us characterize them. For example, each substance absorbs a uniquely different set of infrared frequencies. A plot of the wavelengths absorbed versus the intensities of absorption is called an infrared absorption spectrum. It can be used to identify a compound, because each infrared spectrum is as unique as a set of fingerprints. (See Figure 7.4.) Many substances absorb visible and ultraviolet radiations in unique ways, too, and they have visible and ultraviolet spectra (Figure 7.5).

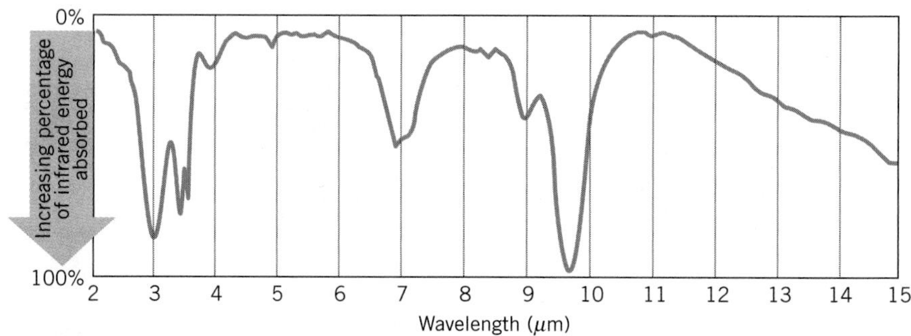

FIG. 7.4 Infrared absorption spectrum of methyl alcohol (also called wood alcohol), the fuel in "canned heat" products such as Sterno. In an infrared spectrum, the usual practice is to show the amount of light absorbed increasing from top to bottom in the graph. Thus, there is a peak in the percentage of light absorbed at about 3 μm. (Spectrum courtesy Sadtler Research Laboratories, Inc., Philadelphia, Pa.)

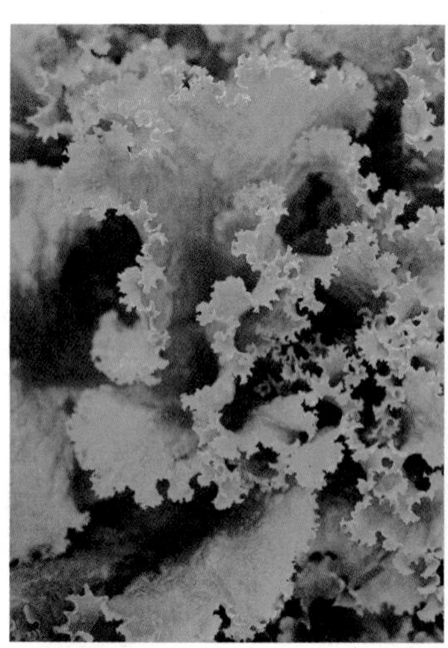

(*a*)

Amount of light absorbed
Violet
Blue
Green
Yellow
Red
400
500
600
700
Wavelength (nm)

(*b*)

FIG. 7.5 Absorption of light by chlorophyll. (*a*) Chlorophyll is the green pigment plants use to harvest solar energy for photosynthesis. (*b*) In this visible absorption spectrum of chlorophyll, the percentage of light absorbed increases from bottom to top. Thus, there is a peak in the light absorbed at about 420 nm and another at about 660 nm. This means the pigment strongly absorbs blue-violet and red light. The green color we *see* is the light that's *not* absorbed. It's composed of the wavelengths of visible light that are reflected. (Our eyes are most sensitive to green, so we don't notice the yellow components of the reflected light.) *((a) Gary Braasch/Stone/Getty Images.)*

Electromagnetic radiation can be viewed as a stream of photons

When an electromagnetic wave passes an object, the oscillating electric and magnetic fields may interact with it, as ocean waves interact with a buoy in a harbor. A tiny charged particle placed in the path of the wave will be yanked back and forth by the oscillating electric and magnetic fields. For example, when a radio wave strikes an antenna, electrons within the antenna begin to bounce up and down, creating an alternating current which can be detected and decoded electronically. Because the wave exerts a force on the antenna's electrons and moves them through a distance, work is done. Thus, as energy is lost by the source of the wave (the radio transmitter) energy is gained by the electrons in the antenna.

A series of groundbreaking experiments showed that classical physics does not correctly describe energy transfer by electromagnetic radiation. In 1900 a German physicist named Max Planck (1858–1947) proposed that electromagnetic radiation can be viewed as a stream of tiny packets or **quanta** of energy that were later called **photons.** Each photon travels at the speed of light. Planck proposed, and Albert Einstein (1879–1955) confirmed, that *the energy of a photon of electromagnetic radiation is proportional to the radiation's frequency,* not to its intensity or brightness as had been believed up to that time. (See Facets of Chemistry 7.1.)

The energy of one photon is called one **quantum** of energy.

$$\text{Energy of a photon} = E = h\nu \tag{7.2}$$

In this expression, h is a proportionality constant that we now call **Planck's constant.** Note that Equation 7.2 relates two representations of electromagnetic radiation. The left-hand side of the equation deals with a property of particles (energy per photon); the right-hand side deals with a property of waves (the frequency). Quantum theory unites the two representations, so we can use whichever representation of electromagnetic radiation is convenient for describing experimental results. For example, in describing the photoelectric effect (Facets of Chemistry 7.1), we represent radiation as a stream of particles. When describing

The value of Planck's constant is 6.626×10^{-34} J s. It has units of energy (joules) multiplied by time (seconds).

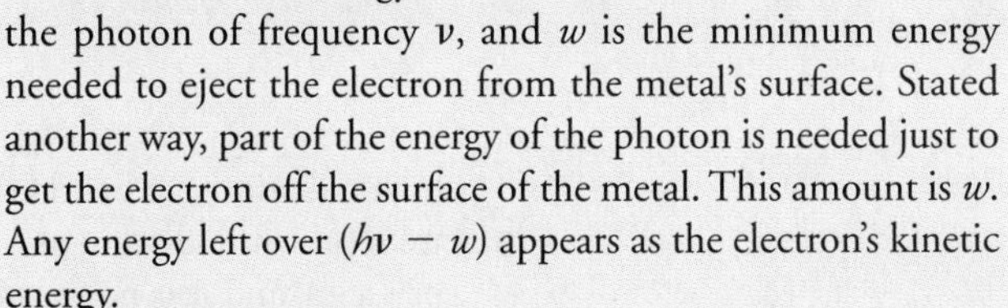

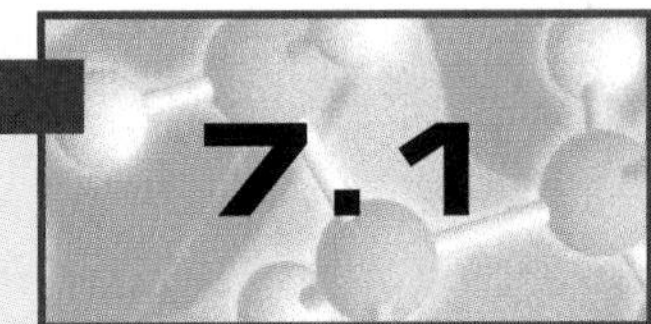

Photoelectricity and Its Applications

One of the earliest clues to the relationship between the frequency of light and its energy was the discovery of the photoelectric effect. In the latter part of the nineteenth century, it was found that certain metals acquired a positive charge when they were illuminated by light. Apparently, light is capable of kicking electrons out of the surface of the metal.

When this phenomenon was studied in detail, it was discovered that electrons could be made to leave a metal's surface only if the frequency of the incident radiation was above some minimum value, which was named the threshold frequency. This threshold frequency differs for different metals, depending on how tightly the metal atom holds onto electrons. Above the threshold frequency, the kinetic energy of the emitted electron increases with increasing frequency of the light. Interestingly, however, its kinetic energy does not depend on the intensity of the light. In fact, if the frequency of the light is below the minimum frequency, no electrons are observed at all, no matter how bright the light is. To physicists of that time, this was very perplexing because they believed the energy of light was related to its brightness. The explanation of the phenomenon was finally given by Albert Einstein in the form of a very simple equation.

$$\text{KE} = h\nu - w$$

where KE is the kinetic energy of the electron that is emitted, $h\nu$ is the energy of the photon of frequency ν, and w is the minimum energy needed to eject the electron from the metal's surface. Stated another way, part of the energy of the photon is needed just to get the electron off the surface of the metal. This amount is w. Any energy left over ($h\nu - w$) appears as the electron's kinetic energy.

Besides its important theoretical implications, the photoelectric effect has many practical applications. For example, automatic "electric eye" door openers use this phenomenon by sensing the interruption of a light beam caused by the person wishing to use the door. The phenomenon is also responsible for photoconduction by certain substances that are used in light meters in cameras and other devices. The production of sound in motion pictures was first made possible by incorporating a strip along the edge of the film (called the sound track) that causes the light passing through it to fluctuate in intensity according to the frequency of the sound that's been recorded. A photocell converts this light to a varying electric current that is amplified and played through speakers in the theater. Even the sensitivity of photographic film to light is related to the release of photoelectrons within tiny grains of silver bromide that are suspended in a coating on the surface of the film.

FACETS OF CHEMISTRY 7.2

Electromagnetic Fields and Their Possible Physiological Effects

Over the past few years you may have heard reports of possible dangers associated with living near high-voltage power lines or operating electrical equipment. What has caused public concern is the fear that 60 Hz electromagnetic radiation emitted by electricity passing through wires might be affecting the health of those nearby. Such fears have been fueled by news media reports of increased cancer rates among groups receiving strong exposure to this radiation, although the actual evidence supporting a relationship between exposure and cancer is weak and even partly contradictory.

When an oscillating electric current with a certain frequency passes through a wire it emits radiation with that same frequency. In fact, that's how radio and TV stations broadcast their signals—by pulsing an electric current through a transmitting antenna. Ordinary household AC (alternating current) electricity has a frequency of 60 Hz, and weak electromagnetic signals are emitted by all wires that carry it. This radiation is most intense when the voltage is highest, as in the lines that carry electricity over long distances between power plants and cities.

The energy possessed by photons of 60 Hz electromagnetic radiation is extremely small (you might try the calculation), so small that it cannot affect the bonds that hold molecules together or even cause heating effects the way microwaves do. How, then, can they affect the activities of cells? The answer might be in the weak pulsating electric and magnetic fields induced in the body by this radiation.

Strong electromagnetic fields have been shown to affect the rate of bone growth as well as the amounts of various proteins produced in cells. Experiments have also revealed that cells exposed to an electric field hold onto calcium ions more than cells not exposed. Other experiments have demonstrated that weak magnetic fields increase the uptake of calcium ions in cells that have been exposed to a substance that triggers cell division. It is believed that this additional calcium increases the tendency of these cells to divide. Since cancer growth depends on the rate of cell division, these results suggest one way cancers might be promoted by such radiation.

Despite laboratory evidence, the connection between electromagnetic radiation and cancer remains inconclusive. Although it is at least *possible* that cancer can be induced by electromagnetic fields, a lot of additional research will be necessary to pin down the answer.

High-voltage power lines such as these emit low levels of 60 Hz electromagnetic radiation (also called *ultralow frequency*, or ULF, radiation). *(Martin Heitner/Stock Connection Worldwide/NewsCom.)*

how radiation can bend around small obstacles and fan out after passing through pinholes, we represent radiation as a wave phenomenon. Electromagnetic radiation is not a stream of particles, and it is not a wave, it's a combination of both that is difficult to describe. However, we can say that in some experiments radiation is best described as a wave and in other experiments the particle explanation works best.

Planck's and Einstein's discovery was really quite surprising. If a particular event requiring energy, such as photosynthesis in green plants, is initiated by the absorption of light, it is the frequency of the light that is important, not its intensity or brightness. This makes sense when we view light as a stream of photons, since "high frequency" is associated with higher energy of the photons, while higher intensity is associated with greater numbers of photons. But it makes no sense at all when we view radiation as a wave phenomenon.

Brighter light delivers more photons; higher frequency light delivers more energetic photons.

The idea that electromagnetic radiation can be represented as either a stream of photons or a wave is a cornerstone of the quantum theory. Physicists were able to use the concept of photons to understand many experimental results that classical physics simply had no explanation for. The success of the quantum theory in describing radiation paved the way for a second startling realization: electrons, like radiation, could be represented as either waves or particles. We now turn our attention to the first experimental evidence that led to our modern quantum mechanical model of atomic structure: the existence of discrete lines in atomic spectra.

7.2 ATOMIC LINE SPECTRA ARE EVIDENCE THAT ELECTRONS IN ATOMS HAVE QUANTIZED ENERGIES

Each spectrum described in Figure 7.3 is called a **continuous spectrum** because it contains a continuous unbroken distribution of light of *all* colors. It is formed when the light from an object that's been heated to a very high temperature (such as the filament in an electric lightbulb) is split by a prism and displayed on a screen. A rainbow after a summer shower appears as a continuous spectrum that most people have seen. In this case, tiny water droplets in the air spread out the colors contained in sunlight.

A rather different kind of spectrum is observed if we examine the light that is given off when an *electric discharge*, or spark, passes through a gas such as hydrogen. The electric discharge is an electric current that *excites,* or energizes, the atoms of the gas. More specifically, the electric current transfers energy to the electrons in the atoms raising them to **excited states.** The atoms then emit the absorbed energy in the form of light as the electrons return to lower energy states. When a narrow beam of this light is passed through a prism, as shown in Figure 7.6, we do *not* see a continuous spectrum. Instead, only a few colors are observed, displayed as a series of individual lines. This series of lines is called the element's **atomic spectrum** or **emission spectrum.** Figure 7.7 shows the visible portions of the atomic spectra of two common elements, sodium and hydrogen, and how they compare with a continuous spectrum. Notice that the spectra of these elements are quite different. In fact, each element has its own unique atomic spectrum that is as characteristic as a fingerprint, and can be used to produce fireworks as in the photo on page 250.

Atoms of an element can also be excited by adding them to the flame of a Bunsen burner.

An emission spectrum is also called a **line spectrum** because the light corresponding to the individual emissions appears as lines on the screen.

Characteristic emissions from these substances add the color to modern fireworks: strontium for red; sodium for yellow; barium for green; copper for blue; charcoal for gold and burning titanium, aluminum or magnesium for silver and white.

A simple pattern of lines in the spectrum of hydrogen suggests a simple explanation for atomic spectra

The first success in explaining atomic spectra quantitatively came with the study of the spectrum of hydrogen. This is the simplest element, since its atoms have only one electron, and it produces the simplest spectrum with the fewest lines.

The atomic spectrum of hydrogen actually consists of several series of lines. One series is in the visible region of the electromagnetic spectrum and is shown in Figure 7.7*c*. Another series is in the ultraviolet region, and the rest are in the infrared. In 1885, J. J. Balmer found an equation that was able to give the wavelengths of the lines in the visible portion of the spectrum. This was soon extended to a more general equation, called the **Rydberg equation,** that could be used to calculate the wavelengths of *all* the spectral lines of hydrogen.

Rydberg equation: $$\frac{1}{\lambda} = R_H\left(\frac{1}{n_1^2} - \frac{1}{n_2^2}\right)$$

The symbol λ stands for the wavelength, R_H is the Rydberg constant (109,678 cm^{-1}), and n_1 and n_2 are variables whose values are whole numbers that range from 1 to ∞. The only

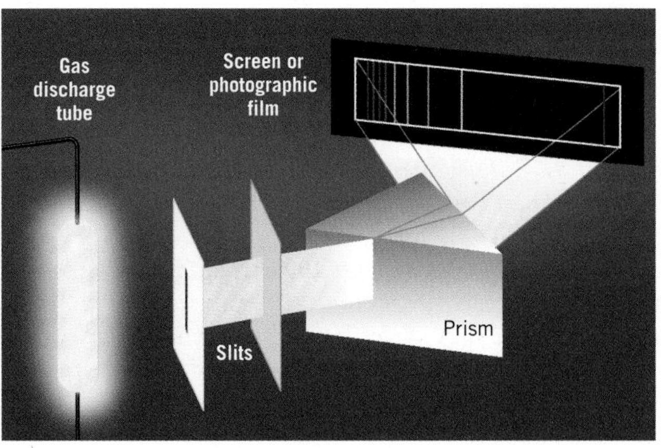

FIG. 7.6 Production and observation of an atomic spectrum. Light emitted by excited atoms is formed into a narrow beam by the slits. It then passes through a prism, which divides the light into relatively few narrow beams with frequencies that are characteristic of the particular element that's emitting the light. When these beams fall on a screen, a series of lines is observed, which is why the spectrum is also called a *line spectrum.*

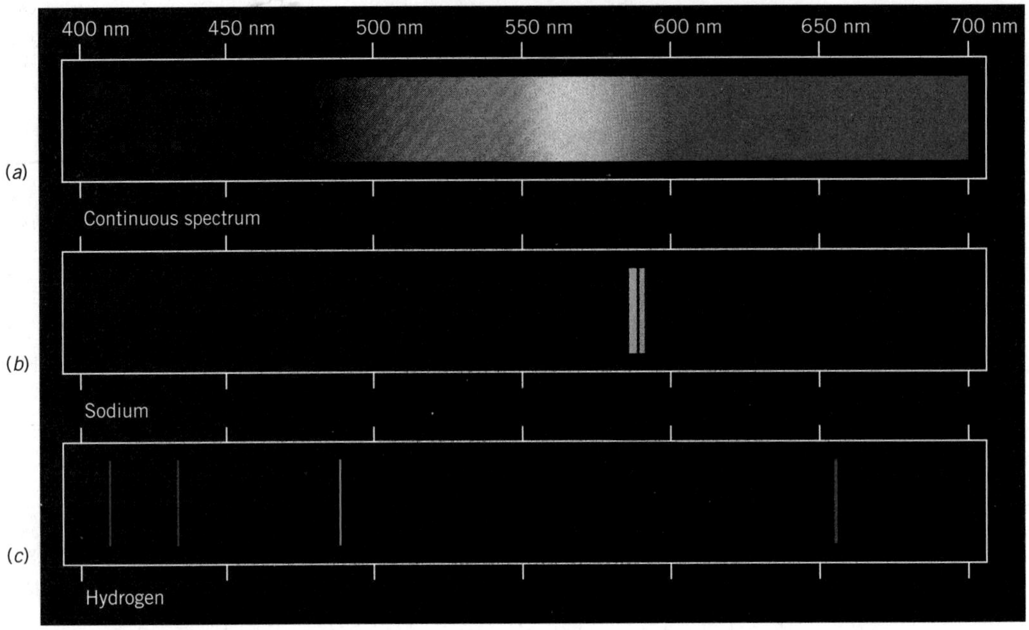

FIG. 7.7 **Continuous and atomic emission spectra.** (*a*) The continuous visible spectrum produced by the sun or an incandescent lamp. (*b*) The atomic spectrum emission produced by sodium. The emission spectrum of sodium actually contains more than 90 lines in the visible region. The two brightest lines are shown here. All the others are less than 1% as bright as these. (*c*) The atomic spectrum (line spectrum) produced by hydrogen. There are only four lines in this visible spectrum. They vary in brightness by only a factor of five, so they are all shown.

restriction is that the value of n_2 must be larger than n_1. (This assures that the calculated wavelength has a positive value.) Thus, if $n_1 = 1$, acceptable values of n_2 are 2, 3, 4, . . . , ∞. The Rydberg constant is an *empirical constant*, which means its value was chosen so that the equation gives values for λ that match the ones determined experimentally. The use of the Rydberg equation is straightforward, as illustrated in the following example.

EXAMPLE 7.3

Calculating the Wavelength of a Line in the Hydrogen Spectrum

The lines in the visible portion of the hydrogen spectrum are called the Balmer series, for which $n_1 = 2$ in the Rydberg equation. Calculate, to four significant figures, the wavelength in nanometers of the spectral line in this series for which $n_2 = 4$.

ANALYSIS: The Rydberg equation will be used for this calculation. The values for n_1 and n_2 are clearly given and the answer is calculated.

SOLUTION: To solve this problem, we substitute values into the Rydberg equation, which will give us $1/\lambda$. Taking the reciprocal will then give the wavelength. As usual, we must be careful with the units.

Substituting $n_1 = 2$ and $n_2 = 4$ into the Rydberg equation gives

$$\begin{aligned}\frac{1}{\lambda} &= 109{,}678 \text{ cm}^{-1}\left(\frac{1}{2^2} - \frac{1}{4^2}\right) \\ &= 109{,}678 \text{ cm}^{-1}\left(\frac{1}{4} - \frac{1}{16}\right) \\ &= 109{,}678 \text{ cm}^{-1}\,(0.2500 - 0.0625) \\ &= 109{,}678 \text{ cm}^{-1}\,(0.1875) \\ &= 2.0565 \times 10^4 \text{ cm}^{-1}\end{aligned}$$

Taking the reciprocal gives the wavelength in centimeters.

$$\lambda = \frac{1}{2.0565 \times 10^4\ \text{cm}^{-1}}$$
$$= 4.8626 \times 10^{-5}\ \text{cm}$$

Finally, we convert to nanometers.

$$\lambda = 4.8626 \times 10^{-5}\ \cancel{\text{cm}} \times \frac{10^{-2}\ \cancel{\text{m}}}{1\ \cancel{\text{cm}}} \times \frac{1\ \text{nm}}{10^{-9}\ \cancel{\text{m}}}$$
$$= 486.3\ \text{nm}$$

Note that we kept one extra significant figure until the end and then rounded to the desired four significant figures.

IS THE ANSWER REASONABLE? Besides double-checking the arithmetic, we can note that the wavelength falls within the visible region of the spectrum, 400 to 700 nm. We can also check the answer against the experimental spectrum of hydrogen in Figure 7.7*c*. This wavelength corresponds to the turquoise line in the hydrogen spectrum.

Practice Exercise 4: Calculate the wavelength in micrometers, μm, of radiation expected when $n_1 = 4$ and $n_2 = 6$. Report your result to three significant figures. (Hint: The values of n_1 and n_2 are used to calculate the term in parentheses first.)

Practice Exercise 5: Calculate the wavelength in nanometers of the spectral line in the visible spectrum of hydrogen for which $n_1 = 2$ and $n_2 = 3$. What color is this line?

The discovery of the Rydberg equation was both exciting and perplexing. The fact that the wavelength of any line in the hydrogen spectrum can be calculated by a simple equation involving just one constant and the reciprocals of the squares of two whole numbers is remarkable. What is there about the behavior of the electron in the atom that could account for such simplicity?

The energy of electrons in atoms is quantized

Earlier you saw that there is a simple relationship between the frequency of light and its energy, $E = h\nu$. Because excited atoms emit light of only certain specific frequencies, it must be true that only certain characteristic energy changes are able to take place within the atoms. For instance, in the spectrum of hydrogen there is a red line (see Figure 7.7*c*) that has a wavelength of 656.4 nm and a frequency of 4.567×10^{14} Hz. The energy of a photon of this light is 3.026×10^{-19} J. Whenever a hydrogen atom emits red light, the frequency of the light is always precisely 4.567×10^{14} Hz and the energy of the atom decreases by *exactly* 3.026×10^{-19} J, never more and never less. Atomic spectra, then, tell us that *when an excited atom loses energy, not just any arbitrary amount can be lost.* The same is true if the atom gains energy.

How is it that atoms of a given element always undergo exactly the same specific energy changes? The answer seems to be that in an atom an electron can have only certain definite amounts of energy and no others. We say that the electron is restricted to certain **energy levels,** and that the energy of the electron is **quantized.**

The energy of an electron in an atom might be compared to the energy of the tortoise in the exhibit shown in Figure 7.8*b*. The tortoise trapped inside the zoo exhibit can only be "stable" on one of the ledges, so it has certain specific amounts of potential energy as determined by the "energy levels" of the various ledges. If the tortoise is raised to a higher ledge, its potential energy is increased. When it drops to a lower ledge, its potential energy decreases. If the tortoise tries to occupy heights between the ledges, it immediately falls to the lower ledge. Therefore, the energy changes for the tortoise are restricted to the differences in potential energy between the ledges.

Any potential energy allowed: energy values are *continuous*

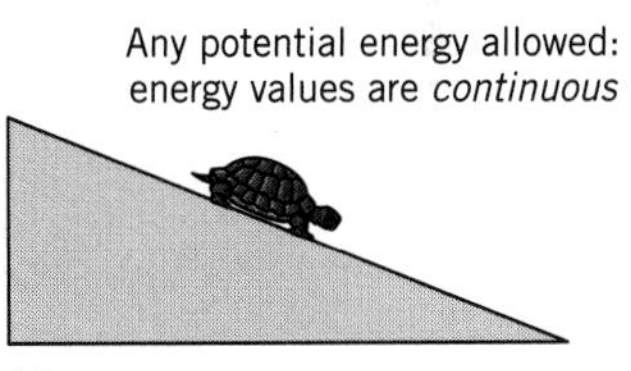

(*a*)

Potential energy restricted: energy values are *discrete*

(*b*)

FIG. 7.8 Continuous and discrete energies. (*a*) The tortoise is free to move to any height less than the height of the hill. Its potential energy can take on any value between the maximum (at the hill top) and the minimum (at the bottom). Similarly, the energy of a free electron can take on any value. (*b*) The tortoise trapped inside a zoo exhibit is found only at three heights: at the bottom (the lowest, or *ground state*), middle, or top ledges. The potential energy of the tortoise at rest is quantized. Similarly, the energy of the electron trapped inside an atom is restricted to certain values, which correspond to the various energy levels in an atom.

So it is with an electron in an atom. The electron can only have energies corresponding to the set of electron energy levels in the atom. When the atom is supplied with energy (by an electric discharge, for example), an electron is raised from a low-energy level to a higher one. When the electron drops back, energy equal to the difference between the two levels is released and emitted as a photon. Because only certain energy jumps can occur, only certain frequencies of light can appear in the spectrum.

The existence of specific energy levels in atoms, as implied by atomic spectra, forms the foundation of all theories about electronic structure. Any model of the atom that attempts to describe the positions or motions of electrons must also account for atomic spectra.

The Bohr model explains the simple pattern of lines seen in the spectrum of hydrogen

Niels Bohr won the 1922 Nobel Prize in Physics for his work on atomic structure.

Classical physical laws, such as those discovered by Isaac Newton, place no restrictions on the sizes or energies of orbits.

The first theoretical model of the hydrogen atom that successfully accounted for the Rydberg equation was proposed in 1913 by Niels Bohr (1885–1962), a Danish physicist. In his model, Bohr likened the electron moving around the nucleus to a planet circling the sun. He suggested that the electron moves around the nucleus along fixed paths, or orbits. His model broke with the classical laws of physics by placing restrictions on the sizes of the orbits and the energy that the electron could have in a given orbit. This ultimately led Bohr to an equation that described the energy of the electron in the atom. The equation includes a number of physical constants such as the mass of the electron, its charge, and Planck's constant. It also contains an integer, n, that Bohr called a **quantum number.** Each of the orbits is identified by its value of n. When all the constants are combined, Bohr's equation becomes

Bohr's equation for the energy actually is

$$E = -\frac{2\pi^2 me^4}{n^2h^2}$$

where m is the mass of the electron, e is the charge on the electron, n is the quantum number, and h is Planck's constant. Therefore, in Equation 7.3,

$$b = -\frac{2\pi^2 me^4}{h^2} = 2.18 \times 10^{-18}\text{ J}.$$

$$E = \frac{-b}{n^2} \tag{7.3}$$

where E is the energy of the electron and b is the combined constant (its value is 2.18×10^{-18} J). The allowed values of n are whole numbers that range from 1 to ∞ (i.e., n could equal 1, 2, 3, 4, . . ., ∞). From this equation the energy of the electron in any particular orbit could be calculated.

Because of the negative sign in Equation 7.3, the lowest (most negative) energy value occurs when $n = 1$, which corresponds to the *first Bohr orbit.* The lowest energy state of an atom is the most stable one and is called the **ground state.** For hydrogen, the ground state occurs when its electron has $n = 1$. According to Bohr's theory, this orbit brings the electron closest to the nucleus. Conversely, an atom with $n = \infty$ would correspond to an "unbound" electron that had escaped from the nucleus. Such an electron has an energy of zero in Bohr's theory. The negative sign in Equation 7.3 ensures that any electron with a finite value of n has a lower energy than an unbound electron. Thus, energy is released when a free electron is bound to a proton to form a hydrogen atom.

When a hydrogen atom absorbs energy, as it does when an electric discharge passes through it, the electron is raised from the orbit having $n = 1$ to a higher orbit, to $n = 2$ or $n = 3$ or even higher. The hydrogen atom is now in an excited state. These higher orbits are less stable than the lower ones, so the electron quickly drops to a lower orbit. When this happens, energy is emitted in the form of light (see Figure 7.9). Since the energy of the electron in a given orbit is fixed, a drop from one particular orbit to another, say, from $n = 2$ to $n = 1$, always releases the same amount of energy, and the frequency of the light emitted because of this change is always precisely the same.

The success of Bohr's theory was in its ability to account for the Rydberg equation. When the atom emits a photon, an electron drops from a higher initial energy E_{high} to a lower final energy E_{low}. If the initial quantum number of the electron is n_{high} and the final quantum number is n_{low}, then the energy change, calculated as a positive quantity, is

$$\Delta E = E_{\text{high}} - E_{\text{low}} = \left(\frac{-b}{n_{\text{high}}^2}\right) - \left(\frac{-b}{n_{\text{low}}^2}\right)$$

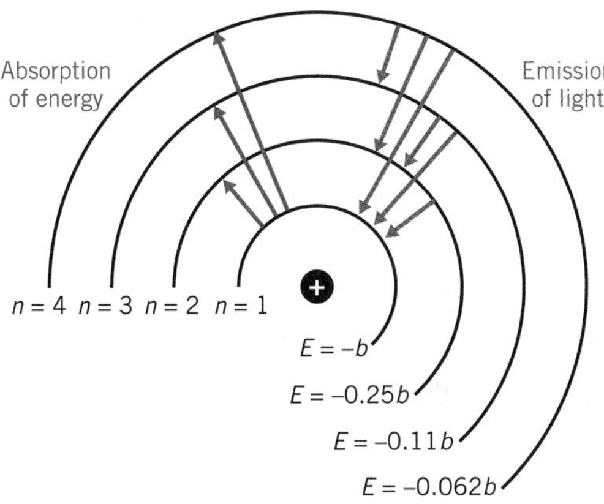

FIG. 7.9 Absorption of energy and emission of light by the hydrogen atom. When the atom absorbs energy, the electron is raised to a higher energy level. When the electron falls to a lower energy level, light of a particular energy and frequency is emitted.

This can be rearranged to give

$$\Delta E = b\left(\frac{1}{n_{\text{low}}^{2}} - \frac{1}{n_{\text{high}}^{2}}\right) \qquad \text{with } n_{\text{high}} > n_{\text{low}}$$

By combining Equations 7.1 and 7.2, the relationship between the energy "ΔE" of a photon and its wavelength λ is

$$\Delta E = \frac{hc}{\lambda} = hc\left(\frac{1}{\lambda}\right)$$

Substituting and solving for $1/\lambda$ give

$$\frac{1}{\lambda} = \frac{b}{hc}\left(\frac{1}{n_{\text{low}}^{2}} - \frac{1}{n_{\text{high}}^{2}}\right) \qquad \text{with } n_{\text{high}} > n_{\text{low}}$$

Notice how closely this equation derived from Bohr's theory matches the Rydberg equation, which was obtained solely from the experimentally measured atomic spectrum of hydrogen. Equally satisfying is that the combination of constants, b/hc, has a value of 109,730 cm^{-1}, which differs by only 0.05% from the experimentally derived value of R_{H} in the Rydberg equation.

The solar system model of the atom was the way that Niels Bohr's model was presented to the public. Today this simple picture of the atom makes a nice corporate logo, but the idea of an atom with electrons orbiting a nucleus as planets orbit a sun has been replaced by the wave mechanical model.

The Bohr model fails for atoms with more than one electron

Bohr's model of the atom was both a success and a failure. By calculating the energy changes that occur between energy levels, Bohr was able to account for the Rydberg equation and, therefore, for the atomic spectrum of hydrogen. However, the theory was not able to explain quantitatively the spectra of atoms with more than one electron, and all attempts to modify the theory to make it work met with failure. Gradually, it became clear that Bohr's picture of the atom was flawed and that another theory would have to be found. Nevertheless, the concepts of quantum numbers and fixed energy levels were important steps forward.

7.3 Electrons Have Properties of Both Particles and Waves

Bohr's efforts to develop a theory of electronic structure were doomed from the very beginning because the classical laws of physics—those known in his day—simply do not apply to objects as small as the electron. Classical physics fails for atomic particles because matter is not really as our physical senses perceive it. When bound inside an atom, electrons behave not like solid particles, but instead like waves. This idea was proposed in 1924 by a young French graduate student, Louis de Broglie.

All the objects that had been studied by scientists until the time of Bohr were large and massive in comparison with the electron, so no one had detected the limits of classical physics.

De Broglie was awarded a Nobel prize in 1929.

FIG. 7.10 Diffraction of water waves on the surface of a pond. As the waves cross, the amplitudes increase where the waves are in phase and cancel where they are out of phase. *(Mandy Collins/Alamy Images.)*

Gigantic waves, called rogue waves, with heights up to 100 ft have been observed in the ocean and are believed to be formed when a number of wave sets moving across the sea become in phase simultaneously.

In Section 7.1 you learned that light waves are characterized by their wavelengths and their frequencies. The same is true of matter waves. De Broglie suggested that the wavelength of a matter wave, λ is given by the equation

$$\lambda = \frac{h}{mv} \tag{7.4}$$

where h is Planck's constant, m is the particle's mass, and v is its velocity. Notice that this equation allows us to connect a wave property, wavelength, with particle properties, mass and velocity. We may describe the electron either as a particle or a wave, and the de Broglie relationship provides a link between the two descriptions.

When first encountered, the concept of a particle of matter behaving as a wave rather than as a solid object is difficult to comprehend. This book certainly seems solid enough, especially if you drop it on your toe! The reason for the book's apparent solidity is that in de Broglie's equation (Equation 7.4) the mass appears in the denominator. This means that heavy objects have extremely short wavelengths. The peaks of the matter waves for heavy objects are so close together that the wave properties go unnoticed and can't even be measured experimentally. But tiny particles with very small masses have much longer wavelengths, so their wave properties become an important part of their overall behavior.

Diffraction provides evidence that electrons have wave properties

Perhaps by now you've begun to wonder if there is any way to *prove* that matter has wave properties. Actually, these properties can be demonstrated by a phenomenon that you have probably witnessed. When raindrops fall on a quiet pond, ripples spread out from where the drops strike the water, as shown in Figure 7.10. When two sets of ripples cross, there are places where the waves are *in phase*, which means that the peak of one wave coincides with the peak of the other. At these points the intensities of the waves add and the height of the water is equal to the sum of the heights of the two crossing waves. At other places the crossing waves are *out of phase*, which means the peak of one wave occurs at the trough of the other. In these places the intensities of the waves cancel. This reinforcement and cancellation of wave intensities, referred to, respectively, as *constructive* and *destructive interference*, is a phenomenon called **diffraction.** It is examined more closely in Figure 7.11. Note how diffraction creates characteristic **interference fringes** when waves pass through adjacent pinholes or reflect off closely spaced grooves. You have seen interference fringes yourself if you've ever noticed the rainbow of colors that shine from the surface of a compact disc (Figure 7.12). When white light, which contains all the visible wavelengths,

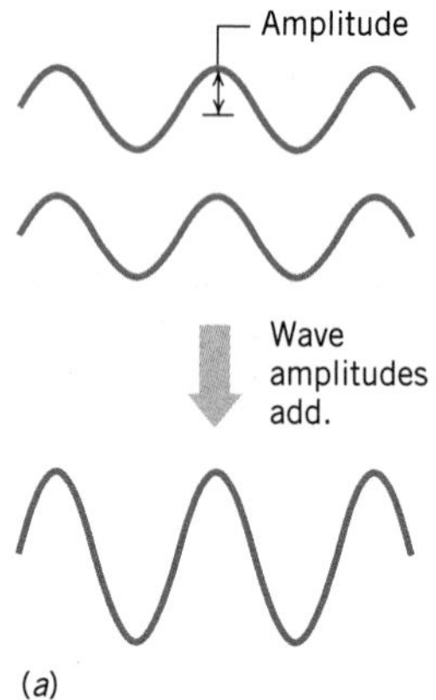

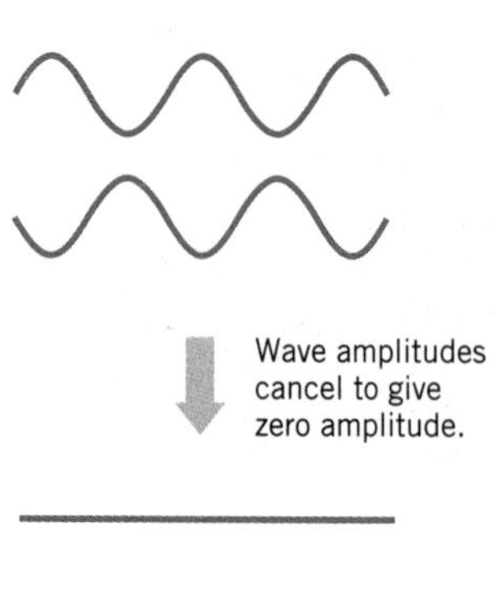

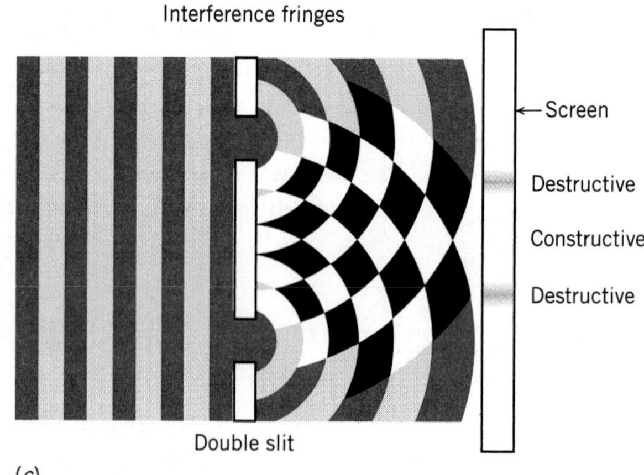

FIG. 7.11 Constructive and destructive interference. (*a*) Waves in phase produce constructive interference and an increase in intensity. (*b*) Waves out of phase produce destructive interference and yield cancellation of intensity. (*c*) Light waves passing through two pinholes fan out and interfere with each other, producing an interference pattern characteristic of waves.

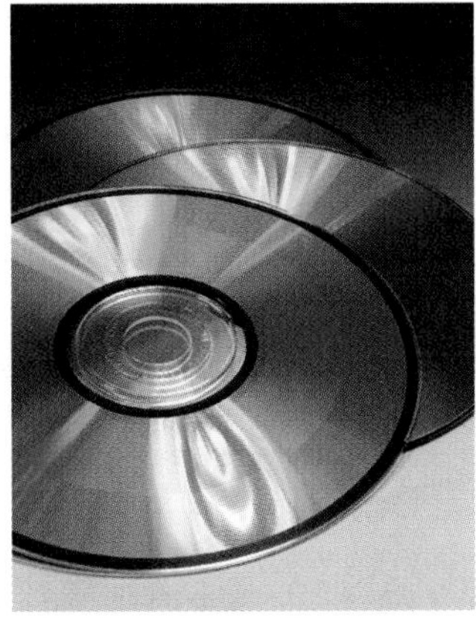

FIG. 7.12 Diffraction of light from a compact disc. Colored interference fringes are produced by the diffraction of reflected light from the closely spaced grooves on the surface of a compact disc. *(John Paul Endress/Corbis Stock Market.)*

is reflected from the closely spaced "grooves" on the CD, it is divided into many individual light beams. For a given angle between the incoming and reflected light, the light waves experience interference with each other for all wavelengths (colors) except for one wavelength which is reinforced. Our eye sees the wavelength of light that is reinforced and as the angle changes the wavelengths that are reinforced change. The result is a rainbow of colors reflected from the CD.

Diffraction is a phenomenon that can only be explained as a property of waves, and we have seen how it can be demonstrated with water waves and light waves. Experiments can also be done to show that electrons, protons, and neutrons experience diffraction, which demonstrates their wave nature (see Figure 7.13). In fact, electron diffraction is the principle on which the electron microscope is based (see Facets of Chemistry 7.3).

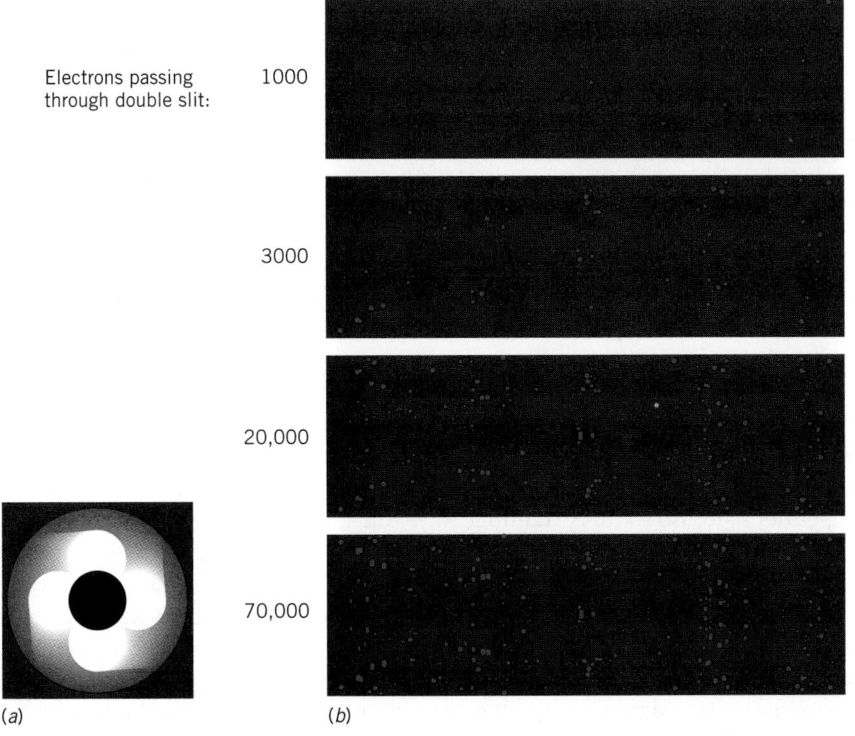

FIG. 7.13 Experimental evidence of wave behavior in electrons. (*a*) An electron diffraction pattern collected by reflecting a beam of electrons from crystalline silicon. (Semiconductor Surface Physics Group, Queens University.) (*b*) Electrons passing *one at a time* through a double slit. Each spot shows an electron impact on a detector. As more and more electrons are passed through the slits, interference fringes are observed.

Bound electrons have quantized energies because they behave like standing waves

Before we can discuss how electron waves behave in atoms, we need to know a little more about waves in general. There are basically two kinds of waves, **traveling waves** and **standing waves**. On a lake or ocean the wind produces waves whose crests and troughs

FACETS OF CHEMISTRY 7.3

The Electron Microscope

The usefulness of a microscope in studying small specimens is limited by its ability to distinguish between closely spaced objects. We call this ability the *resolving power* of the microscope. Through optics, it is possible to increase the magnification and thereby increase the resolving power, but only within limits. These limits depend on the wavelength of the light that is used. Objects with diameters less than the wavelength of the light cannot be seen in detail. Since the smallest wavelength of visible light is about 400 nm, objects smaller than this can't be clearly seen with a microscope that uses visible light.

The electron microscope uses electron waves to "see" very small objects. De Broglie's equation, $\lambda = h/mv$, suggests that if an electron, proton, or neutron has a very high velocity, its wavelength will be very small. In the electron microscope, electrons are accelerated to high speeds across high-voltage electrodes. This gives electron waves with typical wavelengths of about 0.006 to 0.001 nm that strike the sample and are then focused magnetically (using "magnetic lenses") onto a fluorescent screen where they form a visible image. Because of certain difficulties, the actual resolving power of the instrument is quite a bit less than the wavelength of the electron waves—generally on the order of 6 to 1 nm. Some high resolution electron microscopes, however, are able to reveal the shadows of individual atoms in very thin specimens through which the electron beam passes.

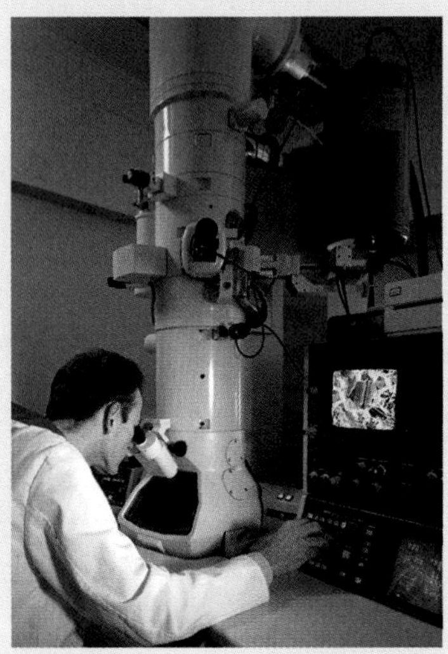

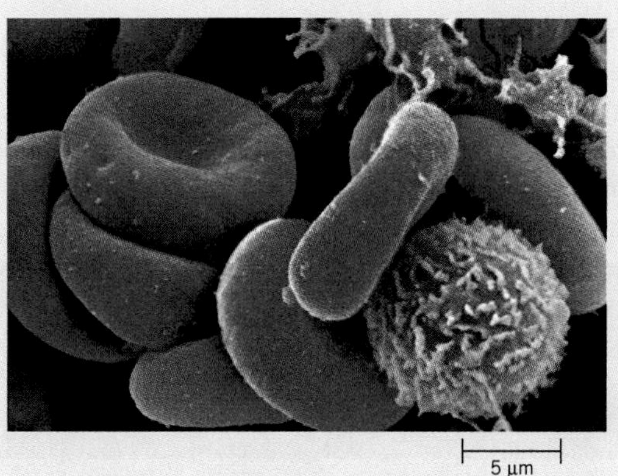

A modern electron microscope operated by trained technicians is used to obtain electron-micrographs such as the one shown here depicting red and white blood cells. *(Brand X/Superstock; Yorgos Nikas/Stone/Getty Images.)*

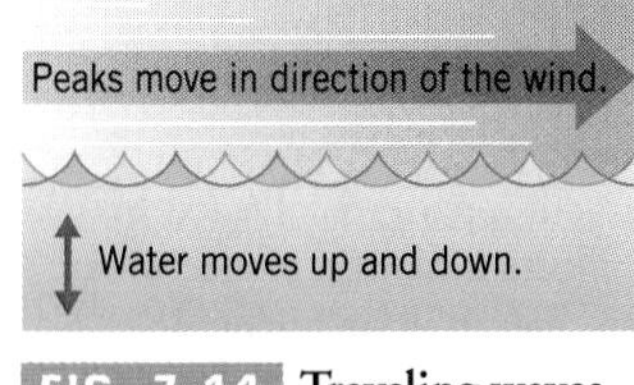

FIG. 7.14 Traveling waves.

move across the water's surface, as shown in Figure 7.14. The water moves up and down while the crests and troughs travel horizontally in the direction of the wind. These are examples of **traveling waves.**

A more important kind of wave for us is the standing wave. An example is the vibrating string of a guitar. When the string is plucked, its center vibrates up and down while the ends, of course, remain fixed. The crest, or point of maximum amplitude of the wave, occurs at one position. At the ends of the string are points of zero amplitude, called **nodes,** and their positions are also fixed. A **standing wave,** then, is one in which the crests and nodes do not change position. One of the interesting things about standing waves is that they lead naturally to "quantum numbers." Let's see how this works using the guitar as an example.

As you know, many notes can be played on a guitar string by shortening its effective length with a finger placed at frets along the neck of the instrument. But even without shortening the string, we can play a variety of notes. For instance, if the string is touched momentarily at its midpoint at the same time it is plucked, the string vibrates as shown in Figure 7.15 and produces a tone an octave higher. The wave that produces this higher tone has a wavelength exactly half of that formed when the untouched string is plucked. In Figure 7.15 we see that other wavelengths are possible, too, and each gives a different note.

Notes played this way are called harmonics.

If you examine Figure 7.15, you will see that there are some restrictions on the wavelengths that can exist. Not just any wavelength is possible because the nodes at either end of the string are in fixed positions. The only waves that can occur are those for which a half-wavelength is repeated *exactly* a whole number of times. Expressed another way, the

FIG. 7.15 **Standing waves on a guitar string.**

length of the string is a whole-number multiple of half-wavelengths. In a mathematical form we could write this as

$$L = n\left(\frac{\lambda}{2}\right)$$

where L is the length of the string, λ is a wavelength (therefore, $\lambda/2$ is half the wavelength), and n is an integer. Rearranging this to solve for the wavelength gives

$$\lambda = \frac{2L}{n} \tag{7.5}$$

We see that the waves that are possible are determined quite naturally by a set of whole numbers (similar to quantum numbers).

We are now in a position to demonstrate how quantum theory unites wave and particle descriptions to build a simple but accurate model of a bound electron. Let's look at an electron that is confined to a wire of length L. To keep things simple, let's assume that the wire is infinitely thin, so that the electron can only move in straight lines along the wire. The wire is clamped in place at either end, and its ends cannot move up or down.

Notes played on a guitar rely on standing waves. The ends of the strings correspond to nodes of the standing waves. Different notes can be played by shortening the effective lengths of the strings with fingers placed along the neck of the instrument. *(© Index Stock.)*

First, let's consider a classical particle model: the "bead on a wire" model shown in Figure 7.16*a*. The bead can slide in either direction along the wire, like a bead on an abacus. If the electron's mass is m and its velocity is v, its kinetic energy is given by

$$E = \frac{1}{2}mv^2$$

The bead can have any velocity, even zero, so the energy E can have any value, even zero. No position on the wire is any more favorable than any other, and the bead is equally likely to be found anywhere on the wire. There is no reason why the bead's position and velocity cannot be known simultaneously.

Now consider a classical wave along the wire, Figure 7.16*b*. It is exactly like the guitar string we looked at in Figure 7.15. The ends of the wire are clamped in place, so there *must* be a whole number of peaks and troughs along the wire. The wavelength is restricted to values calculated by Equation 7.5. We can see that the quantum number, n, is just the number of peaks and troughs along the wave. It has integer values (1, 2, 3, . . .) because you can't have half a peak or half a trough.

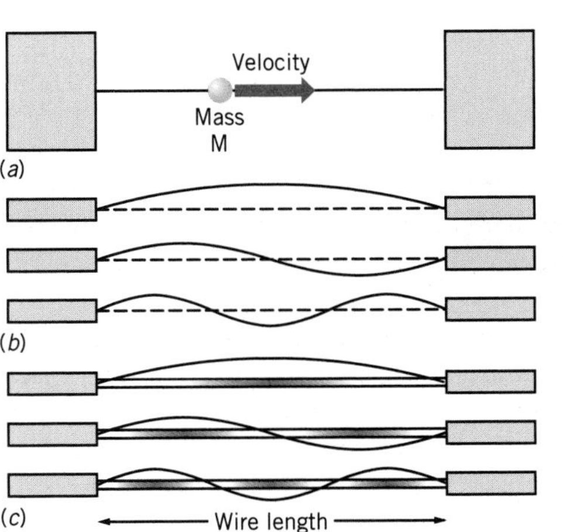

FIG. 7.16 **Three models of an electron on an infinitely thin wire of length L.** (*a*) A classical model of the electron as a bead that can slide along the wire. Any energy is possible, even zero, and the exact position and velocity of the bead can be known simultaneously. (*b*) A classical model of the electron as a standing wave on a wire. An integer number of peaks and troughs (n) is required. The wavelength is restricted to values given by Equation 7.5. (*c*) A quantum mechanical model of the electron on a wire obtained by uniting model (*a*) with model (*b*), using the de Broglie relationship (Equation 7.4). The shaded areas indicate most probable positions for the electron. Energy is quantized and is never zero.

Now let's use the de Broglie relation, Equation 7.4, to unite these two classical models. Our goal is to derive an expression for the energy of an electron trapped on the wire. Notice that to calculate the kinetic energy of the bead on a wire, we need to know its velocity, a particle property. The de Broglie relation lets us relate particle velocities with wavelengths. Rearranging Equation 7.4, we have

$$v = \frac{h}{m\lambda}$$

and inserting this into the equation for kinetic energy gives

$$E = \frac{1}{2}m\left(\frac{h}{m\lambda}\right)^2$$
$$= \frac{h^2}{2m\lambda^2}$$

This equation will give us the energy of the electron from its wavelength. If we substitute in Equation 7.5, which gives the wavelength of the standing wave in terms of the wire length L and the quantum number n, we have

$$E = \frac{n^2h^2}{8mL^2} \tag{7.6}$$

This equation has a number of profound implications. The fact that the electron's energy depends on an integer, n, means that *only certain energy states are allowed.* The allowed states are plotted on the energy level diagram shown in Figure 7.17. The lowest value of n is 1, so the lowest energy level (the ground state) is $E_1 = h^2/8mL^2$. Energies lower than this are not allowed, so the energy cannot be zero! This indicates that the electron will always have some residual kinetic energy. The electron is never at rest. This is true for the electron trapped in a wire and it is also true for an electron trapped in an atom. Thus, *quantum theory resolves the collapsing atom paradox.*

Note that the spacing between energy levels is proportional to $1/L^2$. This means that when the wire is made longer, the energy levels become more closely spaced. In general, *the more room an electron has to move in, the smaller the spacings between its energy levels.* Chemical reactions sometimes change the way that electrons are confined in molecules. This causes changes in the wavelengths of light the reacting mixture absorbs. This is why color changes sometimes occur during chemical changes.

Electron waves are represented by wave functions

The wave that corresponds to the electron is called a **wave function.** The wave function is usually represented by the symbol ψ (Greek letter psi). The wave function can be used to describe the shape of the electron wave and its energy. The wave function is not an

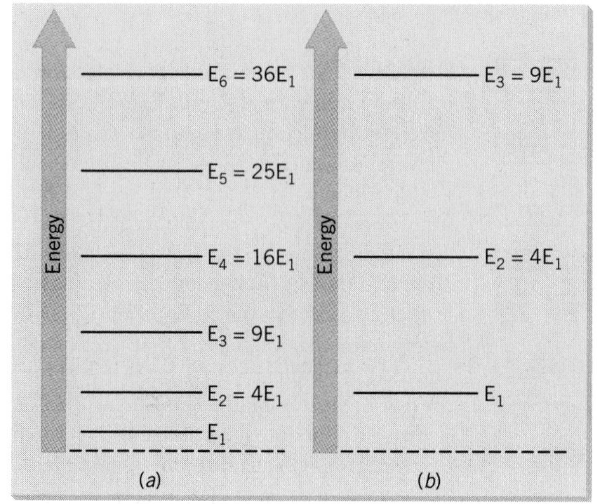

FIG. 7.17 Energy level diagram for the electron on a wire model. (*a*) A long wire, with $L = 2$ nm, and (*b*) a short wire, with $L = 1$ nm. Notice how the energy levels become more closely spaced when the electron has more room to move. Also notice that the energy in the ground state is not zero.

oscillation of the wire, like a guitar wave, nor is it an electromagnetic wave. The wave's amplitude at any given point can be related to the probability of finding the electron there.

The electron waves shown in Figure 7.16*c* show that, unlike the bead on a wire model, the electron is more likely to be found at some places on the wire than others. For the ground state, with $n = 1$, the electron is most likely to be found in the center of the wire. Where the amplitude is zero, for example, at the ends of the wire or the center of the wire in the $n = 2$ state, there is a zero probability of finding the electron! Points where the amplitude of the electron wave is zero are called **nodes.** Notice that the higher the quantum number n, the more nodes the electron wave has and, from Equation 7.6, the more energy the electron has. It is generally true that *the more nodes an electron wave has, the higher its energy.*

The probability of finding an electron at a given point is proportional to the amplitude of the electron wave squared. Thus, peaks and troughs in the electron wave indicate places where there is the greatest buildup of negative charge.

Electron waves in atoms are called orbitals

In 1926 Erwin Schrödinger (1887–1961), an Austrian physicist, became the first scientist to successfully apply the concept of the wave nature of matter to an explanation of electronic structure. His work and the theory that developed from it are highly mathematical. Fortunately, we need only a qualitative understanding of electronic structure, and the main points of the theory can be understood without all the math.

Schrödinger won a Nobel prize in 1933 for his work. The equation he developed that gives electronic wave functions and energies is known as *Schrödinger's equation.* The equation is extremely difficult to solve. Even an approximate solution of the equation for large molecules can require hours or days of supercomputer time.

Schrödinger developed an equation that can be solved to give wave functions and energy levels for electrons trapped inside atoms. Wave functions for electrons in atoms are called **orbitals.** Not all of the energies of the waves are different, but most are. *Energy changes within an atom are simply the result of an electron changing from a wave pattern with one energy to a wave pattern with a different energy.*

Electron waves are described by the term *orbital* to differentiate them from the notion of *orbits,* which was part of the Bohr model of the atom.

We will be interested in two properties of orbitals, their energies and their shapes. Their energies are important because when an atom is in its most stable state (its *ground state*), the atom's electrons have waveforms with the lowest possible energies. The shapes of the wave patterns (i.e., where their amplitudes are large and where they are small) are important because the theory tells us that the amplitude of a wave at any particular place is related to the likelihood of finding the electron there. This will be important when we study how and why atoms form chemical bonds to each other.

"Most stable" almost always means "lowest energy."

In much the same way that the characteristics of a wave on a one-dimensional guitar string can be related to a single integer, wave mechanics tells us that the three-dimensional electron waves (orbitals) can be characterized by a set of *three* integer quantum numbers, n, ℓ, and m_ℓ. In discussing the energies of the orbitals, it is usually most convenient to sort the orbitals into groups according to these quantum numbers.

The principal quantum number, n

The quantum number $\boldsymbol{n}$ is called the **principal quantum number,** and all orbitals that have the same value of n are said to be in the same **shell.** The values of n can range from $n = 1$ to $n = \infty$. The shell with $n = 1$ is called the *first shell,* the shell with $n = 2$ is the *second shell,* and so forth. The various shells are also sometimes identified by letters, beginning (for no significant reason) with K for the first shell ($n = 1$).

The term shell comes from an early notion that atoms could be thought of as similar to onions, with the electrons being arranged in layers around the nucleus.

The principal quantum number is related to the size of the electron wave (i.e., how far the wave effectively extends from the nucleus). The higher the value of n, the larger is the electron's average distance from the nucleus. This quantum number is also related to the energy of the orbital. As n increases, the energies of the orbitals also increase.

Bohr's theory took into account only the principal quantum number n. His theory worked fine for hydrogen because hydrogen just happens to be the one element in which all orbitals having the same value of n also have the same energy. Bohr's theory failed for atoms other than hydrogen, however, because when the atom has more than one electron, orbitals with the same value of n can have different energies.

Bohr was fortunate to have used the element hydrogen to develop his model of the atom. If he had chosen a different element, his model would not have worked.

The secondary quantum number, ℓ

The **secondary quantum number,** $\boldsymbol{\ell}$**,** divides the shells into smaller groups of orbitals called **subshells.** The value of n determines which values of ℓ are allowed. For a given n, ℓ can range from $\ell = 0$ to $\ell = (n - 1)$. Thus, when $n = 1$, $(n - 1) = 0$, so the only value of ℓ that's allowed is zero. This means that when $n = 1$, there is only one subshell (the shell and subshell are really identical). When $n = 2$, ℓ can have values of 0 or 1. (The maximum value

ℓ is also called the **azimuthal quantum number** and the **orbital angular momentum number.**

Relationship between n and ℓ

Value of n	Values of ℓ
1	0
2	0, 1
3	0, 1, 2
4	0, 1, 2, 3
5	0, 1, 2, 3, 4
n	0, 1, 2, ..., $(n - 1)$

The number of subshells in a given shell equals the value of n for that shell. For example, when $n = 3$, there are three subshells.

of $\ell = n - 1 = 2 - 1 = 1$.) This means that when $n = 2$, there are two subshells. One has $n = 2$ and $\ell = 0$, and the other has $n = 2$ and $\ell = 1$. The relationship between n and the allowed values of ℓ are summarized in the table in the margin.

Subshells could be identified by their value of ℓ. However, to avoid confusing numerical values of n with those of ℓ, a letter code is normally used to specify the value of ℓ.

Value of ℓ	0	1	2	3	4	5	...
Letter designation	*s*	*p*	*d*	*f*	*g*	*h*	...

To designate a particular subshell, we write the value of its principal quantum number followed by the letter code for the subshell. For example, the subshell with $n = 2$ and $\ell = 1$ is the $2p$ subshell; the subshell with $n = 4$ and $\ell = 0$ is the $4s$ subshell. Notice that because of the relationship between n and ℓ, every shell has an s subshell ($1s$, $2s$, $3s$, etc.). All the shells except the first have a p subshell ($2p$, $3p$, $4p$, etc.). All but the first and second shells have a d subshell ($3d$, $4d$, etc.); and so forth.

Practice Exercise 6: How many subshells are there in each of the first six shells of an atom? (Hint: Count the subshells based on sets of groups representing *s*, *p*, *d*, and *f* orbitals.)

Practice Exercise 7: What subshells would be found in the shells with $n = 3$ and $n = 4$?

The secondary quantum number determines the shape of the orbital, which we will examine more closely later. Except for the special case of hydrogen, which has only one electron, the value of ℓ also affects the energy. This means that in atoms with two or more electrons, the subshells within a given shell differ slightly in energy, with the energy of the subshell increasing with increasing ℓ. Therefore, within a given shell, the s subshell is lowest in energy, p is the next lowest, followed by d, then f, and so on. For example,

$$4s < 4p < 4d < 4f$$

— increasing energy →

The magnetic quantum number, m_ℓ

Spectroscopists used m_ℓ to explain additional lines that appear in atomic spectra when atoms emit light while in a magnetic field, which explains how this quantum number got its name.

The third quantum number, m_ℓ, is known as the **magnetic quantum number.** Its value indicates individual orbitals within a subshell. Also, its values are related to the way the individual orbitals are oriented relative to each other in space. As with ℓ, there are restrictions as to the possible values of m_ℓ, which can range from $+\ell$ to $-\ell$. When $\ell = 0$, m_ℓ can have only the value 0 because $+0$ and -0 are the same. An s subshell, then, has just a single orbital. When $\ell = 1$, the possible values of m_ℓ are $+1$, 0, and -1. A p subshell therefore has three orbitals: one with $\ell = 1$ and $m_\ell = 1$, another with $\ell = 1$ and $m_\ell = 0$, and a third with $\ell = 1$ and $m_\ell = -1$. Similarly, we find that a d subshell has five orbitals and an f subshell has seven orbitals. The numbers of orbitals in the subshells are easy to remember because they follow a simple arithmetic progression.

s	*p*	*d*	*f*	...
1	3	5	7	...

The whole picture

The relationships among all three quantum numbers are summarized in Table 7.1. In addition, the relative energies of the subshells in an atom containing two or more electrons are depicted in Figure 7.18. Several important features should be noted. First, observe that each orbital on this energy diagram is indicated by a separate circle—one for an s subshell, three for a p subshell, and so forth. Second, notice that all the orbitals of a given subshell have the *same* energy. Third, note that, in going upward on the energy scale, the spacing between successive shells decreases as the number of subshells increases. This leads to some overlapping of shells having different values of n. For instance, the $4s$ subshell is lower in energy than the $3d$ subshell, $5s$ is lower than $4d$, and $6s$ is lower than $5d$. In addition, the $4f$ subshell is below the $5d$ subshell and $5f$ is below $6d$.

TABLE 7.1 Summary of Relationships among the Quantum Numbers n, ℓ, and m_ℓ

Value of n	Value of ℓ	Values of m_ℓ	Subshell	Number of Orbitals
1	0	0	1*s*	1
2	0	0	2*s*	1
	1	−1, 0, 1	2*p*	3
3	0	0	3*s*	1
	1	−1, 0, 1	3*p*	3
	2	−2, −1, 0, 1, 2	3*d*	5
4	0	0	4*s*	1
	1	−1, 0, 1	4*p*	3
	2	−2, −1, 0, 1, 2	4*d*	5
	3	−3, −2, −1, 0, 1, 2, 3	4*f*	7

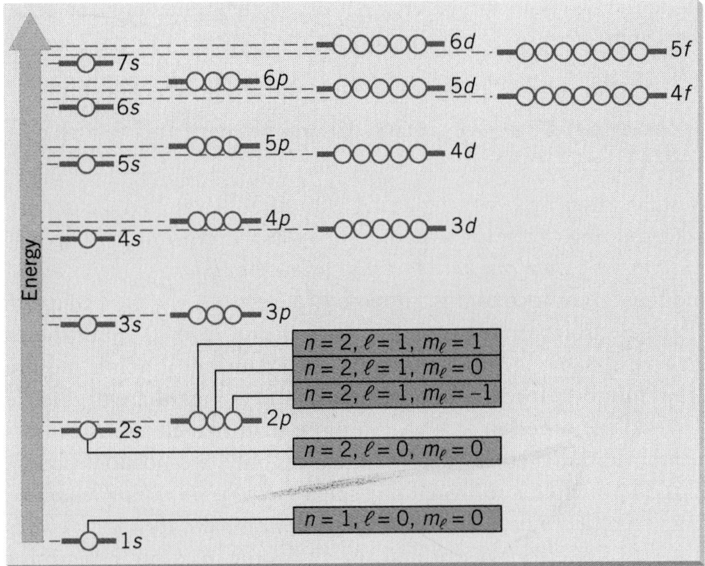

FIG. 7.18 Approximate energy level diagram for atoms with two or more electrons. The quantum numbers associated with the orbitals in the first two shells are also shown.

We will see shortly that Figure 7.18 is very useful for predicting the electronic structures of atoms. Before discussing this, however, we must study another very important property of the electron, a property called spin. Electron spin gives rise to a fourth quantum number.

In an atom, an electron can take on many different energies and waveshapes, each of which is called an orbital that is identified by a set of values for n, ℓ, and m_ℓ. When the electron wave possesses a given set of n, ℓ, and m_ℓ, we say the electron "occupies the orbital" with that set of quantum numbers.

7.4 Electron Spin Affects the Distribution of Electrons among Orbitals in Atoms

Earlier it was stated that an atom is in its most stable state (its ground state) when its electrons have the lowest possible energies. This occurs when the electrons "occupy" the lowest energy orbitals that are available. But what determines how the electrons "fill" these orbitals? Fortunately, there are some simple rules that can help. These govern both the maximum number of electrons that can be in a particular orbital and how orbitals with the same energy become filled. One important factor that influences the distribution of electrons is the phenomenon known as *electron spin.*

Electrons behave like tiny charges that can spin in one of two directions

When a beam of atoms with an odd number of electrons is passed through an uneven magnetic field, the beam is split in two, as shown in Figure 7.19. The splitting occurs because the electrons within the atoms interact with the magnetic field in two different ways. The electrons behave like tiny magnets, and they are attracted to one or the other of the poles depending on their orientation. This can be explained by imagining that an electron spins around its axis, like a toy top. A moving charge creates a moving electric field, which in

FIG. 7.19 **The discovery of electron spin.** In this classic experiment by Stern and Gerlach, a beam of atoms with an odd number of electrons is passed through an uneven magnetic field, created by magnet pole faces of different shapes. The beam splits in two, indicating that the electrons in the atoms behave as tiny magnets, which are attracted to one or the other of the poles depending on their orientation. "Electron spin" was proposed to account for the two possible orientations for the electron's magnetic field.

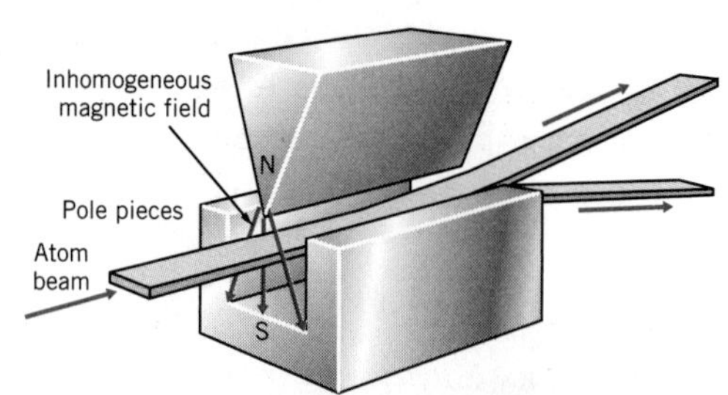

Electrons don't actually spin. We've seen that electrons are not simply particles. But the magnetic properties of electrons are just what we'd see if the electron were a spinning charged particle, and it is useful to picture the electron as spinning.

turn creates a magnetic field. The spinning electrical charge of the electron creates its own magnetic field. This **electron spin** could occur in two possible directions, which accounted for the two beams.

Electron spin gives us a fourth quantum number for the electron, called the **spin quantum number, m_s,** which can take on two possible values: $m_s = +\frac{1}{2}$ or $m_s = -\frac{1}{2}$, corresponding to the two beams in Figure 7.19. The actual values of m_s and the reason they are not integers aren't very important to us, but the fact that there are *only* two values is very significant.

No two electrons in an atom have identical sets of quantum numbers

Pauli received the 1945 Nobel Prize in Physics for his discovery of the exclusion principle.

In 1925 an Austrian physicist, Wolfgang Pauli (1900–1958), expressed the importance of electron spin in determining electronic structure. The **Pauli exclusion principle** states that *no two electrons in the same atom can have identical values for all four of their quantum numbers.* To understand the significance of this, suppose two electrons were to occupy the 1*s* orbital of an atom. Each electron would have $n = 1$, $\ell = 0$, and $m_\ell = 0$. Since these three quantum numbers are the same for both electrons, the exclusion principle requires that their fourth quantum numbers (their spin quantum numbers) be different; one electron must have $m_s = +\frac{1}{2}$ and the other, $m_s = -\frac{1}{2}$. No more than two electrons can occupy the 1*s* orbital of the atom simultaneously because there are only two possible values of m_s. Thus the Pauli exclusion principle is really telling us that *the maximum number of electrons in any orbital is two*, and that *when two electrons are in the same orbital, they must have opposite spins.*

The limit of two electrons per orbital also limits the maximum electron populations of the shells and subshells. For the subshells we have

Remember that a shell is a group of orbitals with the same value of *n*. A subshell is a group of orbitals with the same values of *n* and ℓ.

Subshell	Number of Orbitals	Maximum Number of Electrons
s	1	2
p	3	6
d	5	10
f	7	14

The maximum electron population per shell is shown below.

Shell	Subshells	Maximum Shell Population	
1	1*s*	2	
2	2*s* 2*p*	8	(2 + 6)
3	3*s* 3*p* 3*d*	18	(2 + 6 + 10)
4	4*s* 4*p* 4*d* 4*f*	32	(2 + 6 + 10 + 14)

This trend shows that the maximum electron population of a shell is $2n^2$.

Atoms with unpaired electrons are weakly attracted to magnets

The electron can spin in either of two directions in the presence of an external magnetic field.

We have seen that when two electrons occupy the same orbital they must have different values of m_s. When this occurs, we say that the spins of the electrons are *paired*, or simply that the electrons are *paired.* Such pairing leads to the cancellation of the magnetic effects of the electrons because the north pole of one electron magnet is opposite the south pole of the other. Atoms with more electrons that spin in one direction than in the other are said to contain *unpaired* electrons. For these atoms, the magnetic effects do not cancel and the atoms themselves become tiny magnets that can be attracted to an external magnetic field. This weak attraction of a substance containing unpaired electrons to a magnet indicates that the material is **paramagnetic.** Substances in which all the electrons are paired are not attracted to a magnet and are said to be **diamagnetic.**

Diamagnetic substances are actually weakly repelled by a magnetic field.

Paramagnetism and diamagnetism are measurable properties that provide experimental verification of the presence or absence of unpaired electrons in substances. In addition, the quantitative measurement of the strength of the attraction of a paramagnetic substance toward a magnetic field permits the calculation of the number of unpaired electrons in its atoms, molecules, or ions.

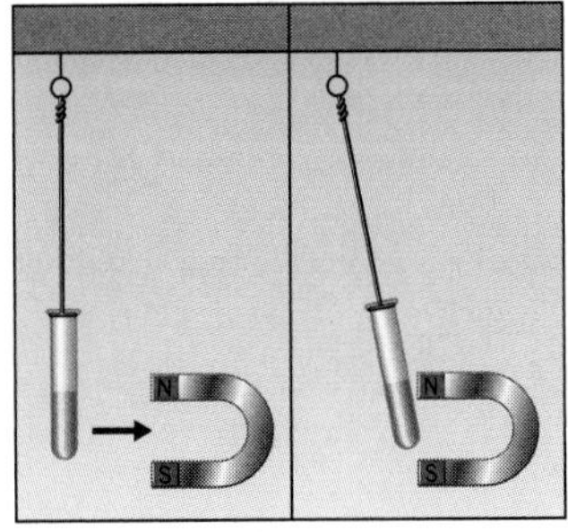

A paramagnetic substance is attracted to a magnetic field.

7.5 THE GROUND STATE ELECTRON CONFIGURATION IS THE LOWEST ENERGY DISTRIBUTION OF ELECTRONS AMONG ORBITALS

The distribution of electrons among the orbitals of an atom is called the atom's **electronic structure** or **electron configuration.** This is something very useful to know about an element because the arrangement of electrons in the outer parts of an atom, which is determined by its electron configuration, controls the chemical properties of the element.

We are interested in the ground state electron configurations of the elements. This is the configuration that yields the lowest energy for an atom and can be predicted for many of the elements by the use of the energy level diagram in Figure 7.18 and application of the Pauli exclusion principle. To see how we go about this, let's begin with the simplest atom of all, hydrogen.

Hydrogen has an atomic number, Z, equal to 1, so a neutral hydrogen atom has one electron. In its ground state this electron occupies the lowest energy orbital that's available, which is the 1*s* orbital. To indicate symbolically the electron configuration we list the subshells that contain electrons and indicate their electron populations by appropriate superscripts. Thus the electron configuration of hydrogen is written as

$$\text{H} \quad 1s^1$$

Another way of expressing electron configurations that we will sometimes find useful is the **orbital diagram.** In it, a circle will represent each orbital and arrows will be used to indicate the individual electrons, head up for spin in one direction and head down for spin in the other. The orbital diagram for hydrogen is simply

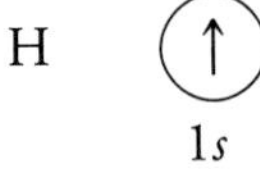

It doesn't matter whether the arrow points up or down. The energy of the electron is the same whether it has one spin or the other. For consistency, when an orbital is half-filled, we will show the electron with an arrow that points up.

Electron configurations are built up by filling lowest energy orbitals first

To arrive at the electron configuration of an atom of another element, we imagine that we begin with a hydrogen atom and then add one proton after another (plus whatever neutrons are also needed) until we obtain the nucleus of the atom of interest. As we proceed, we also must add electrons, one at a time to the lowest available orbital, until we have added enough electrons to give the neutral atom of the element in question. This process for obtaining the electronic structure of an atom is known as the **aufbau principle,** the word *aufbau* being German for "building up."

How s orbitals fill

Let's look at the way this works for helium, which has $Z = 2$. This atom has two electrons, both of which are permitted to occupy the 1*s* orbital. The electron configuration and orbital diagram of helium can now be written as

$$\text{He} \quad 1s^2 \qquad \text{or} \qquad \text{He} \quad \underset{1s}{(\uparrow\downarrow)}$$

Notice that the orbital diagram shows that both electrons in the 1*s* orbital are paired.

We can proceed in the same fashion to predict successfully the electron configurations of most of the elements in the periodic table. For example, the next two elements in the table are lithium, Li ($Z = 3$), and beryllium, Be ($Z = 4$), which have three and four electrons, respectively. For each of these, the first two electrons enter the $1s$ orbital with their spins paired. The Pauli exclusion principle tells us that the $1s$ subshell is filled with two electrons, and Figure 7.18 shows that the orbital of next lowest energy is the $2s$, which can also hold up to two electrons. Therefore, the third electron of lithium and the third and fourth electrons of beryllium enter the $2s$. We can represent the electronic structures of lithium and beryllium as

Li $1s^22s^1$ or Li (↑↓) (↑)
$1s$ $2s$

Be $1s^22s^2$ or Be (↑↓) (↑↓)
$1s$ $2s$

Filling *p* orbitals

After beryllium comes boron, B ($Z = 5$). Referring to Figure 7.18, we see that the first four electrons of this atom complete the $1s$ and $2s$ subshells, so the fifth electron must be placed into the $2p$ subshell.

B $1s^22s^22p^1$

In the orbital diagram for boron, the fifth electron can be put into any one of the $2p$ orbitals—which one doesn't matter because they are all of equal energy.

B (↑↓) (↑↓) (↑)( )( )
$1s$ $2s$ $2p$

Notice, however, that when we give this orbital diagram we show *all* of the orbitals of the $2p$ subshell even though two of them are empty.

Next we come to carbon, which has six electrons. As before, the first four electrons complete the $1s$ and $2s$ orbitals. The remaining two electrons go in the $2p$ subshell to give

C $1s^2\,2s^2\,2p^2$

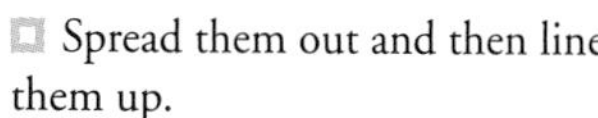

Spread them out and then line them up.

Now, however, to write the orbital diagram we have to make a decision as to where to put the two p electrons. (At this point you may have an unprintable suggestion! But try to bear up. It's really not all that bad.) To make this decision, we apply **Hund's rule,** which states that *when electrons are placed in a set of orbitals of equal energy, they are spread out as much as possible to give as few paired electrons as possible.* Both theory and experiment have shown that if we follow this rule, we obtain the electron configuration with the lowest energy. For carbon, it means that the two p electrons are in separate orbitals and their spins are in the same direction.[1]

C (↑↓) (↑↓) (↑)(↑)( )
$1s$ $2s$ $2p$

It doesn't matter which two orbitals are shown as occupied. Any of these are okay for the ground state of carbon.

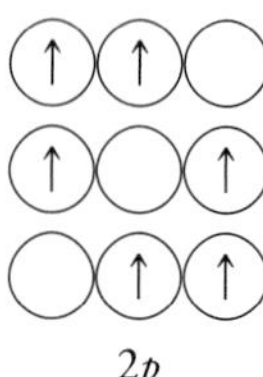

Applying the Pauli exclusion principle and Hund's rule, we can now complete the electron configurations and orbital diagrams for the rest of the elements of the second period.

[1] Hund's rule gives us the *lowest* energy (ground state) distribution of electrons among the orbitals. However, configurations such as

C (↑↓) (↑↓) (↑)(↓)( )
C (↑↓) (↑↓) (↑↓)( )( )
$1s$ $2s$ $2p$

are not impossible; it is just that neither of them corresponds to the lowest energy distribution of electrons in the carbon atom.

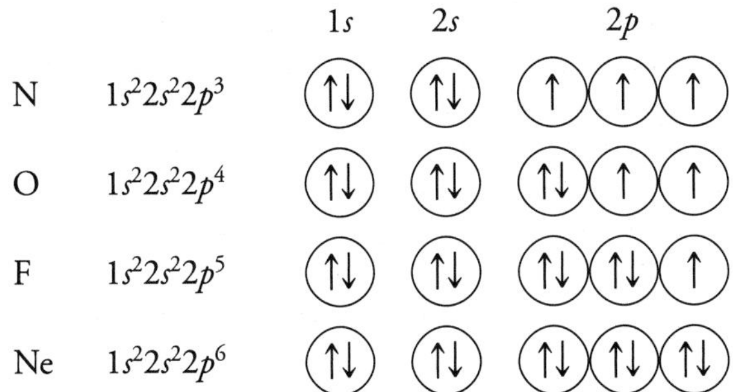

We can continue to predict electron configurations in this way, using Figure 7.18 as a guide to tell us which subshells become occupied and in what order. For instance, after completing the $2p$ subshell at neon, Figure 7.18 predicts that the next two electrons enter the $3s$, followed by the filling of the $3p$. Then we find the $4s$ is lower in energy than the $3d$, so it is filled first. Next, the $3d$ is completed before we go on to fill the $4p$, and so forth.

You might be thinking that you'll have to consult Figure 7.18 whenever you need to write down the ground state electron configuration of an element. However, as you will see in the next section, all the information contained in that figure is also contained in the periodic table.

7.6 ELECTRON CONFIGURATIONS EXPLAIN THE STRUCTURE OF THE PERIODIC TABLE

In Chapter 2 you learned that when Mendeleev constructed his periodic table, elements with similar chemical properties were arranged in vertical columns called groups. Later work led to the expanded version of the periodic table we use today. The basic structure of this table is one of the strongest empirical supports for the quantum theory, which we have been using to predict electron configurations, and it also permits us to use the periodic table itself as a device for predicting electron configurations.

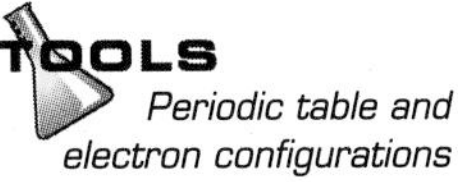

The periodic table is a guide for predicting electron configurations

Consider, for example, the way the table is laid out (Figure 7.20). On the left there is a block of *two* columns shown in blue; on the right there is a block of *six* columns shown in pink; in the center there is a block of *ten* columns shown in yellow, and below the table there are two rows consisting of *fourteen* elements each shown in gray. These numbers—2, 6, 10, and 14—are *precisely* the numbers of electrons that the quantum theory tells us can occupy *s*, *p*, *d*, and *f* subshells, respectively. In fact, *we can use the structure of the periodic table to predict the filling order of the subshells when we write the electron configuration of an element.*

This would be an amazing coincidence if the theory were wrong!

To use the periodic table to predict electron configurations, we follow the aufbau principle as before. We start with hydrogen and then move through the table row after row

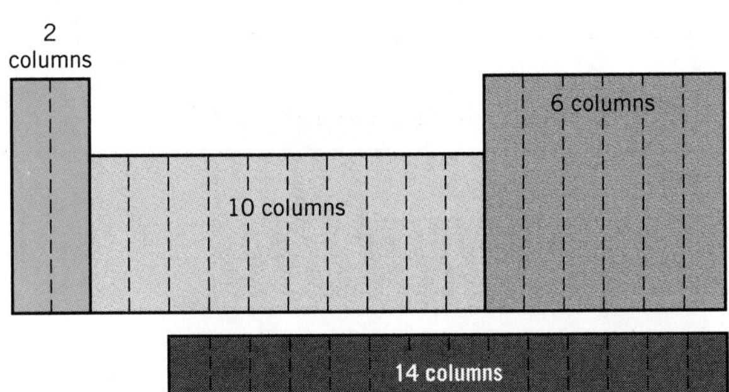

FIG. 7.20 The overall column structure of the periodic table. The table is naturally divided into regions of 2, 6, 10, and 14 columns, which are the numbers of electrons that can occupy *s*, *p*, *d*, and *f* subshells.

until we reach the element of interest, noting as we go along which regions of the table we pass through. For example, consider the element calcium. Using Figure 7.20, we obtain the electron configuration

$$\text{Ca} \qquad 1s^2 2s^2 2p^6 3s^2 3p^6 4s^2$$

To see how this configuration relates to the periodic table, refer to the inside front cover of the book.

Filling periods 1, 2, and 3

Notice that the first period has just two elements, H and He. Starting with hydrogen and passing through this period, two electrons are added to the atom. These enter the $1s$ subshell and are written as $1s^2$. Next we move across the second period, where the first two elements, Li and Be, are in the block of two columns. The two electrons for Li and Be enter the $2s$ subshell and are written as $2s^2$. Then we move to the block of six columns, and as we move across this region in the second row we fill the $2p$ subshell with six electrons, writing $2p^6$. Now we go to the third period where we first pass through the block of two columns, filling the $3s$ subshell with two electrons ($3s^2$), and then through the block of six columns, filling the $3p$ with six electrons ($3p^6$).

Filling periods 4 and 5

Next, we move to the fourth period. To get to calcium in the example above we step through two elements in the two-column block, filling the $4s$ subshell with two more electrons. Up to calcium we have only filled s and p subshells. Figure 7.21 shows that after calcium the $3d$ subshell will be filled. Ten electrons are then added for the ten elements in the first row of the transition elements. After the $3d$ subshell we complete the fourth period by filling the $4p$ subshell with six electrons. The fifth period fills the $5s$, $4d$, and $5p$ subshells in sequence with 2, 10, and 6 electrons, respectively.

Filling periods 6 and 7

The inner transition elements (f subshells) begin to fill in the sixth and seventh periods. Period 6 fills the $6s$, $4f$, $5d$, and $6p$ subshells with 2, 14, 10, and 6 electrons, respectively. The reason for this sequence is that, in general, the $6s$ subshell has the lowest energy. Subshells $4f$, $5d$, and $6p$ have increasingly greater energies in this period. There are many irregularities in this sequence, however, and Appendix A should be consulted for the correct electron configurations. The seventh period fills the $7s$, $5f$, $6d$, and $7p$ subshells. Once again, the sequence is based on the energies of the subshells and there are many irregularities that make it advisable to consult Appendix A for the correct electron configurations.

Notice that in the above patterns the s and p subshells always have the same number as the period they are in. The d subshells always have a number that is one less than the period they are in. Finally, the f subshells are always two less than the period in which they reside.

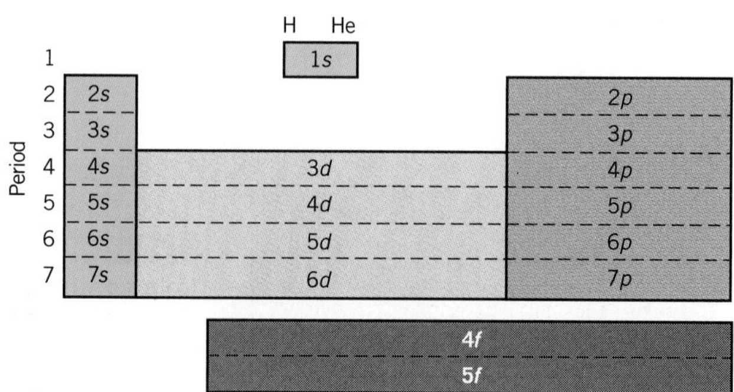

FIG. 7.21 Arrangement of subshells in the periodic table. This format illustrates the sequence for filling subshells.

EXAMPLE 7.4
Predicting Electron Configurations

What is the electron configuration of (a) Mn and (b) Bi?

ANALYSIS: We use the periodic table as a tool to tell us which subshells become filled. It is best if you can do this without referring to Figure 7.21.

SOLUTION: (a) To get to manganese, we cross the following regions of the table, with the results indicated.

Period 1	Fill the 1*s* subshell
Period 2	Fill the 2*s* and 2*p* subshells
Period 3	Fill the 3*s* and 3*p* subshells
Period 4	Fill the 4*s* and then move five places in the 3*d* region

The electron configuration of Mn is therefore

$$\text{Mn} \quad 1s^2 2s^2 2p^6 3s^2 3p^6 4s^2 3d^5$$

This configuration is correct as is. In elements with electrons in *d* and *f* orbitals, we can also group all subshells of the same shell together. For manganese, this gives

$$\text{Mn} \quad 1s^2 2s^2 2p^6 3s^2 3p^6 3d^5 4s^2$$

We'll see later that writing the configuration with this way is convenient when working out ground state electronic configurations for ions.

(b) To get to bismuth, we fill the following subshells:

Period 1	Fill the 1*s*
Period 2	Fill the 2*s* and 2*p*
Period 3	Fill the 3*s* and 3*p*
Period 4	Fill the 4*s*, 3*d*, and 4*p*
Period 5	Fill the 5*s*, 4*d*, and 5*p*
Period 6	Fill the 6*s*, 4*f*, 5*d*, and then add three electrons to the 6*p*

This gives

$$\text{Bi} \quad 1s^2 2s^2 2p^6 3s^2 3p^6 4s^2 3d^{10} 4p^6 5s^2 4d^{10} 5p^6 6s^2 4f^{14} 5d^{10} 6p^3$$

Grouping subshells with the same value of *n* gives

$$\text{Bi} \quad 1s^2 2s^2 2p^6 3s^2 3p^6 3d^{10} 4s^2 4p^6 4d^{10} 4f^{14} 5s^2 5p^6 5d^{10} 6s^2 6p^3$$

Again, either of these configurations is correct.

IS THE ANSWER REASONABLE? We can count the number of electrons to be sure we have 25 for Mn and 83 for Bi and none have been left out or added. If the textbook is available we can look at Appendix A to check our configurations.

Practice Exercise 8: Use the periodic table to predict the electron configurations of (a) Mg, (b) Ge, (c) Cd, and (d) Gd. Group subshells of the same shell together. (Hint: Recall which areas of the periodic table represent *s, p, d,* and *f* electrons.)

Practice Exercise 9: Describe in your own words how to use the periodic table to write the electron configuration of an element.

Practice Exercise 10: Use the periodic table to predict the electron configurations of (a) O, S, Se and (b) P, N, Sb. What is the same about all the elements in part (a), and the elements in part (b)?

Practice Exercise 11: Draw orbital diagrams for (a) Na, (b) S, and (c) Fe. (Hint: Recall how we indicate paired electron spins.)

Practice Exercise 12: Use orbital diagrams to determine how many unpaired electrons are in each of these elements: (a) Mg, (b) Ge, (c) Cd, and (d) Gd.

Electron configurations can be abbreviated using noble gas core configurations

When we consider the chemical reactions of atoms, our attention is usually focused on the distribution of electrons in the **outer shell** of the atom (the occupied shell with the largest value of n). This is because the **outer electrons** (those in the outer shell) are the ones that are exposed to other atoms when the atoms react. The inner electrons of an atom, called the **core electrons,** are buried deep within the atom and normally do not play a role when chemical bonds are formed.

Because we are interested primarily in the electrons of the outer shell, we often write electron configurations in an abbreviated, or shorthand, form. To illustrate, let's consider the elements sodium and magnesium. Their full electron configurations are

$$\text{Na} \quad 1s^22s^22p^63s^1$$
$$\text{Mg} \quad 1s^22s^22p^63s^2$$

The outer electrons of both atoms are in the $3s$ subshell and each has the core configuration $1s^22s^22p^6$, which is the same as that of the noble gas neon.

To write the *shorthand configuration* for an element we indicate what the core is by placing in brackets the symbol of the noble gas whose electron configuration is the same as the core configuration. This is followed by the configuration of the outer electrons for the particular element. The noble gas used is almost always the one that occurs at the end of the period preceding the period containing the element whose configuration we wish to represent. Thus, for sodium and magnesium we write

$$\text{Na} \quad [\text{Ne}]\ 3s^1$$
$$\text{Mg} \quad [\text{Ne}]\ 3s^2$$

Even with the transition elements, we need only concern ourselves with the outermost shell and the d subshell just below. For example, iron has the configuration

$$\text{Fe} \quad 1s^22s^22p^63s^23p^63d^64s^2$$

Only the $4s$ and $3d$ electrons play a role in the chemistry of iron. In general, the electrons below the outer s and d subshells of a transition element—the *core* electrons—are relatively unimportant. In every case these core electrons have the electron configuration of a noble gas. Therefore, for iron the core is $1s^22s^22p^63s^23p^6$, which is the same as the electron configuration of argon. The abbreviated configuration of iron is therefore written as

$$\text{Fe} \quad [\text{Ar}]\ 3d^64s^2$$

EXAMPLE 7.5
Writing Shorthand Electron Configurations

What is the shorthand electron configuration of manganese? Draw the orbital diagram for manganese using the shorthand configuration.

ANALYSIS: As usual, the periodic table serves as our tool for deriving the electron configuration. For the shorthand configuration of an element, we write, in brackets, the symbol for the noble gas that is at the end of the preceding period, followed by the electron configuration of the occupied subshells that exist beyond the noble gas core. For the orbital diagram, we use a circle to represent each of the orbitals in a populated subshell. The orbitals are then populated with the required number of electrons, using arrows as appropriate, following Hund's rule.

SOLUTION: Manganese is in Period 4. The preceding noble gas is argon, Ar, so to write the abbreviated configuration we write the symbol for argon in brackets followed by the electron configuration that exists beyond the argon core. We can obtain this by noting that to get to Mn in Period 4, we first cross the "*s* region" by adding two electrons to the 4*s* subshell, and then go five steps into the "*d* region" where we add five electrons to the 3*d* subshell. Therefore, the shorthand electron configuration for Mn is

$$\text{Mn} \quad [\text{Ar}]\ 4s^2 3d^5$$

Placing the electrons that are in the highest shell farthest to the right gives

$$\text{Mn} \quad [\text{Ar}]\ 3d^5 4s^2$$

To draw the orbital diagram, we simply distribute the electrons in the 3*d* and 4*s* orbitals following Hund's rule. This gives

Mn [Ar] (↑)(↑)(↑)(↑)(↑) (↑↓)
3*d* 4*s*

Notice that each of the 3*d* orbitals is half-filled. The atom contains five unpaired electrons and is paramagnetic.

IS THE ANSWER REASONABLE? From the left, count the groups needed to reach Mn. These should equal seven, the number of electrons in the orbitals beyond Ar as written above. The orbital diagram must have seven electrons and must obey Hund's rule.

Practice Exercise 13: Can an element with an even atomic number be paramagnetic? (Hint: Try writing the orbital diagrams of a few of the transition elements in Period 4.)

Practice Exercise 14: Write shorthand configurations and abbreviated orbital diagrams for (a) P and (b) Sn. Where appropriate, place the electrons that are in the highest shell farthest to the right. How many unpaired electrons does each of these atoms have?

Chemical properties of the representative elements depend on valence shell electron configurations

You learned that Mendeleev constructed the periodic table by placing elements with similar properties in the same group. We are now ready to understand the reason for these similarities in terms of the electronic structures of atoms. It seems reasonable, therefore, that elements with similar properties should have similar outer shell electron configurations. This is, in fact, exactly what we observe. For example, let's look at the alkali metals of Group IA. Going by our rules, we obtain the following configurations.

Li	[He] $\mathbf{2s^1}$	or	$1s^2\mathbf{2s^1}$
Na	[Ne] $\mathbf{3s^1}$	or	$1s^22s^22p^6\mathbf{3s^1}$
K	[Ar] $\mathbf{4s^1}$	or	$1s^22s^22p^63s^23p^6\mathbf{4s^1}$
Rb	[Kr] $\mathbf{5s^1}$	or	$1s^22s^22p^63s^23p^63d^{10}4s^24p^6\mathbf{5s^1}$
Cs	[Xe] $\mathbf{6s^1}$	or	$1s^22s^22p^63s^23p^63d^{10}4s^24p^64d^{10}5s^25p^6\mathbf{6s^1}$

Each of these elements has just one outer shell electron that is in an *s* subshell. We know that when they react, the alkali metals each lose one electron to form ions with a charge of 1+. For each, the electron that is lost is this outer *s* electron, and the electron configuration of the ion that is formed is the same as that of the preceding noble gas.

Li^+	$1s^2$	He	$1s^2$
Na^+	$1s^22s^22p^6$	Ne	$1s^22s^22p^6$
K^+	$1s^22s^22p^63s^23p^6$	Ar	$1s^22s^22p^63s^23p^6$

etc.

If you write the electron configurations of the members of any of the groups in the periodic table, you will find the same kind of similarity among the configurations of the outer shell electrons. The differences are in the value of the principal quantum number of the outer electrons.

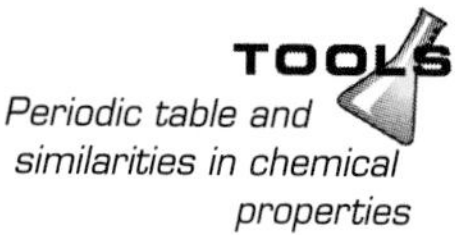

Periodic table and similarities in chemical properties

For the representative elements (those in the longer columns), the only electrons that are normally important in controlling chemical properties are the ones in the outer shell. This outer shell is known as the **valence shell,** and it is always the occupied shell with the largest value of n. The electrons in the valence shell are called **valence electrons.** (The term *valence* comes from the study of chemical bonding and relates to the combining capacity of an element, but that's not important here.)

For the representative elements it is very easy to determine the electron configuration of the valence shell by using the periodic table. *The valence shell always consists of just the s and p subshells that we encounter crossing the period that contains the element in question.* Thus, to determine the valence shell configuration of sulfur, a Period 3 element, we note that to reach sulfur in Period 3 we place two electrons into the $3s$ and four electrons into the $3p$. The valence shell configuration of sulfur is therefore

$$\text{S} \quad 3s^2 3p^4$$

EXAMPLE 7.6
Writing Valence Shell Configurations

Predict the electron configuration of the valence shell of arsenic ($Z = 33$).

ANALYSIS: To determine the number of s and p electrons in the highest shell we write the electron configuration grouping the subshells in each shell together. The last subshell must be either an s or p subshell. If it is an s subshell, the valence shell consists of the s electrons. If the last subshell is a p subshell, the valence shell includes both the s and p electrons in that shell.

SOLUTION: To reach arsenic in Period 4, we add electrons to the $4s$, $3d$, and $4p$ subshells. But the $3d$ is not part of the fourth shell and therefore not part of the valence shell, so all we need be concerned with are the electrons in the $4s$ and $4p$ subshells. This gives us the valence shell configuration of arsenic,

$$\text{As} \quad 4s^2 4p^3$$

IS THE ANSWER REASONABLE? First of all, only s and p electrons can be valence electrons. We can then count backward from As to the start of the period counting only the p and then s electrons. These add up to the numbers above, indicating the answer is correct.

Practice Exercise 15: Give an example of a valence shell with more than eight electrons. If that is not possible, explain why. (Hint: Are there any elements where d shell electrons are valence shell electrons?)

Practice Exercise 16: What is the valence shell electron configuration of (a) Se, (b) Sn, and (c) I?

Configurations for transition and rare earth elements are sometimes unexpected

The rules you've learned for predicting electron configurations work most of the time, but not always. Appendix A gives the electron configurations of all of the elements as determined experimentally. Close examination reveals that there are quite a few exceptions to the rules. Some of these exceptions are significant for us because they occur with common elements.

Two important exceptions are for chromium and copper. Following the rules, we would expect the configurations to be

$$\text{Cr} \quad [\text{Ar}]\, 3d^4 4s^2$$
$$\text{Cu} \quad [\text{Ar}]\, 3d^9 4s^2$$

However, the actual electron configurations, determined experimentally, are

$$\text{Cr} \quad [\text{Ar}]\ 3d^5 4s^1$$
$$\text{Cu} \quad [\text{Ar}]\ 3d^{10} 4s^1$$

The corresponding orbital diagrams are

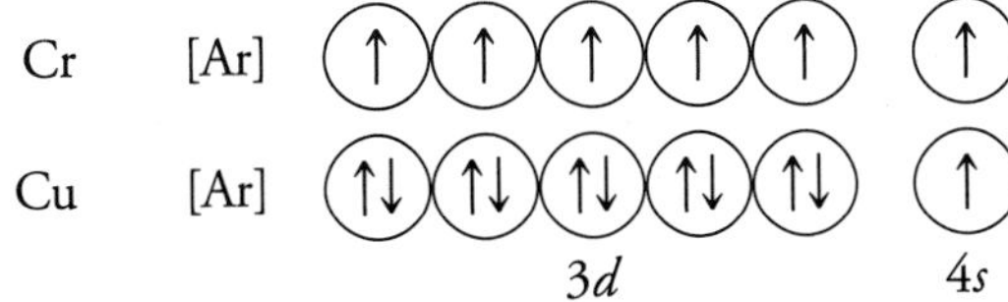

Notice that for chromium, an electron is "borrowed" from the $4s$ subshell to give a $3d$ subshell that is exactly half-filled. For copper the $4s$ electron is borrowed to give a completely filled $3d$ subshell. A similar thing happens with silver and gold, which have filled $4d$ and $5d$ subshells, respectively.

$$\text{Ag} \quad [\text{Kr}]\ 4d^{10} 5s^1$$
$$\text{Au} \quad [\text{Xe}]\ 4f^{14} 5d^{10} 6s^1$$

Apparently, half-filled and filled subshells (particularly the latter) have some special stability that makes such borrowing energetically favorable. This subtle but nevertheless important phenomenon affects not only the ground state configurations of atoms but also the relative stabilities of some of the ions formed by the transition elements. Similar irregularities occur among the lanthanide and the actinide elements.

7.7 QUANTUM THEORY PREDICTS THE SHAPES OF ATOMIC ORBITALS

To picture what electrons are doing within the atom, we are faced with imagining an object that behaves like a particle in some experiments and like a wave in others. There is nothing in our experience that is comparable. Fortunately we can still think of the electron as a particle in the usual sense by using the statistical probability of the electron being found at a particular place. We can then use quantum mechanics to mathematically connect the particle and wave representations of the electron. Even though we may have trouble imagining an object that can be represented both ways, mathematics describes its behavior very accurately.

Describing the electron's position in terms of statistical probability is based on more than simple convenience. The German physicist Werner Heisenberg showed mathematically that it is impossible to measure with complete precision both a particle's velocity and position at the same instant. To measure an electron's position or velocity, we have to bounce another particle off it. Thus, the very act of making the measurement changes the electron's position and velocity. We cannot determine both exact position and exact velocity simultaneously, no matter how cleverly we make the measurements. This was Heisenberg's famous **uncertainty principle.** The theoretical limitations on measuring speed and position are not significant for large objects. For small particles such as the electron, however, these limitations prevent us from ever knowing or predicting where in an atom an electron will be at a particular instant, so we speak of probabilities instead.

The uncertainty principle is often stated mathematically as

$$\Delta x = \frac{h}{4\pi m}\left(\frac{1}{\Delta v}\right)$$

where Δx is the minimum uncertainty in the particle's location, h is Planck's constant, m is the mass of the particle, and Δv is the minimum uncertainty in the particle's velocity. Notice that we can generally measure the particle's location more precisely if the particle is heavier. Notice also that the greater the uncertainty in the velocity, the smaller the uncertainty in the particle's location.

Wave mechanics views the probability of finding an electron at a given point in space as equal to the square of the amplitude of the electron wave (given by the square of the wave function, ψ^2) at that point. It seems quite reasonable to relate probability to amplitude, or intensity, because where a wave is intense its presence is strongly felt. The amplitude is squared because, mathematically, the amplitude can be either positive or negative, but probability only makes sense if it is positive. Squaring the amplitude assures us that the probabilities will be positive. We need not be very concerned about this point, however.

The notion of electron probability leads to two very important and frequently used concepts. One is that an electron behaves as if it were spread out around the nucleus in a sort of **electron cloud**. Figure 7.22*a* is a *dot-density diagram* that illustrates the way

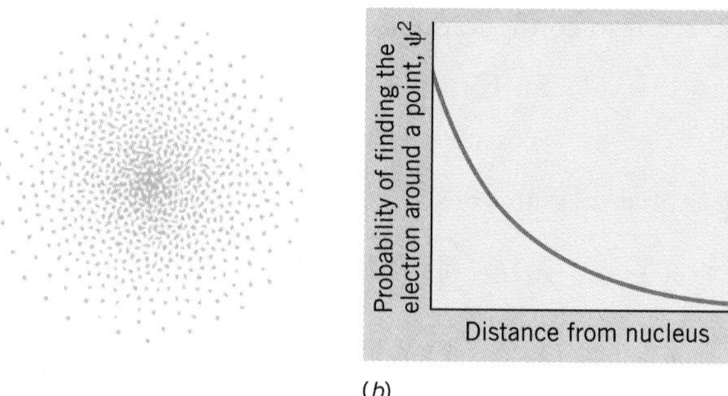

FIG. 7.22 Electron distribution in a 1*s* orbital. (*a*) A dot-density diagram that illustrates the electron probability distribution for a 1*s* electron. (*b*) A graph that shows how the probability of finding the 1*s* electron around a given point, ψ^2, decreases as the distance from the nucleus increases.

The amplitude of an electron wave is described by a *wave function,* which is usually given the symbol ψ (the Greek letter psi). The probability of finding the electron in a given location is given by ψ^2.

the probability of finding the electron varies in space for a 1*s* orbital. In those places where the dot density is large (i.e., where there are large numbers of dots per unit volume), the amplitude of the wave is large and the probability of finding the electron is also large. Figure 7.22*b* shows how the electron probability for a 1*s* orbital varies as we move away from the nucleus. As you might expect, the probability of finding the electron close to the nucleus is large and decreases with increasing distance from the nucleus.

The other important concept that stems from the notion that the electron probability varies from place to place is **electron density,** which relates to how much of the electron's charge is packed into a given volume. In regions of high probability there is a high concentration of electrical charge (and mass) and the electron density is large; in regions of low probability, the electron density is small.

Remember that an electron confined to a tiny space no longer behaves much like a particle. It's more like a cloud of negative charge. Like clouds made of water vapor, the density of the cloud varies from place to place. In some places the cloud is dense; in others the cloud is thinner and may be entirely absent. This is a useful picture to keep in mind as you try to visualize the shapes of atomic orbitals.

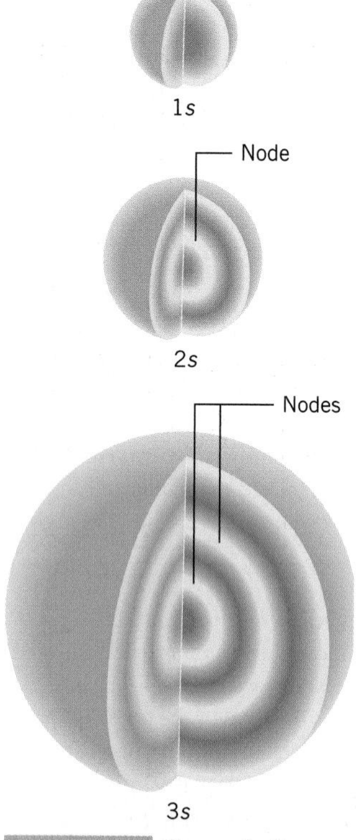

FIG. 7.23 Size variations among *s* orbitals. The orbitals become larger as the principal quantum number, *n*, becomes larger.

s orbitals are spherical; *p* orbitals have two lobes

In looking at the way the electron density distributes itself in atomic orbitals, we are interested in three things: the *shape* of the orbital, its *size,* and its *orientation* in space relative to other orbitals.

The electron density in an orbital doesn't end abruptly at some particular distance from the nucleus. It gradually fades away. Therefore, to define the size and shape of an orbital, it is useful to picture some imaginary surface enclosing, say, 90% of the electron density of the orbital, and on which the probability of finding the electron is everywhere the same. For the 1*s* orbital in Figure 7.22, we find that if we go out a given distance from the nucleus in *any* direction, the probability of finding the electron is the same. This means that all the points of equal probability lie on the surface of a sphere, so we can say that the shape of the orbital is spherical. In fact, all *s* orbitals are spherical. As suggested earlier, their sizes increase with increasing *n*. This is illustrated in Figure 7.23. Notice that beginning with the 2*s* orbital, there are certain places where the electron density drops to zero. These are spherical shaped nodes of the *s* orbital electron waves. It is interesting that electron waves have nodes just like the waves on a guitar string. For *s* orbital electron waves, however, the nodes consist of imaginary spherical *surfaces* on which the electron density is zero.

The *p* orbitals are quite different from *s* orbitals, as shown in Figure 7.24. Notice that the electron density is equally distributed in two regions on opposite sides of the nucleus. Figure 7.24*a* illustrates the two "lobes" of *one* 2*p* orbital. Between the lobes is a **nodal plane**—an imaginary flat surface on which every point has an electron density of zero. The size of the *p* orbitals also increases with increasing *n* as illustrated by the cross section

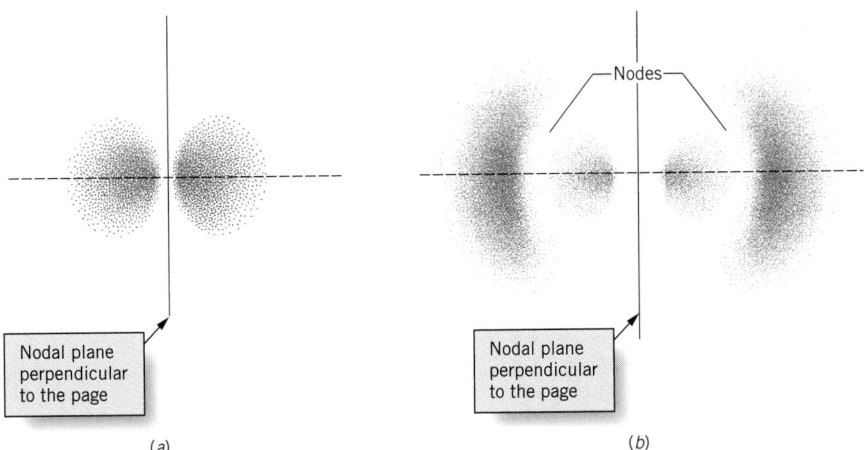

FIG. 7.24 Distribution of electron density in *p* orbitals. (*a*) Dot-density diagram that represents a cross section of the probability distribution in a 2*p* orbital. There is a nodal plane between the two lobes of the orbital. (*b*) Cross section of a 3*p* orbital. Note the nodes in the electron density that are in addition to the nodal plane passing through the nucleus.

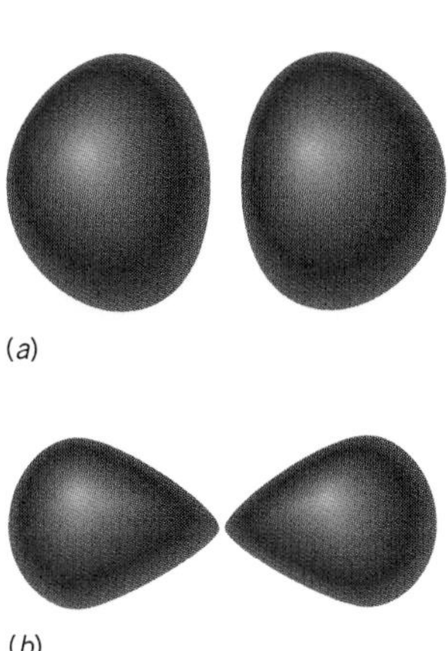

FIG. 7.25 Representations of the shapes of *p* orbitals. (*a*) Shape of a surface of constant probability for a 2*p* orbital. (*b*) A simplified representation of a *p* orbital that emphasizes the directional nature of the orbital.

of a 3*p* orbital in Figure 7.24*b*. The 3*p* and higher *p* orbitals have additional nodes besides the nodal plane that passes through the nucleus.

Figure 7.25*a* illustrates the shape of a surface of constant probability for a 2*p* orbital. Often chemists will simplify this shape by drawing two "balloons" connected at the nucleus and pointing in opposite directions as shown in Figure 7.25*b*. Both representations emphasize the point that a *p* orbital has two equal-sized lobes that extend in opposite directions along a line that passes through the nucleus.

Orbitals in a *p* subshell are oriented at 90° to each other

As you've learned, a *p* subshell consists of three orbitals of equal energy. Wave mechanics tells us that the lines along which the orbitals have their maximum electron densities are oriented at 90° angles to each other, corresponding to the axes of an imaginary *xyz* coordinate system (Figure 7.26). For convenience in referring to the individual *p* orbitals they are often labeled according to the axis along which they lie. The *p* orbital concentrated along the *x* axis is labeled p_x, and so forth.

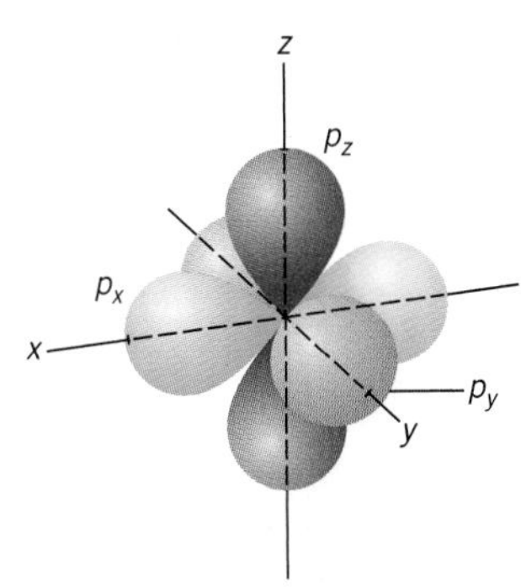

FIG. 7.26 The orientations of the three *p* orbitals in a *p* subshell. Because the directions of maximum electron density lie along lines that are mutually perpendicular, like the axes of an *xyz* coordinate system, it is convenient to label the orbitals p_x, p_y, and p_z.

Four of the five *d* orbitals in a *d* subshell have the same shape

The shapes of the *d* orbitals, illustrated in Figure 7.27, are a bit more complex than are those of the *p* orbitals. Because of this, and because there are five orbitals in a *d* subshell, we haven't attempted to draw all of them at the same time on the same set of coordinate axes. Notice that four of the five *d* orbitals have the same shape and consist of four lobes

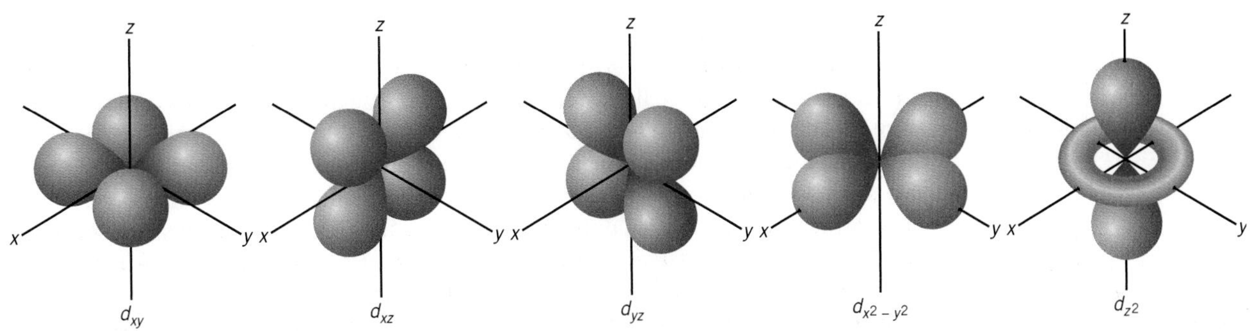

FIG. 7.27 The shapes and directional properties of the five *d* orbitals of a *d* subshell.

The f orbitals are even more complex than the d orbitals, but we will have no need to discuss their shapes.

of electron density. These four d orbitals each have two perpendicular nodal planes that intersect the nucleus. These orbitals differ only in their orientations around the nucleus (their labels come from the mathematics of wave mechanics). The fifth d orbital, labeled d_{z^2}, has two lobes that point in opposite directions along the z axis plus a doughnut-shaped ring of electron density around the center that lies in the x–y plane. The two nodes for the d_{z^2} orbital are conic surfaces whose peaks meet at the nucleus. We will see that the d orbitals are important in the formation of chemical bonds in certain molecules, and that their shapes and orientations are important in understanding the properties of the transition metals, which will be discussed in Chapter 21.

The shapes of the f orbitals are more complex than those of the d orbitals, having more lobes and a variety of shapes. Use of f orbitals for bonding is not important for this course. However, we should note that each f orbital has three nodal planes or surfaces.

7.8 Atomic properties correlate with an atom's electron configuration

There are many chemical and physical properties that vary in a more or less systematic way according to an element's position in the periodic table. For example, in Chapter 2 we noted that the metallic character of the elements increases from top to bottom in a group and decreases from left to right across a period. In this section we discuss several physical properties of the elements that have an important influence on chemical properties. We will see how these properties correlate with an atom's electron configuration, and because electron configuration is also related to the location of an element in the periodic table, we will study their periodic variations as well.

Effective nuclear charge is the positive charge "felt" by outer electrons

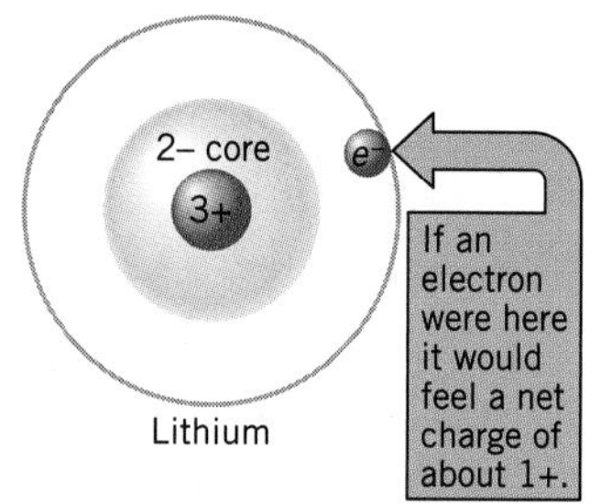

FIG. 7.28 Effective nuclear charge. If the 2− charge of the $1s^2$ core of lithium were 100% effective at shielding the $2s$ electron from the nucleus, the valence electron would feel an effective nuclear charge of only about 1+.

Many of an atom's properties are determined by the amount of positive charge felt by the atom's outer electrons. Except for hydrogen, this positive charge is always *less* than the full nuclear charge, because the negative charge of the electrons in inner shells partially offsets, or "neutralizes," the positive charge of the nucleus.

To gain a better understanding of this, consider the element lithium, which has the electron configuration $1s^22s^1$. The core electrons ($1s^2$), which lie beneath the valence shell ($2s^1$), are tightly packed around the nucleus and for the most part lie between the nucleus and the electron in the outer shell. This core has a charge of 2− and it surrounds a nucleus that has a charge of 3+. When the outer $2s$ electron "looks toward" the center of the atom, it "sees" the 3+ charge of the nucleus reduced to only about 1+ because of the intervening 2− charge of the core. In other words, the 2− charge of the core effectively neutralizes two of the positive charges of the nucleus, so the net charge that the outer electron feels, which we call the **effective nuclear charge,** is only about 1+. This is illustrated in an overly simplified way in Figure 7.28.

An electron spends very little time between the nucleus and another electron in the same shell, so it shields that other electron poorly.

Although electrons in inner shells shield the electrons in outer shells quite effectively from the nuclear charge, electrons in the *same* shell are much less effective at shielding each other. For example, in the element beryllium ($1s^22s^2$) each of the electrons in the outer $2s$ orbital is shielded quite well from the nuclear charge by the inner $1s^2$ core, but one $2s$ electron doesn't shield the other $2s$ electron very well at all. This is because electrons in the same shell are at about the same average distance from the nucleus, and in attempting to stay away from each other they only spend a very small amount of time one below the other, which is what's needed to provide shielding. Since electrons in the same shell hardly shield each other at all from the nuclear charge, *the effective nuclear charge felt by the outer electrons is determined primarily by the difference between the charge on the nucleus and the charge on the core.* With this as background, let's examine some properties controlled by the effective nuclear charge.

Atomic and ionic sizes increase with increasing *n* and decreasing effective nuclear charge

The wave nature of the electron makes it difficult to define exactly what we mean by the "size" of an atom or ion. As we've seen, the electron cloud doesn't simply stop at some particular distance from the nucleus; instead it gradually fades away. Nevertheless, atoms and ions do behave in many ways as though they have characteristic sizes. For example, in a whole host of hydrocarbons, ranging from methane (CH_4, natural gas) to octane (C_8H_{18}, in gasoline) to many others, the distance between the nuclei of carbon and hydrogen atoms is virtually the same. This would suggest that carbon and hydrogen have the same relative sizes in each of these compounds.

The C—H distance in most hydrocarbons is about 110 pm (110×10^{-12} m).

Experimental measurements reveal that the diameters of atoms range from about 1.4×10^{-10} to 5.7×10^{-10} m. Their radii, which is the usual way that size is specified, range from about 7.0×10^{-11} to 2.9×10^{-10} m. Such small numbers are difficult to comprehend. A million carbon atoms placed side by side in a line would extend a little less than 0.2 mm, or about the diameter of the period at the end of this sentence.

The sizes of atoms and ions are rarely expressed in meters because the numbers are so cumbersome. Instead, a unit is chosen that makes the values easier to comprehend. A unit that scientists have traditionally used is called the **angstrom** (symbolized **Å**), which is defined as

$$1 \text{ Å} = 1 \times 10^{-10} \text{ m}$$

The angstrom is named after Anders Jonas Ångström (1814–1874), a Swedish physicist who was the first to measure the wavelengths of the four most prominent lines of the hydrogen spectrum.

However, the angstrom is not an SI unit, and in many current scientific journals, atomic dimensions are in picometers, or sometimes in nanometers (1 pm $= 10^{-12}$ m and 1 nm $= 10^{-9}$ m). In this book, we will normally express atomic dimensions in picometers, but because much of the scientific literature has these quantities in angstroms, you may someday find it useful to remember the conversions:

$$1 \text{ Å} = 100 \text{ pm}$$
$$1 \text{ Å} = 0.1 \text{ nm}$$

Atomic size varies periodically

The variations in atomic radii within the periodic table are illustrated in Figure 7.29. Here we see that atoms generally become larger going from top to bottom in a group, and they become smaller going from left to right across a period. To understand these variations we must consider two factors. One is the value of the principal quantum number of the valence electrons, and the other is the effective nuclear charge felt by the valence electrons.

Large atoms are found in the lower left of the periodic table, and small atoms are found in the upper right.

Periodic trends in atomic size

Going from top to bottom within a group, the effective nuclear charge felt by the outer electrons remains nearly constant, while the principal quantum number of the valence shell increases. For example, consider the elements of Group IA. For lithium, the valence shell configuration is $2s^1$; for sodium, it is $3s^1$; for potassium, it is $4s^1$; and so forth. For each of these elements, the core has a negative charge that is one less than the nuclear charge, so the valence electron of each experiences a nearly constant effective nuclear charge of about 1+. However, as we descend the group, the value of *n* for the valence shell increases, and as you learned earlier, the larger the value of *n*, the larger is the orbital. Therefore, the atoms become larger as we go down a group simply because the orbitals containing the valence electrons become larger. This same argument applies whether the valence shell orbitals are *s* or *p*.

Moving from left to right across a period, electrons are added to the same shell. The orbitals holding the valence electrons all have the *same* value of *n*. In this case we have to examine the variation in the effective nuclear charge felt by the valence electrons.

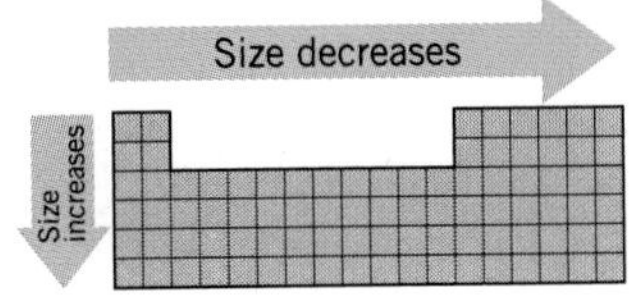

As we move from left to right across a period, the nuclear charge increases; the outer shells of the atoms become more populated, but the inner core remains the same. For example, from lithium to fluorine the nuclear charge increases from 3+ to 9+. The core ($1s^2$) stays the same, however. As a result, the outer electrons feel an increase in positive charge (i.e., effective nuclear charge) that causes them to be drawn inward, and thereby causes the sizes of the atoms to decrease.

Across a row of transition elements or inner transition elements, the size variations are less pronounced than among the representative elements. This is because the outer shell

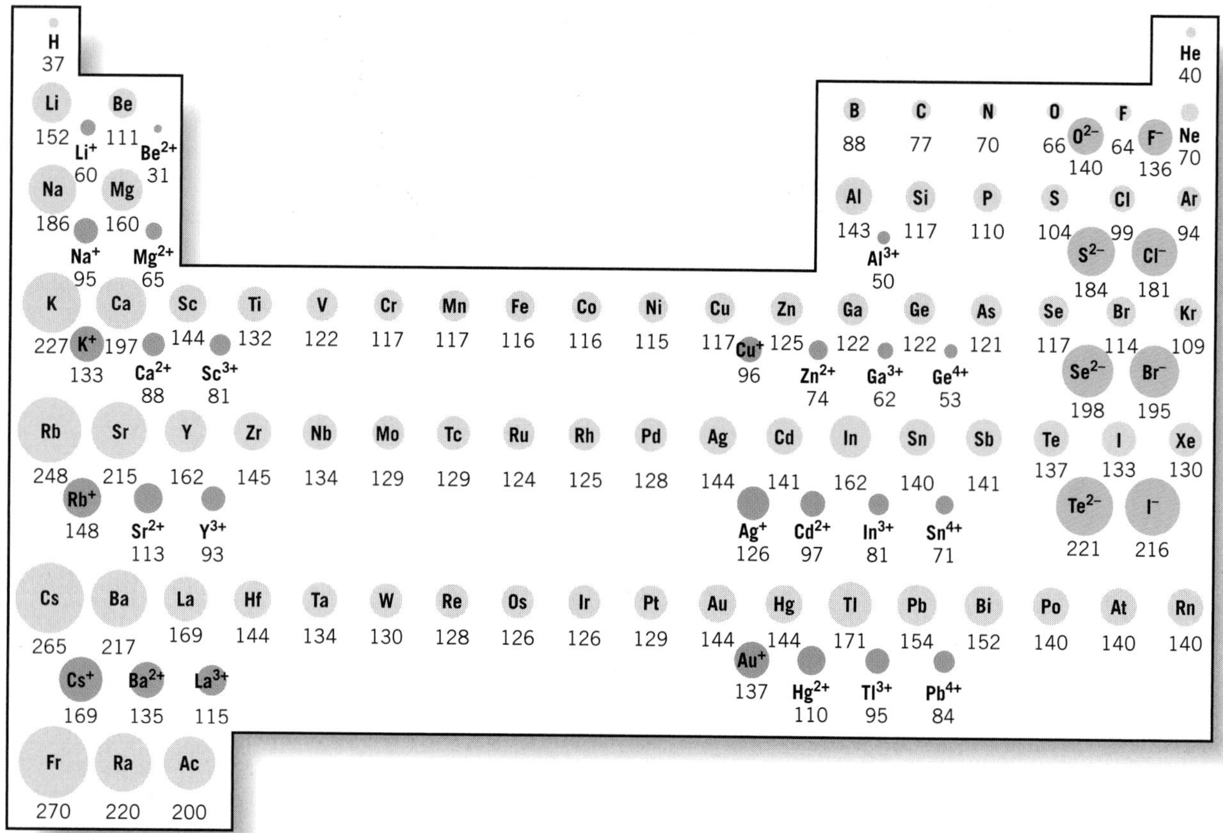

FIG. 7.29 **Variation in atomic and ionic radii in the periodic table.** Values are in picometers.

configuration remains essentially the same while an inner shell is filled. From atomic numbers 21 to 30, for example, the outer electrons occupy the 4*s* subshell while the underlying 3*d* subshell is gradually completed. The amount of shielding provided by the addition of electrons to this inner 3*d* level is greater than the amount of shielding that would occur if the electrons were added to the outer shell, so the effective nuclear charge felt by the outer electrons increases more gradually. As a result, the decrease in size with increasing atomic number is also more gradual.

Ion sizes show the same trends as atom sizes, but anions are larger and cations are smaller than their parent atoms

Figure 7.29 also illustrates how sizes of the ions compare with those of the neutral atoms. As you can see, when atoms gain or lose electrons to form ions, rather significant size changes take place. The reasons are easy to understand and remember.

When electrons are added to an atom, the mutual repulsions between them increase. This causes the electrons to push apart and occupy a larger volume. Therefore, *negative ions are always about 1.5 to 2 times larger than the atoms from which they are formed* (Figure 7.30).

Adding electrons creates an ion that is larger than the neutral atom; removing electrons produces an ion that is smaller than the neutral atom.

When electrons are removed from an atom, the electron–electron repulsions decrease, which allows the remaining electrons to be pulled closer together around the nucleus. Therefore, *positive ions are always smaller than the atoms from which they are formed.* As Figure 7.29 shows, cations often are only 1/2 to 2/3 the size of their parent atom. This is also illustrated in Figure 7.30 for the elements lithium and iron. For lithium, removal of the outer 2*s* electron completely empties the valence shell and exposes the smaller $1s^2$ core. When a metal is able to form more than one positive ion, the sizes of the ions decrease as the amount of positive charge on the ion increases. To form the Fe^{2+} ion, an iron atom loses its outer 4*s* electrons. To form the Fe^{3+} ion, an additional electron is lost from the 3*d* subshell that lies beneath the 4*s*. Comparing sizes, we see that the radius of an iron atom

Li $1s^22s^1$
Li^+ $1s^2$

Fe [Ar] $3d^64s^2$
Fe^{2+} [Ar] $3d^6$
Fe^{3+} [Ar] $3d^5$

is 116 pm whereas the radius of the Fe^{2+} ion is 76 pm. Removing yet another electron to give Fe^{3+} decreases electron–electron repulsions in the *d* subshell and gives the Fe^{3+} ion a radius of 64 pm.

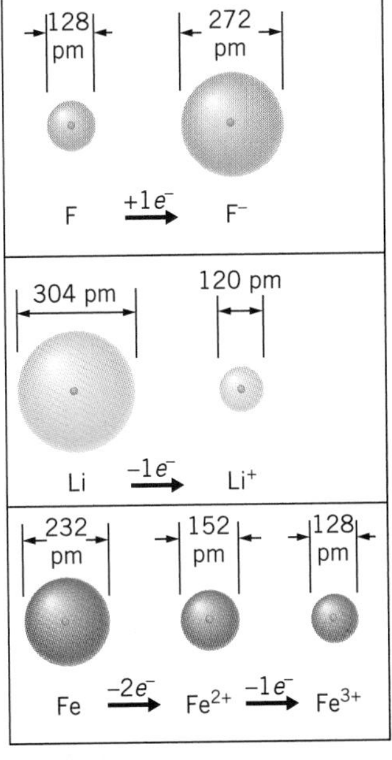

FIG. 7.30 Changes in size when atoms gain or lose electrons to form ions. Adding electrons leads to an increase in the size of the particle, as illustrated for fluorine. Removing electrons leads to a decrease in the size of the particle, as shown for lithium and iron.

Practice Exercise 17: Use the periodic table to choose the largest atom or ion in each set. (Hint: Recall that electrons repel each other.)
(a) Ge, Te, Se, Sn (c) Cr, Cr^{2+}, Cr^{3+}
(b) C, F, Br, Ga (d) O, O^{2-}, S, S^{2-}

Practice Exercise 18: Use the periodic table to determine the smallest atom or ion in each group.
(a) Si, Ge, As, P (c) Db, W, Tc, Fe
(b) Fe^{2+}, Fe^{3+}, Fe (d) Br^-, I^-, Cl^-

Energy changes are associated with the gain or loss of electrons by atoms

The ionization energy is the energy required to remove an electron from an atom or ion

The **ionization energy** (abbreviated **IE**) *is the energy required to remove an electron from an isolated, gaseous atom or ion in its ground state.* For an atom of an element *X*, it is the increase in potential energy associated with the change

$$X(g) \longrightarrow X^+(g) + e^-$$

In effect, the ionization energy is a measure of how much work is required to pull an electron from an atom, so it reflects how tightly the electron is held by the atom. Usually, the ionization energy is expressed in units of kilojoules per mole (kJ/mol), so we can also view it as the energy needed to remove one mole of electrons from one mole of gaseous atoms.

Table 7.2 gives the ionization energies of the first 12 elements. As you can see, atoms with more than one electron have more than one ionization energy. These correspond to the stepwise removal of electrons, one after the other. Lithium, for example, has three ionization energies because it has three electrons. Removing the outer 2s electrons from

TABLE 7.2 Successive Ionization Energies for Hydrogen through Magnesium (kJ/mol)[a]

	1st	2nd	3rd	4th	5th	6th	7th	8th
H	1312							
He	2372	5250						
Li	520	7297	11,810					
Be	899	1757	14,845	21,000				
B	800	2426	3659	25,020	32,820			
C	1086	2352	4619	6221	37,820	47,260		
N	1402	2855	4576	7473	9442	53,250	64,340	
O	1314	3388	5296	7467	10,987	13,320	71,320	84,070
F	1680	3375	6045	8408	11,020	15,160	17,860	92,010
Ne	2080	3963	6130	9361	12,180	15,240	—	—
Na	496	4563	6913	9541	13,350	16,600	20,113	25,666
Mg	737	1450	7731	10,545	13,627	17,995	21,700	25,662

[a]Note the sharp increase in ionization energy when crossing the "staircase," indicating that the last of the valence electrons has been removed.

one mole of isolated lithium atoms to give one mole of gaseous lithium ions, Li^+, requires 520 kJ; so the *first ionization energy* of lithium is 520 kJ/mol. The second IE of lithium is 7297 kJ/mol, and corresponds to the process

$$Li^+(g) \longrightarrow Li^{2+}(g) + e^-$$

This involves the removal of an electron from the now-exposed 1*s* core of lithium. Removal of the third (and last) electron requires the third IE, which is 11,810 kJ/mol. In general, successive ionization energies always increase because each subsequent electron is being pulled away from an increasingly more positive ion, and that requires more work.

Ionization energies are additive. For example,

$$Li(g) \longrightarrow Li^+(g) + e^- \qquad IE_1 = 520 \text{ kJ}$$

$$Li^+(g) \longrightarrow Li^{2+}(g) + e^- \qquad IE_2 = 7297 \text{ kJ}$$

$$Li(g) \longrightarrow Li^{2+}(g) + 2e^- \qquad IE_{total} = IE_1 + IE_2 = 7817 \text{ kJ}$$

It is often helpful to remember that the trends in IE are just the opposite of the trends in atomic size within the periodic table; when size increases, IE decreases.

Larger atoms have lower ionization energies

Within the periodic table there are trends in the way IE varies that are useful to know and to which we will refer in later discussions. We can see these by examining a graph that shows how the first ionization energy varies with an element's position in the table, which is shown in Figure 7.31. Notice that the elements with the largest ionization energies are the nonmetals in the upper right of the periodic table, and that those with the smallest ionization energies are the metals in the lower left of the table. In general, then, the following trends are observed.

Periodic trends in ionization energy

Ionization energy generally increases from bottom to top within a group and increases from left to right within a period. Overall the ionization energy increases from the lower left corner of the periodic table to the upper right corner. This is usually referred to as a **diagonal relationship.**

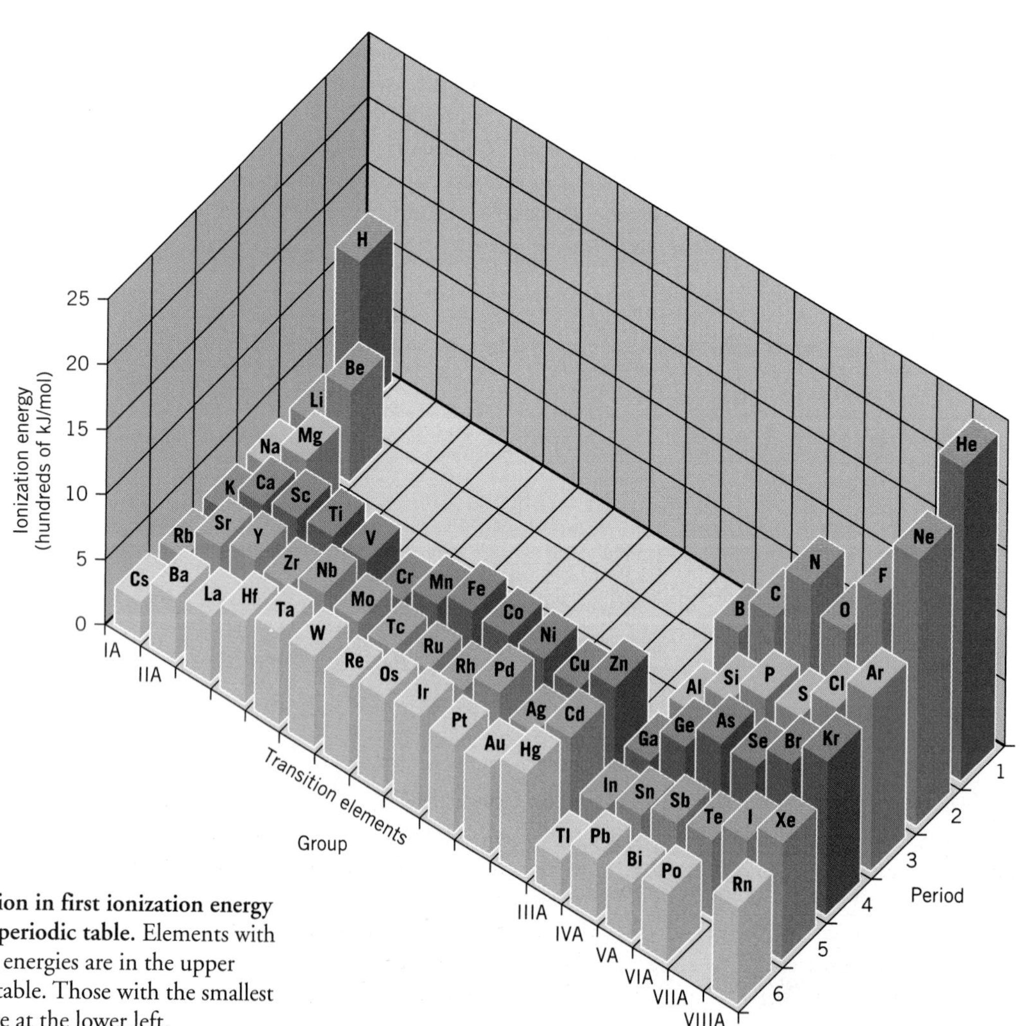

FIG. 7.31 Variation in first ionization energy with location in the periodic table. Elements with the largest ionization energies are in the upper right of the periodic table. Those with the smallest ionization energies are at the lower left.

The same factors that affect atomic size also affect ionization energy. As the value of n increases going down a group, the orbitals become larger and the outer electrons are farther from the nucleus. Electrons farther from the nucleus are bound less tightly, so IE decreases from top to bottom. Of course, this is just the same as saying that it increases from bottom to top.

As you can see, there is a gradual overall increase in IE as we move from left to right across a period, although the horizontal variation of IE is somewhat irregular (see Facets of Chemistry 7.4). The reason for the overall trend is the increase in effective nuclear charge felt by the valence electrons as we move across a period. As we've seen, this draws the valence electrons closer to the nucleus and leads to a decrease in atomic size as we move from left to right. But the increasing effective nuclear charge also causes the valence electrons to be held more tightly, which makes it more difficult to remove them.

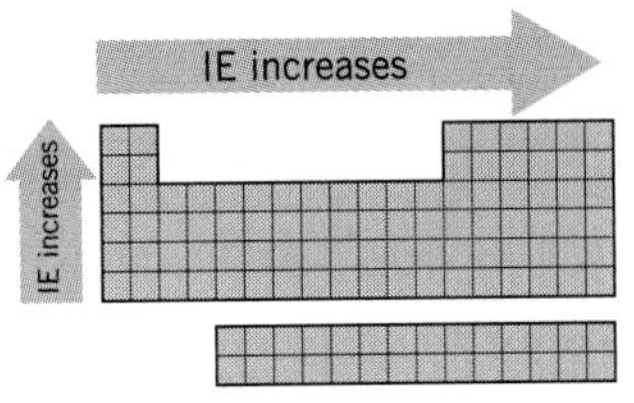

The results of these trends place elements with the largest IE in the upper right-hand corner of the periodic table. It is very difficult to cause these atoms to lose electrons. In the lower left-hand corner of the table are elements that have loosely held valence electrons. These elements form positive ions relatively easily, as you learned in Chapter 2.

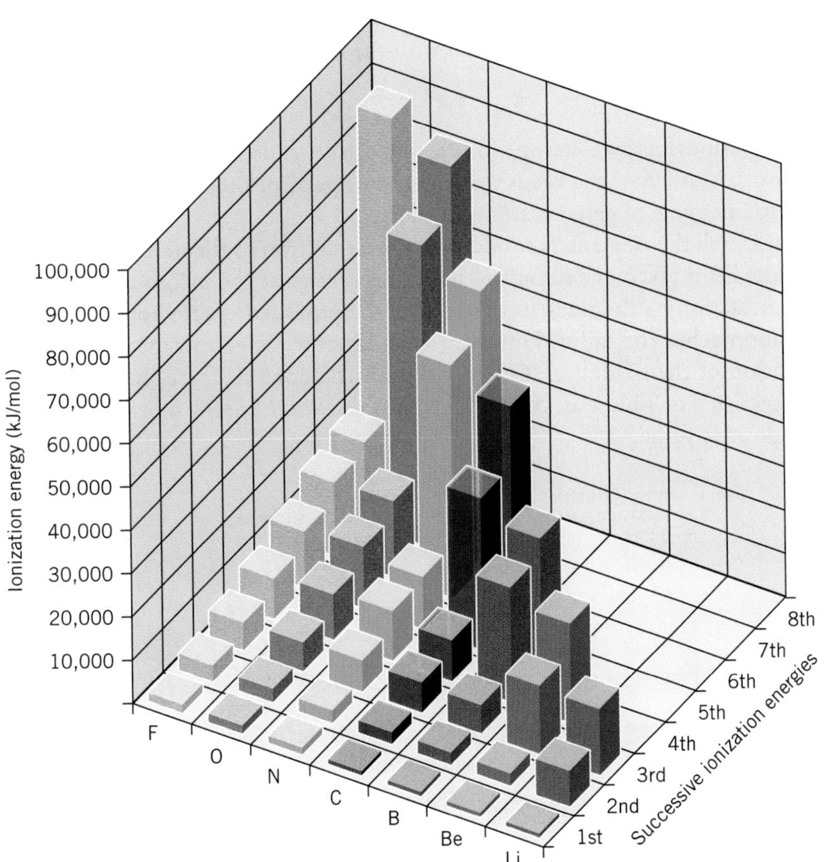

FIG. 7.32 **Variations in successive ionization energies for the elements lithium through fluorine.**

Noble gas configurations are extremely stable

Table 7.2 shows that, for a given element, successive ionization energies increase gradually until the valence shell is emptied. Then a very much larger increase in IE occurs when core electrons are removed. This is illustrated graphically in Figure 7.32 for the Period 2 elements lithium through fluorine. For lithium, we see that the first electron (the 2*s* electron) is removed rather easily, but the second and third electrons, which come from the 1*s* core, are much more difficult to dislodge. For beryllium, the large jump in IE occurs after two electrons (the two 2*s* electrons) are removed. In fact, for all of these elements, the big jump in IE happens when the core electrons are removed.

The data displayed in Figure 7.32 suggest that although it may be moderately difficult to empty the valence shell of an atom, it is *extremely* difficult to remove the core electrons that have the noble gas configuration. As you will learn, this is one of the factors that influences the number of positive charges on ions formed by the representative metals.

Practice Exercise 19: Use the periodic table to select the atom with the most positive value for its first ionization energy, IE: (a) Na, Sr, Be, Rb; (b) B, Al, C, Si. (Hint: Use the diagonal relationships of the IE values within the periodic table.)

Practice Exercise 20: Use Table 7.2 to determine which of the following is expected to have the most positive ionization energy, IE: (a) Na^+, Mg^+, H, C^{2+}; (b) Ne, F, Mg^{2+}, Li^+.

Electron affinity is energy released or absorbed when a particle gains an electron

The **electron affinity** (abbreviated **EA**) *is the potential energy change associated with the addition of an electron to a gaseous atom or ion in its ground state.* For an element *X*, it is the change in potential energy associated with the process

$$X(g) + e^- \longrightarrow X^-(g)$$

As with ionization energy, electron affinities are usually expressed in units of kilojoules per mole, so we can also view the EA as the energy change associated with adding one mole of electrons to one mole of gaseous atoms or ions.

For nearly all the elements, the addition of one electron to the neutral atom is exothermic, and the EA is given as a negative value. This is because the incoming electron experiences an attraction to the nucleus that causes the potential energy to be lowered as the electron approaches the atom. However, when a second electron must be added, as in the formation of the oxide ion, O^{2-}, work must be done to force the electron into an already negative ion. This is an endothermic process where energy must be added and the EA has a positive value.

Change	EA (kJ/mol)
$O(g) + e^- \longrightarrow O^-(g)$	−141
$O^-(g) + e^- \longrightarrow O^{2-}(g)$	+844
$O(g) + 2e^- \longrightarrow O^{2-}(g)$	+703 (net)

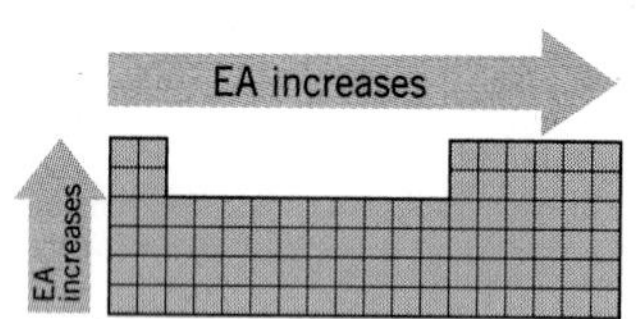

FIG. 7.33 Variation of electron affinity (as an exothermic quantity) within the periodic table.

Notice that more energy is absorbed adding an electron to the O^- ion than is released by adding an electron to the O atom. Overall, the formation of an isolated oxide ion leads to a net increase in potential energy (so we say its formation is *endothermic*). The same applies to the formation of *any* negative ion with a charge larger than 1−.

The electron affinities of the representative elements are given in Table 7.3, and we see that periodic trends in electron affinity roughly parallel those for ionization energy. (See Facets of Chemistry 7.4 for a discussion of some of the irregularities in the trends.)

Although there are some irregularities, overall, the electron affinities of the elements become more *exothermic* going from left to right across a period and from bottom to top in a group (Figure 7.33).

This shouldn't be surprising, because a valence shell that loses electrons easily (low IE) will have little attraction for additional electrons (small EA). On the other hand, a valence shell that holds its electrons tightly will also tend to bind an additional electron tightly.

TABLE 7.3 Electron Affinities of the Representative Elements (kJ/mol)

IA	IIA	IIIA	IVA	VA	VIA	VIIA
H						
−73						
Li	Be	B	C	N	O	F
−60	+238	−27	−122	~ +9	−141	−328
Na	Mg	Al	Si	P	S	Cl
−53	+230	−44	−134	−72	−200	−348
K	Ca	Ga	Ge	As	Se	Br
−48	+155	−30	−120	−77	−195	−325
Rb	Sr	In	Sn	Sb	Te	I
−47	+167	−30	−121	−101	−190	−295
Cs	Ba	Tl	Pb	Bi	Po	At
−45	+50	−30	−110	−110	−183	−270

FACETS OF CHEMISTRY 7.4

Irregularities in the Periodic Variations in Ionization Energy and Electron Affinity

The variation in first ionization energy across a period is not a smooth one, as seen in the graph for the elements in period 2. The first irregularity occurs between Be and B, where the IE increases from Li to Be but then decreases from Be to B. This happens because there is a change in the nature of the subshell from which the electron is being removed. For Li and Be, the electron is removed from the 2*s* subshell, but at boron the first electron comes from the higher energy 2*p* subshell where it is not bound so tightly.

Another irregularity occurs between nitrogen and oxygen. For nitrogen, the electron that's removed comes from a singly occupied orbital. For oxygen, the electron is taken from an orbital that already contains an electron. We can diagram this as follows:

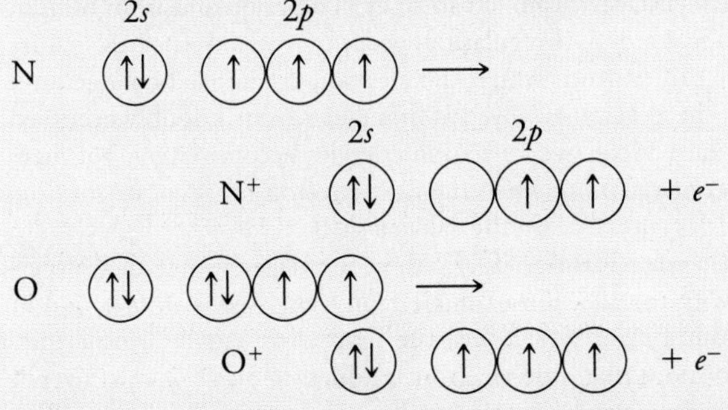

For oxygen, repulsions between the two electrons in the *p* orbital that's about to lose an electron help the electron leave. This "help" is absent for the electron that's about to leave the *p* orbital of nitrogen. As a result, it is not as difficult to remove one electron from an oxygen atom as it is to remove one electron from a nitrogen atom.

As with ionization energy, there are irregularities in the periodic trends for electron affinity. For example, the Group IIA elements have little tendency to acquire electrons because their outer shell *s* orbitals are filled. The incoming electron must enter a higher energy *p* orbital. We also see that the EA for elements in Group VA are either endothermic or only

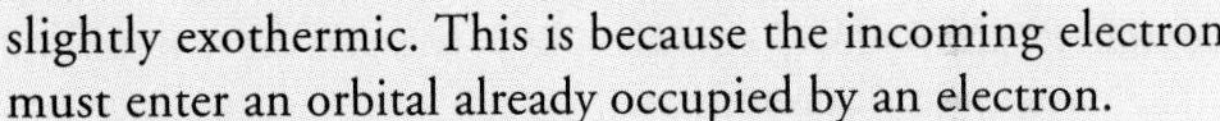

slightly exothermic. This is because the incoming electron must enter an orbital already occupied by an electron.

One of the most interesting irregularities occurs between periods 2 and 3 among the nonmetals. In any group, the element in period 2 has a less exothermic electron affinity than the element below it. The reason seems to be the small size of the nonmetal atoms of period 2, which are among the smallest elements in the periodic table. Repulsions between the many electrons in the small valence shells of these atoms leads to a lower than expected attraction for an incoming electron and a less exothermic electron affinity than the element below in period 3.

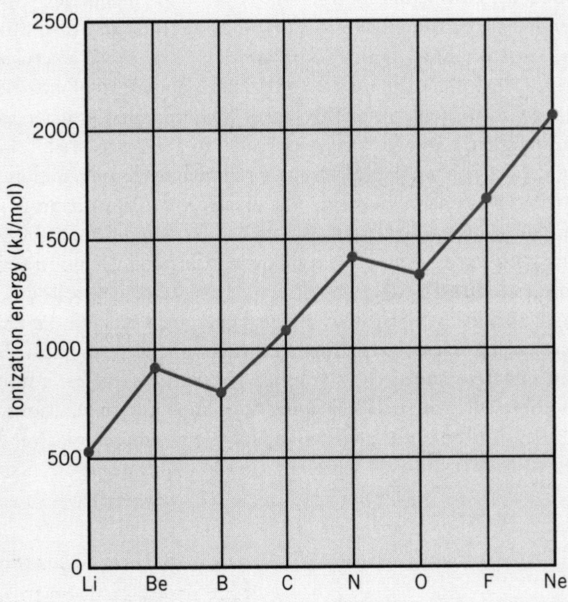

Variation in IE for the Period 2 elements, Li through Ne.

SUMMARY

Electromagnetic Radiation. Electromagnetic radiation, or light energy, travels through space at a constant speed of 3.00×10^8 m s^{-1} in the form of waves. The **wavelength,** λ, and **frequency,** ν, of the wave are related by the equation $\lambda\nu = c$, where c is the **speed of light.** The SI unit for frequency is the **hertz** (**Hz,** 1 Hz = 1 s^{-1}). Light also behaves as if it consists of small packets of energy called **photons** or **quanta.** The energy delivered by a photon is proportional to the frequency of the light, and is given by the equation $E = h\nu$, where h is **Planck's constant.** White light is composed of all the frequencies visible to the eye and can be split into a **continuous spectrum.** Visible light represents only a small portion of the entire **electromagnetic spectrum,** which also includes **X rays, ultraviolet, infrared, microwaves,** and **radio** and **TV** waves.

Atomic Spectra. The occurrence of **line spectra** tells us that atoms can emit energy only in discrete amounts and suggests that the energy of the electron is **quantized;** that is, the electron is restricted to certain specific **energy levels** in an atom. Niels Bohr recognized this and, although his theory was later shown to be incorrect, he was the first to propose a model that was able to account for the **Rydberg equation.** Bohr was the first to introduce the idea of **quantum numbers.**

Matter Waves. The wave behavior of electrons and other tiny particles, which can be demonstrated by **diffraction** experiments, was suggested by de Broglie. Schrödinger applied wave theory to the atom and launched the theory we call **wave mechanics** or **quantum mechanics.** This theory tells us that electron waves in atoms are **standing waves** whose crests and **nodes** are stationary. Each standing wave, or **orbital,** is characterized by three quantum numbers, n, ℓ, and m_ℓ (**principal, secondary,** and **magnetic quantum numbers,** respectively). **Shells** are designated by n (which can range from 1 to ∞), **subshells** by ℓ (which can range from 0 to $n - 1$), and orbitals within subshells by m_ℓ (which can range from $-\ell$ to $+\ell$).

Electron Configurations. The electron has magnetic properties that are explained in terms of spin. The **spin quantum number,** m_s, can have values of $+\frac{1}{2}$ or $-\frac{1}{2}$. The **Pauli exclusion principle** limits orbitals to a maximum population of two electrons with **paired spins.** Substances with unpaired electrons are **paramagnetic** and are attracted weakly to a magnetic field. Substances with only paired electrons are **diamagnetic** and are slightly repelled by a magnetic field. The **electron configuration** of an element in its **ground state** is obtained by filling orbitals beginning with the 1s subshell and following the Pauli exclusion principle and **Hund's rule** (which states that electrons spread out as much as possible in orbitals of equal energy). The periodic table serves as a guide in predicting electron configurations. **Abbreviated configurations** show subshell populations outside a noble gas **core. Valence shell configurations** show the populations of subshells in the **outer shell** of an atom of the representative elements. Sometimes we represent electron configurations using **orbital diagrams.** Unexpected configurations occur for chromium and copper because of the extra stability of half-filled and filled subshells, respectively.

Orbital Shapes. The **Heisenberg uncertainty principle** says we cannot know exactly the position and velocity of an electron both at the same instant. Consequently, wave mechanics describes the probable locations of electrons in atoms. In each orbital the electron is conveniently viewed as an **electron cloud** with a varying **electron density.** All s orbitals are spherical; each p orbital consists of two lobes with a **nodal plane** between them. A p subshell has three p orbitals whose axes are mutually perpendicular and point along the x, y, and z axes of an imaginary coordinate system centered at the nucleus. Four of the five d orbitals in a d subshell have the same shape, with four lobes of electron density each. The fifth has two lobes of electron density pointing in opposite directions along the z axis and a ring of electron density in the x–y plane.

Atomic Properties. The amount of positive charge felt by the valence electrons of an atom is the **effective nuclear charge.** This is less than the actual nuclear charge because core electrons partially shield the valence electrons from the full positive charge of the nucleus. **Atomic radii** depend on the value of n of the valence shell orbitals and the effective nuclear charge experienced by the valence electrons. These radii are expressed in units of picometers or nanometers, or an older unit called the **angstrom (Å),** where 1 Å = 100 pm = 0.1 nm. Atomic radii decrease from left to right in a period and from bottom to top in a group in the periodic table. Negative ions are larger than the atoms from which they are formed; positive ions are smaller than the atoms from which they are formed.

Ionization energy (IE) is the energy needed to remove an electron from an isolated gaseous atom, molecule, or ion in its ground state; it is endothermic. The first ionization energies of the elements increase from left to right in a period and from bottom to top in a group. (Irregularities occur in a period when the nature of the orbital from which the electron is removed changes and when the electron removed is first taken from a doubly occupied p orbital.) Successive ionization energies become larger, but there is a very large jump when the next electron must come from the noble gas core beneath the valence shell.

Electron affinity (EA) is the potential energy change associated with the addition of an electron to a gaseous atom or ion in its ground state. For atoms, the first EA is usually exothermic. When more than one electron is added to an atom, the overall potential energy change is endothermic. In general, electron affinity becomes more exothermic from left to right in a period and from bottom to top in a group. (However, the EA of second period nonmetals is less exothermic than for the nonmetals of the third period. Irregularities across a period occur when the electron being added must enter the next higher energy subshell and when it must enter a half-filled p subshell.)

TOOLS FOR PROBLEM SOLVING

Below we list the tools you have learned in this chapter. Notice that only two of them are related to numerical calculations. The others are conceptual tools that we use in analyzing properties of substances in terms of the underlying structure of matter. Review all these tools and refer to them, if necessary, when working on the Review Questions and Problems that follow.

Wavelength–frequency relationship *(page 253)* Use this equation to convert between frequency and wavelength.

$$c = \frac{\lambda}{\nu}$$

Energy of a photon *(page 257)* Use the following equation to calculate the energy carried by a photon of frequency ν. Also, ν can be calculated if E is known.

$$E = h\nu = h\frac{c}{\lambda}$$

Periodic table We will use the periodic table as a tool for many purposes. In this chapter you learned to use the periodic table as an aid in writing electron configurations (page *275*) of the elements and as a tool to correlate an element's location in the table to similarities in chemical properties (page 280).

Periodic trends in atomic and ionic size *(pages 285, 286)* The periodic table helps us predict relative sizes of atoms and ions.

Periodic trends in ionization energy *(page 288)* Trends are used to compare the ease with which atoms of the elements lose electrons.

Periodic trends in electron affinity *(page 290)* Trends help to compare the tendency of atoms or ions to gain electrons.

QUESTIONS, PROBLEMS, AND EXERCISES

Answers to problems whose numbers are printed in color are given in Appendix B. More challenging problems are marked with asterisks. ILW = Interactive Learningware solution is available at www.wiley.com/college/brady. OH = an Office Hours video is available for this problem.

REVIEW QUESTIONS

Electromagnetic Radiation

7.1 In general terms, why do we call light *electromagnetic radiation*?

7.2 In general, what does the term *frequency* imply? What is meant by the term *frequency of light*? What symbol is used for it, and what is the SI unit (and symbol) for frequency?

7.3 What is meant by the term *wavelength* of light? What symbol is used for it?

7.4 Sketch a picture of a wave and label its wavelength and its amplitude.

7.5 Which property of light waves is a measure of the brightness of the light? Which specifies the color of the light? Which is related to the energy of the light?

7.6 Arrange the following regions of the electromagnetic spectrum in order of increasing wavelength (i.e., shortest wavelength → longest wavelength): microwaves, TV, X rays, ultraviolet, visible, infrared, gamma rays.

7.7 What wavelength range is covered by the *visible spectrum*?

7.8 Arrange the following colors of visible light in order of increasing wavelength: orange, green, blue, yellow, violet, red.

7.9 Write the equation that relates the wavelength and frequency of a light wave. (Define all symbols used.)

7.10 How is the frequency of a particular type of radiation related to the energy associated with it? (Give an equation, defining all symbols.)

7.11 What is a photon?

7.12 Show that the energy of a photon is given by the equation $E = hc/\lambda$.

7.13 Examine each of the following pairs and state which of the two has the higher *energy*.
(a) microwaves and infrared
(b) visible light and infrared
(c) ultraviolet light and X rays
(d) visible light and ultraviolet light

7.14 What is a quantum of energy?

Atomic Spectra

7.15 What is an atomic spectrum? How does it differ from a continuous spectrum?

7.16 What fundamental fact is implied by the existence of atomic spectra?

Bohr Atom and the Hydrogen Spectrum

7.17 Describe Niels Bohr's model of the structure of the hydrogen atom.

7.18 In qualitative terms, how did Bohr's model account for the atomic spectrum of hydrogen?

7.19 What is the "ground state"?

7.20 In what way was Bohr's theory a success? How was it a failure?

Wave Nature of Matter

7.21 How does the behavior of very small particles differ from that of the larger, more massive objects that we meet in everyday life? Why don't we notice this same behavior for the larger, more massive objects?

7.22 Describe the phenomenon called diffraction. How can this be used to demonstrate that de Broglie's theory was correct?

7.23 What experiment could you perform to determine whether a beam was behaving as a wave or as a stream of particles?

7.24 What is *wave/particle duality*?

7.25 What is the difference between a *traveling wave* and a *standing wave*?

7.26 What is the collapsing atom paradox?

7.27 How does quantum mechanics resolve the collapsing atom paradox?

Electron Waves in Atoms

7.28 What are the names used to refer to the theories that apply the matter–wave concept to electrons in atoms?

7.29 What is the term used to describe a particular waveform of a standing wave for an electron?

7.30 What are the two properties of orbitals in which we are most interested? Why?

Quantum Numbers

7.31 What are the allowed values of the principal quantum number?

7.32 What is the value for n for (a) the K shell and (b) the M shell?

7.33 Why does every shell contain an s subshell?

7.34 How many orbitals are found in (a) an s subshell, (b) a p subshell, (c) a d subshell, and (d) an f subshell?

7.35 If the value of m_ℓ for an electron in an atom is 2, could another electron in the same subshell have $m_\ell = -3$?

7.36 Suppose an electron in an atom has the following set of quantum numbers: $n = 2, \ell = 1, m_\ell = 1, m_s = +\frac{1}{2}$. What set of quantum numbers is impossible for another electron in this same atom?

Electron Spin

7.37 What physical property of electrons leads us to propose that they spin like a toy top?

7.38 What is the name of the magnetic property exhibited by atoms that contain unpaired electrons?

7.39 What is the Pauli exclusion principle? What effect does it have on the populating of orbitals by electrons?

7.40 What are the possible values of the spin quantum number?

Electron Configuration of Atoms

7.41 What do we mean by the term *electronic structure?*

7.42 Within any given shell, how do the energies of the s, p, d, and f subshells compare?

7.43 What fact about the energies of subshells was responsible for the apparent success of Bohr's theory about electronic structure?

7.44 How do the energies of the orbitals belonging to a given subshell compare?

7.45 Give the electron configurations of the elements in Period 2 of the periodic table.

7.46 Give the correct electron configurations of (a) Cr and (b) Cu.

7.47 What is the correct electron configuration of silver?

7.48 How are the electron configurations of the elements in a given group similar? Illustrate your answer by writing shorthand configurations for the elements in Group VIA.

7.49 Define the terms *valence shell* and *valence electrons.*

Shapes of Atomic Orbitals

7.50 Why do we use probabilities when we discuss the position of an electron in the space surrounding the nucleus of an atom?

7.51 Sketch the approximate shape of (a) a $1s$ orbital and (b) a $2p$ orbital.

7.52 How does the size of a given type of orbital vary with n?

7.53 How are the p orbitals of a given p subshell oriented relative to each other?

7.54 What is a *nodal plane*?

7.55 What is a spherical node?

7.56 How many nodal planes does a p orbital have? How many does a d orbital have?

7.57 On appropriate coordinate axes, sketch the shape of the following d orbitals: (a) d_{xy}, (b) $d_{x^2-y^2}$, and (c) d_{z^2}.

Atomic and Ionic Size

7.58 What is the meaning of *effective nuclear charge?* How does the effective nuclear charge felt by the outer electrons vary going down a group? How does it change as we go from left to right across a period?

7.59 In what region of the periodic table are the largest atoms found? Where are the smallest atoms found?

7.60 Going from left to right in the periodic table, why are the size changes among the transition elements more gradual than those among the representative elements?

Ionization Energy

7.61 Define ionization energy. Why are ionization energies of atoms and positive ions endothermic quantities?

7.62 For oxygen, write a reaction for the change associated with (a) its first ionization energy and (b) its third ionization energy.

7.63 Explain why ionization energy increases from left to right in a period and decreases from top to bottom in a group.

7.64 Why is an atom's second ionization energy always larger than its first ionization energy?

7.65 Why is the fifth ionization energy of carbon so much larger than its fourth?

7.66 Why is the first ionization energy of aluminum less than the first ionization energy of magnesium?

7.67 Why does phosphorus have a larger first ionization energy than sulfur?

Electron Affinity

7.68 Define *electron affinity*.

7.69 For sulfur, write an equation for the change associated with (a) its first electron affinity and (b) its second electron affinity. How should they compare?

7.70 Why does Cl have a more exothermic electron affinity than F? Why does Br have a less exothermic electron affinity than Cl?

7.71 Why is the second electron affinity of an atom always endothermic?

7.72 How is electron affinity related to effective nuclear charge? On this basis, explain the relative magnitudes of the electron affinities of oxygen and fluorine.

REVIEW PROBLEMS

Electromagnetic Radiation

7.73 What is the frequency in hertz of blue light having a wavelength of 430 nm?

OH **7.74** Ultraviolet light with a wavelength of more than 280 nm has little germicidal value. What is the frequency that corresponds to this wavelength?

7.75 A certain substance strongly absorbs infrared light having a wavelength of 6.85 μm. What is the frequency of this light in hertz?

7.76 The sun emits many wavelengths of light. The brightest light is emitted at about 0.48 μm. What frequency does this correspond to?

7.77 Ozone protects the earth's inhabitants from the harmful effects of ultraviolet light arriving from the sun. This shielding is a maximum for UV light having a wavelength of 295 nm. What is the frequency in hertz of this light?

7.78 The meter is defined as the length of the path light travels in a vacuum during the time interval of 1/299,792,458 of a second. The standards body recommends use of light from a helium–neon laser for realizing the meter. The light from the laser has a wavelength of 632.99139822 nm. What is the frequency of this light, in hertz?

7.79 In New York City, radio station WCBS broadcasts its FM signal at a frequency of 101.1 megahertz (MHz). What is the wavelength of this signal in meters?

7.80 Sodium vapor lamps are often used in residential street lighting. They give off a yellow light having a frequency of 5.09×10^{14} Hz. What is the wavelength of this light in nanometers?

7.81 There has been some concern in recent times about possible hazards to people who live very close to high-voltage electric power lines. The electricity in these wires oscillates at a frequency of 60 Hz, which is the frequency of any electromagnetic radiation that they emit. What is the wavelength of this radiation in meters? What is it in kilometers?

7.82 An X-ray beam has a frequency of 1.50×10^{18} Hz. What is the wavelength of this light in nanometers and in picometers?

7.83 Calculate the energy in joules of a photon of red light having a frequency of 4.0×10^{14} Hz. What is the energy of one mole of these photons?

7.84 Calculate the energy in joules of a photon of green light having a wavelength of 560 nm.

Atomic Spectra

7.85 In the spectrum of hydrogen, there is a line with a wavelength of 410.3 nm. (a) What color is this line? (b) What is its frequency? (c) What is the energy of each of its photons?

7.86 In the spectrum of sodium, there is a line with a wavelength of 589 nm. (a) What color is this line? (b) What is its frequency? (c) What is the energy of each of its photons?

OH **7.87** Use the Rydberg equation to calculate the wavelength in nanometers of the spectral line of hydrogen for which $n_2 = 6$ and $n_1 = 3$. Would we be expected to see the light corresponding to this spectral line? Explain your answer.

7.88 Use the Rydberg equation to calculate the wavelength in nanometers of the spectral line of hydrogen for which $n_2 = 5$ and $n_1 = 2$. Would we be expected to see the light corresponding to this spectral line? Explain your answer.

7.89 Calculate the wavelength of the spectral line produced in the hydrogen spectrum when an electron falls from the tenth Bohr orbit to the fourth. In which region of the electromagnetic spectrum (UV, visible, or infrared) is the line?

7.90 Calculate the energy in joules and the wavelength in nanometers of the spectral line produced in the hydrogen spectrum when an electron falls from the fourth Bohr orbit to the first. In which region of the electromagnetic spectrum (UV, visible, or infrared) is the line?

Quantum Numbers

7.91 What is the letter code for a subshell with (a) $\ell = 1$ and (b) $\ell = 3$?

7.92 What is the value of ℓ for (a) an *f* orbital and (b) a *d* orbital?

7.93 Give the values of n and ℓ for the following subshells: (a) 3*s* and (b) 5*d*.

7.94 Give the values of n and ℓ for the following subshells: (a) 4*p* and (b) 6*f*.

7.95 For the shell with $n = 6$, what are the possible values of ℓ?

7.96 In a particular shell, the largest value of ℓ is 7. What is the value of n for this shell?

7.97 What are the possible values of m_ℓ for a subshell with (a) $\ell = 1$ and (b) $\ell = 3$?

7.98 If the value of ℓ for an electron in an atom is 5, what are the possible values of m_ℓ that this electron could have?

7.99 If the value of m_ℓ for an electron in an atom is -4, what is the smallest value of ℓ that the electron could have? What is the smallest value of n that the electron could have?

7.100 How many orbitals are there in an *h* subshell ($\ell = 5$)? What are their values of m_ℓ?

OH **7.101** Give the complete set of quantum numbers for all of the electrons that could populate the $2p$ subshell of an atom.

7.102 Give the complete set of quantum numbers for all of the electrons that could populate the $3d$ subshell of an atom.

7.103 In an antimony atom, how many electrons have $\ell = 1$? How many electrons have $\ell = 2$ in an antimony atom?

7.104 In an atom of barium, how many electrons have (a) $\ell = 0$ and (b) $m_\ell = 1$?

Electron Configurations of Atoms

7.105 Give the electron configurations of (a) S, (b) K, (c) Ti, and (d) Sn.

7.106 Write the electron configurations of (a) As, (b) Cl, (c) Ni, and (d) Si.

7.107 Which of the following atoms in their ground states are expected to be paramagnetic: (a) Mn, (b) As, (c) S, (d) Sr, and (e) Ar?

7.108 Which of the following atoms in their ground states are expected to be diamagnetic: (a) Ba, (b) Se, (c) Zn, and (d) Si?

ILW OH **7.109** How many unpaired electrons would be found in the ground state of (a) Mg, (b) P, and (c) V?

7.110 How many unpaired electrons would be found in the ground state of (a) Cs, (b) S, and (c) Ni?

7.111 Write the shorthand electron configurations for (a) Ni, (b) Cs, (c) Ge, (d) Br, and (e) Bi.

7.112 Write the shorthand electron configurations for (a) Al, (b) Se, (c) Ba, (d) Sb, and (e) Gd.

7.113 Draw complete orbital diagrams for (a) Mg and (b) Ti.

7.114 Draw complete orbital diagrams for (a) As and (b) Ni.

7.115 Draw orbital diagrams for the shorthand configurations of (a) Ni, (b) Cs, (c) Ge, and (d) Br.

7.116 Draw orbital diagrams for the shorthand configurations of (a) Al, (b) Se, (c) Ba, and (d) Sb.

7.117 What is the value of *n* for the valence shells of (a) Sn, (b) K, (c) Br, and (d) Bi?

7.118 What is the value of *n* for the valence shells of (a) Al, (b) Se, (c) Ba, and (d) Sb?

7.119 Give the configuration of the valence shell for (a) Na, (b) Al, (c) Ge, and (d) P.

7.120 Give the configuration of the valence shell for (a) Mg, (b) Br, (c) Ga, and (d) Pb.

7.121 Draw the orbital diagram for the valence shell for (a) Na, (b) Al, (c) Ge, and (d) P.

7.122 Draw the orbital diagram for the valence shell of (a) Mg, (b) Br, (c) Ga, and (d) Pb.

Atomic Properties

7.123 If the core electrons were 100% effective at shielding the valence electrons from the nuclear charge and the valence electrons provided no shielding for each other, what would be the effective nuclear charge felt by a valence electron in (a) Na, (b) S, and (c) Cl?

7.124 If the core electrons were 100% effective at shielding the valence electrons from the nuclear charge and the valence electrons provided no shielding for each other, what would be the effective nuclear charge felt by a valence electron in (a) Mg, (b) Si, and (c) Br?

7.125 Choose the larger atom in each pair: (a) Mg or S; (b) As or Bi.

7.126 Choose the larger atom in each pair: (a) Al or Ar; (b) Tl or In.

7.127 Choose the largest atom among the following: Ge, As, Sn, and Sb.

7.128 Place the following in order of increasing size: N^{3-}, Mg^{2+}, Na^+, Ne, F^-, and O^{2-}.

7.129 Choose the larger particle in each pair: (a) Na or Na^+; (b) Co^{3+} or Co^{2+}; (c) Cl or Cl^-.

OH **7.130** Choose the larger particle in each pair: (a) S or S^{2-}; (b) Al^{3+} or Al; (c) Au^+ or Au^{3+}.

7.131 Choose the atom with the larger first ionization energy in each pair: (a) B or N; (b) Se or S; (c) Cl or Ge.

7.132 Choose the atom with the larger first ionization energy in each pair: (a) Li or Rb ; (b) Al or F; (c) F or C.

7.133 Choose the atom with the more exothermic electron affinity in each pair: (a) I or Br; (b) Ga or As.

7.134 Choose the atom with the more exothermic electron affinity in each pair: (a) S or As; (b) Si or N.

7.135 Use the periodic table to select the element in the following list for which there is the largest difference between the second and third ionization energies: Na, Mg, Al, Si, P, Se, and Cl.

7.136 Use the periodic table to select the element in the following list for which there is the largest difference between the fourth and fifth ionization energies: Na, Mg, Al, Si, P, Se, and Cl.

ADDITIONAL EXERCISES

7.137 The human ear is sensitive to sound ranging from 20 to 20,000 Hz. The speed of sound is 330 m/s in air, and 1500 m/s under water. What is the longest and the shortest wavelength that can be heard (a) in air and (b) under water?

7.138 Microwaves are used to heat food in microwave ovens. The microwave radiation is absorbed by moisture in the food. This heats the water, and as the water becomes hot, so does the food. How many photons having a wavelength of 3.00 mm would have to be absorbed by 1.00 g of water to raise its temperature by 1.00 °C?

7.139 In the spectrum of hydrogen, there is a line with a wavelength of 410.3 nm. Use the Rydberg equation to calculate the value of *n* for the higher energy Bohr orbit involved in the emission of this light. Assume the value of *n* for the lower energy orbit equals 2.

7.140 Calculate the wavelength in nanometers of the shortest wavelength of light emitted by a hydrogen atom.

7.141 Which of the following electronic transitions could lead to the emission of light from an atom?

$$1s \longrightarrow 4p \longrightarrow 3d \longrightarrow 5f \longrightarrow 4d \longrightarrow 2p$$

7.142 A neon sign is a gas discharge tube in which electrons traveling from the cathode to the anode collide with neon atoms in the tube and knock electrons off of them. As electrons return to the neon ions and drop to lower energy levels, light is given off. How fast would an electron have to be moving to eject an electron from an atom of neon, which has a first ionization energy equal to 2080 kJ mol^{-1}?

OH ****7.143** How many grams of water could have its temperature raised by 5.0 °C by a mole of photons that have a wavelength of (a) 600 nm and (b) 300 nm?

OH ****7.144** It has been found that when the chemical bond between chlorine atoms in Cl_2 is formed, 328 kJ is released per mole of Cl_2 formed. What is the wavelength of light that would be required to break chemical bonds between chlorine atoms?

7.145 Calculate the wavelengths of the lines in the spectrum of hydrogen that result when an electron falls from a Bohr orbit with (a) $n = 5$ to $n = 1$, (b) $n = 4$ to $n = 2$, and (c) $n = 6$ to $n = 4$. In which regions of the electromagnetic spectrum are these lines?

OH **7.146** What, if anything, is wrong with the following electron configurations for atoms in their ground states?

(a) $1s^2 2s^1 2p^3$ (c) $1s^2 2s^2 2p^4$
(b) [Kr] $3d^7 4s^2$ (d) [Xe] $4f^{14} 5d^8 6s^1$

7.147 Suppose students gave the following orbital diagrams for the 2s and 2p subshell in the ground state of an atom. What, if anything, is wrong with them? Are any of these electron distributions impossible?

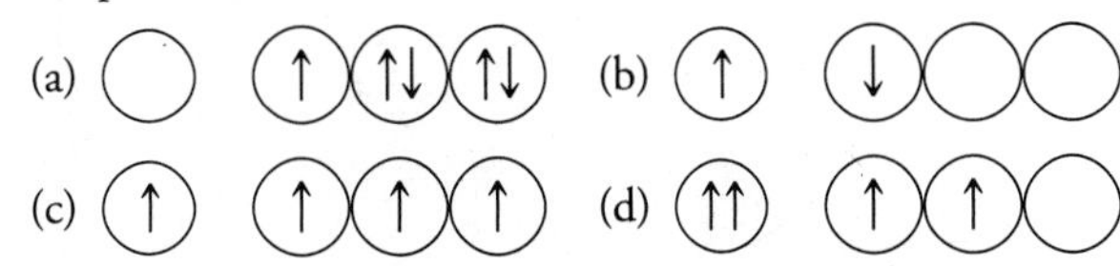

7.148 How many electrons are in p orbitals in an atom of germanium?

7.149 What are the quantum numbers of the electrons that are lost by an atom of iron when it forms the ion Fe^{2+}?

****7.150** The removal of an electron from the hydrogen atom corresponds to raising the electron to the Bohr orbit that has $n = \infty$. On the basis of this statement, calculate the ionization energy of hydrogen in units of (a) joules per atom and (b) kilojoules per mole.

7.151 Use orbital diagrams to illustrate what happens when an oxygen atom gains two electrons. On the basis of what you have learned about electron affinities and electron configurations, why is it extremely difficult to place a third electron on the oxygen atom?

7.152 From the data available in this chapter, determine the ionization energy of (a) F^-, (b) O^-, and (c) O^{2-}. Are any of these energies exothermic?

7.153 For an oxygen atom, which requires more energy, the addition of two electrons or the removal of one electron?

EXERCISES IN CRITICAL THINKING

7.154 Our understanding of the quantum mechanical atom has been developing since the early 1900s. Has quantum mechanics had any effect on the daily lives of people?

7.155 When a copper atom loses an electron to become a Cu^+ ion, what are the possible quantum numbers of the electron that was lost?

7.156 The "red shift" of spectral features of distant stars is used to estimate their relative velocity compared to the earth. Using the information in this chapter, and other reference sources, explain how this is done.

7.157 Placing a small piece of an element from Group IA in water results in increasingly rapid and violently spectacular reactions as we progress from lithium down to cesium. What information in this chapter makes this behavior understandable?

7.158 In this chapter we saw that atoms in an excited state emit light when their electrons relax to lower energy levels. Lasers also emit certain wavelengths of light. What are the similarities and differences in how laser light is produced compared to atomic spectra?

7.159 In this chapter we saw that atoms in an excited state emit light when their electrons relax to lower energy levels. Light emitting diodes, LEDs, can be designed to emit light of different colors. How does the way light is produced in an LED differ from an excited state atom?

CHEMICAL BONDING: GENERAL CONCEPTS 8

Children learn that they can build complex structures by linking together blocks such as these. Molecules in chemical compounds are likewise built by linking together atoms of various kinds. In this chapter we begin our study of the nature of the forces, called chemical bonds, that bind atoms to each other in compounds. Such knowledge is important because the kinds and strengths of chemical bonds determine the chemical properties of substances. *(Ted Horowitz/Corbis)*

CHAPTER OUTLINE

THIS CHAPTER IN CONTEXT In earlier chapters we examined the nature of some ionic and molec lar substances. For example, you learned that ionic compounds, such as ordinary table salt, consist of electrica charged particles (ions) that bind to each other by electrostatic forces. In water, ionic compounds dissociate ar undergo the kinds of reactions we discussed in Chapter 4. Molecular compounds, on the other hand, are compose of atoms that bind to each other by sharing electrons, which causes them to undergo different kinds of reactions. T understand such differences in chemical properties, we need to know more about the nature of the attractions, whic we call **chemical bonds,** that exist between atoms in compounds. Therefore, our principal goal in this chapter is t gain some insight into the reasons why certain combinations of atoms prefer electron transfer and the formation ions (leading to *ionic bonding*) while other combinations bind by electron sharing (leading to *covalent bonding*).

As with electronic structure, models of chemical bonding have evolved, and in this chapter we introduce you t relatively simple theories. Although more complex theories exist (some of which we will explore in Chapter 9), th basic concepts you will study in this chapter still find many useful applications in modern chemical thought.

8.1 ELECTRON TRANSFER LEADS TO THE FORMATION OF IONIC COMPOUNDS

In Chapter 2 you learned that ionic compounds are formed when metals react with nonmetals. Among the examples discussed was sodium chloride, table salt. You learned that when this compound is formed from its elements, each sodium atom loses one electron to form a sodium ion, Na^+, and each chlorine atom gains one electron to become a chloride ion, Cl^-.

$$Na \longrightarrow Na^+ + e^-$$
$$Cl + e^- \longrightarrow Cl^-$$

Once formed, these ions become tightly packed together, as illustrated in Figure 8.1, because their opposite charges attract. *This attraction between positive and negative ions in an ionic compound is what we call an* **ionic bond.**

The reason Na^+ and Cl^- ions attract each other is easy to understand. But *why* are electrons transferred between these and other atoms? *Why* does sodium form Na^+ and not Na^- or Na^{2+}? And *why* does chlorine form Cl^- instead of Cl^+ or Cl^{2-}? To answer such questions we have to consider factors that are related to the potential energy of the system of reactants and products. This is because *for any stable compound to form from its elements, there must be a net lowering of the potential energy.* In other words, the reaction must be exothermic.

Keep in mind the relationship between potential energy changes and endothermic and exothermic processes:

endothermic ⇔ increase in PE
exothermic ⇔ decrease in PE

The lattice energy enables ionic compounds to form

Let's begin by examining the energy change associated with the exchange of electrons between sodium atoms and chlorine atoms. If we deal with a collection of gaseous atoms, we can use the ionization energy (IE) and electron affinity (EA) of sodium and chlorine, respectively. Working on a mole basis, we have

$Na(g) \longrightarrow Na^+(g) + e^-$	+495.4 kJ	(IE of sodium)
$Cl(g) + e^- \longrightarrow Cl^-(g)$	−348.8 kJ	(EA of chlorine)
Net	+146.6 kJ	

This calculation tells us that forming gaseous sodium and chloride ions from gaseous sodium and chlorine atoms requires a substantial increase in the potential energy. In fact, if the IE and EA were the only energy changes involved, ionic sodium chloride would not form from sodium and chlorine. So where does the stability of the compound come from? The answer is seen if we examine a quantity called the lattice energy.

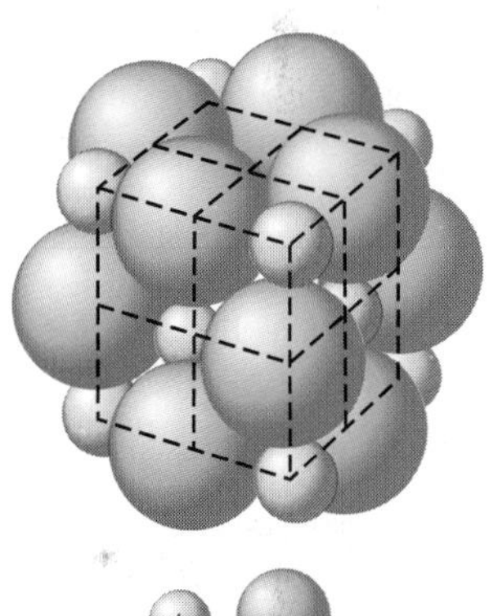

FIG. 8.1 **Packing of ions in NaCl.** In NaCl electrostatic forces hold the ions in place in the solid. Those forces constitute ionic bonds.

In the calculation above, we looked at the formation of gaseous ions, but salt is not a gas; it's a solid in which the ions are packed together in a way that maximizes the attractions between oppositely charged ions. Imagine, now, pulling these ions away from each other to form a gas of ions. This process would require a lot of work and would lead to a large increase in the potential energies of the ions. The **lattice energy** *is the energy required to completely separate the ions in one mole of a solid compound from each other to form a cloud of gaseous ions.* For sodium chloride, the process associated with the lattice energy is pictured in Figure 8.2 and in equation form can be represented as

$$NaCl(s) \longrightarrow Na^+(g) + Cl^-(g)$$

The energy associated with this change (the *lattice energy* of sodium chloride) has been both measured and calculated to be +787.0 kJ mol^{-1}. The positive sign means that it takes 787.0 kJ to separate the ions of one mole of NaCl. *It also means that if we bring together a mole of Na^+ and Cl^- ions from the gaseous state into one mole of crystalline NaCl, 787 kJ will be released.* If we now include the lattice energy (with its sign reversed because we're changing the direction of the change) along with the IE and EA, we have

The name lattice energy comes from the word *lattice*, which is used to describe the regular pattern of ions or atoms in a crystal. Lattice energies are endothermic and so are given positive signs.

	$Na(g) \longrightarrow Na^+(g) + e^-$	+495.4 kJ	(IE of sodium)
	$Cl(g) + e^- \longrightarrow Cl^-(g)$	−348.8 kJ	(EA of chlorine)
	$Na^+(g) + Cl^-(g) \longrightarrow NaCl(s)$	−787.0 kJ	(−lattice energy)
Net	$Na(g) + Cl(g) \longrightarrow NaCl(s)$	−640.4 kJ	

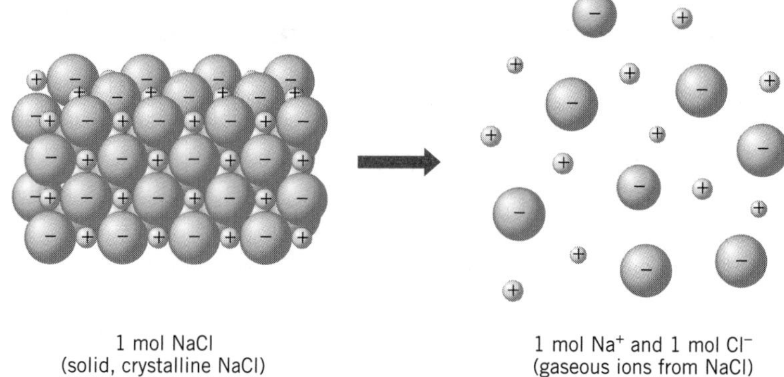

FIG. 8.2 Lattice energy of NaCl. The lattice energy is equal to the amount of energy needed to separate the ions in one mole of an ionic compound. For NaCl, the process requires converting a mole of crystalline NaCl into two moles of ions (1 mol Na^+ and 1 mol Cl^-). The amount of energy absorbed equals 787 kJ.

Thus, the release of energy equivalent to the lattice energy provides a large net lowering of the potential energy as solid NaCl is formed. We can also say that *it is the lattice energy that provides the stabilization necessary for the formation of NaCl.* Without it, the solid compound could not exist.

At this point, you may be wondering about our starting point in these energy calculations, gaseous sodium atoms and chlorine atoms. In nature, sodium is a solid metal and chlorine consists of gaseous Cl_2 molecules. A complete analysis of the energy changes has to take this into account, and we do this in Facets of Chemistry 8.1. Our overall conclusion doesn't change, however. *For any ionic compound, the chief stabilizing influence is the lattice energy, which when released is large enough to overcome the net energy input required to form the ions from the elements.*

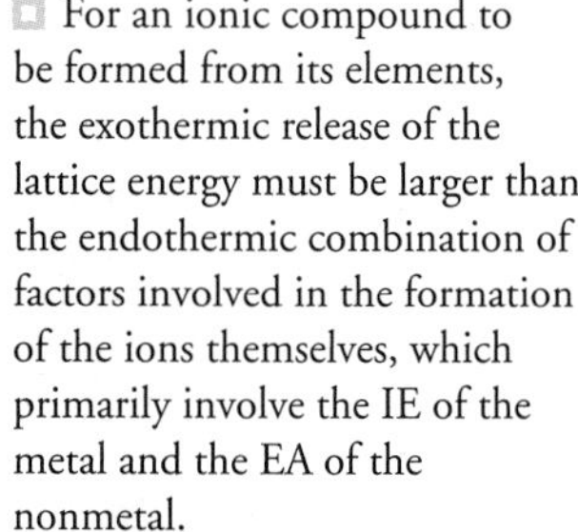

For an ionic compound to be formed from its elements, the exothermic release of the lattice energy must be larger than the endothermic combination of factors involved in the formation of the ions themselves, which primarily involve the IE of the metal and the EA of the nonmetal.

Lattice energy depends on ionic size and charge

The lattice energies of some ionic compounds are given in Table 8.1. As you can see, they are all very large endothermic quantities. Their magnitudes depend on a number of factors, including the sizes of the ions and their charges.[1] In general, as the ions become smaller, the lattice energy increases; smaller ions allow the charges to get closer together

TABLE 8.1 Lattice Energies of Some Ionic Compounds

Compound	Ions	Lattice Energy (kJ mol^{-1})
LiCl	Li^+ and Cl^-	845
NaCl	Na^+ and Cl^-	787
KCl	K^+ and Cl^-	709
LiF	Li^+ and F^-	1033
$CaCl_2$	Ca^{2+} and Cl^-	2258
$AlCl_3$	Al^{3+} and Cl^-	5492
CaO	Ca^{2+} and O^{2-}	3401
Al_2O_3	Al^{3+} and O^{2-}	15,916

[1] The potential energy of two particles with charges q_1 and q_2 separated by a distance r is

$$E = \frac{q_1 q_2}{kr}$$

where k is a proportionality constant. In an ionic solid, q_1 and q_2 have opposite signs, so E is calculated to be a negative quantity. If r increases, E becomes smaller and therefore less negative. As r approaches infinity, corresponding to complete separation of the ions, E approaches zero. Therefore, the more negative the value of E for a pair of ions in the solid, the larger is the amount of energy needed to separate the ions. In a rough way, the size of the lattice energy parallels the magnitude of E. When the charges q_1 and q_2 become larger, E becomes more negative and the lattice energy becomes larger. Similarly, when r becomes smaller, corresponding to smaller ions, E also becomes more negative and the lattice energy becomes larger.

FACETS OF CHEMISTRY 8.1

Calculating the Lattice Energy

In the main body of the text, we described the pivotal role played by the lattice energy in the formation of an ionic compound. But how can we possibly measure the amount of energy required to completely vaporize an ionic compound to give gaseous ions? Actually, we can't measure the lattice energy directly, but we can use Hess's law and some other experimental data to calculate the lattice energy indirectly.

In Chapter 6 you learned that the enthalpy change for a process is the same regardless of the path we follow from start to finish. With this in mind, we can construct a set of alternate paths from the free elements to the solid ionic compound. This is called a Born–Haber cycle after the scientists who were the first to use it to calculate lattice energies, and is shown in the accompanying figure for the formation of sodium chloride. You may recognize the figure as an enthalpy diagram.

We begin with the free elements, sodium and chlorine. The direct path at the bottom left has as its enthalpy change the heat of formation of NaCl, ΔH_f°.

$$Na(s) + \tfrac{1}{2}Cl_2(g) \longrightarrow NaCl(s) \qquad \Delta H_f^\circ = -411.3 \text{ kJ}$$

The alternative path is divided into a number of steps. The first two steps, both of which have ΔH° values that can be measured experimentally, are endothermic. They change Na(*s*) and $Cl_2(g)$ into gaseous atoms, Na(*g*) and Cl(*g*). The next two steps change these atoms to ions, first by the endothermic ionization energy (IE) of Na followed by the exothermic electron affinity (EA) of Cl. This brings us to the gaseous ions, $Na^+(g)$ and $Cl^-(g)$. Notice that at this point, if we add all the energy changes, the ions are at a considerably higher energy than the reactants. If these were the only energy terms involved in the formation of NaCl, the heat of formation would be endothermic and the compound would be unstable; it could not be formed by direct combination of the elements.

The last step on the right finally brings us to solid NaCl and corresponds the negative of the lattice energy. (Remember, the lattice energy is defined as the energy needed to *separate* the ions; in the last step, we are *bringing the ions together* to form the solid.) To make the net energy changes the same along both paths, the energy released when the ions condense to form the solid must equal −787.0 kJ. Therefore, the calculated lattice energy of NaCl must be +787.0 kJ mol^{-1}.

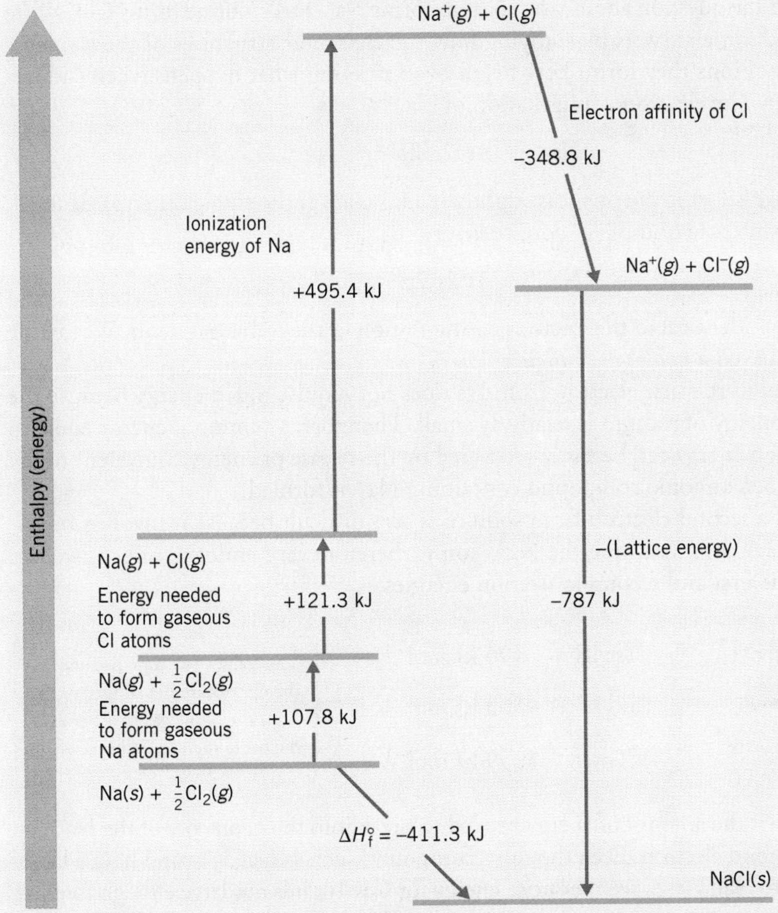

Born–Haber cycle for sodium chloride. The path at the lower left labeled ΔH_f° leads directly to NaCl(*s*). The upper path involves the formation of gaseous atoms from the elements, then the formation of gaseous ions from the atoms, and finally the condensation of Na^+ and Cl^- ions to give solid NaCl. The final step releases energy equivalent to the lattice energy. Both the upper and lower paths lead from the elements to solid NaCl and yield the same net energy change.

which makes them more difficult to pull apart. For example, the lattice energy for LiCl is larger than that for NaCl, reflecting the smaller size of the lithium ion.

The lattice energy also becomes larger as the amount of charge on the ions increases, because more highly charged ions attract each other more strongly. Thus, salts of Ca^{2+} have larger lattice energies than comparable salts of Na^+, and those containing Al^{3+} have even larger lattice energies.

Besides affecting the ability of ionic compounds to form, lattice energies are also important in determining the solubilities of ionic compounds in water and other solvent. We will explore this topic in Chapter 12.

Energy factors determine that metals form cations and nonmetals form anions
We are now in a position to understand why metals tend to form positive ions and nonmetals tend to form negative ions. Metals, at the left of the periodic table, are elements with small ionization energies and electron affinities. Relatively little energy is needed to remove electrons from them to produce positive ions. Nonmetals, at the upper right of the periodic table, have large ionization energies and generally exothermic electron affinities. It is quite difficult to remove electrons from these elements, but sizable amounts of energy are released when they gain electrons. On an energy basis, therefore, it is least "expensive" to form a cation from a metal and an anion from a nonmetal, so it is relatively easy for the energy-lowering effect of the lattice energy to exceed the net energy-raising effect of the ionization energy and electron affinity. In fact, metals combine with nonmetals to form ionic compounds simply because ionic bonding is favored energetically over other types whenever atoms with small ionization energies combine with atoms that have large exothermic electron affinities.

The stability of the noble gas configuration can control which ions are possible

Earlier we raised the question about why sodium forms Na^+ and chlorine forms Cl^-. We're now able to find some answers by studying how the electronic structures of the elements affect the kinds of ions they form. Let's begin by examining what happens when sodium loses an electron. The electron configuration of Na is

$$\text{Na} \quad 1s^2 2s^2 2p^6 3s^1$$

The electron that is lost is the one least tightly held, which is the single outer 3*s* electron. The electronic structure of the Na^+ ion, then, is

$$\text{Na}^+ \quad 1s^2 2s^2 2p^6$$

Notice that this is identical to the electron configuration of the noble gas neon. We say the Na^+ ion has achieved a *noble gas configuration.*

The removal of the first electron from Na does not require much energy because the first ionization energy of sodium is relatively small. Therefore, an input of energy equal to the first ionization energy can be easily recovered by the release of energy equivalent to the lattice energy when an ionic compound containing Na^+ is formed.

Removal of a second electron from sodium is *very* difficult because it involves breaking into the $2s^2 2p^6$ core. Forming the Na^{2+} ion is therefore very endothermic, as we can see by adding the first and second ionization energies.

$$\text{Na}(g) \longrightarrow \text{Na}^+(g) + e^- \qquad \text{1st IE} = 496 \text{ kJ mol}^{-1}$$
$$\text{Na}^+(g) \longrightarrow \text{Na}^{2+}(g) + e^- \qquad \text{2nd IE} = \underline{4563 \text{ kJ mol}^{-1}}$$
$$\text{Total} \quad 5059 \text{ kJ mol}^{-1}$$

This value is so large because the electron removed is from the noble gas core beneath the outer shell of sodium.

Notice how large is the amount of energy needed to break into the neon core of the Na^+ ion to remove the second electron. Even though a compound such as $NaCl_2$ would have a larger lattice energy than NaCl (e.g., see the lattice energy for $CaCl_2$), it is not large enough to make the formation of the compound exothermic. As a result, $NaCl_2$ cannot form. Similar situations exist for other compounds of sodium, so when sodium forms a cation, electron loss stops once the Na^+ ion is formed and a noble gas electron configuration is reached.

Similar situations exist for other metals, too. For example, the first two electrons to be removed from a calcium atom come from the 4*s* valence shell.

$$\text{Ca} \quad 1s^2 2s^2 2p^6 3s^2 3p^6 4s^2$$
$$\text{Ca}^{2+} \quad 1s^2 2s^2 2p^6 3s^2 3p^6$$

The energy needed to accomplish this can be recovered by the release of energy equivalent to the lattice energy when a calcium compound forms.[2] Electron loss ceases at this point, however, because of the huge amount of energy needed to break into the noble gas core. As a result, a calcium atom loses just two electrons when it reacts.

In the case of sodium and calcium, we find that the large ionization energy of the noble gas core just below their outer shells limits the number of electrons they lose, so the ions that are formed have noble gas electron configurations.

Nonmetals also tend to have noble gas configurations when they form anions. For example, when a chlorine atom reacts, it gains one electron.

$$\text{Cl} \quad 1s^22s^22p^63s^23p^5$$
$$\text{Cl}^- \quad 1s^22s^22p^63s^23p^6$$

At this point, we have a noble gas configuration (that of argon). Electron gain ceases, because if another electron were to be added, it would have to enter an orbital in the next higher shell, which is very energetically unfavorable. Similar arguments apply to the other nonmetals as well.

For calcium:

1st IE = 590 kJ/mol
2nd IE = 1146 kJ/mol
3rd IE = 4940 kJ/mol

Because it is so difficult to break into the noble gas core, it's convenient to think of it as being very stable.

The octet rule can sometimes provide a guide to the ions formed by elements

In the preceding discussion you learned that a balance of energy factors causes many atoms to form ions that have a noble gas electron configuration. Historically, this is expressed in the form of a generalization: *When they form ions, atoms of most of the representative elements tend to gain or lose electrons until they have obtained a configuration identical to that of the nearest noble gas.* Because all the noble gases except helium have outer shells with eight electrons, this rule has become known as the **octet rule,** which can be stated as follows: *Atoms tend to gain or lose electrons until they have achieved an outer shell that contains an* **octet of electrons** *(eight electrons).*

TOOLS *Electron configurations of ions of the representative elements*

Many cations do not obey the octet rule

The octet rule, as applied to ionic compounds, really works well only for the cations of the Group IA and IIA metals and for the anions of the nonmetals. It does not work so well for the transition metals and post-transition metals (the metals that follow a row of transition metals).

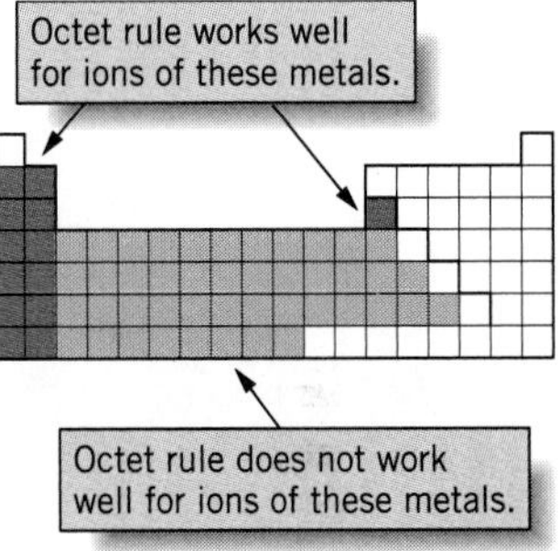
Octet rule works well for ions of these metals.
Octet rule does not work well for ions of these metals.

To obtain the correct electron configurations of the cations of these metals, we apply the following rules:

Obtaining the electron configuration of an ion

1. The first electrons to be lost by an atom or ion are always those from the shell with the largest value of n (i.e., the outer shell).
2. As electrons are removed from a given shell, they come from the highest-energy occupied subshell first, before any are removed from a lower-energy subshell. Within a given shell, the energies of the subshells vary as follows: $s < p < d < f$. This means that f is emptied before d, which is emptied before p, which is emptied before s.

TOOLS *Order in which electrons are lost from an atom*

Applying these rules to the metals of Groups IA and IIA also gives the correct electron configurations.

Let's look at two examples.

Tin (a post-transition metal) forms two ions, Sn^{2+} and Sn^{4+}. The electron configurations are

$$\text{Sn} \quad [\text{Kr}]\ 4d^{10}5s^25p^2$$
$$\text{Sn}^{2+} \quad [\text{Kr}]\ 4d^{10}5s^2$$
$$\text{Sn}^{4+} \quad [\text{Kr}]\ 4d^{10}$$

[2] Compounds of Ca^+ don't exist because the lattice energies of compounds containing Ca^{2+} are so large that they drive the loss of another electron from the Ca^+ ion. In other words, $CaCl_2$ is more stable than CaCl, so the latter never forms.

Notice that the Sn^{2+} ion is formed by the loss of the higher energy $5p$ electrons first. Then, further loss of the two $5s$ electrons gives the Sn^{4+} ion. However, neither of these ions has a noble gas configuration.

For the transition elements, the first electrons lost are the s electrons of the outer shell. Then, if additional electrons are lost, they come from the underlying d subshell. An example is iron, which forms the ions Fe^{2+} and Fe^{3+}. The element iron has the electron configuration

$$\text{Fe} \qquad [\text{Ar}]\ 3d^6 4s^2$$

Iron loses its $4s$ electrons fairly easily to give Fe^{2+}, which has the electron configuration

$$\text{Fe}^{2+} \qquad [\text{Ar}]\ 3d^6$$

The Fe^{3+} ion results when another electron is removed, this time from the $3d$ subshell.

$$\text{Fe}^{3+} \qquad [\text{Ar}]\ 3d^5$$

Iron is able to form Fe^{3+} because the $3d$ subshell is close in energy to the $4s$, so it is not very difficult to remove the third electron. Notice once again that the first electrons to be removed come from the shell with the largest value of n (the $4s$ subshell). Then, after this shell is emptied, the next electrons are removed from the shell below.

Because so many of the transition elements are able to form ions in a way similar to that of iron, the ability to form more than one positive ion is usually cited as one of the characteristic properties of the transition elements. Frequently, one of the ions formed has a 2+ charge, which arises from the loss of the two outer s electrons. Ions with larger positive charges result when additional d electrons are lost. Unfortunately, it is not easy to predict exactly which ions can form for a given transition metal, nor is it simple to predict their relative stabilities with respect to oxidation or reduction.

EXAMPLE 8.1
Writing Electron Configurations of Ions

How do the electron configurations change (a) when a nitrogen atom forms the N^{3-} ion and (b) when an antimony atom forms the Sb^{3+} ion?

ANALYSIS: For the nonmetals, you've learned that the octet rule does work, so the ion that is formed by nitrogen will have a noble gas configuration.

Antimony is a post-transition element, so we don't expect its cation to obey the octet rule. To determine the electron configuration of Sb^{3+}, our tool will be the rules describing the order in which electrons are lost. The rules tell us that when a cation is formed, electrons are removed first from the outer shell of the atom (the shell with the largest value of the principal quantum number, n). Within a given shell, electrons are always removed first from the subshell highest in energy.

SOLUTION: (a) The electron configuration for nitrogen is

$$\text{N} \qquad [\text{He}]\ 2s^2 2p^3$$

To form N^{3-}, three electrons are gained. These enter the $2p$ subshell because it is the lowest available energy level. Filling the $2p$ subshell completes the octet; the configuration for the ion is therefore

$$\text{N}^{3-} \qquad [\text{He}]\ 2s^2 2p^6$$

(b) Let's begin with the ground state electron configuration for antimony.

$$\text{Sb} \qquad [\text{Kr}]\ 4d^{10} 5s^2 5p^3$$

To form the Sb^{3+} ion, three electrons must be removed. These will come from the outer shell, which has $n = 5$. Within this shell, the energies of the subshells increase in the order $s < p < d < f$. Therefore, the $5p$ subshell is higher in energy than the $5s$, so all three electrons are removed from the $5p$. This gives

$$Sb^{3+} \quad [Kr]\ 4d^{10}5s^2$$

ARE THE ANSWERS REASONABLE? In Chapter 2 you learned to use the periodic table to figure out the charges on anions of the nonmetals. For nitrogen, we would take three steps to the right to get to the nearest noble gas, neon. The electron configuration we obtained for N^{3-} is that of neon, so our answer should be correct.

For antimony, we had to remove three electrons, which completely emptied the $5p$ subshell. That's also good news, because ions do not tend to have partially filled s or p subshells (although partially filled d subshells are not uncommon for the transition metals). If we had taken the electrons from any other subshells, the Sb^{3+} ion would have had a partially filled $5p$ subshell.

EXAMPLE 8.2
Writing Electron Configurations of Ions

What is the electron configuration of the V^{3+} ion? Give the orbital diagram for the ion.

ANALYSIS: To obtain the electron configuration of an ion, always begin with the electron configuration of the neutral atom. In this case we will then remove three electrons to obtain the electron configuration of the ion. We have to keep in mind that the electrons are lost first from the occupied shell with highest n.

SOLUTION: The electron configuration of vanadium is

$$V \quad \underbrace{1s^22s^22p^63s^23p^6}_{\text{argon core}}3d^34s^2$$

Notice that we've written the configuration showing the outer shell $4s$ electrons farthest to the right. To form the V^{3+} cation, three electrons must be removed from the neutral atom. The first two come from the $4s$ subshell and the third comes from the $3d$. This means we won't have to take any from the $3s$ or $3p$ subshells, so the argon core will remain intact. Therefore, let's rewrite the electron configuration in abbreviated form.

$$V \quad [Ar]\ 3d^34s^2$$

Removing the three electrons gives

$$V^{3+} \quad [Ar]\ 3d^2$$

To form the orbital diagram, we show all five orbitals of the $3d$ subshell and then spread the two electrons out with spins unpaired. This gives

$$V^{3+} \quad [Ar]\ \underbrace{(\uparrow)(\uparrow)(\ )(\ )(\ )}_{3d}$$

IS THE ANSWER REASONABLE? First, we check that we've written the correct electron configuration of vanadium, which we have. (A quick count of the electrons gives 23, which is the atomic number of vanadium.) We've also taken electrons away from the atom following the rules, so the electron configuration of the ion seems okay. Finally, we remembered to show all five orbitals of the $3d$ subshell, even though only two of the them are occupied.

Practice Exercise 1: What is wrong with the following electron configuration of the In^+ ion? What should the electron configuration be?

$$In^+ \quad 1s^22s^22p^63s^23p^63d^{10}4s^24p^64d^{10}5s^15p^1$$

(Hint: Check the rules that tell us the order in which electrons are lost by an atom or ion.)

Practice Exercise 2: How do the electron configurations change when a chromium atom forms the following ions: (a) Cr^{2+}, (b) Cr^{3+}, and (c) Cr^{6+}?

Practice Exercise 3: How are the electron configurations of S^{2-} and Cl^- related?

8.2 LEWIS SYMBOLS HELP KEEP TRACK OF VALENCE ELECTRONS

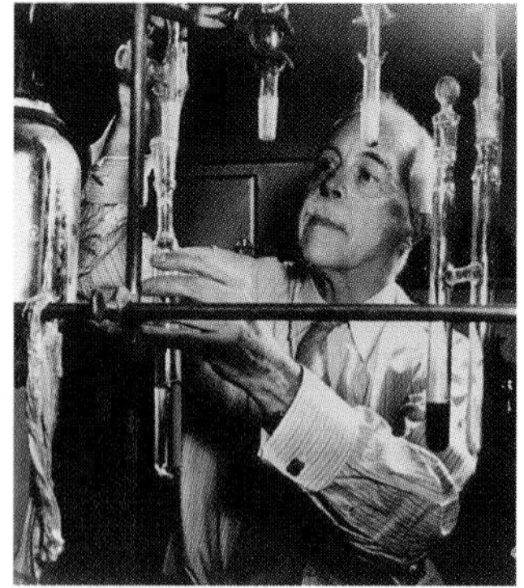

Gilbert N. Lewis, chemistry professor at the University of California, helped develop theories of chemical bonding. In 1916, he proposed that atoms form bonds by sharing pairs of electrons between them. *(Lawrence Berkeley National/Photo Researchers.)*

In the last section you saw how the valence shells of atoms change when electrons are transferred during the formation of ions. We will soon study how many atoms share their valence electrons with each other when they form covalent bonds. In these discussions it is useful to be able to keep track of valence electrons. To help us do this, we use a simple bookkeeping device called Lewis symbols, named after their inventor, a famous American chemist, G. N. Lewis (1875–1946).

To draw the **Lewis symbol** for an element, we write its chemical symbol surrounded by a number of dots (or some other similar mark), which represent the atom's valence electrons. For example, the element lithium, which has one valence electron in its 2*s* subshell, has the Lewis symbol

Li·

In fact, each element in Group IA has a similar Lewis symbol, because each has only one valence electron. The Lewis symbols for all of the Group IA metals are

Li· Na· K· Rb· Cs·

The Lewis symbols for the eight A group elements of Period 2 are[3]

Group	IA	IIA	IIIA	IVA	VA	VIA	VIIA	VIIIA
Symbol	Li·	·Be·	·Ḃ·	·Ċ·	·N:	·O:	·F:	:Ne:

The elements below each of these in their respective groups have identical Lewis symbols except, of course, for the chemical symbol of the element. Notice that when an atom has more than four valence electrons, the additional electrons are shown to be paired with others. Also notice that *for the representative elements, the group number is equal to the number of valence electrons* when the North American convention for numbering groups in the periodic table is followed.

This is one of the advantages of the North American convention for numbering groups in the periodic table.

EXAMPLE 8.3 Writing Lewis Symbols

What is the Lewis symbol for arsenic?

ANALYSIS: We need to know the number of valence electrons, which we can obtain from the group number. Then we distribute the electrons (dots) around the chemical symbol.

SOLUTION: The symbol for arsenic is As and we find it in Group VA. The element therefore has five valence electrons. The first four are placed around the symbol for arsenic as follows:

·Ȧs·

[3] For beryllium, boron, and carbon, the number of unpaired electrons in the Lewis symbol doesn't agree with the number predicted from the atom's electron configuration. Boron, for example, has two electrons paired in its 2*s* orbital and a third electron in one of its 2*p* orbitals; therefore, there is actually only one unpaired electron in a boron atom. The Lewis symbols are drawn as shown, however, because when beryllium, boron, and carbon form bonds, they *behave* as if they have two, three, and four unpaired electrons, respectively.

The fifth electron is paired with one of the first four. This gives

$$\cdot\dot{\underset{\cdot}{\mathrm{As}}}:$$

The location of the fifth electron doesn't really matter, so equally valid Lewis symbols are

$$\cdot\ddot{\underset{\cdot}{\mathrm{As}}}\cdot \quad \text{or} \quad :\dot{\underset{\cdot}{\mathrm{As}}}\cdot \quad \text{or} \quad \cdot\dot{\underset{\cdot\cdot}{\mathrm{As}}}\cdot$$

IS THE ANSWER REASONABLE? There's not much to check here. Have we got the correct chemical symbol? Yes. Do we have the right number of dots? Yes.

Although we will use Lewis symbols mostly to follow the fate of valence electrons in covalent bonds, they can also be used to describe what happens during the formation of ions. For example, when a sodium atom reacts with a chlorine atom, the electron transfer can be depicted as

$$\mathrm{Na}\cdot + \cdot\ddot{\underset{\cdot\cdot}{\mathrm{Cl}}}: \longrightarrow \mathrm{Na}^+ + \left[:\ddot{\underset{\cdot\cdot}{\mathrm{Cl}}}:\right]^-$$

The valence shell of the sodium atom is emptied, so no dots remain. The outer shell of chlorine, which formerly had seven electrons, gains one to give a total of eight. The brackets are drawn around the chloride ion to show that all eight electrons are the exclusive property of the Cl^- ion.

We can diagram a similar reaction between calcium and chlorine atoms.

$$:\ddot{\underset{\cdot\cdot}{\mathrm{Cl}}}\cdot \quad \cdot\mathrm{Ca}\cdot \quad \cdot\ddot{\underset{\cdot\cdot}{\mathrm{Cl}}}: \longrightarrow \mathrm{Ca}^{2+} + 2\left[:\ddot{\underset{\cdot\cdot}{\mathrm{Cl}}}:\right]^-$$

EXAMPLE 8.4
Using Lewis Symbols

Use Lewis symbols to diagram the reaction that occurs between sodium and oxygen atoms to give Na^+ and O^{2-} ions.

ANALYSIS: For electrical neutrality, the formula will be Na_2O, so we know we will use two sodium atoms and one oxygen atom. The sodiums will lose one electron each to give Na^+ and the oxygen will gain two electrons to give O^{2-}.

SOLUTION: First let's draw the Lewis symbols for Na and O.

$$\mathrm{Na}\cdot \qquad \cdot\ddot{\underset{\cdot}{\mathrm{O}}}:$$

It takes two electrons to complete the octet around oxygen. Each Na supplies one. Therefore,

$$\mathrm{Na}\cdot \; \cdot\ddot{\underset{\cdot}{\mathrm{O}}}: \; \cdot\mathrm{Na} \longrightarrow 2\mathrm{Na}^+ + \left[:\ddot{\underset{\cdot\cdot}{\mathrm{O}}}:\right]^{2-}$$

Notice that we have put brackets around the oxide ion.

IS THE ANSWER REASONABLE? We have accounted for all the valence electrons (an important check), the net charge is the same on both sides of the arrow (the equation is balanced), and we've placed the brackets around the oxide ion to emphasize that the octet belongs exclusively to that ion.

Practice Exercise 4: Use Lewis symbols to diagram the formation of CaI_2 from Ca and I atoms. (Hint: Begin by determining how many electrons are gained or lost by each atom.)

Practice Exercise 5: Diagram the reaction between magnesium and oxygen atoms to give Mg^{2+} and O^{2-} ions.

8.3 Covalent bonds are formed by electron sharing

Most of the substances we encounter in our daily lives are not ionic. Instead, they are composed of electrically neutral molecules. The chemical bonds that bind the atoms to each other in such molecules are electrical in nature, but they arise from the sharing of electrons rather than by electron transfer.

Forming a covalent bond lowers the potential energy

Earlier we saw that for ionic bonding to occur, the energy-lowering effect of the lattice energy must be greater than the combined net energy-raising effects of the ionization energy (IE) and electron affinity (EA). Many times this is not possible, particularly when the ionization energies of all the atoms involved are large. This happens, for example, when nonmetals combine with each other to form molecules. In such cases, nature uses a different way to lower the energy—electron sharing.

Let's look at what happens when two hydrogen atoms join to form an H_2 molecule (Figure 8.3). As the two atoms approach each other, the electron of each atom begins to feel the attraction of both nuclei. This causes the electron density around each nucleus to shift toward the region between the two atoms. Therefore, as the distance between the nuclei decreases, there is an increase in the probability of finding either electron near either nucleus. In effect, as the molecule is formed, each of the hydrogen atoms in the H_2 molecule acquires a share of two electrons.

In the H_2 molecule, the buildup of electron density between the two atoms attracts both nuclei and pulls them together. Being of the same charge, however, the two nuclei also repel each other, as do the two electrons. In the molecule that forms, therefore, the atoms are held at a distance at which all these attractions and repulsions are balanced. Overall, the nuclei are kept from separating, and the net force of attraction produced by sharing the pair of electrons is called a **covalent bond.**

Every covalent bond is characterized by two quantities, the average distance between the nuclei held together by the bond, and the amount of energy needed to separate the two atoms to produce neutral atoms again. In the hydrogen molecule, the attractive forces pull the nuclei to a distance of 75 pm, and this distance is called the **bond length** (or sometimes the **bond distance**). Because a covalent bond holds atoms together, work must be done (energy must be supplied) to separate them. The amount of energy needed to "break" the bond (or the energy released when the bond is formed) is called the **bond energy.**

As the distance between the nuclei and the electron cloud that lies between them decreases, the potential energy decreases.

Figure 8.4 shows how the potential energy changes when two hydrogen atoms come together to form H_2. We see that the minimum potential energy occurs at a bond distance of 75 pm, and that 1 mol of hydrogen molecules is more stable than 2 mol of hydrogen atoms by 435 kJ. In other words, the bond energy of H_2 is 435 kJ/mol.

Pairing of electrons occurs when a covalent bond forms

Before joining, each of the separate hydrogen atoms has one electron in a 1*s* orbital. When these electrons are shared, the 1*s* orbital of each atom is, in a sense, filled. Because the electrons now share the same space, they become paired as required by the Pauli exclusion

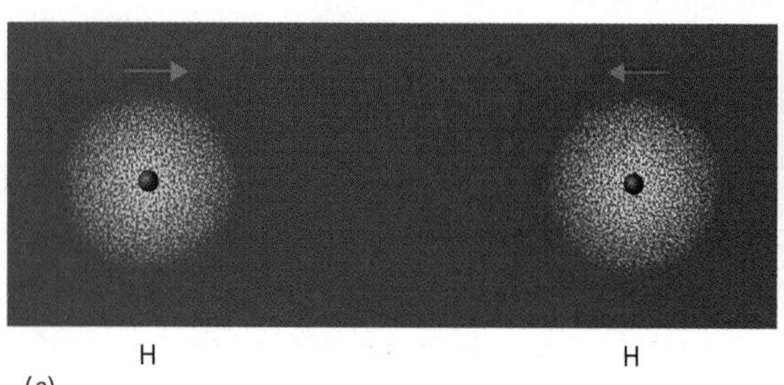

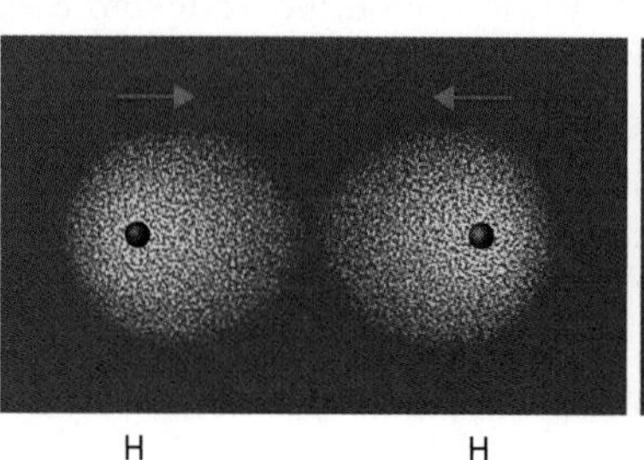

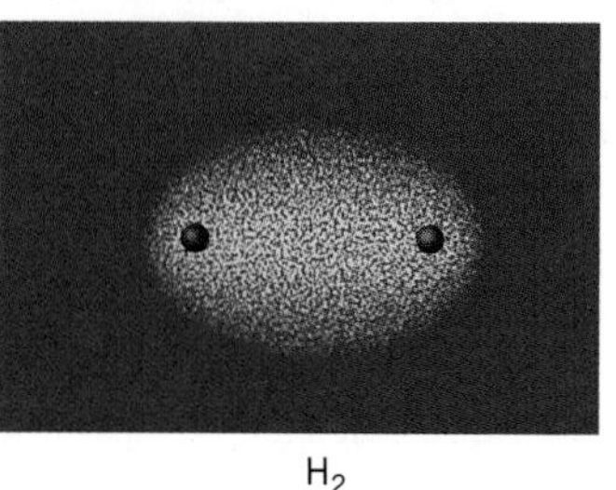

FIG. 8.3 Formation of a covalent bond between two hydrogen atoms. (*a*) Two H atoms separated by a large distance. (*b*) As the atoms approach each other, their electron densities are pulled into the region between the two nuclei. (*c*) In the H_2 molecule, the electron density is concentrated between the nuclei. Both electrons in the bond are distributed over both nuclei.

FACETS OF CHEMISTRY

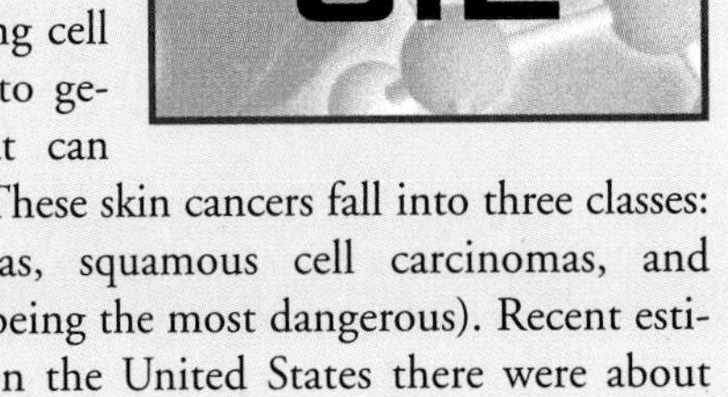

Sunlight and Skin Cancer

The ability of light to provide the energy for chemical reactions enables life to exist on our planet. Green plants absorb sunlight and, with the help of chlorophyll, convert carbon dioxide and water into carbohydrates (e.g., sugars and cellulose), which are essential constituents of the food chain. However, not all effects of sunlight are so beneficial.

As you know, light packs energy that's proportional to its frequency, and if the photons that are absorbed by a substance have enough energy, they can rupture chemical bonds and initiate chemical reactions. Light that is able to do this has frequencies in the UV region of the electromagnetic spectrum, and the sunlight bombarding the earth contains substantial amounts of UV radiation. Fortunately, a layer of ozone (O_3) in the stratosphere, a region of the atmosphere extending from about 45 to 55 km altitude, absorbs most of the incoming UV, protecting life on the surface. However, some UV radiation does get through, and the part of the spectrum of most concern is called "UV-B" with wavelengths between 280 and 320 nm.

What makes UV-B so dangerous is its ability to affect the DNA in our cells. (The structure of DNA and its replication are discussed in Chapter 22.) Absorption of UV radiation causes constituents of the DNA, called *pyrimidine bases,* to undergo reactions that form bonds between them. This causes transcription errors when the DNA replicates during cell division, giving rise to genetic mutations that can lead to skin cancers. These skin cancers fall into three classes: basal cell carcinomas, squamous cell carcinomas, and melanomas (the last being the most dangerous). Recent estimates indicate that in the United States there were about 500,000 cases of the first, 100,000 cases of the second, and 27,600 cases of the third. It has also been estimated that more than 90% of the skin cancers are due to absorption of UV-B radiation.

In recent years concern has grown over the apparent depletion of the ozone layer in the stratosphere caused by the release of gases called chlorofluorocarbons (CFCs), which have been widely used in refrigerators and air conditioners. Some scientists have estimated a substantial increase in the rate of skin cancer caused by increased amounts of UV-B reaching the earth's surface due to this ozone depletion.

Dawn of a new day brings the risk of skin cancer to those particularly susceptible. Fortunately, understanding the risk allows us to protect ourselves with clothing and sunblock creams. *(Taxi/Getty Images.)*

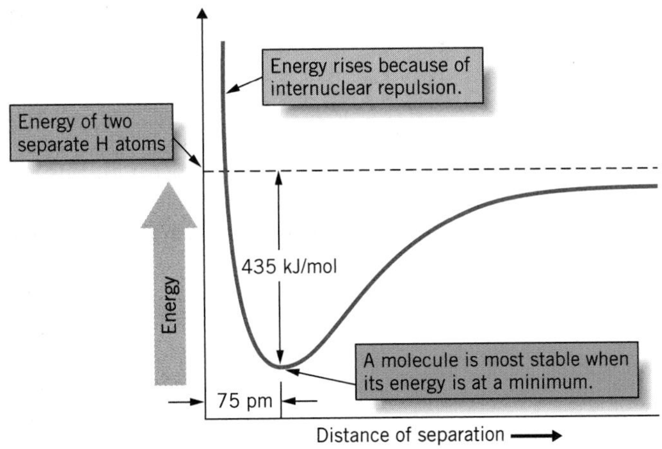

FIG. 8.4 Changes in the potential energies of two hydrogen atoms as they form H_2. The energy of the molecule reaches a minimum when there is a balance between the attractions and repulsions.

principle; that is, m_s is $+\frac{1}{2}$ for one of the electrons and $-\frac{1}{2}$ for the other. In general, the electrons involved almost always become paired when atoms form covalent bonds. In fact, a covalent bond is sometimes referred to as an **electron pair bond.**

In Chapter 7 you learned that when two electrons occupy the same orbital, and therefore share the same space, their spins must be paired. The pairing of electrons is an important part of the formation of a covalent bond.

Lewis symbols are often used to keep track of electrons in covalent bonds. The electrons that are shared between two atoms are shown as a pair of dots placed between the symbols for the bonded atoms. The formation of H_2 from hydrogen atoms, for example, can be depicted as

$$H\cdot + H\cdot \longrightarrow H{:}H$$

Because the electrons are shared, each H atom is considered to have two electrons.

H:H

(Colored circles emphasize that two electrons can be counted around each of the H atoms.)

For simplicity, the electron pair in a covalent bond is usually depicted as a single dash. Thus, the hydrogen molecule is represented as

H—H

A formula such as this, which is drawn with Lewis symbols, is called a **Lewis formula** or **Lewis structure.** It is also called a **structural formula** because it shows which atoms are present in the molecule *and* how they are attached to each other.

Covalent bonding often follows the octet rule

You have seen that when a nonmetal atom forms an anion, electrons are gained until the *s* and *p* subshells of its valence shell are completed. The tendency of a nonmetal atom to finish with a completed valence shell, usually consisting of eight electrons, also influences the number of electrons the atom tends to acquire by sharing, and it thereby affects the number of covalent bonds the atom forms.

As you will see, it is useful to remember that hydrogen atoms form only one covalent bond.

Hydrogen, with just one electron in its 1*s* orbital, completes its valence shell by obtaining a share of just one electron from another atom, so a hydrogen atom forms just one covalent bond. When this other atom is hydrogen, the H_2 molecule is formed.

Octet rule and covalent bonding

Many atoms form covalent bonds by sharing enough electrons to give them complete *s* and *p* subshells in their outer shells. This is the noble gas configuration mentioned earlier and is the basis of the octet rule described in Section 8.1. As applied to covalent bonding, the **octet rule** can be stated as follows: *When atoms form covalent bonds, they tend to share sufficient electrons so as to achieve an outer shell having eight electrons.*

Often, the octet rule can be used to explain the number of covalent bonds an atom forms. This number normally equals the number of electrons the atom must acquire to have a total of eight (an octet) in its outer shell. For instance, the halogens (Group VIIA) all have seven valence electrons. The Lewis symbol for a typical member of this group, chlorine, is

$\cdot\ddot{\underset{\cdot\cdot}{\text{Cl}}}\!:$

We can see that only one electron is needed to complete its octet. Of course, chlorine can actually gain this electron and become a chloride ion. This is what it does when it forms an ionic compound such as sodium chloride (NaCl). But when chlorine combines with another nonmetal, the complete transfer of an electron is not energetically favorable. Therefore, in forming such compounds as HCl or Cl_2, chlorine gets the one electron it needs by forming a covalent bond.

$$\text{H}\cdot + \cdot\ddot{\underset{\cdot\cdot}{\text{Cl}}}\!: \longrightarrow \text{H}\!:\!\ddot{\underset{\cdot\cdot}{\text{Cl}}}\!:$$

$$:\!\ddot{\underset{\cdot\cdot}{\text{Cl}}}\cdot + \cdot\ddot{\underset{\cdot\cdot}{\text{Cl}}}\!: \longrightarrow :\!\ddot{\underset{\cdot\cdot}{\text{Cl}}}\!:\!\ddot{\underset{\cdot\cdot}{\text{Cl}}}\!:$$

The HCl and Cl_2 molecules can also be represented using dashes for the bonds.

$$\text{H}-\ddot{\underset{\cdot\cdot}{\text{Cl}}}\!: \qquad \text{and} \qquad :\!\ddot{\underset{\cdot\cdot}{\text{Cl}}}-\ddot{\underset{\cdot\cdot}{\text{Cl}}}\!:$$

There are many nonmetals that form more than one covalent bond. For example, the three most important elements in biochemical systems are carbon, nitrogen, and oxygen.

$$\cdot\dot{\underset{\cdot}{\text{C}}}\cdot \qquad \cdot\dot{\underset{\cdot\cdot}{\text{N}}}\cdot \qquad \cdot\dot{\underset{\cdot\cdot}{\text{O}}}\!:$$

The simplest hydrogen compounds of these elements are methane, CH_4, ammonia, NH_3, and water, H_2O. Their Lewis structures are

```
  H          H          H
  ..         ..         ..
H:C:H      H:N:H      H:O:
  ..         ..         ..
  H
```

or

```
    H            H           H
    |            |           |
H — C — H    H — N — H   H — O:
    |            ..          ..
    H
```

methane ammonia water

Multiple bonds consist of two or more pairs of electrons

The bond produced by the sharing of *one* pair of electrons between two atoms is called a **single bond.** So far, these have been the only kind we've discussed. There are, however, many molecules in which more than a single pair of electrons are shared between two atoms. For example, we can diagram the formation of the bonds in CO_2 as follows.

$$:\dot{\underset{..}{O}}\cdot \leftrightarrow \cdot\dot{\underset{\cdot}{C}}\cdot \leftrightarrow \cdot\ddot{\underset{\cdot}{O}}: \longrightarrow :\underset{..}{O}::C::\ddot{O}:$$

The central carbon atom shares two of its electrons with each of the oxygen atoms, and each oxygen shares two electrons with carbon. The result is the formation of two **double bonds.** Notice that in the Lewis formula, both of the shared electron pairs are placed between the symbols for the two atoms joined by the double bond. Once again, if we circle the valence shell electrons that "belong" to each atom, we see that each has an octet.

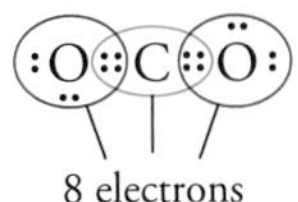

The locations of the unshared pairs of electrons around the oxygen are unimportant. Two equally valid Lewis structures for CO_2 are

$$:\ddot{O}{=}C{=}\ddot{O}: \quad \text{and} \quad \underset{..}{\ddot{O}}{=}C{=}\underset{..}{\ddot{O}}$$

The Lewis structure for CO_2, using dashes, is

$$:\underset{..}{O}{=}C{=}\ddot{O}:$$

Sometimes three pairs of electrons are shared between two atoms. The most abundant gas in the atmosphere, nitrogen, occurs in the form of diatomic molecules, N_2. As we've just seen, the Lewis symbol for nitrogen is

$$\cdot\dot{\underset{\cdot}{N}}:$$

and each nitrogen atom needs three electrons to complete its octet. When the N_2 molecule is formed, each of the nitrogen atoms shares three electrons with the other.

$$:\dot{\underset{\cdot}{N}}\cdot \leftrightarrow \cdot\dot{\underset{\cdot}{N}}: \longrightarrow :N:::N:$$

The result is called a **triple bond.** Again, notice that we place all three electron pairs of the bond between the two atoms. We count all of these electrons as though they belong to both of the atoms. Each nitrogen therefore has an octet.

8 electrons 8 electrons

$$:N:::N:$$

The triple bond is usually represented by three dashes, so the bonding in the N_2 molecule is normally shown as

$$:N{\equiv}N:$$

More complex molecules can contain single, double, and/or triple bonds

All of the bonds in a molecule don't have to be of the same kind. Often we find single, double, and even triple bonds in the same molecule. For example, consider propylene, the raw material in the manufacture of polypropylene, a plastic used to form containers and in making fibers for textiles, carpet, and rope.

```
        H
        |
    H—C—C=C—H
        |  |  |
        H  H  H
```

In this molecule there are both single and double bonds between carbon atoms, as well as the C—H single bonds. Notice that each carbon atom forms four covalent bonds to neighboring atoms, thereby completing its octet. The formation of four covalent bonds is a characteristic property of carbon and is a feature of almost all organic compounds.

Another example is acetic acid, $HC_2H_3O_2$. The shape of the molecule was illustrated in Figure 4.8 (page 137), showing the single hydrogen atom that is capable of ionizing in the formation of H_3O^+. The Lewis structures of acetic acid and the acetate ion are

```
    H  :O:               ⎡    H  :O:     ⎤⁻
    |   ‖                ⎢    |   ‖      ⎥
H—C—C—Ö—H                ⎢ H—C—C—Ö:      ⎥
    |                    ⎢    |          ⎥
    H                    ⎣    H          ⎦
  acetic acid                acetate ion
```

Acetic acid is but one of a large number of substances known as **organic acids** or **carboxylic acids.** In general, their structures are characterized by the presence of the **carboxyl group,** $—CO_2H$.

```
     :O:
      ‖
   —C—Ö—H
carboxyl group
```

All of the organic acids you will encounter in this course are weak acids.

8.4 Covalent bonds can have partial charges at opposite ends

When two identical atoms form a covalent bond, as in H_2 or Cl_2, each atom has an equal share of the bond's electron pair. The electron density at both ends of the bond is the same, because the electrons are equally attracted to both nuclei. However, when different kinds of atoms combine, as in HCl, one nucleus usually attracts the electrons in the bond more strongly than the other.

The result of unequal attractions for the bonding electrons is an unbalanced distribution of electron density within the bond. For example, chlorine attracts electrons in a bond more strongly than hydrogen does. In the HCl molecule, therefore, the electron cloud is pulled more tightly around the Cl, and that end of the molecule experiences a slight buildup of negative charge. The electron density that shifts toward the chlorine is removed from the hydrogen, which causes the hydrogen end to acquire a slight positive charge. These charges are less than full 1+ and 1− charges and are called **partial charges,** which are usually indicated by the lowercase Greek letter delta, δ (see Figure 8.5). Partial charges can also be indicated on Lewis structures. For example,

```
H—C̈l:
δ+  δ−
```

A bond that carries partial positive and negative charges on opposite ends is called a **polar covalent bond,** or often simply a **polar bond** (the word *covalent* is understood). The term *polar* comes from the notion of *poles* of equal but opposite charge at either end of the bond. Because *two poles* of electric charge are involved, the bond is said to be an **electric dipole.**

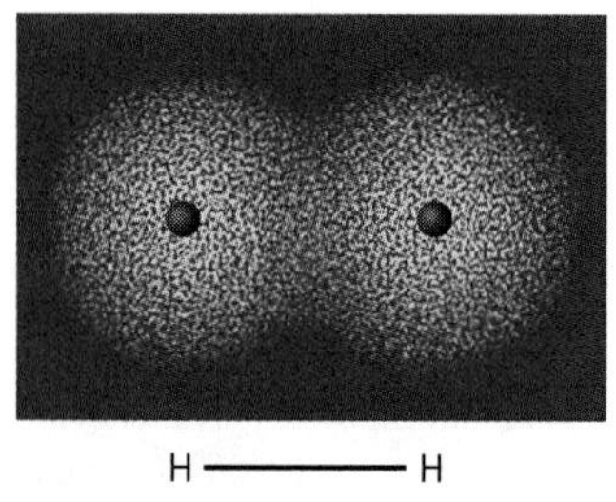

H —— H

(*a*)

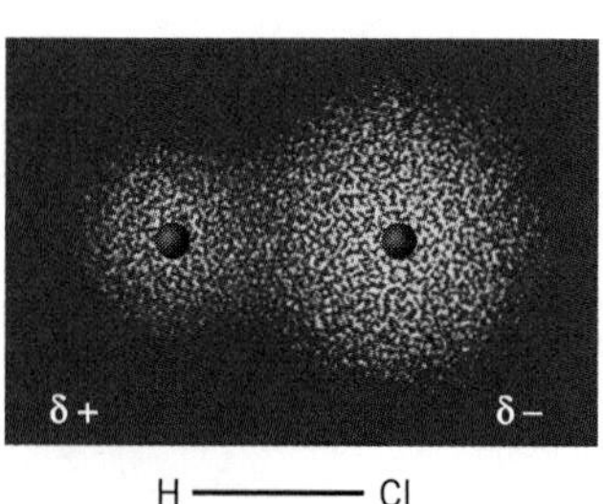

H —— Cl

(*b*)

FIG. 8.5 Nonpolar and polar covalent bonds. (*a*) The electron density of the electron pair in the bond is spread evenly between the two H atoms in H_2, which gives a nonpolar covalent bond. (*b*) In HCl, the electron density of the bond is pulled more tightly around the Cl end of the molecule, causing that end of the bond to become slightly negative. At the same time, the opposite end of the bond becomes slightly positive. The result is a polar covalent bond.

The polar bond in HCl causes the molecule as a whole to have opposite charges on either end, so the HCl molecule as a whole is an electric dipole. We say that HCl is a **polar molecule.** The magnitude of its polarity is expressed quantitatively by its **dipole moment** (symbol μ), which is equal to the amount of charge on either end of the molecule, q, multiplied by the distance between the charges, r.

$$\mu = q \times r \qquad (8.1)^4$$

Table 8.2 lists the dipole moments and bond lengths for some diatomic molecules. The dipole moments are reported in *debye* units (symbol D), where 1 D = 3.34×10^{-30} C m (coulomb × meter).

By separate experiments, it is possible to measure both μ and r (which corresponds to the bond length in a diatomic molecule such as HCl). Knowledge of μ and r allows calculation of the amount of charge on opposite ends of the dipole. For HCl, such calculations show that q equals 0.17 electronic charge units, which means the hydrogen carries a charge of $+0.17e$ and the chlorine a charge of $-0.17e$.

The electronic charge unit is represented by the symbol e. We use the symbol e^- to stand for an electron.

One of the main reasons we are concerned about whether a molecule is polar or not is because many physical properties, such as melting point and boiling point, are affected by it. This is because polar molecules attract each other more strongly than do nonpolar molecules. The positive end of one polar molecule attracts the negative end of another. The strength of the attraction depends both on the amount of charge on either end of the molecule and on the distance between the charges; in other words, it depends on the molecule's dipole moment.

TABLE 8.2 Dipole Moments and Bond Lengths for Some Diatomic Molecules[a]

Compound	Dipole Moment (D)	Bond Length (pm)
HF	1.83	91.7
HCl	1.09	127
HBr	0.82	141
HI	0.45	161
CO	0.11	113
NO	0.16	115

[a]*Source*: National Institute of Standards and Technology.

EXAMPLE 8.5
Calculating the Charge on the End of a Polar Molecule

The HF molecule has a dipole moment of 1.83 D and a bond length of 91.7 pm. What is the amount of charge, in electronic charge units, on either end of the bond?

ANALYSIS: The tool that relates the quantities in this problem is Equation 8.1. To answer the problem correctly, we will have to be especially careful of the units.

SOLUTION: We will solve Equation 8.1 for q.

$$q = \frac{\mu}{r}$$

The debye unit, D, equals 3.34×10^{-30} C m, so the dipole moment of HF is

$$\mu = 1.83 \times (3.34 \times 10^{-30} \text{ C m}) = 6.11 \times 10^{-30} \text{ C m}$$

The SI prefix p (pico) means $\times 10^{-12}$, so the bond length $r = 91.7 \times 10^{-12}$ m. Substituting in the equation above gives

$$q = \frac{6.11 \times 10^{-30} \text{ C } \cancel{\text{m}}}{91.7 \times 10^{-12} \cancel{\text{m}}} = 6.66 \times 10^{-20} \text{ C}$$

[4] In this case, we're using the symbol q to mean electric charge, not heat as in the preceding chapter. Because the number of letters in the alphabet is limited, it's not uncommon in science for the same letter to be used to stand for different quantities. This usually doesn't present a problem as long as the symbol is defined in the context in which it is used.

The amount of charge on an electron (i.e., an electronic charge unit) equals 1.602×10^{-19} C, which we can express as

$$1\ e = 1.602 \times 10^{-19}\ \text{C}$$

The value of q in electronic charge units is therefore

$$q = 6.66 \times 10^{-20}\ \cancel{\text{C}} \left(\frac{1\ e}{1.602 \times 10^{-19}\ \cancel{\text{C}}} \right) = 0.416e$$

As in HCl, the hydrogen carries the positive charge, so the charge on the hydrogen end of the molecule is $+0.416e$ and the charge on the fluorine end is $-0.416e$.

IS THE ANSWER REASONABLE? If we look over the units, we see they cancel correctly, so that gives us confidence we've done the calculation correctly. The fact that our answer is between zero and one electronic charge unit, and therefore a partial electrical charge, further suggests we've solved the problem correctly.

Practice Exercise 6: The chlorine end of the chlorine monoxide molecule carries a charge of $+0.167e$. The bond length is 154.6 pm. Calculate the dipole moment of the molecule in debye units. (Hint: Be sure to convert the charge to coulombs.)

Practice Exercise 7: Although isolated Na^+ and Cl^- ions are unstable, these ions can exist in the gaseous state as *ion pairs*. An ion pair consists of an NaCl unit in which the bond length is 236 pm. The dipole moment of the ion pair is 9.00 D. What are the actual amounts of charge on the sodium and chlorine atoms in this NaCl pair? What percentage of full 1+ and 1− charges are these? (This is the *percentage ionic character* in the NaCl pair.)

Linus Pauling contributed greatly to our understanding of chemical bonding. He was the winner of two Nobel prizes, in 1954 for Chemistry and in 1962 for Peace. *(Roger Ressmeyer/Corbis Images.)*

The noble gases are assigned electronegativities of zero and are omitted from the figure.

Electronegativity expresses an atom's attraction for electrons in a bond

The degree to which a covalent bond is polar depends on the difference in the abilities of the bonded atoms to attract electrons. The greater the difference, the more polar the bond, and the more the electron density is shifted toward the atom that attracts electrons more.

The term that we use to describe the attraction an atom has for the electrons in a bond is called **electronegativity.** In HCl, for example, chlorine is *more electronegative* than hydrogen. This causes the electron pair of the covalent bond to spend more of its time around the more electronegative atom, which is why the Cl end of the bond acquires a partial negative charge.

The first scientist to develop numerical values for electronegativity was Linus Pauling (1901–1994). He observed that polar bonds have a bond energy larger than would be expected if the opposite ends of the bonds were electrically neutral. Pauling reasoned that the extra bond energy arises because of the attraction between the partial charges on opposite ends of the bond. By estimating the extra bond energy, he was able to develop a scale of electronegativities for the elements. Other scientists have used different approaches to measuring electronegativities, with similar results.

A set of numerical values for the electronegativities of the elements is shown in Figure 8.6. These data are useful because the *difference* in electronegativity provides an estimate of the degree of polarity of a bond. For instance, the data tell us fluorine is more electronegative than chlorine, so we expect HF to be more polar than HCl. (This is confirmed by the larger dipole moment of the HF molecule.) In addition, the relative magnitudes of the electronegativities indicate which ends of a bond carry the partial positive and negative charges. Thus, hydrogen is less electronegative than either fluorine or chlorine, so in both of these molecules the hydrogen bears the partial positive charge.

$$\underset{\delta+}{\text{H}}-\underset{\delta-}{\ddot{\underset{..}{\text{F}}}}: \qquad \underset{\delta+}{\text{H}}-\underset{\delta-}{\ddot{\underset{..}{\text{Cl}}}}:$$

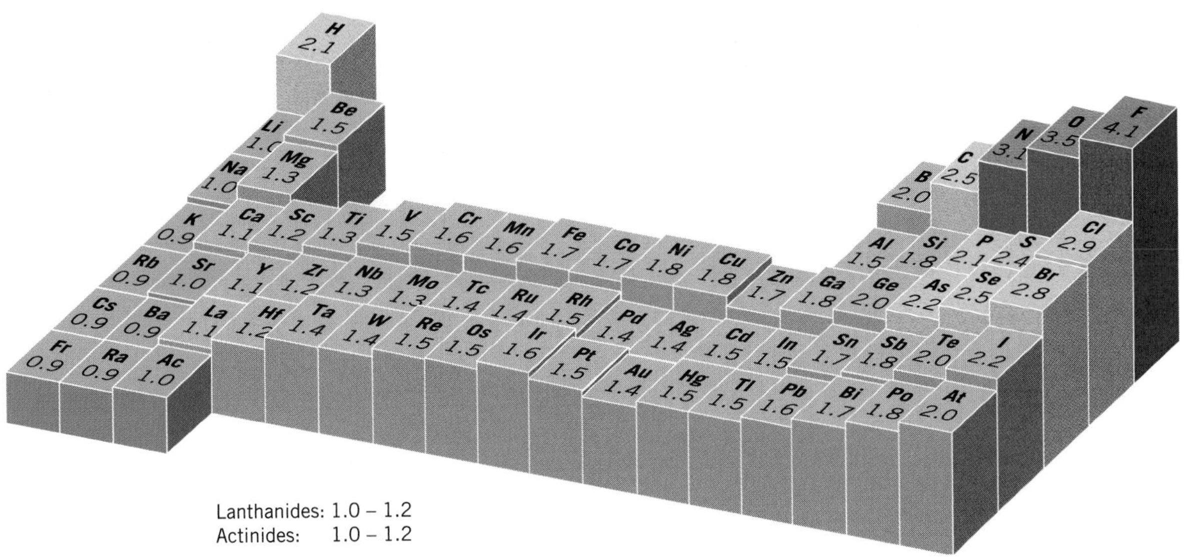

FIG. 8.6 The electronegativities of the elements.

Practice Exercise 8: Bromine and chlorine form a molecular substance with the formula BrCl. Is the bond polar? If so, which atom carries the negative charge? (Hint: Compare electronegativities.)

Practice Exercise 9: For each of the following bonds, choose the atom that carries the partial negative charge: (a) P—Br, (b) Si—Cl, and (c) S—Cl.

By studying electronegativity values and their differences we find that there is no sharp dividing line between ionic and covalent bonding. Ionic bonding and *nonpolar covalent bonding* simply represent the two extremes. A bond is mostly ionic when the difference in electronegativity between two atoms is very large; the more electronegative atom acquires essentially complete control of the bonding electrons. In a **nonpolar covalent bond,** there is no difference in electronegativity, so the pair of bonding electrons is shared equally.

$$Cs^+ \left[:\ddot{\underset{..}{F}}:\right]^- \qquad :\ddot{\underset{..}{F}}:\ddot{\underset{..}{F}}:$$

"bonding pair" held exclusively by fluorine — bonding pair shared equally

The degree to which the bond is polar, which we might think of as the amount of **ionic character** of the bond, varies in a continuous way with changes in the electronegativity difference (Figure 8.7). The bond becomes more than 50% ionic when the electronegativity difference exceeds approximately 1.7.

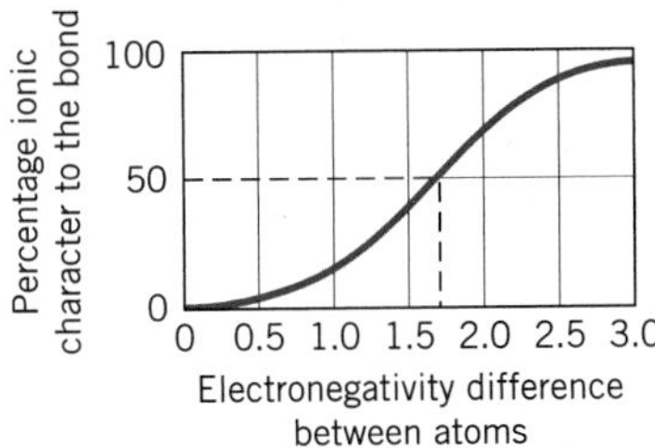

FIG. 8.7 Variation in the percentage ionic character of a bond with electronegativity difference. The bond becomes about 50% ionic when the electronegativity difference equals 1.7, which means that the atoms in the bond carry a partial charge of approximately ±0.5 units.

Electronegativity follows trends within the periodic table

An examination of Figure 8.6 reveals trends in electronegativity within the periodic table; *electronegativity increases from bottom to top in a group, and from left to right in a period.* Notice that the trends follow those for ionization energy (IE). This is because an atom that has a small IE will lose an electron more easily than an atom with a large IE, just as an atom with a small electronegativity will lose its share of an electron pair more readily than an atom with a large electronegativity.

Elements located in the same region of the table (for example, the nonmetals) have similar electronegativities, which means that if they form bonds with each other, the electronegativity differences will be small and the bonds will be more covalent than ionic. On the other hand, if elements from widely separated regions of the table combine, large electronegativity differences occur and the bonds will be predominantly ionic. This is what happens, for example, when an element from Group IA or Group IIA reacts with a nonmetal from the upper right-hand corner of the periodic table.

It is found that the electronegativity is proportional to the average of the ionization energy and the electron affinity of an element.

8.5 THE REACTIVITIES OF METALS AND NONMETALS CAN BE RELATED TO THEIR ELECTRONEGATIVITIES

In addition to enabling us to determine the polarity of covalent bonds, electronegativities can also help us understand chemical properties, particularly those that involve an atom's tendency to gain or lose electrons. Thus, there are parallels between an element's electronegativity and its **reactivity**—its tendency to undergo redox reactions. In this section we will examine some of these trends.

Reactivities of metals relate to their ease of oxidation

Reactivity refers in general to the tendency of a substance to react with something. The *reactivity of a metal* refers to its tendency specifically to undergo *oxidation.*

In nearly every compound containing a metal, the metal exists in a positive oxidation state. Therefore, for a metal, *reactivity* relates to how easily the metal is oxidized. For example, a metal like sodium, which is very easily oxidized, is said to be very reactive, whereas a metal like platinum, which is very difficult to oxidize, is said to be unreactive.

There are several ways to compare how easily metals are oxidized. In Chapter 5 we saw that by comparing the abilities of metals to displace each other from compounds we are able to establish their relative ease of oxidation. This was the basis for the activity series (Table 5.2).

As useful as the activity series is in predicting the outcome of certain redox reactions, it is difficult to remember in detail. Furthermore, often it is sufficient just to know approximately where an element stands in relation to others in a broad range of reactivity. This is where the periodic table can be especially useful to us once again, because there are trends and variations in reactivity within the periodic table that are simple to identify and remember.

Figure 8.8 illustrates how the ease of oxidation (reactivity) of metals varies in the periodic table. In general, these trends roughly follow the variations in electronegativity, with the metal being less easily oxidized as its electronegativity increases. You might expect this, because electronegativity is a measure of how strongly the atom of an element attracts electrons when combining with an atom of a different element. The more strongly the atom attracts electrons, the more difficult it is to oxidize. This relationship between reactivity and electronegativity is only approximate, however, because many other factors affect the stability of the compounds that are formed.

In Figure 8.8, we see that the metals that are most easily oxidized are found at the far left in the periodic table. These are elements with very low electronegativities. The metals in Group IA, for example, are so easily oxidized that all of them react with water to liberate hydrogen. Because of their reactivity toward moisture and oxygen, they have no useful applications that require exposure to the atmosphere, so we rarely encounter them as free metals. The same is true of the heavier metals in Group IIA, calcium through barium. These elements also react with water to liberate hydrogen. In Figure 8.6 we see that electronegativity decreases going down a group, which explains why the heavier elements in Group IIA are more reactive than those at the top of the group.

TOOLS *Trends in the reactivity of metals in the periodic table*

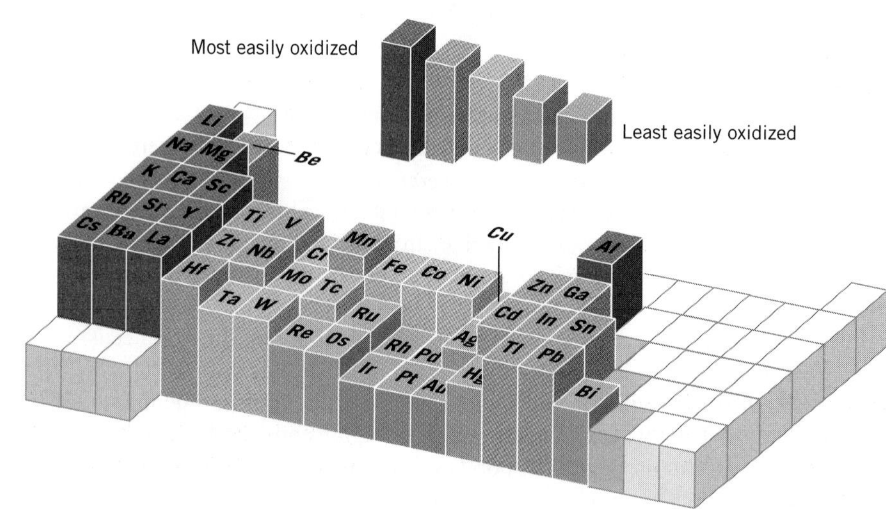

FIG. 8.8 The variation of the ease of oxidation of metals with position in the periodic table.

In Figure 8.8 we can also locate the metals that are the most difficult to oxidize. They occur for the most part among the heavier transition elements in the center of the periodic table, where we find the very unreactive elements platinum and gold—metals used to make fine jewelry. Their bright luster and lack of any tendency to corrode in air or water combine to make them particularly attractive for this purpose. This same lack of reactivity also is responsible for their industrial uses. Gold, for example, is used to coat the electrical contacts in low-voltage circuits found in microcomputers, because even small amounts of corrosion on more reactive metals would be sufficient to impede the flow of electricity so much as to make the devices unreliable.

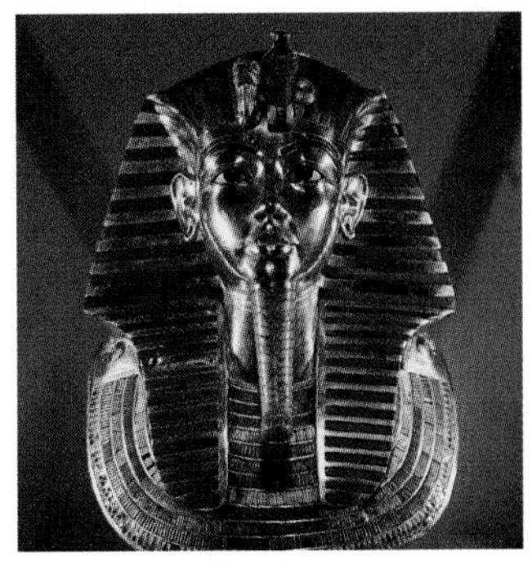

The funeral mask of Tut'ankhamun, almost 3,300 years old, shows no sign of age. It is made of gold with glass eyes and lapis lazuli eyebrows and eyelashes. *(Roger Wood/Corbis Images.)*

The oxidizing power of nonmetals is related to their electronegativities

The reactivity of a metal is determined by its ease of oxidation, and therefore its ability to serve as a reducing agent. *For a nonmetal, reactivity is usually gauged by its ability to serve as an oxidizing agent.* This ability also varies according to the element's electronegativity. Nonmetals with high electronegativities have strong tendencies to acquire electrons and are therefore strong oxidizing agents. In parallel with changes in electronegativities in the periodic table, *the oxidizing abilities of nonmetals increase from left to right across a period and from bottom to top in a group.* Thus, the most powerful oxidizing agent is fluorine, followed closely by oxygen, both in the upper right-hand corner of the periodic table.

TOOLS *Trends in the reactivity of nonmetals in the periodic table*

Single replacement reactions occur among the nonmetals, just as with the metals (which you studied in Chapter 5). For example, heating a metal sulfide in oxygen causes the sulfur to be replaced by oxygen. The displaced sulfur then combines with additional oxygen to give sulfur dioxide. The equation for a typical reaction is

$$2CuS(s) + 3O_2(g) \longrightarrow 2CuO(s) + 2SO_2(g)$$

Displacement reactions are especially evident among the halogens, where a particular halogen in its elemental form will oxidize the *anion* of any halogen below it in Group VIIA, as illustrated in the margin. Thus, F_2 will oxidize Cl^-, Br^-, and I^-. However, Cl_2 will only oxidize Br^- and I^-, and Br_2 will only oxidize I^-.

Fluorine:

$$F_2 + 2Cl^- \longrightarrow 2F^- + Cl_2$$
$$F_2 + 2Br^- \longrightarrow 2F^- + Br_2$$
$$F_2 + 2I^- \longrightarrow 2F^- + I_2$$

Chlorine:

$$Cl_2 + 2Br^- \longrightarrow 2Cl^- + Br_2$$
$$Cl_2 + 2I^- \longrightarrow 2Cl^- + I_2$$

Bromine:

$$Br_2 + 2I^- \longrightarrow 2Br^- + I_2$$

8.6 DRAWING LEWIS STRUCTURES IS A NECESSARY SKILL

Lewis structures are very useful in chemistry because they give us a relatively simple way to describe the structures of molecules. As a result, much chemical reasoning is based on them. Also, as you will learn in the next chapter, we can use the Lewis structure to make reasonably accurate predictions about the shape of a molecule.

In Section 8.3 you saw Lewis structures for a variety of molecules that obey the octet rule. Examples included CO_2, Cl_2, and N_2. The octet rule is not always obeyed, however. For instance, there are some molecules in which one or more atoms must have more than an octet in the valence shell. Examples are PCl_5 and SF_6, whose Lewis structures are

```
      :Cl:                 :F:
  :Cl—  |  —Cl:       :F.   |   .F:
       P                 \  |  /
      / \                   S
  :Cl.    .Cl:           /  |  \
                      :F.   |   .F:
                           :F:
```

Lewis structures just describe which atoms are bonded to each other and the kinds of bonds involved. Thus, the Lewis structure for water can be drawn as H—Ö—H, but it does not mean the water molecule is linear, with all the atoms in a straight line. Actually, water isn't linear; the two O—H bonds form an angle of about 104°.

In these molecules the formation of more than four bonds to the central atom requires that the central atom have a share of more than eight electrons. With the exception of Period 2 elements like carbon and nitrogen, most nonmetals can have more than an octet of electrons in the outer shell.

There are also some molecules (but not many) in which the central atom behaves as though it has less than an octet. The most common examples involve compounds of beryllium and boron.

$$\cdot \text{Be} \cdot + 2 \cdot \ddot{\underset{\cdot\cdot}{\text{Cl}}}: \longrightarrow \; :\ddot{\underset{\cdot\cdot}{\text{Cl}}}-\text{Be}-\ddot{\underset{\cdot\cdot}{\text{Cl}}}:$$

four electrons around Be

$$\cdot \dot{\text{B}} \cdot + 3 \cdot \ddot{\underset{\cdot\cdot}{\text{Cl}}}: \longrightarrow \; :\ddot{\underset{\cdot\cdot}{\text{Cl}}}-\underset{}{\overset{:\ddot{\text{Cl}}:}{\overset{|}{\text{B}}}}-\ddot{\underset{\cdot\cdot}{\text{Cl}}}:$$

six electrons around B

Although Be and B sometimes have less than an octet, the elements in Period 2 never *exceed* an octet. The reason is because their valence shells, having $n = 2$, can hold a maximum of only 8 electrons. (This explains why the octet rule works so well for atoms of carbon, nitrogen, and oxygen.) However, elements in periods below Period 2, such as phosphorus and sulfur, sometimes do exceed an octet, because their valence shells can hold more than 8 electrons. For example, the valence shell for elements in Period 3, for which $n = 3$, can hold a maximum of 18 electrons, and the valence shell for Period 4 elements, which have *s*, *p*, *d*, and *f* subshells, can hold as many as 32 electrons.

TOOLS

Method for drawing Lewis structures

A simple procedure enables us to draw Lewis structures

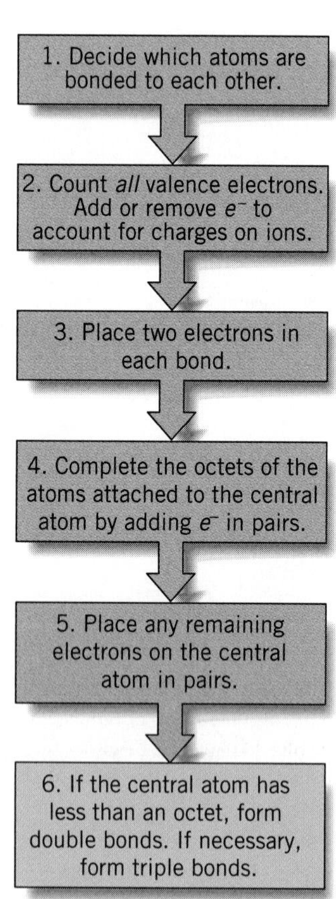

1. Decide which atoms are bonded to each other.
2. Count *all* valence electrons. Add or remove e^- to account for charges on ions.
3. Place two electrons in each bond.
4. Complete the octets of the atoms attached to the central atom by adding e^- in pairs.
5. Place any remaining electrons on the central atom in pairs.
6. If the central atom has less than an octet, form double bonds. If necessary, form triple bonds.

FIG. 8.9 Summary of steps in writing a Lewis structure. If you follow these steps, you will obtain a Lewis structure in which the octet rule is obeyed by the maximum number of atoms.

Figure 8.9 outlines a series of steps that provides a systematic method for writing Lewis structures. The first is to decide which atoms are bonded to each other, so that we know where to put the dots or dashes. This is not always a simple matter. Many times the formula suggests the way the atoms are arranged because the central atom, which is usually the least electronegative one, is usually written first. Examples are CO_2 and ClO_4^-, which have the following *skeletal structures* (i.e., arrangements of atoms):

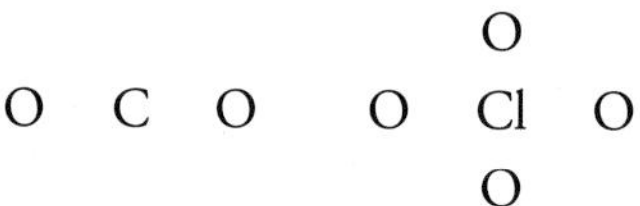

Sometimes, obtaining the skeletal structure is not quite so simple, especially when more than two elements are present. Some generalizations are possible, however. For example, the skeletal structure of nitric acid, HNO_3, is

```
            O
H    O    N    O        (correct)
```

rather than one of the following.

```
        O
    O   N   O     or     H   O   O   N   O      (incorrect)
        H
```

Nitric acid is an oxoacid (Section 4.4), and it happens that the hydrogen atoms that can be released from molecules of oxoacids are always bonded to oxygen atoms, which are in turn bonded to the third nonmetal atom. Therefore, recognizing HNO_3 as the formula of an oxoacid allows us to predict that the three oxygen atoms are bonded to the nitrogen, and the hydrogen is bonded to one of the oxygens. (It is also useful to remember that hydrogen forms only one bond, so we would not choose it to be a central atom.)

There are times when no reasonable basis can be found for choosing a particular skeletal structure. If you must make a guess, choose the most symmetrical arrangement of atoms, because it has the greatest chance of being correct.

After you've decided on the skeletal structure, the next step is to count all of the *valence electrons* to find out how many dots must appear in the final formula. Using the periodic table, locate the groups in which the elements in the formula occur to determine the number of valence electrons contributed by each atom. If the structure you wish to draw is that of an ion, *add one additional valence electron for each negative charge or remove a valence electron for each positive charge.* Some examples are

SO_3	Sulfur (Group VIA) contributes $6e^-$.	$1 \times 6 = 6e^-$
	Each oxygen (Group VIA) contributes $6e^-$.	$3 \times 6 = 18e^-$
		Total $24e^-$
ClO_4^-	Chlorine (Group VIIA) contributes $7e^-$.	$1 \times 7 = 7e^-$
	Each oxygen (Group VIA) contributes $6e^-$.	$4 \times 6 = 24e^-$
	Add $1e^-$ for the 1− charge.	$+1e^-$
		Total $32e^-$
NH_4^+	Nitrogen (Group VA) contributes $5e^-$.	$1 \times 5 = 5e^-$
	Each hydrogen (Group IA) contributes $1e^-$.	$4 \times 1 = 4e^-$
	Subtract $1e^-$ for the 1+ charge.	$-1e^-$
		Total $8e^-$

After we have determined the number of valence electrons, we place them into the skeletal structure in pairs following the steps outlined in Figure 8.9. Let's look at some examples of how we go about this.

EXAMPLE 8.6
Drawing Lewis Structures

What is the Lewis structure of the chloric acid molecule, $HClO_3$?

ANALYSIS: The first step is to select a reasonable skeletal structure. Because the substance is an oxoacid, we can expect the hydrogen to be bonded to an oxygen, which in turn is bonded to the chlorine. The other two oxygens would also be bonded to the chlorine. This gives

```
     O
H  O Cl  O
```

After this, we follow the procedure outlined in Figure 8.9.

SOLUTION: The total number of valence electrons is 26 ($1e^-$ from H, $6e^-$ from each O, and $7e^-$ from Cl). To distribute the electrons, we start by placing a pair of electrons in each bond, because we know that there must be at least one pair of electrons between each pair of atoms.

```
     O
     ..
H:O:Cl:O
```

This has used $8e^-$, so we still have $18e^-$ to go. Next, we work on the atoms surrounding the chlorine (which is the central atom in this structure). No additional electrons are needed around the H, because $2e^-$ are all that can occupy its valence shell. Therefore, we next complete the octets of the oxygens, which uses 16 more electrons.

```
      ..
     :O:
  ..  ..  ..
H:O:Cl:O:
  ..      ..
```

The valence shell of hydrogen contains only the 1*s* subshell, which can hold a maximum of two electrons. This means hydrogen can have a share of only two electrons and can form just one covalent bond.

We have now used a total of $24e^-$, so there are two electrons left. "Leftover" electrons are always placed on the central atom in pairs (the Cl atom, in this case). This gives

```
      ..
     :O:
  ..  ..  ..
H:O:Cl:O:
  ..  ..  ..
```

which we can also write as follows, using dashes for the electron pairs in the bonds.

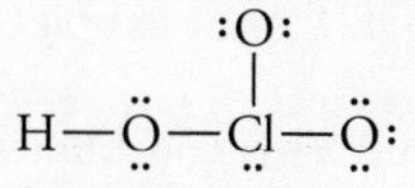

```
          ..
         :O:
          |
    ..    |    ..
  H—O—Cl—O:
    ..   ..    ..
```

The chlorine and the three oxygens have octets, and the valence shell of hydrogen is complete with $2e^-$, so we are finished.

IS THE ANSWER REASONABLE? The most common error is to have either too many or too few valence electrons in the structure, so that's always the best place to begin your check. Doing this will confirm that the number of e^- is correct.

EXAMPLE 8.7
Drawing Lewis Structures

Draw the Lewis structure for the SO_3 molecule.

ANALYSIS: Sulfur is less electronegative than oxygen and it is written first in the formula, so we expect it to be the central atom, surrounded by the three O atoms. This gives the skeletal structure

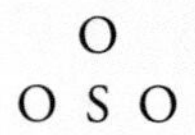

```
  O
O S O
```

From here we proceed to count valence electrons and follow the appropriate steps in entering them into the structure.

SOLUTION: The total number of electrons in the formula is 24 ($6e^-$ from the sulfur, plus $6e^-$ from each oxygen). We begin to distribute the electrons by placing a pair in each bond. This gives

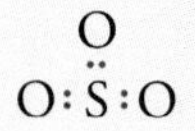

```
   O
   ..
O:S:O
```

We have used $6e^-$, so there are $18e^-$ left. We next complete the octets around the oxygens, which uses the remaining electrons.

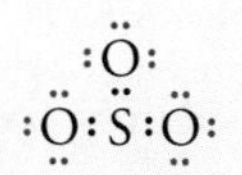

```
     ..
    :O:
 ..  ..  ..
:O:S:O:
 ..      ..
```

At this point all the electrons have been placed into the structure, but we see that the sulfur still lacks an octet. We cannot simply add more dots because the total must be 24. Therefore, according to the last step of the procedure in Figure 8.9, we have to create a multiple bond. To do this we move a pair of electrons that we have shown to belong solely to an oxygen into a sulfur–oxygen bond so that it can be counted as belonging to both the oxygen *and* the sulfur. In other words, we place a double bond between sulfur and one of the oxygens. It doesn't matter which oxygen we choose for this honor.

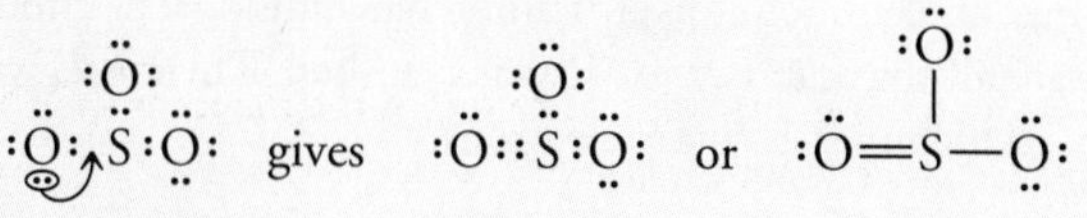

```
    ..              ..                 ..
   :O:             :O:                :O:
 ..  ..  ..     ..  ..  ..             |      ..
:O:↷S:O:  gives  :O::S:O:   or   :O=S—O:
  ..     ..                ..      ..     ..
```

Notice that each atom has an octet.

IS THE ANSWER REASONABLE? The key step in completing the structure is recognizing what we have to do to obtain an octet around the sulfur. We have to add more electrons to the valence shell of sulfur, but without removing them from any of the oxygen atoms. By forming the double bond, we accomplish this. A quick check also confirms that we've placed exactly the correct number of valence electrons into the structure.

EXAMPLE 8.8
Drawing Lewis Structures

What is the Lewis structure for the ion IF_4^-?

ANALYSIS: We can anticipate that iodine will be the central atom, so our skeletal structure is

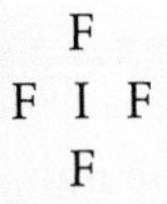

```
  F
F I F
  F
```

Next, we count valence electrons, remembering to add an extra electron to account for the negative charge. Then we distribute the electrons in pairs following the usual procedure.

SOLUTION: The iodine and fluorine atoms are in Group VIIA and each contribute 7 electrons, for a total of $35e^-$. The negative charge requires one additional electron to give a total of $36e^-$.

First we place $2e^-$ into each bond, and then we complete the octets of the fluorine atoms. This uses 32 electrons.

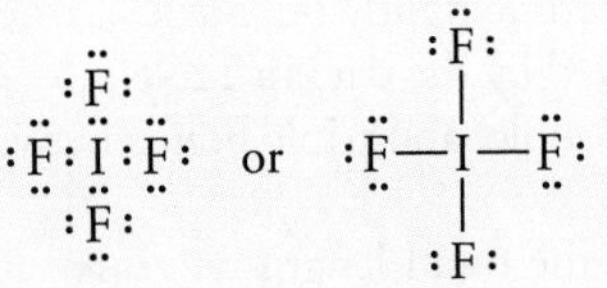

```
    ..                  ..
   :F:                 :F:
 .. .. ..               |
:F:I:F:     or    :F̤̈—I—F̤̈:
 .. .. ..               |
   :F:                 :F:
    ..                  ..
```

There are four electrons left, and according to step 5 in Figure 8.9 they are placed on the central atom as *pairs* of electrons. This gives

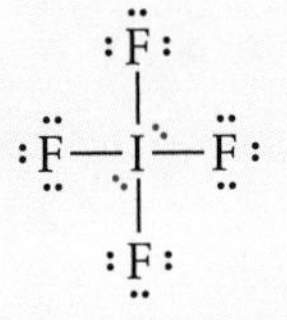

```
      ..
     :F:
      |
  :F̤̈—Ï̤—F̤̈:
      |
     :F:
      ..
```

The last step is to add brackets around the formula and write the charge outside as a superscript.

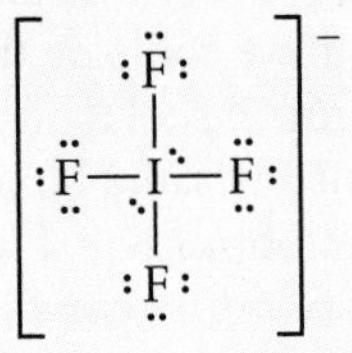

```
 ┌          ┐ −
 │    ..    │
 │   :F:    │
 │    |     │
 │:F̤̈—Ï̤—F̤̈: │
 │    |     │
 │   :F:    │
 │    ..    │
 └          ┘
```

IS THE ANSWER REASONABLE? We can recount the valence electrons, which tells us we have the right number of them, and all are in the Lewis structure. Each fluorine atom has an octet, which is proper. Notice that we have placed the "leftover" electrons onto the central atom. This gives iodine more than an octet, but that's okay because iodine is not a Period 2 element.

Practice Exercise 10: Predict a reasonable skeletal structure for $H_2PO_4^-$ and determine the number of valence electrons that should be in its Lewis structure. (Hint: It's an ion derived from an oxoacid.)

Practice Exercise 11: Predict reasonable skeletal structures for SO_2, NO_3^-, $HBrO_3$, and H_3AsO_4.

Practice Exercise 12: How many valence electrons should appear in the Lewis structures of SO_2, SeO_4^{2-}, and NO^+?

Practice Exercise 13: Draw Lewis structures for OF_2, NH_4^+, SO_2, NO_3^-, ClF_3, and $HClO_4$.

Formal charges help select correct Lewis structures

Lewis structures are meant to describe how atoms share electrons in chemical bonds. Such descriptions are theoretical explanations or predictions that relate to the forces that hold molecules and polyatomic ions together. But, as you learned in Chapter 1, a theory is only as good as the observations on which it is based, so to have confidence in a theory about chemical bonding, we need to have a way to check it. We need experimental observations that relate to the description of bonding.

Two properties that are related to the number of electron pairs shared between two atoms are *bond length,* the distance between the nuclei of the bonded atoms, and *bond energy,* the energy required to separate the bonded atoms to give neutral particles. For example, we mentioned in Section 8.3 that measurements have shown the H_2 molecule has a bond length of 75 pm and a bond energy of 435 kJ/mol, which means that it takes 435 kJ to break the bonds of 1 mol of H_2 molecules to give 2 mol of hydrogen atoms.

A single bond has a bond order of 1, a double bond a bond order of 2, and a triple bond a bond order of 3.

Comparing bonds between the same elements, the bond length and bond energy depend on the **bond order,** which is defined as *the number of pairs of electrons shared between two atoms.* The bond order is a measure of the amount of electron density in the bond, and the greater the electron density, the more tightly the nuclei are held and the more closely they are drawn together. This is illustrated by the data in Table 8.3, which gives typical bond lengths and bond energies for single, double, and triple bonds between carbon atoms. In summary:

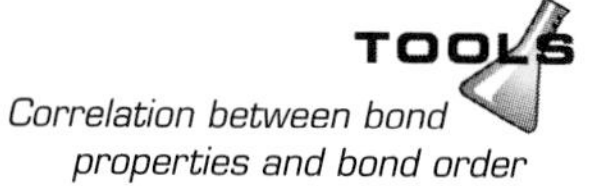

Correlation between bond properties and bond order

> As the bond order increases, the bond length decreases and the bond energy increases, provided we are comparing bonds between the same elements.

With this as background, let's examine the Lewis structure of sulfuric acid, drawn according to the procedure given in Figure 8.9.

```
          :Ö:
           |
  H—Ö—S—Ö—H        (Structure I)
           |
          :Ö:
```

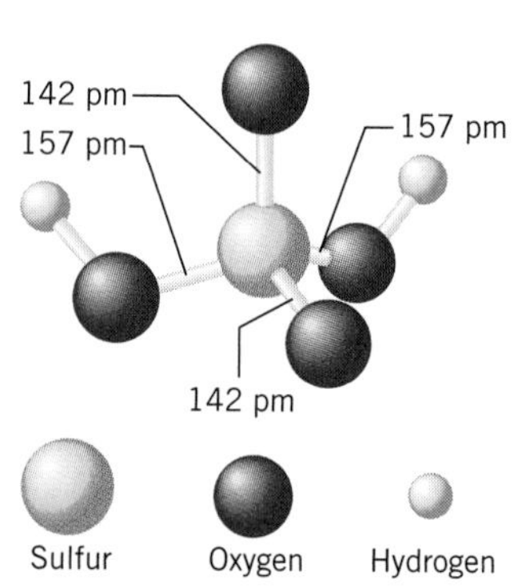

FIG. 8.10 The structure of sulfuric acid in the vapor state. Notice the difference in the sulfur–oxygen bond lengths.

It obeys the octet rule, and there doesn't seem to be any need to attempt to write any other structures for it. But a problem arises if we compare the predicted bond lengths with those found experimentally. In our Lewis structure, all four sulfur-oxygen bonds are single bonds, which means they should have about the same bond lengths. However, experimentally it has been found that the bonds are not of equal length, as illustrated in Figure 8.10. The S—O bonds are shorter than the S—OH bonds, which means they must have a larger bond order. Therefore, we need to modify our Lewis structure to make it conform to reality.

Because sulfur is in Period 3, its valence shell has 3*s*, 3*p*, and 3*d* subshells, which together can accommodate more than eight electrons. Therefore, sulfur is able to form more than four bonds, so we are allowed to increase the bond order in the S—O bonds by moving electron pairs to create sulfur–oxygen double bonds as shown below.

```
          :Ö:                       :Ö
           |                         ‖
  H—Ö—S—Ö—H    gives    H—Ö—S—Ö—H        (Structure II)
           |                         ‖
          :Ö:                       :Ö
```

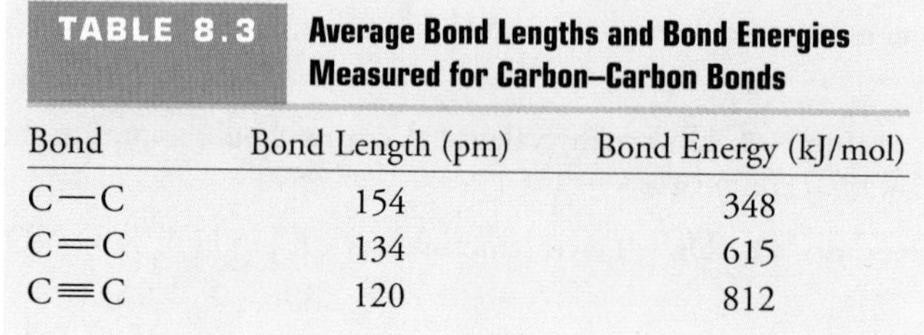

TABLE 8.3 Average Bond Lengths and Bond Energies Measured for Carbon–Carbon Bonds

Bond	Bond Length (pm)	Bond Energy (kJ/mol)
C—C	154	348
C═C	134	615
C≡C	120	812

Now we have a Lewis structure that better fits experimental observations because the sulfur–oxygen double bonds are expected to be shorter than the sulfur–oxygen single bonds. Because this second Lewis structure agrees better with the actual structure of the molecule, it is the *preferred* Lewis structure, even though it violates the octet rule.

Formal charges are apparent charges on atoms

Are there any criteria that we could have applied that would have allowed us to predict that the second Lewis structure for H_2SO_4 is better than the one with only single bonds, even though it seems to violate the octet rule unnecessarily? To answer this question, let's take a closer look at the two Lewis structures we've drawn.

In Structure I, there are only single bonds between the sulfur and oxygen atoms. If the electrons in the bonds are shared equally by S and O, then each atom "owns" half of the electron pair, or the equivalent of one electron. In other words, the four single bonds place the equivalent of four electrons in the valence shell of the sulfur. An isolated single atom of sulfur, however, has six valence electrons, so in Structure I the sulfur has two electrons *less* than it does as just an isolated atom. Thus, at least in a bookkeeping sense, it would appear that if sulfur obeyed the octet rule in H_2SO_4, it would have a charge of 2+. This *apparent* charge on the sulfur atom is called its **formal charge.**

Notice that in defining formal charge, we've stressed the word "apparent." *The formal charge arises because of the bookkeeping we've done and should not be confused with whatever the actual charge is on an atom in the molecule.* (The situation is somewhat similar to the oxidation numbers you learned to assign in Chapter 5, which are artificial charges assigned according to a set of rules.) Here's how formal charges are assigned.

The actual charges on the atoms in a molecule are determined by the relative electronegativities of the atoms.

Calculating the formal charge on an atom

Step 1. Write down the number of valence electrons in an isolated atom of the element.

Step 2. Using the Lewis structure, add up the valence electrons that "belong to" the atom in the molecule or ion, and then subtract this total from the value in Step 1.

In performing the calculation in Step 2, electrons in bonds are divided equally between the two atoms, while unshared electrons are assigned exclusively to the atom on which they reside. For example, for Structure I above, we have

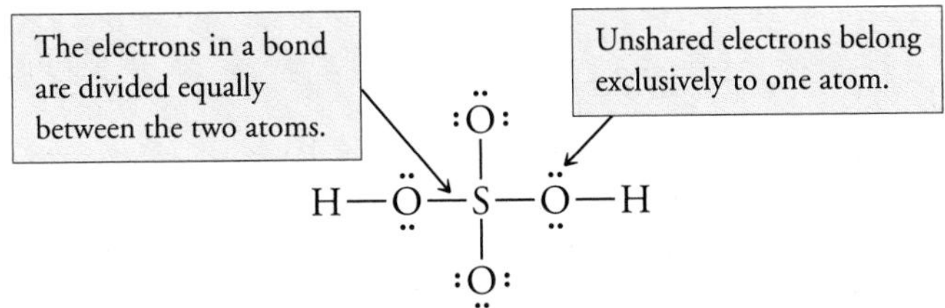

Therefore, the calculation of formal charge is summarized by the following equation:

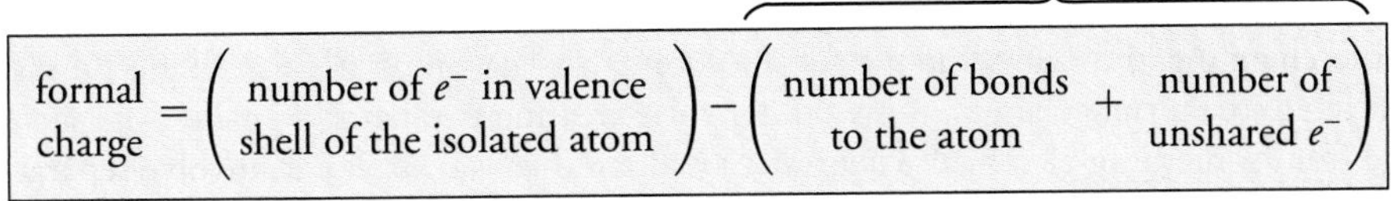

$$\text{formal charge} = \left(\begin{array}{c}\text{number of } e^- \text{ in valence}\\ \text{shell of the isolated atom}\end{array}\right) - \left(\begin{array}{c}\text{number of bonds}\\ \text{to the atom}\end{array} + \begin{array}{c}\text{number of}\\ \text{unshared } e^-\end{array}\right) \quad (8.2)$$

For example, for the sulfur in Structure I, we get

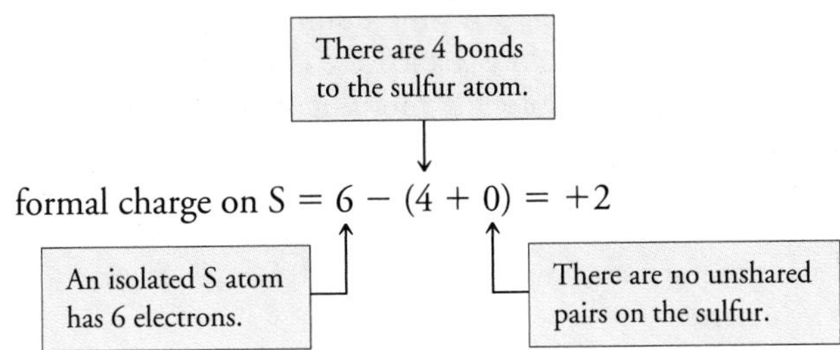

Let's also calculate the formal charges on the hydrogen and oxygen atoms in Structure I. An isolated H atom has one electron. In Structure I each H has one bond and no unshared electrons. Therefore,

$$\text{Formal charge on H} = 1 - (1 + 0) = 0$$

In Structure I, we also see that there are two kinds of oxygens to consider. An isolated oxygen atom has six electrons, so we have, for the oxygens also bonded to hydrogen,

$$\text{Formal charge} = 6 - (2 + 4) = 0$$

and for the oxygens not bonded to hydrogen,

$$\text{Formal charge} = 6 - (1 + 6) = -1$$

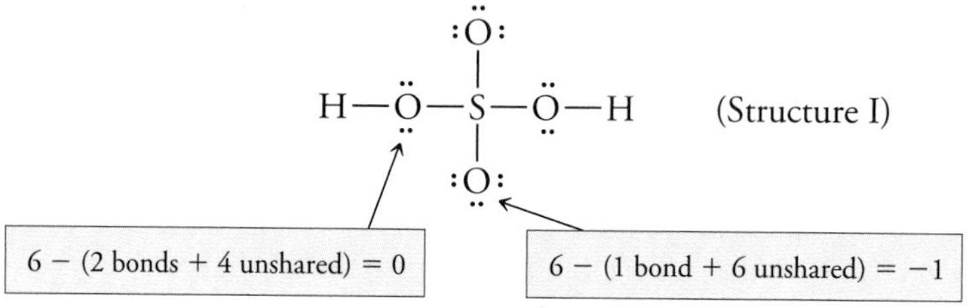

```
            ..
           :O:
            |
      ..    |    ..
   H—O—S—O—H        (Structure I)
      ..    |    ..
            |
           :O:
            ..
```

6 − (2 bonds + 4 unshared) = 0

6 − (1 bond + 6 unshared) = −1

Nonzero formal charges are indicated in a Lewis structure by placing them in circles alongside the atoms, as shown below.

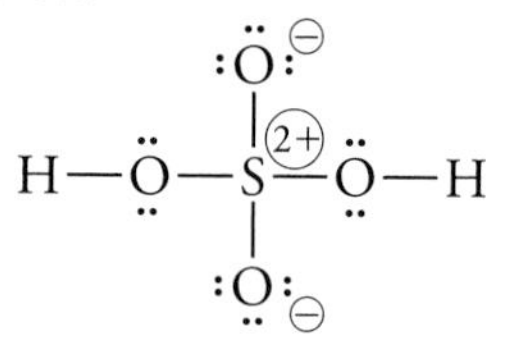

```
            ..
           :O:⊖
            |
      ..    |(2+) ..
   H—O—S—O—H
      ..    |    ..
            |
           :O:
            .. ⊖
```

Notice that the sum of the formal charges in the molecule is zero. It is useful to remember that, in general, *the formal charges in any Lewis structure add up to the charge on the particle.*

Now let's look at the formal charges in Structure II. For sulfur we have

$$\text{Formal charge on S} = 6 - (6 + 0) = 0$$

so the sulfur has no formal charge. The hydrogens and the oxygens that are also bonded to H are the same in this structure as before, so they have no formal charges. And finally, the oxygens that are not bonded to hydrogen have

$$\text{Formal charge} = 6 - (2 + 4) = 0$$

These oxygens also have no formal charges.

Now let's compare the two structures side by side.

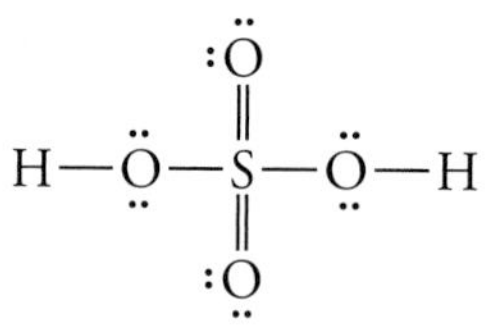

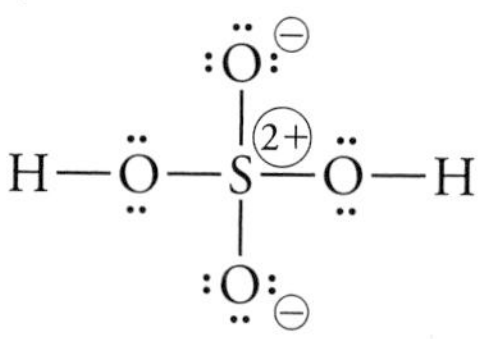

```
            ..                       ..
           :O                       :O:⊖
            ||                       |
      ..    ||   ..            ..    |(2+) ..
   H—O—S—O—H              H—O—S—O—H
      ..    ||   ..            ..    |    ..
            ||                       |
           :O                       :O:
            ..                       .. ⊖
```

Imagine changing the one with the double bonds to the one with the single bonds. According to the formal charges, this would involve creating two pairs of positive–negative charge from something electrically neutral. Stated another way, it would involve separating negative charges from positive charges, and this would require an increase in the potential energy. Our conclusion is that the singly bonded structure on the right has a higher potential energy than the one with the double bonds. In general, the lower the potential energy of a molecule, the more stable it is. Therefore, the lower energy structure with the double bonds is, in principle, the more stable structure, so it is preferred over the one with only single bonds. This now gives us a rule that we can use in selecting the best Lewis structures for a molecule or ion:

TOOLS

Selecting the best Lewis structure

> When several Lewis structures are possible, the one with formal charges closest to zero is the most stable and is preferred.

EXAMPLE 8.9
Selecting Lewis Structures Based on Formal Charges

A student drew three Lewis structures for the nitric acid molecule:

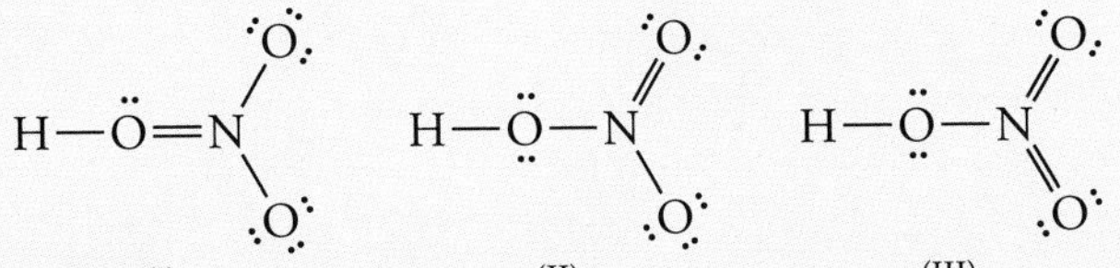

(I) (II) (III)

Which one is preferred?

ANALYSIS: When we have to select among several Lewis structures to find the best one, the tool is the procedure for assigning formal charges. As a rule, the structure with the fewest formal charges will be the best structure. We have to be careful, however, that we don't select a structure in which an atom is assigned more electrons than its valence shell can actually hold. Such a structure must be eliminated from consideration.

SOLUTION: Except for hydrogen, all the atoms in the molecule are from Period 2, and therefore can have a maximum of eight electrons in their valence shells. (Period 2 elements *never* exceed an octet because their valence shells have only *s* and *p* subshells and can accommodate a maximum of eight electrons.) Scanning the structures, we see that I and II show octets around both N and O. However, the nitrogen in Structure III has 5 bonds to it, which require 10 electrons. Therefore, this structure is not acceptable and can be eliminated immediately. Our choice is then between Structures I and II. Let's calculate formal charges on the atoms in each of them.

In Structure III, the formal charges are zero on each of the atoms, but this cannot be the "preferred structure" because the nitrogen atom has too many electrons in its valence shell.

Structure I

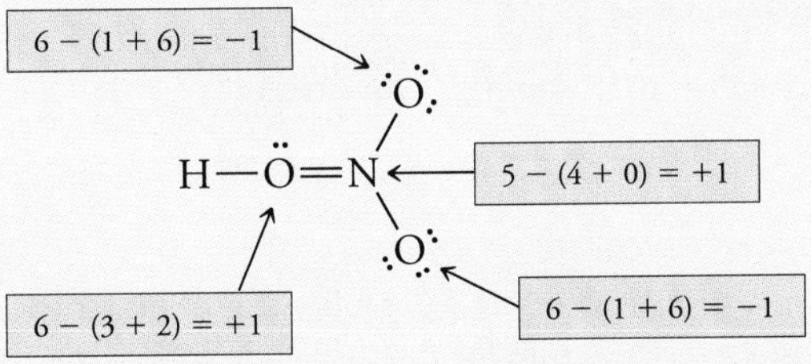

$6 - (1 + 6) = -1$

$5 - (4 + 0) = +1$

$6 - (1 + 6) = -1$

$6 - (3 + 2) = +1$

Structure II

$6 - (2 + 4) = 0$

$5 - (4 + 0) = +1$

$6 - (1 + 6) = -1$

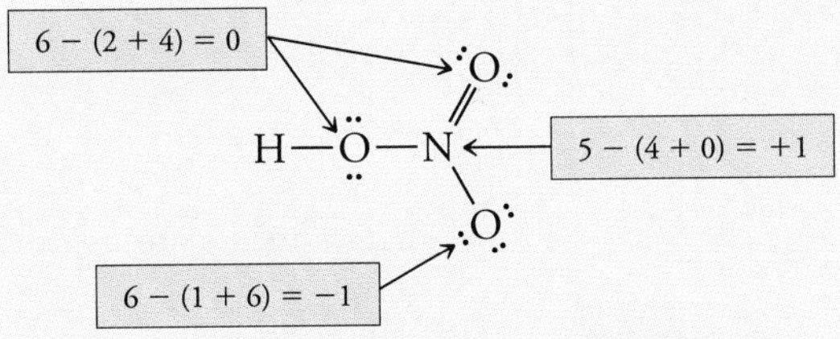

Next, we place the formal charges on the atoms in the structures.

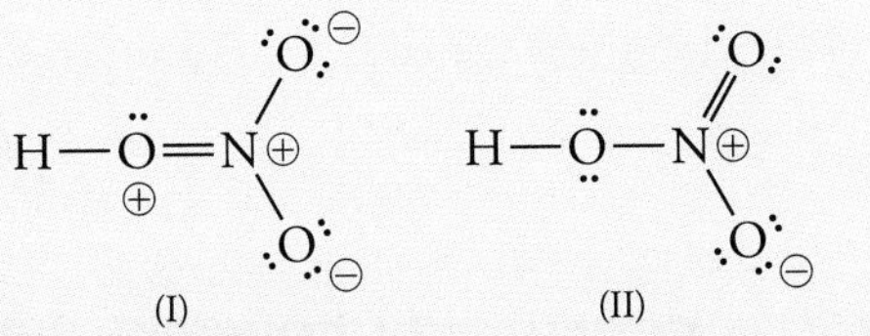

(I) (II)

Because Structure II has fewer formal charges than Structure I, it is the lower-energy, preferred Lewis structure for HNO_3.

IS THE ANSWER REASONABLE? One simple check we can do is to add up the formal charges in each structure. The sum must equal the net charge on the particle, which is zero for HNO_3. Adding formal charges gives zero for each structure, so we can be confident we've assigned them correctly. This gives us confidence in our answer, too.

EXAMPLE 8.10

Selecting Lewis Structures Based on Formal Charges

Two structures can be drawn for BCl_3, as shown below

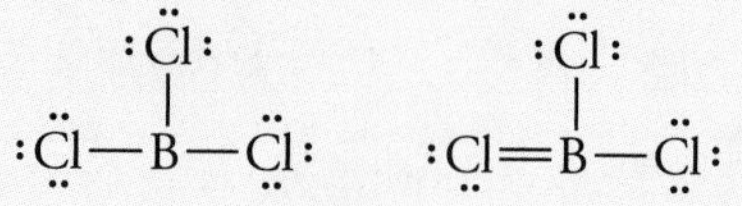

```
        :C̈l:                  :C̈l:
         |                     |
   :C̈l—B—C̈l:            :Cl=B—C̈l:
                           ¨       ¨
        (I)                   (II)
```

Why is the one that violates the octet rule preferred?

ANALYSIS: We're asked to select between Lewis structures, which tells us that we have to consider formal charges. We'll assign them and then see if we can answer the question.

SOLUTION: Assigning formal charges gives

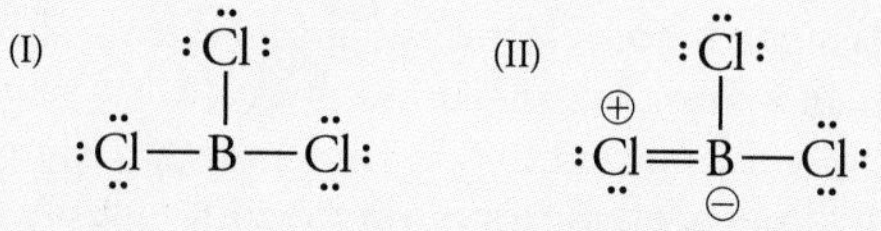

```
(I)     :C̈l:           (II)      :C̈l:
         |                ⊕       |
   :C̈l—B—C̈l:            :Cl=B—C̈l:
                                 ⊖
```

In Structure I all the formal charges are zero. In Structure II, two of the atoms have formal charges, so this alone would argue in favor of Structure I. There is another argument in favor as well. Notice that the formal charges in Structure II place the positive charge on the more electronegative chlorine atom and the negative charge on the less electronegative boron. If charges could form in the molecule, they certainly would not be expected to form in this way. Therefore, there are two factors that make the structure with the double bond unfavorable, so we usually write the Lewis structure for BCl_3 as shown in Structure I.

IS THE ANSWER REASONABLE? We've assigned the formal charges correctly, and our reasoning seems sound, so we appear to have answered the question adequately.

Practice Exercise 14: A student drew the following Lewis structure for the sulfite ion, SO_3^{2-}. Is this the best Lewis structure for the ion? (Hint: Negative formal charges should be on the more electronegative atoms.)

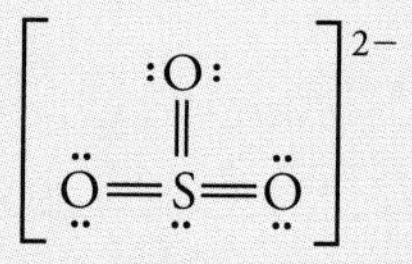

```
 [     :O:     ] 2−
 [     ‖       ]
 [  Ö=S=Ö      ]
```

Practice Exercise 15: Assign formal charges to the atoms in the following Lewis structures.

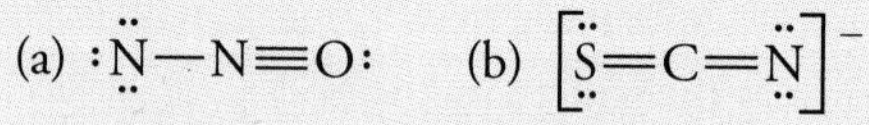

(a) $:\ddot{N}-N\equiv O:$ (b) $[\ddot{S}=C=\ddot{N}]^-$

Practice Exercise 16: Select the preferred Lewis structure for (a) SO_2, (b) $HClO_3$, and (c) H_3PO_4.

Both electrons in a coordinate covalent bond come from the same atom

Often we use Lewis structures to follow the course of chemical reactions. For example, we can diagram the reaction of hydrogen ion combining with a water molecule to form the hydronium ion, which occurs in aqueous solutions of acids.

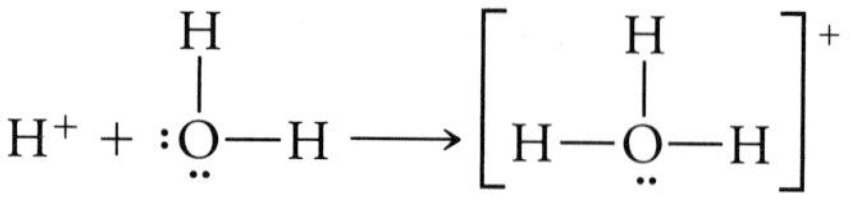

```
          H              [    H    ]+
          |              [    |    ]
H⁺ + :Ö—H  ⟶           [ H—Ö—H  ]
```

The formation of the bond between H^+ and H_2O follows a different path than the covalent bonds we discussed earlier in this chapter. For instance, when two H atoms combine to form H_2, each atom brings one electron to the bond.

$$\text{H}\cdot + \cdot\text{H} \longrightarrow \text{H—H}$$

But in the formation of H_3O^+, both of the electrons that become shared between the H^+ and the O originate on the oxygen atom of the water molecule. *This type of bond, in which both electrons of the shared pair come from just one of the two atoms, is called a* **coordinate covalent bond.**

Although we can make a distinction about the origin of the electrons shared in the bond, once the bond is formed a coordinate covalent bond is really the same as any other covalent bond. In other words, we can't tell where the electrons in the bond came from *after* the bond has been formed. In the H_3O^+ ion, for example, all three O—H bonds are identical once they've been formed.

The concept of a coordinate covalent bond is helpful in explaining what happens to atoms in a chemical reaction. For example, when ammonia is mixed with boron trichloride, an exothermic reaction takes place and the compound NH_3BCl_3 is formed in which there is a boron–nitrogen bond. Using Lewis structures, we can diagram this reaction as follows.

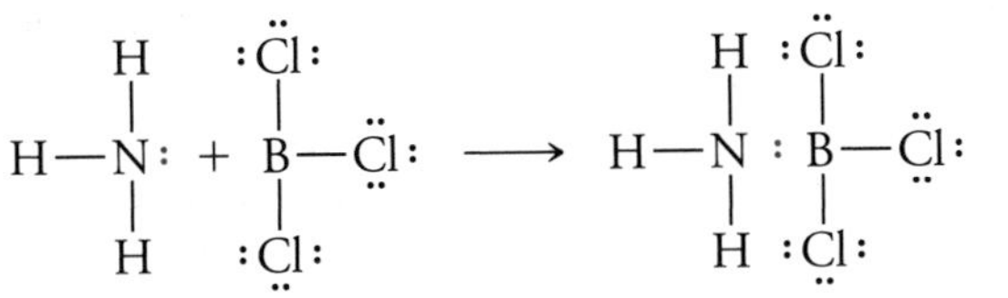

$$\text{H}_3\text{N}: + \text{BCl}_3 \longrightarrow \text{H}_3\text{N}:\text{BCl}_3$$

All electrons are alike, of course. We are using different colors for them so we can see where the electrons in the bond came from.

In the reaction, we might say that "the boron forms a coordinate covalent bond with the nitrogen of the ammonia molecule."

An arrow sometimes is used to represent the donated pair of electrons in a coordinate covalent bond. The direction of the arrow indicates the direction in which the electron pair is donated, in this case from the nitrogen to the boron.

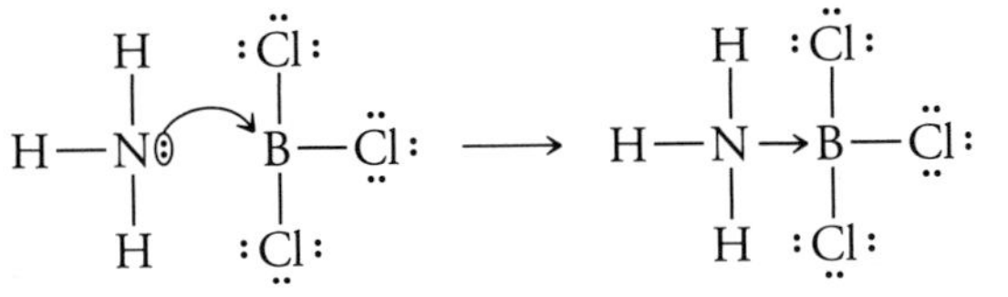

$$\text{H}_3\text{N}: + \text{BCl}_3 \longrightarrow \text{H}_3\text{N}\rightarrow\text{BCl}_3$$

Compounds like BCl_3NH_3, which are formed by simply joining two smaller molecules, are sometimes called **addition compounds.**

Practice Exercise 17: Use Lewis structures to show how the formation of NH_4^+ from NH_3 and H^+ involves formation of a coordinate covalent bond. How does this bond differ from the other NH bonds in NH_4^+? (Hint: You need to keep in mind the definition of a coordinate covalent bond.)

Practice Exercise 18: Use Lewis structures to explain how the reaction between hydroxide ion and hydrogen ion involves the formation of a coordinate covalent bond.

8.7 Resonance Applies When a Single Lewis Structure Fails

There are some molecules and ions for which we cannot write Lewis structures that agree with experimental measurements of bond length and bond energy. An example is the formate ion, CHO_2^-. Following the usual steps, we would write its Lewis structure as

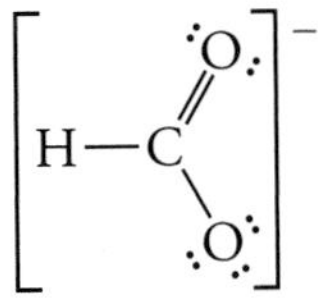

$$\left[\text{H—C}\begin{matrix}\text{//O}\\ \text{\\O}\end{matrix}\right]^-$$

Formate ion is formed by neutralizing formic acid, an organic acid with the structure

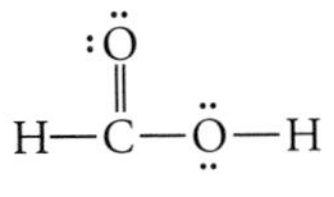

$$\text{H—C(=O)—O—H}$$

The name formic acid comes from *formica*, the Latin word for ant. The one shown here is a fire ant. Formic acid is the substance that causes the stinging sensation in bites from this creature. *(J. H. Robinson/Photo Researchers.)*

This structure suggests that one carbon–oxygen bond should be longer than the other, but experiment shows that they are identical with lengths that are about halfway between the expected values for a single bond and a double bond. The Lewis structure doesn't match the experimental evidence, and there's no way to write one that does. It would require showing all of the electrons in pairs and, at the same time, showing 1.5 pairs of electrons in each carbon–oxygen bond.

The way we get around problems like this is through the use of a concept called **resonance.** We view the actual structure of the molecule or ion, which we cannot draw satisfactorily, as a composite, or average, of a number of Lewis structures that we can draw. For example, for formate we write

No atoms have been moved; the electrons have just been redistributed.

where we have simply shifted electrons around in going from one structure to the other. The bond between the carbon and a particular oxygen is depicted as a single bond in one structure and as a double bond in the other. The average of these is 1.5 bonds (halfway between a single and a double bond), which is in agreement with the experimental bond lengths. These two Lewis structures are called **resonance structures** or **contributing structures.** The actual structure of the ion, which we can't draw, is called a **resonance hybrid** of these two resonance structures. The double-ended arrow is used to show that we are drawing resonance structures and implies that the true hybrid structure is a composite of the two resonance structures.[5]

TOOLS

Determining resonance structures

Equivalent choices for double bond locations lead to resonance structures

There is a simple way to determine when resonance should be applied to Lewis structures. If you find that you must move electrons to create one or more double bonds while following the procedures developed in the previous section, the number of resonance structures is equal to the number of equivalent choices for the locations of the double bonds. For example, in drawing the Lewis structure for the NO_3^- ion, we reach the stage

A double bond must be created to give the nitrogen an octet. Since it can be placed in any one of three locations, there are three resonance structures for this ion.

The three oxygens in NO_3^- are said to be equivalent; that is, they are all alike in their chemical environment. Each oxygen is bonded to a nitrogen atom that's attached to two other oxygen atoms.

Notice that each structure is the same, except for the location of the double bond.

[5] The term *resonance* is often misleading to the beginning student. The word itself suggests that the actual structure flip-flops back and forth between the two structures shown. This is *not* the case! A mule, which is the *hybrid* offspring of a donkey and a horse, isn't a donkey one minute and a horse the next! Although it may have characteristics of both parents, a mule is a mule. A *resonance hybrid* also has characteristics of its "parents," but it never has the exact structure of any of them.

In the nitrate ion, the extra bond that moves around from one structure to another is divided among all three bond locations. Therefore, the average bond order in the N—O bonds is expected to be $1\frac{1}{3}$, or 1.33.[6]

EXAMPLE 8.11
Drawing Resonance Structures

Use formal charges to show that resonance applies to the preferred Lewis structure for the sulfite ion, SO_3^{2-}. Draw the resonance structures and determine the average bond order of the S—O bonds.

ANALYSIS: Following our usual procedure we obtain the Lewis structure

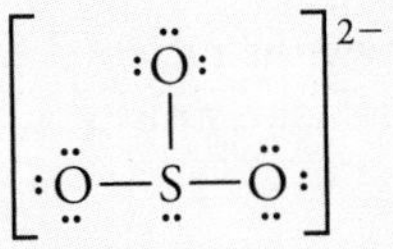

All of the valence electrons have been placed into the structure and we have octets around all of the atoms, so it doesn't seem that we need the concept of resonance. However, the question refers to the "preferred" structure, which suggests that we are going to have to assign formal charges and determine what the preferred structure is. Then we can decide whether the concept of resonance will apply.

SOLUTION: When we assign formal charges, we get

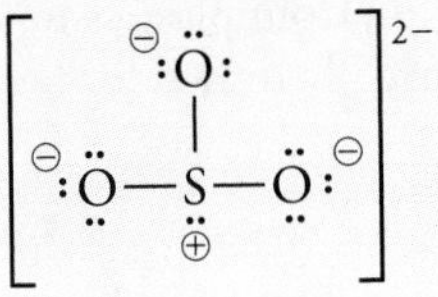

We can obtain a better Lewis structure if we can reduce the number of formal charges. This can be accomplished by moving an unshared pair from one of the oxygens into an S—O bond, thereby forming a double bond. Let's do this using the oxygen at the left.

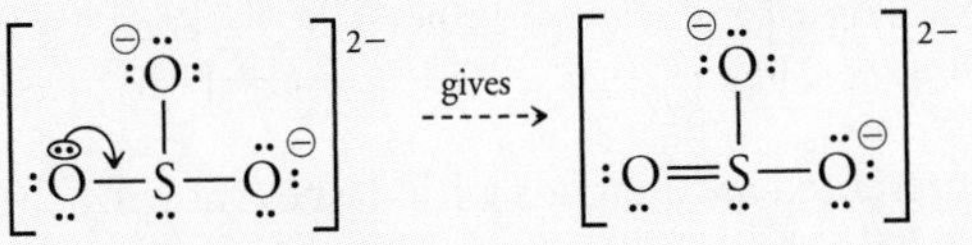

However, we could have done this with any of the three S—O bonds, so there are three equivalent choices for the location of the double bond. Therefore, there are three resonance structures.

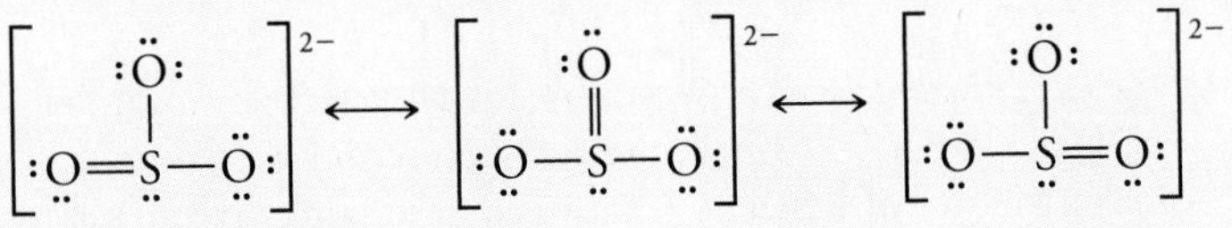

As with the nitrate ion, we expect an average bond order of 1.33.

IS THE ANSWER REASONABLE? If we can answer "yes" to the following questions, the problem is solved correctly: Have we counted valence electrons correctly? Have we properly placed the electrons into the skeletal structure? Do the formal charges we've calculated add up to the charge on the SO_3^{2-} ion? Have we correctly determined the number of equivalent positions for the double bond? Have we computed the average bond order correctly?

[6] For resonance structures, the average bond order can be calculated by adding up the total number of bonds and dividing by the number of equivalent positions. In the NO_3^- ion, we have a total of four bonds (two single bonds and a double bond) distributed over three equivalent positions, so the bond order is 4/3 = 1 1/3.

Practice Exercise 19: The phosphate ion has the following Lewis structure, where we've used formal charges to obtain the best structure.

```
[       :O:       ]3−
         ‖
   :Ö—P—Ö:
         |
        :O:
        ¨        ]
```

How many resonance structures are there for this ion? (Hint: How many equivalent positions are there for the double bond?)

Practice Exercise 20: Draw the resonance structures for HCO_3^-.

Practice Exercise 21: Determine the preferred Lewis structure for the bromate ion, BrO_3^-, and, if appropriate, draw resonance structures.

Resonance is used to explain the stability of some molecules and ions

One of the benefits that a molecule or ion derives from existing as a resonance hybrid is that its total energy is lower than that of any one of its resonance structures. A particularly important example of this occurs with the compound benzene, C_6H_6. This is a flat, hexagonal ring-shaped molecule (Figure 8.11) with a basic structure that appears in many important organic molecules, ranging from plastics to amino acids.

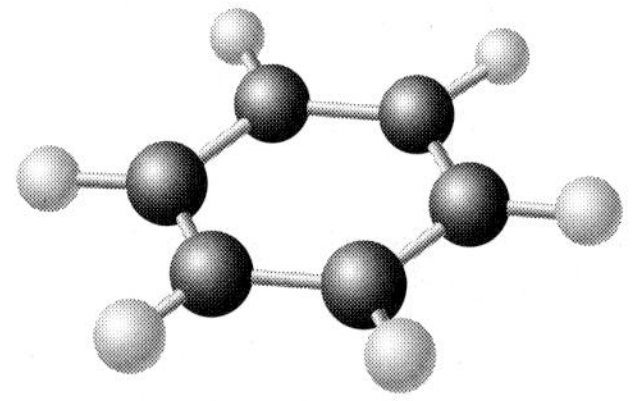

FIG. 8.11 Benzene. The molecule has a planar hexagonal structure.

Two resonance structures are usually drawn for benzene.

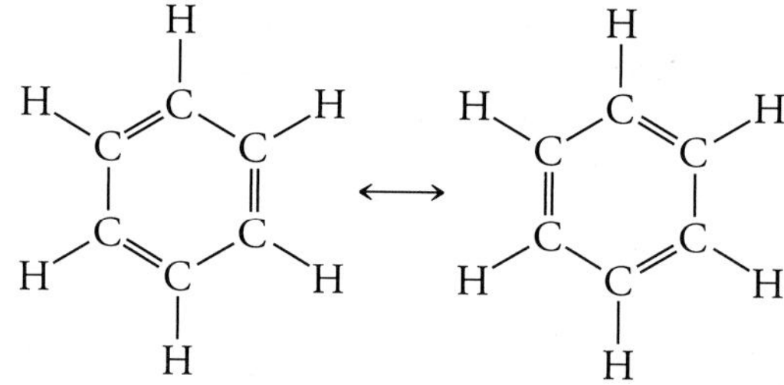

These are generally represented as hexagons with dashes showing the locations of the double bonds. It is assumed that at each apex of the hexagon there is a carbon bonded to a hydrogen as well as to the adjacent carbon atoms.

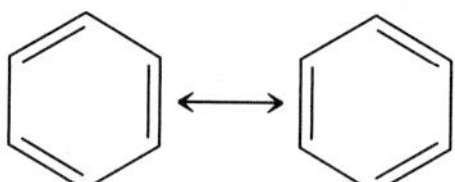

Polystyrene plastic, a common polymer, contains benzene rings joined to alternating carbon atoms in a long hydrocarbon chain. When a gas is blown into melted polystyrene, the lightweight product called Styrofoam is formed. Styrofoam is used to make a wide variety of packaging products for consumer goods. *(Courtesy Polystyrene Packaging Council.)*

Usually, the actual structure of benzene (that of its resonance hybrid) is represented as a hexagon with a circle in the center. This is intended to show that the electron density of the three extra bonds is evenly distributed around the ring.

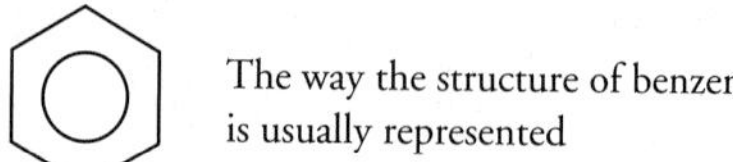

The way the structure of benzene is usually represented

Although the individual resonance structures for benzene show double bonds, the molecule does not react like other organic molecules that have true carbon–carbon double bonds. The reason appears to be that the resonance hybrid is considerably more stable than either of the resonance forms. In fact, it has been calculated that the actual structure of the benzene molecule is more stable than either of its resonance structures by approximately 146 kJ/mol. This extra stability achieved through resonance is called the **resonance energy.**

SUMMARY

Ionic Bonding. In ionic compounds, the forces of attraction between positive and negative ions are called **ionic bonds.** The formation of ionic compounds by electron transfer is favored when atoms of low ionization energy react with atoms of high electron affinity. The chief stabilizing influence in the formation of ionic compounds is the release of the **lattice energy,** which is the energy required to completely separate the ions of an ionic compound. When atoms of the elements in Groups IA and IIA as well as the nonmetals form ions, they usually gain or lose enough electrons to achieve a noble gas electron configuration. Transition elements lose their outer *s* electrons first, followed by loss of *d* electrons from the shell below the outer shell. Post-transition metals lose electrons from their outer *p* subshell first, followed by electrons from the outer *s* subshell.

Covalent Bonding. Electron sharing between atoms occurs when electron transfer is energetically too "expensive." Shared electrons attract the positive nuclei, and this leads to a lowering of the potential energy of the atoms as a covalent bond forms. Electrons generally become paired when they are shared. An atom tends to share enough electrons to complete its valence shell. Except for hydrogen, the valence shell usually holds eight electrons, which forms the basis of the octet rule. The **octet rule** states that atoms of the representative elements tend to acquire eight electrons in their outer shells when they form bonds. **Single, double,** and **triple bonds** involve the sharing of one, two, and three pairs of electrons, respectively, between two atoms. Boron and beryllium often have less than an octet in their compounds. Atoms of the elements of Period 2 cannot have more than an octet because their outer shells can hold only eight electrons. Elements in Periods 3, 4, 5, and 6 can exceed an octet if they form more than four bonds.

Bond energy (the energy needed to separate the bonded atoms) and **bond length** (the distance between the nuclei of the atoms connected by the bond) are two experimentally measurable quantities that can be related to the number of pairs of electrons in the bond. For bonds between atoms of the same elements, bond energy increases and bond length decreases as the **bond order** increases.

Electronegativity and Polar Bonds. The attraction an atom has for the electrons in a bond is called the atom's **electronegativity.** When atoms of different electronegativities form a bond, the electrons are shared unequally and the bond is **polar,** with **partial positive** and **partial negative** charges at opposite ends. This causes the bond to be an electric **dipole.** In a **polar molecule,** such as HCl, the product of the charge at either end multiplied by the distance between the charges gives the **dipole moment,** μ. When the two atoms have the same electronegativity, the bond is **nonpolar.** The extent of polarity of the bond depends on the electronegativity difference between the two bonded atoms. When the electronegativity difference is very large, ionic bonding results. A bond is approximately 50% ionic when the electronegativity difference is 1.7. In the periodic table, electronegativity increases from left to right across a period and from bottom to top in a group.

Reactivity and Electronegativity. The **reactivity** of metals is related to the ease with which they are oxidized (lose electrons); for nonmetals it is related to the ease with which they are reduced (gain electrons). Metals with low electronegativities lose electrons easily, are good reducing agents, and tend to be very reactive. The most reactive metals are located in Groups IA and IIA, and their ease of oxidation increases going down the group. For nonmetals, the higher the electronegativity, the stronger is their ability to serve as oxidizing agents. The strongest oxidizing agent is fluorine. Among the halogens, oxidizing strength decreases from fluorine to iodine.

Lewis Symbols and Lewis Structures. Lewis symbols are a bookkeeping device used to keep track of valence electrons in ionic and covalent bonds. The **Lewis symbol** of an element consists of the element's chemical symbol surrounded by a number of dots equal to the number of valence electrons. In the **Lewis structure** for an ionic compound, the Lewis symbol for the anion is enclosed in brackets (with the charge written outside) to show that all the electrons belong entirely to the ion. The Lewis structure for a molecule or polyatomic ion uses pairs of dots between chemical symbols to represent shared pairs of electrons. The electron pairs in covalent bonds usually are represented by dashes; one dash equals two electrons. The following procedure is used to draw the Lewis structure: (1) decide on the skeletal structure (remember that the least electronegative atom is usually the central atom and is usually first in the formula); (2) count all the valence electrons, taking into account the charge, if any; (3) place a pair of electrons in each bond; (4) complete the octets of atoms other than the central atom (but remember that hydrogen can only have two electrons); (5) place any leftover electrons on the central atom in pairs; (6) if the central atom still has *less than* an octet, move electron pairs to make double or triple bonds.

Formal Charges. The **formal charge** assigned to an atom in a Lewis structure (which usually differs from the actual charge on the atom) is calculated as the difference between the number of valence electrons of an isolated atom of the element and the number of electrons that "belong" to the atom because of its bonds to other atoms and its unshared valence electrons. The sum of the formal charges always equals the net charge on the molecule or ion. The most stable (lowest energy) Lewis structure for a molecule or ion is the one with formal charges closest to zero. This is usually the preferred Lewis structure for the particle.

Coordinate Covalent Bonding. For bookkeeping purposes, we sometimes single out a covalent bond whose electron pair originated from one of the two bonded atoms. An arrow is sometimes used to indicate the donated pair of electrons. Once formed, a coordinate covalent bond is no different from any other covalent bond.

Resonance. Two or more atoms in a molecule or polyatomic ion are *chemically equivalent* if they are attached to the same kinds of atoms or groups of atoms. Bonds to chemically equivalent atoms must be the same; they must have the same bond length and the same bond energy, which means they must involve the sharing of the same number of electron pairs. Sometimes the Lewis structures we draw suggest that the bonds to chemically equivalent atoms are not the same. Typically, this occurs when it is necessary to form multiple bonds during the drawing of a Lewis structure. When alternatives exist for the location of a multiple bond among two or more equivalent atoms, then each possible Lewis structure is actually a **resonance structure** or **contributing structure,** and we draw them all. In drawing resonance structures, the relative locations of the nuclei must be identical in all. Remember that none of the resonance structures corresponds to a real molecule, but their composite—the **resonance hybrid**—does approximate the actual structure of the molecule or ion.

TOOLS FOR PROBLEM SOLVING

In this chapter you learned to apply the following concepts as tools in solving problems. Study each one carefully so that you know what each is used for. When faced with solving a problem, recall what each tool does and consider whether it will be helpful in finding a solution. This will aid you in selecting the tools you need.

Electron configurations of ions of representative elements *(page 303)* Metals in Groups IA and IIA and the nonmetals obey the octet rule when they form ions. Use this knowledge to derive the electron configurations of ions of these elements.

Order in which electrons are lost from an atom *(page 303)* Electrons are lost first from the shell with largest n. For a given shell, electrons are lost from subshells in the following order: f before d before p before s. Use this knowledge to obtain electron configurations of ions of the transition and post-transition metals.

Lewis symbols *(page 306)* Lewis symbols are a bookkeeping device that we use to keep track of valence electrons in atoms and ions. For a neutral atom of the representative elements, the Lewis symbol consists of the atomic symbol surrounded by dots equal in number to the group number.

Octet rule and covalent bonding *(page 310)* The octet rule helps us construct Lewis structures for covalently bonded molecules. Elements in Period 2 never exceed an octet in their valence shells.

Dipole moment *(page 313)* Dipole moments are a measure of the polarity of molecules, so they can be used to compare molecular polarity. The dipole moment (μ) of a diatomic molecule is calculated as the charge on an end of the molecule, q, multiplied by the bond length, r. Dipole moments are expressed in debye units.

$$\mu = q \times r$$

Periodic trends in electronegativity *(page 315)* The trends revealed in Figure 8.6 allow us to use the locations of elements in the periodic table to estimate the degree of polarity of bonds and to estimate which of two atoms in a bond is the most electronegative.

Trends in the reactivity of metals in the periodic table *(page 316)* A knowledge of where the most reactive and least reactive metals are located in the periodic table gives a qualitative feel for how reactive a metal is by locating it in the periodic table.

Trends in the reactivity of nonmetals in the periodic table *(page 317)* The periodic table correlates the position of a nonmetal with its strength as an oxidizing agent. Oxidizing ability increases from left to right across a period and from bottom to top in a group.

Method for drawing Lewis structures *(page 318)* The method described in Figure 8.9 yields Lewis structures in which the maximum number of atoms obey the octet rule.

Correlation between bond properties and bond order *(page 322)* The correlations allow us to compare experimental covalent bond properties (bond energy and bond length) with those predicted by theory.

Formal charges *(page 323)* Use formal charges to select the best Lewis structure for a molecule or polyatomic ion. The best structure is usually the one with the fewest formal charges. Assign formal charges to atoms as follows.

$$\begin{matrix}\text{Formal} \\ \text{charge}\end{matrix} = \begin{pmatrix}\text{number of } e^- \text{ in valence} \\ \text{shell of the isolated atom}\end{pmatrix} - \begin{pmatrix}\text{number of bonds} \\ \text{to the atom}\end{pmatrix} + \begin{matrix}\text{number of} \\ \text{unshared } e^-\end{matrix}\Bigg)$$

Selecting the best Lewis structure *(page 324)* The structure having the smallest formal charges is preferred. Be sure none of the atoms in the structure appears to have more electrons than permitted by its location in the periodic table.

Method for determining resonance structures *(page 328)* By distributing multiple bonds over equivalent atoms in a molecule we obtain a better description of the bonding in the molecule. We also know that when resonance structures can be drawn, a molecule or ion will be more stable than any of the individual resonance structures. Expect resonance structures when there is more than one option for assigning the location of a double bond.

QUESTIONS, PROBLEMS, AND EXERCISES

Answers to problems whose numbers are printed in color are given in Appendix B. More challenging problems are marked with asterisks. ILW = Interactive Learningware solution is available at www.wiley.com/college/brady. OH = an Office Hours video is available for this problem.

REVIEW QUESTIONS

Ionic Bonding

8.1 What must be true about the change in the total potential energy of a collection of atoms for a stable compound to be formed from the elements?

8.2 What is an *ionic bond?*

8.3 How is the tendency to form ionic bonds related to the IE and EA of the atoms involved?

8.4 Define the term *lattice energy*. In what ways does the lattice energy contribute to the stability of ionic compounds?

8.5 Magnesium forms the ion Mg^{2+}, but not the ion Mg^{3+}. Why?

8.6 Why doesn't chlorine form the ion Cl^{2-}?

8.7 Why do many of the transition elements in Period 4 form ions with a 2+ charge?

8.8 If we were to compare the first, second, third, and fourth ionization energies of aluminum, between which pair of successive ionization energies would there be the largest difference? (Refer to the periodic table in answering this question.)

8.9 In each of the following pairs of compounds, which would have the larger lattice energy? (a) CaO or Al_2O_3, (b) BeO or SrO, (c) NaCl or NaBr.

Lewis Symbols

8.10 The Lewis symbol for an atom only accounts for electrons in the valence shell of the atom. Why are we not concerned with the other electrons?

8.11 Which of these Lewis symbols is incorrect?

(a) :Ȯ: (b) ·C̈l· (c) :N̈e: (d) :Sb:

Electron Sharing

8.12 In terms of the potential energy change, why doesn't ionic bonding occur when two nonmetals react with each other?

8.13 Describe what happens to the electron density around two hydrogen atoms as they come together to form an H_2 molecule.

8.14 What happens to the energy of two hydrogen atoms as they approach each other? What happens to the spins of the electrons?

8.15 Is the formation of a covalent bond endothermic or exothermic?

8.16 What factors control the bond length in a covalent bond?

Covalent Bonding and the Octet Rule

8.17 What is the *octet rule?* What is responsible for it?

8.18 How many covalent bonds are normally formed by (a) hydrogen, (b) carbon, (c) oxygen, (d) nitrogen, and (e) chlorine?

8.19 Why do Period 2 elements never form more than four covalent bonds? Why are Period 3 elements able to exceed an octet?

8.20 Define (a) *single bond,* (b) *double bond,* and (c) *triple bond.*

8.21 The Lewis structure for hydrogen cyanide is H—C≡N:. Draw circles enclosing electrons to show that carbon and nitrogen obey the octet rule.

8.22 Why doesn't hydrogen obey the octet rule? How many covalent bonds does a hydrogen atom form?

8.23 Use Lewis structures to show the ionization of the following organic acid (a weak acid) in water. If necessary, refer to Figure 4.11 on page 142.

$$CH_3-CH_2-\overset{\overset{:\ddot{O}:}{\|}}{C}-\ddot{\underset{..}{O}}H$$

8.24 The compound below is called an amine. It is a weak base and undergoes ionization in water following a path similar to that of ammonia. Use Lewis structures to diagram the reaction of this amine with water. If necessary, refer to Figure 4.12 on page 143.

$$CH_3-\underset{..}{\overset{\overset{H}{|}}{N}}-CH_2-CH_3$$

8.25 Use Lewis structures to diagram the reaction between the acid in Question 8.23 and the base in Question 8.24. If necessary, refer to Figure 4.17 on page 151.

Polar Bonds and Electronegativity

8.26 What is a polar covalent bond?

8.27 Define *dipole moment* in the form of an equation. What is the value of the *debye* (with appropriate units)?

8.28 Define *electronegativity.* On what basis did Pauling develop his scale of electronegativities?

8.29 Which element has the highest electronegativity? Which is the second most electronegative element?

8.30 Which elements are assigned electronegativities of zero? Why?

8.31 Among the following bonds, which are more ionic than covalent?

(a) Si—O, (b) Ba—O, (c) Se—Cl, (d) K—Br

8.32 If an element has a low electronegativity, is it likely to be a metal or a nonmetal? Explain your answer.

Electronegativity and the Reactivities of the Elements

8.33 When we say that aluminum is more *reactive* than iron, which kind of reaction of these elements are we describing?

8.34 In what groups in the periodic table are the most reactive metals found? Where do we find the least reactive metals?

8.35 How is the electronegativity of a metal related to its reactivity?

8.36 Arrange the following metals in their approximate order of reactivity (most reactive first, least reactive last) based on their locations in the periodic table: (a) iridium, (b) silver, (c) calcium, and (d) iron.

8.37 Complete and balance equations for the following. If no reaction occurs, write "N.R."
(a) $KCl + Br_2 \longrightarrow$
(b) $NaI + Cl_2 \longrightarrow$
(c) $KCl + F_2 \longrightarrow$
(d) $CaBr_2 + Cl_2 \longrightarrow$
(e) $AlBr_3 + F_2 \longrightarrow$
(f) $ZnBr_2 + I_2 \longrightarrow$

8.38 In each pair, choose the better oxidizing agent.
(a) O_2 or F_2
(b) As_4 or P_4
(c) Br_2 or I_2
(d) P_4 or S_8
(e) Se_8 or Cl_2
(f) As_4 or S_8

Failure of the Octet Rule

8.39 How many electrons are in the valence shells of (a) Be in $BeCl_2$, (b) B in BCl_3, and (c) H in H_2O?

8.40 What is the minimum number of electrons that would be expected to be in the valence shell of As in $AsCl_5$?

8.41 Nitrogen and arsenic are in the same group in the periodic table. Arsenic forms both $AsCl_3$ and $AsCl_5$, but with chlorine, nitrogen only forms NCl_3. On the basis of the electronic structures of N and As, explain why this is so.

Bond Length and Bond Energy

8.42 Define *bond length* and *bond energy*.

8.43 Define *bond order*. How are bond energy and bond length related to bond order? Why are there these relationships?

8.44 The energy required to break the H—Cl bond to give H^+ and Cl^- ions would not be called the H—Cl bond energy. Why?

Formal Charge

8.45 What is the definition of *formal charge*?

8.46 How are formal charges used to select the best Lewis structure for a molecule? What is the basis for this method of selection?

8.47 What are the formal charges on the atoms in the HCl molecule? What are the actual charges on the atoms in this molecule? (Hint: See Section 8.4.) Are formal charges the same as actual charges?

Resonance

8.48 Why is the concept of resonance needed?

8.49 What is a *resonance hybrid*? How does it differ from the resonance structures drawn for a molecule?

8.50 Draw the resonance structures of the benzene molecule. Why is benzene more stable than one would expect if the ring contained three carbon–carbon double bonds?

8.51 Polystyrene plastic consists of a long chain of carbon atoms in which every other carbon is attached to a benzene ring. The ring is attached by replacing a hydrogen of benzene with a bond to the carbon chain. In the chain, carbons not attached to other carbons are bonded to hydrogen atoms. Sketch a portion of a polystyrene molecule that contains five benzene rings.

Coordinate Covalent Bonds

8.52 What is a *coordinate covalent bond*?

8.53 Once formed, how (if at all) does a coordinate covalent bond differ from an ordinary covalent bond?

8.54 BCl_3 has an incomplete valence shell. Use Lewis structures to show how it could form a coordinate covalent bond with a water molecule.

REVIEW PROBLEMS

Electron Configurations of Ions

8.55 Explain what happens to the electron configurations of Mg and Br when they react to form magnesium bromide.

8.56 Describe what happens to the electron configurations of lithium and nitrogen when they react to form lithium nitride.

ILW OH **8.57** What are the electron configurations of the Pb^{2+} and Pb^{4+} ions?

8.58 What are the electron configurations of the Bi^{3+} and Bi^{5+} ions?

8.59 Write the abbreviated electron configuration of the Mn^{3+} ion. How many unpaired electrons does the ion contain?

8.60 Write the abbreviated electron configuration of the Co^{3+} ion. How many unpaired electrons does the ion contain?

Lewis Symbols

8.61 Write Lewis symbols for the following atoms: (a) Si, (b) Sb, (c) Ba, (d) Al, and (e) S.

8.62 Write Lewis symbols for the following atoms: (a) K, (b) Ge, (c) As, (d) Br, and (e) Se.

8.63 Use Lewis symbols to diagram the reactions between (a) Ca and Br, (b) Al and O, and (c) K and S.

OH **8.64** Use Lewis symbols to diagram the reactions between (a) Mg and S, (b) Mg and Cl, and (c) Mg and N.

Dipole Moments:

8.65 Use the data in Table 8.2 (page 313) to calculate the amount of charge on the oxygen and nitrogen in the nitrogen monoxide molecule, expressed in electronic charge units ($e = 1.60 \times 10^{-19}$ C). Which atom carries the positive charge?

8.66 The molecule bromine monofluoride has a dipole moment of 1.42 D and a bond length of 176 pm. Calculate the charge on the ends of the molecule, expressed in electronic charge units ($e = 1.60 \times 10^{-19}$ C). Which atom carries the positive charge?

8.67 The dipole moment of HF is 1.83 D and the bond length is 91.7 pm. Calculate the amount of charge (in electronic charge units) on the hydrogen and the fluorine atoms in the HF molecule.

OH **8.68** In the vapor state, cesium and fluoride ions pair to give CsF formula units that have a bond length of 0.255 nm and a dipole moment of 7.88 D. What is the actual charge on the cesium and fluorine atoms in CsF? What percentage of full 1+ and 1− charges is this?

Bond Energy

8.69 How many grams of water could have its temperature raised from 25 °C (room temperature) to 100 °C (the boiling point of water) by the amount of energy released in the formation of 1 mol of H_2 from hydrogen atoms? The bond energy of H_2 is 435 kJ/mol.

8.70 How much energy, in joules, is required to break the bond in *one* chlorine molecule? The bond energy of Cl_2 is 242.6 kJ/mol.

8.71 The reason there is danger in exposure to high-energy radiation (e.g., ultraviolet and X rays) is that the radiation can rupture chemical bonds. In some cases, cancer can be caused by it. A carbon–carbon single bond has a bond energy of approximately 348 kJ per mole. What wavelength of light is required to provide sufficient energy to break the C—C bond? In which region of the electromagnetic spectrum is this wavelength located?

8.72 A mixture of H_2 and Cl_2 is stable, but a bright flash of light passing through it can cause the mixture to explode. The light causes Cl_2 molecules to split into Cl atoms, which are highly reactive. What wavelength of light is necessary to cause the Cl_2 molecules to split? The bond energy of Cl_2 is 242.6 kJ per mole.

Covalent Bonds and the Octet Rule

8.73 Use Lewis structures to diagram the formation of (a) Br_2, (b) H_2O, and (c) NH_3 from neutral atoms.

OH **8.74** Chlorine tends to form only one covalent bond because it needs just one electron to complete its octet. What are the Lewis structures for the simplest compound formed by chlorine with (a) nitrogen, (b) carbon, (c) sulfur, and (d) bromine?

8.75 Use the octet rule to predict the formula of the simplest compound formed from hydrogen and (a) selenium, (b) arsenic, and (c) silicon. (Remember, however, that the valence shell of hydrogen can hold only two electrons.)

8.76 What would be the formula for the simplest compound formed from (a) phosphorus and chlorine, (b) carbon and fluorine, and (c) iodine and chlorine?

Electronegativity

OH **8.77** Use Figure 8.6 to choose the atom in each of the following bonds that carries the partial positive charge: (a) N—S, (b) Si—I, (c) N—Br, and (d) C—Cl.

8.78 Use Figure 8.6 to choose the atom that carries the partial negative charge in each of the following bonds: (a) Hg—I, (b) P—I, (c) Si—F, and (d) Mg—N.

8.79 Which of the bonds in Problem 8.77 is the most polar?

8.80 Which of the bonds in Problem 8.78 is the least polar?

Drawing Lewis Structures

8.81 What are the expected skeletal structures for (a) $SiCl_4$, (b) PF_3, (c) PH_3, and (d) SCl_2?

8.82 What are the expected skeletal structures for (a) HIO_3, (b) H_2CO_3, (c) HCO_3^-, and (d) PCl_4^+?

8.83 How many dots must appear in the Lewis structures of (a) $SiCl_4$, (b) PF_3, (c) PH_3, and (d) SCl_2?

8.84 How many dots must appear in the Lewis structures of (a) HIO_3, (b) H_2CO_3, (c) HCO_3^-, and (d) PCl_4^+?

ILW **8.85** Draw Lewis structures for (a) $AsCl_4^+$, (b) ClO_2^-, (c) HNO_2, and (d) XeF_2.

8.86 Draw Lewis structures for (a) TeF_4, (b) ClF_5, (c) PF_6^-, and (d) XeF_4.

8.87 Draw Lewis structures for (a) $SiCl_4$, (b) PF_3, (c) PH_3, and (d) SCl_2.

8.88 Draw Lewis structures for (a) HIO_3, (b) H_2CO_3, (c) HCO_3^-, and (d) PCl_4^+.

8.89 Draw Lewis structures for (a) carbon disulfide and (b) cyanide ion.

8.90 Draw Lewis structures for (a) selenium trioxide and (b) selenium dioxide.

8.91 Draw Lewis structures for (a) AsH_3, (b) $HClO_2$, (c) H_2SeO_3, and (d) H_3AsO_4.

OH **8.92** Draw Lewis structures for (a) NO^+, (b) NO_2^-, (c) $SbCl_6^-$, and (d) IO_3^-.

8.93 Draw the Lewis structure for (a) CH_2O (the central atom is carbon, which is attached to two hydrogens and an oxygen) and (b) $SOCl_2$ (the central atom is sulfur, which is attached to an oxygen and two chlorines).

8.94 Draw Lewis structures for (a) $GeCl_4$, (b) CO_3^{2-}, (c) PO_4^{3-}, and (d) O_2^{2-}.

Formal Charge

8.95 Assign formal charges to each atom in the following structures:

(a) $\mathrm{H{-}\ddot{\underset{..}{O}}{-}\ddot{\underset{..}{Cl}}{-}\ddot{\underset{..}{O}}\!:}$

(b) $\mathrm{\ddot{\underset{..}{O}}{=}\overset{:\ddot{O}:\,|}{S}{-}\ddot{\underset{..}{O}}\!:}$

(c) $\mathrm{\ddot{\underset{..}{O}}{=}\ddot{S}{-}\ddot{\underset{..}{O}}\!:}$

8.96 Assign formal charges to each atom in the following structures:

(a) $\mathrm{:\ddot{\underset{..}{Cl}}{-}\overset{:O:\,\|}{N}{-}\ddot{\underset{..}{O}}\!:}$

(b) $\mathrm{:\ddot{\underset{..}{F}}{-}\underset{|\,:\underset{..}{F}:}{\overset{:\ddot{F}:\,|}{N}}{-}\ddot{\underset{..}{O}}\!:}$

(c) $\mathrm{:\ddot{\underset{..}{F}}{-}\underset{|\,:\underset{..}{O}:}{\overset{:\ddot{O}:\,|}{S}}{-}\ddot{\underset{..}{F}}\!:}$

ILW **8.97** Draw the Lewis structure for $HClO_4$ according to the procedure described in Figure 8.9. Assign formal charges to each atom in the formula. Determine the preferred Lewis structure for this compound.

8.98 Draw the Lewis structure for SO_2Cl (sulfur bonded to two Cl and one O). Assign formal charges to each atom. Determine the preferred Lewis structure for this molecule.

8.99 Below are two structures for $BeCl_2$. Give two reasons why the one on the left is the preferred structure.

$\mathrm{:\ddot{\underset{..}{Cl}}{-}Be{-}\ddot{\underset{..}{Cl}}\!:}$ $\qquad$ $\mathrm{:\ddot{\underset{..}{Cl}}{-}Be{=}\ddot{Cl}\!:}$

8.100 The following are two Lewis structures that can be drawn for phosgene, a substance that has been used as a war gas.

```
   :O:              O:
    |               ||
:Cl—C=Cl:       :Cl—C—Cl:
```

Which is the better Lewis structure? Why?

Resonance

8.101 Draw the resonance structures for CO_3^{2-}. Calculate the average C—O bond order.

ILW **8.102** Draw all the resonance structures for the N_2O_4 molecule and determine the average N—O bond order. The skeletal structure of the molecule is

```
O   O
 N N
O   O
```

OH **8.103** How should the N—O bond lengths compare in the NO_3^- and NO_2^- ions?

8.104 Arrange the following in order of increasing C—O bond length: CO, CO_3^{2-}, CO_2, and HCO_2^- (formate ion, page 327).

8.105 The Lewis structure of CO_2 was given as

```
O=C=O
```

but two other resonance structures can also be drawn for it. What are they? On the basis of formal charges, why are they not preferred structures?

8.106 Use formal charges to establish the preferred Lewis structures for the ClO_3^- and ClO_4^- ions. Draw resonance structures for both ions and determine the average Cl—O bond order in each. Which of these ions would be expected to have the shorter Cl—O bond length?

Coordinate Covalent Bonds

8.107 Use Lewis structures to show that the hydronium ion, H_3O^+, can be considered to be formed by the creation of a coordinate covalent bond between H_2O and H^+.

8.108 Use Lewis structures to show that the reaction

$$BF_3 + F^- \longrightarrow BF_4^-$$

involves the formation of a coordinate covalent bond.

ADDITIONAL EXERCISES

8.109 Use data from the tables of ionization energies and electron affinities on pages 287 and 291 to calculate the energy changes for the following reactions.

$$Na(g) + Cl(g) \longrightarrow Na^+(g) + Cl^-(g)$$
$$Na(g) + 2Cl(g) \longrightarrow Na^{2+}(g) + 2Cl^-(g)$$

Approximately how many times larger would the lattice energy of $NaCl_2$ have to be compared to the lattice energy of NaCl for $NaCl_2$ to be more stable than NaCl?

8.110 Changing 1 mol of Mg(*s*) and 1/2 mol of $O_2(g)$ to gaseous atoms requires a total of approximately 150 kJ of energy. The first and second ionization energies of magnesium are 737 and 1450 kJ/mol, respectively; the first and second electron affinities of oxygen are −141 and +844 kJ/mol, respectively, and the standard heat of formation of MgO(*s*) is −602 kJ/mol. Construct an enthalpy diagram similar to the one in Facets of Chemistry 8.1 and use it to calculate the lattice energy of magnesium oxide. How does the lattice energy of MgO compare with that of NaCl? What might account for the difference?

***8.111** Use an enthalpy diagram to calculate the lattice energy of $CaCl_2$ from the following information. Energy needed to vaporize one mole of Ca(*s*) is 192 kJ. For calcium, 1st IE = 589.5 kJ mol^{-1}, 2nd IE = 1146 kJ mol^{-1}. Electron affinity of Cl is −348 kJ mol^{-1}. Bond energy of Cl_2 is 242.6 kJ per mole of Cl—Cl bonds. Standard heat of formation of $CaCl_2$ is −795 kJ mol^{-1}.

***8.112** Use an enthalpy diagram and the following data to calculate the electron affinity of bromine. Standard heat of formation of NaBr is −360 kJ mol^{-1}. Energy needed to vaporize one mole of $Br_2(l)$ to give $Br_2(g)$ is 31 kJ mol^{-1}. The energy needed to change 1 mol Na(*s*) to 1 mol Na(*g*) is 107.8 kJ. The first ionization energy of Na is 495.4 kJ mol^{-1}. The bond energy of Br_2 is 192 kJ per mole of Br—Br bonds. The lattice energy of NaBr is 743.3 kJ mol^{-1}.

8.113 In many ways, tin(IV) chloride behaves more like a covalent molecular species than as a typical ionic chloride. Draw the Lewis structure for the tin(IV) chloride molecule.

8.114 In each pair, choose the one with more polar bonds. (Use the periodic table to answer the question.)

(a) PCl_3 or $AsCl_3$ (c) $SiCl_4$ or SCl_2
(b) SF_2 or GeF_4 (d) SrO or SnO

8.115 How many electrons are in the outer shell of the Zn^{2+} ion?

8.116 The Lewis structure for carbonic acid (formed when CO_2 dissolves in water) is usually given as

```
        O:
        ||
H—O—C—O—H
```

What is wrong with the following structures?

```
          :O:                    :O:
           |                      |
(a) H—O—C—O—H        (b) H—O=C—O—H

          :O:
           |
(c) H—O=C=O—H
```

8.117 Are the following Lewis structures considered to be resonance structures? Explain. Which is the more likely structure for $POCl_3$?

```
    :Cl:           :Cl:
     |              |
:Cl—P=O        :Cl—O=P
     |              |
    :Cl:           :Cl:
```

8.118 Assign formal charges to all the atoms in the following Lewis structure of hydrazoic acid, HN_3.

```
H—N≡N—N:
```

Suggest a lower energy resonance structure for this molecule.

8.119 Assign formal charges to all the atoms in the Lewis structure

```
        :O—H
         |
   :O—As—O—H
         |
       H—O:
```

Suggest a lower energy Lewis structure for this molecule.

8.120 The inflation of an "air bag" when a car experiences a collision occurs by the explosive decomposition of sodium azide, NaN_3, which yields nitrogen gas that inflates the bag. The following resonance structures can be drawn for the azide ion, N_3^-. Identify the best and worst of them.

$$\left[:\ddot{N}=N=\ddot{N}:\right]^- \longleftrightarrow \left[N\equiv N-\ddot{\underset{..}{N}}:\right]^-$$

$$\updownarrow \qquad\qquad\qquad \updownarrow$$

$$\left[:N\equiv N=\ddot{N}:\right]^- \longleftrightarrow \left[:\ddot{\underset{..}{N}}-N-\ddot{\underset{..}{N}}:\right]^-$$

8.121 How should the sulfur–oxygen bond lengths compare for the species SO_3, SO_2, SO_3^{2-}, and SO_4^{2-}?

8.122 What is the most reasonable Lewis structure for S_2Cl_2?

***8.123** There are two acids that have the formula HCNO. Which of the following skeletal structures are most likely for them? Justify your answer.

H C O N   H N O C   H O C N

H C N O   H N C O   H O N C

***8.124** What wavelength of light, if absorbed by a hydrogen molecule, could cause the molecule to split into the ions H^+ and H^-? (The data required are available in this and previous chapters.)

8.125 In the vapor state, ion pairs of KF can be identified. The dipole moment of such a pair is measured to be 8.59 D and the K—F bond length is found to be 217 pm. Is the K—F bond 100% ionic? If not, what percent of full 1+ and 1− charges do the K and F atoms carry, respectively?

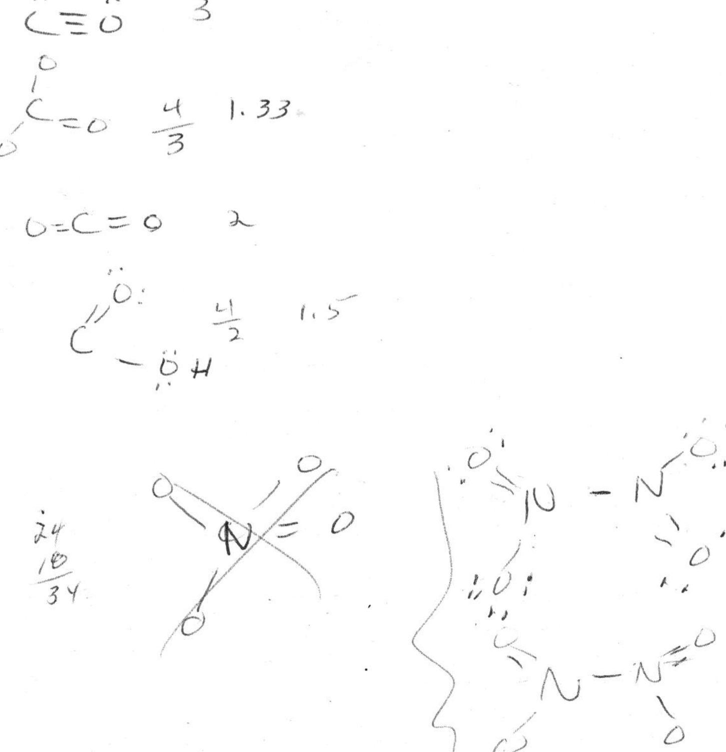

EXERCISES IN CRITICAL THINKING

8.126 What is the average bond energy of a C—C covalent bond? What wavelength of light provides enough energy to break such a bond? Using this information, explain why unfiltered sunlight is damaging to the skin.

8.127 One way of estimating the electronegativity of an atom is to use an average of its ionization energy and electron affinity. Why would these two quantities be related to electronegativity?

8.128 The attractions between molecules of a substance can be associated with the size of the molecule's dipole moment. Explain why this is so.

8.129 The positive end of the dipole in a water molecule is not located on an atom. Explain why this happens and suggest other simple molecules that show the same effect.

8.130 In describing the structures of molecules we use Lewis structures, formal charges, and experimental evidence. Rank these in terms of importance in deciding on the true structure of a molecule, and defend your choice.

CHEMICAL BONDING AND MOLECULAR STRUCTURE

9

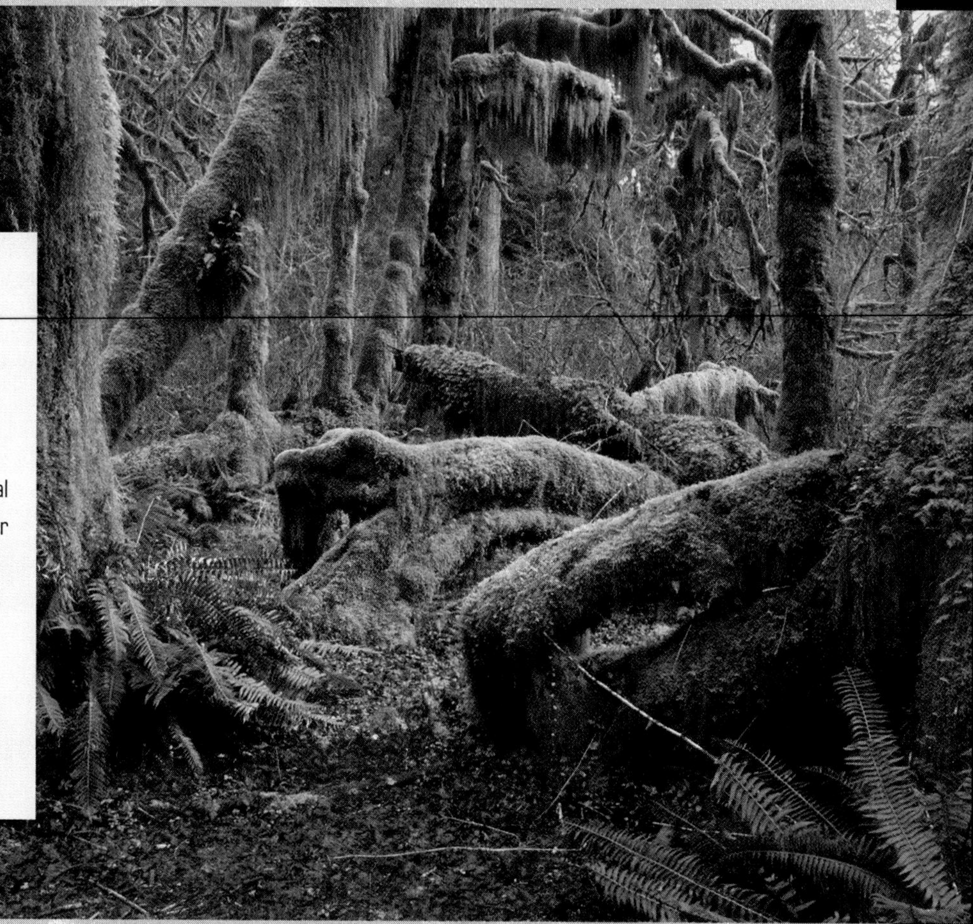

All of the variety of living things found in this rain forest, and elsewhere on earth, depends on the three-dimensional shapes of molecules and the requirement that certain molecules fit precisely together for biochemical reactions to occur. In this chapter we will study the kinds of shapes that molecules have and the way the electronic structures of atoms influence the chemical bonds that determine molecular geometry. *(Steve Satushek/Iconica/Getty Images)*

CHAPTER OUTLINE

9.1 Molecules are three-dimensional with shapes that are built from five basic arrangements

9.2 Molecular shapes are predicted using the VSEPR model

9.3 Molecular symmetry affects the polarity of molecules

9.4 Valence bond theory explains bonding as an overlap of atomic orbitals

9.5 Hybrid orbitals are used to explain experimental molecular geometries

9.6 Hybrid orbitals can be used to describe multiple bonds

9.7 Molecular orbital theory explains bonding as constructive interference of atomic orbitals

9.8 Molecular orbital theory uses delocalized orbitals to describe molecules with resonance structures

THIS CHAPTER IN CONTEXT The structure of an ionic compound, such as NaCl, is controlled primarily by the sizes of the ions and their charges. The attractions between the ions have no preferred directions, so if an ionic compound is melted, this structure is lost and the array of ions collapses into a jumbled liquid state. Molecular substances are quite different, however. Molecules have three-dimensional shapes that are determined by the relative orientations of their covalent bonds, and this structure is maintained regardless of whether the substance is a solid, a liquid, or a gas.

Many of the properties of a molecule depend on the three-dimensional arrangement of its atoms. For example, the functioning of enzymes, which are substances that affect how fast biochemical reactions occur, requires that there be a very precise fit between one molecule and another. Even slight alterations in molecular geometry can destroy this fit and deactivate the enzyme, which in turn prevents the biochemical reaction involved from occurring. Similarly, the structures of polymer molecules in plastics have a strong influence on the properties of materials made from them.

In this chapter we will explore the topic of molecular geometry and study theoretical models that allow us to explain, and in some cases predict, the shapes of molecules. We will also examine theories that explain, in terms of wave mechanics and the electronic structures of atoms, *why* covalent bonds form and *why* they are so highly directional in nature. You will find the knowledge gained here helpful in later discussions of physical properties of substances such as melting points and boiling points.

9.1 MOLECULES ARE THREE-DIMENSIONAL WITH SHAPES THAT ARE BUILT FROM FIVE BASIC ARRANGEMENTS

We live in a three-dimensional world made up of three-dimensional molecules. However, the Lewis structures we've been using to describe the bonding in molecules do not convey any information about the shapes of molecules; they simply describe which atoms are bonded to each other. Our goal now is to examine theories that predict molecular shapes and explain covalent structures in terms of quantum theory. We'll begin by studying some of the kinds of shapes molecules have.

Molecular shape only becomes a question when there are at least three atoms present. If there are only two, there is no doubt as to how they are arranged; one is just alongside the other. But when there are three or more atoms in a molecule we find that its shape is often built from just one or another of five basic geometrical structures.

Basic molecular shapes

Linear molecules

In a **linear molecule** the atoms lie in a straight line. When the molecule has three atoms, the angle formed by the covalent bonds, which we call the **bond angle,** equals 180° as illustrated below.

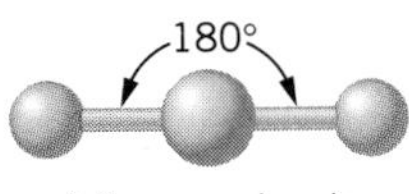

A linear molecule

Planar triangular molecules

A **planar triangular molecule** is one in which three atoms are located at the corners of a triangle and are bonded to a fourth atom that lies in the center of the triangle.

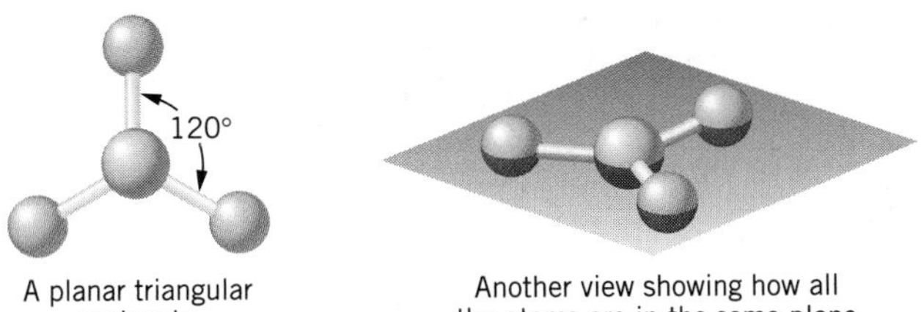

A planar triangular molecule

Another view showing how all the atoms are in the same plane

In this molecule, all four atoms lie in the same plane and the bond angles are all equal to 120°.

As you study these structures, you should try hard to visualize them in three dimensions. You should also learn how to sketch them in a way that conveys the three-dimensional information.

Tetrahedral molecules

A **tetrahedron** is a four-sided geometric figure shaped like a pyramid with triangular faces. A **tetrahedral molecule** is one in which four atoms, located at the vertices of a tetrahedron, are bonded to a fifth atom in the center of the structure.

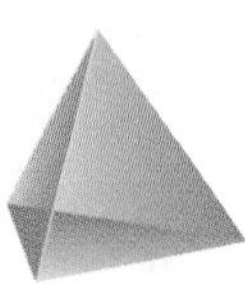

A tetrahedron

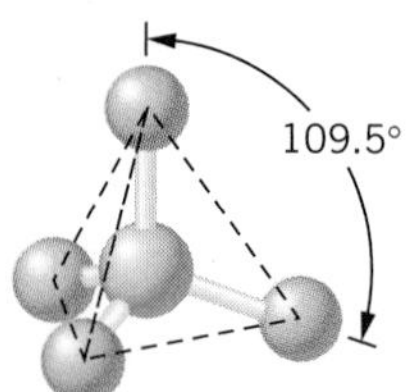

A tetrahedral molecule

All the bond angles in a tetrahedral molecule are the same and are equal to 109.5°.

Trigonal bipyramidal molecules

A **trigonal bipyramid** consists of two *trigonal pyramids* (pyramids with triangular faces) that share a common base. In a **trigonal bipyramidal molecule,** the central atom is located in the middle of the triangular plane shared by the upper and lower trigonal pyramids and is bonded to five atoms that are at the vertices of the figure.

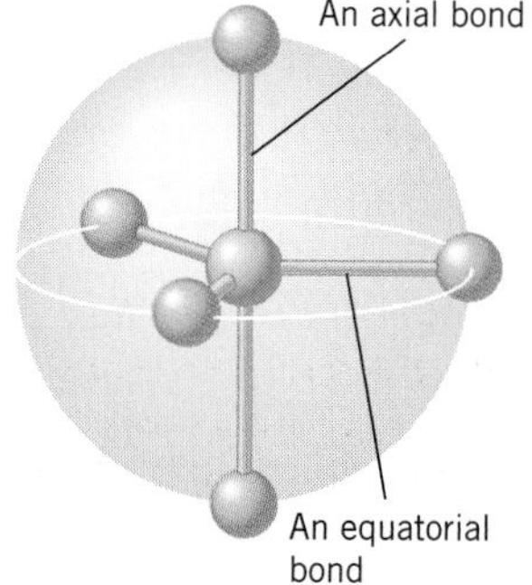

FIG. 9.1 Axial and equatorial bonds in a trigonal bipyramidal molecule.

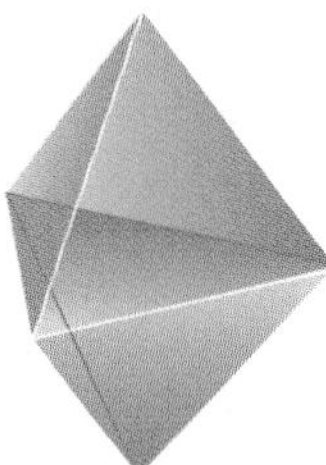

A trigonal bipyramid

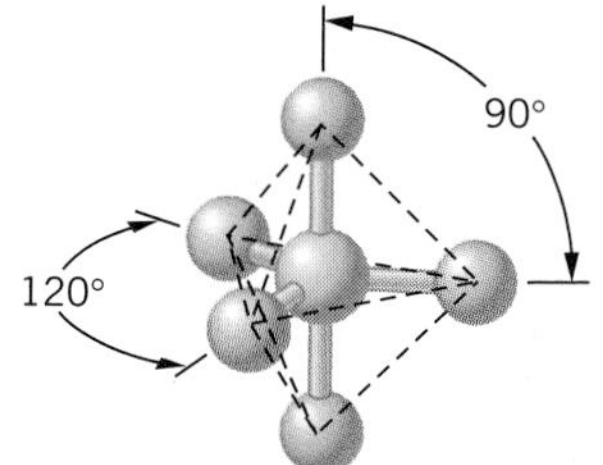

A trigonal bipyramidal molecule

In this type of molecule, not all the bonds are equivalent. If we imagine the trigonal bipyramid centered inside a sphere similar to the earth, as illustrated in Figure 9.1, the atoms in the triangular plane are located around the equator. The bonds to these atoms are called **equatorial bonds.** The angle between two equatorial bonds is 120°. The two vertical bonds pointing along the north and south axis of the sphere are 180° apart and are called **axial bonds.** The bond angle between an axial bond and an equatorial bond is 90°.

A simplified representation of a trigonal bipyramid is illustrated in Figure 9.2. The equatorial triangular plane is sketched as it would look tilted, so we're looking at it from its edge. The axial bonds are represented as lines pointing up and down. To add more three-dimensional character, notice that the bond pointing down appears to be partially hidden by the triangular plane in the center.

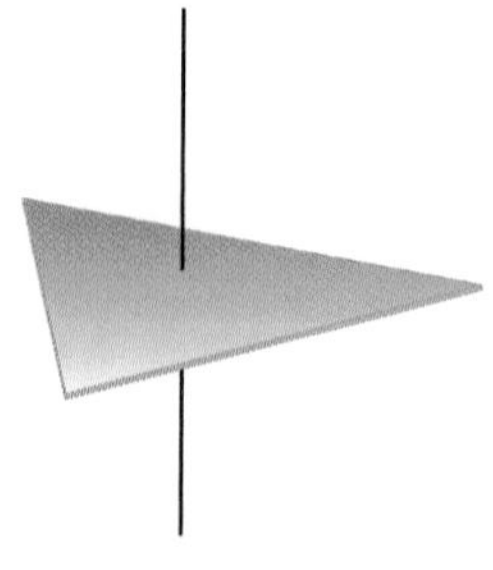

FIG. 9.2 A simplified way of drawing a trigonal bipyramid. To form a trigonal bipyramidal molecule, atoms would be attached at the corners of the triangle in the center and at the ends of the bonds that extend vertically up and down.

Octahedral molecules

An **octahedron** is an eight-sided figure, which you might think of as two *square pyramids* sharing a common square base. The octahedron has only six vertices, and in an **octahedral molecule** we find an atom in the center of the octahedron bonded to six other atoms at the vertices.

All the bonds in an octahedral molecule are equivalent, with angles between adjacent bonds equal to 90°.

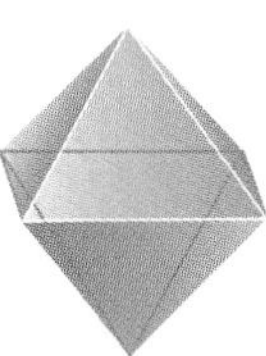

An octahedron

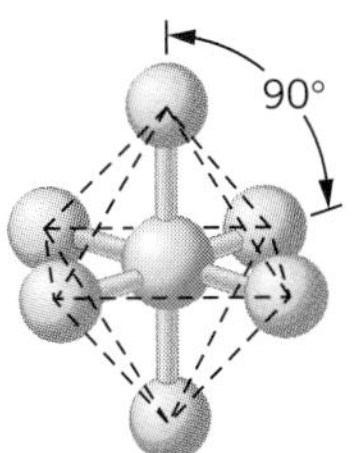

An octahedral molecule

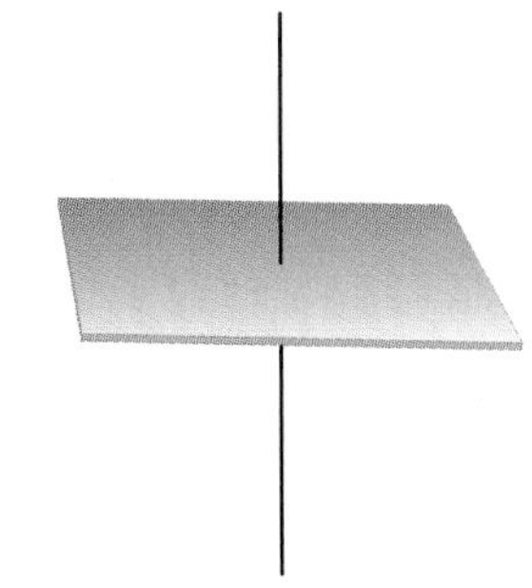

FIG. 9.3 A simplified way of drawing an octahedron. To form an octahedral molecule, atoms would be attached at the corners of the square in the center and at the ends of the bonds that extend vertically up and down.

A simplified representation of an octahedron is shown in Figure 9.3. The square plane in the center of the octahedron, when drawn in perspective and viewed from its edge, looks like a parallelogram. The bonds to the top and bottom of the octahedron are shown as vertical lines. Once again, the bond pointing down is drawn so it appears to be partially hidden by the square plane in the center.

9.2 MOLECULAR SHAPES ARE PREDICTED USING THE VSEPR MODEL

A useful theoretical model should explain known facts, and it should be capable of making accurate predictions. The **valence shell electron pair repulsion model** (called the **VSEPR model,** for short) is remarkably successful at both and is also conceptually simple. The model is based on the following idea:

> Groups of electrons in the valence shell of an atom repel each other and will position themselves in the valence shell so that they are as far apart as possible.

We will use the term **electron domain** to describe regions in space where groups of valence shell electrons can be found. We will consider two types of domains:

An *electron domain* can be a bond, a lone pair, or an unpaired electron. Some prefer to call the VSEPR model (or VSEPR theory) the *electron domain model.*

- **Bonding domains** contain electron pairs that are involved in bonds between pairs of atoms. *All of the electrons within a given single, double, or triple bond are considered to be in the same bonding domain.* A double bond (containing 4 electrons) will occupy more space than a single bond (with only 2 electrons) but all electrons in a bond occupy the same region in space, so they all belong to the same bonding domain.
- **Nonbonding domains** contain valence electrons that are associated with *a single atom.* A nonbonding domain is either an unshared pair of valence electrons (called a **lone pair)** or a single unpaired electron (found in molecules with an odd number of valence electrons).

Figure 9.4 shows the orientations assumed by different numbers of electron domains which permit them to minimize repulsions by remaining as far apart as possible. Notice that when atoms are attached to these electron domains, molecules are formed having the shapes described in the preceding section.

Lewis structures permit us to apply the VSEPR model

When we speak of the shape of a molecule, we are referring to the arrangement of some number of atoms around a central atom. To apply the VSEPR model in predicting shape, we have to know how many electron domains are in the valence shell of the central atom. This is where Lewis structures are helpful.

Consider the $BeCl_2$ molecule. On page 318 we gave its Lewis structure as

$$:\ddot{\underset{..}{Cl}}-Be-\ddot{\underset{..}{Cl}}:$$

Because there are three atoms, the molecule is either linear or nonlinear; that is, the atoms lie in a straight line, or they form some angle less than 180°.

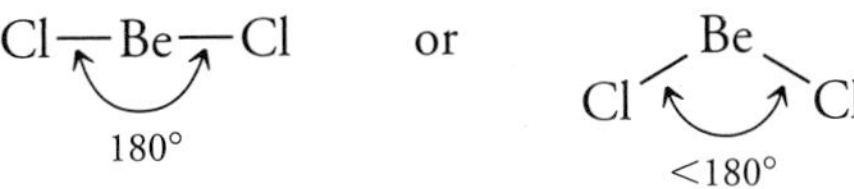

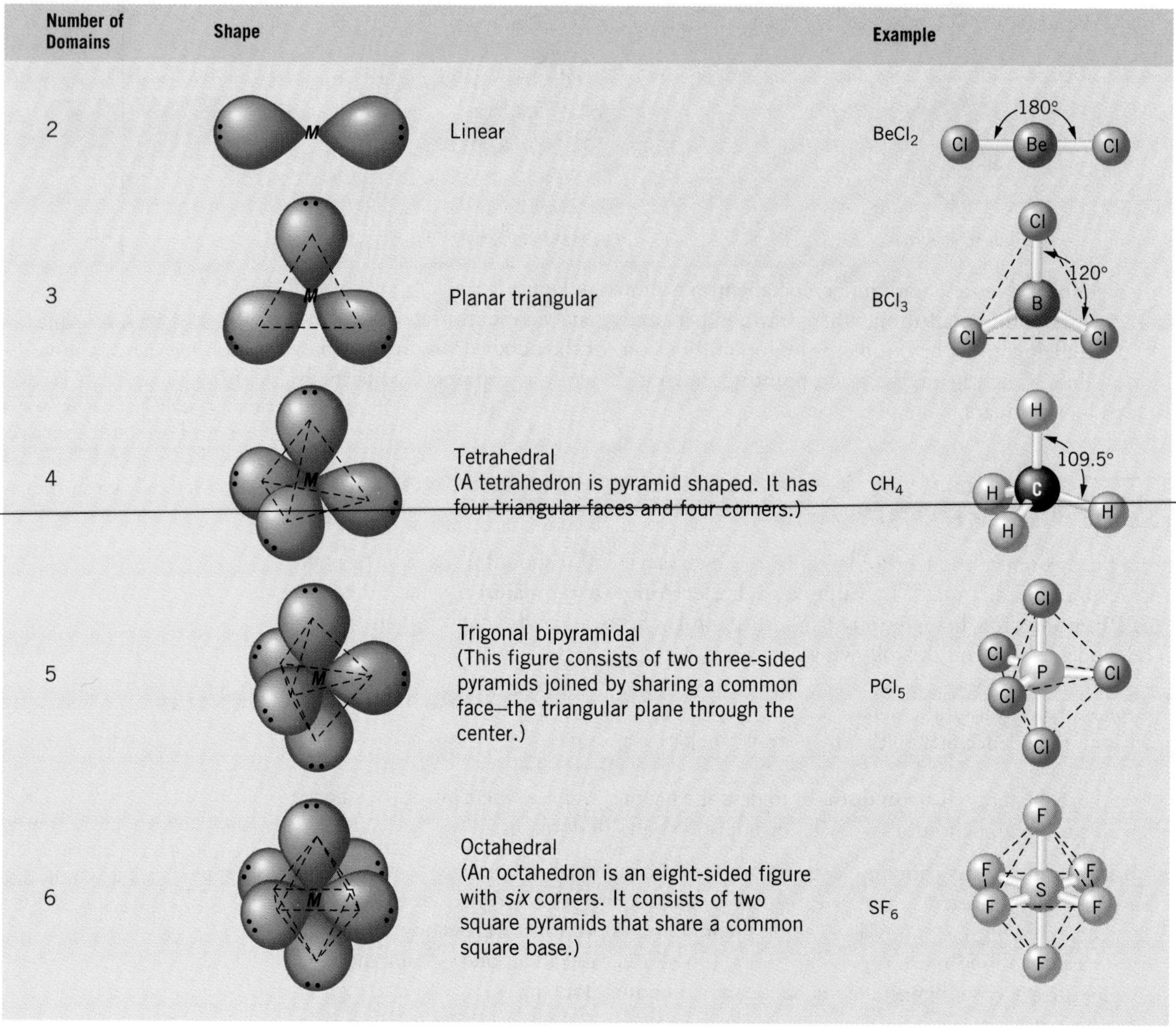

Number of Domains	Shape		Example	
2		Linear	$BeCl_2$	180°
3		Planar triangular	BCl_3	120°
4		Tetrahedral (A tetrahedron is pyramid shaped. It has four triangular faces and four corners.)	CH_4	109.5°
5		Trigonal bipyramidal (This figure consists of two three-sided pyramids joined by sharing a common face—the triangular plane through the center.)	PCl_5	
6		Octahedral (An octahedron is an eight-sided figure with *six* corners. It consists of two square pyramids that share a common square base.)	SF_6	

FIG. 9.4 Shapes expected for different numbers of electron domains around a central atom, *M*. Each lobe represents an electron domain.

electron domains

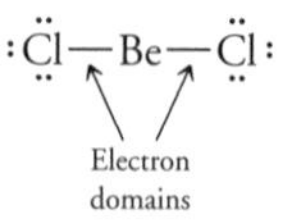

$:\ddot{Cl}-Be-\ddot{Cl}:$

Electron domains

To decide on the structure, we begin by counting the number of electron domains in the valence shell of the Be atom. In this molecule Be forms two single bonds to Cl atoms, each of which corresponds to a domain, so Be has two bonding domains in its valence shell. In Figure 9.4, we see that when there are two electron domains in the valence shell of an atom, minimum repulsion occurs if they are on opposite sides of the nucleus, pointing in opposite directions. We can represent this as

to suggest the approximate locations of the electron clouds of the valence shell electron pairs. In order for the electrons to be in the Be—Cl bonds, the Cl atoms must be placed where the electrons are; the result is that we predict that a $BeCl_2$ molecule should be linear.

$$Cl—Be—Cl$$

In fact, this is the shape of $BeCl_2$ molecules in the vapor state.

EXAMPLE 9.1
Predicting Molecular Shapes

Carbon tetrachloride was once used as a cleaning fluid until it was discovered that it causes liver damage if absorbed by the body. What is the shape of the molecule?

ANALYSIS: The primary tool for solving this kind of problem is the Lewis structure, which we will draw following the procedure in Chapter 8. First, however, we need the chemical formula. Applying the rules of nomenclature in Chapter 2 gives CCl_4. After we draw the Lewis structure, we can count the number of electron domains around the central atom. Finally, we'll use the VSEPR model as a tool to deduce the structure of the molecule.

SOLUTION: Following the procedure in Figure 8.9, the Lewis structure of CCl_4 is

```
        ..
       :Cl:
   ..   |   ..
  :Cl — C — Cl:
   ..   |   ..
       :Cl:
        ..
```

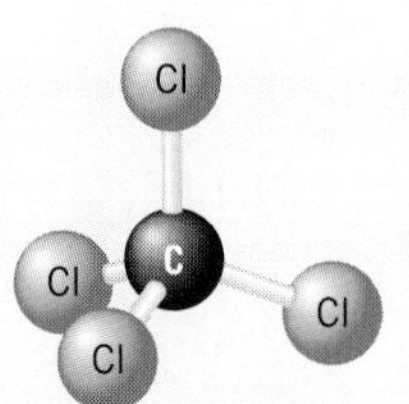

There are four bonds, each corresponding to a bonding domain around the carbon. According to Figure 9.4, the domains can be farthest apart when arranged tetrahedrally, so the molecule is expected to be tetrahedral. (This is, in fact, its structure.)

IS THE ANSWER REASONABLE? The answer depends critically on the Lewis structure, so be sure to check that you've constructed it correctly. Once we're confident in the Lewis structure the rest is straightforward. The arrangement of domains gives us the arrangement of Cl atoms around the C atom.

Practice Exercise 1: What is the shape of the SeF_6 molecule? (Hint: If necessary, refer to Figure 8.9 on page 318.)

Practice Exercise 2: What shape is expected for the $SbCl_5$ molecule?

Nonbonding domains affect the shape of a molecule

Some molecules have a central atom with one or more nonbonding domains, consisting of unshared electron pairs (lone pairs) or unpaired valence electrons. These nonbonding domains affect the geometry of the molecule. An example is $SnCl_2$.

```
   ..   ..   ..
  :Cl — Sn — Cl:
   ..        ..
```

The electronegativity difference between tin and chlorine is only 1.1, which means tin–chlorine bonds have a significant degree of covalent character. Many compounds of tin are molecular, especially those of Sn^{IV}. $SnCl_2$ is another example of a molecule that behaves as though it has less than an octet around the central atom.

There are *three* domains around the tin atom, two bonding domains plus a nonbonding domain (the lone pair). According to Figure 9.4, the domains are farthest apart when at the corners of a triangle. For the moment, let's ignore the chlorine atoms and concentrate on how the electron domains are arranged.

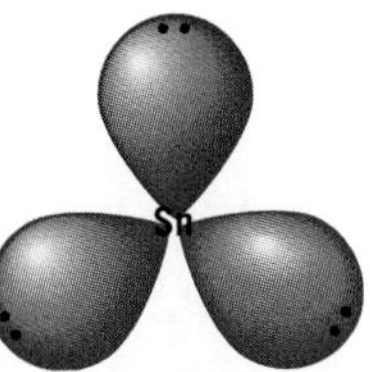

We can see the shape of the molecule, now, by placing the two Cl atoms where two of the domains are.

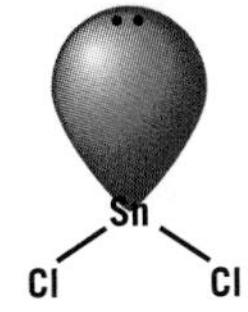

We can't describe this molecule as triangular, even though that is how the domains are arranged. *Molecular shape describes the arrangement of atoms, not the arrangement of domains.* Therefore, we describe the shape of the $SnCl_2$ molecule as being **nonlinear** or **bent** or **V-shaped.**

Notice that when there are three domains around the central atom, *two* different molecular shapes are possible. If all three are bonding domains, a molecule with a planar triangular shape is formed, as shown for BCl_3 in Figure 9.4. If one of the domains is nonbonding, as in $SnCl_2$, the arrangement of the atoms in the molecule is said to be nonlinear. The predicted shapes of both, however, are *derived* by first noting the triangular arrangement of domains around the central atom and *then* adding the necessary number of atoms.

Molecules with four domains have shapes derived from a tetrahedron

There are many molecules with four electron pairs (an octet) in the valence shell of the central atom. When these electron pairs are used to form four bonds, as in methane (CH_4), the resulting molecule is tetrahedral (Figure 9.4). There are many examples, however, where nonbonding domains are also present. For example,

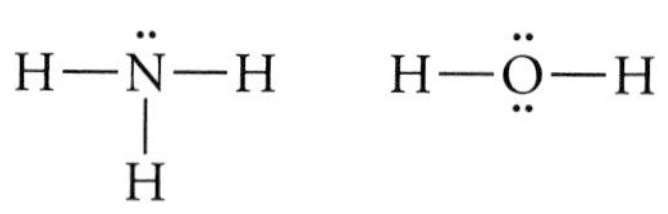

$$\mathrm{H{-}\ddot{N}{-}H}\ \text{(with a third H bonded below N)} \qquad \mathrm{H{-}\ddot{\underset{\cdot\cdot}{O}}{-}H}$$

one lone pair   two lone pairs

Figure 9.5 shows how the nonbonding domains affect the shapes of molecules of this type.

With one nonbonding domain, the central atom is at the top of a pyramid with three atoms at the corners of the triangular base. The resulting structure is said to be **trigonal pyramidal.** When there are two nonbonding domains in the tetrahedron, the three atoms of the molecule (the central atom plus the two atoms bonded to it) do not lie in a straight line, so the structure is described as nonlinear or bent.

Note once again that in describing the shape of the molecule, we look at how the atoms are arranged and ignore the nonbonding domains.

Number of Bonding Domains	Number of Nonbonding Domains	Structure	
4	0	109.5°	**Tetrahedral** (example, CH_4) All bond angles are 109.5°.
3	1		**Trigonal pyramidal** (pyramid shaped) (example, NH_3)
2	2		**Nonlinear, bent** (example, H_2O)

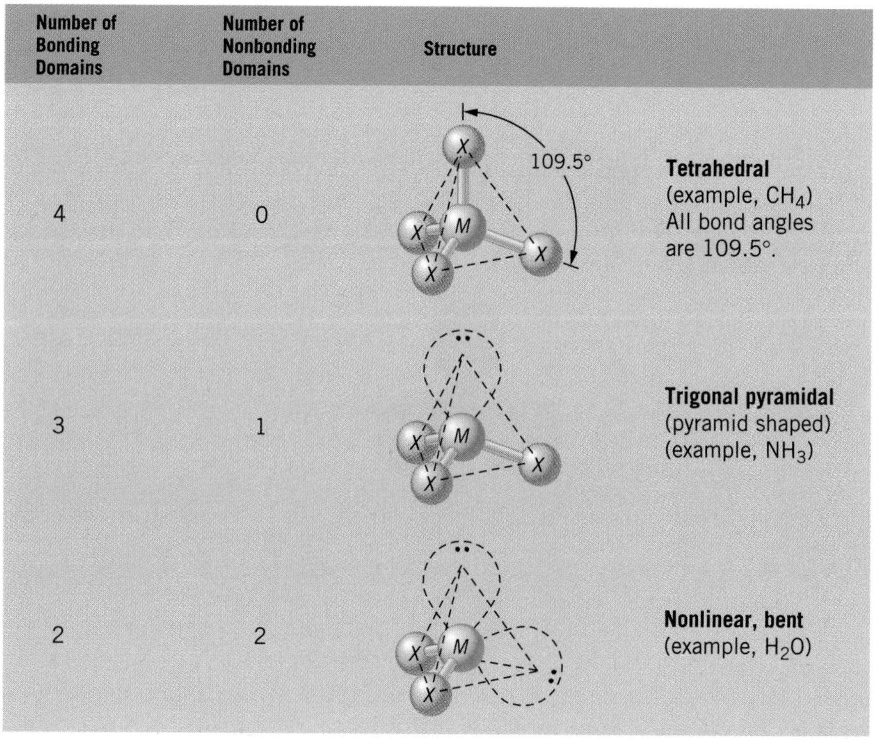

FIG. 9.5 Molecular shapes with four domains around the central atom. The molecules MX_4, MX_3, and MX_2 shown here all have four domains arranged tetrahedrally around the central atom. The shapes of the molecules and their descriptions are derived from the way the X atoms are arranged around M, ignoring the nonbonding domains.

Bonding domain

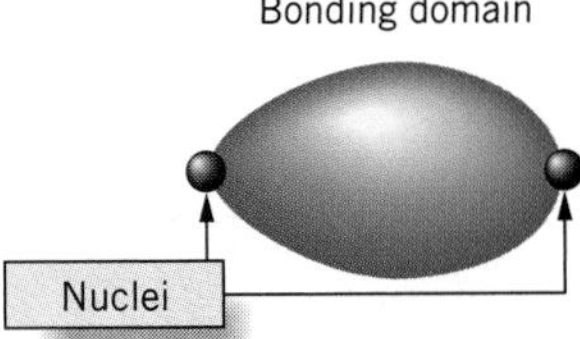

Nonbonding domain

Nucleus

FIG. 9.6 Relative sizes of bonding and nonbonding domains.

Number of Bonding Domains	Number of Nonbonding Domains	Structure	
5	0	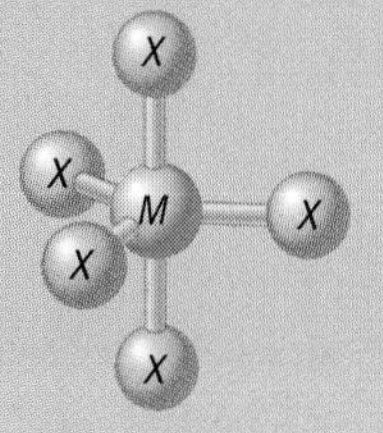	**Trigonal bipyramidal** (example, PCl_5)
4	1	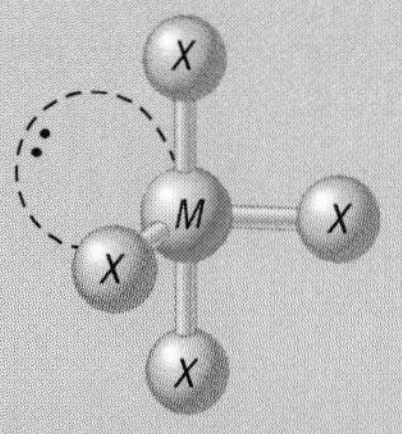	**Distorted tetrahedral** 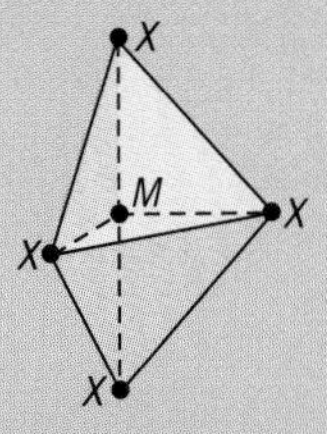 (example, SF_4)
3	2	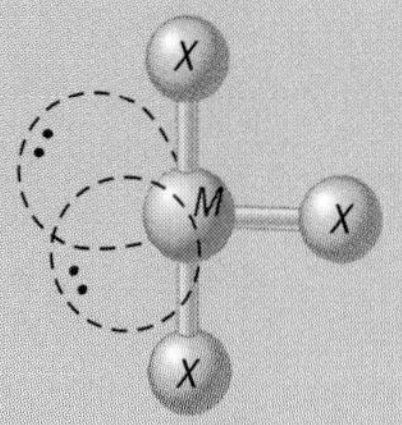	**T-shaped** 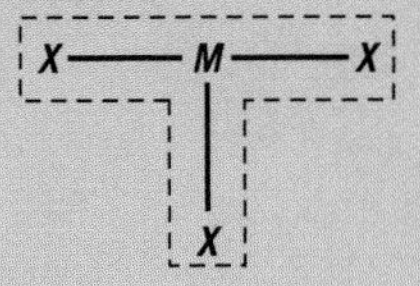 (example, ClF_3)
2	3	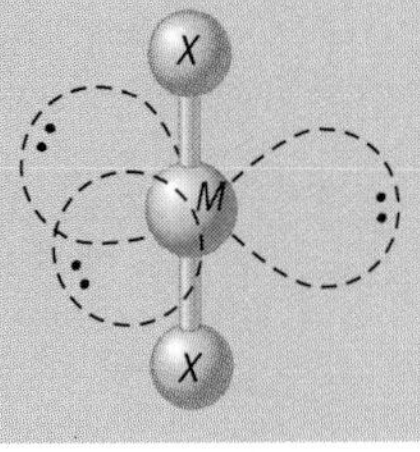	**Linear** (example, I_3^-)

FIG. 9.7 Molecular shapes with five domains around the central atom. Four different molecular structures are possible, depending on the number of nonbonding domains around the central atom *M*.

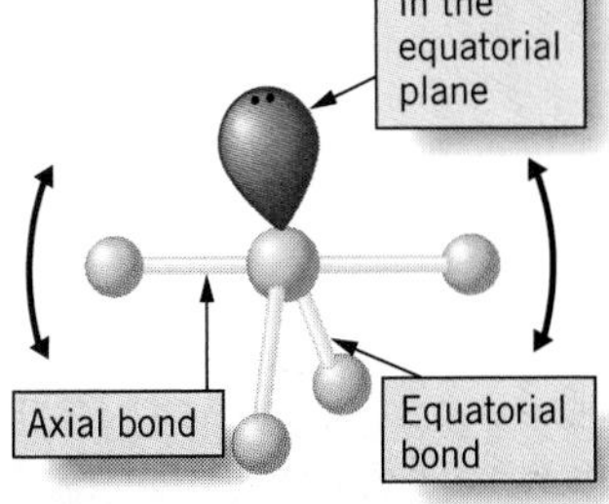

FIG. 9.8 This distorted tetrahedron is sometimes described as a *seesaw structure.* The origin of this description can be seen if we tip the structure over so it stands on the two atoms in the equatorial plane, with the nonbonding domain pointing up.

Molecules with five domains have shapes derived from a trigonal bipyramid

When five domains are present around the central atom, they are directed toward the vertices of a trigonal bipyramid. Molecules such as PCl_5 have this geometry, as shown in Figure 9.4, and are said to have a **trigonal bipyramidal** shape.

In the trigonal bipyramid, nonbonding domains always occupy positions in the *equatorial plane* (the triangular plane through the center of the molecule). This is because nonbonding domains, which have a positive nucleus only at one end, are larger than bonding domains, as illustrated in Figure 9.6. The larger nonbonding domains are less crowded in the equatorial plane, where they have just two closest neighbors at 90°, than they would be in an axial position, where they would have three closest neighbors at 90°.

Figure 9.7 shows the kinds of geometries that we find for different numbers of nonbonding domains in the trigonal bipyramid. When there is only one nonbonding domain, as in SF_4, the structure is described as a **distorted tetrahedron** or **seesaw** in which the central atom lies along one edge of a four-sided figure (Figure 9.8).

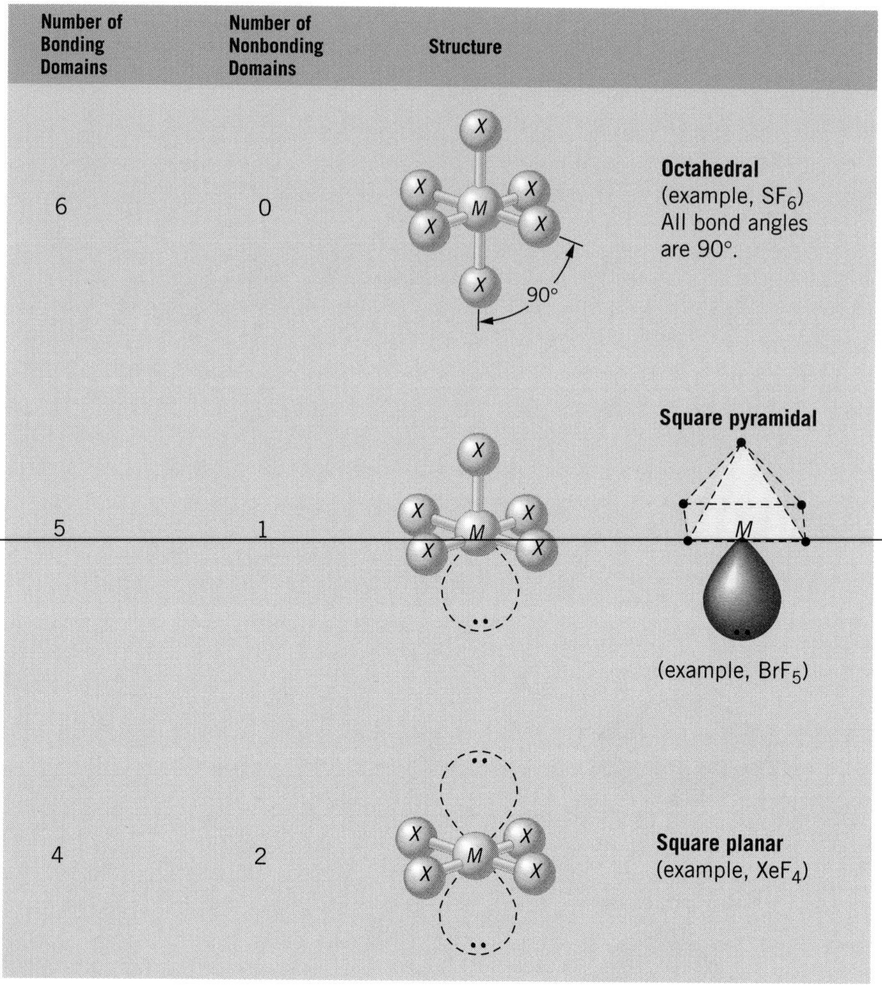

Number of Bonding Domains	Number of Nonbonding Domains	Structure	
6	0	X, M, 90°	**Octahedral** (example, SF_6) All bond angles are 90°.
5	1	X, M	**Square pyramidal** M (example, BrF_5)
4	2	X, M	**Square planar** (example, XeF_4)

FIG. 9.10 Molecular shapes with six domains around the central atom. Although more are theoretically possible, only three different molecular shapes are observed, depending on the number of nonbonding domains around the central atom.

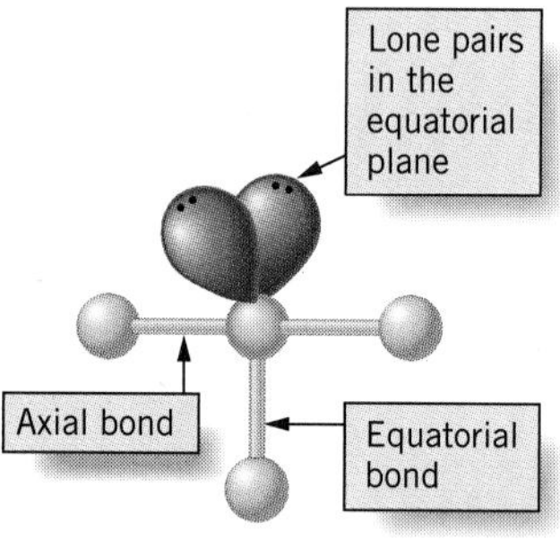

FIG. 9.9 A molecule with two nonbonding domains in the equatorial plane of the trigonal bipyramid. When tipped over, the molecule looks like a T, so it is called *T-shaped.*

With two nonbonding domains in the equatorial plane, the molecule is **T-shaped** (shaped like the letter T, Figure 9.9), and when there are three nonbonding domains in the equatorial plane, the molecule is **linear.**

No common molecule or ion with six domains around the central atom has more than two nonbonding domains.

Molecules with six domains have shapes derived from an octahedron

Finally, we come to molecules or ions that have six domains around the central atom. When all are in bonds, as in SF_6, the molecule is octahedral (Figure 9.4). When one nonbonding domain is present the molecule or ion has the shape of a **square pyramid,** and when two nonbonding domains are present they take positions on opposite sides of the nucleus and the molecule or ion has a **square planar** structure. These shapes are shown in Figure 9.10.

EXAMPLE 9.2
Predicting the Shapes of Molecules and Ions

Do we expect the ClO_2^- ion to be linear?

ANALYSIS: To predict the shape, the first step is to draw the Lewis structure for the ClO_2^- ion. Then we can count the number of domains around the central atom. This will determine which of the basic geometries describes the orientations of the domains. We sketch this shape and then attach the two oxygens. Finally, we ignore any nonbonding domains to arrive at the description of the shape of the ion. In other words, in this last step we note how the *atoms* are arranged, not how the domains are arranged.

SOLUTION: Following the usual procedure, we obtain

$$\left[:\ddot{\underset{..}{O}}-\ddot{\underset{..}{Cl}}-\ddot{\underset{..}{O}}:\right]^-$$

There are four domains around the chlorine: two bonding and two nonbonding. Four domains (according to the theory) are always arranged tetrahedrally. This gives

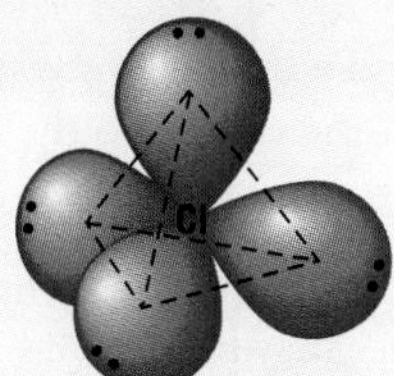

Now we add the two oxygens. It doesn't matter which locations in the tetrahedron we choose because all the bond angles are equal.

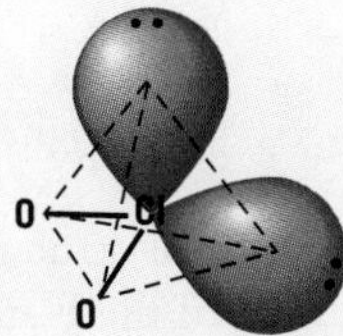

We see that the O—Cl—O angle is less than 180°, so the ion is expected to be nonlinear.

For a molecule with three atoms, there are only two ways they can be arranged, either in a straight line (linear) or in a nonlinear arrangement.

IS THE ANSWER REASONABLE? Here are questions we have to answer to check our work. First, has the Lewis structure been drawn correctly? Have we counted the domains correctly? Have we selected the correct orientation of the domains? And finally, have we correctly described the structure obtained by adding the two oxygen atoms? Our answer to each of these questions is "Yes," so we can be confident our answer is right.

EXAMPLE 9.3
Predicting the Shapes of Molecules and Ions

Xenon is one of the noble gases, and is generally quite unreactive. In fact, it was long believed that all the noble gases were totally unable to form compounds. It came as quite a surprise, therefore, when it was discovered that some compounds could be made. One of these is xenon difluoride. What would you expect the geometry of xenon difluoride to be, linear or nonlinear?

ANALYSIS: First, we use the rules of nomenclature from Chapter 2 to obtain the formula, which is XeF_2. From here, the procedure is the same as before. First we write the compound's Lewis structure. Then we count domains around the central atom to determine which of the basic geometries forms the basis for the structure. Next, we sketch the structure and attach the fluorine atoms to two of the domains. Finally, we ignore any nonbonding domains and observe how the atoms are arranged to describe the shape of the molecule.

SOLUTION: The outer shell of xenon, of course, has a noble gas configuration, which contains 8 electrons. Each fluorine has 7 valence electrons. Using this information we obtain the following Lewis structure for XeF_2.

$$:\ddot{\underset{..}{F}}-\underset{..}{\overset{..\,..}{Xe}}-\ddot{\underset{..}{F}}:$$

Next we count domains around xenon; there are five of them, three nonbonding and two bonding domains. When there are five domains, they are arranged in a trigonal bipyramid.

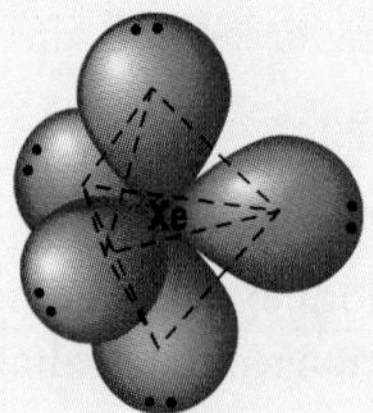

Now we must add the fluorine atoms. In a trigonal bipyramid, the nonbonding domains always occur in the equatorial plane through the center, so the fluorines go on the top and bottom. This gives

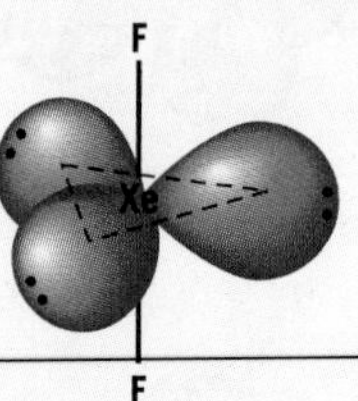

The three atoms, F—Xe—F, are arranged in a straight line, so the molecule is linear.

IS THE ANSWER REASONABLE? Is the Lewis structure correct? Yes. Have we selected the correct basic geometry? Yes. Have we attached the F atoms to the Xe correctly? Yes. All is in order, so the answer is correct.

EXAMPLE 9.4
Predicting the Shapes of Molecules and Ions

The Lewis structure for the very poisonous gas hydrogen cyanide, HCN, is

H—C≡N:

Is the HCN molecule linear or nonlinear?

ANALYSIS: We already have the Lewis structure, so we count electron domains and proceed as before. The critical link in this problem is remembering that *all the electron pairs in a given bond belong to the same bonding domain.*

SOLUTION: There are two bonding domains around the carbon, one for the triple bond with nitrogen, and one for the single bond with hydrogen.

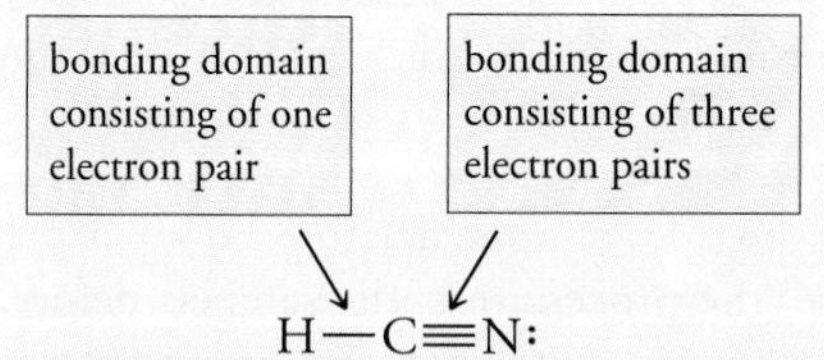

Therefore, we expect the two bonds to locate themselves 180° apart, yielding a linear HCN molecule.

IS THE ANSWER REASONABLE? Have we correctly counted bonding domains? Yes. We can expect our answer to be correct.

Practice Exercise 3: The first known compound of the noble gas argon is HArF. What shape is expected for the HArF molecule? (Hint: Remember, argon is a noble gas with eight electrons in its valence shell.)

Practice Exercise 4: What shape is expected for the I_3^- ion?

Practice Exercise 5: What shape is expected for the XeF_4 molecule?

Practice Exercise 6: Predict the shapes of SO_3^{2-}, CO_3^{2-}, XeO_4, and OF_2.

9.3 MOLECULAR SYMMETRY AFFECTS THE POLARITY OF MOLECULES

The dipole moment of a molecule is a property that can be determined experimentally, and when this is done, an interesting observation is made. There are many molecules that have no dipole moment even though they contain bonds that are polar. Stated differently, they are nonpolar molecules, even though they have polar bonds. The reason for this can be seen if we examine the key role that molecular structure plays in determining molecular polarity.

For a diatomic molecule such as HCl, the polar bond causes the molecule as a whole to be polar. Similarly, H_2 is nonpolar because the H—H bond is nonpolar. For molecules that contain more than two atoms, however, we have to consider the combined effects of all the bonds. Sometimes, when all the atoms attached to the central atom are the same, the effects of the individual polar bonds cancel and the molecule as a whole is nonpolar. Some examples are shown in Figure 9.11. In this figure the dipoles associated with the bonds themselves—the **bond dipoles**—are shown as arrows crossed at one end, $\longmapsto$. The arrowhead indicates the negative end of the bond dipole and the crossed end corresponds to the positive end.

Any molecule composed of just two atoms that differ in electronegativity must be polar because the bond is polar.

The CO_2 molecule is symmetric. Both bonds are identical, so each bond dipole is of the same magnitude. Because CO_2 is a linear molecule, these bond dipoles point in opposite directions and work against each other. The net result is that their effects cancel, and CO_2 is nonpolar. Although it isn't so easy to visualize, the same thing also happens in BCl_3 and CCl_4. In each of these molecules, the influence of one bond dipole is canceled by the effects of the others.

Bond dipoles can be treated as vectors, and the polarity of a molecule is predicted by taking the vector sum of the bond dipoles. If you've studied vectors before, this may help your understanding.

Perhaps you've noticed that the structures of the molecules in Figure 9.11 correspond to three of the basic shapes that we used to derive the shapes of molecules. Molecules with the remaining two structures, trigonal bipyramidal and octahedral, also are nonpolar if all the atoms attached to the central atom are the same. All of the basic shapes are "balanced," or **symmetric**,[1] if all of the domains and groups attached to them are identical. Examples

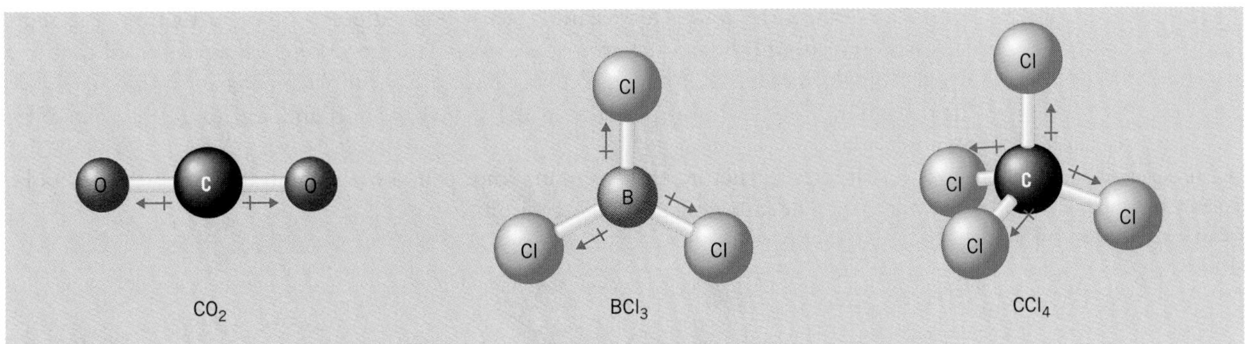

FIG. 9.11 In symmetric molecules such as these, the bond dipoles cancel to give nonpolar molecules.

[1] Symmetry is a more complex subject than we present it here. When we describe the symmetry properties of a molecule, we are specifying the various ways the molecule can be turned and otherwise manipulated while leaving the molecule looking exactly as it appeared before the manipulation. For example, imagine the BCl_3 molecule in Figure 9.11 being rotated 120° around an axis, perpendicular to the page, that passes through the B atom. Performing this rotation leaves the molecule looking just as it did before the rotation. This rotation axis is a symmetry property of BCl_3.

Intuitively, we can recognize when an object possesses symmetry elements such as rotation axes (as well as other symmetry properties we haven't mentioned). Comparing objects, we can usually tell when one is more symmetric than another, and in our discussions in this chapter we rely on this qualitative sense of symmetry.

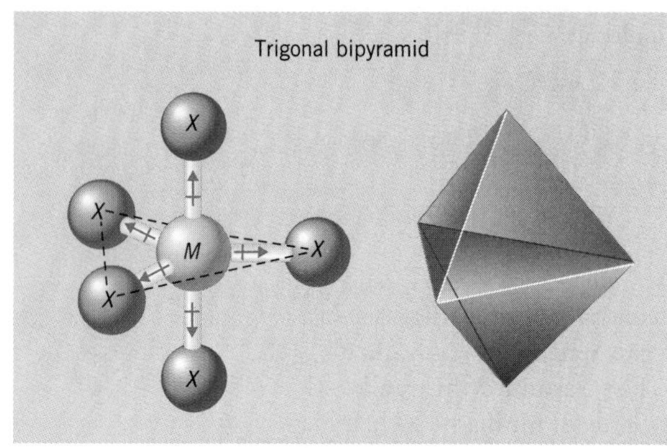

(a)

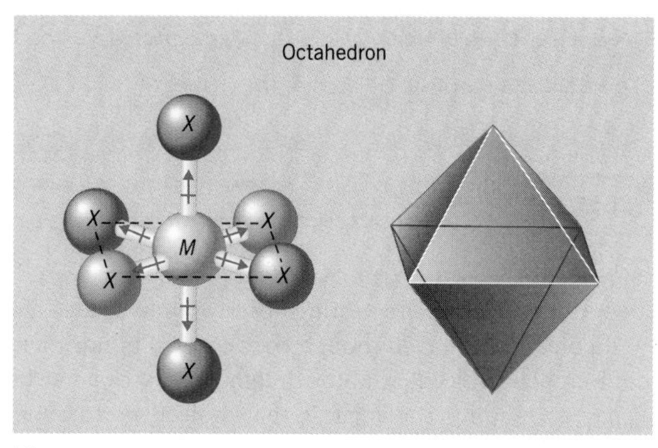

(b)

FIG. 9.12 Cancellation of bond dipoles in trigonal bipyramidal and octahedral molecules. (*a*) A trigonal bipyramidal molecule, MX_5, in which the central atom M is bonded to five identical atoms X. The set of three bond dipoles in the triangular plane in the center (in blue) cancel, as do the linear set of dipoles (red). Overall, the molecule is nonpolar. (*b*) An octahedral molecule MX_6 in which the central atom is bonded to six identical atoms. This molecule contains three linear sets of bond dipoles. Cancellation occurs for each set, so the molecule is nonpolar overall.

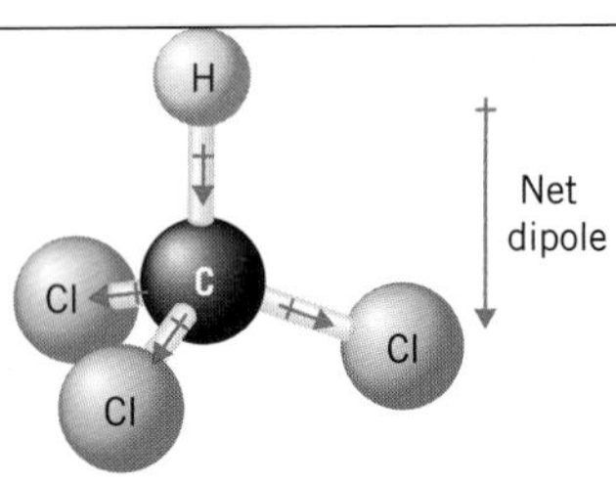

FIG. 9.13 Bond dipoles in the chloroform molecule, $CHCl_3$. Because C is slightly more electronegative than H, the C—H bond dipole points toward the carbon. The small C—H bond dipole actually adds to the effects of the C—Cl bond dipoles. All the bond dipoles are additive, and this causes $CHCl_3$ to be a polar molecule.

are shown in Figure 9.12. The trigonal bipyramidal structure can be viewed as a planar triangular set of atoms (shown in blue) plus a pair of atoms arranged linearly (shown in red). All the bond dipoles in the planar triangle cancel, as do the two dipoles of the bonds arranged linearly, so the molecule is nonpolar overall. Similarly, we can look at the octahedral molecule as consisting of three linear sets of bond dipoles. Cancellation of bond dipoles occurs in each set, so overall the octahedral molecule is also nonpolar.

*If all the atoms attached to the central atom are not the same, or if there are lone pairs in the valence shell of the central atom, the molecule is **usually** polar.* For example, in $CHCl_3$, one of the atoms in the tetrahedral structure is different from the others. The C—H bond is less polar than the C—Cl bonds, and the bond dipoles do not cancel (Figure 9.13). An "unbalanced" structure such as this is said to be **dissymmetric.**

Two familiar molecules that have lone pairs in the valence shells of their central atoms are shown in Figure 9.14. Here the bond dipoles are oriented in such a way that their effects do not cancel. In water, for example, each bond dipole points partially in the same direction, toward the oxygen atom. As a result, the bond dipoles partially add to give a net dipole moment for the molecule. The same thing happens in ammonia where three bond dipoles point partially in the same direction and add to give a polar NH_3 molecule.

The lone pairs also influence the polarity of a molecule, but we will not explore this any further here.

Not every structure that contains lone pairs on the central atom produces polar molecules. The following are two exceptions.

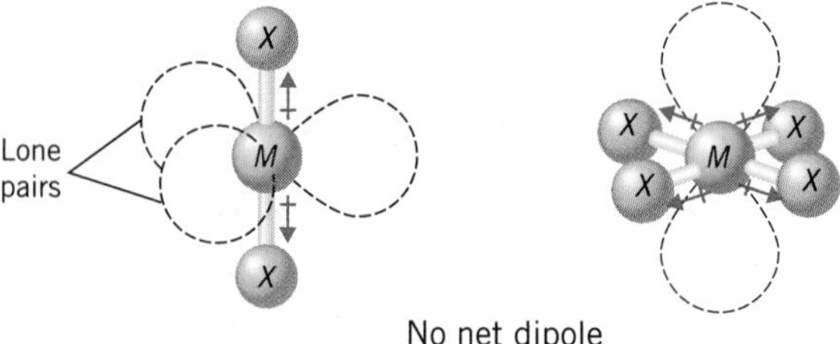

In the first case, we have a pair of bond dipoles arranged linearly, just as they are in CO_2. In the second, the bonded atoms lie at the corners of a square, which can be viewed as two linear sets of bond dipoles. *If the atoms attached to the central atom are the same*, cancellation of bond dipoles is bound to occur and produce nonpolar molecules. This means that molecules such as linear XeF_2 and square planar XeF_4 are nonpolar.

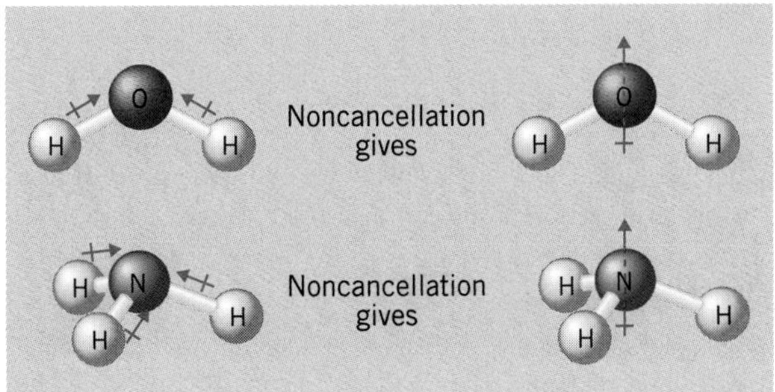

FIG. 9.14 **When nonbonding domains occur on the central atom, the bond dipoles usually do not cancel, and polar molecules result.**

In Summary

- A molecule will be nonpolar if (a) the bonds are nonpolar or (b) there are no lone pairs in the valence shell of the central atom and all the atoms attached to the central atom are the same.
- A molecule in which the central atom has lone pairs of electrons will usually be polar, with the two exceptions described above.

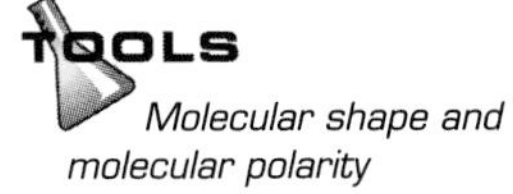

Molecular shape and molecular polarity

On the basis of the preceding discussions, let's see how we can predict whether molecules are expected to be polar or nonpolar.

EXAMPLE 9.5
Predicting Molecular Polarity

Do we expect the phosphorus trichloride molecule to be polar or nonpolar?

ANALYSIS: As before, the first step is to convert the name of the compound into a chemical formula. Our tool is the set of rules given in Chapter 2, which gives the formula PCl_3. Next, we'll use electronegativities as a tool to determine whether the bonds in the molecule are polar. If they're not, the molecule will be nonpolar regardless of its structure. If the bonds are polar, we then need to determine the molecular structure. Our tools will be the procedure for drawing the Lewis structure and the VSEPR model. On the basis of the molecular structure, we can then decide whether the bond dipoles cancel.

SOLUTION: The electronegativities of the atoms (P = 2.1, Cl = 2.9) tell us that the individual P—Cl bonds will be polar. Therefore, to predict whether or not the molecule is polar, we need to know its shape. First we draw the Lewis structure following our usual procedure.

```
       ..
      :Cl:
       |
  ..   ..  ..
 :Cl—P—Cl:
  ..   ..  ..
```

There are four domains around the phosphorus, so they should be arranged tetrahedrally. This means that the PCl_3 molecule should have a trigonal pyramidal shape, as shown in the margin. Because of the structure, the bond dipoles do not cancel, and we expect the molecule to be polar.

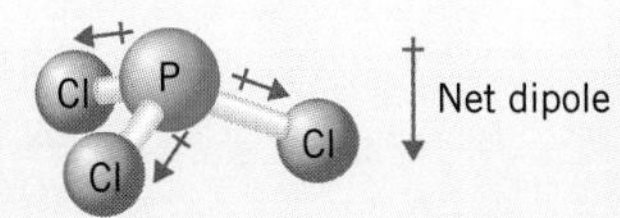

IS THE ANSWER REASONABLE? There's not much to do here except to carefully check the Lewis structure to be sure it's correct and to check that we've applied the VSEPR theory correctly.

EXAMPLE 9.6
Predicting Molecular Polarity

Would you expect the molecule HCN to be polar or nonpolar?

ANALYSIS AND SOLUTION: Our analysis proceeds as in the preceding example. To begin, we have polar bonds because carbon is slightly more electronegative than hydrogen and nitrogen is slightly more electronegative than carbon. The Lewis structure of HCN is

$$\text{H—C}\equiv\text{N:}$$

There are two domains around the central carbon atom, so we expect a linear shape. However, the two bond dipoles do not cancel. One reason is that they are not of equal magnitude, which we know because the difference in electronegativity between C and H is 0.4, whereas the difference in electronegativity between C and N is 0.6. The other reason is because both bond dipoles point in the same direction, from the atom of low electronegativity to the one of high electronegativity. This is illustrated in the margin. Notice that the bond dipoles add to give a net dipole moment for the molecule.

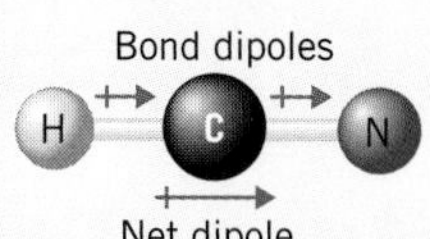

IS THE ANSWER REASONABLE? As in the preceding example, we can check that we've obtained the correct Lewis structure and applied the VSEPR theory correctly, which we have. We can also check to see whether we've used the electronegativities to reach the right conclusions, which we have. We can be confident in our conclusions.

Practice Exercise 7: Is the sulfur tetrafluoride molecule polar or nonpolar? (Hint: Use the VSEPR model to sketch the shape of the molecule.)

Practice Exercise 8: Which of the following molecules would you expect to be polar? (a) SF_6, (b) SO_2, (c) $BrCl$, (d) AsH_3, (e) CF_2Cl_2

9.4 VALENCE BOND THEORY EXPLAINS BONDING AS AN OVERLAP OF ATOMIC ORBITALS

So far we have described the bonding in molecules using Lewis structures. Lewis structures, however, tell us nothing about *why* covalent bonds are formed or *how* electrons manage to be shared between atoms. Nor does the VSEPR model explain *why* electrons group themselves into domains as they do. Thus, we must look beyond these simple models to understand more fully the covalent bond and the factors that determine molecular geometry.

There are fundamentally two theories of covalent bonding that have evolved based on quantum theory, the **valence bond theory** (or **VB theory,** for short) and the **molecular orbital theory (MO theory).** They differ principally in the way they construct a theoretical model of the bonding in a molecule. The valence bond theory imagines individual atoms, each with its own orbitals and electrons, coming together to form the covalent bonds of the molecule. The molecular orbital theory doesn't concern itself with *how* the molecule is formed. It just views a molecule as a collection of positively charged nuclei surrounded in some way by electrons that occupy a set of *molecular orbitals*, in much the same way that the electrons in an atom occupy *atomic orbitals*. (In a sense, MO theory would look at an atom as if it were a special case—a molecule having only one positive center, instead of many.)

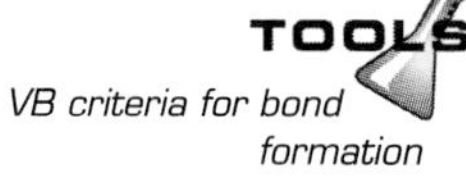

VB criteria for bond formation

According to VB theory, *a bond between two atoms is formed when* ***two electrons*** *with their* ***spins paired*** *are shared by two* ***overlapping*** *atomic orbitals, one orbital from each of the atoms joined by the bond.* By **overlap of orbitals** we mean that portions of two atomic orbitals from different atoms share the same space.

An important part of the theory, as suggested by the bold italic type above, is that only *one* pair of electrons, with paired spins, can be shared by two overlapping orbitals. This

FIG. 9.15 The formation of the hydrogen molecule according to valence bond theory.

electron pair becomes concentrated in the region of overlap and helps "cement" the nuclei together, so the amount that the potential energy is lowered when the bond is formed is determined in part by the extent to which the orbitals overlap. Therefore, *atoms tend to position themselves so that the maximum amount of orbital overlap occurs because this yields the minimum potential energy and therefore the strongest bonds.*

The way VB theory views the formation of a hydrogen molecule is shown in Figure 9.15. As the two atoms approach each other, their $1s$ orbitals begin to overlap and merge as the electron pair spreads out over both orbitals, thereby giving the H—H bond. The description of the bond in H_2 provided by VB theory is essentially the same as that discussed in Chapter 8.

Now let's look at the HF molecule, which is a bit more complex than H_2. Following the usual rules we can write its Lewis structure as

$$\mathrm{H{-}\ddot{\underset{..}{F}}\!:}$$

and we can diagram the formation of the bond as

$$\mathrm{H\cdot + \cdot\ddot{\underset{..}{F}}\!: \longrightarrow H{-}\ddot{\underset{..}{F}}\!:}$$

Our Lewis symbols suggest that the H—F bond is formed by the pairing of electrons, one from hydrogen and one from fluorine. To explain this according to VB theory, we must have two half-filled orbitals, one from each atom, that can be joined by overlap. (They must be half-filled, because we can't place more than two electrons into the bond.) To see clearly what must happen, it is best to look at the orbital diagrams of the valence shells of hydrogen and fluorine.

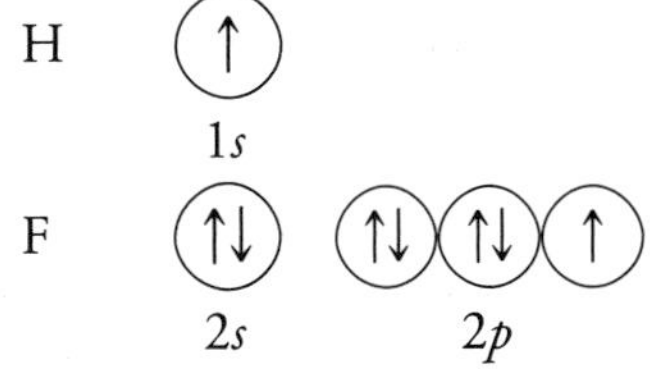

The requirements for bond formation are met by overlapping the half-filled $1s$ orbital of hydrogen with the half-filled $2p$ orbital of fluorine; there are then two orbitals plus two electrons whose spins can adjust so they are paired. The formation of the bond is illustrated in Figure 9.16.

The overlap of orbitals provides a means for sharing electrons, thereby allowing each atom to complete its valence shell. It is sometimes convenient to indicate this using orbital diagrams. For example, the diagram below shows how the fluorine atom completes its $2p$ subshell by acquiring a share of an electron from hydrogen.

F (in HF) (↑↓) (↑↓)(↑↓)(↑↓) (Colored arrow is the H electron.)
2s 2p

Notice that in both the Lewis and VB descriptions of the formation of the H—F bond, the atoms' valence shells are completed.

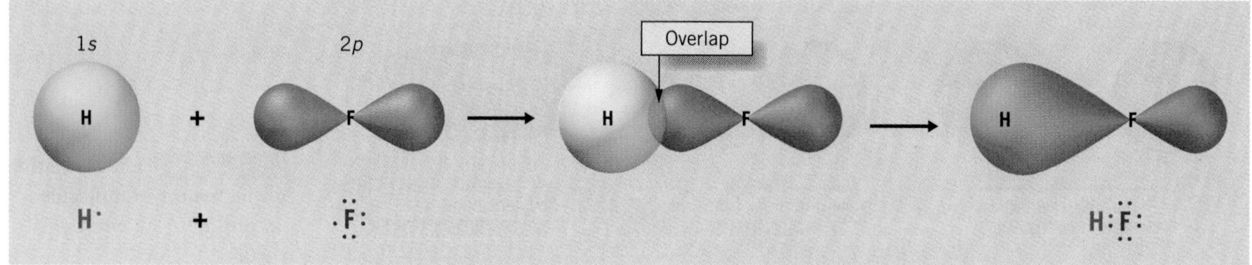

FIG. 9.16 The formation of the hydrogen fluoride molecule according to valence bond theory. For clarity, only the half-filled $2p$ orbital of fluorine is shown. The other $2p$ orbitals of fluorine are filled and cannot participate in bonding.

Valence bond theory can explain bond angles

H_2S is the compound that gives rotten eggs their foul odor.

Let's look now at a more complex molecule, hydrogen sulfide, H_2S. Experiments have shown that it is a nonlinear molecule in which the H—S—H bond angle is about 92°.

$$\underset{92^\circ}{\text{H}\diagup\overset{\text{S}}{}\diagdown\text{H}}$$

Using Lewis symbols, we would diagram the formation of H_2S as

$$2\text{H}\cdot + \cdot\overset{\cdot\cdot}{\underset{\cdot\cdot}{\text{S}}}\cdot \longrightarrow \text{H}-\overset{\cdot\cdot}{\underset{\cdot\cdot}{\text{S}}}-\text{H}$$

If necessary, review the procedure for drawing orbital diagrams on page 273–275 in Chapter 7.

Our Lewis symbols suggest that each H—S bond is formed by the pairing of two electrons, one from H and one from S. Applying this to VB theory, each bond requires the overlap of two half-filled orbitals, one on H and one on S. Therefore, forming *two* H—S bonds in H_2S will require *two* half-filled orbitals on sulfur to form bonds to two separate H atoms. To clearly see what happens, let's look at the orbital diagrams of the valence shells of hydrogen and sulfur.

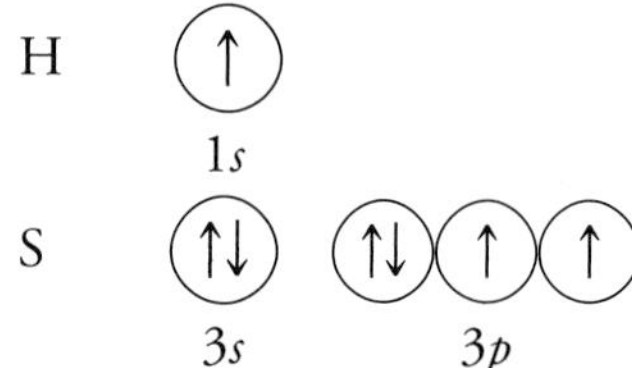

Sulfur has two $3p$ orbitals that each contain only one electron. Each of these can overlap with the $1s$ orbital of a hydrogen atom, as shown in Figure 9.17. This overlap completes the $3p$ subshell of sulfur because each hydrogen provides one electron.

S (in H_2S) (↑↓) (↑↓)(↑↓)(↑↓) (Colored arrows are H electrons.)
3s 3p

In VB theory, two orbitals from different atoms never overlap simultaneously with opposite ends of the same p orbital.

In Figure 9.17, notice that when the $1s$ orbital of a hydrogen atom overlaps with a p orbital of sulfur, the best overlap occurs when the hydrogen atom lies along the axis of the p orbital. Because p orbitals are oriented at 90° to each other, the H—S bonds are expected to be at this angle, too. Therefore, the predicted bond angle is 90°. This is very close to the actual bond angle of 92° found by experiment. Thus, the VB theory requirement for maximum overlap quite nicely explains the geometry of the hydrogen sulfide molecule. Also, notice that both the Lewis and VB descriptions of the formation of the H—S bonds account for the completion of the atoms' valence shells. Therefore, *a Lewis structure can be viewed, in a very qualitative sense, as a shorthand notation for the valence bond description of a molecule.*

Other kinds of orbital overlaps are also possible. For example, according to VB theory the bonding in the fluorine molecule, F_2, occurs by the overlap of two $2p$ orbitals, as shown

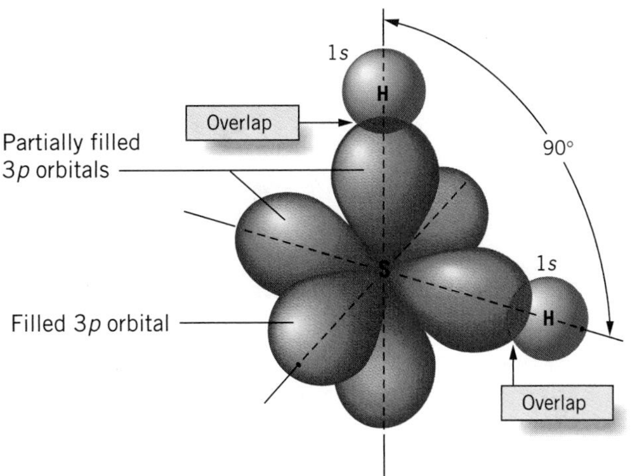

FIG. 9.17 Bonding in H_2S. We expect the hydrogen 1*s* orbitals to position themselves so that they can best overlap with the two partially filled 3*p* orbitals of sulfur, which gives a predicted bond angle of 90°. The experimentally measured bond angle of 92° is very close to the predicted angle.

in Figure 9.18. The formation of the other diatomic molecules of the halogens, all of which are held together by single bonds, could be similarly described.

Practice Exercise 9: Use the principles of VB theory to explain the bonding in HCl. Give the orbital diagram for chlorine in the HCl molecule and indicate the orbital that shares the electron with one from hydrogen. Sketch the orbital overlap that gives rise to the H—Cl bond. (Hint: Remember that a half-filled orbital on each atom is required to form the covalent bond.)

Practice Exercise 10: The phosphine molecule, PH_3, has a trigonal pyramidal shape with H—P—H bond angles equal to 93.7°. Give the orbital diagram for phosphorus in the PH_3 molecule and indicate the orbitals that share electrons with those from hydrogen. On a set of *xyz* coordinate axes, sketch the orbital overlaps that give rise to the P—H bonds.

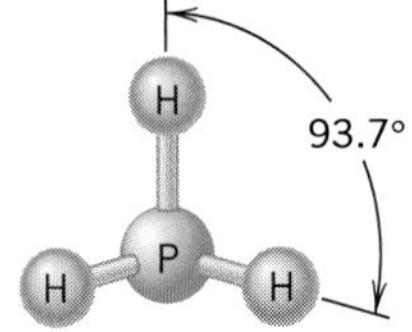

9.5 HYBRID ORBITALS ARE USED TO EXPLAIN EXPERIMENTAL MOLECULAR GEOMETRIES

There are many molecules for which the simple VB theory described above fails to account for the correct shape. For example, there are no simple atomic orbitals that are oriented so they point toward the corners of a tetrahedron, yet there are many tetrahedral molecules. Therefore, to explain the bonds in molecules such as CH_4 we must study the way atomic

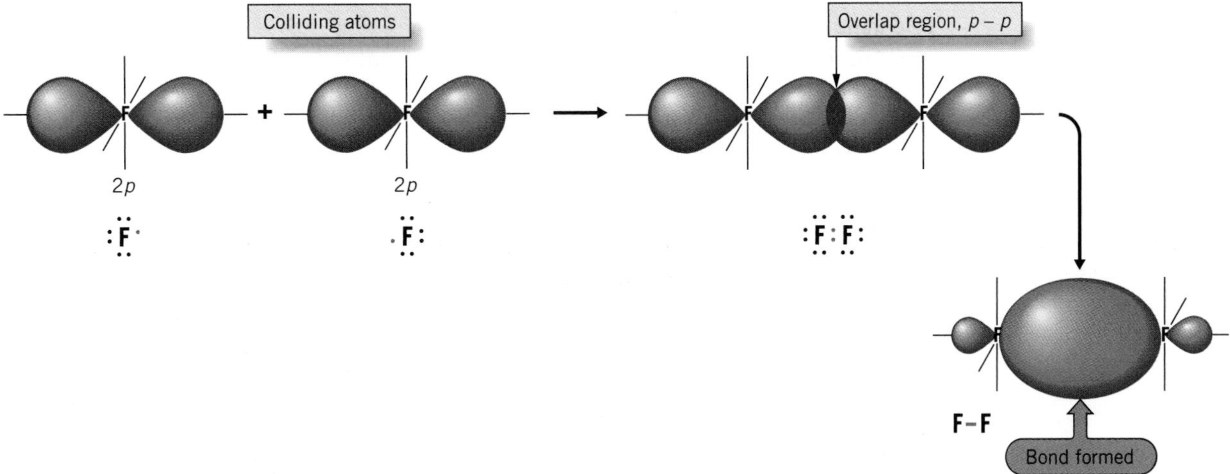

FIG. 9.18 Bonding in the fluorine molecule according to valence bond theory. The two completely filled *p* orbitals on each fluorine atom are omitted, for clarity.

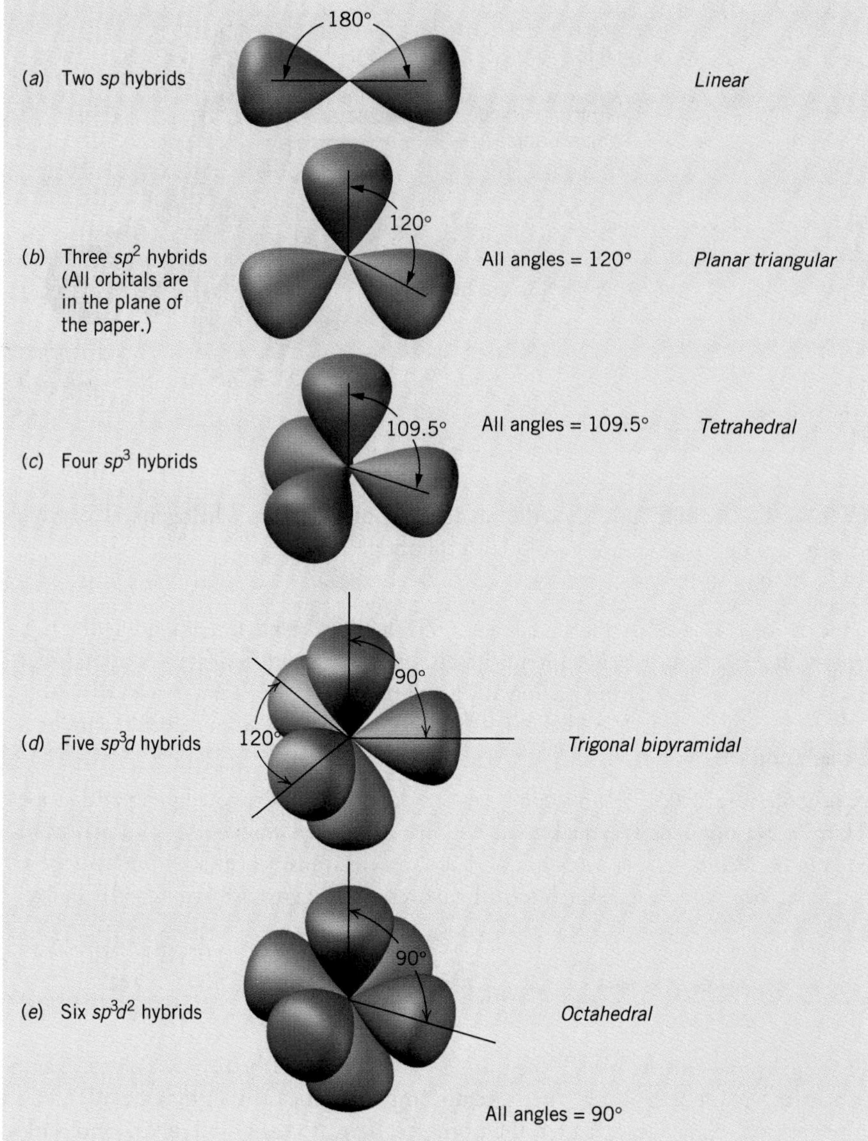

FIG. 9.19 Directional properties of hybrid orbitals formed from *s*, *p*, and *d* atomic orbitals. (*a*) *sp* hybrid orbitals oriented at 180° to each other. (*b*) sp^2 hybrid orbitals formed from an *s* orbital and two *p* orbitals. The angle between them is 120°. (*c*) sp^3 hybrid orbitals formed from an *s* orbital and three *p* orbitals. The angle between them is 109.5°. (*d*) sp^3d hybrid orbitals formed from an *s* orbital, three *p* orbitals, and a *d* orbital. The orbitals point toward the vertices of a trigonal bipyramid. (*e*) sp^3d^2 hybrid orbitals formed from an *s* orbital, three *p* orbitals, and two *d* orbitals. The orbitals point toward the vertices of an octahedron.

orbitals *of the same atom* combine with each other to produce new orbitals with the correct orientations when bonds are formed.

Atomic orbitals (*s*, *p*, and *d*) are the basic building blocks of chemical bonds. To account for the variety of molecular shapes, we often must blend two or more of these basic building blocks to form **hybrid atomic orbitals.**[2] The new orbitals have new shapes and new directional properties, and they can be overlapped to give structures that have bond angles that match those found by experiment. It's important to realize that hybrid orbitals are part of the valence bond *theory.* They can't be directly observed in an experiment. We use them to describe molecular structures that have been determined experimentally.

Figure 9.19 shows the approximate shapes and orientations of five important kinds of hybrid orbitals. You may recognize them as being the same as the orientations of electron domains predicted by the VSEPR model. This is as it should be, of course, because both theories are used to describe the same molecular structures; if they didn't agree with each other, we would have to discard the one that gave incorrect results.

[2] Mathematically, hybrid orbitals are formed by the addition and subtraction of the wave functions for the basic atomic orbitals. This process produces a new set of wave functions corresponding to the hybrid orbitals. The hybrid orbital wave functions describe the shape and directional properties of the orbitals.

In identifying hybrid orbitals, we specify which kinds of pure atomic orbitals, as well as the number of each, that are mixed to form the hybrids. Thus, hybrid orbitals labeled sp^3d^2 are formed by blending one *s* orbital, three *p* orbitals, and two *d* orbitals. *The total number of hybrid orbitals in a set is equal to the number of basic atomic orbitals used to form them.* Therefore, a set of sp^3d^2 hybrids consists of six orbitals, whereas a set of sp^3 orbitals consists of four orbitals.

Figure 9.20 illustrates how an *s* and *p* orbital mix to form a set of two *sp* hybrid orbitals. Notice that each of the hybrid orbitals has the same shape; each has one large lobe and another much smaller one. The large lobe extends farther from the nucleus than either the *s* or *p* orbital from which the hybrid was formed. This allows the hybrid orbital to overlap more effectively with an orbital on another atom when a bond is formed. Therefore, hybrid orbitals form stronger, more stable bonds than would be possible if just simple atomic orbitals were used. In fact, the strength of these bonds is in good agreement with bond strengths that are determined experimentally.

In general, the greater the overlap of two orbitals, the stronger is the bond. At a given distance between nuclei, the greater "reach" of a hybrid orbital gives better overlap than either an *s* or *p* orbital.

Another point to notice in Figure 9.20 is that the large lobes of the two *sp* hybrid orbitals point in opposite directions; that is, they are 180° apart. If bonds are formed by the overlap of these hybrids with orbitals of other atoms, the other atoms will occupy positions on opposite sides of this central atom. Let's look at a specific example, the linear beryllium hydride molecule, BeH_2, as it would be formed in the gas phase.[3]

The orbital diagram for the valence shell of beryllium is

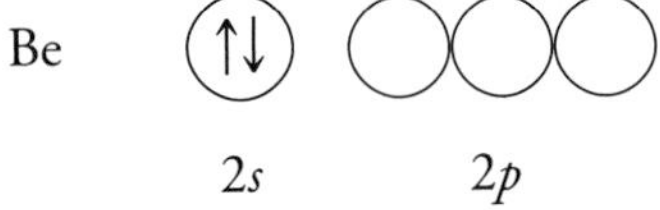
Be
2s
2p

Notice that the 2*s* orbital is filled and the three 2*p* orbitals are empty. For bonds to form at a 180° angle between beryllium and the two hydrogen atoms, two conditions must be met: (1) the two orbitals that beryllium uses to form the Be—H bonds must point in opposite directions, and (2) each of the beryllium orbitals must contain only one electron. In satisfying these requirements, the electrons of the beryllium atom become unpaired and the resulting half-filled *s* and *p* atomic orbitals become hybridized.

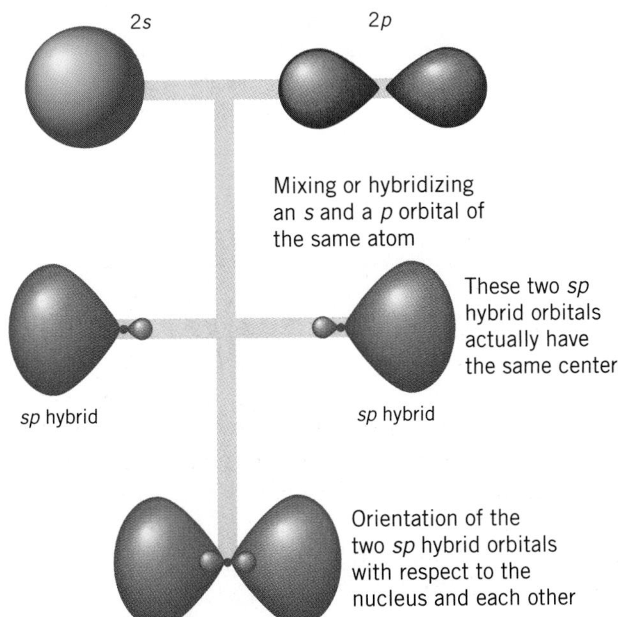
2s
2p
Mixing or hybridizing an s and a p orbital of the same atom
These two sp hybrid orbitals actually have the same center.
sp hybrid
sp hybrid
Orientation of the two sp hybrid orbitals with respect to the nucleus and each other

FIG. 9.20 Formation of *sp* hybrid orbitals. Mixing of the 2*s* and 2*p* atomic orbitals produces a pair of *sp* hybrid orbitals. The large lobes of these orbitals point in opposite directions.

[3] In the solid state, BeH_2 has a complex structure not consisting of simple BeH_2 molecules.

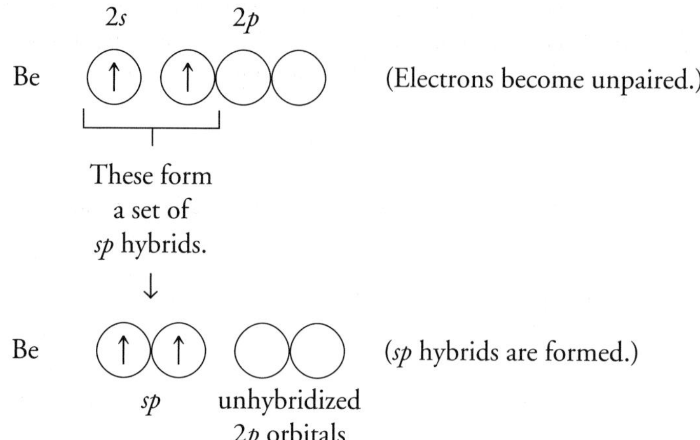

Now the 1*s* orbitals of the hydrogen atoms can overlap with the *sp* hybrids of beryllium to form the bonds, as shown in Figure 9.21. Because the two *sp* hybrid orbitals of beryllium are identical in shape and energy, the two Be—H bonds are alike except for the directions in which they point, and we say that the bonds are *equivalent*. Since the bonds point in opposite directions, the linear geometry of the molecule is also explained. The orbital diagram for beryllium in this molecule is

Be (in BeH_2) (↑↓)(↑↓) ( )( ) (Colored arrows are H electrons.)

sp unhybridized 2*p* orbitals

Even if we had not known the shape of the BeH_2 molecule, we could have obtained the same bonding picture by applying the VSEPR model first. The Lewis structure for BeH_2 is H:Be:H, with two bonding domains around the central atom, so the molecule is linear. Once the shape is known, we can apply the VB theory to explain the bonding in

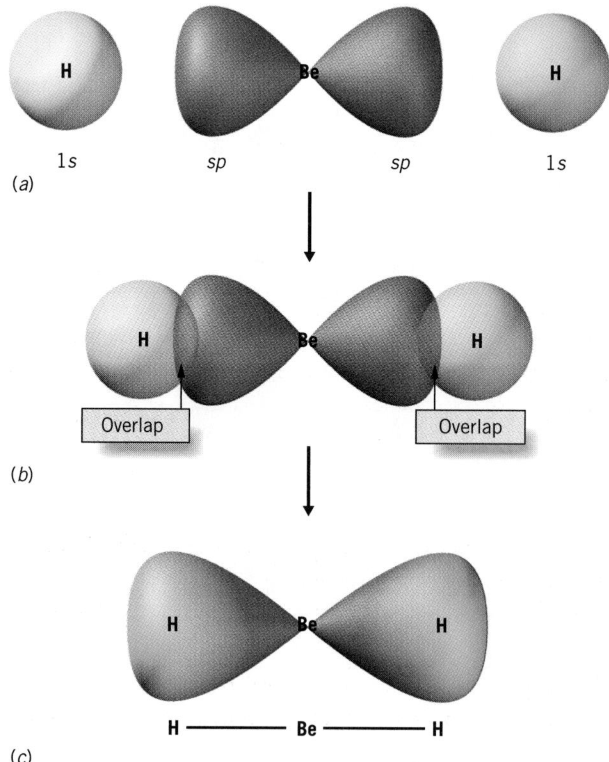

FIG. 9.21 Bonding in BeH_2 according to valence bond theory. Only the larger lobe of each *sp* hybrid orbital is shown. (*a*) The two hydrogen 1*s* orbitals approach the pair of *sp* hybrid orbitals of beryllium. (*b*) Overlap of the hydrogen 1*s* orbitals with the *sp* hybrid orbitals. (*c*) A representation of the distribution of electron density in the two Be—H bonds after they have been formed.

terms of orbital overlaps. Thus, the VB theory and VSEPR model complement each other well. The VSEPR model allows us to predict geometry in a simple way, and once the geometry is known, it is relatively easy to analyze the bonding in terms of VB theory.

EXAMPLE 9.7
Explaining Bonding with Hybrid Orbitals

Methane, CH_4, is a tetrahedral molecule. How is this explained in terms of valence bond theory?

ANALYSIS: No pure atomic orbitals have the correct orientations to form a tetrahedral molecule, so we expect hybrid orbitals will be used. Our tool is Figure 9.19, which permits us to select which hybrids are appropriate based on the geometry of the molecule. Then we can use the orbital diagram of carbon to follow the changes leading to bond formation.

SOLUTION: The tetrahedral structure of the molecule suggests that sp^3 hybrid orbitals are involved in bonding. Let's examine the valence shell of carbon.

C (↑↓) (↑)(↑)( )
2*s* 2*p*

To form four C—H bonds, we need four half-filled orbitals. Unpairing the electrons in the 2*s* and moving one to the vacant 2*p* orbital satisfies this requirement. Then we can hybridize all the orbitals to give the desired sp^3 set.

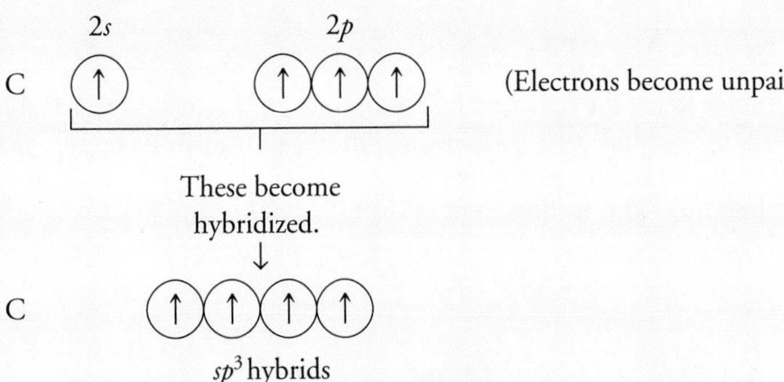

Then we form the four bonds to hydrogen 1*s* orbitals.

C (in CH_4) (↑↓)(↑↓)(↑↓)(↑↓) (Colored arrows are H electrons.)
sp^3

This is illustrated in Figure 9.22.

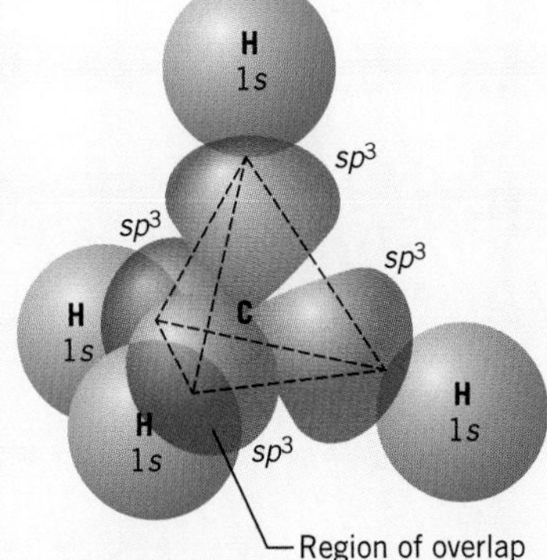

FIG. 9.22 Formation of the bonds in methane. Each bond results from the overlap of a hydrogen 1*s* orbital with an sp^3 hybrid orbital on the carbon atom.

IS THE ANSWER REASONABLE? The positions of the hydrogen atoms around the carbon give the correct tetrahedral shape for the molecule. The Lewis structure for CH_4 shows four bonding domains around the carbon atom, which is consistent with the idea of four sp^3 orbitals overlapping with hydrogen 1*s* orbitals to form four bonds.

Practice Exercise 11: The BCl_3 molecule has a planar triangular shape. What kind of hybrid orbitals does boron use in this molecule? Use orbital diagrams to explain how the bonds are formed. (Hint: Which kind of hybrid orbitals are oriented correctly to give this molecular shape?)

Practice Exercise 12: In the gas phase, beryllium fluoride exists as linear molecules. Which kind of hybrid orbitals does Be use in this compound? Use orbital diagrams to explain how the bonds are formed.

In methane, carbon forms four single bonds with hydrogen atoms by using sp^3 hybrid orbitals. In fact, carbon uses these same kinds of orbitals in all of its compounds in which it is bonded to four other atoms by single bonds. This makes the tetrahedral orientation of atoms around carbon one of the primary structural features of organic compounds, and organic chemists routinely think in terms of "tetrahedral carbon."

In the alkane series of hydrocarbons (compounds with the general formula C_nH_{2n+2}, page 61), carbon atoms are bonded to other carbon atoms. An example is ethane, C_2H_6.

```
    H   H
    |   |
H — C — C — H
    |   |
    H   H
   ethane
```

In this molecule, the carbons are bonded together by the overlap of sp^3 hybrid orbitals (Figure 9.23). One of the most important characteristics of this bond is that the overlap of the orbitals in the C—C bond is hardly affected at all if one portion of the molecule rotates relative to the other around the bond axis. Such rotation, therefore, is said to occur freely and permits different possible relative orientations of the atoms in the molecule. These different relative orientations are called **conformations.** With complex molecules, the number of possible conformations is enormous. For example, Figure 9.24 illustrates three of the large number of possible conformations of the pentane molecule, C_5H_{12}, one of the low molecular weight organic compounds in gasoline.

VSEPR model and hybridization

We can use the VSEPR model to predict hybridization

We've seen that if we know the structure of a molecule, we can make a reasonable guess as to the kind of hybrid orbitals that the central atom uses to form its bonds. Because the VSEPR model works so well in predicting geometry, we can use it to help us obtain VB descriptions of bonding. This is illustrated in the following example.

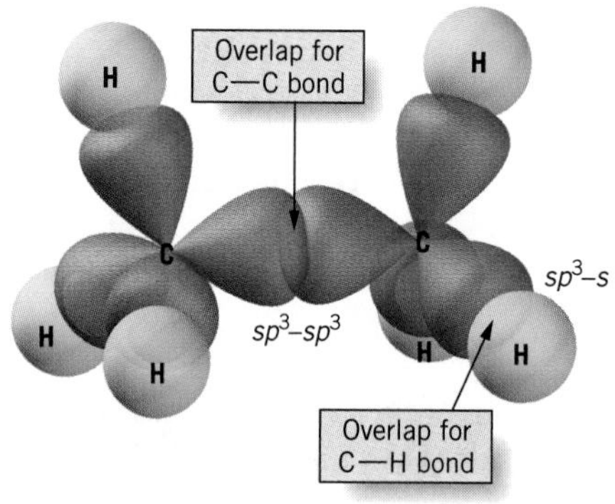

(a)

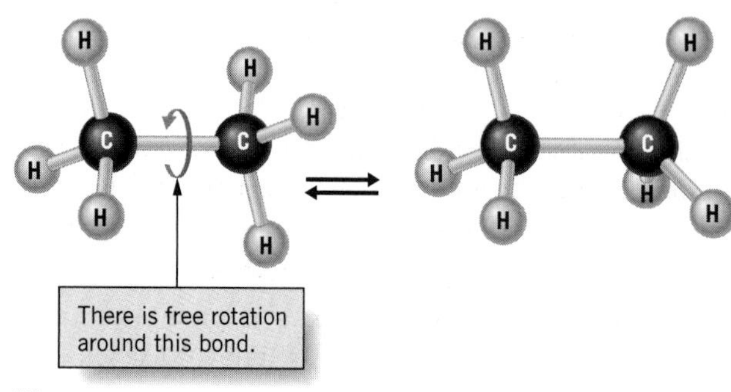

(b)

FIG. 9.23 The bonds in the ethane molecule. (*a*) Overlap of orbitals. (*b*) The degree of overlap of the sp^3 orbitals in the carbon–carbon bond is not appreciably affected by the rotation of the two CH_3— groups relative to each other around the bond.

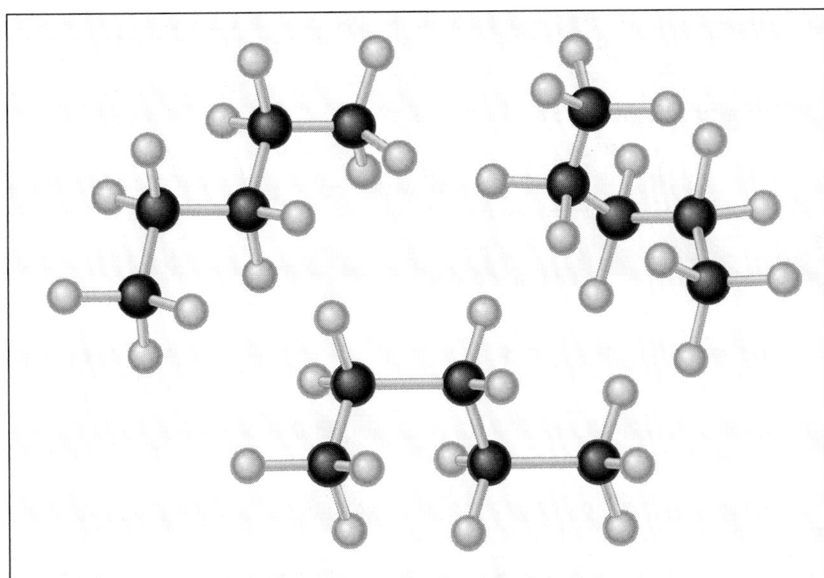

FIG. 9.24 **Three of the many conformations of the atoms in the pentane molecule, C_5H_{12}.** Free rotation around single bonds makes these different conformations possible.

EXAMPLE 9.8
Using the VSEPR Model to Predict Hybridization

Predict the shape of the sulfur hexafluoride molecule and describe the bonding in the molecule in terms of valence bond theory.

ANALYSIS: The rules of nomenclature tell us the chemical formula is SF_6. Based on the formula, we will write a Lewis structure for the molecule. This will permit us to determine its geometry. Then we will examine Figure 9.19 to select the hybrid orbitals that fit this geometry. Finally, we will write the orbital diagram for sulfur to determine which orbitals are used to form the hybrids.

SOLUTION: Following the procedure discussed earlier, the Lewis structure for SF_6 is

```
        :F:
   :F.  |  .F:
      \ | /
        S
      / | \
   :F'  |  'F:
        :F:
```

The VSEPR model tells us the molecule should be octahedral, and referring to Figure 9.19, we find that the hybrid set that fits this structure is sp^3d^2.

As before, let's examine the valence shell of sulfur.

S (↑↓) (↑↓)(↑)(↑)
3s 3p

To form six bonds to fluorine atoms we need six half-filled orbitals, but we show only four orbitals altogether. However, sp^3d^2 orbitals tell us we need to look for d orbitals to include in the set of hybrids.

An isolated sulfur atom has electrons only in its 3s and 3p subshells, so these are the only ones we usually show in the orbital diagram. But the third shell also has a d subshell, which is empty in a sulfur atom. Therefore, let's rewrite the orbital diagram to show the vacant 3d subshell.

Sulfur is able to exceed an octet because of the availability of d orbitals in its valence shell.

S (↑↓) (↑↓)(↑)(↑)
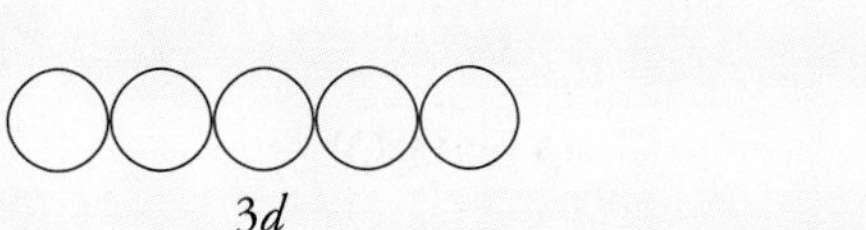
3s 3p 3d

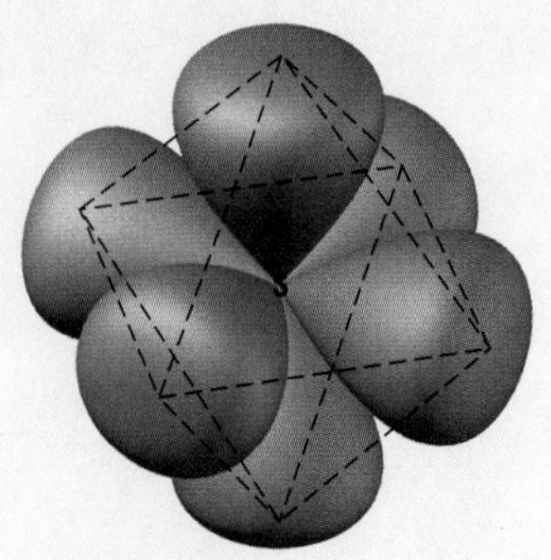

FIG. 9.25 The sp^3d^2 hybrid orbitals of sulfur in SF_6.

Unpairing all of the electrons to give six half-filled orbitals, followed by hybridization, gives the required set of half-filled sp^3d^2 orbitals (Figure 9.25).

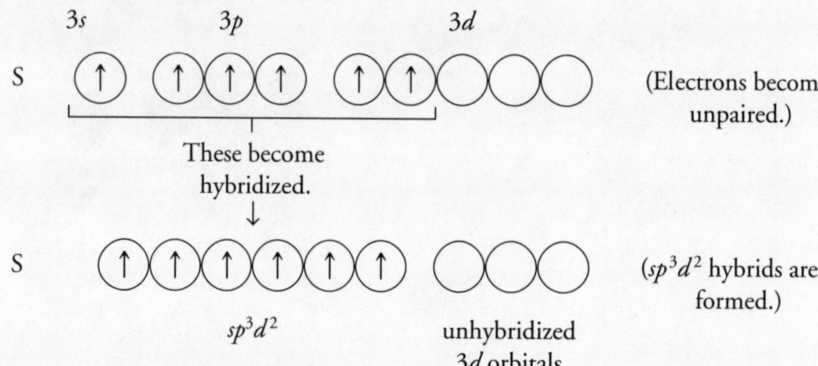

Finally, the six S—F bonds are formed by overlap of the half-filled $2p$ orbitals of the fluorine atoms with these half-filled sp^3d^2 hybrids.

S (in SF_6) (↑↓)(↑↓)(↑↓)(↑↓)(↑↓)(↑↓) ( )( )( ) (Colored arrows are F electrons.)

sp^3d^2 unhybridized 3d orbitals

IS THE ANSWER REASONABLE? The fact that all the parts fit together so well to account for the structure of SF_6 makes us feel that our explanation of the bonding is reasonable.

Practice Exercise 13: What kind of hybrid orbitals are expected to be used by the central atom in PCl_5? (Hint: Which hybrid orbitals have the same geometry as the molecule?)

Practice Exercise 14: What kind of hybrid orbitals are expected to be used by the central atom in SiH_4? Use orbital diagrams to describe the bonding in the molecule.

Practice Exercise 15: Use the VSEPR model to predict the shape of the $AsCl_5$ molecule and then describe the bonding in the molecule using valence bond theory.

Hybrid orbitals can be used to describe molecules with nonbonding domains

Methane is a tetrahedral molecule with sp^3 hybridization of the orbitals of carbon and H—C—H bond angles that are each equal to 109.5°. In ammonia, NH_3, the H—N—H bond angles are 107°, and in water the H—O—H bond angle is 104.5°. Both NH_3 and H_2O have H—X—H bond angles that are close to the bond angles expected for a molecule whose central atom has sp^3 hybrids. The use of sp^3 hybrids by oxygen and nitrogen, therefore, is often used to explain the geometry of H_2O and NH_3.

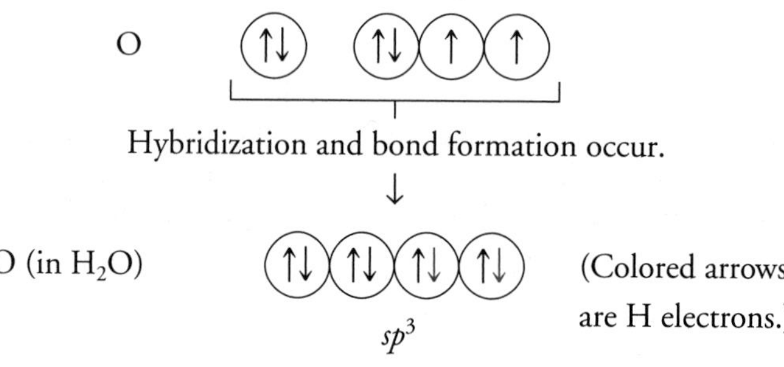

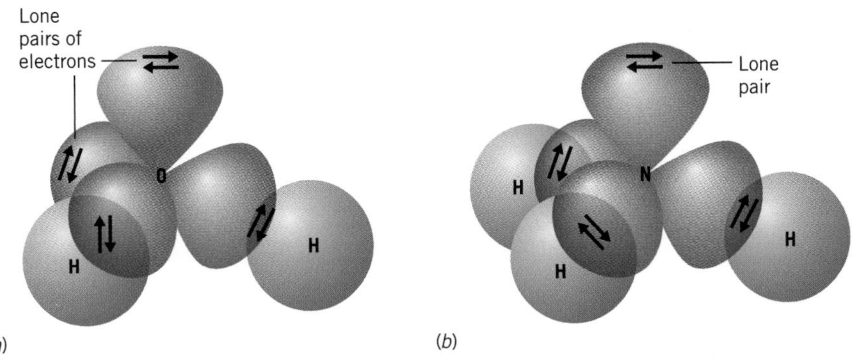

FIG. 9.26 **Hybrid orbitals can hold lone pairs of electrons.** (*a*) In water, two lone pairs on oxygen are held in sp^3 hybrid orbitals. (*b*) Ammonia has one lone pair in an sp^3 hybrid orbital.

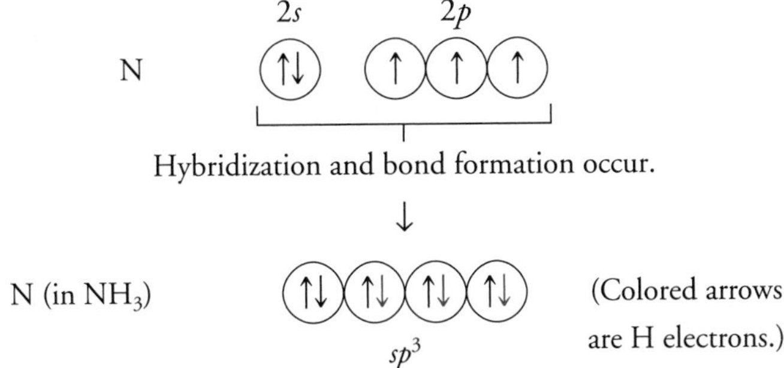

According to these descriptions, not all of the hybrid orbitals of the central atom must be used for bonding. Lone pairs of electrons can be accommodated in them too, as illustrated in Figure 9.26. In fact, putting the lone pair on the nitrogen in an sp^3 hybrid orbital gives a geometry that agrees well with the experimentally determined structure of the ammonia molecule.

EXAMPLE 9.9
Explaining Bonding with Hybrid Orbitals

Use valence bond theory to explain the bonding in the SF_4 molecule.

ANALYSIS: Once again, our primary tool in answering the question is the Lewis structure, which will tell us the number of electron domains around the central atom. We'll use this information to select the kinds of hybrid orbitals used by sulfur in the molecule. Then we can construct the orbital diagrams as before.

SOLUTION: Let's begin by constructing the Lewis structure for the molecule. Following the usual procedure, we obtain

```
      ..
     :F:
      |
  ..  |  ..
 :F—S—F:
  ..  ..  ..
      |
     :F:
      ..
```

The VSEPR model predicts that the electron pairs around the sulfur should be in a trigonal bipyramidal arrangement, and the only hybrids that fit this geometry are sp^3d. To see how they are formed, we look at the valence shell of sulfur, including the vacant 3*d* subshell.

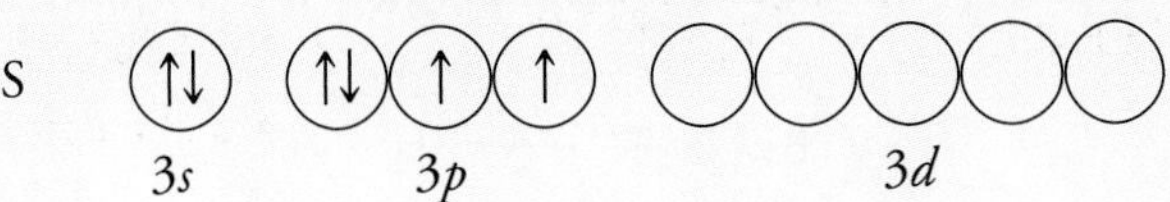

To form the four bonds to fluorine atoms, we need four half-filled orbitals, so we unpair the electrons in one of the filled orbitals. This gives

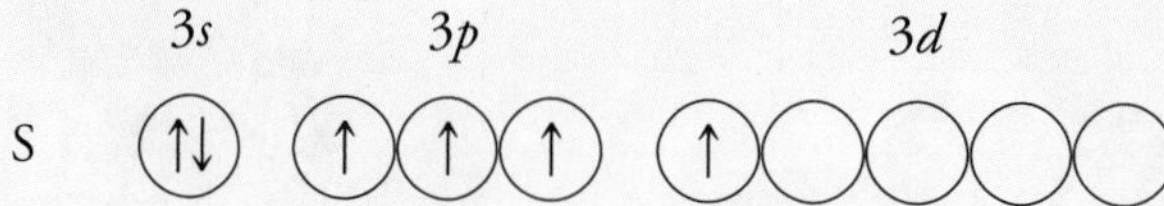

Next, we form the hybrid orbitals. In doing this, we use all the valence shell orbitals that have electrons in them.

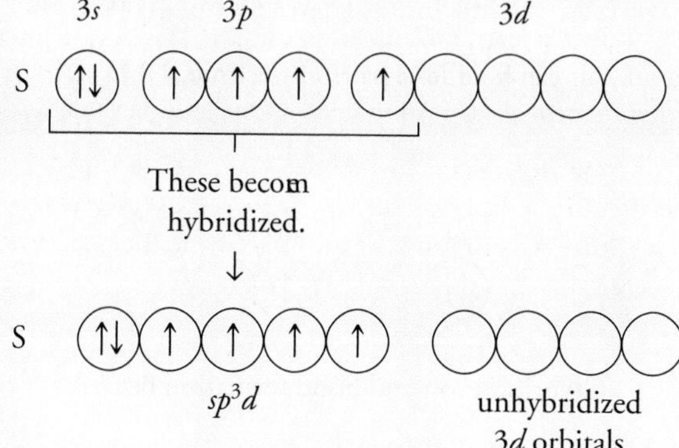

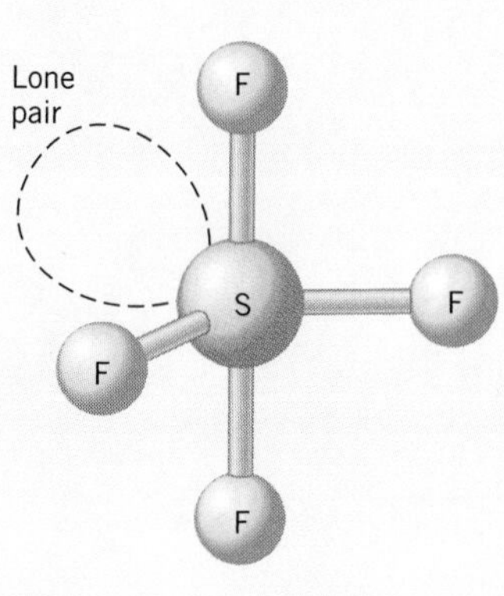

The structure of SF_4

Now, four S—F bonds can be formed by overlap of half-filled $2p$ orbitals of fluorine with the sp^3d hybrid orbitals of sulfur.

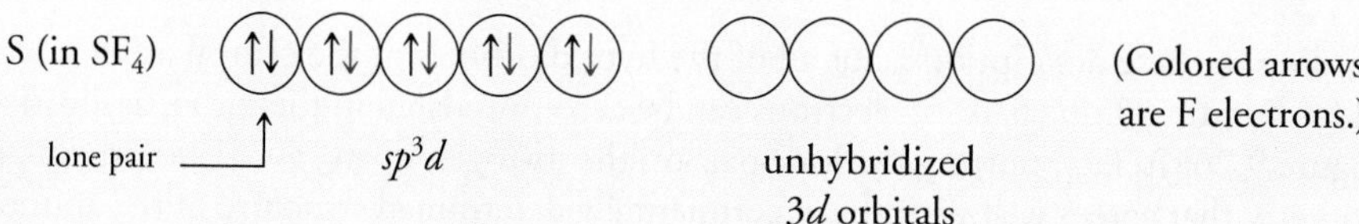

IS THE ANSWER REASONABLE? Counting electrons in the Lewis structure gives a total of 34, which is how many electrons are in the valence shells of one sulfur and four fluorine atoms, so the Lewis structure appears to be correct. The rest of the answer flows smoothly, so the bonding description appears to be reasonable.

Practice Exercise 16: What kind of orbitals are used by Xe in the XeF_4 molecule? (Hint: An Xe atom has eight valence electrons.)

Practice Exercise 17: What kind of hybrid orbitals would we expect the central atom to use for bonding in (a) PCl_3 and (b) ClF_3?

Hybrid orbitals can be used to explain the formation of coordinate covalent bonds

In Section 8.6 we defined a coordinate covalent bond as one in which both of the shared electrons are provided by just one of the joined atoms. Such a bond is formed when boron trifluoride, BF_3, combines with an additional fluoride ion to form the tetrafluoroborate ion, BF_4^-.

$$BF_3 + F^- \longrightarrow \underset{\text{tetrafluoroborate ion}}{BF_4^-}$$

We can diagram this reaction as follows:

```
   :F:                   :F:    -
:F:B  + [:F:]-  ---> [ :F:B:F: ]
   :F:                   :F:
```

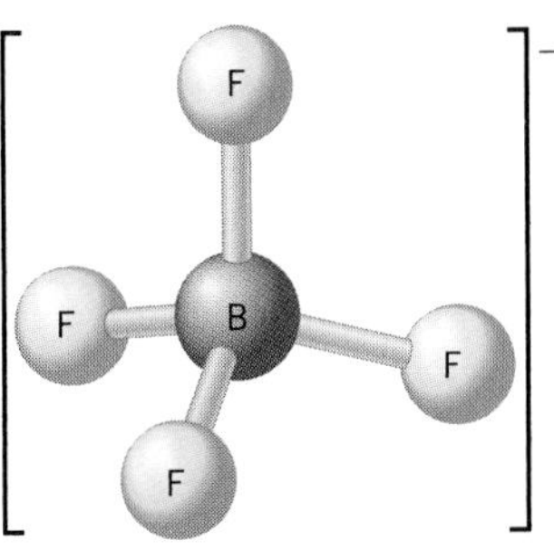

FIG. 9.27 **The tetrahedral structure of the BF_4^- ion.**

As we mentioned previously, the coordinate covalent bond is really no different from any other covalent bond once it has formed. The distinction is made *only* for bookkeeping purposes. One place where such bookkeeping is useful is in keeping track of the orbitals and electrons used when atoms bond together.

The VB theory requirements for bond formation—two overlapping orbitals sharing two paired electrons—can be satisfied in two ways. One, as we have already seen, is by the overlapping of two half-filled orbitals. This gives an "ordinary" covalent bond. The other is by overlapping one filled orbital with one empty orbital. The atom with the filled orbital donates the shared pair of electrons, and a coordinate covalent bond is formed.

The structure of the BF_4^- ion, which the VSEPR model predicts to be tetrahedral (Figure 9.27), can be explained as follows. First we examine the orbital diagram for boron.

	$2s$	$2p$
B	(↑↓)	(↑)( )( )

To form four bonds, we need four hybrid orbitals arranged tetrahedrally around the boron, so we expect boron to use sp^3 hybrids. Notice we spread the electrons out over the hybrid orbitals as much as possible.

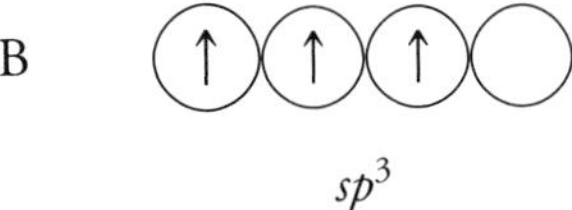

Boron forms three ordinary covalent bonds with fluorine atoms plus one coordinate covalent bond with a fluoride ion.

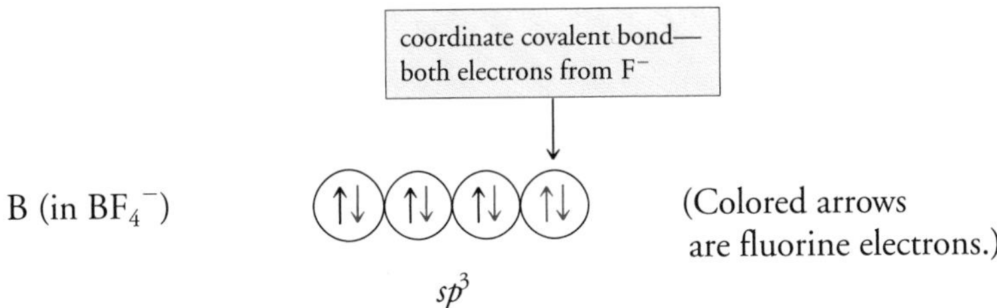

Practice Exercise 18: If we assume nitrogen uses sp^3 hybrid orbitals in NH_3, use valence bond theory to account for the formation of NH_4^+ from NH_3 and H^+. (Hint: Which atom donates the pair of electrons in the formation of the bond between H^+ and NH_3?)

Practice Exercise 19: What is the shape of the PCl_6^- ion? What hybrid orbitals are used by phosphorus in PCl_6^-? Draw the orbital diagram for phosphorus in PCl_6^-.

9.6 Hybrid Orbitals Can Be Used to Describe Multiple Bonds

The types of orbital overlap that we have described so far produce bonds in which the electron density is concentrated most heavily between the nuclei of the two atoms along an imaginary line that joins their centers. Any bond of this kind, whether formed from the overlap of s orbitals, p orbitals, or hybrid orbitals (Figure 9.28), is called a **sigma bond** (or **σ bond**).

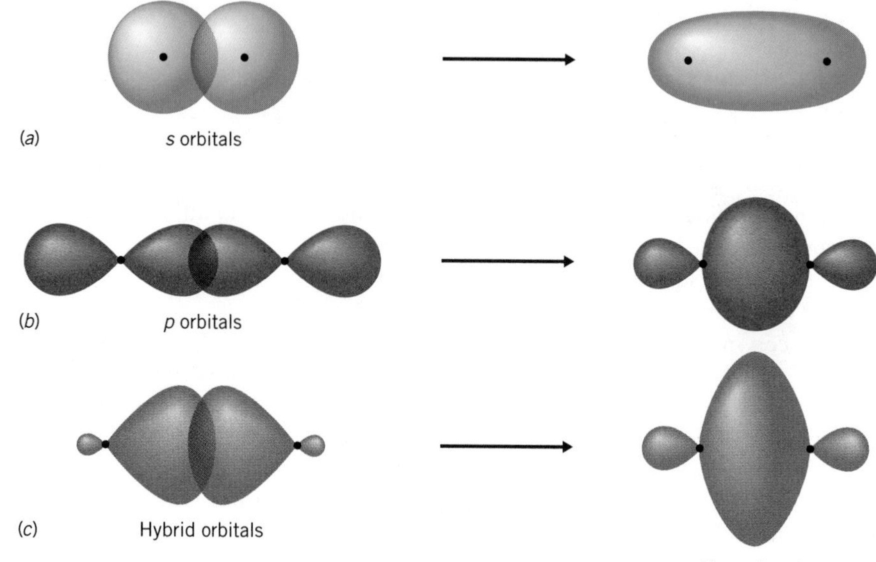

FIG. 9.28 Formation of σ bonds. Sigma bonds concentrate electron density along the line between the two atoms joined by the bond. (*a*) From the overlap of *s* orbitals. (*b*) From the end-to-end overlap of *p* orbitals. (*c*) From the overlap of hybrid orbitals.

Another way that *p* orbitals can overlap is shown in Figure 9.29. This produces a bond in which the electron density is divided between two separate regions that lie on opposite sides of an imaginary line joining the two nuclei. This kind of bond is called a **pi bond** (or **π bond**). Notice that a π bond, like a *p* orbital, consists of two parts, and each part makes up just half of the π bond; it takes *both* of them to equal *one* π bond. The formation of π bonds allows atoms to form double and triple bonds.

A double bond consists of a sigma bond and a pi bond

A hydrocarbon that contains a double bond is ethylene (also called ethene), C_2H_4. It has the Lewis structure

```
H       H
 \     /
  C=C
 /     \
H       H
```

ethylene

The molecule is planar and each carbon atom lies in the center of a triangle surrounded by three other atoms (two H and one C atom). A planar triangular arrangement of bonds suggests that carbon uses sp^2 hybrid orbitals. Therefore, let's look at the distribution of electrons among the orbitals that carbon has available in its valence shell, assuming sp^2 hybridization.

Overlap

Bond axis

Overlap

One π bond consisting of two separate regions of electron density

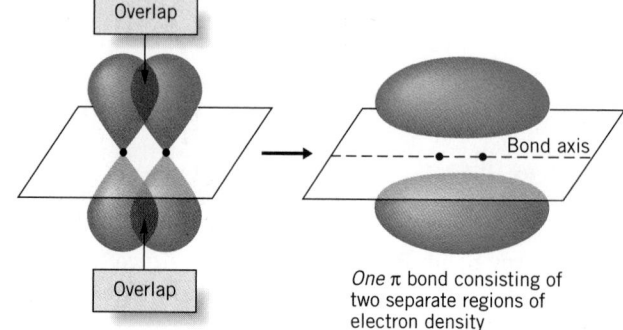

FIG. 9.29 Formation of a π bond. Two *p* orbitals overlap sideways instead of end-to-end. The electron density is concentrated in two regions on opposite sides of the bond axis.

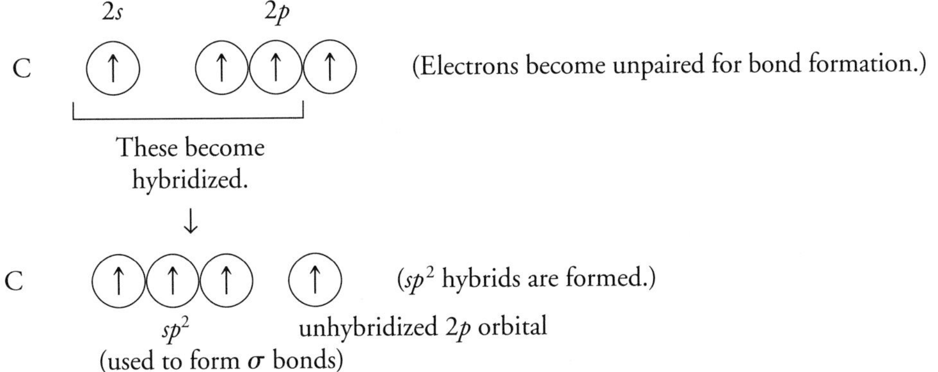

Notice that the carbon atom has an unpaired electron in an unhybridized $2p$ orbital. This p orbital is oriented perpendicular to the triangular plane of the sp^2 hybrid orbitals, as shown in Figure 9.30. Now we can see how the molecule goes together.

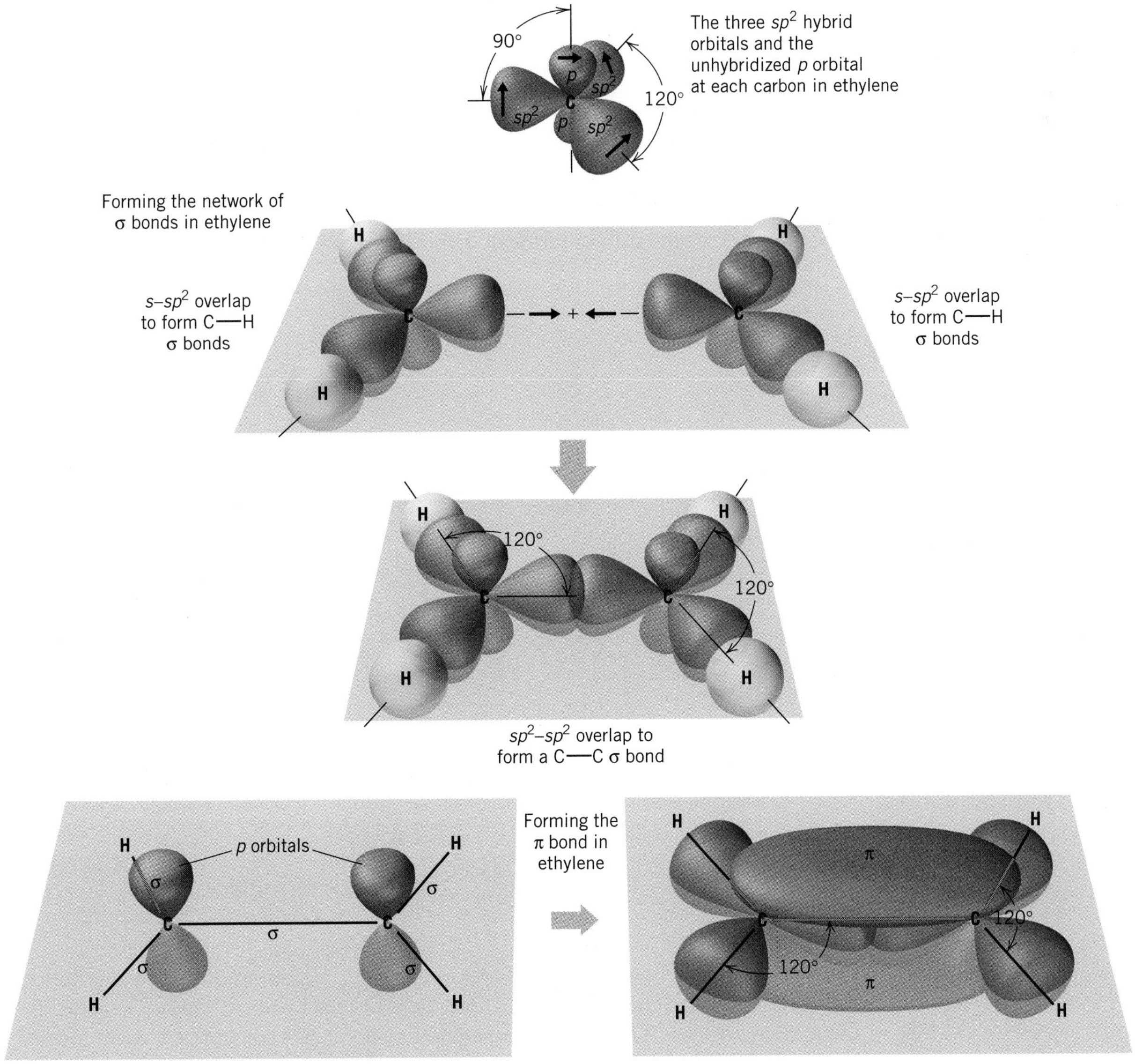

FIG. 9.30 **The carbon–carbon double bond.**

The basic framework of the molecule is determined by the formation of σ bonds. Each carbon uses two of its sp^2 hybrids to form σ bonds to hydrogen atoms. The third sp^2 hybrid on each carbon is used to form a σ bond between the two carbon atoms, thereby accounting for one of the two bonds of the double bond. Finally, the remaining unhybridized $2p$ orbitals, one from each carbon atom, overlap to produce a π bond, which accounts for the second bond of the double bond.

The double bond is a little like a hot dog on a bun. The sigma bond is like the hot dog, and the π bond is like the two parts of the bun.

This description of the bonding in C_2H_4 accounts for one of the most important properties of double bonds: rotation of one portion of the molecule relative to the rest around the axis of the double bond occurs only with great difficulty. The reason for this is illustrated in Figure 9.31. We see that as one CH_2 group is rotated relative to the other around the carbon–carbon bond, the unhybridized p orbitals become misaligned and can no longer overlap effectively. This destroys the π bond. In effect, then, rotation around a double bond involves bond breaking, which requires more energy than is normally available to molecules at room temperature. As a result, rotation around the axis of a double bond usually doesn't take place.

Restricted rotation around double bonds affects the properties of organic and biochemical molecules.

In almost every instance, a double bond consists of a σ bond and a π bond. Another example is the compound formaldehyde (the substance used as a preservative for biological specimens and as an embalming fluid). The Lewis structure of this compound is

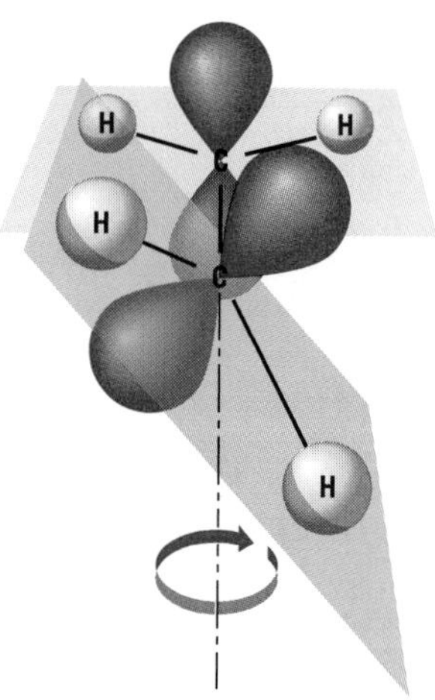

FIG. 9.31 Restricted rotation around a double bond. As the CH_2 group closest to us rotates relative to the one at the rear, the unhybridized p orbitals become misaligned, as shown here. This destroys the overlap and breaks the π bond. Bond breaking requires a lot of energy, more than is available to the molecule through the normal bending and stretching of its bonds at room temperature. Because of this, rotation about the double bond axis is hindered or "restricted."

$$\begin{array}{l} H \\ \quad \backslash \\ \quad\; C{=}\ddot{\underset{..}{O}} \\ \quad / \\ H \end{array}$$

formaldehyde

As with ethylene, the carbon forms sp^2 hybrids, leaving an unpaired electron in an unhybridized p orbital.

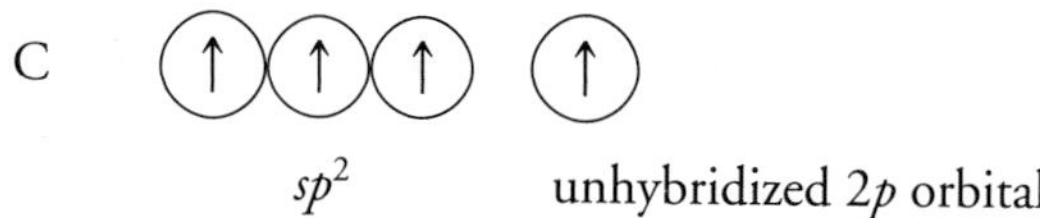

The oxygen can also form sp^2 hybrids, with electron pairs in two of them and an unpaired electron in the third. This means that the remaining unhybridized p orbital also has an unpaired electron.

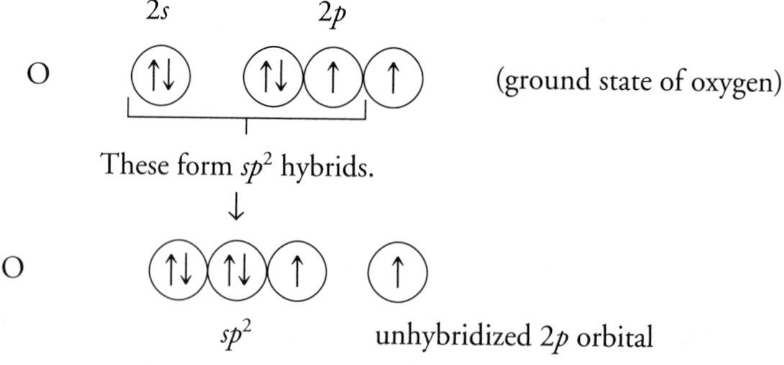

The two filled sp^2 hybrids on the oxygen become lone pairs on the oxygen atom in the molecule.

Figure 9.32 shows how the carbon, hydrogen, and oxygen atoms form the molecule. As before, the basic framework of the molecule is formed by the σ bonds. These determine the molecular shape. The carbon–oxygen double bond also contains a π bond formed by the overlap of the unhybridized p orbitals.

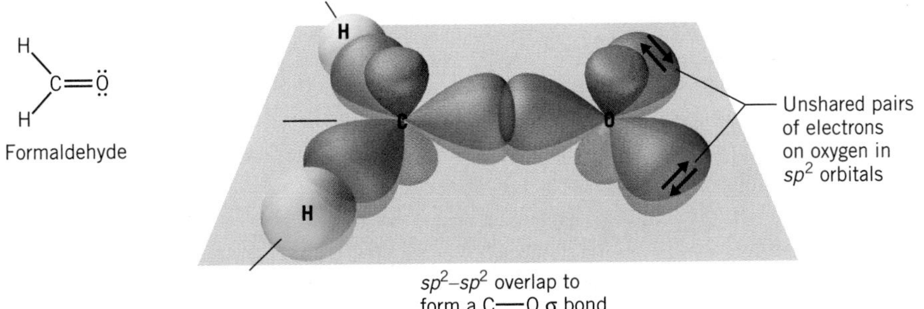

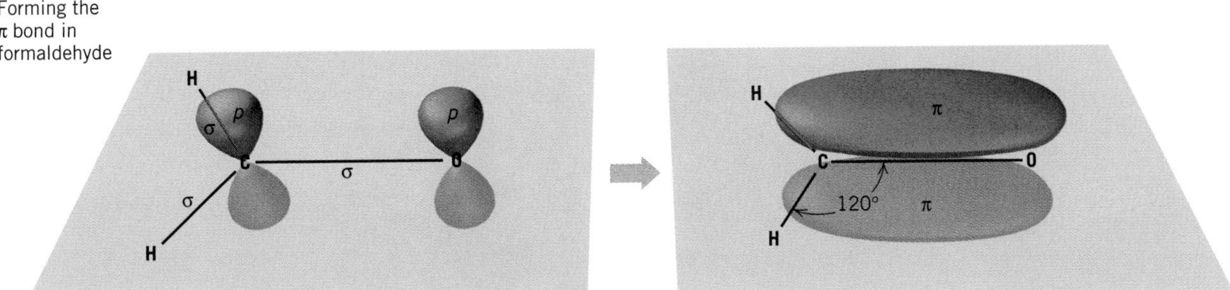

FIG. 9.32 Bonding in formaldehyde. The carbon–oxygen double bond consists of a σ bond and a π bond. The σ bond is formed by overlap of sp^2 hybrid orbitals. Overlap of unhybridized p orbitals on the two atoms gives the π bond.

A triple bond consists of a sigma bond and two pi bonds

An example of a molecule containing a triple bond is ethyne, also known as acetylene, C_2H_2 (a gas used as a fuel for welding torches).

$$\mathrm{H{-}C{\equiv}C{-}H}$$

ethyne
(acetylene)

In the linear acetylene molecule, each carbon needs two hybrid orbitals to form two σ bonds—one to a hydrogen atom and one to the other carbon atom. These can be provided by mixing the $2s$ and one of the $2p$ orbitals to form sp hybrids. To help us visualize the bonding, we will imagine that there is an xyz coordinate system centered at each carbon atom and that it is the $2p_z$ orbital that becomes mixed in the hybrid orbitals.

We label the orbitals p_x, p_y, and p_z just for convenience; they are really all equivalent.

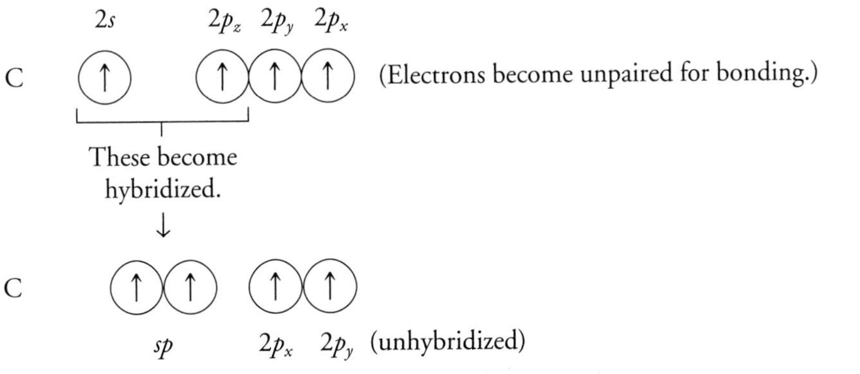

Figure 9.33 shows how the molecule is formed. The sp orbitals point in opposite directions and are used to form the σ bonds. The unhybridized $2p_x$ and $2p_y$ orbitals are perpendicular to the C—C bond axis and overlap sideways to form two separate π bonds that surround the C—C σ bond. Notice that we now have three pairs of electrons in three bonds—one σ bond and two π bonds—whose electron densities are concentrated in different places. Also notice that the use of sp hybrid orbitals for the σ bonds allows us to explain the linear arrangement of atoms in the molecule.

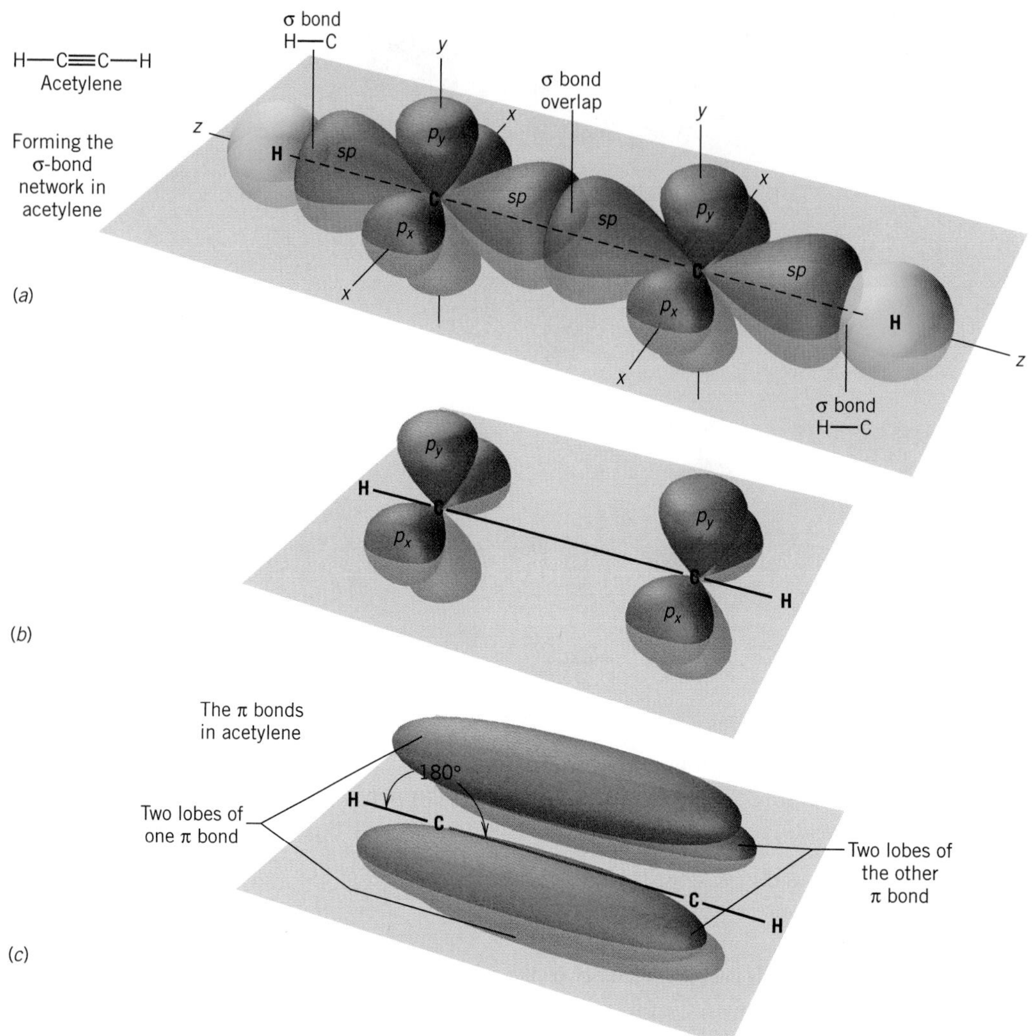

FIG. 9.33 The carbon–carbon triple bond in acetylene. (*a*) The *sp* hybrid orbitals on the carbon atoms are used to form sigma bonds to the hydrogen atoms and to each other. This accounts for one of the three bonds between the carbon atoms. (*b*) Sideways overlap of unhybridized $2p_x$ and $2p_y$ orbitals of the carbon atoms produces two π bonds. (*c*) The two π bonds in acetylene after they've formed surround the σ bond.

Similar descriptions can be used to explain the bonding in other molecules that have triple bonds. Figure 9.34, for example, shows how the nitrogen molecule, N_2, is formed. In it, too, the triple bond is composed of one σ bond and two π bonds.

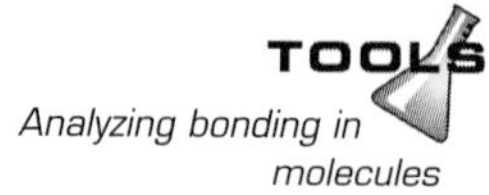

A Brief Summary

On the basis of the preceding discussion, we can make some observations that are helpful in applying the valence bond theory to a variety of molecules.

1. The basic molecular framework of a molecule is determined by the arrangement of its σ bonds.
2. Hybrid orbitals are used by an atom to form its σ bonds and to hold lone pairs of electrons.
3. The number of hybrid orbitals needed by an atom in a structure equals the number of atoms to which it is bonded *plus* the number of lone pairs of electrons in its valence shell.
4. When there is a double bond in a molecule, it consists of one σ bond and one π bond.
5. When there is a triple bond in a molecule, it consists of one σ bond and two π bonds.

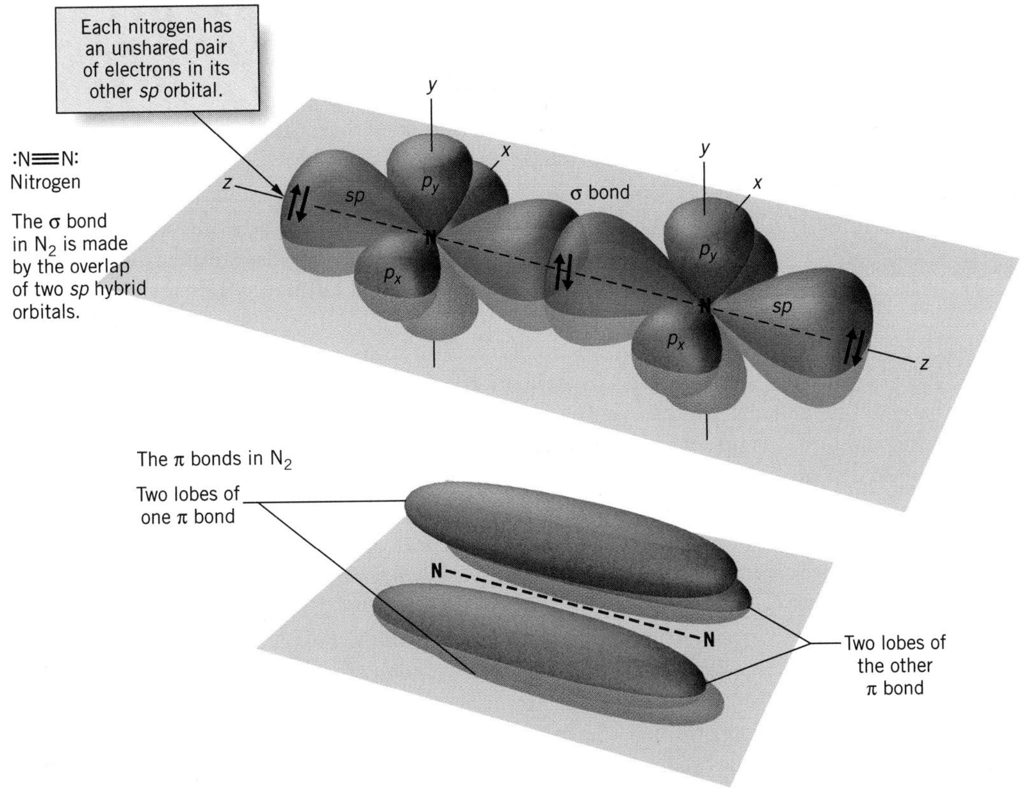

FIG. 9.34 Bonding in nitrogen. The triple bond in nitrogen, N_2, is formed like the triple bond in acetylene. A sigma bond is formed by overlap of *sp* hybrid orbitals. The two unhybridized 2*p* orbitals on each nitrogen atom overlap to give the two π bonds. On each nitrogen, there is a lone pair of electrons in the *sp* hybrid orbital that's not used to form the sigma bond.

Practice Exercise 20: Consider the molecule below. What kind of hybrid orbitals are used by atoms 1, 2, and 3? How many sigma bonds and pi bonds are in the molecule? (Hint: Study the brief summary above.)

```
       ①
        ↘
        :O:  H   H   H
         ‖   |   |   |
     H—C—C—C=C—H
            |
         ②  H    ③
```

(Atom ① is the O; atom ② is the second C (bonded to H above and below); atom ③ is the third C.)

Practice Exercise 21: Consider the molecule below. What kind of hybrid orbitals are used by atoms 1, 2, and 3? How many sigma bonds and pi bonds are in the molecule?

```
      ①       H   H   H   ③
               |   |   |
    H—C≡C—C=C—N—H
                      ..
            ②
```

(Atom ① is the second C of the C≡C; atom ② is the third C; atom ③ is the N.)

9.7 MOLECULAR ORBITAL THEORY EXPLAINS BONDING AS CONSTRUCTIVE INTERFERENCE OF ATOMIC ORBITALS

Molecular orbital theory takes the view that a molecule is similar to an atom in one important respect. Both have energy levels that correspond to various orbitals that can be populated by electrons. In atoms, these orbitals are called atomic orbitals; in molecules, they are called **molecular orbitals.** (We shall frequently call them **MOs.**)

In most cases, the actual shapes and energies of molecular orbitals cannot be determined exactly. Nevertheless, reasonably good estimates of their shapes and energies can be obtained by combining the electron waves corresponding to the atomic orbitals of the atoms that make up the molecule. In forming molecular orbitals, these waves interact by constructive and destructive interference just like other waves we've seen. Their intensities are either added or subtracted when the atomic orbitals overlap.

The number of MOs formed is always equal to the number of atomic orbitals that are combined.

Figure 9.35 illustrates the formation of molecular orbitals by the overlap of two 1*s* orbitals. Notice that the *two* 1*s* atomic orbitals combine to give *two* MOs. In one MO, the intensities of the electron waves add between the nuclei, which gives a buildup of electron density that helps hold the nuclei near each other. Such an MO is called a **bonding molecular orbital.** *Electrons in bonding MOs tend to stabilize a molecule.* In the other MO, cancellation of the electron waves reduces the electron density between the nuclei, which allows the nuclei to repel each other strongly. This is an **antibonding molecular orbital.** *Antibonding MOs tend to destabilize a molecule when occupied by electrons.*

Both MOs in Figure 9.35 have their maximum electron density on an imaginary line that passes through the two nuclei, giving them properties of sigma bonds. MOs like this are also designated as sigma (σ), with a subscript showing which atomic orbitals make up the MO. An asterisk indicates which is an antibonding MO. Thus, the bonding and antibonding MOs formed by overlap of 1*s* orbitals are symbolized as σ_{1s} and σ_{1s}^* respectively.

Bonding MOs are lower in energy than antibonding MOs formed from the same atomic orbitals, as shown in Figure 9.35. When electrons populate molecular orbitals, they fill the lower energy, bonding MOs first. The rules that apply to filling MOs are the same as those for filling atomic orbitals: *Electrons spread out over molecular orbitals of equal energy (Hund's rule) and two electrons can only occupy the same orbital if their spins are paired.* When filling the MOs, we also have to be sure we've accounted for all of the valence electrons of the separate atoms.

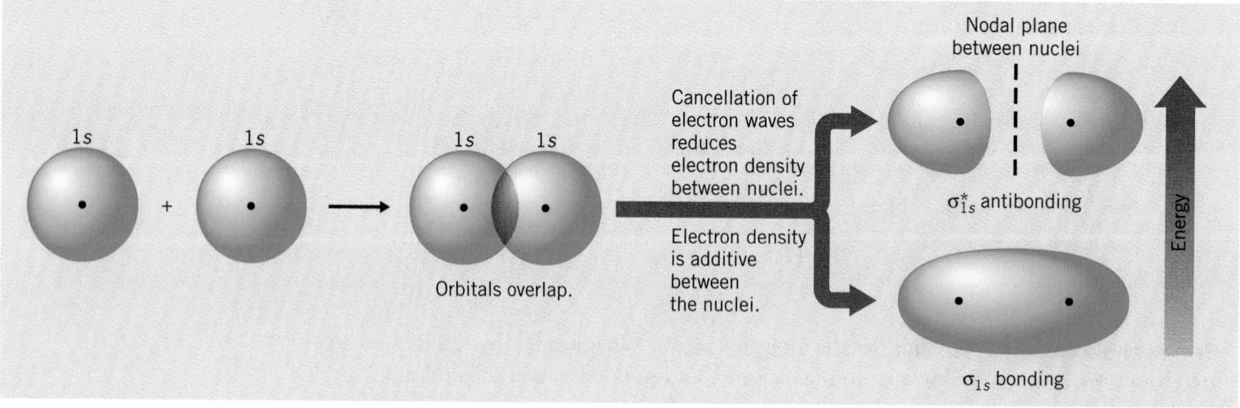

FIG. 9.35 Interaction of 1*s* atomic orbitals to produce bonding and antibonding molecular orbitals. These are σ-type orbitals because the electron density is concentrated along the imaginary line that passes through both nuclei. The antibonding orbital has a nodal plane between the nuclei where the electron density drops to zero.

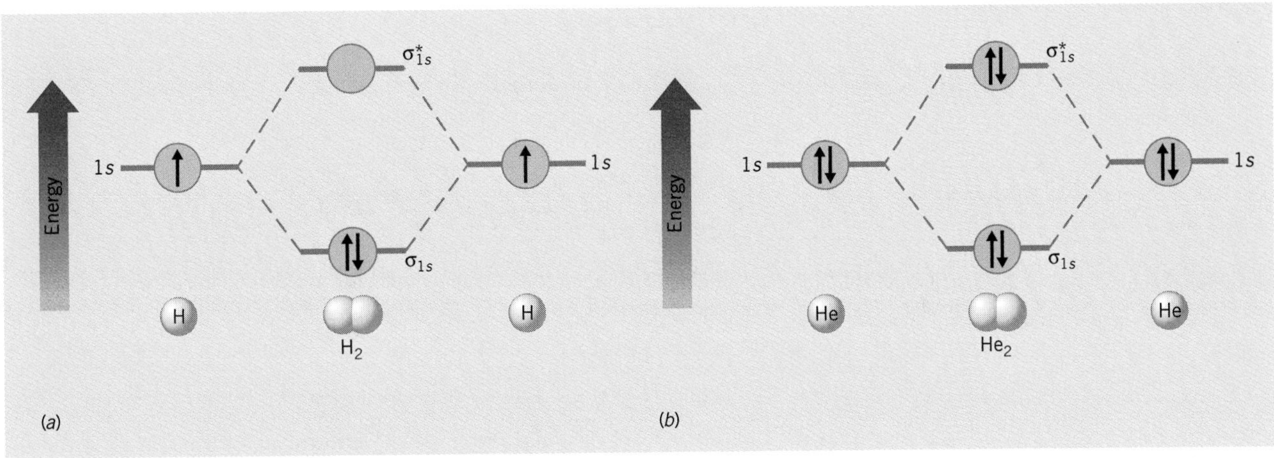

FIG. 9.36 **Molecular orbital descriptions of H_2 and He_2.** (*a*) Molecular orbital energy level diagram for H_2. (*b*) Molecular orbital energy level diagram for He_2.

Molecular orbital theory can explain why some molecules exist and others do not

Figure 9.36*a* is an MO energy level diagram for H_2. The energies of the separate 1*s* atomic orbitals are indicated at the left and right; those of the molecular orbitals are shown in the center. The H_2 molecule has two electrons, and both can be placed in the σ_{1s} orbital. The shape of this bonding orbital, shown in Figure 9.35, should be familiar. It's the same as the shape of the electron cloud that we described using the valence bond theory.

Next, let's consider what happens when two helium atoms come together. Why can't a stable molecule of He_2 be formed? Figure 9.36*b* is the energy diagram for He_2. Notice that both bonding and antibonding orbitals are filled. In situations such as this there is a net destabilization because the antibonding MO is raised in energy more than the bonding MO is lowered, relative to the orbitals of the separated atoms. This means the total energy of He_2 is larger than that of two separate He atoms, so the "molecule" is unstable and immediately comes apart.

In general, the effects of **antibonding electrons** (those in antibonding MOs) cancel the effects of an equal number of bonding electrons, and molecules with equal numbers of bonding and antibonding electrons are unstable. If we remove an antibonding electron from He_2 to give He_2^+, there is a net excess of bonding electrons, and the ion should be capable of existence. In fact, the emission spectrum of He_2^+ can be observed when an electric discharge is passed through a helium-filled tube, which shows that He_2^+ is present during the electric discharge. However, the ion is not very stable and cannot be isolated.

Bond order is related to the difference in the number of electron pairs in bonding and antibonding orbitals

The concept of **bond order** was introduced in Section 8.6 where it was defined as the number of pairs of electrons shared between two atoms. To translate the MO description into these terms, we compute the bond order as follows:

$$\text{Bond order} = \frac{(\text{number of bonding } e^-) - (\text{number of antibonding } e^-)}{2 \text{ electrons/bond}}$$

For the H_2 molecule, we have

$$\text{Bond order} = \frac{2-0}{2} = 1$$

A bond order of 1 corresponds to a single bond. For He_2 we have

$$\text{Bond order} = \frac{2-2}{2} = 0$$

A bond order of zero means there is no bond, so the He_2 molecule is unable to exist. For the He_2^+ ion, which is able to form, the calculated bond order is

$$\text{Bond order} = \frac{2-1}{2} = 0.5$$

Notice that the bond order does not have to be a whole number. In this case, it indicates a bond character equivalent to about half a bond.

Molecular orbital theory successfully predicts the properties of second period diatomics

The outer shell of a Period 2 element (Li through Ne) consists of 2*s* and 2*p* subshells. When atoms of this period bond to each other, the atomic orbitals of these subshells interact strongly to produce molecular orbitals. The 2*s* orbitals, for example, overlap to form σ_{2s} and σ_{2s}^* molecular orbitals having essentially the same shapes as the σ_{1s} and σ_{1s}^* MOs, respectively. Figure 9.37 shows the shapes of the bonding and antibonding MOs produced

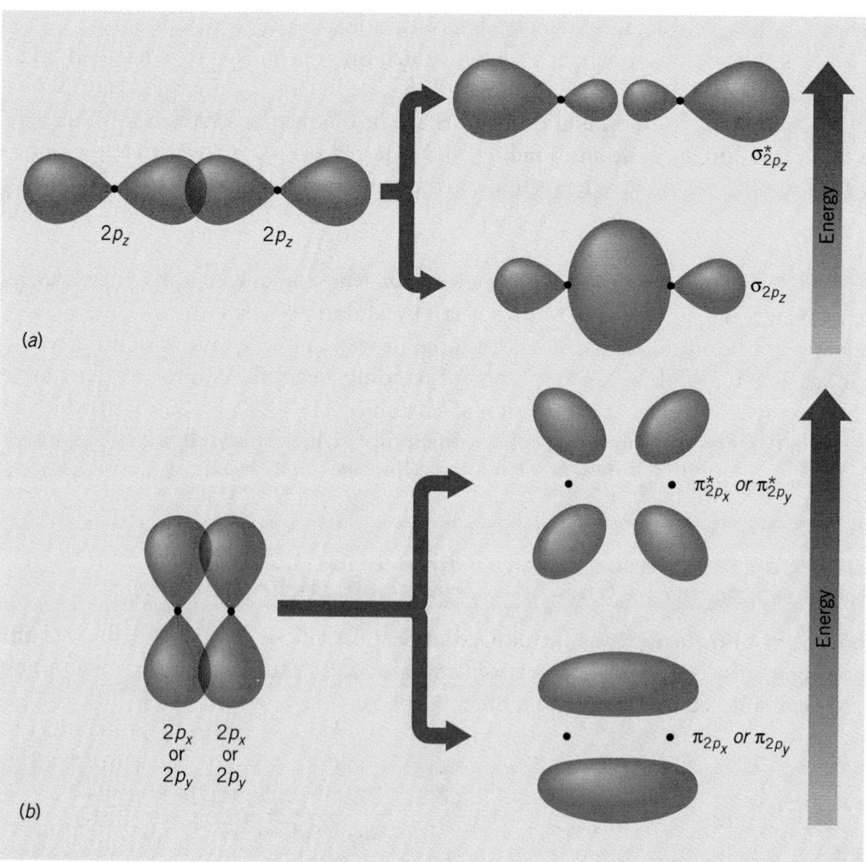

FIG. 9.37 Formation of molecular orbitals by the overlap of *p* orbitals. (*a*) Two $2p_z$ orbitals that point at each other give bonding and antibonding σ-type MOs. (*b*) Perpendicular to the $2p_z$ orbitals are $2p_x$ and $2p_y$ orbitals that overlap to give two sets of bonding and antibonding π-type MOs.

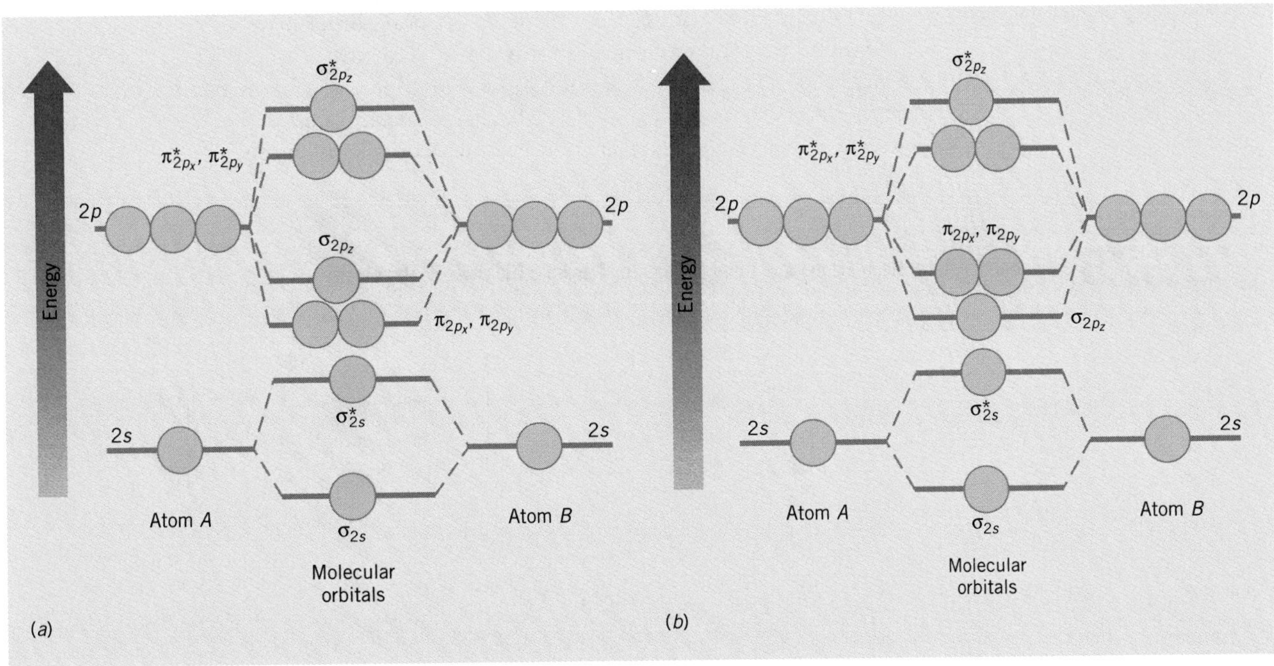

FIG. 9.38 Approximate relative energies of molecular orbitals in second period diatomic molecules. (*a*) Li_2 through N_2. (*b*) O_2 through Ne_2.

when the $2p$ orbitals overlap. If we label those that point toward each other as $2p_z$, a set of bonding and antibonding MOs are formed that we can label as σ_{2p_z} and $\sigma^*_{2p_z}$. The $2p_x$ and $2p_y$ orbitals, which are perpendicular to the $2p_z$ orbitals, overlap sideways to give π-type molecular orbitals. They are labeled π_{2p_x} and $\pi^*_{2p_x}$, and π_{2p_y} and $\pi^*_{2p_y}$, respectively.

The approximate relative energies of the MOs formed from the second shell atomic orbitals are shown in Figure 9.38. Notice that from Li to N, the energies of the π_{2p_x} and π_{2p_y} orbitals are lower than the energy of the σ_{2p_z}. Then from O to Ne, the energies of the two levels are reversed. The reasons for this are complex and beyond the scope of this book.

Using Figure 9.38, we can predict the electronic structures of diatomic molecules of Period 2. These *MO electron configurations* are obtained using the same rules that are applied to the filling of atomic orbitals in atoms.

How Electrons Fill Molecular Orbitals

1. Electrons fill the lowest energy orbitals that are available.
2. No more than two electrons, with spins paired, can occupy any orbital.
3. Electrons spread out as much as possible, with spins unpaired, over orbitals that have the same energy.

Applying these rules to the valence electrons of Period 2 atoms gives the MO electron configurations shown in Table 9.1. Let's see how well MO theory performs by examining data that are available for these molecules.

According to Table 9.1, MO theory predicts that molecules of Be_2 and Ne_2 should not exist at all because they have bond orders of zero. In beryllium vapor and in gaseous neon, no evidence of Be_2 or Ne_2 has ever been found. MO theory also predicts that diatomic molecules of the other Period 2 elements should exist because they all have bond orders greater than zero. These molecules have, in fact, been observed. Although lithium, boron, and carbon are complex solids under ordinary conditions, they can be vaporized. In the vapor, molecules of Li_2, B_2, and C_2 can be detected. Nitrogen, oxygen, and fluorine, as you know, are gaseous elements that exist as N_2, O_2, and F_2.

In Table 9.1, we also see that the predicted bond order increases from boron to carbon to nitrogen and then decreases from nitrogen to oxygen to fluorine. As the bond order increases, the *net* number of bonding electrons increases, so the bonds should become stronger and the bond lengths shorter. The *experimentally measured* bond energies and bond lengths given in Table 9.1 agree with these predictions quite nicely.

TABLE 9.1 Molecular Orbital Populations and Bond Orders for Period 2 Diatomic Molecules[a]

Energy ↑	Li_2	Be_2	B_2	C_2	N_2	Energy ↑	O_2	F_2	Ne_2
$\sigma^*_{2p_z}$	○	○	○	○	○	$\sigma^*_{2p_z}$	○	○	(↑↓)
$\pi^*_{2p_x}, \pi^*_{2p_y}$	○○	○○	○○	○○	○○	$\pi^*_{2p_x}, \pi^*_{2p_y}$	(↑)(↑)	(↑↓)(↑↓)	(↑↓)(↑↓)
σ_{2p_z}	○	○	○	○	(↑↓)	π_{2p_x}, π_{2p_y}	(↑↓)(↑↓)	(↑↓)(↑↓)	(↑↓)(↑↓)
π_{2p_x}, π_{2p_y}	○○	○○	(↑)(↑)	(↑↓)(↑↓)	(↑↓)(↑↓)	σ_{2p_z}	(↑↓)	(↑↓)	(↑↓)
σ^*_{2s}	○	(↑↓)	(↑↓)	(↑↓)	(↑↓)	σ^*_{2s}	(↑↓)	(↑↓)	(↑↓)
σ_{2s}	(↑↓)	(↑↓)	(↑↓)	(↑↓)	(↑↓)	σ_{2s}	(↑↓)	(↑↓)	(↑↓)
Number of Bonding Electrons	2	2	4	6	8		8	8	8
Number of Antibonding Electrons	0	2	2	2	2		4	6	8
Bond Order	1	0	1	2	3		2	1	0
Bond Energy (kJ/mol)	110	—	300	612	953		501	129	—
Bond Length (pm)	267	—	158	124	109		121	144	—

[a] Although the order of the energy levels corresponding to the σ_{2p_z}, and the π bonding MOs become reversed at oxygen, either sequence would yield the same result—a triple bond for N_2, a double bond for O_2, and a single bond for F_2.

Molecular orbital theory is particularly successful in explaining the electronic structure of the oxygen molecule. Experiments show that O_2 is paramagnetic (it's weakly attracted to a magnet) and that the molecule contains two unpaired electrons. In addition, the bond length in O_2 is about what is expected for an oxygen–oxygen double bond. These data cannot be explained by valence bond theory. For example, if we write a Lewis structure for O_2 that shows a double bond and also obeys the octet rule, all the electrons appear in pairs.

:Ö::Ö: (not acceptable based on experimental evidence because all electrons are paired)

On the other hand, if we show the unpaired electrons, the structure has only a single bond and doesn't obey the octet rule.

:Ö:Ö: (not acceptable based on experimental evidence because of the O—O single bond)

Although MO theory handles easily the bonding situations that VB theory has trouble with, MO theory loses the simplicity of VB theory. For even quite simple molecules, MO theory is too complicated to make predictions without extensive calculations.

With MO theory, we don't have any of these difficulties. By applying Hund's rule, the two electrons in the π^* orbitals of O_2 spread out over these orbitals with their spins unpaired because both orbitals have the same energy (see Table 9.1). The electrons in the two antibonding π^* orbitals cancel the effects of two electrons in the two bonding π orbitals, so the net bond order is 2 and the bond is effectively a double bond.

Bonding of heteronuclear diatomic molecules is also explained by MO theory

As molecules become more complex, the simple application of MO theory becomes much more difficult. This is because it is necessary to consider the relative energies of the individual atomic orbitals as well as the orientations of the orbitals relative to those on other atoms. Nevertheless, we can take a brief look at the MO descriptions of a couple of diatomic molecules to see what happens when both atoms in the molecule are not the same. Such molecules are said to be **heteronuclear**.

The MO description of HF is similar to that of valence bond theory

When we consider the possible interaction of the orbitals of different atoms to form molecular orbitals, the first factor we have to consider is the relative energies of the orbitals. This is because orbitals interact most effectively when they are of about equal energy; the greater the difference in energy between the orbitals, the less the orbitals interact, and the more the orbitals behave like simple atomic orbitals.

In HF, the 1*s* orbital of hydrogen is higher in energy than either the 2*s* or 2*p* subshell of fluorine, but it is closest in energy to the 2*p* subshell (Figure 9.39). Taking the *z* axis as the internuclear axis, the hydrogen 1*s* orbital overlaps with the $2p_z$ orbital of fluorine to give bonding and antibonding σ type orbitals, as illustrated in Figure 9.40. The $2p_x$ and $2p_y$ orbitals of fluorine, however, have no orbitals on hydrogen with which to interact, so they are unchanged when the molecule is formed. These two orbitals are said to be **nonbonding orbitals** because they are neither bonding nor antibonding; they have no effect on the stability of the molecule.

In the MO description of HF, we have a pair of electrons in the bonding MO formed by the overlap of the hydrogen 1*s* orbital with the fluorine $2p_z$ orbital. Earlier we saw that valence bond theory explains the bond in HF in the same way, as a pair of electrons shared between the hydrogen 1*s* orbital and a fluorine 2*p* orbital.

The MO energy diagram for CO is similar to that of Period 2 homonuclear diatomics

A **homonuclear diatomic molecule** is one in which both atoms are of the same element. Examples are N_2 and O_2. Carbon monoxide is a heteronuclear molecule, but both atoms are from Period 2 and we expect the orbitals of the second shell to be the ones used to form

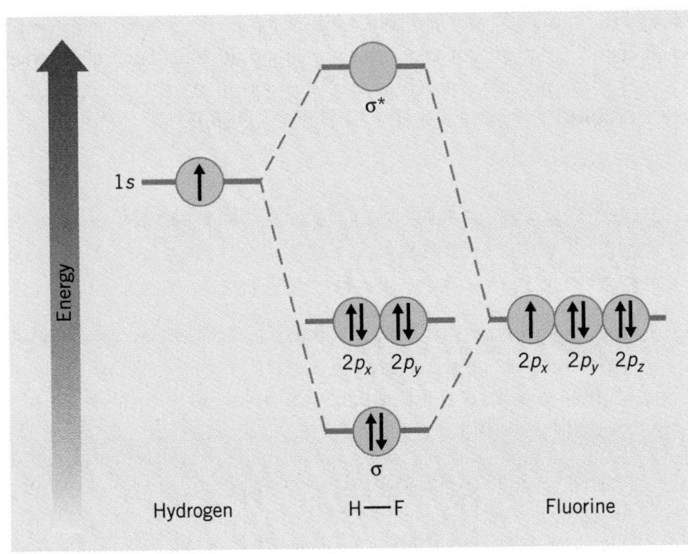

FIG. 9.39 **Molecular orbital energy diagram for HF.** Only the 1*s* orbital of hydrogen and the 2*p* orbitals of fluorine are shown.

The 2*s* orbital of fluorine is so much lower in energy than the 2*p* subshell that we don't need to consider its interaction with the hydrogen 1*s* orbital. That's why it isn't shown in the energy diagram.

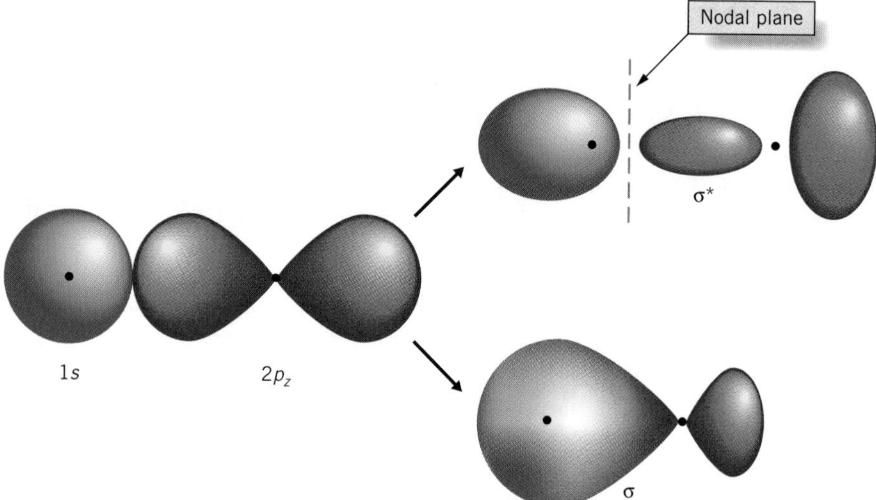

FIG. 9.40 Formation of σ and σ^* orbitals in HF. Notice that the antibonding σ^* orbital has a nodal plane between the nuclei, which effectively removes electron density from the region between the nuclei and, if occupied, leads to destabilization of the molecule.

the MOs. The orbital overlaps are similar to those of the homonuclear diatomics of Period 2, so the energy diagram resembles the one shown in Figure 9.38*a*.

Because the outer shell electrons of oxygen experience a larger effective nuclear charge than those of carbon, the oxygen orbitals will be somewhat lower in energy. This is shown in Figure 9.41. There's a total of 10 valence electrons (4 from carbon and 6 from oxygen) to distribute among the MOs of the molecule. When we do this, there are 8 bonding electrons and 2 antibonding electrons, so the net bond order is 3, corresponding to a triple bond. As expected, it consists of a σ bond and two π bonds.

Practice Exercise 22: The molecular orbital energy level diagram for the cyanide ion, CN^-, is similar to that of the Period 2 homonuclear diatomics. Sketch the energy diagram for the ion and indicate the electron population of the MOs. What is the bond order in the ion? How does this agree with the bond order predicted from the Lewis structure of the ion? (Hint: How many valence electrons are there in the ion?)

Practice Exercise 23: The MO energy level diagram for the nitrogen monoxide molecule is essentially the same as that shown in Table 9.1 for O_2, except the oxygen orbitals are slightly lower in energy than the corresponding nitrogen orbitals. Sketch the energy diagram for nitrogen monoxide and indicate which MOs are populated. Calculate the bond order for the molecule. (Hint: Make adjustments to Figure 9.38*b*.)

9.8 MOLECULAR ORBITAL THEORY USES DELOCALIZED ORBITALS TO DESCRIBE MOLECULES WITH RESONANCE STRUCTURES

One of the least satisfying aspects of the way valence bond theory explains chemical bonding is the need to write resonance structures for certain molecules and ions. For example, consider benzene, C_6H_6. As you learned earlier, this molecule has the shape of a ring whose resonance structures can be written as

The MO description of bonding in this molecule is as follows: The basic structure of the molecule is determined by the sigma-bond framework, which requires that the carbon atoms use sp^2 hybrid orbitals. This allows each carbon to form three σ bonds (two to other C atoms and one to an H atom). Each carbon atom is left with a half-filled unhybridized p orbital perpendicular to the plane of the ring. These p orbitals overlap to give

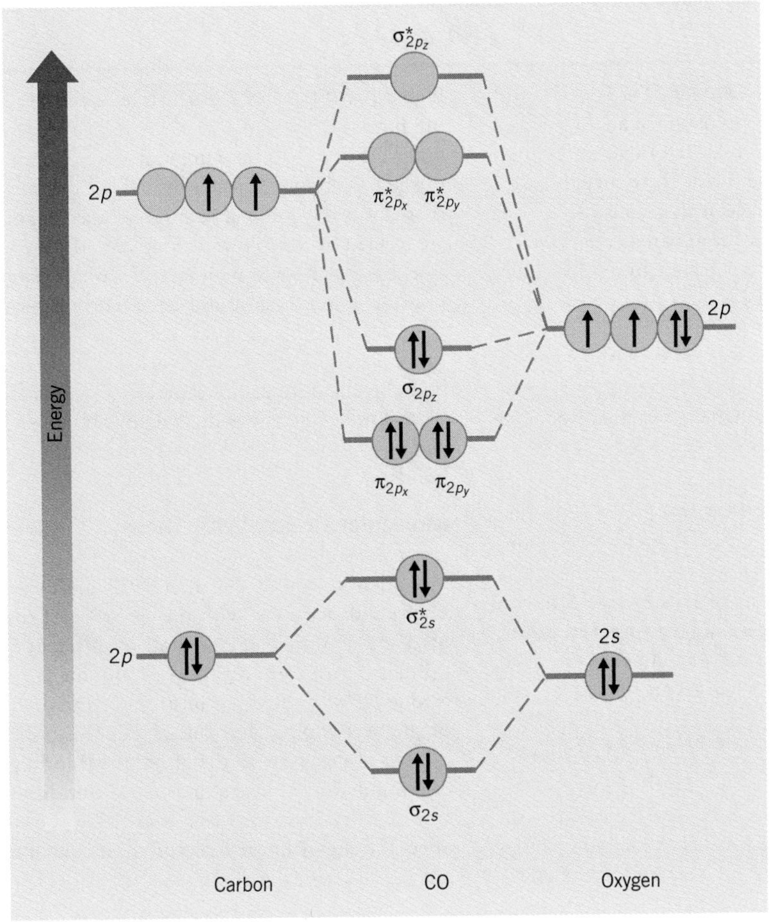

FIG. 9.41 Approximate molecular orbital energy diagram for carbon monoxide. The oxygen orbitals are lower in energy than the corresponding carbon orbitals. The net bond order for CO is 3.

a delocalized π-electron cloud (a **delocalized molecular orbital**) that looks something like two doughnuts with the sigma-bond framework sandwiched between them (Figure 9.42). The delocalized nature of the pi electrons is the reason we usually represent the structure of benzene as

One of the special characteristics of delocalized bonds is that they make a molecule or ion more stable than it would be if it had localized bonds. In Section 8.7 this was described in terms of *resonance energy*. In the molecular orbital theory, we no longer speak of resonance; instead, we refer to the electrons as being delocalized. The extra stability that is associated with this delocalization is therefore described, in the language of MO theory, as the **delocalization energy.**

Functionally, the terms *resonance energy* and *delocalization energy* are the same; they just come from different approaches to bonding theory.

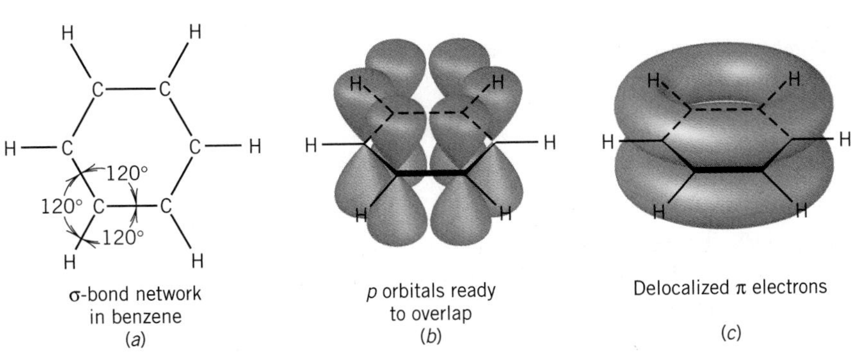

FIG. 9.42 Benzene. (*a*) The σ-bond framework. All atoms lie in the same plane. (*b*) The unhybridized *p* orbitals at each carbon prior to side-to-side overlap. (*c*) The double doughnut-shaped electron cloud formed by the delocalized π electrons.

SUMMARY

Molecular Shapes and VSEPR Theory. The structures of most molecules can be described in terms of one or another of five basic geometries: **linear, planar triangular, tetrahedral, trigonal bipyramidal,** and **octahedral.** The **VSEPR theory** predicts molecular geometry by assuming that **electron domains**—regions of space that contain bonding electrons, unpaired valence electrons, or lone pairs—stay as far apart as possible from each other, while staying as close as possible to the central atom. Figures 9.4, 9.5, 9.7, and 9.10 illustrate the structures obtained with different numbers of groups of electrons in the valence shell of the central atom in a molecule or ion and with different numbers of lone pairs and attached atoms. The correct shape of a molecule or polyatomic ion can usually be predicted from the Lewis structure.

Molecular Shape and Molecular Polarity. A molecule that contains identical atoms attached to a central atom will be nonpolar if there are no lone pairs of electrons in the central atom's valence shell. It will be polar if lone pairs are present, except in two cases: (1) when there are three lone pairs and two attached atoms, and (2) when there are two lone pairs and four attached atoms. If all the atoms attached to the central atom are not alike, the molecule will usually be polar.

Valence Bond (VB) Theory. According to VB theory, a covalent bond is formed between two atoms when an atomic orbital on one atom **overlaps** with an atomic orbital on the other and a pair of electrons with paired spins is shared between the overlapping orbitals. In general, the better the overlap of the orbitals, the stronger the bond. A given atomic orbital can only overlap with one other orbital on a different atom, so a given atomic orbital can only form one bond with an orbital on one other atom.

Hybrid Atomic Orbitals. **Hybrid orbitals** are formed by mixing pure *s*, *p*, and *d* orbitals. Hybrid orbitals overlap better with other orbitals than the pure atomic orbitals from which they are formed, so bonds formed by hybrid orbitals are stronger than those formed by ordinary atomic orbitals. **Sigma bonds** (σ bonds) are formed by the following kinds of orbital overlap: *s–s*, *s–p*, end-to-end *p–p*, and overlap of hybrid orbitals. Sigma bonds allow free rotation around the bond axis. The side-by-side overlap of *p* orbitals produces a **pi bond** (π bond). Pi bonds do not permit free rotation around the bond axis because such a rotation involves bond breaking. In complex molecules, the basic molecular framework is built with σ bonds. A double bond consists of one σ bond and one π bond. A triple bond consists of one σ bond and two π bonds.

Molecular Orbital (MO) Theory. This theory begins with the supposition that molecules are similar to atoms, except they have more than one positive center. They are treated as collections of nuclei and electrons, with the electrons of the molecule distributed among **molecular orbitals** of different energies. Molecular orbitals can spread over two or more nuclei, and can be considered to be formed by the constructive and destructive interference of the overlapping electron waves corresponding to the atomic orbitals of the atoms in the molecule. **Bonding MOs** concentrate electron density between nuclei; **antibonding MOs** remove electron density from between nuclei. **Nonbonding MOs** do not affect the energy of the molecule. The rules for the filling of MOs are the same as those for atomic orbitals. The ability of MO theory to describe **delocalized orbitals** avoids the need for resonance theory. Delocalization of bonds leads to a lowering of the energy by an amount called the **delocalization energy** and produces more stable molecular structures.

TOOLS FOR PROBLEM SOLVING

In this chapter you learned to apply the following concepts as tools in solving problems related to chemical bonding and molecular structure. Study each one carefully so that you know what each is used for. When faced with solving a problem, recall what each tool does and consider whether it will be helpful in finding a solution. This will aid you in selecting the tools you need.

Basic molecular shapes *(pages 339–341)* You need an understanding of the five basic geometries discussed. Practice drawing them and be sure you know their names.

VSEPR model *(page 341)* Electron groups repel each other and arrange themselves in the valence shell of an atom to yield minimum repulsions, which is what determines the shape of the molecule. This tool serves as the foundation for understanding the VSEPR model.

Molecular shape and polarity *(page 351)* We can use molecular shape to determine whether a molecule will be polar or nonpolar. Refer to the summary on page 351.

Criteria for bond formation according to VB theory *(page 352)* A bond requires overlap of two orbitals sharing two electrons with paired spins. Both orbitals can be half-filled, or one can be filled and the other empty. We use these criteria to establish which orbitals atoms use when bonds are formed.

The VSEPR model *(page 341)* **and orientations of hybrid orbitals** *(page 356)* These tools are interrelated. Lewis structures permit us to use the VSEPR model to predict molecular shape, which then allows us to select the correct hybrid orbitals for the valence bond description of bonding. After forming the Lewis structure, we determine the number of domains, from which we derive the structure of the molecule or ion. A convenient way of doing this is to describe the VSEPR structure symbolically. In doing this we represent the central atom by *M*, the atoms attached to the central atom by *X*, and lone pairs by *E*. We can then signify the number of bonding and nonbonding domains around *M* as a formula MX_nE_m, where *n* is the number of bonding domains and *m* is the number of nonbonding domains. The resulting formula is related to the structure of the molecule or ion, and to the hybrid orbitals used by the central atom, as shown below. Practice sketching the structures, associating them with the appropriate generalized formula MX_nE_m, and using the structures to select the appropriate set of hybrid orbitals.

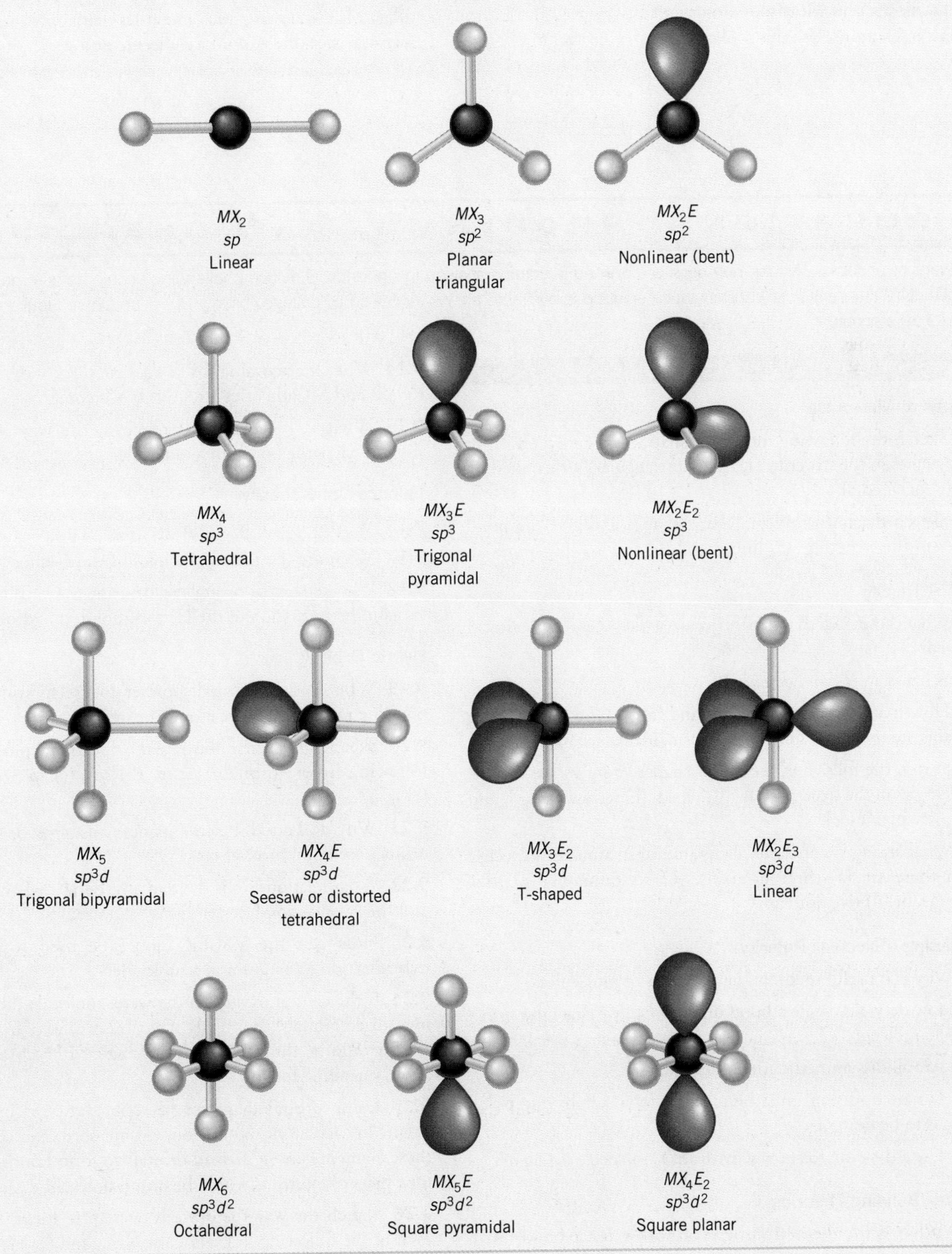

Criteria for determining numbers of σ and π bonds and hybridization *(page 370)* The Lewis structure for a polyatomic molecule lets us apply these criteria to determine how many σ and π bonds are between atoms and the kind of hybrid orbitals each atom uses. Remember that the shape of the molecule is determined by the framework of σ bonds, with π bonds used in double and triple bonds.

Calculating bond order in MO theory *(page 373)*

$$\text{Bond order} = \frac{(\text{number of bonding } e^-) - (\text{number of antibonding } e^-)}{2 \text{ electrons/bond}}$$

How electrons fill molecular orbitals *(page 375)* Electrons populate MOs following the same rules that apply to atomic orbitals in an atom. Use this tool to obtain the correct distribution of electrons over the MOs of a molecule or ion.

QUESTIONS, PROBLEMS, AND EXERCISES

Answers to problems whose numbers are printed in color are given in Appendix B. More challenging problems are marked with asterisks. ILW = Interactive Learningware solution is available at www.wiley.com/college/brady. OH = an Office Hours video is available for this problem.

REVIEW QUESTIONS

Shapes of Molecules

9.1 Sketch the following molecular shapes and give the various bond angles in the structure: (a) planar triangular, (b) tetrahedral, and (c) octahedral.

9.2 Sketch the following molecular shapes and give the bond angles in the structure: (a) linear, and (b) trigonal bipyramidal.

VSEPR Theory

9.3 What is the underlying principle on which the VSEPR model is based?

9.4 What is an *electron domain?*

9.5 How many bonding domains and how many nonbonding domains are there in a molecule of formaldehyde, HCHO?

9.6 Sketch the following molecular shapes and give the various bond angles in the structures: (a) T-shaped, (b) seesaw-shaped, and (c) square pyramidal.

9.7 What arrangements of domains around an atom are expected when there are (a) three domains, (b) six domains, (c) four domains, or (d) five domains?

Predicting Molecular Polarity

9.8 Why is it useful to know the polarities of molecules?

9.9 How do we indicate a bond dipole when we draw the structure of a molecule?

9.10 Are all dissymmetric molecules polar?

9.11 What condition must be met if a molecule having polar bonds is to be nonpolar?

9.12 Use a drawing to show why the SO_2 molecule is polar.

Modern Bonding Theories

9.13 What is the theoretical basis of both valence bond (VB) theory and molecular orbital (MO) theory?

9.14 What shortcomings of Lewis structures and VSEPR theory do VB and MO theories attempt to overcome?

9.15 What is the main difference in the way VB and MO theories view the bonds in a molecule?

Valence Bond Theory

9.16 What is meant by *orbital overlap*?

9.17 What are the main principles of the valence bond theory?

9.18 Use sketches of orbitals to describe how VB theory would explain the formation of the H—Br bond in hydrogen bromide.

Hybrid Orbitals

9.19 Why do atoms usually prefer to use hybrid orbitals for bonding rather than pure atomic orbitals?

9.20 Sketch figures that illustrate the directional properties of the following hybrid orbitals: (a) sp, (b) sp^2, (c) sp^3, (d) sp^3d, and (e) sp^3d^2.

9.21 Why do Period 2 elements never use sp^3d or sp^3d^2 hybrid orbitals for bond formation?

9.22 What relationship is there, if any, between Lewis structures and the valence bond descriptions of molecules?

9.23 How can the VSEPR model be used to predict the hybridization of an atom in a molecule?

9.24 If the central oxygen in the water molecule did not use sp^3 hybridized orbitals (or orbitals of any other kind of hybridization), what would be the expected bond angle in H_2O (assuming no angle-spreading force)?

9.25 Using orbital diagrams, describe how sp^3 hybridization occurs in each atom: (a) carbon, (b) nitrogen, and (c) oxygen. If these elements use sp^3 hybrid orbitals to form bonds, how many lone pairs of electrons would be found on each?

9.26 Sketch the way the orbitals overlap to form the bonds in each of the following: (a) CH_4, (b) NH_3, and (c) H_2O. (Assume the central atom uses hybrid orbitals.)

9.27 We explained the bond angles of 107° in NH_3 by using sp^3 hybridization of the central nitrogen atom. Had the original unhybridized p orbitals of the nitrogen been used to overlap with $1s$ orbitals of each hydrogen, what would have been the H—N—H bond angles? Explain.

9.28 Using sketches of orbitals and orbital diagrams, describe sp^2 hybridization at (a) boron and (b) carbon.

9.29 What two basic shapes have hybridizations that include d orbitals?

Coordinate Covalent Bonds and VB Theory

9.30 The ammonia molecule, NH_3, can combine with a hydrogen ion, H^+ (which has an empty $1s$ orbital), to form the ammonium ion, NH_4^+. (This is how ammonia can neutralize an acid and therefore function as a base.) Sketch the geometry of the ammonium ion, indicating the bond angles.

9.31 How does the geometry around B and O change in the following reaction? How does the hybridization of each atom change?

```
   H        :C̈l:                  H  :C̈l:
   |         |   ..               |   |    ..
H—Ö:   +    B—C̈l:   ⟶       H—Ö—B—C̈l:
   ..        |                    ..  |    ..
            :C̈l:                    :C̈l:
```

Multiple Bonds and Hybrid Orbitals

9.32 How do σ and π bonds differ?

9.33 Why can free rotation occur easily around a σ-bond axis but not around a π-bond axis?

9.34 Using sketches, describe the bonds and bond angles in ethylene, C_2H_4.

9.35 Sketch the way the bonds form in acetylene, C_2H_2.

9.36 How does VB theory treat the benzene molecule? (Draw sketches describing the orbital overlaps and the bond angles.)

Molecular Orbital Theory

9.37 Why is the higher-energy MO in H_2 called an antibonding orbital?

9.38 Using a sketch, describe the two lowest energy MOs of H_2 and their relationship to their parent atomic orbitals.

9.39 Explain why He_2 does not exist but H_2 does.

9.40 How does MO theory account for the paramagnetism of O_2?

9.41 On the basis of MO theory, explain why Li_2 molecules can exist but Be_2 molecules cannot. Could the ion Be_2^+ exist?

9.42 What are the bond orders in (a) O_2^+, (b) O_2^-, and (c) C_2^+?

9.43 What relationship is there between bond order and bond energy?

9.44 Sketch the shapes of the π_{2p_y} and $\pi^*_{2p_y}$ MOs.

9.45 What is a *delocalized MO?*

9.46 What problem encountered by VB theory does MO theory avoid by delocalized bonding?

9.47 Draw the representation of the benzene molecule that indicates its delocalized π system.

9.48 What effect does delocalization have on the stability of the electronic structure of a molecule?

9.49 What is delocalization energy? How is it related to resonance energy?

REVIEW PROBLEMS

9.50 Predict the shapes (a) NH_2^-, (b) CO_3^{2-}, (c) IF_3, (d) Br_3^-, and (e) GaH_3.

9.51 Predict the shapes of (a) SF_3^+, (b) NO_3^-, (c) SO_4^{2-}, (d) O_3, and (e) N_2O.

ILW **9.52** Predict the shapes of (a) FCl_2^+, (b) AsF_5, (c) AsF_3, (d) SbH_3, and (e) SeO_2.

9.53 Predict the shapes of (a) TeF_4, (b) $SbCl_6^-$, (c) NO_2^-, (d) PCl_4^+, and (e) PO_4^{3-}.

9.54 Predict the shapes of (a) IO_4^-, (b) ICl_4^-, (c) TeF_6, (d) SiO_4^{4-}, and (e) ICl_2^-.

9.55 Predict the shapes of (a) CS_2, (b) BrF_4^-, (c) ICl_3, (d) ClO_3^-, and (e) SeO_3.

9.56 Which of the following has a shape described by the figure below? (a) IO_4^-, (b) ICl_4^-, (c) $SnCl_4$, (d) BrF_4^+

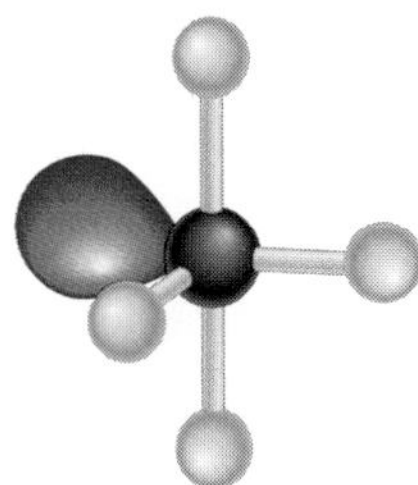

9.57 Which of the following has a shape described by the figure below? (a) BrF_3, (b) PF_3, (c) NO_3^-, (d) SCl_3^-

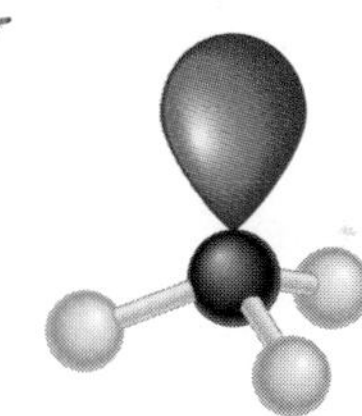

9.58 Acetylene, a gas used in welding torches, has the Lewis structure H—C≡C—H. What would you expect the H—C—C bond angle to be in this molecule?

OH **9.59** Ethylene, a gas used to ripen tomatoes artificially, has the Lewis structure

```
   H   H
   |   |
H—C=C—H
```

What would you expect the H—C—H and H—C—C bond angles to be in this molecule? (Caution: Don't be fooled by the way the structure is drawn here.)

9.60 Predict the bond angle for each of the following. (a) Cl_2O, (b) H_2O, (c) SO_2, (d) I_3^-, (e) NH_2^-

9.61 Predict the bond angle for each of the following. (a) HOCl, (b) PH_2^-, (c) OCN^-, (d) O_3, (e) SnF_2

Predicting Molecular Polarity

ILW **9.62** Which of the following molecules would be expected to be polar? (a) HBr, (b) $POCl_3$, (c) CH_2O, (d) $SnCl_4$, (e) $SbCl_5$

9.63 Which of the following molecules would be expected to be polar? (a) PBr_3, (b) SO_3, (c) $AsCl_3$, (d) ClF_3, (e) BCl_3

9.64 Which of the following molecules or ions would be expected to have a net dipole moment? (a) ClNO, (b) XeF_3^+, (c) $SeBr_4$, (d) NO, (e) NO_2

9.65 Which of the following molecules or ions would be expected to have a net dipole moment? (a) H_2S, (b) BeH_2, (c) SCN^-, (d) CN^-, (e) $BrCl_3$

OH **9.66** Explain why SF_6 is nonpolar, but SF_5Br is polar.

9.67 Explain why CH_3Cl is polar, but CCl_4 is not.

Valence Bond Theory

9.68 Hydrogen selenide is one of nature's most foul-smelling substances. Molecules of H_2Se have H—Se—H bond angles very close to 90°. How would VB theory explain the bonding in H_2Se? Use sketches of orbitals to show how the bonds are formed. Illustrate with appropriate orbital diagrams as well.

OH **9.69** Use sketches of orbitals to show how VB theory explains the bonding in the F_2 molecule. Illustrate with appropriate orbital diagrams as well.

Hybrid Orbitals

9.70 Use orbital diagrams to explain how the beryllium chloride molecule is formed. What kind of hybrid orbitals does beryllium use in this molecule?

9.71 Use orbital diagrams to describe the bonding in (a) tin tetrachloride and (b) antimony pentachloride. Be sure to indicate hybrid orbital formation.

OH **9.72** Draw Lewis structures for the following and use the geometry predicted by the VSEPR model to determine what kind of hybrid orbitals the central atom uses in bond formation: (a) ClO_3^-, (b) SO_3, and (c) OF_2.

9.73 Draw Lewis structures for the following and use the geometry predicted by the VSEPR model to determine what kind of hybrid orbitals the central atom uses in bond formation: (a) $SbCl_6^-$, (b) $BrCl_3$, and (c) XeF_4.

9.74 Use the VSEPR model to help you describe the bonding in the following molecules according to VB theory: (a) arsenic trichloride and (b) chlorine trifluoride. Use orbital diagrams for the central atom to show how hybridization occurs.

9.75 Use the VSEPR model to help you describe the bonding in the following molecules according to VB theory: (a) antimony pentafluoride and (b) selenium dichloride. Use orbital diagrams for the central atom to show how hybridization occurs.

Coordinate Covalent Bonds and VB Theory

9.76 Use orbital diagrams to show that the bonding in SbF_6^- involves the formation of a coordinate covalent bond.

9.77 What kind of hybrid orbitals are used by tin in $SnCl_6^{2-}$? Draw the orbital diagram for Sn in $SnCl_6^{2-}$. What is the geometry of $SnCl_6^{2-}$?

Multiple Bonding and Valence Bond Theory

9.78 A nitrogen atom can undergo sp^2 hybridization when it becomes part of a carbon–nitrogen double bond, as in $H_2C{=}NH$.

(a) Using a sketch, show the electron configuration of sp^2 hybridized nitrogen just before the overlapping occurs to make the double bond.
(b) Using sketches (and the analogy to the double bond in C_2H_4), describe the two bonds of the carbon–nitrogen double bond.
(c) Describe the geometry of $H_2C{=}NH$ (using a sketch that shows all expected bond angles).

9.79 A nitrogen atom can undergo *sp* hybridization and then become joined to carbon by a triple bond to give the structural unit —C≡N:. This triple bond consists of one σ bond and two π bonds.

(a) Write the orbital diagram for *sp* hybridized nitrogen as it would look before any bonds form.
(b) Using the carbon–carbon triple bond as the analogy, and drawing pictures to show which atomic orbitals overlap with which, show how the three bonds of the triple bond in —C≡N: form.
(c) Again using sketches, describe all the bonds in hydrogen cyanide, H—C≡N:
(d) What is the likeliest H—C—N bond angle in HCN?

9.80 Tetrachloroethylene, a common dry-cleaning solvent, has the formula C_2Cl_4. Its structure is

```
:Cl:        :Cl:
    \      /
     C == C
    /      \
:Cl:        :Cl:
```

Use the electron domain and VB theories to describe the bonding in this molecule. What are the expected bond angles?

9.81 Phosgene, $COCl_2$, was used as a war gas during World War I. It reacts with moisture in the lungs of its victims to form CO_2 and gaseous HCl, which cause the lungs to fill with fluid. Phosgene is a simple molecule having the structure

```
          :Cl:
         /
:O == C
         \
          :Cl:
```

Describe the bonding in this molecule using VB theory.

9.82 What kind of hybrid orbitals do the numbered atoms use in the following molecule?

```
     H        (2)   :O: <-(4)
     |          \    ||           H
 H — C — C ≡ C — C — C = C
     |  ^       ^         |    \
(1)  H        (3)         H     H
```

OH **9.83** What kinds of bonds (σ, π) are found in the numbered bonds in the following molecule?

```
     H        (2)   :O:  (4)
     |          \    ||           H
 H — C — C ≡ C — C — C = C
     |                |    \
(1)  H         (3)    H     H
```

Molecular Orbital Theory

ILW OH **9.84** Use the MO energy diagram to predict which in each pair has the greater bond energy: (a) O_2 or O_2^+, (b) O_2 or O_2^-, and (c) N_2 or N_2^+.

9.85 Assume that in the NO molecule the molecular orbital energy level sequence is similar to that for O_2. What happens to the NO bond length when an electron is removed from NO to give NO^+?

9.86 In each of the following pairs, which substance has the longer bond length? (a) N_2 or N_2^+, (b) NO or NO^+, (c) O_2 or O_2^-

9.87 Which of the following molecules or ions are paramagnetic? (a) O_2^+, (b) O_2, (c) O_2^-, (d) NO, (e) N_2.

9.88 Construct the MO energy level diagram for the OH molecule assuming it is similar to that for HF. How many electrons are in (a) bonding MOs and (b) nonbonding MOs? What is the net bond order in the molecule?

9.89 If boron and nitrogen were to form a molecule with the formula BN, what would its MO energy level diagram look like, given that the energies of the $2p$ orbitals of nitrogen are lower than those of boron? If Figure 9.38*a* applies, would the molecule be paramagnetic or diamagnetic? What is the net bond order in the molecule?

ADDITIONAL EXERCISES

OH **9.90** Formaldehyde has the Lewis structure

```
      H
      |    ..
  H—C=O
           ..
```

What would you predict its shape to be?

9.91 The molecule XCl_3 is pyramidal. In which group in the periodic table is element *X* found? If the molecule were planar triangular, in which group would *X* be found? If the molecule were T-shaped, in which group would *X* be found? Why is it unlikely that element *X* is in Group VI?

9.92 Antimony forms a compound with hydrogen that is called stibine. Its formula is SbH_3 and the H—Sb—H bond angles are 91.3°. Which kinds of orbitals does Sb most likely use to form the Sb—H bonds, pure *p* orbitals or hybrid orbitals? Explain your reasoning.

9.93 Describe the changes in molecular geometry and hybridization that take place during the following reactions:

(a) $BF_3 + F^- \longrightarrow BF_4^-$

(b) $PCl_5 + Cl^- \longrightarrow PCl_6^-$

(c) $ICl_3 + Cl^- \longrightarrow ICl_4^-$

(d) $PCl_3 + Cl_2 \longrightarrow PCl_5$

(e) $C_2H_2 + H_2 \longrightarrow C_2H_4$

9.94 Which one of the following five diagrams best represents the structure of $BrCl_4^+$?

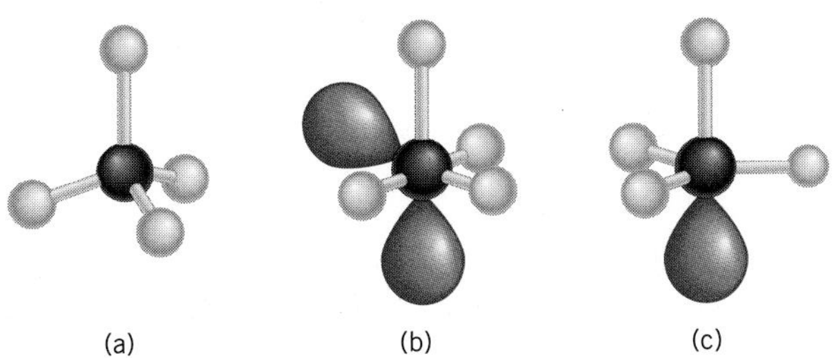

(a) (b) (c)

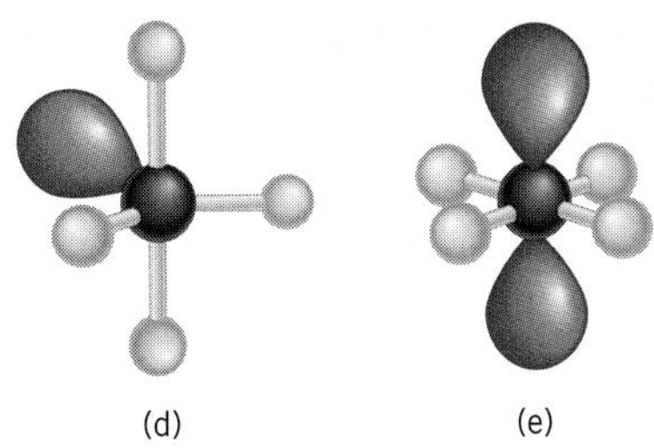

(d) (e)

OH **9.95** Cyclopropane is a triangular molecule with C—C—C bond angles of 60°. Explain why the σ bonds joining carbon atoms in cyclopropane are weaker than the carbon–carbon σ bonds in the noncyclic propane.

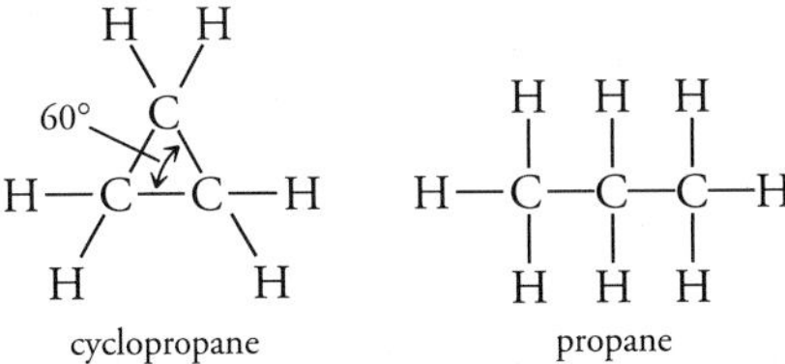

9.96 Phosphorus trifluoride, PF_3, has F—P—F bond angles of 97.8°.

(a) How would VB theory use hybrid orbitals to explain these data?

(b) How would VB theory use unhybridized orbitals to account for these data?

(c) Do either of these models work very well?

9.97 A six-membered ring of carbons can hold a double bond but not a triple bond. Explain.

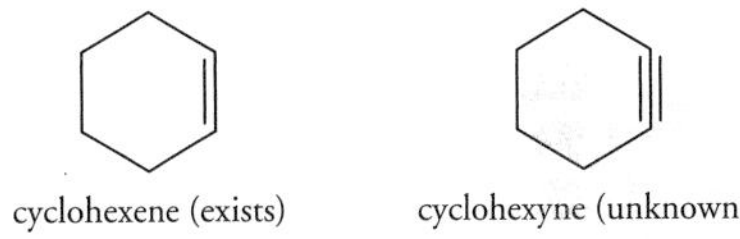

cyclohexene (exists) cyclohexyne (unknown)

***9.98** There exists a hydrocarbon called butadiene, which has the molecular formula C_4H_6 and the structure

```
            H       H
             \     /
      H       C=C
       \     /     \
        C=C         H
       /     \
      H       H
```

The C=C bond lengths are 134 pm (about what is expected for a carbon–carbon double bond), but the C—C bond length in this molecule is 147 pm, which is shorter than a normal C—C single bond. The molecule is planar (i.e., all the atoms lie in the same plane).

(a) What kind of hybrid orbitals do the carbon atoms use in this molecule to form the carbon–carbon bonds?

(b) Between which pairs of carbon atoms do we expect to find sideways overlap of *p* orbitals (i.e., π-type *p*–*p* overlap)?

(c) On the basis of your answer to part (b), do you expect to find localized or delocalized π bonding in the carbon chain in this molecule?

(d) Based on your answer to part (c), explain why the center carbon–carbon bond is shorter than a carbon–carbon single bond.

***9.99** *The more electronegative are the atoms bonded to the central atom, the less are the repulsions between the electron pairs in the bonds.* On the basis of this statement, predict the most probable structure for the molecule PCl_3F_2. Do we expect the molecule to be polar or nonpolar?

***9.100** *A lone pair of electrons in the valence shell of an atom has a larger effective volume than a bonding electron pair. Lone pairs therefore repel other electron pairs more strongly than do bonding pairs.* On the basis of these statements, describe how the bond angles in TeF_4 and BrF_4^- deviate from those found in a trigonal bipyramid and an octahedron, respectively. Sketch the molecular shapes of TeF_4 and BrF_4^- and indicate these deviations on your drawing.

***9.101** *The two electron pairs in a double bond repel other electron pairs more than the single pair of electrons in a single bond.* On the basis of this statement, which bond angles should be larger in SO_2Cl_2, the O—S—O bond angles or the Cl—S—Cl bond angles? (In the molecule, sulfur is bonded to two oxygen atoms and two chlorine atoms. *Hint*: Assign formal charges and work with the best Lewis structure for the molecule.)

***9.102** *A hybrid orbital does not distribute electron density symmetrically around the nucleus of an atom. Therefore, a lone pair in a hybrid orbital contributes to the overall polarity of a molecule.* On the basis of these statements and the fact that NH_3 is a very polar molecule and NF_3 is a nearly nonpolar molecule, justify the notion that the lone pair of electrons in each of these molecules is held in an sp^3 hybrid orbital.

9.103 In a certain molecule, a *p* orbital overlaps with a *d* orbital as shown below. Which kind of bond is formed, σ or π? Explain your choice.

9.104 If we take the internuclear axis in a diatomic molecule to be the *z* axis, what kind of *p* orbital (p_x, p_y, or p_z) on one atom would have to overlap with a d_{xz} orbital on the other atom to give a pi bond?

9.105 The peroxynitrite ion, $OONO^-$, is a potent toxin formed in cells affected by diseases such as diabetes or atherosclerosis. Peroxynitrite ion can oxidize and destroy biomolecules crucial for the survival of the cell.

(a) Give the O—O—N and O—N—O bond angles in peroxynitrite ion.
(b) What is the hybridization of the N atom in peroxynitrite ion?
(c) Suggest why the peroxynitrite ion is expected to be much less stable than the nitrate ion, NO_3^-.

EXERCISES IN CRITICAL THINKING

9.106 Five basic molecular shapes were described for simple molecular structures containing a central atom bonded to various numbers of surrounding atoms. Can you suggest additional possible structures? Provide arguments about the likelihood that these other structures might actually exist.

9.107 Compare and contrast the concepts of *delocalization* and *resonance*.

9.108 Why doesn't a carbon–carbon quadruple bond exist?

9.109 What might the structure of the iodine heptafluoride molecule be? If you can think of more than one possible structure, which is likely to be of lowest energy based on the VSEPR model?

9.110 The F—F bond is weaker than the Cl—Cl bond. How might the lone pairs on the atoms in the molecules be responsible for this?

9.111 Molecular orbital theory predicts the existence of antibonding molecular orbitals. How do antibonding electrons affect the stability in a molecule?

9.112 The structure of the diborane molecule, B_2H_6, is sometimes drawn as

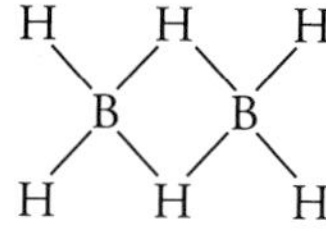

There are not enough valence electrons in the molecule to form eight single bonds, which is what the structure implies. Assuming that the boron atoms use sp^3 hybrid orbitals, suggest a way that hydrogen 1*s* orbitals can be involved in forming delocalized molecular orbitals that bridge the two boron atoms. Use diagrams to illustrate your answer. What would be the average bond order in the bridging bonds?

CHAPTERS 7-9

BRINGING IT TOGETHER

Once again you have an opportunity to test your understanding of concepts, your knowledge of scientific terms, and your problem-solving skills. Read through the following questions carefully, and answer each as fully as possible. When necessary, review topics that give you difficulty. When you are able to answer these questions correctly, you are ready to study the next group of chapters.

1. What are the three principal particles that make up the atom? On the atomic mass scale, what are their approximate masses? What are their electrical charges?

2. A beam of green light has a wavelength of 500 nm. What is the frequency of this light? What is the energy, in joules, of one photon of this light? What is the energy, in joules, of one mole of photons of this light? Would blue light have more or less energy per photon than this light?

3. Arrange the following kinds of electromagnetic radiation in order of increasing frequency: X rays, blue light, radio waves, gamma rays, microwaves, red light, infrared light, ultraviolet light.

4. What is a *continuous spectrum*? How does it differ from an *atomic spectrum*?

5. What experimental evidence is there that matter has wavelike properties?

6. What is the difference between a *traveling wave* and a *standing wave*? What is a *node*?

7. How is the energy of an electron related to the number of nodes in its electron wave?

8. What is a *wave function*? What Greek letter is usually used to represent a wave function? What word do we use to refer to an electron wave in an atom?

9. What are the quantum numbers of the electrons in the valence shells of (a) sulfur, (b) strontium, (c) lead, (d) bromine, and (e) boron?

10. If a given shell has $n = 4$, which kinds of subshells (*s*, *p*, etc.) does it have? What is the maximum number of electrons that could populate this shell?

11. Use the periodic table to predict the electron configurations of (a) tin, (b) germanium, (c) silicon, (d) lead, and (e) nickel.

12. Give the electron configurations of the ions (a) Pb^{2+}, (b) Pb^{4+}, (c) S^{2-}, (d) Fe^{3+}, and (e) Zn^{2+}.

13. What causes an atom, molecule, or ion to be paramagnetic? Which of the ions in the preceding question are paramagnetic? What term describes the magnetic properties of the others?

14. Give the shorthand electron configurations of (a) Ni, (b) Cr, (c) Sr, (d) Sb, and (e) Po.

15. Define *ionization energy* and *electron affinity*. In terms of these properties, which kinds of elements tend to react to form ionic compounds?

16. In general, the second ionization energy of an atom is larger than the first, the third is larger than the second, and so on. Why?

17. Which of the following elements has the largest difference between its second and third ionization energy? Explain your choice.

(a) Li (b) Be (c) B (d) C

18. Which of the following processes are endothermic?

(a) $P^-(g) + e^- \longrightarrow P^{2-}(g)$

(b) $Fe^{3+}(g) + e^- \longrightarrow Fe^{2+}(g)$

(c) $Cl(g) + e^- \longrightarrow Cl^-(g)$

(d) $S(g) + 2e^- \longrightarrow S^{2-}(g)$

19. Sketch the shape of (a) an *s* orbital, (b) a *p* orbital, and (c) the $3d_{xz}$ orbital.

20. What is meant by the term *electron density*?

21. Give orbital diagrams for the valence shells of selenium and thallium.

22. Which ion would be larger? (a) Fe^{2+} or Fe^{3+}, (b) O^- or O^{2-}

23. Which of the following pairs of elements would be expected to form ionic compounds? (a) Br and F, (b) H and P, (c) Ca and F.

24. Use Lewis symbols to diagram the reaction of calcium with sulfur to form CaS.

25. Draw Lewis structures for (a) SbH_3, (b) IF_3, (c) $HClO_2$, (d) C_2^{2-}, (e) AsF_5, (f) O_2^{2-}, (g) HCO_3^-, (h) TeF_6, and (i) HNO_3.

26. Use the VSEPR theory to predict the shapes of (a) $SbCl_3$, (b) IF_5, (c) AsH_3, (d) BrF_2, and (e) OF_2.

27. What kinds of hybrid orbitals are used by the central atom in each of the species in the preceding question?

28. Referring to your answers to Questions 25 and 26, which of the following molecules would be nonpolar? SbH_3, IF_3, AsF_5, $SbCl_3$, OF_2

29. The oxalate ion has the following arrangement of atoms.

```
O   O
 C C
O   O
```

Draw all of its resonance structures.

30. What is meant by the term *overlap of orbitals?*

31. What are *sigma bonds*? What are *pi bonds*? How are sigma and pi bonds used to explain the formation of double and triple bonds?

32. Some resonance structures that can be drawn for carbon dioxide are shown below.

$:\ddot{O}{=}C{=}\ddot{O}:$	$:O{\equiv}C{-}\ddot{\underset{..}{O}}:$	$:\ddot{\underset{..}{O}}{-}C{\equiv}O:$
I	II	III

Assign formal charges to the atoms in these structures. Explain why Structure I is the preferred structure.

33. Ozone, O_3, consists of a chain of three oxygen atoms.

(a) Draw the two resonance structures for ozone that obey the octet rule.

(b) Based on your answer to (a), is the molecule linear or nonlinear?

(c) Assign formal charges to the atoms in the resonance structures you have drawn in part (a).

(d) On the basis of your answers to (b) and (c), explain why ozone is a polar molecule even though it is composed of three atoms that have identical electronegativities.

34. Why, on the basis of formal charges and relative electronegativities, is it more reasonable to expect the structure of $POCl_3$ to be the one on the left rather than the one on the right?

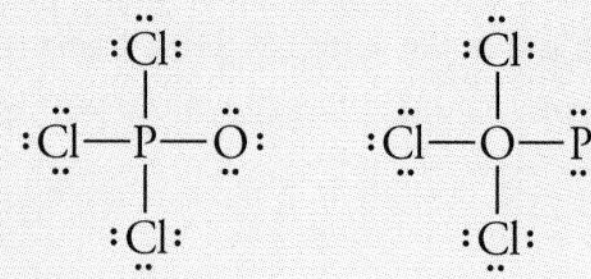

Is either of these the "best" Lewis structure that can be drawn for this molecule?

35. A certain element X was found to form three compounds with chlorine having the formulas XCl_2, XCl_4, and XCl_6. One of its oxides has the formula XO_3, and X reacts with sodium to form the compound Na_2X.
   (a) Is X a metal or a nonmetal?
   (b) In which group in the periodic table is X located?
   (c) In which periods in the periodic table could X possibly be located?
   (d) Draw Lewis structures for XCl_2, XCl_4, XCl_6, and XO_3. (Where possible, follow the octet rule.) Which has multiple bonding?
   (e) What do we expect the molecular structures of XCl_2, XCl_4, XCl_6, and XO_3 to be? Which are polar molecules?
   (f) The element X also forms the oxide XO_2. Draw a Lewis structure for XO_2 that obeys the octet rule.
   (g) Assign formal charges to the atoms in the Lewis structures for XO_2 and XO_3 drawn for parts (d) and (f).
   (h) What kinds of hybrid orbitals would X use for bonding in XCl_4 and XCl_6?
   (i) If X were to form a compound with aluminum, what would be its formula?
   (j) Which compound of X would have the more ionic bonds, Na_2X or MgX?
   (k) If X were in Period 5, what would be the electron configuration of its valence shell?

36. Where in the periodic table are the very reactive metals located? Where are the least reactive ones located?

37. What are bonding and antibonding molecular orbitals? How do they differ in shape and energy?

38. What is a *delocalized molecular orbital*? How does molecular orbital theory avoid the concept of resonance?

39. Predict the shapes of the following molecules and ions:
   (a) PF_3 (c) PF_6^-
   (b) PF_4^+ (d) PF_5

40. Which of the substances in the preceding question have a net dipole moment?

41. According to the VSEPR model, which of the following best illustrates the structure of the $AsCl_3^{2-}$ ion?

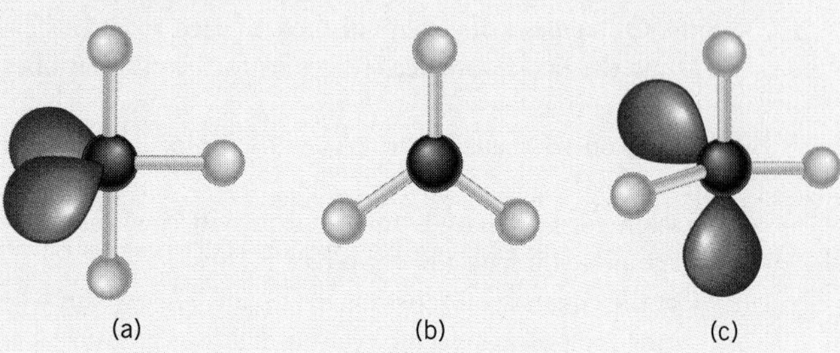

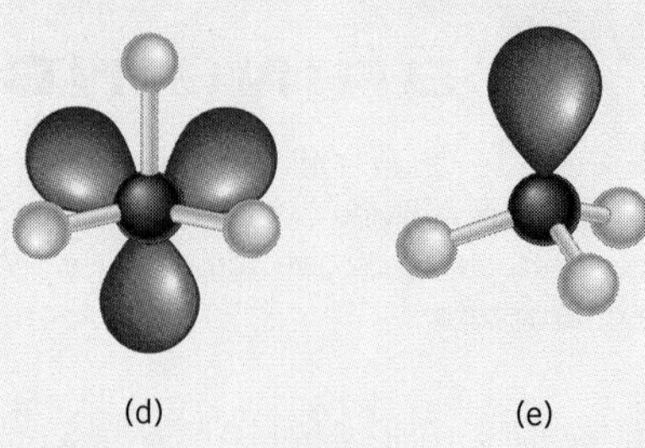

42. Describe how molecular orbital theory explains the bonding in the oxygen molecule.

43. Consider the following statements: (1) Fe^{2+} is easily oxidized to Fe^{3+}, and (2) Mn^{2+} is difficult to oxidize to Mn^{3+}. On the basis of the electron configurations of the ions, explain the difference in ease of oxidation.

44. For each of the following pairs of compounds, which has the larger lattice energy: (a) MgO or NaCl, (b) MgO or BeO, (c) NaI or NaF, and (d) MgO or CaS? Explain your choices.

45. The melting point of Al_2O_3 is much higher than the melting point of NaCl. On the basis of lattice energies, explain why this is so.

46. Why is the change in atomic size, going from one element to the next in a period, smaller among the transition elements than among the representative elements?

47. The VSEPR model predicts the structure below for a certain molecule. Which kind of hybrid orbitals does the central atom in the molecule use to form its covalent bonds?

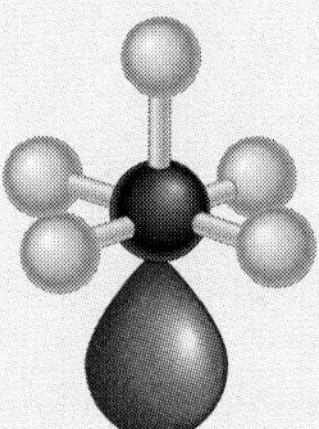

48. Which kind of bond, σ or π, is produced by the overlap of d orbitals pictured below?

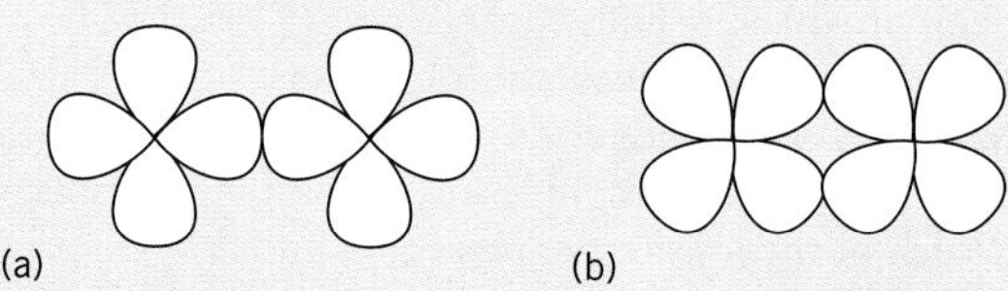

49. The following is the chemical structure of acetaminophen, the painkiller in Tylenol.

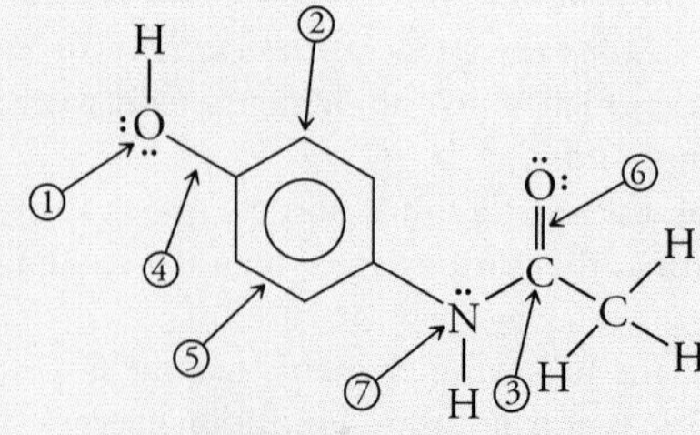

What kinds of hybrid orbitals are used by atoms 1, 2, 3, and 7? How many σ and π bonds are in bonds 4 and 6? What is the average bond order of bond 5?

10 PROPERTIES OF GASES

A cool summer morning provides the ideal conditions for hot-air balloon enthusiasts. Propane heaters inflate each balloon, no matter what shape the creative designers choose. Balloonists then control these brightly colored craft using the properties of gases introduced in this chapter. *(Raymond Watt/ Albuquerque International Balloon Fiesta, Inc.)*

CHAPTER OUTLINE

HIS CHAPTER IN CONTEXT In the preceding chapters we've discussed the chemical properties of ariety of different substances. We've also studied the kinds of forces (chemical bonds) that hold molecular and ic substances together and their three-dimensional structures. In fact, it is the nature of the chemical bonds that tates chemical properties and it is the five simple structures that are used repeatedly to create fantastic molecules ch as DNA and modern plastics. But this is only part of the description of matter as we know it. Another impor- it topic concerns the interaction of one molecule with another without a chemical reaction occurring. These interactions are *physical properties* that are macroscopic manifestations of atomic-scale interactions between molecules, atoms, and ions. These interactions cause phospholipids to arrange themselves into a cell membrane. They also hold DNA together in the famous double helix while allowing the DNA to "unzip" to replicate in the life process. These interactions also explain something as simple as why your pancake syrup is thick when cold but thins out when warmed. The next few chapters will reveal the scientific principles behind why matter acts as it does, even in the absence of a chemical reaction. With this chapter we begin a systematic study of the physical properties of materials, including the factors that govern the behavior of gases, liquids, and solids. We study gases first because they are the easiest to understand and their behavior will help explain some of the properties of liquids and solids in the next chapter.

We live in a mixture of gases called the earth's atmosphere. Through everyday experience, you have become familiar with many of the properties that gases have. Our goal in this chapter is to refine this understanding in terms of the physical laws that govern the way gases behave. You will also learn how this behavior is interpreted in terms of the way we view gases at a molecular level. In our discussions we will describe how the energy concepts, first introduced to you in Chapter 6, provide an explanation of the gas laws. Finally, you will learn how a close examination of gas properties furnishes clues about molecular size and the attractions that exist between molecules.

10.1 FAMILIAR PROPERTIES OF GASES CAN BE EXPLAINED AT THE MOLECULAR LEVEL

For a long time, early scientists didn't recognize the existence of gases as examples of matter. Of course, we now understand that gases are composed of chemical substances that exist in one of the three common states of matter. The reason for the early confusion is that the physical properties of gases differ so much from those of liquids and solids. Consider water, for example. We can see and feel it as a liquid, but it seems to disappear when it evaporates and surrounds us as water vapor. With this in mind, let's examine some of the properties of gaseous substances to look for clues that suggest the nature of gases when viewed at a molecular level.

Air is roughly 21% O_2 and 79% N_2, but it has traces of several other gases.

The most common gas familiar to people is air. (Actually, air is a mixture, but that doesn't matter much when it comes to the physical properties of gases.) Because you've grown up surrounded by gas, you already are aware of many properties that gases have. Let's look at two of them.

- You can wave your hand through air with little resistance. (Compare that with waving your hand through a tub filled with water.)
- The air in a bottle has little weight to it, so if a bottle of air is submerged under water and released, it quickly bobs to the surface.

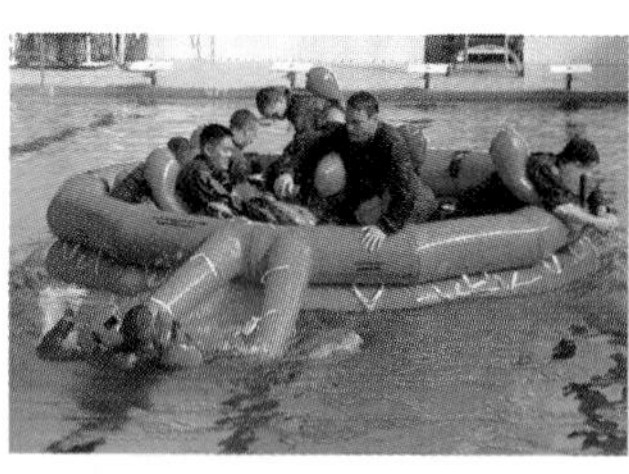

Because air has so little weight for a given volume, it makes things filled with air float, much to the relief of the occupants of an inflatable life raft. *(Courtesy U.S. Air Force photo by K.L. Kibrell.)*

Both of these observations suggest that a given volume of air doesn't have much matter in it. (We can express this by saying that air has a low density.) What else do you know about gases?

- Gases can be compressed. Inflating a tire involves pushing more and more air into the same container (the tire). This behavior is a lot different from liquids; you can't squeeze more water into an already filled bottle.
- Gases exert a pressure. Whenever you inflate a balloon you have an experience with gas pressure, and the "feel" of a balloon suggests that the pressure acts equally in all directions.
- The pressure of a gas depends on *how much* gas is confined. The *more air* you pump into a tire, the greater the pressure.
- Gases fill completely any container into which they're placed. You've never heard of half a bottle of air. If you put air in a container, it expands and fills the container's entire volume. (This is certainly a lot different than the behavior of liquids and solids.)
- Gases mix freely and quickly with each other. You've experienced this when you've smelled the perfume of someone passing by. The vapors of the person's perfume mix with and spread through the air.

- The pressure of a gas rises when its temperature is increased. That's why there's the warning "Do Not Incinerate" printed on aerosol cans. A sealed can, if made too hot, is in danger of exploding from the increased pressure.

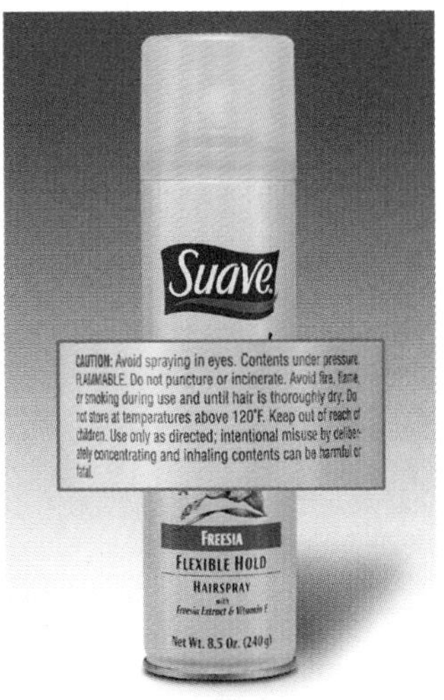

Aerosol cans carry a warning about subjecting them to high temperatures because the internal pressure can become large enough to cause them to explode. *(Andy Washnik.)*

Properties suggest a molecular model

The simple qualitative observations about gases described above suggest what gases must be like when viewed at a molecular level (Figure 10.1). The fact that there's so little matter in a given volume suggests that there is a lot of space between the individual molecules, especially when compared to liquids or solids. This would also explain why gases can be so easily compressed—squeezing a gas simply removes some of the empty space.

It also seems reasonable to believe that the molecules of a gas are moving around fairly rapidly. How else could we explain the travel of the molecules of a perfume so quickly through the air? Furthermore, if gas molecules didn't move, gravity would cause them to settle to the bottom of a container (which they don't do). And if gas molecules are moving, some must be colliding with the walls of the container, and the force of these tiny collisions would explain the pressure a gas exerts. It also explains why adding more gas increases the pressure; the more gas in the container, the more collisions with the walls, and the higher the pressure.

Finally, the fact that gas pressure rises with increasing temperature suggests that the molecules move faster with increasing temperature, because faster molecules would exert greater forces when they collide with the walls.

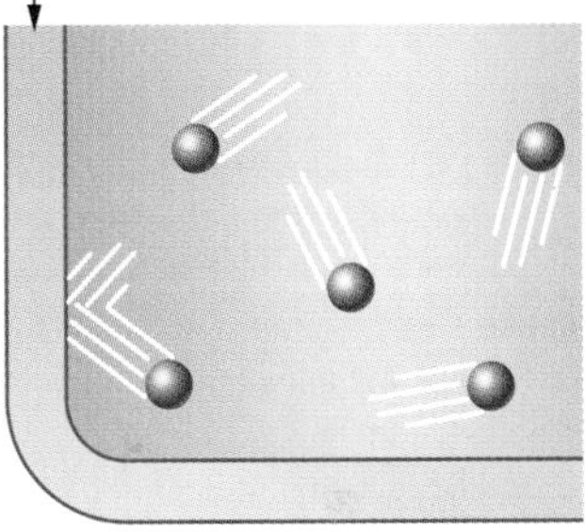

FIG. 10.1 A gas viewed at the molecular level. Simple qualitative observations of the properties of gases lead us to conclude that a gas is composed of widely spaced molecules that are in constant motion. Collisions of molecules with the walls produce tiny forces that, when taken all together, are responsible for the gas pressure.

10.2 PRESSURE IS A MEASURED PROPERTY OF GASES

As we discussed in Chapter 6, **pressure** *is force per unit area*, calculated by dividing the force by the area over which the force acts.

$$\text{Pressure} = \frac{\text{force}}{\text{area}}$$

The earth exerts a gravitational force on everything with mass that is on it or near it. What we call the *weight* of an object, like a book, is simply our measure of the force it exerts because gravity acts on it.

The pressure of the atmosphere is measured with a barometer

Earth's gravity pulls on the air mass of the atmosphere, causing it to cover the earth's surface like an invisible blanket. The molecules in the air collide with every object the air contacts, and by doing so, produce a pressure we call the *atmospheric pressure.*

At any particular location on the earth, the atmospheric pressure acts equally in all directions—up, down, and sideways. In fact, it presses against our bodies with a surprising amount of force, but we don't really feel it because the fluids in our bodies push back with equal pressure. We can observe atmospheric pressure, however, if we pump the air from a collapsible container such as the can in Figure 10.2. Before air is removed, the walls of the can experience atmospheric pressure equally inside and out. When some air is pumped out, however, the pressure inside decreases, making the atmospheric pressure outside of the can greater than the pressure inside. The net inward pressure is sufficiently great to make the can crumple.

To measure atmospheric pressure we use a device called a **barometer**. The simplest type is the *Torricelli barometer*[1] (Figure 10.3), which consists of a glass tube sealed at one end, 80 cm or more in length. To set up the apparatus, the tube is filled with mercury, capped,

[1] In 1643 Evangelista Torricelli, an Italian mathematician, suggested an experiment, later performed by a colleague, that demonstrated that atmospheric pressure determines the height to which a fluid will rise in a tube inverted over the same liquid. This concept led to the development of the Torricelli barometer, which is named in his honor.

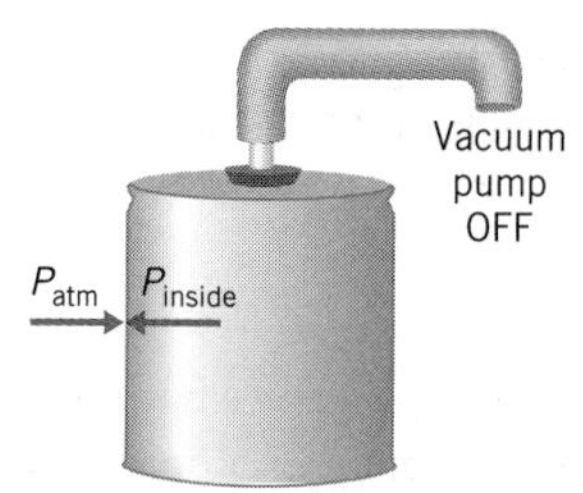

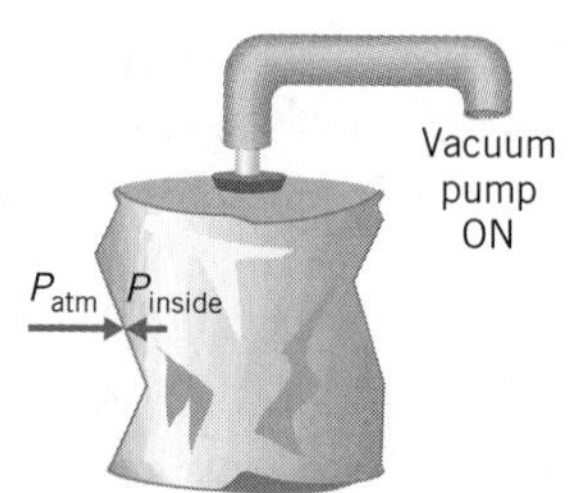

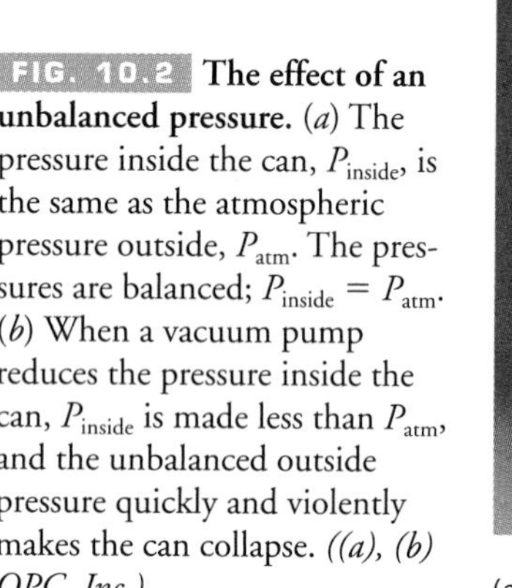

(a)

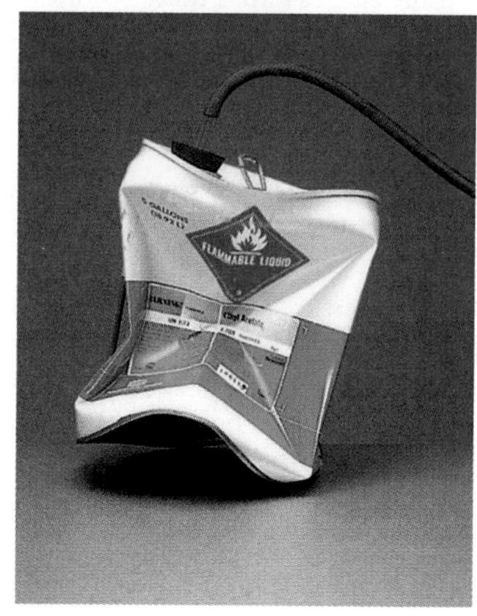

(b)

FIG. 10.2 The effect of an unbalanced pressure. (*a*) The pressure inside the can, P_{inside}, is the same as the atmospheric pressure outside, P_{atm}. The pressures are balanced; $P_{inside} = P_{atm}$. (*b*) When a vacuum pump reduces the pressure inside the can, P_{inside} is made less than P_{atm}, and the unbalanced outside pressure quickly and violently makes the can collapse. *((a), (b) OPC, Inc.)*

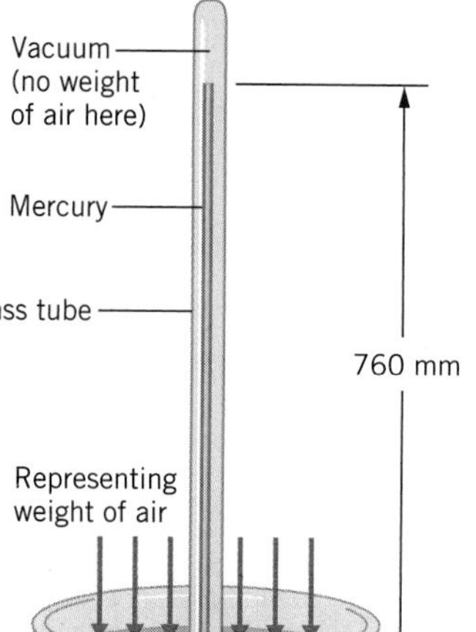

FIG. 10.3 A Torricelli barometer. The apparatus is also called a mercury barometer. The height of the mercury column inside the tube is directly proportional to the atmospheric pressure. In the United States, weather reports often give the height of the mercury column in inches.

inverted, and then its capped end is immersed in a dish of mercury. When the cap is removed, some mercury runs out, but not all.[2] Atmospheric pressure, pushing on the surface of the mercury in the dish, holds most of the mercury in the tube. Opposing the atmosphere is the downward pressure caused by the weight of the mercury still inside the tube. When the two pressures become equal, no more mercury can run out, but a space inside the tube above the mercury level has been created having essentially no atmosphere; it's a *vacuum.*

The height of the mercury column, measured from the surface of the mercury in the dish, is directly proportional to atmospheric pressure. On days when the atmospheric pressure is high, more mercury is forced from the dish into the tube and the height of

[2] Today, mercury is kept as much as possible in closed containers. Although atoms of mercury do not readily escape into the gaseous state, mercury vapor is a dangerous poison.

the column increases. When the atmospheric pressure drops, during an approaching storm, for example, some mercury flows out of the tube and the height of the column decreases. Most people live where this height fluctuates between 730 and 760 mm.

Units of pressure include the pascal, atmosphere, and torr

At sea level, the height of the mercury column in a barometer fluctuates around a value of 760 mm. Some days it's a little higher, some a little lower, depending on the weather. The average pressure at sea level has long been used by scientists as a standard unit of pressure. The **standard atmosphere (atm)** was originally defined as the pressure needed to support a column of mercury 760 mm high measured at 0 °C.[3]

In the SI, the unit of pressure is the **pascal,** symbolized **Pa.** In SI units, the pascal is the ratio of force in *newtons* (N, the SI unit of force) to area in meters squared,

$$1 \text{ Pa} = \frac{1 \text{ N}}{1 \text{ m}^2} = 1 \text{ N m}^{-2}$$

It's a very small pressure; 1 Pa is approximately the pressure exerted by the weight of a lemon spread over an area of 1 m^2.

To bring the standard atmosphere unit in line with other SI units, it has been redefined in terms of the pascal as follows.

$$1 \text{ atm} = 101{,}325 \text{ Pa (exactly)}$$

A unit of pressure related to the pascal is the **bar,** which is defined as 100 kPa. Consequently, one bar is slightly smaller than one standard atmosphere (1 bar = 0.9869 atm). You may have heard the **millibar** unit (1 mb = 10^{-3} bar) used in weather reports describing pressures inside storms such as hurricanes. For example, the lowest atmospheric pressure at sea level ever observed in the Atlantic basin was 882 mb during Hurricane Wilma on October 19, 2005. The storm later weakened but still caused extensive damage as it crossed Florida.

For ordinary laboratory work, the pascal (or kilopascal) is not a conveniently measured unit. Usually we use a unit of pressure called the **torr** (named after Torricelli). The torr is defined as 1/760th of 1 atm.

$$1 \text{ torr} = \tfrac{1}{760} \text{ atm}$$

$$1 \text{ atm} = 760 \text{ torr (exactly)}$$

The torr is very close to the pressure that is able to support a column of mercury 1 mm high. In fact, the *millimeter of mercury* (abbreviated *mm Hg*) is often itself used as a pressure unit. Except when the most exacting measurements are being made, it is safe to use the relationship

$$1 \text{ torr} = 1 \text{ mm Hg}$$

Practice Exercise 1: Using the margin table above, determine the atmospheric pressure in pounds per square inch and inches of mercury when the barometer reads 730 mm Hg. (Hint: Recall your tools for converting units.)

Practice Exercise 2: The second lowest barometric pressure ever recorded at sea level in the western hemisphere was 888 mb during Hurricane Gilbert in 1988. What was the pressure in pascals and in torr?

[3] Because any metal, including mercury, expands or contracts as the temperature increases or decreases, the height of the mercury column varies with temperature (just like in a thermometer). Therefore, the definition of the standard atmosphere required that the temperature at which the mercury height is measured be specified.

In English units, one atmosphere of pressure is 14.7 lb in.$^{-2}$. This means that at sea level, each square inch of your body is experiencing a force of nearly 15 pounds.

Below are values of the standard atmosphere (atm) expressed in different pressure units. Studying the table will give you a feel for the sizes of the different units.

760 torr
101,325 Pa
101.325 kPa
1.013 bar
1013 mb
14.7 lb in.$^{-2}$
1.034 kg cm^{-2}

A satellite photo of Hurricane Wilma when it reached Category 5 strength, with sustained winds of 175 miles per hour in the eye wall. *(Terra MODIS data acquired by direct broadcast at the University of South Florida [Judd Taylor] Image processed at the "University of Wisconsin-Madison [Liam Gumley]".)*

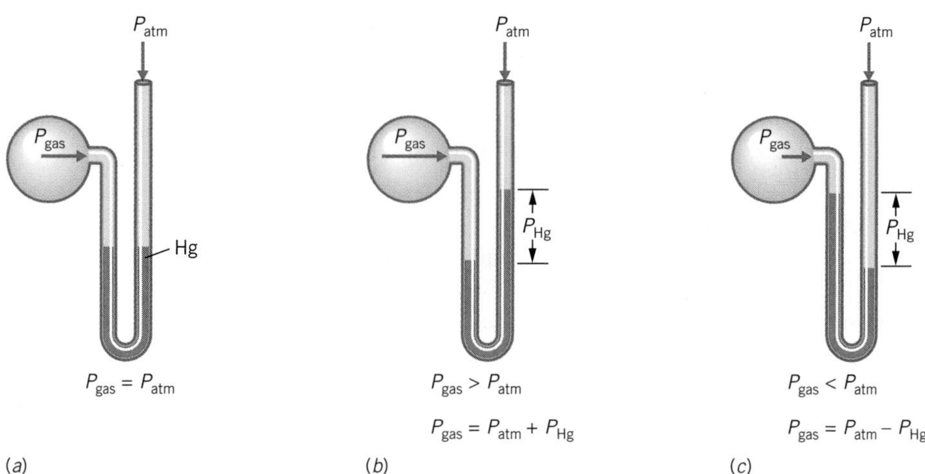

FIG. 10.4 **An open-end manometer.** The difference in the heights of the mercury in the two arms equals the pressure difference, in torr, between the atmospheric pressure, P_{atm}, and the pressure of the trapped gas, P_{gas}.

Manometers measure the pressures of trapped gas samples

Gases used as reactants or formed as products in chemical reactions are kept from escaping by using closed glassware. To measure pressures inside such vessels, a **manometer** is used. Two types are common, open-end and closed-end manometers.

Advantages of mercury over other liquids are its low reactivity, its low melting point, and particularly its very high density, which permits short manometer tubes.

An **open-end manometer** consists of a U-tube partly filled with a liquid, usually mercury (see Figure 10.4). One arm of the U-tube is open to the atmosphere; the other is exposed to a container of some trapped gas. In Figure 10.4*a* the pressure in the flask is equal to the atmospheric pressure and the mercury levels are equal. Figure 10.4*b* illustrates a flask containing a gas at a pressure that is greater than the atmospheric pressure. In Figure 10.4*c*, the mercury level is higher in the arm connected to the container of gas, indicating that the pressure of the atmosphere must be higher than the gas pressure. When mercury is the liquid in the tube, the difference in the heights in the two arms, represented here as P_{Hg}, is equal to the difference between the pressure of the gas and the pressure of the atmosphere. By measuring P_{Hg} in millimeters, the value equals the pressure difference in torr. For the situation shown in Figure 10.4*b*, we would calculate the pressure of the trapped gas as

$$P_{gas} = P_{atm} + P_{Hg} \qquad (\text{when } P_{gas} > P_{Hg})$$

For the situation shown in Figure 10.4*c* we note that the pressure in the flask must be less than the atmospheric pressure or

$$P_{gas} = P_{atm} - P_{Hg} \qquad (\text{when } P_{gas} < P_{Hg})$$

Example 10.1 illustrates how the open-end manometer is used.

EXAMPLE 10.1
Measuring the Pressure of a Gas Using a Manometer

A student collected a gas in an apparatus connected to an open-end manometer, as illustrated in the figure in the margin. The difference in the heights of the mercury in the two columns was 10.2 cm and the atmospheric pressure was measured to be 756 torr. What was the pressure of the gas in the apparatus?

ANALYSIS: From the preceding discussion, we know we will use the atmospheric pressure and either add to it or subtract from it the difference in heights of the mercury columns, expressed in pressure units of torr. But which should we do? In a problem of this type, it is best to use some common sense (something that will help a lot in working problems involving gases).

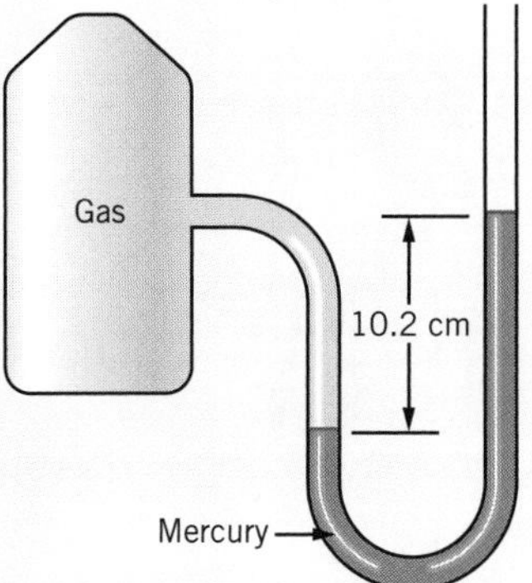

When we look at the diagram of the apparatus, we see that the mercury is pushed up into the arm of the manometer that's open to the air. Common sense tells us that the pressure of the gas inside must be larger than the pressure of the air outside. Therefore, we will add the pressure difference to the atmospheric pressure.

Finally, before we can do the arithmetic, we must be sure the pressure difference is calculated in torr. We know that 1 torr is equivalent to 1 mm Hg, but the difference in heights was measured in centimeters. Therefore, we have to convert 10.2 cm to millimeters.

SOLUTION: You probably know that 1 cm equals 10 mm. If so, that makes the conversion of the pressure difference easy:

$$10.2 \text{ cm Hg} = 102 \text{ mm Hg}$$

If you weren't sure of the conversion, you could always go back to the definitions of the SI prefixes. The two equalities we need are: 1 cm = 10^{-2} m and 1 mm = 10^{-3} m. Ratios of these equalities provide the factor labels for the conversion as shown next.

$$10.2 \, \cancel{\text{cm}} \times \frac{10^{-2} \, \cancel{\text{m}}}{1 \, \cancel{\text{cm}}} \times \frac{1 \text{ mm}}{10^{-3} \, \cancel{\text{m}}} = 102 \text{ mm}$$

Either way, we find the difference in pressures to be 102 mm Hg, which is equivalent to 102 torr.

To find the gas pressure, we add 102 torr to the atmospheric pressure.

$$\begin{aligned} P_{\text{gas}} &= P_{\text{atm}} + P_{\text{Hg}} \\ &= 756 \text{ torr} + 102 \text{ torr} \\ &= 858 \text{ torr} \end{aligned}$$

The gas pressure is 858 torr.

IS THE ANSWER REASONABLE? We can check our conversion of cm to mm by recalling that for the same measurement, the number of millimeters should always be larger than the measurement expressed in centimeters. We can also double-check that we've done the right *kind* of arithmetic (adding or subtracting). Look at the apparatus again. If the gas pressure were lower than atmospheric pressure, it would appear as though the atmosphere was pushing the mercury higher in the same way that the heavier child on a seesaw pushes his lighter friend higher. That's not what we see here. It almost looks like the gas is trying to push the mercury out of the manometer, so we conclude that the gas pressure must be higher than atmospheric pressure. That's exactly what we've found in our calculation, so we can feel confident we've solved the problem correctly.

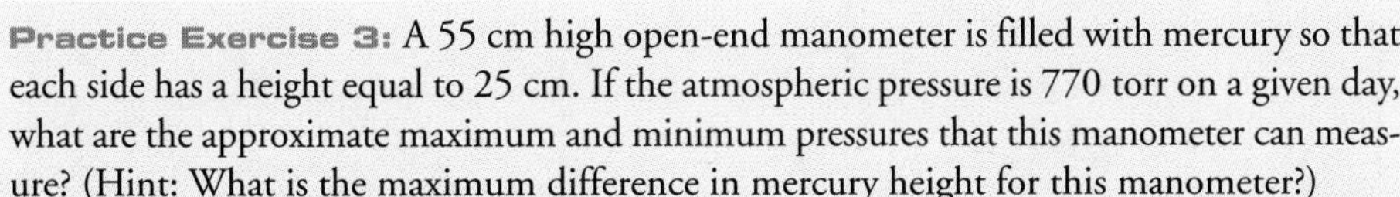

Practice Exercise 3: A 55 cm high open-end manometer is filled with mercury so that each side has a height equal to 25 cm. If the atmospheric pressure is 770 torr on a given day, what are the approximate maximum and minimum pressures that this manometer can measure? (Hint: What is the maximum difference in mercury height for this manometer?)

Practice Exercise 4: In another experiment, it was found that the mercury level in the arm of the manometer attached to the container of gas was 10.7 cm higher than in the arm open to the air (P_{atm} = 770 torr). What was the pressure of the gas?

A closed-end manometer avoids the need to measure atmospheric pressure

A **closed-end manometer** (Figure 10.5) is made by sealing the arm that will be farthest from the gas sample and then filling the closed arm completely with mercury. When the gas pressures to be measured are expected to be small, the filled arm can be made short, making the entire apparatus compact. In this design, the mercury is pushed to the top of the closed arm when the open arm of the manometer is exposed to the atmosphere. When connected to a gas at a low pressure, however, the mercury level in the sealed arm will drop,

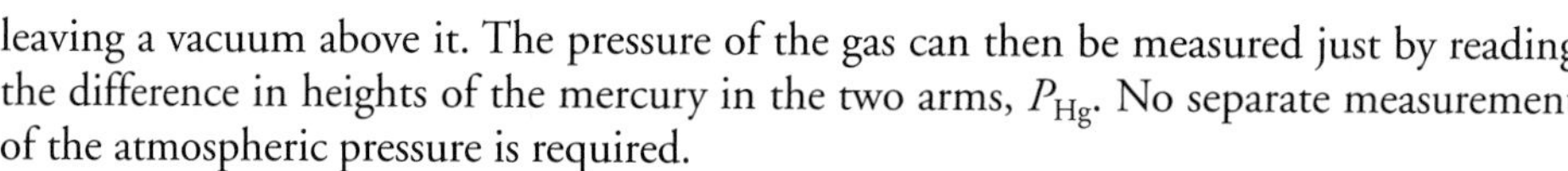

leaving a vacuum above it. The pressure of the gas can then be measured just by reading the difference in heights of the mercury in the two arms, P_{Hg}. No separate measurement of the atmospheric pressure is required.

10.3 THE GAS LAWS SUMMARIZE EXPERIMENTAL OBSERVATIONS

Earlier, we examined some properties of gases that are familiar to you. Our discussion was only qualitative, however, and now that we've discussed pressure and its units, we are ready to examine gas behavior quantitatively.

There are four variables that affect the properties of a gas—pressure, volume, temperature, and the amount of gas. In this section, we will study situations in which the amount of gas (measured by either grams or moles) remains constant and observe how gas samples respond to changes in pressure, volume, and temperature.

At constant temperature for a fixed amount of gas, the volume is inversely proportional to the pressure

When you inflate a bicycle tire with a hand pump, you squeeze air into a smaller volume and increase its pressure. Packing molecules into a smaller space causes an increased number of collisions with the walls, and because these collisions are responsible for the pressure, the pressure increases (Figure 10.6).

Robert Boyle, an English scientist (1627–1691), performed experiments to determine quantitatively how the volume of a fixed amount of gas varies with pressure. Because the volume is also affected by the temperature, he held that variable constant. A graph of typical data collected in his experiments is shown in Figure 10.7*a* and demonstrates that *the volume of a given amount of gas held at constant temperature varies inversely with the applied pressure.* Mathematically, this can be expressed as

$$V \propto \frac{1}{P} \qquad \text{(temperature and amount of gas held constant)}$$

where V is the volume and P is the pressure. This relationship between pressure and volume is now called **Boyle's law** or the **pressure–volume law.**

In the expression above, the proportionality sign, $\propto$, can be removed by introducing a proportionality constant, C.

$$V = \frac{1}{P} \times C$$

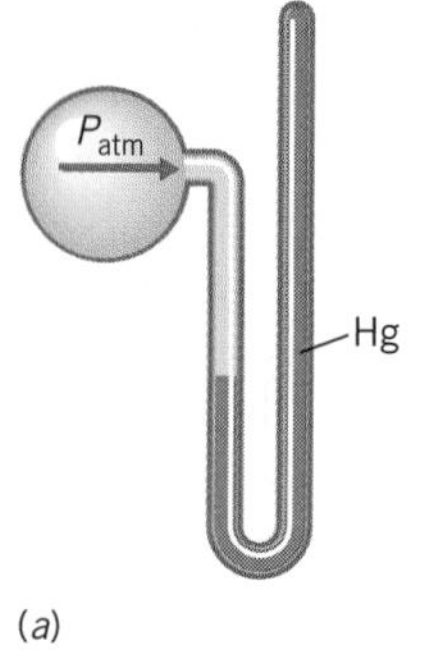

(*a*)

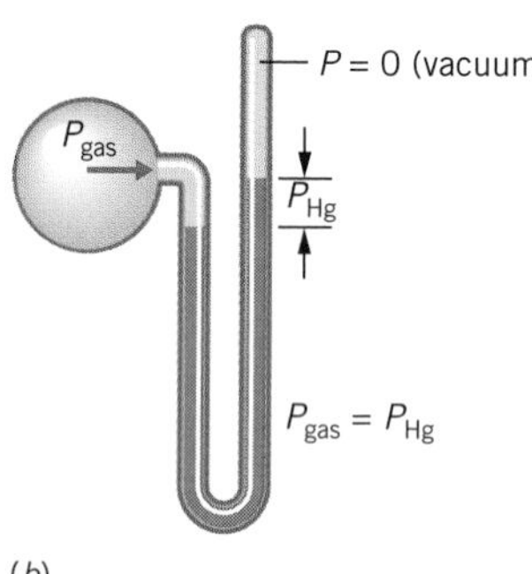

(*b*)

FIG. 10.5 A closed-end manometer for measuring gas pressures less than 1 atm or 760 torr. (*a*) When constructed, the tube is fully evacuated and then mercury is allowed to enter the tube to completely fill the closed arm. (*b*) When the tube is connected to a bulb containing a gas at low pressure, the difference in the mercury heights (P_{Hg}) equals the pressure, in torr, of the trapped gas, P_{gas}.

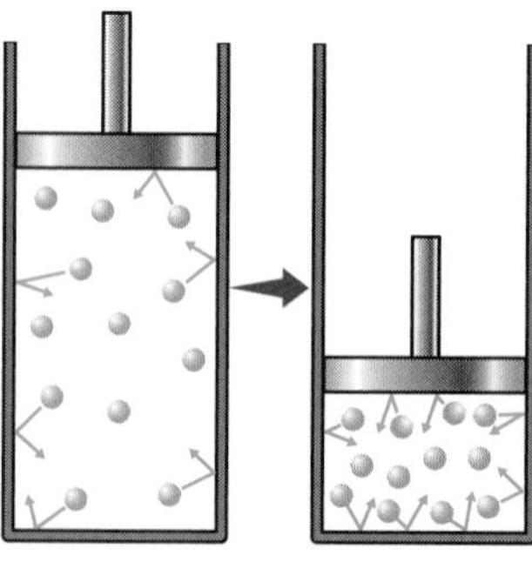

FIG. 10.6 Compressing a gas increases its pressure. A molecular view of what happens when a gas is squeezed into a smaller volume. By crowding the molecules together, the number of collisions with a given area of the walls increases, which causes the pressure to rise.

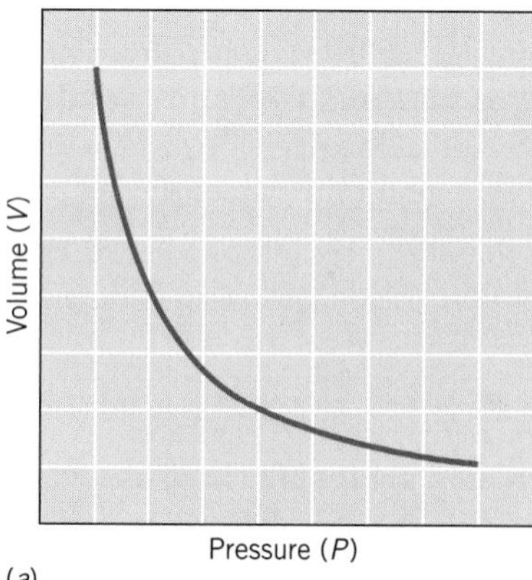

(*a*)

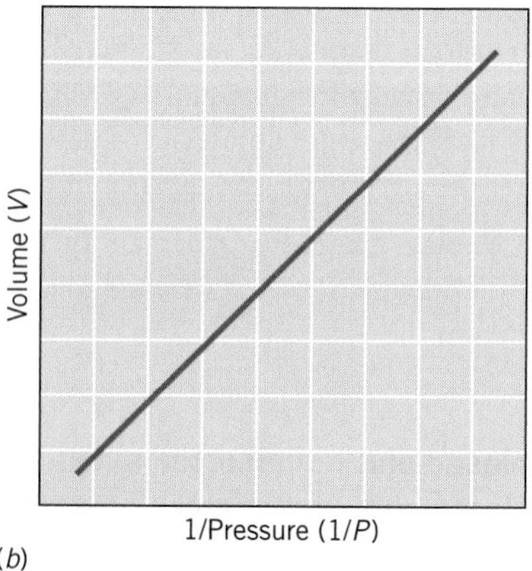

(*b*)

FIG. 10.7 The variation of volume with pressure at constant temperature for a fixed amount of gas. (*a*) A typical graph of volume versus pressure, showing that as the pressure increases, the volume decreases. (*b*) A straight line is obtained when volume is plotted against $1/P$, which shows that $V \propto 1/P$.

Rearranging gives

$$PV = C$$

This equation tells us, for example, that if at constant temperature the gas pressure is doubled the gas volume must be cut in half so that the product $P \times V$ doesn't change.

What is remarkable about Boyle's discovery is that *this relationship is essentially the same for all gases at temperatures and pressures usually found in the laboratory.*

An ideal gas would obey Boyle's law perfectly

When very precise measurements are made, it's found that Boyle's law doesn't quite work. This is especially a problem when the pressure of the gas is very high or when the gas is at a temperature where it's on the verge of changing to a liquid. Although real gases do not *exactly* obey Boyle's law, or any of the other gas laws that we'll study, it is often useful to imagine a hypothetical gas that would. We call such a hypothetical gas an *ideal gas.* An **ideal gas** *would obey the gas laws exactly over all temperatures and pressures.* A real gas behaves more and more like an ideal gas as its pressure decreases and its temperature increases. Most gases we work with in the lab can be treated as ideal gases unless we're dealing with extremely precise measurements.

At constant pressure for a fixed amount of gas, the volume increases with increasing temperature

In 1787 a French chemist and mathematician named Jacques Alexandre Charles became interested in hot-air ballooning, which at the time was becoming popular in France. His new interest led him to study what happens to the volume of a sample of gas when the temperature is varied, keeping the pressure constant.

Jacques Alexandre César Charles (1746–1823), a French scientist, had a keen interest in hot air balloons. He was the first to inflate a balloon with hydrogen.

When data from experiments such as his are plotted, a graph like that shown in Figure 10.8 is obtained. Here the volume of the gas is plotted against the temperature in degrees Celsius. The colored points correspond to typical data, and the lines are drawn to most closely fit the data. Each line represents data collected for a different sample. Because all gases eventually become liquids if cooled sufficiently, the solid portions of the lines correspond to temperatures at which measurements are possible; at lower temperatures the gas liquefies. However, if the lines are extrapolated (reasonably extended) back to a point where the volume of the gas would become zero if it didn't condense, all the lines meet at the same temperature, −273.15 °C. Especially significant is the fact that this exact same behavior is exhibited by all gases; when plots of volume versus temperature are extrapolated to zero volume, the temperature axis is always crossed at −273.15 °C. This point represents the temperature at which all gases, if they did not condense, would have a volume of zero, and below which they would have a negative volume. Negative volumes are impossible,

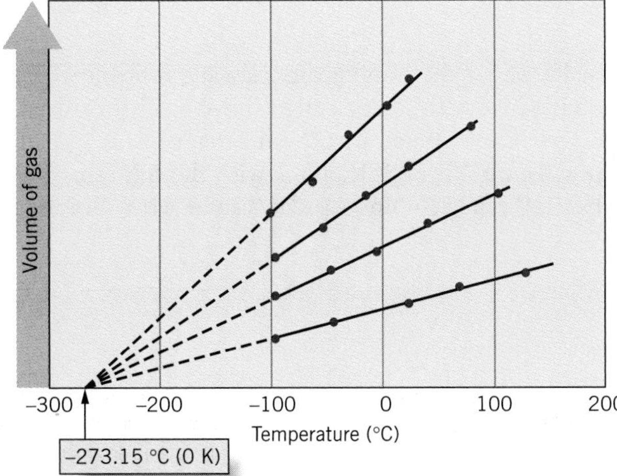

FIG. 10.8 Charles' law plots. Each line shows how the gas volume changes with temperature for a different size sample of the same gas.

of course, so it was reasoned that −273.15 °C must be nature's coldest temperature, and it was called **absolute zero.**

As you learned earlier, absolute zero corresponds to the zero point on the Kelvin temperature scale, and to obtain a Kelvin temperature, we add 273.15 °C to the Celsius temperature.[4]

$$T_K = t_C + 273.15$$

For most purposes, we will need only three significant figures, so we can use the following approximate relationship.

$$T_K = t_C + 273$$

The straight lines in Figure 10.8 suggest that at constant pressure, the volume of a gas is directly proportional to its temperature, provided the temperature is expressed in kelvins. This became known as **Charles' law** (or the **temperature–volume law)** and is expressed mathematically as

$$V \propto T \qquad \text{(pressure and amount of gas held constant)}$$

The value of C' depends on the size and pressure of the gas sample.

Using a different proportionality constant, C', we can write

$$V = C'T \qquad \text{(pressure and amount of gas held constant)}$$

At constant volume for a fixed amount of gas, the pressure is proportional to the absolute temperature

Joseph Louis Gay-Lussac (1778–1850), a French scientist, was a codiscoverer of the element boron.

The French scientist Joseph Louis Gay-Lussac studied how the pressure and temperature of a fixed amount of gas at constant volume are related. (Such conditions exist, for example, when a gas is confined in a vessel with rigid walls, like an aerosol can.) The relationship that he established, called **Gay-Lussac's law** or the **pressure–temperature law,** states that *the pressure of a fixed amount of gas held at constant volume is directly proportional to the Kelvin temperature.* Thus,

$$P \propto T \qquad \text{(volume and amount of gas held constant)}$$

Using still another constant of proportionality, Gay-Lussac's law becomes

We're using different symbols for the various gas law constants because they are different for each law.

$$P = C''T \qquad \text{(volume and amount of gas held constant)}$$

The combined gas law brings together the other gas laws we've studied

The three gas laws we've just examined can be brought together into a single equation known as the **combined gas law,** which states that *the ratio PV/T is a constant for a fixed amount of gas.*

$$\frac{PV}{T} = \text{constant} \qquad \text{(for a fixed amount of gas)}$$

Usually, we use the combined gas law in problems where we know some given set of conditions of temperature, pressure, and volume (for a fixed amount of gas), and wish to find out how one of these variables will change when the others are changed. If we label the initial conditions of *P, V,* and *T* with the subscript 1 and the final conditions with the subscript 2, the combined gas law can be written in the following useful form.

TOOLS Combined gas law

$$\frac{P_1V_1}{T_1} = \frac{P_2V_2}{T_2} \qquad (10.1)$$

[4]In Chapter 1 we presented this equation as $T_K = (t_C + 273.15\ °C)\ (1\ K/1\ °C)$ to emphasize unit cancellation. Operationally, however, we just add 273.15 to the Celsius temperature to obtain the Kelvin temperature (or we just add 273 if three significant figures are sufficient).

In applying this equation, T must always be in kelvins. The pressure and volume can have any units, but whatever the units are on one side of the equation, they must be the same on the other side.

It is simple to show that Equation 10.1 contains each of the other gas laws as special cases. Boyle's law, for example, applies when the temperature is constant. Under these conditions, T_1 equals T_2 and temperature cancels from the equation. This leaves us with

$$P_1V_1 = P_2V_2 \qquad \text{(when } T_1 \text{ equals } T_2\text{)}$$

which is one way to write Boyle's law. Similarly, under conditions of constant pressure, P_1 equals P_2 and the pressure cancels, so Equation 10.1 reduces to

$$V_1/T_1 = V_2/T_2 \qquad \text{(when } P_1 \text{ equals } P_2\text{)}$$

This, of course, is another way of writing Charles' law. Under the constant-volume conditions required by Gay-Lussac's law, V_1 equals V_2, and Equation 10.1 reduces to

$$P_1/T_1 = P_2/T_2 \qquad \text{(when } V_1 \text{ equals } V_2\text{)}$$

EXAMPLE 10.2
Using the Combined Gas Law

An ordinary incandescent lightbulb contains a tungsten filament, which becomes white hot (about 2500 °C) when electricity is passed through it. To prevent the filament from rapidly vaporizing, the bulb is filled to a low pressure with the inert (unreactive) gas argon. Suppose a 12.0 L cylinder containing compressed argon at a pressure of 57.8 atm measured at 25 °C is to be used to fill electric lightbulbs, each with a volume of 158 mL, to a pressure of 3.00 torr at 20 °C. How many of these lightbulbs could be filled by the argon in the cylinder?

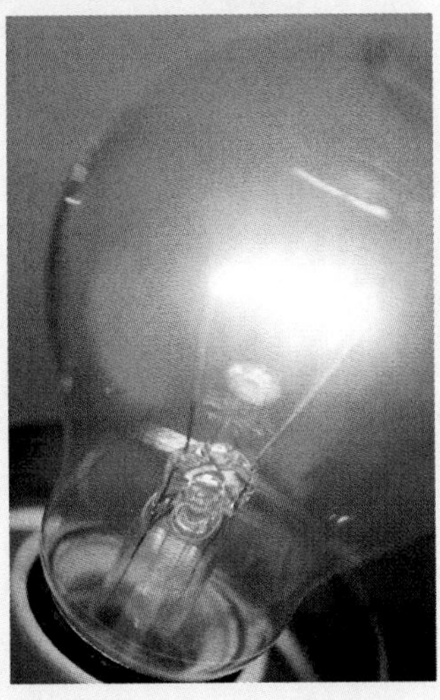

(PhotoDisc, Inc./Getty Images.)

ANALYSIS: What do we need to know to figure out how many lightbulbs can be filled? If we knew the total volume of gas with a pressure of 3.00 torr at 20 °C, we could just divide by the volume of one lightbulb; the result is the number of lightbulbs that can be filled. So, our main problem is determining what volume the argon in the cylinder will occupy when its pressure is reduced to 3.00 torr and its temperature is lowered to 20 °C. Because the total amount of argon isn't changing, we can use the combined gas law for the calculation.

The initial conditions will be 12.0 L of argon at 57.8 atm and 25 °C. The final pressure and temperature will be 3.00 torr and 20 °C. To apply the combined gas law, the pressure units will have to be the same, so we can convert 57.8 atm to torr. The temperatures will have to be in kelvins, so we add 273 to the Celsius temperatures. We're now ready to perform the math.

SOLUTION: We'll begin by converting 57.8 atm to torr. We know that 1 atm = 760 torr, so

$$57.8 \cancel{\text{atm}} \times \frac{760 \text{ torr}}{1 \cancel{\text{atm}}} = 4.39 \times 10^4 \text{ torr}$$

Now we can assemble the data, again using "1" for the initial conditions and "2" for the final ones.

	Initial (1)	Final (2)
P	4.39×10^4 torr	3.00 torr
V	12.0 L	?
T	298 K (25 + 273)	293 K (20 + 273)

Now we have to use the combined gas law.

$$\frac{P_1V_1}{T_1} = \frac{P_2V_2}{T_2}$$

To obtain the final volume, we solve for V_2. We'll do this and then rearrange the equation slightly.

ratio of temperatures

$$V_2 = V_1 \times \frac{P_1}{P_2} \times \frac{T_2}{T_1}$$

ratio of pressures

Notice that for V_2 to have the same units as V_1, the units in numerator and denominator of both ratios must cancel. That's the reason we had to convert atmospheres to torr for P_1. Let's now substitute values from our table of data.

$$V_2 = 12.0 \text{ L} \times \frac{4.39 \times 10^4 \cancel{\text{torr}}}{3.00 \cancel{\text{torr}}} \times \frac{293 \cancel{\text{K}}}{298 \cancel{\text{K}}}$$
$$= 1.73 \times 10^5 \text{ L}$$

Before we can divide by the volume of one lightbulb (158 mL), we have to convert the total volume to milliliters.

$$1.73 \times 10^5 \cancel{\text{L}} \times \frac{1000 \text{ mL}}{1 \cancel{\text{L}}} = 1.73 \times 10^8 \text{ mL}$$

The volume per lightbulb gives us the relationship

$$1 \text{ lightbulb} \Leftrightarrow 158 \text{ mL argon}$$

which we can use to find the number of lightbulbs that can be filled.

$$1.73 \times 10^8 \cancel{\text{mL argon}} \times \frac{1 \text{ lightbulb}}{158 \cancel{\text{mL argon}}} = 1.09 \times 10^6 \text{ lightbulbs}$$

That's 1.09 million lightbulbs! As you can see, there's not much argon in each one.

IS THE ANSWER REASONABLE? Looking at our setup, one way we can round the numbers to estimate the answer is

$$V_2 = 12.0 \text{ L} \times \frac{4.5 \times 10^4 \cancel{\text{torr}}}{3 \cancel{\text{torr}}} \times \frac{300 \cancel{\text{K}}}{300 \cancel{\text{K}}}$$

The estimated answer is $V_2 = 18 \times 10^4$ L, which is very close to our calculator answer. We can multiply the liters by 10^3 to get 18×10^7 mL and then we will "round" the 158 mL/bulb to 18×10^1 mL/bulb. Dividing the two gives us 1×10^6 bulbs. This again agrees quite well with the calculated answer.

To determine whether we've set up the combined gas law properly, we check to see if the pressure and temperature ratios move the volume in the right direction. Going from the initial to final conditions, the pressure *decreases* from 4.39×10^4 torr to 3.00 torr, so the volume should *increase* a lot. The pressure ratio we used is much larger than 1, so multiplying by it should increase the volume, which agrees with what we expect. Next, look at the temperature change; a *drop in temperature* should tend to *decrease* the volume, so we should be multiplying by a ratio that's less than 1. The ratio 293/298 is less than one, so that ratio is correct, too.

(An observant student might note that in the end 12 L of argon must remain in the cylinder. However, we can divide 12,000 mL by 158 to get an answer of approximately 75 bulbs. Subtracting 75 from 1.09×10^6 still leaves us, when rounded, with 1.09×10^6 lightbulbs.)

Notice that we use our knowledge of how gases behave to determine whether we've done the correct arithmetic. If you learn to do this, you can catch your mistakes.

Practice Exercise 5: Use the combined gas law to determine by what factor the pressure of an ideal gas must change if the Kelvin temperature is doubled and the volume is tripled. (Hint: Sometimes it is easier to assume a starting set of temperature, volume, and pressure readings and then apply the conditions of the problem.)

Practice Exercise 6: A sample of nitrogen has a volume of 880 mL and a pressure of 740 torr. What pressure will change the volume to 870 mL at the same temperature?

Practice Exercise 7: What will be the final pressure of a sample of nitrogen with a volume of 950 m^3 at 745 torr and 25.0 °C if it is heated to 60.0 °C and given a final volume of 1150 m^3?

10.4 GAS VOLUMES CAN BE USED IN SOLVING STOICHIOMETRY PROBLEMS

At constant *T* and *P*, gas volumes are in whole-number ratios in a gas-phase reaction

When scientists studied reactions between gases quantitatively, they made an interesting discovery. If the volumes of the reacting gases, as well as the volumes of gaseous products, are measured under the same conditions of temperature and pressure, the volumes are in simple whole-number ratios. For example, hydrogen gas reacts with chlorine gas to give gaseous hydrogen chloride. Beneath the names in the following equation are the relative volumes with which these gases interact (at the same *T* and *P*).

$$\underset{\text{1 volume}}{\text{hydrogen}} + \underset{\text{1 volume}}{\text{chlorine}} \longrightarrow \underset{\text{2 volumes}}{\text{hydrogen chloride}}$$

What this means is that if we were to use 1.0 L of hydrogen, it would react with 1.0 L of chlorine and produce 2.0 L of hydrogen chloride. If we were to use 10.0 L of hydrogen, all the other volumes would be multiplied by 10 as well.

Similar simple, whole-number ratios by volume are observed when hydrogen combines with oxygen to give water, which is a gas above 100 °C.

$$\underset{\text{2 volumes}}{\text{hydrogen}} + \underset{\text{1 volume}}{\text{oxygen}} \longrightarrow \underset{\text{2 volumes}}{\text{water (gaseous)}}$$

Notice that the reacting volumes, *measured under identical temperatures and pressures,* are in ratios of simple, whole numbers.[5]

Observations such as those above led Gay-Lussac to formulate his **law of combining volumes,** which states that *when gases react at the same temperature and pressure, their combining volumes are in ratios of simple whole numbers.* Much later, it was learned that these "simple whole numbers" are the coefficients of the balanced equations for the reactions.

At a given temperature and pressure, equal volumes of gas contain the same number of molecules

The observation that gases react in whole-number volume ratios led Amedeo Avogadro to conclude that, at the same *T* and *P*, equal volumes of gases must have identical numbers of molecules. Today, we know that "equal numbers of *molecules*" is the same as "equal numbers of *moles,*" so Avogadro's insight, now called **Avogadro's principle,** is expressed as follows. *When measured at the same temperature and pressure, equal volumes of gases contain equal numbers of moles.* A corollary to Avogadro's principle is that *the volume of a gas is directly proportional to its number of moles, n.*

Amedeo Avogadro (1776–1856), an Italian scientist, helped to put chemistry on a quantitative basis.

$$V \propto n \qquad \text{(at constant } T \text{ and } P\text{)}$$

The standard molar volume is the volume occupied by one mole of gas at STP
Avogadro's principle implies that the volume occupied by one mole of *any* gas—its *molar volume*—must be identical for all gases under the same conditions of pressure and temperature. To compare the molar volumes of different gases, scientists agreed to use 1 atm and 273.15 K (0 °C) as the **standard conditions of temperature and pressure,** or **STP** for short.

[5]The great French chemist Antoine Laurent Lavoisier (1743–1794) was the first to observe the volume relationships of this particular reaction. In his 1789 textbook, *Elements of Chemistry,* he wrote that the formation of water from hydrogen and oxygen requires that two volumes of hydrogen be used for every volume of oxygen. Lavoisier was unable to extend the study of this behavior of hydrogen and oxygen to other gas reactions because he was beheaded during the French Revolution. (See Michael Laing, *The Journal of Chemical Education,* February 1998, page 177.)

TABLE 10.1 Molar Volumes of Some Gases

Gas	Formula	Molar Volume (L)
Helium	He	22.398
Argon	Ar	22.401
Hydrogen	H_2	22.410
Nitrogen	N_2	22.413
Oxygen	O_2	22.414
Carbon dioxide	CO_2	22.414

If we measure the molar volumes for a variety of gases at STP, we find that the values fluctuate somewhat because the gases are not "ideal." Some typical values are shown in Table 10.1 and if we were to examine the data for many gases, we would find, an average of around 22.4 L per mole. This value is taken to be the molar volume of an *ideal gas* at STP and is now called the **standard molar volume** of a gas.

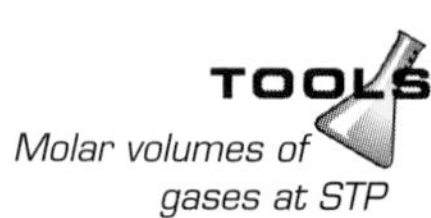

Molar volumes of gases at STP

For an ideal gas at STP:

$$1 \text{ mol gas} \Leftrightarrow 22.4 \text{ L gas}$$

Avogadro's principle was a remarkable advance in our understanding of gases. His insight enabled chemists for the first time to determine the formulas of gaseous elements.[6]

Gas volumes can be used in solving stoichiometry problems

For reactions involving gases, Avogadro's principle lets us use a new kind of stoichiometric equivalency, one between *volumes* of gases. Earlier, for example, we noted the following reaction and its gas volume relationships.

$$\underset{\text{2 volumes}}{2H_2(g)} + \underset{\text{1 volume}}{O_2(g)} \longrightarrow \underset{\text{2 volumes}}{2H_2O(g)}$$

The recognition that equivalencies in gas *volumes* are numerically the same as those for numbers of moles of gas in reactions involving gases simplifies many calculations.

Provided we are dealing with gas volumes measured at the same temperature and pressure, we can write the following stoichiometric equivalencies.

2 volumes $H_2(g) \Leftrightarrow$ 1 volume $O_2(g)$	just as	2 mol $H_2 \Leftrightarrow$ 1 mol O_2
2 volumes $H_2(g) \Leftrightarrow$ 2 volumes $H_2O(g)$	just as	2 mol $H_2 \Leftrightarrow$ 2 mol H_2O
1 volume $O_2(g) \Leftrightarrow$ 2 volumes $H_2O(g)$	just as	1 mol $O_2 \Leftrightarrow$ 2 mol H_2O

Relationships such as these can greatly simplify stoichiometry problems, as we see in Example 10.3.

EXAMPLE 10.3
Stoichiometry of Reactions of Gases

How many liters of hydrogen, measured at STP, are needed to combine exactly with 1.50 L of nitrogen, also measured at STP, to form ammonia?

ANALYSIS: The problem states that the reactants are at STP, standard temperature (273 K) and pressure (760 torr). Therefore all the gases in this example are at the same temperature and

[6]Suppose that hydrogen chloride, for example, is correctly formulated as HCl, not as H_2Cl_2 or H_3Cl_3 or higher, and certainly not as $H_{0.5}Cl_{0.5}$. Then the only way that *two* volumes of hydrogen chloride could come from just *one* volume of hydrogen and *one* of chlorine is if each particle of hydrogen and each of chlorine were to consist of *two* atoms of H and Cl, respectively, H_2 and Cl_2. If these particles were single-atom particles, H and Cl, then one volume of H and one volume of Cl could give only *one* volume of HCl, not two. Of course, if the initial assumption were incorrect so that hydrogen chloride is, say, H_2Cl_2 instead of HCl, then hydrogen would be H_4 and chlorine would be Cl_4. The extension to larger subscripts works in the same way.

pressure. That makes the ratios by volume the same as the ratio by moles, which means that the ratios by volume are the same as the ratios of the coefficients of the balanced equation.

We can restate the problem as

$$1.50 \text{ L } N_2 \Leftrightarrow ? \text{ L } H_2$$

We're dealing with a chemical reaction, so we need the chemical equation.

$$3H_2(g) + N_2(g) \longrightarrow 2NH_3(g)$$

Avogadro's principle gives us the equivalency we need to perform the calculation.

$$3 \text{ volumes } H_2 \Leftrightarrow 1 \text{ volume } N_2$$

SOLUTION: The volume of hydrogen is given in liters, so we need to express the volume equivalency in those units as well.

$$3 \text{ L } H_2 \Leftrightarrow 1 \text{ L } N_2$$

We then multiply the given amount, 1.50 L N_2, by a conversion factor made from the equivalency above:

$$\text{Volume of } H_2 = 1.50 \cancel{\text{L } N_2} \times \frac{3 \text{ L } H_2}{1 \cancel{\text{L } N_2}}$$
$$= 4.50 \text{ L } H_2$$

IS THE ANSWER REASONABLE? The volume of H_2 needed is three times the volume of N_2, and 3×1.5 equals 4.5, so the answer is correct. Remember, however, that the simplicity of this problem arises because the volumes are at the same temperature and pressure.

EXAMPLE 10.4
Stoichiometry Calculations when Gases Are Not at the Same *T* and *P*

Nitrogen monoxide, a pollutant released by automobile engines, is oxidized by molecular oxygen to give the reddish-brown gas nitrogen dioxide, which gives smog its characteristic color. The equation is

$$2NO(g) + O_2(g) \longrightarrow 2NO_2(g)$$

How many milliliters of O_2, measured at 20 °C and 755 torr, are needed to react with 180 mL of NO, measured at 45 °C and 720 torr?

ANALYSIS: Once again, we have a stoichiometry problem, but this one is more complicated than in the preceding example because the gases are not at the same temperature and pressure. The way to resolve this difficulty is to make the temperature and pressure the same for both gases.

We can use the combined gas law to find what volume the NO would occupy if it were at the same temperature and pressure as the O_2. Then we can use the coefficients of the equation to find the volume of O_2.

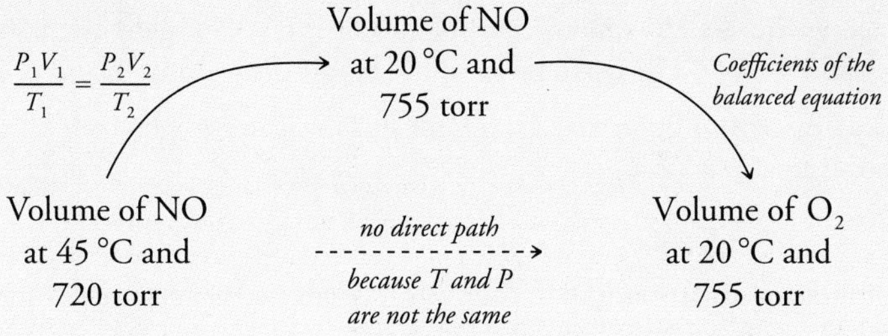

$\frac{P_1V_1}{T_1} = \frac{P_2V_2}{T_2}$

Volume of NO at 20 °C and 755 torr

Coefficients of the balanced equation

Volume of NO at 45 °C and 720 torr

no direct path - - - - → *because T and P are not the same*

Volume of O_2 at 20 °C and 755 torr

SOLUTION: The first step is to use the combined gas law applied to the given volume of NO, so let's set up the data as usual.

	Initial (1)	Final (2)
P	720 torr	755 torr
V	180 mL	?
T	318 K (45 °C + 273)	293 K (20 °C+ 273)

Solving the combined gas law for V_2 gives

$$V_2 = V_1 \times \left(\frac{P_1}{P_2}\right) \times \left(\frac{T_2}{T_1}\right)$$

Next we substitute values.

$$V_2 = 180 \text{ mL} \times \left(\frac{720 \cancel{\text{torr}}}{755 \cancel{\text{torr}}}\right) \times \left(\frac{293 \cancel{\text{K}}}{318 \cancel{\text{K}}}\right)$$

$$= 158 \text{ mL NO}$$

Now we have the volume of NO *at the same temperature and pressure as the oxygen*. This lets us use the coefficients of the equation to establish the equivalency

$$2 \text{ mL NO} \Leftrightarrow 1 \text{ mL } O_2$$

and apply it to find the volume of O_2 required for the reaction.

$$158 \cancel{\text{mL NO}} \times \frac{1 \text{ mL } O_2}{2 \cancel{\text{mL NO}}} = 79.0 \text{ mL } O_2$$

IS THE ANSWER REASONABLE? For the first part of the calculation, we can check to see whether the pressure and temperature ratios move the volume in the right direction. The pressure is increasing (702 torr $\longrightarrow$ 755 torr), so that should have a lowering effect on the volume. The pressure ratio is smaller than 1, so it is having the proper effect. The temperature is dropping (318 K$\longrightarrow$293 K), so this change should also have a volume lowering effect. The temperature ratio is smaller than 1, so it is also having the correct effect. The volume of NO at 20 °C and 755 torr is probably correct. We can also observe that the two ratios are only slightly less than 1.0 and we expect the answer to be close to the given volume, which it is.

The check of the second part of the calculation is simple. According to the equation, the volume of O_2 required should be half the volume of NO, which it is, so the final answer seems to be okay.

Practice Exercise 8: Methane burns according to the following equation.

$$CH_4(g) + 2O_2(g) \longrightarrow CO_2(g) + 2H_2O(g)$$

The combustion of 4.50 L of CH_4 consumes how many liters of O_2, both volumes measured at 25 °C and 740 torr? (Hint: Recall Avogadro's principle concerning the number of molecules in a fixed volume of gas at a given temperature and pressure.)

Practice Exercise 9: How many liters of air (air is 21% oxygen) are required for the combustion of 6.75 L of CH_4? (Assume air and CH_4 are at the same T and P.)

Practice Exercise 10: Butane (C_4H_{10}) is the fuel in cigarette lighters. It burns in oxygen according to the equation

$$2C_4H_{10}(g) + 13O_2(g) \longrightarrow 8CO_2(g) + 10H_2O(g)$$

How many milliliters of O_2 at 35 °C and 725 torr are needed to react completely with 75.0 mL of C_4H_{10} measured at 45 °C and 760 torr?

10.5 THE IDEAL GAS LAW RELATES P, V, T, AND THE NUMBER OF MOLES OF GAS, n

In our discussion of the combined gas law, we noted that the ratio PV/T equals a constant for a fixed amount of gas. However, the value of this "constant" is actually proportional to the number of moles of gas, n, in the sample.[7]

To create an equation even more general than the combined gas law, therefore, we can write

$$\frac{PV}{T} \propto n$$

We can replace the proportionality symbol with an equals sign by including another proportionality constant.

$$\frac{PV}{T} = n \times \text{constant}$$

This new constant is given the symbol ***R*** and is called the **universal gas constant.** We can now write the combined gas law in a still more general form called the **ideal gas law.**

$$\frac{PV}{T} = nR$$

An ideal gas would obey this law exactly over all ranges of the gas variables. The equation, sometimes called the **equation of state for an ideal gas,** is usually rearranged and written as follows.

Sometimes this equation is called the *universal gas law.*

Ideal Gas Law (Equation of State for an Ideal Gas)

$$PV = nRT \tag{10.2}$$

Equation 10.2 tells us how the four important variables for a gas, P, V, n, and T, are related. If we know the values of three, we can calculate the fourth. In fact, Equation 10.2 tells us that if values for three of the four variables are fixed for a given gas, *the fourth can only have one value.* We can define the *state* of a given gas simply by specifying any three of the four variables.

If n, P, and T in Equation 10.2 are known, for example, then V can have *only one value.*

To use the ideal gas law, we have to know the value of the universal gas constant, R. To calculate it, we use the standard conditions of pressure and temperature, and we use the standard molar volume, which sets n equal to 1 mol, the number of moles of the sample. We still have to decide what units to use for pressure and volume, and the value of R differs with these choices. Our choices for this chapter are to express volumes in *liters* and pressures in *atmospheres.* Thus for one molar volume at STP, $n = 1$ mol, $V = 22.4$ L, $P = 1.00$ atm, and $T = 273$ K. Using these values lets us calculate R as follows.

$$R = \frac{PV}{nT} = \frac{(1.00 \text{ atm})(22.4 \text{ L})}{(1.00 \text{ mol})(273 \text{ K})}$$

$$= 0.0821 \frac{\text{atm L}}{\text{mol K}}$$

Or, arranging the units in a commonly used order,

$$R = 0.0821 \text{ L atm mol}^{-1} \text{ K}^{-1}$$

More precise measurements give $R = 0.082057$ L atm mol^{-1} K^{-1}.

To use this value of R in working problems, we have to be sure to express volumes in liters and pressures in atmospheres. And, of course, temperatures must be expressed in kelvins.

[7] In the problems we worked earlier, we were able to use the combined gas law expressed as

$$\frac{P_1V_1}{T_1} = \frac{P_2V_2}{T_2}$$

because the amount of gas remained fixed.

EXAMPLE 10.5
Using the Ideal Gas Law

In Example 10.2, we described filling a 158 mL lightbulb with argon at a temperature of 20 °C and a pressure of 3.00 torr. How many grams of argon are in the lightbulb under these conditions?

ANALYSIS: In gas law calculations we usually have the possibility of either using the combined gas law equation (used if we are changing from one set of P, V, and T to another set of P, V, and T) or the ideal gas law equation (used if we know three of the four variables P, V, n, and T). In this case we see that the tool to accomplish this is the ideal gas law equation, $PV = nRT$. We will solve it for n and substitute values for P, V, R, and T. Once we have the number of moles of argon, a simple moles-to-grams conversion using the atomic mass will give us the mass of Ar in grams.

In performing the calculation, we will have to be sure to convert the volume to liters, pressure to atmospheres, and the temperature to kelvins so the units will cancel correctly.

SOLUTION: To use $R = 0.0821 \text{ L atm mol}^{-1} \text{ K}^{-1}$, we must have V in liters, P in atmospheres, and T in kelvins. Gathering the data and making the necessary unit conversions as we go, we have

$$P = 3.95 \times 10^{-3} \text{ atm} \qquad \text{from } 3.00 \text{ torr} \times \frac{1 \text{ atm}}{760 \text{ torr}}$$

$$V = 0.158 \text{ L} \qquad \text{from } 158 \text{ mL}$$

$$T = 293 \text{ K} \qquad \text{from } (20 \text{ °C} + 273)$$

Solving the ideal gas law for n gives us

$$n = \frac{PV}{RT}$$

Substituting the proper values of P, V, R, and T into this equation gives

$$n = \frac{(3.95 \times 10^{-3} \text{ atm})(0.158 \text{ L})}{(0.0821 \text{ L atm mol}^{-1} \text{K}^{-1})(293 \text{ K})}$$

$$= 2.59 \times 10^{-5} \text{ mol Ar}$$

The atomic mass of Ar is 39.95, so 1 mol Ar = 39.95 g Ar. A conversion factor made from this relationship lets us convert "mol Ar" into "g Ar."

$$2.59 \times 10^{-5} \text{ mol Ar} \times \frac{39.95 \text{ g Ar}}{1 \text{ mol Ar}} = 1.04 \times 10^{-3} \text{ g Ar}$$

Thus, the lightbulb contains only about one-thousandth of a gram of argon.

IS THE ANSWER REASONABLE? As in other problems we approximate the answer by rounding all values to one significant figure to get:

$$n = \frac{(4 \times 10^{-3} \text{ atm})(0.2 \text{ L})}{(0.1 \text{ L atm mol}^{-1} \text{ K}^{-1})(300 \text{ K})} = \frac{0.8 \times 10^{-3} \text{ mol}}{30} = 2.66 \times 10^{-5} \text{ mol}$$

For ease of calculation we round this answer to 2.5×10^{-5} mol and multiply by 40 g Ar per mol Ar to get 1×10^{-3} grams, very close to our calculated value. An alternate rounding of the moles to 3×10^{-5} will give an answer of 1.2×10^{-3} grams, equally close to our original answer.

Another view tells us we can also expect that the amount of argon in the bulb will also be quite small, since one mole occupies 22.4 L at STP; the 158 mL bulb has a much smaller volume than 22.4 L so our final answer also makes sense. As a final check, we can look to be sure the units cancel correctly, and we see that they do. (Whenever you're working a problem where there is unit cancellation, be sure to check to be sure the units do cancel as they are supposed to.)

Practice Exercise 11: Dry ice, $CO_2(s)$, can be made by allowing pressurized $CO_2(g)$ to expand rapidly. If 35% of the expanding $CO_2(g)$ ends up as $CO_2(s)$, how many grams of dry ice can be made from a tank of $CO_2(g)$ that has a volume of 6.0 cubic feet with a gauge pressure of 2000 pounds per square inch (PSIG) at 22 °C? (Hint: The information in the margin table on page 393 will help set up the conversions needed. PSIG is the pressure above the prevailing atmospheric pressure.)

Practice Exercise 12: How many grams of argon were in the 12.0 L cylinder of argon used to fill the lightbulbs described in Example 10.2? The pressure of the argon was 57.8 atm and the temperature was 25 °C.

Molar mass can be calculated from measurements of *P, V, T*, and mass

When a chemist makes a new compound, its molar mass is usually determined to help establish its chemical identity. In general, *to determine the molar mass of a compound experimentally, we need to find two pieces of information about a given sample: the mass of the sample and the number of moles of the substance in the sample.* Once we have mass and moles for the same sample, we simply divide the number of grams by the number of moles to find the molar mass. For instance, if we had a sample weighing 6.40 g and found that it also contained 0.100 mol of the substance, the molar mass would be

TOOLS *Determination of molar mass from ideal gas law*

$$\frac{6.40\ \text{g}}{0.100\ \text{mol}} = 64.0\ \text{g mol}^{-1}$$

If the compound is a gas, its molar mass can be found using experimental values of pressure, volume, temperature, and sample mass. The *P, V, T* data allow us to calculate the number of moles using the ideal gas law (as in Example 10.5). Once we know the number of moles of gas and the mass of the gas sample, the molar mass is obtained by taking the ratio of *grams to moles.*

Recall that when the molecular mass of a substance is expressed in units of grams per mole, the quantity is called the *molar mass.*

EXAMPLE 10.6
Determining the Molar Mass of a Gas

As part of a rock analysis, a student added hydrochloric acid to a rock sample and observed a fizzing action, indicating a gas was being evolved (see the figure in the margin). The student collected a sample of the gas in a 0.220 L gas bulb until its pressure reached 0.757 atm at a temperature of 25.0 °C. The sample weighed 0.299 g. What is the molar mass of the gas? What kind of compound was the likely source of the gas?

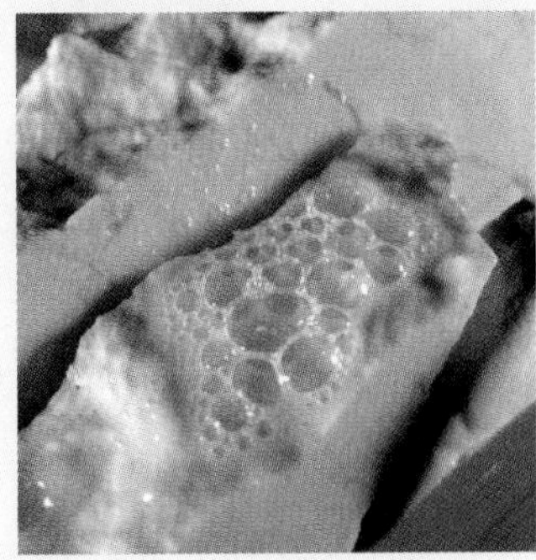

Hydrochloric acid reacting with a rock sample. *(Andy Washnik.)*

ANALYSIS: The strategy for finding the molar mass was described above. We use the *P, V, T* data to calculate the number of moles of gas in the sample. To do this, we solve the ideal gas law for *n* and then substitute values, being sure the units will cancel correctly. Then we divide the given mass by the number of moles to find the molar mass.

SOLUTION: Pressure is already in atmospheres and the volume is in liters, but we must convert degrees Celsius into kelvins. Gathering our data, we have

$$P = 0.757\ \text{atm} \qquad V = 0.220\ \text{L} \qquad T = 298\ \text{K} \quad (25.0\ ^\circ\text{C} + 273)$$

Next, we solve the ideal gas law ($PV = nRT$) for the number of moles, *n*.

$$n = \frac{PV}{RT}$$

Now we can substitute the data for *P, V*, and *T*, along with the value of *R*. This gives

$$n = \frac{(0.757\ \cancel{\text{atm}})(0.220\ \cancel{\text{L}})}{(0.0821\ \cancel{\text{L}}\ \cancel{\text{atm}}\ \text{mol}^{-1}\ \cancel{\text{K}^{-1}})(298\ \cancel{\text{K}})}$$
$$= 6.81 \times 10^{-3}\ \text{mol}$$

The molar mass is obtained from the ratio of grams to moles.

$$\text{Molar mass} = \frac{0.299 \text{ g}}{6.81 \times 10^{-3} \text{ mol}} = 43.9 \text{ g mol}^{-1}$$

We now know the measured molar mass is 43.9, but what gas could this be? Looking back on our discussions in Chapter 4, what gases do we know are given off when a substance reacts with acids? On page 152 we find some options; the gas might be H_2S, HCN, CO_2, or SO_2. Using atomic masses to calculate their molar masses we get

H_2S	34 g mol^{-1}	CO_2	44 g mol^{-1}
HCN	27 g mol^{-1}	SO_2	64 g mol^{-1}

The only gas with a molar mass close to 43.9 is CO_2, and that gas would be evolved if we treat a carbonate with an acid. The rock probably contains a carbonate compound. (Limestone and marble are examples of such minerals.)

ARE THE ANSWERS REASONABLE? If our value for n is correct, then the rest of the calculation is most likely okay too. Let's round the data and estimate n as

$$n = \frac{(3/4 \text{ atm})(0.2 \text{ L})}{(0.1 \text{ L atm mol}^{-1} \text{ K}^{-1})(300 \text{ K})} = \frac{0.15 \text{ atm L}}{30 \text{ L atm mol}^{-1}} = 0.005 \text{ mol}$$

This is reasonably close to the 6.81×10^{-3} mol calculated above. Therefore our molar mass is most likely correct. The rest of the reasoning seems sound and CO_2 from a carbonate compound seems to be reasonable.

Practice Exercise 13: A glass bulb is found to have a volume of 544.23 mL. The mass of the glass bulb filled with air is 735.6898 g. The bulb is then flushed with a gaseous organic compound. The bulb, now filled with the organic gas, weighs 735.6220 g. The measurements were made at STP. What is the molar mass of the organic gas? (Hint: Calculate the difference in molar masses. The average molar mass of air is 28.8 g/mol air.)

Practice Exercise 14: The label on a cylinder of a noble gas became illegible, so a student allowed some of the gas to flow into an evacuated glass bulb with a volume of 300.0 mL until the pressure was 685 torr. The mass of the glass bulb increased by 1.45 g; its temperature was 27.0 °C. What is the molar mass of this gas? Which of the Group VIIIA gases was it?

Gas densities depend on molar masses

Because one mole of any gas occupies the same volume at a particular pressure and temperature, the mass contained in that volume depends on the molar mass of the gas. Consider, for example, one-mole samples of O_2 and CO_2 at STP (Figure 10.9). Each sample occupies a volume of 22.4 L. The oxygen sample has a mass of 32.0 g while the carbon dioxide sample has a mass of 44.0 g. If we calculate the densities of the gases, we find the density of CO_2 is larger than that of O_2.

$$d_{O_2} = \frac{32.0 \text{ g}}{22.4 \text{ L}} = 1.43 \text{ g L}^{-1} \qquad d_{CO_2} = \frac{44.0 \text{ g}}{22.4 \text{ L}} = 1.96 \text{ g L}^{-1}$$

Recall from Chapter 1 that density is the ratio of mass to volume. For liquids and solids, we usually use units of g mL^{-1} (or g cm^{-3}), but because gases have such low densities, units of g L^{-1} give numbers that are easier to comprehend.

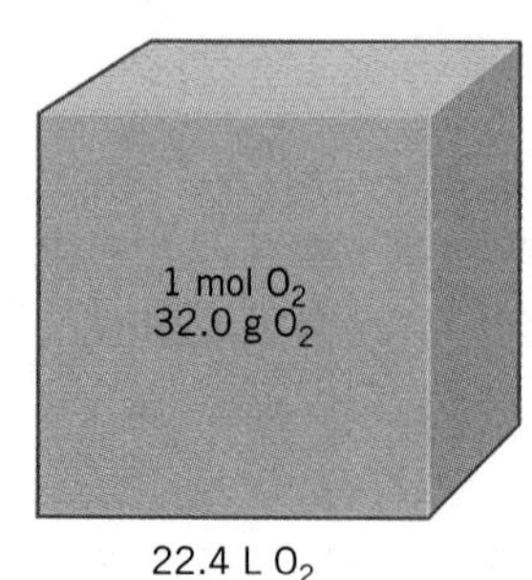

FIG. 10.9 One-mole samples of O_2 and CO_2 at STP. Each sample occupies 22.4 L, but the O_2 weighs 32.0 g whereas the CO_2 weighs 44.0 g. The CO_2 has more mass per unit volume than the O_2 and has the higher density.

Because the volume of a gas is affected by temperature and pressure, the density of a gas changes as these variables change. Gases become less dense as their temperatures rise, which is why hot air balloons are able to float (see chapter opening photo); the less dense hot air inside the balloon floats in the more dense cool air that surrounds it. Gases also become more dense as their pressures increase because increasing the pressure packs more molecules into the same space. To calculate the density of a gas at conditions other than STP, we use the ideal gas law, as illustrated in Example 10.7.

EXAMPLE 10.7
Calculating the Density of a Gas

One procedure used to separate the isotopes of uranium to obtain material to construct a nuclear weapon employs a uranium compound with the formula UF_6. The compound boils at about 56 °C, so at 100 °C it is a gas. What is the density of UF_6 at 100 °C if the pressure of the gas is 740 torr? (Assume the gas contains the mix of uranium isotopes commonly found in nature.)

ANALYSIS: In our discussion above, you learned that if the gas were at STP, we could calculate the density by dividing the mass of one mole of the gas by its molar volume, 22.4 L. We can do a similar calculation here if we first calculate the molar volume at 100 °C and 740 torr. To do this, we can use the ideal gas law, setting n equal to 1.00 mol and converting the pressure to atmospheres and the temperature to kelvins.

SOLUTION: The pressure is 740 torr, so to convert to atmospheres, we use the relationship between the atmosphere and torr.

$$740\ \cancel{\text{torr}} \times \frac{1\ \text{atm}}{760\ \cancel{\text{torr}}} = 0.974\ \text{atm}$$

The temperature is 100 °C, which converts to 373 K, and we said we would use $n = 1.00$ mol. To find the molar volume, we solve the ideal gas law for V.

$$V = \frac{nRT}{P}$$

Now we substitute values.

$$V = \frac{1.00\ \cancel{\text{mol}} \times 0.0821\ \text{L}\ \cancel{\text{atm}}\ \cancel{\text{mol}^{-1}}\ \cancel{\text{K}^{-1}} \times 373\ \cancel{\text{K}}}{0.974\ \cancel{\text{atm}}} = 31.4\ \text{L}$$

At these conditions of temperature and pressure, the molar volume is 31.4 L per mole. To calculate the density, we need one more piece of data, the molar mass of UF_6. Adding up the atomic masses, we get a molar mass of 352.0 g mol^{-1}. The density is therefore

$$d_{UF_6} = \frac{352.0\ \text{g}}{31.4\ \text{L}} = 11.2\ \text{g L}^{-1}$$

IS THE ANSWER REASONABLE? At STP, the density would be equal to 352 g ÷ 22.4 L = 15.7 g L^{-1}. The pressure is almost 1 atm, but the temperature is quite a bit larger than 273 K. Gases expand when heated, so a liter of the gas at the higher temperature will have less UF_6 in it. That means the density will be lower at the higher temperature, and our answer agrees with this analysis. It is probably correct.

Practice Exercise 15: Radon, a radioactive gas, is formed in one step of the natural radioactive decay sequence of U-235 to Pb-207. Radon usually escapes harmlessly through the soil to the atmosphere. When the soil is frozen or saturated with water the only escape route is through cracks in the basements of houses and other buildings. In order to detect radon in a residence, would you place the sensor in the attic, ground floor living area, or the basement? Justify your answer. (Hint: Compare the approximate density of air from data given in Practice Exercise 13 to the density of radon.)

Practice Exercise 16: Sulfur dioxide is a gas that has been used in commercial refrigeration, but not in residential refrigeration because it is toxic. If your refrigerator used SO_2, you could be injured if it developed a leak. What is the density of SO_2 gas measured at −20 °C and a pressure of 96.5 kPa?

Gas densities can be used to calculate molar masses

One of the ways we can use the density of a gas is to determine the molar mass. To do this, we also need to know the temperature and pressure at which the density was measured. Example 10.8 illustrates the reasoning and calculation involved.[8]

EXAMPLE 10.8
Calculating the Molar Mass from Gas Density

A liquid sold under the name Perclene is used as a dry cleaning solvent. It has an empirical formula CCl_2 and a boiling point of 121 °C. When vaporized, the gaseous compound has a density of 4.93 g L^{-1} at 785 torr and 150 °C. What is the molar mass of the compound and what is its molecular formula?

ANALYSIS: Earlier you saw that for gases, we can calculate the molar mass if we have *P, V,* and *T* data for a weighed sample of the substance. We use the *P, V, T* data with the ideal gas law to calculate *n*, and then divide the number of grams by the number of moles. But in this problem we seem to be given values only for *P* and *T.* How will we obtain the other data we need (*V* and grams of gas)? The answer comes from what the density tells us. A value of 4.93 g L^{-1} translates to 4.93 g per liter, which means that if we had 1.00 L of the gas, it would have a mass of 4.93 g. Thus, the density gives the other two data items we need.

$$1.00 \text{ L gas} \Leftrightarrow 4.93 \text{ g gas}$$

After we've calculated the molar mass, we find the molecular formula following the procedure discussed in Chapter 3. We divide the molar mass by the empirical formula mass to find the factor by which we multiply the subscripts in the empirical formula to obtain the molecular formula.

SOLUTION: To calculate the number of moles of gas in 1.00 L, we use the ideal gas law, making unit conversions as needed. The data are

$$T = 423 \text{ K} \qquad \text{From } (150\ ^\circ\text{C} + 273)$$

$$P = 1.03 \text{ atm} \qquad \text{From } 785 \cancel{\text{torr}} \times \frac{1 \text{ atm}}{760 \cancel{\text{torr}}}$$

$$V = 1.00 \text{ L}$$

[8] From the ideal gas law we can derive an equation from which we could calculate the molar mass directly from the density. If we let the mass of gas equal *g*, we could calculate the number of moles, *n*, by the ratio

$$n = \frac{g}{\text{molar mass}}$$

Substituting into the ideal gas law gives

$$PV = nRT = \frac{gRT}{\text{molar mass}}$$

Solving for molar mass, we have

$$\text{Molar mass} = \frac{gRT}{PV} = \left(\frac{g}{V}\right) \times \frac{RT}{P}$$

The quantity *g/V* is the ratio of mass to volume, which is the density *d*, so making this substitution gives

$$\text{Molar mass} = \frac{dRT}{P}$$

This equation could also be used to solve the problem in Example 10.8 by substituting values for *d, R, T,* and *P.*

Solving the ideal gas law for n and substituting values, we have

$$n = \frac{PV}{RT} = \frac{(1.03\ \text{atm})(1.00\ \text{L})}{(0.0821\ \text{L atm mol}^{-1}\text{K}^{-1})(423\ \text{K})}$$
$$= 2.97 \times 10^{-2}\ \text{mol}$$

Now we can calculate the molar mass by dividing the mass, 4.93 g, by the number of moles, 2.97×10^{-2} mol.

$$\frac{4.93\ \text{g}}{2.97 \times 10^{-2}\ \text{mol}} = 166\ \text{g mol}^{-1}$$

The molar mass of the compound is thus 166 g mol^{-1}.

The empirical formula mass that we calculate from the empirical formula, CCl_2, is 82.9. We now divide the molar mass by this value to see how many times CCl_2 occurs in the molecular formula.

$$\frac{166}{82.9} = 2.00$$

To find the molecular formula, we multiply the subscripts of the empirical formula by 2.

$$\text{Molecular formula} = C_{1\times 2}Cl_{2\times 2} = C_2Cl_4$$

(This is the formula for a compound commonly called tetrachloroethylene, which is indeed used as a dry cleaning fluid.)

ARE THE ANSWERS REASONABLE? Sometimes we don't have to do any arithmetic to see that an answer is almost surely correct. The fact that the molar mass we calculated from the gas density is evenly divisible by the empirical formula mass suggests we've worked the problem correctly.

Practice Exercise 17: A gaseous compound of phosphorus and fluorine with an empirical formula of PF_2 has a density of 5.60 g L^{-1} at 23.0 °C and 750 torr. Determine the molecular formula of this compound. (Hint: Calculate the molar mass from the density.)

Practice Exercise 18: A compound composed of only carbon and hydrogen has a density of 5.55 g/L at 40.0 °C and 1.25 atm. What is the molar mass of the compound? What are the possible combinations of C and H that add up to that molar mass? Using the information in Section 2.6, determine which of your formulas is most likely the correct one.

The ideal gas law can be used in stoichiometry calculations

Many chemical reactions either consume or give off gases. The ideal gas law can be used to relate the volumes of such gases to the amounts of other substances involved in the reaction, as illustrated by the following example.

EXAMPLE 10.9
Calculating the Volume of a Gaseous Product Using the Ideal Gas Law

An important chemical reaction in the manufacture of Portland cement is the high temperature decomposition of calcium carbonate to give calcium oxide and carbon dioxide. Suppose a 1.25 g sample of calcium carbonate is decomposed by heating. How many milliliters of carbon dioxide gas will be evolved if the volume will be measured at 740 torr and 25 °C?

ANALYSIS: The first thing to do is determine the reactants and products of the reaction and to balance the equation. Using our rules for naming compounds we can determine that calcium carbonate is $CaCO_3$, carbon dioxide is CO_2, and calcium oxide is CaO. The balanced equation is

$$CaCO_3(s) \longrightarrow CaO(s) + CO_2(g)$$

This balanced equation tells us

$$1 \text{ mol } CaCO_3 \Leftrightarrow 1 \text{ mol } CO_2$$

Now, we will calculate the *moles* of $CaCO_3$, which equals the number of moles of CO_2. We will use this value for n in the ideal gas law equation to find the volume of CO_2.

SOLUTION: The formula mass of $CaCO_3$ is 100.1, so

$$\text{moles of } CaCO_3 = 1.25 \cancel{\text{ g } CaCO_3} \times \frac{1 \text{ mol } CaCO_3}{100.1 \cancel{\text{ g } CaCO_3}}$$

$$= 1.25 \times 10^{-2} \text{ mol } CaCO_3$$

Because $1 \text{ mol } CaCO_3 \Leftrightarrow 1 \text{ mol } CO_2$,

$$n = 1.25 \times 10^{-2} \text{ mol } CO_2$$

Before we use n in the ideal gas law equation, we must convert the given pressure and temperature into the units required by R.

$$P = 740 \cancel{\text{ torr}} \times \frac{1 \text{ atm}}{760 \cancel{\text{ torr}}} = 0.974 \text{ atm} \qquad T = 298 \text{ K } (25.0 \text{ °C} + 273)$$

By rearranging the ideal gas law equation we obtain

$$V = \frac{nRT}{P}$$

$$= \frac{(1.25 \times 10^{-2} \cancel{\text{ mol}})(0.0821 \text{ L} \cancel{\text{ atm}} \cancel{\text{ mol}^{-1}} \cancel{\text{K}^{-1}})(298 \cancel{\text{K}})}{0.974 \cancel{\text{ atm}}}$$

$$= 0.314 \text{ L} = 314 \text{ mL}$$

The reaction will yield 314 mL of CO_2 at the conditions specified.

IS THE ANSWER REASONABLE? We round all of the numbers in our calculation to one significant figure to get

$$V = \frac{(1 \times 10^{-2} \cancel{\text{ mol}})(0.1 \text{ L} \cancel{\text{ atm}} \cancel{\text{ mol}^{-1}} \cancel{\text{K}^{-1}})(300 \cancel{\text{K}})}{1 \cancel{\text{ atm}}} = 0.3 \text{ L} \quad \text{or} \quad 300 \text{ mL}$$

This answer is close to what we calculated above and we may assume the calculation is correct. We also note that all units cancel to leave the desired liter units for volume.

Practice Exercise 19: Carbon disulfide is an extremely flammable substance. It can be ignited by any small spark or even a very hot surface such as a steam pipe. When 10.0 g of CS_2 is burned in excess oxygen, how many liters of CO_2 and of SO_2 are formed at 28 °C and 880 torr? (Hint: Treat this as an ordinary stoichiometry problem but write the balanced equation first.)

Practice Exercise 20: In one lab, the gas collecting apparatus used a glass bulb with a volume of 250 mL. How many grams of $CaCO_3(s)$ need to be heated to prepare enough $CO_2(g)$ to fill this bulb to a pressure of 738 torr at a temperature of 23 °C?

10.6 IN A MIXTURE EACH GAS EXERTS ITS OWN PARTIAL PRESSURE

So far in our discussions we've dealt with only pure gases. However, gas mixtures, such as the air we breathe, are quite common. In general, gas mixtures obey the same laws as pure gases, so Boyle's law applies equally to both pure oxygen and to air. There are times, though, when we must be concerned with the composition of a gas mixture, such as when we are concerned with a pollutant in the atmosphere. In these cases, the variables affected by the composition of a gas mixture are the numbers of moles of each component and the contribution each component makes to the total observed pressure. Because gases mix completely, all the components of the mixture occupy the same volume—that of the container holding them. Furthermore, the temperature of each gaseous component is the same as the temperature of the entire mixture. Therefore, in a gas mixture, each of the components has the same volume and the same temperature.

Each gas in a mixture exerts its own pressure which contributes to the total pressure

In a mixture of nonreacting gases such as air, each gas contributes to the total pressure in proportion to the fraction (by moles) in which it is present (see Figure 10.10). This contribution to the total pressure is called the **partial pressure** of the gas. It is the pressure the gas would exert if it were the only gas in a container of the same size at the same temperature.

Under ordinary temperatures and pressures, nitrogen and oxygen in air do not react.

The general symbol we will use for the partial pressure of gas A is P_A. For a particular gas, the formula of the gas may be put into the subscript, as in P_{O_2}. What John Dalton discovered about partial pressures is now called **Dalton's law of partial pressures:** *The total pressure of a mixture of gases is the sum of their individual partial pressures.* In equation form, the law is

$$P_{\text{total}} = P_A + P_B + P_C + \cdots \tag{10.3}$$

In dry CO_2-free air at STP, for example, P_{O_2} is 159.12 torr, P_{N_2} is 593.44 torr, and P_{Ar} is 7.10 torr. These partial pressures add up to 759.66 torr, just 0.34 torr less than 760 torr or 1.00 atm. The remaining 0.34 torr is contributed by several trace gases, including other noble gases.

Practice Exercise 21: At 20 °C a 1.00 liter flask is filled with 10.0 g of Ar, 10.0 g of N_2, and 10.0 g of O_2. What are the partial pressures of each gas and what is the total pressure in the flask? (Hint: Start by calculating the moles of each gas present.)

Practice Exercise 22: Suppose a tank of oxygen-enriched air prepared for scuba diving has a volume of 17.00 L and a pressure of 237.0 atm at 25 °C. How many grams of oxygen are present if all the other gases in the tank exert a combined partial pressure of 115.0 atm?

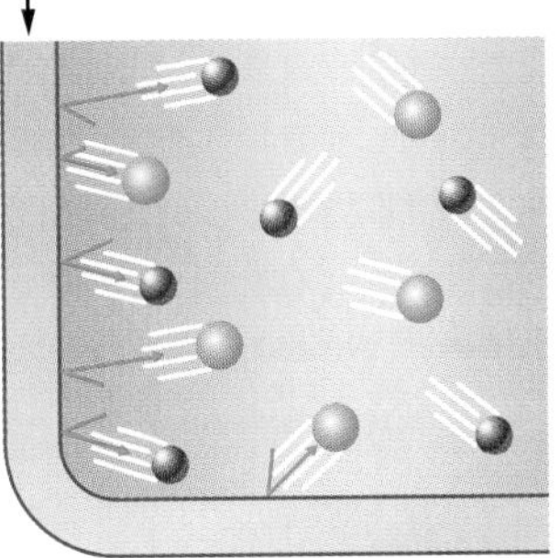

FIG. 10.10 Partial pressures viewed at the molecular level. In a mixture of two gases, A and B, both collide with the walls of the container and thereby contribute their partial pressures to the total pressure.

Collecting gases over water

When gases that do not react with water are prepared in the laboratory, they can be trapped over water by an apparatus like that shown in Figure 10.11. Because of the way the gas is collected, it is saturated with water vapor. (We say the gas is "wet.") Water vapor in a mixture of gases has a partial pressure like that of any other gas.

The vapor present in the space above *any* liquid always contains some of the liquid's vapor, which exerts its own pressure called the liquid's **vapor pressure**. Its value for any given

Even the mercury in a barometer has a tiny vapor pressure—0.0012 torr at 20 °C, which is much too small to affect readings of barometers and the manometers studied in this chapter.

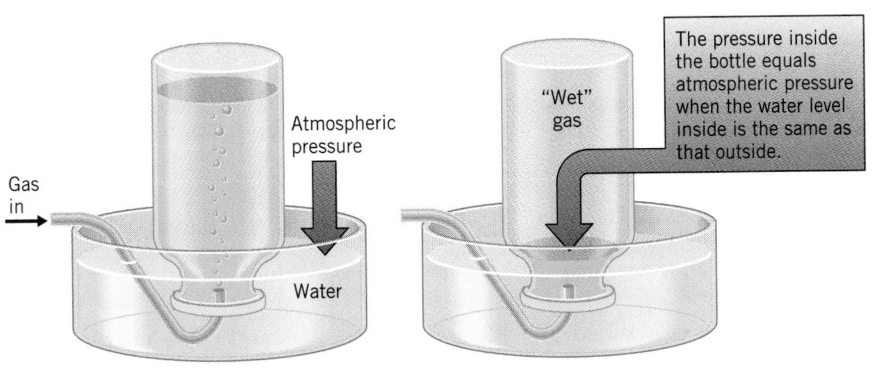

FIG. 10.11 Collecting a gas over water. As the gas bubbles through the water, water vapor goes into the gas, so the total pressure inside the bottle includes the partial pressure of the water vapor at the temperature of the water.

TABLE 10.2 Vapor Pressure of Water at Various Temperatures

Temperature (°C)	Vapor Pressure (torr)	Temperature (°C)	Vapor Pressure (torr)
0	4.579	50	92.51
5	6.543	55	118.0
10	9.209	60	149.4
15	12.79	65	187.5
20	17.54	70	233.7
25	23.76	75	289.1
30	31.82	80	355.1
35	41.18	85	433.6
37[a]	47.07	90	525.8
40	55.32	95	633.9
45	71.88	100	760.0

[a] Human body temperature.

substance depends only on the temperature. The vapor pressures of water at different temperatures, for example, are given in Table 10.2. A more complete table is in Appendix C.4.

If we have adjusted the height of the collecting jar so the water level inside matches that outside, the total pressure of the trapped gas equals the atmospheric pressure, so the value for P_{total} is obtained from the laboratory barometer. We thus calculate P_{gas}, which is the pressure that the gas would exert if it were dry (i.e., without water vapor in it) and inside the same volume that was used to collect it.[9]

EXAMPLE 10.10
Collecting a Gas over Water

A sample of oxygen is collected over water at 20 °C and a pressure of 738 torr. Its volume is 310 mL. (a) What is the partial pressure of the oxygen? (b) What would be its volume when dry at STP?

ANALYSIS: There are two parts to this problem, so before we get started, let's determine which of the gas laws will be our tools for solving them. It seems pretty clear that for part (a) we will need Dalton's law of partial pressures. We have O_2 collected over water, so we know the container holds both O_2 and $H_2O(g)$. For part (b), we have to determine how the volume of the O_2 collected will change when the conditions change to those of STP. We're not changing the amount of O_2. The amount of O_2 is fixed (constant); it just undergoes changes in P, V, and T. To work problems of this kind, our tool is the combined gas law.

To solve part (a), we will have to subtract the vapor pressure of water from the total pressure of the gas mixture (738 torr). That's all there is to it.

To solve part (b), we substitute into the combined gas law after solving it for V_2. Our starting conditions will be 310 mL, 20 °C, and the pressure obtained in part (a). The final conditions will have a temperature of 0 °C and a pressure of 760 torr.

SOLUTION: To calculate the partial pressure of the oxygen, we use Dalton's law. We will need the vapor pressure of water at 20 °C, which we find in Table 10.2 to be 17.5 torr.

$$P_{O_2} = P_{total} - P_{water\ vapor}$$

$$= 738 \text{ torr} - 17.5 \text{ torr} = 720 \text{ torr}$$

The answer to part (a) is that the partial pressure of O_2 is 720 torr.

[9] If the water levels are not the same inside the flask and outside, a correction has to be calculated and applied to the room pressure to obtain the true pressure in the flask. For example, if the water level is higher inside the flask than outside, the pressure in the flask is lower than atmospheric pressure. The difference in levels is in millimeters of *water*, so this has to be converted to the equivalent in millimeters of mercury using the fact that the pressure exerted by a column of fluid is inversely proportional to the fluid's density.

For part (b), we'll begin by assembling the data.

	Initial (1)	Final (2)
P	720 torr (which is P_{O_2})	760 torr (standard pressure)
V	310 mL	?
T	293 K (20.0 °C + 273)	273 K (standard temperature)

We use these in the combined gas law equation:

$$\frac{P_1V_1}{T_1} = \frac{P_2V_2}{T_2}$$

Solving for V_2 and rearranging the equation a bit we have

$$V_2 = V_1 \times \left(\frac{P_1}{P_2}\right) \times \left(\frac{T_2}{T_1}\right)$$

Now we can substitute values and calculate V_2.

$$V_2 = 310 \text{ mL} \times \left(\frac{720 \cancel{\text{torr}}}{760 \cancel{\text{torr}}}\right) \times \left(\frac{273 \cancel{\text{K}}}{293 \cancel{\text{K}}}\right)$$
$$= 274 \text{ mL}$$

Thus, when the water vapor is removed from the gas sample, the dry oxygen will occupy a volume of 274 mL at STP.

ARE THE ANSWERS REASONABLE? We know the pressure of the dry O_2 will be less than that of the wet gas, so the answer to part (a) seems reasonable. To check part (b), we can see if the pressure and temperature ratios move the volume in the right direction. The pressure is increasing (720 torr ⟶ 760 torr), which should tend to lower the volume. The pressure ratio above will do that. The temperature change (293 K ⟶ 273 K) should also lower the volume, and once again, the temperature ratio above will have that effect. Our answer to part (b) is probably okay.

Practice Exercise 23: A 2.50 L sample of methane was collected over water at 30 °C until the pressure in the flask was 775 torr. A small amount of $CaSO_4(s)$ was then added to the flask to absorb the water vapor [forming $CaSO_4 \cdot 2H_2O(s)$]. What is the pressure inside the flask once all the water is absorbed? Assume that the addition of $CaSO_4(s)$ absorbed all the water and did not change the volume of the flask. How many moles of $CH_4(g)$ have been collected? (Hint: Find the partial pressure of water at 30 °C.)

Practice Exercise 24: Suppose you prepared a sample of nitrogen and collected it over water at 15 °C at a total pressure of 745 torr and a volume of 310 mL. Find the partial pressure of the nitrogen and the volume it would occupy at STP.

The composition of a mixture can be expressed in mole fractions or mole percents

One of the useful ways of describing the composition of a mixture is in terms of the *mole fractions* of the components. The **mole fraction** *is the ratio of the number of moles of a given component to the total number of moles of all components.* Expressed mathematically, the mole fraction of substance A in a mixture of $A, B, C, ..., Z$ substances is

The concept of mole fraction applies to any uniform mixture in any physical state—gas, liquid, or solid.

$$X_A = \frac{n_A}{n_A + n_B + n_C + n_D + \cdots + n_Z} \quad (10.4)$$

where X_A is the mole fraction of component A, and $n_A, n_B, n_C, \ldots, n_Z$ are the numbers of moles of each component, $A, B, C, \ldots, Z$, respectively. The sum of all mole fractions for a mixture must always equal 1.

You can see in Equation 10.4 both numerator and denominator have the same units (moles), so they cancel. As a result, a mole fraction has no units. Nevertheless, always remember the definition: a mole fraction stands for the ratio of *moles* of one component to the total number of *moles* of all components.

Sometimes the mole fraction composition of a mixture is expressed on a percentage basis; we call it a **mole percent (mol%).** The mole percent is obtained by multiplying the mole fraction by 100 mol%.

Mole fractions of gases are related to partial pressures

Partial pressure data can be used to calculate the mole fractions of individual gases in a gas mixture because the number of *moles* of each gas is directly proportional to its partial pressure. We can demonstrate this as follows. The partial pressure, P_A, for any one gas, A, in a gas mixture with a total volume V at a temperature T is found by the ideal gas law equation, $PV = nRT$. So to calculate the number of moles of A present, we have

$$n_A = \frac{P_A V}{RT}$$

For any particular gas mixture at a given temperature, the values of V, R, and T are all constants, making the ratio V/RT a constant, too. We can therefore simplify the previous equation by using C to stand for V/RT. In other words, we can write

$$n_A = P_A C$$

The result is the same as saying that *the number of moles of a gas in a mixture of gases is directly proportional to the partial pressure of the gas.* The constant C is the same for all gases in the mixture. So by using different letters to identify individual gases, we can let $P_B C$ stand for n_B, $P_C C$ stand for n_C, and so on in Equation 10.4. Thus,

$$X_A = \frac{P_A C}{P_A C + P_B C + P_C C + \cdots + P_Z C}$$

The constant, C, can be factored out and canceled, so

$$X_A = \frac{P_A}{P_A + P_B + P_C + \cdots + P_Z}$$

The denominator is the sum of the partial pressures of all the gases in the mixture, but this sum equals the total pressure of the mixture (Dalton's law of partial pressures). Therefore, the previous equation simplifies to

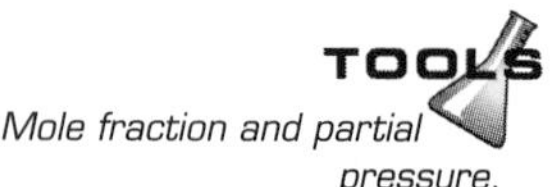

Mole fraction and partial pressure.

$$X_A = \frac{P_A}{P_{\text{total}}} \tag{10.5}$$

Thus, the mole fraction of a gas in a gas mixture is simply the ratio of its partial pressure to the total pressure. Equation 10.5 also gives us a simple way to calculate the partial pressure of a gas in a gas mixture when we know its mole fraction.

EXAMPLE 10.11
Using Mole Fractions to Calculate Partial Pressures

Suppose a mixture of oxygen and nitrogen is prepared in which there are 0.200 mol O_2 and 0.500 mol N_2. If the total pressure of the mixture is 745 torr, what are the partial pressures of the two gases in the mixture?

ANALYSIS: In this problem we know the composition of the mixture and the total pressure. It is important to remember that we can calculate the partial pressure of a gas if we know its mole fraction in the mixture and the total pressure. To solve the problem we will first calculate the mole fraction of each gas and then apply Equation 10.5.

SOLUTION: The mole fractions are calculated as follows:

$$X_{O_2} = \frac{\text{moles of } O_2}{\text{moles of } O_2 + \text{moles of } N_2}$$
$$= \frac{0.200 \text{ mol}}{0.200 \text{ mol} + 0.500 \text{ mol}}$$
$$= \frac{0.200 \cancel{\text{mol}}}{0.700 \cancel{\text{mol}}} = 0.286$$

Similarly, for N_2 we have[10]

$$X_{N_2} = \frac{0.500 \cancel{\text{mol}}}{0.200 \cancel{\text{mol}} + 0.500 \cancel{\text{mol}}} = 0.714$$

We can now use Equation 10.5 to calculate the partial pressure. Solving the equation for partial pressure, we have

$$P_{O_2} = X_{O_2} P_{\text{total}}$$
$$= 0.286 \times 745 \text{ torr}$$
$$= 213 \text{ torr}$$
$$P_{N_2} = X_{N_2} P_{\text{total}}$$
$$= 0.714 \times 745 \text{ torr}$$
$$= 532 \text{ torr}$$

Thus, the partial pressure of O_2 is 213 torr and the partial pressure of N_2 is 532 torr.

ARE THE ANSWERS REASONABLE? There are three things we can check here. First, the mole fractions add up to 1.000, which they must. Second, the partial pressures add up to 745 torr, which equals the given total pressure. Third, the mole fraction of N_2 is somewhat more than twice that for O_2, so its partial pressure should be somewhat more than twice that of O_2. Examining the answers, we see this is true, so our answers should be correct.

Practice Exercise 25: Sulfur dioxide and oxygen react according to the equation

$$2SO_2(g) + O_2(g) \longrightarrow 2SO_3(g)$$

If 50.0 g of $SO_2(g)$ is added to a flask resulting in a pressure of 0.750 atm, what will be the total pressure in the flask when a stoichiometric amount of oxygen is added? (Hint: This problem gives you more information than is needed.)

Practice Exercise 26: Suppose a mixture containing 2.15 g H_2 and 34.0 g NO has a total pressure of 2.05 atm. What are the partial pressures of both gases in the mixture?

Practice Exercise 27: What is the mole fraction and the mole percent of oxygen in exhaled air if P_{O_2} is 116 torr and P_{total} is 760 torr?

10.7 EFFUSION AND DIFFUSION IN GASES LEADS TO GRAHAM'S LAW

If you've ever walked past a restaurant and found your mouth watering after smelling the aroma of food, you've learned firsthand about diffusion! **Diffusion** is the spontaneous mixing of the molecules of one gas, like those of the food aromas, with molecules of another gas, like the air outside the restaurant. (See Figure 10.12*a*.) **Effusion,** on the other hand,

[10]Notice that the sum of the mole fractions (0.286 + 0.714) equals 1.000. In fact, we could have obtained the mole fraction of nitrogen with less calculation by subtracting the mole fraction of oxygen from 1.00.

$$X_{O_2} + X_{N_2} = 1.000$$
$$X_{N_2} = 1.000 - X_{O_2}$$
$$= 1.000 - 0.286 = 0.714$$

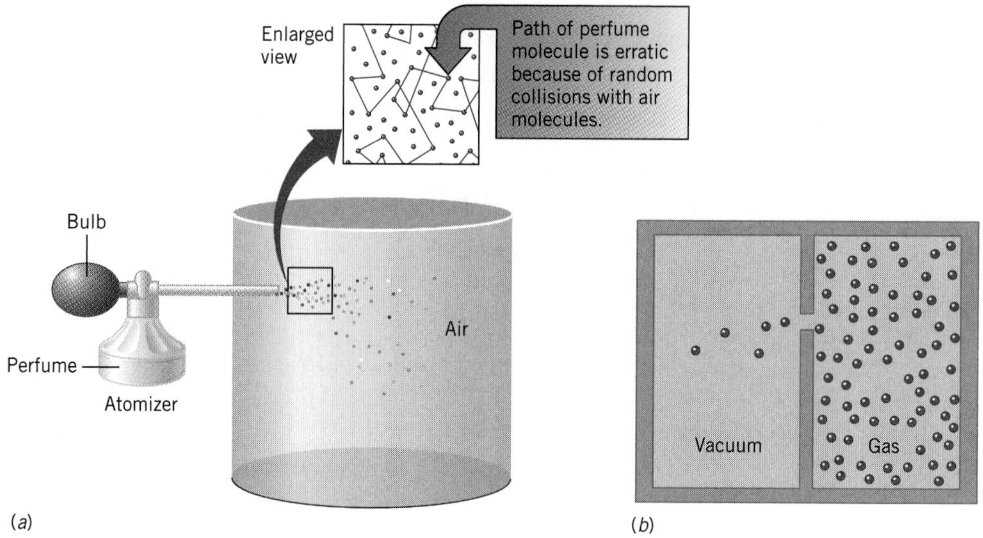

Enlarged view

Path of perfume molecule is erratic because of random collisions with air molecules.

Bulb

Air

Perfume

Atomizer

Vacuum

Gas

(*a*) (*b*)

FIG. 10.12 Spontaneous movements of gases. (*a*) Diffusion. (*b*) Effusion.

is the gradual movement of gas molecules through a very tiny hole into a vacuum (Figure 10.12*b*). The rates at which both of these processes occur depends on the speeds of gas molecules; the faster the molecules move, the more rapidly diffusion and effusion occur.

The Scottish chemist Thomas Graham (1805–1869) studied the rates of diffusion and effusion of a variety of gases through porous clay pots and through small apertures. Comparing different gases at the same temperature and pressure, Graham found that their rates of effusion were inversely proportional to the square roots of their densities. This relationship is now called **Graham's law.**

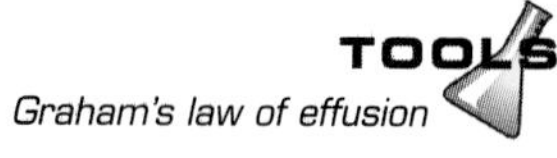

$$\text{Effusion rate} = \frac{k}{\sqrt{d}} \quad \left(\begin{array}{c}\text{when compared at the}\\ \text{same } T \text{ and } P\end{array}\right)$$

By taking the ratio, the proportionality constant, *k*, cancels from numerator and denominator.

Graham's law is usually used in comparing the rates of effusion of different gases, so the proportionality constant can be eliminated and an equation can be formed by writing the ratio of effusion rates.

$$\frac{\text{effusion rate }(A)}{\text{effusion rate }(B)} = \frac{\sqrt{d_B}}{\sqrt{d_A}} = \sqrt{\frac{d_B}{d_A}} \tag{10.6}$$

Earlier you saw that the density of a gas is directly proportional to its molar mass. Therefore, we can re-express Equation 10.6 as follows.

$$\frac{\text{effusion rate }(A)}{\text{effusion rate }(B)} = \frac{\sqrt{d_B}}{\sqrt{d_A}} = \sqrt{\frac{M_B}{M_A}} \tag{10.7}$$

where M_A and M_B are the molar masses of gases *A* and *B*.

Molar masses also affect the rates at which gases undergo diffusion. Gases with low molar masses diffuse (and effuse) more rapidly than gases with high molar masses. Thus, hydrogen with a molar mass of 2 will diffuse more rapidly than methane, CH_4, with a molar mass of 16.

EXAMPLE 10.12
Using Graham's Law

At a given temperature and pressure, which effuses more rapidly and by what factor, ammonia or hydrogen chloride?

ANALYSIS: A gas effusion problem requires the use of Graham's law, which says that the gas with the smaller molar mass will effuse more rapidly. So we need to find the molar masses. Then the relative rates of effusion are found by using them in Equation 10.7.

FACETS OF CHEMISTRY

Effusion and Nuclear Energy

The fuel used in almost all nuclear reactors is uranium, but only one of its naturally occurring isotopes, ^{235}U, can be easily split to yield energy. Unfortunately, this isotope is present in a very low concentration (about 0.72%) in naturally occurring uranium. Most of the element as it is mined consists of the more abundant isotope ^{238}U. Therefore, before uranium can be fabricated into fuel elements, it must be enriched to a ^{235}U concentration of about 2 to 5 percent. Enrichment requires that the isotopes be separated, at least to some degree.

Separating the uranium isotopes is not feasible by chemical means because the chemical properties of both isotopes are essentially identical. Instead, a method is required that is based on the very small difference in the masses of the isotopes. As it happens, uranium forms a compound with fluorine, UF_6, that is easily vaporized at a relatively low temperature. The UF_6 gas thus formed consists of two kinds of molecules, $^{235}UF_6$ and $^{238}UF_6$, with molecular masses of 349 and 352, respectively. Because of their different masses, their rates of effusion are slightly different; $^{235}UF_6$ effuses 1.0043 times faster than $^{238}UF_6$. Although the difference is small, it is sufficient to enable enrichment, provided the effusion is carried out over and over again enough times. In fact, it takes over 1400 separate effusion chambers arranged one after another to achieve the necessary level of enrichment.

SOLUTION: The molar masses are 17.03 for NH_3 and 36.46 for HCl, so we know that NH_3, with its smaller molar mass, effuses more rapidly than HCl. The ratio of the effusion rates is given by

$$\frac{\text{effusion rate } (NH_3)}{\text{effusion rate } (HCl)} = \sqrt{\frac{\mathcal{M}_{HCl}}{\mathcal{M}_{NH_3}}}$$

$$= \sqrt{\frac{36.46}{17.03}} = 1.463$$

We can rearrange the result as

$$\text{Effusion rate } (NH_3) = 1.463 \times \text{effusion rate } (HCl)$$

Thus, ammonia effuses 1.463 times more rapidly than HCl under the same conditions.

IS THE ANSWER REASONABLE? The only quick check is to be sure that the arithmetic tells us ammonia with its lower molar mass effuses more rapidly than the HCl, and that's what our result tells us.

Practice Exercise 28: Bromine has two isotopes with masses of 78.9 and 80.9 (to three significant figures), respectively. What is the expected ratio of the rate of effusion of Br-81 compared to Br-79? (Hint: Note that the example above gives the equation for the rate of effusion of ammonia compared to hydrogen chloride.)

Practice Exercise 29: The hydrogen halide gases all have the same general formula, *HX*, where *X* can be F, Cl, Br, or I. If HCl(*g*) effuses 1.88 times more rapidly than one of the others, which hydrogen halide is the other, HF, HBr, or HI?

10.8 THE KINETIC MOLECULAR THEORY EXPLAINS THE GAS LAWS

Scientists in the nineteenth century, who already knew the gas laws, wondered what had to be true about all gases to account for their conformity to a common set of gas laws. The **kinetic molecular theory of gases** provided an answer. We introduced some of its ideas in Chapter 6, and at the beginning of this chapter we described a number of observations that suggest what gases must be like when viewed at a molecular level. Let's look more closely now at the kinetic molecular theory to see how well it explains the behavior of gases.

The theory, often called simply the kinetic theory of gases, consisted of a set of postulates that describe the makeup of an ideal gas. Then the laws of physics and statistics were applied to see whether the observed gas laws could be predicted from the model. The results were splendidly successful.

Postulates of the Kinetic Theory of Gases

1. A gas consists of an extremely large number of very tiny particles that are in constant, random motion.
2. The gas particles themselves occupy a net volume so small in relation to the volume of their container that their contribution to the total volume can be ignored.
3. The particles often collide in perfectly elastic collisions[11] with themselves and with the walls of the container, and they move in straight lines between collisions neither attracting nor repelling each other.

The particles are assumed to be so small that they have no dimensions at all. They are essentially points in space.

In summary, the model pictures an ideal gas as a collection of constantly moving, extremely small, billiard balls that continually bounce off each other and the walls of their container, and so exert a net pressure on the walls (as described in Figure 10.1, page 391). The gas particles are assumed to be so small that their individual volumes can be ignored, so an ideal gas is effectively all empty space.

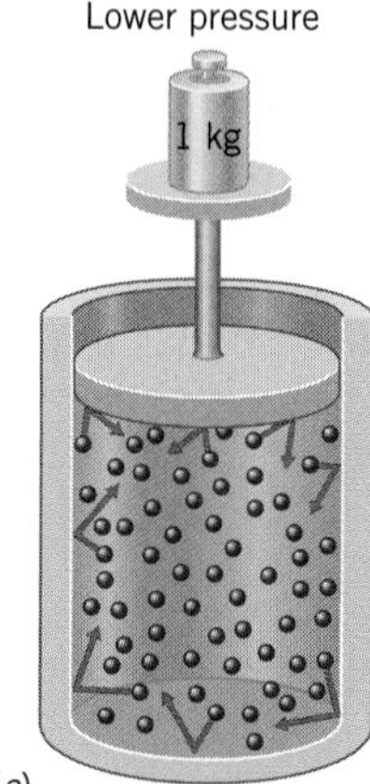

(a)

(b)

FIG. 10.13 The kinetic theory and the pressure–volume law (Boyle's law). When the gas volume is made smaller in going from (*a*) to (*b*), the number of collisions per second with each unit area of the container's walls increases. Therefore, the pressure increases.

The gas laws are predicted by the kinetic theory

According to the model, gases are mostly empty space. As we noted earlier, this explains why gases, unlike liquids and solids, can be compressed so much (squeezed to smaller volumes). It also explains why we have gas laws for gases, and *the same laws for all gases*, but not comparable laws for liquids or solids. The chemical identity of the gas does not matter, because gas molecules do not touch each other except when they collide, and there are extremely weak interactions, if any, between them.

We cannot go over the mathematical details, but we can describe some of the ways in which the laws of physics and the model of an ideal gas account for the gas laws.

Kinetic molecular theory relates temperature to average kinetic energy

The greatest triumph of the kinetic theory came with its explanation of gas temperature, which we discussed in Section 6.2. What the calculations showed was that the product of gas pressure and volume, PV, is proportional to the average kinetic energy of the gas molecules.

$$PV \propto \text{average molecular KE}$$

But from the experimental study of gases, culminating in the equation of state for an ideal gas, we have another term to which PV is proportional, namely, the Kelvin temperature of the gas.

$$PV \propto T$$

(We know what the proportionality constant here is, namely, nR, because by the ideal gas law, PV equals nRT.) With PV proportional *both* to T and to the "average molecular KE," then it must be true that the temperature of a gas is proportional to the average molecular KE.

$$T \propto \text{average molecular KE}$$

Kinetic theory explains the pressure–volume law (Boyle's law)

Using the model of an ideal gas, physicists were able to demonstrate that gas pressure is the net effect of innumerable collisions made by gas particles with the walls of the container. Let's imagine that one wall of a gas container is a movable piston that we can push in (or pull out) and so change the volume (see Figure 10.13). If we reduce the volume by one-half, we double the number of molecules per unit volume. This would double the number of collisions per second with each unit area of the wall and therefore double the pressure.

[11]In *perfectly elastic* collisions, no energy is lost by friction as the colliding objects deform momentarily.

Thus, cutting the volume in half forces the pressure to double, which is exactly what Boyle discovered:

$$P \propto \frac{1}{V} \quad \text{or} \quad V \propto \frac{1}{P}$$

Kinetic theory explains the pressure–temperature law (Gay-Lussac's law)
As you learned earlier, the kinetic theory tells us that increasing the temperature increases the average velocity of gas particles. At higher velocities, the particles strike the container's walls more frequently and with greater force. If we don't change the volume, the *area* being struck remains the same, so the force per unit area (the pressure) must increase. Thus, the kinetic theory explains how the pressure of a fixed amount of gas is proportional to temperature (at constant volume), which is the pressure–temperature law of Gay-Lussac.

Kinetic theory explains the temperature–volume law (Charles' law)
We've just seen that the kinetic theory predicts that increasing the temperature of a gas should increase the pressure if the volume doesn't change. But suppose we wished to keep the pressure constant when we raised the temperature. We could only do this if we allowed the gas to expand. Therefore, a gas expands with increasing T in order to keep P constant, which is another way of saying that V is proportional to T at constant P. Thus, the kinetic theory explains Charles' law.

Kinetic theory explains Dalton's law of partial pressures
The law of partial pressures is actually evidence for that part of the third postulate in the kinetic theory that pictures particles of an ideal gas moving in straight lines between collisions, neither attracting nor repelling each other (see Figure 10.14). By not interacting with each other, the molecules act *independently,* so each gas behaves as though it were alone in the container. Only if the particles of each gas do act independently can the partial pressures of the gases add up in a simple way to give the total pressure.

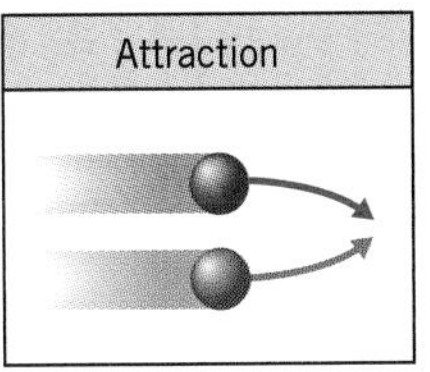

(*a*)

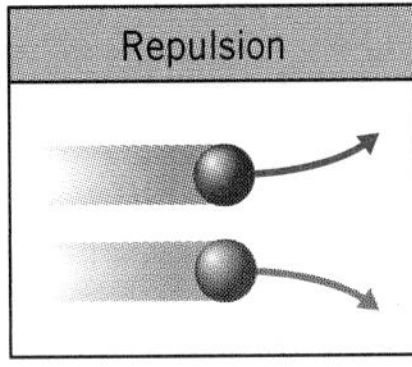

(*b*)

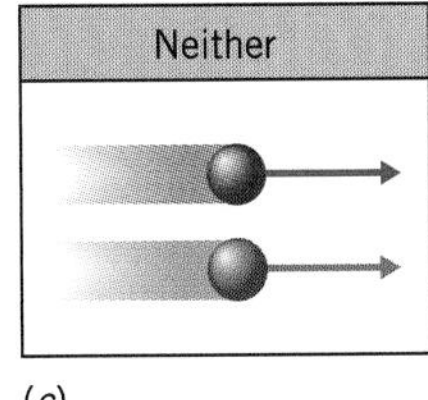

(*c*)

FIG. 10.14 Gas molecules act independently when they neither attract nor repel each other. Gas molecules would not travel in straight lines if they attracted each other as in (*a*) or repelled each other as in (*b*). They would have to travel farther between collisions with the walls and, therefore, would collide with the walls less frequently. This would affect the pressure. Only if the molecules traveled in straight lines with no attractions or repulsions, as in (*c*), would their individual pressures not be influenced by near misses or by collisions between the molecules.

Kinetic theory explains Graham's law of effusion
The key conditions of Graham's law are that the rates of effusion of two gases with different molecular masses must be compared at the same pressure and temperature and under conditions where the gas molecules do not hinder each other. When two gases have the same temperature, their particles have identical average molecular kinetic energies. Using subscripts 1 and 2 to identify two gases with molecules having different masses m_1 and m_2, we can write that at a given temperature

$$\overline{\text{KE}_1} = \overline{\text{KE}_2}$$

where the bar over KE signifies "average."

For a single molecule, its kinetic energy is $\text{KE} = \frac{1}{2}mv^2$. For a large collection of molecules of the same substance, the average kinetic energy is $\overline{\text{KE}} = \frac{1}{2}m\overline{v^2}$, where $\overline{v^2}$ is the *average of the velocities squared* (called the *mean square* velocity). We have not extended the "average" notation (the bar) over the mass because all the molecules of a given substance have the same mass (the average of their masses is just the mass).

Once again comparing two gases, 1 and 2, we take $\overline{v_1^2}$ and $\overline{v_2^2}$ to be the average of the velocities squared of their molecules. If both gases are at the same temperature, we have

$$\overline{\text{KE}_1} = \frac{1}{2}m_1\overline{v_1^2} = \frac{1}{2}m_2\overline{v_2^2} = \overline{\text{KE}_2}$$

Now let's rearrange the previous equation to get the ratio of $\overline{v^2}$ terms.

$$\frac{\overline{v_1^2}}{\overline{v_2^2}} = \frac{m_2}{m_1}$$

Next, we'll take the square root of both sides. When we do this, we obtain a ratio of quantities called the **root mean square** (abbreviated **rms**) **speeds**, which we will represent as $(\overline{v_1})_{\text{rms}}$ and $(\overline{v_2})_{\text{rms}}$.

$$\frac{(\overline{v_1})_{\text{rms}}}{(\overline{v_2})_{\text{rms}}} = \sqrt{\frac{m_2}{m_1}}$$

Suppose we have two molecules with speeds of 6 and 10 m s^{-1}. The average speed is $\frac{1}{2}(6 + 10) = 8$ m s^{-1}. The rms speed is obtained by squaring each speed, averaging the squared values, and then taking the square root of the result. Thus,

$$\overline{v_{rms}} = \sqrt{\tfrac{1}{2}(6^2 + 10^2)} = 8.2 \text{ m s}^{-1}$$

The rms speed, $(\overline{v})_{rms}$, is not actually the same as the average speed of the gas molecules, but instead represents the speed of a molecule that would have the average kinetic energy. (The difference is subtle, and the two averages do not differ by much, as noted in the margin.)

For any substance, the mass of an individual molecule is proportional to the molecular mass. Representing a molecular mass of a gas by M, we can restate this as $m \propto M$. The proportionality constant is the same for all gases. (It's in grams per atomic mass unit when we express m in atomic mass units.) When we take a ratio of two molecular masses, the constant cancels anyway, so we can write

$$\frac{(\overline{v_1})_{rms}}{(\overline{v_2})_{rms}} = \sqrt{\frac{m_2}{m_1}} = \sqrt{\frac{M_2}{M_1}}$$

Notice that the preceding equation tells us that the rms speed of the molecules is inversely proportional to the square root of the molecular mass. This means that *at a given temperature, molecules of a gas with a high molecular mass move more slowly, on average, than molecules of a gas with a low molecular mass.*

As you might expect, fast moving molecules will find an opening in the wall of a container more often than slow moving molecules, so they will effuse faster. Therefore, the rate of effusion of a gas is proportional to the average speed of its molecules, and it is also proportional to $1/\sqrt{M}$.

$$\text{Effusion rate} \propto (\overline{v})_{rms} \propto \frac{1}{\sqrt{M}}$$

Let's use k as the proportionality constant. This gives

$$\text{Effusion rate} = \frac{k}{\sqrt{M}}$$

Comparing two gases, 1 and 2, and taking a ratio of effusion rates to cause k to cancel, we have

$$\frac{\text{effusion rate (gas 1)}}{\text{effusion rate (gas 2)}} = \sqrt{\frac{M_2}{M_1}}$$

This is the way we expressed Graham's law in Equation 10.7. Thus, still another gas law supports the model of an ideal gas.

Kinetic theory predicts an absolute zero

The kinetic theory found that the temperature is proportional to the average kinetic energy of the molecules.

$$T \propto \text{average molecular KE} \propto \tfrac{1}{2}m\overline{v^2}$$

If the average molecular KE becomes zero, the temperature must also become zero. But mass (m) cannot become zero, so the only way that the average molecular KE can be zero is if v goes to zero. A particle cannot move any slower that it does at a dead standstill, so if the particles stop moving entirely, the substance is as cold as anything can get. It's at absolute zero.[12]

[12]Actually, even at 0 K, there must be some slight motion. It's required by the Heisenberg uncertainty principle, which says (in one form) that it's impossible to know precisely both the speed and the location of a particle simultaneously. (If one knows the speed, then there's uncertainty in the location, for example.) If the molecules were actually dead still at absolute zero, there would be no uncertainty in their speed. But then the uncertainty in their *position* would be infinitely great. We would not know where they were! But we do know; they're in this or that container. Thus some uncertainty in speed must exist to have less uncertainty in position and so locate the sample!

10.9 REAL GASES DON'T OBEY THE IDEAL GAS LAW PERFECTLY

According to the ideal gas law, the ratio PV/T equals a product of two constants, nR. But, experimentally, for real gases PV/T is actually not quite a constant. When we use experimental values of P, V, and T for a real gas, such as O_2, to plot actual values of PV/T as a function of P, we get the curve shown in Figure 10.15. The *horizontal* line at $PV/T = 1$ in Figure 10.15 is what we should see if PV/T were truly constant over all values of P, as it would be for an ideal gas.

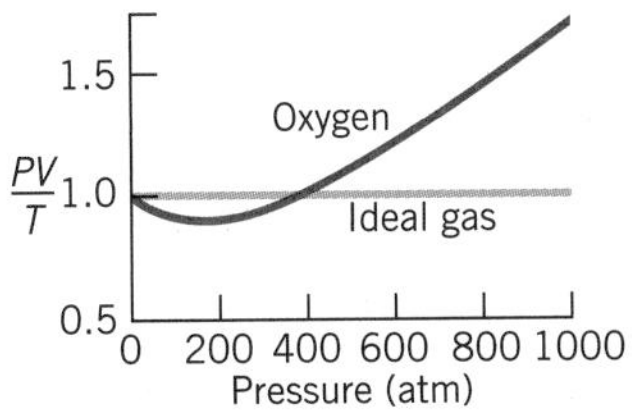

FIG. 10.15 Deviation from the ideal gas law. A graph of PV/T versus P for an ideal gas is equivalent to plotting a constant versus P. The graph must be a straight line, as shown, because the ideal gas law equation tells us that $PV/T = nR$ (a product of constants). The same plot for oxygen, a real gas, is not a straight line, showing that O_2 is not "ideal."

A real gas, like oxygen, deviates from ideal behavior for two important reasons. First, the model of an ideal gas assumes that gas molecules are infinitesimally small—that the individual molecules have no volume. But real molecules do take up some space. (If all of the kinetic motions of the gas molecules ceased and the molecules settled, you could imagine the net space that the molecules would occupy in and of themselves.) Second, in an ideal gas there would be no attractions between molecules, but in a real gas molecules do experience weak attractions toward each other.

At room temperature and atmospheric pressure, most gases behave nearly like an ideal gas, also for two reasons. First, the space between the molecules is so large that the volume occupied by the molecules themselves is insignificant. By doubling the pressure, we are able to squeeze the gas into very nearly half the volume. Second, the molecules are moving so rapidly and are so far apart that the attractions between them are hardly felt. As a result, the gas behaves almost as though there are no attractions.

Deviations from ideal behavior are felt most when the gas is at very high pressure and when the temperature is low. Raising the pressure can reduce only the empty space between the molecules, not the volume of the individual particles themselves. At high pressure, the space taken by the molecules themselves is a significant part of the total volume, so doubling the pressure cannot halve the total volume. As a result, the actual volume of a real gas is larger than expected for an ideal gas, and the ratio PV/T is larger than if the gas were ideal. We see this for O_2 at the right side of the graph in Figure 10.15.

The attractive forces between molecules reveal themselves by causing the pressure of a real gas to be slightly lower than that expected for an ideal gas. The attractions cause the paths of the molecules to bend whenever they pass near each other (Figure 10.16). Because the molecules are not traveling in straight lines, as they would in an ideal gas, they have to travel farther between collisions with the walls. As a result, the molecules of a real gas don't strike the walls as frequently as they would if the gas were ideal, and this reduced frequency of collision translates to a reduced pressure. Thus, the ratio PV/T is *less* than that for an ideal gas, particularly where the problem of particle volume is least, at lower pressures. The curve for O_2 in Figure 10.15, therefore, dips at lower pressures.

The van der Waals equation corrects for deviations from ideal behavior

Many attempts have been made to modify the equation of state of an ideal gas to get an equation that better fits the experimental data for individual real gases. One of the more successful efforts was that of J. D. van der Waals. He found ways to correct the measured values of P and V to give better fits of the data to the general gas law equation. The result of his derivation is called the *van der Waals equation of state for a real gas.* Let's take a brief look at how van der Waals made corrections to measured values of P and V to obtain expressions that fit the ideal gas law.

J. D. van der Waals (1837–1923), a Dutch scientist, won the 1910 Nobel Prize in Physics.

As you know, if a gas were ideal, it would obey the equation

$$P_{\text{ideal}}V_{\text{ideal}} = nRT$$

But for a real gas, using the measured pressure, P_{meas}, and measured volume, V_{meas},

$$P_{\text{meas}}V_{\text{meas}} \neq nRT$$

The reason is because P_{meas} is smaller than P_{ideal} (as a result of attractive forces between real gas molecules), and because V_{meas} is larger than V_{ideal} (because real molecules do take up

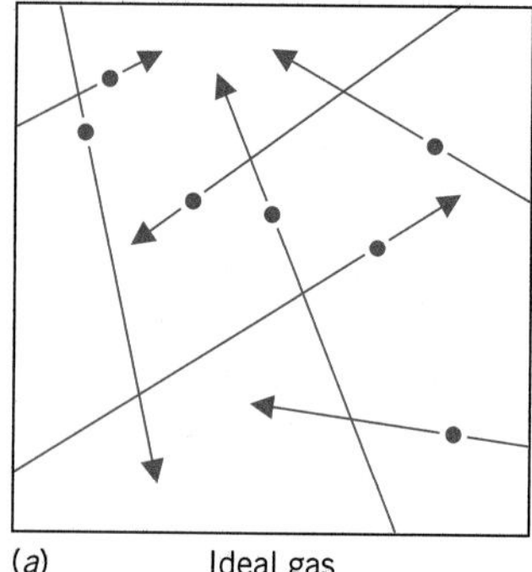
(a) Ideal gas

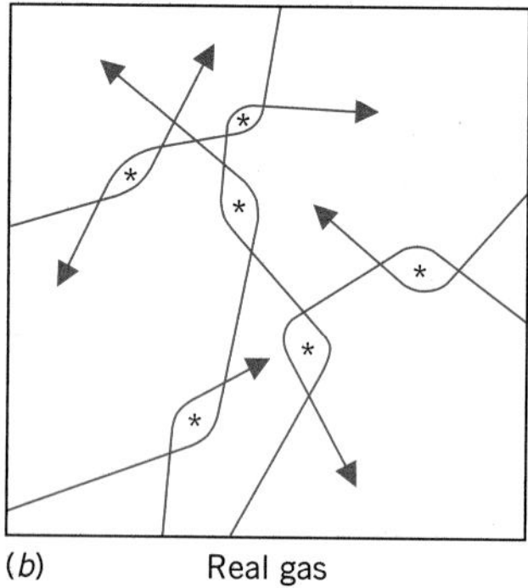
(b) Real gas

FIG. 10.16 The effect of attractive forces on the pressure of a real gas. (*a*) In an ideal gas, the molecules would travel in straight lines. (*b*) In a real gas, the paths curve as one molecule passes close to another because the molecules attract each other. Asterisks indicate the points at which molecules come close to each other. Because of the curved paths, molecules of a real gas take longer to reach the walls between collisions, which reduces the collision frequency and so slightly lowers the pressure.

some space). Therefore, to get the pressure and volume to obey the ideal gas law, we have to *add* something to the measured pressure and *subtract* something from the measured volume. That's exactly what van der Waals did. Here's his equation.

the measured pressure ↓ ; the measured volume ↓

$$\left(P_{\text{meas}} + \frac{n^2a}{V^2}\right)(V_{\text{meas}} - nb) = nRT \tag{10.8}$$

↗ correction to bring measured P up to the pressure an ideal gas would exert

↖ correction to reduce measured V to the volume an ideal gas would have

The constants a and b are called *van der Waals constants* (see Table 10.3). They are determined for each real gas by carefully measuring P, V, and T under varying conditions. Then trial calculations are made to figure out what values of the constants give the best matches between the observed data and the van der Waals equation.

Notice that the constant a involves a correction to the pressure term of the ideal gas law, so the size of a would indicate something about attractions between molecules. Larger values of a mean stronger attractive forces between molecules. Thus, the most easily liquefied substances, like water and ethyl alcohol, have the largest values of the van der Waals constant a, suggesting relatively strong attractive forces between their molecules.

The constant b helps to correct for the volume occupied by the molecules themselves, so the size of b indicates something about the sizes of particles in the gas. Larger values of b mean larger molecular sizes. Looking at data for the noble gases in Table 10.3, we see that as the atoms become larger from helium through xenon, the values of b also become larger. In the next chapter we'll continue the study of factors that control the physical state of a substance, particularly attractive forces and their origins.

TABLE 10.3 Van der Waals Constants

Substance	a (L^2 atm mol^{-2})	b (L mol^{-1})
Noble Gases		
Helium, He	0.03421	0.02370
Neon, Ne	0.2107	0.01709
Argon, Ar	1.345	0.03219
Krypton, Kr	2.318	0.03978
Xenon, Xe	4.194	0.05105
Other Gases		
Hydrogen, H_2	0.02444	0.02661
Oxygen, O_2	1.360	0.03183
Nitrogen, N_2	1.390	0.03913
Methane, CH_4	2.253	0.04278
Carbon dioxide, CO_2	3.592	0.04267
Ammonia, NH_3	4.170	0.03707
Water, H_2O	5.464	0.03049
Ethyl alcohol, C_2H_5OH	12.02	0.08407

SUMMARY

Barometers, Manometers, and Pressure Units. Atmospheric pressure is measured with a **barometer** in which a pressure of one **standard atmosphere** (1 **atm**) will support a column of mercury 760 mm high. This is a pressure of 760 **torr.** By definition, 1 atm = 101,325 **pascals** (**Pa**) and 1 **bar** = 100 kPa. **Manometers,** both open end and closed end, are used to measure the pressure of trapped gases.

Gas Laws. An **ideal gas** is a hypothetical gas that obeys the gas laws exactly over all ranges of pressure and temperature. Real gases exhibit ideal gas behavior most closely at low pressures and high temperatures, which are conditions remote from those that liquefy a gas.

Boyle's Law (Pressure–Volume Law). For a fixed amount of gas at constant temperature, volume varies inversely with pressure: $V \propto 1/P$. A useful form of the equation is $P_1V_1 = P_2V_2$.

Charles' Law (Temperature–Volume Law). For a fixed amount of gas at constant pressure, volume varies directly with the Kelvin temperature: $V \propto T$, or $V_1/V_2 = T_1/T_2$.

Gay-Lussac's Law (Temperature–Pressure Law). For a fixed amount of gas at constant volume, pressure varies directly with Kelvin temperature: $P \propto T$, or $P_1/P_2 = T_1/T_2$.

Avogadro's Principle. Equal volumes of gases contain equal numbers of moles when compared at the same temperature and pressure. At **STP**, 1 mol of an ideal gas occupies a volume of 22.4 L.

Combined Gas Law. PV divided by T for a given gas sample is a constant: $PV/T = C$, or $P_1V_1/T_1 = P_2V_2/T_2$.

Ideal Gas Law. $PV = nRT$. When P is in atmospheres and V is in liters, the value of R is 0.0821 L atm mol^{-1} K^{-1} (T being, as usual, in kelvins).

Gay-Lussac's Law of Combining Volumes. When measured at the same temperature and pressure, the volumes of gases consumed and produced in chemical reactions are in the same ratios as their coefficients.

Mole Fraction. The mole fraction X_A of a substance A equals the ratio of the number of moles of A, n_A, to the total number of moles, n_{total}, of all the components of a mixture:

$$X_A = \frac{n_A}{n_{total}}$$

Dalton's Law of Partial Pressures. The total pressure of a mixture of gases is the sum of the partial pressures of the individual gases:

$$P_{total} = P_A + P_B + P_C + \cdots$$

In terms of mole fractions, $P_A = X_AP_{total}$ and $X_A = P_A/P_{total}$.

Graham's Law of Effusion. The rate of effusion of a gas varies inversely with the square root of its density (or the square root of its molecular mass) at constant pressure and temperature. Comparing different gases at the same temperature and pressure,

$$\frac{\text{effusion rate }(A)}{\text{effusion rate }(B)} = \sqrt{\frac{d_B}{d_A}} = \sqrt{\frac{M_B}{M_A}}$$

Kinetic Theory of Gases. An ideal gas consists of a large number of particles, each having essentially zero volume, that are in constant, chaotic, random motion, traveling in straight lines with no attractions or repulsions between them. When the laws of physics and statistics are applied to this model, and the results compared with the ideal gas law, the Kelvin temperature of a gas is found to be proportional to the average kinetic energy of the gas particles. Pressure is the result of forces of collision of the particles with the container's walls.

Real Gases. Because individual gas particles do have real volumes and because small forces of attraction do exist between them, real gases do not exactly obey the gas laws. The van der Waals equation of state for a real gas makes corrections for the volume of the gas molecules and for the attractive forces between gas molecules. The van der Waals constant a provides a measure of the attractive forces between molecules, whereas the constant b gives a measure of the relative size of the gas molecules.

Reaction Stoichiometry: A Summary. With the study in this chapter of the stoichiometry of reactions involving gases, we have completed our study of the tools needed for the calculations of all variations of reaction stoichiometry. The critical link in all such calculations is the set of coefficients given by the balanced equation, which provides the stoichiometric equivalencies needed to convert from the number of moles of one substance to the numbers of moles of any of the others in the reaction. To use the coefficients requires that all the calculations must funnel through *moles.* Whether we start with grams of some compound in a reaction, with the molarity of its solution plus a volume, or with P–V–T data for a gas in the reaction, *we must get the essential calculation into moles.* After applying the coefficients, we can then move back to any other kind of unit we wish. The flowchart below summarizes what we have been doing.

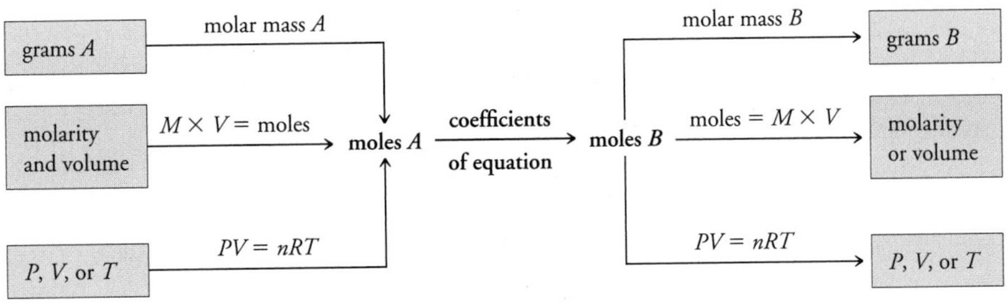

grams A —(molar mass A)→ moles A
molarity and volume —($M \times V$ = moles)→ moles A
P, V, or T —($PV = nRT$)→ moles A
moles A —(**coefficients of equation**)→ moles B
moles B —(molar mass B)→ grams B
moles B —(moles = $M \times V$)→ molarity or volume
moles B —($PV = nRT$)→ P, V, or T

The labels on the arrows of the flowchart suggest the basic tools. Formula masses or molecular masses get us from grams to moles or from moles to grams. Molarity and volume data move us from concentration to moles or back. With P–V–T data we can find moles or, knowing moles of a gas, we can calculate P, V, or T, as long as the other two are known.

TOOLS FOR PROBLEM SOLVING

The compilation below lists the concepts that you've learned in this chapter which can be applied as tools in solving problems. Study each one carefully so that you know what each is used for. When faced with solving a problem, recall what each tool does and consider whether it will be helpful in finding a solution. This will aid you in selecting the tools you need. If necessary, refer to this table when working on the Review Exercises that follow.

Combined gas law *(page 398)*,

$$\frac{P_1V_1}{T_1} = \frac{P_2V_2}{T_2}$$

This law applies when the amount of gas is constant and we are asked how one of the variables (P, V, or T) changes when we change two of the others. When the amount of gas and one of the variables (P, V, or T) are constant, the problem reduces to one involving Boyle's, Charles', or Gay-Lussac's law.

Molar volumes *(page 402)*. For one mole of any gas at STP

$$1 \text{ mol} \Leftrightarrow 22.4 \text{ L}$$

Ideal gas law *(page 405)*,

$$PV = nRT$$

This law applies when any three of the four variables P, V, T, or n, are known and we wish to calculate the value of the fourth.

Determination of molar mass: Molar mass M can be determined in a variety of ways from equations used in this chapter.

From ideal gas law *(page 407)*,

$$\text{molar mass} = \frac{g}{V}\frac{RT}{P} = d\frac{RT}{P}$$

From effusion rates *(Rearrangement of Equation 10.7 page 418)*,

$$M_B = M_A\left(\frac{\text{effusion rate }(A)}{\text{effusion rate }(B)}\right)^2$$

From mole fractions *(Rearrangement of Equation 10.4 page 415)*,

$$M_A = \frac{g_A}{X_A \times n_{\text{total}}}$$

Dalton's law of partial pressures *(page 413)*,

$$P_{\text{total}} = P_A + P_B + P_C + \cdots$$

We use this law to calculate the partial pressure of one gas in a mixture of gases. This requires the total pressure and either the partial pressures of the other gases or their mole fractions. If the partial pressures are known, their sum is the total pressure. When a gas is collected over water, this law is used to obtain the partial pressure of the collected gas in the "wet" gas mixture: $P_{\text{total}} = P_{water} + P_{gas}$.

Mole fractions *(pages 415 and 416)*,

$$X_A = \frac{n_A}{n_{\text{total}}} = \frac{P_A}{P_{\text{total}}}$$

Given the composition of a gas mixture, we can calculate the mole fraction of a component. The mole fraction can then be used to find the partial pressure of the component given the total pressure. If the total pressure and partial pressure of a component are known, we can calculate the mole fraction of the component.

Graham's law of effusion *(page 418)*,

$$\frac{\text{effusion rate }(A)}{\text{effusion rate }(B)} = \sqrt{\frac{d_B}{d_A}} = \sqrt{\frac{M_B}{M_A}}$$

This law allows us to calculate relative rates of effusion of gases (A and B in this equation). It also allows us to calculate molecular masses from relative rates of effusion.

QUESTIONS, PROBLEMS, AND EXERCISES

Answers to problems whose numbers are printed in color are given in Appendix B. More challenging problems are marked with asterisks. ILW = Interactive Learningware solution is available at www.wiley.com/college/brady. OH = an Office Hours video is available for this problem.

REVIEW QUESTIONS

Concept of Pressure; Manometers and Barometers

10.1 If you get jabbed by a pencil, why does is hurt so much more if it's with the sharp point rather than the eraser? Explain in terms of the concepts of force and pressure.

10.2 Write expressions that could be used to form conversion factors to convert between:
(a) kilopascal and atm
(b) torr and mm Hg
(c) torr and atm
(d) torr and pascal
(e) bar and pascal
(f) bar and atm

10.3 At 20 °C the density of mercury is 13.6 g mL^{-1} and that of water is 1.00 g mL^{-1}. At 20 °C, the vapor pressure of mercury is 0.0012 torr and that of water is 18 torr. Give and explain two reasons why water would be an inconvenient fluid to use in a Torricelli barometer.

10.4 What is the advantage of using a closed-end manometer, rather than an open-end one, when measuring the pressure of a trapped gas?

Gas Laws

10.5 Express the following gas laws in equation form: (a) temperature–volume law (Charles' law), (b) temperature–pressure law (Gay-Lussac's law), (c) pressure–volume law (Boyle's law), and (d) combined gas law.

10.6 Which of the four important variables in the study of the physical properties of gases are assumed to be held constant in each of the following laws? (a) Boyle's law, (b) Charles' law, (c) Gay-Lussac's law, (d) combined gas law

10.7 What is meant by an *ideal gas*? Under what conditions does a real gas behave most like an ideal gas?

10.8 State the ideal gas law in the form of an equation. What is the value of the gas constant in units of L atm mol^{-1} K^{-1}?

10.9 State Dalton's law of partial pressures in the form of an equation.

10.10 Define *mole fraction.* How is the partial pressure of a gas related to its mole fraction and the total pressure?

10.11 Consider the diagrams below that illustrate three mixtures of gases *A* and *B*. If the total pressure of each mixture is 1.00 atm, which of the drawings corresponds to a mixture in which the partial pressure of *A* equals 0.600 atm? What are the partial pressures of *A* in the other mixtures? What are the partial pressures of *B*?

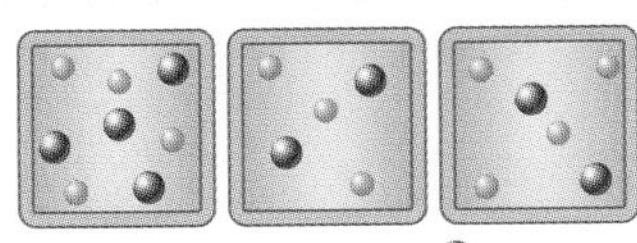

10.12 What is the difference between *diffusion* and *effusion*? State Graham's law in the form of an equation.

Kinetic Theory of Gases

10.13 Describe the model of a gas proposed by the kinetic theory of gases.

10.14 If the molecules of a gas at constant volume are somehow given a lower average kinetic energy, what two measurable properties of the gas will change and in what direction?

10.15 Explain *how* raising the temperature of a gas causes it to expand at constant pressure. (Describe how the model of an ideal gas connects the increase in temperature to the gas expansion.)

10.16 Explain in terms of the kinetic theory *how* raising the temperature of a confined gas makes its pressure increase.

10.17 How does the kinetic theory explain the existence of an absolute zero, 0 K?

10.18 Which of the following gases has the largest value of $(\overline{v})_{rms}$ at 25 °C? (a) N_2, (b) CO_2, (c) NH_3, or (d) HBr

10.19 How would you expect the rate of effusion of a gas to depend on (a) the pressure of the gas and (b) the temperature of the gas?

Real Gases

10.20 Which postulates of the kinetic theory are not strictly true, and why?

10.21 A small value for the van der Waals constant *a* suggests something about the molecules of the gas. What?

10.22 Which of the molecules below has the larger value of the van der Waals constant *b*? Explain your choice.

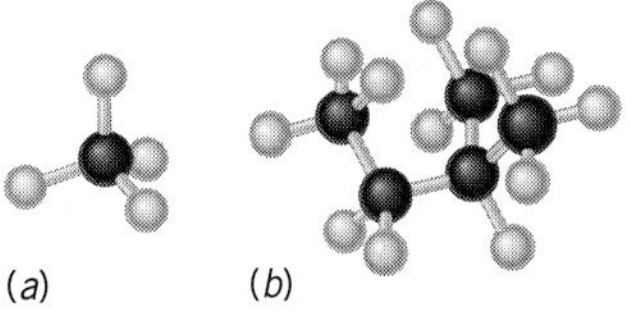

10.23 Under the same conditions of *T* and *V*, why is the pressure of a real gas less than the pressure the gas would exert if it were ideal? At a given *T* and *P*, why is the volume of a real gas larger than it would be if the gas were ideal?

10.24 Suppose we have a mixture of helium and argon. On average, which atoms are moving faster at 25 °C, and why?

REVIEW PROBLEMS

Pressure Unit Conversions

OH **10.25** Carry out the following unit conversions: (a) 1.26 atm to torr, (b) 740 torr to atm, (c) 738 torr to mm Hg, and (d) 1.45 × 10^3 Pa to torr.

10.26 Carry out the following unit conversions: (a) 0.625 atm to torr, (b) 825 torr to atm, (c) 62 mm Hg to torr, and (d) 1.22 kPa to bar.

10.27 What is the pressure in torr of each of the following?
(a) 0.329 atm (summit of Mt. Everest, the world's highest mountain)
(b) 0.460 atm (summit of Mt. Denali, the highest mountain in the United States)

10.28 What is the pressure in atm of each of the following? (These are the values of the pressures exerted individually by N_2, O_2, and CO_2, respectively, in typical inhaled air.)
(a) 595 torr (b) 160 torr (c) 0.300 torr

Manometers and Barometers

10.29 An open-end manometer containing mercury was connected to a vessel holding a gas at a pressure of 720 torr. The atmospheric pressure was 765 torr. Sketch a diagram of the apparatus showing the relative heights of the mercury in the two arms of the manometer. What is the difference in the heights of the mercury expressed in centimeters?

10.30 An open-end manometer containing mercury was connected to a vessel holding a gas at a pressure of 820 torr. The atmospheric pressure was 750 torr. Sketch a diagram of the apparatus showing the relative heights of the mercury in the two arms of the manometer. What is the difference in the heights of the mercury expressed in centimeters?

10.31 An open-end mercury manometer was connected to a flask containing a gas at an unknown pressure. The mercury in the arm open to the atmosphere was 65 mm higher than the mercury in the arm connected to the flask. The atmospheric pressure was 748 torr. What was the pressure of the gas in the flask (in torr)?

10.32 An open-end mercury manometer was connected to a flask containing a gas at an unknown pressure. The mercury in the arm open to the atmosphere was 82 mm lower than the mercury in the arm connected to the flask. The atmospheric pressure was 752 torr. What was the pressure of the gas in the flask (in torr)?

10.33 Suppose that in a closed-end manometer the mercury in the closed arm was 12.5 cm higher than the mercury in the arm connected to a vessel containing a gas. What is the pressure of the gas expressed in torr?

10.34 Suppose a gas is in a vessel connected to both an open-end and a closed-end manometer. The difference in heights of the mercury in the closed-end manometer was 236 mm, while in the open-end manometer the mercury level in the arm open to the atmosphere was 512 mm below the level in the arm connected to the vessel. Calculate the atmospheric pressure and sketch a diagram of the apparatus.

Gas Laws for a Fixed Amount of Gas

10.35 A gas has a volume of 255 mL at 725 torr. What volume will the gas occupy at 365 torr if the temperature of the gas doesn't change?

10.36 A bicycle pump has a barrel that is 75.0 cm long (about 30 in.). If air is drawn into the pump at a pressure of 1.00 atm during the upstroke, how long must the downstroke be, in centimeters, to raise the pressure of the air to 5.50 atm (approximately the pressure in the tire of a 10-speed bike)? Assume no change in the temperature of the air.

10.37 A gas has a volume of 3.86 L at a temperature of 45 °C. What will the volume of the gas be if its temperature is raised to 80 °C while its pressure is kept constant?

OH 10.38 A balloon has a volume of 2.50 L indoors at a temperature of 22 °C. If the balloon is taken outdoors on a cold day when the air temperature is −15 °C (5 °F), what will its volume be in liters? Assume constant air pressure within the balloon.

10.39 A sample of a gas has a pressure of 850 torr at 285 °C. To what Celsius temperature must the gas be heated to double its pressure if there is no change in the volume of the gas?

10.40 Before taking a trip, you check the air in a tire of your automobile and find it has a pressure of 45 lb in.$^{-2}$ on a day when the air temperature is 10 °C (50 °F). After traveling some distance, you find that the temperature of the air in the tire has risen to 43 °C (approximately 110 °F). What is the air pressure in the tire at this higher temperature, expressed in units of lb in.$^{-2}$?

ILW 10.41 A sample of helium at a pressure of 740 torr and in a volume of 2.58 L was heated from 24.0 to 75.0 °C. The volume of the container expanded to 2.81 L. What was the final pressure (in torr) of the helium?

10.42 When a sample of neon with a volume of 648 mL and a pressure of 0.985 atm was heated from 16.0 to 63.0 °C, its volume became 689 mL. What was its final pressure (in atm)?

10.43 What must be the new volume of a sample of nitrogen (in L) if 2.68 L at 745 torr and 24.0 °C is heated to 375.0 °C under conditions that let the pressure change to 760 torr?

10.44 When 280 mL of oxygen at 741 torr and 18.0 °C was warmed to 33.0 °C, the pressure became 760 torr. What was the final volume (in mL)?

10.45 A sample of argon with a volume of 6.18 L, a pressure of 761 torr, and a temperature of 20.0 °C expanded to a volume of 9.45 L and a pressure of 373 torr. What was its final temperature in °C?

10.46 A sample of a refrigeration gas in a volume of 455 mL, at a pressure of 1.51 atm, and at a temperature of 25.0 °C was compressed into a volume of 220 mL with a pressure of 2.00 atm. To what temperature (in °C) did it have to change?

Ideal Gas Law

10.47 What would be the value of the gas constant R in units of $mL\ torr\ mol^{-1}\ K^{-1}$?

10.48 The SI generally uses its base units to compute constants involving derived units. The SI unit for volume, for example, is the cubic meter, called the *stere*, because the meter is the base unit of length. We learned about the SI unit of pressure, the pascal, in this chapter. The temperature unit is the kelvin. Calculate the value of the gas constant in the SI units $m^3\ Pa\ mol^{-1}\ K^{-1}$.

10.49 What volume in liters does 0.136 g of O_2 occupy at 20.0 °C and 748 torr?

10.50 What volume in liters does 1.67 g of N_2 occupy at 22.0 °C and 756 torr?

10.51 What pressure (in torr) is exerted by 10.0 g of O_2 in a 2.50 L container at a temperature of 27 °C?

10.52 If 12.0 g of water is converted to steam in a 3.60 L pressure cooker held at a temperature of 108 °C, what pressure would be produced?

10.53 A sample of carbon dioxide has a volume of 26.5 mL at 20.0 °C and 624 torr. How many grams of CO_2 are in the sample?

10.54 Methane is formed in landfills by the action of certain bacteria on buried organic matter. If a sample of methane collected from a landfill has a volume of 250 mL at 750 torr and 27 °C, how many grams of methane are in the sample?

10.55 To three significant figures, calculate the density in g L^{-1} of the following gases at STP: (a) C_2H_6 (ethane), (b) N_2, (c) Cl_2, and (d) Ar.

10.56 To three significant figures, calculate the density in g L^{-1} of the following gases at STP: (a) Ne, (b) O_2, (c) CH_4 (methane), and (d) CF_4.

OH 10.57 What density (in g L^{-1}) does oxygen have at 24.0 °C and 742 torr?

10.58 At 748.0 torr and 20.65 °C, what is the density of argon (in g L^{-1})?

ILW 10.59 A chemist isolated a gas in a glass bulb with a volume of 255 mL at a temperature of 25.0 °C and a pressure (in the bulb) of 10.0 torr. The gas weighed 12.1 mg. What is the molecular mass of the gas?

10.60 Boron forms a variety of unusual compounds with hydrogen. A chemist isolated 6.3 mg of one of the boron hydrides in a glass bulb with a volume of 385 mL at 25.0 °C and a bulb pressure of 11 torr.

(a) What is the molecular mass of this hydride?

(b) Which of the following is likely to be its molecular formula, BH_3, B_2H_6, or B_4H_{10}?

10.61 At 22.0 °C and a pressure of 755 torr, a gas was found to have a density of 1.13 g L^{-1}. Calculate its molecular mass.

10.62 A gas was found to have a density of 0.08747 mg mL^{-1} at 17.0 °C and a pressure of 760 torr. What is its molecular mass? Can you tell what the gas most likely is?

Stoichiometry of Reactions of Gases

10.63 How many milliliters of oxygen are required to react completely with 175 mL of butane if the volumes of both gases are measured at the same temperature and pressure?

OH 10.64 How many milliliters of O_2 are consumed in the complete combustion of a sample of hexane if the reaction produces 855 mL of CO_2? Assume all gas volumes are measured at the same temperature and pressure.

ILW 10.65 Propylene, C_3H_6, reacts with hydrogen under pressure to give propane, C_3H_8:

$$C_3H_6(g) + H_2(g) \longrightarrow C_3H_8(g)$$

How many liters of hydrogen (at 740 torr and 24 °C) react with 18.0 g of propylene?

10.66 Nitric acid is formed when NO_2 is dissolved in water.

$$3NO_2(g) + H_2O(l) \longrightarrow 2HNO_3(aq) + NO(g)$$

How many milliliters of NO_2 at 25 °C and 752 torr are needed to form 12.0 g of HNO_3?

10.67 How many milliliters of O_2 measured at 27 °C and 654 torr are needed to react completely with 16.8 mL of CH_4 measured at 35 °C and 725 torr?

10.68 How many milliliters of H_2O vapor, measured at 318 °C and 735 torr, are formed when 33.6 mL of NH_3 at 825 torr and 127 °C react with oxygen according to the following equation?

$$4NH_3(g) + 3O_2(g) \longrightarrow 2N_2(g) + 6H_2O(g)$$

10.69 Calculate the maximum number of milliliters of CO_2, at 745 torr and 27 °C, that could be formed in the combustion of carbon monoxide if 300 mL of CO at 683 torr and 25 °C is mixed with 150 mL of O_2 at 715 torr and 125 °C.

10.70 A mixture of ammonia and oxygen is prepared by combining 300 mL of NH_3 (measured at 750 torr and 28 °C) with 220 mL of O_2 (measured at 780 torr and 50 °C). How many milliliters of N_2 (measured at 740 torr and 100 °C) could be formed if the following reaction occurs?

$$4NH_3(g) + 3O_2(g) \longrightarrow 2N_2(g) + 6H_2O(g)$$

Dalton's Law of Partial Pressures

OH 10.71 A 1.00 L container was filled by pumping into it 1.00 L of N_2 at 200 torr, 1.00 L of O_2 at 150 torr, and 1.00 L of He at 300 torr. All volumes and pressures were measured at the same temperature. What was the total pressure inside the container after the mixture was made?

10.72 A mixture of N_2, O_2, and CO_2 has a total pressure of 740 torr. In this mixture the partial pressure of N_2 is 120 torr and the partial pressure of O_2 is 400 torr. What is the partial pressure of the CO_2?

ILW 10.73 A 22.4 L container at 0 °C contains 0.30 mol N_2, 0.20 mol O_2, 0.40 mol He, and 0.10 mol CO_2. What are the partial pressures of each of the gases?

10.74 A 0.200 mol sample of a mixture of nitrogen and carbon dioxide with a total pressure of 840 torr is exposed to solid calcium oxide which reacts with the carbon dioxide to form solid calcium carbonate. Assume the reaction goes to completion. After the reaction was complete, the pressure of the gas had dropped to 320 torr. How many moles of CO_2 were in the original mixture? (Assume no change in volume or temperature.)

10.75 A sample of carbon monoxide was prepared and collected over water at a temperature of 20 °C and a total pressure of 754 torr. It occupied a volume of 268 mL. Calculate the partial pressure of the CO in torr as well as its dry volume (in mL) under a pressure of 1.00 atm and 20 °C.

10.76 A sample of hydrogen was prepared and collected over water at 25 °C and a total pressure of 742 torr. It occupied a volume of 288 mL. Calculate its partial pressure (in torr) and what its dry volume would be (in mL) under a pressure of 1.00 atm at 25 °C.

10.77 What volume of "wet" methane would you have to collect at 20.0 °C and 742 torr to be sure that the sample contains 244 mL of dry methane (also at 742 torr)?

10.78 What volume of "wet" oxygen would you have to collect if you need the equivalent of 275 mL of dry oxygen at 1.00 atm?

The atmospheric pressure in the lab is 746 torr, and the oxygen is to be collected over water at 15.0 °C.

Graham's Law

10.79 Under conditions in which the density of CO_2 is 1.96 g L^{-1} and that of N_2 is 1.25 g L^{-1}, which gas will effuse more rapidly? What will be the ratio of the rates of effusion of N_2 to CO_2?

10.80 Arrange the following gases in order of increasing rate of diffusion at 25 °C: Cl_2, C_2H_4, SO_2.

10.81 Uranium hexafluoride is a white solid that readily passes directly into the vapor state. (Its vapor pressure at 20.0 °C is 120 torr.) A trace of the uranium in this compound—about 0.7%—is uranium-235, which can be used in a nuclear power plant. The rest of the uranium is essentially uranium-238, and its presence interferes with these applications for uranium-235. Gas effusion of UF_6 can be used to separate the fluoride made from uranium-235 and the fluoride made from uranium-238. Which hexafluoride effuses more rapidly? By how much? (You can check your answer by reading Facets of Chemistry 10.1 on page 419.)

OH **10.82** An unknown gas *X* effuses 1.65 times faster than C_3H_8. What is the molecular mass of gas *X*?

ADDITIONAL EXERCISES

10.83 One of the oldest units for atmospheric pressure is lb $in.^{-2}$ (pounds per square inch, or *psi*). Calculate the numerical value of the standard atmosphere in these units to three significant figures. Calculate the mass in pounds of a uniform column of water 33.9 ft high having an area of 1.00 $in.^2$ at its base. (Use the following data: density of mercury = 13.6 g mL^{-1}; density of water = 1.00 g mL^{-1}; 1 mL = 1 cm^3; 1 lb = 454 g; 1 in. = 2.54 cm.)

***10.84** A typical automobile has a weight of approximately 3500 lb. If the vehicle is to be equipped with tires, each of which will contact the pavement with a "footprint" that is 6.0 in. wide by 3.2 in. long, what must the gauge pressure of the air be in each tire? (Gauge pressure is the amount that the gas pressure exceeds atmospheric pressure. Assume that atmospheric pressure is 14.7 lb $in.^{-2}$.)

10.85 Suppose you were planning to move a house by transporting it on a large trailer. The house has an estimated weight of 45.6 tons (1 ton = 2000 lb). The trailer is expected to weigh 8.3 tons. Each wheel of the trailer will have tires inflated to a gauge pressure of 85 psi (which is actually 85 psi above atmospheric pressure). If the area of contact between a tire and the pavement can be no larger than 100 $in.^2$ (10 in. × 10 in.), what is the minimum number of wheels the trailer must have? (Remember, tires are mounted in multiples of two on a trailer. Assume that atmospheric pressure is 14.7 psi.)

***10.86** The motion picture *Titanic* described the tragedy of the collision of the ocean liner of the same name with an iceberg in the North Atlantic. The ship sank soon after the collision on April 14, 1912, and now rests on the sea floor at a depth of 12,468 ft. Recently, the wreck was explored by the research vessel *Nautile,* which has successfully recovered a variety of items from the debris field surrounding the sunken ship. Calculate the pressure in atmospheres and pounds per square inch exerted on the hull of the *Nautile* as it explores the seabed surrounding the *Titanic.* (Hint: The height of a column of liquid required to exert a given pressure is inversely proportional to the liquid's density. Seawater has a density of approximately 1.025 g mL^{-1}; mercury has a density of 13.6 g mL^{-1}; 1 atm = 14.7 lb $in.^{-2}$.)

***10.87** Two flasks (which we will refer to as Flask 1 and Flask 2) are connected to each other by a U-shaped tube filled with an oil having a density of 0.826 g mL^{-1}. The oil level in the arm connected to Flask 2 is 16.24 cm higher than in the arm connected to Flask 1. Flask 1 is also connected to an open-end mercury manometer. The mercury level in the arm open to the atmosphere is 12.26 cm higher than the level in the arm connected to Flask 1. The atmospheric pressure is 0.827 atm. What is the pressure of the gas in Flask 2 expressed in torr? (See the hint given in the preceding exercise.)

OH ***10.88** A bubble of air escaping from a diver's mask rises from a depth of 100 ft to the surface where the pressure is 1.00 atm. Initially, the bubble has a volume of 10.0 mL. Assuming none of the air dissolves in the water, how many times larger is the bubble just as it reaches the surface. Use your answer to explain why scuba divers constantly exhale as they slowly rise from a deep dive. (The density of seawater is approximately 1.025 g mL^{-1}; the density of mercury is 13.6 g mL^{-1}.)

***10.89** In a diesel engine, the fuel is ignited when it is injected into hot compressed air, heated by the compression itself. In a typical high-speed diesel engine, the chamber in the cylinder has a diameter of 10.7 cm and a length of 13.4 cm. On compression, the length of the chamber is shortened by 12.7 cm (a "5-inch stroke"). The compression of the air changes its pressure from 1.00 to 34.0 atm. The temperature of the air before compression is 364 K. As a result of the compression, what will be the final air temperature (in K and °C) just before the fuel injection?

***10.90** Early one cool (60.0 °F) morning you start on a bike ride with the atmospheric pressure at 14.7 lb $in.^{-2}$ and the tire gauge pressure at 50.0 lb $in.^{-2}$. (Gauge pressure is the amount that the pressure exceeds atmospheric pressure.) By late afternoon, the air had warmed up considerably, and this plus the heat generated by tire friction sent the temperature inside the tire to 104 °F. What will the tire gauge now read, assuming that the volume of the air in the tire and the atmospheric pressure have not changed?

OH **10.91** The range of temperatures over which an automobile tire must be able to withstand pressure changes is roughly −50 to 120 °F. If a tire is filled to 35 lb $in.^{-2}$ at −50 °F (on a cold day in Alaska, for example), what will be the pressure in the tire (in the same pressure units) on a hot day in Death Valley when the temperature is 120 °F? (Assume that the volume of the tire does not change.)

10.92 Chlorine reacts with sulfite ion to give sulfate ion and chloride ion. How many milliliters of Cl_2 gas measured at 25 °C and 734 torr are required to react with all the SO_3^{2-} in 50.0 mL of 0.200 *M* Na_2SO_3 solution?

10.93 A common laboratory preparation of hydrogen on a small scale uses the reaction of zinc with hydrochloric acid. Zinc chloride is the other product.

(a) Write the balanced equation for the reaction.

(b) If 12.0 L of H_2 at 760 torr and 20.0 °C is wanted, how many grams of zinc are needed, in theory?

(c) If the acid is available as 8.00 *M* HCl, what is the minimum volume of this solution (in milliliters) required to produce the amount of H_2 described in part (b)?

***10.94** In an experiment designed to prepare a small amount of hydrogen by the method described in the preceding exercise, a student was limited to using a gas-collecting bottle with a maximum capacity of 335 mL. The method involved collecting the hydrogen over water. What are the minimum number of grams of Zn and the minimum number of milliliters of 6.00 *M* HCl needed to produce the *wet* hydrogen that can exactly fit this collecting bottle at 740 torr and 25.0 °C?

10.95 Carbon dioxide can be made in the lab by the reaction of hydrochloric acid with calcium carbonate.

$$CaCO_3(s) + 2HCl(aq) \longrightarrow CaCl_2(aq) + H_2O(l) + CO_2(g)$$

How many milliliters of dry CO_2 at 20.0 °C and 745 torr can be prepared from a mixture of 12.3 g of $CaCO_3$ and 185 mL of 0.250 *M* HCl?

OH **10.96** A mixture was prepared in a 500 mL reaction vessel from 300 mL of O_2 (measured at 25 °C and 740 torr) and 400 mL of H_2 (measured at 45 °C and 1250 torr). The mixture was ignited and the H_2 and O_2 reacted to form water. What was the final pressure inside the reaction vessel after the reaction was over if the temperature was held at 120 °C?

10.97 A student collected 18.45 mL of H_2 over water at 24 °C. The water level inside the collection apparatus was 8.5 cm higher than the water level outside. The barometric pressure was 746 torr. How many grams of zinc had to react with HCl(*aq*) to produce the H_2 that was collected?

10.98 A mixture of gases is prepared from 87.5 g of O_2 and 12.7 g of H_2. After the reaction of O_2 and H_2 is complete, what is the total pressure of the mixture if its temperature is 160 °C and its volume is 12.0 L? What are the partial pressures of the gases remaining in the mixture?

10.99 A sample of an unknown gas with a mass of 3.620 g was made to decompose into 2.172 g of O_2 and 1.448 g of S. Prior to the decomposition, this sample occupied a volume of 1120 mL at 750 torr and 25.0 °C.

(a) What is the percentage composition of the elements in this gas?

(b) What is the empirical formula of the gas?

(c) What is its molecular formula?

***10.100** A sample of a new antimalarial drug with a mass of 0.2394 g was made to undergo a series of reactions that changed all of the nitrogen in the compound into N_2. This gas had a volume of 18.90 mL when collected over water at 23.80 °C and a pressure of 746.0 torr. At 23.80 °C, the vapor pressure of water is 22.110 torr.

(a) Calculate the percentage of nitrogen in the sample.

(b) When 6.478 mg of the compound was burned in pure oxygen, 17.57 mg of CO_2 and 4.319 mg of H_2O were obtained. What are the percentages of C and H in this compound? Assuming that any undetermined element is oxygen, write an empirical formula for the compound.

(c) The molecular mass of the compound was found to be 324. What is its molecular formula?

***10.101** In one analytical procedure for determining the percentage of nitrogen in unknown compounds, weighed samples are made to decompose to N_2, which is collected over water at known temperatures and pressures. The volumes of N_2 are then translated into grams and then into percentages.

(a) Show that the following equation can be used to calculate the percentage of nitrogen in a sample having a mass of *W* grams when the N_2 has a volume of *V* mL and is collected over water at t_C °C at a total pressure of *P* torr. The vapor pressure of water occurs in the equation as $P^\circ_{H_2O}$

$$\text{Percentage N} = 0.04489 \times \frac{V(P - P^\circ_{H_2O})}{W(273 + t_C)}$$

(b) Use this equation to calculate the percentage of nitrogen in the sample described in the preceding exercise.

10.102 The odor of a rotten egg is caused by hydrogen sulfide, H_2S. Most people can detect it at a concentration of 0.15 ppb (parts per billion), meaning 0.15 L of H_2S in 10^9 L of space. A typical student lab is 40 × 20 × 8 ft.

(a) At STP, how many liters of H_2S could be present in a typical lab to have a concentration of 0.15 ppb?

(b) How many milliliters of 0.100 *M* Na_2S would be needed to generate the amount of H_2S in part (a) by the following reaction with hydrochloric acid?

$$Na_2S(aq) + 2HCl(aq) \longrightarrow H_2S(g) + 2NaCl(aq)$$

EXERCISES IN CRITICAL THINKING

10.103 Firefighters advise that you get out of a burning building by keeping close to the floor. We learned that carbon dioxide and most other hazardous compounds are more dense than air and they should settle to the floor. What other facts do we know that make the firefighter's advice correct?

10.104 Carbon dioxide is implicated in global warming. Propose ways to control or perhaps decrease the carbon dioxide content in our air. Rank each proposal based on its feasibility.

10.105 Methane is another gas implicated in the global warming problem. It has been proposed that much of the methane in the atmosphere is from ruminating cows. Evaluate that suggestion and come up with other possible sources of methane.

10.106 Why does the barometric pressure decrease when a rainstorm approaches? What was the lowest barometric pressure ever measured on the earth?

10.107 What gases are responsible for photochemical smog? What chemical reactions are involved? Propose a solution to the problem and estimate the cost to implement it.

10.108 The ozone hole is caused by gases released from ice crystals that melt as spring returns to the south pole. What issues are involved with the ozone hole and how does that affect you?

INTERMOLECULAR ATTRACTIONS AND THE PROPERTIES OF LIQUIDS AND SOLIDS

1

Heated to high temperatures within the earth, lava bursts forth from a volcano's caldera on Hawaii. In addition to the lava, high temperatures and pressures inside the earth contribute to the formation of important minerals such as diamonds. The extreme properties of molten lava and diamond formation can be understood based on the principles in this chapter. *(Joanna McCarthy/Photographers Choice/Getty Images.)*

CHAPTER OUTLIN

THIS CHAPTER IN CONTEXT In the preceding chapter we studied the physical properties of gases, and we observed that all gases behave pretty much alike, regardless of their chemical composition. This is especially so at low pressures and high temperatures, which allows us to use one set of gas laws to describe the behavior of *any* gas. However, when we compare substances in their liquid or solid states (their *condensed states*), the situation is quite different. When a substance is a liquid or a solid, its particles are packed closely together and the forces between them, which we call *intermolecular forces*, are quite strong. Chemical composition and molecular structure play an important role in determining the strengths of such forces, and this causes different substances to behave quite differently from each other when they are liquids or solids.

In this chapter, we focus our attention on the properties of liquids and solids. We begin our study by looking at the basic differences among the states of matter in terms of both common observable properties and the way the states of matter differ at the molecular level. In this chapter we will also examine the different kinds and relative strengths of intermolecular forces. You will learn how they are related to molecular composition and structure, and how intermolecular forces influence a variety of familiar physical properties of liquids, such as boiling points and ease of evaporation. And by studying the energy changes associated with changes of states (for example, evaporation or condensation), you will become familiar with the forces that affect practical applications ranging from evaporative air-conditioning to weather prediction.

11.1 GASES, LIQUIDS, AND SOLIDS DIFFER BECAUSE INTERMOLECULAR FORCES DEPEND ON THE DISTANCES BETWEEN MOLECULES

There are differences among gases, liquids, and solids that are immediately obvious and familiar to everyone. For example, any gas will expand to fill whatever volume is available to it, even if it has to mix with other gases to do so. Liquids and solids, however, retain a constant volume when transferred from one container to another. A solid, such as an ice cube, also keeps its shape, but a liquid such as soda conforms to the shape of whatever bottle or glass we put it in.

In Chapter 10 you learned that gases are easily compressed. Liquids and solids, on the other hand, are nearly *incompressible*, which means their volumes change very little when they are subjected to high pressures. Properties such as the ones we've described can be understood in terms of the way the particles are distributed in the three states of matter, which is summarized in Figure 11.1.

Intermolecular forces depend on distance

If you've ever played with magnets you know that their mutual attraction weakens rapidly as the distance between them increases. Intermolecular attractions are similarly affected by the distance between molecules, rapidly becoming weaker as the distance between the molecules increases.

In gases, the molecules are so far apart that intermolecular attractions are almost negligible, so differences between the attractive forces hardly matter. As a result, chemical composition has little effect on the properties of a gas. But in a liquid or a solid, the molecules are close together and the attractions are strong. Differences among these attractions caused by differences in chemical makeup are greatly amplified, so the properties of liquids and solids depend quite heavily on chemical composition.

The closer two molecules are, the more strongly they attract each other.

11.2 INTERMOLECULAR ATTRACTIONS INVOLVE ELECTRICAL CHARGES

Inter*molecular forces** (the attractions *between* molecules) are always much weaker than the attractions between atoms *within* molecules (intra*molecular forces,** which are the *chemical bonds* that hold molecules together). In a molecule of HCl, for example,

Observable Properties

Gases are easily compressed, but expand spontaneously to fill whatever container they are in.

Liquids retain their volume when placed into a container, but conform to the shape of the container. They are fluid and are able to flow. Liquids are nearly incompressible.

Solids keep their shape and volume *and* are virtually incompressible. They often have crystalline shapes.

Gas

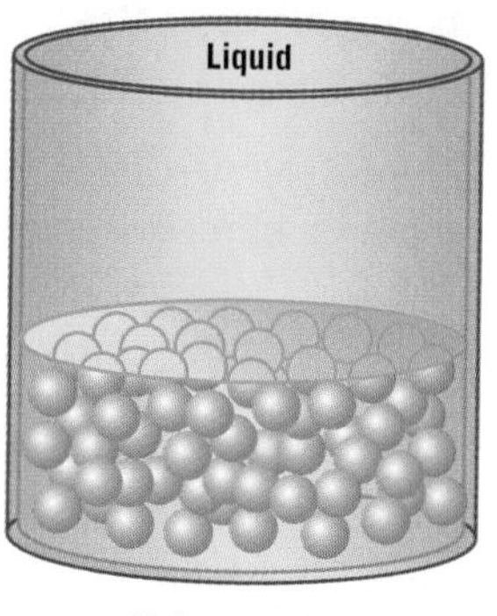

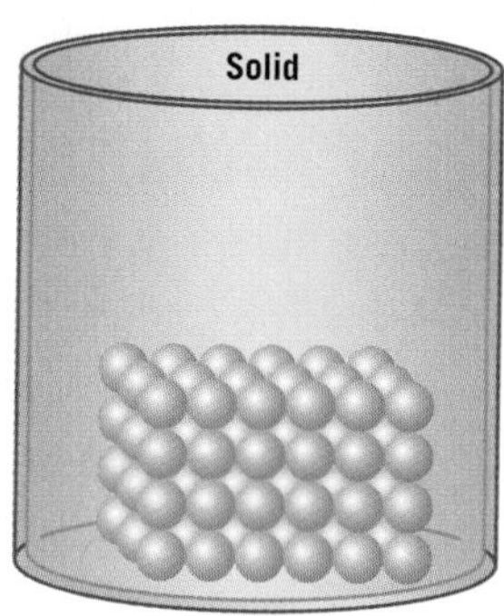

Molecular View

Widely spaced molecules with much empty space between them. Motion is random and with very weak attractions between the molecules.

Molecules tightly packed but with little order. They are able to move past each other with little difficulty. Intermolecular attractive forces are relatively strong.

Molecules tightly packed and highly ordered. Very strong attractions between molecules hold them in place so they are effectively locked in position.

FIG. 11.1 General properties of gases, liquids, and solids. Properties can be understood in terms of how tightly the molecules are packed together and the strengths of the intermolecular attractions between them.

the H and Cl atoms are held very tightly to each other by a covalent bond, and it is the strength of this bond that affects the *chemical properties* of HCl. The strength of the chemical bond also keeps the molecule intact as it moves about. When a particular chlorine atom moves, the hydrogen atom bonded to it is forced to follow along (see Figure 11.2). Attractions between neighboring HCl molecules, in contrast, are much weaker. In fact, they are only about 4% as strong as the covalent bond in HCl. These weaker attractions are what determine the *physical properties* of liquid and solid HCl.

There are several kinds of intermolecular attractions, which are discussed in this section. They all have something in common; *they arise from attractions between opposite electrical charges.* Collectively, they are called **van der Waals forces,** after J. D. van der Waals who studied the nonideal behavior of real gases.

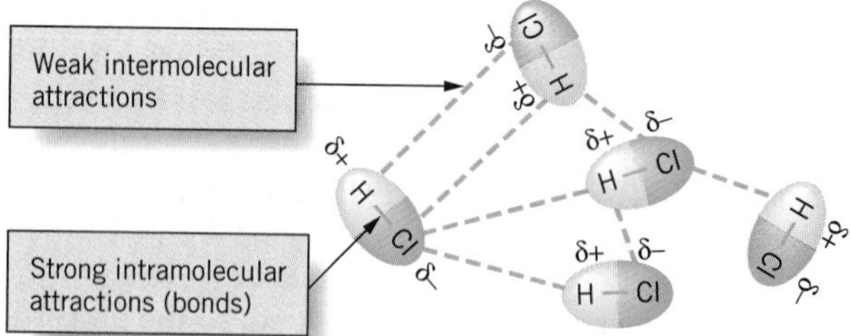

FIG. 11.2 Attractions within and between hydrogen chloride molecules. Strong *intramolecular* attractions (chemical bonds) exist between H and Cl atoms within HCl molecules. These attractions control the chemical properties of HCl. Weaker *intermolecular* attractions exist between neighboring HCl molecules. The intermolecular attractions control the physical properties of this substance.

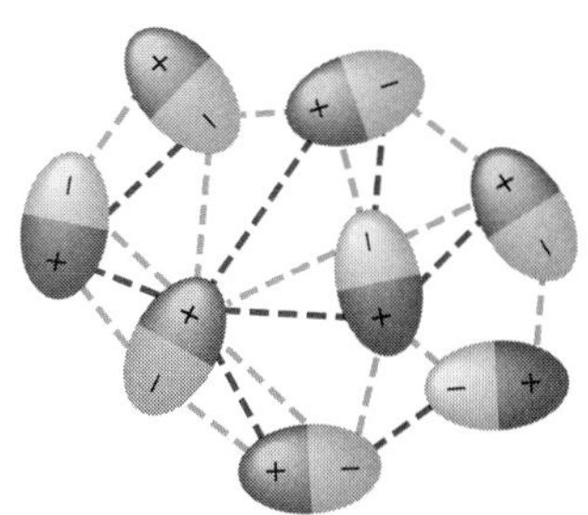
Attractions (- -) are greater than repulsions (- -), so the molecules feel a net attraction to each other.

FIG. 11.3 **Dipole–dipole attractions.** Attractions between polar molecules occur because the molecules tend to align themselves so that opposite charges are near each other and like charges are as far apart as possible. The alignment is not perfect because the molecules are constantly moving and colliding.

Dipole–dipole attractions

Polar molecules, such as HCl, have a partial positive charge at one end and a partial negative charge at the other. Because unlike charges attract, polar molecules tend to line up so the positive end of one dipole is near the negative end of another. Thermal energy (molecular kinetic energy), however, causes the molecules to collide and become disoriented, so the alignment isn't perfect. Nevertheless, there is still a net attraction between them (see Figure 11.3). We call this kind of intermolecular force a **dipole–dipole attraction.** Because collisions lead to substantial misalignment of the dipoles and because the attractions are only between partial charges, dipole–dipole forces are much weaker than covalent bonds, being only about 1–4% as strong. Dipole–dipole attractions fall off rapidly with distance, with the energy required to separate a pair of dipoles being proportional to $1/d^3$, where d is the distance between the dipoles.

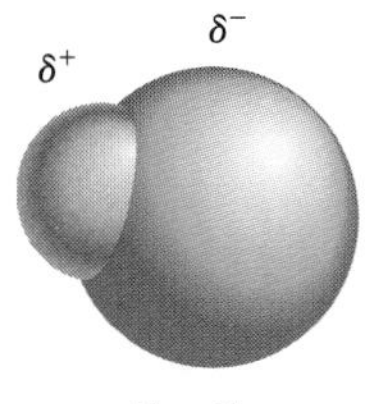
δ^+
δ^-
H—Cl

Hydrogen bonds

When hydrogen is covalently bonded to a very small, highly electronegative atom (principally, fluorine, oxygen, or nitrogen), a particularly strong type of dipole–dipole attraction occurs that's called **hydrogen bonding.** Hydrogen bonds are exceptionally strong because F—H, O—H, and N—H bonds are very polar, and because the partial charges can get quite close because they are concentrated on very small atoms. Typically, a hydrogen bond is about five to ten times stronger than other dipole–dipole attractions.

Hydrogen bonds in water and biological systems

Most substances become more dense when they change from a liquid to a solid. Not so with water. In liquid water, the molecules experience hydrogen bonds that continually break and re-form as the molecules move around (Figure 11.4*b*). As water begins to freeze, however, the molecules become locked in place, and each water molecule participates in four hydrogen bonds (Figure 11.4*c*). The resulting structure occupies a larger volume than the same amount of liquid water, so ice is less dense than the liquid. Because of this, ice cubes and icebergs float in the more dense liquid. The expansion of freezing water is capable of cracking a car's engine block, which is one reason we add antifreeze to a car's cooling system. Ice formation is also responsible for erosion, causing rocks to split where water has seeped into cracks. And in northern cities, freezing water breaks up pavement, creating potholes in the streets.

A hydrogen bond is not a covalent bond. In water there are oxygen–hydrogen covalent bonds within H_2O molecules and hydrogen bonds between H_2O molecules.

Hydrogen bonding is especially important in biological systems because many molecules in our bodies contain N—H and O—H bonds. Examples are proteins and DNA. Proteins are made up mostly (in some cases, entirely) of long chains of amino acids, linked head to tail to form polypeptides.

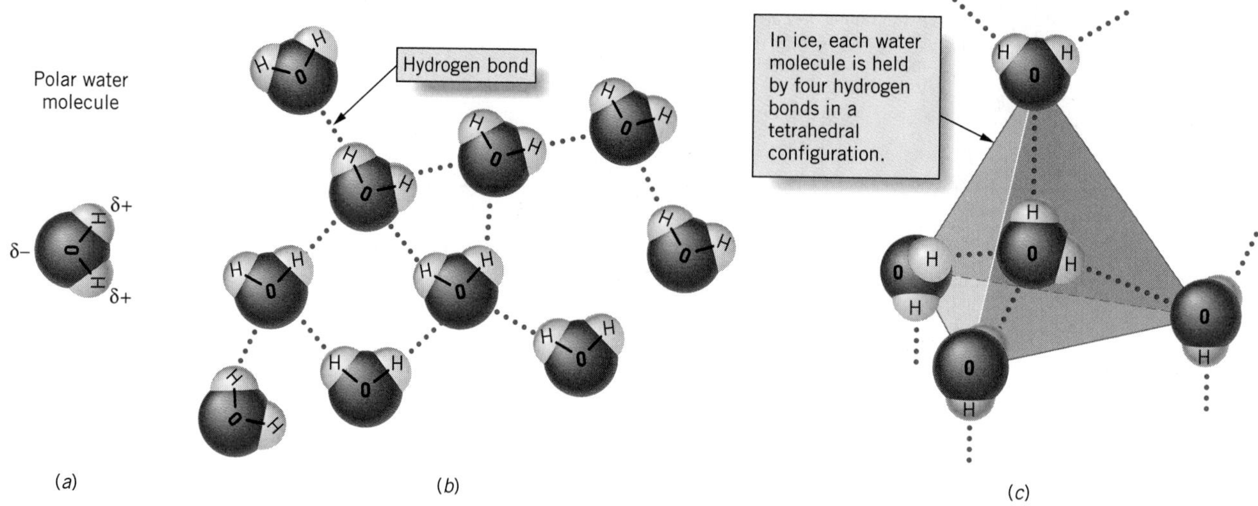
Polar water molecule
δ+
δ−
δ+
Hydrogen bond
In ice, each water molecule is held by four hydrogen bonds in a tetrahedral configuration.
(a)
(b)
(c)

FIG. 11.4 Hydrogen bonding in water. (*a*) The polar water molecule. (*b*) Hydrogen bonding (dotted lines) produces strong attractions between water molecules in the liquid. (*c*) Hydrogen bonding between water molecules in ice, where each water molecule is held by four hydrogen bonds in a tetrahedral configuration.

Polypeptides are examples of *polymers*, which are large molecules made by linking together many smaller units called *monomers* (in this case, amino acids). An amino acid contains both a carboxyl group and an amine group, NH_2. An example is glycine, NH_2CH_2COOH.

Part of a polypeptide chain is shown below.

```
      H      O ¦ H       O ¦ H      O  H
      |      ‖ ¦ |       ‖ ¦ |      ‖  |
---N—CHC—N—CHC—N—CHC—N—CHC---
         |   ¦       |   ¦      |           |
         H   ¦      CH3  ¦      H        CH2C6H5
```

One amino acid segment of a polypeptide

Hydrogen bonding between N—H units in one part of the chain and polar C=O groups in another part help determine the shape of the protein, which greatly influences its biological function. Hydrogen bonding is also responsible for the double helix structure of DNA, which carries our genetic information. This structure is illustrated in Figure 11.5.

London forces

Even nonpolar substances experience intermolecular attractions, as evidenced by the ability of the noble gases and nonpolar molecules such as Cl_2 and CH_4 to condense to liquids, and then crystallize into solids, when cooled to very low temperatures. In such liquids or solids attractions between their particles must exist to cause them to cling together.

In 1930 Fritz London, a German physicist, explained how the particles in even nonpolar substances can experience intermolecular attractions. He noted that in any atom or molecule the electrons are constantly moving. If we could examine such motions in two neighboring particles, we would find that the movement of electrons in one influences the movement of electrons in the other. Because electrons repel each other, as an electron of one particle gets near the other particle, electrons on the second particle are pushed away. This happens continually as the electrons move around, so to some extent, the electron density in both particles flickers back and forth in a synchronous fashion. This is illustrated in Figure 11.6, which depicts a series of instantaneous views of the electron density. Notice that *at any given moment the electron density of a particle can be unsymmetrical,* with more negative charge on one side than on the other. For that particular instant, the particle is a dipole, and we call it a momentary dipole or **instantaneous dipole.**

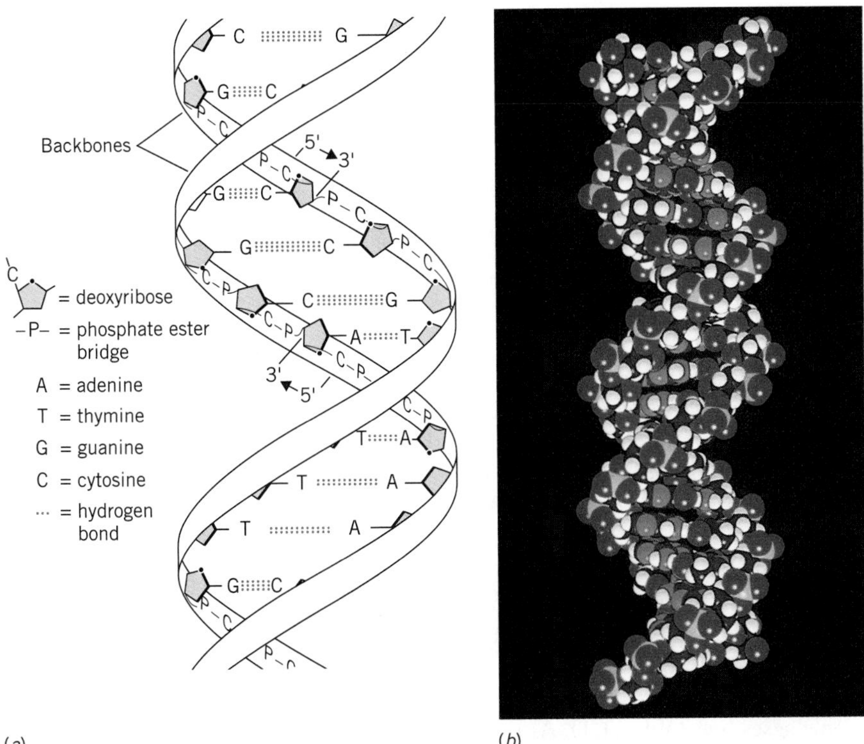

FIG. 11.5 Hydrogen bonding holds the DNA double helix together. (*a*) A schematic drawing in which the hydrogen bonds between the two strands are indicated by dotted lines. The legend to the left of the structure describes the various components of the DNA molecule. (*b*) Phosphorous atoms are shown in blue and they help us trace the DNA "backbone" in this short section of the double helix. *(Nelson Max/Peter Arnold, Inc.)*

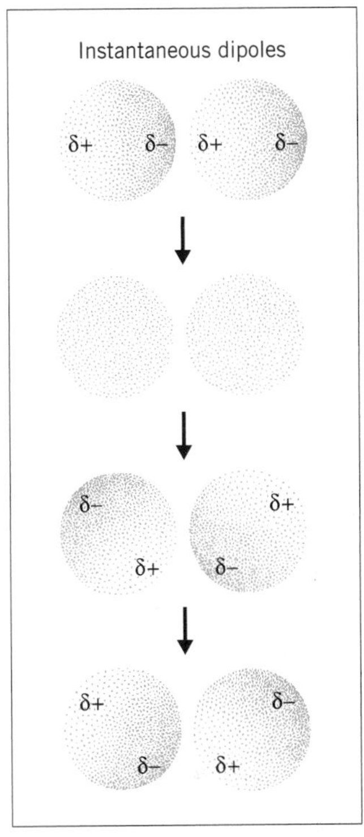

FIG. 11.6 Instantaneous "frozen" views of the electron density in two neighboring particles. Attractions occur between the instantaneous dipoles while they exist.

As an instantaneous dipole forms in one particle, it causes the electron density in its neighbor to become unsymmetrical, too. As a result, this second particle also becomes a dipole. We call it an **induced dipole** because it is caused by, or *induced* by, the formation of the first dipole. Because of the way the dipoles are formed, they always have the positive end of one near the negative end of the other, so there is a dipole–dipole attraction between them. It is a very short-lived attraction, however, because the electrons keep moving; the dipoles vanish as quickly as they form. But, in another moment, the dipoles will reappear in a different orientation and there will be another brief dipole–dipole attraction. In this way the short-lived dipoles cause momentary tugs between the particles. When averaged over a period of time, there is a net, overall attraction. It tends to be relatively weak, however, because the attractive forces are only "turned on" part of the time.

The momentary dipole–dipole attractions that we've just discussed are called *instantaneous dipole–induced dipole attractions*, to distinguish them from the kind of permanent dipole–dipole attractions that exist without interruption in polar substances like HCl. They are also called **London dispersion forces** (or simply **London forces** or **dispersion forces**).

London forces exist between all molecules and ions. Although they are the only kind of attraction possible between nonpolar molecules, London forces also contribute significantly to the total intermolecular attraction between polar molecules, where they are present in addition to the regular dipole–dipole attractions. London forces even occur between oppositely charged ions, but their effects are relatively weak compared to ionic attractions. London forces contribute little to the net overall attractions between ions and are often ignored.

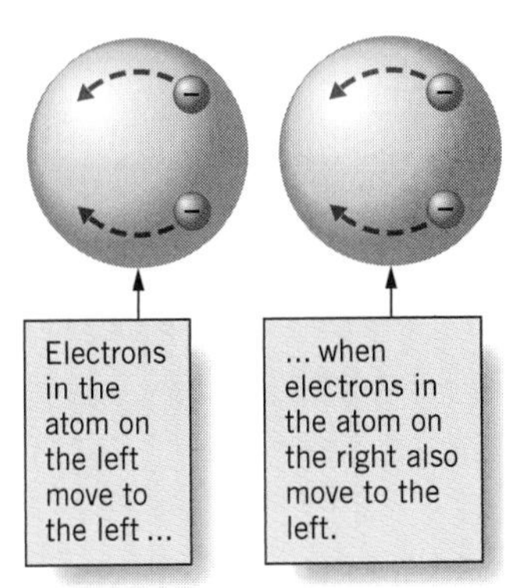

The strengths of London forces

To compare the strengths of intermolecular attractions, a property we can use is boiling point. As we will explain in more detail later in this chapter, the higher the boiling point, the stronger are the attractions between molecules in the liquid.

TABLE 11.1 Boiling Points of the Halogens and Noble Gases

Group VIIA	Boiling Point (°C)	Group VIIIA	Boiling Point (°C)
F_2	−188.1	He	−268.6
Cl_2	−34.6	Ne	−245.9
Br_2	58.8	Ar	−185.7
I_2	184.4	Kr	−152.3
		Xe	−107.1
		Rn	−61.8

London forces decrease very rapidly as the distance between particles increases. The energy required to separate particles held by London forces varies as $1/d^6$, where d is the distance between the particles.

The strengths of London forces are found to depend chiefly on three factors. One is the **polarizability** of the electron cloud of a particle, which is a measure of the ease with which the electron cloud is distorted, and thus is a measure of the ease with which the instantaneous and induced dipoles can form. In general, *as the volume of the electron cloud increases, its polarizability also increases.* When an electron cloud is large, the outer electrons are generally not held very tightly by the nucleus (or nuclei, if the particle is a molecule). This causes the electron cloud to be "mushy" and rather easily deformed, so instantaneous dipoles and induced dipoles form without much difficulty (see Figure 11.7). As a result, particles with large electron clouds experience stronger London forces than do similar particles with small electron clouds.

The effects of size can be seen if we compare the boiling points of the halogens and the noble gases (see Table 11.1). As the atoms become larger, the boiling points increase, reflecting increasingly stronger intermolecular attractions (stronger London forces).

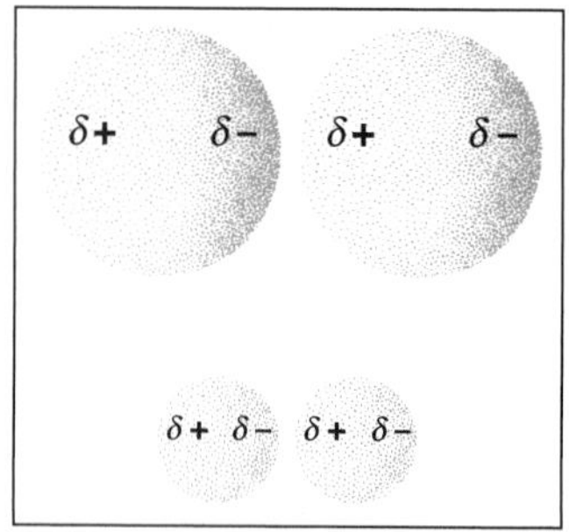

FIG. 11.7 Effect of molecular size on the strengths of London dispersion forces. A large electron cloud is more easily deformed than a small one, so in a large molecule the charges on opposite ends of an instantaneous dipole are larger than in a small molecule. Large molecules therefore experience stronger London forces than small molecules.

A second factor that affects the strengths of London forces is the number of atoms in a molecule. For molecules containing the same elements, London forces increase with the number of atoms, as illustrated by the hydrocarbons (see Table 11.2). As the number of atoms increases, there are more places along their lengths where instantaneous dipoles can develop and lead to London attractions (Figure 11.8). Even if the strength of attraction at each location is about the same, the *total* attraction experienced between the longer molecules is greater.[1]

The third factor that affects the strengths of London forces is molecular shape. Even with molecules that have the same number of the same kinds of atoms, those that have compact shapes experience weaker London forces than long, chainlike molecules (Figure 11.9). Presumably, because of the compact shape of the $(CH_3)_4C$ molecule, the individual hydrogens on neighboring molecules cannot interact with each other as effectively as those on the chainlike molecule.

TABLE 11.2 Boiling Points of Some Hydrocarbons[a]

Molecular Formula	Boiling Point at 1 atm (°C)
CH_4	−161.5
C_2H_6	−88.6
C_3H_8	−42.1
C_4H_{10}	−0.5
C_5H_{12}	36.1
C_6H_{14}	68.7
⋮	⋮
$C_{10}H_{22}$	174.1
⋮	⋮
$C_{22}H_{46}$	327

[a] The molecules of each hydrocarbon in this table have carbon chains of the type C—C—C—C—etc.; that is, one carbon follows another.

[1] The effect of large numbers of atoms on the total strengths of London forces can be compared to the bond between loop and hook layers of the familiar product Velcro. Each loop-to-hook attachment is not very strong, but when large numbers of them are involved, the overall bond between Velcro layers is quite strong.

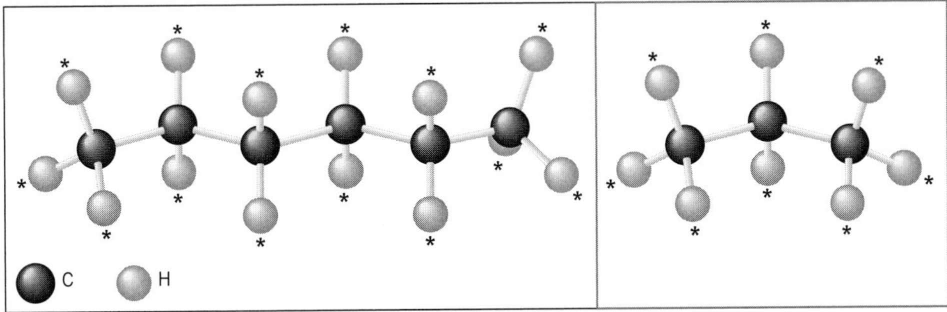

FIG. 11.8 **The number of atoms in a molecule affects London forces.** The C_6H_{14} molecule, left, has more sites (indicated by asterisks, *) along its chain where it can be attracted to other molecules nearby than does the shorter C_3H_8 molecule, right. As a result, the boiling point of C_6H_{14} (hexane, 68.7 °C) is higher than that of C_3H_8 (propane, −42.1 °C).

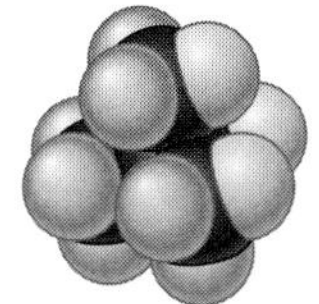

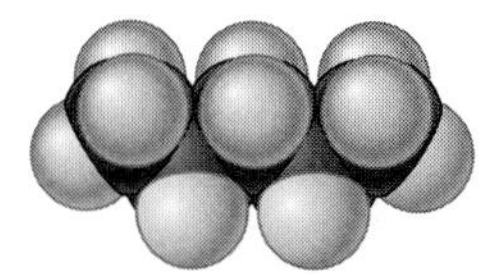

FIG. 11.9 **Molecular shape affects the strengths of London forces.** Shown are two molecules with the formula C_5H_{12}. Not all hydrogen atoms can be seen in these space-filling models. The neopentane molecule, $(CH_3)_4C$, has a more compact shape than the *n*-pentane molecule, $CH_3CH_2CH_2CH_2CH_3$. In the more compact structure, the H atoms cannot interact with those on neighboring molecules as well as the H atoms in the long chainlike structure, so overall the intermolecular attractions are weaker between the more compact molecules

Ion–dipole and ion–induced dipole forces of attraction

In addition to the attractions that exist between neutral molecules, which we discussed above, there are also forces that arise when ions interact with molecules. For example, ions are able to attract the charged ends of polar molecules to give **ion–dipole attractions.** This occurs in water, for example, when ionic compounds dissolve to give hydrated ions. Cations become surrounded by water molecules that are oriented with the negative ends of their dipoles pointing toward the cation. Similarly, anions attract the positive ends of water dipoles. This is illustrated in Figure 11.10. These same interactions can persist into the solid state as well. For example, aluminum chloride crystallizes from water as a hydrate with the formula $AlCl_3 \cdot 6H_2O$. In it the Al^{3+} ion is surrounded by water molecules at the vertices of an octahedron, as illustrated in Figure 11.11. They are held there by ion–dipole attractions.

Ions are also capable of distorting nearby electron clouds, thereby creating dipoles in neighboring particles (like molecules of a solvent, or even other ions). This leads to **ion–induced dipole** attractions, which can be quite strong because the charge on the ion doesn't flicker on and off like the instantaneous charges responsible for ordinary London dispersion forces.

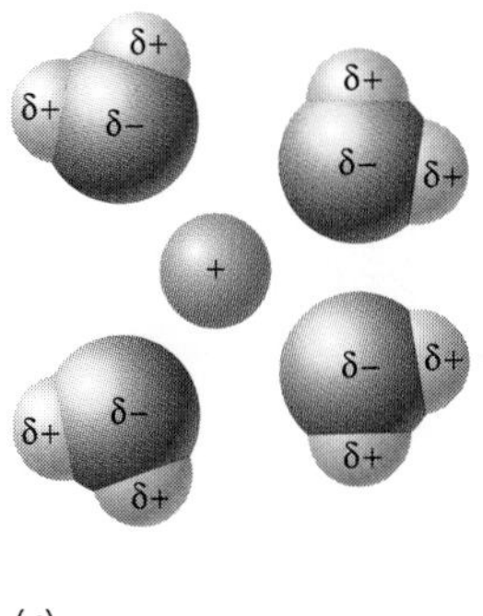

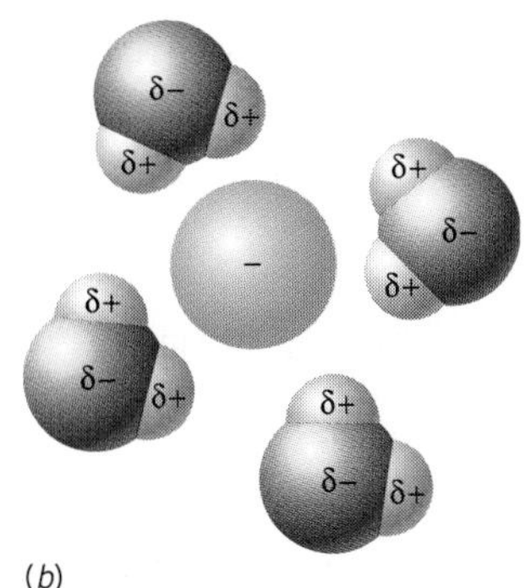

FIG. 11.10 **Ion–dipole attractions.** Here we see the attractions between water molecules and positive and negative ions. (*a*) The negative ends of water dipoles surround a cation and are attracted to the ion. (*b*) The positive ends of water molecules surround an anion, which gives a net attraction.

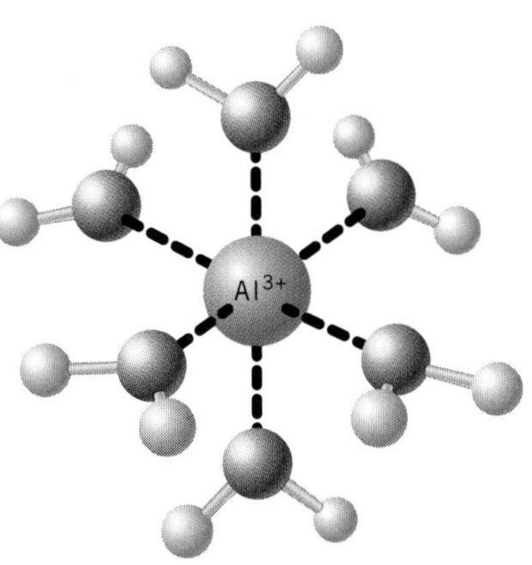

FIG. 11.11 **Ion–dipole attractions hold water molecules in a hydrate.** Water molecules are arranged at the vertices of an octahedron around an aluminum ion in $AlCl_3 \cdot 6H_2O$.

TOOLS
Intermolecular attractions: Hydrogen bonding, dipole–dipole forces, and London forces

TABLE 11.3 Summary of Intermolecular Attractions

Intermolecular Attraction	Types of Substances That Exhibit Attraction	Strength Relative to a Covalent Bond
Dipole–dipole attractions	Occurs between molecules that have permanent dipoles (i.e., polar molecules)	1–5%
Hydrogen bonding	Occurs when molecules contain N—H and O—H bonds	5–10%
London dispersion forces	All atoms, molecules, and ions experience these kinds of attractions. They are present in all substances.	Depends on sizes and shapes of molecules. For large molecules, the cumulative effect of many weak attractions can lead to a large net attraction.
Ion–dipole attractions	Occurs when ions interact with polar molecules	~10%; depends on ion charge and polarity of molecule
Ion–induced dipole attractions	Occurs when an ion creates a dipole in a neighboring particle, which may be a molecule or another ion	Variable, depends on the charge on the ion and the polarizability of its neighbor

Estimating the effects of intermolecular forces

In this section we have described a number of different types of intermolecular attractive forces and the kinds of substances in which they occur (see the summary in Table 11.3). With this knowledge, you should now be able to make some estimate of the nature and relative strengths of intermolecular attractions if you know the molecular structure of a substance. This will enable you to understand and sometimes predict how the physical properties of different substances compare. For example, we've already mentioned that boiling point is a property that depends on the strengths of intermolecular attractions. By being able to compare intermolecular forces in different substances, we can sometimes predict how their boiling points compare. This is illustrated in Example 11.1.

EXAMPLE 11.1
Using Relative Attractive Forces to Predict Properties

Below are structural formulas of ethanol (ethyl alcohol) and propylene glycol (a compound used as a nontoxic antifreeze). Which of these compounds would be expected to have the higher boiling point?

```
    H  H              H  H   H
    |  |              |  |   |
 H—C—C—OH         H—C—C—C—OH
    |  |              |  |   |
    H  H              H  OH  H
   ethanol          propylene glycol
```

ANALYSIS: We know that boiling points are related to the strengths of intermolecular attractions—the stronger the attractions, the higher the boiling point. Therefore, if we can determine which compound has the stronger intermolecular attractions, we can answer the question. Let's decide which kinds of attractions are present and then try to determine their relative strengths.

SOLUTION: We know that both substances will experience London forces, because they are present between *all* molecules. London forces become stronger as molecules become larger, so the London forces should be stronger in propylene glycol.

Looking at the structures, we see that both contain —OH groups (one in ethanol and two in propylene glycol). This means we can expect that there will be hydrogen bonding in both liquids. Because there are more —OH groups per molecule in propylene glycol than in ethanol, we might reasonably expect that there are more opportunities for the ethylene glycol molecules to participate in hydrogen bonding. This would make the hydrogen bonding forces greater in propylene glycol.

Our analysis tells us that both kinds of attractions are stronger in propylene glycol than in ethanol, so propylene glycol should have the higher boiling point.

IS THE ANSWER REASONABLE? There's not much we can do to check our answer other than to review the reasoning, which is sound. (We could also check a reference book, where we would find that the boiling point of ethanol is 78.5 °C and the boiling point of propylene glycol is 188.2 °C!)

Practice Exercise 1: List the following in order of their boiling points from lowest to highest. (a) $Ca(OH)_2$, $CH_3CH_2CH_2CH_2CH_3$, CH_3CH_2OH; (b) $CH_3CH_2NH_2$, $CH_3{-}O{-}CH_3$, $HOCH_2CH_2CH_2CH_2OH$. (Hint: Determine what types of intermolecular attractive forces are important for each molecule.)

Practice Exercise 2: Propylamine and trimethylamine have the same molecular formula, C_3H_9N, but quite different structures, as shown below. Which of these substances is expected to have the higher boiling point? Why?

$$CH_3{-}CH_2{-}CH_2{-}NH_2$$

propylamine

$$\begin{array}{c} CH_3 \\ | \\ H_3C{-}N{-}CH_3 \end{array}$$

trimethylamine

11.3 INTERMOLECULAR FORCES AND TIGHTNESS OF PACKING AFFECT THE PHYSICAL PROPERTIES OF LIQUIDS AND SOLIDS

Earlier we briefly described some properties of liquids and solids. We continue here with a more in-depth discussion, and we'll start by examining two properties that depend mostly on how tightly packed the molecules are, namely, *compressibility* and *diffusion*. Other properties depend much more on the strengths of intermolecular attractive forces, properties such as *retention of volume or shape, surface tension*, the ability of a liquid to *wet* a surface, the *viscosity* of a liquid, and a solid's or liquid's *tendency to evaporate.*

Compressibility and diffusion depend primarily on tightness of packing

Compressibility

The **compressibility** of a substance is a measure of its ability to be forced into a smaller volume. Gases are highly compressible because the molecules are far apart. However, in a liquid or solid most of the space is taken up by the molecules, and there is very little empty space into which to crowd other molecules. As a result, it is very difficult to compress liquids or solids to a smaller volume by applying pressure, so we say that these states of matter are nearly **incompressible.** This is a useful property. When you "step on the brakes" of a car, for example, you rely on the incompressibility of the brake fluid to transmit the pressure you apply with your foot to the brake shoes on the wheels. The incompressibility of liquids is also the foundation of the engineering science of *hydraulics*, which uses fluids to transmit forces that lift or move heavy objects.

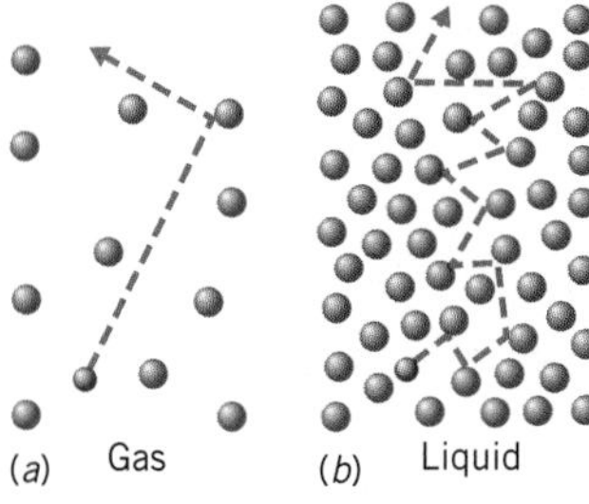

FIG. 11.12 Diffusion in a gas and a liquid viewed at the molecular level. (*a*) Diffusion in a gas is rapid because relatively few collisions occur between widely spaced molecules. (*b*) Diffusion in a liquid is slow because of many collisions between closely spaced particles.

Diffusion

Diffusion occurs much more rapidly in gases than in liquids, and hardly at all in solids. In gases, molecules diffuse rapidly because they travel relatively long distances between collisions, as illustrated in Figure 11.12. In liquids, however, a given molecule suffers many collisions as it moves about, so it takes longer to move from place to place and diffusion is much slower. Diffusion in solids is almost nonexistent at room temperature because the particles of a solid are held tightly in place. At high temperatures, though, the particles of a solid sometimes have enough kinetic energy to jiggle their way past each other, and diffusion can occur slowly. Such high temperature solid-state diffusion is used to make electronic devices, like transistors.

Most physical properties depend primarily on the strengths of intermolecular attractions

Retention of volume and shape

In gases, intermolecular attractions are too weak to prevent the molecules from moving apart to fill an entire vessel, so a gas will conform to the shape and volume of its container. In liquids and solids, however, the attractions are much stronger and are able to hold the particles closely together. As a result, liquids and solids keep the same volume regardless of the size of their container. In a solid, the attractions are even stronger than in a liquid. They hold the particles more or less rigidly in place, so a solid retains its shape when moved from one container to another.

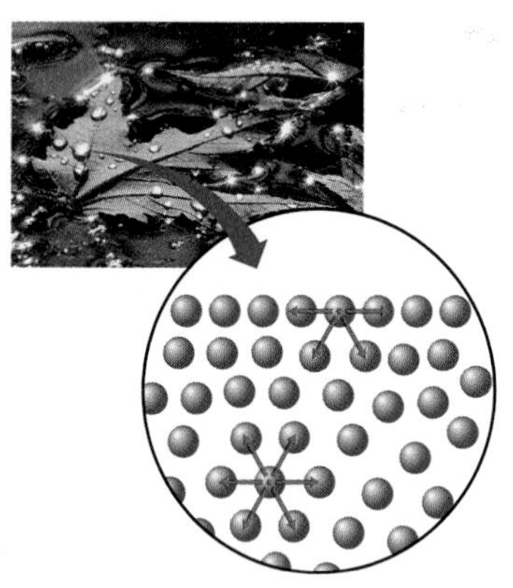

FIG. 11.13 Surface tension and intermolecular attractions. In water, as in other liquids, molecules at the surface are surrounded by fewer molecules than those below the surface. As a result, surface molecules experience fewer attractions than molecules within the liquid. *(Pat O'Hara/Stone/Getty Images.)*

Surface tension

A property that is especially evident for liquids is *surface tension*, which is related to the tendency of a liquid to seek a shape that yields the minimum surface area. For a given volume, the shape with the minimum surface area is a sphere—it's a principle of solid geometry. This is why raindrops tend to be little spheres.

To understand surface tension, we need to examine why molecules would prefer to be within a liquid rather than at its surface. In Figure 11.13 we see that a molecule *within* the liquid is surrounded by densely packed molecules on all sides, whereas one at the *surface* has neighbors beside and below it, but none above. As a result, a surface molecule is attracted to fewer neighbors than one within the liquid. With this in mind, let's imagine how we might change an interior molecule to one at the surface. To accomplish this, we would have to pull away some of the surrounding molecules. Because there are intermolecular attractions, removing neighbors requires work; so there's an increase in potential energy involved. This leads to the conclusion that *a molecule at the surface has a higher potential energy than a molecule in the bulk of the liquid.*

In general, a system becomes more stable when its potential energy decreases. For a liquid, reducing its surface area (and thereby reducing the number of molecules at the surface) lowers its potential energy. The lowest energy is achieved when the liquid has the smallest surface area possible (namely, a spherical shape). In more accurate terms, then, *the* **surface tension** *of a liquid is proportional to the energy needed to expand its surface area.*

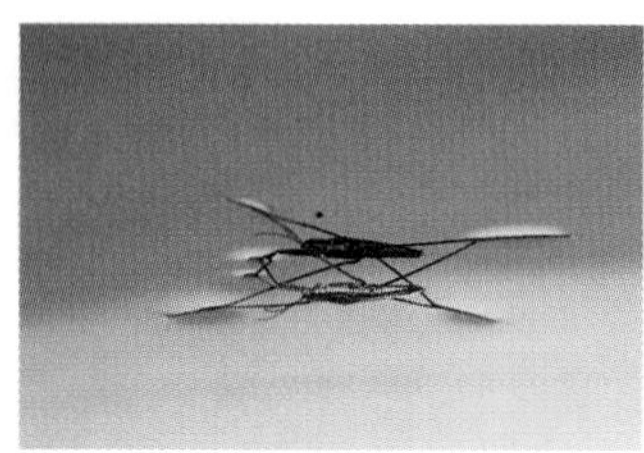

An insect called a *water strider*, shown here, is able to walk on water because of the liquid's surface tension, which causes the water to behave as though it has a skin that resists piercing by the insect's legs. *(Hermann Eisenbeiss/Photo Researchers.)*

The tendency of a liquid to spontaneously acquire a minimum surface area explains many common observations. For example, surface tension causes the sharp edges of glass tubing to become rounded when the glass is softened in a flame, an operation called "fire polishing." Surface tension is also what allows us to fill a water glass above the rim, giving the surface a rounded appearance (Figure 11.14). The surface behaves as if it has a thin, invisible "skin" that lets the water in the glass pile up, trying to assume a spherical shape. Gravity, of course, works in opposition, tending to pull the water down. If too much water is added to the glass, the gravitational force finally wins and the skin breaks; the water overflows. If you push on the surface of a liquid, it resists expansion and pushes back, so the surface "skin" appears to resist penetration. This is what enables certain insects to "walk on water," as illustrated in the photo in the margin.

Surface tension is a property that varies with the strengths of intermolecular attractions. Liquids with strong intermolecular attractive forces have large differences in potential energy between their interior and surface molecules, and have large surface tensions.

Not surprisingly, water's surface tension is among the highest known (comparisons being at the same temperature); its intermolecular forces are hydrogen bonds, the strongest kind of dipole–dipole attraction. In fact, the surface tension of water is roughly three times that of gasoline, which consists of relatively nonpolar hydrocarbon molecules able to experience only London forces.

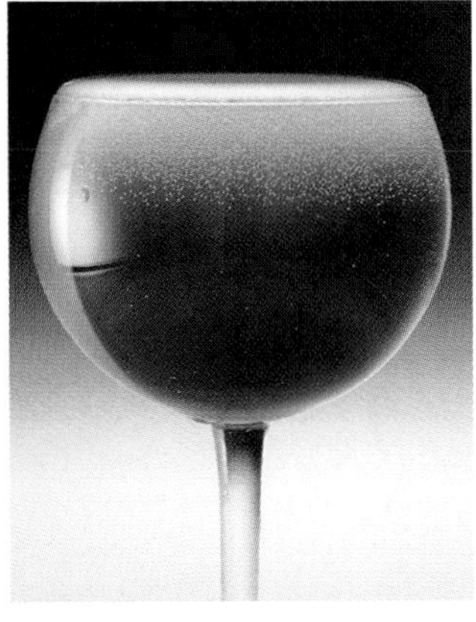

FIG. 11.14 Surface tension in a liquid. Surface tension allows a glass to be filled with water above the rim. *(Michael Watson.)*

Wetting of a surface by a liquid

A property we associate with liquids, especially water, is their ability to wet things. **Wetting** is the spreading of a liquid across a surface to form a thin film. Water wets clean glass, such as the windshield of a car, by forming a thin film over the surface of the glass (see Figure 11.15*a*). Water won't wet a greasy windshield, however. Instead, on greasy glass water forms tiny beads (see Figure 11.15*b*).

For wetting to occur, the intermolecular attractions between the liquid and the surface must be of about the same strength as the attractions within the liquid itself. Such a rough equality exists when water touches clean glass. This is because the glass surface contains lots of oxygen atoms to which water molecules can form hydrogen bonds. As a result, part of the energy needed to expand the water's surface area when wetting occurs is recovered by the formation of hydrogen bonds to the glass surface.

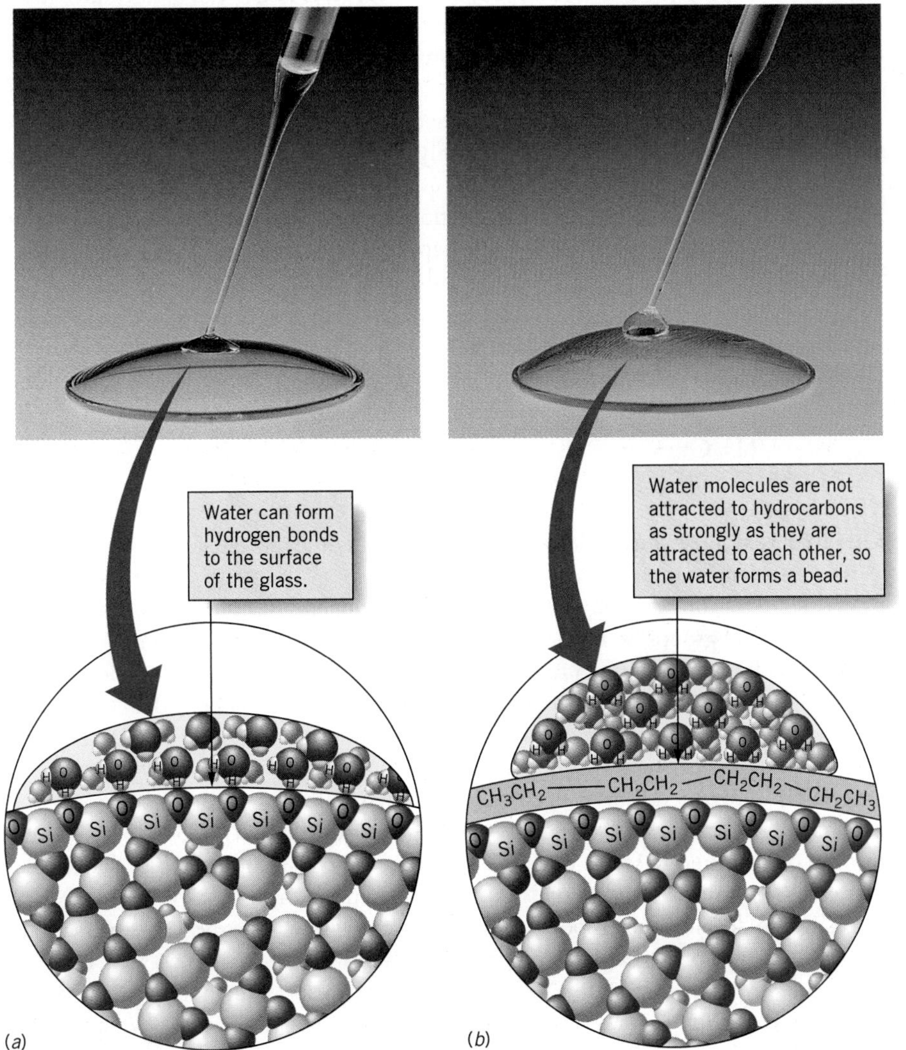

Water can form hydrogen bonds to the surface of the glass.
Water molecules are not attracted to hydrocarbons as strongly as they are attracted to each other, so the water forms a bead.
CH_3CH_2—CH_2CH_2—CH_2CH_2—CH_2CH_3
(a)
(b)

FIG. 11.15 Intermolecular attractions affect the ability of water to wet a surface. (*a*) Water wets a clean glass surface because the surface contains many oxygen atoms to which water molecules can form hydrogen bonds. (*b*) If the surface has a layer of grease, to which water molecules are only weakly attracted, the water doesn't wet it. The water resists spreading and forms a bead instead. *((a), (b) Michael Watson.)*

Glass is characterized by a vast network of silicon–oxygen bonds.

When the glass is coated by a film of oil or grease, the surface exposed to the water drop becomes oil and grease and is now composed of relatively nonpolar molecules (Figure 11.15*b*). These attract other molecules (including water) largely by London forces, which are weak compared with hydrogen bonds. Therefore, the attractions *within* liquid water are much stronger than the attractions *between* water molecules and the greasy surface. The weak water-to-grease London forces can't overcome the hydrogen bonding within liquid water, so the water doesn't spread out; it forms beads instead.

One of the reasons why detergents are used for such chores as doing laundry or washing floors is that detergents contain chemicals called **surfactants** that drastically lower the surface tension of water. This makes the water "wetter," which allows the detergent solution to spread more easily across the surface to be cleaned.

When a liquid has a low surface tension, like gasoline, we know that it has weak intermolecular attractions, and such a liquid easily wets solid surfaces. The weak attractions between molecules in gasoline, for example, are readily replaced by attractions to almost any surface, so gasoline easily spreads to a thin film. If you've ever spilled a little gasoline, you have experienced firsthand that it doesn't bead.

Viscosity

As everybody knows, syrup flows less readily or is more resistant to flow than water (both at the same temperature). Flowing is a change in the *form* of the liquid, and such resistance to a change in form is called the liquid's **viscosity.** We say that syrup is more *viscous* than water. The concept of viscosity is not confined to liquids, however, although it is with liquids that the property is most commonly associated. Solid things, even rock, also yield to forces acting to change their shapes, but normally do so only gradually and imperceptibly. Gases also have viscosity, but they respond almost instantly to form-changing forces.

Viscosity has been called the "internal friction" of a material. It is influenced both by intermolecular attractions and by molecular shape and size. For molecules of similar size, we find that as the strengths of the intermolecular attractions increase, so does the viscosity. For example, consider acetone (nail polish remover) and ethylene glycol (automotive antifreeze), each of which contains 10 atoms.

$$\underset{\text{acetone}}{H_3C{-}\overset{\overset{\displaystyle O}{\|}}{C}{-}CH_3} \qquad \underset{\text{ethylene glycol}}{HO{-}CH_2{-}CH_2{-}OH}$$

> If you've ever spilled a little acetone, you know that it flows very easily. In contrast, ethylene glycol has an "oily" thickness to it and flows more slowly.

Ethylene glycol is more viscous than acetone, and looking at the molecular structures, it's easy to see why. Acetone contains a polar carbonyl group ($>C{=}O$), so it experiences dipole–dipole attractions as well as London forces. Ethylene glycol, on the other hand, contains two $-OH$ groups, so in addition to London forces, ethylene glycol molecules also participate in hydrogen bonding (a much stronger interaction than ordinary dipole–dipole forces). Strong hydrogen bonding in ethylene glycol makes it more viscous than acetone.

Molecular size and the ability of molecules to tangle with each other is another major factor in determining viscosity. The long, floppy, entangling molecules in heavy machine oil (almost entirely a mixture of long chain, nonpolar hydrocarbons), plus the London forces in the material, give it a viscosity roughly 600 times that of water at 15 °C. Vegetable oils, like the olive oil or corn oil used to prepare salad dressings, consist of molecules that are also large but generally nonpolar. Olive oil is roughly 100 times more viscous than water.

> As the temperature drops, molecules move more slowly and intermolecular forces become more effective at restraining flow.

Viscosity also depends on temperature; as the temperature drops, the viscosity increases. When water, for example, is cooled from its boiling point to room temperature, its viscosity increases by over a factor of three. The increase in viscosity with cooling is why operators of vehicles use a "light," thin (meaning less viscous) motor oil during subzero weather.

Evaporation and sublimation are affected by intermolecular attractions

One of the most important physical properties of liquids and solids is their tendency to undergo a change of state from liquid to gas or from solid to gas. For liquids, the change is called **evaporation.** For solids, which can also change directly to the gaseous state by evaporation without going through the liquid state, we use a special term, **sublimation.** Solid carbon dioxide is commonly called *dry ice* because it doesn't melt. Instead, at atmospheric pressure it *sublimes,* changing directly to gaseous CO_2. Naphthalene, the ingredient in some brands of moth flakes, is another substance that can sublime and seemingly disappear.

To understand evaporation and sublimation, we have to examine the motions of molecules. In a solid or liquid, molecules are not stationary; they bounce around, colliding with their neighbors. At a given temperature, there is *exactly the same* distribution of kinetic energies in a liquid or a solid as there is in a gas, which means that Figure 6.4 on page 212 applies to liquids and solids, too. This figure tells us that at a given temperature a small fraction of the molecules have very large kinetic energies and therefore very high velocities. When one of these high velocity molecules is at the surface and is moving outward fast enough, it can escape the attractions of its neighbors and enter the vapor state. We say the molecule has left by evaporation (or by sublimation, if the substance is a solid).

Naphthalene sublimes when heated and the vapor condenses directly to a solid when it encounters a cool surface. Here, beautiful flaky naphthalene crystals have been formed on the bottom of a flask containing ice water. *(Michael Watson.)*

Evaporation produces a cooling effect

One of the things we notice about the evaporation of a liquid is that it produces a cooling effect. You've experienced this if you've stepped out of a shower and been chilled by the air. The evaporation of water from your body produced this effect. In fact, our bodies use the evaporation of perspiration to maintain a constant body temperature.

We can see why liquids become cool during evaporation by examining Figure 11.16, which illustrates the kinetic energy distribution in a liquid at a particular temperature. A marker along the horizontal axis shows the minimum kinetic energy needed by a molecule to escape the attractions of its neighbors. Only molecules with kinetic energies equal to or greater than the minimum can leave the liquid. Others may begin to leave, but before they can escape they slow to a stop and then fall back. Notice in Figure 11.16 that the minimum kinetic energy needed to escape is much larger than the average, which means that when molecules evaporate they carry with them large amounts of kinetic energy. As a result, the average kinetic energy of the molecules left behind decreases. (You might think of this as being similar to removing people taller than 6 ft from a large class of students. When this is done, the average height of those who are left is less.) Because the Kelvin temperature of the remaining liquid is directly proportional to the now lower average kinetic energy, the temperature is lower; in other words, evaporation causes the liquid that remains to be cooler.

The rate of evaporation depends on surface area, temperature, and strengths of intermolecular attractions

Later in this chapter we are going to be concerned about the *rate of evaporation* of a liquid. There are several factors that control this. You are probably already aware of one of them—the surface area of the liquid. Because evaporation occurs from the liquid's surface and not from within, it makes sense that as the surface area is increased, more molecules are able to escape

To understand how temperature and intermolecular forces affect the rate of evaporation, we must compare evaporation rates from the same size surface area. In this discussion, therefore, "rate of evaporation" means "rate of evaporation *per unit surface area.*"

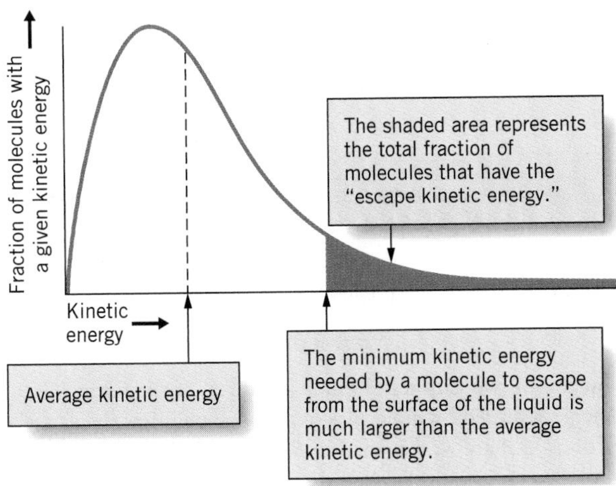
Fraction of molecules with a given kinetic energy
Kinetic energy
The shaded area represents the total fraction of molecules that have the "escape kinetic energy."
Average kinetic energy
The minimum kinetic energy needed by a molecule to escape from the surface of the liquid is much larger than the average kinetic energy.

FIG. 11.16 Cooling of a liquid by evaporation. Molecules that are able to escape from the liquid have kinetic energies larger than the average. When they leave, the average kinetic energy of the molecules left behind is less, so the temperature is lower.

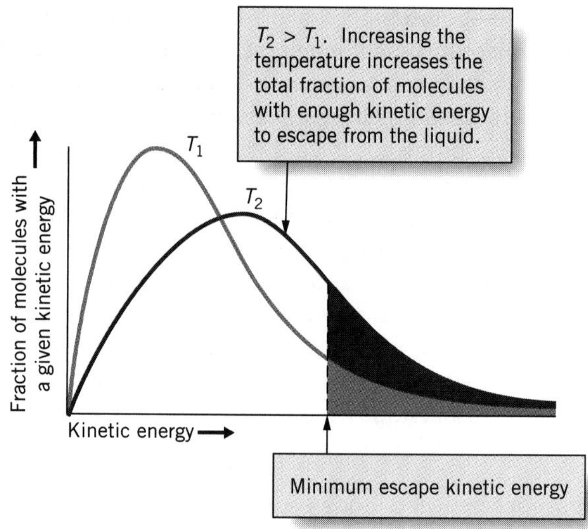

FIG. 11.17 Effect of increasing the temperature on the rate of evaporation of a liquid. At the higher temperature, the total fraction of molecules with enough kinetic energy to escape is larger, so the rate of evaporation is larger.

and the liquid evaporates more quickly. For liquids having the same surface area, the rate of evaporation depends on two factors, namely, temperature and the strengths of intermolecular attractions. Let's examine each of them separately.

TOOLS
Effect of temperature on rate of evaporation

The influence of temperature on evaporation rate is no surprise; you already know that hot water evaporates faster than cold water. The reason can be seen by studying Figure 11.17, which shows kinetic energy distributions for the *same* liquid at two temperatures. Notice two important features of the figure. First, the same minimum kinetic energy is needed for the escape of molecules at both temperatures. This minimum is determined by the kinds of attractive forces between the molecules, and is independent of temperature. Second, the shaded area of the curve represents the *total* fraction of molecules having kinetic energies equal to or greater than the minimum. At the higher temperature, the total fraction is larger, which means that at the higher temperature a greater total fraction has the ability to evaporate. As you might expect, when more molecules have the needed energy, more evaporate in a unit of time. Therefore, *the rate of evaporation per unit surface area of a given liquid is greater at a higher temperature.*

The effect of intermolecular attractions on evaporation rate can be seen by studying Figure 11.18. Here we have kinetic energy distributions for two *different* liquids—call them

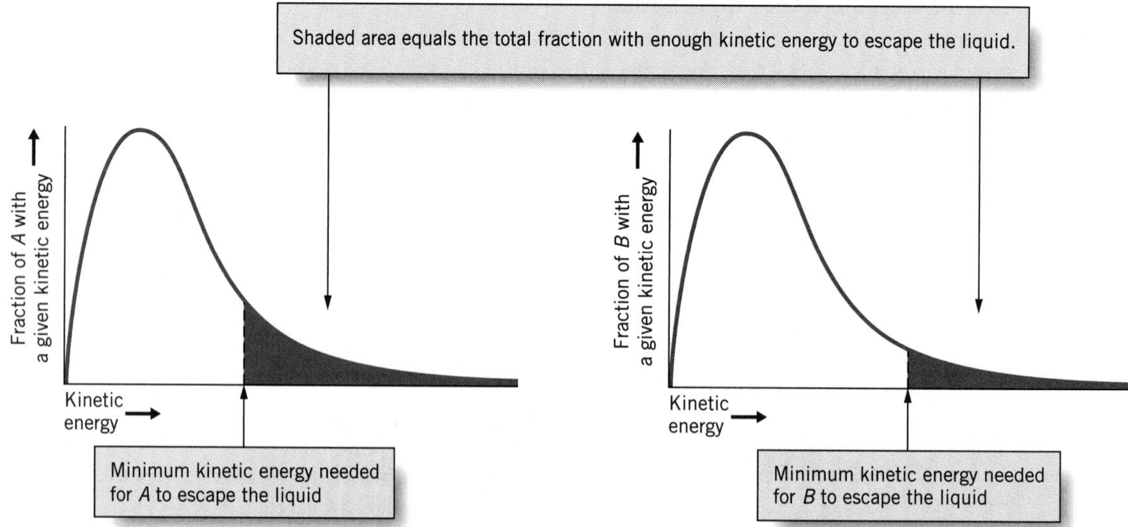

FIG. 11.18 Kinetic energy distribution in two different liquids, *A* and *B*, at the same temperature. The minimum kinetic energy required by molecules of *A* to escape is less than that for *B* because the intermolecular attractions in *A* are weaker than in *B*. This causes *A* to evaporate faster than *B*.

A and *B*—both at the same temperature. In liquid *A*, the attractive forces are weak; they might be of the London type, for example. As we see, the minimum kinetic energy needed by *A* molecules to escape is not very large because they are not attracted very strongly to each other. In liquid *B*, the intermolecular attractive forces are much stronger; they might be hydrogen bonds, for instance. Molecules of *B*, therefore, are held more tightly to each other at the liquid's surface and must have a higher kinetic energy to evaporate. As you can see from the figure, the total fraction of molecules with enough energy to evaporate is greater for *A* than for *B*, which means that *A* evaporates faster than *B*. In general, then, *the weaker the intermolecular attractive forces, the faster is the rate of evaporation at a given temperature.* You are probably also aware of this phenomenon. At room temperature, for example, nail polish remover [acetone, $(CH_3)_2CO$], whose molecules experience weak dipole–dipole and London forces of attraction, evaporates faster than water, whose molecules feel the effects of much stronger hydrogen bonds.

TOOLS
Intermolecular forces and rate of evaporation

11.4 CHANGES OF STATE LEAD TO DYNAMIC EQUILIBRIA

A **change of state** occurs when a substance is transformed from one physical state to another. Evaporation of a liquid and sublimation of a solid are two examples. Others are the melting of a solid such as ice and the freezing of a liquid such as water.

In chemistry (unless otherwise indicated), when we use the term *equilibrium*, we always mean *dynamic equilibrium.*

One of the important features about changes of state is that, at any particular temperature, they always tend toward a condition of *dynamic equilibrium*. We introduced the concept of dynamic equilibrium on page 142 with an example of a system at chemical equilibrium. The same general principles apply to a physical equilibrium, such as that between a liquid and its vapor. Let's see how such an equilibrium is established.

When a liquid is placed in an empty container, it immediately begins to evaporate and molecules of the substance begin to collect in the space above the liquid (see Figure 11.19*a*). As they fly around in the vapor, the molecules collide with each other, with the walls of the container, and with the surface of the liquid itself. Those that strike the liquid's surface tend to stick because their kinetic energies become scattered among the surface molecules. This change, which involves vapor molecules changing to the liquid state, is called **condensation.**

Initially, when the liquid is first introduced into the container, the rate of evaporation is high, but the rate of condensation is very low because there are few molecules in the vapor state. As vapor molecules accumulate, the rate of condensation increases. This continues until the rate at which molecules are condensing becomes equal to the rate at which they are evaporating (Figure 11.19*b*). From that moment on, the number of molecules in the vapor will remain constant, because over a given period of time the

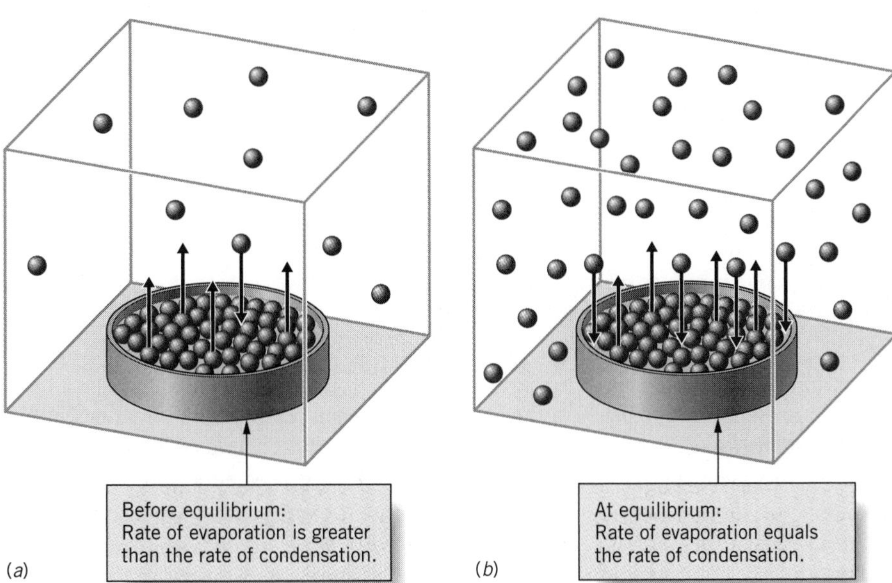
Before equilibrium:
Rate of evaporation is greater than the rate of condensation.
(*a*)
At equilibrium:
Rate of evaporation equals the rate of condensation.
(*b*)

FIG. 11.19 Evaporation of a liquid into a sealed container. (*a*) The liquid has just begun to evaporate into the container. The rate of evaporation is greater than the rate of condensation. (*b*) A dynamic equilibrium is reached when the rate of evaporation equals the rate of condensation. In a given time period, the number of molecules entering the vapor equals the number that leave, so there is no net change in the number of gaseous molecules.

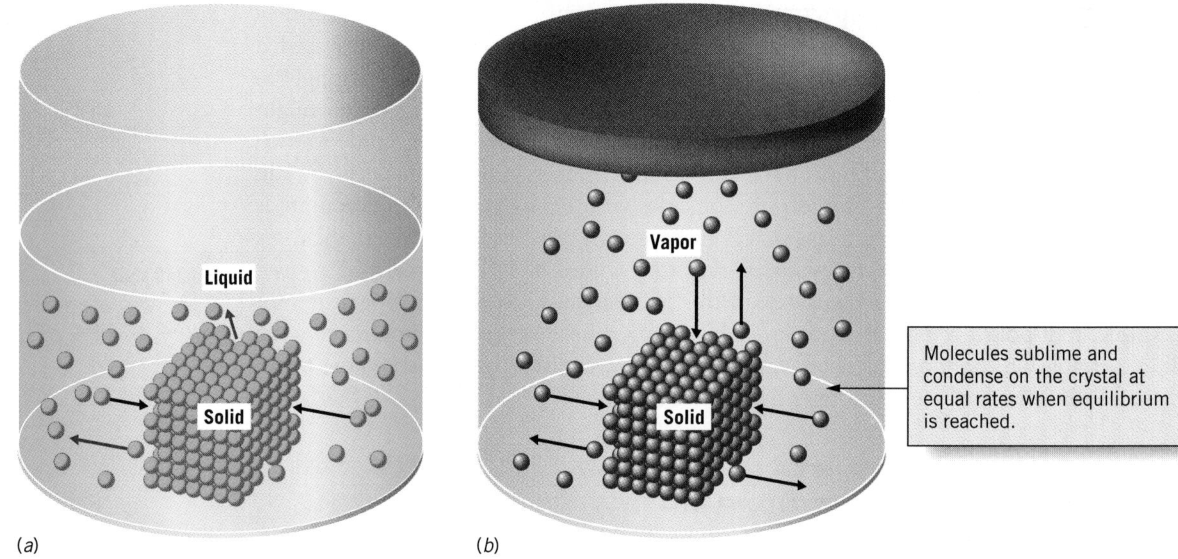

FIG. 11.20 **Solid–liquid and solid–vapor equilibria.** (*a*) As long as no heat is added or removed, melting (red arrows) and freezing (black arrows) occur at equal rates and the number of particles in the solid remains constant. (*b*) Equilibrium is established when molecules evaporate from the solid at the same rate as they condense from the vapor.

A vapor–liquid equilibrium is possible only in a closed container. When the container is open, vapor molecules drift away and the liquid might completely evaporate.

number that enters the vapor is the same as the number that leaves. At this point we have a condition of *dynamic equilibrium,* one in which two opposing effects, evaporation and condensation, are occurring at equal rates.

Similar equilibria are also reached in melting and sublimation. At a temperature called the **melting point,** a solid begins to change to a liquid as heat is added. At this temperature a dynamic equilibrium can exist between molecules in the solid and those in the liquid. Molecules leave the solid and enter the liquid at the same rate as molecules leave the liquid and join the solid (Figure 11.20*a*). As long as no heat is added or removed from such a solid–liquid equilibrium mixture, melting and freezing occur at equal rates. For sublimation, the situation is exactly the same as in the evaporation of a liquid into a sealed container (see Figure 11.20*b*). After a few moments, the rates of sublimation and condensation become the same and equilibrium is established.

11.5 Vapor Pressures of Liquids and Solids Are Controlled by Temperature and Intermolecular Attractions

A liquid with a high vapor pressure at a given temperature is said to be **volatile.**

When a liquid evaporates, the molecules that enter the vapor exert a pressure called the **vapor pressure.** From the very moment a liquid begins to evaporate into the vapor space above it, there is a vapor pressure. If the evaporation is taking place inside a sealed container, this pressure grows until finally equilibrium is reached. Once the rates of evaporation and condensation become equal, the concentration of molecules in the vapor remains constant and the vapor exerts a constant pressure. This final pressure is called the **equilibrium vapor pressure of the liquid.** In general, when we refer to the *vapor pressure,* we really mean the equilibrium vapor pressure.

Two factors determine the equilibrium vapor pressure

TOOLS
Factors that affect vapor pressure

Figure 11.21 shows plots of equilibrium vapor pressure versus temperature for a few liquids. From these graphs we see that both a liquid's temperature and its chemical composition are the major factors affecting its vapor pressure. Once we have selected a particular liquid, however, only the temperature matters. The reason is that the vapor pressure of a given liquid is a function solely of its rate of evaporation *per unit area of the liquid's surface.* When this rate is large, a large concentration of molecules in the vapor state is necessary to establish

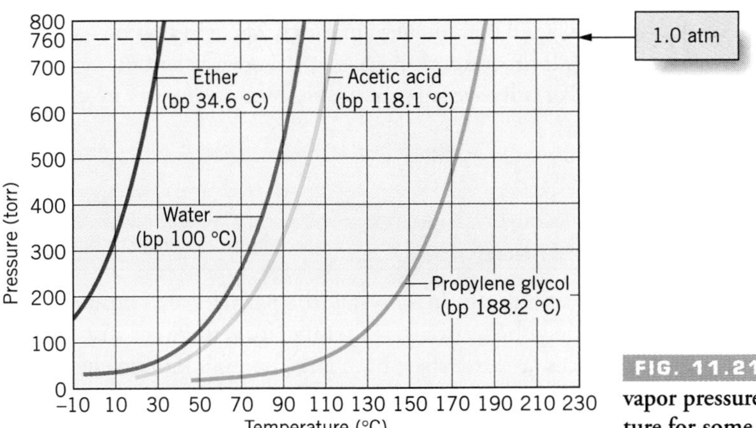

FIG. 11.21 Variation of vapor pressure with temperature for some common liquids.

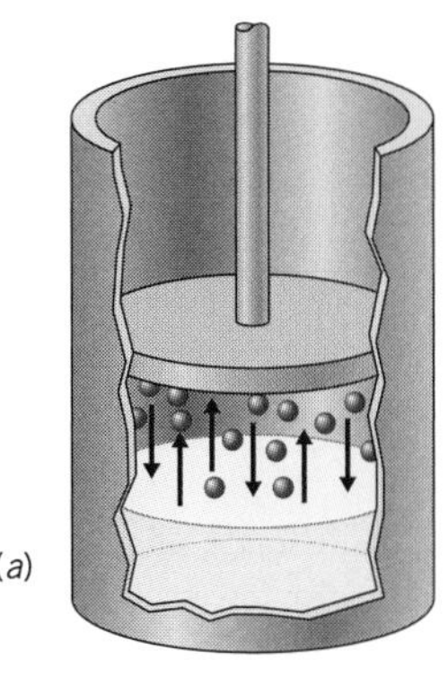

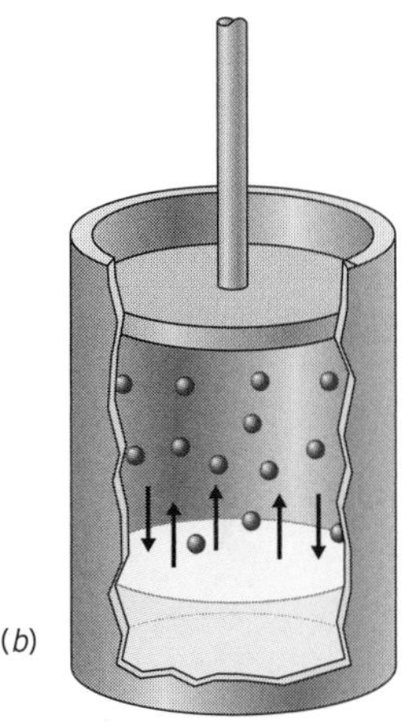

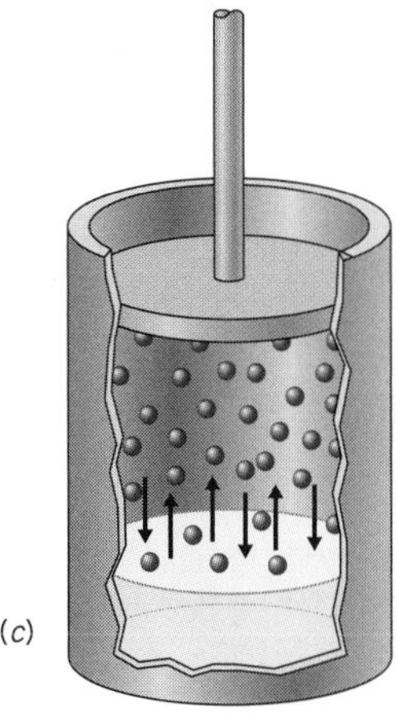

FIG. 11.22 Effect of a volume change on the vapor pressure of a liquid. (*a*) Equilibrium exists between liquid and vapor. (*b*) The volume is increased, which upsets the equilibrium and causes the pressure to drop. The rate of condensation is now less than the rate of evaporation, which hasn't changed. (*c*) After more liquid has evaporated, equilibrium is restored and the rates of condensation and evaporation are again equal and the vapor pressure has returned to its initial value.

equilibrium, which is another way of saying that the vapor pressure is relatively high when the evaporation rate is high. As the temperature of a given liquid increases, so does its rate of evaporation and so does its equilibrium vapor pressure.

As chemical composition changes in going from one liquid to another, the strengths of intermolecular attractions change. If the attractions increase, the rates of evaporation at a given temperature decrease, and the vapor pressures decrease. These data on relative vapor pressures tell us that, of the four liquids in Figure 11.21, intermolecular attractions are strongest in propylene glycol, next strongest in acetic acid, third strongest in water, and weakest in ether. Thus, *we can use vapor pressures as indications of relative strengths of the attractive forces in liquids.*

Some factors do not affect the vapor pressure

An important fact about vapor pressure is that *its magnitude doesn't depend on the* ***total*** *surface area of the liquid, nor on the volume of the liquid in the container, nor on the volume of the container itself, just as long as some liquid remains when equilibrium is reached.* The reason is because none of these factors affects the rate of evaporation *per unit surface area.*

Increasing the *total* surface area does increase the *total* rate of evaporation, but the larger area is also available for condensation; so the rate at which molecules return to the liquid also increases. The rates of both evaporation and condensation are thus affected equally, and no change occurs to the equilibrium vapor pressure.

Adding more liquid to the container can't affect the equilibrium either because evaporation occurs from the *surface*. Having more molecules in the bulk of the liquid does not change what is going on at the surface.

To understand why the vapor pressure doesn't depend on the *size* of the vapor space, consider a liquid in equilibrium with its vapor in a cylinder with a movable piston, as illustrated in Figure 11.22*a*. Withdrawing the piston (Figure 11.22*b*) increases the volume of the vapor space; as the vapor expands, the pressure it exerts becomes less, so there's a momentary drop in the pressure. The molecules of the vapor, being more spread out now, no longer strike the surface as frequently, so the rate of condensation also decreases. The rate of evaporation hasn't changed, however, so for a moment the system is not at equilibrium and the substance is evaporating faster than it is condensing (Figure 11.22*b*). This condition prevails, changing more liquid into vapor, until the concentration of molecules in the vapor has risen enough to make the condensation rate again equal to the evaporation rate (Figure 11.22*c*). At this point the vapor pressure has returned to its original value. Therefore, the net result of expanding the space above the liquid is to change more liquid into vapor, but it does not affect the equilibrium vapor pressure. Similarly, we expect that reducing the volume of the vapor space above the liquid will also not affect the equilibrium vapor pressure.

Practice Exercise 3: Considering Figure 11.22, in which direction should the piston be moved to decrease the number of molecules in the gas phase? (Hint: Consider what must happen to re-establish equilibrium after the piston is moved.)

Practice Exercise 4: Suppose a liquid is in equilibrium with its vapor in a piston–cylinder apparatus like that in Figure 11.22. If the piston is pushed in a short way and the system is allowed to return to equilibrium, what will have happened to the *total number* of molecules in both the liquid and the vapor?

Solids also have vapor pressures

In many solids, such as NaCl, the attractive forces are so strong that virtually no particles have enough kinetic energy to escape at room temperature, so essentially no evaporation occurs. Their vapor pressures at room temperature are virtually zero.

Solids have vapor pressures just as liquids do. In a crystal, the particles are constantly jiggling around, bumping into their neighbors. At a given temperature there is a distribution of kinetic energies, so some particles at the surface have large enough kinetic energies to break away from their neighbors and enter the vapor state. When particles in the vapor collide with the crystal, they can be recaptured, so condensation can occur too. Eventually, the concentration of particles in the vapor reaches a point where the rate of sublimation equals the rate of condensation, and a dynamic equilibrium is established. The pressure of the vapor that is in equilibrium with the solid is called the **equilibrium vapor pressure of the solid.** As with liquids, this equilibrium vapor pressure is usually referred to simply as the vapor pressure. Like that of a liquid, the vapor pressure of a solid is determined by the strengths of the attractive forces between the particles and by the temperature.

11.6 Boiling occurs when a liquid's vapor pressure equals atmospheric pressure

If you were asked to check whether a pot of water was boiling, what would you look for? The answer, of course, is *bubbles*. When a liquid boils, large bubbles usually form at many places on the inner surface of the container and rise to the top. If you were to place a thermometer into the boiling water, you would find that the temperature remains constant, regardless of how you adjust the flame under the pot. A hotter flame just makes the water bubble faster, but it doesn't raise the temperature. *Any pure liquid remains at a constant temperature while it is boiling,* a temperature that's called the liquid's *boiling point.*

On the top of Mt. Everest, the world's tallest peak, water boils at only 69 °C.

If you measure the boiling point of water in Philadelphia, New York, or any place else that is nearly at sea level, your thermometer will read 100 °C or very close to it. However, if you try this experiment in Denver, Colorado, you will find that water boils at about 95 °C. Denver, at a mile above sea level, has a lower atmospheric pressure, so we find that the boiling point depends on the atmospheric pressure.

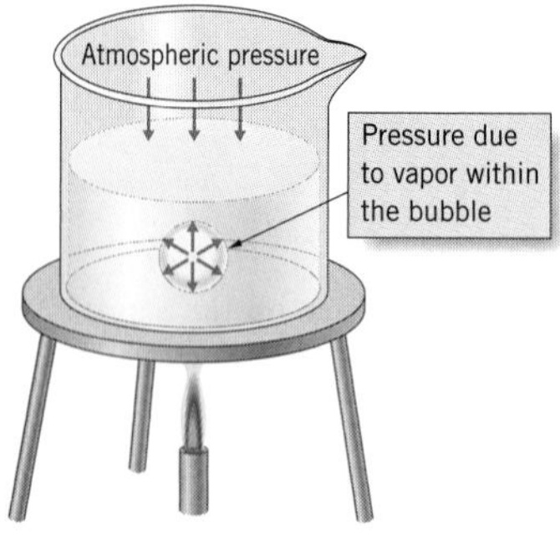

FIG. 11.23 A liquid at its boiling point. The pressure of the vapor within a bubble in a boiling liquid pushes the liquid aside against the opposing pressure of the atmosphere. Bubbles can't form unless the vapor pressure of the liquid is at least equal to the pressure of the atmosphere.

These observations raise some interesting questions. Why do liquids boil? And why does the boiling point depend on the pressure of the atmosphere? The answers become apparent when we realize that inside the bubbles of a boiling liquid is the *liquid's vapor,* not air. When water boils, the bubbles contain water vapor (steam); when alcohol boils, the bubbles contain alcohol vapor. As a bubble grows, liquid evaporates into it, and the pressure of the vapor pushes the liquid aside, making the size of the bubble increase (see Figure 11.23). Opposing the bubble's internal vapor pressure, however, is the pressure of the atmosphere pushing down on the top of the liquid, attempting to collapse the bubble. The only way the bubble can exist and grow is for the vapor pressure within it to equal (maybe just slightly exceed) the pressure exerted by the atmosphere. In other words, bubbles of vapor cannot even form until the temperature of the liquid rises to a point at which the liquid's vapor pressure equals the atmospheric pressure. Thus, in scientific terms, the **boiling point** is defined as *the temperature at which the vapor pressure of the liquid is equal to the prevailing atmospheric pressure.*

Now we can easily understand why water boils at a lower temperature in Denver than it does in New York City. Because the atmospheric pressure is lower in Denver, the water there doesn't have to be heated to as high a temperature to make its vapor pressure equal to the atmospheric pressure. The lower temperature of boiling water at places with high altitudes, like Denver, makes it necessary to cook foods longer. At the other extreme, a pressure cooker is a device that increases the pressure over the boiling water and thereby raises the boiling point. At the higher temperature, foods cook more quickly.

To make it possible to compare the boiling points of different liquids, chemists have chosen 1 atm as the reference pressure. The boiling point of a liquid at 1 atm is called its **normal boiling point.** (If a boiling point is reported without also mentioning the pressure at which it was measured, we assume it to be the normal boiling point.) Notice in Figure 11.21, page 449, that we can find the normal boiling points of ether, water, acetic acid, and propylene glycol by noting the temperatures at which their vapor pressure curves cross the 1 atm pressure line.

Boiling point is affected by intermolecular attractions

Earlier we mentioned that the boiling point is a property whose value depends on the strengths of the intermolecular attractions in a liquid. When the attractive forces are strong, the liquid has a low vapor pressure at a given temperature, so it must be heated to a high temperature to bring its vapor pressure up to atmospheric pressure. High boiling points therefore result from strong intermolecular attractions, so we often use normal boiling point data to assess relative intermolecular attractions among different liquids. (In fact, we did this in solving Example 11.1.)

The effects of intermolecular attractions on boiling point are easily seen by examining Figure 11.24, which gives the plots of the boiling points versus period numbers for some families of binary hydrogen compounds. Notice, first, the gradual increase in boiling point for the hydrogen compounds of the Group IVA elements (CH_4 through GeH_4). These compounds are composed of nonpolar tetrahedral molecules. The boiling points increase from CH_4 to GeH_4 simply because the molecules become larger and their electron clouds become more polarizable, which leads to an increase in the strengths of the London forces.

When we look at the hydrogen compounds of the other nonmetals, we find the same trend from Period 3 through Period 5. Thus, for three compounds of the Group VA series, PH_3, AsH_3, and SbH_3, there is a gradual increase in boiling point, corresponding again to the increasing strengths of London forces. Similar increases occur for the three Group VIA compounds (H_2S, H_2Se, and H_2Te) and for the three Group VIIA compounds (HCl, HBr, and HI). Significantly, however, the Period 2 members of each of these series (NH_3, H_2O, and HF) have much higher boiling points than might otherwise be expected. The reason is that each is involved in hydrogen bonding, which is a much stronger attraction than London forces.

TOOLS *Boiling ponts are related to intermolecular attractive forces*

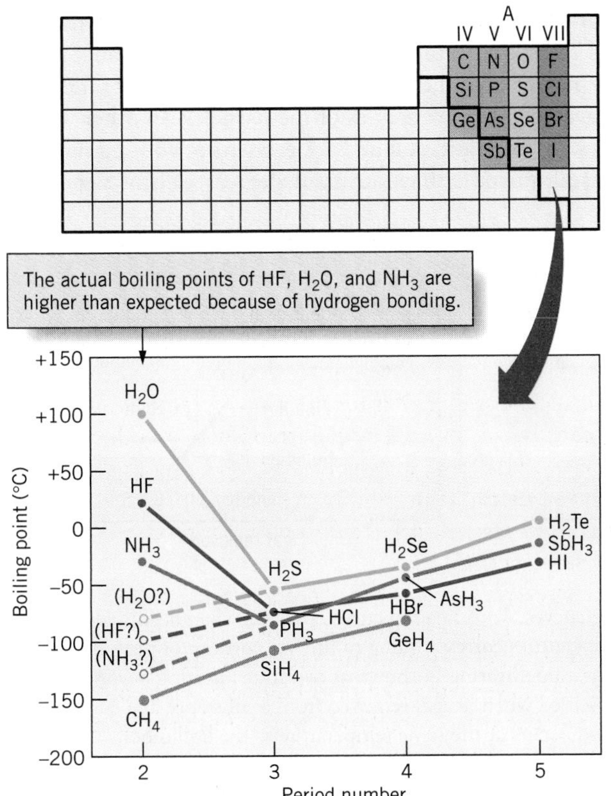

FIG. 11.24 Effects of intermolecular attractions on boiling point. Boiling points of the hydrogen compounds of elements of Groups IVA, VA, VIA, and VIIA of the periodic table. The dashed lines lead to hypothetical boiling points, if hydrogen bonding did not exist, for HF, H_2O, and NH_3.

One of the most interesting and far-reaching consequences of hydrogen bonding is that it causes water to be a liquid, rather than a gas, at temperatures near 25 °C. If it were not for hydrogen bonding, water would have a boiling point somewhere near −80 °C and could not exist as a liquid except at still lower temperatures. At such low temperatures it is unlikely that life as we know it could have developed.

Practice Exercise 5: The Dead Sea is approximately 1300 ft below sea level and its barometric pressure is approximately 830 torr. Will the boiling point of water be elevated from 100 °C by (a) less than 10 °C, (b) 10 °C to 25 °C, (c) 25 °C to 50 °C, or (d) above 50 °C? (Hint: Extrapolate from Figure 11.21.)

Practice Exercise 6: The atmospheric pressure at the top of Mt. McKinley in Alaska, 3.85 miles above sea level, is 330 torr. Use Figure 11.21 to estimate the boiling point of water at the top of the mountain.

11.7 Energy Changes Occur During Changes of State

When a liquid or solid evaporates or a solid melts, there are increases in the distances between the particles of the substance. Particles that normally attract each other are forced apart, increasing their potential energies. Such energy changes affect our daily lives in many ways, especially the energy changes associated with the changes in state of water, changes which even control the weather on our planet. To study these energy changes, let's begin by examining how the temperature of a substance varies as it is heated.

Heating curves and cooling curves reveal changes in kinetic and potential energy

When a solid or liquid is heated, the volume expands only slightly, so there are only small changes in the average distance between the particles. This means that very small changes in potential energy take place, so almost all the heat added goes to increasing the kinetic energy.

Figure 11.25*a* illustrates the way the temperature of a substance changes as we add heat to it *at a constant rate*, starting with the solid and finishing with the gaseous state of the substance. The graph is sometimes called a **heating curve** for the substance.

First, let's look at the portions of the graph that slope upward. These occur where we are increasing the temperature of the solid, liquid, and gas phases. Because temperature is related to average kinetic energy, nearly all of the heat we add in these regions of the heating curve goes to increasing the average kinetic energies of the particles. In other words, the added heat makes the particles go faster and collide with each other with more force. In addition, the slopes of the rising portions have units of degrees Celsius per joule.

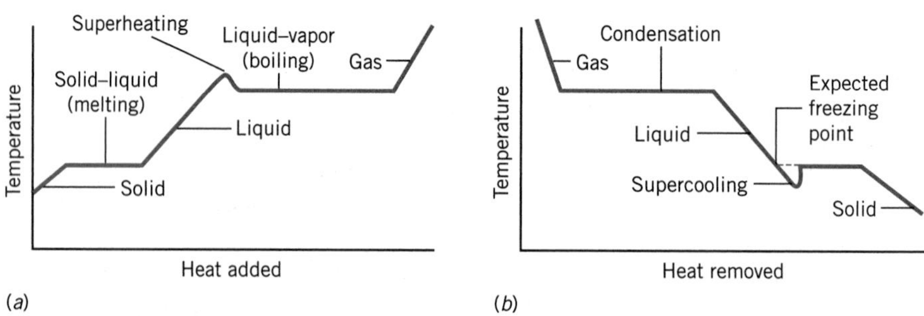

FIG. 11.25 Heating and cooling curves. (*a*) A heating curve observed when heat is added to a substance at a constant rate. The temperatures corresponding to the flat portions of the curve occur at the melting point and boiling point. Superheating in shown as continued heating beyond the boiling point. (*b*) A cooling curve observed when heat is removed from a substance at a constant rate. Condensation of vapor to a liquid occurs at the same temperature as the liquid boils. Supercooling is seen here as the temperature of the liquid dips below its freezing point (the same temperature as its melting point). Once a tiny crystal forms, the temperature rises to the freezing point.

This is the reciprocal of the heat capacity, meaning that the greater the slope, the lower the heat capacity. Gases have lower heat capacities than liquids and solids, therefore the heating of the gas phase has the largest slope.

In those portions of the heating curve where the temperature remains constant, the average kinetic energy of the particles is not changing. This means that all the heat being added must go to increase the *potential energies* of the particles. During melting, the particles held rigidly in the solid begin to separate slightly as they form the mobile liquid phase. The potential energy increase accompanying this process equals the amount of heat input during the melting process. During boiling, there is an even greater increase in the distance between the molecules. Here they go from the relatively tight packing in the liquid to the widely spaced distribution of molecules in the gas. This gives rise to an even larger increase in the potential energy, which we see as a longer flat region on the heating curve during the boiling of the liquid.

The opposite of a heating curve is a **cooling curve** (see Figure 11.25*b*). Here we start with a gas and gradually cool it—remove heat from it at a constant rate—until we have reached a solid.

Superheating and supercooling

Looking at Figure 11.25 again we notice two unusual features, one on each curve. There is a small "blip" on the heating curve near the transition from the liquid to a gas. A similar feature occurs when a liquid is cooled to a solid. These "blips" represent the phenomena of superheating and supercooling. Superheating occurs when the liquid is heated above the boiling point without boiling. If disturbed, a **superheated liquid** will erupt with a shower of vapor and liquid. Many people have discovered this effect when heating their favorite beverage in a microwave oven. When cooling a liquid it is possible to decrease the temperature below the freezing point without solidification occurring, creating a supercooled liquid. Once again if the supercooled liquid is disturbed, very rapid crystallization occurs. Some commercial products take advantage of supercooling to provide an instant heat source for minor injuries since the crystallization process often evolves large quantities of heat.

Supercooling of a vapor is also possible. Condensation of supercooled water vapor onto solid surfaces leads to dew in warm weather and frost in freezing weather.

Molar heats of fusion, vaporization, and sublimation

Because phase changes occur at constant temperature and pressure, the potential energy changes associated with melting and vaporization can be expressed as enthalpy changes. Usually, enthalpy changes are expressed on a "per mole" basis and are given special names to identify the kind of change involved. For example, using the word **fusion** instead of "melting," *the* **molar heat of fusion, ΔH_{fusion},** *is the heat absorbed by one mole of a solid when it melts to give a liquid at the same temperature and pressure.* Similarly, *the* **molar heat of vaporization, $\Delta H_{\text{vaporization}}$,** *is the heat absorbed when one mole of a liquid is changed to one mole of vapor at a constant temperature and pressure.* Finally, *the* **molar heat of sublimation, $\Delta H_{\text{sublimation}}$,** *is the heat absorbed by one mole of a solid when it sublimes to give one mole of vapor, once again at a constant temperature and pressure.* The values of ΔH for fusion, vaporization, and sublimation are all positive because the phase change in each case is endothermic, being accompanied by a net increase in potential energy.

These are also called *enthalpies* of fusion, vaporization, and sublimation.

Examples of the influence of these energy changes on our daily lives abound. For example, you've added ice to a drink to keep it cool because as the ice melts, it absorbs heat (its heat of fusion). Your body uses the heat of vaporization of water to cool itself through the evaporation of perspiration. During the summer, ice cream trucks carry dry ice because the sublimation of CO_2 absorbs its heat of sublimation and keeps the ice cream cold. And perhaps most importantly, weather on our planet is driven by the heat of vaporization of water which serves to convert solar energy into the energy of winds and storms. For example, over oceans large storms such as hurricanes rely on a continued supply of warm moist air produced by rapid evaporation of H_2O from tropical waters. Continual condensation in the high clouds forms rain and supplies the energy needed to feed the storm's winds.

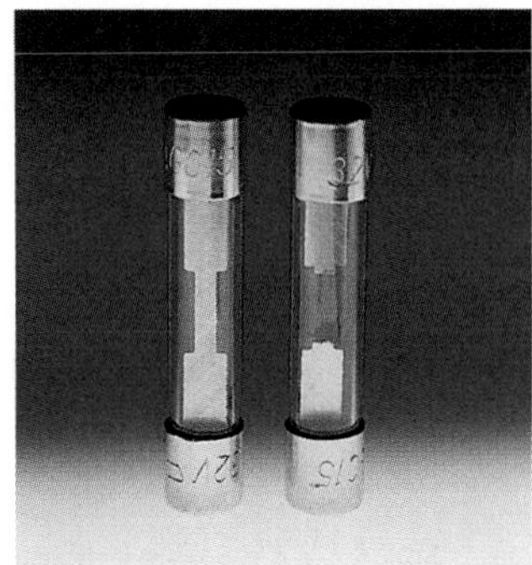

Fusion means melting. The thin metal band in an electrical fuse becomes hot as electricity passes through it. It protects a circuit by melting if too much current is drawn. On the right we see a fuse that has done its job. *(Michael Watson.)*

Solidification of a liquid to a crystal, condensation of a gas to a liquid, or deposition of a gas as a solid are simply the reverse of fusion, vaporization, and sublimation processes. Therefore the heat of crystallization is equal to the heat of fusion but it has the opposite algebraic sign. Similarly the heats of condensation and deposition have the opposite signs of their counterparts.

Since heat is released when liquids condense or when gases become solids or liquids, that heat can be put to practical use. For example, a supersaturated solution of ammonium nitrate can be caused to crystallize inside a plastic bag. The heat liberated is used to treat some sports injuries with this "heat pack." Similarly, meteorologists can often use the "dew point" of the atmosphere to predict overnight low temperatures. The dew point is the temperature at which moisture in the air begins to condense. If the nighttime temperature decreases to the dew point, the heat released by the condensation of water keeps the temperature from falling much further.

FACETS OF CHEMISTRY 11.1

Determining Heats of Vaporization

The way the vapor pressure varies with temperature, which was described in Section 11.5 and Figure 11.21, depends on the heat of vaporization of a substance. The relationship, however, is not a simple proportionality. Instead, it involves *natural logarithms*, which are logarithms to the base *e* as compared to the more familiar base-10 logarithms. With modern calculators the logarithm function can be applied with the press of the *ln* key for natural logarithms and the *log* key for base-10 logarithms. The reverse process of taking an "antilogarithm" uses an *inv* or *2nd* function key with the *ln* or *log* keys on most calculators.

Rudolf Clausius (1822–1888), a German physicist, and Benoit Clapeyron (1799–1864), a French engineer, used the principles of thermodynamics (a subject that's discussed in Chapter 18) to derive the following equation that relates the vapor pressure, heat of vaporization (ΔH_{vap}), and temperature

$$\ln P = \frac{-\Delta H_{vap}}{RT} + C \qquad (1)$$

The quantity ln P is the natural logarithm of the vapor pressure, R is the gas constant expressed in energy units ($R = 8.314$ J mol^{-1} K^{-1}), T is the absolute temperature, and C is a constant. Scientists call this the **Clausius–Clapeyron equation.**

The Clausius–Clapeyron equation provides a convenient graphical method for determining heats of vaporization from experimentally measured vapor pressure–temperature data. To see this, let's rewrite the equation as follows.

$$\ln P = \left(\frac{-\Delta H_{vap}}{R}\right)\frac{1}{T} + C$$

Recall from algebra that a straight line is represented by the general equation

$$y = mx + b$$

where x and y are variables, m is the slope, and b is the intercept of the line with the y axis. In this case, we can make the substitutions

$$y = \ln P \qquad x = \frac{1}{T} \qquad m = \frac{-\Delta H_{vap}}{R} \qquad b = C$$

Therefore, we have

$$\begin{array}{ccccc} \ln P & = & \left(\dfrac{-\Delta H_{vap}}{R}\right) & \dfrac{1}{T} & + \; C \\ \Updownarrow & & \Updownarrow & \Updownarrow & \Updownarrow \\ y & = & m & x & + \; b \end{array}$$

Thus, a graph of ln P versus $1/T$ should give a straight line that has a slope equal to $-\Delta H_{vap}/R$. Such straight line relationships are illustrated in Figure 1 in which experimental data are plotted for water, acetone, and ethanol. From the graphs in Figure 1, the calculated values of ΔH_{vap} are as follows: for water, 43.9 kJ mol^{-1}; for acetone, 32.0 kJ mol^{-1}; and for ethanol, 40.5 kJ mol^{-1}.

Using Equation 1 above, a "two point" form of the Clausius–Clapeyron equation can be derived that can be used to calculate ΔH_{vap} if the vapor pressure is known at two different temperatures. This equation is

$$\ln\frac{P_1}{P_2} = \frac{\Delta H_{vap}}{R}\left(\frac{1}{T_2} - \frac{1}{T_1}\right) \qquad (2)$$

If we know the value of the heat of vaporization, Equation 2 can also be used to calculate the vapor pressure at some particular temperature (say, P_2 at a temperature T_2) if we already know the vapor pressure P_1 at a temperature T_1.

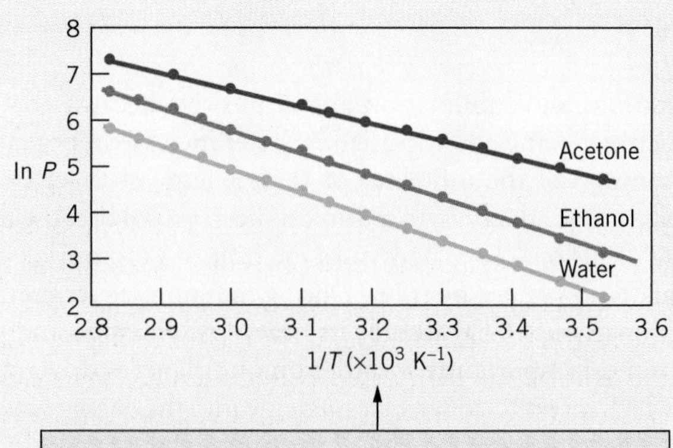

Figure 1 A graph showing plots of ln P versus $1/T$ for acetone, ethanol, and water.

TABLE 11.4 Some Typical Heats of Vaporization

Substance	$\Delta H_{vaporization}$(kJ mol^{-1})	Type of Attractive Force
H_2O	+43.9	Hydrogen bonding and London
NH_3	+21.7	Hydrogen bonding and London
HCl	+15.6	Dipole–dipole and London
SO_2	+24.3	Dipole–dipole and London
F_2	+5.9	London
Cl_2	+10.0	London
Br_2	+15.0	London
I_2	+22.0	London
CH_4	+8.16	London
C_2H_6	+15.1	London
C_3H_8	+16.9	London
C_6H_{14}	+30.1	London

Energy changes are related to intermolecular attractions

When a liquid evaporates or a solid sublimes, the particles go from a situation in which the attractive forces are very strong to one in which the attractive forces are so small they can almost be ignored. Therefore, the values of $\Delta H_{vaporization}$ and $\Delta H_{sublimation}$ give us directly the energy needed to separate molecules from each other. We can examine such values to obtain reliable comparisons of the strengths of intermolecular attractions.

TOOLS *Intermolecular attractive forces are related to heats of vaporization and sublimation*

In Table 11.4, notice that the heats of vaporization of water and ammonia are very large, which is just what we would expect for hydrogen-bonded substances. By comparison, CH_4, a nonpolar substance composed of atoms of similar size, has a very small heat of vaporization. Note also that polar substances such as HCl and SO_2 have fairly large heats of vaporization compared with nonpolar substances. For example, compare HCl with Cl_2. Even though Cl_2 contains two relatively large atoms, and therefore would be expected to have larger London forces than HCl, the HCl has the larger $\Delta H_{vaporization}$. This must be due to dipole–dipole attractions between polar HCl molecules—attractions that are absent in nonpolar Cl_2.

The stronger the attractions, the more the potential energy will increase when the molecules become separated, and the larger will be the value of ΔH.

Heats of vaporization also reflect the factors that control the strengths of London forces. For example, the data in Table 11.4 show the effect of chain length on the intermolecular attractions between hydrocarbons; as the chain length increases from one carbon in CH_4 to six carbons in C_6H_{14}, the heat of vaporization also increases, showing that the London forces also increase. Similarly, the heats of vaporization of the halogens in Table 11.4 show that the strengths of London forces increase as the electron clouds of the particles become larger.

11.8 Changes in a Dynamic Equilibrium Can Be Analyzed using Le Châtelier's Principle

Throughout this chapter, we have studied various dynamic equilibria. One example was the equilibrium that exists between a liquid and its vapor in a closed container. You learned that when the temperature of the liquid is increased in this system, its vapor pressure also increases. Let's briefly review why this occurs.

Initially, the liquid is in equilibrium with its vapor, which exerts a certain pressure. When the temperature is increased, equilibrium no longer exists because evaporation occurs more rapidly than condensation. Eventually, as the concentration of molecules in the vapor increases, *the system reaches a new equilibrium* in which there is more vapor and a little less liquid. The greater concentration of molecules in the vapor causes a larger pressure.

The way a liquid's equilibrium vapor pressure responds to a temperature change is an example of a general phenomenon. *Whenever a dynamic equilibrium is upset by some disturbance, the system changes in a way that will, if possible, bring the system back to equilibrium again.* It's also important to understand that in the process of regaining equilibrium, the system undergoes a net change. Thus, when the temperature of a liquid is raised, there is some net conversion of liquid into vapor as the system returns to equilibrium. When the new equilibrium is reached, the amount of liquid and the amount of vapor are not the same as they were before.

Throughout the remainder of this book, we will deal with many kinds of equilibria, both chemical and physical. It would be very time-consuming and sometimes very difficult to carry out a detailed analysis each time we wish to know the effects of some disturbance on an equilibrium system. Fortunately, there is a relatively simple and fast method for predicting the effect of a disturbance, one based on a principle proposed in 1888 by a brilliant French chemist, Henry Le Châtelier (1850–1936).

Le Châtelier's Principle. When a dynamic equilibrium in a system is upset by a disturbance, the system responds in a *direction* that tends to counteract the disturbance and, if possible, restore equilibrium.

Let's see how we can apply Le Châtelier's principle to a liquid–vapor equilibrium that is subjected to a temperature increase. We cannot increase a temperature, of course, without adding heat. Thus, *the addition of heat is really the disturbing influence when a temperature is increased.* So let's incorporate "heat" as a member of the equation used to represent the liquid–vapor equilibrium.

$$\text{Heat} + \text{liquid} \rightleftharpoons \text{vapor} \qquad (11.1)$$

Recall that we use double arrows, $\rightleftharpoons$, to indicate a dynamic equilibrium in an equation. They imply opposing changes happening at equal rates. Evaporation is endothermic, so the heat is placed on the left side of Equation 11.1 to show that heat is absorbed by the liquid when it changes to the vapor, and that heat is released when the vapor condenses to a liquid.

Le Châtelier's principle tells us that when we add heat to raise the temperature of the equilibrium system, the system will try to adjust in a way that absorbs some of the added heat. This can happen if some liquid evaporates, because vaporization is endothermic. When liquid evaporates, the amount of vapor increases and causes the pressure to rise. Thus, we have reached the correct conclusion in a very simple way, namely, that heating a liquid must increase its vapor pressure.

We often use the term **position of equilibrium** to refer to the relative amounts of the substances on opposite sides of the double arrows in an equilibrium expression such as Equation 11.1. Thus, we can think of how a disturbance affects the position of equilibrium. For example, increasing the temperature increases the amount of vapor and decreases the amount of liquid, and we say *the position of equilibrium has shifted*; in this case, it has shifted in the direction of the vapor, or it has *shifted to the right.* In using Le Châtelier's principle, it is often convenient to think of a disturbance as "shifting the position of equilibrium" in one direction or another in the equilibrium equation.

FIG. 11.26 Crystals of table salt. The size of the tiny cubic sodium chloride crystals can be seen in comparison with a penny. *(The Photo Works.)*

Practice Exercise 7: Use Le Châtelier's principle to predict how a temperature increase will affect the vapor pressure of a solid. (Hint: Solid + heat $\rightleftharpoons$ vapor.)

Practice Exercise 8: Designate whether each of the following physical processes is exothermic or endothermic. Can any of them be exothermic for some substances and endothermic for others? Boiling, melting, condensing, subliming, freezing

11.9 Crystalline Solids Have an Ordered Internal Structure

When many substances freeze, or when they separate as a solid from a solution, they tend to form crystals that have highly regular features. For example, Figure 11.26 is a photograph of crystals of sodium chloride—ordinary table salt. Notice that each particle is very nearly a perfect little cube. Whenever a solution of NaCl is evaporated, the crystals that form have edges that intersect at 90° angles. Thus, cubes are the norm for NaCl.

Crystals in general tend to have flat surfaces that meet at angles that are characteristic of the substance. The regularity of these surface features reflects the high degree of order among the particles that lie within the crystal. This is true whether the particles are atoms, molecules, or ions.

A high degree of regularity is the principal feature that makes solids different from liquids. A liquid lacks this long-range repetition of structure because the particles in a liquid are jumbled and disorganized as they move about.

Crystal structures are described by lattices and unit cells

Any repetitive pattern has a symmetrical aspect about it, whether it be a wallpaper design or the orderly packing of particles in a crystal (Figure 11.27). For example, we can easily recognize certain repeating distances between the elements of the pattern, and we can see that the lines along which the elements of the pattern repeat are at certain angles to each other.

To concentrate on the symmetrical features of a repeating structure, it is convenient to describe it in terms of a set of points that have the same repeat distances as the structure, arranged along lines oriented at the same angles. Such a pattern of points is called a **lattice,** and when we apply it to describe the packing of particles in a solid, we often call it a **crystal lattice.**

In a crystal, the number of particles is enormous. If you could imagine being at the center of even the tiniest crystal, you would find that the particles go on as far as you can see in every direction. Describing the positions of all these particles or their lattice points is impossible and, fortunately, unnecessary. All we need to do is describe the repeating unit of the lattice, which we call the *unit cell.* To see this, and to gain an insight into the usefulness of the lattice concept, let's begin in two dimensions.

In Figure 11.28 we see a two-dimensional *square lattice,* which means the lattice points lie at the corners of squares. The repeating unit of the lattice, its **unit cell,** is indicated in the drawing. If we began with this unit cell, we could produce the entire lattice by moving it repeatedly left and right and up and down by distances equal to its edge length. In this sense, all the properties of a lattice are contained in the properties of its unit cell.

An important fact about lattices is that the same lattice can be used to describe many different designs or structures. For example, in Figure 11.28*b*, we see a design formed by associating a pink heart with each lattice point. Using a square lattice, we could form any number of designs just by using different design elements (for example, a rose or a diamond) or by changing the lengths of the edges of the unit cell. *The only requirement is that the same design element must be associated with each lattice point.* In other words, if there is a rose at one lattice point, then there must be a rose at all the other lattice points.

Extending the lattice concept to three dimensions is straightforward. Illustrated in Figure 11.29 is a *simple cubic* (also called a **primitive cubic**) lattice, the simplest and most symmetrical three-dimensional lattice. Its unit cell, the **simple cubic unit cell,** is a cube with lattice points only at its eight corners. Figure 11.29*c* shows the packing of atoms in a crystal

A wallpaper design

Packing of atoms in a crystal

FIG. 11.27 Symmetry among repetitive patterns. A wallpaper design and particles arranged in a crystal each show a repeating pattern of structural units. The pattern can be described by the distances between the repeating units and the angles along which the repetition of structure occurs.

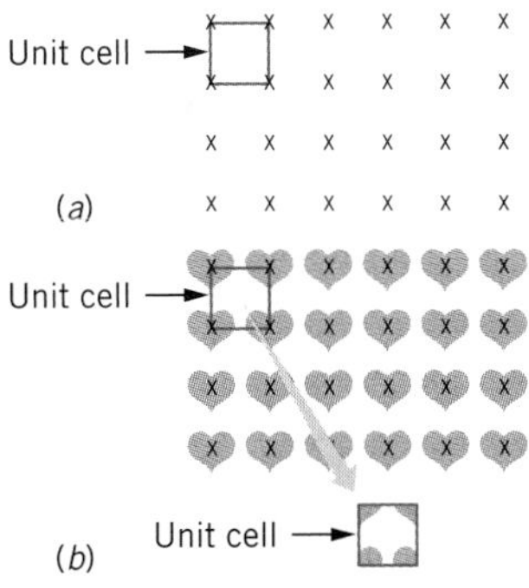

FIG. 11.28 A two-dimensional lattice. (*a*) A simple square lattice, for which the unit cell is a square with lattice points at the corners. (*b*) A wallpaper pattern formed by associating a design element (pink heart) with each lattice point. The X centered on each heart corresponds to a lattice point. The unit cell contains portions of a heart at each corner.

(a) (b) (c)

FIG. 11.29 A three-dimensional simple cubic lattice. (*a*) A simple cubic unit cell showing the locations of the lattice points. (*b*) A portion of a simple cubic lattice built by stacking simple cubic unit cells. (*c*) The crystal structure of polonium having a simple cubic lattice with identical atoms at the lattice points. Only a portion of each atom lies within this particular unit cell.

of polonium that crystallizes in a simple cubic lattice as well as the unit cell for that substance.[2] Notice that when the unit cell is "carved out" of the crystal, we find only part of an atom (1/8th of an atom, actually) at each corner. The rest of each atom resides in adjacent unit cells. Because the unit cell has eight corners, if we put all the corner pieces together we would obtain one complete atom. Thus, we conclude that this unit cell contains just one atom.

$$8\ \cancel{\text{corners}} \times \frac{1/8\ \text{atom}}{\cancel{\text{corner}}} = 1\ \text{atom}$$

As with the two-dimensional lattice, we could use the same simple cubic lattice to describe the structures of many different substances. The *sizes* of the unit cells would vary because the sizes of atoms vary, but the essential symmetry of the stacking would be the same in them all. This fact about lattices makes it possible to describe limitless numbers of different compounds with just a small set of three-dimensional lattices. In fact, it has been shown mathematically that there are only 14 different three-dimensional lattices possible, which means that all the chemical substances that can exist must form crystals with one or another of these 14 lattice types.

There are three cubic lattices

In addition to simple cubic, two other cubic lattices are possible: face-centered cubic and body-centered cubic. The **face-centered cubic** (abbreviated **fcc**) **unit cell** has lattice points (and therefore, identical particles) at each of its eight corners plus another in the center of each face, as shown in Figure 11.30. Many common metals—copper, silver, gold, aluminum, and lead, for example—form crystals that have face-centered cubic lattices. Each of these metals has the same *kind* of lattice, but the *sizes* of their unit cells differ because the sizes of the atoms differ (see Figure 11.31).

The **body-centered cubic (bcc) unit cell** has lattice points at each corner plus one in the center of the cell, as illustrated in Figure 11.32. The body-centered cubic lattice is also common among a number of metals; examples are chromium, iron, and platinum. Again, these are substances with the same *kind* of lattice, but the dimensions of the lattices vary because of the *different sizes* of the particular atoms.

Not all unit cells are cubic. Some have edges of different lengths or edges that intersect at angles other than 90°. Although you should be aware of the existence of other unit cells and lattices, we will limit the remainder of our discussion to cubic lattices and their unit cells.

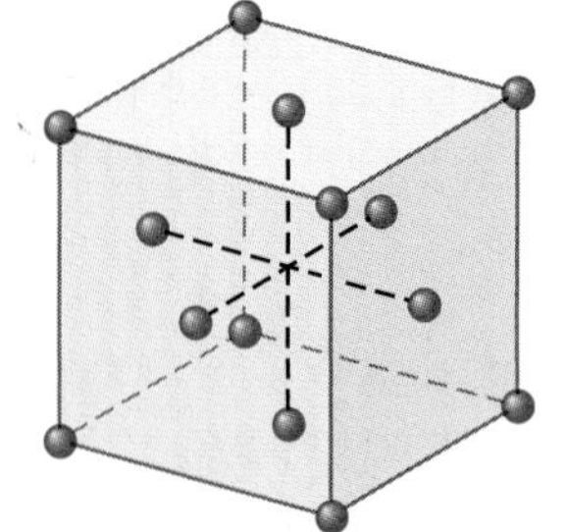

FIG. 11.30 A face-centered cubic unit cell. Lattice points are found at each of the eight corners and in the center of each face.

Many compounds crystallize with cubic lattices

We have seen that a number of metals have cubic lattices. The same is true for many compounds. Figure 11.33, for example, is a view of a portion of a sodium chloride crystal.

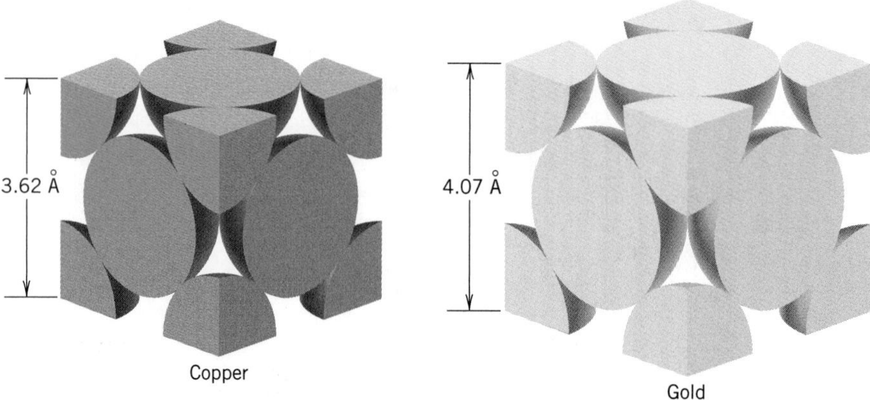

FIG. 11.31 Unit cells for copper and gold. These metals both crystallize in a face-centered cubic structure with similar face centered cubic unit cells. The atoms are arranged in the same way, but their unit cells have edges of different lengths because the atoms are of different sizes (1 Å = 1×10^{-10} m).

[2] Polonium is the only element known to crystallize with a simple cubic lattice. Some compounds also form simple cubic lattices.

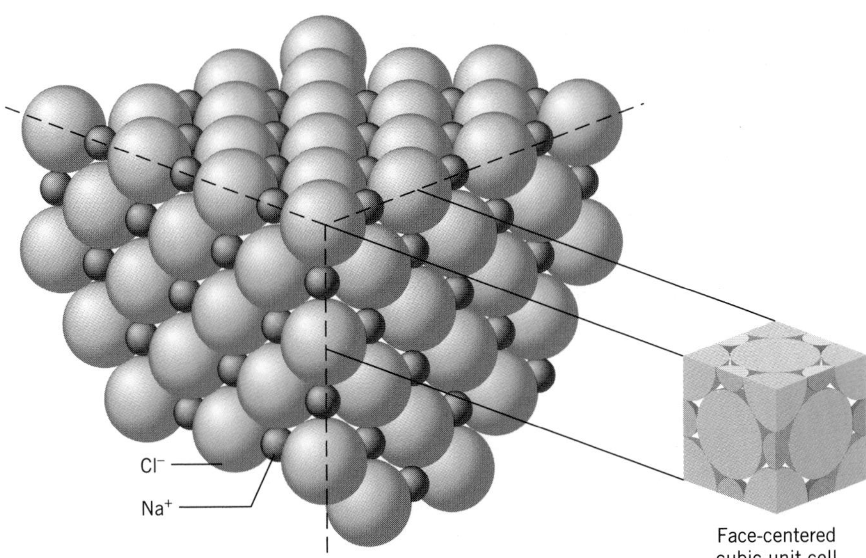

FIG. 11.33 The packing of ions in a sodium chloride crystal. Chloride ions are shown here to be associated with the lattice points of a face-centered cubic unit cell, with the sodium ions placed between the chloride ions.

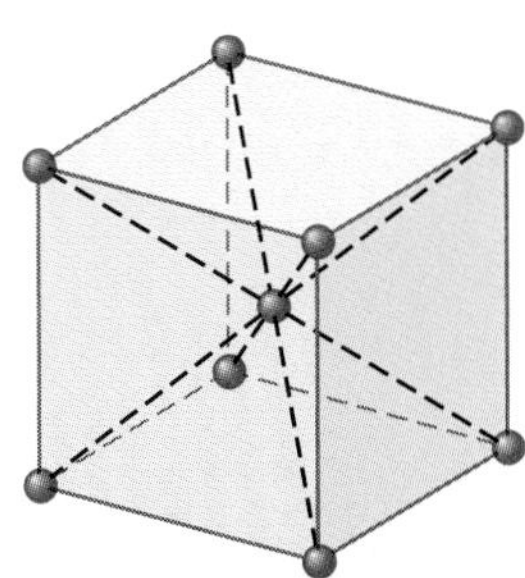

FIG. 11.32 A body-centered cubic unit cell. Lattice points are located at each of the eight corners and in the center of the unit cell.

The Cl^- ions (green) are shown at the lattice points that correspond to a face-centered cubic unit cell. The smaller gray spheres represent Na^+ ions. Notice that they fill the spaces between the Cl^- ions. If we look at the locations of identical particles (Cl^-, for example) we find them at lattice points that describe a face-centered cubic structure. Thus, sodium chloride is said to have a face-centered cubic lattice, and the cubic shape of this lattice is what accounts for the cubic shape of a sodium chloride crystal.

Many of the alkali halides (Group IA–VIIA compounds), such as NaBr and KCl, crystallize with fcc lattices that have the same arrangement of ions as found in NaCl. In fact, this arrangement of ions is so common that it's called the **rock salt structure** (rock salt is the mineral name of NaCl). Because sodium bromide and potassium chloride both have the same kind of lattice as sodium chloride, Figure 11.33 also could be used to describe their unit cells. The *sizes* of their unit cells are different, however, because K^+ is a larger ion than Na^+, and Br^- is larger than Cl^-.

Other examples of cubic unit cells are shown in Figures 11.34 and 11.35. The structure of cesium chloride in Figure 11.34 is simple cubic, although at first glance it may appear to be body centered. This is because in a crystal lattice, identical chemical units must be at each lattice point. In CsCl, Cs^+ ions are found at the corners, but not in the center, so the Cs^+ ions describe a simple cubic unit cell.

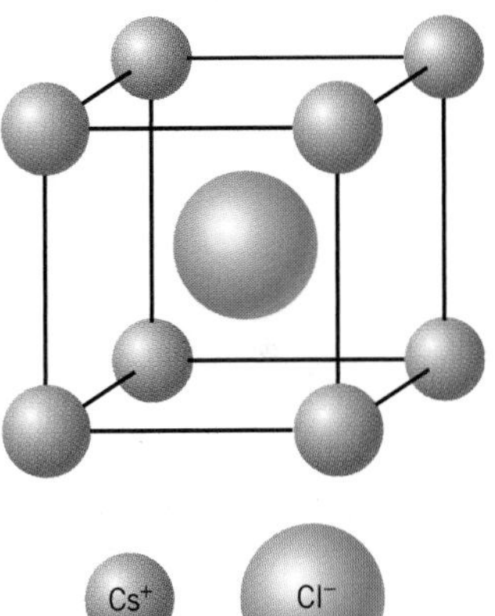

FIG. 11.34 The unit cell for cesium chloride, CsCl. The chloride ion is located in the center of the unit cell. The ions are not shown full-size to make it easier to see their locations in the unit cell.

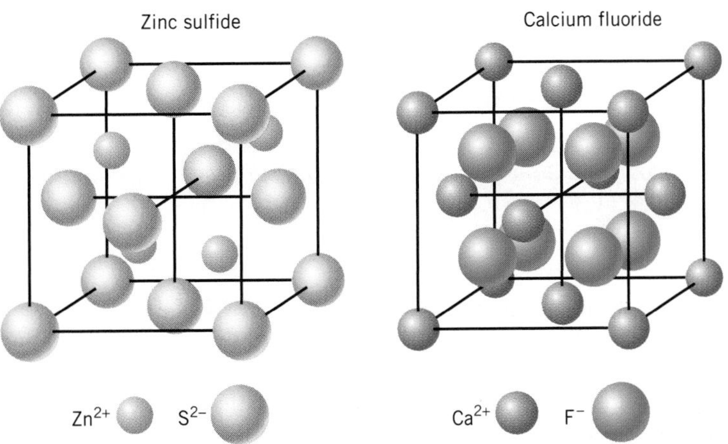

FIG. 11.35 Crystal structures based on the face-centered cubic lattice. Both zinc sulfide, ZnS, and calcium fluoride, CaF_2, have crystal structures that fit a face-centered cubic lattice. In ZnS, the sulfide ions are shown at the fcc lattice sites with the four zinc ions entirely within the unit cell. In CaF_2, the calcium ions are at the lattice points with the eight fluorides entirely within the unit cell. The ions are not shown full-size to make it easier to see their locations in the unit cells.

Both zinc sulfide and calcium fluoride in Figure 11.35 have face-centered cubic unit cells that differ from that for sodium chloride, which illustrates once again how the same basic kind of lattice can be used to describe a variety of chemical structures.

Stoichiometry affects the packing of atoms in a unit cell

At this point, you may wonder why a compound crystallizes with a particular structure. Although this is a complex issue, at least one factor is the stoichiometry of the substance. Because the crystal is made up of a huge number of identical unit cells, the stoichiometry within the unit cell must match the overall stoichiometry of the compound. Let's see how this applies to sodium chloride.

EXAMPLE 11.2
Counting Atoms or Ions in a Unit Cell

How many sodium and chloride ions are there in the unit cell of sodium chloride?

ANALYSIS: To answer this question, we have to look closely at the unit cell of sodium chloride. The critical link is realizing that when the unit cell is carved out of the crystal, it encloses *parts of ions*, so we have to determine how many *whole* sodium and chloride ions can be constructed from the pieces within a given unit cell.

SOLUTION: Let's look at an "exploded" view of the NaCl unit cell, shown in Figure 11.36. We see that we have parts of chloride ions at the corners and in the center of each face. Let's add the parts.

For chloride:

$$8 \text{ corners} \times \tfrac{1}{8}\text{Cl}^- \text{ per corner} = 1 \text{ Cl}^-$$

$$6 \text{ faces} \times \tfrac{1}{2}\text{Cl}^- \text{ per face} = 3 \text{ Cl}^-$$

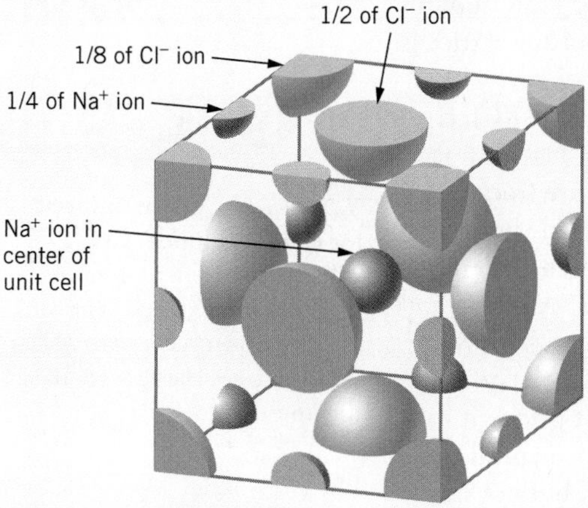

FIG. 11.36 An exploded view of the unit cell of sodium chloride.

For the sodium ions, we have parts along each of the 12 edges plus one whole Na^+ ion in the center of the unit cell. Let's add them.

For sodium:

$$12 \text{ edges} \times \tfrac{1}{4}\text{Na}^+ \text{ per edge} = 3 \text{ Na}^+$$

$$1 \text{ Na}^+ \text{ in the center} = 1 \text{ Na}^+$$

$$\text{Total} = 4 \text{ Na}^+$$

Thus, in one unit cell, there are four chloride ions and four sodium ions.

IS THE ANSWER REASONABLE? The ratio of the ions is 4 to 4, which is the same as 1 to 1. That's the ratio of the ions in NaCl, so the answer is correct.

Practice Exercise 9: How many calcium ions and how many fluoride ions are in the unit cell of calcium fluoride, CaF_2? (Hint: see Figure 11.35.)

Practice Exercise 10: What is the ratio of the ions in the unit cell of cesium chloride? See Figure 11.34.

The calculation in Example 11.2 shows why NaCl can have the crystal structure it does; the unit cell has the proper ratio of cations to anions. It also shows why a compound such as $CaCl_2$ could *not* crystallize with the same kind of unit cell as NaCl. The sodium chloride structure demands a 1-to-1 ratio of cation to anion, so it could not be used by $CaCl_2$ (which has a 1-to-2 cation-to-anion ratio).

Efficiency of packing also can affect how atoms are arranged in a solid

For many solids, particularly metals, the type of crystal structure formed is controlled by maximizing the number of neighbors that surround a given atom. The more neighbors an atom has, the greater are the number of interatomic attractions and the greater is the energy lowering when the solid forms. Structures that achieve the maximum density of packing are known as **closest-packed structures,** and there are two of them that are only slightly different. To visualize how these are produced, let's look at ways to pack spheres of identical size.

Figure 11.37*a* illustrates a layer of blue spheres packed as tightly as possible. Notice that each sphere is touched by six others in this layer. When we add a second layer, each sphere (red) rests in a depression formed by three spheres in the first layer, as illustrated in Figures 11.37*b* and *c*.

The difference between the two closest-packed structures lies in the relative orientations of the spheres in the first layer and those that form the third layer. In Figure 11.38*a* the green spheres in the third layer each lie in a depression between red spheres that is directly above a depression between blue spheres in the first layer. This kind of packing is called **cubic closest packing,** abbreviated **ccp,** because when viewed from a different perspective, the atoms are located at positions corresponding to a face-centered cubic lattice. Figure 11.38*b* describes the other closest-packed structure in which a green sphere in the third layer rests in a depression between red spheres and directly above a blue sphere in the first layer. This arrangement of spheres is called **hexagonal closest packing,** abbreviated **hcp**.

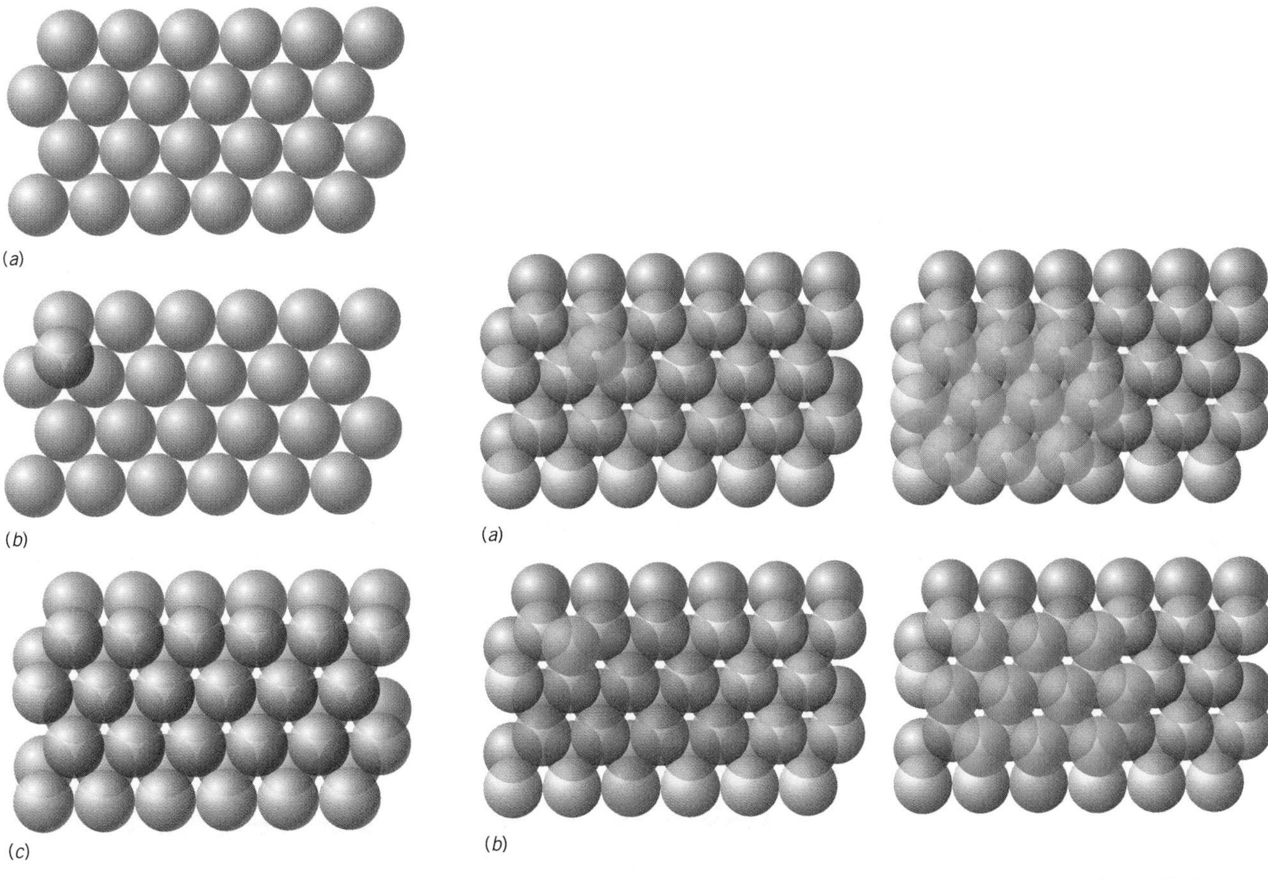

FIG. 11.37 **Packing of spheres.** (*a*) One layer of closely packed spheres. (*b*) A second layer is started by placing a sphere (colored red) in a depression formed between three spheres in the first layer. (*c*) A second layer of spheres shown slightly transparent so we can see how the atoms are stacked over the first layer.

FIG. 11.38 **Closest-packed structures.** (*a*) Cubic closest packing of spheres. (*b*) Hexagonal closest packing of spheres. In both (*a*) and (*b*) the left diagram illustrates the position of one atom on the third layer and the right diagram shows the third layer partially complete. Notice that there are subtle differences between the two modes of packing.

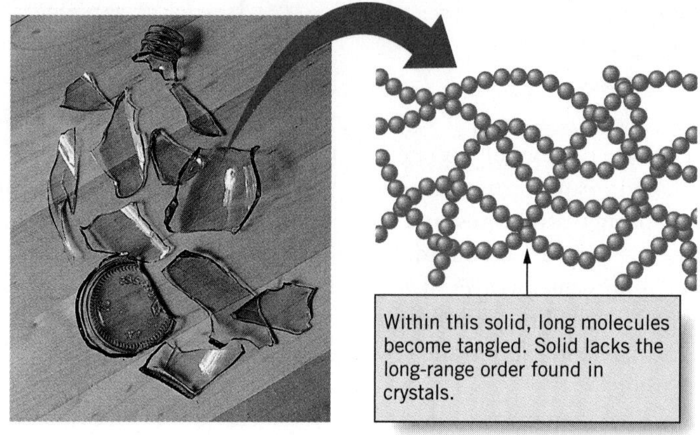

Within this solid, long molecules become tangled. Solid lacks the long-range order found in crystals.

FIG. 11.39 Glass is a noncrystalline solid. When glass breaks, the pieces have sharp edges, but their surfaces are not flat planes. This is because in an amorphous solid like glass, long molecules (much simplified here for clarity) are tangled and disorganized, so there is no long-range order characteristic of a crystal. *(Robert Capece.)*

In the hcp structure, the layers alternate in an A-B-A-B. . . pattern, where A stands for the orientations of the first, third, fifth, etc. layers, and B stands for the orientations of the second, fourth, sixth, etc. layers. Thus, the spheres in the third, fifth, seventh, etc. layers are directly above those in the first, while spheres in the fourth, sixth, eighth, etc. layers are directly above those in the second. In the ccp structure, there is an A-B-C-A-B-C. . . pattern. The first layer is oriented like the fourth, the second like the fifth, and the third like the sixth.

Both the ccp and hcp structures yield very efficient packing of identically sized atoms. In both structures, each atom is in contact with 12 neighboring atoms: 6 atoms in its own layer, 3 atoms in the layer below, and 3 atoms in the layer above. Metals that crystallize with the ccp structure include copper, silver, gold, aluminum, and lead. Metals with the hcp structure include titanium, zinc, cadmium, and magnesium.

Not all solids are crystalline

If a cubic salt crystal is broken, the pieces still have flat faces that intersect at 90° angles. If you shatter a piece of glass, on the other hand, the pieces often have surfaces that are not flat. Instead, they tend to be smooth and curved (see Figure 11.39). This behavior illustrates a major difference between crystalline solids, such as NaCl, and noncrystalline solids, also called **amorphous solids,** such as glass.

The word *amorphous* is derived from the Greek word *amorphos,* which means "without form." Amorphous solids do not have the kinds of long-range repetitive internal structures that are found in crystals. In some ways their structures, being jumbled, are more like liquids than solids. Examples of amorphous solids are ordinary glass and many plastics. In fact, the word **glass** is often used as a general term to refer to any amorphous solid.

As suggested in Figure 11.39, substances that form amorphous solids often consist of long, chainlike molecules that are intertwined in the liquid state somewhat like long strands of cooked spaghetti. To form a crystal from the melted material, these long molecules would have to become untangled and line up in specific patterns. But as the liquid cools, the molecules slow down. Unless the liquid is cooled extremely slowly, the molecular motion decreases too rapidly for the untangling to take place, and the substance solidifies with the molecules still intertwined. As a result, amorphous solids are sometimes described as **supercooled liquids,** a term suggesting the kind of structural disorder found in liquids.

11.10 X-Ray Diffraction Is Used to Study Crystal Structures

When atoms in a crystal are bathed in X rays, they absorb some of the radiation and then emit it again in all directions. In effect, each atom becomes a tiny X-ray source. If we look at radiation from two such atoms (Figure 11.40), we find that the X rays emitted are in phase in some directions but out of phase in others. In Chapter 7 you learned that constructive (in-phase) and destructive (out-of-phase) interferences create a phenomenon called *diffraction.* X-Ray diffraction by crystals has enabled many scientists to win Nobel prizes by determining the structures of extremely complex compounds in a particularly elegant way.

Out of phase
Emitting atoms
In phase

FIG. 11.40 Diffraction of X rays from atoms in a crystal. X rays emitted from atoms are in phase in some directions and out of phase in other directions.

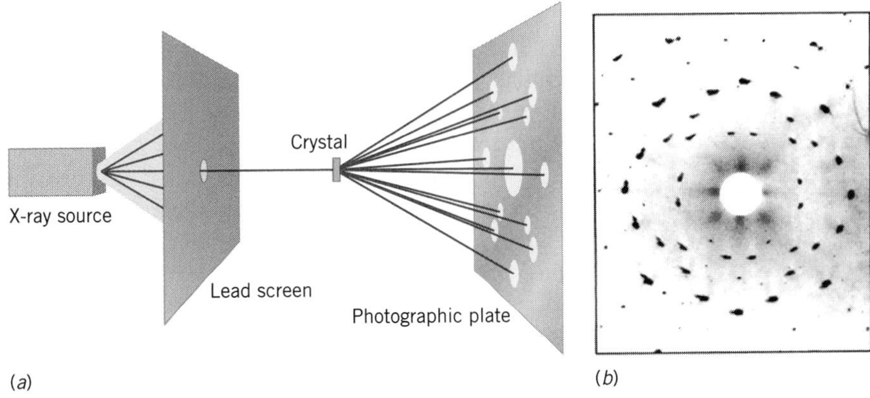

FIG. 11.41 **X-Ray diffraction.** (*a*) The production of an X-ray diffraction pattern. (*b*) An X-ray diffraction pattern produced by sodium chloride recorded on photographic film. *(Visuals Unlimited.)*

In a crystal, there are enormous numbers of atoms, evenly spaced throughout the lattice. When the crystal is bathed in X rays, intense beams are diffracted because of constructive interference, and they appear only in specific directions. In other directions, no X rays appear because of destructive interference. When the X rays coming from the crystal fall on photographic film, the diffracted beams form a **diffraction pattern** (see Figure 11.41). The film is darkened only where the X rays strike.[3]

In 1913, the British physicist William Henry Bragg and his son William Lawrence Bragg discovered that just a few variables control the appearance of an X-ray diffraction pattern. These are shown in Figure 11.42, which illustrates the conditions necessary to obtain constructive interference of the X rays from successive layers of atoms (planes of atoms) in a crystal. A beam of X rays having a wavelength λ strikes the layers at an angle θ. Constructive interference causes an intense diffracted beam to emerge at the same angle θ. The Braggs derived an equation, now called the **Bragg equation,** relating λ, θ, and the distance between the planes of atoms, d,

$$n\lambda = 2d \sin \theta \qquad (11.2)$$

where n is a whole number. The Bragg equation is the basic tool used by scientists in the study of solid structures.

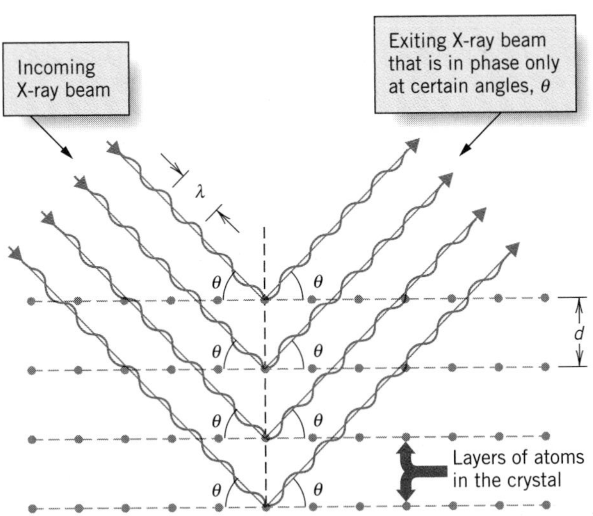

FIG. 11.42 **Diffraction of X rays from successive layers of atoms in a crystal.** The layers of atoms are separated by a distance d. The X rays of wavelength λ enter and emerge at an angle θ relative to the layers of the atoms. For the emerging beam of X rays to have any intensity, the condition $n\lambda = 2d \sin \theta$ must be fulfilled, where n is a whole number.

[3] Modern X-ray diffraction instruments use electronic devices to detect and measure the angles and intensities of the diffracted X rays.

To determine the structure of a crystal, the angles θ at which diffracted X-ray beams emerge from a crystal are measured. These angles are used to calculate the distances between the various planes of atoms in the crystal. The calculated interplanar distances are then used to work backward to deduce where the atoms in the crystal must be located so that layers of atoms are indeed separated by these distances. This is not a simple task, and some sophisticated mathematics as well as computers are needed to accomplish it. The efforts, however, are well rewarded because the calculations give the locations of atoms within the unit cell and the distances between them. This information, plus a lot of chemical "common sense," is used by chemists to arrive at the shapes and sizes of the molecules in the crystal. Example 11.3 below provides a very simple illustration of how such data are used.

X-Ray diffraction has had a profound impact on the study of biochemical molecules. For example, the general shape of DNA molecules, the chemicals of genes, was deduced by using X-ray diffraction. Today, X-ray diffraction continues to be one of the tools used by biochemists to determine the structures of complex proteins and enzymes.

EXAMPLE 11.3
Using Crystal Structure Data to Calculate Atomic Sizes

X-Ray diffraction measurements reveal that copper crystallizes with a face-centered cubic lattice in which the unit cell length is 3.62 Å (see Figure 11.31). What is the radius of a copper atom expressed in angstroms and in picometers?

ANALYSIS: In Figure 11.31, we see that copper atoms are in contact along a diagonal (the dashed line below) that runs from one corner of a face to another corner.

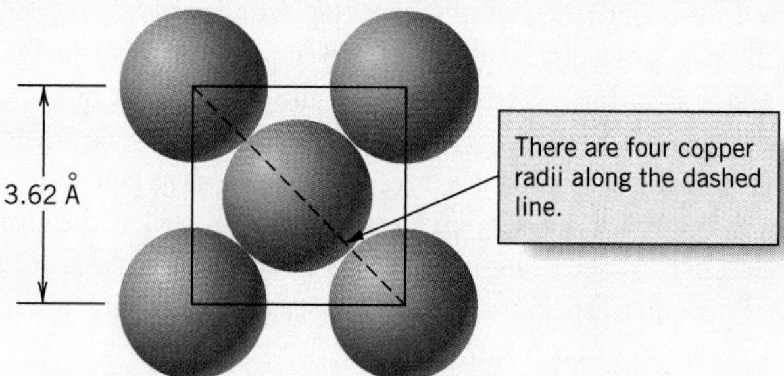
3.62 Å
There are four copper radii along the dashed line.

By geometry, we can calculate the length of this diagonal, which equals four times the radius of a copper atom. Once we calculate the radius in angstrom units we can convert to picometers using the relationships

$$1\ \text{Å} = 1 \times 10^{-10}\ \text{m}$$
$$1\ \text{pm} = 1 \times 10^{-12}\ \text{m}$$

SOLUTION: From geometry, the length of the diagonal is $\sqrt{2}$ times the length of the edge of the unit cell.

$$\text{Diagonal} = \sqrt{2} \times (3.62\ \text{Å}) = 5.12\ \text{Å}$$

If we call the radius of the copper atom r_{Cu}, then the diagonal equals $4 \times r_{Cu}$. Therefore,

$$4 \times r_{Cu} = 5.12\ \text{Å}$$
$$r_{Cu} = 1.28\ \text{Å}$$

The calculated radius of the copper atom is 1.28 Å.

Next we convert this to picometers.

$$1.28\ \cancel{\text{Å}} \times \frac{1 \times 10^{-10}\ \cancel{\text{m}}}{1\ \cancel{\text{Å}}} \times \frac{1\ \text{pm}}{1 \times 10^{-12}\ \cancel{\text{m}}} = 128\ \text{pm}$$

IS THE ANSWER REASONABLE? It's difficult to get an intuitive feel for the sizes of atoms, so we should be careful to check the calculation. The length of the diagonal seems about right; it is longer than the edge of the unit cell. If we look again at the diagram above, we can see that along the diagonal there are four copper radii. The rest of the arithmetic is okay, so our answer is correct.

11.11 Physical Properties of Solids Are Related to Their Crystal Types

Solids exhibit a wide range of properties. Some, such as diamond, are very hard whereas others, such as ice and naphthalene (moth flakes), are relatively soft. Some, such as salt crystals, have high melting points, whereas others, such as candle wax, melt at low temperatures. And some conduct electricity but others are nonconducting. Physical properties such as these depend on the kinds of particles in the solid as well as on the strengths of attractive forces holding the solid together. Even though we can't make exact predictions about such properties, some generalizations do exist. In discussing them, it is convenient to divide crystals into four types: ionic, molecular, covalent, and metallic.

Ionic crystals have cations and anions at lattice sites

Ionic crystals have ions at the lattice sites and the binding between them is mainly electrostatic, which is essentially nondirectional. As a result, the kind of lattice formed is determined mostly by the relative sizes of the ions and their charges. When the crystal forms, the ions arrange themselves to maximize attractions and minimize repulsions.

Because electrostatic forces are strong, ionic crystals tend to be hard. They also tend to have high melting points because the ions have to be given a lot of kinetic energy to enable them to break free of the lattice and enter the liquid state. The forces between ions can also be used to explain the brittle nature of many ionic compounds. For example, when struck by a hammer, a salt crystal shatters into many small pieces. A view at the atomic level reveals how this could occur (Figure 11.43). The slight movement of a layer of ions within an ionic crystal suddenly places ions of the *same* charge next to each other, and for that instant there are large repulsive forces that split the solid.

In the solid state, ionic compounds do not conduct electricity because the charges present are not able to move. When melted, however, ionic compounds are good conductors of electricity. Melting frees the electrically charged ions to move.

Molecular crystals have neutral molecules at lattice sites

Molecular crystals are solids in which the lattice sites are occupied either by atoms (as in solid argon or krypton) or by molecules (as in solid CO_2, SO_2, or H_2O). If the molecules of such solids are relatively small, the crystals tend to be soft and have low melting points because the particles in the solid experience relatively weak intermolecular attractions. In crystals of argon, for example, the attractive forces are exclusively London forces. In SO_2, which is composed of polar molecules, there are dipole–dipole attractions as well as London forces. And in water crystals (ice) the molecules are held in place primarily by strong hydrogen bonds. Molecular compounds do not conduct electricity in either the solid or liquid state because they are unable to transport electrical charges.

Molecular crystals are soft because little effort is needed to separate the particles or cause them to move past each other.

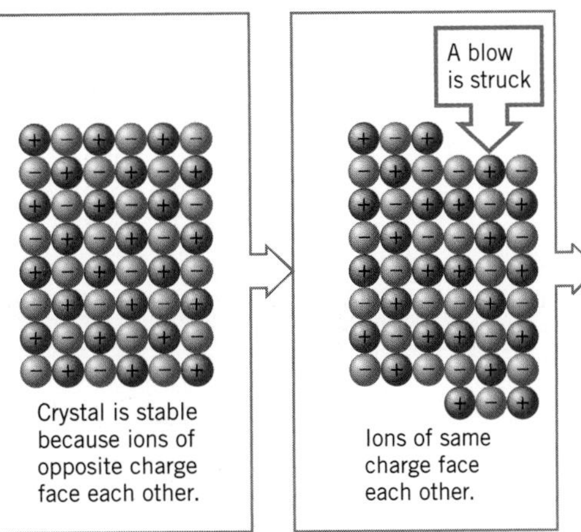
A blow is struck
Crystal is stable because ions of opposite charge face each other.
Ions of same charge face each other.

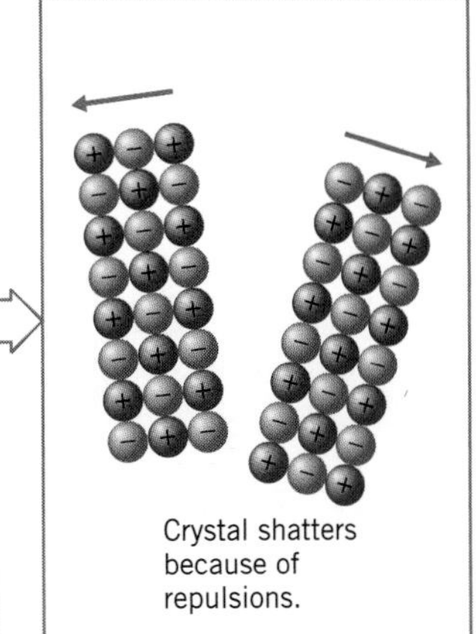
Crystal shatters because of repulsions.

FIG. 11.43 An ionic crystal shatters when struck. In this microview we see that striking an ionic crystal causes some of the layers to shift. This can bring ions of like charge face-to-face. The repulsions between the ions can then force parts of the crystal apart, causing the crystal to shatter. *(Andy Washnik.)*

Covalent crystals have atoms at lattice sites covalently bonded to other atoms

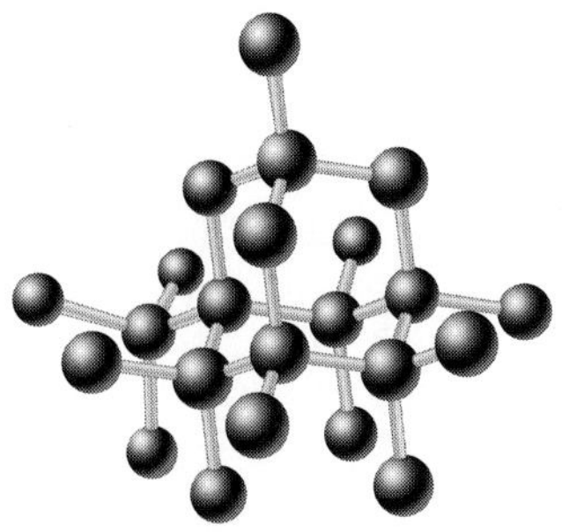

FIG. 11.44 The structure of diamond. Each carbon atom is covalently bonded to four others at the corners of a tetrahedron. This is just a tiny portion of a diamond, of course; the structure extends throughout the entire diamond crystal.

Covalent crystals are solids in which lattice positions are occupied by atoms that are covalently bonded to other atoms at neighboring lattice sites. The result is a crystal that is essentially one gigantic molecule. These solids are sometimes called **network solids** because of the interlocking network of covalent bonds extending throughout the crystal in all directions. A typical example is diamond (see Figure 11.44). Covalent crystals tend to be very hard and to have very high melting points because of the strong attractions between covalently bonded atoms. Other examples of covalent crystals are quartz (SiO_2, found in some types of sand) and silicon carbide (SiC, a common abrasive used in sandpaper). Covalent crystals are poor conductors of electricity, although some, such as silicon, are semiconductors.

Metallic crystals have cations at lattice sites surrounded by mobile electrons

Metallic crystals have properties that are quite different from those of the other three types. Metallic crystals conduct heat and electricity well, and they have the luster characteristically associated with metals. A number of different models have been developed to explain metallic crystals. One of the simplest models views the lattice positions of a metallic crystal as being occupied by *positive ions* (nuclei plus core electrons). Surrounding them is a "cloud" of electrons formed by the valence electrons, which extends throughout the entire solid (see Figure 11.45). The electrons in this cloud belong to no single positive ion,

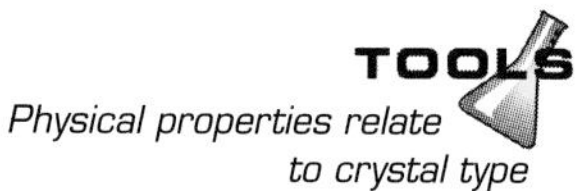

Physical properties relate to crystal type

TABLE 11.5 Types of Crystals

Crystal Type	Particles Occupying Lattice Sites	Type of Attractive Force	Typical Examples	Typical Properties
Ionic	Positive and negative ions	Attractions between ions of opposite charge	NaCl, $CaCl_2$, $NaNO_3$	Relatively hard; brittle; high melting points; nonconductors of electricity as solids, but conduct when melted
Molecular	Atoms or molecules	Dipole–dipole attractions, London forces, hydrogen bonding	HCl, SO_2, N_2, Ar, CH_4, H_2O	Soft; low melting points; nonconductors of electricity in both solid and liquid states
Covalent (network)	Atoms	Covalent bonds between atoms	Diamond, SiC, (silicon carbide), SiO_2 (sand, quartz)	Very hard; very high melting points; nonconductors of electricity
Metallic	Positive ions	Attractions between positive ions and an electron cloud that extends throughout the crystal	Cu, Ag, Fe, Na, Hg	Range from very hard to very soft; melting points range from high to low; conduct electricity in both solid and liquid states; have characteristic luster

FIG. 11.45 The "electron sea" model of a metallic crystal. In this highly simplified view of a metallic solid, metal atoms lose valence electrons to the solid as a whole and exist as positive ions surrounded by a mobile "cloud" of electrons.

but rather to the crystal as a whole. Because the electrons aren't localized on any one atom, they are free to move easily, which accounts for the high electrical conductivity of metals. The electrons can also carry kinetic energy rapidly through the solid, so metals are also good conductors of heat. This model explains the luster of metals, too. When light shines on the metal, the loosely held electrons vibrate easily and readily re-emit the light with essentially the same frequency and intensity.

Some metals, like tungsten, have very high melting points (mp = 3422 °C). Others, such as sodium (mp = 97.8 °C) and mercury, which is a liquid at room temperature, have quite low melting points. To some extent, the melting point depends on the charge of the positive ions in the metallic crystal. The Group IA metals have just one valence electron, so their cores are cations with a 1+ charge, which are only weakly attracted to the "electron cloud" that surrounds them. Atoms of the Group IIA metals, however, form ions with a 2+ charge. These are attracted more strongly to the surrounding electron sea, so the Group IIA metals have higher melting points than their neighbors in Group IA. Metals with very high melting points, like tungsten, must have very strong attractions between their atoms, which suggests that there probably is some covalent bonding between them as well.

The different ways of classifying crystals and a summary of their general properties are given in Table 11.5.

EXAMPLE 11.4
Identifying Crystal Types from Physical Properties

The metal osmium, Os, forms an oxide with the formula OsO_4. The soft crystals of OsO_4 melt at 40 °C, and the resulting liquid does not conduct electricity. To which crystal type does solid OsO_4 probably belong?

ANALYSIS: You might be tempted to suggest that the compound is ionic simply because it is formed from a metal and a nonmetal. However, the properties of the compound are inconsistent with its being ionic. Therefore, we have to consider that there may be exceptions to the generalization discussed earlier about metal–nonmetal compounds. If so, what do the properties of OsO_4 suggest about its crystal type?

SOLUTION: The characteristics of the OsO_4 crystals—softness and low melting point—suggest that solid OsO_4 is a molecular solid and that it contains molecules of OsO_4. This is further supported by the fact that liquid OsO_4 does not conduct electricity, which is evidence for the lack of ions in the liquid.

IS THE ANSWER REASONABLE? There's not much we can do to check ourselves here except to review our analysis.

Practice Exercise 11: Stearic acid is an organic acid that has a chain of 18 carbon atoms. It is a soft solid with a melting point of 70 °C. What crystal type best describes this compound? (Hint: Determine the dominant attractive forces that cause stearic acid to be a solid at room temperature.)

Practice Exercise 12: Boron nitride, which has the empirical formula BN, melts at 2730 °C and is almost as hard as a diamond. What is the probable crystal type for this compound?

Practice Exercise 13: Crystals of elemental sulfur are easily crushed and melt at 113 °C to give a clear yellow liquid that does not conduct electricity. What is the probable crystal type for solid sulfur?

11.12 PHASE DIAGRAMS GRAPHICALLY REPRESENT PRESSURE–TEMPERATURE RELATIONSHIPS

Sometimes it is useful to know under what combinations of temperature and pressure a substance will be a liquid, a solid, or a gas, or the conditions of temperature and pressure that produce an equilibrium between any two phases. A simple way to determine this is to

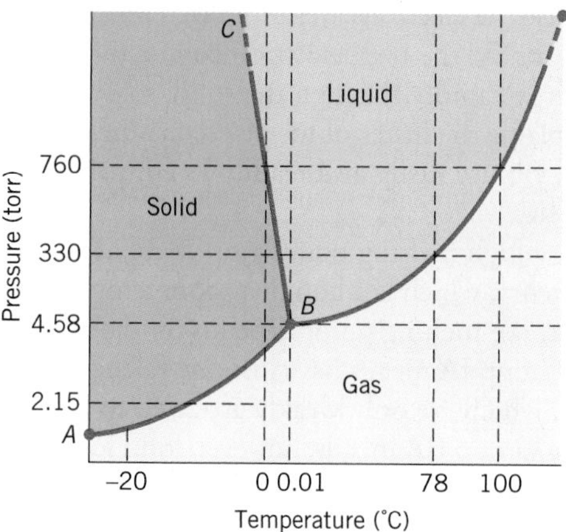

FIG. 11.46 The phase diagram for water, distorted to emphasize certain features. Temperatures and pressures corresponding to the dashed lines on the diagram are referred to in the text discussion.

use a **phase diagram**—a graphical representation of the pressure–temperature relationships that apply to the equilibria between the phases of the substance.

Figure 11.46 is the phase diagram for water. On it, there are three lines that intersect at a common point. Points on these lines correspond to temperatures and pressures at which equilibria between phases can exist. For example, line *AB* is the vapor pressure curve for the solid (ice). Every point on this line gives a temperature and a pressure at which ice and its vapor are able to coexist in equilibrium.

The melting point and boiling point can be read directly from the phase diagram.

Line *BD* is the vapor pressure curve for liquid water. It gives the temperatures and pressures at which the liquid and vapor are able to coexist in equilibrium. Notice that when the temperature is 100 °C, the vapor pressure is 760 torr. Therefore, this diagram also tells us that water boils at 100 °C when the pressure is 1 atm (760 torr), because that is the temperature at which the vapor pressure equals 1 atm.

The solid–vapor equilibrium line, *AB,* and the liquid–vapor line, *BD,* intersect at a common point, *B.* Because this point is on both lines, there is equilibrium between all three phases at the same time.

$$\begin{array}{ccc} & \text{vapor} & \\ \swarrow\!\!\nearrow & & \nwarrow\!\!\searrow \\ \text{liquid} & \rightleftharpoons & \text{solid} \end{array}$$

In the SI, the triple point of water is used to define the Kelvin temperature of 273.16 K.

The temperature and pressure at which this triple equilibrium occurs define the **triple point** of the substance. For water, the triple point occurs at 0.01 °C and 4.58 torr. Every known chemical substance except helium has its own characteristic triple point, which is controlled by the balance of intermolecular forces in the solid, liquid, and vapor.

Line *BC,* which extends upward from the triple point, is the solid–liquid equilibrium line or *melting point line.* It gives temperatures and pressures at which the solid and the liquid are able to be in equilibrium. At the triple point, the melting of ice occurs at +0.01 °C (and 4.58 torr); at 760 torr, melting occurs very slightly lower, at 0 °C. Thus, we can tell that *increasing the pressure on ice lowers its melting point.*

The effect of pressure on the melting point of ice can be predicted using Le Châtelier's principle and the knowledge that a given mass of liquid water occupies *less volume* than the same mass of ice (i.e., liquid water is more dense than ice). Consider an equilibrium that is established between ice and liquid water at 0 °C and 1 atm in an apparatus like that shown in Figure 11.47,

$$H_2O(s) \rightleftharpoons H_2O(l)$$

If the piston is forced in slightly, the pressure increases. According to Le Châtelier's principle the system should respond, if possible, in a way that reduces the pressure. This can happen if some of the ice melts, so the ice–liquid mixture won't require as much space. Then the molecules won't push as hard against each other and the walls, and the pressure

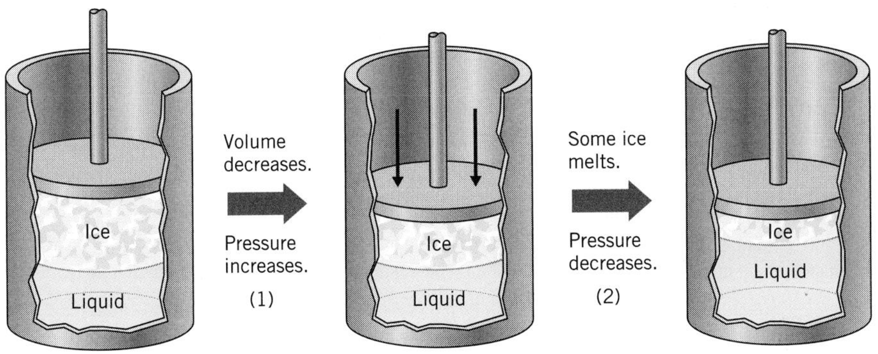

FIG. 11.47 The effect of pressure on the equilibrium $H_2O(s) \rightleftharpoons H_2O(l)$. (1) Pushing down on the piston decreases the volume of both the ice and liquid water by a small amount and increases the pressure. (2) Some of the ice melts, producing the more dense liquid. As the total volume of ice and liquid water decreases, the pressure drops and equilibrium is restored.

will drop. Thus, a pressure increasing disturbance to the system favors a volume decreasing change, which corresponds to the melting of some ice.

Now, suppose we have ice at a pressure just below the solid–liquid line, *BC*. If, *at constant temperature,* we raise the pressure to a point just above the line, the ice will melt and become a liquid. This could only happen if the melting point becomes lower as the pressure is raised.

Water is very unusual. Almost all other substances have melting points that increase with increasing pressure as illustrated by the phase diagram for carbon dioxide (see Figure 11.48). For CO_2 the solid–liquid line slants to the right (it slanted to the left for water). Also notice that carbon dioxide has a triple point that's above 1 atm. At atmospheric pressure, the only equilibrium that can be established is between solid carbon dioxide and its vapor. At a pressure of 1 atm, this equilibrium occurs at a temperature of −78 °C. This is the temperature of dry ice, which sublimes at atmospheric pressure at −78 °C.

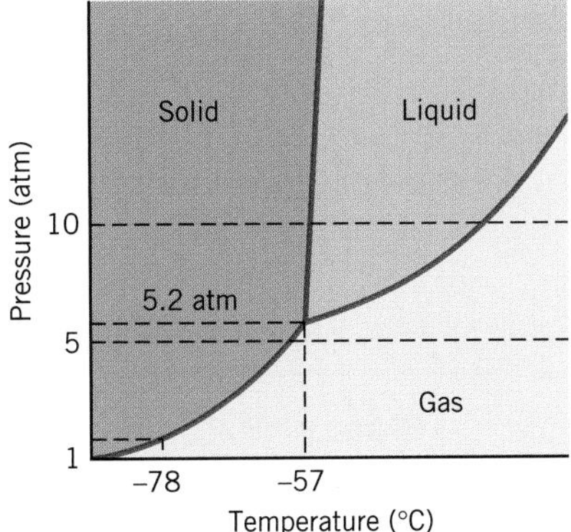

FIG. 11.48 The phase diagram for carbon dioxide.

Single phase regions can be identified in a phase diagram

Besides specifying phase equilibria, the three intersecting lines on a phase diagram serve to define regions of temperature and pressure at which only a single phase can exist. For example, between lines *BC* and *BD* in Figure 11.46 are temperatures and pressures at which water exists as a liquid without being in equilibrium with either vapor or ice. At 760 torr, water is a liquid anywhere between 0 °C and 100 °C. For instance, we are told by the diagram that we can't have ice with a temperature of 25 °C if the pressure is 760 torr (which, of course, you already knew; ice never has a temperature of 25 °C). The diagram also says that we can't have water vapor with a pressure of 760 torr when the temperature is 25 °C (which, again, you already knew; the temperature has to be taken to 100 °C for the vapor pressure to reach 760 torr). Instead, we are told by the phase diagram that the *only* phase for pure water at 25 °C and 1 atm is the liquid. Below 0 °C at 760 torr, water is a solid; above 100 °C at 760 torr, water is a vapor. On the phase diagram for water, the phases that can exist in the different temperature–pressure regions are marked.

EXAMPLE 11.5
Interpreting a Phase Diagram

What phase would we expect for water at 0 °C and 4.58 torr?

ANALYSIS: The words, "What phase. . .," as well as the specified temperature and pressure suggest that we refer to the phase diagram of water (Figure 11.46).

SOLUTION: First we find 0 °C on the temperature axis of the phase diagram of water. Then we move upward until we intersect a line corresponding to 4.58 torr. This intersection occurs in the "Solid" region of the diagram. At 0 °C and 4.58 torr, then, water exists as a solid.

IS THE ANSWER REASONABLE? We've seen that the freezing point of water increases slightly when we lower the pressure, so below 1 atm, water should still be a solid at 0 °C. That agrees with the answer we obtained from the phase diagram.

EXAMPLE 11.6
Interpreting a Phase Diagram

What phase changes occur if water at 0 °C is gradually compressed from a pressure of 2.15 torr to 800 torr?

ANALYSIS: Asking about "what phase changes occur" suggests once again that we use the phase diagram of water (Figure 11.46).

SOLUTION: According to the phase diagram, at 0 °C and 2.15 torr, water exists as a gas (water vapor). As the vapor is compressed, we move upward along the 0 °C line until we encounter the solid–vapor line. Here, an equilibrium will exist as compression gradually transforms the gas into solid ice. Once all the vapor has frozen, further compression raises the pressure and we continue the climb along the 0 °C line until we next encounter the solid–liquid line at 760 torr. As further compression takes place, the solid will gradually melt. After all the ice has melted, the pressure will continue to climb while the water remains a liquid. At 800 torr and 0 °C, the water will be liquid.

IS THE ANSWER REASONABLE? There's not too much we can do to check all this except to take a fresh look at the phase diagram. We do expect that above 760 torr the melting point of ice will be less than 0 °C, so at 0 °C and 800 torr we can anticipate that water will be a liquid.

Practice Exercise 14: The equilibrium line from point B to D in Figure 11.46 is present in another figure in this chapter. Identify what that line represents. (Hint: A review of the other figures will reveal the nature of the line.)

Practice Exercise 15: What phase changes will occur if water at −20 °C and 2.15 torr is heated to 50 °C under constant pressure?

Practice Exercise 16: What phase will water be in if it is at a pressure of 330 torr and a temperature of 50 °C?

Above the critical temperature only one phase is possible

For water (Figure 11.46), the vapor pressure line for the liquid, which begins at point *B*, terminates at point *D*, which is known as the **critical point.** The temperature and pressure at *D* are called the **critical temperature,** T_c, and **critical pressure,** P_c. Above the critical temperature, a distinct liquid phase cannot exist, *regardless of the pressure.*

Figure 11.49 illustrates what happens to a substance as it approaches its critical point. In Figure 11.49*a*, we see a liquid in a container with some vapor above it. We can distinguish

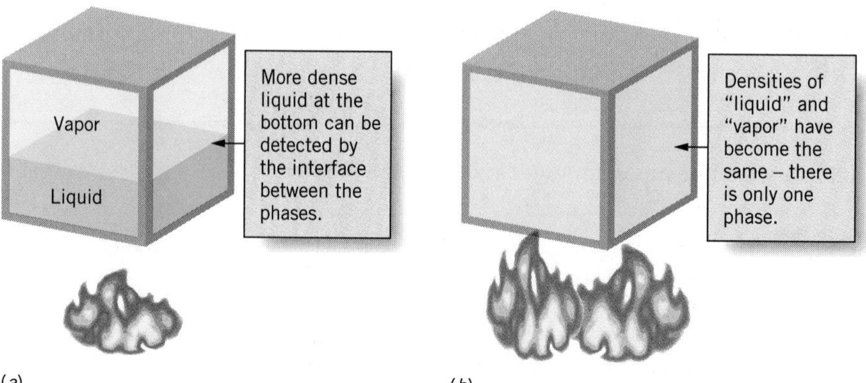

FIG. 11.49 Changes that are observed when a liquid is heated in a sealed container. (*a*) Below the critical temperature. (*b*) Above the critical temperature.

FACETS OF CHEMISTRY 11.2

Decaffeinated Coffee and Supercritical Carbon Dioxide

Many people prefer to avoid caffeine, yet still enjoy a cup of coffee. For them, decaffeinated coffee is just the thing. To satisfy this demand, coffee producers remove caffeine from the coffee beans before roasting them. Several methods have been used, some of which use solvents such as methylene chloride (CH_2Cl_2) or ethyl acetate ($CH_3CO_2C_2H_5$) to dissolve the caffeine. Even though only trace amounts of these solvents remain after the coffee beans are dried, there are those who would prefer not to have any such chemicals in their coffee. And that's where carbon dioxide comes into the picture.

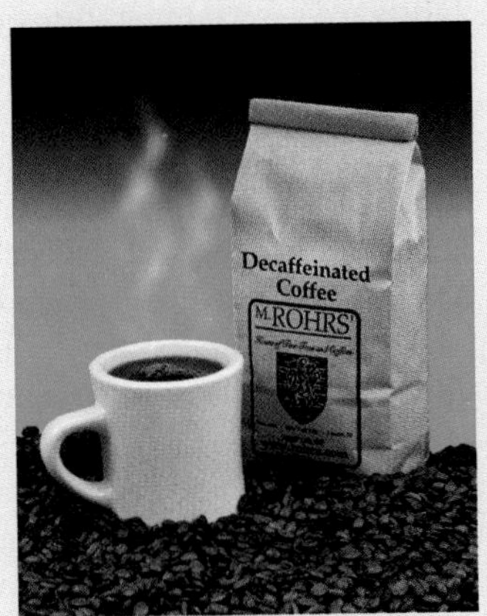

(Andy Washnik.)

It turns out that supercritical carbon dioxide is an excellent solvent for many organic substances, including caffeine. To make it, gaseous CO_2 is heated to a temperature above its critical temperature of 31 °C, typically to about 80 °C. It is then compressed to about 200 atm. This gives it a density near that of a liquid, but with some properties of a gas. The fluid has a very low viscosity and readily penetrates coffee beans that have been softened with steam, drawing out the water and caffeine. After several hours, the CO_2 has removed as much as 97% of the caffeine, and the fluid containing the water and caffeine is then drawn off. When the pressure of the supercritical CO_2 solution is reduced, the CO_2 turns to a gas and the water and caffeine separate. The caffeine is recovered and sold to beverage or pharmaceutical companies. Meanwhile, the pressure over the coffee beans is also reduced and the beans are warmed to about 120 °C, causing residual CO_2 to evaporate. Because CO_2 is not a toxic gas, any small traces of CO_2 that remain are totally harmless.

Decaffeination of coffee is not the only use of supercritical CO_2. It is also used to extract the essential flavor ingredients in spices and herbs for use in a variety of products. As with coffee, using supercritical CO_2 as a solvent completely avoids any potential harm that might be caused by small residual amounts of other solvents.

between the two phases because they have different densities, which causes them to bend light differently. This allows us to see the interface, or surface, between the more dense liquid and the less dense vapor. If this liquid is now heated, two things happen. First, more liquid evaporates. This causes an increase in the number of molecules per cubic centimeter of vapor which, in turn, causes the density of the vapor to increase. Second, the liquid expands (just like mercury does in a thermometer). This means that a given mass of liquid occupies more volume, so its density decreases. As the temperature of the liquid and vapor continue to increase, the vapor density rises and the liquid density falls; they approach each other. Eventually the densities become equal, and a separate liquid phase no longer exists; everything is the same (see Figure 11.49*b*). The highest temperature at which a liquid phase still exists is the critical temperature, and the pressure of the vapor at this temperature is the critical pressure. A substance that has a temperature above its critical temperature and a density near its liquid density is described as a **supercritical fluid.** Supercritical fluids have some unique properties that make them excellent solvents, and one that is particularly useful is supercritical carbon dioxide, which is used as a solvent to decaffeinate coffee.

This interface between liquid and gas is called the **meniscus.**

The values of the critical temperature and critical pressure are unique for every chemical substance and are controlled by the intermolecular attractions (see Table 11.6). Notice

TABLE 11.6 Some Critical Temperatures and Pressures

Compound	T_c (°C)	P_c (atm)
Water	374.1	217.7
Ammonia	132.5	112.5
Carbon dioxide	31	72.9
Ethane (C_2H_6)	32.2	48.2
Methane (CH_4)	−82.1	45.8
Helium	−267.8	2.3

that liquids with strong intermolecular attractions, like water, tend to have high critical temperatures. Under pressure, the strong attractions between the molecules are able to hold them together in a liquid state even when the molecules are jiggling about violently at an elevated temperature. In contrast, substances with weak intermolecular attractions, such as methane and helium, have low critical temperatures. For these substances, even the small amounts of kinetic energy possessed by the molecules at low temperatures is sufficient to overcome the intermolecular attractions and prevent the molecules from sticking together as a liquid, despite being held close together under high pressure.

At room temperature, some gases will liquefy and others will not

When a gaseous substance has a temperature below its critical temperature, it is capable of being liquefied by compressing it. For example, carbon dioxide is a gas at room temperature (approximately 25 °C). This is below its critical temperature of 31 °C. If the $CO_2(g)$ is gradually compressed, a pressure will eventually be reached that lies on the liquid–vapor curve for CO_2, and further compression will cause the CO_2 to liquefy. In fact, that's what happens when a CO_2 fire extinguisher is filled; the CO_2 that's pumped in is a liquid under a high pressure. If you shake a filled CO_2 fire extinguisher, you can feel the liquid sloshing around inside, provided the temperature of the fire extinguisher is below 31 °C (88 °F). When the fire extinguisher is used, a valve releases the pressurized CO_2, which rushes out to extinguish the fire.

On a very hot day, when the temperature is in the 90s, a filled CO_2 fire extinguisher won't give the sensation that it's filled with a liquid. At such temperatures, the CO_2 is in a supercritical state and no separate liquid phase exists.

Gases such as O_2 and N_2, which have critical temperatures far below 0 °C, can never be liquids at room temperature. When they are compressed, they simply become high-pressure gases. To make liquid N_2 or O_2, the gases must be made very cold as well as being compressed to high pressures.

SUMMARY

Physical Properties: Gases, Liquids, and Solids. Gases expand to fill the entire volume of a container. Liquids and solids retain a constant volume if transferred from one container to another. Solids also retain a constant shape. These characteristics are related to how tightly packed the particles are and to the relative strengths of the intermolecular attractions in the different states of matter. Most physical properties depend primarily on intermolecular attractions. In gases, these attractions are weak because the molecules are so far apart. They are much stronger in liquids and solids, where the particles are packed together tightly.

Intermolecular Attractions. Polar molecules attract each other by **dipole–dipole attractions,** which arise because the positive end of one dipole attracts the negative end of another. Nonpolar molecules are attracted to each other by **London dispersion forces,** which are **instantaneous dipole–induced dipole attractions.** London forces are present between all particles, including atoms, polar and nonpolar molecules, and ions. Among different substances, London forces increase with an increase in size of a particle's electron cloud; they also increase with increasing chain length among molecules such as the hydrocarbons. Compact molecules experience weaker London forces than similar long chain molecules. For large molecules, the cumulative effect of large numbers of weak London force interactions can be quite strong and outweigh other intermolecular attractions. **Hydrogen bonding,** a special case of dipole–dipole attractions, occurs between molecules in which hydrogen is covalently bonded to a small, very electronegative atom—principally, nitrogen, oxygen, or fluorine. Hydrogen bonding is much stronger than the other types of intermolecular attractions. **Ion–dipole attractions** occur when ions interact with polar substances. **Ion–induced dipole attractions** result when an ion creates a dipole in a neighboring molecule or ion.

General Properties of Liquids and Solids. Properties that depend mostly on closeness of packing of particles are **compressibility** (or the opposite, incompressibility) and **diffusion.** Diffusion is slow in liquids and almost nonexistent in solids at room temperature. Properties that depend mostly on the strengths of intermolecular attractions are **retention of volume and shape, surface tension,** and **ease of evaporation. Surface tension** is related to the energy needed to expand a liquid's surface area. A liquid can **wet** a surface if its molecules are attracted to the surface about as strongly as they are attracted to each other. **Evaporation** of liquids and solids is endothermic and produces a cooling effect. The overall rate of evaporation increases with increasing surface area. The rate of evaporation from a given surface area of a liquid increases with increasing temperature, and with decreasing intermolecular attractions. Evaporation of a solid is called **sublimation.**

Changes of State. Changes from one physical state to another, such as melting, vaporization, or sublimation, can occur as dynamic equilibria. In a **dynamic equilibrium,** opposing processes occur continually at equal rates, so over time there is no apparent change in the composition of the system. For liquids and solids, equilibria are established when vaporization occurs in a sealed container. A solid is in equilibrium with its liquid at the melting point.

Vapor Pressures. When the rates of evaporation and condensation of a liquid are equal, the vapor above the liquid exerts a pressure called the **equilibrium vapor pressure** (or more commonly, just the **vapor pressure**). The vapor pressure is controlled by the rate of evaporation *per unit surface area.* When the intermolecular attractive forces are large, the rate of evaporation is small and the vapor pressure is small. Vapor pressure increases with increasing temperature because the rate of evaporation increases as the temperature rises. The vapor pressure is independent of the *total* surface area of the liquid. Solids have vapor pressures just as liquids do.

Boiling Point. A substance boils when its vapor pressure equals the prevailing atmospheric pressure. The **normal boiling point** of a liquid is the temperature at which its vapor pressure equals 1 atm. Substances with high boiling points have strong intermolecular attractions.

Energy Changes Associated with Changes of State. On a **heating curve,** flat portions correspond to phase changes in which the heat added changes the potential energies of the particles without changing their average kinetic energy. **Superheating** sometimes occurs when a liquid is heated above its boiling point. On a **cooling curve, supercooling** sometimes occurs when the temperature of the liquid drops below the freezing point of the substance. The enthalpy changes for melting, vaporization of a liquid, and sublimation are the **molar heat of fusion,** ΔH_{fusion}, the **molar heat of vaporization,** $\Delta H_{\text{vaporization}}$, and the **molar heat of sublimation,** $\Delta H_{\text{sublimation}}$, respectively. They are all endothermic and are related in size as follows: $\Delta H_{\text{fusion}} < \Delta H_{\text{vaporization}} < \Delta H_{\text{sublimation}}$. The sizes of these enthalpy changes are large for substances with strong intermolecular attractive forces.

Le Châtelier's Principle. When the equilibrium in a system is upset by a disturbance, the system changes in a direction that minimizes the disturbance and, if possible, brings the system back to equilibrium. By this principle, we find that raising the temperature favors an endothermic change. Decreasing the volume favors a change toward a more dense phase.

Crystalline Solids. Crystalline solids have highly ordered arrangements of particles within them, which can be described in terms of repeating three-dimensional arrays of points called **lattices.** The simplest portion of a lattice is its **unit cell.** Many structures can be described with the same lattice by associating different units (atoms, molecules, or ions) to lattice points and by changing the dimensions of the unit cell. Three cubic unit cells are possible—**simple cubic, face-centered cubic,** and **body-centered cubic.** Sodium chloride and many other alkali metal halides crystallize in the **rock salt structure,** which contains four formula units per unit cell. Two modes of closest packing of atoms are **cubic closest packing (ccp)** and **hexagonal closest packing (hcp).** The ccp structure has an A-B-C-A-B-C. . . alternating stacking of layers of spheres; the hcp structure has an A-B-A-B-. . . stacking of layers. **Amorphous** solids lack the internal structure of crystalline solids. Glass is an amorphous solid and is sometimes called a **supercooled liquid.**

X-Ray Diffraction. Information about crystal structures is obtained experimentally from **X-ray diffraction patterns** produced by constructive and destructive interference of X rays scattered by atoms. Distances between planes of atoms in a crystal can be calculated by the Bragg equation, $n\lambda = 2d \sin \theta$, where n is a whole number, λ is the wavelength of the X rays, d is the distance between planes of atoms producing the diffracted beam, and θ is the angle at which the diffracted X-ray beam emerges relative to the planes of atoms producing the diffracted beam.

Crystal Types. Crystals can be divided into four general types: **ionic, molecular, covalent,** and **metallic.** Their properties depend on the kinds of particles within the lattice and on the attractions between the particles, as summarized in Table 11.5.

Phase Diagrams. Temperatures and pressures at which equilibria can exist between phases are given graphically in a **phase diagram.** The three equilibrium lines intersect at the **triple point.** The liquid–vapor line terminates at the **critical point.** At the **critical temperature,** a liquid has a vapor pressure equal to its **critical pressure.** Above the critical temperature a liquid phase cannot be formed; the single phase that exists is called a **supercritical fluid.** The equilibrium lines also divide a phase diagram into temperature–pressure regions in which a substance can exist in just a single phase. Water is different from most substances in that its melting point decreases with increasing pressure.

TOOLS FOR PROBLEM SOLVING

The concepts that you've learned in this chapter are collected below. They can be applied as tools in solving problems. Study each one carefully so that you know what each is used for. When faced with solving a problem, recall what each tool does and consider whether it will be helpful in finding a solution. This will aid you in selecting the tools you need. If necessary, refer to this table when working on the Review Problems that follow.

Relationship between intermolecular forces and molecular structure *(page 433)* From molecular structure, we can determine whether a molecule is polar or not and whether it has N—H or O—H bonds. This lets us predict and compare the strengths of intermolecular attractions. You should be able to identify when dipole–dipole, London, and hydrogen bonding occurs. See the summary Table 11.3 on page *440.*

Factors that affect rates of evaporation and vapor pressure *(pages 446, 447, and 448)* They allow us to compare the relative rates of evaporation based on temperature and the strengths of intermolecular forces in substances. They also allow us to compare the strengths of intermolecular forces based on the relative magnitudes of vapor pressures at a given temperature.

Boiling points of substances *(page 451)* They allow us to compare the strengths of intermolecular forces in substances based on their boiling points.

Enthalpy changes during phase changes *(pages 453 and 455)* They allow us to compare the strengths of intermolecular forces in substances based on relative values of $\Delta H_{\text{vaporization}}$ and $\Delta H_{\text{sublimation}}$.

Le Châtelier's principle *(page 456)* Le Châtelier's principle enables us to predict the direction in which the position of equilibrium is shifted when a dynamic equilibrium is upset by a disturbance. You should be able to predict how the position of equilibrium between phases is affected by temperature and pressure changes.

Bragg equation *(page 463)*

$$n\lambda = 2d \sin \theta$$

Using the wavelength of X rays and angles at which X rays are diffracted from a crystal, the distances between planes of atoms can be calculated.

Unit cell structures for simple cubic, face-centered cubic, and body-centered cubic lattices *(page 458)* By knowing the arrangements of atoms in these unit cells, we can use the dimensions of the unit cell to calculate atomic radii and other properties.

Properties of crystal types *(page 466)* By examining certain physical properties of a solid (hardness, melting point, electrical conductivity in the solid and liquid state), we can often predict the nature of the particles that occupy lattice sites in the solid and the kinds of attractive forces between them.

Phase diagram *(page 467)* We use a phase diagram to identify temperatures and pressures at which equilibrium can exist between phases of a substance, and to identify conditions under which only a single phase can exist.

QUESTIONS, PROBLEMS, AND EXERCISES

Answers to problems whose numbers are printed in color are given in Appendix B. More challenging problems are marked with asterisks. ILW = Interactive Learningware solution is available at www.wiley.com/college/brady. OH = an Office Hours video is available for this problem.

REVIEW QUESTIONS

Comparisons among the States of Matter

11.1 Why are the intermolecular attractive forces stronger in liquids and solids than they are in gases?

11.2 Compare the behavior of gases, liquids, and solids when they are transferred from one container to another.

11.3 For a given substance, how do the intermolecular attractive forces compare in its gaseous, liquid, and solid states?

Intermolecular Attractions

11.4 Which kinds of attractive forces, intermolecular or intramolecular, are responsible for chemical properties? Which kind are responsible for physical properties?

11.5 Describe *dipole–dipole attractions.*

11.6 What are *London forces*? How are they affected by the sizes of the atoms in a molecule? How are they affected by the number of atoms in a molecule? How are they affected by the shape of a molecule?

11.7 Define *polarizability.* How does this property affect the strengths of London forces?

11.8 Which nonmetals, besides hydrogen, are most often involved in hydrogen bonding? Why these and not others?

11.9 Which is expected to have the higher boiling point, C_8H_{18} or C_4H_{10}? Explain your choice.

11.10 Ethanol and dimethyl ether have the same molecular formula, C_2H_6O. Ethanol boils at 78.4 °C, whereas dimethyl ether boils at −23.7 °C. Their structural formulas are

CH_3CH_2OH	CH_3OCH_3
ethanol	dimethyl ether

Explain why the boiling point of the ether is so much lower than the boiling point of ethanol.

11.11 How do the strengths of covalent bonds and dipole–dipole attractions compare? How do the strengths of ordinary dipole–dipole attractions compare with the strengths of hydrogen bonds?

11.12 For each pair, in which compound are the ion–induced dipole attractions stronger? (a) CaO or CaS, (b) MgO or Al_2O_3

General Properties of Liquids and Solids

11.13 Name two physical properties of liquids and solids that are controlled primarily by how tightly packed the particles are. Name three that are controlled mostly by the strengths of the intermolecular attractions.

11.14 Why does diffusion occur more slowly in liquids than in gases? Why does diffusion occur extremely slowly in solids?

11.15 On the basis of kinetic theory, would you expect the rate of diffusion in a liquid to increase or decrease as the temperature is increased? Explain your answer.

11.16 What is *surface tension*? Why do molecules at the surface of a liquid behave differently from those within the interior?

11.17 Which liquid is expected to have the larger surface tension at a given temperature, CCl_4 or H_2O? Explain your answer.

11.18 What does *wetting* of a surface mean? What is a *surfactant*? What is its purpose and how does it function?

11.19 Polyethylene plastic consists of long chains of carbon atoms, each of which is also bonded to hydrogens, as shown below:

```
       H   H   H   H   H   H   H   H
       |   |   |   |   |   |   |   |
  ...—C—C—C—C—C—C—C—C—...
       |   |   |   |   |   |   |   |
       H   H   H   H   H   H   H   H
```

Water forms beads when placed on a polyethylene surface. Why?

11.20 The structural formula for glycerol is

```
     H    H    H
     |    |    |
  H—C——C——C—H
     |    |    |
    OH   OH   OH
```

Would you expect this liquid to wet glass surfaces? Explain your answer.

11.21 On the basis of what happens on a molecular level, why does evaporation lower the temperature of a liquid?

11.22 On the basis of the distribution of kinetic energies of the molecules of a liquid, explain why increasing the liquid's temperature increases the rate of evaporation.

11.23 How is the rate of evaporation of a liquid affected by increasing the surface area of the liquid? How is the rate of evaporation affected by the strengths of intermolecular attractive forces?

11.24 During the cold winter months, snow often disappears gradually without melting. How is this possible? What is the name of the process responsible for this phenomenon?

Changes of State and Equilibrium

11.25 What terms do we use to describe the following changes of state?
(a) solid $\longrightarrow$ gas, (b) liquid $\longrightarrow$ gas, (c) gas $\longrightarrow$ liquid,
(d) solid $\longrightarrow$ liquid, (e) liquid $\longrightarrow$ solid

11.26 When a molecule escapes from the surface of a liquid by evaporation, it has a kinetic energy that's much larger than the average KE. Why is it likely that after being in the vapor for a while its kinetic energy will be much less? If this molecule collides with the surface of the liquid, is it likely to bounce out again?

11.27 Why does a molecule of a vapor that collides with the surface of a liquid tend to be captured by the liquid, even if the incoming molecule has a large kinetic energy?

11.28 When an equilibrium is established in the evaporation of a liquid into a sealed container, we refer to it as a *dynamic equilibrium*. Why?

11.29 Viewed at a molecular level, what is happening when a dynamic equilibrium is established between the liquid and solid forms of a substance? What is the temperature called at which there is an equilibrium between a liquid and a solid?

11.30 Is it possible to establish an equilibrium between a solid and its vapor? Explain.

Vapor Pressure

11.31 Define *equilibrium vapor pressure.* Why do we call the equilibrium involved a *dynamic equilibrium*?

11.32 Explain why changing the volume of a container in which there is a liquid–vapor equilibrium has no effect on the equilibrium vapor pressure.

11.33 Why doesn't a change in the surface area of a liquid cause a change in the equilibrium vapor pressure?

11.34 What effect does increasing the temperature have on the equilibrium vapor pressure of a liquid? Why?

11.35 Why does moisture condense on the outside of a cool glass of water in the summertime?

11.36 Why do we feel more uncomfortable in humid air at 90 °F than in dry air at 90 °F?

Boiling Points of Liquids

11.37 Define *boiling point* and *normal boiling point.*

11.38 Why does the boiling point vary with atmospheric pressure?

11.39 Mt. Kilimanjaro in Tanzania is the tallest peak in Africa (19,340 ft). The normal barometric pressure at the top of this mountain is about 345 torr. At what Celsius temperature would water be expected to boil there? (See Figure 11.21.)

11.40 When liquid ethanol begins to boil, what is present inside the bubbles that form?

11.41 The radiator cap of an automobile engine is designed to maintain a pressure of approximately 15 $lb/in.^2$ above normal atmospheric pressure. How does this help prevent the engine from "boiling over" in hot weather?

11.42 Butane, C_4H_{10}, has a boiling point of -0.5 °C (which is 31 °F). Despite this, liquid butane can be seen sloshing about inside a typical butane lighter, even at room temperature. Why isn't the butane boiling inside the lighter at room temperature?

11.43 Why does H_2S have a lower boiling point than H_2Se? Why does H_2O have a much higher boiling point than H_2S?

11.44 An H—F bond is more polar than an O—H bond, so HF forms stronger hydrogen bonds than H_2O. Nevertheless, HF has a lower boiling point than H_2O. Explain why this is so.

Energy Changes That Accompany Changes of State

11.45 Below is a cooling curve for one mole of a substance.

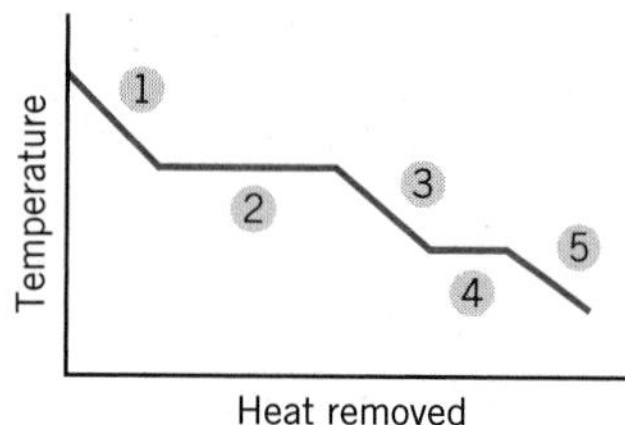

(a) On which portions of this graph do we find the average kinetic energy of the molecules of the substance changing?
(b) On which portions of this graph is the amount of heat removed related primarily to a lowering of the potential energy of the molecules?
(c) Which portion of the graph corresponds to the release of the heat of vaporization?
(d) Which portion of the graph corresponds to the release of the heat of fusion?
(e) Which is larger, the heat of fusion or the heat of vaporization?
(f) On the graph, indicate the melting point of the solid.
(g) On the graph, indicate the boiling point of the liquid.
(h) On the drawing, indicate how supercooling of the liquid would affect the graph.

11.46 Why is $\Delta H_{\text{vaporization}}$ larger than ΔH_{fusion}? How does $\Delta H_{\text{sublimation}}$ compare with $\Delta H_{\text{vaporization}}$? Explain your answer.

11.47 Would the "heat of condensation," $\Delta H_{\text{condensation}}$, be exothermic or endothermic?

11.48 Hurricanes can travel for thousands of miles over warm water, but they rapidly lose their strength when they move over a large land mass or over cold water. Why?

11.49 Ethanol (grain alcohol) has a molar heat of vaporization of 39.3 kJ mol^{-1}. Ethyl acetate, a common solvent, has a molar heat of vaporization of 32.5 kJ mol^{-1}. Which of these substances has the larger intermolecular attractions?

11.50 A burn caused by steam is much more serious than one caused by the same amount of boiling water. Why?

11.51 Arrange the following substances in order of their increasing values of $\Delta H_{\text{vaporization}}$: (a) HF, (b) CH_4, (c) CF_4, and (d) HCl.

Le Châtelier's Principle

11.52 State Le Châtelier's principle in your own words.

11.53 What do we mean by the *position of equilibrium*?

11.54 Use Le Châtelier's principle to predict the effect of adding heat in the equilibrium: solid + heat $\rightleftharpoons$ liquid.

11.55 Use Le Châtelier's principle to explain why lowering the temperature lowers the vapor pressure of a solid.

Crystalline Solids and X-Ray Diffraction

11.56 What is the difference between a crystalline solid and an amorphous solid?

11.57 What is a *lattice*? What is a *unit cell*?

11.58 What relationship is there between a crystal lattice and its unit cell?

11.59 The diagrams below illustrate typical arrangements of paving bricks in a patio or driveway. Sketch the unit cells that correspond to these patterns of bricks.

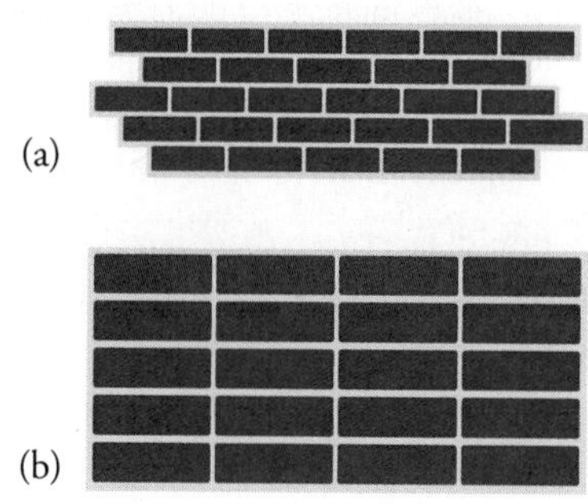

11.60 Below is illustrated the way the atoms of two different elements are packed in a certain solid. The nuclei occupy positions at the corners of a cube. Is this cube the unit cell for this substance? Explain your answer.

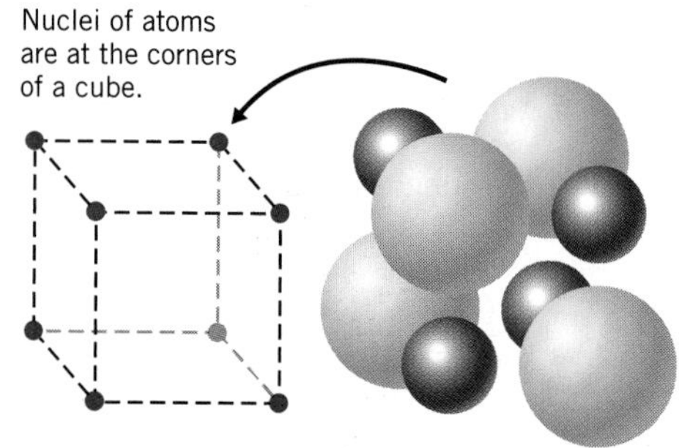

11.61 Make a sketch of a layer of sodium ions and chloride ions in a NaCl crystal. Indicate how the ions are arranged in a face-centered cubic pattern, regardless of whether we place lattice points at the Cl^- ions or Na^+ ions.

11.62 How do the crystal structures of copper and gold differ? In what way are they similar? On the basis of the locations of the elements in the periodic table, what kind of crystal structure would you expect for silver?

11.63 What kind of lattice does zinc sulfide have? What kind of lattice does calcium fluoride have?

11.64 Only 14 different kinds of crystal lattices are possible. How can this be true, considering the fact that there are millions of different chemical compounds that are able to form crystals?

11.65 Write the Bragg equation and define the symbols.

11.66 Why can't $CaCl_2$ or $AlCl_3$ form crystals with the same structure as NaCl?

Crystal Types

11.67 What kinds of particles are located at the lattice sites in a metallic crystal?

11.68 What kinds of attractive forces exist between particles in (a) molecular crystals, (b) ionic crystals, and (c) covalent crystals?

11.69 Why are covalent crystals sometimes called *network solids*?

Amorphous Solids

11.70 What does the word *amorphous* mean?

11.71 What is an *amorphous solid*? Compare what happens when crystalline and amorphous solids are broken into pieces.

Phase Diagrams

11.72 For most substances, the solid is more dense than the liquid. Use Le Châtelier's principle to explain why the melting point of such substances should *increase* with increasing pressure. Sketch the phase diagram for such a substance, being sure to have the solid–liquid equilibrium line slope in the correct direction.

11.73 Define *critical temperature* and *critical pressure.*

11.74 What is a *supercritical fluid*? Why is supercritical CO_2 used to decaffeinate coffee?

11.75 What phases of a substance are in equilibrium at the triple point?

11.76 Why doesn't CO_2 have a normal boiling point?

11.77 At room temperature, hydrogen can be compressed to very high pressures without liquefying. On the other hand, butane becomes a liquid at high pressure (at room temperature). What does this tell us about the critical temperatures of hydrogen and butane?

REVIEW PROBLEMS

Intermolecular Attractions and Molecular Structure

11.78 Which liquid evaporates faster at 25 °C, diethyl ether (an anesthetic) or butanol (a solvent used in the preparation of shellac and varnishes)? Both have the molecular formula $C_4H_{10}O$, but their structural formulas are different, as shown below.

$CH_3CH_2CH_2CH_2OH$ (butanol) $CH_3CH_2-O-CH_2CH_3$ (diethyl ether)

OH **11.79** Which compound should have the higher vapor pressure at 25 °C, butanol or diethyl ether? Which should have the higher boiling point?

11.80 What kinds of intermolecular attractive forces (dipole–dipole, London, hydrogen bonding) are present in the following substances?
(a) HF (b) PCl_3 (c) SF_6 (d) SO_2

11.81 What kinds of intermolecular attractive forces are present in the following substances?

(a) $CH_3-\overset{\overset{\displaystyle O}{\|}}{C}-OH$ (b) H_2S (c) SO_3 (d) CH_3NH_2

OH **11.82** Consider the compounds $CHCl_3$ (chloroform, an important solvent that was once used as an anesthetic) and $CHBr_3$ (bromoform, which has been used as a sedative). Compare the strengths of their dipole–dipole attractions and the strengths of their London forces. Their boiling points are 61 °C and 149 °C, respectively. For these compounds, which kinds of attractive forces (dipole–dipole or London) are more important in determining their boiling points? Justify your answer.

11.83 Carbon dioxide does not liquefy at atmospheric pressure, but instead forms a solid that sublimes at −78 °C. Nitrogen dioxide forms a liquid that boils at 21 °C at atmospheric pressure. How do these data support the statement that CO_2 is a linear molecule whereas NO_2 is nonlinear?

11.84 Which should have the higher boiling point, ethanol (CH_3CH_2OH, found in alcoholic beverages) or ethanethiol (CH_3CH_2SH, a foul-smelling liquid found in the urine of rabbits that have feasted on cabbage)?

11.85 How do the strengths of London forces compare in $CO_2(l)$ and $CS_2(l)$? Which of these is expected to have the higher boiling point? (Check your answer by referring to the *Handbook of Chemistry and Physics*, which is available in your school library.)

OH **11.86** Below are the vapor pressures of some relatively common chemicals measured at 20 °C. Arrange these substances in order of increasing intermolecular attractive forces.

Benzene, C_6H_6	80 torr
Acetic acid, $HC_2H_3O_2$	11.7 torr
Acetone, C_3H_6O	184.8 torr
Diethyl ether, $C_4H_{10}O$	442.2 torr
Water	17.5 torr

11.87 The boiling points of some common substances are given here. Arrange these substances in order of increasing strengths of intermolecular attractions.

Ethanol, C_2H_5OH	78.4 °C
Ethylene glycol, $C_2H_4(OH)_2$	197.2 °C
Water	100 °C
Diethyl ether, $C_4H_{10}O$	34.5 °C

Energy Changes That Accompany Changes of State

OH **11.88** The molar heat of vaporization of water at 25 °C is +43.9 kJ mol^{-1}. How many kilojoules of heat would be required to vaporize 125 mL (0.125 kg) of water?

11.89 The molar heat of vaporization of acetone, C_3H_6O, is 30.3 kJ mol^{-1} at its boiling point. How many kilojoules of heat would be liberated by the condensation of 5.00 g of acetone?

11.90 Suppose 45.0 g of water at 85 °C is added to 105.0 g of ice at 0 °C. The molar heat of fusion of water is 6.01 kJ mol^{-1}, and the specific heat of water is 4.18 J/g °C. On the basis of these data, (a) what will be the final temperature of the mixture and (b) how many grams of ice will melt?

11.91 A cube of solid benzene (C_6H_6) at its melting point and weighing 10.0 g is placed in 10.0 g of water at 30 °C. Given that the heat of fusion of benzene is 9.92 kJ mol^{-1}, to what temperature will the water have cooled by the time all of the benzene has melted?

Crystalline Solids and X-Ray Diffraction

11.92 How many zinc and sulfide ions are present in the unit cell of zinc sulfide? (See Figure 11.35.)

11.93 How many copper atoms are within the face-centered cubic unit cell of copper? (Hint: See Figure 11.31 and add up all the *parts* of atoms in the fcc unit cell.)

OH ILW **11.94** The atomic radius of nickel is 1.24 Å. Nickel crystallizes in a face-centered cubic lattice. What is the length of the edge of the unit cell expressed in angstroms and in picometers?

11.95 Silver forms face-centered cubic crystals. The atomic radius of a silver atom is 144 pm. Draw the face of a unit cell with the nuclei of the silver atoms at the lattice points. The atoms are in contact along the diagonal. Calculate the length of an edge of this unit cell.

11.96 Potassium ions have a radius of 133 pm, and bromide ions have a radius of 195 pm. The crystal structure of potassium bromide is the same as for sodium chloride. Estimate the length of the edge of the unit cell in potassium bromide.

11.97 The unit cell edge in sodium chloride has a length of 564.0 pm. The sodium ion has a radius of 95 pm. What is the *diameter* of a chloride ion?

OH **11.98** Calculate the angles at which X rays of wavelength 229 pm will be observed to be defracted from crystal planes spaced (a) 1000 pm apart and (b) 250 pm apart. Assume $n = 1$ for both calculations.

11.99 Calculate the interplanar spacings (in picometers) that correspond to defracted beams of X rays at $\theta = 20.0°$, 27.4°, and 35.8°, if the X rays have a wavelength of 141 pm. Assume that $n = 1$.

11.100 Cesium chloride forms a simple cubic lattice in which Cs^+ ions are at the corners and a Cl^- ion is in the center (see Figure 11.34). The cation–anion contact occurs along the *body diagonal* of the unit cell. (The body diagonal starts at one corner and then runs through the center of the cell to the opposite corner.) The length of the edge of the unit cell is 412.3 pm. The Cl^- ion has a radius of 181 pm. Calculate the radius of the Cs^+ ion.

11.101 Rubidium chloride has the rock salt structure. Cations and anions are in contact along the edge of the unit cell, which is 658 pm long. The radius of the chloride ion is 181 pm. What is the radius of the Rb^+ ion?

Crystal Types

11.102 Tin(IV) chloride, $SnCl_4$, has soft crystals with a melting point of −30.2 °C. The liquid is nonconducting. What type of crystal is formed by $SnCl_4$?

11.103 Elemental boron is a semiconductor, is very hard, and has a melting point of about 2250 °C. What type of crystal is formed by boron?

11.104 Columbium is another name for one of the elements. This element is shiny, soft, and ductile. It melts at 2468 °C, and the solid conducts electricity. What kind of solid does columbium form?

11.105 Elemental phosphorus consists of soft white "waxy" crystals that are easily crushed and melt at 44 °C. The solid does not conduct electricity. What type of crystal does phosphorus form?

11.106 Indicate which type of crystal (ionic, molecular, covalent, metallic) each of the following would form when it solidifies. (a) Br_2, (b) LiF, (c) MgO, (d) Mo, (e) Si, (f) PH_3, (g) NaOH

11.107 Indicate which type of crystal (ionic, molecular, covalent, metallic) each of the following would form when it solidifies: (a) O_2, (b) H_2S, (c) Pt, (d) KCl, (e) Ge, (f) $Al_2(SO_4)_3$, (g) Ne

Phase Diagrams

11.108 Sketch the phase diagram for a substance that has a triple point at −15.0 °C and 0.30 atm, melts at −10.0 °C at 1 atm, and has a normal boiling point of 90 °C.

11.109 Based on the phase diagram of the preceding problem, below what pressure will the substance undergo sublimation? How does the density of the liquid compare with the density of the solid?

11.110 According to Figure 11.48, what phase(s) should exist for CO_2 at (a) −60 °C and 6 atm, (b) −60 °C and 2 atm, (c) −40 °C and 10 atm, and (d) −57 °C and 5.2 atm?

OH **11.111** Looking at the phase diagram for CO_2 (Figure 11.48), how can we tell that solid CO_2 is more dense than liquid CO_2?

ADDITIONAL EXERCISES

11.112 Make a list of *all* of the attractive forces that exist in solid Na_2SO_3.

OH **11.113** Calculate the mass of water vapor present in 10.0 L of air at 20 °C if the relative humidity is 75%.

11.114 Should acetone molecules be attracted to water molecules more strongly than to other acetone molecules? Explain your answer. The structure of acetone is shown below.

```
    H  O  H
    |  ‖  |
 H—C—C—C—H
    |     |
    H     H
```

OH **11.115** Acetic acid has a heat of fusion of 10.8 kJ mol^{-1} and a heat of vaporization of 24.3 kJ mol^{-1}.

$$HC_2H_3O_2(s) \longrightarrow HC_2H_3O_2(l) \quad \Delta H_{fusion} = 10.8 \text{ kJ mol}^{-1}$$

$$HC_2H_3O_2(l) \longrightarrow HC_2H_3O_2(g) \quad \Delta H_{vaporization} = 24.3 \text{ kJ mol}^{-1}$$

Use Hess's law to estimate the value for the heat of sublimation of acetic acid, in kilojoules per mole.

***11.116** Melting point is sometimes used as an indication of the extent of covalent bonding in a compound—the higher the melting point, the more ionic the substance. On this basis, oxides of metals seem to become less ionic as the charge on the metal ion increases. Thus, Cr_2O_3 has a melting point of 2266 °C whereas CrO_3 has a melting point of only 196 °C. The explanation often given is similar in some respects to explanations of the variations in the strengths of certain intermolecular attractions given in this chapter. Provide an explanation for the greater degree of electron sharing in CrO_3 as compared with Cr_2O_3.

***11.117** When warm moist air sweeps in from the ocean and rises over a mountain range, it expands and cools. Explain how this cooling is related to the attractive forces between gas molecules. Why does this cause rain to form? When the air drops down the far side of the range, its pressure rises as it is compressed. Explain why this causes the air temperature to rise. How does the humidity of this air compare with the air that originally came in off the ocean? Now, explain why the coast of California is lush farmland, whereas valleys (such as Death Valley) that lie to the east of the tall Sierra Nevada are arid and dry.

OH ***11.118** Gold crystallizes in a face-centered cubic lattice. The edge of the unit cell has a length of 407.86 pm. The density of gold is 19.31 g cm^{-3}. Use these data and the atomic mass of gold to calculate the value of Avogadro's number.

11.119 Gold crystallizes with a face-centered cubic unit cell with an edge length of 407.86 pm. Calculate the atomic radius of gold in units of picometers.

*11.120 Calculate the amount of empty space (in pm^3) in simple cubic, body-centered cubic, and face-centered cubic unit cells if the lattice points are occupied by identical atoms with a diameter of 1.00 pm. Which of these structures gives the most efficient packing of atoms.

11.121 Silver has an atomic radius of 144 pm. What would be the density of silver (in g cm^{-3}) if it were to crystallize in (a) a simple cubic lattice, (b) a body-centered cubic lattice, and (c) a face-centered cubic lattice? The actual density of silver is 10.6 g cm^{-3}. Which cubic lattice does silver have?

*11.122 Potassium chloride crystallizes with the rock salt structure. When bathed in X rays, the layers of atoms corresponding to the surfaces of the unit cell produce a diffracted beam of X rays at an angle of 12.8°. Calculate the density of KCl.

11.123 Why do clouds form when the humid air of a weather system called a *warm front* encounters the cool, relatively dry air of a *cold front*?

EXERCISES IN CRITICAL THINKING

11.124 Supercritical CO_2 is used to decaffeinate coffee. Propose other uses for supercritical fluids.

11.125 Freshly precipitated crystals are usually very small. Over time the crystals tend to grow larger. How can we use the concept of dynamic equilibrium to explain this phenomenon?

11.126 What are some "everyday" applications of Le Châtelier's principle? For example, we turn up the heat in an oven to cook a meal faster.

11.127 Lubricants, oils, greases, etc. are very important in everyday life. Explain how a lubricant works in terms of intermolecular forces.

11.128 Galileo's thermometer is a tube of liquid that has brightly colored glass spheres that float or sink depending on the temperature. Using your knowledge from the past two chapters, explain all of the processes that are involved in this thermometer.

11.129 Use the Clausius–Clapeyron equation to plot the vapor pressure curve of a gas that has a heat of vaporization of 21.7 kJ mol^{-1} and a boiling point of 48 °C. Compare your results to the other vapor pressure curves in this chapter.

11.130 Will the near weightless environment of the international space station have any effect on intermolecular forces and chemical reactions? What types of chemical reactions may benefit from a weightless environment?

PROPERTIES OF SOLUTIONS; MIXTURES OF SUBSTANCES AT THE MOLECULAR LEVEL

12

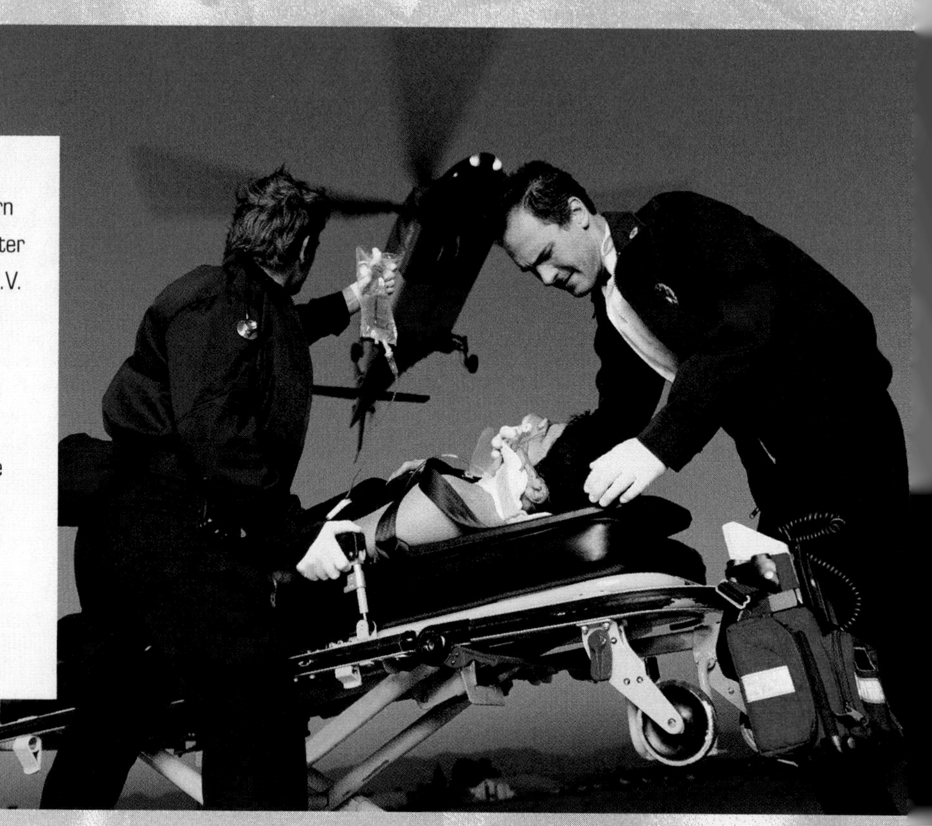

Life-saving intravenous fluids are as indispensable to modern medicine as the rapid helecopter transport to the hospital. An I.V. helps maintain body fluids and also provides a quick way to administer drug solutions to a patient. These solutions are critical to patient care and are best understood by looking at the molecular level of the solution process presented in this chapter. *(Mike Powell/Riser/Getty Images)*

CHAPTER OUTLIN

12.1 Substances mix spontaneously when there is no energy barrier to mixing

12.2 Heats of solution come from unbalanced intermolecular attractions

12.3 A substance's solubility changes with temperature

12.4 Gases become more soluble at higher pressures

12.5 Molarity changes with temperature; molality, weight percentages, and mole fractions do not

12.6 Solutes lower the vapor pressure of a solvent

12.7 Solutions have lower melting points and higher boiling points than pure solvents

12.8 Osmosis is a flow of solvent through a semipermeable membrane due to unequal concentrations

12.9 Ionic solutes affect colligative properties differently from nonionic solutes

THIS CHAPTER IN CONTEXT In Chapter 4 we discussed solutions as a medium for carry chemical reactions. Our focus at that time was the kinds of reactions that take place in aqueous solution. chapter we will examine how the addition of a solute affects the *physical properties* of the mixture. For the c we will see that changes in these physical properties can be used to determine molar masses. The expe chemist uses the properties of various mixtures to dissolve or precipitate substances in order to purify them daily lives we add ethylene glycol (antifreeze) to a car's radiator to protect it against freezing and overheating since water has a lower freezing point and a higher boiling point when such a solute is dissolved in it.

Here we will study not only aqueous solutions, but solutions involving other solvents as well. We will concentrate on liquid solutions, although solutions can also be gaseous or solid. In fact, all gases mix completely at the molecular level, so the gas mixtures we studied in Chapter 10 were gaseous solutions. Solid solutions of metals are called alloys and include such materials as brass and bronze.

12.1 SUBSTANCES MIX SPONTANEOUSLY WHEN THERE IS NO ENERGY BARRIER TO MIXING

There is a wide variation in the ability of liquids to dissolve different solutes. For example, water and gasoline do not dissolve in each other, but water and alcohol do. Similarly, salt will dissolve in water, but not in liquid hydrocarbons such as paint thinner. To understand the reasons for these differences, we must examine the factors that drive solution formation as well as those that inhibit it.

In Chapter 10 you learned that all gases mix spontaneously to form homogeneous mixtures (i.e., solutions). If two gases are placed in separate compartments of a container such as that in Figure 12.1, and the movable partition between them is removed, the molecules will begin to intermingle. The random motions of the molecules will cause them to diffuse one into the other until a uniform mixture is achieved.

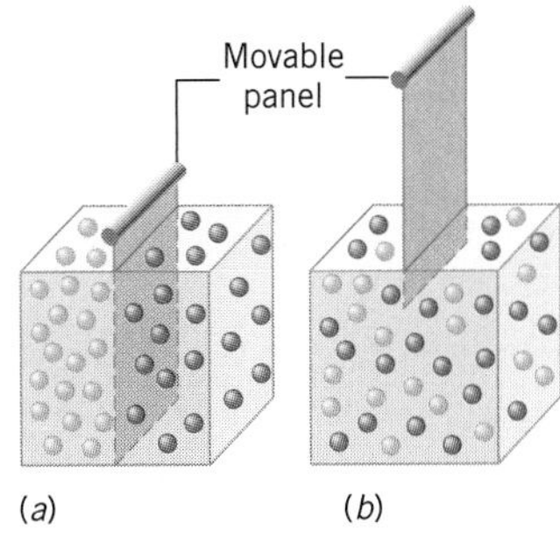

FIG. 12.1 Mixing of gases. When two gases, initially in separate compartments (*a*), suddenly find themselves in the same container (*b*), they mix spontaneously.

The spontaneous mixing of gases illustrates one of nature's strong "driving forces" for change. *A system, left to itself, will tend toward the most probable state.*[1] At the instant we remove the partition, the container holds two separate gas samples, in contact but unmixed. This represents a highly improbable state because of the natural motions of the molecules. The vastly more probable distribution is one in which the molecules are thoroughly mixed, so formation of the gaseous solution involves a transition from a highly improbable state to a highly probable one.

Attractive forces affect the ability of solutions to form

The drive to attain the most probable state favors the formation of *any* solution. What limits the ability of most substances to mix completely, however, are intermolecular forces of attraction. Such attractions are negligible in gases, so regardless of the chemical makeup of the molecules, the forces are unable to prevent them from mixing. That's why all gases spontaneously form solutions with each other. In liquids and solids, however, the situation is much different because intermolecular attractions are so much stronger.

For a liquid solution to form, there must be a balance among the attractive forces so the natural tendency of particles to intermingle can proceed. In other words, the attractive forces between molecules within the solvent and between molecules within the solute must be about as strong as attractions between solute and solvent molecules. Let's look at two examples, mixtures of water with benzene and with ethanol.

```
          H
          |
 H     C     H               H   H
   \ // \  /                 |   |
    C     C              H—C—C—O—H
    |     ||                 |   |
    C     C                  H   H
   / \\ / \
 H     C     H
       |
       H
```

benzene (nonpolar)

ethanol (contains a polar OH group)

Water and benzene (C_6H_6) are insoluble in each other. In water, there are strong hydrogen bonds between the molecules; in benzene the molecules attract each other by relatively weak London forces and are not able to form hydrogen bonds. Suppose that we

[1] In Chapter 18 this driving force will be called *entropy*.

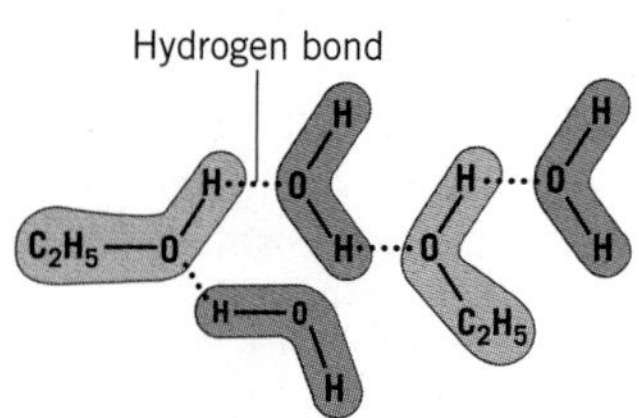

FIG. 12.2 Hydrogen bonds in aqueous ethanol. Ethanol molecules form hydrogen bonds (···) to water molecules.

did manage to disperse water molecules in benzene. As they move about, the water molecules would occasionally encounter each other. Because the water molecules attract each other so much more strongly than they attract benzene molecules, hydrogen bonds would cause water molecules to stick together at each such encounter. This would continue to happen until all the water was in a separate phase. Thus, a solution of water in benzene would not be stable and would gradually separate into two phases. We say water and benzene are **immiscible,** meaning they are mutually insoluble.

Water and ethanol (C_2H_5OH), on the other hand, are **miscible;** they are soluble in all proportions. This is because water and ethanol molecules can form hydrogen bonds with each other that are nearly equivalent to those in the separate pure liquids (Figure 12.2). Mixing these molecules offers little resistance, so they are able to mingle relatively freely.

Like dissolves like rule

Observations like those for benzene–water and ethanol–water mixtures led to a generalization often called the **"like dissolves like" rule:** when solute and solvent have molecules "like" each other in polarity, they tend to form a solution. When solute and solvent molecules are quite different in polarity, solutions of any appreciable concentration do not form. The rule has long enabled chemists to use chemical composition and molecular structure to predict the likelihood of two substances dissolving in each other.

The solubility of solids depends on the relative strengths of intermolecular attractions

The "like dissolves like" principle also applies to the solubility of solids in liquid solvents. Polar solvents tend to dissolve polar and ionic compounds, whereas nonpolar solvents tend to dissolve nonpolar compounds.

Figure 12.3 depicts a section of a crystal of NaCl in contact with water. The dipoles of water molecules orient themselves so that the negative ends of some point toward Na^+ ions and the positive ends of others point at Cl^- ions. In other words, *ion–dipole* attractions

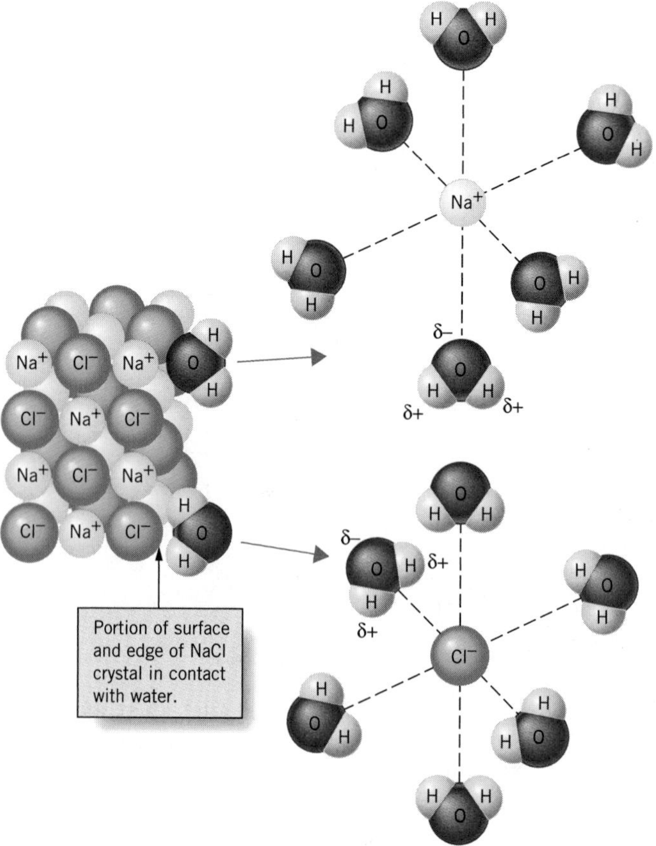

FIG. 12.3 Hydration of ions. Hydration involves a complex redirection of forces of attraction and repulsion. Before this solution forms, water molecules are attracted only to each other, and Na^+ and Cl^- ions have only each other in the crystal to be attracted to. In the solution, the ions have water molecules to take the places of their oppositely charged counterparts; in addition, water molecules are attracted to ions even more than they are to other water molecules.

occur that tend to tug and pull ions from the crystal. At the corners and edges of the crystal, ions are held by fewer neighbors within the solid and so are more readily dislodged than those elsewhere on the crystal's surface. As water molecules dislodge these ions, new corners and edges are exposed, and the crystal continues to dissolve.

Water molecules collide everywhere along the crystal surface, but *successful* collisions—those that dislodge ions—are more likely to occur at corners and edges.

As they become free, the ions become completely surrounded by water molecules (also shown in Figure 12.3). The phenomenon is called the **hydration** of ions. The *general* term for the surrounding of a solute particle by solvent molecules is **solvation,** so hydration is just a special case of solvation. Ionic compounds are able to dissolve in water when the attractions between water dipoles and ions overcome the attractions of the ions for each other within the crystal.

Similar events explain why solids composed of polar molecules, like those of sugar, dissolve in water (see Figure 12.4). Attractions between the solvent and solute dipoles help to dislodge molecules from the crystal and bring them into solution. Again we see that "like dissolves like"; a polar solute dissolves in a polar solvent.

Sugar molecules such as this have polar OH groups that interact with water by hydrogen bonding.

The same reasoning explains why nonpolar solids like wax are soluble in nonpolar solvents such as benzene. Wax is a solid mixture of long chain hydrocarbons, held together by London forces. The attractions between benzene molecules are also London forces, of comparable strength, so molecules of the wax can easily be dispersed among those of the solvent.

When intermolecular attractive forces within solute and solvent are sufficiently different, the two do not form a solution. For example, ionic solids or very polar molecular solids (like sugar) have strong attractions between their particles which cannot be overcome by attractions to molecules of a nonpolar solvent such as benzene.

12.2 HEATS OF SOLUTION COME FROM UNBALANCED INTERMOLECULAR ATTRACTIONS

Because intermolecular attractive forces are important when liquids and solids are involved, the formation of a solution is inevitably associated with energy exchanges. The total energy absorbed or released when a solute dissolves in a solvent at constant pressure to make a solution is called the **molar enthalpy of solution,** or usually just the **heat of solution, ΔH_{soln}.**

Energy is required to separate the particles of solute and also those of the solvent and make them spread out to make room for each other. This step is *endothermic,* because we must overcome the attractions between molecules to spread the particles out. But once the particles come back together as a *solution,* the attractive forces between approaching solute and solvent particles yield a decrease in the system's potential energy, and this is an *exothermic* change. The enthalpy of solution, ΔH_{soln}, is simply the net result of these two opposing enthalpy contributions.

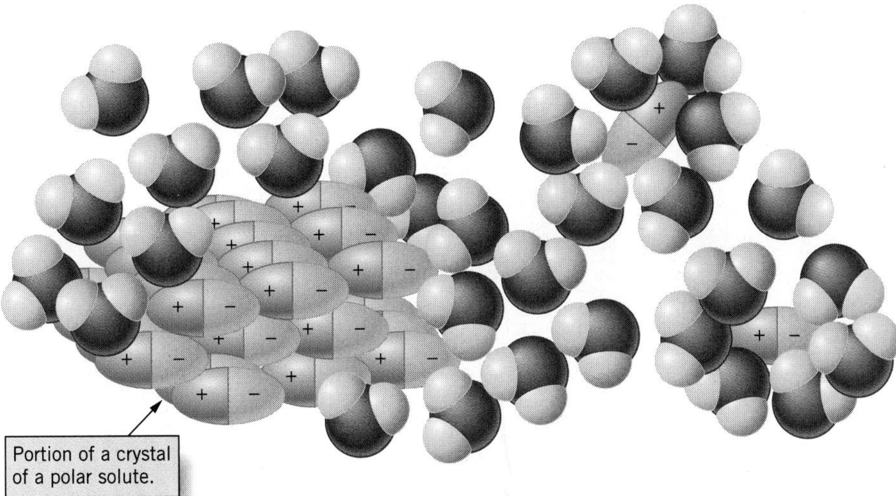

FIG. 12.4 Hydration of a polar molecule. A polar molecule of a molecular compound (such as the sugar glucose) can trade the forces of attraction it experiences for other molecules of its own kind for forces of attraction to molecules of water in an aqueous solution.

The heat of solution for a solid is the lattice energy plus the solvation energy

Because enthalpy is a state function, the magnitude of ΔH_{soln} doesn't depend on the path we take from the separated solute and solvent to the solution. For a solid dissolving in a liquid, it is convenient to imagine a two-step path.

Step 1. *Vaporize the solid to form individual solute particles.* The particles are molecules for molecular substances and ions for ionic compounds. The energy absorbed is the lattice energy of the solid.

Step 2. *Bring the separated gaseous solute particles into the solvent to form the solution.* This step is exothermic, and the enthalpy change when the particles from one mole of solute are dissolved in the solvent is called the **solvation energy.** If the solvent is water, the solvation energy can also be called the **hydration energy.**

The enthalpy diagram showing these steps for potassium iodide is given in Figure 12.5. Step 1 corresponds to the lattice energy of KI, which is represented by the thermochemical equation

$$\mathrm{KI}(s) \longrightarrow \mathrm{K}^+(g) + \mathrm{I}^-(g) \qquad \Delta H = +632 \text{ kJ}$$

Step 2 corresponds to the hydration energy of gaseous K^+ and I^- ions.

$$\mathrm{K}^+(g) + \mathrm{I}^-(g) \longrightarrow \mathrm{K}^+(aq) + \mathrm{I}^-(aq) \qquad \Delta H = -619 \text{ kJ}$$

The *enthalpy of solution* is obtained from the sum of the equations for Steps 1 and 2 and is the enthalpy change when one mole of crystalline KI dissolves in water (corresponding to the direct path in Figure 12.5).

$$\mathrm{KI}(s) \longrightarrow \mathrm{K}^+(aq) + \mathrm{I}^-(aq) \qquad \Delta H_{soln} = +13 \text{ kJ}$$

The value of ΔH_{soln} indicates that the solution process is endothermic for KI, in agreement with the observation that when KI is added to water and the mixture is stirred, it becomes cool as the KI dissolves.

Small percentage errors in very large numbers can cause huge percentage changes in the *differences* between such numbers.

Table 12.1 provides a comparison between values of ΔH_{soln} obtained by the method described above and values obtained by direct measurements. The agreement between calculated and measured values doesn't seem particularly impressive, but this is partly because lattice and hydration energies are not precisely known and partly because the model used in our analysis is evidently too simple. Notice, however, that when "theory" predicts

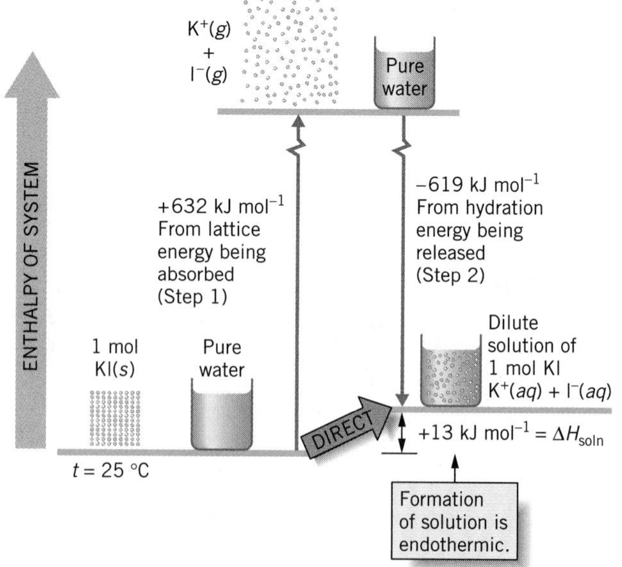

Lattice energy:	$\mathrm{KI}(s) \longrightarrow \mathrm{K}^+(g) + \mathrm{I}^-(g)$	$\Delta H = +632$ kJ
Hydration energy:	$\mathrm{K}^+(g) + \mathrm{I}^-(g) \longrightarrow \mathrm{K}^+(aq) + \mathrm{I}^-(aq)$	$\Delta H = -619$ kJ
Total:	$\mathrm{KI}(s) \longrightarrow \mathrm{K}^+(aq) + \mathrm{I}^-(aq)$	$\Delta H_{soln} = +13$ kJ

FIG. 12.5 Enthalpy diagram for the heat of solution of one mole of potassium iodide. Adding the lattice energy to the hydration energy gives a positive value for ΔH_{soln}, indicating the solution process is endothermic.

TABLE 12.1 Lattice Energies, Hydration Energies, and Heats of Solution for Some Group IA Metal Halides

Compound	Lattice Energy (kJ mol^{-1})	Hydration Energy (kJ mol^{-1})	ΔH_{soln}[a] Calculated[b] ΔH_{soln} (kJ mol^{-1})	ΔH_{soln}[a] Measured ΔH_{soln} (kJ mol^{-1})
LiCl	+833	−883	−50	−37.0
NaCl	+766	−770	−4	+3.9
KCl	+690	−686	+4	+17.2
LiBr	+787	−854	−67	−49.0
NaBr	+728	−741	−13	−0.602
KBr	+665	−657	+8	+19.9
KI	+632	−619	+13	+20.33

[a] Heats of solution refer to the formation of extremely dilute solutions.
[b] Calculated ΔH_{soln} = lattice energy + hydration energy.

relatively large heats of solution, the experimental values are also relatively large, and that both values have the same sign (except for NaCl). Notice also that the variations among the values follow the same trends when we compare the three chloride salts—LiCl, NaCl, and KCl—or the three bromide salts—LiBr, NaBr, and KBr.

Solution of a liquid in another liquid can be modeled as a three-step process

To consider heats of solution when liquids dissolve in liquids, it's useful to imagine a three-step path going from the initial to the final state (see Figure 12.6). We will designate one liquid as the solute and the other as the solvent.

Step 1. *Expand the solute liquid.* First, we imagine that the molecules of one liquid are moved apart just far enough to make room for molecules of the other liquid. Because we have to overcome forces of attraction, this step increases the system's potential energy and so is *endothermic.*

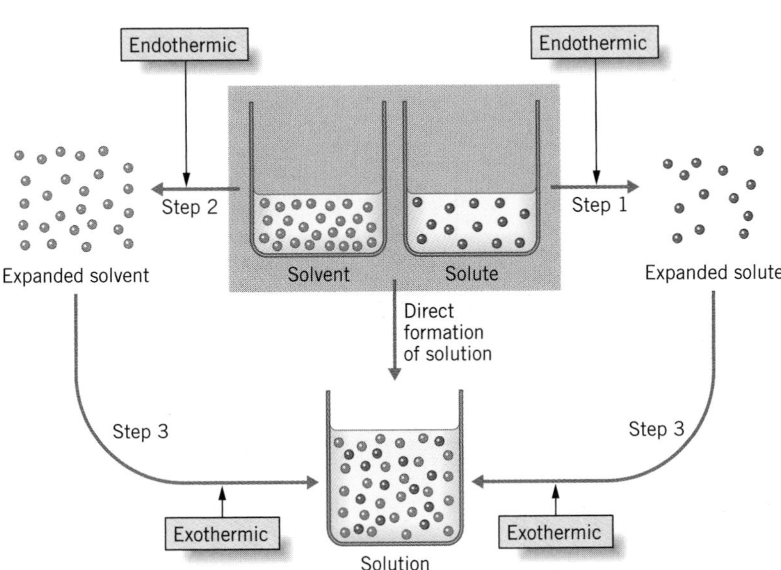

FIG. 12.6 Enthalpy of solution for the mixing of two liquids. To analyze the enthalpy change for the formation of a solution of two liquids, we can imagine the hypothetical steps shown here. **Step 1.** The molecules of the liquid designated as the solute move apart slightly to make room for the solvent molecules, an endothermic process. **Step 2.** The molecules of the solvent are made to take up a larger volume to make room for the solute molecules, which is also an endothermic change. **Step 3.** The expanded samples of solute and solvent spontaneously intermingle, their molecules also attracting each other making the step exothermic.

Step 2. *Expand the solvent liquid.* The second step is like the first, but is done to the other liquid (solvent). On an enthalpy diagram (Figure 12.7) we have climbed two energy steps and have both the solvent and the solute in their slightly "expanded" conditions.

Step 3. *Mix the expanded liquids.* The third step brings the molecules of the expanded solvent and solute together to form the solution. Because the molecules of the two liquids experience mutual forces of attraction, bringing them together lowers the system's potential energy, so Step 3 is exothermic. The value of ΔH_{soln} will, again, be the net energy change for these steps.

Ideal solutions have ΔH_{soln} equal to zero

The enthalpy diagram in Figure 12.7 shows the case when the sum of the energy inputs for Steps 1 and 2 is equal to the energy released in Step 3, so the overall value of ΔH_{soln} is zero. This is very nearly the case when we make a solution of benzene and carbon tetrachloride. Attractive forces between molecules of benzene are almost exactly the same as those between molecules of CCl_4, or between molecules of benzene and those of CCl_4. If all such intermolecular forces were identical, the net ΔH_{soln} would be exactly zero, and the resulting solution would be called an **ideal solution.** Be sure to notice the difference between an *ideal solution* and an *ideal gas*. In an ideal gas, there are no attractive forces. In an ideal solution, there are attractive forces, but they all have the same magnitude.

For most liquids that are mutually soluble, ΔH_{soln} is not zero. Instead, heat is either given off or absorbed. For example, acetone and water are liquids that form a solution exothermically (ΔH_{soln} is negative). With these liquids, the third step releases more energy than the sum of the first two chiefly because molecules of water and acetone attract each other more strongly than acetone molecules attract each other. This is because water molecules can form hydrogen bonds to acetone molecules in the solution, but acetone molecules cannot form hydrogen bonds to other acetone molecules in the pure liquid.

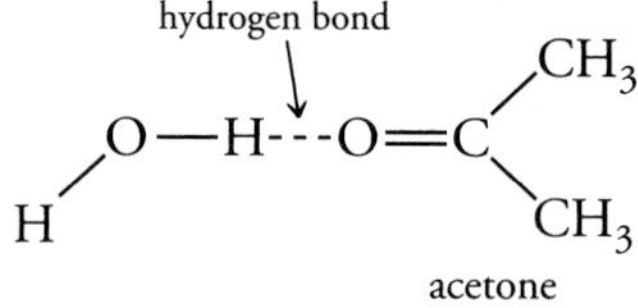

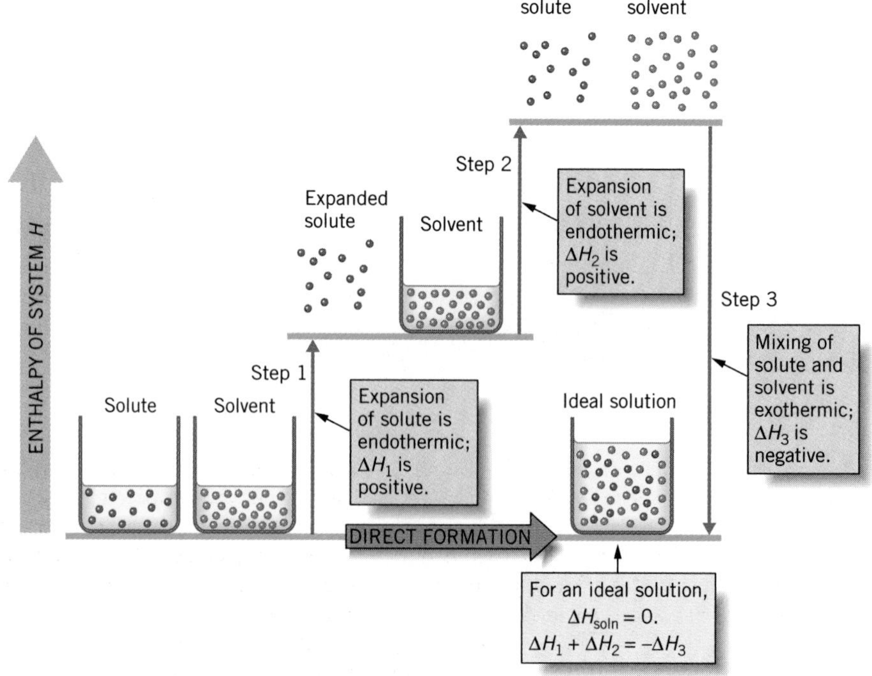

FIG. 12.7 Enthalpy changes in the formation of an ideal solution. The three-step and the direct-formation paths both start and end at the same place with the same enthalpy outcome. The sum of the positive ΔH values for the two endothermic steps, 1 and 2, numerically equals the negative ΔH value for the exothermic step, 3. The net ΔH for the formation of an ideal solution is therefore zero.

Ethanol and hexane form a solution endothermically. In this case, the release of energy in the third step is not enough to compensate for the energy demands of Steps 1 and 2, and the solution cools as it forms. Ethanol molecules attract each other more strongly than they can attract hexane molecules. Hexane molecules cannot push their way into ethanol without breaking up some of the hydrogen bonding between ethanol molecules.

$CH_3CH_2CH_2CH_2CH_2CH_3$
hexane

Gas solubility can be understood using a simple molecular model

Unlike the case for solid and liquid solutes, only very weak attractions exist between gas molecules, so the energy required to "expand the solute" is negligible. The heat absorbed or released when a gas dissolves in a liquid has essentially two contributions, as shown in Figure 12.8:

1. *Energy is absorbed to open "pockets" in the solvent that can hold gas molecules.* The solvent must be expanded slightly to accommodate the molecules of the gas. This requires a small energy input since attractions between solvent molecules must be overcome. Water is a special case; it already contains open holes in its network of loose hydrogen bonds around room temperature. For water, very little energy is required to create pockets that can hold gas molecules.
2. *Energy is released when gas molecules enter the pockets.* Intermolecular attractions between the gas molecules and the surrounding solvent molecules lower the total energy, and energy is released as heat. The stronger the attractions are, the more heat is released. Water is capable of forming hydrogen bonds with some gases, while organic solvents often can't. A larger amount of heat is released when a gas molecule is placed in the pocket in water than in organic solvents.

These factors lead to two generalizations. *Heats of solution for gases in organic solvents are often endothermic* because the energy required to open up pockets is greater than the energy released by attractions formed between the gas and solvent molecules. *Heats of solution for gases in water are often exothermic* because water already contains pockets to hold the gas molecules, and energy is released when water and gas molecules attract each other.

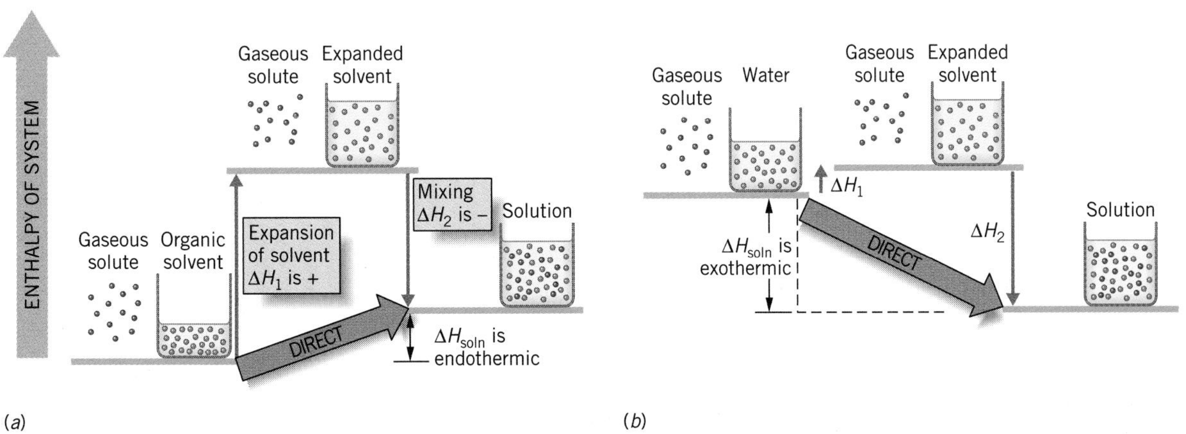

FIG. 12.8 A molecular model of gas solubility. *(a)* A gas dissolves in an organic solvent. Energy is absorbed to open "pockets" in the solvent that can hold the gas molecules. In the second step, energy is released when the gas molecules enter the pockets where they are attracted to the solvent molecules. Here the solution process is shown to be endothermic. *(b)* At room temperature, water's loose network of hydrogen bonds already contains pockets that can accommodate gas molecules, so little energy is needed to prepare the solvent to accept the gas. In the second step, energy is released as the gas molecules take their places in the pockets where they experience attractions to the water molecules. In this case, the solution process is exothermic.

12.3 A SUBSTANCE'S SOLUBILITY CHANGES WITH TEMPERATURE

By "solubility" we mean the mass of solute that forms a *saturated* solution with a given mass of solvent at a specified temperature. The units often are grams of solute per 100 g of solvent. In such a solution there is an equilibrium between the undissolved solute and the solute dissolved in the solution.

$$\text{Solute}_{\text{undissolved}} \rightleftharpoons \text{Solute}_{\text{dissolved}}$$

(Solute contacts the saturated solution.) (Solute is in the saturated solution.)

As long as the temperature is held constant, the concentration of the solute in the solution remains the same. If the temperature of the mixture changes, however, this equilibrium tends to be upset and either more solute will dissolve or some will precipitate. To analyze how temperature affects solubility we can use Le Châtelier's principle, which we introduced in Chapter 11. Recall that this principle tells us that if a system at equilibrium is disturbed, the system will change in a direction that counteracts the disturbance and returns the system to equilibrium (if it can).

To increase the temperature of a solution, heat (energy) is added. When solute dissolves in a solvent, heat is absorbed or evolved. As in Section 11.8, heat is the factor through which Le Châtelier's principle is applied. Let's consider a common situation in which energy is absorbed when additional solute is dissolved in an already saturated solution.

$$\text{Solute}_{\text{undissolved}} + \text{energy} \rightleftharpoons \text{Solute}_{\text{dissolved}}$$

According to Le Châtelier's principle, when we add heat energy to raise the temperature, the system responds by consuming some of the energy we've added. This causes the equilibrium to "shift to the right." In other words, more solute dissolves and when equilibrium is re-established, there is more solute dissolved in the solution. Thus, *when dissolving more solute in a saturated solution is endothermic, raising the temperature increases the solubility of the solute.* This is a common situation for solids dissolving in liquid solvents. How much the solubility is affected by temperature varies widely, as seen in Figure 12.9

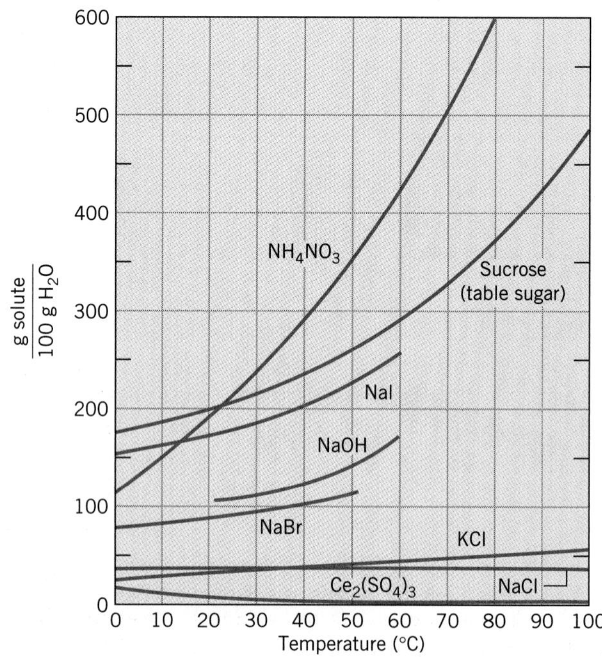

FIG. 12.9 Solubility in water versus temperature for several substances. Most substances become more soluble when the temperature of the solution is increased, but the amount of this increased solubility varies considerably.

TABLE 12.2 Solubilities of Common Gases in Water[a]

Gas	Temperature 0 °C	20 °C	50 °C	100 °C
Nitrogen, N_2	0.0029	0.0019	0.0012	0
Oxygen, O_2	0.0069	0.0043	0.0027	0
Carbon dioxide, CO_2	0.335	0.169	0.076	0
Sulfur dioxide, SO_2	22.8	10.6	4.3	1.8[b]
Ammonia, NH_3	89.9	51.8	28.4	7.4[c]

[a] Solubilities are in grams of solute per 100 g of water when the gaseous space over the liquid is saturated with the gas and the total pressure is 1 atm.
[b] Solubility at 90 °C.
[c] Solubility at 96 °C.

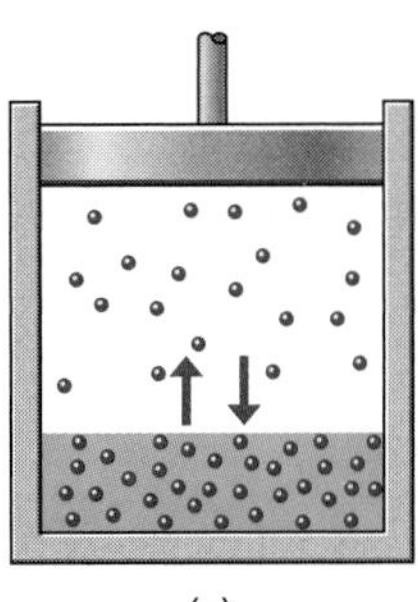

(*a*)

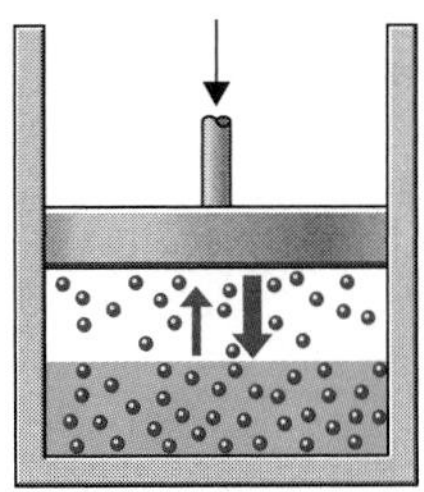

(*b*)

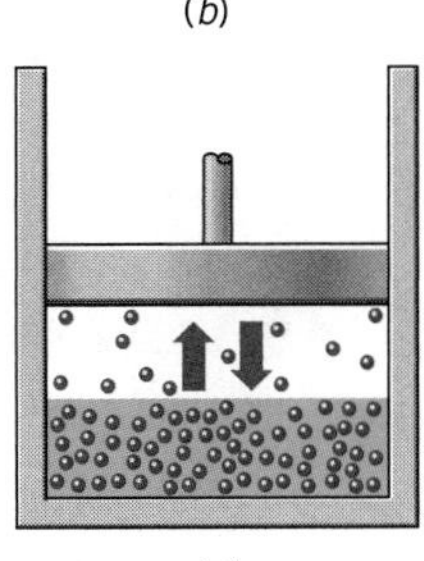

(*c*)

FIG. 12.10 How pressure increases the solubility of a gas in a liquid. (*a*) At some specific pressure, equilibrium exists between the vapor phase and the solution. (*b*) An increase in pressure upsets the equilibrium. More gas molecules are dissolving than are leaving the solution. (*c*) More gas has dissolved and equilibrium is restored.

Some solutes, such as cerium(III) sulfate, $Ce_2(SO_4)_3$, become *less* soluble with increasing temperature (Figure 12.9). Energy must be *released* from a saturated solution of $Ce_2(SO_4)_3$ to make more solute dissolve. For its equilibrium equation, we must show "energy" on the right side because heat is released when more of the solute dissolves into a saturated solution.

$$\text{Solute}_{\text{undissolved}} \rightleftharpoons \text{Solute}_{\text{dissolved}} + \text{energy}$$

Temperature affects the solubility of a gas in a liquid

Table 12.2 gives data for the solubilities of several common gases in water at different temperatures, but all under 1 atm of pressure. In water, gases are usually more soluble at colder temperatures. But the solubility of gases, like other solubilities, can increase or decrease with temperature, depending on the gas and the solvent. For example, the solubilities of H_2, N_2, CO, He, and Ne actually rise with rising temperature in common organic solvents like carbon tetrachloride, toluene, and acetone.

12.4 Gases Become More Soluble at Higher Pressures

The solubility of a gas in a liquid increases with increasing pressure. To understand this at the molecular level, imagine the following equilibrium established in a closed container fitted with a movable piston (Figure 12.10*a*).

$$\text{gas} + \text{solvent} \rightleftharpoons \text{solution} \tag{12.1}$$

If the piston is pushed down (Figure 12.10*b*), the gas is compressed and its pressure increases. This causes the concentration of the gas molecules over the solution to increase, so the rate at which the gas dissolves is now greater than the rate at which it leaves the solution. Eventually, equilibrium is re-established when the concentration of the gas in the solution has increased enough to make the rate of escape equal to the rate at which the gas dissolves (Figure 12.10*c*). At this point, the concentration of the gas in the solution is larger than before.

The effect of pressure on the solubility of a gas can also be explained by Le Châtelier's principle. In this case, the disturbance is an increase in the pressure of the gas above the solution. How could the system counteract the pressure increase? The answer is, by having more gas dissolve in the solution. In this way, the pressure of the gas is reduced and the concentration of the gas in the solution is increased. In other words, increasing the pressure of the gas will cause the gas to become more soluble.

Bottled carbonated beverages fizz when the cap is opened because the sudden drop in pressure causes a sudden drop in gas solubility. *(Andy Washnik.)*

Henry's law relates gas solubility to pressure

William Henry (1774–1836), an English chemist, first reported the relationship between gas solubility and pressure.

Figure 12.11 shows how the solubility in water of oxygen and nitrogen vary with pressure. The straight lines on the graph indicate that the concentration of the gas is directly proportional to its pressure above the solution. This is expressed quantitatively by **Henry's law** (also called the **Pressure-Solubility law**) which states that *the concentration of a gas in a liquid at any given temperature is directly proportional to the partial pressure of the gas over the solution.*

TOOLS
Henry's law

$$C_{\text{gas}} = k_{\text{H}}P_{\text{gas}} \qquad (T \text{ is constant})$$

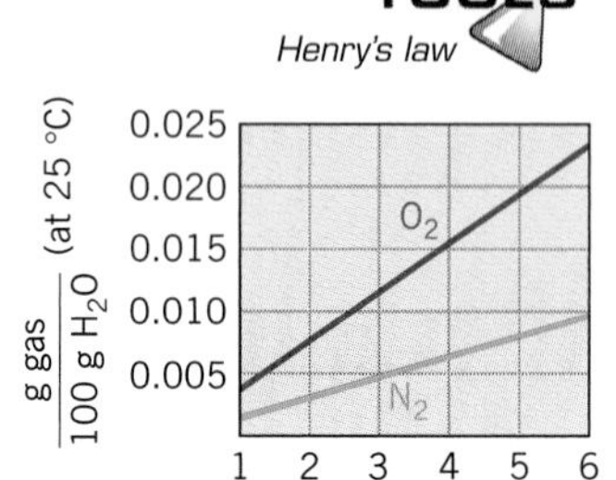

FIG. 12.11 Solubility in water versus pressure for two gases.

where C_{gas} is the concentration of the gas and P_{gas} is the partial pressure of the gas above the solution. The proportionality constant, k_{H}, called the Henry's law constant, is unique to each gas. The equation is true only at low concentrations and pressures and for gases that do not react with the solvent.

An alternate (and commonly used) form of Henry's law is

$$\frac{C_1}{P_1} = \frac{C_2}{P_2} \qquad (12.2)$$

where the subscripts 1 and 2 refer to initial and final conditions, respectively. By taking the ratio, the Henry's law constant cancels.

EXAMPLE 12.1
Using Henry's Law

At 20 °C the solubility of N_2 in water is 0.0150 g L^{-1} when the partial pressure of nitrogen is 580 torr. What will be the solubility of N_2 in water at 20 °C when its partial pressure is 800 torr? Calculate your answer to three significant figures.

ANALYSIS: This problem deals with the effect of gas pressure on gas solubility, so Henry's law applies. We use this law in its form given by Equation 12.2 because it lets us avoid having to know or calculate the Henry's law constant.

SOLUTION: Let's gather the data first.

$$C_1 = 0.0150 \text{ g L}^{-1} \qquad C_2 = ?$$
$$P_1 = 580 \text{ torr} \qquad P_2 = 800 \text{ torr}$$

Using Equation 12.2, we have

$$\frac{0.0150 \text{ g L}^{-1}}{580 \text{ torr}} = \frac{C_2}{800 \text{ torr}}$$

Solving for C_2,

$$C_2 = 0.0207 \text{ g L}^{-1}$$

The solubility under the higher pressure is 0.0207 g L^{-1}.

IS THE ANSWER REASONABLE? In relationship to the initial concentration, the size of the answer makes sense because Henry's law tells us to expect a greater solubility at the higher pressure.

Practice Exercise 1: At room temperature and pressure a hydrogen sulfide solution can be made by bubbling $H_2S(g)$ into water; the result is a 0.1 molar solution. Since $H_2S(g)$ is more dense than air we can assume that a layer of pure $H_2S(g)$ at atmospheric pressure covers the solution. What is the value of Henry's law constant, and is the solubility of $H_2S(g)$ much greater or smaller than other gases mentioned in this section? If so, suggest a reason for the difference. (Hint: Assume 20 °C and 1.0 atm for room temperature and pressure.)

Practice Exercise 2: How many grams of nitrogen and oxygen are dissolved in 100 g of water at 20 °C when the water is saturated with air? At 1 atm pressure, the solubility of oxygen in water is 0.00430 g O_2/100 g H_2O, and the solubility of nitrogen in water is 0.00190 g N_2/100 g H_2O. In pure, dry air, P_{N_2} equals 593 torr and P_{O_2} equals 159 torr. Calculate your answers to three significant figures.

Gases with polar molecules or that react with the solvent are more soluble in water

The gases sulfur dioxide, ammonia, and, to a lesser extent, carbon dioxide are far more soluble in water than are oxygen or nitrogen (see Table 12.2). Part of the reason is that SO_2, NH_3, and CO_2 molecules have polar bonds and sites of partial charge that attract water molecules, forming hydrogen bonds to help hold the gases in solution. Ammonia molecules, in addition, not only can accept hydrogen bonds from water (O—H···N) but also can donate them through their N—H bonds (N—H···O).

The more soluble gases also react with water to some extent as the following chemical equilibria show.

$$CO_2(aq) + H_2O \rightleftharpoons H_2CO_3(aq) \rightleftharpoons H^+(aq) + HCO_3^-(aq)$$

$$SO_2(aq) + H_2O \rightleftharpoons H^+(aq) + HSO_3^-(aq)$$

$$NH_3(aq) + H_2O \rightleftharpoons NH_4^+(aq) + OH^-(aq)$$

The forward reactions contribute to the higher concentrations of the gases in solution, as compared to gases such as O_2 and N_2 that do not react with water at all. Gaseous sulfur trioxide is very soluble in water because it reacts *quantitatively* (i.e., completely) with water to form sulfuric acid.[2]

$$SO_3(g) + H_2O \longrightarrow H_2SO_4(aq)$$

12.5 Molarity Changes with Temperature; Molality, Weight Percentages, and Mole Fractions Do Not

In Section 4.6 you learned that for stoichiometry *molar concentration* or *molarity*, mol L^{-1}, is a convenient unit of concentration because it lets us measure out moles of a solute simply by measuring volumes of solution. For studying physical properties, however, molarity is not preferred because the molarity of a solution varies slightly with temperature. Most liquids expand slightly when heated, so a given solution will have a larger volume, and therefore a lower ratio of moles to volume, as its temperature is raised. For this reason, temperature-insensitive concentration units are used. The most common are *percentage by mass, molality*, and *mole fraction* (or *mole percent*).

Percent concentration

Concentrations of solutions are often expressed as a **percentage by mass** (sometimes called a *percent by weight*), which gives grams of solute per 100 grams of solution. Percentage by

[2] The "concentrated sulfuric acid" of commerce has a concentration of 93 to 98% H_2SO_4, or roughly 18 *M*. This solution takes up water avidly and very exothermically, removing moisture even out of humid air itself (and so must be kept in stoppered bottles). Because it is dense, oily, sticky, and highly corrosive, it must be handled with extreme care. Safety goggles and gloves must be worn when dispensing concentrated sulfuric acid. When making a more dilute solution, *always pour (slowly with stirring) the concentrated acid into the water.* If water is poured onto concentrated sulfuric acid, it can layer on the acid's surface, because the density of the acid is so much higher than water's (1.8 g mL^{-1} vs. 1.0 g mL^{-1} for water). At the interface, such intense heat can be generated that the steam thereby created could explode from the container, spattering acid around.

mass is indicated by % (w/w), where the "w" stands for "weight." To calculate a percentage by mass from solute and solution masses, we can use the following formula:

$$\text{Percentage by mass} = \frac{\text{mass of solute}}{\text{mass of solution}} \times 100\%$$

For example, a solution labeled "0.85% (w/w) NaCl," is one in which the ratio of solute to solution is 0.85 g of NaCl to 100 g of NaCl solution. Often, the "(w/w)" is omitted.[3]

Percentages could also be called *parts per hundred.* Other similar expressions of concentration are *parts per million* (ppm) and *parts per billion* (ppb), where 1 ppm equals 1 g of component in 10^6 g of the mixture and 1 ppb equals 1 g of component in 10^9 g of the mixture.

EXAMPLE 12.2
Using Percent Concentrations

Seawater is typically 3.5% sea salt and has a density of about 1.03 g mL^{-1}. How many grams of sea salt would be needed to prepare enough seawater solution to completely fill a 62.5 L aquarium?

ANALYSIS: We need the number of grams of sea salt in 62.5 L of seawater solution. We can write the problem as

$$62.5 \text{ L soln} \Leftrightarrow ?\text{ g sea salt}$$

To solve the problem, we'll need a relationship that links seawater and sea salt. The percent concentration is the link we need. If we assume that the percent is a percentage by mass, we can write "3.5% sea salt" as follows:

$$3.5 \text{ g sea salt} \Leftrightarrow 100 \text{ g soln}$$

We now need a link between grams of solution and liters of solution. Density relates the mass of a substance to its volume, so we can write a density of 1.03 g mL^{-1} as

$$1.03 \text{ g soln} \Leftrightarrow 1.00 \text{ mL soln}$$

If we convert 62.5 L solution into milliliters of solution, we can use this relationship to obtain the grams of solution. We can then use the percent to convert grams of solution into grams of sea salt.

SOLUTION: The hard work has been done. We just need to assemble the information so that the units cancel properly.

$$62.5 \cancel{\text{L soln}} \times \frac{1000 \cancel{\text{mL soln}}}{1 \cancel{\text{L soln}}} \times \frac{1.03 \cancel{\text{g soln}}}{1.00 \cancel{\text{mL soln}}} \times \frac{3.5 \text{ g sea salt}}{100 \cancel{\text{g soln}}} = 2.2 \times 10^3 \text{ g sea salt}$$

IS THE ANSWER REASONABLE? If seawater is about 4% sea salt, 100 g of seawater should contain about 4 g of salt. A liter of seawater weighs about 1000 g, so it would contain about 40 g of salt. We have about 60 L of seawater, which would contain 60 × 40 g of salt, or 2400 g of salt. This is not too far from our answer, so the answer seems reasonable.

[3] Clinical laboratories sometimes report concentrations as percent by mass/volume, using the symbol "% (w/v)":

$$\text{Percentage by mass/volume} = \frac{\text{mass of solute (g)}}{\text{volume of solution (mL)}} \times 100\%$$

For example, a solution that is 4% (w/v) $SrCl_2$ contains 4 g of $SrCl_2$ dissolved in 100 mL of solution. Notice that with a percentage by mass, we can use any units for the mass of the solute and mass of the solution, as long as they are the same for both. With percentage by mass/volume, we *must* use grams for the mass of solute and milliliters for the volume of solution. If a percent concentration does not include a (w/v) or (w/w) designation, we assume that it is a percentage by mass.

Practice Exercise 3: What volume of water at 20 °C ($d = 0.9982$ g cm^{-3}) is needed to dissolve 45.0 g of sucrose to make a 10% (w/w) solution? (Hint: you need the mass of water to solve this problem.)

Practice Exercise 4: How many grams of NaBr are needed to prepare 250 g of 1.00% (w/w) NaBr solution in water? How many grams of water are needed? How many milliliters are needed, given that the density of water at room temperature is 0.988 g mL^{-1}?

Practice Exercise 5: Hydrochloric acid can be purchased from chemical supply houses as a solution that is 37% (w/w) HCl. What mass of this solution contains 7.5 g of HCl?

Molal concentration

The number of moles of solute per kilogram of *solvent* is called the **molal concentration** or the **molality** of a solution. The usual symbol for molality is ***m***.

$$\text{Molality} = m = \frac{\text{mol of solute}}{\text{kg of solvent}}$$

For example, if we dissolve 0.500 mol of sugar in 1.00 kg of water, we have a 0.500 m solution of sugar. We would not need a volumetric flask to prepare the solution, because we weigh the solvent. Some important physical properties of a solution are related in a simple way to its molality, as we'll soon see.

It is important not to confuse *molarity* with *molality*.

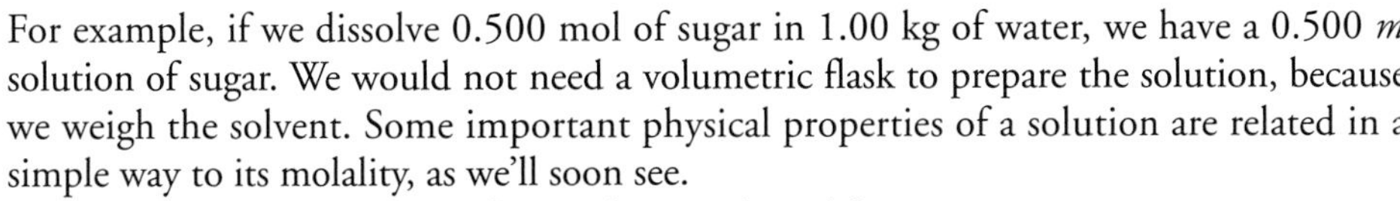

$$\text{Molality} = m = \frac{\text{mol of solute}}{\text{kg of } \textit{solvent}} \qquad \text{Molarity} = M = \frac{\text{mol of solute}}{\text{L of } \textit{solution}}$$

Be sure to notice that molality is defined per kilogram of *solvent*, not kilogram of solution.

As we pointed out at the beginning of this section, the molarity of a solution changes slightly with temperature, but molality does not. Therefore, molality is more convenient in experiments involving temperature changes.

When water is the solvent, a solution's molarity approaches its molality as the solution becomes more dilute. In very dilute solutions, 1 L of *solution* is nearly 1 L of water, which has a mass close to 1 kg. Under these conditions, the ratio of moles per liter (molarity) is very nearly the same as the ratio of moles per kilogram (molality).

This only applies when the solvent is water. For other solvents molarity and molality have quite different values for the same solution.

EXAMPLE 12.3 Calculation to Prepare a Solution of a Given Molality

An experiment calls for a 0.150 m solution of sodium chloride in water. How many grams of NaCl would have to be dissolved in 500.0 g of water to prepare a solution of this molality?

ANALYSIS: *Molality* means moles of solute per kilogram of solvent. The concentration 0.150 m means that the solution must contain 0.150 mol of NaCl for every 1.000 kg (1000 g) of water. We can write

$$0.150 \text{ mol NaCl} \Leftrightarrow 1000 \text{ g } H_2O$$

To calculate the number of moles of NaCl needed for 500.0 g of H_2O, we use this relationship as a conversion factor. We can then use the molar mass of NaCl to convert moles of NaCl into grams of NaCl.

SOLUTION:

$$500.0 \cancel{\text{g H}_2\text{O}} \times \frac{0.150 \cancel{\text{mol NaCl}}}{1000 \cancel{\text{g H}_2\text{O}}} \times \frac{58.44 \text{ g NaCl}}{1 \cancel{\text{mol NaCl}}} = 4.38 \text{ g NaCl}$$

When 4.38 g of NaCl is dissolved in 500 g of H_2O, the concentration is 0.150 m NaCl.

IS THE ANSWER REASONABLE? We'll round the formula mass of NaCl to 60, so 0.15 mol would weigh about 9 g. A 0.15 m solution would contain about 9 g NaCl in 1000 g of water. If we used only 500 g of water, we would need half as much salt, or about 4.5 g. Our answer is close to this, so it seems reasonable.

Practice Exercise 6: If you prepare a solution by dissolving 44.00 g of Na_2SO_4 in 250.0 g of water, what is the molality of the solution? Is the molarity of this solution numerically larger or smaller than its molality? [Hint: Recall the definition of molality and molarity (see above and Section 4.6).]

Practice Exercise 7: Water freezes at a lower temperature when it contains solutes. To study the effect of methanol on the freezing point of water, we might begin by preparing a series of solutions of known molalities. Calculate the number of grams of methanol (CH_3OH) needed to prepare a 0.250 m solution, using 2.000 kg of water.

Practice Exercise 8: What mass of a 0.853 molal solution of $Fe(NO_3)_3$ is needed to obtain (a) 0.0200 mol of $Fe(NO_3)_3$, (b) 0.0500 mol of Fe^{3+} ions, and (c) 0.00300 mol of nitrate ions?

Mole fraction (mole percent)

Mole fraction and mole percent are important tools that were discussed in Chapter 10 (page 415), so for the sake of completeness we will just review the definitions. The mole fraction X_A of a substance A is given by

Mole fraction

$$X_A = \frac{n_A}{n_A + n_B + n_C + n_D + \cdots + n_Z}$$

where n_A, n_B, n_C, . . ., n_Z are the numbers of moles of each component, A, B, C, . . ., Z, respectively. The sum of all mole fractions for a mixture must always equal 1. The mole percent is obtained by multiplying the mole fraction by 100%.

Conversions among concentration units

There are times when we need to relate concentrations expressed in different units. The following example shows that all we need to convert concentrations are the definitions of the units.

EXAMPLE 12.4
Finding Molality from Mass Percent

What is the molality of 10.0% (w/w) aqueous NaCl?

ANALYSIS: The main tools used in performing conversions among concentration units are the definitions of molality and mass percent. The question asks us to go from

$$\frac{10.0 \text{ g NaCl}}{100.0 \text{ g NaCl soln}} \quad \text{to} \quad \frac{? \text{ mol NaCl}}{1 \text{ kg water}}$$

We have two conversions to perform:

- In the numerator, we must convert 10.0 g of NaCl into moles of NaCl. The conversion equality is 58.44 g NaCl = 1 mol NaCl.
- In the denominator, we must convert 100.0 grams of NaCl *solution* to kilograms of *water*. If we subtract the mass of NaCl (10.0 g) from the mass of the NaCl solution (100.0 g), we see that 100.0 g of NaCl solution contains 90.0 g of water. We can write

$$100.0 \text{ g NaCl soln} \Leftrightarrow 90.0 \text{ g water}$$

- We can then convert grams of water to kilograms.

SOLUTION: We write the linking relationships as conversion factors, arranging them so that units cancel properly:

$$\frac{10.0 \text{ g NaCl}}{100.0 \text{ g NaCl soln}} \times \frac{1 \text{ mol NaCl}}{58.44 \text{ g NaCl}} \times \frac{100.0 \text{ g NaCl soln}}{90.0 \text{ g water}} \times \frac{1000 \text{ g water}}{1 \text{ kg water}} = 1.90\ m \text{ NaCl}$$

A 10.0% NaCl solution is also 1.90 molal.

IS THE ANSWER REASONABLE? A convenient rounding of the numbers in our calculation to one significant figure might be:

$$\frac{10 \text{ g NaCl}}{100 \text{ g NaCl soln}} \times \frac{1 \text{ mol NaCl}}{50 \text{ g NaCl}} \times \frac{100 \text{ g NaCl soln}}{100 \text{ g water}} \times \frac{1000 \text{ g water}}{1 \text{ kg water}} = 2\ m \text{ NaCl}$$

This is close to our calculated value and we conclude we were correct. Notice that rounding the 58.44 down to 50 compensates for rounding 90 up to 100 so that our estimate is closer to the correct number. We do not need a calculator for this estimate. First cancel five zeros from the numerator numbers and five zeros from the denominator numbers. This leaves you with 10 divided by 5, which equals 2.

Practice Exercise 9: A bottle on the stockroom shelf reads 50% sodium hydroxide solution. What is the molality of this solution? [Hint: Recall that the % means (w/w) percentage.]

Practice Exercise 10: A certain sample of concentrated hydrochloric acid is 37.0% HCl. Calculate the molality of this solution.

Another kind of calculation is to find the molarity of a solution from its mass percent. This cannot be done without the density of the solution, as the next example shows.

EXAMPLE 12.5
Finding Molarity from Mass Percent

A certain supply of concentrated hydrochloric acid has a concentration of 36.0% HCl. The density of the solution is 1.19 g mL^{-1}. Calculate the molar concentration of HCl.

ANALYSIS: Recall the definitions of percentage by mass and molarity. We must perform the following conversion:

$$\frac{36.0 \text{ g HCl}}{100 \text{ g HCl soln}} \quad \text{to} \quad \frac{? \text{ mol HCl}}{1 \text{ L HCl soln}}$$

We have two conversions to perform:

- The numerators show that we need a grams-to-moles conversion for the solute, HCl. The link between grams and moles is the molar mass (36.46 g mol^{-1} for HCl).
- The denominators tell us that we must carry out a grams-to-liters conversion for the HCl solution, which is why we need the solution's density. Remember from Chapter 1 that *density is the link between a substance's mass and volume.* We can write

$$1.19 \text{ g HCl soln} \Leftrightarrow 1.00 \text{ mL HCl soln}$$

To complete the conversion we'll have to change milliliters to liters.

SOLUTION: We can do the conversions in either order. Here, we've converted the numerator first:

$$\frac{36.0 \text{ g HCl}}{100 \text{ g HCl soln}} \times \frac{1 \text{ mol HCl}}{36.46 \text{ g HCl}} \times \frac{1.19 \text{ g HCl soln}}{1.00 \text{ mL HCl soln}} \times \frac{1000 \text{ mL HCl soln}}{1 \text{ L HCl soln}} = 11.7\ M \text{ HCl}$$

So 36.0% HCl is also 11.8 M HCl.

IS THE ANSWER REASONABLE? We can estimate the answer in our head. First cancel the 36 g HCl in the numerator and the 36.46 g HCl in the denominator to give 1.0 (shown in red below). Now we divide 1000 in the numerator by 100 in the denominator to get 10 (shown in blue). All that is left is $1.19 \times 10 = 11.9\ M$ as shown below.

$$\frac{36.0 \text{ g HCl}}{100 \text{ g HCl soln}} \times \frac{1 \text{ mol HCl}}{36.46 \text{ g HCl}} \times \frac{1.19 \text{ g HCl soln}}{1.00 \text{ mL HCl soln}} \times \frac{1000 \text{ mL HCl soln}}{1 \text{ L HCl soln}} = 11.9\ M \text{ HCl}$$

This is very close to our calculator answer and we can assume we did the calculation correctly.

Practice Exercise 11: Hydrobromic acid can be purchased as 40.0% HBr. The density of this solution is 1.38 g mL^{-1}. What is the molar concentration of HBr in this solution?

Practice Exercise 12: One gram of $Al(NO_3)_3$ is dissolved in 1.00 liter of water at 20 °C. The density of water at this temperature is 0.9982 g cm^{-3} and the density of the resulting solution is 0.9989 g cm^{-3}. Calculate the molarity and molality of this solution. (Hint: remember that density can be used to convert mass to volume.)

Practice Exercise 12 illustrated that when the concentration of an aqueous solution is very low, one gram per liter for example, the molality and the molarity are very close to the same numerical value. As a result we can conveniently interchange molality and molarity when aqueous solutions are very dilute.

12.6 Solutes Lower the Vapor Pressure of a Solvent

After the Greek *kolligativ*, depending on number and not on nature.

The physical properties of solutions to be studied in this and succeeding sections are called **colligative properties,** because they depend mostly on the relative *populations* of particles in mixtures, not on their chemical identities. In this section, we examine the effects of solutes on the vapor pressures of solvents in liquid solutions.

Francois Marie Raoult (1830–1901) was a French scientist.

All liquid solutions of **nonvolatile** solutes (solutes that have no tendency to evaporate) have lower vapor pressures than their pure solvents. The vapor pressure of such a solution is proportional to how much of the solution actually consists of the solvent. This proportionality is given by **Raoult's law** (also known as the **Vapor Pressure–Concentration law**) which says that the vapor pressure of the solution, $P_{solution}$, equals the mole fraction of the solvent, $X_{solvent}$, multiplied by its vapor pressure when pure, $P°_{solvent}$. In equation form, Raoult's law is expressed as follows.

Raoult's law equation

$$P_{solution} = X_{solvent} P°_{solvent}$$

Because of the form of this equation, a plot of $P_{solution}$ versus $X_{solvent}$ should be *linear* at all concentrations when the system obeys Raoult's law (see Figure 12.12).

Notice that the mole fraction in Raoult's law refers to the solvent, not the solute. Usually we're more interested in the effect of the *solute's* mole fraction concentration, X_{solute}, on the vapor pressure. We can show that the change in vapor pressure, ΔP, is directly proportional to the mole fraction of solute, X_{solute}, as follows.[4]

$$\Delta P = X_{solute} \, P^{\circ}_{solvent} \tag{12.3}$$

The change in vapor pressure equals the mole fraction of the solute times the solvent's vapor pressure when pure.

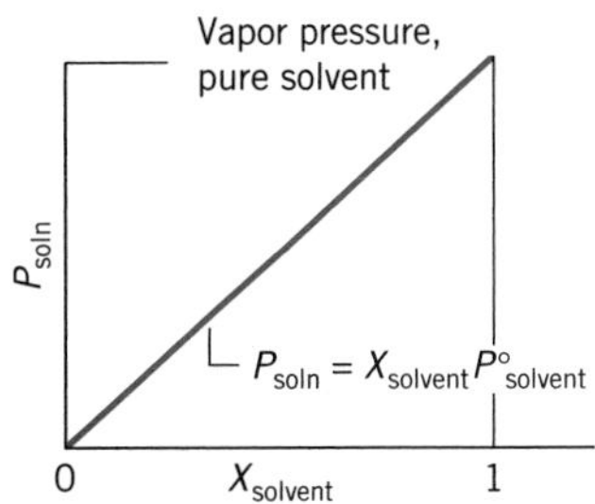

FIG. 12.12 A Raoult's law plot. When the vapor pressure of a solution is plotted against the mole fraction of the solvent, the result is a straight line.

EXAMPLE 12.6
Calculating with Raoult's Law

Carbon tetrachloride has a vapor pressure of 100 torr at 23 °C. This solvent can dissolve candle wax, which is essentially nonvolatile. Although candle wax is a mixture, we can take its molecular formula to be $C_{22}H_{46}$ (molar mass 311 g mol^{-1}). What is the vapor pressure at 23 °C of a solution prepared by dissolving 10.0 g of wax in 40.0 g of CCl_4 (molar mass 154 g mol^{-1})?

ANALYSIS: This problem involves the vapor pressure of a solution and so we need Equation 12.3. We need the mole fraction of the *solute* to use the equation, so we must first calculate the total number of moles of solute and solvent.

SOLUTION: First we calculate the moles of each component.

For CCl_4 $$40.0 \text{ g CCl}_4 \times \frac{1 \text{ mol CCl}_4}{154 \text{ g CCl}_4} = 0.260 \text{ mol CCl}_4$$

For $C_{22}H_{46}$ $$10.0 \text{ g C}_{22}\text{H}_{46} \times \frac{1 \text{ mol C}_{22}\text{H}_{46}}{311 \text{ g C}_{22}\text{H}_{46}} = 0.0322 \text{ mol C}_{22}\text{H}_{46}$$

The total number of moles = 0.292 mol

Now, we can calculate the mole fraction of the solute, $C_{22}H_{46}$.

$$X_{C_{22}H_{46}} = \frac{0.0322 \text{ mol}}{0.292 \text{ mol}} = 0.110$$

The amount that the vapor pressure is lowered, ΔP, will be this particular mole fraction, 0.110, times the vapor pressure of pure CCl_4 (100 torr), calculated by Equation 12.3.

$$\Delta P = 0.110 \times 100 \text{ torr} = 11.0 \text{ torr}$$

The presence of the wax in the CCl_4 lowers the vapor pressure of the CCl_4 by 11.0 torr from 100 to 89 torr.

[4] To derive Equation 12.3, note that the value of ΔP is simply the following difference.

$$\Delta P = P^{\circ}_{solvent} - P_{solution}$$

The mole fractions for our two-component system, $X_{solvent}$ and X_{solute}, must add up to 1; it's the nature of mole fractions.

$$X_{solvent} = 1 - X_{solute}$$

We now insert this expression for $X_{solvent}$ into the Raoult's law equation.

$$P_{solution} = X_{solvent} \, P^{\circ}_{solvent}$$
$$P_{solution} = (1 - X_{solute})P^{\circ}_{solvent} = P^{\circ}_{solvent} - X_{solute} \, P^{\circ}_{solvent}$$

So, by rearranging terms,

$$X_{solute} \, P^{\circ}_{solvent} = P^{\circ}_{solvent} - P_{solution} = \Delta P$$

This result can be rearranged again to give Equation 12.3.

IS THE ANSWER REASONABLE? Raoult's law implies that *the vapor pressure of a solvent in a solution must be less than that of the pure solvent.* Our answer does show a lowering of the vapor pressure of CCl_4 in the wax solution. We could also apply the Raoult's law equation directly; we multiply the mole fraction of the *solvent* by the vapor pressure of the solvent. The mole fraction of the solvent must be 1 minus 0.11 (the other mole fraction) or 0.89. Taking 0.89 times 100 torr gives 89 torr, which is our answer.

Practice Exercise 13: Dibutyl phthalate, $C_{16}H_{22}O_4$ (molecular mass 278 g mol^{-1}), is an oil sometimes used to soften plastic articles. Its vapor pressure is negligible around room temperature. What is the vapor pressure, at 20 °C, of a solution of 20.0 g of dibutyl phthalate in 50.0 g of octane, C_8H_{18} (molecular mass 114 g mol^{-1})? The vapor pressure of pure octane at 20 °C is 10.5 torr. (Hint: Calculate mole fractions first.)

Practice Exercise 14: Acetone (molar mass 58.1 g mol^{-1}) has a vapor pressure at a given temperature of 162 torr. How many grams of the nonvolatile stearic acid (molar mass 284.5 g mol^{-1}) must be added to 156 g of acetone to decrease its vapor pressure to 150 torr?

A molecular explanation can be provided for Raoult's law

How does a nonvolatile solute lower the vapor pressure of the solvent in a solution? Consider the following analogy. Imagine a crowded stadium, filled to capacity. The stadium has separate restrooms for men and women. Men will be entering and leaving the men's restrooms at a certain rate. The rate will depend on the number of men in the stadium. If 50% of the people in the stadium are men, we expect the men's restrooms to be twice as busy as it would be if only 25% of the people were men. In other words, increasing the concentration of women in the stadium lowers the amount of traffic in and out of the men's rooms. Increasing the number of women in the stadium also lowers the number of men occupying the men's restrooms, on average.

In our analogy, the stadium and the men's restrooms represent the solution and vapor phases. The men represent the solvent molecules and the women represent nonvolatile solute. Adding a nonvolatile solute to a solution will lower the evaporation rate of the solvent, just as increasing the fraction of women in the stadium will decrease traffic to the men's restrooms. Increasing the concentration of solute lowers the number of solvent molecules in the vapor phase, on average, just as increasing the fraction of women in the stadium lowers the number of men occupying the men's rooms. If the number of solvent molecules in the vapor phase is lowered, the partial pressure of the solvent above the solution is lowered, as well. The effect is shown in Figure 12.13.

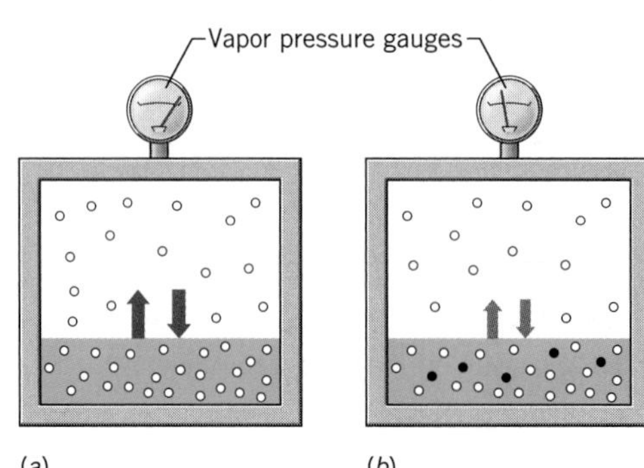

FIG. 12.13 Effect of a nonvolatile solute on the vapor pressure of a solvent. (*a*) Equilibrium between a pure solvent and its vapor. With a high number of solvent molecules in the liquid phase, the rates of evaporation and condensation are relatively high. (*b*) In the solution, some of the solvent molecules have been replaced with solute molecules. There are fewer solvent molecules available to evaporate from solution. The evaporation rate is lower. When equilibrium is established, there are fewer molecules in the vapor. The vapor pressure of the solution is less than that of the pure solvent.

How can we predict the vapor pressure of a solution that contains more than one volatile component?

When two (or more) components of a liquid solution can evaporate, the vapor contains molecules of each substance. Each volatile component contributes its own partial pressure to the total pressure. By Raoult's law, the partial pressure of a particular component is directly proportional to the component's mole fraction in the solution. By Dalton's law of partial pressures, the total vapor pressure will be the sum of the partial pressures. To calculate these partial pressures, we use the Raoult's law equation for each component. When component A is present in a mole fraction of X_A, its partial pressure (P_A) is this fraction of its vapor pressure when pure, namely, P_A°.

$$P_A = X_A P_A^\circ$$

Remember, P_A and P_B here are the *partial* pressures as calculated by Raoult's law.

And, by the same argument, P_B, the partial pressure of component B, is

$$P_B = X_B P_B^\circ$$

The total vapor pressure of the solution of liquids A and B is then, by Dalton's law of partial pressures, the sum of P_A and P_B.

$$P_{\text{total}} = X_A P_A^\circ + X_B P_B^\circ$$

Notice that this equation contains the Raoult's law equation as a special case. If one component, say, component B, is nonvolatile, it has no vapor pressure (P_B° is zero) so the $X_B P_B^\circ$ term drops out, leaving the Raoult's law equation.

EXAMPLE 12.7
Calculating the Vapor Pressure of a Solution of Two Volatile Liquids

Acetone is a solvent for both water and molecular liquids that do not dissolve in water, like benzene. At 20 °C, acetone has a vapor pressure of 162 torr. The vapor pressure of water at 20 °C is 17.5 torr. Assuming that the mixture obeys Raoult's law, what would be the vapor pressure of a solution of acetone and water with 50.0 mol% of each?

ANALYSIS: To find P_{total} we need to calculate the individual partial pressures and then add them.

SOLUTION: A concentration of 50.0 mol% corresponds to a mole fraction of 0.500, so

$$P_{\text{acetone}} = 0.500 \times 162 \text{ torr} = 81.0 \text{ torr}$$

$$P_{\text{water}} = 0.500 \times 17.5 \text{ torr} = \underline{8.75 \text{ torr}}$$

$$P_{\text{total}} = 89.8 \text{ torr}$$

IS THE ANSWER REASONABLE? The vapor pressure of the solution (89.8 torr) has to be much higher than that of pure water (17.5 torr), because of the volatile acetone, but much less than that of pure acetone (162 torr), because of the high mole fraction of water. The answer seems reasonable.

Practice Exercise 15: At 20 °C, the vapor pressure of cyclohexane, a hydrocarbon solvent, is 66.9 torr and that of toluene (another solvent) is 21.1 torr. What is the vapor pressure of a solution of the two at 20 °C when the mole fraction of toluene is 0.250? (Hint: Recall that the sum of the all mole fractions in a mixture must add up to 1.00.)

Practice Exercise 16: Using the information from the exercise above, calculate the expected vapor pressure of a mixture of cyclohexane and toluene that consists of 100.0 grams of each solvent.

Only ideal solutions obey Raoult's law exactly. The vapor pressures of components over real solutions are sometimes higher or sometimes lower than Raoult's law would predict. These differences between ideal and real solution behavior are quite useful. They can tell us whether the attractive forces in the pure solvents, before mixing, are stronger or weaker than the attractive forces in the mixture. Comparing experimental vapor pressures with the predictions of Raoult's law provides a simple way to compare the relative strengths of attractions between molecules in solution. When the attractions between unlike molecules in the mixture are strongest, the experimental vapor pressure will be lower than calculated with Raoult's law. Conversely, when the attractions between identical molecules in the pure solvents are strongest, the experimental vapor pressure is higher than Raoult's law would predict.

12.7 Solutions Have Lower Melting Points and Higher Boiling Points than Pure Solvents

The lowering of the vapor pressure produced by the presence of a nonvolatile solute affects both the boiling and freezing points of a solution. This is illustrated for water in Figure 12.14.

The solid blue lines in the figure correspond to the three equilibrium lines in the phase diagram for pure water, which we discussed in Section 11.12. Adding a nonvolatile solute lowers the vapor pressure of the solution, giving a new liquid–vapor equilibrium line for the solution, which is shown as the red line connecting points *A* and *B*.

When the solution freezes, the solid that forms is pure ice; there is no solute within the ice crystals. This is because the highly ordered structure of the solid doesn't allow solute molecules to take the place of water molecules. As a result, both pure water and the solution

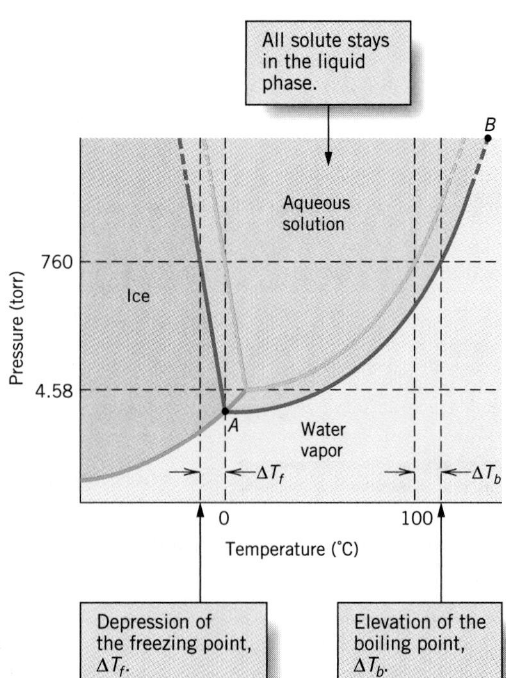

FIG. 12.14 Phase diagram for water and an aqueous solution. Phase diagram for pure water (blue curves) and for an aqueous solution of a nonvolatile solute (red curves).

TABLE 12.3 Molal Boiling Point Elevation and Freezing Point Depression Constants

Solvent	BP (°C)	K_b (°C m^{-1})	MP (°C)	K_f (°C m^{-1})
Water	100	0.51	0	1.86
Acetic acid	118.3	3.07	16.6	3.57
Benzene	80.2	2.53	5.45	5.07
Chloroform	61.2	3.63	—	—
Camphor	—	—	178.4	37.7
Cyclohexane	80.7	2.69	6.5	20.0

BP = normal boiling point; MP = normal melting point.

have the same solid–vapor equilibrium line on the phase diagram. Point A on the diagram is at the intersection of the liquid–vapor and solid–vapor equilibrium lines for the solution and represents the new triple point for the solution. Rising from this triple point is the solid–liquid equilibrium line for the solution, shown in red.

If we look at where the solid–liquid and liquid–vapor lines cross the 1 atm (760 torr) pressure line, we can find the normal freezing and boiling points. For water (the blue lines), the freezing point is 0 °C and the boiling point is 100 °C. Notice that the solid–liquid line for the solution meets the 760 torr line at a temperature *below* 0 °C. In other words, the freezing point of the solution is below that of pure water. The amount by which the freezing point is lowered is called the **freezing point depression,** and is given the symbol ΔT_f. Similarly, we can see that the liquid–vapor line for the solution crosses the 760 torr line at a temperature *above* 100 °C, so the solution boils at a higher temperature than pure water. The amount by which the boiling point is raised is called the **boiling point elevation.** It is given the symbol ΔT_b.

Recall that the normal boiling and freezing points are those at a pressure of 1 atm.

There are many practical applications of freezing point depression and boiling point elevation. For example, antifreeze solutions are added to the radiator of a car to prevent freezing in winter and boiling over in summer. Similarly, salt is spread on icy roads to cause the ice to melt and salt is mixed with ice to decrease the temperature in home ice cream machines.

ΔT_f and ΔT_b are related to the molality of the solute

Freezing point depression and boiling point elevation are both colligative properties. The magnitudes of ΔT_f and ΔT_b are proportional to the relative populations of solute and solvent molecules. Molality (m) is the preferred concentration expression (rather than mole fraction or percent by mass) because of the resulting simplicity of the equations relating ΔT to concentration. The following equations work reasonably well only for dilute solutions, however.

$$\Delta T_f = K_f m \tag{12.4}$$

$$\Delta T_b = K_b m \tag{12.5}$$

TOOLS
Freezing point depression, boiling point elevation, determination of molar mass

K_f and K_b are proportionality constants and are called, respectively, the **molal freezing point depression constant** and the **molal boiling point elevation constant.** The values of both K_f and K_b are characteristic of each solvent (see Table 12.3). The units of each constant are °C m^{-1}. Thus, the value of K_f for a given solvent corresponds to the number of degrees of freezing point lowering for each molal unit of concentration. The K_f for water is 1.86 °C m^{-1}. A 1.00 m solution in water freezes at 1.86 °C *below* the normal freezing point of 0 °C, or at −1.86 °C. A 2.00 m solution should freeze at −3.72 °C. (We say "should" but systems are seldom this ideal.) Similarly, because K_b for water is 0.51 °C m^{-1}, a 1.00 m aqueous solution at 1 atm pressure boils at (100 + 0.51) °C or 100.51 °C, and a 2.00 m solution should boil at 101.02 °C.

EXAMPLE 12.8
Estimating a Freezing Point Using a Colligative Property

Estimate the freezing point of a solution made from 10.0 g of urea, $CO(NH_2)_2$ (molar mass 60.06 g mol^{-1}), and 125 g of water.

ANALYSIS: Equation 12.4 relates concentration to freezing point depression. To use the equation, we must first calculate the molality of the solution.

SOLUTION: Molality is the ratio of moles of solute to kilograms of solvent, so we'll convert grams of urea to moles.

$$10.0 \text{ g CO(NH}_2)_2 \times \frac{1 \text{ mol CO(NH}_2)_2}{60.06 \text{ g CO(NH}_2)_2} = 0.166 \text{ mol CO(NH}_2)_2$$

This is the number of moles of $CO(NH_2)_2$ in 125 g or 0.125 kg of water, so the molality is given by

$$\text{Molality} = \frac{0.166 \text{ mol}}{0.125 \text{ kg}} = 1.33\ m$$

For our estimate we must next use Equation 12.4 although it is most reliably used only for dilute solutions. Table 12.3 tells us that K_f for water is 1.86 °C m^{-1}.

$$\Delta T_f = K_f m = (1.86 \text{ °C } m^{-1})(1.33\ m)$$
$$= 2.47 \text{ °C}$$

The solution should freeze at 2.47 °C below 0 °C, or at −2.47 °C.

IS THE ANSWER REASONABLE? For every unit of molality, the freezing point must be depressed by about 2 °C. The molality of this solution is between 1 m and 2 m, so we expect the freezing point depression to be between about 2 °C and 4 °C. It is.

Practice Exercise 17: At what temperature will a 10% aqueous solution of sugar ($C_{12}H_{22}O_{11}$) boil? (Hint: Recall that the boiling point elevation is a change in temperature.)

Practice Exercise 18: How many grams of glucose (molar mass 180.2 g mol^{-1}) must be dissolved in 250 g of water to raise the boiling point to 102.36 °C?

Freezing point depression and boiling point elevation can be used to determine molar masses

We have described freezing point depression and boiling point elevation as *colligative* properties; they depend on the relative *numbers* of particles, not on their kinds. Because the effects are proportional to molal concentrations, experimentally measured values of ΔT_f or ΔT_b can be useful for calculating the molar masses of unknown solutes, as illustrated in the following example.

EXAMPLE 12.9
Calculating a Molar Mass from Freezing Point Depression Data

A solution made by dissolving 5.65 g of an unknown molecular compound in 110.0 g of benzene froze at 4.39 °C. What is the molar mass of the solute?

ANALYSIS: *To calculate the molar mass, we need to know two things about the same sample, the number of moles and the number of grams.* Dividing grams by moles gives the number of grams per mole, which is the molar mass. In this example, we've been given the number of grams, and we have to use the remaining data to calculate the number of moles.

We can use the value of ΔT_f (the difference between the freezing point of pure benzene and that of the solution) along with the value of K_f for benzene from Table 12.3 to calculate the molality of the solution. This gives the ratio of moles of solute to kilograms of benzene, which we can then use as a conversion factor to calculate the number of moles dissolved in the 110.0 g of benzene.

SOLUTION: Table 12.3 tells us that the melting point of benzene is 5.45 °C and that the value of K_f for benzene is 5.07 °C m^{-1}. The amount of freezing point depression is

$$\Delta T_f = 5.45\ °\text{C} - 4.39\ °\text{C} = 1.06\ °\text{C}$$

We now use Equation 12.4 to find the molality of the solution.

$$\Delta T_f = K_f m$$

$$\text{Molality} = \frac{\Delta T_f}{K_f} = \frac{1.06\ \cancel{°\text{C}}}{5.07\ \cancel{°\text{C}}\ m^{-1}} = 0.209\ m$$

This means that for every kilogram of benzene in the solution, there are 0.209 mol of solute. But we have only 110.0 g or 0.1100 kg of benzene. So the actual number of moles of solute in the given solution is found by

$$0.1100\ \cancel{\text{kg benzene}} \times \frac{0.209\ \text{mol solute}}{1\ \cancel{\text{kg benzene}}} = 0.0230\ \text{mol solute}$$

We can now obtain the molar mass. There are 5.65 g of solute per 0.0230 mol of solute:

$$\frac{5.65\ \text{g}}{0.0230\ \text{mol}} = 246\ \text{g mol}^{-1}$$

The mass of one mole of the solute is 246 g.

IS THE ANSWER REASONABLE? A common mistake to avoid is using the given value of the freezing point, 4.39 °C, in Equation 12.4, instead of calculating ΔT_f. A check shows that we *have* calculated ΔT_f correctly. The ratio of ΔT_f to K_f is about 1/5, or 0.2 *m*, which corresponds to a ratio of 0.2 mol solute per 1 kg of benzene. The solution prepared has only 0.1 kg of solvent, so to have the same ratio, the amount of solute must be about 0.02 mol (2×10^{-2} mol). The sample weighs about 5 g, and dividing 5 g by 2×10^{-2} mol gives 2.5×10^2 g/mol, or 250 g/mol. This is very close to the answer we obtained, so we can feel confident it's correct.

Practice Exercise 19: A solution made by dissolving 3.46 g of an unknown compound in 85.0 g of benzene froze at 4.13 °C. What is the molar mass of the compound? (Hint: Recall the equation for calculating moles of a substance to find the key relationship.)

Practice Exercise 20: A mixture is prepared that is 5.0% (w/w) of an unknown substance mixed with naphthalene (molar mass 128.2 g mol^{-1}). The freezing point of the mixture is found to be 77.3 °C. What is the molar mass of the unknown substance? (The melting point of naphthalene is 80.2 °C and K_f = 6.9 °C.)

12.8 Osmosis Is a Flow of Solvent through a Semipermeable Membrane due to Unequal Concentrations

In living things, membranes of various kinds keep mixtures and solutions organized and separated. Yet some substances have to be able to pass through membranes in order that nutrients and products of chemical work can be distributed correctly. These membranes, in other words, must have a selective *permeability*. They must keep some substances from going through while letting others pass. Such membranes are said to be *semipermeable.*

The degree of permeability varies with the kind of membrane. Cellophane, for example, is permeable to water and small solute particles—ions or molecules—but impermeable to very large molecules, like those of starch or proteins. Special membranes can even be prepared that are permeable only to water, not to any solutes.

Depending on the kind of membrane separating solutions of different concentration, two similar phenomena, *dialysis* and *osmosis*, can be observed. Both are functions of the relative populations of the particles of the dissolved materials, so they are colligative properties.

When a membrane is able to let both water and *small* solute particles through, such as the membranes in living systems, the process is called **dialysis,** and the membrane is called a *dialyzing membrane.* It does not permit huge molecules through, like those of proteins and starch. Artificial kidney machines use dialyzing membranes to help remove the smaller molecules of wastes from the blood while letting the blood retain its large protein molecules.

Osmosis involves the movement of solvent across a membrane

When a semipermeable membrane will let only solvent molecules get through, this movement is called *osmosis,* and the special membrane needed to observe it is called an **osmotic membrane.**

When **osmosis** occurs, there's a net shift of solvent across the membrane from the more dilute solution (or pure solvent) into the more concentrated solution. This happens because there is a tendency toward the equalization of concentrations between the two solutions in contact with one another across the membrane. The rate of passage of solvent molecules through the membrane into the more concentrated solution is greater than their rate of passage in the opposite direction, presumably because at the surface of the membrane the solvent concentration is greater in the more dilute solution (Figure 12.15*a*). This leads to a gradual net flow of water through the membrane into the more concentrated solution.

We observe an effect similar to osmosis if two solutions with unequal concentrations of a nonvolatile solute are placed in a sealed container (Figure 12.15*b*). The rate of evaporation from the more dilute solution is greater than that of the more concentrated solution, but the rate of return to each is the same because both solutions are in contact with the same gas phase. As a result, neither solution is in equilibrium with the vapor. In the dilute solution, molecules of solvent are evaporating faster than they're condensing. But in the concentrated solution, just the opposite occurs; more water molecules return to the solution than leave it. Therefore, over time there is a gradual net transfer of solvent from the dilute solution into the more concentrated one until they both achieve the same concentration.

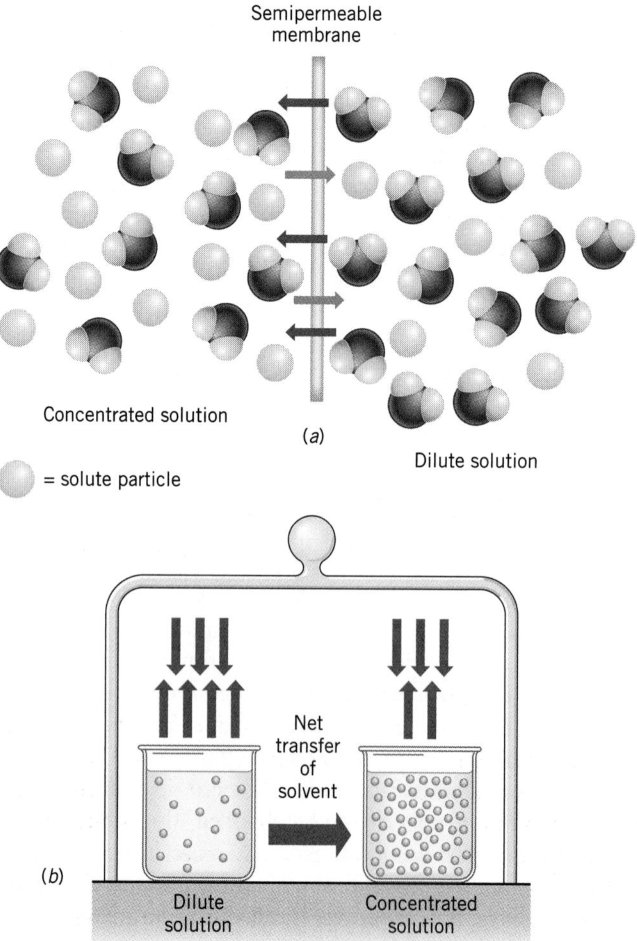

FIG. 12.15 Principles at work in osmosis. (*a*) Osmosis. Solvent molecules pass more frequently from the more dilute solution into the more concentrated one, as indicated by the arrows. This leads to a gradual transfer of solvent from the less concentrated solution into the more concentrated one. (*b*) Because the two solutions have unequal vapor pressures, there is a gradual net transfer of solvent from the more dilute solution into the more concentrated one.

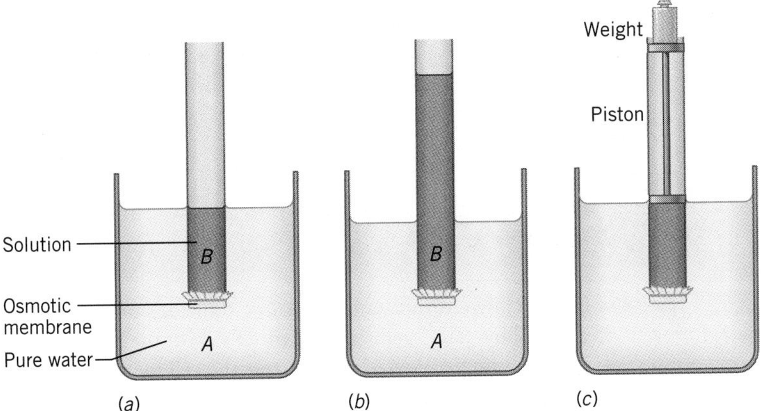

FIG. 12.16 Osmosis and osmotic pressure. (*a*) Initial conditions. A solution, *B*, is separated from pure water, *A*, by an osmotic membrane; no osmosis has yet occurred. (*b*) After a while, the volume of fluid in the tube has increased visibly. Osmosis has taken place. (*c*) A back-pressure is needed to prevent osmosis. The amount of back-pressure is the osmotic pressure of the solution.

Applying pressure to a solution can prevent osmosis

An osmosis experiment is illustrated in Figure 12.16. Initially, we have a solution (*B*) in a tube fitted with an osmotic membrane that dips into a container of pure water (*A*). As time passes, the volume of liquid in the tube increases as solvent molecules transfer into the solution. In Figure 12.16*b*, the net transport of water into the solution has visibly increased the volume.

The weight of the rising fluid column in Figure 12.16*b* provides a push or opposing pressure that makes it increasingly more difficult for molecules of water to enter the solution. Eventually, this pressure becomes sufficient to stop the osmosis. If we apply further pressure, as illustrated in Figure 12.16*c*, we can force enough water back through the membrane to restore the system to its original condition. The exact opposing pressure needed to prevent any osmotic flow *when one of the liquids is pure solvent* is called the **osmotic pressure** of the solution.

Notice that the term "osmotic pressure" uses the word *pressure* in a novel way. Apart from osmosis, a solution does not "have" a special pressure called osmotic pressure. What the solution has is a *concentration* that can generate the occurrence of osmosis and the associated osmotic pressure under the right circumstances. Then, in proportion to the solution's concentration, a specific back-pressure is required to prevent osmosis. By *exceeding* this back-pressure, osmosis can be reversed. *Reverse osmosis* is widely used to purify seawater, both on ocean-going ships and in remote locations where fresh water is unavailable.

Reverse osmosis systems are available at home centers for home use. They can be installed to remove impurities and foul tastes from drinking water. Some bottled water available on supermarket shelves contains water that's been purified by reverse osmosis.

The symbol for osmotic pressure is the Greek capital pi, Π. In a dilute aqueous solution, Π is proportional both to temperature, T, and molar concentration of the solute in the solution, M:

$$\Pi \propto MT$$

The proportionality constant turns out to be the gas constant, R, so for a *dilute* aqueous solution we can write

$$\Pi = MRT \tag{12.6}$$

Of course, M is the ratio mol/L, which we can write as n/V, where n is the number of moles and V is the volume in liters. If we replace M with n/V in Equation 12.6, and rearrange terms, we have an equation for osmotic pressure identical in form to the ideal gas law.

$$\Pi V = nRT \tag{12.7}$$

Equation 12.7 is the *van't Hoff equation for osmotic pressure.*

Osmotic pressure is of tremendous importance in biology and medicine. Cells are surrounded with membranes that restrict the flow of salts but allow water to pass through freely. To maintain a constant amount of water, the osmotic pressure of solutions on either side of the cell membrane must be identical. For example, a solution that is 0.9% (w/v) NaCl has the same osmotic pressure as the contents of red blood cells, and red blood cells bathed in this solution can maintain their normal water content. The solution is said to be isotonic with red blood cells. Blood plasma is an **isotonic solution.**

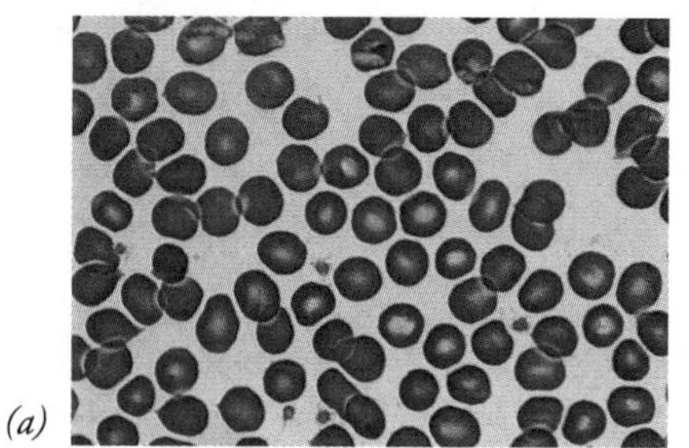
(a)

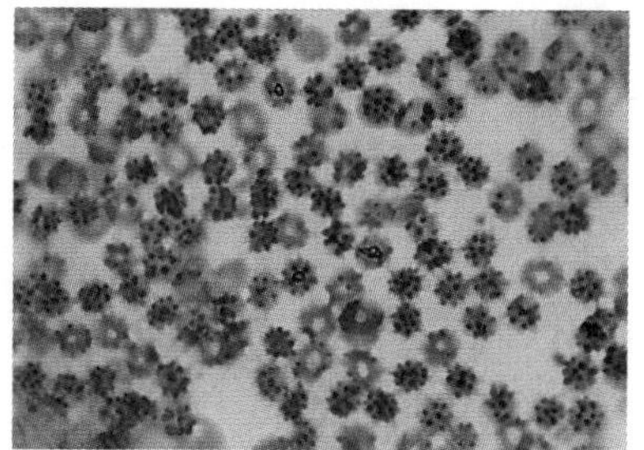
(b)

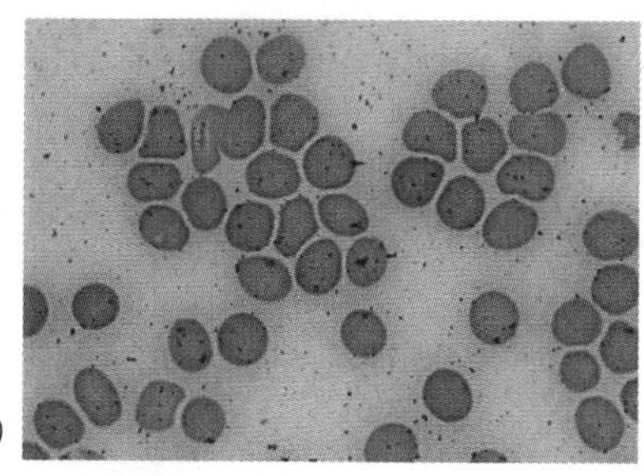
(c)

FIG. 12.17 Effects of isotonic, hypertonic, and hypotonic solutions on red blood cells. *(a)* In an isotonic solution (0.85% NaCl by mass), solutions on either side of the membrane have the same osmotic pressure, and there is no net flow of water across the cell membrane. *(b)* In this hypertonic solution (5.0% NaCl by mass), water flows from areas of lower salt concentration (inside the cell) to higher concentration (the hypertonic solution), causing the cell to dehydrate. *(c)* In this hypotonic solution (0.1% NaCl by mass), water flows from areas of lower salt concentration (the hypotonic solution) to higher concentration (inside the cell). The cells swell and burst. *((a), (b), and (c) Dennis Strete/Fundamental Photographs.)*

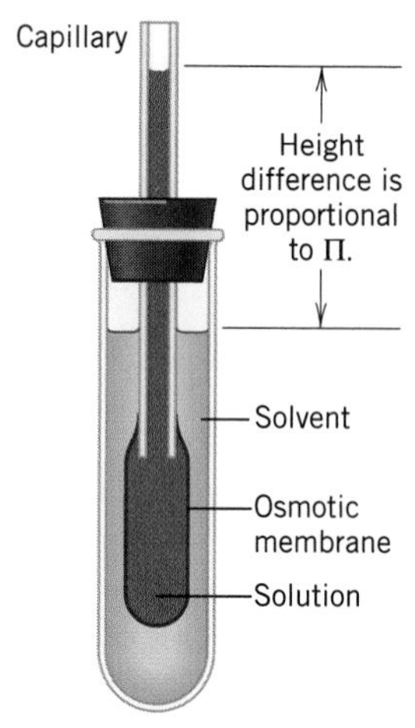

FIG. 12.18 Simple osmometer. When solvent moves into the solution by osmosis, the level of the solution in the capillary rises. The height reached can be related to the osmotic pressure of the solution.

If the cell is placed in a solution with a salt concentration higher than the concentration within the cell, osmosis causes water to flow out of the cell. Such a solution is said to be **hypertonic.** The cell shrinks and dehydrates, and eventually dies. This process kills freshwater fish and plants that are washed out to sea.

On the other hand, water will flow into the cell if it is placed into a solution with an osmotic pressure that is much lower than the osmotic pressure of the cell's contents. Such a solution is called a **hypotonic solution.** A cell placed in distilled water, for example, will swell and burst. If you've ever tried to put in a pair of contact lenses with tap water instead of an isotonic saline solution, you've experienced cell damage from a hypotonic solution. The effects of isotonic, hypotonic, and hypertonic solutions on cells are shown in Figure 12.17.

Obviously, the measurement of osmotic pressure can be very important in preparing solutions that will be used to culture tissues or to administer medicines intravenously. Osmotic pressures can be measured by an instrument called an *osmometer,* illustrated and explained in Figure 12.18. Osmotic pressures can be very high, even in dilute solutions, as Example 12.10 shows.

EXAMPLE 12.10
Calculating an Osmotic Pressure

A very dilute solution, 0.00100 *M* sugar in water, is separated from pure water by an osmotic membrane. What osmotic pressure in torr develops at 25 °C?

ANALYSIS: This is nothing more than an application of Equation 12.6. However, we must be sure to use $R = 0.0821$ L atm mol^{-1} K^{-1}, otherwise the units won't work out correctly.

SOLUTION: Substituting into Equation 12.6, being sure to use the temperature in kelvins so the units cancel correctly,

$$\Pi = (0.00100 \text{ mol L}^{-1})(0.0821 \text{ L atm mol}^{-1} \text{ K}^{-1})(298 \text{ K})$$
$$= 0.0245 \text{ atm}$$

In torr we have

$$\Pi = 0.0245 \text{ atm} \times \frac{760 \text{ torr}}{1 \text{ atm}} = 18.6 \text{ torr}$$

The osmotic pressure of 0.00100 *M* sugar in water is 18.6 torr.

IS THE ANSWER REASONABLE? A 1 *M* solution should have an osmotic pressure of *RT.* Rounding the gas law constant to 0.08 and the temperature to 300, the osmotic pressure of a 1 *M* solution should be about 24 atm. The osmotic pressure of an 0.001 *M* solution should be 1/1000 of this, or about 0.024 atm, which is fairly close to what we got before we converted to torr.

Practice Exercise 21: What is the osmotic pressure, in mm Hg and mm water, of a protein solution when 5.00 g of the protein (molar mass 230,000 g mol^{-1}) is used to prepare 100.0 mL of an aqueous solution at 4.0 °C? (Hint: The height of a column of liquid supported by any pressure is inversely proportional to the density of the liquid. The equivalence to use is 1 torr = 13.6 mm H_2O.)

Practice Exercise 22: An aqueous solution of glucose ($C_6H_{12}O_6$) has an osmotic pressure of 42.5 torr at 25 °C. How many grams of glucose are in 125 mL of this solution?

In the preceding example, you saw that a 0.00100 *M* solution of sugar has an osmotic pressure of 18.6 torr, which is equivalent to 18.6 mm Hg. This pressure is sufficient to support a column of the solution (which is mostly water) roughly 25 cm or 10 in. high. If the solution had been 100 times as concentrated, 0.100 *M* sugar—still relatively dilute—the height of the column supported would be roughly 25 m or over 80 ft!

Molar mass can be estimated from osmotic pressure measurements

An osmotic pressure measurement taken of a dilute solution can be used to determine the molar concentration of the solute, regardless of its chemical composition. Such data along with the mass of the solute in the solution permits us to calculate the molar mass.

Determination of molar mass by osmotic pressure is much more sensitive than determination by freezing point depression or boiling point elevation. The following example illustrates the method.

EXAMPLE 12.11 Calculating Molar Mass from Osmotic Pressure

An aqueous solution with a volume of 100 mL and containing 0.122 g of an unknown molecular compound has an osmotic pressure of 16.0 torr at 21.0 °C. What is the molar mass of the solute?

ANALYSIS: As we noted in Example 12.9, to determine a molar mass we need to measure two quantities for the same sample, its mass and the number of moles. Then, the molar mass is the ratio of grams to moles. We're given the number of *grams* of the solute, so we need to find the number of *moles* equivalent to this mass in order to compute the ratio.

We can use Equation 12.6 to find the molarity of the solute from the osmotic pressure and the temperature. In the calculation, we have to remember to use $R = 0.0821$ L atm mol^{-1} K^{-1}. Then we can use the given volume of the solution (in liters) and the molarity to calculate the number of moles of solute. This is how many moles are in the 0.122 g of solute. Finally, we can calculate the molar mass by dividing grams of solute by moles of solute.

SOLUTION: First, the solution's molarity, *M*, is calculated using the osmotic pressure. This pressure (Π), 16.0 torr, corresponds to 0.0211 atm. The temperature, 20.1 °C, corresponds to 294 K (273 + 21). Using Equation 12.6,

$$16.0 \text{ torr} \times \frac{1 \text{ atm}}{760 \text{ torr}} = 0.0211 \text{ atm}$$

$$\Pi = MRT$$

$$0.0211 \text{ \cancel{atm}} = (M)(0.0821 \text{ L \cancel{atm} mol}^{-1} \text{ \cancel{K^{-1}}})(294 \text{ \cancel{K}})$$

$$M = 8.74 \times 10^{-4} \text{ mol L}^{-1}$$

The 0.122 g of solute is in 100 mL or 0.100 L of solution, not in a whole liter. So the number of moles of solute corresponding to 0.122 g is found by

$$\text{Moles of solute} = 0.100 \text{ \cancel{L soln}} \times \frac{8.74 \times 10^{-4} \text{ mol}}{1 \text{ \cancel{L soln}}}$$

$$= 8.74 \times 10^{-5} \text{ mol}$$

The molar mass of the solute is the number of grams of solute per mole of solute:

$$\text{Molar mass} = \frac{0.122 \text{ g}}{8.74 \times 10^{-5} \text{ mol}} = 1.40 \times 10^3 \text{ g mol}^{-1}$$

IS THE ANSWER REASONABLE? In using Equation 12.6, it is essential that the pressure be in atmospheres when using $R = 0.0821$ L atm mol^{-1} K^{-1}. If all the units cancel correctly (and they do), we've computed the molarity correctly. We've also changed the volume of the solution to liters, so calculating the moles of solute seems to be okay. The moles of solute is approximately 10×10^{-5}, or 1×10^{-4}. Dividing the mass, roughly 0.12 g, by 1×10^{-4} mol gives 0.12×10^{4} g per mol or 1.2×10^{3} g per mol, which is close to the value we found.

Practice Exercise 23: Estimate the molecular mass of a protein when 0.137 g of the protein, dissolved in 100.0 mL of water at 4 °C, supports a column of water that is 6.45 cm high. The density of mercury is 13.6 g ml^{-1} (Hint: The answer should be very large. Don't forget to convert centimeters of water to atmospheres of pressure.)

Practice Exercise 24: A solution of a carbohydrate prepared by dissolving 72.4 mg in 100 mL of solution has an osmotic pressure of 25.0 torr at 25.0 °C. What is the molecular mass of the compound?

12.9 IONIC SOLUTES AFFECT COLLIGATIVE PROPERTIES DIFFERENTLY FROM NONIONIC SOLUTES

The molal freezing point depression constant for water is 1.86 °C m^{-1}. So you might think that a 1.00 m solution of NaCl would freeze at −1.86 °C. Instead, it freezes at −3.37 °C. This greater depression of the freezing point by the salt, which is almost twice 1.86 °C, is not hard to understand if we remember that colligative properties depend on the concentrations of *particles.* We know that NaCl(s) dissociates into two ions in water.

$$NaCl(s) \longrightarrow Na^{+}(aq) + Cl^{-}(aq)$$

If the ions are truly separated, 1.00 m NaCl actually has a concentration of dissolved solute particles of 2.00 m, twice the given molal concentration. Theoretically, 1.00 m NaCl should freeze at $2 \times (-1.86\ °C)$ or −3.72 °C. (Why it actually freezes a little higher than this, at −3.37 °C, will be discussed shortly.)

If we made up a solution of 1.00 m $(NH_4)_2SO_4$, we would have to consider the following dissociation.

$$(NH_4)_2SO_4(s) \longrightarrow 2NH_4^{+}(aq) + SO_4^{2-}(aq)$$

One mole of $(NH_4)_2SO_4$ can give a total of 3 mol of ions (2 mol of NH_4^{+} and 1 mol of SO_4^{2-}). We would expect the freezing point of a 1 m solution of $(NH_4)_2SO_4$ to be $3 \times (-1.86\ °C) = -5.58\ °C$.

When we want to roughly *estimate* a colligative property of a solution of an electrolyte, we recalculate the solution's molality using an assumption about the way the solute dissociates or ionizes.

EXAMPLE 12.12
Estimating the Freezing Point of a Salt Solution

Estimate the freezing point of aqueous 0.106 m $MgCl_2$, assuming that it dissociates completely.

ANALYSIS: The equation that relates a change in freezing temperature to molality is Equation 12.4.

$$\Delta T_f = K_f m$$

From Table 12.3, K_f for water is 1.86 °C m^{-1}. We cannot simply use 0.106 m as the molality because $MgCl_2$ is an ionic compound that dissociates in water. We must use the total molality of ions calculated from the equation for the dissociation.

SOLUTION: When $MgCl_2$ dissolves in water it breaks up as follows.

$$MgCl_2(s) \longrightarrow Mg^{2+}(aq) + 2Cl^-(aq)$$

Because 1 mol of $MgCl_2$ gives 3 mol of ions, the effective (assumed) molality of ions in the solution is three times 0.106 *m*.

$$\text{Effective molality} = (3)(0.106\ m) = 0.318\ m$$

Now we can use Equation 12.4.

$$\Delta T_f = (1.86\ °C\ \cancel{m^{-1}})(0.318\ \cancel{m})$$
$$= 0.591\ °C$$

The freezing point is depressed below 0.000 °C by 0.591 °C, so we calculate that this solution freezes at −0.591 °C.

IS THE ANSWER REASONABLE? The molality, as recalculated, is roughly 0.3, so 3/10th of 1.86 (call it 2) is 0.6. After we add the unit, °C, and subtract from 0 °C we get −0.6 °C, which is close to the answer.

Practice Exercise 25: Calculate the freezing point of aqueous 0.237 *m* LiCl on the assumption that it is 100% dissociated. Calculate what its freezing point would be if the percent dissociation were 0%. (Hint: Recall the relationship developed in Section 12.7.)

Practice Exercise 26: Determine the freezing point depression of aqueous solutions of $MgSO_4$ that are (a) 0.1 *m*, (b) 0.01 *m*, and (c) 0.001 *m*. Which of these could be measured with a laboratory thermometer that has markings at one °C intervals?

Percent ionization can be estimated from colligative property measurements

Experiments show that neither the 1.00 *m* NaCl nor the 1.00 *m* $(NH_4)_2SO_4$ solutions described earlier in this section freeze quite as low as calculated. Our assumption that an electrolyte separates *completely* into its ions is incorrect. Some oppositely charged ions exist as very closely associated pairs, called **ion pairs,** which behave as single "molecules" (Figure 12.19). Clusters larger than two ions probably also exist. The formation of ion pairs and clusters makes the actual *particle* concentration in a 1.00 *m* NaCl solution somewhat less than 2.00 *m*. As a result, the freezing point depression of 1.00 *m* NaCl is not quite as large as calculated on the basis of 100% dissociation.

As solutions of electrolytes are made more and more *dilute,* the observed and calculated freezing points come closer and closer together. At greater dilutions, the **association** (coming together) of ions is less a complication because the ions can be farther apart. So the solutes behave increasingly as if they were 100% separated into their ions at ever higher dilutions.

"Association" is the opposite of dissociation. It is the coming together of particles to form larger particles.

(a)

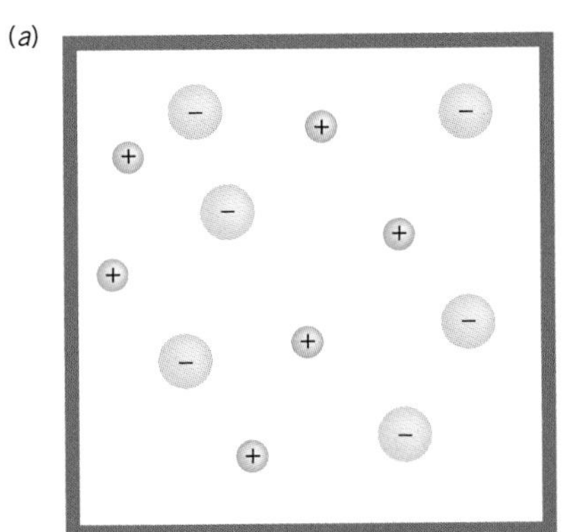

(b)

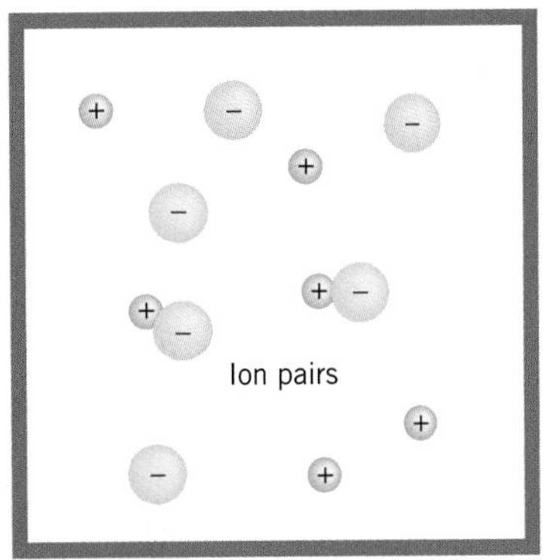

FIG. 12.19 Ion pairs in a solution of NaCl. (*a*) If NaCl were completely dissociated in water, the Na^+ and Cl^- ions would be totally independent. (*b*) Interionic attractions cause some ions to group together as ion pairs, which reduces the total number of independent particles in the solution. In this diagram, two ion pairs are shown.

Chemists compare the degrees of dissociation of electrolytes at different dilutions by a quantity called the **van't Hoff factor, *i*.** It is the ratio of the observed freezing point depression to the value calculated on the assumption that the solute dissolves as a nonelectrolyte.

$$i = \frac{(\Delta T_f)_{\text{measured}}}{(\Delta T_f)_{\text{calculated as a nonelectrolyte}}}$$

The hypothetical van't Hoff factor, i, is 2 for NaCl, KCl, and $MgSO_4$, which break up into two ions on 100% dissociation. For K_2SO_4, the theoretical value of i is 3 because one K_2SO_4 unit gives 3 ions. The actual van't Hoff factors for several electrolytes at different dilutions are given in Table 12.4. Notice that with decreasing concentration (that is, at higher dilutions) the experimental van't Hoff factors agree better with their corresponding hypothetical van't Hoff factors.

TABLE 12.4 Van't Hoff Factors versus Concentration

	Van't Hoff Factor, i			
	Molal Concentration ($mol_{salt}\ kg_{water}^{-1}$)			Value of i if 100% Dissociation Occurred
Salt	0.1	0.01	0.001	
NaCl	1.87	1.94	1.97	2.00
KCl	1.85	1.94	1.98	2.00
K_2SO_4	2.32	2.70	2.84	3.00
$MgSO_4$	1.21	1.53	1.82	2.00

The increase in the percentage dissociation that comes with greater dilution is not the same for all salts. In going from concentrations of 0.1 to 0.001 m, the increase in percentage dissociation of KCl, as measured by the change in i, is only about 7%. But for K_2SO_4, the increase for the same dilution is about 22%, a difference caused by the anion, SO_4^{2-}. It has twice the charge as the anion in KCl and so the SO_4^{2-} ion attracts K^+ more strongly than can Cl^-. Hence, letting an ion of 2− charge and an ion of 1+ charge get farther apart by dilution has a greater effect on their acting independently than giving ions of 1− and 1+ charge more room. When *both* cation and anion are doubly charged, the improvement in percent dissociation with dilution is even greater. We can see from Table 12.4 that there is an almost 50% increase in the value of i for $MgSO_4$ as we go from a 0.1 to a 0.001 m solution.

Colligative properties provide evidence for clustering of solute particles

Some molecular solutes produce *weaker* colligative effects than their molal concentrations would lead us to predict. These unexpectedly weak colligative properties are often evidence that solute molecules are clustering or associating in solution. For example, when dissolved in benzene, benzoic acid molecules associate as **dimers.** They are held together by hydrogen bonds, indicated by the dotted lines in the following.

Di-signifies two, so a dimer is the result of the combination of two single molecules.

$$2\,C_6H_5{-}\overset{\overset{\displaystyle O}{\|}}{C}{-}O{-}H \rightleftharpoons C_6H_5{-}C\begin{matrix} \nearrow O\cdots H{-}O \searrow \\ \searrow O{-}H\cdots O \nearrow \end{matrix}C{-}C_6H_5$$

benzoic acid — benzoic acid dimer

C_6H_5— means

```
   H     H
    \   /
    C—C
   //   \\
H—C      C—
   \    /
    C=C
   /   \
  H     H
```

Because of association, the depression of the freezing point of a 1.00 m solution of benzoic acid in benzene is only about one-half the calculated value. By forming a dimer, benzoic acid has an effective molecular mass that is twice as much as normally calculated. The larger effective molecular mass reduces the molal concentration of particles by half, and the effect on the freezing point depression is reduced by one-half.

SUMMARY

Solutions. Polar or ionic solutes generally dissolve well in polar solvents such as water. Nonpolar, molecular solutes dissolve well in nonpolar solvents. These observations are behind the **"like dissolves like" rule.** Nature's driving force for establishing the more statistically probable mixed state plus intermolecular attractions are the major factors in the formation of a solution. When both solute and solvent are nonpolar, nature's tendency toward the more probable mixed state dominates because intermolecular attractive forces are weak. Ion–dipole attractions and the **solvation** or **hydration** of dissolved species are major factors in forming solutions of ionic or polar solutes in something polar, like water.

Heats of Solution. The **molar enthalpy of solution** of a solid in a liquid, also called the **heat of solution,** is the net of the **lattice energy** and the **solvation energy** (or, when water is the solvent, the **hydration energy**). The lattice energy is the increase in the potential energy of the system required to separate the molecules or ions of the solute from each other. The solvation energy corresponds to the potential energy lowering that occurs in the system from the subsequent attractions of those particles to the solvent molecules as intermingling occurs.

When liquids dissolve in liquids, an **ideal solution** forms if the potential energy increase needed to separate the molecules equals the energy lowering as the separated molecules come together. Usually, some net energy exchange occurs, however, and few solutions are ideal.

When a gas dissolves in a gas, the energy cost to separate the particles is virtually zero, because they're already separated. They remain separated in the gas mixture, so the net enthalpy of solution is very small if not zero. When a gas dissolves in an organic solvent, the solution process is often endothermic. When a gas dissolves in water, the solution process is often exothermic.

Pressure and Gas Solubility. At pressures not too much different from atmospheric pressure, the solubility of a gas in a liquid is directly proportional to the partial pressure of the gas—**Henry's law.**

Concentration Expressions. To study and use colligative properties, a solution's concentration ideally is expressed either in mole fractions or as a **molal concentration,** or **molality.** The **molality** of a solution, ***m*,** is the ratio of the number of moles of solute to the kilograms of *solvent* (not solution, but solvent). **Mass fractions** and **mass percents** are concentration expressions used often for reagents when direct information about number of moles of solutes is not considered important. Mole fraction, molality, and mass percent are temperature-independent concentration expressions.

Colligative Properties. **Colligative properties** are those that depend on the ratio of the particles of the solute to the molecules of the solvent. These properties include the **lowering of the vapor pressure,** the **depression of a freezing point,** the **elevation of a boiling point,** and the **osmotic pressure** of a solution.

According to **Raoult's law,** the vapor pressure of a solution of a nonvolatile (molecular) solute is the vapor pressure of the pure solvent times the mole fraction of the solvent. An alternative expression of this law is that the *change* in vapor pressure caused by the solute equals the mole fraction of the solute times the pure solvent's vapor pressure. When the components of a solution are volatile liquids, then Raoult's law calculations find the *partial pressures* of the vapors of the individual liquids. The sum of the partial pressures equals the total vapor pressure of the solution.

An *ideal solution* is one that would obey Raoult's law exactly and in which all attractions between molecules are equal. An ideal solution has $\Delta H_{soln} = 0$. Solutions of liquids that form exothermically usually have vapor pressures lower than predicted by Raoult's law. Those that form endothermically have vapor pressures higher than predicted.

In proportion to its molal concentration, a solute causes a **freezing point depression** and a **boiling point elevation.** The proportionality constants are the **molal freezing point depression constant** and the **molal boiling point elevation constant,** and they differ from solvent to solvent. Freezing point and boiling point data from a solution made from known masses of both solute and solvent can be used to calculate the molecular mass of a solute.

When a solution is separated from the pure solvent (or from a less concentrated solution) by an *osmotic membrane*, **osmosis** occurs, which is the net flow of solvent into the more concentrated solution. The back-pressure required to prevent osmosis is called the solution's **osmotic pressure,** which, in a dilute aqueous solution, is proportional to the product of the Kelvin temperature and the molar concentration. The proportionality constant is R, the ideal gas constant, and the equation relating these variables is $\Pi = MRT$.

When the membrane separating two different solutions is a *dialyzing membrane*, it permits **dialysis,** the passage not only of solvent molecules but also of small solute molecules or ions. Only very large molecules are denied passage.

Colligative Properties of Electrolytes. Because an electrolyte releases more ions in solution than indicated by the molal (or molar) concentration, a solution of an electrolyte has more pronounced colligative properties than a solution of a molecular compound at the same concentration. The dissociation of a strong electrolyte approaches 100% in very dilute solutions, particularly when both ions are singly charged.

The ratio of the value of a particular colligative property (e.g., ΔT_f) to the value expected if there were no dissociation is the **van't Hoff factor, *i*.** A solute whose formula unit breaks into two ions, like NaCl or $MgSO_4$, would have a van't Hoff factor of 2 if it were 100% dissociated. Observed van't Hoff factors are less than the theoretical values because of formation of **ion pairs,** brought about by **association** of ions in solution. The van't Hoff factors approach those corresponding to 100% dissociation only as the solutions are made more and more dilute.

TOOLS FOR PROBLEM SOLVING

The compilation below lists the tools you have learned in this chapter that are applicable to problem solving. Review them if necessary, and refer to them when working on the Review Problems that follow.

"Like dissolves like" rule *(page 482)* This rule uses the polarity and strengths of attractive forces along with chemical composition and structure to predict whether two substances can form a solution.

Henry's law *(page 490)* The solubility of a gas at a given pressure can be determined from its solubility at another pressure by using the Henry's law equation.

$$C_{gas} = k_H P_{gas}$$

Mass fraction; mass percent *(page 492)* These concentration terms are one way to express concentration and they can be used as conversion factors to calculate the mass of solution needed to deliver a desired mass of solute.

$$\text{mass fraction} = \frac{\text{mass of solute}}{\text{mass of solution}}$$

$$\text{mass percent} = \%(\text{w/w}) = \frac{\text{mass of solute}}{\text{mass of solution}} \times 100\%$$

Molal concentration *(page 493)* This is a temperature-independent concentration unit that is used in experiments where temperature may be varied.

$$\text{molality} = \frac{\text{moles of solute}}{\text{kg of solvent}}$$

This equation also relates the mass of solvent to the moles of solute. Molality is also used in freezing point depression and boiling point elevation experiments to determine molar masses.

Mole fraction; mole percent *(page 494)* This is a temperature-independent concentration unit used with Raoult's law.

$$\text{mole fraction} = X_A = \frac{n_A}{n_{total}}$$

$$\text{mole percent} = \frac{n_A}{n_{total}} \times 100\%$$

Raoult's law *(page 496)* This law summarizes the effect of a non-volatile solute on the vapor pressure of a solution.

$$P_{solution} = X_{solvent} P^\circ_{solvent}$$

Raoult's law for a mixture of two volatile solvents, *A* and *B*, is

$$P_{solution} = X_A P_A^\circ + X_B P_B^\circ$$

Equations for freezing point depression and boiling point elevation *(page 501)*

$$\Delta T_f = K_f m \qquad \Delta T_b = K_b m$$

These equations are used to estimate the decrease in freezing point and increase in boiling point of solutions. These equations are also used to estimate the molar mass of non-dissociating solutes.

Equation for osmotic pressure *(page 505)*

$$\Pi V = nRT \qquad \Pi = MRT$$

The osmotic pressure equation is similar to the ideal gas law equation. It is used to calculate molecular masses from osmotic pressure and concentration data. In this form, remember that R must be 0.0821 L atm mol^{-1} K^{-1}.

QUESTIONS, PROBLEMS, AND EXERCISES

Answers to problems whose numbers are printed in color are given in Appendix B. More challenging problems are marked with asterisks. ILW = Interactive Learningware solution is available at www.wiley.com/college/brady. OH = an Office Hours video is available for this problem.

REVIEW QUESTIONS

Why Solutions Form

12.1 Why do two gases spontaneously mix when they are brought into contact?

12.2 When substances form liquid solutions, what two factors are involved in determining the solubility of the solute in the solvent?

12.3 Methanol, $CH_3—O—H$, and water are miscible in all proportions. What does this mean? Explain how the O—H unit in methanol contributes to this.

12.4 Hexane (C_6H_{12}) and water are immiscible. What does this mean? Explain why they are immiscible in terms of structural features of their molecules and the forces of attraction between them.

12.5 Explain how ion–dipole forces help to bring potassium chloride into solution in water.

12.6 Explain why potassium chloride will not dissolve in carbon tetrachloride, CCl_4.

Heat of Solution

12.7 The value of ΔH_{soln} for a soluble compound is, say, +26 kJ mol^{-1}, and a nearly saturated solution is prepared in an insulated container (e.g., a coffee cup calorimeter). Will the system's temperature increase or decrease as the solute dissolves?

12.8 Referring to the preceding question, which value for this compound would be numerically larger, its lattice energy or its hydration energy?

12.9 Which would be expected to have the larger hydration energy, Al^{3+} or Li^{+}? Why? (Both ions are about the same size.)

12.10 Suggest a reason why the value of ΔH_{soln} for a gas such as CO_2, dissolving in water, is negative.

12.11 The value of ΔH_{soln} for the formation of an acetone–water solution is negative. Explain this in general terms that discuss intermolecular forces of attraction.

12.12 The value of ΔH_{soln} for the formation of an ethanol–hexane solution is positive. Explain this in general terms that involve intermolecular forces of attraction.

12.13 When a certain solid dissolves in water, the solution becomes cool. Is ΔH_{soln} for this solute positive or negative? Explain your reasoning. Is the solubility of this substance likely to increase or decrease with increasing temperature? Explain your answer using Le Châtelier's principle.

12.14 If the value of ΔH_{soln} for the formation of a mixture of two liquids A and B is zero, what does this imply about the relative strengths of A–A, B–B, and A–B intermolecular attractions?

Temperature and Solubility

12.15 If a saturated solution of NH_4NO_3 at 70 °C is cooled to 10 °C, how many grams of solute will separate if the quantity of the solvent is 100 g? (Use data in Figure 12.9.)

12.16 Anglers know that on hot summer days, the largest fish will be found in deep sinks in lake bottoms, where the water is coolest. Use the temperature dependence of oxygen solubility in water to explain why.

Pressure and Solubility

12.17 What is Henry's law?

12.18 Mountain streams often contain fewer living things than equivalent streams at sea level. Give one reason why this might be true in terms of oxygen solubilities at different pressures.

12.19 Why is ammonia so much more soluble in water than is nitrogen? Would you expect hydrogen chloride gas to have a high or low solubility in water? Explain your answers to both questions.

12.20 Why does a bottled carbonated beverage fizz when you take the cap off?

Expressions of Concentration

12.21 Write the definition for each of the following concentration units: mole fraction, mole percent, molality, percent by mass.

12.22 How does the molality of a solution vary with increasing temperature? How does the molarity of a solution vary with increasing temperature?

12.23 Suppose a 1.0 m solution of a solute is made using a solvent with a density of 1.15 g/mL. Will the molarity of this solution be numerically larger or smaller than 1.0? Explain.

Colligative Properties

12.24 What specific fact about a physical property of a solution must be true to call it a colligative property?

12.25 What kinds of data would have to be obtained to find out if a solution of two miscible liquids is almost exactly an ideal solution?

12.26 When octane is mixed with methanol, the vapor pressure of the octane over the solution is higher than what we would calculate using Raoult's law. Why? Explain the discrepancy in terms of intermolecular attractions.

12.27 Explain why a nonvolatile solute dissolved in water makes the system have (a) a higher boiling point than water and (b) a lower freezing point than water.

12.28 Why do we call dialyzing and osmotic membranes *semipermeable*? What is the opposite of *permeable*?

12.29 What is the key difference between dialyzing and osmotic membranes?

12.30 At a molecular level, explain why in osmosis there is a net migration of solvent from the side of the membrane less concentrated in solute to the side more concentrated in solute.

12.31 Two glucose solutions of unequal molarity are separated by an osmotic membrane. Which solution will *lose* water, the one with the higher or the lower molarity?

12.32 Which aqueous solution has the higher osmotic pressure, 10% glucose, $C_6H_{12}O_6$, or 10% sucrose, $C_{12}H_{22}O_{11}$? (Both are molecular compounds.)

12.33 When a solid is *associated* in a solution, what does this mean? What difference does it make to expected colligative properties?

12.34 What is the difference between a *hypertonic* solution and a *hypotonic* solution?

Colligative Properties of Electrolytes

12.35 Why are colligative properties of solutions of ionic compounds usually more pronounced than those of solutions of molecular compounds of the same molalities?

12.36 What is the van't Hoff factor? What is its expected value for all nondissociating molecular solutes? If its measured value is slightly larger than 1.0, what does this suggest about the solute? What is suggested by a van't Hoff factor of approximately 0.5?

12.37 Which aqueous solution, if either, is likely to have the higher boiling point, 0.50 *m* NaI or 0.50 *m* Na_2CO_3?

REVIEW PROBLEMS

Heat of Solution

12.38 Consider the formation of a solution of aqueous potassium chloride. Write the thermochemical equations for (a) the conversion of solid KCl into its gaseous ions and (b) the subsequent formation of the solution by hydration of the ions. The lattice energy of KCl is -690 kJ mol^{-1}, and the hydration energy of the ions is -686 kJ mol^{-1}. Calculate the enthalpy of solution of KCl in kJ mol^{-1}.

OH **12.39** For an ionic compound dissolving in water, $\Delta H_{soln} = -50$ kJ mol^{-1} and the hydration energy is -890 kJ mol^{-1}. Estimate the lattice energy of the ionic compound.

Henry's Law

OH **12.40** The solubility of methane, the chief component of natural gas, in water at 20 °C and 1.0 atm pressure is 0.025 g L^{-1}. What is its solubility in water at 1.5 atm and 20 °C?

12.41 At 740 torr and 20 °C, nitrogen has a solubility in water of 0.018 g L^{-1}. At 620 torr and 20 °C, its solubility is 0.015 g L^{-1}. Show that nitrogen obeys Henry's law.

12.42 If the solubility of a gas in water is 0.010 g L^{-1} at 25 °C with the partial pressure of the gas over the solution at 1.0 atm, predict the solubility of the gas at the same temperature but at double the pressure.

12.43 If 100.0 mL of water is shaken with oxygen gas at 1.0 atm, it will dissolve 0.0039 g O_2. Estimate the Henry's law constant for oxygen gas in water.

Expressions of Concentration

12.44 What is the molality of NaCl in a solution that is 3.000 *M* NaCl, with a density of 1.07 g mL^{-1}?

12.45 A solution of acetic acid, CH_3COOH, has a concentration of 0.143 *M* and a density of 1.00 g mL^{-1}. What is the molality of this solution?

12.46 What is the molal concentration of glucose, $C_6H_{12}O_6$, a sugar found in many fruits, in a solution made by dissolving 24.0 g of glucose in 1.00 kg of water? What is the mole fraction of glucose in the solution? What is the mass percent of glucose in the solution?

12.47 If you dissolved 11.5 g of NaCl in 1.00 kg of water, what would be its molal concentration? What are the mass percent NaCl and the mole percent NaCl in the solution? The volume of this solution is virtually identical to the original volume of the 1.00 kg of water. What is the molar concentration of NaCl in this solution? What would have to be true about any solvent for one of its dilute solutions to have essentially the same molar and molal concentrations?

12.48 A solution of ethanol, CH_3CH_2OH, in water has a concentration of 1.25 *m*. Calculate the mass percent of ethanol.

12.49 A solution of NaCl in water has a concentration of 19.5%. Calculate the molality of the solution.

OH **12.50** A solution of NH_3 in water is at a concentration of 7.50% by mass. Calculate the mole percent NH_3 in the solution. What is the molal concentration of the NH_3?

12.51 An aqueous solution of isopropyl alcohol, C_3H_8O, rubbing alcohol, has a mole fraction of alcohol equal to 0.250. What is the percent by mass of alcohol in the solution? What is the molality of the alcohol?

ILW **12.52** Sodium nitrate, $NaNO_3$, is sometimes added to tobacco to improve its burning characteristics. An aqueous solution of $NaNO_3$ has a concentration of 0.363 *m*. Its density is 1.0185 g mL^{-1}. Calculate the molar concentration of $NaNO_3$ and the mass percent of $NaNO_3$ in the solution. What is the mole fraction of $NaNO_3$ in the solution?

12.53 In an aqueous solution of sulfuric acid, the concentration is 1.89 mol% of acid. The density of the solution is 1.0645 g mL^{-1}. Calculate the following: (a) the molal concentration of H_2SO_4, (b) the mass percent of the acid, and (c) the molarity of the solution.

Raoult's Law

OH **12.54** At 25 °C, the vapor pressure of water is 23.8 torr. What is the vapor pressure of a solution prepared by dissolving 65.0 g of $C_6H_{12}O_6$ (a nonvolatile solute) in 150 g of water? (Assume the solution is ideal.)

12.55 The vapor pressure of water at 20 °C is 17.5 torr. A 35% solution of the nonvolatile solute ethylene glycol, $C_2H_4(OH)_2$, in water is prepared. Estimate the vapor pressure of the solution.

12.56 At 25 °C the vapor pressures of benzene (C_6H_6) and toluene (C_7H_8) are 93.4 and 26.9 torr, respectively. A solution is made by mixing 35.0 g of benzene and 65.0 g of toluene. At what applied pressure, in torr, will this solution boil at 25 °C?

12.57 Pentane (C_5H_{12}) and heptane (C_7H_{16}) are two hydrocarbon liquids present in gasoline. At 20 °C, the vapor pressure of pentane is 420 torr and the vapor pressure of heptane is 36.0 torr. What will be the total vapor pressure (in torr) of a solution prepared by mixing equal masses of the two liquids?

***12.58** Benzene and toluene help get good engine performance from lead-free gasoline. At 40 °C, the vapor pressure of benzene is 180 torr and that of toluene is 60 torr. Suppose you wished to prepare a solution of these liquids that will have a total vapor pressure of 96 torr at 40 °C. What must be the mole percent concentrations of each in the solution?

*12.59 The vapor pressure of pure methanol, CH_3OH, at 30 °C is 160 torr. How many grams of the nonvolatile solute glycerol, $C_3H_5(OH)_3$, must be added to 100 g of methanol to obtain a solution with a vapor pressure of 140 torr?

12.60 A solution containing 8.3 g of a nonvolatile nondissociating substance dissolved in 1 mol of chloroform, $CHCl_3$, has a vapor pressure of 511 torr. The vapor pressure of pure $CHCl_3$ at the same temperature is 526 torr. Calculate (a) the mole fraction of the solute, (b) the number of moles of solute in the solution, and (c) the molecular mass of the solute.

12.61 At 21.0 °C, a solution of 18.26 g of a nonvolatile, nonpolar compound in 33.25 g of ethyl bromide, C_2H_5Br, had a vapor pressure of 336.0 torr. The vapor pressure of pure ethyl bromide at this temperature is 400.0 torr. What is the molecular mass of the compound?

Freezing Point Depression and Boiling Point Elevation

12.62 How many grams of sucrose ($C_{12}H_{22}O_{11}$) are needed to lower the freezing point of 100 g of water by 3.00 °C?

12.63 To make sugar candy a concentrated sucrose solution is boiled until the temperature reaches 270 °F. What is the molality and what is the mole fraction of sucrose in this mixture?

OH 12.64 A solution of 12.00 g of an unknown nondissociating compound dissolved in 200.0 g of benzene freezes at 3.45 °C. Calculate the molecular mass of the unknown.

12.65 A solution of 14 g of a nonvolatile, nondissociating compound in 1.0 kg of benzene boils at 81.7 °C. Calculate the molecular mass of the unknown.

ILW 12.66 What are the molecular mass and molecular formula of a nondissociating molecular compound whose empirical formula is C_4H_2N if 3.84 g of the compound in 500 g of benzene gives a freezing point depression of 0.307 °C?

12.67 Benzene reacts with hot concentrated nitric acid dissolved in sulfuric acid to give chiefly nitrobenzene, $C_6H_5NO_2$. A by-product is often obtained, which consists of 42.86% C, 2.40% H, and 16.67% N (by mass). The boiling point of a solution of 5.5 g of the by-product in 45 g of benzene was 1.84 °C higher than that of benzene. (a) Calculate the empirical formula of the by-product. (b) Calculate a molecular mass of the by-product and determine its molecular formula.

Osmotic Pressure

ILW 12.68

(a) Show that the following equation is true.

$$\text{Molar mass of solute} = \frac{(\text{grams of solute})RT}{\Pi V}$$

(b) An aqueous solution of a compound with a very high molecular mass was prepared in a concentration of 2.0 g L^{-1} at 25 °C. Its osmotic pressure was 0.021 torr. Calculate the molecular mass of the compound.

OH 12.69 A saturated solution is made by dissolving 0.400 g of a polypeptide (a substance formed by joining together in a chainlike fashion some number of amino acids) in water to give 1.00 L of solution. The solution has an osmotic pressure of 3.74 torr at 27 °C. What is the approximate molecular mass of the polypeptide?

Colligative Properties of Electrolyte Solutions

12.70 The vapor pressure of water at 20 °C is 17.5 torr. What would be the vapor pressure at 20 °C of a solution made by dissolving 23.0 g of NaCl in 100 g of water? (Assume complete dissociation of the solute and an ideal solution.)

12.71 How many grams of $AlCl_3$ would have to be dissolved in 150 mL of water to give a solution that has a vapor pressure of 38.7 torr at 35 °C? Assume complete dissociation of the solute and ideal solution behavior. (At 35 °C, the vapor pressure of pure water is 42.2 torr.)

OH 12.72 What is the osmotic pressure, in torr, of a 2.0% solution of NaCl in water when the temperature of the solution is 25 °C?

12.73 Below are the concentrations of the most abundant ions in seawater.

Ion	Molality
Chloride	0.566
Sodium	0.486
Magnesium	0.055
Sulfate	0.029
Calcium	0.011
Potassium	0.011
Bicarbonate	0.002

Use these data to estimate the osmotic pressure of seawater at 25 °C in units of atmospheres. What is the minimum pressure in atm needed to desalinate seawater by reverse osmosis?

12.74 What is the expected freezing point of a 0.20 m solution of $CaCl_2$? (Assume complete dissociation.)

12.75 The freezing point of a 0.10 m solution of mercury(I) nitrate is approximately −0.27 °C. Show that these data suggest that the formula of the mercury(I) ion is Hg_2^{2+}.

Interionic Attractions and Colligative Properties

12.76 The van't Hoff factor for the solute in 0.100 m $NiSO_4$ is 1.19. What would this factor be if the solution behaved as if it were 100% dissociated?

12.77 What is the expected van't Hoff factor for K_2SO_4 in an aqueous solution, assuming 100% dissociation?

12.78 A 0.118 m solution of LiCl has a freezing point of −0.415 °C. What is the van't Hoff factor for this solute at this concentration?

12.79 What is the approximate osmotic pressure of a 0.118 m solution of LiCl at 10 °C? Express the answer in torr. (Use the data in the preceding problem.)

ADDITIONAL EXERCISES

*12.80 The "bends" is a medical emergency caused by the formation of tiny bubbles in the blood vessels of divers who rise too quickly to the surface from a deep dive. The origin of the problem is seen in the calculations in this problem. At 37 °C (normal body temperature), the solubility of N_2 in water is 0.015 g L^{-1} when its pressure over the solution is 1 atm. Air is approximately 78 mol% N_2. How many moles of N_2 are dissolved per liter of blood (essentially an aqueous solution) when a diver inhales air at a pressure of 1 atm? How many moles of N_2 dissolve per liter of blood when the diver is submerged to a depth of approximately 100 ft, where the total pressure of the air being breathed is 4 atm? If the diver suddenly surfaces, how many milliliters of N_2 gas, in the form of tiny bubbles, are released into the bloodstream from each liter of blood (at 37 °C and 1 atm)?

OH **12.81** The vapor pressure of a mixture of 400 g of carbon tetrachloride and 43.3 g of an unknown compound is 137 torr at 30 °C. The vapor pressure of pure carbon tetrachloride at 30 °C is 143 torr, while that of the pure unknown is 85 torr. What is the approximate molecular mass of the unknown?

***12.82** An experiment calls for the use of the dichromate ion, $Cr_2O_7^{2-}$, in sulfuric acid as an oxidizing agent for isopropyl alcohol, C_3H_8O. The chief product is acetone, C_3H_6O, which forms according to the following equation.

$$3C_3H_8O + Na_2Cr_2O_7 + 4H_2SO_4 \longrightarrow 3C_3H_6O + Cr_2(SO_4)_3 + Na_2SO_4 + 7H_2O$$

(a) The oxidizing agent is available only as sodium dichromate dihydrate. What is the minimum number of grams of sodium dichromate dihydrate needed to oxidize 21.4 g of isopropyl alcohol according to the balanced equation?
(b) The amount of acetone actually isolated was 12.4 g. Calculate the percentage yield of acetone.
(c) The reaction produces a volatile by-product. When a sample of it with a mass of 8.654 mg was burned in oxygen, it was converted into 22.368 mg of carbon dioxide and 10.655 mg of water, the sole products. (Assume that any unaccounted for element is oxygen.) Calculate the percentage composition of the by-product and determine its empirical formula.
(d) A solution prepared by dissolving 1.338 g of the by-product in 115.0 g of benzene had a freezing point of 4.87 °C. Calculate the molecular mass of the by-product and write its molecular formula.

OH **12.83** What is the osmotic pressure in torr of a 0.010 *M* aqueous solution of a molecular compound at 25 °C?

***12.84** Ethylene glycol, $C_2H_6O_2$, is used in some antifreeze mixtures. Protection against freezing to as low as −40 °F is sought.
(a) How many moles of solute are needed per kilogram of water to ensure this protection?
(b) The density of ethylene glycol is 1.11 g mL^{-1}. To how many milliliters of solute does your answer to part (a) correspond?
(c) Calculate the number of quarts of ethylene glycol that should be mixed with each quart of water to get the desired protection.

12.85 The osmotic pressure of a dilute solution of a slightly soluble *polymer* (a compound composed of large molecules formed by linking many smaller molecules together) in water was measured using the osmometer in Figure 12.18. The difference in the heights of the liquid levels was determined to be 1.26 cm at 25 °C. Assume the solution has a density of 1.00 g mL^{-1}. (a) What is the osmotic pressure of the solution in torr? (b) What is the molarity of the solution? (c) At what temperature would the solution be expected to freeze? (d) On the basis of the results of these calculations, explain why freezing point depression cannot be used to determine the molecular masses of compounds composed of very large molecules.

12.86 A solution of ethanol, C_2H_5OH, in water has a concentration of 4.613 mol L^{-1}. At 20 °C, its density is 0.9677 g mL^{-1}. Calculate the following: (a) the molality of the solution and (b) the percent concentration of the alcohol. The density of ethanol is 0.7893 g mL^{-1} and the density of water is 0.9982 g mL^{-1} at 20 °C.

12.87 Consider an aqueous 1.00 *m* solution of Na_3PO_4, a compound with useful detergent properties.
(a) Calculate the boiling point of the solution on the assumption that Na_3PO_4 does not ionize at all in solution.
(b) Do the same calculation assuming that the van't Hoff factor for Na_3PO_4 reflects 100% dissociation into its ions.
(c) The 1.00 *m* solution boils at 100.183 °C at 1 atm. Calculate the van't Hoff factor for the solute in this solution.

EXERCISES IN CRITICAL THINKING

12.88 A certain organic substance is soluble in solvent *A* but it is insoluble in solvent *B*. If solvents *A* and *B* are miscible, will the organic compound be soluble in a mixture of *A* and *B*? What additional information is needed to answer this question?

12.89 The situation described in the previous exercise is actually quite common. How might it be used to purify the organic compound?

12.90 Compile and review all of the methods discussed for the determination of molecular masses. Assess which methods are the most reliable, which are the most sensitive, and which are the most convenient to use.

12.91 Having had some laboratory experience by now, evaluate whether preparation of a 0.25 *molar* solution is easier or more difficult than preparing a 0.25 *molal* solution. What experiments require the use of *molal* concentrations?

12.92 This chapter focused on the physical description of osmosis and its use in determining molar masses. What other practical uses are there for osmosis?

12.93 What are the chemical and physical processes that lead to oxygen depletion and the possibility of large-scale fish kills?

12.94 Some gases exhibit an effect that can be described as dissolving in a solid. What gases "dissolve" to a significant extent in what metals? Why is this property of practical interest?

CHAPTERS 10–12

BRINGING IT TOGETHER

Here is another chance for you to test your understanding of concepts, your knowledge of scientific terms, and your skills at problem solving. Read through the following questions carefully, and answer each as fully as possible. Review topics when necessary. When you are able to answer these questions correctly, you are ready to go on to the next group of chapters.

1. A 15.5 L sample of neon at 25.0 °C and a pressure of 748 torr is kept at 25.0 °C as it is allowed to expand to a final volume of 25.4 L. What is the final pressure?
2. An 8.95 L sample of nitrogen at 25.0 °C and 1.00 atm is compressed to a volume of 0.895 L and a pressure of 5.56 atm. What must its final temperature be?
3. A mixture of propane and air will explode if it is heated to 466 °C. If a 20.0 L sample of such a mixture originally at 25.0 °C and 1.00 atm is to be detonated by the heat that is generated by compression alone, what will the pressure of the mixture be at 466 °C if its volume is to be 1.00 L?
4. A sample of oxygen-enriched air with a volume of 12.5 L at 25.0 °C and 1.00 atm consists of 45.0% (v/v) oxygen and 55.0% (v/v) nitrogen What are the partial pressures of oxygen and nitrogen (in torr) in this sample after it has been warmed to a temperature of 37.0 °C and is still at a final volume of 12.5 L?
5. If a gas in a cylinder pushes back a piston against a constant opposing pressure of 3.0×10^5 pascals and undergoes a volume change of 0.50 m^3, how much work will the gas do, expressed in joules?
6. What is the formula mass of a gaseous element if 6.45 g occupies 1.92 L at 745 torr and 25.0 °C? Which element is it?
7. What is the formula mass of a gaseous element if at room temperature it effuses through a pinhole 2.16 times as rapidly as xenon? Which element is it?
8. Briefly and qualitatively explain how the model of an ideal gas, as described by the kinetic theory of gases, explains the following.
   (a) Boyle's law
   (b) Graham's law
   (c) Charles' law
   (d) the meaning of gas temperature
   (e) pressure–temperature law
   (f) absolute zero
9. Which has a higher value of the van der Waals constant *a*, a gas whose molecules are polar or one whose molecules are nonpolar? Explain.
10. What is the van der Waals constant *b* used to correct for, and in what way is this correction accomplished?
11. How many milliliters of dry CO_2, measured at STP, could be evolved in the reaction between 20.0 mL of 0.100 *M* $NaHCO_3$ and 30.0 mL of 0.0800 *M* HCl?
12. How many milliliters of Cl_2 gas, measured at 25 °C and 740 torr, are needed to react with 10.0 mL of 0.10 *M* NaI if the I^- is oxidized to IO_3^- and Cl_2 is reduced to Cl^-?
13. Potassium hypobromite, KOBr, converts ammonia to nitrogen by the following reaction.

$$3KOBr + 2NH_3 \longrightarrow N_2 + 3KBr + 3H_2O$$

To prepare 475 mL of dry N_2, when measured at 24.0 °C and 738 torr, what is the minimum number of grams of KOBr required?

14. Hydrogen peroxide, H_2O_2, is decomposed by potassium permanganate according to the following reaction.

$$5H_2O_2 + 2KMnO_4 + 3H_2SO_4 \longrightarrow 5O_2 + 2MnSO_4 + K_2SO_4 + 8H_2O$$

What is the minimum number of milliliters of 0.125 *M* $KMnO_4$ required to prepare 375 mL of dry O_2 when the gas volume is measured at 22.0 °C and 738 torr?

15. One way to make chlorine is to let manganese dioxide, MnO_2, react with hydrochloric acid according to the following equation.

$$4HCl + MnO_2 \longrightarrow Cl_2 + MnCl_2 + 2H_2O$$

What is the minimum volume (in mL) of 6.44 *M* HCl needed to prepare 525 mL of dry chlorine when the gas is obtained at 24.0 °C and 742 torr?

16. A sample of 248 mL of wet nitrogen gas was collected over water at a total gas pressure of 736 torr and a temperature of 21.0 °C. (The vapor pressure of water at 21.0 °C is 18.7 torr.) The nitrogen was produced by the reaction of sulfamic acid, HNH_2SO_3, with 425 mL of a solution of sodium nitrite according to the following equation.

$$NaNO_2 + HNH_2SO_3 \longrightarrow N_2 + NaHSO_4 + H_2O$$

Calculate what must have been the molar concentration of the sodium nitrite.

17. What two factors are principally responsible for the differences in the behavior of gases and liquids?
18. If the ideal gas law worked well for all substances at all temperatures and pressures, what volume would 1.00 mol of water vapor occupy at 25 °C and 1.00 atm? What volume does 1.00 mol of water actually occupy under these conditions?
19. Which properties of liquids and solids are controlled chiefly by the closeness of the packing of molecules in these states? Which properties are determined chiefly by the strengths of the intermolecular attractions?
20. Consider the molecule $POCl_3$, in which phosphorus is the central atom and is bonded to an oxygen atom and three chlorine atoms.
    (a) Draw the Lewis structure of $POCl_3$ and predict its geometry.
    (b) Is the molecule polar or nonpolar? Explain.
    (c) What kinds of attractive forces would be present between $POCl_3$ molecules in the liquid?
21. What kinds of attractive forces, including chemical bonds, would be present between the particles in the following?
    (a) $H_2O(l)$
    (b) $CCl_4(l)$
    (c) $CH_3OH(l)$
    (d) $BrCl(l)$
    (e) $NaCl(s)$
    (f) $Na_2SO_4(s)$
22. What is a change of state? What terms are used to describe the energy changes associated with the change (a) solid $\longrightarrow$ liquid, (b) solid $\longrightarrow$ gas, and (c) liquid $\longrightarrow$ gas?
23. What is a *dynamic equilibrium*? In terms of Le Châtelier's principle and the "equation"

$$\text{Liquid} + \text{heat} \rightleftharpoons \text{vapor}$$

explain why raising the temperature of a liquid increases the liquid's equilibrium vapor pressure.

24. Can a solid have a vapor pressure? How would the vapor pressure of a solid vary with temperature?

25. Trimethylamine, $(CH_3)_3N$, is a substance responsible in part for the smell of fish. It has a boiling point of 3.5 °C and a molecular weight of 59.1. Dimethylamine, $(CH_3)_2NH$, has a similar odor and boils at a slightly higher temperature, 7 °C, even though it has a somewhat lower molecular mass (45.1). How can this be explained in terms of the kinds of attractive forces between their molecules?

26. Methanol, CH_3OH, commonly known as wood alcohol, has a boiling point of 64.7 °C. Methylamine, a fishy-smelling chemical found in herring brine, has a boiling point of −6.3 °C. Ethane, a hydrocarbon present in petroleum, has a boiling point of −88 °C.

```
     H               H   H              H   H
     |               |   |              |   |
 H—C—O—H         H—C—N—H          H—C—C—H
     |               |                  |   |
     H               H                  H   H

  methanol        methylamine          ethane
 b.p. 64.7 °C    b.p. −6.3 °C        b.p. −88 °C
```

Each has nearly the same molecular mass. Account for the large differences in their boiling points in terms of the attractive forces between their molecules.

27. Based on what you've learned in these chapters, explain the following.
(a) A breeze cools you when you're perspiring.
(b) Droplets of water form on the outside of a glass of cold soda on a warm, humid day.
(c) You feel more uncomfortable on a warm, humid day than on a warm, dry day.
(d) The origin of the energy in a violent thunderstorm.
(e) Clouds form as warm, moist air flows over a mountain range.

28. How do the magnitudes of ΔH_{fusion}, $\Delta H_{\text{vaporization}}$, and $\Delta H_{\text{sublimation}}$ compare for a given substance?

29. Make sketches of (a) a face-centered cubic unit cell, (b) a body-centered cubic unit cell, and (c) a simple cubic unit cell. Which type of unit cell does NaCl have?

30. Aluminum has a density of 2.70 g cm^{-3} and crystallizes in a face-centered cubic lattice. Use these and other data to calculate the atomic radius of an aluminum atom.

31. What is the difference between the closest packed structures identified as ccp and hcp? In each of these structures, how many atoms are in contact with any given atom?

32. Tin tetraiodide (stannic iodide) has the formula SnI_4. It forms soft, yellow to reddish crystals that melt at about 143 °C. What kind of solid does SnI_4 form? What kind of bonding occurs in SnI_4?

33. A certain compound has the formula MCl_2. Crystals of the compound melt at 772 °C and give a liquid that is electrically conducting. What kind of crystal does this compound form?

34. What general properties are expected of covalent crystals?

35. Silicon dioxide, SiO_2, forms very hard crystals that melt at 1610 °C to yield a liquid that does not conduct electricity. What crystal type does SiO_2 form?

36. Sketch the phase diagram for a substance that has a triple point at 25 °C and 100 torr, a normal boiling point of 150 °C, and a melting point at 1 atm of 27 °C. Is the solid more dense or less dense than the liquid? Where on the curve would the critical temperature and critical pressure be? What phase would exist at 30 °C and 10.0 torr?

37. How many grams of 4.00% (w/w) solution of KOH in water are needed to neutralize completely the acid in 10.0 mL 0.256 M H_2SO_4?

38. Calculate the molar concentration of 15.00% (w/w) Na_2CO_3 solution at 20.0 °C given that its density is 1.160 g mL^{-1}.

39. The solubility of pure oxygen in water at 20.0 °C and 760 torr is 4.30×10^{-2} g O_2 per liter of H_2O. When air is in contact with water and the air pressure is 585 torr at 20 °C, how many grams of oxygen from the air dissolve in 1.00 L of water? The average concentration of oxygen in the air is 21.1% (v/v).

40. Compound A is a white solid with a high melting point. When it melts, it conducts electricity. In which solvent is it likely to be more soluble, water or gasoline? Explain.

41. Compound XY is an ionic compound that dissociates as it dissolves in water. The lattice energy of XY is −600 kJ mol^{-1}. The hydration energy of its ions is −610 kJ mol^{-1}.
(a) Write the thermochemical equations for the two steps in the formation of a solution of XY in water.
(b) Write the sum of these two equations in the form of a thermochemical equation, showing the net ΔH.
(c) Draw an enthalpy diagram for the formation of this solution.

42. A 0.270 M KOH solution has a density of 1.01 g mL^{-1}. Calculate the percent concentration (w/w) of KOH.

43. A 5.30 M solution of glycerol in water has a density of 1.11 g mL^{-1}. Calculate the percent concentration by mass of glycerol ($C_3H_8O_3$) and the mole fraction of glycerol present.

44. At 20 °C a 40.00% (v/v) solution of ethyl alcohol, C_2H_5OH, in water, has a density of 0.9369 g mL^{-1}. The density of pure ethyl alcohol at this temperature is 0.7907 g mL^{-1} and that of water is 0.9982 g mL^{-1}.
(a) Calculate the molar concentration and the molal concentration of C_2H_5OH in this solution.
(b) Calculate the mole fraction and mole percent of C_2H_5OH in this solution.
(c) The vapor pressure of ethyl alcohol at 20 °C is 41.0 torr and that of water is 17.5 torr. If the 40.00% (v/v) solution were ideal, what would be the vapor pressure of each component over the solution?

45. Estimate the boiling point of 1.0 molal $Al(NO_3)_3$, assuming that it dissociates entirely into Al^{3+} and NO_3^- ions in solution.

46. Squalene is an oil found chiefly in shark liver oil but also present in low concentrations in olive oil, wheat germ oil, and yeast. A qualitative analysis disclosed that its molecules consist entirely of carbon and hydrogen. When a sample of squalene with a mass of 0.5680 g was burned in pure oxygen, there was obtained 1.8260 g of carbon dioxide and 0.6230 g of water.
(a) Calculate the empirical formula of squalene.
(b) When 0.1268 g of squalene was dissolved in 10.50 g of molten camphor, the freezing point of this solution was 177.3 °C. (The melting point of pure camphor is 178.4 °C, and its molal freezing point depression constant is 37.7 °C kg $camphor^{-1}$). Calculate the molar mass of squalene and determine its molecular formula.

13 KINETICS: THE STUDY OF RATES OF REACTION

This photo shows how speed adds excitement to the movie *Mission Impossible III*. Wreckage of a high-speed chase and the explosion that the hero, Tom Cruise, barely escapes as he dashes to safety are evidence of fast chemical reactions. Kinetics is the name that describes studies of fast (and slow) chemical reactions that we will learn about in this chapter. *(Stephen Vaughan/Paramount Pictures/The Kobal Collection, Ltd.)*

CHAPTER OUTLINE

THIS CHAPTER IN CONTEXT Chemical reactions run at many different speeds. Some, such as the rusting of iron or the breakdown of plastics in the environment, take place very slowly. Others, like the combustion of gasoline or the explosion of gunpowder, occur very quickly. **Chemical kinetics** is the study of the speeds (or *rates*) of chemical reactions. On a practical level, it is concerned with factors that affect the speeds of reactions and how reaction speeds can be controlled. This is essential in industry, where synthetic reactions must take place at controlled speeds. If a reaction takes weeks or months to occur, it may not be economically feasible; if it occurs too quickly or uncontrollably, it may not be safe to carry out. For the consumer, studies on rates of decomposition allow a manufacturer to reliably determine the shelf life, or expiration date, of a product or drug. At a more fundamental level, a study of the speed of a reaction often gives clues that lead to an understanding about *how*, at a molecular level, reactants change into products. Understanding the reaction at this level of detail often allows even finer control of the reaction's speed, and suggests ways to modify the reaction to produce new types of products, or to improve the reaction's yield by preventing undesirable side reactions from occurring.

13.1 FIVE FACTORS AFFECT REACTION RATES

The **rate of reaction** for a given chemical change is the speed with which the reactants disappear and the products form. It is measured by the amount of products produced or reactants consumed per unit time. Usually this is done by monitoring the concentrations of the reactants or products over time, as the reaction runs (see Figure 13.1).

Before we take up the quantitative aspects of reaction rates, let's look qualitatively at factors that can make a reaction run faster or slower. There are five principal factors that influence reaction rates.

1. Chemical nature of the reactants
2. Ability of the reactants to come in contact with each other
3. Concentrations of the reactants
4. Temperature
5. Availability of rate-accelerating agents called *catalysts*

Chemical nature of the reactants

Bonds break and new bonds form during reactions. The most fundamental differences among reaction rates, therefore, lie in the reactants themselves, in the inherent tendencies of their atoms, molecules, or ions to undergo changes in chemical bonds. Some reactions are fast by nature and others are slow (Figure 13.2). Because sodium atoms lose electrons so easily, for example, a freshly exposed surface of metallic sodium tarnishes almost instantly when exposed to air and moisture. Under identical conditions, potassium also reacts with air and moisture, but the reaction is much faster because potassium atoms lose electrons more easily than sodium atoms.

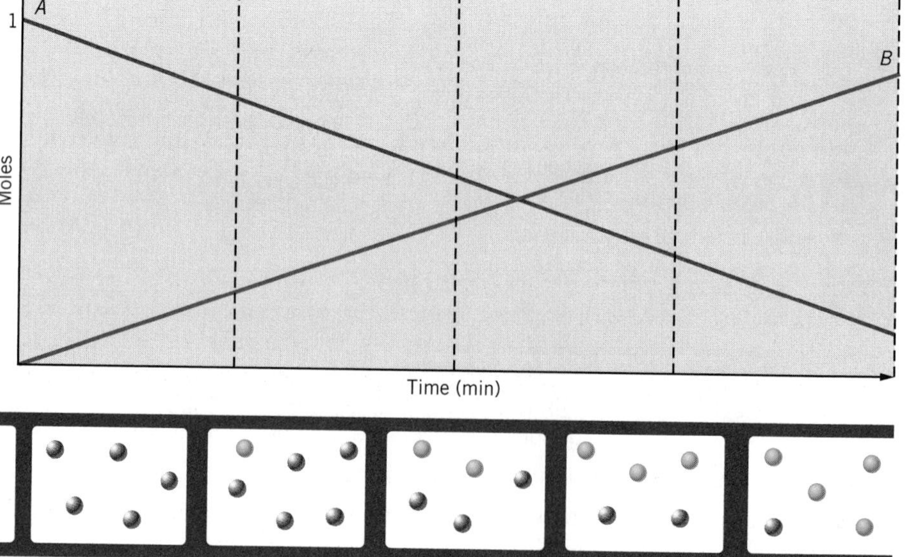

FIG. 13.1 Reaction rates are measured by monitoring concentration changes over time. The progress of the reaction $A \longrightarrow B$. Note that the number of A molecules (in red) decreases with time, as the number of B molecules (in blue) increases. The steeper the concentration versus time curves are, the faster the rate of reaction is. The filmstrip represents the relative numbers of molecules of A and B at each time.

(a)

(b)

FIG. 13.2 The chemical nature of reactants affects reaction rates. (*a*) Sodium loses electrons easily, so it reacts quickly with water. (*b*) Potassium loses electrons even more easily than sodium, so its reaction with water is explosively fast. *(Fundamental Photographs)*

Ability of the reactants to meet

Most reactions involve two or more reactants whose particles (atoms, ions, or molecules) must collide with each other for the reaction to occur. This is why reactions are so often carried out in liquid solutions or in the gas phase, states in which the particles are able to intermingle at a molecular level and collide with each other easily.

Reactions in which all the reactants are in the same phase are called **homogeneous reactions.** Examples include the neutralization of sodium hydroxide by hydrochloric acid when both are dissolved in water, and the explosive gas-phase reaction of gasoline vapor with oxygen that can occur when the two are mixed in the right proportions. (An *explosion* is an extremely rapid reaction that quickly generates hot expanding gases.)

When the reactants are present in different phases—for example, when one is a gas and the other a liquid or a solid—the reaction is called a **heterogeneous reaction.** An example is the combustion of coal, in which solid carbon combines with gaseous oxygen. In a heterogeneous reaction, the reactants are able to meet only at the interface between the phases, so *the area of contact between the phases is a major factor in determining the rate of the reaction.* This area is controlled by the sizes of the particles of the reactants. By pulverizing a solid, the total surface area can be hugely increased (Figure 13.3). This maximizes contact between the atoms, ions, or molecules in the solid state with those in a different phase.

Although heterogeneous reactions are important, they are very complex and difficult to analyze. In this chapter, therefore, we'll focus mostly on homogeneous systems.

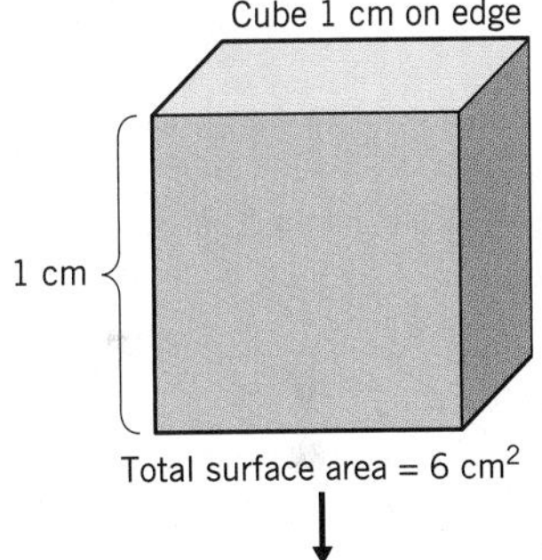

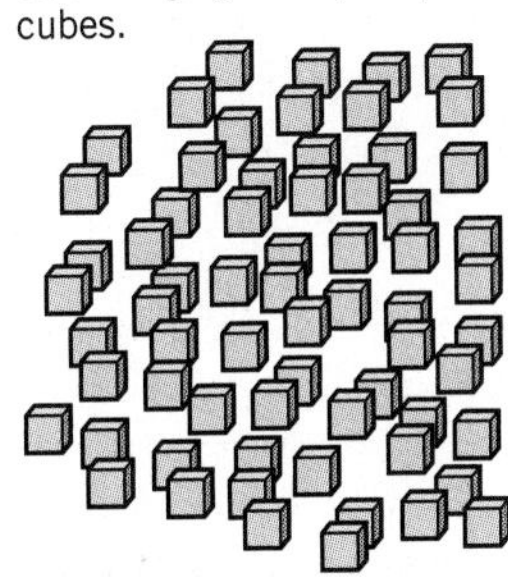

FIG. 13.3 Effect of crushing a solid. When a single solid is subdivided into much smaller pieces, the total surface area on all of the pieces becomes very large.

Concentrations of the reactants

The rates of both homogeneous and heterogeneous reactions are affected by the concentrations of the reactants. For example, wood burns relatively quickly in air but extremely rapidly in pure oxygen. It has been estimated that if air were 30% oxygen instead of 20%, it would not be possible to put out forest fires. Even red hot steel wool, which only sputters and glows in air, bursts into flame when thrust into pure oxygen (see Figure 13.4).

Temperature of the system

Almost all chemical reactions occur faster at higher temperatures than they do at lower temperatures. You may have noticed, for example, that insects move more slowly when the air is cool. An insect is a cold-blooded creature, which means that its body temperature is determined by the temperature of its surroundings. As the air cools, insects cool, and so the rates of their chemical metabolism slow down, making insects sluggish.

Presence of catalysts

Catalysts are substances that increase the rates of chemical reactions without being used up. They affect every moment of our lives. Catalysts are used in the chemical industry to make

FIG. 13.4 Effect of concentration on rate. Steel wool, after being heated to redness in a flame, burns spectacularly when dropped into pure oxygen. *(OPC, Inc.)*

gasoline, plastics, fertilizers, and other products that have become virtual necessities in our lives. In our bodies, substances called enzymes serve as catalysts for biochemical reactions. By making enzymes available or not, a cell is able to direct our body chemistry by controlling which chemical reactions can occur rapidly.

13.2 RATES OF REACTION ARE MEASURED BY MONITORING CHANGE IN CONCENTRATION OVER TIME

The qualitative factors we covered in the previous section can also be described quantitatively. To do this we need to express reaction rates in mathematical terms. Let's start with the concept of **rate,** which always implies a ratio in which a unit of time is in the denominator. Suppose, for example, that you have a job with a pay rate of ten dollars per hour. Because *per* can be translated as *divided by*, your rate of pay can be written as a fraction (abbreviating hour as hr).

$$\text{Rate of pay} = 10 \text{ dollars per hour} = \frac{10 \text{ dollars}}{1 \text{ hr}} = 10 \text{ dollars hr}^{-1}$$

where we've expressed 1/hr as hr^{-1}.

When chemical reactions occur, the concentrations of reactants decrease as they are used up, while the concentrations of the products increase as they form. So one way to describe a reaction's rate is to pick one substance in the reaction's equation and describe its change in concentration per unit of time. The result is the rate of the reaction *with respect to that substance.* Remembering that we always take "final minus initial," the rate of reaction with respect, say, to substance X is

$$\text{Rate with respect to } X = \frac{(\text{conc. of } X \text{ at time } t_{\text{final}} - \text{conc. of } X \text{ at time } t_{\text{initial}})}{(t_{\text{final}} - t_{\text{initial}})}$$

$$= \frac{\Delta(\text{conc. of } X)}{\Delta t}$$

Molarity (mol/L) is normally the concentration unit, and the second (s) is the most often used unit of time. Therefore, the units for reaction rates are most frequently the following.

$$\frac{\text{mol/L}}{\text{s}}$$

Because 1/L and 1/s can also be written as L^{-1} and s^{-1}, the units for a reaction rate can be expressed as mol L^{-1} s^{-1}. For instance, if the concentration of one product of a reaction increases by 0.50 mol/L each second, the rate of its formation is 0.50 mol L^{-1} s^{-1}. Similarly, if the concentration of a reactant decreases by 0.20 mol/L per second, its rate of reaction is 0.20 mol L^{-1} s^{-1}. By convention, reaction rate is reported as a positive value whether something increases or decreases in concentration.

Relative rates of reaction depend on the coefficients in the equation

When we know the value of a reaction rate with respect to one substance, the coefficients of the reaction's balanced equation can be used to find the rates with respect to the other substances. For example, in the combustion of propane,

$$C_3H_8(g) + 5O_2(g) \longrightarrow 3CO_2(g) + 4H_2O(g)$$

five moles of O_2 *must* be consumed per unit of time for each mole of C_3H_8 used in the same time. Therefore, in this reaction oxygen *must* react five times faster than propane in units of mol L^{-1} s^{-1}. Similarly, CO_2 forms three times faster than C_3H_8 reacts and H_2O four times faster. The magnitudes of the rates relative to each other are thus in the same relationship as the coefficients in the balanced equation.

EXAMPLE 13.1

Relationships of Rates within a Reaction

Butane, the fuel in cigarette lighters, burns in oxygen to give carbon dioxide and water. If, in a certain experiment, the butane concentration is decreasing at a rate of 0.20 mol L^{-1} s^{-1}, what is the rate at which the oxygen concentration is decreasing, and what are the rates at which the product concentrations are increasing?

ANALYSIS: As always we need a balanced chemical equation. Butane, C_4H_{10}, burns in oxygen to give CO_2 and H_2O according to the equation

$$2C_4H_{10}(g) + 13O_2(g) \longrightarrow 8CO_2(g) + 10H_2O(g)$$

We now need to relate the rate in terms of oxygen and the products to the given rate in terms of butane. The chemical equation is the tool that links amounts of these substances to the amount of butane. The magnitudes of the rates relative to each other are in the same relationship as the coefficients in the balanced equation.

SOLUTION: For oxygen

$$\frac{0.20 \text{ mol } C_4H_{10}}{\text{L s}} \times \frac{13 \text{ mol } O_2}{2 \text{ mol } C_4H_{10}} = \frac{1.3 \text{ mol } O_2}{\text{L s}}$$

Oxygen is reacting at a rate of 1.3 mol L^{-1} s^{-1}. For CO_2 and H_2O, we have similar calculations.

$$\frac{0.20 \text{ mol } C_4H_{10}}{\text{L s}} \times \frac{8 \text{ mol } CO_2}{2 \text{ mol } C_4H_{10}} = \frac{0.80 \text{ mol } CO_2}{\text{L s}}$$

$$\frac{0.20 \text{ mol } C_4H_{10}}{\text{L s}} \times \frac{10 \text{ mol } H_2O}{2 \text{ mol } C_4H_{10}} = \frac{1.0 \text{ mol } H_2O}{\text{L s}}$$

Therefore,

$$\text{Rate of formation of } CO_2 = 0.80 \text{ mol } L^{-1} s^{-1}$$
$$\text{Rate of formation of } H_2O = 1.0 \text{ mol } L^{-1} s^{-1}$$

ARE THE ANSWERS REASONABLE? If what we've calculated is correct, then the ratio of the numerical values of the last two rates, namely 0.80 to 1.0, should check out to be the same as the ratio of the corresponding coefficients in the chemical equation, namely, 8 to 10 (the same as 0.8 to 1.0).

Practice Exercise 1: The iodate ion reacts with sulfite ions in the reaction

$$IO_3^- + 3SO_3^{2-} \longrightarrow I^- + 3SO_4^{2-}$$

At what rate are the iodide and sulfate ions being produced if the sulfite ion is disappearing at a rate of 2.4×10^{-4} mol L^{-1} s^{-1}? (Hint: Recall the names of polyatomic ions.)

Practice Exercise 2: Hydrogen sulfide burns in oxygen to form sulfur dioxide and water. If sulfur dioxide is being formed at a rate of 0.30 mol L^{-1} s^{-1}, what are the rates of disappearance of hydrogen sulfide and oxygen?

Because the rates of reaction of reactants and products are all related, it doesn't matter which species we pick to follow concentration changes over time. For example, to study the decomposition of hydrogen iodide, HI, into H_2 and I_2,

$$2HI(g) \longrightarrow H_2(g) + I_2(g)$$

it is easiest to monitor the I_2 concentration because it is the only colored substance in the reaction. As the reaction proceeds, purple iodine vapor forms, and there are instruments that

TABLE 13.1 Data, at 508 °C, for the Reaction $2HI(g) \longrightarrow H_2(g) + I_2(g)$

Concentration of HI (mol L^{-1})	Time (s)
0.100	0
0.0716	50
0.0558	100
0.0457	150
0.0387	200
0.0336	250
0.0296	300
0.0265	350

allow us to relate the intensity of the color to the iodine concentration. Then, once we know the rate of formation of iodine, we also know the rate of formation of hydrogen. It's the same because the coefficients of H_2 and I_2 are the same. And the rate of disappearance of HI, which has a coefficient of 2 in the equation, is twice as fast as the rate of formation of I_2.

Most reactions slow down as reactants are used up

A reaction rate is generally not constant throughout the reaction but commonly changes as the reactants are used up. This is because the rate usually depends on the concentrations of the reactants, and these change as the reaction proceeds. For example, Table 13.1 contains data for the decomposition of hydrogen iodide at a temperature of 508 °C. The data, which show the changes in molar HI concentration over time, are plotted in Figure 13.5. Notice that the molar HI concentration drops fairly rapidly during the first 50 seconds of the reaction, which means that the initial rate is relatively fast. However, later, in the interval between 300 s and 350 s, the concentration changes by only a small amount, so the rate has slowed considerably. Thus, the steepness of the curve at any moment reflects the rate of the reaction; the steeper the curve, the higher is the rate.

When we use the term "reaction rate" we mean the instantaneous rate, unless we state otherwise.

The rate at which the HI is being consumed at any particular moment is called the **instantaneous rate.** The instantaneous rate can be determined from the slope (or tangent) of the curve measured at the time we have chosen. The slope, which can be read off the graph, is the ratio (expressed positively) of the change in concentration to the change in time. In Figure 13.5, for example, the rate of the decomposition of hydrogen iodide is determined for a time 100 seconds from the start of the reaction. After the tangent to the

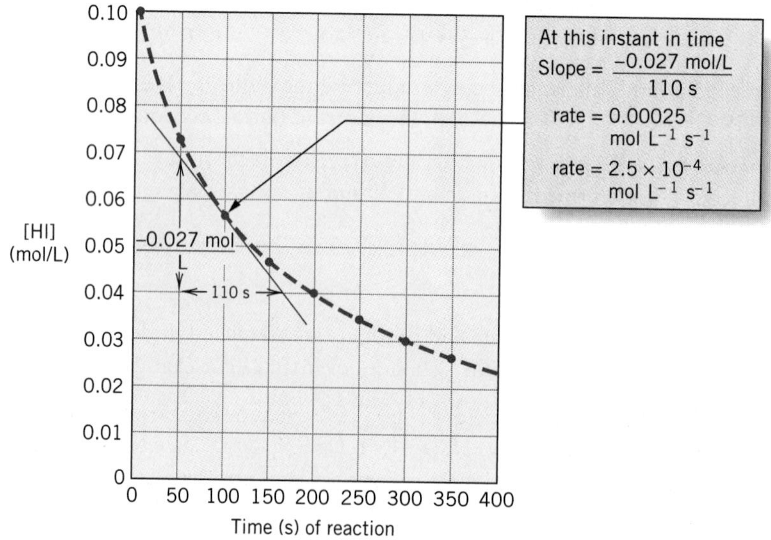

FIG. 13.5 Effect of time on concentration. The data for this plot of the change in the concentration of HI with time for the reaction

$$2HI(g) \longrightarrow H_2(g) + I_2(g)$$

at 508 °C are taken from Table 13.1. The slope is negative because we're measuring the *disappearance* of HI. But when its value is used as a rate of reaction, we express the rate as positive, as we do all rates of reaction.

curve is drawn, we measure the concentration change (a decrease of 0.027 mol/L) and the time change (110 s) from the graph. Because the rate is based on a *decreasing* concentration, we use a minus sign for the *equation* that describes this rate so that the rate itself will be a positive quantity. We use square brackets to signify concentrations specifically in moles per liter; [HI] thus means the molar concentration of HI.

$$\text{Rate}_{(\text{with respect to HI})} = -\left(\frac{[\text{HI}]_{\text{final}} - [\text{HI}]_{\text{initial}}}{t_{\text{final}} - t_{\text{initial}}}\right) = -\left(\frac{-0.027 \text{ mol/L}}{110 \text{ s}}\right)$$

$$\text{Rate}_{(\text{with respect to HI})} = 2.5 \times 10^{-4} \text{ mol L}^{-1} \text{ s}^{-1}$$

Thus, at this moment in the reaction, the rate with respect to HI is 2.5×10^{-4} mol L^{-1} s^{-1}. In the following example, we'll use this technique to obtain the *initial instantaneous rate* of the reaction, that is, the instantaneous rate of reaction at time zero.

EXAMPLE 13.2
Estimating the Initial Rate of a Reaction

For the experimental data shown in Figure 13.5, what is the initial rate of the reaction

$$2\text{HI}(g) \longrightarrow \text{H}_2(g) + \text{I}_2(g)$$

at 508 °C, with respect to HI?

ANALYSIS: The question asks about the initial rate, which is the instantaneous rate of the reaction at time zero. Using Figure 13.5, we can draw a tangent line to the curve showing HI concentration as a function of time at time zero. The instantaneous rate will be the slope of the tangent line. Remember that the slope of a straight line can be calculated from the coordinates of any two points (x_1, y_1) and (x_2, y_2) using the equation

$$\text{Slope} = \frac{y_2 - y_1}{x_2 - x_1}$$

SOLUTION: Remembering that a tangent line to a curve touches the curve at only one point, we can draw the line as shown in the graph below. To more precisely determine the slope of the tangent line, we should choose two points that are as far apart as possible. The point on the curve (0 s, 0.10 mol/L) and the intersection of the tangent line with the time axis (130 s, 0 mol/L) are widely separated:

$$\text{Slope} = \frac{0.10 \text{ mol/L} - 0.00 \text{ mol/L}}{0 \text{ s} - 130 \text{ s}} = -7.7 \times 10^{-4} \text{ mol L}^{-1} \text{ s}^{-1}$$

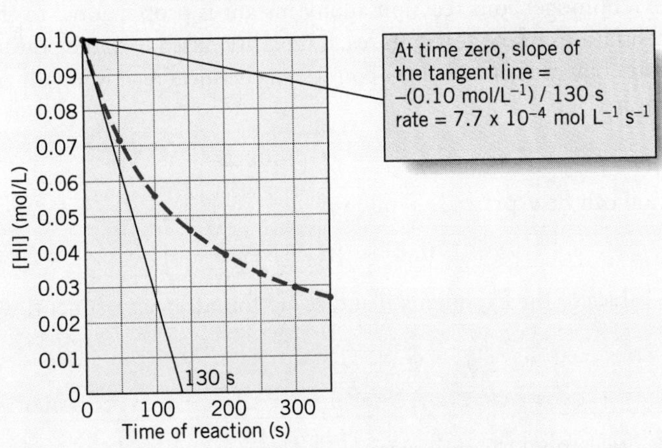

The slope is negative because the concentration of HI is decreasing as time increases. Rates are positive quantities, so we can report the initial rate of reaction as 7.7×10^{-4} mol L^{-1} s^{-1}. Since the tangent line might be drawn a number of different ways, the time difference is uncertain by more than ten seconds and the rate can be reported simply as 8×10^{-4} mol L^{-1} s^{-1}.

The average rate is the slope of a line connecting two points on a concentration versus time graph. The instantaneous rate is the slope of a tangent line at a single point. The average and instantaneous rates are quite different from each other.

IS THE ANSWER REASONABLE? The instantaneous rate at time zero ought to be slightly larger than the *average rate* between zero and 50 seconds. We can compute the average rate directly from data in Table 13.1 by selecting a pair of concentrations at two different times.

$$\text{Slope} = \frac{0.0716 \text{ mol/L} - 0.100 \text{ mol/L}}{50 \text{ s} - 0 \text{ s}} = -5.7 \times 10^{-4} \text{ mol L}^{-1} \text{ s}^{-1}$$

Thus, the average rate from 0 to 50 s is 5.7×10^{-4} mol L^{-1} s^{-1}. As expected, this is slightly less than the instantaneous rate at time zero, 8×10^{-4} mol L^{-1} s^{-1}.

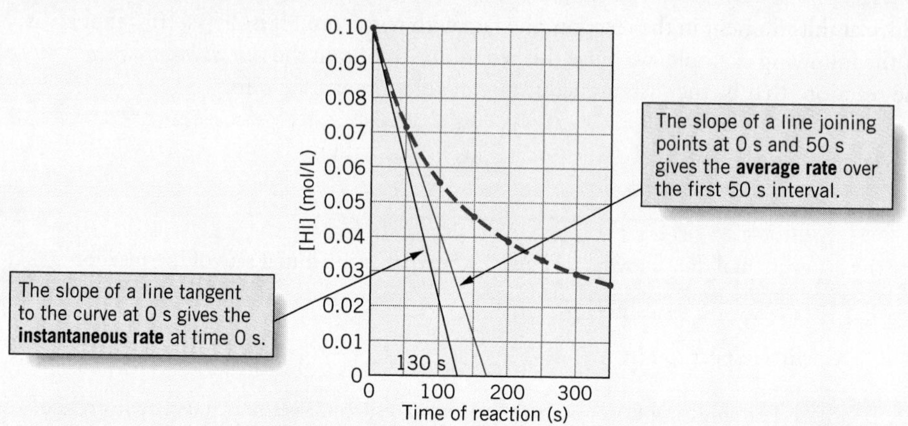

Practice Exercise 3: Use the graph in Figure 13.5 to estimate the rate of reaction with respect to HI 2.00 minutes after the start of the reaction. (Hint: You need to have time in seconds units to find the point where the tangent should be drawn.)

Practice Exercise 4: Use the graph in Figure 13.5 to estimate the rate of reaction with respect to HI 250 seconds after the start of the reaction.

13.3 RATE LAWS GIVE REACTION RATE AS A FUNCTION OF REACTANT CONCENTRATIONS

Thus far we have focused on a rate with respect to *one* component of a reaction. We'll now broaden our focus to consider a rate expression that includes all reactants.

The rate of a homogeneous reaction at any instant is proportional to the product of the molar concentrations of the reactants, each molarity raised to some power or exponent that has to be found by experiment. Let's consider a chemical reaction with an equation of the following form.

$$A + B \longrightarrow \text{products}$$

Its rate of reaction can be expressed as follows.

$$\text{Rate} \propto [A]^m[B]^n \tag{13.1}$$

As we said, the values of the exponents n and m are found by experiment, which we'll go into shortly.

Rate laws relate reaction rates and concentrations

The proportionality symbol, $\propto$, in Equation 13.1 can be replaced by an equals sign if we introduce a proportionality constant, k, which is called the **rate constant** for the reaction. This gives Equation 13.2.

$$\text{Rate} = k[A]^m[B]^n \tag{13.2}$$

Equation 13.2 is called the **rate law** for the reaction of A with B. Once we have found values for k, n, and m, the rate law allows us to calculate the rate of the reaction at any set of known values of concentrations. Consider, for example, the following reaction.

$$H_2SeO_3 + 6I^- + 4H^+ \longrightarrow Se + 2I_3^- + 3H_2O$$

Its rate law is of the form

$$\text{Rate} = k[H_2SeO_3]^x[I^-]^y[H^+]^z$$

The exponents have been found experimentally to be the following for the initial rate of this reaction (i.e., the rate when the reactants are first combined).

$$x = 1, y = 3, \text{ and } z = 2$$

At 0 °C, k equals 5.0×10^5 L^5 mol^{-5} s^{-1}. (We have to specify the temperature because k varies with it.) Substituting the exponents and the value of k into the rate law equation gives the rate law for the reaction.

$$\text{Rate} = (5.0 \times 10^5 \text{ L}^5 \text{ mol}^{-5} \text{ s}^{-1})[H_2SeO_3][I^-]^3[H^+]^2 \qquad \text{(at 0 °C)}$$

We can calculate the rate of the reaction at 0 °C for any set of concentrations of H_2SeO_3, I^-, and H^+ using this rate law.

The value of k depends on the particular reaction being studied as well as the temperature at which the reaction occurs.

The units of the rate constant are such that the calculated rate will have the units mol L^{-1} s^{-1}.

EXAMPLE 13.3
Calculating Reaction Rate from the Rate Law

In the stratosphere, molecular oxygen (O_2) can be broken into two oxygen atoms by ultraviolet radiation from the sun. When one of these oxygen atoms strikes an ozone (O_3) molecule in the stratosphere, the ozone molecule is destroyed, and two oxygen molecules are created:

$$O(g) + O_3(g) \longrightarrow 2O_2(g)$$

This reaction is part of the natural cycle of ozone destruction and creation in the stratosphere. What is the rate of ozone destruction *for this reaction alone* at an altitude of 25 km, if the rate law for the reaction is

$$\text{Rate} = 4.15 \times 10^5 \text{ L mol}^{-1} \text{ s}^{-1}[O_3][O]$$

and the reactant concentrations at 25 km are the following: $[O_3] = 1.2 \times 10^{-8}$ M and $[O] = 1.7 \times 10^{-14}$ M?

ANALYSIS: The tool we use for calculating the rate is the rate law. Because we already know the rate law, the answer to this question is merely a matter of substituting the given molar concentrations into this law.

SOLUTION: To see how the units work out, let's write all of the concentration values as well as the rate constant's units in fraction form.

$$\text{Rate} = \frac{4.15 \times 10^5 \cancel{\text{L}}}{\cancel{\text{mol}} \text{ s}} \times \left(\frac{1.2 \times 10^{-8} \cancel{\text{mol}}}{\cancel{\text{L}}}\right) \times \left(\frac{1.7 \times 10^{-14} \text{ mol}}{\text{L}}\right)$$

Performing the arithmetic and canceling the units, we see that

$$\text{Rate} = \frac{8.5 \times 10^{-17} \text{ mol}}{\text{L s}} = 8.5 \times 10^{-17} \text{ mol L}^{-1} \text{ s}^{-1}$$

IS THE ANSWER REASONABLE? There's obviously no simple check. Multiplying the powers of ten for the rate constant and the concentrations together reassures us that the rate is of the correct order of magnitude, and we can see that at least the answer has the correct units for a reaction rate.

Practice Exercise 5: The rate law for the reaction $2NO(g) + 2H_2(g) \longrightarrow N_2(g) + 2H_2O(g)$ is

$$\text{Rate} = k[NO]^2[H_2]$$

If the rate of reaction is 7.86×10^{-3} mol L^{-1} s^{-1} when the concentrations of NO and H_2 are both 2×10^{-6} mol L^{-1} (a) what is the value of the rate constant and (b) what are the units for the rate constant? (Hint: Note the units of the reaction rate.)

Practice Exercise 6: The rate law for the decomposition of HI to I_2 and H_2 is

$$\text{Rate} = k[HI]^2$$

Figure 13.5 shows that at 508 °C, the rate of the reaction of HI was found to be 2.5×10^{-4} mol L^{-1} s^{-1} when the HI concentration was 0.0558 *M* (see Figure 13.5). (a) What is the value of *k*? (b) What are the units of *k*?

You cannot predict the rate law for a reaction from the overall balanced equation for the reaction

Although a rate law's exponents are generally unrelated to the chemical equation's coefficients, they sometimes are the same by coincidence, as is the case in the decomposition of hydrogen iodide.

$$2HI(g) \longrightarrow H_2(g) + I_2(g)$$

The rate law, as we've said, is

$$\text{Rate} = k[HI]^2$$

The exponent of [HI] in the rate law, namely 2, happens to match the coefficient of HI in the overall chemical equation, but *there is no way we could have predicted this match without experimental data.* Therefore, *never* simply assume the exponents and the coefficients are the same; it's a trap that many students fall into.

An exponent in a rate law is called the **order of the reaction**[1] with respect to the corresponding reactant. For instance, the decomposition of gaseous N_2O_5 into NO_2 and O_2,

$$2N_2O_5(g) \longrightarrow 4NO_2(g) + O_2(g)$$

has the rate law

$$\text{Rate} = k[N_2O_5]$$

When the exponent on a concentration term is equal to 1, it is usually omitted.

The exponent of $[N_2O_5]$ is 1, so the reaction rate is said to be *first order* in N_2O_5. The rate law for the decomposition of HI has an exponent of 2 for the HI concentration, so its reaction rate is *second order* in HI. The rate law

$$\text{Rate} = k[H_2SeO_3][I^-]^3[H^+]^2$$

describes a reaction rate that is first order with respect to H_2SeO_3, third order with respect to I^-, and second order with respect to H^+.

The **overall order of reaction** is the sum of the orders with respect to each reactant in the rate law. The decomposition of N_2O_5 is a **first-order reaction,** and the decomposition of HI is a **second-order reaction.** The overall order for the reaction with H_2SeO_3 above is $1 + 3 + 2 = 6$.

The exponents in a rate law are usually small whole numbers, but fractional and negative exponents are occasionally found. A negative exponent means that the concentration term really belongs in the denominator, which means that as the concentration of the species increases, the rate of reaction decreases.

[1] The reason for describing the *order* of a reaction is to take advantage of a great convenience; namely, the mathematics involved in the treatment of the data is the same for all reactions having the same order. We will not go into this very deeply, but you should be familiar with this terminology; it's often used to describe the effects of concentration on reaction rates.

There are even **zero-order reactions.** They have reaction rates that are independent of the concentration of any reactant. Zero-order reactions usually involve a small amount of a catalyst that is saturated with reactants. This is rather like the situation in a crowded supermarket with only a single checkout lane open. It doesn't matter how many people join the line; the line will move at the same rate no matter how many people are standing in it. An example of a zero-order reaction is the elimination of ethyl alcohol in the liver. Regardless of the blood alcohol level, the rate of alcohol removal by the body is constant, because the number of available catalyst molecules present in the liver is constant. Another zero-order reaction is the decomposition of gaseous ammonia into H_2 and N_2 on a hot platinum surface. The rate at which ammonia decomposes is the same, regardless of its concentration in the gas. The rate law for a zero-order reaction is simply

$$\text{Rate} = k$$

where the rate constant k has units of mol L^{-1} s^{-1}. The rate constant depends on the amount, quality, and available surface area of the catalyst. For example, forcing ammonia through hot platinum powder (with a high surface area) would cause it to decompose faster than simply passing it over a hot platinum surface.

Practice Exercise 7: The following reaction

$$BrO_3^- + 3SO_3^{2-} \longrightarrow Br^- + 3SO_4^{2-}$$

has the rate law

$$\text{Rate} = k[BrO_3^-][SO_3^{2-}]$$

What is the order of the reaction with respect to each reactant? What is the overall order of the reaction? (Hint: Recall that a concentration with no exponent has, in effect, an exponent of 1.)

Practice Exercise 8: A certain reaction has an experimental rate law that is found to be second order in Cl_2 and first order in NO. Write the rate law for this reaction.

The order of a reaction must be determined experimentally

We've mentioned several times that the exponents in the rate law of an overall reaction must be determined experimentally. *This is the only way to know for sure what the exponents are.* To determine the exponents, we study how changes in concentration affect the rate of the reaction. For example, consider again the following hypothetical reaction.

$$A + B \longrightarrow \text{products}$$

Suppose, further, that the data in Table 13.2 have been obtained in a series of five experiments. We know the form of the rate law for the reaction will be

$$\text{Rate} = k[A]^m[B]^n$$

TABLE 13.2 Concentration–Rate Data for the Hypothetical Reaction $A + B \longrightarrow$ products

Experiment	Initial Concentrations $[A]$ (mol L^{-1})	Initial Concentrations $[B]$ (mol L^{-1})	Initial Rate of Formation of Products (mol L^{-1} s^{-1})
1	0.10	0.10	0.20
2	0.20	0.10	0.40
3	0.30	0.10	0.60
4	0.30	0.20	2.40
5	0.30	0.30	5.40

The values of m and n can be discovered by looking for patterns in the rate data given in the table. *One of the easiest ways to reveal patterns in data is to form ratios of results using different sets of conditions.* Because this technique is quite generally useful, let's look in some detail at how it is applied to the problem of finding the rate law exponents.

For experiments 1, 2, and 3 in Table 13.2, the concentration of B has been held constant at 0.10 M. Any change in the rate for these first three experiments must be due to the change in $[A]$. The rate law tells us that when the concentration of B is held constant, the rate must be proportional to $[A]^m$, so if we take the ratio of rate laws for experiments 2 and 1, we obtain

$$\frac{\text{Rate}_2}{\text{Rate}_1} = \frac{k[A]_2^m[B]_2^n}{k[A]_1^m[B]_1^n} = \frac{k}{k}\left(\frac{[A]_2}{[A]_1}\right)^m\left(\frac{[B]_2}{[B]_1}\right)^n$$

For experiments 1 and 2, the left side of this equation is

$$\frac{\text{Rate}_2}{\text{Rate}_1} = \frac{0.40 \cancel{\text{mol L}^{-1}\text{s}^{-1}}}{0.20 \cancel{\text{mol L}^{-1}\text{s}^{-1}}} = 2.0$$

and on the right side of the equation the identical concentrations of B and the rate constant k cancel to give

$$\left(\frac{[A]_2}{[A]_1}\right)^m = \left(\frac{0.20 \cancel{\text{mol L}^{-1}}}{0.10 \cancel{\text{mol L}^{-1}}}\right)^m = 2.0^m$$

so doubling $[A]$ in going from experiment 1 to experiment 2 doubles the rate, and the relationship reduces to $2.0 = 2.0^m$. For each unique combination of experiments 1, 2, and 3, we have

$$2.0 = 2.0^m \qquad \text{(for experiments 2 and 1)}$$
$$3.0 = 3.0^m \qquad \text{(for experiments 3 and 1)}$$
$$1.5 = 1.5^m \qquad \text{(for experiments 3 and 2)}$$

The only value of m that makes all of these equations true is $m = 1$. Therefore, this reaction must be first order with respect to A.

A similar method will give us the exponent on $[B]$. In the final three experiments, the concentration of B changes while the concentration of A is held constant. This time it is the concentration of B that affects the rate. Taking the ratio of rate laws for experiments 4 and 3, we have

$$\frac{\text{Rate}_4}{\text{Rate}_3} = \frac{k\,[A]_4^m\,[B]_4^n}{k\,[A]_3^m\,[B]_3^n}$$

that, after cancelling the identical concentrations of A and the rate constant, k, becomes

$$\frac{\text{Rate}_4}{\text{Rate}_3} = \left(\frac{[B]_4}{[B]_3}\right)^n$$

For each unique combination of experiments 3, 4, and 5, we have

$$4.0 = 2.0^n \qquad \text{(for experiments 4 and 3)}$$
$$9.0 = 3.0^n \qquad \text{(for experiments 5 and 3)}$$
$$2.25 = 1.5^n \qquad \text{(for experiments 5 and 4)}$$

The only value of n that makes all of these equations true is $n = 2$, so the reaction must be second order with respect to B.

Having determined the exponents for the concentration terms, we now know that the rate law for the reaction must be

$$\text{Rate} = k[A]^1[B]^2$$

To calculate the value of k, we substitute rate and concentration data into the rate law for any one of the sets of data.

TABLE 13.3 Relationship between the Order of a Reaction and Changes in Concentration and Rate

Factor by Which the Concentration Is Changed	Factor by Which the Rate Changes	Exponent on the Concentration Term in the Rate Law
2	Rate	0
3	is	0
4	unchanged	0
2	$2 = 2^1$	1
3	$3 = 3^1$	1
4	$4 = 4^1$	1
2	$4 = 2^2$	2
3	$9 = 3^2$	2
4	$16 = 4^2$	2
2	$8 = 2^3$	3
3	$27 = 3^3$	3
4	$64 = 4^3$	3

$$k = \frac{\text{rate}}{[A]^1[B]^2}$$

Using the data from the first set in Table 13.2,

$$k = \frac{0.20 \text{ mol L}^{-1} \text{ s}^{-1}}{(0.10 \text{ mol L}^{-1})(0.10 \text{ mol L}^{-1})^2}$$

$$= \frac{0.20 \text{ mol L}^{-1} \text{ s}^{-1}}{0.0010 \text{ mol}^3 \text{ L}^{-3}}$$

After canceling such units as we can, the value of k with the net units is

$$k = 2.0 \times 10^2 \text{ L}^2 \text{ mol}^{-2} \text{ s}^{-1}$$

Practice Exercise 9: Use the data from the other four experiments (Table 13.2) to calculate k for this reaction. What do you notice about the values of k? (Hint: Don't forget the exponents in the rate law.)

Practice Exercise 10: Use the rate law determined above to describe what will happen to the reaction rate under the following conditions: (a) the concentration of B is tripled, (b) the concentration of A is tripled, (c) the concentration of A is tripled and the concentration of B is halved.

Table 13.3 summarizes the reasoning used to determine the order with respect to each reactant from experimental data.

EXAMPLE 13.4
Determining the Exponents of a Rate Law

Sulfuryl chloride, SO_2Cl_2, is used to manufacture the antiseptic chlorophenol. The following data were collected on the decomposition of SO_2Cl_2 at a certain temperature.

$$SO_2Cl_2(g) \longrightarrow SO_2(g) + Cl_2(g)$$

Initial Concentration of SO_2Cl_2 (mol L^{-1})	Initial Rate of Formation of SO_2 (mol L^{-1} s^{-1})
0.100	2.2×10^{-6}
0.200	4.4×10^{-6}
0.300	6.6×10^{-6}

What are the rate law and the value of the rate constant for this reaction?

ANALYSIS: The first step is to write the general form of the expected rate law so we can see which exponents have to be determined. Then we study the data to see how the rate changes when the concentration is changed by a certain factor.

SOLUTION: We expect the rate law to have the form

$$\text{Rate} = k[SO_2Cl_2]^x$$

Let's examine the data from the first two experiments. Notice that when we double the concentration from 0.100 *M* to 0.200 *M*, the initial rate doubles (from 2.2×10^{-6} mol L^{-1} s^{-1} to 4.4×10^{-6} mol L^{-1} s^{-1}). If we look at the first and third, we see that when the concentration triples (from 0.100 *M* to 0.300 *M*), the rate also triples (from 2.2×10^{-6} mol L^{-1} s^{-1} to 6.6×10^{-6} mol L^{-1} s^{-1}). This behavior tells us that the reaction must be first order in the SO_2Cl_2 concentration. The rate law is therefore

We could also use experiments 2 and 3. From the second to the third, the rate increases by the same factor, 1.5, as the concentration, so by these data, too, the reaction must be first order.

$$\text{Rate} = k[SO_2Cl_2]^1$$

To evaluate *k*, we can use any of the three sets of data. Choosing the first,

$$k = \frac{\text{rate}}{[SO_2Cl_2]^1} = \frac{2.2 \times 10^{-6} \cancel{\text{mol L}^{-1}}\ \text{s}^{-1}}{0.100 \cancel{\text{mol L}^{-1}}} = 2.2 \times 10^{-5}\ \text{s}^{-1}$$

IS THE ANSWER REASONABLE? We should get the same value of *k* by picking any other pair of values. With the last pair of data, at an initial molar concentration of SO_2Cl_2 of 0.300 mol L^{-1} and an initial rate of 6.6×10^{-6} mol L^{-1} s^{-1}, we calculate *k* again to be 2.2×10^{-5} s^{-1}.

EXAMPLE 13.5
Determining the Exponents of a Rate Law

The following data were measured for the reduction of nitric oxide with hydrogen.

$$2NO(g) + 2H_2(g) \longrightarrow N_2(g) + 2H_2O(g)$$

Initial Concentrations (mol L^{-1})		Initial Rate of Formation of H_2O
[NO]	[H_2]	(mol L^{-1} s^{-1})
0.10	0.10	1.23×10^{-3}
0.10	0.20	2.46×10^{-3}
0.20	0.10	4.92×10^{-3}

What is the rate law for the reaction?

ANALYSIS: This time we have two reactants. To see how their concentrations affect the rate we must vary only one concentration at a time. Therefore, we choose two experiments in which the concentration of one reactant doesn't change and examine the effect of a change in the concentration of the other reactant. Then we repeat the procedure for the second reactant.

SOLUTION: We expect the rate law to have the form

$$\text{Rate} = k[NO]^m[H_2]^n$$

Let's look at the first two experiments. Here the concentration of NO remains the same, so the rate is being affected by the change in the H_2 concentration. When we double the H_2 concentration, the rate doubles, so the reaction is first order with respect to H_2. This means $n = 1$.

Next, we need to pick two experiments in which the H_2 concentration doesn't change. Working with the first and third, we see that [NO] doubles and the rate increases by a factor of $4.92/1.23 = 4.00$. When doubling the concentration of a species quadruples the rate, the reaction is second order in that species, so $m = 2$.

Therefore, the rate law for the reaction is

$$\text{Rate} = k[\text{NO}]^2[\text{H}_2]$$

IS THE ANSWER REASONABLE? The only data that we haven't used as a pair are the data for the second and third reactions. The value of [NO] increases by 2 in going from the second to the third set of data, so this should multiply the rate by 2^2 or 4, if we've found the right exponents. But the value for $[H_2]$ halves at the same time, so this should take a rate that is otherwise four times as large and cut it by a factor of $(1/2)^1$ or in half. The net effect, then, is to make the rate of the third reaction two times as large as that of the second reaction, which is the observed rate change.

Practice Exercise 11: The following reaction is investigated to determine its rate law.

$$2\text{NO}(g) + 2\text{H}_2(g) \longrightarrow \text{N}_2(g) + 2\text{H}_2\text{O}(g)$$

Experiments yielded the following results.

Initial Concentrations (mol L^{-1}) [NO]	[H_2]	Initial Rate of Formation of N_2 (mol L^{-1} s^{-1})
0.40×10^{-4}	0.30×10^{-4}	1.0×10^{-8}
0.80×10^{-4}	0.30×10^{-4}	4.0×10^{-8}
0.80×10^{-4}	0.60×10^{-4}	8.0×10^{-8}

(a) Show that these data yield the same rate law as in the preceding Example. (b) What is the value of the rate constant? (c) What are the units for the rate constant? (Hint: Identify the two experiments where only the [NO] changes and the two experiments where only the $[H_2]$ varies.)

Practice Exercise 12: Ordinary sucrose (table sugar) reacts with water in an acidic solution to produce two simpler sugars, glucose and fructose, that have the same molecular formulas.

$$\underset{\text{sucrose}}{\text{C}_{12}\text{H}_{22}\text{O}_{11}} + \text{H}_2\text{O} \longrightarrow \underset{\text{glucose}}{\text{C}_6\text{H}_{12}\text{O}_6} + \underset{\text{fructose}}{\text{C}_6\text{H}_{12}\text{O}_6}$$

In a series of experiments, the following data were obtained.

Initial Sucrose Concentration (mol L^{-1})	Rate of Formation of Glucose (mol L^{-1} s^{-1})
0.10	6.17×10^{-5}
0.20	1.23×10^{-4}
0.50	3.09×10^{-4}

(a) What is the order of the reaction with respect to sucrose? (b) What is the value of the rate constant, with its units?

Practice Exercise 13: A certain reaction has the following equation: $A + B \longrightarrow C + D$. Experiments yielded the following results.

Initial Concentrations (mol L^{-1})		Initial Rate of Formation of C (mol L^{-1} s^{-1})
$[A]$	$[B]$	
0.40	0.30	1.00×10^{-4}
0.60	0.30	2.25×10^{-4}
0.80	0.60	1.60×10^{-3}

(a) What is the rate law for the reaction? (b) What is the value of the rate constant? (c) What are the units for the rate constant? (d) What is the overall order of this reaction?

13.4 INTEGRATED RATE LAWS GIVE CONCENTRATION AS A FUNCTION OF TIME

The rate law tells us how the speed of a reaction varies with the concentrations of the reactants. Often, however, we are more interested in how the concentrations change over time. For instance, if we were preparing some compound, we might want to know how long it will take for the reactant concentrations to drop to some particular value, so we can decide when to isolate the products.

The relationship between the concentration of a reactant and time can be derived from a rate law using calculus. By summing or "integrating" the instantaneous rates of a reaction from the start of the reaction until some specified time t, we can obtain **integrated rate laws** that quantitatively give concentration as a function of time. The form of the integrated rate law depends on the order of the reaction. The mathematical expressions that relate concentration and time in complex reactions can be complicated, so we will concentrate on using integrated rate laws for a few simple first- and second-order reactions.

The natural logarithm of concentration is related linearly with time for first-order reactions

A reaction that is first order has a rate law of the type

$$\text{Rate} = k[A]$$

Using calculus[2] the following equation can be derived that relates the concentration of A and time.

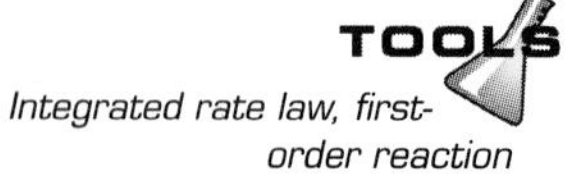

Integrated rate law, first-order reaction

$$\ln \frac{[A]_0}{[A]_t} = kt \tag{13.3}$$

[2] For a first-order reaction, the integrated rate law is obtained by calculus as follows. The instantaneous rate of change of the reactant A is given as

$$\text{Rate} = \frac{-d[A]}{dt} = k[A]$$

This can be rearranged to

$$\frac{d[A]}{[A]} = -k\,dt$$

Next, we integrate between $t = 0$ and $t = t$ as the concentration of A changes from $[A]_0$ to $[A]_t$.

$$\int_{[A]_0}^{[A]_t} \frac{d[A]}{[A]} = \int_0^t -k\,dt$$

$$\ln[\text{A}]_t - \ln[\text{A}]_0 = -kt$$

Using the properties of logarithms, this can be rearranged to give

$$\ln \frac{[A]_0}{[A]_t} = kt$$

The symbol "ln" means natural logarithm. The expression to the left of the equals sign is the natural logarithm of the ratio of $[A]_0$ (the initial concentration of A at $t = 0$) to $[A]_t$ (the concentration of A at a time t after the start of the reaction). We take advantage of a property of logarithms that allows us to write the ratio as a difference in logarithms.

$$\ln[A]_0 - \ln[A]_t = kt$$

We can take the antilogarithm of both sides of Equation 13.3 and rearrange it to obtain the concentration at time t directly as a function of time. Taking the antilogarithm and rearranging algebraically gives[3]

$$[A]_t = [A]_0\, e^{-kt} \qquad (13.4)$$

where e is the base of the system of natural logarithms ($e = 2.718\ldots$). Equation 13.4 shows that the concentration of A decays (decreases) exponentially with time. Calculations can use Equation 13.3 or 13.4. When both of the concentrations are known, it is easiest to use Equation 13.3; when we wish to calculate either $[A]_0$ or $[A]_t$ it may be easier to use Equation 13.4.

□ $[A]_t$ decreases exponentially because the product kt increases with time, but its negative value becomes more negative with time. As the exponent of e becomes a larger negative number the value of the expression becomes smaller.

EXAMPLE 13.6
Concentration–Time Calculations for First-Order Reactions

Dinitrogen pentoxide is not very stable. In the gas phase or dissolved in a nonaqueous solvent, like carbon tetrachloride, it decomposes by a first-order reaction into dinitrogen tetroxide and molecular oxygen.

$$2N_2O_5 \longrightarrow 2N_2O_4 + O_2$$

The rate law is

$$\text{Rate} = k[N_2O_5]$$

At 45 °C, the rate constant for the reaction in carbon tetrachloride is $6.22 \times 10^{-4}\ s^{-1}$. If the initial concentration of N_2O_5 in a carbon tetrachloride solution at 45 °C is 0.500 M, what will its concentration be after exactly one hour?

ANALYSIS: We're dealing with a first-order reaction and the relationship between concentration and time, so the tool we have to apply is either Equation 13.3 or 13.4. Specifically, we have to solve for an unknown concentration. The easiest form of the equation to use when one of the unknowns is a concentration term is Equation 13.4. In performing the calculation, we have to remember that the unit of k involves seconds, not hours, so we must convert the given 1 hr into seconds (1 hr = 3600 s).

SOLUTION: Let's begin by listing the data.

$$[N_2O_5]_0 = 0.500\ M \qquad [N_2O_5]_t = ?\ M$$
$$k = 6.22 \times 10^{-4}\ s^{-1} \qquad t = 3600\ s$$

Using Equation 13.4,[4]

$$\begin{aligned}[N_2O_5]_t &= [N_2O_5]_0\, e^{-kt} \\ &= (0.500\ M) \times e^{-(6.22 \times 10^{-4} s^{-1}) \times 3600\ s} \\ &= (0.500\ M) \times e^{-2.24} \\ &= (0.500\ M) \times 0.11 \\ &= 0.055\ M\end{aligned}$$

□ Calculating $e^{-2.24}$ is a simple operation using a scientific calculator. In most cases it is the inverse of the ln function.

[3] Because of the nature of logarithms, if $\ln x = y$, then $e^{\ln x} = e^y$. But $e^{\ln x} = x$, so $x = e^y$. A similar relationship exists for common (base 10) logarithms. If $\log x = y$, then $10^{\log x} = x = 10^y$.

[4] There are special rules for significant figures for logarithms and antilogarithms. In writing the logarithm of a quantity, the number of digits written *after the decimal point* equals the number of significant figures in the quantity. Raising e to the -2.24 power is the same as taking the antilogarithm of -2.24. Because the quantity -2.24 has two digits after the decimal, the antilogarithm, 0.1064. . ., must be rounded to 0.11 to show just two significant figures.

After one hour, the concentration of N_2O_5 will have dropped to 0.055 *M*.

The calculation could also have been done using Equation 13.3. We would begin by solving for the concentration ratio, substituting values for k and t.

$$\ln\left(\frac{[N_2O_5]_0}{[N_2O_5]_t}\right) = (6.22 \times 10^{-4}\ s^{-1}) \times 3600\ s$$

To take the antilogarithm (antiln), we raise e to the 2.24 power.

$$\text{antiln}\left[\ln\left(\frac{[N_2O_5]_0}{[N_2O_5]_t}\right)\right] = \frac{[N_2O_5]_0}{[N_2O_5]_t}$$

$$\begin{aligned}\text{antiln}\ (2.24) &= e^{2.24} \\ &= 9.4 \text{ (rounding to 2 significant figures)}\end{aligned}$$

This means that

$$\frac{[N_2O_5]_0}{[N_2O_5]_t} = 9.4$$

Now we can substitute the known concentration, $[N_2O_5]_0 = 0.500\ M$. This gives

$$\frac{0.500\ M}{[N_2O_5]_t} = 9.4$$

Solving for $[N_2O_5]_t$ gives

$$\begin{aligned}[N_2O_5]_t &= \frac{0.500\ M}{9.4} \\ &= 0.053\ M\end{aligned}$$

The answers obtained by the two methods differ slightly because of "rounding errors." You can see that using Equation 13.4 is much easier for working this particular problem.

IS THE ANSWER REASONABLE? Notice that the final concentration of N_2O_5 is *less* than its initial concentration. You'd know that you made a huge mistake if the calculated final concentration was larger than the initial 0.500 *M* because reactants are used up by reactions. Also, we obtained essentially the same answer using both equations, so we can be confident it's correct. (If you're faced with a problem like this, you don't have to do it both ways; we just wanted to show that either equation could be used.)

Practice Exercise 14: When designing a consumer product it is desirable that it have a two year shelf life. Often this means that the active ingredient in the product should not decrease by more than 5% in two years. If the reaction is first order, what rate constant must the decomposition reaction of the active ingredient have? (Hint: What are the initial and final percentages of active ingredient?)

Practice Exercise 15: In Practice Exercise 12, the reaction of sucrose with water in an acidic solution was described.

$$\underset{\text{sucrose}}{C_{12}H_{22}O_{11}} + H_2O \longrightarrow \underset{\text{glucose}}{C_6H_{12}O_6} + \underset{\text{fructose}}{C_6H_{12}O_6}$$

The reaction is first order with a rate constant of $6.2 \times 10^{-5}\ s^{-1}$ at 35 °C, when the H^+ concentration is 0.10 *M*. Suppose, in an experiment, the initial sucrose concentration was 0.40 *M*. (a) What will its concentration be after exactly 2 hours? (b) How many minutes will it take for the concentration of sucrose to drop to 0.30 *M*?

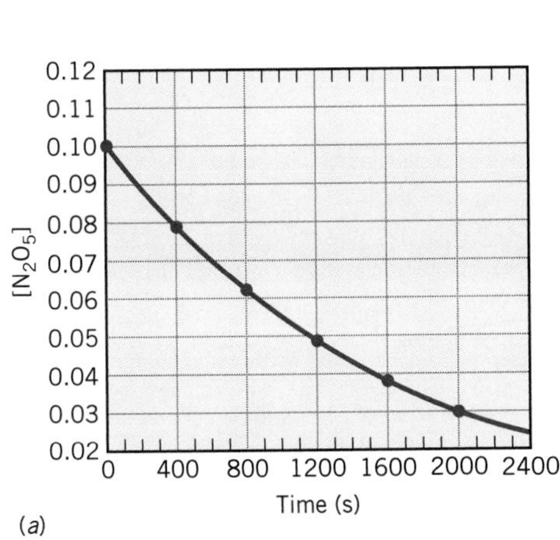

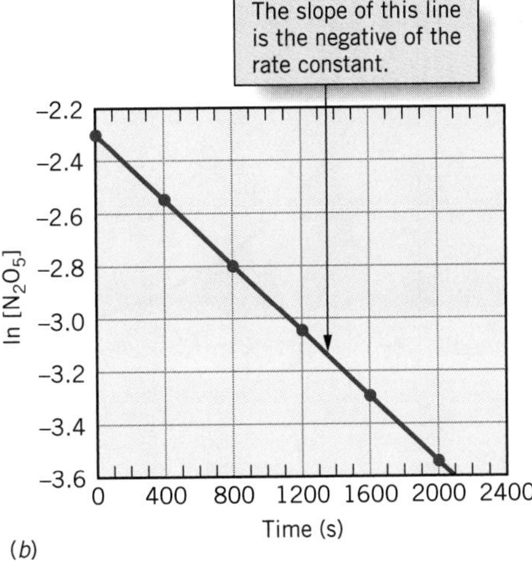

FIG. 13.6 The decomposition of N_2O_5. (*a*) A graph of concentration versus time for the decomposition at 45 °C. (*b*) A straight line is obtained if the logarithm of the concentration is plotted versus time. The slope of this line equals the negative of the rate constant for the reaction.

The rate constant can be determined graphically

Using the properties of logarithms,[5] Equation 13.3 can be rewritten in a form that corresponds to the equation for a straight line.

$$\begin{array}{ccccc} \ln[A]_t & = & -kt & + & \ln[A]_0 \\ \updownarrow & & \updownarrow\updownarrow & & \updownarrow \\ y & = & mx & + & b \end{array}$$

The equation for a straight line is usually written

$$y = mx + b$$

where x and y are variables, m is the slope, and b is the intercept of the line with the y axis.

A plot of the values of $\ln[A]_t$ (vertical axis) versus values of t (horizontal axis) should give a straight line that has a slope equal to $-k$. Such a plot is illustrated in Figure 13.6 for the decomposition of N_2O_5 into N_2O_4 and O_2 in the solvent carbon tetrachloride.

The half-life of a reactant is a measure of its speed of reaction

The *half-life* of a reactant is a convenient way to describe how fast it reacts, particularly for a first-order process. A reactant's **half-life, $t_{1/2}$,** is the amount of time required for half of the reactant to disappear. A rapid reaction has a short half-life because half of the reactant disappears quickly. The equations for half-lives depend on the order of the reaction.

When a reaction, overall, is first order, the half-life of the reactant can be obtained from Equation 13.3 by setting $[A]_t$ equal to one-half of $[A]_0$.

$$[A]_t = \tfrac{1}{2}[A]_0$$

Substituting $\frac{1}{2}[A]_0$ for $[A]_t$ and $t_{1/2}$ for t in Equation 13.3, we have

$$\ln \frac{[A]_0}{\frac{1}{2}[A]_0} = kt_{1/2}$$

Noting that the left-hand side of the equation simplifies to ln 2, and solving the equation for $t_{1/2}$, we have

$$t_{1/2} = \frac{\ln 2}{k} \qquad (13.5)$$

Because ln 2 equals 0.693, Equation 13.5 is sometimes written

$$t_{1/2} = \frac{0.693}{k}$$

Half-life

Because k is a constant for a given reaction, the half-life is also a constant for any particular first-order reaction (at any given temperature). Remarkably, in other words, *the half-life of a first-order reaction is not affected by the initial concentration of the reactant.* This can be illustrated by one of the most common first-order events in nature, the change that radioactive isotopes undergo during radioactive "decay." In fact, you have probably heard the term *half-life* used in reference to the life spans of radioactive substances.

[5] The logarithm of a quotient, $\ln \frac{a}{b}$, can be written as the difference, $\ln a - \ln b$.

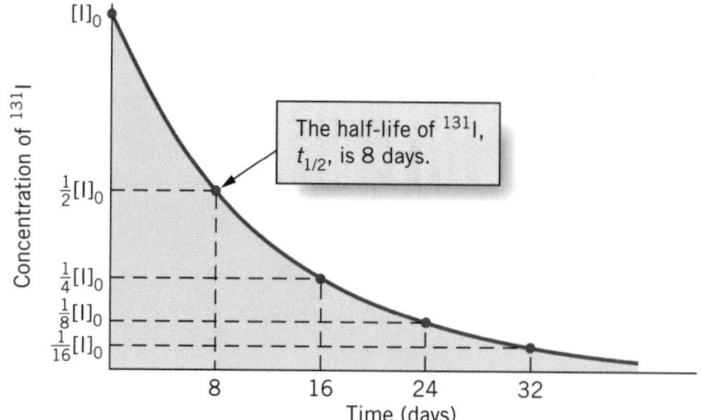

FIG. 13.7 First-order radioactive decay of iodine-131. The initial concentration of the isotope is represented by $[I]_0$.

$^{131}_{53}I \rightarrow {}^{131}_{54}Xe + {}^{0}_{-1}e$

Iodine-131, an unstable, radioactive isotope of iodine, undergoes a nuclear reaction whereby it emits a beta particle (${}^{0}_{-1}e$) and changes into a stable isotope of xenon.[6] The intensity of the radiation decreases, or *decays,* with time (see Figure 13.7). Notice that the time it takes for the first half of the ^{131}I to disappear is 8 days. Then, during the next 8 days half of the remaining ^{131}I disappears, and so on. Regardless of the initial amount, it takes 8 days for half of that amount of ^{131}I to disappear, which means that the half-life of ^{131}I is a constant.

EXAMPLE 13.7
Half-life Calculations

Suppose a patient is given a certain amount of iodine-131 as part of a diagnostic procedure for a thyroid disorder. Given that the half-life of radioactive iodine-131 is 8.0 days, what fraction of the initial iodine-131 would be present in a patient after 24 days if none of it were eliminated through natural body processes?

ANALYSIS: We've learned that radioactive iodine-131 decays by a first-order process with a constant half-life. A period of 24 days is exactly three 8.0 day half-lives. Therefore, let's apply the half-life concept three times.

SOLUTION: If we take the fraction initially present to be 1, we can set up a table

The fraction remaining after n half-lives is $(\frac{1}{2})^n$, or simply $\frac{1}{2^n}$.

Half-life	0	1	2	3
Fraction	1	$\frac{1}{2}$	$\frac{1}{4}$	$\frac{1}{8}$

Half of the iodine-131 is lost in the first half-life, half of that disappears in the second half-life, and so on. Therefore, the fraction remaining after three half-lives is $\frac{1}{8}$.

IS THE ANSWER REASONABLE? We could also have solved the problem using the integrated first-order rate law, Equation 13.3. We'll need the first order rate constant, k, which we can obtain from the half-life by rearranging Equation 13.5:

$$k = \frac{\ln 2}{t_{1/2}} = \frac{0.693}{8.0 \text{ days}} = 0.0866 \text{ day}^{-1}$$

Then we can use Equation 13.3 to compute the fraction $\frac{[A]_0}{[A]_t}$

$$\ln \frac{[A]_0}{[A]_t} = kt = (0.0866 \text{ day}^{-1})(24.0 \text{ day}) = 2.08$$

[6] Iodine-131 is used in the diagnosis of thyroid disorders. The thyroid gland is a small organ located just below the Adam's apple and astride the windpipe. It uses iodide ion to make a hormone, so when a patient is given a dose of $^{131}I^-$ mixed with nonradioactive I^-, both ions are taken up by the thyroid gland. The change in (temporary) radioactivity of the gland is a measure of thyroid activity.

Taking the antilogarithm of both sides, we have

$$\frac{[A]_0}{[A]_t} = e^{2.08} = 8.0$$

The initial concentration, $[A]_0$, is 8.0 times as large as the concentration after 24.0 days, so the fraction remaining after 24 days is $\frac{1}{8}$, which is exactly what we obtained much more simply above.

Practice Exercise 16: In Practice Exercise 12, the reaction of sucrose with water was found to be first order with respect to sucrose. The rate constant under the conditions of the experiments was $6.17 \times 10^{-4}\ s^{-1}$. Calculate the value of $t_{1/2}$ for this reaction in minutes. How many minutes would it take for three-quarters of the sucrose to react? (Hint: What fraction of the sucrose remains?)

Practice Exercise 17: From the answer to Practice Exercise 14, determine the half-life of an active ingredient that has a shelf life of 2.00 years.

Carbon-14 dating determines the age of organic substances

Carbon-14 is a radioactive isotope that is formed in small amounts in the upper atmosphere by the action of cosmic rays on nitrogen atoms. Once formed, the carbon-14 diffuses into the lower atmosphere. It becomes oxidized to carbon dioxide and enters the earth's biosphere by means of photosynthesis. Carbon-14 thus becomes incorporated into plant substances and into the materials of animals that eat plants. As the carbon-14 decays, more is ingested by the living thing. The net effect is an overall equilibrium involving carbon-14 in the global system. As long as the plant or animal is alive, its ratio of carbon-14 atoms to carbon-12 atoms is constant. At death, an organism's remains have as much carbon-14 as they can ever have, and they now slowly lose this carbon-14 by decay. The decay is a first-order process with a rate independent of the *number* of original carbon atoms. The ratio of carbon-14 to carbon-12, therefore, can be related to the years that have elapsed between the time of death and the time of the measurement. The critical assumption in carbon-14 dating is that the steady-state availability of carbon-14 from the atmosphere has remained largely unchanged over the period for which measurements are valid.[7]

In contemporary biological samples the ratio $^{14}C/^{12}C$ is about 1.2×10^{-12}. Thus, each fresh 1.0 g sample of biological carbon in equilibrium with the $^{14}CO_2$ of the atmosphere has a ratio of 5.8×10^{10} atoms of carbon-14 to 4.8×10^{22} atoms of carbon-12. The ratio decreases by a factor of 2 for each half-life period of ^{14}C (5730 years).

In all dating experiments the amounts of sample are extremely small and extraordinary precautions must be taken to avoid contaminating specimens with "modern" materials.

The dating of an object makes use of the fact that radioactive decay is a first-order process. If we let r_0 stand for the $^{14}C/^{12}C$ ratio at the time of death of the carbon-containing species and r_t stand for the $^{14}C/^{12}C$ ratio now, after the elapse of t years, we can substitute into Equation 13.3 to obtain

$$\ln \frac{r_0}{r_t} = kt \qquad (13.6)$$

Willard F. Libby won the Nobel Prize in Chemistry in 1960 for his discovery of the carbon-14 method for dating ancient objects.

where k is the rate constant for the decay (the *decay constant* for ^{14}C) and t is the elapsed time. We can obtain the rate constant from the half-life of ^{14}C using Equation 13.5.

$$\ln 2 = kt_{1/2}$$

[7] The available atmospheric pool of carbon-14 atoms fluctuates somewhat with the intensities of cosmic ray showers, with slow, long-term changes in the earth's magnetic field, and with the huge injections of carbon-12 into the atmosphere from the large-scale burning of coal and petroleum in the 1900s. To reduce the uncertainties in carbon-14 dating, results of the method have been corrected against dates made by tree-ring counting. For example, an uncorrected carbon-14 dating of a Viking site at L'anse aux Meadows, Newfoundland, gave a date of AD 895 ± 30. When corrected, the date of the settlement became AD 997, almost exactly the time indicated in Icelandic sagas for Leif Eriksson's landing at "Vinland," now believed to be the L'anse aux Meadows site.

Substituting 5730 yr for $t_{1/2}$ and solving for k gives $k = 1.21 \times 10^{-4}\ \text{yr}^{-1}$. We can now substitute this value into Equation 13.6 to give

$$\ln \frac{r_0}{r_t} = (1.21 \times 10^{-4}\ \text{yr}^{-1})t \tag{13.7}$$

Equation 13.7 can be used to calculate the age of a once-living object if its current $^{14}C/^{12}C$ ratio can be measured.

EXAMPLE 13.8
Calculating the Age of an Object by ^{14}C Dating

Using a device similar to a mass spectrometer, a sample of an ancient wooden object was found to have a ratio of ^{14}C to ^{12}C equal to 3.3×10^{-13}. What is the age of the object?

ANALYSIS: This is a straightforward calculation that involves using Equation 13.7. We simply substitute values.

SOLUTION: The contemporary ratio of ^{14}C to ^{12}C was given earlier as 1.2×10^{-12}. This corresponds to r_0 in Equation 13.7. Substituting into Equation 13.7 gives

$$\ln \frac{1.2 \times 10^{-12}}{3.3 \times 10^{-13}} = (1.21 \times 10^{-4}\ \text{yr}^{-1})t$$

$$\ln(3.6) = (1.21 \times 10^{-4}\ \text{yr}^{-1})t$$

Solving for t gives an age of 1.1×10^4 years (11,000 years).

IS THE ANSWER REASONABLE? We've been told that the object is ancient, so 11,000 years old seems to make sense. (Also, if you had substituted incorrectly into Equation 13.7, the answer would have been negative, and that certainly doesn't make sense!)

Practice Exercise 18: The ^{14}C content of an ancient piece of wood was found to be one-eighth of that in living trees. How many years old is this piece of wood ($t_{1/2} = 5730$ for ^{14}C)? (Hint: Recall the relationship between the integrated rate equation and half-life.)

Practice Exercise 19: When using carbon-14 dating, samples that have decayed less than 5% and those that have decayed more than 95% may have unacceptably large uncertainties. With that information, what are the upper and lower limits of dates before present, BP, that can be determined?

The reciprocal of the concentration is related linearly to time for second-order reactions

For simplicity, we will only consider a second-order reaction with a rate law of the following type.

$$\text{Rate} = k[B]^2$$

The relationship between concentration and time for a reaction with such a rate law is given by Equation 13.8, an equation that is quite different from that for a first-order reaction.

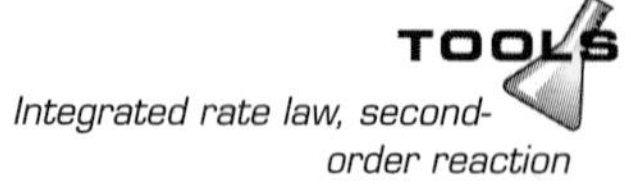

Integrated rate law, second-order reaction

$$\frac{1}{[B]_t} - \frac{1}{[B]_0} = kt \tag{13.8}$$

$[B]_0$ is the initial concentration of B and $[B]_t$ is the concentration at time t. The next example illustrates how Equation 13.8 is applied to calculations.

EXAMPLE 13.9
Concentration–Time Calculations for Second-Order Reactions

Nitrosyl chloride, NOCl, decomposes slowly to NO and Cl_2.

$$2NOCl \longrightarrow 2NO + Cl_2$$

The rate law shows that the rate is second order in NOCl.

$$\text{Rate} = k[NOCl]^2$$

The rate constant k equals 0.020 L mol^{-1} s^{-1} at a certain temperature. If the initial concentration of NOCl in a closed reaction vessel is 0.050 M, what will the concentration be after 30 minutes?

ANALYSIS: We're given a rate law and so can see that it is for a second-order reaction and has the simple form to which our study is limited. We must calculate $[NOCl]_t$, the molar concentration of NOCl, after 30 minutes (1800 s). Our tool for doing this is Equation 13.8.

SOLUTION: Let's begin by tabulating the data.

$$[NOCl]_0 = 0.050\ M \qquad [NOCl]_t = ?\ M$$
$$k = 0.020\ \text{L mol}^{-1}\ \text{s}^{-1} \qquad t = 1800\ \text{s}$$

The equation we wish to substitute into is

$$\frac{1}{[NOCl]_t} - \frac{1}{[NOCl]_0} = kt$$

Making the substitutions gives

$$\frac{1}{[NOCl]_t} - \frac{1}{0.050\ \text{mol L}^{-1}} = (0.020\ \text{L mol}^{-1}\ \text{s}^{-1}) \times (1800\ \text{s})$$

Solving for $1/[NOCl]_t$ gives

$$\frac{1}{[NOCl]_t} - 20\ \text{L mol}^{-1} = 36\ \text{L mol}^{-1}$$
$$\frac{1}{[NOCl]_t} = 56\ \text{L mol}^{-1}$$

Taking the reciprocals of both sides gives us the value of $[NOCl]_t$.

$$[NOCl]_t = \frac{1}{56\ \text{L mol}^{-1}} = 0.018\ \text{mol L}^{-1} = 0.018\ M$$

The molar concentration of NOCl has decreased from 0.050 M to 0.018 M after 30 minutes.

IS THE ANSWER REASONABLE? The concentration of NOCl has decreased, so the answer appears to be reasonable.

Practice Exercise 20: For the reaction in the preceding example, determine how many minutes it would take for the NOCl concentration to drop from 0.040 M to 0.010 M. (Hint: In solving Equation 13.8 time must be a positive value.)

Practice Exercise 21: A sample of nitrosyl chloride was collected for analysis at 10:35 am. At 3:15 pm the same day the sample was analyzed and was found to contain 0.00035 M NOCl. What was the concentration of NOCl at the time the sample was collected?

The second-order rate constant also can be determined graphically

The rate constant k for a second-order reaction, one with a rate following Equation 13.8, can be determined graphically by a method similar to that used for a first-order reaction. We can rearrange Equation 13.8 so that it corresponds to an equation for a straight line.

$$\begin{array}{ccccc} \dfrac{1}{[B]_t} & = & kt & + & \dfrac{1}{[B]_0} \\ \updownarrow & & \updownarrow\updownarrow & & \updownarrow \\ y & = & mx & + & b \end{array}$$

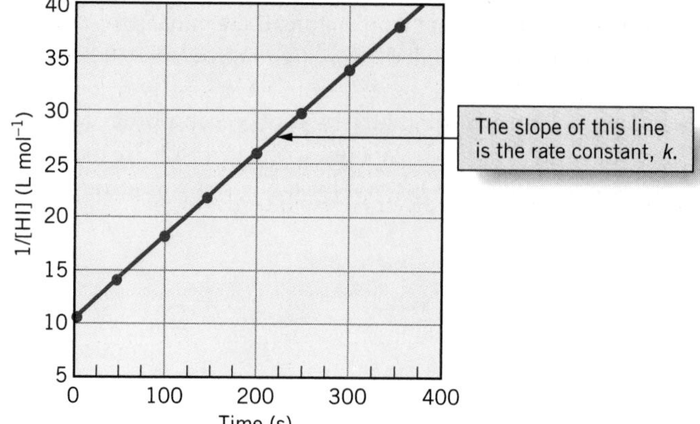

FIG. 13.8 Second-order kinetics. A graph of 1/[HI] versus time for the data in Table 13.1.

When a reaction is second order, then, a plot of $1/[B]_t$ versus t should yield a straight line having a slope k. This is illustrated in Figure 13.8 for the decomposition of HI, using data in Table 13.1.

Half-lives of second-order reactions depend on concentration

The half-life of a second-order reaction *does* depend on initial reactant concentrations. We can see this by examining Figure 13.5 (page 524), which follows the decomposition of gaseous HI, a second-order reaction. The reaction begins with a hydrogen iodide concentration of 0.10 *M*. After 125 seconds, the concentration of HI drops to 0.050 *M*, so 125 s is the observed half-life when the initial concentration of HI is 0.10 *M*. If we then take 0.050 *M* as the next "initial" concentration, we find that it takes 250 seconds (at a *total* elapsed time of 375 seconds) to drop to 0.025 *M*. If we cut the initial concentration in half, from 0.10 *M* to 0.05 *M*, the half-life doubles, from 125 to 250 seconds.

It can be shown that for a second-order reaction of the type we're studying, the half-life is inversely proportional to the initial concentration of the reactant. The half-life is related to the rate constant by Equation 13.9.

$$t_{1/2} = \frac{1}{k \times (\text{initial concentration of reactant})} \tag{13.9}$$

EXAMPLE 13.10
Half-life Calculations

The reaction $2HI(g) \longrightarrow H_2(g) + I_2(g)$ has the rate law, Rate $= k[HI]^2$, with $k = 0.079$ L mol^{-1} s^{-1} at 508 °C. What is the half-life for this reaction at this temperature when the initial HI concentration is 0.10 *M*?

ANALYSIS: The rate law tells us that the reaction is second order. To calculate the half-life, we need to use Equation 13.9.

SOLUTION: The initial concentration is 0.10 mol L^{-1}; $k = 0.079$ L mol^{-1} s^{-1}. Substituting these values into Equation 13.9 gives

$$t_{1/2} = \frac{1}{(0.079 \text{ L mol}^{-1} \text{ s}^{-1})(0.10 \text{ mol L}^{-1})}$$
$$= 1.3 \times 10^2 \text{ s}$$

IS THE ANSWER REASONABLE? To estimate the answer we round the 0.079 to 0.1. The estimated answer is $\frac{1}{0.1 \times 0.1} = 100$. This is close to our calculated value. In addition we check that the units cancel to leave only the seconds units. Both of these checks support our answer.

Practice Exercise 22: The reaction $2NO_2 \longrightarrow 2NO + O_2$ is second order with respect to NO_2. If the initial concentration of $NO_2(g)$ is 6.54×10^{-4} mol L^{-1}, what is the rate constant if the initial reaction rate is 4.42×10^{-7} mol L^{-1} s^{-1}? What is the half-life of this system? (Hint: Start by setting up and solving the rate law before the integrated equation.)

Practice Exercise 23: Suppose that the value of $t_{1/2}$ for a certain reaction was found to be independent of the initial concentration of the reactants. What could you say about the order of the reaction? Justify your answer.

13.5 REACTION RATE THEORIES EXPLAIN EXPERIMENTAL RATE LAWS IN TERMS OF MOLECULAR COLLISIONS

In Section 13.1 we mentioned that nearly all reactions proceed faster at higher temperatures. As a rule, the reaction rate increases by a factor of about 2 or 3 for each 10 °C increase in temperature, although the actual amount of increase differs from one reaction to another. Temperature evidently has a strong effect on reaction rate. To understand why, we need to develop theoretical models that explain our observations. One of the simplest models is called *collision theory.*

Reaction rate is related to the number of effective collisions per second between reactant molecules

The basic postulate of **collision theory** is that the rate of a reaction is proportional to the number of *effective* collisions per second among the reactant molecules. An *effective collision* is one that actually gives product molecules. Anything that can increase the frequency of effective collisions should, therefore, increase the rate.

The kinetic theory provides insights for reaction rate theory.

One of the several factors that influences the number of effective collisions per second is *concentration.* As reactant concentrations increase, the number of collisions per second of all types, including effective collisions, cannot help but increase. We'll return to the significance of concentration in Section 13.7.

Not *every* collision between reactant molecules actually results in a chemical change. We know this because the reactant atoms or molecules in a gas or a liquid undergo an enormous number of collisions per second with each other. If every collision were effective, all reactions would be over in an instant. *Only a very small fraction of all the collisions can really lead to a net change.* Why is this so?

Molecular orientation is important

In most reactions, when two reactant molecules collide they must be oriented correctly for a reaction to occur. For example, the reaction represented by the following equation.

$$2NO_2Cl \longrightarrow 2NO_2 + Cl_2$$

appears to proceed by a two-step mechanism. One step involves the collision of an NO_2Cl molecule with a chlorine atom.

$$NO_2Cl + Cl \longrightarrow NO_2 + Cl_2$$

The orientation of the NO_2Cl molecule when hit by the Cl atom is important (see Figure 13.9). The poor orientation shown in Figure 13.9*a* cannot result in the formation of Cl_2 because the two Cl atoms are not being brought close enough together for a new Cl—Cl bond to form as an N—Cl bond breaks. Figure 13.9*b* shows the necessary orientation if the collision of NO_2Cl and Cl is to effectively lead to products.

A minimum molecular kinetic energy is required

At the start of the reaction described by Figure 13.5, only about one of every billion billion (10^{18}) collisions leads to a net chemical reaction. In each of the other collisions, the reactant molecules just bounce off each other.

Not all collisions, even those correctly oriented, are energetic enough to result in products, and this is the major reason that only a small percentage of all collisions actually lead to chemical change. The colliding particles must carry into the collision a certain minimum combined molecular kinetic energy, called the **activation energy, E_a.** In a successful collision, activation energy changes over to potential energy as the particles hit each other

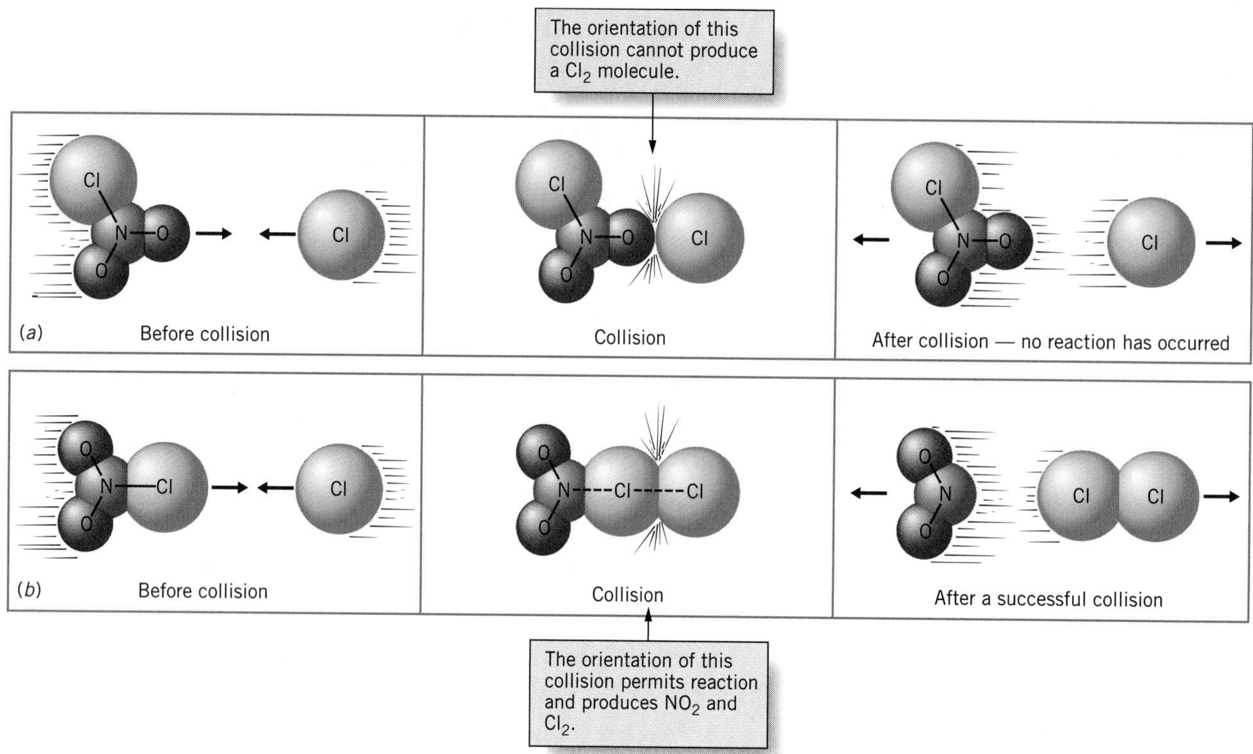
The orientation of this collision cannot produce a Cl_2 molecule.
Cl
N
O
O
Cl
(a)
Before collision
Collision
After collision — no reaction has occurred
(b)
Before collision
Collision
After a successful collision
The orientation of this collision permits reaction and produces NO_2 and Cl_2.

FIG. 13.9 **The importance of molecular orientation during a collision in a reaction.** The key step in the decomposition of NO_2Cl to NO_2 and Cl_2 is the collision of a Cl atom with a NO_2Cl molecule. (*a*) A poorly oriented collision. (*b*) An effectively oriented collision.

and chemical bonds become reorganized into those of the products. For most chemical reactions, the activation energy is quite large, and only a small fraction of all well-oriented, colliding molecules have it.

We can understand the existence of activation energy by studying in detail what actually takes place during a collision. For old bonds to break and new bonds to form, the atomic nuclei within the colliding particles must get close enough together. The molecules on a collision course must, therefore, be moving with a combined kinetic energy great enough to overcome the natural repulsions between electron clouds. Otherwise, the molecules simply veer away or bounce apart. Only fast molecules with large kinetic energies can collide with enough collision energy to enable their nuclei and electrons to overcome repulsions and thereby reach the positions required for the bond breaking and bond making that the chemical change demands.

Rising temperature increases reaction rates

With the concept of activation energy, we can now explain why the rate of a reaction increases so much with increasing temperature. We'll use the two curves in Figure 13.10, each corresponding to a different temperature for the same mixture of reactants. Each curve is a plot of the different *fractions* of all collisions (vertical axis) that have particular values of kinetic energy of collision (horizontal axis). (The total area under a curve then represents the total number of collisions, because all of the fractions must add up to this total.) Notice what happens to the plots when the temperature is increased; the maximum point shifts to the right and the curve flattens somewhat. However, *a modest increase in temperature generally does not affect the reaction's activation energy.* Within reason, the activation energy of a reaction is not affected by a change in temperature. In other words, as the curve flattens and shifts to the right with an increase in temperature, the value of E_a stays the same.

The shaded areas under the curves in Figure 13.10 represent the sum of all those fractions of the total collisions that equal or exceed the activation energy. This sum—we could call it the *reacting fraction*—is relatively much greater at the higher temperature than at the lower temperature because a significant fraction of the curve shifts beyond the activation

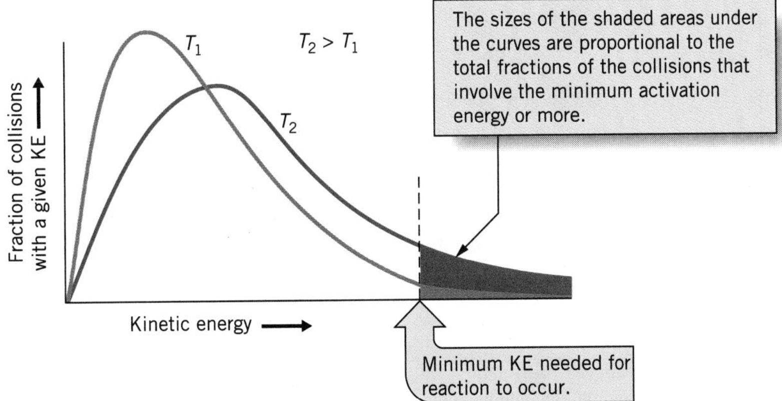

Fraction of collisions with a given KE
T_1
$T_2 > T_1$
T_2
The sizes of the shaded areas under the curves are proportional to the total fractions of the collisions that involve the minimum activation energy or more.
Kinetic energy
Minimum KE needed for reaction to occur.

FIG. 13.10 Kinetic energy distributions for a reaction mixture at two different temperatures.

energy in even a modest change to a higher temperature. In other words, at the higher temperature, a much greater fraction of the collisions occurring each second results in a chemical change, so the reactants disappear faster at the higher temperature.

On the molecular scale we can write an equation that summarizes the three factors involved in the collision theory as

$$\text{Reaction rate (molecules L}^{-1}\text{ s}^{-1}) = N \times f_{\text{orientation}} \times f_{\text{KE}}$$

where N represents the collisions per second per liter of the mixture, approximately $10^{27}\ \text{s}^{-1}$. The two other terms represent the fraction of collisions with the correct orientation, $f_{\text{orientation}}$, and the fraction of collisions with the required total kinetic energy, f_{KE}. To convert this to the laboratory scale rate of mol $\text{L}^{-1}\ \text{s}^{-1}$, we divide the equation by Avogadro's number.

$$\text{Reaction rate (mol L}^{-1}\text{ s}^{-1}) = \frac{\text{reaction rate (}\cancel{\text{molecules}}\text{ L}^{-1}\text{ s}^{-1})}{6.02 \times 10^{23}\ (\cancel{\text{molecules}}\text{ mol}^{-1})}$$

The fraction of molecules, f_{KE}, having or exceeding the activation energy, E_a, is given by the expression

$$\ln f_{\text{KE}} = \frac{-E_a}{RT}$$

The transition state is the arrangement of atoms at the top of the activation energy "hill"

Transition state theory is used to explain in detail what happens when reactant molecules come together in a collision. Most often, those in a head-on collision slow down, stop, and then fly apart unchanged. When a collision does cause a reaction, the particles that separate are those of the products. Regardless of what happens to them, however, as the molecules on a collision course slow down, their total kinetic energy decreases as it changes into potential energy (PE). It's like the momentary disappearance of the kinetic energy of a tennis ball when it hits the racket. In the deformed racket and ball, this energy becomes potential energy, which soon changes back to kinetic energy as the ball takes off in a new direction.

Potential energy diagrams summarize energy changes during the course of a reaction

To visualize the relationship between activation energy and the development of total potential energy we sometimes use a *potential energy diagram* (see Figure 13.11). The vertical axis represents changes in *potential* energy as the kinetic energy of the colliding particles changes over to this form. The horizontal axis is called the **reaction coordinate,** and it represents the extent to which the reactants have changed to the products. It helps us follow the path taken by the reaction as reactant molecules come together and change into product molecules. Activation energy appears as a potential energy "hill" or barrier between the reactants and products. Only colliding molecules, properly oriented, that can deliver kinetic energy into potential energy at least as large as E_a are able to climb over the hill and produce products.

We can use a potential energy diagram to follow the progress of both an unsuccessful and a successful collision (see Figure 13.12). As two reactant molecules collide, we say that they begin to climb the potential energy barrier as they slow down and experience the

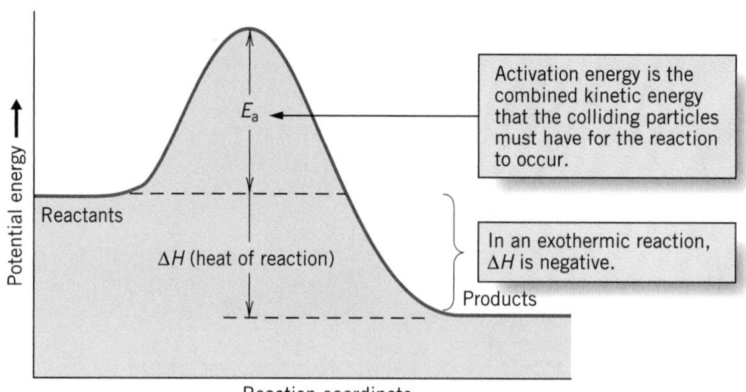

FIG. 13.11 Potential energy diagram for an exothermic reaction.

conversion of their kinetic energy into potential energy. But if their combined initial kinetic energies are equivalent to a potential energy that is less than E_a, the molecules are unable to reach the top of the hill (Figure 13.12*a*). Instead, they fall back toward the reactants. They bounce apart chemically unchanged with their original total kinetic energy; no net reaction has occurred. On the other hand, if the combined kinetic energy of the colliding molecules equals or exceeds E_a, and if the molecules are oriented properly, they are able to pass over the activation energy barrier and form product molecules (Figure 13.12*b*).

Potential energy diagrams also show the heat of reaction

A reaction's potential energy diagram, such as that of Figure 13.11, helps us to visualize the *heat of reaction*, ΔH, a concept introduced in Chapter 6. It's the difference between the potential energy of the products and the potential energy of the reactants. Figure 13.11 is for an *exothermic* reaction because the products have a *lower* potential energy than the reactants. In such a system, the net decrease in potential energy appears as an increase in the molecular kinetic energy of the emerging product molecules. The temperature of the system increases during an exothermic reaction because the average molecular kinetic energy of the system increases.

A potential energy diagram for an endothermic reaction is shown in Figure 13.13. Now the products have a *higher* potential energy than the reactants and, in terms of the heat of reaction, a net input of energy is needed to form the products. Endothermic reactions produce a cooling effect as they proceed because there is a net conversion of molecular kinetic energy to potential energy. As the *total* molecular kinetic energy decreases, the *average* molecular kinetic energy decreases as well, and the temperature drops.

Notice that E_a for an endothermic process is invariably greater than (or it might be equal to) the heat of reaction. If ΔH is both positive and *high*, E_a must also be high, making such reactions very slow. However, for an exothermic reaction (ΔH is negative) we cannot tell from ΔH how large E_a is. It could be high, making for a slow reaction despite its being exothermic. If E_a is low, the reaction would be rapid and all its heat would appear quickly.

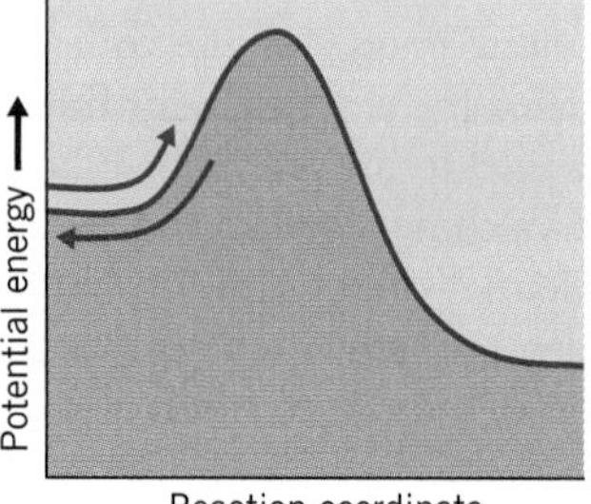

(*a*) An unsuccessful collision; the colliding molecules separate unchanged.

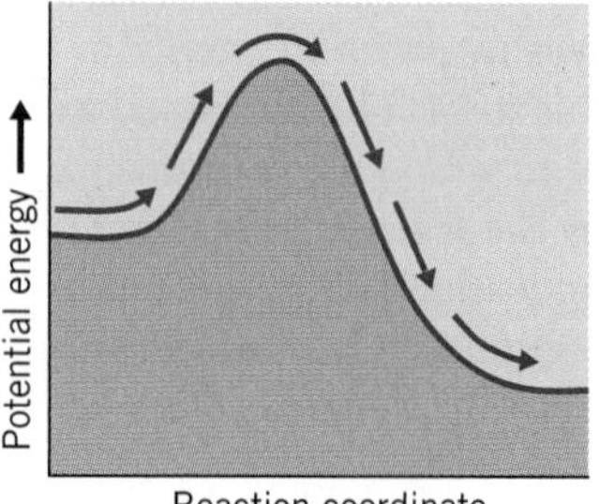

(*b*) A successful collision; the activation energy barrier is crossed and the products are formed.

FIG. 13.12 The difference between an unsuccessful and a successful collision.

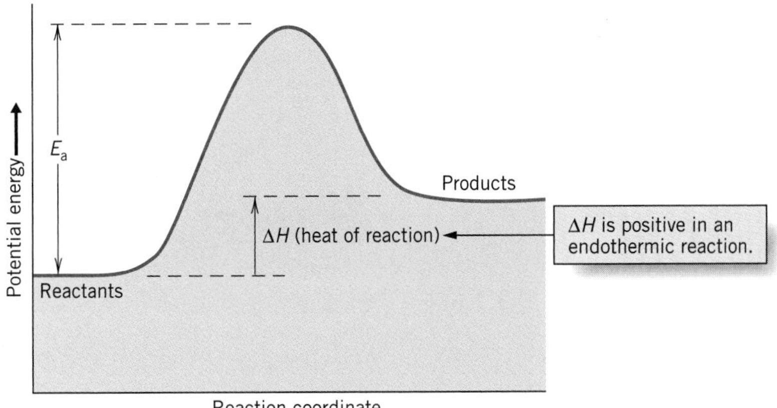

FIG. 13.13 A potential energy diagram for an endothermic reaction.

In Chapter 6 we saw that when the direction of a reaction is reversed, the sign given to the enthalpy change, ΔH, is reversed. In other words, a reaction that is exothermic in the forward direction *must* be endothermic in the reverse direction, and vice versa. This might seem to suggest that reactions are generally reversible. Many are, but if we look again at the energy diagram for a reaction that is exothermic in the forward direction (Figure 13.11), it is obvious that in the opposite direction the reaction is endothermic *and must have a significantly higher activation energy* than the forward reaction. What differs most for the forward and reverse directions is the relative height of the activation energy barrier (see Figure 13.14).

One of the main reasons for studying activation energies is that they provide information about what actually occurs during an effective collision. For example, in Figure 13.9*b* on page 544, we described a way that NO_2Cl could react successfully with a Cl atom during a collision. During this collision, there is a moment when the N—Cl bond is partially broken and the new Cl—Cl bond is partially formed. This brief moment during a successful collision is called the reaction's **transition state.** The potential energy of the transition state corresponds to the high point on the potential energy diagram (see Figure 13.15). The unstable chemical species that momentarily exists at this instant, O_2N---Cl---Cl, with its partially formed and partially broken bonds, is called the **activated complex.**

The size of the activation energy tells us about the relative importance of bond breaking and bond making during the formation of the activated complex. A very high activation energy suggests, for instance, that bond *breaking* contributes very heavily to the formation of the activated complex because bond breaking is an energy-absorbing process. On the other hand, a low activation energy may mean that bonds of about equal strength are being both broken and formed simultaneously.

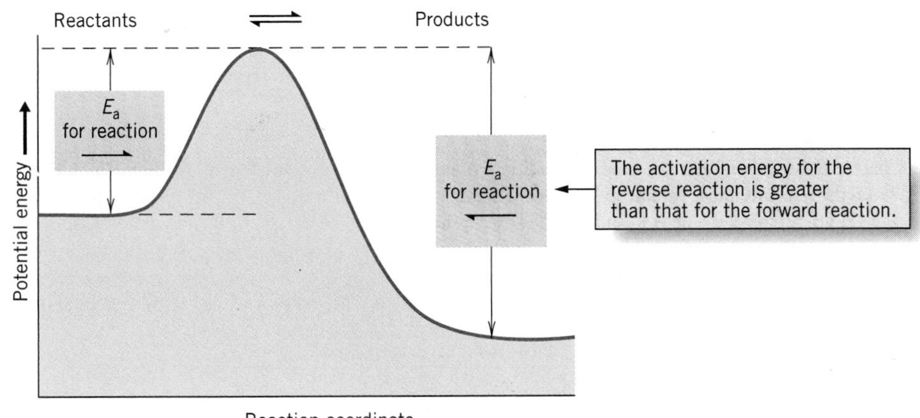

FIG. 13.14 Activation energy barrier for the forward and reverse reactions.

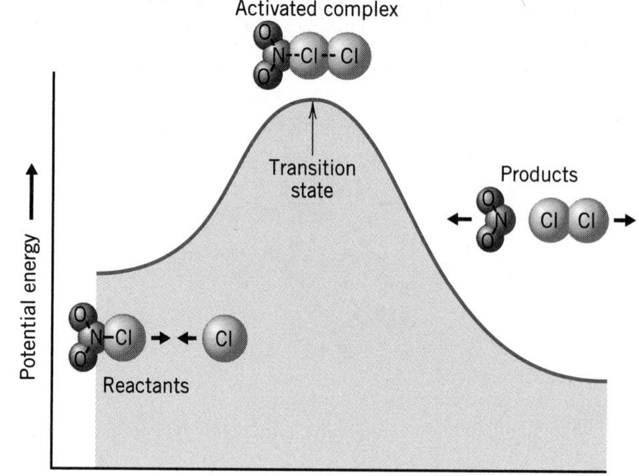

FIG. 13.15 Transition state and the activated complex. Formation of an activated complex in the reaction between NO_2Cl and Cl.
$NO_2Cl + Cl \longrightarrow NO + Cl_2$

13.6 Activation Energies Are Measured by Fitting Experimental Data to the Arrhenius Equation

We've noted that the activation energy is a useful quantity to know because its value can provide clues to the relative importance of bond breaking and bond making during the formation of the activated complex. Determining the value of E_a is accomplished by observing how temperature affects the value of the rate constant, k.

The activation energy is linked to the rate constant by a relationship discovered in 1889 by Svante Arrhenius, whose name you may recall from our discussion of electrolytes and acids and bases in Chapter 4. The usual form of the **Arrhenius equation** is

Arrhenius equation

$$k = A\,e^{-E_a/RT} \tag{13.10}$$

where k is the rate constant, e is the base of the natural logarithm system, and T is the Kelvin temperature. A is a proportionality constant sometimes called the **frequency factor** or the **pre-exponential factor.** R is the gas constant, which we'll express in our study of kinetics in energy units, namely, R equals 8.314 J mol^{-1} K^{-1}.[8]

The activation energy can be determined graphically

Equation 13.10 is normally used in its logarithmic form. If we take the natural logarithm of both sides, we obtain

$$\ln k = \ln A - E_a/RT$$

Let's rewrite the equation as

$$\ln k = \ln A - (E_a/R) \times (1/T) \tag{13.11}$$

[8] The units of R given here are actually SI units, namely, the joule (J), the mole (n), and the kelvin (K). To calculate R in these units we need to go back to the defining equation for the universal gas law, rearranging terms.

$$R = PV/nT$$

In Chapter 10 we learned that the standard conditions of pressure and temperature are 1 atm and 273.15 K; we expressed the standard molar volume in liters, namely 22.414 L. But 1 atm equals 1.01325×10^5 N m^{-2}, where N is the SI unit of force, the newton, and m is the meter, the SI unit of length. So m^2 is area given in SI units. From Chapter 10, the ratio of force to area given by N m^{-2} is called the pascal, Pa, and that force times distance or N m defines one unit of energy in the SI and is called the joule, J. In the SI, volume must be expressed as m^3, to employ the SI unit of length to define volume, and 1 L equals 10^{-3} m^3. So now we can calculate R in SI units.

$$R = \frac{(1.01325 \times 10^5 \text{ N m}^{-2}) \times (22.414 \times 10^{-3} \text{ m}^3)}{(1 \text{ mol} \times 273.15 \text{ K})}$$
$$= 8.314 \text{ N m mol}^{-1} \text{ K}^{-1} = 8.314 \text{ J mol}^{-1} \text{ K}^{-1}$$

We know that the rate constant k varies with the temperature T, which also means that the quantity $\ln k$ varies with the quantity $(1/T)$. These two quantities, namely, $\ln k$ and $1/T$, are variables, so Equation 13.11 is in the form of an equation for a straight line.

$$\begin{array}{ccccccc} \ln k & = & \ln A & + & (-E_a/R) & \times & (1/T) \\ \updownarrow & & \updownarrow & & \updownarrow & & \updownarrow \\ y & = & b & + & m & & x \end{array}$$

To determine the activation energy, we can make a graph of $\ln k$ versus $1/T$, measure the slope of the line, and then use the relationship

$$\text{Slope} = -E_a/R$$

to calculate E_a. Example 13.11 illustrates how this is done.

EXAMPLE 13.11

Determining Energy of Activation Graphically

Consider again the decomposition of NO_2 into NO and O_2. The equation is

$$2NO_2(g) \longrightarrow 2NO(g) + O_2(g)$$

The following data were collected for the reaction.

Rate Constant, k ($L\ mol^{-1}\ s^{-1}$)	Temperature (°C)
7.8	400
10	410
14	420
18	430
24	440

Use the graphical method to determine the activation energy for the reaction in kilojoules per mole.

ANALYSIS: Equation 13.11 is the tool that applies. However, the use of rate data to determine the activation energy graphically requires that we plot $\ln k$, not k, versus the *reciprocal* of the *Kelvin* temperature, so we have to convert the given data into $\ln k$ and $1/T$ before we can construct the graph.

SOLUTION: To illustrate, using the first set of data, the conversions are

$$\ln k = \ln(7.8) = 2.05$$

$$\frac{1}{T} = \frac{1}{(400 + 273)\text{ K}} = \frac{1}{673\text{ K}}$$

$$= 1.486 \times 10^{-3}\text{ K}^{-1}$$

We are carrying extra "significant figures" for the purpose of graphing the data. The remaining conversions give the table below. Then we plot $\ln k$ versus $1/T$ as shown on the next page.

$\ln k$	$1/T$ (K^{-1})
2.05	1.486×10^{-3}
2.30	1.464×10^{-3}
2.64	1.443×10^{-3}
2.89	1.422×10^{-3}
3.18	1.403×10^{-3}

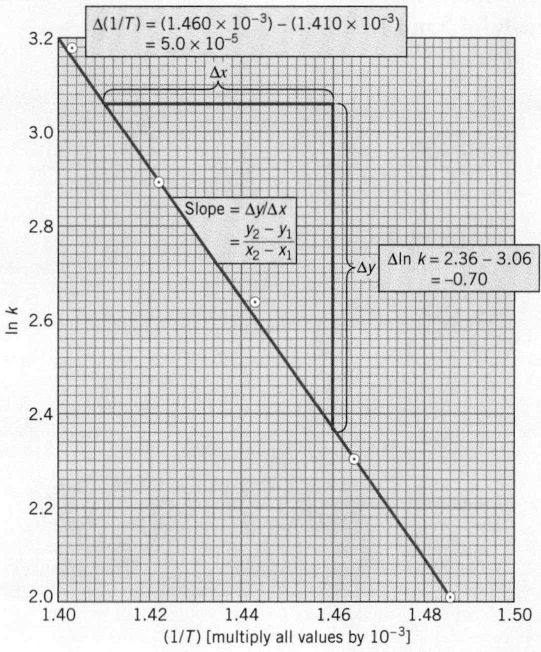

The slope of the curve is obtained as the ratio

$$\text{Slope} = \frac{\Delta(\ln k)}{\Delta(1/T)}$$

$$= \frac{-0.70}{5.0 \times 10^{-5}\ \text{K}^{-1}}$$

$$= -1.4 \times 10^{4}\ \text{K} = -E_a/R$$

After changing signs and solving for E_a we have

$$E_a = (8.314\ \text{J mol}^{-1}\ \text{K}^{-1})(1.4 \times 10^{4}\ \text{K})$$

$$= 1.2 \times 10^{5}\ \text{J mol}^{-1}$$

$$= 1.2 \times 10^{2}\ \text{kJ mol}^{-1}$$

IS THE ANSWER REASONABLE? Activation energies must always have a positive sign and our result is positive. In addition, a check of the units shows that they cancel to give us the correct kJ mol^{-1}. We could try a different pair of points on the same graph to check our work.

The activation energy can be calculated from rate constants measured at two temperatures

If the activation energy and the rate constant at a particular temperature are known, the rate constant at another temperature can be calculated using the following relationship, which can be derived from Equation 13.11,

Arrhenius equation, alternate form

$$\ln\left(\frac{k_2}{k_1}\right) = \frac{-E_a}{R}\left(\frac{1}{T_2} - \frac{1}{T_1}\right) \qquad (13.12)$$

This equation can also be used to calculate the activation energy from rate constants measured at two different temperatures. However, the graphical method discussed earlier gives more precise values of E_a.

EXAMPLE 13.12
Calculating the Rate Constant at a Particular Temperature

The reaction $2NO_2 \longrightarrow 2NO + O_2$ has an activation energy of 111 kJ mol^{-1}. At 400 °C, $k = 7.8$ L mol^{-1} s^{-1}. What is the value of k at 430 °C?

ANALYSIS: We know the activation energy and k at one temperature. We will need to use Equation 13.12 as our tool to obtain k at the other temperature. Since the logarithm term contains the ratio of the rate constants, we will solve for the value of this ratio, substitute the known value of k, and then solve for the unknown k.

SOLUTION: Let's begin by writing Equation 13.12.

$$\ln\left(\frac{k_2}{k_1}\right) = \frac{-E_a}{R}\left(\frac{1}{T_2} - \frac{1}{T_1}\right)$$

Organizing the data gives us the following table.

	k (L mol^{-1} s^{-1})	T (K)
1	7.8	400 + 273 = 673 K
2	?	430 + 273 = 703 K

We must use $R = 8.314$ J mol^{-1} K^{-1} and express E_a in joules ($E_a = 1.11 \times 10^5$ J mol^{-1}). Next, we substitute values into the right side of the equation and solve for $\ln(k_2/k_1)$.

$$\ln\left(\frac{k_2}{k_1}\right) = \frac{-1.11 \times 10^5 \cancel{\text{J mol}^{-1}}}{8.314 \cancel{\text{J mol}^{-1}} \text{ K}^{-1}}\left(\frac{1}{703 \text{ K}} - \frac{1}{673 \text{ K}}\right)$$

$$= (-1.34 \times 10^4 \cancel{\text{K}})(-6.34 \times 10^{-5} \cancel{\text{K}^{-1}})$$

Therefore,

$$\ln\left(\frac{k_2}{k_1}\right) = 0.850$$

Taking the antilog gives the ratio of k_2 to k_1.

$$\frac{k_2}{k_1} = e^{0.850} = 2.34$$

Solving for k_2,

$$k_2 = 2.34k_1$$

Substituting the value of k_1 from the data table gives

$$k_2 = 2.34\ (7.8 \text{ L mol}^{-1} \text{ s}^{-1})$$
$$= 18 \text{ L mol}^{-1} \text{ s}^{-1}$$

IS THE ANSWER REASONABLE? Although no simple check exists, we have at least found that the value of k for the higher temperature is greater than it is for the lower temperature, as it should be.

Practice Exercise 24: The rate constant is directly proportional to the reaction rate if the same reactant concentrations are used. When determining the stability of a consumer product, less than 5% should decompose in two years at 25 °C. What temperature should we set our oven to if we want to see that same 5% decomposition in one week? Assume the activation energy was previously determined to be 154 kJ mol^{-1}. (Hint: All the information is here to apply the Arrhenius equation.)

Practice Exercise 25: The reaction $CH_3I + HI \longrightarrow CH_4 + I_2$ was observed to have rate constants $k = 3.2$ L mol^{-1} s^{-1} at 350 °C and $k = 23$ L mol^{-1} s^{-1} at 400 °C. (a) What is the value of E_a expressed in kJ mol^{-1}? (b) What would be the rate constant at 300 °C?

13.7 EXPERIMENTAL RATE LAWS CAN BE USED TO SUPPORT OR REJECT PROPOSED MECHANISMS FOR A REACTION

A balanced equation generally describes only a net overall change. Usually, however, the net change is the result of a series of simple reactions that are not at all evident from the equation. Consider, for example, the combustion of propane, C_3H_8.

$$C_3H_8(g) + 5O_2(g) \longrightarrow 3CO_2(g) + 4H_2O(g)$$

Anyone who has ever played billiards knows that this reaction simply cannot occur in a single, simultaneous collision between one propane molecule and five oxygen molecules. Just getting only three balls to come together with but one "click" on a flat, two-dimensional surface is extremely improbable. How unlikely it must be, then, for the *simultaneous* collision in three-dimensional space of six reactant molecules, one of which must be C_3H_8 and the other five O_2. Instead, the combustion of propane proceeds very rapidly by a series of much more probable steps, involving colliding chemical species of fleeting existence. *The series of individual steps that add up to the overall observed reaction is called the* **mechanism of a reaction.** Information about reaction mechanisms is one of the dividends paid by the study of rates.

Each individual step in a reaction mechanism is a simple chemical reaction called an *elementary process.* An **elementary process** is a reaction involving collisions between molecules. As you will soon see, its rate law can be written from its own chemical equation, using coefficients as exponents for the concentration terms without requiring experiments to determine the exponents. For most reactions, the individual elementary processes cannot actually be observed; instead, we only see the net reaction. Therefore, the mechanism a chemist writes is really a *theory* about what occurs step by step as the reactants are changed to the products.

Because the individual steps in a mechanism usually cannot be observed directly, devising a mechanism for a reaction requires some ingenuity. However, we can immediately tell whether a proposed mechanism is feasible. *The overall rate law derived from the mechanism must agree with the observed rate law for the overall reaction.*

The rate law for an elementary process can be predicted from the chemical equation for the process

Consider the following elementary process that involves collisions between two identical molecules leading directly to the products shown.

$$2NO_2 \longrightarrow NO_3 + NO \qquad (13.13)$$

$$\text{Rate} = k[NO_2]^x$$

How can we predict the value of the exponent x? Suppose the NO_2 concentration were doubled. There would now be *twice* as many individual NO_2 molecules and *each* would have *twice* as many neighbors with which to collide. The number of NO_2-to-NO_2 collisions per second would doubly double—in other words, increase by a factor of 4. This would cause the rate to increase by a factor of 4, which is 2^2. Earlier we saw that when doubling a concentration leads to a fourfold increase in the rate, the concentration of that reactant is raised to the second power in the rate law. Thus, if Equation 13.13 represents an elementary process, its rate law should be

We double the number of NO_2 molecules and double the number each can collide with, so the collision frequency increases by a factor of 4.

$$\text{Rate} = k[NO_2]^2$$

Notice that the exponent in the rate law for this elementary process is the same as the coefficient in the chemical equation. Similar analyses for other types of elementary processes lead to similar observations and the following statement:

> The exponents in the rate law for an elementary process are equal to the coefficients of the reactants in the chemical equation for that elementary process.

Remember that this rule applies only to *elementary processes.* If all we know is the balanced equation for the overall reaction, the only way we can find the exponents of the rate law is by doing experiments.

The rate law for the slowest step in a mechanism should agree with the experimental rate law

How does the ability to predict the rate law of an elementary process help chemists predict reaction mechanisms? To answer this question, let's look at two reactions and what are believed to be their mechanisms. (There are many other, more complicated systems, and Facets of Chemistry 13.1 describes one type, the free radical chain reaction, that is particularly important.)

FACETS OF CHEMISTRY 13.1

Free Radicals, Explosions, Octane Ratings, Aging and Health

A **free radical** is a very reactive species that contains one or more unpaired electrons. Examples are chlorine atoms formed when a Cl_2 molecule absorbs a photon (light) of the appropriate energy:

$$Cl_2 + \text{light energy } (h\nu) \longrightarrow 2Cl\cdot$$

(A dot placed next to the symbol of an atom or molecule represents an unpaired electron and indicates that the particle is a free radical.) The reason free radicals are so reactive is because of the tendency of electrons to become paired through the formation of either ions or covalent bonds.

Free radicals are important in many gaseous reactions, including those responsible for the production of photochemical smog in urban areas. Reactions involving free radicals have useful applications, too. For example, many plastics are made by reactions that take place by mechanisms that involve free radicals. In addition, free radicals play a part in one of the most important processes in the petroleum industry, *thermal cracking.* This reaction is used to break C—C and C—H bonds in long chain hydrocarbons to produce the smaller molecules that give gasoline a higher octane rating. An example is the formation of free radicals in the thermal cracking reaction of butane. When butane is heated to 700–800 °C, one of the major reactions that occurs is

$$CH_3{-}CH_2{:}CH_2{-}CH_3 \xrightarrow{\text{heat}} CH_3CH_2\cdot + CH_3CH_2\cdot$$

The central C—C bond of butane is shown here as a pair of dots, **:**, rather than the usual dash. When the bond is broken, the electron pair is divided between the two free radicals that are formed. This reaction produces two ethyl radicals, $CH_3CH_2\cdot$.

Free radical reactions tend to have high initial activation energies because chemical bonds must be broken to form the radicals. Once the free radicals are formed, however, reactions in which they are involved tend to be very rapid.

Free Radical Chain Reactions

In many cases, a free radical reacts with a reactant molecule to give a product molecule plus another free radical. Reactions that involve such a step are called **chain reactions.**

Many explosive reactions are chain reactions involving free radical mechanisms. One of the most studied reactions of this type is the formation of water from hydrogen and oxygen. The elementary processes involved can be described according to their roles in the mechanism.

The reaction begins with an **initiation step** that gives free radicals.

$$H_2 + O_2 \xrightarrow{\text{hot surface}} 2OH\cdot \qquad \text{(initiation)}$$

The chain continues with a **propagation step,** which produces the product plus another free radical.

$$OH\cdot + H_2 \longrightarrow H_2O + H\cdot \qquad \text{(propagation)}$$

The reaction of H_2 and O_2 is explosive because the mechanism also contains **branching steps.**

$$\left.\begin{array}{l} H\cdot + O_2 \longrightarrow OH\cdot + O\cdot \\ O\cdot + H_2 \longrightarrow OH\cdot + H\cdot \end{array}\right] \quad \text{branching}$$

Thus, the reaction of one H· with O_2 leads to the net production of two OH· plus an O·. Every time an H· reacts with oxygen, then, there is an increase in the number of free radicals in the system. The free radical concentration grows rapidly, and the reaction rate becomes explosively fast.

Chain mechanisms also contain **termination steps,** which remove free radicals from the system. In the reaction of H_2 and O_2, the wall of the reaction vessel serves to remove H·, which tends to halt the chain process.

$$2H\cdot \xrightarrow{\text{wall}} H_2$$

Free Radicals and Aging

Direct experimental evidence also exists for the presence of free radicals in functioning biological systems. These highly reactive species play many roles, but one of the most interesting is their apparent involvement in the aging process. One theory suggests that free radicals attack protein molecules in collagen. Collagen is composed of long strands of fibers of proteins and is found throughout the body, especially in the flexible tissues of the lungs, skin, muscles, and blood vessels. Attack by free radicals seems to lead to cross-linking between these fibers, which stiffens them and makes them less flexible. The most readily observable result of this is the stiffening and hardening of the skin that accompanies aging or too much sunbathing.

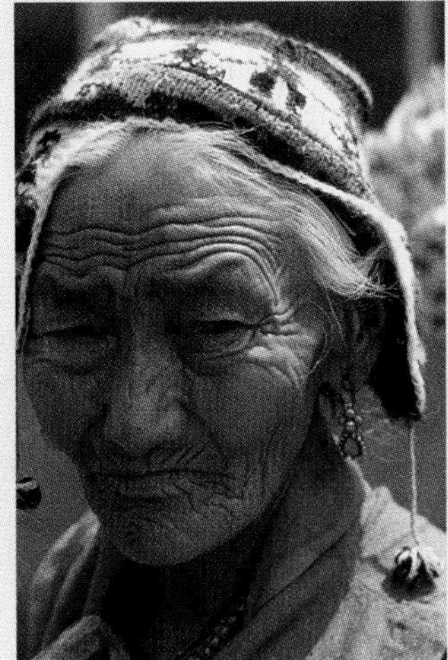

People exposed to sunlight over long periods, like this woman from Nepal (a small country between India and Tibet), tend to develop wrinkles because ultraviolet radiation causes changes in their skin. *(Alison Wright/Corbis)*

Free Radicals and Health

Around 1987 evidence began to surface identifying a stable free radical, nitrogen monoxide (nitric oxide, NO), as a key inorganic molecule controlling a variety of biological functions, from the chemiluminescent flash of a firefly to beneficial effects in the human body. Amyl nitrate and nitroglycerine have been used medicinally since the early 1900s, and NO is apparently the metabolic product that makes these substances effective drugs. Currently, at an estimated rate of 3000 research papers per year, the discoveries of the functions of NO have become major medical milestones of the 21st century. For example, this very simple, diatomic molecule can be used to treat high blood pressure, angina, pulmonary hypertension, breathing problems in newborn babies, erectile dysfunction, and even Alzheimer's and Parkinson's disease. Discovery of the effects of NO and its possible therapeutic applications led to the 1998 Nobel Prize in Medicine honoring Robert Furchgott, Louis Ignarro, and Ferid Murad.

First, consider the gaseous reaction

$$2NO_2Cl \longrightarrow 2NO_2 + Cl_2 \tag{13.14}$$

Experimentally, the rate is first order in NO_2Cl, so the rate law is

$$\text{Rate} = k[NO_2Cl] \qquad (\textit{experimental})$$

The first question we might ask is, Could the overall reaction (Equation 13.14) occur in a single step by the collision of two NO_2Cl molecules? The answer is no, because then it would be an elementary process and the rate law predicted for it would include a squared term, $[NO_2Cl]^2$. But the experimental rate law is first order in NO_2Cl. So the predicted and experimental rate laws don't agree, and we must look further to find the mechanism of the reaction.

On the basis of chemical intuition and other information that we won't discuss here, chemists believe the actual mechanism of the reaction in Equation 13.14 is the following two-step sequence of elementary processes.

$$NO_2Cl \longrightarrow NO_2 + Cl$$
$$NO_2Cl + Cl \longrightarrow NO_2 + Cl_2$$

The Cl atom formed here is called a *reactive intermediate.* We never actually observe the Cl because it reacts so quickly.

Notice that when the two reactions are added, the *intermediate*, Cl, drops out and we obtain the net overall reaction given in Equation 13.14. *Being able to add the elementary processes and thus to obtain the overall reaction is another major test of a mechanism.*

In any multistep mechanism, one step is usually much slower than the others. In this mechanism, for example, it is believed that the first step is slow and that once a Cl atom forms, it reacts very rapidly with another NO_2Cl molecule to give the final products.

The final products of a multistep reaction cannot appear faster than the products of the slow step, so the slow step in a mechanism is called the **rate-determining step** or the **rate-limiting step.** In the two-step mechanism above, then, the first reaction is the rate-determining step because the final products can't be formed faster than the rate at which Cl atoms form.

The rate-determining step is similar to a slow worker on an assembly line. The production rate depends on how quickly the slow worker works, regardless of how fast the other workers are. The factors that control the speed of the rate-determining step therefore also control the overall rate of the reaction. This means that *the rate law for the rate-determining step is directly related to the rate law for the overall reaction.*

Because the rate-determining step is an elementary process, we can predict its rate law from the coefficients of its reactants. The coefficient of NO_2Cl in its relatively slow breakdown to NO_2 and Cl is 1. Therefore, the rate law predicted for the first step is

$$\text{Rate} = k[NO_2Cl] \qquad (\textit{predicted})$$

Notice that the predicted rate law derived for the two-step mechanism agrees with the experimentally measured rate law. Although this doesn't *prove* that the mechanism is correct, it does provide considerable support for it. From the standpoint of kinetics, therefore, the mechanism is reasonable.

The second reaction mechanism that we will study is that of the following gas-phase reaction.

$$2NO + 2H_2 \longrightarrow N_2 + 2H_2O \tag{13.15}$$

The experimentally determined rate law is

$$\text{Rate} = k[NO]^2[H_2] \qquad (\textit{experimental})$$

We can quickly tell from this rate law that Equation 13.15 could *not* itself be an elementary process. If it were, the exponent for $[H_2]$ would have to be 2. Obviously, a mechanism involving two or more steps must be involved.

A chemically reasonable mechanism that yields the correct form for the rate law consists of the following two steps.

$$2NO + H_2 \longrightarrow N_2O + H_2O \qquad (\text{slow})$$
$$N_2O + H_2 \longrightarrow N_2 + H_2O \qquad (\text{fast})$$

One test of the mechanism, as we said, is that the two equations must add to give the correct overall equation; and they do. Further, the chemistry of the second step has actually been observed in separate experiments. N_2O is a known compound, and it does react with H_2 to give N_2 and H_2O. Another test of the mechanism involves the coefficients of NO and H_2 in the predicted rate law for the first step, the supposed rate-determining step.

$$\text{Rate} = k[NO]^2[H_2] \qquad (\textit{predicted})$$

This rate equation does match the experimental rate law, but there is still a serious flaw in the proposed mechanism. If the postulated slow step actually describes an elementary process, it would involve the simultaneous collision between three molecules, two NO and one H_2. A three-way collision is so unlikely that if it were really involved in the mechanism, the overall reaction would be extremely slow. Reaction mechanisms seldom include elementary processes that involve more than two-body or **bimolecular collisions.**

Chemists believe the reaction in Equation 13.15 proceeds by the following three-step sequence of bimolecular elementary processes.

$$2NO \rightleftharpoons N_2O_2 \qquad (\text{fast})$$
$$N_2O_2 + H_2 \longrightarrow N_2O + H_2O \qquad (\text{slow})$$
$$N_2O + H_2 \longrightarrow N_2 + H_2O \qquad (\text{fast})$$

In this mechanism the first step is proposed to be a rapidly established equilibrium in which the unstable intermediate N_2O_2 forms in the forward reaction and then quickly decomposes into NO by the reverse reaction. The rate-determining step is the reaction of N_2O_2 with H_2 to give N_2O and a water molecule. The third step is the reaction mentioned above. Once again, notice that the three steps add to give the net overall change.

The occurrence of N_2O_2 as an intermediate in the proposed mechanism can only be surmised. The compound is never present at a detectable concentration because, as supposed, it's too unstable.

Since the second step is rate determining, the rate law for the reaction should match the rate law for this step. We predict this to be

$$\text{Rate} = k[N_2O_2][H_2] \tag{13.16}$$

However, the experimental rate law does not contain the species N_2O_2. Therefore, we must find a way to express the concentration of N_2O_2 in terms of the reactants in the overall reaction. To do this, let's look closely at the first step of the mechanism, which we view as a reversible reaction.

The rate in the forward direction, in which NO is the reactant, is

$$\text{Rate (forward)} = k_f[NO]^2$$

The rate of the reverse reaction, in which N_2O_2 is the reactant, is

$$\text{Rate (reverse)} = k_r[N_2O_2]$$

Recall that in a dynamic equilibrium forward and reverse reactions occur at equal rates.

If we view this as a dynamic equilibrium, then the rate of the forward and reverse reactions are equal, which means that

$$k_f[NO]^2 = k_r[N_2O_2] \tag{13.17}$$

Since we would like to eliminate N_2O_2 from the rate law in Equation 13.16, let's solve Equation 13.17 for $[N_2O_2]$.

$$[N_2O_2] = \frac{k_f}{k_r}[NO]^2$$

Substituting into the rate law in Equation 13.16 yields

$$\text{Rate} = k\left(\frac{k_f}{k_r}\right)[NO]^2[H_2]$$

Combining all the constants into one (k') gives

$$\text{Rate} = k'[NO]^2[H_2] \qquad (\textit{predicted})$$

There are many reactions that do not follow simple first- or second-order rate laws and have mechanisms far more complex than those studied in this section. Even so, their more complex kinetics still serve as clues to their complex set of elementary processes.

Now the rate law derived from the mechanism matches the rate law obtained experimentally. The three-step mechanism does appear to be reasonable on the basis of kinetics.

The procedure we have worked through here applies to many reactions that proceed by mechanisms involving sequential steps. Steps that precede the rate-determining step are considered to be rapidly established equilibria involving unstable intermediates.

A proposed mechanism must always account for the experimental rate law
Although chemists may devise other experiments to help prove or disprove the correctness of a mechanism, one of the strongest pieces of evidence is the experimentally measured rate law for the overall reaction. No matter how reasonable a particular mechanism may appear, if its elementary processes cannot yield a predicted rate law that matches the experimental one, the mechanism is wrong and must be discarded.

Practice Exercise 26: Select the reactions below that may be elementary processes. For those not selected explain why they are not likely to be elementary processes.

(a) $2N_2O_5 \longrightarrow 2N_2O_4 + O_2$
(b) $NO + O_3 \longrightarrow NO_2 + O_2$
(c) $2NO + H_2 \longrightarrow N_2O + H_2O$
(d) $C_3H_8(g) + 5O_2(g) \longrightarrow 3CO_2(g) + 4H_2O(g)$
(e) $C_{12}H_{22}O_{11} + H_2O \longrightarrow C_6H_{12}O_6 + C_6H_{12}O_6$
(f) $3H_2 + N_2 \longrightarrow 2NH_3$

(Hint: How many molecules are likely to collide at exactly the same time?)

Practice Exercise 27: Ozone, O_3, reacts with nitric oxide, NO, to form nitrogen dioxide and oxygen.

$$NO + O_3 \longrightarrow NO_2 + O_2$$

This is one of the reactions involved in the formation of photochemical smog. If this reaction occurs in a single step, what is the expected rate law for the reaction?

Practice Exercise 28: The mechanism for the decomposition of NO_2Cl is

$$NO_2Cl \longrightarrow NO_2 + Cl$$
$$NO_2Cl + Cl \longrightarrow NO_2 + Cl_2$$

What would the predicted rate law be if the second step in the mechanism were the rate-determining step?

13.8 CATALYSTS CHANGE REACTION RATES BY PROVIDING ALTERNATIVE PATHS BETWEEN REACTANTS AND PRODUCTS

A **catalyst** is a substance that changes the rate of a chemical reaction without itself being used up. In other words, all of the catalyst added at the start of a reaction is present chemically unchanged after the reaction has gone to completion. The action caused by a catalyst is called **catalysis.** Broadly speaking, there are two kinds of catalysts. *Positive catalysts* speed up reactions, and *negative catalysts,* usually called *inhibitors,* slow reactions down. After this, when we use "catalyst" we'll mean positive catalyst, the usual connotation.

Although the catalyst is not part of the overall reaction, it does participate by changing the mechanism of the reaction. The catalyst provides a path to the products that has a rate-determining step with a lower activation energy than that of the uncatalyzed reaction (see Figure 13.16). Because the activation energy along this new route is lower, a greater fraction of the collisions of the reactant molecules have the minimum energy needed to react, so the reaction proceeds faster.

Catalysts can be divided into two groups—**homogeneous catalysts,** which exist in the same phase as the reactants, and **heterogeneous catalysts,** which exist in a separate phase.

Homogeneous catalysts are in the same phase as the reactants

An example of homogeneous catalysis is found in the now outdated *lead chamber process* for manufacturing sulfuric acid. To make sulfuric acid by this process, sulfur is burned to give SO_2, which is then oxidized to SO_3. The SO_3 is dissolved in water as it forms to give H_2SO_4.

In the modern process for making sulfuric acid, the contact process, vanadium(V) oxide, V_2O_5, is a heterogeneous catalyst that promotes the oxidation of sulfur dioxide to sulfur trioxide.

$$S + O_2 \longrightarrow SO_2$$

$$SO_2 + \tfrac{1}{2}O_2 \longrightarrow SO_3$$

$$SO_3 + H_2O \longrightarrow H_2SO_4$$

Unassisted, the second reaction, oxidation of SO_2 to SO_3, occurs slowly. In the lead chamber process, the SO_2 is combined with a mixture of NO, NO_2, air, and steam in large lead-lined reaction chambers. The NO_2 readily oxidizes the SO_2 to give NO and SO_3. The NO is then reoxidized to NO_2 by oxygen.

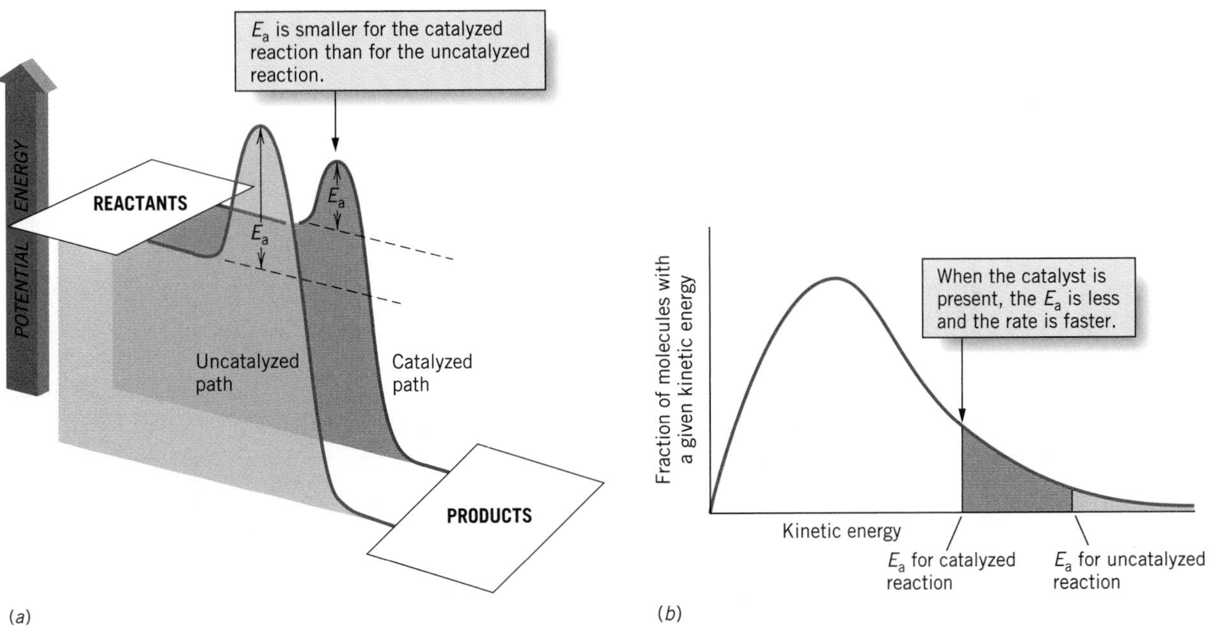

FIG. 13.16 Effect of a catalyst on a reaction. (*a*) The catalyst provides an alternative, low-energy path from the reactants to the products. (*b*) A larger fraction of molecules have sufficient energy to react when the catalyzed path is available.

The NO_2 is regenerated in the second reaction and so is recycled over and over. Thus, only small amounts of it are needed in the reaction mixture to do an effective catalytic job.

$$NO_2 + SO_2 \longrightarrow NO + SO_3$$

$$NO + \tfrac{1}{2}O_2 \longrightarrow NO_2$$

The NO_2 serves as a catalyst by being an oxygen carrier and by providing a low-energy path for the oxidation of SO_2 to SO_3. Notice, as must be true for any catalyst, the NO_2 is regenerated; it has not been permanently changed.

Heterogeneous catalysts are in a separate phase from the reactants

Adsorption means that molecules bind to a surface.

A heterogeneous catalyst is commonly a solid, and it usually functions by promoting a reaction on its surface. One or more of the reactant molecules are adsorbed onto the surface of the catalyst where an interaction with the surface increases their reactivity. An example is the synthesis of ammonia from hydrogen and nitrogen by the Haber process.

$$3H_2 + N_2 \longrightarrow 2NH_3$$

The reaction takes place on the surface of an iron catalyst that contains traces of aluminum and potassium oxides. It is thought that hydrogen molecules and nitrogen molecules dissociate while being held on the catalytic surface. The hydrogen atoms then combine with the nitrogen atoms to form ammonia. Finally, the completed ammonia molecule breaks away, freeing the surface of the catalyst for further reaction. This sequence of steps is illustrated in Figure 13.17.

Heterogeneous catalysts are used in many important commercial processes. The petroleum industry uses heterogeneous catalysts to crack hydrocarbons into smaller fragments and then re-form them into the useful components of gasoline (see Figure 13.18). The availability of such catalysts allows refineries to produce gasoline, jet fuel, or heating oil from crude oil in any ratio necessary to meet the demands of the marketplace.

A vehicle that uses unleaded gasoline is equipped with a catalytic converter (Figure 13.19) designed to lower the concentrations of exhaust gas pollutants, such as carbon monoxide, unburned hydrocarbons, and nitrogen oxides. The catalysts are nanometer size particles of platinum, ruthenium, and rhodium dispersed in a honeycomb of a high temperature ceramic. The large ratio of surface area to mass enables the catalytic converter to react large volumes of exhaust efficiently. Air is introduced into the exhaust stream that then passes over a catalyst that adsorbs CO, NO, and O_2. The NO dissociates into N and O atoms, and the O_2 also dissociates into atoms. Pairing of nitrogen

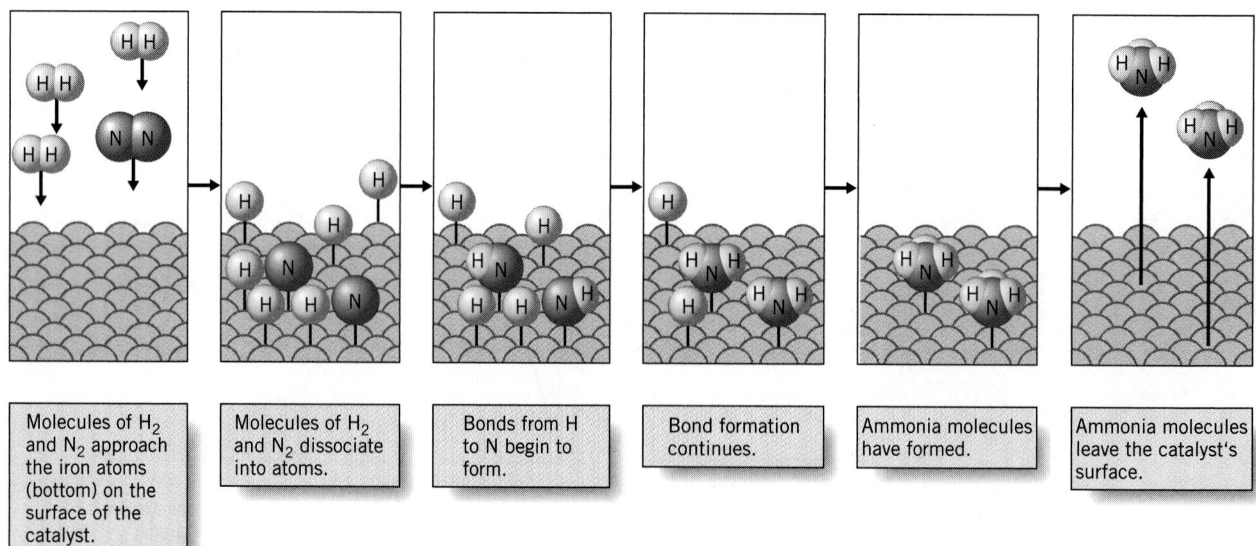

Molecules of H_2 and N_2 approach the iron atoms (bottom) on the surface of the catalyst.
Molecules of H_2 and N_2 dissociate into atoms.
Bonds from H to N begin to form.
Bond formation continues.
Ammonia molecules have formed.
Ammonia molecules leave the catalyst's surface.

FIG. 13.17 The Haber process. Catalytic formation of ammonia molecules from hydrogen and nitrogen occurs on the surface of a catalyst.

(*a*)

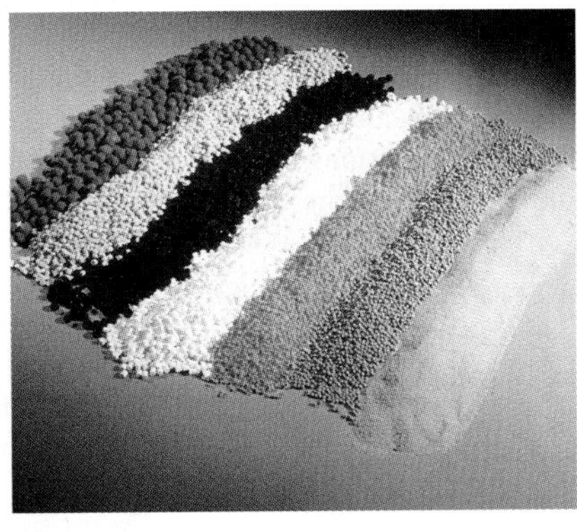

(*b*)

FIG. 13.18 Catalysts are very important in the petroleum industry. (*a*) Catalytic cracking towers at a Standard Oil refinery. (*b*) A variety of catalysts are available as beads, powders, or in other forms for various refinery operations. *(Courtesy American Petroleum Institute/Courtesy Englehard Corporation)*

atoms then produces N_2, and oxidation of CO by oxygen atoms produces CO_2. Unburned hydrocarbons are also oxidized to CO_2 and H_2O. The catalysts in catalytic converters are deactivated or "poisoned" by lead-based octane boosters like tetraethyl lead [$Pb(C_2H_5)_4$]. "Leaded" gasoline was finally banned in 1995. Leaded gasoline also posed an environmental hazard from the lead emitted in automobile exhaust.

The poisoning of catalysts is also a major problem in many industrial processes. Methyl alcohol (methanol, CH_3OH), for example, is a promising fuel that can be made from coal and steam by the reaction

$$\text{C (from coal)} + H_2O \longrightarrow CO + H_2$$

followed by

$$CO + 2H_2 \longrightarrow CH_3OH$$

A catalyst for the second step is copper(I) ion held in solid solution with zinc oxide. However, traces of sulfur, a contaminant in coal, must be avoided because sulfur reacts with the catalyst and destroys its catalytic activity.

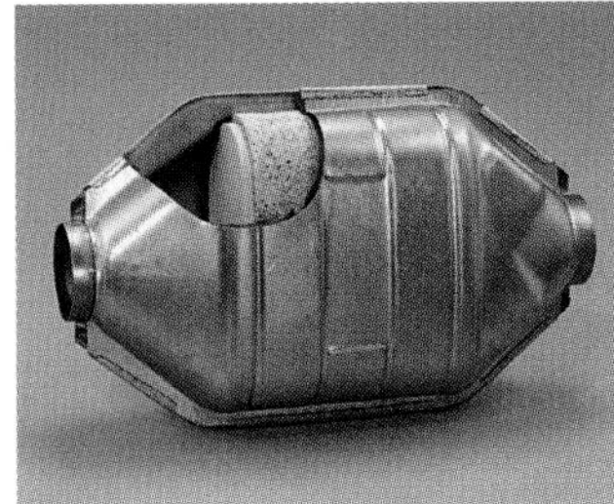

FIG. 13.19 A modern catalytic converter of the type used in about 80% of new cars. Part of the converter has been cut away to reveal the porous ceramic material that serves as the support for the catalyst. *(Courtesy AC Spark Plug)*

Enzymes are biological catalysts

In living systems, complex protein-based molecules called **enzymes** catalyze almost every reaction that occurs in living cells. Enzymes contain a specially shaped area called an active site that lowers the energy of the transition state of the reaction being catalyzed. This causes the reaction rate to increase significantly. Many poisons have been shown to work by blocking important enzyme systems. Heavy metals bind to sulfur-containing groups and distort the active site. Molecular modeling is used to design drug molecules to have optimum shapes to fit enzymatic active sites.

SUMMARY

Reaction Rates. The speeds or **rates** of reactions are controlled by five factors: (1) the nature of the reactants, (2) the ability of reactants to meet, (3) the concentrations of the reactants, (4) the temperature, and (5) the presence of catalysts. The rates of **heterogeneous reactions** are determined largely by the area of contact between the phases; the rates of **homogeneous reactions** are determined by the concentrations of the reactants. The rate is measured by monitoring the change in reactant or product concentrations with time.

$$\text{Rate} = \Delta(\text{concentration})/\Delta(\text{time})$$

In any chemical reaction, the rates of formation of products and the rates of disappearance of reactants are related by the coefficients of the balanced overall chemical equation.

Rate Laws. The **rate law** for a reaction relates the reaction rate to the molar concentrations of the reactants. The rate is proportional to the product of the molar concentrations of the reactants, each raised to an appropriate power. These exponents must be determined by experiments in which the concentrations are varied and the effects of the variations on the rate are measured. The proportionality constant, ***k***, is called the **rate constant.** Its value depends on temperature but not on the concentrations of the reactants. The sum of the exponents in the rate law is the **order** (or overall order) of the reaction.

Concentration and Time. Equations exist that relate the concentration of a reactant at a given time t to the initial concentration and the rate constant. The time required for half of a reactant to disappear is the **half-life, $t_{1/2}$**. For a first-order reaction, the half-life is a constant that depends only on the rate constant for the reaction; it is independent of the initial concentration. The half-life for a second-order reaction is inversely proportional both to the initial concentration of the reactant and to the rate constant.

Theories of Reaction Rate. According to **collision theory,** the rate of a reaction depends on the number per second of **effective collisions** of the reactant particles, which is only an extremely small fraction of the total number of collisions per second. This fraction is so small partly because the reactant molecules must be suitably oriented, but mostly because the colliding molecules must jointly possess a minimum molecular kinetic energy called the **activation energy, E_a**. As the temperature increases, a larger fraction of the collisions have this necessary energy, making more collisions effective each second and the reaction faster.

Transition state theory visualizes how the energies of molecules and the orientations of their nuclei interact as they collide. In this theory, the energy of activation is viewed as an energy barrier on the reaction's potential energy diagram. The *heat of reaction* is the net potential energy difference between the reactants and the products. In reversible reactions, the values of E_a for both the forward and reverse reactions can be identified on an energy diagram. The species at the high point on an energy diagram is the **activated complex** and is said to be in the **transition state.**

Determining the Activation Energy. The **Arrhenius equation** lets us see how changes in activation energy and temperature affect a rate constant. The Arrhenius equation also lets us determine E_a either graphically or by a calculation using the appropriate form of the Arrhenius equation. The calculation requires two rate constants determined at two temperatures. The graphical method uses more values of rate constants at more temperatures and thus usually yields more accurate results. The activation energy and the rate constant at one temperature can be used to calculate the rate constant at another temperature.

Reaction Mechanisms. The detailed sequence of elementary processes that lead to the net chemical change is the **mechanism** of the reaction. Since intermediates usually cannot be detected, the mechanism is a theory. Support for a mechanism comes from matching the predicted rate law for the mechanism with the rate law obtained from experimental data. For the **rate-determining step** or for any **elementary process** the corresponding rate law has exponents equal to the coefficients in the balanced equation for the elementary process.

Catalysts. **Catalysts** are substances that change a reaction rate but are not consumed by the reaction. Negative catalysts inhibit reactions. Positive catalysts provide alternative paths for reactions for which at least one step has a smaller activation energy than the uncatalyzed reaction. **Homogeneous catalysts** are in the same phase as the reactants. **Heterogeneous catalysts** provide a path of lower activation energy by having a surface on which the reactants are adsorbed and react. Catalysts in living systems are called **enzymes.**

TOOLS FOR PROBLEM SOLVING

In this chapter you learned to apply the following concepts as tools in solving problems dealing with aspects of chemical kinetics. Study each tool carefully so that you know what each is used for. When faced with solving a problem, recall what each tool does and consider whether it will be helpful in finding a solution. This will aid you in selecting the tools you need. Remember that at times tools from previous chapters will be needed along with the new ones in this chapter.

Rate law of a reaction *(page 526)* A rate law allows us to calculate the rate of reaction for a given set of reactant concentrations. It also serves as a guide in devising reasonable reaction mechanisms. Exponents for the concentrations in the rate law are always determined experimentally as described on page 529. Rate constants determined at different temperatures are used in the Arrhenius equation described below.

$$\text{Rate} = k[A]^n[B]^m$$

Integrated first-order rate law *(page 534)* For a first-order reaction with known k, this equation is used when we need to calculate the concentration of a reactant at some specified time after the start of the reaction. We could also calculate the time required for the concentration to drop to some specified value. This equation is also used for carbon-14 dating of organic materials.

$$\ln \frac{[A]_0}{[A]_t} = kt$$

Integrated second-order rate law *(page 540)* For a second-order reaction of the form, Rate $= k[B]^2$, with known k, this equation is used to calculate the concentration of a reactant at some specified time after the start of the reaction, or the time required for the concentration to drop to some specified value.

$$\frac{1}{[B]_t} - \frac{1}{[B]_0} = kt$$

Half-lives of a first-order reaction *(page 537)* This equation relates the rate constant to the half-life, $t_{1/2}$, for first-order reactions. Use of half-lives can be a convenient alternative to the integrated first-order rate law to determine the concentration of a reactant after it has reacted for a whole number of number of half-lives. The amount of reactant left after n half-lives is equal to $(1/2)^n$.

$$\ln 2 = kt_{1/2}$$

Arrhenius equation *(pages 548 and 550)* This equation relates the rate constant, k, to the activation energy, E_a, and temperature. Activation energies are determined by measuring rate constants at a variety of temperatures and graphically analyzing the data (see page 549). This equation is also used to determine the shelf life of a wide variety of consumer products.

$$k = A\,e^{-E_a/RT} \qquad \ln\left(\frac{k_2}{k_1}\right) = \frac{-E_a}{R}\left(\frac{1}{T_2} - \frac{1}{T_1}\right)$$

QUESTIONS, PROBLEMS, AND EXERCISES

Answers to problems whose numbers are printed in color are given in Appendix B. More challenging problems are marked with asterisks. ILW = Interactive Learningware solution is available at www.wiley.com/college/brady. OH = an Office Hours video is available for this problem.

REVIEW QUESTIONS

Factors That Affect Reaction Rate

13.1 Give an example from everyday experience of (a) a very fast reaction, (b) a moderately fast reaction, and (c) a slow reaction.

13.2 Suppose we compared two reactions, one requiring the simultaneous collision of three molecules and the other requiring a collision between two molecules. From the standpoint of statistics, and all other factors being equal, which reaction should be faster? Explain your answer.

13.3 How does an instantaneous rate of reaction differ from an average rate of reaction?

13.4 Explain how the initial instantaneous rate of reaction can be determined from experimental concentration versus time data.

13.5 What is a *homogeneous reaction*? What is a *heterogeneous reaction*? Give examples.

13.6 Why are chemical reactions usually carried out in solution?

13.7 What is the major factor that affects the rate of a heterogeneous reaction?

13.8 How does particle size affect the rate of a heterogeneous reaction? Why?

13.9 The rate of hardening of epoxy glue depends on the amount of hardener that is mixed into the glue. What factor affecting reaction rates does this illustrate?

13.10 A Polaroid instant photograph develops faster if it's kept warm than if it is exposed to cold. Why?

13.11 Insects have no way of controlling their body temperatures like mammals do. In cool weather, they become sluggish and move less quickly. How can this be explained using the principles developed in this chapter?

13.12 On the basis of what you learned in Chapter 11, why do foods cook faster in a pressure cooker than in an open pot of boiling water?

13.13 Persons who have been submerged in very cold water and who are believed to have drowned sometimes can be revived. On the other hand, persons who have been submerged in warmer water for the same length of time have died. Explain this in terms of factors that affect the rates of chemical reactions.

Concentration and Rate; Rate Laws

13.14 What are the units of reaction rate?

13.15 What are the units of the rate constant for (a) a first-order reaction, (b) a second-order reaction, and (c) a zero-order reaction?

13.16 How does the dependence of reaction rate on concentration differ between a zero-order and a first-order reaction?

13.17 Is there any way of using the coefficients in the balanced overall equation for a reaction to predict with certainty what the exponents are in the rate law?

13.18 If the concentration of a reactant is doubled and the reaction rate is unchanged, what must be the order of the reaction with respect to that reactant?

13.19 If the concentration of a reactant is doubled and the reaction rate doubles, what must be the order of the reaction with respect to that reactant?

13.20 If the concentration of a reactant is doubled, by what factor will the rate increase if the reaction is second order with respect to that reactant?

13.21 In an experiment, the concentration of a reactant was tripled. The rate increased by a factor of 27. What is the order of the reaction with respect to that reactant?

13.22 Biological reactions usually involve the interaction of an enzyme with a *substrate*, the substance that actually undergoes the chemical change. In many cases, the rate of reaction depends on the concentration of the enzyme but is independent of the substrate concentration. What is the order of the reaction with respect to the substrate in such instances?

13.23 A reaction has the following rate law:

$$\text{Rate} = k[A]^2[B][C]$$

What are the units of the rate constant, k?

Concentration and Time, Half-lives

13.24 How is the half-life of a first-order reaction affected by the initial concentration of the reactant?

13.25 How is the half-life of a second-order reaction affected by the initial reactant concentration?

13.26 Derive the equations for $t_{1/2}$ for first- and second-order reactions from Equations 13.3 and 13.8, respectively.

13.27 The integrated rate law for a zero-order reaction is

$$[A]_t - [A]_0 = -kt$$

Derive an equation for the half-life of a zero-order reaction.

Effect of Temperature on Rate

13.28 What is the basic postulate of collision theory?

13.29 What two factors influence the effectiveness of molecular collisions in producing chemical change?

13.30 In terms of the kinetic theory, why does an increase in temperature increase the reaction rate?

13.31 Draw the potential energy diagram for an endothermic reaction. Indicate on the diagram the activation energy for both the forward and reverse reactions. Also indicate the heat of reaction.

13.32 Explain, in terms of the law of conservation of energy, why an endothermic reaction leads to a cooling of the reaction mixture (provided heat cannot enter from outside the system).

13.33 Draw a potential energy diagram for an exothermic reaction and indicate on the diagram the location of the transition state.

13.34 The decomposition of carbon dioxide,

$$CO_2 \longrightarrow CO + O$$

has a very large activation energy of approximately 460 kJ mol^{-1}. Explain why this is consistent with a mechanism that involves the breaking of a C=O bond.

Reaction Mechanisms

13.35 What is the definition of an *elementary process?* How are elementary processes related to the mechanism of a reaction?

13.36 What is a *rate-determining step?*

13.37 In what way is the rate law for a reaction related to the rate-determining step?

13.38 A reaction has the following mechanism.

$$2NO \longrightarrow N_2O_2$$
$$N_2O_2 + H_2 \longrightarrow N_2O + H_2O$$
$$N_2O + H_2 \longrightarrow N_2 + H_2O$$

What is the net overall change that occurs in this reaction? Identify any intermediates in the reaction.

13.39 If the reaction $NO_2 + CO \longrightarrow NO + CO_2$ occurs by a one-step collision process, what would be the expected rate law for the reaction? The actual rate law is Rate $= k[NO_2]^2$. Could the reaction actually occur by a one-step collision between NO_2 and CO? Explain.

13.40 Oxidation of NO to NO_2—one of the reactions in the production of smog—appears to involve carbon monoxide. A possible mechanism is

$$CO + \cdot OH \longrightarrow CO_2 + H\cdot$$
$$H\cdot + O_2 \longrightarrow HOO\cdot$$
$$HOO\cdot + NO \longrightarrow \cdot OH + NO_2$$

(The formulas with dots represent extremely reactive species with unpaired electrons and are called *free radicals.*) Write the net chemical equation for the reaction.

13.41 Show that the following two mechanisms give the same net overall reaction.

Mechanism 1

$$OCl^- + H_2O \longrightarrow HOCl + OH^-$$
$$HOCl + I^- \longrightarrow HOI + Cl^-$$
$$HOI + OH^- \longrightarrow H_2O + OI^-$$

Mechanism 2

$$OCl^- + H_2O \longrightarrow HOCl + OH^-$$
$$I^- + HOCl \longrightarrow ICl + OH^-$$
$$ICl + 2OH^- \longrightarrow OI^- + Cl^- + H_2O$$

13.42 The experimental rate law for the reaction

$$NO_2 + CO \longrightarrow CO_2 + NO$$

is rate $= k[NO_2]^2$. If the mechanism is

$$2NO_2 \longrightarrow NO_3 + NO \qquad \text{(slow)}$$
$$NO_3 + CO \longrightarrow NO_2 + CO_2 \qquad \text{(fast)}$$

show that the predicted rate law is the same as the experimental rate law.

Catalysts

13.43 How does a catalyst increase the rate of a chemical reaction?

13.44 What is a *homogeneous catalyst*? How does it function, in general terms?

13.45 What is the difference in meaning between the terms *adsorption* and *absorption*? (If necessary, use a dictionary.) Which one applies to heterogeneous catalysts?

13.46 What does the catalytic converter do in the exhaust system of an automobile? Why should leaded gasoline not be used in cars equipped with catalytic converters?

REVIEW PROBLEMS

Measuring Rates of Reaction

13.47 The following data were collected at a certain temperature for the decomposition of sulfuryl chloride, SO_2Cl_2, a chemical used in a variety of organic syntheses.

$$SO_2Cl_2 \longrightarrow SO_2 + Cl_2$$

Time (min)	$[SO_2Cl_2]$ (mol L^{-1})
0	0.1000
100	0.0876
200	0.0768
300	0.0673
400	0.0590
500	0.0517
600	0.0453
700	0.0397
800	0.0348
900	0.0305
1000	0.0267
1100	0.0234

Make a graph of concentration versus time and determine the rate of formation of SO_2 at $t = 200$ minutes and $t = 600$ minutes.

13.48 The following data were collected for the decomposition of acetaldehyde, CH_3CHO (used in the manufacture of a variety of chemicals including perfumes, dyes, and plastics), into methane and carbon monoxide. The data were collected at a temperature of 530 °C.

$$CH_3CHO \longrightarrow CH_4 + CO$$

$[CH_3CHO]$ (mol L^{-1})	Time (s)
0.200	0
0.153	20
0.124	40
0.104	60
0.090	80
0.079	100
0.070	120
0.063	140
0.058	160
0.053	180
0.049	200

Make a graph of concentration versus time and determine the rate of reaction of CH_3CHO after 60 seconds and after 120 seconds.

13.49 In the reaction $3H_2 + N_2 \longrightarrow 2NH_3$, how does the rate of disappearance of hydrogen compare to the rate of disappearance of nitrogen? How does the rate of appearance of NH_3 compare to the rate of disappearance of nitrogen?

OH **13.50** For the reaction $2A + B \longrightarrow 3C$, it was found that the rate of disappearance of B was 0.30 mol L^{-1} s^{-1}. What were the rate of disappearance of A and the rate of appearance of C?

13.51 In the combustion of hexane (a low-boiling component of gasoline),

$$2C_6H_{14}(g) + 19O_2(g) \longrightarrow 12CO_2(g) + 14H_2O(g)$$

it was found that the rate of reaction of C_6H_{14} was 1.20 mol L^{-1} s^{-1}.
(a) What was the rate of reaction of O_2?
(b) What was the rate of formation of CO_2?
(c) What was the rate of formation of H_2O?

13.52 At a certain moment in the reaction

$$2N_2O_5 \longrightarrow 4NO_2 + O_2$$

N_2O_5 is decomposing at a rate of 2.5×10^{-6} mol L^{-1} s^{-1}. What are the rates of formation of NO_2 and O_2?

Rate Laws for Reactions

13.53 Estimate the rate of the reaction

$$H_2SeO_3 + 6I^- + 4H^+ \longrightarrow Se + 2I_3^- + 3H_2O$$

given the rate law for the reaction at 0 °C is

$$\text{Rate} = (5.0 \times 10^5\ L^5\ mol^{-5}\ s^{-1})[H_2SeO_3][I^-]^3[H^+]^2$$

and the reactant concentrations are $[H_2SeO_3] = 2.0 \times 10^{-2}\ M$, $[I^-] = 2.0 \times 10^{-3}\ M$, and $[H^+] = 1.0 \times 10^{-3}\ M$.

13.54 Estimate the rate of the reaction

$$H^+(aq) + OH^-(aq) \longrightarrow H_2O(l)$$

given that the rate law for the reaction is

$$\text{Rate} = (1.3 \times 10^{11}\ L\ mol^{-1}\ s^{-1})[OH^-][H^+]$$

for neutral water where $[H^+] = 1.0 \times 10^{-7}\ M$ and $[OH^-] = 1.0 \times 10^{-7}\ M$.

OH **13.55** The oxidation of NO (released in small amounts in the exhaust of automobiles) produces the brownish-red gas NO_2, which is a component of urban air pollution.

$$2NO + O_2 \longrightarrow 2NO_2$$

The rate law for the reaction is Rate $= k[NO]^2[O_2]$. At 25 °C, $k = 7.1 \times 10^9\ L^2\ mol^{-2}\ s^{-1}$. What would be the rate of the reaction if $[NO] = 0.0010$ mol L^{-1} and $[O_2] = 0.034$ mol L^{-1}?

13.56 The rate law for the decomposition of N_2O_5 is

$$\text{Rate} = k[N_2O_5]$$

If $k = 1.0 \times 10^{-5}\ s^{-1}$, what is the reaction rate when the N_2O_5 concentration is 0.0010 mol L^{-1}?

13.57 The following data were collected for the reaction

$$M + N \longrightarrow P + Q$$

Initial Concentrations (mol L^{-1})		Initial rate of reaction
$[M]$	$[N]$	(mol L^{-1} s^{-1})
0.010	0.010	2.5×10^{-3}
0.020	0.010	5.0×10^{-3}
0.020	0.030	4.5×10^{-2}

What is the rate law for the reaction? What is the value of the rate constant (with correct units)?

13.58 Cyclopropane, C_3H_6, is a gas used as a general anesthetic. It undergoes a slow molecular rearrangement to propylene.

cyclopropane $\longrightarrow$ $H-CH_2-CH=CH_2$ (H–C(H)(H)–C(H)=C(H)–H)

cyclopropane propylene

At a certain temperature, the following data were obtained relating concentration and rate.

Initial Concentration of C_3H_6 (mol L^{-1})	Rate of Formation of Propylene (mol L^{-1} s^{-1})
0.050	2.95×10^{-5}
0.100	5.90×10^{-5}
0.150	8.85×10^{-5}

What is the rate law for the reaction? What is the value of the rate constant, with correct units?

13.59 The reaction of iodide ion with hypochlorite ion, OCl^- (the active ingredient in a "chlorine bleach" such as Clorox), follows the equation $OCl^- + I^- \longrightarrow OI^- + Cl^-$. It is a rapid reaction that gives the following rate data.

Initial Concentrations (mol L^{-1})		Rate of Formation of Cl^- (mol L^{-1} s^{-1})
$[OCl^-]$	$[I^-]$	
1.7×10^{-3}	1.7×10^{-3}	1.75×10^4
3.4×10^{-3}	1.7×10^{-3}	3.50×10^4
1.7×10^{-3}	3.4×10^{-3}	3.50×10^4

What is the rate law for the reaction? Determine the value of the rate constant with its correct units.

13.60 The formation of small amounts of nitric oxide, NO, in automobile engines is the first step in the formation of smog. As noted in Problem 13.55, nitric oxide is readily oxidized to nitrogen dioxide by the reaction

$$2NO(g) + O_2(g) \longrightarrow 2NO_2(g)$$

The following data were collected in a study of the rate of this reaction.

Initial Concentrations (mol L^{-1})		Rate of Formation of NO_2 (mol L^{-1} s^{-1})
$[O_2]$	[NO]	
0.0010	0.0010	7.10
0.0040	0.0010	28.4
0.0040	0.0030	255.6

What is the rate law for the reaction? What is the rate constant with its correct units?

ILW **13.61** At a certain temperature the following data were collected for the reaction $2ICl + H_2 \longrightarrow I_2 + 2HCl$

Initial Concentrations (mol L^{-1})		Initial Rate of Formation of I_2 (mol L^{-1} s^{-1})
[ICl]	$[H_2]$	
0.10	0.10	0.0015
0.20	0.10	0.0030
0.10	0.0500	0.00075

Determine the rate law and the rate constant (with correct units) for the reaction.

13.62 The following data were obtained for the reaction of $(CH_3)_3CBr$ with hydroxide ion at 55 °C.

$$(CH_3)_3CBr + OH^- \longrightarrow (CH_3)_3COH + Br^-$$

Initial Concentrations (mol L^{-1})		Initial Rate of Formation of $(CH_3)_3COH$ (mol L^{-1} s^{-1})
$[(CH_3)_3CBr]$	$[OH^-]$	
0.10	0.10	1.0×10^{-3}
0.20	0.10	2.0×10^{-3}
0.30	0.10	3.0×10^{-3}
0.10	0.20	1.0×10^{-3}
0.10	0.30	1.0×10^{-3}

What is the rate law for the reaction? What is the value of the rate constant (with correct units) at this temperature?

Concentration and Time

13.63 Data for the decomposition of SO_2Cl_2 according to the equation $SO_2Cl_2(g) \longrightarrow SO_2(g) + Cl_2(g)$ were given in Problem 13.47. Show graphically that these data fit a first-order rate law. Graphically determine the rate constant for the reaction.

13.64 For the data in Problem 13.48, decide graphically whether the reaction is first or second order. Determine the rate constant for the reaction described in that problem.

ILW OH **13.65** The decomposition of SO_2Cl_2 described in Problem 13.47 has a first-order rate constant $k = 2.2 \times 10^{-5}$ s^{-1} at 320 °C. If the initial SO_2Cl_2 concentration in a container is 0.0040 M, what will its concentration be (a) after 1.00 hour and (b) after 1.00 day?

13.66 If it takes 75.0 min for the concentration of a reactant to drop to 20% of its initial value in a first-order reaction, what is the rate constant for the reaction in the units min^{-1}?

13.67 The concentration of a drug in the body is often expressed in units of milligrams per kilogram of body weight. The initial dose of a drug in an animal was 25.0 mg/kg body weight. After 2.00 hours, this concentration had dropped to 15.0 mg/kg body weight. If the drug is eliminated metabolically by a first-order process, what is the rate constant for the process in units of min^{-1}?

13.68 In the preceding problem, what must the initial dose of the drug be in order for the drug concentration 3.00 hours afterward to be 5.0 mg/kg body weight?

13.69 The decomposition of hydrogen iodide follows the equation $2HI(g) \longrightarrow H_2(g) + I_2(g)$. The reaction is second order and has a rate constant equal to 1.6×10^{-3} L mol^{-1} s^{-1} at 700 °C. If the initial concentration of HI in a container is 3.4×10^{-2} M, how many minutes will it take for the concentration to be reduced to 8.0×10^{-4} M?

13.70 The second-order rate constant for the decomposition of HI at 700 °C was given in the preceding problem. At 2.5×10^3 minutes after a particular experiment had begun, the HI concentration was equal to 4.5×10^{-4} mol L^{-1}. What was the initial molar concentration of HI in the reaction vessel?

Half-lives

13.71 The half-life of a certain first-order reaction is 15 minutes. What fraction of the original reactant concentration will remain after 2.0 hours?

13.72 Strontium-90 has a half-life of 28 years. How long will it take for all of the strontium-90 presently on the earth to be reduced to 1/32nd of its present amount?

13.73 Using the graph from Problem 13.47, determine the time required for the SO_2Cl_2 concentration to drop from 0.100 mol L^{-1} to 0.050 mol L^{-1}. How long does it take for the concentration to drop from 0.050 mol L^{-1} to 0.025 mol L^{-1}? What is the order of this reaction? (Hint: How is the half-life related to concentration?)

13.74 Using the graph from Problem 13.48, determine how long it takes for the CH_3CHO concentration to decrease from 0.200 mol L^{-1} to 0.100 mol L^{-1}. How long does it take the concentration to drop from 0.100 mol L^{-1} to 0.050 mol L^{-1}? What is the order of this reaction? (Hint: How is the half-life related to concentration?)

Radiological Dating

***13.75** A 500 mg sample of rock was found to have 2.45×10^{-6} mol of potassium-40 ($t_{1/2} = 1.3 \times 10^9$ yr) and 2.45×10^{-6} mol of argon-40. How old was the rock? (What assumption is made about the origin of the argon-40?)

***13.76** If a rock sample was found to contain 1.16×10^{-7} mol of argon-40, how much potassium-40 would also have to be present for the rock to be 1.3×10^9 years old?

ILW **13.77** A tree killed by being buried under volcanic ash was found to have a ratio of carbon-14 atoms to carbon-12 atoms of 4.8×10^{-14}. How long ago did the eruption occur?

13.78 A wooden door lintel from an excavated site in Mexico would be expected to have what ratio of carbon-14 to carbon-12 atoms if the lintel is 9.0×10^3 yr old?

Calculations Involving the Activation Energy

13.79 The following data were collected for a reaction.

Rate Constant (L mol^{-1} s^{-1})	Temperature (°C)
2.88×10^{-4}	320
4.87×10^{-4}	340
7.96×10^{-4}	360
1.26×10^{-3}	380
1.94×10^{-3}	400

Determine the activation energy for the reaction in kJ mol^{-1} both graphically and by calculation using Equation 13.12. For the calculation of E_a, use the first and last sets of data in the table above.

13.80 Rate constants were measured at various temperatures for the reaction $HI(g) + CH_3I(g) \longrightarrow CH_4(g) + I_2(g)$. The following data were obtained.

Rate Constant (L mol^{-1} s^{-1})	Temperature (°C)
1.91×10^{-2}	205
2.74×10^{-2}	210
3.90×10^{-2}	215
5.51×10^{-2}	220
7.73×10^{-2}	225
1.08×10^{-1}	230

Determine the activation energy in kJ mol^{-1} both graphically and by calculation using Equation 13.12. For the calculation of E_a, use the first and last sets of data in the table above.

ILW OH **13.81** The decomposition of NOCl, $2NOCl \longrightarrow 2NO + Cl_2$ has $k = 9.3 \times 10^{-5}$ L mol^{-1} s^{-1} at 100 °C and $k = 1.0 \times 10^{-3}$ L mol^{-1} s^{-1} at 130 °C. What is E_a for this reaction in kJ mol^{-1}? Use the data at 100 °C to calculate the frequency factor.

13.82 The conversion of cyclopropane, an anesthetic, to propylene (see Problem 13.58) has a rate constant $k = 1.3 \times 10^{-6}$ s^{-1} at 400 °C and $k = 1.1 \times 10^{-5}$ s^{-1} at 430 °C.
(a) What is the activation energy in kJ mol^{-1}?
(b) What is the value of the frequency factor, A, for this reaction?
(c) What is the rate constant for the reaction at 350 °C?

13.83 The decomposition of N_2O_5 has an activation energy of 103 kJ mol^{-1} and a frequency factor of 4.3×10^{13} s^{-1}. What is the rate constant for this decomposition at (a) 20 °C and (b) 100 °C?

13.84 At 35 °C, the rate constant for the reaction

$$\underset{\text{sucrose}}{C_{12}H_{22}O_{11}} + H_2O \longrightarrow \underset{\text{glucose}}{C_6H_{12}O_6} + \underset{\text{fructose}}{C_6H_{12}O_6}$$

is $k = 6.2 \times 10^{-5}$ s^{-1}. The activation energy for the reaction is 108 kJ mol^{-1}. What is the rate constant for the reaction at 45 °C?

ADDITIONAL EXERCISES

13.85 For the reaction and data given in Problem 13.48, make a graph of concentration versus time for the *formation* of CH_4. What are the rates of formation of CH_4 at $t = 40$ s and $t = 100$ s?

13.86 The age of wine can be determined by measuring the trace amount of radioactive tritium, ^3H, present in a sample. Tritium is formed from hydrogen in water vapor in the upper atmosphere by cosmic bombardment, so all naturally occurring water contains a small amount of this isotope. Once the water is in a bottle of wine, however, the formation of additional tritium from the water is negligible, so the tritium initially present gradually diminishes by a first-order radioactive decay with a half-life of 12.5 years. If a bottle of wine is found to have a tritium concentration that is 0.100 that of freshly bottled wine (i.e., $[^3H]_t = 0.100[^3H]_0$), what is the age of the wine?

OH **13.87** Carbon-14 dating can be used to estimate the age of formerly living materials because the uptake of carbon-14 from carbon dioxide in the atmosphere stops once the organism dies. If tissue samples from a mummy contain about 81.0% of the carbon-14 expected in living tissue, how old is the mummy? The half-life for decay of carbon-14 is 5730 years.

*13.88 What percentage of cesium chloride made from cesium-137 ($t_{1/2}$=30 yr; beta emitter) remains after 150 yr? What *chemical* product forms?

*13.89 One of the reactions that occurs in polluted air in urban areas is $2NO_2(g) + O_3(g) \longrightarrow N_2O_5(g) + O_2(g)$. It is believed that a species with the formula NO_3 is involved in the mechanism, and the observed rate law for the overall reaction is Rate = $k[NO_2][O_3]$. Propose a mechanism for this reaction that includes the species NO_3 and is consistent with the observed rate law.

OH *13.90 Suppose a reaction occurs with the mechanism

(1) $2A \rightleftharpoons A_2$ (fast)

(2) $A_2 + E \longrightarrow B + C$ (slow)

in which the first step is a very rapid reversible reaction that can be considered to be essentially an equilibrium (forward and reverse reactions occurring at the same rate) and the second is a slow step.

(a) Write the rate law for the forward reaction in step (1).
(b) Write the rate law for the reverse reaction in step (1).
(c) Write the rate law for the rate-determining step.
(d) What is the chemical equation for the net reaction that occurs in this chemical change?
(e) Use the results of parts (a) and (b) to rewrite the rate law of the rate-determining step in terms of the concentrations of the reactants in the overall balanced equation for the reaction.

OH 13.91 The decomposition of urea, $(NH_2)_2CO$, in 0.10 M HCl follows the equation

$$(NH_2)_2CO(aq) + 2H^+(aq) + H_2O \longrightarrow 2NH_4^+(aq) + CO_2(g)$$

At 60 °C, $k = 5.84 \times 10^{-6}$ min^{-1} and at 70 °C, $k = 2.25 \times 10^{-5}$ min^{-1}. If this reaction is run at 80 °C starting with a urea concentration of 0.0020 M, how many minutes will it take for the urea concentration to drop to 0.0012 M?

13.92 Show that for a reaction that obeys the general rate law

$$\text{Rate} = k[A]^n$$

a graph of log(rate) versus log[A] should yield a straight line with a slope equal to the order of the reaction. For the reaction in Problem 13.47, measure the rate of the reaction at t = 150, 300, 450, and 600 s. Then graph log(rate) versus log[SO_2Cl_2] and determine the order of the reaction with respect to SO_2Cl_2.

OH *13.93 It was mentioned that the rates of many reactions approximately double for each 10 °C rise in temperature. Assuming a starting temperature of 25 °C, what would the activation energy be, in kJ mol^{-1}, if the rate of a reaction were to be twice as large at 35 °C?

*13.94 The development of a photographic image on film is a process controlled by the kinetics of the reduction of silver halide by a developer. The time required for development at a particular temperature is inversely proportional to the rate constant for the process. Below are published data on development times for Kodak's Tri-X film using Kodak D-76 developer. From these data, estimate the activation energy (in units of kJ mol^{-1}) for the development process. Also estimate the development time at 15 °C.

Temperature (°C)	Development Time (minutes)
18	10
20	9
21	8
22	7
24	6

13.95 The rate at which crickets chirp depends on the ambient temperature, because crickets are cold-blooded insects whose body temperature follows the temperature of their environment. It has been found that the Celsius temperature can be estimated by counting the number of chirps in 8 seconds and then adding 4. In other words, t_C = (number of chirps in 8 seconds) + 4.

(a) Calculate the number of chirps in 8 seconds for temperatures of 20, 25, 30, and 35 °C.
(b) The number of chirps per unit of time is directly proportional to the rate constant for a biochemical reaction involved in the cricket's chirp. On the basis of this assumption, make a graph of ln(chirps in 8 s) versus ($1/T$). Calculate the activation energy for the biochemical reaction involved.
(c) How many chirps would a cricket make in 8 seconds at a temperature of 40 °C?

*13.96 The cooking of an egg involves the denaturation of a protein called albumen. The time required to achieve a particular degree of denaturation is inversely proportional to the rate constant for the process. This reaction has a high activation energy, E_a = 418 kJ mol^{-1}. Calculate how long it would take to cook a traditional three-minute egg on top of Mt. McKinley in Alaska on a day when the atmospheric pressure there is 355 torr.

*13.97 The following question is based on Facets of Chemistry 13.1. The reaction of hydrogen and bromine appears to follow the mechanism

$$Br_2 \xrightarrow{h\nu} 2Br\cdot$$

$$Br\cdot + H_2 \longrightarrow HBr + H\cdot$$

$$H\cdot + Br_2 \longrightarrow HBr + Br\cdot$$

$$2Br\cdot \longrightarrow Br_2$$

(a) Identify the initiation step in the mechanism.
(b) Identify any propagation steps.
(c) Identify the termination step.
(d) The mechanism also contains the reaction

$$H\cdot + HBr \longrightarrow H_2 + Br\cdot$$

How does this reaction affect the rate of formation of HBr?

EXERCISES IN CRITICAL THINKING

13.98 Provide three examples of ordinary occurrences that mimic a reaction mechanism and have a rate-limiting step.

13.99 Can a reaction have a negative activation energy? Explain your response.

13.100 Assume you have a three-step mechanism. Would the potential energy diagram have three peaks? If so, how would you distinguish the rate-limiting step?

13.101 What range of ages can C-14 dating reliably determine?

13.102 Why are initial reaction rates used to determine rate laws?

13.103 If a reaction is reversible (i.e., the products can react to re-form the reactants) what would the rate law look like?

***13.104** The ozone layer protects us from high energy radiation. Description of the ozone layer is a kinetics problem concerning the formation and destruction of ozone to produce a steady state. How can we write this mathematically assuming that the processes are elementary reactions?

***13.105** How would you measure an extremely fast reaction?

CHEMICAL EQUILIBRIUM: GENERAL CONCEPTS

14

One of the chemicals used in a dry chemical fire extinguisher, like the one being used here to extinguish a car fire, is sodium bicarbonate, $NaHCO_3$. Its decomposition when heated produces carbon dioxide, which helps to smother the flames. The decomposition of sodium bicarbonate involves a heterogeneous chemical equilibrium and is one of the reactions discussed in this chapter. *(Sean Murphy/Stone/Getty Images, Inc)*

CHAPTER OUTLINE

14.1 Dynamic equilibrium is achieved when the rates of forward and reverse processes become equal

14.2 A law relating equilibrium concentrations can be derived from the balanced chemical equation for a reaction

14.3 Equilibrium laws for gaseous reactions can be written in terms of either concentrations or pressures

14.4 Heterogeneous equilibria involve reaction mixtures with more than one phase

14.5 When K is large, the position of equilibrium lies toward the products

14.6 Le Châtelier's principle tells us how a chemical equilibrium responds when disturbed

14.7 Equilibrium concentrations can be used to determine equilibrium constants, and vice versa

THIS CHAPTER IN CONTEXT In Chapter 13 you learned that most chemical reactions slow down as the reactants are consumed and the products form. The focus there was learning about the factors that affect how *fast* the reactant and product concentrations change. In this and the following four chapters we will turn our attention to the ultimate fate of chemical systems, *dynamic chemical equilibrium*—the situation that exists when the concentrations cease to change.

You have already encountered the concept of a dynamic equilibrium several times earlier in this book. Recall that such an equilibrium is established when two opposing processes occur at equal rates. In a liquid–vapor equilibrium, for example, the processes are evaporation and condensation. In a chemical equilibrium, such as the ionization of a weak acid, they are the forward and reverse reactions represented by the chemical equation.

In this chapter we will study the equilibrium condition both qualitatively and quantitatively. Understanding the factors that affect equilibrium systems is important not only in chemistry but in other sciences as well. They apply to biology because living cells must control the concentrations of substances within them in order to survive. And they are fundamental in understanding chemical reactions that affect many current environmental problems, including global warming, acid rain, and stratospheric ozone depletion.

We begin our discussion of equilibrium here by examining principles that apply to chemical systems in general. In Chapters 15 through 17 we will focus on equilibria in aqueous solutions. Then, in Chapter 18 we will tie together the concept of chemical equilibrium with the energetics of chemical change and the drive toward the most probable state.

14.1 DYNAMIC EQUILIBRIUM IS ACHIEVED WHEN THE RATES OF FORWARD AND REVERSE PROCESSES BECOME EQUAL

Let's review the process by which a system reaches equilibrium. For this purpose, we'll examine the reaction for the decomposition of gaseous N_2O_4 into NO_2.

$$N_2O_4(g) \longrightarrow 2NO_2(g)$$

When we first begin, there are no molecules of NO_2, so the only reaction taking place is the decomposition of N_2O_4. As the concentration of N_2O_4 decreases, its rate of decomposition slows, as illustrated in Figure 14.1. At the same time, the concentration of NO_2 rises, which is also shown in Figure 14.1. When NO_2 molecules collide, they are able to re-form N_2O_4, so while the rate of decomposition of N_2O_4 slows, the rate at which NO_2 molecules combine increases. This corresponds to the reverse reaction in the equation above. Eventually, the rate of the reverse reaction becomes equal to the rate of the forward reaction, at which point the concentrations cease to change. N_2O_4 is

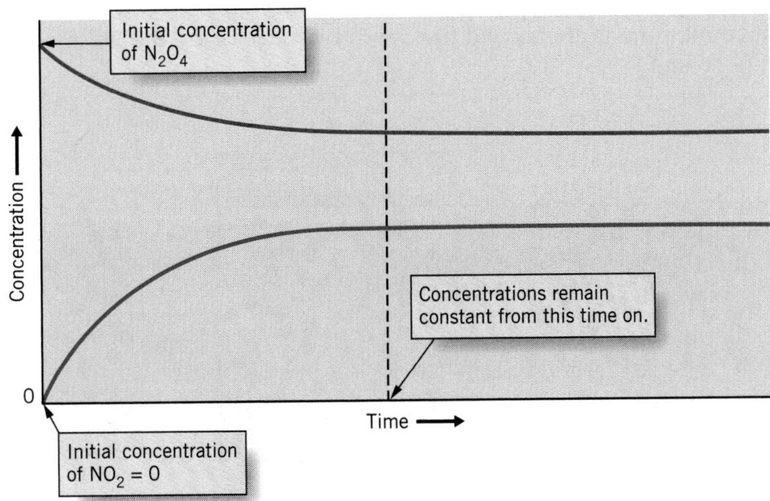

FIG. 14.1 The approach to equilibrium. In the decomposition of $N_2O_4(g)$ into $NO_2(g)$,

$$N_2O_4(g) \rightleftharpoons 2NO_2(g)$$

the concentrations of the N_2O_4 and NO_2 change relatively fast at first. As time passes, the concentrations change more and more slowly. When equilibrium is reached, the concentrations of N_2O_4 and NO_2 no longer change with time; they remain constant.

decomposing as fast as it's being formed, and we've reached *dynamic equilibrium*. As you learned earlier, when we write the equation to describe the equilibrium, we use double arrows ($\rightleftharpoons$).

$$N_2O_4(g) \rightleftharpoons 2NO_2(g) \tag{14.1}$$

Most chemical systems reach a state of dynamic equilibrium, given enough time. Sometimes, though, this equilibrium is extremely difficult (or even impossible) to detect, because in some reactions the amounts of either the reactants or products present at equilibrium are virtually zero. For instance, in water vapor at room temperature there are no detectable amounts of either H_2 or O_2 from the equilibrium

$$2H_2O(g) \rightleftharpoons 2H_2(g) + O_2(g)$$

Water molecules are so stable that we can't detect whether any of them decompose. Even in cases such as this, however, it is often *convenient* to assume that an equilibrium does exist.

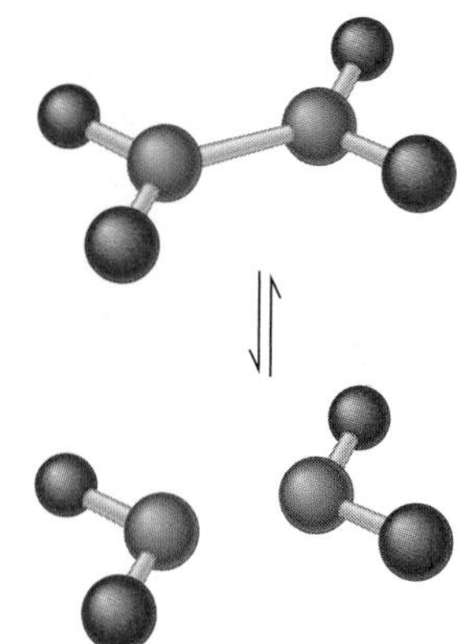

The equilibrium between N_2O_4 and NO_2.

The meaning of reactants and products in a chemical equilibrium

For the forward reaction in Equation 14.1, N_2O_4 is the reactant and NO_2 is the product. But the reverse reaction is also occurring, where NO_2 is the reactant and N_2O_4 is the product. When the system is at equilibrium, the usual definitions of reactant and product don't make a lot of sense. In these situations, when we use the term *reactants*, we simply mean those substances written to the left of the double arrows. Similarly, we use the term *products* to mean the substances to the right of the double arrows.

For the equilibrium $A \rightleftharpoons B$, the reaction read from left to right is the *forward reaction*; the reaction read from right to left is the *reverse reaction*.

$A \longrightarrow B$ (forward reaction)
$A \longleftarrow B$ (reverse reaction)

Remember, even after equilibrium has been reached, both the forward and reverse reactions continue to occur.

Closed systems reach the same equilibrium concentrations whether we start with reactants or with products

The composition of an equilibrium mixture is found to be independent of whether we begin the reaction from the "reactant side" or the "product side." For example, suppose we set up the two experiments shown in Figure 14.2. In the first one-liter flask we place 0.0350 mol N_2O_4. Since no NO_2 is present, some N_2O_4 must decompose for the mixture to reach equilibrium, so the reaction in Equation 14.1 will proceed in the forward direction (i.e., from left to right). When equilibrium is reached, we find the concentration of N_2O_4 has dropped to 0.0292 mol L^{-1} and the concentration of NO_2 has become 0.0116 mol L^{-1}.

In the second one-liter flask we place 0.0700 mol of NO_2 (*precisely* the amount of NO_2 that would form if 0.0350 mol of N_2O_4—the amount placed in the first flask—decomposed completely). In this second flask there is no N_2O_4 present initially, so NO_2 molecules must combine, following the reverse reaction (right to left) in Equation 14.1, to give enough N_2O_4 for equilibrium. When we measure the concentrations at equilibrium in the second flask, we find, once again, 0.0292 mol L^{-1} of N_2O_4 and 0.0116 mol L^{-1} of NO_2.

Stoichiometrically, 0.0700 mol of NO_2 could be formed from 0.0350 mol of N_2O_4, since the ratio of NO_2 to N_2O_4 is 2:1.

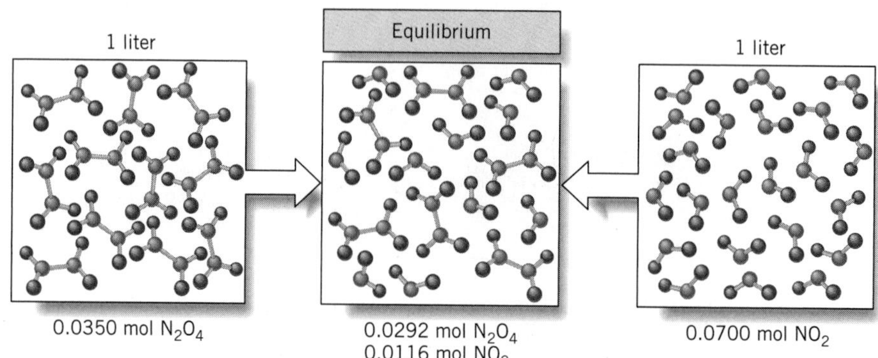

FIG. 14.2 Reaction reversibility for the equilibrium $N_2O_4(g) \rightleftharpoons 2NO_2(g)$. The same equilibrium composition is reached from either the forward or reverse direction, provided the overall system composition is the same. Because pure NO_2 is brown, and pure N_2O_4 is colorless, the amber color of the equilibrium mixture indicates that both species are present at equilibrium.

We see here that the same equilibrium composition is reached whether we begin with pure NO_2 or pure N_2O_4, as long as the *total* amount of nitrogen and oxygen to be divided between these two substances is the same. Similar observations apply to other chemical systems as well, which leads to the following generalization.

> For a given *overall* system composition, we always reach the same equilibrium concentrations whether equilibrium is approached from the forward or reverse direction.

There are times when it is very convenient to be able to imagine the approach to equilibrium starting from the product side of the equation.

14.2 A LAW RELATING EQUILIBRIUM CONCENTRATIONS CAN BE DERIVED FROM THE BALANCED CHEMICAL EQUATION FOR A REACTION

For any chemical system at equilibrium, there exists a simple relationship among the molar concentrations of the reactants and products. To see this, let's consider the gaseous reaction of hydrogen with iodine to form hydrogen iodide.

$$H_2(g) + I_2(g) \rightleftharpoons 2HI(g)$$

Figure 14.3 shows the results of several experiments measuring equilibrium amounts of each gas, starting with different amounts of the reactants and product in a 10.0 L reaction vessel. When equilibrium is reached, the amounts of H_2, I_2, and HI are different for each experiment, as are their molar concentrations. This isn't particularly surprising, but what is amazing is that the relationship among the concentrations is very simple and can actually be predicted (once we've learned how) from the balanced equation for the reaction.

The molar concentrations are obtained by dividing the number of moles of each substance by the volume, 10.0 L.

For each experiment in Figure 14.3, if we square the molar concentration of HI at equilibrium and then divide this by the product of the equilibrium molar

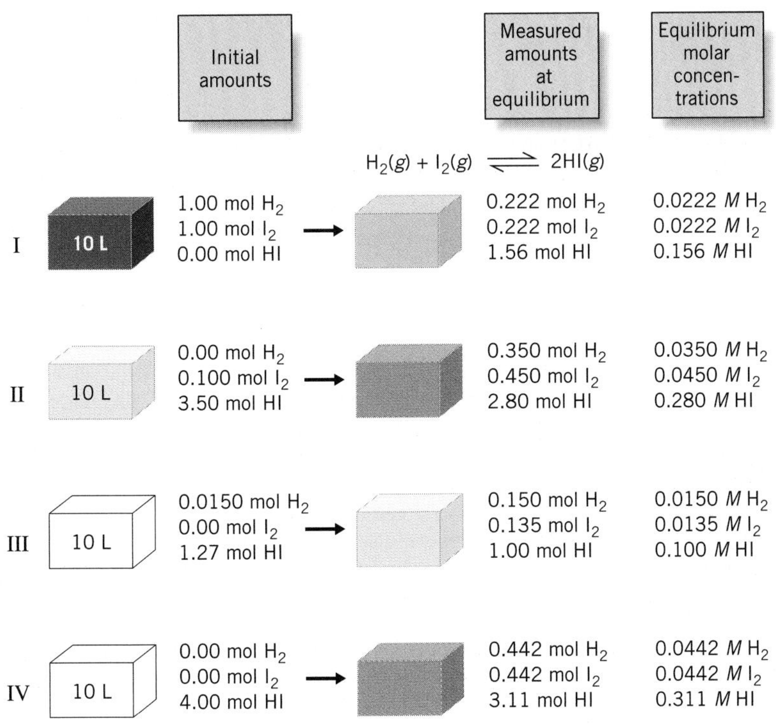

FIG. 14.3 Four experiments to study the equilibrium among H_2, I_2, and HI gases. Different amounts of the reactants and product are placed in a 10.0 L reaction vessel at 440 °C where the gases established the equilibrium

$$H_2(g) + I_2(g) \rightleftharpoons 2HI(g)$$

When equilibrium is reached, different amounts of reactants and products remain in each experiment, which gives different equilibrium concentrations.

concentrations of H_2 and I_2, we obtain the same numerical value. This is shown in Table 14.1 where we have once again used square brackets around formulas as symbols for molar concentrations.

The fraction used to calculate the values in the last column of Table 14.1,

$$\frac{[HI]^2}{[H_2][I_2]}$$

is called the **mass action expression.**[1] The origin of this term isn't important; just consider it a name we use to refer to this fraction. The numerical value of the mass action expression is called the **reaction quotient,** and is often symbolized by the letter ***Q***. For this reaction, we can write

$$Q = \frac{[HI]^2}{[H_2][I_2]}$$

This equation applies for any set of H_2, I_2, and HI concentrations, whether we are at equilibrium or not.

In Table 14.1, notice that when H_2, I_2, and HI are in dynamic equilibrium at 440 °C, Q is equal to essentially the same constant value of 49.5. In fact, if we repeated the experiments in Figure 14.3 over and over again starting with different amounts of H_2, I_2, and HI, we would still obtain the same reaction quotient, provided the systems had reached equilibrium and the temperature was 440 °C. Therefore, for this reaction at equilibrium we can write

$$\frac{[HI]^2}{[H_2][I_2]} = 49.5 \qquad \text{(at equilibrium at 440 °C)} \qquad (14.2)$$

This relationship is called the **equilibrium law** for the system. Significantly, it tells us that for a mixture of these three gases to be at equilibrium at 440 °C, the value of the mass action expression (the reaction quotient) must equal 49.5. If the reaction quotient has any other value, then the gases are not in equilibrium at this temperature. The constant 49.5, which characterizes this equilibrium system, is called the **equilibrium constant.** The equilibrium constant is usually symbolized by K_c (the subscript c because we write the mass action expression using molar concentrations). Thus, we can state the equilibrium law as follows:

$$\frac{[HI]^2}{[H_2][I_2]} = K_c = 49.5 \qquad \text{(at 440 °C)} \qquad (14.3)$$

TABLE 14.1 Equilibrium Concentrations and the Mass Action Expression

Experiment	Equilibrium Concentrations (mol L^{-1}) $[H_2]$	$[I_2]$	$[HI]$	$\frac{[HI]^2}{[H_2][I_2]}$
I	0.0222	0.0222	0.156	$(0.156)^2/(0.0222)(0.0222) = 49.4$
II	0.0350	0.0450	0.280	$(0.280)^2/(0.0350)(0.0450) = 49.8$
III	0.0150	0.0135	0.100	$(0.100)^2/(0.0150)(0.0135) = 49.4$
IV	0.0442	0.0442	0.311	$(0.311)^2/(0.0442)(0.0442) = 49.5$
				Average = 49.5

[1] The mass action expression is derived using thermodynamics, which we'll discuss in Chapter 18. Technically, each concentration in the mass action expression should be divided by its standard state value before inserting it into the expression. This makes the value of any mass action expression unitless. *For substances in solution,* the standard state is an effective molar concentration of 1 *M*, so we would have to divide each concentration by 1 *M* (1 mol L^{-1}) which makes no difference in the numerical value of the mass action expression as long as all concentrations are molarities. *For gases,* the standard state is 1 bar, so we would have to divide each gas concentration by its concentration at that pressure, or each gas partial pressure by 1 bar. For simplicity, we'll leave the standard state values out of our mass action expressions. Later we'll see that these values are lumped into the numerical value of the expression.

It is often useful to think of an equilibrium law such as Equation 14.3 as a *condition* that must be met for equilibrium to exist.

For chemical equilibrium to exist in a reaction mixture, the reaction quotient Q must be equal to the equilibrium constant, K_c.

As you've probably noticed, we have repeatedly mentioned the temperature when referring to the value of K_c. This is because the value of the equilibrium constant changes when the temperature changes. Thus, if we had performed the experiments in Figure 14.3 at a temperature other than 440 °C, we would have obtained a different value for K_c.

An important fact about the mass action expression and the equilibrium law is that it can *always* be predicted from the balanced chemical equation for the reaction. For example, for the general chemical equation

$$dD + eE \rightleftharpoons fF + gG$$

where D, E, F, and G represent chemical formulas and d, e, f, and g are their coefficients, the mass action expression is

$$\frac{[F]^f[G]^g}{[D]^d[E]^e}$$

The exponents in the mass action expression are the same as the stoichiometric coefficients in the balanced equation.

The condition for equilibrium in this reaction is given by the equation

$$\frac{[F]^f[G]^g}{[D]^d[E]^e} = K_c$$

where the only concentrations that satisfy the equation are *equilibrium concentrations.*

Notice that in writing the mass action expression the molar concentrations of the products are always placed in the numerator and those of the reactants appear in the denominator. Also note that after being raised to appropriate powers the concentration terms are *multiplied,* not added.

In general, it is necessary to specify the temperature when giving a value of K_c, because K_c changes when the temperature changes. For example,

$$CH_4(g) + H_2O(g) \rightleftharpoons CO(g) + 3H_2(g)$$

$K_c = 1.78 \times 10^{-3}$ at 800 °C
$K_c = 4.68 \times 10^{-2}$ at 1000 °C
$K_c = 5.67$ at 1500 °C

Notice that even though we cannot predict the *rate law* from the balanced overall equation, we can predict the *equilibrium law.*

The equilibrium law

EXAMPLE 14.1
Writing the Equilibrium Law

Most of the hydrogen produced in the United States is derived from methane in natural gas, using the forward reaction of the equilibrium

$$CH_4(g) + H_2O(g) \rightleftharpoons CO(g) + 3H_2(g)$$

What is the equilibrium law for this reaction?

ANALYSIS: The equilibrium law sets the mass action expression equal to the equilibrium constant. To form the mass action expression, we place the concentrations of the products in the numerator and the concentrations of the reactants in the denominator. The coefficients in the equation become exponents on the concentrations.

SOLUTION: The equilibrium law is

$$\frac{[CO][H_2]^3}{[CH_4][H_2O]} = K_c$$

IS THE ANSWER REASONABLE? Check to see that the products are on top of the fraction and that the reactants are on the bottom. Also check the exponents and be sure they are the same as the coefficients in the balanced chemical equation. Notice that we omit writing the exponent when it is equal to 1.

Practice Exercise 1: The equilibrium law for a reaction is

$$\frac{[NO_2]^4}{[N_2O_3]^2[O_2]} = K_c$$

Write the chemical equation for the equilibrium. (Hint: In the equilibrium law, remember where the reactant and product concentrations go and what the exponents mean.)

Practice Exercise 2: Write the equilibrium law for each of the following:
(a) $2H_2(g) + O_2(g) \rightleftharpoons 2H_2O(g)$
(b) $CH_4(g) + 2O_2(g) \rightleftharpoons CO_2(g) + 2H_2O(g)$

The rule that we always write the concentrations of the products in the numerator of the mass action expression and the concentrations of the reactants in the denominator is not required by nature. It is simply a convention chemists have agreed on. Certainly, if the mass action expression is equal to a constant, its reciprocal is also equal to a constant (let's call it K_c')

$$\frac{[HI]^2}{[H_2][I_2]} = K_c \qquad \frac{[H_2][I_2]}{[HI]^2} = \frac{1}{K_c} = K_c'$$

The value in having a set rule for constructing the mass action expression is that we don't have to specify the mass action expression when we give the equilibrium constant for a reaction. For example, suppose we're told that at a particular temperature $K_c = 10.0$ for the reaction

$$2NO_2(g) \rightleftharpoons N_2O_4(g)$$

From the chemical equation, we can write the correct mass action expression and the correct equilibrium law.

$$K_c = \frac{[N_2O_4]}{[NO_2]^2} = 10.0$$

The balanced chemical equation contains all the information we need to write the equilibrium law.

Equilibrium laws can be combined, scaled, and reversed

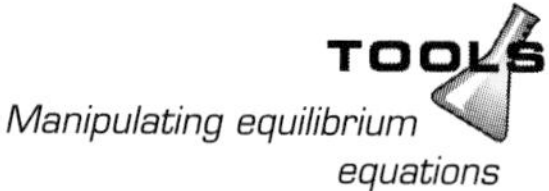

Manipulating equilibrium equations

Sometimes it is useful to be able to combine chemical equilibria to obtain the equation for some other reaction of interest. In doing this, we perform various operations such as reversing an equation, multiplying the coefficients by some factor, and adding the equations to give the desired equation. In our discussion of thermochemistry, you learned how such manipulations affect ΔH values. Some different rules apply to changes in the mass action expressions and equilibrium constants.

Changing the Direction of an Equilibrium

When the direction of an equation is reversed, the new equilibrium constant is the reciprocal of the original. You have just seen this in the discussion above. As another example, when we reverse the equilibrium

$$PCl_3 + Cl_2 \rightleftharpoons PCl_5 \qquad K_c = \frac{[PCl_5]}{[PCl_3][Cl_2]}$$

we obtain

$$PCl_5 \rightleftharpoons PCl_3 + Cl_2 \qquad K_c' = \frac{[PCl_3][Cl_2]}{[PCl_5]}$$

The mass action expression for the second reaction is the reciprocal of that for the first, so K_c' equals $1/K_c$.

Multiplying the Coefficients by a Factor

When the coefficients in an equation are multiplied by a factor, the equilibrium constant is raised to a power equal to that factor. For example, suppose we multiply the coefficients of the equation

$$PCl_3 + Cl_2 \rightleftharpoons PCl_5 \qquad K_c = \frac{[PCl_5]}{[PCl_3][Cl_2]}$$

by 2. This gives

$$2PCl_3 + 2Cl_2 \rightleftharpoons 2PCl_5 \qquad K_c'' = \frac{[PCl_5]^2}{[PCl_3]^2[Cl_2]^2}$$

Comparing mass action expressions, we see that $K_c'' = K_c^2$.

Adding Chemical Equilibria

When chemical equilibria are added, their equilibrium constants are multiplied. For example, suppose we add the following two equations.

We have numbered the equilibrium constants just to distinguish one from the other.

$$2N_2 + O_2 \rightleftharpoons 2N_2O \qquad K_{c1} = \frac{[N_2O]^2}{[N_2]^2[O_2]}$$

$$2N_2O + 3O_2 \rightleftharpoons 4NO_2 \qquad K_{c2} = \frac{[NO_2]^4}{[N_2O]^2[O_2]^3}$$

$$2N_2 + 4O_2 \rightleftharpoons 4NO_2 \qquad K_{c3} = \frac{[NO_2]^4}{[N_2]^2[O_2]^4}$$

If we multiply the mass action expression for K_{c1} by that for K_{c2}, we obtain the mass action expression for K_{c3}.

$$\frac{\cancel{[N_2O]^2}}{[N_2]^2[O_2]} \times \frac{[NO_2]^4}{\cancel{[N_2O]^2}[O_2]^3} = \frac{[NO_2]^4}{[N_2]^2[O_2]^4}$$

Therefore, $K_{c1} \times K_{c2} = K_{c3}$.

Practice Exercise 3: At 25 °C, $K_c = 7.0 \times 10^{25}$ for the reaction

$$2SO_2(g) + O_2(g) \rightleftharpoons 2SO_3(g)$$

What is the value of K_c for the reaction $SO_3(g) \rightleftharpoons SO_2(g) + \frac{1}{2}O_2(g)$? (Hint: What do you have to do with the original equation to obtain the new equation?)

Practice Exercise 4: At 25 °C, the following reactions have the equilibrium constants noted to the right of their equations.

$$2CO(g) + O_2(g) \rightleftharpoons 2CO_2(g) \qquad K_c = 3.3 \times 10^{91}$$

$$2H_2(g) + O_2(g) \rightleftharpoons 2H_2O(g) \qquad K_c = 9.1 \times 10^{80}$$

Use these data to calculate K_c for the reaction

$$H_2O(g) + CO(g) \rightleftharpoons CO_2(g) + H_2(g)$$

14.3 EQUILIBRIUM LAWS FOR GASEOUS REACTIONS CAN BE WRITTEN IN TERMS OF EITHER CONCENTRATIONS OR PRESSURES

When all the reactants and products are gases, we can formulate mass action expressions in terms of partial pressures as well as molar concentrations. This is possible because the molar concentration of a gas is proportional to its partial pressure. This comes from the ideal gas law,

$$PV = nRT$$

Solving for P gives

$$P = \left(\frac{n}{V}\right)RT$$

The quantity n/V has units of mol/L and is simply the molar concentration. Therefore, we can write

If you double the molar concentration of a gas without changing its temperature or volume, you double its pressure.

$$P = (\text{molar concentration}) \times RT \qquad (14.4)$$

This equation applies whether the gas is by itself in a container or part of a mixture. In the case of a gas mixture, P is the partial pressure of the gas.

The relationship expressed in Equation 14.4 lets us write the mass action expression for reactions between gases either in terms of molarities or partial pressures. However, when we make a switch we can't expect the numerical values of the equilibrium constants to be the same, so we use two different symbols for K. When molar concentrations are used, we use the symbol K_c. When partial pressures are used, then $\boldsymbol{K_P}$ is the symbol. For example, the equilibrium law for the reaction of nitrogen with hydrogen to form ammonia

$$N_2(g) + 3H_2(g) \rightleftharpoons 2NH_3(g)$$

can be written in either of the following two ways

$$\frac{[NH_3]^2}{[N_2][H_2]^3} = K_c \qquad \left(\begin{array}{l}\text{because molar concentrations} \\ \text{are used in the mass action} \\ \text{expression}\end{array}\right)$$

$$\frac{P_{NH_3}^2}{P_{N_2}P_{H_2}^3} = K_P \qquad \left(\begin{array}{l}\text{because partial pressures are} \\ \text{used in the mass action} \\ \text{expression}\end{array}\right)$$

The equilibrium molar concentrations can be used to calculate K_c, whereas the equilibrium partial pressures can be used to calculate K_P.

EXAMPLE 14.2
Writing Expressions for K_P

Most of the world's supply of methanol, CH_3OH, is produced by the following reaction.

$$CO(g) + 2H_2(g) \rightleftharpoons CH_3OH(g)$$

What is the expression for K_P for this equilibrium?

ANALYSIS: For K_P we use partial pressures in the mass action expression. We put the equilibrium partial pressures of the products in the numerator and the equilibrium partial pressures of the reactants in the denominator. The coefficients in the equation become exponents on the pressures.

SOLUTION: The expression for K_P for the reaction is

$$K_P = \frac{P_{CH_3OH}}{(P_{CO})(P_{H_2})^2}$$

IS THE ANSWER REASONABLE? Check to see that the mass action expression gives products over reactants, and not the other way around. Check each exponent against the coefficients in the balanced chemical equation.

Practice Exercise 5: Write the equilibrium law in terms of partial pressures for the following reaction. (Hint: Keep in mind how the mass action expression is written.)

$$2N_2(g) + O_2(g) \rightleftharpoons 2N_2O(g)$$

Practice Exercise 6: Using partial pressures, write the equilibrium law for the reaction

$$H_2(g) + I_2(g) \rightleftharpoons 2HI(g)$$

A simple expression relates K_P and K_c

For some reactions K_P is equal to K_c, but for many others the two constants have different values. It is therefore desirable to have a way to calculate one from the other.

Converting between K_P and K_c uses the relationship between partial pressure and molarity. Equation 14.4 can be used to change K_P to K_c by substituting

$$(\text{molar concentration}) \times RT$$

for the partial pressure of each gas in the mass action expression for K_P. Similarly, K_c can be changed to K_P by solving Equation 14.4 for the molar concentrations, and then substituting the result, P/RT, into the appropriate expression for K_c. This sounds like a lot of work, and it is. Fortunately, there is a general equation, which can be derived from these relationships, that we can use to make these conversions simply.

TOOLS
Converting between K_P and K_c

$$K_P = K_c(RT)^{\Delta n_g} \tag{14.5}$$

In this equation, the value of Δn_g is equal to the change in the *number of moles of gas* in going from the reactants to the products.

$$\Delta n_g = (\text{moles of } \textit{gaseous} \text{ products}) - (\text{moles of } \textit{gaseous} \text{ reactants})$$

We use the coefficients of the balanced equation for the reaction to calculate the numerical value of Δn_g. For example, the equation

$$N_2(g) + 3H_2(g) \rightleftharpoons 2NH_3(g) \tag{14.6}$$

tells us that two moles of NH_3 are formed when one mole of N_2 and three moles of H_2 react. In other words, two moles of gaseous product are formed from a total of four moles of gaseous reactants. That's a decrease of two moles of gas, so Δn_g for this reaction equals -2.

For some reactions, the value of Δn_g is equal to zero. An example is the decomposition of HI.

$$2HI(g) \rightleftharpoons H_2(g) + I_2(g)$$

If we take the coefficients to mean moles, there are two moles of gas on each side of the equation, so $\Delta n_g = 0$. Because (RT) raised to the zero power is equal to 1, $K_P = K_c$.

EXAMPLE 14.3
Converting between K_P and K_c

At 500 °C, the reaction between N_2 and H_2 to form ammonia

$$N_2(g) + 3H_2(g) \rightleftharpoons 2NH_3(g)$$

has $K_c = 6.0 \times 10^{-2}$. What is the numerical value of K_P for this reaction?

ANALYSIS: The tool we need to convert between K_P and K_c is Equation 14.5.

$$K_P = K_c(RT)^{\Delta n_g}$$

In the discussion above, we saw that $\Delta n_g = -2$ for this reaction. All we need now are appropriate values of R and T. The temperature, T, must be expressed in kelvins. (When used to stand for temperature, a capital letter T in an equation always means the absolute temperature.) Next we must choose an appropriate value for R. Referring back to Equation 14.4, if the partial pressures are expressed in atm and the concentration in mol L^{-1}, the only value of R that includes all of these units (L, mol, atm, and K) is $R = 0.0821$ L atm mol^{-1} K^{-1}, and this is the *only* value of R that can be used in Equation 14.5.

SOLUTION: Assembling the data, then, we have

$$K_c = 6.0 \times 10^{-2} \qquad \Delta n_g = -2$$
$$T = (500 + 273)\ \text{K} = 773\ \text{K} \qquad R = 0.0821\ \text{L atm mol}^{-1}\ \text{K}^{-1}$$

Substituting these into the equation for K_P gives

$$\begin{aligned} K_P &= (6.0 \times 10^{-2}) \times [(0.0821) \times (773)]^{-2} \\ &= (6.0 \times 10^{-2})/(63.5)^2 \\ &= 1.5 \times 10^{-5} \end{aligned}$$

In this case, K_P has a numerical value quite different from that of K_c.

IS THE ANSWER REASONABLE? In working these problems, check to be sure you have used the correct value of R and that the temperature is expressed in kelvins. Notice that if Δn_g is negative, as it is in this reaction, the expression $(RT)^{\Delta n_g}$ will be less than one. That should make K_P smaller than K_c, which is consistent with our results.

1 mol $NaHCO_3$
38.9 cm^3

$$\text{Molarity} = \frac{1 \text{ mol NaHCO}_3}{0.0389 \text{ L}} = 25.7 \text{ mol/L}$$

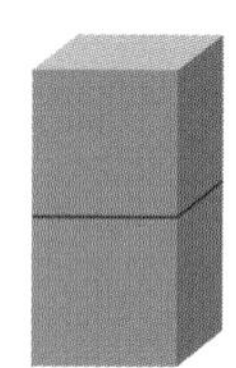

2 mol $NaHCO_3$
77.8 cm^3

$$\text{Molarity} = \frac{2 \text{ mol NaHCO}_3}{0.0778 \text{ L}} = 25.7 \text{ mol/L}$$

FIG. 14.4 The concentration of a substance in the solid state is a constant. Doubling the number of moles also doubles the volume, but the *ratio* of moles to volume remains the same.

Practice Exercise 7: Nitrous oxide, N_2O, is a gas used as an anesthetic; it is sometimes called "laughing gas." This compound has a strong tendency to decompose into nitrogen and oxygen following the equation

$$2N_2O(g) \rightleftharpoons 2N_2(g) + O_2(g)$$

but the reaction is so slow that the gas appears to be stable at room temperature (25 °C). The decomposition reaction has $K_c = 7.3 \times 10^{34}$. What is the value of K_P for this reaction at 25 °C? (Hint: Be careful in calculating Δn_g.)

Practice Exercise 8: Methanol, CH_3OH, is a promising fuel that can be synthesized from carbon monoxide and hydrogen according to the reaction

$$CO(g) + 2H_2(g) \rightleftharpoons CH_3OH(g)$$

For this reaction at 200 °C, $K_P = 3.8 \times 10^{-2}$. Do you expect K_P to be larger or smaller than K_c? Calculate the value of K_c at this temperature.

14.4 HETEROGENEOUS EQUILIBRIA INVOLVE REACTION MIXTURES WITH MORE THAN ONE PHASE

In a **homogeneous reaction**—or a **homogeneous equilibrium**—all of the reactants and products are in the same phase. Equilibria among gases are homogeneous because all gases mix freely with each other, so a single phase exists. There are also many equilibria in which reactants and products are dissolved in the same liquid phase.

When more than one phase exists in a reaction mixture, we call it a **heterogeneous reaction.** A common example is the combustion of wood, in which a solid fuel reacts with gaseous oxygen. Another is the thermal decomposition of sodium bicarbonate (baking soda), which occurs when the compound is sprinkled on a fire.

$$2NaHCO_3(s) \longrightarrow Na_2CO_3(s) + H_2O(g) + CO_2(g)$$

Safety-minded cooks keep a box of baking soda nearby because this reaction makes it an excellent fire extinguisher for fats or oils that have caught fire. The fire is smothered by the products of the reaction.

If $NaHCO_3$ is placed in a sealed container so that no CO_2 or H_2O can escape, the gases and solids come to a heterogeneous equilibrium.

$$2NaHCO_3(s) \rightleftharpoons Na_2CO_3(s) + H_2O(g) + CO_2(g)$$

Following our usual procedure, we can write the equilibrium law for this reaction as

$$\frac{[Na_2CO_3(s)][H_2O(g)][CO_2(g)]}{[NaHCO_3(s)]^2} = K$$

However, the equilibrium law for reactions involving pure liquids and solids can be written in an even simpler form. This is because the concentration of a pure liquid or solid is unchangeable at a given temperature. *For any pure liquid or solid, the ratio of amount of substance to volume of substance is a constant.* For example, if we had a 1 mole crystal of $NaHCO_3$, it would occupy a volume of 38.9 cm^3. Two moles of $NaHCO_3$ would occupy twice this volume, 77.8 cm^3 (Figure 14.4), but the *ratio* of moles to liters (i.e., the molar concentration) remains the same.

Similar reasoning shows that the concentration of Na_2CO_3 in pure solid Na_2CO_3 is a constant, too. This means that the equilibrium law now has three constants, K plus two of the concentration terms. It makes sense to combine all of the numerical constants together.

$$[H_2O(g)][CO_2(g)] = \frac{K[NaHCO_3(s)]^2}{[Na_2CO_3(s)]} = K_c$$

The equilibrium law for a heterogeneous reaction is written without concentration terms for pure solids or liquids.

Equilibrium laws for heterogeneous reactions

Equilibrium constants that are given in tables represent all of the constants combined.[2]

EXAMPLE 14.4
Writing the Equilibrium Law for a Heterogeneous Reaction

The air pollutant sulfur dioxide can be removed from a gas mixture by passing the gases over calcium oxide. The equation is

$$CaO(s) + SO_2(g) \rightleftharpoons CaSO_3(s)$$

Write the equilibrium law for the reaction.

ANALYSIS: For a heterogeneous equilibrium, we do not include solids or pure liquids in the mass action expression. Therefore, we exclude CaO and $CaSO_3$ and include only SO_2.

SOLUTION: The equilibrium law is simply

$$\frac{1}{[SO_2(g)]} = K_c$$

IS THE ANSWER REASONABLE? There's not much to check here except to review the reasoning, which appears to be okay.

Practice Exercise 9: Write the equilibrium law for the following reaction. (Hint: Follow the reasoning above.)

$$NH_3(g) + HCl(g) \rightleftharpoons NH_4Cl(s)$$

Practice Exercise 10: Write the equilibrium law for the following heterogeneous reactions.
(a) $2Hg(l) + Cl_2(g) \rightleftharpoons Hg_2Cl_2(s)$
(b) $Na(s) + H_2O(l) \rightleftharpoons NaOH(aq) + H_2(g)$
(c) Dissolving solid Ag_2CrO_4 in water: $Ag_2CrO_4(s) \rightleftharpoons 2Ag^+(aq) + CrO_4^{2-}(aq)$
(d) $CaCO_3(s) + H_2O(l) + CO_2(aq) \rightleftharpoons Ca^{2+}(aq) + 2HCO_3^-(aq)$

14.5 When K Is Large, the Position of Equilibrium Lies Toward the Products

Whether we work with K_P or K_c, a bonus of always writing the mass action expression with the product concentrations in the numerator is that the size of the equilibrium constant gives us a measure of the *position of equilibrium* (how far the reaction proceeds toward completion when equilibrium is reached). For example, the reaction

$$2H_2(g) + O_2(g) \rightleftharpoons 2H_2O(g)$$

has $K_c = 9.1 \times 10^{80}$ at 25 °C. For an equilibrium between these gases,

$$K_c = \frac{[H_2O]^2}{[H_2]^2[O_2]} = \frac{9.1 \times 10^{80}}{1}$$

[2] Thermodynamics handles heterogeneous equilibria in a more elegant way by expressing the mass action expression in terms of "effective concentrations," or **activities**. In doing this, thermodynamics *defines* the *activity* of any pure liquid or solid as equal to 1, which means terms involving such substances drop out of the mass action expression.

By writing K_c as a fraction, $(9.1 \times 10^{80})/1$, we see that the numerator of the mass action expression must be enormous compared with the denominator, which means that the concentration of H_2O is huge compared with the concentrations of H_2 and O_2. At equilibrium, therefore, almost all of the hydrogen and oxygen atoms in the system are found in H_2O molecules and very few are present in H_2 and O_2. Thus, the enormous value of K_c tells us that the position of equilibrium in this reaction lies far toward the right and that the reaction of H_2 and O_2 goes essentially to completion.

Actually, you would need about 200,000 L of water vapor at 25 °C just to find one molecule of O_2 and two molecules of H_2.

The reaction between N_2 and O_2 to give NO

$$N_2(g) + O_2(g) \rightleftharpoons 2NO(g)$$

has a very small equilibrium constant; $K_c = 4.8 \times 10^{-31}$ at 25 °C. The equilibrium law for the reaction is

$$\frac{[NO]^2}{[N_2][O_2]} = 4.8 \times 10^{-31} = \frac{4.8}{10^{31}}$$

Here the denominator is huge compared with the numerator, so the concentrations of N_2 and O_2 must be very much larger than the concentration of NO. This means that in a mixture of N_2 and O_2 at this temperature, the amount of NO that is formed is negligible. The reaction hardly proceeds at all toward completion before equilibrium is reached, and the position of equilibrium lies far toward the left.

In air at 25 °C, the equilibrium concentration of NO *should be* about 10^{-17} mol/L. It is usually higher because NO is formed in various reactions, such as those responsible for air pollution caused by automobiles.

The relationship between the equilibrium constant and the position of equilibrium can thus be summarized as follow (also see Figure 14.5):

TOOLS *Significance of the magnitude of K*

When *K* is very large	The reaction proceeds far toward completion. The position of equilibrium lies far to the right, toward the products.
When $K \approx 1$	The concentrations of reactants and products are nearly the same at equilibrium. The position of equilibrium lies approximately midway between reactants and products.
When *K* is very small	Extremely small amounts of products are formed. The position of equilibrium lies far to the left, toward the reactants.

Notice that we have omitted the subscript for *K* in this summary. The same qualitative predictions about the extent of reaction apply whether we use K_P or K_c.

One of the ways that we can use equilibrium constants is to compare the extents to which two or more reactions proceed to completion. Take care in making such

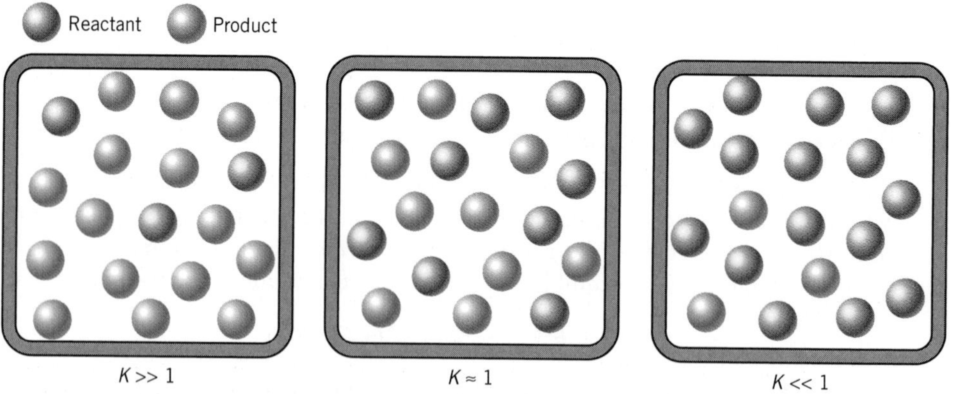

FIG. 14.5 The magnitude of *K* and the position of equilibrium. For the reaction

$$\text{Reactant} \rightleftharpoons \text{Product}$$

a large amount of product and very little reactant are in the reaction mixture at equilibrium when *K* is very large ($K \gg 1$), so we say *the position of equilibrium lies to the right*. When $K \approx 1$, approximately equal amounts of reactant and product are present at equilibrium. When $K \ll 1$, the reaction mixture contains a large amount of reactant at equilibrium and very little product, so we say *the position of equilibrium lies to the left*.

comparisons, however, because unless the values of K are greatly different, the comparison is valid only if both reactions have the same number of reactant and product molecules appearing in their balanced chemical equations.

Practice Exercise 11: Suppose a mixture contained equal concentrations of H_2, Br_2, and HBr. Given that the reaction $H_2(g) + Br_2(g) \rightleftharpoons 2HBr(g)$ has $K_c = 1.4 \times 10^{-21}$, will the reaction proceed to the left or right in order to reach equilibrium? (Hint: In the mass action expression for this reaction, how should the size of the numerator compare with that of the denominator when the system reaches equilibrium?)

Practice Exercise 12: Which of the following reactions will tend to proceed farthest toward completion?

(a) $H_2(g) + Br_2(g) \rightleftharpoons 2HBr(g)$ $\quad K_c = 1.4 \times 10^{-21}$
(b) $2NO(g) \rightleftharpoons N_2(g) + O_2(g)$ $\quad K_c = 2.1 \times 10^{30}$
(c) $2BrCl \rightleftharpoons Br_2 + Cl_2$ (in CCl_4 solution) $\quad K_c = 0.145$

14.6 Le Châtelier's Principle Tells Us How a Chemical Equilibrium Responds When Disturbed

In Section 14.7 you will see that it is possible to perform calculations that tell us what the composition of an equilibrium system is. However, many times we really don't need to know exactly what the equilibrium concentrations are. Instead, we may want to know what actions we should take to control the relative amounts of the reactants or products at equilibrium. For instance, if we were designing gasoline engines, we would like to know what could be done to minimize the formation of nitrogen oxide pollutants. Or, if we were preparing ammonia, NH_3, by the reaction of N_2 with H_2, we might want to know how to maximize the yield of NH_3.

Le Châtelier's principle, introduced in Chapter 11, provides us with the means for making qualitative predictions about changes in chemical equilibria. It does this in much the same way that it allows us to predict the effects of outside influences on equilibria that involve physical changes, such as liquid–vapor equilibria. Recall that **Le Châtelier's principle** states that *if an outside influence upsets an equilibrium, the system undergoes a change in a direction that counteracts the disturbing influence and, if possible, returns the system to equilibrium.* Let's examine the kinds of "outside influences" that affect chemical equilibria.

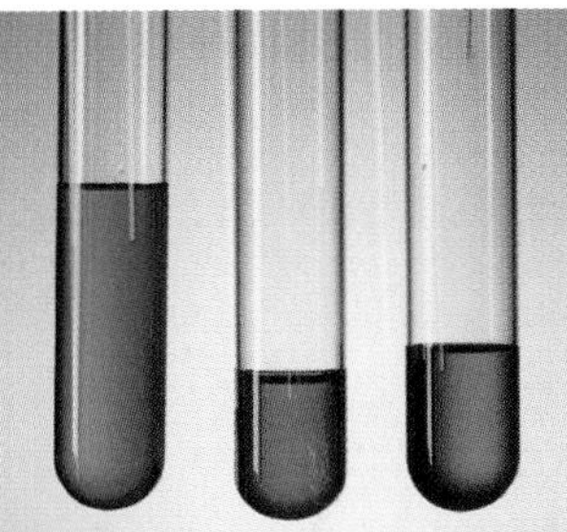

FIG. 14.6 The effect of concentration changes on the position of equilibrium. The solution in the center contains a mixture of blue $Cu(H_2O)_4^{2+}$ and yellow $CuCl_4^{2-}$, so it has a blue-green color. At the right is some of the same solution after the addition of concentrated HCl. It has a more pronounced green color because the equilibrium is shifted toward $CuCl_4^{2-}$. At the left is some of the original solution after the addition of water. It is blue because the equilibrium has shifted toward $Cu(H_2O)_4^{2+}$. *(Michael Watson.)*

Adding a reactant or product changes Q

For a homogeneous system at equilibrium, adding or removing a reactant or product changes the value of Q so that it no longer equals K. The equilibrium is upset, and according to Le Châtelier's principle the system should change in a direction that opposes the disturbance we've introduced. If we've added a substance, the reaction should proceed in a direction to remove some of it; if we've removed a substance, the reaction should proceed in a direction to replace it.

As an example, let's study the equilibrium between two ions of copper.

$$\underset{\text{blue}}{Cu(H_2O)_4^{2+}(aq)} + 4Cl^-(aq) \rightleftharpoons \underset{\text{yellow}}{CuCl_4^{2-}(aq)} + 4H_2O(l)$$

$Cu(H_2O)_4^{2+}$ and $CuCl_4^{2-}$ are called *complex ions.* Some of the interesting properties of complex ions of metals are discussed in Chapter 21.

As noted, $Cu(H_2O)_4^{2+}$ is blue and $CuCl_4^{2-}$ is yellow. Mixtures of the two have an intermediate color and therefore appear blue-green, as illustrated in Figure 14.6, center.

Suppose we add chloride ion to an equilibrium mixture of these copper ions. The system can remove some Cl^- by reacting it with $Cu(H_2O)_4^{2+}$. This gives more $CuCl_4^{2-}$ (Figure 14.6, right), and we say that the equilibrium has "shifted to the right" or "shifted toward the products." In the new position of equilibrium, there is less $Cu(H_2O)_4^{2+}$ and more $CuCl_4^{2-}$ and uncombined H_2O. There is also more Cl^-, because not all that we add reacts. In other words, in this new position of equilibrium, *all* the concentrations have changed in a way that causes Q to become equal to K_c. Similarly, the position of equilibrium is shifted to the left when we add water to the mixture (Figure 14.6, left). The system is able to get rid of some of the H_2O by reaction with $CuCl_4^{2-}$, so more of the blue $Cu(H_2O)_4^{2+}$ is formed.

If we were able to remove a reactant or product, the position of equilibrium would also be changed. For example, if we add Ag^+ to a solution that contains both copper ions in equilibrium, we see an enhancement of the blue color.

$$\underset{\text{blue}}{Cu(H_2O)_4^{2+}(aq)} + 4Cl^-(aq) \rightleftharpoons \underset{\text{yellow}}{CuCl_4^{2-}(aq)} + 4H_2O$$

$$\Big\downarrow +Ag^+$$

$$Ag^+(aq) + Cl^-(aq) \rightleftharpoons AgCl(s)$$

As the Ag^+ reacts with Cl^- to form insoluble AgCl in the second reaction, the equilibrium in the first reaction shifts to the left to replace some of the Cl^-, which also produces more blue $Cu(H_2O)_4^{2+}$. The position of equilibrium shifts to replace the substance that is removed.

TOOLS *Le Châtelier's principle; adding or removing a reactant or product*

> The position of equilibrium shifts in a way to remove reactants or products that have been added, or to replace reactants or products that have been removed.

Changing the volume in gaseous reactions will sometimes upset the equilibrium

Changing the volume of a mixture of reacting gases changes molar concentrations and partial pressures, so we expect volume to have some effect on the position of equilibrium. Let's consider the equilibrium:

$$3H_2(g) + N_2(g) \rightleftharpoons 2NH_3(g)$$

If we reduce the volume of the reaction mixture, we expect the pressure to increase. The system can oppose the pressure change if it is able to reduce the number of molecules of gas, because fewer molecules of gas exert a lower pressure. If the reaction proceeds to the right, two NH_3 molecules appear when four molecules (one N_2 and three H_2) disappear. Therefore, this equilibrium is shifted to the right when the volume of the reaction mixture is reduced.

Now let's look at the equilibrium:

$$H_2(g) + I_2(g) \rightleftharpoons 2HI(g)$$

If this reaction proceeds in either direction, there is no change in the number of molecules of gas. This reaction, then, cannot respond to pressure changes, so changing the volume of the reaction vessel has virtually no effect on the equilibrium.

The simplest way to analyze the effects of a volume change on an equilibrium system involving gases is to count the number of molecules of gaseous substances on both sides of the equation.

TOOLS *Le Châtelier's principle; changing the volume of a gaseous reaction*

> Reducing the volume of a gaseous reaction mixture causes the position of equilibrium to shift in a direction that decreases the number of molecules of gas. No change in the equilibrium will occur if a volume change cannot affect the number of molecules of gas.

As a final note here, *moderate pressure changes have negligible effects on reactions involving only liquids or solids.* Substances in these states are virtually incompressible, and reactions involving them have no way to counteract pressure changes.

Changing the temperature involves adding or removing heat

When the system is heated, the reaction will shift in the direction that absorbs heat. If the reaction is endothermic, it will run in the forward direction to remove the added energy. For example, the melting of ice absorbs heat and is endothermic:

$$H_2O(s) \rightleftharpoons H_2O(l) \qquad \Delta H^\circ = +6 \text{ kJ (at 0 °C)}$$

If you cup your hands around a glass of ice and water at 0 °C, the heat you've added to the system shifts the equilibrium to the right, and some of the ice melts. If we include energy as a reactant in the equation, we could write

$$\text{heat} + H_2O(s) \rightleftharpoons H_2O(l)$$

Adding a reactant drives the reaction in the forward direction, with energy behaving like any other reactant in this case.

Removing some amount of reactant should shift the equilibrium to the left. We expect that removing energy from this system would have the same effect. Putting the glass in the freezer causes some of the water to freeze. Some of the energy is removed, so the equilibrium shifts toward the reactant side of the equation.

This same kind of analysis was used in Chapter 12 to predict how solubility changes with temperature.

The same arguments can be used for equations that describe chemical changes. For instance, the reaction to produce NH_3 from N_2 and H_2 is exothermic; heat is released when NH_3 molecules are formed ($\Delta H_f^\circ = -46.19$ kJ/mol from Table 6.2). Since the reaction releases energy, we can include energy as a product in the equilibrium equation:

$$3H_2(g) + N_2(g) \rightleftharpoons 2NH_3(g) + \text{heat}$$

It is easy to predict what will happen when the reaction mixture is heated or cooled. Heating the mixture adds energy, so the system will counter by shifting to the left. The added energy will be absorbed, and at the same time NH_3 will decompose to produce more H_2 and N_2. Cooling the mixture removes energy, so the equilibrium shifts to the right. H_2 and N_2 will react to produce more NH_3, and energy will be released to compensate for energy removed by cooling.

Increasing the temperature shifts a reaction in a direction that produces an endothermic (heat-absorbing) change.

Decreasing the temperature shifts a reaction in a direction that produces an exothermic (heat-releasing) change.

TOOLS *Le Châtelier's principle; raising or lowering the temperature*

This gives us a way to experimentally determine whether a reaction is exothermic or endothermic, without using a thermometer. Earlier we described the equilibrium involving complex ions of copper. The effect of temperature on this equilibrium is demonstrated in Figure 14.7. When the reaction mixture is heated, we see from the color change that the equilibrium shifts toward the products. Therefore, the reaction must be endothermic.

When heat is added to an equilibrium mixture, it is added to all of the substances present (reactants *and* products). As the system returns to equilibrium, the net reaction that occurs is the one that is endothermic.

Changing the temperature changes *K* for a reaction

Changes in concentrations or volume can shift the position of equilibrium, but *they do not change the equilibrium constant.* However, changing the temperature causes the position of equilibrium to shift because it changes the value of *K*. The enthalpy change of the reaction is the critical factor. Let's look once again at the equilibrium law for the industrial synthesis of ammonia.

Temperature is the only factor that can change *K* for a given reaction.

$$3H_2(g) + N_2(g) \rightleftharpoons 2NH_3(g) \qquad \Delta H^\circ = -46.19 \text{ kJ}$$

$$\frac{[NH_3]^2}{[H_2]^3[N_2]} = K_c$$

FIG. 14.7 **The effect of temperature on the equilibrium $Cu(H_2O)_4^{2+} + 4Cl^- \rightleftharpoons CuCl_4^{2-} + 4H_2O$.** In the center is an equilibrium mixture of the two complexes. When the solution is cooled in ice (left), the equilibrium shifts toward the blue $Cu(H_2O)_4^{2+}$. When heated in boiling water (right), the equilibrium shifts toward $CuCl_4^{2-}$. This behavior indicates that the reaction is endothermic in the forward direction. *(Michael Watson.)*

We'll see how to quantitatively relate K and temperature in Chapter 18.

Because the reaction is exothermic, increasing the temperature shifts the equilibrium to the left. The concentration of NH_3 decreases while the concentrations of N_2 and H_2 increase. Therefore, the numerator of the mass action expression becomes *smaller* and the denominator becomes *larger*. This gives a smaller reaction quotient and therefore a smaller value of K_c.

TOOLS

Le Châtelier's principle; changing temperature changes K

Increasing the temperature of an exothermic reaction makes its equilibrium constant smaller.

Increasing the temperature of an endothermic reaction makes its equilibrium constant larger.

Catalysts don't appear in the chemical equation for a reaction, so they don't appear in the mass action expression, either.

Catalysts have no effect on the position of equilibrium

Recall that catalysts are substances that affect the speeds of chemical reactions without actually being used up. Catalysts do not, however, affect the position of equilibrium in a system. The reason is that a catalyst affects both the forward and reverse reactions equally. Both are speeded up to the same degree, so adding a catalyst to a system has no net effect on the system's equilibrium composition. The catalyst's only effect is to bring the system to equilibrium faster.

Adding an inert gas at constant volume has no effect on the position of equilibrium

A change in volume is not the only way to change the pressure in an equilibrium system of gaseous reactants and products. The pressure can also be changed by keeping the volume the same and adding another gas. If this gas cannot react with any of the gases already present (i.e., if the added gas is *inert* toward the substances in equilibrium), the concentrations of the reactants and products won't change. The concentrations will continue to satisfy the equilibrium law and the reaction quotient will continue to equal K_c, so there will be no change in the position of equilibrium.

EXAMPLE 14.5

Application of Le Châtelier's Principle

The reaction $N_2O_4(g) \rightleftharpoons 2NO_2(g)$ is endothermic, with $\Delta H^\circ = +56.9$ kJ. How will the amount of NO_2 at equilibrium be affected by (a) adding N_2O_4, (b) lowering the pressure by increasing the volume of the container, (c) raising the temperature, and (d) adding a catalyst to the system? Which of these changes will alter the value of K_c?

ANALYSIS: We are applying various types of disturbances to an equilibrium mixture of NO_2 and N_2O_4. The tool we use to judge the effects of such disturbances is Le Châtelier's principle. As we discussed above, we expect the equilibrium to shift to the left or right to counteract the disturbance in most cases.

SOLUTION: (a) Adding N_2O_4 will cause the equilibrium to shift to the right—in a direction that will consume some of the added N_2O_4. The amount of NO_2 will increase.

(b) When the pressure in the system drops, the system responds by producing more molecules of gas, which will tend to raise the pressure and partially offset the change. Since more gas molecules are formed if some N_2O_4 decomposes, the amount of NO_2 at equilibrium will increase.

(c) Because the reaction is endothermic, we write the equation showing heat as a reactant:

$$\text{heat} + N_2O_4(g) \rightleftharpoons 2NO_2(g)$$

Raising the temperature is accomplished by adding heat, so the system will respond by absorbing heat. This means that the equilibrium will shift to the right and the amount of NO_2 at equilibrium will increase.

(d) A catalyst causes a reaction to reach equilibrium more quickly, but it has no effect on the position of chemical equilibrium. Therefore, the amount of NO_2 at equilibrium will not be affected.

Finally, the *only* change that alters K is the temperature change. Raising the temperature (adding heat) will increase K_c for this endothermic reaction.

ARE THE ANSWERS REASONABLE? There's not much we can do here other than review our reasoning. Doing so will reveal we've answered the questions correctly.

Practice Exercise 13: Consider the equilibrium $PCl_5(g) + 4H_2O(g) \rightleftharpoons H_3PO_4(l) + 5HCl(g)$. How will the amount of H_3PO_4 at equilibrium be affected by decreasing the volume of the container holding the reaction mixture? (Hint: Notice that this is a heterogeneous equilibrium.)

Practice Exercise 14: Consider the equilibrium $PCl_3(g) + Cl_2(g) \rightleftharpoons PCl_5(g)$, for which $\Delta H^\circ = -88$ kJ. How will the amount of Cl_2 at equilibrium be affected by (a) adding PCl_3, (b) adding PCl_5, (c) raising the temperature, and (d) decreasing the volume of the container? How (if at all) will each of these changes affect K_P for the reaction?

14.7 EQUILIBRIUM CONCENTRATIONS CAN BE USED TO DETERMINE EQUILIBRIUM CONSTANTS, AND VICE VERSA

You have seen that the magnitude of an equilibrium constant gives us some feel for the extent to which the reaction proceeds at equilibrium. Sometimes, however, it is necessary to have more than merely a qualitative knowledge of equilibrium concentrations. This requires that we be able to use the equilibrium law for purposes of calculation.

Equilibrium calculations for gaseous reactions can be performed using either K_P or K_c, but for reactions in solution we must use K_c. Whether we deal with concentrations or partial pressures, however, the same basic principles apply.

Overall, we can divide equilibrium calculations into two main categories:

1. Calculating equilibrium constants from known equilibrium concentrations or partial pressures.
2. Calculating one or more equilibrium concentrations or partial pressures using the known value of K_c or K_P.

Equilibrium concentrations can be used to calculate K_c

One way to determine the value of K_c is to carry out the reaction, measure the concentrations of reactants and products after equilibrium has been reached, and then use the equilibrium values in the equilibrium law to compute K_c. As an example, let's look again at the decomposition of N_2O_4.

$$N_2O_4(g) \rightleftharpoons 2NO_2(g)$$

In Section 14.1, we saw that if 0.0350 mol of N_2O_4 is placed into a 1 liter flask at 25 °C, the concentrations of N_2O_4 and NO_2 at equilibrium are

$$[N_2O_4] = 0.0292 \text{ mol/L}$$

$$[NO_2] = 0.0116 \text{ mol/L}$$

To calculate K_c for the reaction, we substitute the equilibrium concentrations into the mass action expression of the equilibrium law.

$$\frac{[NO_2]^2}{[N_2O_4]} = K_c$$

$$\frac{(0.0116)^2}{(0.0292)} = K_c$$

Performing the arithmetic gives

$$K_c = 4.61 \times 10^{-3}$$

Concentration tables are an aid in analyzing and solving equilibrium problems

Only rarely are we given all the equilibrium concentrations required to calculate the equilibrium constant. Usually, we have information about the composition of a reaction mixture when it is first prepared, as well as some additional data that we can use to figure out what the equilibrium concentrations are. To help organize our thinking, we will set up a **concentration table** beneath the chemical equation in which we keep track of the concentrations of the substances involved in the reaction as it proceeds toward equilibrium. The table has three rows, which we label *Initial Concentrations, Changes in Concentrations*, and *Equilibrium Concentrations*. To be useful, all entries in the table *must* have units of molar concentration (mol L^{-1}). Let's look at how the data for each row is obtained.

TOOLS Concentration table

Remember, all entries in the concentration table must be molarities. If you use moles instead, you may end up with incorrect answers.

Initial Concentrations

The initial concentrations are those present in the reaction mixture when it's prepared; *we imagine that no reaction occurs until everything is mixed.* Often, the statement of a problem will give the initial molar concentrations, and these are then entered into the table under the appropriate formulas. In other cases, the number of moles of reactants or products dissolved in a specified volume are given and we must then calculate the molar concentrations. If the amount or concentration of a reactant or product is not mentioned in the problem, we will assume none was added to the reaction mixture initially, so its initial concentration is entered as zero.

Changes in Concentrations

If the reaction mixture is not at equilibrium when it's prepared, the chemical reaction must occur (either to the left or right) to bring the system to equilibrium. When this happens, the concentrations change. We will use a positive sign to indicate that a concentration increases and a negative sign to show a decrease in concentration.

The changes in concentrations *always* occur in the same ratio as the coefficients in the balanced equation. For example, if we were dealing with the equilibrium

$$3H_2(g) + N_2(g) \rightleftharpoons 2NH_3(g)$$

and found that the $N_2(g)$ concentration decreases by 0.10 M during the approach to equilibrium, the entries in the "change" row would be as follows:

	$3H_2(g)$	+ $N_2(g)$	$\rightleftharpoons$ $2NH_3(g)$
Changes in Concentrations:	$-3 \times (0.10\ M)$	$-1 \times (0.10\ M)$	$+2 \times (0.10\ M)$
	↓	↓	↓
	$-0.30\ M$	$-0.10\ M$	$+0.20\ M$

In constructing the "change" row, be sure the reactant concentrations all change in the same direction, and that the product concentrations all change in the opposite direction. If the concentrations of the reactants decrease, all the entries for the reactants in the "change" row should have a minus sign, and all the entries for the products should be positive. In this example, N_2 reacts with H_2 (so their concentrations both decrease) and form NH_3 (so its concentration increases).

Equilibrium Concentrations

These are the concentrations of the reactants and products when the system finally reaches equilibrium. We obtain them by adding the changes in concentrations to the initial concentrations.

$$\begin{pmatrix}\text{Initial} \\ \text{concentration}\end{pmatrix} + \begin{pmatrix}\text{change in} \\ \text{concentration}\end{pmatrix} = \begin{pmatrix}\text{equilibrium} \\ \text{concentration}\end{pmatrix}$$

The equilibrium concentrations are the only quantities that satisfy the equilibrium law.

The concentration table is one of our most useful tools for analyzing and setting up equilibrium problems. As you will see, we will use it in almost every problem. The following example illustrates how we can use it in calculating an equilibrium constant.

EXAMPLE 14.6
Calculating K_c from Equilibrium Concentrations

At a certain temperature, a mixture of H_2 and I_2 was prepared by placing 0.200 mol of H_2 and 0.200 mol of I_2 into a 2.00 liter flask. After a period of time the equilibrium

$$H_2(g) + I_2(g) \rightleftharpoons 2HI(g)$$

was established. The purple color of the I_2 vapor was used to monitor the reaction, and from the decreased intensity of the purple color it was determined that, at equilibrium, the I_2 concentration had dropped to 0.020 mol L^{-1}. What is the value of K_c for this reaction at this temperature?

ANALYSIS: The first step in any equilibrium problem is to write the balanced chemical equation and the related equilibrium law. The equation is already given, and the equilibrium law corresponding to it is

$$\frac{[HI]^2}{[H_2][I_2]} = K_c$$

To calculate the value of K_c we must substitute the *equilibrium concentrations* of H_2, I_2, and HI into the mass action expression. But what are they? We have been given only one directly, the value of $[I_2]$. To obtain the others, we have to do some reasoning. Our tool for this will be a concentration table, constructed following the procedure described above. Once we've filled in the last row of the table, we can substitute the values for the equilibrium concentrations into the equilibrium law and calculate K_c.

SOLUTION: To construct the concentration table, we begin by entering the data given in the statement of the problem. For the initial concentrations, we have to calculate the ratio of moles to liters for both the H_2 and I_2, (0.200 mol/2.00 L) = 0.100 *M*. Because no HI was placed into the reaction mixture, its initial concentration is set to zero. These quantities are shown in colored type in the first row. The problem statement also gives us the equilibrium concentration of I_2, so we enter this in the table in the last row (also shown in colored type). The values in regular type are then derived as described below.

	$H_2(g)$	+	$I_2(g)$	$\rightleftharpoons$	$2HI(g)$
Initial concentrations (*M*)	0.100		0.100		0.000
Changes in concentrations (*M*)	−0.080		−0.080		+2(0.080)
Equilibrium concentrations (*M*)	0.020		0.020		0.160

Changes in Concentrations. We have been given both the initial and equilibrium concentrations of I_2, so by difference we can calculate the change for I_2 (−0.080 *M*). The other changes are then calculated from the mole ratios specified in the chemical equation. Because H_2 and I_2 have the same coefficients (i.e., 1), their changes are equal. The coefficient of HI is 2, so its change must be twice that of I_2. Reactant concentrations are decreasing, so their changes are negative; the product concentration increases, so its change is positive.

The changes in concentrations are controlled by the stoichiometry of the reaction. If we can find one of the changes, we can calculate the others from it by using the coefficients of the balanced equation.

Equilibrium Concentrations. For H_2 and HI, we just add the change to the initial value.

Now we can substitute the equilibrium concentrations into the mass action expression and calculate K_c.

$$K_c = \frac{(0.160)^2}{(0.020)(0.020)} = 64$$

IS THE ANSWER REASONABLE? First, carefully examine the equilibrium law. As always, products must be placed in the numerator and reactants in the denominator. Be sure that each exponent corresponds to the correct coefficient in the balanced chemical equation.

Next, check the concentration table to be sure all entries are molar concentrations. Note that the initial concentration of HI is zero. The change for HI must be positive, because the HI concentration cannot become smaller than zero. The positive value in the table agrees with this assessment. Notice, also, that the changes for both reactants have the same sign, which is opposite that of the product. *This relationship among the signs of the changes is always true and serves as a useful check when you construct a concentration table.*

Finally, we can check to be sure we've entered each of the equilibrium concentrations into the mass action expression correctly. At this point we can be confident we've set up the problem correctly. To be sure of the answer, redo the calculation on your calculator.

Practice Exercise 15: In a particular experiment, it was found that when $O_2(g)$ and $CO(g)$ were mixed and reacted according to the equation

$$2CO(g) + O_2(g) \rightleftharpoons 2CO_2(g)$$

the O_2 concentration had decreased by 0.030 mol L^{-1} when the reaction reached equilibrium. How had the concentrations of CO and CO_2 changed? (Hint: How are the changes in concentrations related to one another as the reaction approaches equilibrium?)

Practice Exercise 16: An equilibrium was established for the reaction

$$CO(g) + H_2O(g) \rightleftharpoons CO_2(g) + H_2(g)$$

at 500 °C. (This is an industrially important reaction for the preparation of hydrogen.) At equilibrium, the following concentrations were found in the reaction vessel: [CO] = 0.180 *M*, $[H_2O]$ = 0.0411 *M*, $[CO_2]$ = 0.150 *M*, and $[H_2]$ = 0.200 *M*. What is the value of K_c for this reaction?

Practice Exercise 17: A student placed 0.200 mol of $PCl_3(g)$ and 0.100 mol of $Cl_2(g)$ into a 1.00 liter container at 250 °C. After the reaction

$$PCl_3(g) + Cl_2(g) \rightleftharpoons PCl_5(g)$$

came to equilibrium it was found that the flask contained 0.120 mol of PCl_3.

(a) What were the initial concentrations of the reactants and product?
(b) By how much had the concentrations changed when the reaction reached equilibrium?
(c) What were the equilibrium concentrations?
(d) What is the value of K_c for the reaction at that temperature?

Equilibrium concentrations can be calculated using K_c

In the simplest calculation of this type, all but one of the equilibrium concentrations are known, as illustrated in Example 14.7.

EXAMPLE 14.7
Using K_c to Calculate Concentrations at Equilibrium

The reversible reaction

$$CH_4(g) + H_2O(g) \rightleftharpoons CO(g) + 3H_2(g)$$

has been used as a commercial source of hydrogen. At 1500 °C, an equilibrium mixture of the gases was found to have the following concentrations: [CO] = 0.300 *M*, $[H_2]$ = 0.800 *M*, and $[CH_4]$ = 0.400 *M*. At 1500 °C, K_c = 5.67 for the reaction. What was the equilibrium concentration of $H_2O(g)$ in the mixture?

ANALYSIS AND SOLUTION: This is really a very simple problem. The first step, once we have the chemical equation for the equilibrium, is to write the equilibrium law for the reaction.

$$K_c = \frac{[CO][H_2]^3}{[CH_4][H_2O]}$$

The equilibrium constant and all of the equilibrium concentrations except that for H_2O are known, so we substitute these values into the equilibrium law and solve for the unknown quantity.

$$5.67 = \frac{(0.300)(0.800)^3}{(0.400)[H_2O]}$$

Solving for $[H_2O]$ gives

$$[H_2O] = \frac{(0.300)(0.800)^3}{(0.400)(5.67)}$$

$$= \frac{0.154}{2.27}$$

$$= 0.0678\ M$$

IS THE ANSWER REASONABLE? We can always check problems of this kind by substituting all of the equilibrium concentrations into the mass action expression to see if we get the original equilibrium constant back. We have

$$K_c = \frac{[CO][H_2]^3}{[CH_4][H_2O]} = \frac{(0.300)(0.800)^3}{(0.400)(0.0678)} = 5.66$$

Rounding off the H_2O concentration caused a slight error in K_c, but it is consistent with the equilibrium constant given in the problem.

Practice Exercise 18: The decomposition of N_2O_4 at 25 °C,

$$N_2O_4(g) \rightleftharpoons 2NO_2(g)$$

has $K_c = 4.61 \times 10^{-3}$. A 2.00 L vessel contained 0.0466 mol N_2O_4 at equilibrium. What was the concentration of NO_2 in the vessel? (Hint: What was the concentration of N_2O_4 in the container at equilibrium?)

Practice Exercise 19: Ethyl acetate, $CH_3CO_2C_2H_5$, is an important solvent used in lacquers, adhesives, the manufacture of plastics, and even as a food flavoring. It is produced from acetic acid and ethanol by the reaction

$$\underset{\text{acetic acid}}{CH_3CO_2H(l)} + \underset{\text{ethanol}}{C_2H_5OH(l)} \rightleftharpoons CH_3CO_2C_2H_5(l) + H_2O(l)$$

At 25 °C, $K_c = 4.10$ for this reaction. In a reaction mixture, the following equilibrium concentrations were observed: $[CH_3CO_2H] = 0.210\ M$, $[H_2O] = 0.00850\ M$, and $[CH_3CO_2C_2H_5] = 0.910\ M$. What was the concentration of C_2H_5OH in the mixture?

Acetic acid has the structure

```
   H  O
   |  ‖
H—C—C—O—H
   |
   H
```

where the hydrogen released by ionization is indicated in red. The formula of the acid is written either as CH_3CO_2H (which emphasizes its molecular structure) or as $HC_2H_3O_2$ (which emphasizes that the acid is monoprotic).

Equilibrium concentrations can be calculated using K_c and initial concentrations

A more complex type of calculation involves the use of initial concentrations and K_c to compute equilibrium concentrations. Although some of these problems can be so complicated that a computer is needed to solve them, we can learn the general principles involved by working on simple calculations. Even these, however, require a little applied algebra. This is where the concentration table can be especially helpful.

EXAMPLE 14.8
Using K_c to Calculate Equilibrium Concentrations

The reaction

$$CO(g) + H_2O(g) \rightleftharpoons CO_2(g) + H_2(g)$$

has $K_c = 4.06$ at 500 °C. If 0.100 mol of CO and 0.100 mol of $H_2O(g)$ are placed in a 1.00 liter reaction vessel at this temperature, what are the concentrations of the reactants and products when the system reaches equilibrium?

ANALYSIS: The key to solving this kind of problem is recognizing that at equilibrium the mass action expression must equal K_c.

$$\frac{[CO_2][H_2]}{[CO][H_2O]} = 4.06 \qquad \text{(at equilibrium)}$$

We must find values for the concentrations that satisfy this condition. Our tool in setting up the calculation will be the concentration table.

The problem gives us information about the initial concentrations, which we'll use to establish values for the first line of the table. If we knew how much the concentrations change, we could calculate the equilibrium concentrations. There are no data that lets us calculate directly what any of the changes are, so we will represent them algebraically as unknowns. Combining the changes with the initial concentrations will give us algebraic expressions for the equilibrium concentrations. The relationship among the equilibrium concentrations is established by substituting them into the mass action expression.

The quantities representing the equilibrium concentrations must satisfy the equation given by the equilibrium law.

To see how all this works, let's begin by building the concentration table.

SOLUTION: To build the table, we need quantities to enter into the "initial concentrations," "changes in concentrations," and "equilibrium concentrations" rows.

Initial Concentrations. The initial concentrations of CO and H_2O are each 0.100 mol/1.00 L = 0.100 *M*. Since no CO_2 or H_2 are initially placed into the reaction vessel, their initial concentrations both are zero.

Changes in Concentrations. Some CO_2 and H_2 must form for the reaction to reach equilibrium. This also means that some CO and H_2O must react. But how much? If we knew the answer, we could calculate the equilibrium concentrations. Therefore, the changes in concentration are our unknown quantities.

Let us allow x to be equal to the number of moles per liter of CO that react. The change in the concentration of CO is then $-x$ (it is negative because the change decreases the CO concentration). Because CO and H_2O react in a 1:1 mole ratio, the change in the H_2O concentration is also $-x$. Since one mole each of CO_2 and H_2 are formed when one mole of CO reacts, the CO_2 and H_2O concentrations each increase by x (their changes are $+x$).

We could just as easily have chosen x to be the number of mol/L of H_2O that reacts or the number of mol/L of CO_2 or H_2 that forms. There's nothing special about having chosen CO to define x.

Equilibrium Concentrations. We obtain the equilibrium concentrations as

$$\begin{pmatrix}\text{Equilibrium} \\ \text{concentration}\end{pmatrix} = \begin{pmatrix}\text{initial} \\ \text{concentration}\end{pmatrix} + \begin{pmatrix}\text{change in} \\ \text{concentration}\end{pmatrix}$$

Here is the completed concentration table.

	$CO(g)$ +	$H_2O(g)$ $\rightleftharpoons$	$CO_2(g)$ +	$H_2(g)$
Initial concentrations (*M*)	0.100	0.100	0.0	0.0
Changes in concentrations (*M*)	$-x$	$-x$	$+x$	$+x$
Equilibrium concentrations (*M*)	$0.100 - x$	$0.100 - x$	x	x

The last line in the table tells us the equilibrium CO and H_2O concentrations are equal to the number of moles per liter that were present initially minus the number of moles per liter that react. The equilibrium concentrations of CO_2 and H_2 equal the number of moles per liter of each that forms, since no CO_2 or H_2 is present initially.

Next, we substitute the quantities from the "equilibrium concentrations" row into the mass action expression and solve for x.

$$\frac{(x)(x)}{(0.100 - x)(0.100 - x)} = 4.06$$

which we can write as

$$\frac{x^2}{(0.100 - x)^2} = 4.06$$

In this example we can solve the equation for x most easily by taking the square root of both sides.

$$\sqrt{\frac{x^2}{(0.100 - x)^2}} = \frac{x}{(0.100 - x)} = \sqrt{4.06} = 2.01$$

Taking the square root of 4.06 actually gives two values, $+2.01$ and -2.01. However, the negative root leads to a negative concentration, which doesn't make sense. Therefore, the remainder of the calculation is performed using the positive root.

Clearing fractions gives

$$x = 2.01(0.100 - x)$$
$$x = 0.201 - 2.01x$$

Collecting terms in x gives

$$x + 2.01x = 0.201$$
$$3.01x = 0.201$$
$$x = 0.0668$$

Now that we know the value of x, we can calculate the equilibrium concentrations from the last row of the table.

$$[CO] = 0.100 - x = 0.100 - 0.0668 = 0.033\ M$$
$$[H_2O] = 0.100 - x = 0.100 - 0.0668 = 0.033\ M$$
$$[CO_2] = x = 0.0668\ M$$
$$[H_2] = x = 0.0668\ M$$

IS THE ANSWER REASONABLE? First, we should check to see that all concentrations are positive numbers. They are. As in the preceding example, we can check the answer by substituting the equilibrium concentrations we've found into the mass action expression and evaluate the reaction quotient. If our answers are correct, Q should equal K_c. Let's do this.

$$Q = \frac{(0.0668)^2}{(0.033)^2} = 4.1$$

Rounding K_c to two significant figures gives 4.1, so the calculated concentrations satisfy the equilibrium law.

EXAMPLE 14.9
Using K_c to Calculate Equilibrium Concentrations

In the preceding example it was stated that the reaction

$$CO(g) + H_2O(g) \rightleftharpoons CO_2(g) + H_2(g)$$

has $K_c = 4.06$ at 500 °C. Suppose 0.0600 mol each of CO and H_2O are mixed with 0.100 mol each of CO_2 and H_2 in a 1.00 L reaction vessel. What will the concentrations of all the substances be when the mixture reaches equilibrium at that temperature?

ANALYSIS: We will proceed in much the same way as in the preceding example. However, this time determining the algebraic signs of x will not be quite so simple because none of the initial concentrations is zero. The best way to determine the algebraic signs is to use the initial concentrations to calculate the initial reaction quotient. Then we can compare Q to K_c. By reasoning we will figure out which way the reaction must proceed to make Q equal to K_c.

SOLUTION: The equilibrium law for the reaction is

$$\frac{[CO_2][H_2]}{[CO][H_2O]} = K_c$$

Let's use the initial concentrations, shown in the first row of the concentration table below, to determine the initial value of the reaction quotient.

$$Q_{\text{initial}} = \frac{(0.100)(0.100)}{(0.0600)(0.0600)} = 2.78 < K_c$$

As indicated, Q_{initial} is less than K_c, so the system is not at equilibrium. To reach equilibrium Q must become larger, which requires an increase in the concentrations of CO_2 and H_2 as the reaction proceeds. This means that for CO_2 and H_2, the change must be positive, and, for CO and H_2O, the change must be negative.

Here is the completed concentration table.

	$CO(g)$	+	$H_2O(g)$	$\rightleftharpoons$	$CO_2(g)$	+	$H_2(g)$
Initial concentrations (M)	0.0600		0.0600		0.100		0.100
Change in concentrations (M)	$-x$		$-x$		$+x$		$+x$
Equilibrium concentrations (M)	$0.0600 - x$		$0.0600 - x$		$0.100 + x$		$0.100 + x$

Substituting equilibrium quantities into the mass action expression in the equilibrium law gives us

$$\frac{(0.100 + x)^2}{(0.0600 - x)^2} = 4.06$$

Taking the square root of both sides yields

$$\frac{0.100 + x}{0.0600 - x} = 2.01$$

To solve for x we first multiply each side by $(0.0600 - x)$ to obtain

$$0.100 + x = 2.01(0.0600 - x)$$
$$0.100 + x = 0.121 - 2.01x$$

Collecting terms in x to one side and the constants to the other gives

$$x + 2.01x = 0.121 - 0.100$$
$$3.01x = 0.021$$
$$x = 0.0070$$

Now we can calculate the equilibrium concentrations:

$$[CO] = [H_2O] = (0.0600 - x) = 0.0600 - 0.0070 = 0.0530\ M$$
$$[CO_2] = [H_2] = (0.100 + x) = 0.100 + 0.0070 = 0.107\ M$$

IS THE ANSWER REASONABLE? As a check, let's evaluate the reaction quotient using the calculated equilibrium concentrations.

$$Q = \frac{(0.107)^2}{(0.0530)^2} = 4.08$$

This is acceptably close to the value of K_c. (That it is not *exactly* equal to K_c is because of the rounding of answers during the calculations.)

In each of the preceding two examples, we were able to calculate the answer directly by taking the square root of both sides of the algebraic equation obtained by substituting equilibrium concentrations into the mass action expression. Such direct calculations are not always possible, however, as illustrated in the next example.

EXAMPLE 14.10
Using K_c to Calculate Equilibrium Concentrations

At a certain temperature, $K_c = 4.50$ for the reaction

$$N_2O_4(g) \rightleftharpoons 2NO_2(g)$$

If 0.300 mol of N_2O_4 is placed into a 2.00 L container at that temperature, what will be the equilibrium concentrations of both gases?

ANALYSIS: As in the preceding example, at equilibrium the mass action expression must be equal to K_c.

$$\frac{[NO_2]^2}{[N_2O_4]} = 4.50$$

We will need to find algebraic expressions for the equilibrium concentrations and substitute them into the mass action expression. To obtain these, we set up the concentration table for the reaction.

SOLUTION:

Initial Concentrations. The initial concentration of N_2O_4 is 0.300 mol/2.00 L = 0.150 *M*. Since no NO_2 was placed in the reaction vessel, its initial concentration is 0.000 *M*.

Changes in Concentrations. There is no NO_2 in the reaction mixture, so we know its concentration must increase. This means the N_2O_4 concentration must decrease as some of the NO_2 is formed. Let's allow x to be the number of moles per liter of N_2O_4 that reacts, so the change in the N_2O_4 concentration is $-x$. Because of the stoichiometry of the reaction, the NO_2 concentration must increase by $2x$, so its change in concentration is $+2x$.

Equilibrium Concentrations. As before, we add the change to the initial concentration in each column to obtain expressions for the equilibrium concentrations.

Here is the completed concentration table.

	$N_2O_4(g)$ $\rightleftharpoons$	$2NO_2(g)$
Initial concentrations (*M*)	0.150	0.000
Changes in concentrations (*M*)	$-x$	$+2x$
Equilibrium concentrations (*M*)	$0.150 - x$	$2x$

Where the changes in concentrations are unknown, it is convenient to let the coefficients of x be the same as the coefficients in the balanced equation. This ensures they are in the correct ratio.

Now we substitute the equilibrium quantities into the mass action expression.

$$\frac{(2x)^2}{(0.150 - x)} = 4.50$$

or

$$\frac{4x^2}{(0.150 - x)} = 4.50 \qquad (14.7)$$

This time the left side of the equation is not a perfect square, so we cannot just take the square root of both sides as in Example 14.9. However, because the equation involves terms in x^2, x, and a constant, we can use the quadratic formula to obtain the value of x. Recall that for a quadratic equation of the form

$$ax^2 + bx + c = 0$$

$$x = \frac{-b \pm \sqrt{b^2 - 4ac}}{2a}$$

Expanding Equation 14.7 above gives

$$\begin{aligned} 4x^2 &= 4.50(0.150 - x) \\ &= 0.675 - 4.50x \end{aligned}$$

Arranging terms in the standard order gives

$$4x^2 + 4.50x - 0.675 = 0$$

Therefore, the quantities we will substitute into the quadratic formula are as follows: $a = 4$, $b = 4.50$, and $c = -0.675$. Making these substitutions gives

$$x = \frac{-4.50 \pm \sqrt{(4.50)^2 - 4(4)(-0.675)}}{2(4)}$$

$$= \frac{-4.50 \pm \sqrt{31.05}}{8}$$

$$= \frac{-4.50 \pm 5.57}{8}$$

Because of the $\pm$ term, there are two values of x that satisfy the equation, $x = 0.134$ and $x = -1.26$. However, only the first value, $x = 0.134$, makes any sense chemically. Using this value, the equilibrium concentrations are

$$[N_2O_4] = 0.150 - 0.134 = 0.016\ M$$

$$[NO_2] = 2(0.134) = 0.268\ M$$

Notice that if we had used the negative root, -1.26, the equilibrium concentration of NO_2 would be negative. Negative concentrations are impossible, so $x = -1.26$ is not acceptable *for chemical reasons.* In general, whenever you use the quadratic equation in a chemical calculation, one root will be satisfactory and the other will lead to answers that are nonsense.

IS THE ANSWER REASONABLE? Once again, we can evaluate the reaction quotient using the calculated equilibrium values. When we do this, we obtain $Q = 4.49$, which is acceptably close to the value of K_c given.

Practice Exercise 20: During an experiment, 0.200 mol of H_2 and 0.200 mol of I_2 were placed into a 1.00 liter vessel where the reaction

$$H_2(g) + I_2(g) \rightleftharpoons 2HI(g)$$

came to equilibrium. For this reaction, $K_c = 49.5$ at the temperature of the experiment. What were the equilibrium concentrations of H_2, I_2, and HI? (Hint: The quadratic formula will not be necessary.)

Practice Exercise 21: In an experiment, 0.200 mol H_2 and 0.100 mol I_2 were placed in a 1.00 L vessel where the following equilibrium was established:

$$H_2(g) + I_2(g) \rightleftharpoons 2HI(g)$$

For this reaction, $K_c = 49.5$ at the temperature of the experiment. What were the equilibrium concentrations of H_2, I_2, and HI?

Equilibrium problems can be much more complex than the ones we have just discussed. However, sometimes you can make assumptions that simplify the problem so that an approximate solution can be obtained.

Equilibrium calculations can usually be simplified when K_c is very small

Many chemical reactions have equilibrium constants that are either very large or very small. For example, most weak acids have very small values for K_c. Therefore, only very tiny amounts of products form when these weak acids react with water.

When the K_c for a reaction is very small, the position of equilibrium lies far to the left, toward the reactants. Usually this permits us to simplify the calculations considerably, as shown in the next example.

EXAMPLE 14.11
Simplifying Equilibrium Calculations for Reactions with Small K_c

Hydrogen, a potential fuel, is found in great abundance in water. Before the hydrogen can be used as a fuel, however, it must be separated from the oxygen; the water must be split into H_2 and O_2. One possibility is thermal decomposition, but this requires very high temperatures. Even at 1000 °C, $K_c = 7.3 \times 10^{-18}$ for the reaction

$$2H_2O(g) \rightleftharpoons 2H_2(g) + O_2(g)$$

If at 1000 °C the H_2O concentration in a reaction vessel is set initially at 0.100 *M*, what will the H_2 concentration be when the reaction reaches equilibrium?

ANALYSIS: We know that at equilibrium

$$\frac{[H_2]^2[O_2]}{[H_2O]^2} = 7.3 \times 10^{-18}$$

As usual, our tool is the concentration table. After we've set up the table, we substitute equilibrium concentrations into the equilibrium law. The equilibrium concentrations will be expressions that contain the unknown quantity *x*. We can solve the equilibrium law for *x* and use the concentration table to relate *x* to the H_2 concentration that we're trying to calculate.

SOLUTION:

Initial Concentrations. The initial concentration of H_2O is 0.100 *M*; those of H_2 and O_2 are both 0.0 *M*.

Changes in Concentrations. We know the changes must be in the same ratio as the coefficients in the balanced equation, so we place *x*'s in this row with coefficients equal to those in the chemical equation. Because there are no products present initially, their changes must be positive and the change for the water must be negative.

The complete concentration table is

	$2H_2O(g)$ $\rightleftharpoons$	$2H_2(g)$ +	$O_2(g)$
Initial concentrations (*M*)	0.100	0.0	0.0
Changes in concentrations (*M*)	$-2x$	$+2x$	$+x$
Equilibrium concentrations (*M*)	$0.100 - 2x$	$2x$	x

When we substitute the equilibrium quantities into the mass action expression we get

$$\frac{(2x)^2x}{(0.100 - 2x)^2} = 7.3 \times 10^{-18}$$

$(2x)^2x = (4x^2)x = 4x^3$

or

$$\frac{4x^3}{(0.100 - 2x)^2} = 7.3 \times 10^{-18}$$

This is a *cubic equation* (one term involves x^3) and can be rather difficult to solve unless we can simplify it. In this instance we are able to do so because the very small value of K_c tells us that hardly any of the H_2O will decompose. Whatever the actual value of *x*, we know *in advance* that it is going to be very small. This means that 2*x* will also be small, so when this tiny value is subtracted from 0.100, the result will still be very, very close to 0.100. We will make the assumption, then, that the term in the denominator will be essentially unchanged from 0.100 by subtracting 2*x* from it; that is, we will assume that $0.100 - 2x \approx 0.100$.

Even before we solve the problem, we know that hardly any H_2 and O_2 will be formed, because K_c is so small.

This assumption greatly simplifies the math. We now have

$$\frac{4x^3}{(0.100 - 2x)^2} \approx \frac{4x^3}{(0.100)^2} = 7.3 \times 10^{-18}$$

$$4x^3 = (0.0100)(7.3 \times 10^{-18}) = 7.3 \times 10^{-20}$$

$$x^3 = 1.8 \times 10^{-20}$$

$$x = \sqrt[3]{1.8 \times 10^{-20}}$$

$$= 2.6 \times 10^{-7}$$

Notice that the value of x that we've obtained is indeed very small. If we double it and subtract the answer from 0.100, we still get 0.100 when we round to the correct number of significant figures (the third decimal place).

Always be sure to check your assumptions when solving a problem of this kind.

$$0.100 - 2x = 0.100 - 2(2.6 \times 10^{-7}) = 0.09999948$$

$$= 0.100 \qquad \text{(rounded correctly)}$$

This check verifies that our assumption was valid. Finally, we have to obtain the H_2 concentration. Our table gives

$$[H_2] = 2x$$

Therefore,

$$[H_2] = 2(2.6 \times 10^{-7}) = 5.2 \times 10^{-7}\,M$$

IS THE ANSWER REASONABLE? The equilibrium constant is very small for this reaction (7.3×10^{-18}), so we expect the amount of product at equilibrium to be small as well. The very low concentration of H_2 seems reasonable. The value of x was positive, which means that there was a decrease in reactant concentrations and an increase in product concentrations as we would expect if only reactants were originally present. Finally, we can check the calculation by seeing if the mass action expression is equal to the equilibrium constant:

$$\frac{[H_2]^2[O_2]}{[H_2O]^2} = \frac{(5.2 \times 10^{-7})^2(2.6 \times 10^{-7})}{0.100^2} = 7.0 \times 10^{-18}$$

The calculated value differs from the original equilibrium constant only because the value of x was rounded off when we took the cube root.

Certain situations suggest that simplification of the math will work

The simplifying assumption made in the preceding example is valid because a very small number is subtracted from a much larger one. We could also have neglected x (or $2x$) if it were a very small number that was being added to a much larger one. Remember that you can only neglect an x that's *added* or *subtracted*; you can never drop an x that occurs as a multiplying or dividing factor. Some examples are

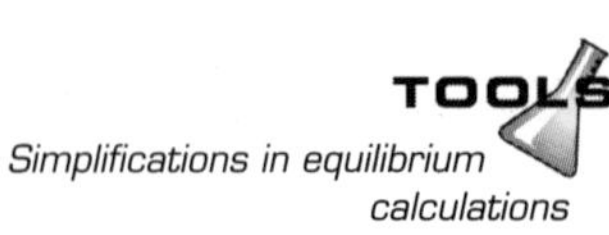

Simplifications in equilibrium calculations

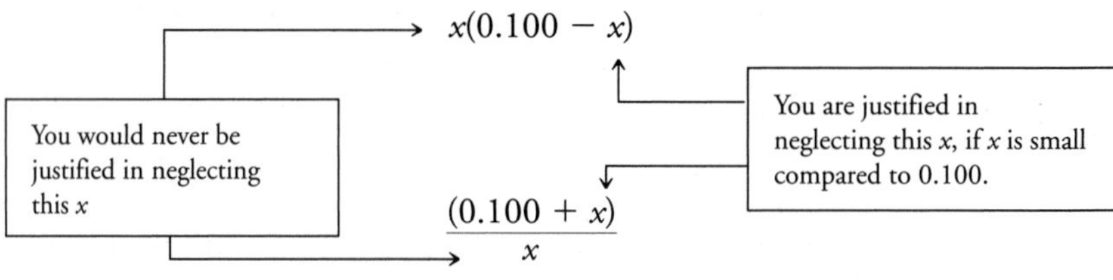

As a rule of thumb, you can expect that these simplifying assumptions will be valid if the concentration from which x is subtracted, or to which x is added, is at least 1000 times greater than K. For instance, in the preceding example, $2x$ was subtracted from 0.100. Since 0.100 is much larger than $1000 \times (7.3 \times 10^{-18})$ we expect the assumption $0.100 - 2x \approx 0.100$ to be valid. However, even though the simplifying assumption is expected to be valid, always check to see if it really is after finishing the calculation. If the assumption proves invalid, then some other way to solve the algebra must be found.

Practice Exercise 22: At 25 °C, the reaction $2NH_3(g) \rightleftharpoons N_2(g) + 3H_2(g)$ has $K_c = 2.3 \times 10^{-9}$. If 0.041 mol NH_3 is placed in a 1.00 L container, what will the concentrations of N_2 and H_2 be when equilibrium is established? (Hint: Review the math beneath the concentration table in the preceding example.)

Practice Exercise 23: In air at 25 °C and 1 atm, the N_2 concentration is 0.033 *M* and the O_2 concentration is 0.00810 *M*. The reaction

$$N_2(g) + O_2(g) \rightleftharpoons 2NO(g)$$

has $K_c = 4.8 \times 10^{-31}$ at 25 °C. Taking the N_2 and O_2 concentrations given above as initial values, calculate the equilibrium NO concentration that should exist in our atmosphere from this reaction at 25 °C.

SUMMARY

Dynamic Equilibrium. When the forward and reverse reactions in a chemical system occur at equal rates, a dynamic equilibrium exists and the concentrations of the reactants and products remain constant. For a given overall chemical composition, the amounts of reactants and products that are present at equilibrium are the same regardless of whether the equilibrium is approached from the direction of pure "reactants," pure "products," or any mixture of them. (In a chemical equilibrium, the terms *reactants* and *products* do not have the usual significance because the reaction is proceeding in both directions simultaneously. Instead, we use *reactants* and *products* simply to identify the substances on the left- and right-hand sides, respectively, of the equation for the equilibrium.)

The Equilibrium Law. The **mass action expression** is a fraction. The concentrations of the products, raised to powers equal to their coefficients in the chemical equation, are multiplied together in the numerator. The denominator is constructed in the same way from the concentrations of the reactants raised to powers equal to their coefficients. The numerical value of the mass action expression is the **reaction quotient, *Q*.** At equilibrium, the reaction quotient is equal to the **equilibrium constant, K_c.** If partial pressures of gases are used in the mass action expression, K_P is obtained. The magnitude of the equilibrium constant is roughly proportional to the extent to which the reaction proceeds to completion when equilibrium is reached. Equilibrium equations can be manipulated by multiplying the coefficients by a common factor, by changing the direction of the reaction, and by adding two or more equilibria. The rules given in the description of the *Tools for Problem Solving* below apply.

Relating K_P to K_c. The values of K_P and K_c are only equal if the same number of moles of gas are represented on both sides of the chemical equation. When the number of moles of gas are different, K_P is related to K_c by the equation $K_P = K_c(RT)^{\Delta n_g}$. Remember to use $R = 0.0821$ L atm mol^{-1} K^{-1} and T = absolute temperature. Also, be careful to calculate Δn_g as the difference between the number of moles of *gaseous* products and the number of moles of *gaseous* reactants in the balanced equation.

Heterogeneous Equilibria. An equilibrium involving substances in more than one phase is a **heterogeneous equilibrium.** The mass action expression for a heterogeneous equilibrium omits concentration terms for pure liquids and/or pure solids.

Le Châtelier's Principle. This principle states that *when an equilibrium is upset, a chemical change occurs in a direction that opposes the disturbing influence and brings the system to equilibrium again.* Adding a reactant or a product causes a reaction to occur that uses up part of what has been added. Removing a reactant or a product causes a reaction that replaces part of what has been removed. Increasing the pressure (by reducing the volume) drives a reaction in the direction of the fewer number of moles of gas. Pressure changes have virtually no effect on equilibria involving only solids and liquids. Raising the temperature causes an equilibrium to shift in an endothermic direction. The value of *K* increases with increasing temperature for reactions that are endothermic in the forward direction. A change in temperature is the only factor that changes *K*. Addition of a catalyst or an inert gas has no effect on an equilibrium.

Equilibrium Calculations. The initial concentrations in a chemical system are controlled by the person who combines the chemicals at the start of the reaction. The changes in concentration are determined by the stoichiometry of the reaction. Only equilibrium concentrations satisfy the equilibrium law. When these are used, the value of the mass action expression, *Q*, is equal to K_c. When a change in concentration is expected to be very small compared to the initial concentration, the change may be neglected and the algebraic equation derived from the equilibrium law can be simplified. In general, this simplification is valid if the initial concentration is at least 1000 times larger than *K*.

TOOLS FOR PROBLEM SOLVING

In this chapter you learned to apply the following concepts as tools in solving problems dealing with aspects of chemical equilibrium. Study each tool carefully so that you know what each is used for. When faced with solving a problem, recall what each tool does and consider whether it will be helpful in finding a solution. This will aid you in selecting the tools you need.

The approach to equilibrium *(page 571)* There are times when it is very helpful to remember that the same equilibrium composition is reached regardless of whether it is approached from the direction of the reactants or from the direction of the products. Some equilibrium problems can be greatly simplified if we first imagine the reaction going to completion (converting all the reactants to products) and then approach the equilibrium from the direction of the products.

The equilibrium law *(page 573)* To solve most chemical equilibrium problems we need the equation for the equilibrium and the equilibrium law. The knowledge of how to construct the equilibrium law from the chemical equation is an essential tool for dealing with such problems. An equation of the form

$$dD + eE \rightleftharpoons fF + gG$$

has the equilibrium law

$$\frac{[F]^f[G]^g}{[D]^d[E]^e} = K_c$$

Remember that for gaseous reactions, partial pressures can be used in place of concentrations, in which case the mass action expression equals K_P. In general, K_P does not equal K_c.

Manipulating equilibrium equations *(pages 574 through 575)* There are occasions when it is necessary to modify an equation for an equilibrium, or combine two or more chemical equilibria. The tools discussed here are used to obtain the new equilibrium constants for the new equations.

- When two equations are added, we multiply their *K*s to obtain the new *K*.
- When an equation is multiplied by a factor n to obtain a new equation, we raise its *K* to the power n to obtain the *K* for the new equation.
- When an equation is reversed, we take the reciprocal of its *K* to obtain the new *K*.

Converting between K_P and K_c *(page 577)* In Chapter 18 you will learn how to calculate equilibrium constants from thermodynamic data. These calculations give K_P for gaseous reactions. To change to K_c values, we use the equation

$$K_P = K_c(RT)^{\Delta n_g}$$

Remember to use $R = 0.0821$ L atm mol^{-1} K^{-1}.

Equilibrium laws for heterogeneous reactions *(page 579)* Being able to write the equilibrium law for a heterogeneous reaction is a tool you will need in later chapters when we deal with solubility equilibria. Remember that pure solids and liquids do not appear in the mass action expression.

Magnitude of *K* *(page 580)* Use this tool to gain a rough estimate of the position of equilibrium.

- When *K* is very large, the position of equilibrium lies far to the right (toward the products).
- When *K* is very small, the position of equilibrium lies far to the left (toward the reactants).

Le Châtelier's principle This tool lets us predict how disturbing influences shift the position of equilibrium. Factors to consider are as follows:

- Adding or removing a reactant or product *(page 582)*
- Changing the volume for gaseous reactions *(page 582)*
- Changing the temperature *(pages 583, 584)*

Catalysts or inert gases have no effect on the position of equilibrium.

Concentration table *(page 586)* This is a tool you will use in almost all equilibrium calculations. Remember the following points when constructing the table:

- All entries in the table must have units of molarity (mol L^{-1}).
- Any reactant or product for which an initial concentration or amount is not given in the statement of the problem is assigned an initial concentration of zero.

- Any substance with an initial concentration of zero must have a positive change in concentration when the reaction proceeds to equilibrium.
- The changes in concentration are in the same ratio as the coefficients in the balanced equation. When the changes are unknown, the coefficients of x can be the same as the coefficients in the balanced equation.
- Only quantities in the last row (Equilibrium Concentrations) satisfy the equilibrium law.

Simplifications in equilibrium calculations *(page 596)* When the initial reactant concentrations are larger than $1000 \times K$, they will change only slightly as the reaction approaches equilibrium. You can therefore expect to be able to neglect the change when it is being added to or subtracted from an initial concentration. This tool is especially useful when working problems in which K is very small and the initial conditions are not far from the final position of equilibrium. *If you use these simplifying approximations, be sure to check their validity after obtaining an answer.*

QUESTIONS, PROBLEMS, AND EXERCISES

Answers to problems whose numbers are printed in color are given in Appendix B. More challenging problems are marked with asterisks. ILW = Interactive Learningware solution is available at www.wiley.com/college/brady. OH = an Office Hours video is available for this problem.

REVIEW QUESTIONS

General

14.1 Sketch a graph showing how the concentrations of the reactants and products of a typical chemical reaction vary with time during the course of the reaction. Assume no products are present at the start of the reaction. Indicate on the graph where the system has reached equilibrium.

14.2 What meanings do the terms *reactants* and *products* have when describing a chemical equilibrium?

Mass Action Expression, K_P, and K_c

14.3 What is an *equilibrium law*? How is the term *reaction quotient* defined?

14.4 Under what conditions does the reaction quotient equal K_c?

14.5 Suppose for the reaction $A \longrightarrow B$ the value of Q is less than K_c. Which way does the reaction have to proceed to reach equilibrium, in the forward or reverse direction?

14.6 When a chemical equation and its equilibrium constant are given, why is it not necessary to also specify the form of the mass action expression?

14.7 At 225 °C, $K_P = 6.3 \times 10^{-3}$ for the reaction

$$CO(g) + 2H_2(g) \rightleftharpoons CH_3OH(g)$$

Would we expect this reaction to go nearly to completion?

14.8 Here are some reactions and their equilibrium constants.
(a) $2CH_4(g) \rightleftharpoons C_2H_6(g) + H_2(g) \quad K_c = 9.5 \times 10^{-13}$
(b) $CH_3OH(g) + H_2(g) \rightleftharpoons CH_4(g) + H_2O(g) \quad K_c = 3.6 \times 10^{20}$
(c) $H_2(g) + Br_2(g) \rightleftharpoons 2HBr(g) \quad K_c = 2.0 \times 10^{9}$
Arrange these reactions in order of their increasing tendency to go toward completion.

Converting between K_P and K_c

14.9 State the equation relating K_P to K_c and define all terms. Which is the only value of R that can be properly used in the equation?

14.10 Use the ideal gas law to show that the partial pressure of a gas is directly proportional to its molar concentration. What is the proportionality constant?

Heterogeneous Equilibria

14.11 What is the difference between a *heterogeneous equilibrium* and a *homogeneous equilibrium*?

14.12 Why do we omit the concentrations of pure liquids and solids from the mass action expression of heterogeneous reactions?

14.13 Consider the following equilibrium.

$$2NaHCO_3(s) \rightleftharpoons Na_2CO_3(s) + CO_2(g) + H_2O(g)$$

If you were converting between K_P and K_c, what value of Δn_g would you use?

Le Châtelier's Principle

14.14 State Le Châtelier's principle in your own words.

14.15 Explain, using its effect on the reaction quotient, why adding a reactant to the following equilibrium shifts the position of equilibrium to the right.

$$PCl_3(g) + Cl_2(g) \rightleftharpoons PCl_5(g)$$

14.16 Halving the volume of a gas doubles its pressure. Using the reaction quotient corresponding to K_P, explain why halving the volume shifts the following equilibrium to the left.

$$N_2O_4(g) \rightleftharpoons 2NO_2(g)$$

14.17 How will the value of K_P for the following reactions be affected by an increase in temperature?
(a) $CO(g) + 2H_2(g) \rightleftharpoons CH_3OH(g) \quad \Delta H° = -18$ kJ
(b) $N_2O(g) + NO_2(g) \rightleftharpoons 3NO(g) \quad \Delta H° = +155.7$ kJ
(c) $2NO(g) + Cl_2(g) \rightleftharpoons 2NOCl(g) \quad \Delta H° = -77.07$ kJ

14.18 Why doesn't a catalyst affect the position of equilibrium in a chemical reaction?

REVIEW PROBLEMS

Equilibrium Laws for K_P and K_c

14.19 Write the equilibrium law for each of the following reactions in terms of molar concentrations:

(a) $2PCl_3(g) + O_2(g) \rightleftharpoons 2POCl_3(g)$

(b) $2SO_3(g) \rightleftharpoons 2SO_2(g) + O_2(g)$

(c) $N_2H_4(g) + 2O_2(g) \rightleftharpoons 2NO(g) + 2H_2O(g)$

(d) $N_2H_4(g) + 6H_2O_2(g) \rightleftharpoons 2NO_2(g) + 8H_2O(g)$

(e) $SOCl_2(g) + H_2O(g) \rightleftharpoons SO_2(g) + 2HCl(g)$

14.20 Write the equilibrium law for each of the following gaseous reactions in terms of molar concentrations.

(a) $3Cl_2(g) + NH_3(g) \rightleftharpoons NCl_3(g) + 3HCl(g)$

(b) $PCl_3(g) + PBr_3(g) \rightleftharpoons PCl_2Br(g) + PClBr_2(g)$

(c) $NO(g) + NO_2(g) + H_2O(g) \rightleftharpoons 2HNO_2(g)$

(d) $H_2O(g) + Cl_2O(g) \rightleftharpoons 2HOCl(g)$

(e) $Br_2(g) + 5F_2(g) \rightleftharpoons 2BrF_5(g)$

14.21 Write the equilibrium law for the reactions in Problem 14.19 in terms of partial pressures.

14.22 Write the equilibrium law for the reactions in Problem 14.20 in terms of partial pressures.

OH **14.23** Write the equilibrium law for each of the following reactions in aqueous solution.

(a) $Ag^+(aq) + 2NH_3(aq) \rightleftharpoons Ag(NH_3)_2{}^+(aq)$

(b) $Cd^{2+}(aq) + 4SCN^-(aq) \rightleftharpoons Cd(SCN)_4{}^{2-}(aq)$

14.24 Write the equilibrium law for each of the following reactions in aqueous solution.

(a) $HClO(aq) + H_2O \rightleftharpoons H_3O^+(aq) + ClO^-(aq)$

(b) $CO_3{}^{2-}(aq) + HSO_4{}^-(aq) \rightleftharpoons HCO_3{}^-(aq) + SO_4{}^{2-}(aq)$

Manipulating Equilibrium Equations

14.25 At 25 °C, $K_c = 1 \times 10^{-85}$ for the reaction

$$7IO_3{}^-(aq) + 9H_2O + 7H^+(aq) \rightleftharpoons I_2(aq) + 5H_5IO_6(aq)$$

What is the value of K_c for the following reaction?

$$I_2(aq) + 5H_5IO_6(aq) \rightleftharpoons 7IO_3{}^-(aq) + 9H_2O + 7H^+(aq)$$

14.26 Use the following equilibria

$$2CH_4(g) \rightleftharpoons C_2H_6(g) + H_2(g) \qquad K_c = 9.5 \times 10^{-13}$$

$$CH_4(g) + H_2O(g) \rightleftharpoons CH_3OH(g) + H_2(g) \qquad K_c = 2.8 \times 10^{-21}$$

to calculate K_c for the reaction

$$2CH_3OH(g) + H_2(g) \rightleftharpoons C_2H_6(g) + 2H_2O(g)$$

14.27 Write the equilibrium law for each of the following reactions in terms of molar concentrations:

(a) $H_2(g) + Cl_2(g) \rightleftharpoons 2HCl(g)$

(b) $\frac{1}{2}H_2(g) + \frac{1}{2}Cl_2(g) \rightleftharpoons HCl(g)$

How does K_c for reaction (a) compare with K_c for reaction (b)?

14.28 Write the equilibrium law for the reaction

$$2HCl(g) \rightleftharpoons H_2(g) + Cl_2(g)$$

How does K_c for this reaction compare with K_c for reaction (a) in the preceding problem?

Converting between K_P and K_c

14.29 A 345 mL container holds NH_3 at a pressure of 745 torr and a temperature of 45 °C. What is the molar concentration of ammonia in the container?

14.30 In a certain container at 145 °C the concentration of water vapor is 0.0200 *M*. What is the partial pressure of H_2O in the container?

14.31 For which of the following reactions does $K_P = K_c$?

(a) $2H_2(g) + C_2H_2(g) \rightleftharpoons C_2H_6(g)$

(b) $N_2(g) + O_2(g) \rightleftharpoons 2NO(g)$

(c) $2NO(g) + O_2(g) \rightleftharpoons 2NO_2(g)$

OH **14.32** For which of the following reactions does $K_P = K_c$?

(a) $CO_2(g) + H_2(g) \rightleftharpoons CO(g) + H_2O(g)$

(b) $PCl_3(g) + Cl_2(g) \rightleftharpoons PCl_5(g)$

(c) $N_2O_4(g) \rightleftharpoons 2NO_2(g)$

ILW **14.33** The reaction $CO(g) + 2H_2(g) \rightleftharpoons CH_3OH(g)$ has $K_P = 6.3 \times 10^{-3}$ at 225 °C. What is the value of K_c at that temperature?

14.34 The reaction $HCO_2H(g) \rightleftharpoons CO(g) + H_2O(g)$ has $K_P = 1.6 \times 10^6$ at 400 °C. What is the value of K_c for the reaction at that temperature?

14.35 The reaction $N_2O(g) + NO_2(g) \rightleftharpoons 3NO(g)$ has $K_c = 4.2 \times 10^{-4}$ at 500 °C. What is the value of K_P at this temperature?

14.36 One possible way of removing NO from the exhaust of a gasoline engine is to cause it to react with CO in the presence of a suitable catalyst.

$$2NO(g) + 2CO(g) \rightleftharpoons N_2(g) + 2CO_2(g)$$

At 300 °C, the reaction has $K_c = 2.2 \times 10^{59}$. What is K_P at 300 °C?

14.37 At 773 °C the reaction

$$CO(g) + 2H_2(g) \rightleftharpoons CH_3OH(g)$$

has $K_c = 0.40$. What is K_P at that temperature?

14.38 The reaction $COCl_2(g) \rightleftharpoons CO(g) + Cl_2(g)$ has $K_P = 4.6 \times 10^{-2}$ at 395 °C. What is K_c at that temperature?

Heterogeneous Equilibria

14.39 Calculate the molar concentration of water in (a) 18.0 mL of H_2O, (b) 100.0 mL of H_2O, and (c) 1.00 L of H_2O. Assume that the density of water is 1.00 g/mL.

14.40 The density of sodium chloride is 2.164 g cm^{-3}. What is the molar concentration of NaCl in a 12.0 cm^3 sample of pure NaCl? What is the molar concentration of NaCl in a 25.0 g sample of pure NaCl?

14.41 Write the equilibrium law corresponding to K_c for each of the following heterogeneous reactions.

(a) $2C(s) + O_2(g) \rightleftharpoons 2CO(g)$

(b) $2NaHSO_3(s) \rightleftharpoons Na_2SO_3(s) + H_2O(g) + SO_2(g)$

(c) $2C(s) + 2H_2O(g) \rightleftharpoons CH_4(g) + CO_2(g)$

(d) $CaCO_3(s) + 2HF(g) \rightleftharpoons CaF_2(s) + H_2O(g) + CO_2(g)$

(e) $CuSO_4 \cdot 5H_2O(s) \rightleftharpoons CuSO_4(s) + 5H_2O(g)$

14.42 Write the equilibrium law corresponding to K_c for each of the following heterogeneous reactions.

(a) $CaCO_3(s) + SO_2(g) \rightleftharpoons CaSO_3(s) + CO_2(g)$

(b) $AgCl(s) + Br^-(aq) \rightleftharpoons AgBr(s) + Cl^-(aq)$

(c) $Cu(OH)_2(s) \rightleftharpoons Cu^{2+}(aq) + 2OH^-(aq)$

(d) $Mg(OH)_2(s) \rightleftharpoons MgO(s) + H_2O(g)$

(e) $3CuO(s) + 2NH_3(g) \rightleftharpoons 3Cu(s) + N_2(g) + 3H_2O(g)$

OH **14.43** The heterogeneous reaction

$$2HCl(g) + I_2(s) \rightleftharpoons 2HI(g) + Cl_2(g)$$

has $K_c = 1.6 \times 10^{-34}$ at 25 °C. Suppose 0.100 mol of HCl and solid I_2 is placed in a 1.00 L container. What will be the equilibrium concentrations of HI and Cl_2 in the container?

14.44 At 25 °C, $K_c = 360$ for the reaction

$$AgCl(s) + Br^-(aq) \rightleftharpoons AgBr(s) + Cl^-(aq)$$

If solid AgCl is added to a solution containing 0.10 *M* Br^-, what will be the equilibrium concentrations of Br^- and Cl^-?

Le Châtelier's Principle

14.45 How will the position of equilibrium in the following reaction

$$\text{heat} + CH_4(g) + 2H_2S(g) \rightleftharpoons CS_2(g) + 4H_2(g)$$

be affected by:

(a) Adding $CH_4(g)$?

(b) Adding $H_2(g)$?

(c) Removing $CS_2(g)$?

(d) Decreasing the volume of the container?

(e) Increasing the temperature?

OH **14.46** The reaction $CO(g) + 2H_2(g) \rightleftharpoons CH_3OH(g)$ has $\Delta H° = -18$ kJ. How will the amount of CH_3OH present at equilibrium be affected by the following?

(a) Adding CO(*g*).

(b) Removing $H_2(g)$.

(c) Decreasing the volume of the container.

(d) Adding a catalyst.

(e) Increasing the temperature.

14.47 Consider the equilibrium

$$N_2O(g) + NO_2(g) \rightleftharpoons 3NO(g) \qquad \Delta H° = +155.7 \text{ kJ}$$

In which direction will this equilibrium be shifted by the following changes?

(a) Adding N_2O.

(b) Removing NO_2.

(c) Adding NO.

(d) Increasing the temperature of the reaction mixture.

(e) Adding helium gas to the reaction mixture at constant volume.

(f) Decreasing the volume of the container at constant temperature.

14.48 Consider the equilibrium

$$2NO(g) + Cl_2(g) \rightleftharpoons 2NOCl(g)$$

for which $\Delta H° = -77.07$ kJ. How will the amount of Cl_2 at equilibrium be affected by the following?

(a) Removing NO(*g*).

(b) Adding NOCl(*g*).

(c) Raising the temperature.

(d) Decreasing the volume of the container.

Equilibrium Calculations

14.49 At a certain temperature, $K_c = 0.18$ for the equilibrium

$$PCl_3(g) + Cl_2(g) \rightleftharpoons PCl_5(g)$$

Suppose a reaction vessel at that temperature contained these three gases at the following concentrations: $[PCl_3] = 0.0420$ *M*, $[Cl_2] = 0.0240$ *M*, $[PCl_5] = 0.00500$ *M*.

(a) Is the system in a state of equilibrium?

(b) If not, in which direction will the reaction have to proceed to get to equilibrium?

14.50 At 460 °C, the reaction

$$SO_2(g) + NO_2(g) \rightleftharpoons NO(g) + SO_3(g)$$

has $K_c = 85.0$. A reaction flask at 460 °C contains these gases at the following concentrations: $[SO_2] = 0.00250$ *M*, $[NO_2] = 0.00350$ *M*, $[NO] = 0.0250$ *M*, $[SO_3] = 0.0400$ *M*.

(a) Is the reaction at equilibrium?

(b) If not, in which way will the reaction have to proceed to arrive at equilibrium?

OH **14.51** At a certain temperature, the reaction

$$CO(g) + 2H_2(g) \rightleftharpoons CH_3OH(g)$$

has $K_c = 0.500$. If a reaction mixture at equilibrium contains 0.180 *M* CO and 0.220 *M* H_2, what is the concentration of CH_3OH?

14.52 At a certain temperature $K_c = 64$ for the reaction

$$N_2(g) + 3H_2(g) \rightleftharpoons 2NH_3(g)$$

Suppose it was found that an equilibrium mixture of these gases contained 0.360 *M* NH_3 and 0.0192 *M* N_2. What was the concentration of H_2 in the mixture?

14.53 At 773 °C, a mixture of CO(*g*), $H_2(g)$, and $CH_3OH(g)$ was allowed to come to equilibrium. The following equilibrium concentrations were then measured: $[CO] = 0.105$ *M*, $[H_2] = 0.250$ *M*, $[CH_3OH] = 0.00261$ *M*. Calculate K_c for the reaction

$$CO(g) + 2H_2(g) \rightleftharpoons CH_3OH(g)$$

14.54 Ethylene, C_2H_4, and water react under appropriate conditions to give ethanol. The reaction is

$$C_2H_4(g) + H_2O(g) \rightleftharpoons C_2H_5OH(g)$$

An equilibrium mixture of these gases at a certain temperature had the following concentrations: $[C_2H_4] = 0.0148$ *M*, $[H_2O] = 0.0336$ *M*, and $[C_2H_5OH] = 0.180$ *M*. What is the value of K_c?

ILW **14.55** At high temperature, 2.00 mol of HBr was placed in a 4.00 L container where it decomposed to give the equilibrium

$$2HBr(g) \rightleftharpoons H_2(g) + Br_2(g)$$

At equilibrium the concentration of Br_2 was measured to be 0.0955 *M*. What is K_c for the reaction at that temperature?

14.56 A 0.050 mol sample of formaldehyde vapor, CH_2O, was placed in a heated 500 mL vessel and some of it decomposed. The reaction is

$$CH_2O(g) \rightleftharpoons H_2(g) + CO(g)$$

At equilibrium, the $CH_2O(g)$ concentration was 0.066 mol L^{-1}. Calculate the value of K_c for this reaction.

14.57 The reaction $NO_2(g) + NO(g) \rightleftharpoons N_2O(g) + O_2(g)$ reached equilibrium at a certain high temperature. Originally, the reaction vessel contained the following initial concentrations: $[N_2O] = 0.184\ M$, $[O_2] = 0.377\ M$, $[NO_2] = 0.0560\ M$, and $[NO] = 0.294\ M$. The concentration of the NO_2, the only colored gas in the mixture, was monitored by following the intensity of the color. At equilibrium, the NO_2 concentration had become 0.118 *M*. What is the value of K_c for the reaction at that temperature?

14.58 At 25 °C, 0.0560 mol O_2 and 0.020 mol N_2O were placed in a 1.00 L container where the following equilibrium was then established.

$$2N_2O(g) + 3O_2(g) \rightleftharpoons 4NO_2(g)$$

At equilibrium, the NO_2 concentration was 0.020 *M*. What is the value of K_c for this reaction?

14.59 At 25 °C, $K_c = 0.145$ for the following reaction in the solvent CCl_4.

$$2BrCl \rightleftharpoons Br_2 + Cl_2$$

If the initial concentration of BrCl in the solution is 0.050 *M*, what will the equilibrium concentrations of Br_2 and Cl_2 be?

14.60 At 25 °C, $K_c = 0.145$ for the following reaction in the solvent CCl_4.

$$2BrCl \rightleftharpoons Br_2 + Cl_2$$

If the initial concentrations of Br_2 and Cl_2 are each 0.0250 *M*, what will their equilibrium concentrations be?

14.61 The equilibrium constant, K_c, for the reaction

$$SO_3(g) + NO(g) \rightleftharpoons NO_2(g) + SO_2(g)$$

was found to be 0.500 at a certain temperature. If 0.240 mol of SO_3 and 0.240 mol of NO are placed in a 2.00 L container and allowed to react, what will be the equilibrium concentration of each gas?

14.62 For the reaction in the preceding problem, a reaction mixture is prepared in which 0.120 mol NO_2 and 0.120 mol of SO_2 are placed in a 1.00 L vessel. After the system reaches equilibrium, what will be the equilibrium concentrations of all four gases? How do these equilibrium values compare to those calculated in Problem 14.61? Account for your observation.

14.63 At a certain temperature the reaction

$$CO(g) + H_2O(g) \rightleftharpoons CO_2(g) + H_2(g)$$

has $K_c = 0.400$. Exactly 1.00 mol of each gas was placed in a 100.0 L vessel and the mixture underwent reaction. What was the equilibrium concentration of each gas?

14.64 At 25 °C, $K_c = 0.145$ for the following reaction in the solvent CCl_4.

$$2BrCl \rightleftharpoons Br_2 + Cl_2$$

If the initial concentrations of each substance in a solution are 0.0400 *M*, what will their equilibrium concentrations be?

ILW 14.65 The reaction $2HCl(g) \rightleftharpoons H_2(g) + Cl_2(g)$ has $K_c = 3.2 \times 10^{-34}$ at 25 °C. If a reaction vessel contains initially 0.0500 mol L^{-1} of HCl and then reacts to reach equilibrium, what will be the concentrations of H_2 and Cl_2?

14.66 At 200 °C, $K_c = 1.4 \times 10^{-10}$ for the reaction

$$N_2O(g) + NO_2(g) \rightleftharpoons 3NO(g)$$

If 0.200 mol of N_2O and 0.400 mol NO_2 are placed in a 4.00 L container, what would the NO concentration be if this equilibrium were established?

ILW 14.67 At 2000 °C, the decomposition of CO_2,

$$2CO_2(g) \rightleftharpoons 2CO(g) + O_2(g)$$

has $K_c = 6.4 \times 10^{-7}$. If a 1.00 L container holding 1.0×10^{-2} mol of CO_2 is heated to 2000 °C, what will be the concentration of CO at equilibrium?

14.68 At 500 °C, the decomposition of water into hydrogen and oxygen,

$$2H_2O(g) \rightleftharpoons 2H_2(g) + O_2(g)$$

has $K_c = 6.0 \times 10^{-28}$. How many moles of H_2 and O_2 are present at equilibrium in a 5.00 L reaction vessel at that temperature if the container originally held 0.015 mol H_2O?

14.69 At a certain temperature, $K_c = 0.18$ for the equilibrium

$$PCl_3(g) + Cl_2(g) \rightleftharpoons PCl_5(g)$$

If 0.026 mol of PCl_5 is placed in a 2.00 L vessel at that temperature, what will the concentration of PCl_3 be at equilibrium?

14.70 At 460 °C, the reaction

$$SO_2(g) + NO_2(g) \rightleftharpoons NO(g) + SO_3(g)$$

has $K_c = 85.0$. Suppose 0.100 mol of SO_2, 0.0600 mol of NO_2, 0.0800 mol of NO, and 0.120 mol of SO_3 are placed in a 10.0 L container at that temperature. What will the concentrations of all the gases be when the system reaches equilibrium?

14.71 At a certain temperature, $K_c = 0.500$ for the reaction

$$SO_3(g) + NO(g) \rightleftharpoons NO_2(g) + SO_2(g)$$

If 0.100 mol SO_3 and 0.200 mol NO are placed in a 2.00 L container and allowed to come to equilibrium, what will the NO_2 and SO_2 concentrations be?

14.72 At 25 °C, $K_c = 0.145$ for the following reaction in the solvent CCl_4.

$$2BrCl \rightleftharpoons Br_2 + Cl_2$$

A solution was prepared with the following initial concentrations: $[BrCl] = 0.0400\ M$, $[Br_2] = 0.0300\ M$, and $[Cl_2] = 0.0200\ M$. What will their equilibrium concentrations be?

***14.73** At a certain temperature, $K_c = 4.3 \times 10^5$ for the reaction

$$HCO_2H(g) \rightleftharpoons CO(g) + H_2O(g)$$

If 0.200 mol of HCO_2H is placed in a 1.00 L vessel, what will be the concentrations of CO and H_2O when the system reaches equilibrium? (Hint: The same equilibrium concentrations are reached whether equilibrium is approached from the left or the right.)

5 ACIDS AND BASES: A SECOND LOOK

The economic health of a country is often reflected in the activity of the stock and commodity exchanges. Chemicals, including pharmaceuticals, agrochemicals and home-care products, make up a large component of the economic activity of many countries. Acids and bases described in this chapter are a major component of the chemical industry. *(© AP/Wide World Photos)*

HAPTER OUTLINE

15.1 Brønsted–Lowry acids and bases exchange protons

15.2 Strengths of Brønsted acids and bases follow periodic trends

15.3 Lewis acids and bases involve coordinate covalent bonds

15.4 Elements and their oxides demonstrate acid–base properties

15.5 pH is a measure of the acidity of a solution

15.6 Strong acids and bases are fully dissociated in solution

CHAPTER IN CONTEXT In Chapter 4 we introduced you to compounds that we call acids and Many common substances, from household products such as vinegar and ammonia, to biologically important unds such as amino acids, are conveniently classified as either an acid or a base. The significant property that such classifications useful is that *acids react with bases*. In fact, this is such a useful relationship that the se concept has been expanded far beyond the limited Arrhenius definition we discussed in Chapter 4. We urn to acid–base chemistry to learn about these broader and often more useful views.

the same time, we will study how trends in the strengths of acids and bases correlate with the periodic table. use the principles developed in previous chapters to explain why one acid should be stronger than another. lity to judge relative acid strengths is an important skill to bring to the study of organic chemistry. In the next we will assign a numerical value that helps us more precisely compare strengths of acids and bases.

FIG. 15.1 The reaction of gaseous HCl with gaseous NH_3. As each gas escapes from its concentrated aqueous solution and mingles with the other, a cloud of microcrystals of NH_4Cl forms above the bottles. *(Andy Washnik.)*

15.1 BRØNSTED-LOWRY ACIDS AND BASES EXCHANGE PROTONS

In our earlier discussion, an acid was described as a substance that produces H_3O^+ in water, whereas a base gives OH^-. An acid–base neutralization, according to Arrhenius, is a reaction in which an acid and a base combine to produce water and a salt. However, many reactions resemble neutralizations without involving H_3O^+, OH^-, or even H_2O. For example, when open bottles of concentrated hydrochloric acid and concentrated aqueous ammonia are placed side by side, a white cloud forms when the vapors from the two bottles mix (see Figure 15.1). The cloud consists of tiny crystals of ammonium chloride which form when ammonia and hydrogen chloride gases, escaping from the open bottles, mix in air and react.

$$NH_3(g) + HCl(g) \longrightarrow NH_4Cl(s)$$

What's interesting is that this is the same net reaction that occurs when an aqueous solution of ammonia (a base) is neutralized by an aqueous solution of hydrogen chloride (an acid). Yet, the gaseous reaction doesn't fit the description of an acid–base neutralization according to the Arrhenius definition because there's no water involved.

If we look at both the aqueous and gaseous reactions, they do have something in common. Both involve the transfer of a *proton* (a hydrogen ion, H^+) from one particle to another.[1] In water, where HCl is completely ionized, the transfer is from H_3O^+ to NH_3, as we discussed on page 150. The ionic equation is

$$NH_3(aq) + H_3O^+(aq) + Cl^-(aq) \longrightarrow \underbrace{NH_4^+(aq) + Cl^-(aq)}_{\text{The ions of } NH_4Cl} + H_2O$$

In the gas phase, the proton is transferred directly from the HCl molecule to the NH_3 molecule.

Electron pair in the bond becomes a lone pair on Cl as the chloride ion is formed.

$$H-\underset{H}{\overset{H}{N}}: + H-\ddot{\underset{..}{Cl}}: \longrightarrow \left[H-\underset{H}{\overset{H}{N}}-H\right]^+ + \left[:\ddot{\underset{..}{Cl}}:\right]^-$$

Electron pair on N binds to a proton (H^+) as the proton separates from the electron pair of the H—Cl bond.

We are using arrows to show how electrons shift and rearrange during the formation of the ions. The electron pair on the nitrogen atom of the NH_3 molecule binds to H^+ as it is removed from the electron pair in the H—Cl bond. The electron pair in the H—Cl bond shifts entirely to the Cl as the Cl^- ion is formed.

As ammonium ions and chloride ions form, they attract each other, gather, and settle as crystals of ammonium chloride.

The Brønsted-Lowry concept views acid-base reactions as proton transfers

Johannes Brønsted (1879–1947), a Danish chemist, and Thomas Lowry (1874–1936), a British scientist, realized that the important event in most acid–base reactions is simply the transfer of a proton from one species to another. Therefore, they redefined *acids* as species that donate protons and *bases* as species that accept protons. The heart of the *Brønsted–Lowry concept of acids and bases* is that *acid–base reactions are proton transfer reactions.* The definitions are therefore very simple.

[1] When the single electron is removed from a hydrogen atom, what remains is just the nucleus of the atom, which is a proton. Therefore, a hydrogen ion, H^+, consists of a proton, and the terms *proton* and *hydrogen ion* are often used interchangeably.

Brønsted–Lowry Definitions of Acids and Bases[2]

An **acid** is a proton donor.
A **base** is a proton acceptor.

Brønsted-Lowry acids and bases

Accordingly, hydrogen chloride is an acid because when it reacts with ammonia, HCl molecules donate protons to NH_3 molecules. Similarly, ammonia is a base because NH_3 molecules accept protons.

Even when water is the solvent, chemists use the Brønsted–Lowry definitions more often than those of Arrhenius. Thus, the reaction between hydrogen chloride and water to form hydronium ion (H_3O^+) and chloride ion (Cl^-), which is another proton transfer reaction, is clearly a Brønsted–Lowry acid–base reaction. Molecules of HCl are the acid in this reaction, and water molecules are the base. HCl molecules collide with water molecules and protons transfer during the collisions.

water + hydrogen chloride → collision "complex" of proper orientation and energy for proton transfer → hydronium ion + chloride ion

Conjugate acids and bases differ by a single proton

Under the Brønsted–Lowry view, it is useful to consider any acid–base reaction as a chemical equilibrium, having both a forward and a reverse reaction. We first encountered a chemical equilibrium in a discussion of weak acids on page 142. Let's examine it again in the light of the Brønsted–Lowry definitions, using formic acid, $HCHO_2$, as an example. Formic acid is a weak acid, so we represent its ionization as a chemical equilibrium in which water is not just a solvent but also a chemical reactant, a proton acceptor.[3]

$$HCHO_2(aq) + H_2O \rightleftharpoons H_3O^+(aq) + CHO_2^-(aq)$$

H—O—C(=O)—H
formic acid ($HCHO_2$)
Only the H in red is available in an acid–base reaction.

$^-$O—C(=O)—H
formate ion (CHO_2^-)

In the forward reaction, a formic acid molecule donates a proton to the water molecule and changes to a formate ion, CHO_2^- (see Figure 15.2*a*). Thus $HCHO_2$ behaves as a **Brønsted acid,** a **proton donor.** Because water accepts this proton from $HCHO_2$, water behaves as a **Brønsted base,** a **proton acceptor.**

Now let's look at the reverse reaction (see Figure 15.2*b*). In it, H_3O^+ behaves as a Brønsted acid because it donates a proton to the CHO_2^- ion. The CHO_2^- ion behaves as a Brønsted base by accepting the proton.

The equilibrium involving $HCHO_2$, H_2O, H_3O^+, and CHO_2^- is typical of proton transfer equilibria in general, in that we can identify *two* acids (e.g., $HCHO_2$ and H_3O^+) and *two* bases (e.g., H_2O and CHO_2^-). Notice that in the aqueous formic acid equilibrium, the acid on the right of the arrows (H_3O^+) is formed from the base on the left (H_2O), and the base on the right (CHO_2^-) is formed from the acid on the left ($HCHO_2$).

[2] Although Brønsted and Lowry are both credited with defining acids and bases in terms of proton transfer, Brønsted carried the concepts further. For the sake of brevity, we will often use the terms *Brønsted acid* and *Brønsted base* when referring to the substances involved in proton transfer reactions.

[3] Just as we have represented acetic acid as $HC_2H_3O_2$, placing the ionizable or acidic hydrogen first (as we do in HCl or HNO_3), so we give the formula of formic acid as $HCHO_2$ and the formula of the formate ion as CHO_2^-. As the chemical structure shows, however, the acidic hydrogen in formic acid resides on O, not on C. Many chemists prefer to write formic acid as HCO_2H or HCOOH.

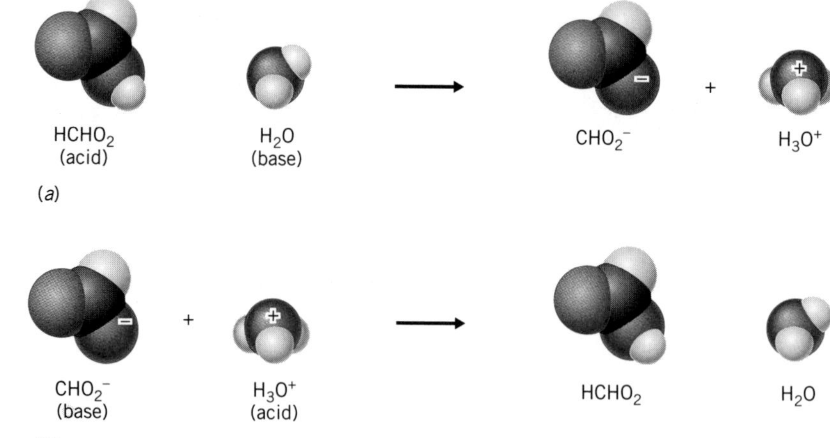

FIG. 15.2 Brønsted acids and bases in aqueous formic acid. (*a*) Formic acid transfers a proton to a water molecule. $HCHO_2$ is the acid and H_2O is the base. (*b*) When hydronium ion transfers a proton to the CHO_2^- ion, H_3O^+ is the acid and CHO_2^- is the base.

TOOLS *Conjugate acids and bases differ by one proton, H^+*

Two substances that differ from each other only by one proton are referred to as a **conjugate acid–base pair.** Thus, H_3O^+ and H_2O are such a pair; they are alike except that H_3O^+ has one more proton than H_2O. One member of the pair is called the **conjugate acid** because it is the proton donor of the two. The other member is the **conjugate base,** because it is the pair's proton acceptor. We say that *H_3O^+ is the conjugate acid of H_2O,* and *H_2O is the conjugate base of H_3O^+*. Notice that *the acid member of the pair has one more H^+ than the base member.*

Notice how the conjugate acid always has one more H^+ than the conjugate base.

The pair $HCHO_2$ and CHO_2^- is the other conjugate acid–base pair in the aqueous formic acid equilibrium. $HCHO_2$ has one more H^+ than CHO_2^-, so the conjugate acid of CHO_2^- is $HCHO_2$; the conjugate base of $HCHO_2$ is CHO_2^-. One way to highlight the two members of a conjugate acid–base pair in an equilibrium equation is to connect them by a line.

TOOLS *A Brønsted–Lowry acid–base equilibrium has two acid–base pairs*

$$\overbrace{\underset{\text{acid}}{HCHO_2} + \underbrace{\underset{\text{base}}{H_2O} \rightleftharpoons \underset{\text{acid}}{H_3O^+}}_{\text{conjugate pair}} + \underset{\text{base}}{CHO_2^-}}^{\text{conjugate pair}}$$

In any Brønsted–Lowry acid–base equilibrium, there are invariably *two* conjugate acid–base pairs. It is important that you learn how to pick them out of an equation by inspection and to write them from formulas.

EXAMPLE 15.1
Determining the Formulas of Conjugate Acids and Bases

What is the conjugate base of nitric acid, HNO_3, and what is the conjugate acid of the hydrogen sulfate ion, HSO_4^-?

ANALYSIS: The first tool in this chapter says that members of any conjugate acid–base pair differ by one H^+, with the member having the greater number of hydrogens being the acid. We are asked to find the conjugate base of HNO_3, so HNO_3 must be the acid member of the pair. To find the formula of the base, we remove one H^+ from the acid, HNO_3. This is equivalent to removing one hydrogen from the acid and decreasing its positive charge by one unit (or, increasing the negative charge by one unit).

We are also asked to find the conjugate acid of HSO_4^-, so HSO_4^- must be the base member of an acid–base pair. To find the formula of the acid, we add one H^+ to the base, which is equivalent to adding one hydrogen to the formula of the base and increasing its positive charge (or decreasing its negative charge) by one unit.

SOLUTION: Removing one H^+ (both the atom and the charge) from HNO_3 leaves NO_3^-. The nitrate ion, NO_3^-, is thus the conjugate base of HNO_3. Adding an H^+ to HSO_4^- gives its conjugate acid, H_2SO_4. (Notice that the charge goes from 1− to zero because we've added the positively charged H^+.)

ARE THE ANSWERS REASONABLE? As a check, we can quickly compare the two formulas in each pair.

$$\begin{array}{ll} HNO_3 & NO_3^- \\ H_2SO_4 & HSO_4^- \end{array}$$

In each case, the formula on the right has one less H^+ than the one on the left, so it is the conjugate base. We've answered the question correctly.

Practice Exercise 1: Which of the following are conjugate acid–base pairs? Describe why the others are not true conjugate acid–base pairs. (a) H_3PO_4 and $H_2PO_4^-$, (b) HI and H^+, (c) NH_2^- and NH_3, (d) HNO_2 and NH_4^+, (e) CO_3^{2-} and CN^-, (f) HPO_4^{2-} and $H_2PO_4^-$ (Hint: Recall that conjugate acid–base pairs must differ by one H^+.)

Practice Exercise 2: Write the formula of the conjugate base for each of the following Brønsted acids. (a) H_2O, (b) HI, (c) HNO_2, (d) H_3PO_4, (e) $H_2PO_4^-$, (f) HPO_4^{2-}, (g) H_2S, (h) NH_4^+

Practice Exercise 3: Write the formula of the conjugate acid for each of the following Brønsted bases. (a) HO_2^-, (b) SO_4^{2-}, (c) CO_3^{2-}, (d) CN^-, (e) NH_2^-, (f) NH_3, (g) $H_2PO_4^-$, (h) HPO_4^{2-}

EXAMPLE 15.2
Identifying Conjugate Acid-Base Pairs in a Brønsted–Lowry Acid-Base Reaction

The anion of sodium hydrogen sulfate, HSO_4^-, reacts as follows with the phosphate ion, PO_4^{3-}.

$$HSO_4^-(aq) + PO_4^{3-}(aq) \longrightarrow SO_4^{2-}(aq) + HPO_4^{2-}(aq)$$

Identify the two conjugate acid–base pairs.

Sodium hydrogen sulfate (also called sodium bisulfate) is used in the manufacture of certain kinds of cement and to clean oxide coatings from metals.

ANALYSIS: There are two things to look for in identifying the conjugate acid–base pairs in an equation. One is our tool that reminds us that the members of a conjugate pair are alike except for the number of hydrogens and charge. The second is the tool that the members of each pair must be on opposite sides of the arrow in the Brønsted–Lowry acid–base equation. In each pair, of course, the acid is the one with the greater number of hydrogens.

SOLUTION: Two of the formulas in the equation contain "PO_4," so they must belong to the same conjugate pair. The one with the greater number of hydrogens, HPO_4^{2-}, must be the Brønsted acid, and the other, PO_4^{3-}, must be the Brønsted base. Therefore, one conjugate acid–base pair is HPO_4^{2-} and PO_4^{3-}. The other two ions, HSO_4^- and SO_4^{2-}, belong to the second conjugate acid–base pair; HSO_4^- is the conjugate acid and SO_4^{2-} is the conjugate base.

conjugate pair

$$\underset{\text{acid}}{HSO_4^-(aq)} + \underset{\text{base}}{PO_4^{3-}(aq)} \longrightarrow \underset{\text{base}}{SO_4^{2-}(aq)} + \underset{\text{acid}}{HPO_4^{2-}(aq)}$$

conjugate pair

IS THE ANSWER REASONABLE? A check satisfies us that we have fulfilled the requirements that each conjugate pair has one member on one side of the arrow and the other member on the opposite side of the arrow and that the members of each pair differ from each other by one (and *only* one) H^+.

Practice Exercise 4: Sodium cyanide solution, when poured into excess hydrochloric acid, releases hydrogen cyanide as a gas. The reaction is

$$NaCN(aq) + HCl(aq) \longrightarrow HCN(g) + NaCl(aq)$$

Identify the conjugate acid–base pairs in this reaction. (Hint: It may be more obvious if the spectator ions are removed.)

Practice Exercise 5: One kind of baking powder contains sodium bicarbonate and calcium dihydrogen phosphate. When water is added, a reaction occurs by the following net ionic equation.

$$HCO_3^-(aq) + H_2PO_4^-(aq) \longrightarrow H_2CO_3(aq) + HPO_4^{2-}(aq)$$

Identify the two Brønsted acids and the two Brønsted bases in this reaction. (The H_2CO_3 decomposes to release CO_2, which causes the cake batter to rise.)

Practice Exercise 6: When some of the strong cleaning agent "trisodium phosphate" is mixed with household vinegar, which contains acetic acid, the following equilibrium is one of the many that are established. (The position of equilibrium lies to the right.) Identify the pairs of conjugate acids and bases.

$$PO_4^{3-}(aq) + HC_2H_3O_2(aq) \rightleftharpoons HPO_4^{2-}(aq) + C_2H_3O_2^-(aq)$$

Amphoteric substances can behave as either acids or bases

Some molecules or ions are able to function either as an acid or as a base, depending on the kind of substance mixed with them. For example, in its reaction with hydrogen chloride, water behaves as a *base* because it *accepts* a proton from the HCl molecule.

$$\underset{\text{base}}{H_2O} + \underset{\text{acid}}{HCl(g)} \longrightarrow H_3O^+(aq) + Cl^-(aq)$$

On the other hand, water behaves as an *acid* when it reacts with the weak base ammonia.

$$\underset{\text{acid}}{H_2O} + \underset{\text{base}}{NH_3(aq)} \rightleftharpoons NH_4^+(aq) + OH^-(aq)$$

Here, H_2O *donates* a proton to NH_3 in the forward reaction.

A substance that can be either an acid or a base depending on the other substance present is said to be **amphoteric.** Another term is **amphiprotic** to stress that the *proton* donating or accepting ability is of central concern.

From the Greek *amphoteros*, "partly one and partly the other."

Amphoteric or amphiprotic substances may be either molecules or ions. For example, anions of acid salts, such as the bicarbonate ion of baking soda, are amphoteric. The HCO_3^- ion can either donate a proton to a base or accept a proton from an acid. Thus, toward the hydroxide ion, the bicarbonate ion is an acid; it donates its proton to OH^-.

$$\underset{\text{acid}}{HCO_3^-(aq)} + \underset{\text{base}}{OH^-(aq)} \longrightarrow CO_3^{2-}(aq) + H_2O$$

Toward hydronium ion, however, HCO_3^- is a base; it accepts a proton from H_3O^+.

$$\underset{\text{base}}{HCO_3^-(aq)} + \underset{\text{acid}}{H_3O^+(aq)} \longrightarrow H_2CO_3(aq) + H_2O$$

[Recall that H_2CO_3 (carbonic acid) almost entirely decomposes to $CO_2(g)$ and water as it forms.]

Practice Exercise 7: Which of the following are amphoteric and which are not? Provide reasons for your decisions. (a) $H_2PO_4^-$, (b) HPO_4^{2-}, (c) H_2S, (d) H_3PO_4, (e) NH_4^+, (f) H_2O, (g) HI, (h) HNO_2 (Hint: Amphoteric substances must be able to provide an H^+ and also react with an H^+.)

Practice Exercise 8: The anion of sodium monohydrogen phosphate, Na_2HPO_4, is amphoteric. Using H_3O^+ and OH^-, write net ionic equations that illustrate this property.

15.2 STRENGTHS OF BRØNSTED ACIDS AND BASES FOLLOW PERIODIC TRENDS

Brønsted acids and bases have differing abilities to lose or gain protons. In this section we examine how these abilities can be compared and how we can anticipate differences according to the locations of key elements in the periodic table.

Acid and base strengths are measured relative to a standard

When we speak of the strength of a Brønsted acid, we are referring to its ability to donate a proton to a base. We measure this by determining the extent to which the reaction of the acid with the base proceeds toward completion—the more complete the reaction, the stronger the acid. To compare the strengths of a series of acids, we have to select some reference base, and because we are interested most in reactions in aqueous media, that base is usually water (although other reference bases could also be chosen).

In Chapter 4 we discussed strong and weak acids from the Arrhenius point of view, and much of what we said there applies when the same acids are studied using the Brønsted–Lowry concept. Thus, acids such as HCl and HNO_3 react completely with water to give H_3O^+ because they are strong proton donors. Hence, we classify them as *strong Brønsted acids.* On the other hand, acids such as HNO_2 (nitrous acid) and $HC_2H_3O_2$ (acetic acid) are much weaker proton donors. Their reactions with water are far from complete and we classify them as *weak acids.*

In a similar manner, the relative strengths of Brønsted *bases* are assigned according to their abilities to accept and bind protons. Once again, to compare strengths, we have to choose a standard acid. Because water is amphiprotic, it can also serve as the standard acid. Substances that are powerful proton acceptors, such as the oxide ion, react completely and are considered to be *strong Brønsted bases.*

$$O^{2-} + H_2O \xrightarrow{100\%} 2OH^-$$

Weaker proton acceptors, such as ammonia, undergo incomplete reactions with water; we classify them *weak bases.*

Hydronium ion and hydroxide ion are the strongest acid and base that can exist in the presence of water

Both HCl and HNO_3 are very powerful proton donors. When placed in water they react completely, losing their protons to water molecules to yield H_3O^+ ions. Representing them by the general formula HA, we have

$$\underset{\text{acid}}{HA} + \underset{\text{base}}{H_2O} \xrightarrow{100\%} \underset{\text{acid}}{H_3O^+} + \underset{\text{base}}{A^-}$$

Because both reactions go to completion, we really can't tell which of the two, HCl or HNO_3, is actually the better proton donor (stronger acid). This would require a reference base less willing than water to accept protons. In water, both HCl and HNO_3 are converted quantitatively to another acid, H_3O^+. The conclusion, therefore, is that *H_3O^+ is the strongest acid we will ever find in an aqueous solution,* because stronger acids react completely with water to give H_3O^+.

A similar conclusion is reached regarding hydroxide ion. We noted that the strong Brønsted base O^{2-} reacts completely with water to give OH^-. Another very powerful proton acceptor is the amide ion, NH_2^-, which also reacts completely with water.

$$\underset{\text{base}}{NH_2^-} + \underset{\text{acid}}{H_2O} \xrightarrow{100\%} \underset{\text{acid}}{NH_3} + \underset{\text{base}}{OH^-}$$

Using water as the reference acid, we can't tell which is the better proton acceptor, O^{2-} or NH_2^-, because both react completely, being replaced by another base, OH^-. Therefore, we can say that *OH^- is the strongest base we will ever find in an aqueous solution,* because stronger bases react completely with water to give OH^-.

Comparing the acid–base strengths of conjugate pairs

As we noted earlier, the chemical equation for the equilibrium present in an aqueous solution of a weak acid actually shows two acids. One is almost always stronger than the other, and the position of the equilibrium tells us which of the two acids is stronger. Let's see how this works using the acetic acid equilibrium, in which the position of equilibrium lies to the left.

$$\underset{\text{acid}}{HC_2H_3O_2(aq)} + \underset{\text{base}}{H_2O} \rightleftharpoons \underset{\text{acid}}{H_3O^+(aq)} + \underset{\text{base}}{C_2H_3O_2^-(aq)}$$

The two Brønsted acids in this equilibrium are $HC_2H_3O_2$ and H_3O^+, and it's helpful to think of them as competing with each other in donating protons to acceptors. The fact that nearly all potential protons stay on the $HC_2H_3O_2$ molecules, and only a relative few spend their time on the H_3O^+ ions, means that the *hydronium ion is a better proton donor than the acetic acid molecule.* Thus, the hydronium ion is a stronger Brønsted acid than acetic acid, and we inferred this relative acidity from the position of equilibrium.

The acetic acid equilibrium also has two bases, $C_2H_3O_2^-$ and H_2O. Both compete for any available protons. But at equilibrium, most of the protons originally carried by acetic acid are still found on $HC_2H_3O_2$ molecules; relatively few are joined to H_2O in the form of H_3O^+ ions. This means that *acetate ions must be more effective than water molecules at obtaining and holding protons from proton donors.* This is the same as saying that the acetate ion is a stronger base than the water molecule. So this illustrates a relative basicity that we are able to infer from the position of the acetic acid equilibrium.

Notice what our discussion of the two conjugate pairs in the acetic acid equilibrium has brought out. Both the weaker of the two acids and the weaker of the two bases are found on the same side of the equation, which is the side favored by the position of equilibrium.

TOOLS

Relative strengths of conjugate acids and bases are related to the position of equilibrium

The position of an acid–base equilibrium favors the weaker acid and base.

$$\underset{\text{weaker acid}}{HC_2H_3O_2(aq)} + \underset{\text{weaker base}}{H_2O} \rightleftharpoons \underset{\text{stronger acid}}{H_3O^+(aq)} + \underset{\text{stronger base}}{C_2H_3O_2^-(aq)}$$

← *Position of equilibrium lies to the left, in favor of the weaker acid and base.*

There's a reciprocal relationship within a conjugate acid–base pair

One aid in predicting the relative strengths of acids and bases is the existence of a reciprocal relationship.

TOOLS

A strong Brønsted acid has a weak conjugate base

The stronger a Brønsted acid is, the weaker is its conjugate base.

To illustrate, recall that $HCl(g)$ is a very strong Brønsted acid; it's 100% ionized in a dilute aqueous solution.

$$HCl(g) + H_2O \xrightarrow{100\%} H_3O^+(aq) + Cl^-(aq)$$

As we explained in Chapter 4, we don't write double equilibrium arrows for the ionization of a strong acid. Not doing so with HCl is another way of saying that the chloride ion, the conjugate base of HCl, must be a very weak Brønsted base. Even in the presence of H_3O^+, a very strong proton donor, chloride ions aren't able to win protons. So HCl, the strong acid, has a particularly weak conjugate base, Cl^-.

There's a matching reciprocal relationship.

The weaker a Brønsted acid is, the stronger is its conjugate base.

TOOLS
A strong Brønsted base has a weak conjugate acid

Consider, for example, the conjugate pair OH^- and O^{2-}. The hydroxide ion is the conjugate acid and the oxide ion is the conjugate base. But the hydroxide ion must be an *extremely* weak Brønsted acid; in fact, we've known it so far only as a base. Given the extraordinary weakness of OH^- as an acid, its conjugate base, the oxide ion, must be an exceptionally strong base. And as you've already learned, oxide ion is such a strong base that its reaction with water is 100% complete. That's why we don't write double equilibrium arrows in the equation for the reaction.

$$\underset{\text{base}}{O^{2-}} + \underset{\text{acid}}{H_2O} \xrightarrow{100\%} \underset{\text{acid}}{OH^-} + \underset{\text{base}}{OH^-}$$

The very strong base O^{2-} has a very weak conjugate acid, OH^-.

An amphoteric substance will act as a base if reacted with an acid but it will act as an acid if mixed with a base. You might ask, if two amphoteric substances are mixed together which will act as the acid and which as the base? The obvious answer is that the stronger acid will act as the acid and the other will be the base. To discover the stronger acid, consider the following reaction

$$H_2S(aq) + HCO_3^-(aq) \rightleftharpoons HS^-(aq) + H_2CO_3(aq)$$

where the position of equilibrium lies to the left (with the reactants). This is interpreted as meaning that $HCO_3^-(aq)$ is a weaker base than $HS^-(aq)$. Therefore, when we mix the two together the reaction will be

$$HS^-(aq) + HCO_3^-(aq) \longrightarrow H_2S(aq) + CO_3^{2-}(aq)$$

because the stronger base, $HS^-(aq)$, will remove the H^+ from the weaker base HCO_3^-.

EXAMPLE 15.3
Using Reciprocal Relationships to Predict Equilibrium Positions

In the reaction below, will the position of equilibrium lie to the left or the right, given the fact that acetic acid is known to be a stronger acid than the hydrogen sulfite ion?

$$HSO_3^-(aq) + C_2H_3O_2^-(aq) \rightleftharpoons HC_2H_3O_2(aq) + SO_3^{2-}(aq)$$

ANALYSIS: We just learned that the position of an acid–base equilibrium favors the weaker acid and base. That's the tool we need to solve this problem. We have to identify which acid and base make up the weaker set. When we do this, we'll have discovered which substances make up the stronger set, and we can then predict where the position of equilibrium will lie.

SOLUTION: We'll write the equilibrium equation using the given fact about the relative strengths of acetic acid and the hydrogen sulfite ion to start writing labels.

$$\underset{\text{weaker acid}}{HSO_3^-(aq)} + C_2H_3O_2^-(aq) \rightleftharpoons \underset{\text{stronger acid}}{HC_2H_3O_2(aq)} + SO_3^{2-}(aq)$$

Now we'll use our tool about the reciprocal relationships to label the two bases. The stronger acid must have the weaker conjugate base; the weaker acid must have the stronger conjugate base.

$$\underset{\text{weaker acid}}{HSO_3^-(aq)} + \underset{\text{weaker base}}{C_2H_3O_2^-(aq)} \rightleftharpoons \underset{\text{stronger acid}}{HC_2H_3O_2(aq)} + \underset{\text{stronger base}}{SO_3^{2-}(aq)}$$

Finally, because the position of equilibrium favors the weaker acid and base, the position of equilibrium lies to the left.

IS THE ANSWER REASONABLE? There are two things we can check. First, both of the weaker conjugates should be on the same side of the equation. They are, so that suggests we've made the correct assignments. Second, the reaction will proceed farther in the direction of the weaker acid and base, so that places the position of equilibrium on the left, which agrees with our answer.

Practice Exercise 9: Given that HSO_4^- is a stronger acid than HPO_4^{2-}, what is the chemical reaction if solutions containing those ions are mixed together? (Hint: One of these must be an acid and the other a base.)

Practice Exercise 10: Given that HSO_4^- is a stronger acid than HPO_4^{2-}, determine whether the substances on the left of the arrows or those on the right are favored in the following equilibrium.

$$HSO_4^-(aq) + PO_4^{3-}(aq) \rightleftharpoons SO_4^{2-}(aq) + HPO_4^{2-}(aq)$$

Periodic trends exist in the strengths of binary acids

Many of the binary compounds between hydrogen and nonmetals, which we may represent by HX, H_2X, H_3X, etc., are acidic and are called **binary acids.** HCl is a common example, but Table 15.1 lists the others that are acids in water. The strong acids are marked by asterisks.

TABLE 15.1 Acidic Binary Compounds of Hydrogen and Nonmetals[a]

Group VIA	
(H_2O)	
H_2S	Hydrosulfuric acid
H_2Se	Hydroselenic acid
H_2Te	Hydrotelluric acid
Group VIIA	
HF	Hydrofluoric acid
*HCl	Hydrochloric acid
*HBr	Hydrobromic acid
*HI	Hydriodic acid

[a] The *names* are for the aqueous solutions of these compounds. Strong acids are marked with asterisks.

The relative strengths of binary acids correlate with the periodic table in two ways.

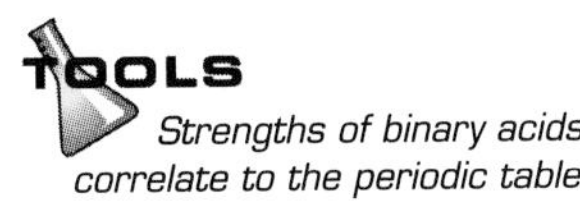

TOOLS
Strengths of binary acids correlate to the periodic table

> The strengths of the binary acids increase from left to right within the same period.
>
> The strengths of binary acids increase from top to bottom within the same group.

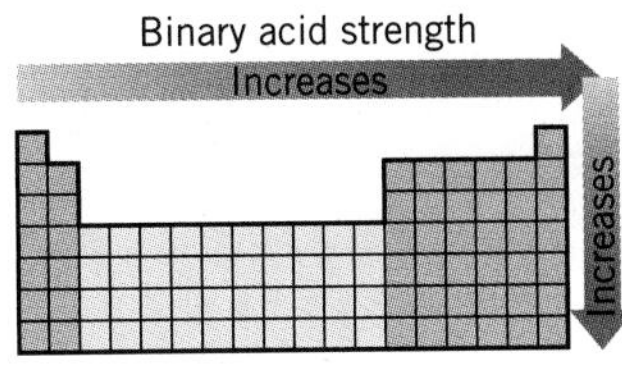

First, as we go left to right within a period, the increase in electronegativities causes the corresponding H—X bonds to become more polar, making the partial positive charge ($\delta+$) on H greater. This makes it easier for the hydrogen to separate as H^+, so the molecule becomes a better proton donor. For example, as we go from S to Cl in Period 3, the electronegativity increases and we find that HCl is a stronger acid than H_2S. A similar increase in electronegativity occurs going left to right in Period 2, from O to F, and HF is a stronger acid than H_2O.

Second, binary acids become stronger from the top of a group to the bottom. Among the binary acids of the halogens, for example, the following is the order of relative acidity.[4]

$$HF < HCl < HBr < HI$$

Thus, HF is the weakest acid in the series, and HI is the strongest. The identical trend occurs in the series of the binary acids of Group VIA elements, having formulas of the general type H_2X. These trends in acidity are opposite what we would expect on the basis of trends in electronegativities, which tell us that the H—F bond is more polar than the H—I bond and that the O—H bond is more polar than the H—S bond.

In understanding the proton donating ability of an acid, one of the most important factors to consider is the strength of the H—X bond. Breaking this bond is essential for the hydrogen to separate as an H^+ ion, so anything that contributes to variations in bond strength will also impact variations in acid strength.

In general, small atoms tend to form stronger bonds than large atoms. Moving horizontally within a period, atomic size varies relatively little, so the strengths of the H—X bonds are nearly the same. As a result, the most significant influence on acid strength is variations in the polarity of the H—X bonds. Descending a group, however, there is a significant increase in atomic size from one element to the next, which is accompanied by a rapid decrease in the H—X bond strength. Apparently, this more than compensates for the decrease in the polarity of the H—X bonds, and the molecules become better able to release protons as we go down a group. The net effect of the two opposing factors, therefore, is an increase in the strengths of the binary acids as we go from top to bottom in a group.

Practice Exercise 11: Order the following groups of acids from the weakest to the strongest. (a) HI, HF, HBr, (b) HCl, PH_3, H_2S, (c) H_2Te, H_2O, H_2Se, (d) AsH_3, HBr, H_2Se, (e) HI, PH_3, H_2Se (Hint: These are all binary acids.)

Practice Exercise 12: Using *only* the periodic table, choose the stronger acid of each pair. (a) H_2Se or HBr, (b) H_2Se or H_2Te, (c) H_2O or H_2S

Trends exist in the strengths of oxoacids

Acids composed of hydrogen, oxygen, and some other element are called **oxoacids** (see Table 15.2). Those that are strong acids in water are marked in the table by asterisks.

[4] We compare acid strengths by the acid's ability to donate a proton to a particular base. As we noted earlier, for strong acids such as HCl, HBr, and HI, water is too strong a proton acceptor to permit us to see differences among their proton-donating abilities. All three of these acids are completely ionized in water and so appear to be of equal strength, a phenomenon called the *leveling effect* (the differences are obscured or *leveled* out). To compare the acidities of these acids, a solvent that is a weaker proton acceptor than water (HF or $HC_2H_3O_2$, for example) has to be used.

TABLE 15.2 Some Oxoacids of Nonmetals and Metalloids[a]

Group IVA	Group VA	Group VIA	Group VIIA
H_2CO_3 Carbonic acid	*HNO_3 Nitric acid		HFO Hypofluorous acid
	HNO_2 Nitrous acid		
	H_3PO_4 Phosphoric acid	*H_2SO_4 Sulfuric acid	*$HClO_4$ Perchloric acid
	H_3PO_3 Phosphorous acid[b]	H_2SO_3 Sulfurous acid[c]	*$HClO_3$ Chloric acid
			$HClO_2$ Chlorous acid
			HClO Hypochlorous acid
	H_3AsO_4 Arsenic acid	*H_2SeO_4 Selenic acid	*$HBrO_4$ Perbromic acid[d]
	H_3AsO_3 Arsenous acid	H_2SeO_3 Selenous acid	*$HBrO_3$ Bromic acid
			HIO_4 Periodic acid (H_5IO_6)[e]
			HIO_3 Iodic acid

[a]Strong acids are marked with asterisks.
[b]Phosphorous acid, despite its formula, is only a diprotic acid.
[c]Hypothetical. An aqueous solution actually contains just dissolved sulfur dioxide, $SO_2(aq)$.
[d]Pure perbromic acid is unstable; a dihydrate is known.
[e]H_5IO_6 is formed from $HIO_4 + 2H_2O$.

A feature common to the structures of all oxoacids is the presence of O—H groups bonded to some central atom. For example, the structures of two oxoacids of the Group VIA elements are

$$\begin{array}{c} \mathrm{O} \\ \| \\ \mathrm{H{-}O{-}S{-}O{-}H} \\ \| \\ \mathrm{O} \end{array} \qquad \begin{array}{c} \mathrm{O} \\ \| \\ \mathrm{H{-}O{-}Se{-}O{-}H} \\ \| \\ \mathrm{O} \end{array}$$

H_2SO_4
sulfuric acid

H_2SeO_4
selenic acid

When an oxoacid ionizes, the hydrogen that's lost as an H^+ comes from the same kind of bond in every instance, specifically, an O—H bond. The "acidity" of such a hydrogen, meaning the ease with which it's released as H^+, is determined by how the group of atoms attached to the oxygen affects the polarity of the O—H bond. If this group of atoms makes the O—H bond more polar, it will cause the H to come off more easily as H^+ and thereby increase the acidity of the molecule.

$$\overset{}{G}{-}\overset{\delta-}{\mathrm{O}}{-}\overset{\delta+}{\mathrm{H}}$$

If the group of atoms, *G*, attached to the O—H group is able to draw electron density from the O atom, the O will pull electron density from the O—H bond, thereby making the bond more polar.

It turns out that there are two principal factors that determine how the polarity of the O—H bond is affected. One is the electronegativity of the central atom in the oxoacid and the other is the number of oxygens attached to the central atom.

The electronegativity of the central atom affects the acidity of an oxoacid

To study the effects of the electronegativity of the central atom, we must compare oxoacids having the same number of oxygens. When we do this, we find that as the electronegativity of the central atom increases, the oxoacid becomes a better proton donor (i.e., a stronger acid). The following diagram illustrates the effect.

$$-\overset{|}{\underset{|}{X}}-\overset{\delta-}{O}-\overset{\delta+}{H}$$

As the electronegativity of X increases, electron density is drawn away from O, which draws electron density away from the O—H bond. This makes the bond more polar and makes the molecule a better proton donor.

Because electronegativity increases from bottom to top within a group and from left to right within a period, we can make the following generalization.

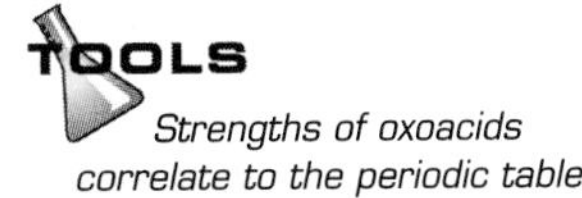

Strengths of oxoacids correlate to the periodic table

> When the central atoms of oxoacids hold the same number of oxygen atoms, the acid strength increases from bottom to top within a group and from left to right within a period.

In Group VIA, for example, H_2SO_4 is a stronger acid than H_2SeO_4 because sulfur is more electronegative than selenium. Similarly, among the halogens, acid strength increases for acids with the formula HXO_4 as follows

$$HIO_4 < HBrO_4 < HClO_4$$

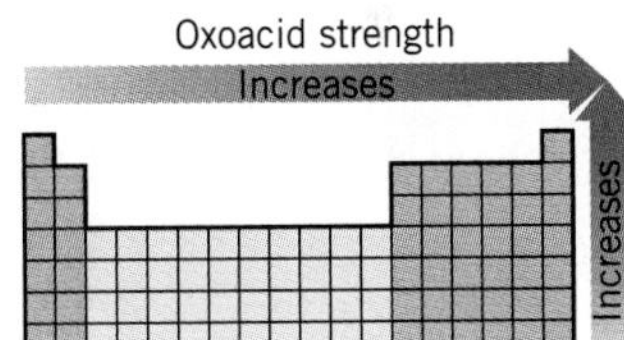

Going from left to right within Period 3, we can compare the acids H_3PO_4, H_2SO_4, and $HClO_4$, where we find the following order of acidities.

$$H_3PO_4 < H_2SO_4 < HClO_4$$

Practice Exercise 13: Which is the stronger acid? (a) $HClO_3$ or $HBrO_3$, (b) H_3PO_4 or H_2SO_4 (Hint: Note that each pair has the same number of oxygen atoms.)

Practice Exercise 14: In each pair indicate the weaker acid. (a) H_3PO_4 or H_3AsO_4, (b) HIO_4 or H_2TeO_4

The number of oxygens affects the acidity of an oxoacid

Comparing oxoacids with the same central atom, we find that as the number of *lone oxygens* increases, the oxoacid becomes a better proton donor. (A *lone oxygen* is one that is bonded only to the central atom and not to a hydrogen.) Thus, comparing HNO_3 with HNO_2, we find that HNO_3 is the stronger acid. To understand why, let's look at their molecular structures.

$$\overset{\delta-}{O}{=}N{-}O{-}H \quad < \quad (\overset{\delta-}{O})(\overset{\delta-}{O})N{-}O{-}H$$

nitrous acid nitric acid

In an oxoacid, lone oxygens pull electron density away from the central atom, which increases the central atom's ability to draw electron density away from the O—H bond. It's as though the lone oxygens make the central atom more electronegative. Therefore, the more lone oxygens that are attached to a central atom, the more polar will be the O—H bonds of the acid and the stronger will be the acid. Thus, the two lone oxygens in HNO_3 produce a greater effect than the one lone oxygen in HNO_2, so HNO_3 is the stronger acid.

Oxygen is a very electronegative element and has a strong tendency to pull electron density away from any atom to which it is attached in an oxoacid.

Similar effects are seen among other oxoacids, as well. For the oxoacids of chlorine, for instance, we find this trend.

$$HClO < HClO_2 < HClO_3 < HClO_4$$

Comparing their structures, we have

$$:\ddot{\underset{..}{Cl}}-\ddot{\underset{..}{O}}-H \qquad \ddot{O}=\ddot{\underset{..}{Cl}}-\ddot{\underset{..}{O}}-H \qquad \ddot{O}=Cl(=\ddot{O}:)-\ddot{\underset{..}{O}}-H \qquad \ddot{O}=Cl(=\ddot{O}:)(=\ddot{O}:)-\ddot{\underset{..}{O}}-H$$

$HClO$ $\qquad$ $HClO_2$ $\qquad$ $HClO_3$ $\qquad$ $HClO_4$

Usually, the formula for hypochlorous acid is written HOCl. We've written it as HClO here to make it easier to follow the trend in acid strengths among the oxoacids of chlorine.

This leads to another generalization.

Strengths of oxo acids depend on lone oxygens

For a given central atom, the acid strength of an oxoacid increases with the number of oxygens held by the central atom.

The ability of lone oxygens to affect acid strength extends to organic compounds as well. For example, compare the molecules below.

$$H-CH_2-CH_2-O-H \qquad H-CH_2-C(=O)-O-H$$

ethanol $\qquad$ acetic acid

In water, ethanol (ethyl alcohol) is not acidic at all. Replacing the two hydrogens on the carbon adjacent to the OH group with an oxygen, however, yields acetic acid. The greater ability of oxygen to pull electron density from the carbon produces a greater polarity of the O—H bond, which is one factor that causes acetic acid to be a better proton donor than ethanol.

Delocalization of negative charge to the lone oxygens affects the basicity of the anion of an oxoacid

Earlier we noted the reciprocal relationship between the strength of an acid and that of its conjugate base. For oxoacids, the lone oxygens play a part in determining the basicity of the anion formed in the ionization reaction. Consider the acids $HClO_3$ and $HClO_4$.

$$H-O-Cl(=O)=O \qquad H-O-Cl(=O)(=O)=O$$

$HClO_3$ $\qquad$ $HClO_4$

From our earlier discussion, we expect $HClO_4$ to be a stronger acid than $HClO_3$, which it is. Ionizations of their protons yield the anions ClO_3^- and ClO_4^-.

$$\left[O-Cl(=O)=O\right]^- \qquad \left[O-Cl(=O)(=O)=O\right]^-$$

ClO_3^- $\qquad$ ClO_4^-

The concept of delocalization of electrons to provide stability was presented in Section 9.8.

In an **oxoanion** (an anion formed from an oxoacid) the negative charge is delocalized over the oxygens. In the ClO_3^- ion, the single negative charge is spread over three oxygens, so each carries a charge of about $\frac{1}{3}-$. This is not a *formal* charge obtained by the rules on page 322. We're only saying that the charge of $1-$ is delocalized over three oxygens to give each a charge of $\frac{1}{3}-$. By the same reasoning, in the ClO_4^- ion each oxygen carries a charge of about $\frac{1}{4}-$. The smaller negative charge on the oxygens in ClO_4^- makes this ion less able than ClO_3^- to attract H^+ ions from H_3O^+, so ClO_4^- is a weaker base than ClO_3^-. Thus, the anion of the stronger acid is the weaker base.

Practice Exercise 15: In each pair, select the stronger acid: (a) HIO_3 or HIO_4, (b) H_2TeO_3 or H_2TeO_4, (c) H_3AsO_3 or H_3AsO_4. (Hint: This problem focuses on the effect of oxygen atoms in acids.)

Practice Exercise 16: In each pair select the weaker acid: (a) H_2SO_4 or $HClO_4$, (b) H_3AsO_4 or H_2SO_4.

Other groups also affect the strength of organic acids

Acetic acid and most organic acids are characterized by the —COOH functional group. The delocalization of the electron on the $—COO^-$ stabilizes the anion after the H^+ ionizes. If other electronegative groups such as halogens are bonded to carbon atoms near the —COOH group, they increase the strength of the acid. As a result, chloroacetic acid is a stronger acid than acetic acid. Dichloroacetic acid and trichloroacetic acids are increasingly stronger acids.

$$CH_3COOH < CH_2ClCOOH < CHCl_2COOH < CCl_3COOH$$

Addition of each chlorine effectively withdraws more electron density from the —O—H bond, resulting in a weaker bond and a stronger acid.

15.3 LEWIS ACIDS AND BASES INVOLVE COORDINATE COVALENT BONDS

In our preceding discussions, acids and bases have been characterized by their tendency to lose or gain protons. However, there are many reactions not involving proton transfer that have properties we associate with acid–base reactions. For example, if gaseous SO_3 is passed over solid CaO, a reaction occurs in which $CaSO_4$ forms.

Remember, a coordinate covalent bond is just like any other covalent bond once it has formed. By using this term, we are following the origin of the electron pair that forms the bond.

$$CaO(s) + SO_3(g) \longrightarrow CaSO_4(s) \qquad (15.1)$$

If these reactants are dissolved in water first, they react to form $Ca(OH)_2$ and H_2SO_4, and when their solutions are mixed the following reaction takes place.

$$Ca(OH)_2(aq) + H_2SO_4(aq) \longrightarrow CaSO_4(s) + 2H_2O \qquad (15.2)$$

The same two initial reactants, CaO and SO_3, form the same ultimate product, $CaSO_4$. It certainly seems that if Reaction 15.2 is an acid–base reaction, we should be able to consider Reaction 15.1 to be an acid–base reaction, too. But there are no protons being transferred, so our definitions require further generalizations. These were provided by G. N. Lewis, after whom Lewis symbols are named.

Lewis Definitions of Acids and Bases

1. A **Lewis acid** is any ionic or molecular species that can accept a pair of electrons in the formation of a coordinate covalent bond.
2. A **Lewis base** is any ionic or molecular species that can donate a pair of electrons in the formation of a coordinate covalent bond.
3. **Neutralization** is the formation of a coordinate covalent bond between the donor (base) and the acceptor (acid).

Examples of Lewis acid–base reactions

The reaction between BF_3 and NH_3 illustrates a Lewis acid–base neutralization. The reaction is exothermic because a bond is formed between N and B, with the nitrogen donating an electron pair and the boron accepting it.

A similar reaction was described in Section 8.6 between BCl_3 and ammonia.

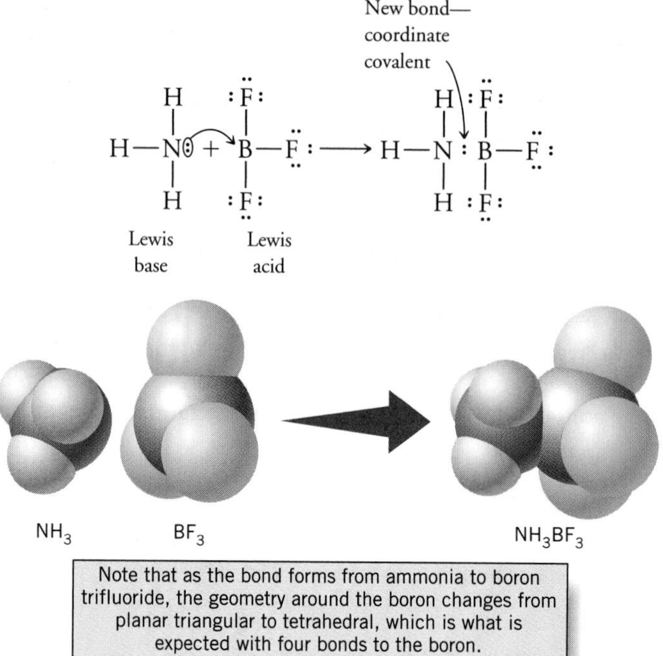

Note that as the bond forms from ammonia to boron trifluoride, the geometry around the boron changes from planar triangular to tetrahedral, which is what is expected with four bonds to the boron.

Compounds like BF_3NH_3, which are formed by simply joining two smaller molecules, are called *addition compounds.*

The ammonia molecule thus acts as a *Lewis base*. The boron atom in BF_3, having only six electrons in its valence shell and needing two more to achieve an octet, accepts the pair of electrons from the ammonia molecule. Hence, BF_3 is functioning as a *Lewis acid.*

As this example illustrates, Lewis bases are substances that have completed valence shells *and unshared pairs of electrons* (e.g., NH_3, H_2O, and O^{2-}). A Lewis acid, on the other hand, can be a substance with an incomplete valence shell, such as BF_3 or H^+.

A substance can also be a Lewis acid even when it has a central atom with a complete valence shell. This works when the central atom has a double bond that, by the shifting of an electron pair to an adjacent atom, can make room for an incoming pair of electrons from a Lewis base. Carbon dioxide is an example. When carbon dioxide is bubbled into aqueous sodium hydroxide, the gas is instantly trapped as the bicarbonate ion.

$$CO_2(g) + OH^-(aq) \longrightarrow HCO_3^-(aq)$$

Lewis acid–base theory represents the movement of electrons in this reaction as follows.

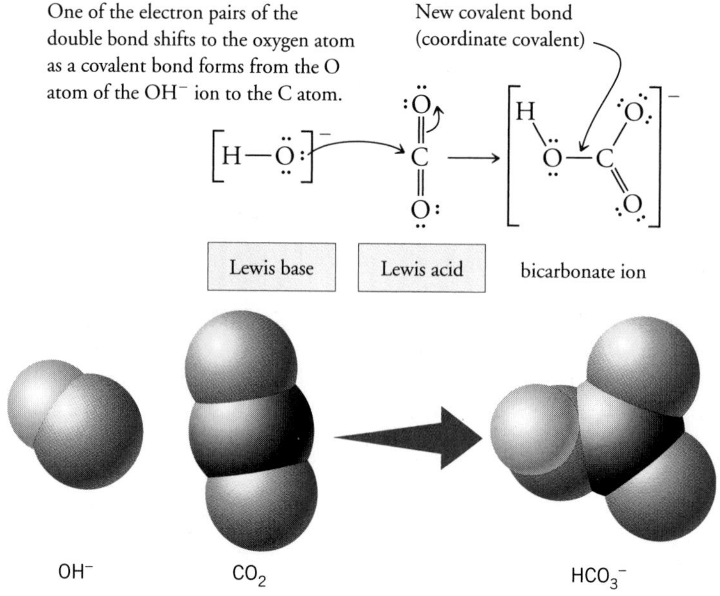

The donation of an electron pair *from* the oxygen of the OH^- ion produces a bond, so the OH^- ion is the Lewis base. The carbon atom of the CO_2 accepts the electron pair, so CO_2 is the Lewis acid.

Lewis acids can also be substances that have valence shells capable of holding more electrons. For example, consider the reaction of sulfur dioxide as a Lewis acid with oxide ion as a Lewis base to make the sulfite ion. This reaction occurs when gaseous sulfur dioxide, an acidic anhydride, mingles with solid calcium oxide, a basic anhydride, to give calcium sulfite, $CaSO_3$.

$$SO_2(g) + CaO(s) \longrightarrow CaSO_3(s)$$

Let's see how electrons relocate as the sulfite ion forms. We use one of the two resonance structures of SO_2 and one of the three such structures for the sulfite ion.

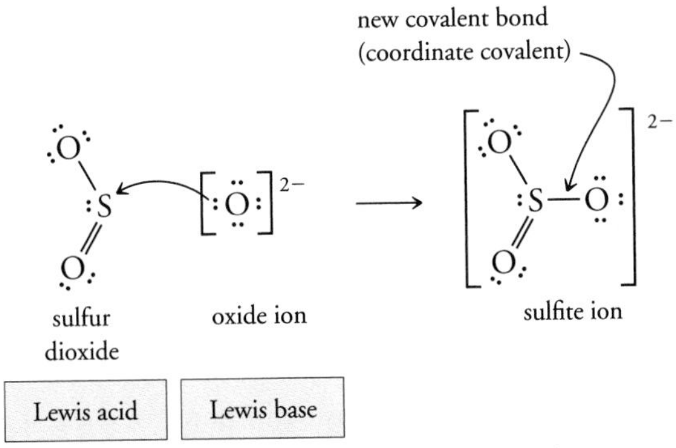

new covalent bond
(coordinate covalent)

sulfur dioxide

oxide ion

sulfite ion

Lewis acid

Lewis base

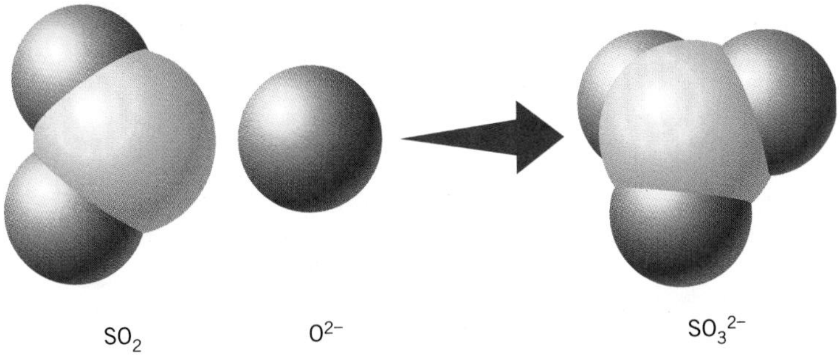

SO_2 O^{2-} SO_3^{2-}

The two very electronegative oxygens attached to the sulfur in SO_2 give the sulfur a substantial positive partial charge, which induces the formation of the coordinate covalent bond from the oxide ion to the sulfur. In this case, sulfur can accommodate more than an octet in its valence shell, so relocation of electron pairs is not necessary.

Table 15.3 summarizes the kinds of substances that behave as Lewis acids and bases. Study it and then work on the practice exercises on the next page.

TABLE 15.3 Types of Substances That Are Lewis Acids and Bases

Lewis Acids
Molecules or ions with incomplete valence shells (e.g., BF_3, H^+)
Molecules or ions with complete valence shells, but with multiple bonds that can be shifted to make room for more electrons (e.g., CO_2)
Molecules or ions that have central atoms capable of holding additional electrons (usually, atoms of elements in Period 3 and below) (e.g., SO_2)
Lewis Bases
Molecules or ions that have unshared pairs of electrons and that have complete valence shells (e.g., O^{2-}, NH_3)

Practice Exercise 17: Identify the Lewis acids and bases in each aqueous reaction.
(a) $NH_3 + H^+ \rightleftharpoons NH_4^+$
(b) $(CH_3)_2O + BCl_3 \rightleftharpoons (CH_3)_2OBCl_3$
(c) $Ag^+ + 2NH_3 \rightleftharpoons Ag(NH_3)_2^+$
(Hint: Draw Lewis structures of the reactants.)

Practice Exercise 18: (a) Is the fluoride ion more likely to behave as a Lewis acid or a Lewis base? Explain. (b) Is the $BeCl_2$ molecule more likely to behave as a Lewis acid or a Lewis base? Explain. (c) Is the SO_3 molecule more likely to behave as a Lewis acid or a Lewis base? Explain.

A Brønsted acid–base reaction is an exchange of a Lewis acid between two Lewis bases

Reactions involving the transfer of a proton can be analyzed by either the Brønsted or Lewis definitions of acids and bases. Consider, for example, the reaction between hydronium ion and ammonia in aqueous solution.

$$H_3O^+ + NH_3 \longrightarrow H_2O + NH_4^+$$

Applying the Brønsted definitions, we have an acid, H_3O^+, reacting with a base, NH_3, to give the corresponding conjugates: H_2O being the conjugate base of H_3O^+ and NH_4^+ being the conjugate acid of NH_3.

To interpret this reaction using the Lewis definitions, we view it as the movement of a Lewis acid (the proton, H^+) from a weaker Lewis base (H_2O) to a stronger Lewis base (NH_3). In other words, we view the hydronium ion as the "neutralization" product of the Lewis base H_2O with the Lewis acid H^+. To emphasize that it is the proton that is transferring, we might write the equation as

$$H_2O{-}H^+ + NH_3 \longrightarrow H_2O + H^+{-}NH_3$$

We can diagram this as follows using Lewis structures:

Electron pair in the OH bond shifts to the oxygen as the water molecule is formed.

$$\left[\text{H}_2\ddot{\text{O}}{-}\text{H}\right]^+ + :\text{NH}_3 \longrightarrow \text{H}{-}\ddot{\text{O}}{:}{-}\text{H} + \left[\text{NH}_4\right]^+$$

The electron pair of the nitrogen captures H^+ (a Lewis acid) from the oxygen of the hydronium ion.

The molecular view of the reaction is

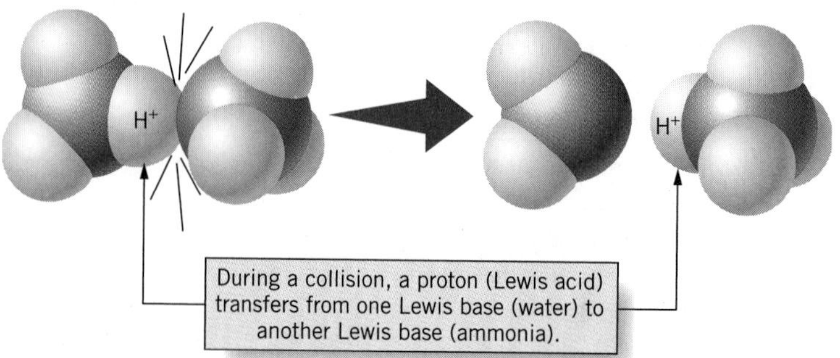
H+
H+
During a collision, a proton (Lewis acid) transfers from one Lewis base (water) to another Lewis base (ammonia).

In general, all Brønsted–Lowry acid–base reactions can be analyzed from the Lewis point of view following this same pattern, with the position of equilibrium favoring the proton being attached to the stronger Lewis base.

A word of caution is in order here. In analyzing proton transfer reactions, it is wise to stick to one interpretation or the other (Brønsted–Lowry or Lewis). Usually we will use the Brønsted–Lowry approach because it is useful to think in terms of conjugate acid–base pairs. However, if you switch to the Lewis interpretation, don't try to apply Brønsted–Lowry terminology at the same time; it won't work and you'll only get confused.

15.4 ELEMENTS AND THEIR OXIDES DEMONSTRATE ACID–BASE PROPERTIES

The elements most likely to form acids are the nonmetals in the upper right-hand corner of the periodic table. Those most likely to form basic hydroxides are similarly grouped in one general location in the table, among the metals, particularly those in Groups IA (alkali metals) and IIA (alkaline earth metals). Thus, elements can themselves be classified according to their abilities to be involved in acids or bases.

In general, the experimental basis for classifying an element according to its abilities to form an acid or base depends on how its *oxide* behaves toward water. Earlier you learned that metal oxides like Na_2O and CaO are called *basic anhydrides* ("anhydride" means "without water") because they react with water to form hydroxides.

$$Na_2O + H_2O \longrightarrow 2NaOH \qquad \text{sodium hydroxide}$$

$$CaO + H_2O \longrightarrow Ca(OH)_2 \qquad \text{calcium hydroxide}$$

The reactions of metal oxides with water are really reactions of their oxide ions, which take H^+ from molecules of H_2O, leaving ions of OH^-.

Many metal oxides are insoluble in water, so their oxide ions are unable to take H^+ ions from H_2O molecules. Many are able to react with acids, which attack the oxide ions in the solid. Iron(III) oxide, for example, reacts with acid as follows.

$$Fe_2O_3(s) + 6H^+(aq) \longrightarrow 2Fe^{3+}(aq) + 3H_2O$$

This is a common method for removing rust from iron in industrial processes.

Nonmetal oxides are usually *acidic anhydrides;* those that react with water give acidic solutions. Typical examples of the formation of acids from nonmetal oxides are the following reactions.

$$SO_3(g) + H_2O \longrightarrow H_2SO_4(aq) \qquad \text{sulfuric acid}$$

$$N_2O_5(g) + H_2O \longrightarrow 2HNO_3(aq) \qquad \text{nitric acid}$$

$$CO_2(g) + H_2O \longrightarrow H_2CO_3(aq) \qquad \text{carbonic acid}$$

Hydrated metal ions can behave as weak acids

When an ionic compound dissolves in water, molecules of the solute gather around the ions and we say the ions are *hydrated.* Within a hydrated cation, the metal ion behaves as a Lewis acid, binding to the partial negative charge on the oxygens of the surrounding water molecules, which serve as Lewis bases. Hydrated metal ions themselves tend to be Brønsted acids because of the equilibrium shown below. For simplicity, the equation represents a metal ion as a *mono*hydrate, namely, $M(H_2O)^{n+}$ with a net positive charge of $n+$ (n being 1, 2, or 3, depending on M).

$$M(H_2O)^{n+} + H_2O \rightleftharpoons MOH^{(n-1)+} + H_3O^+$$

In other words, hydrated metal ions tend to be proton donors in water. Let's see why.

The positive charge on the metal ion attracts the water molecule and draws electron density from the O—H bonds, causing them to become more polar as shown in the margin. This increases the partial positive charge on H and weakens the O—H bond with respect to the transfer of H^+ to another nearby water molecule in the formation of a hydronium ion, which we can illustrate as follows.

The force of attraction between an ion and water molecules is often strong enough to persist when the solvent water evaporates. Solid *hydrates* crystallize from solution, for example, $CuSO_4 \cdot 5H_2O$ (page 56).

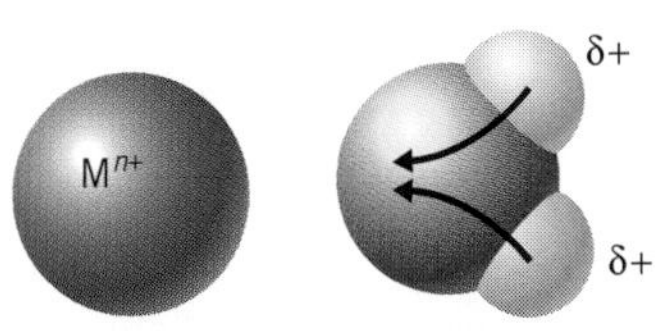

The positive charge on the metal ion pulls electron density away from the H atoms of the water molecule.

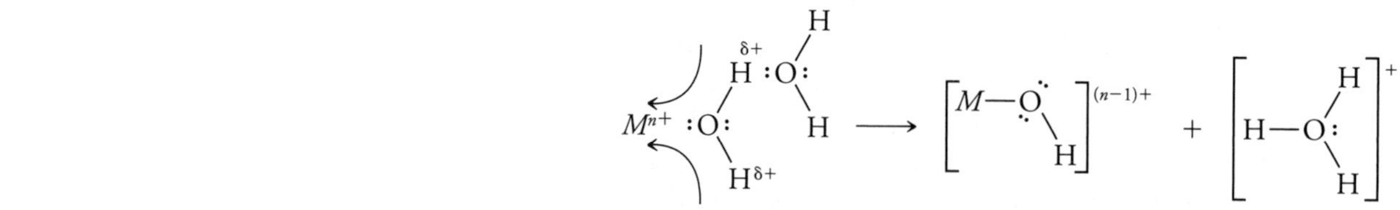

$$M^{n+} :\ddot{O}(H^{\delta+})\cdots H^{\delta+}:\ddot{O}(H)H \longrightarrow \left[M-\ddot{O}-H\right]^{(n-1)+} + \left[H-\ddot{O}(H)-H\right]^{+}$$

Electron density is reduced in the O—H bonds of water (indicated by the curved arrows) by the positive charge of the metal ion, thereby increasing the partial positive charge on the hydrogens. This promotes the transfer of H^+ to a water molecule.

The degree to which metal ions produce acidic solutions depends on the amount of charge on the cation and the cation's size. As the cation's charge increases, the polarizing effect is increased, which thereby favors the release of H^+. This means that highly charged metal ions ought to produce more acidic solutions than ions of low charge, and that is generally the case.

The reason the size of the cation also affects its acidity is that when the cation is small, the positive charge is highly concentrated. A highly concentrated positive charge is better able to pull electrons from an O—H bond than a positive charge that is more spread out. Therefore, for a given positive charge, the smaller the cation, the more acidic are its solutions.

Both size and amount of charge can be considered together by referring to a metal ion's *positive charge density*, the ratio of the positive charge to the volume of the cation (its ionic volume).

$$\text{Charge density} = \frac{\text{ionic charge}}{\text{ionic volume}}$$

The higher the positive charge density, the more effective the metal ion is at drawing electron density from the O—H bond and the more acidic is the hydrated cation.

Very small cations with large positive charges have large positive charge densities and tend to be quite acidic. An example is the hydrated aluminum ion, Al^{3+}. The hexahydrate, $Al(H_2O)_6{}^{3+}$, is one of several of this cation's hydrated forms that are present in an aqueous solution of an aluminum salt. This ion is acidic in water because of the equilibrium

$$[Al(H_2O)_6]^{3+}(aq) + H_2O \rightleftharpoons [Al(H_2O)_5(OH)]^{2+}(aq) + H_3O^+(aq)$$

The equilibrium, while not actually *strongly* favoring the products, does produce enough hydronium ion so that a 0.1 M solution of $AlCl_3$ in water has about the same concentration of hydronium ions as a 0.1 M solution of acetic acid, roughly 1×10^{-3} M.

The acidities of metal ions follow periodic trends

Within the periodic table, atomic size increases down a group and decreases from left to right in a period. Cation sizes follow these same trends, so within a given group, the cation of the metal at the top of the group has the smallest volume and the largest charge density. Therefore, hydrated metal ions at the top of a group in the periodic table are the most acidic within the group.

The cations of the Group IA metals (Li^+, Na^+, K^+, Rb^+, or Cs^+), with charges of just 1+, have little tendency to increase the H_3O^+ concentration in an aqueous solution.

Within Group IIA, the Be^{2+} cation is very small and has sufficient charge density to cause the hydrated ion to be a weak acid. The other cations of Group IIA (Mg^{2+}, Ca^{2+}, Sr^{2+}, Ba^{2+}) have charge densities that become progressively smaller as we go down the group. Although their hydrated ions all generate some hydronium ion in water, the amount is negligible.

Some transition metal ions are also acidic, especially those with charges of 3+. For example, solutions containing salts of Fe^{3+} and Cr^{3+} tend to be acidic because their ions in solution exist as $Fe(H_2O)_6{}^{3+}$ and $Cr(H_2O)_6{}^{3+}$, respectively, and undergo the same ionization reaction as does the $Al(H_2O)_6{}^{3+}$ ion discussed above.

Acid–base properties of metal oxides are influenced by the oxidation number of the metal

Not all metal oxides are basic. As the oxidation number (or charge) on a metal ion increases, the metal ion becomes more *acidic*; it becomes a better electron pair acceptor. For metal hydrates, we've seen that this causes electron density to be pulled from the OH bonds of water molecules, causing the hydrate itself to become a weak proton donor. The increasing acidity of metal ions with increasing charge also affects the basicity of their oxides.

When the positive charge on a metal is small, the oxide tends to be basic, as we've seen for oxides such as Na_2O and CaO. With ions having a 3+ charge, the oxides are less basic and begin to take on acidic properties as well; they become amphoteric. (Recall that a substance is *amphoteric* if it is capable of reacting as either an acid or a base.) Aluminum oxide is an example; it is able to react with both acids and bases. It has basic properties when it dissolves in acid.

$$Al_2O_3(s) + 6H^+(aq) \longrightarrow 2Al^{3+}(aq) + 3H_2O$$

As noted earlier, the hydrated aluminum ion has six water molecules surrounding it, so in an acidic solution the aluminum exists primarily as $Al(H_2O)_6^{3+}$.

Aluminum oxide exhibits acidic properties when it dissolves in a base. One way to write the equation for the reaction is

$$Al_2O_3(s) + 2OH^-(aq) \longrightarrow 2AlO_2^-(aq) + H_2O$$

Actually, in basic solution the formula of the aluminum-containing species is more complex than this and is better approximated by $Al(H_2O)_2(OH)_4^-$. Note that the difference between the two formulas is just the number of water molecules involved in the formation of the ion.

$$Al(H_2O)_2(OH)_4^- \quad \text{is equivalent to} \quad AlO_2^- + 4H_2O$$

However we write the formula for the aluminum-containing ion, it is an anion, not a cation.

When the metal is in a very high oxidation state, the oxide becomes acidic. Chromium(VI) oxide, CrO_3, is an example. When dissolved in water, the resulting solution is quite acidic and is called chromic acid. One of the principal species in the solution is H_2CrO_4, which is a strong acid that is more than 95% ionized. The acid forms salts containing the chromate ion, CrO_4^{2-}.

15.5 pH IS A MEASURE OF THE ACIDITY OF A SOLUTION

There are literally thousands of weak acids and bases, and they vary widely in how weak they are. Both the acetic acid in vinegar and the carbonic acid in pressurized soda water, for example, are classified as weak. Yet carbonic acid is only about 3% as strong an acid as acetic acid. To study such differences quantitatively, we need to explore acid–base equilibria in greater depth. This requires that we first discuss a particularly important equilibrium that exists in *all* aqueous solutions, namely, the ionization of water itself.

Water is a very weak electrolyte

Using sensitive instruments, pure water is observed to weakly conduct electricity, indicating the presence of very small concentrations of ions. They arise from the very slight self-ionization, or *autoionization,* of water itself, represented by the following equilibrium equation.

$$H_2O + H_2O \rightleftharpoons H_3O^+ + OH^-$$

The forward reaction requires a collision of two H_2O molecules.

We've omitted the usual (*aq*) following the symbols for ions in water for the sake of simplicity.

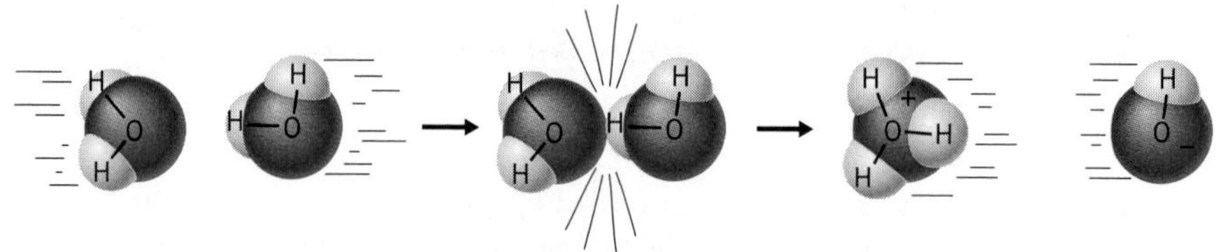

Its equilibrium law, following the procedures developed in Chapter 14, is

$$\frac{[H_3O^+][OH^-]}{[H_2O]^2} = K_c$$

In pure water, and even in dilute aqueous solutions, the molar concentration of water is essentially a constant, with a value of 55.6 *M*. Therefore, the term $[H_2O]^2$ in the denominator is a constant which we can combine with K_c.

$$[H_3O^+][OH^-] = K_c \times [H_2O]^2$$

The product of the two constants, $K_c \times [H_2O]^2$, must also be a constant. Because of the importance of the autoionization equilibrium, it is given the special symbol $\boldsymbol{K_w}$ and is called the **ion product constant of water.**

$$[H_3O^+][OH^-] = K_w$$

Often, for convenience, we omit the water molecule that carries the hydrogen ion and write H^+ in place of H_3O^+. The equilibrium equation for the autoionization of water then simplifies as follows.

$$H_2O \rightleftharpoons H^+ + OH^-$$

The equation for K_w based on this is likewise simplified.

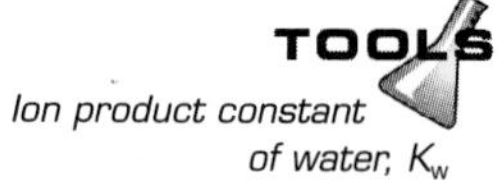

Ion product constant of water, K_w

$$[H^+][OH^-] = K_w \qquad (15.3)$$

In pure water, the concentrations of H^+ and OH^- produced by the autoionization are equal because the ions are formed in equal numbers. It's been found that the concentrations have the following values at 25 °C.

$$[H^+] = [OH^-] = 1.0 \times 1.0^{-7} \text{ mol L}^{-1}$$

Therefore, at 25 °C,

$$K_w = (1.0 \times 10^{-7}) \times (1.0 \times 10^{-7})$$

$$K_w = 1.0 \times 10^{-14} \qquad \text{(at 25 °C)} \qquad (15.4)$$

As with other equilibrium constants, the value of K_w varies with temperature (see Table 15.4). But, for simplicity, we will generally deal with systems at 25 °C, so we'll usually not specify the temperature each time. The value of K_w at 25 °C ($K_w = 1.0 \times 10^{-14}$) is so important that it should be learned (memorized).

Solutes can affect $[H^+]$ and $[OH^-]$, but they can't alter K_w

Water's autoionization takes place in *any* aqueous solution, but because of the effects of other solutes, the molar concentrations of H^+ and OH^- may not be equal. Nevertheless, their product, K_w, is the same. Thus, although Equations 15.3 and 15.4 were derived for pure water, they also apply to dilute aqueous solutions. The significance of this must be emphasized. *In any aqueous solution, the product of* $[H^+]$ *and* $[OH^-]$ *equals* K_w, *although the two molar concentrations may not actually equal each other.*

TABLE 15.4 K_w at Various Temperatures

Temperature (°C)	K_w
0	1.5×10^{-15}
10	3.0×10^{-15}
20	6.8×10^{-15}
25	1.0×10^{-14}
30	1.5×10^{-14}
37[a]	2.5×10^{-14}
40	3.0×10^{-14}
50	5.5×10^{-14}
60	9.5×10^{-14}

[a] Normal body temperature 98.6°F

Criteria for acidic, basic, and neutral solutions

One of the consequences of the autoionization of water is that *in any aqueous solution, there are always both H_3O^+ and OH^- ions, regardless of what solutes are present.* This means that in a solution of the acid HCl there is some OH^-, and in a solution of the base NaOH, there is some H_3O^+. So what criteria do we use to determine whether a solution is neutral, acidic, or basic?

Because of the autoionization of water, even the most acidic solution has some OH^-, and even the most basic solution has some H_3O^+.

A **neutral solution** is one in which the molar concentrations of H_3O^+ and OH^- are equal; neither ion is present in a greater concentration than the other. An **acidic solution** is one in which some solute has made the molar concentration of H_3O^+ greater than that of OH^-. On the other hand, a **basic solution** exists when the molar concentration of OH^- exceeds that of H_3O^+. We therefore define acidic and basic solutions in terms of the *relative* molarities of H_3O^+ and OH^-.

Neutral solution	$[H_3O^+] = [OH^-]$
Acidic solution	$[H_3O^+] > [OH^-]$
Basic solution	$[H_3O^+] < [OH^-]$

EXAMPLE 15.4 Finding $[H^+]$ from $[OH^-]$ or Finding $[OH^-]$ from $[H^+]$

In a sample of blood at 25 °C, $[H^+] = 4.6 \times 10^{-8}$ *M*. Find the molar concentration of OH^-, and decide if the sample is acidic, basic, or neutral.

ANALYSIS: To reach a decision, we need to know how $[H^+]$ and $[OH^-]$ compare. The tool to use is the relationship between the concentrations of $[H^+]$ and $[OH^-]$. We know that the values of $[H^+]$ and $[OH^-]$ are *always* related to each other and to K_w at 25 °C as follows.

$$1.0 \times 10^{-14} = [H^+][OH^-]$$

If we know one concentration, we can *always* find the other.

SOLUTION: We substitute the given value of $[H^+]$ into this equation and solve for $[OH^-]$.

$$1.0 \times 10^{-14} = (4.6 \times 10^{-8})[OH^-]$$

Solving for $[OH^-]$, and remembering that its units are mol L^{-1} or *M*,

$$[OH^-] = \frac{1.0 \times 10^{-14}}{4.6 \times 10^{-8}} M = 2.2 \times 10^{-7} M$$

When we compare $[H^+]$ equaling 4.6×10^{-8} *M* with $[OH^-]$ equaling 2.2×10^{-7} *M*, we see that $[OH^-] > [H^+]$. Our answer, then, is that the blood is slightly basic.

ARE THE ANSWERS REASONABLE? We know $K_w = 1.0 \times 10^{-14} = [H^+][OH^-]$. So one check is to note that if $[H^+]$ is slightly *less* than 1×10^{-7}, $[OH^-]$ will have to be slightly *more* than 1×10^{-7}, as it is. (Be careful in comparing numbers when the exponent on 10 is negative. A value of 10^{-7} is larger than 10^{-8}.)

Practice Exercise 19: Commercial, concentrated, hydrochloric acid has a concentration of 12 moles per liter. If the molarity of hydronium ions is assumed to also be 12 *M*, what is the concentration of hydroxide ions? (Hint: Use the relationships just discussed.)

Practice Exercise 20: An aqueous solution of sodium bicarbonate, $NaHCO_3$, has a molar concentration of hydroxide ion of 7.8×10^{-6} *M*. What is the molar concentration of hydrogen ion? Is the solution acidic, basic, or neutral?

The pH concept provides a logarithmic scale of acidity

In most solutions of weak acids and bases, the molar concentrations of H^+ and OH^- are very small, like those in Example 15.4. When you tried to compare the two values in that example, namely, 4.6×10^{-8} *M* and 2.2×10^{-7} *M*, you had to look in four places, two before the multiplication symbol and the two negative exponents. There's an easier approach involving only one number. A Danish chemist, S. P. L. Sørenson (1868–1939), suggested it.

To make comparisons of small values of $[H^+]$ easier, Sørenson defined a quantity that he called the **pH** of the solution as follows.

Definition of pH

$$pH = -\log [H^+] \qquad (15.5)^5$$

The properties of logarithms let us rearrange Equation 15.5 as follows.

The expression $[H^+] = 10^{-pH}$ corresponds to taking the antilogarithm of the negative of the pH, i.e., $[H^+] = \text{antilog}(-pH)$

$$[H^+] = 10^{-pH} \qquad (15.6)$$

Equation 15.5 can be used to calculate the pH of a solution if its molar concentration of H^+ is known. On the other hand, if the pH is known and we wish to calculate the molar concentration of hydrogen ion, we apply Equation 15.6. We will illustrate these calculations with examples shortly.

The logarithmic definition of pH, Equation 15.5, has proved to be so useful that it has been adapted to quantities other than $[H^+]$. Thus, for any quantity *X*, we may define a term, p*X*, in the following way.

Defining the p-function, pX

$$pX = -\log X \qquad (15.7)$$

For example, to express small concentrations of hydroxide ion, we can define the **pOH** of a solution as

$$pOH = -\log [OH^-]$$

Similarly, for K_w we can define $\mathbf{pK_w}$ as follows.

$$pK_w = -\log K_w$$

The numerical value of pK_w at 25 °C equals $-\log (1.0 \times 10^{-14})$ or $-(-14.00)$. Thus,

$$pK_w = 14.00 \qquad \text{(at 25 °C)}$$

[5] In this equation and similar ones, like Equation 15.7, the logarithm of only the numerical part of the bracketed term is taken. The physical units must be mol L^{-1}, but they are set aside for the calculation.

A useful relationship among pH, pOH, and pK_w can be derived from Equation 15.3 that defines K_w.

$$[H^+][OH^-] = K_w \qquad \text{(Equation 15.3)}$$

By the properties of logarithms,

$$\log [H^+] + \log [OH^-] = \log K_w$$

We next multiply the terms on both sides by -1.

$$-\log [H^+] + -\log [OH^-] = -\log K_w$$

But each of these terms is in the form of pX (Equation 15.7), so

$$pH + pOH = pK_w = 14.00 \qquad \text{(at 25 °C)} \tag{15.8}$$

Relationship of pH and pOH to pK_w

This tells us that in an aqueous solution of any solute at 25 °C, the sum of pH and pOH is 14.00.

Neutral pH is 7.00 only for 25 °C. At other temperatures a neutral solution still has $[H^+] = [OH^-]$, but because K_w changes with temperature, so do the values of $[H^+]$ and $[OH^-]$. This causes the pH of the neutural solution to differ from 7.00. For example, at normal body temperature (37 °C), a neutral solution has $[H^+] = [OH^-] = 1.6 \times 10^{-7}$, so pH = 6.80.

Acidic, basic, and neutral solutions are identified by their pH

One meaning attached to pH is that it is a measure of the acidity of a solution. Hence, we may define *acidic*, *basic*, and *neutral* in terms of pH values. In pure water, or in any solution that is *neutral*,

$$[H^+] = [OH^-] = 1.0 \times 10^{-7}\ M$$

Therefore, by Equation 15.5, *the pH of a neutral solution at 25 °C is 7.00.*[6]

An *acidic solution* is one in which $[H^+]$ is larger than $10^{-7}\ M$ and so has a pH *less* than 7.00. Thus, *as a solution's acidity increases, its pH decreases.*

A *basic solution* is one in which the value of $[H^+]$ is less than $10^{-7}\ M$ and so has a pH that is *greater* than 7.00. *As a solution's acidity decreases, its pH increases.* These simple relationships may be summarized as follows. At 25 °C:

pH $<$ 7.00	**Acidic solution**
pH $>$ 7.00	**Basic solution**
pH = 7.00	**Neutral solution**

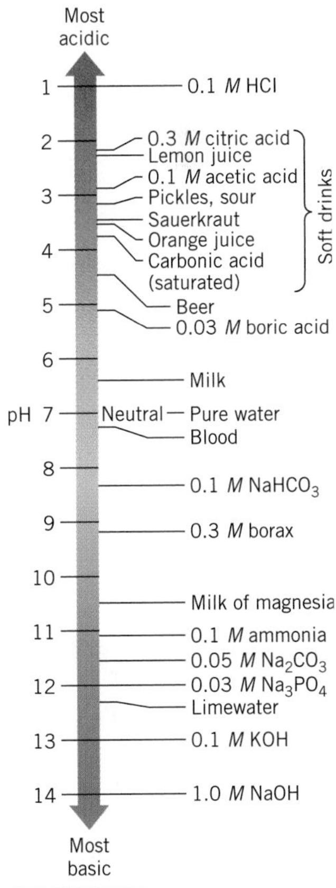

FIG. 15.3 The pH scale.

The pH values for some common substances are given on a pH scale in Figure 15.3.

One of the deceptive features of pH is how much the hydrogen ion concentration changes with a relatively small change in pH. Thus, a change of one pH unit corresponds to a 10-fold change in the hydrogen ion concentration. For example, by Equation 15.5, at pH 6.0 the hydrogen ion concentration is 1×10^{-6} mol L^{-1}. At pH 5.0, it's 1×10^{-5} mol L^{-1}, or 10 times greater. That's a major difference.

The pH concept is almost never used when a pH value would be a negative number, namely, with solutions having hydrogen ion concentrations greater than 1 M. For example, in a solution where the value of $[H^+]$ is 2 M, the "2" is a simple enough number; we can't simplify it further by using its corresponding pH value. When $[H^+]$ is 2.00 M, the pH, calculated by Equation 15.5, would be $-\log(2.00)$ or -0.301. There is nothing wrong with negative pH values, but they offer no advantage over the actual value of $[H^+]$.

[6] Recall that the rule for significant figures in logarithms is that *the number of* ***decimal places*** *in the logarithm of a number equals the number of significant figures in the number.* For example, 3.2×10^{-5} has just two significant figures. The logarithm of that number, displayed on a pocket calculator, is -4.494850022. We write the logarithm, when correctly rounded, as -4.49.

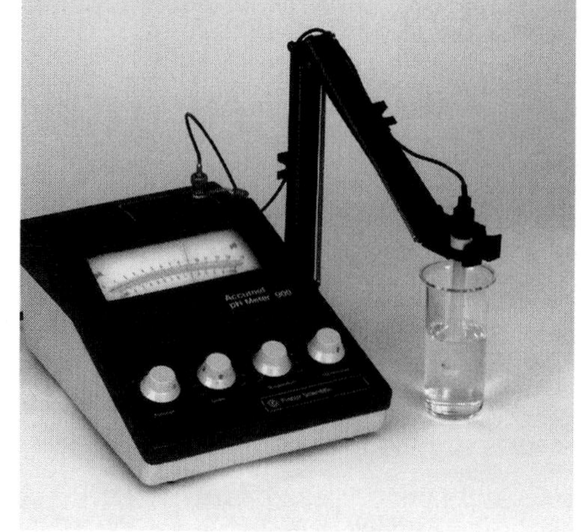

FIG. 15.4 A pH meter. A special combination electrode that is sensitive to hydrogen ion concentration is dipped into the solution to be tested. After the instrument has been calibrated using a solution of known pH, the electrode is dipped into the solution to be tested and the pH is read from the meter. *(Courtesy Hanna Instruments.)*

pH can be measured or estimated

One of the remarkable things about pH is that it can be easily measured using an instrument called a pH meter (Figure 15.4). An electrode system sensitive to the hydrogen ion concentration in a solution is first dipped into a solution of known pH to calibrate the instrument. Once calibrated, the apparatus can then be used to measure the pH of any other solution simply by immersing the electrode into it. Most modern pH meters are able to determine pH values to within ± 0.01 pH units, and research-grade instruments are capable of even greater precision in the pH range of 0.0 to 14.0.

Another less precise method of obtaining the pH uses acid–base indicators. As you learned in Chapter 4, these are dyes whose colors in aqueous solution depend on the acidity of the solution. Table 15.5 gives several examples. Indicators change color over a narrow range of pH values, and Figure 15.5 shows the colors of some indicators at opposite ends of their color change ranges. Facets of Chemistry 15.1 describes some applications of indicators to measuring pH.

TABLE 15.5 Common Acid–Base Indicators

Indicator	Approximate pH Range over Which the Color Changes	Color Change (Lower to Higher pH)
Methyl green	0.2–1.8	Yellow to blue
Thymol blue	1.2–2.8	Yellow to blue
Methyl orange	3.2–4.4	Red to yellow
Ethyl red	4.0–5.8	Colorless to red
Methyl purple	4.8–5.4	Purple to green
Bromocresol purple	5.2–6.8	Yellow to purple
Bromothymol blue	6.0–7.6	Yellow to blue
Phenol red	6.4–8.2	Yellow to red/violet
Litmus	4.7–8.3	Red to blue
Cresol red	7.0–8.8	Yellow to red
Thymol blue	8.0–9.6	Yellow to blue
Phenolphthalein	8.2–10.0	Colorless to pink
Thymolphthalein	9.4–10.6	Colorless to blue
Alizarin yellow R	10.1–12.0	Yellow to red
Clayton yellow	12.2–13.2	Yellow to amber

FIG. 15.5 Colors of some common acid–base indicators. *(Andy Washnik.)*

FACETS OF CHEMISTRY

Swimming Pools, Aquariums, and Flowers

Something these have in common is the need for a proper pH. The water in swimming pools is more than just water; its a dilute mix of chemicals that prevent the growth of bacteria and stabilize the pool lining. For the pool chemistry to be properly balanced, the optimum pH range is between 7.2 and 7.6. Aquariums have to have their pH controlled because fish are very sensitive to how acidic or basic the water is. Depending on the species of fish in the tank, the pH should be between 6.0 and 7.6, and if it drifts much outside this range, the fish will die.

The pH can also affect the availability of substances that plants need to grow. A pH of from 6 to 7 is best for most plants because most nutrients are more soluble when the soil is slightly acidic than when it is neutral or slightly basic. If the soil pH is too high, metal ions such as iron, manganese, and boron that plants need will precipitate and not be available in the groundwater. A very low pH is not good either. If the pH drops to between 4 and 5, metal ions that are toxic to many plants become released as their compounds become more soluble. A low pH also inhibits the growth of certain beneficial bacteria that are needed to decompose organic matter in the soil and release nutrients, especially nitrogen.

There are chemicals that can be added to swimming pools, aquariums, and soil that will raise or lower the pH, but to know how to use them, it's necessary to measure the pH. Commercial test kits are available to enable you to do this, and they involve the use of acid–base indicators. Kits for testing the pH of pool water (Figure 1) use *phenol red* indicator, which changes color over the pH range 6.4–8.2. This places the intermediate color of the indicator right in the middle of the desired range of 7.0–7.6. To use the kit, a sample of pool water is placed into the plastic cylinder and 5 drops of the indicator solution are added. After shaking the mixture to ensure uniform mixing, the color is compared to the standards to gauge the pH. It can then be decided whether chemicals have to be added to either raise or lower the pH.

Similar test kits are available to test the water in an aquarium, except that the indicator used is bromothymol blue, which changes color over a pH range of 6.0–7.6. This generally matches the desired pH range for the water in the aquarium. Soil test kits, such as that shown in Figure 2, use a "universal indicator," which is a mixture of indicators that enable estimation of the pH over a wider range. A sample of the soil to be tested is placed in the plastic apparatus that comes with the kit. Water is added along with a tablet of the indicator. The mixture is shaken and then the color of the water is compared with the color chart. The color that most closely matches the color of the solution gives the estimated pH.

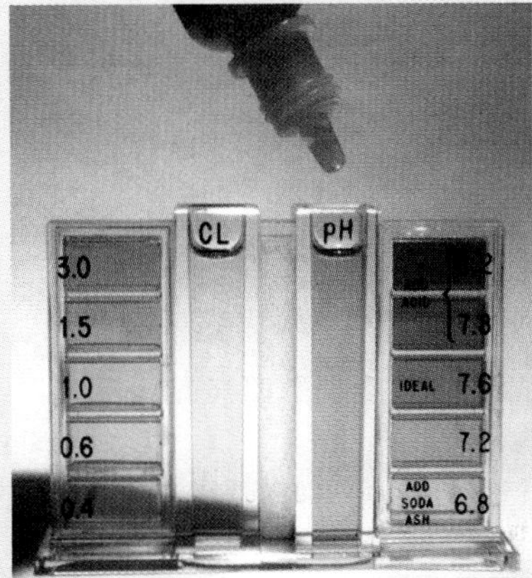

FIG. 1 Swimming pool test kit. This apparatus is used to test for both pH and chlorine concentration. The color of the indicator on the right tells us the pH of the pool water is approximately 7.6, which is slightly basic. *(Larry Stepanowicz/ Fundamental Photographs.)*

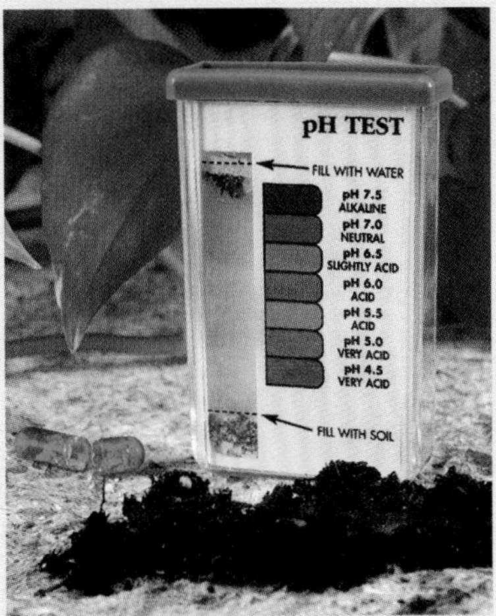

FIG. 2 Testing soil pH. The indicator reveals that the soil sample being tested has a pH of approximately 6.5, which is slightly acidic. *(Andy Washnik.)*

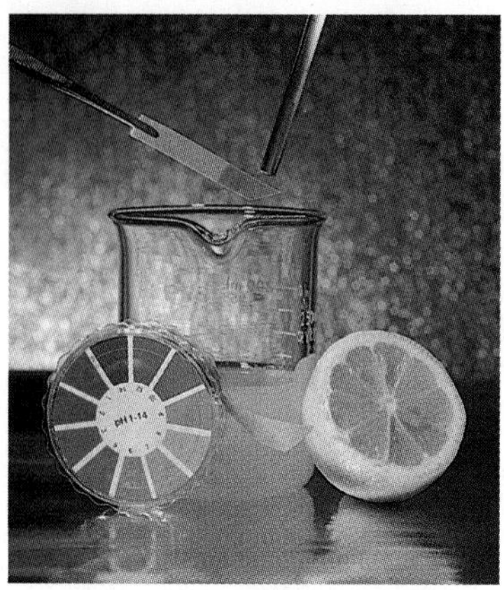

FIG. 15.6 Using a pH test paper. The color of this Hydrion test strip changed to orange when a drop of the lemon juice solution in the beaker was placed on it. According to the color code, the pH of the solution is closer to 3 than to 5. *(Andy Washnik.)*

pH test papers are available that are impregnated with one or more indicator dyes. To obtain a rough idea of the pH, a drop of the solution to be tested is touched to a strip of the test paper and the resulting color is compared with a color code. Some commercial test papers (e.g., Hydrion) are impregnated with several dyes, with their vials carrying the color code (see Figure 15.6).

Litmus paper, commonly found in chemistry labs, is often used qualitatively to test whether a solution is acidic or basic. It consists of porous paper impregnated with litmus dye, made either red or blue by exposure to acid or base, and then dried. Below pH 4.7, litmus is red and above pH 8.3 it is blue. The transition for the color change occurs over a pH range of 4.7 to 8.3 with the center of the change at about 6.5, very nearly a neutral pH. To test whether a solution is acidic, a drop of it is placed on blue litmus paper. If the dye turns pink, the solution is acidic. Similarly, to test if the solution is basic, a drop is touched to red litmus paper. If the dye turns blue, the solution is basic.

pH calculations

Let's now look at some examples of pH calculations using Equations 15.5 and 15.6. These are calculations you will be performing often in Section 15.6 and throughout Chapter 16, so be sure you become proficient at them. The first example calculates a pH from a value either of $[H^+]$ or of $[OH^-]$. Another finds the value of $[H^+]$ or $[OH^-]$ given a pH.

EXAMPLE 15.5
Calculating pH and pOH from $[H^+]$

Because rain washes pollutants out of the air, the lakes in many parts of the world have undergone pH changes. In a New England state, the water in one lake was found to have a hydrogen ion concentration of 3.2×10^{-5} mol L^{-1}. What are the calculated pH and pOH values of the lake's water? Is the water acidic or basic?

ANALYSIS: If we know the value of $[H^+]$, Equation 15.5 is the tool we use to find the pH. The calculation is straightforward.

SOLUTION: First, let's write the equation we will use.

$$pH = -\log [H^+] \qquad \text{(Equation 15.5)}$$

Be sure to notice that *common* logs (base 10 logs) are used in pH calculations. Don't use the *natural* log function on your scientific calculator by mistake.

Next, we simply make the substitution for $[H^+]$.

$$pH = -\log (3.2 \times 10^{-5})$$

Using a calculator to find the logarithm of 3.2×10^{-5} gives the value -4.49 (note the negative sign). To find the pH, we change its algebraic sign.

$$pH = -(-4.49) = 4.49$$

The pH is less than 7.00, so the lake's water is acidic (too acidic, in fact, for game fish to survive).

Once we know the pH, finding the pOH is most easily done by using the fact that the sum of pH and pOH equals 14.00. Therefore, the pOH of the lake water is

$$\begin{aligned} pOH &= 14.00 - 4.49 \\ &= 9.51 \end{aligned}$$

ARE THE ANSWERS REASONABLE? The value of $[H^+]$ is between 1×10^{-5} and 1×10^{-4}. Using these in Equation 15.5 yields pH values of 5 and 4, respectively, so the pH has to be between 5 and 4, which it is. Therefore, the pOH must be between 9 and 10, as it is.

Practice Exercise 21: Water draining from old coal and mineral mines often has pH values of 4.0 or lower. The cause is thought to stem from reactions of ground water and oxygen with iron pyrite, FeS. What is the pOH, $[H^+]$, and $[OH^-]$ concentrations in an acid mine drainage sample that has a pH of 4.25? (Hint: Recall that pH, pOH, $[H^+]$, and $[OH^-]$ are all interrelated using K_w and pK_w.)

Practice Exercise 22: A carbonated beverage was found to have a hydrogen ion concentration of 3.67×10^{-4} mol L^{-1}. What are the calculated pH and pOH values of the beverage? Is it acidic, basic, or neutral?

EXAMPLE 15.6
Calculating pH from $[OH^-]$

What is the pH of a sodium hydroxide solution at 25 °C in which the hydroxide ion concentration equals 0.0026 *M*?

ANALYSIS: There are two ways to solve this problem. We could use the tool concerning the K_w expression ($K_w = [H^+][OH^-] = 1 \times 10^{-14}$) and the given value of the hydroxide ion concentration to find $[H^+]$, and then calculate the pH using Equation 15.5. The second way would be to use the tool defining pOH to calculate the pOH from the hydroxide ion concentration ($pOH = -\log [OH^-]$) and then subtract the pOH from 14.00 to find the pH. The second path requires less effort, so let's proceed that way.

SOLUTION: First, we substitute the $[OH^-]$, which equals 0.0026 *M*, into the equation for pOH.

$$\begin{aligned} pOH &= -\log [OH^-] \\ &= -\log (0.0026) \\ &= -(-2.59) = 2.59 \end{aligned}$$

Then we subtract this pOH value from 14.00 to find the pH.

$$\begin{aligned} pH &= 14.00 - 2.59 \\ &= 11.41 \end{aligned}$$

Notice that the pH in this basic solution is well above 7.

IS THE ANSWER REASONABLE? The molar concentration of hydroxide ion is between 0.001 (or 10^{-3}) and 0.01 (or 10^{-2}), so the pOH must be between 3 and 2, which it is. Thus, the pH ought to be between 11 and 12.

Practice Exercise 23: The molarity of OH^- in the water in which a soil sample has soaked overnight is 1.47×10^{-9} mol L^{-1}. What is the pH of the solution? (Hint: Recall the relationships between pH, pOH, $[H^+]$, and $[OH^-]$.)

Practice Exercise 24: Soil that is close to neutral is often best for growing plants. Based on the above exercise would you recommend adding lime (CaO) or aluminum sulfate $[Al_2(SO_4)_3]$? Recall the discussion in Section 15.4 concerning metal ions.

EXAMPLE 15.7
Calculating $[H^+]$ from pH

"Calcareous soil" is soil rich in calcium carbonate. The pH of the moisture in such soil generally ranges from just over 7 to as high as 8.3. After one particular soil sample was soaked in water, the pH of the water was measured to be 8.14. What value of $[H^+]$ corresponds to a pH of 8.14? Is the soil acidic or basic?

ANALYSIS: The tool we use to calculate $[H^+]$ from pH is Equation 15.6.

$$[H^+] = 10^{-pH}$$

To convert pH into $[H^+]$ we use the 10^x function on the calculator. This is often the inverse or 2nd function of the log key.

SOLUTION: We substitute the pH value, 8.14, into the equation above, which gives

$$[H^+] = 10^{-8.14}$$

Because the pH has two digits following the decimal point, we obtain two significant figures in the hydrogen ion concentration. Therefore, the answer, correctly rounded is

$$[H^+] = 7.2 \times 10^{-9}\ M$$

Because the pH > 7 and $[H^+] < 1 \times 10^{-7}\ M$, the soil is basic.

IS THE ANSWER REASONABLE? The given value of 8.14 is between 8 and 9, so at a pH of 8.14, the hydrogen ion concentration must lie between 10^{-8} and 10^{-9} mol L^{-1}. And that's what we found.

Practice Exercise 25: What are the pH and pOH values of the concentrated hydrochloric acid mentioned in Practice Exercise 19? (Hint: The correct answers may be surprising.)

Practice Exercise 26: Find the values of $[H^+]$ and $[OH^-]$ that correspond to each of the following values of pH. State whether each solution is acidic or basic.

(a) 2.90 (the approximate pH of lemon juice)
(b) 3.85 (the approximate pH of sauerkraut)
(c) 10.81 (the pH of milk of magnesia, a laxative)
(d) 4.11 (the pH of orange juice, on the average)
(e) 11.61 (the pH of dilute, household ammonia)

15.6 Strong Acids and Bases Are Fully Dissociated in Solution

Many solutes affect the pH of an aqueous solution. In this section we examine how strong acids and bases behave and how to calculate the pH of their solutions. Weak acids and bases have similar effects, which we will discuss in the next chapter.

Calculating the pH of dilute solutions of strong acids and bases takes into account their complete dissociation

If necessary, review the list of strong acids on page 140.

By now you know that strong acids and bases are considered to be 100% dissociated in an aqueous solution. This makes calculating the concentrations of H^+ and OH^- in their solutions a relatively simple task.

When the solute in a solution is a strong monoprotic acid, such as HCl or HNO_3, we expect to obtain one mole of H^+ for every mole of the acid in the solution. Thus, a 0.010 *M* solution of HCl contains 0.010 mol L^{-1} of H^+ and a 0.0020 *M* solution of HNO_3 contains 0.0020 mol L^{-1} of H^+. (In this chapter we will not consider strong diprotic acids, such as H_2SO_4, because only the first step in their ionization is complete. In solutions of H_2SO_4, for example, only about 10% of the HSO_4^- ions are further ionized to SO_4^{2-} ions and hydrogen ions.)

To calculate the pH of a solution of a strong monoprotic acid, we use the molarity of the H^+ obtained from the stated molar concentration of the acid. Thus, the 0.010 *M* HCl solution mentioned above has $[H^+] = 0.010\ M$, from which we calculate the pH to be equal to 2.00.

For strong bases, calculating the OH^- concentration is similarly straightforward. A 0.050 *M* solution of NaOH contains 0.050 mol L^{-1} of OH^- because the base is fully dissociated and each mole of NaOH releases one mole of OH^- when it dissociates. For bases such as $Ba(OH)_2$ we have to recognize that two moles of OH^- are released by each mole of the base.

$$Ba(OH)_2(s) \longrightarrow Ba^{2+}(aq) + 2OH^-(aq)$$

Therefore, if a solution contained 0.010 mol $Ba(OH)_2$ per liter, the concentration of OH^- would be 0.020 *M*. Of course, once we know the OH^- concentration we can calculate pOH, from which we can calculate the pH. The following example illustrates the kinds of calculations we've just described.

EXAMPLE 15.8
Calculating the pH, pOH, and $[OH^-]$ for Solutions of Strong Acids or Strong Bases

Calculate the values of pH, pOH, and $[OH^-]$ for the following solutions: (a) 0.020 *M* HCl, (b) 0.00035 *M* $Ba(OH)_2$.

ANALYSIS: The first step in calculating the pH of a solution is an examination of the solute (or solutes) that are present. There are several questions you need to ask yourself. Are they strong or weak electrolytes? Are they acids or bases? How will they affect the pH? The answers to these questions will determine how you proceed next.

In this example, we expect the solutes to be strong electrolytes because that's what we've been discussing. However, let's take a look at them. The solute in part (a) is HCl, which you should recognize as one of the strong acids. Therefore, we expect it to be 100% ionized. From each mole of HCl, we expect one mole of H^+, so we will use the molar concentration of HCl to obtain $[H^+]$, from which we can calculate the pH, pOH, and $[OH^-]$.

The solute in part (b), $Ba(OH)_2$, is a soluble metal hydroxide, which is also a strong electrolyte. From each mole of $Ba(OH)_2$, there are *two* moles of OH^- liberated. We'll use the molarity of the $Ba(OH)_2$ to calculate the hydroxide ion concentration. From $[OH^-]$, we can calculate pOH and then pH.

SOLUTION: (a) In 0.020 *M* HCl, $[H^+] = 0.020\ M$. Therefore,

$$\begin{aligned} pH &= -\log(0.020) \\ &= 1.70 \end{aligned}$$

Thus in 0.020 *M* HCl, the pH is 1.70. The pOH is $(14.00 - 1.70) = 12.30$. To find $[OH^-]$, we can use this value of pOH.

$$\begin{aligned} [OH^-] &= 10^{-pOH} \\ &= 10^{-12.30} \\ &= 5.0 \times 10^{-13}\ M \end{aligned}$$

Notice how much smaller $[OH^-]$ is in this acidic solution than it is in pure water. This makes sense, because the solution is quite acidic.

(b) As noted, for each mole of $Ba(OH)_2$ we obtain two moles of OH^-. Therefore,

$$\begin{aligned} [OH^-] &= 2 \times 0.00035\ M \\ &= 0.00070\ M \\ pOH &= -\log(0.00070) \\ &= 3.15 \end{aligned}$$

Thus the pOH of this solution is 3.15, and the $pH = (14.00 - 3.15) = 10.85$.

ARE THE ANSWERS REASONABLE? In part (a), the molarity of H^+ is between 0.01 *M* and 0.1 *M*, or between 10^{-2} *M* and 10^{-1} *M*. So the pH must be between 1 and 2, as we found. In part (b), the molarity of OH^- is between 0.0001 and 0.001, or between 10^{-4} and 10^{-3}. Hence, the pOH must be between 3 and 4, which it is.

Practice Exercise 27: Calculate the $[H^+]$, pH, and pOH in 0.0050 *M* HNO_3. (Hint: HNO_3 is a strong acid.)

Practice Exercise 28: Calculate the pOH, $[H^+]$, and pH in a solution made by weighing 1.20 g of KOH and dissolving it in sufficient water to make 250.0 mL of solution.

Practice Exercise 29: Rhododendrons are shrubs that produce beautiful flowers in the springtime. They only grow well in soil that has a pH that is 5.5 or slightly lower. What is the hydrogen ion concentration in the soil moisture if the pH is 5.5?

Acidic or basic solutes suppress the ionization of water

In the preceding calculations, we have made a critical and correct assumption, namely, that the autoionization of water contributes negligibly to the total $[H^+]$ in a solution of an acid and to the total $[OH^-]$ in a solution of a base.[7] Let's take a closer look at this.

In a solution of an acid, there are actually *two* sources of H^+. One is from the ionization of the acid solute itself and the other is from the autoionization of water. Thus,

$$[H^+]_{total} = [H^+]_{from\ solute} + [H^+]_{from\ H_2O}$$

A similar equation applies to the total OH^- concentration in a solution of a base.

$[OH^-]_{total} = [OH^-]_{from\ solute} + [OH^-]_{from\ H_2O}$

Except in very dilute solutions of acids, the amount of H^+ contributed by the water ($[H^+]_{from\ H_2O}$) is small compared to the amount of H^+ contributed by the solute ($[H^+]_{from\ solute}$). For instance, in Example 15.8 we saw that in 0.020 *M* HCl the molarity of OH^- was 5.0×10^{-13} *M*. The only source of OH^- in this acidic solution is from the autoionization of water, and the amounts of OH^- and H^+ *formed by the autoionization of water* must be equal. Therefore, $[H^+]_{from\ H_2O}$ also equals 5.0×10^{-13} *M*. If we now look at the total $[H^+]$ for this solution, we have

$$[H^+]_{total} = \underset{\text{(from HCl)}}{0.020\ M} + \underset{\text{(from } H_2O)}{5.0 \times 10^{-13}\ M}$$

$$= 0.020\ M \quad \text{(rounded correctly)}$$

In any solution of an acid, the autoionization of water is suppressed by the H^+ furnished by the solute. It's simply an example of Le Châtelier's principle. If we look at the autoionization reaction, we can see that if some H^+ is provided by an external source (an acidic solute, for example), the position of equilibrium will be shifted to the left.

$$H_2O \rightleftharpoons H^+ + OH^-$$

Adding H^+ from a solute causes the position of equilibrium to shift to the left.

As the results of our calculation have shown, the concentrations of H^+ and OH^- *from the autoionization reaction* are reduced well below their values in a neutral solution (1.0×10^{-7} *M*). Therefore, except for *very* dilute solutions (10^{-6} *M* or less), we will assume that all of the H^+ in the solution of an acid comes from the solute. Similarly, *we'll assume that in a solution of a base, all the* OH^- *comes from the dissociation of the solute.*

SUMMARY

Brønsted–Lowry Acids and Bases. A **Brønsted–Lowry acid** (or more simply a **Brønsted acid**) is a proton donor; a **Brønsted base** is a proton acceptor. According to the Brønsted–Lowry approach, an acid–base reaction is a proton transfer event. In an equilibrium involving a Brønsted acid and base, there are two **conjugate acid–base pairs.** The members of any given pair differ from each other by only one H^+. A substance that can be either an acid or a base, depending on the nature of the other reactant, is **amphoteric** or, with emphasis on proton transfer reactions, **amphiprotic.**

Lewis Acids and Bases. A **Lewis acid** accepts a pair of electrons from a **Lewis base** in the formation of a coordinate covalent bond. Lewis bases often have filled valence shells and must have at least one unshared electron pair. Lewis acids have an incomplete valence shell that can accept an electron pair, have double bonds that allow electron pairs to be moved to make room for an incoming electron pair from a Lewis base, or have valence shells that can accept more than an octet of electrons.

Relative Acidities and the Periodic Table. **Binary acids** contain only hydrogen and another nonmetal. Their strengths increase from top to bottom within a group and from left to right across a period. **Oxoacids,** which contain oxygen atoms in addition to hydrogen and another element, increase in strength as the number of oxygen atoms on the same central atom increases. Delocalization of negative charge enhances the stability of oxoacid anions, making them weaker bases, and as a result, their conjugate acids are correspondingly stronger. Oxoacids having the same number of oxygens generally increase in strength as the central atom moves from bottom to top within a group and from left to right across a period.

Acid–Base Properties of the Elements and Their Oxides. Oxides of metals are basic anhydrides when the charge on the ion is small. Those of the Group IA and IIA metals neutralize acids and tend to react with water to form soluble metal hydroxides. The hydrates of metal ions tend to be proton donors when the positive charge density of the metal ion is itself sufficiently high, as it is when the ion has a 3+ charge. The small beryllium ion, Be^{2+}, forms a weakly acidic hydrated ion. Metal oxides become more acidic as the oxidation number of the metal becomes larger. Aluminum oxide is amphoteric, dissolving in both acids and bases. Chromium(VI) oxide is acidic, forming chromic acid when it dissolves in water.

The Autoionization of Water and the pH Concept. Water reacts with itself to produce small amounts of H_3O^+ (often abbreviated H^+) and OH^- ions. The concentrations of these ions both in pure water *and* in dilute aqueous solutions are related by the expression

$$[H^+][OH^-] = K_w = 1.0 \times 10^{-14} \qquad \text{(at 25 °C)}$$

K_w is the **ion product constant of water.** In pure water,

$$[H^+] = [OH^-] = 1.0 \times 10^{-7}\,M$$

The **pH** of a solution is a measure of the acidity of a solution and is normally measured with a pH meter. As the pH decreases, the acidity increases. The defining equation for pH is $pH = -\log [H^+]$. In exponential form, this relationship between $[H^+]$ and pH is given by $[H^+] = 10^{-pH}$.

Comparable expressions can be used to describe low OH^- ion concentrations in terms of **pOH** values: $pOH = -\log [OH^-]$ and $[OH^-] = 10^{-pOH}$. At 25 °C, pH + pOH = 14.00. A solution is acidic if its pH is less than 7.00 and basic if its pH is greater than 7.00. A neutral solution has a pH of 7.00.

Solutions of Strong Acids and Strong Bases. In calculating the pH of solutions of strong acids and bases, we assume that they are 100% ionized. The autoionization of water contributes a negligible amount to the $[H^+]$ in a solution of an acid. It also contributes a negligible amount to the $[OH^-]$ in a solution of a base.

TOOLS FOR PROBLEM SOLVING

In this chapter you learned to apply the following concepts as tools in solving problems dealing with aspects of acid–base properties and equilibria. Study each tool carefully so that you know what each is used for. When faced with solving a problem, recall what each tool does and consider whether it will be helpful in finding a solution. This will aid you in selecting the tools you need. Remember that at times tools from previous chapters will be needed along with the new ones in this chapter.

Brønsted–Lowry conjugate acids and bases *(page 607)* The Brønsted–Lowry acid–base equilibrium is based on conjugate acid–base pairs. Every conjugate acid–base pair consists of an acid and a base that has lost one proton (H^+) *(page 608)* The conjugate acid is always on the opposite side of an equation from its conjugate base, and the Brønsted–Lowry acid–base equilibrium is typically written with two sets of conjugate acid–base pairs *(page 608)*. Finally we can use the position of equilibrium to determine the relative strengths of conjugate acids and bases *(pages 612 and 613)*.

Periodic trends in strengths of binary acids *(page 615)* You can use the trends to predict the relative acidities of X—H bonds, both for the binary hydrides themselves and for molecules that contain X—H bonds. In general these acids increase in strength going from left to right in a period and from top to bottom of a group in the periodic table.

Periodic trends in strengths of oxoacids *(pages 617, 618)* You can use the trends to predict the relative acidities of oxoacids according to the nature of the central nonmetal as well as the number of oxygens attached to a given nonmetal. The principles involved also let you compare acidities of compounds containing different electronegative elements.

Lewis acids and bases *(page 619)* A Lewis acid is an electron pair acceptor and a Lewis base is an electron pair donor.

Ion product constant for water *(page 626)* Use this equation to calculate $[H^+]$ if you know $[OH^-]$, and vice versa. Be sure you have learned the value of K_w.

$$[H^+][OH^-] = K_w$$

Defining equations for p*X* *(page 628)* Use the equations of this type to calculate pH and pOH from $[H^+]$ and $[OH^-]$, respectively. You also use them to calculate $[H^+]$ and $[OH^-]$ from pH and pOH, respectively.

$$pX = -\log X \qquad \text{and} \qquad X = 10^{-pX}$$

Relationship between pH and pOH *(page 629)* This relationship is used to calculate pH if you know pOH, and vice versa.

$$pH + pOH = pK_w = 14.00 \qquad \text{(at 25°C)}$$

QUESTIONS, PROBLEMS, AND EXERCISES

Answers to problems whose numbers are printed in color are given in Appendix B. More challenging problems are marked with asterisks. ILW = Interactive Learningware solution is available at www.wiley.com/college/brady. OH = an Office Hours video is available for this problem.

REVIEW QUESTIONS

Brønsted–Lowry Acids and Bases

15.1 How is a *Brønsted acid* defined? How is a *Brønsted base* defined?

15.2 How are the formulas of the members of a conjugate acid–base pair related to each other? Within the pair, how can you tell which is the acid?

15.3 Is H_2SO_4 the conjugate acid of SO_4^{2-}? Explain your answer.

15.4 What is meant by the term *amphoteric*? Give two chemical equations that illustrate the amphoteric nature of water.

15.5 Define the term *amphiprotic*.

Trends in Acid–Base Strengths

15.6 Within the periodic table, how do the strengths of the binary acids vary from left to right across a period? How do they vary from top to bottom within a group?

15.7 Astatine, atomic number 85, is radioactive and does not occur in appreciable amounts in nature. On the basis of what you have learned in this chapter, answer the following.
(a) How would the acidity of HAt compare to HI?
(b) How would the acidity of $HAtO_3$ compare with $HBrO_3$?

15.8 Explain why nitric acid is a stronger acid than nitrous acid.

$HO{-}NO_2$	$HO{-}NO$
nitric acid	nitrous acid

15.9 Explain why H_2S is a stronger acid than H_2O.

15.10 Which is the stronger Brønsted base, $CH_3CH_2O^-$ or $CH_3CH_2S^-$? What is the basis for your selection?

15.11 Explain why $HClO_4$ is a stronger acid than H_2SeO_4.

15.12 The position of equilibrium in the equation below lies far to the left. Identify the conjugate acid–base pairs. Which of the two acids is stronger?

$$HOCl(aq) + H_2O \rightleftharpoons H_3O^+(aq) + OCl^-(aq)$$

15.13 Consider the following: CO_3^{2-} is a weaker base than hydroxide ion, and HCO_3^- is a stronger acid than water. In the equation below, would the position of equilibrium lie to the left or to the right? Justify your answer.

$$CO_3^{2-}(aq) + H_2O \rightleftharpoons HCO_3^-(aq) + OH^-(aq)$$

15.14 Acetic acid, $HC_2H_3O_2$, is a weaker acid than nitrous acid, HNO_2. How do the strengths of the bases $C_2H_3O_2^-$ and NO_2^- compare?

15.15 Nitric acid, HNO_3, is a very strong acid. It is 100% ionized in water. In the reaction below, would the position of equilibrium lie to the left or to the right?

$$NO_3^-(aq) + H_2O \rightleftharpoons HNO_3(aq) + OH^-(aq)$$

15.16 $HClO_4$ is a stronger proton donor than HNO_3, but in water both acids appear to be of equal strength; they are both 100% ionized. Why is this so? What solvent property would be necessary in order to distinguish between the acidities of these two Brønsted acids?

15.17 Formic acid, $HCHO_2$, and acetic acid, $HC_2H_3O_2$, are classified as weak acids, but in water $HCHO_2$ is more fully ionized than $HC_2H_3O_2$. However, if we use liquid ammonia as a solvent for these acids, they both appear to be of equal strengths; both are 100% ionized in liquid ammonia. Explain why this is so.

15.18 Which of the molecules below is expected to be the stronger Brønsted acid? Why?

```
    H   O          F   O
    |  //          |  //
H—C—C         F—C—C
    |   \          |   \
    H    O—H       F    O—H
```

15.19 In which of the molecules below is the hydrogen printed in color the more acidic hydrogen? Explain your choice.

```
     Cl  O          Cl  O
     |  //          |  //
Cl—C—C         H—C—C
     |   \          |   \
     Cl   O—H       Cl   O—H
```

Lewis Acids and Bases

15.20 Define *Lewis acid* and *Lewis base*.

15.21 Explain why the addition of a proton to a water molecule to give H_3O^+ is a Lewis acid–base reaction.

15.22 Methylamine has the formula CH_3NH_2 and the structure

```
    H   H
    |   |
H—C—N:
    |   |
    H   H
```

Use Lewis structures to illustrate the reaction of methylamine with boron trifluoride.

15.23 Use Lewis structures to show the Lewis acid–base reaction between CO_2 and H_2O to give H_2CO_3. Identify the Lewis acid and the Lewis base in the reaction.

15.24 Explain why the oxide ion, O^{2-}, can function as a Lewis base but not as a Lewis acid.

15.25 The molecule SbF_5 is able to function as a Lewis acid. Explain why it is able to be a Lewis acid.

15.26 In the reaction of calcium with oxygen to form calcium oxide, each calcium gives a pair of electrons to an oxygen atom. Why isn't this viewed as a Lewis acid–base reaction?

15.27 Boric acid is very poisonous and is used in ant bait (to kill ant colonies) and to poison cockroaches. It is a weak acid with a formula often written as H_3BO_3, although it is better written as $B(OH)_3$. It functions not as a Brønsted acid, but as a Lewis acid. Using Lewis structures, show how $B(OH)_3$ can bind to a water molecule and cause the resulting product to be a weak Brønsted acid.

Acid–Base Properties of the Elements and Their Oxides

15.28 Suppose that a new element was discovered. Based on the discussions in this chapter, what properties (both physical and chemical) might be used to classify the element as a metal or a nonmetal?

15.29 If the oxide of an element dissolves in water to give an acidic solution, is the element more likely to be a metal or a nonmetal?

15.30 Many chromium salts crystallize as hydrates containing the ion $Cr(H_2O)_6^{3+}$. Solutions of these salts tend to be acidic. Explain why.

15.31 Which ion is expected to give the more acidic solution, Fe^{2+} or Fe^{3+}? Why?

15.32 Ions of the alkali metals have little effect on the acidity of a solution. Why?

15.33 What acid is formed when the following oxides react with water?
(a) SO_3, (b) CO_2, (c) P_4O_{10}

15.34 Consider the following oxides: CrO, Cr_2O_3, and CrO_3.
(a) Which is most acidic?
(b) Which is most basic?
(c) Which is most likely to be amphoteric?

15.35 Write equations for the reaction of Al_2O_3 with (a) a strong acid, and (b) a strong base.

Ionization of Water and the pH Concept

15.36 Write the chemical equation for the autoionization of water and the equilibrium law for K_w.

15.37 How are acidic, basic, and neutral solutions in water defined (a) in terms of $[H^+]$ and $[OH^-]$ and (b) in terms of pH?

15.38 At 25 °C, how are the pH and pOH of a solution related to each other?

15.39 Explain how acids and bases suppress the ionization of water, often called the common ion effect.

15.40 Explain the leveling effect of water.

REVIEW PROBLEMS

Brønsted Acids and Bases

OH **15.41** Write the formula for the conjugate acid of each of the following.
(a) F^-
(b) N_2H_4
(c) C_5H_5N
(d) O_2^{2-}
(e) $HCrO_4^-$

15.42 Write the formula for the conjugate base of each of the following.
(a) NH_2OH
(b) HSO_3^-
(c) HCN
(d) H_5IO_6
(e) HNO_2

15.43 Identify the conjugate acid–base pairs in the following reactions.
(a) $HNO_3 + N_2H_4 \rightleftharpoons NO_3^- + N_2H_5^+$
(b) $NH_3 + N_2H_5^+ \rightleftharpoons NH_4^+ + N_2H_4$
(c) $H_2PO_4^- + CO_3^{2-} \rightleftharpoons HPO_4^{2-} + HCO_3^-$
(d) $HIO_3 + HC_2O_4^- \rightleftharpoons IO_3^- + H_2C_2O_4$

15.44 Identify the conjugate acid–base pairs in the following reactions.
(a) $HSO_4^- + SO_3^{2-} \rightleftharpoons HSO_3^- + SO_4^{2-}$
(b) $S^{2-} + H_2O \rightleftharpoons HS^- + OH^-$
(c) $CN^- + H_3O^+ \rightleftharpoons HCN + H_2O$
(d) $H_2Se + H_2O \rightleftharpoons HSe^- + H_3O^+$

Trends in Acid–Base Strengths

15.45 Choose the stronger acid: (a) HBr or HCl; (b) H_2O or HF; (c) H_2S or HBr. Give your reasons.

15.46 Choose the stronger acid: (a) H_2S or H_2Se; (b) H_2Te or HI; (c) PH_3 or NH_3. Give your reasons.

15.47 Choose the stronger acid: (a) HOCl or $HClO_2$; (b) H_2SeO_4 or H_2SeO_3. Give your reasons.

15.48 Choose the stronger acid: (a) HIO_3 or HIO_4; (b) H_3AsO_4 or H_3AsO_3. Give your reasons.

OH **15.49** Choose the stronger acid: (a) $HClO_3$ or HIO_3; (b) HIO_2 or $HClO_3$; (c) H_2SeO_3 or $HBrO_4$. Give your reasons.

15.50 Choose the stronger acid: (a) H_3AsO_4 or H_3PO_4; (b) H_2CO_3 or HNO_3; (c) H_2SeO_4 or $HClO_4$. Give your reasons.

Lewis Acids and Bases

15.51 Use Lewis symbols to diagram the reaction

$$NH_2^- + H^+ \longrightarrow NH_3$$

Identify the Lewis acid and Lewis base in the reaction.

15.52 Use Lewis symbols to diagram the reaction

$$BF_3 + F^- \longrightarrow BF_4^-$$

Identify the Lewis acid and Lewis base in the reaction.

15.53 Beryllium chloride, $BeCl_2$, exists in the solid as a polymer composed of long chains of $BeCl_2$ units arranged as indicated on p. 640. The formula of the chain can be represented as $(BeCl_2)_n$, where n is a large number. Use Lewis structures to show how the reaction $n\text{BeCl}_2 \longrightarrow (\text{BeCl}_2)_n$ is a Lewis acid–base reaction.

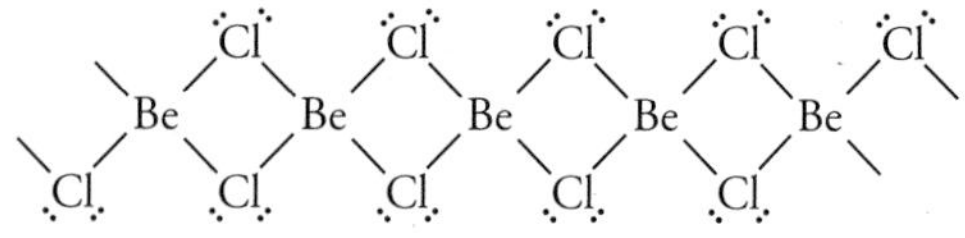

15.54 Aluminum chloride, $AlCl_3$, forms molecules with itself with the formula Al_2Cl_6. Its structure is

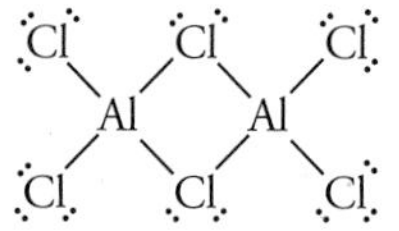

Use Lewis structures to show how the reaction $2AlCl_3 \longrightarrow Al_2Cl_6$ is a Lewis acid–base reaction.

OH **15.55** Use Lewis structures to diagram the reaction

$$CO_2 + H_2O \longrightarrow H_2CO_3$$

Identify the Lewis acid and Lewis base in this reaction.

15.56 Use Lewis structures to diagram the reaction

$$CO_2 + O^{2-} \longrightarrow CO_3^{2-}$$

Identify the Lewis acid and Lewis base in this reaction.

15.57 Use Lewis structures to show how the following reaction can be viewed as the displacement of one Lewis base by another Lewis base from a Lewis acid. Identify the two Lewis bases and the Lewis acid.

$$NH_2^- + H_2O \longrightarrow NH_3 + OH^-$$

OH **15.58** Use Lewis structures to show how the following reaction involves the transfer of a Lewis base from one Lewis acid to another. Identify the two Lewis acids and the two Lewis bases.

$$CO_3^{2-} + SO_2 \longrightarrow CO_2 + SO_3^{2-}$$

Autoionization of Water and pH

15.59 Deuterium oxide, D_2O, ionizes like water. At 20 °C its K_w or ion product constant, analogous to that of water, is 8.9×10^{-16}. Calculate $[D^+]$ and $[OD^-]$ in deuterium oxide at 20 °C. Calculate also the pD and the pOD.

15.60 At the temperature of the human body, 37 °C, the value of K_w is 2.5×10^{-14}. Calculate $[H^+]$, $[OH^-]$, pH, and pOH of pure water at that temperature. What is the relationship between pH, pOH, and K_w at that temperature? Is water neutral at that temperature?

15.61 Calculate the $[H^+]$, pH, and pOH in each of the following solutions in which the hydroxide ion concentrations are
(a) 0.0068 M
(b) 6.4×10^{-5} M
(c) 1.6×10^{-8} M
(d) 8.2×10^{-12} M

15.62 Calculate the $[OH^-]$, pH, and pOH in each of the following solutions in which the H^+ concentrations are
(a) 3.5×10^{-7} M
(b) 0.0017 M
(c) 2.5×10^{-11} M
(d) 7.9×10^{-2} M

OH **15.63** A certain brand of beer had a H^+ concentration equal to 1.9×10^{-5} mol L^{-1}. What is the pH of the beer?

15.64 A soft drink was put on the market with $[H^+] = 1.4 \times 10^{-5}$ mol L^{-1}. What is its pH?

15.65 Calculate the molar concentrations of H^+ and OH^- in solutions that have the following pH values.
(a) 8.14
(b) 2.56
(c) 11.25
(d) 13.28
(e) 6.70

15.66 Calculate the molar concentrations of H^+ and OH^- in solutions that have the following pOH values.
(a) 12.27
(b) 6.14
(c) 10.65
(d) 4.28
(e) 3.76

15.67 The interaction of water droplets in rain with carbon dioxide that is naturally present in the atmosphere causes rainwater to be slightly acidic because CO_2 is an acidic anhydride. As a result, pure clean rain has a pH of about 5.7. What are the hydrogen ion and hydroxide ion concentrations in this rainwater?

15.68 "Acid rain" forms when rain falls through air polluted by oxides of sulfur and nitrogen, which dissolve to form acids such as H_2SO_3, H_2SO_4, and HNO_3. Trees and plants are affected if the acid rain has a pH of 3.5 or lower. What is the hydrogen ion concentration in acid rain that has a pH of 3.16? What is the pH of a solution having twice your calculated hydrogen ion concentration?

Solutions of Strong Acids and Bases

15.69 What is the concentration of H^+ in 0.00065 M HNO_3? What is the pH of the solution? What is the OH^- concentration in the solution?

15.70 What is the concentration of H^+ in 0.031 M $HClO_4$? What is the pH of the solution? What is the OH^- concentration in the solution? By how much does the pH change if the concentration of H^+ is doubled?

ILW **15.71** A sodium hydroxide solution is prepared by dissolving 6.0 g NaOH in 1.00 L of solution. What is the molar concentration of OH^- in the solution? What is the pOH and the pH of the solution? What is the hydrogen ion concentration in the solution?

OH **15.72** A solution was made by dissolving 0.837 g $Ba(OH)_2$ in 100 mL final volume. What is the molar concentration of OH^- in the solution? What are the pOH and the pH? What is the hydrogen ion concentration in the solution?

15.73 A solution of $Ca(OH)_2$ has a measured pH of 11.60. What is the molar concentration of the $Ca(OH)_2$ in the solution? What is the molar concentration of $Ca(OH)_2$ if the solution is diluted so that the pH is 10.60?

15.74 A solution of HCl has a pH of 2.50. How many grams of HCl are there in 250 mL of this solution? How many grams of HCl are in 250 mL of an HCl solution that has twice the pH?

15.75 How many milliliters of 0.0100 M KOH are needed to completely neutralize the HCl in 300 mL of a hydrochloric acid solution that has a pH of 2.25?

15.76 It was found that 25.20 mL of an HNO_3 solution is needed to react completely with 300 mL of a LiOH solution that has a pH of 12.05. What is the molarity of the HNO_3 solution?

15.77 In a 0.0020 *M* solution of NaOH, how many moles per liter of OH^- come from the ionization of water?

15.78 In a certain solution of HCl, the ionization of water contributes 3.4×10^{-11} moles per liter to the H^+ concentration. What is the total H^+ concentration in the solution?

ADDITIONAL EXERCISES

15.79 What is the formula of the conjugate acid of dimethylamine, $(CH_3)_2NH$? What is the formula of its conjugate base?

15.80 Suppose 10.0 mL of HCl gas at 25 °C and 734 torr is bubbled into 250 mL of pure water. What will be the pH of the resulting solution, assuming all the HCl dissolves in the water?

15.81 Suppose the HCl described in the preceding exercise is bubbled into 200 mL of a solution of NaOH that has a pH of 10.50. What will the pH of the resulting solution be?

15.82 Write equations that illustrate the amphiprotic nature of the bicarbonate ion.

15.83 Hydrogen peroxide is a stronger Brønsted acid than water. (a) Explain why this is so. (b) Is an aqueous solution of hydrogen peroxide acidic or basic?

15.84 Hydrazine, N_2H_4, is a weaker Brønsted base than ammonia. In the following reaction, would the position of equilibrium lie to the left or to the right? Justify your answer.

$$N_2H_5^+ + NH_3 \rightleftharpoons N_2H_4 + NH_4^+$$

15.85 Identify the two Brønsted acids and two Brønsted bases in the reaction

$$NH_2OH + CH_3NH_3^+ \rightleftharpoons NH_3OH^+ + CH_3NH_2$$

15.86 In the reaction in the preceding exercise, the position of equilibrium lies to left. Identify the stronger acid in each of the conjugate pairs in the reaction.

***15.87** How would you expect the degree of ionization of $HClO_3$ to compare in the solvents $H_2O(l)$ and $HF(l)$? The reactions are

$$HClO_3 + H_2O \rightleftharpoons H_3O^+ + ClO_3^-$$

$$HClO_3 + HF \rightleftharpoons H_2F^+ + ClO_3^-$$

Justify your answer.

***15.88** Suppose 38.0 mL of 0.000200 *M* HCl is added to 40.0 mL of 0.000180 *M* NaOH. What will be the pH of the final mixture?

***15.89** What is the pH of a 3.0×10^{-7} *M* solution of HCl?

***15.90** How many milliliters of 0.10 *M* NaOH must be added to 200 mL of 0.010 *M* HCl to give a mixture with a pH of 3.00?

15.91 Milk of magnesia is a suspension of magnesium hydroxide in water. Although $Mg(OH)_2$ is relatively insoluble, a small amount does dissolve in the water, which makes the mixture slightly basic and gives it a pH of 10.08. How many grams of $Mg(OH)_2$ are actually dissolved in 100 mL of milk of magnesia?

***15.92** A 1.0 *M* solution of acetic acid has a pH of 2.37. What percentage of the acetic acid is ionized in the solution?

EXERCISES IN CRITICAL THINKING

15.93 Are all Arrhenius acids Brønsted acids? Lewis acids? Give examples if they are not. Give a reasoned explanation if they are.

15.94 How could you determine whether HBr is a stronger acid than HI?

15.95 What happens to the pH of a solution as it is heated? Does that mean that the autoionization of water is exothermic or endothermic?

15.96 Can the pH of a solution ever have a negative value? If so give an example of a situation where the pH has a negative value.

15.97 Alcohols are organic compounds that have an —OH group. Are alcohols acids or bases? Sugars have an —OH group on almost every carbon atom. Are sugars acids or bases? Phenol is a benzene ring with an —OH group. Is phenol an acid or base?

15.98 Acid rain, acid mine drainage, and acid leaching of metals from soils are important environmental considerations. What do these topics refer to and how do they affect you as a person?

EQUILIBRIA IN SOLUTIONS OF WEAK ACIDS AND BASES

16

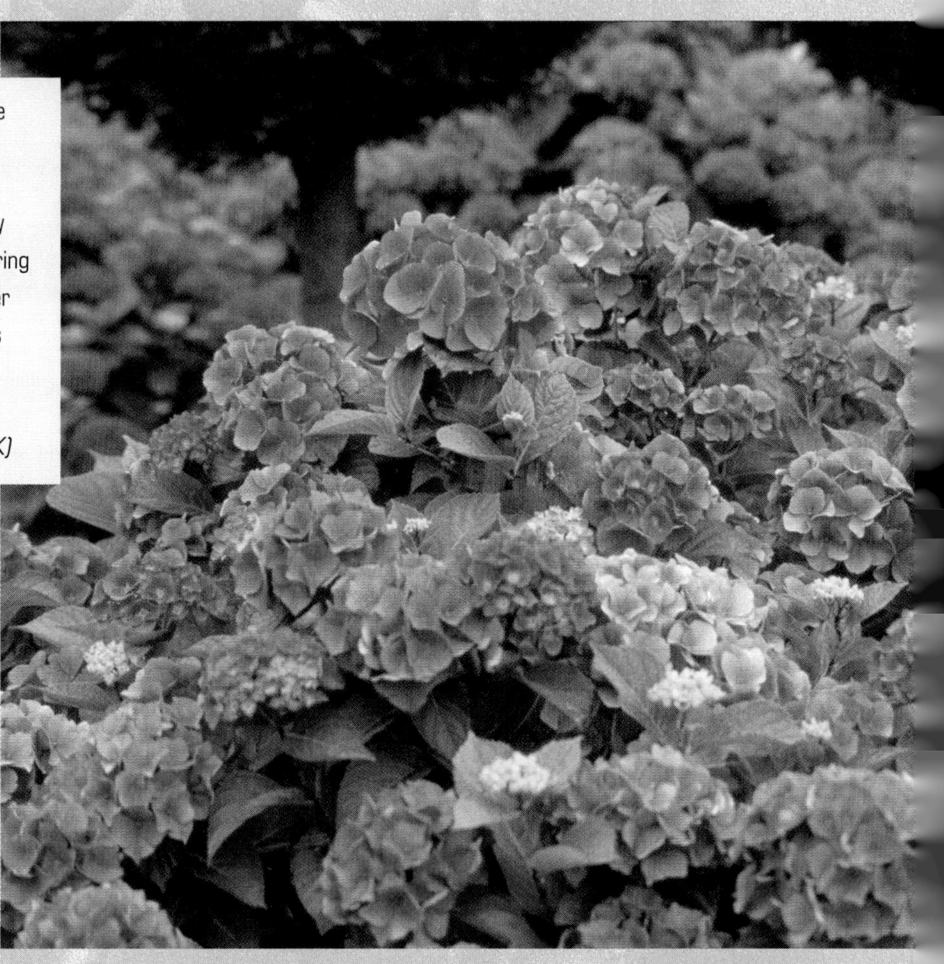

Interesting gardens often have flowering plants with different colors. In this photo all of the plants are hydrangias but they are growing in soils with differing levels of acidity. In this chapter we will find methods chemists use to precisely control the acidity of solutions and soils. *(age fotostock/SUPERSTOCK)*

CHAPTER OUTLINE

THIS CHAPTER IN CONTEXT In Chapters 14 and 15 we discussed the general principles of c equilibrium and how to deal with calculating the hydrogen and hydroxide ion concentrations and pH of solu strong acids and bases. Our goal in this chapter is to bring these concepts together as we examine the c equilibria that exist in solutions of weak acids and bases. There are literally thousands of such substances, r which are found in the environment or in biological systems as a consequence of the natural acidity or basicity of biological molecules.

The principles we develop here have applications not only in traditional chemistry labs, but also in labs that investigate environmental, forensic, and biochemical problems. High-technology laboratories interested in materials science and nanotechnology often use the principles of acid–base equilibria. The consumer oriented industries that make products such as cosmetics, foods, beverages, and cleaning chemicals all employ chemists who are aware that control of pH is very important in safe and effective consumer products.

16.1 IONIZATION CONSTANTS CAN BE DEFINED FOR WEAK ACIDS AND BASES

As you've learned, weak acids and bases are incompletely ionized in water and exist in solution as molecules that are in equilibria with the ions formed by their reactions with water. To deal quantitatively with these equilibria it is essential that you be able to write correct chemical equations for the equilibrium reactions, from which you can then obtain the corresponding correct equilibrium laws. Fortunately, there is a pattern that applies to the way these substances react. *Once you've learned how to write the correct equation for one acid or base, you can write the correct equation for any other.*

A weak acid reacts with water to give its conjugate base and hydronium ion

In aqueous solutions, all weak acids behave the same way. They are Brønsted acids and, therefore, proton donors. Some examples are $HC_2H_3O_2$, HSO_4^-, and NH_4^+. In water, these participate in the following equilibria.

$$HC_2H_3O_2 + H_2O \rightleftharpoons H_3O^+ + C_2H_3O_2^-$$

$$HSO_4^- + H_2O \rightleftharpoons H_3O^+ + SO_4^{2-}$$

$$NH_4^+ + H_2O \rightleftharpoons H_3O^+ + NH_3$$

Notice that, in each case, the acid reacts with water to give H_3O^+ and the corresponding conjugate base. We can represent these reactions in a general way using HA to stand for the formula of the acid.

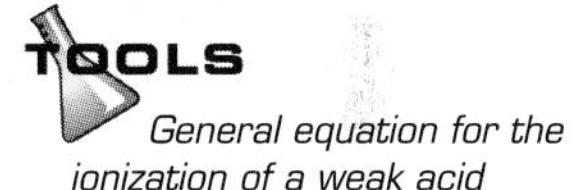

General equation for the ionization of a weak acid

$$HA + H_2O \rightleftharpoons H_3O^+ + A^- \tag{16.1}$$

As you can see from the equations above, HA does not have to be electrically neutral; it can be a molecule such as $HC_2H_3O_2$, a negative ion such as HSO_4^-, or a positive ion such as NH_4^+. (Of course, the actual charge on the conjugate base will then depend on the charge on the parent acid.)

Following the procedure developed in Chapter 14, we can also write a general equation for the equilibrium law for Reaction 16.1, using K_c' to stand for the equilibrium constant.

$$K_c' = \frac{[H_3O^+][A^-]}{[HA][H_2O]}$$

In our discussion of the autoionization of water in Chapter 15, we noted that in dilute aqueous solutions $[H_2O]$ can be considered a constant, so it can be combined with K_c' to give a new equilibrium constant. Doing this gives

$$K_c' \times [H_2O] = \frac{[H_3O^+][A^-]}{[HA]} = K_a$$

The new constant $\boldsymbol{K_a}$ is called an **acid ionization constant.** Abbreviating H_3O^+ as H^+, the equation for the ionization of the acid can be simplified as

Some call K_a the *acid dissociation constant.*

$$HA \rightleftharpoons H^+ + A^-$$

from which the expression for K_a is obtained directly.

$$K_a = \frac{[H^+][A^-]}{[HA]} \tag{16.2}$$

You should learn how to write the chemical equation for the ionization of a weak acid and be able to write the equilibrium law corresponding to its K_a. This is illustrated in Example 16.1.

EXAMPLE 16.1

Writing the K_a Expression for a Weak Acid

Nitrous acid, HNO_2, is a weak acid that's formed in the stomach when nitrite food preservatives encounter stomach acid. There has been some concern that this acid may form carcinogenic products by reacting with proteins. Write the chemical equation for the equilibrium ionization of HNO_2 in water and the appropriate K_a expression.

The meat products shown here contain nitrite ion as a preservative. *(Andy Washnik.)*

ANALYSIS: To solve this problem, we have to think about what happens when the acid reacts with water so we can construct the chemical equation. Once we have the equation, we use it to construct the equilibrium law, which will consist of the mass action expression set equal to K_a.

SOLUTION: When HNO_2 reacts with water, the products will be hydronium ion, H_3O^+, and the conjugate base. The conjugate base of HNO_2 is NO_2^-, so the equation for the ionization reaction is

$$HNO_2 + H_2O \rightleftharpoons H_3O^+ + NO_2^-$$

In the K_a expression, we leave out the H_2O that appears on the left:

$$K_a = \frac{[H_3O^+][NO_2^-]}{[HNO_2]}$$

For simplicity, we usually represent H_3O^+ as H^+, so we can write the expression as

$$K_a = \frac{[H^+][NO_2^-]}{[HNO_2]}$$

Alternatively, we could have written the simplified equation for the ionization reaction, from which the K_a expression above is obtained directly.

$$HNO_2 \rightleftharpoons H^+ + NO_2^-$$

IS THE ANSWER REASONABLE? Our K_a expression has products multiplied in the numerator and reactants in the denominator as required. Also required is that terms within brackets must be multiplied or divided and they are (there are no addition, +, or subtraction, −, signs where they don't belong). We are now confident we have solved the problem correctly.

Practice Exercise 1: For each of the following acids, write the equation for its ionization in water and the appropriate expression for K_a: (a) $HC_2H_3O_2$, (b) $(CH_3)_3NH^+$, and (c) H_3PO_4. (Hint: Determine the conjugate base for each of these acids.)

Practice Exercise 2: For each of the following acids, write the equation for its ionization in water and the appropriate expression for K_a: (a) $HCHO_2$, (b) $(CH_3)_2NH_2^+$, and (c) $H_2PO_4^-$.

For weak acids, values of K_a are usually quite small and can be conveniently represented in a logarithmic form similar to pH. Thus, we can define the **pK_a** of an acid as

Also, K_a = antilog($-$pK_a) or $K_a = 10^{-pK_a}$.

$$pK_a = -\log K_a$$

The strength of a weak acid is determined by its value of K_a; the larger the K_a, the stronger and more fully ionized the acid. Because of the negative sign in the defining equation for pK_a, the stronger the acid, the *smaller* is its value of pK_a. The values of K_a and pK_a for some typical weak acids are given in Table 16.1. A more complete list is located in Appendix C.7.

TABLE 16.1 K_a and pK_a Values for Weak Monoprotic Acids at 25 °C

Name of Acid	Formula	K_a	pK_a
Iodic acid	HIO_3	1.7×10^{-1}	0.77
Chloroacetic acid	$HC_2H_2O_2Cl$	1.36×10^{-3}	2.87
Nitrous acid	HNO_2	7.1×10^{-4}	3.15
Hydrofluoric acid	HF	6.8×10^{-4}	3.17
Cyanic acid	$HOCN$	3.5×10^{-4}	3.46
Formic acid	$HCHO_2$	1.8×10^{-4}	3.74
Barbituric acid	$HC_4H_3N_2O_3$	9.8×10^{-5}	4.01
Acetic acid	$HC_2H_3O_2$	1.8×10^{-5}	4.74
Hydrazoic acid	HN_3	1.8×10^{-5}	4.74
Butanoic acid	$HC_4H_7O_2$	1.52×10^{-5}	4.82
Propanoic acid	$HC_3H_5O_2$	1.34×10^{-5}	4.87
Hypochlorous acid	$HOCl$	3.0×10^{-8}	7.52
Hydrocyanic acid (*aq*)	HCN	6.2×10^{-10}	9.21
Phenol	HC_6H_5O	1.3×10^{-10}	9.89
Hydrogen peroxide	H_2O_2	1.8×10^{-12}	11.74

The values of K_a for strong acids are very large and are not tabulated. For many strong acids, K_a values have not been measured.

EXAMPLE 16.2
Interpreting pK_a and Finding K_a

A certain acid was found to have a pK_a equal to 4.88. Is this acid stronger or weaker than acetic acid? What is the value of K_a for the acid?

ANALYSIS: We can compare the acid strengths by comparing their pK_a values. The larger the pK_a, the weaker the acid. Finding K_a from pK_a involves the same kind of calculation as finding $[H^+]$ from pH, which you learned to do in Chapter 15.

SOLUTION: From Table 16.1, the pK_a of acetic acid equals 4.74. The acid referred to in the problem has pK_a equal to 4.88. Because the acid has a larger pK_a than acetic acid, it is a weaker acid.

To find K_a from pK_a, we use an equation similar to that for finding $[H^+]$ from pH, namely

$$K_a = 10^{-pK_a}$$

Substituting,

$$K_a = 10^{-4.88} = 1.3 \times 10^{-5}$$

On most calculators this is the inverse (or 2nd) operation of the *log* function.

IS THE ANSWER REASONABLE? As a quick check, we can compare the K_a values. For acetic acid, $K_a = 1.8 \times 10^{-5}$. This is larger than 1.3×10^{-5}, which tells us that acetic acid is the stronger acid. That agrees with our conclusion based on the pK_a values.

Practice Exercise 3: Use Table 16.1 to find all the acids that are stronger than acetic acid and weaker than formic acid. (Hint: It may be easier to focus on the K_a values since they are directly related to acid strength.)

Practice Exercise 4: Two acids, H*A* and H*B*, have pK_a values of 3.16 and 4.14, respectively. Which is the stronger acid? What are the K_a values for the acids?

A weak base reacts with water to give its conjugate acid and hydroxide ion

As with weak acids, all weak bases behave in a similar manner in water. They are weak Brønsted bases and are therefore proton acceptors. Examples are ammonia, NH_3, and acetate ion, $C_2H_3O_2^-$. Their reactions with water are

$$NH_3 + H_2O \rightleftharpoons NH_4^+ + OH^-$$

$$C_2H_3O_2^- + H_2O \rightleftharpoons HC_2H_3O_2 + OH^-$$

Notice that, in each instance, the base reacts with water to give OH^- and the corresponding conjugate acid. We can also represent these reactions by a general equation. If we represent the base by the symbol *B*, the reaction is

TOOLS

General equation for the ionization of a weak base

$$B + H_2O \rightleftharpoons BH^+ + OH^- \qquad (16.3)$$

(As with weak acids, B does not have to be electrically neutral.) This yields the equilibrium law

$$K_c' = \frac{[BH^+][OH^-]}{[B][H_2O]}$$

As with solutions of weak acids, the quantity $[H_2O]$ in the denominator is effectively a constant that can be combined with K_c' to give a new constant that we call the **base ionization constant, K_b.**

K_b is also called the *base dissociation constant.*

$$K_b = \frac{[BH^+][OH^-]}{[B]} \qquad (16.4)$$

EXAMPLE 16.3
Writing the K_b Expression for a Weak Base

Hydrazine is a rocket fuel and is used in the manufacture of semiconductor devices.

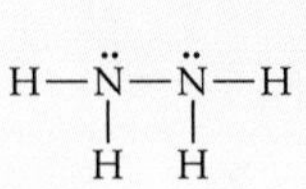

H—N̈—N̈—H (with H on each N) hydrazine N_2H_4

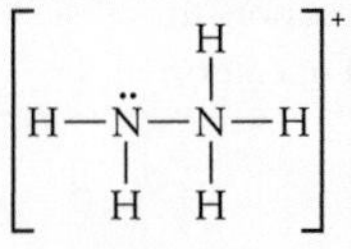

[H—N̈—N(H)—H (with H on each N)]$^+$ hydrazinium ion $N_2H_5^+$

Hydrazine, N_2H_4, is a weak base. It is a poisonous substance that's sometimes formed when a "chlorine bleach," which contains hypochlorite ion, is added to an aqueous solution of ammonia. Write the equation for the reaction of hydrazine with water and write the expression for its K_b.

ANALYSIS: To solve this problem, we have to think about what happens when the base reacts with water so we can correctly construct the chemical equation. Once we have the equation, we use it to construct the equilibrium law, which will consist of the mass action expression set equal to K_b.

SOLUTION: When N_2H_4 reacts with water, the products will be OH^- and the conjugate acid. The conjugate acid of N_2H_4 is $N_2H_5^+$, which we obtain by adding a proton (H^+) to N_2H_4. Therefore, the equation for the ionization reaction in water is[1]

$$N_2H_4 + H_2O \rightleftharpoons N_2H_5^+ + OH^-$$

[1] When a proton is accepted by a weak base that contains a nitrogen atom, such as NH_3, N_2H_4, or $(CH_3)_2NH$, the proton binds to a lone pair of electrons on a nitrogen atom of the base. With water serving as a Brønsted acid, we can diagram the reactions of these bases as follows. (The curved arrows indicate how a lone pair extracts the H^+ from the H_2O molecule and how the electron pair in the O—H bond becomes a lone pair in the OH^- ion.)

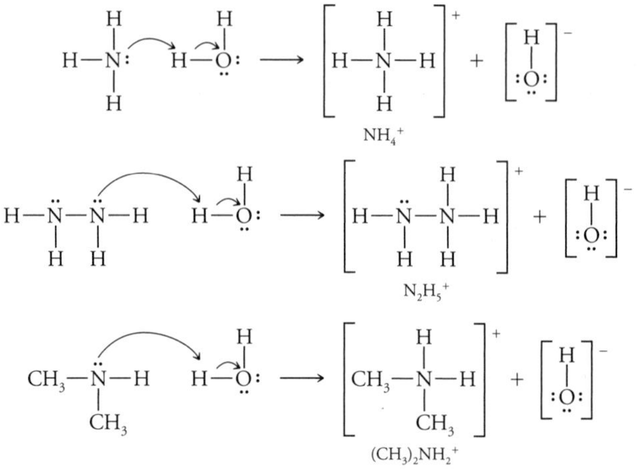

$$NH_3 + H\text{—}\ddot{O}\text{—}H \longrightarrow [NH_4]^+ + [:\ddot{O}\text{—}H]^-$$

NH_4^+

$$N_2H_4 + H\text{—}\ddot{O}\text{—}H \longrightarrow [N_2H_5]^+ + [:\ddot{O}\text{—}H]^-$$

$N_2H_5^+$

$$(CH_3)_2NH + H\text{—}\ddot{O}\text{—}H \longrightarrow [(CH_3)_2NH_2]^+ + [:\ddot{O}\text{—}H]^-$$

$(CH_3)_2NH_2^+$

In the expression for K_b we omit the H_2O, so the equilibrium law is

$$K_b = \frac{[N_2H_5^+][OH^-]}{[N_2H_4]}$$

IS THE ANSWER REASONABLE? The most important test is to be sure that products are in the numerator and reactants in the denominator and that all operations are multiplication or division (no addition or subtraction should appear here). Comparing our answer to Equations 16.3 and 16.4, we see they fit the correct pattern for a weak base, so our answers are correct.

EXAMPLE 16.4
Writing the K_b Expression for a Weak Base

Solutions of chlorine bleach such as Clorox contain the hypochlorite ion, OCl^-, which is a weak base. Write the chemical equation for the reaction of OCl^- with water and the appropriate expression for K_b for this anion.

ANALYSIS: This is similar to the preceding example. We have to be sure we have the correct formulas for the reactants and products of the equation.

SOLUTION: The reaction of hypochlorite ion as a base will yield its conjugate acid plus OH^-. We obtain the formula for the conjugate acid of OCl^- by adding one H^+ to the anion, which gives HOCl. Therefore, the equation for the equilibrium is

$$OCl^- + H_2O \rightleftharpoons HOCl + OH^-$$

As before, we omit H_2O from the expression for K_b.

$$K_b = \frac{[HOCl][OH^-]}{[OCl^-]}$$

IS THE ANSWER REASONABLE? We can check that we followed the rules for writing equilibrium constants, with the reactants in the denominator and products in the numerator. We also can compare our answers to Equations 16.3 and 16.4. The answers are okay.

Practice Exercise 5: For each of the following bases, write the equation for its ionization in water and the appropriate expression for K_b. (Hint: See footnote 1.) (a) $(CH_3)_3N$ (trimethylamine), (b) SO_3^{2-} (sulfite ion), (c) NH_2OH (hydroxylamine)

Practice Exercise 6: Write the ionization reaction, and the K_b expression, when each of the following amphiprotic substances acts as a base. (a) HSO_4^-, (b) $H_2PO_4^-$, (c) HPO_4^{2-}, (d) HCO_3^-, (e) HSO_3^-

H
|
H—N̈—O—H

hydroxylamine, NH_2OH

Because the K_b values for weak bases are usually small numbers, the same kind of logarithmic notation is often used to represent their equilibrium constants. Thus, $\mathbf{p}K_b$ is defined as

$$pK_b = -\log K_b$$

Also, K_b = antilog($-pK_b$) or $K_b = 10^{-pK_b}$.

Table 16.2 lists some molecular bases and their corresponding values of K_b and pK_b. A more complete list is located in Appendix C.7.

The product of K_a and K_b equals K_w for an acid–base conjugate pair

Formic acid, $HCHO_2$ (a substance partly responsible for the sting of a fire ant), is a typical weak acid that ionizes according to the equation

$$HCHO_2 + H_2O \rightleftharpoons H_3O^+ + CHO_2^-$$

TABLE 16.2 K_b **and** pK_b **Values for Weak Molecular Bases at 25 °C**

Name of Base	Formula	K_b	pK_b
Butylamine	$C_4H_9NH_2$	5.9×10^{-4}	3.23
Methylamine	CH_3NH_2	4.4×10^{-4}	3.36
Ammonia	NH_3	1.8×10^{-5}	4.74
Strychnine	$C_{21}H_{22}N_2O_2$	1.0×10^{-6}	6.00
Hydrazine	N_2H_4	9.6×10^{-7}	6.02
Morphine	$C_{17}H_{19}NO_3$	7.5×10^{-7}	6.13
Hydroxylamine	$HONH_2$	6.6×10^{-9}	8.18
Pyridine	C_5H_5N	1.5×10^{-9}	8.82
Aniline	$C_6H_5NH_2$	4.1×10^{-10}	9.39

Formic acid and formate ion have the structures

H—C(=O)—O—H
formic acid, $HCHO_2$
(Acidic hydrogen shown in red.)

H—C(=O)—O$^-$
formate ion, CHO_2^-

Often the formula for formic acid is written HCOOH to emphasize its molecular structure. In this chapter we will write the formulas for acids with the acidic hydrogens first in the formula, which is why we've given the formula for formic acid as $HCHO_2$.

As you've seen, we write its K_a expression as

$$K_a = \frac{[H^+][CHO_2^-]}{[HCHO_2]}$$

The conjugate base of formic acid is the formate ion, CHO_2^-, and when a solute that contains this ion (e.g., $NaCHO_2$) is dissolved in water, the solution is slightly basic. In other words, the formate ion is a weak base in water and participates in the equilibrium

$$CHO_2^- + H_2O \rightleftharpoons HCHO_2 + OH^-$$

The K_b expression for this base is

$$K_b = \frac{[HCHO_2][OH^-]}{[CHO_2^-]}$$

There is an important relationship between the equilibrium constants for this acid–base pair: the product of K_a times K_b equals K_w. We can see this by multiplying the mass action expressions.

$$K_a \times K_b = \frac{[H^+]\cancel{[CHO_2^-]}}{\cancel{[HCHO_2]}} \times \frac{\cancel{[HCHO_2]}[OH^-]}{\cancel{[CHO_2^-]}} = [H^+][OH^-] = K_w$$

In fact, this same relationship exists for *any* acid–base conjugate pair.

For *any* acid–base conjugate pair:

Relationship between K_a *and* K_b

$$K_a \times K_b = K_w \qquad (16.5)$$

Another useful relationship, which can be derived by taking the negative logarithm of each side of Equation 16.5, is

Relationship between pK_a *and* pK_b

$$pK_a + pK_b = pK_w = 14.00 \qquad \text{(at 25 °C)} \qquad (16.6)$$

There are some important consequences of the relationship expressed in Equation 16.5. One is that it is not necessary to tabulate both K_a and K_b for the members of an acid–base pair; if one K is known, the other can be calculated. For example, the K_a for $HCHO_2$ and the K_b for NH_3 will be found in most tables of acid–base equilibrium constants, but these tables usually will not contain the equilibrium constants for the ions that are the conjugates. Thus, tables usually will not contain the K_b for CHO_2^- or the K_a for NH_4^+. If we need them for a calculation, we can calculate them using Equation 16.5.

Tables of ionization constants usually give values for only the molecular member of an acid–base pair.

Practice Exercise 7: The methylammonium ion $CH_3NH_3^+$ has a K_a of 2.3×10^{-11}. What is the K_b for the base methylamine? (Hint: Recall that K_a and K_b are inversely proportional to each other.)

Practice Exercise 8: The value of K_a for $HCHO_2$ is 1.8×10^{-4}. What is the value of K_b for the CHO_2^- ion?

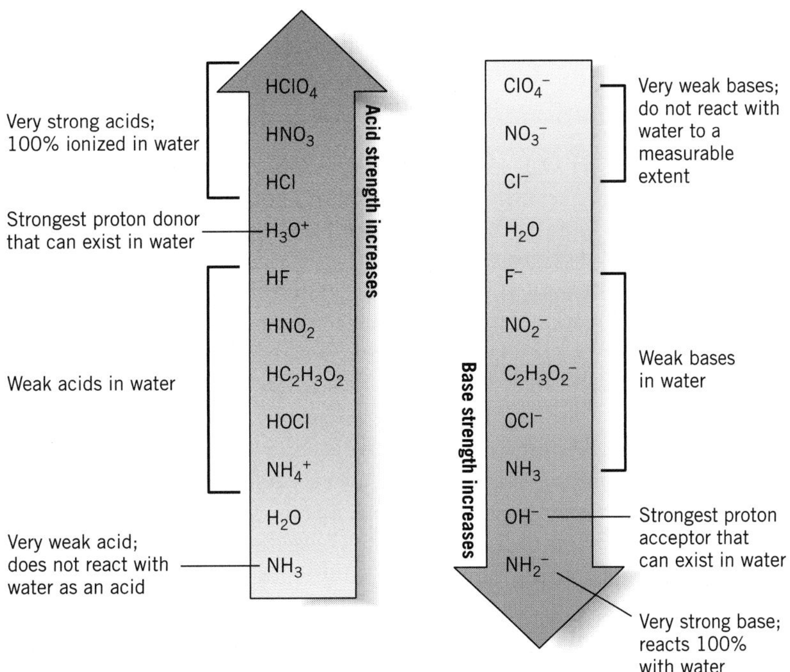

FIG. 16.1 The relative strengths of conjugate acid–base pairs. The stronger the acid, the weaker is its conjugate base. The weaker the acid, the stronger is its conjugate base. Very strong acids are 100% ionized and their conjugate bases do not react with water to any measurable extent.

Another interesting and useful observation is that *there is an inverse relationship between the strengths of the acid and base members of a conjugate pair.* This is illustrated graphically in Figure 16.1. Because the product of K_a and K_b is a constant, the larger the value of K_a is, the smaller is the value of K_b. In other words, *the stronger the conjugate acid, the weaker is its conjugate base* (a fact that we noted in Chapter 15 in our discussion of the strengths of Brønsted acids and bases). We will say more about this relationship in Section 16.3.

16.2 Calculations Can Involve Finding or Using K_a and K_b

Our goal in this section is to develop a general strategy for dealing quantitatively with the equilibria of weak acids and bases in water. Generally, these calculations fall into two categories. The first involves calculating the value of K_a or K_b from the initial concentration of the acid or base and the measured pH of the solution (or some other data about the equilibrium composition of the solution). The second involves calculating equilibrium concentrations given K_a or K_b and initial concentrations.

Initial concentrations and equilibrium data can be used to calculate K_a or K_b

In problems of this type, our *first* goal is to obtain numerical values for *all* the equilibrium concentrations that are needed to evaluate the mass action expression in the definition of K_a or K_b. (The reason, of course, is that at equilibrium the reaction quotient, Q, equals the equilibrium constant.) We are usually given the molar concentration of the acid or base as it would appear on the label of a bottle containing the solution. Also provided is information from which we can obtain directly at least one of the equilibrium concentrations. Thus, we might be given the measured pH of the solution, which provides an estimate of the equilibrium concentration of H^+. (Equilibrium is achieved rapidly in these solutions, so when we measure the pH, the value obtained can be used to calculate the equilibrium concentration of H^+.) Alternatively, we might be given the **percentage ionization** of the acid or base, which we define as follows:

$$\text{Percentage ionization} = \frac{\text{moles ionized per liter}}{\text{moles available per liter}} \times 100\% \tag{16.7}$$

Let's look at some examples that illustrate how to determine K_a and K_b from the kind of data mentioned.

EXAMPLE 16.5
Calculating K_a and pK_a from pH

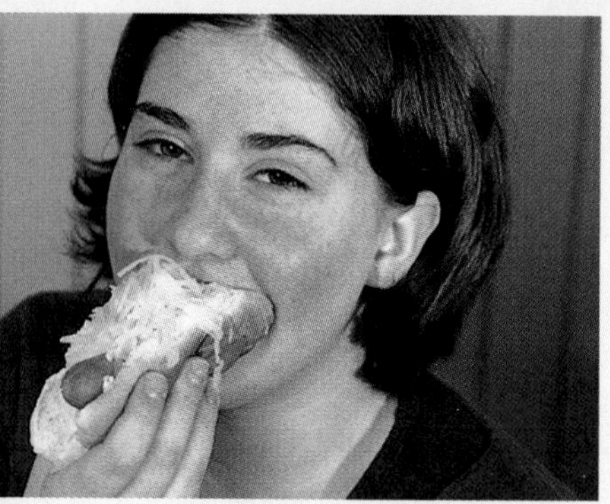

For some, nothing goes better on a hot dog than sauerkraut and mustard. Lactic acid gives sauerkraut its sour taste. *(Coco McCoy/Rainbow.)*

Lactic acid ($HC_3H_5O_3$), which is present in sour milk, also gives sauerkraut its tartness. It is a monoprotic acid. In a 0.100 *M* solution of lactic acid, the pH is 2.44 at 25 °C. Calculate the K_a and pK_a for lactic acid at that temperature.

ANALYSIS: In any problem involving the ionization of a weak acid or base, the first step is to write the chemical equation. From the equation we can write the correct equilibrium law and perform any necessary stoichiometric reasoning. You've learned how to write equations for such reactions, so all we need to know is whether the solute is an acid or a base. We're told it's an acid, so we'll begin by writing the equilibrium equation, and then the K_a expression.

We're given the pH of the solution, so we can use this to calculate the H^+ concentration. This will be the *equilibrium* $[H^+]$; as noted above, equilibrium is reached rapidly in solutions of acids and bases, so when a pH is measured, it is the pH at equilibrium. Once we know the $[H^+]$, we will perform some reasoning using a concentration table of the type we developed in Chapter 14 to figure out the rest of the equilibrium concentrations. After we know them all, we will substitute them into the mass action expression to calculate K_a. (It's important to remember that values that satisfy the K_a expression are *equilibrium* values only.)

SOLUTION: We know the solute is an acid, and we know the general equation for the ionization of an acid, which we apply to the solute in question.

$$HC_3H_5O_3 \rightleftharpoons H^+ + C_3H_5O_3^- \qquad K_a = \frac{[H^+][C_3H_5O_3^-]}{[HC_3H_5O_3]}$$

At this point it will help to set up an equilibrium table as introduced in Chapter 14. To obtain the initial concentrations we see that the only solute is the weak acid (0.100 *M*), so the only source of the ions, H^+ and $C_3H_5O_3^-$, is the ionization of the acid. Their concentrations are listed initially as zero. (Remember, in a solution of an acid, it is safe to ignore the small amount of H^+ contributed by the autoionization of water.) These entries are shown in black.

We now find the $[H^+]$ from pH. This gives us the *equilibrium* value of $[H^+]$.

Remember, the measured pH *always* gives the equilibrium concentration of H^+.

$$[H^+] = 10^{-2.44}$$
$$= 0.0036\ M$$

Because the ions are formed in a 1 to 1 ratio, $[C_3H_5O_3^-] = [H^+]$, so we now know that $[C_3H_5O_3^-] = 0.0036\ M$. These concentrations are equilibrium values and are entered as the red numbers in the table. Since the initial concentration added to the corresponding change gives us the equilibrium concentration, we can deduce that changes for H^+ and $C_3H_5O_3^-$ are 0.0036 and the change for $HC_3H_5O_3$ must be −0.0036. Finally we can calculate the equilibrium concentration of $HC_3H_5O_3$ as shown in boldface in the table.

	$HC_3H_5O_3$	$\rightleftharpoons$	H^+	+	$C_3H_5O_3^-$
Initial concentrations (*M*)	0.100		0		0
Changes in concentrations caused by the ionization (*M*)	−0.0036		+0.0036		+0.0036
Final concentrations at equilibrium (*M*)	**(0.100 − 0.0036) = 0.096 (correctly rounded)**		0.0036		0.0036

The last row of data contains the equilibrium concentrations that we now use to calculate K_a. We simply substitute them into the K_a expression.

$$K_a = \frac{(3.6 \times 10^{-3})(3.6 \times 10^{-3})}{0.096} = 1.4 \times 10^{-4}$$

Thus the acid ionization constant for lactic acid is 1.4×10^{-4}. To find pK_a, we take the negative logarithm of K_a.

$$pK_a = -\log K_a = -\log(1.4 \times 10^{-4}) = 3.85$$

IS THE ANSWER REASONABLE? Weak acids have small ionization constants, so the value we obtained for K_a seems to be reasonable. *We should also check the entries in the concentration table to be sure they are reasonable.* For example, the "changes" for the ions are both positive, meaning both of their concentrations are increasing. This is the way it *must* be because neither ion can have a concentration less than zero. The changes are caused by the ionization reaction, so they both have to change in the same direction and must be equal to each other. Also, we have the concentration of the molecular acid decreasing, as it should if the ions are being formed by the ionization.

EXAMPLE 16.6

Calculating K_b and pK_b from pH

Methylamine, CH_3NH_2, is a weak base and one of several substances that give herring brine its pungent odor. In 0.100 M CH_3NH_2, only 6.4% of the base is ionized. What are K_b and pK_b of methylamine?

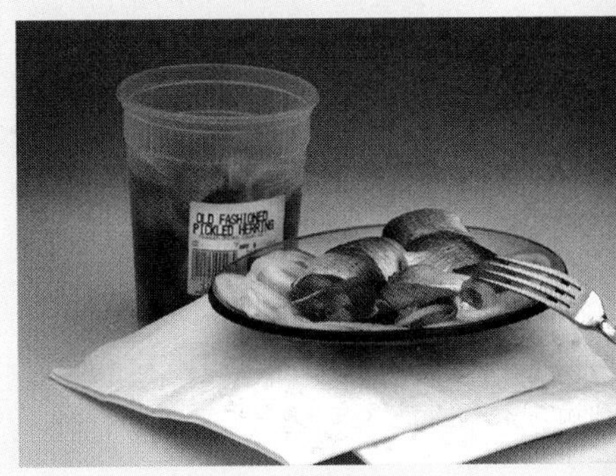

Methylamine is responsible in part for the fishy aroma of pickled herring. *(Andy Washnik.)*

ANALYSIS: In this problem we've been given the percentage ionization of the base. We will use this to calculate the equilibrium concentrations of the ions in the solution. After we know these, solving the problem follows the same path as in the preceding example.

SOLUTION: The first step is to write the chemical equation for the equilibrium and the equilibrium law. To write the chemical equation we need the formula for the conjugate acid of CH_3NH_2, which is $CH_3NH_3^+$ (we've added one H^+ to the nitrogen atom of the base to obtain the conjugate acid). Therefore, applying the general equation for the ionization of a weak base given on page 646, the chemical equation and equilibrium law are

$$CH_3NH_2(aq) + H_2O \rightleftharpoons CH_3NH_3^+(aq) + OH^-(aq)$$

$$K_b = \frac{[CH_3NH_3^+][OH^-]}{[CH_3NH_2]}$$

We can start constructing the equilibrium table at this point with the chemical equation and the appropriate rows and columns. We also enter the initial concentrations as shown in black. The initial concentration means those concentrations before any reaction takes place. Therefore the ions are listed as zeros.

The percentage ionization tells us that 6.4% of the CH_3NH_2 has reacted. Therefore, the number of moles per liter of the base that has ionized at equilibrium in this solution is:

To find 6.4% of 0.100 M the percentage must be converted to the fraction 0.064 first, as shown.

$$\text{Moles per liter of } CH_3NH_2 \text{ ionized} = 0.064 \times 0.100\ M = 0.0064\ M$$

This value represents the *decrease* in the concentration of CH_3NH_2, which we record in the "change" row of the concentration table as −0.0064. We can now use the change in $[CH_3NH_2]$ to determine the concentrations of the other species. These are entered in red. The final equilibrium concentrations are calculated from the initial concentrations and the changes we calculated. They are shown in boldface in the table.

	H_2O + CH_3NH_2	$\rightleftharpoons$ $CH_3NH_3^+$	+ OH^-
Initial concentrations (M)	0.100	0	0
Changes in concentrations caused by the ionization (M)	−0.0064	+0.0064	+0.0064
Final concentrations at equilibrium (M)	**(0.100 − 0.0064) = 0.094 (properly rounded)**	**0.0064**	**0.0064**

Now we can calculate K_b by substituting equilibrium quantities into the mass action expression.

$$K_b = \frac{(0.0064)(0.0064)}{(0.094)} = 4.4 \times 10^{-4}$$

and

$$pK_b = -\log (4.4 \times 10^{-4}) = 3.36$$

IS THE ANSWER REASONABLE? First, the value of K_b is small, so that suggests we've probably worked the problem correctly. We can also check the "Change" row to be sure the algebraic signs are correct. Both ions' initial concentrations start out at zero, so they must both increase, which they do. The change for the CH_3NH_2 is opposite in sign to that for the ions, and that's as it should be. Therefore, we appear to have set up the problem correctly.

Practice Exercise 9: Salicylic acid reacts with acetic acid to form aspirin, acetylsalicylic acid. A 0.200 molar solution of salicylic acid has a pH of 1.836. Calculate the K_a and pK_a of salicylic acid. (Hint: Convert the pH to $[H^+]$ first.)

$$CH_3—CH_2—CH_2—\overset{O}{\overset{\|}{C}}—O—H$$

butyric acid
$HC_4H_7O_2$

Practice Exercise 10: When butter turns rancid, its foul odor is mostly that of butyric acid, $HC_4H_7O_2$, a weak monoprotic acid similar to acetic acid in structure. In a 0.0100 *M* solution of butyric acid at 20 °C, the acid is 4.0% ionized. Calculate the K_a and pK_a of butyric acid at that temperature.

Practice Exercise 11: Few substances are more effective in relieving intense pain than morphine. Morphine is an alkaloid—an alkali-like compound obtained from plants—and alkaloids are all weak bases. In 0.010 *M* morphine, the pOH is 3.90. Calculate the K_b and pK_b for morphine. (You don't need to know the formula for morphine, just that it's a base. Use whatever symbol you want to write the equation.)

Equilibrium concentrations can be calculated from K_a (or K_b) and initial concentrations

These calculations are not especially difficult providing you start off on the right foot. The key to success is examining the statement of the problem carefully so you can construct the correct chemical equation, which will then lead you to write the equilibrium law that's required to obtain the solution. With this in mind, let's take an overview of the "landscape" to develop a method for selecting the correct approach to the problem.

Almost any problem in which you are given a value of K_a or K_b falls into one of three categories: (1) the aqueous solution contains a weak acid as its *only* solute, (2) the solution contains a weak base as its *only* solute, or (3) the solution contains *both* a weak acid and its conjugate base. The approach we take for each of these conditions is described below and is summarized in Figure 16.2.

Conditions 1 and 2 apply to solutions of weak molecular acids and bases and to solutions of salts that contain an ion that is an acid or a base. Condition 3 applies to solutions called buffers, which are discussed in Section 16.5.

1. If the solution contains only a weak acid as the solute, then the problem *must* be solved using K_a. This means that the correct chemical equation for the problem is the ionization of the weak acid. It also means that if you are given the K_b for the acid's conjugate base, you will have to calculate the K_a in order to solve the problem.
2. If the solution contains only a weak base, the problem *must* be solved using K_b. The correct chemical equation is for the ionization of the weak base. If the problem gives you K_a for the base's conjugate acid, then you must calculate K_b to solve the problem.
3. Solutions that contain two solutes, one a weak acid and the other its conjugate base, have some special properties that we will discuss later. (Such solutions are called buffers.) To work problems for these kinds of mixtures, we can use *either* K_a or K_b—it doesn't matter which we use, because we will obtain the same answers either way. However, if we elect to use K_a, then the chemical equation we use must be for the ionization of

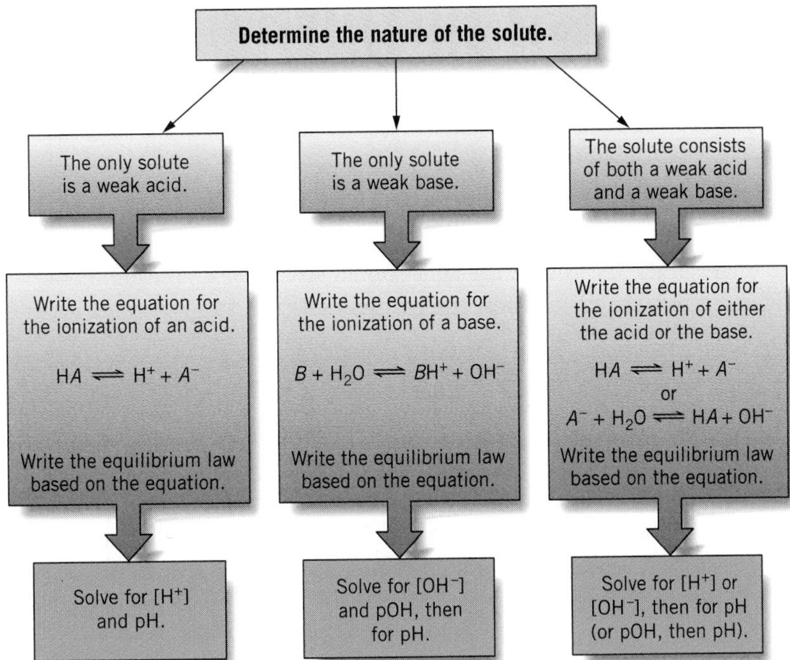

FIG. 16.2 Determining how to proceed in acid–base equilibrium problems. The nature of the solute species determines how the problem is approached. Following this diagram will get you started in the right direction.

the acid. If we decide to use K_b, then the correct chemical equation is for the ionization of the base. Usually, the choice of whether to use K_a or K_b is made on the basis of which constant is most readily available.

When you begin to solve a problem, decide which of the three conditions above applies. If necessary, keep Figure 16.2 handy to guide you in your decision, so that you select the correct path to follow.

Simplifications can usually be made in acid–base equilibrium calculations

In Chapter 14 you learned that when the equilibrium constant is small, it is frequently possible to make simplifying assumptions that greatly reduce the algebraic effort in obtaining equilibrium concentrations. For many acid–base equilibrium problems such simplifications are particularly useful.

Let's consider a solution of 1.0 *M* acetic acid, $HC_2H_3O_2$, for which $K_a = 1.8 \times 10^{-5}$. What is involved in determining the equilibrium concentrations in the solution?

To answer this question, we begin with the chemical equation for the equilibrium. Because the only solute in the solution is the weak acid, we must use K_a and therefore write the equation for the ionization of the acid. Let's use the simplified version.

$$HC_2H_3O_2 \rightleftharpoons H^+ + C_2H_3O_2^-$$

$$CH_3{-}\overset{\overset{\displaystyle O}{\|}}{C}{-}O{-}H$$

acetic acid

$HC_2H_3O_2$

The equilibrium law is

$$K_a = \frac{[H^+][C_2H_3O_2^-]}{[HC_2H_3O_2]} = 1.8 \times 10^{-5}$$

We will take the given concentration, 1.0 *M*, to be the initial concentration of $HC_2H_3O_2$ (i.e., the concentration of the acid before any ionization has occurred). This concentration will drop slightly as the acid ionizes and the ions form. If we let x be the amount of acetic acid that ionizes per liter, then the $HC_2H_3O_2$ concentration will decrease by x (its change will be $-x$) and the H^+ and $C_2H_3O_2^-$ concentrations will each increase by x (their changes will be $+x$). We can now construct the following concentration table.

The initial concentrations of the ions are set equal to zero because none of them have been supplied by a solute.

	$HC_2H_3O_2 \rightleftharpoons$	H^+ +	$C_2H_3O_2^-$
Initial concentrations (*M*)	1.0	0	0
Changes in concentrations caused by the ionization (*M*)	$-x$	$+x$	$+x$
Concentrations at equilibrium (*M*)	$1.0 - x$	x	x

The values in the last row of the table should satisfy the equilibrium law. When they are substituted into the mass action expression we obtain

$$\frac{(x)(x)}{1.0-x} = 1.8 \times 10^{-5}$$

This equation involves a term in x^2, and it could be solved using the quadratic formula. However, this and many other similar calculations involving weak acids and bases can be simplified. Let's look at the reasoning.

The equilibrium constant, 1.8×10^{-5}, is quite small, so we can anticipate that very little of the acetic acid will be ionized at equilibrium. This means x will be very small, so we make the approximation $(1.0 - x) \approx 1.0$. Replacing $1.0 - x$ with 1.0 yields the equation

We are assuming that when x is subtracted from 1.0 and *the result is rounded to the correct number of significant figures*, the answer will round to 1.0.

$$\frac{x^2}{1.0} = 1.8 \times 10^{-5}$$

Solving for x gives $x = 0.0042$ *M*. We see that the value of x is indeed negligible compared to 1.0 *M* (i.e., if we subtract 0.0042 *M* from 1.0 *M* and round correctly, we obtain 1.0 *M*).

Notice that when we make this approximation, *the initial concentration of the acid is used as if it were the equilibrium concentration.* The approximation is valid when the equilibrium constant is small and the concentrations of the solutes are reasonably high—conditions that will apply to most situations you will encounter in this chapter. In Section 16.4 we will discuss the conditions under which the approximation is not valid.

Let's look at some examples that illustrate typical acid–base equilibrium problems.

EXAMPLE 16.7

Calculating the Values of [H⁺] and pH for a Solution of a Weak Acid from Its K_a Value

A student planned an experiment that would use 0.10 *M* propionic acid, $HC_3H_5O_2$. Calculate the value of $[H^+]$ and the pH for this solution. For propionic acid, $K_a = 1.34 \times 10^{-5}$.

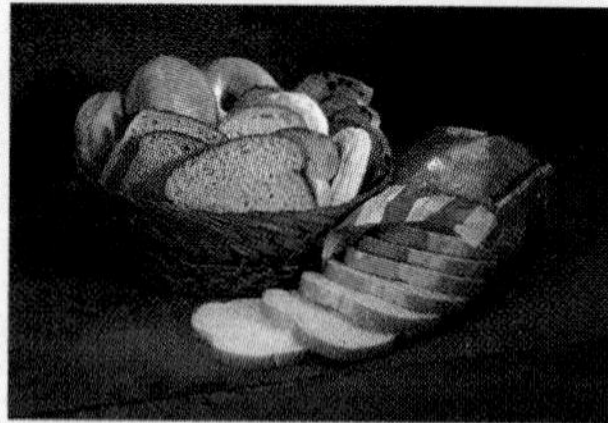

The calcium salt of propionic acid, calcium propionate, is used as a preservative in baked products. *(Andy Washnik.)*

ANALYSIS: First, we note that the only solute in the solution is a weak acid, so we know we will have to use K_a and write the equation for the ionization of the acid.

$$HC_3H_5O_2 \rightleftharpoons H^+ + C_3H_5O_2^-$$

$$K_a = \frac{[H^+][C_3H_5O_2^-]}{[HC_3H_5O_2]}$$

The initial concentration of the acid is 0.10 *M*, and the initial concentrations of the ions are both 0 *M*. Then we construct the concentration table.

SOLUTION: All concentrations in the table are in moles per liter.

	$HC_3H_5O_2 \rightleftharpoons$	H^+ +	$C_3H_5O_2^-$
Initial concentrations (*M*)	0.10	0	0
Changes in concentrations caused by the ionization (*M*)	$-x$	$+x$	$+x$
Final concentrations at equilibrium (*M*)	$(0.10 - x)$ ≈ 0.10	x	x

Notice that the equilibrium concentrations of H^+ and $C_3H_5O_2^-$ are the same; they are represented by x. Anticipating that x will be very small, we make the simplifying approximation $(0.10 - x) \approx 0.10$, so we take the equilibrium concentration of $HC_3H_5O_2$ to be 0.10 M. Substituting these quantities into the K_a expression gives

$$K_a = \frac{[H^+][C_3H_5O_2^-]}{[HC_3H_5O_2]} = \frac{(x)(x)}{(0.10 - x)} \approx \frac{(x)(x)}{(0.10)} = 1.34 \times 10^{-5}$$

$CH_3{-}CH_2{-}\overset{O}{\overset{\|}{C}}{-}O{-}H$

propionic acid
$HC_3H_5O_2$

Solving for x yields

$$x = 1.2 \times 10^{-3}$$

Because $x = [H^+]$,

$$[H^+] = 1.2 \times 10^{-3}\ M$$

Finally, we calculate the pH

$$\begin{aligned} pH &= -\log(1.2 \times 10^{-3}) \\ &= 2.92 \end{aligned}$$

IS THE ANSWER REASONABLE? First, we check to be sure our assumption was reasonable, and it is ($0.10 - 0.0012 = 0.10$ when rounded correctly). Also we see that the calculated pH is less than 7. This tells us that the solution is acidic, which it should be for a solution of an acid. Also, the pH is higher than it would be if the acid were strong. (A 0.10 M solution of a strong acid would have $[H^+] = 0.10\ M$ and pH = 1.0). If we wish to further check the accuracy of the calculation, we can substitute the calculated equilibrium concentrations into the mass action expression. If the calculated quantities are correct, the reaction quotient should equal K_a. Let's do the calculation.

$$\frac{[H^+][C_3H_5O_2^-]}{[HC_3H_5O_2]} = \frac{(x)(x)}{(0.10)} = \frac{(1.2 \times 10^{-3})^2}{0.10} = 1.4 \times 10^{-5} \approx K_a$$

The check works, so we know we have done the calculation correctly.

Practice Exercise 12: Boric acid, H_3BO_3, behaves as a monoprotic acid with a $K_a = 5.8 \times 10^{-10}$. Calculate $[H^+]$ and the pH of a 0.050 M solution of H_3BO_3. (Hint: Consider using an assumption to solve this problem.)

Practice Exercise 13: Nicotinic acid, $HC_2H_4NO_2$, is a B vitamin. It is also a weak acid with $K_a = 1.4 \times 10^{-5}$. Calculate $[H^+]$ and the pH of a 0.050 M solution of $HC_2H_4NO_2$.

EXAMPLE 16.8
Calculating the pH of a Solution and the Percentage Ionization of the Solute

A solution of hydrazine, N_2H_4, has a concentration of 0.25 M. What is the pH of the solution, and what is the percentage ionization of the hydrazine? Hydrazine has $K_b = 9.6 \times 10^{-7}$.

ANALYSIS: Hydrazine must be a weak base, because we have its value of K_b (the "b" in K_b tells us this is a *base* ionization constant). Since hydrazine is the only solute in the solution, we will have to write the equation for the ionization of a weak base and then set up the K_b expression. The conjugate acid of N_2H_4 has one additional H^+, so its formula is $N_2H_5^+$. We need this formula so we can write the correct chemical equation.

SOLUTION: We begin with the chemical equation and the K_b expression.

$$N_2H_4 + H_2O \rightleftharpoons N_2H_5^+ + OH^-$$

$$K_b = \frac{[N_2H_5^+][OH^-]}{[N_2H_4]}$$

Let's once again construct the concentration table. The only sources of $N_2H_5^+$ and OH^- are the ionization of the N_2H_4, so their initial concentrations are both zero. They will form in equal amounts, so we let their changes in concentration equal $+x$. The concentration of N_2H_4 will decrease by x, so its change is $-x$.

	H_2O +	N_2H_4	$\rightleftharpoons$	$N_2H_5^+$	+ OH^-
Initial concentrations (M)		0.25		0	0
Changes in concentrations caused by the ionization (M)		$-x$		$+x$	$+x$
Final concentrations at equilibrium (M)		$(0.25 - x)$ ≈ 0.25		x	x

Because K_b is so small, $[N_2H_4] \approx 0.25\ M$. (As before, we assume the initial concentration will be effectively the same as the equilibrium concentration, an approximation that we expect to be valid.) Substituting into the K_b expression,

$$\frac{(x)(x)}{0.25 - x} \approx \frac{(x)(x)}{0.25} = 9.6 \times 10^{-7}$$

Solving for x gives $x = 4.9 \times 10^{-4}$. This value represents the hydroxide ion concentration, from which we can calculate the pOH.

$$\text{pOH} = -\log(4.9 \times 10^{-4}) = 3.31$$

The pH of the solution can then be obtained from the relationship

$$\text{pH} + \text{pOH} = 14.00$$

Thus,

$$\text{pH} = 14.00 - 3.31 = 10.69$$

To calculate the percentage ionization, we need to use Equation 16.7 on page 649. This requires that we know the number of moles per liter of the base that has ionized. From the concentration table, we see that this value is also equal to x, the amount of N_2H_4 that ionizes per liter (i.e., the change in the N_2H_4 concentration). Therefore,

$$\text{Percentage ionization} = \frac{4.9 \times 10^{-4}\ M}{0.25\ M} \times 100\%$$
$$= 0.20\%$$

The base is 0.20% ionized.

IS THE ANSWER REASONABLE? First, we can quickly check to see whether the value of x satisfies our assumption, and it does $(0.25 - 0.00049 = 0.25)$. The value of the hydroxide ion concentration is small and the pH indicates a basic solution. Additionally the small concentration of hydroxide ions agrees with the small percentage ionization we calculated. We could also use our equilibrium concentrations to calculate K_b as in the previous example.

Practice Exercise 14: Aniline, $C_6H_5NH_2$, is a precursor for many dyes used to color fabrics, known as aniline dyes. Aniline has a K_b of 4.1×10^{-10}. What is the pOH of a 0.025 molar solution of aniline? (Hint: Calculate the OH^- concentration.)

Practice Exercise 15: Pyridine, C_5H_5N, is a foul-smelling liquid for which $K_b = 1.5 \times 10^{-9}$. What is the pH of a 0.010 M aqueous solution of pyridine?

Practice Exercise 16: Phenol is an acidic organic compound for which $K_a = 1.3 \times 10^{-10}$. What is the pH of a 0.15 M solution of phenol in water?

16.3 SALT SOLUTIONS ARE NOT NEUTRAL IF THE IONS ARE WEAK ACIDS OR BASES

When we study solutions of salts, we find that many have a neutral pH. Examples are NaCl or KNO_3. However, not all salt solutions behave this way. Some, such as $NaC_2H_3O_2$, are slightly basic while others, such as NH_4Cl, are slightly acidic. To understand this behavior, we have to examine how the ions of a salt can affect the pH of a solution.

In Section 16.1 you saw that weak acids and bases are not limited to molecular substances. For instance, on page 643 NH_4^+ was given as an example of a weak acid, and on page 645 $C_2H_3O_2^-$ was cited as an example of a weak base. Ions such as these always come to us in compounds in which there is both a cation *and* an anion. Therefore, to place NH_4^+ in water, we need a salt such as NH_4Cl, and to place $C_2H_3O_2^-$ in water, we need a salt such as $NaC_2H_3O_2$.

Because a salt contains two ions, the pH of its solution can potentially be affected by either the cation or the anion, or perhaps even both. Therefore, we have to consider *both* ions if we wish to predict the effect of a salt on the pH of a solution.

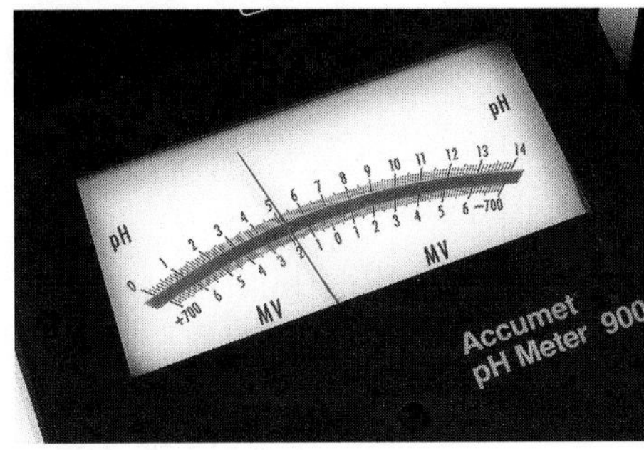

FIG. 16.3 Ammonium ion as a weak acid. The measured pH of a solution of the salt NH_4Cl is less than 7, which indicates the solution is acidic. *(Andy Washnik.)*

Cations can be acids

If the cation of a salt is able to influence the pH of a solution, it does so by behaving as a weak acid. Not all cations are acidic, however, so let's look at the possibilities.

Conjugate acids of molecular bases are weak acids

The ammonium ion, NH_4^+, is the conjugate acid of the molecular base, NH_3. We have already learned that NH_4^+, supplied for example by NH_4Cl, is a weak acid. As seen in Figure 16.3, solutions of this salt have a pH lower than 7. The equation for the reaction as an acid is

$$NH_4^+(aq) + H_2O \rightleftharpoons NH_3(aq) + H_3O^+(aq)$$

The reaction of a salt with water to give either an acidic or basic solution is sometimes called **hydrolysis.** Hydrolysis means *reaction with water.*

Writing this in a simplified form, along with the K_a expression, we have

$$NH_4^+(aq) \rightleftharpoons NH_3(aq) + H^+(aq) \qquad K_a = \frac{[NH_3][H^+]}{[NH_4^+]}$$

Because the K_a values for ions are seldom tabulated, we would usually expect to calculate the K_a value using the relationship: $K_a \times K_b = K_w$. In Table 16.2 the K_b for NH_3 is given as 1.8×10^{-5}. Therefore,

$$K_a = \frac{K_w}{K_b} = \frac{1.0 \times 10^{-14}}{1.8 \times 10^{-5}} = 5.6 \times 10^{-10}$$

Another example is the hydrazinium ion, $N_2H_5^+$, which is also a weak acid.

$$N_2H_5^+(aq) \rightleftharpoons N_2H_4(aq) + H^+(aq) \qquad K_a = \frac{[N_2H_4][H^+]}{[N_2H_5^+]}$$

The tabulated K_b for N_2H_4 is 1.7×10^{-6}. From this, the calculated value of K_a equals 5.9×10^{-9}.

These examples illustrate a general phenomenon:

Identification of acidic cations

> Cations that are the conjugate acids of molecular bases tend to be weakly acidic.

Cations that are not conjugate acids of molecular bases are generally metal ions such as Na^+, K^+, and Ca^{2+}. In Chapter 15 we learned that the cations of the Group IA metals are very weak acids and are unable to affect the pH of a solution. Except for Be^{2+}, the cations of Group IIA also do not affect pH.

Anions can be bases

When a Brønsted acid loses a proton, its conjugate base is formed. Thus Cl^- is the conjugate base of HCl, and $C_2H_3O_2^-$ is the conjugate base of $HC_2H_3O_2$.

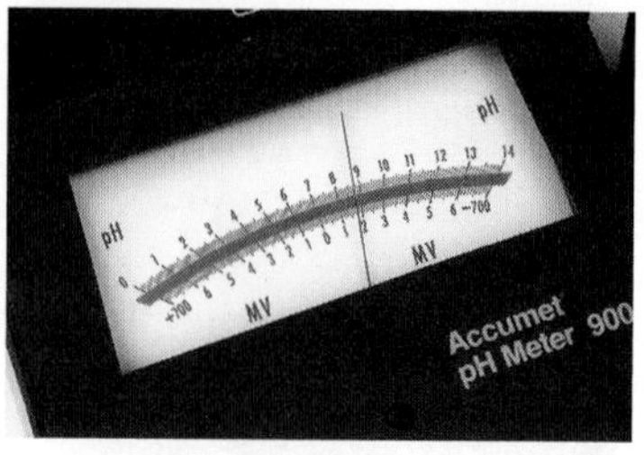

FIG. 16.4 Acetate ion is a weak base in water. The pH of a solution of $KC_2H_3O_2$ is greater than 7, which shows that the solution is basic. Acetate ion reacts with water to yield small amounts of hydroxide ion.

$$C_2H_3O_2^- + H_2O \rightleftharpoons HC_2H_3O_2 + OH^-$$

(Andy Washnik.)

Acid	*Base*
HCl	Cl^-
$HC_2H_3O_2$	$C_2H_3O_2^-$

Although both Cl^- and $C_2H_3O_2^-$ are bases, only the latter affects the pH of an aqueous solution. Why?

Earlier you learned that there is an inverse relationship between the strengths of an acid and its conjugate base—the stronger the acid, the weaker is the conjugate base. Therefore, when an acid is *extremely strong,* as in the case of HCl or other "strong" acids that are 100% ionized (such as HNO_3), the conjugate base is *extremely weak* — too weak to affect in a measurable way the pH of a solution. Consequently, we have the following generalization:

> The anion of a strong acid is too weak a base to influence the pH of a solution.

Acetic acid is much weaker than HCl, as evidenced by its value of K_a (1.8×10^{-5}). Because acetic acid is a weak acid, its conjugate base is much stronger than Cl^- and we can expect it to affect the pH (Figure 16.4). We can calculate the value of K_b for $C_2H_3O_2^-$ from the K_a for $HC_2H_3O_2$, also with the equation $K_a \times K_b = K_w$.

$$K_b = \frac{1.0 \times 10^{-14}}{1.8 \times 10^{-5}} = 5.6 \times 10^{-10}$$

This leads to another conclusion:

TOOLS *Identification of basic anions*

> The anion of a weak acid is a weak base and can influence the pH of a solution. It will tend to make the solution basic.

Acid–base properties of a salt can be predicted

To decide if any given salt will affect the pH of an aqueous solution, we must examine each of its ions and see what it alone might do. There are four possibilities:

1. If neither the cation nor the anion can affect the pH, the solution should be neutral.
2. If only the cation of the salt is acidic, the solution will be acidic.
3. If only the anion of the salt is basic, the solution will be basic.
4. If a salt has a cation that is acidic and an anion that is basic, the pH of the solution is determined by the *relative* strengths of the acid and base based on the K_a and K_b of the ions.

Let's work some examples to show how we use these generalizations.

EXAMPLE 16.9 Predicting the Effect of a Salt on the pH of a Solution

Will a solution of NaOCl, an ingredient in many common bleaching and disinfecting agents, be acidic, basic, or neutral? Will a solution of $Ca(NO_3)_2$, a compound used in fireworks and fertilizers, be acidic, basic, or neutral?

ANALYSIS: Each solute is a salt, so we assume it to be 100% dissociated in water.

$$NaOCl(s) \xrightarrow{H_2O} Na^+(aq) + OCl^-(aq)$$

$$Ca(NO_3)_2(s) \xrightarrow{H_2O} Ca^{2+}(aq) + 2NO_3^-(aq)$$

To answer the question for each solution, we take each ion, in turn, and examine its effect on the acidity of the solution.

SOLUTION: *For NaOCl:* The Na^+ ion is an ion of a Group IA metal and is *not* acidic. Therefore, it does not affect the pH of the solution. We might say Na^+ has a "neutral" effect on the pH. Next, we need to determine whether the anion might be basic. To decide, we need to know about the strength of its conjugate acid, which we obtain by adding H^+ to OCl^- to give HOCl. The list of strong acids is short, and HOCl is not on it, so we can expect that it is a weak acid. If HOCl is a weak acid, then OCl^- is a weak base and its presence should tend to make the solution basic.

Of the two ions in the salt, one is basic and the other is neutral. Therefore, a solution of NaOCl will be basic.

The only time an anion might be acidic is if it is from a partially neutralized polyprotic acid, e.g., HSO_4^-. Because OCl^- doesn't have a hydrogen, it can't be a Brønsted acid.

For $Ca(NO_3)_2$: The Ca^{2+} ion is an ion of a Group IIA metal, so it is not acidic. It will have no effect on the pH. The anion, NO_3^-, comes from the acid HNO_3, which is on our list of strong acids. Therefore, NO_3^- is an extremely weak base and will also have no effect on the pH. Our conclusion, therefore, is that a solution of $Ca(NO_3)_2$ will be neutral.

ARE THE ANSWERS REASONABLE? All we can do here is check our reasoning, which appears to be sound.

Practice Exercise 17: Are solutions of (a) $NaNO_2$, (b) KCl, and (c) NH_4Br acidic, basic, or neutral? (Hint: Write the ions for these compounds. What are the conjugate acids and bases of these ions?)

Practice Exercise 18: Are solutions of (a) $NaNO_3$, (b) KOCl, and (c) NH_4NO_3 acidic, basic, or neutral?

EXAMPLE 16.10
Calculating the pH of a Salt Solution

What is the pH of a 0.10 *M* solution of NaOCl? For HOCl, $K_a = 3.0 \times 10^{-8}$.

ANALYSIS: Problems such as this are just like the other acid–base equilibrium problems you have learned to solve. We proceed as follows: (1) We determine the nature of the solute: Is it a weak acid, is it a weak base, or are both a weak acid and weak base present? (2) We write the appropriate chemical equation and equilibrium law. (3) We proceed with the solution.

The solute is a salt, which is dissociated into the ions Na^+ and OCl^-. The analysis we performed in Example 16.9 tells us that only OCl^- can affect the pH. It is the conjugate *base* of HOCl, so for the purposes of problem solving, the active solute species is a weak base. On page 652 you learned that when the *only* solute is a weak base, we must use the K_b expression to solve the problem. This is where we begin the solution to the problem.

SOLUTION: We write the chemical equation for the equilibrium ionization of the base, OCl^-, and its K_b expression.

If you had trouble writing the correct chemical equation for this equilibrium, you should review Section 16.1.

$$OCl^- + H_2O \rightleftharpoons HOCl + OH^- \qquad K_b = \frac{[HOCl][OH^-]}{[OCl^-]}$$

The data provided in the problem gives the K_a for HOCl. But we can easily calculate K_b because $K_a \times K_b = K_w$.

$$K_b = \frac{K_w}{K_a} = \frac{1.0 \times 10^{-14}}{3.0 \times 10^{-8}} = 3.3 \times 10^{-7}$$

Now let's set up the concentration table. The only sources of HOCl and OH^- are the reaction of the OCl^-, so their concentrations will each increase by *x* and the concentration of OCl^- will decrease by *x*.

	H_2O +	OCl^-	$\rightleftharpoons$	$HOCl$	+	OH^-
Initial concentrations (M)		0.10		0		0
Changes in concentrations caused by the ionization (M)		$-x$		$+x$		$+x$
Final concentrations at equilibrium (M)		$(0.10 - x) \approx 0.10$		x		x

At equilibrium, the concentrations of HOCl and OH^- are the same, x.

$$[HOCl] = [OH^-] = x$$

Because K_b is so small, $[OCl^-] \approx 0.10\ M$. (Once again, we expect x to be so small that we are able to use the initial concentration of the base as if it were the equilibrium concentration.) Substituting into the K_b expression gives

$$\frac{(x)(x)}{0.10-x} \approx \frac{(x)(x)}{0.10} = 3.3 \times 10^{-7}$$

$$x = 1.8 \times 10^{-4}\ M$$

This value of x represents the OH^- concentration, from which we can calculate the pOH and then the pH.

$$[OH^-] = 1.8 \times 10^{-4}\ M$$

$$pOH = -\log(1.8 \times 10^{-4}) = 3.74$$

$$pH = 14.00 - pOH$$

$$= 14.00 - 3.74$$

$$= 10.26$$

The pH of this solution is 10.26.

IS THE ANSWER REASONABLE? First we check our assumption and find it is correct $(0.10 - 0.00018 = 0.10)$. Also, from the nature of the salt, we expect the solution to be basic. The calculated pH corresponds to a basic solution, so the answer seems to be reasonable. You've also seen that we can check the accuracy of the answer by substituting the calculated equilibrium concentrations into the mass action expression.

$$\frac{[HOCl][OH^-]}{[OCl^-]} = \frac{(x)(x)}{0.10} = \frac{(1.8 \times 10^{-4})^2}{0.10} = 3.2 \times 10^{-7}$$

The result is quite close to the given value of K_a, so our answers are correct.

EXAMPLE 16.11
Calculating the pH of a Salt Solution

What is the pH of a 0.20 M solution of hydrazinium chloride, N_2H_5Cl? Hydrazine, N_2H_4, is a weak base with $K_b = 9.6 \times 10^{-7}$.

ANALYSIS: This problem is quite similar to the preceding one. Looking over the statement of the problem, we should realize that N_2H_5Cl is a salt composed of $N_2H_5^+$ (the conjugate acid of N_2H_4) and Cl^-. Since N_2H_4 is a weak base, we expect the $N_2H_5^+$ ion to be a weak acid and thereby affect the pH of the solution. On the other hand, Cl^- is the conjugate base of HCl (a strong acid) and is too weak to influence the pH. Therefore, the only active solute species is the acid, $N_2H_5^+$, which means that to solve the problem we must write the equation for the ionization of the acid and use the K_a expression.

SOLUTION: We will begin with the simplified chemical equation for the equilibrium and write the K_a expression.

$$N_2H_5^+ \rightleftharpoons H^+ + N_2H_4 \qquad K_a = \frac{[H^+][N_2H_4]}{[N_2H_5^+]}$$

The problem has given us K_b for N_2H_4, but we need K_a for $N_2H_5^+$. We obtain this by solving the equation $K_a \times K_b = K_w$ for K_a.

$$K_a = \frac{K_w}{K_b} = \frac{1.0 \times 10^{-14}}{9.6 \times 10^{-7}} = 1.0 \times 10^{-8}$$

Now we set up the concentration table. The initial concentrations of H^+ and N_2H_4 are both set equal to zero; there is no strong acid to give H^+ in the solution and no N_2H_4 is present before the reaction of water with $N_2H_5^+$. Next, we indicate that the concentration of $N_2H_5^+$ decreases by x and the concentrations of H^+ and N_2H_4 both increase by x.

	$N_2H_5^+ \rightleftharpoons$	H^+ +	N_2H_4
Initial concentrations (M)	0.20	0	0
Changes in concentrations caused by the ionization (M)	$-x$	$+x$	$+x$
Final concentrations at equilibrium (M)	$(0.20 - x)$ ≈ 0.20	x	x

At equilibrium, equal amounts of H^+ and N_2H_4 are present, and their equilibrium concentrations are each equal to x.

$$[H^+] = [N_2H_4] = x$$

We also assume that $[N_2H_5^+] \approx 0.20\ M$ and then substitute quantities into the mass action expression.

$$\frac{(x)(x)}{(0.20 - x)} = \frac{(x)(x)}{0.20} = 1.0 \times 10^{-8}$$

$$x = 4.5 \times 10^{-5}\ M$$

Since $x = [H^+]$, the pH of the solution is

$$\text{pH} = -\log(4.5 \times 10^{-5})$$
$$= 4.35$$

IS THE ANSWER REASONABLE? The assumption holds true ($0.20 - 0.000045 = 0.20$). The active solute species in the solution is a weak acid and the calculated pH is less than 7, so the answer seems reasonable. Check the accuracy yourself by substituting equilibrium concentrations into the mass action expression.

Practice Exercise 19: What is the pH of a solution when 25.0 g of methylammonium chloride, CH_3NH_3Cl, is dissolved in enough water to make 500 mL of solution? (Hint: What is the molarity of the CH_3NH_3Cl solution?)

Practice Exercise 20: What is the pH of a 0.10 M solution of $NaNO_2$?

Practice Exercise 21: If 500 mL of a 0.20 M solution of ammonia is mixed with 500 mL of a 0.20 M solution of HBr, what is the pH of the resulting solution of NH_4Br?

Acidity of solutions that contain both the salt of a weak acid and a weak base can be predicted

There are many salts whose ions are both able to affect the pH of the salt solution. Whether or not the salt has a net effect on the pH now depends on the relative strengths of its ions in functioning one as an acid and the other as a base. If they are matched in their respective strengths, the salt has no net effect on pH. In ammonium acetate, for example, the ammonium ion is an acidic cation and the acetate ion is a basic anion. However, the K_a of NH_4^+ is 5.6×10^{-10} and the K_b of $C_2H_3O_2^-$ just happens to be the same,

If *neither* the cation nor anion are able to affect the pH, the salt solution will be neutral (provided no other acidic or basic solutes are present).

5.6×10^{-10}. The cation tends to produce H^+ ions to the same extent that the anion tends to produce OH^-. So in aqueous ammonium acetate, $[H^+] = [OH^-]$, and the solution has a pH of 7.

Consider, now, ammonium formate, NH_4CHO_2. The formate ion, CHO_2^-, is the conjugate base of the weak acid, formic acid, so it is a Brønsted base. Its K_b is 5.6×10^{-11}. Comparing this value to the (slightly larger) K_a of the ammonium ion, 5.6×10^{-10}, we see that NH_4^+ is slightly stronger as an acid than the formate ion is as a base. As a result, a solution of ammonium formate is slightly acidic. We are not concerned here about calculating a pH, only in predicting if the solution is acidic, basic, or neutral.

EXAMPLE 16.12
Predicting How a Salt Affects the pH of Its Solution

Will an aqueous solution that is 0.20 *M* NH_4F be acidic, basic, or neutral?

ANALYSIS: This is a salt in which the cation is a weak acid (it's the conjugate acid of a weak base, NH_3) and the anion is a weak base (it's the conjugate base of a weak acid, HF). The question, then, is, "How do the two ions compare in their abilities to affect the pH of the solution?" We have to calculate their respective K_a and K_b to compare their strengths.

SOLUTION: The K_a of NH_4^+ (calculated from the K_b for NH_3) is 5.6×10^{-10}. Similarly, the K_b of F^- is 1.5×10^{-11} (calculated from the K_a for HF, 6.8×10^{-4}). Comparing the two equilibrium constants, we see that the acid (NH_4^+) is stronger than the base (F^-), so we expect the solution to be slightly acidic.

IS THE ANSWER REASONABLE? We can use the reciprocal relationship between the strengths of the members of conjugate acid–base pairs as a qualitative check. From their respective ionization constants, HF is more fully ionized than NH_3. Therefore, the extent to which F^- reacts with water is *less* than NH_4^+. In other words, F^- is weaker as a base than NH_4^+ is as an acid, so the solution should be slightly acidic.

Practice Exercise 22: Will an aqueous solution of ammonium cyanide, NH_4CN, be acidic, basic, or neutral? (Hint: These anions and cations are conjugates of an acid and a base listed in Tables 16.1 and 16.2.)

Practice Exercise 23: In the discussion above it was shown that an aqueous solution of ammonium acetate, $NH_4C_2H_3O_2$, will be neutral. Will a solution containing an equal number of moles of potassium acetate, $KC_2H_3O_2$, and ammonium nitrate, NH_4NO_3, also be neutral? Justify your answer.

16.4 SIMPLIFICATIONS FAIL FOR SOME EQUILIBRIUM CALCULATIONS

In the preceding two sections we used initial concentrations of solutes as though they were equilibrium concentrations when we performed calculations. This is only an approximation, as we discussed on page 654, but it is one that works most of the time. Unfortunately, it does not work in all cases, so now that you have learned the basic approach to solving equilibrium problems, we will examine those conditions under which simplifying approximations do and do not work. We will also study how to solve problems when the approximations cannot be used.

When simplifying assumptions fail other methods work

When a weak acid, HA, ionizes in water, its concentration is reduced as the ions form. If we let x represent the amount of acid that ionizes per liter, the equilibrium concentration becomes

$$[HA]_{\text{equilib}} = [HA]_{\text{initial}} - x$$

and because the equilibrium constants are small we can often assume that x is very small and

$$[HA]_{\text{equilib}} \approx [HA]_{\text{initial}}$$

Even when x is up to 5% (or one twentieth) of the initial concentration of HA our assumption will still work very well[2]. It is not difficult to show that for x to be less than or equal to ±5% of $[HA]_{\text{initial}}$, $[HA]_{\text{initial}}$ must be greater than or equal to 400 times the value of K_a.

When simplifications work

Simplifications work when $[HA]_{\text{initial}} \geq 400 \times K_a$.

For the equilibrium problems in the preceding sections, this condition was fulfilled, so our simplifications were valid. We now want to study what to do when $[HA]_{\text{initial}} < 400 \times K_a$ (i.e., when we cannot justify simplifying the algebra).

The quadratic equation is one solution

When the algebraic equation obtained by substituting quantities into the equilibrium law is a quadratic equation, we can use the quadratic formula to obtain the solution. This is illustrated by the following example.

EXAMPLE 16.13
Using the Quadratic Formula in Equilibrium Problems

Chloroacetic acid, $HC_2H_2O_2Cl$, is used as an herbicide and in the manufacture of dyes and other organic chemicals. It is a weak acid with $K_a = 1.4 \times 10^{-3}$. What is the pH of a 0.010 M solution of $HC_2H_2O_2Cl$?

ANALYSIS: Before we begin the solution, we check to see whether we can use our usual simplifying approximation. For the simplification to work, the initial concentration of the acid (0.010 M) must be larger than $400 \times K_a$, so let's compute this quantity first.

$$400 \times K_a = 400\,(1.4 \times 10^{-3}) = 0.56\,M$$

The initial concentration of $HC_2H_2O_2Cl$ is *less than* 0.56 M so we know the simplification does not work.

```
   H  O
   |  ||
H—C—C—O—H
   |
   Cl
```
chloroacetic acid

SOLUTION: Let's begin by writing the chemical equation and the K_a expression. (We know we must use K_a because the only solute is the acid.)

$$HC_2H_2O_2Cl \rightleftharpoons H^+ + C_2H_2O_2Cl^-$$

$$K_a = \frac{[H^+][C_2H_2O_2Cl^-]}{[HC_2H_2O_2Cl]} = 1.4 \times 10^{-3}$$

The initial concentration of the acid will be reduced by an amount x as it undergoes ionization to form the ions. From this, let's build the concentration table.

[2] Neglecting x if it is within 5% of the initial concentration of the acid or base will introduce some error into our results. However, there is a similar or greater error introduced by using molar concentrations rather than activities (effective concentrations of ions in solution) in our calculations. Activities are generally covered in higher level chemistry courses. The overall effect is that our calculated pH results are not exact and can be in error by as much as 0.02 pH units..

	$HC_2H_2O_2Cl \rightleftharpoons$	H^+ +	$C_2H_2O_2Cl^-$
Initial concentrations (M)	0.010	0	0
Changes in concentrations (M)	$-x$	$+x$	$+x$
Equilibrium concentrations (M)	$0.010 - x$	x	x

Substituting equilibrium concentrations into the equilibrium law gives

$$\frac{(x)(x)}{(0.010 - x)} = 1.4 \times 10^{-3}$$

We solved a quadratic equation almost exactly the same as this in Section 14.7 so we need only show an abbreviated solution here. The quadratic formula is

$$x = \frac{-b \pm \sqrt{b^2 - 4ac}}{2a}$$

Rearranging our equation to follow the general form gives

$$x^2 + (1.4 \times 10^{-3})x - (1.4 \times 10^{-5}) = 0$$

and we make the substitutions $a = 1$, $b = 1.4 \times 10^{-3}$, and $c = -1.4 \times 10^{-5}$. Entering these into the quadratic formula gives

$$x = \frac{-1.4 \times 10^{-3} \pm \sqrt{(1.4 \times 10^{-3})^2 - 4(1)(-1.4 \times 10^{-5})}}{2(1)}$$

Because of the ± sign we obtain two values for x, but as you saw in Chapter 14, only one of them makes any sense. Here are the two values:

$$x = 3.1 \times 10^{-3}\,M \qquad \text{and} \qquad x = -4.5 \times 10^{-3}\,M$$

We know that x cannot be negative because that would give negative concentrations for the ions (which is impossible), so we must choose the first value as the correct one. This yields the following equilibrium concentrations.

$$[H^+] = 3.1 \times 10^{-3}\,M$$
$$[C_2H_2O_2Cl^-] = 3.1 \times 10^{-3}\,M$$
$$[HC_2H_2O_2Cl] = 0.010 - 0.0031 = 0.007\,M$$

Finally, we calculate the pH of the solution.

$$\text{pH} = -\log(3.1 \times 10^{-3}) = 2.51$$

IS THE ANSWER REASONABLE? We have no assumptions to check. Our concentrations are small and the pH represents an acidic solution expected of an acid. As before, a quick check can be performed by substituting the calculated equilibrium concentrations into the mass action expression.

(Note: If we made the usual assumption, the result would have been $[H^+] = 3.7 \times 10^{-3}\,M$ which is a 20% error in the hydrogen ion concentration!)

Notice that, indeed, x is not negligible compared to the initial concentration, so the simplifying approximation would not have been valid.

Practice Exercise 24: Calculate the pH of a 0.0010 M solution of dimethylamine, $(CH_3)_2NH$, for which $K_b = 9.6 \times 10^{-4}$. (Hint: Check to see if you can use any simplifications to solve this problem.)

Practice Exercise 25: Calculate, using the quadratic equation and the usual simplification, the pH of a 0.030 M solution of hydrofluoric acid, HF, for which $K_a = 6.8 \times 10^{-4}$. Is there a significant difference between the results?

Practice Exercise 26: Calculate the pH of a 0.0010 *M* solution of methylamine, CH_3NH_2, for which $K_b = 4.4 \times 10^{-4}$.

Another solution is the successive approximations method

If you worked your way through the discussion of the use of the quadratic equation, you'll surely be better able to appreciate the *method of successive approximations.* It's not only much faster, particularly with a scientific calculator, but just as accurate. The procedure is outlined in Figure 16.5; follow the diagram as we apply the method below.

If we were to attempt to use the simplifying approximation in the preceding Example, we would obtain the following:

$$\frac{x^2}{0.010} = 1.4 \times 10^{-3}$$

for which we obtain the solution $x = 3.7 \times 10^{-3}$. We will call this our *first approximation* to a solution to the equation

$$\frac{(x)(x)}{(0.010 - x)} = 1.4 \times 10^{-3}$$

Let's put $x = 3.7 \times 10^{-3}$ into the term in the denominator, $(0.010 - x)$, and recalculate x. This gives us

$$\frac{x^2}{(0.010 - 0.0037)} = 1.4 \times 10^{-3}$$

or,

$$x^2 = (0.006) \times 1.4 \times 10^{-3}$$

Taking only the positive root,

$$x = 2.9 \times 10^{-3}$$

Notice that this value of x is much closer to the value calculated using the quadratic equation (which gave $x = 3.1 \times 10^{-3}$). Now we'll call $x = 2.9 \times 10^{-3}$ our *second approximation,* and repeat the process. This gives

$$\frac{x^2}{(0.010 - 0.0029)} = 1.4 \times 10^{-3}$$

$$x^2 = (0.007) \times 1.4 \times 10^{-3}$$
$$x = 3.1 \times 10^{-3}$$

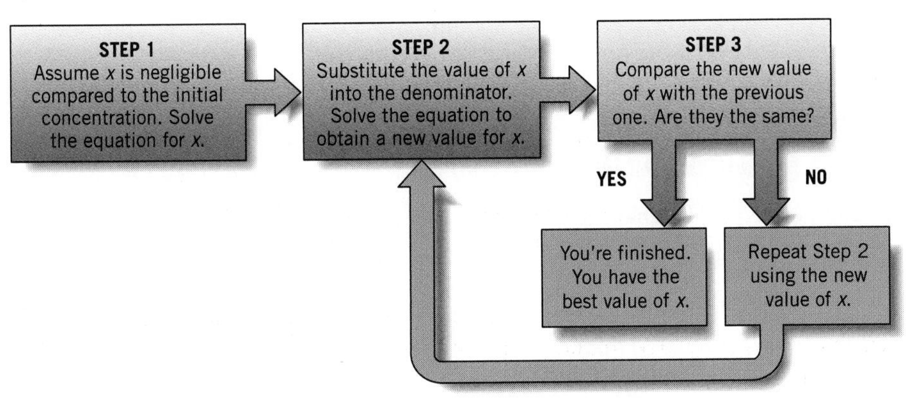
STEP 1
Assume x is negligible compared to the initial concentration. Solve the equation for x.
STEP 2
Substitute the value of x into the denominator. Solve the equation to obtain a new value for x.
STEP 3
Compare the new value of x with the previous one. Are they the same?
YES
NO
You're finished. You have the best value of x.
Repeat Step 2 using the new value of x.

TOOLS *Method of successive approximations*

FIG. 16.5 The method of successive approximations. Following the steps outlined here leads to a rapid solution of equilibrium problems when the usual simplifications fail.

This is the identical value obtained using the quadratic equation. Notice also that the *change* in x between the two approximations grew smaller. The second approximation gave a smaller correction than the first. Each succeeding approximation differs from the preceding one by smaller and smaller amounts. We stop the calculation when the difference between two approximations is insignificant. Try this approach by reworking Practice Exercise 26 and solving for $[OH^-]$ in the 0.0010 M CH_3NH_2 solution using the method of successive approximations.[3]

16.5 BUFFERS ENABLE THE CONTROL OF pH

Many chemical and biological systems are quite sensitive to pH. For example, if the pH of your blood were to change from what it should be, within the range of 7.35 to 7.42, either to 7.00 or to 8.00, you would die. Thus, a change in pH can produce unwanted effects, and systems that are sensitive to pH must be protected from the H^+ or OH^- that might be formed or consumed by some reaction. *Buffers* are mixtures of solutes that accomplish this. The solution containing this mix of solutes is said to be *buffered* or it is described as a *buffer solution.*

Buffers contain a weak acid and a weak base

A **buffer** contains solutes that enable it to resist large changes in pH when small amounts of either strong acid or strong base are added to it. Ordinarily, the buffer consists of two solutes, one providing a weak Brønsted acid and the other a weak Brønsted base. Usually, the acid and base represent a conjugate pair. If the acid is molecular, then the conjugate base is *supplied by a soluble salt of the acid.* For example, a common buffer system consists of acetic acid plus sodium acetate, with the salt's acetate ion serving as the Brønsted base. In your blood, carbonic acid (H_2CO_3, a weak diprotic acid) and the bicarbonate ion (HCO_3^-, its conjugate base) serve as one of the buffer systems used to maintain a remarkably constant pH in the face of the body's production of organic acids by metabolism. Another common buffer consists of the weakly acidic cation, NH_4^+, supplied by a salt like NH_4Cl, and its conjugate base, NH_3.

One important point about buffers is the distinction between keeping a solution at a particular pH and keeping it neutral—at a pH of 7. Although it is certainly possible to prepare a buffer to work at pH 7, buffers can be made that will work around any pH value throughout the pH scale.

A buffer works by neutralizing small additions of strong acid or base

TOOLS
Reactions in a buffer when H^+ or OH^- is added

To work, a buffer must be able to neutralize either a strong acid or strong base that is added to it. This is precisely what the weak base and weak acid components of the buffer do. Consider, for example, a buffer composed of acetic acid, $HC_2H_3O_2$, and acetate ion, $C_2H_3O_2^-$, supplied by a salt such as $NaC_2H_3O_2$. If we add extra H^+ to the buffer (from a strong acid) the acetate ion (the weak conjugate base) can react with it as follows.

$$H^+(aq) + C_2H_3O_2^-(aq) \longrightarrow HC_2H_3O_2(aq)$$

Not all the added H^+ is neutralized, so the pH is lowered a little. Soon we'll see how much it changes.

Thus, the added H^+ changes some of the buffer's Brønsted base, $C_2H_3O_2^-$, to its conjugate (weak) acid, $HC_2H_3O_2$. This reaction prevents a large buildup of H^+ that would otherwise be caused by the addition of the strong acid.

A similar response occurs when a strong base is added to the buffer. The OH^- from the strong base will react with some $HC_2H_3O_2$.

$$HC_2H_3O_2(aq) + OH^-(aq) \longrightarrow C_2H_3O_2^-(aq) + H_2O$$

Here the added OH^- changes some of the buffer's Brønsted acid, $HC_2H_3O_2$, into its conjugate base, $C_2H_3O_2^-$. This prevents a buildup of OH^-, which would otherwise cause

[3] Many modern handheld calculators have "solver" functions that enable the user to solve equilibrium problems such as those discussed here without having to make approximations. You might wish to check the instruction manual for your calculator to see if it has these capabilities. You should also check with your instructor to be sure you are allowed to use this kind of calculator on an exam.

FACETS OF CHEMISTRY 16.1

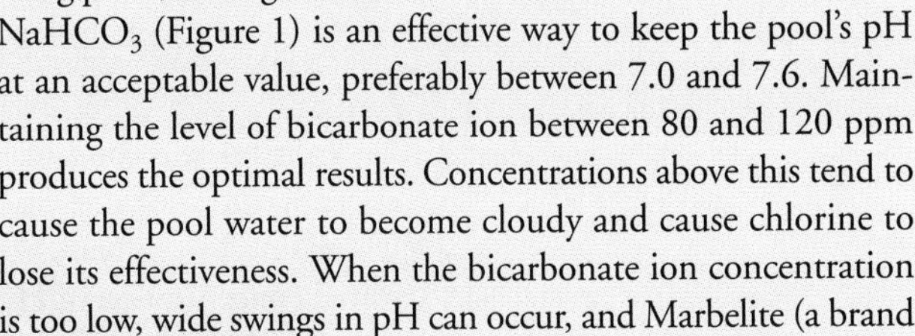

Swimming Pools and Buffers

The water in a swimming pool is a highly dilute solution of chemicals that prevent the growth of bacteria and help stabilize the pool lining. These substances can affect the pH of the pool water, making it unpleasant for swimmers. Therefore, the pH must be monitored and controlled, and control of pH is what buffers do.

Although most buffer systems consist of two separate species that react with H^+ or OH^-, the bicarbonate ion is an example of a single ion that is able to serve both functions. The reactions are

$$HCO_3^-(aq) + H^+(aq) \longrightarrow H_2CO_3(aq)$$

$$HCO_3^-(aq) + OH^-(aq) \longrightarrow H_2O + CO_3^{2-}(aq)$$

Because sodium bicarbonate is nontoxic and because the pH of a solution of the salt is close to 7.0, there are many practical applications of the HCO_3^- buffer. For swimming pools, adding nontoxic $NaHCO_3$ (Figure 1) is an effective way to keep the pool's pH at an acceptable value, preferably between 7.0 and 7.6. Maintaining the level of bicarbonate ion between 80 and 120 ppm produces the optimal results. Concentrations above this tend to cause the pool water to become cloudy and cause chlorine to lose its effectiveness. When the bicarbonate ion concentration is too low, wide swings in pH can occur, and Marbelite (a brand of high-strength polymer cement) and plaster walls will become etched, metals can corrode, the pool's walls and floor can stain, the water can turn green, and the water can cause eyes to burn.

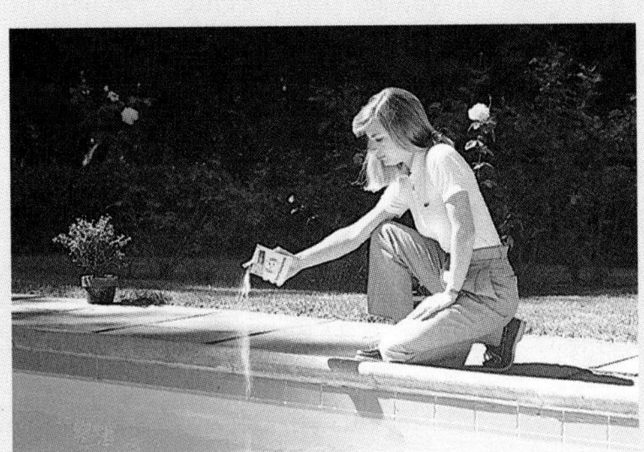

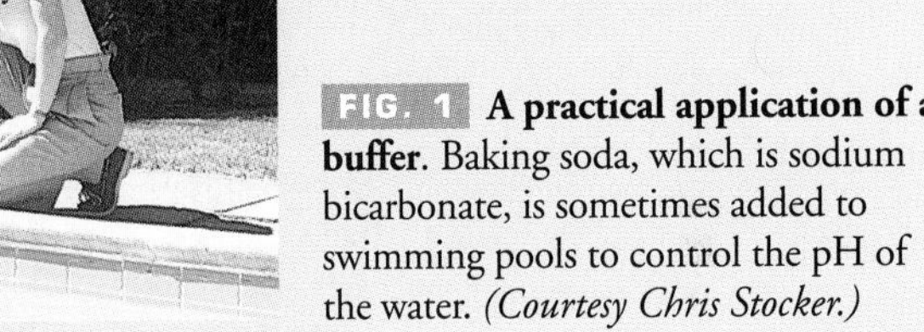

FIG. 1 A practical application of a buffer. Baking soda, which is sodium bicarbonate, is sometimes added to swimming pools to control the pH of the water. *(Courtesy Chris Stocker.)*

a large change in the pH. Thus, one member of a buffer team neutralizes H^+ that might get into the solution, and the other member neutralizes OH^-. *Understanding buffers is an important tool for chemists to use in applications ranging from the protocols of a research project to designing a consumer product.*

Practice Exercise 27: Acetic acid, $HC_2H_3O_2$, and sodium acetate, $NaC_2H_3O_2$ (this provides the acetate ion, $C_2H_3O_2^-$), can be used to make an "acetate" buffer. Does the acetate ion or the acetic acid concentration increase when a strong acid is added to the buffer? Is it the acetate ion or acetic acid that decreases when a strong base is added to the buffer? Explain your answers. (Hint: Which of the buffer components will react with HCl? Which will react with NaOH?)

Practice Exercise 28: For a buffer composed of NH_3 and NH_4^+ (from NH_4Cl), write chemical equations that show what happens when (a) a small amount of strong acid is added, and (b) a small amount of strong base is added.

Calculating the pH of a buffer solution

Calculating the pH of a buffer mixture follows the same procedures we employed in Section 16.2 with a few small changes. The following example illustrates the principles involved.

EXAMPLE 16.14
Calculating the pH of a Buffer

To study the effects of a weakly acidic medium on the rate of corrosion of a metal alloy, a student prepared a buffer solution containing both 0.11 *M* $NaC_2H_3O_2$ and 0.090 *M* $HC_2H_3O_2$. What is the pH of the solution?

ANALYSIS: The buffer solution contains *both* the weak acid $HC_2H_3O_2$ and its conjugate base $C_2H_3O_2^-$. Earlier we noted that when both solute species are present we can use *either* K_a or K_b to perform calculations, whichever is handy. In our tables we find $K_a = 1.8 \times 10^{-5}$ for

$HC_2H_3O_2$, so the simplest approach is to use the equation for the ionization of the acid. We will also be able to use the simplifying approximations developed earlier; these always work for buffers.

SOLUTION: As usual, we begin with the chemical equation and the expression for K_a.

$$HC_2H_3O_2 \rightleftharpoons H^+ + C_2H_3O_2^- \qquad K_a = \frac{[H^+][C_2H_3O_2^-]}{[HC_2H_3O_2]} = 1.8 \times 10^{-5}$$

Let's set up the concentration table so we can proceed carefully. We will take the initial concentrations of $HC_2H_3O_2$ and $C_2H_3O_2^-$ to be the values given in the problem statement. There's no H^+ present from a strong acid, so we set that concentration equal to zero. If the initial concentration of H^+ is zero, its concentration must increase on the way to equilibrium, so under H^+ in the change row we enter $+x$. The other changes follow from that. Here's the completed table.

	$HC_2H_3O_2$	$\rightleftharpoons$	H^+	+	$C_2H_3O_2^-$
Initial concentrations (M)	0.090		0		0.11
Changes in concentrations (M)	$-x$		$+x$		$+x$
Equilibrium concentrations (M)	$(0.090 - x) \approx 0.090$		x		$(0.11 + x) \approx 0.11$

For buffer solutions the quantity x will be very small, so it is safe to make the simplifying approximations. What remains, then, is to substitute the quantities from the last row of the table into the K_a expression.

$$\frac{(x)(0.11 + x)}{(0.090 - x)} \approx \frac{(x)(0.11)}{(0.090)} = 1.8 \times 10^{-5}$$

Solving for x gives us

$$x = \frac{(0.090) \times 1.8 \times 10^{-5}}{(0.11)} = 1.5 \times 10^{-5}$$

Because x equals $[H^+]$, we now have $[H^+] = 1.5 \times 10^{-5}\ M$. Then we calculate pH:

$$\text{pH} = -\log(1.5 \times 10^{-5}) = 4.82$$

Thus, the pH of the buffer is 4.82.

Notice how small x is compared to the initial concentrations. The simplification was valid.

IS THE ANSWER REASONABLE? Again, we check our assumptions and find they worked ($0.090 - 0.000015 = 0.090$ and $0.11 + 0.000015 = 0.11$). The pK_a of acetic acid is 4.74 and our pH is within one pH unit of that value. In fact, the ratio of the conjugate acid and conjugate base concentrations is close to 1.0 and our pH is similarly close to the pK_a. Also, we can check the answer in the usual way by substituting our calculated equilibrium values into the mass action expression. Let's do it.

$$\frac{[H^+][C_2H_3O_2^-]}{[HC_2H_3O_2]} = \frac{(1.5 \times 10^{-5})(0.11)}{(0.090)} = 1.8 \times 10^{-5}$$

The reaction quotient equals K_a, so the values we've obtained are correct equilibrium concentrations.

Practice Exercise 29: Calculate the pH of the buffer solution in the preceding example by using the K_b for $C_2H_3O_2^-$. Be sure to write the chemical equation for the equilibrium as the reaction of $C_2H_3O_2^-$ with water. Then use the chemical equation as a guide in setting up the equilibrium expression for K_b. (Hint: If you work the problem correctly, you should obtain the same answer as above.)

Practice Exercise 30: One liter of buffer is made by dissolving 100.0 grams of acetic acid, $HC_2H_3O_2$, and 100.0 grams of sodium acetate, $NaC_2H_3O_2$, in enough water to make one liter. What is the pH of the solution?

Le Châtelier's principle explains the "common ion effect"

If the solution in the preceding example had contained only acetic acid with a concentration of 0.090 *M*, the calculated $[H^+]$ would have been 1.3×10^{-3} *M*, considerably higher than that of the buffer which also contains 0.11 *M* $C_2H_3O_2^-$. The effect of adding sodium acetate, a substance containing $C_2H_3O_2^-$ ion, to a solution of acetic acid is to suppress the ionization of the acid—it's an example of Le Châtelier's principle. Suppose, for example, we had established the equilibrium

$$HC_2H_3O_2 \rightleftharpoons H^+ + C_2H_3O_2^-$$

According to Le Châtelier's principle, if we add $C_2H_3O_2^-$, the reaction will proceed in a direction to remove some of it. This will shift the equilibrium to the left, thereby reducing the concentration of H^+.

In this example, acetate ion is said to be a **common ion,** in the sense that it is *common* to both the acetic acid equilibrium and to the salt we added, sodium acetate. The suppression of the ionization of acetic acid by addition of the common ion is referred to as the **common ion effect.** We will encounter this phenomenon again in Chapter 17 when we consider the solubilities of salts.

Simplifications are permitted in buffer calculations

There are two useful simplifications that we can use in working buffer calculations. The first is the one we made in Example 16.14:

> Because the initial concentrations are so close to the equilibrium concentrations in the buffer mixture, we can use *initial* concentrations of both the weak acid and its conjugate base as though they were equilibrium values.

A further simplification can be made because the mass action expression contains the ratio of the molar concentrations (in units of moles per liter) of the acid and conjugate base. Let's enter these units for the acid and its conjugate base into the mass action expression. For an acid *HA*,

$$K_a = \frac{[H^+][A^-]}{[HA]} = \frac{[H^+](\text{mol } A^- \cancel{L^{-1}})}{(\text{mol } HA \cancel{L^{-1}})} = \frac{[H^+](\text{mol } A^-)}{(\text{mol } HA)} \qquad (16.8)$$

Notice that the units L^{-1} cancel from the numerator and denominator. This means that for a given acid–base pair, $[H^+]$ is determined by the *mole* ratio of conjugate base to conjugate acid; we don't *have* to use molar concentrations.

> *For buffer solutions* ***only,*** we can use either molar concentrations or moles in the K_a (or K_b) expression to express the amounts of the members of the conjugate acid–base pair (but we must use the same units for each member of the pair).

A further consequence of the relationship derived above is that the pH of a buffer should not change if the buffer is diluted. Dilution changes the volume of a solution but it does not change the number of moles of the solutes, so their mole *ratio* remains constant and so does $[H^+]$.

Buffers can be prepared having a desired pH

The hydrogen ion concentration of a buffer is controlled by both K_a and the ratio of concentrations (or ratio of moles) of the members of the acid–base pair. This can be seen by rearranging Equation 16.8 to solve for $[H^+]$:

$$[H^+] = K_a \frac{[HA]}{[A^-]} \tag{16.9}$$

or

$$[H^+] = K_a \frac{\text{mol } HA}{\text{mol } A^-} \tag{16.10}$$

Buffers are most effective when the mole ratio of acid to base is nearly one, in other words, when $[H^+]=K_a$. Therefore, if we want to prepare a buffer that works well near some specified pH, we look for an acid with a K_a as close to the desired pH as possible. Usually, this means selecting an acid with a pK_a within ± 1 of the desired pH. For experiments in biology, the toxicity of the members of the acid–base pair must also be considered, and that often narrows the choices considerably.

EXAMPLE 16.15
Preparing a Buffer Solution to Have a Predetermined pH

A solution buffered at a pH of 5.00 is needed in an experiment. Can we use acetic acid and sodium acetate to make it? If so, how many moles of $NaC_2H_3O_2$ must be added to 1.0 L of a solution that contains 1.0 mol $HC_2H_3O_2$ to prepare the buffer?

ANALYSIS: There are two parts to this problem. To answer the first, we must check the pK_a of acetic acid to see if it is in the desired range of $pH = pK_a \pm 1$. If it is, we can convert pH to $[H^+]$ and then use Equation 16.10 to calculate the necessary mole ratio. Once we have this, we can proceed to calculate the number of moles of $C_2H_3O_2^-$ needed, and then the number of moles of $NaC_2H_3O_2$.

SOLUTION: Because we want the pH to be 5.00, the pK_a of the selected acid should be 5.00 ± 1, meaning the pK_a should be between 4.00 and 6.00. Because $K_a = 1.8 \times 10^{-5}$ for acetic acid, $pK_a = -\log(1.8 \times 10^{-5}) = 4.74$. So the pK_a of acetic acid falls in the desired range, and acetic acid can be used together with the acetate ion to make the buffer.

Next, we use Equation 16.10 to find the mole ratio of solutes.

$$[H^+] = K_a \times \frac{\text{mol } HC_2H_3O_2}{\text{mol } C_2H_3O_2^-}$$

First, let's solve for the mole ratio.

$$\frac{\text{mol } HC_2H_3O_2}{\text{mol } C_2H_3O_2^-} = \frac{[H^+]}{K_a}$$

The desired pH = 5.00, so $[H^+] = 1.0 \times 10^{-5}$; also $K_a = 1.8 \times 10^{-5}$. Substituting gives

$$\frac{\text{mol } HC_2H_3O_2}{\text{mol } C_2H_3O_2^-} = \frac{1.0 \times 10^{-5}}{1.8 \times 10^{-5}} = 0.56$$

This is the *mole* ratio of the buffer components we want. The solution we are preparing contains 1.0 mol $HC_2H_3O_2$, so the number of moles of acetate ion required is

$$\text{moles } C_2H_3O_2^- = \frac{1.0 \text{ mol } HC_2H_3O_2}{0.56}$$
$$= 1.8 \text{ moles } C_2H_3O_2^-$$

For each mole of $NaC_2H_3O_2$ there is one mole of $C_2H_3O_2^-$, so to prepare the solution we need 1.8 mol $NaC_2H_3O_2$.

IS THE ANSWER REASONABLE? A 1-to-1 mole ratio of $C_2H_3O_2^-$ to $HC_2H_3O_2$ would give $pH = pK_a = 4.74$. The desired pH of 5.00 is slightly more basic than 4.74, so we can expect that the amount of conjugate base needed should be larger than the amount of conjugate acid. Our answer of 1.8 mol of $NaC_2H_3O_2$ appears to be reasonable.

Practice Exercise 31: From Table 16.1 select an acid that, along with its sodium salt, can be used to make a buffer that has a pH of 5.25. If you have 500.0 mL of a 0.200 *M* solution of that acid, how many grams of the corresponding sodium salt do you have to dissolve to obtain the desired pH? (Hint: There is more than one correct answer to this problem. The first step is to determine the ratio of molarities of the conjugate acid and conjugate base.)

Practice Exercise 32: A chemist needed an aqueous buffer with a pH of 3.90. Would formic acid and its salt, sodium formate, make a good pair for this purpose? If so, what mole ratio of the acid, $HCHO_2$, to the anion of this salt, CHO_2^-, is needed? How many grams of $NaCHO_2$ would have to be added to a solution that contains 0.10 mol $HCHO_2$?

If you take a biology course, you're likely to run into a logarithmic form of Equation 16.9 called the **Henderson–Hasselbalch equation.** This is obtained by taking the negative logarithm of both sides of Equation 16.9 and rearranging the term involving the concentrations.

$$\mathrm{pH} = \mathrm{p}K_a + \log\frac{[A^-]}{[HA]} \tag{16.11}$$

In most buffers used in the life sciences, the anion A^- comes from a salt in which the cation has a charge of 1+, such as Na*A*, and the acid is monoprotic. With these as conditions, the equation is sometimes written

$$\mathrm{pH} = \mathrm{p}K_a + \log\frac{[\text{salt}]}{[\text{acid}]} \tag{16.12}$$

For practice, you may wish to apply this equation to buffer problems at the end of the chapter.

We can calculate the change in pH when strong acid or base is added to a buffer

Earlier we described how a buffer is able to neutralize small amounts of either strong acid or base. We can use this knowledge now to calculate how much the pH will change.

EXAMPLE 16.16
Calculating How a Buffer Resists Changes in pH

How much will the pH change if 0.020 mol of HCl is added to a buffer solution that was made by dissolving 0.12 mol of NH_3 and 0.095 mol of NH_4Cl in 250 mL of water?

ANALYSIS: This problem will require two calculations. The first is the pH of the original buffer. The second is the pH of the mixture after the HCl has been added. For the second calculation, we have to determine how the acid changes the amounts of NH_3 and NH_4^+. This requires that we examine how the buffer functions. In this case, the HCl (a strong acid) will supply H^+ that will react completely with NH_3 (the conjugate base of the buffer pair), changing it to NH_4^+.

For both calculations, the pH is determined by the mole ratio of the members of the acid–base pair. To calculate the pH we will be able to use the moles of NH_3 and NH_4^+ directly in the mass action expression. To set up the equilibrium law we can use either the K_a for NH_4^+ or the K_b for NH_3. Since K_b is tabulated, we will use it and write the equation for the ionization of the base.

SOLUTION: We begin with the chemical equation and the K_b expression.

$$NH_3 + H_2O \rightleftharpoons NH_4^+ + OH^- \qquad K_b = \frac{[NH_4^+][OH^-]}{[NH_3]} = 1.8 \times 10^{-5}$$

The solution also contains 0.095 mol of Cl^-, of course, but this ion is not involved in the equilibrium.

The solution contains 0.12 mol of NH_3 and 0.095 mol of NH_4^+ from the complete dissociation of the salt NH_4Cl. We can enter these quantities into the mass action expression in place of concentrations and solve for $[OH^-]$.

$$1.8 \times 10^{-5} = \frac{\text{mol } NH_4^+ \times [OH^-]}{\text{mol } NH_3} = \frac{(0.095)[OH^-]}{0.12}$$

solving for $[OH^-]$ gives

$$[OH^-] = 2.3 \times 10^{-5}$$

To calculate the pH, we obtain pOH and subtract it from 14.00.

$$\text{pOH} = -\log(2.3 \times 10^{-5}) = 4.64$$
$$\text{pH} = 14.00 - 4.64 = 9.36$$

This is the pH before we add any HCl.

Next, we consider the reaction that takes place when we add the HCl to the buffer. The 0.020 mol of HCl is completely ionized, so we're adding 0.020 mol H^+. This will react as follows.

The equilibrium constant for this reaction is 1.8×10^{10}, so the reaction goes almost to completion. Nearly all the added H^+ reacts with NH_3.

$$NH_3(aq) + H^+(aq) \longrightarrow NH_4^+(aq)$$

The 0.020 mol of H^+ will react with 0.020 mol of NH_3 to form 0.020 mol of NH_4^+. This causes the number of moles of NH_3 to *decrease* by 0.020 mol and the number of moles of NH_4^+ to *increase* by 0.020 mol. After addition of the acid, we have

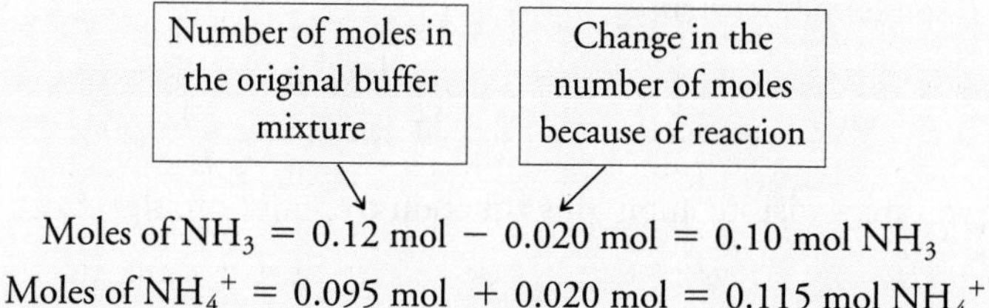

$$\text{Moles of } NH_3 = 0.12 \text{ mol} - 0.020 \text{ mol} = 0.10 \text{ mol } NH_3$$
$$\text{Moles of } NH_4^+ = 0.095 \text{ mol} + 0.020 \text{ mol} = 0.115 \text{ mol } NH_4^+$$

We now use these new amounts of NH_3 and NH_4^+ to calculate the new pH of the buffer solution. First we calculate $[OH^-]$.

$$1.8 \times 10^{-5} = \frac{(0.115)[OH^-]}{0.10}$$

Solving for $[OH^-]$ gives

$$[OH^-] = 1.6 \times 10^{-5}$$

To calculate the pH, we obtain pOH and subtract it from 14.00.

$$\text{pOH} = -\log(1.6 \times 10^{-5}) = 4.80$$
$$\text{pH} = 14.00 - 4.80 = 9.20$$

This is the pH after addition of the acid. We're asked for the change in pH, so we take the difference between the two values.

$$\text{Change in pH} = 9.36 - 9.20 = 0.16$$

Thus, the pH has dropped by 0.16 pH units.

IS THE ANSWER REASONABLE? In general, we can assume that changes in pH should be small (less than 1.0 pH unit); otherwise, there's no simple way to estimate how *much* the pH will change when H^+ or OH^- is added to the buffer. However, we can anticipate the *direction* of the change. Adding H^+ to a buffer will lower the pH somewhat and adding OH^- will raise it. In this example, the buffer is absorbing H^+, so its pH should drop, and our calculations agree. If our calculated pH had been higher than the original, we might expect to find errors in the calculation of the final values for the number of moles of NH_3 and NH_4^+.

Practice Exercise 33: How much will the pH change if we add 0.15 mol NaOH to 1.00 L of a buffer that contains 1.00 mol $HC_2H_3O_2$ and 1.00 mol $NaC_2H_3O_2$? (Hint: Will the NaOH react with $HC_2H_3O_2$ or $NaC_2H_3O_2$? Write the equation for the reaction.)

Practice Exercise 34: A buffer is prepared by mixing 50.0 g of NH_3 and 50.0 g of NH_4Cl in 500 mL of solution. What is the pH of this buffer and what will the pH change to if 5.00 g of HCl is then added to the mixture?

If you look back over the calculations in the preceding example, you can get some feel for how effective buffers can be at preventing large swings in pH. Notice that we've added enough HCl to react with approximately 20% of the NH_3 in the solution, but the pH changed by only 0.16 pH units. If we were to add this same amount of HCl to 250 mL of pure water, it would give a solution with $[H^+] = 0.080\ M$. Such a solution would have a pH of 1.10, which is quite acidic. However, the buffer mixture was still basic, with a pH of 9.20, even after addition of the acid. That's pretty remarkable!

16.6 Polyprotic Acids Ionize in Two or More Steps

Polyprotic acids ionize stepwise

Until now our discussion of weak acids has focused entirely on equilibria involving monoprotic acids. There are, of course, many acids capable of supplying more than one H^+ per molecule. Recall that these are called polyprotic acids. Examples include sulfuric acid, H_2SO_4, carbonic acid, H_2CO_3, and phosphoric acid, H_3PO_4. These acids undergo ionization in a series of steps, each of which releases one proton. For weak polyprotic acids, such as H_2CO_3 and H_3PO_4, each step is an equilibrium. Even sulfuric acid, which we consider a strong acid, is not completely ionized. Loss of the first proton to yield the HSO_4^- ion is complete, but the loss of the second proton is incomplete and involves an equilibrium. In this section, we will focus our attention on weak polyprotic acids as well as solutions of their salts.

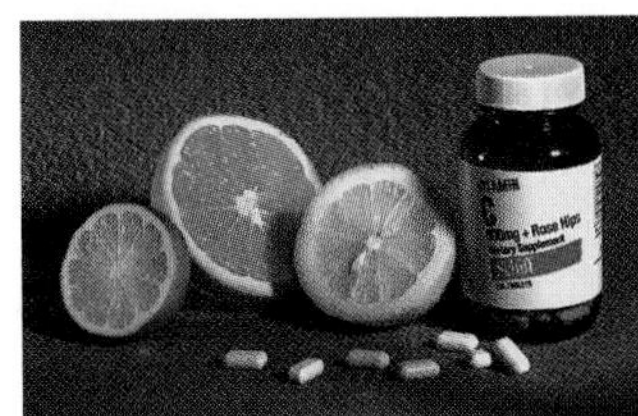

Vitamin C. Many fruits and vegetables contain ascorbic acid (vitamin C), a weak diprotic acid with the formula $H_2C_6H_6O_6$. *(Andy Washnik.)*

Let's begin with the weak diprotic acid H_2CO_3. In water, the acid ionizes in two steps, each of which is an equilibrium that transfers an H^+ ion to a water molecule.

$$H_2CO_3 + H_2O \rightleftharpoons H_3O^+ + HCO_3^-$$
$$HCO_3^- + H_2O \rightleftharpoons H_3O^+ + CO_3^{2-}$$

As usual, we can use H^+ in place of H_3O^+ and simplify these equations to give

$$H_2CO_3 \rightleftharpoons H^+ + HCO_3^-$$
$$HCO_3^- \rightleftharpoons H^+ + CO_3^{2-}$$

Each step has its own ionization constant, K_a, which we identify as K_{a_1} for the first step and K_{a_2} for the second. For carbonic acid,

$$K_{a_1} = \frac{[H^+][HCO_3^-]}{[H_2CO_3]} = 4.5 \times 10^{-7}$$

$$K_{a_2} = \frac{[H^+][CO_3^{2-}]}{[HCO_3^-]} = 4.7 \times 10^{-11}$$

Notice that each ionization makes a contribution to the total molar concentration of H^+, and one of our goals here is to relate the K_a values and the concentration of the acid to $[H^+]$. At first glance, this seems to be a formidable task, but certain simplifications are justified that make the problem relatively simple to solve.

Simplifications in calculations involving polyprotic acids

The principal factor that simplifies calculations involving many polyprotic acids is the large differences between successive ionization constants. Notice that for H_2CO_3, K_{a_1} is much larger than K_{a_2} (they differ by a factor of nearly 10,000). Similar differences between K_{a_1} and K_{a_2} are observed for many diprotic acids. One reason is that an H^+ is lost much more easily from the neutral H_2A molecule than from the HA^- ion. The stronger attraction of the opposite charges inhibits the second ionization. Typically, K_{a_1} is greater than K_{a_2} by a factor of between 10^4 and 10^5, as the data in Table 16.3 show. For a triprotic acid, such as phosphoric acid, H_3PO_4, the second acid ionization constant is similarly greater than the third.

Because K_{a_1} is so much larger than K_{a_2}, virtually all the H^+ in a solution of the acid comes from the first step in the ionization. In other words,

$$[H^+]_{\text{total}} = [H^+]_{\text{first step}} + [H^+]_{\text{second step}}$$

$$[H^+]_{\text{first step}} \gg [H^+]_{\text{second step}}$$

Therefore, we make the approximation that

$$[H^+]_{\text{total}} \approx [H^+]_{\text{first step}}$$

This means that as far as calculating the H^+ concentration is concerned, we can treat the acid as though it were a monoprotic acid and ignore the second step in the ionization. In addition we have

$$[HCO_3^-]_{\text{total}} = [HCO_3^-]_{\text{first step}} - [CO_3^{2-}]_{\text{second step}}$$

However, since $[CO_3^{2-}]_{\text{second step}} = [H^+]_{\text{second step}}$ and $[H^+]_{\text{second step}}$ is very small we can say that

$$[HCO_3^-]_{\text{total}} \approx [HCO_3^-]_{\text{first step}}$$

It is therefore reasonable to conclude that

$$[H^+]_{\text{first step}} = [HCO_3^-]_{\text{first step}}$$

Example 16.17 illustrates how these relationships are applied.

Additional K_a values of polyprotic acids are located in Appendix C.7.

TABLE 16.3 Acid Ionization Constants for Polyprotic Acids

		Acid Ionization Constant for Successive Ionizations (25 °C)		
Name	Formula	K_{a_1}	K_{a_2}	K_{a_3}
Carbonic acid	H_2CO_3	4.5×10^{-7}	4.7×10^{-11}	
Hydrosulfuric acid	$H_2S(aq)$	9.5×10^{-8}	1×10^{-19}	
Phosphoric acid	H_3PO_4	7.1×10^{-3}	6.3×10^{-8}	4.5×10^{-13}
Arsenic acid	H_3AsO_4	5.6×10^{-3}	1.7×10^{-7}	4.0×10^{-12}
Sulfuric acid	H_2SO_4	Large	1.0×10^{-2}	
Selenic acid	H_2SeO_4	Large	1.2×10^{-2}	
Telluric acid	H_6TeO_6	2×10^{-8}	1×10^{-11}	
Sulfurous acid	H_2SO_3	1.2×10^{-2}	6.6×10^{-8}	
Selenous acid	H_2SeO_3	4.5×10^{-3}	1.1×10^{-8}	
Tellurous acid	H_2TeO_3	3.3×10^{-3}	2.0×10^{-8}	
Ascorbic acid (vitamin C)	$H_2C_6H_6O_6$	6.8×10^{-5}	2.7×10^{-12}	
Oxalic acid	$H_2C_2O_4$	5.6×10^{-2}	5.4×10^{-5}	
Citric acid (18 °C)	$H_3C_6H_5O_7$	7.1×10^{-4}	1.7×10^{-5}	6.3×10^{-6}

EXAMPLE 16.17
Calculating the Concentrations of Solute Species in a Solution of a Polyprotic Acid

Calculate the concentrations of all the species produced in the ionization of 0.040 *M* H_2CO_3 as well as the pH of the solution.

ANALYSIS: As in any equilibrium problem, we begin with the chemical equations and the K_a expressions:

$$H_2CO_3 \rightleftharpoons H^+ + HCO_3^- \qquad K_{a_1} = \frac{[H^+][HCO_3^-]}{[H_2CO_3]} = 4.5 \times 10^{-7}$$

$$HCO_3^- \rightleftharpoons H^+ + CO_3^{2-} \qquad K_{a_2} = \frac{[H^+][CO_3^{2-}]}{[HCO_3^-]} = 4.7 \times 10^{-11}$$

We will want to calculate the concentrations of H^+, HCO_3^-, and CO_3^{2-}, and the concentration of H_2CO_3 that remains at equilibrium. As noted in the preceding discussion, the problem is simplified considerably by assuming that the second reaction occurs to a negligible extent compared to the first. This assumption (which we will justify later in the problem) allows us to calculate the H^+ and HCO_3^- concentrations as though the solute were a monoprotic acid. This type of problem is one we've worked on in Section 16.2.

To obtain $[CO_3^{2-}]$, we will have to use the second step in the ionization. Once again, the large difference between K_{a_1} and K_{a_2} permits some simplifications. They involve relationships we have already examined.

$$[H^+]_{\text{equilib}} \approx [H^+]_{\text{formed in first step}}$$

and

$$[HCO_3^-]_{\text{equilib}} \approx [HCO_3^-]_{\text{formed in first step}}$$

Keep in mind that for a given solution, the values of H^+ used in the expressions for K_{a_1} and K_{a_2} are identical. At equilibrium there is only *one* equilibrium H^+ concentration.

SOLUTION: Treating H_2CO_3 as though it were a monoprotic acid yields a problem similar to many others we worked earlier in this chapter. We know some of the H_2CO_3 ionizes; let's call this amount *x*. If *x* moles per liter of H_2CO_3 ionizes, then *x* moles per liter of H^+ and HCO_3^- are formed, and the concentration of H_2CO_3 is reduced by *x*.

	H_2CO_3 $\rightleftharpoons$	H^+ +	HCO_3^-
Initial concentrations (*M*)	0.040	0	0
Changes in concentrations (*M*)	$-x$	$+x$	$+x$
Equilibrium concentrations (*M*)	$(0.040 - x)$ ≈ 0.040	x	x

Substituting into the expression for K_{a_1} gives

$$\frac{x^2}{(0.040 - x)} \approx \frac{x^2}{0.040} = 4.5 \times 10^{-7}$$

Solving for *x* gives $x = 1.3 \times 10^{-4}$, so $[H^+] = [HCO_3^-] = 1.3 \times 10^{-4}$ *M*. From this we can now calculate the pH.

$$\text{pH} = -\log(1.3 \times 10^{-4}) = 3.89$$

Now let's calculate the concentration of CO_3^{2-}. We obtain this from the expression for K_{a_2}.

$$K_{a_2} = \frac{[H^+][CO_3^{2-}]}{[HCO_3^-]} = 4.7 \times 10^{-11}$$

In our analysis, we concluded that we can use the approximations

$$[H^+]_{\text{equilib}} \approx [H^+]_{\text{first step}}$$
$$[HCO_3^-]_{\text{equilib}} \approx [HCO_3^-]_{\text{first step}}$$

Substituting the values obtained above gives us

$$\frac{(1.3 \times 10^{-4})[CO_3^{2-}]}{1.3 \times 10^{-4}} = 4.7 \times 10^{-11}$$

$$[CO_3^{2-}] = 4.7 \times 10^{-11} = K_{a_2}$$

Let's summarize the results. At equilibrium, we have

$$[H_2CO_3] = 0.040\ M$$

$$[H^+] = [HCO_3^-] = 1.3 \times 10^{-4}\ M \qquad \text{(and pH = 3.89)}$$

$$[CO_3^{2-}] = 4.7 \times 10^{-11}\ M$$

ARE THE ANSWERS REASONABLE? First, we check our simplifying assumptions. The initial concentration was assumed to be much larger than the hydrogen ion concentration and it is $(0.040 - 0.00013 = 0.040)$. The second approximation was that the amount of CO_3^{2-} formed would be very small in relation to the HCO_3^- (and also H^+) concentration calculated in the first step. This assumption is true too $(1.3 \times 10^{-4} - 4.7 \times 10^{-11} = 1.3 \times 10^{-4})$. We can also check that the hydrogen ions added in the second step are inconsequential $(1.3 \times 10^{-4} + 4.7 \times 10^{-11} = 1.3 \times 10^{-4})$.

We've checked our assumptions and they are okay. The pH is what we expect from an acid. If more verification is needed you can always perform a quick check of the answers by substituting them into the appropriate mass action expressions:

$$\frac{(1.3 \times 10^{-4})(1.3 \times 10^{-4})}{0.040} = 4.2 \times 10^{-7}, \text{ which is very close to } K_{a_1}$$

$$\frac{(1.3 \times 10^{-4})(4.7 \times 10^{-11})}{(1.3 \times 10^{-4})} = 4.7 \times 10^{-11}, \text{ which equals } K_{a_2}$$

One last thing, the problem asked for all species produced in the ionization of H_2CO_3. Two other substances are in the reaction mixture, water and hydroxide ions. A rigorous solution would also include those species.

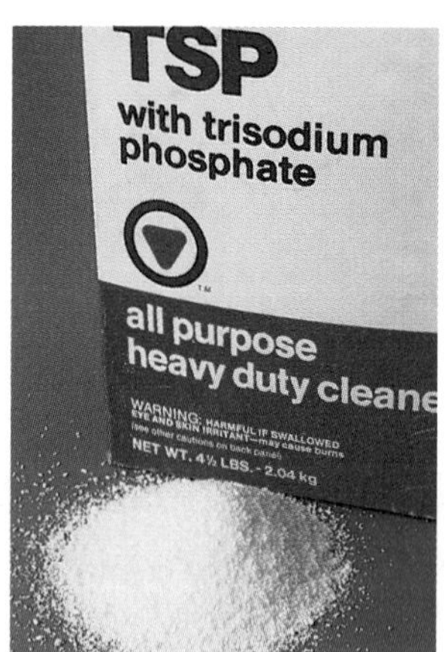

The salt Na_3PO_4 is sold under the name trisodium phosphate or TSP. Solutions of the salt in water are quite basic and are used as an aid in cleaning grime from painted surfaces. Some states restrict the sale of TSP because the phosphate ion it contains presents a pollution problem by promoting the growth of algae in lakes. *(Coco McCoy/Rainbow.)*

Practice Exercise 35: Write the three ionization steps for phosphoric acid, H_3PO_4, and write the K_a expressions for each step. (Hint: Recall that one proton is ionized in each step.)

Practice Exercise 36: Ascorbic acid (vitamin C) is a diprotic acid, $H_2C_6H_6O_6$. See Table 16.3. Calculate $[H^+]$, pH, and $[C_6H_6O_6^{2-}]$ in a 0.10 M solution of ascorbic acid.

One of the most interesting observations we can derive from the preceding example is that *in a solution that contains a polyprotic acid* ***as the only solute,*** *the concentration of the ion formed in the second step of the ionization equals* K_{a_2}.

Salts of polyprotic acids give basic solutions

You learned earlier that the pH of a salt solution is controlled by whether the salt's cation, anion, or both are able to react with water. For simplicity, the salts of polyprotic acids that we will discuss here will be limited to those containing *nonacidic* cations, such as sodium or potassium ion. In other words, we will study salts in which the anion alone is basic and thereby affects the pH of a solution.

A typical example of a salt of a polyprotic acid is sodium carbonate, Na_2CO_3. It is a salt of H_2CO_3, and the salt's carbonate ion is a Brønsted base that is responsible for *two* equilibria that furnish OH^- ion and thereby affect the pH of the solution. These equilibria and their corresponding expressions for K_b are as follows:

$$CO_3^{2-}(aq) + H_2O \rightleftharpoons HCO_3^-(aq) + OH^-(aq) \qquad (16.13)$$

$$K_{b_1} = \frac{[HCO_3^-][OH^-]}{[CO_3^{2-}]}$$

$$HCO_3^-(aq) + H_2O \rightleftharpoons H_2CO_3(aq) + OH^-(aq) \qquad (16.14)$$

$$K_{b_2} = \frac{[H_2CO_3][OH^-]}{[HCO_3^-]}$$

The calculations required to determine the concentrations of the species involved in these equilibria are very much like those involving weak diprotic acids. The difference is that the chemical reactions here are those of bases instead of acids. In fact, *the simplifying assumptions in our calculations will be almost identical to those for weak polyprotic acids.*

First, let's compare the K_b values for the successive equilibria. To obtain these constants, we will use the relationship that $K_a \times K_b = K_w$ along with the values of K_{a_1} and K_{a_2} for H_2CO_3.

In general, K_b values for anions of polyprotic acids are not tabulated. When needed, they are calculated from the appropriate K_a values for the acids.

Notice that in the equilibrium in which CO_3^{2-} is the base (Equation 16.13), the conjugate acid is HCO_3^-. Therefore, to calculate K_b for CO_3^{2-} (which we called K_{b_1}), we must use the K_a for HCO_3^-, which corresponds to K_{a_2} for carbonic acid. Thus,

$$K_{b_1} = \frac{K_w}{K_{a_2}} = \frac{1.0 \times 10^{-14}}{4.7 \times 10^{-11}} = 2.1 \times 10^{-4}$$

Similarly, in the equilibrium in which HCO_3^- is the base (Equation 16.14), the conjugate acid is H_2CO_3. Therefore, to calculate K_b for HCO_3^- (which we called K_{b_2}), we must use the K_a for H_2CO_3, which is K_{a_1} for carbonic acid.

$$K_{b_2} = \frac{K_w}{K_{a_1}} = \frac{1.0 \times 10^{-14}}{4.5 \times 10^{-7}} = 2.2 \times 10^{-8}$$

Remember: K_{b_1} comes from K_{a_2}, and K_{b_2} comes from K_{a_1}.

Now we can compare the two K_b values. For CO_3^{2-}, $K_{b_1} = 2.1 \times 10^{-4}$, and for the (much) weaker base, HCO_3^-, K_{b_2} is 2.2×10^{-8}. Thus CO_3^{2-} has a K_b nearly 10,000 times that of HCO_3^-. This means that the reaction of CO_3^{2-} with water (Equation 16.13) generates far more OH^- than the reaction of HCO_3^- (Equation 16.14). The contribution of the latter to the total pool of OH^- is relatively so small that we can safely ignore it. The simplification here is very much the same as in our treatment of the ionization of weak polyprotic acids.

$$[OH^-]_{total} = [OH^-]_{formed\ in\ first\ step} + [OH^-]_{formed\ in\ second\ step}$$

$$[OH^-]_{formed\ in\ first\ step} \gg [OH^-]_{formed\ in\ second\ step}$$

$$[OH^-]_{total} \approx [OH^-]_{formed\ in\ first\ step}$$

Thus, if we wish to calculate the pH of a solution of a basic anion of a polyprotic acid, *we may work exclusively with K_{b_1} and ignore any further reactions.* Let's study an example of how this works.

EXAMPLE 16.18

Calculating the pH of a Solution of a Salt of a Weak Diprotic Acid

What is the pH of a 0.15 *M* solution of Na_2CO_3?

ANALYSIS: We can ignore the reaction of HCO_3^- with water, so the only relevant equilibrium is

$$CO_3^{2-} + H_2O \rightleftharpoons HCO_3^- + OH^- \qquad K_{b_1} = \frac{[HCO_3^-][OH^-]}{[CO_3^{2-}]} = 2.1 \times 10^{-4}$$

(The value of K_{b_1} was calculated in the discussion preceding this example.) Also, the initial concentration of CO_3^{2-} (0.15 *M*) is larger than $100 \times K_{b_1}$, so we can safely use 0.15 *M* as the equilibrium concentration of CO_3^{2-}.

SOLUTION: Some of the CO_3^{2-} will react; we'll let this be x mol L^{-1}. The concentration table, then, is as follows.

Some detergents that use sodium carbonate as their caustic ingredient. Detergents are best able to dissolve grease and grime if their solutions are basic. In these products, the basic ingredient is sodium carbonate. The salt Na_2CO_3 is relatively nontoxic and the carbonate ion it contains is a relatively strong base. The carbonate ion also serves as a water softener by precipitating Ca^{2+} ions as insoluble $CaCO_3$. *(Paul Silverman/Fundamental Photographs.)*

	CO_3^{2-} + H_2O $\rightleftharpoons$	HCO_3^- +	OH^-
Initial concentrations (*M*)	0.15	0	0
Changes in concentrations caused by the ionization (*M*)	$-x$	$+x$	$+x$
Equilibrium concentrations (*M*)	$(0.15 - x) \approx 0.15$	x	x

Now we're ready to insert the values in the last row of the table into the equilibrium expression for the carbonate ion.

$$\frac{[HCO_3^-][OH^-]}{[CO_3^{2-}]} = \frac{(x)(x)}{0.15 - x} \approx \frac{(x)(x)}{0.15} = 2.1 \times 10^{-4}$$

$$x = [OH^-] = 5.6 \times 10^{-3} \text{ mol L}^{-1}$$

and

$$pOH = -\log(5.6 \times 10^{-3}) = 2.25$$

Finally, we calculate the pH by subtracting pOH from 14.00.

$$pH = 14.00 - 2.25 = 11.75$$

Thus, the pH of 0.15 *M* Na_2CO_3 is calculated to be 11.75. Once again we see that an aqueous solution of a salt composed of a basic anion and a neutral cation is basic.

IS THE ANSWER REASONABLE? Our assumption is reasonable since 0.0056 is less than 5% of 0.015 (0.15− 0.0056 = 0.14). Also, the pH we obtained corresponds to a basic solution, which is what we expect when the anion of the salt reacts as a weak base. The answer seems reasonable.

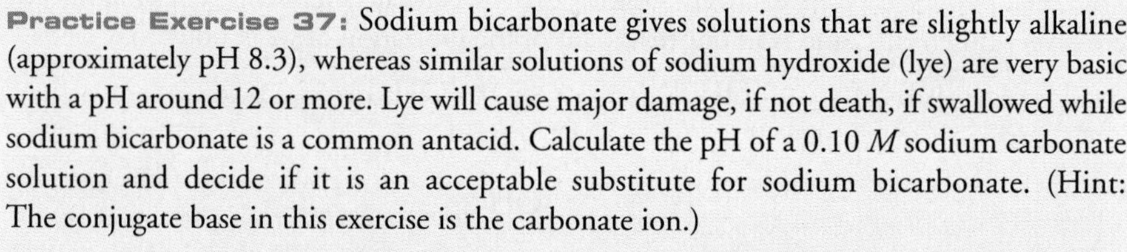
Practice Exercise 37: Sodium bicarbonate gives solutions that are slightly alkaline (approximately pH 8.3), whereas similar solutions of sodium hydroxide (lye) are very basic with a pH around 12 or more. Lye will cause major damage, if not death, if swallowed while sodium bicarbonate is a common antacid. Calculate the pH of a 0.10 *M* sodium carbonate solution and decide if it is an acceptable substitute for sodium bicarbonate. (Hint: The conjugate base in this exercise is the carbonate ion.)

Practice Exercise 38: What is the pH of a 0.20 *M* solution of Na_2SO_3 at 25 °C? For the diprotic acid, H_2SO_3, $K_{a_1} = 1.2 \times 10^{-2}$ and $K_{a_2} = 6.6 \times 10^{-8}$.

Practice Exercise 39: Reasoning by analogy from what you learned about solutions of weak polyprotic acids, what is the molar concentration of H_2SO_3 in a 0.010 *M* solution of Na_2SO_3?

16.7 Acid–Base Titrations Have Sharp Changes in pH at the Equivalence Point

An acid–base titration. Phenolphthalein is often used as an indicator to detect the end point. *(Peter Lerman.)*

In Chapter 4 we studied the overall procedure for an acid–base titration, and we saw how titration data can be used in various stoichiometric calculations. In performing this procedure, the titration is halted at the **end point** when a change in color of an indicator occurs. Ideally, this end point should occur at the **equivalence point,** when stoichiometrically equivalent amounts of acid and base have combined. To obtain this ideal result, selecting an appropriate indicator requires foresight. We will understand this better by studying how the pH of the solution being titrated changes with the addition of titrant.

When the pH of a solution at different stages of a titration is plotted against the volume of titrant added, we obtain a **titration curve.** The values of pH in these plots can be measured using a pH meter during the titration, or it can be calculated by the procedures studied in this chapter.

We will use the calculation method, and also demonstrate that all titration curves can be described by four calculations. Two calculations describe single points on the curve, the starting point and the equivalence point. The other two calculations are simple limiting reactant calculations for mixtures (a) between the start and the equivalence point and (b) after the equivalence point.

Titration of a strong acid by a strong base uses four equations

Titration curves have four regions

The titration of HCl(*aq*) with a standardized NaOH solution illustrates the titration of a strong acid by a strong base. The molecular and net ionic equations are

$$HCl(aq) + NaOH(aq) \longrightarrow NaCl(aq) + H_2O$$

$$H^+(aq) + OH^-(aq) \longrightarrow H_2O$$

Let's consider what happens to the pH of a solution, initially 25.00 mL of 0.2000 *M* HCl, as small amounts of the titrant, 0.2000 *M* NaOH, are added. We will calculate the pH of the resulting solution at various stages of the titration, retaining only two significant figures, and plot the values against the volume of titrant.

The *titrant* is the solution being slowly added from a buret to a solution in the receiving flask.

At the start, before any titrant has been added, the receiving flask contains only 0.2000 *M* HCl. Because this is a strong acid, we know that

$$[H^+] = [HCl] = 0.2000\ M$$

So the initial pH is

$$pH = -\log(0.2000) = 0.70$$

After the start but before the equivalence point we need to determine the concentration of the excess reactant in a simple limiting reactant problem. Let's first calculate the amount of HCl initially present in 25.00 mL of 0.2000 *M* HCl.

$$25.00\text{ mL HCl soln} \times \frac{0.2000\text{ mol HCl}}{1000\text{ mL HCl soln}} = 5.000 \times 10^{-3}\text{ mol HCl}$$

Despite the high precision assumed for the molarity and the volume of the titrant, we will retain only two significant figures in the calculated pH. Precision higher than this is hard to obtain and seldom sought in actual lab work.

Now suppose we add 10.00 mL of 0.2000 *M* NaOH from the buret. The moles of NaOH added is

$$10.00\text{ mL NaOH soln} \times \frac{0.2000\text{ mol NaOH}}{1000\text{ mL NaOH soln}} = 2.000 \times 10^{-3}\text{ mol NaOH}$$

We can see that we have more moles of HCl than NaOH (and the mole ratio is 1:1) so that the base neutralizes 2.000×10^{-3} mol of HCl, and the amount of HCl remaining is

$$(5.000 \times 10^{-3} - 2.000 \times 10^{-3})\text{ mol HCl} = 3.000 \times 10^{-3}\text{ mol HCl remaining}$$

To obtain the concentration we divide by the total volume of the solution that is now (25.00 + 10.00) mL = 35.00 mL = 0.03500 L. The $[H^+]$ is

$$[H^+] = \frac{3.000 \times 10^{-3}\text{ mol}}{0.03500\text{ L}} = 8.571 \times 10^{-2}\ M$$

$$[H^+] = \frac{M_AV_A - M_BV_B}{V_A + V_B}$$

The corresponding pH is 1.07.

We can repeat this calculation for as many different volumes as we care to choose as long as the volume of NaOH is less than the equivalence point volume. A spreadsheet will allow us to quickly calculate the pH of large numbers of data points, but only a few are needed to see the nature of the curve.

At the equivalence point we can calculate the volume of NaOH added as

$$M_{HCl} \times V_{HCl} = M_{NaOH} \times V_{NaOH}$$

$$V_{NaOH}(\text{at eq pt}) = (0.2000\ M_{HCl})(25.00\text{ mL HCl})/(0.2000\ M\text{ NaOH}) = 25.00\text{ mL}$$

TABLE 16.4 Titration of 25.00 mL of 0.2000 *M* HCl with 0.2000 *M* NaOH

Initial Volume of HCl (mL)	Initial Amount of HCl (mol)	Volume of NaOH added (mL)	Amount of NaOH (mol)	Amount of Excess Reagent (mol)	Total Volume of Solution (mL)	Molarity of Ion in Excess (mol L^{-1})	pH
25.00	5.000×10^{-3}	0.00	0.000	5.000×10^{-3} (H^+)	25.00	0.2000 (H^+)	0.70
25.00	5.000×10^{-3}	10.00	2.000×10^{-3}	3.000×10^{-3} (H^+)	35.00	8.571×10^{-2} (H^+)	1.07
25.00	5.000×10^{-3}	20.00	4.000×10^{-3}	1.000×10^{-3} (H^+)	45.00	2.222×10^{-2} (H^+)	1.65
25.00	5.000×10^{-3}	24.00	4.800×10^{-3}	2.000×10^{-4} (H^+)	49.00	4.082×10^{-3} (H^+)	2.39
25.00	5.000×10^{-3}	24.90	4.980×10^{-3}	2.000×10^{-5} (H^+)	49.90	4.000×10^{-4} (H^+)	3.40
25.00	5.000×10^{-3}	24.99	4.998×10^{-3}	2.000×10^{-6} (H^+)	49.99	4.000×10^{-5} (H^+)	4.40
25.00	5.000×10^{-3}	25.00	5.000×10^{-3}	0	50.00	0	7.00
25.00	5.000×10^{-3}	25.01	5.002×10^{-3}	2.000×10^{-6} (OH^-)	50.01	3.999×10^{-5} (OH^-)	9.60
25.00	5.000×10^{-3}	25.10	5.020×10^{-3}	2.000×10^{-5} (OH^-)	50.10	3.992×10^{-4} (OH^-)	10.60
25.00	5.000×10^{-3}	26.00	5.200×10^{-3}	2.000×10^{-4} (OH^-)	51.00	3.922×10^{-3} (OH^-)	11.59
25.00	5.000×10^{-3}	50.00	1.000×10^{-2}	5.000×10^{-3} (OH^-)	75.00	6.667×10^{-2} (OH^-)	12.82

At this point we have exactly neutralized all of the acid with base and neither HCl nor NaOH are in excess. The solution contains only NaCl and we have already observed that NaCl solutions are neutral with a pH = 7.00. The pH at the equivalence point of all strong acid–strong base titrations is 7.00.

The pH of all strong acid–strong base titrations is 7.00 at the equivalence point.

After the equivalence point we will have an excess of base. As we did before, we calculate the moles of acid and the moles of base. Now the moles of base will be larger and we subtract the moles of acid to find the excess moles of base. This is divided by the total volume to obtain the [OH^-] from which the pH is calculated.

The equivalence point in a titration can be found using a pH meter by plotting pH versus volume of base added and noting the sharp rise in pH. If a pH meter is used, we don't have to use an indicator.

$$[OH^-] = \frac{M_B V_B - M_A V_A}{V_A + V_B}$$

Table 16.4 shows the results of calculating the pH of the solution in the receiving flask after the addition of further small volumes of the titrant. Figure 16.6 shows a plot of the pH of the solution in the receiving flask versus the volume of the 0.2000 *M* NaOH solution added—the *titration curve* for the titration of a strong acid with a strong base. Notice how the pH of the solution increases slowly until we are very close to the equivalence point, pH = 7.00. With the addition of a very small amount of base, the curve rises very sharply and then, almost as suddenly, levels off again to reflect just a gradual increase in pH.

Titration of a weak acid by a strong base uses four equations

The calculations for the titration of a weak acid by a strong base are a bit more complex than when both the acid and base are strong. This is because we have to consider the equilibria involving the weak acid and its conjugate base. As an example, let's do the calculations and

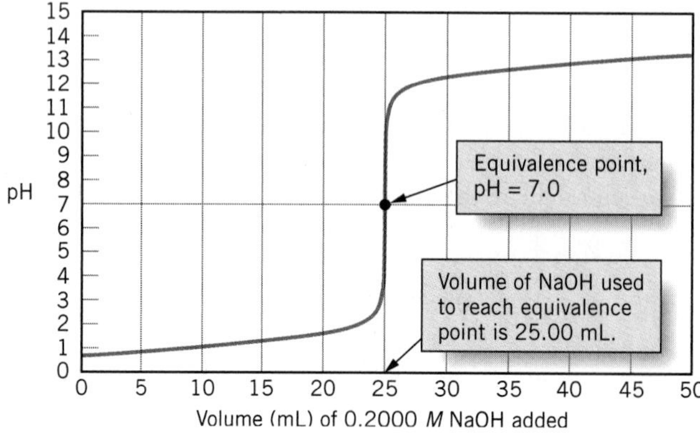

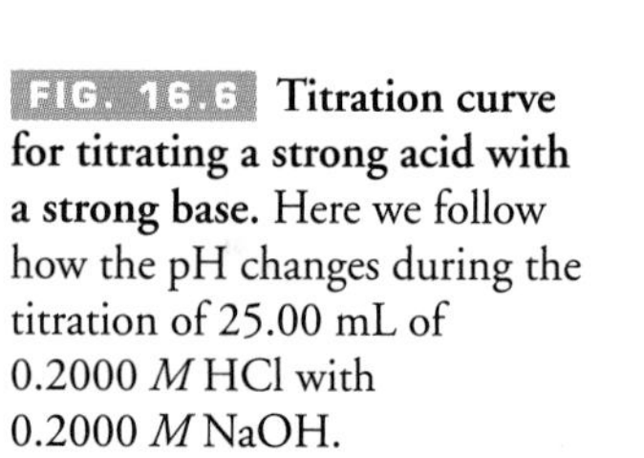
FIG. 16.6 Titration curve for titrating a strong acid with a strong base. Here we follow how the pH changes during the titration of 25.00 mL of 0.2000 *M* HCl with 0.2000 *M* NaOH.

draw a titration curve for the titration of 25.00 mL of 0.2000 M $HC_2H_3O_2$ with 0.2000 M NaOH. The same four points are calculated but they use different equations.

The molecular and net ionic equations we will use are

$$HC_2H_3O_2(aq) + NaOH(aq) \longrightarrow NaC_2H_3O_2(aq) + H_2O$$

$$HC_2H_3O_2(aq) + OH^-(aq) \longrightarrow C_2H_3O_2^-(aq) + H_2O$$

Because the concentrations of both solutions are the same, and because the acid and base react in a one-to-one mole ratio, we know that we will need exactly 25.00 mL of the base to reach the equivalence point. With this as background, let's look at what is involved in calculating the pH at various points along the titration curve.

At the start the solution is simply a solution of the weak acid $HC_2H_3O_2$. We must use K_a to calculate the pH. We have seen that in many situations the simplifying assumptions hold and the calculations are summarized as

$$K_a = \frac{[H^+][C_2H_3O_2^-]}{[HC_2H_3O_2]} = \frac{(x)(x)}{(0.2000 - x)} \approx \frac{(x)(x)}{(0.2000)} = 1.8 \times 10^{-5}$$

$[H^+] = \sqrt{K_a C_{HA}}$

$$x^2 = 3.6 \times 10^{-6} \quad \text{(rounded)}$$

$$x = [H^+] = 1.9 \times 10^{-3}$$

This result gives us a pH of 2.72. This is the pH before any NaOH is added.

Between the start and the equivalence point, as we add NaOH to the $HC_2H_3O_2$, the chemical reaction produces $C_2H_3O_2^-$, so the solution contains both $HC_2H_3O_2$ and $C_2H_3O_2^-$ (it is a buffer solution). Our equilibrium law can be rearranged to read

$$[H^+] = \frac{K_a[HC_2H_3O_2]}{[C_2H_3O_2^-]} = \frac{K_a(\text{mol } HC_2H_3O_2)}{(\text{mol } C_2H_3O_2^-)}$$

Now all we need to do is solve a limiting reactant calculation to determine the moles of acetic acid, $HC_2H_3O_2$, left after each addition of NaOH and at the same time calculate the moles of acetate ion, $C_2H_3O_2^-$, formed.

Let's see what happens when we add 10.00 mL of NaOH solution. The initial moles of acetic acid are

$$0.2000\ M\ HC_2H_3O_2 \times 0.02500\text{ L} = 5.000 \times 10^{-3}\text{ mol } HC_2H_3O_2$$

The moles of NaOH added are

$$0.2000\ M\ NaOH \times 0.01000\text{ L} = 2.000 \times 10^{-3}\text{ mol NaOH}$$

The moles of acetic acid left are

$$5.000 \times 10^{-3}\text{ mol} - 2.000 \times 10^{-3}\text{ mol} = 3.000 \times 10^{-3}\text{ mol } HC_2H_3O_2$$

From the chemical equations above, the moles of NaOH added are stoichiometrically equal to the moles of acetate ion formed. We therefore have 2.000×10^{-3} mol $C_2H_3O_2^-$. Entering the values for the K_a and the moles of acetic acid and acetate ions we get

$$[H^+] = \frac{(1.8 \times 10^{-5})(3.000 \times 10^{-3}\text{ mol } HC_2H_3O_2)}{(2.000 \times 10^{-3}\text{ mol } C_2H_3O_2^-)} = 2.7 \times 10^{-5}\ M$$

$[H^+] = K_a \dfrac{M_A V_A - M_B V_B}{M_B V_B}$

and the pH is 4.57.

At the equivalence point, all the $HC_2H_3O_2$ has reacted and the solution contains the salt $NaC_2H_3O_2$ (i.e., we have a mixture of the ions Na^+ and $C_2H_3O_2^-$).

We began with 5.000×10^{-3} mol of $HC_2H_3O_2$, so we now have 5.000×10^{-3} mol of $C_2H_3O_2^-$ (in 0.05000 L). The concentration of $C_2H_3O_2^-$ is

$$[C_2H_3O_2^-] = \frac{5.000 \times 10^{-3}\text{ mol}}{0.05000\text{ L}} = 0.1000\ M$$

Because $C_2H_3O_2^-$ is a base, we have to use K_b and write the chemical equation for the reaction of a base with water.

$$C_2H_3O_2^-(aq) + H_2O \rightleftharpoons HC_2H_3O_2(aq) + OH^-(aq)$$

$$K_b = \frac{[OH^-][HC_2H_3O_2]}{[C_2H_3O_2^-]} = \frac{K_w}{K_a} = 5.6 \times 10^{-10}$$

Substituting into the mass action expression gives

$$\frac{[OH^-][HC_2H_3O_2]}{[C_2H_3O_2^-]} = \frac{(x)(x)}{0.1000 - x} \approx \frac{(x)\,(x)}{0.1000} = 5.6 \times 10^{-10}$$

$$x^2 = 5.6 \times 10^{-11}$$

$$x = [OH^-] = 7.5 \times 10^{-6}$$

$[OH^-] = \sqrt{K_b \dfrac{M_A V_A}{V_A + V_B}}$

It is interesting that although we describe the overall reaction in the titration as *neutralization*, at the equivalence point the solution is slightly basic, not neutral.

The pOH calculates to be 5.12. Finally, the pH is (14.00 − 5.12) or 8.88. The pH at the equivalence point in this titration is 8.88, somewhat basic because the solution contains the Brønsted base, $C_2H_3O_2^-$. *In the titration of any weak acid with a strong base, the pH at the equivalence point will be greater than 7.*

After the equivalence point, the additional OH^- now shifts the following equilibrium to the left:

$$C_2H_3O_2^-(aq) + H_2O \rightleftharpoons HC_2H_3O_2(aq) + OH^-(aq)$$

The production of OH^- *by this route* is thus suppressed as we add more and more NaOH solution. The only source of OH^- that affects the pH is from the base added after the equivalence point. From here on, therefore, the pH calculations become identical to those for the last half of the titration curve for HCl and NaOH shown above. Each point is just a matter of calculating the additional number of moles of NaOH, calculating the new final volume, taking their ratio (to obtain the molar concentration), and then calculating pOH and pH.

$[OH^-] = \dfrac{M_B V_B - M_A V_A}{V_A + V_B}$

Table 16.5 summarizes the titration data, and the titration curve is given in Figure 16.7. Once again, notice how the change in pH occurs slowly at first, then shoots up through the equivalence point, and finally looks just like the last half of the curve in Figure 16.6.

Practice Exercise 40: When a weak acid such as acetic acid, $HC_2H_3O_2$, is titrated with KOH what are ALL of the species in solution at (a) the equivalence point, (b) at the start of the titration, (c) after the equivalence point, and (d) before the equivalence point? For each part list the species in order from highest concentration to lowest. (Hint: H_2O should be the first item in all your lists and either $[H^+]$ or $[OH^-]$ should be the last.)

Practice Exercise 41: Suppose we titrate 20.0 mL of 0.100 *M* $HCHO_2$ with 0.100 *M* NaOH. Calculate the pH (a) before any base is added, (b) when half of the $HCHO_2$ has been neutralized, (c) after a total of 15.0 mL of base has been added, and (d) at the equivalence point.

Practice Exercise 42: Suppose 30.0 mL of 0.15 *M* NaOH is added to 50.0 mL of 0.20 *M* $HCHO_2$. What will be the pH of the resulting solution?

TABLE 16.5 Titration of 25.00 mL of 0.2000 *M* $HC_2H_3O_2$ with 0.2000 *M* NaOH

Volume of Base Added (mL)	Molar Concentration of Species in Parentheses	pH
0.00	1.9×10^{-3} (H^+)	2.72
10.00	2.7×10^{-5} (H^+)	4.57
24.90	7.2×10^{-8} (H^+)	7.14
24.99	7.1×10^{-9} (H^+)	8.14
25.00	7.5×10^{-6} (OH^-)	8.88
25.01	4.0×10^{-5} (OH^-)	9.60
25.10	4.0×10^{-4} (OH^-)	10.60
26.00	3.9×10^{-3} (OH^-)	11.59
35.00	3.3×10^{-2} (OH^-)	12.52

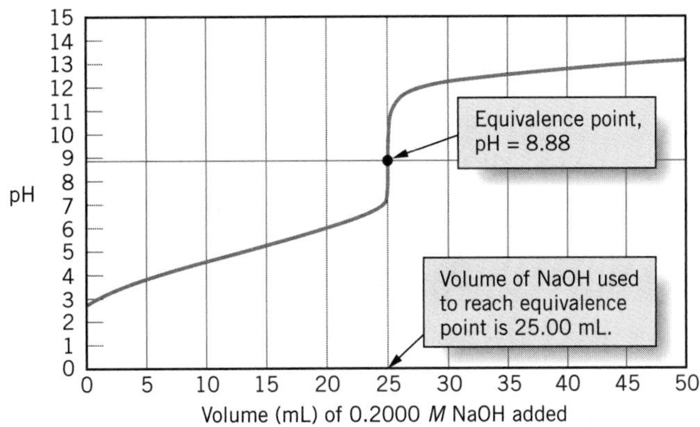

FIG. 16.7 Titration curve for titrating a weak acid with a strong base. In this titration, we follow the pH as 25.00 mL of 0.2000 *M* acetic acid is titrated with 0.2000 *M* NaOH.

Weak base–strong acid titrations also use four equations

The calculations involved here are nearly identical to those of the weak acid–strong base titration. We will review what is required but will not actually perform the calculations.

If we titrate 25.00 mL of 0.2000 *M* NH_3 with 0.2000 *M* HCl, the molecular and net ionic equations are

$$NH_3(aq) + HCl(aq) \longrightarrow NH_4Cl(aq)$$

$$NH_3(aq) + H^+(aq) \longrightarrow NH_4^+(aq)$$

Before the titration begins, the solution is simply a solution of the weak base, NH_3. Since the only solute is a base, we must use K_b to calculate the pH.

During the titration but before the equivalence point, as we add HCl to the NH_3, the neutralization reaction produces NH_4^+, so the solution contains both NH_3 and NH_4^+ (it is a buffer solution). We do a limiting reactant calculation to determine the moles of ammonia left and the moles of ammonium ions produced by the addition of HCl. We use the equilibrium law to perform the calculation as we did with the weak acid.

At the equivalence point, all the NH_3 has reacted and the solution contains the salt NH_4Cl. We calculate the concentration of ammonium ions and then use the techniques we developed to determine the pH of the conjugate acid of a weak base.

After the equivalence point, the H^+ introduced by further addition of HCl has nothing with which to react, so it causes the solution to become more and more acidic. The concentration of H^+ calculated from the excess of HCl is used to calculate the pH.

Figure 16.8 illustrates the titration curve for this system.

Titration curves for diprotic acids show two equivalence points

When a weak diprotic acid such as ascorbic acid (vitamin C) is titrated with a strong base, there are two protons to be neutralized and there are two equivalence points. Provided that

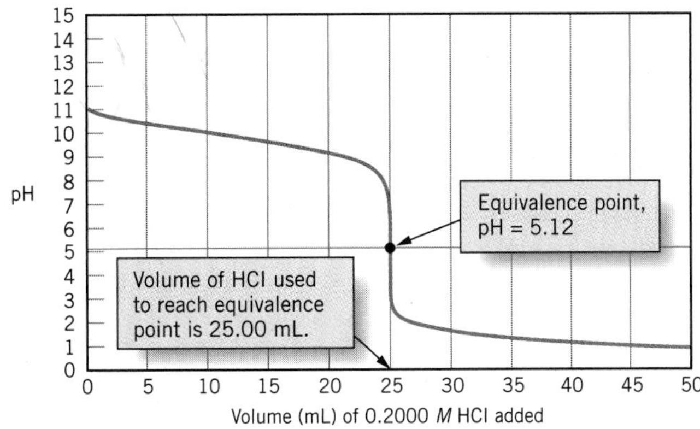

FIG. 16.8 Titration curve for the titration of a weak base with a strong acid. Here we follow the pH as 25.00 mL of 0.2000 *M* NH_3 is titrated with 0.2000 *M* HCl.

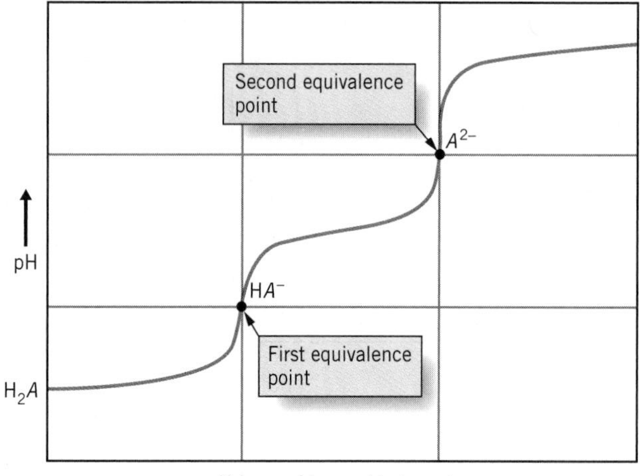

FIG. 16.9 Titration of a diprotic acid, H_2A, by a strong base. As each equivalence point is reached, there is a sharp rise in the pH.

the values of K_{a_1} and K_{a_2} differ by several powers of 10, the neutralization takes place stepwise and the resulting titration curve shows two sharp increases in pH. We won't perform the calculations here, because they're complex, but instead just look at the general shape of the titration curve, which is shown in Figure 16.9.

Acid–base indicators are weak acids

Most dyes that work as acid–base indicators are also weak acids. Therefore, let's represent an indicator by the formula H*In*. In its un-ionized state, H*In* has one color. Its conjugate base, In^-, has a different color, the more strikingly different the better. In solution, the indicator is involved in a typical acid–base equilibrium:

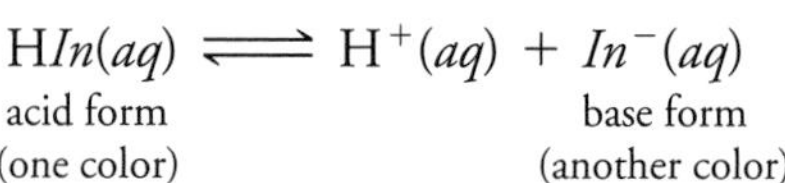

$$\underset{\substack{\text{acid form} \\ \text{(one color)}}}{HIn(aq)} \rightleftharpoons H^+(aq) + \underset{\substack{\text{base form} \\ \text{(another color)}}}{In^-(aq)}$$

The corresponding acid ionization constant, K_{In}, is given by

$$K_{In} = \frac{[H^+][In^-]}{[HIn]}$$

In a strongly acidic solution, when the H^+ concentration is high, the equilibrium is shifted to the left and most of the indicator exists in its "acid form." Under these conditions, the color we observe is that of H*In*. If the solution is made basic, the H^+ concentration drops and the equilibrium shifts to the right, toward In^-, and the color we observe is that of the "base form" of the indicator.

The *observed* change in color for an indicator actually occurs gradually over a range of pH values. This is because of the human eye's limited ability to discern color changes. Sometimes a change of as much as 2 pH units is needed for some indicators—more for litmus—before the eye notices the color change. This is why tables of acid–base indicators, such as Table 15.5 on page 630, provide approximate pH ranges for the color changes.

How acid-base indicators work

In a typical acid–base titration, you've seen that as we pass the equivalence point, there is a sudden and large change in the pH. For example, in the titration of HCl with NaOH described earlier, the pH just one-half drop before the equivalence point (when 24.97 mL of the base has been added) is 3.92. Just one drop later (when 25.03 mL of base has been added) we have passed the equivalence point and the pH has risen to 10.08. This large swing in pH (from 3.92 to 10.08) causes a sudden shift in the position of equilibrium for the indicator, and we go from a condition where most of the indicator is in its acid form to a condition in which most is in the base form. This is observed visually as a change in color from that of the acid form to that of the base form.

SUMMARY

Acid and Base Ionization Constants. A weak acid HA ionizes according to the general equation

$$HA + H_2O \rightleftharpoons H_3O^+ + A^-$$

or more simply,

$$HA \rightleftharpoons H^+ + A^-$$

The equilibrium constant is called the **acid ionization constant, K_a** (sometimes called an *acid dissociation constant*).

$$K_a = \frac{[H^+][A^-]}{[HA]}$$

A weak base B ionizes by the general equation

$$B + H_2O \rightleftharpoons BH^+ + OH^-$$

The equilibrium constant is called the **base ionization constant, K_b** (sometimes called a *base dissociation constant*).

$$K_b = \frac{[OH^-][BH^+]}{[B]}$$

The smaller the values of K_a (or K_b), the weaker are the substances as Brønsted acids (or bases).

Another way to compare the relative strengths of acids or bases is to use the negative logarithms of K_a and K_b, called **pK_a** and **pK_b**, respectively.

$$pK_a = -\log K_a \qquad pK_b = -\log K_b$$

The *smaller* the pK_a or pK_b, the *stronger* is the acid or base. For a conjugate acid–base pair,

$$K_a \times K_b = K_w$$

and

$$pK_a + pK_b = 14.00 \qquad \text{(at 25 °C)}$$

Equilibrium Calculations—Determining K_a and K_b. The values of K_a and K_b can be obtained from initial concentrations of the acid or base and either the pH of the solution or the percentage ionization of the acid or base. The measured pH gives the equilibrium value for $[H^+]$. The **percentage ionization** is defined as

$$\text{Percentage ionization} = \frac{\text{amount ionized}}{\text{amount available}} \times 100\%$$

Equilibrium Calculations—Determining Equilibrium Concentrations when K_a or K_b Is Known. Problems fall into one of three categories: (1) the only solute is a weak acid (we must use the K_a expression), (2) the only solute is a weak base (we must use the K_b expression), and (3) the solution contains both a weak acid and its conjugate base (we can use either K_a or K_b).

When the initial concentration of the acid (or base) is larger than 400 times the value of K_a (or K_b), it is safe to use initial concentrations of acid or base as though they were equilibrium values in the mass action expression. When this approximation cannot be used, we can use the quadratic formula.

Ions as Acids or Bases. A solution of a salt is acidic if the cation is acidic but the anion is neutral. Metal ions with high charge densities generally are acidic, but those of Groups IA and IIA (except Be^{2+}) are not. Cations such as NH_4^+, which are the conjugate acids of weak *molecular* bases, are themselves proton donors and are acidic.

When the anion of a salt is the conjugate base of a *weak* acid, the anion is a weak base. Anions of strong acids, such as Cl^- and NO_3^-, are such weak Brønsted bases that they cannot affect the pH of a solution.

If the salt is derived from a weak acid *and* a weak base, its net effect on pH has to be determined on a case by case basis by determining which of the two ions is the stronger.

Buffers. A solution that contains both a weak acid and a weak base (usually an acid–base conjugate pair) is called a **buffer,** because it is able to absorb $[H^+]$ from a strong acid or $[OH^-]$ from a strong base without suffering large changes in pH. For the general acid–base pair, HA and A^-, the following reactions neutralize H^+ and OH^-.

When H^+ is added to the buffer: $A^- + H^+ \longrightarrow HA$

When OH^- is added to the buffer:

$$HA + OH^- \longrightarrow A^- + H_2O$$

The pH of a buffer is controlled by the ratio of weak acid to weak base, expressed either in terms of molarities or moles.

$$[H^+] = K_a \times \frac{[HA]}{[A^-]} = K_a \times \frac{\text{moles } HA}{\text{moles } A^-}$$

Because the $[H^+]$ is determined by the mole ratio of HA to A^-, dilution does not change the pH of a buffer. The **Henderson–Hasselbalch equation,**

$$pH = pK_a + \log \frac{[A^-]}{[HA]}$$

can be used to calculate the pH directly from the pK_a of the acid and the concentrations of the conjugate acid, HA, and base, A^-.

In performing buffer calculations, the usually valid simplifications are

$$[HA]_{\text{equilib}} \approx [HA]_{\text{initial}}$$

$$[A^-]_{\text{equilib}} \approx [A^-]_{\text{initial}}$$

The value of $[A^-]_{\text{initial}}$ is found from the molarity of the *salt* of the weak acid in the buffer. Once $[H^+]$ is found by this calculation, the pH is calculated.

Buffer calculations can also be performed using K_b for the weak base component. Similar equations apply, the principal difference being that it is the OH^- concentration that is calculated. Thus, using K_b we have

$$[OH^-] = K_b \times \frac{[\text{conjugate base}]}{[\text{conjugate acid}]}$$

$$= K_b \times \frac{\text{moles conjugate base}}{\text{moles conjugate acid}}$$

A solution is buffered when the pK_a of the acid member of the buffer pair lies within ±1 unit of the solution's pH.

Equilibria in Solutions of Polyprotic Acids. A polyprotic acid has a K_a value for the ionization of each of its hydrogen ions. Successive values of K_a often differ by a factor of 10^4 to 10^5. This allows us to calculate the pH of a solution of a polyprotic acid by using just the value of K_{a_1}. If the polyprotic acid is the only solute, the anion formed in the second step of the ionization, A^{2-}, has a concentration equal to K_{a_2}.

The anions of weak polyprotic acids are bases that react with water in successive steps, the last of which has the molecular polyprotic acid as a product. For a diprotic acid, $K_{b_1} = K_w/K_{a_2}$ and $K_{b_2} = K_w/K_{a_1}$. Usually, $K_{b_1} \gg K_{b_2}$, so virtually all the OH^- produced in the solution comes from the first step. The pH of the solution can be calculated using just K_{b_1} and the reaction

$$A^{2-} + H_2O \rightleftharpoons HA^- + OH^-$$

Acid–Base Titrations and Indicators. Neutralization titrations can be experimentally measured using a pH meter or they can be calculated using the principles developed in this chapter. Both methods yield the same result, a titration curve. The titration curve contains important information about the chemical system including the equivalence point and equilibrium constant data.

Each calculated titration curve has four distinct calculations:

1. The starting pH of a solution that contains only an acid or base.
2. The pH values for titrant volumes between the start and the equivalence points.
3. The pH at the equivalence point.
4. The pH values for titrant volumes past the equivalence point.

A graph of the pH values versus the volume of titrant gives us a titration curve. Titration curves show a sharp change in pH at the equivalence point. Titration curves also have distinct differences in shape when strong acid–strong base titrations are compared to weak acid–weak base titrations.

Acid–base indicators are weak acids in which the conjugate acid and base have different colors. The sudden change in pH at the equivalence point causes a rapid shift from one color of the indicator to the other. If matched to the equivalence point, a pH indicator will change color abruptly when a titration reaches the equivalence point.

TOOLS FOR PROBLEM SOLVING

The concepts that you've learned in this chapter can be applied as tools in solving problems. Study each one carefully so that you know what each is used for. When faced with solving a problem, recall what each tool does and consider whether it will be helpful in finding a solution. This will aid you in selecting the tools you need. If necessary, refer to this table when working on the Review Exercises and Review Problems that follow.

General equations for the ionization of a weak acid *(page 643)* These are the two general equations for the ionization of weak acids in aqueous solution. We use these as a general outline for the reaction from which we can construct the correct K_a equation

$$HA + H_2O \rightleftharpoons H_3O^+ + A^- \quad \text{or} \quad HA \rightleftharpoons H^+ + A^-$$

General equation for the ionization of a weak base *(page 646)* This is the general equation for the ionization of a weak base in aqueous solution. We use this as a general outline for the reaction from which we can construct the correct K_b equation

$$B + H_2O \rightleftharpoons BH^+ + OH^-$$

Inverse relationships of acid and base constants *(page 648)* These two equations are used to convert between K_a and K_b or between pK_a and pK_b values

$$K_a \times K_b = K_w \quad \text{and} \quad pK_a + pK_b = pK_w = 14.00$$

Percentage ionization *(page 649)* Use this equation to calculate the percentage ionization from the initial concentration of acid (or base) and the change in the concentrations of the ions. If the percentage ionization is known, you can calculate the change in the concentration of the acid or base and then use that information to calculate the K_a (or K_b).

$$\text{Percentage ionization} = \frac{\text{moles per liter ionized}}{\text{moles per liter available}} \times 100\%$$

Identification of acidic cations and basic anions *(pages 657 and 658)* We use the concepts developed here to determine whether a salt solution is acidic, basic, or neutral. This is also the first step in calculating the pH of a salt solution.

Criterion for applying simplifications in acid–base equilibrium calculations *(page 663)* We use this to determine whether initial concentrations can be used as though they are equilibrium values in the mass action expression when working acid–base equilibrium problems. If the condition fails, then we have to use the quadratic equation or the method of successive approximations *(page 665)*.

$$[X]_{\text{initial}} > 400 \times K$$

Reactions when H^+ or OH^- are added to a buffer *(page 666)* These reactions determine how the concentrations of conjugate acid and base change when a strong acid or strong base is added to a buffer. Adding H^+ decreases $[A^-]$ and increases $[HA]$; adding OH^- decreases $[HA]$ and increases $[A^-]$.

Ionization of polyprotic acids *(page 673)* Polyprotic acids have more than one ionizable proton in their formulas. Each proton will ionize sequentially and have a form similar to the monoprotic acid ionization tool above.

Four stages of a titration curve *(page 678)* A titration curve is a graph of the pH as a function of the volume of titrant added to a sample. To plot such a curve, there are four different stages of calculation. They are (1) the starting point, (2) the stage from the start to the equivalence point, (3) the equivalence point, and (4) after the equivalence point.

How acid-base indicators work *(page 684)* Acid-base indicators are weak acids that have one color for the conjugate acid and a different color for the conjugate base. Whether the indicator is in the conjugate acid or conjugate base form depends on the pH. The rapid, large change in pH at a titration equivalence point results in a distinct color change that signals a titration end point.

QUESTIONS, PROBLEMS, AND EXERCISES

Answers to problems whose numbers are printed in color are given in Appendix B. More challenging problems are marked with asterisks. ILW = Interactive Learningware solution is available at www.wiley.com/college/brady. OH = an Office Hours video is available for this problem.

REVIEW QUESTIONS

Acid and Base Ionization Constants: K_a, K_b, pK_a, and pK_b

16.1 Write the general equation for the ionization of a weak acid in water. Give the equilibrium law corresponding to K_a.

16.2 Write the chemical equation for the ionization of each of the following weak acids in water. (Some are polyprotic acids; for these write only the equation for the first step in the ionization.)

(a) HNO_2 (c) $HAsO_4^{2-}$
(b) H_3PO_4 (d) $(CH_3)_3NH^+$

16.3 For each of the acids in Question 16.2, write the appropriate K_a expression.

16.4 Write the general equation for the ionization of a weak base in water. Give the equilibrium law corresponding to K_b.

16.5 Write the chemical equation for the ionization of each of the following weak bases in water.

(a) $(CH_3)_3N$ (c) NO_2^-
(b) AsO_4^{3-} (d) $(CH_3)_2N_2H_2$

16.6 For each of the bases in Question 16.5, write the appropriate K_b expression.

16.7 The pK_a of HCN is 9.21 and that of HF is 3.17. Which is the stronger Brønsted base, CN^- or F^-?

16.8 Write the structural formulas for the conjugate acids of the following:

(a) $CH_3-CH_2-\ddot{N}(CH_3)-H$ (b) pyridine ($C_5H_5N:$) (c) $H-\ddot{O}-\ddot{N}(H)-H$

16.9 Write the structural formulas for the conjugate bases of the following:

(a) $C_6H_5-NH_3^+$ (b) $[CH_3-N(CH_3)_2-H]^+$ (c) $[H-\ddot{N}(H)-N(H)(H)-H]^+$

16.10 How is percentage ionization defined? Write the equation.

16.11 If a weak acid is 1.2% ionized in a 0.22 *M* solution, what is the molar concentration of H^+ in the solution?

16.12 What criterion do we use to determine whether or not the equilibrium concentration of an acid or base will be effectively the same as its initial concentration when we calculate the pH of the solution?

16.13 For which of the following are we permitted to make the assumption that the equilibrium concentration of the acid or base is the same as the initial concentration when we calculate the pH of the solution specified?

(a) 0.020 *M* $HC_2H_3O_2$ (c) 0.002 *M* N_2H_4
(b) 0.10 *M* CH_3NH_2 (d) 0.050 *M* $HCHO_2$

Acid–Base Properties of Salt Solutions

16.14 Aspirin is acetylsalicylic acid, a monoprotic acid whose K_a value is 3.27×10^{-4}. Does a solution of the sodium salt of aspirin in water test acidic, basic, or neutral? Explain.

16.15 The K_b value of the oxalate ion, $C_2O_4^{2-}$, is 1.9×10^{-10}. Is a solution of $K_2C_2O_4$ acidic, basic, or neutral? Explain.

16.16 Consider the following compounds and suppose that 0.5 *M* solutions are prepared of each: NaI, KF, $(NH_4)_2SO_4$, KCN, $KC_2H_3O_2$, $CsNO_3$, and KBr. Write the *formulas* of those that have solutions that are (a) acidic, (b) basic, and (c) neutral.

16.17 Will an aqueous solution of $AlCl_3$ turn litmus red or blue? Explain.

16.18 A solution of hydrazinium acetate is slightly acidic. Without looking at the tables of equilibrium constants, is K_a for acetic acid larger or smaller than K_b for hydrazine? Justify your answer.

16.19 When ammonium nitrate is added to a suspension of magnesium hydroxide in water, the $Mg(OH)_2$ dissolves. Write a net ionic equation to show how this occurs.

Buffers

16.20 Write ionic equations that illustrate how each pair of compounds can serve as a buffer pair.

(a) H_2CO_3 and $NaHCO_3$ (the "carbonate" buffer in blood)

(b) NaH_2PO_4 and Na_2HPO_4 (the "phosphate" buffer inside body cells)

(c) NH_4Cl and NH_3

16.21 Bicarbonate ion is able to act as a buffer all by itself. Write chemical equations that show how this ion reacts with (a) H^+ and (b) OH^-.

Ionization of Polyprotic Acids

16.22 When sulfur dioxide, an air pollutant from the burning of sulfur-containing coal or oil, dissolves in water, an acidic solution is formed that can be viewed as containing sulfurous acid, H_2SO_3.

$$H_2O + SO_2(g) \rightleftharpoons H_2SO_3(aq)$$

Write the expression for K_{a_1} and K_{a_2} for sulfurous acid.

16.23 Citric acid, found in citrus fruits, is a triprotic acid, $H_3C_6H_5O_7$. Write chemical equations for the three-step ionization of this acid in water and the corresponding K_a expressions.

16.24 What simplifying assumptions do we usually make in working problems involving the ionization of polyprotic acids? Why are they usually valid? Under what conditions do they fail?

Salts of Polyprotic Acids

16.25 Write the equations for the chemical equilibria that exist in solutions of (a) Na_2SO_3, (b) Na_3PO_4, and (c) $K_2C_4H_4O_6$.

16.26 What simplifying assumptions do we usually make in working problems involving equilibria of salts of polyprotic acids? Why are they usually valid? Under what conditions do they fail?

Titrations and Acid–Base Indicators

16.27 Define the terms *equivalence point* and *end point* as they apply to an acid–base titration.

16.28 When a formic acid solution is titrated with sodium hydroxide, will the solution be acidic, neutral, or basic at the equivalence point?

16.29 When a solution of hydrazine is titrated by hydrochloric acid, will the solution be acidic, neutral, or basic at the equivalence point?

16.30 Qualitatively, describe how an acid–base indicator works. Why do we want to use a minimum amount of indicator in a titration?

16.31 If you use methyl orange in the titration of $HC_2H_3O_2$ with NaOH, will the end point of the titration correspond to the equivalence point? Justify your answer.

REVIEW PROBLEMS

Acid and Base Ionization Constants: K_a, K_b, pK_a, and pK_b

16.32 The K_a for HF is 6.8×10^{-4}. What is the K_b for F^-?

16.33 The barbiturate ion, $C_4H_3N_2O_3^-$, has $K_b = 1.0 \times 10^{-10}$. What is K_a for barbituric acid?

16.34 Lactic acid, $HC_3H_5O_3$, is responsible for the sour taste of sour milk. At 25 °C, its $K_a = 1.4 \times 10^{-4}$. What is the K_b of its conjugate base, the lactate ion, $C_3H_5O_3^-$?

OH **16.35** Iodic acid, HIO_3, has a pK_a of 0.77. (a) What is the formula and the K_b of its conjugate base? (b) Is its conjugate base a stronger or a weaker base than the acetate ion?

Equilibrium Calculations

16.36 A 0.20 *M* solution of a weak acid, *HA*, has a pH of 3.22. What is the percentage ionization of the acid? What is the value of K_a for the acid?

16.37 If a weak base is 0.030% ionized in 0.030 *M* solution, what is the pH of the solution? What is the value of K_b for the base?

OH **16.38** Periodic acid, HIO_4, is an important oxidizing agent and a moderately strong acid. In a 0.10 *M* solution, $[H^+] = 3.8 \times 10^{-2}$ mol L^{-1}. Calculate the K_a and pK_a for periodic acid.

16.39 Chloroacetic acid, $HC_2H_2O_2Cl$, is a stronger monoprotic acid than acetic acid. In a 0.10 *M* solution, the pH is 1.96. Calculate the K_a and pK_a for chloroacetic acid.

ILW **16.40** Ethylamine, $CH_3CH_2NH_2$, has a strong, pungent odor similar to that of ammonia. Like ammonia, it is a Brønsted base. A 0.10 *M* solution has a pH of 11.86. Calculate the K_b and pK_b for ethylamine. What is the percentage ionization of ethylamine in the solution?

16.41 Hydroxylamine, $HONH_2$, like ammonia, is a Brønsted base. A 0.15 *M* solution has a pH of 10.12. What is the K_b and pK_b for hydroxylamine? What is the percentage ionization of the $HONH_2$?

ILW **16.42** What are the concentrations of all the solute species in 0.150 *M* lactic acid, $HC_3H_5O_2$? What is the pH of the solution? This acid has $K_a = 1.4 \times 10^{-4}$.

16.43 What are the concentrations of all the solute species in a 1.0 *M* solution of hydrogen peroxide, H_2O_2? What is the pH of the solution? For H_2O_2, $K_a = 1.8 \times 10^{-12}$.

16.44 Codeine, a cough suppressant extracted from crude opium, is a weak base with a pK_b of 5.79. What will be the pH of a 0.020 *M* solution of codeine? (Use *Cod* as a symbol for codeine.)

16.45 Pyridine, C_5H_5N, is a bad-smelling liquid that is a weak base in water. Its pK_b is 8.82. What is the pH of a 0.20 *M* aqueous solution of the compound?

16.46 A solution of acetic acid has a pH of 2.54. What is the concentration of acetic acid in this solution?

16.47 How many moles of NH_3 must be dissolved in water to give 500 mL of solution with a pH of 11.22?

Equilibrium Calculations when Simplifications Fail

16.48 What is the pH of a 0.0050 *M* solution of sodium cyanide?

16.49 What is the pH of a 0.020 *M* solution of chloroacetic acid, for which $K_a = 1.36 \times 10^{-3}$?

16.50 The compound *para*-aminobenzoic acid (PABA) is a powerful sun screening agent whose salts were once used widely in sun tanning and screening lotions. The parent acid, which we may symbolize as H-*Paba*, is a weak acid with a pK_a of 4.92 (at 25 °C). What is the $[H^+]$ and pH of a 0.030 *M* solution of the acid?

16.51 Barbituric acid, $HC_4H_3N_2O_3$ (which we will abbreviate H-*Bar*), was discovered by the Nobel prize–winning organic chemist Adolph von Baeyer and named after his friend, Barbara. It is the parent compound of widely used sleeping drugs, the barbiturates. Its pK_a is 4.01. What is the $[H^+]$ and pH of a 0.020 *M* solution of H-*Bar*?

Acid–Base Properties of Salt Solutions

16.52 Calculate the pH of 0.20 *M* NaCN. What is the concentration of HCN in the solution?

16.53 Calculate the pH of 0.40 *M* KNO_2. What is the concentration of HNO_2 in the solution?

ILW **16.54** Calculate the pH of 0.15 *M* CH_3NH_3Cl. For methylamine, CH_3NH_2, $K_b = 4.4 \times 10^{-4}$.

16.55 Calculate the pH of 0.10 *M* hydrazinium chloride, N_2H_5Cl.

16.56 A 0.18 *M* solution of the sodium salt of nicotinic acid (also known pharmaceutically as niacin) has a pH of 9.05. What is the value of K_a for nicotinic acid?

16.57 A weak base *B* forms the salt *B*HCl, composed of the ions BH^+ and Cl^-. A 0.15 *M* solution of the salt has a pH of 4.28. What is the value of K_b for the base *B*?

OH ***16.58** Liquid chlorine bleach is really nothing more than a solution of sodium hypochlorite, NaOCl, in water. Usually, the concentration is approximately 5% NaOCl by weight. Use this information to calculate the approximate pH of a bleach solution, assuming no other solutes are in the solution except NaOCl. (Assume the bleach has a density of 1.0 g/mL.)

***16.59** The conjugate acid of a molecular base has the hypothetical formula, BH^+, which has a pK_a of 5.00. A solution of a salt of this cation, *BHY*, tests slightly basic. Will the conjugate acid of Y^-, H*Y*, have a pK_a greater than 5.00 or less than 5.00? Explain.

Buffers

16.60 What is the pH of a solution that contains 0.15 *M* $HC_2H_3O_2$ and 0.25 *M* $C_2H_3O_2^-$? Use $K_a = 1.8 \times 10^{-5}$ for $HC_2H_3O_2$.

16.61 Rework the preceding problem using the K_b for the acetate ion. (Be sure to write the proper chemical equation and equilibrium law.)

OH **16.62** A buffer is prepared containing 0.25 *M* NH_3 and 0.45 *M* NH_4^+. Calculate the pH of the buffer using the K_b for NH_3.

16.63 Calculate the pH of the buffer in the preceding problem using the K_a for NH_4^+.

16.64 Suppose 25.0 mL of 0.10 *M* HCl is added to a 250 mL portion of a buffer composed of 0.25 *M* NH_3 and 0.20 *M* NH_4Cl. By how much will the concentrations of the NH_3 and NH_4^+ ions change after the addition of the strong acid?

16.65 A student added 100 mL of 0.10 *M* NaOH to 250 mL of a buffer that contained 0.15 *M* $HC_2H_3O_2$ and 0.25 *M* $C_2H_3O_2^-$. By how much did the concentrations of $HC_2H_3O_2$ and $C_2H_3O_2^-$ change after the addition of the strong base?

16.66 By how much will the pH change if 0.025 mol of HCl is added to 1.00 L of the buffer in Problem 16.60?

16.67 By how much will the pH change if 25.0 mL of 0.20 *M* NaOH is added to 500 mL of the buffer in Problem 16.60?

16.68 By how much will the pH change if 0.040 mol of HCl is added to 1.00 L of the buffer in Problem 16.62?

16.69 By how much will the pH change if 35 mL of 0.10 *M* KOH is added to 200 mL of the buffer in Problem 16.62?

16.70 How many grams of sodium acetate, $NaC_2H_3O_2$, would have to be added to 1.0 L of 0.15 *M* acetic acid (pK_a 4.74) to make the solution a buffer for pH 4.00?

16.71 How many grams of sodium formate, $NaCHO_2$, would have to be dissolved in 1.0 L of 0.12 *M* formic acid (pK_a 3.74) to make the solution a buffer for pH 3.50?

16.72 Suppose 30.00 mL of 0.100 *M* HCl is added to an acetate buffer prepared by dissolving 0.100 mol of acetic acid and 0.110 mol of sodium acetate in 100 mL of solution. What are the initial and final pH values? What would be the pH if the same amount of HCl solution were added to 100 mL of pure water?

16.73 How many milliliters of 0.15 *M* HCl would have to be added to the original 100 mL of the buffer described in Problem 16.72 to make the pH decrease by 0.05 pH unit? How many milliliters of the same HCl solution would, if added to 100 mL of pure water, make the pH decrease by 0.05 pH unit?

Solutions of Polyprotic Acids

ILW **16.74** Calculate the concentrations of all the solute species in a 0.15 *M* solution of ascorbic acid (vitamin C). What is the pH of the solution?

16.75 Tellurium, in the same family as sulfur, forms an acid analogous to sulfuric acid and called telluric acid. It exists, however, as H_6TeO_6 (which looks like the formula $H_2TeO_4 + 2H_2O$). It is a diprotic acid with $K_{a_1} = 2 \times 10^{-8}$ and $K_{a_2} = 1 \times 10^{-11}$. Calculate the concentrations of H^+, $H_5TeO_6^-$, and $H_4TeO_6^{2-}$ in a 0.25 *M* solution of H_6TeO_6. What is the pH of the solution?

OH **16.76** Calculate the concentrations of all of the solute species involved in the equilibria in a 2.0 *M* solution of H_3PO_4. Calculate the pH of the solution.

16.77 What is the pH of a 0.25 *M* solution of arsenic acid, H_3AsO_4? In this solution, what are the concentrations of $H_2AsO_4^-$ and $HAsO_4^{2-}$?

***16.78** Phosphorous acid, H_3PO_3, is actually a diprotic acid for which $K_{a_1} = 3.0 \times 10^{-2}$ and $K_{a_2} = 1.6 \times 10^{-7}$. What are the values of $[H^+]$, $[H_2PO_3^-]$, and $[HPO_3^{2-}]$ in a 1.0 *M* solution of H_3PO_3? What is the pH of the solution?

*16.79 What is the pH of a 0.20 *M* solution of oxalic acid, $H_2C_2O_4$?

Solutions of Salts of Polyprotic Acids

16.80 Calculate the pH of 0.24 *M* Na_2SO_3. What are the concentrations of HSO_3^- and H_2SO_3 in the solution?

16.81 Calculate the pH of 0.33 *M* K_2CO_3. What are the concentrations of HCO_3^- and H_2CO_3 in the solution?

16.82 Sodium citrate, $Na_3C_6H_5O_7$, is used as an anticoagulant in the collection of blood. What is the pH of a 0.10 *M* solution of this salt?

16.83 What is the pH of a 0.25 *M* solution of sodium oxalate, $Na_2C_2O_4$?

*16.84 What is the pH of a 0.50 *M* solution of Na_3PO_4? In this solution, what are the concentrations of HPO_4^{2-}, $H_2PO_4^-$, and H_3PO_4?

*16.85 The pH of a 0.10 *M* Na_2CO_3 solution is adjusted to 12.00 using a strong base. What is the concentration of CO_3^{2-} in this solution?

Acid–Base Titrations

16.86 When 50 mL of 0.050 *M* formic acid, $HCHO_2$, is titrated with 0.050 *M* sodium hydroxide, what is the pH at the equivalence point? (Be sure to take into account the change in volume during the titration.) Select a good indicator for this titration from Table 15.5.

OH 16.87 When 25 mL of 0.12 *M* aqueous ammonia is titrated with 0.12 *M* hydrobromic acid, what is the pH at the equivalence point? Select a good indicator for this titration from Table 15.5.

16.88 What is the pH of a solution prepared by mixing 25.0 mL of 0.180 *M* $HC_2H_3O_2$ with 40.0 mL of 0.250 *M* NaOH?

16.89 What is the pH of a solution prepared by mixing exactly 25.0 mL of 0.200 *M* $HC_2H_3O_2$ with 15.0 mL of 0.400 *M* KOH?

*16.90 For the titration of 75.00 mL of 0.1000 *M* acetic acid with 0.1000 *M* NaOH, calculate the pH (a) before the addition of any NaOH solution, (b) after 25.00 mL of the base has been added, (c) after half of the $HC_2H_3O_2$ has been neutralized, and (d) at the equivalence point.

*16.91 For the titration of 50.00 mL of 0.1000 *M* ammonia with 0.1000 *M* HCl, calculate the pH (a) before the addition of any HCl solution, (b) after 20.00 mL of the acid has been added, (c) after half of the NH_3 has been neutralized, and (d) at the equivalence point.

ADDITIONAL EXERCISES

16.92 Calculate the percentage ionization of acetic acid in solutions having concentrations of 1.0 *M*, 0.10 *M*, and 0.010 *M*. How does the percentage ionization of a weak acid change as the acid becomes more dilute?

16.93 What is the pH of a solution that is 0.100 *M* in HCl and also 0.125 *M* in $HC_2H_3O_2$? What is the concentration of acetate ion in this solution?

16.94 A solution is prepared by mixing 300 mL of 0.500 *M* NH_3 and 100 mL of 0.500 *M* HCl. Assuming that the volumes are additive, what is the pH of the resulting mixture?

16.95 A solution is prepared by dissolving 15.0 g of pure $HC_2H_3O_2$ and 25.0 g of $NaC_2H_3O_2$ in 750 mL of solution (the final volume). (a) What is the pH of the solution? (b) What would the pH of the solution be if 25.00 mL of 0.25 *M* NaOH were added? (c) What would the pH be if 25.0 mL of 0.40 *M* HCl were added to the original 750 mL of buffer solution?

*16.96 For an experiment involving what happens to the growth of a particular fungus in a slightly acidic medium, a biochemist needs 250 mL of an acetate buffer with a pH of 5.12. The buffer solution has to be able to hold the pH to within ±0.10 pH unit of 5.12 even if 0.0100 mol of NaOH or 0.0100 mol of HCl enters the solution.

(a) What is the minimum number of grams of acetic acid and of sodium acetate dihydrate that must be used to prepare the buffer?

(b) Describe the buffer by giving its molarity in acetic acid and its molarity in sodium acetate.

(c) What is the pH of an unbuffered solution made by adding 0.0100 mol of NaOH to 250 mL of pure water?

(d) What is the pH of an unbuffered solution made by adding 0.0100 mol of HCl to 250 mL of pure water?

16.97 Predict whether the pH of 0.120 *M* NH_4CN is greater than, less than, or equal to 7.00. Give your reasons.

*16.98 What is the pH of a 4.5×10^{-2} *M* solution of ammonium acetate, $NH_4C_2H_3O_2$?

*16.99 What is the approximate freezing point of a 0.50 *M* solution of dichloroacetic acid, $HC_2HO_2Cl_2$ ($K_a = 5.0 \times 10^{-2}$). Assume the density of the solution is 1.0 g/mL.

*16.100 How many milliliters of ammonia gas measured at 25 °C and 740 torr must be dissolved in 250 mL of 0.050 *M* HNO_3 to give a solution with a pH of 9.26?

*16.101 The hydrogen sulfate ion, HSO_4^-, is a moderately strong Brønsted acid with a K_a of 1.0×10^{-2}.

(a) Write the chemical equation for the ionization of the acid and give the appropriate K_a expression.

(b) What is the value of $[H^+]$ in 0.010 *M* HSO_4^- (furnished by the salt, $NaHSO_4$)? Do NOT make simplifying assumptions.

(c) What is the calculated $[H^+]$ in 0.010 *M* HSO_4^-, obtained by using the usual simplifying assumption?

(d) How much error is introduced by incorrectly using the simplifying assumption?

16.102 Some people who take megadoses of ascorbic acid will drink a solution containing as much as 6.0 g of ascorbic acid dissolved in a glass of water. Assuming the volume to be 250 mL, calculate the pH of the solution.

*16.103 For the titration of 25.00 mL of 0.1000 *M* HCl with 0.1000 *M* NaOH, calculate the pH of the reaction mixture after each of the following total volumes of base have been added to the original solution. (Remember to take into account the change in total volume.) Construct a graph showing the titration curve for this experiment.

(a) 0 mL	(d) 24.99 mL	(g) 25.10 mL
(b) 10.00 mL	(e) 25.00 mL	(h) 26.00 mL
(c) 24.90 mL	(f) 25.01 mL	(i) 50.00 mL

16.104 Below is a diagram illustrating a mixture HF and F^- in an aqueous solution. For this mixture, does pH equal pK_a for HF? Explain. Describe how the number of HF molecules and F^- ions will change after three OH^- ions are added. How will the number of HF and F^- change if two H^+ ions are added? Explain your answers by using chemical equations.

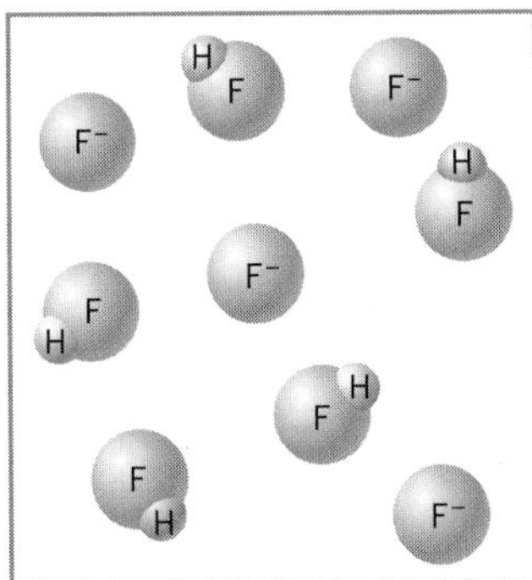

EXERCISES IN CRITICAL THINKING

16.105 Suppose that when doing a titration, a mistake causes an error of 10% in the concentration of hydronium ions. What is the error in terms of pH? Is the pH error the same for a 1.0×10^{-3} *M* $[H^+]$ solution as compared to a 5.0×10^{-6} *M* solution?

16.106 In the 1950s it was discovered that lakes in the northeastern United States had declining fish populations due to increased acidity. In order to address the problem of lake acidification, make a list of what you need to know in order to start addressing the problem.

16.107 Where are buffers found in everyday consumer products? Propose reasons why a manufacturer would use a buffer in a given product of your choice.

16.108 Why must the acid used for a buffer have a pK_a within one pH unit of the pH of the buffer? What happens if that condition is not met?

16.109 What conjugate acid–base pairs are used to buffer (a) over the counter drugs, (b) foods, (c) cosmetics, and (d) shampoos?

16.110 Your blood at 37 °C needs to be maintained within a narrow pH range of 7.35 to 7.45 to maintain optimal health. What possible conjugate acid–base pairs are present in blood that can buffer the blood and keep it within this range?

16.111 Develop a list of the uses of phosphoric acid in various consumer products.

16.112 Investigate the salts of phosphoric acid (i.e., those that contain the anions $H_2PO_4^-$, HPO_4^{2-}, and PO_4^{3-}) to find out how they are used in consumer products.

SOLUBILITY AND SIMULTANEOUS EQUILIBRIA 17

Small marine organisms called corals extract calcium ions and carbonate ions from sea water and precipitate calcium carbonate to form their shells. Over time, these small shells become the major structural feature of coral reefs, such as the one shown here. The equilibria involved in the solubility of metal salts constitute the major focus of this chapter. *(Scott Tuason/ Image Quest Marine)*

CHAPTER OUTLIN

THIS CHAPTER IN CONTEXT In the preceding chapters you learned how the principles of equilit can be applied to aqueous solutions of acids and bases. In this chapter we extend these principles to aqu reactions that involve the formation and dissolving of precipitates (a topic we introduced in Chapter 4). Many reactions are common in the world around us. For example, groundwater rich in carbon dioxide dissolves dep of calcium carbonate, producing vast underground caverns, and as the remaining calcium-containing solution gra evaporates, stalactites and stalagmites form. Within living organisms, precipitation reactions form the hard ca carbonate shells of clams, oysters, and coral as well as the unwanted calcium oxalate and calcium phosphate der we call kidney stones. Also, dilute acids in our mouths promote the dissolving of tooth enamel, which is comp of a mineral made up of calcium phosphate and calcium hydroxide.

In this chapter you will learn how we can calculate the solubilities of "insoluble" salts in water and in other solutions, and how the formation of substances called complex ions can affect solubilities. The concepts we develop here can tell us when precipitates will form and when they will dissolve, and we can use these in a practical lab setting in the separation of metal ions for chemical analysis.

17.1 An Insoluble Salt Is in Equilibrium with the Solution Around It

The equilibrium constant for an "insoluble" salt is called the solubility product constant, K_{sp}

None of the salts we described in Chapter 4 as being insoluble are *totally* insoluble. For example, the solubility rules tell us that AgCl is "insoluble," but if some solid AgCl is placed in water, a very small amount does dissolve. Once the solution has become saturated, the following equilibrium is established between the undissolved AgCl and its ions in the solution.

$$AgCl(s) \rightleftharpoons Ag^+(aq) + Cl^-(aq)$$

(In a saturated solution of AgCl)

Recall that when salts dissolve, they dissociate essentially completely, so the equilibrium is between the solid and the ions that are in the solution.

This is a heterogeneous equilibrium because it involves a solid reactant (AgCl) in equilibrium with ions in aqueous solution. Using the procedure developed in Section 14.4 (page 579), we write the equilibrium law as follows, omitting the solid from the mass action expression.

$$[Ag^+][Cl^-] = K_{sp} \quad (17.1)$$

TOOLS Solubility product constant, K_{sp}

The equilibrium constant, K_{sp}, is called the **solubility product constant** (because the system is a *solubility* equilibrium and the constant equals a *product* of ion concentrations).

It's important that you understand the distinction between solubility and solubility product. The *solubility* of a salt is the amount of the salt that dissolves in a given amount of solvent to give a saturated solution. The *solubility product* is the product of the molar concentrations of the ions in the saturated solution, raised to appropriate powers (see below).

The solubilities of salts change with temperature, so a value of K_{sp} applies only at the temperature at which it was determined. Some typical K_{sp} values are in Table 17.1 and in Appendix C.

In solubility equilibria, the reaction quotient is called the ion product

In preceding chapters we described the value of the mass action expression as the reaction quotient, Q. For simple solubility equilibria like those discussed in this section, the mass action expression is a product of ion concentrations raised to appropriate powers, so Q is often called the **ion product** of the salt. Thus, for AgCl,

$$\text{Ion product} = [Ag^+][Cl^-] = Q$$

At any dilution of a salt throughout the range of possibilities for an *unsaturated* solution, there will be varying values for the ion concentrations and, therefore, for Q. However, Q acquires a constant value, K_{sp}, in a *saturated* solution. When a solution is less than saturated, the value of Q is less than K_{sp}. Thus, we can use the numerical value of Q for a given solution as a test for saturation by comparing it to the value of K_{sp}.

Many salts produce more than one of a given ion per formula unit when they dissociate, and this introduces exponents into the ion product expression. For example, when silver chromate, Ag_2CrO_4, precipitates (Figure 17.1), it enters into the following solubility equilibrium.

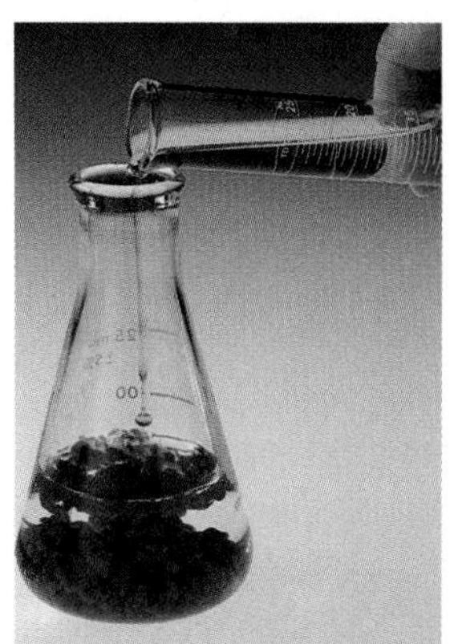

FIG. 17.1 Silver chromate. When sodium chromate is added to a solution of silver nitrate, deep red "insoluble" silver chromate, Ag_2CrO_4, precipitates. *(Michael Watson.)*

$$Ag_2CrO_4(s) \rightleftharpoons 2Ag^+(aq) + CrO_4^{2-}(aq)$$

The equilibrium law is obtained following the procedure we developed in Chapter 14, using coefficients as exponents in the mass action expression.

TABLE 17.1 Solubility Product Constants

Type	Salt		Ions of Salt	K_{sp} (25 °C)
Halides	CaF_2	$\rightleftharpoons$	$Ca^{2+} + 2F^-$	3.9×10^{-11}
	$AgCl$	$\rightleftharpoons$	$Ag^+ + Cl^-$	1.8×10^{-10}
	$AgBr$	$\rightleftharpoons$	$Ag^+ + Br^-$	5.0×10^{-13}
	AgI	$\rightleftharpoons$	$Ag^+ + I^-$	8.3×10^{-17}
	PbF_2	$\rightleftharpoons$	$Pb^{2+} + 2F^-$	3.6×10^{-8}
	$PbCl_2$	$\rightleftharpoons$	$Pb^{2+} + 2Cl^-$	1.7×10^{-5}
	$PbBr_2$	$\rightleftharpoons$	$Pb^{2+} + 2Br^-$	2.1×10^{-6}
	PbI_2	$\rightleftharpoons$	$Pb^{2+} + 2I^-$	7.9×10^{-9}
Hydroxides	$Al(OH)_3$	$\rightleftharpoons$	$Al^{3+} + 3OH^-$	3×10^{-34} [a]
	$Ca(OH)_2$	$\rightleftharpoons$	$Ca^{2+} + 2OH^-$	6.5×10^{-6}
	$Fe(OH)_2$	$\rightleftharpoons$	$Fe^{2+} + 2OH^-$	7.9×10^{-16}
	$Fe(OH)_3$	$\rightleftharpoons$	$Fe^{3+} + 3OH^-$	1.6×10^{-39}
	$Mg(OH)_2$	$\rightleftharpoons$	$Mg^{2+} + 2OH^-$	7.1×10^{-12}
	$Zn(OH)_2$	$\rightleftharpoons$	$Zn^{2+} + 2OH^-$	3.0×10^{-16} [b]
Carbonates	Ag_2CO_3	$\rightleftharpoons$	$2Ag^+ + CO_3^{2-}$	8.1×10^{-12}
	$MgCO_3$	$\rightleftharpoons$	$Mg^{2+} + CO_3^{2-}$	3.5×10^{-8}
	$CaCO_3$	$\rightleftharpoons$	$Ca^{2+} + CO_3^{2-}$	4.5×10^{-9} [c]
	$SrCO_3$	$\rightleftharpoons$	$Sr^{2+} + CO_3^{2-}$	9.3×10^{-10}
	$BaCO_3$	$\rightleftharpoons$	$Ba^{2+} + CO_3^{2-}$	5.0×10^{-9}
	$CoCO_3$	$\rightleftharpoons$	$Co^{2+} + CO_3^{2-}$	1.0×10^{-10}
	$NiCO_3$	$\rightleftharpoons$	$Ni^{2+} + CO_3^{2-}$	1.3×10^{-7}
	$ZnCO_3$	$\rightleftharpoons$	$Zn^{2+} + CO_3^{2-}$	1.0×10^{-10}
Chromates	Ag_2CrO_4	$\rightleftharpoons$	$2Ag^+ + CrO_4^{2-}$	1.2×10^{-12}
	$PbCrO_4$	$\rightleftharpoons$	$Pb^{2+} + CrO_4^{2-}$	1.8×10^{-14} [d]
Sulfates	$CaSO_4$	$\rightleftharpoons$	$Ca^{2+} + SO_4^{2-}$	2.4×10^{-5}
	$SrSO_4$	$\rightleftharpoons$	$Sr^{2+} + SO_4^{2-}$	3.2×10^{-7}
	$BaSO_4$	$\rightleftharpoons$	$Ba^{2+} + SO_4^{2-}$	1.1×10^{-10}
	$PbSO_4$	$\rightleftharpoons$	$Pb^{2+} + SO_4^{2-}$	6.3×10^{-7}
Oxalates	CaC_2O_4	$\rightleftharpoons$	$Ca^{2+} + C_2O_4^{2-}$	2.3×10^{-9}
	MgC_2O_4	$\rightleftharpoons$	$Mg^{2+} + C_2O_4^{2-}$	8.6×10^{-5}
	BaC_2O_4	$\rightleftharpoons$	$Ba^{2+} + C_2O_4^{2-}$	1.2×10^{-7}
	FeC_2O_4	$\rightleftharpoons$	$Fe^{2+} + C_2O_4^{2-}$	2.1×10^{-7}
	PbC_2O_4	$\rightleftharpoons$	$Pb^{2+} + C_2O_4^{2-}$	2.7×10^{-11}

[a] Alpha form. [b] Amorphous form. [c] Calcite form. [d] At 10 °C.

$$[Ag^+]^2[CrO_4^{2-}] = K_{sp}$$

Thus, the ion product contains the ion concentrations raised to powers equal to the number of ions released per formula unit. This means that to obtain the correct ion product expression you have to know the formulas of the ions that make up the salt. In other words, you have to realize that Ag_2CrO_4 is composed of Ag^+ and the polyatomic anion CrO_4^{2-}. If necessary, review the list of polyatomic ions in Table 2.8 on page 69.

Practice Exercise 1: Write the equation for the equilibrium involved in the solubility of barium phosphate, $Ba_3(PO_4)_2$, and write the equilibrium law corresponding to K_{sp}. (Hint: Remember that it's a heterogeneous equilibrium.)

Practice Exercise 2: What are the ion product expressions for the following salts? (a) barium oxalate, (b) silver sulfate

K_{sp} can be determined from molar solubilities

Molar solubility

One way to determine the value of K_{sp} for a salt is to measure its solubility—how much of the salt is required to give a saturated solution in a specified amount of solution. It is useful to express this as the **molar solubility,** which equals *the number of moles of salt dissolved in one liter of its **saturated** solution.* The molar solubility can be used to calculate the K_{sp} under the assumption that all of the salt that dissolves is 100% dissociated into the ions implied in the salt's formula.[1]

EXAMPLE 17.1
Calculating K_{sp} from Solubility Data

Silver bromide, AgBr, is the light-sensitive compound used in nearly all photographic film. The solubility of AgBr in water was measured to be 1.3×10^{-4} g L^{-1} at 25 °C. Calculate K_{sp} for AgBr at that temperature.

ANALYSIS: As usual, we begin with the chemical equation for the equilibrium, from which we construct the equilibrium law (here, the expression for K_{sp}).

$$AgBr(s) \rightleftharpoons Ag^+(aq) + Br^-(aq)$$

$$K_{sp} = [Ag^+][Br^-]$$

To calculate K_{sp}, we need the concentrations of the ions expressed in moles per liter. We can obtain these from the molar solubility—the number of moles of AgBr dissolved per liter. Therefore, the first step will be to change 1.3×10^{-4} g L^{-1} to moles per liter.

For problems dealing with solubility, the concentration table is an especially useful tool. In setting up the table, it is helpful if we imagine the formation of the saturated solution to occur stepwise. First, we will look at the composition of the solvent into which the salt will be placed. Does it contain any of the ions involved in the equilibrium? If it doesn't, the initial concentrations will be set equal to zero. However, if the solvent contains a solute that is a source of one of the ions in the equilibrium, we will use its concentration as the initial concentration of that ion.

When the salt dissolves, the concentrations of the ions increase, so the entries in the "change" row will be positive and will have values determined by the formula of the salt. We will obtain these from the molar solubility.

The entries in the last row, which are the equilibrium values, are obtained by adding the initial concentrations to the changes. Once we have the equilibrium ion concentrations, we substitute them into the ion product expression to calculate K_{sp}.

SOLUTION: First, we calculate the number of moles of AgBr dissolved per liter.

$$\text{Molar solubility} = \frac{1.3 \times 10^{-4}\ \cancel{\text{g AgBr}}}{1.00\ \text{L soln}} \times \frac{1.00\ \text{mol AgBr}}{187.77\ \cancel{\text{g AgBr}}} = 6.9 \times 10^{-7}\ \text{mol L}^{-1}$$

Now we can begin to set up the concentration table, which is shown on the next page. Notice first that there are no entries in the "reactant" column. This is because AgBr is a solid and doesn't appear in the mass action expression.

Initial Concentrations In the first row, under the formulas of the ions, we enter the initial concentrations of Ag^+ and Br^-. *Remember, these are the concentrations of the ions present in the solvent before any of the AgBr dissolves.* In this case the solvent is pure water. Neither Ag^+ nor Br^- are present in pure water, so we set the initial concentrations equal to zero.

[1] This assumption works reasonably well for slightly soluble salts made up of singly charged ions, like silver bromide. For simplicity, and to illustrate the nature of calculations involving solubility equilibria, we will work on the assumption that *all* salts behave as though they are 100% dissociated. This is not entirely true, especially for salts of multiply charged ions, so the accuracy of our calculations is limited. We discussed the reasons for the incomplete dissociation of salts in Section 12.9.

Changes in Concentrations In the "change" row, we enter data on how the concentrations of the $Ag^+(aq)$ and $Br^-(aq)$ change when the AgBr dissolves. Because dissolving the salt always increases these concentrations, these entries are always positive. They are also related to each other by the stoichiometry of the dissociation reaction. Thus, when AgBr dissolves, Ag^+ and Br^- ions are released in a 1:1 ratio. So, when 6.9×10^{-7} mol of AgBr dissolves (per liter), 6.9×10^{-7} mol of Ag^+ ion and 6.9×10^{-7} mol of Br^- ion go into solution. The concentrations of these species thus *change* (increase) by those amounts.

Equilibrium Concentrations As usual, we obtain the equilibrium values by adding the "initial concentrations" to the "changes."

There are no entries in the column under AgBr(*s*) because this substance does not appear in the K_{sp} expression.

	$AgBr(s)$	$\rightleftharpoons$	$Ag^+(aq)$	+	$Br^-(aq)$
Initial concentrations (*M*)	No entries in this column		0		0
Changes in concentrations when AgBr dissolves (*M*)			$+6.9 \times 10^{-7}$		$+6.9 \times 10^{-7}$
Equilibrium concentrations (*M*)			6.9×10^{-7}		6.9×10^{-7}

We now substitute the equilibrium ion concentrations into the K_{sp} expression.

The K_{sp} we calculated from the solubility data here differs by only 4% from the value in Table 17.1.

$$\begin{aligned} K_{sp} &= [Ag^+][Br^-] \\ &= (6.9 \times 10^{-7})(6.9 \times 10^{-7}) \\ &= 4.8 \times 10^{-13} \end{aligned}$$

The K_{sp} of AgBr is thus calculated to be 4.8×10^{-13} at 25 °C.

IS THE ANSWER REASONABLE? We've divided 1.3×10^{-4} by a number that's approximately 200, which would give a value of about 6.5×10^{-7}, so our molar solubility seems reasonable. (Also, we know AgBr has a very low solubility in water, so we expect the molar solubility to be very small.) The reasoning involved in the change row also makes sense; the number of moles of ions formed per liter must each equal the number of moles of AgBr that dissolve. Finally, if we round 6.9×10^{-7} to 7×10^{-7} and square it, we obtain $49 \times 10^{-14} = 4.9 \times 10^{-13}$. Our answer of 4.8×10^{-13} seems to be okay.

EXAMPLE 17.2
Calculating K_{sp} from Molar Solubility Data

The molar solubility of silver chromate, Ag_2CrO_4, in water is 6.7×10^{-5} mol L^{-1} at 25 °C. What is K_{sp} for Ag_2CrO_4?

ANALYSIS: We begin with the equilibrium equation and K_{sp} expression.

$$Ag_2CrO_4(s) \rightleftharpoons 2Ag^+(aq) + CrO_4^{2-}(aq) \qquad K_{sp} = [Ag^+]^2[CrO_4^{2-}]$$

Once again, we use the concentration table as our tool for analyzing the concentrations.

The solute is pure water, so neither Ag^+ nor CrO_4^{2-} are present before the salt dissolves; their initial concentrations are zero. For the "change" row, we have to be careful to take into account the formula of the salt. In a liter of water, when 6.7×10^{-5} mol of Ag_2CrO_4 dissolves, we obtain 6.7×10^{-5} mol of CrO_4^{2-} and $2 \times (6.7 \times 10^{-5}$ mol) of Ag^+. With this information, we can fill in the "initial" and "change" rows.

SOLUTION: As usual, we add the values in the "initial" and "change" rows to obtain the equilibrium concentrations of the ions.

	$Ag_2CrO_4(s) \rightleftharpoons$ $2Ag^+(aq)$	+ $CrO_4^{2-}(aq)$
Initial concentrations (*M*)	0	0
Changes in concentrations when Ag_2CrO_4 dissolves (*M*)	$+[2 \times (6.7 \times 10^{-5})] = +1.3 \times 10^{-4}$	$+6.7 \times 10^{-5}$
Equilibrium concentrations (*M*)	1.3×10^{-4}	6.7×10^{-5}

Substituting the equilibrium concentrations into the mass action expression for K_{sp} gives

$$K_{sp} = (1.3 \times 10^{-4})^2(6.7 \times 10^{-5})$$
$$= 1.1 \times 10^{-12}$$

So the K_{sp} of Ag_2CrO_4 at 25 °C is calculated to be 1.1×10^{-12}.

IS THE ANSWER REASONABLE? The critical step in solving the problem correctly is placing the correct quantities in the "initial" and "change" rows. The Ag_2CrO_4 is dissolved in water, so the initial concentrations of the ions must be zero; these entries are okay. For the "changes," it's important to remember that the quantities are related to each other by the coefficients in the chemical equation. That means the change for Ag^+ must be twice as large as the change for CrO_4^{2-}. Studying the table, we see that we've done this correctly. We can also see that we've added the "change" to the "initial" values correctly, and that we've performed the proper arithmetic in evaluating the mass action expression.

Practice Exercise 3: The solubility of thallium(I) iodide, TlI, in water at 20 °C is 5.9×10^{-3} g L^{-1}. Using this fact, calculate K_{sp} for TlI on the assumption that it is 100% dissociated in the solution. (Hint: What is the molar solubility of TlI?)

Practice Exercise 4: One liter of water is able to dissolve 2.15×10^{-3} mol of PbF_2 at 25 °C. Calculate the value of K_{sp} for PbF_2.

Now let's introduce a complication. Let's suppose that the aqueous system into which we're dissolving a slightly soluble salt is not pure water, but instead is a solution of a solute that provides one of the ions of the salt.

EXAMPLE 17.3
Calculating K_{sp} from Molar Solubility Data

The molar solubility of $PbCl_2$ in a 0.10 *M* NaCl solution is 1.7×10^{-3} mol L^{-1} at 25 °C. Calculate the K_{sp} for $PbCl_2$.

ANALYSIS: Again, we start by writing the equation for the equilibrium and the K_{sp} expression.

$$PbCl_2(s) \rightleftharpoons Pb^{2+}(aq) + 2Cl^-(aq) \qquad K_{sp} = [Pb^{2+}][Cl^-]^2$$

In this problem, the $PbCl_2$ is being dissolved not in pure water but in 0.10 *M* NaCl, which contains 0.10 *M* Cl^-. (It also contains 0.10 *M* Na^+, but that's not important here because Na^+ doesn't affect the equilibrium and so doesn't appear in the K_{sp} expression.) The initial concentration of Pb^{2+} is zero because none is in solution to begin with. The *initial* concentration of Cl^-, however, is 0.10 *M*.

For the change row, we note that when 1.7×10^{-3} mole of $PbCl_2$ dissolves per liter, the Pb^{2+} concentration increases by 1.7×10^{-3} *M* and the Cl^- concentration increases by $2 \times (1.7 \times 10^{-3}$ *M*). With these data we build the concentration table and evaluate the ion product to obtain K_{sp}.

SOLUTION: Here is the completed concentration table based on the analysis above.

	$PbCl_2(s) \rightleftharpoons$	$Pb^{2+}(aq)$	+	$2Cl^-(aq)$
Initial concentrations (*M*)		0		0.10 (because the solvent is 0.10 *M* NaCl)
Changes in concentrations when $PbCl_2$ dissolves (*M*)		$+1.7 \times 10^{-3}$		$+[2 \times (1.7 \times 10^{-3})] = +3.4 \times 10^{-3}$
Equilibrium concentrations (*M*)		1.7×10^{-3}		$0.10 + 0.0034 = 0.10$ (rounded correctly)

Substituting the equilibrium concentrations into the K_{sp} expression gives

$$K_{sp} = (1.7 \times 10^{-3})(0.10)^2$$
$$= 1.7 \times 10^{-5}$$

The K_{sp} of $PbCl_2$ is calculated to be 1.7×10^{-5} at 25 °C.

IS THE ANSWER REASONABLE? The solvent here is 0.10 *M* NaCl, so we have Cl^- in the mixture before any $PbCl_2$ dissolves. We check the "initial" row to be sure we've entered 0.10 *M* under Cl^-, and we have. Next, we check to be sure we've taken the stoichiometry of the equilibrium equation into account when placing values into the "change" row. We've done this correctly, too, because the change for Cl^- is twice the change for Pb^{2+}. Finally, we can check to be sure we've performed the correct arithmetic on the equilibrium concentrations. We have, so our answer should be correct.

Practice Exercise 5: The molar solubility of Ag_2SO_4 in a solution containing 28.4 g Na_2SO_4 per liter is 4.3×10^{-3} *M*. What is K_{sp} for Ag_2SO_4? (Hint: What is the molarity of the Na_2SO_4 solution?)

Practice Exercise 6: At 25 °C, the molar solubility of $CoCO_3$ in a 0.10 *M* Na_2CO_3 solution is 1.0×10^{-9} mol L^{-1}. Calculate the value of K_{sp} for $CoCO_3$.

Practice Exercise 7: The molar solubility of PbF_2 in a 0.10 *M* $Pb(NO_3)_2$ solution at 25 °C is 3.1×10^{-4} mol L^{-1}. Calculate the value of K_{sp} for PbF_2.

Molar solubility can be calculated from K_{sp}

Besides calculating K_{sp} from solubility information, we can also compute solubilities from values of K_{sp}. The following examples illustrate the calculations.[2]

EXAMPLE 17.4
Calculating Molar Solubility from K_{sp}

What is the molar solubility of AgCl in pure water at 25 °C?

ANALYSIS: To solve this problem, we need the chemical equation, the equilibrium law, and the value of K_{sp} for AgCl (which we obtain from Table 17.1). The relevant equations are

$$AgCl(s) \rightleftharpoons Ag^+(aq) + Cl^-(aq) \qquad K_{sp} = [Ag^+][Cl^-] = 1.8 \times 10^{-10}$$

[2] As noted earlier, these calculations ignore the fact that the salt that dissolves is not truly 100% dissociated into the ions implied in the salt's formula. This is particularly a problem with the salts of multiply charged ions, so the solubilities of such salts, when calculated from their K_{sp} values, must be taken as rough estimates. In fact, many calculations involving K_{sp} give values that are merely estimates.

Now we can build a concentration table. First, we see that the solvent is pure water, so neither Ag^+ nor Cl^- ion is present in solution at the start; their *initial* concentrations are zero.

Next, we turn to the "change" row. If we knew what the changes were, we could calculate the equilibrium concentrations and figure out the molar solubility. But we don't know the changes, so we will have to find them algebraically. To do this, *we will define our unknown* x *as the molar solubility of the salt*—the number of moles of AgCl that dissolves in one liter. Because 1 mol of AgCl yields 1 mol Ag^+ and 1 mol Cl^-, the concentration of each of these ions increases by x (their changes are each $+x$).

SOLUTION: Our concentration table is as follows.

	$AgCl(s) \rightleftharpoons Ag^+(aq)$	+	$Cl^-(aq)$
Initial concentrations (*M*)	0		0
Changes in concentrations when AgCl dissolves (*M*)	$+x$		$+x$
Equilibrium concentrations (*M*)	x		x

By defining x as the molar solubility, the coefficients of x in the "change" row are the same as the coefficients of Ag^+ and Cl^- in the equation for the equilibrium.

We make the substitutions into the K_{sp} expression, using equilibrium quantities from the last row of the table:

$$(x)(x) = 1.8 \times 10^{-10}$$
$$x = 1.3 \times 10^{-5}$$

The calculated molar solubility of AgCl in water at 25 °C is 1.3×10^{-5} mol L^{-1}.

IS THE ANSWER REASONABLE? The solvent is water, so the initial concentrations are zero. If x is the molar solubility, then the changes in the concentrations of Ag^+ and Cl^- when the salt dissolves must also be equal to x. We can also check the algebra, which we've done correctly, so the answer seems to be okay.

EXAMPLE 17.5
Calculating Molar Solubility from K_{sp}

Calculate the molar solubility of lead iodide in water from its K_{sp} at 25 °C.

ANALYSIS: First, we need the formula for lead iodide, which is PbI_2 (obtained from the rules of nomenclature in Chapter 2). To set up the problem, we begin with the equation for the equilibrium and the K_{sp} expression. We obtain K_{sp} from Table 17.1.

$$PbI_2(s) \rightleftharpoons Pb^{2+}(aq) + 2I^-(aq) \qquad K_{sp} = [Pb^{2+}][I^-]^2 = 7.9 \times 10^{-9}$$

The solvent is water, so the initial concentrations of the ions are zero. As before, we will define x as the molar solubility of the salt. By doing this, the coefficients of x in the "change" row will be identical to the coefficients of the ions in the chemical equation. This assures us that the changes in the concentrations are in the correct mole ratio.

SOLUTION: Here is the concentration table.

	$PbI_2(s)$	$\rightleftharpoons$	$Pb^{2+}(aq)$	+	$2I^-(aq)$
Initial concentrations (*M*)			0		0
Changes in concentrations when PbI_2 dissolves (*M*)			$+x$		$+2x$
Equilibrium concentrations (*M*)			x		$2x$

A precipitate of yellow PbI_2 forms when the two colorless solutions, one with sodium iodide and the other of lead(II) nitrate, are mixed. *(Lawrence Migdale/Photo Researchers.)*

Substituting equilibrium quantities into the K_{sp} expression gives

$$K_{sp} = (x)(2x)^2 = 4x^3 = 7.9 \times 10^{-9}$$
$$x^3 = 2.0 \times 10^{-9}$$
$$x = 1.3 \times 10^{-3}$$

Thus, the molar solubility of PbI_2 is calculated to be 1.3×10^{-3} mol L^{-1}.

IS THE ANSWER REASONABLE? We check the entries in the table. The solvent is water, so the initial concentrations are equal to zero. By setting x equal to the moles per liter of PbI_2 that dissolves (i.e., the molar solubility), the coefficients of x in the "change" row have to be the same as the coefficients of Pb^{2+} and Cl^- in the chemical equation. The entries in the "change" row are okay. In performing the algebra, notice that when we square $2x$, we get $4x^2$. Multiplying $4x^2$ by x gives $4x^3$. After that, the rest is straightforward.

Practice Exercise 8: Calculate the molar solubility of Ag_3PO_4 in water. Its $K_{sp} = 2.8 \times 10^{-18}$. (Hint: Review the algebra in the preceding example.)

Practice Exercise 9: What is the calculated molar solubility in water at 25 °C of (a) AgBr and (b) Ag_2CO_3?

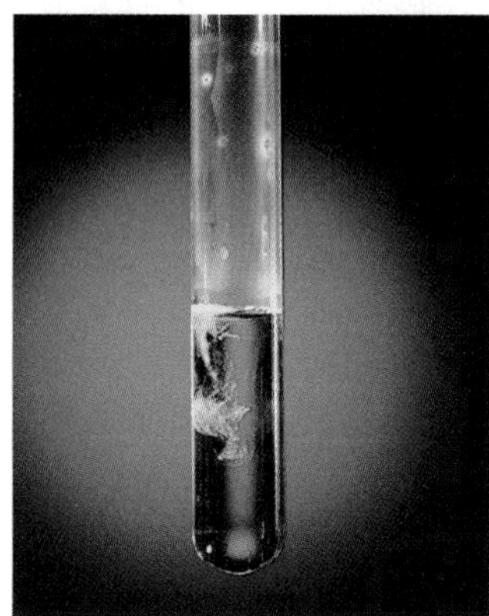

FIG. 17.2 The common ion effect. The test tube shown here initially held a saturated solution of NaCl, where the equilibrium $NaCl(s) \rightleftharpoons Na^+(aq) + Cl^-(aq)$ had been established. Addition of a few drops of concentrated HCl, containing a high concentration of the common ion Cl^-, forced the equilibrium to shift to the left. This caused some white crystals of solid NaCl to precipitate. *(Michael Watson.)*

A salt is less soluble if the solvent already contains an ion of the salt

Suppose that we stir some lead chloride (a compound having a low solubility) with water long enough to establish the following equilibrium.

$$PbCl_2(s) \rightleftharpoons Pb^{2+}(aq) + 2Cl^-(aq)$$

If we now add a concentrated solution of a soluble lead compound, such as $Pb(NO_3)_2$, the increased concentration of Pb^{2+} in the $PbCl_2$ solution will drive the position of equilibrium to the left, causing some $PbCl_2$ to precipitate. The phenomenon is simply an application of Le Châtelier's principle, the net result being that $PbCl_2$ is less soluble in a solution that contains Pb^{2+} from another source than it is in pure water. The same effect is produced if a concentrated solution of a soluble chloride salt such as NaCl is added to the saturated $PbCl_2$ solution. The added Cl^- will drive the equilibrium to the left, reducing the amount of dissolved $PbCl_2$.

The phenomenon described above is an example of the *common ion effect,* which we described in Chapter 16. In this case, Pb^{2+} is the common ion when we add $Pb(NO_3)_2$ and Cl^- is the common ion when we add NaCl. Figure 17.2 shows how even a relatively soluble salt, NaCl, can be forced out of its saturated solution simply by adding concentrated hydrochloric acid, which serves as a source of the common ion, Cl^-. The common ion effect can dramatically lower the solubility of a salt, as the next example demonstrates.

EXAMPLE 17.6
Calculations Involving the Common Ion Effect

What is the molar solubility of PbI_2 in a 0.10 *M* NaI solution?

ANALYSIS: We begin with the chemical equation for the equilibrium, the appropriate K_{sp} expression, and the value of K_{sp} obtained from Table 17.1.

$$PbI_2(s) \rightleftharpoons Pb^{2+}(aq) + 2I^-(aq) \qquad K_{sp} = [Pb^{2+}][I^-]^2 = 7.9 \times 10^{-9}$$

As usual, our tool to analyze what happens is the concentration table. As before, we imagine that we are adding the PbI_2 to a solvent into which it dissolves. This time, however, the solvent isn't water; it's a solution of NaI, which contains one of the ions of the salt PbI_2. Therefore, our initial concentrations will not both be zero. The solvent doesn't contain any lead compound, so the initial concentration of Pb^{2+} is equal to zero, but the solvent does contain 0.10 *M* NaI, which is completely dissociated and yields 0.10 *M* Na^+ and 0.10 *M* I^-. The initial concentration of I^- is therefore 0.10 *M*. These are the values we place in the "initial" row.

Next, we let x be the molar solubility of PbI_2. When x mol of PbI_2 dissolves per liter, the concentration of Pb^{2+} changes by $+x$ and that of I^- changes by twice as much, or $+2x$. Finally, the equilibrium concentrations are obtained by summing the initial concentrations and the changes.

SOLUTION: Here is the concentration table.

$PbI_2(s)$	$\rightleftharpoons$	$Pb^{2+}(aq)$	+	$2I^-(aq)$
Initial concentrations (*M*)		0		0.10
Changes in concentrations when PbI_2 dissolves (*M*)		$+x$		$+2x$
Equilibrium concentrations (*M*)		x		$0.10 + 2x$

Substituting equilibrium values into the K_{sp} expression gives

$$K_{sp} = (x)(0.10 + 2x)^2 = 7.9 \times 10^{-9}$$

Just a brief inspection reveals that solving this expression for x will be difficult if we cannot simplify the math. Fortunately, a simplification is possible, because the small value of K_{sp} for PbI_2 tells us that the salt has a very low solubility. This means very little of the salt will dissolve, so x (or even $2x$) will be quite small. Let's assume that $2x$ will be much smaller than 0.10. If this is so, then

$$0.10 + 2x \approx 0.10 \qquad \text{(assuming } 2x \text{ is negligible compared to 0.10)}$$

Substituting 0.10 *M* for the I^- concentration gives

$$K_{sp} = (x)(0.10)^2 = 7.9 \times 10^{-9}$$

$$x = \frac{7.9 \times 10^{-9}}{(0.10)^2}$$

$$= 7.9 \times 10^{-7}\ M$$

Thus, the molar solubility of PbI_2 in 0.10 *M* NaI solution is calculated to be 7.9×10^{-7} *M*.

IS THE ANSWER REASONABLE? Check the entries in the table. The initial concentrations come from the solvent, which contains no Pb^{2+} but does contain 0.10 *M* I^-. By letting x equal the molar solubility, the coefficients of x in the "change" row equal the coefficients in the equation for the equilibrium. *We should also check to see if our simplifying assumption is valid.* Notice that $2x$, which equals 1.6×10^{-6}, is indeed vastly smaller than 0.10, just as we anticipated. (If we add 1.6×10^{-6} to 0.10 and round correctly, we obtain 0.10.)

Practice Exercise 10: Calculate the molar solubility of AgI in 0.20 *M* CaI_2 solution. Compare the answer to the calculated molar solubility of AgI in pure water. (Hint: What is the I^- concentration in 0.20 *M* CaI_2?)

Practice Exercise 11: Calculate the molar solubility of $Fe(OH)_3$ in a solution where the OH^- concentration is initially 0.050 *M*. Assume the dissociation of $Fe(OH)_3$ is 100%.

A Mistake to Avoid

The most common mistake that students make with problems like Example 17.6 is to use the coefficient of an ion in the solubility equilibrium at the wrong moment in the calculation. The coefficient of I^- in the PbI_2 equilibrium is 2. The mistake is to use this 2 to double the *initial* concentration of I^-. However, the *initial* concentration of I^- was

READ THESE TWO PARAGRAPHS!

provided not by PbI_2 but by NaI. When 0.10 mol of NaI dissociates, it gives 0.10 mol of I^-, not 2×0.10 mol. *The coefficients in the equation for the equilibrium are only used to obtain the quantities in the "change" row.*

To avoid mistakes, it is useful to always view the formation of the final solution as a two-step process. You begin with a solvent into which the "insoluble solid" will be placed. In some problems, the solvent may be pure water, in which case the initial concentrations of the ions will be zero. In other problems, like Example 17.6, the solvent will be a *solution* that contains a common ion. When this is so, first decide what the concentration of the common ion is and enter that value into the "initial concentration" row of the table. Next, imagine that the solid is added to the solvent and a little of it dissolves. *The amount that dissolves is what gives the values in the "change" row.* These entries must be in the same ratio as the coefficients in the equilibrium, which is accomplished if we let x be the molar solubility of the salt. Then the coefficients of x will be the same as the coefficients of the ions in the chemical equation for the equilibrium.

In Example 17.5 (page 699) we found that the molar solubility of PbI_2 in *pure* water is 1.3×10^{-3} M. In water that contains 0.10 M NaI (Example 17.6), the solubility of PbI_2 is 7.9×10^{-7} M, well over a thousand times smaller. As we said, the common ion effect can cause huge reductions in the solubilities of sparingly soluble compounds by shifting the equilibrium toward the formation of the solid.

We can use K_{sp} to determine if a precipitate will form in a solution

If we know the anticipated concentrations of the ions of a salt in a solution, we can use the value of K_{sp} for the salt to predict whether or not a precipitate should form. This is because the computed value for Q (the ion product) can tell us whether a solution is unsaturated, saturated, or supersaturated. *For a precipitate of a salt to form, the solution must be supersaturated.*

If the solution is unsaturated, its ion concentrations are less than required for saturation, Q is less than K_{sp}, and no precipitate should form. For a saturated solution, we have Q equal to K_{sp}, and no precipitate will form. (If a precipitate were to form, the solution would become unsaturated and the precipitate would redissolve.) If the solution is supersaturated, the ion concentrations exceed those required for saturation and Q is larger than K_{sp}. Only in this last instance should we expect a precipitate to form. This can be summarized as follows.

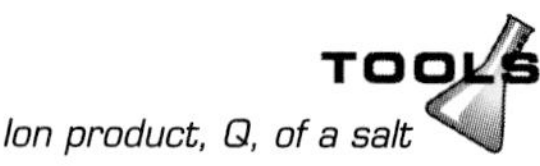

Ion product, Q, of a salt

Precipitate will form	$Q > K_{sp}$	(supersaturated)
No precipitate will form	$Q = K_{sp}$	(saturated)
	$Q < K_{sp}$	(unsaturated)

Let's look at some sample calculations.

EXAMPLE 17.7
Predicting whether a Precipitate Will Form

Suppose we wish to prepare 0.500 L of a solution containing 0.0075 mol of NaCl and 0.075 mol of $Pb(NO_3)_2$. Knowing from the solubility rules that the chloride of Pb^{2+} is "insoluble," we are concerned that a precipitate of $PbCl_2$ might form. Will it?

ANALYSIS: Our tool for answering this question is the criteria described above, which tell us a precipitate will only form if the computed Q for the solution is larger than K_{sp}. Therefore, our first task is to compute the value of the ion product (Q) appropriate for $PbCl_2$ *using the concentrations of the ions in the solution to be prepared.* This means we need the correct form of the ion product, which we can obtain by writing the solubility equilibrium and the K_{sp} expression that applies to a saturated solution of $PbCl_2$.

$$PbCl_2(s) \rightleftharpoons Pb^{2+}(aq) + 2Cl^-(aq) \qquad K_{sp} = [Pb^{2+}][Cl^-]^2$$

From Table 17.1, $K_{sp} = 1.7 \times 10^{-5}$. Also, we have to convert moles to molarity, because these are the quantities that we have to substitute into the ion product expression.

SOLUTION: The planned solution would have the following molar concentrations.

$$[Pb^{2+}] = \frac{0.075 \text{ mol}}{0.500 \text{ L}} = 0.15\ M \qquad [Cl^-] = \frac{0.0075 \text{ mol}}{0.500 \text{ L}} = 0.015\ M$$

We use these values to compute the ion product for $PbCl_2$ in the solution.

$$Q = [Pb^{2+}][Cl^-]^2 = (0.15)(0.015)^2 = 3.4 \times 10^{-5}$$

This value of Q is larger than the K_{sp} of $PbCl_2$, 1.7×10^{-5}, which means that a precipitate of $PbCl_2$ is likely to form if we attempt to prepare this solution.

It is usually difficult to prevent the extra salt from precipitating out of a supersaturated solution. (Sodium acetate is a notable exception. Supersaturated solutions of this salt are easily made.)

IS THE ANSWER REASONABLE? We can double-check that the molarities of the ions in the planned solution are correct. Then, we check the calculation of Q. All seems to be in order, so our answer is correct.

Practice Exercise 12: Will a precipitate of $CaSO_4$ form in a solution if the Ca^{2+} concentration is 0.0025 M and the SO_4^{2-} concentration is 0.030 M? (Hint: What is the form of the ion product for $CaSO_4$?)

Practice Exercise 13: Will a precipitate form in a solution containing 3.4×10^{-4} M CrO_4^{2-} and 4.8×10^{-5} M Ag^+?

EXAMPLE 17.8
Predicting whether a Precipitate Will Form

What possible precipitate might form by mixing 50.0 mL of 1.0×10^{-4} M NaCl with 50.0 mL of 1.0×10^{-6} M $AgNO_3$? Will it form? (Assume the volumes are additive.)

ANALYSIS: In this problem we are being asked, in effect, whether a metathesis reaction will occur between NaCl and $AgNO_3$. We should be able to use the solubility rules in Chapter 4 as a tool to predict whether this *might* occur. If the solubility rules suggest a precipitate, we can then calculate the ion product for the compound using the concentrations of the ions in the final solution. If this ion product exceeds K_{sp} for the salt, then a precipitate is expected.

To calculate Q correctly requires that we use the concentrations of the ions *after the solutions have been mixed.* Therefore, before computing the ion product, we must first take into account the fact that mixing the solutions dilutes each of the solutes. Dilution problems were covered in Chapter 4, where you learned that the tool for working such calculations is Equation 4.6 (page 158).

To obtain the solution to this problem we must bring together tools from two different chapters.

SOLUTION: Let's begin by writing the equation for the potential metathesis reaction between NaCl and $AgNO_3$. Our tool is the solubility rules (page 147), which helps us determine whether each product is soluble or insoluble.

$$NaCl(aq) + AgNO_3(aq) \longrightarrow AgCl(s) + NaNO_3(aq)$$

The solubility rules indicate that we expect a precipitate of AgCl. But are the concentrations of Ag^+ and Cl^- actually high enough?

In the original solutions, the 1.0×10^{-6} M $AgNO_3$ contains 1.0×10^{-6} M Ag^+ and the 1.0×10^{-4} M NaCl contains 1.0×10^{-4} M Cl^-. What are the concentrations of these ions after dilution? To determine these we use the equation that applies to all dilution problems involving molarity (Equation 4.6, page 158).

$$V_{dil} \cdot M_{dil} = V_{conc} \cdot M_{conc} \qquad \text{(Equation 4.6)}$$

Solving for M_{dil} gives

$$M_{dil} = \frac{V_{conc}M_{conc}}{V_{dil}}$$

The initial volumes of the more concentrated solutions are 50.0 mL and when the two solutions are combined, the final total volume is 100 mL. Therefore,

$$[Ag^+]_{final} = \frac{(50.0 \text{ mL})(1.0 \times 10^{-6}\ M)}{100.0 \text{ mL}} = 5.0 \times 10^{-7}\ M$$

$$[Cl^-]_{final} = \frac{(50.0 \text{ mL})(1.0 \times 10^{-4}\ M)}{100.0 \text{ mL}} = 5.0 \times 10^{-5}\ M$$

Now we use these to calculate Q for AgCl, which we can obtain from the dissociation reaction of the salt.

$$AgCl(s) \rightleftharpoons Ag^+(aq) + Cl^-(aq)$$
$$Q = [Ag^+][Cl^-]$$

Substituting the concentrations computed above gives

$$Q = (5.0 \times 10^{-7})(5.0 \times 10^{-5}) = 2.5 \times 10^{-11}$$

In Table 17.1, the K_{sp} for AgCl is given as 1.8×10^{-10}. Notice that Q is *smaller* than K_{sp}, which means that the final solution is unsaturated in AgCl and a precipitate will *not* form.

IS THE ANSWER REASONABLE? There are several things we should check here. They include writing the equation for the metathesis, calculating the concentrations of the ions after dilution (we've doubled the volume, so the concentrations are halved), and setting up the correct ion product. All appear to be correct, so the answer should be okay.

Practice Exercise 14: What precipitate might be expected if we pour together 100.0 mL of $1.0 \times 10^{-3}\ M\ Pb(NO_3)_2$ and 100.0 mL of $2.0 \times 10^{-3}\ M\ MgSO_4$? Will some form? (Assume that the volumes are additive.) (Hint: What tools will you need to solve the problem?)

Practice Exercise 15: What precipitate might be expected if we pour together 50.0 mL of 0.10 $M\ Pb(NO_3)_2$ and 20.0 mL of 0.040 M NaCl? Will some form? (Assume that the volumes are additive.)

17.2 SOLUBILITY EQUILIBRIA OF METAL OXIDES AND SULFIDES INVOLVE REACTION WITH WATER

Aqueous equilibria involving insoluble oxides and sulfides are more complex than those we've considered so far because of reactions of the anions with the solvent, water.

When a metal oxide dissolves in water, it does so by reacting with water instead of by a simple dissociation of ions that remain otherwise unchanged. Sodium oxide, for example, consists of Na^+ and O^{2-} ions, and it readily dissolves in water. The solution, however, does not contain the oxide ion, O^{2-}. Instead, the hydroxide ion forms. The equation for the reaction is

$$Na_2O(s) + H_2O \longrightarrow 2NaOH(aq)$$

This actually involves the reaction of oxide ions with water as the crystals of Na_2O break up.

$$O^{2-}(s) + H_2O \longrightarrow 2OH^-(aq)$$

The oxide ion is simply too powerful a base to exist in water at any concentration worthy of experimental note. We can understand why from the extraordinarily high (estimated) value of K_b for O^{2-}, 1×10^{22}. Thus there is no way to supply *oxide ions* to an aqueous solution in order to form an insoluble metal oxide directly. Oxide ions react with water, instead, to generate the hydroxide ion. When an insoluble metal *oxide* instead of an insoluble metal *hydroxide* does precipitate from a solution, it forms because the specific metal ion is able to react with OH^-, extract O^{2-}, and leave H^+ (or H_2O) in the solution. The silver ion, for example, precipitates as brown silver oxide, Ag_2O, when OH^- is added to aqueous silver salts (Figure 17.3).

$$2Ag^+(aq) + 2OH^-(aq) \longrightarrow Ag_2O(s) + H_2O$$

The reverse of this reaction, written as an equilibrium, corresponds to the solubility equilibrium for Ag_2O.

$$Ag_2O(s) + H_2O \rightleftharpoons 2Ag^+(aq) + 2OH^-(aq)$$

FIG. 17.3 Silver oxide. A brown, mudlike precipitate of silver oxide forms as sodium hydroxide solution is added to a solution of silver nitrate. *(Michael Watson.)*

Metal sulfides behave like oxides in their solubility equilibria

When we shift from oxygen to sulfur in Group VIA and consider metal sulfides, we find many similarities to oxides. One is that the sulfide ion, S^{2-}, like the oxide ion, is such a strong Brønsted base that it does not exist in any ordinary aqueous solution. The sulfide ion has not been detected in an aqueous solution even in the presence of 8 *M* NaOH where one might think that the reaction

$$OH^- + HS^- \longrightarrow H_2O + S^{2-}$$

could generate some detectable S^{2-}. An 8 *M* NaOH solution is at a concentration well outside the bounds of the "ordinary." Thus, Na_2S, like Na_2O, dissolves in water *by reacting with it*, not by releasing an otherwise unchanged divalent anion, S^{2-}.

$$Na_2S(s) + H_2O \longrightarrow 2Na^+(aq) + HS^-(aq) + OH^-(aq)$$

As with Na_2O, water reacts with the sulfide ion as the crystals break apart and the ions enter the solution.

$$S^{2-}(s) + H_2O \longrightarrow HS^-(aq) + OH^-(aq)$$

The sulfides of most metals have quite low solubilities in water. Simply bubbling hydrogen sulfide gas into an aqueous solution of any one of a number of metal ions—Cu^{2+}, Pb^{2+}, and Ni^{2+}, for example—causes their sulfides to precipitate. Many of these have distinctive colors (Figure 17.4) that can be used to help identify which metal ion is in solution.

In the laboratory, the qualitative analysis of metal ions uses the precipitation of metal sulfides to separate some metal ions from others.

To write the solubility equilibrium for an insoluble metal sulfide we have to take into account the reaction of the sulfide ion with water. For example, for CuS the equilibrium is best written as

$$CuS(s) + H_2O \rightleftharpoons Cu^{2+}(aq) + HS^-(aq) + OH^-(aq) \qquad (17.2)$$

This equation yields the ion product $[Cu^{2+}][HS^-][OH^-]$, and the solubility product constant for CuS is expressed by the equation.

$$K_{sp} = [Cu^{2+}][HS^-][OH^-]$$

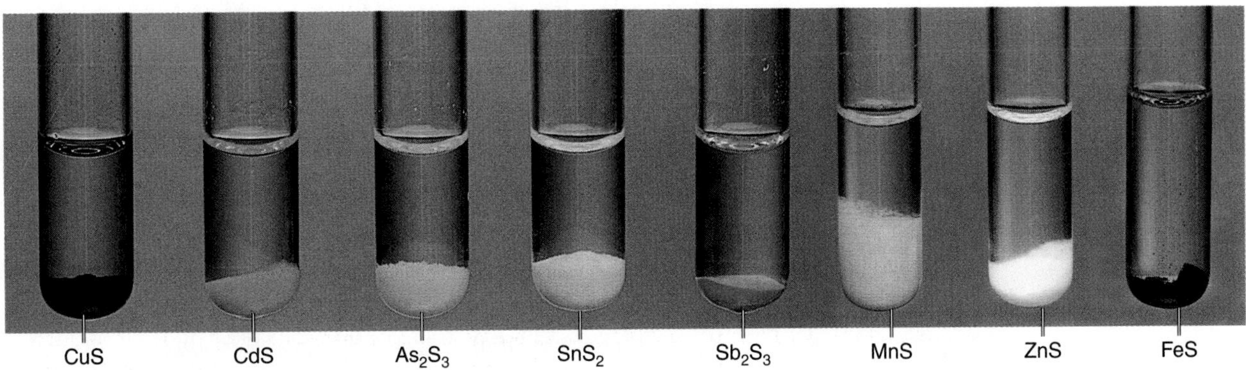

FIG. 17.4 The colors of some metal sulfides. *(OPC, Inc.)*

Values of K_{sp} of this form for a number of metal sulfides are given in the last column in Table 17.2. Notice particularly how much the K_{sp} values vary—from 2×10^{-53} to 3×10^{-11}, a spread of a factor of 10^{42}.

Acid-insoluble sulfides are so insoluble they don't dissolve in acid

Many metal sulfides are able to react with acid and thereby dissolve. An example is zinc sulfide.

$$ZnS(s) + 2H^+(aq) \longrightarrow Zn^{2+}(aq) + H_2S(aq)$$

However, some metal sulfides, referred to as the *acid-insoluble sulfides,* have K_{sp} values so low that they do not dissolve in acid. The cations in this group can be precipitated from the other cations simply by bubbling hydrogen sulfide into a sufficiently acidic solution that contains several metal ions.

When the solution is acidic, we have to treat the solubility equilibria differently. In acid, HS^- and OH^- would be neutralized, leaving their conjugate acids, H_2S and H_2O. Under these conditions, the equation for the solubility equilibrium becomes

$$CuS(s) + 2H^+(aq) \rightleftharpoons Cu^{2+}(aq) + H_2S(aq)$$

Acid solubility product, K_{spa}

This changes the mass action expression for the solubility product equilibrium, which we will now call the **acid solubility product, K_{spa}.** The *a* in the subscript indicates that the medium is acidic.

$$K_{spa} = \frac{[Cu^{2+}][H_2S]}{[H^+]^2}$$

Table 17.2 also gives K_{spa} values for the metal sulfides. Notice that all K_{spa} values are 10^{21} larger than the K_{sp} values. Metal sulfides are clearly vastly more soluble in dilute acid than in water. Yet several—the acid-insoluble sulfides—are so insoluble that even the most soluble of them, SnS, barely dissolves, even in moderately concentrated acid. So there are two families of sulfides, the *acid-insoluble sulfides* and the acid-soluble ones, otherwise

TABLE 17.2 Metal Ions Separable by Selective Precipitation of Sulfides[a]

Metal Ion	Sulfide	K_{spa}	K_{sp}
Acid-Insoluble Sulfides			
Hg^{2+}	HgS (black form)	2×10^{-32}	2×10^{-53}
Ag^+	Ag_2S	6×10^{-30}	6×10^{-51}
Cu^{2+}	CuS	6×10^{-16}	6×10^{-37}
Cd^{2+}	CdS	3×10^{-7}	3×10^{-28}
Pb^{2+}	PbS	3×10^{-7}	3×10^{-28}
Sn^{2+}	SnS	1×10^{-5}	1×10^{-26}
Base-Insoluble Sulfides (Acid-Soluble Sulfides)			
Zn^{2+}	α-ZnS	3×10^{-4}	3×10^{-25}
	β-ZnS	3×10^{-2}	3×10^{-23}
Co^{2+}	CoS	5×10^{-1}	5×10^{-22}
Ni^{2+}	NiS	4×10^{1}	4×10^{-20}
Fe^{2+}	FeS	6×10^{2}	6×10^{-19}
Mn^{2+}	MnS (pink form)	3×10^{10}	3×10^{-11}
	MnS (green form)	3×10^{7}	3×10^{-14}

[a]Data are for 25 °C. See R.J. Meyers, *J. Chem. Ed.*, vol. 63, 1986, p. 689.

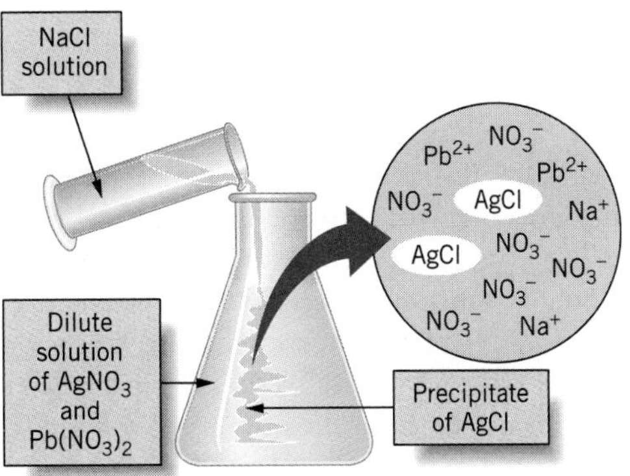

FIG. 17.5 Selective precipitation. When dilute sodium chloride is added to a - solution containing both Ag^+ ions and Pb^{2+} ions (both dissolved as their nitrate salts), the less soluble AgCl precipitates before the more soluble $PbCl_2$. Precipitation of the AgCl is nearly complete before any $PbCl_2$ begins to form. *(Andy Washnik.)*

known as the *base-insoluble sulfides*. As we discuss in the next section, the differing solubilities of their sulfides in acid provide a means of separating the two classes of metal ions from each other.

17.3 METAL IONS CAN BE SEPARATED BY SELECTIVE PRECIPITATION

Selective precipitation means causing one metal ion to precipitate while holding another in solution. Often this is possible because of large differences in the solubilities of salts that we would generally consider to be insoluble. For example, the K_{sp} values for AgCl and $PbCl_2$ are 1.8×10^{-10} and 1.7×10^{-5}, respectively, which gives them molar solubilities in water of 1.3×10^{-5} *M* for AgCl and 1.6×10^{-2} *M* for $PbCl_2$. In terms of molar solubilities, lead chloride is approximately 1200 times more soluble in water than AgCl. If we had a solution containing both 0.10 *M* Pb^{2+} and 0.10 *M* Ag^+ and began adding Cl^-, AgCl would precipitate first (Figure 17.5). In fact, we can calculate that before any $PbCl_2$ starts to precipitate, the concentration of Ag^+ will have been reduced to 1.4×10^{-8} *M*. Thus, nearly all the silver is removed from the solution without precipitating any of the lead, and the ions are effectively separated. All we need to do is find a way of adjusting the concentration of the Cl^- to achieve the separation.

Selective precipitation of metal sulfides is accomplished by controlling the pH

The large differences in K_{sp} values between the acid-insoluble and the base-insoluble metal sulfides make it possible to separate the corresponding cations from each other when they are in the same solution. The sulfides of the acid-insoluble cations are selectively precipitated by hydrogen sulfide from a solution kept at a pH that keeps the other cations in solution. A solution saturated in H_2S is used, for which the molarity of H_2S is 0.1 *M*.

Let's work an example to show how we can calculate the pH needed to allow the selective precipitation of two metal cations as their sulfides. We will use Cu^{2+} and Ni^{2+} ions to represent a cation from each class.

EXAMPLE 17.9
Selective Precipitation of the Sulfides of Acid-Insoluble Cations from Base-Insoluble Cations

Over what range of hydrogen ion concentrations (and pH) is it possible to separate Cu^{2+} from Ni^{2+} when both metal ions are present in a solution at a concentration of 0.010 *M* and the solution is made saturated in H_2S (where $[H_2S] = 0.1$ *M*)?

ANALYSIS: We must work with the chemical equilibria for the metal sulfides and their associated equilibrium expressions (the equations for their K_{spa}).

$$CuS(s) + 2H^+(aq) \rightleftharpoons Cu^{2+}(aq) + H_2S(aq) \qquad K_{spa} = \frac{[Cu^{2+}][H_2S]}{[H^+]^2} = 6 \times 10^{-16}$$

$$NiS(s) + 2H^+(aq) \rightleftharpoons Ni^{2+}(aq) + H_2S(aq) \qquad K_{spa} = \frac{[Ni^{2+}][H_2S]}{[H^+]^2} = 4 \times 10^{1}$$

The K_{spa} values (from Table 17.2) tell us that NiS is much more soluble in an acidic solution than CuS. Therefore, we want to make the H^+ concentration large enough to prevent NiS from precipitating, but small enough that CuS does precipitate.

The problem reduces to two questions. The first is, "What hydrogen ion concentration would be needed to keep the Cu^{2+} *in solution*?" (The answer to this question will give us the *upper limit* on $[H^+]$; we would really want a lower H^+ concentration so CuS *will* precipitate.) The second question is, "What is the hydrogen ion concentration just before NiS precipitates?" The answer to this will be the *lower limit* on $[H^+]$. At any lower H^+ concentration, NiS will precipitate, so we want an H^+ concentration equal to or larger than this value. Once we know these limits, we know that any hydrogen ion concentration in between them will permit CuS to precipitate but retain Ni^{2+} in solution.

SOLUTION: We will find the upper limit first. If CuS *does not precipitate,* the Cu^{2+} concentration will be the given value, 0.010 *M*, so we substitute this along with the H_2S concentration (0.1 *M*) into the expression for K_{spa}.

$$K_{spa} = \frac{[Cu^{2+}][H_2S]}{[H^+]^2} = \frac{(0.010)(0.1)}{[H^+]^2} = 6 \times 10^{-16}$$

Now we solve for $[H^+]$.

$$[H^+]^2 = \frac{(0.010)(0.1)}{6 \times 10^{-16}} = 2 \times 10^{12}$$

$$[H^+] = 1 \times 10^{6}\ M$$

If we could make $[H^+] = 1 \times 10^6$ *M*, we could prevent CuS from forming. However, it isn't possible to have 10^6 or a *million* moles of H^+ per liter! What the calculated $[H^+]$ tells us, therefore, is that *no matter how acidic the solution is, we cannot prevent* CuS *from precipitating when we saturate the solution with* H_2S. (You can now see why CuS is classed as an "acid-insoluble sulfide.")

To obtain the lower limit, we calculate the $[H^+]$ required to give an equilibrium concentration of Ni^{2+} equal to 0.010 *M*. If we keep the value of $[H^+]$ *equal to or larger* than this value, then NiS will be prevented from precipitating. The calculation is exactly like the one above. First, we substitute values into the K_{spa} expression.

$$K_{spa} = \frac{[Ni^{2+}][H_2S]}{[H^+]^2} = \frac{(0.010)(0.1)}{[H^+]^2} = 4 \times 10^{1}$$

Once again, we solve for $[H^+]$.

$$[H^+]^2 = \frac{(0.010)(0.1)}{4 \times 10^{1}}$$

$$[H^+] = 5 \times 10^{-3}\ M$$

$$pH = 2.3$$

If we maintain the pH of the solution of 0.010 *M* Cu^{2+} and 0.010 *M* Ni^{2+} at 2.3 or lower (more acidic), as we make the solution saturated in H_2S, virtually all the Cu^{2+} will precipitate as CuS, but all the Ni^{2+} will stay in solution.

IS THE ANSWER REASONABLE? The K_{spa} values certainly tell us that CuS is quite insoluble in acid and that NiS is a lot more soluble, so from that standpoint the results seem reasonable. There's not much more we can do to check the answer, other than to review the entire set of calculations.

Practice Exercise 16: Suppose a solution contains 0.0050 *M* Co^{2+} and is saturated with H_2S (with $[H_2S] = 0.1\ M$). Would CoS precipitate if the solution has a pH of 3.5? (Hint: What relationship do we use to determine whether a precipitate is going to form in a solution?)

Practice Exercise 17: Consider a solution containing Hg^{2+} and Fe^{2+}, both at molarities of 0.010 *M*. It is to be saturated with H_2S. Calculate the highest pH that this solution could have that would keep Fe^{2+} in solution while causing Hg^{2+} to precipitate as HgS.

In actual lab work, separation of the acid-insoluble sulfides from the base-insoluble ones is performed with $[H^+] \approx 0.3\ M$ which corresponds to a pH of about 0.5. This ensures that the base-insoluble sulfides will not precipitate. The calculation above also demonstrated why NiS can be classified as a "base-insoluble sulfide." If the solution is *basic* when it is made saturated in H_2S, the pH will surely be larger than 2.3 and NiS will precipitate.

Selective precipitation can be applied to metal carbonates

The principles of selective precipitation by the control of pH apply to any system where the anion is that of a weak acid. The metal carbonates are examples, with many being quite insoluble in water (see Table 17.1). The dissociation of magnesium carbonate in water, for example, involves the following equilibrium and K_{sp} equations.

$$MgCO_3(s) \rightleftharpoons Mg^{2+}(aq) + CO_3^{2-}(aq) \qquad K_{sp} = 3.5 \times 10^{-8}$$

For strontium carbonate, the equations are

$$SrCO_3(s) \rightleftharpoons Sr^{2+}(aq) + CO_3^{2-}(aq) \qquad K_{sp} = 9.3 \times 10^{-10}$$

Would it be possible to separate the magnesium ion from the strontium ion by taking advantage of the difference in K_{sp} of their carbonates?

We can do so if we can control the carbonate ion concentration. Because the carbonate ion is a relatively strong Brønsted base, its control is available, indirectly, by adjusting the pH of the solution. This is because the hydrogen ion is one of the species in each of the following equilibria.[3]

$$H_2CO_3(aq) \rightleftharpoons H^+(aq) + HCO_3^-(aq) \qquad K_{a_1} = 4.5 \times 10^{-7}$$

$$HCO_3^-(aq) \rightleftharpoons H^+(aq) + CO_3^{2-}(aq) \qquad K_{a_2} = 4.7 \times 10^{-11}$$

You can see that if we increase the hydrogen ion concentration, both equilibria will shift to the left, in accordance with Le Châtelier's principle, and this will reduce the concentration of the carbonate ion. The value of $[CO_3^{2-}]$ thus decreases with decreasing pH. On the other hand, if we decrease the hydrogen ion concentration, making the solution more basic, we will cause the two equilibria to shift to the right. The concentration of the carbonate ion thus increases with increasing pH.

For the calculations ahead, it will be useful to combine the two hydrogen carbonate equilibria into one overall equation that relates the molar concentrations of carbonic acid, carbonate ion, and hydrogen ion. So we first add the two.

$$\begin{array}{rcl} H_2CO_3(aq) & \rightleftharpoons & H^+(aq) + HCO_3^-(aq) \\ HCO_3^-(aq) & \rightleftharpoons & H^+(aq) + CO_3^{2-}(aq) \\ \hline H_2CO_3(aq) & \rightleftharpoons & 2H^+(aq) + CO_3^{2-}(aq) \end{array} \qquad (17.3)$$

[3] The situation involving aqueous carbonic acid is complicated by the presence of dissolved CO_2, which we could represent in an equation as $CO_2(aq)$. In fact, this is how most of the dissolved CO_2 exists, namely, as $CO_2(aq)$, not as $H_2CO_3(aq)$. But we may use $H_2CO_3(aq)$ as a surrogate or stand-in for $CO_2(aq)$, because the latter changes smoothly to the former on demand. The following two successive equilibria involving carbonic acid exist in a solution of aqueous CO_2.

$$CO_2(aq) + H_2O \rightleftharpoons H_2CO_3(aq) \rightleftharpoons H^+(aq) + HCO_3^-(aq)$$

The value of K_{a_1} cited here for $H_2CO_3(aq)$ is really the product of the equilibrium constants of these two equilibria.

Recall from Chapter 14 that when we add two equilibria to get a third, the equilibrium constant of the latter is the product of the two equilibria that are combined. Thus, for Equation 17.3 the equilibrium constant, which we'll symbolize as K_a, is obtained as follows.

TOOLS

Combined K_a expression for a diprotic acid

$$K_a = K_{a_1} \times K_{a_2} = \frac{[H^+]^2[CO_3^{2-}]}{[H_2CO_3]} \tag{17.4}$$

$$= (4.5 \times 10^{-7}) \times (4.7 \times 10^{-11})$$

$$= 2.1 \times 10^{-17}$$

We can safely use Equation 17.4 *only* when two of the three concentrations are known.

With this K_a value we may now study how to find the pH range within which Mg^{2+} and Sr^{2+} can be separated by taking advantage of their difference in K_{sp} values. To perform such a separation we would saturate a solution with CO_2, which provides H_2CO_3 through the rapidly established equilibrium

$$CO_2(aq) + H_2O \rightleftharpoons H_2CO_3(aq)$$

Thus, a solution that contains 0.030 M CO_2 has an effective H_2CO_3 concentration of 0.030 M.

EXAMPLE 17.10

Separation of Metal Ions by the Selective Precipitation of Their Carbonates

A solution contains magnesium nitrate and strontium nitrate, each at a concentration of 0.10 M. Carbon dioxide is to be bubbled in to make the solution saturated in $CO_2(aq)$, approximately 0.030 M. What pH range would make it possible for the carbonate of one cation to precipitate but not that of the other?

ANALYSIS: There are really two parts to this. First, what is the range in values of $[CO_3^{2-}]$ that allow one carbonate to precipitate but not the other? Second, given this range, what are the values of the solution's pH that produce this range in $[CO_3^{2-}]$ values?

To answer the first question, we will use the K_{sp} values of the two carbonate salts and their molar concentrations to find the CO_3^{2-} concentrations in their saturated solutions. To keep the more soluble carbonate from precipitating, the CO_3^{2-} concentration must be at or below that required for saturation. For the less soluble of the two, we must have a CO_3^{2-} concentration larger than in a saturated solution so that it will precipitate.

To answer the second question, we will use the combined K_a expression for H_2CO_3 to find the $[H^+]$ that gives the necessary CO_3^{2-} concentrations obtained in answering the first question. Having these $[H^+]$, it is then a simple matter to convert them to pH values.

SOLUTION: The range in $[CO_3^{2-}]$ values is obtained by using the K_{sp} values of the two carbonates.

$$K_{sp} = [Mg^{2+}][CO_3^{2-}] = 3.5 \times 10^{-8}$$

$$K_{sp} = [Sr^{2+}][CO_3^{2-}] = 9.3 \times 10^{-10}$$

On the basis of the K_{sp} values, we see that $MgCO_3$ is the more soluble of the two. Let's find the $[CO_3^{2-}]$ values required to give saturated solutions of these two salts.

For Magnesium Carbonate:

$$[CO_3^{2-}] = \frac{K_{sp}}{[Mg^{2+}]} = \frac{3.5 \times 10^{-8}}{0.10} = 3.5 \times 10^{-7}\ M$$

For Strontium Carbonate:

$$[CO_3^{2-}] = \frac{K_{sp}}{[Sr^{2+}]} = \frac{9.3 \times 10^{-10}}{0.10} = 9.3 \times 10^{-9}\ M$$

For separation, we have to keep the $[CO_3^{2-}]$ less than or equal to 3.5×10^{-7} M and larger than 9.3×10^{-9} M.

$[CO_3^{2-}] > 9.3 \times 10^{-9}\,M$ (to precipitate $SrCO_3$)

$[CO_3^{2-}] \leq 3.5 \times 10^{-7}\,M$ (to prevent $MgCO_3$ from precipitating)

The second phase of our calculation now asks what values of $[H^+]$ correspond to the calculated limits on $[CO_3^{2-}]$. For this we will use the K_a expression given by Equation 17.4. First we solve for the square of $[H^+]$, using the molarity of the dissolved CO_2, 0.030 *M*, as the value of $[H_2CO_3]$. This gives us

$$[H^+]^2 = K_a \times \frac{[H_2CO_3]}{[CO_3^{2-}]} = 2.1 \times 10^{-17} \times \frac{0.030}{[CO_3^{2-}]}$$

Because $CO_2(aq)$ can rapidly combine with water to give $H_2CO_3(aq)$, the value of $[H_2CO_3]$ is taken to be that of $[CO_2(aq)]$, namely, 0.030 *M*.

We will use this equation for each of the two boundary values of $[CO_3^{2-}]$ to calculate the corresponding two values of $[H^+]^2$. Once we have them, the steps to values of $[H^+]$ and pH are easy.

For Magnesium Carbonate: To *prevent* the precipitation of $MgCO_3$, $[CO_3^{2-}]$ must be no higher than 3.5×10^{-7} *M*, so $[H^+]^2$ must not be less than

$$[H^+]^2 = 2.1 \times 10^{-17} \times \frac{0.030}{3.5 \times 10^{-7}} = 1.8 \times 10^{-12}$$

$$[H^+] = 1.3 \times 10^{-6}\,M$$

This corresponds to a pH of 5.89. At a higher (more basic) pH, magnesium carbonate precipitates. Thus, pH $\leq$ 5.89 prevents precipitation of $MgCO_3$.

For Strontium Carbonate: To have a saturated solution of $SrCO_3$ with $[Sr^{2+}] = 0.10\,M$, the value of $[CO_3^{2-}]$ would be 9.3×10^{-9} *M*, as we calculated above. This corresponds to a value of $[H^+]^2$ found as follows.

$$[H^+]^2 = 2.1 \times 10^{-17} \times \frac{0.030}{9.3 \times 10^{-9}} = 6.8 \times 10^{-11}$$

$$[H^+] = 8.2 \times 10^{-6}\,M$$

$$\text{pH} = 5.09$$

To cause $SrCO_3$ to precipitate, the $[CO_3^{2-}]$ would have to be higher than 9.3×10^{-9} *M*, and that would require that $[H^+]$ be *less than* 8.2×10^{-6} *M*. If $[H^+]$ were less than 8.2×10^{-6} *M*, the pH would have to be higher than 5.09. Thus, to cause $SrCO_3$ to precipitate from the given solution, pH > 5.09.

In summary, when the pH of the given solution is kept above 5.09 and less than or equal to 5.89, Sr^{2+} will precipitate as $SrCO_3$ but Mg^{2+} will remain dissolved.

IS THE ANSWER REASONABLE? The only way to be sure of the answer is to go back over the reasoning and the calculations. However, there is a sign that the answer is probably correct. Notice that the two K_{sp} values are not vastly different—they differ by just a factor of about 40. Therefore, it's not surprising that to achieve separation we would have to keep the pH within a rather narrow range (from 5.09 to 5.89).

Practice Exercise 18: The K_{sp} for barium oxalate, BaC_2O_4, is 1.2×10^{-7} and for oxalic acid, $H_2C_2O_4$, $K_{a_1} = 5.6 \times 10^{-2}$ and $K_{a_2} = 5.4 \times 10^{-5}$. If a solution containing 0.10 *M* $H_2C_2O_4$ and 0.050 *M* $BaCl_2$ is prepared, what must the minimum H^+ concentration be in the solution to prevent the formation of a BaC_2O_4 precipitate? (Hint: Work with a combined K_a expression for $H_2C_2O_4$.)

Practice Exercise 19: A solution contains calcium nitrate and nickel nitrate, each at a concentration of 0.10 *M*. Carbon dioxide is to be bubbled in to make its concentration equal 0.030 *M*. What pH range would make it possible for the carbonate of one cation to precipitate but not that of the other?

FIG. 17.6 The complex ion of Cu^{2+} and water. A solution containing copper sulfate has a blue color because it contains the complex ion $Cu(H_2O)_4^{2+}$. *(Andy Washnik.)*

Recall that a *Lewis base* is an electron pair donor in the formation of a coordinate covalent bond.

17.4 COMPLEX IONS PARTICIPATE IN EQUILIBRIA IN AQUEOUS SOLUTIONS

Metal ions can combine with anions or neutral molecules to form complex ions

In our previous discussions of metal-containing compounds, we left you with the impression that the only kinds of bonds in which metals are ever involved are ionic bonds. For some metals, like the alkali metals of Group IA, this is close enough to the truth to warrant no modifications. But for many other metal ions, especially those of the transition metals and the post-transition metals, it is not. This is because the ions of many of these metals are able to behave as Lewis acids (i.e., as electron pair acceptors in the formation of coordinate covalent bonds). Thus, by participating in Lewis acid–base reactions they become *covalently* bonded to other atoms. Copper(II) ion is a typical example.

In aqueous solutions of copper(II) salts, like $CuSO_4$ or $Cu(NO_3)_2$, the copper is not present as simple Cu^{2+} ions. Instead, each Cu^{2+} ion becomes bonded to four water molecules to give a pale blue ion with the formula $Cu(H_2O)_4^{2+}$ (see Figure 17.6). We call this species a **complex ion** because it is composed of a number of simpler species (i.e., it is *complex*, not simple). The chemical equation for the formation of the $Cu(H_2O)_4^{2+}$ ion is

$$Cu^{2+} + 4H_2O \longrightarrow Cu(H_2O)_4^{2+}$$

which can be diagrammed using Lewis structures as follows.

$$4\;\mathrm{H{-}\ddot{O}{:}{-}H} \quad + \quad \mathrm{Cu^{2+}} \longrightarrow \left[\mathrm{Cu}(\leftarrow\!\mathrm{\ddot{O}H_2})_4\right]^{2+}$$

$Cu(H_2O)_4^{2+}$

As you can see in this analysis, the Cu^{2+} ion accepts pairs of electrons from the water molecules, so Cu^{2+} is a Lewis acid and the water molecules are each Lewis bases.

The number of complex ions formed by metals, especially the transition metals, is enormous, and the study of the properties, reactions, structures, and bonding in complex ions like $Cu(H_2O)_4^{2+}$ has become an important specialty within chemistry. We will provide a more complete discussion of them in Chapter 21. For now, we will introduce you to some of the terminology that we use in describing these substances.

A Lewis base that attaches itself to a metal ion is called a **ligand** (from the Latin *ligare*, meaning "to bind"). Ligands can be neutral molecules with unshared pairs of electrons (like H_2O), or they can be anions (like Cl^- or OH^-). The atom in the ligand that actually provides the electron pair is called the **donor atom,** and the metal ion is the **acceptor.** The result of combining a metal ion with one or more ligands is a *complex ion*, or simply just a *complex*. The word "complex" avoids problems when the particle formed is electrically neutral, as sometimes happens. Compounds that contain complex ions are generally referred to as **coordination compounds** because the bonds in a complex ion can be viewed as coordinate covalent bonds. Sometimes the complex itself is called a *coordination complex.*

FIG. 17.7 The complex ion of Cu^{2+} and ammonia. Ammonia molecules displace water molecules from $Cu(H_2O)_4^{2+}$ (left) to give the deep blue $Cu(NH_3)_4^{2+}$ ion (right). *(Andy Washnik.)*

In an aqueous solution, the formation of a complex ion is really a reaction in which water molecules are replaced by other ligands. Thus, when NH_3 is added to a solution of copper ion, the water molecules in the $Cu(H_2O)_4^{2+}$ ion are replaced, one after another, by molecules of NH_3 until the deeply blue complex $Cu(NH_3)_4^{2+}$ is formed (Figure 17.7). Each successive reaction is an equilibrium, so the entire chemical system involves many species and is quite complicated. Fortunately, when the ligand concentration is *large* relative to that of the metal ion, the concentrations of the intermediate complexes are very small and we can work only with the *overall* reaction for the formation of the final complex. Our study of complex ion equilibria will be limited to these situations. The equilibrium equation for the formation of $Cu(NH_3)_4^{2+}$, therefore, can be written as though the complex forms in one step.

$$Cu(H_2O)_4^{2+}(aq) + 4NH_3(aq) \rightleftharpoons Cu(NH_3)_4^{2+}(aq) + 4H_2O$$

We will simplify this equation for the purposes of dealing quantitatively with the equilibria by omitting the water molecules. (It's safe to do this because the concentration of H_2O in aqueous solutions is taken to be effectively a constant and need not be included in mass action expressions.) In simplified form, we write the equilibrium above as follows:

$$Cu^{2+}(aq) + 4NH_3(aq) \rightleftharpoons Cu(NH_3)_4^{2+}(aq)$$

We have two goals here: to study such equilibria themselves and to learn how they can be used to influence the solubilities of metal ion salts.

According to Le Châtelier's principle, when the concentration of ammonia is high, the position of equilibrium in this reaction is shifted far to the right, so effectively all complex ions with water molecules are changed to those with ammonia molecules.

Formation constants reflect the stabilities of complex ions

When the chemical equation for the equilibrium is written so that the complex ion is the product, the equilibrium constant for the reaction is called the **formation constant, K_{form}.** The equilibrium law for the formation of $Cu(NH_3)_4^{2+}$ in the presence of excess NH_3, for example, is

TOOLS *Formation constants of complexes*

$$\frac{[Cu(NH_3)_4^{2+}]}{[Cu^{2+}][NH_3]^4} = K_{form}$$

Sometimes this equilibrium constant is called the **stability constant.** The larger its value, the greater is the concentration of the complex at equilibrium, and so the more stable is the complex.

Table 17.3 provides several more examples of complex ion equilibria and their associated equilibrium constants. (Additional examples are in Appendix C, Table C.6.) Notice that the most stable complex in the table, $Co(NH_3)_6^{3+}$, has, as you would expect, the largest value of K_{form}.

Instability constants are the inverse of formation constants

Some chemists prefer to describe the relative stabilities of complex ions differently. The *inverses* of formation constants are cited and are called **instability constants, K_{inst}.** This approach focuses attention on the *breakdown* of the complex, not its formation. Therefore,

TABLE 17.3 Formation Constants and Instability Constants for Some Complex Ions

Ligand	Equilibrium	K_{form}	K_{inst}
NH_3	$Ag^+ + 2NH_3 \rightleftharpoons Ag(NH_3)_2^+$	1.6×10^7	6.3×10^{-8}
	$Co^{2+} + 6NH_3 \rightleftharpoons Co(NH_3)_6^{2+}$	5.0×10^4	2.0×10^{-5}
	$Co^{3+} + 6NH_3 \rightleftharpoons Co(NH_3)_6^{3+}$	4.6×10^{33}	2.2×10^{-34}
	$Cu^{2+} + 4NH_3 \rightleftharpoons Cu(NH_3)_4^{2+}$	1.1×10^{13}	9.1×10^{-14}
	$Hg^{2+} + 4NH_3 \rightleftharpoons Hg(NH_3)_4^{2+}$	1.8×10^{19}	5.6×10^{-20}
F^-	$Al^{3+} + 6F^- \rightleftharpoons AlF_6^{3-}$	1×10^{20}	1×10^{-20}
	$Sn^{4+} + 6F^- \rightleftharpoons SnF_6^{2-}$	1×10^{25}	1×10^{-25}
Cl^-	$Hg^{2+} + 4Cl^- \rightleftharpoons HgCl_4^{2-}$	5.0×10^{15}	2.0×10^{-16}
Br^-	$Hg^{2+} + 4Br^- \rightleftharpoons HgBr_4^{2-}$	1.0×10^{21}	1.0×10^{-21}
I^-	$Hg^{2+} + 4I^- \rightleftharpoons HgI_4^{2-}$	1.9×10^{30}	5.3×10^{-31}
CN^-	$Fe^{2+} + 6CN^- \rightleftharpoons Fe(CN)_6^{4-}$	1.0×10^{24}	1.0×10^{-24}
	$Fe^{3+} + 6CN^- \rightleftharpoons Fe(CN)_6^{3-}$	1.0×10^{31}	1.0×10^{-31}

the associated equilibrium equation is written as the reverse of the formation of the complex. The equilibrium for the copper–ammonia complex would be written as follows, for example:

$$Cu(NH_3)_4^{2+}(aq) \rightleftharpoons Cu^{2+}(aq) + 4NH_3(aq)$$

The equilibrium constant for this equilibrium is called the *instability constant.*

Among the complexes shown in Table 17.3, notice that Co^{3+} forms the most stable complex with NH_3 whereas Co^{2+} forms the least stable one.

$$K_{inst} = \frac{[Cu^{2+}][NH_3]^4}{[Cu(NH_3)_4^{2+}]} = \frac{1}{K_{form}}$$

Notice that K_{inst} is the reciprocal of K_{form}. The K_{inst} is called an *instability* constant because the larger its value, the more *unstable* the complex is. The data in the last column of Table 17.3 show this. The least stable complex in the table, $Co(NH_3)_6^{2+}$, has the largest value of K_{inst}.

17.5 Complex Ion Formation Increases the Solubility of a Salt

The silver halides are extremely insoluble salts, as we've learned. The K_{sp} of AgBr at 25 °C, for example, is only 5.0×10^{-13}. In a saturated aqueous solution, the concentration of each ion of AgBr is only 7.1×10^{-7} mol L^{-1}. Suppose that we start with a saturated solution in which undissolved AgBr is present and equilibrium exists. Now suppose that we begin to add aqueous ammonia to the system. Because NH_3 molecules bind strongly to silver ions, they begin to form $Ag(NH_3)_2^+$ with the trace amount of Ag^+ ions initially in solution. The reaction is

$$\textit{Complex ion equilibrium:}\quad Ag^+(aq) + 2NH_3(aq) \rightleftharpoons Ag(NH_3)_2^+(aq)$$

Because the forward reaction removes *uncomplexed* Ag^+ ions from solution, it upsets the solubility equilibrium:

$$\textit{Solubility equilibrium:}\quad AgBr(s) \rightleftharpoons Ag^+(aq) + Br^-(aq)$$

By withdrawing uncomplexed Ag^+ ions, the ammonia causes the solubility equilibrium to shift to the right to generate more Ag^+ ions from AgBr(*s*). The way the two equilibria are related can be viewed as follows.

$$AgBr(s) \rightleftharpoons Ag^+(aq) + Br^-(aq)$$
$$+$$
$$2NH_3(aq)$$
$$\Updownarrow$$
$$Ag(NH_3)_2^+(aq)$$

Addition of ammonia removes free silver ion from the solution, causing the solubility equilibrium to shift to the right.

Thus, adding ammonia causes more AgBr(*s*) to dissolve—an example of Le Châtelier's principle at work. Our example illustrates a general phenomenon:

> The solubility of a slightly soluble salt increases when one of its ions can be changed to a soluble complex ion.

To analyze what happens, let's put the two equilibria together.

$$\textit{Complex ion equilibrium:}\quad Ag^+(aq) + 2NH_3(aq) \rightleftharpoons Ag(NH_3)_2^+(aq)$$
$$\textit{Solubility equilibrium:}\quad AgBr(s) \rightleftharpoons Ag^+(aq) + Br^-(aq)$$
$$\textit{Sum of equilibria:}\quad AgBr(s) + 2NH_3(aq) \rightleftharpoons Ag(NH_3)_2^+(aq) + Br^-(aq)$$

The equilibrium constant for the net overall reaction is written in the usual way. The term for [AgBr(*s*)] is omitted because it refers to a solid and so has a constant value.

$$K_c = \frac{[Ag(NH_3)_2^+][Br^-]}{[NH_3]^2}$$

FACETS OF CHEMISTRY 17.1

No More Soap Scum—Complex Ions and Solubility

A problem that has plagued homeowners with hard water—water that contains low concentrations of divalent cations, especially Ca^{2+}—is the formation of insoluble deposits of "soap scum" as well as "hard water spots" on surfaces such as shower tiles, shower curtains, and bathtubs. These deposits form when calcium ions interact with large anions in soap to form precipitates and also when hard water that contains bicarbonate ion evaporates, causing precipitation of calcium carbonate, $CaCO_3$.

$$Ca^{2+}(aq) + 2HCO_3^-(aq) \longrightarrow CaCO_3(s) + CO_2(g) + H_2O$$

A variety of consumer products are sold that contain ingredients intended to prevent these precipitates from forming, and they accomplish this by forming complex ions with calcium ions, which has the effect of increasing the solubilities of the soap scum and calcium carbonate deposits. A principal ingredient in these products is an organic compound called *ethylenediaminetetraacetic* acid (mercifully abbreviated *EDTA*). The structure of the compound, which is also abbreviated as H_4EDTA to emphasize that it contains four acidic hydrogens which are part of carboxyl groups, is

$$(HO{-}\overset{O}{\overset{\|}{C}}{-}CH_2)_2N{-}CH_2{-}CH_2{-}N(CH_2{-}\overset{O}{\overset{\|}{C}}{-}OH)_2$$

H_4EDTA

Acidic hydrogens are shown in blue, donor atoms are shown in red.

A metal ion surrounded octahedrally by all six donor atoms if H_4EDTA loses all of the acidic hydrogens.

This molecule is an excellent complex-forming ligand; it contains a total of *six* donor atoms (in red) that can bind to a metal ion, enabling the ligand to wrap itself around the metal ion as illustrated below. (The usual colors are used to identify the various elements in the ligand.)

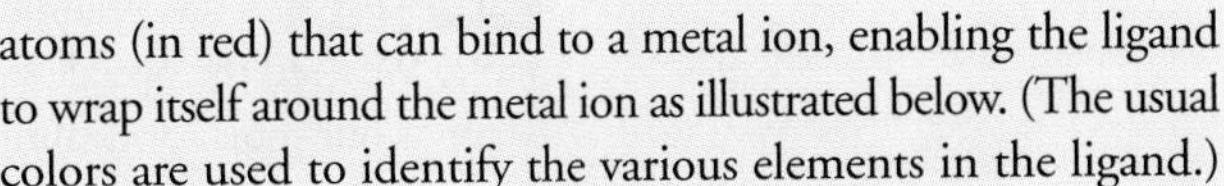

One consumer product, called Clean Shower (Figure 1), contains H_4EDTA along with substances called surfactants. When the EDTA binds with calcium ions, it releases just two H^+ ions

$$H_4EDTA(aq) + Ca^{2+}(aq) \longrightarrow CaH_2EDTA(aq) + 2H^+(aq)$$

The H^+ combine with anions of the soap to form neutral organic compounds called fatty acids that are usually not water-soluble. The surfactants in the product, however, enable the fatty acids to dissolve, preventing formation of precipitates. The Clean Shower product is sprayed on the wet walls after you take a shower, forming the products described above. The next time you take a shower, the water washes away the soluble products, keeping soap scum from building up and keeping the shower walls clean.

FIG. 1 The product Clean Shower contains EDTA, which prevents the formation of soap scum and hard water spots in showers and on bathtubs. *(Andy Washnik.)*

Because we have added two equations to obtain a third equation, K_c for this expression is the product of K_{form} and K_{sp}. The values for K_{form} for $Ag(NH_3)_2^+$ and K_{sp} for AgBr are known, so by multiplying the two, we find the overall value of K_c for the silver bromide–ammonia system.

Recall that when equilibria are added, their equilibrium constants are multiplied.

$$K_c = (1.6 \times 10^7)(5.0 \times 10^{-13})$$
$$= 8.0 \times 10^{-6}$$

This approach thus gives us a way to calculate the solubility of a sparingly soluble salt when a substance able to form a complex with the metal ion is put into its solution. The next example shows how this works.

EXAMPLE 17.11
Calculating the Solubility of a Slightly Soluble Salt in the Presence of a Ligand

How many moles of AgBr can dissolve in 1.0 L of 1.0 *M* NH_3?

ANALYSIS: A few preliminaries have to be done before we can take advantage of a concentration table. We need the overall equation and its associated equation for the equilibrium constant, which serve as our tools for dealing with these kinds of problems. The overall equilibrium is

$$AgBr(s) + 2NH_3(aq) \rightleftharpoons Ag(NH_3)_2^+(aq) + Br^-(aq)$$

The equation for K_c is

$$K_c = \frac{[Ag(NH_3)_2^+][Br^-]}{[NH_3]^2} = 8.0 \times 10^{-6} \quad \text{(as calculated earlier)}$$

Now let's prepare the concentration table. To do this, we imagine we are adding solid AgBr to the ammonia solution.

Before any reaction takes place, the concentration of NH_3 is 1.0 *M* and the concentration of $Ag(NH_3)_2^+$ is zero. The concentration of Br^- is also zero, because there is none of it in the ammonia solution.

If we define x as the molar solubility of the AgBr in the solution, then the concentrations of $Ag(NH_3)_2^+$ and Br^- both increase by x and the concentration of ammonia decreases by $2x$ (because of the coefficient of NH_3 in the equation).

The equilibrium values are obtained as usual by adding the initial and change rows. However, another comment is in order here. Letting the concentration of Br^- equal that of $Ag(NH_3)_2^+$ is valid because *and only because* K_{form} is such a large number. Essentially *all* Ag^+ ions that do dissolve from the insoluble AgBr are changed to the complex ion. There are relatively few uncomplexed Ag^+ ions in the solution, so the number of $Ag(NH_3)_2^+$ and Br^- ions in the solution are very nearly the same.

SOLUTION: Here is the concentration table, built using the reasoning above.

	$AgBr(s) + 2NH_3(aq)$	$\rightleftharpoons Ag(NH_3)_2^+(aq)$	$+ Br^-(aq)$
Initial concentrations (*M*)	1.0	0	0
Changes in concentrations caused by NH_3 (*M*)	$-2x$	$+x$	$+x$
Equilibrium concentrations (*M*)	$(1.0 - 2x)$	x	x

We've done all the hard work. Now all we need to do is substitute the values in the last row of the concentration table into the equation for K_c:

$$K_c = \frac{(x)(x)}{(1.0 - 2x)^2} = 8.0 \times 10^{-6}$$

This can be solved by taking the square root of both sides, which gives

$$\frac{x}{(1.0 - 2x)} = \sqrt{8.0 \times 10^{-6}} = 2.8 \times 10^{-3}$$

Solving for x and rounding to the correct number of significant figures gives

$$x = 2.8 \times 10^{-3}$$

Because we've defined x as the molar solubility of AgBr in the ammonia solution, we can say that 2.8×10^{-3} mol of AgBr dissolves in 1.0 L of 1.0 *M* NH_3. (This is not very much, of course, but in contrast, only 7.1×10^{-7} mol of AgBr dissolves in 1.0 L of pure water. Thus AgBr is nearly 4000 times more soluble in the 1.0 *M* NH_3 than in pure water.)

IS THE ANSWER REASONABLE? Everything depends on the reasoning used in preparing the concentration table, particularly letting $-2x$ represent the change in the concentration of NH_3 as a result of the presence of AgBr and the formation of the complex. Additionally, we can see that the maximum solubility of AgBr if all the NH_3 were complexed is 0.5 *M* while the minimum solubility if the NH_3 had no effect is the solubility of AgBr in distilled water (7×10^{-7} *M*). Our result is between these two limits, so the answer is reasonable.

Practice Exercise 20: Calculate the solubility of silver chloride in 0.10 *M* NH_3 and compare it with its solubility in pure water. Refer to Table 17.1 for the K_{sp} for AgCl. [Hint: Assume all the Ag^+ from the AgCl becomes incorporated into $Ag(NH_3)_2^+$.]

Practice Exercise 21: How many moles of NH_3 have to be added to 1.0 L of water to dissolve 0.20 mol of AgCl? The complex ion $Ag(NH_3)_2^+$ forms.

SUMMARY

Solubility Equilibria for Salts. The mass action expression for a solubility equilibrium of a salt is called the **ion product.** It equals the product of the molar concentrations of its ions, each raised to a power equal to the subscript of the ion in the salt's formula. At a given temperature, the value of the ion product, Q, in a *saturated* solution of the salt equals a constant called the **solubility product constant,** or K_{sp}.

For solubility equilibria, the *common ion effect* is the ability of an ion of a soluble salt to suppress the solubility of a sparingly soluble compound that has the same (the "common") ion.

If soluble salts are mixed together in the same solution, a cation of one and an anion of another will precipitate if the ion product, Q, exceeds the solubility product constant of the salt formed from them.

Selective Precipitation. Metal sulfides are vastly more soluble in an acidic solution than in water. To express their solubility equilibria in acid, the **acid solubility constant** or K_{spa} is used. Several metal sulfides have such low values of K_{spa} that they are insoluble even at low pH. These **acid-insoluble sulfides** can thus be selectively separated from the **base-insoluble sulfides** by making the solution both quite acidic as well as saturated in H_2S. By adjusting the pH, salts of other weak acids, like the carbonate salts, can also be selectively precipitated.

Complex Ions of Metals. Coordination compounds contain **complex ions** (also called **complexes** or **coordination complexes**), formed from a metal ion and a number of ligands. **Ligands** are Lewis bases and may be electrically neutral or negatively charged. Water and ammonia are common neutral ligands. The equilibrium constant for the formation of a complex in the presence of an excess of ligand is called the **formation constant,** K_{form}, of the complex. The larger the value of K_{form}, the more stable is the complex. Salts whose cations form stable complexes, like Ag^+ salts whose cation forms a stable complex with ammonia, $Ag(NH_3)_2^+$, are made more soluble when the ligand is present.

TOOLS FOR PROBLEM SOLVING

In this chapter you learned to apply the following concepts as tools in solving problems dealing with aspects of the solubility of salts. Study each tool carefully so that you know what each is used for. When faced with solving a problem, recall what each tool does and consider whether it will be helpful in finding a solution. This will aid you in selecting the tools you need.

Solubility product constant, K_{sp} *(page 693)* We use K_{sp} to calculate the molar solubility of a salt either in pure water or in a solution that contains a common ion. Comparing K_{sp} with the ion product of a potential precipitating salt permits us to decide whether a precipitate will form.

Molar solubility *(page 695)* This is used for calculating the value of K_{sp} for a salt. When calculating the solubility of a salt from K_{sp}, we let x represent the molar solubility in the concentration table.

Ion product, Q, of a salt *(page 702)* Compare Q with K_{sp} to determine whether a precipitate will form:

$Q > K_{sp}$ a precipitate forms

$Q \leq K_{sp}$ a precipitate does not form

Acid solubility product constant, K_{spa} *(page 706)* Use K_{spa} data to calculate the solubility of a metal sulfide at a given pH. We can use K_{spa} data for two or more metal sulfides to calculate the pH at which one will selectively precipitate from a solution saturated in H_2S.

Combined K_a expression for a diprotic acid *(page 710)* Combining expressions for K_{a_1} and K_{a_2} for a diprotic acid, H_2A, yields the equation

$$\frac{[H^+]^2[A^{2-}]}{[H_2A]} = K_a = K_{a_1} \times K_{a_2}$$

This equation can be used if (and only if) we know two of the three quantities in the combined mass action expression. This is a useful tool when working problems involving selective precipitation of salts of diprotic acids where the concentration of the anion A^{2-} is controlled by adjusting the pH of the solution.

Formation constants (stability constants) of complexes *(page 713)* We can use K_{form} values to make judgments concerning the relative stabilities of complexes. Along with K_{sp} values, we can use formation constants of complexes to determine the solubility of a sparingly soluble salt in a solution containing a ligand that's able to form a complex with the cation of the salt.

QUESTIONS, PROBLEMS, AND EXERCISES

Answers to problems whose numbers are printed in color are given in Appendix B. More challenging problems are marked with asterisks. ILW = Interactive Learningware solution is available at www.wiley.com/college/brady. OH = an Office Hours video is available for this problem.

REVIEW QUESTIONS

Solubility Products

17.1 What is the difference between an *ion product* and an *ion product constant*?

17.2 Use the equilibrium below to demonstrate why the K_{sp} expression does not include the concentration of $Ba_3(PO_4)_2$ in the denominator.

$$Ba_3(PO_4)_2(s) \rightleftharpoons 3Ba^{2+}(aq) + 2PO_4^{3-}(aq)$$

17.3 What is the *common ion effect*? How does Le Châtelier's principle explain it? Use the solubility equilibrium for AgCl to illustrate the common ion effect.

17.4 If sodium acetate is added to a solution of acetic acid, the pH increases. Explain how this is an example of the common ion effect.

17.5 With respect to K_{sp}, what conditions must be met if a precipitate is going to form in a solution?

17.6 What limits the accuracy and reliability of solubility calculations based on K_{sp} values?

Selective Precipitations

17.7 Potassium oxide is readily soluble in water, but the resulting solution contains essentially no oxide ion. Explain, using an equation, what happens to the oxide ion.

17.8 What chemical reaction takes place when solid sodium sulfide is dissolved in water? Write the chemical equation.

17.9 Consider cobalt(II) sulfide.
(a) Write its solubility equilibrium and K_{sp} equation for a saturated solution in water. (Remember, there is no free sulfide ion in the solution.)
(b) Write its solubility equilibrium and K_{spa} equation for a saturated solution in aqueous acid.

17.10 Use Le Châtelier's principle to explain how adjusting the pH enables the control of the concentration of $C_2O_4^{2-}$ in a solution of oxalic acid, $H_2C_2O_4$.

17.11 Suppose you wished to control the PO_4^{3-} concentration in a solution of phosphoric acid by controlling the pH of the solution. If you assume you know the H_3PO_4 concentration, what combined equation would be useful for that purpose?

Simultaneous and Complex Ion Equilibria

17.12 A solution of $MgBr_2$ can be changed to a solution of $MgCl_2$ by adding AgCl(*s*) and stirring the mixture well. In terms of the equilibria involved, explain how this happens.

17.13 On the basis of Le Châtelier's principle, explain how the addition of solid NH_4Cl to a beaker containing solid $Mg(OH)_2$ in contact with water is able to cause the $Mg(OH)_2$ to dissolve. Write equations for *all* the chemical equilibria that exist in the solution after the addition of the NH_4Cl.

17.14 Using Le Châtelier's principle, explain how the addition of aqueous ammonia dissolves silver chloride. If HNO_3 is added after the AgCl has dissolved in the NH_3 solution, it causes AgCl to re-precipitate. Explain why.

17.15 For $PbCl_3^-$, $K_{form} = 2.5 \times 10^1$. If a solution containing this complex ion is diluted with water, $PbCl_2$ precipitates. Write the equations for the equilibria involved and use them together with Le Châtelier's principle to explain how this happens.

REVIEW PROBLEMS

Solubility Products

17.16 Write the K_{sp} expressions for each of the following compounds.
(a) CaF_2
(b) Ag_2CO_3
(c) $PbSO_4$
(d) $Fe(OH)_3$
(e) PbI_2
(f) $Cu(OH)_2$

17.17 Write the K_{sp} expressions for each of the following compounds.
(a) AgI
(b) Ag_3PO_4
(c) $PbCrO_4$
(d) $Al(OH)_3$
(e) $ZnCO_3$
(f) $Zn(OH)_2$

Determining K_{sp}

17.18 In water, the solubility of lead(II) chloride is 0.016 *M*. Use that information to calculate the value of K_{sp} for $PbCl_2$.

OH 17.19 A student evaporated 100.0 mL of a saturated BaF_2 solution and found the solid BaF_2 she recovered weighed 0.132 g. From those data, calculate K_{sp} for BaF_2.

17.20 Barium sulfate is so insoluble that it can be swallowed without significant danger, even though Ba^{2+} is toxic. At 25 °C, 1.00 L of water dissolves only 0.00245 g of $BaSO_4$. Calculate K_{sp} for $BaSO_4$.

17.21 A student found that a maximum of 0.800 g silver acetate is able to dissolve in 100.0 mL of water. What is the molar solubility and the K_{sp} for the salt?

17.22 It was found that the molar solubility of $BaSO_3$ in 0.10 *M* $BaCl_2$ is 8.0×10^{-6} *M*. What is the value of K_{sp} for $BaSO_3$?

17.23 A student prepared a saturated solution of $CaCrO_4$ and found that when 156 mL of the solution was evaporated, 0.649 g of $CaCrO_4$ was left behind. What is the value of K_{sp} for the salt?

17.24 At 25 °C, the molar solubility of silver phosphate is 1.8×10^{-5} mol L^{-1}. Calculate K_{sp} for the salt.

17.25 The molar solubility of barium phosphate in water at 25 °C is 1.4×10^{-8} mol L^{-1}. What is the value of K_{sp} for the salt?

Using K_{sp} to Calculate Solubilities

17.26 What is the molar solubility of $PbBr_2$ in water?

17.27 What is the molar solubility of Ag_2CrO_4 in water?

17.28 Calculate the molar solubility of Ag_2CO_3 in water. (Ignore the reaction of the CO_3^{2-} ion with water.)

17.29 Calculate the molar solubility of PbF_2 in water.

17.30 At 25 °C, the value of K_{sp} for LiF is 1.7×10^{-3}, and that for BaF_2 is 1.7×10^{-6}. Which salt, LiF or BaF_2, has the larger molar solubility in water? Calculate the solubility of each in units of mol L^{-1}.

17.31 At 25 °C, the value of K_{sp} for AgCN is 2.2×10^{-16} and that for $Zn(CN)_2$ is 3×10^{-16}. In terms of grams per 100 mL of solution, which salt is the more soluble in water?

17.32 A salt whose formula is *MX* has a K_{sp} equal to 3.2×10^{-10}. Another sparingly soluble salt, MX_3, must have what value of K_{sp} if the molar solubilities of the two salts are to be identical?

17.33 A salt having a formula of the type M_2X_3 has $K_{sp} = 2.2 \times 10^{-20}$. Another salt, M_2X, has to have what K_{sp} value if M_2X has twice the molar solubility of M_2X_3?

17.34 Calcium sulfate is found in plaster. At 25 °C the value of K_{sp} for $CaSO_4$ is 2.4×10^{-5}. What is the calculated molar solubility of $CaSO_4$ in water?

17.35 Chalk is $CaCO_3$, and at 25 °C its $K_{sp} = 4.5 \times 10^{-9}$. What is the molar solubility of $CaCO_3$? How many grams of $CaCO_3$ dissolve in 100 mL of aqueous solution? (Ignore the reaction of CO_3^{2-} with water.)

Common Ion Effect

ILW 17.36 Copper(I) chloride has $K_{sp} = 1.9 \times 10^{-7}$. Calculate the molar solubility of copper(I) chloride in (a) pure water, (b) 0.0200 *M* HCl solution, (c) 0.200 *M* HCl solution, and (d) 0.150 *M* $CaCl_2$ solution.

17.37 Gold(III) chloride has $K_{sp} = 3.2 \times 10^{-25}$. Calculate the molar solubility of gold(III) chloride in (a) pure water, (b) 0.010 *M* HCl solution, (c) 0.010 *M* $MgCl_2$ solution, and (d) 0.010 *M* $Au(NO_3)_3$ solution.

17.38 Calculate the molar solubility of $Mg(OH)_2$ in a solution that is basic with a pH of 12.50.

17.39 Calculate the molar solubility of $Al(OH)_3$ in a solution that is slightly basic with a pH of 9.50.

17.40 What is the highest concentration of Pb^{2+} that can exist in a solution of 0.10 *M* HCl?

17.41 Will lead(II) bromide be less soluble in 0.10 *M* $Pb(C_2H_3O_2)_2$ or in 0.10 *M* NaBr?

17.42 Calculate the molar solubility of Ag_2CrO_4 at 25 °C in (a) 0.200 *M* $AgNO_3$ and (b) 0.200 *M* Na_2CrO_4. For Ag_2CrO_4 at 25 °C, $K_{sp} = 1.2 \times 10^{-12}$.

17.43 What is the molar solubility of $Mg(OH)_2$ in (a) 0.20 *M* NaOH and (b) 0.20 *M* $MgSO_4$? For $Mg(OH)_2$, $K_{sp} = 7.1 \times 10^{-12}$.

17.44 How much will the percent ionization of the acetic acid change in 0.500 L of 0.10 *M* $HC_2H_3O_2$ if 0.050 mol of solid $NaC_2H_3O_2$ is added? (Assume no change in the volume of the solution.) How much will the pH change?

17.45 How much will the percent ionization of the acetic acid change in 0.500 L of 0.10 *M* $HC_2H_3O_2$ if 0.025 mol of gaseous HCl is dissolved in the solution? (Assume no change in volume.) How will the pH of the solution change?

*17.46 In an experiment 2.20 g of NaOH(*s*) is added to 250 mL of 0.10 *M* $FeCl_2$ solution. What mass of $Fe(OH)_2$ will be formed? What will the molar concentration of Fe^{2+} be in the final solution?

*17.47 Suppose that 1.75 g of NaOH(*s*) is added to 250 mL of 0.10 *M* $NiCl_2$ solution. What mass, in grams, of $Ni(OH)_2$ will be formed? What will be the pH of the final solution? For $Ni(OH)_2$, $K_{sp} = 6 \times 10^{-16}$.

17.48 What is the molar solubility of $Fe(OH)_2$ in a buffer that has a pH of 9.50?

17.49 What is the molar solubility of $Ca(OH)_2$ in (a) 0.10 *M* $CaCl_2$ and (b) 0.10 *M* NaOH?

Precipitation

17.50 Does a precipitate of $PbCl_2$ form when 0.0150 mol of $Pb(NO_3)_2$ and 0.0120 mol of NaCl are dissolved in 1.00 L of solution?

17.51 Silver acetate, $AgC_2H_3O_2$, has $K_{sp} = 2.3 \times 10^{-3}$. Does a precipitate form when 0.015 mol of $AgNO_3$ and 0.25 mol of $Ca(C_2H_3O_2)_2$ are dissolved in a total volume of 1.00 L of solution?

ILW **17.52** Does a precipitate of $PbBr_2$ form if 50.0 mL of 0.0100 *M* $Pb(NO_3)_2$ is mixed with (a) 50.0 mL of 0.0100 *M* KBr and (b) 50.0 mL of 0.100 *M* NaBr?

17.53 Would a precipitate of silver acetate form if 22.0 mL of 0.100 *M* $AgNO_3$ were added to 45.0 mL of 0.0260 *M* $NaC_2H_3O_2$? For $AgC_2H_3O_2$, $K_{sp} = 2.3 \times 10^{-3}$.

17.54 Both AgCl and AgI are very sparingly soluble salts, but the solubility of AgI is much less than that of AgCl, as can be seen by their K_{sp} values. Suppose that a solution contains both Cl^- and I^- with $[Cl^-] = 0.050\ M$ and $[I^-] = 0.050\ M$. If solid $AgNO_3$ is added to 1.00 L of this mixture (so that no appreciable change in volume occurs), what is the value of $[I^-]$ when AgCl first begins to precipitate?

17.55 Suppose that Na_2SO_4 is added gradually to 100.0 mL of a solution that contains both Ca^{2+} ion (0.15 *M*) and Sr^{2+} ion (0.15 *M*). (a) What will the Sr^{2+} concentration be (in mol L^{-1}) when $CaSO_4$ just begins to precipitate? (b) What percentage of the strontium ion has precipitated when $CaSO_4$ just begins to precipitate?

17.56 Suppose 50.0 mL each of 0.0100 *M* solutions of NaBr and $Pb(NO_3)_2$ are poured together. Does a precipitate form? If so, calculate the molar concentrations of the ions at equilibrium.

17.57 If a solution of 0.10 *M* Mn^{2+} and 0.10 *M* Cd^{2+} is gradually made basic, what will the concentration of Cd^{2+} be when $Mn(OH)_2$ just begins to precipitate? Assume no change in the volume of the solution.

Selective Precipitation

ILW OH **17.58** What value of $[H^+]$ and what pH permits the selective precipitation of the sulfide of just one of the two metal ions in a solution that has a concentration of 0.010 *M* Pb^{2+} and 0.010 *M* Co^{2+}?

17.59 What pH would yield the maximum separation Mn^{2+} from Sn^{2+} in a solution that is 0.010 *M* in Mn^{2+}, 0.010 *M* in Sn^{2+}, and saturated in H_2S? (Assume the green form of MnS in Table 17.2.)

17.60 What range of pH would permit the selective precipitation of Cu^{2+} as $Cu(OH)_2$ from a solution that contains 0.10 *M* Cu^{2+} and 0.10 *M* Mn^{2+}? For $Mn(OH)_2$, $K_{sp} = 1.6 \times 10^{-13}$ and for $Cu(OH)_2$, $K_{sp} = 4.8 \times 10^{-20}$.

17.61 Kidney stones often contain insoluble calcium oxalate, CaC_2O_4, which has $K_{sp} = 2.3 \times 10^{-9}$. Calcium oxalate is considerably less soluble than magnesium oxalate, MgC_2O_4, which has $K_{sp} = 8.6 \times 10^{-5}$. Suppose a solution contained both Ca^{2+} and Mg^{2+} at a concentration of 0.10 *M*. What pH would be required to achieve maximum separation of these ions by precipitation of CaC_2O_4 if the solution also contains oxalic acid, $H_2C_2O_4$ at a concentration of 0.10 *M*? For $H_2C_2O_4$, $K_{a_1} = 5.6 \times 10^{-2}$ and $K_{a_2} = 5.4 \times 10^{-5}$.

Complex Ion Equilibria

17.62 Write the chemical equilibria and equilibrium laws that correspond to K_{form} for the following complexes:
(a) $CuCl_4^{2-}$
(b) AgI_2^-
(c) $Cr(NH_3)_6^{3+}$

17.63 Write the chemical equilibria and equilibrium laws that correspond to K_{form} for the following complexes:
(a) $Ag(S_2O_3)_2^{3-}$
(b) $Zn(NH_3)_4^{2+}$
(c) SnS_3^{2-}

OH **17.64** Write equilibria that correspond to K_{form} for each of the following complex ions and write the equations for K_{form}.
(a) $Co(NH_3)_6^{3+}$ (b) HgI_4^{2-} (c) $Fe(CN)_6^{4-}$

17.65 Write the equilibria that are associated with the equations for K_{form} for each of the following complex ions. Write also the equations for the K_{form} of each.
(a) $Hg(NH_3)_4^{2+}$ (b) SnF_6^{2-} (c) $Fe(CN)_6^{3-}$

17.66 For $PbCl_3^-$, $K_{form} = 2.5 \times 10^1$. Use this information plus the K_{sp} for $PbCl_2$ to calculate K_c for the reaction

$$PbCl_2(s) + Cl^-(aq) \rightleftharpoons PbCl_3^-(aq)$$

17.67 The overall formation constant for $Ag(CN)_2^-$ equals 5.3×10^{18}, and the K_{sp} for AgCN equals 1.2×10^{-16}. Calculate K_c for the reaction

$$AgCN(s) + CN^-(aq) \rightleftharpoons Ag(CN)_2^-(aq)$$

17.68 How many grams of solid NaCN have to be added to 1.2 L of water to dissolve 0.11 mol of $Fe(OH)_3$ in the form of $Fe(CN)_6^{3-}$? Use data as needed from Tables 17.1 and 17.3. (For simplicity, ignore the reaction of CN^- ion with water.)

17.69 In photography, unexposed silver bromide is removed from film by soaking the film in a solution of sodium thiosulfate, $Na_2S_2O_3$. Silver ion forms a soluble complex with thiosulfate ion, $S_2O_3^{2-}$, that has the formula $Ag(S_2O_3)_2^{3-}$, and formation of the complex causes the AgBr in the film to dissolve. The $Ag(S_2O_3)_2^{3-}$ complex has $K_{form} = 2.0 \times 10^{13}$. How many grams of AgBr ($K_{sp} = 5.0 \times 10^{-13}$) will dissolve in 125 mL of 1.20 *M* $Na_2S_2O_3$ solution?

17.70 Silver iodide is very insoluble and can be difficult to remove from glass apparatus, but it forms a relatively stable complex ion, AgI_2^- ($K_{form} = 1 \times 10^{11}$), that makes AgI fairly soluble in a solution containing I^-. When a solution containing the AgI_2^- ion is diluted with water, AgI precipitates. Explain why this happens in terms of the equilibria that are involved. How many grams of AgI will dissolve in 100 mL of 1.0 *M* KI solution?

17.71 Silver forms a sparingly soluble cyanide salt, AgCN, for which $K_{sp} = 1.2 \times 10^{-16}$. It also forms a soluble cyanide complex, $Ag(CN)_2^-$, for which $K_{form} = 5.3 \times 10^{18}$. How many grams of solid KCN must be added to 100 mL of water to dissolve 0.020 mol AgCN? For simplicity, ignore the reaction of CN^- with water. Be sure to include all the cyanide that's added to the solution.

17.72 The formation constant for $Ag(CN)_2^-$ equals 5.3×10^{18}. Use the data in Table 17.1 to determine the molar solubility of AgI in 0.010 *M* KCN solution.

17.73 Suppose that some dipositive cation, M^{2+}, is able to form a complex ion with a ligand, *L*, by the following equation.

$$M^{2+} + 2L \rightleftharpoons M(L)_2^{2+}$$

The cation also forms a sparingly soluble salt, MCl_2. In which of the following circumstances would a given quantity of ligand be more able to bring larger quantities of the salt into solution? Explain and justify the calculation involved.

(a) $K_{form} = 1 \times 10^2$ and $K_{sp} = 1 \times 10^{-15}$

(b) $K_{form} = 1 \times 10^{10}$ and $K_{sp} = 1 \times 10^{-20}$

17.74 The molar solubility of $Zn(OH)_2$ in 1.0 *M* NH_3 is 5.7×10^{-3} mol L^{-1}. Determine the value of the instability constant of the complex ion, $Zn(NH_3)_4^{2+}$. Ignore the reaction, $NH_3 + H_2O \rightleftharpoons NH_4^+ + OH^-$.

OH **17.75** Calculate the molar solubility of $Cu(OH)_2$ in 2.0 *M* NH_3. (For simplicity, ignore the reaction of NH_3 as a base.)

ADDITIONAL EXERCISES

***17.76** How many milliliters of 0.10 *M* HCl would have to be added to 100 mL of a saturated solution of $PbCl_2$ in contact with 50.0 g of solid $PbCl_2$ to reduce the Pb^{2+} concentration to 0.0050 *M*? (Don't forget to take into account the combined volumes of the two solutions.)

17.77 Magnesium hydroxide, $Mg(OH)_2$, found in milk of magnesia, has a solubility of 7.05×10^{-3} g L^{-1} at 25 °C. Calculate K_{sp} for $Mg(OH)_2$.

17.78 Does iron(II) sulfide dissolve in 8 *M* HCl? Perform the calculations that prove your answer.

17.79 As noted earlier, milk of magnesia is an aqueous suspension of $Mg(OH)_2$. If we assume that besides the $Mg(OH)_2$ the only other component is water, use K_{sp} to estimate the pH of milk of magnesia.

***17.80** Suppose that 25.0 mL of 0.10 *M* HCl is added to 1.000 L of saturated $Mg(OH)_2$ in contact with more than enough $Mg(OH)_2(s)$ to react with all the HCl. After reaction has ceased, what will the molar concentration of Mg^{2+} be? What will the pH of the solution be?

OH ***17.81** Solid $Mn(OH)_2$ is added to a solution of 0.100 *M* $FeCl_2$. After reaction, what will be the molar concentrations of Mn^{2+} and Fe^{2+} in the solution? What will be the pH of the solution? For $Mn(OH)_2$, $K_{sp} = 1.6 \times 10^{-13}$.

***17.82** Suppose that 50.0 mL of 0.12 *M* $AgNO_3$ is added to 50.0 mL of 0.048 *M* NaCl solution. (a) What mass of AgCl would form? (b) Calculate the final concentrations of all of the ions in the solution that is in contact with the precipitate. (c) What percentage of the Ag^+ ions have precipitated?

***17.83** A sample of hard water was found to have 278 ppm Ca^{2+} ion. Into 1.00 L of this water, 1.00 g of Na_2CO_3 was dissolved. What is the new concentration of Ca^{2+} in parts per million? (Assume that the addition of Na_2CO_3 does not change the volume, and assume that the density of the aqueous solutions involved are all 1.00 g mL^{-1}.)

***17.84** What value of $[H^+]$ and what pH would allow the selective separation of the carbonate of just one of the two metal ions in a solution that has a concentration of 0.010 *M* La^{3+} and 0.010 *M* Pb^{2+}? For $La_2(CO_3)_3$, $K_{sp} = 4.0 \times 10^{-34}$; for $PbCO_3$, $K_{sp} = 7.4 \times 10^{-14}$. A saturated solution of CO_2 in water has a concentration of H_2CO_3 equal to 3.3×10^{-2} *M*.

***17.85** When solid NH_4Cl is added to a suspension of $Mg(OH)_2(s)$, some of the $Mg(OH)_2$ dissolves.

(a) Write equations for *all* the chemical equilibria that exist in the solution after the addition of the NH_4Cl.

(b) How many moles of NH_4Cl must be added to 1.0 L of a suspension of $Mg(OH)_2$ to dissolve 0.10 mol of $Mg(OH)_2$?

(c) What is the pH of the solution after the 0.10 mol of $Mg(OH)_2$ has dissolved in the solution containing the NH_4Cl?

***17.86** After solid $CaCO_3$ was added to a slightly basic solution, the pH was measured to be 8.50. What was the molar solubility of $CaCO_3$ in the solution?

17.87 In modern construction, walls and ceilings are constructed of "drywall," which consists of plaster sandwiched between sheets of heavy paper. Plaster is composed of calcium sulfate, $CaSO_4 \cdot 2H_2O$. Suppose you had a leak in a water pipe that was dripping water on a drywall ceiling 1/2 in. thick at a rate of 2.00 L per day. Use the K_{sp} of calcium sulfate to estimate how many days it would take to dissolve a hole 1.0 cm in diameter. Assume the density of the plaster is 0.97 g cm^{-3}.

17.88 What is the molar solubility of $Fe(OH)_3$ in water? (Hint: Don't forget the self-ionization of water.)

***17.89** In Example 17.11 (page 716) we say, "There are relatively few uncomplexed Ag^+ ions in the solution." Calculate the molar concentration of Ag^+ ion actually left after the complex forms as described in Example 17.11.

***17.90** What are the concentrations of Pb^{2+}, Br^-, and I^- in an aqueous solution that's in contact with both PbI_2 and $PbBr_2$?

***17.91** Will a precipitate form in a solution made by dissolving 1.0 mol of $AgNO_3$ and 1.0 mol $HC_2H_3O_2$ in 1.0 L of solution? For $AgC_2H_3O_2$, $K_{sp} = 2.3 \times 10^{-3}$ and for $HC_2H_3O_2$, $K_a = 1.8 \times 10^{-5}$.

***17.92** How many grams of solid sodium acetate must be added to 0.200 L of a solution containing 0.200 *M* $AgNO_3$ and 0.10 *M* nitric acid to cause silver acetate to begin to precipitate? For $HC_2H_3O_2$, $K_a = 1.8 \times 10^{-5}$ and for $AgC_2H_3O_2$, $K_{sp} = 2.3 \times 10^{-3}$.

***17.93** How many grams of solid potassium fluoride must be added to 200 mL of a solution that contains 0.20 *M* $AgNO_3$ and 0.10 *M* acetic acid to cause silver acetate to begin to precipitate? For HF, $K_a = 6.8 \times 10^{-4}$; for $HC_2H_3O_2$, $K_a = 1.8 \times 10^{-5}$; for $AgC_2H_3O_2$, $K_{sp} = 2.3 \times 10^{-3}$.

***17.94** What is the molar solubility of $Mg(OH)_2$ in 0.10 *M* NH_3 solution? Remember that NH_3 is a weak base.

***17.95** If 100 mL of 2.0 *M* NH_3 is added to 400 mL of a solution containing 0.10 *M* Mn^{2+} and 0.10 *M* Sn^{2+}, what minimum number of grams of HCl would have to be added to the mixture to prevent $Mn(OH)_2$ from precipitating? For $Mn(OH)_2$, $K_{sp} = 1.6 \times 10^{-13}$. Assume that virtually all the tin is precipitated as $Sn(OH)_2$ by the reaction

$$Sn^{2+}(aq) + 2NH_3(aq) + 2H_2O \longrightarrow Sn(OH)_2(s) + 2NH_4^+(aq)$$

***17.96** What is the molar concentration of Cu^{2+} ion in a solution prepared by mixing 0.50 mol of NH_3 and 0.050 mol of $CuSO_4$ in 1.00 L of solution? For NH_3, $K_b = 1.8 \times 10^{-5}$; for $Cu(OH)_2$, $K_{sp} = 4.8 \times 10^{-20}$; and for $Cu(NH_3)_4^{2+}$, $K_{form} = 1.1 \times 10^{13}$.

***17.97** On the basis of the K_{sp} of $Al(OH)_3$, what would be the pH of a mixture consisting of solid $Al(OH)_3$ mixed with pure water? (Assume 100% dissociation of the aluminum hydroxide.)

17.98 You are given a sample containing NaCl(*aq*) and NaBr(*aq*), both with concentrations of 0.020 *M*. Some of the figures below represent a series of snapshots of what molecular level views of the sample would show as a 0.200 *M* $Pb(NO_3)_2(aq)$ solution is slowly added. In these figures, Br^- is red-brown and Cl^- is green. Also, spectator ions are not shown, nor are ions whose concentrations are much less than the other ions. The K_{sp} constants for $PbCl_2$ and $PbBr_2$ are 1.7×10^{-5} and 2.1×10^{-6}, respectively.

(1) Arrange the figures in a time sequence to show what happens as the lead nitrate solution is added, excluding any that do not "make sense."

(2) Explain why you excluded any figures that did not belong in the observed time sequence.

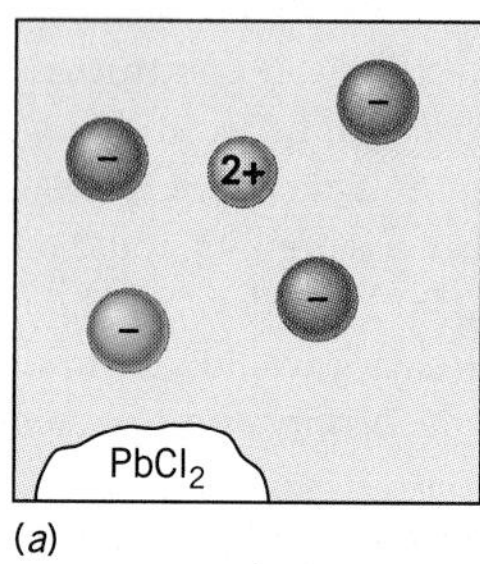

(a)

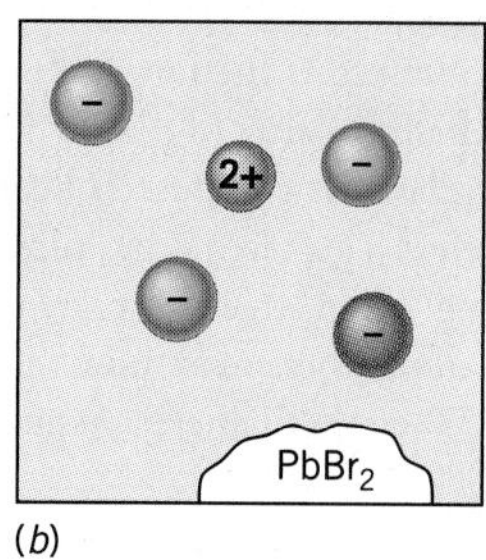

(b)

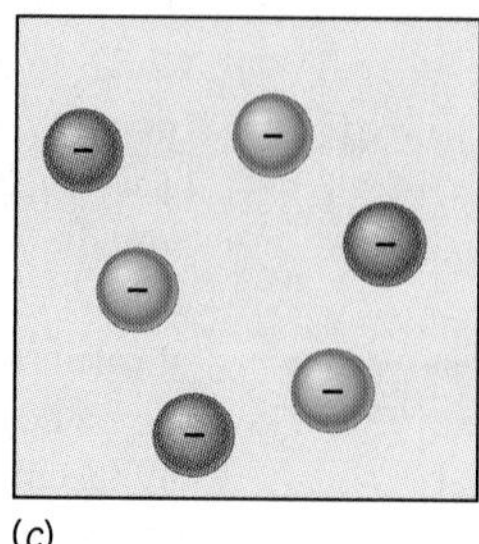
(c)

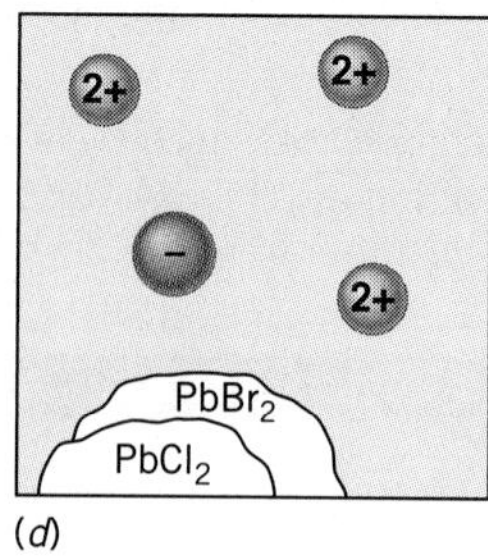

(d)

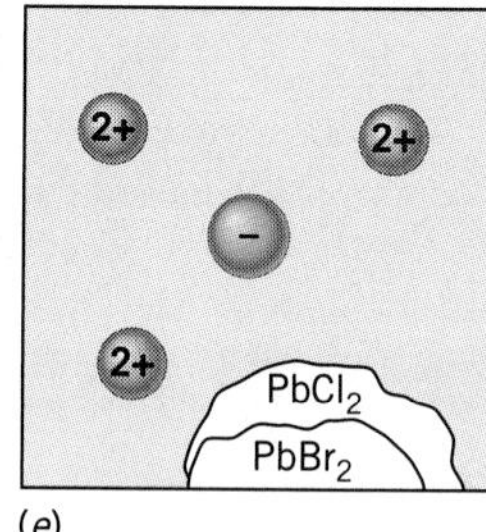

(e)

EXERCISES IN CRITICAL THINKING

17.99 Consider mercury(II) sulfide, HgS, which has a solubility product of 2×10^{-53}. Suppose some solid HgS was added to 1.0 L of water. How many ions of Hg^{2+} and S^{2-} are present in the water when equilibrium is reached? If your answer is accurate, is there a true equilibrium between HgS(*s*) and the ions in the solution? Explain your answer.

17.100 If aqueous ammonia is added gradually to a solution of copper sulfate, a pale blue precipitate forms that then dissolves to give a deep blue solution. Describe the chemical reactions that take place during these changes.

17.101 From a practical standpoint, can you effectively separate Pb^{2+} and Sr^{2+} ions by selective precipitation of their sulfates? Support your conclusions using calculations.

17.102 A salt with the formula of MX_2 is slightly soluble in water. Estimate the minimum reliable mass you can determine on your laboratory balance. Use that estimate to determine the smallest value of the solubility product of MX_2 that could be determined by evaporating a liter of a solution that is saturated in MX_2.

17.103 In older textbooks the solubility equilibrium for lead(II) sulfide was written as $PbS(s) \rightleftharpoons Pb^{2+}(aq) + S^{2-}(aq)$ with $K_{sp} = 3.0 \times 10^{-28}$. Calculate the solubility of lead sulfide using K_{sp} (assuming no reaction of S^{2-} with water) and with K_{spa}. Is there a difference between the two answers? Discuss the results of your calculations.

17.104 Suppose two silver wires, one coated with silver chloride and the other coated with silver bromide, are placed in a beaker containing pure water. Over time, what if anything will happen to the compositions of the coatings on the two wires? Justify your answer.

CHAPTERS 13–17

BRINGING IT TOGETHER

Once again you have an opportunity to test your understanding of concepts, your knowledge of scientific terms, and your skills at solving chemistry problems. Read through the following questions carefully, and answer each as fully as possible. Review topics when necessary. When you are able to answer these questions correctly, you are ready to go on to the next group of chapters.

1. List the factors that affect the speed of a chemical reaction.

2. At 25° C and $[OH^-] = 1.00\ M$, the reaction

$$I^-(aq) + OCl^-(aq) \longrightarrow OI^-(aq) + Cl^-(aq)$$

has the following rate law

$$\text{Rate} = (0.60\ \text{L mol}^{-1}\ \text{s}^{-1})[I^-][OCl^-]$$

Calculate the rate of the reaction when $[OH^-] = 1.00\ M$ and

(a) $[I^-] = 0.0100\ M$ and $[OCl^-] = 0.0200\ M$

(b) $[I^-] = 0.100\ M$ and $[OCl^-] = 0.0400\ M$

3. A reaction has the stoichiometry: $3A + B \longrightarrow C + D$. The following data were obtained for the initial rate of formation of C at various concentrations of A and B.

Initial Concentrations		Initial Rate of Formation
$[A]$	$[B]$	of C (mol L^{-1} s^{-1})
0.010	0.010	2.0×10^{-4}
0.020	0.020	8.0×10^{-4}
0.020	0.010	8.0×10^{-4}

(a) What is the rate law for the reaction?

(b) What is the value of the rate constant?

(c) What is the rate at which C is formed if $[A] = 0.017\ M$ and $[B] = 0.033\ M$?

4. If the concentration of a particular reactant is doubled and the rate of the reaction is cut in half, what must be the order of the reaction with respect to that reactant?

5. Organic compounds that contain large proportions of nitrogen and oxygen tend to be unstable and are easily decomposed. Hexanitroethane, $C_2(NO_2)_6$, decomposes according to the equation

$$C_2(NO_2)_6 \longrightarrow 2NO_2 + 4NO + 2CO_2$$

The reaction in CCl_4 as a solvent is first order with respect to $C_2(NO_2)_6$. At 70 °C, $k = 2.41 \times 10^{-6}\ s^{-1}$ and at 100 °C $k = 2.22 \times 10^{-4}\ s^{-1}$.

(a) What is the half-life of $C_2(NO_2)_6$ at 70 °C? What is the half-life at 100 °C?

(b) If 0.100 mol of $C_2(NO_2)_6$ is dissolved in CCl_4 at 70 °C to give 1.00 L of solution, what will be the $C_2(NO_2)_6$ concentration after 500 minutes?

(c) What is the value of the activation energy of this reaction, expressed in kilojoules?

(d) What is the reaction's rate constant at 120 °C?

6. Radioactive strontium-90, ^{90}Sr, has a half-life of 28 years.

(a) What fraction of a sample of ^{90}Sr will remain after three half-lives?

(b) What fraction of a sample of ^{90}Sr will remain after 168 years?

(c) If the amount of ^{90}Sr remaining in a sample is only one-sixteenth of the amount originally present, how many years has the sample been undergoing radioactive decay?

(d) If the amount of ^{90}Sr remaining in a sample is only one-sixth of the amount originally present, how many years has the sample been undergoing radioactive decay?

7. The reaction $2A + 2B \longrightarrow M + N$ has this rate law: Rate $= k[A]^2$. At 25 °C, $k = 1.0 \times 10^{-4}$ L mol^{-1} s^{-1}. If the initial concentrations of A and B are 0.250 M and 0.150 M, respectively

(a) What is the half-life of the reaction?

(b) What will be the concentrations of A and B after 30 minutes?

8. Define *reaction mechanism, rate-determining step,* and *elementary process.*

9. The decomposition of ozone, O_3, is believed to occur by the two-step mechanism

$$\begin{array}{rcll} O_3 & \longrightarrow & O_2 + O & \text{(slow)} \\ O + O_3 & \longrightarrow & 2O_2 & \text{(fast)} \\ \hline 2O_3 & \longrightarrow & 3O_2 & \text{(net reaction)} \end{array}$$

If this is the mechanism, what is the reaction's rate law?

10. Draw a diagram showing how the potential energy varies during an exothermic reaction. Identify the activation energy for both the forward and reverse reactions. Also, identify the heat of reaction.

11. How does a heterogeneous catalyst increase the rate of a chemical reaction?

12. One possible mechanism for the decomposition of ethane, C_2H_6, into ethylene, C_2H_4, and hydrogen,

$$C_2H_6 \longrightarrow C_2H_4 + H_2$$

includes the following steps.

(1) $C_2H_6 \longrightarrow 2CH_3\cdot$

(2) $CH_3\cdot + C_2H_6 \longrightarrow CH_4 + C_2H_5\cdot$

(3) $C_2H_5\cdot \longrightarrow C_2H_4 + H\cdot$

(4) $H\cdot + C_2H_6 \longrightarrow C_2H_5\cdot + H_2$

(5) $H\cdot + C_2H_5\cdot \longrightarrow C_2H_6$

(a) Which steps initiate the reaction?

(b) Which are propagation steps?

(c) Which is a termination step?

13. Write the appropriate mass action expression, using molar concentrations, for these reactions.

(a) $NO_2(g) + N_2O(g) \rightleftharpoons 3NO(g)$

(b) $CaSO_3(s) \rightleftharpoons CaO(s) + SO_2(g)$

(c) $NiCO_3(s) \rightleftharpoons Ni^{2+}(aq) + CO_3^{2-}(aq)$

14. At a certain temperature, the reaction $2HF(g) \rightleftharpoons 2HF(g) + F_2(g)$ has $K_c = 1 \times 10^{-13}$. Does this reaction proceed far toward completion when equilibrium is reached? If 0.010 mol HF was placed in a 1.00 L container and the system was permitted to come to equilibrium, what would be the concentrations of H_2 and F_2 in the container?

15. At 100°C, the reaction $2NO_2(g) \rightleftharpoons N_2O_4(g)$ has $K_P = 6.5 \times 10^{-2}$. What is the value of K_c at that temperature?

16. At 1000°C, the reaction $NO_2(g) + SO_2(g) \rightleftharpoons NO(g) + SO_3(g)$ has $K_c = 3.60$. If 0.100 mol NO_2 and 0.100 mol SO_2 are placed in a 5.00 L container and allowed to react, what will all the concentrations be when equilibrium is reached? What will the new equilibrium concentrations be if 0.010 mol NO and 0.010 mol SO_3 are added to the original equilibrium mixture?

17. For the reaction in the preceding question, $\Delta H° = -41.8$ kJ. How will the equilibrium concentration of NO be affected if
(a) More NO_2 is added to the container?
(b) Some SO_3 is removed from the container?
(c) The temperature of the reaction mixture is raised?
(d) Some SO_2 is removed from the mixture?
(e) The pressure of the gas mixture is lowered by expanding the volume to 10.0 L?

18. At 60°C, $K_w = 9.5 \times 10^{-14}$. What is the pH of pure water at that temperature? Why can we say that the water is neither acidic nor basic?

19. At 25 °C, the water in a natural pool of water in one of the western states was found to contain hydroxide ions at a concentration of 4.7×10^{-7} g OH^- per liter. Calculate the pH of the water and state if it is acidic, basic, or neutral.

20. Which is the stronger acid, H_3AsO_3 or H_3AsO_4? How can one tell without a table of weak and strong acids?

21. Which is the stronger acid, H_2S or H_2Te?

22. What are the conjugate acids of (a) HSO_3^- and (b) N_2H_4?

23. What are the conjugate bases of (a) HSO_3^-, (b) N_2H_4, and (c) $C_5H_5NH^+$?

24. Identify the conjugate acid–base pairs in the reaction

$$CH_3NH_2 + NH_4^+ \rightleftharpoons CH_3NH_3^+ + NH_3$$

25. What is the definition of an *amphoteric substance*? What is a *Lewis acid?* What is a *Lewis base*?

26. Use Lewis structures to diagram the reaction between the Lewis base OH^- and the Lewis acid SO_3.

27. *X*, *Y*, and *Z* are all nonmetallic elements in the same period of the periodic table where they occur, left to right, in the order given. Which would be a stronger binary acid than the binary acid of *Y*, the binary acid of *X* or the binary acid of *Z*? Explain.

28. The first antiseptic to be used in surgical operating rooms was phenol, C_6H_5OH, a weak acid and a potent bactericide. A 0.550 *M* solution of phenol in water was found to have a pH of 5.07.
(a) Write the chemical equation for the equilibrium involving C_6H_5OH in the solution.
(b) Write the equilibrium law corresponding to K_a for C_6H_5OH.
(c) Calculate the values of K_a and pK_a for phenol.
(d) Calculate the values of K_b and pK_b for the phenoxide ion, $C_6H_5O^-$.

29. The pK_a of saccharin, $HC_7H_3SO_3$, a sweetening agent, is 11.68.
(a) What is the pK_b of the saccharinate ion, $C_7H_3SO_3^-$?
(b) Does a solution of sodium saccharinate in water have a pH of 7, or is the solution acidic or basic? If the pH is not 7, calculate the pH of a 0.010 *M* solution of sodium saccharinate in water.

30. At 25 °C the value of K_b for codeine, a pain-killing drug, is 1.6×10^{-6}. Calculate the pH of a 0.0115 *M* solution of codeine in water.

31. Methylamine, CH_3NH_2, is a weak base. Write the chemical equation for the equilibrium that occurs in an aqueous solution of this solute. Write the equilibrium law corresponding to K_b for CH_3NH_2.

32. The pK_b of methylamine, CH_3NH_2, is 3.43. Calculate the pK_a of its conjugate acid, $CH_3NH_3^+$.

33. Ascorbic acid, $H_2C_6H_6O_6$, is a diprotic acid usually known as vitamin C. For this acid, pK_{a_1} is 4.10 and pK_{a_2} is 11.79. When 125 mL of a solution of ascorbic acid was evaporated to dryness, the residue of pure ascorbic acid had a mass of 3.12 g.
(a) Calculate the molar concentration of ascorbic acid in the solution before it was evaporated.
(b) Calculate the pH of the solution and the molar concentration of the ascorbate ion, $C_6H_6O_6^{2-}$, before the solution was evaporated.

34. What ratio of molar concentrations of sodium acetate to acetic acid can buffer a solution at a pH of 4.50?

35. Write a chemical equation for the reaction that would occur in a buffer composed of sodium acetate and acetic acid if
(a) Some HCl were added.
(b) Some NaOH were added.

36. If 0.020 mol of NaOH were added to 0.500 L of a sodium acetate–acetic acid buffer that contains 0.10 *M* $NaC_2H_3O_2$ and 0.15 *M* $HC_2H_3O_2$, by how many pH units will the pH of the buffer change?

37. A biology experiment requires the use of a nutrient fluid buffered at a pH of 4.85, and 625 mL of the solution is needed. It has to be buffered to be able to hold the pH to within ±0.10 pH unit of 5.00 even if 5.00×10^{-3} mol of OH^- or 5.00×10^{-3} mol of H^+ ion enter the buffer.
(a) Using tabulated data, pick the best acid and its sodium salt that could be used to prepare the solution.
(b) Calculate the minimum number of grams of the pure acid and its salt that are needed to prepare the buffer solution.
(c) What are the molar concentrations of the acid and of its salt in the solution?

38. How would each of the following aqueous solutions test, acidic, basic, or neutral? (Assume that each is at least 0.2 *M*.) (a) potassium nitrate, (b) chromium(III) chloride, (c) ammonium iodide, (d) potassium dihydrogen phosphate.

39. When 50.00 mL of an acid with a concentration of 0.115 *M* (for which $pK_a = 4.87$) is titrated with 0.100 *M* NaOH, what is the pH at the equivalence point? What would be a good indicator for the titration?

40. Calculate the pH of a 0.050 *M* solution of sodium ascorbate, $Na_2C_6H_6O_6$. For ascorbic acid, $H_2C_6H_6O_6$, $K_{a_1} = 7.9 \times 10^{-5}$ and $K_{a_2} = 1.6 \times 10^{-12}$.

41. When 25.0 mL of 0.100 *M* NaOH was added to 50.0 mL of a 0.100 *M* solution of a weak acid, HX, the pH of the mixture reached a value of 3.56. What is the value of K_a for the weak acid?

42. How many grams of solid NaOH would have to be added to 0.100 L of a 0.100 M solution of NH_4Cl to give a mixture with a pH of 9.26?

43. The molar solubility of silver chromate, Ag_2CrO_4, in water is 6.7×10^{-5} M. What is K_{sp} for Ag_2CrO_4?

44. What is the pH of a saturated solution of magnesium hydroxide?

45. What is the solubility of iron(II) hydroxide in grams per liter if the solution is buffered to a pH of 10.00?

46. Suppose 30.0 mL of 0.100 M $Pb(NO_3)_2$ is added to 20.0 mL of 0.500 M KI.

    (a) How many grams of PbI_2 will be formed?

    (b) What will the molar concentrations of all the ions be in the mixture after equilibrium has been reached?

47. How many moles of NH_3 must be added to 1.00 L of solution to dissolve 1.00 g of $CuCO_3$? For $CuCO_3$, $K_{sp} = 2.3 \times 10^{-10}$. Ignore hydrolysis of CO_3^{2-}, but consider the formation of the complex ion, $Cu(NH_3)_4^{2+}$.

48. Over what pH range must a solution be buffered to achieve a selective separation of the carbonates of barium, $BaCO_3$ ($K_{sp} = 5.0 \times 10^{-9}$), and lead, $PbCO_3$ ($K_{sp} = 7.4 \times 10^{-14}$)? The solution is initially 0.010 M in Ba^{2+} and 0.010 M in Pb^{2+}.

49. A solution containing 0.10 M Pb^{2+} and 0.10 M Ni^{2+} is to be saturated with H_2S. What range of pH values could the solution have so that when the procedure is completed one of the ions remains in solution while the other is precipitated as its sulfide?

50. A solution that contains 0.10 M Fe^{2+} and 0.10 M Sn^{2+} is maintained at a pH of 3.00 while H_2S is gradually added to it. What will be the concentration of Sn^{2+} in the solution when FeS just begins to precipitate?

51. A metal sulfide MS has a value of K_{sp} of 4.0×10^{-29}. (a) What is the value of K_{spa} for this compound? (b) Calculate the molar solubility of MS in 0.30 M HCl.

THERMODYNAMICS 1

A perfume serves its purpose because its aroma drifts spontaneously through the air as molecules of the perfume mix with those of the atmosphere. Such spontaneous events are predicted by thermodynamics. In this chapter you will learn about the factors that determine whether or not a process is spontaneous. *(Vincent Besnault/Sygma/Corbis)*

CHAPTER OUTLIN

THIS CHAPTER IN CONTEXT The chemistry we observe in our world is controlled both by w happen and what *cannot.* For example, hydrocarbon fuels such as octane (C_8H_{18}) *can* burn, forming CO_2 a and releasing heat. When combustion reactions are started, they proceed *spontaneously* (i.e., on their own, further assistance), and we use them to provide energy to move vehicles, generate electricity, and heat hon offices. On the other hand, if we mix CO_2 and H_2O, there is nothing we can do to entice them to react sponta to form hydrocarbons. These reactions *cannot* take place on their own. If they could, our problems with fossil fuel supplies and greenhouse gas production could easily be solved.

Observations like those described above raise the fundamental question, "What determines whether or not a chemical reaction is possible when substances are combined?" The answer to this question is found in the study of thermodynamics, which expands on topics introduced in Chapter 6—energy changes in chemical reactions. As you will see, not only will we be able to use thermodynamics to determine the possibility of reaction, but we will find another explanation of chemical equilibrium and another way of finding equilibrium constants.

18.1 INTERNAL ENERGY CAN BE TRANSFERRED AS HEAT OR WORK, BUT IT CANNOT BE CREATED OR DESTROYED

Chemical thermodynamics is the study of the role of energy in chemical change and in determining the behavior of materials. It is based on a few laws that summarize centuries of experimental observation. Each law is a statement about relationships between energy, heat, work, and temperature. Because the laws manifest themselves in so many different ways, and underlie so many different phenomena, there are many alternative but equivalent ways to state them. For ease of reference, the laws are identified by number and are called the first law, the second law, and the third law.

Thermo implies heat, *dynamics* implies movement.

The **first law of thermodynamics,** which was discussed in Chapter 6, states that internal energy may be transferred as heat or work but it cannot be created or destroyed. This law, you recall, serves as the foundation for Hess's law, which we used in our computations involving enthalpy changes in Chapter 6. Let's review the first law in more detail.

Recall that the **internal energy** of a system, which is given the symbol ***E***, is the system's total energy—the sum of all the kinetic and potential energies of its particles. For a chemical reaction, a change in the internal energy, ΔE, is defined as

$$\Delta E = E_{\text{products}} - E_{\text{reactants}}$$

Expanding gases can do work. In a steam engine, such as the one powering this locomotive, heat is absorbed by water to turn it into high temperature, high pressure steam (q is positive). The steam then expands, pushing pistons and causing the locomotive to move. In the expansion, the steam does work and w is negative. *(Richard A. Cooke III/Stone/Getty Images.)*

Thus, ΔE is positive if energy flows into a system and negative if energy flows out.

The first law of thermodynamics considers two ways by which energy can be exchanged between a system and its surroundings. One is by the absorption or release of heat, which is given the symbol q. The other involves **work, *w*.** If *work is done on a system,* as in the compression of a gas, the system gains and stores energy. Conversely, if *the system does work on the surroundings,* as when a gas expands and pushes a piston, the system loses some energy by changing part of its potential energy to kinetic energy which is transferred to the surroundings. The first law of thermodynamics expresses the net change in energy mathematically by the equation

$$\Delta E = q + w$$

ΔE = (heat input) + (work input)

In Chapter 6 you learned that a positive sign on an energy change indicates energy gained by the system, whereas a negative sign indicates energy lost by the system. Thus, when. . .

q is (+)	Heat is absorbed by the system.
q is (−)	Heat is released by the system.
w is (+)	Work is done on the system.
w is (−)	Work is done by the system.

In Chapter 6 you also learned that ΔE is a state function, which means that its value does not depend on how a change from one state to another is carried out. On the other hand, q and w are not state functions.

When a chemical system expands, it does pressure-volume work on the surroundings

There are two kinds of work that chemical systems can do (or have done on them) that are of concern to us. One is electrical work, which is examined in the next chapter and

FIG. 18.1 Work being done on a gas. When a gas is *compressed* by an external pressure, w is positive because work is done on the gas. Because V decreases when the gas is compressed, the negative sign on ΔV assures that w will be positive for the compression of a gas. *(Lawrence Manning/Corbis Images.)*

will be discussed there. The other is work associated with the expansion or contraction of a system under the influence of an external pressure. An example is the work you perform on a gas when you compress it to fill a tire (Figure 18.1). Such "pressure–volume" or P–V work was discussed in Chapter 6 where it was shown that this kind of work is given by the equation

$$w = -P\Delta V$$

where P is the *external pressure* on the system.

If P–V work is the only kind of work involved in a chemical change, the equation for ΔE takes the form

$$\Delta E = q + (-P\Delta V) = q - P\Delta V$$

In Chapter 6 we also showed that when a reaction takes place in a container whose volume cannot change, the entire energy change must appear as heat. Therefore, ΔE is called the heat at constant volume (q_v)

$$\Delta E = q_v$$

Practice Exercise 1: Molecules of an ideal gas have no intermolecular attractions and therefore undergo no change in potential energy on expansion of the gas. If the expansion is also at constant temperature, there is no change in the kinetic energy, so the isothermal (constant temperature) expansion of an ideal gas has $\Delta E = 0$. Suppose such a gas expands at constant temperature from a volume of 1.0 L to 12.0 L against a constant opposing pressure of 14.0 atm. In units of L atm, what are q and w for this change? (Hints: Is the system doing work, or is work done on the system? What must be the sum of q and w in this case?)

Practice Exercise 2: If a gas is compressed under adiabatic conditions (allowing no transfer of heat to or from the surroundings) by application of an external pressure, the temperature of the gas increases. Why?

Gases become hot when compressed. Diesel engines are used to power large trucks and other heavy equipment such as this diesel locomotive. In the cylinders of a diesel engine, air is compressed rapidly to very small volumes, raising the temperature to the point where fuel ignites spontaneously when injected into it. *(John Griffin/The Image Works.)*

Enthalpy is a convenient state function for studying heats of reaction under constant pressure

Rarely do we carry out reactions in containers of fixed volume. Usually, reactions take place in containers open to the atmosphere where they are exposed to a constant pressure. To study heats of reactions under constant pressure conditions, enthalpy was invented. Recall that the **enthalpy, H,** is defined by the equation

$$H = E + PV$$

In Section 6.5 we showed that at constant pressure, the **enthalpy change, ΔH,** is equal to q_p

$$\Delta H = q_p$$

where q_p is the **heat of reaction at constant pressure.** Thus, the value of ΔH for a system equals the heat at constant pressure.

The pressure-volume work is the difference between ΔE and ΔH for a chemical reaction

In Chapter 6 we noted that ΔE and ΔH are not equal. They differ by the pressure–volume work, $-P\Delta V$.

$$\Delta E - \Delta H = -P\Delta V$$

The only time ΔE and ΔH differ by a significant amount is when gases are formed or consumed in a reaction, and even then they do not differ by much. To calculate

The ΔV values for reactions involving only solids and liquids are very tiny, so ΔE and ΔH for the reactions are nearly the same size.

ΔE from ΔH (or ΔH from ΔE), we must have a way of calculating the pressure–volume work.

If we assume the gases in the reaction behave as ideal gases, then we can use the ideal gas law. Solved for V this is

$$V = \frac{nRT}{P}$$

A volume change can therefore be expressed as

$$\Delta V = \Delta\left(\frac{nRT}{P}\right)$$

For a change at constant pressure and temperature we can rewrite this as

$$\Delta V = \Delta n\left(\frac{RT}{P}\right)$$

Thus, when the reaction occurs, the volume change is caused by a change in the number of moles of *gas*. Of course, in chemical reactions not all reactants and products need be gases, so to be sure we compute Δn correctly, let's express the change in the number of moles of gas as Δn_{gas}. It's defined as

$$\Delta n_{\text{gas}} = (n_{\text{gas}})_{\text{products}} - (n_{\text{gas}})_{\text{reactants}}$$

The $P\Delta V$ product is, therefore,

$$P\Delta V = P \cdot \Delta n_{\text{gas}}\left(\frac{RT}{P}\right) = \Delta n_{\text{gas}} RT$$

Substituting into the equation for ΔH gives

$$\Delta H = \Delta E + \Delta n_{\text{gas}} RT \tag{18.1}$$

The following example illustrates how small the difference is between ΔE and ΔH.

EXAMPLE 18.1
Conversion between ΔE and ΔH

The decomposition of calcium carbonate in limestone is used industrially to make carbon dioxide.

$$CaCO_3(s) \longrightarrow CaO(s) + CO_2(g)$$

The reaction is endothermic and has $\Delta H° = +571$ kJ. What is the value of $\Delta E°$ for this reaction?

ANALYSIS: The problem asks us to convert between ΔE and ΔH, so the tool we need to use is Equation 18.1. The superscript ° on ΔH and ΔE tells us the temperature is 25 °C (standard temperature for measuring heats of reaction). It is also important to remember that in calculating Δn_{gas}, we count just the numbers of moles of gas. Also, because we wish to calculate the term nRT in units of kilojoules, we will use $R = 8.314$ J mol^{-1} K^{-1}.

SOLUTION: The equation we need is

$$\Delta H = \Delta E + \Delta n_{\text{gas}} RT$$

Solving for ΔE and applying the superscript ° gives

$$\Delta E° = \Delta H° - \Delta n_{\text{gas}} RT$$

To calculate Δn_{gas} we take the coefficients in the equation to represent numbers of moles. Therefore, there is one mole of gas among the products and no moles of gas among the reactants, so

$$\Delta n_{\text{gas}} = 1 - 0 = 1$$

The temperature is 298 K, and $R = 8.314 \text{ J mol}^{-1} \text{ K}^{-1}$. Substituting,

$$\begin{aligned}\Delta E^\circ &= +571 \text{ kJ} - (1 \cancel{\text{mol}})(8.314 \text{ J} \cancel{\text{mol}^{-1}} \cancel{\text{K}^{-1}})(298 \cancel{\text{K}}) \\ &= +571 \text{ kJ} - 2.48 \text{ kJ} \\ &= +569 \text{ kJ}\end{aligned}$$

Notice that ΔH° and ΔE° differ by only 2 kJ, which is approximately 0.4%.

IS THE ANSWER REASONABLE? A gas is formed in the reaction, so the system must expand and push against the opposing pressure of the atmosphere when the reaction occurs at constant pressure. Energy must be supplied to accomplish this. At constant volume this expansion would not be necessary; the pressure would simply increase. Therefore, decomposing the $CaCO_3$ should require more energy at constant pressure (ΔH°) than at constant volume (ΔE°) by an amount equal to the work done pushing back the atmosphere, and we see that ΔH° is larger than ΔE°, so the answer is reasonable.

Practice Exercise 3: Calculate the difference, in kilojoules, between ΔE and ΔH for the following exothermic reaction at 45 °C. Which is more exothermic, ΔE or ΔH? (Hint: Note the temperature.)

$$2N_2O(g) + 3O_2(g) \longrightarrow 4NO_2(g)$$

Practice Exercise 4: The reaction

$$CaO(s) + 2HCl(g) \longrightarrow CaCl_2(s) + H_2O(g)$$

has $\Delta H^\circ = -217.1$ kJ. Calculate ΔE° for this reaction. What is the percentage difference between ΔE° and ΔH°?

18.2 A SPONTANEOUS CHANGE IS A CHANGE THAT CONTINUES WITHOUT OUTSIDE INTERVENTION

Now that we've reviewed the way thermodynamics treats energy changes, we turn our attention to one of our main goals in studying this subject—finding relationships among the factors that control whether events are spontaneous. By **spontaneous change** we mean one that occurs by itself, without continuous outside assistance. Examples are water flowing over a waterfall and the melting of ice cubes in a cold drink on a warm day. These are events that proceed on their own.

Some spontaneous changes occur very rapidly. An example is the detonation of a stick of dynamite, or the exposure of photographic film. Other spontaneous events, such as the rusting of iron or the erosion of stone, occur slowly and many years may pass before a change is noticed. Still others occur at such an extremely slow rate under ordinary conditions that they appear not to be spontaneous at all. Gasoline–oxygen mixtures appear perfectly stable indefinitely at room temperature because under those conditions they react very slowly. However, if heated, their rate of reaction increases tremendously and they react explosively.

Each day we also witness events that are obviously *not* spontaneous. We may pass by a pile of bricks in the morning and later in the day find that they have become a brick wall. We know from experience that the wall didn't get there by itself. A pile of bricks becoming a brick wall is *not* spontaneous; it requires the intervention of a bricklayer. Similarly, the decomposition of water into hydrogen and oxygen is not spontaneous. We see water all the time and we know that it's stable. Nevertheless, we can cause water to decompose by passing an electric current through it in a process called *electrolysis* (Figure 18.2.)

$$2H_2O(l) \xrightarrow{\textit{electrolysis}} 2H_2(g) + O_2(g)$$

This decomposition will continue, however, only as long as the electric current is maintained. As soon as the supply of electricity is cut off, the decomposition ceases. This example demonstrates the difference between spontaneous and nonspontaneous changes. Once

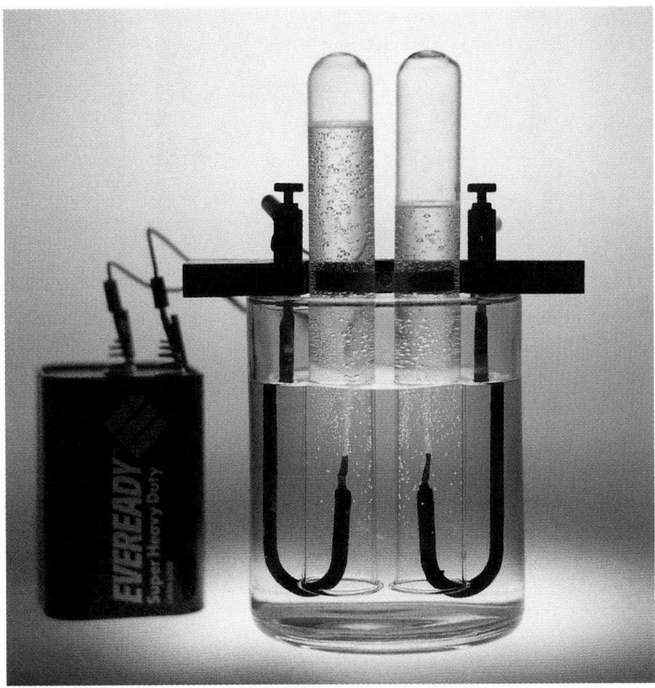

FIG. 18.2 The electrolysis of water produces H_2 and O_2 gases. It is a nonspontaneous change that only continues as long as electricity is supplied. *(Charles D. Winters/Photo Researchers.)*

a spontaneous event begins, it has a tendency to continue until it is finished. A nonspontaneous event, on the other hand, can continue only as long as it receives some sort of outside assistance.

Nonspontaneous changes have another common characteristic. They are able to occur only when accompanied by some spontaneous change. For example, a bricklayer consumes food, and a series of spontaneous biochemical reactions then occur that supply the necessary muscle power to build a wall. Similarly, the nonspontaneous electrolysis of water requires some sort of spontaneous mechanical or chemical change to generate the needed electricity. In short, *all nonspontaneous events occur at the expense of spontaneous ones.* Everything that happens can be traced, either directly or indirectly, to spontaneous changes.

The driving of nonspontaneous reactions to completion by linking them to spontaneous ones is an important principle in biochemistry.

Reaction rate affects the apparent spontaneity of reactions

The reaction of gasoline vapor with oxygen mentioned above illustrates an important observation about spontaneous changes. Even though the reaction between these substances has a strong tendency to occur, and is therefore spontaneous, the *rate of the reaction* at room temperature is so slow that the mixture appears to be stable. In other words, the reaction *appears* to be nonspontaneous because its rate is so slow. There are many reactions in nature that are spontaneous but occur at such a slow rate that they aren't observed. Biochemical reactions often fall into this category. Without the presence of a catalyst (an enzyme), these reactions are so slow that effectively they do not occur. Living systems control their chemical reactions by selectively making enzymes available when spontaneous reactions or their products are needed.

The direction of spontaneous change is often but not always in the direction of lower energy

What determines the direction of spontaneous change? Let's begin by examining some everyday events such as those depicted in Figure 18.3. When iron rusts, heat is released, so the reaction lowers the internal energy of the system. Similarly, the chemical substances in the gasoline–oxygen mixture lose chemical energy by evolving heat as the gasoline burns to produce CO_2 and H_2O.

We might be tempted to conclude (as some nineteenth century chemists did) that spontaneous events occur in the direction of lowest energy, so that energy lowering is a

FIG. 18.3 **Three common spontaneous events.** Iron rusts, fuel burns, and an ice cube melts at room temperature. *(left: George B. Diebold/Corbis Images; center: Lowell J. Georgia/Photo Resarchers; right: Susumu Sato/Corbis Images).*

"driving force" behind spontaneous change. For example, we may argue that when we drop a book, it falls to the floor because that lowers its potential energy. Since a change that lowers the potential energy of a system can be said to be exothermic, we can state this factor another way—*exothermic changes have a tendency to proceed spontaneously.*

Most, but not all, chemical reactions that are exothermic occur spontaneously.

Yet if this is so, how do we explain the third photograph in Figure 18.3? The melting of ice at room temperature is clearly a spontaneous process. But ice absorbs heat from the surroundings as it melts. This absorbed heat gives the water from the melted ice cube a higher internal energy than the original ice had. This is an example of a spontaneous but endothermic process. There are many other examples of spontaneous endothermic processes: the evaporation of water from a lake, the expansion of carbon dioxide gas into a vacuum, or the operation of a chemical "cold pack."

In Chapter 12 we discussed the solution process, and we said that one of the principal driving forces in the formation of a solution is the fact that the mixed state is much more probable than the unmixed state. It turns out that arguments like this can be used to explain the direction of any spontaneous process.

18.3 Spontaneous Processes Tend to Proceed from States of Lower Probability to States of Higher Probability

In Chapter 6 you learned that when a hot object is placed in contact with a colder one, heat will flow spontaneously from the hot object to the colder one. But why? Energy can be conserved no matter which direction the heat flows.

To analyze what happens in the spontaneous flow of heat, let's imagine a situation in which we have two objects in contact with one another, one with a high temperature and the other with a low temperature. Because of the relationship between average kinetic energy and temperature, we expect the high temperature object to have many rapidly moving molecules, whereas the low temperature object will have more slow speed molecules. With these objects in contact, how likely is it that no energy will be transferred between them?

Where the objects touch, many of the molecular collisions will involve fast-moving "hot" molecules and slow-moving "cold" ones. In such a collision, it is very unlikely that the fast molecule will gain kinetic energy at the expense of the slow one. Instead, the fast molecule will lose kinetic energy and slow down, while the slow molecule will gain kinetic

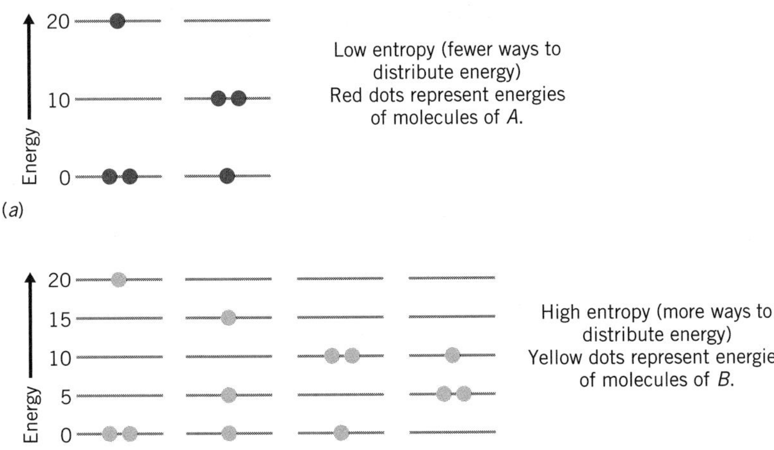

FIG. 18.4 A positive value for ΔS means an increase in the number of ways energy can be distributed among a system's molecules. Consider the reaction $A \longrightarrow B$, where A can take on energies that are multiples of 10 energy units, and B can take on energies that are multiples of 5 units. Suppose that the total energy of the reacting mixture is 20 units. (*a*) There are 2 ways to distribute 20 units of energy among 3 molecules of A. (*b*) There are 4 ways to distribute 20 units of energy among 3 molecules of B. The entropy of B is higher than the entropy of A because there are more ways to distribute the same amount of energy in B than in A.

energy and speed up. Over time, the result of many such collisions is that the hot object cools and the cool object warms. While this happens, the combined kinetic energy of the two objects becomes distributed over all the molecules in the system.

What we learn here is that heat flows because of the probable outcome of intermolecular collisions. Viewed on a larger scale, the uniform distribution of kinetic energy between the objects, caused by the flow of heat, is much more probable than a situation in which no heat flow occurs, which illustrates the role of probability in determining the direction of a spontaneous change. *Spontaneous processes tend to proceed from states of low probability to states of higher probability.* The higher probability states are those that allow more options for distributing energy among the molecules, so we can also say that *spontaneous processes tend to disperse energy.*

Entropy is a measure of the number of equivalent ways to distribute energy in the system

Because statistical probability is so important in determining the outcome of chemical and physical events, thermodynamics defines a quantity, called **entropy** (symbol ***S***), that describes the number of equivalent ways that energy can be distributed in a system. The greater the number of energetically equivalent versions there are of the system, the larger is its statistical probability, and therefore the larger the value of the entropy (Figure 18.4).

The greater the statistical probability of a particular state, the greater is the entropy.

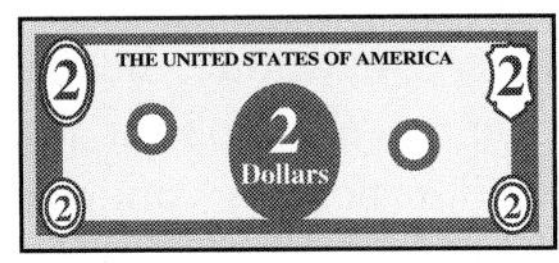

(*a*) Two ways to count out \$2 with paper money

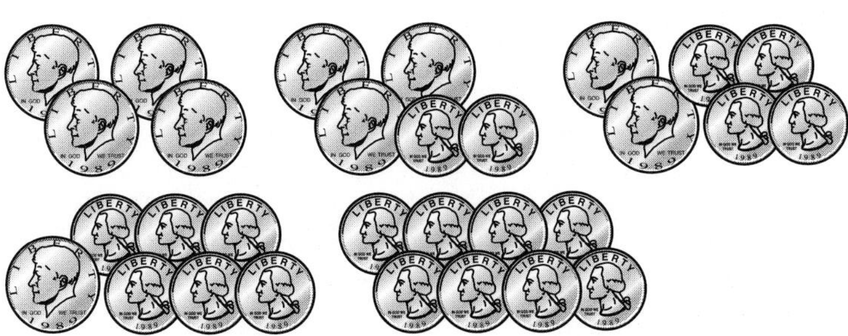

(*b*) Five ways to count out \$2 with 50¢ and 25¢ coins

Entropy. If energy were money, entropy would describe the number of different ways of counting it out. For example, there are only two ways of counting out \$2 using American paper money. But there are five ways of counting out \$2 using 50-cent and 25-cent coins. We could say that the "entropy" of a system that dealt in coins was higher than that of a system that dealt only in paper money.

In chemistry, we usually deal with systems that contain very large numbers of particles and it is usually impractical to count the number of ways that the particles can be arranged to produce a system with a particular energy. Fortunately, we will not need to do so. The entropy of the system can be related to experimental heat and temperature measurements.

Like enthalpy, entropy is a state function. It depends only on the state of the system, so an **entropy change, ΔS,** is independent of the path from start to finish. As with other thermodynamic quantities, ΔS is defined as "final minus initial" or "products minus reactants." Thus

$$\Delta S = S_{\text{final}} - S_{\text{initial}}$$

or, for a chemical system,

$$\Delta S = S_{\text{products}} - S_{\text{reactants}}$$

As you can see, when S_{final} is larger than S_{initial} (or when S_{products} is larger than $S_{\text{reactants}}$), the value of ΔS is positive. A positive value for ΔS means an increase in the number of energy-equivalent ways the system can be produced, and we have seen that this kind of change tends to be spontaneous. This leads to a general statement about entropy:

> Any event that is accompanied by an increase in the entropy of the system will have a *tendency* to occur spontaneously.

An increase in freedom of molecular motions corresponds to an increase in entropy

It is often possible to predict whether ΔS is positive or negative for a particular change. This is because several factors influence the magnitude of the entropy in predictable ways.

TOOLS
Factors affecting entropy: volume changes

Volume

For gases, the entropy increases with increasing volume, as illustrated in Figure 18.5. Below we see a gas confined to one side of a container, separated from a vacuum by a removable partition. Let's suppose the partition could be pulled away in an instant, as shown in Figure 18.5*b*. Now we find a situation in which all the molecules of the gas are at one end of a larger container. There are many more possible ways that the total kinetic energy can be distributed among the molecules in the larger volume. That makes the configuration in Figure 18.5*b* extremely unlikely. The gas expands spontaneously to achieve a more probable (higher entropy) particle distribution.

TOOLS
Factors affecting entropy: temperature changes

Temperature

The entropy is also affected by the temperature; the higher the temperature, the larger is the entropy. For example, when a substance is a solid at absolute zero, its particles are essentially motionless. There is relatively little kinetic energy, and so there are few ways to distribute kinetic energy among the particles; thus, the entropy of the solid is relatively low (Figure 18.6*a*).

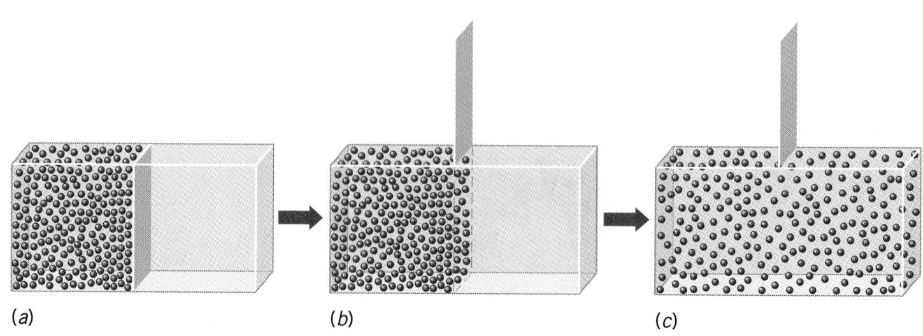

FIG. 18.5 The expansion of a gas into a vacuum. (*a*) A gas in a container separated from a vacuum by a partition. (*b*) The gas at the moment the partition is removed. (*c*) The gas expands to achieve a more probable (higher entropy) particle distribution.

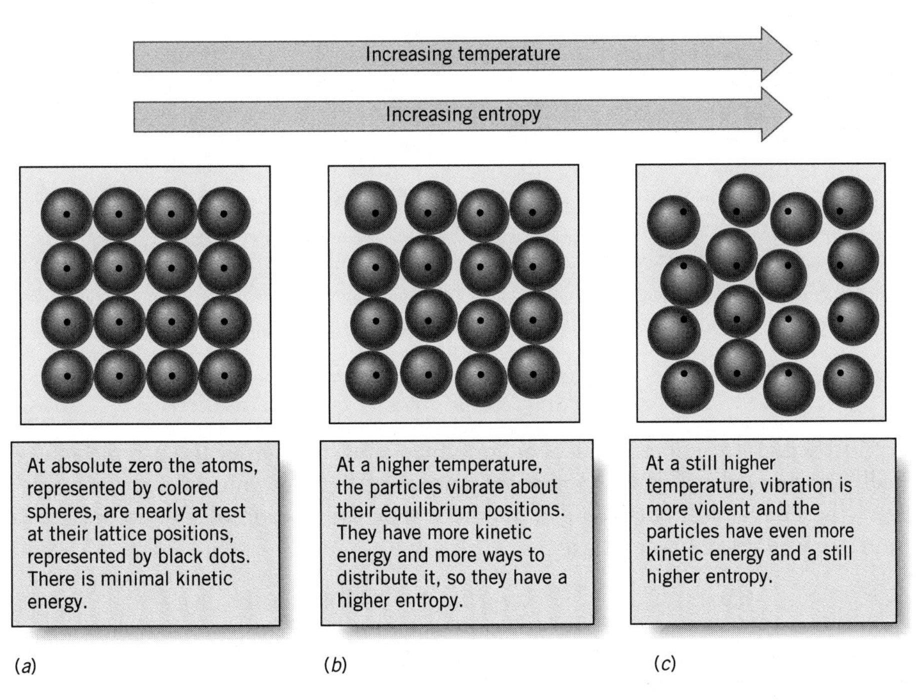

FIG. 18.6 Variation of entropy with temperature. (*a*) At absolute zero the atoms, represented by colored circles, rest at their equilibrium lattice positions, represented by black dots. Entropy is relatively low. (*b*) At a higher temperature, the particles vibrate about their equilibrium positions and there are more different ways to distribute kinetic energy among the molecules. Entropy is higher than in (*a*). (*c*) At a still higher temperature, vibration is more violent and at any instant the particles are found in even more arrangements. Entropy is higher than in (*b*).

If some heat is added to the solid, the kinetic energy of the particles increases along with the temperature. This causes the particles to move and vibrate within the crystal, so at a particular moment (pictured in Figure 18.6*b*) the particles are not found exactly at their lattice sites. There is more kinetic energy than at the lower temperature, and there are more ways to distribute it among the molecules, so the entropy is larger. If the temperature is raised further, the particles are given even more kinetic energy with an even larger number of possible ways to distribute it, causing the solid to have a still higher entropy (Figure 18.6*c*).

Physical state

TOOLS *Factors affecting entropy: physical state*

One of the major factors that affects the entropy of a system is its physical state, which is demonstrated in Figure 18.7. Suppose that the diagrams represent ice, water, and steam at the same temperature. There is greater freedom of molecular movement in water than in ice at the same temperature, and so there are more ways to distribute kinetic energy among the molecules of liquid water than there are in ice. The water molecules in steam are free to move through the entire container. They are able to distribute their kinetic energies in a very large number of ways. In general, therefore, there are many more possible ways to distribute kinetic energy among gas molecules than there are in liquids and solids. In fact, a gas has such a large entropy compared with a liquid or solid that changes which produce gases from liquids or solids are almost always accompanied by increases in entropy.

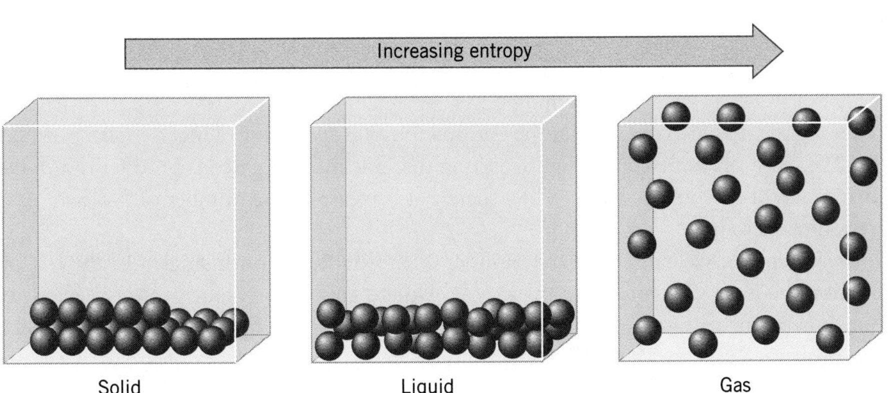

FIG. 18.7 Comparison of the entropies of the solid, liquid, and gaseous states of a substance. The crystalline solid has a very low entropy. The liquid has a higher entropy because its molecules can move more freely and there are more ways to distribute kinetic energy among them. All the particles are still found at one end of the container. The gas has the highest entropy because the particles are randomly distributed throughout the entire container, so there are many, many ways to distribute kinetic energy among the molecules.

FIG. 18.8 Entropy is affected by number of particles. Adding additional particles to a system increases the number of ways that energy can be distributed in the system, so with all other things being equal, a reaction that produces more particles will have a positive value of ΔS.

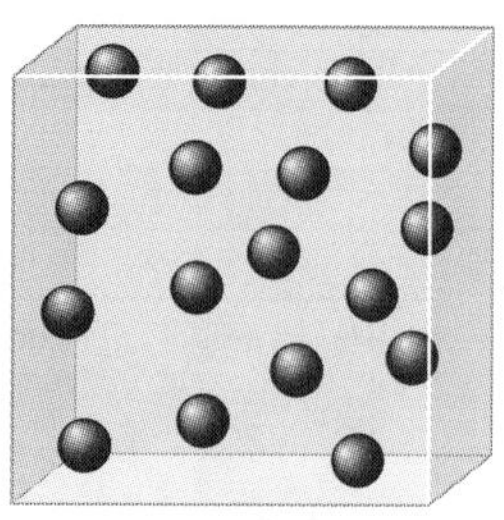
Lower entropy

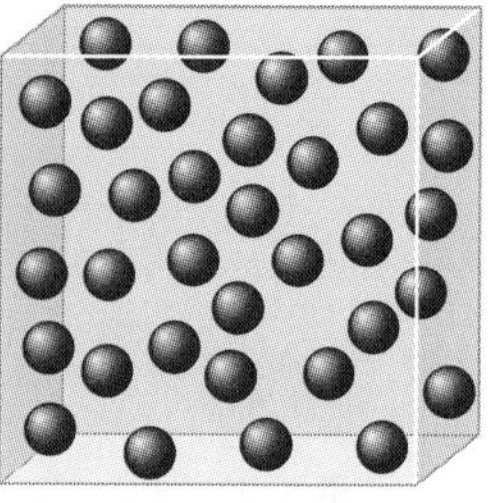
Higher entropy

As you will learn soon, absolute values for S can be obtained. This is entirely different from E and H, where we can only determine differences (i.e., ΔE and ΔH).

The sign of ΔS for a chemical reaction can sometimes be predicted by examining the chemical equation

When a chemical reaction produces or consumes gases, the sign of its entropy change is usually easy to predict. This is because the entropy of a gas is so much larger than that of either a liquid or solid. For example, the thermal decomposition of sodium bicarbonate produces two gases, CO_2 and H_2O.

$$2NaHCO_3(s) \xrightarrow{\text{heat}} Na_2CO_3(s) + CO_2(g) + H_2O(g)$$

Because the amount of gaseous products is larger than the amount of gaseous reactants, we can predict that the entropy change for the reaction is positive. On the other hand, the reaction

$$CaO(s) + SO_2(g) \longrightarrow CaSO_3(s)$$

(which can be used to remove sulfur dioxide from a gas mixture) has a negative entropy change.

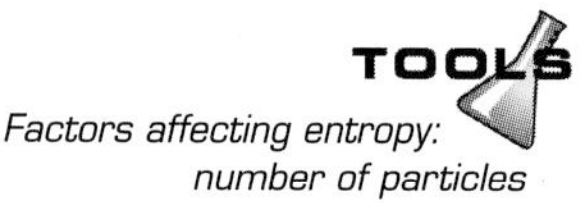

Factors affecting entropy: number of particles

As the number of particles increases, the number of ways of distributing energy also increases

For chemical reactions, another major factor that affects the sign of ΔS is a change in the total number of molecules as the reaction proceeds. When more molecules are produced during a reaction, more ways of distributing the energy among the molecules are possible. *When all other things are equal, reactions that increase the number of particles in the system tend to have a positive entropy change,* as shown in Figure 18.8.

EXAMPLE 18.2
Predicting the Sign of ΔS

Predict the algebraic sign of ΔS for the reactions:

(a) $2NO_2(g) \longrightarrow N_2O_4(g)$
(b) $C_3H_8(g) + 5O_2(g) \longrightarrow 3CO_2(g) + 4H_2O(g)$

ANALYSIS: The tools we use in analyzing questions such as these are the factors that affect the entropy of a system. Thus, we look for changes in the number of moles of gas and changes in the number of particles on going from reactants to products.

SOLUTION: In reaction (a) we are forming fewer, more complex molecules (N_2O_4) from simpler ones (NO_2). Since we are forming fewer molecules, there are fewer ways to distribute energy among them, which means that the entropy must be decreasing. Therefore, ΔS must be negative. We reach the same conclusion by noting that one mole of gaseous product is formed from two moles of gaseous reactant. When there is a decrease in the number of moles of gas, the reaction tends to have a negative ΔS.

For reaction (b), we can count the number of molecules on both sides. On the left of the equation we have six molecules; on the right there are seven. There are more ways to distribute kinetic energy among seven molecules than among six, so for reaction (b), we expect ΔS to be positive. We reach the same conclusion by counting the number of moles of gas on both sides of the equation. On the left there are 6 moles of gas; on the right there are 7 moles of gas. Because the number of moles of gas is increasing, we expect ΔS to be positive.

ARE THE ANSWERS REASONABLE? The only check we can perform here is to carefully review our reasoning. It appears sound, so the answers appear to be correct.

Practice Exercise 5: Would you expect ΔS to be positive or negative for the following?

$$Ag^{+}(aq) + Cl^{-}(aq) \longrightarrow AgCl(s)$$

(Hint: How is the freedom of movement of the ions affected?)

Practice Exercise 6: Predict the sign of the entropy change for (a) the condensation of steam to liquid water and (b) the sublimation of a solid.

Practice Exercise 7: Predict the sign of ΔS for the following reactions:

(a) $2SO_2(g) + O_2(g) \longrightarrow 2SO_3(g)$
(b) $CO(g) + 2H_2(g) \longrightarrow CH_3OH(g)$

Practice Exercise 8: What is the expected sign of ΔS for the following reactions? Justify your answers.

(a) $2H_2(g) + O_2(g) \longrightarrow 2H_2O(l)$
(b) $N_2(g) + 3H_2(g) \longrightarrow 2NH_3(g)$
(c) $Ca(OH)_2(s) \xrightarrow{H_2O} Ca^{2+}(aq) + 2OH^{-}(aq)$

18.4 ALL SPONTANEOUS PROCESSES INCREASE THE TOTAL ENTROPY OF THE UNIVERSE

You have learned that enthalpy and entropy are two factors that affect the spontaneity of a physical or chemical event. Sometimes they work together to favor a change, as in the combustion of gasoline where heat is given off (an exothermic change) and large volumes of gases are formed (an entropy increase).

In many situations, the enthalpy and entropy changes oppose each other, as in the melting of ice. Melting absorbs heat and is endothermic, which tends to make the process nonspontaneous. But the greater freedom of motion of the molecules that accompanies melting has the opposite effect and tends to make the change spontaneous.

When the enthalpy and entropy changes conflict, temperature becomes a critical factor that can influence the direction in which the change is spontaneous. For example, consider a mixture of solid ice and liquid water. If we attempt to raise the temperature of the mixture to 25 °C, all the solid will melt. At 25 °C the change *solid* $\longrightarrow$ *liquid* is spontaneous. On the other hand, if we attempt to cool the mixture to −25 °C, freezing occurs, so at −25 °C the opposite change (*liquid* $\longrightarrow$ *solid*) is spontaneous. *Thus, there are actually three factors that can influence spontaneity: the enthalpy change, the entropy change, and the temperature.* The balance between these factors comes into focus through the second law of thermodynamics.

The second law of thermodynamics states that all real processes increase the total entropy of the universe

One of the most far-reaching observations in science is incorporated into the **second law of thermodynamics,** which states, in effect, that *whenever a spontaneous event takes place in our universe, the total entropy of the universe increases* ($\Delta S_{total} > 0$). Notice that the increase in entropy that's referred to here is for the *total* entropy of the *universe* (system *plus* surroundings), not just the system alone. This means that a system's entropy can decrease, just as long as there is a larger increase in the entropy of the surroundings so that the *overall* entropy change is positive. Because everything that happens relies on spontaneous changes of some sort, the entropy of the universe is constantly rising.

Now let's examine more closely the total entropy change for the universe. As we've suggested, this quantity equals the sum of the entropy change for the system plus the entropy change for the surroundings.

$$\Delta S_{\text{total}} = \Delta S_{\text{system}} + \Delta S_{\text{surroundings}}$$

It can be shown that the entropy change for the surroundings is equal to the heat transferred *to* the surroundings *from* the system, $q_{\text{surroundings}}$, divided by the Kelvin temperature, T, at which it is transferred.

$$\Delta S_{\text{surroundings}} = \frac{q_{\text{surroundings}}}{T}$$

The law of conservation of energy requires that the heat transferred to the surroundings equals the negative of the heat added to the system, so we can write

$$q_{\text{surroundings}} = -q_{\text{system}}$$

In our study of the first law of thermodynamics we saw that for changes at constant temperature and pressure, $q_{\text{system}} = \Delta H$ for the system. By substitutions, therefore, we arrive at the relationship

$$\Delta S_{\text{surroundings}} = \frac{-\Delta H_{\text{system}}}{T}$$

and the entropy change for the entire universe becomes

$$\Delta S_{\text{total}} = \Delta S_{\text{system}} - \frac{\Delta H_{\text{system}}}{T}$$

By rearranging the right side of this equation, we obtain

$$\Delta S_{\text{total}} = \frac{T\Delta S_{\text{system}} - \Delta H_{\text{system}}}{T}$$

Now let's multiply both sides of the equation by T to give

$$T\Delta S_{\text{total}} = T\Delta S_{\text{system}} - \Delta H_{\text{system}}$$

or

$$T\Delta S_{\text{total}} = -(\Delta H_{\text{system}} - T\Delta S_{\text{system}})$$

Because ΔS_{total} must be positive for a spontaneous change, the quantity in parentheses, $(\Delta H_{\text{system}} - T\Delta S_{\text{system}})$, must be negative. Therefore, we can state that for a change to be spontaneous,

$$\Delta H_{\text{system}} - T\Delta S_{\text{system}} < 0 \qquad (18.2)$$

The origin of *free* in *free energy* is discussed in Section 18.7.

Equation 18.2 gives us a way of examining the balance between ΔH, ΔS, and temperature in determining the spontaneity of an event, and it becomes convenient at this point to introduce another thermodynamic state function. It is called the **Gibbs free energy, G,** named to honor one of America's greatest scientists, Josiah Willard Gibbs (1839–1903). (It's called **free energy** because it is related, as we will see later, to the maximum energy in a change that is "free" or "available" to do useful work.) The Gibbs free energy is defined as

$$G = H - TS \qquad (18.3)$$

so for changes at constant T and P, the **Gibbs free energy change, ΔG,** becomes

$$\Delta G = \Delta H - T\Delta S \qquad (18.4)$$

Because G is defined entirely in terms of state functions, it is also a state function. This means that

$$\Delta G = G_{\text{final}} - G_{\text{initial}} \qquad (18.5)$$

By comparing Equations 18.3 and 18.5, we arrive at the special importance of the free energy change:

> *At constant temperature and pressure, a change can only be spontaneous if it is accompanied by a decrease in the free energy of the system.*

TOOLS ΔG as a predictor of spontaneity

In other words, for a change to be spontaneous, G_{final} must be less than $G_{initial}$ and ΔG must be negative. With this in mind, we can now examine how ΔH, ΔS, and T are related in determining spontaneity.

Reactions that occur with a free energy decrease are sometimes said to be **exergonic.** Those that occur with a free energy increase are sometimes said to be **endergonic.**

When ΔH is negative and ΔS is positive, the process will be spontaneous

The combustion of octane (a component of gasoline),

$$2C_8H_{18}(l) + 25O_2(g) \longrightarrow 16CO_2(g) + 18H_2O(g)$$

is a very exothermic reaction. There is also a large increase in entropy because the number of particles in the system increases and large volumes of gases are formed. For this change, ΔH is negative and ΔS is positive, both of which favor spontaneity. Let's analyze how this affects the sign of ΔG.

$$\Delta H \text{ is negative } (-)$$

$$\Delta S \text{ is positive } (+)$$

$$\Delta G = \Delta H - T\Delta S$$

$$= (-) - [T(+)]$$

Notice that regardless of the Kelvin temperature, which must be a positive number, ΔG will be negative. This means that regardless of the temperature, such a change must be spontaneous. In fact, once started, fires will continue to consume all available fuel or oxygen at nearly any temperature because combustion reactions are always spontaneous (see Figure 18.9*a*).

When ΔH is positive and ΔS is negative, the process is not spontaneous

When a change is endothermic and is accompanied by a lowering of the entropy, both factors work against spontaneity.

(*a*)  (*b*)  (*c*) 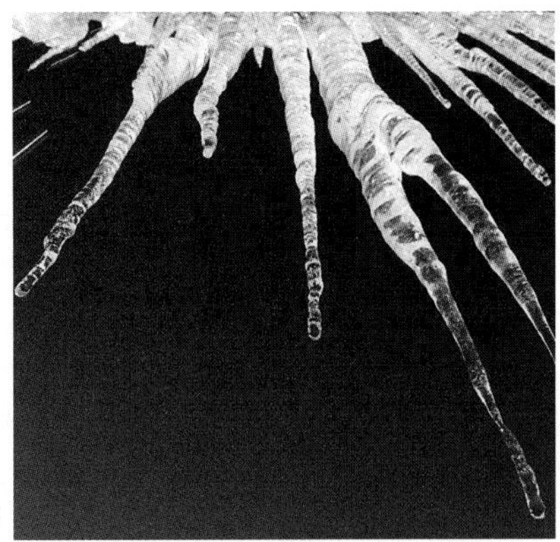

FIG. 18.9 Process spontaneity can be predicted if ΔH, ΔS, and T are known. (*a*) When ΔH is negative and ΔS is positive, as in any combustion reaction, the reaction is spontaneous at any temperature. (*b*) When ΔH is positive and ΔS is negative, the process is not spontaneous. Left to themselves, ash, carbon dioxide, and water will not spontaneously combine to form wood. (*c*) When ΔH and ΔS have the same sign, temperature determines whether the process is spontaneous or not. Water spontaneously becomes ice below 0 °C, and ice spontaneously melts into liquid water above 0 °C. *(left: Corbis images; center: Andrea Pistolesi/The Image Bank/Getty Images; right: John Berry/The Image Works)*

ΔH is positive (+)

ΔS is negative (−)

$$\Delta G = \Delta H - T\Delta S$$
$$= (+) - [T(-)]$$

Now, no matter what the temperature is, ΔG will be positive and the change must be nonspontaneous. An example would be carbon dioxide and water coming back together to form wood and oxygen again in a fire (Figure 18.9*b*).[1] If you saw such a thing happen on a film, experience would tell you that the film was being played backward.

When ΔH and ΔS have the same sign, temperature determines spontaneity

When ΔH and ΔS have the same algebraic sign, the temperature becomes the determining factor in controlling spontaneity. If ΔH and ΔS are both positive, then

$$\Delta G = (+) - [T(+)]$$

Thus, ΔG is the difference between two positive quantities, ΔH and $T\Delta S$. This difference will only be negative if the term $T\Delta S$ is larger in magnitude than ΔH, and this will only be true when the temperature is high. In other words, *when ΔH and ΔS are both positive, the change will be spontaneous at high temperature but not at low temperature.* A change noted above is the melting of ice.

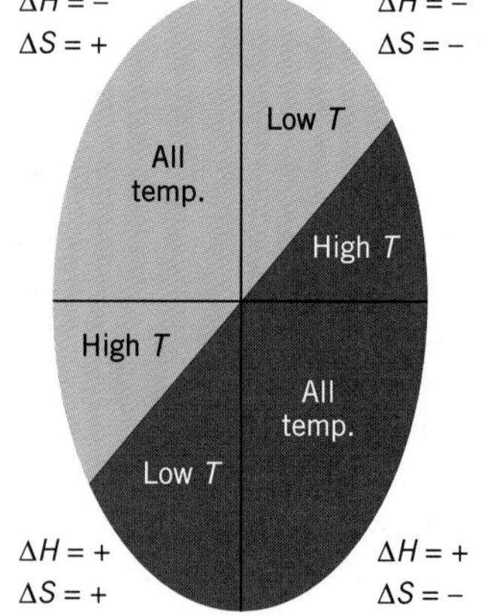

FIG. 18.10 Summary of the effects of the signs of ΔH and ΔS on spontaneity as a function of temperature. When ΔH and ΔS have the same sign, spontaneity is determined by the temperature.

$$H_2O(s) \longrightarrow H_2O(l)$$

This is a change that is endothermic and also accompanied by an increase in entropy. We know that at high temperatures (above 0 °C) melting is spontaneous, but at low temperatures (below 0 °C) it is not.

For similar reasons, when ΔH and ΔS are both negative, ΔG will be negative (and the change spontaneous) only when the temperature is low.

$$\Delta G = (-) - [T(-)]$$

Only when the negative value of ΔH is larger in magnitude than the negative value of $T\Delta S$ will ΔG be negative. Such a change is only spontaneous at low temperatures. An example is the freezing of water (see Figure 18.9*c*).

$$H_2O(l) \longrightarrow H_2O(s)$$

This is an exothermic change that is accompanied by a decrease in entropy; it is only spontaneous at low temperatures (i.e., below 0 °C).

Figure 18.10 summarizes the effects of the signs of ΔH and ΔS on ΔG, and hence on the spontaneity of physical and chemical events.

18.5 THE THIRD LAW OF THERMODYNAMICS MAKES EXPERIMENTAL MEASUREMENT OF ABSOLUTE ENTROPIES POSSIBLE

Earlier we described how the entropy of a substance depends on temperature, and we noted that at absolute zero the order within a crystal is a maximum and the entropy is a minimum. The **third law of thermodynamics** goes one step further by stating: *At absolute zero the entropy of a perfectly ordered pure crystalline substance is zero.*

$$S = 0 \quad \text{at} \quad T = 0\text{ K}$$

[1] Plants can manufacture wood from carbon dioxide and water that they take in, but the reaction by itself is *not* spontaneous. Plant cells manufacture wood by coupling the nonspontaneous reaction to a complex series of reactions with negative values of ΔG, so that the entire chain of reactions *taken together* is spontaneous overall. The same trick is used to drive nonspontaneous processes forward in human cells. The high negative free energy change from the breakdown of sugar and other nutrients is coupled with the nonspontaneous synthesis of complex proteins from simple starting materials, driving cell growth and making life possible.

TABLE 18.1 Standard Entropies of Some Typical Substances at 298.15 K

Substance	$S°$ (J mol^{-1} K^{-1})	Substance	$S°$ (J mol^{-1} K^{-1})
$Ag(s)$	42.55	$H_2O(g)$	188.7
$AgCl(s)$	96.2	$H_2O(l)$	69.96
$Al(s)$	28.3	$HCl(g)$	186.7
$Al_2O_3(s)$	51.00	$HNO_3(l)$	155.6
$C(s, \text{graphite})$	5.69	$H_2SO_4(l)$	157
$CO(g)$	197.9	$HC_2H_3O_2(l)$	160
$CO_2(g)$	213.6	$Hg(l)$	76.1
$CH_4(g)$	186.2	$Hg(g)$	175
$CH_3Cl(g)$	234.2	$K(s)$	64.18
$CH_3OH(l)$	126.8	$KCl(s)$	82.59
$CO(NH_2)_2(s)$	104.6	$K_2SO_4(s)$	176
$CO(NH_2)_2(aq)$	173.8	$N_2(g)$	191.5
$C_2H_2(g)$	200.8	$NH_3(g)$	192.5
$C_2H_4(g)$	219.8	$NH_4Cl(s)$	94.6
$C_2H_6(g)$	229.5	$NO(g)$	210.6
$C_2H_5OH(l)$	161	$NO_2(g)$	240.5
$C_8H_{18}(l)$	466.9	$N_2O(g)$	220.0
$Ca(s)$	41.4	$N_2O_4(g)$	304
$CaCO_3(s)$	92.9	$Na(s)$	51.0
$CaCl_2(s)$	114	$Na_2CO_3(s)$	136
$CaO(s)$	40	$NaHCO_3(s)$	102
$Ca(OH)_2(s)$	76.1	$NaCl(s)$	72.38
$CaSO_4(s)$	107	$NaOH(s)$	64.18
$CaSO_4 \cdot \frac{1}{2}H_2O(s)$	131	$Na_2SO_4(s)$	149.49
$CaSO_4 \cdot 2H_2O(s)$	194.0	$O_2(g)$	205.0
$Cl_2(g)$	223.0	$PbO(s)$	67.8
$Fe(s)$	27	$S(s)$	31.9
$Fe_2O_3(s)$	90.0	$SO_2(g)$	248.5
$H_2(g)$	130.6	$SO_3(g)$	256.2

Because we know the point at which entropy has a value of zero, it is possible by *experimental measurement* and calculation to determine the total amount of entropy that a substance has at temperatures above 0 K. If the entropy of one mole of a substance is determined at a temperature of 298 K (25 °C) and a pressure of 1 atm, we call it the **standard entropy, $S°$**. Table 18.1 lists the standard entropies for a number of substances.[2] (Notice that entropy has the dimensions of energy/temperature (i.e., joules per kelvin); this is explained in Facets of Chemistry 18.1).

25 °C and 1 atm are the same standard conditions we used in our discussion of $\Delta H°$ in Chapter 6.

Once we have the entropies of a variety of substances, we can calculate the **standard entropy change, $\Delta S°$**, for chemical reactions in much the same way as we calculated $\Delta H°$ in Chapter 6.

$$\Delta S° = (\text{sum of } S° \text{of the products}) - (\text{sum of } S° \text{ of the reactants}) \quad (18.6)$$

If the reaction we are working with happens to correspond to the formation of 1 mol of a compound from its elements, then the $\Delta S°$ that we calculate can be referred to as the **standard entropy of formation, $\Delta S_f°$**. Values of $\Delta S_f°$ are not tabulated, however; if we need them for some purpose, we must calculate them from tabulated values of $S°$.

This is simply a Hess's law type of calculation. Note, however, that elements have nonzero $S°$ values, which must be included in the bookkeeping.

[2] In our earlier discussions of standard states (Chapter 6) we defined the *standard pressure* as 1 atm. This was the original pressure unit used by thermodynamicists. In the SI, however, the recognized unit of pressure is the pascal (Pa), not the atmosphere. After considerable discussion, the SI adopted the *bar* as the standard pressure for thermodynamic quantities: 1 bar = 10^5 Pa. One bar differs from one atmosphere by only 1.3%, and for thermodynamic quantities that we deal with in this text, their values at 1 atm and at 1 bar differ by an insignificant amount. Since most available thermodynamic data are still specified at 1 atm rather than 1 bar, we shall continue to use 1 atm for the standard pressure.

FACETS OF CHEMISTRY 18.1

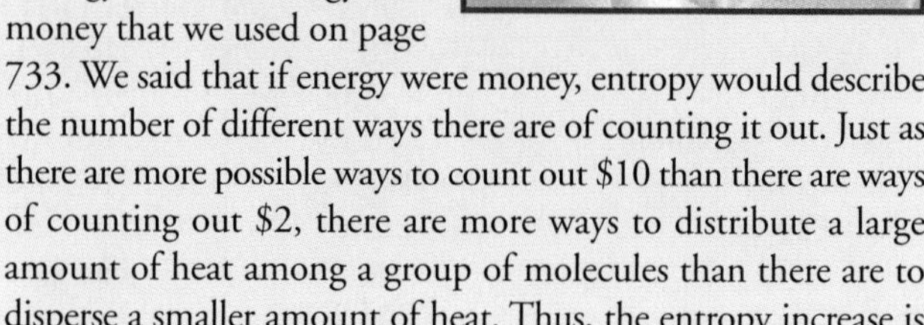

Why the Units of Entropy Are Energy/Temperature

Entropy is a state function, just like enthalpy, so the value of ΔS doesn't depend on the "path" that is followed during a change. In other words, we can proceed from one state to another in any way we like and ΔS will be the same. If we choose a *reversible path* in which just a slight alteration in the system can change the direction of the process, we can measure ΔS directly. For example, at 25 °C an ice cube will melt and there is nothing we can do at that temperature to stop it. This change is *nonreversible* in the sense described above, so we couldn't use this path to measure ΔS. At 0 °C, however, it is simple to stop the melting process and reverse its direction. At 0 °C the melting of ice is a reversible process.

If we set up a change so that it is reversible, we can calculate the entropy change as $\Delta S = q/T$, where q is the heat added to the substance and T is the temperature at which the heat is added. The more energy we add to the system as heat, the more possible ways there are to distribute it among the molecules. To understand this, let's use the analogy between energy and money that we used on page 733. We said that if energy were money, entropy would describe the number of different ways there are of counting it out. Just as there are more possible ways to count out $10 than there are ways of counting out $2, there are more ways to distribute a large amount of heat among a group of molecules than there are to disperse a smaller amount of heat. Thus, the entropy increase is *directly proportional to the amount of heat added.*

Temperature also affects the size of the entropy increase. To see why, study Figure 18.6. At lower temperatures, the entropy of a material is low, so the change in entropy produced when a given amount of heat is added to the material will be greater than it would have been at higher temperatures. The low temperature material can distribute the energy in more ways than the higher temperature material can. For a given quantity of heat the entropy change is *inversely proportional to the temperature.* Thus, $\Delta S = q/T$, and entropy has units of energy divided by temperature (e.g., $J\ K^{-1}$).

EXAMPLE 18.3
Calculating $\Delta S°$ from Standard Entropies

Urea (a compound found in urine) is manufactured commercially from CO_2 and NH_3. One of its uses is as a fertilizer where it reacts slowly with water in the soil to produce ammonia and carbon dioxide. The ammonia provides a source of nitrogen for growing plants.

$$\underset{\text{urea}}{CO(NH_2)_2(aq)} + H_2O(l) \longrightarrow CO_2(g) + 2NH_3(g)$$

What is the standard entropy change when one mole of urea reacts with water?

ANALYSIS: This problem is a straightforward application of Equation 18.6 as a tool to compute the standard entropy change for the reaction. We'll need the standard entropies $S°$ of each reactant and product. The data we need can be found in Table 18.1 and are collected in the table below.

Substance	$S°$ (J/mol K)
$CO(NH_2)_2(aq)$	173.8
$H_2O(l)$	69.96
$CO_2(g)$	213.6
$NH_3(g)$	192.5

SOLUTION: Applying Equation 18.6, we have

$$\begin{aligned}\Delta S° &= [S°_{CO_2(g)} + 2S°_{NH_3(g)}] - [S°_{CO(NH_2)_2(aq)} + S°_{H_2O(l)}] \\ &= \left[1\ \cancel{mol} \times \left(\frac{213.6\ J}{\cancel{mol}\ K}\right) + 2\ \cancel{mol} \times \left(\frac{192.5\ J}{\cancel{mol}\ K}\right)\right] \\ &\quad - \left[1\ \cancel{mol} \times \left(\frac{173.8\ J}{\cancel{mol}\ K}\right) + 1\ \cancel{mol} \times \left(\frac{69.96\ J}{\cancel{mol}\ K}\right)\right] \\ &= (598.6\ J/K) - (243.8\ J/K) \\ &= 354.8\ J/K\end{aligned}$$

Notice that the unit *mol* cancels in each term, so the units of $\Delta S°$ are joules per kelvin.

Thus, the standard entropy change for this reaction is +354.8 J/K (which we can also write as $+354.8\ J\ K^{-1}$).

IS THE ANSWER REASONABLE? In the reaction, gases are formed from liquid reactants. Since gases have much larger entropies than liquids, we expect $\Delta S°$ to be positive, which agrees with our answer.

Practice Exercise 9: Calculate $\Delta S_f°$ for $NH_3(g)$. (Hint: Write the equation for the reaction.)

Practice Exercise 10: Calculate the standard entropy change, $\Delta S°$, in J K^{-1} for the following:

(a) $CaO(s) + 2HCl(g) \longrightarrow CaCl_2(s) + H_2O(l)$
(b) $C_2H_4(g) + H_2(g) \longrightarrow C_2H_6(g)$

18.6 THE STANDARD FREE ENERGY CHANGE, $\Delta G°$, IS ΔG AT STANDARD CONDITIONS

When ΔG is determined at 25 °C (298 K) and 1 atm, we call it the **standard free energy change, $\Delta G°$**.[3] There are several ways of obtaining $\Delta G°$ for a reaction. One of them is to compute $\Delta G°$ from $\Delta H°$ and $\Delta S°$.

TOOLS

Calculating $\Delta G°$ from $\Delta H°$ and $\Delta S°$

$$\Delta G° = \Delta H° - (298.15\text{ K})\Delta S°$$

Experimental measurement of $\Delta G°$ is also possible, but we will discuss how this is done later.

EXAMPLE 18.4
Calculating $\Delta G°$ from $\Delta H°$ and $\Delta S°$

Calculate $\Delta G°$ for the reaction of urea with water from values of $\Delta H°$ and $\Delta S°$.

$$CO(NH_2)_2(aq) + H_2O(l) \longrightarrow CO_2(g) + 2NH_3(g)$$

ANALYSIS: We can calculate $\Delta G°$ with the equation

$$\Delta G° = \Delta H° - T\Delta S°$$

To calculate $\Delta H°$, we can use Hess's law as a tool with the data in Table 6.2. To obtain $\Delta S°$, we normally would need to do a similar calculation using Equation 18.6 as a tool with data from Table 18.1. However, we already performed this calculation in Example 18.3.

SOLUTION: First we calculate $\Delta H°$ from data in Table 6.2.

$$\begin{aligned}\Delta H° &= [\Delta H°_{f\,CO_2(g)} + 2\Delta H°_{f\,NH_3(g)}] - [\Delta H°_{f\,CO(NH_2)_2(aq)} + \Delta H°_{f\,H_2O(l)}]\\ &= \left[1\,\cancel{\text{mol}} \times \left(\frac{-393.5\text{ kJ}}{\cancel{\text{mol}}}\right) + 2\,\cancel{\text{mol}} \times \left(\frac{-46.19\text{ kJ}}{\cancel{\text{mol}}}\right)\right]\\ &\quad - \left[1\,\cancel{\text{mol}} \times \left(\frac{-319.2\text{ kJ}}{\cancel{\text{mol}}}\right) + 1\,\cancel{\text{mol}} \times \left(\frac{-285.9\text{ kJ}}{\cancel{\text{mol}}}\right)\right]\\ &= (-485.9\text{ kJ}) - (-605.1\text{ kJ})\\ &= +119.2\text{ kJ}\end{aligned}$$

In Example 18.3 we found $\Delta S°$ to be $+354.8$ J K^{-1}. To calculate $\Delta G°$ we also need the Kelvin temperature, which we must express to at least four significant figures to match the number of significant figures in $\Delta S°$. Since standard temperature is *exactly* 25 °C, $T = (25.00 + 273.15)$ K $= 298.15$ K. Also, we must be careful to express $\Delta H°$ and $T\Delta S°$ in the same energy units, so

[3] Sometimes, the temperature is specified as a subscript in writing the symbol for the standard free energy change. For example, $\Delta G°$ can also be written $\Delta G°_{298}$. As you will see later, there are times when it is desirable to indicate the temperature explicitly.

$354.8\ J\ K^{-1} = 0.3548\ kJ\ K^{-1}$

we'll change the units of the entropy change to give $\Delta S° = +0.3548\ kJ\ K^{-1}$. Substituting into the equation for $\Delta G°$,

$$\begin{aligned}\Delta G° &= +119.2\ kJ - (298.15\ \cancel{K})(0.3548\ kJ\ \cancel{K^{-1}}) \\ &= +119.2\ kJ - 105.8\ kJ \\ &= +13.4\ kJ\end{aligned}$$

Therefore, for this reaction, $\Delta G° = +13.4$ kJ.

IS THE ANSWER REASONABLE? We can estimate that the second term will be approximately 100, so the answer should be small, which it is. The answer is therefore reasonable.

Practice Exercise 11: Calculate $\Delta G_f°$ for N_2O_4 from $\Delta H_f°$ and $\Delta S_f°$ for N_2O_4. (Hint: Write the chemical equation.)

Practice Exercise 12: Use the data in Table 6.2 and Table 18.1 to calculate $\Delta G_f°$ for the formation of iron(III) oxide (the iron oxide in rust). The equation for the reaction is

$$4Fe(s) + 3O_2(g) \longrightarrow 2Fe_2O_3(s)$$

In Section 6.8 you learned that it is useful to have tabulated standard heats of formation, $\Delta H_f°$, because they can be used with Hess's law to calculate $\Delta H°$ for many different reactions. Standard free energies of formation, $\Delta G_f°$, can be used in similar calculations to obtain $\Delta G°$.

Using $\Delta G_f°$ values

$$\Delta G° = (\text{sum of } \Delta G_f° \text{ of products}) - (\text{sum of } \Delta G_f° \text{ of reactants}) \qquad (18.7)$$

The $\Delta G_f°$ values for some typical substances are found in Table 18.2. Example 18.5 shows how we can use them to calculate $\Delta G°$ for a reaction.

EXAMPLE 18.5
Calculating $\Delta G°$ from $\Delta G_f°$

Ethanol, C_2H_5OH (also called ethyl alcohol), is made from grain by fermentation and is used as an additive to gasoline to produce a fuel mix called E85 (85% ethanol, 15% gasoline). What is $\Delta G°$ for the combustion of liquid ethanol to give $CO_2(g)$ and $H_2O(g)$?

ANALYSIS: Our tool is Equation 18.7. To use it we will need the balanced chemical equation for the reaction and $\Delta G_f°$ for each reactant and product. The free energy data are available in Table 18.2.

SOLUTION: First, we construct the balanced equation for the reaction.

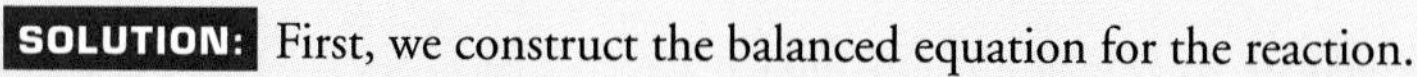

$$C_2H_5OH(l) + 3O_2(g) \longrightarrow 2CO_2(g) + 3H_2O(g)$$

(© AP/Wide World Photos.)

Applying Equation 18.7, we have

$$\Delta G° = [2\Delta G_{f\,CO_2(g)}° + 3\Delta G_{f\,H_2O(g)}°] - [\Delta G_{f\,C_2H_5OH(l)}° + 3\Delta G_{f\,O_2(g)}°]$$

As with $\Delta H_f°$, the $\Delta G_f°$ for any element in its standard state is zero. Therefore, using the data from Table 18.2,

$$\begin{aligned}\Delta G° &= \left[2\ \cancel{mol} \times \left(\frac{-394.4\ kJ}{\cancel{mol}}\right) + 3\ \cancel{mol} \times \left(\frac{-228.6\ kJ}{\cancel{mol}}\right)\right] \\ &\quad - \left[1\ \cancel{mol} \times \left(\frac{-174.8\ kJ}{\cancel{mol}}\right) + 3\ \cancel{mol} \times \left(\frac{0\ kJ}{\cancel{mol}}\right)\right] \\ &= (-1474.6\ kJ) - (-174.8\ kJ) \\ &= -1299.8\ kJ\end{aligned}$$

The standard free energy change for the reaction equals -1299.8 kJ.

IS THE ANSWER REASONABLE? As before, there's no easy way to estimate the answer. To check your answer, be sure you've got the correct algebraic sign for each term.

Practice Exercise 13: Calculate $\Delta G°$ for the reaction of iron(III) oxide with carbon monoxide to give elemental iron and carbon dioxide. (Hint: Be careful writing the chemical equation.)

Practice Exercise 14: Calculate $\Delta G°_{reaction}$ in kilojoules for the following reactions using the data in Table 18.2.

(a) $2NO(g) + O_2(g) \longrightarrow 2NO_2(g)$

(b) $Ca(OH)_2(s) + 2HCl(g) \longrightarrow CaCl_2(s) + 2H_2O(g)$

18.7 ΔG IS THE MAXIMUM AMOUNT OF WORK THAT CAN BE DONE BY A PROCESS

One of the chief uses of spontaneous chemical reactions is the production of useful work. For example, fuels are burned in gasoline or diesel engines to power automobiles and heavy machinery, and chemical reactions in batteries start our autos and run all sorts of modern electronic gadgets, including cellular phones, beepers, and laptop computers.

TABLE 18.2 Standard Free Energies of Formation of Typical Substances at 298.15 K

Substance	$\Delta G_f°$ (kJ mol^{-1})	Substance	$\Delta G_f°$ (kJ mol^{-1})
$Ag(s)$	0	$H_2O(g)$	−228.6
$AgCl(s)$	−109.7	$H_2O(l)$	−237.2
$Al(s)$	0	$HCl(g)$	−95.27
$Al_2O_3(s)$	−1576.4	$HNO_3(l)$	−79.91
$C(s, \text{graphite})$	0	$H_2SO_4(l)$	−689.9
$CO(g)$	−137.3	$HC_2H_3O_2(l)$	−392.5
$CO_2(g)$	−394.4	$Hg(l)$	0
$CH_4(g)$	−50.79	$Hg(g)$	+31.8
$CH_3Cl(g)$	−58.6	$K(s)$	0
$CH_3OH(l)$	−166.2	$KCl(s)$	−408.3
$CO(NH_2)_2(s)$	−197.2	$K_2SO_4(s)$	−1316.4
$CO(NH_2)_2(aq)$	−203.8	$N_2(g)$	0
$C_2H_2(g)$	+209	$NH_3(g)$	−16.7
$C_2H_4(g)$	+68.12	$NH_4Cl(s)$	−203.9
$C_2H_6(g)$	−32.9	$NO(g)$	+86.69
$C_2H_5OH(l)$	−174.8	$NO_2(g)$	+51.84
$C_8H_{18}(l)$	+17.3	$N_2O(g)$	+103.6
$Ca(s)$	0	$N_2O_4(g)$	+98.28
$CaCO_3(s)$	−1128.8	$Na(s)$	0
$CaCl_2(s)$	−750.2	$Na_2CO_3(s)$	−1048
$CaO(s)$	−604.2	$NaHCO_3(s)$	−851.9
$Ca(OH)_2(s)$	−896.76	$NaCl(s)$	−384.0
$CaSO_4(s)$	−1320.3	$NaOH(s)$	−382
$CaSO_4 \cdot \frac{1}{2}H_2O(s)$	−1435.2	$Na_2SO_4(s)$	−1266.8
$CaSO_4 \cdot 2H_2O(s)$	−1795.7	$O_2(g)$	0
$Cl_2(g)$	0	$PbO(s)$	−189.3
$Fe(s)$	0	$S(s)$	0
$Fe_2O_3(s)$	−741.0	$SO_2(g)$	−300.4
$H_2(g)$	0	$SO_3(g)$	−370.4

When chemical reactions occur, however, their energy is not always harnessed to do work. For instance, if gasoline is burned in an open dish, the energy evolved is lost entirely as heat and no useful work is accomplished. Engineers, therefore, seek ways to capture as much energy as possible in the form of work. One of their primary goals is to maximize the efficiency with which chemical energy is converted to work and to minimize the amount of energy transferred unproductively to the environment as heat.

Gas and diesel engines are not very efficient, so most of the energy produced in the combustion of fuel appears as heat, not work. That's why these engines require cooling systems.

Scientists have discovered that the maximum conversion of chemical energy to work occurs if a reaction is carried out under conditions that are said to be thermodynamically reversible. A process is defined as **thermodynamically reversible** if its driving force is opposed by another force that is just the slightest bit weaker, so that the slightest increase in the opposing force will cause the direction of the change to be reversed. An example of a nearly reversible process is illustrated in Figure 18.11, where we have a compressed gas in a cylinder pushing against a piston that's held in place by liquid water above it. If a water molecule evaporates, the external pressure drops slightly and the gas can expand just a bit, performing a small amount of work. Gradually, as one water molecule after another evaporates, the gas inside the cylinder slowly expands and performs work on the surroundings. At any time, however, the process can be reversed by the condensation of a water molecule.[4]

Although we could obtain the maximum work by carrying out a change reversibly, a thermodynamically reversible process requires so many steps that it proceeds at an extremely slow speed. If the work cannot be done at a reasonable rate, it is of little value to us. Our goal, then, is to approach thermodynamic reversibility for maximum efficiency, but to carry out the change at a pace that will deliver work at acceptable rates.

The relationship of useful work to thermodynamic reversibility was illustrated earlier (Section 6.5) in our discussion of the discharge of an automobile battery. Recall that when the battery is shorted with a heavy wrench, no work is done and all the energy appears as heat. In this case there is nothing opposing the discharge, and it occurs in the most thermodynamically irreversible manner possible. However, when the current is passed through a small electric motor, the motor itself offers resistance to the passage of the electricity and the discharge takes place slowly. In this instance, the discharge occurs in a more nearly thermodynamically reversible manner because of the opposition provided by the motor, and a relatively large amount of the available energy appears in the form of the work accomplished by the motor.

The preceding discussion leads naturally to the question: Is there a limit to the amount of the available energy in a reaction that can be harnessed as useful work? The answer to this question is to be found in the Gibbs free energy.

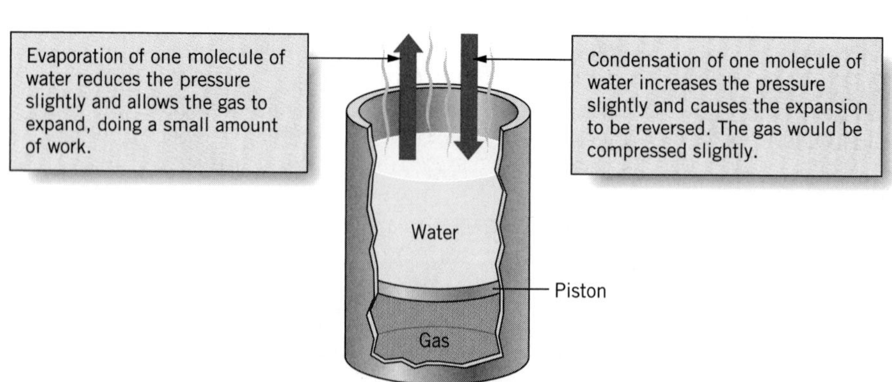

Evaporation of one molecule of water reduces the pressure slightly and allows the gas to expand, doing a small amount of work.
Condensation of one molecule of water increases the pressure slightly and causes the expansion to be reversed. The gas would be compressed slightly.
Water
Piston
Gas

FIG. 18.11 A reversible expansion of a gas. As water molecules evaporate one at a time, the external pressure gradually decreases and the gas slowly expands, performing a small amount of work in each step. The process would be reversed if a molecule of water were to condense into the liquid. The ability of the expansion to be reversed by the slightest increase in the opposing pressure is what makes this a reversible process.

[4] Although we sometimes say that a chemical reaction is "reversible" because it can run in both the forward and reverse directions, we cannot say the reaction is *thermodynamically reversible* unless the concentrations are only infinitesimally different from their equilibrium values as the reaction occurs.

The maximum amount of energy produced by a reaction that can be *theoretically* harnessed as work is equal to ΔG.

This is the energy that need not be lost to the surroundings as heat and is therefore *free* to be used for work. Thus, by determining the value of ΔG, we can find out whether or not a given reaction will be an effective source of useful energy. Also, by comparing the actual amount of work derived from a given system with the ΔG values for the reactions involved, we can measure the efficiency of the system.

EXAMPLE 18.6
Calculating Maximum Work

Calculate the maximum work available, expressed in kilojoules, from the oxidation of 1 mol of octane, $C_8H_{18}(l)$, by oxygen to give $CO_2(g)$ and $H_2O(l)$ at 25 °C and 1 atm.

ANALYSIS: The maximum work is equal to ΔG for the reaction. Standard thermodynamic conditions are specified, so we need to calculate $\Delta G°$. The tool for this is Equation 18.7.

SOLUTION: First we need a balanced equation for the reaction. For the combustion of one mole of C_8H_{18} we have

$$C_8H_{18}(l) + 12\tfrac{1}{2}O_2(g) \longrightarrow 8CO_2(g) + 9H_2O(l)$$

Then we apply Equation 18.7.

$$\Delta G° = [8\Delta G_{f\ CO_2(g)}^{\circ} + 9\Delta G_{f\ H_2O(l)}^{\circ}] - [\Delta G_{f\ C_8H_{18}(l)}^{\circ} + 12.5\Delta G_{f\ O_2(g)}^{\circ}]$$

Referring to Table 18.2 and dropping the canceled mol units,

$$\begin{aligned}\Delta G° &= [8 \times (-394.4)\ \text{kJ} + 9 \times (-237.2)\ \text{kJ}] - [1 \times (+17.3)\ \text{kJ} + 12.5 \times (0)\ \text{kJ}]\\ &= (-5290\ \text{kJ}) - (+17.3\ \text{kJ})\\ &= -5307\ \text{kJ}\end{aligned}$$

Thus, at 25 °C and 1 atm, we can expect no more than 5307 kJ of work from the oxidation of 1 mol of C_8H_{18}.

IS THE ANSWER REASONABLE? Be sure to check the algebraic signs of each of the terms in the calculation.

Practice Exercise 15: Calculate the maximum work that could be obtained from the combustion of 100 g of ethanol, $C_2H_5OH(l)$. How does it compare with the maximum work from the combustion of 100 g of $C_8H_{18}(l)$? Which is the better fuel? (Hint: The work available *per mole* has already been calculated in the worked examples.)

Practice Exercise 16: Calculate the maximum work that could be obtained at 25 °C and 1 atm from the oxidation of 1.00 mol of aluminum by $O_2(g)$ to give $Al_2O_3(s)$. (The oxidation of aluminum to aluminum oxide in booster rockets provides part of the energy that lifts the space shuttle off its launching pad.)

Blastoff of the space shuttle. The large negative heat of formation of Al_2O_3 provides power to the solid booster rockets that lift the space shuttle from its launch pad. *(Courtesy Lockheed Missile and Space Co., Inc.)*

18.8 ΔG IS ZERO WHEN A SYSTEM IS AT EQUILIBRIUM

We have seen that when the value of ΔG for a given change is negative, the change occurs spontaneously. We have also seen that a change is nonspontaneous when ΔG is positive. However, when ΔG is neither positive nor negative, the change is neither spontaneous nor nonspontaneous—the system is in a state of equilibrium. This occurs when ΔG is equal to zero.

> When a system is in a state of dynamic equilibrium,
>
> $$G_{\text{products}} = G_{\text{reactants}} \qquad \text{and} \qquad \Delta G = 0$$

Let's again consider the freezing of water.

$$H_2O(l) \rightleftharpoons H_2O(s)$$

Below 0 °C, ΔG for this change is negative and the freezing is spontaneous. On the other hand, above 0 °C we find that ΔG is positive and freezing is nonspontaneous. When the temperature is exactly 0 °C, $\Delta G = 0$ and an ice–water mixture exists in a condition of equilibrium. As long as heat isn't added or removed from the system, neither freezing nor melting is spontaneous and the ice and liquid water can exist together indefinitely.

No work can be done by a system at equilibrium

We have identified ΔG as a quantity that specifies the amount of work that is available from a system. Since ΔG is zero at equilibrium, the amount of work available is zero also. Therefore, when a system is at equilibrium, no work can be extracted from it. As an example, consider again the common lead storage battery that we use to start our car.

When the battery is fully charged, there are virtually no products of the discharge reaction present. The chemical reactants, however, are present in large amounts. Therefore, the total free energy of the reactants far exceeds the total free energy of products and, since $\Delta G = G_{\text{products}} - G_{\text{reactants}}$, the ΔG of the system has a large negative value. This means that a lot of energy is available to do work. As the battery discharges, the reactants are converted to products and G_{products} gets larger while $G_{\text{reactants}}$ gets smaller; thus ΔG becomes less negative, and less energy is available to do work. Finally, the battery reaches equilibrium. The total free energies of the reactants and the products have become equal, so $G_{\text{products}} - G_{\text{reactants}} = 0$ and $\Delta G = 0$. No further work can be extracted and we say the battery is dead.

Melting points and boiling points can be estimated from ΔH and ΔS

For a phase change such as $H_2O(l) \longrightarrow H_2O(s)$, equilibrium can only exist at one particular temperature at atmospheric pressure. For water, that temperature is 0 °C. Above 0 °C, only liquid water can exist, and below 0 °C all the liquid will freeze to give ice. This yields an interesting relationship between ΔH and ΔS for a phase change. Since $\Delta G = 0$,

$$\Delta G = 0 = \Delta H - T\Delta S$$

Therefore,

$$\Delta H = T\Delta S$$

and

$$\Delta S = \frac{\Delta H}{T} \tag{18.8}$$

Thus, if we know ΔH for the phase change and the temperature at which the two phases coexist, we can calculate ΔS for the phase change. Another interesting relationship that we can obtain is

$$T = \frac{\Delta H}{\Delta S} \tag{18.9}$$

Thus, if we know ΔH and ΔS, we can calculate the temperature at which equilibrium will occur.

EXAMPLE 18.7
Estimating the Equilibrium Temperature for a Phase Change

For the phase change $Br_2(l) \longrightarrow Br_2(g)$, $\Delta H° = +31.0$ kJ mol^{-1} and $\Delta S° = 92.9$ J mol^{-1} K^{-1}. Assuming that ΔH and ΔS are nearly temperature independent, calculate the approximate Celsius temperature at which $Br_2(l)$ will be in equilibrium with $Br_2(g)$ at 1 atm (i.e., the normal boiling point of liquid Br_2).

ANALYSIS: The temperature at which equilibrium exists is given by Equation 18.9,

$$T = \frac{\Delta H}{\Delta S}$$

If we assume that ΔH and ΔS do not depend much on temperature, then we can use $\Delta H°$ and $\Delta S°$ in this equation to approximate the boiling point. That is,

$$T \approx \frac{\Delta H°}{\Delta S°}$$

□ ΔH and ΔS do not change much with changes in temperature. This is because temperature changes affect the enthalpies and entropies of both the reactants and products by about the same amount, so the differences between reactants and products stays fairly constant.

SOLUTION: Substituting the data given in the problem,

$$T \approx \frac{3.10 \times 10^4 \text{ J mol}^{-1}}{92.9 \text{ J mol}^{-1}\text{ K}^{-1}}$$
$$= 334 \text{ K}$$

The Celsius temperature is 334 − 273 = 61 °C. Notice that we were careful to express $\Delta H°$ in joules, not kilojoules, so the units would cancel correctly. It is also interesting that the boiling point we calculated is quite close to the measured normal boiling point of 58.8 °C.

IS THE ANSWER REASONABLE? The $\Delta H°$ value equals 31,000 and the $\Delta S°$ value equals approximately 100, which means the temperature should be about 310 K. Our value, 334 K is not far from that, so the answer is reasonable.

Practice Exercise 17: The heat of vaporization of ammonia is 21.7 kJ mol^{-1} and the boiling point of ammonia is −33.3 °C. Estimate the entropy change for the vaporization of liquid ammonia. (Hint: What algebraic sign do we expect for the entropy change?)

Practice Exercise 18: The heat of vaporization of mercury is 60.7 kJ/mol. For $Hg(l)$, $S° = 76.1$ J mol^{-1} K^{-1} and for $Hg(g)$, $S° = 175$ J mol^{-1} K^{-1}. Estimate the normal boiling point of liquid mercury.

Free energy diagrams for chemical reactions show a minimum in free energy at equilibrium

One way to gain a better understanding of how the free energy changes during a reaction is by studying **free energy diagrams.** As an example, let's study a reaction you've seen before—the decomposition of N_2O_4 into NO_2.

$$N_2O_4(g) \longrightarrow 2NO_2(g)$$

In our discussion of chemical equilibrium in Chapter 14 (page 570), we noted that equilibrium in this system can be approached from *either* direction, with the same equilibrium concentrations being achieved provided we begin with the same overall system composition.

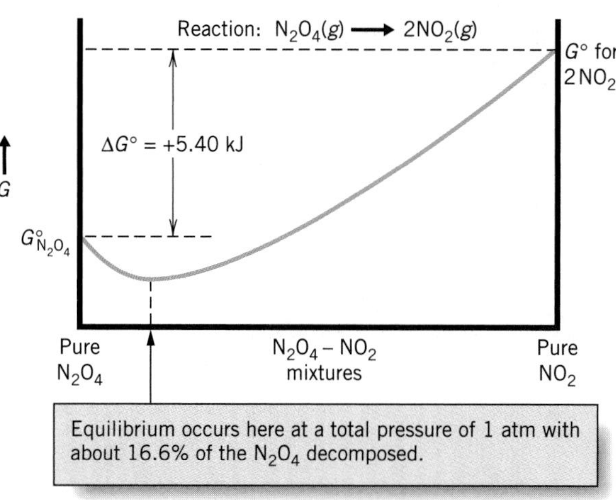

FIG. 18.12 Free energy diagram for the decomposition of $N_2O_4(g)$. The minimum on the curve indicates the composition of the reaction mixture at equilibrium. Because $\Delta G°$ is positive, the position of equilibrium lies close to the reactants. The amount of product that will form by the time the system reaches equilibium will be small.

When a system moves "downhill" on its free energy curve, $G_{\text{final}} < G_{\text{initial}}$ and ΔG is negative. Changes with negative ΔG are spontaneous.

Figure 18.12 shows the free energy diagram for the reaction, which depicts how the free energy changes as we proceed from the reactant to the product. On the left of the diagram we have the free energy of one mole of pure $N_2O_4(g)$, and on the right the free energy of two moles of pure $NO_2(g)$. Points along the horizontal axis represent mixtures of both substances. Notice that in going from reactant (N_2O_4) to product ($2NO_2$), the free energy has a minimum. It drops below that of either pure N_2O_4 or pure NO_2.

Any system will spontaneously seek the lowest point on its free energy curve. If we begin with pure $N_2O_4(g)$, the reaction will proceed from left to right and some $NO_2(g)$ will be formed, because proceeding in the direction of NO_2 leads to a lowering of the free energy. If we begin with pure $NO_2(g)$, a change also will occur. Going downhill on the free energy curve now takes place as the reverse reaction occurs [i.e., $2NO_2(g) \longrightarrow N_2O_4(g)$]. Once the bottom of the "valley" is reached, the system has come to equilibrium. As you learned in Chapter 14, if the system isn't disturbed, the composition of the equilibrium mixture will remain constant. Now we see that the reason is because any change (moving either to the left or right) would require an uphill climb. Free energy increases are not spontaneous, so this doesn't happen.

An important thing to notice in Figure 18.12 is that some reaction takes place spontaneously in the forward direction even though $\Delta G°$ is positive. However, the reaction doesn't proceed far before equilibrium is reached. For comparison, Figure 18.13 shows the shape of the free energy curve for a reaction with a negative $\Delta G°$. We see here that at equilibrium there has been a much greater conversion of reactants to products. Thus, *$\Delta G°$ tells us where the position of equilibrium lies between pure reactants and pure products.*

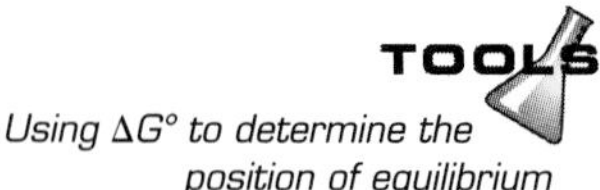

Using $\Delta G°$ to determine the position of equilibrium

$\Delta G°$ and the position of equilibrium

- When $\Delta G°$ is positive, the position of equilibrium lies close to the reactants and little reaction occurs by the time equilibrium is reached. The reaction will appear to be nonspontaneous.
- When $\Delta G°$ is negative, the position of equilibrium lies close to the products and a large amount of products will have formed by the time equilibrium is reached. The reaction will appear to be spontaneous.
- When $\Delta G° = 0$, the position of equilibrium will lie about midway between reactants and products. Substantial amounts of both reactants and products will be present when equilibrium is reached. The reaction will appear to be spontaneous whether we begin with pure reactants or pure products.

$\Delta G°$ can be used to predict the outcome of a chemical reaction

In general, the value of $\Delta G°$ for most reactions is much larger numerically than the $\Delta G°$ for the $N_2O_4 \longrightarrow NO_2$ reaction. In addition, the extent to which a reaction proceeds is very sensitive to the size of $\Delta G°$. If the $\Delta G°$ value for a reaction is reasonably large—about 20 kJ or more—almost no observable reaction will occur when $\Delta G°$ is positive. On the

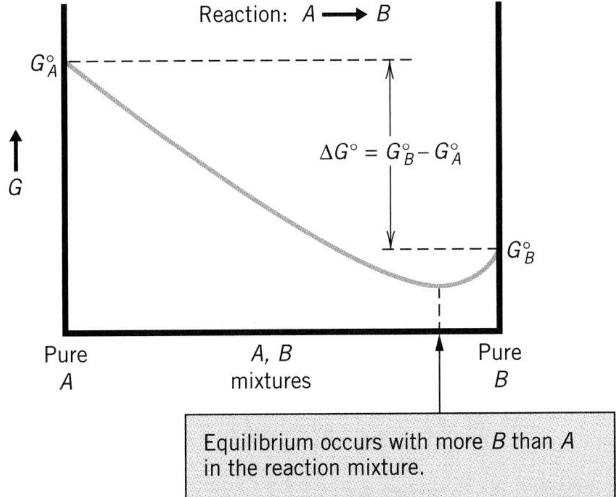

FIG. 18.13 Free energy curve for a reaction having a negative ΔG°. Because G_B° is lower than G_A°, ΔG° is negative for the reaction $A \longrightarrow B$. This causes the position of equilibrium to lie far to the right, near the products. When the system reaches equilibrium, there will be a large amount of products present and little reactants.

other hand, the reaction will go almost to completion if ΔG° is both large and negative.[5] From a practical standpoint, then, *the size and sign of ΔG° serve as indicators of whether an observable spontaneous reaction will occur.*

EXAMPLE 18.8
Using ΔG° as a Predictor of the Outcome of a Reaction

Would we expect to be able to observe the following reaction at 25 °C?

$$NH_4Cl(s) \longrightarrow NH_3(g) + HCl(g)$$

ANALYSIS: To answer this question, our tool is the magnitude and algebraic sign of ΔG°. If ΔG° is reasonably large and positive, the reaction won't be observed. If it is reasonably large and negative, we can expect to see the reaction go nearly to completion.

SOLUTION: First let's calculate ΔG° for the reaction using the data in Table 18.2. The procedure is the same as that discussed earlier.

$$\begin{aligned}\Delta G^\circ &= [\Delta G_f^\circ{}_{NH_3(g)} + \Delta G_f^\circ{}_{HCl(g)}] - [\Delta G_f^\circ{}_{NH_4Cl(s)}] \\ &= [(-16.7\text{ kJ}) + (-95.27\text{ kJ})] - [-203.9\text{ kJ}] \\ &= +91.9\text{ kJ}\end{aligned}$$

Because ΔG° is large and positive, only extremely small amounts of products can form at this temperature.

IS THE ANSWER REASONABLE? Be sure to check the algebraic signs of each of the terms in the calculation.

Practice Exercise 19: Use the data in Table 18.2 to determine whether the reaction

$$SO_2(g) + O_2(g) \longrightarrow SO_3(g) + \tfrac{1}{2}O_2(g)$$

should "occur spontaneously" at 25 °C. (Hint: Be careful with algebraic signs.)

Practice Exercise 20: Use the data in Table 18.2 to determine whether we should expect to see the formation of $CaCO_3(s)$ in the following reaction at 25 °C.

$$CaCl_2(s) + H_2O(g) + CO_2(g) \longrightarrow CaCO_3(s) + 2HCl(g)$$

[5] As we discussed earlier, to actually see a change take place, the speed of a spontaneous reaction must be reasonably fast. For example, the decomposition of the nitrogen oxides into N_2 and O_2 is thermodynamically spontaneous (ΔG° is negative), but their rates of decomposition are so slow that these substances appear to be stable and some are obnoxious air pollutants.

The position of equilibrium changes with temperature because ΔG changes with temperature

So far, we have confined our discussion of the relationship of free energy and equilibrium to a special case, 25 °C. But what about other temperatures? Equilibria certainly can exist at temperatures other than 25 °C, and in Chapter 14 you learned how to apply Le Châtelier's principle to predicting the way temperature affects the position of equilibrium. Now let's see how thermodynamics deals with this.

298 K = 25 °C

You've learned that at 25 °C the position of equilibrium is determined by the difference between the free energy of pure products and the free energy of pure reactants. This difference is given by ΔG°_{298}, where we have now used the subscript "298" to indicate the temperature, 298 K. We define ΔG°_{298} as

$$\Delta G^\circ_{298} = (G^\circ_{\text{products}})_{298} - (G^\circ_{\text{reactants}})_{298}$$

At temperatures other than 25 °C, it is still the difference between the free energies of the products and reactants that determines the position of equilibrium. We might write this as ΔG°_T. Thus, at a temperature other than 25 °C (298 K), we have

$$\Delta G^\circ_T = (G^\circ_{\text{products}})_T - (G^\circ_{\text{reactants}})_T$$

where $(G^\circ_{\text{products}})_T$ and $(G^\circ_{\text{reactants}})_T$ are the total free energies of the pure products and reactants, respectively, at that other temperature.

Next, we must find a way to compute ΔG°_T. Earlier we saw that ΔG° can be obtained from the equation

$$\Delta G^\circ = \Delta H^\circ - (298\text{ K})\Delta S^\circ$$

At a different temperature, T, the equation becomes

$$\Delta G^\circ_T = \Delta H^\circ_T - T\Delta S^\circ_T$$

Now we seem to be getting closer to our goal. If we can compute or estimate the values of ΔH°_T and ΔS°_T for a reaction, we have solved our problem.

The size of ΔG°_T obviously depends very strongly on the temperature—the equation above has temperature as one of its variables. However, as noted in the margin comment on page 749, the magnitudes of the ΔH and ΔS for a reaction are relatively insensitive to the temperature. Therefore, we can use ΔH°_{298} and ΔS°_{298} as reasonable approximations of ΔH°_T and ΔS°_T. This allows us to rewrite the equation for ΔG°_T as

TOOLS
Calculating ΔG° at temperatures other than 25 °C

$$\Delta G^\circ_T \approx \Delta H^\circ_{298} - T\Delta S^\circ_{298} \qquad (18.10)$$

The following examples illustrate how this equation is useful.

EXAMPLE 18.9
ΔG° at Temperatures Other than 25 °C

Earlier we saw that at 25 °C the value of ΔG° for the reaction

$$N_2O_4(g) \longrightarrow 2NO_2(g)$$

has a value of +5.40 kJ. What is the approximate value of ΔG°_T for this reaction at 100 °C?

ANALYSIS: Equation 18.10 is the tool needed to solve the problem. To use it, we need values of ΔH° and ΔS°. The ΔH° for the reaction can be calculated from the data in Table 6.2 on page 236. Here we find the following standard heats of formation.

$N_2O_4(g)$ $\quad \Delta H^\circ_f = +9.67$ kJ/mol
$NO_2(g)$ $\quad \Delta H^\circ_f = +33.8$ kJ/mol

We combine these by a Hess's law calculation to compute ΔH° for the reaction. Next, we compute ΔS° for the reaction using absolute entropy data from Table 18.1.

$N_2O_4(g)$ $\quad S^\circ = 304$ J/mol K
$NO_2(g)$ $\quad S^\circ = 240.5$ J/mol K

Our tool for this calculation is Equation 18.6 on page 741, which is also a Hess's law kind of calculation.

SOLUTION: First, we compute $\Delta H°$ using Hess's law (Equation 6.14, page 238).

$$\Delta H° = [2\Delta H°_{f\ NO_2(g)}] - [\Delta H°_{f\ N_2O_4(g)}]$$
$$= \left[2\ \cancel{mol} \times \left(\frac{33.8\ kJ}{\cancel{mol}}\right)\right] - \left[1\ \cancel{mol} \times \left(\frac{9.67\ kJ}{\cancel{mol}}\right)\right]$$
$$= +57.9\ kJ$$

Next, we use Equation 18.6 to calculate $\Delta S°$.

$$\Delta S° = \left[2\ \cancel{mol} \times \left(\frac{240.5\ J}{\cancel{mol}\ K}\right)\right] - \left[1\ \cancel{mol} \times \left(\frac{304\ J}{\cancel{mol}\ K}\right)\right]$$
$$= +177\ J\ K^{-1}\ \text{or}\ 0.177\ kJ\ K^{-1}$$

The temperature is 100 °C, which is 373 K, so we can call the free energy change $\Delta G°_{373}$. Substituting into Equation 18.10 using $T = 373$ K, we have

$$\Delta G°_{373} \approx (+57.9\ kJ) - (373\ \cancel{K})(0.177\ kJ\ \cancel{K}^{-1})$$
$$\approx -8.1\ kJ$$

Notice that at this higher temperature, the sign of $\Delta G°_T$ has become negative.

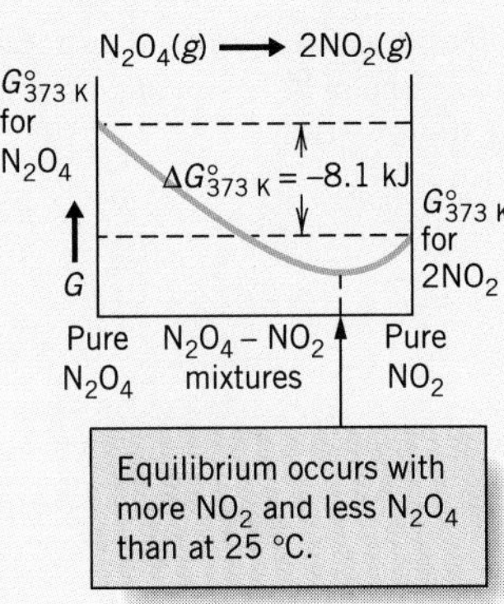

IS THE ANSWER REASONABLE? As usual, we can double-check the algebraic signs to be sure we've performed the calculations correctly. For this problem, however, there's another check we can do.

Let's compare free energy diagrams for this reaction at 25 °C and 100 °C. At 25 °C, $\Delta G°$ is positive and the position of equilibrium lies toward the reactant (Figure 18.12). Because $\Delta G°$ is negative at 100 °C, the free energy diagram will resemble that in Figure 18.13, with the position of equilibrium closer to the product (see the figure in the margin). Raising the temperature has shifted the position of equilibrium of this *endothermic* reaction toward the products, which is just what we would have anticipated by applying Le Châtelier's principle. The answer is therefore reasonable.

Practice Exercise 21: In Examples 18.3 and 18.4 we computed $\Delta S°$ and $\Delta H°$ for the reaction

$$CO(NH_2)_2(aq) + H_2O(l) \longrightarrow CO_2(g) + 2NH_3(g)$$

What is $\Delta G°_T$ for the reaction at 75 °C? (Hint: Be careful with the units.)

Practice Exercise 22: Use the data in Table 18.2 to determine $\Delta G°_{298}$ for the reaction

$$2NaHCO_3(s) \longrightarrow Na_2CO_3(s) + CO_2(g) + H_2O(g)$$

Then calculate the approximate value for $\Delta G°$ for the reaction at 200 °C using the data in Tables 6.2 and 18.1. How does the position of equilibrium for the reaction change as the temperature is increased?

18.9 EQUILIBRIUM CONSTANTS CAN BE ESTIMATED FROM STANDARD FREE ENERGY CHANGES

In the preceding discussion, you learned in a qualitative way that the position of equilibrium in a reaction is determined by the sign and magnitude of $\Delta G°$. You also learned that the direction in which a reaction proceeds depends on where the system composition stands relative to the minimum on the free energy curve. Thus, the reaction will proceed spontaneously in the forward direction only if it will lead to a lowering of the free energy (i.e., if ΔG is negative).

Quantitatively, the relationship between ΔG and $\Delta G°$ is expressed by the following equation, which we will not attempt to justify.

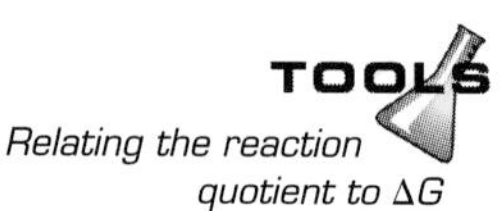

Relating the reaction quotient to ΔG

$$\Delta G = \Delta G° + RT \ln Q \tag{18.11}$$

Here R is the gas constant in appropriate energy units (e.g., 8.314 J mol^{-1} K^{-1}), T is the Kelvin temperature, and ln Q is the natural logarithm of the reaction quotient. For gaseous reactions, Q is calculated using partial pressures expressed in atm[6]; for reactions in solution, Q is calculated from molar concentrations. Equation 18.11 allows us to predict the direction of the spontaneous change in a reaction mixture if we know $\Delta G°$ and the composition of the mixture, as illustrated in Example 18.10.

Recall that in Chapter 14 we defined Q as the *reaction quotient,* the numerical value of the mass action expression.

EXAMPLE 18.10
Determining the Direction of a Spontaneous Reaction

The reaction $2NO_2(g) \rightleftharpoons N_2O_4(g)$ has $\Delta G°_{298} = -5.40$ kJ per mole of N_2O_4. In a reaction mixture, the partial pressure of NO_2 is 0.25 atm and the partial pressure of N_2O_4 is 0.60 atm. In which direction must this reaction proceed to reach equilibrium?

In Chapter 14, you learned that you can also predict the direction of a reaction by comparing Q with K (page 592)

ANALYSIS: Since we know that reactions proceed spontaneously *toward* equilibrium, we are really being asked to determine whether the reaction will proceed spontaneously in the forward or reverse direction. We can use Equation 18.11 to calculate ΔG for the forward reaction. If ΔG is negative, then the forward reaction is spontaneous. However, if the calculated ΔG is positive, the forward reaction is nonspontaneous and it is really the reverse reaction that is spontaneous.

SOLUTION: First, we need the correct form for the mass action expression so we can calculate Q correctly. Expressed in terms of partial pressures, the mass action expression is

$$\frac{P_{N_2O_4}}{P^2_{NO_2}}$$

Therefore, the equation we will use is

$$\Delta G = \Delta G° + RT \ln\left(\frac{P_{N_2O_4}}{P^2_{NO_2}}\right)$$

Next, let's assemble the data:

$$\Delta G°_{298} = -5.40 \text{ kJ mol}^{-1} \qquad T = 298 \text{ K}$$
$$= -5.40 \times 10^3 \text{ J mol}^{-1} \qquad P_{N_2O_4} = 0.60 \text{ atm}$$
$$R = 8.314 \text{ J mol}^{-1}\text{K}^{-1} \qquad P_{NO_2} = 0.25 \text{ atm}$$

Notice we have changed the energy units of $\Delta G°$ to joules so they will be compatible with those calculated using R. The next step is to substitute into the equation for ΔG. In doing this, we leave off the units for the partial pressures because reaction quotients are always unitless quantities.[7] Here is the calculation.[8]

[6] Strictly speaking, we should express pressures in atmospheres or bars depending on which was used to obtain the $\Delta G°$ data. For simplicity, we will use the more familiar pressure unit atm in all our calculations. Because the atmosphere and bar differ by only about 1 percent, any errors that might be introduced will be very small.

[7] For reasons beyond the scope of this text, quantities in mass action expressions used to compute reaction quotients and equilibrium constants are actually ratios. For partial pressures, it's the pressure in atmospheres divided by 1 atm; for concentrations, it's the molarity divided by 1 M. This causes the units to cancel to give a unitless value for Q or K.

[8] As we noted in the footnote on page 535, when taking the logarithm of a quantity, the number of digits *after the decimal point* should equal the number of significant figures in the quantity. Since 0.60 atm and 0.25 atm both have two significant figures, the logarithm of the quantity in square brackets (2.26) is rounded to give two digits after the decimal point.

$$\Delta G = -5.40 \times 10^3 \text{ J mol}^{-1} + (8.314 \text{ J mol}^{-1} \text{ K}^{-1})(298 \text{ K}) \ln \left[\frac{0.60}{(0.25)^2}\right]$$
$$= -5.40 \times 10^3 \text{ J mol}^{-1} + (8.314 \text{ J mol}^{-1} \text{ K}^{-1})(298 \text{ K}) \ln (9.6)$$
$$= -5.40 \times 10^3 \text{ J mol}^{-1} + (8.314 \text{ J mol}^{-1} \text{ K}^{-1})(298 \text{ K})(2.26)$$
$$= -5.40 \times 10^3 \text{ J mol}^{-1} + 5.60 \times 10^3 \text{ J mol}^{-1}$$
$$= +2.0 \times 10^2 \text{ J mol}^{-1}$$

Since ΔG is positive, the forward reaction is nonspontaneous. The reverse reaction is the one that will occur, and to reach equilibrium, some N_2O_4 will have to decompose.

IS THE ANSWER REASONABLE? There is no simple check. However, we can check to be sure the energy units in both terms on the right are the same. Notice that here we have changed the units for $\Delta G°$ to joules to match those of R. Also notice that the temperature is expressed in kelvins, to match the temperature units in R.

Practice Exercise 23: Calculate ΔG for the reaction described in the preceding example if the partial pressure of NO_2 is 0.260 atm and the partial pressure of N_2O_4 is 0.598 atm. Where does the system stand relative to equilibrium? (Hint: What is ΔG for a system at equilibrium?)

Practice Exercise 24: In which direction will the reaction described in the preceding Example proceed to reach equilibrium if the partial pressure of NO_2 is 0.60 atm and the partial pressure of N_2O_4 is 0.25 atm?

The thermodynamic equilibrium constant K can be computed from $\Delta G°$

Earlier you learned that when a system reaches equilibrium the free energy of the products equals the free energy of the reactants and ΔG equals zero. We also know that at equilibrium the reaction quotient equals the equilibrium constant.

$$\text{At equilibrium} \quad \begin{cases} \Delta G = 0 \\ Q = K \end{cases}$$

If we substitute these into Equation 18.11, we obtain

$$0 = \Delta G° + RT \ln K$$

which can be rearranged to give

$$\Delta G° = -RT \ln K \tag{18.12}$$

TOOLS *Determining thermodynamic equilibrium constants*

The equilibrium constant K calculated from this equation is called the **thermodynamic equilibrium constant** and corresponds to K_P for reactions involving gases (with partial pressures expressed in bars) and to K_c for reactions in solution (with concentrations expressed in mol L^{-1}).[9]

Equation 18.12 is useful because it permits us to determine equilibrium constants from either measured or calculated values of $\Delta G°$. As you know, $\Delta G°$ can be determined by a Hess's law type of calculation from tabulated values of $\Delta G_f°$. It also allows us to obtain values of $\Delta G°$ from measured equilibrium constants.

[9] The thermodynamic equilibrium constant requires that gases *always* be included as partial pressures in mass action expressions. The pressure units must be standard pressure units (atmospheres or bars, depending on which convention was used for collecting the $\Delta G°$ data. For a heterogeneous equilibrium involving gases and substances in solution, the mass action expression will mix partial pressures (for the gases) with molarities (for the dissolved substances).

EXAMPLE 18.11
Thermodynamic Equilibrium Constants

The brownish haze often associated with air pollution is caused by nitrogen dioxide, NO_2, a red-brown gas. Nitric oxide, NO, is formed in auto engines and some of it escapes into the air where it is oxidized to NO_2 by oxygen.

$$2NO(g) + O_2(g) \rightleftharpoons 2NO_2(g)$$

The value of K_P for the reaction is 1.7×10^{12} at 25.00 °C. What is $\Delta G°$ for the reaction, expressed in joules per mole? In kilojoules per mole?

ANALYSIS: The tool relating thermodynamic data and equilibrium constants is Equation 18.12.

$$\Delta G° = -RT \ln K_P$$

We'll need the data below to calculate $\Delta G°$. Note that the temperature is given to the nearest hundredth of a degree so we'll use $T = t_C + 273.15$ to compute the Kelvin temperature.

$$R = 8.314 \text{ J mol}^{-1} \text{ K}^{-1}$$
$$T = 298.15 \text{ K}$$
$$K_p = 1.7 \times 10^{12}$$

The units in this calculation give $\Delta G°$ on a per mole (mol^{-1}) basis. This reminds us that we are viewing the coefficients of the reactants and products as representing moles, rather than some other sized quantity.

SOLUTION: Substituting these values into the equation, we have

$$\begin{aligned}\Delta G° &= -(8.314 \text{ J mol}^{-1} \text{ K}^{-1} \times 298.15 \text{ K}) \ln (1.7 \times 10^{12}) \\ &= -(8.314 \text{ J mol}^{-1} \cancel{\text{K}^{-1}} \times 298.15 \cancel{\text{K}}) \times (28.16) \\ &= -6.980 \times 10^4 \text{ J mol}^{-1} \text{ (to four significant figures)}\end{aligned}$$

Expressed in kilojoules, $\Delta G° = -69.80 \text{ kJ mol}^{-1}$.

IS THE ANSWER REASONABLE? The value of K_P tells us that the position of equilibrium lies far to the right, which means $\Delta G°$ must be large and negative. Therefore, the answer, $-69.80 \text{ kJ mol}^{-1}$, seems reasonable.

EXAMPLE 18.12
Thermodynamic Equilibrium Constants

Sulfur dioxide, which is sometimes present in polluted air, reacts with oxygen when it passes over the catalyst in automobile catalytic converters. The product is the very acidic oxide SO_3.

$$2SO_2(g) + O_2(g) \rightleftharpoons 2SO_3(g)$$

For this reaction, $\Delta G° = -1.40 \times 10^2 \text{ kJ mol}^{-1}$ at 25 °C. What is the value of K_P?

To make the units cancel correctly, we have added the unit mol^{-1} to the value of $\Delta G°$. This just emphasizes that the amounts of reactants and products are specified in mole sized quantities.

ANALYSIS: Once again we use the equation

$$\Delta G° = -RT \ln K_P$$

Our data are

$$R = 8.314 \text{ J mol}^{-1} \text{ K}^{-1}$$
$$T = 298 \text{ K}$$
$$\begin{aligned}\Delta G° &= -1.40 \times 10^2 \text{ kJ mol}^{-1} \\ &= -1.40 \times 10^5 \text{ J mol}^{-1}\end{aligned}$$

SOLUTION: To calculate K_P, let's first solve for $\ln K_P$.

$$\ln K_P = \frac{-\Delta G°}{RT}$$

Substituting values gives

$$\ln K_P = \frac{-(-1.40 \times 10^5 \, \cancel{J} \, \cancel{mol^{-1}})}{(8.314 \, \cancel{J} \, \cancel{mol^{-1}} \, \cancel{K^{-1}})(298 \, \cancel{K})}$$
$$= +56.5$$

To calculate K_P, we take the antilogarithm,

$$K_p = e^{56.5}$$
$$= 3 \times 10^{24}$$

Notice that we have expressed the answer to only one significant figure. As discussed earlier, when taking a logarithm, the number of digits written after the decimal place equals the number of significant figures in the number. Conversely, the number of significant figures in the antilogarithm equals the number of digits after the decimal in the logarithm.

IS THE ANSWER REASONABLE? The value of $\Delta G°$ is large and negative, so the position of equilibrium should favor the products. The large value of K_P is therefore reasonable.

Practice Exercise 25: The reaction $N_2(g) + 3H_2(g) \rightleftharpoons 2NH_3(g)$ has $K_P = 6.9 \times 10^5$ at 25.0 °C. Calculate $\Delta G°$ for this reaction in units of kilojoules. (Hint: Be careful with units and significant figures.)

Practice Exercise 26: The reaction $H_2(g) + I_2(g) \rightleftharpoons 2HI(g)$ has $\Delta G° = +3.3$ kJ mol^{-1} at 25.0 °C. What is the value of K_P at that temperature?

18.10 BOND ENERGIES CAN BE ESTIMATED FROM REACTION ENTHALPY CHANGES

You have seen how thermodynamic data allow us to predict the spontaneity of chemical reactions as well as the nature of chemical systems at equilibrium. In addition to these useful and important benefits of the study of thermodynamics, there is a bonus. By studying heats of reaction, and heats of formation in particular, we can obtain fundamental information about the chemical bonds in the substances that react, because the origin of the energy changes in chemical reactions is changes in bond energies.

Recall that the **bond energy** is the amount of energy needed to break a chemical bond to give electrically neutral fragments. It is a useful quantity to know in the study of chemical properties, because during chemical reactions, bonds within the reactants are broken and new ones are formed as the products appear. The first step—bond breaking—is one of the factors that controls the reactivity of substances. Elemental nitrogen, for example, has a very low degree of reactivity, which is generally attributed to the very strong triple bond in N_2. Reactions that involve the breaking of this bond in a single step simply do not occur. When N_2 does react, it is by a stepwise involvement of its three bonds, one at a time.

The heats of formation of the nitrogen oxides are endothermic because of the large bond energy of the N_2 molecule.

Bond energies can be measured or calculated using Hess's law

The bond energies of simple diatomic molecules such as H_2, O_2, and Cl_2 are usually measured *spectroscopically*. A flame or an electric spark is used to excite (energize) the molecules, causing them to emit light. An analysis of the spectrum of emitted light allows scientists to compute the amount of energy needed to break the bond.

For more complex molecules, thermochemical data can be used to calculate bond energies in a Hess's law kind of calculation. We will use the standard heat of formation of methane to illustrate how this is accomplished. However, before we can attempt such a calculation, we must first define a thermochemical quantity that we will call the **atomization energy**, symbolized ΔH_{atom}. This is the amount of energy needed to rupture all the

The kind of bond breaking described here divides the electrons of the bond equally between the two atoms. It could be symbolized as

$$A{:}B \longrightarrow A\cdot + \cdot B$$

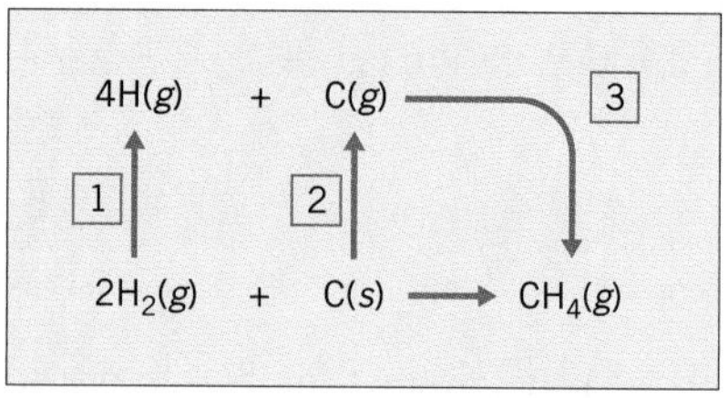

FIG. 18.14 Two paths for the formation of methane from its elements in their standard states. As described in the text, steps 1, 2, and 3 of the upper path involve the formation of gaseous atoms of the elements and the formation of bonds in CH_4. The lower path corresponds to the direct combination of the elements in their standard states to give CH_4. Because ΔH is a state function, the sum of the enthalpy changes along the upper path must equal the enthalpy change for the lower path (ΔH_f°).

chemical bonds in one mole of gaseous molecules to give gaseous atoms as products. For example, the atomization of methane is

$$CH_4(g) \longrightarrow C(g) + 4H(g)$$

and the enthalpy change for the process is ΔH_{atom}. For this particular molecule, ΔH_{atom} corresponds to the total amount of energy needed to break all the C—H bonds in one mole of CH_4; therefore, division of ΔH_{atom} by 4 would give the average C—H bond energy in methane, expressed in kJ mol^{-1}.

Figure 18.14 shows how we can use the standard heat of formation, ΔH_f°, to calculate the atomization energy. Across the bottom we have the chemical equation for the formation of CH_4 from its elements. The enthalpy change for this reaction, of course, is ΔH_f°. In this figure we also can see an alternative three-step path that leads to $CH_4(g)$. One step is the breaking of H—H bonds in the H_2 molecules to give gaseous hydrogen atoms, another is the vaporization of carbon to give gaseous carbon atoms, and the third is the combination of the gaseous atoms to form CH_4 molecules. These changes are labeled 1, 2, and 3 in the figure.

Since ΔH is a state function, the net enthalpy change from one state to another is the same regardless of the path that we follow. This means that the sum of the enthalpy changes along the upper path must be the same as the enthalpy change along the lower path, ΔH_f°. Perhaps this can be more easily seen in Hess's law terms if we write the changes along the upper path in the form of thermochemical equations.

A more complete table of standard heats of formation of gaseous atoms is located in Appendix C.2.

Steps 1 and 2 have enthalpy changes that are called *standard heats of formation of gaseous atoms.* Values for these quantities have been measured for many of the elements, and some are given in Table 18.3. Step 3 is the opposite of atomization, and its enthalpy change will therefore be the negative of ΔH_{atom} (recall that if we reverse a reaction, we change the sign of its ΔH).

(Step 1)	$2H_2(g) \longrightarrow 4H(g)$	$\Delta H_1^\circ = 4\Delta H_{f\,H(g)}^\circ$
(Step 2)	$C(s) \longrightarrow C(g)$	$\Delta H_2^\circ = \Delta H_{f\,C(g)}^\circ$
(Step 3)	$4H(g) + C(g) \longrightarrow CH_4(g)$	$\Delta H_3^\circ = -\Delta H_{atom}$
	$2H_2(g) + C(s) \longrightarrow CH_4(g)$	$\Delta H^\circ = \Delta H_{f\,CH_4(g)}^\circ$

Notice that by adding the first three equations, C(*g*) and 4H(*g*) cancel and we get the equation for the formation of CH_4 from its elements in their standard states. This means that adding the ΔH° values of the first three equations should give ΔH_f° for CH_4.

$$\Delta H_1^\circ + \Delta H_2^\circ + \Delta H_3^\circ = \Delta H_{f\,CH_4(g)}^\circ$$

Let's substitute for ΔH_1°, ΔH_2°, and ΔH_3°, and then solve for ΔH_{atom}. First, we substitute for the ΔH° quantities.

$$4\Delta H_{f\,H(g)}^\circ + \Delta H_{f\,C(g)}^\circ + (-\Delta H_{atom}) = \Delta H_{f\,CH_4(g)}^\circ$$

Next, we solve for $(-\Delta H_{atom})$.

$$-\Delta H_{atom} = \Delta H_{f\,CH_4(g)}^\circ - 4\Delta H_{f\,H(g)}^\circ - \Delta H_{f\,C(g)}^\circ$$

Changing signs and rearranging the right side of the equation just a bit gives

$$\Delta H_{atom} = 4\Delta H_{f\,H(g)}^\circ + \Delta H_{f\,C(g)}^\circ - \Delta H_{f\,CH_4(g)}^\circ$$

Now all we need are values for the ΔH_f° terms on the right side. From Table 18.3 we obtain $\Delta H_{f\,H(g)}^\circ$ and $\Delta H_{f\,C(g)}^\circ$, and the value of $\Delta H_{f\,CH_4(g)}^\circ$ is obtained from Table 6.2. We will round these to the nearest 0.1 kJ/mol.

TABLE 18.3 Standard Heats of Formation of Some Gaseous Atoms from the Elements in Their Standard States

Atom	ΔH_f° per mole of atoms (kJ mol^{-1})[a]
H	217.89
Li	161.5
Be	324.3
B	560
C	716.67
N	472.68
O	249.17
F	79.14
Si	450
P	332.2
S	276.98
Cl	121.47
Br	112.38
I	107.48

[a]All values are positive because forming the gaseous atoms from the elements involves bond breaking and is endothermic.

$$\Delta H^\circ_{f\ H(g)} = +217.9 \text{ kJ/mol}$$
$$\Delta H^\circ_{f\ C(g)} = +716.7 \text{ kJ/mol}$$
$$\Delta H^\circ_{f\ CH_4(g)} = -74.8 \text{ kJ/mol}$$

Substituting these values gives

$$\Delta H_{atom} = 1663.1 \text{ kJ/mol}$$

and division by 4 gives an estimate of the average C—H bond energy in this molecule.

$$\text{Bond energy} = \frac{1663.1 \text{ kJ/mol}}{4}$$
$$= 415.8 \text{ kJ/mol of C—H bonds}$$

This value is quite close to the one in Table 18.4 on page 760, which is an average of the C—H bond energies in many different compounds. The other bond energies in Table 18.4 are also based on thermochemical data and were obtained by similar calculations.

Bond energies can be used to estimate heats of formation

An amazing thing about many covalent bond energies is that they are very nearly the same in many different compounds. This suggests, for example, that a C—H bond is very nearly the same in CH_4 as it is in a large number of other compounds that contain this kind of bond.

Because the bond energy doesn't vary much from compound to compound, we can use tabulated bond energies to estimate the heats of formation of substances. For example, let's calculate the standard heat of formation of methyl alcohol vapor, $CH_3OH(g)$. The structural formula for methanol is

```
      H
      |
  H—C—O—H
      |
      H
```

To perform this calculation, we set up two paths from the elements to the compound, as shown in Figure 18.15. The lower path has an enthalpy change corresponding to $\Delta H^\circ_{f\ CH_3OH(g)}$, while the upper path takes us to the gaseous elements and then through the energy released when the bonds in the molecule are formed. This latter energy can be computed from the bond energies in Table 18.4. As before, the sum of the energy changes along

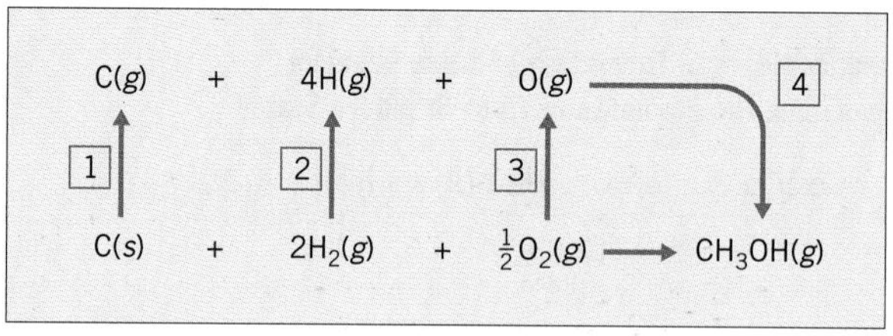

$C(g) \quad + \quad 4H(g) \quad + \quad O(g)$ [4]

[1] [2] [3]

$C(s) \quad + \quad 2H_2(g) \quad + \quad \frac{1}{2}O_2(g) \longrightarrow CH_3OH(g)$

FIG. 18.15 Two paths for the formation of methyl alcohol vapor from its elements in their standard states. The numbered paths are referred to in the discussion.

the upper path must be the same as the energy change along the lower path, and this permits us to compute $\Delta H^\circ_{f\,CH_3OH(g)}$.

Steps 1, 2, and 3 involve the formation of the gaseous atoms from the elements, and their enthalpy changes are obtained from Table 18.3.

$$\Delta H_1^\circ = \Delta H^\circ_{f\,C(g)} = 1 \text{ mol} \times 716.7 \text{ kJ/mol} = 716.7 \text{ kJ}$$

$$\Delta H_2^\circ = 4\Delta H^\circ_{f\,H(g)} = 4 \text{ mol} \times (217.9 \text{ kJ/mol}) = 871.6 \text{ kJ}$$

$$\Delta H_3^\circ = \Delta H^\circ_{f\,O(g)} = 1 \text{ mol} \times (249.2 \text{ kJ/mol}) = 249.2 \text{ kJ}$$

The sum of these values, +1837.5 kJ, is the total energy input (the net ΔH°) for the first three steps.

The formation of the CH_3OH molecule from the gaseous atoms is exothermic because energy is always released when atoms become joined by covalent bonds. In this molecule we can count three C—H bonds, one C—O bond, and one O—H bond. Their formation releases energy equal to their bond energies, which we obtain from Table 18.4.

Bond	Energy (kJ)
3(C—H)	3 × (412 kJ/mol) = 1236
C—O	360
O—H	463

Adding these together gives a total of 2059 kJ. Therefore, ΔH_4° is −2059 kJ (because it is exothermic). Now we can compute the total enthalpy change for the upper path.

TABLE 18.4 Some Average Bond Energies

Bond	Bond Energy (kJ mol^{-1})
C—C	348
C=C	612
C≡C	960
C—H	412
C—N	305
C=N	613
C≡N	890
C—O	360
C=O	743
C—F	484
C—Cl	338
C—Br	276
C—I	238
H—H	436
H—F	565
H—Cl	431
H—Br	366
H—I	299
H—N	388
H—O	463
H—S	338
H—Si	376

$$\Delta H^\circ = (+1837.5 \text{ kJ}) + (-2059 \text{ kJ})$$
$$= -222 \text{ kJ}$$

The value just calculated should be equal to ΔH_f° for $CH_3OH(g)$. For comparison, it has been found experimentally that ΔH_f° for this molecule (in the vapor state) is -201 kJ/mol. At first glance, the agreement doesn't seem very good, but on a relative basis the calculated value (-222 kJ) differs from the experimental one by only about 10%.

SUMMARY

First Law of Thermodynamics. The change in the **internal energy** of a system, ΔE, equals the sum of the heat absorbed by the system, q, and the work done on the system, w. ΔE is a state function, but q and w are not. For pressure–volume work, $w = -P\Delta V$, where P is the external pressure and $\Delta V = V_{\text{final}} - V_{\text{initial}}$. The heat at constant volume, q_v, is equal to ΔE, whereas the heat at constant pressure, q_p, is equal to ΔH. The value of ΔH differs from ΔE by the work expended in pushing back the atmosphere when the change occurs at constant atmospheric pressure. In general, the difference between ΔE and ΔH is quite small. For a chemical reaction, $\Delta H = \Delta E + \Delta n_{\text{gas}}RT$, where Δn_{gas} is the change in the number of moles of *gas* on going from reactants to products.

Spontaneity. A **spontaneous change** occurs without outside assistance, whereas a nonspontaneous change requires continuous help and can occur only if it is accompanied by and linked to some other spontaneous event.

Spontaneity is associated with statistical probability—spontaneous processes tend to proceed from lower probability states to those of higher probability. The thermodynamic quantity associated with the probability of a state is **entropy, S.** Entropy is a measure of the number of energetically equivalent ways a state can be realized. In general, gases have much higher entropies than liquids, which have somewhat higher entropies than solids. Entropy increases with volume for a gas and with the temperature. During a chemical reaction, the entropy tends to increase if the number of molecules increases.

Second Law of Thermodynamics. This law states that the entropy of the universe (system plus surroundings) increases whenever a spontaneous change occurs.

Gibbs Free Energy. The Gibbs free energy change, ΔG, allows us to determine the combined effects of enthalpy and entropy changes on the spontaneity of a chemical or physical change. A change is spontaneous only if the free energy of the system decreases (ΔG is negative). When ΔH and ΔS have the same algebraic sign, the temperature becomes the critical factor in determining spontaneity.

Third Law of Thermodynamics. The entropy of a pure crystalline substance is equal to zero at absolute zero (0 K). **Standard entropies, S°,** are calculated for 25 °C and 1 atm (Table 18.1) and can be used to calculate ΔS° for chemical reactions.

Standard Free Energy Changes. When ΔG is measured at 25 °C and 1 atm, it is the **standard free energy change, ΔG°. Standard free energies of formation, ΔG_f°,** can be used to obtain ΔG° for chemical reactions by a Hess's law type of calculation.

For any system, the value of ΔG is equal to the maximum amount of energy that can be obtained in the form of useful work, which can be obtained only if the change takes place by a **reversible process.** All real changes are irreversible and we always obtain less work than is theoretically available; the rest is lost as heat.

Free Energy and Equilibrium. When a system reaches equilibrium, $\Delta G = 0$ and no useful work can be obtained from it. For a phase change (e.g., solid $\rightleftharpoons$ liquid), ΔS and ΔH are related by the equation $\Delta S = \Delta H/T$, where T is the temperature at which the two phases are in equilibrium.

In chemical reactions, equilibrium occurs at a minimum on the free energy curve part way between pure reactants and pure products. The composition of the equilibrium mixture is determined by where the minimum lies along the reactant $\longrightarrow$ product axis; when it lies close to the products, the proportion of product to reactants is large and the reaction goes far toward completion.

When a reaction has a value of ΔG° that is both large and negative, it will appear to occur spontaneously because a lot of products will be formed by the time equilibrium is reached. If ΔG° is large and positive, it may be difficult to observe any reaction at all because only tiny amounts of products will be formed.

Thermodynamic Equilibrium Constants. The spontaneity of a reaction is determined by ΔG (how the free energy changes with a change in concentration). This is related to the standard free energy change, ΔG°, by the equation $\Delta G = \Delta G^\circ + RT\ln Q$, where Q is the reaction quotient for the system. At equilibrium, $\Delta G^\circ = -RT \ln K$, where $K = K_P$ for gaseous reactions and $K = K_c$ for reactions in solution.

Bond Energies and Heats of Reaction. The bond energy equals the amount of energy needed to break a bond to give neutral fragments. The sum of all the bond energies in a molecule is the **atomization energy, ΔH_{atom},** and, on a mole basis, it corresponds to the energy needed to break one mole of molecules into individual atoms. The heat of formation of a gaseous compound equals the sum of the energies needed to form atoms of the elements that are found in the substance plus the negative of the atomization energy.

TOOLS FOR PROBLEM SOLVING

In this chapter you learned to apply the following concepts as tools in solving problems related to thermodynamics. Study each one carefully so that you know what each is used for. When faced with solving a problem, recall what each tool does and consider whether it will be helpful in finding a solution. This will aid you in selecting the tools you need.

Converting between ΔE and ΔH *(page 729)* Use the following equation when you need to find the difference between ΔE and ΔH, or when converting from one to the other.

$$\Delta H = \Delta E + \Delta n_{\text{gas}}RT$$

Use the coefficients of gaseous reactants and products to calculate Δn_{gas}.

Factors that affect the entropy *(pages 734 through 736)* You can sometimes anticipate the sign of the entropy change for a system because certain factors favor an increase in entropy.

- Volume: Entropy increases with increasing volume.
- Temperature: Entropy increases with increasing temperature.
- Physical state: $S_{\text{gas}} \gg S_{\text{liquid}} > S_{\text{solid}}$. When gases are formed in a reaction, ΔS is almost always positive.
- Number of particles: Entropy increases when the number of particles increases.

Gibbs free energy *(page 738)* The Gibbs free energy is defined in terms of ΔH and ΔS by the equation

$$\Delta G = \Delta H - T\Delta S$$

ΔG as a predictor of spontaneity *(page 739)* You can use the sign of ΔG to tell whether a particular change is spontaneous.

Standard entropies *(page 741)* Calculate the standard entropy change for a reaction using the following equation and standard entropy values from Table 18.1. The calculation is similar to the Hess's law calculation of $\Delta H°$ you learned in Chapter 6.

$$\Delta S° = (\text{sum of } S° \text{ of products}) - (\text{sum of } S° \text{of reactants})$$

Calculating $\Delta G°$ from $\Delta H°$ and $\Delta S°$ *(page 743)* By calculating $\Delta H°$ from data in Table 6.2 and $\Delta S°$ from data in Table 18.1 you can calculate $\Delta G°$ for a reaction. The equation is

$$\Delta G° = \Delta H° - (298.15 \text{ K})\Delta S°$$

Using standard free energies of formation to calculate $\Delta G°$ *(page 744)* When you want $\Delta G°$ for a reaction, use $\Delta G_f°$ data from Table 18.2 and the equation

$$\Delta G° = (\text{sum of } \Delta G_f° \text{of products}) - (\text{sum of } \Delta G_f° \text{of reactants})$$

Using $\Delta G°$ to judge qualitatively the position of equilibrium *(page 750)* The sign and magnitude of $\Delta G°$ can serve as a qualitative indicator of the position of equilibrium and lets us anticipate whether an observable amount of product will be formed in a reaction.

- $\Delta G°$ is large and positive: Position of equilibrium is close to reactants.
- $\Delta G°$ is large and negative: Position of equilibrium is close to products.
- $\Delta G° = 0$: Position of equilibrium lies about midway between reactants and products.

Calculating $\Delta G°$ at temperatures other than 25 °C *(page 752)* When you need $\Delta G°$ at some temperature other than 25 °C, assume $\Delta H°$ and $\Delta S°$ are nearly independent of temperature and use the equation

$$\Delta G_T° \approx \Delta H_{298}° - T\Delta S_{298}°$$

Such a calculation would be needed to estimate an equilibrium constant at a temperature other than 25 °C.

Testing where a reaction stands relative to equilibrium *(page 754)* We can tell whether a particular reaction mixture composition will lead to a spontaneous reaction in the forward direction or the reverse direction, or whether the reaction is at equilibrium, by calculating ΔG by the equation

$$\Delta G = \Delta G° + RT \ln Q$$

- ΔG is negative: The reaction is spontaneous in the forward direction.
- ΔG is positive: The reaction is spontaneous in the reverse direction.
- ΔG is zero: The reaction is at equilibrium.

Thermodynamic equilibrium constants *(page 755)* The following important equation lets us calculate equilibrium constants from thermodynamic data, or calculate $\Delta G°$ from measured equilibrium constants.

$$\Delta G° = -RT \ln K$$

Remember to use $R = 8.314 \text{ J mol}^{-1} \text{ K}^{-1}$, T in kelvins, and that $K = K_P$ for gaseous reactions and K_c for reactions in solution.

QUESTIONS, PROBLEMS, AND EXERCISES

Answers to problems whose numbers are printed in color are given in Appendix B. More challenging problems are marked with asterisks. ILW = Interactive Learningware solution is available at www.wiley.com/college/brady. OH = an Office Hours video is available for this problem.

REVIEW QUESTIONS

First Law of Thermodynamics

18.1 What is the origin of the name *thermodynamics*?

18.2 How is a change in the internal energy defined in terms of the initial and final internal energies?

18.3 What is the algebraic sign of ΔE for an endothermic change? Why?

18.4 State the first law of thermodynamics in words. What equation defines the change in the internal energy in terms of heat and work? Define the meaning of the symbols, including the significance of their algebraic signs.

18.5 Which quantities in the statement of the first law are state functions and which are not?

18.6 Which thermodynamic quantity corresponds to the heat at constant volume? Which corresponds to the heat at constant pressure?

18.7 What are the units of $P\Delta V$ if pressure is expressed in pascals and volume is expressed in cubic meters?

18.8 If there is a decrease in the number of moles of gas during an exothermic chemical reaction, which is numerically larger, ΔE or ΔH? Why?

18.9 Which of the following changes is accompanied by the most negative value of ΔE? (a) A spring is compressed and heated. (b) A compressed spring expands and is cooled. (c) A spring is compressed and cooled. (d) A compressed spring expands and is heated.

Spontaneous Change

18.10 What is a *spontaneous change*? What role does kinetics play in determining the apparent spontaneity of a chemical reaction?

18.11 List five changes that you have encountered recently that occurred spontaneously. List five changes that are nonspontaneous that you have caused to occur.

18.12 Which of the items that you listed in Question 18.11 are exothermic (leading to a lowering of the potential energy) and which are endothermic (accompanied by an increase in potential energy)?

18.13 At constant pressure, what role does the enthalpy change play in determining the spontaneity of an event?

18.14 How do the probabilities of the initial and final states in a process affect the spontaneity of the process?

Entropy

18.15 An instant cold pack purchased in a pharmacy contains a packet of solid ammonium nitrate surrounded by a pouch of water. When the packet of NH_4NO_3 is broken, the solid dissolves in water and a cooling of the mixture occurs because the solution process for NH_4NO_3 in water is endothermic. Explain, in terms of what happens to the molecules and ions, why this mixing occurs spontaneously.

18.16 What is *entropy*?

18.17 How is the entropy of a substance affected by (a) an increase in temperature, (b) a decrease in volume, (c) changing from a liquid to a solid, and (d) dissociating into individual atoms?

18.18 Will the entropy change for each of the following be positive or negative?
(a) Moisture condenses on the outside of a cold glass.
(b) Raindrops form in a cloud.
(c) Gasoline vaporizes in the carburetor of an automobile engine.
(d) Air is pumped into a tire.
(e) Frost forms on the windshield of your car.
(f) Sugar dissolves in coffee.

18.19 On the basis of our definition of entropy, suggest why entropy is a state function.

18.20 State the second law of thermodynamics.

18.21 How can a process have a negative entropy change for the system, and yet still be spontaneous?

Third Law of Thermodynamics and Standard Entropies

18.22 What is the third law of thermodynamics?

18.23 Would you expect the entropy of an alloy (a solution of two metals) to be zero at 0 K? Explain your answer.

18.24 Why does entropy increase with increasing temperature?

18.25 Does glass have $S = 0$ at 0 K? Explain.

Gibbs Free Energy

18.26 What is the equation expressing the change in the Gibbs free energy for a reaction occurring at constant temperature and pressure?

18.27 In terms of the algebraic signs of ΔH and ΔS, under what circumstances will a change be spontaneous:
(a) At all temperatures?

(b) At low temperatures but not at high temperatures?

(c) At high temperatures but not at low temperatures?

18.28 Under what circumstances will a change be nonspontaneous regardless of the temperature?

Free Energy and Work

18.29 How is free energy related to useful work?

18.30 What is a thermodynamically reversible process? How is the amount of work obtained from a change related to thermodynamic reversibility?

18.31 How is the *rate* at which energy is withdrawn from a system related to the amount of that energy which can appear as useful work?

18.32 When glucose is oxidized by the body to generate energy, part of the energy is used to make molecules of ATP (adenosine triphosphate). However, of the total energy released in the oxidation of glucose, only 38% actually goes to making ATP. What happens to the rest of the energy?

18.33 Why are real, observable changes not considered to be thermodynamically reversible processes?

Free Energy and Equilibrium

18.34 In what way is free energy related to equilibrium?

18.35 Considering the fact that the formation of a bond between two atoms is exothermic and is accompanied by an entropy decrease, explain why all chemical compounds decompose into individual atoms if heated to a high enough temperature.

18.36 When a warm object is placed in contact with a cold one, they both gradually come to the same temperature. On a molecular level, explain how this is related to entropy and spontaneity.

18.37 Sketch the shape of the free energy curve for a chemical reaction that has a positive $\Delta G°$. Indicate the composition of the reaction mixture corresponding to equilibrium.

18.38 Many reactions that have large, negative values of $\Delta G°$ are not actually observed to happen at 25 °C and 1 atm. Why?

Thermodynamics and Equilibrium

18.39 Suppose a reaction has a negative $\Delta H°$ and a negative $\Delta S°$. Will more or less product be present at equilibrium as the temperature is raised?

18.40 Write the equation that relates the free energy change to the value of the reaction quotient for a reaction.

18.41 How is the equilibrium constant related to the standard free energy change for a reaction? (Write the equation.)

18.42 What is the value of $\Delta G°$ for a reaction for which $K = 1$?

Bond Energies and Heats of Reaction

18.43 Define the term *atomization energy*.

18.44 Why are the heats of formation of gaseous atoms from their elements endothermic quantities?

18.45 The gaseous C_2 molecule has a bond energy of 602 kJ mol^{-1}. Why isn't the standard heat of formation of $C(g)$ equal to half this value?

REVIEW PROBLEMS

First Law of Thermodynamics

18.46 A certain system absorbs 0.300 kJ of heat and has 0.700 kJ of work performed on it. What is the value of ΔE for the change? Is the overall change exothermic or endothermic?

18.47 The value of ΔE for a certain change is -1455 J. During the change, the system absorbs 812 J of heat. Did the system do work, or was work done on the system? How much work, expressed in joules, was involved?

18.48 Suppose that you were pumping an automobile tire with a hand pump that pushed 24.0 in.^{3} of air into the tire on each stroke, and that during one such stroke the opposing pressure in the tire was 30.0 lb/in.^{2} above the normal atmospheric pressure of 14.7 lb/in.^{2}. Calculate the number of joules of work accomplished during each stroke. (1 L atm = 101.325 J.)

OH **18.49** Consider the reaction between aqueous solutions of baking soda, $NaHCO_3$, and vinegar, $HC_2H_3O_2$.

$$NaHCO_3(aq) + HC_2H_3O_2(aq) \longrightarrow NaC_2H_3O_2(aq) + H_2O(l) + CO_2(g)$$

If this reaction occurs at atmospheric pressure ($P = 1$ atm), how much work, expressed in *L atm*, is done by the system in pushing back the atmosphere when 1.00 mol $NaHCO_3$ reacts at a temperature of 25 °C? (*Hint:* Review the gas laws.)

18.50 Calculate $\Delta H°$ and $\Delta E°$ for the following reactions at 25 °C. (If necessary, refer to the data in Table C.1 in Appendix C.)

(a) $3PbO(s) + 2NH_3(g) \longrightarrow 3Pb(s) + N_2(g) + 3H_2O(g)$

(b) $NaOH(s) + HCl(g) \longrightarrow NaCl(s) + H_2O(l)$

(c) $Al_2O_3(s) + 2Fe(s) \longrightarrow Fe_2O_3(s) + 2Al(s)$

(d) $2CH_4(g) \longrightarrow C_2H_6(g) + H_2(g)$

18.51 Calculate $\Delta H°$ and $\Delta E°$ for the following reactions at 25 °C. (If necessary, refer to the data in Table C.1 in Appendix C.)

(a) $2C_2H_2(g) + 5O_2(g) \longrightarrow 4CO_2(g) + 2H_2O(g)$

(b) $C_2H_2(g) + 5N_2O(g) \longrightarrow 2CO_2(g) + H_2O(g) + 5N_2(g)$

(c) $NH_4Cl(s) \longrightarrow NH_3(g) + HCl(g)$

(d) $(CH_3)_2CO(l) + 4O_2(g) \longrightarrow 3CO_2(g) + 3H_2O(g)$

18.52 The reaction

$$2N_2O(g) \longrightarrow 2N_2(g) + O_2(g)$$

has $\Delta H° = -163.14$ kJ. What is the value of ΔE for the decomposition of 180 g of N_2O at 25 °C? If we assume that ΔH doesn't change appreciably with temperature, what is ΔE for this same reaction at 200 °C?

18.53 A 10.0 L vessel at 20 °C contains butane, $C_4H_{10}(g)$, at a pressure of 2.00 atm. What is the maximum amount of work that can be obtained by the combustion of this butane if the gas is first brought to a pressure of 1 atm and the temperature is brought to

25 °C? Assume the products are also returned to that same temperature and pressure.

Factors That Affect Spontaneity

OH **18.54** Use the data from Table 6.2 to calculate $\Delta H°$ for the following reactions. On the basis of their values of $\Delta H°$, which are favored to occur spontaneously?
(a) $CaO(s) + CO_2(g) \longrightarrow CaCO_3(s)$
(b) $C_2H_2(g) + 2H_2(g) \longrightarrow C_2H_6(g)$
(c) $3CaO(s) + 2Fe(s) \longrightarrow 3Ca(s) + Fe_2O_3(s)$

18.55 Use the data from Table 6.2 to calculate $\Delta H°$ for the following reactions. On the basis of their values of $\Delta H°$, which are favored to occur spontaneously?
(a) $NH_4Cl(s) \longrightarrow NH_3(g) + HCl(g)$
(b) $2C_2H_2(g) + 5O_2(g) \longrightarrow 4CO_2(g) + 2H_2O(g)$
(c) $C_2H_2(g) + 5N_2O(g) \longrightarrow 2CO_2(g) + H_2O(g) + 5N_2(g)$

18.56 What factors must you consider to determine the sign of ΔS for the reaction $2N_2O(g) \longrightarrow 2N_2(g) + O_2(g)$ if it occurs at constant temperature?

18.57 What factors must you consider to determine the sign of ΔS for the reaction $2HI(g) \longrightarrow H_2(g) + I_2(s)$?

18.58 Predict the algebraic sign of the entropy change for the following reactions.
(a) $PCl_3(g) + Cl_2(g) \longrightarrow PCl_5(g)$
(b) $SO_2(g) + CaO(s) \longrightarrow CaSO_3(s)$
(c) $CO_2(g) + H_2O(l) \longrightarrow H_2CO_3(aq)$
(d) $Ni(s) + 2HCl(aq) \longrightarrow H_2(g) + NiCl_2(aq)$

OH **18.59** Predict the algebraic sign of the entropy change for the following reactions.
(a) $I_2(s) \longrightarrow I_2(g)$
(b) $Br_2(g) + 3Cl_2(g) \longrightarrow 2BrCl_3(g)$
(c) $NH_3(g) + HCl(g) \longrightarrow NH_4Cl(s)$
(d) $CaO(s) + H_2O(l) \longrightarrow Ca(OH)_2(s)$

Third Law of Thermodynamics

18.60 Calculate $\Delta S°$ for the following reactions in J K^{-1} from the data in Table 18.1. On the basis of their values of $\Delta S°$, which of these reactions are favored to occur spontaneously?
(a) $N_2(g) + 3H_2(g) \longrightarrow 2NH_3(g)$
(b) $CO(g) + 2H_2(g) \longrightarrow CH_3OH(l)$
(c) $2C_2H_6(g) + 7O_2(g) \longrightarrow 4CO_2(g) + 6H_2O(g)$
(d) $Ca(OH)_2(s) + H_2SO_4(l) \longrightarrow CaSO_4(s) + 2H_2O(l)$
(e) $S(s) + 2N_2O(g) \longrightarrow SO_2(g) + 2N_2(g)$

18.61 Calculate $\Delta S°$ for the following reactions in J K^{-1}, using the data in Table 18.1.
(a) $Ag(s) + \frac{1}{2}Cl_2(g) \longrightarrow AgCl(s)$
(b) $H_2(g) + \frac{1}{2}O_2(g) \longrightarrow H_2O(g)$
(c) $H_2(g) + \frac{1}{2}O_2(g) \longrightarrow H_2O(l)$
(d) $CaCO_3(s) + H_2SO_4(l) \longrightarrow CaSO_4(s) + H_2O(g) + CO_2(g)$
(e) $NH_3(g) + HCl(g) \longrightarrow NH_4Cl(s)$

OH **18.62** Calculate $\Delta S_f°$ for the following compounds in J mol^{-1} K^{-1}.
(a) $C_2H_4(g)$
(b) $CaSO_4{\cdot}2H_2O(s)$
(c) $HC_2H_3O_2(l)$

18.63 Calculate $\Delta S_f°$ for the following compounds in J mol^{-1} K^{-1}.
(a) $Al_2O_3(s)$
(b) $CaSO_4 \cdot \frac{1}{2}H_2O(s)$
(c) $NH_4Cl(s)$

18.64 Nitrogen dioxide, NO_2, an air pollutant, dissolves in rainwater to form a dilute solution of nitric acid. The equation for the reaction is

$$3NO_2(g) + H_2O(l) \longrightarrow 2HNO_3(l) + NO(g)$$

Calculate $\Delta S°$ for the reaction in J K^{-1}.

18.65 Good wine will turn to vinegar if it is left exposed to air because the alcohol is oxidized to acetic acid. The equation for the reaction is

$$C_2H_5OH(l) + O_2(g) \longrightarrow HC_2H_3O_2(l) + H_2O(l)$$

Calculate $\Delta S°$ for the reaction in J K^{-1}.

Gibbs Free Energy

ILW **18.66** Phosgene, $COCl_2$, was used as a war gas during World War I. It reacts with the moisture in the lungs to produce HCl, which causes the lungs to fill with fluid, and CO, which asphyxiates the victim. Both lead ultimately to death. For $COCl_2(g)$, $S° = 284$ J/mol K and $\Delta H_f° = -223$ kJ/mol. Use this information and the data in Table 18.1 to calculate $\Delta G_f°$ for $COCl_2(g)$ in kJ mol^{-1}.

OH **18.67** Aluminum oxidizes rather easily, but forms a thin protective coating of Al_2O_3 that prevents further oxidation of the aluminum beneath. Use the data for $\Delta H_f°$ (Table 6.2) and $S°$ (Table 18.1) to calculate $\Delta G_f°$ for $Al_2O_3(s)$ in kJ mol^{-1}.

18.68 Compute $\Delta G°$ in kJ for the following reactions, using the data in Table 18.2.
(a) $SO_3(g) + H_2O(l) \longrightarrow H_2SO_4(l)$
(b) $2NH_4Cl(s) + CaO(s) \longrightarrow CaCl_2(s) + H_2O(l) + 2NH_3(g)$
(c) $CaSO_4(s) + 2HCl(g) \longrightarrow CaCl_2(s) + H_2SO_4(l)$

18.69 Compute $\Delta G°$ in kJ for the following reactions, using the data in Table 18.2.
(a) $2HCl(g) + CaO(s) \longrightarrow CaCl_2(s) + H_2O(g)$
(b) $2AgCl(s) + Ca(s) \longrightarrow CaCl_2(s) + 2Ag(s)$
(c) $3NO_2(g) + H_2O(l) \longrightarrow 2HNO_3(l) + NO(g)$

18.70 Given the following,

$4NO(g) \longrightarrow 2N_2O(g) + O_2(g) \quad \Delta G° = -139.56$ kJ

$2NO(g) + O_2(g) \longrightarrow 2NO_2(g) \quad \Delta G° = -69.70$ kJ

calculate $\Delta G°$ for the reaction

$$2N_2O(g) + 3O_2(g) \longrightarrow 4NO_2(g)$$

18.71 Given these reactions and their $\Delta G°$ values,

$COCl_2(g) + 4NH_3(g) \longrightarrow CO(NH_2)_2(s) + 2NH_4Cl(s) \quad \Delta G° = -332.0$ kJ

$COCl_2(g) + H_2O(l) \longrightarrow CO_2(g) + 2HCl(g) \quad \Delta G° = -141.8$ kJ

$NH_3(g) + HCl(g) \longrightarrow NH_4Cl(s) \quad \Delta G° = -91.96$ kJ calculate the value of $\Delta G°$ for the reaction

$$CO(NH_2)_2(s) + H_2O(l) \longrightarrow CO_2(g) + 2NH_3(g)$$

Free Energy and Work

OH **18.72** Gasohol is a mixture of gasoline and ethanol (grain alcohol), C_2H_5OH. Calculate the maximum work that could be obtained at 25 °C and 1 atm by burning 1 mol of C_2H_5OH.

$$C_2H_5OH(l) + 3O_2(g) \longrightarrow 2CO_2(g) + 3H_2O(g)$$

18.73 What is the maximum amount of useful work that could possibly be obtained at 25 °C and 1 atm from the combustion of 48.0 g of natural gas, $CH_4(g)$, to give $CO_2(g)$ and $H_2O(g)$?

Free Energy and Equilibrium

18.74 Chloroform, formerly used as an anesthetic and now believed to be a carcinogen (cancer-causing agent), has a heat of vaporization $\Delta H_{vaporization} = 31.4$ kJ mol^{-1}. The change $CHCl_3(l) \longrightarrow CHCl_3(g)$ has $\Delta S° = 94.2$ J mol^{-1} K^{-1}. At what temperature do we expect $CHCl_3$ to boil (i.e., at what temperature will liquid and vapor be in equilibrium at 1 atm pressure)?

OH **18.75** For the melting of aluminum, $Al(s) \longrightarrow Al(l)$, $\Delta H° = 10.0$ kJ mol^{-1} and $\Delta S° = 9.50$ J/mol K. Calculate the melting point of Al. (The actual melting point is 660 °C.)

18.76 Isooctane, an important constituent of gasoline, has a boiling point of 99.3 °C and a heat of vaporization of 37.7 kJ mol^{-1}. What is ΔS (in J mol^{-1} K^{-1}) for the vaporization of 1 mol of isooctane?

18.77 Acetone (nail polish remover) has a boiling point of 56.2 °C. The change $(CH_3)_2CO(l) \longrightarrow (CH_3)_2CO(g)$ has $\Delta H° = 31.9$ kJ mol^{-1}. What is $\Delta S°$ for the change?

Free Energy and Spontaneity of Chemical Reactions

ILW **18.78** Determine whether the following reaction (equation unbalanced) will be spontaneous at 25 °C. (Do we expect appreciable amounts of products to form?)

$$C_2H_4(g) + HNO_3(l) \longrightarrow HC_2H_3O_2(l) + H_2O(l) + NO(g) + NO_2(g)$$

18.79 Which of the following reactions (equations unbalanced) would be expected to be spontaneous at 25 °C and 1 atm?
(a) $PbO(s) + NH_3(g) \longrightarrow Pb(s) + N_2(g) + H_2O(g)$
(b) $NaOH(s) + HCl(g) \longrightarrow NaCl(s) + H_2O(l)$
(c) $Al_2O_3(s) + Fe(s) \longrightarrow Fe_2O_3(s) + Al(s)$
(d) $2CH_4(g) \longrightarrow C_2H_6(g) + H_2(g)$

Thermodynamic Equilibrium Constants

OH **18.80** Calculate the value of the thermodynamic equilibrium constant for the following reactions at 25 °C. (Equations are unbalanced. Refer to the data in Appendix C.)
(a) $PCl_3(g) + O_2(g) \rightleftharpoons POCl_3(g)$
(b) $SO_3(g) \rightleftharpoons SO_2(g) + O_2(g)$

18.81 Calculate the value of the thermodynamic equilibrium constant for the following reactions at 25 °C. (Equations are unbalanced. Refer to the data in Appendix C.)
(a) $N_2H_4(g) + O_2(g) \rightleftharpoons NO(g) + H_2O(g)$
(b) $N_2H_4(g) + H_2O_2(g) \rightleftharpoons NO_2(g) + H_2O(g)$

ILW **18.82** The reaction

$$NO_2(g) + NO(g) \rightleftharpoons N_2O(g) + O_2(g)$$

has $\Delta G^0_{1273} = -9.67$ kJ. A 1.00 L reaction vessel at 1000 °C contains 0.0200 mol NO_2, 0.040 mol NO, 0.015 mol N_2O, and 0.0350 mol O_2. Is the reaction at equilibrium? If not, in which direction will the reaction proceed to reach equilibrium?

18.83 The reaction $CO(g) + H_2O(g) \rightleftharpoons HCHO_2(g)$ has $\Delta G°_{673} = +79.8$ kJ mol^{-1}. If a mixture at 400 °C contains 0.040 mol CO, 0.022 mol H_2O, and 3.8×10^{-3} mol $HCHO_2$ in a 2.50 L container, is the reaction at equilibrium? If not, in which direction will the reaction proceed spontaneously?

18.84 A reaction that can convert coal to methane (the chief component of natural gas) is

$$C(s) + 2H_2(g) \rightleftharpoons CH_4(g)$$

for which $\Delta G° = -50.79$ kJ mol^{-1}. What is the value of K_P for this reaction at 25 °C? Does this value of K_P suggest that studying this reaction as a means of methane production is worthwhile pursuing?

***18.85** One of the important reactions in living cells from which the organism draws energy is the reaction of adenosine triphosphate (ATP) with water to give adenosine diphosphate (ADP) and free phosphate ion.

$$ATP + H_2O \rightleftharpoons ADP + PO_4^{3-}$$

The value of $\Delta G°_{310}$ for the reaction at 37 °C (normal human body temperature) is -33 kJ mol^{-1}. Calculate the value of the equilibrium constant for the reaction at that temperature.

18.86 What is the value of the equilibrium constant for a reaction for which $\Delta G° = 0$? What will happen to the composition of the system if we begin the reaction with the pure products?

18.87 Methanol, a potential replacement for gasoline as an automotive fuel, can be made from H_2 and CO by the reaction

$$CO(g) + 2H_2(g) \rightleftharpoons CH_3OH(g)$$

At 500 K, this reaction has $K_P = 6.25 \times 10^{-3}$. Calculate $\Delta G°_{500}$ for this reaction in units of kilojoules.

Bond Energies and Heats of Reaction

18.88 Use the data in Table 18.4 to compute the approximate atomization energy of NH_3.

18.89 Approximately how much energy would be released during the formation of the bonds in one mole of acetone molecules? Acetone, the solvent usually found in nail polish remover, has the structural formula

```
  H  :O:  H
  |   ‖   |
H—C———C———C—H
  |       |
  H       H
```

OH **18.90** The standard heat of formation of ethanol vapor, $C_2H_5OH(g)$, is -235.3 kJ mol^{-1}. Use the data in Table 18.3 and the average bond energies for C—C, C—H, and O—H bonds to estimate the C—O bond energy in this molecule. The structure of the molecule is

```
   H  H
   |  |   ..
H—C—C—O—H
   |  |   ..
   H  H
```

18.91 The standard heat of formation of ethylene, $C_2H_4(g)$, is $+52.284$ kJ mol^{-1}. Calculate the C═C bond energy in this molecule.

ILW **18.92** Carbon disulfide, CS_2, has the Lewis structure $:\ddot{S}=C=\ddot{S}:$, and for $CS_2(g)$, $\Delta H_f^\circ = +115.3$ kJ mol^{-1}. Use the data in Table 18.3 to calculate the average C═S bond energy in this molecule.

18.93 Gaseous hydrogen sulfide, H_2S, has $\Delta H_f^\circ = -20.15$ kJ mol^{-1}. Use data in Table 18.3 to calculate the average S—H bond energy in this molecule.

18.94 For $SF_6(g)$, $\Delta H_f^\circ = -1096$ kJ mol^{-1}. Use the data in Table 18.3 to calculate the average S—F bond energy in SF_6.

18.95 Use the results of the preceding problem and the data in Table C.2 to calculate the standard heat of formation of $SF_4(g)$. The measured value of ΔH_f° for $SF_4(g)$ is -718.4 kJ mol^{-1}. What is the percentage difference between your calculated value of ΔH_f° and the experimentally determined value?

18.96 Use the data in Tables 18.3 and 18.4 to estimate the standard heat of formation of acetylene, H—C≡C—H, in the gaseous state.

18.97 What would be the approximate heat of formation of CCl_4 vapor at 25 °C and 1 atm?

18.98 Which substance should have the more exothermic heat of formation, CF_4 or CCl_4?

***18.99** Would you expect the value of ΔH_f° for benzene, C_6H_6, computed from tabulated bond energies, to be very close to the experimentally measured value of ΔH_f°? Justify your answer.

ADDITIONAL EXERCISES

18.100 If pressure is expressed in atmospheres and volume is expressed in liters, $P\Delta V$ has units of L atm (liters × atmospheres). In Chapter 10 you learned that 1 atm = 101,325 Pa, and in Chapter 1 you learned that 1 L = 1 dm^3. Use this information to determine the number of joules corresponding to 1 L atm.

18.101 Calculate the work, in joules, done by a gas as it expands at constant temperature from a volume of 3.00 L and a pressure of 5.00 atm to a volume of 8.00 L. The external pressure against which the gas expands is 1.00 atm. (1 atm = 101,325 Pa.)

***18.102** When an ideal gas expands at a constant temperature, $\Delta E = 0$ for the change. Why?

***18.103** When a real gas expands at a constant temperature, $\Delta E > 0$ for the change. Why?

18.104 An ideal gas in a cylinder fitted with a piston expands at constant temperature from a pressure of 5 atm and a volume of 12 L to a final volume of 30 L against a constant opposing pressure of 2.0 atm. How much heat does the gas absorb, expressed in units of L atm (liter × atm)? (Hint: See Problem 18.102)

***18.105** A cylinder fitted with a piston contains 5.00 L of a gas at a pressure of 4.00 atm. The entire apparatus is contained in a water bath to maintain a constant temperature of 25 °C. The piston is released and the gas expands until the pressure inside the cylinder equals the atmospheric pressure outside, which is 1 atm. Assume ideal gas behavior and calculate the amount of work done by the gas as it expands at constant temperature.

***18.106** The experiment described in the preceding problem is repeated, but this time a weight, which exerts a pressure of 2 atm, is placed on the piston. When the gas expands, its pressure drops to 2 atm. Then the weight is removed and the gas is allowed to expand again to a final pressure of 1 atm. Throughout both expansions the temperature of the apparatus was held at a constant 25 °C. Calculate the amount of work performed by the gas in each step. How does the combined total amount of work in this two-step expansion compare with the amount of work done by the gas in the one-step expansion described in the preceding problem? How can even more work be obtained by the expansion of the gas?

18.107 When potassium iodide dissolves in water, the mixture becomes cool. For this change, which is of a larger magnitude, $T\Delta S$ or ΔH?

18.108 The enthalpy of combustion, $\Delta H^\circ_{combustion}$, of oxalic acid, $H_2C_2O_4(s)$, is -246.05 kJ mol^{-1}. Consider the following data:

Substance	ΔH_f° (kJ mol^{-1})	S° (J mol^{-1} K^{-1})
$C(s)$	0	5.69
$CO_2(g)$	−393.5	213.6
$H_2(g)$	0	130.6
$H_2O(l)$	−285.8	69.96
$O_2(g)$	0	205.0
$H_2C_2O_4(s)$	?	120.1

(a) Write the balanced thermochemical equation that describes the combustion of one mole of oxalic acid.

(b) Write the balanced thermochemical equation that describes the formation of one mole of oxalic acid.

(c) Use the information in the table above and the equations in parts (a) and (b) to calculate ΔH_f° for oxalic acid.

(d) Calculate ΔS_f° for oxalic acid and ΔS° for the combustion of one mole of oxalic acid.

(e) Calculate ΔG_f° for oxalic acid and ΔG° for the combustion of one mole of oxalic acid.

OH **18.109** Many biochemical reactions have positive values for ΔG° and seemingly should not be expected to be spontaneous. They occur, however, because they are chemically coupled with other reactions that have negative values of ΔG°. An example is the set of reactions that forms the beginning part of the sequence of reactions involved in the metabolism of glucose, a sugar. Given these reactions and their corresponding ΔG° values,

$$\text{Glucose} + \text{phosphate} \longrightarrow \text{glucose 6-phosphate} + H_2O \qquad \Delta G^\circ = +13.13 \text{ kJ}$$

$$\text{ATP} + H_2O \longrightarrow \text{ADP} + \text{phosphate} \qquad \Delta G^\circ = -32.22 \text{ kJ}$$

calculate ΔG° for the coupled reaction

$$\text{Glucose} + \text{ATP} \longrightarrow \text{glucose 6-phosphate} + \text{ADP}$$

*__18.110__ The reaction

$$2C_4H_{10}(g) + 13O_2(g) \longrightarrow 8CO_2(g) + 10H_2O(g)$$

has $\Delta G° = -5407$ kJ. Determine the value of ΔG_f° for $C_4H_{10}(g)$. Calculate the value of K_c for the reaction at 25 °C.

*__18.111__ At 1500 °C, $K_c = 5.67$ for the reaction

$$CH_4(g) + H_2O(g) \rightleftharpoons CO(g) + 3H_2(g)$$

Calculate the value of ΔG_{1773}° for the reaction at that temperature.

18.112 Given the following reactions and their values of $\Delta G°$, calculate the value of ΔG_f° for $N_2O_5(g)$.

$$2H_2(g) + O_2(g) \longrightarrow 2H_2O(l) \qquad \Delta G° = -474.4 \text{ kJ}$$

$$N_2O_5(g) + H_2O(l) \longrightarrow 2HNO_3(l) \qquad \Delta G° = -37.6 \text{ kJ}$$

$$\tfrac{1}{2}N_2(g) + \tfrac{3}{2}O_2(g) + \tfrac{1}{2}H_2(g) \longrightarrow HNO_3(l) \qquad \Delta G° = -79.91 \text{ kJ}$$

18.113 Ethyl alcohol, C_2H_5OH, has been suggested as an alternative to gasoline as a fuel. In Example 18.5 we calculated $\Delta G°$ for combustion of 1 mol of C_2H_5OH; in Example 18.6 we calculated $\Delta G°$ for combustion of 1 mol of octane. Let's assume that gasoline has the same properties as octane (one of its constituents). The density of C_2H_5OH is 0.7893 g/mL; the density of octane, C_8H_{18}, is 0.7025 g/mL. Calculate the maximum work that could be obtained by burning 1 gallon (3.78 liters) each of C_2H_5OH and C_8H_{18}. On a *volume* basis, which is the better fuel?

*__18.114__ At room temperature (25 °C), the gas ClNO is impure because it decomposes slightly according to the equation

$$2ClNO(g) \rightleftharpoons Cl_2(g) + 2NO(g)$$

The extent of decomposition is about 5%. What is the approximate value of ΔG_{298}° for the reaction at that temperature?

*__18.115__ The reaction

$$N_2O(g) + O_2(g) \rightleftharpoons NO_2(g) + NO(g)$$

has $\Delta H° = -42.9$ kJ and $\Delta S° = -26.1$ J/K. Suppose 0.100 mol of N_2O and 0.100 mol of O_2 are placed in a 2.00 L container at 500 °C and the equilibrium is established. What percentage of the N_2O has reacted? (Note: Assume that ΔH and ΔS are relatively insensitive to temperature, so ΔH_{298}° and ΔS_{298}° are about the same as ΔH_{773}° and ΔS_{773}°, respectively.)

18.116 Use the data in Table 18.3 to calculate the bond energy in the nitrogen molecule and the oxygen molecule.

18.117 The heat of vaporization of carbon tetrachloride, CCl_4, is 29.9 kJ mol^{-1}. Using this information and data in Tables 18.3 and 18.4, estimate the standard heat of formation of liquid CCl_4.

*__18.118__ At 25 °C, 0.0560 mol O_2 and 0.020 mol N_2O were placed in a 1.00 L container where the following equilibrium was then established.

$$2N_2O(g) + 3O_2(g) \rightleftharpoons 4NO_2(g)$$

At equilibrium, the NO_2 concentration was 0.020 M. Calculate the value of $\Delta G°$ for the reaction.

18.119 For the substance $SO_2F_2(g)$, $\Delta H_f^\circ = -858$ kJ mol^{-1}. The structure of the SO_2F_2 molecule is

```
        :Ö
         ‖
   :F̤—S—F̤:
         ‖
        O̤:
```

Use the value of the S—F bond energy calculated in Problem 18.94 and the data in Table C.2 to determine the average S=O bond energy in SO_2F_2 in units of kJ mol^{-1}.

EXERCISES IN CRITICAL THINKING

18.120 On earth, we do not normally find collections of individual subatomic particles, such as protons, neutrons, and electrons. Rather, they are assembled into atoms of various kinds. On the other hand, in the interior of stars individual atoms don't exist. There, an atom would break apart into separate subatomic particles. Explain this in terms of the principles of thermodynamics.

18.121 The average C—H bond energy calculated using the procedure in Section 18.10 is not quite equal to the energy needed to cause the reaction $CH_4(g) \longrightarrow CH_3(g) + H(g)$. Suggest reasons why this is so.

18.122 Discuss the statement: A world near absolute zero would be controlled almost entirely by potential energy.

18.123 If a catalyst were able to affect the position of equilibrium in a reaction, it would be possible to construct a perpetual motion machine (a machine from which energy could be extracted without having to put energy into it). Imagine how such a machine could be made. Why would it violate the first law of thermodynamics?

18.124 At the beginning of this chapter we noted that the reaction of CO_2 with H_2O to form a hydrocarbon fuel is nonspontaneous. According to thermodynamics, where would the position of equilibrium lie for such a reaction. Why?

9 ELECTROCHEMISTRY

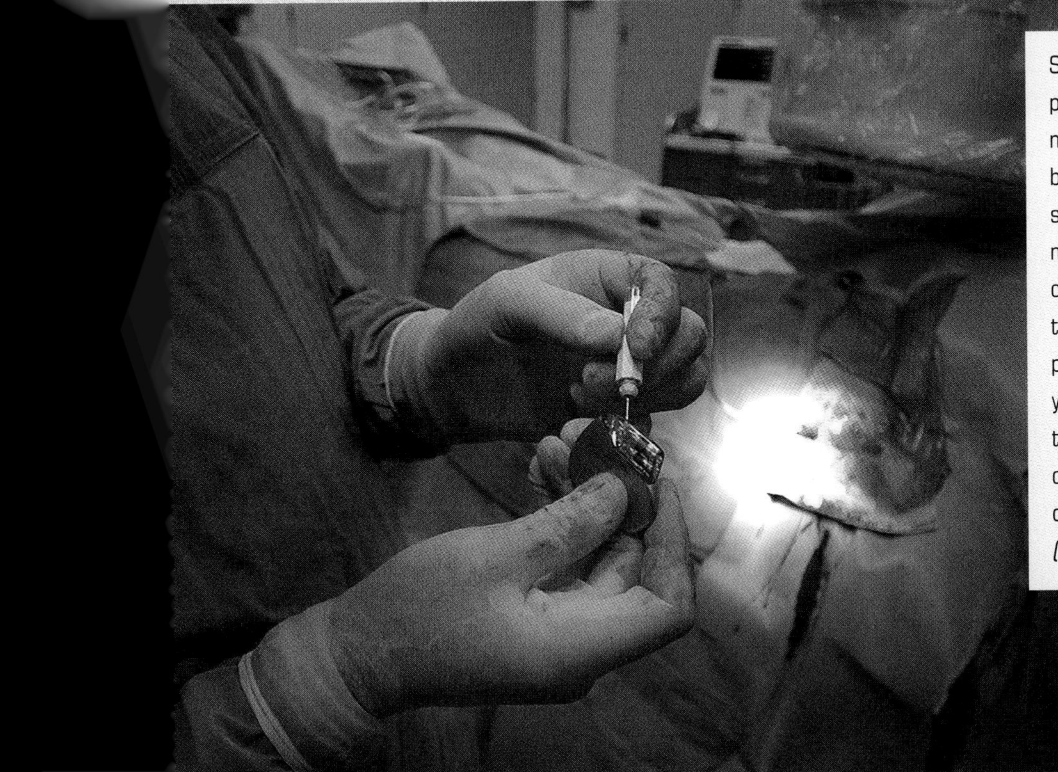

Surgeons implant a heart pacemaker to help this patient maintain a reasonable heart rate by providing a gentle electrical shock to the heart muscle as needed. The most critical component of the pacemaker is the battery that must remain in place and active for up to seven years at a time. Batteries and their design are important parts of the electrochemistry discussed in this chapter. *(© AP/Wide World Photos)*

HAPTER OUTLINE

CHAPTER IN CONTEXT Oxidation and reduction (redox) reactions occur in many chemical s. Examples include our own respiratory system and the complementary photosynthetic system in plants. In there's the toasting of bread, the rusting of iron, the action of bleach on stains, and the production and stion of petroleum that heats us, generates electricity, and moves our cars. In this chapter we will study how ssible to separate the processes of oxidation (electron loss) and reduction (electron gain) and cause them to different physical locations. When we are able to do this, we can use spontaneous redox reactions to produce ity. And by reversing the process, we can use electricity to make nonspontaneous redox reactions happen to e important products by a process called electrolysis.

cause electricity plays a role in these systems, the processes involved are described as **electrochemical s.** The study of such changes is called **electrochemistry.** As you will learn, electrical measurements and the principles of thermodynamics combine to give fundamental information about chemical reactions, such as free energy changes and equilibrium constants. In addition, you will learn some of the practical applications of electrochemistry.

The last section of this chapter describes modern batteries. Development of new, highly efficient, light weight and relatively inexpensive batteries and fuel cells is an area of intensive research. These devices will be important contributors to efforts to conserve fossil fuels.

19.1 GALVANIC CELLS USE REDOX REACTIONS TO GENERATE ELECTRICITY

Batteries have become common sources of portable power for a wide range of consumer products, from cell phones to iPods to laptops and hybrid cars. The energy from a battery comes from a spontaneous redox reaction in which the electron transfer is forced to take place through a wire. The apparatus that provides electricity in this way is called a **galvanic cell,** after Luigi Galvani (1737–1798), an Italian anatomist who discovered that electricity can cause the contraction of muscles. [It is also called a **voltaic cell,** after another Italian scientist, Alessandro Volta (1745–1827), whose inventions led ultimately to the development of modern batteries.]

Galvanic cells form a useful electrical circuit

If a shiny piece of metallic copper is placed into a solution of silver nitrate, a spontaneous reaction occurs. Gradually, a grayish white deposit forms on the copper and the solution itself becomes pale blue as hydrated Cu^{2+} ions enter the solution (see Figure 19.1). The equation is

$$2Ag^{+}(aq) + Cu(s) \longrightarrow Cu^{2+}(aq) + 2Ag(s)$$

Although the reaction is exothermic, no usable energy can be harnessed from it because all the energy is dispersed as heat.

To produce useful *electrical* energy, the two half-reactions involved in the net reaction must be made to occur in separate containers or compartments called **half-cells.** When this is done, electrons must flow through an external circuit to power devices

(a)

(b)

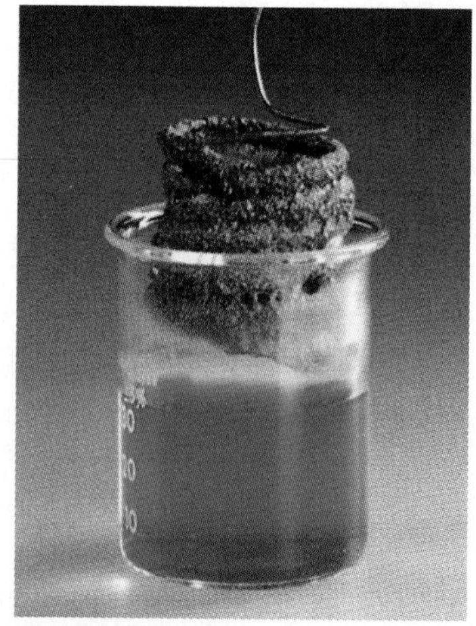
(c)

FIG. 19.1 **Reaction of copper with a solution of silver nitrate.** (*a*) A coil of copper wire stands next to a beaker containing a silver nitrate solution. (*b*) When the copper wire is placed in the solution, copper dissolves, giving the solution its blue color, and metallic silver deposits as glittering crystals on the wire. (*c*) After a while, much of the copper has dissolved and nearly all of the silver has deposited as the free metal. *(Michael Watson.)*

such as a laser pointer, a laptop computer, or your iPod. An apparatus to accomplish this—a **galvanic cell**—is made up of two half-cells, as illustrated in Figure 19.2. On the left, a silver electrode dips into a solution of $AgNO_3$, and, on the right, a copper electrode dips into a solution of $Cu(NO_3)_2$. The two electrodes are connected by an external electrical circuit and the two solutions are connected by a *salt bridge*, the function of which will be described shortly. When the circuit is completed by closing the switch, the reduction of Ag^+ to Ag occurs spontaneously in the half-cell on the left and oxidation of Cu to Cu^{2+} occurs spontaneously in the half-cell on the right. The reaction that takes place in each half-cell is a *half-reaction* of the type you learned to balance by the ion–electron method in Chapter 5. In the silver half-cell, the following half-reaction occurs.

$$Ag^+(aq) + e^- \longrightarrow Ag(s) \qquad \text{(reduction)}$$

When electrons appear as a reactant, the process is reduction; when they appear as a product, it is oxidation.

In the copper half-cell, the half-reaction is

$$Cu(s) \longrightarrow Cu^{2+}(aq) + 2e^- \qquad \text{(oxidation)}$$

When these reactions take place, electrons, left behind by oxidation of the copper, travel through the external circuit to the other electrode where they are transferred to the silver ions, as Ag^+ is reduced to the familiar shiny silver metal.

The cell reaction is the net overall reaction in the cell

The overall reaction that takes place in the galvanic cell is called the **cell reaction.** To obtain it, we combine the individual electrode half-reactions, making sure that the number of electrons gained in one half-reaction equals the number lost in the other. Thus, to obtain the cell reaction we multiply the half-reaction for the reduction of silver by 2 and then add the two half-reactions to obtain the net reaction. (Notice that $2e^-$ appear on each side, and so they cancel.) This is exactly the same as the process we used to balance redox reactions by the ion–electron method described in Section 5.2.

$$\begin{array}{lr} 2Ag^+(aq) + 2e^- \longrightarrow 2Ag(s) & \text{(reduction)} \\ Cu(s) \longrightarrow Cu^{2+}(aq) + 2e^- & \text{(oxidation)} \\ \hline 2Ag^+(aq) + Cu(s) + \cancel{2e^-} \longrightarrow 2Ag(s) + Cu^{2+}(aq) + \cancel{2e^-} & \text{(cell reaction)} \end{array}$$

The two electrons canceled in this reaction are also the number of electrons transferred. These will be important later when we discuss the Nernst equation.

A galvanic cell is composed of two half-cells connected by an external circuit and a salt bridge.

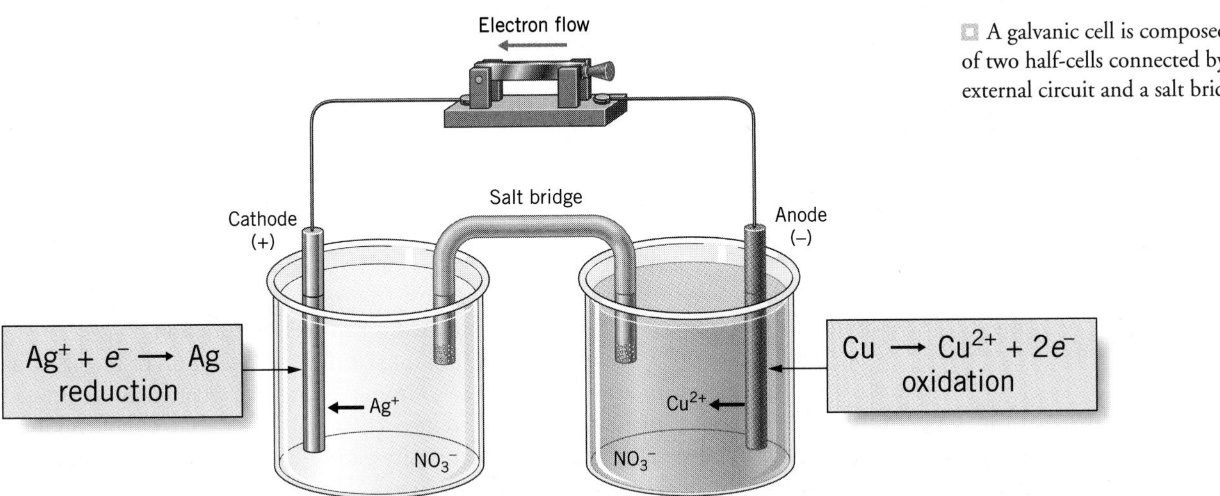

FIG. 19.2 **A galvanic cell.** The cell consists of two half-cells. Oxidation takes place in one half-cell and reduction in the other as indicated by the half-reactions.

Electrodes are named according to the chemical processes that occur at them

The electrodes in electrochemical systems are identified by the names *cathode* and *anode*. The names are *always* assigned according to the nature of the chemical changes that occur at the electrodes. In any electrochemical system:

The **cathode** is the electrode at which reduction (electron gain) occurs.
The **anode** is the electrode at which oxidation (electron loss) occurs.

Thus, in the galvanic cell we've been discussing, the silver electrode is the cathode and the copper electrode is the anode.

Conduction of charge occurs in two ways

In the external circuit of a galvanic cell, electrical charge is transported from one electrode to the other by the movement of *electrons* through the wires. This is called **metallic conduction,** and is how metals in general conduct electricity. In this external circuit, electrons always travel from the anode, where they are left behind by the oxidation process, to the cathode, where they are picked up by the substance being reduced.

In electrochemical cells another kind of electrical conduction also takes place. In a solution that contains ions (or in a molten ionic compound), *electrical charge is carried through the liquid by the movement of ions, not electrons.* The transport of electrical charge by ions is called **electrolytic conduction.**

When the reactions take place in the copper–silver galvanic cell, positive copper ions *enter* the liquid that surrounds the anode while positive silver ions *leave* the liquid that surrounds the cathode (Figure 19.3). For the galvanic cell to work, the solutions in both half-cells must remain electrically neutral. This requires that ions be permitted to enter or leave the solutions. For example, when copper is oxidized, the solution surrounding the anode becomes filled with Cu^{2+} ions, so negative ions are needed to balance their charge. Similarly, when Ag^{+} ions are reduced, NO_3^- ions are left behind in the solution and positive ions are needed to maintain neutrality. The **salt bridge** shown in Figure 19.2 allows the movement of ions required to keep the solutions electrically neutral. The salt bridge is also essential to complete the electrical circuit.

A salt bridge is a tube filled with a solution of a salt composed of ions not involved in the cell reaction. Often KNO_3 or KCl is used. The tube is fitted with porous plugs at each end that prevent the solution from pouring out but at the same time enable the solution in the salt bridge to exchange ions with the solutions in the half-cells.

Often the salt bridge is prepared by saturating a hot agar-agar solution with KNO_3 or KCl. After pouring into a U-shaped tube the agar-agar solidifies on cooling. The salt ions can move but the agar-agar does not flow out of the salt bridge.

During operation of the cell, negative ions can diffuse from the salt bridge into the copper half-cell, or, to a much smaller extent, Cu^{2+} ions can leave the solution and enter the salt bridge. Both processes help keep the copper half-cell electrically neutral. At the silver half-cell, positive ions from the salt bridge can enter or negative NO_3^- ions can, once again to a much smaller extent, leave the half-cell by entering the salt bridge, to keep it electrically neutral.

Without the salt bridge, electrical neutrality could not be maintained and no electrical current could be produced by the cell. Therefore, *electrolytic contact*—contact by means of a solution containing ions—must be maintained for the cell to function.

If we look closely at the overall movement of ions during the operation of the galvanic cell, we find that negative ions (*anions*) move away from the cathode, where they are present in excess, and *toward the anode*, where they are needed to balance the charge of the positive ions being formed. Similarly, we find that positive ions (*cations*) move away from the anode, where they are in excess, and *toward the cathode*, where they can balance the charge of the anions left in excess. In fact, the reason positive ions are called cations and negative ions are called anions is because of the nature of the electrodes toward which they move.

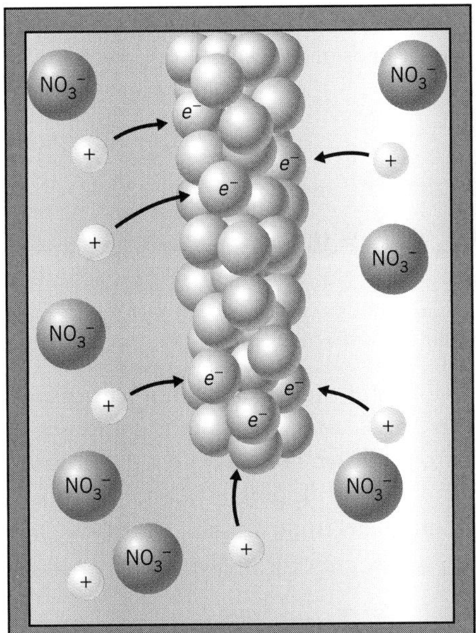

Reduction of silver ions at the cathode extracts electrons from the electrode, so the electrode becomes positively charged.

Anode

Oxidation of copper atoms at the anode leaves electrons behind on the electrode, which becomes negatively charged.

FIG. 19.3 Expanded view of Figure 19.2 to show changes that take place at the anode and cathode in the copper–silver galvanic cell. (Not drawn to scale.) At the anode, Cu^{2+} ions enter the solution when copper atoms are oxidized, leaving electrons behind on the electrode. Unless Cu^{2+} ions move away from the electrode or NO_3^- ions move toward it, the solution around the electrode will become positively charged. At the cathode, Ag^+ ions leave the solution and become silver atoms by acquiring electrons from the electrode surface. Unless more silver ions move toward the cathode or negative ions move away, the solution around the electrode will become negatively charged.

In summary,

> Cations move in the general direction of the cathode.
> Anions move in the general direction of the anode.

Charges on the electrodes come from electron loss and gain

At the anode of the galvanic cell described in Figures 19.2 and 19.3, copper atoms spontaneously leave the electrode and enter the solution as Cu^{2+} ions. The electrons that are left behind give the anode a slight negative charge. (We say the anode has a *negative polarity*.) At the cathode, electrons spontaneously join Ag^+ ions to produce neutral atoms, but the effect is the same as if Ag^+ ions become part of the electrode, so the cathode acquires a slight positive charge. (The cathode has a *positive polarity*.) During the operation of the cell, the amount of positive and negative charge on the electrodes is kept small by the flow of electrons (an electric current) through the external circuit from the anode to the cathode when the circuit is complete. In fact, unless electrons can flow out of the anode and into the cathode, the chemical reactions that occur at their surfaces will cease.

The small difference in charge between the electrodes is formed by the spontaneity of the overall reaction, that is, by the favorable free energy change. Nature's tendency toward electrical neutrality prevents a large buildup of charge on the electrodes and promotes the spontaneous flow of electricity through the external circuit.

Cell notation gives a shorthand description of a galvanic cell

As a matter of convenience, chemists have devised a shorthand way of describing the makeup of a galvanic cell. For example, the copper–silver cell that we have been using in our discussion is represented as follows.

$$Cu(s)|Cu^{2+}(aq)||Ag^{+}(aq)|Ag(s)$$

By convention, in **standard cell notation,** the anode half-cell is specified on the left, with the electrode material of the anode given first. In this case, the anode is copper metal. The single vertical bar represents a *phase boundary*—between the copper electrode and the solution that surrounds it. The double vertical bars represent the two phase boundaries, one at each

Also, notice that for each half-cell, the reactant in the redox half-reaction is given first. In the anode compartment, Cu is the reactant and is oxidized to Cu^{2+}, whereas in the cathode compartment, Ag^+ is the reactant and is reduced to Ag.

end of the salt bridge, which connects the solutions in the two half-cells. On the right, the cathode half-cell is described, with the material of the cathode given last. Thus, the electrodes themselves (copper and silver) are specified at opposite ends of the cell description.

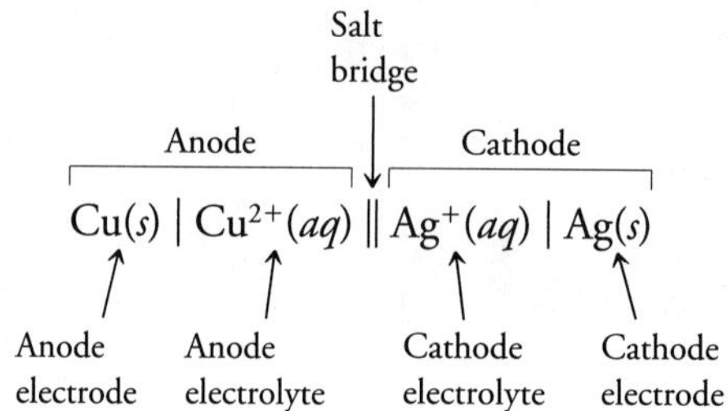

Sometimes, both the oxidized and reduced forms of the reactants in a half-cell are soluble and cannot be used as an electrode. In these cases, an inert electrode composed of platinum or gold is used to provide a site for electron transfer. For example, a galvanic cell can be made using an anode composed of a zinc electrode dipping into a solution containing Zn^{2+}, and a cathode composed of a platinum electrode dipping into a solution containing both Fe^{2+} and Fe^{3+} ions. The cell reaction is

$$2Fe^{3+}(aq) + Zn(s) \longrightarrow 2Fe^{2+}(aq) + Zn^{2+}(aq)$$

The cell notation for this galvanic cell is written as follows.

$$Zn(s)|Zn^{2+}(aq)||Fe^{3+}(aq), Fe^{2+}(aq)|Pt(s)$$

where we have separated the formulas for the two iron ions by a comma. In this cell, the reduction of the Fe^{3+} to Fe^{2+} takes place at the surface of the inert platinum electrode.

EXAMPLE 19.1
Describing Galvanic Cells

The following spontaneous reaction occurs when metallic zinc is dipped into a solution of copper sulfate.

$$Zn(s) + Cu^{2+}(aq) \longrightarrow Zn^{2+}(aq) + Cu(s)$$

Describe a galvanic cell that could take advantage of this reaction. What are the half-cell reactions? What is the standard cell notation? Make a sketch of the cell and label the cathode and anode, the charges on each electrode, the direction of ion flow, and the direction of electron flow.

ANALYSIS: Answering all these questions relies on identifying the anode and cathode in the equation for the cell reaction; this is often the key to solving problems of this type. By definition, the anode is the electrode at which oxidation happens, and the cathode is where reduction occurs. The first step, therefore, is to determine which reactant is oxidized and which is reduced. One way to do this is to divide the cell reaction into half-reactions and balance them by adding electrons. Then, if electrons appear as a product, the half-reaction is oxidation; if the electrons appear as a reactant, the half-reaction is reduction.

SOLUTION: The balanced half-reactions are as follows.

$$Zn(s) \longrightarrow Zn^{2+}(aq) + 2e^-$$
$$Cu^{2+}(aq) + 2e^- \longrightarrow Cu(s)$$

Zinc loses electrons and is oxidized, so it is the anode. The anode half-cell is therefore a zinc electrode dipping into a solution that contains Zn^{2+} [e.g., from dissolved $Zn(NO_3)_2$ or $ZnSO_4$]. Symbolically, the anode half-cell is written with the electrode material at the left of the vertical bar and the oxidation product at the right.

$$Zn(s)|Zn^{2+}(aq)$$

Copper ions gain electrons and are reduced to metallic copper, so the cathode half-cell consists of a copper electrode dipping into a solution containing Cu^{2+} [e.g., from dissolved

$Cu(NO_3)_2$ or $CuSO_4$]. The copper half-cell can be represented as follows, with the electrode material to the right of the vertical bar and the substance reduced on the left.

$$Cu^{2+}(aq)|Cu(s)$$

The standard cell notation places the zinc anode half-cell on the left and the copper cathode half-cell on the right separated by double bars that represent the salt bridge.

$$\underset{\text{anode}}{Zn(s)|Zn^{2+}(aq)}||\underset{\text{cathode}}{Cu^{2+}(aq)|Cu(s)}$$

A sketch of the cell is shown in the margin. The anode always carries a negative charge in a galvanic cell, so the zinc electrode is negative and the copper electrode is positive. Electrons in the external circuit travel from the negative electrode to the positive electrode (i.e., from the Zn anode to the Cu cathode). Anions move toward the anode and cations move toward the cathode as shown.

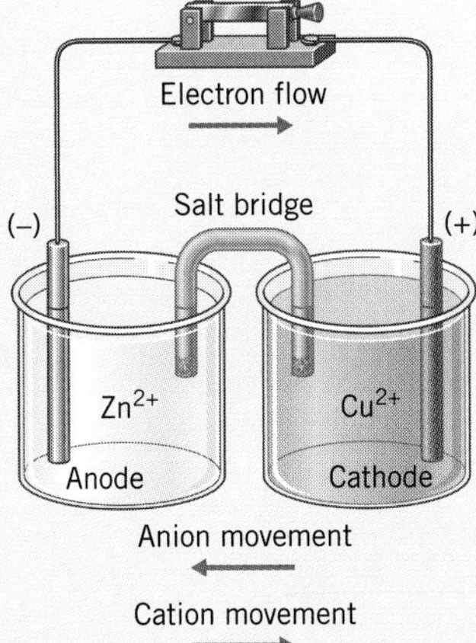

The zinc–copper cell. In the cell notation described in this example, we indicate the anode on the left and the cathode on the right. In this drawing of the apparatus, the anode half-cell is also shown on the left, but it could just as easily be shown on the right, as in Figure 19.2. Be sure you understand that where we place the apparatus on the lab bench doesn't affect which half-cell is the anode and which is the cathode.

ARE THE ANSWERS REASONABLE? All of the answers depend on determining which substance is oxidized and which is reduced, so we check that first. Oxidation is electron loss, and Zn must lose electrons to become Zn^{2+}, so zinc is oxidized and must be the anode. If zinc is the anode, then copper must be the cathode. We can then reason that the oxidation of zinc produces electrons that flow from the anode to the cathode.

Practice Exercise 1: Sketch and label a galvanic cell that makes use of the following spontaneous redox reaction.

$$Mg(s) + Fe^{2+}(aq) \longrightarrow Mg^{2+}(aq) + Fe(s)$$

Write the half-reactions for the anode and cathode. Give the standard cell notation. (Hint: Determine which reactant is being oxidized and which is being reduced.)

Practice Exercise 2: Write the anode and cathode half-reactions for the following galvanic cell. Write the equation for the overall cell reaction.

$$Al(s)|Al^{3+}(aq)||Ni^{2+}(aq)|Ni(s)$$

19.2 CELL POTENTIALS CAN BE RELATED TO REDUCTION POTENTIALS

A galvanic cell has an ability to push electrons through the external circuit. The magnitude of this ability is expressed as a **potential.** Potential is expressed in an electrical unit called the **volt (V),** which is a measure of the amount of energy, in joules, that can be delivered per **coulomb, C,** (the SI unit of charge) as the charges move through the circuit. Thus, a charge flowing under a potential of 1 volt can deliver 1 joule of energy per coulomb.

$$1 \text{ V} = 1 \text{ J/C} \tag{19.1}$$

The potential generated by a galvanic cell has also been called the *electromotive force (emf)*. Modern electrochemistry uses the preferred abbreviations E_{cell} or E°_{cell} to note cell potentials and standard cell potentials, respectively.

The cell potential is the maximum potential produced by a galvanic cell

The voltage or potential of a galvanic cell varies with the amount of charge flowing through the circuit. The *maximum* potential that a given cell can generate is called its **cell potential,** $\boldsymbol{E_{cell}}$, and it depends on the composition of the electrodes, the concentrations of the ions in the half-cells, and the temperature. The standard state for electrochemistry is defined as a system where the temperature is 25 °C, all concentrations are 1.00 *M*, and any gases are at 1.00 atm pressure. When the system is at standard state, the potential of a galvanic cell is the **standard cell potential,** symbolized by $\boldsymbol{E^\circ_{cell}}$.

If charge flows from a cell, some of the cell's voltage is lost overcoming its own internal resistance, and the measured voltage is less than the original E_{cell}.

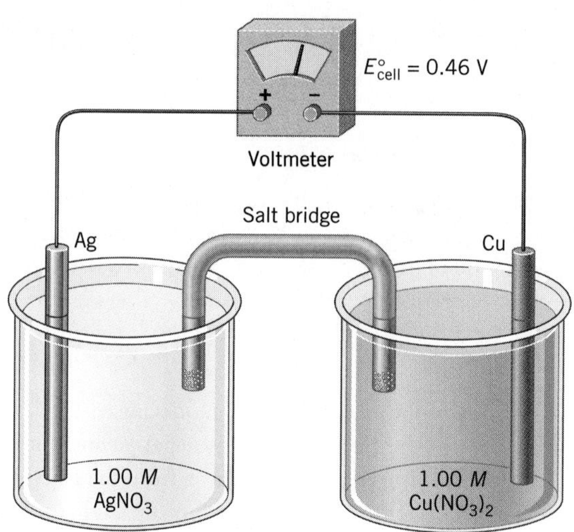

FIG. 19.4 A cell designed to generate the standard cell potential. The concentrations of the Cu^{2+} and Ag^+ ions in the half-cells are 1.00 *M*. It is very important to always connect the negative terminal of the voltmeter to the anode for correct readings.

Cell potentials are rarely larger than a few volts. For example, the standard cell potential for the galvanic cell constructed from silver and copper electrodes shown in Figure 19.4 is only 0.46 V, and one cell in an automobile battery produces only about 2 V. Batteries that generate higher voltages, such as an automobile battery, contain a number of cells arranged in series so that their potentials are additive.

Reduction potentials are a measure of the tendency of reduction half-reactions to occur

It is useful to think of each half-cell as having a certain natural tendency to acquire electrons and proceed as a *reduction*. The magnitude of this tendency is expressed by the half-reaction's **reduction potential.** When measured under standard conditions, namely, 25 °C, concentrations of 1.00 *M* for all solutes, and a pressure of 1 atm, the reduction potential is called the **standard reduction potential.** To represent a standard reduction potential, we will add a subscript to the symbol $E°$ that identifies the substance undergoing reduction. Thus, the standard reduction potential for the half-reaction

Standard reduction potentials are also called *standard electrode potentials.*

$$Cu^{2+}(aq) + 2e^- \longrightarrow Cu(s)$$

is specified as $E^\circ_{Cu^{2+}}$.

When two half-cells are connected to make a galvanic cell, the one with the larger standard reduction potential (the one with the greater tendency to undergo reduction) acquires electrons from the half-cell with the lower standard reduction potential, which is therefore forced to undergo oxidation. The standard cell potential, which is always taken to be a positive number, represents the *difference* between the standard reduction potential of one half-cell and the standard reduction potential of the other. In general, therefore,

TOOLS
Standard reduction potentials are used to calculate E°_{cell}

$$E^\circ_{cell} = \begin{pmatrix}\text{standard reduction potential}\\ \text{of the substance reduced}\end{pmatrix} - \begin{pmatrix}\text{standard reduction potential}\\ \text{of the substance oxidized}\end{pmatrix} \quad (19.2)$$

As an example, let's look at the copper–silver cell. From the cell reaction,

$$2Ag^+(aq) + Cu(s) \longrightarrow 2Ag(s) + Cu^{2+}(aq)$$

we can see that silver ions are reduced and copper is oxidized. If we compare the two possible reduction half-reactions,

$$Ag^+(aq) + e^- \longrightarrow Ag(s)$$
$$Cu^{2+}(aq) + 2e^- \longrightarrow Cu(s)$$

the one for Ag^+ must have a greater tendency to proceed than the one for Cu^{2+}, because it is the silver ion that is actually reduced. This means that the standard reduction potential of Ag^+ must be algebraically larger than the standard reduction potential of Cu^{2+}. In other words, if we knew the values of $E^\circ_{Ag^+}$ and $E^\circ_{Cu^{2+}}$, we could calculate E°_{cell} with Equation 19.2 by subtracting the smaller standard reduction potential (copper) from the larger one (silver).

$$E^\circ_{cell} = E^\circ_{Ag^+} - E^\circ_{Cu^{2+}}$$

Assigning standard reduction potentials requires a reference electrode

Unfortunately there is no way to measure the standard reduction potential of an isolated half-cell. All we can measure is the difference in potential produced when two half-cells are connected. Therefore, to assign numerical values for standard reduction potentials, a reference electrode has been arbitrarily chosen and its standard reduction potential has been defined as *exactly* 0 V. This reference electrode is called the **standard hydrogen electrode** (see Figure 19.5). Gaseous hydrogen at a pressure of 1.00 atm is bubbled over a platinum electrode coated with very finely divided platinum, which provides a large catalytic surface area on which the electrode reaction can occur. This electrode is surrounded by a solution whose temperature is 25 °C and in which the hydrogen ion concentration is 1.00 *M*. The half-cell reaction at the platinum surface, written as a reduction, is

$$2H^+(aq, 1.00\ M) + 2e^- \rightleftharpoons H_2(g, 1.00\ \text{atm}) \qquad E^\circ_{H^+} = 0\ \text{V (exactly)}$$

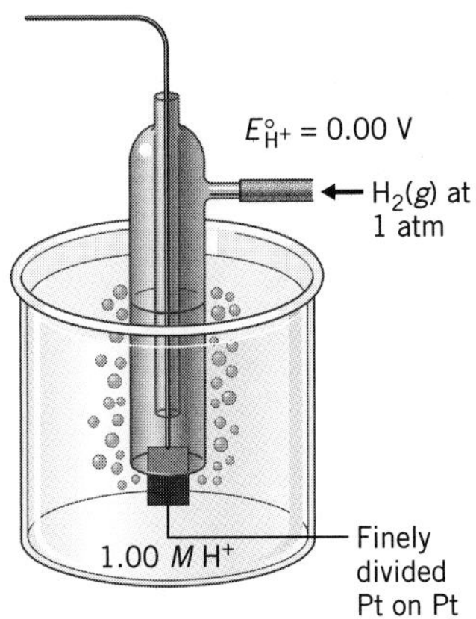

FIG. 19.5 **The hydrogen electrode.** The half-reaction is $2H^+(aq) + 2e^- \rightleftharpoons H_2(g)$.

The double arrows indicate only that the reaction is reversible, not that there is true equilibrium. Whether the half-reaction occurs as reduction or oxidation depends on the standard reduction potential of the half-cell with which it is paired.

Figure 19.6 illustrates the hydrogen electrode connected to a copper half-cell to form a galvanic cell. When we use a voltmeter to measure the potential of the cell, we find that the copper electrode carries a positive charge and the hydrogen electrode a negative charge. Therefore, copper must be the cathode, and Cu^{2+} is reduced to Cu when the cell operates. Similarly, hydrogen must be the anode, and H_2 is oxidized to H^+. The half-reactions and cell reaction, therefore, are

$$Cu^{2+}(aq) + 2e^- \longrightarrow Cu(s) \qquad \text{(cathode)}$$
$$H_2(g) \longrightarrow 2H^+(aq) + 2e^- \qquad \text{(anode)}$$
$$Cu^{2+}(aq) + H_2(g) \longrightarrow Cu(s) + 2H^+(aq) \qquad \text{(cell reaction)}$$[1]

Using Equation 19.2, we can express E°_{cell} in terms of $E^\circ_{Cu^{2+}}$ and $E^\circ_{H^+}$.

$$E^\circ_{cell} = E^\circ_{Cu^{2+}} - E^\circ_{H^+}$$

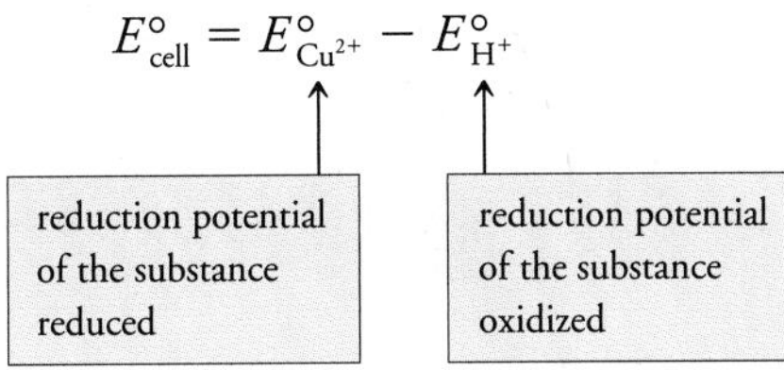

[1]The cell notation for this cell is written as

$$Pt(s), H_2(g)|H^+(aq)||Cu^{2+}(aq)|Cu(s)$$

The notation for the hydrogen electrode (the anode in this case) is shown at the left of the double vertical bars. Although there is a phase boundary between H_2 and Pt, they are shown together to emphasize their simultaneous contact with the solution.

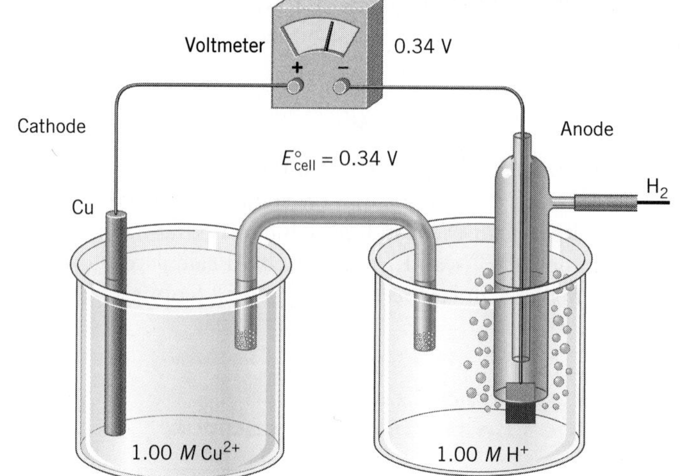

FIG. 19.6 A galvanic cell composed of copper and hydrogen half-cells. The cell reaction is $Cu^{2+}(aq) + H_2(g) \longrightarrow Cu(s) + 2H^+(aq)$.

In a galvanic cell, the measured cell potential is ***always*** taken to be a positive value. This is important to remember.

The measured standard cell potential is 0.34 V and $E^\circ_{H^+}$ equals 0.00 V.[2] Therefore,

$$0.34 \text{ V} = E^\circ_{Cu^{2+}} - 0.00 \text{ V}$$

Relative to the hydrogen electrode, then, the standard reduction potential of Cu^{2+} is +0.34 V. (We have written the value with a plus sign because some standard reduction potentials are negative, as we will see.)

Now let's look at a galvanic cell set up between a zinc electrode and a hydrogen electrode (see Figure 19.7). This time we find that the hydrogen electrode is positive and the zinc electrode is negative, which tells us that the hydrogen electrode is the cathode and the zinc electrode is the anode. This means that hydrogen ion is being reduced and zinc is being oxidized. The half-reactions and cell reaction are therefore

$$2H^+(aq) + 2e^- \longrightarrow H_2(g) \qquad \text{(cathode)}$$
$$Zn(s) \longrightarrow Zn^{2+}(aq) + 2e^- \qquad \text{(anode)}$$
$$\overline{2H^+(aq) + Zn(s) \longrightarrow H_2(g) + Zn^{2+}(aq)} \qquad \text{(cell reaction)}^3$$

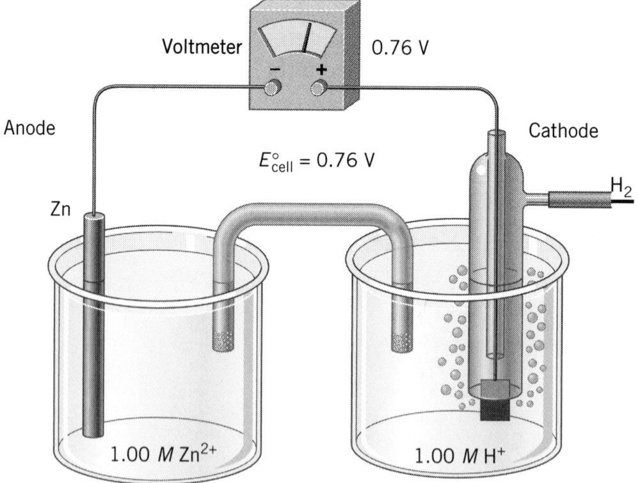

FIG. 19.7 A galvanic cell composed of zinc and hydrogen half-cells. The cell reaction is $Zn(s) + 2H^+(aq) \longrightarrow Zn^{2+}(aq) + H_2(g)$.

[2] The standard reduction potential for the reduction of hydrogen ions is exactly zero. We will be writing its value in mathematical problems with the number of decimal places required to maintain the precision of the opposing electrode potential. In this case we used two decimal places (0.00) to match the two decimal places of the 0.34 V standard cell potential.

[3] This cell is represented as

$$Zn(s)|Zn^{2+}(aq)||H^+(aq)|H_2(g), Pt(s)$$

This time the hydrogen electrode is the cathode and appears at the right of the double vertical bars.

From Equation 19.2, the standard cell potential is given by

$$E^\circ_{cell} = E^\circ_{H^+} - E^\circ_{Zn^{2+}}$$

Substituting into this the measured standard cell potential of 0.76 V and $E^\circ_{H^+} = 0.00$ V, we have

$$0.76 \text{ V} = 0.00 \text{ V} - E^\circ_{Zn^{2+}}$$

which gives

$$E^\circ_{Zn^{2+}} = -0.76 \text{ V}$$

Notice that the standard reduction potential of zinc is negative. A negative standard reduction potential simply means that the substance is not as easily reduced as H^+. In this case, it tells us that Zn is oxidized when it is paired with the hydrogen electrode.

The standard reduction potentials of many half-reactions can be compared to that for the standard hydrogen electrode in the manner described above. Table 19.1 lists values obtained for some typical half-reactions. They are arranged in decreasing order—the

TABLE 19.1 Standard Reduction Potentials at 25 °C

Half-Reaction	E° (volts)
$F_2(g) + 2e^- \rightleftharpoons 2F^-(aq)$	+2.87
$S_2O_8^{2-}(aq) + 2e^- \rightleftharpoons 2SO_4^{2-}(aq)$	+2.01
$PbO_2(s) + HSO_4^-(aq) + 3H^+(aq) + 2e^- \rightleftharpoons PbSO_4(s) + 2H_2O$	+1.69
$2HOCl(aq) + 2H^+(aq) + 2e^- \rightleftharpoons Cl_2(g) + 2H_2O$	+1.63
$MnO_4^-(aq) + 8H^+(aq) + 5e^- \rightleftharpoons Mn^{2+}(aq) + 4H_2O$	+1.51
$PbO_2(s) + 4H^+(aq) + 2e^- \rightleftharpoons Pb^{2+}(aq) + 2H_2O$	+1.46
$BrO_3^-(aq) + 6H^+(aq) + 6e^- \rightleftharpoons Br^-(aq) + 3H_2O$	+1.44
$Au^{3+}(aq) + 3e^- \rightleftharpoons Au(s)$	+1.42
$Cl_2(g) + 2e^- \rightleftharpoons 2Cl^-(aq)$	+1.36
$O_2(g) + 4H^+(aq) + 4e^- \rightleftharpoons 2H_2O$	+1.23
$Br_2(aq) + 2e^- \rightleftharpoons 2Br^-(aq)$	+1.07
$NO_3^-(aq) + 4H^+(aq) + 3e^- \rightleftharpoons NO(g) + 2H_2O$	+0.96
$Ag^+(aq) + e^- \rightleftharpoons Ag(s)$	+0.80
$Fe^{3+}(aq) + e^- \rightleftharpoons Fe^{2+}(aq)$	+0.77
$I_2(s) + 2e^- \rightleftharpoons 2I^-(aq)$	+0.54
$NiO_2(s) + 2H_2O + 2e^- \rightleftharpoons Ni(OH)_2(s) + 2OH^-(aq)$	+0.49
$Cu^{2+}(aq) + 2e^- \rightleftharpoons Cu(s)$	+0.34
$SO_4^{2-}(aq) + 4H^+(aq) + 2e^- \rightleftharpoons H_2SO_3(aq) + H_2O$	+0.17
$AgBr(s) + e^- \rightleftharpoons Ag(s) + Br^-(aq)$	+0.07
$2H^+(aq) + 2e^- \rightleftharpoons H_2(g)$	0
$Sn^{2+}(aq) + 2e^- \rightleftharpoons Sn(s)$	−0.14
$Ni^{2+}(aq) + 2e^- \rightleftharpoons Ni(s)$	−0.25
$Co^{2+}(aq) + 2e^- \rightleftharpoons Co(s)$	−0.28
$PbSO_4(s) + H^+(aq) + 2e^- \rightleftharpoons Pb(s) + HSO_4^-(aq)$	−0.36
$Cd^{2+}(aq) + 2e^- \rightleftharpoons Cd(s)$	−0.40
$Fe^{2+}(aq) + 2e^- \rightleftharpoons Fe(s)$	−0.44
$Cr^{3+}(aq) + 3e^- \rightleftharpoons Cr(s)$	−0.74
$Zn^{2+}(aq) + 2e^- \rightleftharpoons Zn(s)$	−0.76
$2H_2O + 2e^- \rightleftharpoons H_2(g) + 2OH^-(aq)$	−0.83
$Al^{3+}(aq) + 3e^- \rightleftharpoons Al(s)$	−1.66
$Mg^{2+}(aq) + 2e^- \rightleftharpoons Mg(s)$	−2.37
$Na^+(aq) + e^- \rightleftharpoons Na(s)$	−2.71
$Ca^{2+}(aq) + 2e^- \rightleftharpoons Ca(s)$	−2.76
$K^+(aq) + e^- \rightleftharpoons K(s)$	−2.92
$Li^+(aq) + e^- \rightleftharpoons Li(s)$	−3.05

Substances located to the left of the double arrows are *oxidizing agents*, because they become reduced when the reactions proceed in the forward direction. The best oxidizing agents are those most easily reduced, and they are located at the top of the table (e.g., F_2).

Substances located to the right of the double arrows are *reducing agents*; they become oxidized when the reactions proceed from right to left. The best reducing agents are those found at the bottom of the table (e.g., Li).

half-reactions at the top have the greatest tendency to occur as reduction, while those at the bottom have the least tendency to occur as reduction.

EXAMPLE 19.2 Calculating Standard Cell Potentials

We mentioned earlier that the standard cell potential of the silver–copper galvanic cell has a value of +0.46 V. The cell reaction is

$$2Ag^+(aq) + Cu(s) \longrightarrow 2Ag(s) + Cu^{2+}(aq)$$

and we have seen that the standard reduction potential of Cu^{2+}, $E^\circ_{Cu^{2+}}$, is +0.34 V. What is the value of $E^\circ_{Ag^+}$, the standard reduction potential of Ag^+?

ANALYSIS: Since we know the standard potential of the cell and one of the two standard reduction potentials, Equation 19.2 will be our tool to calculate the unknown standard reduction potential. This requires that we identify the substance oxidized and the substance reduced. We can do this by dividing the cell reaction into half-reactions, or we can observe how the oxidation numbers of the reactants change. Recall that if the oxidation number increases algebraically, the substance undergoes oxidation, whereas if the oxidation number decreases, the substance is reduced.

SOLUTION: Silver changes from Ag^+ to Ag; its oxidation number decreases from +1 to 0, so Ag^+ is reduced. Similar reasoning tells us that copper is oxidized from Cu to Cu^{2+}. Therefore, according to Equation 19.2,

$$E^\circ_{cell} = E^\circ_{Ag^+} - E^\circ_{Cu^{2+}}$$

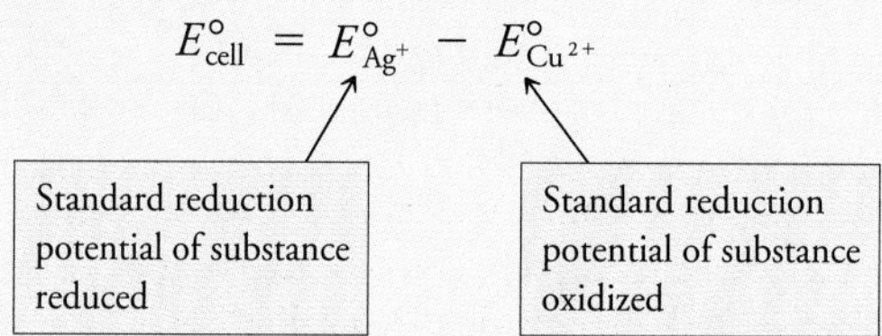

Substituting values for E°_{cell} and $E^\circ_{Cu^{2+}}$,

$$0.46\text{ V} = E^\circ_{Ag^+} - 0.34\text{ V}$$

Then we solve for $E^\circ_{Ag^+}$.

$$E^\circ_{Ag^+} = 0.46\text{ V} + 0.34\text{ V}$$
$$= 0.80\text{ V}$$

The standard reduction potential of silver ion is therefore +0.80 V.

IS THE ANSWER REASONABLE? We know the standard cell potential is the difference between the two standard reduction potentials. The difference between +0.80 V and +0.34 V (subtracting the smaller from the larger) is 0.46 V. Our calculated standard reduction potential for Ag^+ appears to be correct. You can also take a peek at Table 19.1 now for a final check.

Practice Exercise 3: Copper metal and zinc metal will both reduce Ag^+ ions under standard state conditions. Which metal, when used as an electrode in a galvanic cell, will have the larger E°_{cell} under these conditions? (Hint: Write the two possible chemical reactions.)

Practice Exercise 4: The galvanic cell described in Practice Exercise 1 has a standard cell potential of 1.93 V. The standard reduction potential of Mg^{2+} corresponding to the half-reaction $Mg^{2+}(aq) + 2e^- \rightleftharpoons Mg(s)$ is −2.37 V. Calculate the standard reduction potential of iron(II). Check your answer by referring to Table 19.1.

FACETS OF CHEMISTRY

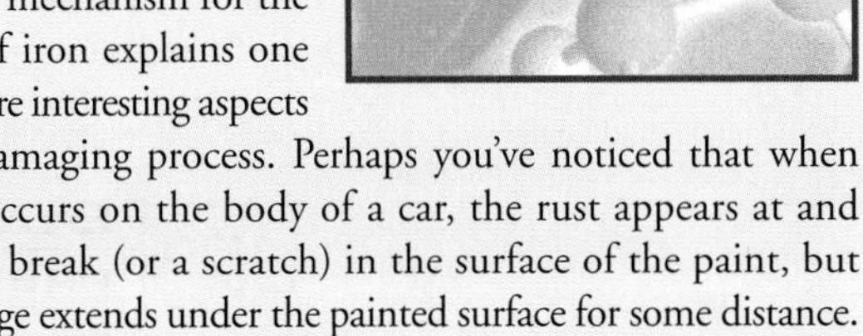

Corrosion of Iron and Cathodic Protection

A problem that has plagued humanity ever since the discovery of methods for obtaining iron and other metals from their ores has been corrosion—the reaction of a metal with substances in the environment. The rusting of iron in particular is a serious problem because iron and steel have so many uses.

The rusting of iron is a complex chemical reaction that involves both oxygen and moisture (see Figure 1). Iron won't rust in pure water that's oxygen free, and it won't rust in pure oxygen in the absence of moisture. The corrosion process is apparently electrochemical in nature, as shown in the accompanying diagram. At one place on the surface, iron becomes oxidized in the presence of water and enters solution as Fe^{2+}.

$$Fe(s) \longrightarrow Fe^{2+}(aq) + 2e^-$$

At this location the iron is acting as an anode.

The electrons that are released when the iron is oxidized travel through the metal to some other place where the iron is exposed to oxygen. This is where reduction takes place (it's a cathodic region on the metal surface), and oxygen is reduced to give hydroxide ion.

$$\tfrac{1}{2}O_2(aq) + H_2O + 2e^- \longrightarrow 2OH^-(aq)$$

The iron(II) ions that are formed at the anodic regions gradually diffuse through the water and eventually contact the hydroxide ions. This causes a precipitate of $Fe(OH)_2$ to form, which is very easily oxidized by O_2 to give $Fe(OH)_3$. This hydroxide readily loses water. In fact, complete dehydration gives the oxide,

$$2Fe(OH)_3 \longrightarrow Fe_2O_3 + 3H_2O$$

When partial dehydration of the $Fe(OH)_3$ occurs, *rust* is formed. It has a composition that lies between that of the hydroxide and that of the oxide, Fe_2O_3, and is usually referred to as a *hydrated oxide*. Its formula is generally represented as $Fe_2O_3 \cdot xH_2O$.

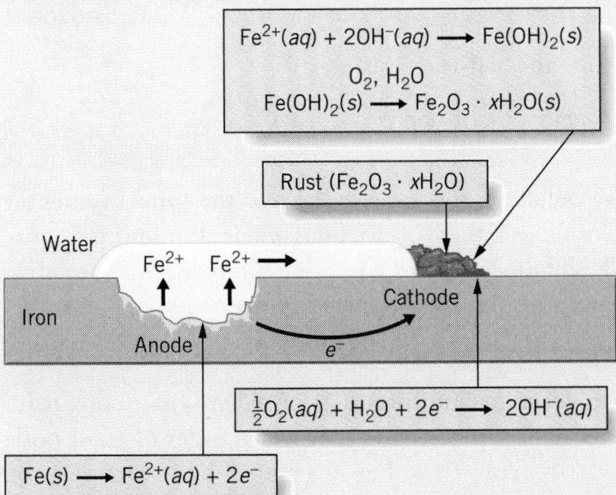

FIG. 1 Corrosion of iron. Iron dissolves in anodic regions to give Fe^{2+}. Electrons travel through the metal to cathodic sites where oxygen is reduced, forming OH^-. The combination of the Fe^{2+} and OH^-, followed by air oxidation, gives rust.

This mechanism for the rusting of iron explains one of the more interesting aspects of this damaging process. Perhaps you've noticed that when rusting occurs on the body of a car, the rust appears at and around a break (or a scratch) in the surface of the paint, but the damage extends under the painted surface for some distance. Apparently, the Fe^{2+} ions that are formed at the anode sites are able to diffuse rather long distances to the hole in the paint, where they finally react with air to form the rust.

Cathodic Protection

One way to prevent the rusting of iron is to coat it with another metal. This is done with "tin" cans, which are actually steel cans that have been coated with a thin layer of tin. However, if the layer of tin is scratched and the iron beneath is exposed, the corrosion is accelerated because iron has a lower reduction potential than tin; the iron becomes the anode in an electrochemical cell and is easily oxidized.

Another way to prevent corrosion is called *cathodic protection*. It involves placing the iron in contact with a metal that is *more easily* oxidized. This causes iron to be a cathode and the other metal to be the anode. If corrosion occurs, iron is protected from oxidation because it is cathodic and the other metal reacts instead.

Zinc is most often used to provide cathodic protection to other metals. For example, zinc sacrificial anodes can be attached to the rudder of a boat (see Figure 2). When the rudder is submerged, the zinc will gradually corrode but the metal of the rudder will not. Periodically, the anodes are replaced to provide continued protection.

Steel objects that must withstand the weather are often coated with a layer of zinc, a process called galvanizing. You've seen this on chain-link fences and metal garbage pails. Even if the steel is exposed through a scratch, it is prevented from being oxidized because it is in contact with a metal that is more easily oxidized.

FIG. 2 Cathodic protection. Before launching, a shiny new zinc anode disk is attached to the bronze rudder of this boat to provide cathodic protection. Over time, the zinc will corrode instead of the less reactive bronze. (The rudder is painted with a special blue paint to inhibit the growth of barnacles.) *(Courtesy James Brady.)*

At this point you may have wondered why the term *cell potential* is used in some places, and *standard cell potential* is used in others. Based on our definitions, the term standard cell potential is used in places where the system is at standard state (i.e., 1.00 *M* concentrations, 1.00 atm pressures, and 25.0 °C). Cell potential is used for *any* set of concentrations, pressures, and temperatures, including the standard conditions. For convenience, our calculations in the next two sections will use standard cell potentials.

19.3 STANDARD REDUCTION POTENTIALS CAN PREDICT SPONTANEOUS REACTIONS

Redox reactions can be predicted by comparing reduction potentials

It's easy to predict the spontaneous reaction between the substances in two half-reactions, at standard state, because we know that *the half-reaction with the more positive reduction potential always takes place as written (namely, as a reduction), while the other half-reaction is forced to run in reverse (as an oxidation).*

EXAMPLE 19.3
Predicting a Spontaneous Reaction

What spontaneous reaction occurs if Cl_2 and Br_2 are added to a solution that contains both Cl^- and Br^-? Assume that the cell is at standard state.

ANALYSIS: We know that in the spontaneous redox reaction, the more easily reduced substance will be the one that undergoes reduction. By assuming we are at standard state we can use the standard reduction potentials for Cl_2 and Br_2 to compare their $E°$ values and determine which is the more easily reduced, and then we will use that information to write the correct "cell reaction." This is the spontaneous reaction, *whether or not it occurs in a galvanic cell.*

SOLUTION: There are two possible reduction half-reactions.

$$Cl_2(g) + 2e^- \longrightarrow 2Cl^-(aq)$$
$$Br_2(aq) + 2e^- \longrightarrow 2Br^-(aq)$$

Referring to Table 19.1, we find that Cl_2 has a more positive standard reduction potential (+1.36 V) than does Br_2 (+1.07 V). This means Cl_2 will be reduced and the half-reaction for Br_2 will be reversed, changing to an oxidation. Therefore, the spontaneous reaction has the following half-reactions.

$$Cl_2(g) + 2e^- \longrightarrow 2Cl^-(aq) \qquad \text{(a reduction)}$$
$$2Br^-(aq) \longrightarrow Br_2(aq) + 2e^- \qquad \text{(an oxidation)}$$

The net reaction is obtained by combining the half-reactions.

$$Cl_2(g) + 2Br^-(aq) \longrightarrow Br_2(aq) + 2Cl^-(aq)$$

Strictly speaking, the $E°$ values only tell us what to expect under standard conditions. However, only when $E°_{cell}$ is small can changes in the concentrations change the direction of the spontaneous reaction.

IS THE ANSWER REASONABLE? We can check to be sure we've read the correct values for $E°_{Cl_2}$ and $E°_{Br_2}$ from Table 19.1, and we can check the half-reactions we used to find the equation for the net reaction. (Experimentally, chlorine does indeed oxidize bromide ion to bromine, a fact used to recover bromine from seawater and natural brine solutions.)

When our cell is at standard state, reactants and products of *spontaneous* redox reactions are easy to spot when standard reduction potentials are listed in order of most positive to least positive (most negative), as in Table 19.1. For *any* pair of half-reactions, the one higher up in the table has the more positive standard reduction potential and occurs as a reduction. The other half-reaction is reversed and occurs as an oxidation. *Therefore, for a spontaneous reaction, the* ***reactants*** *are found on the left side of the higher half-reaction and on the right side of the lower half-reaction. (This is usually, but not always, true of systems that are not at standard state.)*

EXAMPLE 19.4
Predicting the Outcome of Redox Reactions

Predict the reaction that will occur, at 25 °C, when Ni and Fe are added to a solution that is 1.00 *M* in both Ni^{2+} and Fe^{2+}.

ANALYSIS: This system is at standard state and we can use the standard reduction potentials to predict the reaction. The first question we would ask is, "What *possible* reactions could occur?" We have a situation involving possible changes of ions to atoms or of atoms to ions. In other words, the system involves a possible redox reaction, and you've seen we can predict these using data in Table 19.1. One way to do this is to note the relative positions of the half-reactions when arranged as they are in Table 19.1.

$$\boxed{Ni^{2+}(aq)} + 2e^- \rightleftharpoons Ni(s) \qquad E^\circ_{Ni^{2+}} = -0.25$$

$$Fe^{2+}(aq) + 2e^- \rightleftharpoons \boxed{Fe(s)} \qquad E^\circ_{Fe^{2+}} = -0.44$$

In the table, the half-reaction higher up has the more positive (in this case, less negative) standard reduction potential, and will occur as a reduction. As a result, the reactants in the spontaneous reaction are related by the diagonal line that slants from upper left to lower right, as illustrated above. In other words, Ni^{2+} will react with Fe. The products are the substances on the opposite sides of the half-reaction, Ni and Fe^{2+}.

SOLUTION: We've done nearly all the work in our analysis of the problem. All that's left is to write the equation. The reactants are Ni^{2+} and Fe; the products are Ni and Fe^{2+}.

$$Ni^{2+}(aq) + Fe(s) \longrightarrow Ni(s) + Fe^{2+}(aq)$$

The equation is balanced in terms of both atoms and charges, so this is the reaction that will occur in the system specified in the problem.

IS THE ANSWER REASONABLE? Notice that we've predicted the reaction very easily using just the *positions* of the half-reactions relative to each other in the table; we really didn't have to use the values of their standard reduction potentials. We could check ourselves by proceeding as in Example 19.3. In the table, we see that Ni^{2+} has a more positive (less negative) standard reduction potential than Fe^{2+}, so Ni^{2+} is reduced and its half-cell reaction is written just as in Table 19.1. The half-cell reaction for Fe^{2+} in Table 19.1, however, must be reversed; it is Fe that will be oxidized.

$$Ni^{2+}(aq) + 2e^- \longrightarrow Ni(s) \qquad \text{(reduction)}$$

$$Fe(s) \longrightarrow Fe^{2+}(aq) + 2e^- \qquad \text{(oxidation)}$$

$$Ni^{2+}(aq) + Fe(s) \longrightarrow Ni(s) + Fe^{2+}(aq) \qquad \text{(net reaction)}$$

Practice Exercise 5: Based only on the half-reactions in Table 19.1, determine what reaction will occur in each of the following mixtures, at standard state. (a) I_2, I^- and Fe^{2+}, Fe^{3+}, (b) Mg, Mg^{2+} and Cr, Cr^{3+}, (c) Co, Co^{2+} and H_2SO_3, SO_4^{2-}. (Hint: Use Table 19.1 to write the possible half-reactions and, if necessary, determine E°_{cell}.)

Practice Exercise 6: Use the positions of the half-reactions in Table 19.1 to predict the spontaneous reaction when Br^-, SO_4^{2-}, H_2SO_3, and Br_2 are mixed in an acidic solution at standard state.

Practice Exercise 7: From the positions of the respective half-reactions in Table 19.1, predict whether the following reaction will occur if all the ions are 1.0 *M* at 25 °C. If it is not, write the equation for the spontaneous reaction.

$$Ni^{2+}(aq) + 2Fe^{2+}(aq) \longrightarrow Ni(s) + 2Fe^{3+}(aq)$$

Standard reduction potentials predict the cell reaction and standard cell potential of a galvanic cell

We've just seen that we can use standard reduction potentials to predict spontaneous redox reactions. If we intend to use these reactions in a galvanic cell, we can also predict what the standard cell potential will be, as illustrated in the next example.

EXAMPLE 19.5
Predicting the Cell Reaction and Standard Cell Potential of a Galvanic Cell

A typical cell of a lead storage battery of the type used to start automobiles is constructed using electrodes made of lead and lead(IV) oxide (PbO_2) and with sulfuric acid as the electrolyte. The half-reactions and their standard reduction potentials in this system are

$$PbO_2(s) + 3H^+(aq) + HSO_4^-(aq) + 2e^- \rightleftharpoons PbSO_4(s) + 2H_2O \qquad E^\circ_{PbO_2} = 1.69 \text{ V}$$

$$PbSO_4(s) + H^+(aq) + 2e^- \rightleftharpoons Pb(s) + HSO_4^-(aq) \qquad E^\circ_{PbSO_4} = -0.36 \text{ V}$$

What is the cell reaction and what is the standard potential of the cell?

ANALYSIS: Our method for predicting spontaneous reactions specifies that the system should be at standard state. Although a battery is not at standard state, we will assume standard state to make the calculations easier. In the spontaneous cell reaction, the half-reaction with the larger (more positive) standard reduction potential will take place as reduction while the other half-reaction will be reversed and occur as oxidation. The standard cell potential is simply the difference between the two standard reduction potentials, calculated using Equation 19.2.

Remember that half-reactions are combined following the same procedure used in the ion–electron method of balancing redox reactions (Section 5.2).

SOLUTION: PbO_2 has a larger, more positive standard reduction potential than $PbSO_4$, so the first half-reaction will occur in the direction written. The second must be reversed to occur as an oxidation. In the cell, therefore, the half-reactions are

$$PbO_2(s) + 3H^+(aq) + HSO_4^-(aq) + 2e^- \longrightarrow PbSO_4(s) + 2H_2O$$
$$Pb(s) + HSO_4^-(aq) \longrightarrow PbSO_4(s) + H^+(aq) + 2e^-$$

Adding the two half-reactions and canceling electrons gives the cell reaction,

$$PbO_2(s) + Pb(s) + 2H^+(aq) + 2HSO_4^-(aq) \longrightarrow 2PbSO_4(s) + 2H_2O$$

The cell standard potential is obtained by using Equation 19.2.

$$E^\circ_{cell} = (E^\circ \text{ of substance reduced}) - (E^\circ \text{ of substance oxidized})$$

Since the first half-reaction occurs as a reduction and the second as an oxidation,

$$\begin{aligned} E^\circ_{cell} &= E^\circ_{PbO_2} - E^\circ_{PbSO_4} \\ &= (1.69 \text{ V}) - (-0.36 \text{ V}) \\ &= 2.05 \text{ V} \end{aligned}$$

ARE THE ANSWERS REASONABLE? The half-reactions involved in the problem are located in Table 19.1, and their relative positions tell us that PbO_2 will be reduced and that lead will be oxidized. Therefore, we've combined the half-reactions correctly and we can feel confident that we've also applied Equation 19.2 correctly. In addition, the standard cell potential of 2.05 volts tells us that small differences from standard state will still allow us to come to the same conclusion.

EXAMPLE 19.6

Predicting the Cell Reaction and Standard Cell Potential of a Galvanic Cell

At standard state, what would be the cell reaction and the standard cell potential of a galvanic cell employing the following half-reactions?

$$Al^{3+}(aq) + 3e^- \rightleftharpoons Al(s) \qquad E^\circ_{Al^{3+}} = -1.66 \text{ V}$$

$$Cu^{2+}(aq) + 2e^- \rightleftharpoons Cu(s) \qquad E^\circ_{Cu^{2+}} = +0.34 \text{ V}$$

Which half-cell would be the anode?

ANALYSIS: This problem is very similar to the preceding one, so we expect to proceed in essentially the same way.

SOLUTION: Our method for predicting spontaneous reactions indicates that the half-reaction with the more positive standard reduction potential will occur as a reduction; the other will occur as an oxidation. In this cell, then, Cu^{2+} is reduced and Al is oxidized. To obtain the cell reaction, we add the two half-reactions, remembering that the electrons must cancel. This means we must multiply the copper half-reaction by three and the aluminum half-reaction by two.

$$3[Cu^{2+}(aq) + 2e^- \longrightarrow Cu(s)] \qquad \text{(reduction)}$$

$$2[Al(s) \longrightarrow Al^{3+}(aq) + 3e^-] \qquad \text{(oxidation)}$$

$$3Cu^{2+}(aq) + 2Al(s) \longrightarrow 3Cu(s) + 2Al^{3+}(aq) \qquad \text{(cell reaction)}$$

The anode in the cell is aluminum because that is where oxidation takes place (by definition).

To obtain the standard cell potential, we substitute into Equation 19.2.

$$\begin{aligned} E^\circ_{cell} &= E^\circ_{Cu^{2+}} - E^\circ_{Al^{3+}} \\ &= (0.34 \text{ V}) - (-1.66 \text{ V}) \\ &= 2.00 \text{ V} \end{aligned}$$

An important point to notice here is that *although we multiply the half-reactions by factors to make the electrons cancel,* ***we do not multiply the standard reduction potentials by these factors.***[4] To obtain the standard cell potential, we simply subtract one standard reduction potential from the other.

ARE THE ANSWERS REASONABLE? If we locate the half-reactions in Table 19.1, their relative positions tell us we've written the correct equation for the spontaneous reaction. It also means we've identified correctly the substances reduced and oxidized, so we've correctly applied Equation 19.2.

Practice Exercise 8: What are the overall cell reaction and the standard cell potential of a galvanic cell employing the following half-reactions?

$$NiO_2(s) + 2H_2O + 2e^- \rightleftharpoons Ni(OH)_2(s) + 2OH^-(aq) \qquad E^\circ_{NiO_2} = 0.49 \text{ V}$$

$$Fe(OH)_2(s) + 2e^- \rightleftharpoons Fe(s) + 2OH^-(aq) \qquad E^\circ_{Fe(OH)_2} = -0.88 \text{ V}$$

(Hint: Recall that standard cell potentials are based on a spontaneous reaction.)

These are the reactions in an Edison cell, a type of rechargeable storage battery.

Practice Exercise 9: The four substances in the following two half-reactions are placed in the same beaker, at standard state. Write the balanced equation for the spontaneous reaction and determine the standard cell potential if the two reactions are used in a galvanic cell.

$$Cu^{2+}(aq) + 2e^- \rightleftharpoons Cu(s) \qquad E^\circ_{Cu^{2+}} = +0.34 \text{ V}$$

$$Cr^{3+}(aq) + 3e^- \rightleftharpoons Cr(s) \qquad E^\circ_{Cr^{3+}} = -0.74 \text{ V}$$

[4] Reduction potentials are intensive quantities; they have the units volts, which are joules *per coulomb*. The same number of joules are available for each coulomb of charge regardless of the total number of electrons shown in the equation. Therefore, reduction potentials are never multiplied by factors before they are subtracted to give the cell potential.

Practice Exercise 10: What are the overall cell reaction and the standard cell potential of a galvanic cell employing the following half-reactions at standard state?

$$Cr^{3+}(aq) + 3e^- \rightleftharpoons Cr(s) \qquad E^\circ_{Cr^{3+}} = -0.74 \text{ V}$$
$$MnO_4^-(aq) + 8H^+(aq) + 5e^- \rightleftharpoons Mn^{2+}(aq) + 4H_2O \qquad E^\circ_{MnO_4^-} = +1.51 \text{ V}$$

The calculated cell potential can tell us whether a reaction is spontaneous

Because we can predict the spontaneous redox reaction that will take place among a mixture of reactants, it also should be possible to predict whether or not a particular reaction, *as written*, can occur spontaneously. We can do this by calculating the standard cell potential that corresponds to the reaction in question and seeing if the standard potential is *positive*.

In a galvanic cell, the calculated standard cell potential for the spontaneous reaction is always positive. If the calculated standard cell potential is negative, the reaction is spontaneous in the reverse direction.

These generalizations apply under standard conditions: 1 *M* concentrations of all ions, 1 atm pressure for gases, and 25 °C.

For example, to obtain the standard cell potential for a spontaneous reaction in our previous examples, we subtracted the standard reduction potentials in a way that gave a positive answer. Therefore, if we compute the standard cell potential for a particular reaction *based on the way the equation is written* and the standard potential comes out positive, we know the reaction is spontaneous. If the calculated standard cell potential comes out negative, however, the reaction is nonspontaneous. In fact, it is really spontaneous in the opposite direction.

EXAMPLE 19.7 Determining whether a Reaction Is Spontaneous by Using the Calculated Standard Cell Potential

Determine whether the following reactions, at standard state, are spontaneous as written. If a reaction is not spontaneous, write the equation for the reaction that is.

(1) $Cu(s) + 2H^+(aq) \longrightarrow Cu^{2+}(aq) + H_2(g)$

(2) $3Cu(s) + 2NO_3^-(aq) + 8H^+(aq) \longrightarrow 3Cu^{2+}(aq) + 2NO(g) + 4H_2O$

ANALYSIS: Our goal for each reaction will be to calculate the standard cell potential based on the reaction as written. Our tool for spontaneous reactions states that if E°_{cell} is positive, then the reaction is spontaneous. However, if E°_{cell} is negative, then the reaction is not spontaneous as written and reversing the equation will give the spontaneous reaction.

To calculate E°_{cell}, we need to divide the equation into its half-reactions, find the necessary standard reduction potentials in Table 19.1, and then use Equation 19.2 to calculate E°_{cell}. The signs of E°_{cell} will then tell us whether the reactions are spontaneous under standard conditions.

SOLUTION: (1) The half-reactions involved in this reaction are

$$Cu(s) \longrightarrow Cu^{2+}(aq) + 2e^- \qquad \text{(oxidation)}$$
$$2H^+(aq) + 2e^- \longrightarrow H_2(g) \qquad \text{(reduction)}$$

The H^+ is reduced and Cu is oxidized, so Equation 19.2 will take the form

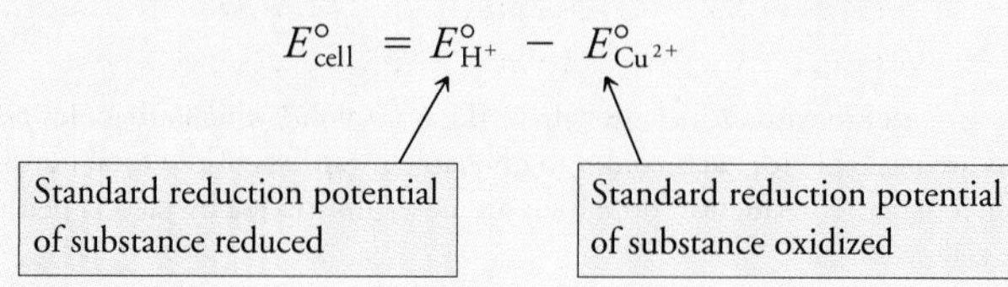

Substituting values from Table 19.1 gives

$$\begin{aligned} E^\circ_{\text{cell}} &= (0.00 \text{ V}) - (0.34 \text{ V}) \\ &= -0.34 \text{ V} \end{aligned}$$

The calculated standard cell potential is negative, so reaction (1) is not spontaneous in the forward direction. The spontaneous reaction is actually the reverse of (1).

$$\underset{\text{Reaction (1) reversed}}{Cu^{2+}(aq) + H_2(g) \longrightarrow Cu(s) + 2H^+(aq)}$$

(2) The half-reactions involved in this equation are

$$\begin{aligned} Cu(s) &\longrightarrow Cu^{2+}(aq) + 2e^- \\ NO_3^-(aq) + 4H^+(aq) + 3e^- &\longrightarrow NO(g) + 2H_2O \end{aligned}$$

The Cu is oxidized while the NO_3^- is reduced. According to Equation 19.2,

$$E^\circ_{\text{cell}} = E^\circ_{NO_3^-} - E^\circ_{Cu^{2+}}$$

Substituting values from Table 19.1 gives

$$\begin{aligned} E^\circ_{\text{cell}} &= (0.96 \text{ V}) - (0.34 \text{ V}) \\ &= +0.62 \text{ V} \end{aligned}$$

Because the calculated standard cell potential is positive, reaction 2 is spontaneous in the forward direction, as written.

Copper dissolves in HNO_3 because it contains the oxidizing agent NO_3^-.

ARE THE ANSWERS REASONABLE? By noting the relative positions of the half-reactions in Table 19.1, you can confirm that we've answered the questions correctly.

Practice Exercise 11: Determine if each of the following reactions, under standard state conditions, is spontaneous.

(a) $Br_2(aq) + 2I^-(aq) \longrightarrow 2Br^-(aq) + I_2(s)$

(b) $MnO_4^-(aq) + 5Ag(aq) + 8H^+(aq) \longrightarrow Mn^{2+}(aq) + 5Ag^+(s) + 4H_2O$

(Hint: Determine the sign of the calculated standard cell potential.)

Practice Exercise 12: Under standard state conditions, which of the following reactions occur spontaneously?

(a) $Br_2(aq) + Cl_2(g) + 2H_2O \longrightarrow 2Br^-(aq) + 2HOCl(aq) + 2H^+(aq)$

(b) $3Zn(s) + 2Cr^{3+}(aq) \longrightarrow 3Zn^{2+}(aq) + 2Cr(s)$

In the previous examples and exercises we have been rigorous in specifying that standard reduction potentials are used when the system is at standard state. If the cell is not at standard state, the cell potential will not be the same as the standard cell potential, but in the majority of cases, the algebraic sign of the calculated cell potential will be the same and we can reach the same conclusions.

19.4 CELL POTENTIALS ARE RELATED TO FREE ENERGY CHANGES

The fact that cell potentials allow us to predict the spontaneity of redox reactions is no coincidence. There is a relationship between the cell potential and the free energy change for a reaction. In Chapter 18 we saw that ΔG for a reaction is a measure of the maximum useful work that can be obtained from a chemical reaction. Specifically, the relationship is

$$-\Delta G = \text{maximum work} \tag{19.3}$$

In an electrical system, work is supplied by the flow of electric charge created by the potential of the cell. It can be calculated from the equation

$$\text{Maximum work} = n\mathcal{F}E_{\text{cell}} \tag{19.4}$$

TOOLS Faraday constant

where n is the number of moles of electrons transferred, $\mathcal{F}$ is a constant called the **Faraday constant** which is equal to the number of coulombs of charge equivalent to 1 mol of electrons (9.65×10^4 coulombs per mole of electrons), and E_{cell} is the potential of the cell in volts. To see that Equation 19.4 gives work (which has the units of energy) we can analyze the units. In Equation 19.1 you saw that 1 volt = 1 joule/coulomb. Therefore,

More precisely, one **Faraday** ($\mathcal{F}$) = 96,485 C.

$$\text{Maximum work} = \cancel{\text{mole } e^-} \times \left(\frac{\cancel{\text{coulombs}}}{\cancel{\text{mole } e^-}}\right) \times \left(\frac{\text{joule}}{\cancel{\text{coulombs}}}\right) = \text{joule}$$

$$\begin{matrix} \Updownarrow & \Updownarrow & \Updownarrow \\ n & \mathcal{F} & E_{\text{cell}} \end{matrix}$$

Combining Equations 19.3 and 19.4 gives us

$$\Delta G = -n\mathcal{F}E_{\text{cell}} \tag{19.5}$$

At standard state we are dealing with the *standard* cell potential, so we can calculate the *standard* free energy change.

TOOLS *Standard free energy change is related to the standard cell potential*

$$\Delta G^\circ = -n\mathcal{F}E^\circ_{\text{cell}} \tag{19.6}$$

Referring back to Chapter 18, if ΔG has a negative value, a reaction will be spontaneous and this corresponds to a positive value of E_{cell}. Up to now we have been careful to predict spontaneity for standard state systems where E°_{cell} is equal to E_{cell}. In Example 19.11 below we will see how to calculate E_{cell} and precisely predict if a reaction is spontaneous.

EXAMPLE 19.8 Calculating the Standard Free Energy Change

Calculate ΔG° for the following reaction, given that its standard cell potential is 0.320 V at 25 °C.

$$NiO_2(s) + 2Cl^-(aq) + 4H^+(aq) \longrightarrow Cl_2(g) + Ni^{2+}(aq) + 2H_2O$$

ANALYSIS: Our tool for solving this problem is Equation 19.6. Taking the coefficients in the equation to stand for *moles*, two moles of Cl^- are oxidized to Cl_2 and two moles of electrons are transferred to the NiO_2 ($n = 2 \text{ mol } e^-$). We will also use the faraday constant, $1\ \mathcal{F} = 96{,}500 \text{ C/mol } e^-$ (recall that the SI abbreviation for coulomb is C).

$$1\ \mathcal{F} = \frac{96{,}500 \text{ C}}{1 \text{ mol } e^-}$$

SOLUTION: Using Equation 19.6, we have

$$\begin{aligned} \Delta G^\circ &= -(2 \cancel{\text{mol } e^-}) \times \left(\frac{9.65 \times 10^4 \cancel{\text{C}}}{1 \cancel{\text{mol } e^-}}\right) \times \left(\frac{0.320 \text{ J}}{\cancel{\text{C}}}\right) \\ &= -6.18 \times 10^4 \text{ J} \\ &= -61.8 \text{ kJ} \end{aligned}$$

IS THE ANSWER REASONABLE? Let's do some approximate arithmetic. The faraday constant equals approximately 100,000, or 10^5. The product $2 \times 0.32 = 0.64$, so ΔG° should be about 0.64×10^5 or 6.4×10^4 J. The answer seems to be okay.

Practice Exercise 13: A certain reaction has an E°_{cell} of 0.107 volts and has a ΔG° of -30.9 kJ. How many electrons are transferred in the reaction? (Hint: See Equation 19.6.)

Practice Exercise 14: Calculate ΔG° for the reactions that take place in the galvanic cells described in Practice Exercises 11 and 12.

Equilibrium constants can be calculated from E°_{cell}

One useful application of electrochemistry is the determination of equilibrium constants. In Chapter 18 you saw that ΔG° is related to the equilibrium constant by the expression

$$\Delta G^\circ = -RT \ln K_c$$

where we have used K_c for the equilibrium constant because electrochemical reactions occur in solution. We've seen in this chapter that ΔG° is also related to E°_{cell}

$$\Delta G^\circ = -n\mathscr{F}E^\circ_{cell}$$

Therefore, E°_{cell} and the equilibrium constant are also related. Equating the right sides of the two equations, we have

$$-n\mathscr{F}E^\circ_{cell} = -RT \ln K_c$$

Solving for E°_{cell} gives[5]

$$E^\circ_{cell} = \frac{RT}{n\mathscr{F}} \ln K_c \qquad (19.7)$$

TOOLS *Equilibrium constants are related to standard cell potentials*

For the units to work out correctly, the value of R must be 8.314 J mol^{-1} K^{-1}, T must be the temperature in kelvins, $\mathscr{F}$ equals 9.65×10^4 C per mole of e^-, and n equals the number of moles of electrons transferred in the reaction.

EXAMPLE 19.9
Calculating Equilibrium Constants from E°_{cell}

Calculate K_c for the reaction in Example 19.8.

ANALYSIS: Equation 19.7 is our tool for solving this problem. We need to collect the terms to insert in this equation to solve it. We need to find E°_{cell}, n, R, T, and $\mathscr{F}$. Remember that T is in kelvins and R must have the appropriate units of J mol^{-1} K^{-1}.

SOLUTION: The reaction in Example 19.8 has $E^\circ_{cell} = 0.320$ V and $n = 2$. The temperature is 25 °C or 298 K. Let's solve Equation 19.7 for ln K_c and then substitute values.

$$\ln K_c = \frac{E^\circ_{cell}\, n\mathscr{F}}{RT}$$

Substituting values and using the relationship that 1 V = 1 J C^{-1},

$$\ln K_c = \frac{0.320 \text{ J C}^{-1} \times 2 \times 9.65 \times 10^4 \text{ C mol}^{-1}}{8.314 \text{ J mol}^{-1} \text{ K}^{-1} \times 298 \text{ K}}$$

$$= 24.9$$

Taking the antilogarithm,

$$K_c = e^{24.9} = 7 \times 10^{10}$$

[5] For historical reasons, Equation 19.7 is sometimes expressed in terms of common logs (base 10 logarithms). Natural and common logarithms are related by the equation

$$\ln x = 2.303 \log x$$

For reactions at 25 °C (298 K), all of the constants (R, T, and $\mathscr{F}$) can be combined with the factor 2.303 to give 0.0592 joules/coulomb. Because joules/coulomb equals volts, Equation 19.7 reduces to

$$E^\circ_{cell} = \frac{0.0592 \text{ V}}{n} \log K_c$$

where n is the number of moles of electrons transferred in the cell reaction as it is written.

IS THE ANSWER REASONABLE? As a rough check, we can look at the magnitude of E°_{cell} and apply some simple reasoning. When E°_{cell} is positive, ΔG° is negative, and in Chapter 18 you learned that when ΔG° is negative, the reaction proceeds far toward completion when equilibrium is reached. Therefore, we expect that K_c will be large, and that agrees with our answer.

A more complete check would require evaluating the fraction used to compute ln K_c. First, we should check to be sure we've substituted correctly into the equation for ln K_c. Next, we could do some approximate arithmetic to check the value of ln K_c. Rounding all numbers to one significant figure we get

$$\ln K_c = \frac{0.3 \text{ J C}^{-1} \times 2 \times 10 \times 10^4 \text{ C mol}^{-1}}{10 \text{ J mol}^{-1}\text{K}^{-1} \times 300 \text{ K}} = \frac{6 \times 10^4}{3000} = 2 \times 10^1 = 20$$

This is close to the 24.9 we calculated above and we are confident the calculation was done correctly.

Practice Exercise 15: The calculated standard cell potential for the reaction

$$Cu^{2+}(aq) + 2Ag(s) \rightleftharpoons Cu(s) + 2Ag^+(aq)$$

is $E^\circ_{cell} = -0.46$ V. Calculate K_c for the reaction as written. Is the reaction spontaneous? If not, what is K_c for the spontaneous reaction? (Hint: The tool described by Equation 19.7 is important here.)

Practice Exercise 16: Use the following half-reactions and the data in Table 19.1 to write the equation for the spontaneous reaction. Write the equilibrium law for the reaction and use the standard cell potential to determine the value of the equilibrium constant. How is this related to the K_{sp} for AgBr?

$$Ag^+(aq) + e^- \longrightarrow Ag(s)$$
$$AgBr(s) + e^- \longrightarrow Ag(s) + Br^-(aq)$$

19.5 CONCENTRATIONS IN A GALVANIC CELL AFFECT THE CELL POTENTIAL

At 25 °C when all of the ion concentrations in a cell are 1.00 *M* and when the partial pressures of any gases involved in the cell reaction are 1.00 atm, the cell potential is equal to the standard potential. When the concentrations or pressures change, however, so does the potential. For example, in an operating cell or battery, the potential gradually drops as the reactants are used up and as the cell reaction approaches its natural equilibrium status. When it reaches equilibrium, the potential has dropped to zero—the battery is dead.

The Nernst equation defines the relationship of cell potential to ion concentrations

The effect of concentration on the cell potential can be obtained from thermodynamics. In Chapter 18, you learned that the free energy change is related to the reaction quotient Q by the equation

$$\Delta G = \Delta G^\circ + RT \ln Q$$

Substituting for ΔG and ΔG° from Equations 19.5 and 19.6 gives

$$-n\mathcal{F}E_{cell} = -n\mathcal{F}E^\circ_{cell} + RT \ln Q$$

Dividing both sides by $-n\mathcal{F}$ gives

Nernst equation

$$E_{cell} = E^\circ_{cell} - \frac{RT}{n\mathcal{F}} \ln Q \qquad (19.8)$$

This equation is commonly known as the **Nernst equation,**[6] named after Walther Nernst, a German chemist and physicist. Notice, if $Q = 1$ then $\ln Q = 0$ and $E_{cell} = E°_{cell}$.

In writing the Nernst equation for a galvanic cell, we will construct the mass action expression (from which we calculate Q) using molar concentrations for ions and partial pressures in atmospheres for gases.[7] Thus, for the following cell using a hydrogen electrode (with the partial pressure of H_2 not necessarily equal to 1 atm) and having the reaction

$$Cu^{2+}(aq) + H_2(g) \longrightarrow Cu(s) + 2H^+(aq)$$

the Nernst equation would be written

$$E_{cell} = E°_{cell} - \frac{RT}{n\mathcal{F}} \ln \frac{[H^+]^2}{[Cu^{2+}]P_{H_2}}$$

This is a heterogeneous reaction, so we have not included the concentration of the solid, Cu(*s*), in the mass action expression.

EXAMPLE 19.10
Calculating the Effect of Concentration on E_{cell}

Suppose a galvanic cell employs the following half-reactions.

$$Ni^{2+}(aq) + 2e^- \rightleftharpoons Ni(s) \qquad E°_{Ni^{2+}} = -0.25 V$$
$$Cr^{3+}(aq) + 3e^- \rightleftharpoons Cr(s) \qquad E°_{Cr^{3+}} = -0.74 V$$

Calculate the cell potential when $[Ni^{2+}] = 1.0 \times 10^{-4}$ *M* and $[Cr^{3+}] = 2.0 \times 10^{-3}$ *M*.

ANALYSIS: Because the concentrations are not 1.00 *M*, we must use the Nernst equation (Equation 19.8) as our tool, but first we need the cell reaction. We need it to determine the number of electrons transferred, *n*, and we need it to determine the correct form of the mass action expression from which we calculate the numerical value of *Q*. We must also note that the reacting system is heterogeneous; both solid metals and a liquid solution of their dissolved ions are involved. We have to remember that a mass action expression does not contain concentration terms for solids, such as Ni and Cr.

SOLUTION: Nickel has the more positive (less negative) standard reduction potential, so its half-reaction will occur as a reduction. This means that chromium will be oxidized. Making electron gain equal to electron loss, the cell reaction is found as follows.

$$3[Ni^{2+}(aq) + 2e^- \longrightarrow Ni(s)] \qquad \text{(reduction)}$$
$$2[Cr(s) \longrightarrow Cr^{3+}(aq) + 3e^-] \qquad \text{(oxidation)}$$
$$3Ni^{2+}(aq) + 2Cr(s) \longrightarrow 3Ni(s) + 2Cr^{3+}(aq) \qquad \text{(cell reaction)}$$

The total number of electrons transferred is six, which means $n = 6$. Now we can write the Nernst equation for the system.

$$E_{cell} = E°_{cell} - \frac{RT}{n\mathcal{F}} \ln \frac{[Cr^{3+}]^2}{[Ni^{2+}]^3}$$

Notice that we've constructed the mass action expression, from which we will calculate the reaction quotient, using the concentrations of the ions raised to powers equal to their coefficients in the net cell reaction, and that we have not included concentration terms for the two solids. This is the procedure we followed for heterogeneous equilibria in Chapter 14.

[6] Using common logarithms instead of natural logarithms and calculating the constants for 25 °C gives another form of the Nernst equation that is sometimes used:

$$E_{cell} = E°_{cell} - \frac{0.0592 \text{ V}}{n} \log Q$$

[7] Because of interionic attractions, ions do not always behave as though their concentrations are equal to their molarities. Strictly speaking, therefore, we should use effective concentrations (called activities) in the mass action expression. Effective concentrations are difficult to calculate, so for simplicity we will use molarities and accept the fact that our calculations are not entirely accurate.

Next we need E°_{cell}. Since Ni^{2+} is reduced,

$$\begin{aligned} E^\circ_{cell} &= E^\circ_{Ni^{2+}} - E^\circ_{Cr^{3+}} \\ &= (-0.25\ V) - (-0.74\ V) \\ &= 0.49\ V \end{aligned}$$

Now we can substitute this value for E°_{cell} along with $R = 8.314$ J mol^{-1} K^{-1}, $T = 298$ K, $n = 6$, $\mathcal{F} = 9.65 \times 10^4$ C mol^{-1}, $[Ni^{2+}] = 1.0 \times 10^{-4}$ *M*, and $[Cr^{3+}] = 2.0 \times 10^{-3}$ *M* into the Nernst equation. This gives

$$\begin{aligned} E_{cell} &= 0.49\ V - \frac{8.314\ J\ \cancel{mol^{-1}}\ \cancel{K^{-1}} \times 298\ \cancel{K}}{6 \times 9.65 \times 10^4\ C\ \cancel{mol^{-1}}} \ln \frac{(2.0 \times 10^{-3})^2}{(1.0 \times 10^{-4})^3} \\ &= 0.49\ V - (0.00428\ V) \ln (4.0 \times 10^6) \\ &= 0.49\ V - (0.00428\ V)(15.20) \\ &= 0.49\ V - 0.0651\ V \\ &= 0.42\ V \end{aligned}$$

The potential of the cell is expected to be 0.42 V.

IS THE ANSWER REASONABLE? There's no simple way to check the answer. However, there are certain important points to consider. First, check that you've combined the half-reactions correctly to calculate E°_{cell} and given the balanced cell reaction, because we need the coefficients of the equation to obtain the correct superscripts in the Nernst equation. Then, be sure you've used the Kelvin temperature, $R = 8.314$ J mol^{-1} K^{-1}, and made the other substitutions correctly.

EXAMPLE 19.11
The Spontaneous Reaction May Be Concentration Dependent

The reaction of tin metal with acid can be written as

$$Sn(s) + 2H^+(aq) \longrightarrow Sn^{2+}(aq) + H_2(g)$$

Calculate the cell potential (a) when the system is at standard state, (b) when the pH is 2.00, and (c) when the pH is 5.00. Assume that $[Sn^{2+}] = 1.00$ *M* and the partial pressure of H_2 is also 1.00 atm.

ANALYSIS: In this reaction the tin metal is the substance that is oxidized and hydrogen ions are reduced. Part (a) is at standard state, and we have used our tool for combining standard reduction potentials to solve problems like that in Examples 19.6 and 19.7. For parts (b) and (c) we must use the Nernst equation as our tool. We determine the number of electrons transferred by noting that tin loses two electrons and each hydrogen gains an electron. Therefore two electrons are transferred from tin to the hydrogen ions. We can also set up *Q* for the Nernst equation by substituting 1.00 for both $[Sn^{2+}]$ and P_{H_2}:

$$Q = \frac{[Sn^{2+}]P_{H_2}}{[H^+]^2} = \frac{1.00}{[H^+]^2}$$

We can also calculate the hydrogen ion concentrations for the pH 2.00 and pH 5.00 solutions as 1.0×10^{-2} *M* and 1.0×10^{-5} *M* respectively.

SOLUTION: For part (a) we find the difference between standard reduction potentials as

$$E^\circ_{cell} = 0.00\ V - (-0.14)\ V = +0.14\ V$$

For parts (b) and (c) we substitute into the Nernst equation

$$E_{cell} = E^\circ_{cell} - \frac{RT}{n\mathscr{F}} \ln \frac{1.00}{[H^+]^2}$$

(b) $$E_{cell} = 0.14\text{ V} - \frac{8.314\text{ J }\cancel{\text{mol}^{-1}}\ \cancel{\text{K}^{-1}} \times 298\ \cancel{\text{K}}}{2 \times 9.65 \times 10^4\text{ C }\cancel{\text{mol}^{-1}}} \ln \frac{1.00}{(1.0 \times 10^{-2})^2}$$

$$E_{cell} = 0.14\text{ V} - 0.12\text{ V} = +0.02\text{ V}$$

(c) $$E_{cell} = 0.14\text{ V} - \frac{8.314\text{ J }\cancel{\text{mol}^{-1}}\ \cancel{\text{K}^{-1}} \times 298\ \cancel{\text{K}}}{2 \times 9.65 \times 10^4\text{ C }\cancel{\text{mol}^{-1}}} \ln \frac{1.00}{(1.0 \times 10^{-5})^2}$$

$$E_{cell} = 0.14\text{ V} - 0.30\text{ V} = -0.16\text{ V}$$

At standard state the reaction is spontaneous. At pH 2.00 it is spontaneous but the potential is a small positive value. At pH 5.00 the reaction is not spontaneous.

ARE THE ANSWERS REASONABLE? The first question to answer is, "Does this make sense?" and indeed it does. Looking at the natural logarithm part of the equation we see that as the $[H^+]$ decreases, the ln term increases and makes a more negative adjustment to the E°_{cell}. Since a decrease in $[H^+]$ is an increase in pH, we expect E_{cell} will decrease as pH increases. We cannot easily estimate natural logarithms so a check of calculations may be easiest if the values are entered into your calculator in the reverse order from the first calculation.

Practice Exercise 17: A galvanic cell is constructed with a copper electrode dipping into a 0.015 *M* solution of Cu^{2+} ions and an electrode made of magnesium immersed in a 2.2×10^{-6} *M* solution of magnesium ions. Write the balanced chemical reaction. What is the cell potential at 25 °C? (Hint: Set up the Nernst equation for the reaction.)

Practice Exercise 18: In a certain zinc–copper cell,

$$Zn(s) + Cu^{2+}(aq) \longrightarrow Zn^{2+}(aq) + Cu(s)$$

the ion concentrations are $[Cu^{2+}] = 0.0100$ *M* and $[Zn^{2+}] = 1.0$ *M*. What is the cell potential at 25 °C?

Experimental cell potentials can be used to determine ion concentrations

One of the principal uses of the relationship between concentration and cell potential is for the measurement of concentrations of redox reactants and products in a galvanic cell. Experimental determination of cell potentials combined with modern developments in electronics has provided a means of monitoring and analyzing the concentrations of all sorts of substances in solution, even some that are not themselves ionic and that are not involved directly in electrochemical changes. In fact, the operation of a pH meter relies on the logarithmic relationship between hydrogen ion concentration and the potential of a special kind of electrode (Figure 19.8 on page 795)

EXAMPLE 19.12
Using the Nernst Equation to Determine Concentrations

To measure the concentration of Cu^{2+} in a large number of samples of water in which the copper ion concentration is expected to be quite small, an electrochemical cell was assembled that consists of a silver electrode, dipping into a 1.00 *M* solution of $AgNO_3$, connected by a salt bridge to a second half-cell containing a copper electrode. The copper half-cell was then filled with one water sample after another, with the cell potential being measured for each sample. In the analysis of one sample, the cell potential at 25 °C was measured to be 0.62 V. The copper electrode was observed to carry a negative charge, so it served as the anode. What was the concentration of copper ion in the sample?

ANALYSIS: In this problem, we've been given the cell potential, E_{cell}, and we can calculate $E°_{cell}$ from the standard reduction potentials in Table 19.1. The unknown quantity is one of the concentration terms in the Nernst equation that we use as our tool for solving this problem.

SOLUTION: The first step is to write the proper equation for the cell reaction, because we need it to compute $E°_{cell}$ and to construct the mass action expression for use in the Nernst equation. Because copper is the anode, it is being oxidized. This also means that Ag^+ is being reduced. Therefore, the equation for the cell reaction is

$$Cu(s) + 2Ag^+(aq) \longrightarrow Cu^{2+}(aq) + 2Ag(s)$$

Two electrons are transferred, so $n = 2$ and the Nernst equation is

$$E_{cell} = E°_{cell} - \frac{RT}{2\mathcal{F}} \ln \frac{[Cu^{2+}]}{[Ag^+]^2}$$

The value of $E°_{cell}$ can be obtained from the tabulated standard reduction potentials in Table 19.1. Following our usual procedure and recognizing that silver ion is reduced,

$$\begin{aligned} E°_{cell} &= E°_{Ag^+} - E°_{Cu^{2+}} \\ &= (0.80 \text{ V}) - (0.34 \text{ V}) \\ &= 0.46 \text{ V} \end{aligned}$$

Now we can substitute values into the Nernst equation and solve for the concentration ratio in the mass action expression.

$$0.62 \text{ V} = 0.46 \text{ V} - \frac{8.314 \text{ J } \cancel{\text{mol}^{-1}} \cancel{\text{K}^{-1}} \times 298 \cancel{\text{K}}}{2 \times 9.65 \times 10^4 \text{ C } \cancel{\text{mol}^{-1}}} \ln \frac{[Cu^{2+}]}{[Ag^+]^2}$$

Solving for $\ln([Cu^{2+}]/[Ag^+]^2)$ gives

$$\ln \frac{[Cu^{2+}]}{[Ag^+]^2} = -12$$

Taking the antilog gives us the value of the mass action expression.

$$\frac{[Cu^{2+}]}{[Ag^+]^2} = 6 \times 10^{-6}$$

Since we know that the concentration of Ag^+ is 1.00 M, we can now solve for the Cu^{2+} concentration.

$$\frac{[Cu^{2+}]}{(1.00)^2} = 6 \times 10^{-6}$$

$$[Cu^{2+}] = 6 \times 10^{-6} \, M$$

IS THE ANSWER REASONABLE? All we can do easily is check to be sure we've written the correct chemical equation, on which all the rest of the solution to the problem rests. Be careful about algebraic signs and that you select the proper value for R and the temperature in kelvins. Also, notice that we first solved for the logarithm of the ratio of concentration terms. Then, after taking the (natural) antilogarithm, we substitute the known value for $[Ag^+]$ and solve for $[Cu^{2+}]$.

As a final point, notice that the Cu^{2+} concentration is indeed very small, and that it can be obtained very easily by simply measuring the potential generated by the electrochemical cell. Determining the concentrations in many samples is also very simple—just change the water sample and measure the potential again.

The ease of such operations and the fact that they lend themselves well to automation and computer analysis make electrochemical analyses especially attractive to scientists.

Practice Exercise 19: A galvanic cell is constructed with a copper electrode dipping into a 0.015 M solution of Cu^{2+} ions and a magnesium electrode immersed in a solution of Mg^{2+} ions. The cell potential is measured as 2.79 volts at 25 °C. What is the concentration of magnesium ions?

$$Mg(s) + Cu^{2+}(aq) \longrightarrow Mg^{2+}(aq) + Cu(s)$$

(Hint: Use the Nernst equation to solve for $[Mg^{2+}]$.)

Practice Exercise 20: In the analysis of two other water samples by the procedure described in Example 19.12, cell potentials (E_{cell}) of 0.57 V and 0.82 V were obtained. Calculate the Cu^{2+} ion concentration in each of these samples.

Practice Exercise 21: A galvanic cell was constructed by connecting a nickel electrode that was dipping into 1.20 M $NiSO_4$ solution to a chromium electrode that was dipping into a solution containing Cr^{3+} at an unknown concentration. The potential of the cell was measured to be 0.552 V, with the chromium serving as the anode. The standard cell potential for the system was determined to be 0.487 V. What was the concentration of Cr^{3+} in the solution of unknown concentration?

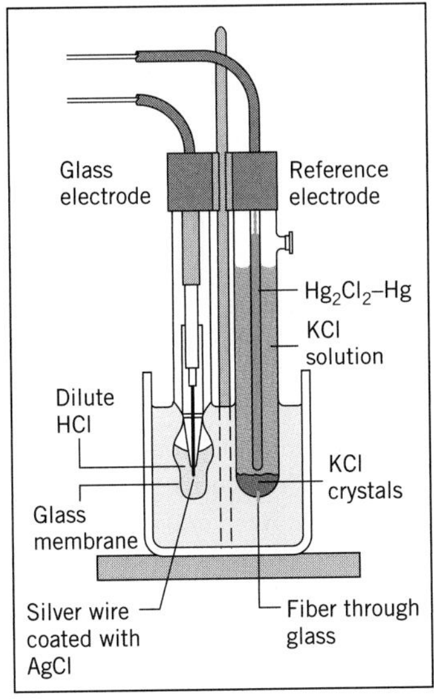

FIG. 19.8 Electrodes used with a pH meter. The electrode on the left is called a *glass electrode*. It contains a silver wire, coated with AgCl, dipping into a dilute solution of HCl. This half-cell has a potential that depends on the difference between the $[H^+]$ inside and outside a thin glass membrane at the bottom of the electrode. On the right is a reference electrode that forms the other half-cell. The galvanic cell formed by the two electrodes produces a potential that is proportional to the pH of the solution into which they are dipped.

Concentration cells consist of two almost identical half-cells

The dependence of cell potential on concentration allows us to construct a galvanic cell from two half-cells composed of the same substances but having different concentrations of the solute species. An example would be a pair of copper electrodes dipping into solutions that have different concentrations of Cu^{2+}, say 0.10 M Cu^{2+} in one and 1.0 M in the other (Figure 19.9). When this cell operates, reactions take place that tend to bring the two Cu^{2+} concentrations toward the same value. Thus, in the half-cell containing 0.10 M Cu^{2+}, copper is oxidized, which adds Cu^{2+} to the more dilute solution. In the other cell, Cu^{2+} is reduced, removing Cu^{2+} from the more concentrated solution. This makes the more dilute half-cell the anode and the more concentrated half-cell the cathode.

$$\underset{\text{anode}}{Cu(s)|Cu^{2+}(0.10\ M)}||\underset{\text{cathode}}{Cu^{2+}(1.0\ M)|Cu(s)}$$

The half-reactions in the spontaneous cell reaction are

$$\begin{aligned} Cu(s) &\longrightarrow Cu^{2+}(0.10\ M) + 2e^- \\ Cu^{2+}(1.0\ M) + 2e^- &\longrightarrow Cu(s) \\ \hline Cu^{2+}(1.0\ M) &\longrightarrow Cu^{2+}(0.10\ M) \end{aligned}$$

The Nernst equation for this cell is

$$E_{cell} = E^\circ_{cell} - \frac{RT}{n\mathscr{F}} \ln \frac{[Cu^{2+}]_{dilute}}{[Cu^{2+}]_{conc}}$$

Because we're dealing with the same substances in the cell, $E^\circ_{cell} = 0$ V (exactly). When the cell operates, $n = 2$, and we'll take $T = 298$ K. Substituting values,

$$\begin{aligned} E_{cell} &= 0\ \text{V} - \frac{8.314\ \text{J}\ \cancel{\text{mol}^{-1}}\ \cancel{\text{K}^{-1}} \times 298\ \cancel{\text{K}}}{2 \times 9.65 \times 10^4\ \text{C}\ \cancel{\text{mol}^{-1}}} \ln\left(\frac{0.10}{1.0}\right) \\ &= 0.030\ \text{V} \end{aligned}$$

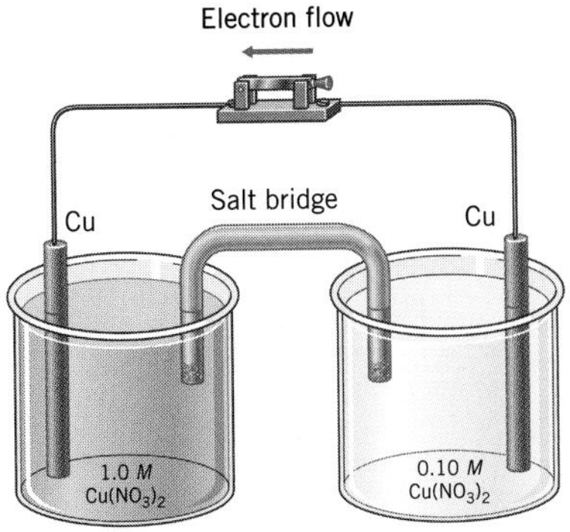

FIG. 19.9 A concentration cell. When the circuit is completed, reactions occur that tend to make the concentrations of Cu^{2+} the same in the two half-cells. Oxidation occurs in the more dilute half-cell, and reduction occurs in the more concentrated one.

In this concentration cell, one solution is ten times more concentrated than the other, yet the potential generated is only 0.03 V. In general, the potential generated by concentration differences are quite small. Yet they are significant in biological systems, where electrical potentials are generated across biological membranes by differences in ion concentrations (e.g., K^+). Membrane potentials are important in processes such as the transmission of nerve impulses.

These small differences in potential also illustrate that if Q, in a system that is not at standard state, is between 0.10 and 10 we can conclude that $E_{cell} \approx E^\circ_{cell}$ and we can generalize the results in Sections 19.2 and 19.3 when predicting spontaneous reactions.

19.6 ELECTROLYSIS USES ELECTRICAL ENERGY TO CAUSE CHEMICAL REACTIONS

In our preceding discussions, we've examined how spontaneous redox reactions can be used to generate electrical energy. We now turn our attention to the opposite process, the use of electrical energy to force nonspontaneous redox reactions to occur.

When electricity is passed through a molten (melted) ionic compound or through a solution of an electrolyte, a chemical reaction occurs that we call **electrolysis.** A typical electrolysis apparatus, called an **electrolysis cell** or **electrolytic cell,** is shown in Figure 19.10. This particular cell contains molten sodium chloride. (A substance undergoing electrolysis must be molten or in solution so its ions can move freely and conduction can occur.) *Inert electrodes*—electrodes that won't react with the molten NaCl—are dipped into the cell and then connected to a source of direct current (DC) electricity.

To perform electrolysis, we must use direct current in which electrons move in only one direction, not in the oscillating, back and forth pattern of alternating current.

The DC source serves as an "electron pump," pulling electrons away from one electrode and pushing them through the external wiring onto the other electrode. The electrode from which electrons are removed becomes positively charged, while the other electrode becomes negatively charged. When electricity starts to flow, chemical changes begin to happen. At the positive electrode, oxidation occurs as electrons are pulled away from negatively charged chloride ions. Because of the nature of the chemical change, therefore, *the positive electrode becomes the anode.* The DC source pumps the electrons through the external electrical circuit to the negative electrode. Here reduction takes place as the electrons are forced onto positively charged sodium ions, so *the negative electrode is the cathode.*

The chemical changes that occur at the electrodes can be described by chemical equations.

$$Na^+(l) + e^- \longrightarrow Na(l) \qquad \text{(cathode)}$$
$$2Cl^-(l) \longrightarrow Cl_2(g) + 2e^- \qquad \text{(anode)}$$

As in a galvanic cell, the overall reaction that takes place in the electrolysis cell is called the *cell reaction.* To obtain it, we add the individual electrode half-reactions together, making sure that the number of electrons gained in one half-reaction equals the number lost in the other.

$$2Na^+(l) + 2e^- \longrightarrow 2Na(l) \qquad \text{(cathode)}$$
$$2Cl^-(l) \longrightarrow Cl_2(g) + 2e^- \qquad \text{(anode)}$$
$$2Na^+(l) + 2Cl^-(l) + \cancel{2e^-} \longrightarrow 2Na(l) + Cl_2(g) + \cancel{2e^-} \qquad \text{(cell reaction)}$$

As you know, table salt is quite stable. It doesn't normally decompose because the reverse reaction, the reaction of sodium and chlorine to form sodium chloride, is highly spontaneous. Therefore, we often write the word *electrolysis* above the arrow in the equation to show that electricity is the driving force for this otherwise nonspontaneous reaction.

$$2Na^+(l) + 2Cl^-(l) \xrightarrow{\text{electrolysis}} 2Na(l) + Cl_2(g)$$

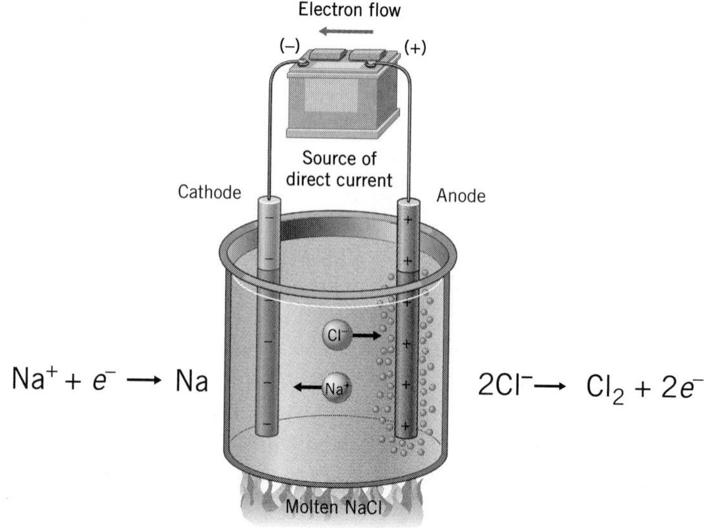

FIG. 19.10 Electrolysis of molten sodium chloride. In this electrolysis cell, the passage of an electric current decomposes molten sodium chloride into metallic sodium and gaseous chlorine. Unless the products are kept apart, they react on contact to re-form NaCl.

Electrolytic and galvanic cells have similarities and differences

In a galvanic cell, the spontaneous cell reaction deposits electrons on the anode and removes them from the cathode. As a result, the anode carries a slight negative charge and the cathode a slight positive charge. In most galvanic cells the reactants must be separated in separate compartments. In an *electrolysis cell*, the situation is reversed. Often the two electrodes are immersed in the same liquid. Also, the oxidation at the anode must be forced to occur, which requires that the anode be positive so it can remove electrons from the reactant at that electrode. On the other hand, the cathode must be made negative so it can force the reactant at the electrode to accept electrons.

Electrolytic Cell	**Galvanic Cell**
Cathode is negative (reduction).	Cathode is positive (reduction).
Anode is positive (oxidation).	Anode is negative (oxidation).
Anode and cathode are often in same compartment.	Anode and cathode are usually in separate compartments.

Even though the charges on the cathode and anode differ between electrolytic cells and galvanic cells, the ions in solution always move in the same direction. In both types of cells, positive ions (cations) move toward the cathode. They are attracted there by the negative charge on the cathode in an electrolysis cell; they diffuse toward the cathode in a galvanic cell to balance the charge of negative ions left behind when ions are reduced. Similarly, negative ions (anions) move toward the anode. They are attracted to the positive anode in an electrolysis cell, and they diffuse toward the anode in a galvanic cell to balance the charge of the positive ions entering the solution.

By agreement among scientists, the names anode and cathode are assigned according to the nature of the reaction taking place at the electrode. If the reaction is oxidation, the electrode is called the anode; if it's reduction, the electrode is called the cathode.

Oxidation and reduction must occur for conduction to continue in an electrolytic cell

The electrical conductivity of a molten salt or a solution of an electrolyte is possible only because of the reactions that take place at the surface of the electrodes. For example, when charged electrodes are dipped into molten NaCl they become surrounded by a layer of ions of the opposite charge. Let's look closely at what happens at one of the electrodes, say the anode (Figure 19.11). Here the positive charge of the

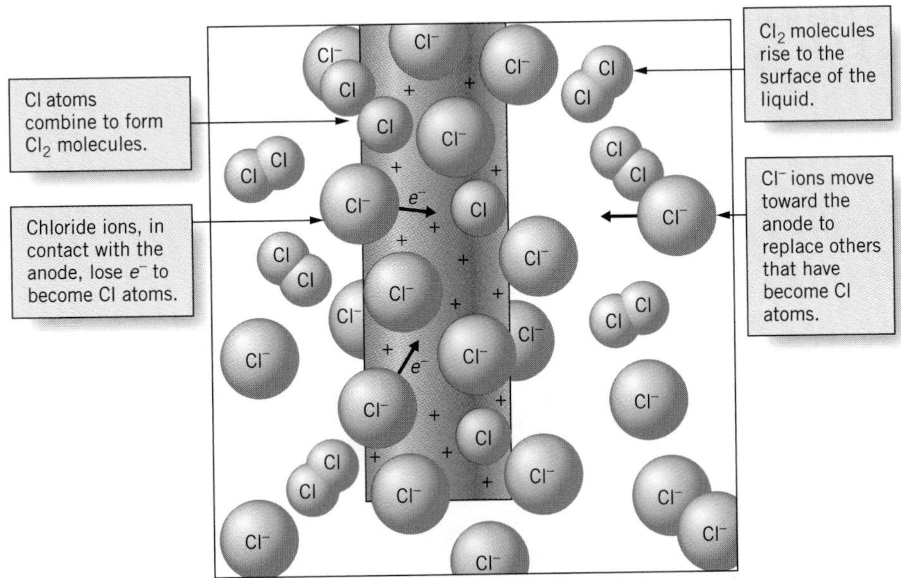

FIG. 19.11 A microscopic view of changes at the anode in the electrolysis of molten NaCl. The positive charge of the electrode attracts a coating of Cl^- ions. At the surface of the electrode, electrons are pulled from the ions, yielding neutral Cl atoms, which combine to form Cl_2 molecules that move away from the electrode and eventually rise to the surface as a gas.

electrode attracts negative Cl^- ions, which form a coating on the electrode's surface. The charge on the anode pulls electrons from the ions, causing them to be oxidized and changing them into neutral Cl atoms that join to become Cl_2 molecules. Because the molecules are neutral, they are not held by the electrode and so move away from the electrode's surface. Their places are quickly taken by negative ions from the surrounding liquid, which tends to leave the surrounding liquid positively charged. Other negative ions from farther away move toward the anode to keep the liquid there electrically neutral. In this way, negative ions gradually migrate toward the anode. By a similar process, positive ions diffuse through the liquid toward the negatively charged cathode where they become reduced.

Cations (positive ions) move toward the cathode and anions (negative ions) migrate toward the anode. This happens in both electrolytic and galvanic cells.

Electrolysis reactions in aqueous solutions can involve oxidation and/or reduction of water

When electrolysis is carried out in an aqueous solution, the electrode reactions are more difficult to predict because at the electrodes there are competing reactions. We have to consider not only the possible oxidation and reduction of the solute, but also the oxidation and reduction of water. For example, consider what happens when electrolysis is performed on a solution of potassium sulfate (Figure 19.12). The products are hydrogen and oxygen. At the cathode, water is reduced, not K^+.

$$2H_2O(l) + 2e^- \longrightarrow H_2(g) + 2OH^-(aq) \qquad \text{(cathode)}$$

At the anode, water is oxidized, not the sulfate ion.

$$2H_2O(l) \longrightarrow O_2(g) + 4H^+(aq) + 4e^- \qquad \text{(anode)}$$

Color changes of an acid–base indicator dissolved in the solution confirm that the solution becomes basic around the cathode, where OH^- is formed, and acidic around the anode, where H^+ is formed (see Figure 19.13). In addition, the gases H_2 and O_2 can be separately collected.

We can understand why these redox reactions happen if we examine the standard reduction potential data from Table 19.1. For example, at the cathode we have the following competing reactions.

$$K^+(aq) + e^- \longrightarrow K(s) \qquad E^\circ_{K^+} = -2.92\text{ V}$$

$$2H_2O(l) + 2e^- \longrightarrow H_2(g) + 2OH^-(aq) \qquad E^\circ_{H_2O} = -0.83\text{ V}$$

Water has a much less negative (and therefore more positive) standard reduction potential than K^+, which means H_2O is much easier to reduce than K^+. As a result, when the

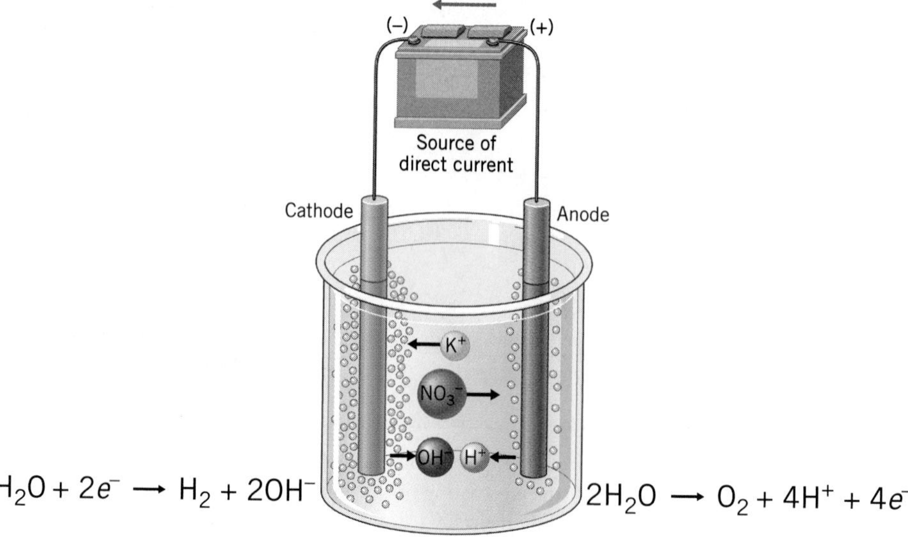

FIG. 19.12 Electrolysis of an aqueous solution of potassium sulfate. The products of the electrolysis are the gases hydrogen and oxygen as shown in the half-reactions for each half-cell.

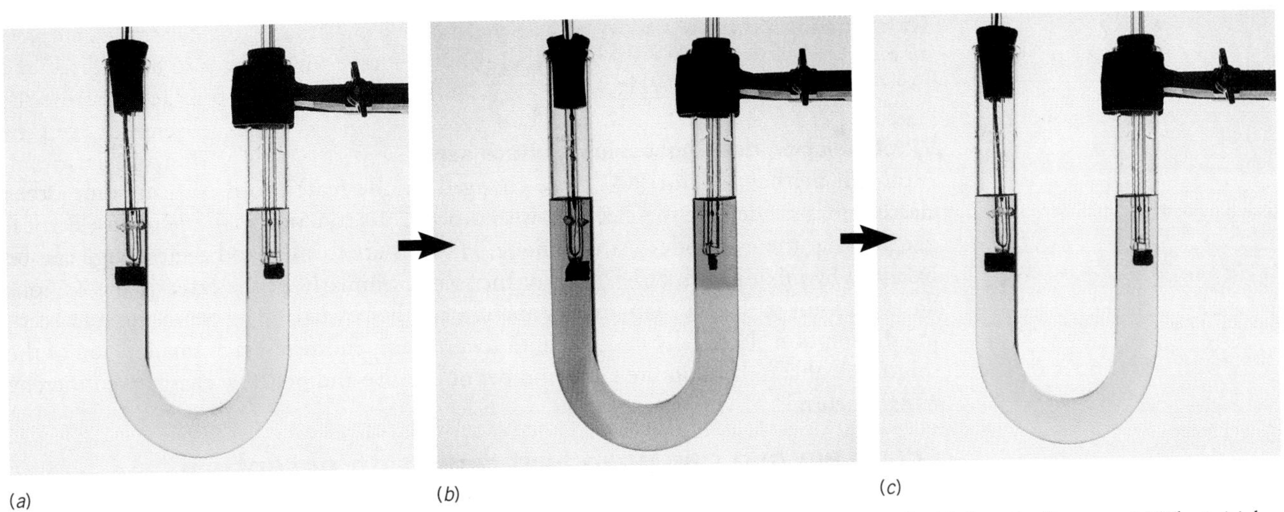

(a) (b) (c)

FIG. 19.13 Electrolysis of an aqueous solution of potassium sulfate in the presence of acid–base indicators. (*a*) The initial yellow color indicates that the solution is neutral (neither acidic nor basic). (*b*) As the electrolysis proceeds, H^+ is produced at the anode (along with O_2) and causes the solution there to become pink. At the cathode, H_2 is evolved and OH^- ions are formed, which turns the solution around that electrode a bluish violet. (*c*) After the electrolysis is stopped and the solution is stirred, the color becomes yellow again as the H^+ and OH^- ions formed by the electrolysis neutralize each other. *(Michael Watson)*

electrolysis is performed the more easily reduced substance is reduced and we observe H_2 being formed at the cathode.

At the anode we have the following possible oxidation half-reactions.

$$2SO_4^{2-}(aq) \longrightarrow S_2O_8^{2-}(aq) + 2e^-$$

$$2H_2O(l) \longrightarrow 4H^+(aq) + O_2(g) + 4e^-$$

In Table 19.1, we find them written in the opposite direction:

$$S_2O_8^{2-}(aq) + 2e^- \longrightarrow 2SO_4^{2-}(aq) \qquad E^\circ_{S_2O_8^{2-}} = +2.01 \text{ V}$$

$$O_2(g) + 4H^+(aq) + 4e^- \longrightarrow 2H_2O(l) \qquad E^\circ_{O_2} = +1.23 \text{ V}$$

The $E°$ values tell us that $S_2O_8^{2-}$ is much more easily reduced than O_2. But if $S_2O_8^{2-}$ is the *more easily reduced*, then the product, SO_4^{2-}, must be *the less easily oxidized*. Stated another way, *the half-reaction with the smaller standard reduction potential is more easily reversed as an oxidation*. As a result, when electrolysis is performed, water is oxidized instead of SO_4^{2-} and we observe O_2 being formed at the anode.

The overall cell reaction for the electrolysis of the K_2SO_4 solution can be obtained as before. Because the number of electrons lost has to equal the number gained, the cathode reaction must occur twice each time the anode reaction occurs once.

$$2 \times [2H_2O(l) + 2e^- \longrightarrow H_2(g) + 2OH^-(aq)] \qquad \text{(cathode)}$$

$$2H_2O(l) \longrightarrow O_2(g) + 4H^+(aq) + 4e^- \qquad \text{(anode)}$$

After adding, we combine the coefficients for water and cancel the electrons from both sides to obtain the cell reaction.

$$6H_2O(l) \longrightarrow 2H_2(g) + O_2(g) + 4H^+(aq) + 4OH^-(aq)$$

Notice that hydrogen ions and hydroxide ions are produced in equal numbers. In Figure 19.13, we see that when the solution is stirred, they combine to form water.

$$6H_2O(l) \longrightarrow 2H_2(g) + O_2(g) + \underbrace{4H^+(aq) + 4OH^-(aq)}_{\downarrow \; H_2O}$$

The net change, then, is

$$2H_2O \xrightarrow{\text{electrolysis}} 2H_2(g) + O_2(g)$$

What function does potassium sulfate serve?
Although neither K^+ nor SO_4^{2-} are changed by the reaction, K_2SO_4 or some other electrolyte is needed for the electrolysis to proceed. Its function is to maintain electrical neutrality at the electrodes. At the anode, H^+ ions are formed and their charge can be balanced by mixing with SO_4^{2-} ions of the solute. Similarly, at the cathode the K^+ ions are able to mix with OH^- ions as they are formed, thereby balancing the charge and keeping the solution electrically neutral. In this way, at any moment, each small region of the solution is able to contain the same number of positive and negative charges and thereby remain neutral.

Often we can use standard reduction potentials to predict electrolysis products

Predicting electrolysis products

Suppose we wished to know what products are expected in the electrolysis of an aqueous solution of copper(II) bromide, $CuBr_2$. Let's examine the possible electrode half-reactions and their respective standard reduction potentials.

At the cathode, possible reactions are the reduction of copper ion and the reduction of water. From Table 19.1,

$$Cu^{2+}(aq) + 2e^- \longrightarrow Cu(s) \qquad E^\circ_{Cu^{2+}} = +0.34\text{ V}$$

$$2H_2O(l) + 2e^- \longrightarrow H_2(g) + 2OH^-(aq) \qquad E^\circ_{H_2O} = -0.83\text{ V}$$

The much more positive standard reduction potential for Cu^{2+} tells us to anticipate that Cu^{2+} will be reduced at the cathode.

At the anode, possible reactions are the oxidation of Br^- and the oxidation of water. The half-reactions are

$$2Br^-(aq) \longrightarrow Br_2(aq) + 2e^-$$

$$2H_2O(l) \longrightarrow O_2(g) + 4H^+(aq) + 4e^-$$

In Table 19.1 they are written as reductions with the following $E°$ values.

$$Br_2(aq) + 2e^- \longrightarrow 2Br^-(aq) \qquad E^\circ_{Br_2} = +1.07\text{ V}$$

$$O_2(g) + 4H^+(aq) + 4e^- \longrightarrow 2H_2O(l) \qquad E^\circ_{O_2} = +1.23\text{ V}$$

The data tell us that O_2 is more easily reduced than Br_2, which means that Br^- is more easily oxidized than H_2O. Therefore, we expect that Br^- will be oxidized at the anode.

In fact, our predictions are confirmed when we perform the electrolysis. The cathode, anode, and net cell reactions are

$$Cu^{2+}(aq) + 2e^- \longrightarrow Cu(s) \qquad \text{(cathode)}$$

$$2Br^-(aq) \longrightarrow Br_2(aq) + 2e^- \qquad \text{(anode)}$$

$$Cu^{2+}(aq) + 2Br^-(aq) \xrightarrow{\text{electrolysis}} Cu(s) + Br_2(aq) \qquad \text{(net reaction)}$$

EXAMPLE 19.13
Predicting the Products in an Electrolysis Reaction

Electrolysis is planned for an aqueous solution that contains a mixture of 0.50 M $ZnSO_4$ and 0.50 M $NiSO_4$. On the basis of standard reduction potentials, what products are expected to be observed at the electrodes? What is the expected net cell reaction?

ANALYSIS: We need to consider the competing reactions at the cathode and the anode. At the cathode, the half-reaction with the most positive (or least negative) standard reduction potential will be the one expected to occur. At the anode, the half-reaction with the *least* *positive* standard reduction potential is the one most easily reversed, and should occur as an oxidation.

SOLUTION: At the cathode, the competing reduction reactions involve the two cations and water. The reactions and their standard reduction potentials are

$$Ni^{2+}(aq) + 2e^- \rightleftharpoons Ni(s) \qquad E° = -0.25\text{ V}$$
$$Zn^{2+}(aq) + 2e^- \rightleftharpoons Zn(s) \qquad E° = -0.76\text{ V}$$
$$2H_2O + 2e^- \rightleftharpoons H_2(g) + 2OH^-(aq) \qquad E° = -0.83\text{ V}$$

The least negative standard reduction potential is that of Ni^{2+}, so we expect that ion to be reduced at the cathode and solid nickel to be formed.

At the anode, the competing oxidation reactions are for water and SO_4^{2-} ion. In Table 19.1, substances oxidized are found on the right side of the half-reactions. The two half-reactions having these as products are

$$S_2O_8^{2-}(aq) + 2e^- \rightleftharpoons 2SO_4^{2-}(aq) \qquad E° = +2.01\text{ V}$$
$$O_2(g) + 4H^+(aq) + 4e^- \rightleftharpoons 2H_2O \qquad E° = +1.23\text{ V}$$

The half-reaction with the least positive $E°$ (the second one here) is most easily reversed as an oxidation, so we expect the oxidation half-reaction to be

$$2H_2O \rightleftharpoons O_2(g) + 4H^+(aq) + 4e^-$$

At the anode, we expect O_2 to be formed.

The predicted net cell reaction is obtained by combining the two expected electrode half-reactions, making the electron loss equal to the electron gain.

$$\begin{array}{rll} 2H_2O & \longrightarrow O_2(g) + 4H^+(aq) + 4e^- & \text{(anode)} \\ 2 \times [Ni^{2+}(aq) + 2e^- & \longrightarrow Ni(s)] & \text{(cathode)} \\ \hline 2H_2O + 2Ni^{2+}(aq) & \longrightarrow O_2(g) + 4H^+(aq) + 2Ni(s) & \text{(net cell reaction)} \end{array}$$

ARE THE ANSWERS REASONABLE? We can check the locations of the half-reactions in Table 19.1 to confirm our conclusions. For the reduction step, the higher up in the table a half-reaction is, the greater its tendency to occur as reduction. Among the competing half-reactions at the cathode, the one for Ni^{2+} is highest, so we expect that Ni^{2+} is the easiest to reduce and Ni(*s*) should be formed at the cathode.

For the oxidation step, the lower down in the table a half-reaction is, the easier it is to reverse and cause to occur as oxidation. On this basis, the oxidation of water is easier than the oxidation of SO_4^{2-}, so we expect H_2O to be oxidized and O_2 to be formed at the anode.

Of course, we could also test our prediction by carrying out the electrolysis experimentally.

Practice Exercise 22: In the electrolysis of an aqueous solution containing Fe^{2+} and I^-, what product do we expect at the anode? (Hint: Write the three oxidation half-reactions possible at the anode.)

Practice Exercise 23: In the electrolysis of an aqueous solution containing both Cd^{2+} and Sn^{2+}, what product do we expect at the cathode? Under what conditions can the other ion be reduced?

Using standard reduction potentials to predict electrolysis doesn't always work

Although we can use standard reduction potentials to predict electrolysis reactions most of the time, there are occasions when standard reduction potentials do not successfully predict electrolysis products. Sometimes concentrations, far from standard conditions, will change the sign of the cell potential. The formation of complex ions can also interfere and produce unexpected results. And sometimes the electrodes themselves are the culprits. For example, in the electrolysis of aqueous NaCl using inert platinum electrodes, we find experimentally that Cl_2 is formed at the anode. Is this what we would have expected? Let's examine the standard reduction potentials of O_2 and Cl_2 to find out.

$$Cl_2(g) + 2e^- \rightleftharpoons 2Cl^-(aq) \qquad E^\circ = +1.36 \text{ V}$$

$$O_2(g) + 4H^+(aq) + 4e^- \rightleftharpoons 2H_2O \qquad E^\circ = +1.23 \text{ V}$$

Michael Faraday (1791–1867), a British scientist and both a chemist and a physicist, made key discoveries leading to electric motors, generators, and transformers. *(Courtesy Edgar Fahs Smith Collection, University of Pennsylvania.)*

Because of its less positive standard reduction potential, we would expect the oxygen half-reaction to be the easier to reverse (with water being oxidized to O_2). Thus, standard reduction potentials predict that O_2 should be formed, but experiment shows that Cl_2 is produced. The nature of the electrode surface and how it interacts with oxygen is part of the answer. Further explanation for why this happens is beyond the scope of this text, but the unexpected result does teach us that we must be cautious in predicting products in electrolysis reactions solely on the basis of standard reduction potentials.

19.7 Stoichiometry of Electrochemical Reactions Involves Electric Current and Time

In about 1833, Michael Faraday discovered that the amount of chemical change that occurs during electrolysis is directly proportional to the amount of electrical charge that is passed through an electrolysis cell. For example, the reduction of copper ion at a cathode is given by the equation

$$Cu^{2+}(aq) + 2e^- \longrightarrow Cu(s)$$

The equation tells us that to deposit one mole of metallic copper requires two moles of electrons. Thus, the half-reaction for an oxidation or reduction relates the amount of chemical substance consumed or produced to the amount of electrons that the electric current must supply. To use this information, however, we must be able to relate it to electrical measurements that can be made in the laboratory.

The SI unit of electric current is the **ampere (A)** and the SI unit of charge is the **coulomb (C).** A coulomb is the amount of charge that passes by a given point in a wire when an electric current of one ampere flows for one second. This means that coulombs are the product of amperes of current multiplied by seconds. Thus

$$1 \text{ coulomb} = 1 \text{ ampere} \times 1 \text{ second}$$

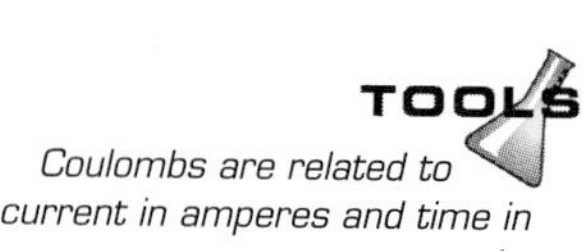

Coulombs are related to current in amperes and time in seconds.

$$1 \text{ C} = 1 \text{ A s}$$

For example, if a current of 4 A flows through a wire for 10 s, 40 C pass by a given point in the wire.

$$(4 \text{ A}) \times (10 \text{ s}) = 40 \text{ A s}$$
$$= 40 \text{ C}$$

As we noted earlier, it has been determined that 1 mol of electrons carries a charge of 9.65×10^4 C, which in honor of Michael Faraday is often called the Faraday constant, $\mathcal{F}$.

$$1 \text{ mol } e^- \Leftrightarrow 9.65 \times 10^4 \text{ C} \qquad \left(\begin{array}{c}\text{to three}\\ \text{significant figures}\end{array}\right)$$

$$1\,\mathcal{F} = 9.65 \times 10^4 \text{ C/mol } e^-$$

Now we have a way to relate laboratory measurements to the amount of chemical change that occurs during an electrolysis. Measuring the current in amperes and the time in seconds allows us to calculate the charge sent through the system in coulombs. From this we can get the amount of electrons (in moles), which we can then use to calculate the amount of chemical change produced. The following examples demonstrate the principles involved for electrolysis, but similar calculations also apply to reactions in galvanic cells.

EXAMPLE 19.14
Calculations Related to Electrolysis

How many grams of copper are deposited on the cathode of an electrolytic cell if an electric current of 2.00 A is run through a solution of $CuSO_4$ for a period of 20.0 min?

ANALYSIS: The balanced half-reaction serves as our tool for relating chemical change to amounts of electricity. The ion being reduced is Cu^{2+}, so the half-reaction is

$$Cu^{2+} + 2e^- \longrightarrow Cu$$

Therefore,

$$1 \text{ mol Cu} \Leftrightarrow 2 \text{ mol } e^-$$

The product of current (in amperes) and time (in seconds) is the tool that will give us charge (in coulombs). We can relate this to the number of moles of electrons by the Faraday constant, another of our tools. Then, from the number of moles of electrons we calculate moles of copper, from which we calculate the mass of copper by using the atomic mass. Here's a diagram of the path to the solution.

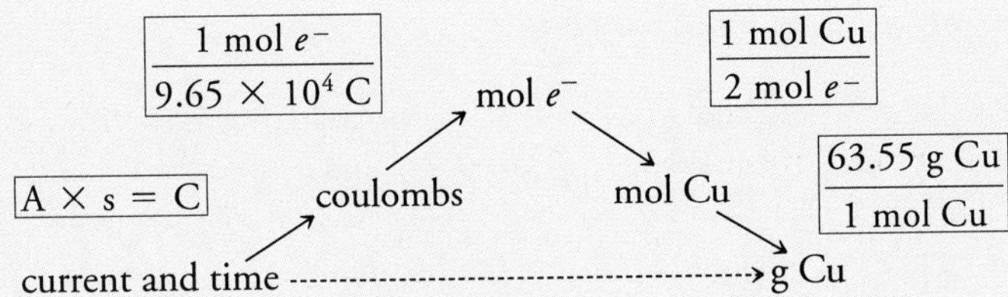

SOLUTION: First we convert minutes to seconds; 20.0 min = 1.20×10^3 s. Then we multiply the current by the time to obtain the number of coulombs (1 A s = 1 C).

$$\begin{aligned}(1.20 \times 10^3 \text{ s}) \times (2.00 \text{ A}) &= 2.40 \times 10^3 \text{ A s} \\ &= 2.40 \times 10^3 \text{ C}\end{aligned}$$

Because $1 \text{ mol } e^- \Leftrightarrow 9.65 \times 10^4$ C,

$$2.40 \times 10^3 \cancel{\text{C}} \times \frac{1 \text{ mol } e^-}{9.65 \times 10^4 \cancel{\text{C}}} = 0.02487 \text{ mol } e^-$$

In stepwise calculations we always carry one or more extra significant figures until the final result to minimize rounding errors.

Next, we use the relationship between mol e^- and mol Cu from the balanced half-reaction along with the atomic mass of copper.

$$0.02487 \cancel{\text{mol } e^-} \times \left(\frac{1 \cancel{\text{mol Cu}}}{2 \cancel{\text{mol } e^-}}\right) \times \left(\frac{63.55 \text{ g Cu}}{1 \cancel{\text{mol Cu}}}\right) = 0.7903 \text{ g Cu}$$

With proper rounding, the electrolysis will deposit 0.790 g of copper on the cathode.

We could have combined all the steps in a single calculation by stringing together the various conversion factors and using the factor-label method to cancel units.

$$2.00 \cancel{\text{A}} \times 20.0 \cancel{\text{min}} \times \frac{60 \cancel{\text{s}}}{1 \cancel{\text{min}}} \times \frac{1 \cancel{\text{mol } e^-}}{9.65 \times 10^4 \cancel{\text{A}} \cancel{\text{s}}} \times \frac{1 \cancel{\text{mol Cu}}}{2 \cancel{\text{mol } e^-}} \times \frac{63.55 \text{ g Cu}}{1 \cancel{\text{mol Cu}}} = 0.790 \text{ g Cu}$$

IS THE ANSWER REASONABLE? As before, we round all numbers to one significant figure to estimate the answer.

$$\text{g Cu} = 2.00 \cancel{\text{A}} \times 20.0 \cancel{\text{min}} \times \frac{60 \cancel{\text{s}}}{1 \cancel{\text{min}}} \times \frac{1 \cancel{\text{mol } e^-}}{10 \times 10^4 \cancel{\text{A}} \cancel{\text{s}}} \times \frac{1 \cancel{\text{mol Cu}}}{2 \cancel{\text{mol } e^-}} \times \frac{60 \text{ g Cu}}{1 \cancel{\text{mol Cu}}}$$

$$= \frac{40 \times 3600}{200000} = \frac{40 \times 0.36}{20} = 0.72 \text{ g Cu}$$

This is very close to our answer and we are even more confident that the calculation was correct when we check the cancellation of units.

EXAMPLE 19.15
Calculations Related to Electrolysis

Electrolysis provides a useful way to deposit a thin metallic coating on an electrically conducting surface. This technique is called electroplating. How much time would it take in minutes to deposit 0.500 g of metallic nickel on a metal object using a current of 3.00 A? The nickel is reduced from the +2 oxidation state.

ANALYSIS: We need an equation for the reduction. Because the nickel is reduced to the free metal from the +2 state, we can write

$$Ni^{2+} + 2e^- \longrightarrow Ni(s)$$

This gives the relationship

$$1 \text{ mol Ni} \Leftrightarrow 2 \text{ mol } e^-$$

We wish to deposit 0.500 g of Ni, which we can convert to moles. Then we can calculate the number of moles of electrons required, which in turn is used with the tool for the Faraday constant to determine the number of coulombs required. Another of our tools tells us that this is the product of amperes and seconds, so we can calculate the time needed to deposit the metal. The calculation can be diagrammed as follows.

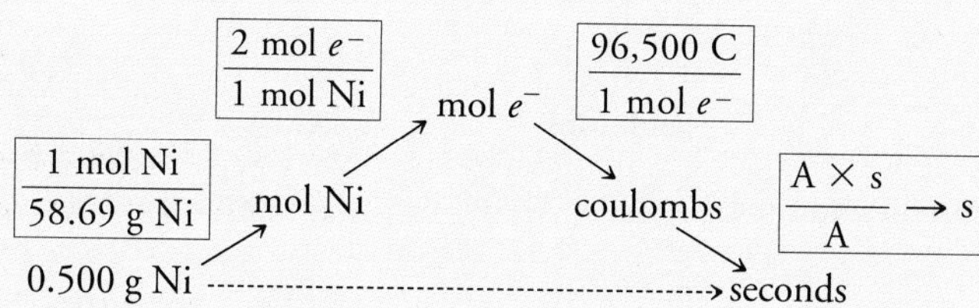

SOLUTION: First, we calculate the number of moles of electrons required (keeping at least one extra significant figure).

$$0.500 \text{ g Ni} \times \left(\frac{1 \text{ mol Ni}}{58.69 \text{ g Ni}}\right) \times \left(\frac{2 \text{ mol } e^-}{1 \text{ mol Ni}}\right) = 0.01704 \text{ mol } e^-$$

Then we calculate the number of coulombs needed.

$$0.01704 \text{ mol } e^- \times \left(\frac{9.65 \times 10^4 \text{ C}}{1 \text{ mol } e^-}\right) = 1.644 \times 10^3 \text{ C} = 1.644 \times 10^3 \text{ A s}$$

Once we've calculated the number of coulombs required, we can calculate the time required if we know the current, or the current needed to perform the electrolysis in a given time.

This tells us that the product of current multiplied by time equals 1.644×10^3 A s. The current is 3.00 A. Dividing 1.644×10^3 A s by 3.00 A gives the time required in seconds, which we then convert to minutes.

$$\left(\frac{1.644 \times 10^3 \text{ A s}}{3.00 \text{ A}}\right) \times \left(\frac{1 \text{ min}}{60 \text{ s}}\right) = 9.133 \text{ min}$$

Properly rounded, this becomes 9.13 min. We could also have combined the calculations in a single string of conversion factors.

$$0.500 \text{ g Ni} \times \frac{1 \text{ mol Ni}}{58.69 \text{ g Ni}} \times \frac{2 \text{ mol } e^-}{1 \text{ mol Ni}} \times \frac{9.65 \times 10^4 \text{ A s}}{1 \text{ mol } e^-} \times \frac{1}{3.00 \text{ A}} \times \frac{1 \text{ min}}{60 \text{ s}} = 9.13 \text{ min}$$

IS THE ANSWER REASONABLE? As in the preceding example, let's round all numbers to one digit:

$$0.500 \text{ g Ni} \times \frac{1 \text{ mol Ni}}{60 \text{ g Ni}} \times \frac{2 \text{ mol } e^-}{1 \text{ mol Ni}} \times \frac{10 \times 10^4 \text{ A s}}{1 \text{ mol } e^-} \times \frac{1}{3.00 \text{ A}} \times \frac{1 \text{ min}}{60 \text{ s}} = \frac{100000 \text{ min}}{3600 \times 3} = \frac{100000 \text{ min}}{10000} = 10 \text{ min}$$

This result, along with the proper cancellation of units, indicates our calculations and setup were correct.

Practice Exercise 24: How many moles of hydroxide ion will be produced at the cathode during the electrolysis of water with a current of 4.00 A for a period of 200 s? The cathode reaction is

$$2e^- + 2H_2O \longrightarrow H_2 + 2OH^-$$

(Hint: The stepwise diagram in Example 19.15 will be helpful.)

Practice Exercise 25: How many minutes will it take for a current of 10.0 A to deposit 3.00 g of gold from a solution of $AuCl_3$?

Practice Exercise 26: What current must be supplied to deposit 3.00 g of gold from a solution of $AuCl_3$ in 20.0 min?

Practice Exercise 27: Suppose the solutions in the galvanic cell depicted in Figure 19.2 (page 771) have a volume of 125 mL and suppose the cell is operated for a period of 1.25 hr with a constant current of 0.100 A flowing through the external circuit. By how much will the concentration of the copper ion increase during that time?

19.8 PRACTICAL APPLICATIONS OF ELECTROCHEMISTRY

Electrochemistry has many applications both in science and in our everyday lives. In this limited space, we can only touch on some of the more common and important examples.

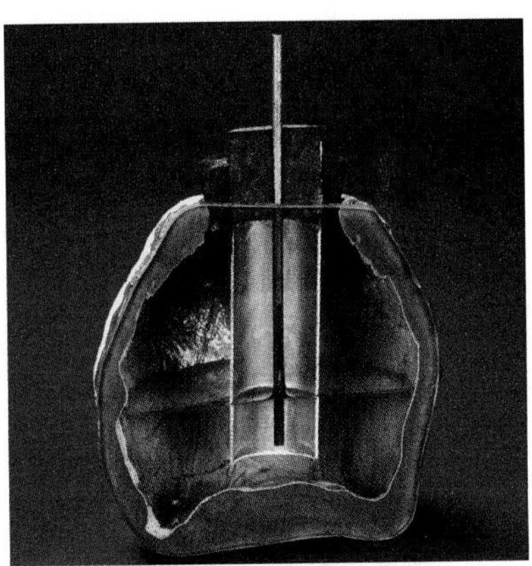

The oldest known electric battery in existence, discovered in 1938 in Baghdad, Iraq, consists of a copper tube surrounding an iron rod. If filled with an acidic liquid such as vinegar, the cell could produce a small electric current. *(Smith College Museum of Ancient Inventions.)*

Batteries are practical examples of galvanic cells

One of the most familiar uses of galvanic cells, popularly called *batteries*, is the generation of portable electrical energy.[8] These devices are classified as being either **primary cells** (cells not designed to be recharged; they are discarded after their energy is depleted) or **secondary cells** (cells designed for repeated use; they are able to be recharged).

The lead storage battery is used in most automobiles

The common **lead storage battery** used to start an automobile is composed of a number of secondary cells, each having a potential of about 2 V, that are connected in series so that their voltages are additive. Most automobile batteries contain six such cells and give about 12 V, but 6, 24, and 32 V batteries are also available.

A typical lead storage battery is illustrated in Figure 19.14. The anode of each cell is composed of a set of lead plates, the cathode consists of another set of plates that hold a coating of PbO_2, and the electrolyte is sulfuric acid. When the battery is discharging the electrode reactions are

$$PbO_2(s) + 3H^+(aq) + HSO_4^-(aq) + 2e^- \longrightarrow PbSO_4(s) + 2H_2O \qquad \text{(cathode)}$$

$$Pb(s) + HSO_4^-(aq) \longrightarrow PbSO_4(s) + H^+(aq) + 2e^- \qquad \text{(anode)}$$

The net reaction taking place in each cell is

$$PbO_2(s) + Pb(s) + \underbrace{2H^+(aq) + 2HSO_4^-(aq)}_{2H_2SO_4} \longrightarrow 2PbSO_4(s) + 2H_2O$$

As the cell discharges, the sulfuric acid concentration decreases, which causes the density of the electrolyte to drop. The state of charge of the battery can be monitored with a **hydrometer,** which consists of a rubber bulb that is used to draw the battery fluid into

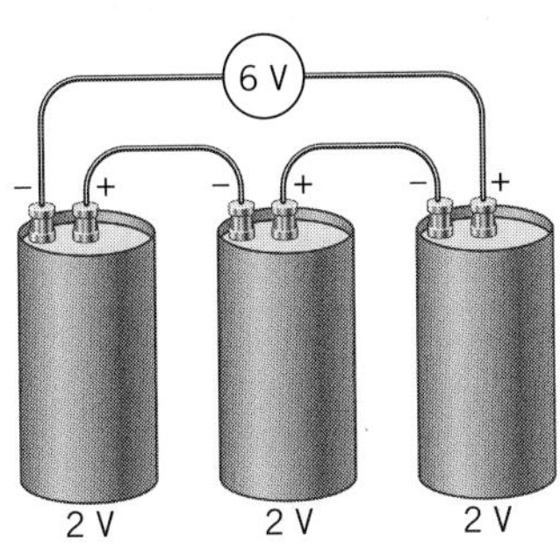

If three 2 volt cells are connected in series, their voltages are additive to provide a total of 6 volts. In today's autos, 12 volt batteries containing six cells are the norm.

[8] Strictly speaking, a cell is a single electrochemical unit consisting of a cathode and an anode. A battery is a collection of cells connected in series.

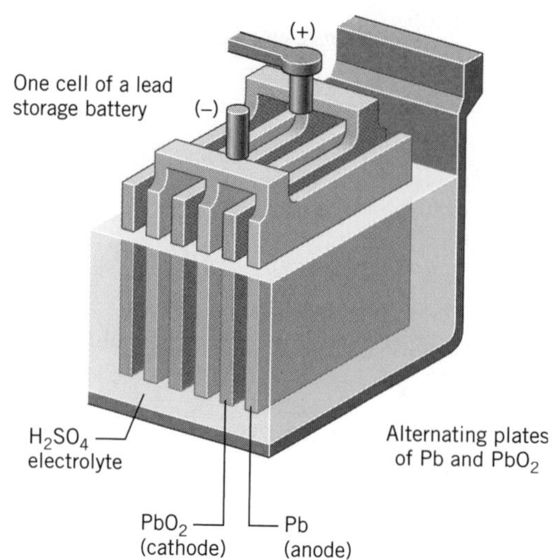

FIG. 19.14 Lead storage battery. A 12 volt lead storage battery, such as those used in most automobiles, consists of six cells like the one shown here. Notice that the anode and cathode each consist of several plates connected together. This allows the cell to produce the large currents necessary to start a car.

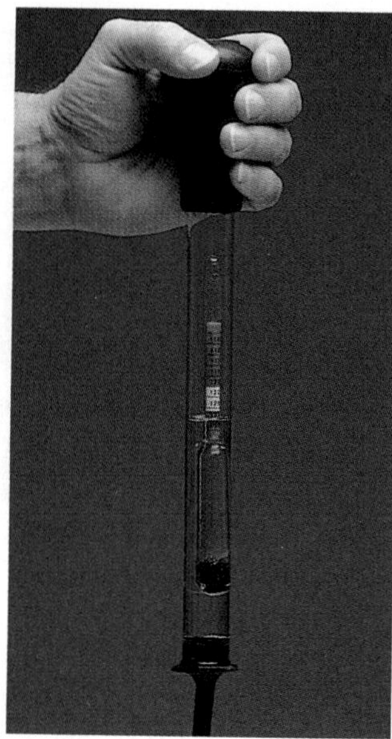

FIG. 19.15 A battery hydrometer. Battery acid is drawn into the glass tube. The depth to which the float sinks is inversely proportional to the concentration of the acid and, therefore, to the state of charge of the battery. *(OPC, Inc.)*

a glass tube containing a float (see Figure 19.15). The depth to which the float sinks is inversely proportional to the density of the liquid—the deeper the float sinks, the lower is the density of the acid and the weaker is the charge on the battery. The narrow neck of the float is usually marked to indicate the state of charge of the battery.

The principal advantage of the lead storage battery is that the cell reactions that occur spontaneously during discharge can be reversed by the application of a voltage from an external source. In other words, the battery can be recharged by electrolysis. The reaction for battery recharge is

$$2PbSO_4(s) + 2H_2O \xrightarrow{\text{electrolysis}} PbO_2(s) + Pb(s) + 2H^+(aq) + 2HSO_4^-(aq)$$

Improper recharging of lead–acid batteries can produce potentially explosive H_2 gas. Most modern lead storage batteries use a lead–calcium alloy as the anode. That reduces the need to have the individual cells vented, and the battery can be sealed, thereby preventing spillage of the corrosive electrolyte.

The zinc–manganese dioxide cell is our familiar dry cell

The ordinary, relatively inexpensive 1.5 V dry cell is the **zinc–manganese dioxide cell,** or **Leclanché cell** (named after its inventor George Leclanché). It is a primary cell used to power flashlights, remote TV, VCR, and DVD controllers, toys, and the like, but it is not really dry (see Figure 19.16). Its outer shell is made of zinc, which serves as the anode. The cathode—the positive terminal of the battery—consists of a carbon (graphite) rod surrounded by a moist paste of graphite powder, manganese dioxide, and ammonium chloride.

The anode reaction is simply the oxidation of zinc.

$$Zn(s) \longrightarrow Zn^{2+}(aq) + 2e^- \qquad \text{(anode)}$$

The cathode reaction is complex, and a mixture of products is formed. One of the major reactions is

$$2MnO_2(s) + 2NH_4^+(aq) + 2e^- \longrightarrow Mn_2O_3(s) + 2NH_3(aq) + H_2O \qquad \text{(cathode)}$$

The ammonia that forms at the cathode reacts with some of the Zn^{2+} produced from the anode to form a complex ion, $Zn(NH_3)_4^{2+}$. Because of the complexity of the cathode half-cell reaction, no simple overall cell reaction can be written.

The alkaline battery is also a primary cell.

A more popular version of the Leclanché battery uses a basic, or *alkaline* electrolyte and is called an **alkaline battery** or **alkaline dry cell.** It too uses Zn and MnO_2 as reactants, but under basic conditions (Figure 19.17). The half-cell reactions are

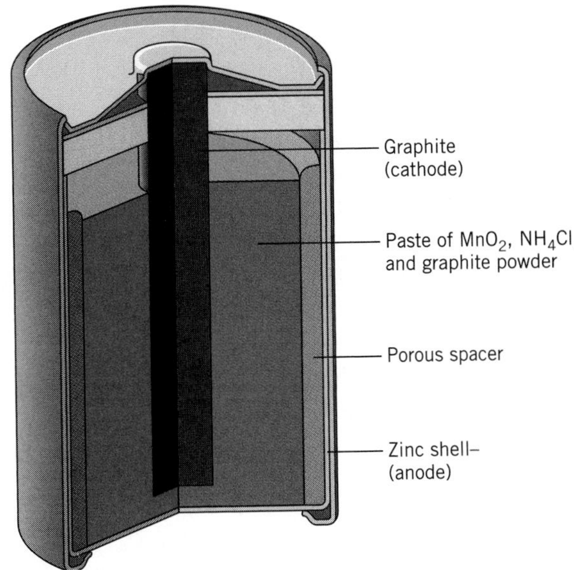

FIG. 19.16 A cutaway view of a zinc–carbon dry cell (Leclanché cell).

$$
\begin{array}{rcll}
Zn(s) + 2OH^-(aq) & \longrightarrow & ZnO(s) + H_2O + 2e^- & \text{(anode)} \\
2MnO_2(s) + H_2O + 2e^- & \longrightarrow & Mn_2O_3(s) + 2OH^-(aq) & \text{(cathode)} \\
\hline
Zn(s) + 2MnO_2(s) & \longrightarrow & ZnO(s) + Mn_2O_3(s) & \text{(net cell reaction)}
\end{array}
$$

and the voltage is about 1.54 V. It has a longer shelf-life and is able to deliver higher currents for longer periods than the less expensive zinc–carbon cell.

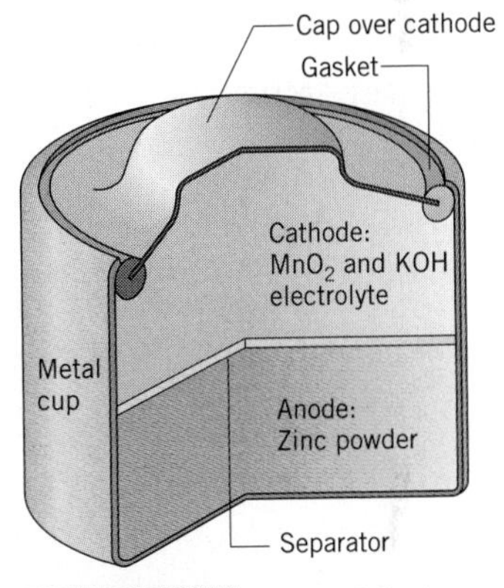

FIG. 19.17 A simplified diagram of an alkaline zinc–carbon dry cell.

The nickel–cadmium storage cell is rechargeable

The **nickel–cadmium storage cell,** or **nicad battery,** is a secondary cell that produces a potential of about 1.4 V, which is slightly lower than that of the zinc–carbon cell. The electrode reactions in the cell during discharge are

$$
\begin{array}{rcll}
Cd(s) + 2OH^-(aq) & \longrightarrow & Cd(OH)_2(s) + 2e^- & \text{(anode)} \\
NiO_2(s) + 2H_2O + 2e^- & \longrightarrow & Ni(OH)_2(s) + 2OH^-(aq) & \text{(cathode)} \\
\hline
Cd(s) + NiO_2(s) + 2H_2O & \longrightarrow & Ni(OH)_2(s) + Cd(OH)_2(s) & \text{(net cell reaction)}
\end{array}
$$

The nickel–cadmium battery can be recharged, in which case the anode and cathode reactions above are reversed to remake the reactants. The battery also can be sealed to prevent leakage, which is particularly important in electronic devices.

Nickel–cadmium batteries work especially well in applications such as portable power tools, CD players, and even electric cars. They have a high **energy density** (available energy per unit volume), they are able to release the energy quickly, and they can be rapidly recharged.

If a battery can supply large amounts of energy and is contained in a small package, it will have a desirably high energy density.

Nickel–metal hydride batteries store more energy than nicad batteries

These secondary cells, which are often referred to as Ni–MH batteries, have been used extensively in recent years to power devices such as cell phones, camcorders, and even electric vehicles. They are similar in many ways to the alkaline nickel–cadmium cells discussed above, except for the anode reactant, which is hydrogen. This is possible because certain metal alloys [e.g., $LaNi_5$ (an alloy of lanthanum and nickel) and Mg_2Ni (an alloy of magnesium and nickel)] have the ability to absorb and hold substantial amounts of hydrogen, and that the hydrogen can be made to participate in reversible electrochemical reactions. The term *metal hydride* is used to describe the hydrogen-holding alloy.

There are compounds of hydrogen with metals such as sodium that actually contain the *hydride ion*, H^-. The metal "hydrides" described here are not of that type.

The cathode in the Ni–MH cell is NiO(OH), a compound of nickel in the +3 oxidation state, and the electrolyte is a solution of KOH. Using the symbol MH to stand for the metal hydride, the reactions in the cell during discharge are

$$
\begin{array}{rcll}
MH(s) + OH^-(aq) & \longrightarrow & M(s) + H_2O + e^- & \text{(anode)} \\
NiO(OH)(s) + H_2O + e^- & \longrightarrow & Ni(OH)_2(s) + OH^-(aq) & \text{(cathode)}
\end{array}
$$

The overall cell reaction is

$$MH(s) + NiO(OH)(s) \longrightarrow Ni(OH)_2(s) + M(s) \qquad E^\circ_{cell} = 1.35\ V$$

When the cell is recharged, the reactions are reversed.

The principal advantage of the Ni–MH cell over the Ni–Cd cell is that it can store about 50% more energy in the same volume. This means, for example, that comparing cells of the same size and weight, a nickel–metal hydride cell can power a laptop computer or a cell phone about 50% longer than a nickel–cadmium cell.

Lithium batteries combine high energy with low weight

Lithium has the most negative standard reduction potential of any metal (Table 19.1), so it has a lot of appeal as an anode material. Furthermore, lithium is a very lightweight metal, so a cell employing lithium as a reactant would also be lightweight. The major problem with using lithium in a galvanic cell is that the metal reacts vigorously with water to produce hydrogen gas and lithium hydroxide.

$$2Li(s) + 2H_2O \longrightarrow 2LiOH(aq) + H_2(g)$$

Therefore, to employ lithium in a galvanic cell scientists had to find a way to avoid aqueous electrolytes. This became possible in the 1970s with the introduction of organic solvents and solvent mixtures that were able dissolve certain lithium salts and thereby serve as electrolytes.

Today's lithium batteries fall into two categories, primary batteries that can be used once and then discarded when fully discharged, and rechargeable cells.

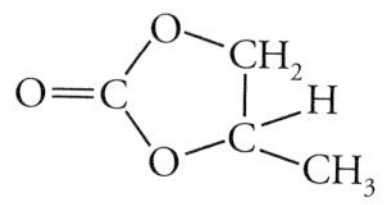

One of the most common nonrechargeable cells is the **lithium–manganese dioxide battery,** which accounts for about 80% of all primary lithium cells. This cell uses a solid lithium anode and a cathode made of heat-treated MnO_2. The electrolyte is a mixture of propylene carbonate and dimethoxyethane (see structures in the margin) containing a dissolved lithium salt such as $LiClO_4$. The cell reactions are as follows (superscripts are the oxidation numbers of the manganese):

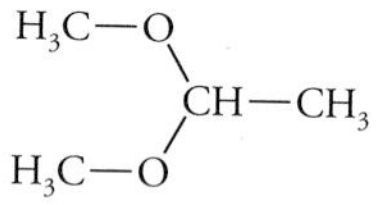

O=C(–O–CH$_2$–)(–O–C(H)(CH$_3$)–)

propylene carbonate

$H_3C{-}O{-}CH(CH_3){-}O{-}CH_3$...

H_3C—O, CH—CH_3, H_3C—O

dimethoxyethane

$$Li \longrightarrow Li^+ + e^- \qquad \text{(anode)}$$

$$Mn^{IV}O_2 + Li^+ + e^- \longrightarrow Mn^{III}O_2(Li^+) \qquad \text{(cathode)}$$

$$Li + Mn^{IV}O_2 \longrightarrow Mn^{III}O_2(Li^+) \qquad \text{(net cell reaction)}$$

This cell produces a voltage of about 3.4 V, which is more than twice that of an alkaline dry cell, and because of the light weight of the lithium, it produces more than twice as much energy for a given weight. These cells are used in applications that require a higher current drain or energy pulses (e.g., photoflash).

Lithium ion cells are rechargeable

Rechargeable lithium batteries found in many cell phones, digital cameras, and laptop computers do not contain metallic lithium. They are called **lithium ion cells** and use lithium ions instead. In fact, the cell's operation doesn't actually involve true oxidation and reduction. Instead, it uses the transport of Li^+ ions through the electrolyte from one electrode to the other accompanied by the transport of electrons through the external circuit to maintain charge balance. Here's how it works.

It was discovered that Li^+ ions are able to slip between layers of atoms in certain crystals such as graphite[9] and $LiCoO_2$ (a process called **intercalation**). When the cell is constructed, it is in its "uncharged" state, with no Li^+ ions between the layers of carbon atoms in the graphite. When the cell is charged (Figure 19.18*a*), Li^+ ions leave $LiCoO_2$ and travel through the electrolyte to the graphite (represented below by the formula C_6).

Pronounced in-*ter*-ca-*la*-tion. Rhymes with *percolation.*

Initial charging:

$$LiCoO_2 + C_6 \longrightarrow Li_{1-x}CoO_2 + Li_xC_6$$

When the cell spontaneously discharges to provide electrical power (Figure 19.18*b*), Li^+ ions move back through the electrolyte to the cobalt oxide while electrons move through

[9] Graphite is one of the forms of elemental carbon and consists of layers of joined benzene-like rings.

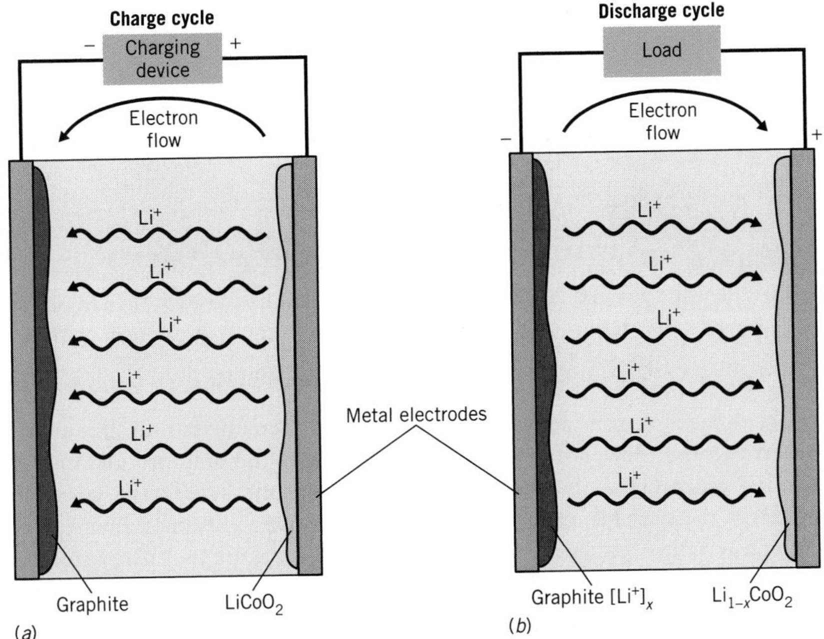

FIG. 19.18 Lithium ion cell. (*a*) During the charging cycle, an external voltage forces electrons through the external circuit and causes lithium ions to travel from the $LiCoO_2$ electrode to the graphite electrode. (*b*) During discharge, the lithium ions spontaneously migrate back to the $LiCoO_2$ electrode, and electrons flow through the external circuit to balance the charge.

the external circuit from the graphite electrode to the cobalt oxide electrode. If we represent the amount of Li^+ transferring by y, the discharge "reaction" is

Discharge:

$$Li_{1-x}CoO_2 + Li_xC_6 \longrightarrow Li_{1-x+y}CoO_2 + Li_{x-y}C_6$$

Thus, the charging and discharging cycles simply sweep Li^+ ions back and forth between the two electrodes, with the electrons flowing through the external circuit to keep the charge in balance.

Fuel cells operate with a continuous supply of reactants

The galvanic cells we've discussed until now can only produce power for a limited time because the electrode reactants are eventually depleted. Fuel cells are different; they are electrochemical cells in which the electrode reactants are supplied continuously and are able to operate without theoretical limit as long as the supply of reactants is maintained. This makes fuel cells an attractive source of power where long-term generation of electrical energy is required.

Figure 19.19 illustrates an early design of a hydrogen–oxygen **fuel cell.** The electrolyte, a hot (~200 °C) concentrated solution of potassium hydroxide in the center

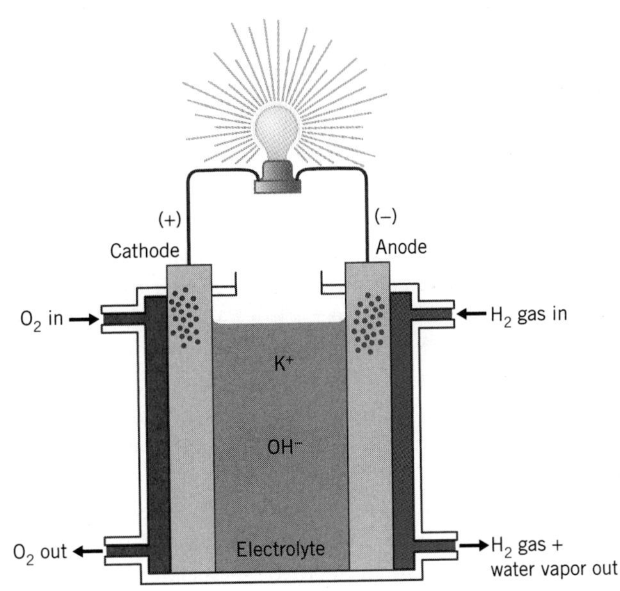

FIG. 19.19 A hydrogen–oxygen fuel cell.

compartment, is in contact with two porous electrodes that contain a catalyst (usually platinum) to facilitate the electrode reactions. Gaseous hydrogen and oxygen under pressure are circulated so as to come in contact with the electrodes. At the cathode, oxygen is reduced.

$$O_2(g) + 2H_2O + 4e^- \longrightarrow 4OH^-(aq) \qquad \text{(cathode)}$$

At the anode, hydrogen is oxidized to water.

$$H_2(g) + 2OH^-(aq) \longrightarrow 2H_2O + 2e^- \qquad \text{(anode)}$$

Part of the water formed at the anode leaves as steam mixed with the circulating hydrogen gas. The net cell reaction, after making electron loss equal to electron gain, is

$$2H_2(g) + O_2(g) \longrightarrow 2H_2O \qquad \text{(net cell reaction)}$$

Hydrogen–oxygen fuel cells are an attractive alternative to gasoline powered engines in part because they are essentially pollution-free—the only product of the reaction is harmless water. Fuel cells are also quite thermodynamically efficient, converting as much as 75% of the available energy to useful work, compared to approximately 25 to 30% for gasoline and diesel engines. Among the major obstacles are the energy costs of generating the hydrogen fuel and problems in providing storage and distribution of the highly flammable H_2.

Electrolysis has many industrial applications

Besides being a useful tool in the chemistry laboratory, electrolysis has many important industrial applications. Here we will briefly examine the chemistry of electroplating and the production of some of our most common chemicals.

This motorcycle sparkles with chrome plating that was deposited by electrolysis. The shiny hard coating of chromium is both decorative and a barrier to corrosion. *(Syracuse Newspapers/ The Image Works.)*

Electroplating deposits metal on a surface

Electroplating, which was mentioned in Examples 19.14 and 19.15, is a procedure in which electrolysis is used to apply a thin (generally 0.03 to 0.05 mm thick) ornamental or protective coating of one metal over another. It is a common technique for improving the appearance and durability of metal objects. For instance, a thin, shiny coating of metallic chromium is applied over steel objects to make them attractive and to prevent rusting.

The exact composition of the electroplating bath varies, depending on the metal to be deposited, and it can affect the appearance and durability of the finished surface. For example, silver deposited from a solution of silver nitrate ($AgNO_3$) does not stick to other metal surfaces very well. However, if it is deposited from a solution of silver cyanide containing $Ag(CN)_2^-$, the coating adheres well and is bright and shiny. Other metals that are electroplated from a cyanide bath are gold and cadmium. Nickel, which can also be applied as a protective coating, is plated from a nickel sulfate solution, and chromium is plated from a chromic acid (H_2CrO_4) solution.

Aluminum is produced from aluminum oxide

Aluminum is a useful but highly reactive metal. It is so difficult to reduce that ordinary metallurgical methods for obtaining it do not work. Early efforts to produce aluminum by electrolysis failed because its anhydrous halide salts (those with no water of hydration) are difficult to prepare and are volatile, tending to evaporate rather than melt. On the other hand, its oxide, Al_2O_3, has such a high melting point (over 2000 °C) that no practical method of melting it could be found.

In 1886, Charles M. Hall discovered that Al_2O_3 dissolves in the molten form of a mineral called cryolite, Na_3AlF_6, to give a conducting mixture with a relatively low melting point from which aluminum could be produced electrolytically. The process was also discovered by Paul Héroult in France at nearly the same time, and today this method for producing aluminum is usually called the **Hall–Héroult process** (see Figure 19.20). Purified aluminum oxide, which is obtained from an ore called *bauxite*, is dissolved in molten cryolite in which the oxide dissociates to give Al^{3+} and O^{2-} ions. At the cathode,

Aluminum is used today as a structural metal, in alloys, and in such products as aluminum foil, electrical wire, window frames, and kitchen pots and pans.

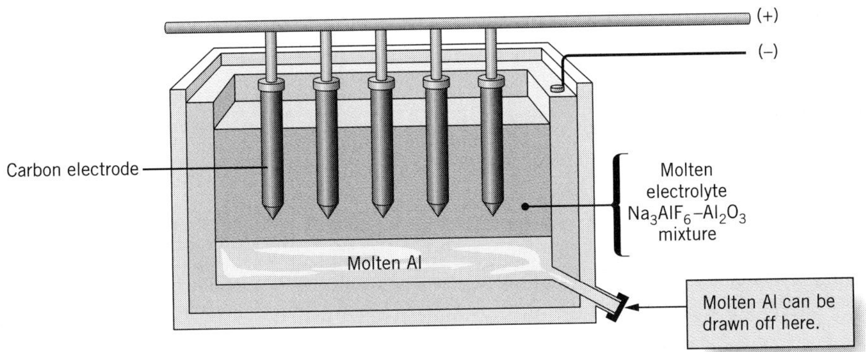

FIG. 19.20 **Production of aluminum by electrolysis.** In the apparatus used to produce aluminum electrolytically by the Hall–Héroult process, Al_2O_3 is dissolved in molten cryolite, Na_3AlF_6. Al^{3+} is reduced to metallic Al and O^{2-} is oxidized to O_2, which reacts with the carbon anodes to give CO_2. Periodically, molten aluminum is drawn off at the bottom of the cell and additional Al_2O_3 is added to the cryolite. The carbon anodes also must be replaced from time to time as they are consumed by their reaction with O_2.

aluminum ions are reduced to produce the free metal, which forms as a layer of molten aluminum below the less dense solvent. At the carbon anodes, oxide ion is oxidized to give free O_2.

$$Al^{3+} + 3e^- \longrightarrow Al(l) \qquad \text{(cathode)}$$
$$2O^{2-} \longrightarrow O_2(g) + 4e^- \qquad \text{(anode)}$$
$$4Al^{3+} + 6O^{2-} \longrightarrow 4Al(l) + 3O_2(g) \qquad \text{(cell reaction)}$$

The oxygen formed at the anode attacks the carbon electrodes (producing CO_2), so the electrodes must be replaced frequently.

The production of aluminum consumes enormous amounts of electrical energy and is therefore very costly, not only in terms of dollars but also in terms of energy resources. For this reason, recycling of aluminum has a high priority as we seek to minimize our use of energy.

Sodium is made by electrolysis of sodium chloride

Sodium is prepared by the electrolysis of molten sodium chloride (see Section 19.6). The metallic sodium and the chlorine gas that form must be kept apart or they will react violently and re-form NaCl. A specialized apparatus called a **Downs cell** accomplishes this separation.

The spectacular reaction of chlorine and sodium is captured in Figure 2.24 on page 62.

Both sodium and chlorine are commercially important. Chlorine is used largely to manufacture plastics such as polyvinyl chloride (PVC), many solvents, and industrial chemicals. A small percentage of the annual chlorine production is used to chlorinate drinking water.

PVC is used to make a large variety of products, from raincoats to wire insulation to pipes and conduits for water and sanitary systems. In the United States, the annual demand for PVC plastics is about 15 billion pounds.

Sodium has been used in the manufacture of tetraethyl lead, an octane booster for gasoline that has been phased out in the United States but which is still used in many other countries. Sodium is also used in the production of energy efficient sodium vapor lamps, which give street lights and other commercial lighting a bright yellow-orange color.

Refining of copper makes it suitable for electrical wire

When copper is first obtained from its ore, it is about 99% pure. The impurities—mostly silver, gold, platinum, iron, and zinc—decrease the electrical conductivity of the copper enough that even 99% pure copper must be further refined before it can be used in electrical wire.

The impure copper is used as the anode in an electrolysis cell that contains a solution of copper sulfate and sulfuric acid as the electrolyte (see Figure 19.21). The cathode is a thin sheet of very pure copper. When the cell is operated at the correct voltage, only copper and impurities more easily oxidized than copper (iron and zinc) dissolve at the anode. The less active metals simply fall off the electrode and settle to the bottom of the container. At the cathode, copper ions are reduced, but the zinc ions and iron ions remain

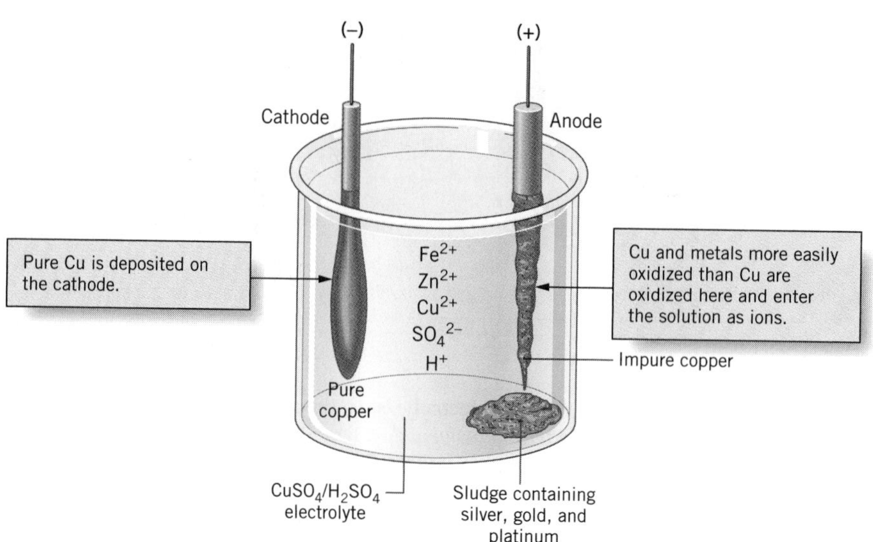

FIG. 19.21 Purification of copper by electrolysis. Impure copper anodes dissolve and pure copper is deposited on the cathodes. Metals less easily oxidized than copper settle to the bottom of the apparatus as "anode mud," while metals less easily reduced than copper remain in solution.

Copper refining provides one-fourth of the silver and one-eighth of the gold produced annually in the United States.

in solution because they are more difficult to reduce than copper. Gradually, the impure copper anode dissolves and the copper cathode, about 99.96% pure, grows larger. The accumulating sludge—called anode mud—is removed periodically, and the value of the silver, gold, and platinum recovered from it virtually pays for the entire refining operation.

Electrolysis of brine produces several important chemicals

One of the most important commercial electrolysis reactions is the electrolysis of concentrated aqueous sodium chloride solutions called **brine.** At the cathode, water is much more easily reduced than sodium ion, so H_2 forms.

$$2H_2O(l) + 2e^- \longrightarrow H_2(g) + 2OH^-(aq) \qquad \text{(cathode)}$$

As we noted earlier, even though water is more easily oxidized than chloride ion, complicating factors at the electrodes actually allow chloride ion to be oxidized instead. At the anode, therefore, we observe the formation of Cl_2.

$$2Cl^-(aq) \longrightarrow Cl_2(g) + 2e^- \qquad \text{(anode)}$$

The net cell reaction is therefore

$$2Cl^-(aq) + 2H_2O(l) \longrightarrow H_2(g) + Cl_2(g) + 2OH^-(aq)$$

If we include the sodium ion, already in the solution as a spectator ion and not involved in the electrolysis directly, we can see why this is such an important reaction.

Copper refining. Copper cathodes, 99.96% pure, are pulled from the electrolytic refining tanks at Kennecott's Utah copper refinery. It takes about 28 days for the impure copper anodes to dissolve and deposit the pure metal on the cathodes. *(Courtesy ASARCO, Inc.)*

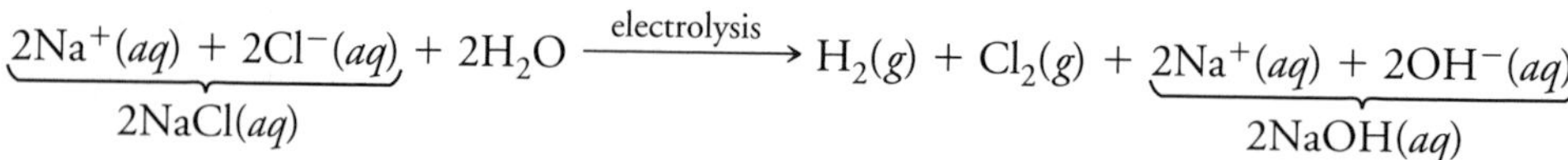

$$\underbrace{2Na^+(aq) + 2Cl^-(aq)}_{2NaCl(aq)} + 2H_2O \xrightarrow{\text{electrolysis}} H_2(g) + Cl_2(g) + \underbrace{2Na^+(aq) + 2OH^-(aq)}_{2NaOH(aq)}$$

Thus, the electrolysis converts inexpensive salt to valuable chemicals: H_2, Cl_2, and NaOH. The hydrogen is used to make other chemicals, including hydrogenated vegetable oils. The chlorine is used for the purposes mentioned earlier. Among the uses of sodium hydroxide, one of industry's most important bases, are the manufacture of soap and paper, the neutralization of acids in industrial reactions, and the purification of aluminum ores.

Sodium hydroxide is commonly known as *lye* or *caustic soda.*

In the industrial electrolysis of brine, it is necessary to capture the H_2 and Cl_2 separately to prevent them from mixing and reacting (explosively). Second, the NaOH from the reaction is contaminated with unreacted NaCl. Third, if Cl_2 is left in the presence of NaOH, the solution becomes contaminated by hypochlorite ion (OCl^-), which forms by the reaction of Cl_2 with OH^-.

$$Cl_2(g) + 2OH^-(aq) \longrightarrow Cl^-(aq) + OCl^-(aq) + H_2O$$

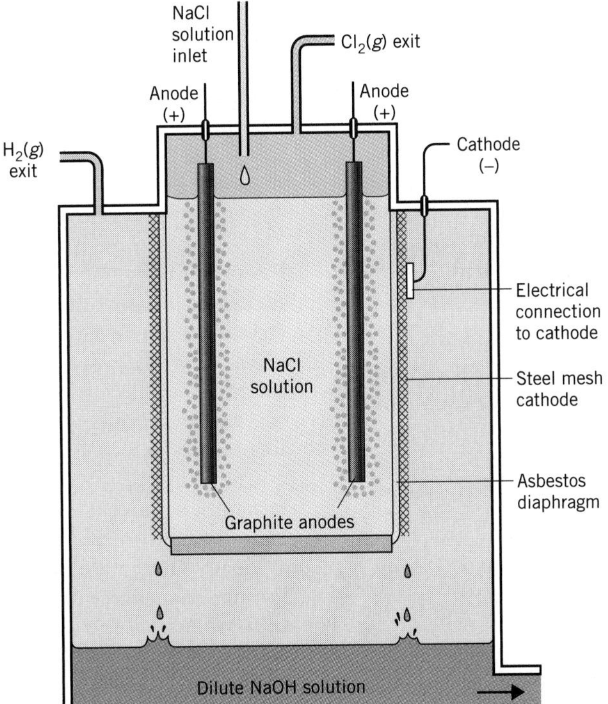

FIG. 19.22 A diaphragm cell used in the commercial production of NaOH by the electrolysis of aqueous NaCl. This is a cross section of a cylindrical cell in which the NaCl solution is surrounded by an asbestos diaphragm supported by a steel mesh cathode. (From J. E. Brady and G. E. Humiston, *General Chemistry: Principles and Structure*, 4th ed. Copyright © 1986, John Wiley & Sons, New York. Used by permission.)

In one manufacturing operation, however, the Cl_2 is not removed as it forms, and its reaction with hydroxide ion is used to manufacture aqueous sodium hypochlorite. For this purpose, the solution is stirred vigorously during the electrolysis so that very little Cl_2 escapes. As a result, a stirred solution of NaCl gradually changes during electrolysis to a solution of NaOCl, a dilute (5%) solution of which is sold as liquid laundry bleach (e.g., Clorox).

Most of the pure NaOH manufactured today is made in an apparatus called a **diaphragm cell.** The design varies somewhat, but Figure 19.22 illustrates its basic features. The cell consists of an iron wire mesh cathode that encloses a porous asbestos shell—the diaphragm. The NaCl solution is added to the top of the cell and seeps slowly through the diaphragm. When it contacts the iron cathode, hydrogen is evolved and is pumped out of the surrounding space. The solution, now containing dilute NaOH, drips off the cell into the reservoir below. Meanwhile, within the cell, chlorine is generated at the anodes dipping into the NaCl solution. Because there is no OH^- in this solution, the Cl_2 can't react to form OCl^- ion and simply bubbles out of the solution and is captured.

SUMMARY

Galvanic Cells. A **galvanic cell** is composed of two **half-cells,** each containing an **electrode** in contact with an electrolyte reactant. A spontaneous redox reaction is thus divided into separate oxidation and reduction half-reactions, with the electron transfer occurring through an external electrical circuit. Reduction occurs at the **cathode;** oxidation occurs at the **anode.** In a galvanic cell, the cathode is positively charged and the anode is negatively charged. The half-cells must be connected electrolytically by a **salt bridge** to complete the electrical circuit, which permits electrical neutrality to be maintained by allowing cations to move toward the cathode and anions toward the anode.

The **potential** (expressed in volts) produced by a cell is equal to the **standard cell potential** when all ion concentrations are 1.00 *M* and the partial pressures of any gases involved equal 1.00 atm and the temperature is 25.0 °C. The **standard cell potential** is the difference between the **standard reduction potentials** of the half-cells. In the spontaneous reaction, the half-cell with the higher standard reduction potential undergoes reduction and forces the other to undergo oxidation. The standard reduction potentials of isolated half-cells can't be measured, but values are assigned by choosing the **hydrogen electrode** as a reference electrode; its standard reduction potential is assigned a value of exactly 0 V. Species more easily reduced than H^+ have positive standard reduction potentials; those less easily reduced have negative standard reduction potentials. Standard reduction potentials can be used to predict the cell reaction. They can also be used to predict spontaneous redox reactions not occurring in galvanic cells and to predict whether or not a given reaction is spontaneous. Often, they can be used to determine the products of electrolysis reactions. Standard reduction potentials are used to calculate E°_{cell}. Since reduction potentials under nonstandard conditions are often numerically close to standard reduction potentials, we can usually use standard reduction potentials to make predictions about reactions with nonstandard conditions. The predictions above about spontaneous reactions, cell reactions, electrolysis products, and reactions outside of galvanic cells are usually valid under nonstandard conditions also.

Thermodynamics and Cell Potentials. The values of ΔG° and K_c for a reaction can be calculated from E°_{cell}. They all involve the **faraday,** $\mathcal{F}$, a constant equal to the number of **coulombs (C)** of charge per mole of electrons ($1\ \mathcal{F} = 96{,}500\ \text{C/mol}\ e^-$). The Nernst equation relates the cell potential to the standard cell potential and the reaction quotient. It allows the cell potential to be calculated for ion concentrations other than 1.00 *M*. The important equations are summarized as

$$\Delta G^\circ = -n\mathcal{F}E^\circ_{cell} = -RT \ln K_c$$

$$E_{cell} = E^\circ_{cell} - \frac{RT}{n\mathcal{F}} \ln Q$$

Electrolysis. In an **electrolytic cell,** a flow of electricity causes an otherwise nonspontaneous reaction to occur. A negatively charged **cathode** causes reduction of one reactant and a positively charged **anode** causes oxidation of another. Ion movement instead of electron transport occurs in the electrolyte. The electrode reactions are determined by which species is most easily reduced and which is most easily oxidized, but in aqueous solutions complex surface effects at the electrodes can alter the natural order. In the electrolysis of water, an electrolyte must be present to maintain electrical neutrality at the electrodes.

Quantitative Aspects of Electrochemical Reactions. The product of current (**amperes**) and time (seconds) gives coulombs. This relationship and the half-reactions that occur at the anode or cathode permit us to relate the amount of chemical change to measurements of current and time.

Practical Galvanic Cells. The **lead storage battery** and the **nickel–cadmium (nicad) battery** are **secondary cells** and are rechargeable. The state of charge of the lead storage battery can be tested with a **hydrometer,** which measures the density of the sulfuric acid electrolyte. The **zinc–manganese dioxide cell** (the **Leclanché cell** or common dry cell) and the common **alkaline battery** (which uses essentially the same reactions as the less expensive dry cell) are **primary cells** and are not rechargeable. The rechargeable **nickel–metal hydride** (Ni–MH) battery uses hydrogen contained in a metal alloy as its anode reactant and has a higher **energy density** than the **nicad battery.** Primary **lithium–manganese dioxide cells** and rechargeable **lithium ion cells** produce a large cell potential and have a very high energy density. Lithium ion cells store and release energy by transferring lithium ions between electrodes where the Li^+ ions are **intercalated** between layers of atoms in the electrode materials. **Fuel cells,** which have high thermodynamic efficiencies, are able to provide continuous power because they consume fuel that can be fed continuously.

Applications of Electrolysis. **Electroplating,** the production of aluminum, the refining of copper, and the electrolysis of molten and aqueous sodium chloride are examples of practical applications of electrolysis.

TOOLS FOR PROBLEM SOLVING

The table below lists the concepts that you've learned in this chapter which can be applied as tools in solving problems. Study each one carefully so that you know what each is used for. When faced with solving a problem, recall what each tool does and consider whether it will be helpful in finding a solution. This will aid you in selecting the tools you need. If necessary, refer to this table when working on the Review Questions and Review Problems that follow.

Electrode reactions *(page 772)* Chemists have agreed to call the electrode at which oxidation occurs the anode. The cathode is the electrode where reduction occurs.

Standard reduction potentials relate to E°_{cell} In a galvanic cell, the difference between two standard reduction potentials equals the standard cell potential *(page 776)*.

$$E^\circ_{cell} = \begin{pmatrix}\text{standard reduction potential}\\ \text{of the substance reduced}\end{pmatrix} - \begin{pmatrix}\text{standard reduction potential}\\ \text{of the substance oxidized}\end{pmatrix}$$

Standard cell potentials are always positive and the reaction that gives us a positive standard cell potential will be a spontaneous reaction at standard state *(page 786)*. Comparing standard reduction potentials also lets us predict the electrode reactions in electrolysis *(page 800)*.

Standard cell potentials are related to thermodynamic quantities The relationship between the standard cell potential, E°_{cell}, and the standard free energy, ΔG°, is $\Delta G^\circ = -n\mathcal{F}E^\circ_{cell}$ *(page 788)*

The relationship between the standard cell potential, E°_{cell}, and the equilibrium constant, K_c, is $E^\circ_{cell} = \frac{RT}{n\mathcal{F}} \ln K_c$ *(page 789)*.

Standard cell potentials are also needed in the Nernst equation (below) to relate concentrations of species in galvanic cells to the cell potential.

Nernst equation *(page 790)* The Nernst equation lets us relate the cell potential, E_{cell}, the standard cell potential, E°_{cell}, and concentration data.

$$E_{cell} = E^\circ_{cell} - \frac{RT}{n\mathcal{F}} \ln Q$$

We can also calculate the concentration of a species in solution from E°_{cell} and a measured value for E_{cell}.

Faraday constant *(page 788)* Besides being a constant in the equations above, the Faraday constant allows us to relate coulombs (obtained from the product of current and time) to moles of chemical change in electrochemical reactions.

$$1\ \mathcal{F} = 9.65 \times 10^4\ \text{C/mol}\ e^-$$

Coulombs *(page 802)* Coulombs, C, can be experimentally determined as the current, in amperes, A, multiplied by time, in seconds. Combined with the faraday, the moles of electrons can be determined.

$$1\ \text{coulomb (C)} = 1\ \text{ampere (A)} \times 1\ \text{second (s)}$$

QUESTIONS, PROBLEMS, AND EXERCISES

Answers to problems whose numbers are printed in color are given in Appendix B. More challenging problems are marked with asterisks. ILW = Interactive Learningware solution is available at www.wiley.com/college/brady. (OH) = an Office Hours video is available for this problem.

REVIEW QUESTIONS

Galvanic Cells

19.1 What is a *galvanic cell*? What is a *half-cell*?

19.2 What is the function of a *salt bridge*?

19.3 In the copper–silver cell, why must the Cu^{2+} and Ag^+ solutions be kept in separate containers?

19.4 Which redox processes take place at the anode and cathode in a galvanic cell? What is the sign of the electrical charges on the anode and cathode in a galvanic cell?

19.5 In a galvanic cell, do electrons travel from anode to cathode, or from cathode to anode? Explain.

19.6 Explain how the movement of the ions relative to the electrodes is the same in both galvanic and electrolytic cells.

19.7 Aluminum will displace tin from solution according to the equation $2Al(s) + 3Sn^{2+}(aq) \longrightarrow 2Al^{3+}(aq) + 3Sn(s)$. What would be the individual half-cell reactions if this were the cell reaction in a galvanic cell? Which metal would be the anode and which the cathode?

Cell Potentials and Reduction Potentials

19.8 What is the difference between a *cell potential* and a *standard cell potential*?

19.9 How are standard reduction potentials combined to give the standard cell potential for a spontaneous reaction?

19.10 Describe the hydrogen electrode. What is the value of its standard reduction potential?

19.11 What do the positive and negative signs of reduction potentials tell us?

19.12 If $E^\circ_{Cu^{2+}}$ had been chosen as the standard reference electrode and had been assigned a potential of 0.00 V, what would the reduction potential of the hydrogen electrode be relative to it?

19.13 If you set up a galvanic cell using metals not found in Table 19.1, what experimental information will tell you which is the anode and which is the cathode in the cell?

Using Standard Reduction Potentials

19.14 Compare Table 5.2 with Table 19.1. What can you say about the basis for the activity series for metals?

19.15 Make a sketch of a galvanic cell for which the cell notation is

$$Fe(s)\ |\ Fe^{3+}(aq)\ ||\ Ag^+(aq)\ |\ Ag(s)$$

(a) Label the anode and the cathode.

(b) Indicate the charge on each electrode.

(c) Indicate the direction of electron flow in the external circuit.

(d) Write the equation for the net cell reaction.

19.16 Make a sketch of a galvanic cell in which inert platinum electrodes are used in the half-cells for the system

$$Pt(s)\ |\ Fe^{2+}(aq),\ Fe^{3+}(aq)\ ||\ Br_2(aq),\ Br^-(aq)\ |\ Pt(s)$$

Label the diagram and indicate the composition of the electrolytes in the two cell compartments. Show the signs of the electrodes and label the anode and cathode. Write the equation for the net cell reaction.

Cell Potentials and Thermodynamics

19.17 Write the equation that relates the standard cell potential to the standard free energy change for a reaction.

19.18 What is the equation that relates the equilibrium constant to the cell potential?

19.19 Show how the equation that relates the equilibrium constant to the cell potential (Equation 19.7) can be derived from the Nernst equation (Equation 19.8).

The Effect of Concentration on Cell Potential

19.20 The cell reaction during the discharge of a lead storage battery is

$$Pb(s) + PbO_2(s) + 2H^+(aq) + 2HSO_4^-(aq) \longrightarrow 2PbSO_4(s) + 2H_2O$$

The standard cell potential is 2.05 V. What is the correct form of the Nernst equation for the reaction at 25 °C?

19.21 What is a *concentration cell*? Why is the E°_{cell} for such a cell equal to zero?

Electrolysis

19.22 What electrical charges do the anode and the cathode carry in an electrolytic cell? What does the term *inert electrode* mean?

19.23 Why must electrolysis reactions occur at the electrodes in order for electrolytic conduction to continue?

19.24 Why must NaCl be melted before it is electrolyzed to give Na and Cl_2? Write the anode, cathode, and overall cell reactions for the electrolysis of molten NaCl.

19.25 Write half-reactions for the oxidation and the reduction of water.

19.26 What happens to the pH of the solution near the cathode and anode during the electrolysis of K_2SO_4? What function does K_2SO_4 serve in the electrolysis of a K_2SO_4 solution?

Stoichiometric Relationships in Electrolysis

19.27 What is a *faraday*? What relationships relate faradays to current and time measurements?

19.28 Using the same current, which will require the greater length of time, depositing 0.10 mol Cu from a Cu^{2+} solution, or depositing 0.10 mol of Cr from a Cr^{3+} solution? Explain your reasoning.

19.29 An electric current is passed through two electrolysis cells connected in series (so the same amount of current passes through each of them). One cell contains Cu^{2+} and the other contains Ag^+. In which cell will the larger number of moles of metal be deposited? Explain your answer.

19.30 An electric current is passed through two electrolysis cells connected in series (so the same amount of current passes through each of them). One cell contains Cu^{2+} and the other contains Fe^{2+}. In which cell will the greater mass of metal be deposited? Explain your answer.

Practical Galvanic Cells

19.31 What are the anode and cathode reactions during the discharge of a lead storage battery? How can a battery produce a potential of 12 V if the cell reaction has a standard potential of only 2 V?

19.32 What are the anode and cathode reactions during the charging of a lead storage battery?

19.33 How is a hydrometer constructed? How does it measure density? Why can a hydrometer be used to check the state of charge of a lead storage battery?

19.34 What reactions occur at the electrodes in the ordinary dry cell?

19.35 What chemical reactions take place at the electrodes in an alkaline dry cell?

19.36 Give the half-cell reactions and the cell reaction that take place in a nicad battery during discharge. What are the reactions that take place during the charging of the cell?

19.37 How is hydrogen held as a reactant in a nickel–metal hydride battery? Write the chemical formula for a typical alloy used in this battery. What is the electrolyte?

19.38 What are the anode, cathode, and net cell reactions that take place in a nickel–metal hydride battery during discharge? What are the reactions when the battery is charged?

19.39 Give two reasons why lithium is such an attractive anode material for use in a battery. What are the problems associated with using lithium for this purpose?

19.40 What are the electrode materials in a typical primary lithium cell? Write the equations for the anode, cathode, and cell reactions.

19.41 What are the electrode materials in a typical lithium ion cell? Explain what happens when the cell is charged. Explain what happens when the cell is discharged.

19.42 Write the cathode, anode, and net cell reaction in a hydrogen–oxygen fuel cell.

19.43 What advantages do fuel cells offer over conventional means of obtaining electrical power by the combustion of fuels?

Applications of Electrolysis

19.44 What is *electroplating*? Sketch an apparatus to electroplate silver.

19.45 Describe the Hall–Héroult process for producing metallic aluminum. What half–reaction occurs at the anode? What half-reaction occurs at the cathode? What is the overall cell reaction?

19.46 In the Hall–Héroult process, why must the carbon anodes be replaced frequently?

19.47 How is metallic sodium produced? What are some uses of metallic sodium? Write equations for the anode and cathode reactions.

19.48 Describe the electrolytic refining of copper. What economic advantages offset the cost of electricity for this process? What chemical reactions occur at the anode and the cathode?

19.49 Describe the electrolysis of aqueous sodium chloride. How do the products of the electrolysis compare for stirred and unstirred reactions? Write chemical equations for the reactions that occur at the electrodes.

REVIEW PROBLEMS

Cell Notation

19.50 Write the half-reactions and the balanced cell reaction for the following galvanic cells.

(a) $Cd(s)|Cd^{2+}(aq)||Au^{3+}(aq)|Au(s)$

(b) $Fe(s)|Fe^{2+}(aq)||Br_2(aq), Br^-(aq)|Pt(s)$

(c) $Cr(s)|Cr^{3+}(aq)||Cu^{2+}(aq)|Cu(s)$

OH **19.51** Write the half-reactions and the balanced cell reaction for the following galvanic cells.

(a) $Zn(s)|Zn^{2+}(aq)||Cr^{3+}(aq)|Cr(s)$

(b) $Pb(s), PbSO_4(s)|HSO_4^-(aq), H^+(aq)||H^+(aq), HSO_4^-(aq)|PbO_2(s), PbSO_4(s)$

(c) $Mg(s)|Mg^{2+}(aq)||Sn^{2+}(aq)|Sn(s)$

19.52 Write the cell notation for the following galvanic cells. For half-reactions in which all the reactants are in solution or are gases, assume the use of inert platinum electrodes.

(a) $NO_3^-(aq) + 4H^+(aq) + 3Fe^{2+}(aq) \longrightarrow 3Fe^{3+}(aq) + NO(g) + 2H_2O$

(b) $Cl_2(g) + 2Br^-(aq) \longrightarrow Br_2(aq) + 2Cl^-(aq)$

(c) $Au^{3+}(aq) + 3Ag(s) \longrightarrow Au(s) + 3Ag^+(aq)$

19.53 Write the cell notation for the following galvanic cells. For half-reactions in which all the reactants are in solution or are gases, assume the use of inert platinum electrodes.

(a) $Cd^{2+}(aq) + Fe(s) \longrightarrow Cd(s) + Fe^{2+}(aq)$

(b) $NiO_2(s) + 4H^+(aq) + 2Ag(s) \longrightarrow Ni^{2+}(aq) + 2H_2O + 2Ag^+(aq)$

(c) $Mg(s) + Cd^{2+}(aq) \longrightarrow Mg^{2+}(aq) + Cd(s)$

Using Reduction Potentials

19.54 For each pair of substances, use Table 19.1 to choose the better reducing agent.

(a) $Sn(s)$ or $Ag(s)$ (c) $Co(s)$ or $Zn(s)$
(b) $Cl^-(aq)$ or $Br^-(aq)$ (d) $I^-(aq)$ or $Au(s)$

19.55 For each pair of substances, use Table 19.1 to choose the better oxidizing agent.

(a) $NO_3^-(aq)$ or $MnO_4^-(aq)$ (c) $PbO_2(s)$ or $Cl_2(g)$
(b) $Au^{3+}(aq)$ or $Co^{2+}(aq)$ (d) $NiO_2(s)$ or $HOCl(aq)$

OH 19.56 Use the data in Table 19.1 to calculate the standard cell potential for each of the following reactions:

(a) $NO_3^-(aq) + 4H^+(aq) + 3Fe^{2+}(aq) \longrightarrow 3Fe^{3+}(aq) + NO(g) + 2H_2O$

(b) $Br_2(aq) + 2Cl^-(aq) \longrightarrow Cl_2(g) + 2Br^-(aq)$

(c) $Au^{3+}(aq) + 3Ag(s) \longrightarrow Au(s) + 3Ag^+(aq)$

19.57 Use the data in Table 19.1 to calculate the standard cell potential for each of the following reactions:

(a) $Cd^{2+}(aq) + Fe(s) \longrightarrow Cd(s) + Fe^{2+}(aq)$

(b) $NiO_2(s) + 4H^+(aq) + 2Ag(s) \longrightarrow Ni^{2+}(aq) + 2H_2O + 2Ag^+(aq)$

(c) $Mg(s) + Cd^{2+}(aq) \longrightarrow Mg^{2+}(aq) + Cd(s)$

19.58 From the positions of the half-reactions in Table 19.1, determine whether the following reactions are spontaneous under standard state conditions.

(a) $2Au^{3+} + 6I^- \longrightarrow 3I_2 + 2Au$

(b) $H_2SO_3 + H_2O + Br_2 \longrightarrow 4H^+ + SO_4^{2-} + 2Br^-$

(c) $3Ca + 2Cr^{3+} \longrightarrow 2Cr + 3Ca^{2+}$

19.59 Use the data in Table 19.1 to determine which of the following reactions should occur spontaneously under standard state conditions.

(a) $Br_2 + 2Cl^- \longrightarrow Cl_2 + 2Br^-$

(b) $3Fe^{2+} + 2NO + 4H_2O \longrightarrow 3Fe + 2NO_3^- + 8H^+$

(c) $Ni^{2+} + Fe \longrightarrow Fe^{2+} + Ni$

ILW 19.60 From the half-reactions below, determine the cell reaction and standard cell potential.

$$BrO_3^- + 6H^+ + 6e^- \rightleftharpoons Br^- + 3H_2O \qquad E^\circ_{BrO_3^-} = 1.44 \text{ V}$$

$$I_2 + 2e^- \rightleftharpoons 2I^- \qquad E^\circ_{I_2} = 0.54 \text{ V}$$

19.61 What is the standard cell potential and the net reaction in a galvanic cell that has the following half-reactions?

$$MnO_2 + 4H^+ + 2e^- \rightleftharpoons Mn^{2+} + 2H_2O \qquad E^\circ_{MnO_2} = 1.23 \text{ V}$$

$$PbCl_2 + 2e^- \rightleftharpoons Pb + 2Cl^- \qquad E^\circ_{PbCl_2} = -0.27 \text{ V}$$

OH 19.62 What will be the spontaneous reaction among H_2SO_3, $S_2O_3^{2-}$, HOCl, and Cl_2? The half-reactions involved are

$$2H_2SO_3 + 2H^+ + 4e^- \rightleftharpoons S_2O_3^{2-} + 3H_2O \qquad E^\circ_{H_2SO_3} = 0.40 \text{ V}$$

$$2HOCl + 2H^+ + 2e^- \rightleftharpoons Cl_2 + 2H_2O \qquad E^\circ_{HOCl} = 1.63 \text{ V}$$

19.63 What will be the spontaneous reaction among Br_2, I_2, Br^-, and I^-?

19.64 Will the following reaction occur spontaneously under standard state conditions?

$$SO_4^{2-} + 4H^+ + 2I^- \longrightarrow H_2SO_3 + I_2 + H_2O$$

Use E°_{cell} calculated from data in Table 19.1 to answer the question.

19.65 Determine whether the reaction

$$S_2O_8^{2-} + Ni(OH)_2 + 2OH^- \longrightarrow 2SO_4^{2-} + NiO_2 + 2H_2O$$

will occur spontaneously under standard state conditions. Use E°_{cell} calculated from the data below to answer the question.

$$NiO_2 + 2H_2O + 2e^- \rightleftharpoons Ni(OH)_2 + 2OH^- \qquad E^\circ_{NiO_2} = 0.49 \text{ V}$$

$$S_2O_8^{2-} + 2e^- \rightleftharpoons 2SO_4^{2-} \qquad E^\circ_{S_2O_8^{2-}} = 2.01 \text{ V}$$

Cell Potentials and Thermodynamics

ILW 19.66 Calculate ΔG° for the following reaction *as written*.

$$2Br^- + I_2 \longrightarrow 2I^- + Br_2$$

OH 19.67 Calculate ΔG° for the reaction

$$2MnO_4^- + 6H^+ + 5HCHO_2 \longrightarrow 2Mn^{2+} + 8H_2O + 5CO_2$$

for which $E^\circ_{cell} = 1.69$ V.

19.68 Given the following half-reactions and their standard reduction potentials, calculate (a) E°_{cell}, (b) ΔG° for the cell reaction, and (c) the value of K_c for the cell reaction.

$$2ClO_3^- + 12H^+ + 10e^- \rightleftharpoons Cl_2 + 6H_2O \qquad E^\circ_{ClO_3^-} = 1.47 \text{ V}$$

$$S_2O_8^{2-} + 2e^- \rightleftharpoons 2SO_4^{2-} \qquad E^\circ_{S_2O_8^{2-}} = 2.01 \text{ V}$$

19.69 Calculate K_c for the system,

$$Ni^{2+} + Co \rightleftharpoons Ni + Co^{2+}$$

Use the data in Table 19.1. Assume $T = 298$ K.

19.70 The system $2AgI + Sn \rightleftharpoons Sn^{2+} + 2Ag + 2I^-$ has a calculated E°_{cell} of -0.015 V. What is the value of K_c for this system?

19.71 Determine the value of K_c at 25 °C for the re action

$$2H_2O + 2Cl_2 \rightleftharpoons 4H^+ + 4Cl^- + O_2$$

The Effect of Concentration on Cell Potential

19.72 The cell reaction

$$NiO_2(s) + 4H^+(aq) + 2Ag(s) \longrightarrow Ni^{2+}(aq) + 2H_2O + 2Ag^+(aq)$$

has $E^\circ_{cell} = 2.48$ V. What will be the cell potential at a pH of 2.00 when the concentrations of Ni^{2+} and Ag^+ are each 0.030 *M*?

19.73 The E°_{cell} is 0.135 V for the reaction

$$3I_2(s) + 5Cr_2O_7^{2-}(aq) + 34H^+ \longrightarrow 6IO_3^-(aq) + 10Cr^{3+}(aq) + 17H_2O$$

What is E_{cell} if $[Cr_2O_7^{2-}] = 0.010\ M$, $[H^+] = 0.10\ M$, $[IO_3^-] = 0.00010\ M$, and $[Cr^{3+}] = 0.0010\ M$?

OH *19.74 A cell was set up having the following reaction.

$$Mg(s) + Cd^{2+}(aq) \longrightarrow Mg^{2+}(aq) + Cd(s) \qquad E^\circ_{cell} = 1.97\ V$$

The magnesium electrode was dipping into a 1.00 *M* solution of $MgSO_4$ and the cadmium electrode was dipping into a solution of unknown Cd^{2+} concentration. The potential of the cell was measured to be 1.54 V. What was the unknown Cd^{2+} concentration?

*19.75 A silver wire coated with AgCl is sensitive to the presence of chloride ion because of the half-cell reaction

$$AgCl(s) + e^- \rightleftharpoons Ag(s) + Cl^- \qquad E^\circ_{AgCl} = 0.2223\ V$$

A student, wishing to measure the chloride ion concentration in a number of water samples, constructed a galvanic cell using the AgCl electrode as one half-cell and a copper wire dipping into 1.00 *M* $CuSO_4$ solution as the other half-cell. In one analysis, the potential of the cell was measured to be 0.0895 V with the copper half-cell serving as the cathode. What was the chloride ion concentration in the water? (Take $E^\circ_{Cu^{2+}} = +0.3419$ V.)

*19.76 At 25 °C, a galvanic cell was set up having the following half-reactions.

$$Fe^{2+}(aq) + 2e^- \rightleftharpoons Fe(s) \qquad E^\circ_{Fe^{2+}} = -0.447\ V$$
$$Cu^{2+}(aq) + 2e^- \rightleftharpoons Cu(s) \qquad E^\circ_{Cu^{2+}} = +0.3419\ V$$

The copper half-cell contained 100 mL of 1.00 *M* $CuSO_4$. The iron half-cell contained 50.0 mL of 0.100 *M* $FeSO_4$. To the iron half-cell was added 50.0 mL of 0.500 *M* NaOH solution. The mixture was stirred and the cell potential was measured to be 1.175 V. Calculate the value of K_{sp} for $Fe(OH)_2$.

*19.77 Suppose a galvanic cell was constructed at 25 °C using a Cu/Cu^{2+} half-cell (in which the molar concentration of Cu^{2+} was 1.00 *M*) and a hydrogen electrode having a partial pressure of H_2 equal to 1 atm. The hydrogen electrode dipped into a solution of unknown hydrogen ion concentration, and the two half-cells were connected by a salt bridge. The precise value of $E^\circ_{Cu^{2+}}$ is +0.3419 V.

(a) Derive an equation for the pH of the solution with the unknown hydrogen ion concentration, expressed in terms of E_{cell} and E°_{cell}.

(b) If the pH of the solution were 5.15, what would be the observed potential of the cell?

(c) If the potential of the cell were 0.645 V, what would be the pH of the solution?

19.78 What is the potential of a concentration cell at 25.0 °C if it consists of silver electrodes dipping into two different solutions of $AgNO_3$, one with a concentration of 0.015 *M* and the other with a concentration of 0.50 *M*?

19.79 What would be the potential of the cell in the preceding problem if the temperature of the cell were 50 °C?

Quantitative Aspects of Electrochemical Reactions

19.80 How many moles of electrons are required to (a) reduce 0.20 mol Fe^{2+} to Fe, (b) oxidize 0.70 mol Cl^- to Cl_2, (c) reduce 1.50 mol Cr^{3+} to Cr, (d) oxidize 1.0×10^{-2} mol Mn^{2+} to MnO_4^-?

19.81 How many moles of electrons are required to (a) produce 5.00 g Mg from molten $MgCl_2$, (b) form 41.0 g Cu from a $CuSO_4$ solution?

ILW 19.82 How many grams of $Fe(OH)_2$ are produced at an iron anode when a basic solution undergoes electrolysis at a current of 8.00 A for 12.0 min?

19.83 How many grams of Cl_2 would be produced in the electrolysis of molten NaCl by a current of 4.25 A for 35.0 min?

ILW 19.84 How many hours would it take to produce 75.0 g of metallic chromium by the electrolytic reduction of Cr^{3+} with a current of 2.25 A?

19.85 How many hours would it take to generate 35.0 g of lead from $PbSO_4$ during the charging of a lead storage battery using a current of 1.50 A? The half-reaction is

$$Pb + HSO_4^- \longrightarrow PbSO_4 + H^+ + 2e^-$$

OH 19.86 How many amperes would be needed to produce 60.0 g of magnesium during the electrolysis of molten $MgCl_2$ in 2.00 hr?

19.87 A large electrolysis cell that produces metallic aluminum from Al_2O_3 by the Hall–Héroult process is capable of yielding 900 lb (409 kg) of aluminum in 24 hr. What current is required?

*19.88 The electrolysis of 250 mL of a brine solution (NaCl) was carried out for a period of 20.00 min with a current of 2.00 A in an apparatus that prevented Cl_2 from reacting with other products of the electrolysis. The resulting solution was titrated with 0.620 *M* HCl. How many milliliters of the HCl solution was required for the titration?

*19.89 An unstirred solution of 2.00 *M* NaCl was electrolyzed for a period of 25.0 min and then titrated with 0.250 *M* HCl. The titration required 15.5 mL of the acid. What was the average current in amperes during the electrolysis?

OH *19.90 A solution of NaCl in water was electrolyzed with a current of 2.50 A for 15.0 min. How many milliliters of Cl_2 gas would be formed if it was collected over water at 25 °C and a total pressure of 750 torr?

*19.91 How many milliliters of dry gaseous H_2, measured at 20 °C and 735 torr, would be produced at the cathode in the electrolysis of dilute H_2SO_4 with a current of 0.750 A for 15.00 min?

Predicting Electrolysis Reactions

19.92 If electrolysis is carried out on an aqueous solution of aluminum sulfate, what products are expected at the electrodes? Write the equation for the net cell reaction.

19.93 If electrolysis is carried out on an aqueous solution of cadmium iodide, what products are expected at the electrodes? Write the equation for the net cell reaction.

OH 19.94 What products would we expect at the electrodes if a solution containing both KBr and $CuSO_4$ were electrolyzed? Write the equation for the net cell reaction.

19.95 What products would we expect at the electrodes if a solution containing both $BaCl_2$ and CuI_2 were electrolyzed? Write the equation for the net cell reaction.

ADDITIONAL EXERCISES

*19.96 A watt is a unit of electrical power and is equal to one joule per second (1 watt = 1 J s^{-1}). How many hours can a calculator drawing 2.0×10^{-3} watt be operated by a mercury battery having a cell potential equal to 1.34 V if a mass of 1.00 g of HgO is available at the cathode? The cell reaction is

$$HgO(s) + Zn(s) \longrightarrow ZnO(s) + Hg(l)$$

*19.97 Suppose that a galvanic cell were set up having the net cell reaction

$$Zn(s) + 2Ag^+(aq) \longrightarrow Zn^{2+}(aq) + 2Ag(s)$$

The Ag^+ and Zn^{2+} concentrations in their respective half-cells initially are 1.00 M, and each half-cell contains 100 mL of electrolyte solution. If this cell delivers current at a constant rate of 0.10 A, what will the cell potential be after 15.00 hr?

*19.98 The value of K_{sp} for AgBr is 5.0×10^{-13}. What will be the potential of a cell constructed of a standard hydrogen electrode as one half-cell and a silver wire coated with AgBr dipping into 0.10 M HBr as the other half-cell? For the Ag/AgBr electrode,

$$AgBr(s) + e^- \rightleftharpoons Ag(s) + Br^-(aq) \qquad E^\circ_{AgBr} = +0.070 \text{ V}$$

*19.99 A student set up an electrolysis apparatus and passed a current of 1.22 A through a 3 M H_2SO_4 solution for 30.0 min. The H_2 formed at the cathode was collected and found to have a volume, over water at 27 °C, of 288 mL at a total pressure of 767 torr. Use the data to calculate the charge on the electron, expressed in coulombs.

*19.100 A hydrogen electrode is immersed in a 0.10 M solution of acetic acid at 25 °C. The electrode is connected to another consisting of an iron nail dipping into 0.10 M $FeCl_2$. What will be the measured potential of this cell? Assume $P_{H_2} = 1.00$ atm.

*19.101 What current would be required to deposit 1.00 m^2 of chrome plate having a thickness of 0.050 mm in 4.50 hr from a solution of H_2CrO_4? The density of chromium is 7.19 g cm^{-3}.

19.102 A solution containing vanadium (chemical symbol V) in an unknown oxidation state was electrolyzed with a current of 1.50 A for 30.0 min. It was found that 0.475 g of V was deposited on the cathode. What was the original oxidation state of the vanadium ion?

*19.103 What masses of H_2 and O_2 in grams would have to react each second in a fuel cell at 110 °C to provide 1.00 kilowatt (kW) of power if we assume a thermodynamic efficiency of 70%? (Hint: Use data in Chapters 6 and 18 to compute the value of ΔG° for the reaction $H_2(g) + \frac{1}{2}O_2(g) \longrightarrow H_2O(g)$ at 110 °C. 1 watt = 1 J s^{-1}.)

*19.104 A Ag/AgCl electrode dipping into 1.00 M HCl has a standard reduction potential of +0.2223 V. The half-reaction is

$$AgCl(s) + e^- \rightleftharpoons Ag(s) + Cl^-(aq)$$

A second Ag/AgCl electrode is dipped into a solution containing Cl^- at an unknown concentration. The cell generates a potential of 0.0478 V, with the electrode in the solution of unknown concentration having a negative charge. What is the molar concentration of Cl^- in the unknown solution?

19.105 Consider the following galvanic cell

$$Ag(s)|Ag^+(0.00030\ M)||Fe^{3+}(0.0011\ M),\ Fe^{2+}(0.040\ M)|Pt(s)$$

Calculate the cell potential. Determine the sign of the electrodes in the cell. Write the equation for the spontaneous cell reaction.

EXERCISES IN CRITICAL THINKING

19.106 In biochemical systems the normal standard state that requires $[H^+] = 1.00\ M$ is not realistic. (a) Which half-reactions in Table 19.1 will have different potentials if pH = 7.00 is defined as the standard state for hydronium ions? (b) What will the new standard reduction potentials be at pH = 7.00 for those reactions? These are called $E^{\circ\prime}_{cell}$ with the prime indicating the potential at pH = 7.00.

19.107 Calculate a new version of Table 19.1 using the lithium half-reaction to define zero. Does this change the results of any problems involving standard cell potentials?

19.108 In Problem 19.79 the potential at 50 °C was calculated. Does the change in molarity of the solutions, due to the change in density of water, have an effect on the potentials?

19.109 There are a variety of methods available for generating electricity. List as many methods as you can think of. Rank each of the methods based on your knowledge of (a) the efficiency of the method and (b) the environmental pollution caused by each method.

19.110 Using the cost of electricity in your area, how much will it cost to produce a case of 24 soda cans, each weighing 0.45 ounces? Assume that alternating current can be converted to direct current with 100% efficiency.

*19.111 Most flashlights use two or more batteries in series. Use the concepts of galvanic cells in this chapter to explain why a flashlight with two new batteries and one "dead" battery will give only a dim light if any light is obtained at all.

*19.112 If two electrolytic cells are placed in series, the same number of electrons must pass through both cells. One student argues you can get twice as much product if two cells are placed in series compared to a single cell and therefore the cost of production (i.e., the cost of electricity) will decrease greatly and profits will increase. Is the student correct? Explain your reasoning based on the principles of electrochemistry.

NUCLEAR REACTIONS AND THEIR ROLE IN CHEMISTRY

2

The nebula RCW49, shown in infrared light in this image from NASA's Spitzer Space Telescope, is a nursery for newborn stars. Nuclear fusion reactions involving hydrogen are responsible for the tremendous amounts of energy given off in stars. Reactions of this type are discussed in this chapter. *(Courtesy NASA/JPL-Caltech/University of Wisconsin-madison)*

CHAPTER OUTLIN

20.1 Mass and energy are conserved in *all* of their forms

20.2 The energy required to break a nucleus into separate nucleons is called the nuclear binding energy

20.3 Radioactivity is an emission of particles and/or electromagnetic radiation by unstable atomic nuclei

20.4 Stable isotopes fall within a "band of stability" on a plot based on numbers of protons and neutrons

20.5 Transmutation is the change of one isotope into another

20.6 How is radiation measured?

20.7 Radionuclides have medical and analytical applications

20.8 Nuclear fission and nuclear fusion release large amounts of energy

THIS CHAPTER IN CONTEXT From the standpoint of chemistry, our interest in the atomic stems primarily from its role in determining the number and energies of an atom's electrons. This is becau the electron distribution in an atom that controls chemical properties. Although the nuclei of most isoto exceptionally stable, many of the elements of interest to us also have one or more isotopes with unstable nuc have unique properties which make them particularly useful in chemistry. These unstable nuclei tend to emit r consisting of particles and/or energy. In the pages ahead you will learn about the different kinds of nuclear ra how this radiation is detected and measured, and how the properties of unstable nuclei can be applied to practical problems.

One of the benefits that comes from studying nuclear transformations is an understanding of the enormous amounts of energy associated with certain nuclear changes. As you study this chapter you will come to appreciate the origin of the energy given off by stars, including our own sun. You will also learn how nuclear reactors work and how nuclear reactions have been applied to produce nuclear weapons.

20.1 MASS AND ENERGY ARE CONSERVED IN ALL OF THEIR FORMS

Changes involving unstable atomic nuclei generally involve large amounts of energy, amounts that are considerably greater than in chemical reactions. To understand how these energy changes arise, we begin our study by re-examining two physical laws that, until this chapter, have been assumed to be separate and independent, namely, the laws of conservation of energy and conservation of mass. They may be safely treated as distinct for chemical reactions but not for nuclear reactions. These two laws, however, are only different aspects of a deeper, more general law.

As atomic and nuclear physics developed in the early 1900s, physicists realized that the mass of a particle cannot be treated as a constant in all circumstances. The mass, m, of a particle depends on the particle's velocity, v, meaning the velocity relative to the observer. A particle's mass is related to this velocity, and to the velocity of light, c, by the following equation.

□ $c = 3.00 \times 10^8$ m s^{-1}, the speed of light.

$$m = \frac{m_0}{\sqrt{1 - (v/c)^2}} \tag{20.1}$$

Notice what happens when v is zero and the particle has no velocity (relative to the observer). The ratio v/c is then zero, the whole denominator reduces to a value of 1, and Equation 20.1 becomes

$$m = m_0$$

This is why the symbol m_0 stands for the particle's *rest mass.*

Rest mass is what we measure in all lab operations, because any object, like a chemical sample, is either at rest (from our viewpoint) or is not moving extraordinarily rapidly. Only as the particle's velocity approaches the speed of light, c, does the v/c term in Equation 20.1 become important. As v approaches c, the ratio v/c approaches 1, and so $[1 - (v/c)^2]$ gets closer and closer to 0. The whole denominator, in other words, approaches a value of 0. If it actually reached 0, then m, which would be $(m_0 \div 0)$, would become infinity. In other words, the mass, m, of the particle moving at the velocity of light would be infinitely great, a physical impossibility. This is why the speed of light is seen as an absolute upper limit on the speed that any particle can approach.

□ Even at $v = 1000$ m s^{-1} (about 2250 mph), the denominator (not rounded) is 0.99999333, or within 7×10^{-4}% of 1.

At the velocities of everyday experience, the mass of anything calculated by Equation 20.1 equals the rest mass to four of five significant figures. The difference cannot be detected by weighing devices. Thus, in all of our normal work, mass appears to be conserved, and the law of conservation of mass functions this way in chemistry.

As a particle's velocity increases, energy is changed to mass

We know that matter cannot appear from nothing, so the extra mass an object acquires as it goes faster must come from the energy supplied to increase the object's velocity. Physicists therefore realized that mass and energy are interconvertible and that in the world of high energy physics, the laws of conservation of mass and conservation of energy are not separate and independent. What emerged was a single, deeper law now called the **law of conservation of mass–energy**.

Law of Conservation of Mass–Energy The sum of all the energy and of all the mass (expressed as an equivalent in energy) in the universe is a constant.

The Einstein equation quantitatively relates rest mass and energy

Albert Einstein, perhaps the most famous physicist of the twentieth century, was able to show that when mass converts to energy, the change in energy, ΔE, is related to the change in rest mass, Δm_0, by the following equation, now called the **Einstein equation.**

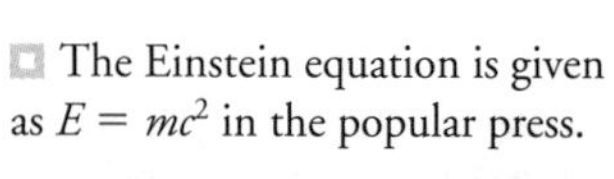

$$\Delta E = \Delta m_0 c^2 \tag{20.2}$$

Again, c is the velocity of light, 3.00×10^8 m s^{-1}.

Because the velocity of light is very large, even if an energy change is enormous, the change in mass, Δm_0, is extremely small. For example, the combustion of methane releases considerable heat per mole.

The Einstein equation is given as $E = mc^2$ in the popular press.

$$CH_4(g) + 2O_2(g) \longrightarrow CO_2(g) + 2H_2O(l) \qquad \Delta H° = -890 \text{ kJ}$$

The release of 890 kJ of heat energy corresponds to a loss of mass, which by the Einstein equation equals a loss of 9.89 ng. This is about 1×10^{-7}% of the mass of 1 mol of CH_4 and 2 mol of O_2. Such a tiny change in mass is not detectable by laboratory balances, so for all practical purposes, mass is conserved. Although the Einstein equation has no direct applications in chemistry, its importance certainly became clear when atomic fission (the breaking apart of heavy atoms to form lighter fragments) was first observed in 1939.

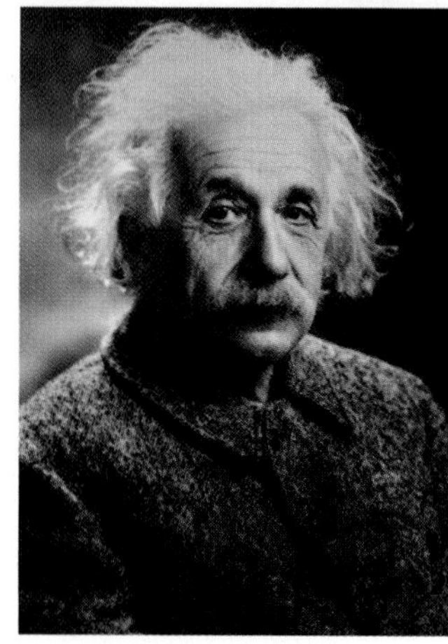

Albert Einstein (1879–1955); Nobel prize, 1921 (physics). *(Photo Researchers.)*

20.2 THE ENERGY REQUIRED TO BREAK A NUCLEUS INTO SEPARATE NUCLEONS IS CALLED THE NUCLEAR BINDING ENERGY

As we will discuss further in Section 20.3, an atomic nucleus is held together by extremely powerful forces of attraction that are able to overcome the repulsions between protons. To break a nucleus into its individual nucleons therefore requires an enormous input of energy. This energy is called the **nuclear binding energy.**

Absorption of the binding energy would produce the individual nucleons—protons and neutrons—that had made up the nucleus. These nucleons would now carry extra mass corresponding to the mass-equivalent of the energy they had absorbed. If we add up their masses, the sum should be larger than the mass of the nucleus from which they had come. And this is exactly what is observed. *For a given atomic nucleus, the sum of the rest masses of all of its nucleons is always a little larger than the actual mass of the nucleus.* The mass difference is called the **mass defect,** and its energy equivalent is the nuclear binding energy.

Keep in mind that nuclear binding energy is not energy actually possessed by the nucleus but is, instead, the energy the nucleus would have to absorb to break apart. Thus, the *higher* the binding energy, the *more stable* is the nucleus.

The nuclear binding energy can be calculated

We can calculate nuclear binding energy using the Einstein equation. Helium-4, for example, has atomic number 2, so its nucleus consists of 4 nucleons (2 protons and 2 neutrons). The rest mass of one helium-4 nucleus is known to be 4.0015061792 u. However, the sum of the rest masses of its four separated nucleons is slightly more, 4.0318827650 u, which we can show as follows. The rest mass of an isolated proton is 1.0072764669 u and that of a neutron is 1.0086649156 u.

$$\text{For 2 protons: } 2 \times 1.0072764669 \text{ u} = 2.0145529338 \text{ u}$$
$$\text{For 2 neutrons: } 2 \times 1.0086649156 \text{ u} = \underline{2.0173298312 \text{ u}}$$
$$\text{Total rest mass of nucleons in } ^4\text{He} = 4.0318827650 \text{ u}$$

The mass defect, the difference between the calculated and measured rest masses for the helium-4 nucleus, is 0.030375858 u, found by

$$\underset{\text{mass of the 4 nucleons}}{4.0318827650 \text{ u}} - \underset{\text{mass of nucleus}}{4.0015061792 \text{ u}} = \underset{\text{mass defect}}{0.0303765858 \text{ u}}$$

Using Einstein's equation, let's calculate to three significant figures the nuclear binding energy that is equivalent to the mass defect for the ^{4}He nucleus. We will round the mass defect to 0.0304 u. To obtain the energy in joules, we have to remember that $1 \text{ J} = 1 \text{ kg m}^2 \text{ s}^{-2}$, so the mass in atomic mass units (u) must be converted to kilograms. The table of constants inside the rear cover of the book gives $1 \text{ u} = 1.6605389 \times 10^{-24}$ g, which equals $1.6605389 \times 10^{-27}$ kg. Substituting into the Einstein equation gives

$$\Delta E = \Delta mc^2 = \underbrace{(0.0304 \cancel{\text{u}}) \times \frac{1.66065389 \times 10^{-27} \text{ kg}}{1 \cancel{\text{u}}}}_{\Delta m \text{ (in kg)}} \times \underbrace{(3.00 \times 10^8 \text{ m s}^{-1})^2}_{c^2}$$

$$= 4.54 \times 10^{-12} \text{ kg m}^2 \text{ s}^{-2}$$

$$= 4.54 \times 10^{-12} \text{ J}$$

There are four nucleons in the helium-4 nucleus, so the binding energy *per* nucleon is (4.54×10^{-12} J/4 nucleons) or 1.14×10^{-12} J/nucleon.

The formation of just one nucleus of ^{4}He releases 4.54×10^{-12} J. If we could make Avogadro's number or 1 mol of ^{4}He nuclei (with a total mass of only 4 g) the net release of energy would be

$$(6.02 \times 10^{23} \text{ nuclei}) \times (4.54 \times 10^{-12} \text{ J/nucleus}) = 2.73 \times 10^{12} \text{ J}$$

This is a huge amount of energy from forming only 4 g of helium. It could keep a 100 watt lightbulb lit for nearly 900 years!

Binding energy per nucleon varies from one element to another

Figure 20.1 shows a plot of binding energies per nucleon versus mass numbers for most of the elements. The curve passes through a maximum at iron-56, which means that the nuclei of iron-56 atoms are the most stable of all. The plot in Figure 20.1, however, does not have a sharp maximum. Thus, a large number of elements with intermediate mass numbers in the broad center of the periodic table include the most stable isotopes in nature.

Nuclei of low mass number have small binding energies per nucleon. Joining two such nuclei to form a heavier nucleus, a process called **nuclear fusion,** leads to a more stable nucleus and a large increase in binding energy per nucleon. This extra energy is released when the two lighter nuclei fuse (join) and is the origin of the energy released in the cores of stars and the detonation of a hydrogen bomb. Nuclear fusion is discussed further in Section 20.8.

As we follow the plot of Figure 20.1 to the highest mass numbers, the nuclei decrease in stability as the binding energies decrease. Among the heaviest atoms, therefore, we might expect to find isotopes that could change to more stable forms by breaking up into lighter nuclei, by undergoing nuclear fission. **Nuclear fission,** also discussed in Section 20.8, is the spontaneous breaking apart of a nucleus to form isotopes of intermediate mass number.

Quite often you'll see the terms *atomic fusion* and *atomic fission* used for nuclear fusion and nuclear fission.

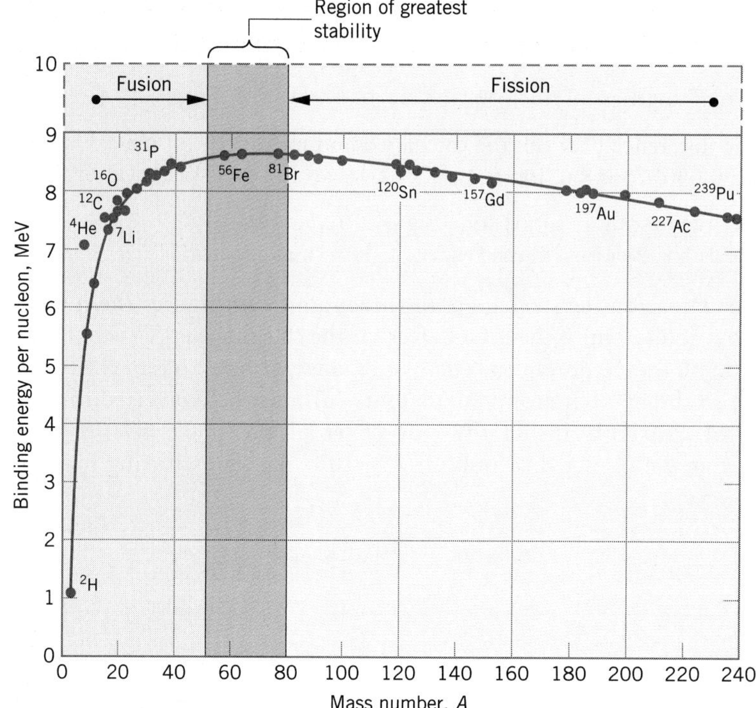

FIG. 20.1 Binding energies per nucleon. The energy unit here is the megaelectron volt or MeV: 1 MeV = 10^6 eV = 1.602×10^{-13} J. *(From D. Halliday and R. Resnick, Fundamentals of Physics, 2nd ed., Revised, 1986. John Wiley & Sons, Inc. Used by permission.)*

20.3 Radioactivity Is an Emission of Particles and/or Electromagnetic Radiation by Unstable Atomic Nuclei

Except for hydrogen, all atomic nuclei have more than one proton, each of which carries a positive charge. Because like charges repel, we might wonder how any nucleus could be stable. Electrostatic forces of attraction and repulsion, such as the kinds present among the ions in a crystal of sodium chloride, are not the only forces at work in the nucleus, however. Protons do, indeed, repel each other electrostatically, but another force, a force of attraction called the *nuclear strong force,* also acts in the nucleus. The nuclear strong force, effective only at very short distances, overcomes the electrostatic force of repulsion between protons, and it binds both protons and neutrons into a nuclear package. Moreover, the neutrons, by helping to keep the protons farther apart, also lessen repulsions between protons.

Adjacent neutrons experience no electrostatic repulsion between each other, only the strong force (of attraction).

One consequence of the difference between the nuclear strong force and the electrostatic force occurs among nuclei that have large numbers of protons but too few intermingled neutrons to dilute the electrostatic repulsions between protons. Such nuclei are often unstable; their nuclei carry more energy than do other arrangements of the same nucleons. To achieve a lower energy and thus more stability, unstable nuclei have a tendency to eject small nuclear fragments, and many simultaneously release high-energy electromagnetic radiation. The stream of particles (or photons) coming from the sample is called **nuclear radiation** or **atomic radiation** and the phenomenon is called **radioactivity.** Isotopes that exhibit this property are said to be **radioactive** and are called **radionuclides.** About 60 of the approximately 350 naturally occurring isotopes are radioactive.

In a sample of a given radionuclide, not all the atoms undergo change at once. The rate at which radiation is emitted (which translates into the intensity of the radiation) depends on the concentration of the isotope in the sample. Over time, as radioactive nuclei change into stable ones, the number of atoms of the radionuclide remaining in the sample decreases, causing the intensity of the radiation to drop, or *decay*. The radionuclide is said to undergo **radioactive decay.**

Naturally occurring atomic radiation consists principally of three kinds: alpha, beta, and gamma radiation, as discussed below.

Alpha radiation is a stream of helium nuclei

Alpha radiation consists of a stream of helium nuclei called **alpha particles** (α particles), symbolized as ${}^{4}_{2}He$, where 4 is the mass number and 2 is the atomic number. The alpha particle bears a charge of 2+, but the charge is omitted from the symbol.

Alpha particles are the most massive of those commonly emitted by radionuclides. When ejected (Figure 20.2), alpha particles move through the atom's electron orbitals, emerging from the atom at speeds of up to one-tenth the speed of light. Their size, however, prevents them from going far. After traveling at most only a few centimeters in air, alpha particles collide with air molecules, lose kinetic energy, pick up electrons, and become neutral helium atoms. Alpha particles cannot penetrate the skin, although enough exposure causes a severe skin burn. If carried in air or on food into the soft tissues of the lungs or the intestinal tract, emitters of alpha particles can cause serious harm, including cancer.

In Chapter 2 you learned that an isotope is identified by writing its mass number, A, as a superscript and its atomic number, Z, as a subscript in front of the chemical symbol, X, as in ${}^{A}_{Z}X$. We use this same notation in representing particles involved in nuclear reactions. For particles that are not atomic nuclei, Z stands for the charge on the particle.

Nuclear equations describe the decay of radioactive nuclei

To symbolize the decay of a nucleus, we construct a **nuclear equation**, which we can illustrate by the alpha decay of uranium-238 to thorium-234.

$$ {}^{238}_{92}U \longrightarrow {}^{234}_{90}Th + {}^{4}_{2}He $$

Unlike chemical reactions, nuclear reactions produce new isotopes, so we need separate rules for balancing nuclear equations.

A nuclear transformation such as the alpha decay of ${}^{238}U$ does not depend on the chemical environment. The same nuclear equation applies whether the uranium is in the form of the free element or in a compound.

Rules for balancing a nuclear equation

1. The sum of the mass numbers on each side of the arrow must be the same.
2. The sum of the atomic numbers (nuclear charge) on each side of the arrow must be the same.

In the nuclear equation for the decay of uranium-238, the atomic numbers balance ($90 + 2 = 92$), and the mass numbers balance ($234 + 4 = 238$). Notice that electrical charges are not shown, even though they are there (initially). The alpha particle, for example, has a charge of 2+. If it was emitted by a neutral uranium atom, then the thorium particle initially has a charge of 2−. These charged particles, however, eventually pick up or lose electrons either from each other or from molecules in the matter through which they travel.

Beta radiation is a stream of electrons

Naturally occurring **beta radiation** consists of a stream of electrons, which in this context are called **beta particles**. In a nuclear equation, the beta particle has the symbol ${}^{0}_{-1}e$, because the electron's mass number is 0 and its charge is 1−. Hydrogen-3 (tritium) is a beta emitter that decays by the following equation.

$$ \underset{\text{tritium}}{{}^{3}_{1}H} \longrightarrow \underset{\text{helium-3}}{{}^{3}_{1}He} + \underset{\text{beta particle (electron)}}{{}^{0}_{-1}e} + \underset{\text{antineutrino}}{\bar{\nu}} $$

Sometimes the beta particle is given the symbol ${}^{0}_{-1}\beta$, or simply ${}^{-}\beta$.

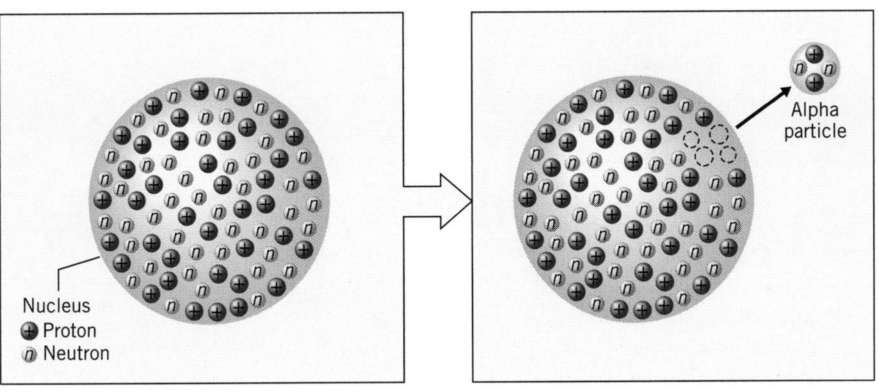

FIG. 20.2 Emission of an alpha particle from an atomic nucleus. Removal of ${}^{4}_{2}He$ from a nucleus decreases the atomic number by 2 and the mass number by 4.

Both the antineutrino (to be described further shortly) and the beta particle come from the atom's nucleus, not its electron shells. We do not think of them as having a prior existence in the nucleus, any more than a photon exists before its emission from an excited atom (see Figure 20.3). Both the beta particle and the antineutrino are created during the decay process in which a neutron is transformed into a proton.

$$\underset{\substack{\text{neutron}\\\text{(in the}\\\text{nucleus)}}}{{}^{1}_{0}n} \longrightarrow \underset{\substack{\text{beta particle}\\\text{(emitted)}}}{{}^{0}_{-1}e} + \underset{\substack{\text{proton}\\\text{(remains in}\\\text{the nucleus)}}}{{}^{1}_{1}p} + \underset{\substack{\text{antineutrino}\\\text{(emitted)}}}{\overline{\nu}}$$

Unlike alpha particles, which are all emitted with the same discrete energy from a given radionuclide, beta particles emerge from a given beta emitter with a continuous spectrum of energies. Their energies vary from zero to some characteristic fixed upper limit for each radionuclide. This fact once gave nuclear physicists considerable trouble, partly because it was an apparent violation of energy conservation. To solve this problem, Wolfgang Pauli proposed in 1927 that beta emission is accompanied by emission of yet another decay particle, this one electrically neutral and almost massless. Enrico Fermi suggested the name *neutrino* ("little neutral one"), but eventually it was named the *antineutrino*, symbolized by $\overline{\nu}$.

An electron is extremely small, so a beta particle is less likely to collide with the molecules of anything through which it travels. Depending on its initial kinetic energy, a beta particle can travel up to 300 cm (about 10 ft) in dry air, much farther than alpha particles. Only the highest energy beta particles can penetrate the skin, however.

Gamma radiation is very high energy electromagnetic radiation

Gamma photons have the symbol ${}^{0}_{0}\gamma$ because they have no charge or mass.

Gamma radiation, which accompanies most nuclear decays, consists of high-energy photons given the symbol ${}^{0}_{0}\gamma$ or, often, simply γ in equations. Gamma radiation is extremely penetrating and is effectively blocked only by very dense materials, like lead.

The emission of gamma radiation involves transitions between energy levels *within* the nucleus. Nuclei have energy levels of their own, much as atoms have orbital energy levels. When a nucleus emits an alpha or beta particle, it sometimes is left in an excited energy state. By the emission of a gamma-ray photon, the nucleus relaxes into a more stable state.

The electron volt is an energy unit

The energy carried by a given radiation is usually described by an energy unit new to our study, the **electron volt,** abbreviated **eV;** 1 eV is the energy an electron receives when accelerated under the influence of 1 volt. It is related to the joule as follows.

$$1 \text{ eV} = 1.602 \times 10^{-19} \text{ J}$$

In the interconversion of mass and energy,

$1\text{eV} \Leftrightarrow 1.783 \times 10^{-36}$ kg
$1\text{MeV} \Leftrightarrow 1.78 \times 10^{-27}$ g
$1 \text{ GeV} \Leftrightarrow 1.783 \times 10^{-24}$ g

As you can see, the electron volt is an extremely small amount of energy, so multiples are commonly used, like the kilo-, mega-, and gigaelectron volt.

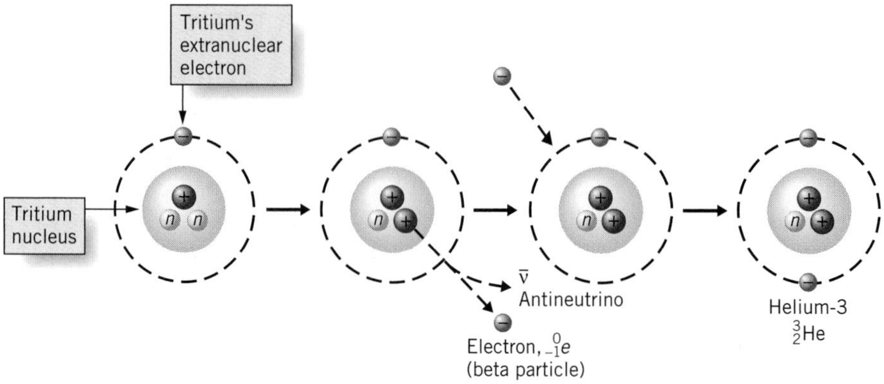

FIG. 20.3 Emission of a beta particle from a tritium nucleus. Emission of a beta particle changes a neutron into a proton. This results in a positively charged ion, which picks up an electron to become a neutral atom.

$$1\ \text{keV} = 10^3\ \text{eV}$$
$$1\ \text{MeV} = 10^6\ \text{eV}$$
$$1\ \text{GeV} = 10^9\ \text{eV}$$

An alpha particle emitted by radium-224 has an energy of 5 MeV. Hydrogen-3 (tritium) emits beta radiation at an energy of 0.05 to 1 MeV. The gamma radiation from cobalt-60, the radiation currently used to kill bacteria and other pests in certain foods, consists of photons with energies of 1.173 MeV and 1.332 MeV.

X rays are high energy electromagnetic radiation

X rays, like gamma rays, consist of high-energy electromagnetic radiation, but their energies are usually less than those of gamma radiation. Although X rays are emitted by some synthetic radionuclides, when needed for medical diagnostic work they are generated by focusing a high-energy electron beam onto a metal target.

X rays used in diagnosis typically have energies of 100 keV or less.

Some radioisotopes emit positrons or neutrons; others capture electrons

Many *synthetic isotopes* (isotopes not found in nature, but synthesized by methods discussed in Section 20.5) emit *positrons,* which are particles with the mass of an electron but a positive instead of a negative charge. A **positron** is a positive beta particle, a positive electron, and its symbol is ${}^{0}_{1}e$. It forms in the nucleus by the conversion of a proton to a neutron (Figure 20.4). Positron emission, like beta emission, is accompanied by a chargeless and virtually massless particle, a *neutrino* (ν), the counterpart of the antineutrino ($\overline{\nu}$) in the realm of antimatter (defined below). Cobalt-54, for example, is a positron emitter and changes to a stable isotope of iron.

The symbol $^{+}\beta$ is sometimes used for the positron.

$$ {}^{54}_{27}\text{Co} \longrightarrow {}^{54}_{26}\text{Fe} + \underset{\text{positron}}{{}^{0}_{1}e} + \underset{\text{neutrino}}{\nu} $$

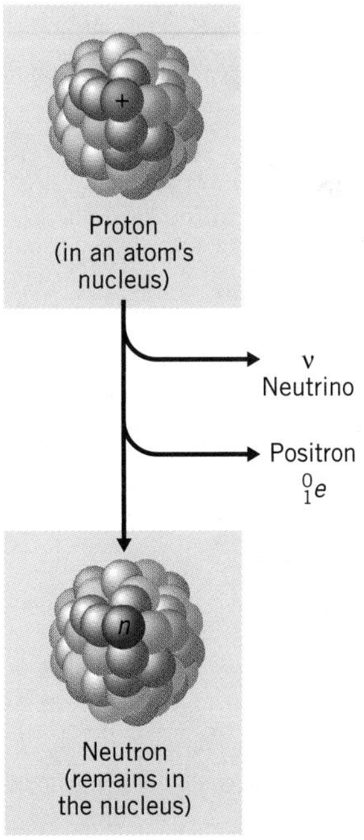

A positron, when emitted, eventually collides with an electron, and the two annihilate each other (Figure 20.5). Their masses change entirely into the energy of two gamma-ray photons called *annihilation radiation photons,* each with an energy of 511 keV.

$$ {}^{0}_{-1}e + {}^{0}_{1}e \longrightarrow 2\,{}^{0}_{0}\gamma $$

Because a positron destroys a particle of ordinary matter (an electron), it is called a particle of antimatter. To be classified as **antimatter,** a particle must have a counterpart among one of ordinary matter, and the two must annihilate each other when they collide. For example, a neutron and an antineutron represent such a pair and annihilate each other when they come in contact.

Neutron emission, another kind of nuclear reaction, does not lead to an isotope of a different element. Krypton-87, for example, decays as follows to krypton-86.

$$ {}^{87}_{36}\text{Kr} \longrightarrow {}^{86}_{36}\text{Kr} + \underset{\text{neutron}}{{}^{1}_{0}n} $$

FIG. 20.4 The emission of a positron replaces a proton by a neutron.

Electron capture, yet another kind of nuclear reaction, is very rare among natural isotopes but common among synthetic radionuclides. For example, an electron can be captured from the orbital electron shell having $n = 1$ or $n = 2$ by a vanadium-50 nucleus, causing it to change to a stable ^{50}Ti nucleus. The transformation is accompanied by the emmision of X rays and a neutrino.

$$ {}^{50}_{23}\text{V} + {}^{0}_{-1}e \xrightarrow{\text{Electron capture}} {}^{50}_{22}\text{Ti} + \text{X rays} + \nu $$

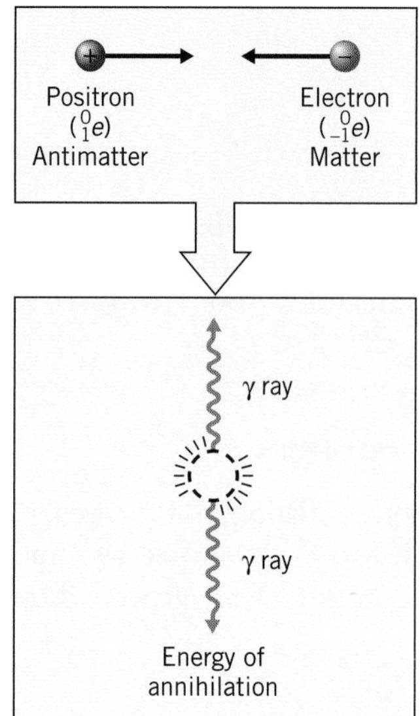

FIG. 20.5 Gamma radiation is produced when a positron and an electron collide. Annihilation of a positron and an electron leads to two gamma-ray photons.

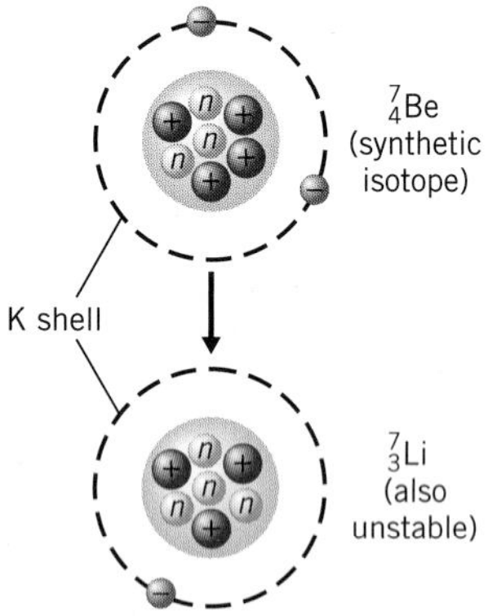

The net effect in the nucleus of electron capture is the conversion of a proton into a neutron (Figure 20.6).

$$^{1}_{1}p + {}^{0}_{-1}e \longrightarrow {}^{1}_{0}n$$

proton (in the nucleus) | electron (captured from the 1*s* orbital) | neutron (in the nucleus)

Electron capture does not change an atom's mass number, only its atomic number. It also leaves a hole in the first or second electron shell, and the atom emits photons of X rays as other orbital electrons drop down to fill the hole. Moreover, the nucleus that has just captured an orbital electron may be in an excited energy state and so can emit a gamma ray photon.

$^{7}_{4}Be$ (synthetic isotope)

K shell

$^{7}_{3}Li$ (also unstable)

FIG. 20.6 Electron capture. Electron capture is the collapse of an orbital electron into the nucleus, and this changes a proton into a neutron. A gap is left in a low-energy electron orbital. When an electron drops from a higher orbital to fill the gap, an X-ray photon is emitted.

A radioactive disintegration series is a sequence of successive nuclear reactions

Sometimes a radionuclide does not decay directly to a stable isotope, but decays instead to another unstable radionuclide. The decay of one radionuclide after another will continue until a stable isotope forms. The sequence of such successive nuclear reactions is called a **radioactive disintegration series.** Four series occur naturally. Uranium-238 is at the head of one of them (Figure 20.7).

The rates of decay of radionuclides vary and are usually described by specifying their half lives, $t_{1/2}$, a topic we studied in Section 13.4. One *half-life* period in nuclear science is

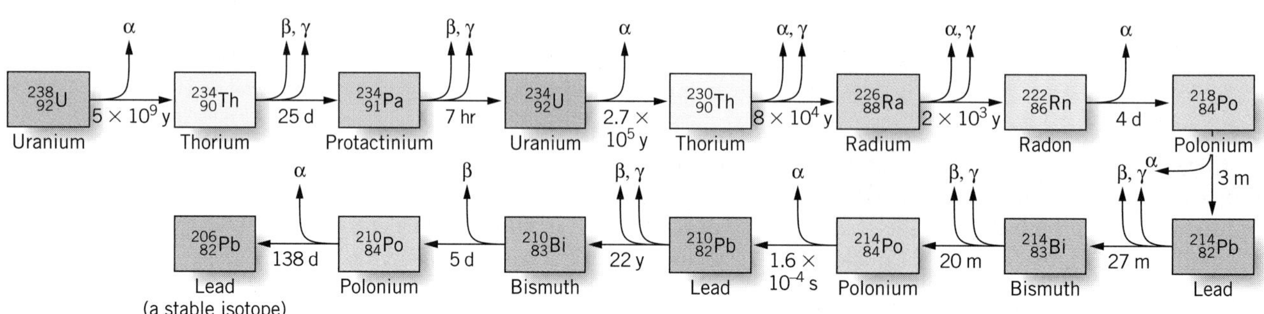

FIG. 20.7 The uranium-238 radioactive disintegration series. The time given beneath each arrow is the half-life period of the preceding isotope (y = year, m = month, d = day, hr = hour, and s = second).

TABLE 20.1 Typical Half-Life Periods

Element	Isotope	Half-Life	Radiations or Mode of Decay
Naturally occurring radionuclides			
Potassium	${}^{40}_{19}\mathrm{K}$	1.3×10^{9} yr	beta, gamma
Tellurium	${}^{123}_{52}\mathrm{Te}$	1.2×10^{13} yr	electron capture
Neodymium	${}^{144}_{60}\mathrm{Nd}$	5×10^{15} yr	alpha
Samarium	${}^{149}_{62}\mathrm{Sm}$	4×10^{14} yr	alpha
Rhenium	${}^{187}_{75}\mathrm{Re}$	7×10^{10} yr	beta
Radon	${}^{222}_{86}\mathrm{Rn}$	3.82 day	alpha
Radium	${}^{226}_{88}\mathrm{Ra}$	1590 yr	alpha, gamma
Thorium	${}^{230}_{90}\mathrm{Th}$	8×10^{4} yr	alpha, gamma
Uranium	${}^{238}_{92}\mathrm{U}$	4.51×10^{9} yr	alpha
Synthetic radionuclides			
Tritium	${}^{3}_{1}\mathrm{T}$	12.26 yr	beta
Oxygen	${}^{15}_{8}\mathrm{O}$	124 s	positron
Phosphorus	${}^{32}_{15}\mathrm{P}$	14.3 day	beta
Technetium[a]	${}^{99m}_{43}\mathrm{Tc}$	6.02 hr	gamma
Iodine	${}^{131}_{53}\mathrm{I}$	8.07 day	beta
Cesium	${}^{137}_{55}\mathrm{Cs}$	30 yr	beta, gamma
Strontium	${}^{90}_{38}\mathrm{Sr}$	28.1 yr	beta
Plutonium	${}^{238}_{94}\mathrm{Pu}$	87.8 yr	alpha
Americium	${}^{243}_{95}\mathrm{Am}$	7.37×10^{3} yr	alpha

[a] The superscript 99m refers to a metastable isotope of technetium, which has a higher energy than technetium-99.

the time it takes for a given sample of a radionuclide to decay to one-half of its initial amount. Radioactive decay is a first-order process, so the period of time taken by one half-life is independent of the initial number of nuclei. The huge variations in the half-lives of several radionuclides are shown in Table 20.1.

EXAMPLE 20.1 Writing a Balanced Nuclear Equation

Cesium-137, ${}^{137}_{55}\mathrm{Cs}$, one of the radioactive wastes from a nuclear power plant or an atomic bomb explosion, emits beta and gamma radiation. Write the nuclear equation for the decay of cesium-137.

Ions of cesium, which is in the same family as sodium, travel in the body to many of the same sites where sodium ions go.

ANALYSIS: We will start with an incomplete equation using the given information. Then we will use the requirements for a balanced nuclear equation as a tool to figure out any other data needed to complete the equation.

SOLUTION: The incomplete nuclear equation is

$${}^{137}_{55}\mathrm{Cs} \longrightarrow {}^{0}_{-1}e + {}^{0}_{0}\gamma + {}^{\_\_}_{\_\_}\_\_$$

Mass number goes here.

Atomic symbol goes here.

Atomic number goes here.

The atomic symbol can be obtained from the table inside the front cover after we have determined the atomic number, Z. Z is found using the fact that the atomic number (55) on the left side of the equation must equal the sum of the atomic numbers on the right side.

$$55 = -1 + 0 + Z$$

$$Z = 56$$

The periodic table tells us that element 56 is Ba (barium). To determine which isotope of barium forms, we recall that the sums of the mass numbers on either side of the equation must also be equal. Letting A be the mass number of the barium isotope,

$$137 = 0 + 0 + A$$

$$A = 137$$

The balanced nuclear equation, therefore, is

$$^{137}_{55}\text{Cs} \longrightarrow {}^{0}_{-1}e + {}^{0}_{0}\gamma + {}^{137}_{56}\text{Ba}$$

IS THE ANSWER REASONABLE? Besides double-checking that element 56 is barium, the answer satisfies the requirements of a nuclear equation, namely, the sums of the mass numbers, 137, are the same on both sides as are the sums of the atomic numbers.

Practice Exercise 1: Marie Curie earned one of her two Nobel prizes for isolating the element radium, which soon became widely used to treat cancer. Radium-226, $^{226}_{88}\text{Ra}$, emits a gamma photon plus a particle to give radon-222. Write a balanced nuclear equation for its decay and identify the particle that's emitted. (Hint: Be sure to balance atomic number and mass.)

Practice Exercise 2: Write the balanced nuclear equation for the decay of strontium-90, a beta emitter. (Strontium-90 is one of the many radionuclides present in the wastes of operating nuclear power plants.)

20.4 STABLE ISOTOPES FALL WITHIN A "BAND OF STABILITY" ON A PLOT BASED ON NUMBERS OF PROTONS AND NEUTRONS

When all known isotopes of each element, both stable and unstable, are arrayed on a plot according to numbers of protons and neutrons, an interesting zone can be defined (Figure 20.8). The two curved lines in the array of Figure 20.8 enclose this zone, called the **band of stability,** within which lie all stable nuclei. (No isotope above element 83, bismuth, is included in Figure 20.8 because none has a *stable* isotope.) Within the band of stability are also some unstable isotopes, because smooth lines cannot be drawn to exclude them.

Any isotope not represented anywhere on the array, inside or outside the band of stability, probably has a half-life too short to permit its detection. For example, an isotope with 50 neutrons and 60 protons would lie well below the band of stability and would likely be extremely unstable. Any attempt to make it would likely be a waste of time and money.

Notice that the band curves slightly upward as the number of protons increases. The curvature means that the *ratio* of neutrons to protons gradually increases from 1:1, a ratio indicated by the straight line in Figure 20.8. The reason is easy to understand. More protons require more neutrons to provide a compensating nuclear strong force and to dilute electrostatic proton–proton repulsions.

Isotopes occurring above and to the left of the band of stability tend to be beta emitters. Isotopes lying below and to the right of the band are positron emitters. The isotopes with atomic numbers above 83 tend to be alpha emitters. Are there any reasons for these tendencies?

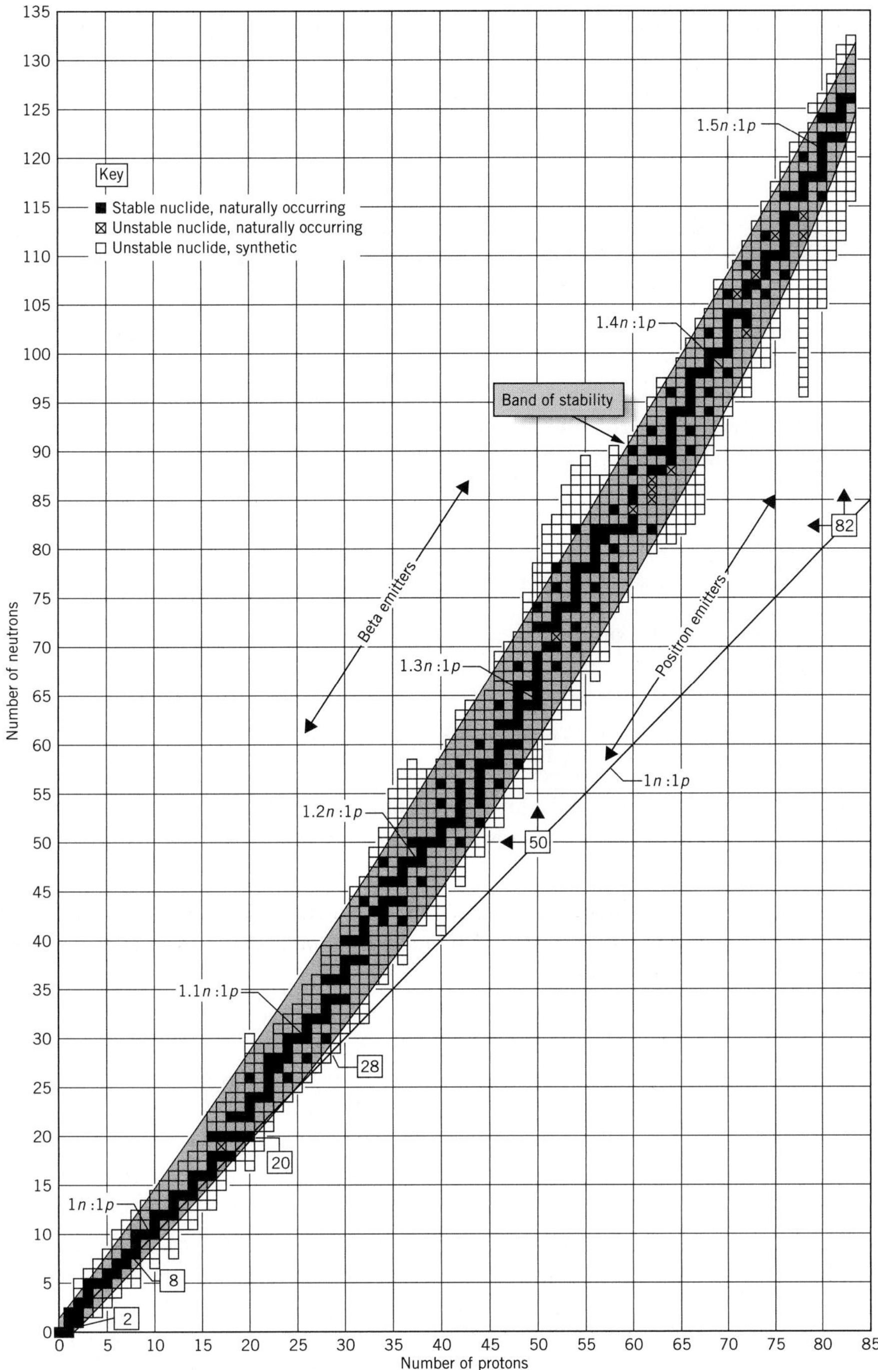

FIG. 20.8 **The band of stability.** Stable nuclei fall within a narrow band in a plot of number of neutrons versus number of protons. Nuclei far outside the band are too unstable to exist.

Alpha Emitters

The alpha emitters, as we said, occur mostly among the radionuclides above atomic number 83. Their nuclei have too many protons, and the most efficient way to lose protons is by the loss of an alpha particle.

Beta Emitters

Beta emitters are generally above the band of stability and so have neutron-to-proton ratios that evidently are too high. By beta decay a nucleus loses a neutron and gains a proton, thus decreasing the ratio.

$$^{1}_{0}n \longrightarrow {}^{1}_{1}p + {}^{\ 0}_{-1}e + \overline{\nu}$$

The proton can also be given the symbol $^{1}_{1}H$ in nuclear equations.

For example, by beta decay fluorine-20 decreases its neutron-to-proton ratio from 11/9 to 10/10.

$$^{20}_{9}\text{F} \longrightarrow {}^{20}_{10}\text{Ne} + {}^{0}_{-1}e + \overline{\nu}$$

$$\frac{\text{neutron}}{\text{proton}} = \frac{11}{9} \qquad \frac{10}{10}$$

The surviving nucleus, that of neon-20, is closer to the center of the band of stability. Figure 20.9, an enlargement of the fluorine part of Figure 20.8, explains this change further. Figure 20.9 also shows how the beta decay of magnesium-27 to aluminum-27 also lowers the neutron-to-proton ratio.

Positron Emitters

In nuclei with too few neutrons to be stable, positron emission increases the neutron-to-proton ratio by changing a proton into a neutron. A fluorine-17 nucleus, for example, increases its neutron-to-proton ratio, improves its stability, and moves into the band of stability by emitting a positron and a neutrino and changing to oxygen-17 (see Figure 20.9).

$$^{17}_{9}\text{F} \longrightarrow {}^{17}_{8}\text{O} + {}^{0}_{1}e + \nu$$

$$\frac{\text{neutron}}{\text{proton}} = \frac{8}{9} \qquad \frac{9}{8}$$

Positron decay by magnesium-23 to sodium-23 produces a similarly favorable shift (also shown in Figure 20.9).

Nuclei with even numbers of protons and neutrons are likely to be stable

Study of the compositions of stable nuclei reveals an interesting relationship: Nature favors even numbers for protons and neutrons. This is summarized by the **odd–even rule**.

Odd–even rule

Odd–Even Rule When the numbers of neutrons and protons in a nucleus are both even, the isotope is far more likely to be stable than when both numbers are odd.

These five stable isotopes, $^{2}_{1}\text{H}$, $^{6}_{3}\text{Li}$, $^{10}_{5}\text{B}$, $^{14}_{7}\text{N}$, and $^{138}_{57}\text{La}$, all have odd numbers of both protons and neutrons.

Of the 264 stable isotopes, only five have odd numbers of both protons and neutrons, whereas 157 have even numbers of both. The rest have an odd number of one nucleon and an even number of the other. To see this, notice in Figure 20.8 how the horizontal lines

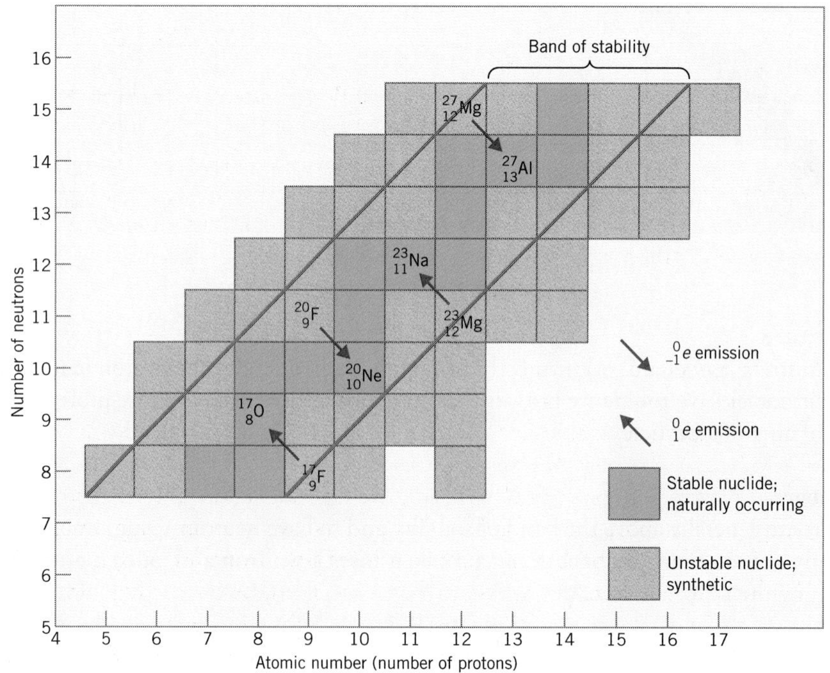

FIG. 20.9 An enlarged section of the band of stability. Beta decay from magnesium-27 and fluorine-20 reduces their neutron-to-proton ratio and moves them closer to the band of stability. Positron decay from magnesium-23 and fluorine-17 increases the ratio and moves those nuclides closer to the band of stability, too.

FACETS OF CHEMISTRY

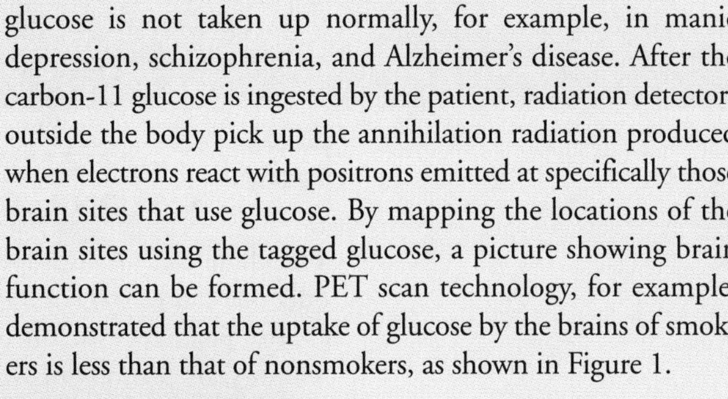
20.1

Positron Emission Tomography (PET)

Positron emitters are used in an important method for studying brain function called the PET scan, for *positron emission tomography*. The technique begins by chemically incorporating positron-emitting radionuclides into molecules, like glucose, that can be absorbed by the brain directly from the blood. It's like inserting radiation generators that act from within the brain rather than focusing X rays or gamma rays from the outside. Carbon-11, for example, is a positron emitter whose atoms can be used in place of carbon-12 atoms in glucose molecules, $C_6H_{12}O_6$. (One way to prepare such glucose is to let a leafy vegetable, Swiss chard, use $^{11}CO_2$ to make the glucose by photosynthesis.)

A PET scan using tiny amounts of carbon-11 glucose detects situations where glucose is not taken up normally, for example, in manic depression, schizophrenia, and Alzheimer's disease. After the carbon-11 glucose is ingested by the patient, radiation detectors outside the body pick up the annihilation radiation produced when electrons react with positrons emitted at specifically those brain sites that use glucose. By mapping the locations of the brain sites using the tagged glucose, a picture showing brain function can be formed. PET scan technology, for example, demonstrated that the uptake of glucose by the brains of smokers is less than that of nonsmokers, as shown in Figure 1.

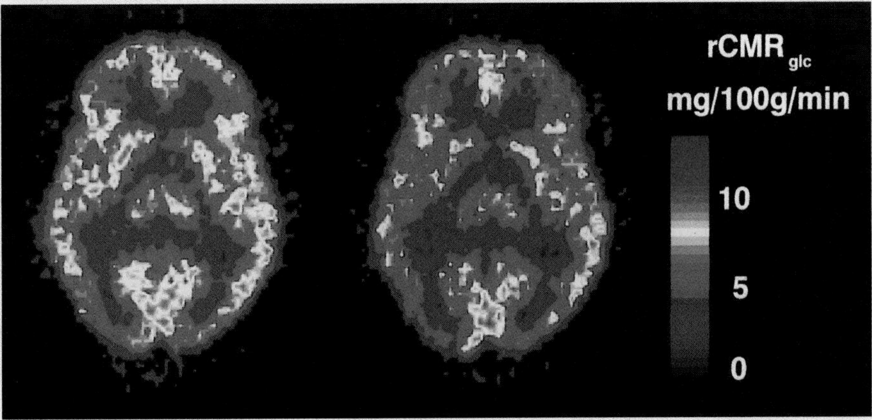
rCMR glc
mg/100g/min
10
5
0
The control, a PET scan of a normal brain.
The PET scan of the brain of a volunteer injected with nicotine.
The color code indicates the rates of glucose metabolism.

FIG. 1 Positron emission tomography (PET) in the study of brain activity. (*Left*) Normal brain. (*Right*) Brain affected by nicotine. The PET scan reveals widespread reduction in the rate of glucose metabolism when nicotine is present. *(Courtesy of E.D. London, National Institute of Drug Abuse.)*

with the largest numbers of dark squares (stable isotopes) most commonly correspond to even numbers of neutrons. Similarly, the vertical lines with the most dark squares most often correspond to even numbers of protons.

The odd–even rule is related to the spins of nucleons. Both protons and neutrons behave as though they spin, like orbital electrons. When two protons or two neutrons have paired spins, meaning the spins are opposite, their combined energy is less than when the spins are unpaired. Only when there are even numbers of protons and neutrons can all the spins be paired and so give the nucleus a lower energy and greater stability. The least stable nuclei tend to be those with both an odd number of protons and an odd number of neutrons.

Isotopes with "magic numbers" are especially stable

Another rule of thumb for nuclear stability is based on *magic numbers* of nucleons. Isotopes with specific numbers of protons or neutrons, the **magic numbers,** are more stable than the rest. The magic numbers of nucleons are 2, 8, 20, 28, 50, 82, and 126, and where they fall is shown in Figure 20.8 (except for magic number 126).

When the numbers of both protons and neutrons are the same magic number, as they are in $^{4}_{2}He$, $^{16}_{8}O$, and $^{40}_{20}Ca$ the isotope is very stable. $^{100}_{50}Sn$ also has two identical magic numbers. Although this isotope of tin is unstable, having a half-life of only several seconds, it is much more stable than nearby radionuclides, whose half-lives are in milliseconds. Thus, although tin-100 lies well outside the band of stability, it is stable enough to be observed. One stable isotope of lead, $^{208}_{82}Pb$, involves two different magic numbers, 82 protons and 126 neutrons.

Magic numbers do not cancel the need for a favorable neutron-to-proton ratio. An atom with 82 protons and 82 neutrons lies far outside the band of stability, and yet 82 is a magic number.

The existence of magic numbers supports the hypothesis that a nucleus has a shell structure with energy levels analogous to electron energy levels. Electron levels, as you already know, are associated with their own special numbers, those that equal the maximum number of electrons allowed in a principal energy level: 2, 8, 18, 32, 50, 72, and 98 (for principal levels with n equal to 1, 2, 3, 4, 5, 6, and 7, respectively). The total numbers of electrons in the atoms of the most chemically stable elements—the noble gases—also make up a special set: 2, 10, 18, 36, 54, and 86 electrons. Thus, special sets of numbers are not unique to nuclei.

20.5 TRANSMUTATION IS THE CHANGE OF ONE ISOTOPE INTO ANOTHER

The change of one isotope into another is called **transmutation,** and radioactive decay is only one cause. Transmutation can also be forced by the bombardment of nuclei with high-energy particles, such as alpha particles from natural emitters, neutrons from atomic reactors, or protons made by stripping electrons from hydrogen. To make them better bombarding missiles, protons and alpha particles can be accelerated in an electrical field (Figure 20.10). This gives them greater energy and enables them to sweep through the target atom's orbital electrons and become buried in its nucleus. Although beta particles can be accelerated, their disadvantage is that they are repelled by a target atom's own electrons.

Transmutation occurs when compound nuclei decay

Compound here refers only to the idea of *combination*, not to a chemical.

Both the energy and the mass of a bombarding particle enter the target nucleus at the moment of capture. The energy of the new nucleus, called a **compound nucleus,** quickly becomes distributed among all of the nucleons, but the nucleus is nevertheless rendered somewhat unstable. To get rid of the excess energy, a compound nucleus generally ejects something (a neutron, proton, or electron) and often emits gamma radiation as well. This leaves a new nucleus of an isotope different than the original target, so a transmutation has occurred overall.

Ernest Rutherford observed the first example of artificial transmutation. When he let alpha particles pass through a chamber containing nitrogen atoms, an entirely new radiation was generated, one much more penetrating than alpha radiation. It proved to be a stream of protons (Figure 20.11). Rutherford was able to show that the protons came from the decay of the compound nuclei of fluorine-18, produced when nitrogen-14 nuclei captured bombarding alpha particles.

The asterisk, *, symbolizes a high-energy nucleus, a *compound nucleus.*

$$\underset{\text{alpha particle}}{{}^{4}_{2}\text{He}} + \underset{\text{nitrogen nucleus}}{{}^{14}_{7}\text{N}} \longrightarrow \underset{\text{fluorine (compound nucleus)}}{{}^{18}_{9}\text{F}^{*}} \longrightarrow \underset{\text{oxygen (a rare but stable isotope)}}{{}^{17}_{8}\text{O}} + \underset{\text{proton (high energy)}}{{}^{1}_{1}p}$$

FIG. 20.10 **Linear accelerator.** In this linear accelerator at Brookhaven National Laboratory on Long Island, New York, protons can be accelerated to just under the speed of light and given up to 33 GeV of energy before they strike their target. (1 GeV = 10^9 eV.) *(Courtesy Brookhaven National Laboratory.)*

Alpha particle ^4_2He

Proton ^1_1H

Nitrogen atom $^{14}_7N$

Fluorine atom $^{18}_9F^*$

Oxygen atom $^{17}_8O$

FIG. 20.11 Transmutation of nitrogen into oxygen. When the nucleus of nitrogen-14 captures an alpha particle it becomes a compound nucleus of fluorine-18. This then expels a proton and becomes the nucleus of oxygen-17.

In the synthesis of alpha particles from lithium-7, protons are used as bombarding particles. The resulting compound nucleus, that of beryllium-8, splits in two.

$$\underset{\text{proton}}{^1_1p} + \underset{\text{lithium}}{^7_3Li} \longrightarrow \underset{\text{beryllium}}{^8_4Be^*} \longrightarrow \underset{\text{alpha particles}}{2\,^4_2He}$$

Compound nuclei can follow various decay modes

A given compound nucleus can be made in a variety of ways. Aluminum-27, for example, forms by any of the following routes.

$$^4_2He + {}^{23}_{11}Na \longrightarrow {}^{27}_{13}Al^*$$

$$^1_1p + {}^{26}_{12}Mg \longrightarrow {}^{27}_{13}Al^*$$

$$\underset{\text{deuteron}}{^2_1H} + {}^{25}_{12}Mg \longrightarrow {}^{27}_{13}Al^*$$

A deuteron is the nucleus of a deuterium atom, just as a proton is the nucleus of a protium (hydrogen) atom.

Each path gives the compound nucleus $^{27}_{13}Al^*$ a different amount of energy. Depending on this energy, different paths of decay are available, and all of the following routes have been observed. They illustrate how the synthesis of such a large number of synthetic isotopes, some stable and others unstable, has been possible.

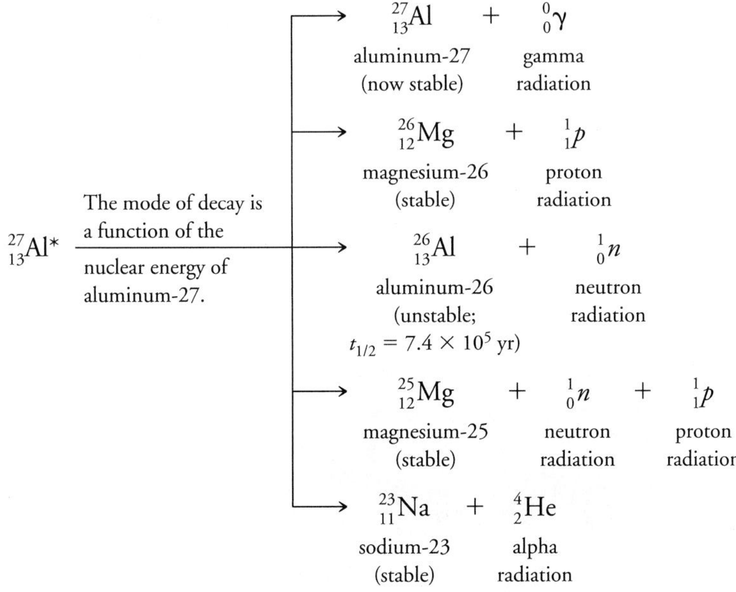

Transmutation is used to produce synthetic elements

Over a thousand isotopes have been made by transmutations. Most do not occur naturally; they appear in the band of stability (Figure 20.8) as open squares, nearly 900 in number. The naturally occurring radioactive isotopes above atomic number 83 all have very long half-lives. Others might have existed, but their half-lives probably were too short to permit them to last into our era. All of the elements beyond neptunium (atomic number 93 and higher, known as the **transuranium elements**) are synthetic. Elements from atomic numbers 93 to 103 complete the actinide series of the periodic table, which starts with element 90, thorium. Beyond this series, elements 104–116 and 118 have also been made.

To make the heaviest elements, bombarding particles larger than neutrons are used, such as alpha particles or the nuclei of heavier atoms. For example, element 110—darmstadtium, Ds—was made when a neutron was ejected from the compound nucleus formed by the fusion of nickel-62 and lead-208.

Element 110 was named after the place of its discovery, Darmstadt, Germany. Element 111 was named in honor of Wilhelm Roentgen, who discovered X rays in 1895.

$$^{62}_{28}\text{Ni} + ^{208}_{82}\text{Pb} \longrightarrow ^{270}_{110}\text{Ds} \longrightarrow ^{269}_{110}\text{Ds} + ^{1}_{0}n$$

Similarly, some atoms of element 111 (roentgenium, Rg) formed when a neutron was lost from the compound nucleus made by bombarding bismuth-209 with nickel-64. Most of these heavy atoms are extremely unstable, with half-lives measured in fractions of milliseconds. An exception is element 114; two isotopes have been detected with half-lives reported to be in seconds.

20.6 How is Radiation Measured?

Atomic radiation is often described as **ionizing radiation** because it creates ions by knocking electrons from molecules in the matter through which the radiation travels. The generation of ions is behind some of the devices for detecting radiation.

The Geiger–Müller tube, one part of a **Geiger counter,** detects beta and gamma radiation having energy high enough to penetrate the tube's window. Inside the tube is a gas under low pressure in which ions form when radiation enters. The ions permit a pulse of electricity to flow, which activates a current amplifier, a pulse counter, and an audible click.

A **scintillation counter** (Figure 20.12) contains a sensor composed of a substance called a *phosphor* that emits a tiny flash of light when struck by a particle of ionizing radiation. These flashes can be magnified electronically and automatically counted.

The darkening of a photographic film exposed to radiation over a period of time is proportional to the total quantity of radiation received. **Film dosimeters** work on this principle (Figure 20.13), and people who work near radiation sources use them. Each person keeps a log of total exposure, and if a predetermined limit is exceeded, the worker must be reassigned to an unexposed workplace.

The rate of radioactive decay is measured in disintegrations per second

The **activity** of a radioactive material is the number of disintegrations per second. The SI unit of activity is the **becquerel (Bq),** and it equals one disintegration per second. A liter

The becquerel is named after Henri Becquerel (1852–1908), the discoverer of radioactivity, who won a Nobel prize in 1903.

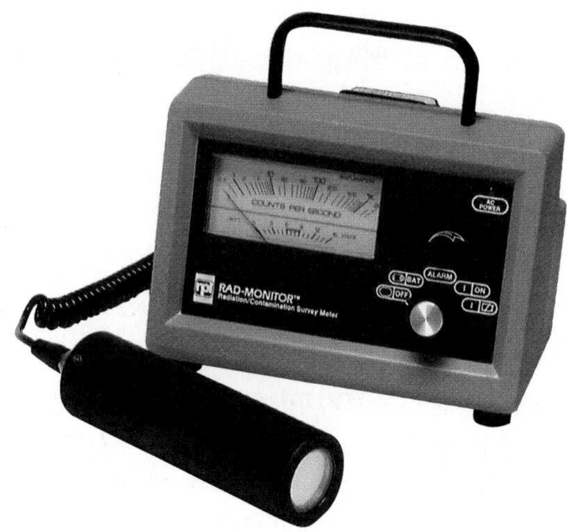

FIG. 20.12 **Scintillation probe.** Energy received from radiations striking the phosphor at the end of the probe is amplified by a photomultiplier unit and sent to the instrument where the intensity of the radiation is displayed on a meter. *(Research Products International Corp.)*

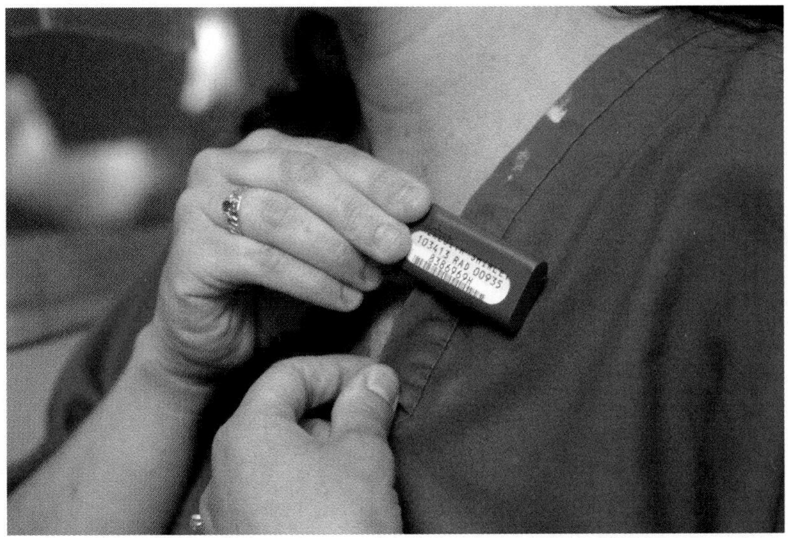

FIG. 20.13 Badge dosimeter. *(Cliff Moore/Photo Researchers, Inc.)*

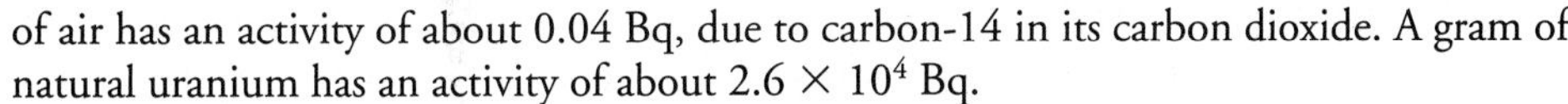

Marie Curie (1867–1934); Nobel prizes 1903 (physics) and 1911 (chemistry). *(Courtesy College of Physicians of Philadelphia.)*

of air has an activity of about 0.04 Bq, due to carbon-14 in its carbon dioxide. A gram of natural uranium has an activity of about 2.6×10^4 Bq.

The **curie (Ci),** named after Marie Curie, the discoverer of radium, is an older unit, being equal to the activity of a 1.0 g sample of radium-226.

$$1 \text{ Ci} = 3.7 \times 10^{10} \text{ disintegrations s}^{-1} = 3.7 \times 10^{10} \text{ Bq} \qquad (20.3)$$

For a sufficiently large sample of a radioactive material, the activity is experimentally found to be proportional to the number of radioactive nuclei, N:

$$\text{Activity} = kN$$

The constant of proportionality, k, is called the **decay constant.** The decay constant is characteristic of the particular radioactive nuclide, and it gives the activity per nuclide in the sample. Since activity is the number of disintegrations per second, and therefore the change in the number of nuclei per second, we can write

$$\text{Activity} = -\frac{\Delta N}{\Delta t} = kN \qquad (20.4)$$

which is called the **law of radioactive decay.**[1] The law shows that radioactive decay is a first-order kinetic process, and the decay constant is really just a first-order rate constant in terms of number of nuclei, rather than concentrations.

Recall from Chapter 13 that the half-life of a first-order reaction is given by Equation 13.5:

$$t_{1/2} = \frac{\ln 2}{k}$$

If we know the half-life of a radioisotope, we can use this relationship to compute its decay constant and also the activity of a known mass of the radioisotope, as Example 20.2 demonstrates.

[1] Note that the minus sign is introduced to make the activity a positive number, since the change in the number of radioactive nuclei ΔN is negative.

EXAMPLE 20.2
Using the Law of Radioactive Decay

Deep space probes such as NASA's Cassini spacecraft need to keep their instruments warm enough to operate effectively. Since solar power is not available in the darkness of deep space, these craft generate heat from the radioactive decay of small pellets of plutonium dioxide. Each pellet is about the size of a pencil eraser and weighs about 2.7 g. If the pellets are pure PuO_2, with the plutonium being ^{238}Pu, what is the activity of one of the fuel pellets, in bequerels?

ANALYSIS: The tool for finding the activity is Equation 20.4. But to use the equation, we'll need the decay constant, k, and the number of plutonium-238 atoms, N, in the fuel pellet.

From Table 20.1, the half-life of ^{238}Pu is 87.8 years. We can rearrange Equation 13.5 to compute the decay constant k from the half-life:

$$k = \frac{\ln 2}{t_{1/2}}$$

We'll want the decay constant in terms of seconds because the becquerel is defined as disintegrations per second. We should therefore convert the half-life into seconds before substituting it into Equation 13.5.

We know that the pellet contains 2.7 g of PuO_2. To calculate N, we'll have to perform the following stoichiometric conversion:

$$2.7 \text{ g PuO}_2 \Leftrightarrow \text{? atoms Pu}$$

The tools here come from Chapter 3. We'll use the molar mass to find moles of PuO_2, the chemical formula to relate this to moles of Pu, and then Avogadro's number to find the number of atoms of Pu. Our strategy is to perform the following conversions:

$$2.7 \text{ g PuO}_2 \longrightarrow \text{mol PuO}_2 \longrightarrow \text{mol Pu} \longrightarrow \text{atoms Pu}$$

We'll need the formula mass of PuO_2 (270 g mol^{-1}) to perform the first conversion. For the second conversion, the formula of PuO_2 tells us that there is 1 mol Pu in 1 mol PuO_2. For the final conversion, recall that 1 mole of atoms is 6.02×10^{23} atoms.

SOLUTION: First, let's compute the decay constant from the half-life. We must convert the half-life into seconds:

$$87.8 \text{ \cancel{years}} \times \frac{365 \text{ \cancel{days}}}{1 \text{ \cancel{year}}} \times \frac{24 \text{ \cancel{hours}}}{1 \text{ \cancel{day}}} \times \frac{60 \text{ \cancel{min}}}{1 \text{ \cancel{hour}}} \times \frac{60 \text{ s}}{1 \text{ \cancel{min}}} = 2.77 \times 10^9 \text{ s}$$

Now we can solve Equation 13.5 for the decay constant:

$$\begin{aligned} k &= \frac{\ln 2}{t_{1/2}} \\ &= \frac{0.693}{2.77 \times 10^9 \text{ s}} \\ &= 2.50 \times 10^{-10} \text{ s}^{-1} \end{aligned}$$

Next, we need the number of ^{238}Pu atoms in the fuel pellet:

$$2.7 \text{ \cancel{g PuO_2}} \times \frac{1 \text{ \cancel{mol PuO_2}}}{270 \text{ \cancel{g PuO_2}}} \times \frac{1 \text{ \cancel{mol Pu}}}{1 \text{ \cancel{mol PuO_2}}} \times \frac{6.02 \times 10^{23} \text{ atoms Pu}}{1 \text{ \cancel{mol Pu}}} = 6.0 \times 10^{21} \text{ atoms Pu}$$

From the law of radioactive decay, Equation 20.4, the activity of the fuel pellet is

$$\text{Activity} = kN = 2.50 \times 10^{-10} \text{ s}^{-1} \times 6.0 \times 10^{21} \text{ atoms Pu} = 1.5 \times 10^{12} \text{ atoms Pu/s}$$

Since the becquerel is defined as the number of disintegrations per second, and each plutonium atom corresponds to one disintegration, we can report the activity as 1.5×10^{12} Bq.

IS THE ANSWER REASONABLE? There is no easy way to check the size of the decay constant beyond checking the arithmetic. The number of Pu atoms in the pellet makes sense, because if we have 2.7 g PuO_2 and the formula mass is 270 g mol^{-1}, we have 1/100th of a mole of PuO_2 and so 1/100 a mole of Pu. We should have (1/100) of 6.02×10^{23} atoms, or 6.02×10^{21} atoms of Pu.

The fuel pellet has about 40 times the activity of a gram of radium-226, as indicated by Equation 20.3, so the activity of ^{238}Pu per gram is about 15 times the activity of ^{226}Ra per gram.

Practice Exercise 3: A 2.00 g sample of a mixture of plutonium with a nonradioactive metal has an activity of 6.22×10^{11} Bq. What is the percentage by mass of plutonium the sample? (Hint: How many atoms of Pu are in the sample?)

Practice Exercise 4: The EPA limit for radon-222 is 6 pCi (picocuries) per liter of air. How many atoms of ^{222}Rn per liter of air will produce that activity?

Other units are used to express amounts of radiation and its effects on tissue

Nuclear radiation can have varying effects depending on the energy of the radiation and its ability to be absorbed. The **gray (Gy)** is the SI unit of *absorbed dose*, and 1 gray corresponds to 1 joule of energy absorbed per kilogram of absorbing material. The **rad** is an older unit of absorbed dose, 1 rad being the absorption of 10^{-2} joule per kilogram of tissue. Thus, 1 Gy equals 100 rad. In terms of danger, if every individual in a large population received 450 rad (4.5 Gy), roughly half of the population would die in 60 days.

The *gray* is named after Harold Gray, a British radiologist. *Rad* stands for **r**adiation **a**bsorbed **d**ose.

The gray is not a good basis for comparing the biological effects of radiation in tissue, because these effects depend not just on the energy absorbed but also on the kind of radiation and the tissue itself. The **sievert (Sv),** the SI unit of *dose equivalent,* was invented to meet this problem. The **rem** is an older unit of dose equivalent, one still used in medicine. Its value is generally taken to equal 10^{-2} Sv. The U.S. government has set guidelines of 0.3 rem per week as the maximum exposure workers may receive. (For comparison, a chest X ray typically involves about 0.007 rem or 7 mrem.)

Rem stands for **r**oentgen **e**quivalent for **m**an, where the *roentgen* is a unit related to X ray and gamma radiation.

Radiation damages living tissue

A whole body dose of 25 rem (0.25 Sv) induces noticeable changes in human blood. A set of symptoms called *radiation sickness* develops at about 100 rem, becoming severe at 200 rem. Among the symptoms are nausea, vomiting, a drop in the white cell count, diarrhea, dehydration, prostration, hemorrhaging, and loss of hair. If each person in a large population absorbed 400 rem, half would die in 60 days. A 600 rem dose would kill everyone in the group in a week. Many workers received at least 400 rem in the moments following the steam explosion that tore apart one of the nuclear reactors at the Ukraine energy park near Chernobyl in 1986.

Radiation produces free radicals

Even small absorbed doses can be biologically harmful. The danger does not lie in the heat energy associated with the dose, which is usually very small. Rather, the harm is in the ability of ionizing radiation to create unstable ions or neutral species with odd (unpaired) electrons, species that can set off other reactions. Water, for example, can interact as follows with ionizing radiation.

$$\mathrm{H{-}\ddot{\underset{\cdot\cdot}{O}}{-}H} \xrightarrow{\text{radiation}} [\mathrm{H{-}\dot{\underset{\cdot\cdot}{O}}{-}H}]^+ + {}^{\ 0}_{-1}e$$

The new cation, $[\mathrm{H{-}\dot{\underset{\cdot\cdot}{O}}{-}H}]^+$, is unstable and one breakup path is

$$[\mathrm{H{-}\dot{\underset{\cdot\cdot}{O}}{-}H}]^+ \longrightarrow \underset{\text{proton}}{\mathrm{H^+}} + \underset{\text{hydroxyl radical}}{:\dot{\underset{\cdot\cdot}{O}}{-}\mathrm{H}}$$

The proton might pick up a stray electron to become a hydrogen atom, H•.

Free radicals are discussed in more detail in Facets of Chemistry 13.1 on page 553.

Both the hydrogen atom and the hydroxyl radical are examples of **free radicals** (often simply called *radicals*), which are neutral or charged particles having one or more unpaired electrons. They are chemically very reactive. What they do once formed depends on the other chemical species nearby, but radicals can set off a series of totally undesirable chemical reactions inside a living cell. This is what makes the injury from absorbed radiation of far greater magnitude than the energy alone could inflict. A dose of 600 rem is a lethal dose in a human, but the same dose absorbed by pure water causes the ionization of only one water molecule in every 36 million.

Background radiation comes from a variety of sources

The presence of radionuclides in nature makes it impossible for us to be free from all exposure to ionizing radiation. Cosmic rays composed of high-energy photons shower on us from the sun and interstellar space. They interact with the air's nitrogen molecules to produce carbon-14, a beta emitter, which enters the food chain via photosynthesis, which converts CO_2 to sugars and starch. From soil and from building stone comes the radiation of radionuclides native to the earth's crust. The top 40 cm of soil holds an average of 1 gram of radium, an alpha emitter, per square kilometer. Naturally occurring potassium-40, a beta emitter, adds its radiation wherever potassium ions are found in the body. The presence of carbon-14 and potassium-40 together produce about 5×10^5 nuclear disintegrations per minute inside an adult human. Radon gas seeps into basements from underground formations. In fact, a little over half of the radiation we experience, on average, is from radon-222 and its decay products.

Diagnostic X rays, both medical and dental, also expose us to ionizing radiation. All these sources produce a combined **background radiation** that averages 360 mrem per person annually in the United States. The averages are roughly 82% from natural radiation and 18% from medical sources.

Radiation can be reduced by shielding and by distance

Gamma radiation and X rays are so powerful that they are effectively shielded only by very dense materials, like lead. Otherwise, one should stay as far from a source as possible, because the intensity of radiation diminishes with the *square* of the distance. This relationship is the **inverse square law,** which can be written mathematically as follows, where *d* is the distance from the source.

$$\text{Radiation intensity} \propto \frac{1}{d^2}$$

When the intensity, I_1, is known at distance d_1, then the intensity I_2 at distance d_2 can be calculated by the following equation.

$$\frac{I_1}{I_2} = \frac{d_2^2}{d_1^2} \qquad (20.5)$$

When the ratio is taken, the proportionality constant cancels, so we don't have to know what its value is.

This law applies only to a small source that radiates equally in all directions, with no intervening shields.

Practice Exercise 5: If the intensity of radiation from a radioactive source is 4.8 units at a distance of 5.0 m, how far from the source would you have to move to reduce the intensity to 0.30 units? (Hint: How does intensity vary with distance?)

Practice Exercise 6: If an operator 10 m from a small source is exposed to 1.4 units of radiation, what will be the intensity of the radiation if he moves to 1.2 m from the source?

20.7 Radionuclides have medical and analytical applications

Because chemical properties depend on the number and arrangement of orbital electrons and not on the specific makeup of nuclei, both radioactive and stable isotopes of an element behave the same chemically. This fact forms the basis for some uses of radionuclides. The chemical and physical properties enable scientists to get radionuclides into place in systems of interest. Then the radiation is exploited for medical or analytical purposes. Tracer analysis is an example.

Tracer analysis is used in medicine to study specific tissues in the body

In **tracer analysis,** the chemical form and properties of a radionuclide enable the system to distribute it to a particular location. The intensity of the radiation then tells something about how that site is working. In the form of the iodide ion, for example, iodine-131 is carried by the body to the thyroid gland, the only user of iodide ion in the body. The gland takes up the iodide ion to synthesize the hormone thyroxine. An underactive thyroid gland is unable to concentrate iodide ion normally and will emit less intense radiation under standard test conditions.

Tracer analyses are also used to pinpoint the locations of brain tumors, which are uniquely able to concentrate the pertechnetate ion, TcO_4^-, made from technetium-99*m*.[2] This strong gamma emitter, which resembles the chloride ion in some respects, is one of the most widely used radionuclides in medicine.

Technetium-99*m* is also used in bone scans to detect and locate bone cancer. Active cancer sites concentrate the Tc, which can be detected using external scanning devices.

Neutron activation analysis is used to detect trace elements

A number of stable nuclei can be changed into emitters of gamma radiation by capturing neutrons, and this makes possible a procedure called **neutron activation analysis**. Neutron capture followed by gamma emission can be represented by the following equation (where *A* is a mass number and *X* is a hypothetical atomic symbol).

$$^{A}X + {}^{1}_{0}n \longrightarrow {}^{(A+1)}X^{*} \longrightarrow {}^{(A+1)}X + {}^{0}_{0}\gamma$$

^{A}X	$^{1}_{0}n$	$^{(A+1)}X^{*}$	$^{(A+1)}X$	$^{0}_{0}\gamma$
isotope of element *X* being analyzed	neutron	compound nucleus (unstable)	more stable form of a new isotope of *X*	gamma radiation

A neutron-enriched compound nucleus emits gamma radiation at its own set of unique frequencies, and these sets of frequencies are now known for each isotope. (Not all isotopes, however, become gamma emitters by neutron capture.) The element can be identified by measuring the specific *frequencies* of gamma radiation emitted. The *concentration* of the element can be determined by measuring the *intensity* of the gamma radiation.

Neutron activation analysis is so sensitive that concentrations as low as $10^{-9}\%$ can be determined. A museum might have a lock of hair of some famous but long dead person suspected of having been slowly murdered by arsenic poisoning. If so, some arsenic would be in the hair, and neutron activation analysis could find it without destroying the specimen of hair.

Radiological dating determines the age of a sample using kinetics

The determination of the age of a geological deposit or an archaeological find by the use of the radionuclides that are naturally present is called **radiological dating.** It is based partly on the premise that the half-lives of radionuclides have been constant throughout the entire geological period. This premise is supported by the finding that half-lives are insensitive to all environmental forces such as heat, pressure, magnetism, or electrical stresses. Radiological dating of archaeological objects by carbon-14 analyses was described in Chapter 13 (page 539).

In geological dating, a pair of isotopes is sought that are related as a "parent" to a "daughter" in a radioactive disintegration series, like the uranium-238 series (Figure 20.7). Uranium-238 (as "parent") and lead-206 (as "daughter") have thus been used as a radiological dating pair of isotopes. The half-life of uranium-238 is very long, a necessary criterion for geological dating. Put simply, after the concentrations of uranium-238

[2] The *m* in technetium-99*m* means that the isotope is in a metastable form. Its nucleus is at a higher energy level than the nucleus in technetium-99, to which technetium-99*m* decays.

and lead-206 are determined in a rock specimen, the *ratio* of the concentrations together with the half-life of uranium-238 is used to calculate the age of the rock.

Probably the most widely used isotopes for dating rock are the potassium-40/argon-40 pair. Potassium-40 is a naturally occurring radionuclide with a half-life nearly as long as that of uranium-238. One of its modes of decay is electron capture, and argon-40 forms.

$$^{40}_{19}\text{K} + {}^{0}_{-1}e \longrightarrow {}^{40}_{18}\text{Ar}$$

The argon produced by the reaction remains trapped within the crystal lattices of the rock and is freed only when the rock sample is melted. How much has accumulated is measured with a mass spectrometer (page 47), and the observed ratio of argon-40 to potassium-40, together with the half-life of the parent, permits the age of the specimen to be estimated. Because the half-lives of uranium-238 and potassium-40 are so long, samples have to be at least 300,000 years old for either of the two parent–daughter isotope pairs to provide reliable results.

For ^{238}U, $t_{1/2} = 4.51 \times 10^9$ yr. For ^{40}K, $t_{1/2} = 1.3 \times 10^9$ yr.

Carbon-14 dating determines the age of organic objects

As discussed in detail in Chapter 13, measurements of the ^{14}C to ^{12}C ratio in an ancient organic sample, such as an object made of wood or bone, permits calculation of the sample's age.

There are two approaches to carbon-14 dating. The older method, introduced by its discoverer, Willard F. Libby (Nobel Prize in Chemistry, 1960), measures the *radioactivity* of a sample taken from the specimen. The radioactivity is proportional to the concentration of carbon-14.

The newer and current method of carbon-14 dating relies on a device resembling a mass spectrometer that is able to separate the atoms of carbon-14 from the other isotopes of carbon (as well as from nitrogen-14) *and count all of them,* not just the carbon-14 atoms that decay. This approach permits the use of smaller samples (0.5–5 mg versus 1–20 g for the Libby method); it works at much higher efficiencies; and it gives more precise dates. Objects of up to 70,000 years old can be dated, but the highest accuracy involves systems no older than 7000 years.

For carbon dating to be accurate, extraordinary precautions must be taken to ensure that specimens are not contaminated by more recent sources of carbon or carbon compounds.

20.8 Nuclear Fission and Nuclear Fusion Release Large Amounts of Energy

Nuclear fission is a process whereby a heavy atomic nucleus splits into two lighter fragments. **Nuclear fusion,** on the other hand, is a process whereby very light nuclei join to form a heavier nucleus. Both processes release large amounts of energy, as we will discuss shortly.

Nuclear fission is initiated by absorption of a neutron by an unstable nucleus

Because of their electrical neutrality, neutrons penetrate an atom's electron cloud relatively easily and so are able to enter the nucleus. Enrico Fermi discovered in the early 1930s that even slow-moving, *thermal neutrons* can be captured. (Thermal neutrons are those whose average kinetic energy puts them in thermal equilibrium with their surroundings at room temperature.) When he directed thermal neutrons at a uranium target, Fermi discovered that several different species of nuclei, all much lighter than uranium, were produced.

Without realizing it, what Fermi had observed was the nuclear fission of one particular isotope, uranium-235, present in small concentrations in naturally occurring uranium. The general reaction can be represented as follows.

$$^{235}_{92}\text{U} + {}^{1}_{0}n \longrightarrow X + Y + b\,{}^{1}_{0}n$$

X and Y can be a large variety of nuclei with intermediate atomic numbers. Over 30 have been identified. The coefficient b has an average value of 2.47, the average number of neutrons produced by fission events. A typical specific fission is

$$^{235}_{92}\text{U} + {}^{1}_{0}n \longrightarrow {}^{236}_{92}\text{U}^* \longrightarrow {}^{94}_{36}\text{Kr} + {}^{139}_{56}\text{Ba} + 3\,{}^{1}_{0}n$$

What actually undergoes fission is the compound nucleus of uranium-236. It has 144 neutrons and 92 protons, giving it a neutron-to-proton ratio of roughly 1.6. Initially, the emerging krypton and barium isotopes have the same ratio, and this is much too high for them. The neutron-to-proton ratio for stable isotopes with 36 to 56 protons is nearer 1.2 to 1.3 (Figure 20.8). Therefore, the initially formed, neutron-rich krypton and barium nuclides promptly eject neutrons, called *secondary neutrons,* that generally have much higher energies than thermal neutrons.

An isotope that can undergo fission after neutron capture is called a **fissile isotope.** The naturally occurring fissile isotope of uranium used in reactors is uranium-235, whose abundance among the uranium isotopes today is only 0.72%. Two other fissile isotopes, uranium-233 and plutonium-239, can be made in nuclear reactors.

Nuclear chain reactions require a critical mass of fissile material

The secondary neutrons released by fission become thermal neutrons as they are slowed by collisions with surrounding materials. They can now be captured by unchanged uranium-235 atoms. Because each fission event produces, on the average, more than two new neutrons, the potential exists for a **nuclear chain reaction** (Figure 20.14). A *chain reaction* is a self-sustaining process whereby products from one event cause one or more repetitions of the process.

If the sample of uranium-235 is small enough, the loss of neutrons to the surroundings is sufficiently rapid to prevent a chain reaction. However, at a certain *critical mass* of uranium-235, about 50 kilograms, this loss of neutrons is insufficient to prevent a sustained reaction. A virtually instantaneous fission of the sample ensues, in other words, an atomic bomb explosion. To trigger an atomic bomb, therefore, two or more subcritical masses of uranium-235 (or plutonium-239) are forced together to form a critical mass.

The energy yield from fission is very large

The binding energy per nucleon (Figure 20.1) in uranium-235 (about 7.6 MeV) is less than the binding energies of the new nuclides (about 8.5 MeV). The net change for a single

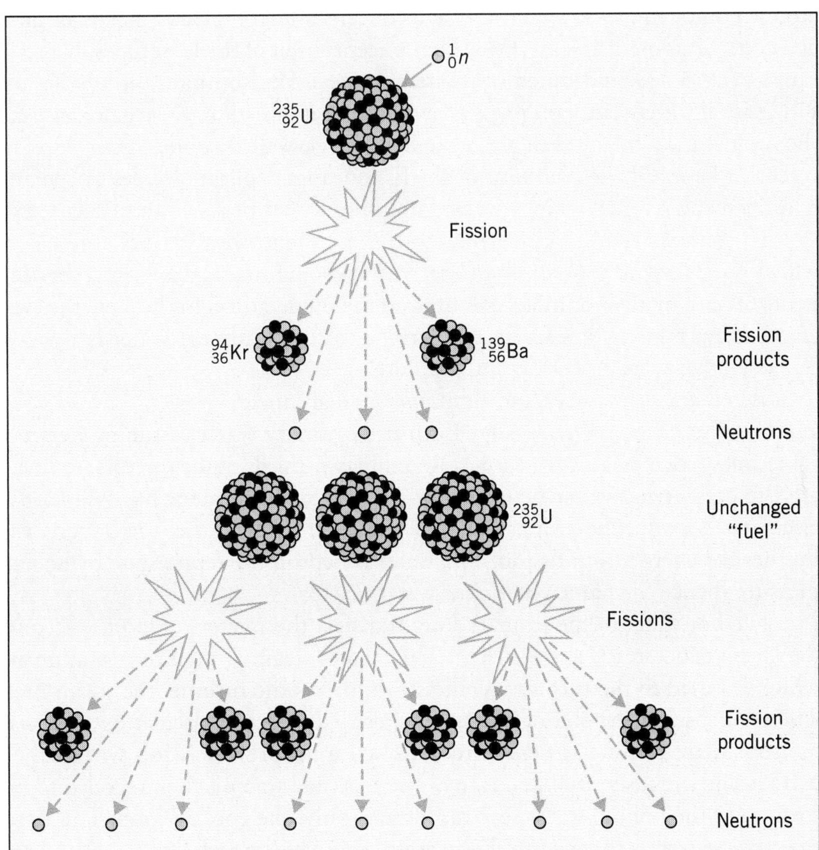

FIG. 20.14 Nuclear chain reaction. Whenever the concentration of a fissile isotope is high enough (at or above the critical mass), the neutrons released by one fission can be captured by enough unchanged nuclei to cause more than one additional fission event. In civilian nuclear reactors, the fissile isotope is too dilute for this to get out of control. In addition, control rods of nonfissile materials that are able to capture excess neutrons can be inserted or withdrawn from the reactor core to make sure that the heat generated can be removed as fast as it forms.

fission event can be calculated as follows (where we intend only *two* significant figures in each result).

Binding energy in krypton-94:

1 MeV = 1.602×10^{-13} J

$$(8.5 \text{ MeV/nucleon}) \times 94 \text{ nucleons} = 800 \text{ MeV}$$

Binding energy in barium-139:

$$(8.5 \text{ MeV/nucleon}) \times 139 \text{ nucleons} = 1200 \text{ MeV}$$

$$\text{Total binding energy of products:} \quad 2000 \text{ MeV}$$

Binding energy in uranium-235:

$$(7.6 \text{ MeV/nucleon}) \times 235 \text{ nucleons} = 1800 \text{ MeV}$$

The difference in total binding energy is (2000 MeV − 1800 MeV), or 200 MeV (3.2×10^{-11} J), which has to be taken as just a rough calculation. This is the energy released by each fission event going by the equation given. The energy produced by the fission of 1 kg (4.25 mole) of uranium-235 is calculated to be roughly 8×10^{13} J, enough to keep a 100 watt lightbulb in energy for 3000 years.

The energy available from 1 kg of uranium-235 is equivalent to the energy of combustion of 3000 tons of soft coal or 13,200 barrels of oil.

Heat from fission reactions can be used to drive electrical generators

Virtually all civilian nuclear power plants throughout the world operate on the same general principles. The energy of fission is used, as heat, either directly or indirectly to increase the pressure of some gas that then drives an electrical generator.

The heart of a nuclear power plant is the *reactor,* where fission takes place in the *fuel core.* The nuclear fuel is generally uranium oxide, enriched to 2–4% in uranium-235 and formed into glasslike pellets. These are housed in long, sealed metal tubes called *cladding.* Bunches of tubes are assembled in spacers that permit a coolant to circulate around the tubes. A reactor has several such assemblies in its fuel core. The coolant carries away the heat of fission.

There is no danger of a nuclear power plant undergoing an atomic bomb explosion. An atomic bomb requires uranium-235 at a concentration of 85% or greater or plutonium-239 at a concentration of at least 93%. The concentration of fissile isotopes in a reactor is in the range of 2 to 4%, and much of the remainder is the common, nonfissile uranium-238. However, if the coolant fails to carry away the heat of fission, the reactor core can melt, and the molten mass might even go through the thick-walled containment vessel in which the reactor is kept. Or the high heat of the fission might split molecules of coolant water into hydrogen and oxygen, which, on recombining, would produce an immense explosion.

To convert secondary neutrons to thermal neutrons, the fuel core has a *moderator,* which is the coolant water itself in nearly all civilian reactors. Collisions between secondary neutrons and moderator molecules heat up the moderator. This heat energy eventually generates steam that enables an electric turbine to run. Ordinary water is a good moderator, but so are heavy water (D_2O) and graphite.

D_2O is deuterium oxide. Deuterium is an isotope of hydrogen, ${}^{2}_{1}H$.

Two main types of reactors dominate civilian nuclear power, the *boiling water reactor* and the *pressurized water reactor.* Both use ordinary water as the moderator and so are sometimes called *light water reactors.* Roughly two-thirds of the reactors in the United States are the pressurized-water type (Figure 20.15). Such a reactor has two loops, and water circulates in both. The primary loop moves water through the reactor core, where it picks up thermal energy from fission. The water is kept in the *liquid* state by being under high pressure (hence the name *pressurized* water reactor).

The hot water in the primary loop transfers thermal energy to the secondary loop at the steam generator (Figure 20.15). This makes steam at high temperature and pressure, which is piped to the turbine. As the steam drives the turbine, the steam pressure drops. The condenser at the end of the turbine, cooled by water circulating from a river or lake or from huge cooling towers, forces a maximum pressure drop within the turbine by condensing the steam to liquid water. The returned water is then recycled to high-pressure steam. (In the boiling water reactor, there is only one coolant loop. The water heated in the reactor itself is changed to the steam that drives the turbine.)

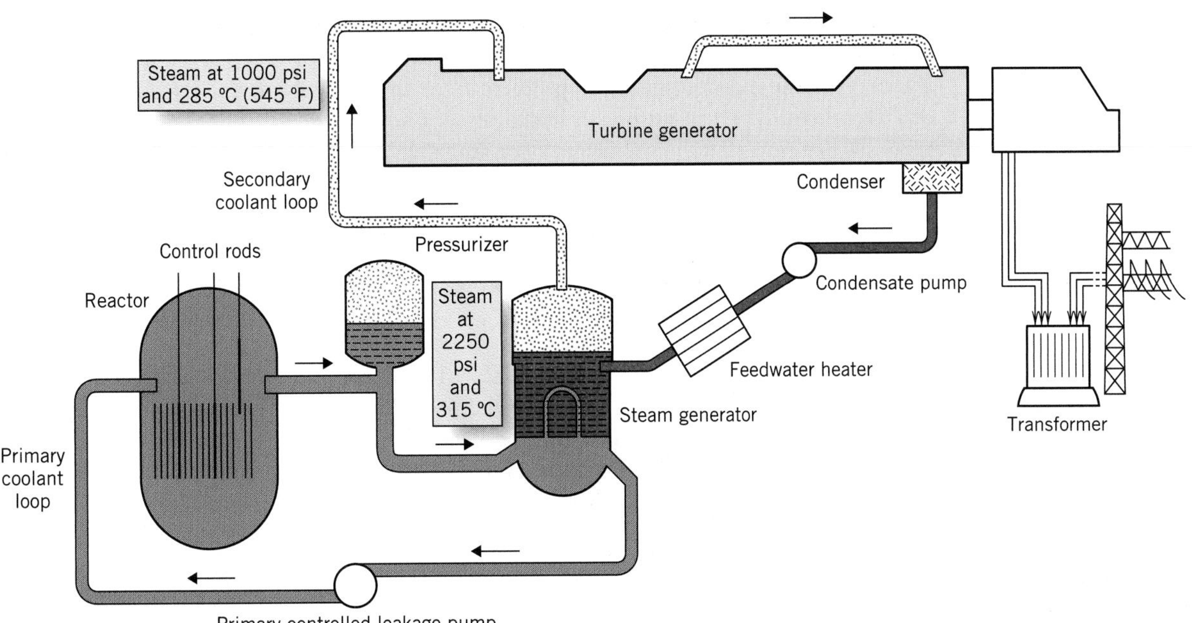

FIG. 20.15 **Pressurized water reactor, the type used in most of the nuclear power plants in the United States.** Water in the primary coolant loop is pumped around and through the fuel elements in the core, and it carries away the heat of the nuclear chain reactions. The hot water delivers its heat to the cooler water in the secondary coolant loop, where steam is generated to drive the turbines. *(Drawing from WASH-1261, U.S. Atomic Energy Commission, 1973.)*

Nuclear power plants produce several types of radioactive waste

Radioactive wastes from nuclear power plants occur as gases, liquids, and solids. The gases are mostly radionuclides of krypton and xenon but, with the exception of xenon-85 ($t_{1/2}$ = 10.4 years), the gases have short half-lives and decay quickly. During decay, they must be contained, and this is one function of the cladding. Other dangerous radionuclides produced by fission include iodine-131, strontium-90, and cesium-137.

Cesium-137 has a half-life of 30 years; that of strontium-90 is 28.1 years. Both are beta emitters.

Iodine-131 must be contained because the human thyroid gland concentrates iodide ion to make the hormone thyroxine. Once the isotope is in the gland, beta radiation from iodine-131 could cause harm, possibly thyroid cancer and possibly impaired thyroid function. An effective countermeasure to iodine-131 poisoning is to take doses of ordinary iodine (as sodium iodide). Statistically, this increases the likelihood that the thyroid will take up stable iodide ion rather than the unstable anion of iodine-131.

Cesium-137 and strontium-90 also pose problems to humans. Cesium is in Group IA together with sodium, so radiating cesium-137 cations travel to some of the same places in the body where sodium ions go. Strontium is in Group IIA with calcium, so strontium-90 cations can replace calcium ions in bone tissue, sending radiation into bone marrow and possibly causing leukemia. The half-lives of cesium-137 and strontium-90, however, are relatively short.

Some radionuclides in wastes are so long-lived that solid reactor wastes must be kept away from all human contact for dozens of centuries, longer than any nation has ever yet endured. Probably the most intensively studied procedure for making solid radioactive wastes secure is to convert them to glasslike or rocklike solids, and bury them deeply within a rock stratum or in a mountain believed to be geologically stable with respect to earthquakes or volcanoes. Finding a location for such a site has been the subject of much scientific and political debate.

Nuclear fusion occurs when light nuclei join to form a heavier nucleus

In Section 20.2 we mentioned that joining, or *fusing,* two light nuclei that lie to the left of the peak in the nuclear binding energy curve in Figure 20.1 leads to a net increase in nuclear binding energy and a corresponding release of energy. The process is called *nuclear fusion,* and the amount of energy released is considerably greater than in fission. Harnessing this energy for peaceful purposes is still a long way off, however, because many immensely difficult scientific and engineering problems remain to be solved.

Deuterium, ^2_1H, an isotope of hydrogen, is a key fuel in all approaches to fusion. It is naturally present as 0.015% of all hydrogen atoms, including those in water. Despite this low percentage, the earth has so much water that the supply of deuterium is virtually limitless.

The fusion reaction most likely to be used in a successful fusion reactor involves fusion of deuterium and another isotope of hydrogen, tritium, ^3_1H.

$$\underset{\text{deuterium}}{^2_1\text{H}} + \underset{\text{tritium}}{^3_1\text{H}} \longrightarrow \underset{\text{helium}}{^4_2\text{He}} + \underset{\text{neutron}}{^1_0n} + 17.6\text{ MeV}$$

This corresponds to an energy yield of 2.82×10^{-12} J for each atom of helium formed, or 1.70×10^9 kJ per mole of He formed. One problem with this reaction is that tritium is radioactive with a relatively short half-life, so it doesn't occur naturally. It can be made in several ways, however, from lithium or even deuterium via other nuclear reactions.

Comparing fission and fusion on a mass basis, fission of one kilogram of ^{235}U yields approximately 8×10^{13} J, whereas forming one kilogram of ^4He by the fusion reaction above yields 4.2×10^{14} J. Therefore, on a mass basis, fusion yields more than five times as much energy as fission. The potential energy yield from fusion is so great that the deuterium in just 0.005 cubic kilometers of ocean would supply the energy needs of the United States for one year.

Thermonuclear fusion uses high temperatures to overcome electrostatic repulsions between nuclei

The central scientific problem with fusion is to get the fusing nuclei close enough for a long enough time that the nuclear strong force (of attraction) can overcome the electrostatic force (of repulsion). As we learned in Section 20.3, the strong force acts over a much shorter range than the electrostatic force. Two nuclei on a collision course, therefore, repel each other virtually until they touch and get into the range of the strong force. The kinetic energies of two approaching nuclei must therefore be very substantial if they are to overcome this electrostatic barrier. Achieving such energies, moreover, must be accomplished by large numbers of nuclei all at once in batch after batch if there is to be any practical generation of electrical power by nuclear fusion. Relatively isolated fusion events achieved in huge accelerators will not do. The only practical way to give batch quantities of nuclei enough energy is to transfer *thermal* energy to them, and so the overall process is called *thermonuclear fusion.* Temperatures required to provide such thermal energy are very high—more than 100 million degrees Celsius!

The atoms whose nuclei we want to fuse must first be stripped of their electrons. Thus, a high energy cost is exacted from the start but, overall, the energy yield will more than pay for it. The product is an electrically neutral, gaseous mixture of nuclei and unattached electrons called a **plasma.** The plasma must then be made so dense that like-charged nuclei are within 2 fm (2×10^{-15} m) of each other, meaning a plasma density of roughly 200 g cm^{-3} as compared with 200 mg cm^{-3} under ordinary conditions. To achieve this, the plasma must be confined at a pressure of several billion atmospheres long enough for the separate nuclei to fuse. The temperature needed is several times the temperature at the center of our sun.

The interior of the sun is at a temperature of approximately 15 million kelvins (15 MK).

Although practical peaceful uses for thermonuclear fusion are still in the distant future, military applications have been around for over 60 years. Thermonuclear fusion is the source of the energy released in the explosion of a hydrogen bomb. The energy needed to trigger the fusion is provided by the explosion of a fission bomb based on either uranium or plutonium.

Fusion reactions are the source of energy in stars

Nature has used thermonuclear fusion since the origin of the universe as the source of energy in stars, where high temperatures (over 15 megakelvins) and huge gravity provide the kinetic energy and high density needed to initiate fusion reactions. The chief process in solar-mass stars like our sun is called the proton–proton cycle:

The Proton–Proton Cycle

$$2\left[{}^1_1\text{H} + {}^1_1\text{H}\right] \longrightarrow 2\,{}^2_1\text{H} + 2\,{}^0_1e + 2\nu$$

$$2\left[{}^1_1\text{H} + {}^2_1\text{H}\right] \longrightarrow 2\,{}^3_2\text{He} + 2\gamma$$

$${}^3_2\text{He} + {}^3_2\text{He} \longrightarrow {}^4_2\text{He} + {}^1_1\text{H} + {}^1_1\text{H}$$

$$\text{Net:}\quad 4\,{}^1_1\text{H} \longrightarrow {}^4_2\text{He} + 2\,{}^0_1e + 2\nu + 2\gamma$$

The positrons produced combine with electrons in the plasma, annihilate each other, and generate additional energy and gamma radiation. Virtually all the neutrinos escape the sun and move into the solar system (and beyond), carrying with them a little less than 2% of the energy generated by the cycle. Not counting the energy of the neutrinos, each operation of one cycle generates 26.2 MeV or 4.20×10^{-12} J, which is equivalent to 2.53×10^{12} J per *mole* of alpha particles produced. This is the source of the solar energy radiated throughout our system, which can continue in this way for another 5 billion years.

Even at a temperature of 15 megakelvins, the rate of energy production per cubic centimeter by fusion in the sun is quite small, only about 10^{-4} J s^{-1} cm^{-3}. (That's thousands of times less than the rate at which a human body generates heat!) But because the sun has such a large volume, the *total* rate of energy production is enormous.

SUMMARY

The Einstein Equation. Mass and energy are interconvertible. The **Einstein equation, $\Delta E = \Delta mc^2$** (where c is the speed of light), lets us calculate one from the other. The total of the energy in the universe and all the mass calculated as an equivalent of energy is a constant, which is the **law of conservation of mass–energy.**

Nuclear Binding Energies. When a nucleus forms from its nucleons, some mass changes into energy. This amount of energy, the **nuclear binding energy,** would be required to break up the nucleus again. The higher the binding energy per nucleon, the more stable is the nucleus.

Radioactivity. The *electrostatic force* by which protons repel each other is overcome in the nucleus by the *nuclear strong force.* The ratio of neutrons to protons is a factor in nuclear stability. By radioactivity, the naturally occurring **radionuclides** adjust their neutron-to-proton ratios, lower their energies, and so become more stable by emitting **alpha** or **beta radiation,** sometimes gamma radiation as well.

The loss of an **alpha particle** leaves a nucleus with four fewer units of mass number and two fewer of atomic number. Loss of a **beta particle** leaves a nucleus with the same mass number and an atomic number one unit higher. **Gamma radiation** lets a nucleus lose some energy without a change in mass or atomic number. Depending on the specific isotope, synthetic radionuclides emit alpha, beta, and gamma radiation. Some emit **positrons** (positive electrons, a form of **antimatter**) that produce gamma radiation by annihilation collisions with electrons. Other synthetic radionuclides decay by **electron capture** and emit **X rays.** Some radionuclides emit neutrons.

Nuclear equations are balanced when the mass numbers and atomic numbers on either side of the arrow respectively balance. The energies of emission are usually described in **electron volts (eV)** or multiples thereof (1 eV = 1.602×10^{-19} J).

A few very long-lived radionuclides in nature, like ^{238}U, are at the heads of **radioactive disintegration series,** which represent the successive decays of "daughter" radionuclides until a stable isotope forms.

Nuclear Stability. Stable nuclides generally fall within a curving band, called the **band of stability,** when all known nuclides are plotted according to their numbers of neutrons and protons. Radionuclides that have too high neutron-to-proton ratios eject beta particles to adjust their ratios downward. Those with neutron-to-proton ratios too low generally emit positrons to change their ratios upward.

Isotopes whose nuclei consist of even numbers of both neutrons and protons are generally much more stable than those with odd numbers of both; this is the **odd–even rule.** Having all neutrons paired and all protons paired is energetically better (more stable) than having any nucleon unpaired. Isotopes with specific numbers of protons or neutrons, the **magic numbers** of 2, 8, 20, 28, 50, 82, and 126, are generally more stable than others.

Transmutation. When a bombardment particle—a proton, deuteron, alpha particle, or neutron—is captured, the resulting **compound nucleus** contains the energy of both the captured particle and its nucleons. The mode of decay of the compound nucleus is a function of its extra energy, not its extra mass. Many radionuclides have been made by these nuclear reactions, including all of the elements from atomic number 93 and higher.

Detecting and Measuring Radiations. Instruments to detect and measure **ionizing radiation—Geiger counters** or **scintillation counters,** for example—take advantage of the ability of radiation to generate ions in air or other matter. Such radiation can harm living tissue by producing **free radicals.**

The **curie (Ci)** and the **becquerel (Bq),** the SI unit, describe how active a source is; 1 Ci = 3.7×10^{10} Bq where 1 Bq = 1 disintegration s^{-1}.

The SI unit of absorbed dose, the **gray (Gy),** is used to describe how much energy is absorbed by a unit mass of absorber; 1 Gy = 1 J kg^{-1}. An older unit, the **rad,** is equal to 0.01 Gy.

The **sievert,** an SI unit, and the **rem,** an older unit, are used to compare doses absorbed by different tissues and caused by different kinds of radiation. A 600 rem whole body dose is lethal.

The normal **background radiation** causes millirem exposures per year. Naturally occurring radon, cosmic rays, radionuclides in soil and stone building materials, medical X rays, and releases from nuclear tests or from nuclear power plants all contribute to this background.

Protection against radiation can be achieved by using dense shields (e.g., lead or thick concrete), by avoiding overuse of radionuclides or X rays in medicine, and by taking advantage of the **inverse square law.** This law tells us that the intensity of radiation decreases with the square of the distance from its source.

Applications. **Tracer analysis** uses small amounts of a radionuclide, which can be detected using devices like the scintillation counter, to follow the path of chemical and biological processes. In **neutron activation analysis,** neutron bombardment causes some elements to become γ emitters. The radiation can be detected and measured, giving the identity and concentration of the activated elements. **Radiological dating** uses the known half-lives of naturally occurring radionuclides to date geological and archeological objects.

Fission and Fusion. Uranium-235, which occurs naturally, and plutonium-239, which can be made from uranium-238, are **fissile isotopes** that serve as the fuel in present-day reactors. When either isotope captures a thermal neutron, the isotope splits in one of several ways to give two smaller isotopes plus energy and more neutrons. The neutrons can generate additional fission events, enabling a nuclear chain reaction. If a critical mass of a fissile isotope is allowed to form, the **nuclear chain reaction** proceeds out of control, and the material detonates as an atomic bomb explosion. *Pressurized water reactors* are the most commonly used fission reactors for power generation, and have two loops of circulating fluids. In the primary loop, water circulates around the reactor core and absorbs the heat of fission. In the secondary loop, water accepts the heat and changes to high-pressure steam, which drives the electrical generator. One major problem with nuclear energy is the storage of radioactive wastes.

Thermonuclear fusion joins two light nuclei to form a heavier nucleus with the release of more energy than nuclear fission. A typical reaction combines ${}^{2}_{1}H$ and ${}^{3}_{1}H$ to give ${}^{4}_{2}He$ and a neutron. High temperatures and pressures are necessary to initiate the fusion reaction. In stars, gravity is able to contain the high temperature **plasma** and allow fusion to occur. In a hydrogen bomb, the reaction is initiated by a fission bomb. Scientific and engineering hurdles must still be overcome before fusion can be a viable peaceful energy source.

TOOLS FOR PROBLEM SOLVING

In this chapter you learned to apply the following concepts as tools in solving problems related to nuclear changes and their applications. Study each tool carefully so that you know what each is used for. When faced with solving a problem, recall what each tool does and consider whether it will be helpful in finding a solution. This will aid you in selecting the tools you need.

The Einstein equation *(page 822)* This equation, $\Delta E = \Delta m_0 c^2$ (or often just $E = mc^2$), is used when you have to relate an amount of mass to its equivalent in energy. Be careful of units in using the equation. To obtain joules, mass must be in units of kilograms and c must have units of m/s (because $1\ J = 1\ kg\ m^2\ s^{-2}$).

Nuclear equations *(page 825)* When you have to write and balance a nuclear equation, remember to apply the following two criteria:

1. The sums of the mass numbers on each side of the arrow must be equal.
2. The sums of the atomic numbers on each side must be the same.

The odd–even rule *(page 832)* This rule allows you to judge and compare the likely stability of nuclei according to their numbers of protons and neutrons.

When the numbers of neutrons and protons in a nucleus are both even, the isotope is far more likely to be stable than when both numbers are odd.

Law of radioactive decay *(page 837)* Use this law to relate the *activity* (in units of disintegrations per second, or Bq) to the *decay constant, k* (a first-order rate constant), and the number of atoms of the radionuclide in the sample, N. The activity is a quantity that can be measured using a Geiger or scintillation counter.

$$\text{Activity} = -\frac{\Delta N}{\Delta t} = kN$$

Half-life of a radionuclide *(page 837)* When you know the half-life of an isotope (which is available in tables), you can calculate the decay constant, k. This is useful when you need to apply the law of radioactive decay (see above).

$$t_{1/2} = \frac{\ln 2}{k}$$

Inverse square law *(page 840)* This simple law lets you compute radiation intensity at various distances from a radioactive source. If the intensity, I_1, is known at distance d_1, then the intensity I_2 at distance d_2 can be calculated by

$$\frac{I_1}{I_2} = \frac{d_2^2}{d_1^2}$$

QUESTIONS, PROBLEMS, AND EXERCISES

Answers to problems whose numbers are printed in color are given in Appendix B. More challenging problems are marked with asterisks. ILW = Interactive Learningware solution is available at www.wiley.com/college/brady. OH = an Office Hours video is available for this problem.

REVIEW QUESTIONS

Conservation of Mass–Energy

20.1 In chemical calculations involving chemical reactions we can regard the law of conservation of mass as a law independent of the law of conservation of energy despite Einstein's union of the two. What fact(s) makes this possible?

20.2 How can we know that the speed of light is the absolute upper limit on the speed of any object?

20.3 State the following.
(a) law of conservation of mass–energy
(b) Einstein equation

20.4 Why isn't the sum of the masses of all nucleons in one nucleus equal to the mass of the actual nucleus?

Radioactivity

20.5 When a substance is described as *radioactive*, what does that mean? Why is the term *radioactive decay* used to describe the phenomenon?

20.6 Three kinds of radiation make up nearly all of the radiation observed from naturally occurring radionuclides. What are they?

20.7 Give the composition of each of the following.
(a) alpha particle
(b) beta particle
(c) positron
(d) deuteron

20.8 Why is the penetrating ability of alpha radiation less than that of beta or gamma radiation?

20.9 With respect to their formation, how do gamma rays and X rays differ?

20.10 How does electron capture generate X rays?

Nuclear Stability

20.11 What data are plotted and what criterion is used to identify the actual band in the band of stability?

20.12 Both barium-123 and barium-140 are radioactive, but which is more likely to have the *longer* half-life? Explain your answer.

20.13 Tin-112 is a stable nuclide but indium-112 is radioactive and has a very short half-life ($t_{1/2}$ = 14 min). What does tin-112 have that indium-112 does not to account for this difference in stability?

20.14 Lanthanum-139 is a stable nuclide but lanthanum-140 is unstable ($t_{1/2}$ = 40 hr). What rule of thumb concerning nuclear stability is involved?

20.15 As the atomic number increases, the neutron-to-proton ratio increases. What does this suggest is a factor in nuclear stability?

20.16 Radionuclides of high atomic number are far more likely to be alpha emitters than those of low atomic number. Offer an explanation for this phenomenon.

20.17 Although lead-164 has two magic numbers, 82 protons and 82 neutrons, this isotope is unknown. Lead-208, however, is known and stable. What problem accounts for the nonexistence of lead-164?

20.18 What decay particle is emitted from a nucleus of low to intermediate atomic number but a relatively high neutron-to-proton ratio? How does the emission of this particle benefit the nucleus?

20.19 What decay particle is emitted from a nucleus of low to intermediate atomic number but a relatively low neutron-to-proton ratio? How does the emission of this particle benefit the nucleus?

20.20 What does electron capture do to the neutron-to-proton ratio in a nucleus, increase it, decrease it, or leave it alone? Which kinds of radionuclides are more likely to undergo this change, those above or those below the band of stability?

Transmutations

20.21 Compound nuclei form and then decay almost at once. What accounts for the instability of a compound nucleus?

20.22 What explains the existence of several decay modes for the compound nucleus aluminum-27?

20.23 Rutherford theorized that a compound nucleus forms when helium nuclei hit nitrogen-14 nuclei. If this compound nucleus decayed by the loss of a neutron instead of a proton, what would be the other product?

Detecting and Measuring Radiations

20.24 What specific property of nuclear radiation is used by the Geiger counter?

20.25 Dangerous doses of radiation can actually involve very small quantities of energy. Explain.

20.26 What units, SI and common, are used to describe each of the following?
(a) the *activity* of a radioactive sample
(b) the *energy* of a particle or of a photon of radiation given off by a nucleus
(c) the amount of energy absorbed by a given mass from a dose of radiation
(d) dose equivalents for comparing biological effects

20.27 A sample giving 3.7×10^{10} disintegrations s^{-1} has what activity in Ci and in Bq?

20.28 Explain the necessity in health sciences for the *sievert.*

Applications of Radionuclides

20.29 Why should a radionuclide used in diagnostic work have a short half-life? If the half-life is too short, what problem arises?

20.30 An alpha emitter is not used in diagnostic work. Why?

20.31 In general terms, explain how neutron activation analysis is used and how it works.

20.32 What is one assumption in the use of the uranium/lead ratio for dating ancient geologic formations?

***20.33** If a sample used for carbon-14 dating is contaminated by air, there is a potentially serious problem with the method. What is it?

20.34 List some of the kinds of radiation that make up our background radiation.

Nuclear Fission and Fusion

20.35 Why is it easier for a nucleus to capture a neutron than a proton?

20.36 What do each of the following terms mean?
(a) thermal neutron (c) fissile isotope
(b) nuclear fission (d) nuclear fusion

20.37 Which fissile isotope occurs in nature?

20.38 What fact about the fission of uranium-235 makes it possible for a *chain reaction* to occur?

20.39 Explain in general terms why fission generates more neutrons than needed to initiate it.

20.40 Why would there be a *subcritical mass* of a fissile isotope? (Why isn't *any* mass of uranium-235 critical?)

20.41 What purpose is served by a *moderator* in a nuclear reactor?

20.42 Why is there no possibility of an atomic bomb explosion from a nuclear power plant?

20.43 Write the nuclear equation for the fusion reaction between deuterium and tritium. Why must tritium be synthesized for this reaction?

20.44 What obstacles make constructing a reactor for controlled nuclear fusion especially difficult?

REVIEW PROBLEMS

Conservation of Mass–Energy

20.45 Calculate the mass equivalent in grams of 1.00 kJ.

***20.46** Calculate the mass in kilograms of a 1.00 kg object when its velocity, relative to us, is (a) 3.00×10^7 m s^{-1}, (b) 2.90×10^8 m s^{-1}, and (c) 2.99×10^8 m s^{-1}. (Notice the progression of these numbers toward the velocity of light, 3.00×10^8 m s^{-1}.)

OH **20.47** Calculate the amount of mass in nanograms that is changed into energy when one mole of liquid water forms by the combustion of hydrogen, all measurements being made at 1 atm and 25 °C. What percentage is this of the total mass of the reactants?

20.48 Show that the mass equivalent to the energy released by the complete combustion of 1 mol of methane (890 kJ) is 9.89 ng.

Nuclear Binding Energies

ILW OH **20.49** Calculate the binding energy in joules per nucleon of the deuterium nucleus, whose mass is 2.0135 u.

20.50 Calculate the binding energy in joules per nucleon of the tritium nucleus, whose mass is 3.01550 u.

Radioactivity

20.51 Complete the following nuclear equations by writing the symbols of the missing particles

(a) $^{211}_{82}\text{Pb} \longrightarrow {}^{0}_{-1}e + \_\_\_\_$

(b) $^{177}_{73}\text{Ta} \xrightarrow{\text{electron capture}} \_\_\_\_$

(c) $^{220}_{86}\text{Rn} \longrightarrow {}^{4}_{2}\text{He} + \_\_\_\_$

(d) $^{19}_{10}\text{Ne} \longrightarrow {}^{0}_{1}e + \_\_\_\_$

20.52 Write the symbols of the missing particles to complete the following nuclear equations.

(a) $^{245}_{96}\text{Cm} \longrightarrow {}^{4}_{2}\text{He} + \_\_\_\_$

(b) $^{146}_{56}\text{Ba} \longrightarrow {}^{0}_{-1}e + \_\_\_\_$

(c) $^{58}_{29}\text{Cu} \longrightarrow {}^{0}_{1}e + \_\_\_\_$

(d) $^{68}_{32}\text{Ge} \xrightarrow{\text{electron capture}} \_\_\_\_$

20.53 Write a balanced nuclear equation for each of the following changes.
(a) alpha emission from plutonium-242
(b) beta emission from magnesium-28
(c) positron emission from silicon-26
(d) electron capture by argon-37

OH **20.54** Write the balanced nuclear equation for each of the following nuclear reactions.
(a) electron capture by iron-55
(b) beta emission by potassium-42
(c) positron emission by ruthenium-93
(d) alpha emission by californium-251

20.55 Write the symbols, including the atomic and mass numbers, for the radionuclides that would give each of the following products.
(a) fermium-257 by alpha emission
(b) bismuth-211 by beta emission
(c) neodymium-141 by positron emission
(d) tantalum-179 by electron capture

20.56 Each of the following nuclides forms by the decay mode described. Write the symbols of the parents, giving both atomic and mass numbers.
(a) rubidium-80 formed by electron capture
(b) antimony-121 formed by beta emission
(c) chromium-50 formed by positron emission
(d) californium-253 formed by alpha emission

20.57 Krypton-87 decays to krypton-86. What other particle forms?

20.58 Write the symbol of the nuclide that forms from cobalt-58 when it decays by electron capture.

Nuclear Stability

OH **20.59** If an atom of potassium-38 had the option of decaying by positron emission or beta emission, which route would it likely take, and why? Write the nuclear equation.

20.60 Suppose that an atom of argon-37 could decay by either beta emission or electron capture. Which route would it likely take, and why? Write the nuclear equation.

20.61 If we begin with 3.00 mg of iodine-131 ($t_{1/2}$ = 8.07 days), how much remains after 6 half-life periods?

20.62 A sample of technetium-99*m* with a mass of 9.00 ng will have decayed to how much of this radionuclide after 4 half-life periods (about 1 day)?

Transmutations

20.63 When vanadium-51 captures a deuteron ($^{2}_{1}\text{H}$), what compound nucleus forms? (Write its symbol.) This particle expels a proton ($^{1}_{1}p$). Write the nuclear equation for the overall change from vanadium-51.

20.64 The alpha-particle bombardment of fluorine-19 generates sodium-22 and neutrons. Write the nuclear equation, including the intermediate compound nucleus.

OH **20.65** Gamma-ray bombardment of bromine-81 causes a transmutation in which a neutron is one product. Write the symbol of the other product.

20.66 Neutron bombardment of cadmium-115 results in neutron capture and the release of gamma radiation. Write the nuclear equation.

20.67 When manganese-55 is bombarded by protons, each ^{55}Mn nucleus releases a neutron. What else forms? Write the nuclear equation.

20.68 Which nuclide forms when sodium-23 is bombarded by alpha particles and the compound nucleus emits a gamma-ray photon?

20.69 The nuclei of which isotope of zinc-70 would be needed as bombardment particles to make nuclei of element 112 from lead-208 if the intermediate compound nucleus loses a neutron?

20.70 Write the symbol of the nuclide whose nuclei would be the target for bombardment by nickel-64 nuclei to produce nuclei of $^{272}_{111}Rg$ after the intermediate compound nucleus loses a neutron.

Detecting and Measuring Radiations

20.71 Suppose that a radiologist who is 2.0 m from a small, unshielded source of radiation receives 2.8 units of radiation. To reduce the exposure to 0.28 units of radiation, to what distance from the source should the radiologist move?

20.72 By what percentage should a radiation specialist increase the distance from a small unshielded source to reduce the radiation intensity by 10.0%?

OH **20.73** If exposure from a distance of 1.60 m gave a worker a dose of 8.4 rem, how far should the worker move away from the source to reduce the dose to 0.50 rem for the same period?

20.74 During work with a radioactive source, a worker was told that he would receive 50 mrem at a distance of 4.0 m during 30 min of work. What would be the received dose if the worker moved closer, to 0.50 m, for the same period?

Law of Radioactive Decay

20.75 Smoke detectors contain a small amount of americium-241, which has a half-life of 1.70×10^5 days. If the detector contains 0.20 mg of ^{241}Am, what is the activity, in becquerels? In microcuries?

20.76 Strontium-90 is a dangerous radioisotope present in fallout produced by nuclear weapons. ^{90}Sr has a half-life of 1.00×10^4 days. What is the activity of 1.00 g of ^{90}Sr, in becquerels? In microcuries?

20.77 Iodine-131 is a radioisotope present in radioactive fallout that targets the thyroid gland. If 1.00 mg of ^{131}I has an activity of 4.6×10^{12} Bq, what is the decay constant for ^{131}I? What is the half-life, in seconds?

20.78 A 10.0 mg sample of thallium-201 has an activity of 7.9×10^{13} Bq. What is the decay constant for ^{201}Tl? What is the half-life of ^{201}Tl, in seconds?

Applications of Radionuclides

OH **20.79** What percentage of cesium chloride made from cesium-137 ($t_{1/2}$ = 30 y; beta emitter) remains after 150 y? What *chemical* product forms?

20.80 A sample of waste has a radioactivity, caused solely by strontium-90 (beta emitter, $t_{1/2}$ = 28.1 yr), of 0.245 Ci g^{-1}. How many years will it take for its activity to decrease to 1.00×10^{-6} Ci g^{-1}?

20.81 A worker in a laboratory unknowingly became exposed to a sample of radiolabeled sodium iodide made from iodine-131 (beta emitter, $t_{1/2}$ = 8.07 days). The mistake was realized 28.0 days after the accidental exposure, at which time the activity of the sample was 25.6×10^{-5} Ci g^{-1}. The safety officer needed to know how active the sample was at the time of the exposure. Calculate that value in curies per gram.

20.82 Technetium-99*m* (gamma emitter, $t_{1/2}$ = 6.02 hr) is widely used for diagnosis in medicine. A sample prepared in the early morning for use that day had an activity of 4.52×10^{-6} Ci. What will its activity be at the end of the day, that is, after 8.00 hr?

20.83 A 0.500 g sample of rock was found to have 2.45×10^{-6} mol of potassium-40 ($t_{1/2} = 1.3 \times 10^9$ yr) and 2.45×10^{-6} mol of argon-40. How old was the rock? (What assumption is made about the origin of the argon-40?)

20.84 If a rock sample was found to contain 1.16×10^{-7} mol of argon-40, how much potassium-40 would also have to be present for the rock to be 1.3×10^9 years old?

20.85 A tree killed by being buried under volcanic ash was found to have a ratio of carbon-14 atoms to carbon-12 atoms of 4.8×10^{-14}. How long ago did the eruption occur?

20.86 A wooden door lintel from an excavated site in Mexico would be expected to have what ratio of carbon-14 to carbon-12 atoms if the lintel is 9.0×10^3 y old?

Nuclear Energy

OH **20.87** Complete the following nuclear equation by supplying the symbol for the other product of the fission.

$$^{235}_{92}U + ^{1}_{0}n \longrightarrow ^{94}_{38}Sr + \_\_\_\_ + 2\,^{1}_{0}n$$

20.88 Both products of the fission in the previous problem are unstable. According to Figures 20.8 and 20.9, what is the most likely way for each of them to decay, by alpha emission, beta emission, or by positron emission? Explain. What are some of the possible fates of the extra neutrons produced by the fission shown in the previous problem?

ADDITIONAL EXERCISES

20.89 What is the nuclear equation for each of the following changes?
(a) beta emission from aluminum-30
(b) alpha emission from einsteinium-252
(c) electron capture by molybdenum-93
(d) positron emission by phosphorus-28

***20.90** Calculate to three significant figures the binding energy in joules per nucleon of the nucleus of an atom of iron-56. The observed mass of one *atom* is 55.9349 u. What information lets us know that no isotope has a *larger* binding energy per nucleon?

***20.91** Calculate to five significant figures the binding energy in joules per nucleon of uranium-235. The observed mass of one *atom* is 235.0439 u.

20.92 Give the nuclear equation for each of these changes.
(a) positron emission by carbon-10
(b) alpha emission by curium-243
(c) electron capture by vanadium-49
(d) beta emission by oxygen-20

***20.93** If a positron is to be emitted spontaneously, how much more *mass* (as a minimum) must an *atom* of the parent have than an *atom* of the daughter nuclide? Explain.

20.94 If a proton and an antiproton were to collide and produce two annihilation photons, what would the wavelength of the photons be, in meters? Which of the decimal multipliers in Table 1.4 (page 12) would be most appropriate for expressing this wavelength?

20.95 There is a gain in binding energy per nucleon when light nuclei fuse to form heavier nuclei. Yet, a tritium atom and a deuterium atom, in a mixture of these isotopes, does not spontaneously fuse to give helium (and energy). Explain why not.

20.96 ^{214}Bi decays to isotope *A* by alpha emission; *A* then decays to *B* by beta emission, which decays to *C* by another beta emission. Element *C* decays to *D* by still another beta emission, and *D* decays by alpha emission to a stable isotope, *E*. What is the proper symbol of element *E*? (Contributed by Prof. W. J. Wysochansky, Georgia Southern University.)

20.97 ^{15}O decays by positron emission with a half-life of 124 s. (a) Give the proper symbol of the product of the decay. (b) How much of a 750 mg sample of ^{15}O remains after 5.0 min of decay? (Contributed by Prof. W. J. Wysochansky, Georgia Southern University.)

20.98 Alpha decay of ^{238}U forms ^{234}Th. What kind of decay from ^{234}Th produces ^{234}Ac? (Contributed by Prof. Mark Benvenuto, University of Detroit—Mercy.)

20.99 A sample of rock was found to contain 2.07×10^{-5} mol of ^{40}K and 1.15×10^{-5} mol of ^{40}Ar. If we assume that all of the ^{40}Ar came from the decay of ^{40}K, what is the age of the rock in years ($t_{1/2} = 1.3 \times 10^9$ years for ^{40}K.)

20.100 The ^{14}C content of an ancient piece of wood was found to be one-eighth of that in living trees. How many years old is this piece of wood ($t_{1/2} = 5730$ years for ^{14}C)?

20.101 Dinitrogen trioxide, N_2O_3, is largely dissociated into NO and NO_2 in the gas phase where there exists the equilibrium, $N_2O_3 \rightleftharpoons NO + NO_2$. In an effort to determine the structure of N_2O_3, a mixture of NO and *NO_2 was prepared containing isotopically labeled N in the NO_2. After a period of time the mixture was analyzed and found to contain substantial amounts of both *NO and *NO_2. Explain how this is consistent with the structure for N_2O_3 being ONONO.

20.102 The reaction $(CH_3)_2Hg + HgI_2 \rightarrow 2CH_3HgI$ is believed to occur through a transition state with the structure

```
H3C        I
    \    .' \
     Hg      Hg
       \   .'  \
        CH3     I
```

If this is so, what should be observed if $(CH_3)_2Hg$ and *HgI_2 are mixed, where the asterisk denotes a radioactive isotope of Hg? Explain your answer.

***20.103** A large, complex piece of apparatus has built into it a cooling system containing an unknown volume of cooling liquid. It is desired to measure the volume of the coolant without draining the lines. To the coolant was added 10.0 mL of methanol whose molecules included atoms of ^{14}C and that had a specific activity of 580 counts per minute per gram (cpm/g), determined using a Geiger counter. The coolant was permitted to circulate to assure complete mixing before a sample was withdrawn that was found to have a specific activity of 29 cpm/g. Calculate the volume of coolant in the system in milliliters. The density of methanol is 0.792 g/mL, and the density of the coolant is 0.884 g/mL.

***20.104** A complex ion of chromium(III) with oxalate ion was prepared from ^{51}Cr-labeled $K_2Cr_2O_7$, having a specific activity of 843 cpm/g (counts per minute per gram), and ^{14}C-labeled oxalic acid, $H_2C_2O_4$, having an specific activity of 345 cpm/g. Chromium-51 decays by electron capture with the emission of gamma radiation, whereas ^{14}C is a pure beta emitter. Because of the characteristics of the beta and gamma detectors, each of these isotopes may be counted independently. A sample of the complex ion was observed to give a gamma count of 165 cpm and a beta count of 83 cpm. From these data, determine the number of oxalate ions bound to each Cr(III) in the complex ion. (Hint: For the starting materials calculate the cpm per mole of Cr and oxalate, respectively.)

20.105 Iodine-131 is used to treat Graves disease, a disease of the thyroid gland. The amount of ^{131}I used depends on the size of the gland. If the dose is 86 microcuries per gram of thyroid gland, how many grams of ^{131}I should be administered to a patient with a thyroid gland weighing 20 g? Assume all the iodine administered accumulates in the thyroid gland.

20.106 The fuel for a thermonuclear bomb (hydrogen bomb) is lithium deuteride, a salt composed of the ions ${}^{6}_{3}Li^{+}$ and ${}^{2}_{1}H^{-}$. Considering the nuclear reaction

$$ {}^{6}_{3}Li + {}^{1}_{1}n \rightarrow {}^{4}_{2}He + {}^{3}_{1}H $$

explain how a ^{235}U fission bomb could serve as a trigger for a fusion bomb. Write appropriate nuclear equations.

20.107 In 2006, the confirmed synthesis of ${}^{294}_{118}Uuo$ (an isotope of element 118) was reported to involve the bombardment of ^{245}Cf with ^{48}Ca. Write an equation for the nuclear reaction, being sure to include any other products of the reaction.

***20.108** Radon, a radioactive noble gas, is an environmental problem in some areas, where it can seep out of the ground and into homes. Exposure to radon-222, an alpha emitter with a half-life of 3.823 days, can increase the risk of lung cancer. At an exposure level of 4 pCi per liter (the level at which the EPA recommends action), the lifetime risk of death from lung cancer due to radon exposure is estimated to be 62 out of 1,000 for current smokers, compared with 73 out of 10,000 for nonsmokers. If the air in a home was analyzed and found to have an activity of 4.1 pCi L^{-1}, how many atoms of ^{222}Rn are there per liter of air?

***20.109** The isotope ^{145}Pr decays by emission of beta particles with an energy of 1.80 MeV each. Suppose a person swallowed, by accident, 1.0 mg of Pr having a specific activity (activity per gram) of 140 Bq g^{-1}. What would be the absorbed dose from ^{145}Pr in units of Gy and rad over a period of 10 minutes? Assume all the beta particles are absorbed by the person's body.

EXERCISES IN CRITICAL THINKING

20.110 A silver wire coated with nonradioactive silver chloride is placed into a solution of sodium chloride that is labeled with radioactive ^{36}Cl. After a while, the AgCl was analyzed and found to contain some ^{36}Cl. How do you interpret the results of this experiment?

20.111 Suppose you were given a piece of cotton cloth and told that it was believed to be 2000 years old. You performed a carbon dating test on a tiny piece of the cloth and your data indicated that it was only 800 years old. If the cloth really was 2000 years old, what factors might account for your results?

20.112 In 2006, the former Soviet spy Alexander Litvinenko was poisoned by the polonium isotope ^{210}Po. He died 23 days after ingesting the isotope, which is an alpha emitter. Find data on the Internet to answer the following: Assuming Litvinenko was fed 1 μg of ^{210}Po and that it became uniformly distributed through the cells in his body, how many atoms of ^{210}Po made their way into each cell in his body? (Assume his body contained the average number of cells found in an adult human.) Also calculate the number of cells affected by the radiation each second, being sure to take into account an estimate of the average number of cells affected by an alpha particle emitted by a ^{210}Po nucleus.

20.113 What would be the formula of the simplest hydrogen compound of element 116? Would a solution of that compound in water be acidic, basic, or neutral? Explain your reasoning.

20.114 Astatine is a halogen. Its most stable isotope, ^{210}At, has a half-life of only 8.3 hours, and only very minute amounts of the element are ever available for study (<0.001 μg). This amount is so small as to be virtually invisible. Describe experiments you might perfom that would tell you whether the silver salt of astatine, AgAt, is insoluble in water.

NONMETALS, METALLOIDS, METALS, AND METAL COMPLEXES 2

A chemical found in many brands of shampoo is EDTA, which is able to form stable complexes with ions found in hard water. These ions, which include Ca^{2+} and Mg^{2+}, can interfere with the action of soaps and some detergents and inhibit the formation of a nice lather, like the one shown here. Complexes of metal ions with EDTA and other substances are studied in detail in this chapter. *(Beauty Photo studio/Age Fotostock America, Inc.)*

CHAPTER OUTLIN

THIS CHAPTER IN CONTEXT In previous chapters you learned many of the concepts that ch have used to develop their understanding of how the elements react with each other and the kinds of com they produce. For example, in our discussions of chemical kinetics, you learned how various factors affect th of reactions, and in our study of thermodynamics you learned how enthalpy and entropy changes affect the p ity of observing chemical changes. Our focus, however, was primarily on the concepts themselves, with exam chemical behavior being used to reinforce and justify them. With these concepts available to us now, we wi our emphasis in the opposite direction and examine some of the physical and chemical properties of the elements and their compounds.

Our intent in this chapter is not to be encyclopedic. Instead, we will concentrate our attention on several aspects of the chemistries of the elements. We will study how the elements occur in nature and explore the methods used to obtain them in their free (uncombined) states. For metals, their extraction from compounds constitutes a vast commercial enterprise under the general heading of *metallurgy.*

In their free states, the nonmetallic elements exist in a broad range of structures, ranging from simple atoms to very complex molecular forms. We will study how the relative tendencies of the nonmetals to form strong σ and π bonds influences the kinds of structures observed.

The last part of the chapter is devoted to an in-depth look at complex ions of metals, a topic first introduced in Chapter 17. These substances, which have applications from food preservatives to catalyzing biochemical reactions, have a variety of structures and colors. We will study the kinds of substances that combine with metals to form complexes with various geometries, how complexes are named, and how the electronic structures and colors of complexes can be explained.

21.1 Nonmetals and Metalloids Are Found as Free Elements and in Compounds

As you learned in Chapter 2, the nonmetals and metalloids are located at the right side of the periodic table (Figure 21.1). As suggested by their positions in the table, metalloids have properties that place them between those of metals and nonmetals. For instance, metalloids have a sheen that gives them the appearance of a metal, and they exhibit weak electrical conductivity (many are semiconductors). Chemically, however, they behave more like nonmetals. For example, metalloids combine with the more active metals in Groups IA and IIA to form compounds that are saltlike, containing anions such as Si^{4-}, As^{3-}, and Te^{2-} (more complex anions are also formed), and they also combine with nonmetals to form molecular compounds in which their oxidation numbers are positive.

A sample of crystalline elemental silicon exhibits a metallic sheen. *(Jeff J. Daly/Fundamental Photographs.)*

Nonmetals occur in compounds and in the free state

Most chemical compounds contain one or more nonmetals. In compounds involving metals, nonmetals occur as either simple anions or in polyatomic anions. Many other compounds are composed of only nonmetals in combinations that range from simple (such as HCl) to very complex (such as DNA).

It is difficult to make general statements about the preparation of the elemental nonmetals. The noble gases, for example, are always found uncombined in nature. The atmosphere is the major source of the noble gases, even though their concentrations in air

H																	He
Li	Be											B	C	N	O	F	Ne
Na	Mg											Al	Si	P	S	Cl	Ar
K	Ca	Sc	Ti	V	Cr	Mn	Fe	Co	Ni	Cu	Zn	Ga	Ge	As	Se	Br	Kr
Rb	Sr	Y	Zr	Nb	Mo	Tc	Ru	Rh	Pd	Ag	Cd	In	Sn	Sb	Te	I	Xe
Cs	Ba	La	Hf	Ta	W	Re	Os	Ir	Pt	Au	Hg	Tl	Pb	Bi	Po	At	Rn
Fr	Ra																

FIG. 21.1 Distribution of nonmetals and metalloids in the periodic table.

are very small. Physical methods are used to separate them from other gases with which they're mixed. Of the noble gases, only helium and radon are not obtained primarily from the atmosphere. Helium is found in gaseous deposits beneath the earth's crust where it has collected after being produced by the capture of electrons by alpha particles (helium nuclei, He^{2+}), which are formed during the radioactive decay of elements such as uranium. Radon itself is radioactive and is produced by the radioactive decay of radium.

Noble gas atoms have completed *s* and *p* subshells in their outer shells and are very unreactive. They have very little tendency to form bonds to other atoms.

$$^{226}Ra \longrightarrow {}^{222}Rn + {}^{4}He$$

Since radon spontaneously decomposes into other elements relatively quickly ($t_{1/2}$ = 3.8 days), it occurs only in minute quantities in nature.

Other nonmetals, while present in many naturally occurring compounds, also are found extensively in the free state as well. For instance, our atmosphere is composed primarily of elemental nitrogen, N_2 (about 80%), and oxygen, O_2 (about 20%). Although nitrogen and oxygen are also found in a vast number of compounds, certainly their most economical source is simply the air itself. Sulfur and carbon are two other elements that occur naturally in both the combined and free states. There are, for instance, many naturally occurring sulfates (for example, $BaSO_4$ and $CaSO_4{\cdot}2H_2O$) and sulfides (FeS_2, CuS, HgS, PbS, and ZnS). In the free state, sulfur has been found in large underground deposits from which it is mined by forcing superheated water under pressure into the sulfur, causing the sulfur to melt. Once molten, the sulfur–water mixture is foamed to the surface using compressed air. Sulfur is also deposited on rock surfaces near volcanic vents and is called *brimstone.*

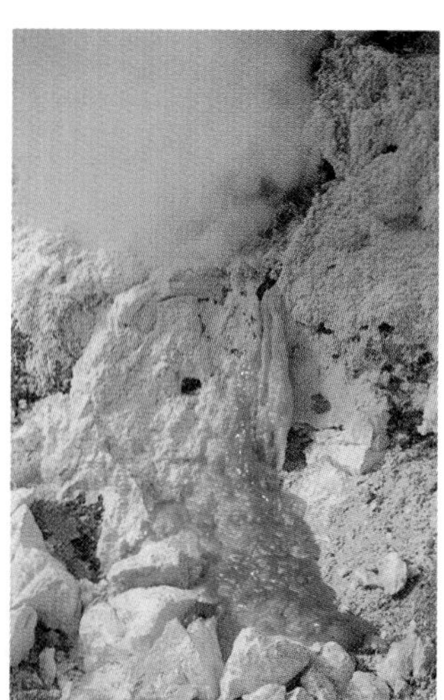

Elemental sulfur in nature. Sulfur is one of the substances released from the earth during volcanic eruptions. Here we see molten sulfur pouring from a vent in the Kawah Ijen volcano in Indonesia. *(© API/Explorer/ Photo Researchers.)*

Turning to carbon, we find that most of its naturally occurring compounds are carbonates, for example, limestone ($CaCO_3$). In the free state most carbon is found in either of its two principal forms, diamond and graphite.

Nonmetals are obtained from compounds by either oxidation or reduction

Because nonmetals combine with each other as well as with metals, no simple generalizations can be made concerning their recovery from compounds. When combined with a metal, the nonmetal is found in a negative oxidation state, so an oxidation must be brought about to generate the free element. For example, the halogens Cl_2, Br_2, and I_2 can be conveniently prepared in the laboratory by reacting one of their salts with an oxidizing agent such as MnO_2 in an acidic solution:

$$2X^- + MnO_2 + 4H^+ \longrightarrow X_2 + Mn^{2+} + 2H_2O$$

where X = Cl, Br, or I.

As you learned in Chapter 19, chlorine is a very important industrial chemical, and vast quantities (approximately 10 million tons annually) are produced by electrolysis of NaCl, both aqueous and molten. Chlorine is used in large amounts in water treatment, in the manufacture of pharmaceuticals, pesticides, and solvents, and in the production of vinyl chloride, which is used to manufacture vinyl plastics.

The halogens themselves can also serve as oxidizing agents in replacement reactions. Since the tendency to acquire electrons (electronegativity) decreases as we proceed downward in a group, the ability of the halogen to serve as an oxidizing agent decreases too. This is seen in their reduction potentials (Table 21.1), which decrease from fluorine to iodine. As a result, a given halogen is a better oxidizing agent than the other halogens below it in Group VIIA and is able to displace them from their binary compounds with metals. Thus

Remember, the larger the reduction potential, the greater is the tendency of the substance to acquire electrons and, therefore, to be an oxidizing agent.

TABLE 21.1 Reduction Potentials of the Halogens

Reaction	$E°$(V)
$F_2(aq) + 2e^- \rightleftharpoons 2F^-(aq)$	2.87
$Cl_2(aq) + 2e^- \rightleftharpoons 2Cl^-(aq)$	1.36
$Br_2(aq) + 2e^- \rightleftharpoons 2Br^-(aq)$	1.07
$I_2(aq) + 2e^- \rightleftharpoons 2I^-(aq)$	0.54

F_2 will displace Cl^-, Br^-, and I^-, while Cl_2 will displace only Br^- and I^- but not F^-, and so on. This is illustrated by the following typical reactions.

For fluorine:

$$F_2 + \begin{Bmatrix} 2NaCl \\ 2NaBr \\ 2NaI \end{Bmatrix} \longrightarrow 2NaF + \begin{Bmatrix} Cl_2 \\ Br_2 \\ I_2 \end{Bmatrix}$$

For chlorine:

$$Cl_2 + \begin{Bmatrix} 2NaBr \\ 2NaI \end{Bmatrix} \longrightarrow 2NaCl + \begin{Bmatrix} Br_2 \\ I_2 \end{Bmatrix}$$

$$Cl_2 + 2NaF \longrightarrow \text{no reaction}$$

For bromine:

$$Br_2 + 2NaI \longrightarrow 2NaBr + I_2$$

$$Br_2 + \begin{Bmatrix} 2NaF \\ 2NaCl \end{Bmatrix} \longrightarrow \text{no reaction}$$

Iodine cannot displace any of the other halogens from their compounds.

The relative oxidizing power of the halogens is used in the commercial preparation of Br_2. Bromine is isolated from seawater and brine solutions pumped from deep wells by passing Cl_2, followed by air, through the liquid. The Cl_2 oxidizes the Br^- to Br_2 and the air sweeps the volatile Br_2 from the solution.

Fluorine, because of its position as the most powerful chemical oxidizing agent, can only be obtained by electrolytic oxidation. This process must be carried out in the absence of water because water is more easily oxidized than the fluoride ion. In an aqueous solution, therefore, the oxidation

$$2H_2O(l) \longrightarrow O_2(g) + 4H^+(aq) + 4e^-$$

will occur in preference to

$$2F^-(aq) \longrightarrow F_2(g) + 2e^-$$

In practice, a molten mixture of KF and HF, which has a lower melting point than KF alone, is electrolyzed, producing H_2 at the cathode and F_2 at the anode.

H^+ is more easily reduced than K^+, which is why H_2 appears at the cathode.

Nonmetals can also be extracted from their compounds by reduction if the nonmetal happens to exist in a positive oxidation state. For instance, elemental phosphorus is produced from a phosphate such as $Ca_3(PO_4)_2$, where it exists in the +5 oxidation state. To obtain the phosphorus, the $Ca_3(PO_4)_2$ is heated to approximately 1500 °C in an electric furnace with a mixture of carbon and SiO_2 (sand).[1]

$$Ca_3(PO_4)_2(s) + 3SiO_2(s) + 5C(s) \xrightarrow{\text{heat}} 3CaSiO_3(l) + 5CO(g) + 2P(g)$$

In this reaction, the SiO_2 is present to combine with the calcium to form calcium silicate, $CaSiO_3$, a compound with a relatively low melting point. Because the reduction of SiO_2 by carbon requires much higher temperatures (>1900 °C), only the phosphorus is reduced to the element at the lower temperatures used in the reaction.

Most metal silicates contain polymeric anions with oxygen bridges between silicon atoms. Silicon dioxide itself has a complex structure consisting of SiO_4 tetrahedra in which each oxygen atom is shared by two silicon atoms.

Metalloids are found in nature in compounds

In most of their compounds, the metalloids are combined with nonmetals, either in a molecular structure such as SiO_2 (found in silica sand) or in an oxoanion such as those found in the silicates (found in many kinds of rocks). In such combinations, the

[1]Phosphorus was first produced in 1669 when alchemist Hennig Brandt heated a mixture of sand and dried urine. He condensed the vapors by passing them through water to give the new element. It was named phosphorus from the Greek *phosphoros* (light bringer) because the solid in a sealed bottle glowed in the dark. The glow is actually the result of slow oxidation of the phosphorus surface by residual oxygen in the container.

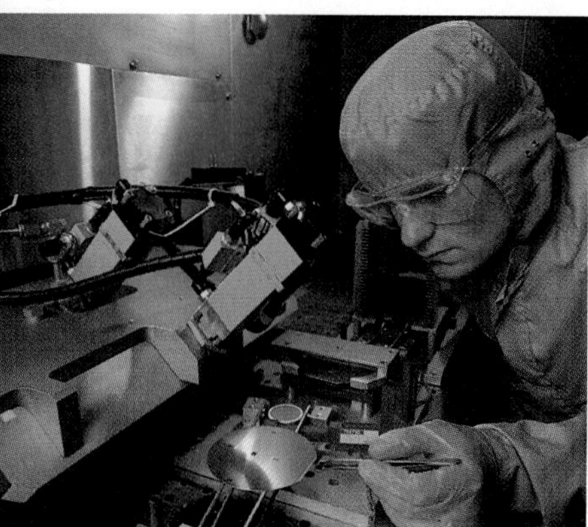

Uses of silicon as a semiconductor. A technician checks a silicon wafer during one stage in the manufacture of computer "chips." The silicon disk contains hundreds of individual chips, each consisting of thousands of electronic components. After all of the individual components have been added, the disk will be cut up to give separate chips that will be incorporated into parts for computer circuits. *(David Parker/Photo Researchers.)*

metalloid has a lower electronegativity than the nonmetal, so the metalloid exists in a positive oxidation state. To obtain the metalloid in its elemental state, therefore, a chemical reduction must be carried out. This is usually accomplished with either hydrogen or carbon as the reducing agent. For example, boron is obtained by passing a mixture of BCl_3 vapor and hydrogen gas over a hot wire, on which the following reaction takes place.

$$2BCl_3(g) + 3H_2(g) \longrightarrow 2B(s) + 6HCl(g)$$

On the other hand, elemental silicon is produced by heating SiO_2 with carbon in an electric furnace, where the reaction

$$SiO_2(s) + C(s) \longrightarrow Si(s) + CO_2(g)$$

occurs once the temperature exceeds approximately 3000 °C (below this temperature the reverse reaction is actually favored).

The remaining metalloids may be obtained from their oxides by heating them with either carbon or hydrogen; for example,

$$GeO_2(s) + C(s) \longrightarrow Ge(s) + CO_2(g)$$
$$GeO_2(s) + 2H_2(g) \longrightarrow Ge(s) + 2H_2O(g)$$

Similarly, we have

$$2As_2O_3(s) + 3C(s) \longrightarrow 4As(s) + 3CO_2(g)$$
$$As_2O_3(s) + 3H_2(g) \longrightarrow 2As(s) + 3H_2O(g)$$

and

$$2Sb_2O_3(s) + 3C(s) \longrightarrow 4Sb(s) + 3CO_2(g)$$
$$Sb_2O_3(s) + 3H_2(g) \longrightarrow 2Sb(s) + 3H_2O(g)$$

There are few aspects of modern life that are not impacted significantly by computer circuits embedded in silicon or germanium.

As you may recall, in very pure form, silicon and germanium have widespread applications in the electronics industry, where they are used in transistors and photoconduction devices. Sophisticated electronic circuits embedded in tiny wafers of silicon make possible a myriad of computer-controlled devices that we've come to take for granted, such as laptop computers; CD, DVD, and MP3 players; handheld computer games; computers that enable automobile engines to manage exhaust emission; cell phones; and digital cameras (and the list goes on).

21.2 Nonmetallic Elements in Their Free States Have Structures of Varying Complexity

Only the noble gases exist in nature as single atoms. All the other nonmetallic elements are found in more complex forms in their free states—some as diatomic molecules and the rest in more complex molecular structures. In this section we will study what these structures are and use bonding theory to understand them. Let's begin by reviewing how atoms form covalent bonds.

You learned in Chapter 9 that atoms are able to share electrons in two basic ways. One is by the formation of σ bonds, which can involve the overlap of *s* orbitals or an end-to-end overlap of *p* orbitals or hybrid orbitals. The second kind of covalent bond is the π bond, which normally requires the sideways overlap of unhybridized *p* orbitals (Figure 21.2). (Pi bonds can also be formed by *d* orbitals, but we do not discuss them in this book.)

Not all atoms have the same tendency to form π bonds. The ability of an atom to form π bonds determines its ability to form multiple bonds, and this in turn greatly affects the kinds of molecular structures that the element produces. One of the most striking illustrations of this is the molecular structures of the elemental nonmetals and metalloids.

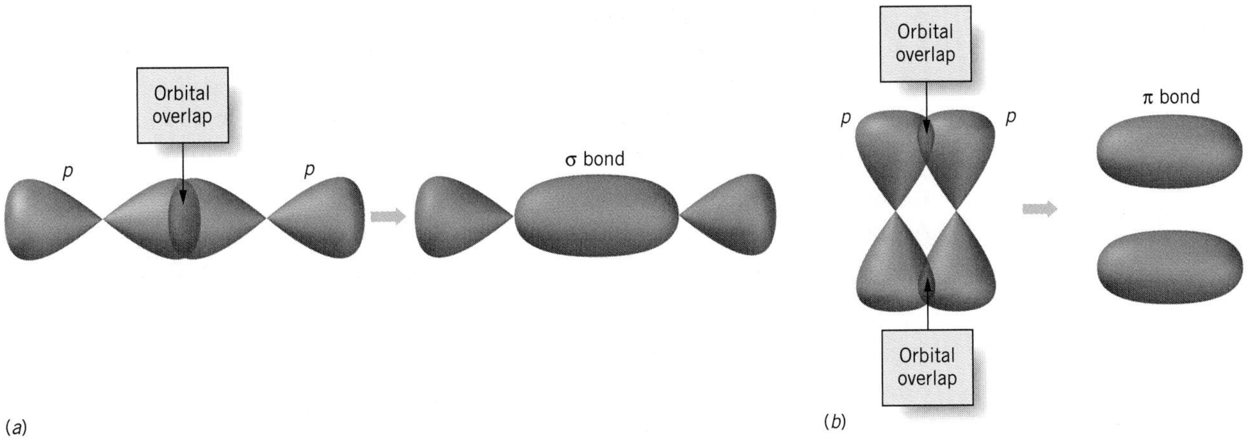

FIG. 21.2 **Formation of sigma and pi bonds.** (*a*) Sigma bonds formed by the head-to-head overlap of *p* orbitals. (*b*) Pi bonds formed by the side-by-side overlap of *p* orbitals.

Nonmetals in Period 2 form multiple bonds relatively easily

One of the controlling factors in determining the complexity of the molecular structures of the nonmetals and metalloids is their ability to form multiple bonds. Small atoms, such as those in Period 2, are able to approach each other closely. As a result, effective sideways overlap of their *p* orbitals can occur, and these atoms form strong π bonds. Therefore, carbon, nitrogen, and oxygen are able to form multiple bonds about as easily as they are able to form single bonds. On the other hand, when the atoms are large—which is the case for atoms from Periods 3, 4, and so on—π-type overlap between their *p* orbitals is relatively ineffective, so π bonds formed by large atoms are relatively weak compared to σ bonds. Therefore, rather than form a double bond consisting of one σ bond and one π bond, these elements prefer to use two separate σ bonds to bond their atoms together. This leads to a useful generalization: *Elements in Period 2 are able to form multiple bonds fairly readily, while elements below them in Periods 3, 4, 5, and 6 have a tendency to prefer single bonds.* Let's see how this affects the structures of the elemental nonmetallic elements.

Oxygen and nitrogen have six and five electrons, respectively, in their valence shells. This means that an oxygen atom needs two electrons to complete its valence shell, and a nitrogen atom needs three. Although a perfectly satisfactory Lewis structure for O_2 can't be drawn, experimental evidence suggests that the oxygen molecule does possess a double bond. The molecular orbital theory, which provides an excellent explanation of the bonding in O_2, also tells us that there is a double bond in the O_2 molecule. The nitrogen molecule, which we discussed earlier, contains a triple bond. Oxygen and nitrogen, because of their small size, are capable of multiple bonding because they are able to form strong π bonds. This allows them to form a sufficient number of bonds with just a single neighbor to complete their valence shells, so they are able to form diatomic molecules.

Oxygen, in addition to forming the stable species O_2 (dioxygen), also can exist in another very reactive molecular form called **ozone,** which has the formula O_3. The structure of ozone can be represented as a resonance hybrid

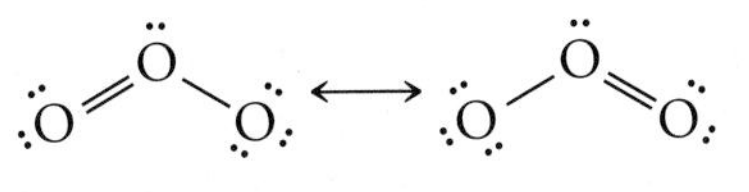

$$\mathrm{O{=}O{-}O} \longleftrightarrow \mathrm{O{-}O{=}O}$$

ΔG_f° for O_3 is $+163$ kJ mol^{-1}, which indicates that the molecule has a strong tendency to decompose to give O_2.

This unstable molecule can be generated by the passage of an electric discharge through ordinary O_2, and the pungent odor of ozone can often be detected in the vicinity of high-voltage electrical equipment. It is also formed in limited amounts in the upper atmosphere by the action of ultraviolet radiation from the sun on O_2. The presence of ozone in the upper atmosphere shields the earth and its creatures from exposure to intense and harmful ultraviolet light from the sun.

The existence of an element in more than one form, either as the result of differences in molecular structure as with O_2 and O_3, or as the result of differences in the packing of molecules in the solid, is a phenomenon called **allotropy.** The different forms of the element are called **allotropes.** Thus, O_2 is one allotrope of oxygen and O_3 is another. Allotropy is not limited to oxygen, as you will soon see.

Carbon forms four allotropes

An atom of carbon, another Period 2 element, has four electrons in its valence shell, so it must share four electrons to complete its octet. There is no way for carbon to form a quadruple bond, so a simple C_2 species is not stable under ordinary conditions. Instead, carbon completes its octet in other ways, leading to four allotropic forms of the element. One of these is **diamond,** in which each carbon atom uses sp^3 hybrid orbitals to form covalent bonds to four other carbon atoms at the corners of a tetrahedron (Figure 21.3*a*).

In its other allotropes, carbon employs sp^2 hybrid orbitals to form ring structures with delocalized π systems covering their surfaces. The most stable form of carbon is **graphite,** which consists of layers of carbon atoms, each composed of many hexagonal "benzene-like" rings fused together in a structure reminiscent of chicken wire.

Graphite is able to serve as a lubricant because the layers of carbon atoms in the structure are able to slide over each other relatively easily.

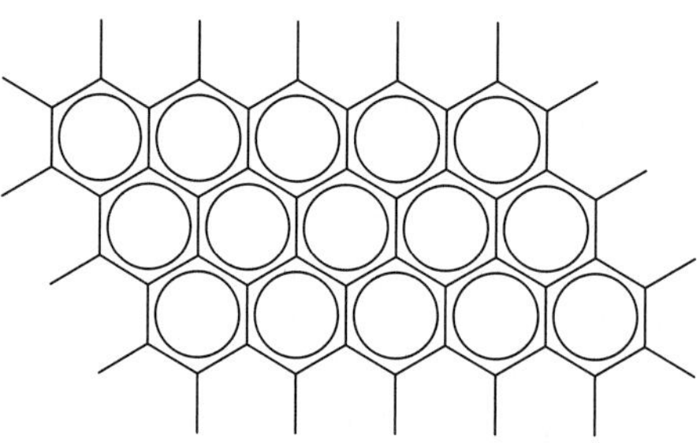

A fragment of a carbon layer in graphite

Another figure showing the layer structure of graphite can be found on the rear cover of the book.

In graphite, these layers are stacked one on top of another, as shown in Figure 21.3*b*. Graphite is an electrical conductor because of the delocalized π electron system that extends across the layers. Electrons can be pumped in at one end of a layer and removed from the other.

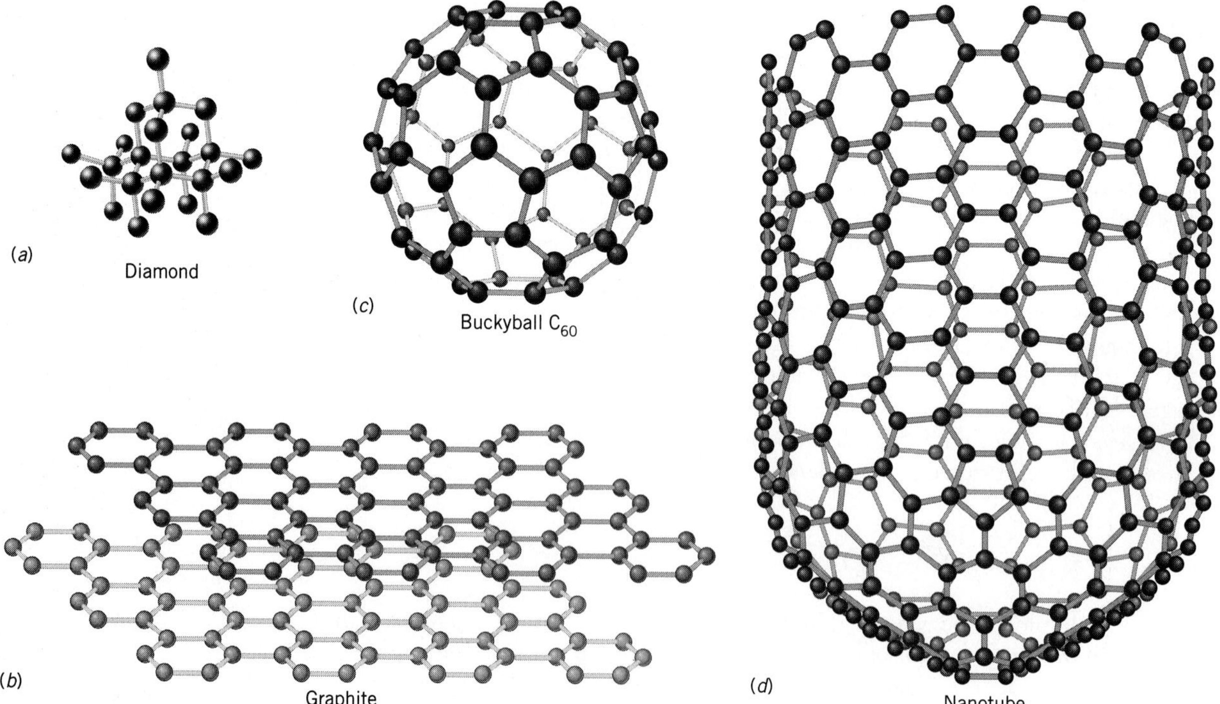

FIG. 21.3 Molecular forms of carbon. (*a*) Diamond. (*b*) Graphite. (*c*) Buckminsterfullerene, or "buckyball," C_{60}. (*d*) A portion of a carbon nanotube showing one closed end.

In 1985, a new form of carbon was discovered that consists of tiny balls of carbon atoms, the simplest of which has the formula C_{60} (Figure 21.3*c*). They were named **fullerenes** and the C_{60} molecule itself was named **buckminsterfullerene** (nickname **buckyball**) in honor of R. Buckminster Fuller, the designer of a type of structure called a geodesic dome. The bonds between carbon atoms in the buckyball occur in a pattern of five- and six-membered rings arranged like the seams in a soccer ball as well as the structural elements of the geodesic dome.

Carbon nanotubes, discovered in 1991, are another form of carbon that is related to the fullerenes. They are formed, along with fullerenes, when an electric arc is passed between carbon electrodes. The nanotubes consist of tubular carbon molecules that we can visualize as rolled up sheets of graphite (with hexagonal rings of carbon atoms). The tubes are capped at each end with half of a spherical fullerene molecule, so a short tube would have a shape like a hot dog. A portion of a carbon nanotube is illustrated in Figure 21.3*d*. Carbon nanotubes have unusual properties that have made them the focus of much research in recent years.

The front cover of the book illustrates the structure and formation of a carbon nanotube.

Weight for weight, carbon nanotubes are about 100 times stronger than stainless steel and about 40 times stronger than the carbon fibers used to make tennis rackets and shafts for golf clubs.

Boron forms a complex structure containing B_{12} units

In Period 2 there is still one element whose structure we have not considered, namely, boron. This element, found in Group IIIA, is quite unlike any of the others, since there is no simple way for it to complete its valence shell. The boron atom has just three electrons in its valence shell. Crystalline boron contains clusters of 12 boron atoms located at the vertices of an icosahedron (a 20-sided geometric figure) as shown in Figure 21.4. In the solid, each of these is also joined to yet another boron atom outside the cluster (Figure 21.4*b*). The electrons available for bonding are therefore delocalized to a large extent over many boron atoms.

The linking together of B_{12} units produces a large three-dimensional covalent solid that is very difficult to break down. As a result, boron is very hard (it is the second hardest element) and has a very high melting point (about 2200 °C).

Nonmetallic elements below Period 2 have structures containing single bonds

In graphite, carbon exhibits multiple bonding, as do nitrogen and oxygen in their molecular forms. As we noted earlier, their ability to do this reflects their ability to form strong π bonds—a requirement for the formation of a double or triple bond. When we move to

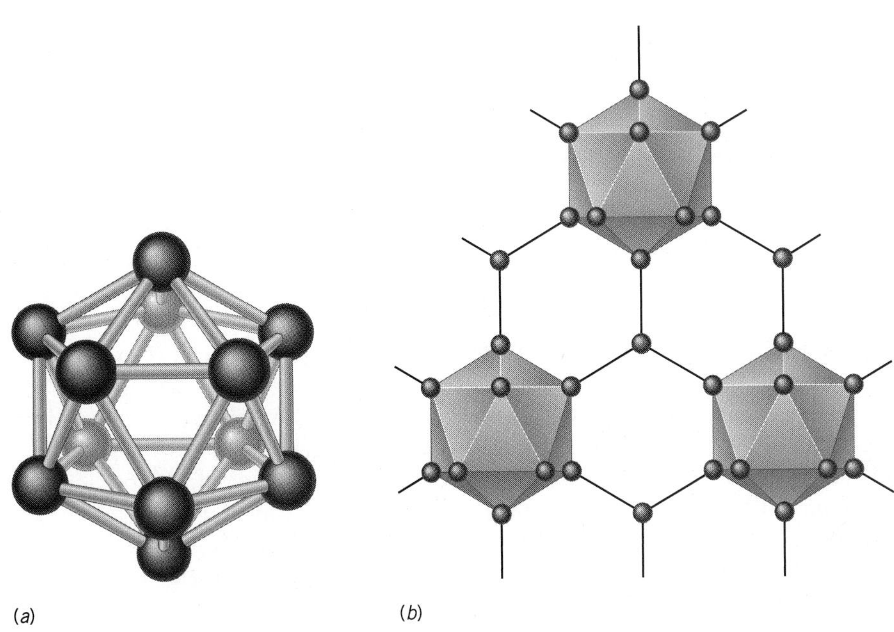

FIG. 21.4 **The structure of elemental boron.** (*a*) The arrangement of 12 boron atoms in a B_{12} cluster. (*b*) The element boron consists of an interconnecting network of B_{12} clusters that produces a very hard and high-melting solid.

the third and successive periods, a different state of affairs exists. Here, we have much larger atoms that are able to form relatively strong σ bonds but much weaker π bonds. Because their π bonds are so weak, these elements prefer single bonds (σ bonds), and the molecular structures of the free elements reflect this.

Elements of Group VIIA

Each of the elements in Group VIIA is diatomic in the free state. This is because π bonding is not necessary in any of their molecular structures. Chlorine, for example, requires just one electron to complete its octet, so it only needs to form one covalent bond with another atom. Therefore, it forms one σ bond to another chlorine atom and is able to exist as diatomic Cl_2. Bromine and iodine form diatomic Br_2 and I_2 molecules for the same reason. The structures of the remaining nonmetals and metalloids are considerably more complex, however.

Elements of Group VIA

Below oxygen in Group VIA is sulfur, which has the Lewis symbol

$$\cdot\ddot{\underset{\cdot\cdot}{S}}\cdot$$

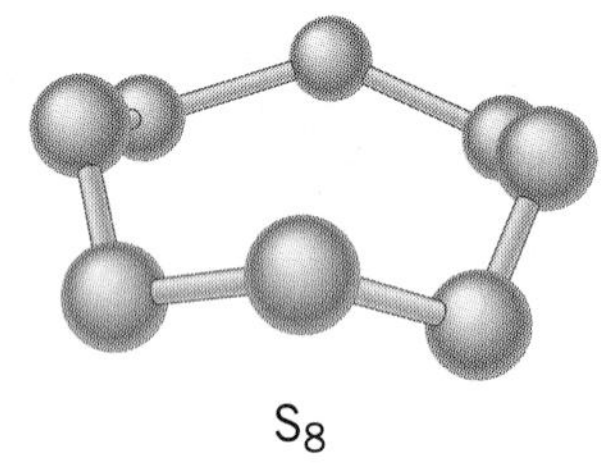

FIG. 21.5 The structure of the puckered S_8 ring.

A sulfur atom requires two more electrons to obtain an octet, so it must form two covalent bonds. But sulfur doesn't form π bonds well to other sulfur atoms; instead, it prefers to form two stronger single bonds to *different* sulfur atoms. Each of these also prefers to bond to two different sulfur atoms, and this gives rise to a

$$-\ddot{\underset{\cdot\cdot}{S}}-\ddot{\underset{\cdot\cdot}{S}}-\ddot{\underset{\cdot\cdot}{S}}-\ddot{\underset{\cdot\cdot}{S}}-\ddot{\underset{\cdot\cdot}{S}}-$$

sequence. Actually, in sulfur's most stable form, called **orthorhombic sulfur,** the sulfur atoms are arranged in an eight-membered ring to give a molecule with the formula S_8 (properly named cyclooctasulfur). The S_8 ring has a puckered crownlike shape, which is illustrated in Figure 21.5. Another allotrope is **monoclinic sulfur,** which also contains S_8 rings that are arranged in a slightly different crystal structure.

Selenium, below sulfur in Group VIA, also forms Se_8 rings in one of its allotropic forms. Both selenium and tellurium also can exist in a gray form in which there are long Se_x and Te_x chains (where the subscript x is a large number).

Elements of Group VA

Like nitrogen, the other elements in Group VA all have five valence electrons. Phosphorus is an example.

$$\cdot\dot{\underset{\cdot\cdot}{P}}\cdot$$

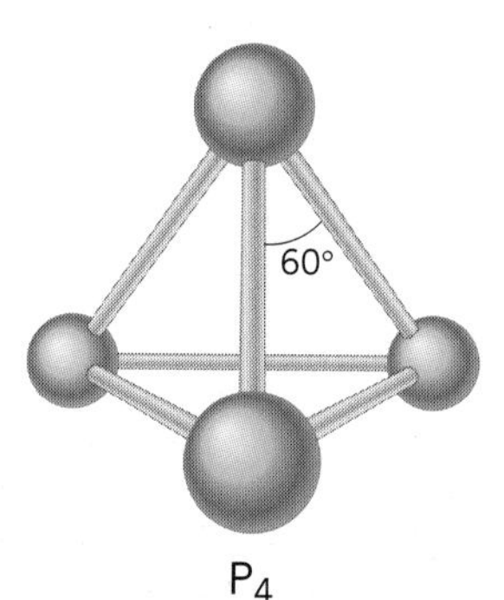

FIG. 21.6 The molecular structure of white phosphorus, P_4. The bond angles of 60° make the phosphorus–phosphorus bonds quite weak and causes the molecule to be very reactive.

The preferred angle between bonds formed by overlap of p orbitals is 90°. Each face of the P_4 tetrahedron is a triangle, however, with 60° angles between edges. This produces less than optimum overlap between the p orbitals in the bonds.

To achieve a noble gas structure, the phosphorus atom must acquire three more electrons. Because there is little tendency for phosphorus to form multiple bonds, as nitrogen does when it forms N_2, the octet is completed by the formation of three single bonds to three *different* phosphorus atoms.

The simplest elemental form of phosphorus is a waxy solid called **white phosphorus** because of its appearance. It consists of P_4 molecules in which each phosphorus atom lies at a corner of a tetrahedron, as illustrated in Figure 21.6. Notice that in this structure each phosphorus is bound to three others. This allotrope of phosphorus is very reactive, partly because of the very small P—P—P bond angle of 60°. At this small angle, the p orbitals of the phosphorus atoms don't overlap very well, so the bonds are weak. As a result, breaking a P—P bond occurs easily. When a P_4 molecule reacts, this bond breaking is the first step, so P_4 molecules are readily attacked by other chemicals, especially oxygen. White phosphorus is so reactive toward oxygen that it ignites and burns spontaneously in air. For this reason, white phosphorus is used in military incendiary devices, and you've probably seen movies in which exploding phosphorus shells produce arching showers of smoking particles.

A second allotrope of phosphorus that is much less reactive is called **red phosphorus.** At the present time, its structure is unknown, although it has been suggested that it contains P_4 tetrahedra joined at the corners as shown in Figure 21.7. Red phosphorus is also used in explosives and fireworks, and it is mixed with fine sand and used on the

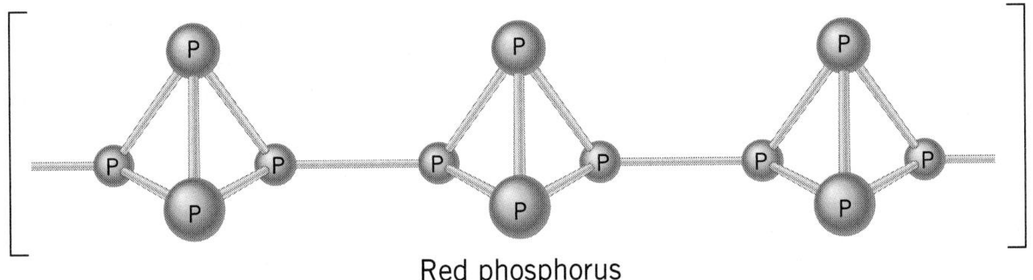

FIG. 21.7 Proposed molecular structure of red phosphorus. Red phosphorus is believed to be composed of long chains of P_4 tetrahedra connected at their corners.

striking surfaces of matchbooks. As a match is drawn across the surface, friction ignites the phosphorus, which then ignites the ingredients in the tip of the match.

A third allotrope of phosphorus is called **black phosphorus,** which is formed by heating white phosphorus at very high pressures. This variety has a layered structure in which each phosphorus atom in a layer is covalently bonded to three others in the same layer. As in graphite, these layers are stacked one atop another, with only weak forces between the layers. As you might expect, black phosphorus has many similarities to graphite.

The elements arsenic and antimony, which are just below phosphorus in Group VA, are also able to form somewhat unstable yellow allotropic forms containing As_4 and Sb_4 molecules, but their most stable forms have a metallic appearance with structures similar to black phosphorus.

Elements of Group IVA

Finally, we look at the heavier nonmetallic elements in Group IVA, silicon and germanium. To complete their octets, each must form four covalent bonds. Unlike carbon, however, they have very little tendency to form multiple bonds, so they don't form allotropes that have a graphite structure. Instead, each of them forms a solid with a structure similar to diamond.

FIG. 21.8 The color of iron oxide. Red Fe_2O_3 gives the clay used by this sculptor its rich color. *(Dan Boler/Stone/Getty Images.)*

21.3 Metals Are Prepared from Compounds by Reduction

When metals form compounds, they almost always lose electrons (become oxidized) and exist in positive oxidation states. Therefore, isolating metals from compounds generally involves reduction. In this section we will look at some of the ways this can be accomplished. Before we do this, however, let's take a brief look at where in nature we find metal compounds.

Metals come from both the earth and the sea

Most metals are reactive enough that they do not occur as free elements in nature. Instead, they are found combined with other elements in compounds. Where we find such compounds depends a great deal on their solubilities in water. For example, you learned that salts of the alkali metals (metals of Group IA) are soluble in water. It is no surprise, therefore, to find Na^+ and K^+ ions in the sea. In fact, the oceans provide a huge storehouse of many minerals, as illustrated by the table in the margin. Sodium and magnesium (as Na^+ and Mg^{2+}) are the metal ions in largest concentration in seawater, partly because their natural abundance among the elements is large,[2] and also because their sulfates, halides, and carbonates are water soluble.

Concentrations of Ions in Seawater

Ion	Molarity
Chloride	0.550
Sodium	0.468
Magnesium	0.055
Sulfate	0.029
Calcium	0.011
Potassium	0.011
Bicarbonate	0.002

Oxygen is the earth's most abundant element, and because it is so reactive it is able to combine with nearly all metals. As you learned when studying the solubility rules in Chapter 4, most metal oxides are insoluble in water (exceptions are the oxides of the alkali metals and some of the oxides of the Group IIA metals). Because of these facts, many metals occur in nature as insoluble oxides found buried in the ground. An example is iron(III) oxide, which gives its rich rust-red color to iron ores and various clay minerals that contain Fe_2O_3 (Figure 21.8).

[2] On the earth, sodium is the fourth most abundant element on an atom basis (approximately 2.6% of the atoms on earth are Na); magnesium is the seventh most abundant (about 1.8%).

As you learned in Chapter 4, nearly all metal carbonates are also insoluble, and the ability of sea creatures, such as clams, oysters, and coral, to extract carbonate ions from seawater to construct their shells accounts for the large formations of carbonate minerals of calcium and magnesium present in many locations around the planet. Over eons of time, the calcium carbonate skeletons of these creatures have built up and then been moved by upheavals in the earth's crust. Such movements have often lifted them to heights far above sea level and at times subjected them to tremendous compressive forces that transformed them into limestone and marble.

Other anions yield insoluble compounds with metal ions as well, which accounts for deposits of metal sulfides, such as copper sulfides (CuS and Cu_2S), lead sulfide (PbS), and calcium phosphate, $Ca_3(PO_4)_2$. The latter mineral is the chief source of phosphate fertilizers, without which farmers could not hope to maintain the large crop harvests required to feed the world's growing population.

In most soils, phosphorus is the limiting nutrient, so phosphate fertilizers are essential to produce good crops.

A few metals, most notably gold and platinum, do occur in the uncombined state in nature. These are metals with very low degrees of reactivity. Although they form many compounds, the metals are found free in nature because their compounds are not particularly stable and are easily decomposed by heat.

Reactive metals are often produced by electrolysis

Recall that for metals, reactivity is a measure of how easily oxidized they are.

The method used to extract a metal from its compounds depends on how "reactive" the metal is. Metals that are easily oxidized, such as the alkali and alkaline earth metals, form compounds that are correspondingly difficult to reduce, and it is difficult and expensive to find chemical reducing agents that are up to the job. For this reason, electrolysis is usually the procedure used. For example, in Chapter 19 we described the elecrolytic production of two reactive metals, sodium and aluminum. Sodium is formed by electrolysis of its molten chloride, whereas aluminum is produced from its oxide by dissolving it in a salt such as Na_3AlF_6 which has a lower melting point.

In an interesting application of Le Châtelier's principle, potassium is usually made by passing sodium vapor over molten potassium chloride at high temperature. Although potassium ions are more difficult to reduce than sodium ions, the equilibrium

$$KCl(l) + Na(g) \rightleftharpoons NaCl(l) + K(g)$$

is shifted to the right because potassium has a higher vapor pressure than sodium. The less volatile sodium condenses as potassium vapor is swept away, thereby gradually shifting the position of equilibrium. The same procedure is used to make metallic rubidium and cesium.

Marble is a favorite stone of sculptors. The type of stone called marble is composed of calcium carbonate, which was deposited by sea creatures long ago and then transformed by heat and pressure caused by movements of the earth's crust. How fortunate for us, because it enabled the great sculptor Michelangelo to create this masterpiece known as *La Pietà*. *(Art Resource.)*

Many metals can be made using a chemical reducing agent

When a metal is of intermediate reactivity, it can be economically extracted from its compounds by reaction with a chemical reducing agent. For example, one of the most common agents used for the reduction of metal oxides is carbon. Tin and lead, for example, can be produced by heating their oxides with carbon, a process called **smelting.**

$$2SnO + C \xrightarrow{\text{heat}} 2Sn + CO_2$$

$$2PbO + C \xrightarrow{\text{heat}} 2Pb + CO_2$$

Carbon is used in large quantities in commercial metallurgy (Section 21.4) because of its abundance and low cost.

Hydrogen is another reducing agent that can be used to liberate metals of moderate chemical activity. For instance tin and lead oxides are also reduced when heated under a stream of H_2.

FACETS OF CHEMISTRY

Polishing Silver—The Easy Way

Using an active metal to reduce a compound of a less active metal has a number of practical applications. One that is handy around the home is using aluminum to remove the tarnish from silver. Generally, silver tarnishes by a gradual reaction with hydrogen sulfide, present in very small amounts in the air. The product of the reaction is silver sulfide, Ag_2S, which is black and forms a dull film over the bright metal. Polishing a silver object using a mild abrasive restores the shine, but it also gradually removes silver from the object as the silver sulfide is rubbed away.

An alternative method, and one that requires little effort, involves lining the bottom of a sink with aluminum foil, adding warm water and detergent (which acts as an electrolyte), and then submerging the tarnished silver object in the detergent–water mixture and placing it in contact with the aluminum (Figure 1). A galvanic cell is established in which the more active aluminum is the anode and the silver object is the cathode. In a short time, the silver sulfide is reduced, restoring the shine and depositing the freed silver metal on the object. As this happens, a small amount of the aluminum foil is oxidized and caused to dissolve. Besides requiring little physical effort, this silver polishing technique doesn't remove silver from the object being polished.

(*a*)

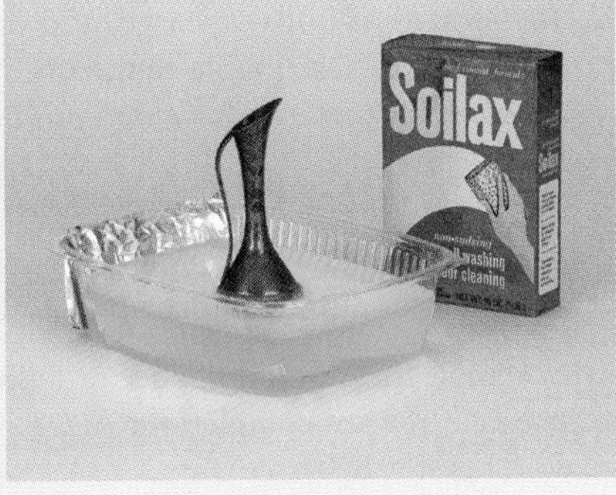

(*b*)

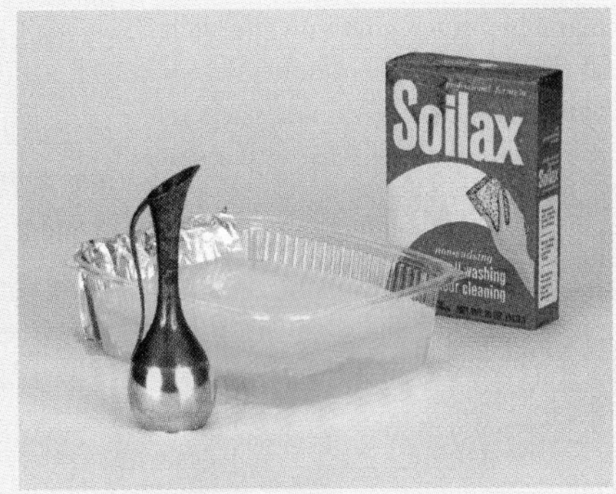

(*c*)

FIG. 1 **Using aluminum to remove tarnish from silver.** (*a*) A badly tarnished silver vase stands next to a container of detergent (Soilax) dissolved in water. The bottom of the container is lined with aluminum foil. (*b*) The vase, shown partly immersed in the detergent, rests on the aluminum foil. (*c*) After a short time, the vase is removed, rinsed with water, and wiped with a soft cloth. Where the vase was immersed in the liquid, much of the tarnish has been reduced to metallic silver. *(Andy Washnik.)*

$$SnO + H_2 \xrightarrow{\text{heat}} Sn + H_2O$$

$$PbO + H_2 \xrightarrow{\text{heat}} Pb + H_2O$$

The use of a more active metal to carry out the reduction is also possible. In Chapter 19 we saw that a galvanic cell could be established between two different metals, for example, Zn and Ag. In that cell the more active reducing agent, Zn, causes the Ag^+ to be reduced. Aluminum was first prepared in 1825 by the reaction of aluminum chloride with the more active metal, potassium.

$$AlCl_3 + 3K \longrightarrow 3KCl + Al$$

Some metal compounds are thermally unstable

Thermal decomposition of a compound involves the conversion of the substance into its elements by heat. Some metal compounds are extremely resistant to such decomposition. For example, magnesium oxide is used to make bricks to line the walls of high-temperature ovens because it has a very high melting point and virtually no tendency to undergo decomposition. On the other hand, mercury(II) oxide decomposes at a relatively low temperature. For example, Joseph Priestley (1733–1804), in his experiments leading to the discovery of oxygen, produced metallic mercury and oxygen from mercury(II) oxide by simply heating it with sunlight focused on the HgO by means of a magnifying glass.

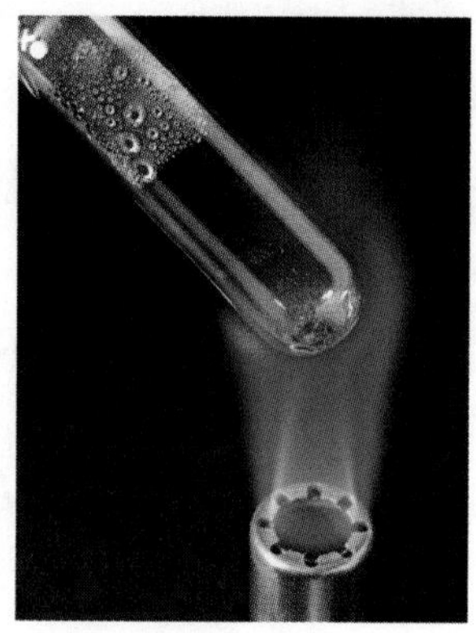

FIG. 21.9 Thermal decomposition of mercury(II) oxide. Heated to a temperature of only 400 °C, HgO turns black and releases oxygen. Drops of mercury metal begin to collect on the walls of the test tube as mercury vapor condenses on the cooler parts of the tube. *(Richard Megna/Fundamental Photographs.)*

In this case, HgO decomposes quite spontaneously at elevated temperatures (Figure 21.9). The equation for the decomposition is.

$$2HgO(s) \longrightarrow 2Hg(l) + O_2(g)$$

The practicality of using a thermal decomposition reaction of this type to produce a free metal depends on the extent to which the reaction proceeds to completion at a given temperature.

In Chapter 18 you learned that at 25 °C the position of equilibrium in a reaction is roughly governed by ΔG°_{298}. If ΔG°_{298} is negative, $K > 1$ and appreciable amounts of products will be expected at equilibrium. Taking ΔG°_T to be the equivalent of ΔG°_{298}, but at a higher temperature, we have the relationship

$$\Delta G^\circ_T = \Delta H^\circ_T - T\Delta S^\circ_T$$

For decomposition, ΔS°_T is positive (decomposition leads to formation of more particles). Most compounds have negative values of ΔH°_f, so we expect the ΔH°_T for the decomposition to be positive, too. That makes ΔG°_T equal to the difference between two positive quantities, ΔH°_T and $T\Delta S^\circ_T$. If ΔH°_T is not *too* positive, the temperature T needed to make ΔG°_T negative is not very large, and thermal decomposition should proceed to a significant degree at a reasonable temperature. *Stated another way, if ΔH°_f for a compound is not too negative, it should be susceptible to thermal decomposition at a reasonably low temperature.*

EXAMPLE 21.1

Calculating the Temperature Required for Thermal Decomposition

Above what temperature would the decomposition of Ag_2O be expected to proceed to an appreciable extent toward completion? At 25 °C, ΔH°_f for Ag_2O is -31.1 kJ mol^{-1}, and $\Delta S^\circ_f = -66.1$ J mol^{-1} K^{-1}.

ANALYSIS: The tool that we will use is Equation 18.10 (page 752), which allows us to estimate ΔG°_T assuming ΔH°_T and ΔS°_T are nearly independent of temperature:

$$\Delta G^\circ_T \approx \Delta H^\circ_{298} - T\Delta S^\circ_{298} \qquad \text{(Equation 18.10)}$$

The decomposition of Ag_2O, $Ag_2O(s) \longrightarrow 2Ag(s) + \frac{1}{2}O_2(g)$, is the reverse of its formation. Therefore, for this reaction at 25 °C we can write

$$\Delta H^\circ_{298} = -\Delta H^\circ_f = +31.1 \text{ kJ mol}^{-1}$$
$$\Delta S^\circ_{298} = -\Delta S^\circ_f = +66.1 \text{ J mol}^{-1}\text{ K}^{-1}$$

Let's calculate the temperature at which $\Delta G^\circ_T = 0$, because above that temperature ΔG°_T will be negative and the decomposition reaction will proceed to a significant extent.

SOLUTION: We will assume, as stated above, that ΔH°_T and ΔS°_T are approximately independent of temperature so that we can use ΔH°_{298} and ΔS°_{298} in the equation for ΔG°_T.

$$\Delta G^\circ_T \approx \Delta H^\circ_{298} - T\Delta S^\circ_{298}$$

When $\Delta G^\circ_T = 0$

$$0 = \Delta H^\circ_{298} - T\Delta S^\circ_{298}$$

Solving for T gives

$$T = \frac{\Delta H^\circ_{298}}{\Delta S^\circ_{298}}$$
$$= \frac{31{,}100 \cancel{\text{J}}\ \cancel{\text{mol}^{-1}}}{66.1 \cancel{\text{J}}\ \cancel{\text{mol}^{-1}}\ \text{K}^{-1}}$$
$$= 470 \text{ K}$$

This corresponds to a Celsius temperature of 197 °C.

Because ΔH°_{298} and ΔS°_{298} are both positive, ΔG°_T will become negative at temperatures above 197 °C. This means that above 197 °C the reaction should become feasible, with much of the Ag_2O undergoing decomposition.

IS THE ANSWER REASONABLE? There is no simple check on this answer, although we can check to be sure that the units cancel properly, and they do. Notice that we were careful to change the units of ΔH°_{298} from kilojoules to joules. If we had not done that, the units would not have canceled properly.

Practice Exercise 1: Above what temperature would Au_2O_3 decompose to give Au and O_2? At 25 °C, $\Delta H^\circ_f = +80.8$ kJ mol^{-1} for Au_2O_3. Also, for Au_2O_3, $S^\circ = 125$ J mol^{-1} K^{-1}; for Au, $S^\circ = 47.7$ J mol^{-1} K^{-1}; for O_2, $S^\circ = 205$ J mol^{-1} K^{-1}. (Hint: Can the absolute temperature be negative?)

Practice Exercise 2: Calculate the temperature required to observe a significant amount of thermal decomposition of MgO. At 25 °C, magnesium oxide has $\Delta H^\circ_f = -601.7$ kJ mol^{-1}; for Mg, $S^\circ = 32.5$ J mol^{-1} K^{-1}, and for O_2, $S^\circ = 205$ J mol^{-1} K^{-1}. For MgO, $S^\circ = 26.9$ J mol^{-1} K^{-1}.

21.4 METALLURGY IS THE SCIENCE AND TECHNOLOGY OF METALS

Metals played an important role in the growth and development of civilization even before the beginning of recorded history. Archaeological evidence indicates that gold was used in making eating utensils and ornaments as early as 3500 BC. Silver was discovered at least as early as 2400 BC, and iron and steel have been used as construction materials since about 1000 BC. Since these earliest times, the methods for obtaining metals from their naturally occurring deposits have evolved, and today they constitute the subject we call metallurgy.

As the title of this section proclaims, **metallurgy** is defined as the science and technology of metals. In modern terms, it is primarily concerned with the procedures and chemical reactions that are used to separate metals from their ores and the preparation of metals for practical use. An **ore** is simply a mineral deposit that has a desirable component in a sufficiently high concentration to make its extraction economically profitable. The economics of the recovery operation, though, is what distinguishes an ore from just another rock. For example, magnesium is found in the mineral *olivine*, which has the formula Mg_2SiO_4. This compound contains over 30% magnesium, but there is no economical way to extract the metal from it. Instead, most magnesium is obtained from seawater, even though its concentration is a mere 0.13%. Although the concentration is much less in the water, a profitable method of obtaining the magnesium has been developed, so seawater is the principal source of magnesium.

Because of the wide variety of sources and the varying nature of the metal compounds in ores, no single method can be applied to the production of all metals. Nevertheless, the sequence of metallurgical processes can be divided into three principal steps.

1. *Concentration.* Ores that contain substantial amounts of impurities, such as rock, must often be treated to concentrate the metal-bearing component. Pretreatment of an ore is also carried out to convert some metal compounds into substances that can be more easily reduced.
2. *Reduction.* The particular procedure employed to reduce the metal compound to give the metal depends on how easily the compound is reduced.

3. *Refining.* Often, during reduction, substantial amounts of impurities become incorporated into the metal. Refining is the process whereby these impurities are removed and the composition of the metal adjusted (alloys formed) to meet the needs of specific applications.

Ores are often treated to enrich the metal-bearing component

When the source of a metal is an ore that is dug from the ground, considerable amounts of sand and dirt are usually mixed in with the ore. To reduce the volume of material that must be processed, the ore is normally concentrated before the metal is separated from it. How this is done depends on the physical and chemical properties of the ore itself, as well as those of the impurities.

Gangue is pronounced "gang."

In some cases, the unwanted rock and sand, called **gangue,** can be removed simply by washing the material with a stream of water. This flushes away the waste and leaves the enriched ore behind. Some iron ores are treated in this way. This procedure also forms the basis for the well-known technique called "panning for gold" that you've probably seen in movies. A sample of sand that might also contain gold is placed into a shallow pan partially filled with water. As the water is swirled around, it washes the less dense sand over the rim of the pan, but leaves any of the more dense bits of gold behind.

Flotation is a method commonly used to enrich the sulfide ores of copper and lead. The ore is crushed, mixed with water, and ground into a souplike slurry which is then transferred to flotation tanks (Figure 21.10) where it is mixed with detergents and oil. The oil adheres to the particles of sulfide ore, but not to the particles of sand and dirt. Air is blown through the mixture and the rising air bubbles become attached to the oil-coated ore particles, bringing them to the surface where they are held in a froth. The detergents in the mixture stabilize the bubbles long enough for the froth and its load of ore particles to be skimmed off. Meanwhile, the sand and dirt settle to the bottom of the tanks and are removed.

Many ores must undergo a second round of pretreatment before the metal can be obtained from them. For example, after enrichment, sulfide ores are usually heated in air. This procedure, called **roasting,** converts the sulfides to oxides which are more conveniently reduced than sulfides. Typical reactions that occur during roasting are

$$2Cu_2S + 3O_2 \longrightarrow 2Cu_2O + 2SO_2$$
$$2PbS + 3O_2 \longrightarrow 2PbO + 2SO_2$$

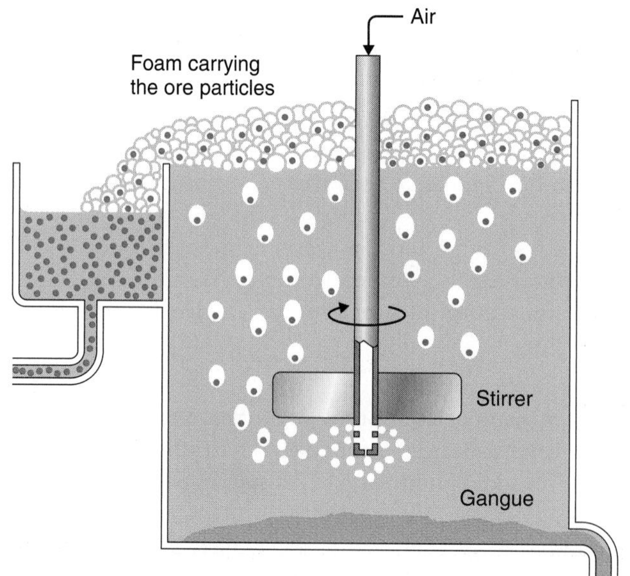

FIG. 21.10 A flotation apparatus.

A by-product of roasting is sulfur dioxide. In years past this was simply released to the atmosphere and was a major source of air pollution. Today we realize that the SO_2 cannot be allowed to escape into the air. One way to remove it from the exhaust gases is to allow it to react with calcium carbonate ($CaCO_3$).

$$CaCO_3(s) + SO_2(g) \longrightarrow CaSO_3(s) + CO_2(g)$$

This method creates another problem, however—the disposal of the solid calcium sulfite. Another way to dispose of the SO_2 is to oxidize it to SO_3. The SO_3 can be converted to sulfuric acid which is then sold.

Aluminum ore, called *bauxite*, must also be pretreated before it can be processed. Bauxite contains aluminum oxide, Al_2O_3, but a number of impurities are also present. To remove the impurities, use is made of aluminum oxide's amphoteric behavior. The ore is mixed with a concentrated sodium hydroxide solution, which dissolves the Al_2O_3.

Aluminum oxide is also called alumina.

$$Al_2O_3(s) + 2OH^-(aq) \longrightarrow 2AlO_2^-(aq) + H_2O$$

The major impurities, however, are insoluble in base, so when the mixture is filtered, the impurities remain on the filter while the aluminum-containing solution passes through. The solution is then neutralized with acid, which precipitates aluminum hydroxide.

$$AlO_2^-(aq) + H^+(aq) + H_2O \longrightarrow Al(OH)_3(s)$$

When the precipitate is heated, water is driven off and the oxide is formed.

$$2Al(OH)_3(s) \xrightarrow{\text{heat}} Al_2O_3(s) + 3H_2O(g)$$

This purified Al_2O_3 then becomes the raw material for the Hall–Héroult process discussed in Section 19.8.

Most metals are obtained from their ores using chemical reducing agents

Except for very reactive metals such as sodium, magnesium, and aluminum, which are reduced using electrolysis (Section 19.8), most metals are formed from their compounds using chemical reduction. A plentiful, and therefore inexpensive, reducing agent that's often used is carbon, which is usually obtained from coal. When coal is heated strongly in the absence of air, volatile components are driven off and **coke** is formed. Coke is composed almost entirely of carbon. Carbon is an effective reducing agent for metal oxides because it combines with the oxygen to form carbon dioxide. For example, after it is roasted, lead oxide is mixed with coke and heated. As noted earlier, this method for reducing a metal oxide is known as *smelting*.

$$2PbO(s) + C(s) \xrightarrow{\text{heat}} 2Pb(l) + CO_2(g)$$

The high thermodynamic stability of CO_2 serves as one driving force for the reaction. The loss of CO_2 and the inability to reach an equilibrium is another.

Copper oxide ores can also be reduced with carbon.

$$2CuO + C \longrightarrow 2Cu + CO_2$$

Coke is made from coal. Coke is made in a battery of side-by-side coke ovens where coal is heated to drive off volatile materials. Here we see a fresh batch of white-hot coke being pushed from one of the ovens into a waiting railroad car. It will be delivered to a blast furnace nearby where it will be used to reduce iron ore to metallic iron. *(David M. Campione/Photo Researchers.)*

This step is unnecessary for some copper sulfide ores if the conditions under which the ore is roasted are properly controlled. For example, heating an ore that contains Cu_2S in air can convert some of the Cu_2S to Cu_2O.

$$2Cu_2S + 3O_2 \longrightarrow 2Cu_2O + 2SO_2$$

At the appropriate time the supply of oxygen is cut off and the mixture of Cu_2S and Cu_2O reacts further to give metallic copper.

$$Cu_2S + 2Cu_2O \longrightarrow 6Cu + SO_2$$

Reduction of iron ore is the first step in making steel

Without question, the most important use of carbon as a reducing agent is in the production of iron and steel. The chemical reactions take place in a huge tower called a **blast furnace**

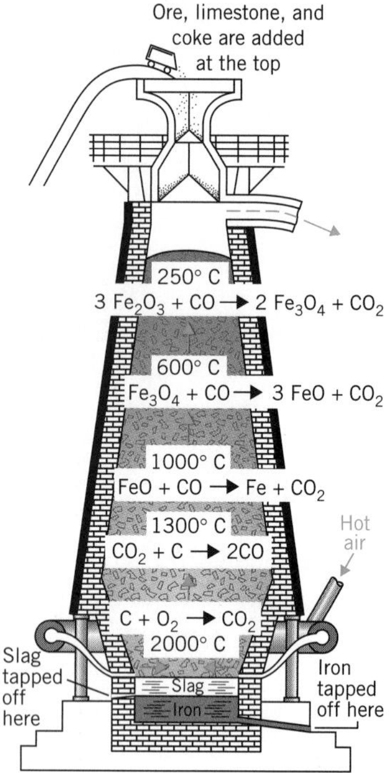

FIG. 21.11 A typical blast furnace for the reduction of iron ore. To make 1 ton of iron requires 1.75 tons of ore, 0.75 ton of coke, and 0.25 ton of limestone.

(see Figure 21.11). Some are as tall as a 15-story building and produce up to 2400 tons of iron a day. They are designed for continuous operation, so the raw materials can be added at the top and molten iron can be tapped off at the bottom. Once started, a typical blast furnace may run continuously for two years or longer before it is worn out and must be rebuilt.

The material put into the top of the blast furnace is called the **charge.** It consists of a mixture of iron ore, limestone, and coke. A typical iron ore consists of an iron oxide (Fe_2O_3, for example) plus impurities of sand and rock. The coke is added to reduce the iron oxide to the free metal. The limestone is added to react with the high-melting impurities to form a **slag,** which has a lower melting point. The slag can then be drained off as a liquid at the base of the furnace.

To understand what happens in the furnace, it is best to begin with the reactions that take place near the bottom. Here, heated air is blown into the furnace where carbon (from the coke) reacts with oxygen to form carbon dioxide.

The blast of hot air is what gives the blast furnace its name.

$$C + O_2 \longrightarrow CO_2$$

The reaction is very exothermic, and the temperature in this part of the furnace rises to nearly 2000 °C. It is the hottest region of the furnace. The hot CO_2 rises and reacts with additional carbon to form carbon monoxide.

$$CO_2 + C \longrightarrow 2CO$$

This reaction is endothermic, which causes the temperature higher up in the furnace to drop to about 1300 °C. As the carbon monoxide rises through the charge, it reacts with the iron oxides and reduces them to the free metal. The reactions are

The active reducing agent in the blast furnace is carbon monoxide.

$$3Fe_2O_3(s) + CO(g) \longrightarrow 2Fe_3O_4(s) + CO_2(g)$$
$$Fe_3O_4(s) + CO(g) \longrightarrow 3FeO(s) + CO_2(g)$$
$$FeO(s) + CO(g) \longrightarrow Fe(l) + CO_2(g)$$

As the charge settles toward the bottom, molten iron trickles down and collects in a well at the base of the furnace.

The high temperature in the furnace also causes the limestone in the reaction mixture to decompose to give calcium oxide.

$$\underset{\text{limestone}}{CaCO_3} \longrightarrow CaO + CO_2$$

The calcium oxide reacts with impurities such as silica (SiO_2) in the sand to form the slag.

The acidic anhydride SiO_2 reacts with oxide ion in CaO to form the silicate ion, SiO_3^{2-}, in $CaSiO_3$. It's a Lewis acid–base reaction.

$$CaO + SiO_2 \longrightarrow \underset{\text{calcium silicate (slag)}}{CaSiO_3}$$

The molten slag also trickles down through the charge. It collects as a liquid layer on top of the more dense molten iron. Periodically, the furnace is tapped and the iron and slag are drawn off. The iron, which still contains some impurities, is called **pig iron.**[3] It is usually treated further to produce steel. The slag itself is a valuable by-product. It is used to make insulating materials and is one of the chief ingredients in the manufacture of Portland cement.

Preparing metals for use is called refining

Before metals can be used, most must be purified, or **refined,** after they are reduced to the metallic state. Sometimes, reduction and purification can take place in a single step. For example, the ore for titanium (a metal used to make aircraft parts) is its oxide, TiO_2. Recovering the metal from the ore is difficult because the metal itself reacts at high temperature with both carbon and nitrogen. The solution is to heat the oxide with carbon

[3] The name pig iron comes from an early method of casting the molten iron into bars for shipment. The molten metal was run through a channel that fed into sand molds. The arrangement looked a little like a litter of pigs feeding from their mother.

in the presence of chlorine gas, which converts the titanium to $TiCl_4$, a volatile compound that is drawn off as a gas.

$$TiO_2(s) + C(s) + 2Cl_2(g) \longrightarrow TiCl_4(g) + CO_2(g)$$

The volatility of the $TiCl_4$ enables it to be removed from any impurities in the ore. After being separated from the CO_2, the $TiCl_4$ is allowed to react with magnesium to form essentially pure titanium metal.

$$TiCl_4 + 2Mg \longrightarrow Ti + 2MgCl_2$$

The purification of nickel also takes advantage of the volatility of one of its compounds. When nickel is heated mildly in the presence of carbon monoxide, it forms a compound called nickel carbonyl, $Ni(CO)_4$.

$$Ni(s) + 4CO(g) \xrightarrow{\text{warm}} \underset{\text{nickel carbonyl}}{Ni(CO)_4(g)}$$

Nickel carbonyl is very toxic.

Vapors of the $Ni(CO)_4$ compound are easily separated from impurities in the nickel, which don't form similarly volatile substances. Later the $Ni(CO)_4$ can be heated to high temperature, causing the substance to decompose and deposit highly purified nickel.[4]

For some metals, refining is accomplished by electrolysis. For example, metallic copper that comes from the smelting process is about 99% pure, but before being used in electrical wiring it is purified electrolytically as described in Chapter 19.

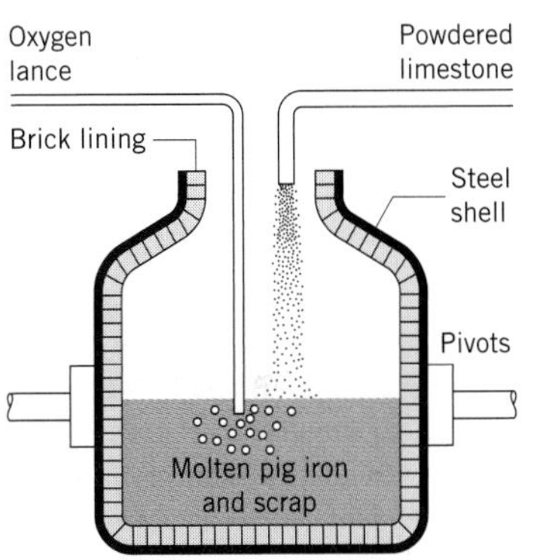

FIG. 21.12 The basic oxygen process for making steel. The reaction vessel is very large, being capable of holding up to 300 tons of steel.

Refining of iron to make steel involves removing impurities and adding other metals

The conversion of pig iron to steel is the most important commercial refining process. The pig iron from a blast furnace consists of about 95% iron, 3 to 4% carbon, and smaller amounts of phosphorus, sulfur, manganese, and other elements. Steel contains much less carbon as well as certain other ingredients in very definite proportions. Converting pig iron to steel, therefore, involves removing the impurities and much of the carbon, and adding other metals in precisely controlled amounts.

Today, most steel is made by a method called the **basic oxygen process.** It uses a large pear-shaped reaction vessel that is mounted on pivots as shown in Figure 21.12. The vessel is lined with an insulating layer of refractory bricks composed of $MgCO_3$ or a mixture of $MgCO_3$ and $CaCO_3$. The charge consists of about 30% scrap iron and scrap steel and about 70% molten pig iron. A tube called an *oxygen lance* is dipped into the charge and pure oxygen is blown through the molten metal. The oxygen rapidly burns off the excess carbon and oxidizes impurities to their oxides. The heat generated also melts the scrap iron. The impurities form a slag with calcium oxide that comes from powdered limestone, which is also added. Finally, other metals are introduced in the proper proportions to give a product with the desired properties. After the steel is ready, the reaction vessel can be tipped to pour out its contents. This method of making steel is very fast. A batch of steel weighing 300 tons can be made in less than an hour.

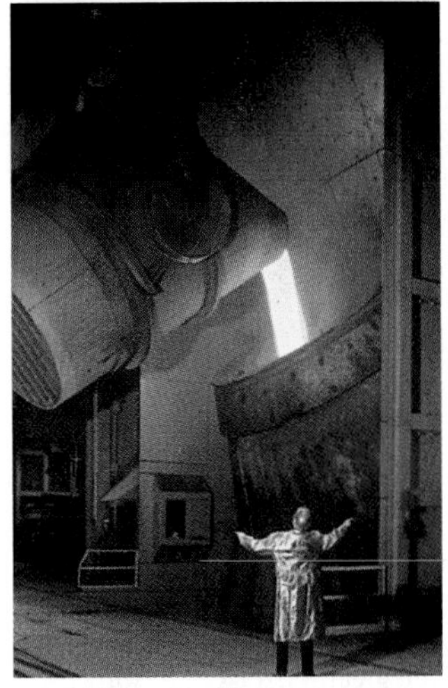

A molten iron charge is added to a basic oxygen furnace. *(James Mejuto Photography.)*

21.5 COMPLEX IONS ARE FORMED BY MANY METALS

In Section 17.4 (page 712), we introduced you to **complex ions** (or more simply, **complexes**) of metals. Recall that these are substances formed when molecules or anions become bonded covalently to metal ions to form more complex species. Two examples that were given were the pale blue $Cu(H_2O)_4^{2+}$ ion and the deep blue $Cu(NH_3)_4^{2+}$ ion. In Chapter 17 we discussed how the formation of complex ions can affect the solubilities of salts, but the importance of these substances reaches far beyond solubility equilibria. The number of complex ions formed by metals, especially the transition metals, is enormous, and the study of the properties, reactions, structures, and bonding in complexes like $Cu(NH_3)_4^{2+}$ has become an important specialty within chemistry. The study of metal

[4] This method of purifying nickel is called the Mond process.

In the space we have available here our goal will simply be to give you a taste of the subject.

complexes even extends into biochemistry. This is because, ultimately, nearly all the metals our bodies require become bound in complexes in order to perform their biochemical functions.

Before we proceed further, let's review some of the basic terminology we will use in our discussions. The molecules or ions that become attached to a metal ion [e.g., the NH_3 molecules in $Cu(NH_3)_4{}^{2+}$] are called **ligands.** Ligands are neutral molecules or anions that contain one or more atoms with at least one unshared electron pair that can be donated to the metal ion in the formation of a metal–ligand bond. The reaction of a ligand with a metal ion is therefore a Lewis acid–base reaction in which the ligand is the Lewis base (electron pair donor) and the metal ion is the Lewis acid (electron pair acceptor).

$$Ag^+ + :NH_3 \longrightarrow [Ag \leftarrow NH_3]^+$$

silver ion (Lewis acid) — ammonia molecule (Lewis base) — complex ion

Coordinate covalent bond between ammonia and the silver ion

Recall that a coordinate covalent bond is one in which both of the shared electrons originate on the same atom. Once formed, of course, a coordinate covalent bond is just like any other covalent bond.

A ligand atom that donates an electron pair to the metal is said to be a **donor atom** and the metal is the **acceptor.** Thus, in the example above, the nitrogen of the ammonia molecule is the donor atom and the silver ion is the acceptor. Because of the way the metal–ligand bond is formed, it can be considered to be a coordinate covalent bond, and compounds that contain metal complexes are often called **coordination compounds.** The complexes themselves are sometimes referred to as **coordination complexes.**

Ligands are Lewis bases

As we've noted, ligands may be either anions or neutral molecules. In either case, *ligands are Lewis bases* and, therefore, contain at least one atom with one or more lone pairs (unshared pairs) of electrons.

Anions that serve as ligands include many simple monatomic ions such as the halide ions (F^-, Cl^-, Br^-, I^-) and the sulfide ion (S^{2-}). Common polyatomic anions that are ligands are nitrite ion ($NO_2{}^-$), cyanide ion (CN^-), hydroxide ion (OH^-), thiocyanate ion (SCN^-), and thiosulfate ion ($S_2O_3{}^{2-}$). (This is really only a small sampling, not a complete list.)

The most common neutral molecule that serves as a ligand is water, and most of the reactions of metal ions in aqueous solutions are actually reactions of their complex ions—ions in which the metal is attached to some number of water molecules. This number isn't always the same, however. Copper(II) ion, for example, forms the complex ion $Cu(H_2O)_4{}^{2+}$ (as we've noted earlier),[5] but cobalt(II) combines with water molecules to form $Co(H_2O)_6{}^{2+}$. Another common neutral ligand is ammonia, NH_3, which has one lone pair of electrons on the nitrogen atom. If ammonia is added to an aqueous solution containing the $Ni(H_2O)_6{}^{2+}$ ion, for example, the color changes dramatically from green to blue as ammonia molecules displace water molecules (see Figure 21.13).

FIG. 21.13 Complex ions of nickel. (*Left*) A solution of nickel chloride, which contains the green $Ni(H_2O)_6{}^{2+}$ ion. (*Right*) Adding ammonia to the nickel chloride solution leads to formation of the blue $Ni(NH_3)_5(H_2O)^{2+}$ ion. *(Andy Washnik.)*

$$\underset{\text{(green)}}{Ni(H_2O)_6{}^{2+}(aq)} + 5NH_3(aq) \longrightarrow \underset{\text{(blue)}}{Ni(NH_3)_5(H_2O)^{2+}(aq)} + 5H_2O$$

Each of the ligands that we have discussed so far is able to use just one atom to attach itself to a metal ion. Such ligands are called **monodentate ligands,** indicating that they have only "one tooth" with which to "bite" the metal ion.

[5] The formula $Cu(H_2O)_4{}^{2+}$ is actually an oversimplification. Copper(II) complexes in water usually have two additional water molecules loosely attached to the Cu^{2+} at a greater distance than the other four ligand atoms. Therefore, the copper(II) complex with water could also be written $Cu(H_2O)_6{}^{2+}$ to indicate the additional water molecules surrounding the Cu^{2+} ion.

There are also many ligands that have two or more donor atoms, and collectively they are referred to as **polydentate ligands.** The most common of these have two donor atoms, so they are called **bidentate ligands.** When they form complexes, *both* donor atoms become attached to the same metal ion. Oxalate ion and ethylenediamine (abbreviated *en* in writing the formula for a complex) are examples of bidentate ligands.

$$^{-}O-\overset{\overset{O}{\|}}{C}-\overset{\overset{O}{\|}}{C}-O^{-} \qquad H_2N-CH_2-CH_2-NH_2$$

oxalate ion   ethylenediamine, en

When these ligands become attached to a metal ion, ring structures are formed as shown below. Complexes that contain such ring structures are called **chelates.**[6]

```
  O       O
   \\    //
    C----C                CH2—CH2
   /      \               /      \
  O        O           H2N        NH2
    ↘    ↙                ↘     ↙
     M^Z+                   M^Z+
an oxalate complex   an ethylenediamine complex
```

Structures like these are important in "complex ion chemistry," as we shall see later in this chapter.

One of the most common polydentate ligands is a compound called ethylenediaminetetraacetic acid, mercifully abbreviated EDTA.

Sometimes the ligand EDTA is abbreviated using small letters (i.e., edta).

```
       :O:                                :O:
        ‖                                  ‖
H—O—C—CH2                          CH2—C—O—H
          \                        /
           N—CH2—CH2—N
          /                        \
H—O—C—CH2                          CH2—C—O—H
        ‖                                  ‖
       :O:                                :O:
```

EDTA (often H_4EDTA)

The H atoms attached to the oxygen atoms are easily removed as protons, which gives an anion with a charge of 4−. The structure of the anion, $EDTA^{4-}$, is shown below with the donor atoms in color.

```
       :O:                                :O:
        ‖                                  ‖
 ⁻O—C—CH2                          CH2—C—O⁻
          \                        /
           N—CH2—CH2—N
          /                        \
 ⁻O—C—CH2                          CH2—C—O⁻
        ‖                                  ‖
       :O:                                :O:
```

$EDTA^{4-}$

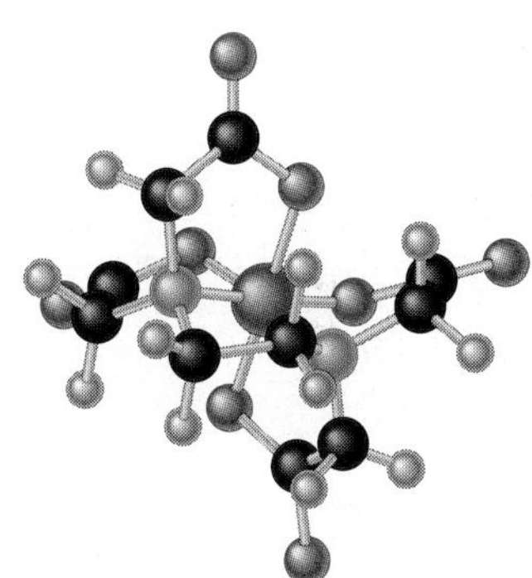

The structure of an EDTA complex. The nitrogen atoms are blue, oxygen is red, carbon is black, and hydrogen is white. The metal ion is in the center of the complex bonded to the two nitrogens and four oxygens.

As you can see, the $EDTA^{4-}$ ion has six donor atoms, and this permits it to wrap itself around a metal ion and form very stable complexes.

EDTA is a particularly useful and important ligand. It is relatively nontoxic, which allows it to be used in small amounts in foods to retard spoilage. If you look at the labels on bottles of salad dressings, for example, you often will find that one of the ingredients is $CaNa_2EDTA$ (calcium disodium EDTA). The $EDTA^{4-}$ available from this salt forms

The *calcium* salt of EDTA is used because the $EDTA^{4-}$ ion would otherwise extract Ca^{2+} ions from bones, and that would be harmful.

[6] The term *chelate* comes from the Greek *chele*, meaning claw. These bidentate ligands grasp the metal ions with two "claws" (donor atoms) somewhat like a crab holds its prey. (Who says scientists have no imagination?)

soluble complex ions with any traces of metal ions that might otherwise promote reactions of the salad oils with oxygen, and thereby lead to spoilage.

Many shampoos contain Na_4EDTA to soften water. The $EDTA^{4-}$ binds to Ca^{2+}, Mg^{2+}, and Fe^{3+} ions, which removes them from the water and prevents them from interfering with the action of detergents in the shampoo (see the photo at the beginning of this chapter). A similar application was described in Facets of Chemistry 17.1 on page 715.

EDTA is also sometimes added in small amounts to whole blood to prevent clotting. It ties up calcium ions, which the clotting process requires. EDTA has even been used as a treatment in cases of poisoning because it can help remove poisonous heavy metal ions, like Pb^{2+}, from the body when they have been accidentally ingested.

Formulas for complexes obey rules

When we write the formula for a complex, we follow rules.

Writing formulas for complexes

1. The symbol for the metal ion is always given first, followed by the ligands.
2. When more than one kind of ligand is present, anionic ligands are written first (in alphabetical order), followed by neutral ligands (also in alphabetical order).
3. The charge on the complex is the algebraic sum of the charge on the metal ion and the charges on the ligands.

For example, the formula of the complex ion of Cu^{2+} and NH_3, which we mentioned earlier, was written $Cu(NH_3)_4{}^{2+}$ with the Cu first followed by the ligands. The charge on the complex is 2+ because the copper ion has a charge of 2+ and the ammonia molecules are neutral. Copper(II) ion also forms a complex ion with four cyanide ions, CN^-, $Cu(CN)_4{}^{2-}$. The metal ion contributes two (+) charges and the four ligands contribute a total of four (−) charges, one for each cyanide ion. The algebraic sum is therefore −2, so the complex ion has a charge of 2−.

Square brackets here do not mean molar concentration. It is usually clear from the context of a discussion whether we intend the brackets to mean molarity.

The formula for a complex ion is often placed within brackets, with the charge *outside*. The two complexes just mentioned would thus be written as $[Cu(H_2O)_4]^{2+}$ and $[Cu(CN)_4]^{2-}$. The brackets emphasize that the ligands are attached to the metal ion and are not free to roam about. One of the many complex ions formed by the chromium(III) ion contains five water molecules and one chloride ion as ligands. To indicate that all are attached to the Cr^{3+} ion, we use brackets and write the complex ion as $[CrCl(H_2O)_5]^{2+}$. When this complex is isolated as a chloride salt, the formula is written $[CrCl(H_2O)_5]Cl_2$, in which $[CrCl(H_2O)_5]^{2+}$ is the cation and so is written first. The formula $[CrCl(H_2O)_5]Cl_2$ clearly shows that five water molecules and a chloride ion are bonded to the chromium ion, and the other two chloride ions are present to provide electrical neutrality for the salt.

In Chapter 2 you learned about hydrates, and one was the beautiful blue hydrate of copper(II) sulfate, $CuSO_4{\cdot}5H_2O$ (see page 56). It was much too early then to make the distinction, but the formula should have been written as $[Cu(H_2O)_4]SO_4{\cdot}H_2O$ to show that four of the five water molecules are held in the crystal as part of the complex ion $[Cu(H_2O)_4]^{2+}$. The fifth water molecule is held in the crystal by being hydrogen bonded to the sulfate ion.

Hexa- means "six."

Many other hydrates of metal salts actually contain complex ions of the metals in which water is the ligand. Cobalt salts like cobalt(II) chloride, for example, crystallize from aqueous solutions as hexahydrates (meaning they contain six H_2O molecules per formula unit of the salt). The compound $CoCl_2{\cdot}6H_2O$ (Figure 21.14) actually is $[Co(H_2O)_6]Cl_2$, and it contains the pink complex $[Co(H_2O)_6]^{2+}$. This ion also gives solutions of cobalt(II) salts a pink color as can be seen in Figure 21.14. Although most hydrates of metal salts contain complex ions, the distinction is seldom made, and it's acceptable to write the formula for these hydrates in the usual fashion, e.g., $CuSO_4{\cdot}5H_2O$ instead of $[Cu(H_2O)_4]SO_4{\cdot}H_2O$.

EXAMPLE 21.2
Writing the Formula for a Complex Ion

Write the formula for the complex ion formed by the metal ion Cr^{3+} and six NO_2^- ions as ligands. Decide whether the complex could be isolated as a chloride salt or a potassium salt, and write the formula for the appropriate salt.

ANALYSIS: The rules on page 874 are the tools we'll use to write the formula of the complex ion correctly. If the complex is a positive ion, it would require a negative ion to form a salt, so it could be isolated as a chloride salt; if the complex is a negative ion, then a potassium salt could form.

SOLUTION: Six NO_2^- ions contribute a total charge of 6−; the metal contributes a charge of 3+. The algebraic sum is (6−) + (3+) = 3−. The formula of the complex ion is therefore $[Cr(NO_2)_6]^{3-}$.

Because the complex is an anion, it requires a cation to form a neutral salt, so the complex could be isolated as a potassium salt, not as a chloride salt. For the salt to be electrically neutral, three K^+ ions are required for each complex ion, $[Cr(NO_2)_6]^{3-}$. The formula of the salt would therefore be $K_3[Cr(NO_2)_6]$. Notice that in writing the formula for the salt, we've specified the cation (K^+) first, followed by the anion.

IS THE ANSWER REASONABLE? Things to check: Have we written the formula with the metal ion first, followed by the ligands? (Yes.) Have we computed the charge on the complex correctly? (Yes.) Does the formula for the salt correspond to an electrically neutral substance? (Yes.) All seems to be okay.

Practice Exercise 3: Write the formula of the complex ion formed by Ag^+ and two thiosulfate ions, $S_2O_3^{2-}$. If we were able to isolate this complex ion as its ammonium salt, what would be the formula for the salt? (Hint: Remember, salts are electrically neutral.)

Practice Exercise 4: Aluminum chloride crystallizes from aqueous solutions as a hexahydrate. Write the formula for the salt and suggest a formula for the complex ion formed by aluminum ion and water.

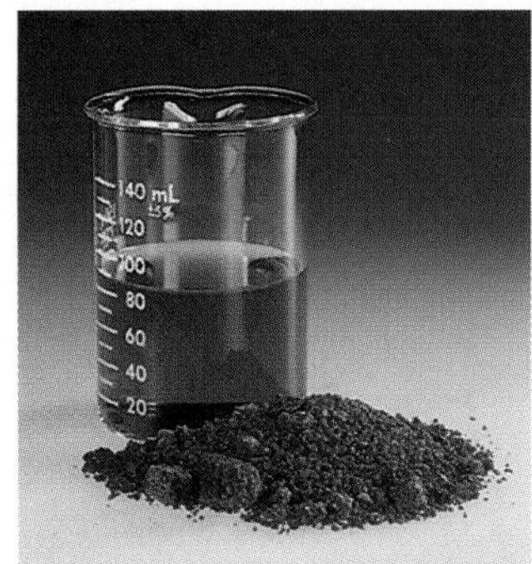

FIG. 21.14 Color of the cobalt(II) ion in hexahydrate salts. The ion $[Co(H_2O)_6]^{2+}$ is pink and gives its color both to crystals and to an aqueous solution of $CoCl_2{\cdot}6H_2O$. *(Michael Watson.)*

The chelate effect leads to extra stability in complexes

An interesting aspect of the complexes formed by ligands such as ethylenediamine and oxalate ion is their stabilities compared to similar complexes formed by monodentate ligands. For example, the complex $[Ni(en)_3]^{2+}$ is considerably more stable than $[Ni(NH_3)_6]^{2+}$, even though both complexes have six nitrogen atoms bound to a Ni^{2+} ion. We can compare them quantitatively by examining their formation constants.

$$Ni^{2+}(aq) + 6NH_3(aq) \rightleftharpoons [Ni(NH_3)_6]^{2+}(aq) \qquad K_{form} = 2.0 \times 10^8$$
$$Ni^{2+}(aq) + 3en(aq) \rightleftharpoons [Ni(en)_3]^{2+}(aq) \qquad K_{form} = 1.4 \times 10^{17}$$

The ethylenediamine complex is more stable than the ammonia complex by a factor of 2×10^9 (2 billion)! This exceptional stability of complexes formed with polydentate ligands is called the **chelate effect,** so named because it occurs in compounds that have these *chelate ring* structures.

Ammonia is a monodentate ligand, and each NH_3 can supply one donor atom to a metal. Ethylenediamine (en) is a bidentate ligand, and each has two nitrogen donor atoms. Therefore, six ammonia molecules and three ethylenediamine molecules supply the same number of donor atoms.

There are two related reasons for the chelate effect, which we can understand best if we examine the ease with which the complexes undergo dissociation once formed. One reason appears to be associated with the probability of the ligand being removed from the vicinity of the metal ion when a donor atom becomes detached. If one end of a bidentate ligand comes loose from the metal, the donor atom cannot wander very far because the other end of the ligand is still attached. There is a high probability that the loose end will become reattached to the metal ion before the other end can let go, so overall the ligand appears to be bound tightly. With a monodentate ligand, however, there is nothing to hold the ligand near the metal ion if it becomes detached. The ligand can easily wander off into the surrounding solution and be lost. As a result, a mono-dentate ligand doesn't behave as if it is as firmly attached to the metal ion as a polydentate ligand.

The second reason is related to the entropy change for the dissociation. In Chapter 18 you learned that when there is an increase in the number of particles in a chemical reaction, the entropy change is positive. Dissociation of both complexes produces more particles, so the entropy change is positive for both of them. However, comparing the two reactions,

$$[Ni(NH_3)_6]^{2+}(aq) \rightleftharpoons Ni^{2+}(aq) + 6NH_3(aq) \qquad K_{inst} = 5.0 \times 10^{-9}$$
$$[Ni(en)_3]^{2+}(aq) \rightleftharpoons Ni^{2+}(aq) + 3en(aq) \qquad K_{inst} = 2.4 \times 10^{-18}$$

Recall that $\Delta G°$ is related to the equilibrium constant. A more favorable $\Delta G°$ for dissociation should yield a larger equilibrium constant, and, indeed, we find the instability constant is larger for the ammonia complex.

we see that dissociation of the ammonia complex gives a net increase of *six* in the number of particles, whereas dissociation of the ethylenediamine complex yields a net increase of *three.* This means the entropy change is more positive for the dissociation of the ammonia complex than for the ethylenediamine complex. The larger entropy change translates into a more favorable $\Delta G°$ for the decomposition, so at equilibrium the ammonia complex should be dissociated to a greater extent. This is, in fact, what is suggested by the values of their instability constants, K_{inst} (recall that $K_{inst} = 1/K_{form}$).

21.6 THE NOMENCLATURE OF METAL COMPLEXES FOLLOWS AN EXTENSION OF THE RULES DEVELOPED EARLIER

The naming of chemical compounds was introduced in Chapter 2 where we discussed the nomenclature system for simple inorganic compounds. This system, revised and kept up to date by the International Union of Pure and Applied Chemistry (IUPAC), has been extended to cover metal complexes. Below are some of the rules that have been developed to name coordination complexes. As you will see, some of the names arrived at by following the rules are difficult to pronounce at first, and may even sound a little odd. However, the primary purpose of this and any other system of nomenclature is to provide a method that gives each unique compound its own unique name and permits us to write the formula of the compound given the name.

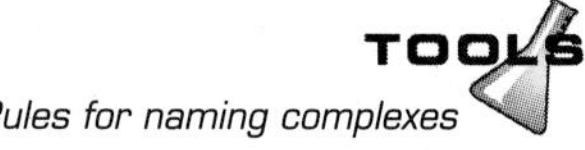

Rules for naming complexes

Rules of Nomenclature for Coordination Complexes

1. **Cationic species are named before anionic species.** This is the same rule that applies to other ionic compounds such as NaCl, where we name the cation first followed by the anion (i.e., sodium chloride).
2. **The names of anionic ligands always end in the suffix -o.**
   (a) Ligands whose names end in *-ide* have this suffix changed to *-o.*

Anion		Ligand
Chloride	Cl^-	chloro-
Bromide	Br^-	bromo-
Cyanide	CN^-	cyano-
Oxide	O^{2-}	oxo-

   (b) Ligands whose names end in *-ite* or *-ate* become *-ito* and *-ato*, respectively.

Anion		Ligand
Carbonate	CO_3^{2-}	carbonato-
Thiosulfate	$S_2O_3^{2-}$	thiosulfato-
Thiocyanate	SCN^-	thiocyanato- (when bonded through sulfur) isothiocyanato- (when bonded through nitrogen)
Oxalate	$C_2O_4^{2-}$	oxalato-
Nitrite	NO_2^-	nitrito- (when bonded through oxygen; written ONO in formula for complex)[a]

[a]An exception to this is when the nitrogen of the NO_2^- ion is bonded to the metal, in which case the ligand is named nitro-.

3. **A neutral ligand is given the same name as the neutral molecule.** By this rule, the molecule ethylenediamine, when serving as a ligand, is called ethylenediamine in the name of the complex. Two very important exceptions to this, however, are water and ammonia. These are named as follows when they serve as ligands.

H_2O aqua- NH_3 ammine - (note the double *m*)

4. **When there is more than one of a particular ligand, their number is specified by the prefixes di- = 2, tri- = 3, tetra- = 4, penta- = 5, hexa- = 6, and so forth. When confusion might result by using those prefixes, the following are used instead: bis- = 2, tris- = 3, tetrakis- = 4.** Following this rule, the presence of two chloride ligands in a complex would be indicated as *dichloro-* (notice, too, the ending on the ligand name). However, if two ethylenediamine ligands are present, use of the prefix *di-* might cause confusion. Someone reading the name might wonder whether diethylenediamine means two ethylenediamine molecules or one molecule of a substance called diethylenediamine. To avoid this problem we place the ligand name in parentheses preceded by *bis*; that is, *bis(ethylenediamine).*

Bis- is employed here so that when the name is used in verbal communication it is clear that the meaning is two ethylenediamine molecules.

5. **As noted earlier, in the *formula* of a complex, the symbol for the metal is written first, followed by those of the ligands. Among the ligands, anionic ligands are written first (in alphabetical order), followed by neutral ligands (also in alphabetical order). In the *name* of the complex, the ligands are named first, in alphabetical order *without regard to charge,* followed by the name of the metal.** For example, suppose we had a complex composed of Co^{3+}, two Cl^- (chloro- ligands), one CN^- (cyano- ligand), and three NH_3 (ammine- ligands). The formula of the electrically neutral complex would be written $[CoCl_2(CN)(NH_3)_3]$. In the name of this complex, the ligands would be specified before the metal as *triamminedichlorocyano-* (*triammine-* for the three NH_3 ligands, *dichloro-* for the two Cl^- ligands, and *cyano-* for the CN^- ligand). Notice that in alphabetizing the names of the ligands, we ignore the prefixes *tri-* and *di-*. Thus *triammine-* is written before *dichloro-* because *ammine-* precedes *chloro-* alphabetically. For the same reason, *dichloro-* is written before *cyano-*. The complete name for the complex is given below under Rule 7.

In the formula, the metal ion appears first, followed by the ligands; in the name, the ligands are specified first, followed by the metal.

The ligands are alphabetized according to the first letter of the name of the ligand, not the first letter of the prefix.

6. **Negative (anionic) complex ions always end in the suffix *-ate.*** This suffix is appended to the English name of the metal atom in most cases. However, if the name of the metal ends in *-ium, -um,* or *-ese*, the ending is dropped and replaced by *-ate.*

Metal	Metal as Named in an Anionic Complex
Aluminum	aluminate
Chromium	chromate
Manganese	manganate
Nickel	nickelate
Cobalt	cobaltate
Zinc	zincate
Platinum	platinate
Vanadium	vanadate

For metals whose symbols are derived from their Latin names, the suffix *-ate* is appended to the Latin stem. (An exception, however, is mercury; in an anion it is named *mercurate.*)

Metal	Stem	Metal as Named in an Anionic Complex
Iron	ferr-	ferrate
Copper	cupr-	cuprate
Lead	plumb-	plumbate
Silver	argent-	argentate
Gold	aur-	aurate
Tin	stann-	stannate

For neutral or positively charged complexes, the metal is *always* specified with the English name for the element, *without any suffix.*

7. **The oxidation state of the metal in the complex is written in Roman numerals within parentheses following the name of the metal.** For example,

As you learned earlier, the charge on the complex is the algebraic sum of the charges on the ligands and the charge on the metal ion.

$[CoCl_2(CN)(NH_3)_3]$ is triamminedichlorocyanocobalt(III)

$[Co(NH_3)_6]^{3+}$ is the hexaamminecobalt(III) ion

$[CuCl_4]^{2-}$ is the tetrachlorocuprate(II) ion

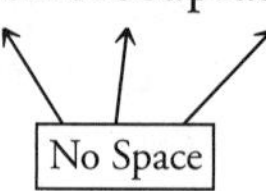

Notice that there are no spaces between the names of the ligands and the name of the metal, and that there is no space between the name of the metal and the parentheses that enclose the oxidation state expressed in Roman numerals.

The following are some additional examples. Study them carefully to see how the rules given above apply. Then try the practice exercises that follow.

Notice, once again, that the alphabetical order of the ligands is determined by the first letter in the name of the ligand, not the first letter in the prefix.

$[Ni(CN)_4]^{2-}$	tetracyanonickelate(II) ion
$K_3[CoCl_6]$	potassium hexachlorocobaltate(III)
$[CoCl_2(NH_3)_4]^+$	tetraamminedichlorocobalt(III) ion
$Na_3[Co(NO_2)_6]$	sodium hexanitrocobaltate(III)
$[Ag(NH_3)_2]^+$	diamminesilver(I) ion
$[Ag(S_2O_3)_2]^{3-}$	dithiosulfatoargentate(I) ion
$[Mn(en)_3]Cl_2$	tris(ethylenediamine)manganese(II) chloride
$[PtCl_2(NH_3)_2]$	diamminedichloroplatinum(II)

Practice Exercise 5: Write the formula for each of the following: (a) hexachlorostannate(IV) ion and (b) ammonium diaquatetracyanoferrate(II). (Hint: Divide each name into its various parts.)

Practice Exercise 6: Name the following compounds: (a) $K_3[Fe(CN)_6]$ and (b) $[CrCl_2(en)_2]_2SO_4$.

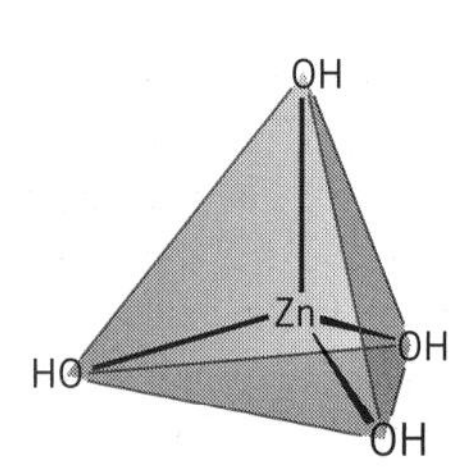

Tetrahedral
$[Zn(OH)_4]^{2-}$

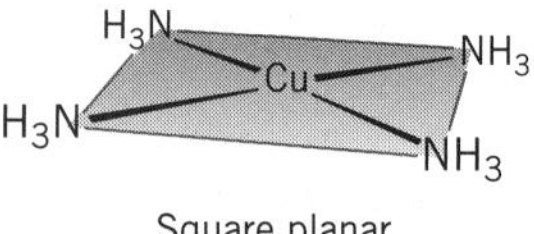

Square planar
$[Cu(NH_3)_4]^{2+}$

FIG. 21.15 Tetrahedral and square planar geometries. These are structures that occur for complexes in which the metal ion has a coordination number of 4. For the copper complex, we are viewing the square planar structure tilted back into the plane of the paper.

21.7 Coordination Number and Structure Are Often Related

One of the most interesting aspects of the study of complexes is the kinds of structures that they form. In many ways, this is related to the **coordination number** of the metal ion, which we define as *the number of donor atoms attached to the metal ion.* For example, in the complex $[Ni(CN)_4]^{2-}$ the nickel is surrounded by the four carbon atoms that belong to the cyanide ions, so the coordination number of Ni^{2+} in the complex is 4. Similarly, the coordination number of Cr^{3+} in the $[Cr(H_2O)_6]^{3+}$ ion is 6, and the coordination number of Ag^+ in $[Ag(NH_3)_2]^+$ is 2.

Sometimes the coordination number isn't immediately obvious from the formula of the complex. For example, you learned that there are many polydentate ligands which contain more than one donor atom that can bind simultaneously to a metal ion. Often, a metal is able to accommodate two or more polydentate ligands to give complexes with formulas such as $[Cr(H_2O)_2(en)_2]^{3+}$ and $[Cr(en)_3]^{3+}$. In each of these examples, the coordination number of the Cr^{3+} is 6. In the $[Cr(en)_3]^{3+}$ ion, there are three ethylenediamine ligands that each supply two donor atoms, for a total of 6, and in $[Cr(H_2O)_2(en)_2]^{3+}$, the two ethylenediamine ligands supply a total of 4 donor atoms and the two H_2O molecules supply another 2, so once again the total is 6.

Ordinarily, the VSEPR model isn't used to predict the structures of transition metal complexes because it can't be relied on to give correct results when the metal has a partially filled *d* subshell.

Structures of complexes depend on coordination number

For metal complexes, there are certain geometries that are usually associated with particular coordination numbers.

Coordination Number 2. Examples are complexes such as $[Ag(NH_3)_2]^+$ and $[Ag(CN)_2]^-$. Usually, these complexes have a linear structure such as

$$[H_3N—Ag—NH_3]^+ \quad \text{and} \quad [NC—Ag—CN]^-$$

(Since the Ag^+ ion has a filled *d* subshell, it behaves as any of the representative elements as far as predicting geometry by VSEPR theory, so these structures are exactly what we would expect based on that theoretical model.)

Coordination Number 4. Two common geometries occur when four ligand atoms are bonded to a metal ion, namely, tetrahedral and square planar. These are illustrated in Figure 21.15. The tetrahedral geometry is usually found with metal ions that have completely filled *d* subshells, such as Zn^{2+}. The complexes $[Zn(NH_3)_4]^{2+}$ and $[Zn(OH)_4]^{2-}$ are examples.

Square planar geometries are observed for complexes of Cu^{2+}, Ni^{2+}, and especially Pt^{2+}. Examples are $[Cu(NH_3)_4]^{2+}$, $[Ni(CN)_4]^{2-}$, and $[PtCl_4]^{2-}$. The most well-studied square planar complexes are those of Pt^{2+}, because they are considerably more stable than the others.

Coordination Number 6. The most common coordination number for complex ions is 6. Examples are $[Al(H_2O)_6]^{3+}$, $[Co(C_2O_4)_3]^{3-}$, $[Ni(en)_3]^{2+}$, and $[Co(EDTA)]^-$. With few exceptions, all complexes with a coordination number of 6 are octahedral. This holds true for those formed from both monodentate and bidentate ligands, as illustrated in Figure 21.16. In describing the shapes of octahedral complexes, most chemists use one of the simplified drawings of the octahedron shown in Figure 21.17.

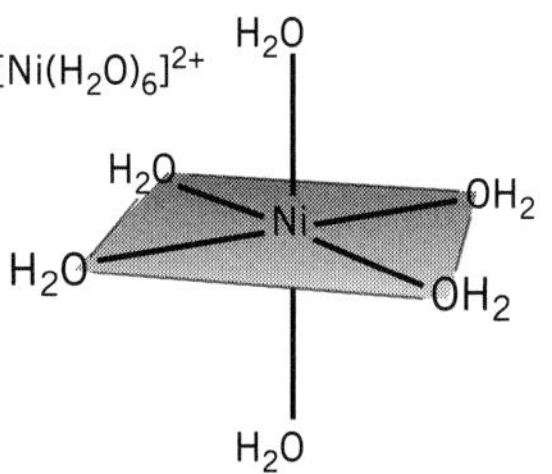

Complex with monodentate ligands

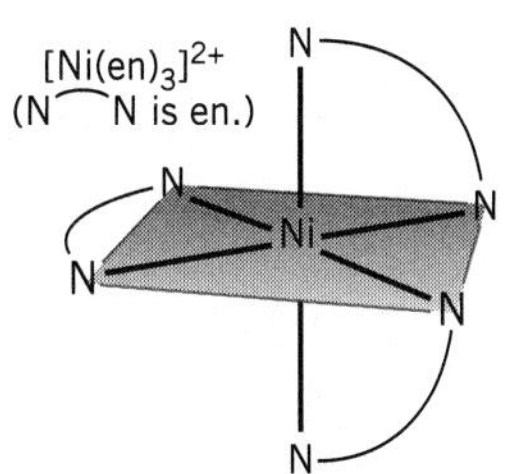

Complex with bidentate ligands

FIG. 21.16 Octahedral complexes. Complexes with this geometry can be formed with either monodentate ligands such as water or with polydentate ligands such as ethylenediamine (en). To simplify the drawing of the ethylenediamine complex, the atoms joining the donor nitrogen atoms in the ligand, $—CH_2—CH_2—$, are represented as the curved line between the N atoms. Also notice that the nitrogen atoms of the bidentate ligand span adjacent positions within the octahedron. This is the case for all polydentate ligands that you will encounter in this book.

Practice Exercise 7: What is the coordination number of the metal ion in $[Cr(C_2O_4)_3]^{3-}$? (Hint: Identify the ligand and sketch its structure.)

Practice Exercise 8: What is the coordination number of the metal ion in (a) $[CoCl_2(C_2O_4)_2]^{3-}$, (b) $[Cr(C_2O_4)_2(en)]^-$, and (c) $[Co(H_2O)_2(en)_2]^{3+}$?

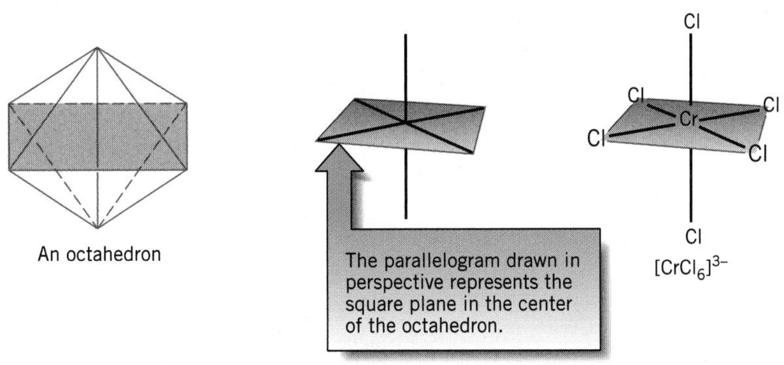

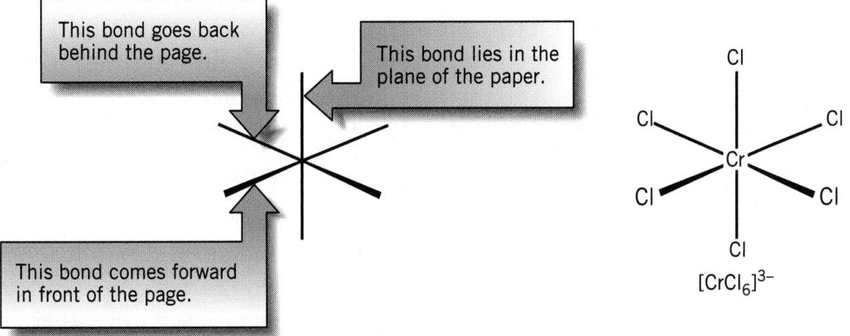

FIG. 21.17 Simplified representations of the octahedral complex, $[CrCl_6]^{3-}$. (*a*) Drawings similar to those you learned to construct in Chapter 9. (*b*) An alternative method of representing the octahedron.

21.8 ISOMERS OF COORDINATION COMPLEXES ARE COMPOUNDS WITH THE SAME FORMULA BUT DIFFERENT STRUCTURES

When you write the chemical formula for a compound, you might be tempted to think that you should also be able to predict exactly what the structure of the molecule or ion is. As you may realize by now, this just isn't possible in many cases. Sometimes we can use simple rules to make reasonable structural guesses, as in the discussion of the drawing of Lewis structures in Chapter 8, but these rules apply only to simple molecules and ions. For more complex substances, there usually is no way of knowing for sure what the structure of the molecule or ion is without performing the necessary experiments to determine the structure.

One of the reasons that structures can't be predicted with certainty from chemical formulas alone is that there are usually many different ways that the atoms in the formula can be arranged. In fact, it is frequently possible to isolate two or more compounds that actually have the same chemical formula. For example, three different solids, each with its own characteristic color and other properties, can be isolated from a solution of chromium(III) chloride. All three have the same overall composition, and in the absence of other data, their formulas are written $CrCl_3 \cdot 6H_2O$. However, experiments have shown that these solids are actually the salts of three different complex ions. Their actual formulas (and colors) are

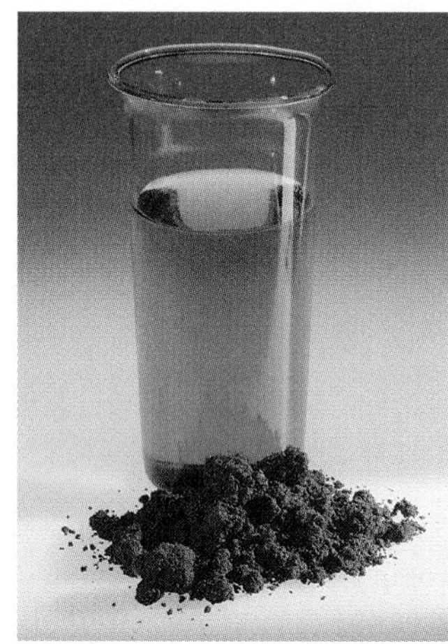

Chromium(III) chloride hexahydrate. The hydrated chromium(III) chloride purchased from chemical supply companies is actually $[CrCl_2(H_2O)_4]Cl \cdot 2H_2O$. Its green color in both the solid state and solution is due to the complex ion $[CrCl_2(H_2O)_4]^+$. *(Michael Watson.)*

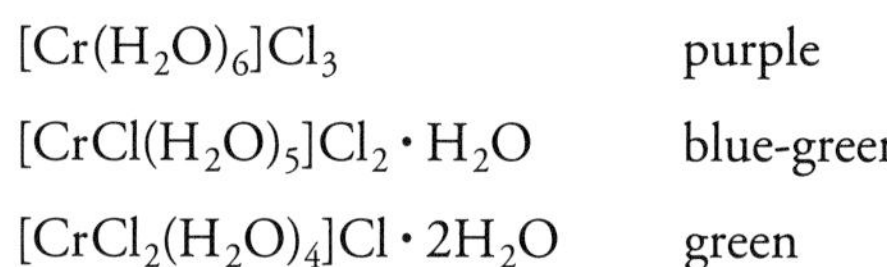

$[Cr(H_2O)_6]Cl_3$	purple
$[CrCl(H_2O)_5]Cl_2 \cdot H_2O$	blue-green
$[CrCl_2(H_2O)_4]Cl \cdot 2H_2O$	green

Even though their overall compositions are the same, each of these substances is a distinct chemical compound with its own unique set of properties.

The existence of two or more compounds, each having the same chemical formula, is known as **isomerism.** In the example above, each salt is said to be an **isomer** of $CrCl_3 \cdot 6H_2O$. For coordination compounds, isomerism can occur in a variety of ways. In the example above, isomers exist because of the different possible ways that the water molecules and chloride ions can be held in the crystals. In one instance, all the water molecules serve as ligands, while in the other two, part of the water is present as water of hydration and some of the chloride is bonded to the metal ion. Another example, which is similar in some respects, is $Cr(NH_3)_5SO_4Br$. This "substance" can be isolated as two isomers.

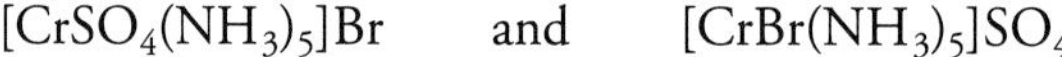

$$[CrSO_4(NH_3)_5]Br \qquad \text{and} \qquad [CrBr(NH_3)_5]SO_4$$

Ag_2SO_4 is soluble but $BaSO_4$ is insoluble. $BaBr_2$ is soluble but AgBr is isoluble.

They can be distinguished chemically by their differing abilities to react with Ag^+ and Ba^{2+}. The first isomer reacts in aqueous solution with Ag^+ to give a precipitate of AgBr, but it doesn't react with Ba^{2+}. This tells us that Br^- exists as a free ion in the solution. It also suggests the SO_4^{2-} is bound to the chromium and is unavailable to react with Ba^{2+} to give insoluble $BaSO_4$.

The second isomer, $[CrBr(NH_3)_5]SO_4$, reacts in solution with Ba^{2+} to give a precipitate of $BaSO_4$, which means there is free SO_4^{2-} in the solution. There is no reaction with Ag^+, however, because Br^- is bonded to the chromium and is not available as free Br^- in the solution. Thus, we see that even though both isomers have the same overall composition, they behave chemically in quite different ways and are therefore distinctly different compounds.

Stereoisomerism relates to how atoms are arranged in space

One of the most interesting kinds of isomerism found among coordination compounds is called **stereoisomerism**, which is defined as *differences among isomers that arise as a result of the various possible orientations of their atoms in space.* In other words, when stereoisomerism exists, we have compounds in which the same atoms are attached to each other, but they differ in the way those atoms are arranged in space relative to one another.

One form of stereoisomerism is called **geometric isomerism;** it is best understood by considering an example. Consider the square planar complexes having the formula $PtCl_2(NH_3)_2$. There are two ways to arrange the ligands around the platinum, as illustrated below. In one isomer, called the **cis isomer,** the chloride ions are *next to each other* and the ammonia molecules are also next to each other. In the other isomer, called the **trans isomer,** identical ligands are *opposite each other*. In identifying (and naming) isomers, *cis* means "on the same side," and *trans* means "on opposite sides."

The isomer on the left, *cis*-$PtCl_2(NH_3)_2$, is the anticancer drug known as *cisplatin*. It is interesting that only the cis isomer of the compound is clinically active against tumors. The trans isomer is totally ineffective.

Cl NH$_3$ / Pt / Cl NH$_3$ and H$_3$N Cl / Pt / Cl NH$_3$

cis isomer — *cis*-diamminedichloroplatinum(II)

trans isomer — *trans*-diamminedichloroplatinum(II)

Geometric isomerism also occurs for octahedral complexes. For example, consider the ions $[CrCl_2(H_2O)_4]^+$ and $[CrCl_2(en)_2]^+$. Both can be isolated as cis and trans isomers.

The structures of the octahedral complexes are being drawn following the method in Figure 21.17*b*.

cis: Cr with H$_2$O (top), H$_2$O, Cl, H$_2$O, Cl, H$_2$O (bottom)

trans: Cr with H$_2$O (top), H$_2$O, Cl, Cl, OH$_2$, H$_2$O (bottom)

$[CrCl_2(H_2O)_4]^+$

Curved lines connecting nitrogen atoms represent —CH$_2$—CH$_2$—, which links the nitrogen atoms in ethylenediamine.

cis: Cr with N, N, N, N, Cl, Cl

trans: Cr with Cl (top), N, N, N, N, Cl (bottom)

$[CrCl_2(en)_2]^+$

N N is en, (ethylenediamine) $NH_2CH_2CH_2NH_2$.

In the cis isomer, the chloride ligands are both on the same side of the metal ion; in the trans isomer, the chloride ligands are on opposite ends of a line that passes through the center of the metal ion.

FIG. 21.18 One difference between left and right hands. A left-hand glove won't fit the right hand. *(Michael Watson.)*

Chirality is a subtle type of isomerism

There is a second kind of stereoisomerism that is much more subtle than geometric isomerism. This occurs when molecules are exactly the same except for one small difference—they are *nonsuperimposable mirror images of each other*. They bear the same relationship to each other as do your left and right hands. (In fact this relationship between such isomers is sometimes referred to by the term *handedness.*)

Although similar in appearance, your two hands are not exactly alike. If you place one hand over the other, palms down, your thumbs point in opposite directions. **Superimposability** is the ability of two objects to fit "one within the other" with no mismatch of parts. Your left and right hands lack this ability, and we say they are *nonsuperimposable*. This is why your right hand doesn't fit into your left-hand glove (Figure 21.18); your right hand does not match *exactly* the space that corresponds to your left hand.

Your hands are also mirror images of each other. If you hold your left hand so it faces a mirror, and then look at its reflection as illustrated in Figure 21.19, you will see that it looks exactly like your right hand. If it were possible to reach "through the looking glass,"

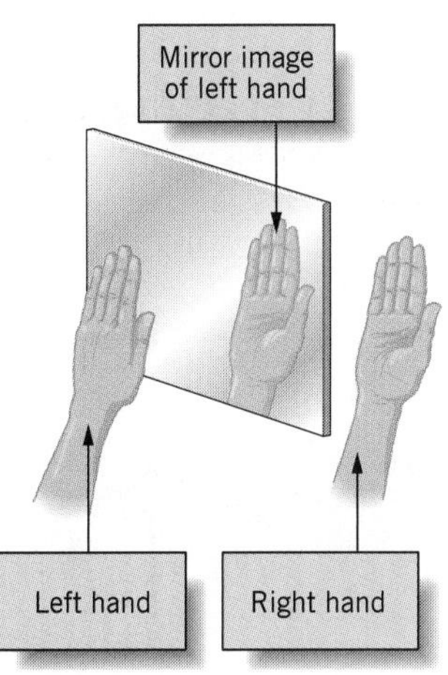

FIG. 21.19 Hands are mirror images of each other. The image of the left hand reflected in the mirror appears the same as the right hand.

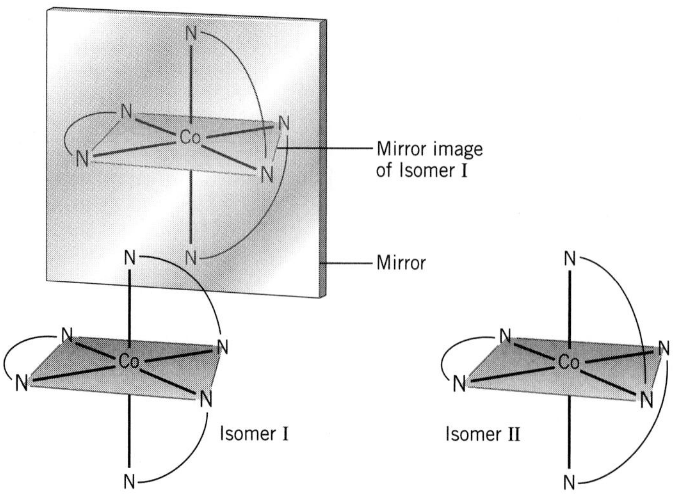

FIG. 21.20 The two isomers of $[Co(en)_3]^{3+}$. Isomer II is constructed as the mirror image of Isomer I. No matter how Isomer II is turned about, it is not superimposable on Isomer I.

Chirality is the technical term for "handedness," meaning objects structurally related to each other as are our left and right hands.

Thumbtacks are not chiral. Thumbtacks and their mirror images are superimposable; they are not chiral and are identical. *(Andy Washnik.)*

your right-hand glove would fit the reflection of your left hand perfectly. Thus, your two hands are *nonsuperimposable mirror images* of each other.

If two objects are nonsuperimposable mirror images of each other, they are not identical and are said to be **chiral.** Your left and right hands have this property and are not exactly alike. A pair of thumbtacks, however, are mirror images of each other and *are* superimposable, so they are identical. Thumbtacks are not chiral.

Some complex ions exhibit chirality

The most common examples of chirality among coordination compounds occur with octahedral complexes that contain two or three bidentate ligands—for instance, $[CoCl_2(en)_2]^+$ and $[Co(en)_3]^{3+}$. For the complex $[Co(en)_3]^{3+}$, the two nonsuperimposable isomers, called **enantiomers,** are shown in Figure 21.20. For the complex $[CoCl_2(en)_2]^+$, only the *cis* isomer is chiral, as described in Figure 21.21.

As you can see, chiral isomers differ in only a very minor way from each other. This difference is so subtle that most of the properties of chiral isomers are identical. They have identical melting points and boiling points, and in nearly all of their reactions, their behavior is exactly alike. The only way that the difference between chiral molecules or ions manifests itself is in the way that they interact with physical or chemical "probes" that also

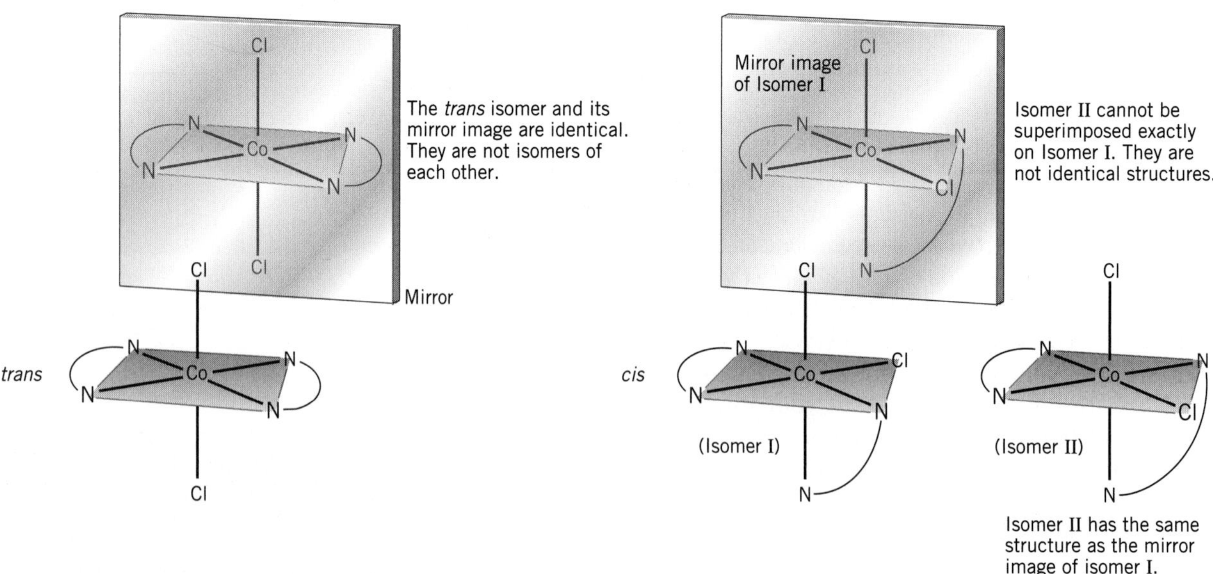

FIG. 21.21 Isomers of the $[CoCl_2(en)_2]^+$ ion. The mirror image of the trans isomer can be superimposed exactly on the original, so the trans isomer is not chiral. The cis isomer (Isomer I) is chiral, however, because its mirror image (Isomer II) cannot be superimposed on the original.

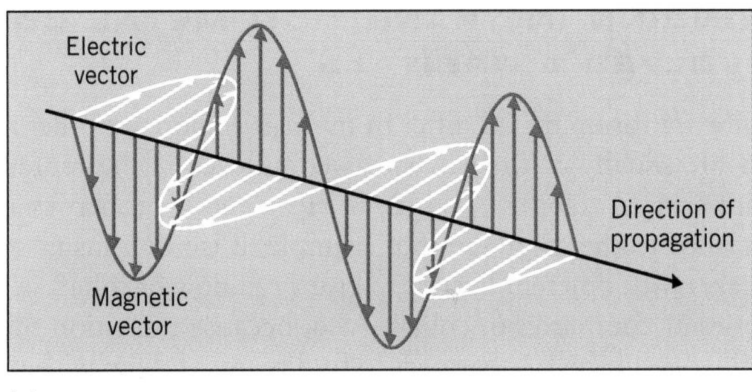
Electric
vector
Direction of
propagation
Magnetic
vector

(*a*)

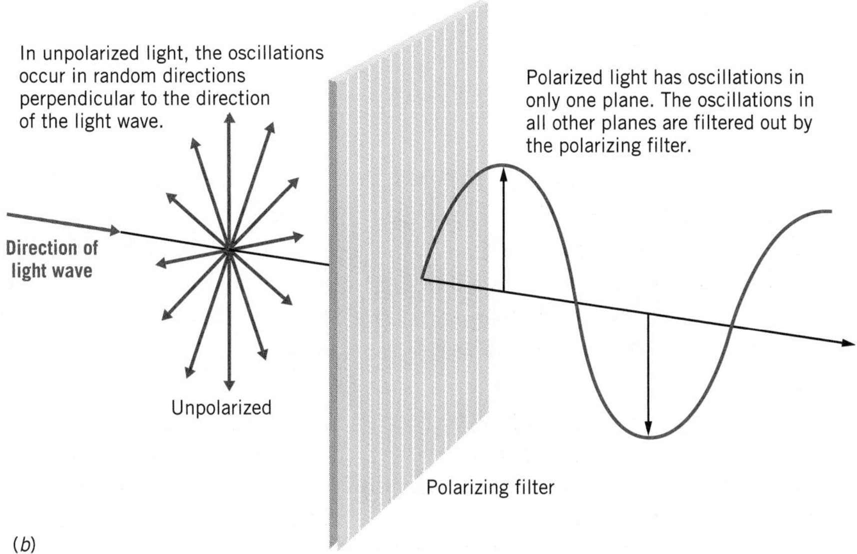
In unpolarized light, the oscillations occur in random directions perpendicular to the direction of the light wave.
Polarized light has oscillations in only one plane. The oscillations in all other planes are filtered out by the polarizing filter.
Direction of light wave
Unpolarized
Polarizing filter

(*b*)

FIG. 21.22 Unpolarized and polarized light. (*a*) Light possesses electric and magnetic components that oscillate perpendicular to the direction of propagation of the light wave. (*b*) In unpolarized light, the electromagnetic oscillations of the photons are oriented at random angles around the axis of propagation of the light wave. The polarizing filter has the effect of preventing oscillations from passing through unless they are in one particular plane. The result is called plane-polarized light.

have a handedness about them. For example, if two reactants are both chiral, then a given isomer of one of them will usually behave slightly differently toward each of the two isomers of the other reactant. This has very profound effects in biochemical reactions, in which nearly all of the molecules involved are chiral.

Chiral isomers can rotate plane-polarized light

One way that chiral isomers differ is in the way they affect polarized light. Light is *electromagnetic radiation* that possesses both electric and magnetic components that behave like vectors. These vectors oscillate in directions perpendicular to the direction in which the light wave is traveling (Figure 21.22*a*). In ordinary light, the oscillations of the electric and magnetic fields of the photons are oriented randomly around the direction of the light beam. In **plane-polarized light,** all the oscillations occur in the same plane (Figure 21.22*b*). Ordinary light can be polarized in several ways. One is to pass it through a special film of plastic, as in a pair of Polaroid sunglasses. This has the effect of filtering out all the vibrations except those that are in one plane (Figure 21.22*b*).

Chiral isomers like those described in Figures 21.20 and 21.21 are said to be **optically active** and have the ability to rotate plane-polarized light, as illustrated in Figure 21.23. Because of this phenomenon, chiral isomers are said to be **optical isomers.**

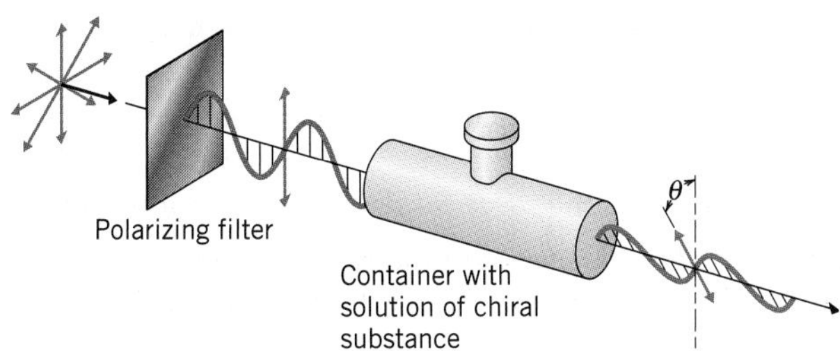
Polarizing filter
Container with
solution of chiral
substance
θ

FIG. 21.23 Rotation of polarized light by a chiral substance. When plane-polarized light passes through a solution of a chiral substance, the plane of polarization is rotated either to the left or to the right. In this illustration, the plane of polarization of the light is rotated to the left by a measurable angle θ (as seen facing the light source).

21.9 Bonding in Transition Metal Complexes Involves *d* Orbitals

Complexes of the transition metals differ from the complexes of other metals in two special ways: (1) they are usually colored, whereas complexes of the representative metals are usually (but not always) colorless, and (2) their magnetic properties are often affected by the ligands attached to the metal ion. For example, it is not unusual for a given metal ion to form complexes with different ligands to give a rainbow of colors, as illustrated in Figure 21.24 for a series of complexes of cobalt. Also, because transition metal ions often have incompletely filled *d* subshells, we expect to find many of them with unpaired *d* electrons, and compounds that contain them should be paramagnetic. But for a given metal ion, the number of unpaired electrons is not always the same from one complex to another. For example, Fe^{2+} has four of its six 3*d* electrons unpaired in the $[Fe(H_2O)_6]^{2+}$ ion, but all of its electrons are paired in the $[Fe(CN)_6]^{4-}$ ion. As a result, the $[Fe(H_2O)_6]^{2+}$ ion is paramagnetic and the $[Fe(CN)_6]^{4-}$ ion is diamagnetic.

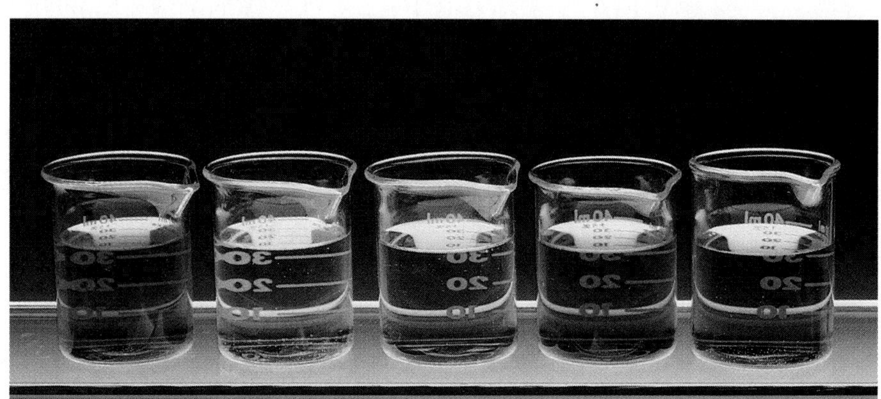

FIG. 21.24 Colors of complex ions depend on the nature of the ligands. Each of these brightly colored solutions contains a complex ion of Co^{3+}. The variety of colors arises because of the different ligands (molecules or anions) that are bonded to the cobalt ion in the complexes. *(Michael Watson.)*

Crystal field theory considers how *d* orbital energies are affected by the ligands

Any theory that attempts to explain the bonding in complex ions must also explain their colors and magnetic properties. One of the simplest theories that does this is the **crystal field theory.** The theory gets its name from its original use in explaining the behavior of transition metal ions in crystals. It was discovered later that the theory works well for transition metal complexes, too.

More complete theories consider the covalent nature of metal–ligand bonding, but crystal field theory nevertheless provides a useful model for explaining the colors and magnetic properties of complexes.

Crystal field theory ignores covalent bonding in complexes. It assumes that the primary stability comes from the electrostatic attractions between the positively charged metal ion and the negative charges of the ligand anions or dipoles. Crystal field theory's unique approach, though, is the way it examines how the negative charges on the ligands affect the energy of the complex by influencing the energies of the *d* orbitals of the metal ion, and this is what we will focus our attention on here. To understand the theory, therefore, it is essential that you know how the *d* orbitals are shaped and especially how they are oriented in space relative to each other. The *d* orbitals were described in Chapter 7, and they are illustrated again in Figure 21.25.

The labels for the *d* orbitals come from the mathematics of quantum mechanics.

First, notice that four of the *d* orbitals have the same shape but point in different directions. These are the $d_{x^2-y^2}$, d_{xy}, d_{xz}, and d_{yz}. Each has four lobes of electron density. The fifth *d* orbital, labeled d_{z^2}, has two lobes that point in opposite directions along the *z* axis plus a small donut-shaped ring of electron density around the center that is concentrated in the *xy* plane.

Of prime importance to us are the *directions* in which the lobes of the *d* orbitals point. Notice that three of them—d_{xy}, d_{xz}, and d_{yz}—point *between* the *x*, *y*, and *z* axes. The other two—the d_{z^2} and $d_{x^2-y^2}$ orbitals—have their maximum electron densities *along* the *x*, *y*, and *z* axes.

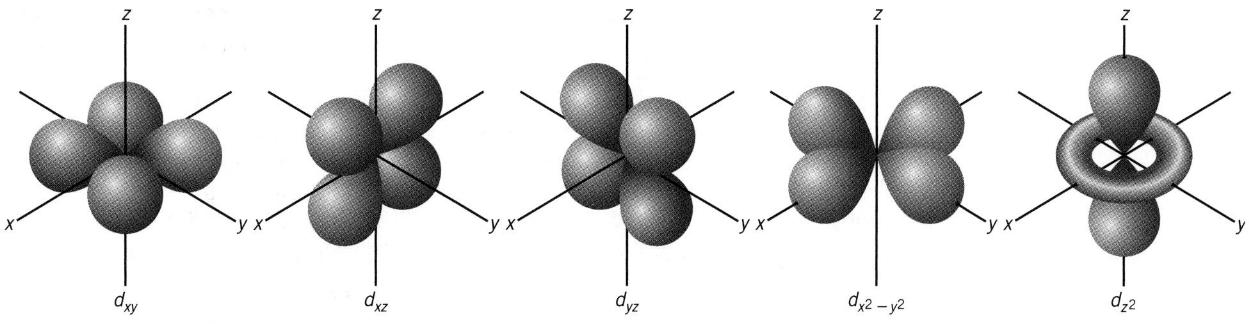

FIG. 21.25 **The shapes and directional properties of the five *d* orbitals of a *d* subshell.**

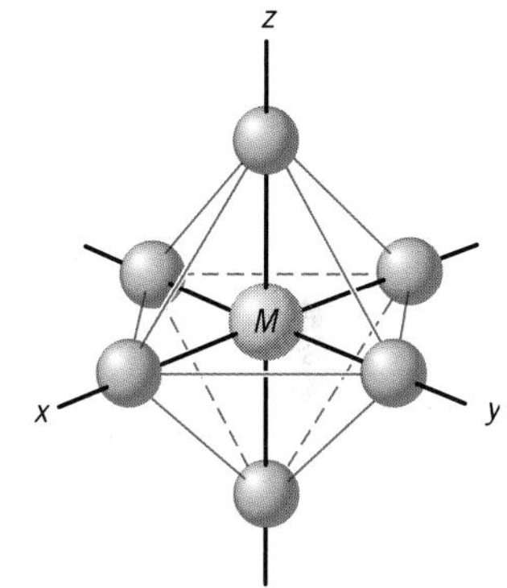

FIG. 21.26 **An octahedral complex ion with ligands that lie along the *x*, *y*, and *z* axes.**

Now let's consider constructing an octahedral complex within this coordinate system. We can do this by bringing ligands in along each of the axes as shown in Figure 21.26. The question we want to answer is, "How do these ligands affect the energies of the *d* orbitals?"

In an isolated atom or ion, all the *d* orbitals of a given *d* subshell have the same energy. Therefore, an electron will have the same energy regardless of which *d* orbital it occupies. In an octahedral complex, however, this is no longer true. If the electron is in the $d_{x^2-y^2}$ or d_{z^2} orbital, it is forced to be nearer the negative charge of the ligands than if it is in a d_{xy}, d_{xz}, or d_{yz} orbital. Since the electron itself is negatively charged and is repelled by the charges of the ligands, the electron's potential energy will be higher in the $d_{x^2-y^2}$ and d_{z^2} orbitals than in a d_{xy}, d_{xz}, or d_{yz} orbital. Therefore, as the complex is formed, the *d* subshell actually splits into *two* new energy levels as shown in Figure 21.27. Here we see that regardless of which orbital the electron occupies, its energy increases because it is repelled by the negative charges of the approaching ligands. However, the electron is repelled *more* (and has a higher energy) if it is in an orbital that points directly at the ligands than if it occupies an orbital that points between them.

In an octahedral complex, the energy difference between the two sets of *d* orbital energy levels is called the **crystal field splitting.** It is usually given the symbol Δ (delta), and its magnitude depends on the following factors:

The nature of the ligand. *Some ligands produce a larger splitting of the energies of the d orbitals than others.* For a given metal ion, for example, cyanide always gives a large value of Δ and F^- always gives a small value. We will have more to say about this later.

The oxidation state of the metal. *For a given metal and ligand, the size of Δ increases with an increase in the oxidation number of the metal.* As electrons are removed from a metal and the charge on the ion becomes more positive, the ion becomes smaller. This

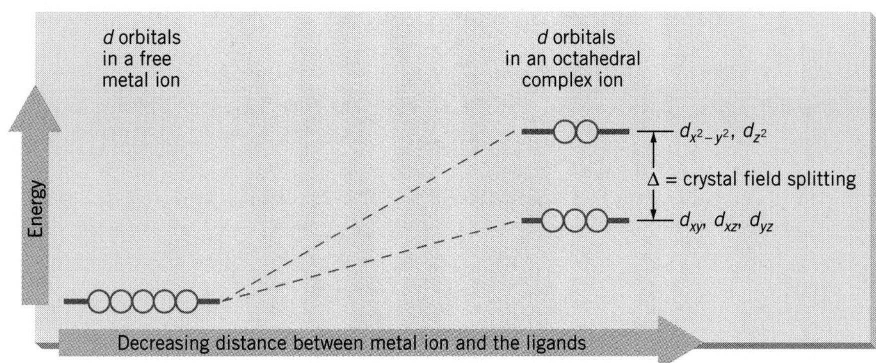

TOOLS *Crystal field splitting pattern for octahedral complexes*

FIG. 21.27 **The changes in the energies of the *d* orbitals of a metal ion as an octahedral complex is formed.** As the ligands approach the metal ion, the *d* orbitals split into two new energy levels.

means that the ligands are attracted to the metal more strongly and they can approach the center of the complex more closely. As a result, they also approach the *d* orbitals along the *x*, *y*, and *z* axes more closely, and thereby cause a greater repulsion. This causes a greater splitting of the two *d* orbital energy levels and a larger value of Δ.

The row in which the metal occurs in the periodic table. *For a given ligand and oxidation state, the size of* Δ increases going down a group. In other words, for a given ligand, an ion of an element in the first row of transition elements has a smaller value of Δ than the ion of a heavier member of the same group. Thus, comparing complexes of Ni^{2+} and Pt^{2+} with the same ligand, we find that the platinum complex has the larger crystal field splitting. The explanation of this is that in the larger ion (e.g., Pt^{2+}), the *d* orbitals are larger and more diffuse and extend farther from the nucleus in the direction of the ligands. This produces a larger repulsion between the ligands and the orbitals that point at them.

The magnitude of Δ is very important in determining the properties of complexes, including the stabilities of oxidation states of the metal ions, the colors of complexes, and their magnetic properties. Let's look at some examples.

Crystal field theory can explain the relative stabilities of oxidation states

Comparing the cations formed by chromium, it is found that the Cr^{2+} ion is very easily oxidized to Cr^{3+}. This is easily explained by crystal field theory. In water, we expect the ions to exist as the complexes $[Cr(H_2O)_6]^{2+}$ and $[Cr(H_2O)_6]^{3+}$, respectively. Let's examine the energies and electron populations of the *d* orbital energy levels in each complex (Figure 21.28).

The element chromium has the electron configuration

$$\text{Cr} \quad [\text{Ar}]\ 3d^5 4s^1$$

Removing two electrons gives the Cr^{2+} ion, and removing three gives Cr^{3+}.

$$\text{Cr}^{2+} \quad [\text{Ar}]\ 3d^4$$

$$\text{Cr}^{3+} \quad [\text{Ar}]\ 3d^3$$

Next, we distribute the *d* electrons among the various *d* orbitals following Hund's rule, but for Cr^{2+} we have to make a choice. Should the electrons all be forced into the lower of the two energy levels, or should they be spread out? From the diagram we see that the fourth electron does not pair with one of the others in the lower energy *d* orbital level. Instead, three electrons populate the lower level and the fourth is in the upper level. We will discuss *why* this happens later, but for now, let's use the two energy diagrams to explain why Cr^{2+} is so easy to oxidize.

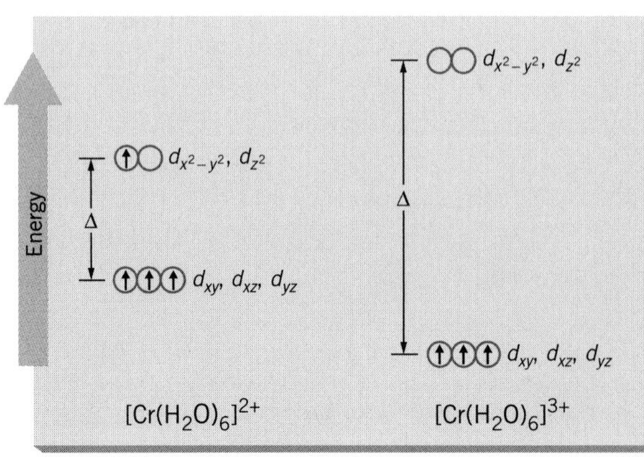

FIG. 21.28 Energy level diagrams for the $[Cr(H_2O)_6]^{2+}$ and $[Cr(H_2O)_6]^{3+}$ ions. The magnitude of Δ is larger for the chromium(III) complex because the Cr^{3+} ion is smaller than the Cr^{2+} ion and the ligands are drawn closer to the metal ion, thereby increasing repulsions felt by the $d_{x^2-y^2}$ and d_{z^2} orbitals.

There are actually two factors that favor the oxidation of Cr(II) to Cr(III). First, the electron that is removed from Cr^{2+} to give Cr^{3+} comes from the *higher* energy level, so oxidizing the chromium(II) *removes* a high-energy electron. The second factor is the effect caused by increasing the oxidation state of the chromium. As we've pointed out, this increases the magnitude of Δ, and as you can see, the energy of the three electrons that remain is lowered. Thus, both the removal of a high-energy electron and the lowering of the energy of the electrons that are left behind help make the oxidation occur, and the $[Cr(H_2O)_6]^{2+}$ ion is very easily oxidized to $[Cr(H_2O)_6]^{3+}$.

Crystal field theory can explain the colors of complex ions

When light is absorbed by an atom, molecule, or ion, the energy of the photon raises an electron from one energy level to another. In many substances, such as sodium chloride, the energy difference between the highest energy populated level and the lowest energy unpopulated level is quite large, so the frequency of a photon that carries the necessary energy lies outside the visible region of the spectrum. The substance appears white because visible light is unaffected; it is reflected unchanged.

Remember, $E = h\nu$. The energy of the photon absorbed determines the frequency (and wavelength) of the absorbed light.

For complex ions of the transition metals, the energy difference between the *d* orbital energy levels is not very large, and photons with frequencies in the visible region of the spectrum are able to raise an electron from the lower energy set of *d* orbitals to the higher energy set. This is shown in Figure 21.29 for the $[Cr(H_2O)_6]^{3+}$ ion.

As you know, white light contains photons of all the frequencies and colors in the visible spectrum. If we shine white light through a solution of a complex, the light that passes through has all the colors *except* those that have been absorbed. It is not difficult to determine what will be seen if we know what colors are being absorbed. All we need is a color wheel like the one shown in Figure 21.30. Across from each other on the color wheel are *complementary colors.* Green-blue is the complementary color to red, and yellow is the complementary color to violet-blue. If a substance absorbs a particular color when bathed in white light, the perceived color of the reflected or transmitted light is the complementary color. In the case of the $[Cr(H_2O)_6]^{3+}$ ion, the light absorbed when the electron is raised from one set of *d* orbitals to the other has a frequency of 5.22×10^{14} Hz, which is the color of yellow light. This is why a solution of the ion appears violet.[7]

Because of the relationship between the energy and frequency of light, we see that the color of the light absorbed by a complex depends on the magnitude of Δ; the larger the size of Δ, the more energy the photon must have and the higher will be the frequency of the absorbed light. For a given metal in a given oxidation state, the size of Δ depends on the ligand. Some ligands give a large crystal field splitting, while others

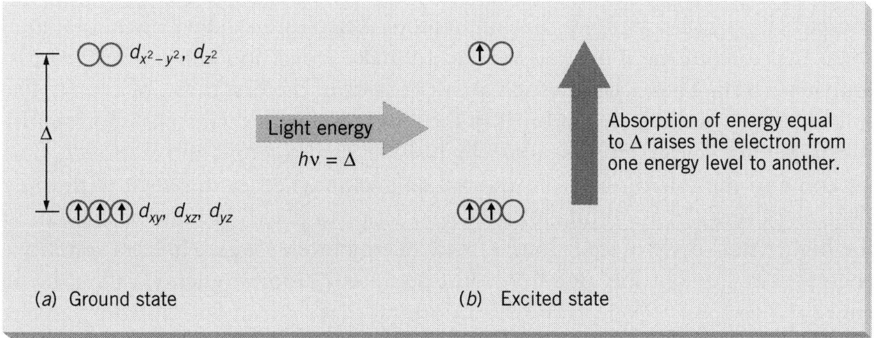

FIG. 21.29 Absorption of a photon by the $[Cr(H_2O)_6]^{3+}$ *complex.* (*a*) The electron distribution in the ground state of the $[Cr(H_2O)_6]^{3+}$ ion. (*b*) Light energy raises an electron from the lower energy set of *d* orbitals to the higher energy set.

[7] The perception of color is actually somewhat more complex than this because of the varying sensitivity of the human eye to various wavelengths. For example, the eye is much more sensitive to green than to red. If a compound reflects both of those colors with equal intensity, it will appear greenish simply because the eye sees green better than it sees red.

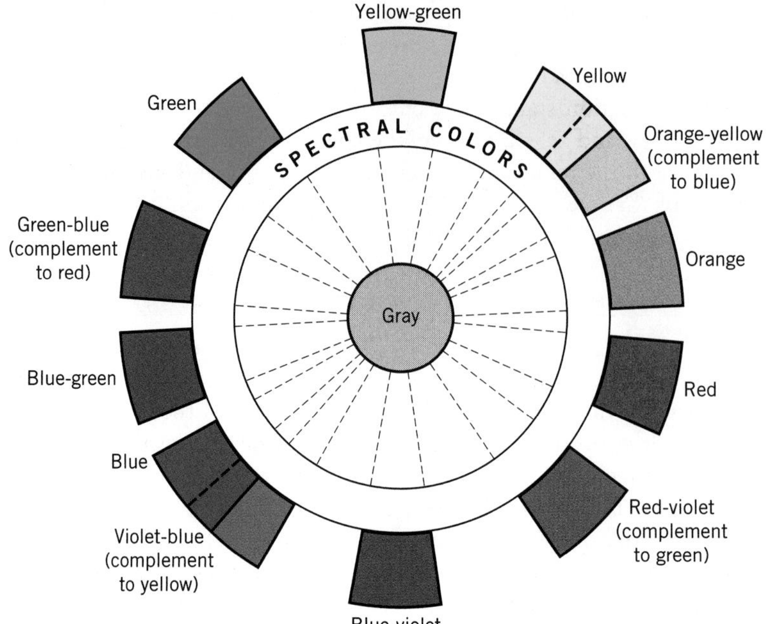

FIG. 21.30 A color wheel. Colors that are across from each other are called complementary colors. When a substance absorbs a particular color, light that is reflected or transmitted has the color of its complement. Thus something that absorbs red light appears green-blue, and vice versa.

give a small splitting. For example, ammonia produces a larger splitting than water, so the complex $[Cr(NH_3)_6]^{3+}$ absorbs light of higher energy and higher frequency than $[Cr(H_2O)_6]^{3+}$. ($[Cr(NH_3)_6]^{3+}$ absorbs blue light and appears yellow.) Because changing the ligand changes Δ, the same metal ion is able to form a variety of complexes with a large range of colors.

The order of the ligands can be determined by measuring the frequencies of the light absorbed by complexes.

A ligand that produces a large crystal field splitting with one metal ion also produces a large Δ in complexes with other metals. For example, cyanide ion is a very effective ligand and always gives a very large Δ, regardless of the metal to which it is bound. Ammonia is less effective than cyanide ion but more effective than water. Thus, ligands can be arranged in order of their effectiveness at producing a large crystal field splitting. This sequence is called the **spectrochemical series.** Such a series containing some common ligands arranged in order of their decreasing strength is

Spectrochemical series

$$CN^- > NO_2^- > en > NH_3 > H_2O > C_2O_4^{2-} > OH^- > F^- > Cl^- > Br^- > I^-$$

For a given metal ion, cyanide ion produces the largest Δ and iodide produces the smallest.

Crystal field theory can explain the magnetic properties of complexes

Let's return to the question of the electron distribution among the *d* orbitals in Cr^{2+} complexes. As you saw above, this ion has four *d* electrons, and we noted that in placing these electrons in the *d* orbitals we had to make a decision about where to place the fourth electron. There's no question about the fate of the first three, of course. They just spread out across the three *d* orbitals in the lower level with their spins unpaired. In other words, we just follow Hund's rule, which we learned to apply in Chapter 7. But when we come to the fourth electron we have to decide whether to pair it with one of the electrons already in a *d* orbital of the lower set or to place it in one of the *d* orbitals of the higher set.[8] If we place it in the lower energy level, we give it extra stability (lower energy), but some of this stability is lost because it requires energy, called the **pairing energy,** *P,* to force the electron into an orbital that's already occupied by an electron. On the other hand, if we place it in the higher level, we are relieved of the burden of pairing the electron, but it also tends to give the electron a higher energy. Thus, for the

[8] We've never had to make this kind of decision before because the energy levels in atoms were always widely spaced. In complex ions, however, the spacing between the two *d* orbital energy levels is fairly small.

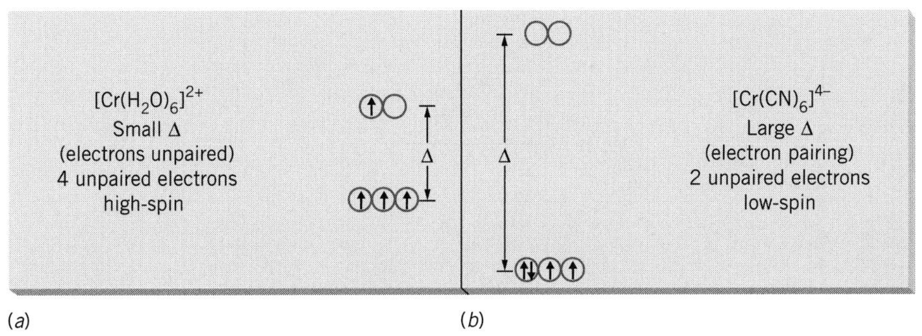

FIG. 21.31 The effect of Δ on the electron distribution in a complex with four *d* electrons. (*a*) When Δ is small, the electrons remain unpaired. (*b*) When Δ is large, the lower energy level accepts all four electrons and two electrons become paired.

fourth electron, "pairing" and "placing" work in opposite directions in the way they affect the energy of the complex.

The critical factor in determining whether the fourth electron enters the lower level and becomes paired, or whether it enters the higher level with the same spin as the other *d* electrons, is the magnitude of Δ. If Δ is larger than the pairing energy *P*, then greater stability is achieved if the fourth electron is paired with one in the lower level. If Δ is small compared to *P*, then greater stability is obtained by spreading the electrons out as much as possible. The complexes $[Cr(H_2O)_6]^{2+}$ and $[Cr(CN)_6]^{4-}$ illustrate this well.

Water is a ligand that does not produce a large Δ, so $P > \Delta$, and minimum pairing of electrons takes place. This explains the energy level diagram for the $[Cr(H_2O)_6]^{2+}$ complex in Figure 21.31*a*. When cyanide is the ligand, however, a very large Δ is obtained, and this leads to pairing of the fourth electron with one in the lower set of *d* orbitals. This is shown in Figure 21.31*b*. It is interesting to note that by measuring the degree of paramagnetism of the two complexes, it can be demonstrated experimentally that $[Cr(H_2O)_6]^{2+}$ has four unpaired electrons and the $[Cr(CN)_6]^{4-}$ ion has just two.

For octahedral chromium(II) complexes, there are two possibilities in terms of the number of unpaired electrons. They contain either four or two, depending on the magnitude of Δ. When there is the maximum number of unpaired electrons, the complex is described as being **high-spin;** when there is the minimum number of unpaired electrons it is described as being **low-spin.** High- and low-spin octahedral complexes are possible when the metal has a d^4, d^5, d^6, or d^7 electron configuration. Let's look at another example—one containing the Fe^{2+} ion, which has the electron configuration

$$Fe^{2+} \quad [Ar]\ 3d^6$$

At the beginning of this section we mentioned that the $[Fe(H_2O)_6]^{2+}$ ion is paramagnetic and has four unpaired electrons, while the $[Fe(CN)_6]^{4-}$ ion is diamagnetic, meaning it has no unpaired electrons. Now we can see why, by referring to Figure 21.32. Water produces a weak splitting and a minimum amount of pairing of electrons. When the six *d* electrons in the Fe^{2+} ion are distributed, one must be paired in the lower level after filling the upper level. The result is four unpaired *d* electrons, and a high-spin complex. Cyanide ion, however, produces a large splitting, so $\Delta > P$. This means that maximum pairing of electrons in the lower level takes place, and a low-spin complex is formed. Six electrons are just the right amount to completely fill all three of these *d* orbitals, and since all the electrons are paired, the complex is diamagnetic.

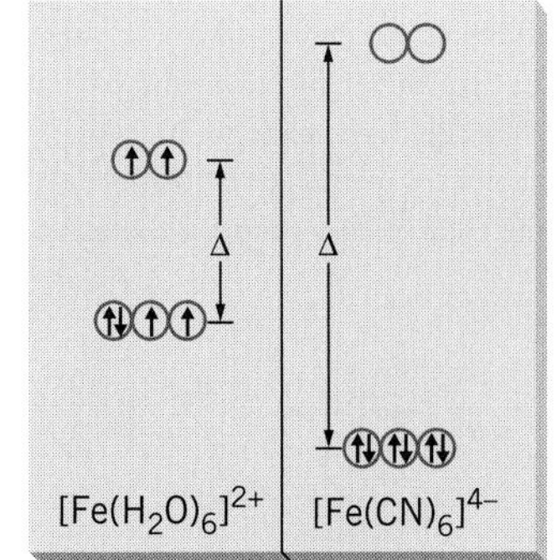

FIG. 21.32 The distribution of *d* electrons in $[Fe(H_2O)_6]^{2+}$ and $[Fe(CN)_6]^{4-}$. The magnitude of Δ for the cyanide complex is much larger than for the water complex. This produces a maximum pairing of electrons in the lower energy set of *d* orbitals in $[Fe(CN)_6]^{4-}$.

Crystal field theory also applies to other geometries

The crystal field theory can be extended to geometries other than octahedral. The effect that changing the structure of the complex has on the energies of the *d* orbitals is to change the splitting pattern.

Square planar complexes show additional splitting of the *d* orbital energies

We can form a square planar complex from an octahedral one by removing the ligands that lie along the *z* axis. As this happens, the ligands in the *xy* plane are able to approach the metal a little closer because they are no longer being repelled by ligands along the

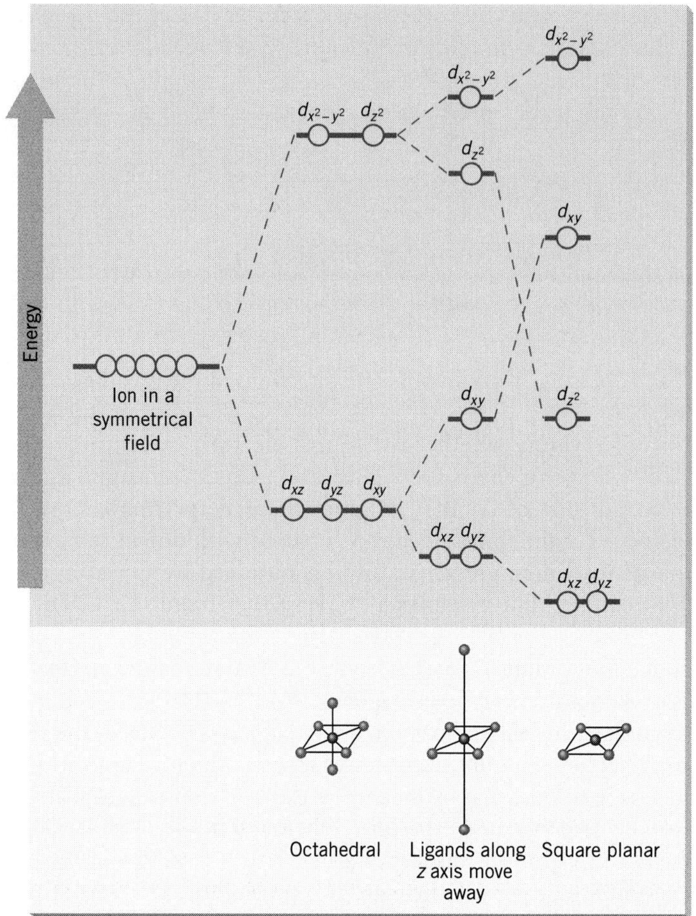

FIG. 21.33 Energies of *d* orbitals in complexes with various structures. The splitting pattern of the *d* orbitals changes as the geometry of the complex changes.

For Figure 21.33, as the energies of the *d* orbitals change, the d_{z^2} orbital drops in energy by the same amount as the $d_{x^2-y^2}$ rises. Similarly, the energies of the d_{xz} and d_{yz} orbitals drop only half as much as the energy of the d_{xy} rises. In this way, if all the *d* orbitals were filled, the changes in geometry would have no effect on the total energy of the complex.

z axis. The effect of these changes on the energies of the *d* orbitals is illustrated in Figure 21.33. The repulsions felt by the *d* orbitals that point in the *z* direction are reduced, so we find that the energies of the d_{z^2}, d_{xz}, and d_{yz} orbitals drop. At the same time, the energies of the orbitals in the *xy* plane feel greater repulsions, so the $d_{x^2-y^2}$ and d_{xy} orbitals rise in energy.

Nickel(II) ion (which has eight *d* electrons) forms a complex with cyanide ion that is square planar and diamagnetic. In this complex the strong field produced by the cyanide ions yields a large energy separation between the d_{xy} and $d_{x^2-y^2}$ orbitals, so that a low-spin complex results. The electron distribution in this complex is illustrated in Figure 21.34.

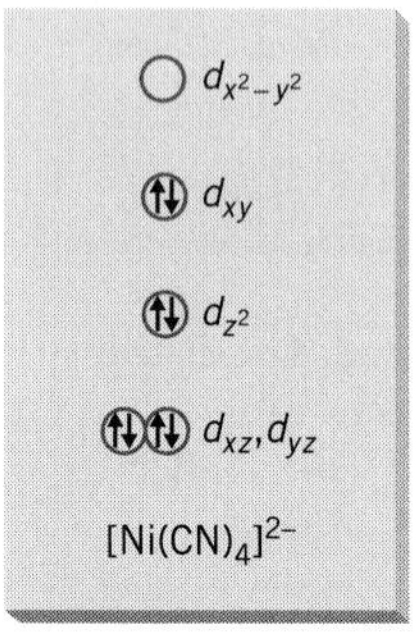

○ $d_{x^2-y^2}$

⇅ d_{xy}

⇅ d_{z^2}

⇅⇅ d_{xz}, d_{yz}

$[Ni(CN)_4]^{2-}$

FIG. 21.34 Distribution of the electrons among the *d* orbitals of nickel in the diamagnetic $[Ni(CN)_4]^{2-}$ ion.

Tetrahedral complexes have a *d* orbital splitting pattern opposite that of octahedral complexes

The splitting pattern for the *d* orbitals in a tetrahedral complex is illustrated in Figure 21.35. Notice that the order of the energy levels is exactly opposite to that in an octahedral complex. In addition, the size of Δ is also much smaller for a tetrahedral complex than for an octahedral one. (Actually, $\Delta_{tet} \approx \frac{4}{9}\,\Delta_{oct}$ for the same metal ion with the same ligands.) This small Δ is always less than the pairing energy, so tetrahedral complexes are always high-spin complexes.

21.10 Metal ions serve critical functions in biological systems

Most of the compounds in our bodies have structures based on carbon as the principal element, and their functions are usually related to the geometries assumed by carbon-containing compounds as well as the breaking and forming of carbon–carbon, carbon–oxygen, and carbon–nitrogen bonds. In Chapter 22 we will discuss some classes of biochemical molecules.

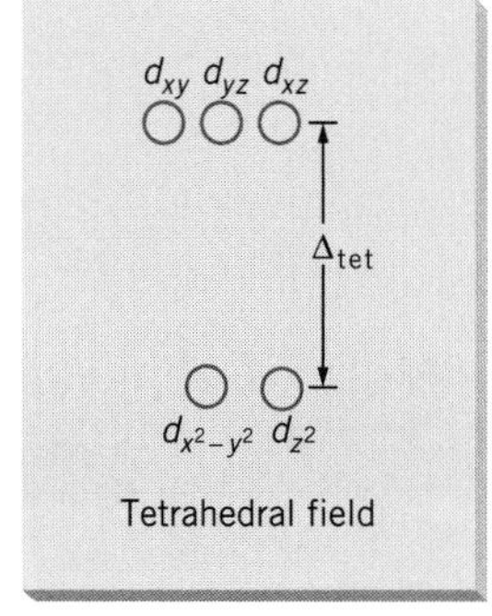

FIG. 21.35 Splitting pattern of the *d* orbitals for a tetrahedral complex. The crystal field splitting for the tetrahedral structure (Δ_{tet}) is smaller than that for the octahedral structure (Δ_{oct}). For complexes with the same ligands, $\Delta_{tet} \approx \frac{4}{9}\Delta_{oct}$.

Our bodies also require certain metal ions in order to operate, and without them life cannot be sustained. For this reason, the study of metals in biological systems has become a very important branch of biochemistry, and a large number of research papers are published annually that deal with this topic. Table 21.2 lists some of the essential metals and the functions fulfilled by their ions.

A few metals, such as sodium and potassium, are found as simple monatomic ions in body fluids. Most metal ions, however, become bound by ligands and do their work as part of metal complexes. As examples, we will look briefly at two metals—iron and cobalt.

Iron is one of the essential elements required by our bodies. We obtain it in a variety of ways in our diets. Iron is involved in oxygen transport in our blood and in retaining oxygen in muscle tissue so that it's available when needed. The iron is present as Fe^{2+} held in a complex in which the basic ligand structure is

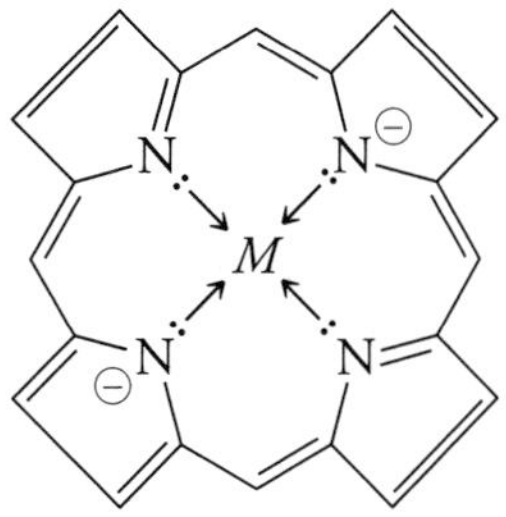

This ligand composition, with its square planar arrangement of nitrogen atoms that bind to a metal ion, is called a *porphyrin structure.* It is the ligand structure in a biologically active unit called heme. Heme is the oxygen-carrying component in the blood protein hemoglobin and in myoglobin, which is found in muscle tissue.

The porphyrin structure is present in chlorophyll, too, where the metal bound in the center is Mg^{2+}. Chlorophyll absorbs sunlight (solar energy), which is used by plants to convert carbon dioxide and water into glucose and oxygen.

In the lungs, O_2 molecules are absorbed by the blood and become bound to Fe^{2+} ions in the heme units of hemoglobin (Figure 21.36). Blood circulation then carries the O_2 to tissues where it is needed, at which time it is released by the Fe^{2+}. One of the important functions of the porphyrin ligand in this process is to prevent the Fe^{2+} from being oxidized by the O_2. (In fact, if the iron is oxidized to Fe^{3+}, it no longer is able to carry O_2.) In muscle tissue, heme units in the protein myoglobin take O_2 from hemoglobin and hold it until it's needed. In this way, muscle tissue is able to store O_2 so that plenty of it is available when the muscle must work hard.

The iron in hemoglobin also binds CO_2 and helps transport it back to the lungs to be exhaled.

Heme units are also present in proteins called cytochromes where the iron is involved in electron transfer reactions that employ the +2 and +3 oxidation states of iron.

TABLE 21.2 Some Biologically Important Metals and their Corresponding Human Body Functions

Metal	Body Function
Na, Ca	Blood pressure and blood coagulation
Fe	Oxygen transport and storage
Ca	Teeth and bone formation
Ca	Urinary stone formation
Zn	Control of pH in blood
Ca, Mg	Muscle contraction
K	Maintenance of stomach acidity
Fe, Cu	Respiration
Cu	Bone health
Ca, Fe, Co	Cell division

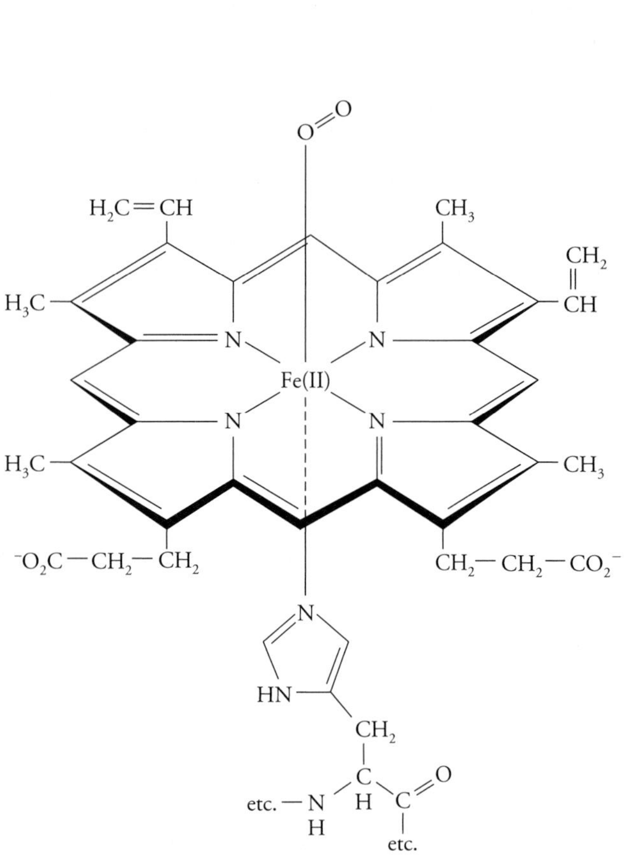

FIG. 21.36 Iron(II) bound to oxygen in heme. The porphyrin ring ligand in heme surrounds an Fe^{2+} ion that binds an O_2 molecule. In its oxygenated form, the Fe^{2+} is octahedrally coordinated in the structure.

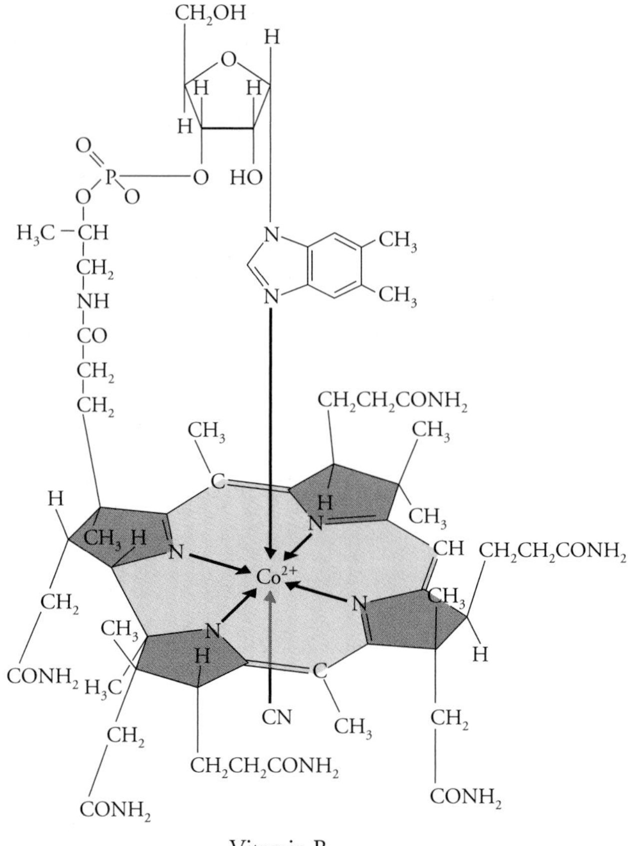

Vitamin B_{12}

FIG. 21.37 The structure of cyanocobalamin. Notice the cobalt ion in the center of the square planar arrangement of nitrogen atoms that are part of the ligand structure. Overall, the cobalt is surrounded octahedrally by donor atoms.

The ability of transition metals to exist in different oxidation states is one reason they are used in biological systems. They can easily take part in oxidation–reduction reactions. Copper(I) and copper(II) ions constitute another pair involved in catalyzing biochemical redox reactions.

A structure similar to heme is found in cyanocobalamin, the form of vitamin B_{12} found in vitamin pills (Figure 21.37). Here, a Co^{2+} ion is held in a square planar ligand structure (called a *corrin ring*) that is slightly different from that found in heme. Certain enzymes (biological catalysts) require cobalamins to function. Vitamin B_{12} is essential in our diets and a deficiency in the vitamin leads to a disease called pernicious anemia.

We have illustrated here just a few examples of the important roles that metal ions play in living systems. They are roles that cannot be fulfilled by other carbon-based compounds, some of which will be described in Chapter 22.

SUMMARY

Obtaining Nonmetallic Elements in Their Free States. Metalloids normally exist in positive oxidation states in compounds and are obtained by reduction. Common reducing agents are carbon and hydrogen. In nature, the noble gases are never found in compounds. Oxygen and nitrogen are obtained from air, and sulfur is found in natural deposits, often deep below the earth's surface. Carbon is found in coal and diamonds. Nonmetals are also found in naturally occurring compounds.

If the nonmetal is combined with a metal, oxidation is used to change it to the free element. The oxidizing strength of the halogens decreases from fluorine to iodine. In Group VIIA, a given halogen (e.g., F_2) is able to oxidize the halide ion below it (e.g., Cl^-), thereby displacing it from its compounds. Fluorine, being the most difficult element to oxidize, is obtained by electrolysis of molten mixtures of HF and KF. Elemental phosphorus is obtained from calcium phosphate by reaction with carbon and SiO_2.

Molecular Structures of the Nonmetals. Elements of Period 2, because of their small size, form strong π bonds. As a result, these elements easily participate in multiple bonding between like atoms, which accounts for diatomic molecules of O_2 and N_2, and the π-bonded structure of graphite. Elements of Periods 3, 4, and 5 are large and their *p* orbitals do not overlap well to form strong π bonds, so these elements prefer single σ bonds between like atoms, which leads to more complex molecular structures.

Different forms of the same element are called **allotropes.** Oxygen exists in two allotropic forms: dioxygen (O_2) and **ozone** (O_3). Carbon forms several allotropes including **diamond, graphite,** C_{60} molecules called **buckminsterfullerene** (one member of the **fullerene** family of structures), and **carbon nanotubes.** Elemental boron consists of B_{12} clusters linked through other boron atoms to give an extremely hard solid. Sulfur forms S_8 molecules that can be arranged in two different allotropic forms. Phosphorus occurs as **white phosphorus** (P_4), **red phosphorus,** and **black phosphorus.** Silicon only forms a diamondlike structure.

Preparation of Metals Most metals occur in compounds where they exist in positive oxidation states and are obtained from their compounds by reduction. Ions of sodium and potassium are found in seawater, along with magnesium. Insoluble carbonate deposits of Ca and Mg arise from the shells of marine animals. Many metals are found in the earth as oxides, sulfides, and phosphates. Active metals such as sodium, magnesium, and aluminum must generally be prepared by electrolysis. Metals of intermediate activity can be obtained using chemical reducing agents such as carbon and hydrogen. Reducing a metal oxide with carbon is called **smelting.** Compounds with small heats of formation tend to be thermally unstable and can be decomposed by heat.

Metallurgy. Usually, when an **ore** is dug from the ground, the metal-bearing component must be enriched by a pretreatment step that removes much of the unwanted **gangue. Flotation** is often used with lead and copper sulfide ores. Sulfide ores are usually **roasted** to convert them to oxides, which are more easily reduced. Aluminum's amphoteric character is exploited in purifying bauxite.

Carbon, in the form of **coke** made from coal, is a common chemical reducing agent because it is plentiful and inexpensive. Metallic iron forms in a **blast furnace** where a charge of iron ore, limestone, and coke reacts in a stream of heated air. Molten iron and **slag** flow to the bottom of the furnace and are periodically tapped. **Steel** is made from this impure iron mostly by the **basic oxygen process.**

Complex Ions of Metals. Coordination compounds contain **complex ions** (also called **complexes** or **coordination complexes**), formed from a metal ion and a number of ligands. **Ligands** are Lewis bases and may be **monodentate, bidentate,** or, in general, **polydentate,** depending on the number of **donor atoms** that they contain. Water is the most common monodentate ligand. Polydentate ligands bind to a metal through two or more donor atoms and yield ring structures called **chelates.** Common bidentate ligands are oxalate ion and ethylenediamine (en); a common polydentate ligand is ethylenediaminetetraacetic acid (EDTA), which has six donor atoms.

In the formula of a complex, the metal is written first, followed by the formulas of the ligands (anions first in alphabetical order followed by neutral ligands in alphabetical order). Brackets are often used to enclose the set of atoms that make up the complex, with the charge on the complex written outside the brackets. The charge on the complex is the algebraic sum of the charges on the metal ion and the charges on the ligands.

Complexes of polydentate ligands are more stable than similar complexes formed with monodentate ligands, partly because a polydentate ligand is less likely to be lost completely if one of its donor atoms becomes detached from the metal ion. The larger positive entropy change for dissociation of complexes with monodentate ligands also favors their dissociation compared with complexes of polydentate ligands. The phenomenon is called the **chelate effect.**

Nomenclature of Complexes. Complexes are named following a set of rules developed by the IUPAC. These are summarized on pages 876 to 878.

Coordination Number and Structure. The **coordination number** of a metal ion in a complex is the number of donor atoms attached to the metal ion. Polydentate ligands supply two or more donor atoms, which must be taken into account when determining the coordination number from the formula of the complex. Geometries associated with common coordination numbers are as follows: for coordination number 2, linear; for coordination number 4, tetrahedral and square planar (especially for Pt^{2+} complexes); and for coordination number 6, octahedral.

Isomers of Coordination Compounds. When two or more distinct compounds have the same chemical formula, they are **isomers** of each other. **Stereoisomers** have the same atoms attached to each other, but the atoms are arranged differently in space. In a **cis isomer,** attached groups of atoms are on the same side of some reference plane through the molecule. In a **trans isomer,** they are on opposite sides. Cis and trans isomerism is a form of **geometric isomerism. Chiral** isomers are exactly the same in every way but one—they are not **superimposable** on their mirror images. These kinds of isomers exist for complexes of the type $M(AA)_3$, where M is a metal ion and AA is a bidentate ligand, and also for complexes of the type *cis*-$M(AA)_2a_2$, where a is a monodentate ligand. Chiral isomers that are related as object to mirror images are said to be **enantiomers.** Because they are able to rotate **plane-polarized light,** they are called **optical isomers.**

Crystal Field Theory. In an octahedral complex, the ligands influence the energies of the *d* orbitals by splitting the *d* subshell into two energy sublevels. The lower energy one consists of the d_{xy}, d_{xz}, and d_{yz} orbitals; the higher energy level consists of the d_{z^2} and $d_{x^2-y^2}$ orbitals. The energy difference between the two new *d* sublevels is the **crystal field splitting,** Δ, and for a given ligand it increases with an increase in the oxidation state of the metal. For a given metal ion, Δ depends on the ligand, and it depends on the period number in which the metal is found.

In the **spectrochemical series,** ligands are arranged in order of their ability to cause a large Δ. Cyanide ion produces the largest crystal field splitting and iodide ion, the smallest. **Low-spin** complexes result when Δ is larger than the **pairing energy**—the energy needed to cause two electrons to become paired in the same orbital. **High-spin** complexes occur when Δ is smaller than the pairing energy. Light of energy equal to Δ is absorbed when an electron is raised from the lower energy set of *d* orbitals to the higher set, and the color of the complex is determined by the colors that remain in the transmitted light. Crystal field theory can also explain the relative stabilities of oxidation states, in many cases.

Different splitting patterns of the *d* orbitals occur for other geometries. The patterns for square planar and tetrahedral geometries are described in Figures 21.33 and 21.35, respectively. The value of Δ for a tetrahedral complex is only about 4/9 that of Δ for an octahedral complex.

Metals in Living Systems. Most metals required by living organisms perform their actions when bound as complex ions. Heme contains Fe^{2+} held in a square planar porphyrin ligand and binds to O_2 in hemoglobin and myoglobin. Heme is also found in cytochromes where it participates in redox reactions involving Fe^{2+} and Fe^{3+}. Vitamin B_{12}, required by the body to prevent the vitamin deficiency disease called pernicious anemia, contains Co^{2+} in a corrin ring structure, which is similar to the porphyrin ring.

TOOLS FOR PROBLEM SOLVING

In this chapter you learned to apply the following concepts as tools in solving problems related to the properties of metal complexes. Study each tool carefully so that you know what each is used for. When faced with solving a problem, recall what each tool does and consider whether it will be helpful in finding a solution. This will aid you in selecting the tools you need.

Rules for writing formulas for complexes *(page 874)* The following rules apply whenever you have to write the formula for a complex ion:

1. The symbol for the metal ion is always given first, followed by the ligands.
2. When more than one kind of ligand is present, anionic ligands are written first (in alphabetical order), followed by neutral ligands (also in alphabetical order).
3. The charge on the complex is the algebraic sum of the charge on the metal ion and the charges on the ligands.

Rules for naming complexes *(pages 876 to 878)* Naming complexes follows rules that are an extension of the rules you learned earlier. You have to learn them and then apply them when you have to name a complex, or write a formula given the name.

Crystal field splitting pattern for an octahedral complex *(page 885)* Figure 21.27 forms the basis for applying the principles of crystal field theory to octahedral complexes. To use it, you need the electron configuration of the metal ion. First write the electron configuration for the metal under consideration. Then remove electrons from the atom starting with the outer *s* subshell first, followed if necessary by electrons from the underlying *d* subshell. For a complex under consideration, set up the splitting diagram and place electrons into the *d* orbitals following Hund's rule. For d^4, d^5, d^6, and d^7 configurations, you may have to decide whether a high- or low-spin configuration is preferred.

Spectrochemical series *(page 888)* Use the location of ligands in the spectrochemical series to compare their effects on the crystal field splitting. The series is

$$CN^- > NO_2^- > en > NH_3 > H_2O > C_2O_4^{2-} > OH^- > F^- > Cl^- > Br^- > I^-$$

QUESTIONS, PROBLEMS, AND EXERCISES

Answers to problems whose numbers are printed in color are given in Appendix B. More challenging problems are marked with asterisks. ILW = Interactive Learningware solution is available at www.wiley.com/college/brady. OH = an Office Hours video is available for this problem.

REVIEW QUESTIONS

Recovery of Nonmetals and Metalloids from Compounds

21.1 What is the major commercial source of N_2 and O_2?

21.2 Why is helium found in underground deposits? Why are only small quantities of radon observed in nature?

21.3 Chlorine, Cl_2, can be made in the lab by the reaction of HCl with $KMnO_4$, with $MnCl_2$ being among the products. Write a balanced net ionic equation for the reaction, keeping in mind that $KMnO_4$ is water soluble.

21.4 Complete the following chemical equations. If no reaction occurs, write N.R.

(a) $Cl_2 + KI \longrightarrow$

(b) $Br_2 + CaF_2 \longrightarrow$

(c) $I_2 + MgCl_2 \longrightarrow$

(d) $F_2 + SrCl_2 \longrightarrow$

21.5 How is Br_2 recovered from seawater?

21.6 Why can't fluorine be produced by the electrolysis of aqueous NaF? What products are formed at inert electrodes in the electrolysis of aqueous NaF?

21.7 Why is a molten mixture of KF and HF used in the production of F_2 by electrolysis rather than molten KF by itself?

21.8 Hydrogen fluoride is a gas that can be liquefied by cooling it to 19.6 °C (just slightly below room temperature). Why can't electrolysis be carried out on liquid HF to form F_2 and H_2?

21.9 Write the chemical equation for the production of elemental phosphorus from calcium phosphate and SiO_2. What is the reducing agent in the reaction? What is the function of the SiO_2? Why isn't SiO_2 reduced to Si in the reaction?

21.10 Under the proper conditions, iodide ion will react with H_2SO_4 to generate I_2 and H_2S. Write a balanced net ionic equation for the reaction.

21.11 Why are metalloids usually recovered from their compounds by chemical reduction of their compounds rather than by oxidation?

21.12 Write chemical equations for

(a) the chemical reduction of BCl_3 with hydrogen.

(b) the production of Si from SiO_2 using carbon as a reducing agent.

(c) the reduction of As_2O_3 with hydrogen.

Molecular Structures of the Nonmetals and Metalloids

21.13 Which of the nonmetals occur in nature in the form of isolated atoms?

21.14 Why are the Period 2 elements able to form much stronger π bonds than the nonmetals of Period 3? Why does a Period 3 nonmetal prefer to form all σ bonds instead of one σ bond and several π bonds?

21.15 Even though the nonmetals of Periods 3, 4, and 5 do not tend to form π bonds between like atoms, each of the halogens is able to form diatomic molecules (Cl_2, Br_2, I_2). Why?

21.16 What are *allotropes*? How do they differ from *isotopes*?

21.17 What are the two allotropes of oxygen?

21.18 Construct the molecular orbital diagram for O_2 and explain why it has two unpaired electrons. What is the net bond order in O_2?

21.19 Draw the Lewis structure for O_3. Is the molecule linear, based on the VSEPR theory? Assign formal charges to the atoms in the Lewis structure. Does this suggest the molecule is polar or nonpolar?

21.20 What beneficial function does ozone serve in the upper atmosphere?

21.21 Describe the structure of diamond. What kind of hybrid orbitals does carbon use to form bonds in diamond? What is the geometry around carbon in this structure?

21.22 Describe the structure of graphite. What kind of hybrid orbitals does carbon use in the formation of the molecular framework of graphite?

21.23 Why does graphite have lubricating properties?

21.24 Describe the C_{60} molecule. What is it called? What name is given to the series of similar substances?

21.25 How is the structure of a carbon nanotube related to the structure of graphite?

21.26 In elemental boron, there are clusters of boron atoms linked through other boron atoms. What is the formula for the boron clusters? What is the shape of a cluster?

21.27 What is the molecular structure of sulfur in its most stable allotropic form?

21.28 Make a sketch that describes the molecular structure of white phosphorus.

21.29 What are the P—P—P bond angles in the P_4 molecule? If phosphorus uses *p* orbitals to form the phosphorus–phosphorus bonds, what bond angle would give the best orbital overlap? On the basis of your answers to these two questions, explain why P_4 is so chemically reactive.

21.30 What structure has been proposed for red phosphorus? How do the reactivities of red and white phosphorus compare?

21.31 What is the molecular structure of black phosphorus? In what way does the structure of black phosphorus resemble that of graphite?

21.32 What is the molecular structure of silicon? Suggest a reason why silicon doesn't form an allotrope that's similar in structure to graphite.

Occurrence of Metals and Recovery of Metals from Compounds

21.33 Which are the three most abundant metals in seawater? Why don't we find large amounts of silver ion in seawater?

21.34 What is the origin of limestone?

21.35 Why are many metals found as oxides and sulfides in the earth?

21.36 Why is carbon used so often as an industrial reducing agent?

21.37 Write equations for the reduction of copper(II) oxide with hydrogen.

21.38 Why isn't thermal decomposition a practical method for obtaining metals such as sodium or magnesium?

21.39 In general, why do compounds tend to decompose at high temperatures but not at low temperatures?

21.40 In terms of thermodynamics, what must be true for us to be able to observe a substantial degree of decomposition of a compound?

21.41 The value of $\Delta G°$ for the decomposition of a metal oxide is negative if the heat of formation of the compound is positive. Why are we able to isolate such compounds at room temperature if they are unstable toward decomposition?

21.42 Many explosives have positive heats of formation. How does this explain why they tend to explode?

Metallurgy

21.43 Use your own words to define *metallurgy*.

21.44 What is an *ore*? What distinguishes an ore from some other potential source of a metal?

21.45 Many rocks are composed of minerals called aluminosilicates. One such mineral is called *orthoclase* and has the formula $KAlSi_3O_8$. Despite their high abundance, aluminosilicates are not considered aluminum ores. What is the probable reason for this?

21.46 What is *gangue*?

21.47 Why can gold be separated from impurities of rock and sand by *panning*?

21.48 Describe the *flotation process*.

21.49 Write chemical equations for the reactions that occur when Cu_2S and PbS are roasted in air. Write a chemical equation to show how SO_2 from the roasting can be kept from being released to the environment. Why might a sulfuric acid plant be located near a plant that roasts sulfide ores?

21.50 Write chemical equations that show how bauxite is purified.

21.51 Why is reduction, rather than oxidation, necessary to extract metals from their compounds?

21.52 Sodium, magnesium, and aluminum are produced by electrolysis instead of by reduction with chemical reducing agents. Why?

21.53 What is *coke*? How is it made?

21.54 Write chemical equations for the reduction of PbO and CuO with carbon.

21.55 Copper(I) sulfide can be converted to metallic copper without adding a reducing agent. Explain this using appropriate chemical equations.

21.56 Why is a blast furnace called a *blast* furnace?

21.57 What is the composition of the charge that's added to a blast furnace?

21.58 Describe the chemical reactions involved in the reduction of Fe_2O_3 that take place in a blast furnace. What is the active reducing agent in the blast furnace?

21.59 What is slag? Write a chemical equation for its formation in a blast furnace. What are some of its uses?

21.60 What does *refining* mean in metallurgy?

21.61 What is the difference between pig iron and steel?

21.62 Describe the basic oxygen process.

Complex Ions of Metals

21.63 The formation of the complex ion $[Cu(H_2O)_4]^{2+}$ is described as a Lewis acid–base reaction. Explain.

(a) What are the formulas of the Lewis acid and the Lewis base in the reaction?
(b) What is the formula of the ligand?
(c) What is the name of the species that provides the donor atom?
(d) What atom is the donor atom, and why is it so designated?
(e) What is the name of the species that is the acceptor?

21.64 To be a ligand, a substance should also be a Lewis base. Explain.

21.65 Why are substances that contain complex ions often called coordination compounds?

21.66 Give the names of two molecules mentioned in the text that are electrically neutral, monodentate ligands.

21.67 Give the formulas of four ions that have 1− charges and are monatomic, monodentate ligands.

21.68 Use Lewis structures to diagram the formation of $Cu(NH_3)_4^{2+}$ and $CuCl_4^{2-}$ ions from their respective components.

21.69 What must be true about the structure of a ligand classified as *bidentate*?

21.70 What is a *chelate*? Use Lewis structures to diagram the way that the oxalate ion, $C_2O_4^{2-}$, functions as a chelating agent.

21.71 How many donor atoms does $EDTA^{4-}$ have?

21.72 Explain how a salt of $EDTA^{4-}$ can retard the spoilage of salad dressing.

21.73 How does a salt of $EDTA^{4-}$ in shampoo make the shampoo work better in hard water?

21.74 The cobalt(III) ion, Co^{3+}, forms a 1:1 complex with $EDTA^{4-}$. What is the net charge, if any, on this complex, and what would be a suitable formula for it (using the symbol EDTA)?

21.75 Which complex is more stable, $[Cr(NH_3)_6]^{3+}$ or $[Cr(en)_3]^{3+}$? Why?

Coordination Number and Structure

21.76 What is a *coordination number*? What structures are generally observed for complexes in which the central metal ion has a coordination number of 4? What is the most common structure observed for coordination number 6?

21.77 Sketch the structure of an octahedral complex that contains only identical monodentate ligands.

21.78 Sketch the structure of the octahedral $[Co(EDTA)]^-$ ion. Remember that donor atoms in a polydentate ligand span adjacent positions in the octahedron.

Isomers of Coordination Compounds

21.79 What are *isomers*?

21.80 Define *stereoisomerism, geometric isomerism, chiral isomers,* and *enantiomers.*

21.81 What are *cis* and *trans* isomers?

21.82 What condition must be fulfilled in order for a molecule or ion to be chiral?

21.83 What are *optical isomers*?

Bonding in Complexes

21.84 On appropriate coordinate axes, sketch and label the five *d* orbitals.

21.85 Which *d* orbitals point *between* the *x*, *y*, and *z* axes? Which point along the coordinate axes?

21.86 Explain why an electron in a $d_{x^2-y^2}$ or d_{z^2} orbital in an octahedral complex will experience greater repulsions because of the presence of the ligands than an electron in a d_{xy}, d_{xz}, or d_{yz} orbital.

21.87 Sketch the *d* orbital energy level diagram for a typical octahedral complex.

21.88 Explain why octahedral cobalt(II) complexes are easily oxidized to cobalt(III) complexes. Sketch the *d* orbital energy diagram and assume a large value of Δ when placing electrons in the *d* orbitals.

21.89 Explain how the same metal in the same oxidation state is able to form complexes of different colors.

21.90 If a complex appears red, what color light does it absorb? What color light is absorbed if the complex appears yellow?

21.91 What does the term *spectrochemical series* mean? How can the order of the ligands in the series be determined?

21.92 What do the terms *low-spin complex* and *high-spin complex* mean?

21.93 For which *d* orbital electron configurations are both high-spin and low-spin complexes possible?

21.94 Indicate by means of a sketch what happens to the *d* orbital electron configuration of the $[Fe(CN)_6]^{4-}$ ion when it absorbs a photon of visible light.

21.95 The complex $[Co(C_2O_4)_3]^{3-}$ is diamagnetic. Sketch the *d* orbital energy level diagram for the complex and indicate the electron populations of the orbitals.

21.96 Consider the complex $[MCl_4(H_2O)_2]^-$ illustrated below on the left. Suppose the structure of the complex is distorted to give the structure on the right, where the water molecules along the *z* axis have moved away from the metal somewhat and the four chloride ions along the *x* and *y* axes have moved closer. What effect will this distortion have on the energy level splitting pattern of the *d* orbitals? Use a sketch of the splitting pattern to illustrate your answer.

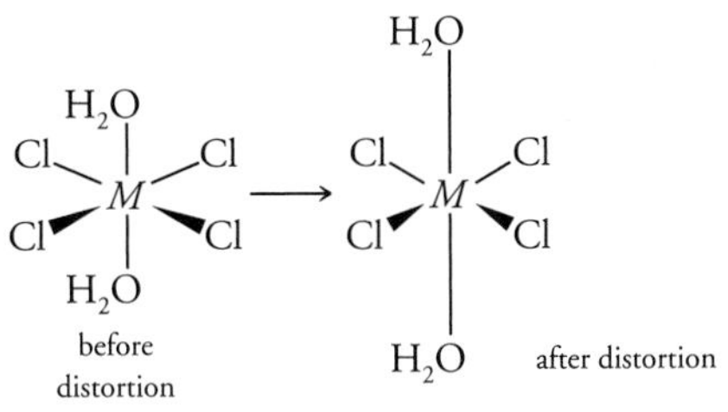

before distortion → after distortion

Metals in Living Systems

21.97 In what ways is the porphyrin structure important in biological systems?

21.98 If a metal ion is held in the center of a porphyrin ring structure, what is its coordination number? (Assume the porphyrin is the only ligand.)

21.99 What function does heme serve in hemoglobin? What does it do in myoglobin?

21.100 How are the ligand ring structures similar in vitamin B_{12} and in heme? What metal is coordinated in cobalamin?

21.101 What are some of the roles played by calcium ion in the body?

REVIEW PROBLEMS

Thermal Stability of Metal Compounds

OH **21.102** Estimate the temperature at which $\Delta G_T^\circ = 0$ for the decomposition of mercury(II) oxide according to the following equation.

$$2HgO(s) \longrightarrow 2Hg(g) + O_2(g)$$

For HgO, $\Delta H_f^\circ = -90.8$ kJ mol^{-1} and $S^\circ = 70.3$ J mol^{-1} K^{-1}; for Hg(*g*), $\Delta H_f^\circ = +61.3$ kJ mol^{-1} and $S^\circ = 175$ J mol^{-1} K^{-1}; for O_2, $S^\circ = 205$ J mol^{-1} K^{-1}.

OH **21.103** From the data below, estimate the temperature at which $K_P = 1$ for the reaction

$$CuO(s) \longrightarrow Cu(s) + \tfrac{1}{2}O_2(g)$$

For CuO, $\Delta H_f^\circ = -155$ kJ mol^{-1}. Absolute entropies: CuO(*s*), 42.6 J mol^{-1} K^{-1}; Cu(*s*), 33.2 J mol^{-1} K^{-1}; and $O_2(g)$, 205 J mol^{-1} K^{-1}.

Complex Ions

21.104 The iron(III) ion forms a complex with six cyanide ions that is often called the ferricyanide ion. What is the net charge on the complex ion, and what is its formula? What is the IUPAC name for the complex?

21.105 The silver ion forms a complex ion with two ammonia molecules. What is the formula of the ion and what is its IUPAC name? Can the complex ion exist as a salt with the sodium ion or with the chloride ion? Write the formula of the possible salt. (Use brackets and parentheses correctly.)

21.106 Write the formula, including its correct charge, for a complex that contains Co^{3+}, two Cl^-, and two ethylenediamine ligands.

OH **21.107** Write the formula, including its correct charge, for a complex that contains Cr^{3+}, two NH_3 ligands, and four NO_2^- ligands.

Naming Complexes

21.108 How would the following molecules or ions be named as ligands when writing the name of a complex ion?

(a) $C_2O_4^{2-}$
(b) S^{2-}
(c) Cl^-
(d) $(CH_3)_2NH$ (dimethylamine)

OH **21.109** How would the following molecules or ions be named as ligands when writing the name of a complex ion?

(a) NH_3
(b) N^{3-}
(c) SO_4^{2-}
(d) $C_2H_3O_2^-$

21.110 Give IUPAC names for each of the following.

(a) $[Ni(NH_3)_6]^{2+}$
(b) $[CrCl_3(NH_3)_3]^-$
(c) $[Co(NO_2)_6]^{3-}$
(d) $[Mn(CN)_4(NH_3)_2]^{2-}$
(e) $[Fe(C_2O_4)_3]^{3-}$

21.111 Give IUPAC names for each of the following.

(a) $[AgI_2]^-$
(b) $[SnS_3]^{2-}$
(c) $[Co(H_2O)_4(en)_2]_2(SO_4)_3$
(d) $[CrCl(NH_3)_5]SO_4$
(e) $K_3[Co(C_2O_4)_3]$

21.112 Write chemical formulas for each of the following.

(a) tetraaquadicyanoiron(III) ion
(b) tetraammineoxalatonickel(II)
(c) diaquatetracyanoferrate(III) ion
(d) potassium hexathiocyanatomanganate(III)
(e) tetrachlorocuprate(II) ion

21.113 Write chemical formulas for each of the following.

(a) tetrachloroaurate(III) ion
(b) bis(ethylenediamine)dinitroiron(III) ion
(c) tetraamminedicarbonatocobalt(III) nitrate
(d) ethylenediaminetetraacetatoferrate(II) ion
(e) diamminedichloroplatinum(II)

Coordination Number and Structure

21.114 In $[FeCl_2(H_2O)_2(en)]$, what is the coordination number of iron ?

OH **21.115** What is the coordination number of nickel in $[Ni(NO_2)_2(C_2O_4)_2]^{4-}$?

21.116 NTA is the abbreviation for nitrilotriacetic acid, a substance that was used at one time in detergents. Its structure is

```
              O
             //
     CH2—C—O—H
    /
   /          O
:N           //
   \ CH2—C—O—H
    \
     CH2—C—O—H
          \\
           O
```

The four donor atoms of the ligand are shown in red. Sketch the structure of an octahedral complex containing the NTA ligand. Assume that two water molecules are also attached to the metal ion and that each oxygen donor atom in the NTA is bonded to a position in the octahedron that is adjacent to the nitrogen of the NTA.

21.117 The compound shown below is called diethylenetriamine and is abbreviated "dien."

$$H_2\ddot{N}—CH_2—CH_2—\ddot{N}H—CH_2—CH_2—\ddot{N}H_2$$

When it bonds to a metal, it is a ligand with three donor atoms.

(a) Which are most likely the donor atoms?

(b) What is the coordination number of cobalt in the complex $Co(dien)_2{}^{3+}$?

(c) Sketch the structure of the complex $[Co(dien)_2]^{3+}$.

(d) Which complex would be expected to be more stable in aqueous solution, $[Co(dien)_2]^{3+}$ or $[Co(NH_3)_6]^{3+}$?

(e) What would be the structure of triethylenetetraamine?

Isomers of Coordination Compounds

21.118 Below are two structures drawn for a complex. Are they actually different isomers, or are they identical? Explain your answer.

```
       Cl                  H2O
H2O    |   OH2      H2O    |   OH2
    \  |  /             \  |  /
       M                   M
    /  |  \             /  |  \
H2O    |   Cl       Cl     |   Cl
      H2O                 H2O
```

Structure I   Structure II

***21.119** Below is a structure for one of the isomers of the complex $[Co(H_2O)_3(dien)]^{3+}$. Are isomers of this complex chiral? Justify your answer.

```
       N
H2O    |   N
    \  |  /
      Co
    /  |  \
H2O    |   N
      H2O
```

N⌒N⌒N represents dien

21.120 Sketch and label the isomers of the square planar complex $[PtBrCl(NH_3)_2]$.

21.121 The complex $[CoCl_3(NH_3)_3]$ can exist in two isomeric forms. Sketch them.

21.122 Sketch the chiral isomers of $[Co(C_2O_4)_3]^{3-}$.

OH **21.123** Sketch the chiral isomers of $[CrCl_2(en)_2]^+$. Is there a nonchiral isomer of the complex?

Bonding in Complexes

21.124 In which complex do we expect to find the larger Δ?

(a) $[Cr(H_2O)_6]^{2+}$ or $[Cr(H_2O)_6]^{3+}$

(b) $[Cr(en)_3]^{3+}$ or $[CrCl_6]^{3-}$

21.125 Arrange the following complexes in order of increasing wavelength of the light absorbed by them: $[Cr(H_2O)_6]^{3+}$, $[CrCl_6]^{3-}$, $[Cr(en)_3]^{3+}$, $[Cr(CN)_6]^{3-}$, $[Cr(NO_2)_6]^{3-}$, $[CrF_6]^{3-}$, $[Cr(NH_3)_6]^{3+}$

OH **21.126** Which complex should be expected to absorb light of the highest frequency, $[Cr(H_2O)_6]^{3+}$, $[Cr(en)_3]^{3+}$, or $[Cr(CN)_6]^{3-}$?

21.127 Which complex should absorb light at the longer wavelength?

(a) $[Fe(H_2O)_6]^{2+}$ or $[Fe(CN)_6]^{4-}$

(b) $[Mn(CN)_6]^{3-}$ or $[Mn(CN)_6]^{4-}$

21.128 In each pair below, which complex is expected to absorb light of the shorter wavelength? Justify your answers.

(a) $[RuCl(NH_3)_5]^{2+}$ or $[FeCl(NH_3)_5]^{2+}$

(b) $[Ru(NH_3)_6]^{2+}$ or $[Ru(NH_3)_6]^{3+}$

21.129 A complex $[CoA_6]^{3+}$ is red. The complex $[CoB_6]^{3+}$ is green. Which ligand, *A* or *B*, produces the larger crystal field splitting, Δ? Explain your answer.

21.130 Referring to the two ligands, *A* and *B*, described in the preceding problem, which complex would be expected to be more easily oxidized, $[CoA_6]^{2+}$ or $[CoB_6]^{2+}$? Explain your answer.

21.131 Referring to the complexes in the preceding two problems, would the color of $[CoA_6]^{2+}$ more likely be red or blue?

21.132 Would the complex $[CoF_6]^{4-}$ more likely be low-spin or high-spin? Could it be diamagnetic?

21.133 Sketch the *d* orbital energy level diagrams for $[Fe(H_2O)_6]^{3+}$ and $[Fe(CN)_6]^{3-}$ and predict the number of unpaired electrons in each.

ADDITIONAL EXERCISES

OH ***21.134** In the decomposition of HgO described in Problem 21.102, what are the equilibrium molar concentrations of Hg(*g*) and $O_2(g)$ above solid HgO at the temperature at which $\Delta G_T^\circ = 0$?

OH **21.135** Estimate K_P at 100, 500, and 1000 °C for the reaction

$$MoO_3(s) \rightleftharpoons Mo(s) + \tfrac{3}{2}O_2(g)$$

given the following data:

	ΔH_f° (kJ mol^{-1})	S° (J mol^{-1} K^{-1})
$MoO_3(s)$	−754.3	78.2
$Mo(s)$	0.0	28.6
$O_2(g)$	0.0	205.0

21.136 Is the complex $[Co(EDTA)]^-$ chiral? Illustrate your answer with sketches.

21.137 The complex $[PtCl_2(NH_3)_2]$ can be obtained as two distinct isomeric forms. Make a model of a tetrahedron and show that if the complex were tetrahedral, two isomers would be impossible.

*21.138 A solution was prepared by dissolving 0.500 g of $CrCl_3 \cdot 6H_2O$ in 100 mL of water. A silver nitrate solution was added and gave a precipitate of AgCl that was filtered from the mixture, washed, dried, and weighed. The AgCl had a mass of 0.538 g.

(a) What is the formula of the complex ion of chromium in the compound?

(b) What is the correct formula for the compound?

(c) Sketch the structure of the complex ion in the compound.

(d) How many different isomers of the complex can be drawn?

*21.139 The compound $Cr_2(NH_3)_6Cl_6$ is a neutral salt in which the cation and anion are both octahedral complex ions. How many isomers (including possible structural and chiral isomers) are there that have this overall composition?

*21.140 The complex $[Co(CN)_6]^{4-}$ is not expected to be perfectly octahedral. Instead, the ligands in the *xy* plane are pulled closer to the Co^{2+} ion while those along the *z* axis move slightly farther away. Using information available in this chapter, explain why the distortion of the octahedral geometry leads to a net lowering of the energy of the complex.

EXERCISES IN CRITICAL THINKING

21.141 In Section 21.2 we assumed that $\Delta H_T^\circ \approx \Delta H_{298}^\circ$. Why is this just an approximation? What factors would cause ΔH_T° to differ from ΔH_{298}°?

21.142 Graphite is a reasonably good conductor in directions parallel to the planes of the carbon atoms, but is a poor conductor in a direction perpendicular to the planes. Why is this so? Would you expect carbon nanotubes to be good conductors of electricity along their length? Explain your answer.

21.143 Considering the fact that unshared electron pairs normally contribute to the polarity of a molecule, is the ozone molecule expected to be polar or nonpolar? How does your answer compare with the answer reached for Question 21.19?

21.144 It was mentioned on page 858 that *d* orbitals are capable of participating in the formation of π bonds. Make a sketch that illustrates how such a bond could be formed between two *d* orbitals and between a *d* and a *p* orbital.

21.145 The two chiral isomers of $[Co(C_2O_4)_3]^{3-}$ (shown below) can be viewed as propellers having either a right- or left-handed twist, respectively. They are identified by the labels Δ and Λ, as indicated. A 50–50 mixture of both isomers is said to be racemic and will not rotate plane-polarized light. Using various laboratory procedures the two isomers present in a racemic mixture can be separated from each other. For this complex, however, a solution of a single isomer is not stable and gradually reverts to a mixture of the two isomers by a process called racemization. Racemization involves the conversion of one isomer to the other until an equilibrium between the two isomers is achieved (i.e., $\Delta \rightleftharpoons \Lambda$). One mechanism proposed for the racemization of isomers of $[Co(C_2O_4)_3]^{3-}$ involves dissociation of an oxalate ion followed by rearrangement and then reattachment of $C_2O_4^{2-}$.

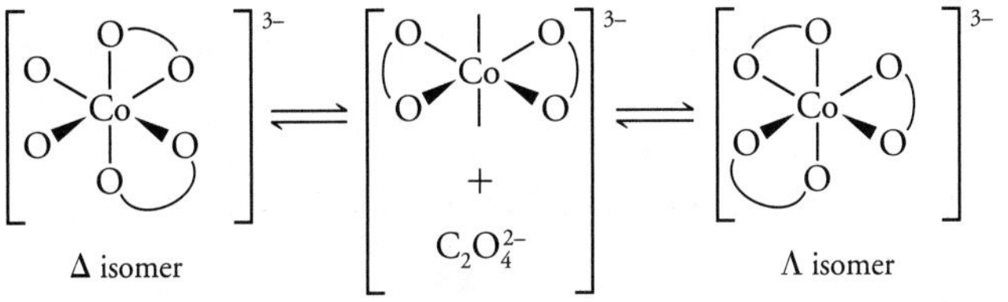

Δ isomer

$C_2O_4^{2-}$

Λ isomer

What experiment could you perform to test whether this is really the mechanism responsible for the racemization of $[Co(C_2O_4)_3]^{3-}$?

ORGANIC COMPOUNDS, POLYMERS, AND BIOCHEMICALS

2

High-speed racing boats often are built with hulls made of a polymer called Kevlar. Because of the polymer's high strength, the hulls can be made thin, thereby reducing weight and increasing speed. How polymers form is one of the topics of this chapter. *(Courtesy Extremeboat.com and Crash.net)*

CHAPTER OUTLIN

22.1 Organic chemistry is the study of carbon compounds

22.2 Hydrocarbons consist of only C and H atoms

22.3 Ethers and alcohols are organic derivatives of water

22.4 Amines are organic derivatives of ammonia

22.5 Organic compounds with carbonyl groups include aldehydes, ketones, and carboxylic acids

22.6 Polymers are composed of many repeating molecular units

22.7 Most biochemicals are organic compounds

22.8 Nucleic acids carry our genetic information

THIS CHAPTER IN CONTEXT A large proportion of the substances we encounter on a daily basis, including foods, fuels, and the many plastics and polymers used to make containers and fabrics, have molecular structures based on atoms of carbon linked to one another by covalent bonds. You've already seen a variety of organic compounds as examples used in discussions throughout this book. They include such substances as hydrocarbons, alcohols, ketones (such as acetone), and organic acids. The number of such compounds is enormous and their study constitutes the branch of chemistry called *organic chemistry,* so named because at one time it was believed that such substances could only be synthesized by living organisms.

At the molecular level of life, nature uses compounds of carbon. The amazing variety of living systems down to the uniqueness of each individual is possible largely because of the properties of this element. We'll take a look at some of carbon's properties in this chapter.

22.1 Organic chemistry is the study of carbon compounds

Organic chemistry is the study of the preparation, properties, identification, and reactions of those compounds of carbon not classified as inorganic. The latter include the oxides of carbon, the bicarbonates and carbonates of metal ions, the metal cyanides, and a handful of other compounds. There are several million known carbon compounds, and all but a very few are classified as organic.

Carbon has some unique properties that enable it to form so many compounds

Sulfur atoms can also form long chains, but they are unable to hold the atoms of any other element strongly at the same time.

What makes the existence of so many organic compounds possible is not just the multivalency of carbon—its atoms always have four bonds in organic compounds. Rather, carbon atoms are unique in their ability to form strong covalent bonds to each other *while at the same time bonding strongly to atoms of other nonmetals, such as H, N, and O.* For example, molecules in the plastic polyethylene have *carbon chains* that are thousands of carbon atoms long with hydrogen atoms attached to each carbon.

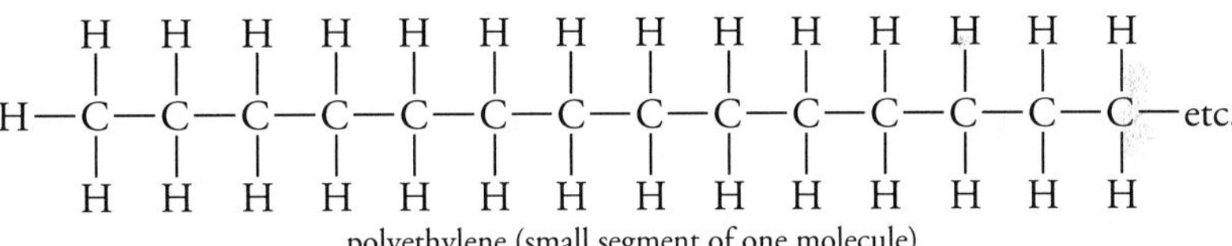

polyethylene (small segment of one molecule)

The longest known sequence of atoms of other members of the carbon family, each also bonded to hydrogens, is eight for silicon, five for germanium, two for tin, and one for lead.

Another reason for the huge number of organic compounds is *isomerism,* which was introduced in Chapter 21 (page 880) where we discussed isomers of complex ions. *Isomers,* recall, are compounds with identical molecular formulas but whose molecules have different structures. Examples are the isomers of butane, C_4H_{10}, and isomers of C_2H_6O (ethyl alcohol and dimethyl ether). See Figure 22.1.

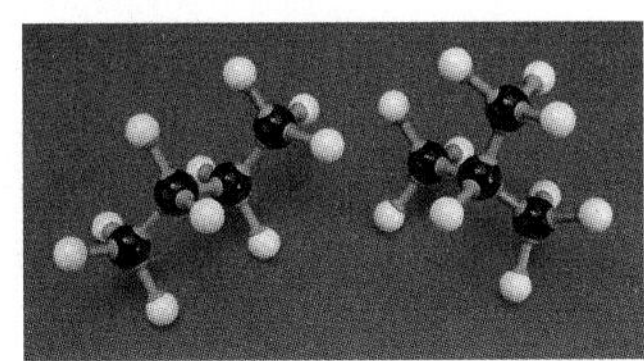

(*a*)

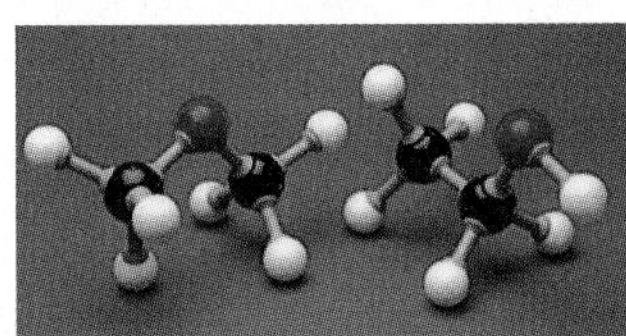

(*b*)

FIG. 22.1 **Isomers.** (*a*) The isomers of C_4H_{10}. Butane is on the left and 2-methylpropane (isobutane) is on the right. (*b*) The isomers of C_2H_6O. Dimethyl ether is on the left and ethanol is on the right. *(Robert Capece.)*

butane	isobutane	ethyl alcohol	dimethyl ether
BP −0.5 °C	BP −11.7 °C	BP 78.5 °C	BP −23 °C

Because the isomers have different structures, the effects of intermolecular attractions are different, which gives rise to differences in boiling points.

Organic compounds can also exhibit chirality, a property we discussed in Chapter 21 on page 881. When a carbon atom is bonded to four different atoms or groups, the carbon

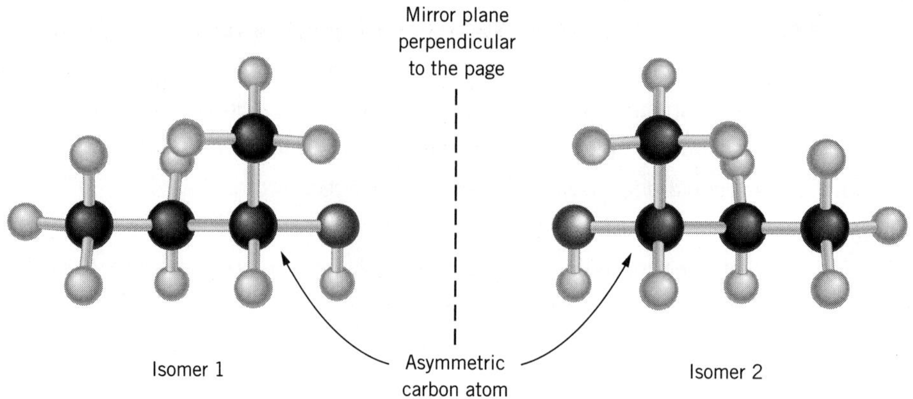

FIG. 22.2 **Chiral isomers of 2-butanol.** Isomer 2 is the mirror image of Isomer 1. No matter how you twist or turn Isomer 1, you cannot get it to fit exactly onto Isomer 2, which means the two isomers are nonsuperimposable mirror images of each other.

is a chiral center. It is called an **asymmetric carbon atom.** An example is 2-butanol, in which the asymmetric carbon, shown in red, is bonded to H, OH, CH_3, and CH_2CH_3.

$$\begin{array}{c} OH \\ | \\ CH_3CHCH_2CH_3 \end{array}$$
2-butanol

The two nonsuperimposable mirror image isomers of this compound are shown in Figure 22.2.

As the number of carbons per molecule increases, the number of possible isomers for any given formula becomes astronomic.

Few of the possible isomers of the compounds with large numbers of carbon atoms have actually been made, but there is nothing except too much crowding within the molecules of some of them to prevent their existence.

Formula	Number of Isomers
C_8H_{18}	18
$C_{10}H_{22}$	75
$C_{20}H_{42}$	366,319
$C_{40}H_{82}$	6.25×10^{13} (estimated)

Organic families are defined by their functional groups

The study of the huge number of organic compounds is manageable because they can be sorted into *organic families* defined by *functional groups.* A major goal of this chapter is to show how they enable us to organize and understand the properties of organic compounds.

Functional groups are small structural units within molecules at which most of the compound's chemical reactions occur (Table 22.1). For example, all *alcohols* have the *alcohol group,* and a molecule of the simplest member of the alcohol family, methyl alcohol, has only one carbon. Ethyl alcohol molecules have two carbons. All members of the family of *carboxylic acids* (organic acids) have the *carboxyl group.* The carboxylic acid with two carbons per molecule is the familiar weak acid, acetic acid.

$$\underset{\text{alcohol group}}{-\overset{|}{\underset{|}{C}}-O-H} \qquad \underset{\text{methyl alcohol}}{H-\overset{H}{\overset{|}{\underset{\underset{H}{|}}{C}}}-O-H} \qquad \underset{\text{carboxyl group}}{-\overset{O}{\overset{\|}{C}}-O-H} \qquad \underset{\text{acetic acid}}{H-\overset{H}{\overset{|}{\underset{\underset{H}{|}}{C}}}-\overset{O}{\overset{\|}{C}}-O-H}$$

One family in Table 22.1, the *alkane* family, has no functional group, just C—C and C—H single bonds. These bonds are virtually nonpolar, because C and H are so alike in electronegativity. Therefore, alkane molecules are the least able of all organic molecules to attract ions or polar molecules. Hence, they do not react, at least at room temperature, with

TABLE 22.1 Some Important Families of Organic Compounds

Family	Characteristic Structural Feature[a]	Example
Hydrocarbons	Only C and H present	
	Families of Hydrocarbons:	
	Alkanes: only single bonds	CH_3CH_3
	Alkenes: C=C	$CH_2{=}CH_2$
	Alkynes: C≡C	HC≡CH
	Aromatic: Benzene ring	(benzene ring)
Ethers	ROR′	CH_3OCH_3
Alcohols	ROH	CH_3CH_2OH
Aldehydes	RCH=O (O double-bonded to C)	$CH_3CH{=}O$
Ketones	RC(=O)R′	$CH_3C({=}O)CH_3$
Carboxylic acids	RC(=O)OH	$CH_3C({=}O)OH$
Esters	RC(=O)OR′	$CH_3C({=}O)OCH_3$
Amines	RNH_2, RNHR′, RNR′R″	CH_3NH_2 CH_3NHCH_3 $CH_3N(CH_3)CH_3$
Amides	RC(=O)—NR′(H)R″(H)	$CH_3C({=}O)NH_2$

[a]R, R′, and R″ represent hydrocarbon groups—*alkyl groups*—defined in the text.

polar or ionic reactants such as strong acids and bases nor with common oxidizing agents, like dichromate or permanganate ion.

Alkanes do react with fluorine, chlorine, bromine, and hot nitric acid. They also burn, a reaction with oxygen.

Condensed structures save time and space

The structural formulas in Table 22.1 are "condensed" because this saves both space and time in writing structures *without sacrificing any structural information.* In condensed structures, C—H bonds are usually "understood." When three H atoms are attached to carbon, they are set alongside the C, as in CH_3, but sometimes H_3C. When a carbon is bonded to two other H atoms, the condensed symbol is usually CH_2, sometimes H_2C. Thus the condensed structure of ethanol is CH_3CH_2OH, and that of dimethyl ether is CH_3OCH_3 or H_3COCH_3.

We have used condensed structures in many of our earlier discussions.

Functional groups can react with polar reactants

When a polar group of atoms, like the OH group or even a halogen atom, is attached to carbon, the molecule has a polar site. It now can attract polar and ionic reactants and undergo chemical changes, but generally at or near just this functional group. This is why compounds of widely varying size but with the same functional group display very similar kinds of reactions.

For example, the reactions of *amines* (organic derivatives of ammonia, discussed in detail in Section 22.4) are similar from amine to amine. Therefore, we need only learn the handful of *kinds* of reactions exhibited by all amines and then adapt this knowledge to a specific example as needed. A special symbol makes this easy.

The symbol R is from the German *radikal.*

In structural formulas R represents an alkane-like group

To focus on the general properties of a functional group, chemists use the symbol R in a structure to represent any and all purely alkane-like groups that do not react with a specific reactant. For example, one type of amine is represented by the formula, R—NH_2. R may be CH_3, CH_3CH_2, $CH_3CH_2CH_2$, and so forth. Because the amines are compounds with an ammonia-like group, they are all Brønsted bases and neutralize acids just as ammonia does. Therefore, we can summarize the reaction of hydrochloric acid with any amine of the R—NH_2 type, regardless of how large the hydrocarbon portion of the molecule is, by the following general equation.

$$R{-}NH_2 + HCl \longrightarrow R{-}NH_3^+ + Cl^-$$

Straight chain

Branched chain

Carbon forms open-chain and ring compounds

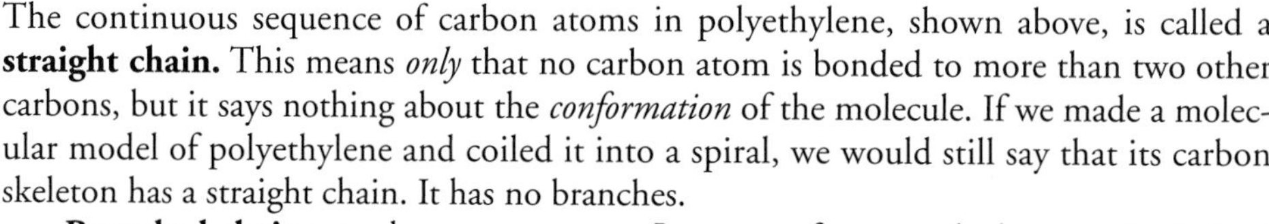

The continuous sequence of carbon atoms in polyethylene, shown above, is called a **straight chain.** This means *only* that no carbon atom is bonded to more than two other carbons, but it says nothing about the *conformation* of the molecule. If we made a molecular model of polyethylene and coiled it into a spiral, we would still say that its carbon skeleton has a straight chain. It has no branches.

Branched chains are also very common. Isooctane, for example, has a *main chain* of five carbon atoms (in black) carrying three CH_3 *branches* (shown in red).

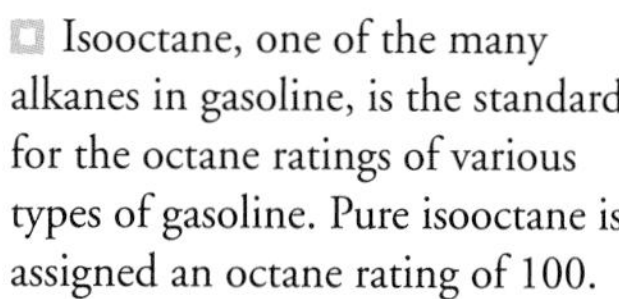

Isooctane, one of the many alkanes in gasoline, is the standard for the octane ratings of various types of gasoline. Pure isooctane is assigned an octane rating of 100.

$$\begin{array}{c} \quad CH_3\;\; CH_3 \\ CH_3\overset{|}{\underset{|}{C}}CH_2\overset{|}{C}HCH_3 \\ CH_3 \end{array}$$

isooctane

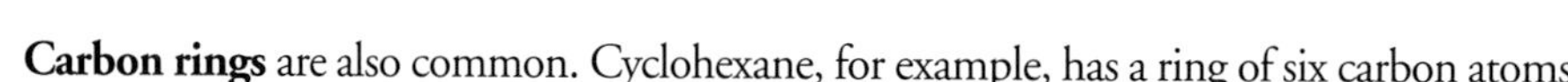

Carbon rings are also common. Cyclohexane, for example, has a ring of six carbon atoms.

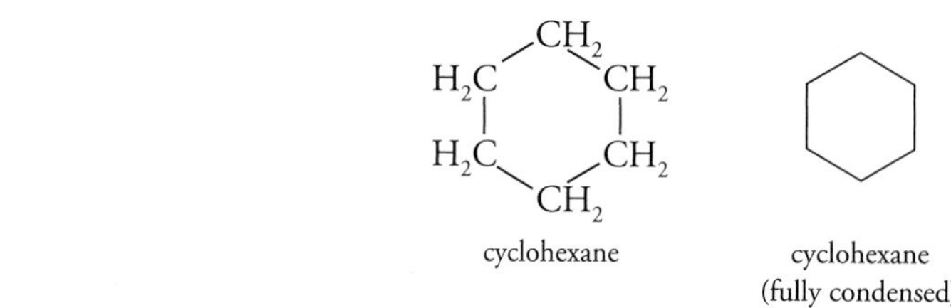

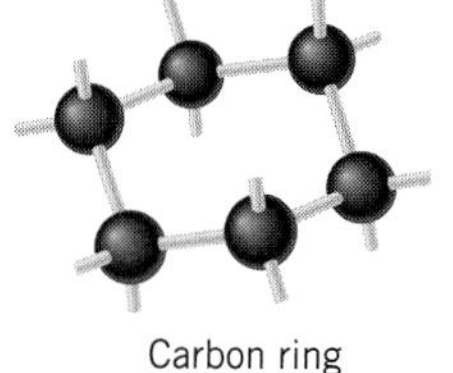

Carbon ring

Just about everything is "understood" in the very convenient, fully condensed structure of cyclohexane. When polygons, like the hexagon, are used to represent rings, the following conventions are used.

Using polygons to represent rings

Using polygons to represent rings

1. C occurs at each corner unless O or N (or another multivalent atom) is explicitly written at a corner.
2. A line connecting two corners is a covalent bond between adjacent ring atoms.
3. Remaining bonds, as required by the covalence of the atom at a corner, are understood to hold H atoms.
4. Double bonds are always explicitly shown.

We can illustrate these rules with the following cyclic compounds.

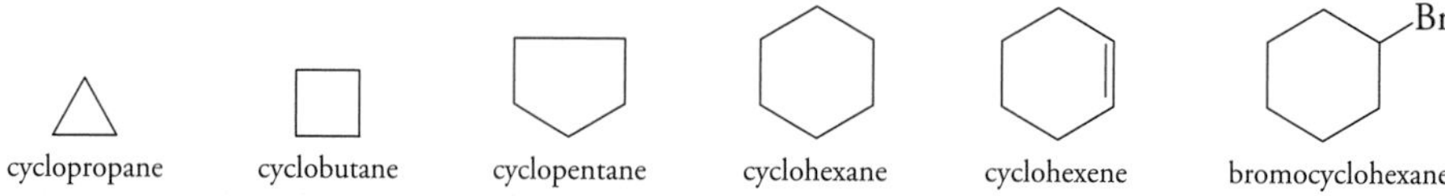

There is no theoretical upper limit to the size of a ring.

Many compounds have **heterocyclic** rings. Their molecules contain an atom other than carbon in a ring position, as in pyrrole, piperidine, and tetrahydropyran.

The tetrahydropyran ring occurs in molecules of sugar.

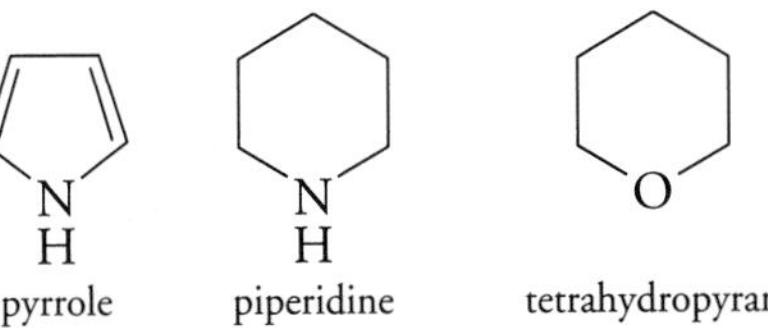

pyrrole   piperidine   tetrahydropyran

The atom in the ring other than carbon is called the *heteroatom*, and any atoms of H or other element bonded to it are not "understood" but always shown.

General principles apply to structure and physical properties

One goal of organic chemistry is to enable the prediction of properties from molecular structure. Functional groups and molecular size determine, for example, whether a compound is soluble in water.

Functional groups affect water solubility

Piperidine and tetrahydropyran (shown below) are both freely soluble in water, whereas cyclohexane is insoluble. The cyclohexane molecule, having only nonpolar C—C and C—H single bonds, is unable to form hydrogen bonds or other polar attractions with water molecules (see page 481). Molecules of piperidine and tetrahydropyran, however, can form hydrogen bonds (shown as "dotted" bonds, ···) with water.

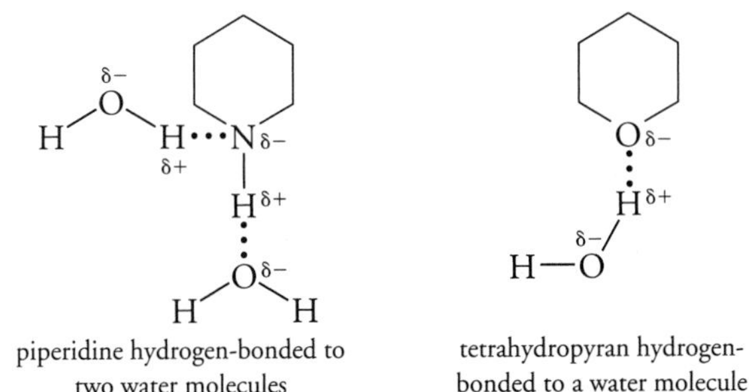

piperidine hydrogen-bonded to two water molecules

tetrahydropyran hydrogen-bonded to a water molecule

Thus organic compounds whose molecules have N and O atoms are more soluble in water than molecules of alkanes of about the same size.

Molecular size affects physical properties

Even with N or O atoms, a compound's solubility in water decreases as the hydrocarbon portion of its molecules becomes larger and larger. The cholesterol molecule, for example, has an OH group, but that group is overwhelmed by the huge hydrocarbon group. So cholesterol is practically insoluble in water.

The insolubility of cholesterol in water accounts for its separation from the bloodstream as heart disease slowly develops.

cholesterol

Cholesterol, however, is correspondingly more soluble in organic solvents such as benzene, chloroform, and acetone.

22.2 HYDROCARBONS CONSIST OF ONLY C AND H ATOMS

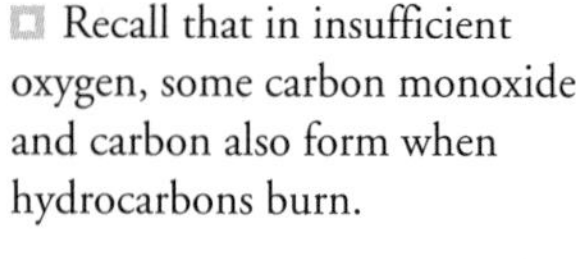

Recall that in insufficient oxygen, some carbon monoxide and carbon also form when hydrocarbons burn.

The addition of H_2 to double or triple bonds is discussed further on page 912.

The alkanes make up just one family of a multifamily group of compounds called the **hydrocarbons** whose molecules consist only of C and H atoms. Besides the alkanes, the hydrocarbons include *alkenes, alkynes,* and *aromatic hydrocarbons* (Table 22.1). All are insoluble in water. All burn, giving carbon dioxide and water if sufficient oxygen is available.

Alkenes and alkynes are called **unsaturated compounds.** Because their molecules have double bonds or triple bonds, respectively, they are able to react with H_2 under appropriate conditions to yield alkanes. The alkanes are **saturated compounds,** compounds with only single bonds, and are unable to react with H_2. The aromatic hydrocarbons are also unsaturated because the carbon atoms of their rings, when represented by simple Lewis structures, also have double bonds. However, they do not react with hydrogen.

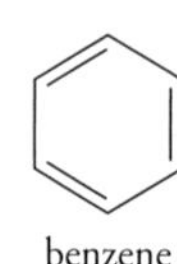

benzene
(Lewis structure)

benzene
(alternative structure)

Most hydrocarbons come from petroleum

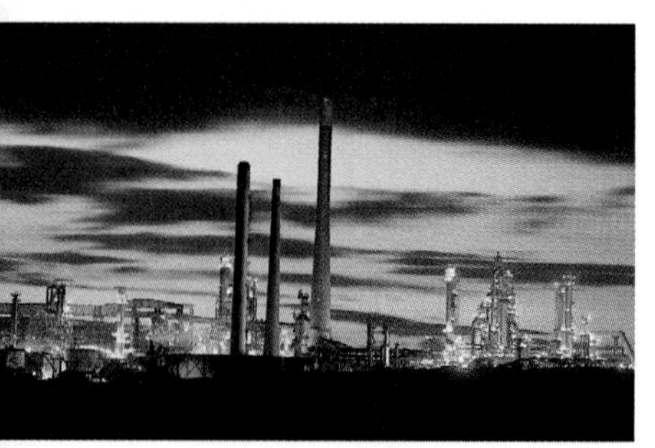

Petroleum refinery. The towers of this petroleum refinery contain catalysts that break up large molecules in hot crude oil to sizes suitable for vehicle engines. *(Martin Bond/Photo Researchers.)*

The *fossil fuels*—coal, petroleum, and natural gas—supply us with virtually all of our hydrocarbons. One of the operations in petroleum refining is to boil crude oil (petroleum freed of natural gas) and selectively condense the vapors between preselected temperature ranges. The liquid collected at each range is called a *fraction*, and the operation is *fractional distillation*. Each fraction is a complex mixture made up almost entirely of hydrocarbons, mostly alkanes. *Gasoline*, for example, is the fraction boiling roughly between 40 and 200 °C. The *kerosene* and *jet fuel* fraction overlaps this range, going from 175 to 325 °C.

The alkanes in gasoline generally have 5 to 10 carbon atoms; those in kerosene, 12 to 18. *Paraffin wax* is part of the nonvolatile residue of petroleum refining, and it consists of alkanes with over 20 carbons per molecule. Low-boiling fractions of crude oil, those having molecules of 4 to 8 carbons, are used as nonpolar solvents.

Alkanes have only carbon–carbon single bonds

All open-chain **alkanes** (those without rings) have the general formula C_nH_{2n+2}, where n equals the number of carbon atoms. Table 22.2 gives the structures, names, and some properties of the first 10 unbranched open-chain alkanes. Their boiling points steadily increase with an increase in the number of atoms, illustrating how London forces become greater with molecular size. The alkanes are generally less dense than water.

TABLE 22.2 Straight-Chain Alkanes

IUPAC Name	Molecular Formula	Structure	Boiling Point (°C)	Melting Point (°C)	Density (g mL^{-1}, 20 °C)
Methane	CH_4	CH_4	−161.5	−182.5	
Ethane	C_2H_6	CH_3CH_3	−88.6	−183.3	
Propane	C_3H_8	$CH_3CH_2CH_3$	−42.1	−189.7	
Butane	C_4H_{10}	$CH_3(CH_2)_2CH_3$	−0.5	−138.4	
Pentane	C_5H_{12}	$CH_3(CH_2)_3CH_3$	36.1	−129.7	0.626
Hexane	C_6H_{14}	$CH_3(CH_2)_4CH_3$	68.7	−95.3	0.659
Heptane	C_7H_{16}	$CH_3(CH_2)_5CH_3$	98.4	−90.6	0.684
Octane	C_8H_{18}	$CH_3(CH_2)_6CH_3$	125.7	−56.8	0.703
Nonane	C_9H_{20}	$CH_3(CH_2)_7CH_3$	150.8	−53.5	0.718
Decane	$C_{10}H_{22}$	$CH_3(CH_2)_8CH_3$	174.1	−29.7	0.730

An IUPAC system of nomenclature applies to organic compounds

The IUPAC rules for naming organic compounds follow a regular pattern. The last syllable in an IUPAC name designates the family of the compound. The names of all saturated hydrocarbons, for example, end in *-ane*. For each family there is a rule for picking out and naming the *parent chain* or *parent ring* within a specific molecule. The compound is then regarded as having *substituents* attached to its parent chain or ring. Let's see how these principles work with naming alkanes.

IUPAC Rules for Naming the Alkanes

IUPAC rules for naming compounds

1. The name ending for all alkanes (and cycloalkanes) is *-ane*.
2. The *parent chain* is the longest continuous chain of carbons in the structure. For example, the branched-chain alkane

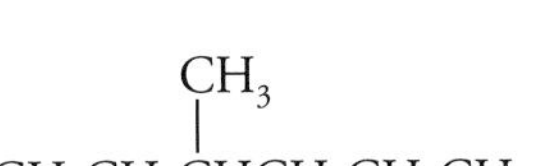

is regarded as being "made" from (derived from) the following parent

$$CH_3CH_2CH_2CH_2CH_2CH_3$$

by replacing a hydrogen atom on the third carbon from the left with CH_3.

$$CH_3CH_2CH(\leftarrow CH_3,\ H\rightarrow)CH_2CH_2CH_3 \longrightarrow CH_3CH_2CH(CH_3)CH_2CH_2CH_3$$

3. A prefix is attached to the name ending, *-ane*, to specify the number of carbon atoms *in the parent chain*. The prefixes through chain lengths of 10 carbons are as follows. The names in Table 22.2 illustrate how the prefixes are used.

meth-	1 C	hex-	6 C
eth-	2 C	hept-	7 C
prop-	3 C	oct-	8 C
but-	4 C	non-	9 C
pent-	5 C	dec-	10 C

The parent chain of our example has six carbons, so the parent is named hexane—*hex* for six carbons and *ane* for being in the alkane family. Therefore, the alkane whose name we are devising is viewed as a *derivative* of this parent, *hexane*.

4. The carbon atoms of the parent chain are numbered starting from whichever end of the chain gives the location of the first branch the lower of two possible numbers. Thus the correct direction for numbering our example is from left to right, not right to left, because this locates the branch (CH_3) at position 3, not position 4.

$$\underset{1\ \ 2\ \ 3\ \ 4\ \ 5\ \ 6}{CH_3CH_2CH(CH_3)CH_2CH_2CH_3} \qquad \underset{6\ \ 5\ \ 4\ \ 3\ \ 2\ \ 1}{CH_3CH_2CH(CH_3)CH_2CH_2CH_3}$$

(correct direction of numbering) (incorrect direction of numbering)

5. Each branch attached to the parent chain is named, so we must now learn the names of some of the alkane-like branches.

Alkyl groups are alkane fragments that replace hydrogen in the parent alkane

Any branch that consists only of carbon and hydrogen and has only single bonds is called an **alkyl group**, and the names of all alkyl groups end in *-yl*. Think of an alkyl group as an alkane minus one of its hydrogen atoms. For example,

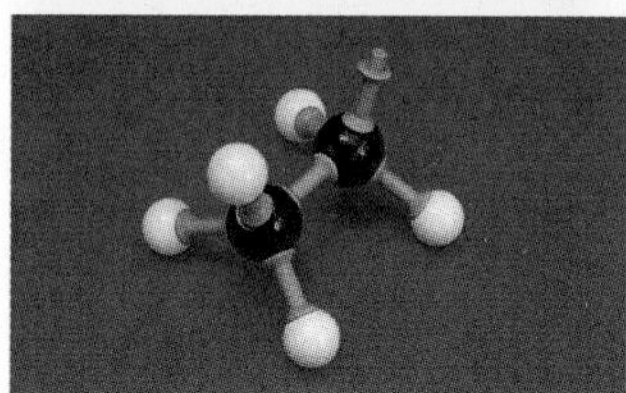

(Robert Capece.)

$$\begin{array}{c} \text{H} \\ | \\ \text{H—C—H} \\ | \\ \text{H} \\ \text{methane} \end{array} \xrightarrow{\text{remove one H}} \begin{array}{c} \text{H} \\ | \\ \text{H—C—} \\ | \\ \text{H} \\ \text{methyl} \end{array} \quad \text{or} \quad CH_3\text{—}$$

$$\begin{array}{c} \text{H H} \\ |\;\;\; | \\ \text{H—C—C—H} \\ |\;\;\; | \\ \text{H H} \\ \text{ethane} \end{array} \xrightarrow{\text{remove one H}} \begin{array}{c} \text{H H} \\ |\;\;\; | \\ \text{H—C—C—} \\ |\;\;\; | \\ \text{H H} \\ \text{ethyl} \end{array} \quad \text{or} \quad CH_3CH_2\text{—}$$

Two alkyl groups can be obtained from propane because the middle position in its chain of three is not equivalent to either of the end positions.

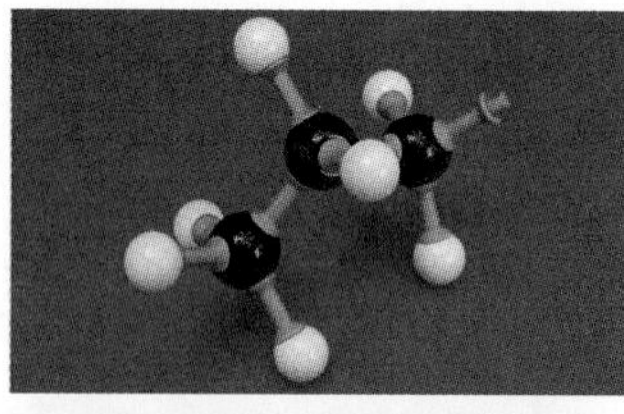

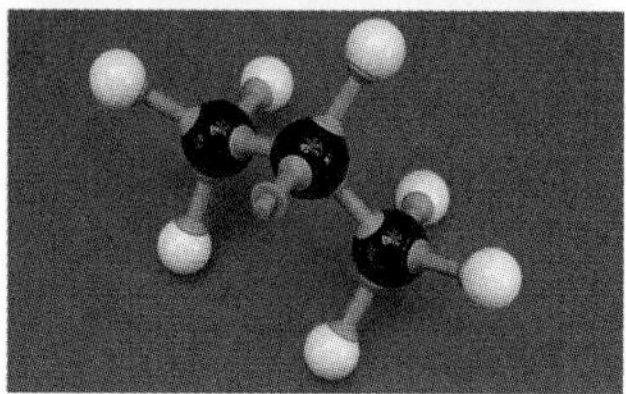

(Robert Capece.)

$$\begin{array}{c} \text{H H H} \\ |\;\;\; |\;\;\; | \\ \text{H—C—C—C—H} \\ |\;\;\; |\;\;\; | \\ \text{H H H} \\ \text{propane} \end{array} \xrightarrow{\text{remove one H}} \begin{array}{c} \text{H H H} \\ |\;\;\; |\;\;\; | \\ \text{H—C—C—C—} \\ |\;\;\; |\;\;\; | \\ \text{H H H} \\ \text{propyl} \end{array} \quad \text{or} \quad CH_3CH_2CH_2\text{—}$$

$$\begin{array}{c} \text{H H H} \\ |\;\;\; |\;\;\; | \\ \text{H—C—C—C—H} \\ |\;\;\; |\;\;\; | \\ \text{H H H} \\ \text{propane} \end{array} \xrightarrow{\text{remove one H}} \begin{array}{c} \text{H}\;\;\;\;\;\;\;\text{H} \\ |\;\;\; |\;\;\; | \\ \text{H—C—C—C—H} \\ |\;\;\; |\;\;\; | \\ \text{H H H} \\ \text{isopropyl} \end{array} \quad \text{or} \quad \begin{array}{c} | \\ CH_3CHCH_3 \end{array}$$

We will not need to know the IUPAC names for any alkyl groups with four or more carbon atoms.

6. The name of each alkyl group is attached to the name of the parent as a prefix, placing its chain location number in front and separating the number from the name by a hyphen. Thus, the original example is named 3-methylhexane.

$$\begin{array}{c} CH_3 \\ | \\ CH_3CH_2CHCH_2CH_2CH_3 \\ \text{3-methylhexane} \end{array}$$

7. When two or more groups are attached to the parent, each is named and located with a number. The names of alkyl substituents are assembled in their alphabetical order. Always use *hyphens* to separate numbers from words. Here is an application.

$$\begin{array}{c} CH_3CH_2 \quad\; CH_3 \\ |\qquad\quad | \\ CH_3CH_2CH_2CHCH_2CHCH_3 \\ 7\;\;\;6\;\;\;5\;\;\;4\;\;3\;\;\;2\;\;1 \\ \text{4-ethyl-2-methylheptane} \end{array}$$

8. When two or more substituents are identical, multiplier prefixes are used: di- (for 2), tri- (for 3), tetra- (for 4), and so forth. The location number of every group must occur in the final name. Always separate a number from another number in a name by a *comma*. For example,

$$\begin{array}{c} CH_3 \quad\; CH_3 \\ |\qquad\quad | \\ CH_3CHCH_2CHCH_2CH_3 \end{array}$$

Correct name: 2,4-dimethylhexane

Incorrect names: 2,4-methylhexane
3,5-dimethylhexane
2-methyl-4-methylhexane
2-4-dimethylhexane

9. When identical groups are on the *same* carbon, the number of this position is repeated in the name. For example,

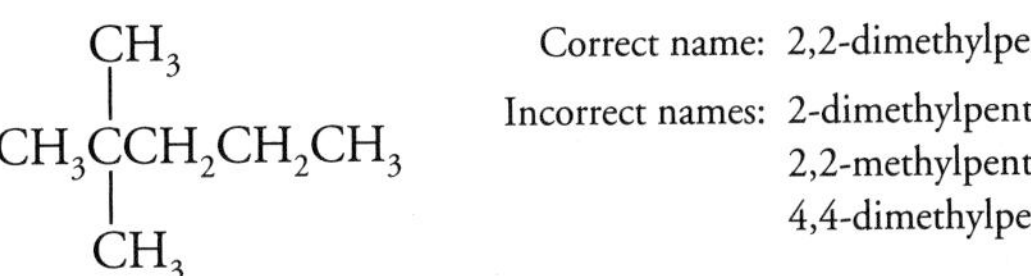

Correct name: 2,2-dimethylpentane

Incorrect names: 2-dimethylpentane
2,2-methylpentane
4,4-dimethylpentane

Common names are still widely used for many compounds. For example, the common name of 2-methylpropane is *isobutane.*

These are not all of the IUPAC rules for alkanes, but they will handle all of our needs.

EXAMPLE 22.1
Using the IUPAC Rules to Name an Alkane

What is the IUPAC name for the following compound?

$$\begin{array}{l} \quad\quad\quad\quad\quad\quad CH_3CH_2 \quad CH_3 \\ \quad\quad\quad\quad\quad\quad\quad | \quad\quad\quad\ | \\ CH_3CH_2CH_2CHCH_2CCH_3 \\ \quad\quad\quad\quad\quad\quad\quad\quad\quad\quad\ | \\ \quad\quad\quad\quad\quad\quad\quad\quad\quad\ CH_3 \end{array}$$

ANALYSIS: The compound is a hydrocarbon, and by studying the structure, we can see that there are only single bonds. Therefore, the substance is an alkane and we must use the IUPAC rules for naming alkanes.

SOLUTION: The ending to the name must be *-ane.* The longest chain is seven carbons long, so the name of the parent alkane is *heptane.* We have to number the chain from right to left in order to reach the first branch with the lower number.

At carbon 2 there are two one-carbon methyl groups. At carbon 4, there is a two-carbon ethyl group. Alphabetically, *ethyl* comes before *methyl,* so we must assemble these names as follows to make the final name. (Names of alkyl groups are alphabetized *before* any prefixes such as di- or tri- are affixed.)

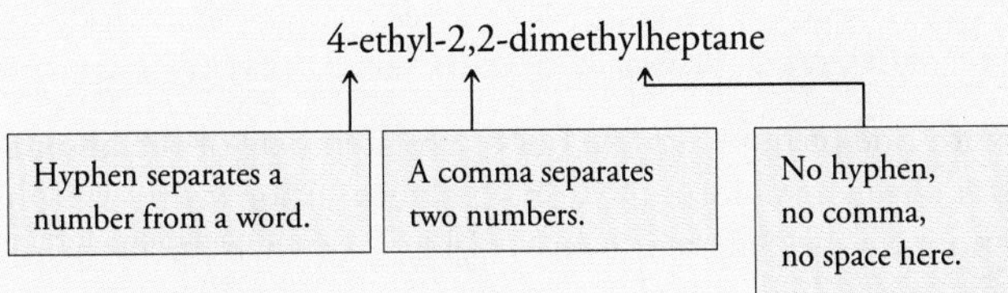
4-ethyl-2,2-dimethylheptane
Hyphen separates a number from a word.
A comma separates two numbers.
No hyphen, no comma, no space here.

IS THE ANSWER REASONABLE? The most common mistake is to pick a shorter chain than the true "parent," a mistake we did not make. Another common mistake is to number the chain incorrectly. One overall check that some people use is to count the number of carbons implied by the name (in our example, 2 + 1 + 1 + 7 = 11) and compare it to the count obtained directly from the structure. If the counts don't match, you know you can't be right.

Practice Exercise 1: A student incorrectly named a compound 1,1,1-trimethylethane. What should the name be? (Hint: Write the structure of 1,1,1-trimethylethane.)

Practice Exercise 2: Write the IUPAC names of the following compounds. In searching for the parent chain, be sure to look for the longest continuous chain of carbons *even if the chain twists and goes around corners.*

(a)
$$\begin{array}{l} CH_3CH_2 \\ \quad\quad\quad \backslash \\ \quad\quad\quad CH{-}CH_3 \\ \quad\quad\quad / \\ CH_2CH_2 \\ | \\ CH_3 \end{array}$$

(b)
$$\begin{array}{l} \quad\quad\quad CH_3 \quad CH_2CH_2CH_3 \\ \quad\quad\quad\quad \backslash \quad\ | \\ \quad\quad\quad\quad CHCHCH_2CH_3 \\ \quad\quad\quad\quad / \\ CH_3CH \\ \quad\quad\ | \\ \quad\quad CH_3 \end{array}$$

(c)

$$
\begin{array}{c}
\quad CH_3 \quad CH_3 \quad CH_3 \\
CH_3CH_2CHCHCHCH_2CHCH_3 \\
CH_3CH_2
\end{array}
$$

Alkanes undergo few chemical reactions

Alkanes, as we have indicated, are generally stable at room temperature toward such different reactants as concentrated sulfuric acid (or any other common acid), concentrated aqueous bases (like NaOH), and even the most reactive metals. Fluorine attacks virtually all organic compounds, including the alkanes, to give mixtures of products. Like all hydrocarbons, the alkanes burn in air to give carbon dioxide and water. *Hot* nitric acid, chlorine, and bromine also react with alkanes. The chlorination of methane, for example, can be made to yield the following compounds.

These are the common names for the chlorinated derivatives of methane, not the IUPAC names.

CH_3Cl	CH_2Cl_2	$CHCl_3$	CCl_4
methyl chloride	methylene chloride	chloroform	carbon tetrachloride

When heated at high temperatures in the absence of air, alkanes "crack," meaning that they break up into smaller molecules. The cracking of methane, for example, yields finely powdered carbon and hydrogen.

$$CH_4 \xrightarrow{\text{high temperature}} C + 2H_2$$

The controlled cracking of ethane gives ethene, commonly called ethylene.

$$\underset{\text{ethane}}{CH_3CH_3} \xrightarrow{\text{high temperature}} \underset{\text{ethene}}{CH_2{=}CH_2} + H_2$$

$HOCH_2CH_2OH$ ethylene glycol

Ethene ("ethylene"), from the cracking of ethane, is one of the most important raw materials in the organic chemicals industry. It is used to make polyethylene plastic as well as ethyl alcohol and ethylene glycol (an antifreeze).

Alkenes and alkynes have double and triple bonds

Hydrocarbons with one or more double bonds are members of the **alkene** family. Open-chain alkenes have the general formula, C_nH_{2n}. Hydrocarbons with triple bonds are in the **alkyne** family and have the general formula, C_nH_{2n-2} (when open-chain).

Alkenes and alkynes, like all hydrocarbons, are insoluble in water and are flammable. The most familiar alkenes are ethene and propene (commonly called ethylene and propylene, respectively), the raw materials for polyethylene and polypropylene, respectively. Ethyne ("acetylene"), an important alkyne, is the fuel for oxyacetylene torches.

The IUPAC accepts both *ethene* and *ethylene* as the name of $CH_2{=}CH_2$. The common name of propene is *propylene*, and other simple alkenes have common names as well.

$CH_2{=}CH_2$	$CH_3CH{=}CH_2$	$HC{\equiv}CH$
ethene (ethylene)	propene (propylene)	ethyne (acetylene)

Alkenes have IUPAC names

The IUPAC rules for the names of alkenes are adaptations of those for alkanes, but with two important differences. First, the parent chain *must include the double bond* even if this means that the parent chain is shorter than another. Second, the parent alkene chain must be *numbered from whichever end gives the first carbon of the double bond the lower of two possible numbers.* This (lower) number, followed by a hyphen, precedes the name of the parent chain, unless there is no ambiguity about where the double bond occurs. The numbers for the locations of branches are not considered in numbering the chain. Otherwise, alkyl groups are named and located as before. Some examples of correctly named alkenes are as follows.

$CH_3CH_2CH{=}CH_2$	$CH_3CH{=}CHCH_3$	$CH_3CH_2CH(CH_3)CH_2CH{=}C(CH_3)CH_3$	(ring structure)
1-butene (not 1,2-butene, not 3-butene)	2-butene	2,5-dimethyl-2-heptene (not 3,6-dimethyl-5-heptene)	cyclohexene

Notice that only one number is used to locate the double bond, the number of the first carbon of the double bond to be reached as the chain is numbered.

Some alkenes have two double bonds and are called *dienes*. Some have three double bonds and are called *trienes*, and so forth. Each double bond has to be located by a number.

$CH_2{=}CHCH{=}CHCH_3$ (1,3-pentadiene) $CH_2{=}CHCH_2CH{=}CH_2$ (1,4-pentadiene) $CH_2{=}CHCH{=}CHCH{=}CH_2$ (1,3,5-hexatriene)

Alkenes exhibit geometric isomerism

As explained in Section 9.6, there is no free rotation at a carbon–carbon double bond (see page 368). Many alkenes, therefore, exhibit **geometric isomerism.** Thus *cis*-2-butene and *trans*-2-butene (see below) are **geometric isomers** of each other. They not only have the same molecular formula, C_4H_8, but also the same skeletons and the same organization of atoms and bonds, namely, $CH_3CH{=}CHCH_3$. The two 2-butenes differ in the *directions* taken by the two CH_3 groups attached at the double bond.

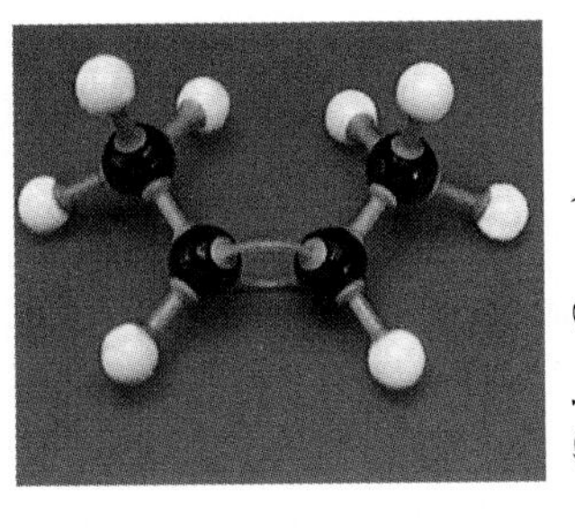

(Robert Capece.)

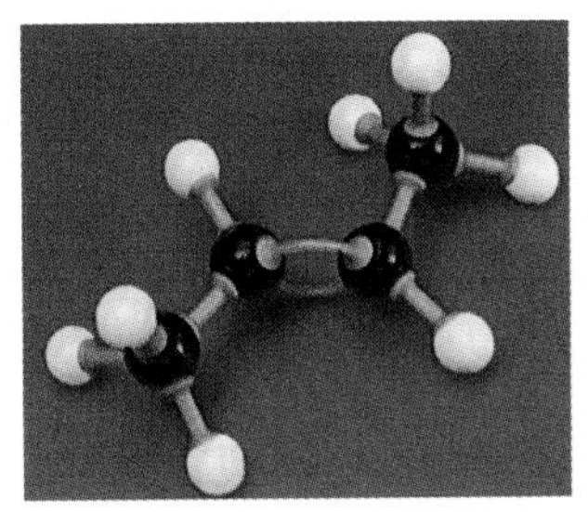

(Robert Capece.)

H₃C and CH₃ on same side: $H_3C(H)C{=}C(H)CH_3$	H and CH₃ opposite: $H(H_3C)C{=}C(CH_3)H$
cis-2-butene BP 3.72 °C	*trans*-2-butene BP 0.88 °C

Because ring structures also lock out free rotation, geometric isomers of ring compounds are possible as well. These two isomers of 1, 2-dimethylcyclopropane are examples.

H_3C … CH_3
***trans* isomer**
BP 28 °C

H_3C … CH_3
***cis* isomer**
BP 37 °C

Cis means "on the same side"; *trans* means "on opposite sides." This difference in orientation gives the two geometric isomers of 2-butene measurable differences in physical properties, as their boiling points show. Because each has a double bond, however, the *chemical* properties of *cis*- and *trans*-2-butene are very similar.

Alkenes undergo addition reactions

Electron-seeking species are naturally attracted to the electron density at the π bond of the double bond, so alkenes react readily with protons provided by strong proton donors. Alkenes thus undergo **addition reactions,** reactions in which the pieces of a reactant become separately attached to the carbons of a double bond. Ethene, for example, readily reacts with hydrogen chloride as follows.

$$CH_2{=}CH_2 + H{-}Cl(g) \longrightarrow Cl{-}CH_2{-}CH_3$$

We say that the hydrogen chloride molecule *adds across the double bond*, with the hydrogen of the HCl bonding to the carbon at one end and the chlorine bonding to the carbon at the other end. A pair of electrons of the π bond move out and take H^+ from HCl, which releases Cl^-. A positive (+) charge is left at one end of the original carbon–carbon double bond as a bond to H forms at the other end.

$$CH_2{=}CH_2 + H{-}Cl \longrightarrow \overset{+}{C}H_2{-}CH_2(H) + Cl^- \longrightarrow Cl{-}CH_2{-}CH_3$$

ethyl carbocation chloroethane

Recall that curved arrows indicate how electron pairs rearrange; they usually are not used to signify the movement of atoms.

A carbocation, having a carbon with only six valence electrons, not an octet, generally has only a fleeting existence.

The result of H^+ transfer is a very unstable cation called a *carbocation*, an ion with a positive charge on carbon. This charged site then quickly attracts the Cl^- ion to give the product, chloroethane.

Another example is 2-butene; like ethene, it too adds HCl.

$$\underset{\text{2-butene}}{CH_3CH{=}CHCH_3} + HCl \longrightarrow \underset{\text{2-chlorobutane}}{CH_3\underset{\displaystyle Cl}{\underset{|}{C}}HCH_2CH_3}$$

Alkene double bonds also add hydrogen bromide, hydrogen iodide, and sulfuric acid ("hydrogen sulfate"). Alkynes undergo similar addition reactions.

A water molecule adds to a double bond if an acid catalyst, like sulfuric acid, is present. The catalyst participates in the first step by donating a proton to one end of the double bond to create a carbocation. This unstable species preferentially attracts what is the most abundant electron-rich species present, namely, the oxygen end of a water molecule.

In the chemical equation, the notation

$$\xrightarrow{HSO_4^-}$$

means that one of the products of the reaction is HSO_4^-. Writing an equation this way lets us focus on the fate of the principal organic reactant(s).

$$\underset{\text{ethene}}{CH_2{=}CH_2} + \underset{\substack{\text{sulfuric acid}\\ \text{(trace; much}\\ H_2O\text{ present)}}}{H{-}OSOH} \xrightarrow{HSO_4^-} \underset{\substack{H_2O \text{ attacks in-}\\ \text{stead of } HSO_4^-}}{\underset{|}{\overset{}{C}}H_2{-}\overset{+}{C}H_2} \longrightarrow \underset{\substack{\text{protonated form}\\ \text{of ethanol}}}{CH_3{-}CH_2{-}\overset{+}{O}H_2} \longrightarrow \underset{\text{ethanol}}{CH_3CH_2OH} + H^+$$

represents recovered catalyst

Note that in the *overall* result, the pieces of the water molecule, H and OH, become attached at the different carbons of the double bond. 2-Butene (either cis or trans) gives a similar reaction.

Other inorganic compounds that add to an alkene double bond are chlorine, bromine, and hydrogen. Chlorine and bromine react rapidly at room temperature. Ethene, for example, reacts with bromine to give 1,2-dibromoethane.

$$\underset{\text{ethene}}{CH_2{=}CH_2} + Br{-}Br \longrightarrow \underset{\text{1,2-dibromoethane}}{\underset{\displaystyle Br}{\underset{|}{C}H_2}{-}\underset{\displaystyle Br}{\underset{|}{C}H_2}}$$

Alkenes undergo hydrogenation by adding hydrogen

The product of the addition of hydrogen to an alkene is an alkane, and the reaction is called *hydrogenation.* It requires a catalyst—powdered platinum, for example—and sometimes a higher temperature and pressure than available under an ordinary room atmosphere. The hydrogenation of 2-butene (cis or trans) gives butane.

$$\underset{\substack{\text{2-butene}\\ (cis \text{ or } trans)}}{CH_3CH{=}CHCH_3} + H{-}H \xrightarrow[\text{catalyst}]{\text{heat, pressure}} CH_3\underset{\displaystyle H}{\underset{|}{C}H}{-}\underset{\displaystyle H}{\underset{|}{C}H}CH_3 \quad \text{or} \quad \underset{\text{butane}}{CH_3CH_2CH_2CH_3}$$

Double bonds react with ozone

Ozone in smog attacks the double bonds in the chlorophyll of green plants, which is able to prevent photosynthesis and so kill the plants.

Ozone (O_3) reacts with anything that has carbon–carbon double or triple bonds, and the reaction breaks the molecules into fragments at each such site, giving a variety of products. Because many important compounds in living systems have alkene double bonds, ozone is a very dangerous material when it is formed in smog.

Aromatic hydrocarbons contain benzene rings

The most common **aromatic compounds** contain the *benzene ring,* a ring of six carbon atoms, each bonded to one H or one other atom or group (in addition to the two adjacent carbons). The benzene ring is represented as a hexagon either with alternating single and double bonds or with a circle. For example,

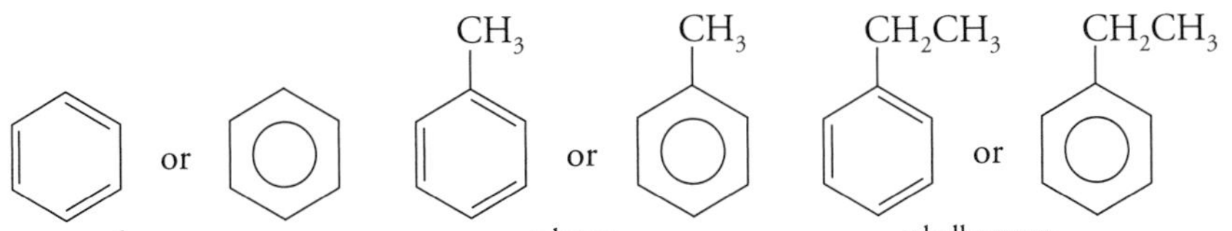

benzene toluene (methylbenzene) ethylbenzene

Hydrocarbons and their oxygen or nitrogen derivatives that are not aromatic are called *aliphatic compounds*.

The circle better represents the delocalized bonds of the benzene ring, which are described in Section 9.8.

In the molecular orbital view of benzene, discussed in Section 9.8, the delocalization of the ring's π electrons strongly stabilizes the ring. This explains why benzene does not easily undergo addition reactions; they would interfere with the delocalization of electron density. Instead, the benzene ring most commonly undergoes **substitution reactions** in which one of the ring H atoms is replaced by another atom or group. For example, benzene reacts with chlorine in the presence of iron(III) chloride to give chlorobenzene, instead of a 1,2-dichloro compound. To dramatize this point, we must use a resonance structure for benzene (see page 378).

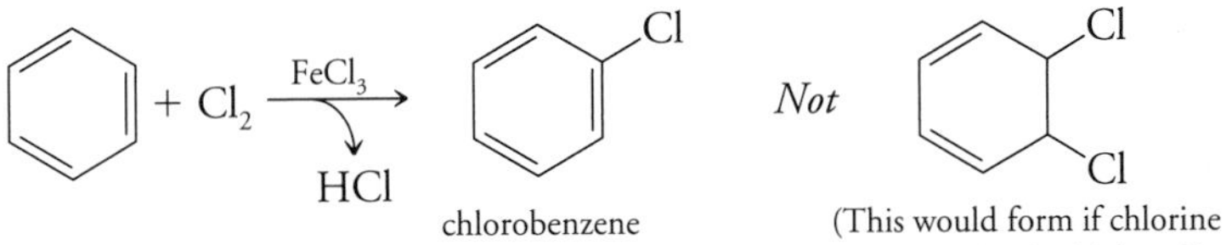

chlorobenzene

(This would form if chlorine *added* to the double bond.)

You can infer that *substitution,* but not addition, leaves intact the closed-circuit, delocalized, and very stable π electron network of the benzene ring.

Provided that a suitable catalyst is present, benzene reacts by substitution with chlorine, bromine, and nitric acid as well as with sulfuric acid. (Recall that Cl_2 and Br_2 readily *add* to alkene double bonds.)

$$C_6H_6 + Br_2 \xrightarrow{FeBr_3 \text{ catalyst}} C_6H_5\text{—}Br + HBr$$

bromobenzene

$$C_6H_6 + HNO_3 \xrightarrow{H_2SO_4 \text{ catalyst}} C_6H_5\text{—}NO_2 + H_2O$$

nitrobenzene

$$C_6H_6 + H_2SO_4 \longrightarrow C_6H_5\text{—}SO_3H + H_2O$$

benzenesulfonic acid

22.3 Ethers and Alcohols Are Organic Derivatives of Water

Ethers have two alkyl groups attached to oxygen

Molecules of **ethers** contain two alkyl groups joined to one oxygen, the two R groups being alike or different. We give only the common names for the following examples.

CH_3OCH_3	$CH_3CH_2OCH_2CH_3$	$CH_3OCH_2CH_3$	$R\text{—}O\text{—}R'$
dimethyl ether	diethyl ether	methyl ethyl ether	ethers
BP −23 °C	BP 34.5 °C	BP 11 °C	(general structure)

Pentane (BP 36.1 °C), like diethyl ether in having no OH group but having nearly the identical number of atoms, has about the same boiling point as diethyl ether.

Diethyl ether is the "ether" once widely used as an anesthetic in surgery.

The contrasting boiling points of alcohols and ethers illustrate the influence of hydrogen bonding. Hydrogen bonds cannot exist between molecules of ethers and the simple ethers have very low boiling points, being substantially lower than those of alcohols of comparable molecular sizes. For example, 1-butanol, $CH_3CH_2CH_2CH_2OH$ (BP 117 °C) boils 83 degrees higher than its structural isomer, diethyl ether (BP 34.5 °C).

Alcohols can be represented by the formula ROH

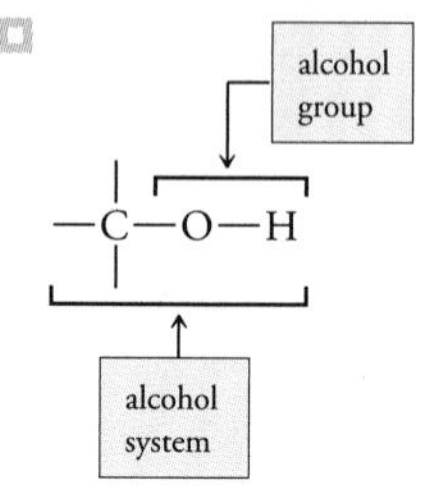

An **alcohol** is any compound with an OH group and three other groups attached to a carbon atom by *single bonds*. Using the symbol R to represent any alkyl group, alcohols have ROH as their general structure.

The four structurally simplest alcohols are the following. (Their common names are in parentheses below their IUPAC names.)

CH_3OH	CH_3CH_2OH	$CH_3CH_2CH_2OH$	$CH_3CH(OH)CH_3$
methanol (methyl alcohol) BP 65 °C	ethanol (ethyl alcohol) BP 78.5 °C	1-propanol (propyl alcohol) BP 97 °C	2-propanol (isopropyl alcohol) BP 82 °C

Ethanol is the alcohol in beverages and is also added to gasoline to make "gasohol" and makes up 85% of the fuel E-85 (the rest is gasoline).

The name ending for an alcohol is *-ol*. It replaces the *-e* ending of the name of the hydrocarbon that corresponds to the parent. The parent chain of an alcohol must be the longest *that includes the carbon bonded to the OH group*. The chain is numbered to give the site of the OH group the lower number regardless of where alkyl substituents occur.

Reactions of alcohols include oxidation, dehydration, and substitution

Ethers are almost as chemically inert as alkanes. They burn (as do alkanes), and they are split apart when boiled in concentrated acids.

The alcohols, in contrast, have a rich chemistry. We'll look at their oxidation and dehydration reactions as well as some substitution reactions.

Oxidation products of alcohols depend on structure

When the carbon atom of the alcohol system, the *alcohol carbon atom,* also is bonded to at least one H, this H can be removed by an oxidizing agent. The H atom of the OH group also leaves, and the two Hs become part of a water molecule. We may think of the oxidizing agent as providing the O atom for H_2O. The number of H atoms left on the original alcohol carbon then determines the *family* of the product.

The oxidation of an alcohol of the RCH_2OH type, with two H atoms on the alcohol carbon, produces first an *aldehyde*, which is further oxidized to a *carboxylic acid*. (The nature of aldehydes and ketones are discussed further in Section 22.5.)

$$RCH_2OH \xrightarrow{\text{oxidation}} \underset{\text{aldehyde}}{RCH{=}O} \xrightarrow{\text{further oxidation}} \underset{\text{carboxylic acid}}{RC({=}O)OH}$$

The net ionic equation for the formation of the aldehyde when dichromate ion is used is

$$3RCH_2OH + Cr_2O_7^{2-} + 8H^+ \longrightarrow 3RCH{=}O + 2Cr^{3+} + 7H_2O$$

The aldehyde group, CH=O, is one of the most easily oxidized, and the carboxyl group, CO_2H, is one of the most oxidation resistant of the functional groups.

Aldehydes are much more easily oxidized than alcohols, so unless the aldehyde is removed from the solution as it forms, it will consume oxidizing agent that has not yet reacted and be changed to the corresponding carboxylic acid.

The oxidation of an alcohol of the R_2CHOH type produces a *ketone*. For example, the oxidation of 2-propanol gives propanone.

$$\underset{\text{2-propanol}}{3CH_3CH(OH)CH_3} + Cr_2O_7^{2-} + 8H^+ \longrightarrow \underset{\text{propanone (acetone)}}{3CH_3C({=}O)CH_3} + 2Cr^{3+} + 7H_2O$$

Ketones (discussed further in Section 22.5) strongly resist oxidation, so a ketone does not have to be removed from the oxidizing agent as it forms.

Alcohols of the type R_3COH have no removable H atom on the alcohol carbon, so they cannot be oxidized in a similar manner.

EXAMPLE 22.2
Alcohol Oxidation Products

What organic product can be made by the oxidation of 2-butanol with dichromate ion? If no oxidation can occur, state so.

ANALYSIS: We first have to look at the structure of 2-butanol.

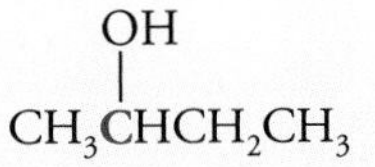

$$\begin{array}{c} \quad OH \\ \quad | \\ CH_3CHCH_2CH_3 \end{array}$$

2-butanol

2-Butanol can be oxidized because it has an H atom on the alcohol carbon (carbon 2). The oxidation results in the detachment of both this H atom and the one joined to the O atom, leaving a double bond to O.

SOLUTION: We carry out the changes that the analysis found. We simply erase the two H atoms that we identified and insert a double bond from C to O. The product is 2-butanone.

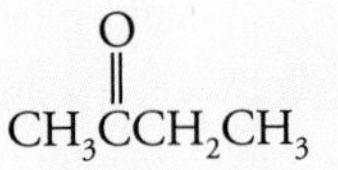

$$\begin{array}{c} \quad O \\ \quad \| \\ CH_3CCH_2CH_3 \end{array}$$

2-butanone (a ketone)

The starting material had two alkyl groups, CH_3 and CH_2CH_3. It doesn't matter what the alkyl groups are, however, because the reaction takes the same course in all cases. All alcohols of the R_2CHOH type can be oxidized to ketones in this way.

IS THE ANSWER REASONABLE? The "skeleton" of heavy atoms, C and O, does not change in this kind of oxidation, and 2-butanone has the same skeleton as 2-butanol. The oxidation also produces a double bond from C to O, as we showed.

Practice Exercise 3: Oxidation of an alcohol gave the following product. What was the formula of the original alcohol? (Hint: How many hydrogens are removed from the alcohol carbon by oxidation?)

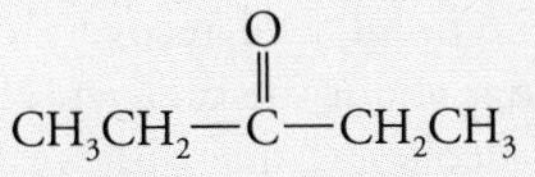

$$\begin{array}{c} O \\ \| \\ CH_3CH_2-C-CH_2CH_3 \end{array}$$

Practice Exercise 4: What are the structures of the products that form by the oxidation of the following alcohols? (a) ethanol, (b) 3-pentanol

Dehydration reactions eliminate the components of water from alcohols

In the presence of a strong acid, like concentrated sulfuric acid, an alcohol molecule can undergo the loss of a water molecule, leaving behind a carbon–carbon double bond. This reaction, called **dehydration,** is one example of an **elimination reaction.** For example,

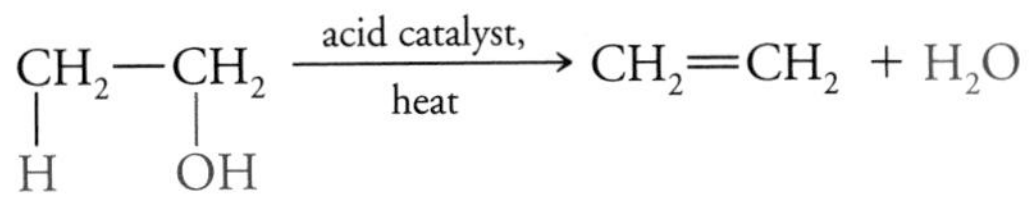

$$\underset{\text{ethanol}}{\begin{array}{cc} CH_2- & CH_2 \\ | & | \\ H & OH \end{array}} \xrightarrow[\text{heat}]{\text{acid catalyst,}} \underset{\text{ethene}}{CH_2=CH_2} + H_2O$$

$$\underset{\text{1-propanol}}{\begin{array}{cc} CH_3CH- & CH_2 \\ | & | \\ H & OH \end{array}} \xrightarrow[\text{heat}]{\text{acid catalyst,}} \underset{\text{propene}}{CH_3CH=CH_2} + H_2O$$

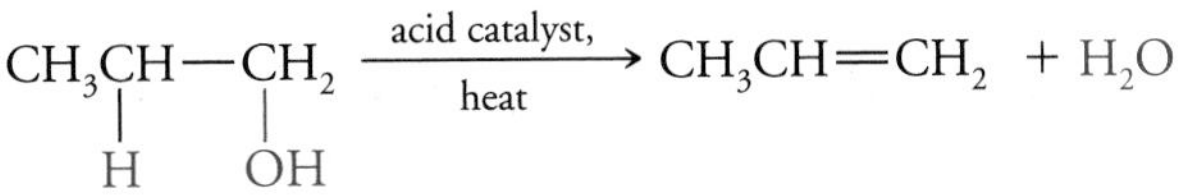

What makes the elimination of water possible is the proton-accepting ability of the O atom of the OH group. Alcohols resemble water in that they react like Brønsted bases toward concentrated strong acids to give an equilibrium mixture involving a protonated form. Ethanol, for example, reacts with (and dissolves in) concentrated sulfuric acid by the following reaction. (H_2SO_4 is written as $H{-}OSO_3H$.)[1]

Remember, the curved arrows in these diagrams show how electrons rearrange during the reaction. They don't indicate the movements of the atoms.

$$CH_3CH_2OH + H{-}OSO_3H \ \text{(catalyst)} \rightleftharpoons CH_3CH_2{-}OH_2^+ + {}^-OSO_3H$$

This bond is now weak.

The organic cation is nothing more than the ethyl derivative of the hydronium ion, $CH_3CH_2OH_2^+$, and *all three bonds to oxygen in this cation are weak*, just like all three bonds to oxygen in H_3O^+.

A water molecule now leaves, taking with it the electron pair that held it to the CH_2 group. The remaining organic species is a carbocation.

$$\underset{\text{protonated form of ethanol}}{CH_3CH_2{-}OH_2^+} \rightleftharpoons \underset{\text{ethyl carbocation}}{CH_3\overset{+}{C}H_2} + H_2O$$

Carbocations, as we said, are unstable. The ethyl carbocation, $CH_3CH_2^+$, loses a proton to become more stable, donating it to some proton acceptor in the medium. The electron pair holding the departing proton stays behind to become the second bond of the new double bond in the product. All carbon atoms now have outer octets. We'll use the HSO_4^- ion (written here as HSO_3O^-) as the proton acceptor.

$$HSO_3O^- + \underset{\text{ethyl carbocation}}{H{-}CH_2{-}\overset{+}{C}H_2} \longrightarrow \underset{\text{ethene}}{CH_2{=}CH_2} + \underset{\text{recovered catalyst}}{HSO_3OH}$$

The last few steps in the dehydration of ethanol—the separation of H_2O, loss of the proton, formation of the double bond, and recovery of the catalyst—probably occur simultaneously. Notice two things about the catalyst, H_2SO_4. First, it works to convert a species with strong bonds, the alcohol, to one with weak bonds strategically located, the protonated alcohol. Second, the catalyst is recovered.

Alcohols participate in substitution reactions

Under acidic conditions, the OH group of an alcohol can be replaced by a halogen atom, using a concentrated hydrohalogen acid. For example,

$$\underset{\text{ethanol}}{CH_3CH_2OH} + HI\ (\text{conc.}) \xrightarrow{\text{heat}} \underset{\text{iodoethane (ethyl iodide)}}{CH_3CH_2I} + H_2O$$

$$\underset{\text{1-propanol}}{CH_3CH_2CH_2OH} + HBr\ (\text{conc.}) \xrightarrow{\text{heat}} \underset{\text{1-bromopropane (propyl bromide)}}{CH_3CH_2CH_2Br} + H_2O$$

[1] The first curved arrow illustrates how an atom attached to one molecule, like the O of the HO group, uses an unshared pair of electrons to pick up an atom of another molecule, like the H of the catalyst. The second curved arrow shows that *both* electrons of the bond from H to O in sulfuric acid remain in the hydrogen sulfate ion that forms. Thus H transfers as H^+.

These reactions, like the earlier reaction between chlorine and benzene, are **substitution reactions.** The first step in each is the transfer of H^+ to the OH of the alcohol to give the protonated form of the OH group.

$$R{-}OH + H^+ \longrightarrow R{-}OH_2{}^+$$

Once again, the acid catalyst works to weaken an important bond. Given the high concentration of halide ion, X^-, it is this species that successfully interacts with $R{-}OH_2{}^+$ to displace OH_2 (water) and yield $R{-}X$.

22.4 Amines Are Organic Derivatives of Ammonia

Amines are organic derivatives of ammonia in which one, two, or three hydrocarbon groups have replaced hydrogens. Examples, together with their common (not IUPAC) names, are

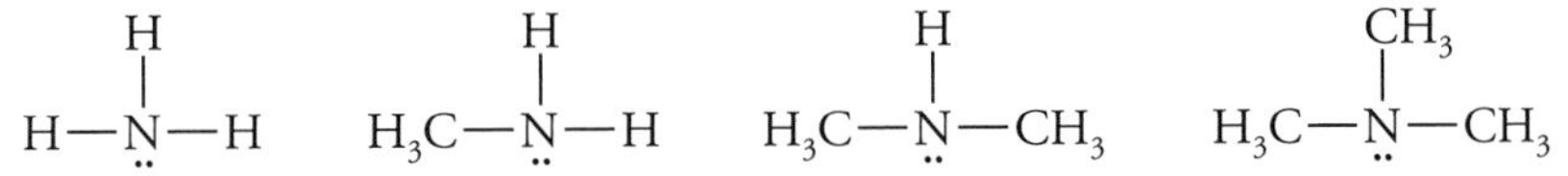

$H{-}\ddot{N}(H){-}H$	$H_3C{-}\ddot{N}(H){-}H$	$H_3C{-}\ddot{N}(H){-}CH_3$	$H_3C{-}\ddot{N}(CH_3){-}CH_3$
ammonia	methylamine	dimethylamine	trimethylamine
BP −33.4 °C	BP −8 °C	BP 8 °C	BP 3 °C

The N—H bond is not as polar as the O—H bond, so amines boil at lower temperatures than alcohols of comparable molecular size. Amines of low molecular mass are soluble in water. Hydrogen bonding between molecules of water and the amine facilitates this.

CH_3CH_2OH, BP 78.5 °C
$CH_3CH_2NH_2$, BP 17 °C
$CH_3CH_2CH_3$, BP −42 °C

The basicity of amines affects their reactions

As we said earlier, amines are Brønsted bases. Behaving like ammonia, amines that do dissolve in water establish an equilibrium in which a low concentration of hydroxide ion exists. For example,

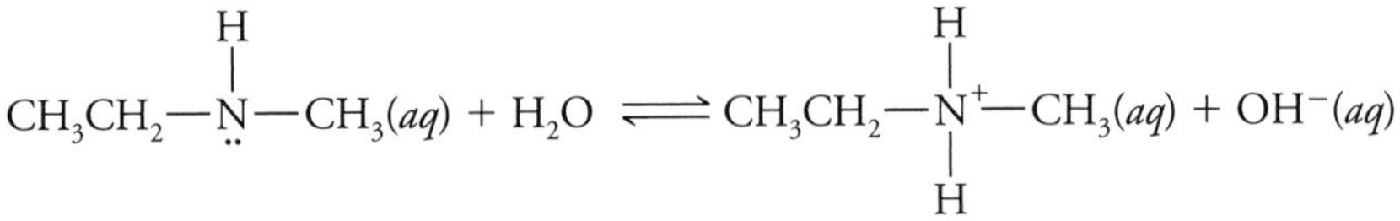

$$\underset{\text{ethylmethylamine}}{CH_3CH_2{-}\ddot{N}(H){-}CH_3(aq)} + H_2O \rightleftharpoons \underset{\text{ethylmethylammonium ion}}{CH_3CH_2{-}\overset{+}{N}H_2{-}CH_3(aq)} + OH^-(aq)$$

As a result, aqueous solutions of amines test basic to litmus and have pH values above 7.

When an amine is mixed with an acid such as hydrochloric acid, the amine and acid react almost quantitatively (i.e., completely). The amine accepts a proton and changes almost 100% into its protonated form. For example,

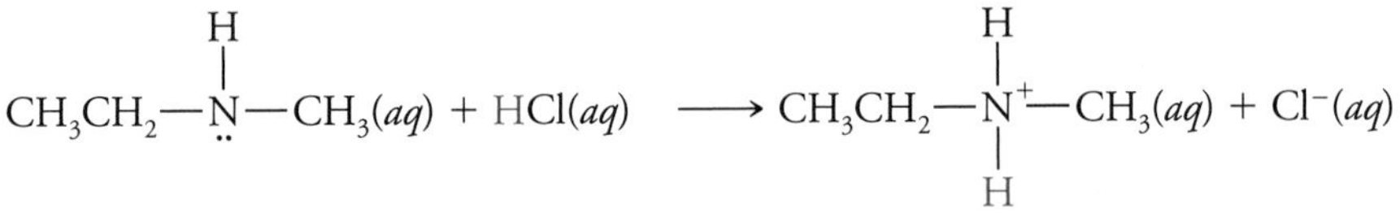

$$\underset{\text{ethylmethylamine}}{CH_3CH_2{-}\ddot{N}(H){-}CH_3(aq)} + HCl(aq) \longrightarrow \underset{\text{ethylmethylammonium ion (a protonated amine)}}{CH_3CH_2{-}\overset{+}{N}H_2{-}CH_3(aq)} + Cl^-(aq)$$

Even water-insoluble amines undergo this reaction, and the resulting salt is much more soluble in water than the original electrically neutral amine.

Many important medicinal chemicals, like quinine, are amines, but they are usually supplied to patients in protonated forms so that the drug can be administered as an aqueous solution, not as a solid. This strategy is particularly important for medicinals that must be given by intravenous drip.

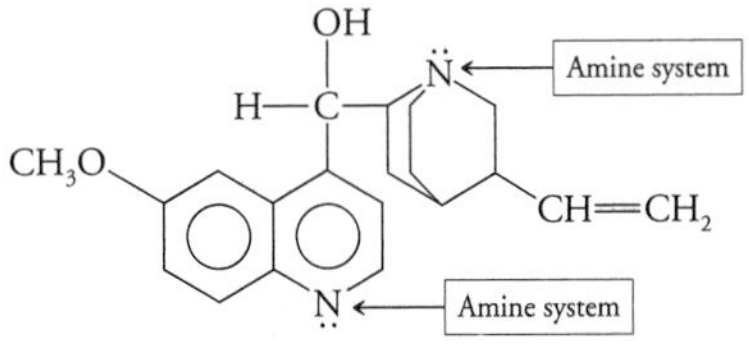

quinine (an antimalarial drug)

Protonated amines are weak Brønsted acids

A protonated amine is a substituted ammonium ion. Like the ammonium ion itself, protonated amines are weak Brønsted acids. They can neutralize strong base. For example,

$$\underset{\text{methylammonium ion}}{CH_3NH_3^+(aq)} + OH^-(aq) \longrightarrow \underset{\text{methylamine}}{CH_3NH_2(aq)} + H_2O$$

This reverses the protonation of an amine and releases the uncharged amine molecule.

22.5 Organic Compounds with Carbonyl Groups Include Aldehydes, Ketones, and Carboxylic Acids

Many organic compounds contain oxygen atoms doubly bonded to a carbon that is also bonded to two other atoms. This grouping of atoms, $>C{=}O$, is called a **carbonyl group,** and it occurs in several organic families. What is attached to the carbon atom in $C{=}O$ determines the specific family.

The carbonyl group is a polar group, and it helps to make compounds containing it much more soluble in water than hydrocarbons of roughly the same molecular size.

Aldehydes and ketones are carbonyl compounds

We've already encountered aldehydes and ketones in our discussion of the oxidation of alcohols. When the carbonyl group binds an H atom plus a hydrocarbon group (or a second H), the compound is an **aldehyde.** When $C{=}O$ is bonded to two hydrocarbon groups at C, the compound is a **ketone.**

O
‖
—C—
carbonyl group

O
‖
—C—H
aldehyde group

O
‖
H—C—H or R—C—H
aldehydes

O
‖
C—C—C
keto group

O
‖
R—C—R′
ketones

The *aldehyde group* is often condensed to CHO, the double bond of the carbonyl group being "understood." The *ketone group* is sometimes condensed to CO.

The carbonyl group occurs widely. As an aldehyde group, it's in the molecules of most sugars, like glucose. Another common sugar, fructose, has the keto group.

Names of aldehydes and ketones have characteristic endings

The IUPAC name ending for an aldehyde is *-al.* The parent chain is the longest chain *that includes the aldehyde group.* Thus, the three-carbon aldehyde is named *propanal,* because "propane" is the name of the three-carbon alkane and the *-e* in propane is replaced by *-al.* The numbering of the chain always starts by assigning the carbon of the aldehyde group position 1. This rule, therefore, makes it unnecessary to include the number locating the aldehyde group in the name, as illustrated by the name 2-methylpropanal.

$HC(=O)H$	$CH_3C(=O)H$	$CH_3CH_2C(=O)H$	$CH_3CH(CH_3)—C(=O)H$
methanal	ethanal	propanal	2-methylpropanal (not 2-methyl-1-propanal)
BP −21 °C	BP 21 °C	BP 49 °C	BP 64 °C

Aldehydes cannot form hydrogen bonds between their own molecules, so they boil at lower temperatures than alcohols of comparable molecular masses.

The name ending for the IUPAC names of ketones is *-one.* The parent chain must include the carbonyl group and be numbered from whichever end reaches the carbonyl carbon first. The number of the ketone group's location must be part of the name whenever there would otherwise be uncertainty.

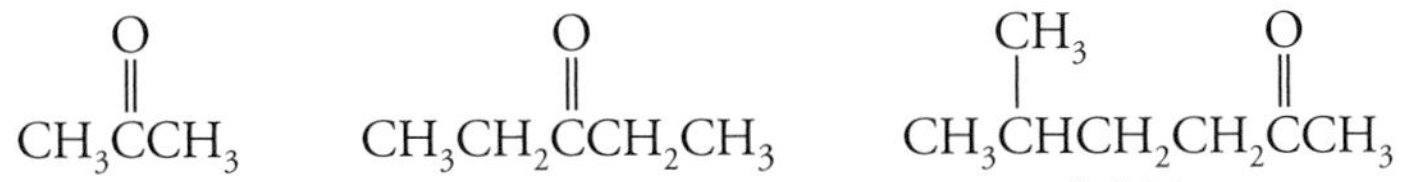

$$\underset{\substack{\text{propanone}\\\text{(acetone)}\\\text{BP 56.5 °C}}}{\mathrm{CH_3\overset{\overset{\displaystyle O}{\|}}{C}CH_3}} \qquad \underset{\substack{\text{3-pentanone}\\\text{BP 101.5 °C}}}{\mathrm{CH_3CH_2\overset{\overset{\displaystyle O}{\|}}{C}CH_2CH_3}} \qquad \underset{\substack{\text{5-methyl-2-hexanone}\\\text{(not 2-methyl-5-hexanone)}\\\text{BP 145 °C}}}{\mathrm{CH_3\overset{\overset{\displaystyle CH_3}{|}}{C}HCH_2CH_2\overset{\overset{\displaystyle O}{\|}}{C}CH_3}}$$

We need not write "2-propanone," because if the carbonyl carbon is anywhere else in a three-carbon chain, the compound is the *aldehyde*, propanal.

Aldehydes and ketones can be made to add hydrogen

Hydrogen is capable of adding across the double bond of the carbonyl group in both aldehydes and ketones. The reaction is just like the addition of hydrogen across the double bond of an alkene and takes place under roughly the same conditions, namely, with a metal catalyst, heat, and pressure. The reaction is called either *hydrogenation* or *reduction*. For example,

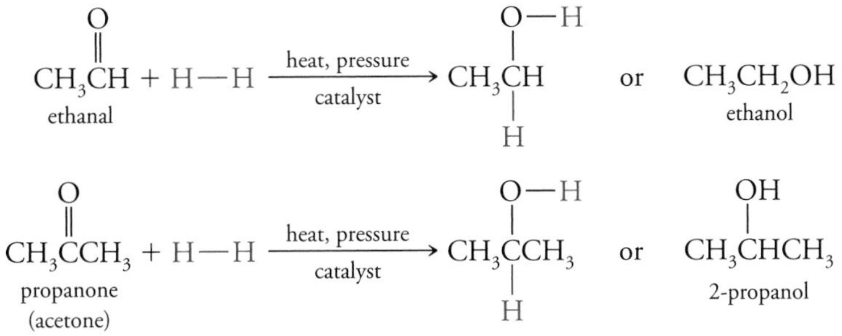

$$\underset{\text{ethanal}}{\mathrm{CH_3\overset{\overset{\displaystyle O}{\|}}{C}H}} + \mathrm{H{-}H} \xrightarrow[\text{catalyst}]{\text{heat, pressure}} \mathrm{CH_3\underset{\underset{\displaystyle H}{|}}{\overset{\overset{\displaystyle O{-}H}{|}}{C}}H} \quad \text{or} \quad \underset{\text{ethanol}}{\mathrm{CH_3CH_2OH}}$$

$$\underset{\substack{\text{propanone}\\\text{(acetone)}}}{\mathrm{CH_3\overset{\overset{\displaystyle O}{\|}}{C}CH_3}} + \mathrm{H{-}H} \xrightarrow[\text{catalyst}]{\text{heat, pressure}} \mathrm{CH_3\underset{\underset{\displaystyle H}{|}}{\overset{\overset{\displaystyle O{-}H}{|}}{C}}CH_3} \quad \text{or} \quad \underset{\text{2-propanol}}{\mathrm{CH_3\overset{\overset{\displaystyle OH}{|}}{C}HCH_3}}$$

The H atoms take up positions at opposite ends of the carbonyl group's double bond, which then becomes a single bond holding an OH group.

Aldehydes are easily oxidized

Aldehydes and ketones are in separate families because of their remarkably different behavior toward oxidizing agents. As we noted on page 914, aldehydes are easily oxidized, but ketones strongly resist oxidation. Even in storage in a bottle, aldehydes are slowly oxidized by the oxygen of the air trapped in the bottle.

Compounds containing a carboxyl group are acids

A **carboxylic acid** carries an OH on the carbon of the carbonyl group.

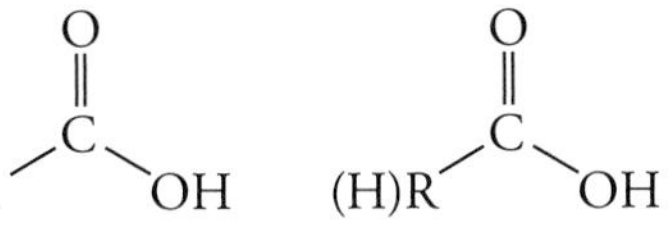

$$\underset{\substack{\text{carboxyl}\\\text{group}}}{-\overset{\overset{\displaystyle O}{\|}}{C}-\mathrm{OH}} \qquad \underset{\substack{\text{carboxylic}\\\text{acids}}}{\mathrm{(H)R}-\overset{\overset{\displaystyle O}{\|}}{C}-\mathrm{OH}}$$

In condensed formulas the **carboxyl group** is often written as CO_2H or COOH.

The name ending of the IUPAC names of carboxylic acids is *-oic acid.* The parent chain must be the longest that includes the carboxyl carbon, which is numbered position 1. The name of the hydrocarbon with the same number of carbons as the parent is then changed by replacing the terminal *-e* with *-oic acid.* For example,

$$\underset{\substack{\text{methanoic acid}\\\text{BP 101 °C}}}{\mathrm{HCO_2H}} \qquad \underset{\substack{\text{ethanoic acid}\\\text{BP 118 °C}}}{\mathrm{CH_3CO_2H}} \qquad \underset{\substack{\text{3-methylbutanoic acid}\\\text{BP 176 °C}}}{\mathrm{CH_3\overset{\overset{\displaystyle CH_3}{|}}{C}HCH_2CO_2H}}$$

Because carboxylic acids have both a lone oxygen and an OH group, their molecules strongly hydrogen bond to each other. Their high boiling points, relative to alcohols of comparable molecular size, reflect this.

The carboxyl group is a weakly acidic group, causing aqueous solutions of the compounds to be weakly acidic. In fact, we've used both formic acid and acetic acid in previous discussions of weak acids. All carboxylic acids, both water soluble and water insoluble, neutralize such bases as the hydroxide, bicarbonate, and carbonate ions. The general equation for the reaction with OH^- is

$$RCO_2H + OH^- \xrightarrow{H_2O} RCO_2^- + H_2O$$

The carboxyl group is present in all of the building blocks of proteins, the amino acids, which are discussed further on page 937.

Esters are derivatives of carboxylic acids

A *derivative* of a carboxylic acid is a compound that can be made from the acid, or which can be changed to the acid by hydrolysis.

Carboxylic acids are used to synthesize two important kinds of derivatives, *esters* and *amides*. In **esters,** the OH of the carboxyl group is replaced by OR. (The H in parentheses means the group attached to the C of the carboxyl group can either be R or H.)

$$\underset{\text{esters}}{(H)R\overset{\overset{O}{\|}}{C}OR'} \quad \text{or} \quad (H)RCO_2R' \quad \text{for example,} \quad \underset{\substack{\text{octyl ethanoate} \\ \text{(fragrance of oranges)}}}{CH_3\overset{\overset{O}{\|}}{C}O(CH_2)_7CH_3}$$

Esters are responsible for many pleasant fragrances in nature.

Ester	Aroma
$HCO_2CH_2CH_3$	Rum
$HCO_2CH_2CH(CH_3)_2$	Raspberry
$CH_3CO_2(CH_2)_4CH_3$	Banana
$CH_3CO_2(CH_2)_2CH(CH_3)_2$	Pear
$CH_3CO_2(CH_2)_7CH_3$	Orange
$CH_3(CH_2)_2CO_2CH_2CH_3$	Pineapple
$CH_3(CH_2)_2CO_2(CH_2)_4CH_3$	Apricot

The IUPAC name of an ester begins with the name of the alkyl group attached to the O atom. This is followed by a separate word, one taken from the name of the parent carboxylic acid but altered by changing *-ic acid* to *-ate.* For example,

$$\underset{\substack{\text{methyl methanoate} \\ \text{BP 31.5 °C}}}{HCO_2CH_3} \qquad \underset{\substack{\text{ethyl ethanoate} \\ \text{BP 77 °C}}}{CH_3CO_2CH_2CH_3} \qquad \underset{\substack{\text{isopropyl 3-methylbutanoate} \\ \text{BP 142 °C}}}{CH_3\overset{\overset{CH_3}{|}}{C}HCH_2CO_2\overset{\overset{CH_3}{|}}{C}HCH_3}$$

One way to prepare an ester is to heat a solution of the parent carboxylic acid and the alcohol in the presence of an acid catalyst. (We'll not go into the details of how this happens.) The following kind of equilibrium forms, but a substantial stoichiometric excess of the alcohol (usually the less expensive reactant) is commonly used to drive the position of the equilibrium toward the ester.

$$\underset{\text{carboxylic acid}}{R\overset{\overset{O}{\|}}{C}OH} + \underset{\text{alcohol}}{HOR'} \underset{\text{heat}}{\overset{H^+ \text{ catalyst}}{\rightleftharpoons}} \underset{\text{ester}}{R\overset{\overset{O}{\|}}{C}OR'} + H_2O$$

For example,

$$\underset{\substack{\text{butanoic acid} \\ \text{(BP 166 °C)}}}{CH_3CH_2CH_2\overset{\overset{O}{\|}}{C}OH} + \underset{\substack{\text{ethanol} \\ \text{(BP 78.5 °C)}}}{HOCH_2CH_3} \underset{\text{heat}}{\overset{H^+ \text{ catalyst}}{\rightleftharpoons}} \underset{\substack{\text{ethyl butanoate} \\ \text{(BP 120 °C)} \\ \text{(fragrance of pineapple)}}}{CH_3CH_2CH_2\overset{\overset{O}{\|}}{C}OCH_2CH_3} + H_2O$$

Esters can be split by reaction with water or base

An ester is hydrolyzed to its parent acid and alcohol when the ester is heated together with a stoichiometric excess of water (plus an acid catalyst). The identical equilibrium as shown above forms, but in ester hydrolysis water is in excess, so the equilibrium shifts to the left to favor the carboxylic acid and alcohol, another illustration of Le Châtelier's principle.

Esters are also split apart by the action of aqueous base, only now the carboxylic acid emerges not as the free acid but as its anion. The reaction is called ester **saponification.** We may illustrate it by the action of aqueous sodium hydroxide on a simple ester, ethyl acetate.

$$\underset{\substack{\text{ethyl ethanoate} \\ \text{(ethyl acetate)}}}{CH_3\overset{\overset{O}{\|}}{C}OCH_2CH_3(aq)} + NaOH(aq) \xrightarrow{\text{heat}} \underset{\substack{\text{ethanoate ion} \\ \text{(acetate ion)}}}{CH_3\overset{\overset{O}{\|}}{C}O^-(aq)} + Na^+(aq) + \underset{\text{ethanol}}{HOCH_2CH_3(aq)}$$

Ester groups abound among the molecules of the fats and oils in our diets. The hydrolysis of their ester groups occurs when we digest them, with an enzyme as the catalyst, not a strong acid. Because this digestion occurs in a region of the intestinal tract where the fluids are slightly basic, the anions of carboxylic acids form.

Amides are amine derivatives of carboxylic acids

Carboxylic acids can also be converted to *amides*, a functional group found in proteins. In **amides,** the OH of the carboxyl group is replaced by trivalent nitrogen, which may also be bonded to any combination of H atoms or hydrocarbon groups.

$$\underset{\text{simple amides}}{\text{(H)R}\overset{\overset{\displaystyle O}{\|}}{\text{C}}\text{NH}_2} \quad \text{or} \quad \text{(H)RCONH}_2, \quad \text{for example,} \quad \underset{\substack{\text{ethanamide}\\\text{(acetamide)}}}{\text{CH}_3\overset{\overset{\displaystyle O}{\|}}{\text{C}}\text{NH}_2}$$

The *simple amides* are those in which the nitrogen bears no hydrocarbon groups, only 2 H atoms. In the place of either or both of these H atoms, however, there can be a hydrocarbon group, and the resulting substance is still in the amide family.

The IUPAC names of the simple amides are devised by first writing the name of the parent carboxylic acid. Then its ending, *-oic acid,* is replaced by *-amide.* For example:

$$\underset{\text{propanamide}}{\text{CH}_3\text{CH}_2\text{CONH}_2} \qquad \underset{\text{pentanamide}}{\text{CH}_3\text{CH}_2\text{CH}_2\text{CH}_2\text{CONH}_2} \qquad \underset{\text{4-methylpentanamide}}{\text{CH}_3\overset{\overset{\displaystyle \text{CH}_3}{|}}{\text{C}}\text{HCH}_2\text{CH}_2\text{CONH}_2}$$

One of the ways to prepare simple amides parallels that of the synthesis of esters, that is, by heating a mixture of the carboxylic acid and an excess of ammonia.

In general:

$$\underset{\text{carboxylic acid}}{\text{R}\overset{\overset{\displaystyle O}{\|}}{\text{C}}\text{OH}} + \underset{\text{ammonia}}{\text{H}{-}\text{NH}_2} \xrightarrow{\text{heat}} \underset{\substack{\text{simple}\\\text{amide}}}{\text{R}\overset{\overset{\displaystyle O}{\|}}{\text{C}}\text{NH}_2} + \text{H}_2\text{O}$$

An example:

$$\underset{\text{ethanoic acid}}{\text{CH}_3\overset{\overset{\displaystyle O}{\|}}{\text{C}}\text{OH}} + \underset{\text{ammonia}}{\text{H}{-}\text{NH}_2} \xrightarrow{\text{heat}} \underset{\text{ethanamide}}{\text{CH}_3\overset{\overset{\displaystyle O}{\|}}{\text{C}}\text{NH}_2} + \text{H}_2\text{O}$$

Amides, like esters, can be hydrolyzed. When simple amides are heated with water, they change back to their parent carboxylic acids and ammonia. Both strong acids and strong bases promote the reaction. As the following equations show, the reaction is the reverse of the formation of an amide.

In general:

$$\underset{\text{simple amide}}{\text{R}\overset{\overset{\displaystyle O}{\|}}{\text{C}}\text{NH}_2} + \text{H}{-}\text{OH} \xrightarrow{\text{heat}} \underset{\text{carboxylic acid}}{\text{R}\overset{\overset{\displaystyle O}{\|}}{\text{C}}\text{OH}} + \text{NH}_3$$

An example:

$$\underset{\text{ethanamide}}{\text{CH}_3\overset{\overset{\displaystyle O}{\|}}{\text{C}}\text{NH}_2} + \text{H}{-}\text{OH} \xrightarrow{\text{heat}} \underset{\text{ethanoic acid}}{\text{CH}_3\overset{\overset{\displaystyle O}{\|}}{\text{C}}\text{OH}} + \text{NH}_3$$

Urea is an important nitrogen fertilizer because it reacts with soil moisture to release ammonia (and carbon dioxide).

$$\underset{\text{urea}}{\text{NH}_2\overset{\overset{\displaystyle O}{\|}}{\text{C}}\text{NH}_2}$$

Amides are not basic like amines

Despite the presence of the NH_2 group in simple amides, the amides are not Brønsted bases like amines or ammonia. This is can be understood by examining the following two resonance structures.

$$\mathrm{R{-}\overset{\overset{\ddot{O}:}{\|}}{C}{-}\ddot{N}{-}H} \longleftrightarrow \mathrm{R{-}\overset{\overset{\ominus:\ddot{O}:}{|}}{C}{=}\overset{\oplus}{N}{-}H}$$

(with H on N in each structure)

Structure 1   Structure 2

Effectively, the lone pair on the nitrogen in Structure 1 becomes partially delocalized onto the oxygen, as shown in Structure 2. This makes the lone pair on the amide nitrogen less available for donation to an H^+ than the corresponding lone pair on the nitrogen of an amine. As a result, the amide nitrogen has little tendency to acquire a proton, so amides are neutral compounds in an acid–base sense.

Practice Exercise 5: Complete the following equation by drawing structural formulas for the products.

$$H_2N{-}CH_2CH_2CH_3\,(aq) + CH_3CH_2{-}\overset{\overset{O}{\|}}{C}{-}OH\,(aq) \longrightarrow$$

(Hint: Identify the functional groups and their properties.)

Practice Exercise 6: Write the structural formula(s) for the principal organic product(s) in the following reactions:

(a) $CH_3CH_2{-}\overset{\overset{O}{\|}}{C}{-}OCH(CH_3)_2 \xrightarrow{OH^-(aq)}$

(b) $CH_3CH_2CH_2OH \xrightarrow[\text{conc.}]{H_2SO_4}$

22.6 POLYMERS ARE COMPOSED OF MANY REPEATING MOLECULAR UNITS

Lightweight synthetic fabrics have made it easier to pack strong tents into remote regions (here, the Wasatch Mountains of Utah). *(Richard Price/Taxi/Getty Images.)*

Nearly all of the compounds that we have studied so far have relatively low molecular masses. Both in nature and in the world of synthetics, however, many substances consist of **macromolecules** made up of hundreds or even thousands of atoms. Synthetics made of macromolecules are examples of how chemists have been able to take very ordinary substances in nature, like coal, oil, air, and water, and make new materials, never seen before, with useful applications. Recording tapes, skis, composites in recreational vehicles and their tires, backpacking gear, and all sorts of other recreational equipment derive strength from macromolecules. Materials that are made into thread and cloth consist of macromolecules, and when paints cure, some of their molecules change into other molecules of enormous sizes.

In nature, substances with macromolecules are almost everywhere you look. Trees and anything made of wood, for example, derive their strength from lignins and cellulose, both consisting of enormous molecules that overlap and intertwine. The proteins and the DNA found in our bodies are also huge molecules. Macromolecules of biological origin will be discussed in Section 22.7; in this section we will look at synthetic macromolecules and some of their uses.

Polymers have structural order

Some macromolecular substances have more structural order than others. A **polymer,** for example, is a macromolecular substance all of whose molecules have a small characteristic structural feature that repeats itself over and over. An example is polypropylene, a polymer with many uses, including dishwasher-safe food containers, indoor–outdoor carpeting, and artificial turf (Figure 22.3). The molecules of polypropylene have the following system.

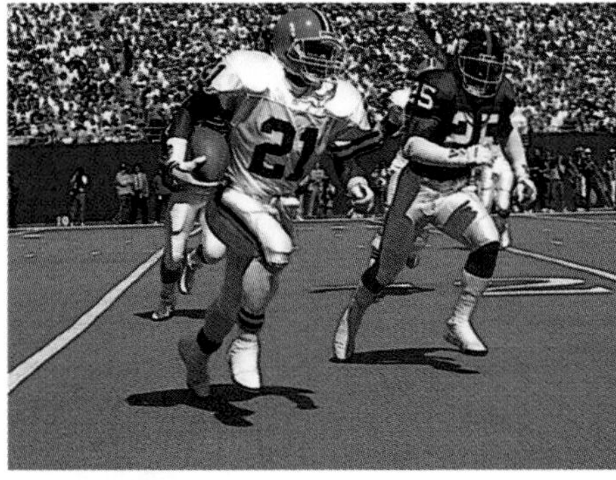

FIG. 22.3 Artificial turf on a football field is made of polypropylene fibers. *(Robert Tringali Jr./Sports Chrome Inc.)*

$$\text{etc.}-CH_2-\underset{\displaystyle CH_3}{\overset{|}{CH}}-CH_2-\underset{\displaystyle CH_3}{\overset{|}{CH}}-CH_2-\underset{\displaystyle CH_3}{\overset{|}{CH}}-CH_2-\underset{\displaystyle CH_3}{\overset{|}{CH}}-CH_2-\underset{\displaystyle CH_3}{\overset{|}{CH}}-\text{etc.}$$

polypropylene

Notice that the polymer consists of a long carbon chain (the polymer's **backbone**) with CH_3 groups attached at periodic intervals.[2]

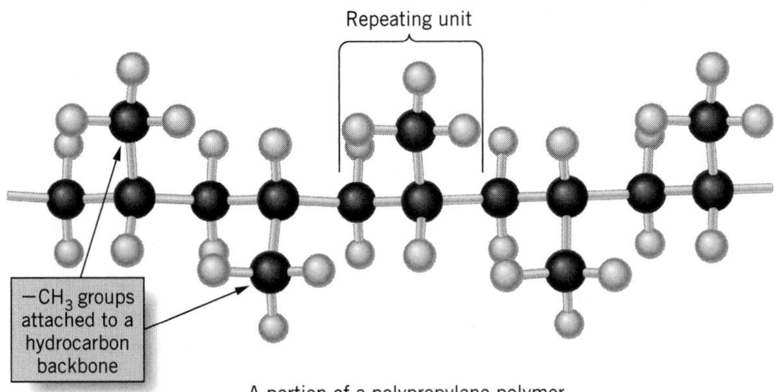

A portion of a polypropylene polymer

If you study this structure, you can see that one structural unit occurs repeatedly (actually thousands of times). In fact, the structure of a polymer is usually represented by the use of only its repeating unit, enclosed in parentheses, with a subscript *n* standing for several thousand units.

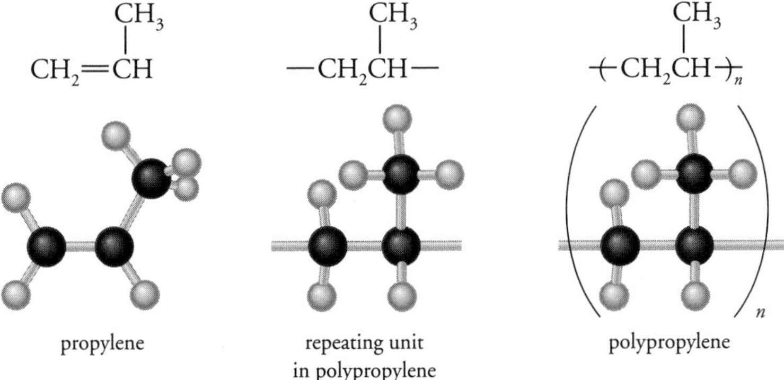

The value of *n* is not a constant for every molecule in a given polymer sample. A polymer, therefore, does not consist of molecules identical in *size*, just identical in *kind*; they have the same repeating unit. Notice that despite the *-ene* ending (which is used in naming alkenes), "polypropylene" has no double bonds. The polymer is named after its starting material, propylene.

The repeating unit of a polymer is contributed by a chemical raw material called a **monomer**. Thus propylene is the monomer for polypropylene. The reaction that makes a polymer out of a monomer is called **polymerization,** and the verb is "to polymerize." Most (but not all) useful polymers are formed from monomers that are considered to be organic compounds.

[2] The drawing of the molecule shows one way the CH_3 groups can be oriented relative to the backbone and each other. Other orientations are also possible, and they affect the physical properties of the polymer.

Some polymers form by addition of monomer units

There are basically two ways that monomers become joined to form polymers. One path involves the simple addition of one monomer unit to another, a process that continues over and over until a very long chain of monomer units is produced. Polymers formed by this process are called **addition polymers.** Polypropylene, discussed above, is an example. A simpler example is *polyethylene*, formed from ethylene ($CH_2{=}CH_2$) monomer units. Under the right conditions and with the aid of a substance called an **initiator**, a pair of electrons in the carbon–carbon double bond of ethylene becomes unpaired. The initiator binds to one carbon, leaving an unpaired electron on the other carbon. The result is a very reactive substance called a *free radical,* which can attack the double bond of another ethylene. In the attack, one of the electron pairs of the double bond becomes unpaired. One of the electrons becomes shared with the unpaired electron of the free radical, forming a bond that joins the two hydrocarbon units together. The other unpaired electron moves to the end of the chain, as illustrated below.

As the free radical approaches the ethylene molecule...

...electrons in one of the bonds unpair and move to opposite ends of the molecule.

initiator—CH₂—CH₂· + H₂C=CH₂ ⟶ initiator—CH₂—CH₂—CH₂—CH₂·

A bond forms between the two hydrocarbon units and the unpaired electron moves to the end of the chain

This process is repeated over and over as a long hydrocarbon chain grows. Eventually the chain becomes terminated and the result is a polyethylene molecule. The molecule is so large that the initiator, which is still present at one end of the chain, is an insignificant part of the whole, so in writing the structure of the polymer, the initiator is generally ignored.

Actually, after ethylene has polymerized, the repeating unit in polyethylene is simply CH_2.

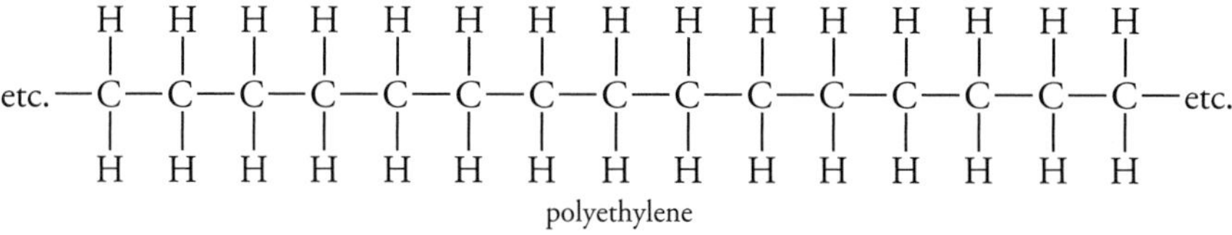

polyethylene

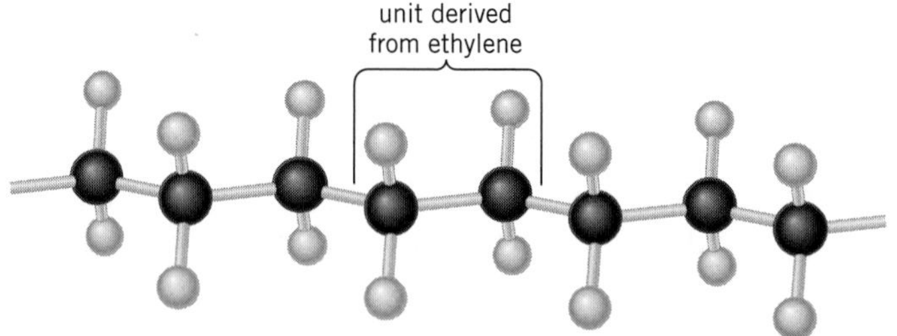

A portion of a polyethylene molecule. Although formed from a monomer with the formula C_2H_4, the actual repeating unit is CH_2.

The process by which the polymer forms has a significant influence on its ultimate structure. For example, the least expensive method for making polyethylene leads to **branching,** which means that polymer chains grow off the main backbone of the molecule as it grows longer (Figure 22.4). Other, more expensive procedures produce molecules without branching, and as we will discuss later, this has a significant effect on the properties of the polymer.

$-CH_2-CH_2-CH_2-CH-CH_2-CH_2-CH-CH_2-CH_2-CH_2-CH_2-CH-CH_2-$

Branches

Main polymer chain

FIG. 22.4 Branching of the polymer chain in polyethylene. Segments of polyethylene grow off the main polymer backbone.

FIG. 22.5 Styrofoam. Made of polystyrene, Styrofoam is widely used as an insulation in construction. *(Michael Ventura/Bruce Coleman, Inc.)*

In addition to polyethylene and polypropylene, another very common addition polymer, called **polystyrene,** is formed by polymerizing styrene,

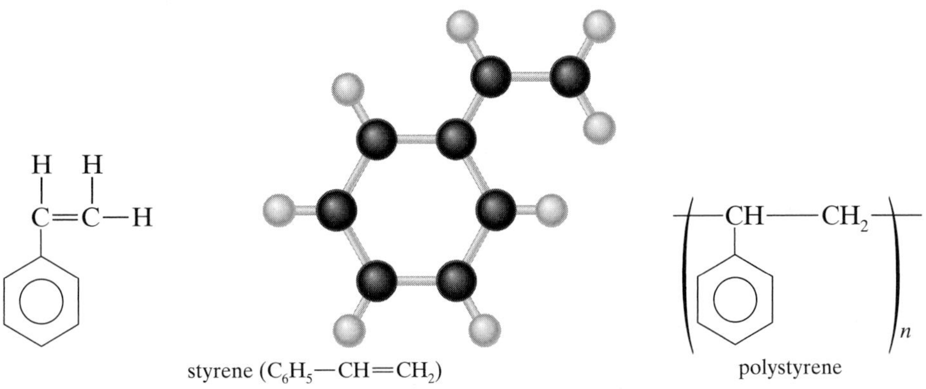

styrene ($C_6H_5-CH=CH_2$) polystyrene

Notice that the monomer is similar to propylene, but with a benzene ring in place of the methyl group ($-CH_3$). Therefore, in the polymer we find C_6H_5- attached to every other atom in the hydrocarbon backbone. Polystyrene has almost as many uses as polyethylene. It's used to make clear plastic drinking glasses, molded car parts, and housings of things like computers and kitchen appliances. Sometimes a gas, like carbon dioxide, is blown through molten polystyrene as it is molded into articles. As the hot liquid congeals, tiny pockets of gas are trapped, and the product is a foamed plastic, like the familiar Styrofoam cups or insulation materials (Figure 22.5).

Hundreds of substances similar to ethylene, propylene, and styrene, as well as their halogen derivatives, have been tested as monomers, and Table 22.3 gives some examples.

A "halogen derivative" is a compound in which one or more halogen atoms substitute for hydrogens in a parent molecule. Thus, CH_3Cl is a halogen derivative of CH_4.

EXAMPLE 22.3
Writing the Formula for an Addition Polymer

Use the information in Table 22.3 to write the structure of the polymer poly(vinyl chloride) showing three repeat units. Write the general formula for the polymer.

ANALYSIS: The polymer we're dealing with is an addition polymer, so we can anticipate that the entire $CH_2=CHCl$ molecule will repeat over and over. The polymer will be formed by opening the double bond, as we saw for ethylene.

SOLUTION: When the double bond opens, bonds to two other monomer units will form.

```
                     H  Cl
                     |  |
other monomer unit—C—C—other monomer unit
                     |  |
                     H  H
```

We need to attach three repeating units to have the answer to the problem.

$$
\begin{array}{cccccccccccc}
 & \text{H} & & \text{Cl} & & \text{H} & & \text{Cl} & & \text{H} & & \text{Cl} \\
 & | & & | & & | & & | & & | & & | \\
- & \text{C} & - & \text{C} & - & \text{C} & - & \text{C} & - & \text{C} & - & \text{C} & - \\
 & | & & | & & | & & | & & | & & | \\
 & \text{H} & & \text{H} & & \text{H} & & \text{H} & & \text{H} & & \text{H}
\end{array}
$$

$\underbrace{\hspace{3em}}_{\text{repeating unit}}$

The general formula for the polymer should indicate the repeating unit occurring n times.

$$
\left(\begin{array}{ccc}
\text{H} & & \text{Cl} \\
| & & | \\
\text{C} & - & \text{C} \\
| & & | \\
\text{H} & & \text{H}
\end{array}\right)_n
$$

IS THE ANSWER REASONABLE? There's not much to check here, other than to be sure that the repeating unit has the same molecular formula as the monomer, which it does.

Practice Exercise 7: Suppose 2-butene were polymerized. Make a sketch showing three repeating units of the monomer in the polymer. (Hint: What is the structure of 2-butene?)

Practice Exercise 8: Write a formula showing three repeating units of the monomer in the polymer Teflon.

TABLE 22.3 Some Addition Polymers Formed from Compounds Related to Ethylene, $CH_2{=}CH_2$

Polymer	Monomer	Uses
Polyethylene	$CH_2{=}CH_2$	Grocery bags, bottles, children's toys, bulletproof vests
Polypropylene	$CH_2{=}CH-CH_3$ (CH₃ on the CH)	Dishwasher-safe plastic kitchenware, indoor–outdoor carpeting, rope
Polystyrene	$C_6H_5-CH{=}CH_2$	Plastic cups, toys, housings for kitchen appliances, Styrofoam insulation.
Poly(vinyl chloride) (PVC)	$CH_2{=}CHCl$	Insulation, credit cards, vinyl siding for houses, bottles, plastic pipe
Poly(tetrafluoroethylene) (Teflon)	$F_2C{=}CF_2$	Nonstick surfaces on cookware, valves
Poly(vinyl acetate) (PVA)	$CH_2{=}C(H)-O-C({=}O)-CH_3$	Latex paint, coatings, glue, molded items
Poly(methyl methacrylate) (Lucite)	$CH_2{=}C(CH_3)-C({=}O)-O-CH_3$	Shatter-resistant windows, coatings, acrylic paints, molded items

Polymers can be formed by condensation reactions

The second way that monomer units can combine to form a polymer is by a process called **condensation,** in which a small molecule is eliminated when the two monomer units become joined. In a simplified way, we can diagram this as follows.

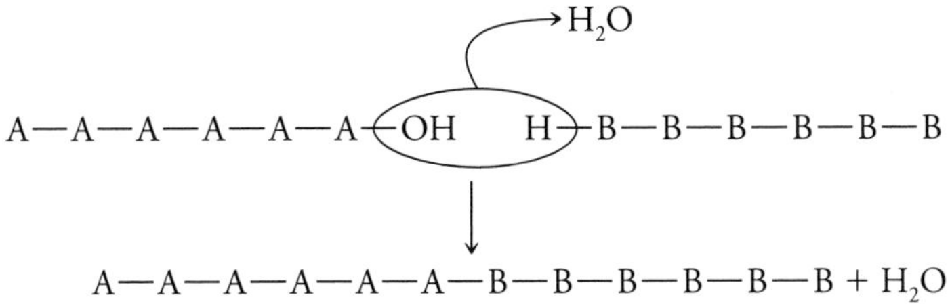

$$\text{A—A—A—A—A—A—OH} \quad \text{H—B—B—B—B—B—B} \;\rightarrow\; \text{A—A—A—A—A—A—B—B—B—B—B—B} + H_2O$$

In this example, an OH group from one molecule combines with an H from another, forming a water molecule. At the same time, the two molecules A and B become joined by a covalent bond. If this can be made to happen at both ends of A and B, long chains are formed and a **condensation polymer** is the result.

The two most familiar condensation polymers are nylon and polyesters. Nylon is formed by combining two different compounds, so it's considered a **copolymer.** The first nylon to be manufactured is called **nylon 6,6** because it forms by combining two compounds each with six carbon atoms.

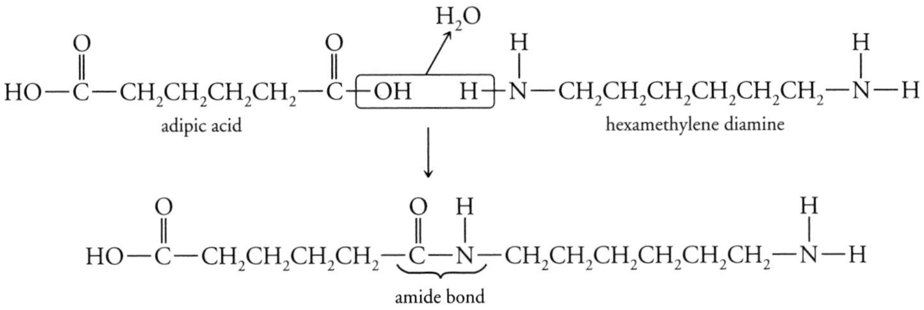

$$\underset{\text{adipic acid}}{HO{-}\overset{O}{\overset{\|}{C}}{-}CH_2CH_2CH_2CH_2{-}\overset{O}{\overset{\|}{C}}{-}OH} \quad \underset{\text{hexamethylene diamine}}{H{-}\overset{H}{\overset{|}{N}}{-}CH_2CH_2CH_2CH_2CH_2CH_2{-}\overset{H}{\overset{|}{N}}{-}H} \xrightarrow{-H_2O}$$

$$HO{-}\overset{O}{\overset{\|}{C}}{-}CH_2CH_2CH_2CH_2{-}\underbrace{\overset{O}{\overset{\|}{C}}{-}\overset{H}{\overset{|}{N}}}_{\text{amide bond}}{-}CH_2CH_2CH_2CH_2CH_2CH_2{-}\overset{H}{\overset{|}{N}}{-}H$$

Notice that adipic acid contains two carboxyl groups (it's called a dicarboxylic acid). The other compound is an amine (diamine, actually, because it has two amine groups). By the elimination of water, the two molecules become joined by a linkage called an *amide bond.* This same linkage is found in proteins, including silk (a fiber nylon was invented to replace) and proteins found in our bodies.

The molecule above with the amide bond still has a carboxyl group on one end and an amine group on the other, so further condensation can occur, ultimately leading to the formation of nylon 6,6.

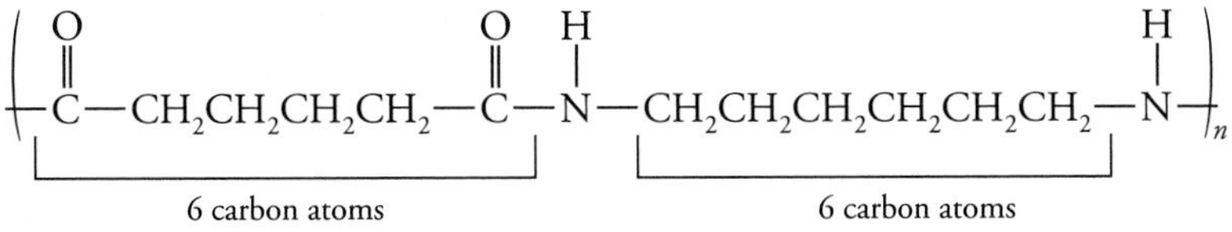

$$\left(\underbrace{\overset{O}{\overset{\|}{C}}{-}CH_2CH_2CH_2CH_2{-}\overset{O}{\overset{\|}{C}}}_{\text{6 carbon atoms}}{-}\underbrace{\overset{H}{\overset{|}{N}}{-}CH_2CH_2CH_2CH_2CH_2CH_2{-}\overset{H}{\overset{|}{N}}}_{\text{6 carbon atoms}}\right)_n$$

nylon 6,6

Nylon was invented in 1940 and became popular as a substitute for silk in women's stockings. It forms strong elastic fibers and is used to make fishing line as well as fibers found in all sorts of clothing and many other products.

An example of a polyester is shown below.

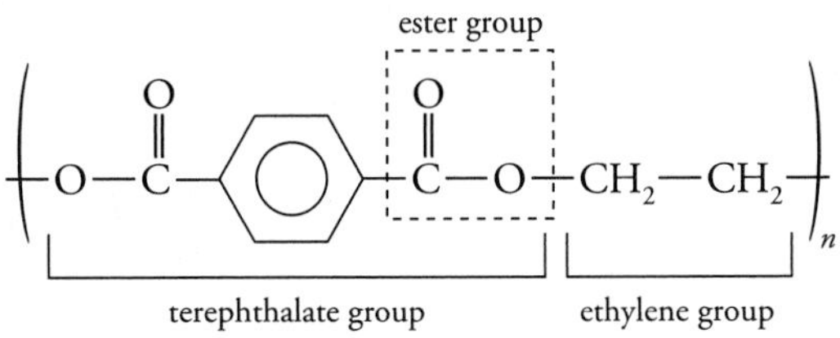

$$\left(\underbrace{O{-}\overset{O}{\overset{\|}{C}}{-}C_6H_4{-}\overbrace{\overset{O}{\overset{\|}{C}}{-}O}^{\text{ester group}}}_{\text{terephthalate group}}{-}\underbrace{CH_2{-}CH_2}_{\text{ethylene group}}\right)_n$$

poly(ethylene terephthalate), also known as PET

It is a copolymer made by condensation polymerization, the first step of which is

$$\underset{\text{dimethyl terephthalate}}{H_3C{-}O{-}\overset{\overset{\large O}{\|}}{C}{-}C_6H_4{-}\overset{\overset{\large O}{\|}}{C}{-}O{-}CH_3} + \underset{\text{ethylene glycol}}{HO{-}CH_2{-}CH_2{-}OH} \longrightarrow H_3C{-}O{-}\overset{\overset{\large O}{\|}}{C}{-}C_6H_4{-}\overset{\overset{\large O}{\|}}{C}{-}O{-}CH_2{-}CH_2{-}OH + CH_3OH$$

Notice that this time the small molecule that's displaced is CH_3OH, methyl alcohol. Continued polymerization ultimately leads to the PET polymer shown above. You probably have heard of this polymer because it also goes by the name *Dacron*.

A variety of starting materials can be used to form different polyesters with a range of properties. Their uses include fibers for fabrics, shatterproof plastic bottles for soft drinks, Mylar for making recording tapes and balloons that don't easily deflate, and shatterproof windows and eyeglasses. See Facets of Chemistry 22.1.

Cross-linking between polymer strands gives increased strength

When natural rubber latex was first discovered, it wasn't particularly useful. You couldn't make rubber tires out of it because when it got hot, the rubber became sticky and would melt; when it became cold, it became brittle and hard. In 1839, Charles Goodyear (of tire fame) discovered that adding sulfur to latex, and then heating it, drastically altered the properties of the rubber. He called his new product **vulcanized rubber.**

What happens when sulfur reacts with latex is that groups of sulfur atoms form bridges, called **cross-links,** between strands of the latex polymer (known technically as polyisoprene). This is shown in Figure 22.6. By linking the strands of polyisoprene together, they are no longer able to slip by each other when hot, so the rubber doesn't melt. The increased strength also prevents the polymer from becoming brittle and easily broken when cold. Cross-linking also gives the polymer a "memory," enabling it to snap back to its original shape when released after being stretched (a property you've experienced using rubber bands). The amount of sulfur used to vulcanize the latex also affects the properties of the finished product. If only a small amount of sulfur is used, the polymer is elastic. But as the amount of sulfur increases, so does the amount of cross-linking, and the product becomes harder and less resilient.

Notice that the polyisoprene in Figure 22.6 has lots of bends and turns in it. When you stretch rubber, you tend to straighten out the polymer strands, but when you release it, the strands snap back to their original shapes.

Cross-linking provides strength to a variety of polymers you may be familiar with. Formica (used in kitchen countertops), epoxy resins (in epoxy glues), and polycarbonates (which have a polyester-like structure) made strong for use in shatterproof eyeglasses are examples. Even the material used to make soft contact lenses is a cross-linked polymer that's capable of absorbing lots of water.

Polymer crystallinity affects physical properties

Beyond chemical stability, physical properties are the features of polymers most sought after. Desirable properties of Teflon, for example, are its chemical inertness and its slipperiness toward just about anything. Nylon isn't eaten by moths—a chemical property, in fact—but its superior strength and its ability to be made into fibers and fabrics of great beauty are what make nylon valuable. Dacron (a polyester) resists mildewing and its fibers are not weakened by mildew like those of cotton. Dacron also has greater strength with lower mass than cotton, and Dacron fibers do not stretch as much, which are properties that account for Dacron's use for making sails.

In many ways, the physical properties of a polymer are related to how the individual polymer strands are able to pack in the solid. For example, earlier we noted that the least expensive way of making polyethylene yields a product that has branching. The branches

(a)

(b)

crosslink to another polymer chain

sulfur crosslinks

crosslink to another polymer chain

FIG. 22.6 Cross-linking of polymer chains in rubber by reaction with sulfur. (*a*) Two strands of polyisoprene molecules. (*b*) Sulfur reacts by opening double bonds in the polymer molecules and forming bridges between adjacent strands.

prevent the molecules from lining up in an orderly fashion, so the molecules twist and intertwine to give an essentially amorphous solid (which means it lacks the kind of order found in crystalline solids). See Figure 22.7*a*. This amorphous product is called *low density polyethylene* (*LDPE*); the polymer molecules have a relatively low molecular mass and the solid has little structural strength. LDPE is the kind of polyethylene used to make the plastic bags grocery stores use to pack your purchases.

HDPE molecules contain approximately 30,000 CH_2 units linked end to end!

Following different methods, polyethylene can be made to form without branching and with molecular masses ranging from 200,000 to 500,000 u. This polymer is called *linear* polyethylene or *high density polyethylene* (*HDPE*). In HDPE, the polymer strands are able to line up alongside each other to produce a large degree of order (and therefore, crystallinity), as illustrated in Figure 22.7*b*. This enables the molecules to form fibers easily, and because the molecules are large and packed so well, the London forces between them are very strong. The result is a strong, tough fiber. DuPont's Tyvek, for example, is made from thin, crystalline HDPE polyethylene fibers randomly oriented and pressed together into a material resembling paper. It is lightweight, strong, and resists water, tears, punctures, and abrasion. Federal Express has been using it for years for envelopes, and builders use it to wrap new construction to prevent water and air intrusion, thereby lowering heating and cooling costs. Tyvek is also made into limited-use protective clothing for use in hazardous environments.

A new house under construction is wrapped with a Tyvek fabric to prevent water and air intrusion, thereby lowering heating and cooling costs. *(DuPont Tyvek Weatherization Systems.)*

UHMWPE molecules contain between 200,000 and 400,000 CH_2 units attached end to end!

Under the right conditions, linear polyethylene molecules can be made extremely long, yielding *ultrahigh molecular weight polyethylene* (*UHMWPE*) with molecules having molecular masses of three to six million. The fibers produced from this polymer are so strong that they are used to make bulletproof vests! Honeywell is producing an oriented polyethylene polymer they call Spectra, which forms flexible fibers that can be woven into a strong, cut-resistant fabric. It is used to make thin, lightweight liners for surgical gloves that resist cuts by scalpels, industrial work gloves, and even sails for sailboats (see Figure 22.8). Mixed with other plastics, it can be molded into strong rigid forms such as helmets for military or sporting applications.

Other good fiber-forming polymers also have long molecules with shapes that permit strong interactions between the individual polymer strands. Nylon, for example, possesses polar carbonyl groups, $>C{=}O$, and $N{-}H$ bonds that form strong hydrogen bonds between the individual molecules.

A carbon atom is at each vertex.

hydrogen bonds

Three strands of nylon 6,6 bound to each other by hydrogen bonds.

(*a*) Two branched polymer chains become twisted with little order.

(*b*) Polymer chains are able to align, producing a tightly packed structure with a large degree of crystallinity.

FIG. 22.7 Amorphous and crystalline polyethylene. (*a*) When branching occurs in LDPE, the polymer strands are not able to become aligned and an amorphous product results. (*b*) Linear HDPE has a high degree of crystallinity, which makes for excellent strong fibers.

Kevlar, another type of nylon, also forms strong hydrogen bonds between polymer molecules and is very crystalline. Its strong fibers are also used to make bulletproof vests. Because the fibers are so strong, they are also used to make thin yet strong hulls of racing boats. This lightweight construction improves speed and performance without sacrificing safety.

22.7 Most Biochemicals Are Organic Compounds

Biochemistry is the systematic study of the chemicals of living systems, their organization into cells, and their chemical interactions. Biochemicals have no life in themselves as isolated compounds, yet life has a molecular basis. Only when chemicals are organized into cells or tissues can interactions occur that enable tissue repair, cell reproduction, the generation of energy, the removal of wastes, and the management of a number of other

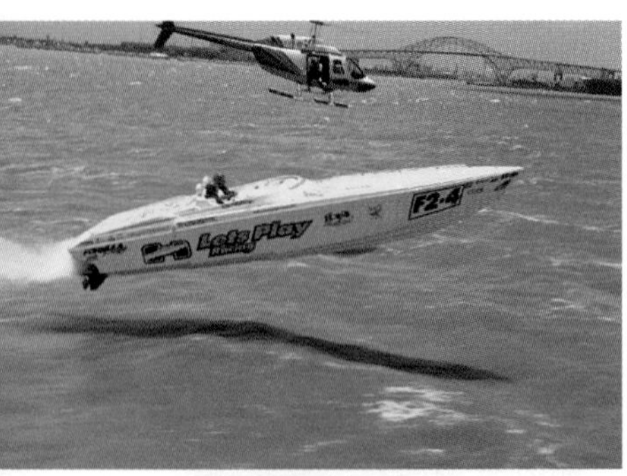

High-speed racing boats often are built with hulls made of Kevlar. Because of the polymer's high strength, the hulls can be made thin, thereby reducing weight and increasing speed. *(Tom Newby/Courtesy of Formula Powerboats.)*

FIG. 22.8 Spectra fibers. Strong, lightweight sails made of Spectra fibers propelled Brad Van Liew to victory in the 2002–2003 "Around Alone" around-the-world yacht race. Van Liew sailed 31,094 miles in seven months to win the race. *(Photo courtesy Brad Van Liew and Tommy Hilfiger. Photo by Billy Black.)*

FACETS OF CHEMISTRY

Bioplastics and Biodegradable Polymers

Almost all the polymers we find incorporated in fabrics and consumer plastics are made from materials derived from petroleum. This leads to two major problems. First, there is only a finite amount of petroleum available on earth, so when we manufacture plastics, we are consuming a precious commodity that can't be renewed. Second, synthetic polymers and plastics are not readily attacked by microorganisms, so when placed in landfills or scattered into the environment, they do not degrade and pose a continuing pollution problem. But changes are in the wind.

DuPont, one of the world's leading chemical producers, is manufacturing a polyester polymer called *Sorona*, a versatile polymer with properties that make it excellent for apparel and upholstery fabrics (Figure 1). It is made by condensation polymerization of propanediol with terephthalic acid.

$$\underset{\text{terephthalic acid}}{HO-\overset{\overset{\displaystyle O}{\|}}{C}-C_6H_4-\overset{\overset{\displaystyle O}{\|}}{C}-\boxed{OH}}\quad \underset{\text{propanediol}}{\boxed{H}O-CH_2-CH_2-CH_2-OH} \xrightarrow{-H_2O}$$

$$\underset{\text{Sorona polymer}}{\left(-O-\overset{\overset{\displaystyle O}{\|}}{C}-C_6H_4-\overset{\overset{\displaystyle O}{\|}}{C}-O-CH_2-CH_2-CH_2-\right)_n}O-$$

DuPont has been obtaining the propanediol from petroleum sources but has developed a genetically engineered microbe to produce it by fermentation of corn sugar. The terephthalic acid, however, is still made from petroleum.

Cargill Dow's *NatureWorks* is a completely bio-based polymer made by polymerization of lactic acid, which is also derived by fermentation of corn sugar. The reaction is

$$\underset{\text{lactic acid}}{HO-\overset{\overset{\displaystyle CH_3}{|}}{CH}-\overset{\overset{\displaystyle O}{\|}}{C}-\boxed{OH}}\quad \underset{\text{lactic acid}}{\boxed{H}O-\overset{\overset{\displaystyle CH_3}{|}}{CH}-\overset{\overset{\displaystyle O}{\|}}{C}-OH} \xrightarrow{-H_2O}$$

$$\underset{\text{NatureWorks polylactide polymer}}{-O\left(-\overset{\overset{\displaystyle CH_3}{|}}{CH}-\overset{\overset{\displaystyle O}{\|}}{C}-O-\right)_n\overset{\overset{\displaystyle CH_3}{|}}{CH}-\overset{\overset{\displaystyle O}{\|}}{C}-O-}$$

The NatureWorks polymer uses no petroleum products and is being used to make fibers for apparel where it is combined with cotton fibers. Other fabric blends for apparel include wool and silk. Clear cold drink cups are also being made with the polymer, which is biodegradable. It's expected that use of the polymer will spread to a variety of packaging materials, too (see Figure 2).

Metabolix, Inc., in Cambridge, Massachusetts, has found a way to avoid the chemical synthesis step in making polymers by using genetic engineering to develop microbes that convert plant matter directly into polymers that are similar to NatureWorks. Depending on the microbe used, a variety of polymers can be produced with the general formula

$$\left[-\overset{\overset{\displaystyle CH_3}{|}}{CH}-(CH_2)_x-\overset{\overset{\displaystyle O}{\|}}{C}-O-\right]_n$$

Different polymer properties are obtained depending on the value of x. At the present time, production of such polymers is at the pilot plant stage.

FIG. 1 Sorona fibers in these fabrics have superior stretch recovery, soft touch, and good dye and print capabilities. *(Anthony Farina/DuPont Media Relations.)*

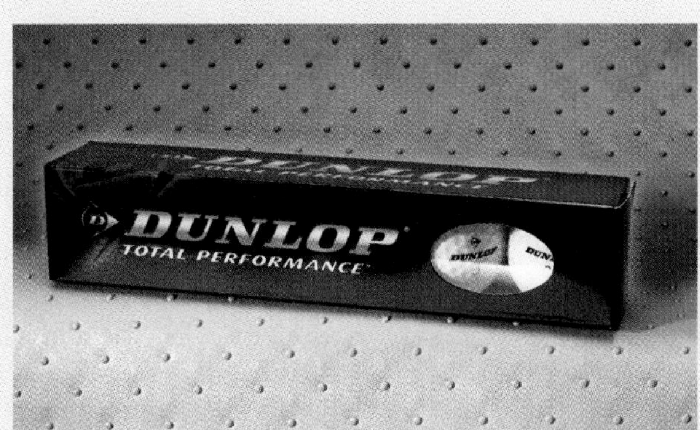

FIG. 2 The plastic film used to package these golf balls is made from NatureWorks polylactide polymer. *(Andy Washnik.)*

functions. The world of living things is composed mostly of organic compounds, including many natural polymers such as proteins, starch, cellulose, and the chemicals of genes.

There are three important requirements for life

A living system requires materials, energy, and information or "blueprints." Our focus in this section will be on the substances that supply them—*carbohydrates, lipids, proteins,* and *nucleic acids*—the basic materials whose molecules, together with water and a few kinds of ions, make up cells and tissues.

Starch, table sugar, and cotton are all *carbohydrates.*

Our brief survey of biochemistry will focus mainly on the *structures* of selected biochemical materials. This is where any study of biochemistry must begin, because the chemical, physical, and biological properties of biochemicals are determined by structure. Where appropriate, we'll describe some of their reactions, particularly with water.

A variety of compounds are required for cells to work. The membranes that enclose all of the cells of your body, for example, are made up mostly of lipid molecules, although molecules of proteins and carbohydrates are also incorporated. Most hormones are either in the lipid or the protein family. Essentially all cellular catalysts—enzymes—are proteins, but many enzyme molecules cannot function without the presence of relatively small molecules of vitamins or certain metal ions.

Lipids include the fats and oils in our diet, like butter, margarine, salad oils, and baking shortenings (e.g., lard). Meat and egg albumin are particularly rich in *proteins.*

Lipids and carbohydrates are also our major sources of the chemical energy we need to function. In times of fasting or starvation, however, the body is also able to draw on its proteins for energy.

The information needed to operate a living system is borne by molecules of nucleic acids. Their structures carry the *genetic code* that instructs cells how to make its proteins, including its enzymes. Hundreds of diseases, like cystic fibrosis and sickle-cell anemia, are caused by defects in the molecular structures of nucleic acids. Viruses, like the AIDS virus, work by taking over the genetic machinery of a cell.

Carbohydrates include sugars, starch, and cellulose

Carbohydrates are naturally occurring polyhydroxyaldehydes or polyhydroxyketones, or else they are compounds that react with water to give these. The carbohydrates include table sugar (sucrose) as well as starch and cellulose.

Most carbohydrates are polymers of simpler units called **monosaccharides.** The most common monosaccharide is glucose, a pentahydroxyaldehyde and probably the most widely occurring structural unit in the entire living world. Glucose is the chief carbohydrate in blood, and it provides the building unit for such important polysaccharides as cellulose and starch. Fructose, a pentahydroxyketone, is produced together with glucose when we digest table sugar. Honey is also rich in fructose.

$$\underset{\text{OH}}{\underset{|}{\text{CH}_2}}\underset{\text{OH}}{\underset{|}{\text{CH}}}-\underset{\text{OH}}{\underset{|}{\text{CH}}}-\underset{\text{OH}}{\underset{|}{\text{CH}}}-\underset{\text{OH}}{\underset{|}{\text{CH}}}-\overset{\text{O}}{\overset{\|}{\text{CH}}}$$

glucose (open-chain form)
a polyhydroxyaldehyde

$$\underset{\text{OH}}{\underset{|}{\text{CH}_2}}\underset{\text{OH}}{\underset{|}{\text{CH}}}-\underset{\text{OH}}{\underset{|}{\text{CH}}}-\underset{\text{OH}}{\underset{|}{\text{CH}}}\overset{\text{O}}{\overset{\|}{\text{C}}}\underset{\text{OH}}{\underset{|}{\text{CH}_2}}$$

fructose (open-chain form)
a polyhydroxyketone

When dissolved in water, the molecules of most carbohydrates exist in a mobile equilibrium involving more than one structure. Glucose, for example, exists as two cyclic forms and one open-chain form in equilibrium in water (see Figure 22.9). The open-chain form, the only one with a free aldehyde group, is represented by less than 0.1% of the solute molecules. Yet, the solute in an aqueous glucose solution still gives the reactions of a polyhydroxyaldehyde. This is possible because the equilibrium between this form and the two cyclic forms shifts to supply more of any of its members when a specific reaction occurs to just one (in accordance with Le Châtelier's principle).

Disaccharides are composed to two monosaccharide units

Carbohydrates whose molecules are split into two monosaccharide molecules by reacting with water are called **disaccharides.** *Sucrose* (table sugar, cane sugar, or beet sugar) is an example, and its hydrolysis gives glucose and fructose.

• rotation at C-1 carbon
• movement of H to form OH

Ring opens here.

Ring opens here.

α-glucose

Open form (polyhydroxyaldehyde)

β-glucose

FIG. 22.9 Structures of glucose. Three forms of glucose are in equilibrium in an aqueous solution. The curved arrows in the open form show how bonds become reoriented as it closes into a cyclic form. Depending on how the CH═O group is turned at the moment of ring closure, the new OH group at C-1 takes up one of two possible orientations, α or β.

oxygen bridge

glucose unit

sucrose

fructose unit

The six-membered rings of the cyclic forms of monosaccharides are not actually *flat* rings. Glucose, for example, has the structure shown below.

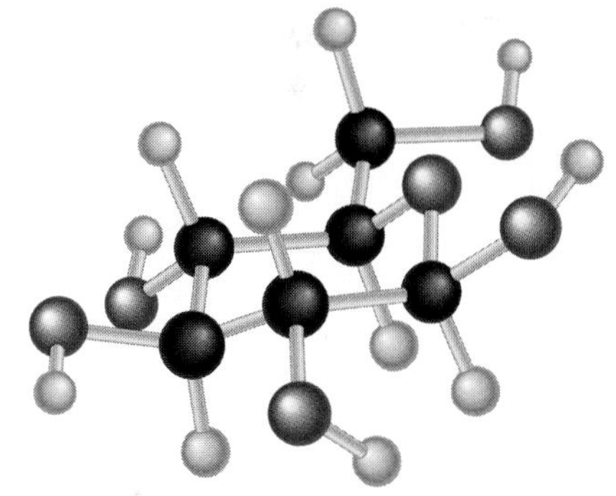

To simplify, let's represent sucrose as Glu—O—Fru, where Glu is a glucose unit and Fru is a fructose unit, both joined by an oxygen bridge, —O—. The hydrolysis of sucrose, the chemical reaction by which we digest it, can thus be represented as follows.[3]

$$\underset{\text{sucrose}}{\text{Glu—O—Fru}} + H_2O \xrightarrow[\text{(hydrolysis)}]{\text{digestion}} \text{glucose} + \text{fructose}$$

Lactose (milk sugar) hydrolyzes to glucose and galactose (Gal), an isomer of glucose, and this is the reaction by which we digest lactose.

$$\underset{\text{lactose}}{\text{Gal—O—Glu}} + H_2O \xrightarrow[\text{(hydrolysis)}]{\text{digestion}} \text{galactose} + \text{glucose}$$

Polysaccharides are polymers of monosaccharides

Some of the most important naturally occurring polymers are the **polysaccharides,** carbohydrates whose molecules involve thousands of monosaccharide units linked to each other by oxygen bridges. They include starch, glycogen, and cellulose. The complete hydrolyses of all three yield only glucose, so their structural differences involve details of the oxygen bridges.

Plants store glucose units for energy needs in molecules of **starch,** found often in seeds and tubers (e.g., potatoes). Starch consists of two kinds of glucose polymers. The structurally simpler kind is *amylose*, which makes up roughly 20% of starch. We may represent its structure as follows, where O is the oxygen bridge linking glucose units.

$$\text{Glu}\left(\text{O—Glu}\right)_n\text{OH}$$

amylose (n is very large)

The average amylose molecule has over 1000 glucose units linked together by oxygen bridges. These are the sites that are attacked and broken when water reacts with amylose

[3] Considerable detail is lost with the simplified structure of sucrose (and other carbohydrates like it to come). We leave the details, however, to other books because we seek a broader view.

during digestion. Molecules of glucose are released and are eventually delivered into circulation in the bloodstream.

$$\text{Amylose} + n\text{H}_2\text{O} \xrightarrow[\text{(hydrolysis)}]{\text{digestion}} n \text{ glucose}$$

The bulk of starch is made up of *amylopectin*, whose molecules are even larger than those of amylose. The amylopectin molecule consists of several amylose molecules linked by oxygen bridges from the end of one amylose unit to a site somewhere along the "chain" of another amylose unit.

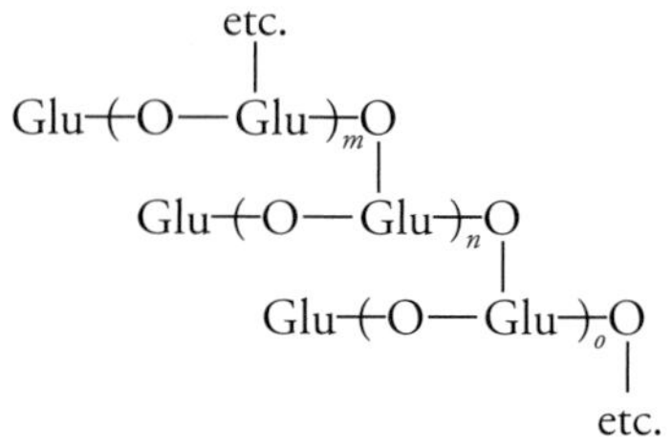

etc.
$\text{Glu}\!-\!(\text{O}\!-\!\text{Glu})_m\!-\!\text{O}$
$\text{Glu}\!-\!(\text{O}\!-\!\text{Glu})_n\!-\!\text{O}$
$\text{Glu}\!-\!(\text{O}\!-\!\text{Glu})_o\!-\!\text{O}$
etc.

amylopectin (*m*, *n*, and *o* are large numbers)

Molecular masses ranging from 50,000 to several million are observed for amylopectin samples from the starches of different plant species. (A molecular mass of 1 million corresponds to about 6000 glucose units.)

Animals store glucose units for energy as **glycogen,** a polysaccharide with a molecular structure very similar to that of amylopectin. When we eat starchy foods and deliver glucose molecules into the bloodstream, any excess glucose not needed to maintain a healthy concentration in the blood is removed from circulation by particular tissues, like the liver and muscles. Liver and muscle cells convert glucose to glycogen. Later, during periods of high energy demand or fasting, glucose units are released from the glycogen reserves so that the concentration of glucose in the blood stays high enough for the needs of the brain and other tissues.

Cellulose is a carbohydrate that humans cannot digest

Cellulose is the chief material in a plant cell wall, and it makes up about 100% of cotton.

Cellulose is a polymer of glucose, much like amylose, but with the oxygen bridges oriented with different geometries. We lack the enzyme needed to hydrolyze its oxygen bridges, so we are unable to use cellulose materials like lettuce for food, only for fiber. Animals that eat grass and leaves, however, have bacteria living in their digestive tracts that convert cellulose into small molecules, which the host organism then appropriates for its own use.

Lipids comprise a family of water-insoluble compounds

Lipids are natural products that are water insoluble but tend to dissolve in nonpolar solvents such as diethyl ether or benzene. The lipid family is huge because the only structural requirement is that lipid molecules be relatively nonpolar with large segments that are entirely hydrocarbon-like. For example, the lipid family includes cholesterol as well as sex hormones, like estradiol and testosterone. You can see from their structures how largely hydrocarbon-like they are.

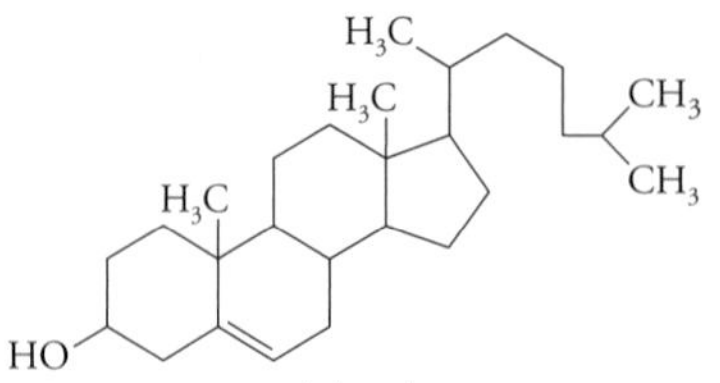

H_3C H_3C CH_3 CH_3 H_3C HO

cholesterol

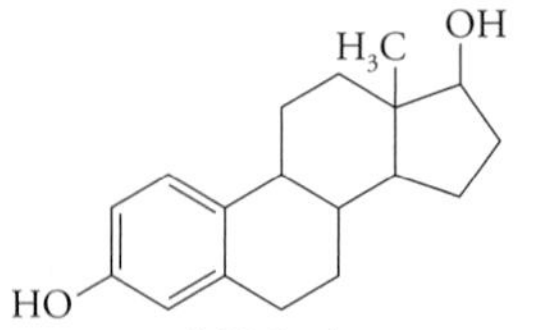

H_3C OH HO

estradiol (a female sex hormone)

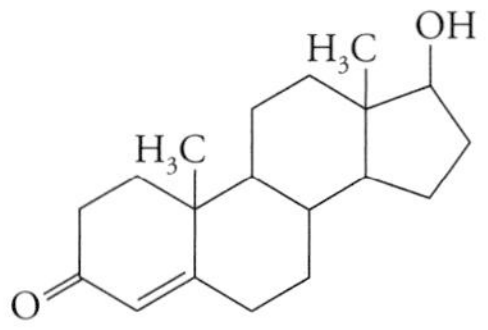

H_3C OH H_3C O

testosterone (a male sex hormone)

Triacylglycerols are esters of glycerol with long-chain carboxylic acids

The lipid family also includes the edible fats and oils in our diets—substances such as olive oil, corn oil, peanut oil, butterfat, lard, and tallow. These are **triacylglycerols,** that is, esters between glycerol, an alcohol with three OH groups, and any three of several long-chain carboxylic acids.

TABLE 22.4 Common Fatty Acids

Fatty Acid	Number of Carbon Atoms	Structure	Melting Point (°C)
Myristic acid	14	$CH_3(CH_2)_{12}CO_2H$	54
Palmitic acid	16	$CH_3(CH_2)_{14}CO_2H$	63
Stearic acid	18	$CH_3(CH_2)_{16}CO_2H$	70
Oleic acid	18	$CH_3(CH_2)_7CH{=}CH(CH_2)_7CO_2H$	4
Linoleic acid	18	$CH_3(CH_2)_4CH{=}CHCH_2CH{=}CH(CH_2)_7CO_2H$	−5
Linolenic acid	18	$CH_3CH_2CH{=}CHCH_2CH{=}CHCH_2CH{=}CH(CH_2)_7CO_2H$	−11

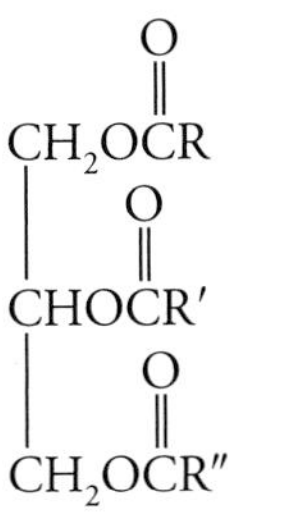

$$\begin{array}{l} CH_2OC(=O)R \\ |\\ CHOC(=O)R' \\ | \\ CH_2OC(=O)R'' \end{array}$$

triacylglycerol (general structural features)

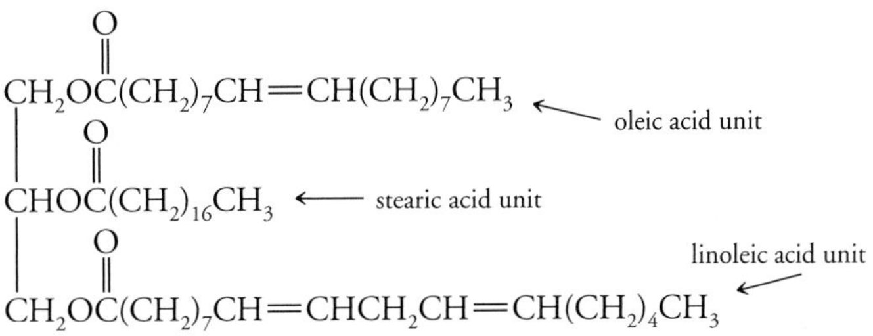

$$\begin{array}{ll} CH_2OC(=O)(CH_2)_7CH{=}CH(CH_2)_7CH_3 & \leftarrow \text{oleic acid unit} \\ | & \\ CHOC(=O)(CH_2)_{16}CH_3 & \leftarrow \text{stearic acid unit} \\ | & \\ CH_2OC(=O)(CH_2)_7CH{=}CHCH_2CH{=}CH(CH_2)_4CH_3 & \leftarrow \text{linoleic acid unit} \end{array}$$

triacylglycerol—typical molecule present in a vegetable oil

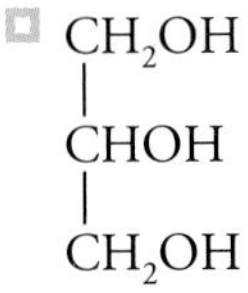

$$\begin{array}{l} CH_2OH \\ | \\ CHOH \\ | \\ CH_2OH \end{array}$$

glycerol

The carboxylic acids used to make triacylglycerols are called **fatty acids** and generally have just one carboxyl group on an unbranched chain with an even number of carbon atoms (Table 22.4). Their long hydrocarbon chains make triacylglycerols mostly like hydrocarbons in physical properties, including insolubility in water. Many fatty acids have alkene groups.

Triacylglycerols obtained from vegetable sources, like olive oil, corn oil, and peanut oil, are called *vegetable oils* and are liquids at room temperature. Triacylglycerols from animal sources, like lard and tallow, are called *animal fats* and are solids at room temperature. The vegetable oils generally have more alkene double bonds per molecule than animal fats, and so are said to be *polyunsaturated.* The double bonds are usually cis, and so the molecules are kinked, making it more difficult for them to nestle close together, experience London forces, and so be in the solid state.

Cooking oils are advertised as *polyunsaturated* because of their several alkene groups per molecule.

Digestion of triacylglycerols involves hydrolysis

We digest the triacylglycerols by hydrolysis, meaning by their reaction with water. Our digestive juices in the upper intestinal tract have enzymes called *lipases* that catalyze these reactions. For example, the complete digestion of the triacylglycerol shown above occurs by the following reaction.

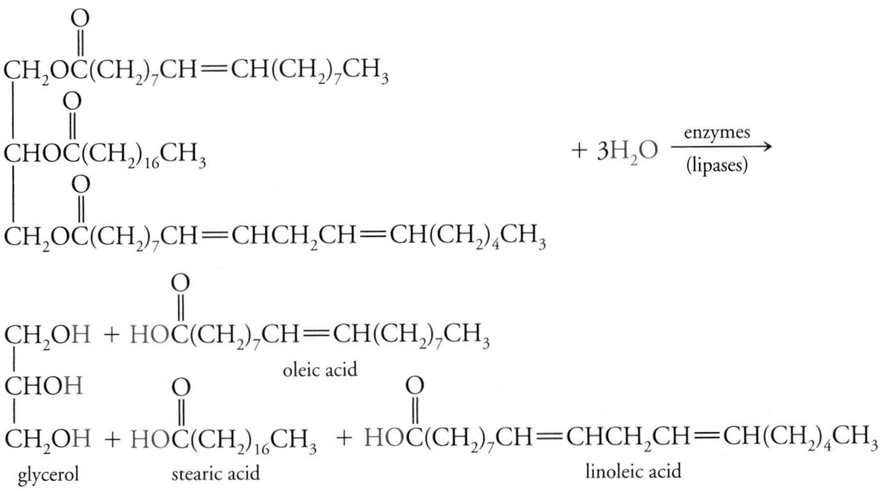

$$\begin{array}{l} CH_2OC(=O)(CH_2)_7CH{=}CH(CH_2)_7CH_3 \\ | \\ CHOC(=O)(CH_2)_{16}CH_3 \\ | \\ CH_2OC(=O)(CH_2)_7CH{=}CHCH_2CH{=}CH(CH_2)_4CH_3 \end{array} + 3H_2O \xrightarrow[\text{(lipases)}]{\text{enzymes}}$$

$$\begin{array}{l} CH_2OH + HOC(=O)(CH_2)_7CH{=}CH(CH_2)_7CH_3 \quad \text{(oleic acid)} \\ | \\ CHOH \\ | \\ CH_2OH + HOC(=O)(CH_2)_{16}CH_3 + HOC(=O)(CH_2)_7CH{=}CHCH_2CH{=}CH(CH_2)_4CH_3 \end{array}$$

glycerol — stearic acid — linoleic acid

Actually, the *anions* of the acids form, because the medium in which lipid digestion occurs is basic.

When the hydrolysis of a triacylglycerol is carried out in the presence of sufficient base so as to release the fatty acids as their anions, the reaction is called *saponification.* The mixture of the salts of long-chain fatty acids is what makes up ordinary *soap.*

Hydrogenation of vegetable oils produces solid fats

Vegetable oils are generally less expensive to produce than butterfat, but because the oils are liquids, few people care to use them as bread spreads. Remember that animal fats, like butterfat, are solids at room temperature, and vegetable oils differ from animal fats only in the number of carbon–carbon double bonds per molecule. Simply adding hydrogen to the double bonds of a vegetable oil, therefore, changes the lipid from a liquid to a solid.

Partial hydrogenation of a vegetable oil can lead to rearrangement of the atoms around the double bonds from cis to trans. The resulting products are known as *trans fats.* Because ingestion of trans fats has been associated with coronary artery disease, food product labels are now required to show the amounts of those fats in the food.

Cell membranes in animals are composed of lipid bilayers

The lipids involved in the structures of cell membranes in animals are not triacylglycerols. Some are diacylglycerols with the third site on the glycerol unit taken up by an attachment to a phosphate unit. This, in turn, is joined to an amino alcohol unit by an ester-like network. The phosphate unit carries one negative charge, and the amino unit has a positive charge. These lipids are called *glycerophospholipids.* Lecithin is one example.

The amino alcohol unit in lecithin is contributed by choline, a cation:

$$HOCH_2CH_2\overset{+}{N}(CH_3)_3$$

Ethanolamine (in its protonated form):

$$HOCH_2CH_2NH_3^+$$

is another amino alcohol that occurs in phospholipids.

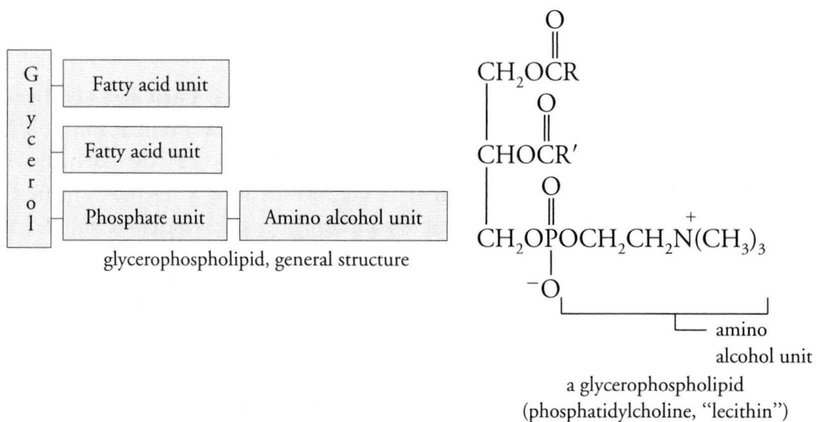

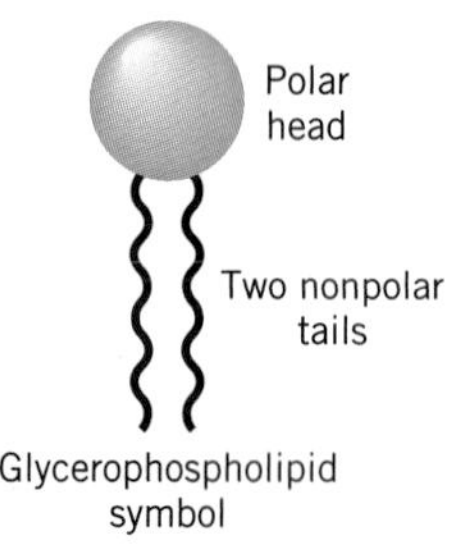

Glycerophospholipid symbol

The glycerophospholipids illustrate that it is possible for lipid molecules to carry polar, even ionic sites, and still not be very soluble in water. The combination of both nonpolar and polar or ionic units within the same molecule enable the glycerophospholipids and similar substances to be major building units for the membranes of animal cells.

The purely hydrocarbon-like portions of a glycerophospholipid molecule (the long R groups contributed by the fatty acid units) are **hydrophobic** ("water fearing"; they avoid water molecules). The portions bearing the electrical charges are **hydrophilic** ("water loving"; they are attracted to water molecules). In an aqueous medium, therefore, the molecules of a glycerophospholipid aggregate in a way that minimizes the exposure of the hydrophobic side chains to water and maximizes contact between the hydrophilic sites and water. These interactions are roughly what take place when glycerophospholipid molecules aggregate to form the *lipid bilayer* membrane of an animal cell (Figure 22.10). The hydrophobic side chains intermingle in the center of the layer where water molecules do not occur. The hydrophilic groups are exposed to the aqueous medium inside and outside of the cell. Not shown in Figure 22.10 are cholesterol and cholesterol ester molecules, which help to stiffen the membranes. Thus cholesterol is essential to the cell membranes of animals.

Neurotransmitters are small molecules that travel between the end of one nerve cell and the surface of the next to transmit the nerve impulse.

Cell membranes also include protein units, which provide several services. Some are molecular recognition sites for molecules such as hormones and neurotransmitters. Others provide channels for the movements of ions, like Na^+, K^+, Ca^{2+}, Cl^-, HCO_3^-, and others, into or out of the cell. Some are channels for the transfer of small organic molecules, like glucose.

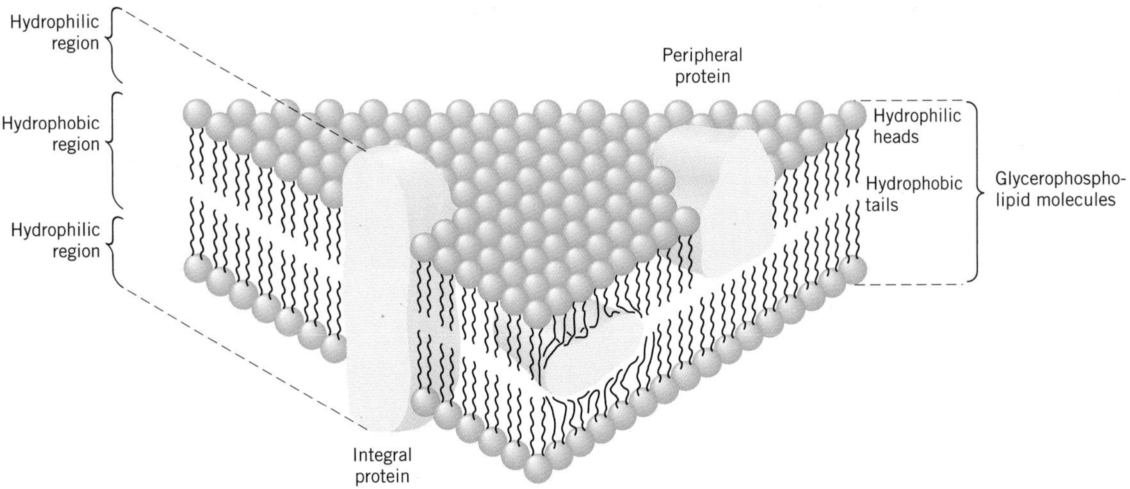

FIG. 22.10 **Animal cell membrane.** The lipid molecules of an animal cell membrane are organized as a bilayer.

Proteins are almost entirely polymers of amino acids

The proteins are a huge family of substances that make up about half of the human body's dry weight. They are found in all cells and in virtually all parts of cells. Proteins serve as enzymes, hormones, and neurotransmitters. They carry oxygen in the bloodstream as well as some of the waste products of metabolism. No other group of compounds has such a variety of functions in living systems.

The dominant structural units of **proteins** are macromolecules called **polypeptides,** which are made from a set of monomers called **α-amino acids.** Most protein molecules include, besides their polypeptide units, small organic molecules or metal ions, and the whole protein lacks its characteristic biological function without these species (Figure 22.11).

The monomer units for polypeptides are a group of about 20 α-amino acids all of which have structural features in common. Some examples of the set of 20 amino acids used to make proteins are given below. The symbol R stands for a structural group, an *amino acid side chain.* All are known by their common names. Each also has a three-letter symbol.

α-position → $^{+}NH_3CH(R)CO^-$ (C=O)

α-amino acid, general structural features

R = H, glycine (Gly)
= CH_3, alanine (Ala)
= CH_2–C_6H_5 (benzene ring), phenylalanine (Phe)
= $CH_2CH_2CO_2H$, glutamic acid (Glu)
= $CH_2CH_2CH_2CH_2NH_2$, lysine (Lys)
= CH_2SH, cysteine (Cys)

The simplest amino acid is aminoacetic acid, or glycine, for which the "side chain" is H.

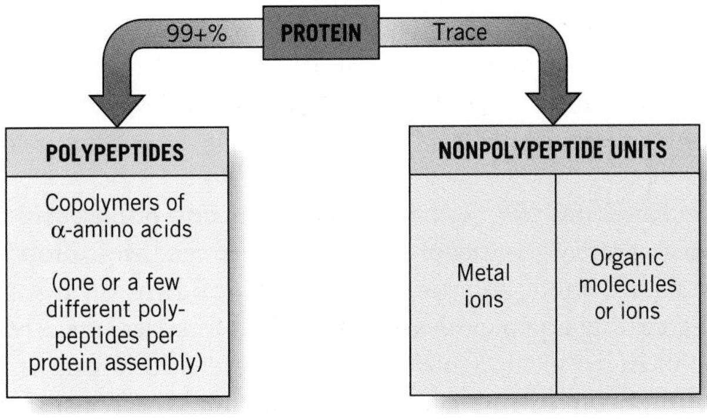

FIG. 22.11 **Components of proteins.** Some proteins consist exclusively of polypeptide molecules, but most also have nonpolypeptide units such as small organic molecules or metal ions, or both.

Glycine, like all of the amino acids in their pure states, exists as a *dipolar ion.* Such an ion results from an internal self-neutralization, by the transfer of a proton from the proton-donating carboxyl group to the proton-accepting amino group.

$$\underset{\text{aminoacetic acid (glycine)}}{NH_2CH_2CO_2H} \longrightarrow \underset{\text{glycine, dipolar ionic form}}{^+NH_3CH_2CO_2^-}$$

Polypeptides are elimination polymers, and they are similar to nylon in some ways. The peptide bond has the same structural features as the bond connecting monomer units in nylon (page 927).

Polypeptides are polymers

Polypeptides are copolymers of the amino acids. The carboxyl group of one amino acid becomes joined to the amino group of another by means of the same kind of carbonyl–nitrogen bond found in amides, but here called the **peptide bond.** Let's see how two amino acids, glycine and alanine, can become linked by a (multistep) splitting out of water.

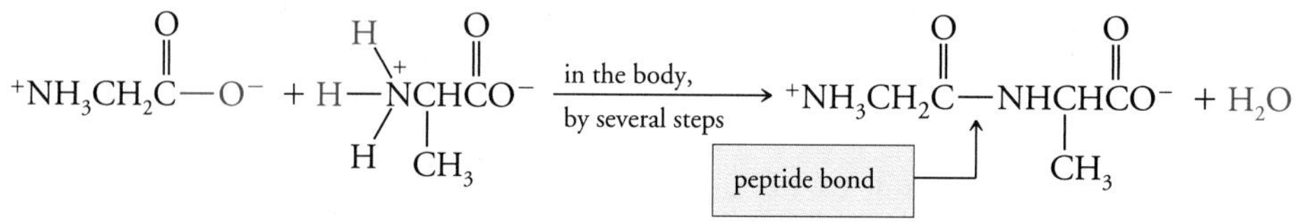

$$\underset{\text{glycine (Gly)}}{^+NH_3CH_2C(=O)-O^-} + \underset{\text{alanine (Ala)}}{H-\overset{+}{N}H_2CH(CH_3)CO^-} \xrightarrow[\text{by several steps}]{\text{in the body,}} \underset{\text{glycylalanine (Gly-Ala)}}{^+NH_3CH_2C(=O)-NHCH(CH_3)C(=O)O^-} + H_2O$$

(peptide bond: the C—N bond between the glycine and alanine units)

The product of the reaction, glycylalanine, is an example of a *dipeptide.* (Notice how the three-letter symbols for the amino acids make up what biochemists sometimes use as the *structural* formulas of such products.)

We could have taken glycine and alanine in different roles and written the equation for the formation of a different dipeptide, Ala-Gly.

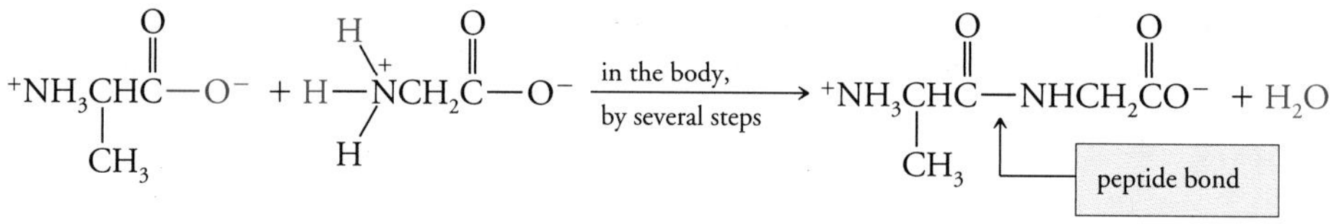

$$\underset{\text{alanine (Ala)}}{^+NH_3CH(CH_3)C(=O)-O^-} + \underset{\text{glycine (Gly)}}{H-\overset{+}{N}H_2CH_2C(=O)-O^-} \xrightarrow[\text{by several steps}]{\text{in the body,}} \underset{\text{alanylglycine (Ala-Gly)}}{^+NH_3CH(CH_3)C(=O)-NHCH_2C(=O)O^-} + H_2O$$

(peptide bond: the C—N bond between the alanine and glycine units)

The artificial sweetener aspartame (NutraSweet) is the methyl ester of a dipeptide.

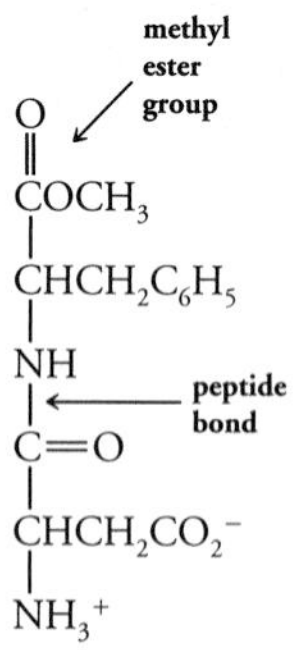

Aspartame: $C(=O)OCH_3$ (methyl ester group) — $CHCH_2C_6H_5$ — NH — (peptide bond) — $C=O$ — $CHCH_2CO_2^-$ — NH_3^+

aspartame

We can think of the formation of these two dipeptides as being like the formation of two 2-letter words from the letters N and O. Taken in one order we get NO; in the other, ON. They have entirely different meanings, yet are made of the same pieces.

Notice that each dipeptide above has a CO_2^- at one end of the chain and a $^+NH_3$ group at the other end. Each end of a dipeptide molecule, therefore, could become involved in the formation of yet another peptide bond involving any of the 20 amino acids. For example, if glycylalanine were to combine with phenylalanine (Phe), two different *tripeptides* could form with the sequences Gly-Ala-Phe or Phe-Gly-Ala. Each of these tripeptides has a CO_2^- at one end and a $^+NH_3$ at the other end. Thus, you can see how very long sequences of amino acid units might be joined together.

As the length of a polypeptide chain grows, the number of possible combinations of amino acids becomes astronomical. For example, with just the three amino acids we've used (Gly, Ala, and Phe), six different polypeptide sequences are possible that differ only in the three side chains, H, CH_3, and $CH_2C_6H_5$, located at the α-carbon atoms.

Proteins usually contain more than one polypeptide chain

Many proteins consist of a single polypeptide. Most proteins, however, involve assemblies of two or more polypeptides. These are identical in some proteins, but in others the aggregating polypeptides are different. Moreover, a relatively small organic molecule may be included in the aggregation, and a metal ion is sometimes present, as well. Thus, the terms "protein" and "polypeptide" are not synonyms. Hemoglobin, for example, has all

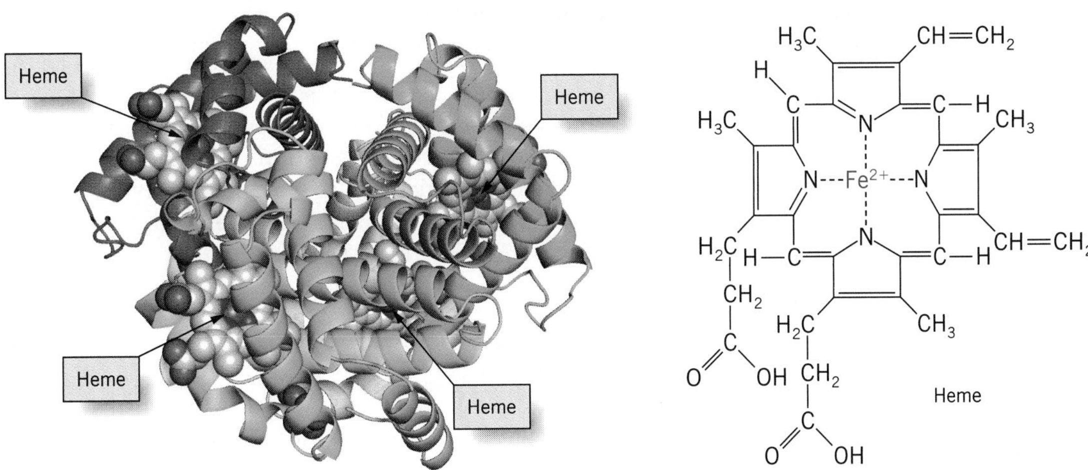

FIG. 22.12 **Hemoglobin.** Its four polypeptide chains, each shown as a different colored ribbon, twist and bend around the four embedded heme units. Each heme unit contains an Fe^{2+} ion in its center which is able to bind to O_2.

of the features just described (Figure 22.12). It is made of four polypeptides—two similar pairs—and four molecules of heme, the organic compound that causes the red color of blood. Heme, in turn, holds an iron(II) ion. The *entire* package is the protein, hemoglobin. If one piece is missing or altered in any way—for example, if iron occurs as Fe^{3+} instead of Fe^{2+}, the substance is not hemoglobin, and it does not transport oxygen in the blood.

Hemoglobin is the oxygen carrier in blood.

Notice in Figure 22.12 how the strands of each polypeptide unit in hemoglobin are coiled and that the coils are kinked and twisted. Such shapes of polypeptides are determined by the amino acid sequence, because the side chains are of different sizes and some are hydrophilic and others are hydrophobic. Polypeptide molecules become twisted and coiled in whatever way minimizes the contact of hydrophobic groups with the surrounding water and maximizes the contacts of hydrophilic groups with water molecules.

The final shape of a protein, called its *native form,* is as critical to its ability to function as anything else about its molecular architecture. For example, just the exchange of one side-chain R group by another changes the shape of hemoglobin and causes a debilitating condition known as sickle-cell anemia.

In one of the subunits of the hemoglobin in those with sickle-cell anemia an isopropyl group, $-CH(CH_3)_2$, is a side chain where a $-CH_2CH_2CO_2H$ side chain should be.

Almost all enzymes are proteins

The catalysts in living cells are called **enzymes,** and virtually all are proteins. Some enzymes require metal ions, such as Mn^{2+}, Co^{2+}, Cu^{2+}, and Zn^{2+}, all of which are on the list of the *trace elements* that must be in a good diet. Some enzymes also require molecules of the B vitamins to be complete enzymes.

Some of our most dangerous poisons work by deactivating enzymes, often those needed for the transmission of nerve signals. For example, the botulinum toxin that causes botulism, a deadly form of food poisoning, deactivates an enzyme in the nervous system. Heavy metal ions, like Hg^{2+} or Pb^{2+}, are poisons because they deactivate enzymes.

Heavy metal ions bond to the HS groups of cysteine side chains in polypeptides.

22.8 Nucleic Acids Carry Our Genetic Information

The enzymes of an organism are made under the chemical direction of a family of compounds called the *nucleic acids.* Both the similarities and the uniqueness of every species as well as every individual member of a species depend on structural features of these compounds.

DNA and RNA are types of nucleic acids

The **nucleic acids** occur as two broad types, namely, **RNA,** or ribonucleic acids, and **DNA,** or deoxyribonucleic acids. DNA is the actual chemical of a gene, the individual unit of heredity and the chemical basis through which we inherit all of our characteristics.

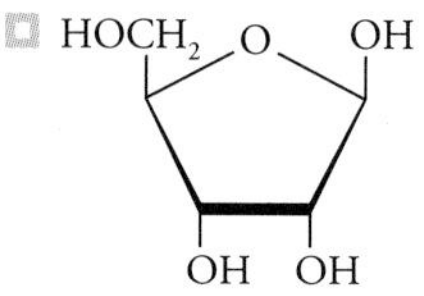

The main chains or "backbones" of DNA molecules consist of alternating units contributed by phosphoric acid and a monosaccharide (see Figure 22.13). In RNA the monosaccharide is ribose (hence the R in RNA). In DNA, the monosaccharide is deoxyribose (*deoxy* means "lacking an oxygen unit"). Thus, both DNA and RNA have the following systems, where *G* stands for *group,* each *G* unit representing a unique nucleic acid side chain or *base.*

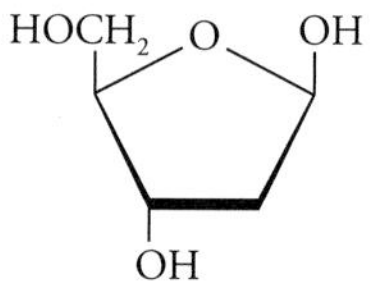

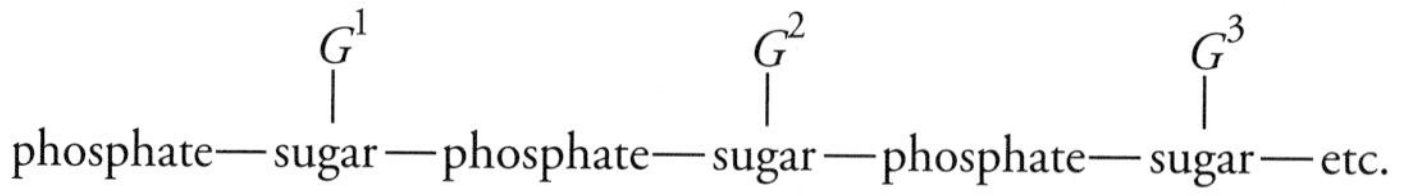

$$\text{phosphate—}\underset{\displaystyle\text{sugar}}{\overset{G^1}{|}}\text{—phosphate—}\underset{\displaystyle\text{sugar}}{\overset{G^2}{|}}\text{—phosphate—}\underset{\displaystyle\text{sugar}}{\overset{G^3}{|}}\text{—etc.}$$

Backbone system in all nucleic acids–many thousands of repeating units long (in DNA, the sugar is deoxyribose; in RNA, the sugar is ribose)

The side chains, *G,* are all heterocyclic amines whose molecular shapes have much to do with their function. Being amines, which are basic in water, they are referred to as the *bases* of the nucleic acids and are represented by single letters—A for adenine, T for thymine, U for uracil, G for guanine, and C for cytosine.

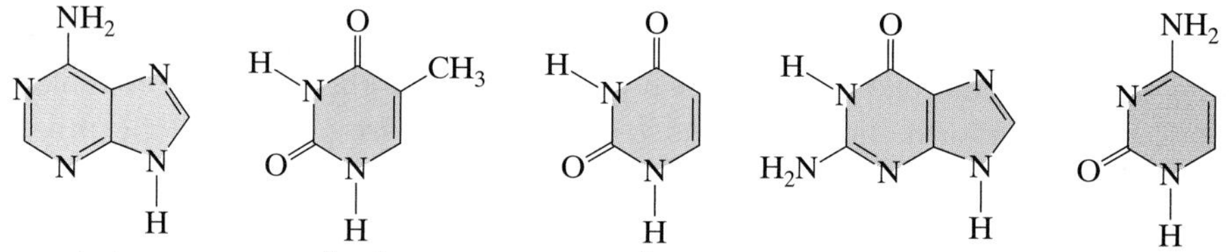

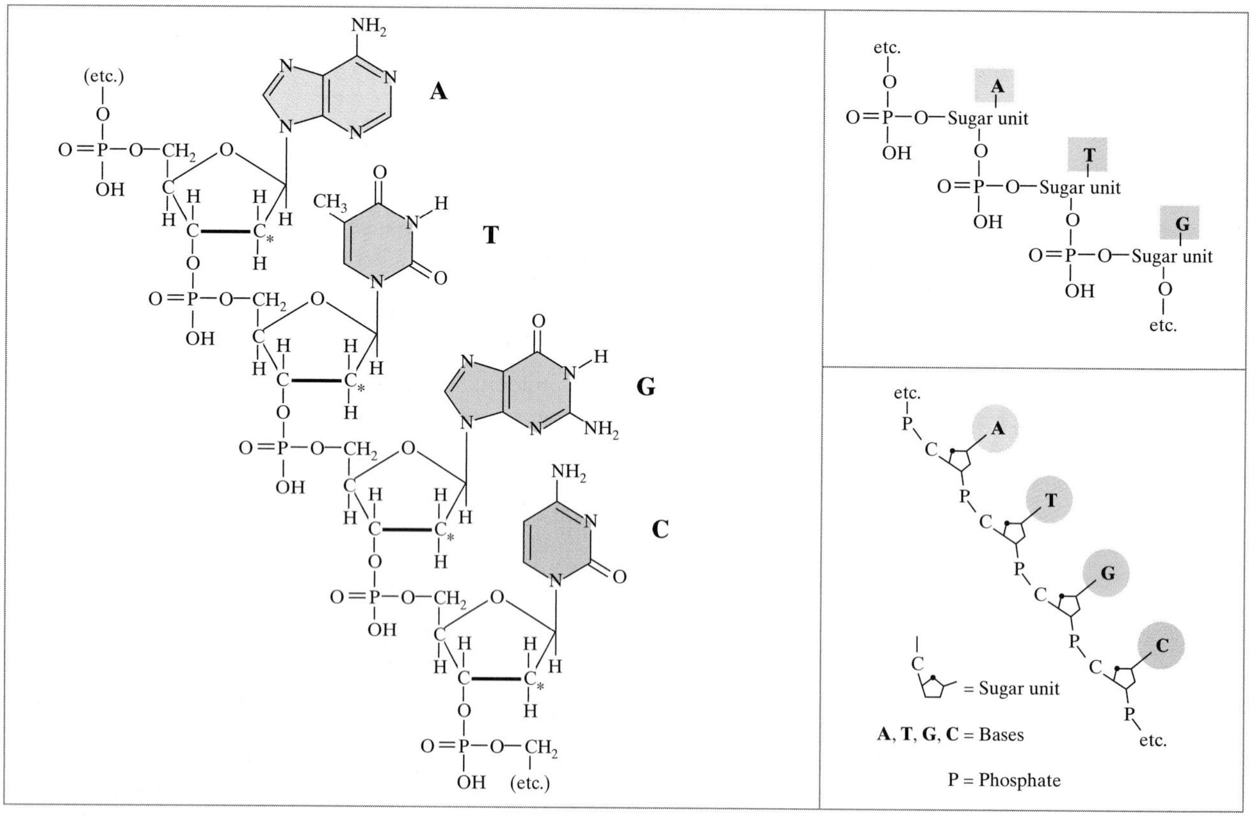

FIG. 22.13 Nucleic acids. A segment of a DNA chain featuring each of the four DNA bases. When the sites marked by asterisks each carry an OH group, the main "backbone" would be that of RNA. In RNA U replaces T. The insets show how simplified versions of a DNA strand can be drawn.

The bases A, T, G, and C occur in DNA, whereas A, U, G, and C are in RNA. These few bases are the "letters" of the genetic alphabet. The messages of all genes are composed with just four letters, A, T, G, and C.

DNA occurs as a double helix

In 1953, F. H. C. Crick of England and J. D. Watson, an American, deduced that DNA occurs in cells as two intertwined, oppositely running strands of molecules coiled like a spiral staircase and called the **DNA double helix** (Figure 22.14). Hydrogen bonds help to hold the two strands side by side, but other factors are also involved.

Crick and Watson shared the 1962 Nobel Prize in Physiology or Medicine with Maurice Wilkins, using X-ray data from Rosalind Franklin to deduce the helical structure of DNA.

The bases have N—H and O═C groups, which enable hydrogen bonds (···) to form between them.

$$\overset{\delta+}{\text{N—H}}\cdots\overset{\delta-}{\text{O}}\text{═C}$$

However, the bases with the best "matching" molecular geometries for maximum hydrogen bonding occur only as particular *pairs* of bases. The functional groups of each pair are in exactly the right locations in their molecules to allow hydrogen bonds between pair members. Adenine (A) pairs with thymine (T) and cytosine (C) pairs with guanine (G) (see Figure 22.15). In DNA, A pairs only with T, never with G or C. Similarly, C pairs only with G, never with A or T. Thus opposite every G on one strand in a DNA double helix, a C occurs on the other. Opposite every A on one strand in DNA, a T is found on the other.

Adenine (A) can also pair with uracil (U), but U occurs in RNA. The A-to-U pairing is an important factor in the work of RNA.

The replication of DNA occurs through base pairing

Prior to cell division, the cell produces duplicate copies of its DNA so that each daughter cell will have a complete set. Such reproductive duplication is called DNA **replication.**

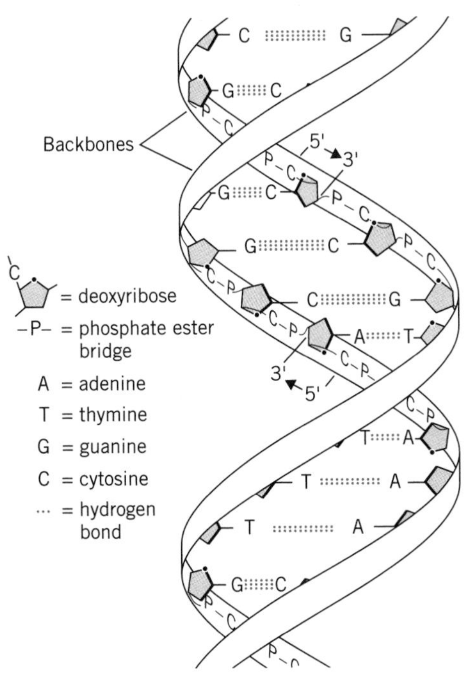

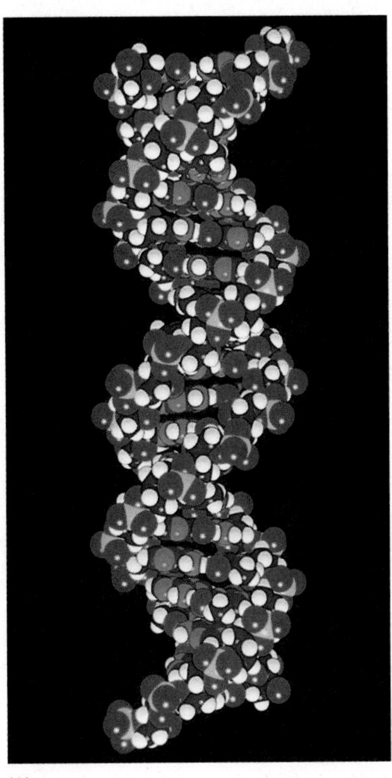

FIG. 22.14 The DNA double helix. (*a*) A schematic drawing in which the hydrogen bonds between the two strands are indicated by dotted lines (*b*) A model of a short section of a DNA double helix. *(Nelson Max/Peter Arnold, Inc.)*

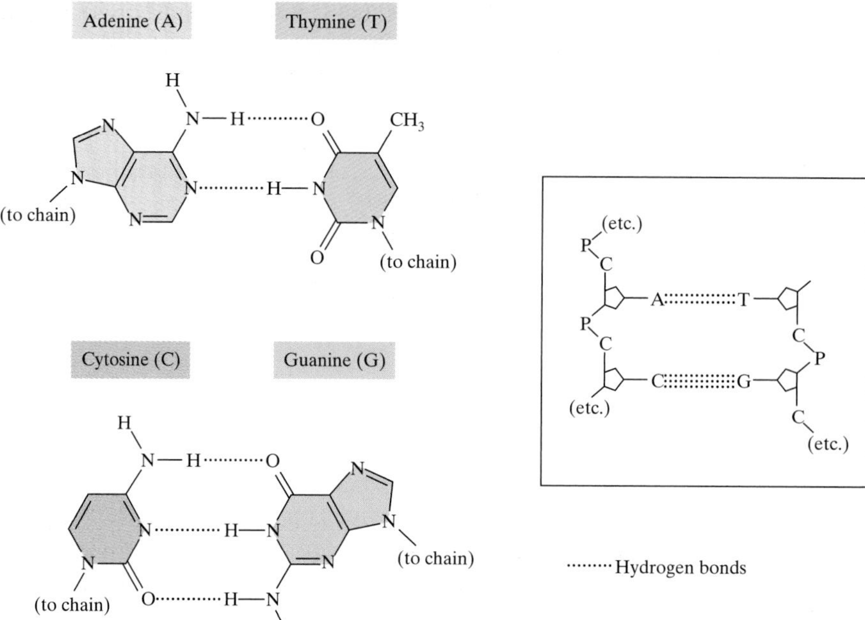

FIG. 22.15 Base pairing in DNA. The hydrogen bonds are indicated by dotted lines.

cytosine unit

a typical nucleotide, one using cytosine as its side chain

The accuracy of DNA replication results from the limitations of the base pairings: A only with T and C only with G (Figure 22.16). The unattached letters, A, T, G, and C, in Figure 22.16 here represent not just the bases but the whole monomer molecules of DNA. These are called *nucleotides* and are molecules made of a phosphate–sugar unit bonded to one particular base. The nucleotides are made by the cell and are present in the cellular "soup."

An enzyme catalyzes each step of replication. As replication occurs, the two strands of the parent DNA double helix separate, and the monomers for new strands assemble along the exposed single strands. Their order of assembling is determined entirely by the specificity of base pairing. For example, a base such as T on a parent strand can accept only a nucleotide with the base A. Two daughter double helices result that are identical to the parent, and each carries one strand from the parent. One new double helix goes to one daughter cell and the second goes to the other.

Genes are made up of segments along a DNA chain

A single human gene has between 1000 and 3000 bases, but they do not occur continuously on a DNA molecule. In multicelled organisms, a gene is neither a whole DNA molecule nor a continuous sequence in one. A single gene consists of the totality of particular *segments* of a DNA chain that, taken together, carry the necessary genetic

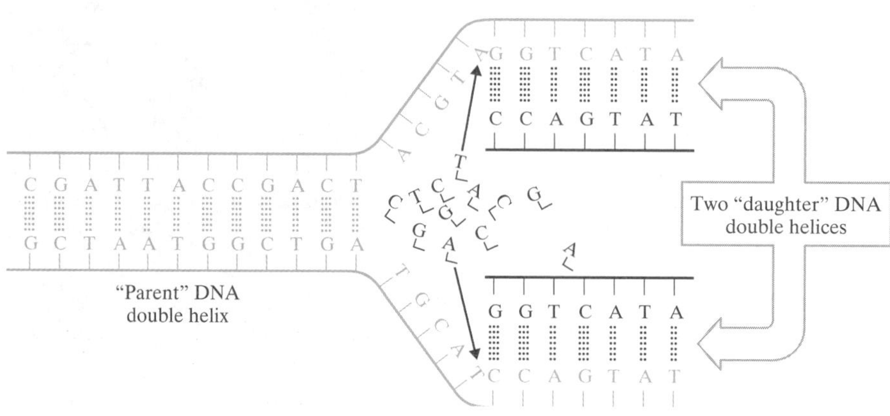

FIG. 22.16 Base pairing and the replication of DNA.

instruction for a particular polypeptide. The individual separated sections making up a gene are called **exons**—a unit that helps to *ex*press a message. The sections of a DNA strand between exons are called **introns**—units that *in*terrupt the gene.

DNA directs the synthesis of polypeptides

Each polypeptide in a cell is made under the direction of its own gene. In broad outline, the steps between a gene and an enzyme occur as follows.

Something like 25,000 genes are needed to account for all of the kinds of proteins in a human body.

DNA	$\xrightarrow{\text{transcription}}$	RNA	$\xrightarrow{\text{translation}}$	Polypeptide
gene	(The genetic message is read off in the cell nucleus and transferred to RNA.)		(The genetic message, now on RNA outside the nucleus, is used to direct the synthesis of a polypeptide.)	

The step labeled **transcription** accomplishes just that—the genetic message, represented by a sequence of bases on DNA, is transcribed into a *complementary* sequence of bases on RNA, but U, not T, is on the RNA. **Translation** means the conversion of the base sequence on RNA into a side-chain sequence on a new polypeptide. It is like translating from one language (the DNA/RNA base sequences) to another language (the polypeptide side-chain sequence).

The cell tells which amino acid unit must come next in a growing polypeptide by means of a code that relates unique groups of nucleic acid bases to specific amino acids. The code, in other words, enables the cell to translate from the four-letter RNA alphabet (A, U, G, and C) to the 20-letter alphabet of amino acids (the amino acid side chains).

Four types of RNA are involved in polypeptide synthesis

Four types of RNA are involved in the connection between gene and polypeptide. One is called *ribosomal* RNA, or rRNA, and it is packaged together with enzymes in small granules called *ribosomes.* Ribosomes are manufacturing stations for polypeptides.

Another RNA is called *messenger* RNA, or mRNA, and it brings the blueprints for a specific polypeptide from the cell nucleus to the manufacturing station (a ribosome). mRNA is the carrier of the genetic message to the polypeptide assembly site.

The mRNA is made from another kind of RNA called *heterogeneous nuclear* RNA, or hnRNA. This is the RNA first made at the direction of a DNA unit, so it contains sections that were specified by both exons and introns (hence, *heterogeneous*).

Little agreement exists on the name *heterogeneous nuclear RNA* among biochemistry references. Some call it pre-RNA; others call it *primary transcript RNA*, or ptRNA.

The last type of RNA is called *transfer* RNA, or tRNA. It has responsibility for picking up the prefabricated parts, the amino acids, and getting them from the cell "soup" to the ribosome.

When some chemical signal tells a cell to make a particular polypeptide, the cell nucleus makes the hnRNA that corresponds to the gene. As we indicated, the base sequence in hnRNA is an exact complement to a base sequence in the DNA being transcribed—both the exon sections of the DNA and the introns. Reactions and enzymes in the cell nucleus then remove those parts of the new hnRNA that were caused by introns. It is as if the message, now on hnRNA, is edited to remove nonsense sections. The result of this editing is messenger RNA.

The mRNA now moves outside the cell nucleus and becomes joined to a ribosome where the enzymes for polypeptide synthesis are found. The polypeptide manufacturing site now awaits the arrival of tRNA molecules bearing amino acid units. All is in readiness for polypeptide synthesis—for *translation.*

The genetic code consists of triplets of bases

To continue with our use of the language of coding, we can describe the genetic message as being written in code words made of three letters—three bases. For example, when the bases G, G, and U occur side by side on an mRNA chain, they together specify the amino acid glycine. Although three other triplets also mean glycine, one of the genetic codes for glycine is GGU. Similarly, GCU is a code "word" for alanine.

Each triplet of bases *as it occurs on mRNA* is called a **codon,** and each codon corresponds to a specific amino acid. Which amino acid goes with which codon constitutes the **genetic code.** One of the most remarkable features of the code is that it is essentially universal. The codon specifying alanine in humans, for example, also specifies alanine in the genetic machinery of bacteria, aardvarks, camels, rabbits, and stinkbugs. Our chemical kinship with the entire living world is profound (even humbling).

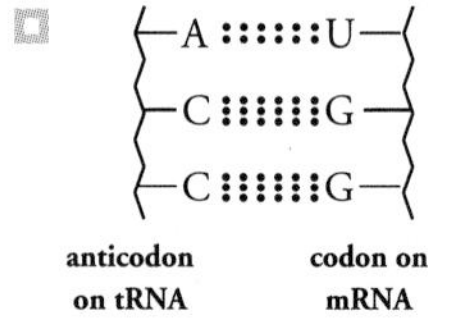

A triplet of bases complementary to a codon occurs on tRNA and is called an **anticodon.** A tRNA molecule, bearing the amino acid corresponding to its anticodon, can line up at a strand of mRNA only where the anticodon "fits" by hydrogen bonding to the matching codon. If the anticodon is ACC, then the A on the anticodon must find a U on the mRNA at the same time that its neighboring two C bases find neighboring G bases on the mRNA strand. Thus the anticodon and codon line up as shown in the margin, where the dotted lines indicate hydrogen bonds.

mRNA codons determine the sequence in which tRNA units can line up, and this sets the sequence in which amino acid units become joined in a polypeptide. Through this overall process, the sequence of bases in DNA are translated into a sequence of amino acids in the polypeptide.

Genetic defects are caused by "errors" in the base sequence in DNA

In 2001 scientists working on the *Human Genome Project*, a major scientific effort to "map" all of the genes of the body by locating where they are on the various chromosomes, announced the sequencing of the human genome. Knowledge of the locations and structures of genes is expected to help cure at least some genetic diseases.

About 2000 diseases are attributed to various kinds of defects in the genetic machinery of cells. With so many steps from gene to polypeptide, we should expect many opportunities for things to go wrong. Suppose, for example, that just one base in a gene were wrong. Then one base on an mRNA strand would be the wrong base—"wrong" meaning with respect to getting the polypeptide we want. This would change every remaining codon. If the next three codons on mRNA, for example, were UCU—GGU—GCU—U—etc., and the first G were deleted, then the sequence would become

UCU—GUG—CUU—etc.

All remaining triplets change! You can imagine how this could lead to an entirely different polypeptide than what we want.

Or suppose that the GGU codon were replaced by, say, a GCU codon so that the UCU—GGU—GCU—etc. becomes

UCU—GCU—GCU—etc.

One amino acid in the resulting polypeptide would not be what we want. Something like this makes the difference, for example, between normal hemoglobin and sickle-cell hemoglobin and the difference between health and a tragic genetic disease.

Atomic radiation, like gamma rays, beta rays, or even X rays, as well as chemicals that mimic radiation can cause genetic defects, too. A stray high-energy photon hitting a gene might cause two opposite bases to fuse together chemically and so make the cell reproductively dead. When it came time for it to divide, the cell could not replicate its genes, could not divide, and that would be the end of it. If this happens to enough cells in a tissue, the organism might be fatally affected.

Viruses take over the genetic machinery in cells

The protein of a virus is a protective overcoat for its nucleic acid, and it provides an enzyme to catalyze a breakthrough by the virus into a host cell.

Viruses are packages of chemicals usually consisting of nucleic acid and protein. Their nucleic acids are able to take over the genetic machinery of the cells of particular host tissues, manufacture more virus particles, and multiply enough to burst the host cell. Because cancer cells divide irregularly, and usually more rapidly than normal cells, some viruses are thought to be among the agents that can cause cancer. For example, strains of human papilloma virus (HPV) are known to cause cervical cancer.

SUMMARY

Organic Compounds. **Organic chemistry** is the study of the covalent compounds of carbon, except for its oxides, carbonates, bicarbonates, and a few others. **Functional groups** attached to nonfunctional and hydrocarbon-like groups are the basis for the classification of organic compounds. Members of a family have the same kinds of chemical properties. Their physical properties are a function of the relative proportions of functional and nonfunctional groups. Opportunities for hydrogen bonding strongly influence the boiling points and water solubilities of compounds with OH or NH groups. Organic compounds exhibit isomerism because often there is more than one possible structure corresponding to a given molecular formula. Molecules containing an **asymmetric carbon atom** exhibit chirality.

Saturated Hydrocarbons. The **alkanes**—saturated compounds found chiefly in petroleum—are the least polar and the least reactive of the organic families. Their carbon skeletons can exist as **straight chains, branched chains,** or **rings.** When they are *open chain,* there is free rotation about single bonds. Rings can include heteroatoms, but **heterocyclic compounds** are not hydrocarbons.

By the IUPAC rules, the names of alkanes end in *-ane.* A prefix denotes the carbon content of the parent chain, the longest continuous chain, which is numbered from whichever end is nearest a branch. The **alkyl groups** are alkanes minus a hydrogen.

Alkanes in general do not react with strong acids or bases or strong redox reactants. At high temperatures, their molecules crack to give H_2 and unsaturated compounds.

Unsaturated Hydrocarbons. The lack of free rotation at a carbon–carbon double bond makes **geometric isomerism** possible. The pair of electrons of a π bond makes alkenes act as Brønsted bases, enabling alkenes to undergo **addition reactions** with hydrogen chloride and water (in the presence of an acid catalyst). Alkenes also add hydrogen (in the presence of a metal catalyst). Bromine and chlorine add to alkenes under mild, uncatalyzed conditions. Strong oxidizing agents, like ozone, readily attack alkenes and break their molecules apart.

Aromatic hydrocarbons, like benzene, do not give the addition reactions shown by alkenes. Instead, they undergo **substitution reactions** that leave the energy lowering, pi-electron network intact. In the presence of suitable catalysts, benzene reacts with chlorine, bromine, nitric acid, and sulfuric acid.

Alcohols and Ethers. The **alcohols,** ROH, and the **ethers,** ROR′, are alkyl derivatives of water. Ethers have almost as few chemical properties as alkanes. Alcohols undergo an **elimination reaction** that splits out H_2O and changes them to alkenes. Concentrated hydrohalogen acids, like HI, react with alcohols by a **substitution reaction,** which replaces the OH group by a halogen. Oxidizing agents convert alcohols of the type RCH_2OH first into aldehydes and then into carboxylic acids. Oxidizing agents convert alcohols of the type R_2CHOH into ketones. Alcohols form esters with carboxylic acids.

Amines. The simple **amines,** RNH_2, as well as the more substituted relatives, RNHR′ and RNR′R″, are organic derivatives of ammonia and thus are weak bases that can neutralize strong acids. The conjugate acids of amines are good proton donors and can neutralize strong bases.

Carbonyl Compounds. **Aldehydes,** RCH═O, among the most easily oxidized organic compounds, are oxidized to carboxylic acids. **Ketones,** RCOR′, strongly resist oxidation. Aldehydes and ketones add hydrogen to make alcohols.

The **carboxylic acids,** RCO_2H, are weak acids but neutralize strong bases. Their anions are Brønsted bases. Carboxylic acids can also be changed to esters by heating them with alcohols.

Esters, RCO_2R', react with water to give back their parent acids and alcohols. When aqueous base is used for hydrolysis, the process is called *saponification* and the products are the salts of the parent carboxylic acids as well as alcohols.

The simple **amides,** $RCONH_2$, can be made from acids and ammonia, and they can be hydrolyzed back to their parent acids and trivalent nitrogen compounds. The amides are not basic.

Polymers. **Polymers,** which are **macromolecules,** are made up of a very large number of atoms in which a small characteristic feature repeats over and over many times. **Polypropylene,** an **addition polymer** that consists of a hydrocarbon **backbone** with a methyl group, CH_3, attached to every other carbon, is formed by **polymerization** of the **monomer** propylene. Polyethylene and polystyrene, which are also addition polymers, are formed from ethylene and styrene, respectively. **Condensation polymers** are formed by elimination of a small molecule such as H_2O or CH_3OH from two monomer units accompanied by the formation of a covalent bond between the monomers. **Nylons** and **polyesters** are **copolymers** because they are formed from two different monomers.

Cross-linking occurs when bridging groups of atoms link polymer chains together. Latex (polyisoprene) can be cross-linked by heating it with sulfur to give **vulcanized rubber.** Polymerization of ethylene can lead to **branching,** which produces an amorphous polymer called low density polyethylene (LDPE). High density polyethylene (HDPE) and ultrahigh molecular weight polyethylene (UHMWPE) are not branched and are more crystalline, which makes them stronger. Nylon's properties are affected by hydrogen bonding between polymer strands.

Carbohydrates. The glucose unit is present in starch, glycogen, cellulose, sucrose, lactose, and simply as a **monosaccharide** in honey. It's the chief "sugar" in blood. As a monosaccharide, glucose is a pentahydroxyaldehyde in one form but it also exists in two cyclic forms. The latter are joined by oxygen bridges in the **disaccharides** and **polysaccharides.** In lactose (milk sugar), a glucose and a galactose unit are joined. In sucrose (table sugar), glucose and fructose units are linked.

The major polysaccharides—**starch** (amylose and amylopectin), glycogen, and cellulose—are all polymers of glucose but with different patterns and geometries of branching. The digestion of the disaccharides and starch requires the hydrolyses of the oxygen bridges to give monosaccharides.

Lipids. Natural products in living systems that are relatively soluble in nonpolar solvents, but not in water, are all in a large group of compounds called **lipids.** They include sex hormones and cholesterol, the triacylglycerols of nutritional importance, and the glycerophospholipids and other phospholipids needed for cell membranes.

The **triacylglycerols** are triesters between glycerol and three of a number of long-chain, often unsaturated **fatty acids.** In the digestion of the triacylglycerols, hydrolysis of the ester groups occurs, and anions of fatty acids form (together with glycerol).

Molecules of glycerophospholipids have large segments that are hydrophobic and others that are hydrophilic. Glycerophospholipid molecules are the major components of cell membranes, where they are arranged in a lipid bilayer.

Amino Acids, Polypeptides, and Proteins. The **amino acids** in nature are mostly compounds with NH_3^+ and CO_2^- groups joined to the same carbon atom (called the alpha position of the amino acid). About 20 amino acids are important to the structures of polypeptides and proteins.

When two amino acids are linked by a **peptide bond** (an amide bond), the compound is a dipeptide. In **polypeptides,** several (sometimes thousands) of regularly spaced peptide bonds occur, involving as many amino acid units. The uniqueness of a polypeptide lies in its chain length and in the sequence of the side-chain groups.

Some **proteins** consist of only a polypeptide, but most proteins involve two or more (sometimes different) polypeptides and often a nonpolypeptide (an organic molecule) and a metal ion.

Nearly all **enzymes** are proteins. Major poisons work by deactivating enzymes.

Nucleic Acids. The formation of a polypeptide with the correct amino acid sequence needed for a given polypeptide is under the control of **nucleic acids.** These are polymers whose backbones are made of a repeating sequence of pentose sugar units. On each sugar unit is a **base,** a heterocyclic amine such as adenine (A), thymine (T), guanine (G), cytosine (C), or uracil (U). In **DNA,** the sugar is deoxyribose, and the bases are A, T, G, and C. In **RNA,** the sugar is ribose and the bases are A, U, G, and C.

DNA exists in cell nuclei as **double helices.** The two strands are associated side by side, with hydrogen bonds occurring between particular pairs of bases. In DNA, A is always paired to T; C is always paired to G. The base U replaces T in RNA, and A can also pair with U. The bases, A, U, G, and C, are the four letters of the genetic alphabet, and the specific combination that codes for one amino acid in polypeptide synthesis is three bases, side by side, on a strand. An individual gene in higher organisms consists of sections of a DNA chain, called **exons,** separated by other sections called **introns.** The **replication** of DNA is the synthesis by the cell of exact copies of the original two strands of a DNA double helix.

Polypeptide Synthesis. The DNA that carries the gene for a particular polypeptide is used to direct the synthesis of a molecule of hnRNA. The hnRNA is then modified by the removal of those segments corresponding to the introns of the DNA. This leaves messenger RNA, mRNA, the carrier of the message of the gene to the ribosomes. The ribosomes contain rRNA (ribosomal RNA). The overall change from gene to mRNA is called **transcription.** mRNA is like an assembly line at a factory (ribosomes) awaiting the arrival of parts—the amino acids. These are borne on transfer RNA molecules, tRNA.

A unit of three bases on mRNA constitutes a **codon** for one amino acid. A tRNA–amino acid unit has an **anticodon** that can pair to a codon on mRNA. In this way a tRNA–amino acid unit "recognizes" only one place on mRNA for delivering the amino acid to the polypeptide developing at the mRNA assembly line. The key that relates codons to specific amino acids is the **genetic code** and it is essentially the same code for all organisms in nature. The involvement of mRNA in directing the synthesis of a polypeptide is called **translation.**

Even small changes or deletions in the sequences of bases on DNA can have large consequences in the form of genetic diseases. Viruses are able to use their own nucleic acids to take over the genetic apparatus of the host tissue.

TOOLS FOR PROBLEM SOLVING

In this chapter you learned to apply the following concepts as tools in solving problems related to organic chemistry. Study each tool carefully so that you know what each is used for. When faced with solving a problem, recall what each tool does and consider whether it will be helpful in finding a solution. This will aid you in selecting the tools you need.

Functional group concept *(page 903)* If you can recognize a functional group in a molecule, you can place the structure into an organic family and so predict the kinds of reactions the compound can give. In particular, study Table 22.1.

Convention in using polygons to represent rings *(page 904)* A carbon atom is understood at each corner; other elements in the ring are explicitly written. Edges of the polygon represent covalent bonds; double bonds are explicitly shown. The remaining bonds, as required by the covalence of the atom at a corner, are understood to hold H atoms.

IUPAC rules of nomenclature *(pages 907, 910, 913, 914, and 917–921)* Each family of compounds has a characteristic name ending:

-ane	alkanes	-al	aldehydes
-ene	alkenes	-one	ketones
-yne	alkynes	-oic acid	carboxylic acids
-ol	alcohols	-oate	esters

Each member of a family has a defined "parent," which is the longest chain that includes the functional group. How the chain is numbered varies with the family, but generally the numbering starts from whichever end of the parent chain reaches the functional group with the lower number. The names of side chains, such as the alkyl groups, are included in the name ahead of the name of the parent, and their locations are designated by the numbers of the carbons at which they are attached to the parent.

QUESTIONS, PROBLEMS, AND EXERCISES

Answers to problems whose numbers are printed in color are given in Appendix B. More challenging problems are marked with asterisks.

REVIEW QUESTIONS

Structural Formulas

22.1 In general terms, what makes possible so many organic compounds?

22.2 What is the condensed structure of R if R—CH_3 represents the compound ethane?

22.3 Which of the following structures are possible, given the numbers of bonds that various atoms can form?
(a) $CH_2CH_2CH_3$
(b) $CH_3{=}CHCH_2CH_3$
(c) $CH_3CH{=}CHCH_2CH_3$

22.4 Write neat, condensed structures of the following.

(a)

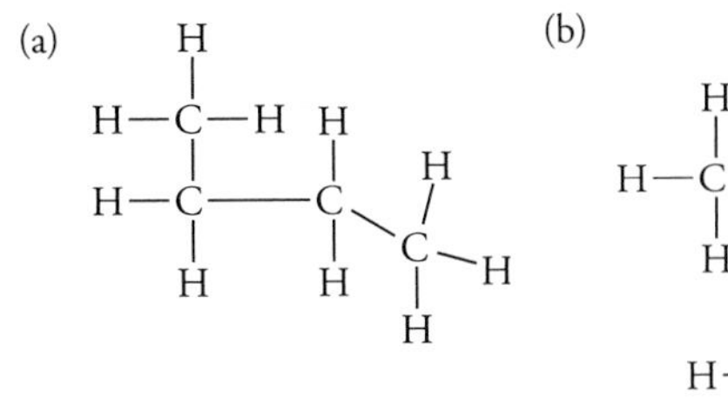

(b)

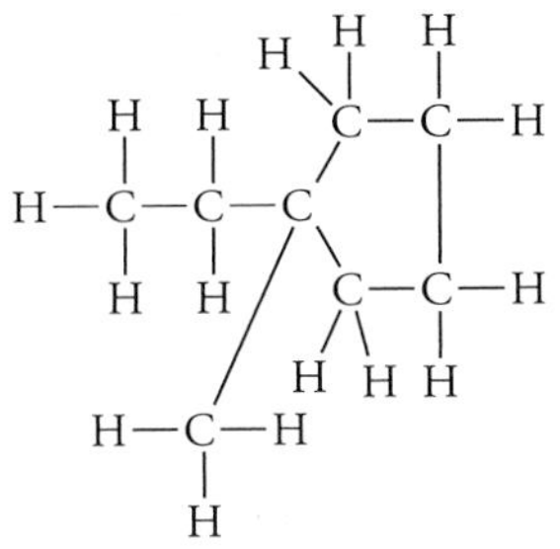

(c)

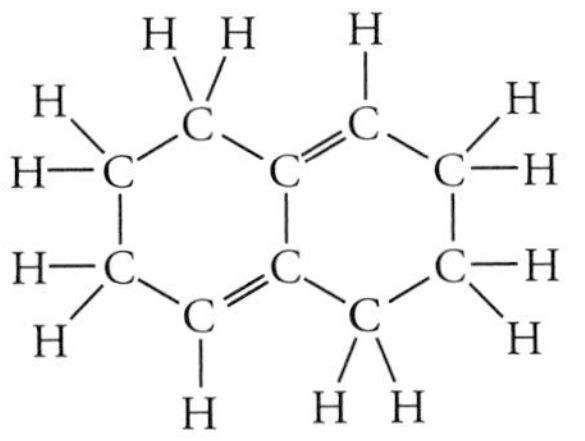

Families of Organic Compounds

22.5 In $CH_3CH_2NH_2$, the NH_2 group is called the *functional* group. In *general terms*, why is it called this?

22.6 In general terms, why do functional groups impart more chemical reactivity to families that have them? Why don't the alkanes display as many reactions as, say, the amines?

Isomers

22.7 What must be true about two substances if they are to be called *isomers* of each other?

22.8 What prevents free rotation about a double bond and so makes geometric isomers possible?

Properties and Structure

22.9 Explain why $CH_3CH_2CH_2OH$ is more soluble in water than $CH_3CH_2CH_2CH_3$.

22.10 Of the two compounds in Question 22.9, which has the higher boiling point? Explain.

22.11 Examine the structures of the following compounds.

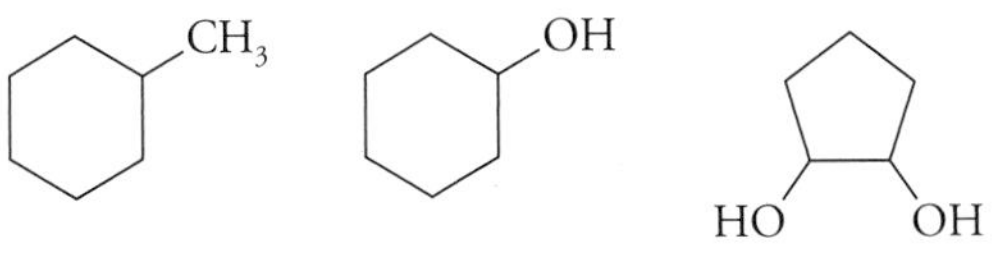

A B C

(a) Which has the highest boiling point? Explain.
(b) Which has the lowest boiling point? Explain.

22.12 Which of the following compounds is more soluble in water? Explain.

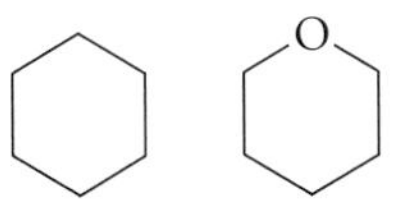

A B

22.13 Which of the following compounds has the higher boiling point? Explain.

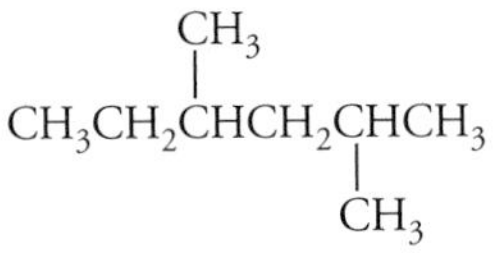

A:
CH_3
$CH_3CH_2CHCH_2CHCH_3$
CH_3

B: $CH_3CH_2CH_2CH_2CH_3$

22.14 Why do aldehydes and ketones have boiling points that are lower than those of their corresponding alcohols?

22.15 Acetic acid boils at 118 °C, higher even than 1-propanol, which boils at 97 °C. How can the higher boiling point of acetic acid be explained?

22.16 Methyl ethanoate has many more atoms than its parent acid, ethanoic acid. Yet methyl ethanoate (BP 59 °C) boils much lower than ethanoic acid (BP 118 °C). How can this be explained?

***22.17** Ethanamide is a solid at room temperature. 1-Aminobutane, which has about the same molecular mass, is a liquid. Explain.

22.18 Amines, RNH_2, do not have as high boiling points as alcohols with comparable numbers of atoms. Why?

Names

22.19 Write the condensed structures of the following compounds.
(a) 2,2-dimethyloctane
(b) 1,3-dimethylcyclopentane
(c) 1,1-diethylcyclohexane
(d) 6-ethyl-5-isopropyl-7-methyl-1-octene
(e) *cis*-2-pentene

22.20 Write condensed structures of the following compounds.
(a) 3-methylbutanal
(b) 4-methyl-2-octanone
(c) 2-chloropropanoic acid
(d) isopropyl ethanoate
(e) 2-methylbutanamide

22.21 Write condensed structures of the following compounds.
(a) 2,3-butanedione
(b) butanedicarboxylic acid
(c) 2-aminopropanal
(d) cyclohexyl 2-methylpropanoate
(e) sodium 2,3-dimethylbutanoate

22.22 "3-Butanol" is not a proper name, but a structure could still be written for it. What is this structure, and what IUPAC name should be used?

22.23 No number is needed to identify the location of the double bond in $CH_3CH{=}CH_2$, propene. Why?

Chemical Properties of Organic Compounds

22.24 What are the products of the complete combustion of any hydrocarbon, assuming an abundant supply of oxygen?

***22.25** Propene is known to react with concentrated sulfuric acid as follows.

$$CH_3CH{=}CH_2 + H_2SO_4 \longrightarrow CH_3\underset{\displaystyle OSO_3H}{\underset{|}{C}}HCH_3$$

The reaction occurs in two steps. Write the equation for the first.

***22.26** 2-Methylpropene reacts with hydrogen chloride as follows.

$$CH_3\overset{\displaystyle CH_3}{\overset{|}{C}}{=}CH_2 + H{-}Cl \longrightarrow CH_3\overset{\displaystyle CH_3}{\overset{|}{\underset{\displaystyle Cl}{\underset{|}{C}}}}CH_3$$

The reaction occurs in two steps. Write the equation of each step.

22.27 The compound 2-butene exists as two isomers, but when both isomers are hydrogenated, the products are identical. Explain.

22.28 Briefly explain how the C—O bond in isopropyl alcohol is weakened when a strong acid is present.

22.29 Which isomer of 1-butanol cannot be oxidized by dichromate ion? Write its structure and IUPAC name.

22.30 A monofunctional organic oxygen compound dissolves in aqueous base but not in aqueous acid. The compound is in which of the families of organic compounds that we studied? Explain.

22.31 A monofunctional organic nitrogen compound dissolves in aqueous hydrochloric acid but not in aqueous sodium hydroxide. What kind of organic compound is it?

22.32 Why are amides neutral while amines are basic?

***22.33** Hydrazine is a Brønsted base but urea does not exhibit basic properties. Offer an explanation.

$$\underset{\text{hydrazine}}{NH_2NH_2} \qquad \underset{\text{urea}}{NH_2\overset{\displaystyle O}{\overset{\|}{C}}NH_2}$$

22.34 Write the equilibrium that is present when $CH_3CH_2CH_2NH_2$ is in an aqueous solution.

22.35 Predict which of the following species can neutralize the hydronium ion in dilute, aqueous HCl at room temperature.

(a) $CH_3\overset{\displaystyle O}{\overset{\|}{C}}O^-$ (c) HO^- (e) $CH_3CH_2CH_2CH_3$

(b) CH_3NH_2 (d) $CH_3\overset{\displaystyle O}{\overset{\|}{C}}NH_2$ (f)  pyrrolidine (five-membered ring containing NH)

22.36 Write the products that can be expected to form in the following situations. If no reaction occurs, write "N.R."
(a) $CH_3CH_2CH_2NH_2 + HBr(aq) \longrightarrow$
(b) $CH_3CH_2CH_2CH_3 + HI(aq) \longrightarrow$
(c) $CH_3CH_2CH_2NH_3^+ + H_3O^+ \longrightarrow$
(d) $CH_3CH_2CH_2NH_3^+ + OH^- \longrightarrow$

22.37 Write the equation for the equilibrium that is present in a solution of propanoic acid and methanol with a trace of strong acid.

Polymers

22.38 What is a *macromolecule*? Name two naturally occurring macromolecular substances.

22.39 What is a *polymer*? Are all macromolecules polymers?

22.40 What is a *monomer*? Draw structures for the monomers used to make (a) polypropylene, (b) poly(tetrafluoroethylene), and (c) poly(vinyl chloride).

22.41 What do we mean by the term *polymer backbone*?

22.42 What is the repeating unit in polypropylene? Write the formula for the polypropylene polymer. Give three uses for polypropylene.

22.43 How do propylene and the repeating unit in polypropylene differ?

22.44 What is the difference between an *addition polymer* and a *condensation polymer*?

22.45 Write the structure of polystyrene showing three of the repeating units. What are three uses for polystyrene plastic?

22.46 What is a *copolymer*?

22.47 Write the structural formula for (a) nylon 6,6 and (b) poly(ethylene terephthalate).

22.48 The structural formula for Kevlar is

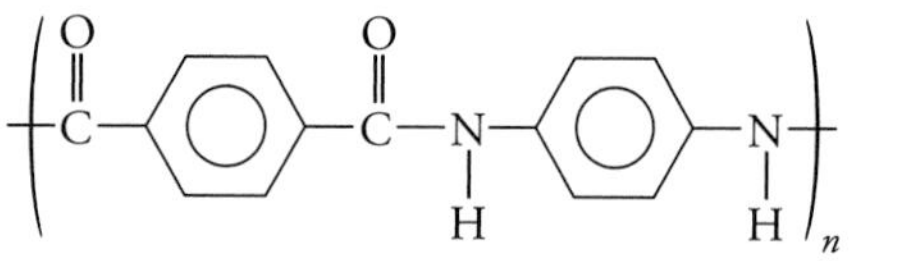

$$\left[\overset{\displaystyle O}{\overset{\|}{C}}{-}C_6H_4{-}\overset{\displaystyle O}{\overset{\|}{C}}{-}\underset{\displaystyle H}{\underset{|}{N}}{-}C_6H_4{-}\underset{\displaystyle H}{\underset{|}{N}}\right]_n$$

Identify the amide bond in the polymer.

22.49 Polycarbonate polymers are polyesters. An example is

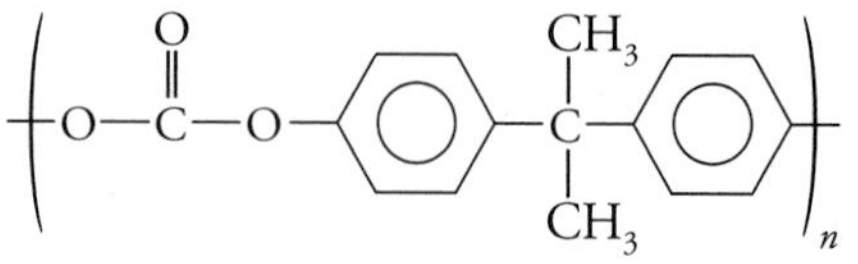

$$\left[O{-}\overset{\displaystyle O}{\overset{\|}{C}}{-}O{-}C_6H_4{-}\overset{\displaystyle CH_3}{\overset{|}{\underset{\displaystyle CH_3}{\underset{|}{C}}}}{-}C_6H_4\right]_n$$

Identify the structural feature that makes it a polyester. Identify the structural feature that makes it a poly*carbonate*.

22.50 What is meant by the term *branching* as applied to polymers? How does branching affect the properties of low density polyethylene?

22.51 What is *cross-linking*? How does it affect the properties of a polymer?

22.52 What is *vulcanized rubber*?

22.53 How does polymer crystallinity affect the physical properties of the polymer?

22.54 Why doesn't low density polyethylene form strong crystalline polymer fibers?

22.55 Why does nylon form strong fibers?

22.56 Show how hydrogen bonding can bind Kevlar polymer chains together. (See Question 22.48.)

22.57 What are some applications of crystalline polymers such as HDPE, UHMWPE, and Kevlar?

Biochemistry

22.58 What is *biochemistry*?

22.59 What are the three fundamental needs for sustaining life and what are the general names for the substances that supply those needs?

Carbohydrates

22.60 How are *carbohydrates* defined?

22.61 What monosaccharide forms when the following polysaccharides are completely hydrolyzed? (a) starch, (b) glycogen, (c) cellulose

22.62 Glucose exists in three forms, only one of which is a polyhydroxyaldehyde. How then can an aqueous solution of any form of glucose undergo the reactions of an aldehyde?

22.63 Name the compounds that form when sucrose is digested.

22.64 The digestion of lactose gives what compounds? Name them.

22.65 The complete hydrolysis of the following compounds gives what product(s)? (Give the names only.)
(a) amylose (b) amylopectin

22.66 Describe the relationships among amylose, amylopectin, and starch.

22.67 Why are humans unable to use cellulose as a source of glucose?

22.68 What function is served by glycogen in the body?

Lipids

22.69 How are *lipids* defined?

22.70 Why are lipids more soluble than carbohydrates in nonpolar solvents?

22.71 Cholesterol is not an ester, yet it is defined as a lipid. Why?

22.72 A product such as corn oil is advertised as "polyunsaturated," but all nutritionally important lipids have unsaturated C=O groups. What does *polyunsaturated* refer to?

***22.73** Is it likely that the following compound could be obtained by the hydrolysis of a naturally occurring lipid? Explain.

$$CH_3(CH_2)_2CH{=}CHCH_2CH{=}CHCH_2CH{=}CH(CH_2)_7CO_2H$$

Amino Acids, Polypeptides, and Proteins

22.74 Describe the specific ways in which the monomers for all polypeptides are (a) alike and (b) different.

22.75 What is the *peptide bond*? How is it similar to the amide bond in nylon?

22.76 Describe the structural way in which two *isomeric* polypeptides would be different.

22.77 Describe the structural ways in which two different polypeptides can differ.

22.78 Why is a distinction made between the terms *polypeptide* and *protein*?

22.79 In general terms, what forces are at work that determine the final *shape* of a polypeptide strand?

22.80 What kind of substance makes up most enzymes?

22.81 The most lethal chemical poisons act on what kinds of substances? Give examples.

Nucleic Acids and Heredity

22.82 In general terms only, how does the body solve the problem of getting a particular amino acid sequence rather than a mixture of randomly organized sequences into a polypeptide?

22.83 In what specific structural way does DNA carry genetic messages?

22.84 What kind of force occurs between the two DNA strands in a double helix?

22.85 How are the two DNA strands in a double helix structurally related?

22.86 In what ways do DNA and RNA differ structurally?

22.87 Which base pairs with (a) A in DNA, (b) A in RNA, and (c) C in DNA or RNA?

22.88 A specific gene in the DNA of a higher organism is said to be *segmented* or *interrupted*. Explain.

22.89 On what kind of RNA are codons found?

22.90 On what kind of RNA are anticodons found?

22.91 What function does each of the following have in polypeptide synthesis?
(a) a ribosome (b) mRNA (c) tRNA

22.92 What kind of RNA forms directly when a cell uses DNA at the start of the synthesis of a particular enzyme? What kind of RNA is made from it?

22.93 The process of *transcription* begins with which nucleic acid and ends with which one?

22.94 The process of *translation* begins with which nucleic acid? What is the end result of translation?

22.95 What is a virus, and (in general terms only) how does it infect a tissue?

22.96 Use the internet to find what genetic engineering is able (in general terms) to accomplish.

REVIEW PROBLEMS

Structural Formulas

22.97 Write full (expanded) structures for each of the following molecular formulas. Remember how many bonds the various kinds of atoms must have. In some you will have to use double or triple bonds. (Hint: A trial-and-error approach will have to be used.)
(a) CH_5N (c) $CHCl_3$ (e) C_2H_2
(b) CH_2Br_2 (d) NH_3O (f) N_2H_4

22.98 Write full (expanded) structures for each of the following molecular formulas. Follow the guidelines of Problem 22.97.

(a) C_2H_6 (c) CH_2O (e) C_3H_3N
(b) CH_2O_2 (d) HCN (f) CH_4O

Families of Organic Compounds

22.99 Name the family to which each compound belongs.

(a) $CH_3CH{=}CH_2$
(b) CH_3CH_2OH
(c) $CH_3CH_2\overset{O}{\overset{\|}{C}}OCH_3$
(d) $CH_3CH_2CH_2\overset{O}{\overset{\|}{C}}OH$
(e) $CH_3CH_2CH_2NH_2$
(f) $HOCH_2CH_2CH_3$

22.100 To what organic family does each compound belong?

(a) $CH_3C{\equiv}CH$
(b) $CH_3CH_2\overset{O}{\overset{\|}{C}}H$
(c) $CH_3\overset{O}{\overset{\|}{C}}CH_2CH_2CH_3$
(d) $CH_3{-}O{-}CH_2CH_3$
(e) $CH_3CH_2NH_2$

$CH_3CH_2\overset{O}{\overset{\|}{C}}NH_2$

22.101 Identify by letter the compounds in Problem 22.99 that are saturated compounds.

22.102 Which parts of Problem 22.100 give the structures of saturated compounds?

22.103 Classify the following compounds as amines or amides. If another functional group occurs, name it, too.

(a) $CH_3CH_2NH_2$
(b) $CH_3CH_2NHCH_3$
(c) $CH_3CH_2\overset{O}{\overset{\|}{C}}NH_2$
(d) $CH_3\overset{O}{\overset{\|}{C}}CH_2NH_2$

22.104 Name the functional groups present in each of the following structures.

(a) 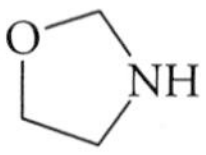 O, NH (five-membered ring)

(b) 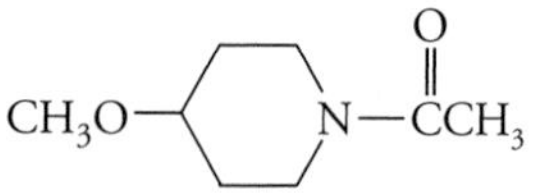 $CH_3O{-}\langle\text{ring}\rangle N{-}\overset{O}{\overset{\|}{C}}CH_3$

(c) O, O, C=O, H (five-membered ring)

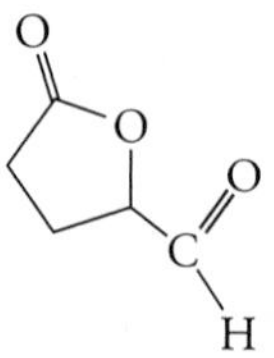

(d) $CH_3O\overset{O}{\overset{\|}{C}}CH_2CH_2\overset{O}{\overset{\|}{C}}OH$

Isomers

22.105 Decide whether the members of each pair are identical, are isomers, or are unrelated.

(a) $CH_3{-}CH_3$ and $CH_3{-}CH_3$ (drawn vertically)

(b) 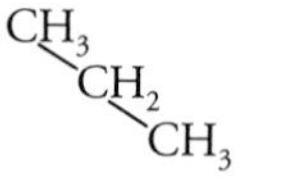 $CH_3CH_2CH_3$ (drawn zigzag) and $CH_3CH_2CH_3$ (drawn bent)

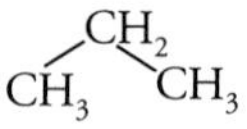

(c) CH_3CH_2OH and $CH_3CH_2CH_2OH$

(d) $CH_3CH{=}CH_2$ and cyclopropane ($CH_2{-}CH_2{-}CH_2$ ring)

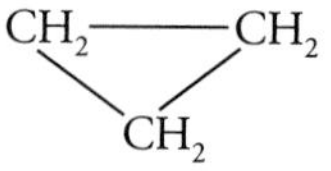

(e) $H{-}\overset{O}{\overset{\|}{C}}{-}CH_3$ and $CH_3{-}\overset{O}{\overset{\|}{C}}{-}H$

(f) $CH_3CH(CH_3)CH_3$ and $CH_3CH(CH_3)CH_3$ (second CH_3 drawn below)

(g) $CH_3CH_2CH_2NH_2$ and $CH_3CH_2{-}NH{-}CH_3$

22.106 Examine each pair and decide if the two are identical, are isomers, or are unrelated.

(a) $CH_3CH_2\overset{O}{\overset{\|}{C}}OH$ and $HO\overset{O}{\overset{\|}{C}}CH_2CH_3$

(b) $H\overset{O}{\overset{\|}{C}}OCH_2CH_3$ and $CH_3CH_2\overset{O}{\overset{\|}{C}}OH$

(c) $H\overset{O}{\overset{\|}{C}}OCH_2CH_2OH$ and $HOCH_2CH_2\overset{O}{\overset{\|}{C}}OH$

(d) $CH_3\overset{O}{\overset{\|}{C}}CH_2CH_3$ and $CH_3CH_2\overset{O}{\overset{\|}{C}}CH_3$

(e) 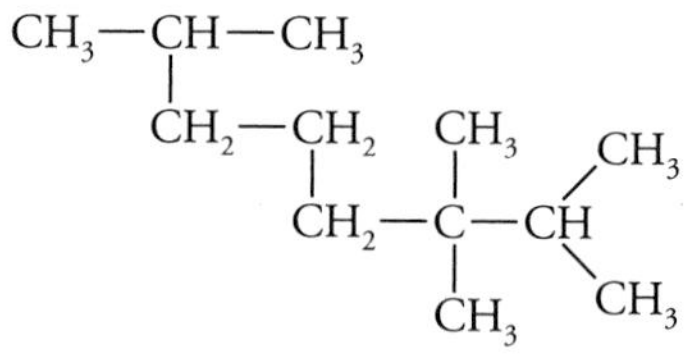 $CH_3{-}CH(CH_2{-}CH_2{-}CH_2{-}C(CH_3)_2{-}CH(CH_3)_2){-}CH_3$

and

$CH_3{-}CH(CH_3){-}CH_2{-}CH_2{-}CH_2{-}C(CH_3)_2{-}CH(CH_3){-}CH_3$

(f) $CH_3{-}NH{-}\overset{O}{\overset{\|}{C}}{-}CH_3$ and $CH_3CH_2\overset{O}{\overset{\|}{C}}NH_2$

(g) $H{-}O{-}O{-}H$ and $H{-}O{-}H$

Names of Hydrocarbons

22.107 Write the IUPAC names of the following hydrocarbons.

(a) $CH_3CH_2CH_2CH_2CH_3$

(b) $CH_3CH_2CH_2CHCH_3$ (with CH_3 on the CH)

(c)
CH_3
$CH_3CHCH_2CHCH_2CH_3$
CH_3

22.108 Write the IUPAC names of the following compounds.

(a)
CH_3
$CH_3CH_2CHCH_2CHCH_3$
CH_3

(b) $CH_3CH_2CH{=}CHCH_2CH_3$

(c)
CH_3
$CH_3CHCH{=}CHCH_3$

Geometric Isomerism

22.109 Write the structures of the cis and trans isomers, *if any,* for the following.

(a) $CH_2{=}CHBr$

(b) $CH_3CH{=}CHCH_2CH_3$

(c)
Br
$CH_3C{=}CHCl$

22.110 Write the structures of the cis and trans isomers, if any, for the following compounds.

(a)
CH_3
$CH_3C{=}CHCH_3$

(b)
CH_3
$CH_3CH{=}CCH_2CH_3$

(c)
Cl Cl
$CH_3CH_2C{=}CCH_2CH_3$

Chemical Reactions of Alkenes

22.111 Write the structures of the products that form when ethylene reacts with each of the following substances by an addition reaction. (Assume that needed catalysts or other conditions are provided.)

(a) H_2 (c) Br_2 (e) $HBr(g)$
(b) Cl_2 (d) $HCl(g)$ (f) H_2O (in acid)

22.112 The isopropyl cation is more stable than the propyl cation, so the former forms almost exclusively when propene undergoes addition reactions with proton-donating species. On the basis of this fact, predict the final products of the reaction of propene with each of the following reagents.

(a) hydrogen chloride (b) hydrogen iodide
(c) water in the presence of an acid catalyst

22.113 Repeat Problem 22.111 using 2-butene.

22.114 Repeat Problem 22.112 using cyclohexene.

22.115 In general terms, why doesn't benzene undergo the same kinds of addition reactions as cyclohexene?

22.116 If benzene were to *add* one mole of Br_2, what would form? What forms, instead, when Br_2, benzene, and an $FeBr_3$ catalyst are heated together? (Write the structures.)

Structures and Properties of Other Functional Groups

22.117 Write condensed structures and the IUPAC names for all of the saturated alcohols with three or fewer carbon atoms per molecule.

22.118 Write the condensed structures and the IUPAC names for all of the possible alcohols with the general formula $C_4H_{10}O$.

22.119 Write the condensed structures of all of the possible ethers with the general formula $C_4H_{10}O$. Following the pattern for the common names of ethers given in the chapter, what are the likely common names of these ethers?

22.120 Write the structures of the isomeric alcohols of the formula $C_4H_{10}O$ that could be oxidized to aldehydes. Write the structure of the isomer that could be oxidized to a ketone.

***22.121** Write the structures of the products of the acid-catalyzed dehydrations of the following compounds

(a) OH (on cyclopentane ring)

(b)
OH
(benzene ring)$-CHCH_3$

(c) (benzene ring)$-CH_2CH_2OH$

22.122 Write the structures of the substitution products that form when the alcohols of Problem 22.121 are heated with concentrated hydriodic acid.

22.123 Write the structures of the aldehydes or ketones that can be prepared by the oxidation of the compounds given in Problem 22.121.

***22.124** Write the structure of the product of the acid-catalyzed dehydration of 2-propanol. Write the mechanism of the reaction.

***22.125** When ethanol is heated in the presence of an acid catalyst, ethene and water form; an elimination reaction occurs. When 2-butanol is heated under similar conditions, *two* alkenes form (although not in equal quantities). Write the structures and IUPAC names of the two alkenes.

***22.126** If the formation of the two alkenes in Problem 22.125 were determined purely by statistics, then the mole ratio of the two alkenes should be in the ratio of what whole numbers? Why?

22.127 Which of the following two compounds could be easily oxidized? Write the structure of the product of the oxidation.

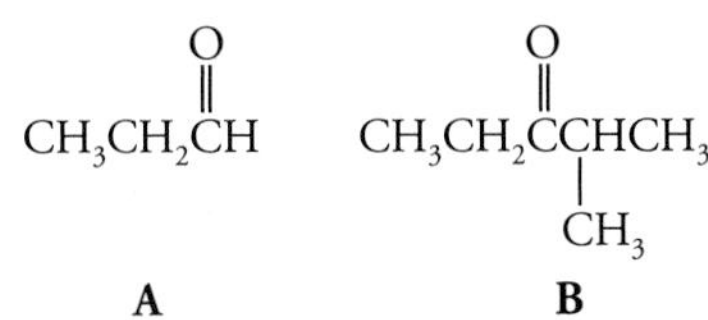

22.128 Which of the following compounds has the chemical property given below? Write the structure of the product of the reaction.

$$\underset{\textbf{A}}{CH_3CH_2CH_2OH} \qquad \underset{\textbf{B}}{CH_3CH_2\overset{O}{\overset{\|}{C}}OH} \qquad \underset{\textbf{C}}{CH_3CH_2\overset{O}{\overset{\|}{C}}H}$$

(a) is easily oxidized
(b) neutralizes NaOH
(c) forms an ester with methyl alcohol
(d) can be oxidized to an aldehyde
(e) can be dehydrated to an alkene

22.129 Write the structures of the products that form in each of the following situations. If no reaction occurs, write "N.R."

(a) $CH_3CH_2CO_2^- + HCl(aq) \longrightarrow$

(b) $CH_3CH_2CO_2CH_3 + H_2O \xrightarrow{\text{heat}}$

(c) $CH_3CH_2CH_2CO_2H + NaOH(aq) \longrightarrow$

22.130 Write the structures of the products that form in each of the following situations. If no reaction occurs, write "N.R."

(a) $CH_3CH_2CO_2H + NH_3 \xrightarrow{\text{heat}}$

(b) $CH_3CH_2CH_2CONH_2 + H_2O \xrightarrow{\text{heat}}$

(c) $CH_3CH_2CO_2CH_3 + NaOH(aq) \xrightarrow{\text{heat}}$

*__22.131__ What organic products with smaller molecules form when the following compound is heated with water?

$$CH_3CH_2\underset{\underset{CH_3CH_2}{|}}{N}\overset{O}{\overset{\|}{C}}CH_3$$

22.132 Which of the following species undergo the reaction specified? Write the structures of the products that form.

$$CH_3CH_2\overset{O}{\overset{\|}{C}}NH_2 \qquad CH_3CH_2CH_2NH_2 \qquad CH_3CH_2NH_3^+$$

(a) neutralizes dilute hydrochloric acid
(b) hydrolyzes (reacts with water)
(c) neutralizes dilute sodium hydroxide

Polymers

22.133 The structure of vinyl acetate is shown below. This compound forms an addition polymer in the same way as ethylene and propylene. Draw a segment of the poly(vinyl acetate) polymer that contains three of the repeating units.

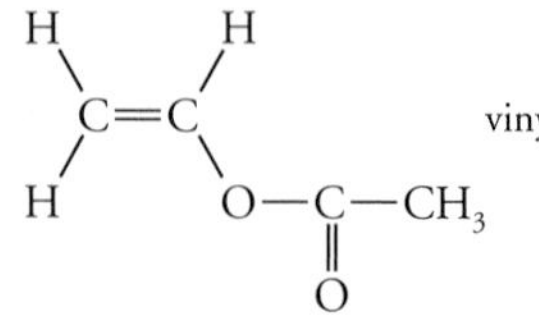

$$\begin{matrix} H & & H \\ & C{=}C & \\ H & & O{-}\overset{}{\underset{\underset{O}{\|}}{C}}{-}CH_3 \end{matrix}$$

vinyl acetate

22.134 If the following two compounds polymerized in the same way that Dacron forms, what would be the repeating unit of their polymer?

$$HO\overset{O}{\overset{\|}{C}}CH_2CH_2\overset{O}{\overset{\|}{C}}OH \qquad HOCH_2CH_2OH$$

22.135 Kodel, a polyester made from the following monomers, is used to make fibers for weaving crease-resistant fabrics.

$$HOCH_2{-}C_6H_{10}{-}CH_2OH \qquad HO\overset{O}{\overset{\|}{C}}{-}C_6H_4{-}\overset{O}{\overset{\|}{C}}OH$$

monomers for Kodel

What is the structure of the repeating unit in Kodel?

22.136 If the following two compounds polymerized in the same way that nylon forms, what would be the repeating unit in the polymer?

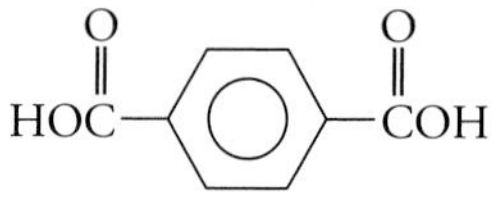

$$HO\overset{O}{\overset{\|}{C}}{-}C_6H_4{-}\overset{O}{\overset{\|}{C}}OH \qquad NH_2CH_2CH_2CH_2CH_2NH_2$$

Lipids

22.137 Write the structure of a triacylglycerol that could be made from palmitic acid, oleic acid, and linoleic acid.

22.138 Write the structures of the products of the complete hydrolysis of the following triacylglycerol.

$$\begin{array}{l} CH_2O\overset{O}{\overset{\|}{C}}(CH_2)_7CH{=}CHCH_2CH{=}CH(CH_2)_4CH_3 \\ | \\ CHO\overset{O}{\overset{\|}{C}}(CH_2)_{12}CH_3 \\ | \\ CH_2O\overset{O}{\overset{\|}{C}}(CH_2)_7CH{=}CH(CH_2)_7CH_3 \end{array}$$

22.139 Write the structure of the triacylglycerol that would result from the complete hydrogenation of the lipid whose structure is given in Problem 22.138.

22.140 If the compound in Problem 22.138 is saponified, what are the products? Give their names and structures.

22.141 What parts of glycerophospholipid molecules provide hydrophobic sites? Hydrophilic sites?

22.142 In general terms, describe the structure of the membrane of an animal cell. What kinds of molecules are present? How are they arranged? What forces keep the membrane intact? What functions are served by the protein components?

Amino Acids and Polypeptides

22.143 Write the structure of the dipeptide that could be hydrolyzed to give two molecules of glycine.

22.144 What is the structure of the tripeptide that could be hydrolyzed to give three molecules of alanine?

22.145 What are the structures of the two dipeptides that could be hydrolyzed to give glycine and phenylalanine?

22.146 Write the structures of the dipolar ionic forms of the amino acids that are produced by the complete digestion of the following compound.

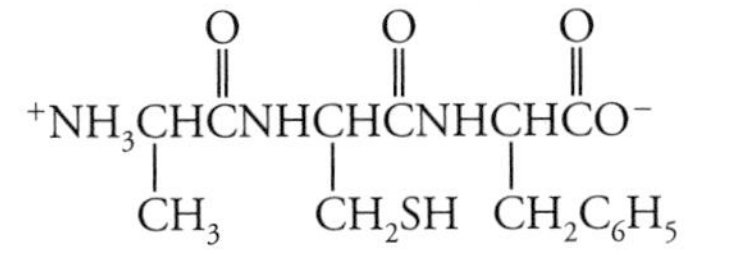

$$^{+}NH_3\underset{CH_3}{CH}\overset{O}{\overset{\|}{C}}NH\underset{CH_2SH}{CH}\overset{O}{\overset{\|}{C}}NH\underset{CH_2C_6H_5}{CH}\overset{O}{\overset{\|}{C}}O^{-}$$

ADDITIONAL EXERCISES

***22.147** The H_2SO_4-catalyzed addition of water to 2-methylpropene could give two isomeric alcohols.
(a) Write their condensed structures.
(b) Actually, only one forms. It is not possible to oxidize this alcohol to an aldehyde or to a ketone having four carbons. Which alcohol, therefore, is the product?
(c) Write the structures of the two possible carbocations that could form when sulfuric acid donates a proton to 2-methylpropene.
(d) One of the two carbocations of part (c) is more stable than the other. Which one is it, and how can you tell?

22.148 An unknown alcohol was either **A** or **B** below. When the unknown was oxidized with an *excess* of strong oxidizing agent, sodium dichromate in acid, the *organic* product isolated was able to neutralize sodium hydroxide.

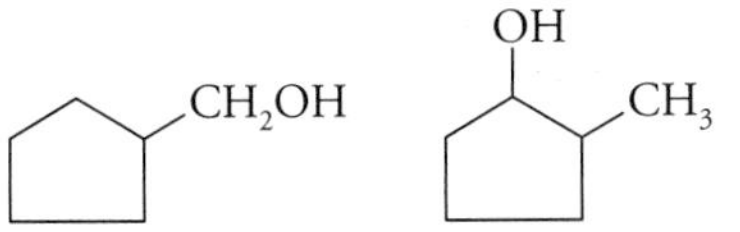

A: cyclopentane ring bearing CH_2OH
B: cyclopentane ring bearing OH and CH_3 on adjacent carbons

(a) Which was the alcohol? Give the reasons for your choice.
(b) Write the balanced net ionic equation for the oxidation.

***22.149** Suggest a reason why trimethylamine, $(CH_3)_3N$, has a *lower* boiling point (BP 3 °C) than dimethylamine, $(CH_3)_2NH$ (BP 8 °C), despite having a larger number of atoms.

***22.150** Write the structures of the organic products that form in each situation. If no reaction occurs, write "N.R."

(a) 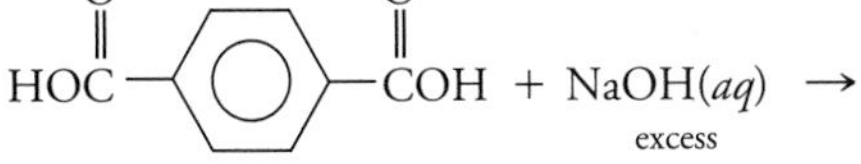 $HO\overset{O}{\overset{\|}{C}}-C_6H_4-\overset{O}{\overset{\|}{C}}OH + NaOH(aq) \xrightarrow[\text{excess}]{}$

(b) $CH_3\overset{O}{\overset{\|}{C}}CH_2CH_2CH_3 + H_2 \xrightarrow[\text{catalyst}]{\text{heat, pressure,}}$

(c) $CH_3\overset{+}{N}H_2CH_2CH_2CH_3 + OH^{-}(aq) \rightarrow$

(d) $CH_3CH_2OCH_2CH_2\overset{O}{\overset{\|}{C}}OCH_3 + H_2O \xrightarrow{\text{heat}}$

(e) 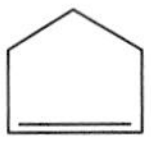 cyclopentene $+ H_2O \xrightarrow[\text{catalyst}]{H_2SO_4}$

(f) 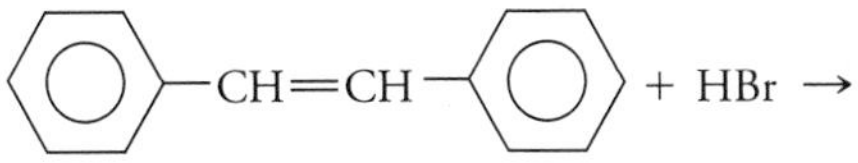 $C_6H_5-CH{=}CH-C_6H_5 + HBr \rightarrow$

***22.151** How many tripeptides can be made from three different amino acids? If the three are glycine (Gly), alanine (Ala), and phenylalanine (Phe), what are the structures of the possible tripeptides? (Hint: Gly-Ala-Phe is one structure.)

***22.152** A 0.5574 mg sample of an organic acid, when burned in pure oxygen, gave 1.181 mg of CO_2 and 0.3653 mg H_2O. Calculate the empirical formula of the acid.

***22.153** When 0.2081 g of an organic acid was dissolved in 50.00 mL of 0.1016 *M* NaOH, it took 23.78 mL of 0.1182 *M* HCl to neutralize the NaOH that was not used up by the sample of the organic acid. Calculate the formula mass from the data given. Is the answer necessarily the *molecular mass* of the organic acid? Explain.

EXERCISES IN CRITICAL THINKING

22.154 The compound CH_2Cl_2 only exists as one isomer. Why does this support the statement that the atoms in the compound are arranged in a tetrahedral rather than a square planar structure?

22.155 The α-carbon atom in an amino acid is a chiral center (except for glycine). Of the two possible enantiomers of these optically active substances, only one is produced naturally. What does this suggest about the mechanism whereby amino acids are synthesized in living creatures?

22.156 Use resonance structures to explain why urea, $(NH_2)_2C{=}O$, is not basic like typical amines.

22.157 Below is the ring structure of a sugar molecule. How many chiral centers does this sugar molecule have? How many isomers are possible?

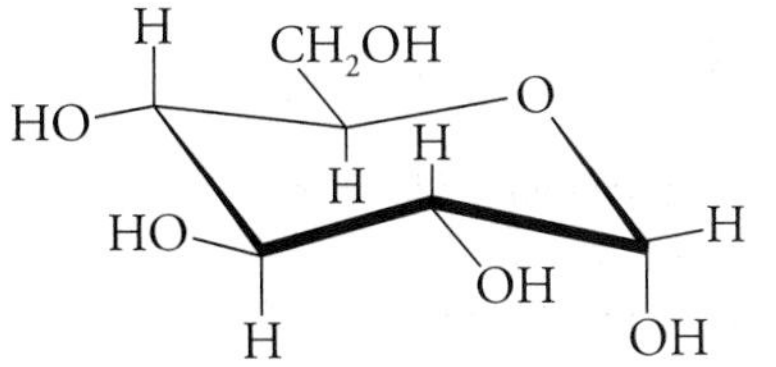

(Ring structure with labels: H, CH_2OH, HO, O, H, H, HO, H, OH, H, OH)

22.158 What would be the repeating unit of the polymer formed by condensation polymerization of the two monomers shown below?

$$CH_3O-\overset{O}{\overset{\|}{C}}-CH_2-CH_2-CH_2-\overset{O}{\overset{\|}{C}}-OCH_3$$

$$H_2N-CH_2-CH_2-NH_2$$

CHAPTERS 18-22

BRINGING IT TOGETHER

Again you have an opportunity to test your understanding of concepts, your knowledge of scientific terms, and your skills at solving chemistry problems. Read through the following questions carefully, and answer each as fully as possible. Review topics when necessary.

1. What is meant by the term *spontaneous change*?
2. Which of these are state functions: E, H, q, S, G, w, T?
3. How is entropy related to statistical probability?
4. What would be the algebraic signs of ΔS for the following reactions?
   (a) $Br_2(l) + Cl_2(g) \longrightarrow 2BrCl(g)$
   (b) $CaO(s) + CO_2(g) \longrightarrow CaCO_3(s)$
5. Which of the following states has the greatest entropy?
   (a) $2H_2O(l)$ (c) $2H_2(l) + O_2(g)$ (e) $4H(g) + 2O(g)$
   (b) $2H_2O(s)$ (d) $2H_2(g) + O_2(g)$
6. Calculate $\Delta S°$ (in J K^{-1}) for the following reactions.
   (a) $H_2O(l) + SO_3(g) \longrightarrow H_2SO_4(l)$
   (b) $2KCl(s) + H_2SO_4(l) \longrightarrow K_2SO_4(s) + 2HCl(g)$
   (c) $C_2H_4(g) + H_2O(g) \longrightarrow C_2H_5OH(l)$
7. Calculate $\Delta G°$ (in kJ) for the following reactions.
   (a) $CaSO_4 \cdot \frac{1}{2}H_2O(s) + \frac{3}{2}H_2O(l) \longrightarrow CaSO_4 \cdot 2H_2O(s)$
   (b) $CH_4(g) + Cl_2(g) \longrightarrow CH_3Cl(g) + HCl(g)$
   (c) $CaSO_4(s) + CO_2(s) \longrightarrow CaCO_3(s) + SO_3(g)$
8. Which of the reactions in the preceding question would appear to be spontaneous?
9. Calculate $\Delta S°$, $\Delta H°$, and $\Delta G_T°$ (at 400 °C) using energy units of joules or kilojoules for these reactions.
   (a) $CaSO_4{\cdot}2H_2O(s) \longrightarrow CaSO_4{\cdot}\frac{1}{2}H_2O(s) + \frac{3}{2}H_2O(g)$
   (b) $NaOH(s) + NH_4Cl(s) \longrightarrow NaCl(s) + NH_3(g) + H_2O(g)$
   (c) $SO_3(g) \longrightarrow SO_2(g) + \frac{1}{2}O_2(g)$
10. For the reaction $3NO(g) \rightleftharpoons NO_2(g) + N_2O(g)$, calculate K_P at 25 °C using values of $\Delta G_f°$ from Table 18.2. Calculate the value of K_c at 25 °C for the reaction.
11. Consider the reaction

$$CH_4(g) + Cl_2(g) \rightleftharpoons CH_3Cl(g) + HCl(g)$$

   (a) Calculate $\Delta G_{473}°$ for the reaction at 200 °C.
   (b) What is the value of K_P for the reaction at 200 °C?
   (c) What is the value of K_c for the reaction at 200 °C?
12. Glycine, one of the important amino acids, has the structure

```
   H  O
   |  ||
H—N—C—C—O—H
   |  |
   H  H
```

   Calculate the atomization energy of the glycine molecule in units of kJ mol^{-1} from the data in Table 18.4.
13. Use bond energies in Table 18.4 to calculate the approximate energy that would be absorbed or given off in the formation of 25.0 g of C_2H_6 by the following reaction in the gas phase.

```
                         H  H
                         |  |
H—C≡C—H + 2H2 ——→ H—C—C—H
                         |  |
                         H  H
```

14. Sketch a diagram of an electrolysis cell in which a concentrated solution of NaCl is undergoing electrolysis.
   (a) Label the cathode and anode, including their charges.
   (b) Write half-reactions for the changes taking place at the electrodes.
   (c) Write a balanced equation for the net cell reaction.
15. Suppose that the electrolysis cell described in the previous question contains 250 mL of brine.
   (a) What will be the pH of the solution if the electrolysis is carried out for 20.0 minutes using a current of 1.00 A?
   (b) How many milliliters of H_2, measured at STP, would be evolved if the cell were operated at 5.00 A for 10.0 minutes?
16. What current would be required to deposit 0.100 g of nickel in 20.0 minutes from a solution of $NiSO_4$?
17. A large electrolysis cell can produce as much as 900 lb of aluminum in 1 day from Al_2O_3. What current is required to accomplish this? (Assume three significant figures.)
18. Sketch a diagram of a galvanic cell consisting of a copper electrode dipping into 1.00 *M* $CuSO_4$ solution and an iron electrode dipping into 1.00 *M* $FeSO_4$ solution.
   (a) Identify the cathode and the anode. Indicate the charge carried by each.
   (b) Write the equation for the net cell reaction.
   (c) Describe the cell by writing its standard cell notation.
   (d) What is the potential of the cell?
19. What is the purpose of a salt bridge in a galvanic cell?
20. Suppose the cell described in Question 18 contains 100 mL of each solution and is operated for a period of 50.0 hr at a constant current of 0.10 A. At the end of that time, what will be the concentrations of Cu^{2+} and Fe^{2+} in their respective solutions? What will the cell potential be at that point?
21. Use data from Table 19.1 to calculate the value of K_c at 25 °C for the reaction

$$O_2(g) + 4Br^-(aq) + 4H^+(aq) \longrightarrow 2Br_2(aq) + 2H_2O$$

22. A galvanic cell was assembled as follows. In one compartment, a copper electrode was immersed in a 1.00 *M* solution of $CuSO_4$. In the other compartment, a manganese electrode was immersed in a 1.00 *M* solution of $MnSO_4$. The potential of the cell was measured to be 1.52 V, with the Mn electrode as the negative electrode. What is the value of $E°$ for the following half-reaction?

$$Mn^{2+}(aq) + 2e^- \rightleftharpoons Mn(s)$$

23. Calculate $E_{cell}°$ for the reaction

$$3Cu(s) + 2NO_3^-(aq) + 8H^+(aq) \longrightarrow 2NO(g) + 4H_2O + 3Cu^{2+}(aq)$$

24. What is the value of K_c for the reaction described in the preceding question?

25. Consider a galvanic cell formed by using the half-reactions

$$NiO_2(s) + 2H_2O + 2e^- \rightleftharpoons Ni(OH)_2(s) + 2OH^-(aq) \quad E° = +0.49 \text{ V}$$

$$PbO_2(s) + H_2O + 2e^- \rightleftharpoons PbO(s) + 2OH^-(aq) \quad E° = +0.25 \text{ V}$$

(a) Write the equation for the spontaneous cell reaction.
(b) Calculate the value of $E°_{cell}$ for the system.
(c) Calculate $\Delta G°$ for the spontaneous cell reaction.

26. A galvanic cell was constructed in which one half-cell consists of a silver electrode coated with silver chloride dipping into a solution that contains chloride ion and the second half-cell consists of a nickel electrode dipping into a solution that contains Ni^{2+}. The half-cell reactions and their reduction potentials are

$$AgCl(s) + e^- \rightleftharpoons Ag(s) + Cl^-(aq) \quad E° = +0.222 \text{ V}$$

$$Ni^{2+}(aq) + 2e^- \rightleftharpoons Ni(s) \quad E° = -0.257 \text{ V}$$

(a) Write the equation for the spontaneous cell reaction.
(b) Calculate $E°_{cell}$ for the cell reaction.
(c) Write the Nernst equation for the cell.
(d) Calculate the cell potential if $[Cl^-] = 0.020\ M$ and $[Ni^{2+}] = 0.10\ M$.
(e) The galvanic cell was used to measure an unknown chloride ion concentration. The Ni^{2+} concentration was 0.200 *M* and the measured cell potential was 0.388 V. What was the chloride ion concentration?

27. Describe how a nickel–metal hydride cell works.
28. Describe how a lithium ion cell works. What does *intercalation* mean?
29. What factors involving the nucleus of helium might be responsible for the large abundance of helium in the universe?
30. What is the symbol for (a) an alpha particle, (b) a beta particle, (c) a positron?
31. What is the rest mass of a particle of gamma radiation?
32. Suppose that the total mass of the reactants in a chemical reaction was 100.00000 g. How many kilojoules of energy would have to evolve from this reaction if the total mass of the products could be no greater than 99.99900 g? If all this energy were used to heat water, how many liters of water could have its temperature raised from 10 °C to 100 °C?
33. Calculate the binding energy in joules per nucleon for the nucleus of the deuterium atom, ${}^{2}_{1}H$. The mass of an atom of deuterium, including its electron, is 2.014102 u.
34. Which nuclear force acts over the longer distance, the electrostatic force or the nuclear strong force?
35. Why is an isotope of hydrogen a better candidate for nuclear fusion than an isotope of helium?
36. When an atom of uranium-238 absorbs a neutron, it can ultimately change to an atom of plutonium-239. Write a nuclear equation for the reaction. If the reaction occurs in two steps, what intermediate nucleus is formed?
37. If an atom of beryllium-7 decays by the capture of a K-electron, into which isotope does it change? Write the equation.
38. An atom of neodymium-144 decays by alpha emission. Write the nuclear equation.
39. Phosphorus-32 decays by beta emission. Write the nuclear equation.
40. The half-life of samarium-149, an alpha emitter, is 4×10^{14} years. The half-life of oxygen-15 is 124 seconds. Assuming that equimolar samples are compared, which is the more intensely radioactive?
41. The half-life of cesium-137 is 30 years. Of an initial 100 g sample, how much cesium-137 will remain after 300 years?
42. What are some "rules of thumb" that can be used to judge if a particular radionuclide might have a long enough half-life to warrant the effort to make it?
43. A particular compound nucleus can form from a variety of combinations of targets and bombarding particles, as in the example of aluminum on page 835. What determines how a compound nucleus breaks up?
44. Strontium-90, a beta emitter, has a half-life of 28.1 yr. If 36.2 mg of ${}^{90}Sr$ were incorporated in the bones of a growing child, how many beta particles would the child absorb from this source in 1.00 day?
45. Which kind of radiation poses a greater health risk, beta or alpha radiation? Why?
46. What reaction, if any, will occur when solutions of KBr(*aq*) and $Cl_2(aq)$ are mixed?
47. Write the equation for the reduction of bismuth(III) oxide by hydrogen.
48. Why can't fluorine be produced by the electrolysis of aqueous NaF? Why can't liquid HF be electrolyzed to give F_2 and H_2?
49. Why doesn't phosphorus form a stable P_2 molecule similar to N_2?
50. Describe the structure of the ozone molecule. What effect does ozone in smog have on the chlorophyll in green plants?
51. What is the molecular structure of diamond? How are the structures of graphite, buckyballs, and carbon nanotubes related?
52. What must be true about the heat of formation of a metal compound if the compound is thermally unstable?
53. Why is carbon used as a reducing agent in industrial metallurgy? Write the equations that take place in a blast furnace during the reduction of Fe_2O_3 to metallic Fe?
54. The reaction

$$Ni(H_2O)_6{}^{2+} + 6NH_3 \longrightarrow Ni(NH_3)_6{}^{2+} + 6H_2O$$

can be described as the displacement of one Lewis base by another. Explain. What is the Lewis acid in the reaction?

55. Sketch the structures of the oxalate ion and ethylenediamine. Identify the donor atoms these ligands use when they form chelate complexes.
56. Sketch the chiral isomers of $[Cr(H_2O)_2(en)_2]^{3+}$. Is there a nonchiral isomer of the complex? What is the name of $[Cr(H_2O)_2(en)_2]Cl_3$?
57. Which complex would absorb light of a shorter wavelength, $[V(H_2O)_6]^{3+}$ or $[V(CN)_6]^{3-}$? Justify your answer.
58. Name the following organic compound.

```
   CH3
   |
CH3CHCH2CH2CH=CCH2CH3
              |
              CH2CH2CH3
```

59. Draw the structure of 2,2-dimethyl-4-ethylheptane.
60. Which of the following compounds (a) neutralizes NaOH, (b) neutralizes HCl, (c) yields an alcohol and an organic acid when hydrolyzed, (d) is easily oxidized to an acid, (e) would

be classified as aromatic, (f) would be oxidized to give a ketone?

(1) $CH_3O{-}\overset{\overset{\displaystyle O}{\|}}{C}CH_2CH_3$

(2) $CH_3CH_2\overset{\overset{\displaystyle O}{\|}}{C}H$

(3) 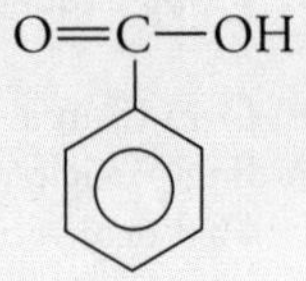 $O{=}C{-}OH$ (attached to a benzene ring)

(4) $CH_3\overset{\overset{\displaystyle OH}{|}}{C}HCH_2CH_3$

(5) 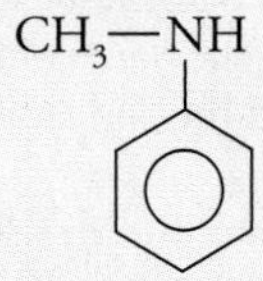 $CH_3{-}NH$ (attached to a benzene ring)

61. What would be the repeating structural unit if a condensation polymer were to form from the following monomers by the elimination of CH_3OH?

$$CH_3O{-}\overset{\overset{\displaystyle O}{\|}}{C}CH_2CH_2\overset{\overset{\displaystyle O}{\|}}{C}{-}OCH_3 \qquad H_2NCH_2CH_2CH_2NH_2$$

62. What effect does cross-linking have on the properties of a polymer?

63. What are the monomer units in (a) starch, (b) a polypeptide, and (c) DNA?

64. How do the water solubilities compare for monosaccharides and lipids?

65. How is the folding of a polypeptide chain affected by the hydrophobic and hydrophilic nature of its side chains?

66. Use data in Table 19.1 to calculate the value of K_{sp} for AgCl.

APPENDIX A

ELECTRON CONFIGURATIONS OF THE ELEMENTS

Atomic Number			Atomic Number			Atomic Number		
1	H	$1s^1$	40	Zr	[Kr] $5s^2\,4d^2$	79	Au	[Xe] $6s^1\,4f^{14}\,5d^{10}$
2	He	$1s^2$	41	Nb	[Kr] $5s^1\,4d^4$	80	Hg	[Xe] $6s^2\,4f^{14}\,5d^{10}$
3	Li	[He] $2s^1$	42	Mo	[Kr] $5s^1\,4d^5$	81	Tl	[Xe] $6s^2\,4f^{14}\,5d^{10}\,6p^1$
4	Be	[He] $2s^2$	43	Tc	[Kr] $5s^2\,4d^5$	82	Pb	[Xe] $6s^2\,4f^{14}\,5d^{10}\,6p^2$
5	B	[He] $2s^2\,2p^1$	44	Ru	[Kr] $5s^1\,4d^7$	83	Bi	[Xe] $6s^2\,4f^{14}\,5d^{10}\,6p^3$
6	C	[He] $2s^2\,2p^2$	45	Rh	[Kr] $5s^1\,4d^8$	84	Po	[Xe] $6s^2\,4f^{14}\,5d^{10}\,6p^4$
7	N	[He] $2s^2\,2p^3$	46	Pd	[Kr] $4d^{10}$	85	At	[Xe] $6s^2\,4f^{14}\,5d^{10}\,6p^5$
8	O	[He] $2s^2\,2p^4$	47	Ag	[Kr] $5s^1\,4d^{10}$	86	Rn	[Xe] $6s^2\,4f^{14}\,5d^{10}\,6p^6$
9	F	[He] $2s^2\,2p^5$	48	Cd	[Kr] $5s^2\,4d^{10}$	87	Fr	[Rn] $7s^1$
10	Ne	[He] $2s^2\,2p^6$	49	In	[Kr] $5s^2\,4d^{10}\,5p^1$	88	Ra	[Rn] $7s^2$
11	Na	[Ne] $3s^1$	50	Sn	[Kr] $5s^2\,4d^{10}\,5p^2$	89	Ac	[Rn] $7s^2\,6d^1$
12	Mg	[Ne] $3s^2$	51	Sb	[Kr] $5s^2\,4d^{10}\,5p^3$	90	Th	[Rn] $7s^2\,6d^2$
13	Al	[Ne] $3s^2\,3p^1$	52	Te	[Kr] $5s^2\,4d^{10}\,5p^4$	91	Pa	[Rn] $7s^2\,5f^2\,6d^1$
14	Si	[Ne] $3s^2\,3p^2$	53	I	[Kr] $5s^2\,4d^{10}\,5p^5$	92	U	[Rn] $7s^2\,5f^3\,6d^1$
15	P	[Ne] $3s^2\,3p^3$	54	Xe	[Kr] $5s^2\,4d^{10}\,5p^6$	93	Np	[Rn] $7s^2\,5f^4\,6d^1$
16	S	[Ne] $3s^2\,3p^4$	55	Cs	[Xe] $6s^1$	94	Pu	[Rn] $7s^2\,5f^6$
17	Cl	[Ne] $3s^2\,3p^5$	56	Ba	[Xe] $6s^2$	95	Am	[Rn] $7s^2\,5f^7$
18	Ar	[Ne] $3s^2\,3p^6$	57	La	[Xe] $6s^2\,5d^1$	96	Cm	[Rn] $7s^2\,5f^7\,6d^1$
19	K	[Ar] $4s^1$	58	Ce	[Xe] $6s^2\,4f^1\,5d^1$	97	Bk	[Rn] $7s^2\,5f^9$
20	Ca	[Ar] $4s^2$	59	Pr	[Xe] $6s^2\,4f^3$	98	Cf	[Rn] $7s^2\,5f^{10}$
21	Sc	[Ar] $4s^2\,3d^1$	60	Nd	[Xe] $6s^2\,4f^4$	99	Es	[Rn] $7s^2\,5f^{11}$
22	Ti	[Ar] $4s^2\,3d^2$	61	Pm	[Xe] $6s^2\,4f^5$	100	Fm	[Rn] $7s^2\,5f^{12}$
23	V	[Ar] $4s^2\,3d^3$	62	Sm	[Xe] $6s^2\,4f^6$	101	Md	[Rn] $7s^2\,5f^{13}$
24	Cr	[Ar] $4s^1\,3d^5$	63	Eu	[Xe] $6s^2\,4f^7$	102	No	[Rn] $7s^2\,5f^{14}$
25	Mn	[Ar] $4s^2\,3d^5$	64	Gd	[Xe] $6s^2\,4f^7\,5d^1$	103	Lr	[Rn] $7s^2\,5f^{14}\,6d^1$
26	Fe	[Ar] $4s^2\,3d^6$	65	Tb	[Xe] $6s^2\,4f^9$	104	Rf	[Rn] $7s^2\,5f^{14}\,6d^2$
27	Co	[Ar] $4s^2\,3d^7$	66	Dy	[Xe] $6s^2\,4f^{10}$	105	Db	[Rn] $7s^2\,5f^{14}\,6d^3$
28	Ni	[Ar] $4s^2\,3d^8$	67	Ho	[Xe] $6s^2\,4f^{11}$	106	Sg	[Rn] $7s^2\,5f^{14}\,6d^4$
29	Cu	[Ar] $4s^1\,3d^{10}$	68	Er	[Xe] $6s^2\,4f^{12}$	107	Bh	[Rn] $7s^2\,5f^{14}\,6d^5$
30	Zn	[Ar] $4s^2\,3d^{10}$	69	Tm	[Xe] $6s^2\,4f^{13}$	108	Hs	[Rn] $7s^2\,5f^{14}\,6d^6$
31	Ga	[Ar] $4s^2\,3d^{10}\,4p^1$	70	Yb	[Xe] $6s^2\,4f^{14}$	109	Mt	[Rn] $7s^2\,5f^{14}\,6d^7$
32	Ge	[Ar] $4s^2\,3d^{10}\,4p^2$	71	Lu	[Xe] $6s^2\,4f^{14}\,5d^1$	110	Ds	[Rn] $7s^2\,5f^{14}\,6d^8$
33	As	[Ar] $4s^2\,3d^{10}\,4p^3$	72	Hf	[Xe] $6s^2\,4f^{14}\,5d^2$	111	Rg	[Rn] $7s^2\,5f^{14}\,6d^9$
34	Se	[Ar] $4s^2\,3d^{10}\,4p^4$	73	Ta	[Xe] $6s^2\,4f^{14}\,5d^3$	112	Uub	[Rn] $7s^2\,5f^{14}\,6d^{10}$
35	Br	[Ar] $4s^2\,3d^{10}\,4p^5$	74	W	[Xe] $6s^2\,4f^{14}\,5d^4$	113	Uut	[Rn] $7s^2\,5f^{14}\,6d^{10}\,7p^1$
36	Kr	[Ar] $4s^2\,3d^{10}\,4p^6$	75	Re	[Xe] $6s^2\,4f^{14}\,5d^5$	114	Uuq	[Rn] $7s^2\,5f^{14}\,6d^{10}\,7p^2$
37	Rb	[Kr] $5s^1$	76	Os	[Xe] $6s^2\,4f^{14}\,5d^6$	115	Uup	[Rn] $7s^2\,5f^{14}\,6d^{10}\,7p^3$
38	Sr	[Kr] $5s^2$	77	Ir	[Xe] $6s^2\,4f^{14}\,5d^7$	116	Uuh	[Rn] $7s^2\,5f^{14}\,6d^{10}\,7p^4$
39	Y	[Kr] $5s^2\,4d^1$	78	Pt	[Xe] $6s^1\,4f^{14}\,5d^9$	118	Uuo	[Rn] $7s^2\,5f^{14}\,6d^{10}\,7p^6$

ANSWERS TO PRACTICE EXERCISES AND SELECTED REVIEW PROBLEMS

CHAPTER 1

Practice Exercises

1.1 meter^3 or m^3. **1.2** $kg\left(\frac{m}{s^2}\right)$ or kg m s^{-2}. **1.3** 187 °C. **1.4** 10 °C, 293 K. **1.5** (a) 42.0 g, (b) 0.857 g/mL, (c) 149 cm. **1.6** (a) 30.0 mL, (b) 54.155 g, (c) 11.3 g, (d) 3.62 ft, (e) 0.48 m^2. **1.7** 11.5 m^2. **1.8** (a) 108 in., (b) 1.25×10^5 cm, (c) 0.0107 ft, (d) 8.59 km L^{-1}. **1.9** $d = 16.5$ g cm^{-3}. The object is not composed of pure gold. **1.10** 647 lb. **1.11** 0.899 g/cm^3. **1.12** 0.0568 cm^3

Review Problems

1.26 (a) 0.01 m, (b) 1000 m, (c) 10^{12} pm, (d) 0.1 m, (e) 0.001 kg, (f) 0.01 g. **1.28** (a)120 °F, (b) 50 °F, (c) −3.61 °C, (d) 9.4 °C, (e) 333 K, (f) 243 K. **1.30** At 39.7 °C, this dog has a fever; the temperature is out of normal canine range. **1.32** 1.0×10^7 K − 2.5×10^7 K, 1.0×10^7 °C − 2.5×10^7 °C, 1.8×10^7 °F − 4.5×10^7 °F. **1.34** −269 °C. **1.36** (a) 4 significant figures, (b) 5 significant figures, (c) 4 significant figures, (d) 2 significant figures, (e) 4 significant figures, (f) 2 significant figures. **1.38** (a) 0.72 m^2, (b) 84.24 kg, (c) 4.19 g/mL, (d) 19.42 g/mL, (e) 857.7 cm^2. **1.40** (a) 11.5 km/h, (b) 8.2×10^6 μg/L, (c) 7.53×10^{-5} kg, (d) 0.1375 L, (e) 25 mL, (f) 3.42×10^{-20} dm. **1.42** (a) 91 cm, (b) 2.3 kg, (c) 2800 mL, (d) 200 mL, (e) 88 km/hr, (f) 80.4 km. **1.44** (a) 7,800 cm^2, (b) 577 km^2, (c) 6.54×10^6 cm^3. **1.46** 4,000 pistachios. **1.48** 90 m/s. **1.50** 1520 mi/hr. **1.52** 5.1×10^{13} mi. **1.54** 11,034 m. **1.56** 0.798 g/mL. **1.58** 31.6 mL. **1.60** 276 g. **1.62** 11 g/mL. **1.64** 0.0709 g/mL

CHAPTER 2

Practice Exercises

2.1 12.3 g Cd. **2.2** Compounds A and D are the same, as are compounds B and C. **2.3** $^{240}_{94}Pu$ 94 electrons. **2.4** $^{35}_{17}Cl$ 17 protons, 17 electrons, and 18 neutrons. **2.5** We can discard the 17 since the 17 tells the number of protons which is information that the symbol "Cl" also provides. In addition, the number of protons equals the number of electrons in a neutral atom, so the symbol "Cl" also indicates the number of electrons. The 35 is necessary to state which isotope of chlorine is in question and therefore the number of neutrons in the atom. **2.6** 26.9814 u. **2.7** 5.2955 times as heavy as carbon. **2.8** 10.8 u. **2.9** (a) 1 Ni, 2 Cl, (b) 1 Fe, 1 S, 4 O, (c) 3 Ca, 2 P, 8 O, (d) 1 Co, 2 N, 12 O, 12 H. **2.10** (a) 2 N nitrogen, 4 H hydrogen, 3 O oxygen, (b) 1 Fe iron, 1 N nitrogen, 4 H hydrogen, 2 S sulfur, 8 O oxygen, (c) 1 Mo molybdenum, 2 N nitrogen, 11 O oxygen, 10 H hydrogen, (d) 6 C carbon, 4 H hydrogen, 1 Cl chlorine, 1 N nitrogen, 2 O oxygen. **2.11** 1 Mg, 2 O, 4 H, and 2 Cl. **2.12** $Mg(OH)_2(s) + 2HCl(aq) \longrightarrow MgCl_2(aq) + 2H_2O$. **2.13** 6 N, 42 H, 2 P, 20 O, 3 Ba, and 12 C. **2.14** $CH_3CH_2CH_2CH_2CH_2CH_2CH_2CH_3$

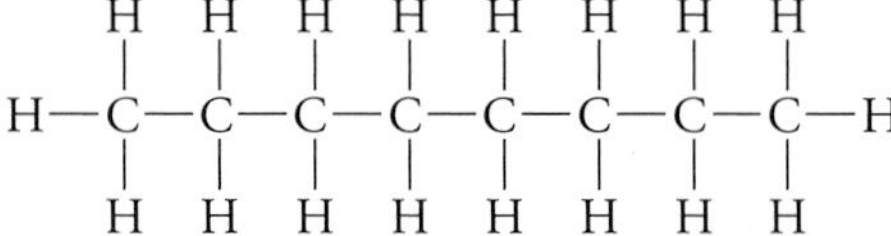

2.15 $CH_3CH_2CH_2CH_2CH_2CH_2CH_2CH_2CH_2CH_3$

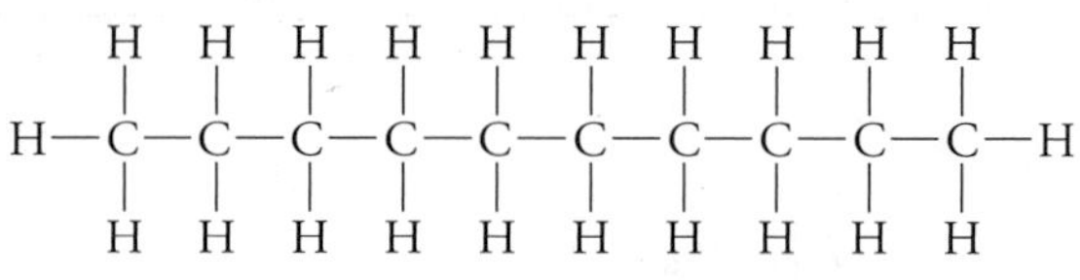

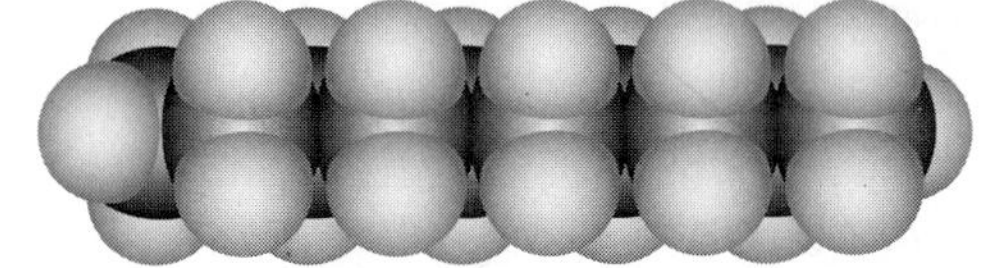

2.16 (a) $CH_3CH_2CH_2OH$

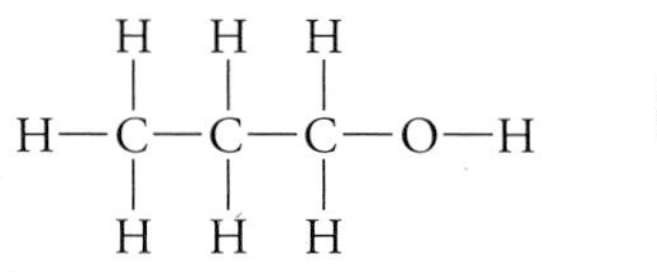

(b) $CH_3CH_2CH_2CH_2OH$

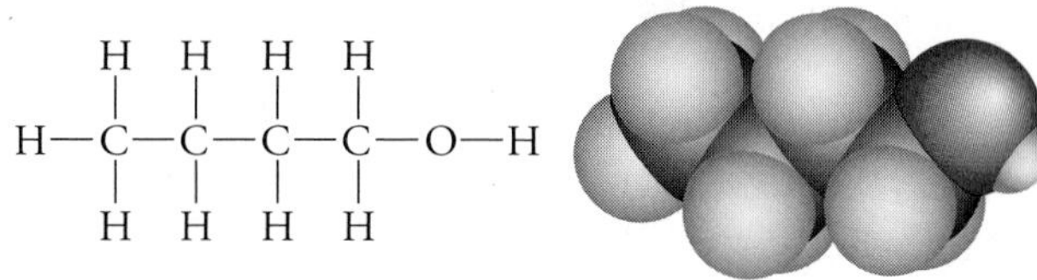

2.17 (a) 26 protons and 26 electrons, (b) 26 protons and 23 electrons, (c) 7 protons and 10 electrons, (d) 7 protons and 7 electrons. **2.18** (a) 8 protons and 8 electrons, (b) 8 protons and 10 electrons, (c) 13 protons and 10 electrons, (d) 13 protons and 13 electrons. **2.19** (a) NaF, (b) Na_2O, (c) MgF_2, (d) Al_4C_3. **2.20** (a) Ca_3N_2, (b) $AlBr_3$, (c) Na_3P, d) CsCl. **2.21** (a) $CrCl_3$ and $CrCl_2$, Cr_2O_3 and CrO, (b) CuCl, $CuCl_2$, Cu_2O and CuO. **2.22** (a) Au_2S and Au_2S_3, Au_3N and AuN, (b) SnS and SnS_2, Sn_3N_2 and Sn_3N_4. **2.23** (a) $KC_2H_3O_2$, (b) $Sr(NO_3)_2$, (c) $Fe(C_2H_3O_2)_3$. **2.24** (a) Na_2CO_3, (b) $(NH_4)_2SO_4$. **2.25** (a) phosphorous trichloride, (b) sulfur dioxide, (c) dichlorine heptaoxide. **2.26** (a) $AsCl_5$, (b) SCl_6, (c) S_2Cl_2. **2.27** (a) K_2O, (b) $BaBr_2$, (c) Na_3N, (d) Al_2S_3. **2.28** (a) aluminum chloride, (b) barium sulfide, (c) sodium bromide, (d) calcium fluoride. **2.29** (a) postassium sulfide, (b) magnesium phosphide, (c) nickel(II) chloride, (d) iron(III) oxide. **2.30** (a) Al_2S_3, (b) SrF_2, (c) TiO_2, (d) Au_2O_3. **2.31** (a) lithium carbonate, (b) iron(III) hydroxide. **2.32** (a) $KClO_3$, (b) $Ni_3(PO_4)_2$. **2.33** diiodine pentaoxide. **2.34** chromium(III) acetate

Review Problems

2.76 $x = 29.3$ g nitrogen. **2.78** 5.54 g ammonia. **2.80** 2.286 g of O. **2.82** 1.008 u. **2.84** 2.01588 u. **2.86** $(0.6917 \times 62.9396\ u) + (0.3083 \times 64.9278\ u) = 63.55\ u$.

2.88

	neutrons	*protons*	*electrons*
(a) Radium-226	138	88	88
(b) ^{206}Pb	124	82	82
(c) Carbon-14	8	6	6
(d) ^{23}Na	12	11	11

2.90 1 Cr, 6 C, 9 H, 6 O. **2.92** $MgSO_4$. **2.94** (a) 2 K, 2 C, 4 O, (b) 2 H, 1 S, 3 O, (c) 12 C, 26 H, (d) 4 H, 2 C, 2 O, (e) 9 H, 2 N, 1 P, 4 O. **2.96** (a) 1 Ni, 2 Cl, 8 O, (b) 1 Cu, 1 C, 3 O, (c) 2 K, 2 Cr, 7 O, (d) 2 C, 4 H, 2 O, (e) 2 N, 9 H, 1 P, 4 O. **2.98** (a) 6 N, 3 O, (b) 4 Na, 4 H, 4 C, 12 O, (c) 2 Cu, 2 S, 18 O, 20 H. **2.100** (a) 6, (b) 3, (c) 27. **2.102** (a) K^+, (b) Br^-, (c) Mg^{2+}, (d) S^{2-}, (e) Al^{3+}. **2.104** (a) NaBr, (b) KI, (c) BaO, (d) $MgBr_2$, (e) BaF_2. **2.106** (a) KNO_3, (b) $Ca(C_2H_3O_2)_2$, (c) NH_4Cl, (d) $Fe_2(CO_3)_3$, (e) $Mg_3(PO_4)_2$. **2.108** (a) PbO and PbO_2, (b) SnO and SnO_2, (c) MnO and Mn_2O_3, (d) FeO and Fe_2O_3, (e) Cu_2O and CuO. **2.110** (a) silicon dioxide, (b) xenon tetrafluoride, (c) tetraphosphorus decaoxide, (d) dichlorine heptaoxide. **2.112** (a) calcium sulfide, (b) aluminum bromide, (c) sodium phosphide, (d) barium arsenide, (e) rubidium sulfide. **2.114** (a) iron(II) sulfide, (b) copper(II) oxide, (c) tin(IV) oxide, (d) cobalt(II) chloride hexahydrate. **2.116** (a) sodium nitrite, (b) potassium permanganate, (c) magnesium sulfate heptahydrate, (d) potassium thiocyanate. **2.118** (a) ionic, chromium(II) chloride, (b) molecular, disulfur dichloride, (c) ionic, ammonium acetate, (d) molecular, sulfur trioxide, (e) ionic, potassium iodate, (f) molecular, tetraphosphorous hexaoxide, (g) ionic, calcium sulfite, (h) ionic, silver cyanide, (i) ionic, zinc(II) bromide, (j) molecular, hydrogen selenide. **2.120** (a) Na_2HPO_4, (b) Li_2Se, (c) $Cr(C_2H_3O_2)_3$, (d) S_2F_{10}, (e) $Ni(CN)_2$, (f) Fe_2O_3, (g) SbF_5. **2.122** (a) $(NH_4)_2S$, (b) $Cr_2(SO_4)_3{\cdot}6H_2O$, (c) SiF_4, (d) MoS_2, (e) $SnCl_4$, (f) H_2Se, (g) P_4S_7. **2.124** diselenium hexasulfide and diselenium tetrasulfide

CHAPTER 3

Practice Exercises

3.1 0.129 mol Al. **3.2** $\pm 7.12 \times 10^{-5}$ mol Si. **3.3** Yes. **3.4** 3.5×10^{18} molecules of sucrose. **3.5** 0.0516 mol Al^{3+}. **3.6** 3.44 mol N atoms. **3.7** 59.5 g Fe. **3.8** 10.5 g Fe. **3.9** 27.9 g Fe. **3.10** 13.04% H, 52.17% C. It is likely that the compound contains another element. **3.11** 36.84% N, 63.16% O, There are no other elements present. **3.12** 30.45% N, 69.55% O.

3.13

N_2O:	63.65% N	36.34% O
NO:	46.68% N	53.32% O
NO_2:	30.45% N	69.55% O
N_2O_3:	36.86% N	63.14% O
N_2O_4:	30.45% N	69.55% O
N_2O_5:	25.94% N	74.06% O

The compound N_2O_3 corresponds to the data in Practice Exercise 3.11. **3.14** NO. **3.15** SO_2. **3.16** Al_2O_3. **3.17** N_2O_5. **3.18** Na_2SO_4. **3.19** C_9H_8O. **3.20** CS_2. **3.21** CH_2O. **3.22** $C_2H_4Cl_2$ and $C_6H_6Cl_6$. **3.23** N_2H_4. **3.24** $AlCl_3(aq) + Na_3PO_4(aq) \longrightarrow AlPO_4(s) + 3NaCl(aq)$. **3.25** $3CaCl_2(aq) + 2K_3PO_4(aq) \longrightarrow Ca_3(PO_4)_2(s) + 6KCl(aq)$. **3.26** 3.38 mol O_2. **3.27** 0.183 mol H_2SO_4. **3.28** 78.5 g Al_2O_3. **3.29** 1.18×10^2 g CO_2. **3.30** 55.0 g CO_2, 34 g HCl remaining. **3.31** 30.01 g NO. **3.32** 36.78 g $HOOCC_6H_4O_2C_2O_3$, 83.5%. **3.33** 30.9 g $HC_2H_3O_2$, 86.1%

Review Problems

3.25 1:2, 2 mol N to 4 mol O. **3.27** 2.59×10^{-3} mole Ta. **3.29** (a) 6 atom C:11 atom H, (b) 12 mole C:11 mole O, (c) 12 atom H:11 atom O, (d) 12 mole H:11 mole O. **3.31** 1.05 mol Bi. **3.33** 4.32 mol Cr.

3.35 (a) $\left(\frac{2 \text{ mol Al}}{3 \text{ mol S}}\right)$ or $\left(\frac{3 \text{ mol S}}{2 \text{ mol Al}}\right)$,
(b) $\left(\frac{3 \text{ mol S}}{1 \text{ mol } Al_2(SO_4)_3}\right)$ or $\left(\frac{1 \text{ mol } Al_2(SO_4)_3}{3 \text{ mol S}}\right)$,
(c) 0.600 mol Al, (d) 3.48 mol S. **3.37** 0.0725 mol N_2, 0.218 mole H_2. **3.39** 0.833 mol CF_4. **3.41** 9.33×10^{23} atoms C. **3.43** 3.76×10^{24} atoms. **3.45** 3.01×10^{23} atoms C–12. **3.47** (a) 75.4 g Fe, (b) 392 g O, (c) 35.1 g Ca. **3.49** 1.30×10^{-10} g K. **3.51** 0.302 mol Ni. **3.53** (a) $NaHCO_3$ 84.0066 g/mol, (b) $(NH_4)_2CO_3$ 96.0858 g/mol, (c) $CuSO_4{\cdot}5H_2O$ 249.685 g/mole, (d) $K_2Cr_2O_7$ 294.1846 g/mole, (e) $Al_2(SO_4)_3$ 342.151 g/mol. **3.55** (a) 388 g $Ca_3(PO_4)_2$, (b) 0.151 g $Fe(NO_3)_3$, (c) 34.9 g C_4H_{10}, (d) 1.39×10^{-4} g $(NH_4)_2CO_3$. **3.57** (a) 0.215 mol $CaCO_3$, (b) 9.16×10^{-11} mol NH_3, (c) 7.94×10^{-2} mol $Sr(NO_3)_2$, (d) 4.31×10^{-8} mol Na_2CrO_4. **3.59** 0.0750 mol Ca, 3.01 g Ca. **3.61** 1.30 mol N, 62.5 g $(NH_4)_2CO_3$. **3.63** 3.43 kg fertilizer. **3.65** Assume one mole total for each of the following.

(a) 19.2% Na	1.68% H	25.8% P	53.3% O
(b) 12.2% N	5.26% H	26.9% P	55.6% O
(c) 62.0% C	10.4% H	27.6% O	
(d) 23.3% Ca	18.6% S	55.7% O	2.34% H
(e) 23.3% Ca	18.6% S	55.7% O	2.34% H.

3.67 Heroin. **3.69** Freon 141b. **3.71** 22.9% P, 77.1% Cl. **3.73** These data are consistent with the experimental values cited in the problem. **3.75** 0.474 g O. **3.77** (a) SCl, (b) CH_2O (c) NH_3, (d) AsO_3, (e) HO. **3.79** $NaTcO_4$. **3.81** CCl_2. **3.83** $C_9H_8O_2$. **3.85** C_2H_6O. **3.87** $C_{19}H_{30}O_2$. **3.89** (a) $Na_2S_4O_6$, (b) $C_6H_4Cl_2$, (c) $C_6H_3Cl_3$. **3.91** $C_{19}H_{30}O_2$. **3.93** HgBr, Hg_2Br_2. **3.95** CHNO, $C_3H_3N_3O_3$. **3.97** 36 mol H. **3.99** $4Fe(s) + 3O_2(g) \longrightarrow 2Fe_2O_3(s)$.

3.101 (a) $Ca(OH)_2 + 2HCl \longrightarrow CaCl_2 + 2H_2O$,
(b) $2AgNO_3 + CaCl_2 \longrightarrow Ca(NO_3)_2 + 2AgCl$,
(c) $Pb(NO_3)_2 + Na_2SO_4 \longrightarrow PbSO_4 + 2NaNO_3$,
(d) $2Fe_2O_3 + 3C \longrightarrow 4Fe + 3CO_2$,
(e) $2C_4H_{10} + 13O_2 \longrightarrow 8CO_2 + 10H_2O$.

3.103 (a) $Mg(OH)_2 + 2HBr \longrightarrow MgBr_2 + 2H_2O$,
(b) $2HCl + Ca(OH)_2 \longrightarrow CaCl_2 + 2H_2O$,
(c) $Al_2O_3 + 3H_2SO_4 \longrightarrow Al_2(SO_4)_3 + 3H_2O$,
(d) $2KHCO_3 + H_3PO_4 \longrightarrow K_2HPO_4 + 2H_2O + 2CO_2$,
(e) $C_9H_{20} + 14O_2 \longrightarrow 9CO_2 + 10H_2O$.

3.105 $2FeCl_3 + SnCl_2 \longrightarrow 2FeCl_2 + SnCl_4$. **3.107** (a) 0.030 mol $Na_2S_2O_3$, (b) 0.24 mol HCl, (c) 0.15 mol H_2O, (d) 0.15 mol H_2O. **3.109** (a) 3.6 g Zn, (b) 22 g Au, (c) 55 g $Au(CN)_2^-$. **3.111** (a) $4P + 5O_2 \longrightarrow P_4O_{10}$, (b) 8.85 g O_2, (c) 14.2 g P_4O_{10}, (d) 3.26 g P. **3.113** 30.28 g HNO_3. **3.115** 0.47 kg O_2. **3.117** (a) Fe_2O_3, (b) 195 g Fe. **3.119** 26.7 g $FeCl_3$. **3.121** 0.913 mg HNO_3. **3.123** 66.98 g $BaSO_4$, 96.22%. **3.125** 88.72%. **3.127** 9.2 g C_7H_8

CHAPTER 4

Practice Exercises

4.1 (a) $FeCl_3(s) \longrightarrow Fe^{3+}(aq) + 3Cl^-(aq)$,
(b) $K_3PO_4(s) \longrightarrow 3K^+(aq) + PO_4^{3-}(aq)$.

4.2 (a) $MgCl_2(s) \longrightarrow Mg^{2+}(aq) + 2Cl^-(aq)$,
(b) $Al(NO_3)_3(s) \longrightarrow Al^{3+}(aq) + 3NO_3^-(aq)$,
(c) $Na_2CO_3(s) \longrightarrow 2Na^+(aq) + CO_3^{2-}(aq)$.

4.3 molecular: $(NH_4)_2SO_4(aq) + Ba(NO_3)_2(aq) \longrightarrow BaSO_4(s) + 2NH_4NO_3(aq)$,
ionic: $2NH_4^+(aq) + SO_4^{2-}(aq) + Ba^{2+}(aq) + 2NO_3^-(aq) \longrightarrow BaSO_4(s) + 2NH_4^+(aq) + 2NO_3^-(aq)$,
net ionic: $Ba^{2+}(aq) + SO_4^{2-}(aq) \longrightarrow BaSO_4(s)$.

4.4 molecular: $CdCl_2(aq) + Na_2S(aq) \longrightarrow CdS(s) + 2NaCl(aq)$,

ionic: $Cd^{2+}(aq) + 2Cl^-(aq) + 2Na^+(aq) + S^{2-}(aq) \longrightarrow CdS(s) + 2Na^+(aq) + 2Cl^-(aq)$,
net ionic: $Cd^{2+}(aq) + S^{2-}(aq) \longrightarrow CdS(s)$.
4.5 $HCHO_2(aq) + H_2O \longrightarrow H_3O^+(aq) + CHO_2^-(aq)$.
4.6 $H_3C_6H_5O_7(s) + H_2O \longrightarrow H_3O^+(aq) + H_2C_6H_5O_7^-(aq)$,
$H_2C_6H_5O_7^-(aq) + H_2O \longrightarrow H_3O^+(aq) + HC_6H_5O_7^{2-}(aq)$,
$HC_6H_5O_7^{2-}(aq) + H_2O \longrightarrow H_3O^+(aq) + C_6H_5O_7^{3-}(aq)$.
4.7 $(C_2H_5)_3N(aq) + H_2O \longrightarrow (C_2H_5)_3NH^+(aq) + OH^-(aq)$.
4.8 $HONH_2(aq) + H_2O \longrightarrow HONH_3^+(aq) + OH^-(aq)$.
4.9 $CH_3NH_2(aq) + H_2O \rightleftharpoons CH_3NH_3^+(aq) + OH^-(aq)$.
4.10 $HNO_2(aq) + H_2O \rightleftharpoons H_3O^+(aq) + NO_2^-(aq)$.
4.11 Sodium arsenate. **4.12** Calcium formate. **4.13** HF: Hydrofluoric acid, sodium salt = sodium fluoride (NaF), HBr: Hydrobromic acid, sodium salt = sodium bromide (NaBr).
4.14 $NaHSO_3$, sodium hydrogen sulfite.
4.15 $H_3PO_4(aq) + NaOH(aq) \longrightarrow NaH_2PO_4(aq) + H_2O$ sodium dihydrogen phosphate,
$NaH_2PO_4(aq) + NaOH(aq) \longrightarrow Na_2HPO_4(aq) + H_2O$ sodium hydrogen phosphate,
$Na_2HPO_4(aq) + NaOH(aq) \longrightarrow Na_3PO_4(aq) + H_2O$ sodium phosphate.
4.16 molecular: $Zn(NO_3)_2(aq) + Ca(C_2H_3O_2)_2(aq) \longrightarrow Zn(C_2H_3O_2)_2(aq) + Ca(NO_3)_2(aq)$,
ionic: $Zn^{2+}(aq) + 2NO_3^-(aq) + Ca^{2+}(aq) + 2C_2H_3O_2^-(aq) \longrightarrow Zn^{2+}(aq) + 2C_2H_3O_2^-(aq) + Ca^{2+}(aq) + 2NO_3^-(aq)$,
net ionic: No reaction.
4.17 (a) molecular: $AgNO_3(aq) + NH_4Cl(aq) \longrightarrow AgCl(s) + NH_4NO_3(aq)$,
ionic: $Ag^+(aq) + NO_3^-(aq) + NH_4^+(aq) + Cl^-(aq) \longrightarrow AgCl(s) + NH_4^+(aq) + NO_3^-(aq)$,
net ionic: $Ag^+(aq) + Cl^-(aq) \longrightarrow AgCl(s)$,
(b) molecular: $Na_2S(aq) + Pb(C_2H_3O_2)_2(aq) \longrightarrow 2NaC_2H_3O_2(aq) + PbS(s)$,
ionic: $2Na^+(aq) + S^{2-}(aq) + Pb^{2+}(aq) + 2C_2H_3O_2^-(aq) \longrightarrow 2Na^+(aq) + 2C_2H_3O_2^-(aq) + PbS(s)$,
net ionic: $S^{2-}(aq) + Pb^{2+}(aq) \longrightarrow PbS(s)$.
4.18 molecular: $2HNO_3(aq) + Ca(OH)_2(aq) \longrightarrow Ca(NO_3)_2(aq) + 2H_2O$,
ionic: $2H^+(aq) + 2NO_3^-(aq) + Ca^{2+}(aq) + 2OH^-(aq) \longrightarrow Ca^{2+}(aq) + 2NO_3^-(aq) + 2H_2O$,
net ionic: $H^+(aq) + OH^-(aq) \longrightarrow H_2O$.
4.19 (a) molecular: $HCl(aq) + KOH(aq) \longrightarrow H_2O + KCl(aq)$,
ionic: $H^+(aq) + Cl^-(aq) + K^+(aq) + OH^-(aq) \longrightarrow H_2O + K^+(aq) + Cl^-(aq)$,
net ionic: $H^+(aq) + OH^-(aq) \longrightarrow H_2O$,
(b) molecular: $HCHO_2(aq) + LiOH(aq) \longrightarrow H_2O + LiCHO_2(aq)$,
ionic: $HCHO_2(aq) + Li^+(aq) + OH^-(aq) \longrightarrow H_2O + Li^+(aq) + CHO_2^-(aq)$,
net ionic: $HCHO_2(aq) + OH^-(aq) \longrightarrow H_2O + CHO_2^-(aq)$,
(c) molecular: $N_2H_4(aq) + HCl(aq) \longrightarrow N_2H_5Cl(aq)$,
ionic: $N_2H_4(aq) + H^+(aq) + Cl^-(aq) \longrightarrow N_2H_5^+(aq) + Cl^-(aq)$,
net ionic: $N_2H_4(aq) + H^+(aq) \longrightarrow N_2H_5^+(aq)$.
4.20 molecular: $CH_3NH_2(aq) + HCHO_2(aq) \longrightarrow CH_3NH_3CHO_2(aq)$,
ionic: $CH_3NH_2(aq) + HCHO_2(aq) \longrightarrow CH_3NH_3^+(aq) + CHO_2^-(aq)$,
net ionic: $CH_3NH_2(aq) + HCHO_2(aq) \longrightarrow CH_3NH_3^+(aq) + CHO_2^-(aq)$.
4.21 molecular: $2HCHO_2(aq) + Co(OH)_2(s) \longrightarrow Co(CHO_2)_2(aq) + 2H_2O$,
ionic: $2HCHO_2(aq) + Co(OH)_2(s) \longrightarrow 2CHO_2^-(aq) + Co^{2+}(aq) + 2H_2O$,
net ionic: $2HCHO_2(aq) + Co(OH)_2(s) \longrightarrow 2CHO_2^-(aq) + Co^{2+}(aq) + 2H_2O$,
4.22 (a) Formic acid, a weak acid will form. Molecular: $KCHO_2(aq) + HCl(aq) \longrightarrow KCl(aq) + HCHO_2(aq)$,
ionic: $K^+(aq) + CHO_2^-(aq) + H^+(aq) + Cl^-(aq) \longrightarrow K^+(aq) + Cl^-(aq) + HCHO_2(aq)$,
net ionic: $CHO_2^-(aq) + H^+(aq) \longrightarrow HCHO_2(aq)$,
(b) Carbonic acid will form and it will further dissociate to water and carbon dioxide:
$CuCO_3(s) + 2H^+(aq) \longrightarrow CO_2(g) + H_2O + Cu^{2+}(aq)$,
molecular: $CuCO_3(s) + 2HC_2H_3O_2(aq) \longrightarrow CO_2(g) + H_2O + Cu(C_2H_3O_2)_2(aq)$,
ionic: $CuCO_3(s) + 2HC_2H_3O_2(aq) \longrightarrow CO_2(g) + H_2O + Cu^{2+}(aq) + 2C_2H_3O_2^-(aq)$,
net ionic: $CuCO_3(s) + 2HC_2H_3O_2(aq) \longrightarrow CO_2(g) + H_2O + Cu^{2+}(aq) + 2C_2H_3O_2^-(aq)$,
c) NR, d) Insoluble nickel hydroxide will precipitate:
$Ni^{2+}(aq) + 2OH^-(aq) \longrightarrow Ni(OH)_2(s)$,
molecular: $NiCl_2(aq) + 2NaOH(aq) \longrightarrow Ni(OH)_2(s) + 2NaCl(aq)$,
ionic: $Ni^{2+}(aq) + 2Cl^-(aq) + 2Na^+(aq) + 2OH^-(aq) \longrightarrow Ni(OH)_2(s) + 2Na^+(aq) + 2Cl^-(aq)$,
net ionic: $Ni^{2+}(aq) + 2OH^-(aq) \longrightarrow Ni(OH)_2(s)$.
4.23 1.53 M HNO_3. **4.24** 0.1837 M NaCl.
4.25 0.0438 mol HCl. **4.26** 143 mL. **4.27** 2.11 g $Sr(NO_3)_2$.
4.28 0.531 g $AgNO_3$. **4.29** 250.0 mL. **4.30** 600 mL of water.
4.31 31.6 mL H_3PO_4. **4.32** 26.8 mL NaOH. **4.33** 0.40 M Fe^{3+}, 1.2 M Cl^-. **4.34** 0.750 M Na^+. **4.35** 0.0449 M $CaCl_2$.
4.36 60.0 mL KOH. **4.37** 0.605 g Na_2SO_4.
4.38 (a) 5.41×10^{-3} mol Ca^{2+}, (b) 5.41×10^{-3} moles Ca^{2+}, (c) 5.41×10^{-3} moles $CaCl_2$, (d) 0.600 g $CaCl_2$, (e) 30.0% $CaCl_2$.
4.39 0.178 M H_2SO_4. **4.40** 0.0220 M HCl, 0.0803%

Review Problems

4.49 (a) ionic:
$2NH_4^+(aq) + CO_3^{2-}(aq) + Mg^{2+}(aq) + 2Cl^-(aq) \longrightarrow 2NH_4^+(aq) + 2Cl^-(aq) + MgCO_3(s)$,
net: $Mg^{2+}(aq) + CO_3^{2-}(aq) \longrightarrow MgCO_3(s)$,
(b) ionic: $Cu^{2+}(aq) + 2Cl^-(aq) + 2Na^+(aq) + 2OH^-(aq) \longrightarrow Cu(OH)_2(s) + 2Na^+(aq) + 2Cl^-(aq)$,
net: $Cu^{2+}(aq) + 2OH^-(aq) \longrightarrow Cu(OH)_2(s)$,
(c) ionic: $3Fe^{2+}(aq) + 3SO_4^{2-}(aq) + 6Na^+(aq) + 2PO_4^{3-}(aq) \longrightarrow Fe_3(PO_4)_2(s) + 6Na^+(aq) + 3SO_4^{2-}(aq)$,
net: $3Fe^{2+}(aq) + 2PO_4^{3-}(aq) \longrightarrow Fe_3(PO_4)_2(s)$,
(d) ionic: $2Ag^+(aq) + 2C_2H_3O_2^-(aq) + Ni^{2+}(aq) + 2Cl^-(aq) \longrightarrow 2AgCl(s) + Ni^{2+}(aq) + 2C_2H_3O_2^-(aq)$,
net: $Ag^+(aq) + Cl^-(aq) \longrightarrow AgCl(s)$.
4.51 $HClO_4(l) + H_2O \longrightarrow H_3O^+(aq) + ClO_4^-(aq)$.
4.53 $N_2H_4(aq) + H_2O \rightleftharpoons N_2H_5^+(aq) + OH^-(aq)$.
4.55 $HNO_2(aq) + H_2O \rightleftharpoons H_3O^+(aq) + NO_2^-(aq)$.
4.57 $H_2CO_3(aq) + H_2O \rightleftharpoons H_3O^+(aq) + HCO_3^-(aq)$,
$HCO_3^-(aq) + H_2O \rightleftharpoons H_3O^+(aq) + CO_3^{2-}(aq)$
4.59 (a) ionic:
$3Fe^{2+}(aq) + 3SO_4^{2-}(aq) + 6K^+(aq) + 2PO_4^{3-}(aq) \longrightarrow Fe_3(PO_4)_2(s) + 6K^+(aq) + 3SO_4^{2-}(aq)$,
net: $3Fe^{2+}(aq) + 2PO_4^{3-}(aq) \longrightarrow Fe_3(PO_4)_2(s)$,
(b) ionic: $3Ag^+(aq) + 3C_2H_3O_2^-(aq) + Al^{3+}(aq) + 3Cl^-(aq) \longrightarrow 3AgCl(s) + Al^{3+}(aq) + 3C_2H_3O_2^-(aq)$,
net: $Ag^+(aq) + Cl^-(aq) \longrightarrow AgCl(s)$.
4.61 molecular: $Na_2S(aq) + Cu(NO_3)_2(aq) \longrightarrow CuS(s) + 2NaNO_3(aq)$,
ionic: $2Na^+(aq) + S^{2-}(aq) + Cu^{2+}(aq) + 2NO_3^-(aq) \longrightarrow CuS(s) + 2Na^+(aq) + 2NO_3^-(aq)$,
net: $Cu^{2+}(aq) + S^{2-}(aq) \longrightarrow CuS(s)$.
4.63 (a), (b), and (d).
4.65 (a) molecular: $Ca(OH)_2(aq) + 2HNO_3(aq) \longrightarrow Ca(NO_3)_2(aq) + 2H_2O$,
ionic: $Ca^{2+}(aq) + 2OH^-(aq) + 2H^+(aq) + 2NO_3^-(aq) \longrightarrow Ca^{2+}(aq) + 2NO_3^-(aq) + 2H_2O$,
net: $H^+(aq) + OH^-(aq) \longrightarrow H_2O$,

(b) molecular: $Al_2O_3(s) + 6HCl(aq) \longrightarrow 2AlCl_3(aq) + 3H_2O$,
ionic: $Al_2O_3(s) + 6H^+(aq) + 6Cl^-(aq) \longrightarrow 2Al^{3+}(aq) + 6Cl^-(aq) + 3H_2O$,
net: $Al_2O_3(s) + 6H^+(aq) \longrightarrow 2Al^{3+}(aq) + 3H_2O$,
(c) molecular: $Zn(OH)_2(s) + H_2SO_4(aq) \longrightarrow ZnSO_4(aq) + 2H_2O$,
ionic: $Zn(OH)_2(s) + 2H^+(aq) + SO_4^{2-}(aq) \longrightarrow Zn^{2+}(aq) + SO_4^{2-}(aq) + 2H_2O$,
net: $Zn(OH)_2(s) + 2H^+(aq) \longrightarrow Zn^{2+}(aq) + 2H_2O$.
4.67 The electrical conductivity would decrease gradually, until one solution had neutralized the other, forming a nonelectrolyte. Once the point of neutralization had been reached, the addition of excess sulfuric acid would cause the conductivity to increase, because sulfuric acid is a strong electrolyte itself.
4.69 (a) $2H^+(aq) + CO_3^{2-}(aq) \longrightarrow H_2O + CO_2(g)$,
(b) $NH_4^+(aq) + OH^-(aq) \longrightarrow NH_3(g) + H_2O$.
4.71 (a) formation of insoluble $Cr(OH)_3$, (b) formation of water, a weak electrolyte. **4.73** (a) molecular:
$3HNO_3(aq) + Cr(OH)_3(s) \longrightarrow Cr(NO_3)_3(aq) + 3H_2O$,
ionic: $3H^+(aq) + 3NO_3^-(aq) + Cr(OH)_3(s) \longrightarrow Cr^{3+}(aq) + 3NO_3^-(aq) + 3H_2O$,
net: $3H^+(aq) + Cr(OH)_3(s) \longrightarrow Cr^{3+}(aq) + 3H_2O$,
(b) molecular: $HClO_4(aq) + NaOH(aq) \longrightarrow NaClO_4(aq) + H_2O$,
ionic: $H^+(aq) + ClO_4^-(aq) + Na^+(aq) + OH^-(aq) \longrightarrow Na^+(aq) + ClO_4^-(aq) + H_2O$,
net: $H^+(aq) + OH^-(aq) \longrightarrow H_2O$,
(c) molecular: $Cu(OH)_2(s) + 2HC_2H_3O_2(aq) \longrightarrow Cu(C_2H_3O_2)_2(aq) + 2H_2O$,
ionic: $Cu(OH)_2(s) + 2HC_2H_3O_2(aq) \longrightarrow Cu^{2+}(aq) + 2C_2H_3O_2^-(aq) + 2H_2O$,
net: $Cu(OH)_2(s) + 2HC_2H_3O_2(aq) \longrightarrow Cu^{2+}(aq) + 2C_2H_3O_2^-(aq) + 2H_2O$,
(d) molecular: $ZnO(s) + H_2SO_4(aq) \longrightarrow ZnSO_4(aq) + H_2O$,
ionic: $ZnO(s) + 2H^+(aq) + SO_4^{2-}(aq) \longrightarrow Zn^{2+}(aq) + SO_4^{2-}(aq) + H_2O$,
net: $ZnO(s) + 2H^+(aq) \longrightarrow Zn^{2+}(aq) + H_2O$.
4.75 (a) molecular: $Na_2SO_3(aq) + Ba(NO_3)_2(aq) \longrightarrow BaSO_3(s) + 2NaNO_3(aq)$,
ionic: $2Na^+(aq) + SO_3^{2-}(aq) + Ba^{2+}(aq) + 2NO_3^-(aq) \longrightarrow BaSO_3(s) + 2Na^+(aq) + 2NO_3^-(aq)$,
net: $Ba^{2+}(aq) + SO_3^{2-}(aq) \longrightarrow BaSO_3(s)$,
(b) molecular: $2HCHO_2(aq) + K_2CO_3(aq) \longrightarrow CO_2(g) + H_2O + 2KCHO_2(aq)$,
ionic: $2HCHO_2(aq) + 2K^+(aq) + CO_3^{2-}(aq) \longrightarrow CO_2(g) + H_2O + 2K^+(aq) + 2CHO_2^-(aq)$,
net: $2HCHO_2(aq) + CO_3^{2-}(aq) \longrightarrow 2CHO_2^-(aq) + CO_2(g) + H_2O$,
(c) molecular: $2NH_4Br(aq) + Pb(C_2H_3O_2)_2(aq) \longrightarrow 2NH_4C_2H_3O_2(aq) + PbBr_2(s)$,
ionic: $2NH_4^+(aq) + 2Br^-(aq) + Pb^{2+}(aq) + 2C_2H_3O_2^-(aq) \longrightarrow 2NH_4^+(aq) + 2C_2H_3O_2^-(aq) + PbBr_2(s)$,
net: $Pb^{2+}(aq) + 2Br^-(aq) \longrightarrow PbBr_2(s)$,
(d) molecular: $2NH_4ClO_4(aq) + Cu(NO_3)_2(aq) \longrightarrow Cu(ClO_4)_2(aq) + 2NH_4NO_3(aq)$,
ionic: $2NH_4^+(aq) + 2ClO_4^-(aq) + Cu^{2+}(aq) + 2NO_3^-(aq) \longrightarrow Cu^{2+}(aq) + 2ClO_4^-(aq) + 2NO_3^-(aq) + 2NH_4^+(aq)$,
net: N.R.
4.77 There are numerous possible answers. One of many possible sets of answers would be,
(a) $NaHCO_3(aq) + HCl(aq) \longrightarrow NaCl(aq) + CO_2(g) + H_2O$,
(b) $FeCl_2(aq) + 2NaOH(aq) \longrightarrow Fe(OH)_2(s) + 2NaCl(aq)$,
(c) $Ba(NO_3)_2(aq) + K_2SO_3(aq) \longrightarrow BaSO_3(s) + 2KNO_3(aq)$,
(d) $2AgNO_3(aq) + Na_2S(aq) \longrightarrow Ag_2S(s) + 2NaNO_3(aq)$,
(e) $ZnO(s) + 2HCl(aq) \longrightarrow ZnCl_2(aq) + H_2O$.
4.79 (a) 1.00 *M* NaOH, (b) 0.577 *M* $CaCl_2$.
4.81 658 mL $NaC_2H_3O_2$. **4.83** (a) 1.46 g NaCl, (b) 16.2 g $C_6H_{12}O_6$, (c) 6.13 g H_2SO_4. **4.85** 0.11 *M* H_2SO_4.
4.87 300 mL. **4.89** 225 mL water. **4.91** (a) 0.0145 mol Ca^{2+}, 0.0290 mol Cl^-, (b) 0.020 mol Al^{3+}, 0.060 mol Cl^-.
4.93 (a) 0.25 *M* Cr^{2+}, 0.50 *M* NO_3^-, (b) 0.10 *M* Cu^{2+}, 0.10 *M* SO_4^{2-}, (c) 0.48 *M* Na^+, 0.16 *M* PO_4^{3-}, (d) 0.15 *M* Al^{3+}, 0.22 *M* SO_4^{2-}. **4.95** 1.0 g $Al_2(SO_4)_3$. **4.97** 12.0 mL $NiCl_2$ soln, 0.36 g $NiCO_3$. **4.99** 0.113 *M* KOH, $KOH(aq) + HCl(aq) \longrightarrow KCl(aq) + H_2O$. **4.101** 0.485 g $Al_2(SO_4)_3$.
4.103 2.00 mL $FeCl_3$ soln, 0.129 g AgCl. **4.105** 13.3 mL $AlCl_3$.
4.107 0.167 *M* Fe^{3+}, 3.67 g Fe_2O_3. **4.109** (a) 8.00×10^{-3} mol AgCl, (b) 0.160 *M* NO_3^-, 0.220 *M* Na^+, 0.060 *M* Cl^-. **4.111** 0.114 *M* HCl.
4.113 2.67×10^{-3} mol $HC_3H_5O_3$. **4.115** $MgSO_4 \cdot 7H_2O$.
4.117 48.40% Pb in the sample. **4.119** 58.6%

CHAPTER 5

Practice Exercises

5.1 Reduced. **5.2** Aluminum is oxidized and is the reducing agent, Chlorine is reduced and is the oxidizing agent. **5.3** +3.
5.4 (a) Ni +2; Cl −1, (b) Mg +2; Ti +4; O −2, (c) K +1; Cr +6; O −2, (d) H +1; P +5, O −2, (e) V +3; C 0; H +1; O −2.
5.5 +8/3. **5.6** $KClO_3$ is reduced and HNO_2 is oxidized.
5.7 Cl_2 is reduced and is the oxidizing agent. $NaClO_2$ is oxidized and is the reducing agent. **5.8** Water, since the oxidation number of oxygen drops from −1 to −2 in the formation of water.
5.9 $2Al(s) + 3Cu^{2+}(aq) \longrightarrow 2Al^{3+}(aq) + 3Cu(s)$.
5.10 $3Sn^{2+} + 16H^+ + 2TcO_4^- \longrightarrow 2Tc^{4+} + 8H_2O + 3Sn^{4+}$.
5.11 $4Cu + 2NO_3^- + 10H^+ \longrightarrow 4Cu^{2+} + N_2O + 5H_2O$.
5.12 $4OH^- + SO_2 \longrightarrow SO_4^{2-} + 2e^- + 2H_2O$.
5.13 $2MnO_4^- + 3C_2O_4^{2-} + 4OH^- \longrightarrow 2MnO_2 + 6CO_3^{2-} + 2H_2O$.
5.14 $Zn \longrightarrow Zn^{2+} + 2e^-$, $2H^+ + 2e^- \longrightarrow H_2$.
5.15 $HNO_3 + 9H^+ + 4Mg \longrightarrow NH_4^+ + 4Mg^{2+} + 3H_2O$.
5.16 (a) molecular: $Mg(s) + 2HCl(aq) \longrightarrow MgCl_2(aq) + H_2(g)$,
ionic: $Mg(s) + 2H^+(aq) + 2Cl^-(aq) \longrightarrow Mg^{2+}(aq) + 2Cl^-(aq) + H_2(g)$,
net ionic: $Mg(s) + 2H^+(aq) \longrightarrow Mg^{2+}(aq) + H_2(g)$,
(b) molecular: $2Al(s) + 6HCl(aq) \longrightarrow 2AlCl_3(aq) + 3H_2(g)$,
ionic: $2Al(s) + 6H^+(aq) + 6Cl^-(aq) \longrightarrow 2Al^{3+}(aq) + 6Cl^-(aq) + 3H_2(g)$;
net ionic: $2Al(s) + 6H^+(aq) \longrightarrow 2Al^{3+}(aq) + 3H_2(g)$.
5.17 $Cu^{2+}(aq) + Mg(s) \longrightarrow Cu(s) + Mg^{2+}(aq)$.
5.18 (a) $2Al(s) + 3Cu^{2+}(aq) \longrightarrow 2Al^{3+}(aq) + 3Cu(s)$,
(b) $Ag(s) + Mg^{2+}(aq) \longrightarrow$ No reaction.
5.19 $2C_{20}H_{42}(s) + 21O_2(g) \longrightarrow 40C(s) + 42H_2O(g)$.
5.20 $2C_4H_{10}(g) + 13O_2(g) \longrightarrow 8CO_2(g) + 10H_2O(g)$.
5.21 $C_2H_5OH(l) + 3O_2(g) \longrightarrow 2CO_2(g) + 3H_2O(g)$.
5.22 $2Sr(s) + O_2(g) \longrightarrow 2SrO(s)$.
5.23 $4Fe(s) + 3O_2(g) \longrightarrow 2Fe_2O_3(s)$.
5.24 0.06100 *M* $H_2C_2O_4$. **5.25** 2.37 g $Na_2S_2O_3$.
5.26 (a) $5Sn^{2+} + 2MnO_4^- + 16H^+ \longrightarrow 5Sn^{4+} + 2Mn^{2+} + 8H_2O$, (b) 0.120 g Sn, (c) 40.0% Sn, (d) 50.7% SnO_2

Review Problems

5.25 (a) S^{2-}: −2, (b) SO_2: S +4, O −2, (c) P_4: P 0, (d) PH_3: P −3, H +1. **5.27** (a) O: −2, Na: +1, Cl: +1, (b) O: −2, Na: +1, Cl: +3, (c) O: −2, Na: +1, Cl: +5, (d) O: −2, Na: +1, Cl: +7.
5.29 The sum of the oxidation numbers should be zero, (a) S: −2, Pb: +2, (b) Cl: −1, Ti: +4, (c) Cs +1, O −1/2, (d) F −1, O +1. **5.31** In the forward direction: Cl_2 is reduced and Cl_2 also oxidized. In the reverse direction: Cl^- is the reducing agent. HOCl must be the oxidizing agent. **5.33** (a) substance reduced (and oxidizing agent): HNO_3, substance oxidized (and reducing agent): H_3AsO_3, (b) substance reduced (and oxidizing agent): HOCl, substance oxidized (and reducing agent): NaI, (c) substance reduced (and oxidizing agent): $KMnO_4$, substance oxidized

(and reducing agent): $H_2C_2O_4$, (d) substance reduced (and oxidizing agent): H_2SO_4, substance oxidized (and reducing agent): Al.
5.35 (a) $OCl^- + 2S_2O_3^{2-} + 2H^+ \longrightarrow S_4O_6^{2-} + Cl^- + H_2O$,
(b) $2NO_3^- + Cu + 4H^+ \longrightarrow 2NO_2 + Cu^{2+} + 2H_2O$,
(c) $3AsO_3^{3-} + IO_3^- \longrightarrow I^- + 3AsO_4^{3-}$,
(d) $Zn + SO_4^{2-} + 4H^+ \longrightarrow Zn^{2+} + SO_2 + 2H_2O$,
(e) $NO_3^- + 4Zn + 10H^+ \longrightarrow 4Zn^{2+} + NH_4^+ + 3H_2O$,
(f) $2Cr^{3+} + 3BiO_3^- + 4H^+ \longrightarrow Cr_2O_7^{2-} + 3Bi^{3+} + 2H_2O$,
(g) $I_2 + 5OCl^- + H_2O \longrightarrow 2IO_3^- + 5Cl^- + 2H^+$,
(h) $2Mn^{2+} + 5BiO_3^- + 14H^+ \longrightarrow 2MnO_4^- + 5Bi^{3+} + 7H_2O$,
(i) $3H_3AsO_3 + Cr_2O_7^{2-} + 8H^+ \longrightarrow 3H_3AsO_4 + 2Cr^{3+} + 4H_2O$,
(j) $2I^- + HSO_4^- + 3H^+ \longrightarrow I_2 + SO_2 + 2H_2O$.
5.37 (a) $2CrO_4^{2-} + 3S^{2-} + 4H_2O \longrightarrow 2CrO_2^- + 3S + 8OH^-$,
(b) $3C_2O_4^{2-} + 2MnO_4^- + 4H_2O \longrightarrow 6CO_2 + 2MnO_2 + 8OH^-$,
(c) $4ClO_3^- + 3N_2H_4 \longrightarrow 4Cl^- + 6NO + 6H_2O$,
(d) $NiO_2 + 2Mn(OH)_2 \longrightarrow Ni(OH)_2 + Mn_2O_3 + H_2O$,
(e) $3SO_3^{2-} + 2MnO_4^- + H_2O \longrightarrow 3SO_4^{2-} + 2MnO_2 + 2OH^-$.
5.39 $4OCl^- + S_2O_3^{2-} + H_2O \longrightarrow 4Cl^- + 2SO_4^{2-} + 2H^+$.
5.41 $O_3 + Br^- \longrightarrow BrO_3^-$.
5.43 (a) $Mn(s) + 2HCl(aq) \longrightarrow MnCl_2(aq) + H_2(g)$,
$Mn(s) + 2H^+(aq) + 2Cl^-(aq) \longrightarrow Mn^{2+}(aq) + 2Cl^-(aq) + H_2(g)$,
$Mn(s) + 2H^+(aq) \longrightarrow Mn^{2+}(aq) + H_2(g)$,
(b) $Cd(s) + 2HCl(aq) \longrightarrow CdCl_2(aq) + H_2(g)$,
$Cd(s) + 2H^+(aq) + 2Cl^-(aq) \longrightarrow Cd^{2+}(aq) + 2Cl^-(aq) + H_2(g)$,
$Cd(s) + 2H^+(aq) \longrightarrow Cd^{2+}(aq) + H_2(g)$,
(c) $Sn(s) + 2HCl(aq) \longrightarrow SnCl_2(aq) + H_2(g)$,
$Sn(s) + 2H^+(aq) + 2Cl^-(aq) \longrightarrow Sn^{2+}(aq) + 2Cl^-(aq) + H_2(g)$,
$Sn(s) + 2H^+(aq) \longrightarrow Sn^{2+}(aq) + H_2(g)$.
5.45 (a) $3Ag(s) + 4HNO_3(aq) \longrightarrow 3AgNO_3(aq) + 2H_2O + NO(g)$,
(b) $Ag(s) + 2HNO_3(aq) \longrightarrow AgNO_3(aq) + H_2O + NO_2(aq)$.
5.47 (a) N.R., (b) $2Cr(s) + 3Pb^{2+}(aq) \longrightarrow 2Cr^{3+}(aq) + 3Pb(s)$,
(c) $2Ag^+(aq) + Fe(s) \longrightarrow 2Ag(s) + Fe^{2+}(aq)$,
(d) $3Ag(s) + Au^{3+}(aq) \longrightarrow Au(s) + 3Ag^+(aq)$.
5.49 Pt, Ru, Tl, Pu.
5.51 $Cd(s) + 2TlCl(aq) \longrightarrow CdCl_2(aq) + 2Tl(s)$.
5.53 (a) $2C_6H_6(l) + 15O_2(g) \longrightarrow 12CO_2(g) + 6H_2O(g)$,
(b) $C_3H_8(g) + 5O_2(g) \longrightarrow 3CO_2(g) + 4H_2O(g)$,
(c) $C_{21}H_{44}(s) + 32O_2(g) \longrightarrow 21CO_2(g) + 22H_2O(g)$.
5.55 (a) $2C_6H_6(l) + 9O_2(g) \longrightarrow 12CO(g) + 6H_2O(g)$,
$2C_3H_8(g) + 7O_2(g) \longrightarrow 6CO(g) + 8H_2O(g)$,
$2C_{21}H_{44}(s) + 43O_2(g) \longrightarrow 42CO(g) + 44H_2O(g)$,
(b) $2C_6H_6(l) + 3O_2(g) \longrightarrow 12C(s) + 6H_2O(g)$,
$C_3H_8(g) + 2O_2(g) \longrightarrow 3C(s) + 4H_2O(g)$,
$C_{21}H_{44}(s) + 11O_2(g) \longrightarrow 21C(s) + 22H_2O(g)$.
5.57 $2CH_3OH(l) + 3O_2(g) \longrightarrow 2CO_2(g) + 4H_2O(g)$.
5.59 $2(CH_3)_2S(g) + 9O_2(g) \longrightarrow 4CO_2(g) + 6H_2O(g) + 2SO_2(g)$.
5.61 (a) $2Zn(s) + O_2(g) \longrightarrow 2ZnO(s)$,
(b) $4Al(s) + 3O_2(g) \longrightarrow 2Al_2O_3(s)$,
(c) $2Mg(s) + O_2(g) \longrightarrow 2MgO(s)$,
(d) $4Fe(s) + 3O_2(g) \longrightarrow 2Fe_2O_3(s)$.
5.63 (a) $IO_3^- + 3SO_3^{2-} \longrightarrow I^- + 3SO_4^{2-}$, (b) 9.55 g Na_2SO_3.
5.65 3.53 g Cu.
5.67 (a) $2MnO_4^- + 5Sn^{2+} + 16H^+ \longrightarrow 2Mn^{2+} + 5Sn^{4+} + 8H_2O$,
(b) 17.4 mL $KMnO_4$.
5.69 (a) 6.38×10^{-3} M I_3^-, (b) 1.01×10^{-3} g SO_2,
(c) 2.10×10^{-3}%, (d) 21 ppm.
5.71 (a) 9.463%, (b) 18.40%.
5.73 (a) 0.02994 g H_2O_2, (b) 2.994% H_2O_2.
5.75 (a) $2CrO_4^{2-} + 3SO_3^{2-} + H_2O \longrightarrow 2CrO_2^- + 3SO_4^{2-} + 2OH^-$,
(b) 0.875 g Cr in the original alloy., (c) 25.4% Cr.
5.77 (a) 5.405×10^{-3} mol $C_2O_4^{2-}$, (b) 0.5999 g $CaCl_2$,
(c) 24.35% $CaCl_2$

CHAPTER 6

Practice Exercises

6.1 55.1 J/°C. **6.2** 5.23 kJ, 1250 cal, 1.25 kcal. **6.3** 12.4 kJ °C^{-1}.
6.4 394 kJ/mol C. **6.5** −58 kJ mol^{-1}. **6.6** 3.7 kJ, 74 kJ/mol.
6.7 $\frac{1}{4}CH_4(g) + \frac{1}{2}O_2(g) \longrightarrow \frac{1}{4}CO_2(g) + \frac{1}{2}H_2O(l)$ $\Delta H = 222.6$ kJ
6.8 $2.500H_2(g) + \frac{2.500}{2}O_2(g) \longrightarrow 2.500H_2O(l)$ $\Delta H = -714.75$ kJ
6.9 $2Cu(s) + O_2(g)$

−310 kJ; −169 kJ; $Cu_2O(s) + \frac{1}{2}O_2(g)$; −141 kJ; $2CuO(s)$

The reaction is exothermic.
6.10 $NO(g) + \frac{1}{2}O_2(g)$

+90.4 kJ; −56.6 kJ; $NO_2(g)$; $\frac{1}{2}N_2(g) + O_2(g)$; +33.8 kJ

The reaction is endothermic.
6.11 $\Delta H = +857.7$ kJ. **6.12** $\Delta H = +99.2$ kJ.
6.13 $\Delta H° = -44.0$ kJ. **6.14** 385 kJ. **6.15** 2.62×10^6 kJ.
6.16 $\frac{1}{2}N_2(g) + 2H_2(g) + \frac{1}{2}Cl_2(g) \longrightarrow NH_4Cl(s)$ $\Delta H_f° = -315.4$ kJ
6.17 $Na(s) + \frac{1}{2}H_2(g) + C(s) + \frac{3}{2}O_2(g) \longrightarrow NaHCO_3(s)$ $\Delta H_f° = -947.7$ kJ/mol
6.18 +800.34 kJ.
6.19 $S(s) + \frac{3}{2}O_2(g) \longrightarrow SO_3(g)$ $\Delta H_f° = -395.2$ kJ/mol
$S(s) + O_2(g) \longrightarrow SO_2(g)$ $\Delta H_f° = -296.9$ kJ/mol
Reverse the first reaction and add the two reactions together to get $SO_3(g) \longrightarrow SO_2(g) + \frac{1}{2}O_2(g)$ $\Delta H_f° = +98.3$ kJ
The answers for the enthalpy of reaction are the same using either method. **6.20** (a) −113.1 kJ, (b) −177.8 kJ

Review Problems

6.41 −17 J. **6.43** +100 J. **6.45** −1320 J, −315 cal.
6.47 (a) 1.67×10^3 J, (b) 1.67×10^3 J, (c) 23.2 J °C^{-1},
(d) 4.64 J g^{-1} °C^{-1}. **6.49** 25.12 J/mol °C. **6.51** −30.4 kJ.
6.53 $HNO_3(aq) + KOH(aq) \longrightarrow KNO_3(aq) + H_2O(l)$, 3.8×10^3 J, 53 kJ/mol.
6.55 (a) $C_3H_8(g) + 5O_2(g) \longrightarrow 3CO_2(g) + 4H_2O(l)$,
(b) 2.22×10^5 J, (c) −222 kJ/mol.
6.57 (a) $2CO(g) + O_2(g) \longrightarrow 2CO_2(g)$, $\Delta H° = -566$ kJ,
(b) −283 kJ/mol. **6.59** 162 kJ of heat are evolved.
6.61 $Ge(s) + O_2(g)$

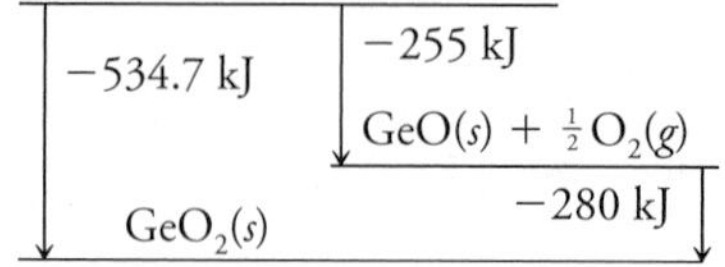

$\Delta H° = -280$ kJ
6.63 $2NO_2(g) \longrightarrow N_2O_4(g)$, $\Delta H° = -57.93$ kJ,
$2NO(g) + O_2(g) \longrightarrow 2NO_2(g)$, $\Delta H° = -113.14$ kJ,
$2NO(g) + O_2(g) \longrightarrow N_2O_4(g)$, $\Delta H° = -171.07$ kJ.
6.65 $\frac{1}{2}Na_2O(s) + HCl(g) \longrightarrow \frac{1}{2}H_2O(l) + NaCl(s)$, $\Delta H° = -253.66$ kJ,
$NaNO_2(s) \longrightarrow \frac{1}{2}Na_2O(s) + \frac{1}{2}NO_2(g) + \frac{1}{2}NO(g)$, $\Delta H° = +213.57$ kJ,
$\frac{1}{2}NO(g) + \frac{1}{2}NO_2(g) \longrightarrow \frac{1}{2}N_2O(g) + \frac{1}{2}O_2(g)$, $\Delta H° = -21.34$ kJ,

$\frac{1}{2}H_2O(l) + \frac{1}{2}O_2(g) + \frac{1}{2}N_2O(g) \longrightarrow HNO_2(l),$
$\Delta H° = -17.18$ kJ,
$HCl(g) + NaNO_2(s) \longrightarrow HNO_2(l) + NaCl(s),$
$\Delta H° = -78.61$ kJ.

6.67 $\frac{1}{2}CaO(s) + \frac{1}{2}Cl_2(g) \longrightarrow \frac{1}{2}CaOCl_2(s)$
$\Delta H° = \frac{1}{2}(-110.9 \text{ kJ}).$
$\frac{1}{2}H_2O(l) + \frac{1}{2}CaOCl_2(s) + NaBr(s) \longrightarrow$
$NaCl(s) + \frac{1}{2}Ca(OH)_2(s) + \frac{1}{2}Br_2(l) \quad \Delta H° = \frac{1}{2}(-60.2 \text{ kJ}).$
$\frac{1}{2}Ca(OH)_2(s) \longrightarrow \frac{1}{2}CaO(s) + \frac{1}{2}H_2O(l) \quad \Delta H° = \frac{1}{2}(+65.1 \text{ kJ}).$
$\frac{1}{2}Cl_2(g) + NaBr(s) \longrightarrow NaCl(s) + \frac{1}{2}Br_2(l)$
$\Delta H° = \frac{1}{2}(-160.1 \text{ kJ}) = -53$ KJ.

6.69 $12NH_3(g) + 21O_2(g) \longrightarrow 12NO_2(g) + 18H_2O(g)$
$\Delta H° = 3(-1132$ kJ),
$12NO_2(g) + 16NH_3(g) \longrightarrow 14N_2(g) + 24H_2O(g)$
$\Delta H° = 2(-2740$ kJ),
$28NH_3(g) + 21O_2(g) \longrightarrow 14N_2(g) + 42H_2O(g)$
$\Delta H° = -8876$ kJ,
Now divide this equation by 7 to get
$4NH_3(g) + 3O_2(g) \longrightarrow 2N_2(g) + 6H_2O(g)$
$\Delta H° = 1/7(-8876 \text{ kJ}) = -1268$ kJ.

6.71 (a) $2C(\textit{graphite}) + 2H_2(g) + O_2(g) \longrightarrow HC_2H_3O_2(l)$
$\Delta H_f° = -487.0$ kJ,
(b) $2C(\textit{graphite}) + \frac{1}{2}O_2(g) + 3H_2(g) \longrightarrow C_2H_5OH(l)$
$\Delta H_f° = -277.63$ kJ,
(c) $Ca(s) + \frac{1}{8}S_8(s) + 3O_2(g) + 2H_2(g) \longrightarrow CaSO_4{\cdot}2H_2O(s)$
$\Delta H_f° = -2021.1$ kJ.

6.73 (a) −196.6 kJ, (b) −177.8 kJ.
6.75 $C_{12}H_{22}O_{11}(s) + 12O_2(g) \longrightarrow 12CO_2(g) + 11H_2O(l)$
$\Delta H_f° = -2021.1 \text{ kJ mol}^{-1} \quad \Delta H°_{combustion} = -5.65 \times 10^3$ kJ/mol.

CHAPTER 7

Practice Exercises

7.1 5.10×10^{14} Hz. **7.2** 3.00×10^{13} Hz. **7.3** 2.874 m. **7.4** 2.63 μm. **7.5** 656.6 nm, red. **7.6** Shell 1 has 1 subshell, shell 2 has 2 subshells, shell 3 has 3 subshells, shell 4 has 4 subshells, shell 5 has 5 subshells, shell 6 has 6 subshells.
7.7 $n = 3$: *s*, *p* and *d* subshells, $n = 4$: *s*, *p*, *d* and *f* subshells.
7.8 (a) Mg: $1s^22s^22p^63s^2$,
(b) Ge: $1s^22s^22p^63s^23p^63d^{10}4s^24p^2$,
(c) Cd: $1s^22s^22p^63s^23p^63d^{10}4s^24p^64d^{10}5s^2$,
(d) Gd: $1s^22s^22p^63s^23p^63d^{10}4s^24p^64d^{10}4f^75s^25p^65d^16s^2$.
7.9 The electron configuration of an element follows the periodic table. The electrons are filled in the order of the periodic table and the energy levels are determined by the row the element is in and the subshell is given by the column, the first two columns are the *s*-block, the last six columns are the *p*-block, the *d*-block has ten columns, and the *f*-block has 14 columns.
7.10 (a) O: $1s^22s^22p^4$, S: $1s^22s^22p^63s^23p^4$, Se: $1s^22s^22p^63s^23p^63d^{10}4s^24p^4$,
(b) P: $1s^22s^22p^63s^23p^3$, N: $1s^22s^22p^3$, Sb: $1s^22s^22p^63s^23p^63d^{10}4s^24p^64d^{10}5s^25p^3$.
The elements have the same number of electrons in the valence shell, and the only differences between the valence shells are the energy levels.

7.11 (a) Na: (↑↓) (↑↓) (↑↓)(↑↓)(↑↓) (↑) ( )( )( ) ( ) ( )( )( )( )( )
1s 2s 2p 3s 3p 4s 3d

(b) S: (↑↓) (↑↓) (↑↓)(↑↓)(↑↓) (↑↓) (↑↓)(↑)(↑) ( ) ( )( )( )( )( )
1s 2s 2p 3s 3p 4s 3d

(c) Fe: (↑↓) (↑↓) (↑↓)(↑↓)(↑↓) (↑↓) (↑↓)(↑↓)(↑↓) (↑↓) (↑↓)(↑)(↑)(↑)(↑)
1s 2s 2p 3s 3p 4s 3d

7.12 (a)
Mg: (↑↓) (↑↓) (↑↓)(↑↓)(↑↓) (↑↓) ( )( )( ) ( ) ( )( )( )( )( ) ( )( )( )
1s 2s 2p 3s 3p 4s 3d 4p
0 unpaired electrons
(b)
Ge: (↑↓) (↑↓) (↑↓)(↑↓)(↑↓) (↑↓) (↑↓)(↑↓)(↑↓) (↑↓) (↑↓)(↑↓)(↑↓)(↑↓)(↑↓) (↑)(↑)( )
1s 2s 2p 3s 3p 4s 3d 4p
2 unpaired electrons
(c)
Cd: (↑↓) (↑↓) (↑↓)(↑↓)(↑↓) (↑↓) (↑↓)(↑↓)(↑↓) (↑↓) (↑↓)(↑↓)(↑↓)(↑↓)(↑↓) (↑↓)(↑↓)(↑↓)
1s 2s 2p 3s 3p 4s 3d 4p
(↑↓) (↑↓)(↑↓)(↑↓)(↑↓)(↑↓)
5s 4d
0 unpaired electrons
(d)
Gd: (↑↓) (↑↓) (↑↓)(↑↓)(↑↓) (↑↓) (↑↓)(↑↓)(↑↓) (↑↓) (↑↓)(↑↓)(↑↓)(↑↓)(↑↓) (↑↓)(↑↓)(↑↓)
1s 2s 2p 3s 3p 4s 3d 4p
(↑↓) (↑↓)(↑↓)(↑↓)(↑↓)(↑↓) (↑↓)(↑↓)(↑↓) (↑↓) (↑)( )( )( )( ) (↑)(↑)(↑)(↑)(↑)(↑)(↑)
5s 4d 5p 6s 5d 4f
8 unpaired electrons

7.13 Yes, Ti, Cr, Fe, Ni and the elements in their groups have even numbers of electrons and are paramagnetic. Additionally, oxygen has eight electrons, but it is paramagnetic since it has two unpaired electrons in the 2*p* orbitals.
7.14 (a) P: $[Ne]3s^23p^3$
[Ne] (↑↓) (↑)(↑)(↑)
3s 3p (3 unpaired electrons),

(b) Sn: $[Kr]4d^{10}5s^25p^2$
[Kr] (↑↓)(↑↓)(↑↓)(↑↓)(↑↓) (↑↓) (↑)(↑)( )
4d 5s 5p (2 unpaired electrons)

7.15 For representative elements the valence shell is defined as the occupied shell with the highest value of *n*. In a ground state atom, only *s* and *p* electrons fit that definition.
7.16 (a) Se: $4s^24p^4$, (b) Sn: $5s^25p^2$, (c) I: $5s^25p^5$.
7.17 (a) Sn, (b) Ga, (c) Cr, (d) S^{2-}.
7.18 (a) P, (b) Fe^{3+}, (c) Fe, (d) Cl^-.
7.19 (a) Be, (b) C. **7.20** (a) Na^+, (b) Mg^{2+}

Review Problems

7.73 6.98×10^{14} Hz. **7.75** 4.38×10^{13} Hz.
7.77 1.02×10^{15} Hz. **7.79** 2.97 m.
7.81 5.0×10^6 m, 5.0×10^3 km.
7.83 2.7×10^{-19} J, 1.6×10^5 J mol^{-1}.
7.85 (a) violet, (b) 7.307×10^{14} Hz, (c) 4.842×10^{-19} J.
7.87 1094 nm, We would not expect to see the light since it is not in the visible region.
7.89 1.737×10^{-6} m, infrared region. **7.91** (a) *p*, (b) *f*.
7.93 (a) $n = 3$, $\ell = 0$, (b) $n = 5$, $\ell = 2$.
7.95 0, 1, 2, 3, 4, 5. **7.97** (a) $m_\ell = 1, 0$, or -1, (b) $m_\ell = 3, 2, 1, 0, -1, -2$, or -3.
7.99 $\ell = 4$ $n = 5$.

7.101

n	ℓ	m_ℓ	m_s
2	1	−1	+1/2
2	1	−1	−1/2
2	1	0	+1/2
2	1	0	−1/2
2	1	+1	+1/2
2	1	+1	−1/2

7.103 21 electrons have $\ell = 1$, 20 electrons have $\ell = 2$.

7.105 (a) S $1s^22s^22p^63s^23p^4$, (b) K $1s^22s^22p^63s^23p^64s^1$, (c) Ti $1s^22s^22p^63s^23p^63d^24s^2$, (d) Sn $1s^22s^22p^63s^23p^63d^{10}4s^24p^64d^{10}5s^25p^2$.
7.107 (a) Mn paramagnetic, (b) As paramagnetic, (c) S paramagnetic, (d) Sr not paramagnetic, (b) Ar not paramagnetic.
7.109 (a) Mg zero unpaired electrons, (b) P three unpaired electrons, (c) V three unpaired electrons.
7.111 (a) Ni $[Ar]3d^84s^2$, (b) Cs $[Xe]6s^1$, (c) Ge $[Ar]3d^{10}4s^24p^2$, (d) Br $[Ar]3d^{10}4s^24p^5$, (e) Bi $[Xe]4f^{14}5d^{10}6s^26p^3$.

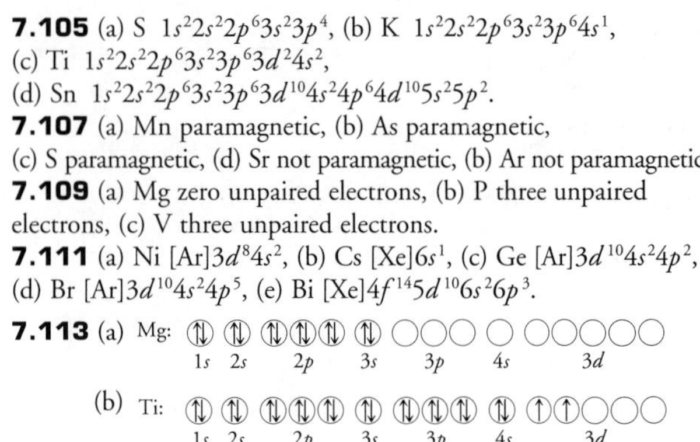

7.113 (a) Mg: (⇅) (⇅) (⇅)(⇅)(⇅) (⇅) ( )( )( ) ( ) ( )( )( )( )( )
1s 2s 2p 3s 3p 4s 3d

(b) Ti: (⇅) (⇅) (⇅)(⇅)(⇅) (⇅) (⇅)(⇅)(⇅) (⇅) (↑)(↑)( )( )( )
1s 2s 2p 3s 3p 4s 3d

7.115

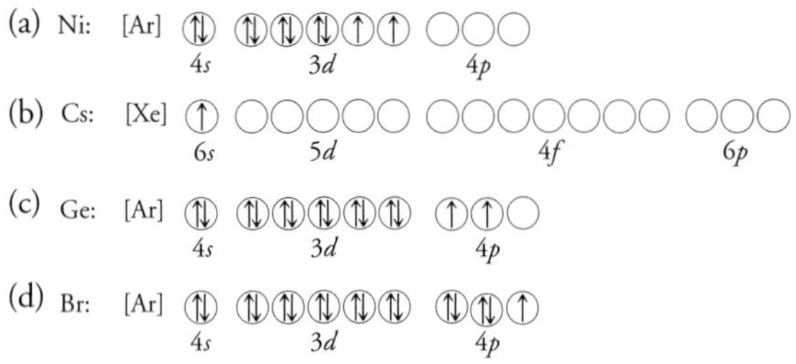

(a) Ni: [Ar] (⇅) (⇅)(⇅)(⇅)(↑)(↑) ( )( )( )
4s 3d 4p

(b) Cs: [Xe] (↑) ( )( )( )( )( ) ( )( )( )( )( )( )( ) ( )( )( )
6s 5d 4f 6p

(c) Ge: [Ar] (⇅) (⇅)(⇅)(⇅)(⇅)(⇅) (↑)(↑)( )
4s 3d 4p

(d) Br: [Ar] (⇅) (⇅)(⇅)(⇅)(⇅)(⇅) (⇅)(⇅)(↑)
4s 3d 4p

7.117 (a) 5, (b) 4, (c) 4, (d) 6.
7.119 (a) $3s^1$, (b) $3s^23p^1$, (c) $4s^24p^2$, (d) $3s^23p^3$.
7.121 (a) Na: (↑)
3s

(b) Al: (⇅) (↑)( )( )
3s 3p

(c) Ge: (⇅) (↑)(↑)( )
4s 4p

(d) P: (⇅) (↑)(↑)(↑)
3s 3p

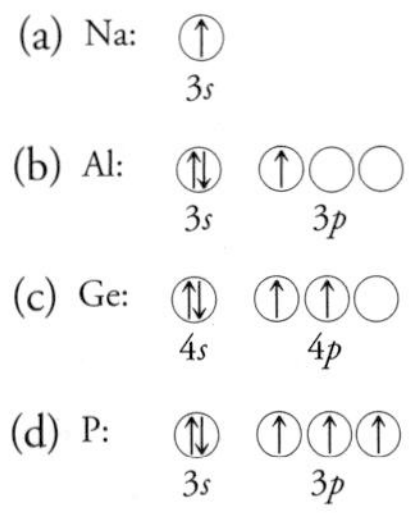

7.123 (a) 1, (b) 6, (c) 7. **7.125** (a) Mg, (b) Bi. **7.127** Sb.
7.129 (a) Na, (b) Co^{2+}, (c) Cl^-. **7.131** (a) N, (b) S, (c) Cl.
7.133 (a) Br, (b) As. **7.135** Mg

CHAPTER 8

Practice Exercises

8.1 There is one electron missing, and it should go into the 5*s* orbital, and the 5*p* orbital should be empty, $1s^12s^22p^63s^23p^63d^{10}4s^24p^64d^{10}5s^2$.
8.2 (a) Cr^{2+}: $[Ar]3d^4$ The 4*s* electron and one 3*d* electron are lost, (b) Cr^{3+}: $[Ar]3d^3$ The 4*s* electron and two 3*d* electrons are lost, (c) Cr^{6+}: [Ar] The 4*s* electron and all of the 3*d* electrons are lost.
8.3 The electron configurations are identical.

8.4 :Ï· ·Ca· ·Ï: ⟶ Ca^{2+} + 2[:Ï:]$^-$

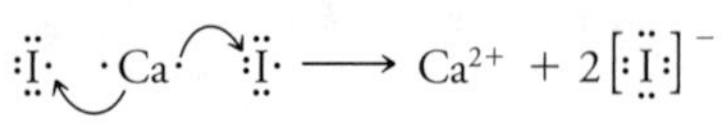

8.5 ·Ö: :Mg· ⟶ Mg^{2+} + [:Ö:]$^{2-}$

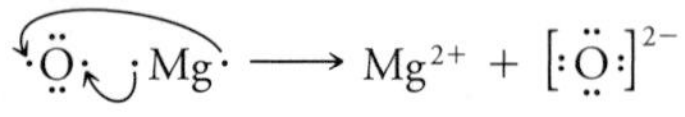

8.6 1.24 D. **8.7** 0.795 e^-, on Na: +0.795 e^-, on Cl: −0.795 e^-, 57.6% positive charge on the Na, 57.6% negative charge on the Cl.
8.8 The bond is polar and the Cl carries the negative charge.
8.9 (a) Br, (b) Cl, (c) Cl.
8.10

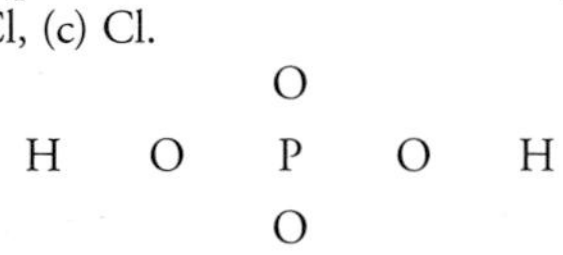

O
H O P O H
O

32 valence electrons.

8.11

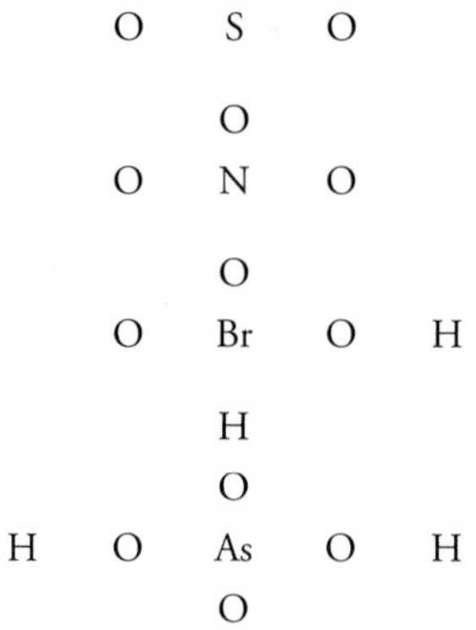

O S O

O
O N O

O
O Br O H

H
O
H O As O H
O

8.12 18 valence electrons, 32 valence electrons, 10 valence electrons.

8.13 :F̤—Ö—F̤:   [H—N(H)(H)—H]$^+$ [:Ö—N(:Ö:)=Ö]$^-$

Ö=S̤—Ö: :F̤—Cl(:F̤:)—F̤:   H—Ö—Cl(:Ö:)(:Ö:)—Ö:

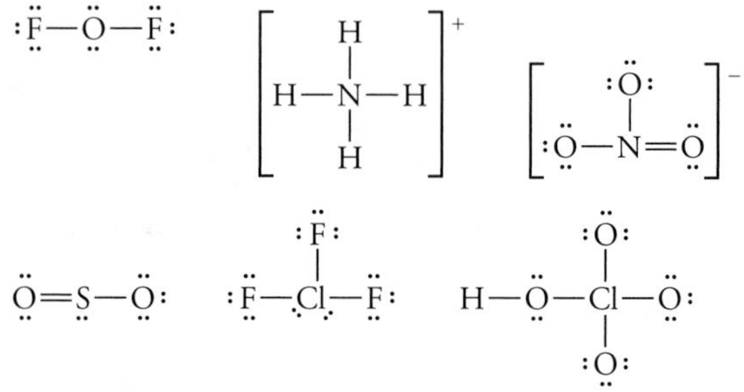

8.14 The negative sign should be on the oxygen, so two of the oxygen atoms should have a single bond and three lone pairs and the sulfur should have one double bond, two single bonds, and a lone pair.

8.15 (a) :N̤(−2)—N(+1)≡O:(+1)   (b) [S̤(0)=C(0)=N(−1)]$^-$

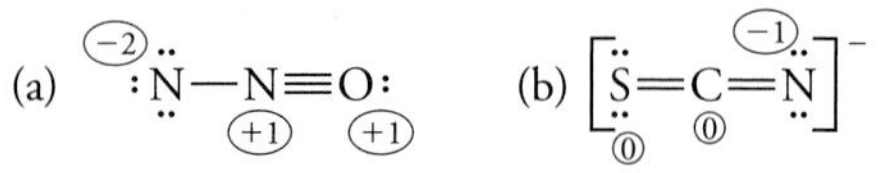

8.16 (a) 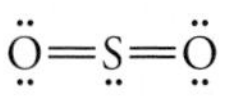 Ö=S̤=Ö

(b) 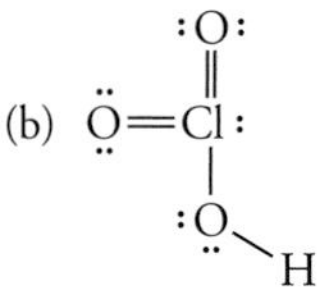 Ö=Cl(=O:)(:)—Ö—H

(c) 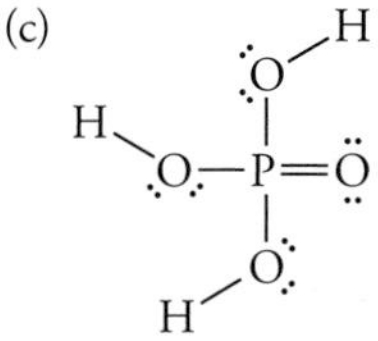 H—Ö—P(=Ö)(—Ö—H)—Ö—H

8.17

H—N̈(H)(H) + H^+ ⟶ [H—N(H)(H)—H]$^+$

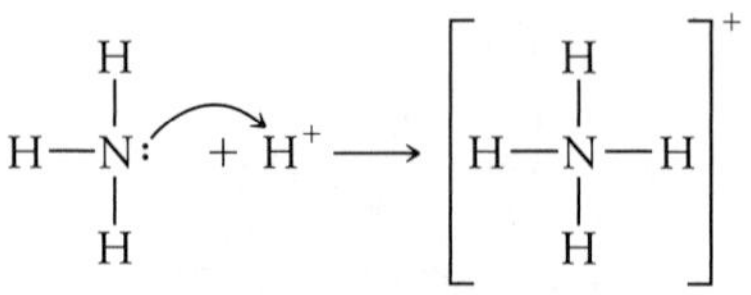

There is no difference between the coordinate covalent bond and the other covalent bonds.

8.18 H—Ö:$^-$ + H^+ ⟶ H—Ö—H

coordinate covalent bond

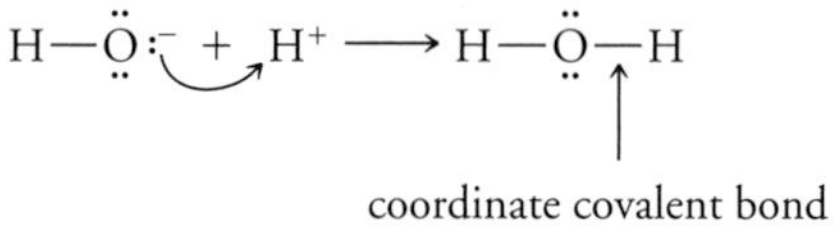

8.19

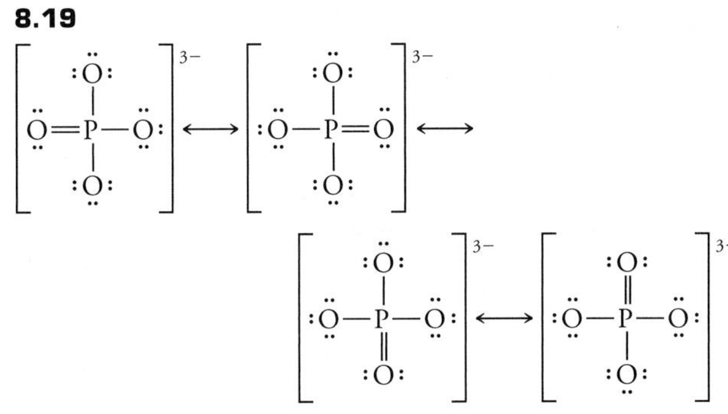

$[:O:-P(-O:)(=O)-O:]^{3-} \longleftrightarrow [:O:-P(-O:)(-O:)=O]^{3-}$

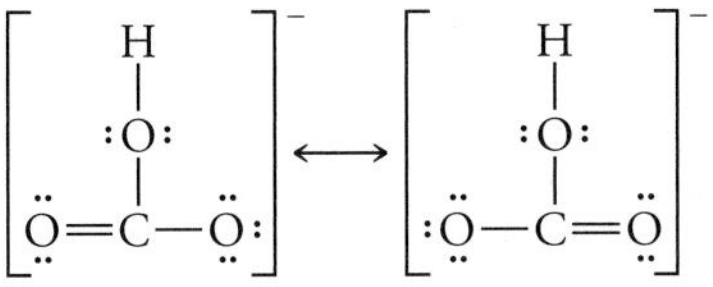

8.20 $[H-O-C(=O)-O:]^- \longleftrightarrow [H-O-C(-O:)=O]^-$

8.21 $[O=Br(=O)-O]^- \longleftrightarrow [O-Br(=O)=O]^- \longleftrightarrow [O=Br(=O)-O]^-$

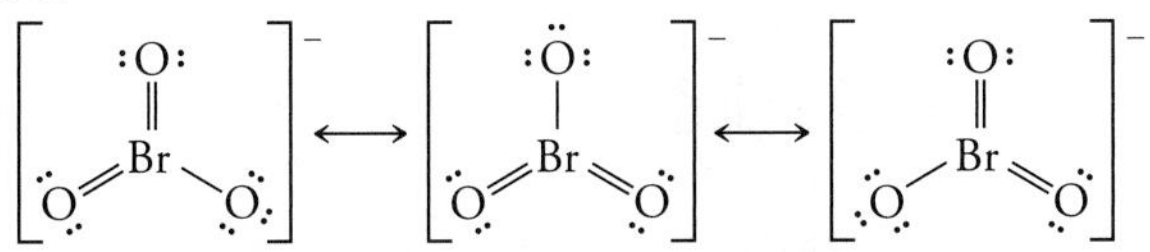

Review Problems

8.55 Magnesium loses two electrons, Bromine gains an electron: To keep the overall change of the formula unit neutral, two Br^- ions combine with one Mg^{2+} ion to form $MgBr_2$.
8.57 $[Xe]4f^{14}5d^{10}6s^2$, $[Xe]4f^{14}5d^{10}$.
8.59 $[Ar]3d^4$ 4 unpaired electrons.
8.61 (a) ·Si· (b) :Sb· (c) ·Ba· (d) ·Al· (e) :S:
8.63
(a) :Br· + Ca· + ·Br: ⟶ Ca^{2+} + $2[:Br:]^-$

(b) ·O· Al ·O· Al ·O· ⟶ $2Al^+$ + $3[:O:]^{2-}$

(c) K· ·S· ·K ⟶ $2K^+$ + $[:S:]^{2-}$

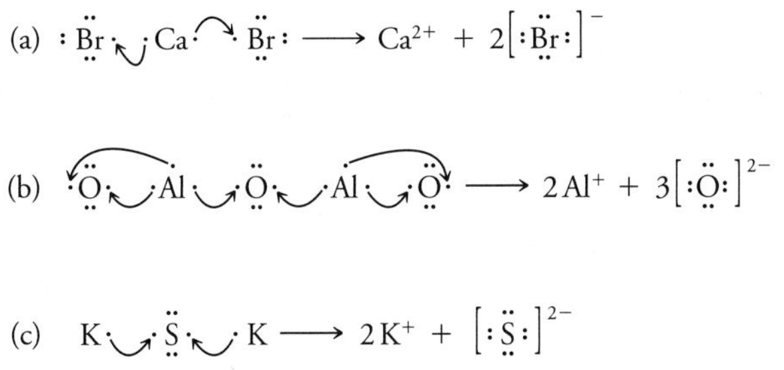

8.65 0.029 e^-, The nitrogen atom is positive. **8.67** 0.42 e^-.
8.69 1.4×10^3 g. **8.71** 344 nm, Ultraviolet region.
8.73
(a) :Br· + ·Br: ⟶ :Br—Br:
(b) 2H· + ·O· ⟶ H—O—H
(c) 3H· + ·N· ⟶ H—N(—H)—H

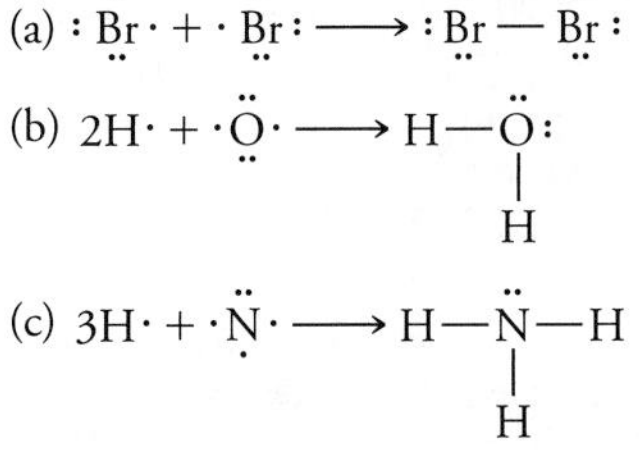

8.75 (a) H_2Se, (b) H_3As, (c) SiH_4. **8.77** (a) S, (b) Si, (c) Br, (d) C. **8.79** N—S.
8.81 (a)

Cl
Cl Si Cl
Cl

(b)

F
F P F

(c)

H
H P H

(d)

Cl S Cl

8.83 (a) 32, (b) 26, (c) 8, (d) 20.
8.85 (a) $[:Cl-As(-Cl:)(-Cl:)-Cl:]^+$ (b) $[:O-Cl-O:]^-$
(c) H—O—N=O (d) :F—Xe—F:
8.87 (a) :Cl—Si(—Cl:)(—Cl:)—Cl: (b) :F—P(—F:)—F:
(c) H—P(—H)—H (d) :Cl—S—Cl:
8.89 (a) S=C=S (b) $[:C\equiv N:]^-$
8.91 (a) H—As(—H)—H (b) H—O—Cl—O:
(c) H—O—Se(=O)—O—H (d) :O—As(—O—H)(—O—H)—O: H
8.93 (a) $H_2C=O$ (b) :Cl—S(=O)—Cl:
8.95 (a) H—O—Cl—O: (b) O=S(—O:)—O: (c) O=S—O:
8.97

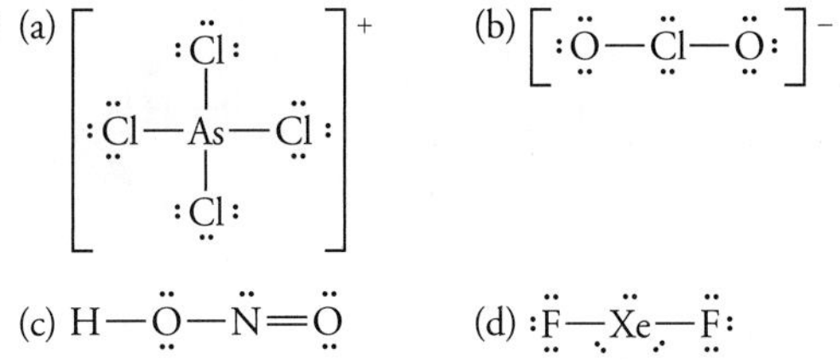

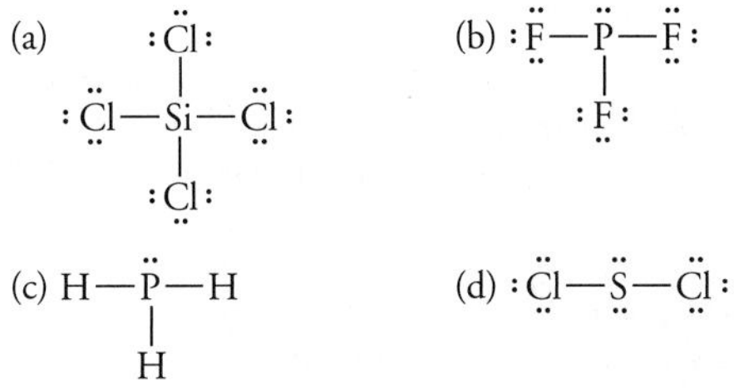

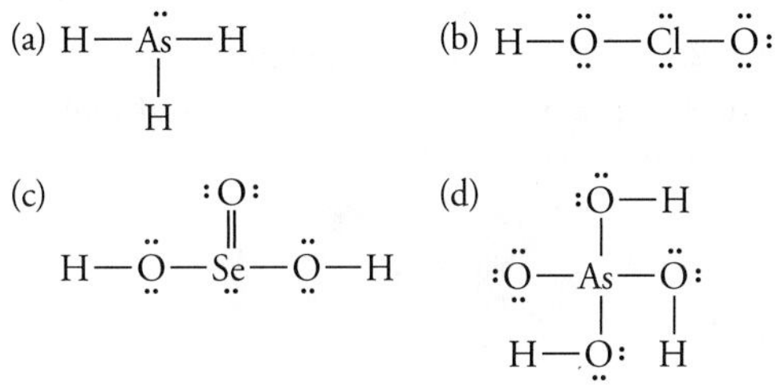

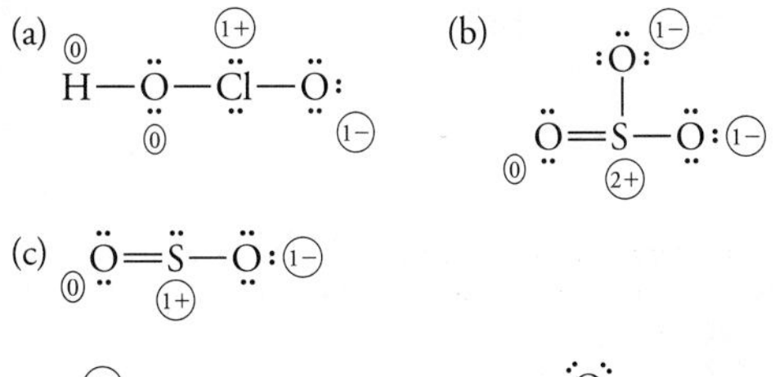

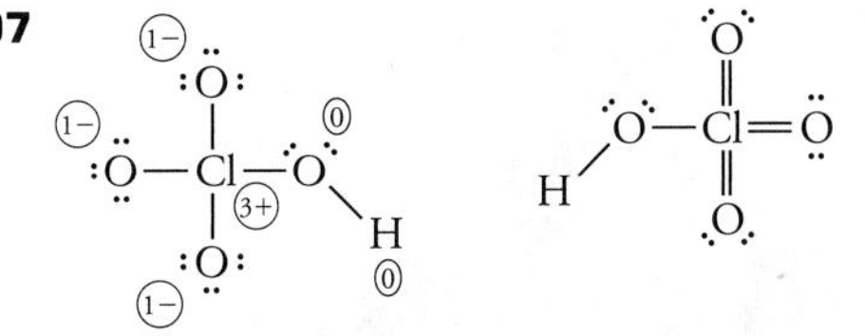

8.99 The formal charges on all of the atoms of the left structure are zero, therefore, the potential energy of this molecule is lower and it is more stable.
8.101 The average bond order is 4/3

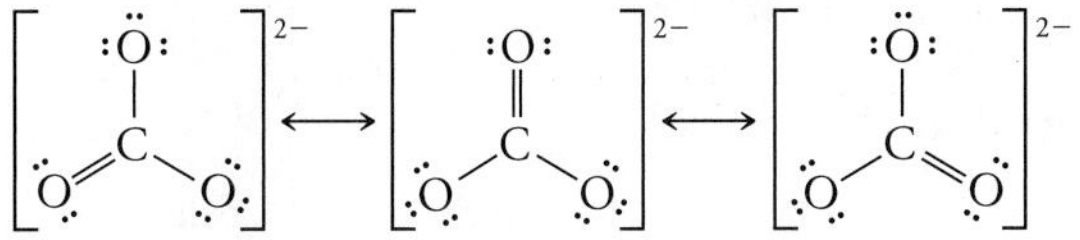

$[O=C(-O)-O]^{2-} \longleftrightarrow [O-C(=O)-O]^{2-} \longleftrightarrow [O-C(-O)=O]^{2-}$

8.103 The N—O bond in NO_2^- should be shorter than that in NO_3^-.

8.105. $:\!O\!\equiv\!C\!-\!\ddot{\underset{..}{O}}\!: \longleftrightarrow :\!\ddot{\underset{..}{O}}\!-\!C\!\equiv\!O\!:$

These are not preferred structures, because in each Lewis diagram, one oxygen atom bears a formal charge of +1 whereas the other bears a formal charge of −1. The structure with the formal charges of zero has a lower potential energy and is more stable.

8.107

$$H-\ddot{\underset{..}{O}}-H \quad H^{+} \longrightarrow \left[\begin{array}{c} H-\ddot{O}-H \\ | \\ H \end{array} \right]^{+}$$

CHAPTER 9

Practice Exercises

9.1 Octahedral shape. **9.2** Trigonal bipyramidal. **9.3** Linear. **9.4** Linear. **9.5** Square planar. **9.6** SO_3^{2-} trigonal pyramidal, CO_3^{2-} planar triangular, XeO_4 tetrahedral, OF_2 bent. **9.7** Polar. **9.8** (a) SF_6 Not polar, (b) SO_2 Polar, (c) BrCl Polar, (d) AsH_3, Polar, (e) CF_2Cl_2 Polar. **9.9** The H—Cl bond is formed by the overlap of the half–filled 1*s* atomic orbital of a H atom with the half–filled 3*p* valence orbital of a Cl atom,

Cl atom in HCl (x = H electron):

(⇅) (⇅)(⇅)(↑x)
3*s* 3*p*

The overlap that gives rise to the H—Cl bond is that of a 1*s* orbital of H with a 3*p* orbital of Cl.

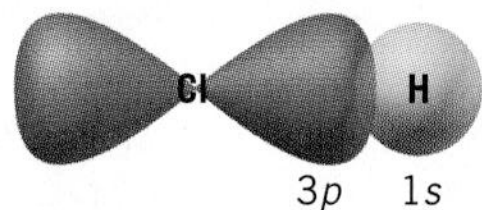

9.10 The half–filled 1*s* atomic orbital of each H atom overlaps with a half–filled 3*p* atomic orbital of the P atom, to give three P–H bonds. This should give a bond angle of 90°,

P atom in PH_3 (x = H electron):

(⇅) (↑x)(↑x)(↑x)
3*s* 3*p*

The orbital overlap that forms the P—H bond combines a 1*s* orbital of hydrogen with a 3*p* orbital of phosphorus:

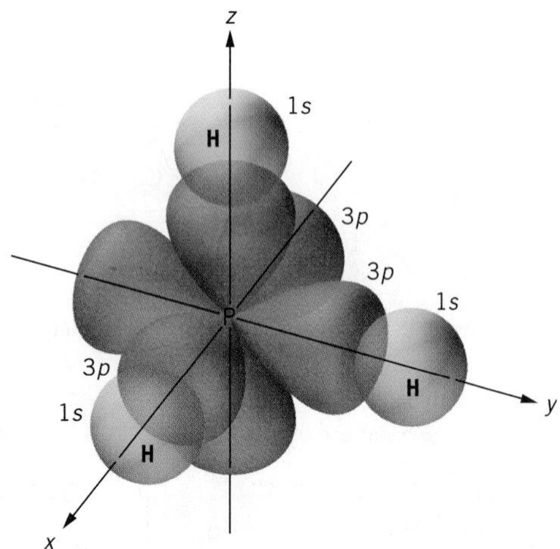

9.11 sp^2, The sp^2 hybrid orbitals on the B, x = Cl electron.

(↑x)(↑x)(↑x) ○
sp^2 2*p*

9.12 *sp*, The *sp* hybrid orbitals on the Be; x = F electron.

(↑x)(↑x) ○○
sp 2*p*

9.13 Since there are five bonding pairs of electrons on the central phosphorus atom, we choose sp^3d hybridization for the P atom. Each of phosphorus' five sp^3d hybrid orbitals overlaps with a 3*p* atomic orbital of a chlorine atom to form a total of five P—Cl single bonds. Four of the 3*d* atomic orbitals of P remain unhybridized.

9.14 sp^3

(↑x)(↑x)(↑x)(↑x)
sp^3

9.15 Trigonal bipyramidal, sp^3d—*p* bonds. **9.16** sp^3d^2. **9.17** (a)sp^3, (b) sp^3d. **9.18** NH_3 is sp^3 hybridized. Three of the electron pairs are use for bonding with the three hydrogens. The fourth pair of electrons is a lone pair of electrons. This pair of electrons is used for the formation of the bond between the nitrogen of NH_3 and the hydrogen ion, H^+.

9.19 Octahedral. sp^3d^2,

P atom in PCl_6 (x = Cl electron):

(↑x)(↑x)(↑x)(↑x)(↑x)(↑x) ○○○
sp^3d^2 3*d*

9.20 atom 1: sp^2, atom 2: sp^3, atom 3: sp^2, 10 σ bonds and 2 π bonds. **9.21** atom 1: *sp*, atom 2: sp^2, atom 3: sp^3, 9 σ bonds and 3 π bonds. **9.22** Bond order: 3 and this does agree with the Lewis structure.

$\sigma^*_{2p_z}$	○
$\pi^*_{2p_x}, \pi^*_{2p_y}$	○○
π_{2p_x}, π_{2p_y}	(⇅)(⇅)
σ_{2p_z}	(⇅)
σ^*_{2s}	(⇅)
σ_{2s}	(⇅)

9.23

$\sigma^*_{2p_z}$	○
$\pi^*_{2p_x}, \pi^*_{2p_y}$	(↑)○
π_{2p_x}, π_{2p_y}	(⇅)(⇅)
σ_{2p_z}	(⇅)
σ^*_{2s}	(⇅)
σ_{2s}	(⇅)

Bond order: 5/2:

Review Problems

9.50 (a) Bent, (b) Planar triangular, (c) T-shaped, (d) Linear, (e) Planar triangular. **9.52** (a) Nonlinear, (b) Trigonal bipyramidal, (c) Trigonal pyramidal, (d) Trigonal pyramidal, (e) Nonlinear. **9.54** (a) Tetrahedral, (b) Square planar, (c) Octahedral, (d) Tetrahedral, (e) Linear. **9.56** BrF_4^-. **9.58** 180°. **9.60** (a) 109.5°, (b) 109.5°, (c) 120°, (d) 180°, (e) 109.5°. **9.62** (a), (b), and (c). **9.64** All are polar. **9.66** In SF_6, although the individual bonds in this substance are polar bonds, the geometry of the bonds is symmetrical which serves to cause the individual dipole moments of the various bonds to cancel one another. In SF_5Br, one of the six bonds has a different polarity so the individual dipole moments of the various bonds do not cancel one another. **9.68** The 1*s* atomic orbitals of the hydrogen atoms overlap with the mutually perpendicular *p* atomic orbitals of the selenium atom.

Se atom in H_2Se (x = H electron):

(⇅) (⇅)(↑x)(↑x)
4*s* 4*p*

9.70

Hybridized Be: (x = a Cl electron)

sp 2p

9.72 (a) sp^3

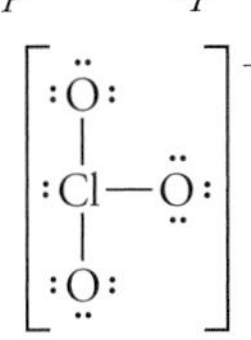

(b) sp^2

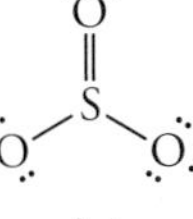

(c) sp^3

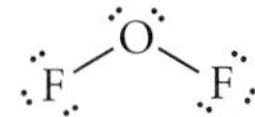

9.74 (a) sp^3,
The hybrid orbital diagram for As: (x = a Cl electron).

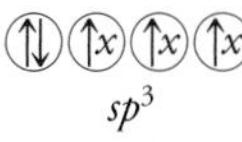

(b) sp^3d,
The hybrid orbital diagram for Cl: (x = a F electron).

sp^3d $3d$

9.76 Sb in SbF_6^-: (xx = an electron pair from the donor F^-).

sp^3d^2

9.78 (a) N in the C═N system,

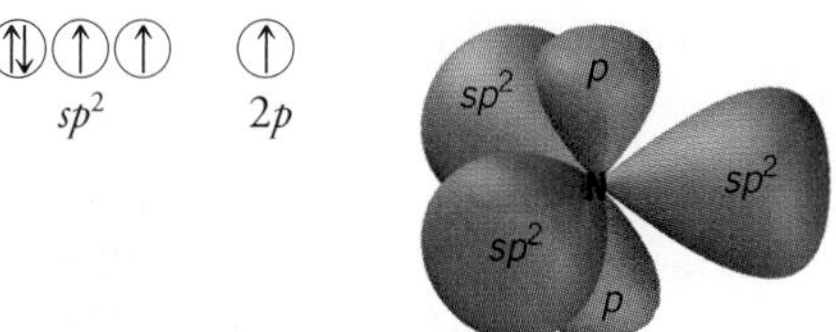

(b) pi bond,

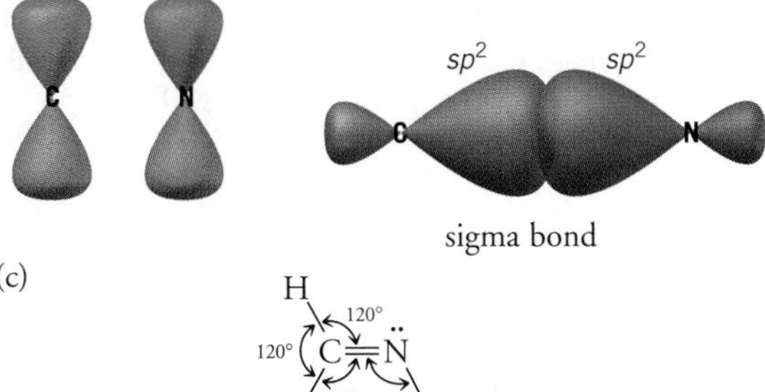

sigma bond

(c)

H, C═N, H, H; bond angles 120°, 120°, 120°, 120°

9.80 Each carbon atom is sp^2 hybridized, and each C—Cl bond is formed by the overlap of an sp^2 hybrid of carbon with a p atomic orbital of a chlorine atom. The C═C double bond consists first of a C—C σ bond formed by "head on" overlap of sp^2 hybrids from each C atom. Secondly, the C═C double bond consists of a side–to–side overlap of unhybridized p orbitals of each C atom, to give one π bond. The molecule is planar, and the expected bond angles are all 120°. **9.82** 1. sp^3, 2. sp, 3. sp^2, 4. sp^2. **9.84** (a) O_2^+, (b) O_2, (c) N_2. **9.86** (a) N_2^+, (b) NO, (c) O_2^-.

9.88

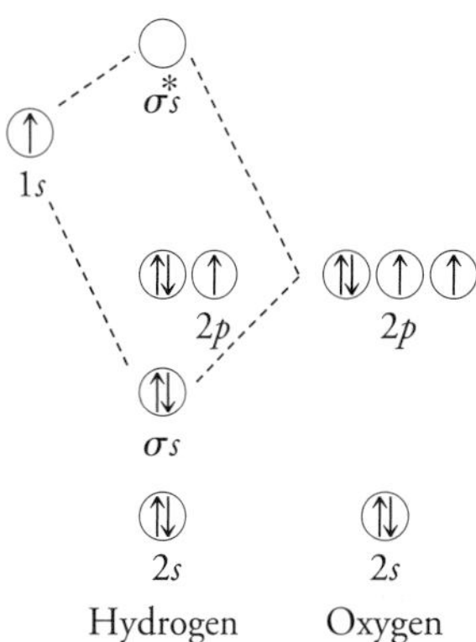

2 electrons in bonding MOs and 3 electrons in nonbonding MOs, Bond order: 1

CHAPTER 10

Practice Exercises

10.1 14.1 psi, 28.7 in. Hg. **10.2** 88,800 Pascal, 666 torr. **10.3** 1270 mm Hg, 270 mm Hg. **10.4** 663 mm Hg. **10.5** 2/3. **10.6** 750 torr. **10.7** 688 torr. **10.8** 9.00 L O_2. **10.9** 64.3 L air. **10.10** 495 mL O_2. **10.11** 15,000 g solid CO_2. **10.12** 1,130 g Ar. **10.13** 26.0 g/mol. **10.14** 132 g mol^{-1}, Xenon. **10.15** Radon is almost eight times denser than air, the sensor should be in the basement. **10.16** 2.94 g L^{-1}. **10.17** P_2F_4. **10.18** 114 g mol^{-1}, 9 C and 6 H, 8 C and 18 H, 7 C and 30 H, 6 C and 42 H, 5 C and 54 H, 4 C and 66 H, 3 C and 78 H, 2 C and 90 H, 1 C and 102 H, 8 C and 18 H is the most likely combination. **10.19** 2.80 L CO_2, 5.60 L SO_2, Total volume = 8.40 L. **10.20** 1.00 g $CaCO_3$. **10.21** P_{Ar} = 6.01 atm, P_{N_2} = 8.59 atm, P_{O_2} = 7.53 atm, P_{total} = 22.13 atm. **10.22** 2713 g O_2. **10.23** 743.18 torr, 0.0983 mol CH_4. **10.24** 732 torr 283 mL. **10.25** 1.125 atm when just added, 0.750 atm when reaction is complete. **10.26** P_{H_2} = 0.996 atm, P_{NO} = 1.05 atm. **10.27** 0.153, 15.3%. **10.28** 0.988. **10.29** HI

Review Problems

10.25 (a) 958 torr, (b) 0.974 atm, (c) 738 mm Hg, (d) 10.9 torr.
10.27 (a) 250 torr, (b) 350 torr.
10.29 4.5 cm Hg,

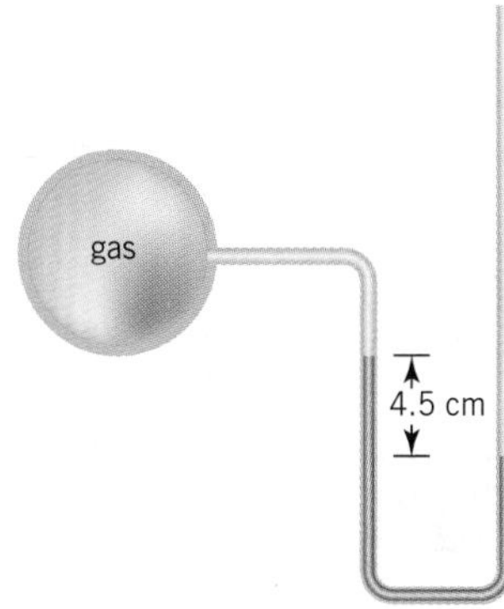

10.31 813 torr. **10.33** 125 torr. **10.35** 507 mL. **10.37** 4.28 L. **10.39** 843 °C. **10.41** 796 torr. **10.43** 5.73 L. **10.45** −53.4 °C. **10.47** 6.24×10^4 mL torr mol^{-1} K^{-1}. **10.49** 0.104 L. **10.51** 2340 torr. **10.53** 0.0398 g. **10.55** (a) 1.34 g L^{-1}, (b) 1.25 g L^{-1}, (c) 3.17 g L^{-1}, (d) 1.78 g L^{-1}. **10.57** 1.28 g/L. **10.59** 88.2 g/mol. **10.61** 27.6 g/mol. **10.63** 1.14×10^3 mL. **10.65** 10.7 L H_2. **10.67** 36.3 mL O_2.

10.69 217 mL. **10.71** 650 torr. **10.73** P_{N_2} = 228 torr, P_{O_2} = 152 torr, P_{He} = 304 torr, P_{CO_2} = 76 torr. **10.75** 736 torr, 260 mL. **10.77** 250 mL. **10.79** N_2, 1.25. **10.81** $^{235}UF_6$ 1.0043

CHAPTER 11

Practice Exercises

11.1 (a) $CH_3CH_2CH_2CH_2CH_3 < CH_3CH_2OH < Ca(OH)_2$, (b) $CH_3—O—CH_3 < CH_3CH_2NH_2 < HOCH_2CH_2CH_2CH_2OH$. **11.2** Propylamine, because of its ability to form hydrogen bonds. **11.3** Pushed in. **11.4** The total number of molecules remains the same. **11.5** (a) less than 10 °C. **11.6** 75 °C. **11.7** The equilibriumwill shift to the right, producing more vapor. **11.8** Boiling, Endothermic, Melting, Endothermic, Condensing, Exothermic, Subliming, Endothermic, Freezing, Exothermic, No, each physical change is always exothermic, or always endothermic as shown. **11.9** 4 Ca^{2+} and 8 F^-. **11.10** 1 to 1. **11.11** Molecular crystal. **11.12** Covalent or network solid. **11.13** Molecular solid. **11.14** Vapor pressure curve, see Fig. 11.21. **11.15** Solid to gas. **11.16** Liquid.

Review Problems

11.78 Diethyl ether. **11.80** (a) London forces, dipole- dipole and hydrogen bonding, (b) London forces and dipole- dipole, (c) London forces, (d) London forces and dipole- dipole. **11.82** Chloroform would be expected to display larger dipole-dipole attractions because it has a larger dipole moment than bromoform. On the other hand, bromoform would be expected to show stronger London forces due to having larger electron clouds which are more polarizable than those of chlorine. Since bromoform in fact has a higher boiling point that chloroform, we must conclude that it experiences stronger intermolecular attractions than chloroform, which can only be due to London forces. Therefore, London forces are more important in determining the boiling points of these two compounds. **11.84** Ethanol. **11.86** ether < acetone < benzene < water < acetic acid. **11.88** 305 kJ. **11.90** (a) 0 °C, (b) 47.9 g. **11.92** 4 Zn^{2+}, 4 S^{2-}. **11.94** 3.51 Å., 351 pm. **11.96** 656 pm. **11.98** (a) 6.57°, (b) 27.3°. **11.100** 176 pm. **11.102** Molecular solid. **11.104** Metallic solid. **11.106** (a)molecular, (b) ionic, (c) ionic, (d) metallic, (e) covalent, (f) molecular, (g) ionic. **11.108**

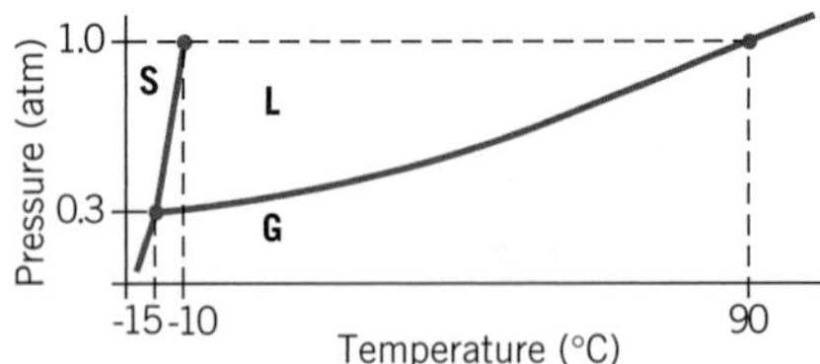

11.110 (a) solid, (b) gas, (c) liquid, (d) solid, liquid, and gas

CHAPTER 12

Practice Exercises

12.1 3.4 g L^{-1} atm^{-1}. Hydrogen sulfide is more soluble in water than nitrogen and oxygen. Hydrogen sulfide reacts with the water to form hydronium ions and $HS^-(aq)$ ions. **12.2** 0.899 mg of O_2, 1.48 mg of N_2. **12.3** 405.7 mL water. **12.4** 2.50 g NaBr, 248 g H_2O, 251 mL H_2O. **12.5** 2.0×10^1 g solution. **12.6** 1.239 *m*, Smaller. **12.7** 16.0 g CH_3OH. **12.8** (a) 28.3 g solution, (b) 70.7 g solution, (c) 1.41 g solution. **12.9** 25 *m* NaOH. **12.10** 16.1 *m*. **12.11** 6.82 M. **12.12** 0.00469 *M* $Al(NO_3)_3$, 0.00470 *m* $Al(NO_3)_3$. **12.13** 9.02 torr. **12.14** 61.2 g stearic acid. **12.15** 55.4 torr. **12.16** 45.0 torr. **12.17** 100.16 °C. **12.18** 209 g glucose. **12.19** 157 g/mol. **12.20** 125 g mol^{-1}. **12.21** 3.75 mm Hg, 51.1 mm H_2O. **12.22** 0.051 g. **12.23** 4.99×10^4 g mol^{-1}. **12.24** 5.38×10^2 g mol^{-1}. **12.25** −0.882 °C, −0.441 °C. **12.26** (a) 0.372 °C, (b) 0.0372 °C, (c) 0.00372 °C, The first freezing point depression could be measured using a laboratory thermometer that can measure 0.1 °C increments.

Review Problems

12.38

(a) $KCl(s) \longrightarrow K^+(g) + Cl^-(g)$, $\Delta H° = +690$ kJ mol^{-1}

(b) $K^+(g) + Cl^-(g) \longrightarrow K^+(aq) + Cl^-(aq)$, $\Delta H° = -686$ kJ mol^{-1}

$KCl(s) \longrightarrow K^+(aq) + Cl^-(aq)$, $\Delta H° = +4$ kJ mol^{-1}

12.40 0.038 g/L. **12.42** 0.020 g L^{-1}. **12.44** 3.35 *m*. **12.46** 0.133 molal, 2.39×10^{-3}, 2.34%. **12.48** = 5.45%. **12.50** 7.89%, 4.76 *m*. **12.52** 0.359 *M* $NaNO_3$, 3.00%, 6.49×10^{-3}. **12.54** 22.8 torr. **12.56** 52.7 torr. **12.58** 70 mol % toluene and 30 mol % benzene. **12.60** (a) 0.029, (b) 2.99×10^{-2} mol soute, (c) 278 g/mol. **12.62** 55.1 g. **12.64** 152 g/mol. **12.66** 127 g/mol, $C_8H_4N_2$.

12.68 (a) $\dfrac{(g) \times (L\ atm\ mol^{-1}K^{-1}) \times (K)}{L \times atm} = \dfrac{g}{mol}$, (b) 1.8×10^6 g/mol. **12.70** 15.3 torr. **12.72** 1.3×10^4 torr. **12.74** −1.1 °C. **12.76** 2. **12.78** 1.89

CHAPTER 13

Practice Exercises

13.1 Rate of production of I^- = 8.0×10^{-5} mol L^{-1} s^{-1}, Rate of production of SO_4^{2-} = 2.4×10^{-4} mol L^{-1} s^{-1}. **13.2** Rate of disappearance of O_2 = 0.45 mol L^{-1} s^{-1}. Rate of disappearance of H_2S = 0.30 mol L^{-1} s^{-1}. **13.3** -2.1×10^{-4} mol L^{-1} s^{-1}. **13.4** 1×10^{-4} mol L^{-1} s^{-1} is correct. **13.5** (a) 9.8×10^{14} L^2 mol^{-2} s^{-1}, (b) L^2 mol^{-2} s^{-1}. **13.6** (a) 8.0×10^{-2} L mol^{-1} s^{-1}, (b) L mol^{-1} s^{-1}. **13.7** order of the reaction with respect to $[BrO_3^-] = 1$, order of the reaction with respect to $[SO_3^{2-}] = 1$, overall order of the reaction = 1 + 1 = 2. **13.8** Rate = $k[Cl_2]^2[NO]$. **13.9** k = 2.0×10^2 L^2 mol^{-2} s^{-1}. Each of the other data sets also gives the same value. **13.10** (a) The rate will increase nine–fold, (b) The rate will increase three–fold, (c) The rate will decrease by three fourths. **13.11** rate = $k[NO]^n [H_2]^m$, (a) rate = $k[NO]^2 [H_2]^1$, (b) k = 2.1×10^5 mol^{-2} L^2 s^{-1}, (c) mol^{-2} L^2 s^{-1}. **13.12** (a) First order with respect to sucrose. (b) 6.17×10^{-4} s^{-1}. **13.13** (a) Rate = $k[A]^2[B]^2$, (b) $k = 6.9 \times 10^{-3}$ L^3 mol^{-3} s^{-1}, (c) L^3 mol^{-3} s^{-1}, (d) 4. **13.14** k = 2.56×10^{-2} yr^{-1}. **13.15** (a) 0.26 *M*, (b) 77 min. **13.16** 18.7 min, 37.4 min. **13.17** 27.0 yr. **13.18** 1.72×10^4 yrs. **13.19** 24,800 years BP, 424 years BP. **13.20** 63 min. **13.21** 4.0×10^{-4} *M*. **13.22** 1.03 L mol^{-1} s^{-1}, $t_{1/2} = 1.48 \times 10^3$ s. **13.23** The reaction is first–order. **13.24** 45 °C. **13.25** a) 1.4×10^2 kJ/mol, b) 0.30 L mol^{-1} s^{-1}. **13.26** (a), (b), and (e) may be elementary processes. Equations (c), (d), and (f) are not elementary processes because they have more than two molecules colliding at one time, and this is very unlikely. **13.27** Rate = $k[NO][O_3]$.

13.28 Rate = $\dfrac{k[NO_2Cl]^2}{[NO_2]}$

Review Problems

13.47

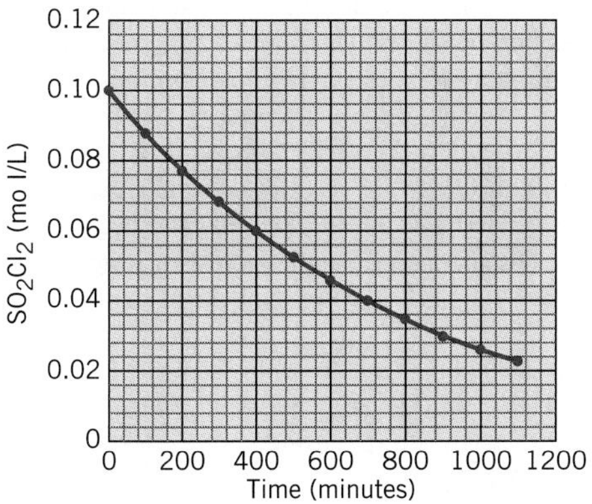

At 200 min: 1×10^{-4} M/s At 600 minutes: 7×10^{-5} M/s.

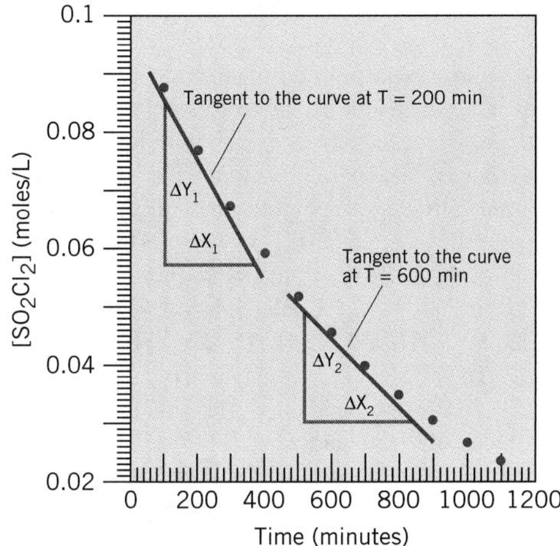

13.49 The rate of disappearance of hydrogen is three times the rate of disappearance of nitrogen. NH_3 appears twice as fast as N_2 disappears. **13.51** (a) rate for O_2 = 11.4 mol L^{-1} s^{-1}, (b) rate for CO_2 = 7.20 mol L^{-1} s^{-1}, (c) rate for H_2O = 8.40 mol L^{-1} s^{-1}. **13.53** rate = 8.0×10^{-11} mol L^{-1} s^{-1}. **13.55** rate = 2.4×10^{2} mol L^{-1} s^{-1}. **13.57** rate = $k \times [M][N]^2$, $k = 2.5 \times 10^3$ L^2 mol^{-2} s^{-1}. **13.59** rate = $k[OCl^-][I^-]$, $k = 6.1 \times 10^9$ L mol^{-1} s^{-1}. **13.61** rate = $k[ICl][H_2]$, $k = 1.5 \times 10^{-1}$ L mol^{-1} s^{-1}. **13.63**

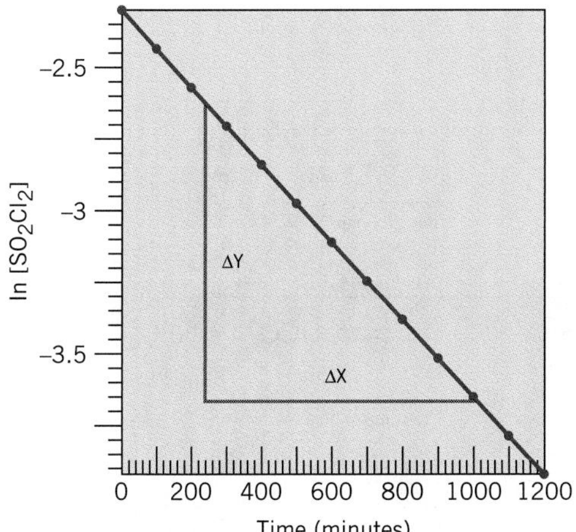

$k = 1.32 \times 10^{-3}$ min^{-1}.
13.65 (a) $x = 3.7 \times 10^{-3}$ M, (b) $x = 6.0 \times 10^{-4}$ M.
13.67 4.26×10^{-3} min^{-1}. **13.69** 1.3×10^4 min. **13.71** 1/256. **13.73** Approximately 500 min, The reaction is first–order in SO_2Cl_2. **13.75** 1.3×10^9 years old. **13.77** 2.7×10^4 years ago. **13.79** $E_a = 7.93 \times 10^4$ J/mol = 79.3 kJ/mol. **13.81** E_a = 99 kJ mol^{-1} and $A = 6.6 \times 10^9$ L mol^{-1} s^{-1}. **13.83** Use equation 13.10: (a) 1.9×10^{-5} s^{-1}, (b) 1.6×10^{-1} s^{-1}

CHAPTER 14

Practice Exercises

14.1 $2N_2O_3 + O_2 \rightleftharpoons 4NO_2$.

14.2 (a) $\frac{[H_2O]^2}{[H_2]^2[O_2]} = K_c$, (b) $\frac{[CO_2][H_2O]^2}{[CH_4][O_2]^2} = K_c$.

14.3 $K_c = 1.2 \times 10^{-13}$. **14.4** $K_c = 1.9 \times 10^5$.

14.5 $K_P = \frac{(P_{N_2O})^2}{(P_{N_2})^2(P_{O_2})}$. **14.6** $K_P = \frac{(P_{HI})^2}{(P_{H_2})(P_{I_2})}$.

14.7 $K_P = 1.8 \times 10^{36}$. **14.8** Smaller K_c = 57.

14.9 $K_c = \frac{1}{[NH_3(g)][HCl(g)]}$.

14.10 (a) $K_c = \frac{1}{[Cl_2(g)]}$,
(b) $K_c = [Na^+ (aq)][OH^- (aq)][H_2 (g)]$,
(c) $K_c = [Ag^+]^2[CrO_4^{2-}]$,
(d) $K_c = \frac{[Ca^{+2}(aq)][HCO_3^-(aq)]^2}{[CO_2(aq)]}$.

14.11 Reaction will proceed to the left. **14.12** Reaction (b). **14.13** There will be no change in the amount of H_3PO_4. **14.14** (a) Decreasing the concentration of Cl_2 at equilibrium, the value of K_P will be unchanged. (b) Increasing the amount of Cl_2 at equilibrium. The value of K_P will be unchanged, (c) Increasing the amount of Cl_2 at equilibrium. Decreasing the value of K_P. (d) Decreasing the concentration of Cl_2 at equilibrium, The value of K_P will be unchanged. **14.15** [CO] decreases by 0.060 mol/L and $[CO_2]$ increases by 0.060 mol/L. **14.16** K_c = 4.06. **14.17** (a) $[PCl_3]$ = 0.200 M, $[Cl_2]$ = 0.100 M, $[PCl_5]$ = 0.000 M, (b) PCl_3 and Cl_2 have decreased by 0.080 M and PCl_5 has increased by 0.080 M. (c) $[PCl_3]$ = 0.120 M. $[PCl_5]$ = 0.080 M $[Cl_2]$ = 0.020 M. (d) K_c = 33. **14.18** $[NO_2] = 1.04 \times 10^{-2}$ M. **14.19** $[C_2H_5OH] = 8.98 \times 10^{-3}$ M. **14.20** $[H_2] = [I_2]$ = 0.044 M, [HI] = 2(0.156) = 0.312 M. **14.21** $[H_2]$ = 0.200 – 0.0934 = 0.107 M, $[I_2]$ = 0.100 – 0.0934 = 0.0066 M, [HI] = 2(0.0934) = 0.1868 M. **14.22** $[N_2] = 6.2 \times 10^{-4}$ M, $[H_2] = 1.9 \times 10^{-3}$ M. **14.23** $[NO] = 1.1 \times 10^{-17}$ M.

Review Problems

14.19 (a) $K_c = \frac{[POCl_3]^2}{[PCl_3]^2[O_2]}$, (b) $K_c = \frac{[SO_2]^2[O_2]}{[SO_3]^2}$,

(c) $K_c = \frac{[NO]^2[H_2O]^2}{[N_2H_4][O_2]^2}$, (d) $K_c = \frac{[NO_2]^2[H_2O]^8}{[N_2H_4][H_2O_2]^6}$,

(e) $K_c = \frac{[SO_2][HCl]^2}{[SOCl_2][H_2O]}$.

14.21 (a) $K_P = \frac{(P_{POCl_3})^2}{(P_{PCl_3})^2(P_{O_2})}$, (b) $K_P = \frac{(P_{SO_2})^2(P_{O_2})}{(P_{SO_3})^2}$,

(c) $K_P = \dfrac{(P_{NO})^2(P_{H_2O})^2}{(P_{N_2H_4})(P_{O_2})^2}$, (d) $K_P = \dfrac{(P_{NO_2})^2(P_{H_2O})^8}{(P_{N_2H_4})(P_{H_2O_2})^6}$,

(e) $K_P = \dfrac{(P_{SO_2})(P_{HCl})^2}{(P_{SOCl_2})(P_{H_2O})}$.

14.23 (a) $K_c = \dfrac{[Ag(NH_3)_2{}^+]}{[Ag^+][NH_3]^2}$, (b) $K_c = \dfrac{[Cd(SCN)_4{}^{2-}]}{[Cd^{2+}][SCN^-]^4}$.

14.25 $K = 1 \times 10^{85}$.

14.27 (a) $K_c = \dfrac{[HCl]^2}{[H_2][Cl_2]}$, (b) $K_c = \dfrac{[HCl]}{[H_2]^{1/2}[Cl_2]^{1/2}}$,

K_c for reaction (b) is the square root of K_c for reaction (a). **14.29** 0.0375 *M*. **14.31** b. **14.33** $K_c = 11$. **14.35** $K_P = 2.7 \times 10^{-2}$. **14.37** $K_P = 5.4 \times 10^{-5}$. **14.39** (a) 55.5 *M*, (b) 55.5 *M*, (c) 55.5 *M*.

14.41 (a) $K_c = \dfrac{[CO]^2}{[O_2]}$, (b) $K_c = [H_2O][SO_2]$,

(c) $K_c = \dfrac{[CH_4][CO_2]}{[H_2O]^2}$, (d) $K_c = \dfrac{[H_2O][CO_2]}{[HF]^2}$,

(e) $K_c = [H_2O]^5$.

14.43 $[HI] = 1.47 \times 10^{-12}\ M$, $[Cl_2] = 7.37 \times 10^{-13}\ M$. **14.45** (a) The system shifts to the right to consume some of the added methane. (b) The system shifts to the left to consume some of the added hydrogen. (c) The system shifts to the right to make some more carbon disulfide. (d) The system shifts to the left to decrease the amount of gaseous moles. (e) The system shifts to the right to absorb some of the added heat. **14.47** (a) right, (b) left, (c) left, (d) right, (e) no effect, (f) left. **14.49** (a) No, (b) To the left. **14.51** $[CH_3OH] = 4.36 \times 10^{-3}\ M$. **14.53** $K_c = 0.398$. **14.55** $K_c = 0.0955$. **14.57** $K_c = 0.915$. **14.59** $[Br_2] = [Cl_2] = 0.011\ M$. **14.61** $[NO_2] = [SO_2] = 0.0497$ mol/L, $[NO] = [SO_3] = 0.0703$ mol/L. **14.63** $[H_2] = [CO_2] = 7.7 \times 10^{-3}\ M$, $[CO] = [H_2O] = 0.0123\ M$. **14.65** $[H_2] = [Cl_2] = 8.9 \times 10^{-19}\ M$. **14.67** $[CO] = 5.0 \times 10^{-4}\ M$. **14.69** $[PCl_5] = 3.0 \times 10^{-5}\ M$. **14.71** $[NO_2] = [SO_2] = 0.0281\ M$. **14.73** $[CO] = [H_2O] = 0.200\ M$.

CHAPTER 15

Practice Exercises

15.1 Conjugate acid base pairs (a), (c), and (f). (b) The conjugate base of HI is I^-. (d) The conjugate base of HNO_2 is NO_2^- and the conjugate base of NH_4^+ is NH_3. (e) The conjugate acid of CO_3^{2-} is HCO_3^- and the conjugate acid of CN^- is HCN. **15.2** (a) OH^-, (b) I^-, (c) NO_2^- (d) $H_2PO_4^-$, (e) HPO_4^{2-}, (f) PO_4^{3-}, (g) HS^- (h) NH_3. **15.3** (a) H_2O_2, (b) HSO_4^-, (c) HCO_3^-, (d) HCN, (e) NH_3 (f) NH_4^+, (g) H_3PO_4 (h) $H_2PO_4^-$. **15.4** HCN and CN^-, HCl and Cl^-. **15.5** Brønsted acids: $H_2PO_4^-(aq)$ and $H_2CO_3(aq)$, Brønsted bases: $HCO_3^-(aq)$ and $HPO_4^{2-}(aq)$.

15.6

conjugate pair

$PO_4^{3-}(aq) + HC_2H_3O_2(aq) \rightleftharpoons HPO_4^{2-}(aq) + C_2H_3O_2^-(aq)$

base acid acid base

conjugate pair

15.7 (a) $H_2PO_4^-$ amphoteric since it can both accept and donate a proton, (b) HPO_4^{2-} amphoteric since it can both accept and donate a proton, (c) H_2S amphoteric since it can both accept and donate a proton, (d) H_3PO_4 not amphoteric: it can only donate protons, (e) NH_4^+ not amphoteric: it can only donate protons, (f) H_2O amphoteric since it can both accept and donate a proton, (g) HI not amphoteric: it can only donate protons, (h) HNO_2 not amphoteric: it can only donate protons.

15.8 $HPO_4^{2-}(aq) + OH^-(aq) \longrightarrow PO_4^{3-}(aq) + H_2O$; $HPO_4^{2-}(aq) + H_3O^+(aq) \longrightarrow H_2PO_4^- + H_2O$. **15.9** $HSO_4^-(aq) + HPO_4^{2-}(aq) \longrightarrow SO_4^{-2}(aq) + H_2PO_4^-(aq)$. **15.10** The substances on the right because they are the weaker acid and base. **15.11** (a) HF < HBr < HI, (b) $PH_3 < H_2S <$ HCl, (c) $H_2O < H_2Se < H_2Te$, (d) $AsH_3 < H_2Se <$ HBr, (e) $PH_3 < H_2Se <$ HI. **15.12** (a) HBr, (b) H_2Te, (c) H_2S. **15.13** (a) $HClO_3$, (b) H_2SO_4. **15.14** (a) H_3AsO_4, (b) H_2TeO_4. **15.15** (a) HIO_4, (b) H_2TeO_4, (c) H_3AsO_4. **15.16** (a) H_2SO_4, (b) H_3AsO_4. **15.17** (a) NH_3 Lewis base, H^+ Lewis acid, (b) CN^- Lewis base, H_2O Lewis acid, (c) Ag^+ Lewis acid, NH_3 Lewis base. **15.18** (a) Fluoride ions have a filled octet of electrons and are likely to behave as Lewis bases, i.e., electron pair donors. (b) $BeCl_2$ is a likely Lewis acid since it has an incomplete shell. The Be atom has only two valence electrons and it can easily accept a pair of electrons. (c) It could reasonably be considered a potential Lewis base since it contains three oxygens, each with lone pairs and partial negative charges. However, it is more effective as a Lewis acid, since the central sulfur bears a significant positive charge. **15.19** $8.3 \times 10^{-16}\ M$. **15.20** $1.3 \times 10^{-9}\ M$, Basic. **15.21** pOH = 9.75, $[H^+] = 5.62 \times 10^{-5}\ M$, $[OH^-] = 1.78 \times 10^{-10}\ M$. **15.22** pH = 3.44, pOH = 10.56, Basic. **15.23** 5.17. **15.24** CaO. **15.25** pH = 1.1, pOH = 15.1. **15.26** (a) $[H^+] = 1.3 \times 10^{-3}\ M$, $[OH^-] = 7.7 \times 10^{-12}\ M$, Acidic. (b) $[H^+] = 1.4 \times 10^{-4}\ M$, $[OH^-] = 7.1 \times 10^{-11}\ M$, Acidic. (c) $[H^+] = 1.5 \times 10^{-11}\ M$, $[OH^-] = 6.7 \times 10^{-4}\ M$, Basic. (d) $[H^+] = 7.8 \times 10^{-5}\ M$, $[OH^-] = 1.3 \times 10^{-10}\ M$, Acidic. (e) $[H^+] = 2.5 \times 10^{-12}\ M$, $[OH^-] = 4.0 \times 10^{-3}\ M$, Basic. **15.27** $[H^+] = 0.005\ M$, pH = 2.30, pOH = 11.70. **15.28** pOH = 1.068, pH = 12.932, $[H^+] = 1.17 \times 10^{-13}\ M$. **15.29** $[H^+] = 3.2 \times 10^{-6}\ M$.

Review Problems

15.41 (a) HF, (b) $N_2H_5^+$, (c) $C_5H_5NH^+$, (d) HO_2^-, (e) H_2CrO_4.

15.43

conjugate pair

(a) $HNO_3 + N_2H_4 \rightleftharpoons N_2H_5^+ + NO_3^-$

acid base acid base

conjugate pair

conjugate pair

(b) $N_2H_5^+ + NH_3 \rightleftharpoons NH_4^+ + N_2H_4$

acid base acid base

conjugate pair

conjugate pair

(c) $H_2PO_4^- + CO_3^{2-} \rightleftharpoons HCO_3^- + HPO_4^{2-}$

acid base acid base

conjugate pair

conjugate pair

(d) $HIO_3 + HC_2O_4^- \rightleftharpoons H_2C_2O_4 + IO_3^-$

acid base acid base

conjugate pair

15.45 (a) HBr, HBr bond is weaker, (b) HF, more electronegative F polarizes and weakens the bond, (c) HBr, larger Br forms a weaker bond with H. **15.47** (a) $HClO_2$, because it has more oxygen atoms, (b) H_2SeO_4, because it has more lone oxygen atoms. **15.49** (a) $HClO_3$, because Cl is more electronegative, (b) $HClO_3$, because the charge is more evenly distributed, (c) $HBrO_4$, because the negative charge is more evenly distributed.
15.51

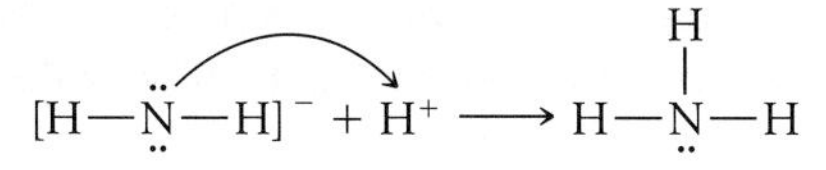

$$[H{-}\ddot{N}{-}H]^- + H^+ \longrightarrow H{-}\underset{..}{\overset{H}{N}}{-}H$$

Lewis bases: NH_2^- Lewis acid: H^+

15.53

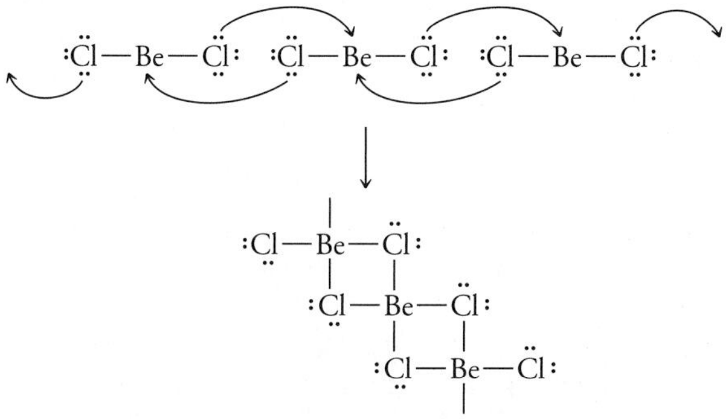

:Cl—Be—Cl: :Cl—Be—Cl: :Cl—Be—Cl:

↓

:Cl—Be—Cl: / :Cl—Be—Cl: / :Cl—Be—Cl: (chain of bridged Be–Cl units)

15.55

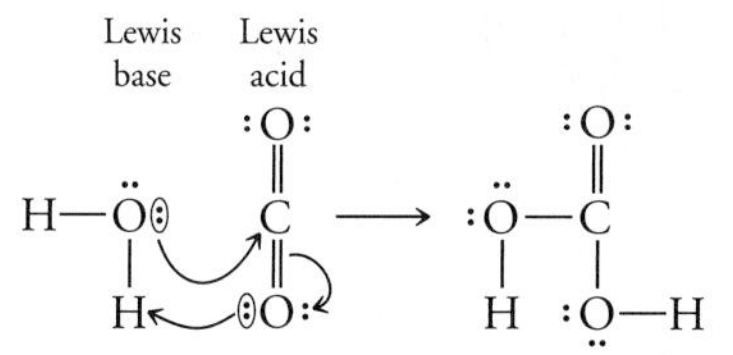

Lewis base Lewis acid

$H{-}\ddot{O}^{\ominus} + C(=O)({-}O^{\ominus}) \longrightarrow H{-}O{-}C(=O){-}O{-}H$

15.57

$$H{-}\overset{H}{N}{:}^{\ominus} + {:}\overset{H}{O}{-}H \longrightarrow H{-}\overset{H}{N}{-}H + {:}\ddot{O}{-}H^{\ominus}$$

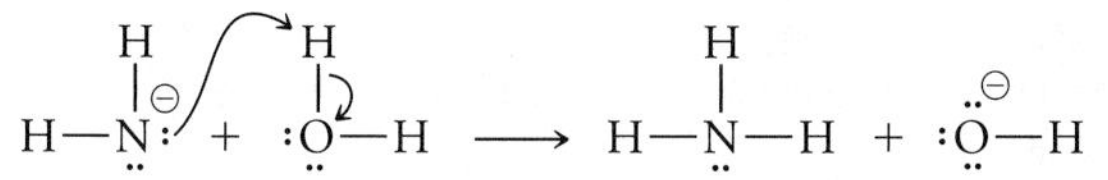

Lewis bases: NH_2^- and OH^-; Lewis acid: H^+

15.59 $[D^+] = [OD^-] = 3.0 \times 10^{-8}\ M$, pD = 7.52, pOD = 7.52. **15.61** (a) $[H^+] = 1.5 \times 10^{-12}\ M$, pH = 11.83, pOH = 2.17, (b) $[H^+] = 1.6 \times 10^{-10}\ M$, pH = 9.81, pOH = 4.19, (c) $[H^+] = 6.3 \times 10^{-7}\ M$, pH = 6.20, pOH = 7.80, (d) $[H^+] = 1.2 \times 10^{-3}\ M$, pH = 2.91, pOH = 11.09. **15.63** 4.72. **15.65** (a) $[H^+] = 7.2 \times 10^{-9}\ M$, $[OH^-] = 1.4 \times 10^{-6}\ M$, (b) $[H^+] = 2.7 \times 10^{-3}\ M$, $[OH^-] = 3.6 \times 10^{-12}\ M$, (c) $[H^+] = 5.6 \times 10^{-12}\ M$, $[OH^-] = 1.8 \times 10^{-3}\ M$, (d) $[H^+] = 5.3 \times 10^{-14}\ M$, $[OH^-] = 1.9 \times 10^{-1}\ M$, (e) $[H^+] = 2.0 \times 10^{-7}\ M$, $[OH^-] = 5.0 \times 10^{-8}\ M$. **15.67** $[H^+] = 2.0 \times 10^{-6}\ M$, $[OH^-] = 5.0 \times 10^{-9}\ M$. **15.69** pH = 3.19, $[OH^-] = 1.55 \times 10^{-11}\ M$. **15.71** 0.15 M OH^-, pOH = 0.82, pH = 13.18, $[H^+] = 6.61 \times 10^{-14}\ M$. **15.73** $2.0 \times 10^{-3}\ M\ Ca(OH)_2$, $2.0 \times 10^{-4}\ M\ Ca(OH)_2$. **15.75** 168 mL 0.0100 M KOH. **15.77** $5 \times 10^{-12}\ M$.

CHAPTER 16

Practice Exercises

16.1 (a) $HC_2H_3O_2 + H_2O \rightleftharpoons H_3O^+ + C_2H_3O_2^-$,

$$K_a = \frac{[H_3O^+][C_2H_3O_2^-]}{[HC_2H_3O_2]},$$

(b) $(CH_3)_3NH^+ + H_2O \rightleftharpoons H_3O^+ + (CH_3)_3N$,

$$K_a = \frac{[H_3O^+][(CH_3)_3N]}{[(CH_3)_3NH^+]},$$

(c) $H_3PO_4 + H_2O \rightleftharpoons H_3O^+ + H_2PO_4^-$,

$$K_a = \frac{[H_3O^+][H_2PO_4^-]}{[H_3PO_4]}.$$

16.2 (a) $HCHO_2 + H_2O \rightleftharpoons H_3O^+ + CHO_2^-$,

$$K_a = \frac{[H_3O^+][CHO_2^-]}{[HCHO_2]},$$

(b) $(CH_3)_2NH_2^+ + H_2O \rightleftharpoons H_3O^+ + (CH_3)_2NH$,

$$K_a = \frac{[H_3O^+][CH_3NH]}{[(CH_3)_2NH_2^+]},$$

(c) $H_2PO_4^- + H_2O \rightleftharpoons H_3O^+ + HPO_4^{2-}$,

$$K_a = \frac{[H_3O^+][HPO_4^{2-}]}{[H_2PO_4^-]}.$$

16.3 Barbituric acid. **16.4** HA is the strongest acid, For HA: $K_a = 6.9 \times 10^{-4}$, For HB: $K_a = 7.2 \times 10^{-5}$.
16.5 (a) $(CH_3)_3N + H_2O \rightleftharpoons (CH_3)_3NH^+ + OH^-$,

$$K_b = \frac{[(CH_3)_3NH^+][OH^-]}{[(CH_3)_3N]},$$

(b) $SO_3^{2-} + H_2O \rightleftharpoons HSO_3^- + OH^-$,

$$K_b = \frac{[HSO_3^-][OH^-]}{[SO_3^{2-}]},$$

(c) $NH_2OH + H_2O \rightleftharpoons NH_3OH^+ + OH^-$,

$$K_b = \frac{[NH_3OH^+][OH^-]}{[NH_2OH]}.$$

16.6 (a) $HSO_4^- + H_2O \rightleftharpoons H_2SO_4 + OH^-$,

$$K_b = \frac{[H_2SO_4][OH^-]}{[HSO_4^-]},$$

(b) $H_2PO_4^- + H_2O \rightleftharpoons H_3PO_4 + OH^-$,

$$K_b = \frac{[H_3PO_4][OH^-]}{[H_2PO_4^-]},$$

(c) $HPO_4^{2-} + H_2O \rightleftharpoons H_2PO_4^- + OH^-$,

$$K_b = \frac{[H_2PO_4^-][OH^-]}{[HPO_4^{2-}]},$$

(d) $HCO_3^- + H_2O \rightleftharpoons H_2CO_3 + OH^-$,

$$K_b = \frac{[H_2CO_3][OH^-]}{[HCO_3^-]},$$

(e) $HSO_3^- + H_2O \rightleftharpoons H_2SO_3 + OH^-$,

$$K_b = \frac{[H_2SO_3][OH^-]}{[HSO_3^-]}.$$

16.7 $K_b = 4.3 \times 10^{-4}$. **16.8** $K_b = 5.6 \times 10^{-11}$.
16.9 $K_a = 1.15 \times 10^{-3}$, $pK_a = 2.938$.
16.10 $K_a = 1.7 \times 10^{-5}$, $pK_a = 4.78$.
16.11 $K_b = 1.6 \times 10^{-6}$, $pK_b = 5.79$.
16.12 $[H^+] = 5.4 \times 10^{-6}\ M$, pH = 5.27.
16.13 $[H^+] = 8.4 \times 10^{-4}\ M$, pH = 3.08.
16.14 pOH = 5.48. **16.15** pH = 8.59. **16.16** pH = 5.36.
16.17 (a) basic, (b) neutral, (c) acidic. **16.18** (a) neutral, (b) basic, (c) acidic. **16.19** pH = 5.39. **16.20** pH = 8.07.
16.21 pH = 5.13. **16.22** basic. **16.23** neutral.

16.24 pH = 10.79. **16.25** Solving using the quadratic formula: pH = 2.38. Solving by simplification pH = 2.35. The difference is a difference of 0.03 pH units. **16.26** pH = 10.68. **16.27** Upon addition of a strong acid, the concentration of $HC_2H_3O_2$ will increase. When a strong base is added, it reacts with the acid to form more of the acetate ion, therefore the concentration of the acetic acid will decrease. **16.28** (a) $H^+ + NH_3 \longrightarrow NH_4^+$, (b) $OH^- + NH_4^+ \longrightarrow H_2O + NH_3$. **16.29** 4.84, the difference is due to rounding errors. **16.30** 4.61. **16.31** Acetic acid buffer, 26.2 g $NaC_2H_3O_2$, or hydrazoic acid buffer, 21.0 g NaN_3, or butanoic acid buffer, 29.6 g $NaC_4H_7O_2$, or propanoic acid buffer, 22.9 g $NaC_3H_5O_2$. **16.32** Yes, 0.692 mol $HCHO_2$ for 1 mol CHO_2^-, 4.7 g $NaCHO_2$. **16.33** 0.13 pH units. **16.34** pH = 9.76, pH = 9.67.

16.35 $H_3PO_4 \rightleftharpoons H^+ + H_2PO_4^- \quad K_a = \dfrac{[H^+][H_2PO_4^-]}{[H_3PO_4]}$,

$H_2PO_4^- \rightleftharpoons H^+ + HPO_4^{2-} \quad K_a = \dfrac{[H^+][HPO_4^{2-}]}{[H_2PO_4^-]}$,

$HPO_4^{2-} \rightleftharpoons H^+ + PO_4^{3-} \quad K_a = \dfrac{[H^+][PO_4^{3-}]}{[HPO_4^{2-}]}$.

16.36 $[H^+] = 2.6 \times 10^{-3}$ *M*, pH = 2.58, $[HC_6H_6O_6^-] = 2.7 \times 10^{-12}$.
16.37 pH = 11.66, It is not a substitute for $NaHCO_3$.
16.38 pH = 10.24. **16.39** $[H_2SO_3] = K_{b2}$ for SO_3^{2-}.
16.40 (a) H_2O, K^+, $HC_2H_3O_2$, H^+, $C_2H_3O_2^-$, and OH^-, $[H_2O] > [K^+] > [C_2H_3O_2^-] > [OH^-] > [HC_2H_3O_2] > [H^+]$, (b) H_2O, $HC_2H_3O_2$, H^+, $C_2H_3O_2^-$, and OH^-, $[H_2O] > [HC_2H_3O_2] > [H^+] > [C_2H_3O_2^-] > [OH^-]$, (c) H_2O, K^+, $HC_2H_3O_2$, H^+, $C_2H_3O_2^-$, and OH^-, $[H_2O] > [K^+] > [OH^-] > [C_2H_3O_2^-] > [HC_2H_3O_2] > [H^+]$, (d) H_2O, K^+, $HC_2H_3O_2$, H^+, $C_2H_3O_2^-$, and OH^-, $[H_2O] > [HC_2H_3O_2] > [K^+] > [C_2H_3O_2^-] > [OH^-] > [H^+]$.
16.41 (a) 2.37, (b) 3.74, (c) 4.22, (d) 8.22. **16.42** pH = 3.66

Review Problems

16.32 1.5×10^{-11}. **16.34** 7.1×10^{-11}. **16.36** 0.30%, $K_a = 1.82 \times 10^{-6}$. **16.38** $K_a = 2.3 \times 10^{-2}$, $pK_a = 1.6$.
16.40 $K_b = 5.6 \times 10^{-4}$ $pK_b = 3.25$, 7.2%.
16.42 $[HC_3H_5O_2] = 0.145$, $[H^+] = 4.5 \times 10^{-3}$, $[C_3H_5O_2^-] = 4.5 \times 10^{-3}$, pH = 2.34. **16.44** 10.26.
16.46 $[HC_2H_3O_2] = 0.47$ *M*. **16.48** pH = 10.44.
16.50 $[H^+] = 6.0 \times 10^{-4}$ *M*, pH = 3.22.
16.52 pH = 11.26, $[HCN] = 1.8 \times 10^{-3}$ *M*.
16.54 pH = 5.72. **16.56** $K_a = 1.4 \times 10^{-5}$.
16.58 pH = 10.67. **16.60** pH = 4.97. **16.62** pH = 9.00.
16.64 $\Delta[NH_4^+] = -0.01$ *M*, $\Delta[NH_3] = -0.03$ *M*.
16.66 −0.12 pH units. **16.68** −0.11 pH units.
16.70 2.2 g $NaCHO_2$. **16.72** $pH_{initial} = 4.788$, $pH_{final} = 4.761$, pH = 1.63. **16.74** $[H_2C_6H_6O_6] \cong 0.15$ *M*, $[H_3O^+] = [HC_6H_6O_6^-] = 3.2 \times 10^{-3}$ *M*, $[C_6H_6O_6^{2-}] = 2.7 \times 10^{-12}$ *M*, $[OH^-] = 3.2 \times 10^{-12}$ *M*, pH = 2.5. **16.76** $[H^+] = [H_2PO_4^-] = 0.12$ *M*, $[H_3PO_4] = 2.0 - 0.12 = 1.9$ *M*, pH = 0.92, $[HPO_4^{2-}] = 6.3 \times 10^{-8}$, $[PO_4^{3-}] = 2.4 \times 10^{-19}$.
16.78 $[H^+] = [H_2PO_3^-] = 0.16$ *M*, $[HPO_3^{2-}] = 1.6 \times 10^{-7}$ *M*, pH = 0.80. **16.80** pH = 10.28, $[HSO_3^-] = 1.9 \times 10^{-4}$, $[H_2SO_3] = 8.3 \times 10^{-13}$. **16.82** pH = 9.10. **16.84** pH = 12.97, $[HPO_4^{2-}] = 9.4 \times 10^{-2}$ *M*, $[H_2PO_4^-] = 1.6 \times 10^{-7}$, $[H_3PO_4^-] = 2.4 \times 10^{-18}$. **16.86** pH = 8.07, Cresol red.
16.88 pH = 12.93. **16.90** (a) 2.8724, (b) 4.444, (c) 4.7447, (d) 8.7236

CHAPTER 17

Practice Exercises

17.1 $Ba_3(PO_4)_2(s) \rightleftharpoons 3Ba^{2+}(aq) + 2PO_4^{3-}(aq)$, $K_{sp} = [Ba^{2+}]^3[PO_4^{3-}]^2$.
17.2 (a) $K_{sp} = [Ba^{2+}][C_2O_4^{2-}]$, (b) $K_{sp} = [Ag^+]^2[SO_4^{2-}]$.
17.3 3.2×10^{-10}. **17.4** 3.98×10^{-8}. **17.5** 1.5×10^{-5}.
17.6 1.0×10^{-10}. **17.7** 3.9×10^{-8}. **17.8** 1.8×10^{-5} *M* Ag_3PO_4.
17.9 (a) 7.1×10^{-7} *M* AgBr, (b) 1.3×10^{-4} *M* Ag_2CO_3.
17.10 2.1×10^{-16} *M* of AgI will dissolve in a 0.20 *M* CaI_2 solution. In pure water, the solubility is 9.1×10^{-9} *M*.
17.11 1.3×10^{-35} *M* of $Fe(OH)_3$. **17.12** A precipitate will form. **17.13** No precipitate will form. **17.14** We expect $PbSO_4(s)$ since nitrates are soluble. A precipitate of $PbSO_4$ is not expected. **17.15** We expect a precipitate of $PbCl_2$ since nitrates are soluble. A precipitate of $PbCl_2$ is not expected.
17.16 No. **17.17** 2.9. **17.18** 0.35 *M* H^+.
17.19 5.40 to 6.13. **17.20** 4.9×10^{-3} *M* in 0.10 *M* NH_3; 1.3×10^{-5} *M* in pure water. **17.21** 4.1 mol NH_3

Review Problems

17.16 (a) $K_{sp} = [Ca^{2+}][F^-]^2$, (b) $K_{sp} = [Ag^+]^2[CO_3^{2-}]$, (c) $K_{sp} = [Pb^{2+}][SO_4^{2-}]$, (d) $K_{sp} = [Fe^{3+}][OH^-]^3$, (e) $K_{sp} = [Pb^{2+}][I^-]^2$, (f) $K_{sp} = [Cu^{2+}][OH^-]^2$.
17.18 1.6×10^{-5}. **17.20** 1.10×10^{-10}. **17.22** 8.0×10^{-7}.
17.24 2.8×10^{-18}. **17.26** 8.1×10^{-3} *M*. **17.28** 1.3×10^{-4} *M*.
17.30 4.1×10^{-2} *M* LiF, 7.5×10^{-3} *M* BaF_2, LiF is more soluble.
17.32 2.8×10^{-18}. **17.34** 4.9×10^{-3} *M*.
17.36 (a) 4.4×10^{-4} *M*, (b) 9.5×10^{-6} *M*, (c) 9.5×10^{-7} *M*, (d) 6.3×10^{-7} *M*. **17.38** 6.9×10^{-9} *M*. **17.40** 1.7×10^{-3} *M*.
17.42 (a) 3.0×10^{-11} mol/L, (b) 1.2×10^{-6} mol/L.
17.44 1.3 %, +1.8 pH units. **17.46** 2.2 g $Fe(OH)_2$, 2.0×10^{-12} *M*. **17.48** 7.9×10^{-7} *M*. **17.50** No precipitate will form. **17.52** (a) No precipitate will form, (b) A precipitate will form. **17.54** 2.3×10^{-8} *M*. **17.56** No precipitate forms.
17.58 $[H^+] = 0.045$ *M*, pH = 1.35. **17.60** pH = 4.8–8.1, $Mn(OH)_2$ will be soluble, but some $Cu(OH)_2$ will precipitate out of solution.

17.62 (a) $Cu^{2+}(aq) + 4Cl^-(aq) \rightleftharpoons CuCl_4^{2-}(aq)$

$$K_{form} = \frac{[CuCl_4^{2-}]}{[Cu^{2+}][Cl^-]^4},$$

(b) $Ag^+(aq) + 2I^-(aq) \rightleftharpoons AgI_2^-(aq) \quad K_{form} = \dfrac{[AgI_2^-]}{[Ag^+][I^-]^2}$,

(c) $Cr^{3+}(aq) + 6NH_3(aq) \rightleftharpoons Cr(NH_3)_6^{3+}(aq)$

$$K_{form} = \frac{[Cr(NH_3)_6^{3+}]}{[Cr^{3+}][NH_3]^6}.$$

17.64 (a) $Co^{3+}(aq) + 6NH_3(aq) \rightleftharpoons Co(NH_3)_6^{3+}(aq)$

$$K_{form} = \frac{[Co(NH_3)_6^{3+}]}{[Co^{3+}][NH_3]^6},$$

(b) $Hg^{2+}(aq) + 4I^-(aq) \rightleftharpoons HgI_4^{2-}(aq)$

$$K_{form} = \frac{[HgI_4^{2-}]}{[Hg^{2+}][I^-]^4},$$

(c) $Fe^{2+}(aq) + 6CN^-(aq) \rightleftharpoons Fe(CN)_6^{4-}(aq)$

$$K_{form} = \frac{[Fe(CN)_6^{4-}]}{[Fe^{2+}][CN^-]^6}.$$

17.66 4.3×10^{-4}. **17.68** 450 g NaCN are required.
17.70 1.9×10^{-4} g AgI. **17.72** 9.5×10^{-3} *M*.
17.74 $K_{inst} = 3.7 \times 10^{-10}$

CHAPTER 18

Practice Exercises

18.1 +154 L atm. **18.2** Energy is added to the system in the form of work. **18.3** −2.64 kJ, ΔE is more exothermic.
18.4 −214.6 kJ, 1.14 %. **18.5** negative.
18.6 (a) negative, (b) positive.
18.7 (a) negative, (b) negative.
18.8 (a) negative, (b) negative, (c) positive.
18.9 -99.1 J K^{-1}.
18.10 (a) −229 J/K, (b) −120.9 J/K.
18.11 +98.3 kJ/mol. **18.12** 1482 kJ.
18.13 −30.3 kJ/mol.
18.14 (a) −69.7 kJ/mol, (b) −120.1 kJ/mol.
18.15 2820 kJ work for $C_2H_5OH(l)$, 4650 kJ work for $C_8H_{18}(l)$, C_8H_{18} is a better fuel on a gram basis.
18.16 788 kJ. **18.17** $+90.5 \text{ J mol}^{-1} \text{ K}^{-1}$.
18.18 614 K (341 °C).
18.19 The reaction should be spontaneous.
18.20 We do not expect to see products formed from reactants.
18.21 −4.3 kJ. **18.22** +32.8 kJ, −29.1 kJ. The equilibrium shifts to products. **18.23** $\Delta G = 0$. The reaction is at equilibrium.
18.24 The reaction will proceed to the right. **18.25** −33 kJ.
18.16 0.26

Review Problems

18.46 +1000 J, Endothermic. **18.48** 121 J.
18.50 (a) $\Delta H° = +24.58$ kJ, $\Delta E° = 19.6$ kJ,
(b) $\Delta H° = -178$ kJ, $\Delta E° = -175$ kJ,
(c) $\Delta H° = 847.6$ kJ, $\Delta E° = \Delta H°$,
(d) $\Delta H° = 65.029$ kJ $\Delta E° = \Delta H°$.
18.52 $\Delta E = -677$ kJ, $\Delta E_{200\,°C} = -683$ kJ.
18.54 (a) −178 kJ spontaneous, (b) −311 kJ spontaneous, (c) +1084.3 kJ not spontaneous. **18.56** number of moles of reactants and products, state of the reactants and products, and complexity of the molecules. **18.58** (a) negative, (b) negative, (c) negative, (d) positive. **18.60** (a) −198.3 J/K not spontaneous, (b) −332.3 J/K not spontaneous, (c) +92.6 J/K spontaneous, (d) +14 J/K spontaneous, (e) +159 J/K spontaneous.
18.62 (a) $-52.8 \text{ J mol}^{-1} \text{ K}^{-1}$, (b) $-868.9 \text{ J mol}^{-1} \text{ K}^{-1}$, (c) $-318 \text{ J mol}^{-1} \text{ K}^{-1}$.
18.64 −269.7 J/K.
18.66 −209 kJ/mol.
18.68 (a) −82.3 kJ, (b) −8.8 kJ, (c) +70.7 kJ.
18.70 +0.16 kJ. **18.72** 1299.8 kJ. **18.74** 333 K.
18.76 $101 \text{ J mol}^{-1} \text{ K}^{-1}$.
18.78 spontaneous.
18.80 (a) $K_P = 10^{263}$, (b) $K_P = 2.90 \times 10^{-25}$.
18.82 The system is not at equilibrium and must shift to the right to reach equilibrium.
18.84 $K_P = 8.000 \times 10^8$. This is a favorable reaction, since the equilibrium lies far to the side favoring products and is worth studying as a method for methane production.
18.86 If $\Delta G° = 0$, $K_c = 1$. If we start with pure products, the value of Q will be infinite (there are zero reactants) and, since $Q > K_c$, the reaction will proceed towards the reactants, i.e., the pure products will decompose to their elements.
18.88 1.16×10^3 kJ/mol. **18.90** 354 kJ/mol.
18.92 577.7 kJ/mol. **18.94** 308.0 kJ/mol.
18.96 85 kJ/mol.
18.98 The heat of formation of CF_4 should be more exothermic than that of CCl_4 because more energy is released on formation of a C—F bond than on formation of a C—Cl bond. Also, less energy is needed to form gaseous F atoms than to form gaseous Cl atoms.

CHAPTER 19

Practice Exercises

19.1 anode: $Mg(s) \longrightarrow Mg^{2+}(aq) + 2e^-$,
cathode: $Fe^{2+}(aq) + 2e^- \longrightarrow Fe(s)$,
cell notation: $Mg(s) \mid Mg^{2+}(aq) \parallel Fe^{2+}(aq) \mid Fe(s)$,

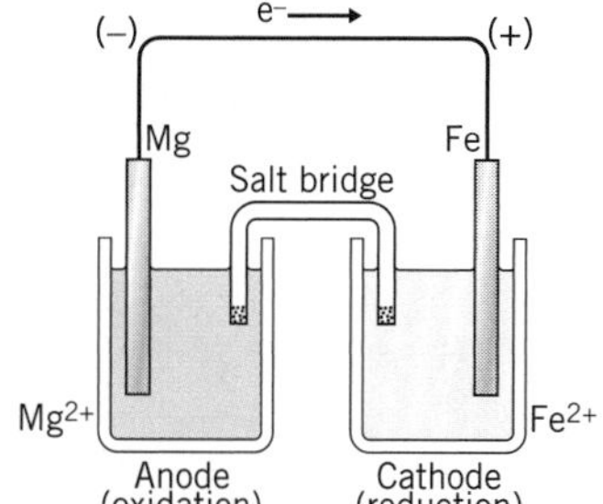

19.2 anode: $Al(s) \longrightarrow Al^{3+}(aq) + 3e^-$,
cathode: $Ni^{2+}(aq) + 2e^- \longrightarrow Ni(s)$,
overall: $3Ni^{2+}(aq) + 2Al(s) \longrightarrow 2Al^{3+}(aq) + 3Ni(s)$.
19.3 Zinc. **19.4** −0.44 V.
19.5 (a) $2I^-(aq) + 2Fe^{3+}(aq) \longrightarrow I_2(s) + 2Fe^{2+}(aq)$
(b) $3Mg(s) + 2Cr^{3+}(aq) \longrightarrow 2Cr(s) + 3Mg^{2+}(aq)$
(c) $Co(s) + 4H^+(aq) + 2SO_4^{2-}(aq) \longrightarrow H_2SO_3(aq) + H_2O + Co^{2+}(aq)$.
19.6 $Br_2(aq) + H_2SO_3(aq) + H_2O \longrightarrow 2Br^-(aq) + SO_4^{2-}(aq) + 4H^+(aq)$.
19.7 Non spontaneous.
$Ni(s) + 2Fe^{3+}(aq) \longrightarrow Ni^{2+}(aq) + 2Fe^{2+}(aq)$.
19.8 $NiO_2(s) + Fe(s) + 2H_2O \longrightarrow Ni(OH)_2(s) + Fe(OH)_2(s)$ $E°_{cell} = 1.37$ V.
19.9 $2Cr(s) + 3Cu^{2+}(aq) \longrightarrow 2Cr^{3+}(aq) + 3Cu(s)$ $E°_{cell} = 1.08$ V.
19.10 $3MnO_4^-(aq) + 24H^+(aq) + 5Cr(s) \longrightarrow 5Cr^{3+}(aq) + 3Mn^{2+}(aq) + 12H_2O$ $E°_{cell} = 2.25$ V.
19.11 (a) Spontaneous, (b) Spontaneous.
19.12 (a) Nonspontaneous, (b) Spontaneous. **19.13** 3 electrons.
19.14 $\Delta G° = -102$ kJ, −343 kJ, +108 kJ and −11.6 kJ, respectively. **19.15** $K_c = 2.7 \times 10^{-16}$, $K = 5.6 \times 10^{16}$.
19.16 $Ag^+(aq) + Br^-(aq) \longrightarrow AgBr(s)$,
$K = \dfrac{1}{[Ag^+][Br^-]}$, 2.2×10^{12}.
19.17 $Cu^{2+}(aq) + Mg(s) \longrightarrow Mg^{2+}(aq) + Cu(s)$, 2.82 V.
19.18 1.04 V. **19.19** 2.9×10^{-5} M Mg^{2+}.
19.20 1.9×10^{-4} M, 6.6×10^{-13} M. **19.21** 6.6×10^{-4} M.
19.22 I_2. **19.23** Tin. **19.24** 8.29×10^{-3} mol OH^-.
19.25 7.33 min. **19.26** 3.67 A. **19.27** + 0.0187 M.

Review Problems

19.50 (a) anode: $Cd(s) \longrightarrow Cd^{2+}(aq) + 2e^-$,
cathode: $Au^{3+}(aq) + 3e^- \longrightarrow Au(s)$,
cell: $3Cd(s) + 2Au^{3+}(aq) \longrightarrow 3Cd^{2+}(aq) + 2Au(s)$,
(b) anode: $Fe(s) \longrightarrow Fe^{2+}(aq) + 2e^-$,
cathode: $Br_2(aq) + 2e^- \longrightarrow 2Br^-(aq)$,
cell: $Fe(s) + Br_2(aq) \longrightarrow Fe^{2+}(aq) + 2Br^-(aq)$,
(c) anode: $Cr(s) \longrightarrow Cr^{3+}(aq) + 3e^-$,
cathode: $Cu^{2+}(aq) + 2e^- \longrightarrow Cu(s)$,
cell: $2Cr(s) + 3Cu^{2+}(aq) \longrightarrow 2Cr^{3+}(aq) + 3Cu(s)$.
19.52 (a) $Pt(s) \mid Fe^{2+}(aq), Fe^{3+}(aq) \parallel NO_3^-(aq), H^+(aq) \mid NO(g) \mid Pt(s)$,
(b) $Pt(s) \mid Cl^-(aq), Cl_2(g) \parallel Br_2(aq), Br^-(aq) \mid Pt(s)$,
(c) $Ag(s) \mid Ag^+(aq) \mid Au^{3+}(aq) \mid Au(s)$.
19.54 (a) Sn(s), (b) $Br^-(aq)$, (c) Zn(s), (d) $I^-(aq)$.
19.56 (a) 0.19 V, (b) −0.29 V, (c) 0.62 V.
19.58 (a) Spontaneous. (b) Spontaneous. (c) Spontaneous.

19.60 $BrO_3^-(aq) + 6I^-(aq) + 6H^+(aq) \longrightarrow 3I_2(s) + Br^-(aq) + 3H_2O$, net reaction. $E^\circ_{cell} = 0.90V$.
19.62 $4HOCl(aq) + 2H^+(aq) + S_2O_3^{2-} \longrightarrow 2Cl_2(g) + H_2O + 2H_2SO_3(aq)$
19.64 Not spontaneous. **19.66** 1.0×10^2 kJ.
19.68 (a) $E^\circ_{cell} = 0.54V$, (b) -5.2×10^2 kJ, (c) 2.1×10^{91}.
19.70 0.31. **19.72** 2.38 V. **19.74** $[Cd^{2+}] = 2.86 \times 10^{-15}$ *M*.
19.76 $K_{sp} = 1.96 \times 10^{-15}$. **19.78** 0.09 V.
19.80 (a) 0.40 mol e^-, (b) 0.70 mol e^-, (c) 4.50 mol e^-, (d) 5.0×10^{-2} mol e^-. **19.82** 2.68 g Fe(OH).
19.84 51.5 hr. **19.86** 66.2 amp. **19.88** 40.1 mL.
19.90 298 mL Cl_2.
19.92 $2H_2O \rightleftharpoons 2H_2(g) + O_2(g)$, $E^\circ = -2.06$ V.
19.94 Br_2 and Cu, $Cu^{2+} + 2Br^- \rightleftharpoons Br_2 + Cu(s)$

CHAPTER 20

Practice Exercises

20.1 ${}^{226}_{88}Ra \longrightarrow {}^{222}_{86}Rn + {}^{4}_{2}He + {}^{0}_{0}\gamma$.
20.2 ${}^{90}_{38}Sr \longrightarrow {}^{90}_{39}Y + {}^{0}_{-1}e$. **20.3** 50.0%.
20.4 1.10×10^5 atoms Rn-222. **20.5** 20 m. **20.6** 100 units

Review Problems

20.45 1.11×10^{-11} g. **20.47** -3.18 ng, 1.09×10^{-5}%.
20.49 1.8×10^{-13} J/nucleon.
20.51 (a) ${}^{211}_{83}Bi$, (b) ${}^{177}_{72}Hf$, (c) ${}^{216}_{84}Po$, (d) ${}^{19}_{9}F$.
20.53 (a) ${}^{242}_{94}Pu \longrightarrow {}^{4}_{2}He + {}^{238}_{92}U$, (b) ${}^{28}_{12}Mg \longrightarrow {}^{0}_{-1}e + {}^{28}_{13}Al$, (c) ${}^{26}_{14}Si \longrightarrow {}^{0}_{1}e + {}^{26}_{13}Al$, (d) ${}^{37}_{18}Ar + {}^{0}_{-1}e \longrightarrow {}^{37}_{17}Cl$.
20.55 (a) ${}^{261}_{102}No$, (b) ${}^{211}_{82}Pb$, (c) ${}^{141}_{61}Pm$, (d) ${}^{179}_{74}W$.
20.57 ${}^{87}_{36}Kr \longrightarrow {}^{86}_{36}Kr + {}^{1}_{0}n$. **20.59** The more likely process is positron emission. ${}^{38}_{19}K \longrightarrow {}^{0}_{1}e + {}^{38}_{18}Ar$. **20.61** 0.0469 mg.
20.63 ${}^{53}_{24}Cr^*$; ${}^{51}_{23}V + {}^{2}_{1}H \longrightarrow {}^{53}_{24}Cr^* \longrightarrow {}^{1}_{1}p + {}^{52}_{23}V$.
20.65 ${}^{80}_{35}Br$. **20.67** ${}^{55}_{26}Fe$; ${}^{55}_{25}Mn + {}^{1}_{1}p \longrightarrow {}^{1}_{0}n + {}^{55}_{26}Fe$.
20.69 ${}^{70}_{30}Zn + {}^{208}_{82}Pb \longrightarrow {}^{278}_{112}Uub \longrightarrow {}^{1}_{0}n + {}^{277}_{112}Uub$.
20.71 6.3 m. **20.73** 6.6 m.
20.75 2.4×10^7 Bq, 6.5×10^2 μCi.
20.77 $k = 1.0 \times 10^{-6}\ s^{-1}$, $t_{1/2} = 6.9 \times 10^5$ s.
20.79 3.1% The chemical product is $BaCl_2$.
20.81 2.84×10^{-3} Ci/g.
20.83 1.3×10^9 years old.
20.85 2.7×10^4 years ago.
20.87 ${}^{235}_{92}U + {}^{1}_{0}n \longrightarrow {}^{94}_{38}Sr + {}^{140}_{54}Xe + 2{}^{1}_{0}n$.

CHAPTER 21

Practice Exercises

21.1 At any temperature. **21.2** 5570 K.
21.3 $[Ag(S_2O_3)_2]^{3-}$, $(NH_4)_3[Ag(S_2O_3)_2]$.
21.4 $AlCl_3{\cdot}6H_2O$, $[Al(H_2O)_6]^{3+}$.
21.5 (a) $[SnCl_6]^{2-}$, (b) $(NH_4)_2[Fe(CN)_4(H_2O)_2]$.
21.6 (a) potassium hexacyanoferrate(III), (b) dichlorobis(ethylenediamine)chromium(III) sulfate.
21.7 Six. **21.8** (a) Six, (b) Six, (c) Six.

Review Problems

21.102 734 K.
21.104 The net charge is -3, and the formula is $[Fe(CN)_6]^{3-}$, hexacyanoferrate(III) ion. **21.106** $[CoCl_2(en)_2]^+$.
21.108 (a) oxalato, (b) sulfido or thio, (c) chloro, (d) dimethylamine. **21.110** (a) hexaamminenickel(II) ion, (b) triamminetrichlorochromate(II) ion, (c) hexanitrocobaltate(III) ion, (d) diamminetetracyanomanganate(II) ion, (e) trioxalatoferrate(III) ion or trisoxalatoferrate(III) ion.
21.112 (a) $[Fe(CN)_2(H_2O)_4]^+$, (b) $[Ni(C_2O_4)(NH_3)_4]$, (c) $[Fe(CN)_4(H_2O)_2]^-$, (d) $K_3[Mn(SCN)_6]$, (e) $[CuCl_4]^{2-}$.
21.114 Six.
21.116

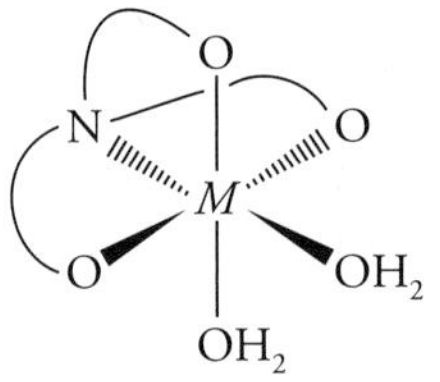

The curved lines represent $-CH_2-\overset{O}{\overset{\|}{C}}-$ groups.
21.118 Since both are the *cis* isomer, they are identical.
21.120

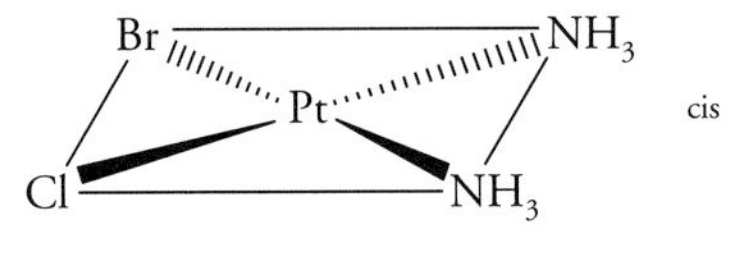

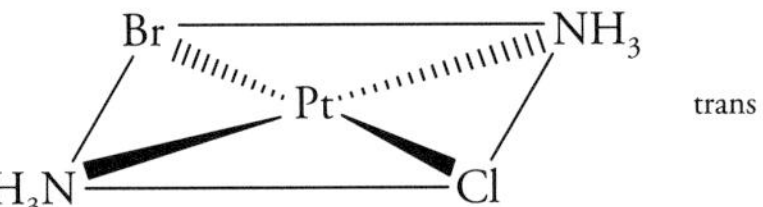

21.122

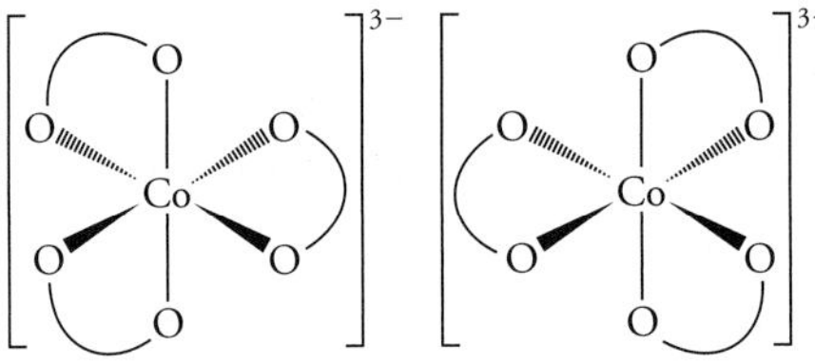

21.124 (a) $[Cr(H_2O)_6]^{3+}$, (b) $[Cr(en)_3]^{3+}$.
21.126 $[Cr(CN)_6]^{3-}$.
21.128 (a) $[RuCl(NH_3)_5]^{3+}$. The value of Δ increases down a group., (b) $[Ru(NH_3)_6]^{3+}$. The value of Δ increases with oxidation state of the metal.
21.130 This is the one with the strongest field ligand, since Co^{2+} is a d^7 ion: CoA_6^{3+}.
21.132 This is a weak field complex of Co^{2+}, and it should be a high–spin d^7 case. It cannot be diamagnetic.

CHAPTER 22

Practice Exercises

22.1 2,2-dimethylpropane.
22.2 (a) 3-methylhexane, (b) 4-ethyl-2,3-dimethylheptane, (c) 5-ethyl-2,4,6-trimethyloctane.
22.3

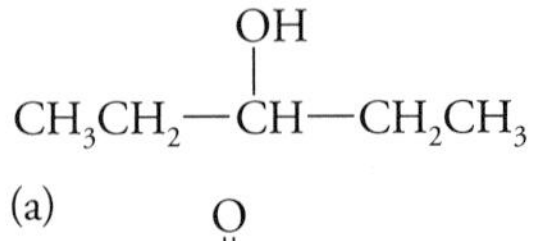

$$CH_3CH_2-\underset{}{\overset{OH}{\overset{|}{C}H}}-CH_2CH_3$$

22.4 (a)

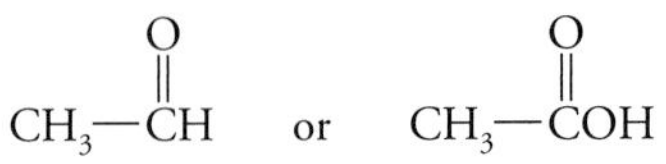

$$CH_3-\overset{O}{\overset{\|}{C}}H \quad \text{or} \quad CH_3-\overset{O}{\overset{\|}{C}}OH$$

(b)

$$CH_3CH_2\overset{O}{\overset{\|}{C}}CH_2CH_3$$

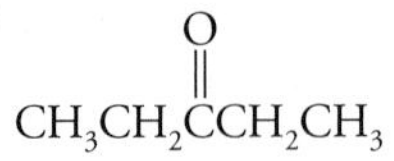

22.5 $CH_3CH_2{-}\overset{O}{\overset{\|}{C}}{-}NHCH_2CH_2CH_3$

22.6 (a) $CH_3CH_2{-}\overset{O}{\overset{\|}{C}}{-}O^- + HOCH(CH_3)_2$

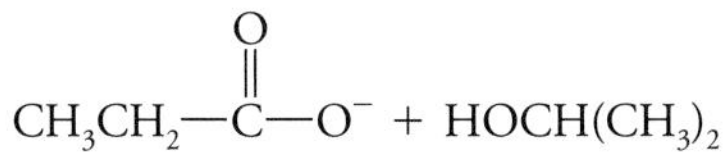

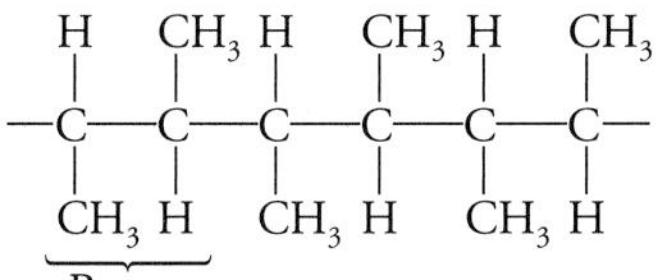

(b) $CH_3CH{=}CH_2$

22.7 $-C(H)(CH_3)-C(CH_3)(H)-C(H)(CH_3)-C(CH_3)(H)-C(H)(CH_3)-C(CH_3)(H)-$ (first two carbons: Repeat unit)

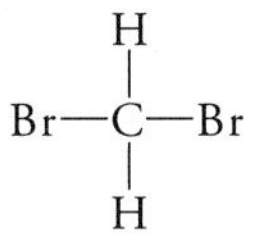

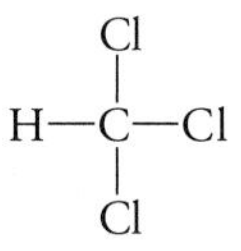

22.8 $-CF_2-CF_2-CF_2-CF_2-CF_2-CF_2-$ (first two carbons: Repeat unit)

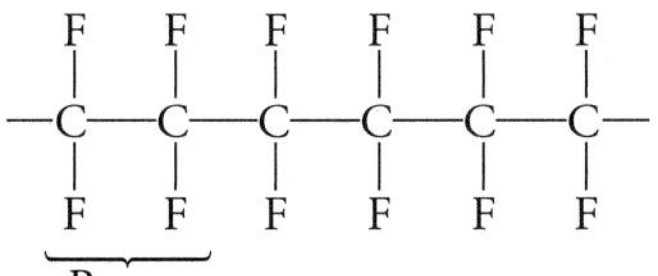

Review Problems

22.97 (a) $H_3C{-}NH_2$ (H—C(H)(H)—N(H)—H) (b) CH_2Br_2 (Br—C(H)(H)—Br)

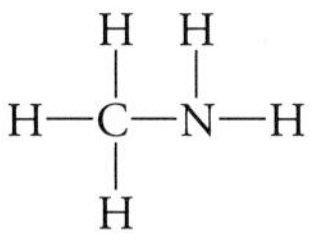

(c) $CHCl_3$ (H—C(Cl)(Cl)—Cl) (d) H—N(H)—O—H

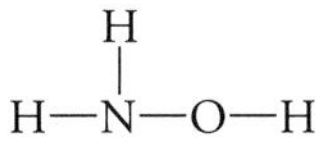

(e) H—C≡C—H (f) H—N(H)—N(H)—H

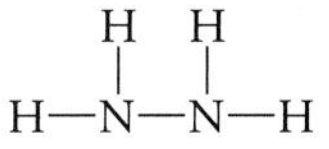

22.99 (a) alkene (b) alcohol (c) ester (d) carboxylic acid (e) amine (f) alcohol.

22.101 b, e, and f.

22.103 (a) amine, (b) amine, (c) amide, (d) amine, ketone.

22.105 (a) identical, (b) identical, (c) unrelated, (d) isomers, (e) identical, (f) identical, (g) isomers.

22.107 (a) pentane, (b) 2-methylpentane, (c) 2,4-dimethylhexane.

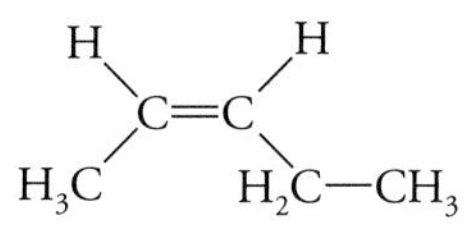

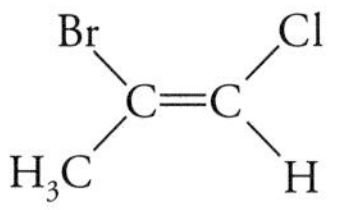

22.109 (a) No isomers,

(b) $H_3C(H)C{=}C(H)(H_2C{-}CH_3)$ trans; $H_3C(H)C{=}C(H)(H_2C{-}CH_3)$ cis

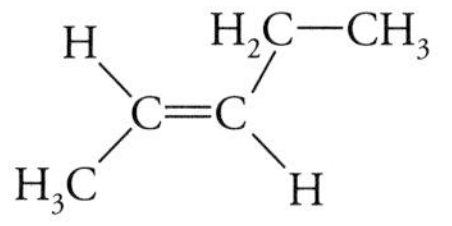

(c) $H_3C(Br)C{=}C(Cl)(H)$ cis; $H_3C(Br)C{=}C(H)(Cl)$ trans

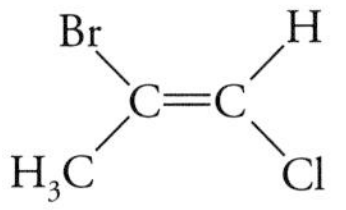

22.111 (a) CH_3CH_3, (b) $ClCH_2CH_2Cl$, (c) $BrCH_2CH_2Br$, (d) CH_3CH_2Cl, (e) CH_3CH_2Br, (f) CH_3CH_2OH.

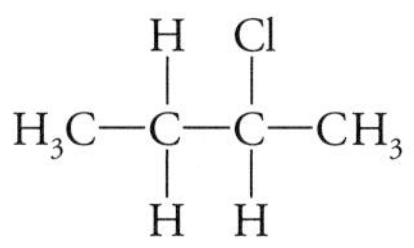

22.113 (a) $CH_3CH_2CH_2CH_3$,

(b) $H_3C{-}CHCl{-}CHCl{-}CH_3$ (c) $H_3C{-}CHBr{-}CHBr{-}CH_3$

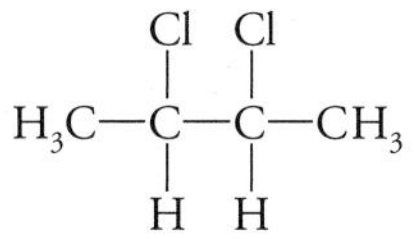

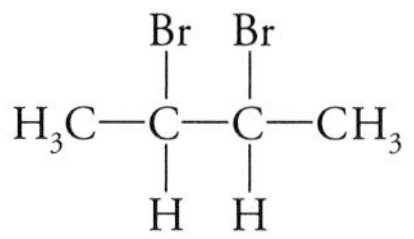

(d) $H_3C{-}CH_2{-}CHCl{-}CH_3$ (e) $H_3C{-}CH_2{-}CHBr{-}CH_3$

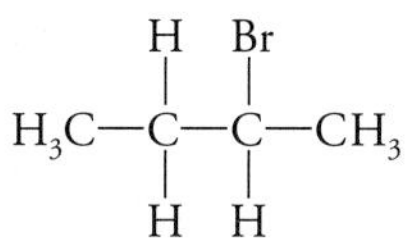

(f) $H_3C{-}CH_2{-}CH(OH){-}CH_3$

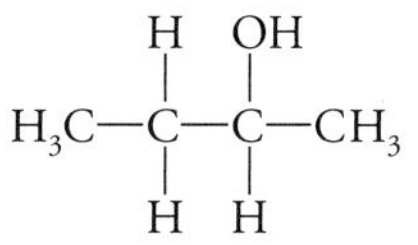

22.115 This sort of reaction would disrupt the π delocalization of the benzene ring. The subsequent loss of resonance energy would not be favorable.

22.117 CH_3OH:
IUPAC name = methanol; common name = methyl alcohol.
CH_3CH_2OH:
IUPAC name = ethanol; common name = ethyl alcohol.
$CH_3CH_2CH_2OH$:
IUPAC name = 1-propanol; common name = propyl alcohol. 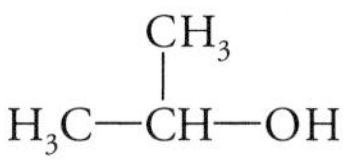
$H_3C{-}CH(CH_3){-}OH$
IUPAC name = 2-propanol; common name = isopropyl alcohol.

22.119 $CH_3CH_2CH_2{-}O{-}CH_3$ methyl propyl ether
$CH_3CH_2{-}O{-}CH_2CH_3$ diethyl ether
$(CH_3)_2CH{-}O{-}CH_3$ methyl 2-propyl ether

22.121 (a) cyclopentene (b) $C_6H_5{-}CH{=}CH_2$

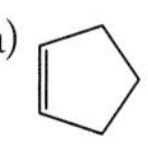

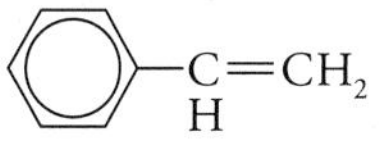

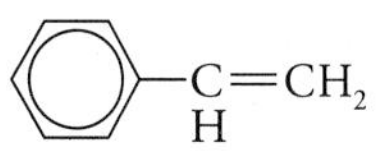

(c) $C_6H_5{-}CH{=}CH_2$

22.123

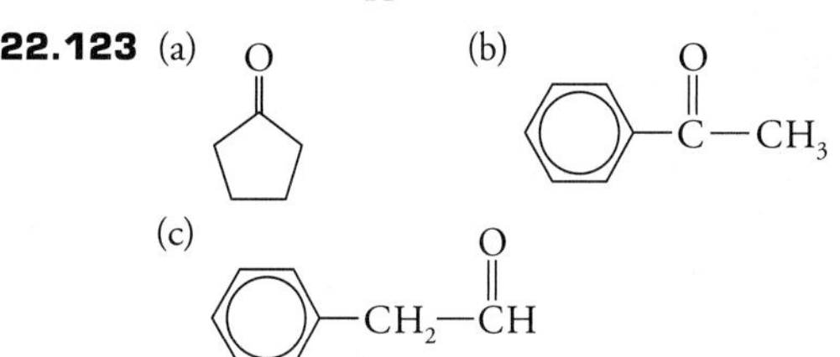

22.125 The elimination of water can result in a C=C double bond in two locations:
$CH_2{=}CHCH_2CH_3$ 1-butene
$CH_3CH{=}CHCH_3$ 2-butene

22.127 The aldehyde is more easily oxidized. The product is:

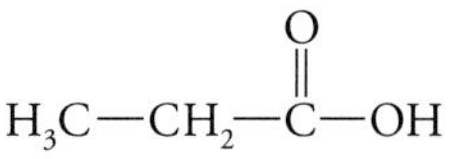

$H_3C{-}CH_2{-}\overset{O}{\overset{\|}{C}}{-}OH$

22.129 (a) $CH_3CH_2CO_2H$, (b) $CH_3CH_2CO_2H + CH_3OH$, (c) $Na^+ + CH_3CH_2CH_2CO_2^- + H_2O$

22.131 $CH_3CO_2H + CH_3CH_2NHCH_2CH_3$

22.133 $-CH(OC(=O)CH_3)-CH_2-CH(OC(=O)CH_3)-CH_2-CH(OC(=O)CH_3)-CH_2-CH(OC(=O)CH_3)-CH_2-$

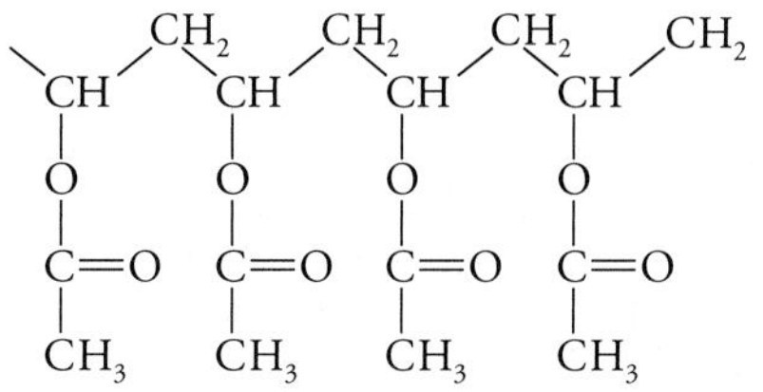

22.135

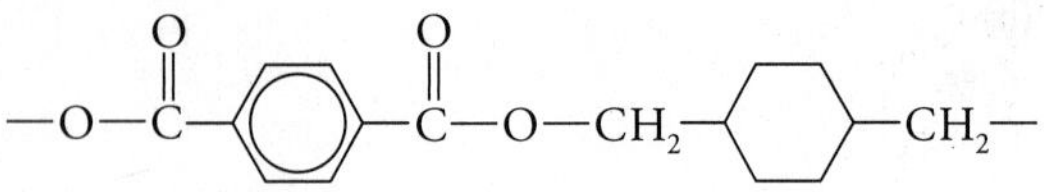

$$-O-\overset{\displaystyle O}{\overset{\|}{C}}-C_6H_4-\overset{\displaystyle O}{\overset{\|}{C}}-O-CH_2-C_6H_{10}-CH_2-$$

22.137

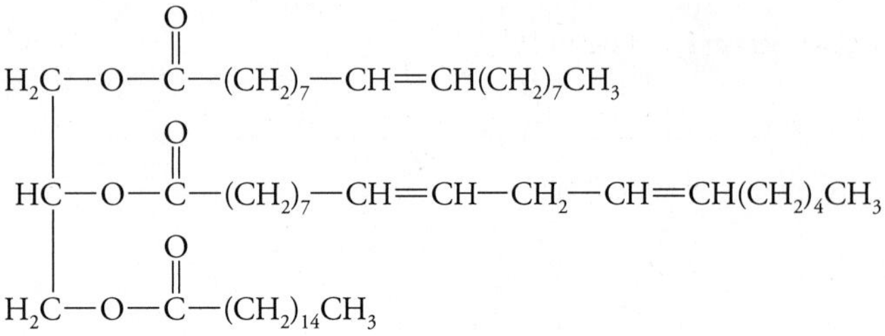

$$H_2C-O-\overset{\displaystyle O}{\overset{\|}{C}}-(CH_2)_7-CH{=}CH(CH_2)_7CH_3$$
$$HC-O-\overset{\displaystyle O}{\overset{\|}{C}}-(CH_2)_7-CH{=}CH-CH_2-CH{=}CH(CH_2)_4CH_3$$
$$H_2C-O-\overset{\displaystyle O}{\overset{\|}{C}}-(CH_2)_{14}CH_3$$

22.139

$$H_2C-O-\overset{\displaystyle O}{\overset{\|}{C}}-(CH_2)_{16}CH_3$$
$$HC-O-\overset{\displaystyle O}{\overset{\|}{C}}-(CH_2)_{12}CH_3$$
$$H_2C-O-\overset{\displaystyle O}{\overset{\|}{C}}-(CH_2)_{16}CH_3$$

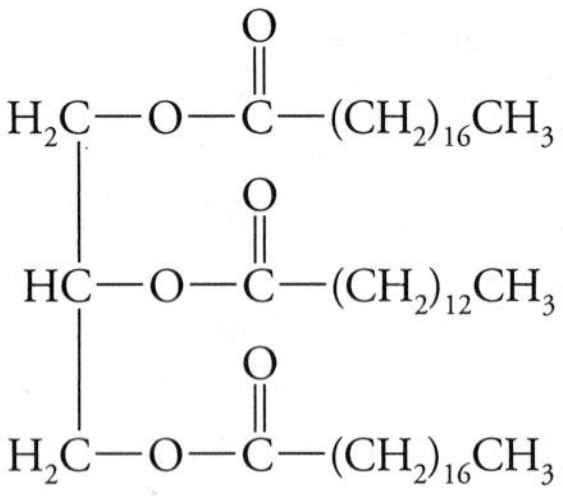

22.141 Hydrophobic sites are composed of fatty acid units.
Hydrophilic sites are composed of charged units.

22.143

$$^{+}H_3N-CH_2-\overset{\displaystyle O}{\overset{\|}{C}}-NH-CH_2-\overset{\displaystyle O}{\overset{\|}{C}}-O^-$$

22.145

$$^{+}H_3N-CH_2-\overset{\displaystyle O}{\overset{\|}{C}}-NH-\underset{\displaystyle CH_2C_6H_5}{\underset{|}{CH}}-\overset{\displaystyle O}{\overset{\|}{C}}-O^- \qquad ^{+}H_3N-\underset{\displaystyle CH_2C_6H_5}{\underset{|}{CH}}-\overset{\displaystyle O}{\overset{\|}{C}}-NH-CH_2-\overset{\displaystyle O}{\overset{\|}{C}}-O^-$$

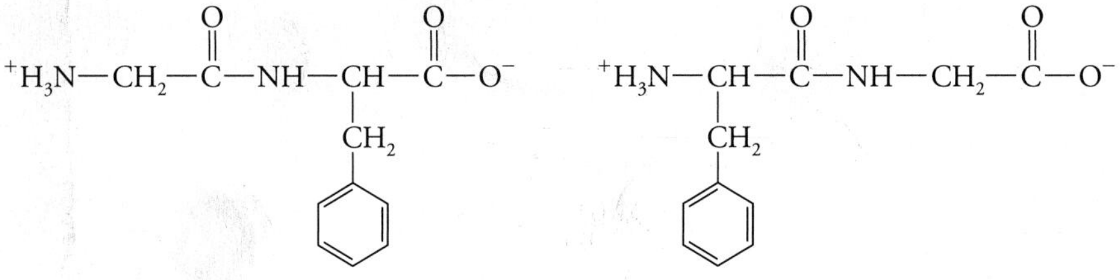

APPENDIX C

TABLES OF SELECTED DATA

TABLE C.1 Thermodynamic Data for Selected Elements, Compounds, and Ions (25 °C)

Substance	ΔH_f°(kJ mol^{-1})	S°(J mol^{-1} K^{-1})	ΔG_f°(kJ mol^{-1})
Aluminum			
Al(*s*)	0	28.3	0
Al^{3+}(*aq*)	−524.7		−481.2
$AlCl_3$(*s*)	−704	110.7	−629
Al_2O_3(*s*)	−1669.8	51.0	−1576.4
$Al_2(SO_4)_3$(*s*)	−3441	239	−3100
Arsenic			
As(*s*)	0	35.1	0
AsH_3(*g*)	+66.4	223	+68.9
As_4O_6(*s*)	−1314	214	−1153
As_2O_5(*s*)	−925	105	−782
H_3AsO_3(*aq*)	−742.2		
H_3AsO_4(*aq*)	−902.5		
Barium			
Ba(*s*)	0	66.9	0
Ba^{2+}(*aq*)	−537.6	9.6	−560.8
$BaCO_3$(*s*)	−1219	112	−1139
$BaCrO_4$(*s*)	−1428.0		
$BaCl_2$(*s*)	−860.2	125	−810.8
BaO(*s*)	−553.5	70.4	−525.1
$Ba(OH)_2$(*s*)	−998.22	−8	−875.3
$Ba(NO_3)_2$(*s*)	−992	214	−795
$BaSO_4$(*s*)	−1465	132	−1353
Beryllium			
Be(*s*)	0	9.50	0
$BeCl_2$(*s*)	−468.6	89.9	−426.3
BeO(*s*)	−611	14	−582
Bismuth			
Bi(*s*)	0	56.9	0
$BiCl_3$(*s*)	−379	177	−315
Bi_2O_3(*s*)	−576	151	−497
Boron			
B(*s*)	0	5.87	0
BCl_3(*g*)	−404	290	−389
B_2H_6(*g*)	+36	232	+87
B_2O_3(*s*)	−1273	53.8	−1194
$B(OH)_3$(*s*)	−1094	88.8	−969
Bromine			
Br_2(*l*)	0	152.2	0
Br_2(*g*)	+30.9	245.4	+3.11
HBr(*g*)	−36	198.5	+53.1
Br^-(*aq*)	−121.55	82.4	−103.96
Cadmium			
Cd(*s*)	0	51.8	0
Cd^{2+}(*aq*)	−75.90	−73.2	−77.61
$CdCl_2$(*s*)	−392	115	−344
CdO(*s*)	−258.2	54.8	−228.4
CdS(*s*)	−162	64.9	−156
$CdSO_4$(*s*)	−933.5	123	−822.6
Calcium			
Ca(*s*)	0	41.4	0
Ca^{2+}(*aq*)	−542.83	−53.1	−553.58
$CaCO_3$(*s*)	−1207	92.9	−1128.8
CaF_2(*s*)	−741	80.3	−1166
$CaCl_2$(*s*)	−795.0	114	−750.2
$CaBr_2$(*s*)	−682.8	130	−663.6
CaI_2(*s*)	−535.9	143	
CaO(*s*)	−635.5	40	−604.2
$Ca(OH)_2$(*s*)	−986.59	76.1	−896.76
$Ca_3(PO_4)_2$(*s*)	−4119	241	−3852
$CaSO_3$(*s*)	−1156		
$CaSO_4$(*s*)	−1433	107	−1320.3
$CaSO_4 \cdot \frac{1}{2}H_2O$(*s*)	−1575.2	131	−1435.2
$CaSO_4 \cdot 2H_2O$(*s*)	−2021.1	194.0	−1795.7
Carbon			
C(*s*, graphite)	0	5.69	0
C(*s*, diamond)	+1.88	2.4	+2.9
CCl_4(*l*)	−134	214.4	−65.3
CO(*g*)	−110.5	197.9	−137.3
CO_2(*g*)	−393.5	213.6	−394.4
CO_2(*aq*)	−413.8	117.6	−385.98
H_2CO_3(*aq*)	−699.65	187.4	−623.08
HCO_3^-(*aq*)	−691.99	91.2	−586.77
CO_3^{2-}(*aq*)	−677.14	−56.9	−527.81
CS_2(*l*)	+89.5	151.3	+65.3
CS_2(*g*)	+117	237.7	+67.2
HCN(*g*)	+135.1	201.7	+124.7
CN^-(*aq*)	+150.6	94.1	+172.4
CH_4(*g*)	−74.848	186.2	−50.79
C_2H_2(*g*)	+226.75	200.8	+209
C_2H_4(*g*)	+52.284	219.8	+68.12
C_2H_6(*g*)	−84.667	229.5	−32.9
C_3H_8(*g*)	−104	269.9	−23
C_4H_{10}(*g*)	−126	310.2	−17.0
C_6H_6(*l*)	+49.0	173.3	+124.3
CH_3OH(*l*)	−238.6	126.8	−166.2
C_2H_5OH(*l*)	−277.63	161	−174.8
$HCHO_2$(*g*)	−363	251	+335
$HC_2H_3O_2$(*l*)	−487.0	160	−392.5
HCHO(*g*)	−108.6	218.8	−102.5
CH_3CHO(*g*)	−167	250	−129

(Continued)

TABLE C.1 Thermodynamic Data for Selected Elements, Compounds, and Ions (25 °C) *(Continued)*

Substance	ΔH_f°(kJ mol^{-1})	S°(J mol^{-1} K^{-1})	ΔG_f°(kJ mol^{-1})
$(CH_3)_2CO(l)$	−248.1	200.4	−155.4
$C_6H_5CO_2H(s)$	−385.1	167.6	−245.3
$CO(NH_2)_2(s)$	−333.19	104.6	−197.2
$CO(NH_2)_2(aq)$	−391.2	173.8	−203.8
$CH_2(NH_2)CO_2H(s)$	−532.9	103.5	−373.4
Chlorine			
$Cl_2(g)$	0	223.0	0
$Cl^-(aq)$	−167.2	56.5	−131.2
$HCl(g)$	−92.30	186.7	−95.27
$HCl(aq)$	−167.2	56.5	−131.2
$HClO(aq)$	−131.3	106.8	−80.21
Chromium			
$Cr(s)$	0	23.8	0
$Cr^{3+}(aq)$	−232		
$CrCl_2(s)$	−326	115	−282
$CrCl_3(s)$	−563.2	126	−493.7
$Cr_2O_3(s)$	−1141	81.2	−1059
$CrO_3(s)$	−585.8	72.0	−506.2
$(NH_4)_2Cr_2O_7(s)$	−1807		
$K_2Cr_2O_7(s)$	−2033.01		
Cobalt			
$Co(s)$	0	30.0	0
$Co^{2+}(aq)$	−59.4	−110	−53.6
$CoCl_2(s)$	−325.5	106	−282.4
$Co(NO_3)_2(s)$	−422.2	192	−230.5
$CoO(s)$	−237.9	53.0	−214.2
$CoS(s)$	−80.8	67.4	−82.8
Copper			
$Cu(s)$	0	33.15	0
$Cu^{2+}(aq)$	+64.77	−99.6	+65.49
$CuCl(s)$	−137.2	86.2	−119.87
$CuCl_2(s)$	−172	119	−131
$Cu_2O(s)$	−168.6	93.1	−146.0
$CuO(s)$	−155	42.6	−127
$Cu_2S(s)$	−79.5	121	−86.2
$CuS(s)$	−53.1	66.5	−53.6
$CuSO_4(s)$	−771.4	109	−661.8
$CuSO_4 \cdot 5H_2O(s)$	−2279.7	300.4	−1879.7
Fluorine			
$F_2(g)$	0	202.7	0
$F^-(aq)$	−332.6	−13.8	−278.8
$HF(g)$	−271	173.5	−273
Gold			
$Au(s)$	0	47.7	0
$Au_2O_3(s)$	+80.8	125	+163
$AuCl_3(s)$	−118	148	−48.5
Hydrogen			
$H_2(g)$	0	130.6	0
$H_2O(l)$	−285.9	69.96	−237.2
$H_2O(g)$	−241.8	188.7	−228.6
$H_2O_2(l)$	−187.6	109.6	−120.3
$H_2Se(g)$	+76	219	+62.3
$H_2Te(g)$	+154	234	+138
Iodine			
$I_2(s)$	0	116.1	0
$I_2(g)$	+62.4	260.7	+19.3
$HI(g)$	+26.6	206	+1.30
Iron			
$Fe(s)$	0	27	0
$Fe^{2+}(aq)$	−89.1	−137.7	−78.9
$Fe^{3+}(aq)$	−48.5	−315.9	−4.7
$Fe_2O_3(s)$	−822.2	90.0	−741.0
$Fe_3O_4(s)$	−1118.4	146.4	−1015.4
$FeS(s)$	−100.0	60.3	−100.4
$FeS_2(s)$	−178.2	52.9	−166.9
Lead			
$Pb(s)$	0	64.8	0
$Pb^{2+}(aq)$	−1.7	10.5	−24.4
$PbCl_2(s)$	−359.4	136	−314.1
$PbO(s)$	−219.2	67.8	−189.3
$PbO_2(s)$	−277	68.6	−219
$Pb(OH)_2(s)$	−515.9	88	−420.9
$PbS(s)$	−100	91.2	−98.7
$PbSO_4(s)$	−920.1	149	−811.3
Lithium			
$Li(s)$	0	28.4	0
$Li^+(aq)$	−278.6	10.3	
$LiF(s)$	−611.7	35.7	−583.3
$LiCl(s)$	−408	59.29	−383.7
$LiBr(s)$	−350.3	66.9	−338.87
$Li_2O(s)$	−596.5	37.9	−560.5
$Li_3N(s)$	−199	37.7	−155.4
Magnesium			
$Mg(s)$	0	32.5	0
$Mg^{2+}(aq)$	−466.9	−138.1	−454.8
$MgCO_3(s)$	−1113	65.7	−1029
$MgF_2(s)$	−1124	79.9	−1056
$MgCl_2(s)$	−641.8	89.5	−592.5
$MgCl_2 \cdot 2H_2O(s)$	−1280	180	−1118
$Mg_3N_2(s)$	−463.2	87.9	−411
$MgO(s)$	−601.7	26.9	−569.4
$Mg(OH)_2(s)$	−924.7	63.1	−833.9
Manganese			
$Mn(s)$	0	32.0	0
$Mn^{2+}(aq)$	−223	−74.9	−228
$MnO_4^-(aq)$	−542.7	191	−449.4
$KMnO_4(s)$	−813.4	171.71	−713.8
$MnO(s)$	−385	60.2	−363
$Mn_2O_3(s)$	−959.8	110	−882.0
$MnO_2(s)$	−520.9	53.1	−466.1

(Continued)

TABLE C.1 Thermodynamic Data for Selected Elements, Compounds, and Ions (25 °C) *(Continued)*

Substance	ΔH_f°(kJ mol^{-1})	S°(J mol^{-1} K^{-1})	ΔG_f°(kJ mol^{-1})
$Mn_3O_4(s)$	−1387	149	−1280
$MnSO_4(s)$	−1064	112	−956
Mercury			
$Hg(l)$	0	76.1	0
$Hg(g)$	+61.32	175	+31.8
$Hg_2Cl_2(s)$	−265.2	192.5	−210.8
$HgCl_2(s)$	−224.3	146.0	−178.6
$HgO(s)$	−90.83	70.3	−58.54
$HgS(s,red)$	−58.2	82.4	−50.6
Nickel			
$Ni(s)$	0	30	0
$NiCl_2(s)$	−305	97.5	−259
$NiO(s)$	−244	38	−216
$NiO_2(s)$			−199
$NiSO_4(s)$	−891.2	77.8	−773.6
$NiCO_3(s)$	−664.0	91.6	−615.0
$Ni(CO)_4(g)$	−220	399	−567.4
Nitrogen			
$N_2(g)$	0	191.5	0
$NH_3(g)$	−46.19	192.5	−16.7
$NH_4^+(aq)$	−132.5	113	−79.37
$N_2H_4(g)$	+95.40	238.4	+159.3
$N_2H_4(l)$	+50.6	121.2	+149.4
$NH_4Cl(s)$	−315.4	94.6	−203.9
$NO(g)$	+90.37	210.6	+86.69
$NO_2(g)$	+33.8	240.5	+51.84
$N_2O(g)$	+81.57	220.0	+103.6
$N_2O_4(g)$	+9.67	304	+98.28
$N_2O_5(g)$	+11	356	+115
$HNO_3(l)$	−173.2	155.6	−79.91
$NO_3^-(aq)$	−205.0	146.4	−108.74
Oxygen			
$O_2(g)$	0	205.0	0
$O_3(g)$	+143	238.8	+163
$OH^-(aq)$	−230.0	−10.75	−157.24
Phosphorus			
$P(s,white)$	0	41.09	0
$P_4(g)$	+314.6	163.2	+278.3
$PCl_3(g)$	−287.0	311.8	−267.8
$PCl_5(g)$	−374.9	364.6	−305.0
$PH_3(g)$	+5.4	210.2	+12.9
$P_4O_6(s)$	−1640		
$POCl_3(g)$	−1109.7	646.5	−1019
$POCl_3(l)$	−1186	26.36	−1035
$P_4O_{10}(s)$	−2984	228.9	−2698
$H_3PO_4(s)$	−1279	110.5	−1119
Potassium			
$K(s)$	0	64.18	0
$K^+(aq)$	−252.4	102.5	−283.3
$KF(s)$	−567.3	66.6	−537.8
$KCl(s)$	−435.89	82.59	−408.3

Substance	ΔH_f°(kJ mol^{-1})	S°(J mol^{-1} K^{-1})	ΔG_f°(kJ mol^{-1})
$KBr(s)$	−393.8	95.9	−380.7
$KI(s)$	−327.9	106.3	−324.9
$KOH(s)$	−424.8	78.9	−379.1
$K_2O(s)$	−361	98.3	−322
$K_2SO_4(s)$	−1433.7	176	−1316.4
Silicon			
$Si(s)$	0	19	0
$SiH_4(g)$	+33	205	+52.3
$SiO_2(s,alpha)$	−910.0	41.8	−856
Silver			
$Ag(s)$	0	42.55	0
$Ag^+(aq)$	+105.58	72.68	+77.11
$AgCl(s)$	−127.0	96.2	−109.7
$AgBr(s)$	−100.4	107.1	−96.9
$AgNO_3(s)$	−124	141	−32
$Ag_2O(s)$	−31.1	121.3	−11.2
Sodium			
$Na(s)$	0	51.0	0
$Na^+(aq)$	−240.12	59.0	−261.91
$NaF(s)$	−571	51.5	−545
$NaCl(s)$	−411.0	72.38	−384.0
$NaBr(s)$	−360	83.7	−349
$NaI(s)$	−288	91.2	−286
$NaHCO_3(s)$	−947.7	102	−851.9
$Na_2CO_3(s)$	−1131	136	−1048
$Na_2O_2(s)$	−510.9	94.6	−447.7
$Na_2O(s)$	−510	72.8	−376
$NaOH(s)$	−426.8	64.18	−382
$Na_2SO_4(s)$	−1384.49	149.49	−1266.83
Sulfur			
$S(s, rhombic)$	0	31.9	0
$SO_2(g)$	−296.9	248.5	−300.4
$SO_3(g)$	−395.2	256.2	−370.4
$H_2S(g)$	−20.6	206	−33.6
$H_2SO_4(l)$	−811.32	157	−689.9
$H_2SO_4(aq)$	−909.3	20.1	−744.5
$SF_6(g)$	−1209	292	−1105
Tin			
$Sn(s,white)$	0	51.6	0
$Sn^{2+}(aq)$	−8.8	−17	−27.2
$SnCl_4(l)$	−511.3	258.6	−440.2
$SnO(s)$	−285.8	56.5	−256.9
$SnO_2(s)$	−580.7	52.3	−519.6
Zinc			
$Zn(s)$	0	41.6	0
$Zn^{2+}(aq)$	−153.9	−112.1	−147.06
$ZnCl_2(s)$	−415.1	111	−369.4
$ZnO(s)$	−348.3	43.6	−318.3
$ZnS(s)$	−205.6	57.7	−201.3
$ZnSO_4(s)$	−982.8	120	−874.5

TABLE C.2 Heats of Formation of Gaseous Atoms from Elements in Their Standard States

Element	ΔH_f°(kJ mol^{-1})*	Element	ΔH_f°(kJ mol^{-1})*
Group IA		*Group IVA*	
H	217.89	C	716.67
Li	161.5	Si	450
Na	107.8	*Group VA*	
K	89.62	N	472.68
Rb	82.0	P	332.2
Cs	78.2		
Group IIA		*Group VIA*	
Be	324.3	O	249.17
Mg	146.4	S	276.98
Ca	178.2	*Group VIIA*	
Sr	163.6	F	79.14
Ba	177.8	Cl	121.47
Group IIIA		Br	112.38
B	560	I	107.48
Al	329.7		

*All values in this table are positive because forming the gaseous atoms from the elements is endothermic: it involves bond breaking.

TABLE C.3 Average Bond Energies

Bond	Bond Energy (kJ mol^{-1})	Bond	Bond Energy (kJ mol^{-1})
C—C	348	C—Br	276
C═C	612	C—I	238
C≡C	960	H—H	436
C—H	412	H—F	565
C—N	305	H—Cl	431
C═N	613	H—Br	366
C≡N	890	H—I	299
C—O	360	H—N	388
C═O	743	H—O	463
C—F	484	H—S	338
C—Cl	338	H—Si	376

TABLE C.4 Vapor Pressure of Water as a Function of Temperature

Temp (°C)	Vapor Pressure (torr)	Temp (°C)	Vapor Pressure (torr)	Temp (°C)	Vapor Pressure (torr)	Temp (°C)	Vapor Pressure (torr)
0	4.58	11	9.84	22	19.8	33	37.7
1	4.93	12	10.5	23	21.1	34	39.9
2	5.29	13	11.2	24	22.4	35	41.2
3	5.68	14	12.0	25	23.8	36	44.6
4	6.10	15	12.8	26	25.2	37	47.1
5	6.54	16	13.6	27	26.7	38	49.7
6	7.01	17	14.5	28	28.3	39	52.4
7	7.51	18	15.5	29	30.0	40	55.3
8	8.04	19	16.5	30	31.8	41	58.3
9	8.61	20	17.5	31	33.7	42	61.5
10	9.21	21	18.7	32	35.7	43	64.8

(Continued)

TABLE C.4 Vapor Pressure of Water as a Function of Temperature *(Continued)*

Temp (°C)	Vapor Pressure (torr)	Temp (°C)	Vapor Pressure (torr)	Temp (°C)	Vapor Pressure (torr)	Temp (°C)	Vapor Pressure (torr)
44	68.3	59	142.6	74	277.2	89	506.1
45	71.9	60	149.4	75	289.1	90	525.8
46	75.6	61	156.4	76	301.4	91	546.0
47	79.6	62	163.8	77	314.1	92	567.0
48	83.7	63	171.4	78	327.3	93	588.6
49	88.0	64	179.3	79	341.0	94	610.9
50	92.5	65	187.5	80	355.1	95	633.9
51	97.2	66	196.1	81	369.7	96	657.6
52	102.1	67	205.0	82	384.9	97	682.1
53	107.2	68	214.2	83	400.6	98	707.3
54	112.5	69	223.7	84	416.8	99	733.2
55	118.0	70	233.7	85	433.6	100	760.0
56	123.8	71	243.9	86	450.9		
57	129.8	72	254.6	87	468.7		
58	136.1	73	265.7	88	487.1		

TABLE C.5 Solubility Product Constants

Salt	K_{sp}	Salt	K_{sp}
Fluorides		***Hydroxides***	
MgF_2	6.6×10^{-9}	$Mg(OH)_2$	7.1×10^{-12}
CaF_2	3.9×10^{-11}	$Ca(OH)_2$	6.5×10^{-6}
SrF_2	2.9×10^{-9}	$Mn(OH)_2$	1.6×10^{-13}
BaF_2	1.7×10^{-6}	$Fe(OH)_2$	7.9×10^{-16}
LiF	1.7×10^{-3}	$Fe(OH)_3$	1.6×10^{-39}
PbF_2	3.6×10^{-8}	$Co(OH)_2$	1×10^{-15}
		$Co(OH)_3$	3×10^{-45}
Chlorides		$Ni(OH)_2$	6×10^{-16}
CuCl	1.9×10^{-7}	$Cu(OH)_2$	4.8×10^{-20}
AgCl	1.8×10^{-10}	$V(OH)_3$	4×10^{-35}
Hg_2Cl_2	1.2×10^{-18}	$Cr(OH)_3$	2×10^{-30}
TlCl	1.8×10^{-4}	Ag_2O	1.9×10^{-8}
$PbCl_2$	1.7×10^{-5}	$Zn(OH)_2$	3.0×10^{-16}
$AuCl_3$	3.2×10^{-25}	$Cd(OH)_2$	5.0×10^{-15}
		$Al(OH)_3$ (alpha form)	3×10^{-34}
Bromides			
CuBr	5×10^{-9}	***Cyanides***	
AgBr	5.0×10^{-13}	AgCN	1.2×10^{-16}
Hg_2Br_2	5.6×10^{-23}	$Zn(CN)_2$	3×10^{-16}
$HgBr_2$	1.3×10^{-19}		
$PbBr_2$	2.1×10^{-6}	***Sulfites***	
		$CaSO_3$	3×10^{-7}
Iodides		Ag_2SO_3	1.5×10^{-14}
CuI	1×10^{-12}	$BaSO_3$	8×10^{-7}
AgI	8.3×10^{-17}		
Hg_2I_2	4.7×10^{-29}	***Sulfates***	
HgI_2	1.1×10^{-28}	$CaSO_4$	2.4×10^{-5}
PbI_2	7.9×10^{-9}	$SrSO_4$	3.2×10^{-7}

(Continued)

TABLE C.5 Solubility Product Constants *(Continued)*

Salt	K_{sp}	Salt	K_{sp}
$BaSO_4$	1.1×10^{-10}	Ag_2CO_3	8.1×10^{-12}
$RaSO_4$	4.3×10^{-11}	Hg_2CO_3	8.9×10^{-17}
Ag_2SO_4	1.5×10^{-5}	$ZnCO_3$	1.0×10^{-10}
Hg_2SO_4	7.4×10^{-7}	$CdCO_3$	1.8×10^{-14}
$PbSO_4$	6.3×10^{-7}	$PbCO_3$	7.4×10^{-14}
Chromates		***Phosphates***	
$BaCrO_4$	2.1×10^{-10}	$Ca_3(PO_4)_2$	2.0×10^{-29}
$CuCrO_4$	3.6×10^{-6}	$Mg_3(PO_4)_2$	6.3×10^{-26}
Ag_2CrO_4	1.2×10^{-12}	$SrHPO_4$	1.2×10^{-7}
Hg_2CrO_4	2.0×10^{-9}	$BaHPO_4$	4.0×10^{-8}
$CaCrO_4$	7.1×10^{-4}	$LaPO_4$	3.7×10^{-23}
$PbCrO_4$	1.8×10^{-14}	$Fe_3(PO_4)_2$	1×10^{-36}
		Ag_3PO_4	2.8×10^{-18}
Carbonates		$FePO_4$	4.0×10^{-27}
$MgCO_3$	3.5×10^{-8}	$Zn_3(PO_4)_2$	5×10^{-36}
$CaCO_3$	4.5×10^{-9}	$Pb_3(PO_4)_2$	3.0×10^{-44}
$SrCO_3$	9.3×10^{-10}	$Ba_3(PO_4)_2$	5.8×10^{-38}
$BaCO_3$	5.0×10^{-9}		
$MnCO_3$	5.0×10^{-10}	***Ferrocyanides***	
$FeCO_3$	2.1×10^{-11}	$Zn_2[Fe(CN)_6]$	2.1×10^{-16}
$CoCO_3$	1.0×10^{-10}	$Cd_2[Fe(CN)_6]$	4.2×10^{-18}
$NiCO_3$	1.3×10^{-7}	$Pb_2[Fe(CN)_6]$	9.5×10^{-19}
$CuCO_3$	2.5×10^{-10}		

TABLE C.6 Formation Constants of Complexes (25 °C)

Complex Ion Equilibrium	K_{form}	Complex Ion Equilibrium	K_{form}
Halide Complexes		$Hg^{2+} + 4NH_3 \rightleftharpoons [Hg(NH_3)_4]^{2+}$	1.8×10^{19}
$Al^{3+} + 6F^- \rightleftharpoons [AlF_6]^{3-}$	1×10^{20}	$Co^{2+} + 6NH_3 \rightleftharpoons [Co(NH_3)_6]^{2+}$	5.0×10^{4}
$Al^{3+} + 4F^- \rightleftharpoons [AlF_4]^-$	2.0×10^{8}	$Co^{3+} + 6NH_3 \rightleftharpoons [Co(NH_3)_6]^{3+}$	4.6×10^{33}
$Be^{2+} + 4F^- \rightleftharpoons [BeF_4]^{2-}$	1.3×10^{13}	$Cd^{2+} + 6NH_3 \rightleftharpoons [Cd(NH_3)_6]^{2+}$	2.6×10^{5}
$Sn^{4+} + 6F^- \rightleftharpoons [SnF_6]^{2-}$	1×10^{25}	$Ni^{2+} + 6NH_3 \rightleftharpoons [Ni(NH_3)_6]^{2+}$	2.0×10^{8}
$Cu^+ + 2Cl^- \rightleftharpoons [CuCl_2]^-$	3×10^{5}	***Cyanide Complexes***	
$Ag^+ + 2Cl^- \rightleftharpoons [AgCl_2]^-$	1.8×10^{5}	$Fe^{2+} + 6CN^- \rightleftharpoons [Fe(CN)_6]^{4-}$	1.0×10^{24}
$Pb^{2+} + 4Cl^- \rightleftharpoons [PbCl_4]^{2-}$	2.5×10^{15}	$Fe^{3+} + 6CN^- \rightleftharpoons [Fe(CN)_6]^{3-}$	1.0×10^{31}
$Zn^{2+} + 4Cl^- \rightleftharpoons [ZnCl_4]^{2-}$	1.6	$Ag^+ + 2CN^- \rightleftharpoons [Ag(CN)_2]^-$	5.3×10^{18}
$Hg^{2+} + 4Cl^- \rightleftharpoons [HgCl_4]^{2-}$	5.0×10^{15}	$Cu^+ + 2CN^- \rightleftharpoons [Cu(CN)_2]^-$	1.0×10^{16}
$Cu^+ + 2Br^- \rightleftharpoons [CuBr_2]^-$	8×10^{5}	$Cd^{2+} + 4CN^- \rightleftharpoons [Cd(CN)_4]^{2-}$	7.7×10^{16}
$Ag^+ + 2Br^- \rightleftharpoons [AgBr_2]^-$	1.7×10^{7}	$Au^+ + 2CN^- \rightleftharpoons [Au(CN)_2]^-$	2×10^{38}
$Hg^{2+} + 4Br^- \rightleftharpoons [HgBr_4]^{2-}$	1×10^{21}		
$Cu^+ + 2I^- \rightleftharpoons [CuI_2]^-$	8×10^{8}	***Complexes with Other Monodentate Ligands***	
$Ag^+ + 2I^- \rightleftharpoons [AgI_2]^-$	1×10^{11}	***Methylamine*** **(CH_3NH_2)**	
$Pb^{2+} + 4I^- \rightleftharpoons [PbI_4]^{2-}$	3×10^{4}	$Ag^+ + 2CH_3NH_2 \rightleftharpoons [Ag(CH_3NH_2)_2]^+$	7.8×10^{6}
$Hg^{2+} + 4I^- \rightleftharpoons [HgI_4]^{2-}$	1.9×10^{30}	***Thiocyanate ion*** **(SCN^-)**	
Ammonia Complexes		$Cd^{2+} + 4SCN^- \rightleftharpoons [Cd(SCN)_4]^{2-}$	1×10^{3}
$Ag^+ + 2NH_3 \rightleftharpoons [Ag(NH_3)_2]^+$	1.6×10^{7}	$Cu^{2+} + 2SCN^- \rightleftharpoons [Cu(SCN)_2]$	5.6×10^{3}
$Zn^{2+} + 4NH_3 \rightleftharpoons [Zn(NH_3)_4]^{2+}$	7.8×10^{8}	$Fe^{3+} + 3SCN^- \rightleftharpoons [Fe(SCN)_3]$	2×10^{6}
$Cu^{2+} + 4NH_3 \rightleftharpoons [Cu(NH_3)_4]^{2+}$	1.1×10^{13}	$Hg^{2+} + 4SCN^- \rightleftharpoons [Hg(SCN)_4]^{2-}$	5.0×10^{21}

(Continued)

TABLE C.6 Formation Constants of Complexes (25 °C) *(Continued)*

Complex Ion Equilibrium	K_{form}	Complex Ion Equilibrium	K_{form}
Hydroxide ion **(OH^-)**		$Ni^{2+} + 3\text{ bipy} \rightleftharpoons [Ni(bipy)_3]^{2+}$	3.0×10^{20}
$Cu^{2+} + 4OH^- \rightleftharpoons [Cu(OH)_4]^{2-}$	1.3×10^{16}	$Co^{2+} + 3\text{ bipy} \rightleftharpoons [Co(bipy)_3]^{2+}$	8×10^{15}
$Zn^{2+} + 4OH^- \rightleftharpoons [Zn(OH)_4]^{2-}$	2×10^{20}	$Mn^{2+} + 3\text{ phen} \rightleftharpoons [Mn(phen)_3]^{2+}$	2×10^{10}
Complexes with Bidentate Ligands*		$Fe^{2+} + 3\text{ phen} \rightleftharpoons [Fe(phen)_3]^{2+}$	1×10^{21}
$Mn^{2+} + 3\text{ en} \rightleftharpoons [Mn(en)_3]^{2+}$	6.5×10^{5}	$Co^{2+} + 3\text{ phen} \rightleftharpoons [Co(phen)_3]^{2+}$	6×10^{19}
$Fe^{2+} + 3\text{ en} \rightleftharpoons [Fe(en)_3]^{2+}$	5.2×10^{9}	$Ni^{2+} + 3\text{ phen} \rightleftharpoons [Ni(phen)_3]^{2+}$	2×10^{24}
$Co^{2+} + 3\text{ en} \rightleftharpoons [Co(en)_3]^{2+}$	1.3×10^{14}	$Co^{2+} + 3C_2O_4{}^{2-} \rightleftharpoons [Co(C_2O_4)_3]^{4-}$	4.5×10^{6}
$Co^{3+} + 3\text{ en} \rightleftharpoons [Co(en)_3]^{3+}$	4.8×10^{48}	$Fe^{3+} + 3C_2O_4{}^{2-} \rightleftharpoons [Fe(C_2O_4)_3]^{3-}$	3.3×10^{20}
$Ni^{2+} + 3\text{ en} \rightleftharpoons [Ni(en)_3]^{2+}$	4.1×10^{17}	***Complexes of Other Polydentate Ligands****	
$Cu^{2+} + 2\text{ en} \rightleftharpoons [Cu(en)_2]^{2+}$	3.5×10^{19}	$Zn^{2+} + EDTA^{4-} \rightleftharpoons [Zn(EDTA)]^{2-}$	3.8×10^{16}
$Mn^{2+} + 3\text{ bipy} \rightleftharpoons [Mn(bipy)_3]^{2+}$	1×10^{6}	$Mg^{2+} + 2NTA^{3-} \rightleftharpoons [Mg(NTA)_2]^{4-}$	1.6×10^{10}
$Fe^{2+} + 3\text{ bipy} \rightleftharpoons [Fe(bipy)_3]^{2+}$	1.6×10^{17}	$Ca^{2+} + 2NTA^{3-} \rightleftharpoons [Ca(NTA)_2]^{4-}$	3.2×10^{11}

*en = ethylenediamine

bipy = bipyridyl

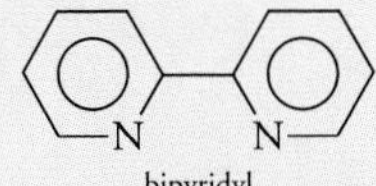

bipyridyl

phen = 1,10-phenanthroline

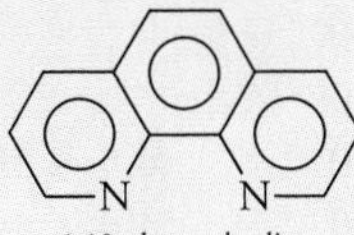

1,10-phenanthroline

$EDTA^{4-}$ = ethylenediaminetetraacetate ion

NTA^{3-} = nitrilotriacetate ion

TABLE C.7 Ionization Constants of Weak Acids and Bases (Alternative Formulas in Parentheses)

Monoprotic Acid	Name	K_a
$HC_2O_2Cl_3$ (Cl_3CCO_2H)	trichloroacetic acid	2.2×10^{-1}
HIO_3	iodic acid	1.69×10^{-1}
$HC_2HO_2Cl_2$ (Cl_2CHCO_2H)	dichloroacetic acid	5.0×10^{-2}
$HC_2H_2O_2Cl$ (ClH_2CCO_2H)	chloroacetic acid	1.36×10^{-3}
HNO_2	nitrous acid	7.1×10^{-4}
HF	hydrofluoric acid	6.8×10^{-4}
HOCN	cyanic acid	3.5×10^{-4}
$HCHO_2$ (HCO_2H)	formic acid	1.8×10^{-4}
$HC_3H_5O_3$ [$CH_3CH(OH)CO_2H$]	lactic acid	1.38×10^{-4}
$HC_4H_3N_2O_3$	barbituric acid	9.8×10^{-5}
$HC_7H_5O_2$ ($C_6H_5CO_2H$)	benzoic acid	6.28×10^{-5}
$HC_4H_7O_2$ ($CH_3CH_2CH_2CO_2H$)	butanoic acid	1.52×10^{-5}
HN_3	hydrazoic acid	1.8×10^{-5}
$HC_2H_3O_2$ (CH_3CO_2H)	acetic acid	1.8×10^{-5}
$HC_3H_5O_2$ ($CH_3CH_2CO_2H$)	propanoic acid	1.34×10^{-5}
$HC_2H_4NO_2$	nicotinic acid (niacin)	1.4×10^{-5}
HOCl	hypochlorous acid	3.0×10^{-8}
HOBr	hypobromous acid	2.1×10^{-9}

(Continued)

TABLE C.7 Ionization Constants of Weak Acids and Bases (Alternative Formulas in Parentheses) *(Continued)*

Monoprotic Acid	Name	K_a		
HCN	hydrocyanic acid	6.2×10^{-10}		
HC_6H_5O	phenol	1.3×10^{-10}		
HOI	hypoiodous acid	2.3×10^{-11}		
H_2O_2	hydrogen peroxide	1.8×10^{-12}		
Polyprotic Acid	**Name**	K_{a_1}	K_{a_2}	K_{a_3}
H_2SO_4	sulfuric acid	large	1.0×10^{-2}	
H_2CrO_4	chromic acid	5.0	1.5×10^{-6}	
$H_2C_2O_4$	oxalic acid	5.6×10^{-2}	5.4×10^{-5}	
H_3PO_3	phosphorous acid	3×10^{-2}	1.6×10^{-7}	
$H_2S(aq)$	hydrosulfuric acid	9.5×10^{-8}	1×10^{-19}	
H_2SO_3	sulfurous acid	1.2×10^{-2}	6.6×10^{-8}	
H_2SeO_4	selenic acid	large	1.2×10^{-2}	
H_2SeO_3	selenous acid	4.5×10^{-3}	1.1×10^{-8}	
H_6TeO_6	telluric acid	2×10^{-8}	1×10^{-11}	
H_2TeO_3	tellurous acid	3.3×10^{-3}	2.0×10^{-8}	
$H_2C_3H_2O_4$ ($HO_2CCH_2CO_2H$)	malonic acid	1.4×10^{-3}	2.0×10^{-6}	
$H_2C_8H_4O_4$	phthalic acid	1.1×10^{-3}	3.9×10^{-6}	
$H_2C_4H_4O_6$	tartaric acid	9.2×10^{-4}	4.3×10^{-5}	
$H_2C_6H_6O_6$	ascorbic acid	6.8×10^{-5}	2.7×10^{-12}	
H_2CO_3	carbonic acid	4.3×10^{-7}	4.7×10^{-11}	
H_3PO_4	phosphoric acid	7.1×10^{-3}	6.3×10^{-8}	4.5×10^{-13}
H_3AsO_4	arsenic acid	5.6×10^{-3}	1.7×10^{-7}	4.0×10^{-12}
$H_3C_6H_5O_7$	citric acid	7.1×10^{-4}	1.7×10^{-5}	6.3×10^{-6}
Weak Base	**Name**	K_b		
$(CH_3)_2NH$	dimethylamine	9.6×10^{-4}		
$C_4H_9NH_2$	butylamine	5.9×10^{-4}		
CH_3NH_2	methylamine	4.4×10^{-4}		
$CH_3CH_2NH_2$	ethylamine	4.3×10^{-4}		
$(CH_3)_3N$	trimethylamine	7.4×10^{-5}		
NH_3	ammonia	1.8×10^{-5}		
$C_{21}H_{22}N_2O_2$	strychnine	1.0×10^{-6}		
N_2H_4	hydrazine	9.6×10^{-7}		
$C_{17}H_{19}NO_3$	morphine	7.5×10^{-7}		
NH_2OH	hydroxylamine	6.6×10^{-9}		
C_5H_5N	pyridine	1.5×10^{-9}		
$C_6H_5NH_2$	aniline	4.1×10^{-10}		
PH_3	phosphine	10^{-28}		

TABLE C.8 Standard Reduction Potentials (25 °C)

$E°$ (Volts)	Half-Cell Reaction
+2.87	$F_2(g) + 2e^- \rightleftharpoons 2F^-(aq)$
+2.08	$O_3(g) + 2H^+(aq) + 2e^- \rightleftharpoons O_2(g) + H_2O$
+2.01	$S_2O_8^{2-}(aq) + 2e^- \rightleftharpoons 2SO_4^{2-}(aq)$
+1.82	$Co^{3+}(aq) + e^- \rightleftharpoons Co^{2+}(aq)$
+1.77	$H_2O_2(aq) + 2H^+(aq) + 2e^- \rightleftharpoons 2H_2O$
+1.695	$MnO_4^-(aq) + 4H^+(aq) + 3e^- \rightleftharpoons MnO_2(s) + 2H_2O$
+1.69	$PbO_2(s) + HSO_4^-(aq) + 3H^+(aq) + 2e^- \rightleftharpoons PbSO_4(s) + 2H_2O$
+1.63	$2HOCl(aq) + 2H^+(aq) + 2e^- \rightleftharpoons Cl_2(g) + 2H_2O$
+1.51	$Mn^{3+}(aq) + e^- \rightleftharpoons Mn^{2+}(aq)$
+1.51	$MnO_4^-(aq) + 8H^+(aq) + 5e^- \rightleftharpoons Mn^{2+}(aq) + 4H_2O$
+1.46	$PbO_2(s) + 4H^+(aq) + 2e^- \rightleftharpoons Pb^{2+}(aq) + 2H_2O$
+1.44	$BrO_3^-(aq) + 6H^+(aq) + 6e^- \rightleftharpoons Br^-(aq) + 3H_2O$
+1.42	$Au^{3+}(aq) + 3e^- \rightleftharpoons Au(s)$
+1.36	$Cl_2(g) + 2e^- \rightleftharpoons 2Cl^-(aq)$
+1.33	$Cr_2O_7^{2-}(aq) + 14H^+(aq) + 6e^- \rightleftharpoons 2Cr^{3+}(aq) + 7H_2O$
+1.24	$O_3(g) + H_2O + 2e^- \rightleftharpoons O_2(g) + 2OH^-(aq)$
+1.23	$MnO_2(s) + 4H^+(aq) + 2e^- \rightleftharpoons Mn^{2+}(aq) + 2H_2O$
+1.23	$O_2(g) + 4H^+(aq) + 4e^- \rightleftharpoons 2H_2O$
+1.20	$Pt^{2+}(aq) + 2e^- \rightleftharpoons Pt(s)$
+1.07	$Br_2(aq) + 2e^- \rightleftharpoons 2Br^-(aq)$
+0.96	$NO_3^-(aq) + 4H^+(aq) + 3e^- \rightleftharpoons NO(g) + 2H_2O$
+0.94	$NO_3^-(aq) + 3H^+(aq) + 2e^- \rightleftharpoons HNO_2(aq) + H_2O$
+0.91	$2Hg^{2+}(aq) + 2e^- \rightleftharpoons Hg_2^{2+}(aq)$
+0.87	$HO_2^-(aq) + H_2O + 2e^- \rightleftharpoons 3OH^-(aq)$
+0.80	$NO_3^-(aq) + 4H^+(aq) + 2e^- \rightleftharpoons 2NO_2(g) + 2H_2O$
+0.80	$Ag^+(aq) + e^- \rightleftharpoons Ag(s)$
+0.77	$Fe^{3+}(aq) + e^- \rightleftharpoons Fe^{2+}(aq)$
+0.69	$O_2(g) + 2H^+(aq) + 2e^- \rightleftharpoons H_2O_2(aq)$
+0.54	$I_2(s) + 2e^- \rightleftharpoons 2I^-(aq)$
+0.49	$NiO_2(s) + 2H_2O + 2e^- \rightleftharpoons Ni(OH)_2(s) + 2OH^-(aq)$
+0.45	$SO_2(aq) + 4H^+(aq) + 4e^- \rightleftharpoons S(s) + 2H_2O$
+0.401	$O_2(g) + 2H_2O + 4e^- \rightleftharpoons 4OH^-(aq)$
+0.34	$Cu^{2+}(aq) + 2e^- \rightleftharpoons Cu(s)$
+0.27	$Hg_2Cl_2(s) + 2e^- \rightleftharpoons 2Hg(l) + 2Cl^-(aq)$
+0.25	$PbO_2(s) + H_2O + 2e^- \rightleftharpoons PbO(s) + 2OH^-(aq)$
+0.2223	$AgCl(s) + e^- \rightleftharpoons Ag(s) + Cl^-(aq)$
+0.172	$SO_4^{2-}(aq) + 4H^+(aq) + 2e^- \rightleftharpoons H_2SO_3(aq) + H_2O$
+0.169	$S_4O_6^{2-}(aq) + 2e^- \rightleftharpoons 2S_2O_3^{2-}(aq)$
+0.16	$Cu^{2+}(aq) + e^- \rightleftharpoons Cu^+(aq)$
+0.15	$Sn^{4+}(aq) + 2e^- \rightleftharpoons Sn^{2+}(aq)$
+0.14	$S(s) + 2H^+(aq) + 2e^- \rightleftharpoons H_2S(g)$
+0.07	$AgBr(s) + e^- \rightleftharpoons Ag(s) + Br^-(aq)$
0 (exactly)	$2H^+(aq) + 2e^- \rightleftharpoons H_2(g)$
−0.13	$Pb^{2+}(aq) + 2e^- \rightleftharpoons Pb(s)$

(Continued)

TABLE C.8 Standard Reduction Potentials (25 °C) *(Continued)*

$E°$ (Volts)	Half-Cell Reaction
−0.14	$Sn^{2+}(aq) + 2e^- \rightleftharpoons Sn(s)$
−0.15	$AgI(s) + e^- \rightleftharpoons Ag(s) + I^-(aq)$
−0.25	$Ni^{2+}(aq) + 2e^- \rightleftharpoons Ni(s)$
−0.28	$Co^{2+}(aq) + 2e^- \rightleftharpoons Co(s)$
−0.34	$In^{3+}(aq) + 3e^- \rightleftharpoons In(s)$
−0.34	$Tl^+(aq) + e^- \rightleftharpoons Tl(s)$
−0.36	$PbSO_4(s) + H^+(aq) + 2e^- \rightleftharpoons Pb(s) + HSO_4^-(aq)$
−0.40	$Cd^{2+}(aq) + 2e^- \rightleftharpoons Cd(s)$
−0.44	$Fe^{2+}(aq) + 2e^- \rightleftharpoons Fe(s)$
−0.56	$Ga^{3+}(aq) + 3e^- \rightleftharpoons Ga(s)$
−0.58	$PbO(s) + H_2O + 2e^- \rightleftharpoons Pb(s) + 2OH^-(aq)$
−0.74	$Cr^{3+}(aq) + 3e^- \rightleftharpoons Cr(s)$
−0.76	$Zn^{2+}(aq) + 2e^- \rightleftharpoons Zn(s)$
−0.81	$Cd(OH)_2(s) + 2e^- \rightleftharpoons Cd(s) + 2OH^-(aq)$
−0.83	$2H_2O + 2e^- \rightleftharpoons H_2(g) + 2OH^-(aq)$
−0.88	$Fe(OH)_2(s) + 2e^- \rightleftharpoons Fe(s) + 2OH^-(aq)$
−0.91	$Cr^{2+}(aq) + e^- \rightleftharpoons Cr(s)$
−1.16	$N_2(g) + 4H_2O + 4e^- \rightleftharpoons N_2O_4(aq) + 4OH^-(aq)$
−1.18	$V^{2+}(aq) + 2e^- \rightleftharpoons V(s)$
−1.216	$ZnO_2^-(aq) + 2H_2O + 2e^- \rightleftharpoons Zn(s) + 4OH^-(aq)$
−1.63	$Ti^{2+}(aq) + 2e^- \rightleftharpoons Ti(s)$
−1.66	$Al^{3+}(aq) + 3e^- \rightleftharpoons Al(s)$
−1.79	$U^{3+}(aq) + 3e^- \rightleftharpoons U(s)$
−2.02	$Sc^{3+}(aq) + 3e^- \rightleftharpoons Sc(s)$
−2.36	$La^{3+}(aq) + 3e^- \rightleftharpoons La(s)$
−2.37	$Y^{3+}(aq) + 3e^- \rightleftharpoons Y(s)$
−2.37	$Mg^{2+}(aq) + 2e^- \rightleftharpoons Mg(s)$
−2.71	$Na^+(aq) + e^- \rightleftharpoons Na(s)$
−2.76	$Ca^{2+}(aq) + 2e^- \rightleftharpoons Ca(s)$
−2.89	$Sr^{2+}(aq) + 2e^- \rightleftharpoons Sr(s)$
−2.90	$Ba^{2+}(aq) + 2e^- \rightleftharpoons Ba(s)$
−2.92	$Cs^+(aq) + e^- \rightleftharpoons Cs(s)$
−2.92	$K^+(aq) + e^- \rightleftharpoons K(s)$
−2.93	$Rb^+(aq) + e^- \rightleftharpoons Rb(s)$
−3.05	$Li^+(aq) + e^- \rightleftharpoons Li(s)$

GLOSSARY

This glossary has the definitions of the key terms that were marked in boldface throughout the chapters plus a few additional terms. The numbers in parentheses that follow the definitions are the numbers of the sections in which the glossary entries received their principal discussions.

A

Absolute Zero: 0 K, −273.15 °C. Nature's lowest temperature. (1.5, 10.3)

Acceptor: A Lewis acid; the central metal ion in a complex ion. (17.4, 21.5)

Accuracy: Freedom from error. The closeness of a measurement to the true value. (1.6)

Acid: *Arrhenius theory:* A substance that produces hydronium ions (hydrogen ions) in water. (4.3)

Brønsted theory: A proton donor. (15.1)

Lewis theory: An electron-pair acceptor. (15.3)

Acid–Base Indicator: A dye with one color in acid and another color in base. (4.3, 4.8)

Acid–Base Neutralization: The reaction of an acid with a base. (4.3, 15.3)

Acid Ionization Constant (K_a):

$$K_a = \frac{[H^+][A^-]}{[HA]} \text{ for the equilibrium,}$$

$$HA \rightleftharpoons H^+ + A^-, \qquad (16.1)$$

Acid Rain: Rain made acidic by dissolved sulfur and nitrogen oxides.

Acid Salt: A salt of a partially neutralized polyprotic acid, for example, $NaHSO_4$ or $NaHCO_3$. (4.4)

Acid Solubility Product: The special solubility product expression for metal sulfides in dilute acid and related to the equation for their dissolving. For a divalent metal sulfide, MS,

$$MS(s) + 2H^+(aq) \longrightarrow M^{2+}(aq) + H_2S(aq)$$

$$K_{spa} = \frac{[M^{2+}][H_2S]}{[H^+]^2} \qquad (17.2)$$

Acidic Anhydride: An oxide that reacts with water to make the solution acidic. (4.3)

Acidic Solution: An aqueous solution in which $[H^+] > [OH^-]$. (15.5)

Actinide Elements (Actinide Series): Elements 90–103. (2.3)

Activated Complex: The chemical species that exists with partly broken and partly formed bonds in the transition state. (13.5)

Activation Energy (E_a): The minimum kinetic energy that must be possessed by the reactants in order to give an effective collision (one that produces products). (13.5)

Activities: Effective concentrations which properly should be substituted into a mass action expression to satisfy the equilibrium law. The activity of a solid is defined as having a value of 1. (14.4)

Activity: For a radioactive material, the number of disintegrations per second. (20.6)

Activity Series: A list of metals in order of their reactivity as reducing agents. (5.4)

Actual Yield: See *Yield, Actual.*

Addition Compound: A molecule formed by the joining of two simpler molecules through formation of a covalent bond (usually a coordinate covalent bond). (8.6)

Addition Polymer: A polymer formed by the simple addition of one monomer unit to another, a process that continues over and over until a very long chain of monomer units is produced. (22.6)

Addition Reaction: The addition of a molecule to a double or triple bond. (22.2)

Adiabatic Change: A change within a system during which no energy enters or leaves the system. (6.3)

Alcohol: An organic compound whose molecules have the OH group attached to tetrahedral carbon. (2.6, 22.3)

Aldehyde: An organic compound whose molecules have the group $-CH{=}O$. (22.5)

Alkali Metals: The Group IA elements (except hydrogen)—lithium, sodium, potassium, rubidium, cesium, and francium. (2.3)

Alkaline Battery (Alkaline Dry Cell): A zinc–manganese dioxide galvanic cell of 1.54 V used commonly in flashlight batteries. (19.8)

Alkaline Earth Metals: The Group IIA elements—beryllium, magnesium, calcium, strontium, barium, and radium. (2.3)

Alkalis: (a) The alkali metals. (2.3) (b) Hydroxides of the alkali metals; strong bases. (4.3)

Alkane: A hydrocarbon whose molecules have only single bonds. (2.6, 22.2)

Alkene: A hydrocarbon whose molecules have one or more double bonds. (22.2)

Alkyl Group: An organic group of carbon and hydrogen atoms related to an alkane but with one less hydrogen atom (e.g., CH_3-, methyl; CH_3CH_2-, ethyl). (22.2)

Alkyne: A hydrocarbon whose molecules have one or more triple bonds. (22.2)

Allotrope: One of two or more forms of an element. (21.2)

Allotropy: The existence of an element in two or more molecular or crystalline forms called allotropes. (21.2)

Alpha Particle (^4_2He): The nucleus of a helium atom. (20.3)

Alpha Radiation: A high-velocity stream of alpha particles produced by radioactive decay. (20.3)

Alum: A double salt with the general formula $M^+M^{3+}(SO_4)_2 \cdot 12H_2O$, such as potassium alum: $KAl(SO_4)_2 \cdot 12H_2O$.

Amalgam: A solution of a metal in mercury.

Amide: An organic compound whose molecules have any one of the following groups: (22.5)

$$\overset{O}{\overset{\|}{-CNH_2}} \qquad \overset{O}{\overset{\|}{-CNHR}} \qquad \overset{O}{\overset{\|}{-CNR_2}}$$

α-Amino Acid: One of about 20 monomers of polypeptides. (22.7)

Amine: An organic compound whose molecules contain the group NH_2, NHR, or NR_2. (22.4)

Amorphous Solid: A noncrystalline solid. A glass. (11.9)

Ampere (A): The SI unit for electric current; one coulomb per second. (19.7)

Amphiprotic Compound: A compound that can act either as a proton donor or as a proton acceptor; an amphoteric compound. (15.1)

Amphoteric Compound: A compound that can react as either an acid or a base. (15.1)

Amplitude: The height of a wave, which is a measure of the wave's intensity. (7.1)

amu: See *Atomic Mass Unit.*

Angstrom (Å): 1 Å $= 10^{-10}$ m $= 100$ pm $= 0.1$ nm. (7.8)

Anhydrous: Without water. (2.5)

Anion: A negatively charged ion. (2.8)

Anode: The positive electrode in a gas discharge tube. The electrode at which oxidation occurs during an electrochemical change. (19.1)

Antibonding Electrons: Electrons that occupy antibonding molecular orbitals. (9.7)

Antibonding Molecular Orbital: A molecular orbital that denies electron density to the space between nuclei and destabilizes a molecule when occupied by electrons. (9.7)

Anticodon: A triplet of bases on a tRNA molecule that pairs to a matching triplet—a codon—on an mRNA molecule during mRNA-directed polypeptide synthesis. (22.8)

Antimatter: Any particle annihilated by a particle of ordinary matter. (20.3)

Aqua Regia: One part concentrated nitric acid and three parts concentrated hydrochloric acid (by volume).

Aqueous Solution: A solution that has water as the solvent.

Aromatic Compound: An organic compound whose molecules have the benzene ring system. (22.2)

Arrhenius Acid: See *Acid.*

Arrhenius Base: See *Base.*

Arrhenius Equation: An equation that relates the rate constant of a reaction to the reaction's activation energy. (13.6)

Association: The joining together of molecules by hydrogen bonds. (12.9)

Asymmetric Carbon Atom: A carbon atom that is bonded to four different groups and which is a chiral center. (22.1)

Atmosphere, Standard (atm): 101,325 Pa. The pressure that supports a column of mercury 760 mm high at 0 °C; 760 torr. (6.5, 10.2)

Atmospheric Pressure: The pressure exerted by the mixture of gases in our atmosphere. (6.5, 10.2)

Atom: A neutral particle having one nucleus; the smallest representative sample of an element. (1.2)

Atomic Mass: The average mass (in u) of the atoms of the isotopes of a given element as they occur naturally. (2.2)

Atomic Mass Unit (u): $1.6605402 \times 10^{-24}$ g; 1/12th the mass of one atom of carbon-12. Sometimes given the symbol amu. (2.2)

Atomic Number: The number of protons in a nucleus. (2.2)

Atomic Radiation: Radiation consisting of particles or electromagnetic radiation given off by radioactive elements. (20.3)

Atomic Spectrum: The line spectrum produced when energized or excited atoms emit electromagnetic radiation. (7.2)

Atomic Weight: See *Atomic Mass.*

Atomization Energy (ΔH_{atom}): The energy needed to rupture all of the bonds in one mole of a substance in the gas state and produce its atoms, also in the gas state. (18.10)

Aufbau Principle: A set of rules enabling the construction of an electron structure of an atom from its atomic number. (7.5)

Average: See *Mean.*

Avogadro's Number (Avogadro's Constant): 6.022×10^{23}; the number of particles or formula units in one mole. (3.1)

Avogadro's Principle: Equal volumes of gases contain equal numbers of molecules when they are at identical temperatures and pressures. (10.4)

Axial Bonds: Covalent bonds oriented parallel to the vertical axis in a trigonal bipyramidal molecule. (9.1)

Azimuthal Quantum Number *(ℓ)*: The quantum number *ℓ*. (See also *Secondary Quantum Number.*) (7.3)

B

Backbone (Polymer): The long chain of atoms in a polymer to which other groups are attached. (22.6)

Background Radiation: The atomic radiation from the natural radionuclides in the environment and from cosmic radiation. (20.6)

Balance: An apparatus for measuring mass. (1.5)

Balanced Equation: A chemical equation that has on opposites sides of the arrow the same number of each atom and the same net charge. (2.5, 3.4)

Band of Stability: The envelope that encloses just the stable nuclides in a plot of all nuclides constructed according to their numbers of neutrons versus their numbers of protons. (20.4)

Bar: The standard pressure for thermodynamic quantities; 1 bar = 10^5 pascals, 1 atm = 101,325 Pa. (6.5, 10.2)

Barometer: An apparatus for measuring atmospheric pressure. (10.2)

Base: *Arrhenius theory:* A substance that releases OH^- ions in water. (4.3)

Brønsted theory: A proton-acceptor. (15.1)

Lewis theory: An electron-pair acceptor. (15.3)

Base Ionization Constant, K_b:

$$K_b = \frac{[BH^+][OH^-]}{[B]} \text{ for the equilibrium,}$$

$$B + H_2O \rightleftharpoons BH^+ + OH^- \qquad (16.1)$$

Base Units: The units of the fundamental measurements of the SI. (1.5)

Basic Anhydride: An oxide that can neutralize acid or that reacts with water to give OH^-. (4.3)

Basic Oxygen Process: A method to convert pig iron into steel. (21.4)

Basic Solution: An aqueous solution in which $[H^+] < [OH^-]$. (15.5)

Battery: One or more galvanic cells arranged to serve as a practical source of electricity.

Becquerel (Bq): 1 disintegration s^{-1}. The SI unit for the activity of a radioactive source. (20.6)

Bent Molecule (V-Shaped Molecule): A molecule that is nonlinear. (9.2)

Beta Particle ($_{-1}^{0}e$): An electron emitted by radioactive decay. (20.3)

Beta Radiation: A stream of electrons produced by radioactive decay. (20.3)

Bidentate Ligand: A ligand that has two atoms that can become simultaneously attached to the same metal ion. (21.5)

Bimolecular Collision: A collision of two molecules. (13.7)

Binary Acid: An acid with the general formula H_nX, where X is a nonmetal. (4.4, 15.2)

Binary Compound: A compound composed of two different elements. (2.8)

Binding Energy, Nuclear: The energy equivalent of the difference in mass between an atomic nucleus and the sum of the masses of its nucleons. (20.2)

Biochemistry: The study of the organic substances in organisms. (22.7)

Biological Catalyst: Biological molecule such as an enzyme that catalyzes a chemical reaction. (13.8)

Black Phosphorus: An allotrope of phosphorus that has a layered structure. (21.2)

Blast Furnace: A structure in which iron ore is reduced to iron. (21.4)

Body-Centered Cubic (bcc) Unit Cell: A unit cell having identical atoms, molecules, or ions at the corners of a cube plus one more particle in the center of the cube. (11.9)

Boiling Point: The temperature at which the vapor pressure of the liquid equals the atmospheric pressure. (11.6)

Boiling Point Elevation: A colligative property of a solution by which the solution's boiling point is higher than that of the pure solvent. (12.7)

Bond Angle: The angle formed by two bonds that extend from the same atom. (9.1)

Bond Dipole: A dipole within a molecule associated with a specific bond. (9.3)

Bond Dissociation Energy: See *Bond Energy.*

Bond Distance: See *Bond Length.*

Bond Energy: The energy needed to break one mole of a particular bond to give electrically neutral fragments. (8.3, 18.10)

Bond Length: The distance between two nuclei that are held together by a chemical bond. (8.3)

Bond Order: The number of electron pairs shared between two atoms. The *net* number of pairs of bonding electrons. (8.6, 9.7)

Bond order = 1/2 × (no. of bonding e^- − no. of antibonding e^-)

Bonding Domain: A region between two atoms that contains one or more electron pairs in bonds and that influences molecular shape. (9.2)

Bonding Electrons: Electrons that occupy bonding molecular orbitals. (9.7)

Bonding Molecular Orbital: A molecular orbital that introduces a buildup of electron density between nuclei and stabilizes a molecule when occupied by electrons. (9.7)

Boundary: The interface between a system and its surroundings across which energy or matter might pass. (6.3)

Boyle's Law: See *Pressure–Volume Law.*

Bragg Equation: $n\lambda = 2d \sin \theta$. The equation used to convert X-ray diffraction data into a crystal structure. (11.10)

Branched-Chain Compound: An organic compound in whose molecules the carbon atoms do not all occur one after another in a continuous sequence. (22.1)

Branching (Polymer): The formation of side chains (branches) along the main backbone of a polymer. (22.6)

Branching Step: A step in a chain reaction that produces more chain-propagating species than it consumes. (Facets of Chemistry 13.1)

Brine: An aqueous solution of sodium chloride, often with other salts. (19.8)

Brønsted Acid: See *Acid.*

Brønsted Base: See *Base.*

Brownian Motion: The random, erratic motions of colloidally dispersed particles in a fluid. (2.6)

Buckminsterfullerene: The C_{60} molecule. Also called buckyball. (21.2)

Buckyball: See *Buckminsterfullerene.*

Buffer: (a) A pair of solutes that can keep the pH of a solution almost constant if either acid or base is added. (b) A solution containing such a pair of solutes. (16.5)

Buffer Capacity: A measure of how much strong acid or strong base is needed to change the pH of a buffer by some specified amount.

Buret: A long tube of glass usually marked in mL and 0.1 mL units and equipped with a stopcock for the controlled addition of a liquid to a receiving flask. (4.8)

By-product: The substances formed by side reactions. (3.6)

C

Calorie (cal): 4.184 J. The energy that will raise the temperature of 1.00 g of water from 14.5 to 15.5 °C. (In popular books on foods, the term *Calorie*, with a capital C, means 1000 cal or 1 kcal.) (6.1)

Calorimeter: An apparatus used in the determination of the heat of a reaction. (6.5)

Calorimetry: The science of measuring the quantities of heat that are involved in a chemical or physical change. (6.5)

Carbohydrates: Polyhydroxyaldehydes or polyhydroxyketones or substances that yield these by hydrolysis and that are obtained from plants or animals. (22.7)

Carbon Nanotube: Tubular carbon molecules that can be visualized as rolled up sheets of graphite (with hexagonal rings of carbon atoms) capped at each end by half of a spherical fullerene molecule. (21.2)

Carbon Ring: A series of carbon atoms arranged in a ring. (22.1)

Carbonyl Group: An organic functional group consisting of a carbon atom joined to an oxygen atom by a double bond; $C{=}O$. (22.5)

Carboxyl Group: $-CO_2H$. (8.3, 22.5)

Carboxylic Acid: An organic compound whose molecules have the carboxyl group $-CO_2H$. (8.3, 22.5)

Catalysis: Rate enhancement caused by a catalyst. (13.8)

Catalyst: A substance that in relatively small proportion accelerates the rate of a reaction without being permanently chemically changed. (13.8)

Catenation: The linking together of atoms of the same element to form chains.

Cathode: The negative electrode in a gas discharge tube. The electrode at which reduction occurs during an electrochemical change. (19.1)

Cathode Ray: A stream of electrons ejected from a hot metal and accelerated toward a positively charged site in a vacuum tube.

Cation: A positively charged ion. (2.8)

Cell Potential, E_{cell}: The potential (voltage) of a galvanic cell when no current is drawn from the cell. (19.2)

Cell Reaction: The overall chemical change that takes place in an electrolytic cell or a galvanic cell. (19.1)

Celsius Scale: A temperature scale on which water freezes at 0 °C and boils at 100 °C (at 1 atm) and that has 100 divisions called Celsius degrees between those two points. (1.5)

Centimeter (cm): 0.01 m. (1.5)

Chain Reaction: A self-sustaining change in which the products of one event cause one or more new events. (Facets of Chemistry 13.1, 20.8)

Change of State: Transformation of matter from one physical state to another. In thermochemistry, any change in a variable used to define the state of a particular system—a change in composition, pressure, volume, or temperature. (11.4)

Charge: The mixture of raw materials added to a blast furnace. (21.4)

Charles' Law: See *Temperature–Volume Law.*

Chelate: A complex ion containing rings formed by polydentate ligands. (21.5)

Chelate Effect: The extra stability found in complexes that contain chelate rings. (21.5)

Chemical Bond: The force of electrical attraction that holds atoms together in compounds. (2.6, 8 Introduction)

Chemical Change: A change that converts substances into other substances; a chemical reaction. (1.3)

Chemical Energy: The potential energy of chemicals that is transferred during chemical reactions. (6.1)

Chemical Equation: A before-and-after description that uses formulas and coefficients to represent a chemical reaction. (2.5)

Chemical Equilibrium: Dynamic equilibrium in a chemical system. (4.3, 14.1)

Chemical Formula: A formula written using chemical symbols and subscripts that describes the composition of a chemical compound or element. (2.5)

Chemical Kinetics: The study of rates of reaction. (13 Introduction)

Chemical Property: The ability of a substance, either by itself or with other substances, to undergo a change into new substances. (1.4)

Chemical Reaction: A change in which new substances (products) form from starting materials (reactants). (1.3)

Chemical Symbol: A formula for an element. (1.3)

Chemical Thermodynamics: See *Thermodynamics.*

Chemistry: The study of the compositions of substances and the ways by which their properties are related to their compositions. (1.1)

Chirality: The "handedness" of an object; the property of an object (like a molecule) that makes it unable to be superimposed onto a model of its own mirror image. (21.8, 22.1)

Cis Isomer: A stereoisomer whose uniqueness is in having two groups on the same side of some reference plane. (21.8, 22.2)

Clausius–Clapeyron Equation: The relationship between the vapor pressure, the temperature, and the molar heat of vaporization of a substance (where C is a constant). (Facets of Chemistry 11.1)

$$\ln P = \frac{\Delta H_{vap}}{RT} + C$$

Closed-End Manometer: See *Manometer.*

Closed System: A system that can absorb or release energy but not mass across the boundary between the system and its surroundings. (6.3)

Closest-Packed Structure: A crystal structure in which atoms or molecules are packed as efficiently as possible. (11.9)

Codon: An individual unit of hereditary instruction that consists of three, side by side, side chains on a molecule of mRNA. (22.8)

Coefficients: Numbers in front of formulas in chemical equations. (2.5)

Coinage Metals: Copper, silver, and gold.

Coke: Coal that has been strongly heated to drive off its volatile components and that is mostly carbon. (21.4)

Collapsing Atom Paradox: The paradox faced by classical physics that predicts a moving electron in an atom should emit energy and spiral into the nucleus. (7 Introduction)

Colligative Property: A property such as vapor pressure lowering, boiling point elevation, freezing point depression, and osmotic pressure whose physical value depends only on the ratio of the numbers of moles of solute and solvent particles and not on their chemical identities. (12.6)

Collision Theory: The rate of a reaction is proportional to the number of effective collisions that occur each second between the reactants. (13.5)

Combined Gas Law: See *Gas Law, Combined.*

Combustion: A rapid reaction with oxygen accompanied by a flame and the evolution of heat and light. (5.5)

Common Ion: The ion in a mixture of ionic substances that is common to the formulas of at least two. (16.5)

Common Ion Effect: The solubility of one salt is reduced by the presence of another having a common ion. (16.5, 17.1)

Competing Reaction: A reaction that reduces the yield of the main product by forming by-products (3.6)

Complex Ion (Complex): The combination of one or more anions or neutral molecules (ligands) with a metal ion. (17.4, 21.5)

Compound: A substance consisting of chemically combined atoms from two or more elements and present in a definite ratio. (1.3)

Compound Nucleus: An atomic nucleus carrying excess energy following its capture of some bombarding particle. (20.5)

Compressibility: Capable of undergoing a reduction in volume under increasing pressure. (10.1, 11.3)

Concentrated Solution: A solution that has a large ratio of the amounts of solute to solvent. (4.1)

Concentration: The ratio of the quantity of solute to the quantity of solution (or the quantity of solvent). (See *Molal Concentration, Molar Concentration, Mole Fraction, Percentage Concentration.*) (4.1)

Concentration Table: A part of the strategy for organizing data needed to make certain calculations, particularly any involving equilibria. (14.7, 16.2)

Conclusion: A statement that is based on what we think about a series of observations. (1.2)

Condensation: The change of a vapor to its liquid or solid state. (11.4)

Condensation Polymer: A polymer formed from monomers by splitting out a small molecule such as H_2O or CH_3OH. (22.6)

Condensation Polymerization: The process of forming a condensation polymer. (22.6)

Conformation: A particular relative orientation or geometric form of a flexible molecule. (9.5)

Conjugate Acid: The species in a conjugate acid–base pair that has the greater number of H^+ units. (15.1)

Conjugate Acid–Base Pair: Two substances (ions or molecules) whose formulas differ by only one H^+ unit. (15.1)

Conjugate Base: The species in a conjugate acid–base pair that has the fewer number of H^+ units. (15.1)

Conservation of Energy, Law of: See *Law of Conservation of Energy.*

Conservation of Mass–Energy, Law of: See *Law of Conservation of Mass–Energy.*

Continuous Spectrum: The electromagnetic spectrum corresponding to the mixture of frequencies present in white light. (7.2)

Contributing Structure: One of a set of two or more Lewis structures used in applying the theory of resonance to the structure of a compound. A resonance structure. (8.7)

Conversion Factor: A ratio constructed from the relationship between two units such as 2.54 cm/1 in., from 1 in. = 2.54 cm. (1.7)

Cooling Curve: A graph showing how the temperature of a substance changes as heat is removed from it at a constant rate as the substance undergoes changes in its physical state. (11.7)

Coordinate Covalent Bond: A covalent bond in which both electrons originated from one of the joined atoms, but otherwise like a covalent bond in all respects. (8.6)

Coordination Compound (Coordination Complex): A complex or its salt. (17.4, 21.5)

Coordination Number: The number of donor atoms that surround a metal ion. (21.7)

Copolymer: A polymer made from two or more different monomers. (22.6)

Core Electrons: The inner electrons of an atom that are not exposed to the electrons of other atoms when chemical bonds form. (7.6)

Corrosion: The slow oxidation of metals exposed to air or water. (5.5)

Coulomb (C): The SI unit of electrical charge; the charge on 6.25×10^{18} electrons; the amount of charge that passes a fixed point of a wire conductor when a current of 1 A flows for 1 s. (19.2, 19.7)

Covalent Bond: A chemical bond that results when atoms share electron pairs. (8.3)

Covalent Crystal (Network Solid): A crystal in which the lattice positions are occupied by atoms that are covalently bonded to the atoms at adjacent lattice sites. (11.11)

Critical Mass: The mass of a fissile isotope above which a self-sustaining chain reaction occurs. (20.8)

Critical Point: The point at the end of a vapor pressure versus temperature curve for a liquid and that corresponds to the critical pressure and the critical temperature. (11.12)

Critical Pressure (P_c): The vapor pressure of a substance at its critical temperature. (11.12)

Critical Temperature (T_c): The temperature above which a substance cannot exist as a liquid regardless of the pressure. (11.12)

Cross-Link: A bridge formed between polymer strands. (22.6)

Crystal Field Splitting (Δ): The difference in energy between sets of *d* orbitals in a complex ion. (21.9)

Crystal Field Theory: A theory that considers the effects of the polarities or the charges of the ligands in a complex ion on the energies of the *d* orbitals of the central metal ion. (21.9)

Crystal Lattice: The repeating symmetrical pattern of atoms, molecules, or ions that occurs in a crystal. (11.9)

Cubic Closest Packing (ccp): Efficient packing of spheres with an A-B-C-A-B-C. . . alternating stacking of layers of spheres. (11.9)

Cubic Meter (m^3): The SI derived unit of volume. (1.5)

Curie (Ci): A unit of activity for radioactive samples, equal to 3.7×10^{10} disintegrations per second. (20.6)

D

Dalton: One atomic mass unit, u.

Dalton's Atomic Theory: Matter consists of tiny, indestructible particles called atoms. All atoms of one element are identical. The atoms of different elements have different masses. Atoms combine in definite ratios by atoms when they form compounds. (2.1)

Dalton's Law of Partial Pressures: See *Partial Pressures, Law of.*

Data: The information (often in the form of physical quantities) obtained in an experiment or other experience or from references. (1.2)

Debye: Unit used to express dipole moments. 1 D = 3.34×10^{-30} C m (coulomb meter). (8.4)

Decay Constant: The first-order rate constant for radioactive decay. (20.6)

Decimal Multipliers: Factors—exponentials of 10 or decimals—that are used to define larger or smaller SI units. (1.5)

Decomposition: A chemical reaction that changes one substance into two or more simpler substances. (1.3)

Dehydration: Removal of water from a substance.

Dehydration Reaction: Formation of a carbon–carbon double bond by removal of the components of water from an alcohol. (22.3)

Deliquescent Compound: A compound able to absorb enough water from humid air to form a concentrated solution.

Delocalization Energy: The difference between the energy a substance would have if its molecules had no delocalized molecular orbitals and the energy it has because of such orbitals. (9.8)

Delocalized Molecular Orbital: A molecular orbital that spreads over more than two nuclei. (9.8)

ΔH_{fusion}: See *Molar Heat of Fusion*

$\Delta H_{sublimation}$: See *Molar Heat of Sublimation*

$\Delta H_{vaporization}$: See *Molar Heat of Vaporization*

Density: The ratio of an object's mass to its volume. (1.8)

Dependent Variable: The experimental variable of a pair of variables whose value is determined by the other, the independent variable.

Derived Unit: Any unit defined solely in terms of base units. (1.5)

Deuterium, ^2_1H: The isotope of hydrogen with a mass number of 2. (20.8)

Diagonal Relationship: Physical or chemical properties that generally vary diagonally from one corner to the other in the periodic table (e.g., ionization energy, electron affinity, and electronegativity). (7.8)

Dialysis: The passage of small molecules and ions, but not species of a colloidal size, through a semipermeable membrane. (12.8)

Diamagnetism: The property experienced by a substance that contains no unpaired electrons whereby the substance is repelled weakly by a magnet. (7.4)

Diamond: A crystalline form of carbon in which each carbon atom is bonded tetrahedrally to four other carbon atoms. (21.2)

Diaphragm Cell: An electrolytic cell used to manufacture sodium hydroxide by the electrolysis of aqueous sodium chloride. (19.8)

Diatomic Substance (Diatomic Molecule): A molecular substance made from two atoms. (2.5)

Diffraction: Constructive and destructive interference by waves. (7.3)

Diffraction Pattern: The image formed on a screen or a photographic film caused by the diffraction of electromagnetic radiation such as visible light or X rays. (11.10)

Diffusion: The spontaneous intermingling of one substance with another. (10.7)

Dilute Solution: A solution in which the ratio of the quantities of solute to solvent is small. (4.1)

Dilution: The process whereby a concentrated solution is made more dilute. (4.6)

Dimensional Analysis: See *Factor-Label Method.*

Dimer: Two monomer units joined by chemical bonds or intramolecular forces. (12.9)

Dipole (Electric): Partial positive and partial negative charges separated by a distance. (8.4)

Dipole–Dipole Attraction: Attraction between molecules that are dipoles. (11.2)

Dipole Moment (μ): The product of the sizes of the partial charges in a dipole multiplied by the distance between them; a measure of the polarity of a molecule. (8.4)

Diprotic Acid: An acid that can furnish two H^+ per molecule. (4.3)

Disaccharide: A carbohydrate whose molecules can be hydrolyzed to two monosaccharides. (22.7)

Dispersion Forces: Another term for London forces. (11.2)

Disproportionation: A redox reaction in which a portion of a substance is oxidized at the expense of the rest, which is reduced.

Dissociation: The separation of preexisting ions when an ionic compound dissolves or melts. (4.2)

Distorted Tetrahedron: A description of a molecule in which the central atom is surrounded by five electron pairs, one of which is a lone pair of electrons. The central atom is bonded to four other atoms. The structure is also said to have a seesaw shape. (9.2)

Dissymmetric: Lacking or deficient in symmetry. In a dissymmetric molecule the effects of the individual bond dipoles do not cancel, causing the molecule as a whole to be polar. (9.3)

DNA: Deoxyribonucleic acid; a nucleic acid that hydrolyzes to deoxyribose, phosphate ion, adenine, thymine, guanine, and cytosine, and that is the carrier of genes. (22.8)

DNA Double Helix: Two oppositely running strands of DNA held in a helical configuration by interstrand hydrogen bonds. (22.8)

Donor Atom: The atom on a ligand that makes an electron pair available in the formation of a complex. (17.4, 21.5)

Double Bond: (a) A covalent bond formed by sharing two pairs of electrons. (8.3) (b) A covalent bond consisting of one sigma bond and one pi bond. (9.6)

Double Replacement Reaction (Metathesis Reaction): A reaction of two salts in which cations and anions exchange partners (e.g., $AgNO_3 + NaCl \longrightarrow AgCl + NaNO_3$). (4.5)

Downs Cell: An electrolytic cell for the industrial production of sodium. (19.8)

Ductility: A metal's ability to be drawn (or stretched) into wire. (2.4)

Dynamic Equilibrium: A condition in which two opposing processes are occurring at equal rates. (4.3, 14.1)

E

ΔE: See *Internal Energy Change.*

Effective Collision: A collision between molecules that is capable of leading to a net chemical change. (13.5)

Effective Nuclear Charge: The net positive charge an outer electron experiences as a result of the partial screening of the full nuclear charge by core electrons. (7.8)

Effusion: The movement of a gas through a very tiny opening into a region of lower pressure. (10.7)

Effusion, Law of (Graham's Law): The rates of effusion of gases are inversely proportional to the square roots of their densities when compared at identical pressures and temperatures.

$$\text{Effusion rate} \propto \frac{1}{\sqrt{d}} \qquad (\text{constant } P \text{ and } T)$$

where d is the gas density. (10.7)

Einstein Equation: $\Delta E = \Delta m_0 c^2$ where ΔE is the energy obtained when a quantity of rest mass, Δm_0, is destroyed, or the energy lost when this quantity of mass is created. (20.1)

Electric Dipole: Two poles of electric charge separated by a distance. (8.4)

Electrochemical Change: A chemical change that is caused by or that produces electricity. (19 Introduction)

Electrochemistry: The study of electrochemical changes. (19 Introduction)

Electrolysis: The production of a chemical change by the passage of electricity through a solution that contains ions or through a molten ionic compound. (19.6)

Electrolysis Cell: An apparatus for electrolysis. (19.6)

Electrolyte: A compound that conducts electricity either in solution or in the molten state. (4.2)

Electrolytic Cell: See *Electrolysis Cell.*

Electrolytic Conduction: The transport of electrical charge by ions. (19.1)

Electromagnetic Spectrum: The distribution of frequencies of electromagnetic radiation among various types of such radiation—microwave, infrared, visible, ultraviolet, X, and gamma rays. (7.1)

Electromagnetic Wave (Electromagnetic Radiation): The successive series of oscillations in the strengths of electrical and magnetic fields associated with light, microwaves, gamma rays, ultraviolet rays, infrared rays, and the like. (7.1)

Electron (e^- or $_{-1}^{0}e$): (a) A subatomic particle with a charge of 1− and mass of 0.0005486 u (9.109383×10^{-28} g) that occurs outside an atomic nucleus.

The particle that moves when an electric current flows. (2.2) (b) A beta particle. (20.3)

Electron Affinity (EA): The energy change (usually expressed in kJ mol^{-1}) that occurs when an electron adds to an isolated gaseous atom or ion. (7.8)

Electron Capture: The capture by a nucleus of an orbital electron and that changes a proton into a neutron in the nucleus. (20.3)

Electron Cloud: Because of its wave properties, an electron's influence spreads out like a cloud around the nucleus. (7.7)

Electron Configuration: The distribution of electrons in an atom's orbitals. (7.5)

Electron Density: The concentration of the electron's charge within a given volume. (7.7)

Electron Domain: A region around an atom where one or more electron pairs are concentrated and which influences the shape of a molecule. (9.2)

Electron Domain Model: See *Valence Shell Electron Pair Repulsion Theory.*

Electron Pair Bond: A covalent bond. (8.3)

Electron Spin: The spinning of an electron about its axis that is believed to occur because the electron behaves as a tiny magnet. (7.4)

Electron Volt (eV): The energy an electron receives when it is accelerated under the influence of 1 V and equal to 1.6×10^{-19} J. (20.3)

Electronegativity: The relative ability of an atom to attract electron density toward itself when joined to another atom by a covalent bond. (8.4)

Electronic Structure: The distribution of electrons in an atom's orbitals. (7.5)

Electroplating: Depositing a thin metallic coating on an object by electrolysis. (19.8)

Element: A substance in which all of the atoms have the same atomic number. A substance that cannot be broken down by chemical reactions into anything that is both stable and simpler. (1.3, 2.2)

Elementary Process: One of the individual steps in the mechanism of a reaction. (13.7)

Elimination Reaction: The loss of a small molecule from a larger molecule as in the elimination of water from an alcohol. (22.3)

Emission Spectrum: See *Atomic Spectrum.*

Empirical Formula: A chemical formula that uses the smallest whole-number subscripts to give the proportions by atoms of the different elements present. (3.3)

Enantiomers: Stereoisomers whose molecular structures are related as an object to its mirror image but that cannot be superimposed. (21.8)

End Point: The moment in a titration when the indicator changes color and the titration is ended. (4.8, 16.7)

Endergonic: Descriptive of a change accompanied by an increase in free energy. (18.4)

Endothermic: Descriptive of a change in which a system's internal energy increases. (6.4)

Energy: Something that matter possesses by virtue of an ability to do work. (6.1)

Energy Density: For a galvanic cell, the ratio of the energy available to the volume of the cell. (19.8)

Energy Level: A particular energy an electron can have in an atom or a molecule. (7.2)

Enthalpy (*H*): The heat content of a system. (6.5, 18.1)

Enthalpy Change (Δ*H*): The difference in enthalpy between the initial state and the final state for some change. (6.5, 18.1)

Enthalpy Diagram: A graphical depiction of enthalpy changes following different paths from reactants to products. (6.7)

Enthalpy of Solution: See *Heat of Solution.*

Entropy (*S*): A thermodynamic quantity related to the number of equivalent ways the energy of a system can be distributed. The greater this number, the more probable is the state and the higher is the entropy. (18.3)

Entropy Change (Δ*S*): The difference in entropy between the initial state and the final state for some change. (18.3)

Enzyme: A catalyst in a living system and that consists of a protein. (13.8, 22.7)

Equation of State of an Ideal Gas: See *Gas Law, Ideal.*

Equatorial Bond: A covalent bond located in the plane perpendicular to the long axis of a trigonal bipyramidal molecule. (9.1)

Equilibrium: See *Dynamic Equilibrium.*

Equilibrium Constant, *K*: The value that the mass action expression has when the system is at equilibrium. (14.2)

Equilibrium Law: The mathematical equation for a particular equilibrium system that sets the mass action expression equal to the equilibrium constant. (14.2)

Equilibrium Vapor Pressure of a Liquid: The pressure exerted by a vapor in equilibrium with its liquid state. (11.5)

Equilibrium Vapor Pressure of a Solid: The pressure exerted by a vapor in equilibrium with its solid state. (11.5)

Equivalence: A relationship between two quantities expressed in different units. (1.8)

Equivalence Point: The moment in a titration when the number of equivalents of the reactant added from a buret equals the number of equivalents of another reactant in the receiving flask. (16.7)

Error in a Measurement: The difference between a measurement and the "true" value we are trying to measure. (1.6)

Ester: An organic compound whose molecules have the ester group. (22.5)

$$-\overset{\overset{\displaystyle O}{\|}}{C}-O-C$$

ester group

Ether: An organic compound in whose molecules two hydrocarbon groups are joined to an oxygen. (22.3)

Ethyl Group: $CH_3CH_2—$. (22.2)

Evaporate: To change from a liquid to a vapor. (11.3)

Exact Number: A number obtained by a direct count or that results by a definition; and that is considered to have an infinite number of significant figures. (1.6)

Excess Reactant: The reactant left over once the limiting reactant is used up. (3.5)

Excited State: A term describing an atom or molecule where all of the electrons are not in their lowest possible energy levels. (7.2)

Exergonic: Descriptive of a change accompanied by a decrease in free energy. (18.4)

Exon: One of a set of sections of a DNA molecule (separated by introns) that, taken together, constitute a gene. (22.8)

Exothermic: Descriptive of a change in which energy leaves a system and enters the surroundings. (6.4)

Expansion Work: See *Pressure–Volume Work.*

Exponential Notation: See *Scientific Notation.*

Extensive Property: A property of an object that is described by a physical quantity whose magnitude is proportional to the size or amount of the object (e.g., mass or volume). (1.4)

F

Face-Centered Cubic (fcc) Unit Cell: A unit cell having identical atoms, molecules, or ions at the corners of a cube and also in the center of each face of the cube. (11.9)

Factor-Label Method: A problem-solving technique that uses the correct cancellation of the units of physical quantities as a guide for the correct setting up of the solution to the problem. (1.7)

Fahrenheit Scale: A temperature scale on which water freezes at 32 °F and boils at 212 °F (at 1 atm) and between which points there are 180 degree divisions called Fahrenheit degrees. (1.5)

Family of Elements: See *Group.*

Faraday ($\mathscr{F}$): One mole of electrons; 9.65×10^4 coulombs. (19.4)

Faraday Constant ($\mathscr{F}$): 9.65×10^4 coulombs/mol e^-. (19.4)

Fatty Acid: One of several long-chain carboxylic acids produced by the hydrolysis (digestion) of a lipid. (22.7)

Film Dosimeter: A device used by people working with radioactive isotopes that records doses of atomic radiation by the darkening of photographic film. (20.6)

First Law of Thermodynamics: A formal statement of the law of conservation of energy. $\Delta E = q + w$. (6.5, 18.1)

First-Order Reaction: A reaction with a rate law in which rate $= k[A]^1$, where A is a reactant. (13.3)

Fissile Isotope: An isotope capable of undergoing fission following neutron capture. (20.8)

Fission: The breaking apart of atomic nuclei into smaller nuclei accompanied by the release of energy, and the source of energy in nuclear reactors. (20.2, 20.8)

Flotation: A method for concentrating sulfide ores of copper and lead by bubbling air through a slurry of oil-coated ore particles. The sulfides, but not soil or other rock particles, stick to the rising air bubbles and collect in the foam at the surface. (21.4)

Force: Anything that can cause an object to change its motion or direction. (1.5)

Formal Charge: The apparent charge on an atom in a molecule or polyatomic ion as calculated by a set of rules. (8.6)

Formation Constant (K_{form}): The equilibrium constant for an equilibrium involving the formation of a complex ion. Also called the stability constant. (17.4)

Formula: See *Chemical Formula.*

Formula Mass: The sum of the atomic masses (in u) of all of the atoms represented in a chemical formula. Often used with units of g mol^{-1} to represent masses of ionic substances. See also *Molar Mass.* (3.1)

Formula Unit: A particle that has the composition given by the chemical formula. (2.7)

Forward Reaction: In a chemical equation, the reaction as read from left to right. (4.3)

Fossil Fuels: Coal, oil, and natural gas.

Free Element: An element that is not combined with another element in a compound. (2.5)

Free Energy: See *Gibbs Free Energy* or *Standard Free Energy Change*

Free Energy Diagram: A plot of the changes in free energy for a multicomponent system versus the composition. (18.8)

Free Radical: An atom, molecule, or ion that has one or more unpaired electrons. (13.7, 22.6)

Freezing Point Depression: A colligative property of a liquid solution by which the freezing point of the solution is lower than that of the pure solvent. (12.7)

Frequency (ν): The number of cycles per second of electromagnetic radiation. (7.1)

Frequency Factor: The proportionality constant, A, in the Arrhenius equation. (13.6)

Fuel Cell: An electrochemical cell in which electricity is generated from the redox reactions of common fuels. (19.8)

Fullerene: An allotrope of carbon made of an extended joining together of five- and six-membered rings of carbon atoms. (21.2)

Functional Group: The group of atoms of an organic molecule that enters into a characteristic set of reactions that are independent of the rest of the molecule. (22.1)

Fusion: (a) Melting. (11.7) (b) The formation of atomic nuclei by the joining together of the nuclei of lighter atoms. (20.2, 20.8)

G

***G*:** See *Gibbs Free Energy.*

ΔG: See *Gibbs Free Energy Change.*

$\Delta G°$: See *Standard Free Energy Change.*

$\Delta G_f°$: See *Standard Free Energy of Formation.*

Galvanic Cell: An electrochemical cell in which a spontaneous redox reaction produces electricity. (19.1)

Gamma Radiation: Electromagnetic radiation with wavelengths in the range of 1 Å or less (the shortest wavelengths of the spectrum). (20.3)

Gangue: The unwanted rock and sand that is separated from an ore. (21.4)

Gas: One of the states of matter. A gas consists of rapidly moving widely spaced atomic or molecular sized particles. (1.4)

Gas Constant, Universal (R): $R = 0.0821$ liter atm mol^{-1} K^{-1} or $R = 8.314$ J mol^{-1} K^{-1} (10.5)

Gas Law, Combined: For a given mass of gas, the product of its pressure and volume divided by its Kelvin temperature is a constant. (10.3)

$$PV/T = \text{a constant}$$

Gas Law, Ideal: $PV = nRT$. (10.5)

Gay-Lussac's Law: See *Pressure–Temperature Law.*

Geiger Counter: A device that detects beta and gamma radiation (20.6)

Genetic Code: The correlation of codons with amino acids. (22.8)

Geometric Isomer: One of a set of isomers that differ only in geometry. (22.2)

Geometric Isomerism: The existence of isomers whose molecules have identical atomic organizations but different geometries; cis-trans isomers. (21.8, 22.2)

Gibbs Free Energy (G): A thermodynamic quantity that relates enthalpy (H), entropy (S), and temperature (T) by the equation: (18.4)

$$G = H - TS$$

Gibbs Free Energy Change (ΔG): The difference given by: (18.4)

$$\Delta G = \Delta H - T\Delta S$$

Glass: Any amorphous solid. (11.9)

Glycogen: A polysaccharide that animals use to store glucose units for energy. (22.7)

Graham's Law: See *Effusion, Law of.*

Gram (g): 0.001 kg. (1.5)

Graphite: The most stable allotrope of carbon, consisting of layers of joined six-membered rings of carbon atoms. (21.2)

Gray (Gy): The SI unit of radiation absorbed dose. (20.6)

$$1 \text{ Gy} = 1 \text{ J kg}^{-1}$$

Greenhouse Effect: The retention of solar energy made possible by the ability of the greenhouse gases (e.g., CO_2, CH_4, H_2O, and the chlorofluorocarbons) to absorb outgoing radiation and reradiate some of it back to earth.

Ground State: The lowest energy state of an atom or molecule. (7.2)

Group: A vertical column of elements in the periodic table. (2.3)

H

ΔH: See *Enthalpy Change.*

ΔH_{atom}: See *Atomization Energy.*

ΔH_c: See *Heat of Combustion.*

$\Delta H°$: See *Standard Heat of Reaction.*

$\Delta H_f°$: See *Standard Heat of Formation.*

ΔH_{fusion}: See *Molar Heat of Fusion.*

ΔH_{soln}: See *Heat of Solution.*

$\Delta H_{sublimation}$: See *Molar Heat of Sublimation.*

$\Delta H_{vaporization}$: See *Molar Heat of Vaporization.*

Half-Cell: That part of a galvanic cell in which either oxidation or reduction takes place. (19.1)

Half-Life ($t_{1/2}$): The time required for a reactant concentration or the mass of a radionuclide to be reduced by half. (13.4)

Half-Reaction: A hypothetical reaction that constitutes exclusively either the oxidation or the reduction half of a redox reaction and in whose equation the correct formulas for all species taking part in the change are given together with enough electrons to give the correct electrical balance. (5.2)

Hall–Héroult Process: A method for manufacturing aluminum by the electrolysis of aluminum oxide in molten cryolite. (19.8)

Halogen Family: Group VIIA in the periodic table—fluorine, chlorine, bromine, iodine, and astatine. (2.3)

Hard Water: Water with dissolved Mg^{2+}, Ca^{2+}, Fe^{2+}, or Fe^{3+} ions at a concentration high enough (above 25 mg L^{-1}) to interfere with the use of soap. (Facets of Chemistry 4.1)

Heat: Energy that flows from a hot object to a cold object as a result of their difference in temperature. (6.1)

Heat Capacity: The quantity of heat needed to raise the temperature of an object by 1 °C. (6.3)

Heat of Combustion: The heat evolved in the combustion of a substance. (6.5)

Heat of Formation, Standard: See *Standard Heat of Formation.*

Heat of Reaction: The heat exchanged between a system and its surroundings when a chemical change occurs in the system. (6.5)

Heat of Reaction at Constant Pressure (q_p): The heat of a reaction in an open system, ΔH. (6.5, 18.1)

Heat of Reaction at Constant Volume (q_v): The heat of a reaction in a sealed vessel, like a bomb calorimeter, ΔE. (6.5)

Heat of Reaction, Standard: See *Standard Heat of Reaction.*

Heat of Solution (ΔH_{soln}): The energy exchanged between the system and its surroundings when one mole of a solute dissolves in a solvent to make a dilute solution. (12.2)

Heating Curve: A graph showing how the temperature of a substance changes as heat is added to it at a constant rate as the substance undergoes changes in its physical state. (11.7)

Henderson–Hasselbalch Equation:

$$\text{pH} = \text{p}K_a + \log \frac{[A^-]_{\text{initial}}}{[HA]_{\text{initial}}} \quad \text{or}$$

$$\text{pH} = \text{p}K_a + \log \frac{[\text{salt}]}{[\text{acid}]}. \quad (16.5)$$

Henry's Law: See *Pressure–Solubility Law.*

Hertz (Hz): 1 cycle s^{-1}; the SI unit of frequency. (7.1)

Hess's Law: For any reaction that can be written in steps, the standard heat of reaction is the same as the sum of the standard heats of reaction for the steps. (6.7)

Hess's Law Equation: For the change,

$$aA + bB + \ldots \longrightarrow nN + mM + \ldots : \quad (6.8)$$

$$\Delta H^\circ = \begin{pmatrix}\text{sum of } \Delta H_f^\circ \text{ of all} \\ \text{of the products}\end{pmatrix} - \begin{pmatrix}\text{sum of } \Delta H_f^\circ \text{ of all} \\ \text{of the reactants}\end{pmatrix}$$

Heterocyclic Compound: A compound whose molecules have rings that include one or more multivalent atoms other than carbon. (22.1)

Heterogeneous Catalyst: A catalyst that is in a different phase than the reactants and onto whose surface the reactant molecules are adsorbed and where they react. (13.8)

Heterogeneous Equilibrium: An equilibrium involving more than one phase. (14.4)

Heterogeneous Mixture: A mixture that has two or more phases with different properties. (1.3)

Heterogeneous Reaction: A reaction in which not all of the chemical species are in the same phase. (13.1, 14.4)

Heteronuclear Molecule: A molecule in which not all atoms are of the same element. (9.7)

Hexagonal Closest Packing (hcp): Efficient packing of spheres with an A-B-A-B-. . . alternating stacking of layers of spheres. (11.9)

High-Spin Complex: A complex ion or coordination compound in which there is the maximum number of unpaired electrons. (21.9)

Homogeneous Catalyst: A catalyst that is in the same phase as the reactants. (13.8)

Homogeneous Equilibrium: An equilibrium system in which all components are in the same phase. (14.4)

Homogeneous Mixture: A mixture that has only one phase and that has uniform properties throughout; a solution. (1.3)

Homogeneous Reaction: A reaction in which all of the chemical species are in the same phase. (13.1, 14.4)

Homonuclear Diatomic Molecule: A diatomic molecule in which both atoms are of the same element. (9.7)

Hund's Rule: Electrons that occupy orbitals of equal energy are distributed with unpaired spins as much as possible among all such orbitals. (7.5)

Hybrid Atomic Orbitals: Orbitals formed by mixing two or more of the basic atomic orbitals of an atom and that make possible more effective overlaps with the orbitals of adjacent atoms than do ordinary atomic orbitals. (9.5)

Hydrate: A compound that contains molecules of water in a definite ratio to other components. (2.5)

Hydrated Ion: An ion surrounded by a cage of water molecules that are attracted by the charge on the ion. (4.2)

Hydration: The development in an aqueous solution of a cage of water molecules about ions or polar molecules of the solute. (12.1)

Hydration Energy: The enthalpy change associated with the hydration of gaseous ions or molecules as they dissolve in water. (12.2)

Hydride: (a) A binary compound of hydrogen. (2.6) (b) A compound containing the hydride ion (H^-).

Hydrocarbon: An organic compound whose molecules consist entirely of carbon and hydrogen atoms. (2.6, 22.2)

Hydrogen Bond: An extra strong dipole–dipole attraction between a hydrogen bound covalently to nitrogen, oxygen, or fluorine and another nitrogen, oxygen, or fluorine atom. (11.2)

Hydrogen Electrode: The standard of comparison for reduction potentials and for which $E^\circ_{H^+}$ has a value of 0.00 V at 25 °C, when P_{H_2} = 1 atm and $[H^+] = 1\ M$ in the reversible half-cell reaction: (19.2)

$$2H^+(aq) + 2e^- \rightleftharpoons H_2(g)$$

Hydrolysis: A reaction with water

Hydrometer: A device for measuring specific gravity. (19.8)

Hydronium Ion: H_3O^+. (4.3)

Hydrophilic Group: A polar molecular unit capable of having dipole–dipole attractions or hydrogen bonds with water molecules. (22.7)

Hydrophobic Group: A nonpolar molecular unit with no affinity for water. (22.7)

Hypertonic Solution: A solution that has a higher osmotic pressure than cellular fluids. (12.8)

Hypothesis: A tentative explanation of the results of experiments. (1.2)

Hypotonic Solution: A solution that has a lower osmotic pressure than cellular fluids. (12.8)

I

Ideal Gas: A hypothetical gas that obeys the gas laws exactly. (10.3)

Ideal Gas Law: $PV = nRT$. (10.5)

Ideal Solution: A hypothetical solution that would obey the vapor pressure–concentration law (Raoult's law) exactly. (12.2)

Immiscible: Mutually insoluble. Usually used to describe liquids that are insoluble in each other. (12.1)

Incompressible: Incapable of losing volume under increasing pressure. (11.3)

Independent Variable: The experimental variable of a pair of variables whose value is first selected and from which the value of the dependent variable then results.

Indicator: A chemical put in a solution being titrated and whose change in color signals the end point. (4.3, 15.5)

Induced Dipole: A dipole created when the electron cloud of an atom or a molecule is distorted by a neighboring dipole or by an ion. (11.2)

Inert Gas: Any of the noble gases—Group VIIIA of the periodic table. Any gas that has virtually no tendency to react. (2.3)

Initiation Step: The step in a chain reaction that produces reactive species that can start chain propagation steps. (Facets of Chemistry 13.1)

Inner Transition Elements: Members of the two long rows of elements below the main body of the periodic table—elements 58–71 and elements 90–103. (2.3)

Inorganic Compound: A compound made from any elements except those compounds of carbon classified as organic compounds. (2.9)

Instability Constant (K_{inst}): The reciprocal of the formation constant for an equilibrium in which a complex ion forms. (17.4)

Instantaneous Dipole: A momentary dipole in an atom, ion, or molecule caused by the erratic movement of electrons. (11.2)

Instantaneous Rate: The rate of reaction at any particular moment during a reaction. (13.2)

Integrated Rate Law: A rate law that relates concentration versus time. (13.4)

Intensive Property: A property whose physical magnitude is independent of the size of the sample, such as density or temperature. (1.4)

Intercalation: The insertion of small atoms or ions between layers in a crystal such as graphite. (19.8)

Interference Fringes: Pattern of light produced by waves that undergo diffraction. (7.3)

Intermolecular Forces (Intermolecular Attractions): Attractions *between* neighboring molecules. (11.2)

Internal Energy (E): The sum of all of the kinetic energies and potential energies of the particles within a system. (6.2, 18.1)

Internal Energy Change (ΔE): The difference in internal energy between the initial state and the final state for some change. (6.2, 6.5)

International System of Units (SI): The successor to the metric system of measurements that retains most of the units of the metric system and their decimal relation-ships but employs new reference standards. (1.5)

Intramolecular Forces: Forces of attraction within molecules; chemical bonds. (11.2)

Intron: One of a set of sections of a DNA molecule that separate the exon sections of a gene from each other. (22.8)

Inverse Square Law: The intensity of a radiation is inversely proportional to the square of the distance from its source. (20.6)

Ion: An electrically charged particle on the atomic or molecular scale of size. (2.7)

Ion–Dipole Attraction: The attraction between an ion and the charged end of a polar molecule. (11.2)

Ion–Electron Method: A method for balancing redox reactions that uses half-reactions. (5.2)

Ion–Induced Dipole Attraction: Attraction between an ion and a dipole induced in a neighboring molecule. (11.2)

Ion Pair: A more or less loosely associated pair of ions in a solution. (12.9)

Ion Product: The mass action expression for the solubility equilibrium involving the ions of a salt and equal to the product of the molar concentrations of the ions, each concentration raised to a power that equals the number of ions obtained from one formula unit of the salt. (17.1)

Ion Product Constant of Water (K_w): $K_w = [H^+][OH^-]$ (15.5)

Ionic Bond: The attractions between ions that hold them together in ionic compounds. (8.1)

Ionic Character: The extent to which a covalent bond has a dipole moment and is polarized. (8.4)

Ionic Compound: A compound consisting of positive and negative ions. (2.7)

Ionic Crystal: A crystal that has ions located at the lattice points. (11.11)

Ionic Equation: A chemical equation in which soluble strong electrolytes are written in dissociated or ionized form. (4.2)

Ionic Reaction: A chemical reaction in which ions are involved. (4.2)

Ionization Energy (IE): The energy needed to remove an electron from an isolated, gaseous atom, ion, or molecule (usually given in units of kJ mol^{-1}). (7.8)

Ionization Reaction: A reaction of chemical particles that produces ions. (4.3)

Ionizing Radiation: Any high-energy radiation—X rays, gamma rays, or radiations from radionuclides—that generates ions as it passes through matter. (20.6)

Isolated System: A system that cannot exchange matter or energy with its surroundings. (6.3)

Isomer: One of a set of compounds that have identical molecular formulas but different structures. (21.8)

Isomerism: The existence of sets of isomers. (21.8)

Isopropyl Group: $(CH_3)_2CH—$. (22.2)

Isotonic Solution: A solution that has the same osmotic pressure as cellular fluids. (12.8)

Isotopes: Atoms of the same element with different atomic masses. Atoms of the same element with different numbers of neutrons in their nuclei. (2.2)

IUPAC Rules: The formal rules for naming substances as developed by the International Union of Pure and Applied Chemistry. (2.3, 2.9)

J

Joule (J): The SI unit of energy. (6.1)

$$1 \text{ J} = 1 \text{ kg m}^2 \text{ s}^{-2}$$

$$4.184 \text{ J} = 1 \text{ cal (exactly)}$$

K

K: See *Kelvin.*

K_a: See *Acid Ionization Constant.*

K_b: See *Base Ionization Constant.*

K_{form}: See *Formation Constant.*

K_{inst}: See *Instability Constant.*

K_{sp}: See *Solubility Product Constant.*

K_{spa}: See *Acid Solubility Product.*

K_w: See *Ion Product Constant of Water.*

K-Capture: See *Electron Capture.*

Kelvin (K): One degree on the Kelvin scale of temperature and identical in size to the Celsius degree. (1.5)

Kelvin Scale: The temperature scale on which water freezes at 273.15 K and boils at 373.15 K and that has 100 degree divisions called kelvins between these points. K = °C + 273.15. (1.5)

Ketone: An organic compound whose molecules have the carbonyl group (C=O) flanked by hydrocarbon groups. (22.5)

Kilocalorie (kcal): 1000 cal. (6.1)

Kilogram (kg): The base unit for mass in the SI and equal to the mass of a cylinder of platinum–iridium alloy kept by the International Bureau of Weights and Measures at Sevres, France. 1 kg = 1000 g. (1.5)

Kilojoule (kJ): 1000 J. (6.1)

Kinetic Energy (KE): Energy of motion. KE = $(1/2)mv^2$. (6.1)

Kinetic Molecular Theory: Molecules of a substance are in constant motion with a distribution of kinetic energies at a given temperature. The average kinetic energy of the molecules is proportional to the Kelvin temperature. (6.2)

Kinetic Molecular Theory of Gases: A set of postulates used to explain the gas laws. A gas consists of an extremely large number of very tiny, very hard particles in constant, random motion. They have negligible volume and, between collisions, experience no forces between themselves. (10.8)

L

Lanthanide Elements: Elements 58–71. (2.3)

Lattice: A symmetrical pattern of points arranged with constant repeat distances along lines oriented at constant angles. (11.9)

Lattice Energy: Energy released by the imaginary process in which isolated ions come together to form a crystal of an ionic compound. (8.1)

Law: A description of behavior (and not an *explanation* of behavior) based on the results of many experiments. (1.2)

Law of Combining Volumes: When gases react at the same temperature and pressure, their combining volumes are in ratios of simple whole numbers. (10.4)

Law of Conservation of Energy: The energy of the universe is constant; it can be neither created nor destroyed but only transferred and transformed. (6.1)

Law of Conservation of Mass: No detectable gain or loss in mass occurs in chemical reactions. Mass is conserved. (2.1)

Law of Conservation of Mass–Energy: The sum of all the mass in the universe and of all of the energy, expressed as an equivalent in mass (calculated by the Einstein equation), is a constant. (20.1)

Law of Definite Proportions: In a given chemical compound, the elements are always combined in the same proportion by mass. (2.1)

Law of Gas Effusion: See *Effusion, Law of.*

Law of Multiple Proportions: Whenever two elements form more than one compound, the different masses of one element that combine with the same mass of the other are in a ratio of small whole numbers. (2.1)

Law of Partial Pressures: See *Partial Pressures, Dalton's Law of.*

Law of Radioactive Decay:

$$\text{Activity} = -\frac{\Delta N}{\Delta t} = kN,$$

where ΔN is the change in the number of radioactive nuclei during the time span Δt, and k is the decay constant. (20.6)

Le Châtelier's Principle: When a system that is in dynamic equilibrium is subjected to a disturbance that upsets the equilibrium, the system undergoes a change that counteracts the disturbance and, if possible, restores the equilibrium. (11.8, 14.6)

Lead Storage Battery: A galvanic cell of about 2 V involving lead and lead(IV) oxide in sulfuric acid. (19.8)

Leclanché Cell: See *Zinc–Manganese Dioxide Cell.*

Lewis Acid: An electron-pair acceptor. (15.3)

Lewis Base: An electron-pair donor. (15.3)

Lewis Structure (Lewis Formula): A structural formula drawn with Lewis symbols and that uses dots and dashes to show the valence electrons and shared pairs of electrons. (8.3)

Lewis Symbol: The symbol of an element that includes dots to represent the valence electrons of an atom of the element. (8.2)

Ligand: A molecule or an anion that can bind to a metal ion to form a complex. (17.4, 21.5)

Like Dissolves Like Rule: Strongly polar and ionic solutes tend to dissolve in polar solvents and nonpolar solutes tend to dissolve in nonpolar solvents. (12.1)

Limiting Reactant: The reactant that determines how much product can form when nonstoichiometric amounts of reactants are used. (3.5)

Line Spectrum: An atomic spectrum. So named because the light emitted by an atom and focused through a narrow slit yields a series of lines when projected on a screen. (7.1)

Linear Molecule: A molecule all of whose atoms lie on a straight line. (9.1, 9.2)

Lipid: Any substance found in plants or animals that can be dissolved in nonpolar solvents. (22.7)

Liquid: One of the states of matter. A liquid consists of tightly packed atomic or molecular sized particles that can move past each other. (1.4)

Liter (L): 1 dm^3. 1 L = 1000 mL = 1000 cm^3. (1.5)

Lithium Ion Cell: A cell in which lithium ions are transferred between the electrodes through an electrolyte, while electrons travel through the external circuit. (19.8)

Lithium–Manganese Dioxide Battery: A battery that uses metallic lithium as the anode and manganese dioxide as the cathode. (19.8)

Localized Bond: A covalent bond in which the bonding pair of electrons is localized between two nuclei. (9.8)

London Forces (Dispersion Forces): Weak attractive forces caused by instantaneous dipole–induced dipole attractions. (11.2)

Lone Pair: A pair of electrons in the valence shell of an atom that is not shared with another atom. An unshared pair of electrons. (9.2)

Low-Spin Complex: A coordination compound or a complex ion with electrons paired as much as possible in the lower energy set of *d* orbitals. (21.9)

M

Macromolecule: A molecule whose molecular mass is very large. (22.6)

Magic Numbers: The numbers 2, 8, 20, 28, 50, 82, and 126, numbers whose significance in nuclear science is that a nuclide in which the number of protons or neutrons equals a magic number has nuclei that are relatively more stable than those of other nuclides nearby in the band of stability. (20.4)

Magnetic Quantum Number (m_ℓ): A quantum number that can have values from $-\ell$ to $+\ell$. (7.3)

Main Group Elements: Elements in any of the A groups in the periodic table. (2.3)

Main Reaction: The desired reaction between the reactants as opposed to competing reactions that give by-products. (3.6)

Malleability: A metal's ability to be hammered or rolled into thin sheets. (2.4)

Manometer: A device for measuring the pressure within a closed system. The two types—*closed end* and *open end*—differ according to whether the operating fluid (e.g., mercury) is exposed at one end to the atmosphere. (10.2)

Mass: A measure of the amount of matter that there is in a given sample. (1.3)

Mass Action Expression: A fraction in which the numerator is the product of the molar concentrations of the products, each raised to a power equal to its coefficient in the equilibrium equation, and the denominator is the product of the molar concentrations of the reactants, each also raised to the power that equals its coefficient in the equation. (For gaseous reactions, partial pressures can be used in place of molar concentrations.) (14.2)

Mass Defect: For a given isotope, it is the mass that changed into energy as the nucleons gathered to form the nucleus, this energy being released from the system. (20.2)

Mass Number: The numerical sum of the protons and neutrons in an atom of a given isotope. (2.2)

Matter: Anything that has mass and occupies space. (1.3)

Mean: The sum of N numerical values divided by N; the average. (1.6)

Measurement: A numerical observation. (1.5)

Mechanism of a Reaction: The series of individual steps (called elementary processes) in a chemical reaction that gives the net, overall change. (13.7)

Melting Point: The temperature at which a substance melts; the temperature at which a solid is in equilibrium with its liquid state. (11.4)

Meniscus: The interface between a liquid and a gas.

Metal: An element or an alloy that is a good conductor of electricity, that has a shiny surface, and that is malleable and ductile; an element that normally forms positive ions and has an oxide that is basic. (2.4, 4.3)

Metallic Conduction: Conduction of electrical charge by the movement of electrons. (19.1)

Metallic Crystal: A solid having positive ions at the lattice positions that are attracted to a "sea of electrons" that extends throughout the entire crystal. (11.11)

Metalloids: Elements with properties that lie between those of metals and nonmetals, and that are found in the periodic table around the diagonal line running from boron (B) to astatine (At). (2.4)

Metallurgy: The science and technology of metals, the procedures and reactions that separate metals from their ores, and the operations that create practical uses for metals. (21.4)

Metathesis Reaction: See *Double Replacement Reaction.*

Meter (m): The SI base unit for length. (1.5)

Methyl Group: CH_3—. (22.2)

Metric Units: A decimal system of units for physical quantities taken over by the SI. (1.5) See also *International System of Units.*

Millibar: 1 mb = 10^{-3} bar. (10.2)

Milliliter (mL): 0.001 L. 1000 mL = 1 L. (1.5)

Millimeter (mm): 0.001 m. 1000 mm = 1 m. (1.5)

Millimeter of Mercury (mm Hg): A unit of measurement that is proportional to pressure; equal to 1/760 atm. 760 mm Hg = 1 atm. 1 mm Hg = 1 torr. (10.2)

Miscible: Mutually soluble. (12.1)

Mixture: Any matter consisting of two or more substances physically combined in no particular proportion by mass. (1.3)

MO Theory: See *Molecular Orbital Theory.*

Model, Theoretical: A picture or a mental construction derived from a set of ideas and assumptions that are imagined to be true because they can be used to explain certain observations and measurements (e.g., the model of an ideal gas). (1.2)

Molal Boiling Point Elevation Constant (K_b): The number of degrees (°C) per unit of molal concentration that a boiling point of a solution is higher than that of the pure solvent. (12.7)

Molal Concentration (m): The number of moles of solute in 1000 g of solvent. (12.5)

Molal Freezing Point Depression Constant (K_f): The number of degrees (°C) per unit of molal concentration that a freezing point of a solution is lower than that of the pure solvent. (12.7)

Molality: The molal concentration. (12.5)

Molar Concentration (M): The number of moles of solute per liter of solution. The molarity of a solution. (4.6)

Molar Enthalpy of Solution: See *Heat of Solution.*

Molar Heat Capacity: The heat that can raise the temperature of 1 mol of a substance by 1 °C; the heat capacity per mole. (6.3)

Molar Heat of Fusion, ΔH_{fusion}: The heat absorbed when 1 mol of a solid melts to give 1 mol of the liquid at constant temperature and pressure. (11.7)

Molar Heat of Sublimation, $\Delta H_{sublimation}$: The heat absorbed when 1 mol of a solid sublimes to give 1 mol of its vapor at constant temperature and pressure. (11.7)

Molar Heat of Vaporization, $\Delta H_{vaporization}$: The heat absorbed when 1 mol of a liquid changes to 1 mol of its vapor at constant temperature and pressure. (11.7)

Molar Mass: The mass of one mole of a substance; the mass in grams equal to the sum of the atomic masses of the atoms in a substance, with units of g mol^{-1}. (3.1)

Molar Solubility: The number of moles of solute required to give 1 L of a saturated solution of the solute. (17.1)

Molar Volume, Standard: The volume of 1 mol of a gas at STP; 22.4 L mol^{-1}. (10.4)

Molarity: See *Molar Concentration.*

Mole (mol): The SI unit for amount of substance; the formula mass in grams of an element or compound; an amount of a chemical substance that contains 6.022×10^{23} formula units. (3.1)

Mole Fraction: The ratio of the number of moles of one component of a mixture to the total number of moles of all components. (10.6)

Mole Percent (mol%): The mole fraction of a component expressed as a percent; mole fraction × 100%. (10.6)

Molecular Compound: A compound consisting of electrically neutral molecules. (2.6)

Molecular Crystal: A crystal that has molecules or individual atoms at the lattice points. (11.11)

Molecular Equation: A chemical equation that gives the full formulas of all of the reactants and products and that is used to plan an actual experiment. (4.2)

Molecular Formula: A chemical formula that gives the actual composition of one molecule. (2.6, 3.3)

Molecular Kinetic Energy: The energy associated with the motions of and within molecules as they fly about, spinning and vibrating. (6.2)

Molecular Mass: The sum of the atomic masses (in u) of all of the atoms represented in a molecular chemical substance; also called the *molecular weight.* May be used with units of g mol^{-1}. See also *Molar Mass.* (3.1)

Molecular Orbital (MO): An orbital that extends over two or more atomic nuclei. (9.7)

Molecular Orbital Theory (MO Theory): A theory about covalent bonds that views a molecule as a collection of positive nuclei surrounded by electrons distributed among a set of bonding, antibonding, and nonbonding orbitals of different energies. (9.4, 9.7)

Molecular Weight: See *Molecular Mass.*

Molecule: A neutral particle composed of two or more atoms combined in a definite ratio of whole numbers. (1.2, 2.6)

Monatomic: A particle consisting of just one atom. (2.9)

Monoclinic Sulfur: An allotrope of sulfur. (21.2)

Monodentate Ligand: A ligand that can attach itself to a metal ion by only one atom. (21.5)

Monomer: A substance of relatively low formula mass that is used to make a polymer. (22.6)

Monoprotic Acid: An acid that can furnish one H^+ per molecule. (4.3)

Monosaccharide: A carbohydrate that cannot be hydrolyzed. (22.7)

N

Negative Charge: A type of electrical charge possessed by certain particles such as the electron. A negative charge is attracted by a positive charge and is repelled by another negative charge. (2.2)

Nernst Equation: An equation relating cell potential and concentration. (19.5)

$$E_{cell} = E^\circ_{cell} - \frac{RT}{n\mathcal{F}} \ln Q$$

Net Ionic Equation: An ionic equation from which spectator ions have been omitted. It is balanced when both atoms and electrical charge balance. (4.2)

Network Solid: See *Covalent Crystal.*

Neutralization, Acid–Base: See *Acid–Base Neutralization.*

Neutral Solution: A solution in which $[H^+] = [OH^-]$. (15.5)

Neutron (n, ${}^1_0n$): A subatomic particle with a charge of zero, a mass of 1.0086649 u (1.674927×10^{-24} g) and that exists in all atomic nuclei except those of the hydrogen-1 isotope. (2.2)

Neutron Activation Analysis: A technique to analyze for trace impurities in a sample by studying the frequencies and intensities of the gamma radiations they emit after they have been rendered radioactive by neutron bombardment of the sample. (20.7)

Neutron Emission: A nuclear reaction in which a neutron is ejected. (20.3)

Nicad Battery: A nickel–cadmium cell. (19.8)

Nickel–Cadmium Storage Cell: A galvanic cell of about 1.4 V involving the reaction of cadmium with nickel(IV) oxide. (19.8)

Nitrogen Family: Group VA in the periodic table—nitrogen, phosphorus, arsenic, antimony, and bismuth. (2.3)

Noble Gases: Group VIIIA in the periodic table—helium, neon, argon, krypton, xenon, and radon. (2.3)

Nodal Plane: A plane that can be drawn to separate opposing lobes of *p*, *d*, and *f* orbitals. (7.7)

Node: A place where the amplitude or intensity of a wave is zero. (7.3)

Nomenclature: The names of substances and the rules for devising names. (2.9)

Nonbonding Domain: A region in the valence shell of an atom that holds an unshared pair of electrons and that influences the shape of a molecule. (9.2)

Nonbonding Molecular Orbital: A molecular orbital that has no net effect on the stability of a molecule when populated with electrons and that is localized on one atom in the molecule. (9.7)

Nonelectrolyte: A compound that in its molten state or in solution cannot conduct electricity. (4.2)

Nonlinear Molecule: A molecule in which the atoms do not lie in a straight line. (9.2)

Nonmetal: A nonductile, nonmalleable, nonconducting element that tends to form negative ions (if it forms them at all) far more readily than positive ions and whose oxide is likely to show acidic properties. (2.4)

Nonmetallic Element: An element without metallic properties; an element with poor electrical conductivity. (2.4)

Nonoxidizing acid: An acid in which the anion is a poorer oxidizing agent than the hydrogen ion (e.g., HCl, H_2SO_4, H_3PO_4). (5.3)

Nonpolar Covalent Bond: A covalent bond in which the electron pair(s) are shared equally by the two atoms. (8.4)

Nonpolar Molecule: A molecule that has no net dipole moment. (9.3)

Nonvolatile: Descriptive of a substance with a high boiling point, a low vapor pressure. and that does not evaporate. (12.6)

Normal Boiling Point: The temperature at which the vapor pressure of a liquid equals 1 atm. (11.6)

Nuclear Binding Energy: See *Binding Energy, Nuclear.*

Nuclear Chain Reaction: A self-sustaining nuclear reaction. (20.8)

Nuclear Equation: A description of a nuclear reaction that uses the special symbols of isotopes, that describes some kind of nuclear transformation or disintegration, and that is balanced when the sums of the atomic numbers on either side of the arrow are equal and the sums of the mass numbers are also equal. (20.3)

Nuclear Fission: See *Fission.*

Nuclear Fusion: See *Fusion.*

Nuclear Radiation: Alpha, beta, or gamma radiation emitted by radioactive nuclei. (20.3)

Nuclear Reaction: A change in the composition or energy of the nuclei of isotopes accompanied by one or more events such as the radiation of nuclear particles or electromagnetic energy, transmutation, fission, or fusion. (13.4, 20.3)

Nucleic Acids: Polymers in living cells that store and translate genetic information and whose molecules hydrolyze to give a sugar unit (ribose from ribonucleic acid, RNA, or deoxyribose from deoxyribonucleic acid, DNA), a phosphate, and a set of four of the five nitrogen-containing, heterocyclic bases (adenine, thymine, guanine, cytosine, and uracil). (22.8)

Nucleon: A proton or a neutron. (2.2)

Nucleus: The hard, dense core of an atom that holds the atom's protons and neutrons. (2.2)

Nylon 6,6: A polymer of a six-carbon dicarboxylic acid and a six-carbon diamine. (22.6)

O

Observation: A statement that accurately describes something we see, hear, taste, feel, or smell. (1.2)

Octahedral Molecule: A molecule in which a central atom is surrounded by six atoms located at the vertices of an imaginary octahedron. (9.1)

Octahedron: An eight-sided figure that can be envisioned as two square pyramids sharing the common square base. (9.1)

Octet (of Electrons): Eight electrons in the valence shell of an atom. (8.1)

Octet Rule: An atom tends to gain or lose electrons until its outer shell has eight electrons. (8.1, 8.3)

Odd–Even Rule: When the numbers of protons and neutrons in an atomic nucleus are both even, the isotope is more likely to be stable than when both numbers are odd. (20.4)

Open System: A system that can exchange both matter and energy with its surroundings. (6.3)

Open-End Manometer: See *Manometer.*

Optical Isomers: Stereoisomers other than geometric (cis–trans) isomers and that include substances that can rotate the plane of plane-polarized light. (21.8)

Orbital: An electron waveform with a particular energy and a unique set of values for the quantum numbers n, ℓ, and m_ℓ. (7.3)

Orbital Diagram: A diagram in which the electrons in an atom's orbitals are represented by arrows to indicate paired and unpaired spins. (7.5)

Order (of a Reaction): The sum of the exponents in the rate law is the *overall* order. Each exponent gives the order of the reaction with respect to a specific reactant. (13.3)

Ore: A substance in the earth's crust from which an element or compound can be extracted at a profit. (21.4)

Organic Acid: An acid that contains the carboxyl group, $-\overset{\overset{\displaystyle O}{\|}}{C}-OH$. (8.3, 22.6)

Organic Chemistry: The study of the compounds of carbon that are not classified as inorganic. (2.6, 22.1)

Organic Compound: Any compound of carbon other than a carbonate, bicarbonate, cyanide, cyanate, carbide, or gaseous oxide. (2.6)

Orthorhombic Sulfur: The most stable allotrope of sulfur, composed of S_8 rings. (21.2)

Osmosis: The passage of solvent molecules, but not those of solutes, through a semipermeable membrane; the limiting case of dialysis. (12.8)

Osmotic Membrane: A membrane that allows passage of solvent, but not solute particles. (12.8)

Osmotic Pressure: The back pressure that would have to be applied to prevent osmosis; one of the colligative properties. (12.8)

Outer Electrons: The electrons in the occupied shell with the largest principal quantum number. An atom's electrons in its valence shell. (7.6)

Outer Shell: The occupied shell in an atom having the highest principal quantum number (n). (7.6)

Overall Order of Reaction: The sum of the exponents on the concentration terms in a rate law. (13.3)

Overlap of Orbitals: A portion of two orbitals from different atoms that share the same space in a molecule. (9.4)

Oxidation: A change in which an oxidation number increases (becomes more positive). A loss of electrons. (5.1)

Oxidation Number: The charge that an atom in a molecule or ion would have if all of the electrons in its bonds belonged entirely to the more electronegative atoms; the oxidation state of an atom. (5.1)

Oxidation State: See *Oxidation Number.*

Oxidation–Reduction Reaction: A chemical reaction in which changes in oxidation numbers occur. (5.1)

Oxidizing Acid: An acid in which the anion is a stronger oxidizing agent than H^+ (e.g., $HClO_4$, HNO_3). (5.3)

Oxidizing Agent: The substance that causes oxidation and that is itself reduced. (5.1)

Oxoacid: An acid that contains oxygen besides hydrogen and another element (e.g., HNO_3, H_3PO_4, H_2SO_4). (4.4, 15.2)

Oxoanion: The anion of an oxoacid (e.g., ClO_4^-, SO_4^{2-}). (15.2)

Oxygen Family: Group VIA in the periodic table—oxygen, sulfur, selenium, tellurium, and polonium. (2.3)

Ozone: A very reactive allotrope of oxygen with the formula O_3. (21.2)

P

Pairing Energy: The energy required to force two electrons to become paired and occupy the same orbital. (21.9)

Paramagnetism: The weak magnetism of a substance whose atoms, molecules, or ions have one or more unpaired electrons. (7.4)

Partial Charge: Charges at opposite ends of a dipole that are fractions of full 1+ or 1− charges. (8.4)

Partial Pressure: The pressure contributed by an individual gas to the total pressure of a gas mixture. (10.6)

Partial Pressure, Law of (Dalton's Law of Partial Pressures): The total pressure of a mixture of gases equals the sum of their partial pressures. (10.6)

Pascal (Pa): The SI unit of pressure equal to 1 newton m^{-2}; 1 atm = 101,325 Pa. (10.2)

Pauli Exclusion Principle: No two electrons in an atom can have the same values for all four of their quantum numbers. (7.4)

Peptide Bond: The amide linkage in molecules of polypeptides. (22.7)

Percentage by Mass (Percentage by Weight): (a) The number of grams of an element combined in 100 g of a compound. (3.3) (b) The number of grams of a substance in 100 g of a mixture or solution. (12.5)

Percentage Composition: A list of the percentages by weight of the elements in a compound. (3.3)

Percentage Concentration: A ratio of the amount of solute to the amount of solution expressed as a percent. (4.1)

Weight/weight: Grams of solute in 100 g of solution.

Weight/volume: Grams of solute in 100 mL of solution.

Volume/volume: Volumes of solute in 100 volumes of solution.

Percentage Ionization: An equation that quantifies the ionization of a substance in solution. (16.2)

$$\text{Percentage ionization} = \frac{\text{amount of substance ionized}}{\text{initial amount of substance}} \times 100\%$$

Percentage Yield: The ratio (taken as a percent) of the mass of product obtained to the mass calculated from the reaction's stoichiometry. (3.6)

Period: A horizontal row of elements in the periodic table. (2.3)

Periodic Table: A table in which symbols for the elements are displayed in order of increasing atomic number and arranged so that elements with similar properties lie in the same column (group). (Inside front cover, 2.3)

pH: $-\log [H^+]$. (15.5)

Phase: A homogeneous region within a sample. (1.3)

Phase Diagram: A pressure–temperature graph on which are plotted the temperatures and the pressures at which equilibrium exists between the states of a substance. It defines regions of T and P in which the solid, liquid, and gaseous states of the substance can exist. (11.12)

Photon: A unit of energy in electromagnetic radiation equal to $h\nu$, where ν is the frequency of the radiation and h is Planck's constant. (7.1)

Photosynthesis: The use of solar energy by a plant to make high-energy molecules from carbon dioxide, water, and minerals. (6.4)

Physical Change: A change that is not accompanied by a change in chemical makeup. (1.3)

Physical Law: A relationship between two or more physical properties of a system, usually expressed as a mathematical equation, that describes how a change in one property affects the others.

Physical Property: A property that can be specified without reference to another substance and that can be measured without causing a chemical change. (1.4)

Physical State: The condition of aggregation of a substance's formula units, whether as a solid, a liquid, or a gas. (1.4)

Pi Bond (π Bond): A bond formed by the sideways overlap of a pair of p orbitals and that concentrates electron density into two separate regions that lie on opposite sides of a plane that contains an imaginary line joining the nuclei. (9.6)

Pig Iron: The impure iron made by a blast furnace. (21.4)

pK_a: $-\log K_a$. (16.1)

pK_b: $-\log K_b$. (16.1)

pK_w: $-\log K_w$. (15.5)

Planar Triangular Molecule: A molecule in which a central atom holds three other atoms located at the corners of an equilateral triangle and that includes the central atom at its center. (9.1)

Plane-Polarized Light: Light in which all the oscillations occur in one plane. (21.8)

Planck's Constant (h): The ratio of the energy of a photon to its frequency; $6.6260755 \times 10^{-34}$ J Hz^{-1}. (7.1)

pOH: $-\log [OH^-]$. (15.5)

Plasma: An electrically neutral, very hot gaseous mixture of nuclei and unattached electrons. (20.8)

Polar Covalent Bond (Polar Bond): A covalent bond in which more than half of the bond's negative charge is concentrated around one of the two atoms. (8.4)

Polar Molecule: A molecule in which individual bond polarities do not cancel and in which, therefore, the centers of density of negative and positive charges do not coincide. (8.4)

Polarizability: A term that describes the ease with which the electron cloud of a molecule or ion is distorted. (11.2)

Polyatomic Ion: An ion composed of two or more atoms. (2.8)

Polydentate Ligand: A ligand that has two or more atoms that can become simultaneously attached to a metal ion. (21.5)

Polymer: A substance consisting of macromolecules that have repeating structural units. (22.6)

Polymerization: A chemical reaction that converts a monomer into a polymer. (22.6)

Polypeptide: A polymer of α-amino acids that makes up all or most of a protein. (22.7)

Polyprotic Acid: An acid that can furnish more than one H^+ per molecule. (4.3)

Polysaccharide: A carbohydrate whose molecules can be hydrolyzed to hundreds of monosaccharide molecules. (22.7)

Polystyrene: An addition polymer of styrene with the following structure. (22.6)

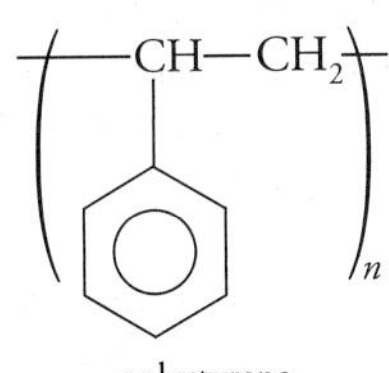

polystyrene

Position of Equilibrium: The relative amounts of the substances on both sides of the double arrows in the equation for an equilibrium. (4.3, 11.8)

Positive Charge: A type of electrical charge possessed by certain particles such as the proton. A positive charge is attracted by a negative charge and is repelled by another positive charge. (2.2)

Positron ($^{1}_{0}p$): A positively charged particle with the mass of an electron. (20.3)

Post-transition Metal: A metal that occurs in the periodic table immediately to the right of a row of transition elements. (2.8)

Potential: See *Volt.*

Potential Energy (PE): Stored energy. (6.1)

Potential Energy Diagram: A diagram indicating the conversion of kinetic energy to potential energy and back again as atoms or molecules collide and then recoil in a chemical reaction. (13.5)

Precipitate: A solid that separates from a solution usually as the result of a chemical reaction. (4.1)

Precipitation Reaction: A reaction in which a precipitate forms. (4.1, 4.5)

Precision: How reproducible measurements are; the fineness of a measurement as indicated by the number of significant figures reported in the physical quantity. (1.6)

Pre-exponential Factor: A number or variable that precedes the exponential part of a number. (13.6)

Pressure: Force per unit area. (6.5, 10.2)

Pressure–Concentration Law: See *Vapor Pressure–Concentration Law.*

Pressure–Solubility Law (Henry's Law): The concentration of a gas dissolved in a liquid at any given temperature is directly proportional to the partial pressure of the gas above the solution. (12.4)

Pressure–Temperature Law (Gay-Lussac's Law): The pressure of a given mass of gas is directly proportional to its Kelvin temperature if the volume is kept constant. $P \propto T$. (10.3)

Pressure–Volume Law (Boyle's Law): The volume of a given mass of a gas is inversely proportional to its pressure if the temperature is kept constant. $V \propto 1/P$. (10.3)

Pressure–Volume Work (*P–V* Work): The energy transferred as work when a system expands or contracts against the pressure exerted by the surroundings. At constant pressure, $w = -P\Delta V$. (6.5)

Primary Cell: A galvanic cell (battery) not designed to be recharged; it is discarded after its energy is depleted. (19.8)

Primitive Cubic Unit Cell: A cubic unit cell that has atoms only at the corners of the cell. (11.9)

Principal Quantum Number (*n*): The quantum number that defines the principal energy levels and that can have values of 1, 2, 3,. . ., ∞. (7.3)

Products: The substances produced by a chemical reaction and whose formulas follow the arrows in chemical equations. (2.5)

Propagation Step: A step in a chain reaction for which one product must serve in a succeeding propagation step as a reactant and for which another (final) product accumulates with each repetition of the step. (Facets of Chemistry 13.1)

Property: A characteristic of matter. (1.4)

Propyl Group: $CH_3CH_2CH_2$—. (22.2)

Protein: A macromolecular substance found in cells that consists wholly or mostly of one or more polypeptides that often are combined with an organic molecule or a metal ion. (22.7)

Proton (^1_1p or $^1_1H^+$): (a) A subatomic particle, with a charge of 1+ and a mass of 1.0072765 u ($1.6726217 \times 10^{-24}$ g) and that is found in atomic nuclei. (2.2) (b) The name often used for the hydrogen ion and symbolized as H^+. (4.3)

Proton Acceptor: A Brønsted base. (15.1)

Proton Donor: A Brønsted acid. (15.1)

Pure Substance: An element or a compound. (1.3)

Q

Qualitative Analysis: The use of experimental procedures to determine what elements are present in a substance. (4.8)

Qualitative Observation: Observations that do not involve numerical information. (1.5)

Quanta: Packets of electromagnetic radiation now commonly called photons. (7.1)

Quantitative Analysis: The use of experimental procedures to determine the percentage composition of a compound or the percentage of a component of a mixture. (4.8)

Quantitative Observation: An observation involving a measurement and numerical information. (1.5)

Quantized: Descriptive of a discrete, definite amount as of *quantized energy.* (7.2)

Quantum: The energy of one photon. (7.1)

Quantum Mechanics: See *Wave Mechanics.*

Quantum Number: A number related to the energy, shape, or orientation of an orbital, or to the spin of an electron. (7.2)

Quantum Theory: The physics of objects that exhibit wave/particle duality.

R

R: See *Gas Constant, Universal.*

Rad (rd): A unit of radiation-absorbed dose and equal to 10^{-5} J g^{-1} or 10^{-2} Gy. (20.6)

Radioactive Decay: The change of a nucleus into another nucleus (or into a more stable form of the same nucleus) by the loss of a small particle or a gamma ray photon. (20.3)

Radioactive Disintegration Series: A sequence of nuclear reactions beginning with a very long-lived radionuclide and ending with a stable isotope of lower atomic number. (20.3)

Radioactivity: The emission of one or more kinds of radiation from an isotope with unstable nuclei. (20.3)

Radiological Dating: A technique for measuring the age of a geologic formation or an ancient artifact by determining the ratio of the concentrations of two isotopes, one radioactive and the other a stable decay product. (13.4, 20.7)

Radionuclide: A radioactive isotope. (20.3)

Raoult's Law: See *Vapor Pressure–Concentration Law.*

Rare Earth Metals: The lanthanides. (2.3)

Rate: A ratio in which a unit of time appears in the denominator, for example, 40 miles hr^{-1} or 3.0 mol L^{-1} s^{-1}. (13.2)

Rate Constant (*k*): The proportionality constant in the rate law; the rate of reaction when all reactant concentrations are 1 *M*. (13.3)

Rate Law: An equation that relates the rate of a reaction to the molar concentrations of the reactants raised to powers. (13.3)

Rate of Reaction: How quickly the reactants disappear and the products form and usually expressed in units of mol L^{-1} s^{-1}. (13.1)

Rate-Determining Step (Rate-Limiting Step): The slowest step in a reaction mechanism. (13.7)

Reactant, Limiting: See *Limiting Reactant.*

Reactants: The substances brought together to react and whose formulas appear before the arrow in a chemical equation. (2.5)

Reaction Coordinate: The horizontal axis of a potential energy diagram of a reaction. (13.5)

Reaction Quotient (*Q*): The numerical value of the mass action expression. See *Mass Action Expression.* (14.2)

Reactivity: A description of the tendency for a substance to undergo reaction. For a metal, it is the tendency to undergo oxidation. (8.5)

Red Phosphorus: A relatively unreactive allotrope of phosphorus. (21.2)

Redox Reaction: An oxidation–reduction reaction. (5.1)

Reducing Agent: A substance that causes reduction and is itself oxidized. (5.1)

Reduction: A change in which an oxidation number decrease (becomes less positive and more negative). A gain of electrons. (5.1)

Reduction Potential: A measure of the tendency of a given half-reaction to occur as a reduction. (19.2)

Refining: The industrial conversion of a compound (ore) containing a desired element into a pure form of the element. (21.4)

Refractory: A high-melting heat-resistant material used to line furnaces and rocket engines, and to shield the space shuttle from the high heat of re-entry. (21.4)

Rem: A dose in rads multiplied by a factor that takes into account the variations that different radiations have in their damage-causing abilities in tissue. (20.6)

Replication: In nucleic acid chemistry, the reproductive duplication of DNA double helices prior to cell division. (22.8)

Representative Element: An element in one of the A groups in the periodic table. (2.3)

Resonance: A concept in which the actual structure of a molecule or polyatomic ion is represented as a composite or average of two or more Lewis structures, which are called the resonance or contributing structures (and none of which has real existence). (8.7)

Resonance Energy: The difference in energy between a substance and its principal resonance (contributing) structure. (8.7)

Resonance Hybrid: The actual structure of a molecule or polyatomic ion taken as a composite or average of the resonance or contributing structures. (8.7)

Resonance Structure: A Lewis structure that contributes to the hybrid structure in resonance-stabilized systems; a contributing structure (8.7)

Reverse Reaction: In a chemical equation, the reaction as read from right to left. (4.3)

Reversible Process: A process that occurs by an infinite number of steps during which the driving force for the change is just barely greater than the force that resists the change. (18.7)

Reversible Reaction: A reaction capable of proceeding in either the forward or reverse direction. (13.5, 14.7, 18.7)

Ring, Carbon: A closed-chain sequence of carbon atoms. (22.1)

RNA: Ribonucleic acid; a nucleic acid that gives ribose, phosphate ion, adenine, uracil, guanine, and cytosine when hydrolyzed. It occurs in several varieties. (22.8)

Roasting: Heating a sulfide ore in air to convert it to an oxide. (21.4)

Rock Salt Structure: The face-centered cubic structure observed for sodium chloride, which is also possessed by crystals of many other compounds. (11.9)

Root Mean Square Speed (rms Speed): The square root of the average of the speeds-squared of the molecules in a substance. (10.8)

Rydberg Equation: An equation used to calculate the wavelengths of all the spectral lines of hydrogen. (7.2)

S

Salt: An ionic compound in which the anion is not OH^- or O^{2-} and the cation is not H^+. (4.2, 4.3)

Salt Bridge: A tube that contains an electrolyte that connects the two half-cells of a galvanic cell. (19.1)

Saponification: The reaction of an organic ester with a strong base to give an alcohol and the salt of the organic acid. (22.5)

Saturated Organic Compound: A compound whose molecules have only single bonds. (22.2)

Saturated Solution: A solution that holds as much solute as it can at a given temperature. A solution in which there is an equilibrium between the dissolved and the undissolved states of the solute. (4.1) (17.1)

Scanning Tunneling Microscope (STM): An instrument that enables the imaging of individual atoms on the surface of an electrically conducting specimen. (2.1)

Scientific Law: See *Law.*

Scientific Method: The observation, explanation, and testing of an explanation by additional experiments. (1.2)

Scientific Notation: The representation of a quantity as a decimal number between 1 and 10 multiplied by 10 raised to a power (e.g., 6.02×10^{23}). (1.5)

Scintillation Counter: A device for measuring nuclear radiation that contains a sensor composed of a substance called a *phosphor* that emits a tiny flash of light when struck by a particle of ionizing radiation. These flashes can be magnified electronically and automatically counted. (20.6)

Second Law of Thermodynamics: Whenever a spontaneous event takes place, it is accompanied by an increase in the entropy of the universe. (18.4)

Second-Order Reaction: A reaction with a rate law of the type: rate $= k[A]^2$ or rate $= k[A][B]$, where A and B are reactants. (13.3)

Secondary Cell: A galvanic cell (battery) designed for repeated use; it is able to be recharged. (19.8)

Secondary Quantum Number (ℓ): The quantum number whose values can be $0, 1, 2, \ldots, (n-1)$, where n is the principal quantum number. (7.3)

Seesaw Shaped Molecule: A description given to a molecule in which the central atom has five electron pairs in its valence shell, one of which is a lone pair and the others are used in bonds to other atoms. See also *Distorted Tetrahedron.* (9.2)

Selective Precipitation: A technique that uses differences in the solubilities of specific salts to separate ions from each other. (17.3)

Semiconductor: A substance that conducts electricity weakly. (2.4)

Shell: All of the orbitals associated with a given value of n (the principal quantum number). (7.3)

SI (International System of Units): The modified metric system adopted in 1960 by the General Conference on Weights and Measures. (1.5)

Side Reaction: A reaction the occurs simultaneously with another reaction (the main reaction) in the same mixture to produce by-products. (3.6)

Sievert (Sv): The SI unit for dose equivalent. The dose equivalent H is calculated from D (the dose in grays), Q (a measure of the effectiveness of the radiation at causing harm), and N (a variable that accounts for other modifying factors). $H = DQN$. (20.6)

Sigma Bond (σ Bond): A bond formed by the head-to-head overlap of two atomic orbitals and in which electron density becomes concentrated along and around the imaginary line joining the two nuclei. (9.6)

Significant Figures (Significant Digits): The digits in a physical measurement that are known to be certain plus the first digit that contains uncertainty. (1.6)

Simple Cubic Unit Cell: See *Primitive Cubic Unit Cell.*

Simplest Formula: See *Empirical Formula.*

Single Bond: A covalent bond in which a single pair of electrons is shared. (8.3)

Single Replacement Reaction: A reaction in which one element replaces another in a compound; usually a redox reaction. (5.4)

Skeletal Structure: A diagram of the arrangement of atoms in a molecule, which is the first step in constructing the Lewis structure. (8.6)

Skeleton Equation: An unbalanced equation showing only the formulas of reactants and products. (5.2)

Slag: A relatively low melting mixture of impurities that forms in a blast furnace or other furnaces used to refine metals. (21.4)

Smelting: A process in which a metal oxide is heated with a reducing agent in order to obtain the free metal. (21.3)

Solid: One of the states of matter. A solid consists of tightly packed atomic or molecular sized particles held rigidly in place. (1.4)

Solubility: The ratio of the quantity of solute to the quantity of solvent in a saturated solution and that is usually expressed in units of (g solute)/(100 g solvent) at a specified temperature. (4.1)

Solubility Product Constant (K_{sp}): The equilibrium constant for the solubility of a salt and that, for a saturated solution, is equal to the product of the molar concentrations of the ions, each raised to a power equal to the number of its ions in one formula unit of the salt. (17.1) See also *Acid Solubility Product.*

Solubility Rules: A set of rules describing salts that are soluble and those that are insoluble. They enable the prediction of the formation of a precipitate in a metathesis reaction. (4.5)

Solute: Something dissolved in a solvent to make a solution. (4.1)

Solution: A homogeneous mixture in which all particles are of the size of atoms, small molecules, or small ions. (1.3, 4.1)

Solvation: The development of a cage-like network of a solution's solvent molecules about a molecule or ion of the solute. (12.1)

Solvation Energy: The enthalpy of the interaction of gaseous molecules or ions of solute with solvent molecules during the formation of a solution. (12.2)

Solvent: A medium, usually a liquid, into which something (a solute) is dissolved to make a solution. (4.1)

***sp* Hybrid Orbital:** A hybrid orbital formed by mixing one *s* and one *p* atomic orbital. The angle between a pair of *sp* hybrid orbitals is 180°. (9.5)

sp^2 Hybrid Orbital: A hybrid orbital formed by mixing one *s* and two *p* atomic orbitals. sp^2 hybrids are planar triangular with the angle between two sp^2 hybrid orbitals being 120°. (9.5)

sp^3 Hybrid Orbital: A hybrid orbital formed by mixing one *s* and three *p* atomic orbitals. sp^3 hybrids point to the corners of a tetrahedron; the angle between two sp^3 hybrid orbitals is 109.5°. (9.5)

sp^3d Hybrid Orbital: A hybrid orbital formed by mixing one *s*, three *p*, and one *d* atomic orbital. sp^3d hybrids point to the corners of a trigonal bipyramid. (9.5)

sp^3d^2 Hybrid Orbital: A hybrid orbital formed by mixing one *s*, three *p*, and two *d* atomic orbitals. sp^3d^2 hybrids point to the corners of an octahedron. (9.5)

Specific Heat (Specific Heat Capacity): The quantity of heat that will raise the temperature of 1 g of a substance by 1 °C, usually in units of cal g^{-1} $°C^{-1}$ or J g^{-1} $°C^{-1}$. (6.3)

Spectator Ion: An ion whose formula appears in an ionic equation identically on both sides of the arrow, that does not participate in the reaction, and that is excluded from the net ionic equation. (4.2)

Spectrochemical Series: A listing of ligands in order of their ability to produce a large crystal field splitting. (21.9)

Speed of Light (c): The speed at which light travels in a vacuum; 3.00×10^8 m s^{-1}. (7.1)

Spin Quantum Number (m_s): The quantum number associated with the spin of a subatomic particle and for the electron can have a value of $+\frac{1}{2}$ or $-\frac{1}{2}$. (7.4)

Spontaneous Change: A change that occurs by itself without outside assistance. (18.2)

Square Planar Molecule: A molecule with a central atom having four bonds that point to the corners of a square. (9.2)

Square Pyramid: A pyramid with four triangular sides and a square base. (9.2)

Stability Constant: See *Formation Constant.*

Stabilization Energy: See *Resonance Energy.*

Standard Atmosphere: See *Atmosphere, Standard*

Standard Cell Notation: A way of describing the anode and cathode half-cells in a galvanic cell. The anode half-cell is specified on the left, with the electrode material of the anode given first and a vertical bar representing the phase boundary between the electrode and the solution. Double bars represent the salt bridge between the half-cells. The cathode half-cell is specified on the right, with the material of the cathode given last. Once again, a single vertical bar represents the phase boundary between the solution and the electrode. (19.1)

Standard Cell Potential ($E°_{cell}$): The potential of a galvanic cell at 25 °C and when all ionic concentrations are exactly 1 *M* and the partial pressures of all gases are 1 atm. (19.2)

Standard Conditions of Temperature and Pressure (STP): Standard reference conditions for gases. 273 K (0 °C) and 1 atm (760 torr). (10.4)

Standard Enthalpy Change ($\Delta H°$): See *Standard Heat of Reaction.*

Standard Enthalpy of Formation ($\Delta H_f°$): See *Standard Heat of Formation.*)

Standard Entropy ($S°$): The entropy of 1 mol of a substance at 25 °C and 1 atm. (18.5)

Standard Entropy Change ($\Delta S°$): The entropy change of a reaction when determined with reactants and products at 25 °C and 1 atm and on the scale of the mole quantities given by the coefficients of the balanced equation. (18.5)

Standard Entropy of Formation ($\Delta S_f°$): The value of $\Delta S°$ for the formation of one mole of a substance from its elements in their standard states. (18.5)

Standard Free Energy Change ($\Delta G°$): $\Delta G° = \Delta H° - T\Delta S°$. (18.6)

Standard Free Energy of Formation ($\Delta G_f°$): The value of $\Delta G°$ for the formation of *one* mole of a compound from its elements in their standard states. (18.6)

Standard Heat of Combustion ($\Delta H_c°$): The enthalpy change for the combustion of one mole of a compound under standard conditions. (6.8)

Standard Heat of Formation ($\Delta H_f°$): The amount of heat absorbed or evolved when one mole of the compound is formed from its elements in their standard states. (6.8)

Standard Heat of Reaction ($\Delta H°$): The enthalpy change of a reaction when determined with reactants and products at 25 °C and 1 atm and on the scale of the mole quantities given by the coefficients of the balanced equation. (6.6)

Standard Hydrogen Electrode: See *Hydrogen Electrode.*

Standard Molar Volume: See *Molar Volume, Standard.*

Standard Reduction Potential $E°_{cell}$: The reduction potential of a half-reaction at 25 °C when all ion concentrations are 1 *M* and the partial pressures of all gases are 1 atm. Also called standard electrode potential. (19.2)

Standard Solution: Any solution whose concentration is accurately known. (4.8)

Standard State: The condition in which a substance is in its most stable form at 25 °C and 1 atm. (6.6, 6.8)

Standing Wave: A wave whose peaks and nodes do not change position. (7.3)

Starch: A polymer of glucose used by plants to store energy. (22.7)

State Function: A quantity whose value depends only on the initial and final states of the system and not on the path taken by the system to get from the initial to the final state. (*P, V, T, H, S,* and *G* are all state functions.) (6.2)

State of a System: The set of specific values of the physical properties of a system—its composition, physical form, concentration, temperature, pressure, and volume. (6.2)

State of Matter: A physical state of a substance: solid, liquid, or gas. See also *Standard State.* (1.4)

Stereoisomerism: The existence of isomers whose structures differ only in spatial orientations (e.g., geometric isomers and optical isomers). (21.8)

Stock System: A system of nomenclature that uses Roman numerals to specify oxidation states. (2.9)

Stoichiometric Equivalence: The ratio by moles between two elements in a formula or two substances in a chemical reaction. (3.2)

Stoichiometry: A description of the relative quantities by moles of the reactants and products in a reaction as given by the coefficients in the balanced equation. (3 Introduction)

Stopcock: A valve on a buret that is used to control the flow of titrant. (4.8)

Stored Energy: See *Potential Energy.*

STP: See *Standard Conditions of Temperature and Pressure.*

Straight-Chain Compound: An organic compound in whose molecules the carbon atoms are joined in one continuous open-chain sequence. (22.1)

Strong Acid: An acid that is essentially 100% ionized in water. A good proton donor. An acid with a large value of K_a. (4.3)

Strong Base: Any powerful proton acceptor. A base with a large value of K_b. A metal hydroxide that dissociates essentially 100% in water. (4.3)

Strong Electrolyte: Any substance that ionizes or dissociates in water to essentially 100%. (4.2, 4.3)

Structural Formula (Lewis Structure): A chemical formula that shows how the atoms of a molecule or polyatomic ion are arranged, to which other atoms they are bonded, and the kinds of bonds (single, double, or triple). (8.3)

Subatomic Particles: Electrons, protons, neutrons, and atomic nuclei. (2.2)

Sublimation: The conversion of a solid directly into a gas without passing through the liquid state. (11.3)

Subscript: In a chemical formula, a number after a chemical symbol, written below the line, and indicating the number of the preceding atoms in the formula (e.g., CH_4). Subscripts are also used to differentiate many variables such as the acid ionization constant (K_a) and the base ionization constant (K_b). (2.5)

Subshell: All of the orbitals of a given shell that have the same value of their secondary quantum number, ℓ. (7.3)

Substance: See *Pure Substance.*

Substitution Reaction: The replacement of an atom or group on a molecule by another atom or group. (22.2, 22.3)

Superconductor: A material in a state in which it offers no resistance to the flow of electricity.

Supercooled Liquid: A liquid at a temperature below its freezing point. An amorphous solid. (11.7)

Supercritical Fluid: A substance at a temperature above its critical temperature. (11.12)

Superheated Liquid: The condition of a substance in its liquid state above its boiling point. (11.7)

Superimposability: A test of structural chirality in which a model of one structure and a model of its mirror image are compared to see if the two could be made to blend perfectly, with every part of one coinciding simultaneously with the parts of the other. (21.8, 22.1)

Supersaturated Solution: A solution that contains more solute than it would hold if the solution were saturated. Supersaturated solutions are unstable and tend to produce precipitates. (4.1)

Surface Tension: A measure of the amount of energy needed to expand the surface area of a liquid. (11.3)

Surfactant: A substance that lowers the surface tension of a liquid and promotes wetting. (11.3)

Surroundings: That part of the universe other than the system being studied and separated from the system by a real or an imaginary boundary. (6.3)

Symmetric: An object is symmetric if it looks the same when rotated, reflected in a mirror, or reflected through a point. (9.3)

System: That part of the universe under study and separated from the surroundings by a real or an imaginary boundary. (6.3, 6.6)

T

$t_{1/2}$**:** See *Half-Life.*

T-Shaped Molecule: A molecule having five electron domains in its valence shell, two of which contain lone pairs. The other three are used in bonds to other atoms. The molecule has the shape of the letter T, with the central atom located at the intersection of the two crossing lines. (9.2)

Tarnishing: See *Corrosion.*

Temperature: A measure of the hotness or coldness of something. A property related to the average kinetic energy of the atoms and molecules in a sample. A property that determines the direction of heat flow—from high temperature to low temperature. (1.5, 6.1)

Temperature–Volume Law (Charles' Law): The volume of a given mass of a gas is directly proportional to its Kelvin temperature if the pressure is kept constant. $V \propto T$. (10.3)

Termination Step: A step in a chain reaction in which a reactive species needed for a chain propagation step disappears without helping to generate more of this species. (Facets of Chemistry 13.1)

Tetrahedral Molecule: A molecule with a central atom bonded to four other atoms located at the corners of an imaginary tetrahedron. (9.1)

Tetrahedron: A four-sided figure with four triangular faces and shaped like a pyramid. (9.1)

Theoretical Model: See *Model, Theoretical.*

Theoretical Yield: The yield of a product calculated from the reaction's stoichiometry. (3.6)

Theory: A tested explanation of the results of many experiments. (1.2)

Thermal Decomposition: The decomposition of a substance caused by heating it. (21.3)

Thermal Energy: The molecular kinetic energy possessed by molecules as a result of the temperature of the sample. Energy that is transferred as heat. (6.1)

Thermal Equilibrium: A condition reached when two or more substances in contact with each other come to the same temperature. (6.2)

Thermal Property: A physical property, like heat capacity or heat of fusion, that concerns a substance's ability to absorb heat without changing chemically.

Thermochemical Equation: A balanced chemical equation accompanied by the value of ΔH° that corresponds to the mole quantities specified by the coefficients. (6.6)

Thermochemistry: The study of the energy changes of chemical reactions. (6 Introduction)

Thermodynamic Equilibrium Constant (K): The equilibrium constant that is calculated from ΔG° (the standard free energy change) for a reaction at T K by the equation, $\Delta G^\circ = RT \ln K$. (18.9)

Thermodynamics (Chemical Thermodynamics): The study of the role of energy in chemical change and in determining the behavior of materials. (6 Introduction, 18.1)

Third Law of Thermodynamics: For a pure crystalline substance at 0 K, $S = 0$. (18.5)

Titrant: The solution added from a buret during a titration. (4.8)

Titration: An analytical procedure in which a solution of unknown concentration is combined slowly and carefully with a standard solution until a color change of some indicator or some other signal shows that equivalent quantities have reacted. Either solution can be the titrant in a buret with the other solution being in a receiving flask. (4.8)

Titration Curve: For an acid–base titration, a graph of pH versus the volume of titrant added. (16.7)

Torr: A unit of pressure equal to 1/760 atm. 1 mm Hg. (10.2)

Tracer Analysis: The use of small amounts of a radioisotope to follow (trace) the course of a chemical or biological change. (20.7)

Transcription: The synthesis of mRNA at the direction of DNA. (22.8)

Trans Isomer: A stereoisomer whose uniqueness lies in having two groups that project on opposite sides of a reference plane. (21.8, 22.2)

Transition Elements: The elements located between Groups IIA and IIIA in the periodic table. (2.3)

Transition Metals: The transition elements. (2.3)

Transition State: The brief moment during an elementary process in a reaction mechanism when the species involved have acquired the minimum amount of potential energy needed for a successful reaction, an amount of energy that corresponds to the high point on a potential energy diagram of the reaction. (13.5)

Transition State Theory: A theory about the formation and breakup of activated complexes. (13.5)

Translation: The synthesis of a polypeptide at the direction of a molecule of mRNA. (22.8)

Transmutation: The conversion of one isotope into another. (20.5)

Transuranium Elements: Elements 93 and higher. (20.5)

Traveling Wave: A wave whose peaks and nodes move. (7.3)

Triacylglycerol: An ester of glycerol and three fatty acids. (22.7)

Trigonal Bipyramid: A six-sided figure made of two three-sided pyramids that share a common face. (9.1)

Trigonal Bipyramidal Molecule: A molecule with a central atom holding five other atoms that are located at the corners of a trigonal bipyramid. (9.1, 9.2)

Trigonal Pyramidal Molecule: A molecule that consists of an atom, situated at the top of a three-sided pyramid, that is bonded to three other atoms located at the corners of the base of the pyramid. (9.2)

Triple Bond: A covalent bond in which three pairs of electrons are shared. (8.3)

Triple Point: The temperature and pressure at which the liquid, solid, and vapor states of a substance can coexist in equilibrium. (11.12)

Triprotic Acid: An acid that can furnish three H^+ ions per molecule. (4.3)

U

u: See *Atomic Mass Unit.*

Ultraviolet Catastrophe: The term given to the fact that classical physics predicts large amounts of ultraviolet radiation should be emitted from heated materials. In fact, very little ultraviolet radiation is produced. (7 Introduction)

Uncertainty: The amount by which a measured quantity deviates from the true or actual value. (1.6)

Uncertainty Principle: There is a limit to our ability to measure a particle's speed and position simultaneously. (7.7)

Unit Cell: The smallest portion of a crystal that can be repeated over and over in all directions to give the crystal lattice. (11.9)

Unit of Measurement: A reference quantity, such as the meter or kilogram, in terms of which the sizes of measurements can be expressed. (1.5)

Universal Gas Constant (*R*): See *Gas Constant, Universal*.

Universe: The system and surroundings taken together. (6.3)

Unsaturated Compound: A compound whose molecules have one or more double or triple bonds. (22.2)

Unsaturated Solution: Any solution with a concentration less than that of a saturated solution of the same solute and solvent. (4.1)

V

V-Shaped Molecule: See *Bent Molecule.*

Vacuum: An enclosed space containing no matter whatsoever. A *partial vacuum* is an enclosed space containing a gas at a very low pressure.

Valence Bond Theory (VB Theory): A theory of covalent bonding that views a bond as being formed by the sharing of one pair of electrons between two overlapping atomic or hybrid orbitals. (9.4)

Valence Electrons: The electrons of an atom in its valence shell that participate in the formation of chemical bonds. (7.6)

Valence Shell: The electron shell with the highest principal quantum number, *n*, that is occupied by electrons. (7.6)

Valence Shell Electron Pair Repulsion Theory (VSEPR Theory): The bonding and nonbonding (lone pair) electron domains in the valence shell of an atom seek an arrangement that leads to minimum repulsions and thereby determine the geometry of a molecule. (9.2)

Van der Waals' Constants: Empirical constants that make the van der Waals' equation conform to the gas law behavior of a real gas. (10.9)

Van der Waals' Equation: An equation of state for a real gas that corrects *V* and *P* for the excluded volume and the effects of intermolecular attractions. (10.9)

Van der Waals' Forces: Attractive forces including dipole–dipole, ion–dipole, and induced dipole forces. (11.2)

Van't Hoff Factor (*i*): The ratio of the observed freezing point depression to the value calculated on the assumption that the solute dissolves as un-ionized molecules. (12.9)

Vapor Pressure: The pressure exerted by the vapor above a liquid (usually referring to the *equilibrium* vapor pressure when the vapor and liquid are in equilibrium with each other). (10.6, 11.5)

Vapor Pressure–Concentration Law (Raoult's Law): The vapor pressure of one component above a mixture of molecular compounds equals the product of its vapor pressure when pure and its mole fraction. (12.6)

Viscosity: A liquid's resistance to flow. (11.3)

Visible Spectrum: That region of the electromagnetic spectrum whose frequencies can be detected by the human eye. (7.1)

Volatile: Descriptive of a liquid that has a low boiling point, a high vapor pressure at room temperature, and therefore evaporates easily. (11.5)

Volt (V): The SI unit of electric potential or emf in joules per coulomb. (19.2)

$$1 \text{ V} = 1 \text{ J C}^{-1}$$

Voltaic Cell: See *Galvanic Cell.*

VSEPR Theory: See *Valence Shell Electron Pair Repulsion Theory.*

Vulcanized Rubber: Rubber that has been treated with a substance such as sulfur that forms cross-links and improves the properties of the rubber. (22.6)

W

Wave: An oscillation that moves outward from a disturbance. (7.1)

Wave Function (ψ): A mathematical function that describes the intensity of an electron wave at a specified location in an atom. The square of the wave function at a particular location specifies the probability of finding an electron there. (7.3)

Wave Mechanics (Quantum Mechanics): A theory of atomic structure based on the wave properties of matter. (7 Introduction)

Wave/Particle Duality: A particle such as the electron behaves like a particle in some experiments and like a wave in others. (7 Introduction)

Wavelength (λ): The distance between crests in the wavelike oscillations of electromagnetic radiations. (7.1)

Weak Acid: An acid with a low percentage ionization in solution; a poor proton donor; an acid with a low value of K_a. (4.3, 16.1)

Weak Base: A base with a low percentage ionization in solution; a poor proton acceptor; a base with a low value of K_b. (4.3, 16.1)

Weak Electrolyte: A substance that has a low percentage ionization or dissociation in solution. (4.3)

Weighing: The operation of measuring the mass of something using a balance. (1.5)

Weight: The force with which something is attracted to the earth by gravity. (1.3)

Weight Percent: See *Percentage by Mass.*

Wetting: The spreading of a liquid across a solid surface. (11.3)

White Phosphorus: A very reactive allotrope of phosphorus consisting of tetrahedral P_4 molecules. (21.2)

Work (*w*): The energy expended in moving an opposing force through some particular distance. Work has units of *force* × *distance*. (6.1, 6.5, 18.1)

X

X Ray: A stream of very high-energy photons emitted by substances when they are bombarded by high-energy beams of electrons or are emitted by radionuclides that have undergone K-electron capture. (20.3)

Y

Yield, Actual: The amount of a product obtained in a laboratory experiment. (3.6)

Yield, Percentage: The ratio, given as a percent, of the quantity of product actually obtained in a reaction to the theoretical yield. (3.6)

Yield, Theoretical: The amount of a product calculated by the stoichiometry of the reaction. (3.6)

Z

Zinc–Manganese Dioxide Dry Cell (Leclanché Cell): A galvanic cell of about 1.5 V involving zinc and manganese dioxide under mildly acidic conditions. (19.8)

Zero-Order Reaction: A reaction that occurs at a constant rate regardless of the concentration of the reactant. (13.3)

INDEX

G

H

L

M

N

Q

R

RELATIONSHIPS AMONG UNITS

(Values in boldface are exact.)

Length

1 in. = **2.54** cm
1 ft = **30.48** cm
1 yd = **91.44** cm
1 mi = **5280** ft
1 ft = **12** in.
1 yd = **36** in.

Volume

1 liq. oz = **29.57353** mL
1 qt = **946.352946** mL
1 gallon = **3.785411784** L
1 gallon = **4** qt = **8** pt
1 qt = **2** pt = **32** liq. oz

Mass

1 oz = **28.349523125** g
1 lb = **453.59237** g
1 lb = **16** oz

Pressure

1 atm = **760** torr
1 atm = **101,325** Pa
1 atm = 14.696 psi (lb/in.^{2})
1 atm = 29.921 in. Hg

Energy

1 cal = **4.184** J
1 ev = 1.6022×10^{-19} J
1 ev/molecule = 96.49 kJ/mol
1 ev/molecule = 23.06 kcal/mol
1 J = 1 kg m^2 s^{-2} = 10^7 erg

PHYSICAL CONSTANTS

Rest mass of electron	$m_e = 5.485799094 \times 10^{-4}$ u ($9.1093821 \times 10^{-28}$ g)
Rest mass of proton	$m_p = 1.0072764668$ u ($1.67262164 \times 10^{-24}$ g)
Rest mass of neutron	$m_n = 1.0086649160$ u ($1.67492721 \times 10^{-24}$ g)
Electronic charge	$e = 1.60217649 \times 10^{-19}$ C
Atomic mass unit	u $= 1.66053878 \times 10^{-24}$ g
Gas constant	R $= 0.0820575$ L atm mol^{-1} K^{-1}
	$= 8.31447$ J mol^{-1} K^{-1}
	$= 1.98721$ cal mol^{-1} K^{-1}
Molar volume, ideal gas	$= 22.4140$ L (at STP)
Avogadro's number	$= 6.0221418 \times 10^{23}$ things/mol
Speed of light in a vacuum	c $= 2.99792458 \times 10^8$ m s^{-1} (Exactly)
Planck's constant	h $= 6.6260690 \times 10^{-34}$ J s
Faraday constant	F $= 9.6485340 \times 10^4$ C mol^{-1}

LABORATORY REAGENTS

(Values are for the average concentrated reagents available commercially.)

Reagent	Percent (w/w)	Mole Solute / Liter Solution	Gram Solute / 100 mL Solution
NH_3	29	15	26
$HC_2H_3O_2$	99.7	17	105
HCl	37	12	44
HNO_3	71	16	101
H_3PO_4	85	15	144
H_2SO_4	96	18	177